AF882604

Vegetational Areas of East Texas

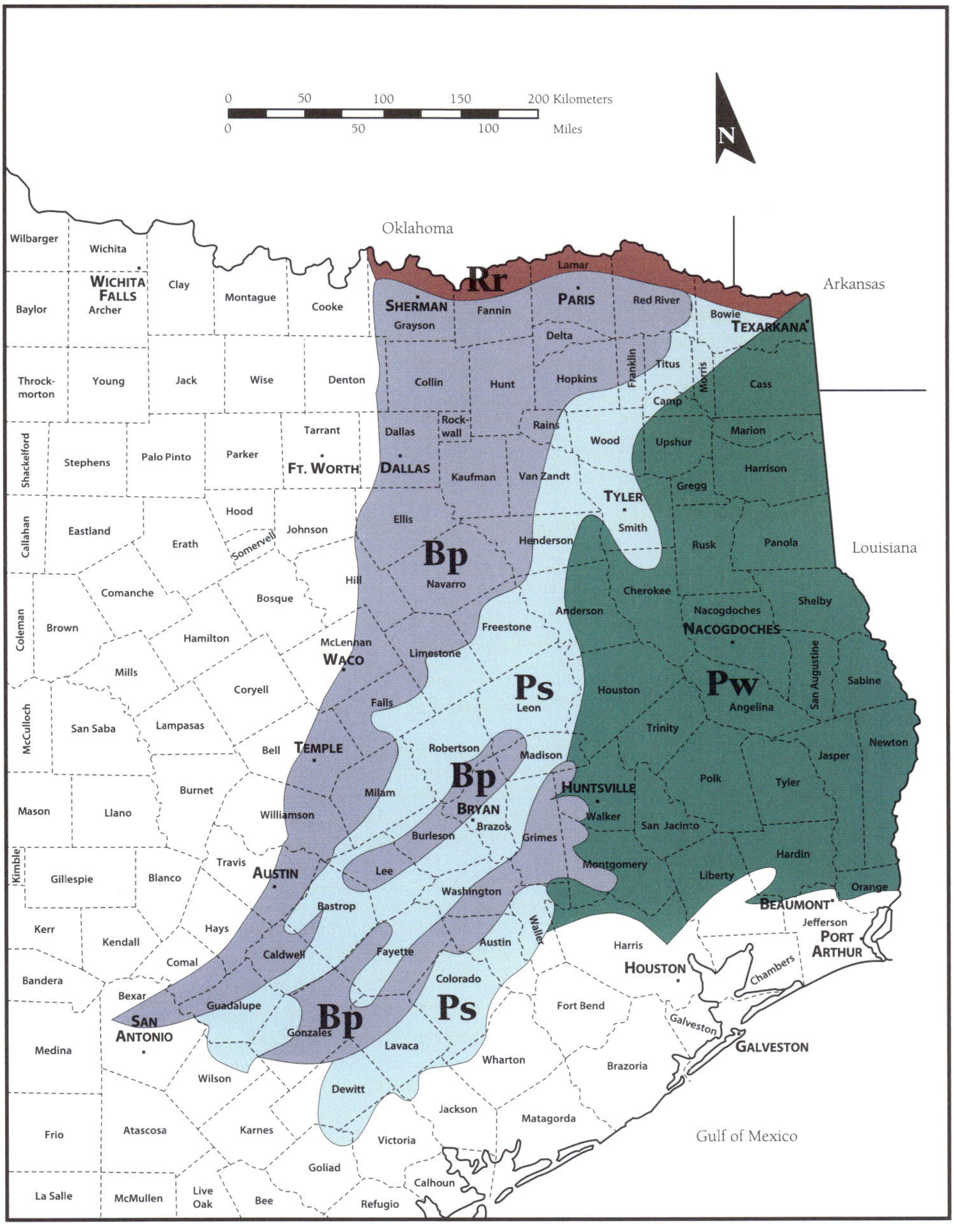

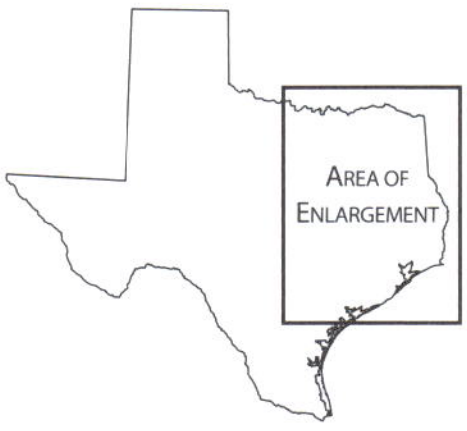

Bp Blackland Prairie
Pw Pineywoods
Ps Post Oak Savannah
Rr Red River Area

Illustrated Flora of East Texas

Illustrated Flora of East Texas
Volume Two

PUBLISHED WITH THE SUPPORT OF

MAJOR BENEFACTORS

SWBC Foundation (Charles Amato)

The Balogh Blessings Fund at Swab Charitable (George and Terri Balogh)

The Cynthia and George Mitchell Foundation

BENEFACTORS

David P. and Patricia Lewis

Warren W. Pruess

Michael and Barbara MacRoberts
in memory of William Walter Shepherd, Cumbria, UK

Sue Tracey

OTHER CONTRIBUTORS

Sonnia Hill

INSTITUTIONAL SUPPORT

Austin College
Botanical Research Institute of Texas

We are extremely grateful to all the donors who made everything possible.

Illustrated Flora of East Texas

George M. Diggs, Jr. | Barney L. Lipscomb | Robert J. George

VOLUME TWO:

Dicots: Acanthaceae–Fabaceae

BRIT PRESS

Fort Worth Botanic Garden | Botanical Research Institute of Texas
Center for Botanical Exploration & Discovery
Fort Worth, Texas

2026

Illustrated Flora of East Texas, Volume Two. Dicots: Acanthaceae–Fabaceae
© 2026 Fort Worth Botanic Garden | Botanical Research Institute of Texas
All rights reserved
First edition 2026
Printed in the South Korea
ISBN: 13-978-1889878-83-6

Creative Commons License CC BY-NC-SA

No part of this book may be reproduced or transmitted in any form by any means, electronic, or mechanical, including photocopying and recording, or by any information storage and retrieval system, except as may be expressly permitted in writing from the publisher. Requests for permission should be addressed to BRIT Press.

Published by BRIT Press, Fort Worth Botanic Garden | Botanical Research Institute of Texas, Fort Worth, Texas, U.S.A.

Cover design
Rebecca Horn, rhorngraphics

Book layout and production
Rebecca Horn, rhorngraphics

Front cover image
Layla Luna, Layla Luna Studio, www.laylaluna.com

Please cite this book as follows:
Diggs, G.M., Jr., B.L. Lipscomb, & R.J. George. 2026. *Illustrated Flora of East Texas, Volume Two. Dicots: Acanthaceae–Fabaceae.* Bot. Misc. 69. BRIT Press, Fort Worth Botanic Garden | Botanical Research Institute of Texas, U.S.A.

Illustrated Flora of East Texas, Volume Two. Dicots: Acanthaceae–Fabaceae by George M. Diggs, Jr., Barney L. Lipscomb, and Robert J. George. Published with support from benefactors, Fort Worth Botanic Garden | Botanical Research Institute of Texas Press, and Austin College.

Distribution of copies by:
BRIT Press, Fort Worth Botanic Garden | Botanical Research Institute of Texas
1700 University Dr.
Fort Worth, Texas 76107-3400, U.S.A.
Telephone: 1.817.332.4441
Fax: 1.817.332.4112
Email: press@brit.org
Websites: britpress.org, shopbritpress.org

BRIT PRESS

Botanical Miscellany, No. 69
ISSN: 0883-1475
15 June 2026

Illustrated Flora of East Texas

Volume Two

IS THE FOURTH PUBLICATION OF THE

Illustrated Texas Floras Project

A COLLABORATIVE PROJECT OF

Austin College Center for Environmental Studies

AND

Botanical Research Institute of Texas

TO PRODUCE ILLUSTRATED FLORISTIC TREATMENTS

DESIGNED TO BE USEFUL TO BOTH BOTANICAL SPECIALISTS

AND A MORE GENERAL AUDIENCE.

AUSTIN COLLEGE

BRIT PRESS

DEDICATIONS

To Monique Dubrule Reed and Linda "Linny" Heagy
Special friends and vested supporters of the Illustrated Texas Floras Project

IN MEMORY
MONIQUE DUBRULE REED (1963–2025)

MONIQUE DUBRULE REED was born in El Paso, Texas, where green is a novelty, so from an early age she found plants fascinating. She earned her undergraduate degree in Floriculture from Texas A&M University (B.S. 1985), as well as an M.S. in Botany (1997). She was the Herbarium Botanist in the Biology Department at A&M from 1985–2012 and worked in the Tracy Herbarium until she retired in 2020. She coordinated undergraduate and graduate laboratories in flowering plant taxonomy and economic botany and served as a botanical information resource for the general public. Her particular interests included the flora of Texas, bog and outcrop plants, cultivated plants, exotic weeds, and education. In 1997, she published the *Manual of the Dicot Flora of Brazos and Surrounding Counties*. She was a co-author of the *Illustrated Flora of East Texas, Volume One: Introduction, Pteridophytes, Gymnosperms, and Monocotyledons* (2006). Monique was not only a knowledgeable botanist, she was an excellent editor, communicator, and a stickler for detail and grammar. Her contributions to Texas botany will inspire future botanists.

—George M. Diggs, Jr., Barney L. Lipscomb, Robert J. George

Monique was immensely gifted in many different fields. Her profession was taxonomic botany (identifying wild plants). She devoted many years to teaching students, sharing her love of plants with the public, and running the herbarium of the Biology Department at Texas A&M. She was a respected authority in her field, having co-authored an award-winning book (*Illustrated Flora of East Texas, Volume One*) and discovered a previously unknown species of wildflower. Many of those she taught and mentored fondly remember Monique's special love of "belly botany"—plants that are so small that one must get down on the ground to see and appreciate them.

—David Reed, Roll Call Tribute, Texas A&M University

IN MEMORY
LINNY HEAGY (1946–2025)

LINNY HEAGY earned a commercial arts degree at John Herron Art Institute of Indiana University and was a professional artist, illustrator, and graphic designer, working extensively on Texas plants. For *Shinners & Mahler's Illustrated Flora of North Central Texas* (1999), she created 226 original line drawings, painted the art for the dust jacket and the frontispiece, and served as creative director/art director for the entire volume. Likewise, for the *Illustrated Flora of East Texas, Volume One*, she produced 28 original illustrations, painted the frontispiece, designed the cover, and served as creative director/art director. In addition, Linny depicted a number of new plant species published in the botanical journal *Sida*. She illustrated *Rare Plants of Texas*, a book on rare and endangered Texas plants, in conjunction with the Texas Parks and Wildlife Department (Poole et al. 2007). She was also commissioned as one of the illustrators for the Flora of North America Project.

—Illustrated Flora of East Texas, Vol. One (2006)

The Maverick Botanical Artist [Linny Heagy] in her own words:
Seems I've always been climbing trees: there was the old cottonwood tree, the apple, plum, and mulberry trees just outside the door of my childhood home in Indiana (that home being two converted railroad box cars set side by side with no inside plumbing), and then there was a large cherry tree with delicious fruit just outside a small Bavarian village where I lived for one year. I guess child curiosity and the urge to travel drew me upward into those monarchs of nature showing me vistas of the biota continuum. However, becoming a scientific botanical illustrator would be a very cicatricial journey.

I had no influences in order to seek a career in the botanical sciences, only my love and curiosity for the natural world. As a young girl I had no idea there were such careers. However, I did have a keen sense of observation and good hand/eye coordination and a natural ability to draw. With the mentoring of my high school art teacher, I secured a scholarship to John Herron Art Institute of Indiana University. I was classically trained in all aspects of the arts: drawing, life drawing, portraiture, anatomy, painting, sculpting, and art history. My major would be in the advertising arts, specializing in graphic design and illustration, specifically fashion illustration. After college, I soon moved to upstate New York to partner with a new advertising agency learning all aspects of creating a business from the ground up. I worked hard and was successful but my urge to travel sent me to Europe for a year. On my return to the States, I moved to Miami setting up my own creative design/illustration studio specializing in fashion illustration and was fashion illustration instructor for Bauder Fashion College teaching fashion designers how to interpret their designs on paper.

Four years later, again the urge to see other parts of North America surfaced. So when I was asked to help set up an art department in Houston, Texas, I said yes. Eventually I settled in the Dallas/Fort Worth area becoming an art director, graphic designer, and fashion illustrator for Neiman Marcus in Dallas.

In 1990 I began to climb my last tree. I left the world of advertising and, finally put myself on that path I didn't take in my salad days so many years ago ... the natural world. While researching a paper for an associates degree in horticultural science, I discovered a treasure: BRIT (Botanical Research Institute of Texas) and got involved with helping create their corporate image and visually launch the Illustrated Texas Floras Project taking on the job of creative director and my first attempts at botanical illustrating (in addition: the book designing, layouts, type specifications, production management, and all that goes with creating a flora). All of my previous career experiences in the advertising arts and fashion world culminated in this commission and I believe we created one of the first floras of its kind. Soon after, I was to illustrate the book, *Rare Plants of Texas*. Eventually, I was recommended by Ted Barkley to be interviewed as scientific botanical illustrator for Flora of North America and was tutored by the well-respected Yevonn Wilson-Ramsey.

With every move I made in my life, one thing I never tired of ... the beauty and diversity of our natural world. I have found contentment in preserving on paper the humble beauty of nature before it vanishes.

While illustrating the *Flora of North Central Texas* for BRIT, I was fortunate to have shared an office with Ruth Ginsburg, archivist and author of the biography *Lloyd Herbert Shinners, By Himself*. I was also honored to have been the designer and producer of the book. One of his thoughts found in the archives has always resonated with me. I quote it here:

> *My life for these years has been a living out of Goethe's words:*
>
> *"Alles Vergangliche ist muse ein Gleichnis. (All that has passed is only a metaphor, an allegory.) I was at once participant and detached observer and student of that in which I participated. It was a process of testing the validity and hope of one of the most romantic of sciences in an age alien and hostile to it. And after it I can declare reaffirmation of faith in quiet scholarship, in scholarly pleasurable labor, in laborious study—in scientia amabilis nostra."*

Linny's additional achievements

1975 Gold Metal Great Menu Award. National Restaurant Association for Cornell Dining Catering Service Menu, Cornell University; Judges: Craig Claiborne, food editor, *New York Times*; Albin Seaberg, author, *"Menu Design, Merchandising and Marketing"*; Janice Garr, food editor, *Restaurant Business Magazine*; Jane Wallace, editor-in-chief, *Institution/Volume Feeding Magazine*; and Henry Bolling, president of The National Restaurant Association. Judging comments: *We reviewed over 1,100 menus. The Gold award was judged on general overall appearance, merchandising or selling power, ease of reading, imagination and originality.*

1981 National Gold ADDY Award, Direct Mail Campaign for Dadeland Merchants Association

1984 New York Art Directors Club 63rd Annual Exhibition Merit Award, Neiman Marcus *German Fortnight* Menu and Recipe book

1986 Dallas Advertising League Merit Award, Four Seasons Hotel, Dallas at Las Colinas brochure

2001 Exhibition of Scientific Botanical Illustrations, The Kochi Prefectural Makino Botanical Garden, Kochi, Japan

2013 Association of Illustrators (AOI) shortlist, Flora of North America, Volume 6, frontispiece illustration of *Dionaea muscipula*

—*Reproduced from the* Flora of North America Newsletter, *2013*

ILLUSTRATED FLORA OF EAST TEXAS
VOLUME TWO
DICOTS: ACANTHACEAE–FABACEAE

A book to Promote Botanical and Environmental Education,
Scientific Exploration, and Conservation in East Texas and Adjacent Areas.

AUTHORS' NOTE

In a flora book this size, it is inevitable that errors and omissions, of all sorts, escape the eyes of authors and editors. Because of the possibility of providing updates and corrections online, we would appreciate corrections, suggestions, or additions from individuals using this book. Such information can be sent to:

Barney Lipscomb, barney@fwbg.org

Also, we hope that this book will spur ongoing interest in the *Illustrated Flora of East Texas*. Plant specimens—particularly county, regional, or state records—would be much appreciated and can be deposited at a number of herbaria in the state including:

Botanical Research Institute of Texas Herbarium (BRIT)
1700 University Drive
Fort Worth, Texas 76107-3400

Stephen F. Austin State University Herbarium (ASTC)
Department of Biology
Biology Building, Room 222
P.O. Box 13003, SFA Station
Nacogdoches, Texas 75962-3033

The core of BRIT's herbarium is the Lloyd H. Shinners Collection in Systematic Botany, begun at Southern Methodist University in Dallas, Texas, in 1944. The Botanical Research Institute of Texas was formed in 1987 to house the transfer of this collection, which moved to Fort Worth in 1991. Areas of geographic concentration encompass Texas, the southeastern United States, Mexico, Central America, South America, and Southeast Asia. The BRIT herbarium holdings are worldwide in scope, and represent most of the earth's plant families.

Founded in 1949, the Stephen F. Austin Herbarium (ASTC) contains nearly 80,000 specimens and is the 4th largest collection in Texas. Personnel include Director Dr. James Van Kley, Curator Imani Harris, and long-time employee Macauley Parker. Although the collection includes specimens from throughout North America, its focus is on the East Texas Pineywoods, where it represents nearly all the native and naturalized species found in the region. The herbarium plays a vital role in teaching, as well as crucial plant identification support to our research in vegetation science. We also entertain loan requests and facilitate specimen exchanges with other herbaria.

The University of Texas Herbarium (TEX) is part of the Billie L. Turner Plant Resources Center (PRC), which houses the combined herbaria of TEX and the Lundell Herbarium (LL). The combined PRC collections include over one million specimens from all parts of the world and represent the 13th largest herbarium in the United States. The two herbaria (TEX and LL) are completely integrated and interfiled, although each sheet is marked as to its herbarium and should be cited as such. The University of Texas Herbarium itself, with close to 700,000 specimens, dates to about 1890 and is especially strong in Texas and Mexican specimens. The research interests of faculty, staff, and graduate students have led to very strong representation of Asteraceae, Fabaceae, and Rhamnaceae, among other families, as well as the Chihuahuan Desert region (Mexico and the United States.) and the mountains and plains of northeastern Mexico. All new specimens received into the Plant Resources Center are now accessioned into TEX.

Online specimens from BRIT and SFA of the East Texas flora can be found through the TORCH data portal: https://portal.torcherbaria.org/portal/index.php. Online specimens from TEX of the East Texas flora can be found through the SERNEC dataportal: https://sernecportal.org/portal/index.php.

TABLE OF CONTENTS	PAGE

ILLUSTRATED FLORA OF EAST TEXAS

ACKNOWLEDGMENTS

The authors would like to express our deep appreciation to the Botanical Research Institute of Texas (BRIT) and to Austin College.

All individuals assisting with the *Illustrated Flora of North Central Texas* are thanked again here. Also, a large number of the individuals acknowledged in volume one of the *Illustrated Flora of East Texas* made contributions that also apply to this volume, and we greatly appreciate their efforts.

The Asteraceae and the Brassicaceae family treatments were contributed by Amanda K. Neill, former director of the BRIT Herbarium. Eric Keith contributed the Fabaceae. A. Michael Powell reviewed the Cactaceae. Monique Reed reviewed final drafts of the Asteraceae, Brassicaceae, and earlier drafts of several other families from 2006–2009.

Several artists contributed original drawings to this volume. First and foremost, we greatly appreciate the original drawings Linny Heagy did for our *Illustrated Flora of North Central Texas* that are included in this volume. Other artists include Leila Ali-Sanchez, Natch Azure, Judith Baker, Gail Dentler, Clair Gaston, Gwen Genius, Robert J. George, Jeannice Gordon, Ellen LeBlanc, Carolyn Patterson, Harper Quill, Maria Thompson, and Amy Welch. The cover painting was done by Layla Luna (Layla Luna Studio, www.laylaluna.com).

Texas county distribution maps were based on USDA PLANTS and Biota of North America Program (BONAP) data and additional herbarium records.

Several individuals deserve special recognition for their contributions: Jason Best, Director of Biodiversity Informatics, Botanical Research Institute of Texas at the Fort Worth Botanic Garden, provided assistance with the Texas county distribution maps. Rebecca Horn (rhorngraphics) provided layout design and production services. The book design follows that which was created by Linny Heagy for *Illustrated Flora North Central Texas* and *Illustrated Flora of East Texas, Volume One.*

Financial support for this volume has come from a number of sources—these are listed on the contributor's page near the beginning of the book.

Preface

This short preface is designed to give a historic overview and context for the Illustrated Texas Floras Project and specifically the *Illustrated Flora of East Texas*, and to provide information on progress toward the completion of that three-volume floristic treatment of the plants of the eastern part of Texas. An extensive introduction to the flora of East Texas is provided in *Illustrated Flora of East Texas, Volume One* (Diggs et al. 2006) and should be referred to as needed.

In the early 1990s, George Diggs, Barney Lipscomb, and Robert "Bob" O'Kennon, while chatting at the then recently formed Botanical Research Institute of Texas (BRIT), discussed the possibility of updating William "Bill" Mahler's *Shinners' Manual of the North Central Texas Flora* (Mahler 1988). Mahler's work was 335 pages, and was so named because it was based on Lloyd Shinners' *Spring Flora of the Dallas-Fort Worth Area, Texas*, first published in 1958.

What began as an effort to update Mahler's work by including additional species then known in the North Central Texas region, ended up as *Shinners and Mahler's Illustrated Flora of North Central Texas* (IFNCT), a fully illustrated flora of 1,636 pages covering 2,223 species. The work evolved over the process of its writing to include extensive introductory material, appendices, illustrations, field work by many individuals, and many more species than were originally known from the region. The project was, from the beginning, a collaboration between the Austin College Center for Environmental Studies and BRIT and eventually was envisioned as the first publication of the Illustrated Texas Floras Project. The stated goal of that project was "to produce illustrated floristic treatments designed to be useful to both botanical specialists and a more general audience."

As the IFNCT evolved this one goal was clearly kept in mind—to make the book useful to not only botanists, but also to other people who were interested in identifying and learning about the plants of the region. The IFNCT is available online in its entirety at https://fwbg.org/science-conservation/brit-press/illustrated-flora-of-north-central-texas-online/.

After the IFNCT was published in 1999, the team decided that the next floristic project undertaken should be East Texas. For the purposes of the project, East Texas (see inside front cover of this volume) is an area of roughly 62,600 square miles (162,200 square kilometers, 40 million acres, or 16.2 million hectares) or about the size of Georgia. This 87-county region stretches from the eastern border of Texas with Arkansas and Louisiana west to the western margin of the Blackland Prairie, and from the Red River border with Oklahoma south nearly to San Antonio. Vegetational areas include the Pineywoods, the Post Oak Savannah, the Blackland Prairie, and the Red River area. It became clear, based on the more than 3,400 species documented to be present, that the flora would be a three-volume work. The species included represent roughly two-thirds of the total Texas flora. Work soon began both in the field, especially in the Big Thicket and along the Red River (two areas especially rich in species), and on an introduction and treatments of the species to be included in Volume 1 (lycophytes, ferns, gymnosperms, and monocots). Volume 1 (Diggs et al. 2006) covered a total of 1,060 species, and represents the contributions of numerous individuals. This work included line drawing illustrations for all species, had nearly 200 color photographs, and in addition had county distribution maps for almost all species. The county distribution maps drew heavily on those by Turner et al. (2003). The treatments in general were more detailed than those in the IFNCT. It also had a 271-page introduction and extensive appendices. Volume 1 in its entirety is available online at https://fwbg.org/science-conservation/brit-press/illustrated-flora-of-east-texas-online/.

Following publication of Volume 1 in 2006, work began on the preparation of taxonomic treatments for Volume 2, which includes the dicot families beginning alphabetically with the Acanthaceae and ending with the Fabaceae. Significant progress was made over the next three years with treatments nearly completed for more than half the families to be included in the volume. However, it then became clear because of financial constraints and institutional priorities that funds were not available to publish the volume in the near future.

At that point in time it was fortunate that a fern enthusiast and generous supporter, Susan Tracey, approached us about the possibility of writing an updated treatment of the ferns and lycophytes of Texas as a whole. As a result, work shifted to the production of *The Ferns and Lycophytes of Texas* (Diggs & Lipscomb 2014). As with the previous floras, this treatment of the ferns and lycophytes was fully illustrated with line drawing illustrations. It also included color photographs, and in addition, Texas county distribution maps as well as distribution maps of the various species throughout the United States and Canada. Much of the material of that volume can be found online at: https://ferns.brit.org/?q=intro. That volume represents the third work of the Illustrated Texas Floras Project.

Amanda K. Neill, former Director of the BRIT Herbarium, began working on the Asteraceae for Volume 2 of the *Illustrated Flora of East Texas* several years before a "nearly-final draft" of the family was made available online in 2019 as the *Asteraceae of East Texas* (Neill 2019). It can be found freely available at https://fwbgstg.wpengine.com/wp-content/uploads/2021/08/Asteraceae_East_Texas_download.pdf. It is a 195-page detailed treatment of the family in East Texas including statewide distribution maps. The Asteraceae is the largest family in East Texas, with 138 genera and 423 species. In 2019, it also seemed possible to move forward with completing Volume 2, and work restarted on writing and updating other family treatments.

Knowledge of both plant taxonomy and how botanical treatments are written have changed a tremendous amount since the beginning of the Illustrated Texas Floras Project. At that time, in the early 1990s, most research was done using physical books and physical copies of scientific journals. Also at that time, molecular studies of plants were in their early stages. In contrast, molecular studies were widely used in the preparation of the *Illustrated Flora of East Texas, Volume 1*, and their influence is even more evident throughout this volume. Insights that were virtually impossible to discover previously are now widely understood thanks to the application of molecular techniques. Thus, the preparation of this second volume of the *Flora of East Texas* has been radically different with one major exception—herbarium specimens are still the crucial ultimate data source and their use is indispensable. Even in the case of herbarium specimens, high resolution scans/digitization of herbarium sheets allows researchers to "examine" specimens remotely. Because of the importance of specimens, BRIT's herbarium is an extremely valuable resource. BRIT's herbarium, which is the largest in the state and among the ten largest in the United States with more than 1.5 million preserved plant specimens, has allowed the Illustrated Texas Floras Project to occur. Another important resource was the *Flora of North America North of Mexico*. An additional important new resource is the work of the Angiosperm Phylogeny Group (APG) and the continual updating of that work (Stevens 2001 onwards). In particular, the Angiosperm Phylogeny Website (Stevens 2001 onwards) is a huge online compendium of botanical knowledge. Family circumscription in general follows APG, the latest published version being APG IV (2016). Family and generic synopses have been drawn from a variety of sources, but the most recent version of *The Plant Book* (Mabberley 2017) has been a major source.

We are optimistic that volume three (Fagaceae through Zygophyllaceae) will be completed in the not-too-distant future. The completion of the three volumes of the *Illustrated Flora of East Texas* is part of the Fort Worth Botanic Garden | Botanical Research Institute of Texas Long Range Strategic Plan. While funding will determine the exact timeline, a number of individuals are committed to writing treatments and seeing the project through to completion.

Abbreviations and Symbols

ABBREVIATIONS/SYMBOLS	MEANING
	family or generic synopsis
	endemic to East Texas
	endemic to Texas
	introduced species, subspecies, or variety (not native to the United States)
⚠	species, subspecies, or variety of conservation concern (e.g., endangered, threatened)
☠	toxic/poisonous plant
⊘	noxious or harmful or potentially harmful exotic plant
$<$	less than
$\leq$	less than or equal to
$>$	more than
$\geq$	more than or equal to
$\pm$	more or less
+	or more (e.g., small tree 2–5+ m tall)
$\times$	times or to indicate hybridization
auct.	Latin: *auctorum,* author
ASTC	herbarium abbreviation for Stephen F. Austin State University, Nacogdoches, TX
BAYLU	herbarium abbreviation for Baylor University Herbarium, Waco, TX
BCNWR	herbarium abbreviation for Balcones Canyonlands National Wildlife Refuge
B.P.	before present
BRCH	herbarium abbreviation for Botanical Research Center Herbarium, Bryan, TX
BRIT	herbarium abbreviation for Botanical Research Institute of Texas, Fort Worth
c	central
ca.	Latin: *circa,* about
cm	centimeter
COA	herbarium abbreviation for City of Austin's Office of Environmental Resource Management
comb. nov.	Latin: *combinatio nova,* new combination of name and epithet
diam.	diameter
dm	decimeter
DUKE	herbarium abbreviation for Duke University, Durham, North Carolina
DUR	herbarium abbreviation for Southeastern Oklahoma State University, Durant, Oklahoma; currently housed at the Botanical Research Institute of Texas, Fort Worth
e	east
e.g.	Latin: *exempli gratia,* for example
ex.	used when an author publishes a new species (or variety or subspecies) based on a name attributed to but not validly published by another author
f.	Latin: *filius,* son; e.g., L. f. indicates the younger Linnaeus
HPC	herbarium abbreviation for Howard Payne University, Brownwood, TX
i.e.	Latin: *id est,* that is
ined.	Latin: *ineditus,* unpublished
m	meter
MICH	herbarium abbreviation for the University of Michigan, Ann Arbor
mm	millimeter

ABBREVIATIONS/SYMBOLS	MEANING
mya	million years ago
n	north
n	= haploid chromosome number
2*n*	= diploid chromosome number
NY	herbarium abbreviation for the New York Botanical Garden, Bronx
nom. illeg.	Latin: *nomen illegitimum,* illegitimate name
of authors, not	used to indicate a name was used in the sense of certain authors, but not in the sense of the author making the combination; technically written as: auct. non
ours	used in descriptions to emphasize character states of East TX species that may be different in other geographic areas
p.p.	Latin: *pro parte,* in part
pers. comm.	personal communication of information to the authors
pers. obs.	personal observation by one of the authors
s	south
sensu	in sense of; used to indicate that a name is used in the sense of one author, not another
sensu lato	in the broad sense, e.g., if a genus is broadly treated to include many species
sensu stricto	in the strict sense, e.g., if a genus is narrowly treated to include few species
SBSC	herbarium abbreviation for Spring Branch Science Center, Spring Branch Independent School District, Houston, TX
SHST	herbarium abbreviation for Sam Houston State University, Huntsville, TX
SMU	herbarium abbreviation for Southern Methodist University Herbarium, now part of the Botanical Research Institute of Texas (BRIT), Fort Worth
s.n.	Latin: *sine numero,* without number
sp.	species
spp.	species (plural)
subsp.	subspecies
TAES	herbarium abbreviation for S.M. Tracy Herbarium, Department of Rangeland Ecology & Management, Texas A&M University, College Station
TAMU	herbarium abbreviation for Biology Department Herbarium, Texas A&M University, College Station
TEX	herbarium abbreviation for University of Texas at Austin (including LL—Lundell Herbarium and RUNYON—Robert Runyon Herbarium)
Univ.	university
UTTYLER	unofficial herbarium abbreviation for University of Texas at Tyler
VDB	herbarium abbreviation for Vanderbilt University Herbarium, currently housed at the Botanical Research Institute of Texas, Fort Worth
var.	variety
w	west

Ranges for measurements, e.g., (10–)12–23(–30) mm long, should be interpreted as "typically 12 to 23 mm long, rarely as little as 10 mm long, rarely as much as 30 mm long"

States are abbreviated using standard, two letter, United States Postal code abbreviations (e.g., TX = Texas, OK = Oklahoma, LA = Louisiana, AR = Arkansas)

Illustrated Flora of East Texas

Lupinus texensis

ACANTHACEAE Juss.

ACANTHUS OR WILD PETUNIA FAMILY

Ours mostly perennial herbs (including 2 aquatics and a vine) or some woody basally and two shrubs, often with squarish stems; foliage often with minute cystoliths (= mineral concretions) on surface (visible as short lines or dots); leaves simple, opposite, sessile to petiolate; leaf blades entire or indistinctly toothed; stipules absent; flowers perfect, axillary or terminal, solitary or in clusters or pedunculate spikes, cymes, or panicles, often with conspicuous bracts (cleistogamous flowers some-times produced in *Ruellia*); sepals 5, united at base or up to 1/3 their length; corollas sympetalous, tubular below, some are 2-lipped and bilaterally symmetrical or nearly radially symmetrical and 4- to 5-lobed, often with a color pattern; stamens 4 (didynamous) or 2 (+ 2 staminodia), inserted on the corolla tube; pistil 1, of 2 carpels; ovary superior; style 1; fruit a 2-valved capsule usually with 2–18 seeds, each seed typically supported by a small hook-like or curved outgrowth (called a retinaculum or jaculator)—the seeds forcefully ejected under certain moisture conditions (fruits usually explosively dehiscent).

A large (ca. 4,320 species in 191 genera—Stevens 2001 onwards); cosmopolitan but largely tropical family of mostly shrubs, with some herbs, twiners, trees, and even aquatics. The main economic value of the family is that a number are used ornamentally, including *Acanthus*, *Aphelandra* (ZEBRA-PLANT), *Fittonia* (NERVE-PLANT), *Justicia brandegeeana* (SHRIMP-PLANT), and *Thunbergia* (CLOCKVINE). The family shares similarities to and is within a lineage (Lamiales) that includes such families as the Bignoniaceae, Lamiaceae, Lentibulariaceae, Orobanchaceae, Scrophulariaceae, and Verbenaceae (Stevens 2001 onwards). However, molecular evidence (McDade et al. 2000b; Scotland & Vollesen 2000) indicates that the family is monophyletic. The Avicenniaceae (a single genus of mainly tropical mangrove trees with pneumatophores (= erect exposed roots involved in gas exchange) is now included in the family by many authorities. We are following Stevens (2001 onwards) and McDade et al. (2008) and are including it in the Acanthaceae. Family name from *Acanthus*, BEAR'S-BREECHES, a genus of 30 mostly spiny xerophytic species of Old World tropical and warm areas; the inspiration of the leaf motif in the capitals of Corinthian columns supposedly originated from *A. spinosus* L. (Greek: *acanthos*, thorn, in reference to the spiny leaves and bracts) (subclass Asteridae—Cronquist; order Lamiales—APG IV)

FAMILY RECOGNITION IN THE FIELD: herbs, 2 shrubs, or 1 vine with opposite, simple, entire or indistinctly toothed leaves; flowers often showy, with sympetalous, 2-lipped and bilaterally symmetrical or nearly radially symmetrical flowers in often bracteate inflorescences; stamens 2 or 4; fruit a capsule usually with seeds on small hook-like structures. In the somewhat similar Lamiaceae, the fruits are 4 one-seeded nutlets, and in the sometimes quite similar Scrophulariaceae, the seeds are not borne on hooks and are usually more than 12.

REFERENCES: Wasshausen 1966, 1998; Long 1970; Daniel 1984a, 2000; Scotland et al. 1995; Manktelow 2000; McDade et al. 2000a, 2000b, 2005; Scotland & Vollesen 2000; Wasshausen & Wood 2004; Wunderlin et al. 2019.

1. Plant a woody shrub to 1 m or more tall; found in tidal flats; leaves leathery; corollas white 4–8 mm long ____ **Avicennia**
1. Plants without the combination of characters above.
   2. Plant a trailing or climbing vine; corollas yellow to orange, cream, or white, with a conspicuous dark purple center; capsules with a beak ca. 1 cm long; seeds not supported by small hook-like retinacula ____ **Thunbergia**
   2. Plant not a vine, either a ± erect herb or small shrub; corollas various, but not as above; capsules not beaked; seeds each supported by small hook-like retinacula.
      3. Corollas nearly radially symmetrical to slightly bilaterally symmetrical, 15–70 mm long; lobes ± equal; stamens 4 or 2.
         4. Corollas reddish to orange or white, lobes 4; stamens 2.
            5. Corollas orange-red; lobes narrowly oblong; flowers borne in long loose spikes; bracts minute, lanceolate, 2–5 mm long; a shrub ca. 1.5 m tall ____ **Anisacanthus**
            5. Corollas usually white (can be pale blue or pink), lobes ovate to broadly oblong; flowers borne in compact strobilus-like spikes; bracts broadly ovate, 15 mm long; herbaceous or subshrubby perennial to 0.3–0.6(–1) m tall ____ **Yeatesia**
         4. Corollas white to blue to purplish or pink with 5 lobes; stamens 4 (+ 1 staminode in *Stenandrium*).
            6. Plants acaulescent; flowers borne in a pedunculate head-like or elongate spike-like inflorescence; anthers 1-celled ____ **Stenandrium**
            6. Plants caulescent; flowers axillary or terminal; anthers 2-celled.
               7. Calyx about 3/4 as long as corollas; anther sacs basally awned or mucronate; leaf blades linear to oblanceolate; corolla tube 7 mm or less long ____ **Dyschoriste**
               7. Calyx < 1/2 as long as corollas; anther sacs basally rounded; leaf blades usually much broader than linear to oblanceolate (except in *R. simplex* and *R. dipteracanthus* with linear to lanceolate leaf blades); corolla tube usually much > 7 mm long ____ **Ruellia**

3. Corollas distinctly bilaterally symmetrical, 6–30 mm long; lobes ± unequal; stamens usually 2 (+ 2 staminodia) (4 only in *Hygrophila lacustris*).
    8. Corolla tube uniformly long, slender, and cylindric, 8.5–19 mm long, 0.9–2 mm wide, at least twice as long as calyx; plants woody at base, plants of dry upland habitats in sw portion of East TX _______ **Justicia** (previously *Siphonoglossa*)
    8. Corolla tube various, usually wider at top, longer or shorter than calyx, 1.5–10 mm long; plants woody or herbaceous at base, plants of dry to aquatic habitats in various portions of East TX.
        9. Plants woody at base from a stout root; plants of dry upland habitats; only in extreme sw part of East TX _______ **Carlowrightia**
        9. Plants mostly herbaceous with fibrous roots; plants of mesic to wet, to aquatic habitats; mostly distributed in various parts of East TX.
            10. Plants of mesic habitats; corollas usually white, lobes subequal _______ **Yeatesia**
            10. Plants of wet to moist or aquatic habitats; corollas white to light blue to purplish or pinkish, often with contrasting markings; lobes unequal.
                11. Corollas 9 mm or less long, the lobes shorter than the tube; stamens 4 or 2 (+ 2 staminodia) _______ **Hygrophila**
                11. Corollas 9–30 mm long, the lobes longer or shorter than the tube; stamens 2.
                    12. Leaves usually long-petioled (petioles (0.5–)2–7 cm or more long); flowers in small axillary clusters, subtended by 2–4 obovate bracts; peduncles < 4 cm long, usually shorter than or equal to the petioles of subtending leaves; corollas with lower lip unlobed _______ **Dicliptera**
                    12. Leaves usually sessile or with shorter petioles; flowers in a slender, 1-sided spike, a dense head-like spike, or in axillary, densely clustered 1–2-flowered inflorescences subtended by small, triangular to linear to subulate bracts or bractioles; peduncles elongate, usually > 4 cm long or much shorter (< 4 mm); corollas with lower lip 3-lobed _______ **Justicia**

ANISACANTHUS Nees DESERT-HONEYSUCKLE, ANISACANTH

A genus of 9 species native to the sw United States, Mexico, and Central America (Mabberley 2017); some are cultivated as ornamentals. It is possibly related to *Carlowrightia* (Daniel 1982; Durkee 1999), a relationship that has been confirmed by molecular evidence (McDade et al. 2000a, 2000b). (Greek: *anisa*, unequal, and *Acanthus*, from *akanthos*, a thorn)

REFERENCES: Hagen 1941; Daniel 1982; Daniel & Henrickson 1982; Henrickson 1986.

Anisacanthus quadrifidus var. **wrightii** (Torr.) Henrickson, (sp.: four-cut; var.: for Charles Wright, 1811–1885, TX collector), FLAME-ACANTHUS, HUMMINGBIRD-BUSH, WRIGHT'S ANISACANTHUS, WRIGHT'S DESERT-HONEYSUCKLE. Shrub to ca. 1.5 m tall; leaf blades to 5 cm long and 2 cm wide; petioles to 1 cm long; flowers single or in pairs in terminal, 1-sided, spike-like inflorescences; calyx ca. 5 mm long; corollas reddish to orange, 3–4 cm long, slightly bilaterally symmetrical, 2-lipped, the upper lip entire or nearly so, the lower lip 3-lobed; corolla tube longer than lobes, ± straight, slender, scarcely dilated at throat; stamens 2; capsules ca. 15 mm long; seeds 2(–4). Used in landscapes and sometimes escaping; Bexar (Turner et al. 2003) and Travis (*Tharp 266* SMU) cos.; adventive in Dallas Co. but not well-established (E. McWilliams, pers. comm.); also Mexico. Jun–Aug. [*A. wrightii* (Torr.) A. Gray] Native of Edwards Plateau, South TX Plains, and n Mexico. This species is attractive to hummingbirds and is now widely cultivated as an ornamental TX native plant; it is winter hardy at least as far north as Grayson Co. (on the OK border). The other three varieties of this species are native to Mexico (Henrickson 1986).

AVICENNIA L. BLACK MANGROVE

A genus of about 8 species (Wunderlin et al. 2019) of the tropics and subtropics. (named for Abu Ali Sina, known in the West as Avicenna, c. 980–1037, a Persian physician, philosopher, and naturalist).

REFERENCE: Wunderlin et al. 2019.

Avicennia germinans (L.) L., (to sprout), BLACK MANGROVE, MANGLE BLANCO. Tree to 20 m in the tropics (a woody stemmed shrub to 1.5 m in our area); leaves opposite, petioles 2–27 mm long, leaf blades to 15 cm long and 4.4 cm wide, oblong or lanceolate to elliptic or obovate, entire, usually grayish-mealy beneath; flowering spikes in axillary or terminal cymes, to 65 mm long and 15 mm wide; calyx 5-lobed, 3–5 mm long, densely appressed-pubescent outside, glabrous inside; corollas white, 5-lobed, 4–8 mm long with yellow center parted to ca. the middle, the tube equaling or shorter than the calyx, the lobes spreading, densely pubescent outside and velvety inside; stamens 4; fruit a dehiscent capsule, pear-shaped or ovate, asymmetric, 1.2–3.2 cm long, 1–2.5 cm in diam., densely gray-pubescent. Pneumatophores (= erect exposed roots) involved in gas exchange present. In lagoons and tidal flats; in East TX known only from Jefferson Co.; coastal U.S. from FL to TX; West Indies; Central America; Africa; Asia; Australia; and Pacific Islands. Summer to fall. [*A. nitida* Jacq.] Native of Africa and Asia.

CARLOWRIGHTIA A. Gray WRIGHTWORT, CARLOWRIGHTIA

Small, suffrutescent (= woody only at base), much-branched perennials; stems straggling-decumbent to ascending; leaves simple, entire; flowers borne all along the stems or in terminal spikes; calyx 5-parted; corollas 4-parted; stamens 2; capsules explosively dehiscent.

A genus of 26 species (8 native to TX) of warm to arid areas from the sw U.S. to Costa Rica (Daniel 2003; Mabberley 2017). It is most diverse in Mexico, with 24 of the species occurring in that country. The key below is modified from Henrickson and Daniel (1979). The corollas of some species superficially resemble those of papilionaceous legumes, with a keel-like lower petal and a banner-like upper petal. (Named for Charles Wright, 1811–1885, botanical collector mainly in TX, Cuba, and his native Connecticut; correspondent of Asa Gray; *Carlo* from Spanish for Charles, Carlos and Wright's surname)
REFERENCES: Henrickson & Daniel 1979; Daniel 1980, 1983, 1984b, 1988, 2003.

1. Corollas white with maroon nerves on each petal lobe; flowers usually borne along entire length of stem; stem pubescence of mainly short (0.1–0.5(–1) mm long), stout, decurved hairs; seeds with entire margins ____ **C. texana**
1. Corollas white except for the upper petal lobe which has a yellow eye surrounded by a maroon ring; flowers mostly in terminal spikes; stem pubescence of both short and long, slender, erect hairs 0.1–1(–1.7) mm long; seeds with crenulate to denticulate margins, the short teeth with minute recurved barbs ____ **C. torreyana**

Carlowrightia texana Henrickson & T.F. Daniel, (of Texas), TEXAS WRIGHTWORT, TEXAS CARLOWRIGHTIA. Strigulose perennial (5–)10–30 cm tall from a much-branched twiggy base; leaves petiolate; leaf blades (2.5–)6–16(–42) mm long, oblong-ovate, broadly ovate to orbicular, entire; calyx 3–5 mm long; corollas 6–7 mm long, white with maroon nerves on lobes; capsules 7.5–11 mm long, glabrous, 4-seeded. Rocky slopes, often disturbed areas; described as a new species by Henrickson and Daniel in 1979; Bexar and Comal cos. (Turner et al. 2003) also Wilson Co. (*Carr & Candee 20969-B* TEX) at the sw corner of East TX; mainly South TX Plains, Edwards Plateau, and Trans-Pecos; AZ, FL, NM, and TX; Mexico and Central America Apr–Nov. Native of the U.S. and Mexico.

Carlowrightia torreyana Wassh., (for John Torrey, 1796–1873, botanist and physician, co-author with Asa Gray, described many w American plants), TORREY'S WRIGHTWORT. Perennial 10–50 cm tall, similar to *C. texana* except where noted in key and below; leaf blades (6–)10–20(–40) mm long, broadly ovate to orbicular reniform; corollas 7–9 mm long, white except for the upper petal lobe which has a yellow eye surrounded by a maroon ring. Rocky slopes and ledges, along creeks, and in plains and chaparral; Bexar (*Lindheimer 7432* SMU) and Comal (Kartesz 2022) cos. at sw margin of East TX; mainly Edwards Plateau; in the U.S. known only from TX. Usually May–Sep. Native of TX and Mexico.

DICLIPTERA Juss. FOLDWING

A genus of ca. 150 species of tropical and warm areas (Mabberley 2017); some are cultivated as ornamentals. (Greek: *diklis*, double-folded, as of doors, etc., and *pteron*, a wing, alluding either to the 2-winged fruit or possibly the involucre)

Dicliptera brachiata Spreng., (branched at right angles), FALSE MINT, BRANCHED FOLDWING. Perennial herb to 0.7(–1) m tall; leaf blades to 12(–15) cm long, (1.5–)2–7 cm wide; petioles (0.5–)2–7 cm or more long; flowers in axillary, short-peduncled to subsessile clusters, the clusters subtended by an involucre of 2–4 floral bracts; bracts obovate, to 7 mm long, 2–5 mm wide; corollas purplish or pinkish (rarely white), zygomorphic, 15–20 mm long including tube, the upper lip entire or shallowly 2–3-lobed, the lower lip usually entire; capsules ovoid, 5–6 mm long, stipitate, apically emarginate with 2–4 seeds; $n = 40$ (Daniel 2000). Moist wooded stream bottoms, wet habitats; throughout much of East TX except for Pineywoods; East Cross Timbers, e Edwards Plateau, and Gulf Prairies and Marshes; se U.S. from VA s to FL w to KS and TX. Jun–Nov. Native of the U.S.

DYSCHORISTE Nees SNAKEHERB, DYSCHORISTE

A genus of ca. 75 species (Wunderlin et al. 2019) of tropical and warm areas. (Greek: *dys*, hard, and *chorist*, separate or asunder, referring to the coherent, hard to separate capsule valves)
REFERENCES: Kobuski 1928; Wasshausen 1998; Henrickson 1999a; Wasshausen & Wood 2003.

Dyschoriste linearis (Torr. & A. Gray) Kuntze var. **linearis,** (linear, narrow with sides nearly parallel), NARROW-LEAF SNAKEHERB, NARROW-LEAF DYSCHORISTE, SNAKEHERB, POLKADOTS. Rhizomatous, ± pubescent, herbaceous perennial; stems erect or partly decumbent, usually 7–30 cm long; leaves sessile, linear to oblanceolate, obtuse; flowers axillary, sessile, shorter than the leaves, with foliaceous bracts; calyx united about 1/4 to nearly 1/2 its length, the lobes 9–13 mm long, long acuminate; corollas lavender, with dark dots in throat, nearly radially symmetrical to slightly zygomorphic, 15–27 mm long; capsules with 2–4 seeds. Rocky or sandy ground; w, and particularly sw part of East TX; throughout much of TX, except for the Panhandle; OK and TX. Apr–Jul, sporadically to Oct. Native of the U.S. and Mexico. The variety occurring in East TX is var. *linearis*. Henrickson (1999a) distinguished var. *sanpatriciensis* Henrickson based mostly on its dense pubescence as oppose to scant pubescence of leaves, stems, inflorescences, and sepals in var. *linearis*, known from Kleberg, San Patricio, Terrell, and Val Verde cos. and n Coahuila, Mexico. However, Turner et al. (2003) included this variety within var. *linearis*.

HYGROPHILA R. Br. SWAMPWEED

Perennial, wet area, submersed to terrestrial herbs; stems decumbent to ascending, rooting at the nodes; leaves opposite, sessile or subsessile, entire, with cystoliths; flowers sessile or nearly so, in axillary clusters or terminal spikes; calyx with 5 segments; corollas with tube slightly longer than limb, zygomorphic, 5-lobed, these contorted in bud; stamens 4 (2 can be staminodial); capsule oblong, with 12–18 seeds, jaculators present.

A mainly tropical genus of both the Old and New worlds with ca. 80 wet area species (Wunderlin et al. 2019); some are cultivated as aquatic ornamentals in aquaria, while others can be weeds in rice fields. The seeds of some are sticky and are dispersed by attaching to the feet of birds (Mabberley 2017). *Hygrophila difformis* (L. f.) Blume (WATER-WISTERIA), native to se Asia and a weed of rice fields, was reported from Florida (Wunderlin et al. 2002). The following key is modified from Angerstein and Lemke (1994) and Wasshausen (1998). (Greek: *hygros*, moist, and *philos*, loving, because of their habitat)

REFERENCES: Angerstein & Lemke 1994; Wasshausen 1998; Wunderlin et al. 2002; Daniel 2013; Wunderlin et al. 2019.

1. Plants terrestrial or emergent; leaf blades 5–12 cm long, 0.5–2.5 cm wide; flowers in axillary clusters along length of stem; stamens 4 ____ **H. lacustris**
1. Plants submersed, emergent, or rarely terrestrial; leaf blades 2–3.5 cm long, 0.4–1 cm wide; flowers in terminal spikes; stamens 2, staminodia 2 ____ **H. polysperma**

Hygrophila lacustris (Schltdl. & Cham.) Nees, (living in ponds or lakes), GULF SWAMPWEED. Weak stemmed, caulescent, perennial herb; stems ± square in cross section, decumbent below, rooting at the nodes, the branches ascending, to ca. 80 cm long; leaves nearly sessile, leaf blades narrowly elliptic to oblong; flowers in sessile axillary clusters along length of stem; corollas white, zygomorphic, the tube ca. 5 mm long, the limb ca. 2.5 mm long; capsule oblong, 8–12 mm long, glabrous; seeds 12–18. Wet areas; mostly in the se portion of East TX; also Gulf Prairies and Marshes; se U.S. from FL and GA w to TX; also West Indies. Jun–Aug. Native of the U.S. Some authors (Wunderlin et al. 2019; Daniel 2013) identify U.S. material of this entity as *H. costata* Nees & T. Nees, a widespread taxon of tropical America. We, however, are following Wasshausen (1998) and Weakly (2024) and assigning U.S. material to *H. lacustris*.

Hygrophila polysperma (Roxb.) T. Anderson, (many-seeded), EAST INDIAN SWAMPWEED, INDIAN SWAMPWEED, EAST INDIAN HYGROPHILA, MIRAMAR WEED. Much-branched, annual herb; stems 0.1–1.5 m long (terrestrial individuals usually 20 cm or less tall; submersed individuals with longer stems); leaves nearly sessile, blade oblong-lanceolate to ovate, the bases slightly connate; flowers in terminal spikes (may appear to be axillary at ends of stems); calyx parted to the middle, the 5 segments linear, pubescent, scarious-ciliate marginally; corollas white, (drying yellowish), the tube ca. 5 mm long, the limb ca. 3 mm long; stamens 2, fertile, 2 staminodial; capsule oblong, 6–7 mm long, pilose distally; seeds more than 15. In rivers and streams; Comal and San Marcos rivers in Comal and Hayes cos., Bexar, Kendall, and Travis cos. (Angerstein & Lemke 1994; Turner et al. 2003; Williams 2019); AL, FL, KY, MS, SC, TX, and VA. Mostly Sep–Oct. [*Justicia polysperma* Roxb.] Native of India and Malaysia. This species was introduced to the U.S. in the 1940s and was first reported from TX in 1994 (Angerstein & Lemke 1994); it was apparently introduced by the aquarium plant industry. This plant is able to easily reproduce vegetatively, with only small fragments growing into new individuals. It has high growth potential, and possibly poses a threat to aquatic ecosystems (Angerstein & Lemke 1994). It is a U.S. federal noxious weed (USDA Natural Resources Conservation Service 2006).

JUSTICIA L. WATER-WILLOW

Ours perennial herbs; leaves simple, entire; flowers in axillary pedunculate spikes or solitary in leaf axils or, on short axillary peduncles 4 mm or less long; calyx lobes 5; corollas 2-lipped, the lower lip 3-lobed; stamens 2; capsules 4-seeded, the 2 valves reflexed at maturity.

A genus of ca. 700 species of herbs and shrubs of tropical and warm areas of the world and temperate North America (Daniel 2011); it is the largest genus in the family (Wasshausen 1998). As treated here, the genus contains a number of species previously treated in *Siphonoglossa* (see Hilsenbeck 1989c, 1990a, 1990b). Based on molecular evidence, within the family, *Justicia* is in a large lineage including *Dicliptera*, *Carlowrightia*, and *Anisacanthus* (McDade et al. 2000a, 2000b). Some species such as *J. brandegeeana* Wassh. & L.B. Sm., SHRIMP-PLANT, (formerly treated in *Beloperone*), are cultivated as ornamentals,

Anisacanthus quadrifidus var. wrightii [HEA]

Avicennia germinans [CO1]

Carlowrightia texana [POW]

Carlowrightia torreyana [LUN]

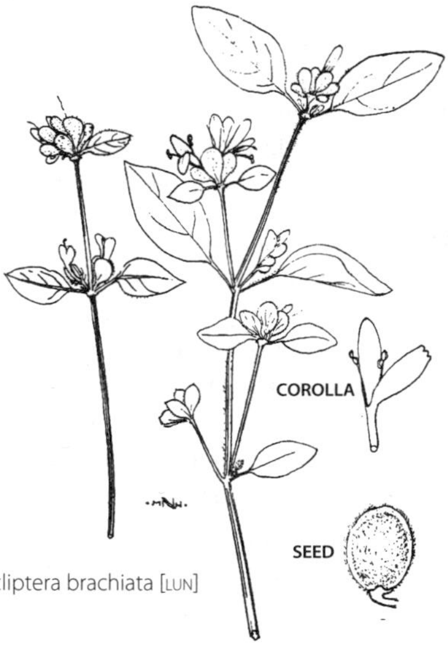

Dicliptera brachiata [LUN]

Dyschoriste linearis var. linearis [LUN]

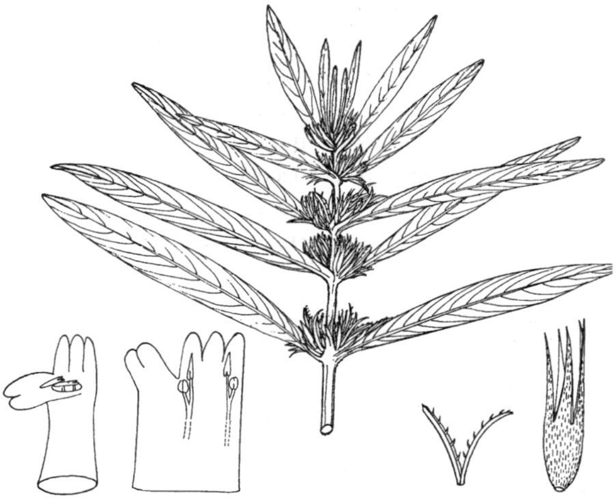

Hygrophila lacustris [LUN]

particularly for their visually striking bracts and flowers. The leaves of a South American species are used as a hallucinogenic snuff by the Yanomamó Indians (Brewer-Carias & Steyermark 1976). (Named for James Justice, 18th century Scottish botanist)
REFERENCES: Graham 1988; Hilsenbeck 1989a, 1989b, 1989c, 1990a, 1990b; Daniel 2011.

1. Bracts conspicuous to 2 cm long, overlapping, arranged in 4 ranks, red to reddish brown ____ **J. brandegeeana**
1. Bracts much smaller, not overlapping, if overlapping then not arranged in ranks, green.
   2. Flowers 2–4 per node, sessile to subsessile; peduncles 4 mm or less long; corollas 12–30 mm long; plants of dry to mesic habitats.
      3. Plants of dry habitats; corolla tube 8.5–19 mm long, cylindric; bracteoles oblanceolate to obovate, 4–16 mm long; calyx lobes not hyaline; calyx subtended only by 2 foliaceous bracteoles; on sw border of East TX ____ **J. pilosella**
      3. Plants of dry to mesic habitats; corolla tube 5–10 mm long, widened at the top; bracteoles linear, 4–19 mm long; calyx subtended by numerous foliaceous bracts, calyx lobes hyaline; in East TX known only from Goliad Co. ____ **J. pacifica**
   2. Flowers in obvious pedunculate spikes; peduncles 6–15 cm long; corollas 9–12 mm long; plants of wet habitats.
      4. Flower spikes dense head-like becoming oblong at maturity; lower lip of corolla arched-recurving; seeds densely warty; plants mostly in w part of East TX ____ **J. americana**
      4. Flower spikes loosely-flowered, elongate, usually spaced along 1 side of the axis of the inflorescence; lower lip of corolla ± flat; seeds smooth; plants mostly in e 2/3 of East TX ____ **J. lanceolata**

Justicia americana (L.) Vahl, (of America), AMERICAN WATER-WILLOW. Plant glabrous, colonial, rhizomatous, producing long leafy shoots late in summer; stems erect, to ca. 1 m tall; leaves sessile or nearly so; leaf blades narrowly oblong-lanceolate to elliptic-lanceolate; flowered portion of spikes to 3 cm long, peduncles to 15 cm long; sepals separate nearly to base, linear-lanceolate, ca. 7 mm long; corollas 10–12 mm long, white or violet with red-purple dots; capsules ca. 12 mm long. Wet ground, margins of streams and ponds; w part of East TX w to mainly Edwards Plateau and Cross Timbers and Prairies; widespread in the e 1/2 of the U.S. from NY s to FL w to NE and TX; se Canada. May–Jun. Native of the U.S. and Canada. This species can form dense colonies, particularly on gravel bars and along stream banks.

Justicia brandegeeana Wassh. & L.B. Sm., (named in honor of T.S. Brandegee, 1843–1925, American botanist), SHRIMP-PLANT. Basally woody perennial herb; stem slightly 4-angled to 1.5 m tall; leaves petiolate, 1–3 cm long, blade 2.5–9 cm long and 2–5 cm wide, ovate or lanceolate-ovate, entire; inflorescence a dense terminal or axillary spike to 10 cm long, bracts broadly ovate, 1–2 cm long, apex obtuse or subacute, base cordate, red to reddish brown, overlapping in 4 ranks; calyx 5-lobed, 6–8 mm long, narrowly lanceolate; corollas white or lavender, tube 1.5–2.5 cm long, upper lip smaller than the lower, subentire, lower lip shallowly 3-lobed, purple spotted on the inner surface; capsule club-shaped, 10–12 mm long, seeds 2–4, ca. 3 mm long, smooth. Disturbed sites, escape from cultivation; known only from Victoria Co. in TX (Singhurst & Holmes 2012); FL and TX; West Indies; Mexico; Central and South America; and Australia. All year. Native of Mexico.

Justicia lanceolata (Chapm.) Small (lanceolate, lance-shaped, referring to the narrow leaves), LANCE-LEAF WATER-WILLOW. Stems erect or spreading, 10–30 cm tall; leaves sessile or nearly so; leaf blades linear to elliptic-lanceolate, to 10 cm long and 3 cm wide, minutely puberulent; spikes axillary, flowered portion to 11.5 cm long, peduncle to 6 cm long, sometimes branched; corollas ca. 9–10 mm long, lavender with darker markings; capsules 10–15 mm long. Wet areas; widespread in e East TX; also Callahan Co. (Stanford 1976) and Tarrant Co. (Turner et al. 2003) further w; mostly se U.S. from IL s to FL and w to OK and TX. Mar–Jun. [*Dianthera lanceolata* (Chapm.) Small, *J. ovata* var. *lanceolata* (Chapm.) R.W. Long] Native of the U.S. Wasshausen (1998) distinguished the western members of this species (including TX populations) as var. *lanceolata* citing differences in blade shape (elliptic-ovate to rhombic-oblong in var. *ovata* and linear to linear-elliptic or elliptic-lanceolate in var. *lanceolata*) and differences in the inflorescences (congested becoming almost head-like in var. *ovata* and more or less secund in var. *lanceolata*). However, Daniel (2011) treated them as species, noting that the differences in them "appear to be as significant as those between other species of the genus in the eastern United States." We are, therefore, following Daniel (2011).

Justicia pacifica (Oerst.) Hemsl., (of the Pacific Ocean, from the collection of the original type specimen, *Adhatoda pacifica* Oerst., in Oaxaca, Mexico), RUNYON'S WATER-WILLOW. Erect to reclining perennial herbs becoming basally woody with age, often with 2 or more stems, spreading to prostrate, to 1 m long; leaves petiolate, 4(–8) mm long, blades ovate to narrowly elliptic to linear, to 12 cm long; flowers in upper leaf axils, congested, on short peduncles to 2 mm long, subtended by numerous foliaceous, ciliate-margined bracts; usually 2–4 per node, bracteole immediately subtending calyx linear to lance-linear, 5–10 mm long; calyx ca. 10 mm long, calyx lobes lance-linear, 3.3–8 mm long, margins hyaline (light yellow), central portion green; corollas pinkish to violet, 12–23 mm long, tube 5–10 mm long, widened at the top, upper lip 2-lobed, lower lip 3-lobed, lobes

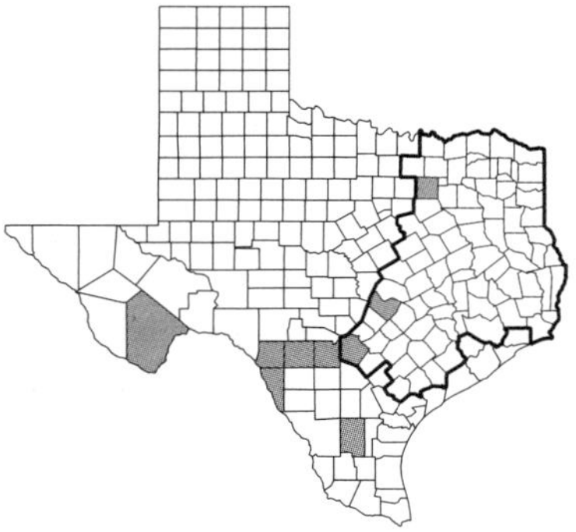

Anisacanthus quadrifidus var. wrightii

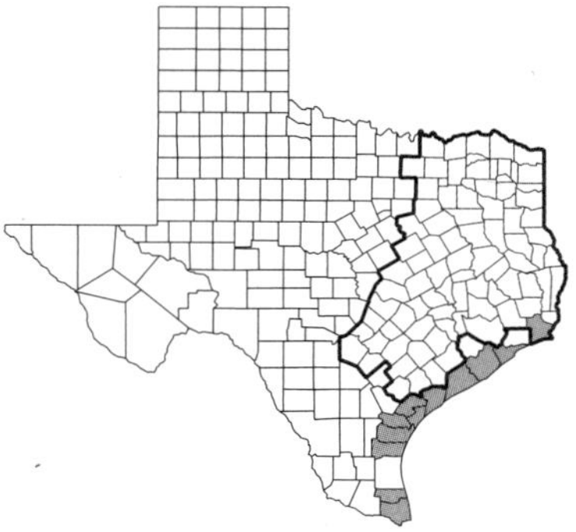

Avicennia germinans

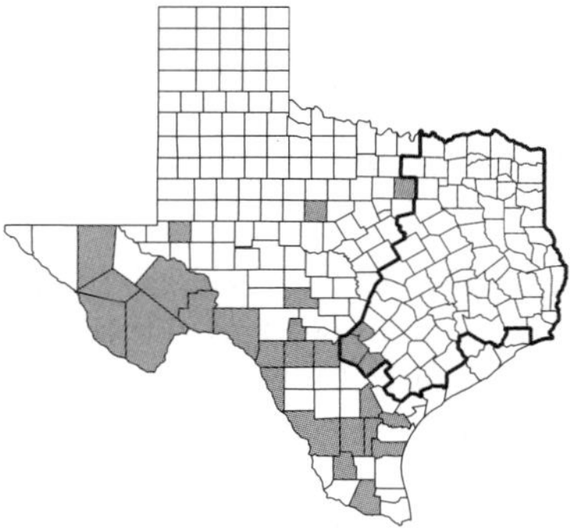

Carlowrightia texana

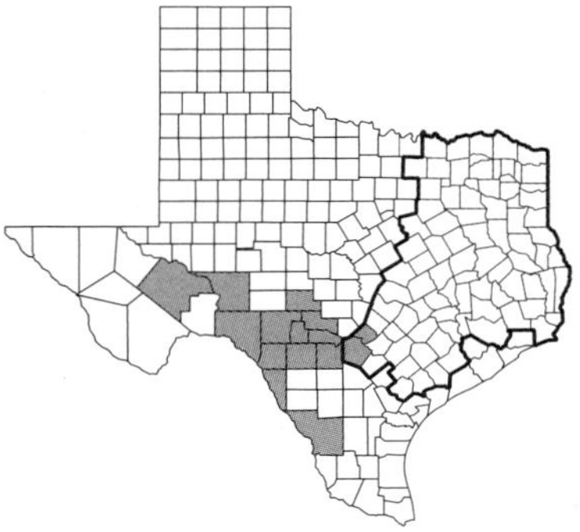

Carlowrightia torreyana

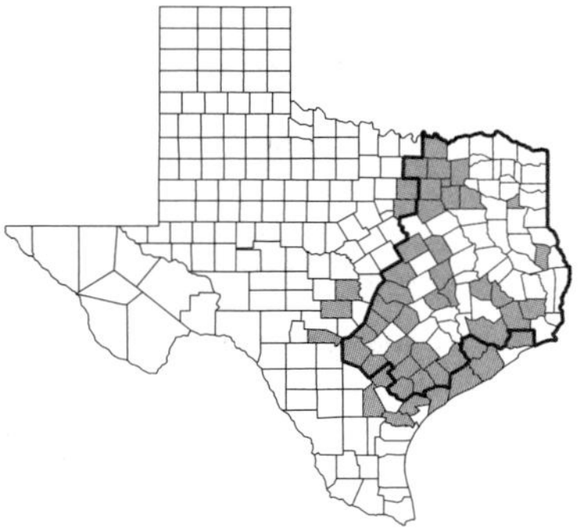

Dicliptera brachiata

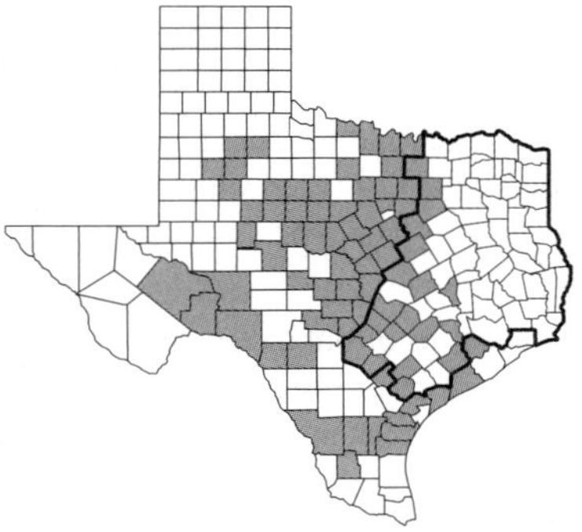

Dyschoriste linearis var. linearis

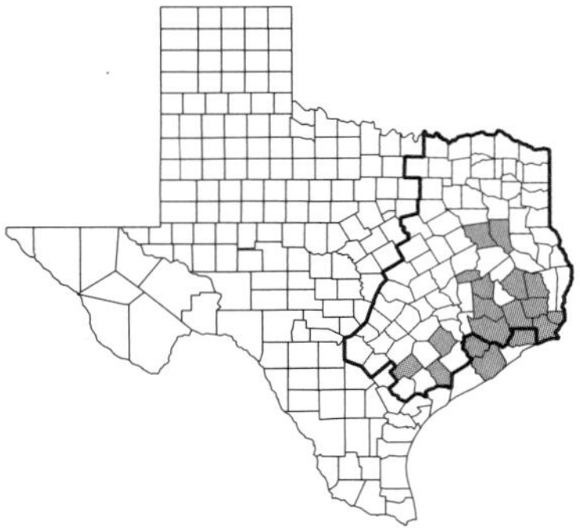

Hygrophila lacustris

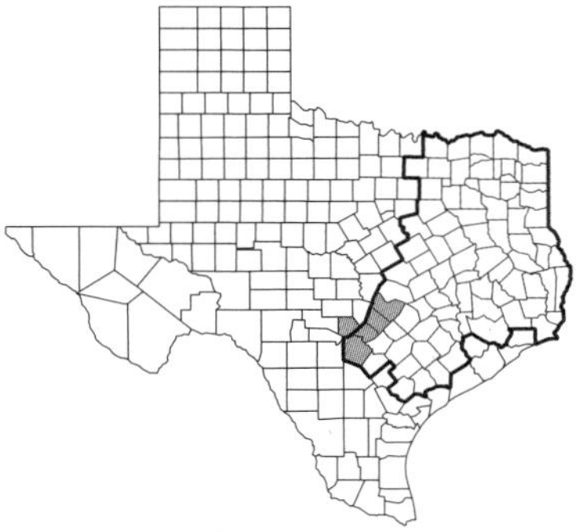

Hygrophila polysperma

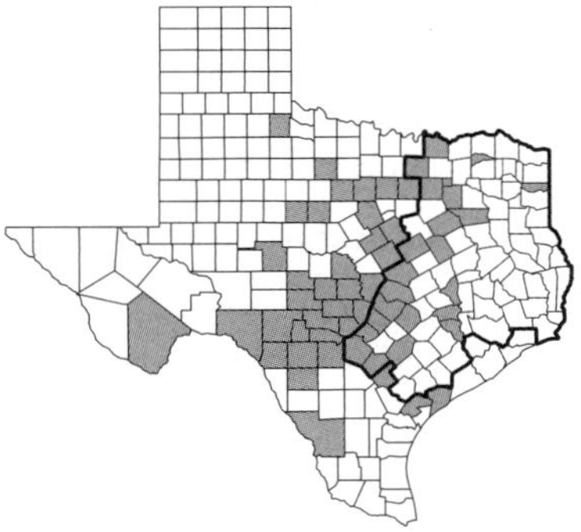

Justicia americana

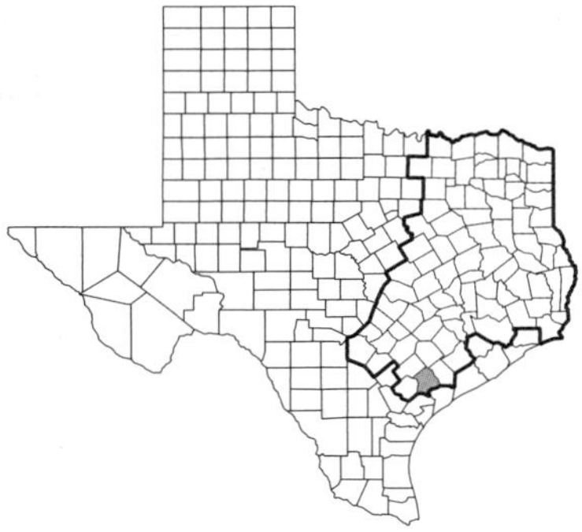

Justicia brandegeeana

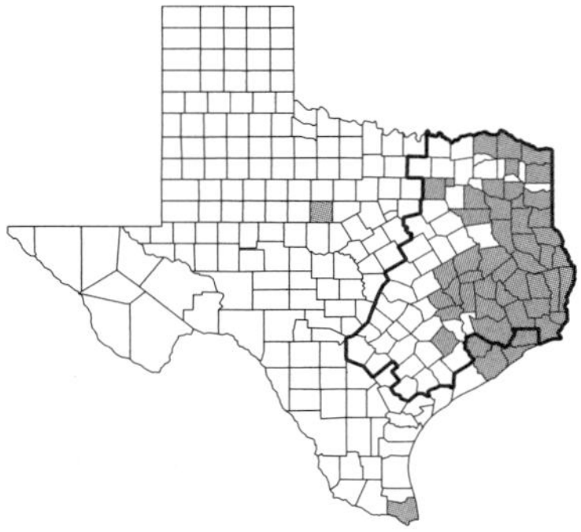

Justicia lanceolata

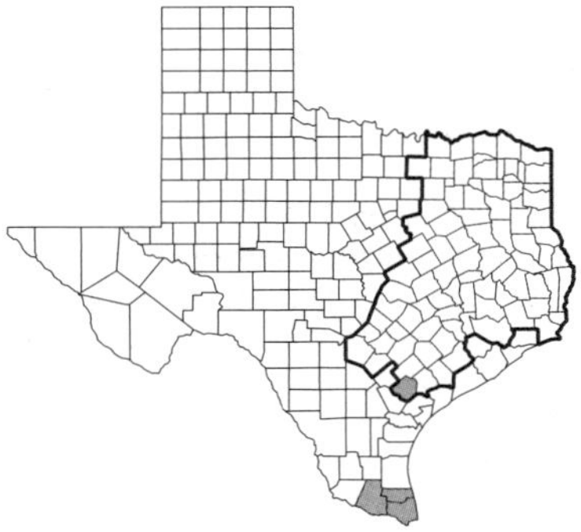

Justicia pacifica

rounded with white markings; capsules ca. 10–15 mm long; seeds smooth. Thorn scrub and evergreen woodland; in East TX known only from Goliad Co. (Turner et al. 2003), also extreme s TX; Mexico. In the U.S. known only from TX. Jul–Dec. [*J. caudata* A. Gray, *J. furcata* Jacq., [nom. illegit.], *J. runyonii* Small] Native of TX and Mexico. Included based on three consecutive collections (*E. Owens 3149*, *3150*, and *3151* (all TAES)) in Goliad Co. at the extreme s boundary of East TX. These collections are from 1940 and no other collections are known.

Justicia pilosella (Nees) Hilsenb., (little pilose—covered with long soft hairs), HAIRY TUBETONGUE, TUBETONGUE, FALSE HONEYSUCKLE, GREGG'S TUBETONGUE. Low suffrutescent perennial; leaves subsessile; leaf blades ovate to oval, to 4 cm long; flowers in upper leaf axils, usually 1–2 per node on peduncles to 4 mm long; floral bracts none; bracteoles subtending the calyx oblanceolate or obovate, 4–16 mm long; calyx 3.5–9 mm long, calyx lobes subulate to lance-subulate, margins not hyaline; corollas lavender to rose or white, to 30 mm long, the tube slender, cylindric, elongate 8.5–19 mm long, the upper lip entire or 2-lobed, the lower lip 3-lobed, much larger, typically with a white spot and purple dots at base; capsules 5–10 mm long; seeds with minute bumps. Rocky, sandy, or grassy areas; in the sw part of East TX; mostly s 2/3 of TX; NM and TX. Apr–Oct. [*Monechma pilosella* Nees, *Siphonoglossa dipteracantha* (Nees) A. Heller, *Siphonoglossa pilosella* (Nees) Torr.] Native of the U.S. and Mexico. In the past, the species, has long been recognized in the genus *Siphonoglossa* (e.g., Wasshausen 1966; Correll & Johnston 1970; Hatch et al. 1990; Diggs et al. 1999), but work by Hilsenbeck (e.g., 1990b) suggests that the species is better placed in *Justicia*, citing chromosomal and flavonoid chemistry similarities between *Justicia* and *Siphonoglossa* (sect. *Pentaloba*), a view that is followed by Jones et al. (1997) and Turner et al. (2003).

RUELLIA L. WILD PETUNIA

Ours glabrous or pubescent perennial herbs, the base sometimes woody; leaves sessile or short-petiolate; leaf blades narrowly lanceolate to ovate-lanceolate or oblong-elliptic; sepals united a short distance at base, acuminate; ours with corollas usually lavender to bluish purple, often with reddish or purple spots down one side, or white (in *R. metziae*), large and showy, opening in the morning, falling by late afternoon (some species also produce cleistogamous flowers); capsules usually explosively dehiscent.

A genus of ca. 300 species (Wunderlin et al. 2019) found mainly in tropical areas of the world, with a number in temperate North America; some are cultivated as ornamentals. Based on molecular evidence, within the family, *Ruellia* is considered closely related to *Hygrophila* and *Dyschoriste* (McDade et al. 2000b). (Named for Jean Ruelle, 1474–1537, early French herbalist)
REFERENCES: Leonard 1927; Tharp & Barkley 1949; Long 1961, 1966, 1971, 1973, 1974, 1975, 1977; Long & Uttal 1962; Turner 1991a; Daniel 1995, 2005, 2013; Wunderlin et al. 2019.

1. Flowers (chasmogamous) terminating the main stem in panicle-like inflorescence on a long peduncle (lateral, few-flowered inflorescences of exclusively cleistogamous flowers also produced early); corollas white or lavender to bluish purple.
   2. Corollas (of chasmogamous flowers) ca. 3–5(–5.5) cm long, lavender to bluish purple; calyx lobes 7–15 mm long in flower.
      3. Leaves mostly caulescent, 2–3 or more leafy nodes per stem; plants 1.5–10 dm tall; flowers in terminal cymes all chasmogamous; capsules of lateral cleistogamous flowers 15–18 mm long.
         4. Plants 1.5–3.5 dm tall; lower internodes with spreading white hairs 1–3 mm long; capsules of lateral cleistogamous flowers 16–18 mm long ____ **R. corzoi**
         4. Plants 3–7 dm (–10) dm tall; lower internodes short pubescent to almost glabrous; capsules of lateral cleistogamous flowers ca. 15 mm long ____ **R. nudiflora**
      3. Leaves mostly basal; rarely more than two leafy nodes per stem; plants 1–3.5 dm tall; flowers in terminal cymes, cleistogamous mixed with chasmogamous; capsules of lateral cleistogamous flowers 10–14 mm long. ____ **R. yucatana**
   2. Corollas (of chasmogamous flowers) (5–)5.5–7 cm long, white to bluish purple; calyx lobes 14–25 mm long in flower.
      5. Corollas ca. 7 cm long, lavender to bluish purple; calyx lobes 22–25 mm long in flower; leaf blades ovate to broadly ovate to subdeltoid ____ **R. occidentalis**
      5. Corollas (5–)5.5–6.5 cm long, usually white (sometimes pale bluish purple); calyx lobes 14–20 mm long in flower; leaf blades oblong-lanceolate to narrowly oblong-ovate ____ **R. metziae**
1. Flowers axillary, sessile or in ± peduncled glomerules or cymes OR at tips of side branches; corollas usually lavender to bluish purple.
   6. Leaf blades linear to linear-lanceolate to lanceolate, mostly 10–20(–30) cm long and 2–3(–4) cm wide; mostly ≥ 10 times long as wide; flowers borne in lax axillary dichotomous cymes; bracts subulate ____ **R. simplex**
   6. Leaf blades wider; lanceolate, oblong, or elliptic to ovate; mostly > 2 cm wide and < 10 cm long; mostly ≤ 10 times long as wide; flowers not borne as above; bracts foliaceous.

Hygrophila polysperma [SID]

Justicia americana [LUN]

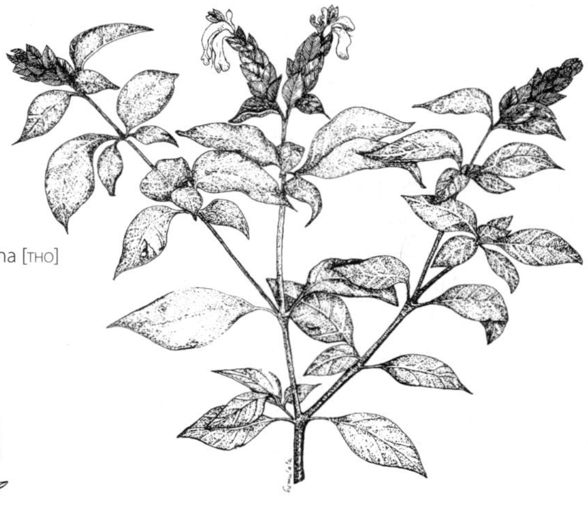

Justicia brandegeeana [THO]

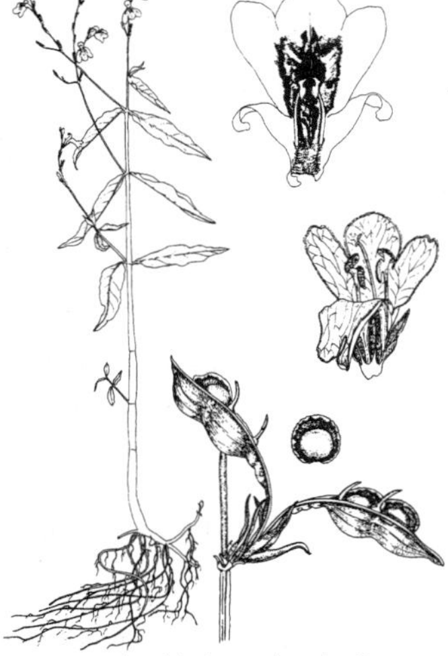

Justicia lanceolata [CO1]

Justicia pacifica [BRT]

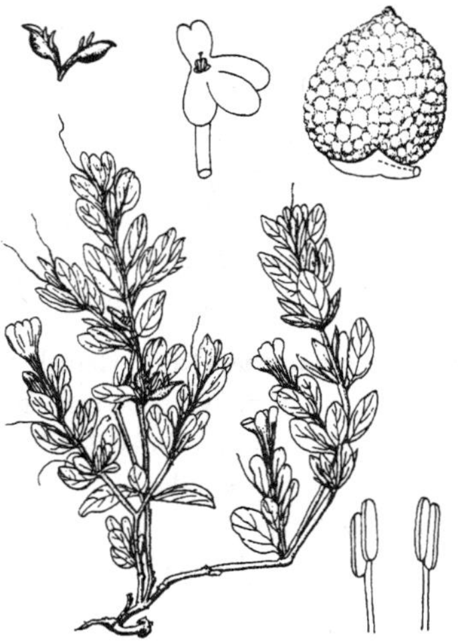

Justicia pilosella [LUN]

7. Flowers (chasmogamous) solitary–3 (or rarely loosely cymose) at ends of usually unbranched peduncles; peduncles 0.2–7 cm long.
 8. Largest leaf blades usually < 30 mm wide; ovary pubescent; calyx lobes narrowly linear, 1–1.2 mm wide, with prolonged linear-acicular tips; peduncles 3–7 cm long). ____ **R. pedunculata**
 8. Largest leaf blades usually > 30 mm wide; ovary glabrous; calyx lobes lanceolate to linear-lanceolate, (1.5–)2–4 mm wide, flat to the tips; peduncle 0.2–3 cm long ____ **R. strepens**
7. Flowers (chasmogamous and cleistogamous) in sessile or subsessile (on very short peduncles) glomerules (or 1–2) from axils of main stem or main leafy branches.
 9. Calyx lobes linear-lanceolate to lanceolate, (1.5–)2–4 mm wide, flat to the tips; flowers mainly cleistogamous ____ **R. strepens**
 9. Calyx lobes narrowly linear, ≤ 1.5 mm wide, the prolonged tips very slender to almost bristle-like; flowers usually chasmogamous, rarely cleistogamous.
  10. Leaves sessile or subsessile (petioles to only 3 mm long), erect-ascending (= staying close to the stem), the leaves near middle of stem with blades usually 5 cm or less long ____ **R. humilis**
  10. Leaves subsessile to distinctly petioled, those on main stem with petioles 2–30 mm long, mostly spreading away from the stem, the leaves near middle of stem with blades usually 5 cm or more long.
   11. Glomerules of flowers few, restricted to the upper nodes; leaf blades linear-lanceolate to lanceolate, usually 20 mm or less wide ____ **R. caroliniensis**
   11. Glomerules of flowers several–many (except 1–2 per axil in *R. dipteracanthsus*), extending well down the stem and branches; leaf blades linear to broadly elliptic or ovate, variable in width, up to 90 mm wide.
    12. Leaf blades relatively narrow, lanceolate to narrowly elliptic to ovate; capsules 10–17 mm long, often glabrous or with long hairs.
     13. Stems erect, simple; capsules 11–17 mm long; plants widespread in e half of East TX ____ **R. caroliniensis**
     13. Stems much branched and spreading; capsules 10–15 mm long; plants known in TX only from Harris Co. ____ **R. dipteracanthus**
    12. Leaf blades relatively wide (to 90 mm), broadly elliptic to ovate; capsules minutely canescent, ca. 11 mm long; species known in East TX primarily from w margin of area ____ **R. drummondiana**

Ruellia caroliniensis (J.F. Gmel.) Steud., (of Carolina), SMALL-FLOWER RUELLIA. Plant to ca. 0.9 m tall, inconspicuously and sparsely pubescent; leaves clearly petiolate; flowers in sessile or very short peduncled glomerules mostly at the upper 1–4 nodes; corollas 2–5 cm long, lavender to lavender-blue. Sandy open woods and open ground; collected by Reverchon in Dallas Co. (Mahler 1988) also McLennan Co. (Hannick et al. 2013) and widespread primarily in the Pineywoods, but also w to Tarrant Co. just w of East TX; widespread in the se portion of the U.S. May–Sep. [*R. caroliniensis* subsp. *ciliosa* (Pursh) R.W. Long, *R. caroliniensis* var. *salicina* Fernald, *R. caroliniensis* var. *semicalva* Fernald, *R. caroliniensis* var. *serrulata* Tharp & F.A. Barkley, *R. ciliosa* Pursh, *R. ciliosa* var. *cinerascens* Fernald] Native of the U.S. We are following Jones et al. (1997), Turner (1991a), and Turner et al. (2003) in not recognizing infraspecific taxa in this species. Turner (1991a) noted that the variation present appears to merely represent "forms of a highly variable *Ruellia caroliniensis*, as amply documented by Long (1971)." Further, Long (1974), who had earlier (1970) recognized a number of infraspecific taxa, noted based on further research that the "value of recognizing infraspecific categories in *R. caroliniensis* appears to be negligible." Long (1971, 1977) indicated that this species is facultatively cleistogamous and can produce both cleistogamous and chasmogamous flowers.

Ruellia corzoi Tharp & F.A. Barkley, (named for Antonio Hernandez Corzo, of the *Instituto Politecnico de Mexico*) CORZO'S WILD PETUNIA. Stems 1.5–3.5 dm; leaf blades 4–6 cm long and 1.5–2.5 cm wide, ovate to obovate; both chasmogamous and cleistogamous flowers present; chasmogamous flowers with calyx lobes 11–14 mm long; corollas ca. 3 cm long, tube 8–9 mm long, throat 11–13 mm long; capsules of cleistogamous flowers 16–18 mm long, 3.5 mm wide. Scrub woodland, caliche soil, rocky open soil, mesic creek sides; in East TX recorded only in Bexar Co., included based on exsiccatae in Wasshausen (1966); South TX Plains, Edwards Plateau, and Trans-Pecos; in the U.S. known only from TX; Mexico. Apr–Nov. Native of the U.S. and Mexico.

Ruellia dipteracanthus (Nees) Hemsl., (from the Greek *di*, two and *pteron*, wing, and *acantha*, spine in reference to the capsule), SHRUBBY WILD PETUNIA. Much branched and spreading perennial herb to 3–4 or more dm long; leaf blades narrowly elliptic-lanceolate, 3–6 cm long and 0.7–1.5 cm wide; flowers solitary or 2 on a very short peduncle in axils; calyx lobes linear-filiform 10–15 mm long; corollas blue-purple 3–4 cm long; capsules 10–15 mm long, glabrous. Disturbed sites, low wet areas; in TX known only in Harris Co. included based on Keith et al. (2017); FL and TX; Mexico. All year. Native of Mexico. Frequently cultivated nearly worldwide and likely escaped elsewhere.

Ruellia drummondiana (Nees) A. Gray, (for its discoverer, Thomas Drummond, 1780–1835, Scottish botanist and collector in North America), DRUMMOND'S RUELLIA, DRUMMOND'S WILD PETUNIA. Plant to ca. 0.75(–1) m tall, densely pubescent with short, spreading to appressed hairs; flowers in short-peduncled, axillary glomerules; corollas ca. 3–4 cm long, lavender. Wooded, riparian, often rocky areas, typically in calcareous soils; w margin of East TX, Bexar and Gonzales cos. (Turner 1991a) north to Dallas Co. (BRIT) and adjacent portions of Edwards Plateau and Cross Timbers and Prairies; endemic to TX. Aug–Sep.

Ruellia humilis Nutt., (dwarf, low-growing), PRAIRIE-PETUNIA, LOW RUELLIA, FRINGE-LEAF RUELLIA, FRINGE-LEAF WILD PETUNIA. Plant to ca. 0.6(–0.8) m tall, usually rather densely short-pilose; leaves sessile or nearly so; leaf blades lanceolate to ovate or broadly elliptic; flowers very few in axils of middle and upper leaves, subsessile; corollas 3–4.5 cm long, lavender to light blue. Prairies and open woods; widespread throughout East TX; also a few scattered locations further w in the state to the Panhandle and Edwards Plateau and s to the n part of the Gulf Prairies and Marshes; ec U.S. from OH s to AL and w to NE and TX. May–Sep. Native of the U.S. While this species is morphologically quite variable, Turner (1991a) pointed out that it is relatively easy to recognize the species based on "its glabrous capsules and nearly sessile leaves." Tharp and Barkley (1949), Wasshausen (1966), and Correll and Johnston (1970) treated the species as having five varieties. Turner (1991a), however, noted that a number of these taxa "appear to be but forms, sometimes populational, of a widespread highly variable var. *humilis*" and Kartesz (2022) does not recognize any infraspecific taxa. Weakley, A.S., and Southeastern Flora Team (2024) states, "This species badly needs taxonomic re-evaluation, looking at variation across its wide and fragmented distribution." We are following Turner (1991a), Jones et al. (1997), and Turner et al. (2003) in recognizing only two varieties, separated as follows by Turner (1991a).

1. Plants mostly 5–20(–30) cm high, the primary stems mostly depauperate, often ascending or recumbent and 1–2 mm in diam.; midstem leaves mostly 0.5–1.5(–2) cm across, mostly sparsely pubescent, the marginal hairs only rarely exceeding 1 mm in length ____ var. **depauperata**
1. Plants mostly 20–60 cm high, the primary stems mostly stiffly erect and 2–5 mm in diam.; midstem leaves mostly 1.5–2.5 cm across, moderately to markedly pubescent, the marginal hairs often 1 mm long or more ____ var. **humilis**

var. **depauperata** Tharp & F.A. Barkley, (impoverished, imperfectly developed, undeveloped). Plants generally depauperate, usually only 5–20(–30) cm tall, smaller and less pubescent than var. *humilis*. Areas of clay soils in gumbo glades and coastal grasslands; s Pineywoods and n Gulf Prairies and Marshes. Turner (1991a) noted that this variety intergrades into var. *humilis* in the s part of the range of var. *humilis* and "is only weakly differentiated and perhaps does not merit recognition."

var. **humilis.** Often, but not always, on alluvial or sandy soils; widespread in East TX, scattered further w. [*R. humilis* Nutt. var. *expansa* Fernald, *R. humilis* Nutt. var. *frondosa* Fernald, *R. humilis* Nutt. var. *longiflora* (A. Gray) Fernald]

Ruellia metziae Tharp, (for Sister Mary Clare Metz, 1907–1998, nun and professor of botany at Our Lady of the Lake College, San Antonio, TX). Plant to ca. 0.6 m tall; leaves petiolate; leaf blades oblong-lanceolate to narrowly ovate-oblong, 3–12 cm long, undulate to toothed; corollas usually white (sometimes pale bluish purple), with tubes 3–5 cm long; cleistogamous flowers sometimes produced early from axillary peduncles only. Limestone outcrops, gravel, calcareous soils, thickets, and open woods; Bexar, McLennan, Travis (BRIT), Bosque, Comal, Grayson, Karnes, and Wilson (Turner et al. 2003), Williamson (Tharp & Barkley 1949) cos.; mainly Edwards Plateau; in the U.S. known only from TX; also Mexico. Apr–Jul. [*R. metziae* Tharp var. *marshii* Tharp & F.A. Barkley, *R. muelleri* Tharp & F.A. Barkley] Native of TX and Mexico. This is the only East TX species that usually has white corollas, and as noted by Turner (1991a), it is a "very distinct taxon, readily distinguished by its low habit and large white flowers with tubes 3–5 cm long."

Ruellia nudiflora (Engelm. & A. Gray) Urb., (naked-flowered), VIOLET RUELLIA, NIGHT-FLOWERING WILD PETUNIA, VIOLET WILD PETUNIA. Stems short pubescent, to almost glabrous, usually 3–7(–10) dm tall; leaves petiolate; corollas purplish blue, 3–5 cm long, tube 15–20 mm long, throat 18–20 mm long; fruit of cleistogamous flowers ca. 15 mm long. Sandy open woods; in East TX widespread in all but the Pineywoods; Edwards Plateau, South TX Plains, and Trans-Pecos; AL, AZ, LA, MS, TX. Apr–May. Native of the U.S. and Mexico. Turner (1991a) studied this species complex and concluded that there are two varieties in TX worthy of formal recognition. He separated the two as follows, but noted that hybridization and intergradation between them can occur.

1. Corollas (of chasmogamous flowers) mostly 4–5 cm long; calyx lobes having glandular hairs with the stalks elongate, 2–5 times as long as the capitate glands; pubescence of capsule not glandular-pubescent throughout, or nearly so ____ var. **nudiflora**
1. Corollas mostly 3–4 cm long; calyx lobes having glandular hairs with the stalks short, about as long as or 1–2 times as long as the capitate glands; pubescence of capsule glandular-pubescent throughout, or nearly so ____ var. **runyonii**

var. **nudiflora.** Areas of clay soils in gumbo glades and coastal grasslands; s and w parts of East TX; widespread primarily in s 1/2 of TX; AL, AR, LA, MS, and TX. [*Dipteracanthus nudiflorus* Engelm. & A. Gray, *R. nudiflora* var. *hispidula* Shinners] var. **runyonii** (Tharp. & F.A. Barkley) B.L. Turner, (for Robert Runyon of Brownsville, a friend of Tharp and Barkley and student of the vegetation of the lower Rio Grande Valley). Calcareous or sandy calcareous soils; Bexar, DeWitt, Goliad, and Wilson (Turner et al. 2003) cos. at sw corner of East TX; mainly South TX Plains. [*R. runyonii* Tharp & F.A. Barkley, *R. runyonii* var. *berlandieri* Tharp & F.A. Barkley]

Ruellia occidentalis (A. Gray) Tharp & F.A. Barkley, (western), WESTERN WILD PETUNIA. Plant 0.3–0.7 m tall, glandular-pubescent; leaves petiolate; corollas ca. 7 cm long, lavender to bluish purple. Dry rocky woods, chaparral, slopes, and along streams; Bexar (Turner et al. 2003), Comal (*Wright 430* GH), Goliad, Travis (BRIT) cos.; mainly South TX Plains; in the U.S. known only from TX; also Mexico. May–Oct. [*R. davisiorum* Tharp & F.A. Barkley, *R. nudiflora* (Engelm. & A. Gray) Urb. var. *occidentalis* (A. Gray) Leonard, *R. strictopaniculata* Tharp & F.A. Barkley, *R. tuberosa* L. var. *occidentalis* A. Gray] Native of TX and Mexico. Turner (1991a) considered *R. occidentalis* to be most closely related to *R. nudiflora*. While *R. davisiorum*, RIO GRANDE WILD PETUNIA, is recognized by some authorities (e.g., Hatch et al. 1990; Correll & Johnston 1070; Kartesz 2015), we are following Turner (1991a; Jones et. al 1997) in including this taxon within *R. occidentalis*. According to Turner (1991a), "Plants with congested panicles and relatively small leaves have been segregated as *R. davisiorum*...but such plants appear to be but forms of a variable *R. occidentalis*." However, Turner (1991a) also noted that field work is needed in the area in southernmost TX where *R. occidentalis* and *R. nudiflora* occur and where the types of *R. davisiorum* and *R. strictopaniculata* were collected.

Ruellia pedunculata Torr. ex A. Gray, (with a flower stalk), STALKED WILD PETUNIA, STALKED RUELLIA. Plant to ca. 0.7 m tall; leaves subsessile or short-petiolate, the petioles only 1–2 mm long; peduncles to ca. 7 cm long; calyx pilose with slender pointed hairs; corollas bluish lavender to purplish, sometimes pale, 3.5–5.5 cm long. Open woods and along streams, sandy soils; Fannin Co. (BRIT) s and e, primarily throughout e portion of East TX; mainly se U.S. from SC s to FL and w to OK and TX and n to IL. May–Sep. [*R. pinetorum* Fernald, *R. pedunculata* subsp. *pinetorum* (Fernald) R.W. Long] Native of the U.S. and Mexico. Turner (1991a) argued that TX plants identified as *R. pinetorum* "are but hybrid derivatives of an occasional cross between *R. pedunculata* and *R. humilis*, at least the characters which are said to mark *R. pinetorum* are found in *R. humilis*, and the hypothetical parents are mostly sympatric, at least in Texas."

Ruellia simplex C. Wright, (simple, undivided, in reference to the stem) LINEAR-LEAF WILD PETUNIA, BRITTON'S WILD PETUNIA, MEXICAN BLUEBELL. Erect perennial herb to 1 m tall, stem decumbent at base; leaves with blades 10–20(–30) cm long, to 2–3(–4) cm wide, lanceolate to linear-lanceolate or linear-elliptic, apex acute; cymes axillary, loose, bracts narrowly lanceolate; calyx lobes subulate, ca. 8 mm long; corollas pale violet to purple, 2.5–4.5 cm long; capsule 10–20 mm long, glabrous. Common in nursery trade, readily escapes cultivation; widely scattered in East TX; widely scattered in South TX Plains; se U.S. from SC s to FL w to TX, also AZ. May–Nov. [*R. brittoniana* Leonard, *R. coerulea* Morong, *R. malacosperma* Greenm., *R. tweediana* Griseb.] Native of e Mexico. The taxonomic and nomenclatural history of *R. simplex* is very complex and convoluted and is best elucidated by Ezcurra and Daniel (2007). We are following the nomenclature of Ezcurra and Daniel (2007) and Wunderlin et al. (2019) in using *Ruellia simplex*.

Ruellia strepens L., (rustling or rattling), LIMESTONE WILD PETUNIA, LIMESTONE RUELLIA, SMOOTH RUELLIA. Stems to ca. 1 m tall, glabrous or pubescent in lines; leaves petiolate; leaf blades ovate- or elliptic-lanceolate, acute; peduncles of 1–3 chasmogamous flowers, leafy-bracted, the bracts lanceolate to elliptic-ovate; corollas pale bluish violet, 3–6 cm long, cleistogamous flowers borne in axillary glomerules. Stream and hillside woods; mostly n and s portions of East TX; Gulf Prairies and Marshes, also Gillespie Co. (Turner et al. 2003) to the w; e U.S. from PA s to GA and w to IA, OK, and TX. Apr–Jun, with solitary, open flowers, also Sep–Oct, with clustered, often smaller, open or usually cleistogamous flowers. [*Dipteracanthus strepens* (L.) Nees, *R. strepens* var. *cleistantha* A. Gray] Native of the U.S. Turner (1991a) argued that this is a widespread and variable entity.

Ruellia yucatana (Leonard) Tharp & F.A. Barkley, (from the collection site as *Ruellia nudiflora* var. *yucatana* in Yucatán, Mexico), YUCATAN WILD RUELLIA. Stems to 35 cm tall, branched from the base, more or less minutely, glandular-pilose and viscid, more so toward the inflorescence; basal leaves radially spreading, clustered from crowded nodes, petioles rarely more than 25 mm long, blade 13 cm long, 8 cm wide, oblong-elliptic to spatulate, apex obtusely rounded, rarely acute, base cuneate, stem leaves similar to basal leaves only smaller; cleistogamous cymes usually several flowered on conspicuous, slender horizontal, upcurved or ascending dichotomous branches toward the base of the plant, chasmogamous flowers in terminal panicles mixed with cleistogamous flowers, calyx lobes 7–13 mm long, linear; corollas of chasmogamous flowers lavender-blue, 3–4 cm long, tube 6–8 mm long; capsules of cleistogamous flowers 10–14 mm long, capsules of chasmogamous flowers 15–17 mm long. In grasslands, caliche hills, chaparral, and on the edge of dense woods; Colorado, Goliad, and Harris (Kartesz 2022) cos.; Edwards Plateau and South TX Plains; AL and TX; Mexico. Mar–Oct. [*R. nudiflora* var. *yucatana* Leonard] Native of TX and Mexico.

Ruellia caroliniensis [STR]

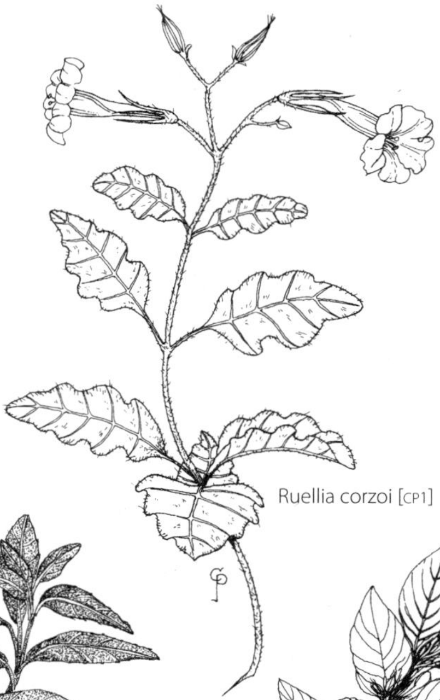

Ruellia corzoi [CP1]

Ruellia dipteracanthus [THO]

Ruellia drummondiana [HEA]

Ruellia humilis [STE]

Ruellia metziae [HEA]

Ruellia nudiflora [LUN]

Ruellia noctiflora (Nees) A. Gray, (night-flowering), NIGHT-FLOWERING WILD PETUNIA. A questionable collection of this species was cited for TX by Tharp and Barkley (1949). We have been unable to find any other documented specimens. The species is known from AL, FL, GA, sw LA, and MS. According to Turner (1991a) "might yet be found in swampy forestland of Newton or Orange counties." If present, it should be easily recognizable by its large (8–9 cm long) white corollas.

STENANDRIUM Nees SHAGGYTUFT

A genus of about 25 species in North America, West Indies, Mexico, Central America, and South America (Wunderlin et al. 2019). (Greek: *stenos* narrow, and *aner* man referring to the narrow stamens)

REFERENCES: Daniel 1984a; Wunderlin et al. 2019.

Stenandrium dulce (Cav.) Nees, (sweet, in reference to the sweet-smelling flowers), SWEET SHAGGYTUFF. Perennial herb to 2 dm tall; leaves in a basal rosette; blade 1.3–10 cm long, 0.5–4 cm wide, ovate, elliptic, or oblanceolate, apex acute to rounded, the base rounded to cuneate or sometimes decurrent along the petiole, upper and lower surfaces pubescent, usually glandular punctuate, petiole 0.5–6.5 cm long; flowers sessile in a head-like or elongated spike to 8.5 cm long on a pubescent to nearly glabrous scape 0.5–20 cm long; bracts imbricate, lanceolate to elliptic or rarely oblanceolate or obovate, foliose, (6–)9–20 mm long; bracteoles lanceolate-subulate to subulate, 3–9 mm long; calyx 4–11 mm long, lobes subulate 4–9 mm long; corollas pink to purple, marked with white within; 1.6–2 cm long, tube 0.9–1.6 cm long, the lobes obovate to narrowly obovate; capsule elliptic, 7–12 mm long, seeds 1–4, 3–4 mm long with long appressed barbed hairs. Wet pinelands, moist soil, river bottoms, low flat woods; included based on a collection (*Rosen & Singhurst 5735* TEX) from Wharton Co.; s TX; AL, FL, NC; and Mexico; Central and South America. All year. Native in FL and TX. No county distribution map provided.

THUNBERGIA Retz. CLOCKVINE

An Old World tropical genus of 100–150 species of climbing or erect herbs and shrubs with bracts enclosing the calyces (Mabberley 2017); a number are cultivated as ornamentals. They differ from many other members in the Acanthaceae in not having cystoliths or retinacula. Molecular evidence (Scotland et al. 1995; McDade 2000b) shows that *Thunbergia* is nearly basal phylogenetically in the Acanthaceae but should be included in the family (Scotland & Vollesen 2000). (Named for Carl Peter Thunberg, 1743–1828, Swedish student and successor of Linnaeus and traveler to Asia and South Africa)

REFERENCES: Schönenberger & Endress 1998; Schönenberger 1999.

Thunbergia alata Bojer ex Sims, (winged), BLACK-EYED-SUSAN-VINE, BLACK-EYED CLOCKVINE. Perennial, herbaceous, trailing or climbing vine, often grown as an annual; stems usually ca. 1 m long; tendrils absent; leaf blades 2.5–8 cm long, ovate to triangular-ovate, apically acute, basally cordate to ± hastate, entire to few-toothed, pubescent; petioles wing-margined, often nearly as long as blades; flowers usually axillary, solitary, on peduncles usually longer than petioles; calyx small, subtended and hidden by 2 conspicuous bracts ca. 15 mm long; corollas large and showy, 2.5–4 cm long, yellow to orange, cream, or white, with a dark reddish purple center "eye," with a slender tube and a spreading 5-lobed limb; lobes of limb rounded distally, not differing greatly in size or shape; stamens 4; capsules 8–10 mm in diam., with a stout beak ca. 1 cm long. Long cultivated in East TX (an 1885 Reverchon collection (cultivated) is known from Dallas—Wasshausen 1966); Bexar and Harris cos. (Wasshausen 1966); also Cameron Co. (BRIT) in extreme s TX; AL, AR, FL, and TX. Mar–Apr. Native of tropical East Africa. The flowers are apparently bee pollinated (Schönenberger 1999).

YEATESIA Small BRACTSPIKE

A genus of 3 species (Mabberley 2017) native from the se U.S. to ne Mexico. (Named for Professor W.S. Yeates, State Geologist of Georgia from the late 1800s until 1908; so named because J.K. Small first encountered the genus in sw Georgia)

REFERENCES: Small 1896; Wasshausen 1966, 1998; Long 1970; Hilsenbeck 1989a.

Yeatesia viridiflora (Nees) Small, (with green flowers—despite the white to pale blue or pink corollas), YELLOW BRACTSPIKE. Herbaceous or subshrubby perennial 0.3–0.6(–1) m tall; leaf blades 8–14 cm long, 2–5 cm wide, apically acuminate, gradually tapered to petiole basally, entire; petiole 1–2 cm long; inflorescences short-peduncled, cylindrical, compact spikes, axillary and terminal, the inflorescences conspicuously subtended by bracts and the flowers by bracteoles, the bracts broadly ovate, to ca. 16 mm long and 9 mm wide, the bracteoles smaller and more linear; calyx deeply 5-parted; corollas nearly equally 4-lobed to slightly zygomorphic, usually white but also pale blue, lavender, or pink, 14–16 mm long, the tube 7–9 mm long, the lobes 5–7 mm long; stamens 2; capsule clavate to clavate-oblong, with stipe-like base, seeds 4 (or fewer if some aborted). Rich bottomlands and mesic mixed forests; Hardin, Jasper, Jefferson, Nacogdoches, Newton (BRIT), Milam, Sabine (Turner et al. 2003), and San

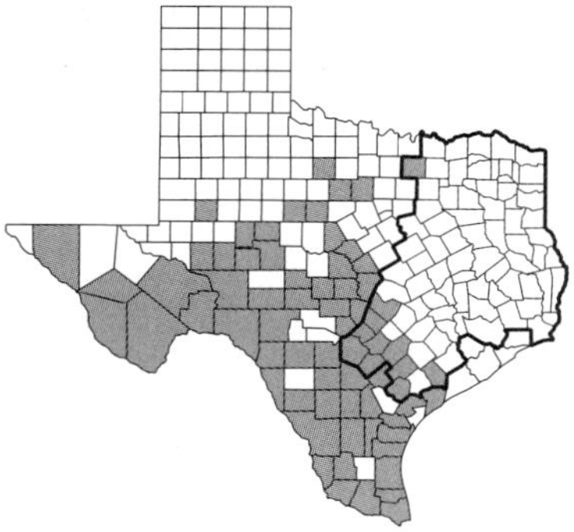

Justicia pilosella

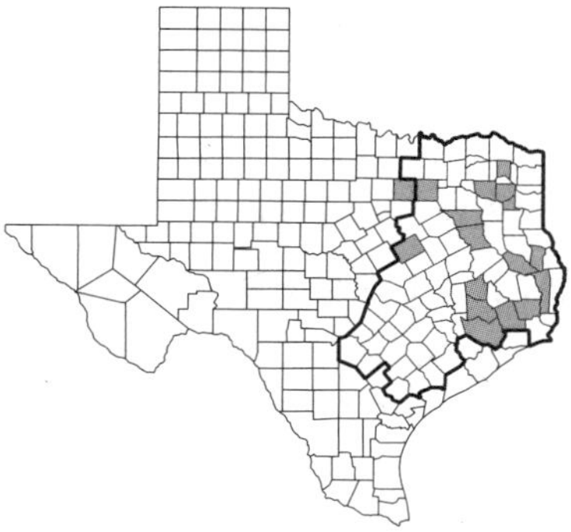

Ruellia caroliniensis

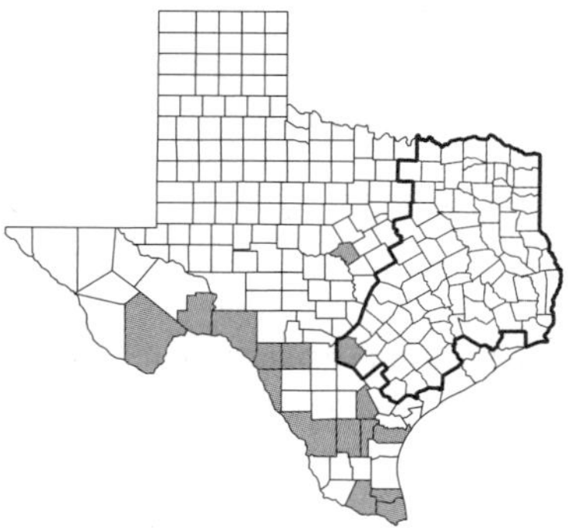

Ruellia corzoi

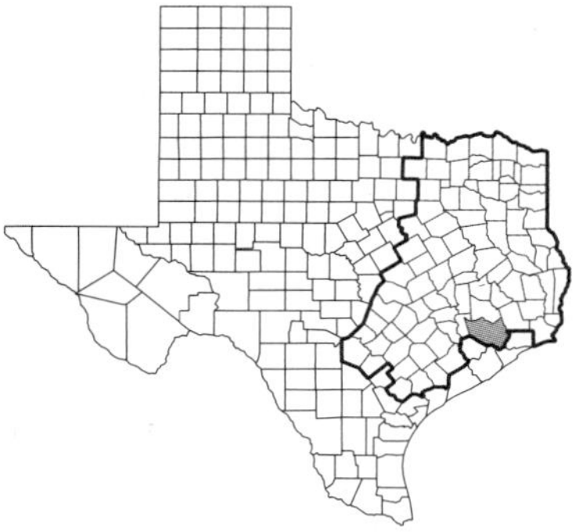

Ruellia dipteracanthus

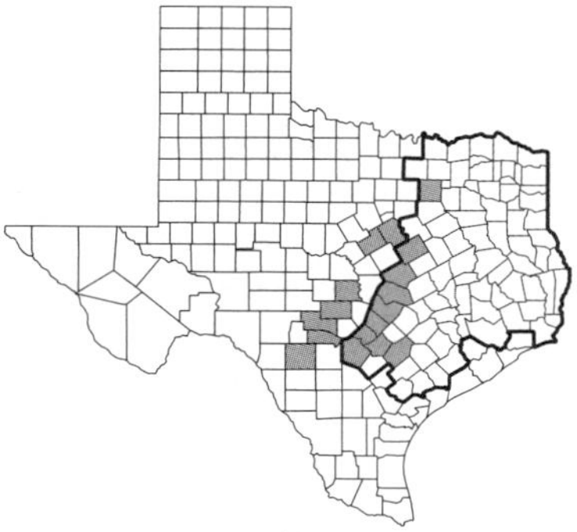

Ruellia drummondiana

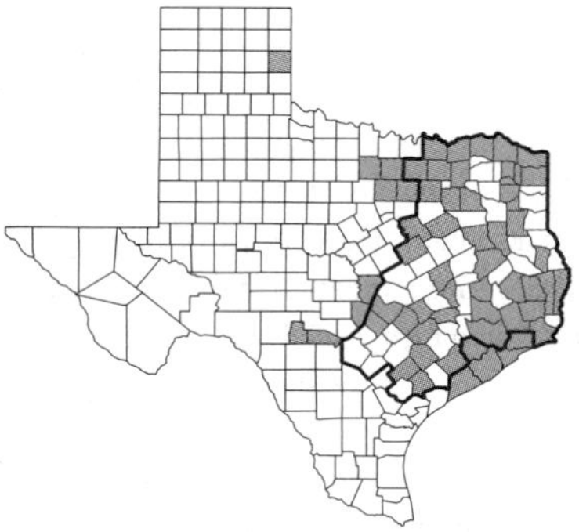

Ruellia humilis

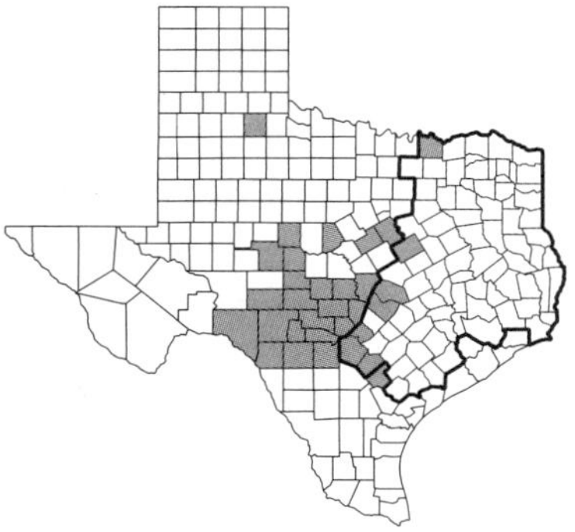

Ruellia metziae

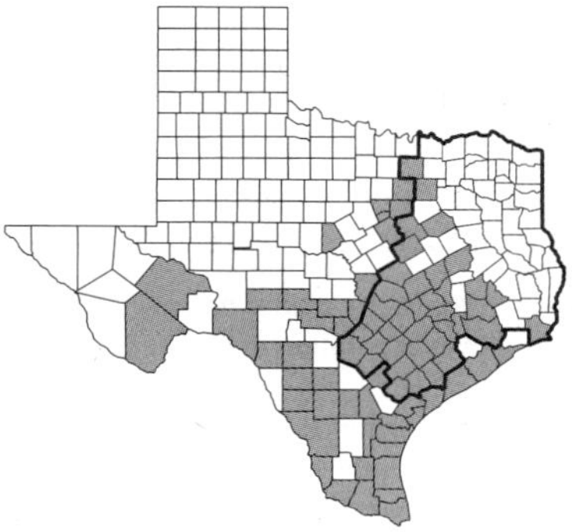

Ruellia nudiflora

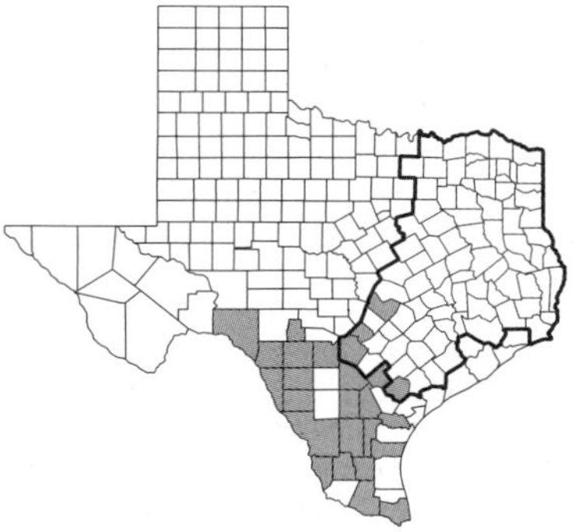

Ruellia occidentalis

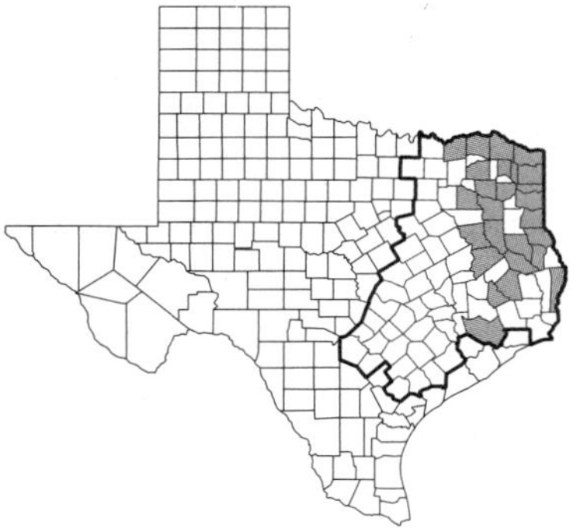

Ruellia pedunculata

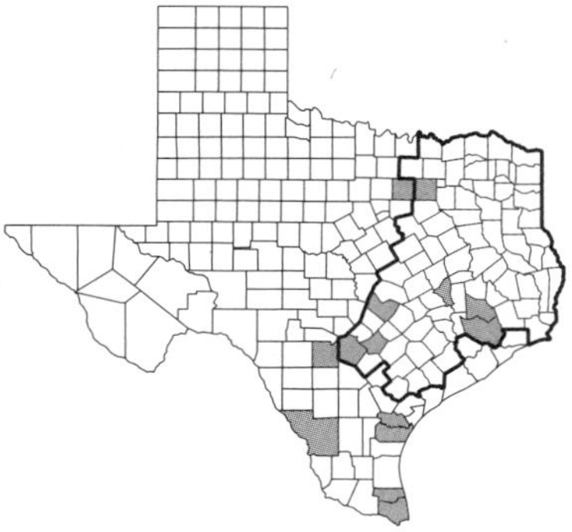

Ruellia simplex

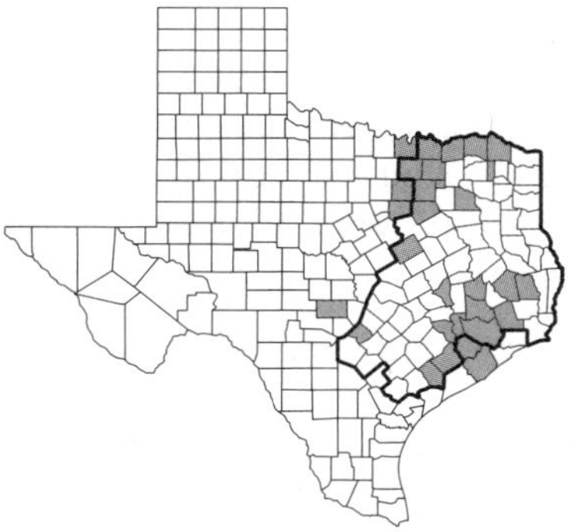

Ruellia strepens

Augustine (Wasshausen 1966), cos.; also Victoria Co. (Wasshausen 1966) in the Gulf Prairies and Marshes; Gulf Coastal Plain from FL and GA w to TX. Jun–Nov. [*Dicliptera viridiflora* (Nees) R.W. Long, *Gatesia laetevirens* A. Gray, *Justicia laetevirens* Vahl, *Rhytiglossa viridiflora* Nees, *Y. laetevirens* (Buckley) Small] Native of the U.S.

AIZOACEAE Martinov

FIG-MARIGOLD OR ICEPLANT FAMILY

Ours annual or perennial succulent herbs; stems prostrate to decumbent, sometimes forming mats; leaves opposite or subopposite; flowers usually solitary(–3), axillary, sessile or pedicellate; calyx lobes 5; petals absent; stamens 5–30; ovaries 2–5 carpellate, superior or half inferior; fruits circumscissile capsules.

A medium-large (ca. 1,880 species in 125 genera—Mabberley 2017), widespread tropical and subtropical family of betalain-containing, succulent herbs or small shrubs centered in s Africa and to a lesser extent in Australia; most have the C4 or CAM types of photosynthesis, physiological adaptations for hot dry conditions; some are cultivated as ornamentals including the PEBBLE PLANTS or LIVING STONES (*Lithops*) that mimic rocks and *Mesembryanthemum* (ICEPLANT), a huge genus sometimes split into over 100 segregate genera; agreement as to the best treatment of that genus has not yet been reached (Stevens 2001 onwards; Klak et al. 2007). *Mollugo* and *Glinus*, treated here in the Molluginaceae following APG IV (2016), were formerly included in the Aizoaceae. The genus *Geocarpon* was also earlier placed in the Aizoaceae and treated in that family by many authors (e.g., Fernald 1950). However, a variety of evidence (e.g., exstipulate leaves, connate sepals, free central placentation, valvular fruit, anthocyanin pigments, molecular data) supports its inclusion in the Caryophyllaceae (Palmer & Steyermark 1950; Bogle et al. 1971; Nienaber 2005; Harbaugh et al. 2010). The Aizoaceae, *sensu stricto*, is distinguished by the "fusion of sepals to form a calyx tube, by perigynous stamen insertion, and by its strong tendency toward leaf succulence" (Boetsch 2002). The Aizoaceae appears related to the Phytolaccaceae and Nyctaginaceae (Stevens 2001 onwards), but its exact relationships within the Caryophyllales remain unclear (Stevens 2001 onwards; Landrum 2002; Walker et al. 2018). Family name from *Aizoon*, a genus of 25 species of the Mediterranean, s Africa, and Australia. (Latin name for an evergreen plant) (subclass Caryophyllidae—Cronquist; order Caryophyllales—APG IV)

FAMILY RECOGNITION IN THE FIELD: the East TX species are succulent, prostrate to decumbent herbs with opposite or subopposite leaves, inconspicuous axillary flowers, and circumscissile capsules; they superficially resemble Portulacaceae (*Portulaca*) but unlike Aizoaceae, our members of that family have alternate or subopposite leaves and flowers with petals and only 2 (vs. 5) sepals.

REFERENCES: Bogle 1970; Bittrich & Hartmann 1988; Hartmann 1993; Behnke & Mabry 1994; Stevens 2001 onwards; Boetsch 2002; Landrum 2002; Klak et al. 2003; Vivrette et al. 2003; Hassan et al. 2005; Thiede et al. 2007.

1. Leaves of each pair equal, the petioles ± absent or short, the blades narrowly oblanceolate to spatulate or narrowly ovate, 1–2.5 cm long; stipules absent; styles 2–5; placentation axile; seeds 30—60; in East TX known only near the coast ____ **Sesuvium**
1. Leaves of each pair unequal, the petioles quite distinct, often as long as the blades or nearly so, the blades broadly obovate to nearly orbicular, to ca. 4(–6.5) cm long; stipules attached to margin of petiole, bidentate, usually papery; styles 1–2; placentation basal; seeds ca. 7; scattered in East TX ____ **Trianthema**

SESUVIUM L. SEA-PURSLANE

Ours annual or perennial, succulent herbs; roots fibrous; stems prostrate to ascending, sometimes forming mats; leaves opposite, those of each pair equal; stipules absent; petiole bases clasping; blades usually flat, spatulate to oblanceolate, obovate, or elliptic-ovate, tapered, the margins entire; flowers usually solitary, axillary, sessile or pedicellate; calyx lobes 5, reddish adaxially, usually hooded near tip; petals absent; stamens 5–30; pistil 2–5-carpellate; ovary half inferior, 2–5-loculed; placentation axile; ovules to 60; styles 2–5, papillate; stigmas 2–5; fruits capsules, ovoid to obconic, membranaceous, the dehiscence circumscissile; seeds to 60.

A genus of 8–14 species (Stevens 2001 onwards; Ferren 2003a) of halophytes of tropical and warm areas around the world, especially along coasts. (Latin: *Sesuvium*, apparently the country of the Sesuvii, a Gallic tribe mentioned by Caesar, the reason for application to this genus unknown—Ferren 2003a)

REFERENCES: Correll & Johnson 1970; Boetsch 2002; Ferren 2003a.

1. Plants annual, not rooting at the nodes; calyx lobes to 2–3 mm long; pedicels usually absent or to 1 mm long; stamens 5; styles and locules 2–3 ____ **S. maritimum**

Ruellia occidentalis [CP1]

Ruellia pedunculata [BB2]

Ruellia simplex [HEA]

Ruellia strepens [BB2]

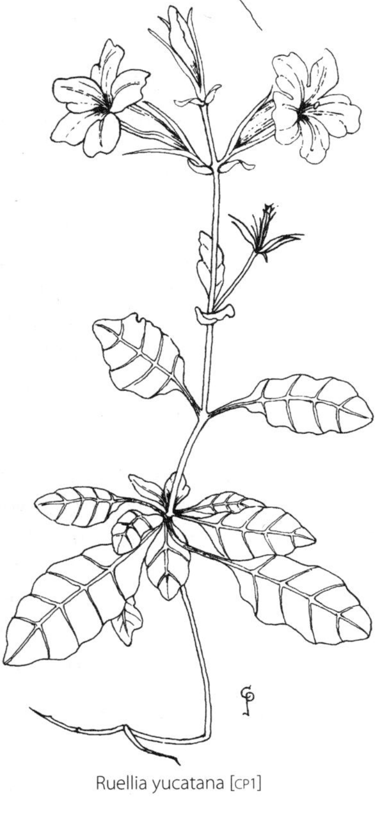

Ruellia yucatana [CP1]

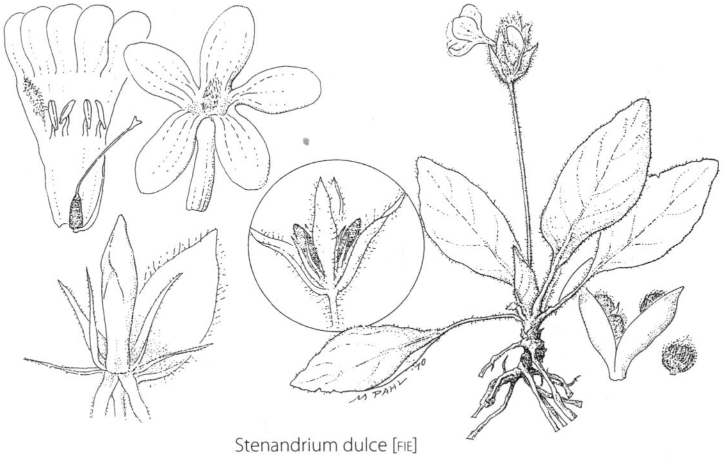

Stenandrium dulce [FIE]

1. Plants perennial, rooting at the nodes; calyx lobes 3–10 mm long; pedicels to 20 mm long; stamens 30; styles and locules 5 ______ **S. portulacastrum**

Sesuvium maritimum (Walter) Britton, Sterns, & Poggenb., (of the sea), SLENDER SEA-PURSLANE. Plant annual, papillate, glabrous, sometimes forming mats; stems prostrate to ascending at tips, 1–4 dm long, not rooting at nodes; petioles short, clasping, blades spatulate to obovate, 1–2.5 cm long, tapered to base; flowers usually solitary; pedicels usually absent or to 1 mm long; calyx lobes pink or purple adaxially, with subapical abaxial appendage, ovate to ovate-oblong; stamens 5; pistil 2–3-carpellate; ovary 2–3-loculed; styles 2–3; capsules ovoid, 4–5 mm long; seeds 30–50, blackish brown, 1 mm or less long, iridescent, smooth. Beaches and sandy areas; Harris Co. (*Rosen & Burnett 6530* TEX), otherwise strictly coastal, se and s TX; NY s to FL and w to TX, also inland in KS, LA, and OK. Flowering all year. [*Pharnaceum maritimum* Walter]

Sesuvium portulacastrum (L.) L., (with flowers like *Portulaca*—purslane), CENCILLA, SHORELINE SEA-PURSLANE. Plant perennial, glabrous, forming mats to 2 m in diam.; stems prostrate, rooting at the nodes; petioles ± absent, blades oblanceolate to elliptic-ovate, to 6 cm long, to 2.5 cm wide, tapered to clasping base; flowers solitary; pedicels to 20 mm long; calyx lobes pink-purple adaxially, with subapical abaxial appendage, ovate to lanceolate, 3–10 mm long, to 6 mm wide; stamens 30; pistil 5-carpellate; ovary 5-loculed; styles 5; capsules conic, 10 mm long, 5–6 mm wide; seeds 30–60, black, 1.2–1.5 mm long, shiny, smooth. Beaches or dunes along the coast, rarely inland; Harris Co. (*Rosen 6182* TEX), otherwise coastal se to s TX and Webb Co. (USDA Plants Database 2019) along the Rio Grande; coastal from PA to FL w to TX, also inland in LA. Flowering all year. [*Portulaca portulacastrum* L.] Ferren (2003a) noted that this species is "a widespread and variable subtropical and tropical species to which many names have been applied, particularly to material collected beyond North America (e.g., Argentina and Brazil)." The species is eaten as a vegetable in some parts of Asia (Mabberley 2017) and is also used as fodder; it is a known halophyte (Rabhi et al. 2012).

Cypselea humifusa Turpin, (sprawling), PANAL, is a glabrous mat-forming annual with opposite leaves, those of each pair unequal. Known from Bossier and Natchitoches parishes (USDA Plants Database 2019) just two parishes e of the TX-LA border. It could therefore be expected in East TX. Molecular data (Hassan et al. 2005) closely link *Cypselea* with *Sesuvium*. This species can be distinguished from other East TX Aizoaceae by its (1–)3 stamens, fringed stipules, very small leaves (blades 3–9 mm long) with distinct petioles often longer than the blades, and seeds ca. 150 per capsule. It is native to the West Indies.

TRIANTHEMA L. HORSE-PURSLANE, SEA-PURSLANE

A genus of 20–32 species (Stevens et al. 2001 onwards; Ferren 2003b) of warm areas of the world, especially Australia; some contain alkaloids and are used medicinally. The genus is in subfamily Sesuvioideae (Hassan et al. 2005). (Greek: *tri*, three, and *anthemon*, flower)

REFERENCES: Ferren 2003b; Hassan et al. 2005.

Trianthema portulacastrum L., (with flowers like *Portulaca*—purslane), DESERT HORSE-PURSLANE, SEA-PURSLANE, LOWLAND-PURSLANE, VERDOLAGA BLANCA. Glabrous, succulent, annual herb branching from the base; branches usually decumbent; leaves subopposite to opposite, in unequal pairs; stipules present, bidentate; petioles sheathing, blades broadly obovate to nearly orbicular, to ca. 4(–6.5) cm long, the margins entire; flowers solitary(–3), axillary, sessile; calyx lobes 5, ca. 2.5 mm long, pinkish purple within, green outside; petals absent; stamens 5–10; carpels 2; ovary superior, but sometimes seemingly almost inferior, 1–2-locular; fruit a capsule ca. 4 mm long with prominent winged appendages at apex, circumscissile near the middle, with ca. 7 seeds. Sandy soils and waste ground, moist or seasonally dry areas; Dallas, Grayson (BRIT!), Gregg (Turner et al. 2003), Brazos, Dewitt, Harris, and Robertson (Kartesz 2015) cos.; mainly South TX Plains and Trans-Pecos; widely scattered in the s 1/2 of the U.S. May–Oct. [*T. monogyna* L. [nom. illeg.], *T. procumbens* Mill.] Ferren (2003b) treats it as native to the U.S. as does USDA Plants Database (2019). However, Boetsch (2002) considers this species to be a pantropical weed, "likely introduced from the Old World tropics" and Yatskievych (2006) likewise considers it an Old World native. Aneja et al. (2000) view it as indigenous to South Africa. We are following these authors in treating it as introduced to the U.S. A single seed is dispersed with the detached (floatable) cap of the capsule, while the other seeds can be dispersed from the bottom portion of the capsule or remain on the parent plant (Ferren 2003b). The leaves are used as a vegetable in Africa and India and the species is sometimes used medicinally; however, it is also reported to be poisonous to livestock (Bogle 1970). It is also considered to be a serious invasive weed in agricultural crops in some areas (e.g., India, Pakistan) (Aneja et al. 2000; Randhawa et al. 2002).

Thunbergia alata [LUN]

Yeatsia viridiflora [LUN]

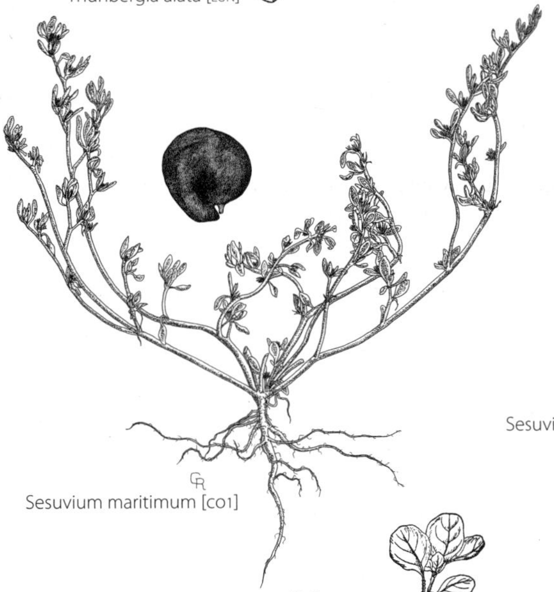

Sesuvium maritimum [CO1]

Sesuvium portulacastrum [CO1]

CAPSULE

Trianthema portulacastrum [GLE]

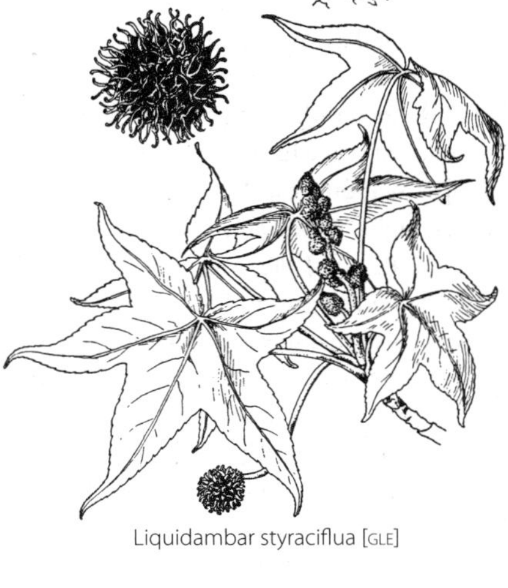

Liquidambar styraciflua [GLE]

ALTINGIACEAE Horan.

SWEET-GUM FAMILY

A small monogeneric family of 15 species of trees (Ickert-Bond & Wen 2013) known from e North America to Central America, w Asia, and e Asia to Malesia. The species are valued worldwide for timber and medicinal extracts or scents (e.g., the fragrant resin known as storax or sometimes called styrax) and locally prized for the roots and bark used in traditional Chinese medicine (Ickert-Bond & Wen 2013). In the past Altingiaceae have usually been included in Hamamelidaceae (e.g., Cronquist 1981; Diggs et al. 1999), sometimes as subfamily Altingioideae; however, molecular evidence now shows that the two families are distinct lineages (Shi et al. 2001; Ickert-Bond & Wen 2006). The group is well known for its Asian-North America disjunct distribution; see the generic discussion of *Campsis* (Bignoniaceae) for a detailed discussion of this biogeographic pattern. (Named for Willem Arnold Alting (1724–1800), the Governor-General of the Dutch East Indies when Noronha, who named the genus, visited Java (Hayne 1830)) (subclass Hamamelidae—Cronquist; order Saxifragales—APG IV)

FAMILY RECOGNITION IN THE FIELD: the only East TX species is a tree with alternate, simple, deeply palmately 5–7-lobed (*star-like* in shape), toothed, fragrant leaves with stipules or at least stipule scars; fruits woody 2-valved capsules (in ours grouped together in stalked *ball-like clusters*).

REFERENCES: Ernst 1963b; Goldblatt & Endress 1977; Endress 1989, 1993b; Meyer 1997; Shi et al. 2001; Ickert-Bond & Wen 2006, 2013.

LIQUIDAMBAR L. SWEETGUM

A genus of 15 species (Ickert-Bond & Wen 2013) of monoecious trees native to e North America to Central America, w Asia (e Mediterranean), and e Asia to Malesia; the species are valuable for timber, for the aromatic balsam (storax) used medicinally and in scents, and as cultivated ornamentals with spectacular fall colors. The genera *Altingia* and *Semiliquidambar* have sometimes been recognized as distinct from *Liquidambar*; however, molecular evidence shows *Altingia* is nested within *Liquidambar*, and that *Semiliquidambar* is an intergeneric hybrid between *Altingia* and *Liquidambar* strongly argues that only one genus be recognized (Ickert-Bond & Wen 2006; Wu et al. 2010; Ickert-Bond & Wen 2013). The Asian *L. orientalis* Mill. is the source of Levant storax, the balm (of Gilead) of the Bible. (According to Fernald (1950), a mongrel name, from Latin: *liquidus*, fluid, and Arabic: *ambar*, amber, in allusion to the fragrant terebinthine juice or gum which exudes from the tree)

REFERENCES: Duncan 1959; Bogle 1986; Hoey & Parks 1994; Ickert-Bond et al. 2005, 2007; Morris et al. 2008; Wu et al. 2010; Ruiz-Sanchez & Ornelas 2014.

Liquidambar styraciflua L., (flowing with storax or gum), SWEETGUM, REDGUM. Monoecious tree to 40 m or more tall, often exuding a gum (chewed by some); small branches often with corky ridges; leaves alternate, simple, deciduous, blades to ca. 18 cm long and 12 cm wide, deeply palmately 5–7-lobed, star-like in shape, toothed, fragrant if crushed, quite showy in fall, yellow to orange, crimson, or purplish, the lobes acuminate, usually serrate; petioles to 12 cm long; flowers unisexual, without petals; staminate flowers in bracteate globose masses on racemose inflorescences to 6 cm long; pistillate flowers in peduncled globose heads that develop into woody globose clusters (2.5 to 3(–4) cm in diam.) of 2-beaked capsules. Wet areas, low or rich woods; widespread mainly in the e part of East TX, but sw to Bastrop and Fayette cos. (Kartesz 2015), and native in Red River drainage w to Lamar Co. (G. Diggs, pers. obs.), widely cultivated and occasionally escaped further w (e.g., Tarrant Co.—Kartesz 2015). Mar–May. SWEETGUM is particularly noted for its strikingly colorful fall foliage. It is also the source of a fragrant gum or balsamic oleo-resin (COPALM BALM, AMERICAN COPALM, AMERICAN STORAX, STORAX) obtained from the inner bark after wounding or gashing; it has a long history of usage including being smoked with tobacco, burned as an incense, used for a chewing gum, as an antiseptic, in soaps and cosmetics, and as a fixative in perfumes, adhesives, and lacquers. It has also been used as a medicine by various Native American groups, including the Aztecs, and was much used by Confederate doctors. Medical uses include treatment of skin diseases, wounds, coughs, ulcers, and dysentery and it is currently still used medicinally in a variety of ways with research ongoing (Standley 1922a; Peattie 1948; Meyer 1997; Moerman 1998; Lingbeck et al. 2015; Mabberley 2017). SWEETGUM is a leading commercial hardwood species in East TX (Cox & Leslie 1991)(see appendix 21 in *Illustrated Flora of East Texas*, Vol. One). Disjunct populations of this species (sometimes previously treated as a different variety or species) occur in the mountains of Mexico and Central America; their occurrence is a relict of a once more widespread distribution with arrival of this and other temperate species in the tropics in the Miocene, with subsequent isolation there due to climatic warming and drying in the late Miocene to early Pleistocene leaving the southern populations "trapped" in the cooler higher elevations (Graham 1999b; Ruiz-Sanchez & Ornelas 2014). It is odd for someone familiar with the flora of the se U.S. to observe such a familiar tree in the floristically very different montane forests of Central America (e.g., Lagos de Montebello, Mexico-Guatemala border, G. Diggs, pers. obs.). [*L. barbata* Stokes, *L. gummifera* Salisb. [nom. illeg.], *L. macrophylla* Oerst., *L. styraciflua* var. *mexicana* Oerst.]

AMARANTHACEAE Juss.

AMARANTH OR PIGWEED FAMILY

Ours annual or perennial, often weedy herbs; leaves opposite in our species (except *Amaranthus* and *Celosia*), simple, entire or nearly so; stipules absent; inflorescences axillary or terminal, of small clusters, head-like, or in spikes or spike-like racemes or compact panicles; flowers very small, each subtended or enclosed by 1–3 scarious (= papery) or prominently scarious-margined bracts (1), bracteoles (2), perfect or unisexual; petals absent; tepals 0–5; stamens usually 5; pistil 1; ovary superior, placentation usually basal; utricle dehiscent or indehiscent.

A medium-sized (ca. 900 species in ca. 65 genera—Robertson & Clemants 2003) family of mostly herbs or shrubs, climbers, or rarely trees of tropical and warm areas of the world with a few in temperate regions; the family includes a number of ornamentals (e.g., *Amaranthus caudatus*—LOVE-LIES-BLEEDING; *Celosia cristata*—COCKSCOMB; *Gomphrena globosa*—GLOBE-AMARANTH) and agricultural weeds as well as grain crops (e.g., *Amaranthus*). The flowers of almost all species are wind-pollinated. As is the case with many members of the Caryophyllales, rather than the typical anthocyanins, the pigments present are of the nitrogen-containing type known as betalains (Cronquist & Thorne 1994). Molecular evidence indicates the family is most closely related to Chenopodiaceae (Downie & Palmer 1994); in fact, authors vary in their treatment of the Amaranthaceae and Chenopodiaceae. While including the Chenopodiaceae in the Amaranthaceae has been considered less so in the past (treated separately in Correll & Johnston 1970; Robertson & Clemants 2003; Turner et al. 2003), recently it has gained more proponents (Stevens 2001 onwards; Mabberley 2017; Meyers et al. 2020; Lesica 2022). However, we are following recent authors (Yatskievych 2006; Hernández-Ledezma et al. 2015; Weakley, A.S., and Southeastern Flora Team 2022) in treating the Chenopodiaceae separate from the Amaranthaceae, rather than as one family. (subclass Caryophyllidae—Cronquist; order Caryophyllales—APG IV)
FAMILY RECOGNITION IN THE FIELD: herbs with simple leaves entire or nearly so and very small apetalous flowers in usually dense inflorescences; flowers subtended by *papery* often colored bracts; *filaments united into a tube or cup*. Chenopodiaceae are similar in flower structure (e.g., very small, lacking petals), but lack scarious bracts and have separate filaments.
REFERENCES: Standley 1917; Reed 1969b; Robertson 1981; Townsend 1993; Behnke & Mabry 1994; Downie & Palmer 1994; Eliasson 1988; Robertson & Clemants 2003.

1. Leaves alternate.
    2. Flowers unisexual (plants either monoecious or dioecious); utricles 1-seeded ____ **Amaranthus**
    2. Flowers perfect; utricle with seeds up to 20 ____ **Celosia**
1. Leaves opposite.
    3. Stems and leaves dense with irregularly branched hairs (at least young portions), sometimes appearing stellate; plants with gray-green foliage and inconspicuous flowers (2–5) in small axillary clusters lacking white or colored bracts ____ **Tidestromia**
    3. Plants without the combination of characters above.
        4. Inflorescence a diffuse panicle up to 20 cm broad; flowers unisexual (occasionally some bisexual flowers present); pistillate flowers subtended by long, dense white hairs much longer than tepals ____ **Iresine**
        4. Inflorescence a dense head or narrow spike (a number of these can be grouped together); flowers perfect, not subtended by long, dense, white hairs much longer than tepals (long hairs sometimes present <u>on</u> tepals).
            5. Inflorescences pedunculate, simple or compound spikes, elongate, at least 2 times as long as wide ____ **Froelichia**
            5. Inflorescences sessile or pedunculate, globose or short cylindric heads < 2 times as long as wide.
                6. Inflorescences sessile or nearly so and axillary (subtended by a single leaf) or pedunculate (without immediately subtending leaf).
                    7. Inflorescences axillary, sessile or nearly so (pedunculate in *A. philoxeroides*); tepals with stiff sharp apex, glabrous to sparsely to densely hairy with barbed or unbarbed straight hairs; interstaminal lobes (pseudostaminodes) present ____ **Alternanthera**
                    7. Inflorescences axillary and sessile; tepals sometimes merely acuminate but without stiff apex, densely hairy with crinkled, unbarbed hairs, sometimes obscuring the tepals; interstaminal lobes (pseudostaminodes) absent.
                        8. Tepals basally united; inflorescences 3–5 mm broad ____ **Guilleminea**
                        8. Tepals separate to base; inflorescences 5–8 mm broad ____ **Gossypianthus**
                6. Inflorescences pedunculate, terminal and/or axillary, with a pair of leaves immediately subtending the head.
                    9. Leaves fleshy; bracteole midribs not crested; inflorescences 7–10 mm broad ____ **Blutaparon**
                    9. Leaves not fleshy; bracteole midrib variously denticulate crested; inflorescences 9–28 mm broad ____ **Gomphrena**

ALTERNANTHERA Forssk. CHAFF-FLOWER

Ours perennials or annual; leaves opposite, entire; inflorescences in head-like spikes (referred to as heads), flowers perfect; tepals usually 5, separate; stamens 5 (without anthers in some cases), the filaments united basally into a tube/cup, with pseudo-staminodes, sometimes called interstaminal lobes (= non anther-bearing appendages from the filament tube) alternating with the stamens; fruit a utricle.

A genus of ca. 80 species (Clemants 2003a) of tropical and warm areas, especially the Americas; some are edible while others, because of their colorful leaves, are used as bedding plants. All 6 members of the genus occurring in East TX are introduced. The following key is adapted from Mears (1977 [1978]) and Clemants (2003a). (Latin: *alternans*, alternating, and *anthera*, an anther, in reference to the alternation of pseudostaminodes and stamens—Clemants 2003a)

REFERENCES: Melville 1958; Mears 1977 (1978); Buckingham 1996; Clemants 2003a.

1. Inflorescences pedunculate, the peduncles ca. 2–7 cm long; plants aquatic or semi-terrestrial; leaves fleshy ____ **A. philoxeroides**
1. Inflorescences sessile or nearly so, the peduncles at most 1 cm long; plants terrestrial in sandy disturbed areas and waste places; leaves not fleshy.
  2. Leaves varicolored, green, red, or yellow; stems varying from prostrate to ascending to erect; anthers 5, oblong; plants escaped from cultivation ____ **A. bettzickiana**
  2. Leaves green; stems prostrate; anthers 3–5, globose; plants weedy, not escaped from cultivation.
    3. Tepals monomorphic (all ± same size), with unbarbed hairs.
      4. Mature utricles exerted between tepals; stems glabrous to pubescent in lines ____ **A. sessilis**
      4. Mature utricles included within tepals; stems with silky pubescence ____ **A. paronychioides**
    3. Tepals dimorphic, tepal hairs barbed with a terminal grappling hook-like barb.
      5. Tepals 5–7 mm long, strongly spinose tipped, sparsely villous; bracts longer than tepals; leaf blades usually as wide as long ____ **A. pungens**
      5. Tepals 3–5 mm long, only slightly spinose tipped, densely villous; bracts ± as long as tepals; leaf blades longer than wide ____ **A. caracasana**

Alternanthera bettzickiana (Regel) G. Nicholson, (named for August Bettzick, a 19th century German gardener), CALICOPLANT, PARROTLEAF. Stems prostrate to ascending or erect; foliage varicolored; heads straw-colored to white; bracts < 1/2 as long as tepals; tepals nearly equal, pilose or hispid; anthers 5. Waste places; included based on mapped location in central part of East TX by Clemants (2003a); only shown for FL in Kartesz (2022) and Weakley, A.S., and Southeastern Flora Team (2022); FL and TX (?). Late winter. If this taxon is present in East TX, it is most likely an ephemeral waif. Native of Brazil. [*Telanthera bettzickiana* Regel, *A. ficoidea* var. *bettzickana* (Regel) Backer, *A. tenella* Colla var. *bettzickiana* (Regel) Veldkamp] Because of its varicolored foliage, this species is widely cultivated in the tropics (Clemants 2003a).

Alternanthera caracasana Kunth, (of Caracas, Venezuela), MAT CHAFF-FLOWER, VERDOLACA DE PUERCO. Stems prostrate, branched, sometimes hirsute; leaves clustered, blades glabrous or sparsely pubescent (chiefly beneath and on margins; tepals awn-tipped, unequal (dimorphic), with numerous minutely barbed hairs. Disturbed areas and waste places; Brazos, Harris, Walker, and Waller (Kartesz 2022) cos., otherwise mainly w part of East TX; s and w TX; widely scattered in se U.S. and throughout sw U.S. Jun–Oct. There are questions concerning the native range of this species; Henrickson (1993) indicated it is native to Central and South America; however, Robertson (1981), in his treatment of the family for the se U.S, considered it native to the s U.S.; Hatch et al. (1991) treated it as native to TX. We are following Clemants (2003) in considering it introduced to the U.S. [*A. peploides* (Humb. & Bonpl. ex Roem. & Schult.) Urb.] Native of Latin America.

Alternanthera paronychioides A. St.-Hil., (resembling *Paronychia*, nailwort or whitlow-wort, Caryophyllaceae). SMOOTH CHAFF-FLOWER, SMOOTH JOYWEED. Stems prostrate; heads white; bracts < 1/2 as long as tepals; tepals nearly equal, the hairs not barbed; anthers 3–5. Waste places, sandy, often wet places; Bastrop, Goliad, Harris (Kartesz 2022), and Travis (BRIT) cos.; mostly Gulf Prairies and Marshes and Trans-Pecos; scattered in the e U.S. Spring–late fall, year-round in very warm areas. Native to the Old World tropics. The specific epithet was originally spelled as *paronichyoides*. The species was reported from se OK (Hoagland & Buthod 2004). [*A. ficoidea* (L.) P. Beauv., *A. polygonoides* of authors, not (L.) R. Br. ex Sweet]

Alternanthera philoxeroides (Mart.) Griseb., (like *Philoxerus*, an Australian genus of Amaranthaceae), ALLIGATOR-WEED. Trailing and mat-forming with erect flowering stems; leaves well-spaced on flowering stems, thick and fleshy, glabrous; heads silvery white; bracts < 1/2 as long as tepals; tepals nearly equal, glabrous. Aquatic or in very wet areas; throughout e 1/2 of East TX and scattered in w 1/2 of East TX; also Tarrant Co. (BRIT) to the w and Gulf Prairies and Marshes; se U.S. from MD s to FL and w to KS and TX, also CA. Mar–Aug. Native of South America. [*Achyranthes philoxeroides* (Mart.) Standl., *Bucholzia philoxeroides* Mart.] This species can be a problematic weed contributing to clogging of aquatic habitats (Buckingham 1996). In Texas,

ALLIGATOR-WEED is considered a "harmful or potentially harmful exotic plant" and it is illegal to release, import, sell, purchase, propagate, or possess this species in the state (Harvey 1998). The species was first reported in the se U.S. in Florida in 1894; it subsequently spread widely and in some areas formed nearly pure stands, outcompeting native vegetation; biological controls (various insects: *Agasicles hygrophila*—flea beetle, *Vogtia malloi*–a moth, and *Amynothrips andersoni* a thrip) have been used (Buckingham 1996); the species apparently does not produce viable seeds in the U.S.—reproduction occurs by plant fragments (Robertson 1981; Clemants 2003).

Alternanthera pungens Kunth (piercing, sharp-pointed), KHAKI-WEED. Stems prostrate; heads straw colored; bracts equaling tepals in length; tepals unequal (dimorphic), strongly awn-tipped, 5–7 mm long, with sparse barbed hairs; anthers 3–5. Waste places; the only TX locality mapped by Clemants (2003) was in the Gulf Prairies and Marshes well to the s of East TX; however, specimens from Bastrop, Brazos, and Travis cos. were found in the BRIT herbarium; also Hidalgo, Llano, Medina, and Wise (BRIT) cos. to the w of East TX; scattered in e coast U.S. states and Gulf Coast. Summer–fall. [*Achyranthes leiantha* (Seub.) Standl.] This species was long (and incorrectly) known as *Alternanthera repens* (L.) G.F. Gmel. Native of South America.

Alternanthera sessilis (L.) R. Br. ex DC., (stalkless), SESSILE JOYWEED. Stems prostrate; heads white; bracts ca. 1/2 as long as tepals; tepals nearly equal, the hairs not barbed; anthers 3–5. Liberty Co. (Davis Hill State Natural Area, *Brown 18040*—first published for TX by Brown & Marcus 1998), also Brazoria Co. (Brown et al. 2007); e U.S. from SC s to FL and w to TX. Summer–early fall. Possibly native of s Asia, now widespread (e.g., Africa, South America). [*Gomphrena sessilis* L.] This species is a U.S. federal noxious weed (USDA Natural Resources Conservation Service 2006). It was reported as new to AR (Reid & Williams 2001).

AMARANTHUS L. GREEN PIGWEED, AMARANTH

Ours annuals with taproots; leaves alternate; inflorescence of dense, terminal or axillary spikes or panicles or small axillary clusters; flowers unisexual, the plants thus monoecious or dioecious; utricle 1-seeded, circumscissile, irregularly splitting or indehiscent; seeds lenticular.

It cannot be stressed enough that *Amaranthus* species can be highly variable as to, branching pattern and plant and leaf size depending on availability of light, water, and nutrients that can make identification difficult. A magnification of 40× or more is often helpful in identification as is true for much of the family.

A genus of ca. 70 species (Mosyakin & Robertson 2003) of primarily tropical, subtropical, and warm temperate areas of the world. As treated here, *Amaranthus* includes the segregates *Acnida* and *Acantochiton*. The seeds of *Amaranthus* species were used as a major "grain" crop (pseudocereal) by Native Americans in pre-Colombian times (e.g., *A. caudatus* L.—INCA-WHEAT, *A. hybridus* L.—especially in Mexico). The protein-rich leaves have been used for millennia and are one of the most common vegetables in the tropics; today several species of *Amaranthus* are often intercropped with beans, maize, and other crops in Mexico (Mapes et al. 1997); some species, however, can accumulate toxic levels of nitrates (Cheatham & Johnston 1995); some are cultivated as ornamentals; others are widespread weeds considered by some to be among the worst annual garden weeds—they grow rapidly and can produce hundreds of thousands of seeds which can survive dormant in the soil for decades (Baumgardt 1982). Other uses by Native Americans include forage for livestock, medicines, face and body paint, and ceremonial items (Cheatham & Johnston 1995; Moerman 1998). Hybridization is common and widespread in the genus (Sauer 1955). Pistillate plants are often necessary for accurate identification of dioecious species (Mosyakin & Robertson 2003). (Greek: *amaranthos*, unfading, because the dry calyx and bracts are persistent) The following key is adapted from Mosyakin and Robertson (2003) and Weakley, A.S., and Southeastern Flora Team (2022).

REFERENCES: Sauer 1955, 1967; Hendrickson 1999; Pratt & Clark 2001; Mosyakin & Robertson 2003.

1. Plants dioecious.
 2. Plants pistillate.
  3. Tepals present and well-developed (usually 5 present, fewer in *A. tuberculatus*), at least the outer tepals > 2 mm long and with a visible midvein).
   4. Tepals 1 or 2 or absent, lanceolate to linear ____ **A. tuberculatus**
   4. Tepals 5, at least the inner spatulate.
    5. Outermost tepal obtuse to subacute (similar to the others), the midvein excurrent slightly or not at all.
     6. Bracts ± equal to tepals; utricle dehiscent (circumscissile); style branches erect; plants widespread in East TX ____ **A. arenicola**
     6. Bracts shorter than tepals; utricle indehiscent; style branches spreading; plants typically on coastal dunes ____ **A. greggii**
    5. Outermost tepal acute or acuminate (dissimilar to the inner obtuse tepals), the midvein excurrent into a rigid point ____ **A. palmeri**

3. Tepals lacking, or rudimentary (often only 1–2 present, these < 1(–2) mm long and lacking a visible midvein).
   7. Utricle with conspicuous and regular longitudinal ridges; bracts mostly > 1.5 mm long, with a stout midrib not far excurrent beyond the bract blade ____ **A. australis**
   7. Utricle smooth or irregularly tuberculate; bracts mostly < 1.5 mm long, with a slender excurrent midrib ____ **A. tuberculatus**

2. Plants staminate (in some cases, pistillate plants may be required for positive identification).
   8. Outer tepals with prominent midribs, usually longer than the inner tepals; bracts > 2 mm long (or 1–2 mm long in *A. tuberculatus*), mostly with prominent midribs.
      9. Outer tepals with apex acute or obtuse; dark midribs with a minute, sharp point.
         10. Leaves usually thin and soft, blade mostly narrowly ovate, obovate, elliptic, or lanceolate, apex subacute to obtuse; plants widespread in East TX ____ **A. arenicola**
         10. Leaves usually thick and coarse, blade ovate, or rhombic-ovate to elliptic, apex rounded-obtuse to emarginate; plants rare in East TX ____ **A. greggii**
      9. Outer tepals with apex acuminate; midribs excurrent as rigid spines.
         11. Bracts ca. 4 mm long, equaling or exceeding the outer tepals ____ **A. palmeri**
         11. Bracts ca. 2 mm long, shorter than the outer tepals ____ **A. tuberculatus**
   8. Outer tepals without prominent midribs, not appreciably longer than the inner tepals; bracts < 2 mm long, the midribs usually not prominent (except sometimes in *A. australis*).
      12. Bracts with moderately prominent midribs; midribs of outer tepals excurrent; utricles straw-colored to brown ____ **A. australis**
      12. Bracts with slender midribs; midribs of outer tepals not excurrent; utricles reddish brown to black ____ **A. tuberculatus**

1. Plants monoecious.
   13. Inflorescences axillary clusters of glomerules (sometimes leafy, terminal spikes also present).
      14. Pistillate flowers usually with only 1 well-developed tepal, sometimes with 1–3 distinctly unequal tepals; utricles regularly or tardily dehiscent ____ **A. californicus**
      14. Pistillate flowers usually with 3 tepals; utricles usually regularly dehiscent (indehiscent in *A. blitum*) OR pistillate flowers usually with (4–)5 tepals; utricles usually indehiscent or tardily dehiscent (regularly dehiscent in *A. blitoides*).
         15. Pistillate flowers usually with 3 tepals; utricles usually regularly dehiscent (indehiscent in *A. blitum*).
            16. Utricles indehiscent; leaf blades usually deeply notched at the tip.
               17. Plants prostrate; leaves (0.75–)1–2(–3.5) cm long; inflorescences axillary cymes (terminal inflorescences reduced or absent) ____ **A. blitum** subsp. **emarginatus**
               17. Plants ascending; leaf blades 2–4(–6) cm long; inflorescences primarily terminal, thin and flexuous ____ **A. blitum** subsp. **pseudogracilis**
            16. Utricles dehiscent; leaf blades obtuse, acuminate, or very shallowly notched at the tip ____ **A. albus**
         15. Pistillate flowers usually with (4–)5 tepals; utricles dehiscent, tardily dehiscent, or indehiscent.
            18. Inflorescence axes thickened, becoming indurate at maturity.
               19. Utricles indehiscent ____ **A. crassipes** var. **crassipes**
               19. Utricles dehiscent ____ **A. sclerpoides**
            18. Inflorescence axes not thickened, not indurate at maturity.
               20. Utricles with regular, circumscissile dehiscence ____ **A. blitoides**
               20. Utricles indehiscent (or tardily and irregularly dehiscent) ____ **A. polygonoides**
   13. Inflorescences terminal spikes or panicles, leafless or nearly so at least in the distal portions (axillary spikes or clusters usually also present).
      21. Utricles indehiscent; tepals of pistillate flowers usually 2–3 (5 in *A. spinosus*); bracts shorter than the tepals.
         22. Stems with paired nodal spines; tepals of pistillate flowers 5 ____ **A. spinosus**
         22. Stems lacking spines; tepals of pistillate flowers 2–3.
            23. Utricles distinctly rugose, equaling or slightly longer than the tepals; terminal inflorescences usually thin and interrupted; leaf blade apex obtuse, rounded or emarginate ____ **A. viridis**
            23. Utricles smooth to faintly rugose (occasionally wrinkled or rugose in dried material), distinctly exceeding the tepals; terminal inflorescences usually thick and dense (or thin and interrupted in some forms of *A. blitum*), leaf blade usually distinctly notched.
               24. Plants ascending; leaf blades 2–4(–6) cm long; inflorescences primarily terminal, thin and flexuous ____ **A. blitum** subsp. **pseudogracilis**
               24. Plants prostrate; leaf (0.75–)1–2(–3.5) cm long; inflorescences axillary cymes (terminal inflorescences reduced or absent) ____ **A. blitum** subsp.. **emarginatus**
      21. Utricles dehiscent; tepals of pistillate flowers usually 5 (3–5 in *A. powellii*); bracts ≥ the tepals (shorter than the tepals in some cultivated forms).
         25. Fully developed inflorescences large and robust, usually brightly colored (red, purple, rarely green); bracts usually not exceeding style branches at maturity; seeds white, ivory, red, or, brown; [cultivated, only weakly naturalized] ____ **A. cruentus**

Alternanthera bettzickiana [MAI]

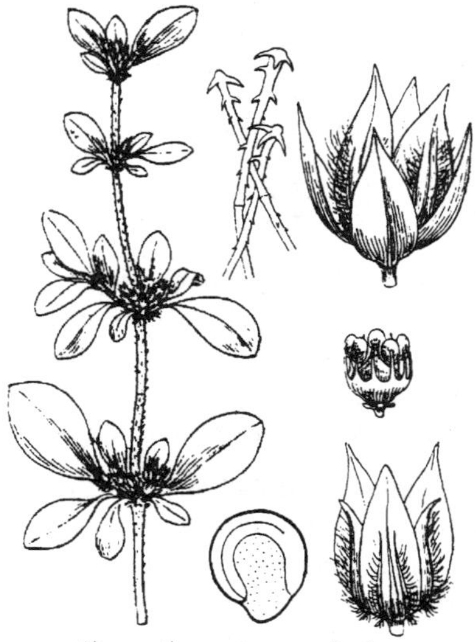

Alternanthera caracasana [FAW]

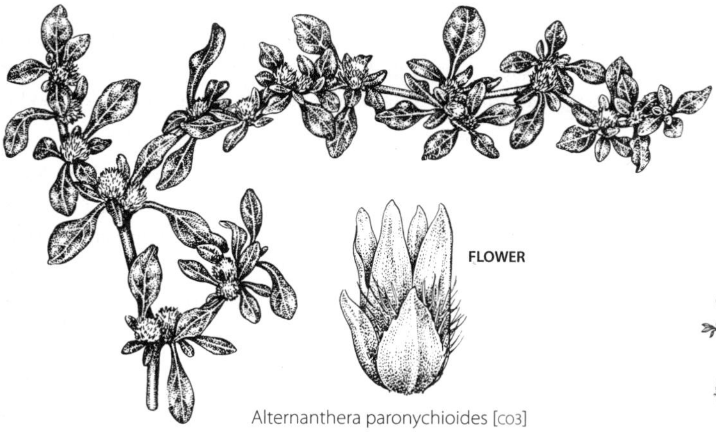

Alternanthera paronychioides [CO3]

Alternanthera philoxeroides [REE]

Alternanthera pungens [MAR]

Alternanthera sessilis [KOO]

Amaranthus albus [REE]

25. Fully developed inflorescences moderately large, usually green (rarely with some whitish or reddish coloration); bracts exceeding the style branches and tepals; seeds brown or black; [wild and weedy].
    26. Tepals of pistillate flowers obtuse, rounded, or slightly notched at the tip; plants rather densely pubescent ____ **A. retroflexus**
    26. Tepals of pistillate flowers acute, acuminate, or aristate at the tip; plants slightly pubescent when young, becoming glabrous or nearly so.
        27. Bracts 2–4 mm long; inflorescences usually soft and lax, with spreading branches ____ **A. hybridus**
        27. Bracts 4–7 mm long; inflorescences usually stiff, with erect branches ____ **A. powellii**

Amaranthus albus L., (white), TUMBLEWEED, TUMBLEWEED AMARANTH, WHITE AMARANTH, TUMBLE PIGWEED. Stems erect, stiff, whitish or pale green, 0.2–1.2 m tall; branches spreading; bracts of pistillate flowers 2 times as long as tepals; staminate and pistillate flowers intermixed; seeds 0.6–0.8 mm in diam., reddish brown, shiny. Disturbed sites; mainly c and w parts of East TX; widespread in TX; throughout the U.S.; s Canada. Aug–Dec. [*A. albus* var. *pubescens* (Uline & W.L. Bray) Fernald, *A. pubescens* (Uline & W.L. Bray) Rydb.] Native distribution difficult to determine.

Amaranthus arenicola I.M. Johnst., (growing in sandy places), TORREY'S AMARANTH, SANDHILL AMARANTH, SAND AMARANTH. Plant erect, to 2(–3) m tall; leaf blades narrowly ovate, obovate, elliptic, or lanceolate; seeds 1–1.3 mm in diam., dark reddish brown, shiny. Sandy areas; mostly in the sw portion of East TX; widespread in TX; native to c and sw Great Plains (Mosyakin & Robertson 2003), but introduced more widely. Jun–Oct. [*A. torreyi* of authors, not (A. Gray) Benth. ex S. Watson] Mosyakin and Robertson (2003) noted that staminate plants of *A. tuberculatus* have often been misidentified as this species.

Amaranthus australis (A. Gray) J.D. Sauer, (southern), SOUTHERN WATER-HEMP, SOUTHERN AMARANTH. Plant erect, 2–3(–9!) m tall; inflorescences mostly terminal spikes to panicles, usually interrupted; tepals of staminate flowers 5; tepals of pistillate flowers absent; seeds 1–1.25 mm in diam., dark reddish brown to black, shiny. Marshy places; widely scattered in East TX; widely scattered in TX; se U.S. from VA s to FL w to TX. May–Aug. [*Acnida australis* A. Gray, *A. cannabina* L. var. *australis* (A. Gray) Uline & W.L. Bray, *A. cuspidata* Bertero ex Spreng.]

Amaranthus blitoides S. Watson, (resembling blithe, a common name for *Chenopodium* species), PROSTRATE PIGWEED, PROSTRATE AMARANTH, MATWEED, MATWEEED AMARANTH, QUELITE MANCHADO. Stems prostrate to rarely ascending, trailing and forming mats, to 1 m long, glabrous or nearly so; leaf blades oblong-oblanceolate; staminate and pistillate flowers intermixed; staminate flowers with 3(–4) tepals; pistillate flowers with (3–)4–5 tepals; seeds 1.3–1.6 mm in diam., black, dull. Disturbed sites, weedy areas; c and w portions of East TX; widespread mainly in w 2/3 of TX; throughout the U.S.; s Canada. Jun–Oct. [*A. graecizans* of authors, not L.]

Amaranthus blitum L., (old generic name for *Chenopodium capitatum* (L.) Ambrosi), PURPLE AMARANTH. Glabrous annual, prostrate to erect 1–6 dm high; petiole equal to 2 times as long as blade, ovate or obovate, apex distinctly notched, mucronate; bracts of pistillate flowers lanceolate, inconspicuous, 0.5 mm long, shorter than tepals; staminate flowers at the tips of the inflorescence; tepals of pistillate flowers 0.8–1.5 mm long, apex broadly acute; utricle exceeding tepals. Fields, roadsides, disturbed habitats; scattered in East TX; Gulf Prairies and Mashes and South TX Plains; widely scattered in e U.S. from NY s to FL and w to IL and TX, also w tier of U.S. states; Central and South America; Eurasia; and Africa. Sep–Oct. [*A. ascendens* Loisel., *A. lividus* L.] Native of the topics. Two varieties in our area.

1. Plants prostrate; leaves (0.75–)1–2(–3.5) cm long; inflorescences axillary cymes (terminal inflorescences reduced or absent); plants widespread in East TX ____ subsp. **emarginatus**
1. Plants ascending; leaf blades 2–4(–6) cm long; inflorescences primarily terminal, thin and flexuous; plants rare in East TX ____ subsp. **pseudogracilis**

Amaranthus blitum L. subsp. **emarginatus** (Uline & W.L. Bray) Carretro, Muñoz Garm., & Pedrol, (subsp.: notched in reference to the leaf apex). Sandy alluvium on banks of rivers, lake shores, and beaches; distribution as in the species.

Amaranthus blitum L. subsp. **pseudogracilis** (Thell.) N. Bayón ex Thell., (subsp.: from Latin: *pseudo*, false and *gracilis*, slender). Weedy in vegetable crops and nurseries; known in TX only in Orange Co. (Costea et al. 2001); FL, LA, NC, and TX (Costea et al. 2001). [*A. blitum* L. var. *pseudogracilis* (Thell.) Costea]

Amaranthus californicus (Moq.) S. Watson, (of California), CALIFORNIA AMARANTH. Stems prostrate, whitish or tinged with pink, 0.1–0.5 m long; petiole 1/2 or less as long as blade; staminate flowers intermixed with pistillate flowers; bracts of pistillate flowers linear, 0.5–1 mm long; utricles subglobose, 1–1.2 mm long; dehiscence regular circumscissile or tardily dehiscent; seeds very dark reddish brown. Waste places other disturbed habitats; known only from Harris Co. (Kartesz 2022); Trans-Pecos and High Plains; scattered in the w 1/2 of the U.S.; Canada. Summer–fall. Native of the U.S.

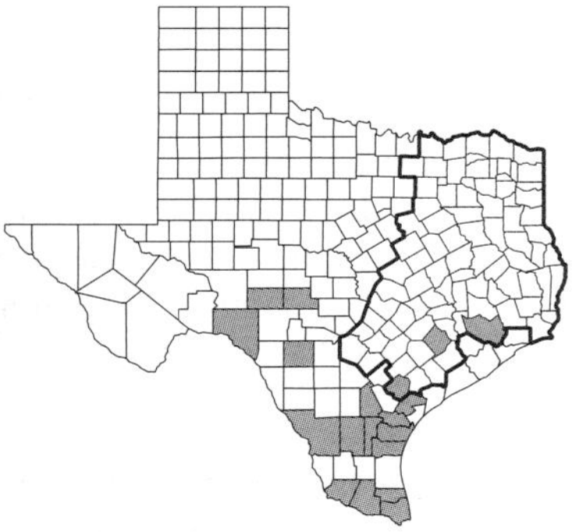

Ruellia yucatana

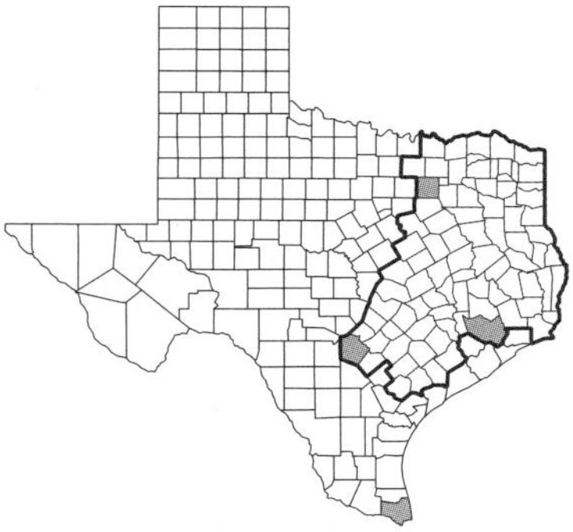

Thunbergia alata

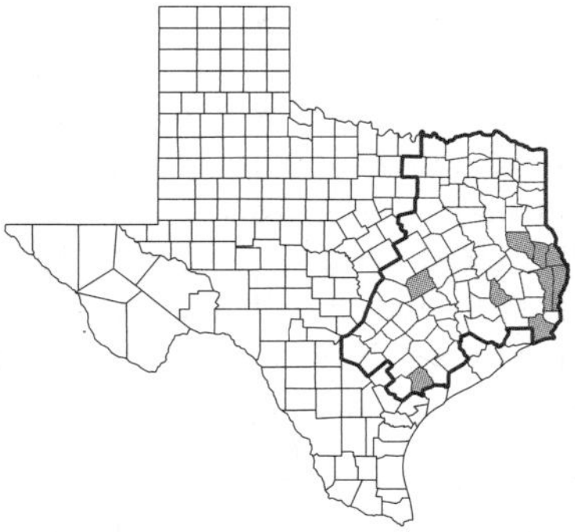

Yeatesia viridiflora

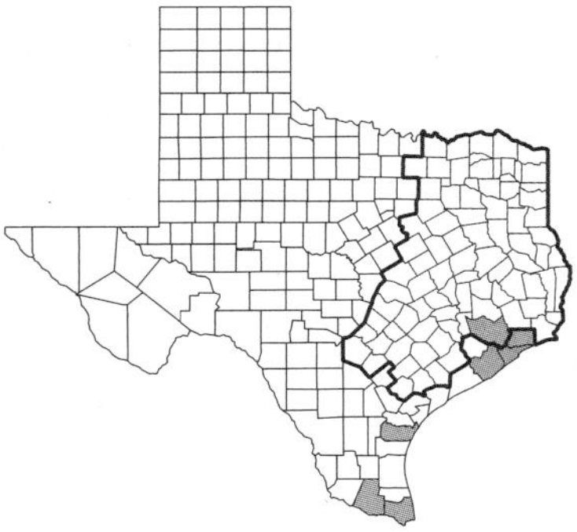

Sesuvium maritimum

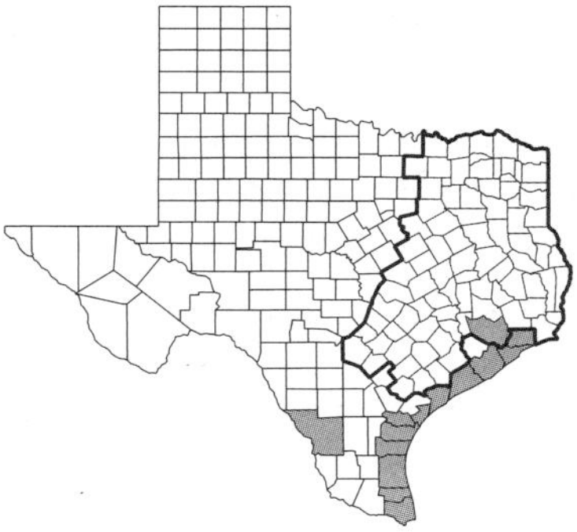

Sesuvium portulacastrum

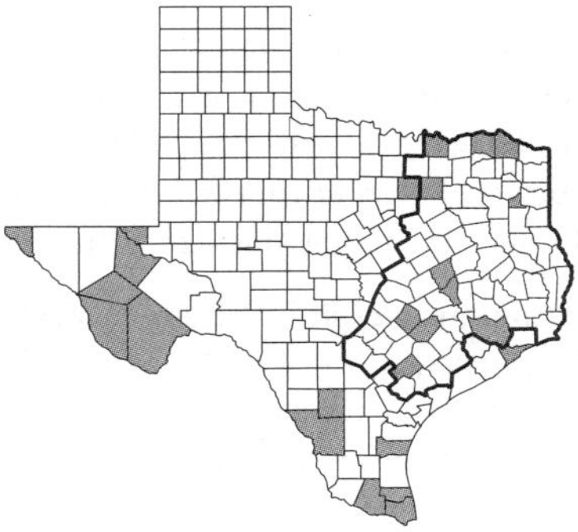

Trianthema portulacastrum

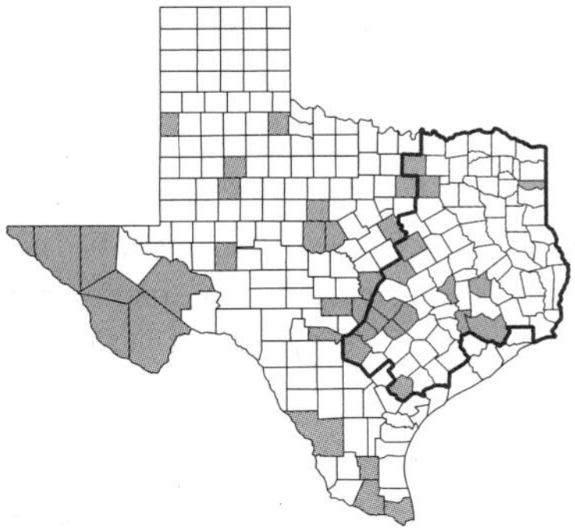

Alternanthera caracasana

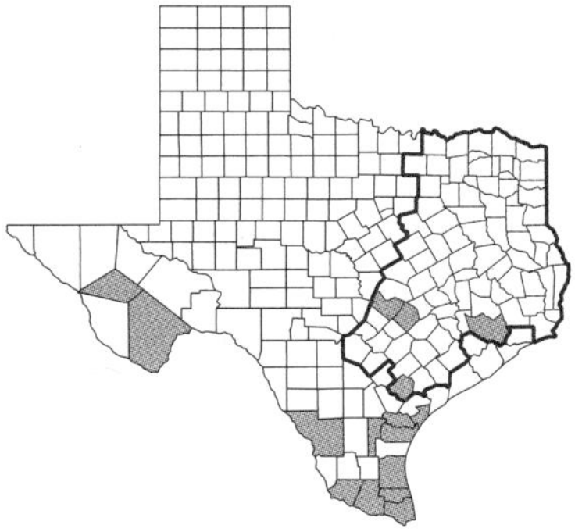

Alternanthera paronychioides

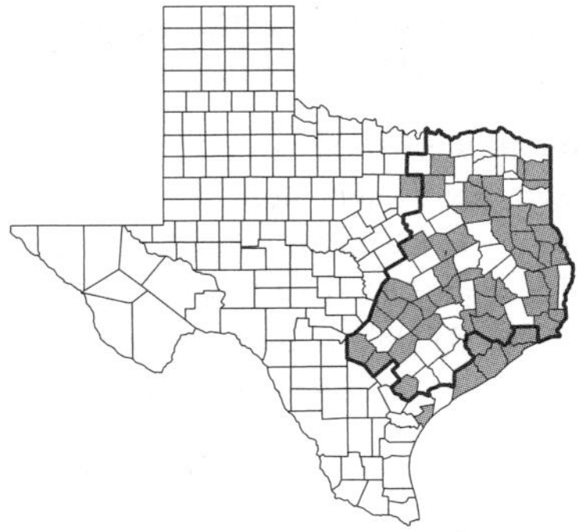

Alternanthera philoxeroides

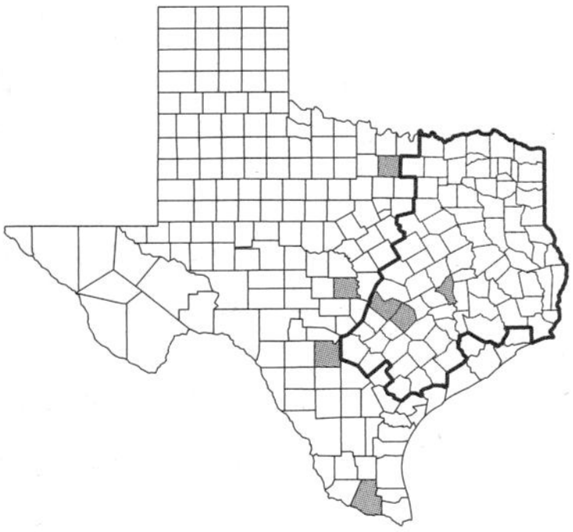

Alternanthera pungens

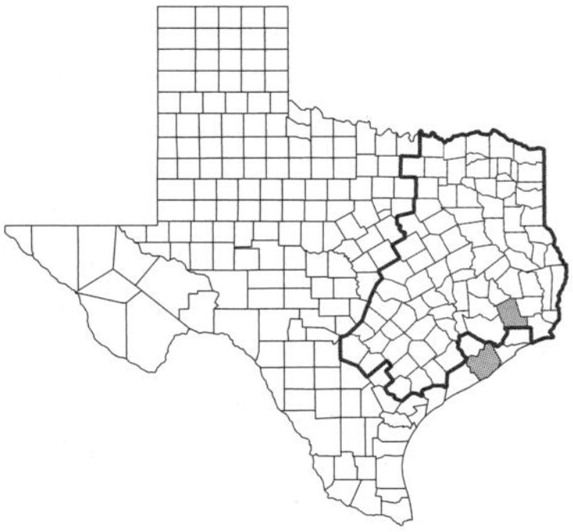

Alternanthera sessilis

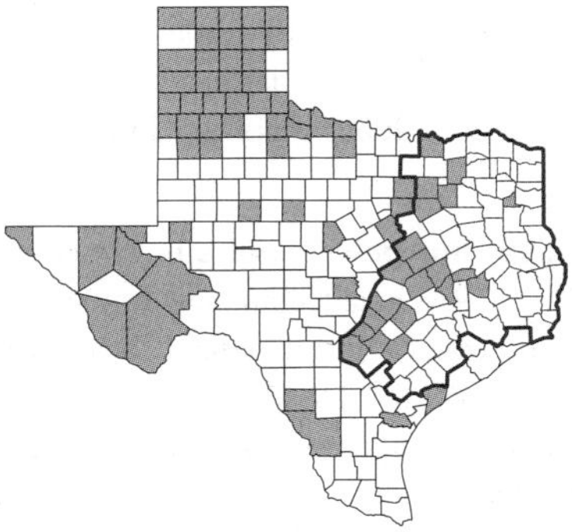

Amaranthus albus

Amaranthus crassipes Schltdl., (with thick stem), CLUBFOOT AMARANTH, SPREADING AMARANTH, TROPICAL SPREADING AMARANTH. Annual with prostrate or weakly ascending stems, 0.1–0.6 m long; leaf blade 1–3 cm long and 0.3–2 cm wide; inflorescence axillary clusters borne from base to apex, 5–9 mm long, the axis becoming thick, hardened, and tortuous; staminate flowers intermixed with pistillate flowers; bracts of pistillate flowers deltate; tepals of pistillate flowers narrowly spatulate, 1.2–2 mm long, becoming hardened; stigma spreading; utricle indehiscent. Open seasonally wet flats, waste places, disturbed habitats; Angelina, Bell, Bexar, McLennan, Travis, and Williamson (Kartesz 2022) cos. on w margin of East TX; South TX Plains and Coastal Prairies and Marshes; AL, AZ, FL, LA, NM, SC, and TX. Summer–fall. Native of tropical America. Henrickson (1999b) distinguished two varieties in this species, a treatment followed by Turner et al. (2003). All East TX collections are of the type variety. The other var. (*A. crassipes* var. *warnockii* (I.M. Johnst.) Henrickson) is known from far w TX and NM.

Amaranthus cruentus L., (blood-red, dull-red), BLOOD AMARANTH, PURPLE AMARANTH, CATERPILLAR AMARANTH. Stems erect, greenish or reddish purple, 0.4–2 m tall; petiole 1/2 as long to equaling blade length, blade 3–15(–20) cm long and 1.5–10(–15) cm wide, ovate, rhombic-ovate to broadly lanceolate; inflorescence dark red to purple, sometimes green; bracts lanceolate to ovate, equaling or slightly longer than tepals, midrib extending into a spine; staminate flowers at the tips of the inflorescence; tepals of pistillate flowers 1–1.6 mm long, narrowly oblong, apex acute; utricles 2–2.5 mm long, smooth or slightly rugose distally. Near cultivated areas; Anderson and Kaufman (Kartesz 2022) cos. in East TX and Webb Co. (Kartesz 2022) in the South TX Plains; scattered in the U.S. Jul–Oct. [*A. hybridus* L. subsp. *cruentus* (L.) Thell.] Mosyakin and Robertson (2003) noted that this species is cultivated as an ornamental and pseudocereal almost worldwide and that it "originated from *A. hybridus* (most probably in cultivation in Central America), with which it shares almost all major morphologic characteristics."

Amaranthus greggii S. Watson, (named for its discoverer Josiah Gregg, American merchant, explorer, and naturalist), GREGG'S AMARANTH, JOSIAH AMARANTH. Stems stout and much-branched at base, 1–10 dm tall; petiole shorter than blade, blade 1.5–4 cm long and 0.5–2 cm wide, ovate or rhombic-ovate to elliptic; thick and course, apex rounded to emarginate with a mucro; bracts of staminate flowers shorter than tepals; bracts of pistillate flowers with short excurrent midrib, shorter than tepals, apex acute; tepals of staminate flowers 3 mm long, apex subacute to obtuse; tepals of pistillate flowers 2–3 mm long, with dark midrib not excurrent, spatulate with mucro, apex subacute to obtuse; utricle 2–3 mm long, equaling tepals. Coastal sand dunes, sea beaches; Bexar, Jefferson, and Travis (Kartesz 2022) cos, typically on coastal dunes; Gulf Prairies and Marshes, also Atascosa Co. in the South TX Plains; LA and TX; endemic to West Gulf Coastal Plain. Jun–Oct. Staminate plants of *A. greggii* can be difficult to distinguish from *A. arenicola*. Both species can be found in coastal areas.

Amaranthus hybridus L., (hybrid), GREEN AMARANTH, GREEN PIGWEED, SMOOTH AMARANTH, SMOOTH PIGWEED, HYBRID AMARANTH, SLENDER PIGWEED. Stems erect, stout, 0.5–1.5(–2.5) m tall; spikes slender, 6–10(–12) mm broad, terminal and axillary; staminate flowers at the tips of the inflorescence; utricle dehiscent; seeds round, 1.1–1.3 mm in diam., black, shiny. Pioneer weed, disturbed areas; widely scattered in East TX; mainly e 1/2 of TX, rare to the w.; throughout the U.S. May–Oct. Native to e North America; now a widespread naturalized weed in such areas as the Mediterranean region, South Africa, Australia, and e Asia (Correll & Johnston 1970; Mosyakin & Robertson 2003).

Amaranthus palmeri S. Watson, (for its discoverer, Edward Palmer, 1831–1911, English-American surgeon and botanist, collected in w U.S. and Mexico), CARELESSWEED, PALMER'S PIGWEED, PALMER'S AMARANTH, REDROOT. Plant erect, 1–2(–3) m tall; branches of inflorescence often very long, densely flowered, cylindrical, half-drooping; bracts of staminate flowers ca. 4 mm long; bracts of pistillate flowers 4–6 mm long including long excurrent midrib; seeds 1–1.3 mm in diam., dark reddish brown. Disturbed sites and weedy areas; nearly throughout TX, often abundant; widespread in the U.S. except Pacific Northwest; Ontario, Canada. Jun–Oct. This now widespread species is thought to have originally been native to the sw U.S. and adjacent Mexico (Mosyakin & Robertson 2003). Implicated in livestock poisoning through the accumulation of nitrates (Kingsbury 1964; Burlage 1968). ☠

Amaranthus polygonoides L., (resembling *Polygonum*—knotweed), TROPICAL AMARANTH, BERLANDIER'S AMARANTH. Stems erect, stiff, 0.15–0.3 m tall, branching from the base; branches spreading, prostrate to ascending; staminate and pistillate flowers intermixed; seeds ca. 1 mm in diam., black, shiny. Disturbed sites; widespread in the s 1/2 of East TX.; widespread in the s 1/2 of TX; FL, NM, SC, and TX. May–Oct. [*A. berlandieri* (Moq.) Uline & W.L. Bray] Native of tropical America. While *A. berlandieri* has often been recognized as a distinct species (e.g., Correll & Johnston 1970; Hatch et al. 1990), Henrickson (1999b) indicated that the evidence does not justify recognizing it as distinct from *A. polygonoides*.

Amaranthus powellii S. Watson, (named for its discoverer John Wesley Powell, 1834–1902, geologist, U.S. Army soldier, explorer, and professor), POWELL'S AMARANTH, GREEN AMARANTH, POWELL'S SMOOTH AMARANTH. Stems erect green or sometimes reddish purple, 0.3–1.5(–2) m; leaf blade rhombic-ovate to broadly lanceolate, 4–8 cm long and 2–3 cm wide; bracts 4–7 mm long, 2–3 times

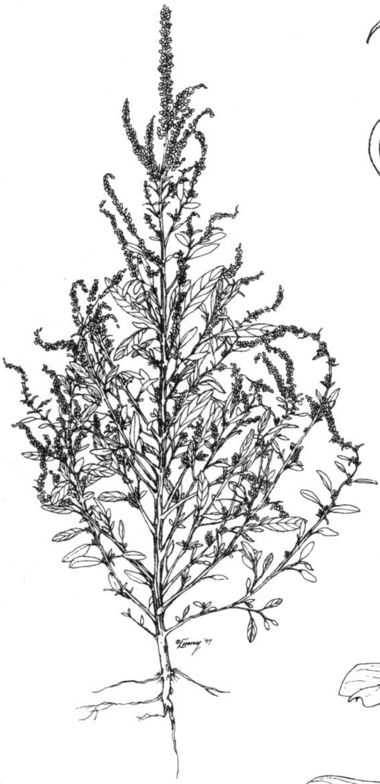

Amaranthus arenicola [HEA]

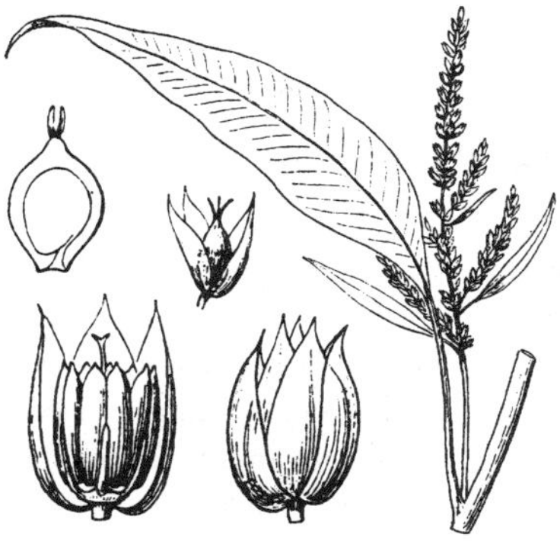

Amaranthus australis [FAW]

Amaranthus blitoides [REE]

Amaranthus blitum [RAD]

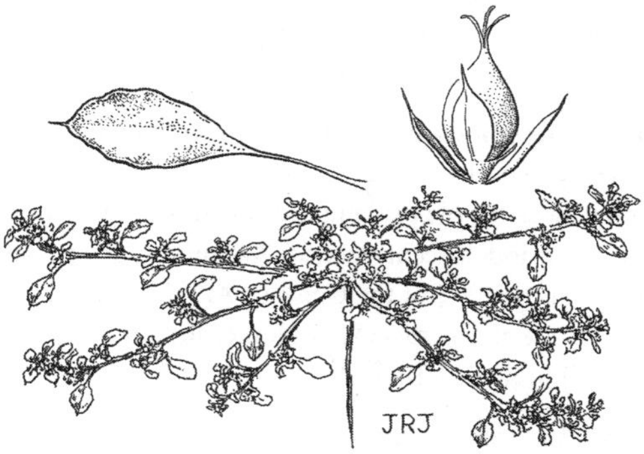

Amaranthus californicus [PN2]

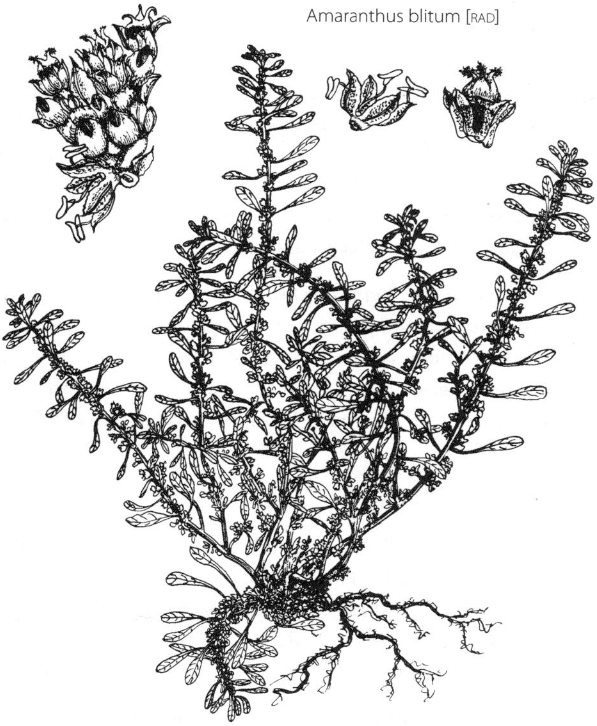

Amaranthus crassipes [CO1]

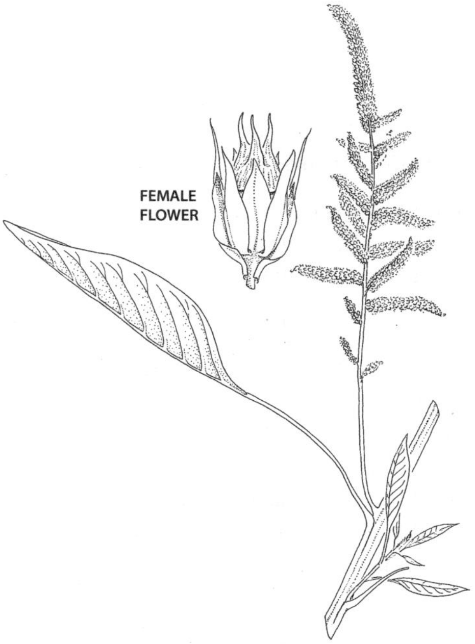

Amaranthus cruentus [HOL]

as long as tepals; staminate flowers at the tips of the inflorescence; utricles equaling or shorter than tepals, dehiscent. Disturbed habitats, roadsides, railroads, banks of streams; scattered sw 1/2 of East TX; Edwards Plateau, Trans-Pecos, and South TX Plains; scattered in e 1/2 of U.S. and throughout w 1/2 of the U.S.; adjacent Canada. Jul–Oct. [*A. bracteosus* Uline & W.L. Bray, *A. retroflexus* L. var. *powellii* (S. Watson) B. Boivin] This species was originally native to the sw U.S. and adjacent Mexico (Mosyakin & Robertson 2003).

Amaranthus retroflexus L., (reflexed or twisted back), RED-ROOT PIGWEED, ROUGH PIGWEED, REDROOT AMARANTH, WILD-BEET AMARANTH, COMMON AMARANTH, GREEN AMARANTH, QUELITE. Plants erect, 0.2–1.5(–2) m tall; leaf blade 2–15 cm long and 1–7 cm wide, ovate to rhombic-ovate; bracts lanceolate to subulate, exceeding tepals, apex acuminate, midrib excurrent, spinose; staminate flowers at the tips of the inflorescence; utricle shorter to nearly equal to tepals, dehiscent; seeds 1–1.3 mm in diam., dark reddish brown, shiny. Disturbed sites; widely scattered in East TX; widespread in TX; throughout the U.S.; much of Canada. May–Oct. Apparently introduced in Texas; first found in the state in 1894 (Mahler 1988). A cosmopolitan weed native to e and c North America (Mosyakin & Robertson 2003). Animals may be poisoned from eating the plants due to the accumulation of large amounts of nitrates (Burlage 1968; Stephens 1980). [*A. retroflexus* L. var. *salicifolius* I.M. Johnst.] ☠

Amaranthus scleropoides Uline & W.L. Bray, (from Greek: *scleros*, hard; *pod*, foot; and *-oides*, resembling, presumably referring to the thickened and hardened inflorescence branches), BONEBRACT AMARANTH, BONEBRACT PIGWEED. Similar to *A. crassipes* var. *crassipes* except utricles regularly circumscissile dehiscent; staminate and pistillate flowers intermixed. Seasonally wet disturbed habitats; sw margin of East TX; Edwards Plateau, South TX Plains, and Trans-Pecos; OK and TX. Jan–Dec. [*A. blitoides* S. Watson var. *scleropoides* (Uline & W.L. Bray) Thell.] A hybrid between this species and *A. crassipes* (*A.* ×*texensis* Henrickson) from the South TX Plains was described by Henrickson (1999b).

Amaranthus spinosus L., (full of spines), SPINY PIGWEED, THORNY AMARANTH, SPINY AMARANTH, QUELITE ESPINSO. Plant erect, 0.2–1.2 m tall; the only species in East TX with distinct spines at the nodes, to 1.5(–2.5) cm long; leaf blades ovate to rhombic-lanceolate; inflorescences simple or compound with terminal staminate spikes and axillary, subglobose mostly pistillate clusters; utricles dehiscent or indehiscent; seeds 0.7–1 mm in diam., black, shiny. Disturbed sites; widespread in East TX; Mainly e 1/2 of TX; throughout the e 1/2 of the U.S. also AZ, CA, and MT; adjacent Canada. May–Oct. Pantropical weed, originally native to lowlands of tropical America (Mosyakin & Robertson 2003). Reported to cause internal mechanical injury and bloat in livestock if ingested (Burlage 1968). ☜

Amaranthus tuberculatus (Moq.) J.D. Sauer, (with tubercles), NUTTALL'S WATER-HEMP, WATER-HEMP, ROUGH-FRUIT WATER HEMP, TALL WATER-HEMP, ROUGH-FRUIT WATER-HEMP. Plant highly variable, glabrous, 0.15–2.5 m tall; spikes often interrupted towards the base into glomerules; seeds ca. 1 mm in diam., dark reddish brown. Low, moist, disturbed sites, fields; invader species; mainly s and w portions of East TX; widely distributed in parts of TX; throughout the midwest and scattered in e and w U.S.; Canada. Late Apr–Oct. [*Acnida tamariscina* of authors, not (Nutt.) Alph. Wood, *A. tuberculata* Moq., *Amaranthus rudis* J.D. Sauer, *A. tamariscinus* of authors, not Nutt.] One of the most abundant native weeds of late summer and fall. In TX this species has gone under the name *Amaranthus rudis* (Hatch et al. 1990; Diggs et al. 1999); however, research (Pratt & Clark 2001) and long-term observation (K.R. Robertson in Mosyakin & Robertson 2003) support treating *A. rudis* as part of *A. tuberculatus*.

Amaranthus viridis L., (green), SLENDER AMARANTH, TROPICAL GREEN AMARANTH. Stems erect 0.2–1 m tall; leaf blades 1–7 cm long and 0.5–5 cm wide, rhombic-ovate or ovate; bracts of pistillate flowers 1 mm long, shorter than tepals; staminate flowers inconspicuous, mostly at the tips of the inflorescence; pistillate flowers and staminate flowers with 3 tepals; utricles ovoid to compressed ovoid, equal to or longer than tepals, prominently or faintly rugose. Fields, railroads, waste areas, disturbed habitats; mostly in the se portion of East TX; Gulf Prairies and Marshes; mostly in se U.S. from VA s to FL and w to OK and TX, also sw U.S. Native to South America. Jul–Oct. [*A. gracilis* Desf. ex Poir.] ☜

BLUTAPARON Raf. SILVERHEAD

A genus of 4 species, mostly coastal, tropical and subtropical America, Asia (Japan), and w Africa (Clemants 2003b). The taxa included here in *Blutaparon* were long considered to be part of the genus *Philoxerus* (e.g., Correll & Johnston 1970). However, that genus is a group of Australasian species closely related to *Gomphrena* and distinct generically from what is now called *Blutaparon* (Mears 1982). (Abridged from old Latin name *Bulutaparon*—Mears 1982)

REFERENCES: Mears 1982; Clemants 2003b.

Blutaparon vermiculare (L.) Mears (worm-like), SILVERHEAD. Perennial (occasionally annual) herb, somewhat succulent; stems mostly decumbent, to 1.8(–2) m long, much-branched; leaves opposite, fleshy, the leaf blades ranging from linear to linear-

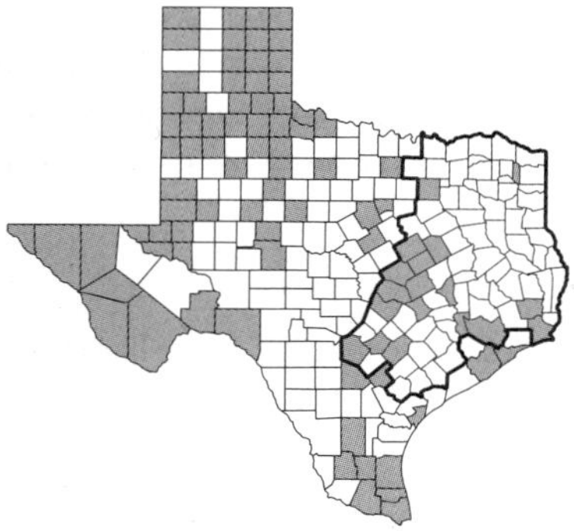

Amaranthus arenicola

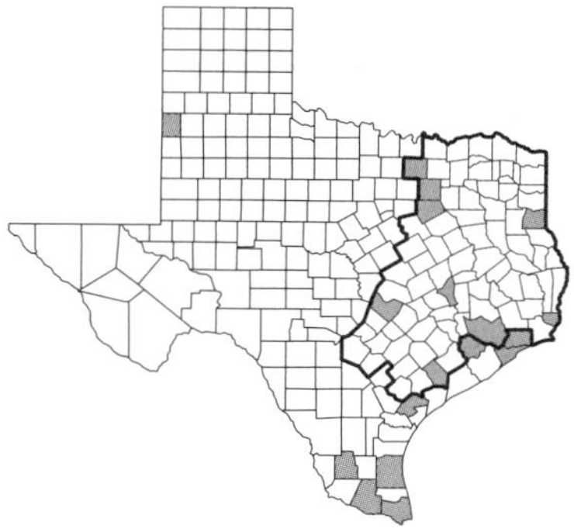

Amaranthus australis

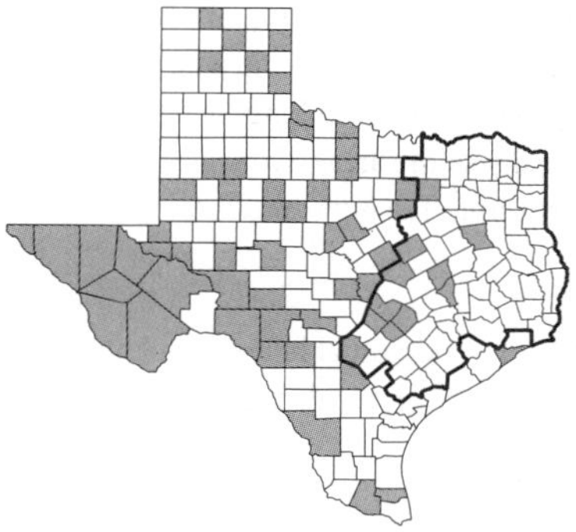

Amaranthus blitoides

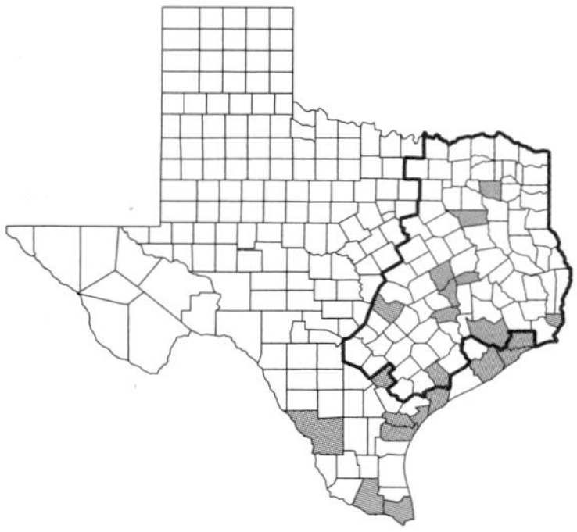

Amaranthus blitum

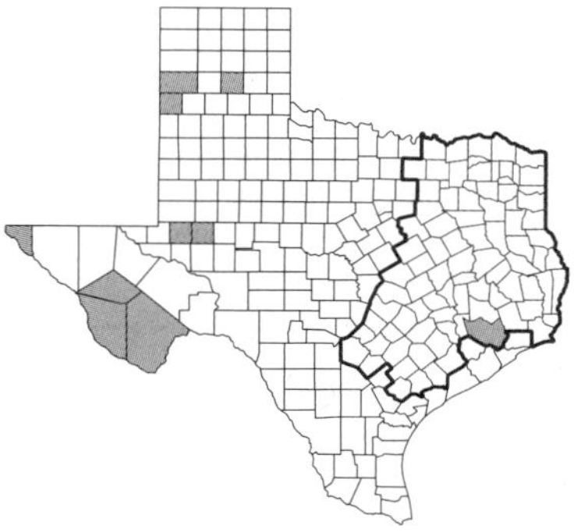

Amaranthus californicus

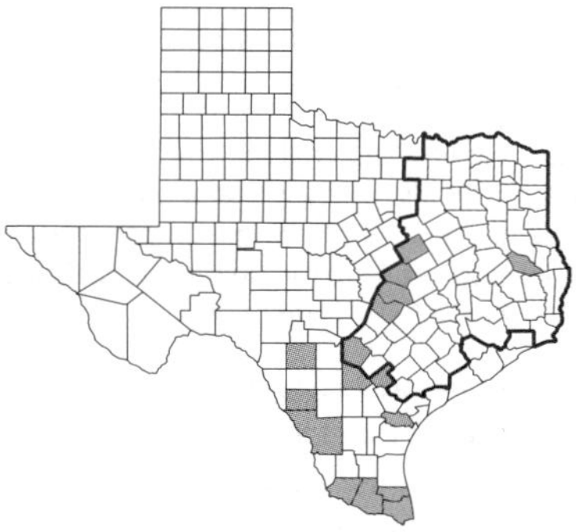

Amaranthus crassipes

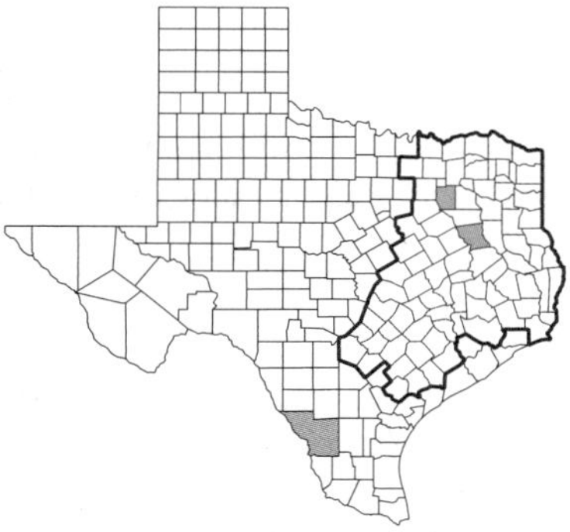

Amaranthus cruentus

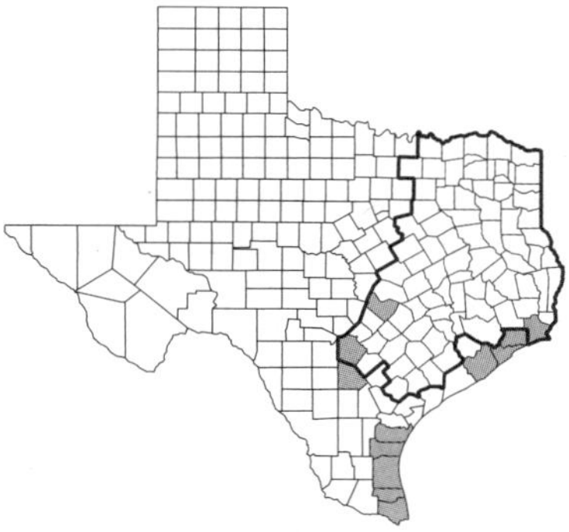

Amaranthus greggii

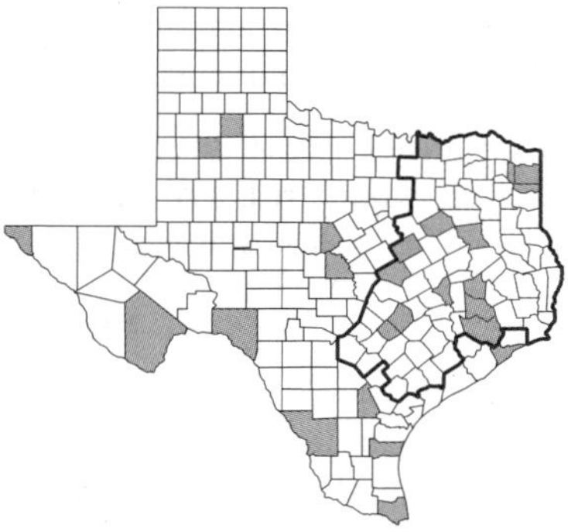

Amaranthus hybridus

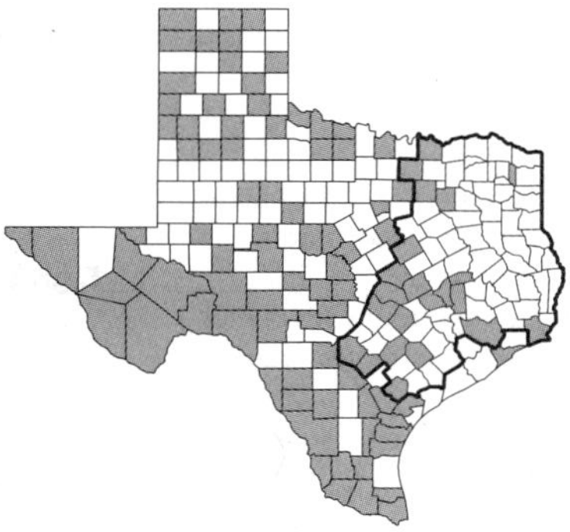

Amaranthus palmeri

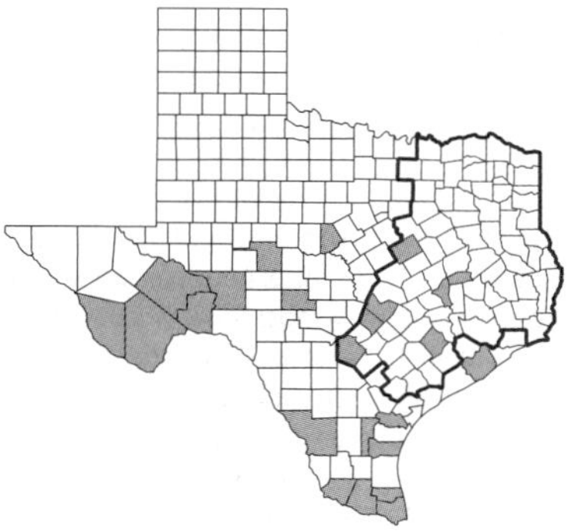

Amaranthus polygonoides

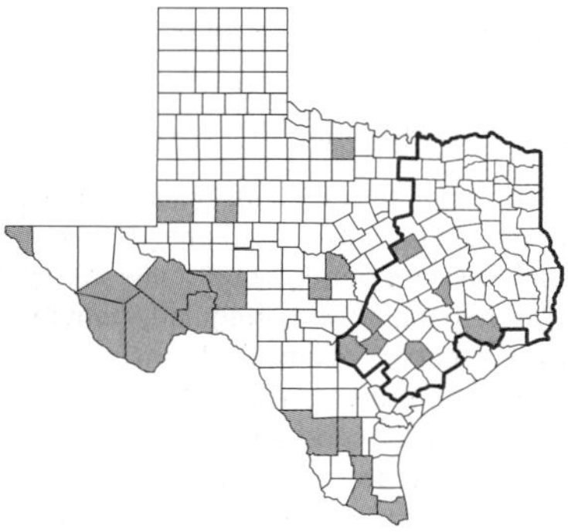

Amaranthus powellii

lanceolate, linear-oblong, or oblanceolate, to 60 mm long and 2–12 mm wide, tapered to a sessile base; inflorescences stalked, subglobose or cylindric heads, densely many-flowered, white or pinkish, 7–30 mm long; flowers perfect with tepals united basally, white, 3–5 mm long, bracteate; stamens 5; utricles indehiscent or splitting irregularly, ca. 1.3 mm long. Sandy areas, dunes, typically saline areas; Gonzales and Jefferson (Kartesz 2022) cos.; mainly coastal in the Gulf Prairies and Marshes; FL, LA, TX. Year-round. [*Gomphrena vermicularis* L., *Philoxerus vermicularis* (L.) R. Br. ex Sm.]

CELOSIA L. COCKSCOMB

Annual or perennial herbs or subshrubs; leaves alternate, petiolate; inflorescences of spikes, these often in panicle-like clusters; flowers perfect; tepals 5, often with a silvery or pinkish tint or variously colored in cultivated forms; stamens 5; utricles with up to 20 seeds.

A genus of ca. 65 species primarily of the tropics and subtropics of the New World, Africa, and Asia (Robertson 2003). The leaves of *C. argentea* are eaten like spinach in some tropical areas, while *C. cristata* L. (COCKSCOMB), with variously colored inflorescences, is widely used as an ornamental; some are used medicinally. (Greek: *keleos*, burning, in reference to the color and or appearance of the inflorescence of *C. cristata*—Robertson 2003)
REFERENCES: Grant 1961, 1962; Robertson 2003.

1. Leaf blades 8–15 cm long, long tapering at base; inflorescences of dense spikes, the individual spikes 13–20 mm in diam.; tepals silvery white or pinkish; style 4 mm long; plants annual, to 1 m tall; seeds 3–8 ____ **C. argentea**
1. Leaf blades 2–7 cm long, abruptly narrowed, truncate or oblique at base; inflorescences of loose spikes, the individual spikes 3–10 mm in diam.; tepals greenish, greenish white or pinkish green; style 1 mm long; plants perennial, 0.3–2 m tall; seeds up to 20 ____ **C. nitidia**

Celosia argentea L., (silvery), SILVER COCKSCOMB, COCKSCOMB, CRESTA DE GALLO. Annual herb; stems erect, to 1 m tall; tepals 6–8 mm long; utricle 4 mm long, with 3–8 seeds. Weedy areas, waste places; Brazos, Harris, Robertson, and Williamson (Kartesz 2022) cos.; otherwise not known from TX; widely scattered in e U.S. from VT s to FL and west to WI and TX, also Utah. Summer–early fall. Perhaps native to India (Robertson 2003).

Celosia nitida Vahl, (shining), WEST INDIAN COCKSCOMB, ALBAHACA. Perennial herb or subshrub; stems erect or clambering, 0.3–2 m tall; tepals 3.5–7 mm long; utricle 4–5 mm long, with up to 20 seeds; in the sw portion of East TX; mainly s TX; FL and TX; Mexico; West Indies and South America. Fall–winter.

Celosia cristata L., (crested), COCKSCOMB, [*C. argentea* L. var. *cristata* (L.) Kuntze]. This annual species, with variously colored inflorescences (pink, red, yellow, purple, white), is widely used as an ornamental. It is known to escape in various parts of the e U.S. including LA. It is easily recognized by the colorful crested, fasciated, or plumose inflorescences.

FROELICHIA Moench SNAKE-COTTON, COTTONWEED

Ours weak perennials or annuals, gray with ± matted pubescence, simple or with long, erect or partly decumbent branches from basal part of plant; leaves few, mostly in basal part of plant, blades oblong or oblanceolate; inflorescences leafless; bracts white or yellowish, unequal, shorter than the perianth; flowers perfect; tepals 5; perianth densely and obviously woolly outside, becoming hardened at maturity, with (0–)1–3 tubercles or spines; stamens with united filaments forming a scarious tube, resembling an additional inner perianth; utricle surrounded by the hardened perianth.

A genus of 18 species of warm areas of the Americas and the Galapagos Islands (McCauley 2004). Hybridization between species is not uncommon. (Named for Joseph Aloys Froelich, 1766–1841, a German botanist)
REFERENCES: McCauley 2004; McCauley & Ballard 2007a, 2007b.

1. Mature perianth with irregularly and deeply cut ("spiny") lateral wings; flowers 2.4–3.8 mm long; stems slender, branched from base (except in depauperate specimens with only 1 slender stalk); floral spikes 3-ranked; bracteoles glabrous, not pubescent distally; stem pubescence grayish-white; widespread and generally weedy ____ **F. gracilis**
1. Mature perianth with irregularly dentate to crenulate lateral wings; flowers 3.5–6 mm long; stems stout with one or more erect to decumbent branches from ground level; floral spikes 5-ranked (rarely 3-ranked); bracteoles sparsely or densely pubescent distally; stem pubescence grayish-white or brownish.
   2. Pseudostaminodes blunt, brown-tinted (often observed as darkened flower tips), 0.5–0.7 mm long, usually not recurved; bracteoles sparsely hairy distally; flowers 3.5–5.4 mm long ____ **F. drummondii**
   2. Pseudostaminodes acute, light colored or pinkish, 0.5–1.3 mm long, usually recurved; bracteoles with small tufts of distinct pubescence distally; flowers 4–6 mm long.
      3. Leaf blades orbicular to broadly elliptic, abaxial surface densely woolly; rare in East TX ____ **F. latifolia**
      3. Leaf blades linear to lanceolate to oblanceolate or oblong, abaxial surface silky; widespread in East TX ____ **F. floridana**

Amaranthus hybridus [GLE]

Amaranthus palmeri [CO1]

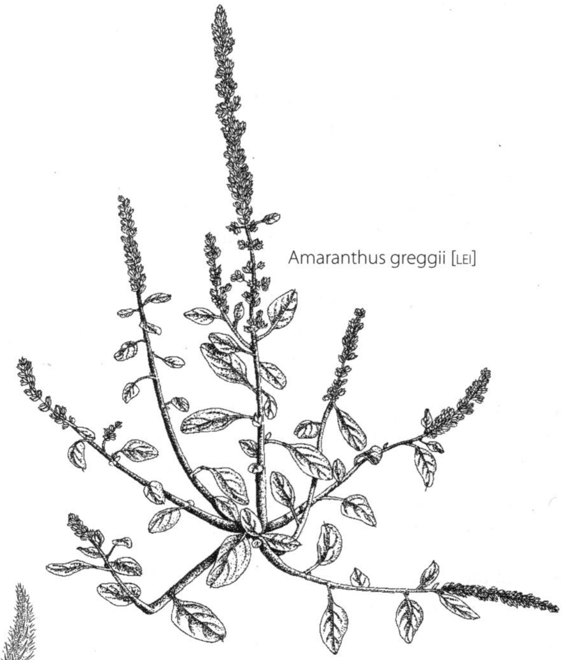

Amaranthus greggii [LEI]

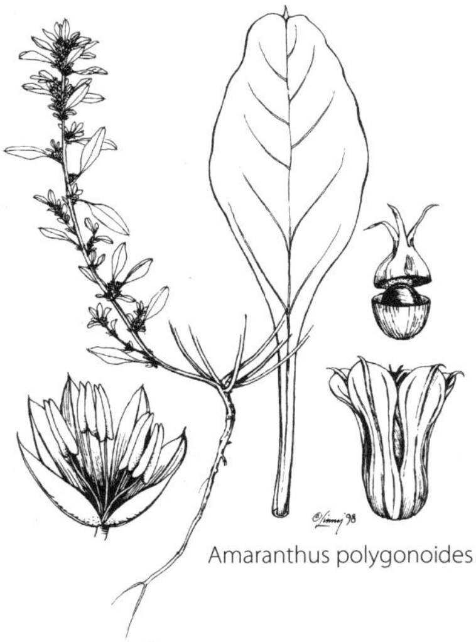

Amaranthus polygonoides [HEA]

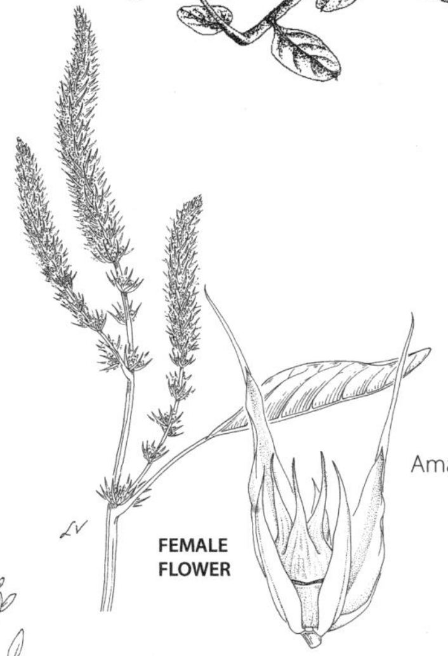

Amaranthus powellii [HOL]

Amaranthus retroflexus [REE]

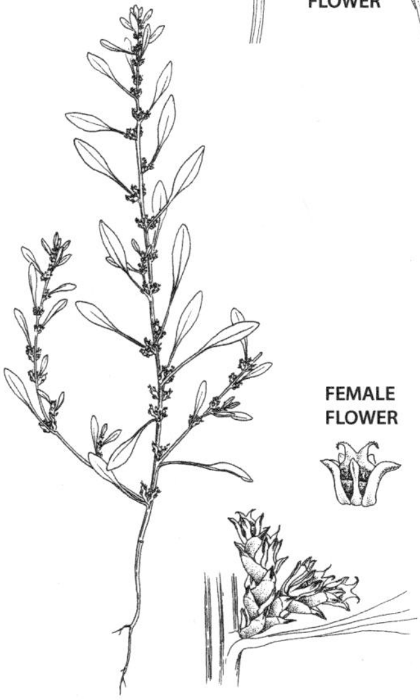

Amaranthus sclerpoides [SID]

Amaranthus spinosus [GLE]

Froelichia drummondii Moq., (for its discoverer Thomas Drummond, 1780–1835, Scottish botanist and collector in North America), DRUMMOND'S SNAKE-COTTON. Leaf blade broadly lanceolate to orbicular; bracteoles glabrous or sparsely pubescent distally; fruiting perianth with irregularly dentate wings, both surfaces of perianth with distinct spines or tubercles. Open sandy prairies, edges of woodlands, and roadsides; in East TX known only from Bexar, Guadalupe (Kartesz 2022), and Wilson (*Nesom 5223* TEX) cos. on the extreme sw margin of East TX; South TX Plains; endemic to TX. Summer.

Froelichia floridana (Nutt.) Moq., (of Florida), FIELD SNAKE-COTTON. Fruiting perianth with narrow to broad, nearly entire to deeply dentate or lacerate lateral wings or crests with one or few tubercles or short spines on each face at base or occasionally up to the middle. Loose sandy soils, often abundant on roadsides and disturbed sites; throughout East TX; mostly in se and c U.S. Jun–Oct. While some authors Kartesz (2022) and Weakley, A.S., and Southeastern Flora Team (2022) recognize two varieties of *Froelichia* (var. *campestris* and var. *floridana*) we are following McCauley and Ballard (2002) and McCauley (2003) in not recognizing vars. *campestris* and *floridana*. [*F. campestris* Small, *F. floridana* var. *campestris* (Small) Fernald] Native of North America and the West Indies. Wind-pollinated and considered a cause of hay fever (Ajilvsgi 1984).

Froelichia gracilis (Hook.) Moq., (graceful), SLENDER SNAKE-COTTON. Fruiting perianth lateral wings spinier than in *F. floridana*, lateral wings of distinct and rather sharp, separate or united, spiny teeth and a tooth at base of each face. Sandy or gravelly ground; mostly in the w 2/3 of East TX; throughout much of TX except far e portion; throughout much of the U.S. from NH s to GA and w to WY and AZ. Jun–Oct. Native of Midwestern U.S. According to McGregor (1986), *F. floridana* and *F. gracilis* often grow together and "putative hybrids between the two are all too frequent."

Froelichia latifolia R.A. McCauley, (with broad leaves), BROAD-LEAF SNAKE-COTTON. Leaf blades orbicular to broadly elliptic; inflorescence 5-ranked, apex often pyramidal and recurved; bracteole pubescent distally; flowers densely woolly, interstaminal lobes acute; lateral wings of perianth irregularly dentate and with distinct spines or tubercles on each face. Open disturbed grasslands and forest edges; Anderson, Bexar, Brazos, Guadalupe, Lavaca, and Leon (McCauley 2004), Bastrop, Dewitt, Goliad, Gonzales, Montgomery, and Wilson (Kartesz 2022) cos.; Gulf Prairies and Marshes; endemic to TX. Jun–Nov.

GOMPHRENA L. GLOBE-AMARANTH

Annual or perennials with prostrate to erect stems, pilose; leaves opposite, blades ovate to obovate, entire; inflorescences immediately subtended by leafy bracts, terminal and/or axillary, pedunculate, heads subglobose to short-cylindric; bract 1; bracteoles (sometimes called bractlets) 2 with crested keel; flowers perfect; tepals 5, perianth woolly, basally fused; stamens 5; filaments fused basally into a tube; ovule 1; style 1; stigmas 2(–3), short and stout or filiform.

A genus of ca. 125 species of tropical and warm areas of the New World and Australia (Mabberley 2017); a number are cultivated as ornamentals. (Modification of Pliny's *Gromphaena*, a species of amaranth)
REFERENCES: Mears 1980; Clemants 2003c.

1. Roots fleshy, spindle-shaped; plants perennial; inflorescence heads 10–12 mm in diam.; plants erect ______ **G. nealleyi**
1. Roots fibrous; plants annual or perennial; inflorescence heads 9–28 mm in diam.; plants prostrate to erect.
   2. Inflorescence heads 20–28 mm in diam.; plants erect, annual ______ **G. globosa**
   2. Inflorescence heads 9–13 mm in diam.; plants prostrate or decumbent, annual or perennial ______ **G. serrata**

Gomphrena globosa L., (globose, spherical), COMMON GLOBE-AMARANTH. Annual erect, stems stout, usually much branched, sometimes simple, to 1 m tall; leaf petiole 0.5–2 cm long, leaf blade apex acute; flowers showy, heads 20–28 mm in diam.; heads variously colored, purple, orange, rose, white, or variegated. Widely cultivated ornamental that sometimes escapes; Brazos, Montgomery and Travis (Kartesz 2022) cos.; extreme s TX and Edwards Plateau; widely scattered in se U.S. from SC s to FL and w to AR and TX, also NY w to IN. Jul–Sep. Native of s Asia. The flowering heads superficially resemble those of CLOVER species.

Gomphrena nealleyi J.M. Coult. & Fisher, (for Greenleaf Cilley Nealley, Texas botanist), NEALLEY'S GLOBE-AMARANTH. Perennial, erect, roots fleshy, stems 1–4 dm tall; leaves sessile, blade apex rounded; heads white or pink, 10–12 mm in diam. Grasslands, sandy soils; mostly in the s portion of East TX; endemic to TX. Winter–summer.

Gomphrena serrata L., (with saw teeth, perhaps in reference to the denticulate bracteole crest), ARRASA-CON-TODO, PROSTRATE GLOBE-AMARANTH. Annual, stems prostrate to decumbent, 1–10 dm long; leaves sessile to petiolate, petiole to 0.6 mm long, blade apex rounded or obtuse; heads white tinged with pink or red, 9–13 mm in diam. Lawns, roadsides, sandy open areas, woodlands; mostly in the s portion of East TX; Coastal Prairies and Marshes; scattered in se U.S. from SC s to FL and w to TX. Year-round. [*G. decumbens* Jacq., *G. dispersa* Standl.] Native of North America, the West Indies, Mexico, and Central America.

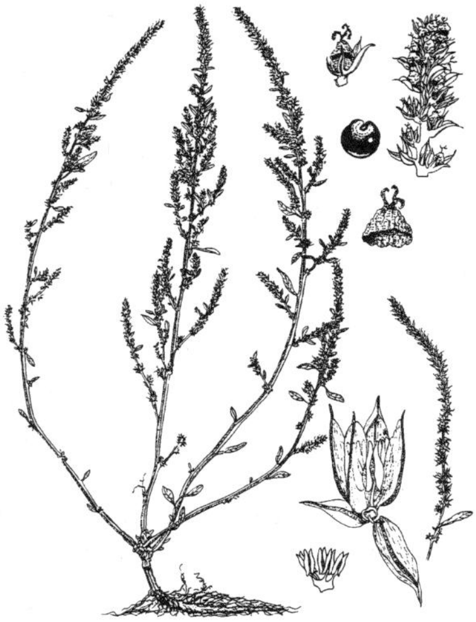

Amaranthus tuberculatus [CO1]

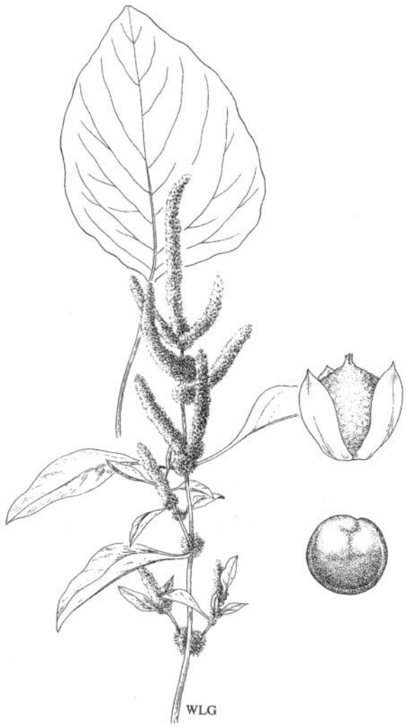

Amaranthus viridis [GLE]

Blutaparon vermiculare [COZ]

Froelichia drummondii [RIC]

Celosia argentea [RAD, WU1]

Celosia nitida [FNA]

Froelichia floridana [GLE, RAD]

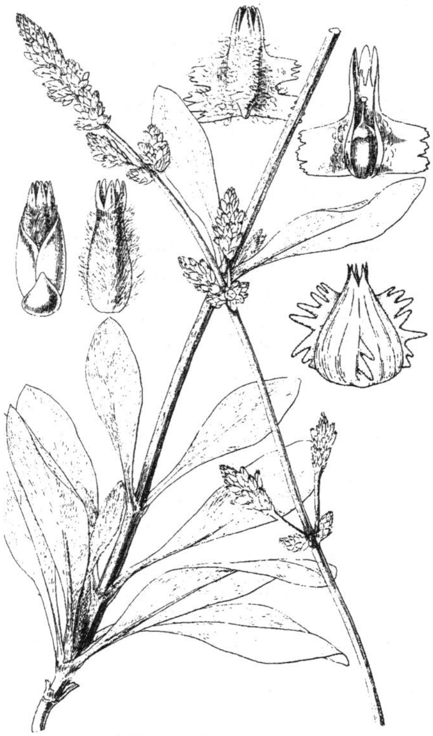

Froelichia gracilis [GLE, IPL]

Gomphrena haageana Klotzsch, (for J.N. Haage, 1826–1878, a seed grower of Erfurt, Germany), HAAGE'S GLOBE-AMARANTH. Perennial with leaf blades oblanceolate to oblong-linear and heads light red with yellow florets, is sometimes cultivated in TX and might be expected to escape; El Paso, Gillespie, and Val Verde (Kartesz 2022) cos. w of East TX.

GOSSYPIANTHUS Hook. COTTON-FLOWER

A genus of 2 species, 1 from OK through TX to adjacent Mexico and in Hispaniola, and the other restricted to Cuba (Henrickson 1987; Clemants 2003d); it has often been included in *Guilleminea*. (Latin: *gossypion*, cotton, and Greek: *anthemon*, flower, presumably in reference to the densely lanate tepals)
REFERENCES: Mears 1967; Henrickson 1987; Clemants 2003d.

Gossypianthus lanuginosus (Poir.) Moq., (woolly). Perennial herb, stems usually procumbent, sometimes ascending, with sparse pilose pubescence to strongly woolly-villous; leaves opposite; rosette leaves with wide-winged petiole without chlorophyll, larger than cauline, to 9.7 cm long, cauline leaves short-petioled, blade oval to obovate, usually 0.5–1.8 cm long; inflorescence an axillary cluster (to 10 mm long) of 6–12 flowers; flowers sessile, 5-merous, perfect; tepals separate, 3-nerved, green between nerves, densely hairy with crinkled hairs; stamens 5, the filament tube free from calyx; style 1; stigmas capitate, bilobed; utricle membranous, indehiscent, 1-seeded; seeds 1 mm long, brown, shiny. Dry soils, sandy, rocky, or disturbed areas. Flowering throughout growing season. This is a highly variable species sometimes divided into several varieties; 2 are recognized here, but are not distinguished on the distribution map.

1. Leaves densely pilose or tomentose, especially beneath; rosette blades lanceolate or oblanceolate to spatulate, usually 4–12 mm wide, obtuse to acute ____ var. **lanuginosus**
1. Leaves nearly glabrous; leaf blades usually linear to lanceolate, usually 3–6 mm wide, acute ____ var. **tenuiflorus**

var. **lanuginosus.** WOOLLY COTTON-FLOWER. Throughout most of TX; KS, NM, OK, and TX. [*Guilleminea lanuginosa* (Poir.) Moq. ex Benth. & Hook. f., *G. lanuginosa* var. *rigidiflora* (Hook.) Mears, *G. lanuginosa* var. *sheldonii* (Uline & W.L. Bray) Mears]

var. **tenuiflorus** (Hook.) Henrickson, (slender-flowered), LANCE-LEAF COTTON-FLOWER. Dallas, Harris, Hunt, and Jackson (Kartesz 2022) cos.; mainly Gulf Prairies and Marshes; AR, OK, and TX. [*Guilleminea lanuginosa* (Poir.) Moq. ex Benth. & Hook. f. var. *tenuiflora* (Hook.) Mears]

GUILLEMINEA Kunth MATWEED

A genus of 5 or 6 species (Pedersen 2000; Clemants 2003e) primarily of South America, but also sw North America and Mexico. Three of the species were described by Pedersen (2000). *Gossypianthus*, which is very similar vegetatively, has sometimes been included in *Guilleminea* (Mears 1967); however, according to Henrickson (1987), the two genera were "so fundamentally different in floral structures that a cogeneric treatment could not be supported." (Named for Antoine Guillemin, 1796–1842, French botanist, author, and explorer—Clemants 2003e)
REFERENCES: Mears 1967; Henrickson 1987; Pedersen 2000; Clemants 2003e.

Guilleminea densa (Willd. ex Roem. & Shult.) Moq. var. **aggregata** Uline & W.L. Bray, (sp.: dense; var.: clustered in a dense mass), SMALL MATWEED. Mat-forming, procumbent perennial herb 0.1–0.8 m in diam.; stems often with conspicuous pubescence, villous to woolly-villous; leaves opposite, petiolate, entire margined, basal and cauline, the basal rosette leaves larger, with winged petioles, the blades usually 2–4 cm long, oblong-elliptic to lanceolate, withering by time of flowering; cauline leaves with petiole to 4 mm long, the blades from ovate to oblong or lanceolate, to 0.4–1(–2.3) cm long; inflorescences small (to 3 mm long) clusters of few flowers, axillary, sessile, white, internodes between individual inflorescences very short producing a very compact mass of several to many inflorescences; bracts and bracteoles membranous; flowers perfect; tepals 5, fused in basal 1/2, the lobes whitish, to 1.5 mm long, densely hairy with crinkled hairs; stamens 5, the filament tube fused with calyx; style 1; utricle 1-seeded to 1.4 mm long. Dry rocky or sandy areas, waste places; Bastrop, Comal, Travis (Kartesz 2022), Bexar (*Ruth 1557* BRIT), and Hopkins (*O'Kennon 10161* BRIT) cos. on sw margin of East TX; mainly Trans-Pecos and scattered elsewhere in w TX; mostly in sw U.S. AZ, CA, CO, NM, OK, and TX. May–Oct. This variety occurs from the sw U.S. s to Oaxaca, Mexico (Henrickson 1987); the type variety is native to South America.

IRESINE P. Browne BLOODLEAF

Ours monoecious or dioecious herbs (occasionally some bisexual flowers present); leaves opposite, petiolate, blades thin, entire or serrulate; inflorescence a panicle of numerous, small flowers, often whitish to straw-colored due to bracts and tepals; flowers unisexual, each subtended by a bract and 2 bracteoles shorter than the tepals; perianth of 5 tepals; pistillate flowers subtended

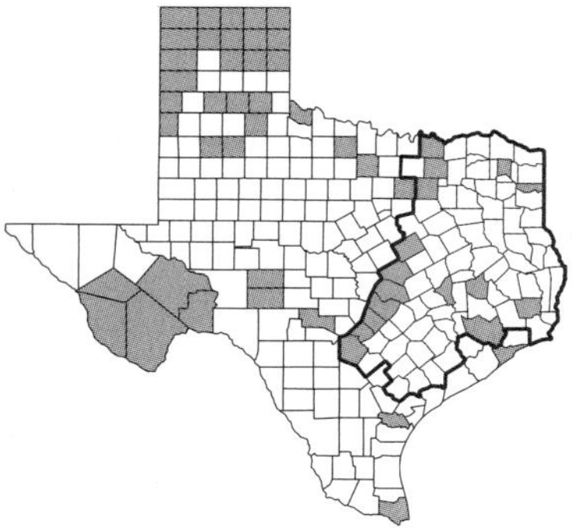
Amaranthus retroflexus

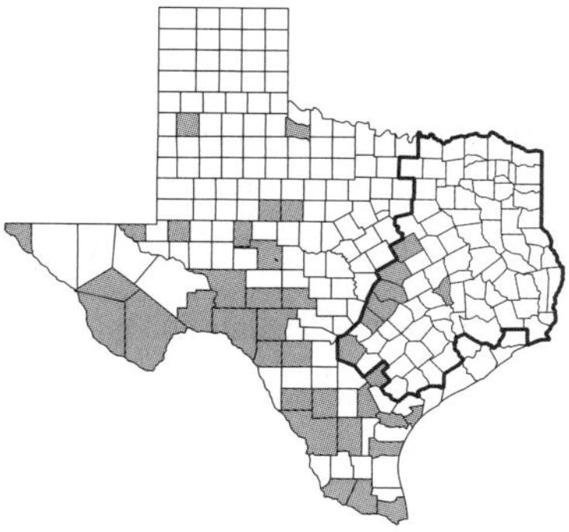
Amaranthus scleropoides

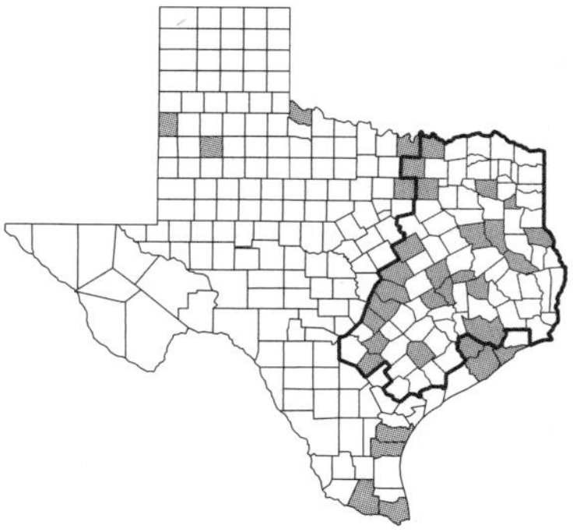
Amaranthus spinosus

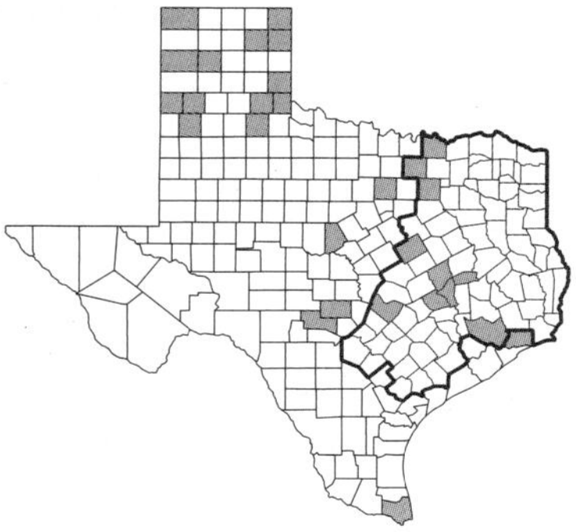
Amaranthus tuberculatus

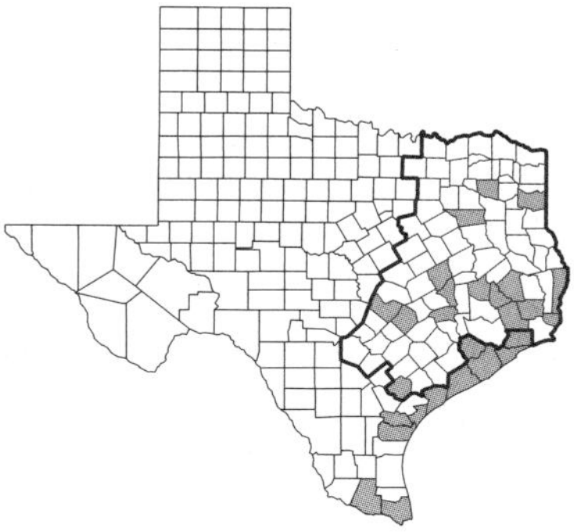
Amaranthus viridis

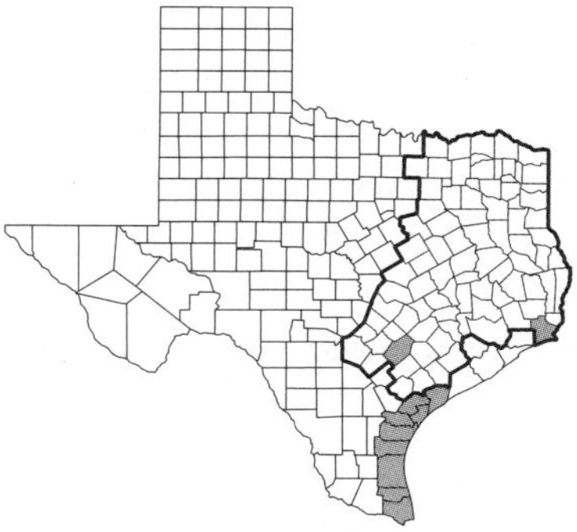
Blutaparon vermiculare

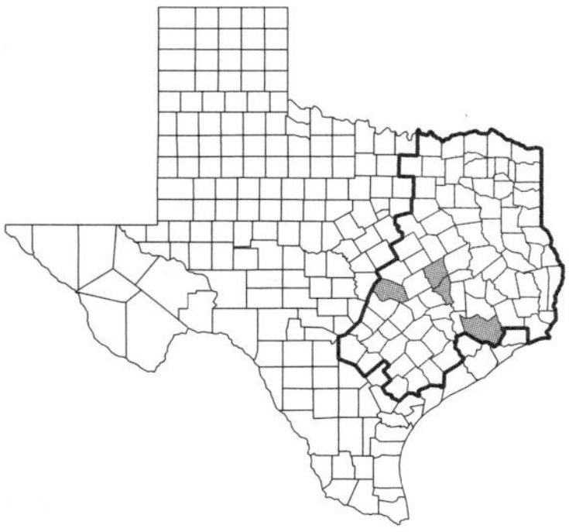
Celosia argentea

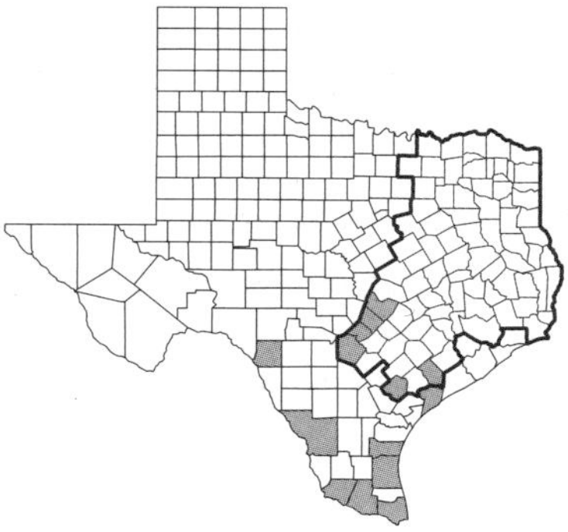
Celosia nitida

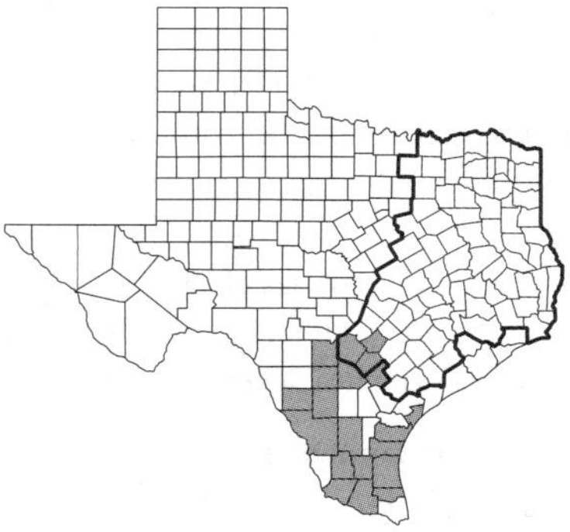
Froelichia drummondii

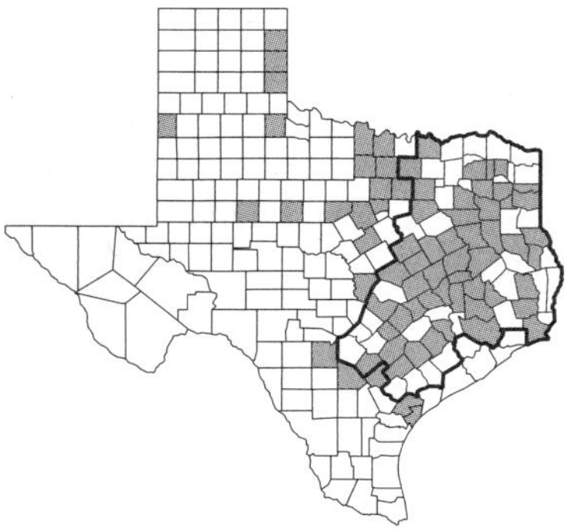
Froelichia floridana

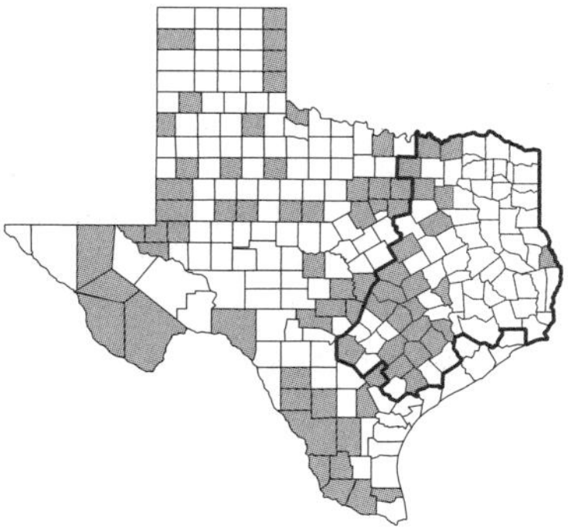
Froelichia gracilis

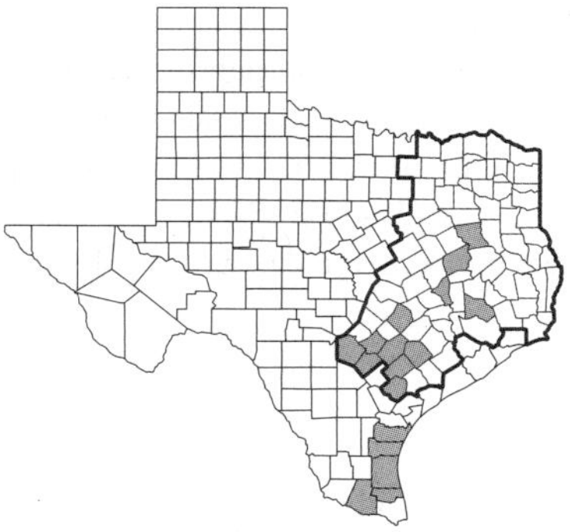
Froelichia latifolia

by conspicuous white wool; staminate flowers essentially glabrous. stamens 5, the filaments fused basally; utricle indehiscent, 1-seeded; seed dark red, shiny.

A genus of ca. 70 species (Clemants 2003f) of tropical and temperate areas, especially the Americas, but also Asia, w Africa, and Australia; some are used medicinally while others are cultivated for their ornamental foliage (e.g., *I. herbstii* Hook., native of South America, with leaves purplish-red or green, with colorful veins). (Greek: *eiresione*, a staff or branch wound with wool, from the calyx, often bearing long hairs)

REFERENCES: Henrickson & Sundberg 1986; Pedersen 2000; Clemants 2003f.

1. Plants annual or weakly perennial from a vertical root; stems usually much-branched, 0.4–3 m long; tepals obtuse or rounded ____ **I. diffusa**
1. Plants perennial with horizontal rhizomes or branching woody rootstocks; stems usually unbranched up to the inflorescence, 0.3–1 m long; tepals acute to acuminate.
   2. Larger leaf blades 4–15 cm long; utricle equaling or longer than tepals of pistillate flower ____ **I. rhizomatosa**
   2. Leaf blades 3–6 cm long; utricle shorter than tepals of pistillate flower; species known only from sw margin of East TX ____ **I. heterophylla**

Iresine diffusa Humb. & Bonpl. ex Willd., (diffuse, spreading), JUBA'S-BUSH. Monoecious annual or weak perennial; stems erect or spreading, sometimes clambering over adjacent plants, 0.4–3 m long; leaf blades ovate to lanceolate; tepals of pistillate flowers 3-nerved, white to straw-colored; utricle shorter than (included within) tepals of pistillate flower. Often in sandy moist areas; in TX known only from McLennan Co. (Hannick et al. 2013); FL, LA, and TX; Mexico; West Indies; Central and South America. Jul–Oct. [*I. celosia* L. [nom. illeg. superfl.]]

Iresine heterophylla Standl., (diversely leaved), STANDLEY BLOODLEAF. Perennial from long slender branching woody rootstocks; stems erect or ascending, 0.5–1 m tall; leaf blades ovate, often asymmetric; tepals of pistillate flowers 3-nerved, whitish to straw-colored; utricle exerted from tepals. Rocky areas; Hays and Travis (Kartesz 2022) cos.; mainly Edwards Plateau and Trans-Pecos; AZ, NM, and TX. Summer.

Iresine rhizomatosa Standl., (very rooted), ROOTSTOCK BLOODLEAF, BLOODLEAF. Dioecious, stoloniferous perennial with slender horizontal rhizomes; stems erect, 0.5–1.5 m tall; leaf blades ovate to ovate-lanceolate to elliptic; tepals faintly 1-nerved, silvery white; utricle included within tepals. Sandy soils, often moist areas; throughout much of East TX except Pineywoods; East Cross Timbers and Edwards Plateau; widespread in the e U.S. from MD s to FL w to KS and TX. Aug–Oct. This species has sometimes been incorrectly referred to as *I. celosioides* (Clemants 2003f).

TIDESTROMIA Standl. HONEYSWEET

A genus of 6 species of sw U.S., n Mexico, and West Indies (Sánchez del Pino & Olvera 2002). (Named for Ivar Frederick Tidestrom, 1864–1956, Swedish-born botanist of sw U.S.)

REFERENCES: Sánchez del Pino & Olvera 2002; Sánchez del Pino & Clemants 2003.

Tidestromia lanuginosa (Nutt.) Standl., (woolly), WOOLLY TIDESTROMIA, ESPANTA VAQUEROS. Annual herb usually to only 15(–50) cm tall and 1 m across; stems usually prostrate to rarely ascending; stems and leaves gray-green to ashy-white or yellowish green, densely pubescent with irregularly branched hairs (sometimes appearing stellate), almost glabrous with age; leaves apparently commonly 3 per node (actually a true alternate leaf and 2 subopposite involucral leaves (Sánchez del Pino & Clemants 2003), obovate to rhombic-ovate or lanceolate, entire, 5–20(–30) mm long; petioles as long as blades or shorter; flowers 2–5 in small axillary clusters (technically dichasia), inconspicuous, perfect; bracts ovate; tepals 5, unequal, 1–3 mm long, yellowish; stamens 5; utricle subglobose. Dry sandy, gravelly, or rocky areas; Brazos, Gregg, Hays, Jefferson, San Jacinto, and Travis (Kartesz 2022) cos.; Gulf Prairies and Marshes, South TX Plains, High Plains, and Trans-Pecos; primarily sw U.S. Mar–Oct. [*T. lanuginosa* (Nutt.) Standl. subsp. *eliassoniana* Sánch. Pino & Flores Olv., *Alternanthera lanuginosa* (Nutt.) Moq.] Infraspecific taxa are sometimes recognized based on pollen and hair characters (Sánchez del Pino & Olvera 2002).

ANACARDIACEAE R. Br.

SUMAC OR CASHEW FAMILY

Ours mostly native subshrubs, shrubs, small trees, or woody vines, often with somewhat milky juice, resin ducts in the bark with resin that often dries black; leaves alternate, with 3(–5) leaflets or once pinnately compound (simple in 1 species in East TX); leaflets entire, toothed, or lobed; flowers small, in terminal or axillary, head-like to loose, open panicles, perfect or often

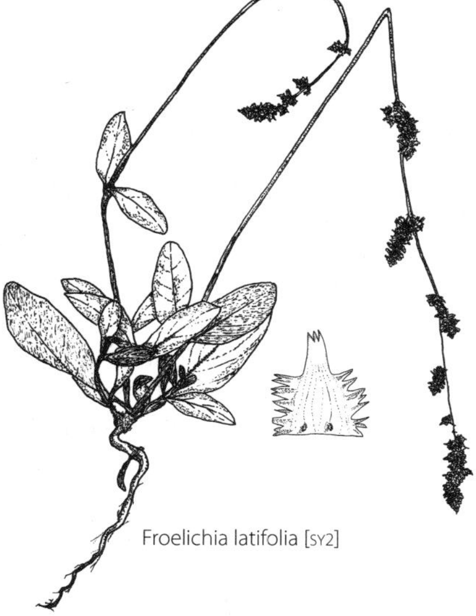

Froelichia latifolia [SY2]

Gomphrena globosa [NIC]

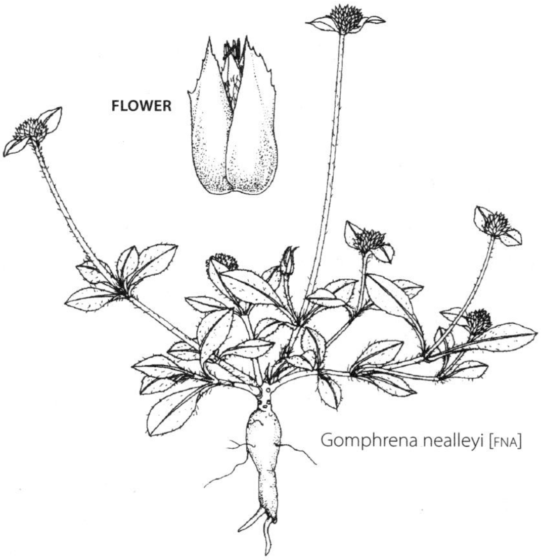

Gomphrena nealleyi [FNA]

Gomphrena serrata [ABM]

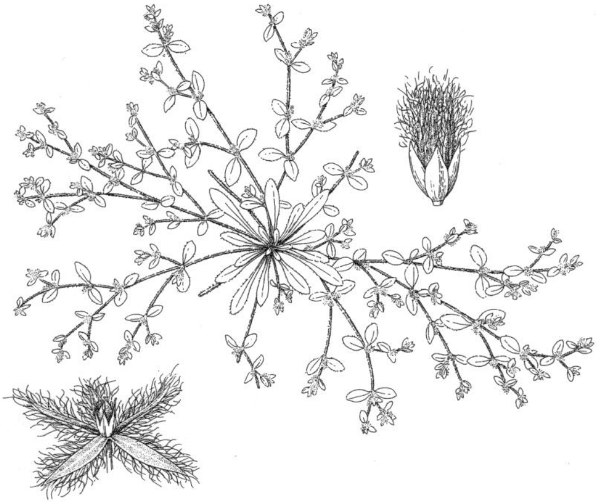

Gossypianthus lanuginosus var. lanuginosus [SID]

Gossypianthus lanuginosus var. tenuiflorus [IPL]

unisexual (plants often dioecious), 5-merous, radially symmetrical; petals white to cream, yellow, or green; stamens 5; pistil 1; fruit a drupe.

A medium-sized (873–985 species in 80 genera) family of the tropics and subtropics with a few temperate species (Stevens 2001 onwards; Mabberley 2017). Economically important taxa include food plants such as *Anacardium occidentale* L. (CASHEW), *Mangifera indica* L. (MANGO), and *Pistacia vera* L. (PISTACHIO), and ornamentals including *Cotinus* (SMOKETREE), *Rhus* (SUMACS), and *Schinus* (PEPPERTREE). A number of species cause contact dermatitis; catechols, resorcinols, and other phenolic compounds, often in mixtures (e.g., urushiol) are involved (Stevens 2001 onwards) (see discussion under *Toxicodendron*). Though temperate *Toxicondendron* is the most known, the majority of poisonous Anacardiaceae are tropical in distribution and little known (Mitchell 1990). In the past, a number of genera with highly reduced flowers and inflorescences have been segregated into separate families (Julianaceae, Podoaceae); however, they are now included in the Anacardiaceae (Stevens 2001 onwards). The family is closely related to the tropical Burseraceae (TORCHWOOD family) (Pell et al. 2024a), which also includes FRANKINCENSE (*Boswellia*) and MYRRH (*Commiphora*). Family name from *Anacardium*, a genus of ca. 11 species of the New World tropics (Pell et al. 2011; Mabberley 2017). (Greek: *ana*, up, and *cardia*, heart, possibly in reference to the large, swollen, fleshy, bright red or yellow, pear-like receptacle above the nut; the receptacle is edible and referred to as the cashew apple) (subclass Rosidae—Cronquist; order Sapindales—APG IV)

FAMILY RECOGNITION IN THE FIELD: subshrubs, shrubs, trees, or woody vines with alternate leaves that either have 3 leaflets or are pinnately compound (simple in 1 species in East TX); flowers small, inconspicuous; fruit a small drupe.

REFERENCES: Barkley 1957, 1961; Brizicky 1962d; Baer 1979; Wannan 2006; Mabberley 2017; Weakley, A.S., and Southeastern Flora Team 2022; Pell et al. 2024a.

1. Leaves simple; flowers small, numerous (mostly sterile, falling to leave sterile pedicels); fruiting panicles with plumose sterile pedicels; styles lateral, unequal; fruits light brown; rare species known only from 2 East TX cos. ____ **Cotinus**
1. Leaves 3(–5)-foliate or pinnately compound; fruiting panicles without plumose sterile pedicels; styles terminal, equal or nearly so; fruits red to purplish, blue, white or yellowish gray; including species common and widespread in East TX.
   2. Leaves usually evenly pinnately compound (without a terminal leaflet); plant an introduced tree ____ **Pistacia**
   2. Leaves usually odd pinnately compound (with a terminal leaflet) OR 3(–5)-foliate; plant a native vine, subshrub, shrub, or small tree.
      3. Fruits red, noticeably pubescent with glandular and eglandular hairs; leaves pinnately compound with > 3 leaflets or palmately 3-foliate (= terminal leaflet sessile); inflorescences terminal (or in *R. aromatica* sometimes appearing axillary); plants without contact poisons ____ **Rhus**
      3. Fruits white to yellowish gray, glabrous or sparingly pubescent with eglandular hairs; leaves usually pinnately 3-foliate (= terminal leaflet stalked), rarely 5-foliate or in 1 species the leaves pinnately compound with 7–15 leaflets; inflorescences axillary; plants with contact poisons ____ **Toxicodendron**

COTINUS Mill. SMOKETREE, SMOKEWOOD, SMOKEBUSH

A genus of ca. 5 species of s Europe to China, sw China, se U.S., and Mexico (Rzedowski & Calderón de Rzedowski 1999; Mabberley 2017). The Mexican species appear to be phylogenetically basal in the genus (Rzedowski & Calderón de Rzedowski 1999). The widespread Old World species, *C. coggygria* Scop. (WIGTREE, EUROPEAN SMOKETREE), is widely cultivated and the wood yields a yellow dye; it can be distinguished from the native species by its smaller (ca. 4–9 cm long), glabrous, often ovate leaf blades. The smoke-like effect of the plumose fruiting pedicels gave rise to the common name. (Greek: *kotinus*, ancient name of the wild olive, used by Pliny for an unidentified shrub of the Apennines, but applied by some pre-Linnean botanists, such as Tournefort, to *C. coggygria*—Brizicky 1962d)

REFERENCES: Baskin & Baskin 1989; Rzedowski & Calderón de Rzedowski 1999.

Cotinus obovatus Raf., (obovate, inversely ovate), AMERICAN SMOKETREE, SMOKEBUSH, CHITTAMWOOD. Usually dioecious shrub or small tree to ca. 12 m tall with strong-smelling sap; wood yellow; leaves simple, blades obovate to elliptic-obovate, 6–17 cm long, 4–9 cm wide, strikingly yellow to orange, red, or scarlet in fall, apically usually obtuse to rounded, with lower surface pubescent; petioles 1.5–6 cm long, often purple or reddish; flowers small, numerous (mostly sterile, falling to leave sterile pedicels), yellowish white to greenish yellow, in loose terminal panicles ca. 10 cm long which can enlarge to ca. 30 cm in fruit; sterile pedicels to 3.5 cm long, plumose with pale purplish or brownish hairs (panicles thus with a pink "smoky" appearance but not as showy as in the introduced *C. coggygria* Scop.); fruits drupes, flattened, kidney-shaped or oblique-oblong, ca. 4 mm long. Rocky woods, limestone outcrops; Hays Co. (Turner et al. 2003) on sw margin of East TX, also Camp Co. (*Busch 110* BRIT); mainly Edwards Plateau in Bandera, Kendall, Kerr, Uvalde (Stanford 1976; Simpson 1988), Blanco, and Medina (Turner et al. 2003) cos.; endemic to sc U.S. in AL, AR, GA, KY, MO, OK, TN, and TX. Apr–May. [*C. americanus* Nutt., *Rhus americana* (Nutt.)

Guilleminea densa [ARF, SID]

Iresine diffusa [JAA]

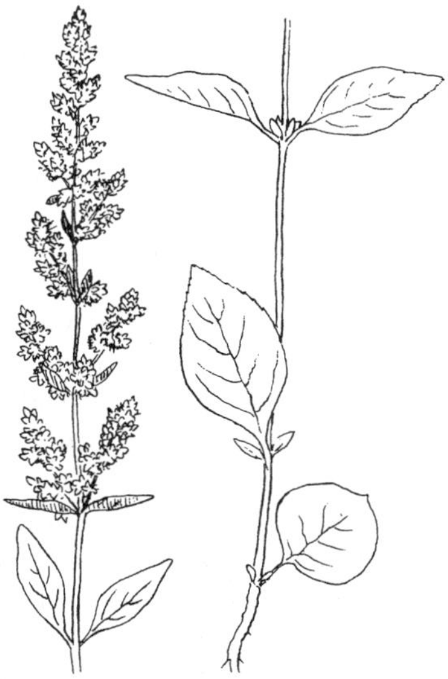

Iresine heterophylla [IV2]

Iresine rhizomatosa [JAA, RAD]

Tidestromia lanuginosa [RYD]

Cotinus obovatus [SA3]

Pistacia chinensis [LUN]

Cotinus obovatus [SA3]

Sudw., *R. cotinoides* Nutt. ex Chapm.] According to Simpson (1988), the TX occurrence of this tree is relictual with most TX populations (in the Edwards Plateau) more than 500 miles from the nearest locality in OK, except for the Camp Co. specimen mentioned above in the ne portion of East TX; the species has an extremely limited and scattered distribution in the e U.S. (Little 1970) and its distribution is considered relictual (Little 1983). The wood yields a yellow dye which was extensively used during the Civil War (Cox & Leslie 1991); the result was that many trees were cut for this purpose (Yatskievych 2006). The striking fall foliage suggests more widespread use as an ornamental.

PISTACIA L. PISTACHIO

A genus of 12 species from the Mediterranean, Asia, s U.S., and Central America (Pell et al. 2011; Mabberley 2017). Species produce resins, gums, oils, and edible seeds and the genus includes the Old World *P. vera* L., PISTACHIO (typically with 5 leaflets), widely cultivated for its edible nuts. One species, *P. mexicana* Kunth [*P. texana* Swingle], TEXAS PISTACHE, AMERICAN PISTACHIO, MEXICAN PISTACHIO, COPALL PISTACHIO, WILD PISTACHIO, LENTISCO, is native to sw TX. Aphids (Fordinae) that form distinctive galls are closely associated with species of *Pistacia*, the sometimes massive, spherical galls produce terpenes that dissuade goats, at least, from eating them; the aphids have diversified in part by occupying different sites on the one plant species (Inbar et al. 2004; Rostás et al. 2013). Fordinae aphids form galls on other Anacardiaceae, interestingly, crown-group Melaphidina (Fordinae) aphids, now on *Rhus*, have been dated to (79–)73.3(–68.3) Mya (Ren et al. 2017). A gall-forming jumping psyllid plant louse, the hemipteran *Calophya*, is notably common on *Schinus*, and other psyllids occur on Anacardiaceae (Burckhardt & Basset 2000; Burckhardt 2005). (Greek *pistake*, pistachio, name for *P. vera*, the edible pistachio)
REFERENCES: Gilman & Watson 1994; McWilliams & Arnold 1998; Pell et al. 2011.

Pistacia chinensis Bunge, (of China), CHINESE PISTACHIO, CHINESE PISTACHE, CHINESE MASTIC TREE. Relatively fast growing, dioecious tree 10–15(–25) m tall; leaves even-pinnately compound (sometimes odd-pinnately compound), becoming strikingly yellowish, orangish, or reddish in fall; leaflets 10–20, 5–10 cm long, 1.2–2 cm wide, base oblique, apex acuminate, entire, paler beneath; inflorescences much-branched panicles; flowers appearing before the leaves, apetalous, reddish to greenish, the staminate in tight clusters 5–7.5 cm long, the pistillate in longer looser groups; fruits single-seeded drupes ca. 5 mm in diam., red to purplish turning bluish. Widely planted as an ornamental; apparently prefers moist well-drained soil, but is heat and drought tolerant and will tolerate a variety of soil conditions (including sandy to black clay); Bexar, Brazos, Dallas, McLennan, and Travis (Kartesz 2022) cos.; Edwards Plateau, also Tarrant Co. (R. O'Kennon, pers. obs.) just w of East TX; widely scattered in AR, AZ, CA, GA, MS, NC, OK, SC, and TX. Spring. Native from Afghanistan to China and the Philippines. While now being widely cultivated, research by McWilliams and Arnold (1998) indicate that the species has the potential of becoming a problematic invasive in East TX. They note that individuals have spread from (apparently by bird dispersal) established landscape plants on the Texas A&M campus, have become naturalized in the local ecosystem, and "in some cases have grown through established plants to top mature *Juniperus virginiana*." The species is wind pollinated (Yu & Lu 2011). The young shoots are used as a Chinese vegetable (Mabberley 2017). The commercial PISTACHIO NUT (*P. vera* L.) is sometimes grafted onto this species (Gilman & Watson 1994).

RHUS L. SUMAC

Shrubs or rarely small trees with long rhizomes, often forming thickets; leaves palmately 3-foliate or pinnately compound, often turning bright red in fall, ours deciduous (except *R. virens*); leaflets entire or toothed, sometimes lobed; fruits red, globose, sometimes somewhat flattened, visibly pubescent. Ours native to the U.S.

A genus of ca. 35 species (not including *Toxicodendron*) (Pell et al. 2011; Mabberley 2017) of temperate and warm areas (Mabberley 2017); *Toxicodendron*, recognized here as a separate genus, is sometimes lumped into *Rhus* as a section or subgenus. According to J. Hennen (pers. comm.), there are rust fungi that attack only *Toxicodendron* (*Pileolaria brevipes* Berk. & Ravenel) and others that attack only *Rhus* (*Pileolaria patzcuarensis* (Holw.) Arthur). In contrast to *Rhus* species, which are all non-poisonous, *Toxicodendron* species are poisonous and can cause contact dermatitis. A number of *Rhus* species form extremely dense and visually striking thickets due to vegetative reproduction via rhizomes. SUMACS often display very early fall foliage color (often strikingly red); this is considered to serve as a "foliar fruit flag" which attracts birds that act as dispersal agents for the fall-ripening fruits (Stiles 1984). Some species are a source of tannin. Native Americans used *Rhus* species medicinally in a number of ways including as a means to induce vomiting, as poultices for skin problems, and the fruits chewed as a breath freshener (Moerman 1998; Yatskievych 2006). (Ancient Greek and Latin name for SUMAC)

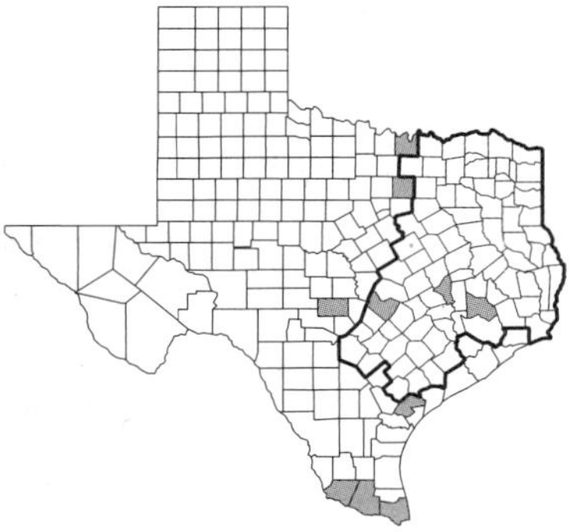

Gomphrena globosa

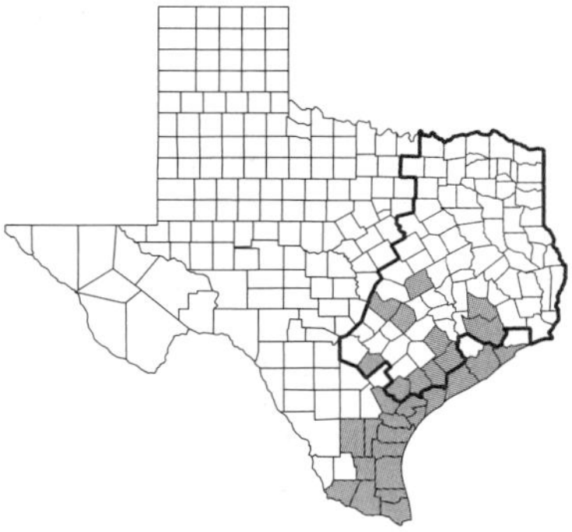

Gomphrena nealleyi

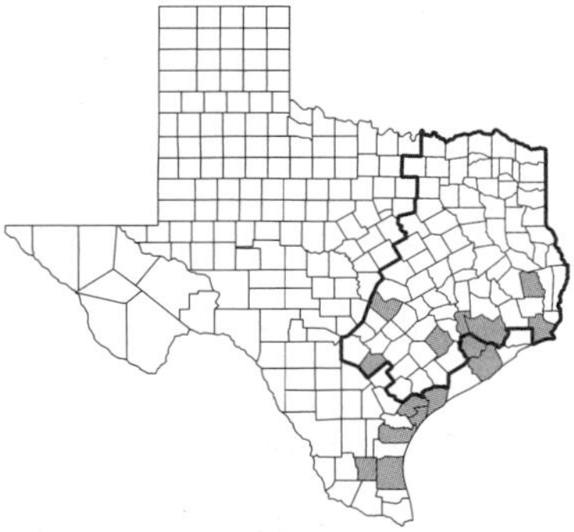

Gomphrena serrata

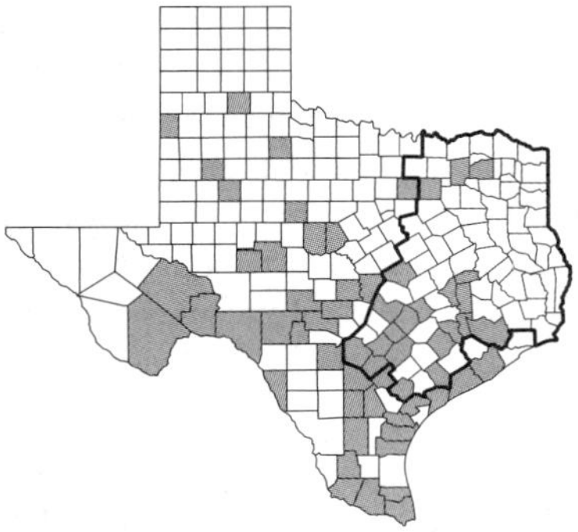

Gossypianthus lanuginosus

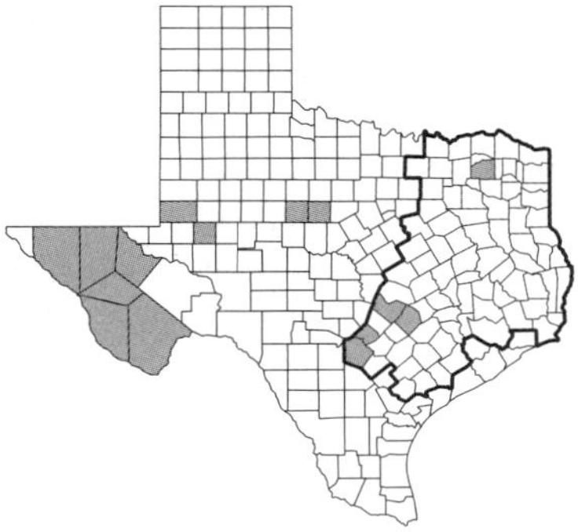

Guilleminea densa

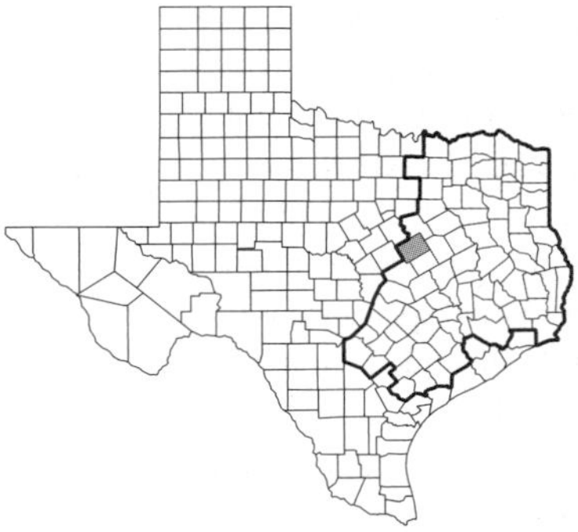

Iresine diffusa

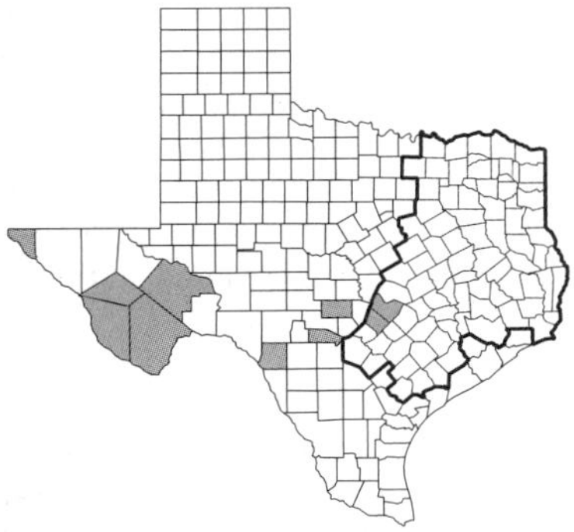

Iresine heterophylla

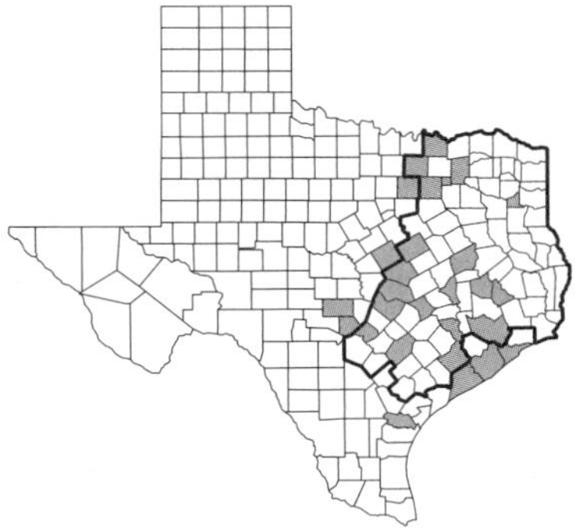

Iresine rhizomatosa

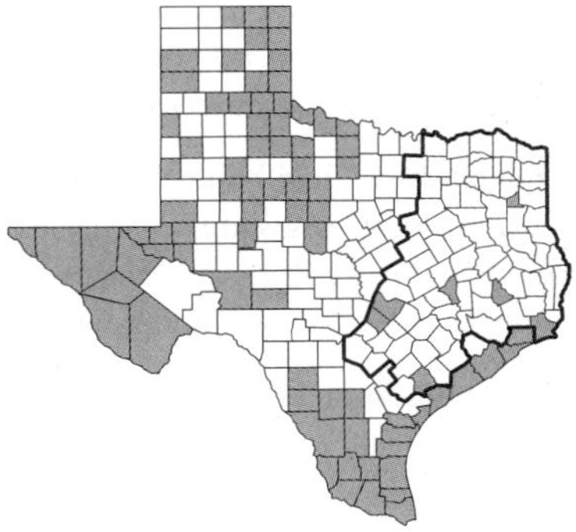

Tidestromia lanuginosa

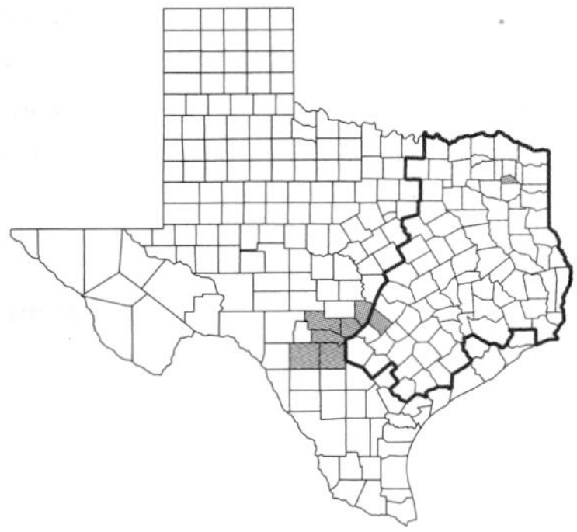

Cotinus obovatus

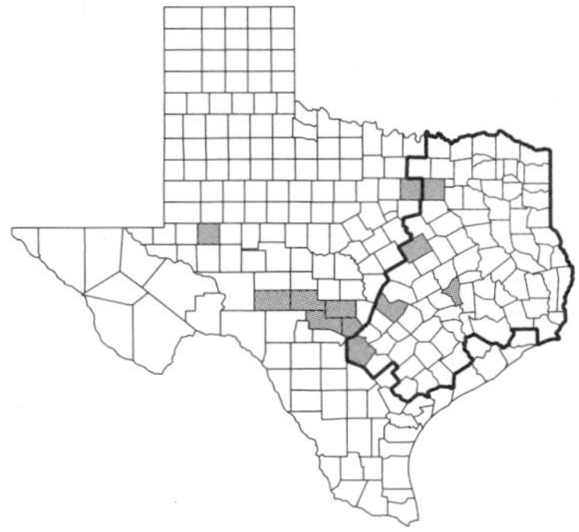

Pistacia chinensis

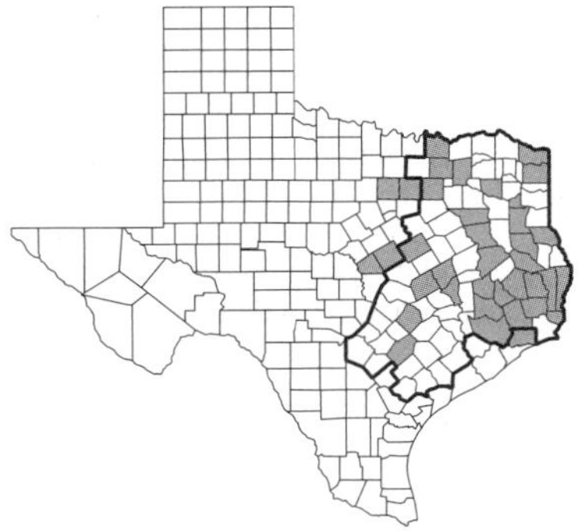

Rhus copallinum var. copallinum

REFERENCES: Barkley 1937; Brizicky 1963a; Young 1978; Turner 1996; Xiaojie et al. 1999; Simpson 2004; Pell et al. 2024b.

1. Leaves palmately compound; leaflets 3 ____ **R. aromatica**
1. Leaves pinnately compound; leaflets 5 or more.
   2. Leaves with rachis completely unwinged AND leaflets entire; leaflets 4–12 cm long ____ see **Toxicodendron vernix**
   2. Leaves with rachis winged OR leaflets toothed OR leaflets 4 cm or less long.
      3. Leaflets (3–)5–9, very small, < 2 cm long and < 6 mm wide; branches often ± spinescent ____ **R. microphylla**
      3. Leaflets 5–31, much > 2 cm long and usually much > 6 mm wide; branches not spinescent.
         4. Leaf rachis unwinged; leaflets sharply toothed or entire.
            5. Leaves deciduous, not coriaceous; leaflets mostly sharply and conspicuously toothed, glabrous and glaucous beneath, to 12 cm long ____ **R. glabra**
            5. Leaves evergreen, coriaceous; leaflets entire, pubescent beneath, soft to the touch, usually 4 cm or less long ____ **R. virens**
         4. Leaf rachis winged; leaflets mostly entire, sometimes serrate (usually apically) ____ **R. copallinum**

Rhus aromatica Aiton, (fragrant), FRAGRANT SUMAC. Shrub to ca. 3 m tall with creeping stem-bases, forming loose clumps; foliage fragrant when bruised; leaves trifoliate; leaflet blades ovate to rhombic or obovate, margins 3–5-lobed, crenate to dentate; flowers in terminal, head-like or irregular, dense cluster-like spikes; petals yellow to pinkish; flowering before leaves appear or with or slightly after unfolding of leaves. Sandy woods and ravines; widespread in East TX; throughout the U.S.; also adjacent Canada. Two varieties occur in East TX. Numerous authors (e.g., Diggs et al. 1999; Weakley, A.S., and Southeastern Flora Team 2022) have recognized the segregate species *Rhus trilobata* Nutt. and its varieties and differentiated it from *Rhus aromatic* base mostly on pubescence and leaf size and shape. According to Pell et al. (2024), *R. trilobata* and many varietal segregates of *Rhus aromatica* display extensive variation in leaf and leaflet size, shape, and pubescence associated with local environmental conditions. They add, "Morphometric analyses of herbarium specimens revealed that in general, leaf traits vary continuously across the continent."

1. Stems glabrous or pilose; lower leaf surfaces puberulent to hairy ____ var. **aromatica**
1. Stems densely pilose; lower leaf surfaces densely pilose ____ var. **pilosissima**

var. **aromatica.** Petioles 0.3–4.3 cm long; leaflet blades 1.2–7.9 cm long, 1–5.6 cm wide. Woodlands, grasslands, hillsides, shrublands; throughout the U.S. and adjacent Canada. Mar–Jun. [*R. aromatica* var. *flabelliformis* Shinners, *R. aromatica* var. *serotina* (Greene) Rehder, *R. aromatica* var. *trilobata* (Nutt.) A. Gray, *R. trilobata* Nutt., *R. trilobata* var. *serotina* (Greene) F.A. Barkley] Because of the numerous synonyms and possible misidentifications, a county level map of TX is not provided.

var. **pilosissima** (Engl.) Shinners, (very hairy). Petioles 0.7–2.6 cm long; leaflet blades 1.2–5.4 mm long, 0.9–4.7 cm wide. Woodlands, hillsides, savannas, shrublands; AZ, CA, CO, KS, NM, OK, and TX; Mexico. Mar–Apr. [*R. aromatica* subsp. *pilosissima* (Engl.) W.A. Weber, *R. trilobata* Nutt. var. *pilosissima* Engl.] Because of the numerous synonyms and possible misidentifications, a county level map of TX is not provided.

Rhus copallinum L., (gummy, resinous), WING-RIB SUMAC, SHINING SUMAC, FLAME-LEAF SUMAC, DWARF SUMAC. Shrub or small tree to 1.5–3(–10) m tall; leaf rachis narrowly to broadly winged; leaflets (5–)7–17(–23) soft-pubescent to nearly glabrous beneath; flowers in dense, subsessile, pyramidal panicles. Sandy woods, hills, and open areas; throughout much of East TX; Edwards Plateau, Cross Timbers and Prairies, and South TX Plains; e 1/2 of the U.S. from ME s to FL and w to WI, IA, AZ, and TX; also adjacent Canada. Jun–Jul. The epithet has sometimes been spelled "copallina," but according to Yatskievych (2006), "the original spelling that Linnaeus used should be retained." Some authors (Turner et al. 2003; Kartesz 2022) do not recognize varieties, but instead treat var. *lanceolata* at the species level. According to Crosswhite (1980), before lemons were readily available, the small fruits of *R. copallinum* were crushed in water to make a tart drink called sumac-ade. Simpson (2004) notes that the tallest sumac species (*R. copallinum*) in Texas is in Upshur County and measures 47 ft (14.3 m) tall, but has a trunk circumference of only 12 inches (30.5 cm).

1. Rachis and its wings on some leaves over 4 mm wide (total distance across rachis and its wings); leaflets 7–17, ovate-lanceolate, 2–4 times as long as wide, scarcely falcate ____ var. **copallinum**
1. Rachis and its wings usually < 3.5 mm wide; leaflets 13–19, linear-lanceolate, 4–9 times as long as wide, usually strongly falcate ____ var. **lanceolata**

var. **copallinum.** Open woods, prairies, riverbanks, savannas, forest edges, disturbed areas, on a wide range of soil types; throughout East TX; e U.S. from ME s to FL and w to WI and s to OK and TX; Canada. (Feb–)Apr–Oct. [*R. copallinum* var. *latifolia* Engl., *R. copallinum* var. *leucantha* (Jacq.) DC., *R. leucantha* Jacq.] According to Pell et al. (2024), "A great deal of leaf diversity exists within var. *copallinum*, some of which may be recognizably distinct locally but which does not warrant recognition at the continental level."

var. **lanceolata** A. Gray, (lance-shaped), prairie sumac, prairie flame-leaf sumac. Open woods, prairies, riverbanks, savannas, forest edges, disturbed areas, usually on limestone; mostly along the w border of East TX, scattered e; AZ, OK, NM, and TX; Mexico. Jul–Nov. [*R. lanceolata* (A. Gray) Britton]

Rhus glabra L., (glabrous, smooth, hairless), smooth sumac, scarlet sumac. Tree-shaped shrub (rarely a small tree) to ca. 3 m tall, the leaves and few branches crowded toward summit of stem; leaflets 9–23, lanceolate to narrowly elliptic, glabrous, glaucous beneath, margins mostly distinctly toothed; flowers in dense, subsessile, ovoid to pyramidal panicles. Sandy or rocky soils, hillside woods, stream banks, fencerows; widespread in East TX; mainly ne quadrant of the state and a few localities in the Panhandle; nearly throughout the U.S. and s Canada. May. A cut-leaved form (less aggressive that the typical plant), f. *laciniata* (Carrière.) B.L. Rob., is sometimes cultivated as an ornamental (Yatskievych 2006) and research has been done on growing the species as a possible source of tannins and oils (Campbell 1984).

Rhus microphylla Engelm., (small-leaved), little-leaf sumac, desert sumac, scrub sumac, small-leaf sumac, correosa, soda-pop-bush. Much-branched, shrub (rarely a small tree) to ca. 5 m tall (usually much smaller); leaflets 5–9, sessile, pilose, very small, < 2 cm long; rachis winged; flowers appearing before (rarely with) the leaves. Dry open areas, limestone mesas; Bexar and Comal cos. on sw margin of East TX and Leon Co. in c part of East TX (Turner et al. 2003); mainly w 1/2 of TX; AZ, NM, OK, and TX. Apr(–May). According to Crosswhite (1980), before lemons were readily available, the small fruits were crushed in water to make a tart drink called sumac-ade.

Rhus virens Lindh. ex A. Gray, (green), evergreen sumac, tobacco sumac, lentisco. Shrub or small tree to 3 m or more tall; leaves evergreen; leaflets 5–9, usually 4 cm or less long, 2 cm or less wide. Limestone areas, bluffs; sw margin of East TX from McLennan s to Bexar (Kartesz 2022) cos.; mainly Edwards Plateau and s Cross Timbers and Prairies w to Trans-Pecos; AZ, NM, and TX; Mexico. Aug–Oct. Turner (1996) treated this and related species; he recognized almost all TX populations of *R. virens* as var. *virens* and recognized two additional varieties, one limited to Mexico (*R. virens* var. *asutralis* B.L. Turner) and the other occurring in the U.S. only in AZ, NM, and the Trans-Pecos of TX (*R. virens* var. *choriophylla* (Wooton & Standl.) L.D. Benson).

TOXICODENDRON Mill. POISON-OAK, POISON-IVY

Poisonous woody vines, shrubs, subshrubs, or small trees; leaves pinnately 3(–5)-foliate or pinnately compound, often turning bright red in fall. Ours native of the U.S.

A genus of 10–15 species of North and South America and e Asia (Weakley, A.S., and Southeastern Flora Team 2022); often included in the genus *Rhus* but rather clearly distinguished (Barkley 1937; Gillis 1971). These species are toxic, causing **severe allergic contact dermatitis** in some individuals. From the standpoint of the plant, the toxins are probably a type of defense against pests since they restrict the growth of disease-causing fungi and deter certain plant-feeding insects (Aguilar-Ortigoza et al. 2003; Aguilar-Ortigoza & Sosa 2004). Physical contact with any part of the plant, exposure to fumes/smoke from burning plants, or contact with pets having touched the plants are common means of exposure. Following a latent period of 12–48 hours after exposure, there is inflammation including reddening of the skin, sometimes accompanied by edematous swelling. This can be followed by the formation of fluid-filled blisters; the fluid cannot spread the dermatitis. Instead, additional spread is from unwashed skin, clothing, or other objects (Lampe 1986). Eating the leaves can result in an internal reaction occasionally known to be fatal (Gillis 1975). Likewise, smoke from burned plants can cause serious internal reactions and is the leading cause of injury among fire-fighters (e.g., lung inflammation, respiratory emergencies) (Anderson 1995; Andersson 1997). According to Gillis (1975), "The poisons may be effective for an indefinite period of time in causing dermatitis. Several hundred year-old herbarium specimens have been known to affect a sensitive person who has handled them!" These reactions are caused by a mixture of resinous phenolic compounds (= have an aromatic ring and several hydroxyl groups) commonly known as urushiols (chemically catechols or more precisely, pentadecylcatechols or heptadecylcatechols), which "penetrate the skin, interact with proteins, and trigger an immune response" (Yatskievych 2006). Urushiols are extremely potent allergens (among the most potent known) and extremely tiny amounts are enough to cause a rash (Anderson 1995). Specifically, in the skin the compounds are "oxidized into quinone derivatives, initiating a complex cell-mediated immune response leading to a blistering rash ... and the formation of immunoglobulin antibodies" (Aguilar-Ortigoza et al. 2003). Because the compounds rapidly bind to the skin and are insoluble in water, rinsing exposed skin with rubbing alcohol (isopropyl) (some alternatively recommend rinsing with Clorox®—mixed one part Clorox® to 9 parts water) or some other organic solvent as soon as possible after contact is recommended. The alcohol should be followed by rinsing extensively with cold water (hot water possibly spreads the urushiols). Subsequently, the area can be washed with soap and water (initially using soap may only spread the toxin). There is some disagreement about the use of soaps; some authorities recommend washing with a strong degreasing soap; however, others indicate that soaps remove natural skin oils which may be protective. A specialized product, now sold as

Tecnu Oak-n-Ivy® Cleanser (originally developed during the cold war to remove nuclear fallout dust), is available for removing the urushiols (Anderson 1995; Stehlin 1996; Andersson 1997; Brooks 1999; Yatskievych 2006; Tecnuextreme.com 2007). All contaminated clothes, tools, etc. should be thoroughly washed with soap or alcohol since the toxins are extremely long-lasting. Gillis (1975) and Frankel (1991) discussed *Toxicodendron* dermatitis. Gillis (1975) also gave several citations documenting the use of POISON-IVY in Native American arrow poisons. The sap of an Asian species, *T. verniciflruum* (Stokes) F.A. Barkley (ORIENTAL LACQUER TREE), is a major source of lacquer; furniture treated with the lacquer can cause dermatitis in sensitive individuals. A number of *Toxicodendron* species, like those in the genus *Rhus*, display very early fall foliage color (often strikingly red); this is considered to serve as a "foliar fruit flag" which attracts birds that act as dispersal agents for the fall-ripening fruits (Stiles 1984). (Latin: *toxicum*, poison, and Greek: *dendron*, tree)

REFERENCES: McNair 1925; Barkley 1937; Brizicky 1963a; Gillis 1971, 1975; Brooks 1977; Baer 1979; Reveal 1990b, 1991; Anderson 1995; Aguilar-Ortigoza et al. 2003; Aguilar-Ortigoza & Sosa 2004; Pell et al. 2024c.

1. Leaflets 7–15; large shrub or small tree; usually in swamps or wet areas; species of extreme se portion of East TX ____ **T. vernix**
1. Leaflets usually 3(–5); vines, subshrubs, or shrubs; usually in habitats of moderate to dry moisture conditions; including species extremely abundant nearly throughout TX.
   2. Plants not climbing (low shrubs), without aerial roots, in sandy soils; leaflets often with broad, blunt apex, sometimes with narrowed apex, margins dentate to deeply lobed, pubescent underneath ____ **T. pubescens**
   2. Plants varying from not climbing (low to tall shrubs) to high-climbing vines, usually with aerial roots, in a variety of soil types; leaflets with apex acute to acuminate, margins entire, serrate, dentate, almost palmatisect or pinnatisect, shallowly to deeply lobed, or sharply toothed, glabrous or pubescent underneath ____ **T. radicans**

Toxicodendron pubescens Mill., (pubescent, downy), EASTERN POISON-OAK. Low, creeping shrub or subshrub, not climbing and without aerial roots; leaflet blades ovate to elliptic, lower leaf surfaces pubescent, margins dentate to deeply lobed. Sandy woods, the soils often low in nutrients; widespread in East TX (Bastrop, Bexar, Bowie, Brazos, Burleson, Caldwell, Dallas, Grayson, Harris, Harrison, Henderson, Houston, Jasper, Lamar, Montague, Montgomery, Polk, Rusk, Smith, Upshur, Walker, Waller, and Wilson cos.—Gillis 1971); also Tarrant Co. (Gillis 1971) just to the w of East TX and scattered to the w; e U.S. from NJ s to FL w to KS, OK, and TX. [*Rhus toxicarium* Salisb. [nom. illeg.], *R. toxicodendron* L., *Toxicodendron toxicodendron* (L.) Britton [nom. illeg.]] Apr–May. Gillis (1971) noted a specimen from Bastrop Co. suggesting a hybrid between *T. pubescens* (as *T. toxicodendron*) and *T. radicans* subsp. *verrucosum*; however, he found "very few instances" of such putative hybrids between the two species and he noted the "relative rarity of intermediate forms . . . where both of these species are frequent, if not common" as a "guiding factor in recognizing them as distinct species." This species causes contact dermatitis.

Toxicodendron radicans (L.) Kuntze, (rooting), POISON-OAK, POISON-IVY, HIEDRA. Low, creeping shrub with aerial roots to erect shrub or low- to high-climbing vine. Leaflet blades narrowly lanceolate, elliptic, ovate, broadly ovate, orbiculate, or deltate, lower leaf surfaces glabrous (pubescent along veins) or pubescent, margins entire to serrate, dentate, or shallowly to deeply lobed, apex acute to acuminate; fruits glabrous or densely pubescent. In a variety of habitats from low woods to forest margins and disturbed areas; widespread in East TX; widespread in TX; throughout most of the U.S.; Canada. Varieties are not distinguished on the county distribution map. Mid-Apr–May(–later). This is the "most common and widespread species in the Anacardiaceae" (Gillis 1971). The varieties of *T. radicans* seem to overlap morphologically and are often difficult to distinguish; according to Yatskievych (2006), there are "many intermediate, transitional, or possibly hybrid plants." Turner et al. (2003) did not recognize any infraspecific taxa and also included *T. pubescens* in *T. radicans*. While we provide a key to the varieties of *T. radicans*, the precise identity of many specimens should be expected to be difficult to determine. Mapped ranges should also be somewhat suspect. This species causes contact dermatitis; prevention is the best approach to avoiding problems with POISON-IVY—contact with the plant should be carefully avoided since an individual's sensitivity can change suddenly and dramatically; the old saying "leaves of three, let it be; leaves of five, let it thrive" is helpful in distinguishing the three leafleted POISON-IVY (and POISON-OAK) from VIRGINIA CREEPER (*Parthenocissus quinquefolia* (L.) Planch.) which has five leaflets; likewise the alternate leaves of *Toxicodendron* can be easily distinguished from the superficially similar, often trifoliate, opposite leaves of BOX ELDER (*Acer negundo* L.). The saying "leaves of three, let it be; berries white, poisonous sight" emphasizes the distinction between the poisonous white-fruited *Toxicodendron* species and the non-toxic red-fruited *Rhus* species. Key to varieties modified from Pell et al. (2024). Varieties are not distinguished on the county map.

1. Leaflet blade margins shallowly to deeply lobed (may be acutely so) to almost palmatisect or pinnatisect; leaflet blades ovate to broadly ovate to deltate.
   2. Leaflet blade margins shallowly to deeply lobed, usually acutely so; leaflet blades ovate to deltate, apex acute; petals abaxially glabrous; fruits glabrous ____ var. **verrucosum**

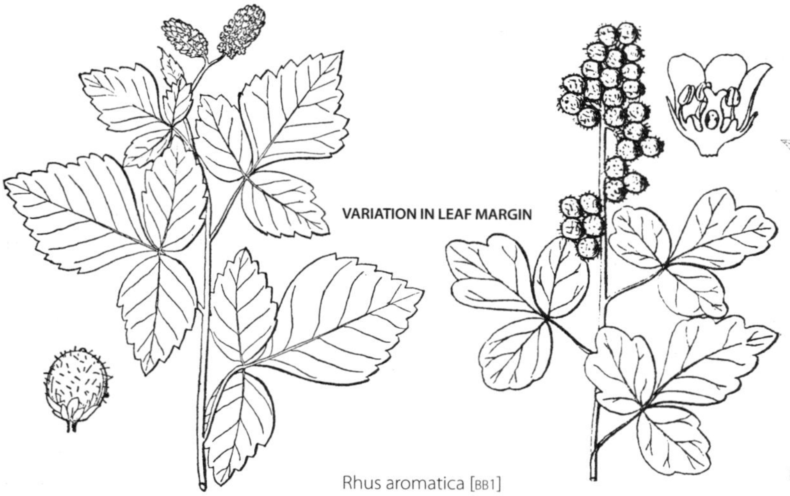

Rhus aromatica [BB1]

Rhus copallinum var. copallinum [SA3]

Rhus copallinum var. lanceolata [SA2]

Rhus glabra [REE]

Rhus microphylla [POW]

Rhus virens [LYN, POW]

Toxicodendron pubescens [REE]

2. Leaflet blade margins deeply lobed to almost palmatisect to pinnatisect, usually rounded; leaflet blades ovate to broadly ovate, apex acute to acuminate; petals abaxially pubescent; fruits pubescent ____ var. **eximium**

1. Leaflet blade margins entire, serrate, dentate, or lobed; leaflet blades lanceolate, orbiculate, broadly elliptic, or ovate.

3. Shrubs or lianas, usually climbing; aerial roots present; leaflet apex acute to acuminate; fruit glabrous or densely hairy ____ var. **radicans**

3. Shrubs, never climbing; aerial roots absent; leaflet apex acuminate; fruit glabrous ____ var. **rydbergii**

var. **eximium** (Greene) F.A. Barkley, (excellent or extraordinary). Shrubs or vines, usually climbing with aerial roots; leaflet blades ovate to broadly ovate, margins deeply lobed to almost palmatisect or pinnatisect, lobes rounded, apex acute to acuminate; fruits pubescent. Canyons, oak woodlands, rocky slopes; in East TX known only from Bexar Co. (Barkley 1961); Edwards Plateau and Trans-Pecos, also just to the w of East TX in Tarrant Co. (Barkley 1961); TX; Mexico. Apr–Sep (dependent on precipitation).

var. **radicans,** POISON-IVY. Shrubs or vines, usually climbing; leaflet blades ovate, sometimes lanceolate to orbiculate, lower leaf surface glabrous (pubescent on veins) or pubescent, margins entire, serrate, dentate, or lobed; fruits glabrous or densely pubescent. Hillsides, open woods, forest edges, disturbed areas; Harris, Montgomery, Walker (Gillis 1971), Anderson, Brazos, Madison, and Washington (Kartesz 2022) cos.; East TX and possibly more widespread in the state; widespread in e U.S. from ME s to FL and w to AR and TX; also se-most Canada. Mar–Aug. [*Rhus radicans* L. var. *littoralis* (Mearns) Deam, *R. toxicodendron* L. var. *microcarpa* Michx., *R. toxicodendron* var. *pubens* Engelm. ex S. Waston, *R. toxicodendron* var. *vulgaris* Michx., *Toxicodendron negundo* Greene, *T. radicans* subsp. *negundo* (Greene) Gillis, *T. radicans* subsp. *pubens* (Engelm. ex S. Watson) Gillis]

var. **rydbergii** (Small ex Rydb.) Erskine, (for Per Axel Rydberg, 1860–1931, first curator of the New York Botanical Garden Herbarium), WESTERN POISON-IVY. Shrubs, never climbing, 0.2–3 m tall; leaflet blades ovate to broadly elliptic, lower leaf surface veins sparsely to densely pubescent, margins entire, serrate or shallowly 1–3-lobed, apex acuminate; fruits glabrous. Hillsides, open woods, forest edges, disturbed areas; Hardin (*S.D. Jones 14621* SAT), Hill (*G.H. Hall 155* BRIT), and Titus (*J.D. Smith 306* BRIT) cos.; Trans-Pecos, Cross Timbers and Prairies, and High Plains; mostly w 1/2 of the U.S., scattered in ne U.S. May–Jul. [*Rhus rydbergii* Small ex Rydb., *R. radicans* L. var. *rydbergii* (Small ex Rydb.) Rehder, *R. toxicodendron* L. var. *rydbergii* (Small ex Rydb.) Garrett, *Toxicodendron rydbergii* (Small ex Rydb.) Greene]

var. **verrucosum** (Scheele) Gillis, (verrucose, warted), POISON-IVY. Shrubs or vines; leaflet blades ovate to deltate, margins shallowly to deeply acutely lobed, apex acute; fruits glabrous. Oak woodlands, rocky slopes, primarily on limestone; scattered in East TX; East TX w to Cross Timbers and Prairies, Edwards Plateau, and Trans-Pecos; OK and TX. [*Rhus verrucosa* Scheele]

Toxicodendron vernix (L.) Kuntze, (varnish), POISON SUMAC, POISON-ELDER, POISON-DOGWOOD. Large shrub or small tree, 2–7.5 m tall; branchlets glabrous; leaflets 4–12 cm long, 2–5 cm wide, entire; flowers in axillary panicles, greenish. Swamps or wet areas; Newton (*Cory 57228* SMU), Polk (Brown et al. 2006a), Hardin, Jasper, Nacogdoches, Sabine, San Augustine, and Shelby (Turner et al. 2003) cos., also cited for vegetational area 4 by Hatch et al. (1990), though we have been unable to confirm its presence in the Blackland Prairie; e U.S. w to MN, IL, and TX; also se Canada. Apr–Jun. [*Rhus vernix* L.] Vegetatively, this species resembles many non-toxic *Rhus* species. It can, however, be easily distinguished by its pinnately compound leaves having entire leaflets and lacking any wing along the rachis; all East TX *Rhus* species with pinnately compound leaves have either toothed leaflets or a winged rachis or in the case of the locally rare *R. virens*, small leaflets 4 cm or less long. This species causes contact dermatitis.

ANNONACEAE Juss.

CUSTARD-APPLE OR ANNONA FAMILY

The Annonaceae is a large family of ca. 2,450–2,500 species in 107–120 genera (Stevens 2001 onwards; Mabberley 2017); it is mainly in tropical areas with *Asimina* extending n to Michigan and Ontario. The family consists of aromatic (due to ethereal oils) trees, shrubs, and lianas, usually with alkaloids, and with 3-merous flowers (unusual except among monocots); they share with the related Magnoliaceae a number of unspecialized flower characters such as numerous free stamens and carpels, laminar stamens, and monocolpate pollen. The family is interesting evolutionarily because it is one of the "magnoliids" (a group of ca. 10,000 species of four related orders—Canellales, Laurales, Magnoliales, Piperales) which are among the approximately 3% of flowering plants near the base of the angiosperm lineage that are neither in the monocots or eudicots (Stevens 2006 [2001 onwards]; APG IV 2016; Shen et al. 2022); it is one of the most diverse of these basal families. Many are important economically with tropical species variously used medicinally or for perfumes or cosmetics (Baumgardt 1982); a number are valued for their edible fruits including *Annona* species (e.g., CUSTARD-APPLE, CHERIMOYA, SOURSOP, and SWEETSOP). Family name from *Annona*, a genus of ca. 100 species of tropical America and Africa. (Name either from Haitian, *anon* or *hanon*, the native name for the widely cultivated *A.*

muricata L.—soursop (Kral 1997), or from Latin: *annona*, yearly produce, crop, grain, food) (subclass Magnoliidae—Cronquist; order Magnoliales—APG IV)

FAMILY RECOGNITION IN THE FIELD: the only two East TX species are shrubs or small trees with alternate, distichous (= 2-ranked) entire, obovate or oblanceolate leaves, hairy *naked buds* (= buds without scales), *3-merous* flowers, and somewhat banana-like fruits.

REFERENCES: Wood 1958; Kessler 1993a; Kral 1997; Johnson 2003; Chatrou et al. 2012; Couvreur et al. 2019.

ASIMINA Adans. AMERICAN PAWPAW

Shrubs or small trees; branches spreading-ascending; buds naked; young twigs with rusty pubescence; leaves alternate, simple, distichous (= in two vertical rows or ranks, not spirally arranged; e.g., leaves occurring in two ranks or rows on opposite sides of a stem), entire, deciduous, obovate to oblanceolate, pinnately-veined, aromatic/malodorous when crushed, without stipules; inflorescences axillary, the flowers usually 1 per axil, appearing before or with the leaves; flowers perfect; receptacle becoming enlarged; sepals 3, early deciduous; petals 6, in two unequal whorls, the outer larger; stamens numerous, with short filaments, in a mass; pistils 3–7(–12), separate; ovaries superior; fruits fleshy berries, unevenly thickly cylindric, usually 3–5 per flower; seeds 3–many per pistil, 1.5–2.8 cm long.

An e U.S. genus of 8 species (Kral 1997; Stevens 2001 onwards); this is one of only 11 genera of trees endemic to e North America (five of these, *Asimina*, *Leitneria*, *Maclura*, *Planera*, and *Taxodium*, occur in East TX) (Little 1983). It is also the northernmost and only extra-tropical genus of a large and otherwise tropical family (Little 1970). *Asimina* is apparently an example of an anachronism (Janzen & Martin 1982; Barlow 2001a, 2001b); the fruits, though now dispersed by some mammals, are thought to have coevolved with dispersal agents that are now extinct; other North American examples of anachronisms include *Maclura pomifera* (BOIS D'ARC, HORSE APPLE; Moraceae) and *Gleditsia triacanthos* (HONEY-LOCUST; Fabaceae). Beetles and flies have been discussed as possible pollinators (Kral 1960a; Willson & Schemske 1980; Norman et al. 1992), with successful fertilization apparently possible only by outcrossing (Yatskievych 2006). (Name through *asiminier* of the French colonists, from the Native American name, *assimin*, an Algonquin word connected to pawpaws—Gordon 2023)

REFERENCES: Nash 1896; Kral 1960a, 1997; Wilbur 1970; Willson & Schemske 1980; Rogstad et al. 1991; Norman et al. 1992; Weakley et al. 2012; Gordon 2023.

1. Leaf blades usually 6–15 cm long (rarely to 20 cm); longest (outer) petals 1–1.3 cm long; peduncle < 1 cm long, with red-brown to tan hairs; plants shrubs to 2(–4) m tall ____ **A. parviflora**
1. Leaf blades 15–30 cm long; longest petals 1.5–2.5 cm long; peduncle 1–2.5 cm long, with dark brown to red-brown hairs; plants large shrub or tree to 12(–15) m tall ____ **A. triloba**

Asimina parviflora (Michx.) Dunal, (small-flowered), SMALL-FLOWER PAWPAW, SMALL-FRUIT PAWPAW, DWARF PAWPAW. Shrubs, often flowering and fruiting when only a few feet tall, resembling a "miniature version of *A. tribloba* in flower, shoot, and leaf" (Kral 1997), usually 2(–4) m or less tall; flowers faintly fetid, often appearing sessile (due to short peduncle); petals maroon, rarely yellow, the inner with basal corrugate nectary zone indistinct; pistils 5–7; fruits 3–6(–7) cm long, greenish yellow. Rich to dry upland woods; s part of East TX and disjunct w to Bastrop and Caldwell (Turner et al. 2003) cos.; also Nacogdoches and Colorado cos. to the n and s (Kartesz 2015) and Chambers Co. in n Gulf Prairies and Marshes (Turner et al. 2003); se U.S. from VA s to FL w to AR and TX. Early spring. [*Orchidocarpum parviflorum* Michx.] Suspected hybrids between this species and *A. triloba* have been reported from AL (Kral 1997).

Asimina triloba (L.) Dunal, (three-lobed), PAWPAW, COMMON PAWPAW. Large shrub or tree to 12(–15) m tall, often root suckering to form colonies; buds naked (= without scales) but protected by pubescence; leaf blades obovate to elliptic-ovate, to 7(–15) cm wide, apically abruptly acuminate to acute, basally cuneate, rusty-pubescent beneath; petioles 5–10 mm long; flowers fetid, yeasty-smelling, strongly protogynous; petals impressed-veiny, greenish yellow when immature, maturing dull purple to maroon, 6, in 2 series, the inner and outer series very unequal, the inner shorter, with a distinct, corrugate nectary zone near the base adaxially; pistils 3–15; fruits to 12(–15) cm long, to ca. 3–4 cm thick, green to dark brown when ripe (in fall), the flesh sweet and edible; seeds 2–10, to 2.8 cm long, flattened. Rich woods and banks of streams; mainly e Pineywoods, also Grayson Co. (*M.D. Bryant s.n.* BRIT!) along the Red River near the mouth of Pawpaw Creek; also Chambers Co. in n Gulf Prairies and Marshes (Turner et al. 2003); with the exception of the Grayson Co. collection, this species is known in TX only from the extreme e part of the state (Turner et al. 2003; USDA Plants Database 2019); widespread in the e 1/2 of the U.S. w to NE, OK, and TX and also Ontario in extreme se Canada. While the Grayson Co. occurrence is possibly "natural" (there are a number of disjuncts in the microclimates of the Red River area), it is also possible that this edible species was transported from further east by Native Americans; the species is known to have been appreciated by Native Americans and explorers, including members of the Lewis and Clark

Expedition (Clark 1806; Gordon 2023). Two streams in n Grayson Co. are named PawPaw Creek. Apr–May. [*Annona triloba* L.] Because of its ability to reproduce vegetatively (clonally), this species can form extensive thickets (Kral 1960a; Rogstad et al. 1991); reproduction by seeds appears to be relatively rare (Willson & Schemske 1980; Norman et al. 1992). The yeasty floral scents are postulated to attract flies and beetles (Goodrich et al. 2006). The fruits are edible (and delicious to some, with a "tropical" flavor) but can cause gastrointestinal problems for certain individuals; they have been called "Indian bananas" (Yatskievych 2006). They are reported to be the largest edible fruit native to North America (Goodrich et al. 2006), can reach nearly a pound in weight (Callaway 1993), have potential as a commercial fruit (Callaway & Callaway 1992), and are being investigated for commercial use (Pomper et al. 2008). The seeds are reported to contain a toxic alkaloid and the plant causes contact dermatitis in a small percentage of people (Peattie 1948; Kingsbury 1964). Yatskievych (2006) noted that most parts of the plant (except the fruit) and especially the bark contain a variety of toxins (alkaloids and fatty acid lactones). Some of these are being investigated "as potential anticancer and antitumor drugs, and also for possible applications as pesticides." The species has few pests, but Zebra swallowtail butterflies (*Protographium marcellus*) utilize the "plant's acetogenins (asimicin affecting mitochondrial electron transport system) against its predators (birds)" (Mabberley 2017). The acetogenins have been investigated for their anti-cancer potential (Nam et al. 2018). The species is apparently unpalatable to deer (Weakley et al. 2012). Strips of the bark were used by Native Americans and pioneers for weaving into cloth and for making fish stringers (Yatskievych 2006).

APIACEAE Lindl. (UMBELLIFERAE Juss.)

CARROT OR PARSLEY FAMILY

Annual or perennial herbs; leaves basal, alternate, or (in *Bowlesia*) opposite; petioles in nearly all species with enlarged, clasping base, blades simple or compound, entire, toothed, lobed, or finely dissected; flowers axillary or terminal, small, in umbels (varying from loose and open to compact and head-like); simple or compound (= branched); compound umbels often subtended by an involucre of leafy bracts; ultimate umbels (= umbellets), often subtended by an involucel of bractlets, are supported by rays and composed of few to many flowers; calyx tube cylindrical to shallowly cup-like, with entire or 5-toothed summit; petals 5, attached at summit of calyx tube, equal or occasionally unequal; stamens 5, attached around a fleshy disk (= stylopodium); pistil 2-carpellate; ovary inferior; styles and stigmas 2; fruit a schizocarp that splits into 2 1-seeded segments (= mericarps); the surfaces where the mericarps are joined is termed the commissure; the outer surface of each mericarp often has 1 dorsal rib, 2 intermediate ribs and 2 lateral ribs adjacent to the commissure.

A large (ca. 3,820 species in ca. 466 genera), nearly cosmopolitan but more commonly n temperate Eurasian and North American family (Plunkett et al. 2018) of mainly herbs or less commonly shrubs or even trees (Mabberley 2017). Partly because of the presence of aromatic oils in secretory canals or oil tubes (= vittae), the family yields many important herbs, spices, and foods including *Anethum* (DILL), *Apium* (CELERY), *Carum* (CARAWAY), *Coriandrum* (CORIANDER, CILANTRO), *Cuminum* (CUMIN), *Daucus* (CARROT), *Foeniculum* (FENNEL), *Pastinaca* (PARSNIP), *Petroselinum* (PARSLEY), and *Pimpinella* (ANISE); it also includes some extremely poisonous plants. To avoid possibly fatal poisoning, no wild member of the family should be eaten without absolute certainty of identification. Some species are also known to cause "phototoxic reactions," also referred to as photodermatitis—contact with the plant can sensitizes skin to ultraviolet light, resulting in sunburn-like effects (Lampe & McCann 1985; Yatskievych 2006). Many species of Apiaceae are used as food by caterpillars of the black swallowtail butterfly (*Papilio polyxenes*); numerous native species (*Berula erecta, Cicuta maculata, Cryptotaenia canadensis, Osmorhiza longistylis, Polytaenia nuttallii, Ptilimnium capillaceum, Sium suave, Spermolepis divaricata*, and *Zizia aurea*) are utilized as well as introduced taxa (DILL, FENNEL, PARSLEY, QUEEN-ANNE'S-LACE, POISON-HEMLOCK). Apiaceae typically produce toxic psoralins (furanocoumarins) to protect themselves from herbivore damage; however, black swallowtail larvae are resistant and actually grow faster in the presence of the chemicals (Scott 1986; Tveten & Tveten 1993). It has long been known that Apiaceae are closely related to Araliaceae and some workers have argued that the two should be combined into a single family (Judd et al. 1994). However, while related, the families are two distinct lineages. Nonetheless, some changes in traditional delimitation have been necessary—the most striking example is that Apiaceae subfamily Hydrocotyloideae is now known to be polyphyletic and cannot be maintained (Downie et al. 2000; Chandler & Plunkett 2004). For example, *Hydrocotyle* and a closely related genus, *Trachymene*, are now recognized in the Araliaceae rather than the Apiaceae. Additionally, in recent years, there have been many advancements in our knowledge of the phylogeny of many groups in the Apiaceae, mostly through molecular research. As a result, there have been many rearrangements at various taxonomic levels. It has been found that the characters often used to identify and categorize Apiaceae are the result of homoplasy (= a shared character between two or more plants that did not arise from a common ancestor). Family name from *Apium*, a temperate genus of ca. 19 species including *A. graveolens* L., CELERY, valued for its edible petioles and fruits (celery "seed"). (Classical Latin name for celery and parsnip) (subclass Rosidae—Cronquist; order Apiales—APG IV)

Family recognition in the field: often aromatic herbs with small, frequently white to green, yellow, or pink flowers usually in *umbels*; leaves alternate and often pinnately compound, sometimes finely dissected, with sheathing petioles; internodes hollow; fruits small, splitting into 2 1-seeded parts.

References: Coulter & Rose 1900; Mathias & Constance 1944–1945, 1961; Mathias 1965; Hiroe 1979; Pimenov & Leonov 1993; Judd et al. 1994; Plunkett et al. 1996; Plunkett et al. 1996 [1997]; Downie et al. 2000; Chandler & Plunkett 2004; Wunderlin et al. 2020; Weakley, A.S., and Southeastern Flora Team 2024.

1. Leaves simple or apparently so (at least the basal in *Thaspium*); leaf blades round to reniform, ovate, lanceolate, or linear, either unlobed or palmately divided 2/3 to base or less, with few, wide lobes.
  2. Leaves reduced to elongate narrowly spatulate to linear, entire, hollow, septate phyllodes, normal blade tissue absent.
    3. Plant rhizomatous; narrowly, spatulate leaves and simple umbels arising from the horizontal rhizome nodes ____ **Lilaeopsis**
    3. Plant caulescent with a woody caudex-like base; leaves linear (rush-like in appearance; umbels compound ____ **Tiedemannia**
  2. Leaves with normal blade tissue.
    4. Leaves perfoliate (= completely surrounding the stem) ____ **Bupleurum**
    4. Leaves not perfoliate.
      5. Leaf blades (at least upper) linear to oblong-lanceolate, often spinose; flowers sessile, numerous, in very dense, ovoid to globose, whitish to bluish heads 5–25 mm in diam. ____ **Eryngium**
      5. Leaf blades round to reniform, broadly ovate, or ovate; flowers not arranged as above.
        6. Leaf margins finely and evenly toothed; basal leaf blades unlobed or with only 1 or 2 lobes; erect perennials from fibrous roots ____ **Thaspium**
        6. Leaf margins nearly entire to slightly scalloped or toothed; leaf blades unlobed or variously lobed; creeping perennials OR prostrate to suberect annuals from slender taproot.
          7. Inflorescences usually long-peduncled; plants glabrous or with simple pubescence; stipules absent; leaves 2–5 per node from a perennial creeping rhizome, rooting at the nodes ____ **Centella**
          7. Inflorescences sessile or nearly so; plants with ± gray, stellate pubescence; stipules present, scarious, lacerate; leaves opposite from a prostrate to suberect slender, taprooted annual ____ **Bowlesia**
1. Leaves compound or decompound or at least the largest deeply lobed; leaf blades triangular-ovate to lanceolate or linear in outline.
  8. Inflorescence a dense "head," globose to elongated and sometimes with a tuft of usually spinose bracts (= coma) exserted apically ____ **Eryngium**
  8. Inflorescence open, loose, not dense, without spiny bracts exserted apically.
    9. Ovaries and fruits bristly or prickly.
      10. Leaves palmately compound with 3–5 leaflets, the leaflets toothed and lanceolate to elliptic, the largest > 1 cm wide, usually much wider ____ **Sanicula**
      10. Leaves pinnately compound or decompound, with numerous segments; ultimate leaf segments < 1 cm wide, often much less.
        11. Umbels with simple, entire basal bracts or bracts absent.
          12. Ultimate leaf segments of upper leaves narrowly linear or thread-like; plants glabrous; mature fruits 1–2 mm long ____ **Spermolepis**
          12. Ultimate leaf segments of upper leaves narrowly lanceolate; plants hispid; mature fruits 3–5 mm long ____ **Torilis**
        11. Umbels with toothed or lobed basal bracts ____ **Daucus**
    9. Ovaries and fruits glabrous or pubescent, but neither bristly nor prickly.
      13. Petals creamy white to golden yellow or green (except white in *Polytaenia albiflora*).
        14. Leaves all basal.
          15. Fruits 5–11 mm long, winged (the lateral ribs of the fruits with corky wings); ultimate leaf segments linear or linear-lanceolate, ≤ 1 mm wide; plants of prairie habitats ____ **Lomatium**
          15. Fruits 3–4 mm long, unwinged; ultimate leaf segments broader, > 1 mm wide; plants of alluvial habitats ____ **Tauschia**
        14. Leaves distributed up the stem.
          16. Ultimate leaf segments long and thread-like (< 0.5 mm wide); plants with strong anise or dill odor.
            17. Plants with strong dill odor; mature fruits with lateral ribs narrowly winged, distinctly different from other ribs ____ **Anethum**
            17. Plants with strong anise/licorice-like odor; mature fruits with the lateral ribs wingless, all the ribs ± the same in appearance ____ **Foeniculum**
          16. Ultimate leaf segments not long and thread-like (> 1 mm wide, often much greater); plants without strong anise or dill odor.
            18. Leaflets entire marginally, without any teeth, unlobed or with 1 or 2 deep lobes; bracts absent below both umbels and umbellets (= secondary or ultimate umbels) ____ **Taenidia**

18. Leaflets usually with at least some teeth and sometimes conspicuously lobed; bracts absent below umbels but bractlets usually present below umbellets (sometimes inconspicuous).
    19. Ultimate segments of upper leaves sharply toothed around the margins, unlobed or with few lobes; fruits 2–6 mm long.
        20. All flowers of umbellets pedicellate, the pedicels 1–4 mm long; fruits winged (the ribs of the fruits with corky wings); roots fibrous ____ **Thaspium**
        20. Central flower of each umbellet sessile or nearly so (pedicel < 0.5 mm long); fruits not winged; roots tuberous-thickened ____ **Zizia**
    19. Ultimate segments of upper leaves toothed at apex or entire, conspicuously lobed; fruits 2–15 mm long.
        21. Fruits dorsally flattened, 5–15 mm long, 4–9.5 mm broad; lateral ribs corky; small calyx teeth present; widespread native species ____ **Polytaenia**
        21. Fruits not dorsally flattened, 2–4 mm long, 1.5–3 mm broad; lateral ribs not corky; calyx teeth absent; rarely escaped introduced cultivar ____ **Petroselinum**

13. Petals white, pink, lavender, maroon, or purple.
    22. Fruits linear with long differentiated beak 20–70 mm long; ovary pubescent ____ **Scandix**
    22. Fruits beakless or nearly so; ovary pubescent or glabrous.
        23. Pedicels 0.3–2 times as long as the ovary when petals have expanded (up to 3 times in fruit).
            24. Umbels small, simple or compound with 1–6 rays; (–10 in *Ammoselinium popei*).
                25. Small bracts below umbellets linear or narrowly lanceolate, acute or small bracts absent.
                    26. Segments of middle and upper leaves obtuse or subacute; plants branching from base; fruits 2–4(–5) mm long ____ **Ammoselinum**
                    26. Segments of middle and upper leaves sharply acute; plants branching in upper part; fruits 8–10 mm long ____ **Trepocarpus**
                25. Small bracts below umbellets oblong-elliptic, obtuse or subacute ____ **Chaerophyllum**
            24. Umbels relatively larger, compound with 6–18 rays.
                27. Involucres of pinnately parted bracts as long as or longer than the rays; fruits not winged ____ **Daucosma**
                27. Involucres absent or of 1–3 minute bracts; fruits winged (the ribs of the fruits with corky wings) ____ **Thaspium**
        23. Pedicels 2–many times as long as the ovary when petals have expanded.
            28. Umbels mostly lateral or axillary, sessile to short-peduncled; fruits ovoid, 1.4–2.3 mm long ____ **Cyclospermum**
            28. Umbels mostly or all terminal, short- or long-peduncled; fruits various.
                29. Bractlets below umbellets oblong-elliptic, obtuse, scarious; flowers pinkish; fruits broadly winged; leaves mostly near base of plant ____ **Vesper**
                29. Bractlets below umbellets absent or slender, acute, not scarious; flowers various; fruits winged or not winged; leaves along stem and/or basal.
                    30. Upper leaves with lanceolate or oblanceolate, usually toothed segments or leaflets.
                        31. Umbels with 2–8 primary rays; ovary elongating, the fruit ellipsoidal.
                            32. Leaves mostly once compound; small bracts below umbellets linear-lanceolate, erect, inconspicuous; pedicels very unequal ____ **Cryptotaenia**
                            32. Leaves 2–3 times compound; small bracts below umbellets lanceolate, reflexed, prominent; pedicels about equal ____ **Osmorhiza**
                        31. Umbels with many (6–30+) primary rays; ovary remaining short, the fruit subglobose to oblong.
                            33. Leaf blades much-dissected, almost parsley-like in appearance; small bracts below umbellets ovate-lanceolate with wide clasping base, abruptly tapered to ± narrow tips ____ **Conium**
                            33. Leaf blades 1–3 times pinnately compound but not much-dissected, not at all parsley-like in appearance; small bracts below umbellets linear to lanceolate, scarcely if at all clasping, not abruptly tapered.
                                34. Leaf blades 1–3 times pinnately compound, the lateral leaflets or lateral leaf segments with distinct stalks.
                                    35. Involucres subtending main umbels absent or of 1–4 entire, inconspicuous bracts; ribs of fruit prominent, corky, obtuse, covering half or more of the fruit surface ____ **Cicuta**
                                    35. Involucres subtending main umbels of numerous usually pinnately divided bracts with filiform segments; ribs of fruit narrow, acute, covering much less than half of the fruit surface ____ **Ammi**
                                34. Leaf blades once pinnately compound (but leaflets can be deeply lobed), the lateral leaflets sessile or nearly so.

36. Leaflets not only toothed but also some often deeply lobed; fruits 1.5–2 mm long, with ribs obscure ____ **Berula**
36. Leaflets finely to coarsely toothed (rarely entire), but not lobed; fruits 2–7 mm long, with ribs prominent.
37. Leaflets often coarsely toothed (teeth to 4 mm or more long); fruits 4–7 mm long; stems not conspicuously corrugated ____ **Oxypolis**
37. Leaflets finely toothed (teeth < 1 mm long); fruits 1.5–3 mm long; stems conspicuously corrugated ____ **Sium**

30. Upper leaves with thread-like to linear-lanceolate, mostly entire segments or some leaves undivided, slender, entire.
38. Leaves once compound (some can be simple), with linear-lanceolate, entire lobes or leaflets.
39. Stem leaves palmately compound; fruits ca. as long as broad with a small narrow beak ____ **Cynosciadium**
39. Stem leaves pinnately compound or pinnatifid or some simple and entire; fruits mostly longer than broad, beakless ____ **Limnosciadium**
38. Leaves 1–3 times compound, the upper with linear-lanceolate to thread-like segments.
40. Petals obtuse or subacute (a few sometimes bilobed), oblong or elliptic, longer than wide; fruits neither constricted into 2 parts nor conspicuously flattened.
41. Petals 0.5–1.3 mm long, white; fruits tuberculate to essentially smooth or scabrous with minute hairs, 1.5–2 mm in diam.; native species widespread in East TX ____ **Spermolepis**
41. Petals (of outer flowers in umbels) 2.2–4 mm long, white, often tinged pink or lavender; fruits smooth, (1.5)2.5–3(–5) mm in diam.; introduced cultivated species rarely escaped in East TX ____ **Coriandrum**
40. Petals with indented or notched apex, obovate or suborbicular, shorter to slightly longer than wide; fruits constricted into 2 parts or conspicuously flattened or neither.
42. Segments of middle and lower leaves linear-lanceolate to thread-like, entire or nearly so; fruits not conspicuously flattened.
43. Petals (of outer flowers in umbels) 0.5–1.3 mm long; fruits not constricted into 2 parts; calyx tube of outer flowers cup-shaped, slightly shorter to much longer than its lobes or teeth ____ **Ptilimnium**
43. Petals (of outer flowers in umbels) 1.6–2.1 mm long; fruits conspicuously constricted into 2 parts; calyx tube of outer flowers subrotate, much shorter than its lobes ____ **Atrema**
42. Segments of middle and lower leaves lanceolate, sharply toothed; fruits conspicuously flattened ____ **Eurytaenia**

AMMI L. BISHOP'S-WEED, BULLWORT

An Old World genus of 3–4 species (Weakley, A.S., and Southeastern Flora Team 2024); *A. visnaga* (L.) Lam. (TOOTH PICK BISHOP'S WEED) has been cultivated since the time of the Assyrians for medical uses, especially in treating angina and asthma. (ancient Greek name used by Dioscorides)

REFERENCES: Mathias & Constance 1961; Yatskievych 2006.

Ammi majus L., (bigger, larger), GREATER AMMI, BISHOP'S-WEED, LARGE BULLWORT. Erect, glabrous, branching annual 0.2–0.8 m tall; superficially resembling *Cicuta*; leaves ternately or pinnately compound or decompound, the ultimate segments narrowly lanceolate, serrate; inflorescences of compound umbels; involucral bracts numerous, usually pinnately divided, the segments filiform; rays 50–60; flowers white; fruits oblong, 1.5–2 mm long, 1 mm or less wide, the ribs narrow, acute, covering much less than half of the fruit surface, unwinged. Waste areas, roadsides; c and s portions of East TX; Edwards Plateau and South TX Plains; scattered in se U.S., also AZ, CA, and OR. Mar–Jun. Native of the Mediterranean. The species is cultivated for the cut-flower trade (Mabberley 2017). The plants, especially the immature fruits, are reported to contain furanocoumarins which can cause photodermatitis in some people (Yatskieych 2006).

AMMOSELINUM Torr. & A. Gray SAND-PARSLEY

Glabrous small annuals branching from base; leaves ternately or ternate-pinnately decompound, the ultimate segments linear to spatulate; flowers inconspicuous, white, in small compound umbels; fruit with prominent ribs, winged or not so.

A genus of 3 species of North America and temperate South America (Mabberley 2017). (Greek: *ammo*, sand, presumably from the habitat, and *selinon*, celery)

REFERENCES: Nesom 2012c, 2024a.

1. Umbels in the axils of stem leaves, ± sessile (appearing simple); fruits 2–2.2 mm long; plants 4–10 cm tall ____ **A. butleri**
1. Umbels axillary or seemingly terminal, on peduncles 2.5–7.5 cm long; fruits 3–4(–5) mm long; plants 8–35(–60) cm tall ____ **A. popei**

Ammoselinum butleri (Engelm. ex S. Watson) J.M. Coult. & Rose, (for George Dexter Butler, 1850–1910, lawyer, teacher, botanist, correspondent of George Engelmann, and one of the early collectors of this species), BUTLER'S SAND-PARSLEY. Ultimate segments rounded to bluntly pointed at tip, sometimes with a minute sharp point, margin and midrib glabrous to slightly scabrous; rays (1–)2–4; umbellets 3–5-flowered; fruits ovoid. Sandy or disturbed ground, fairly common but inconspicuous and overlooked; scattered in East TX; Edwards Plateau and South TX Plains; se U.S. from KY s to AL and w to KS and TX, but absent from FL, GA, and SC. Mar–Apr. Native of North America.

Ammoselinum popei Torr. & A. Gray, (for Clara Maria Pope, active 1760s–1838, British flower painter), PLAINS SAND-PARSLEY. Leaflet margin and midrib prominently scabrous; rays (3–)5–10; umbellets 4–9-flowered; fruits oblong-ovoid. Sandy or gravelly ground; Grayson, Lamar, McLennan, and Travis (Kartesz 2022) cos.; mainly w 2/3 of TX; KS, OK, NM, TN, and TX. Mar–May. Native of North America.

ANETHUM L. DILL

A genus of 4 species of s Europe and sw Asia (Ball 2024a). It is closely related to *Foeniculum* (FENNEL) and hybridization between the two can occur. (From *Anethon*, ancient Greek name of dill, thought to come from *aithein*, blaze, in allusion to the pungent seeds)

REFERENCES: Mathias & Constance 1961; Yatskievych 2006; Ball 2024a.

Anethum graveolens L., (heavy-scented), DILL. Glabrous, strong-scented annual, stems 0.5–1.5 m tall; petioles 1–3 cm long; leaf blades 4–13 cm long, pinnately decompound, ultimate segments 0.5–2 cm long; inflorescence terminal and lateral compound umbels, peduncles 6–12 cm long; rays 15–40, 1.5–3 cm long in flower, 3–7 cm long in fruit; flowers many per umbellet, pedicels unequal, 0.5–6 mm long in flower, 6–9 mm long in fruit; petals yellow; fruits ovate-elliptic, 3–4 mm long, with narrowly winged lateral ribs; similar to *Foeniculum* and reported to hybridize with it. Widely cultivated in TX and reseeds in open ground; Grayson (BRIT), Brazos (Turner et al. 2003), Burleson, Cherokee, and McLennan (Kartesz 2022) cos.; scattered to the s and w of East TX; actually probably more widespread given the ease of its reseeding; scattered in the U.S. and s Canada. May–Summer. Native of sw Asia. This species has been cultivated since at least 400 BC.; the leaves, inflorescences, and fruits are widely used in pickling (e.g., dill pickles) and as a flavoring (Mabberley 2017). Duke (1985) indicated that insects exposed to the dill-derived insecticide parathion alone suffered only 8% mortality; those exposed to the same doses of parathion plus d-carvone or other dill components showed 99% mortality; he raised the question of whether certain naturally occurring plant products can synergistically react with pesticides to produce harmful effects in humans.

ATREMA DC. PRAIRIE-BISHOP

A monotypic genus of sc U.S. (Greek *a-*, without, and *trema*, hole, alluding to unperforated closed commissure of fruits as in *Astoma* and *Bifora*)

REFERENCES: Hardway et al. 2004; Downie et al. 2008; Nesom 2024b.

Atrema americanaum DC., (of America), PRAIRIE-BISHOP. Low, glabrous annual 25–75 cm tall; leaves ternate-pinnately decompound, ultimate segments filiform; involucres 2–8 bracts; rays (4–)5–14; flowers white; fruits subglobose, 2–3 mm long, 2–2.5 mm wide, the two nearly spherical segments separating from one another at maturity. Prairies, rocky slopes, and roadsides, limestone areas; mostly s and w portions of East TX; Edwards Plateau; AL, AR, OK, and TX. May–Jun. [*Bifora americana* (DC.) Benth. & Hook. f. ex S. Watson] Molecular research has shown that *Atrema americanum* is distinct from the Eurasian species (*Bifora testiculata*, *Bifora radians*). Native of s U.S. Speaking of the area of Brazos and surrounding counties (Post Oak Savannah), Reed (2005) noted that "this flower is the last in the temporal sequence: *Nothoscordum*, *Arenaria*, *Atrema* that colors lots and roadsides white in springtime."

Toxicodendren radicans var. radicans [LYN]

Toxicodendron radicans var. verrucosum [LYN]

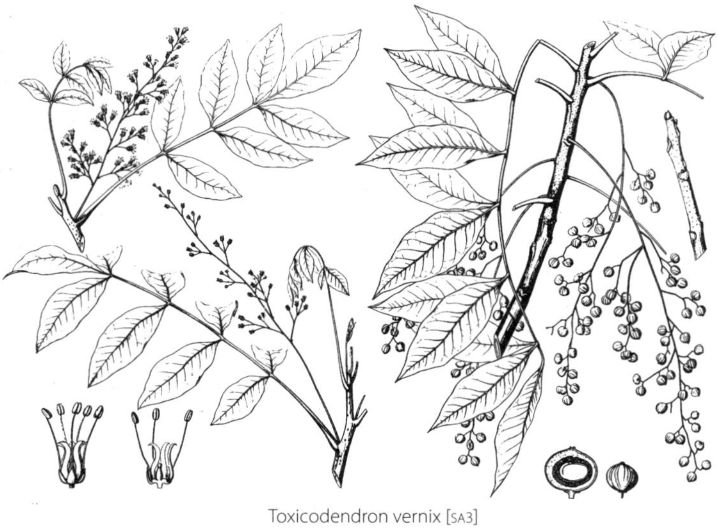
Toxicodendron vernix [SA3]

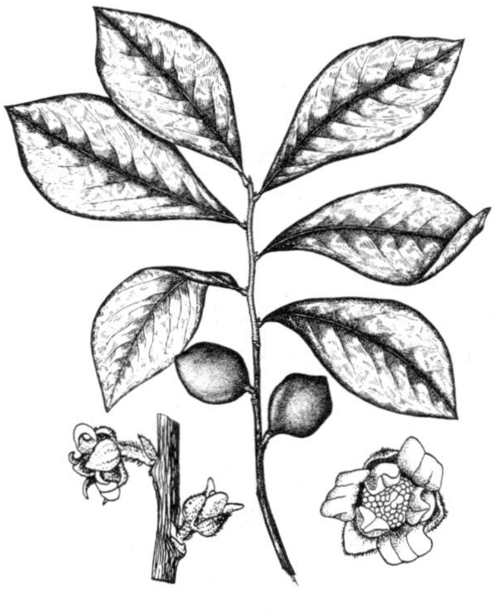
Asimina parviflora [BTT, VIN]

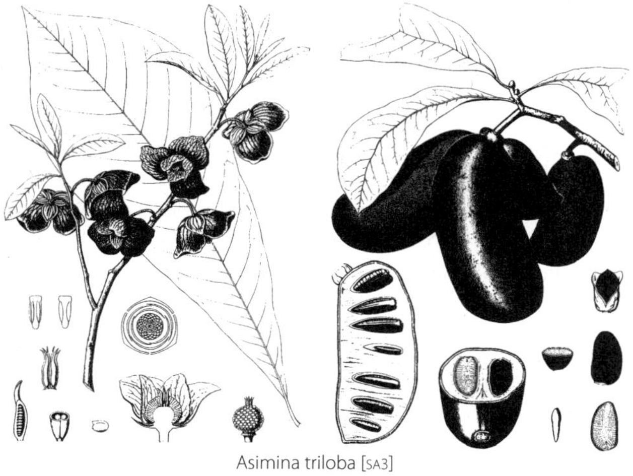
Asimina triloba [SA3]

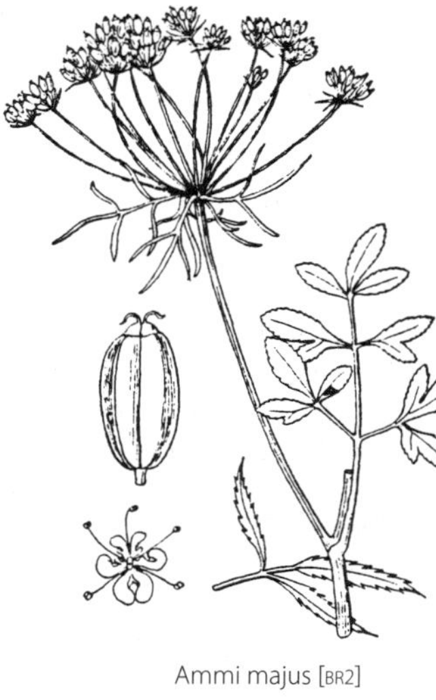
Ammi majus [BR2]

BERULA W.D.J. Koch WATER-PARSNIP, CUT-LEAF WATER-PARSNIP

A genus of 8 species of wet areas of the n temperate zone and e and s Africa (Spalik et al. 2009). Formerly treated as a widespread monotypic genus, molecular research (Spalik & Downie 2006; Spalik et al. 2009) has shown that the taxon in North America is distinct from those in Eurasia. Also, taxa in other genera (*Sium* and *Afrocarum*) were shown to be part of a *Berula* clade in the broad sense and transferred to *Berula*. (Latin name of some aquatic plant)
REFERENCES: Spalik & Downie 2006; Spalik et al. 2009; Nesom 2024c.

Berula incisa (Torr.) G.N. Jones, (cut deeply), STALKY BERULA, WATER-PARSNIP, CUT-LEAF WATER-PARSNIP. Erect or reclining, glabrous perennial; stems 0.2–1 m tall; submerged leaves filiform-dissected, emergent leaves once pinnately compound; leaflets toothed to deeply lobed; umbel from distal nodes; umbellets (6–)10–20-flowered; involucel bractlets 4–8, ovate-lanceolate; flowers white; calyx teeth minute, subulate; fruits oval to orbicular, 1.5–2 mm long. Wet areas; in East TX known only from Travis Co. (Turner et al. 2003); Edwards Plateau and Trans-Pecos; widespread in the w 1/2 of the U.S. Jun–Sep. Native of North America. [*B. erecta* of authors, not (Huds.) Coville, *B. erecta* var. *incisa* (Torr.) Cronquist] It is reported to be extremely poisonous and capable of causing death in cattle (Lewis & Elvin-Lewis 1977).

BOWLESIA Ruiz & Pav. BOWLESIA

A mainly South American genus of 15 species (Mabberley 2017). (Named for William Bowles, 1705–1780, Irish writer on Spanish natural history)
REFERENCE: Mathias & Constance 1965.

Bowlesia incana Ruiz & Pav., (hoary, quite gray), HOARY BOWLESIA. Low, prostrate to suberect annual, ± gray with stellate pubescence; leaves simple, opposite, blades suborbicular in outline, 5–7-lobed, to 3 cm long and 4.5 cm wide, usually much smaller; inflorescence a simple axillary umbel, minute, 2–5-flowered, peduncles sessile to subsessile, flowers minute, white or purplish; fruits sessile to subsessile, ovate to round, 1–2 mm long. Lawn weed, rarely collected but spreading rapidly in some areas; s and w portions of East TX; Cross Timbers and Prairies, Edwards Plateau, Gulf Prairies and Marshes, and South TX Plains; s 1/3 of the U.S. from FL w to CA. Feb–Jun. Native of South America and possibly also s North America; the geographic origin of this species is problematic; it is found in South America and also in the sw U.S. and adjacent Mexico. It was collected in TX in 1828 by Jean Louis Berlandier, but Mathias and Constance (1965) concluded that its North American distribution is probably the result of naturalization from South America; however, Constance (1993) considered it native to California and the sw US.

BUPLEURUM L. THOROW-WAX, THOROUGHWAX

Glabrous annuals with simple, entire, perfoliate, parallel veined leaves and compound umbels; involucres absent; involucels large, ovate; corolla yellow.

A genus of 170–190 species of Eurasia, n Africa, Canary Islands, North America, and s Africa (Neves 2024). The common names THOROW-WAX or THOROUGHWAX, i.e., "throw-wax" (through-grow) result from the perfoliate leaves, the stems apparently growing through the leaves (Mabberley 2017). (Greek: *bous*, an ox, and *pleuron*, a rib, referring to another plant)
REFERENCES: Snogerup & Snogerup 2001; Neves & Watson 2004; Neves 2024.

1. Rays (2–)4–8(–10); fruits smooth ____ **B. rotundifolium**
1. Rays (1–)2–4(–5); fruits rugose ____ **B. lancifolium**

Bupleurum lancifolium Hornem., (lance-leaved), LANCE-LEAF THOROW-WAX. Annual to 50 cm tall; first leaves petiolate, elliptic, soon withering; basal leaves clasping with basal auricles, stem leaf blades ovate to elliptic-ovate, upper leaves perfoliate; peduncles 1.5–3 cm long, umbellets 6–25-flowered; involucel bractlets 5, foliose; fruit rugose. Disturbed sites; in East TX known only from Harris Co. (Shinners 1967); the herbarium sheet (*Miner s.n.*, 1964, SMU), notes that it is a garden weed. We include it here in case it may be encountered again in East TX; also Webb Co. (*Tarvin 65* SMU) with the note: Laredo, Texas; Lake Casa Blanca; AZ, CA, FL, NC, OH, PA, and TX; Europe, n Africa, and the Middle East. It is a Mediterranean species, possibly introduced in bird-seed (Correll & Johnston 1970). Apr–Jul.

Bupleurum rotundifolium L., (round-leaved), ROUND-LEAF THOROUGHWAX, HARE'S-EAR, THOROW-WAX. Annual to 60 cm tall; basal and lower leaves oblong- to obovate-lanceolate, subpetiolate to perfoliate basally; upper leaves ovate to suborbicular, perfoliate; peduncles 2–7 cm long; involucel bractlets of 5–6 conspicuous, broadly ovate to obovate, acuminate bractlets, extending well beyond the flowers; umbellets 5–15-flowered; fruits smooth. Open, weedy areas; Brazos, Dallas, Fannin, Grayson, McLennan,

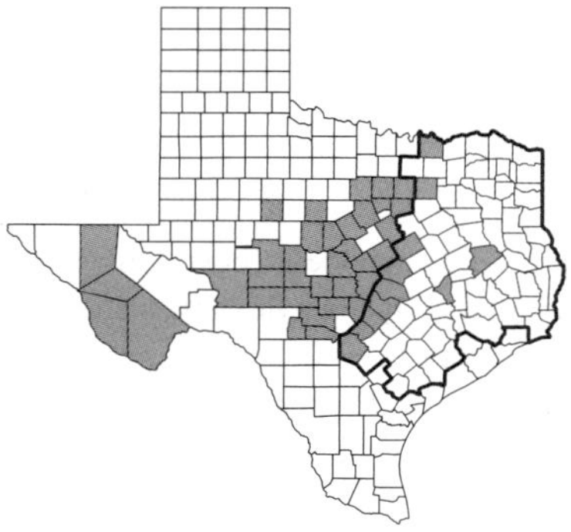

Rhus copallinum var. lanceolata

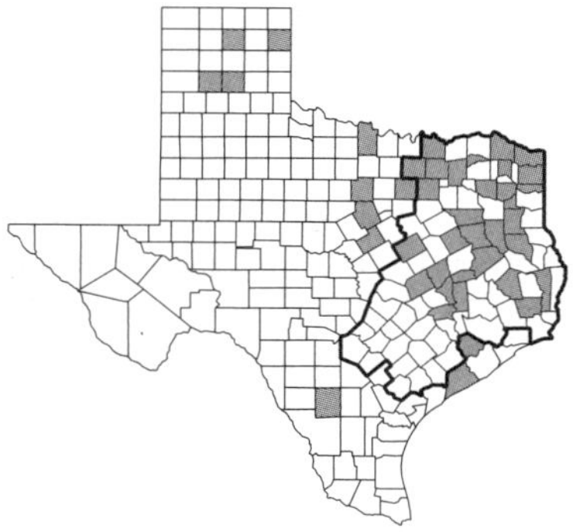

Rhus glabra

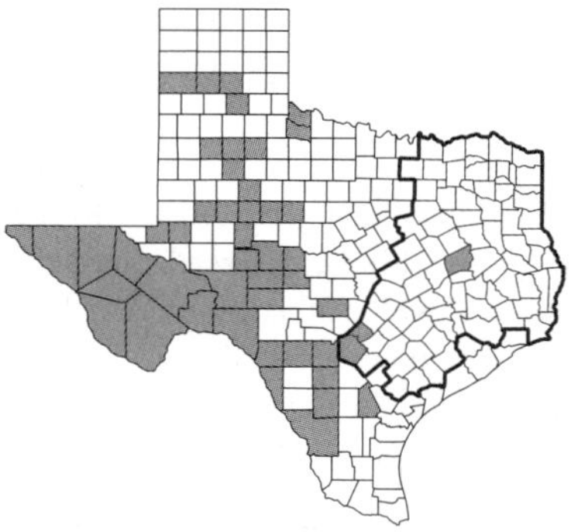

Rhus microphylla

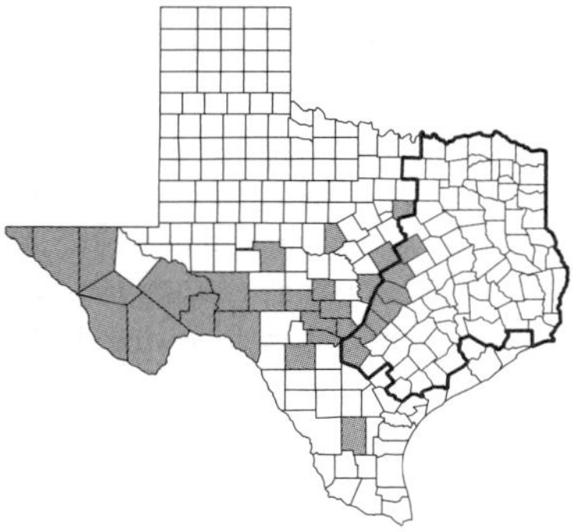

Rhus virens

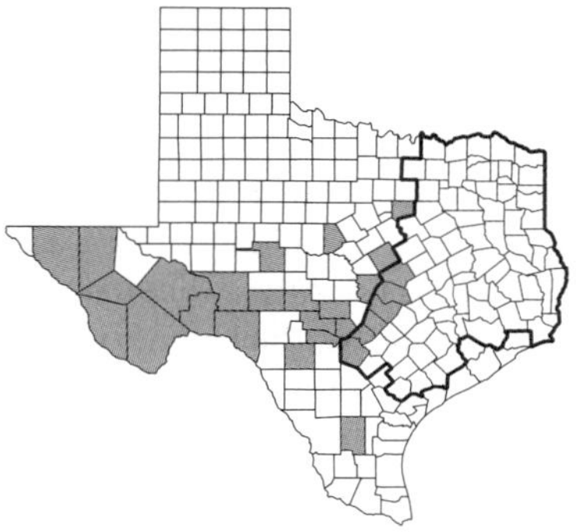

Rhus virens var. virens

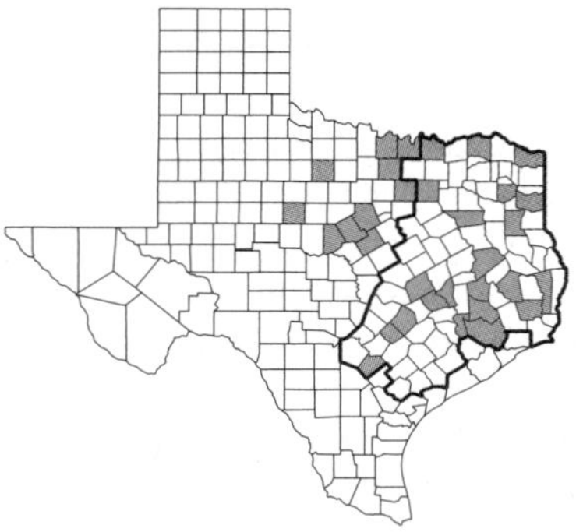

Toxicodendron pubescens

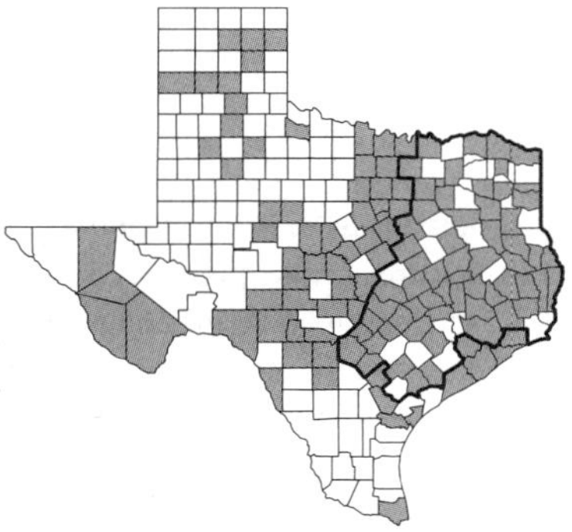

Toxicodendron radicans

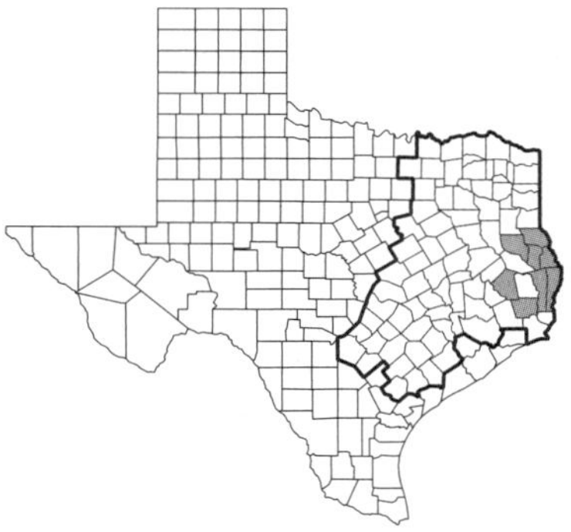

Toxicodendron vernix

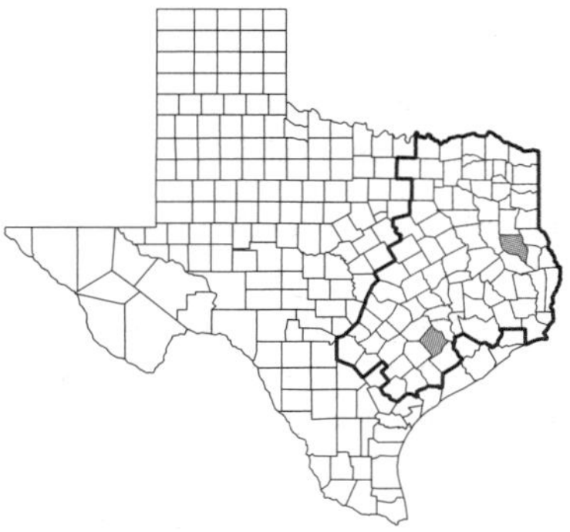

Asimina parviflora

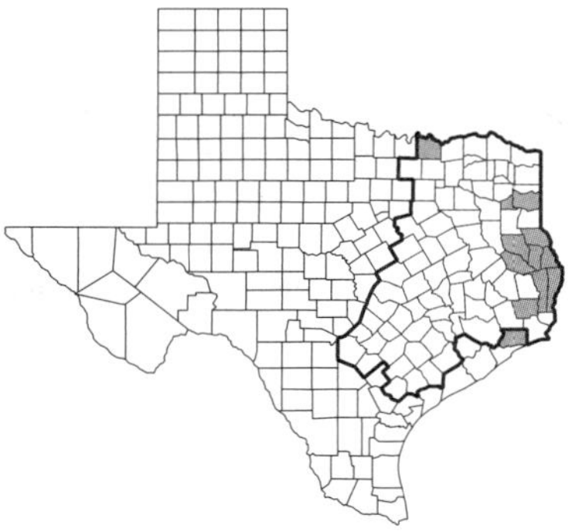

Asimina triloba

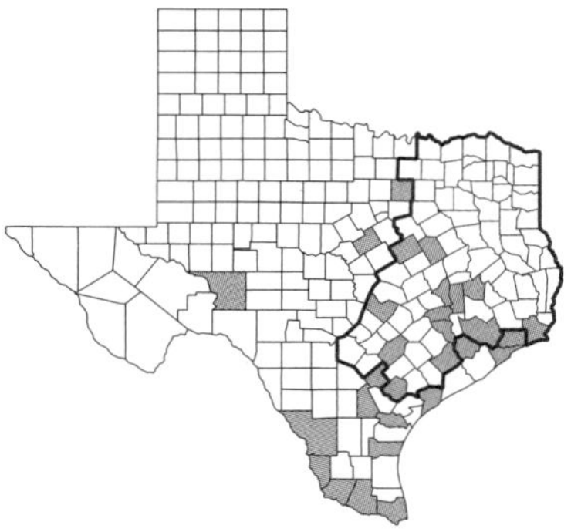

Ammi majus

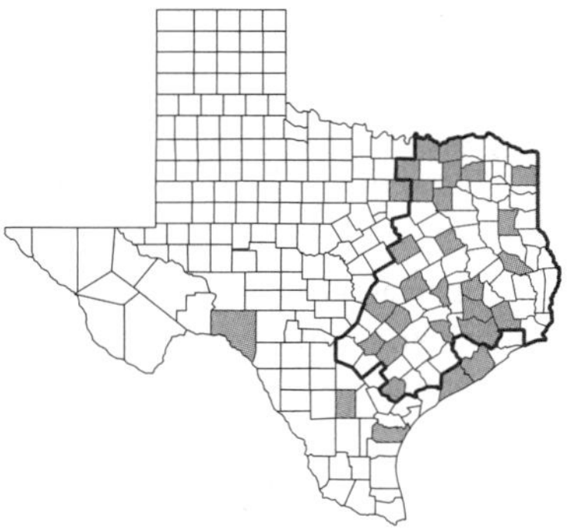

Ammoselinum butleri

Travis, and Washington (Kartesz 2022) cos.; adjacent Cross Timbers and Prairies; scattered in the e 1/2 of the U.S. also AZ; adjacent Canada. Apr–Jun. Native of the Mediterranean region.

CENTELLA L. SPADELEAF

A cosmopolitan genus of ca. 540 species (Nesom 2024d). Molecular work (Chandler & Plunkett 2004) has clarified relationships with the Apiaceae. *Centella* was previously treated in Apiaceae subfamily Hydrocotyloideae (now a subfamily in Araliaceae) with 2 other genera occurring in East TX, *Bowlesia* (currently in the Apiaceae) and *Hydrocotyle* (now treated in the Araliaceae); however, the morphological resemblances leading to this classification are currently thought to be the result of convergent evolution. *Centella* is now considered to be part of a small group of genera including *Mackinlaya* (5 species of Malesia and the w Pacific; also previously in the Araliaceae) which appear related and are placed in the apioid subfamily Mackinlayoideae (Stevens 2001 onwards). (Possibly diminutive of *centem*, a hundred, from the rounded, coin-like leaf blades of some species)
REFERENCES: Fernald 1940a; Schubert & van Wyk 1995; Stevens 2001 onwards; Chandler & Plunkett 2004; Nesom 2024d.

Centella asiatica (L.) Urb., (of Asia), ASIATIC SPADELEAF, SPADELEAF. Acaulescent perennial with creeping rhizomes; leaves arising from the rhizome nodes in clusters of 2–5, simple, leaf blades ovate to broadly ovate, basally cordate to truncate, petiole 0.3–30 cm long; umbels simple, 1–5 per node, 1–4(–5)-flowered; flowers white or rose-tinged; fruits 2.5–4 mm long, suborbicular with 3 primary and 2 secondary ribs, reticulate. Edges of streams and other wet places; widespread in s portions of East TX; also scattered to the s and w; se U.S. from NJ s to FL w to TX, also OH and OR. May–Sep. [*C. erecta* (L. f.) Fernald] Authors vary on the treatment of *Centella* in North America. Weakley, A.S., and Southeastern Flora Team (2024) treats the entity in the se U.S. as *C. erecta* and native to North America (see also Alqahtani et al. 2017), while Kartesz (2022), Wunderlin et al. (2020), and Nesom (2024d) treat it as *C. asiatica* with a pantropical distribution. Until further research comes to light, we are following the latter authors. *Centella asiatica* is sometimes eaten or taken as a tea, it is also said to be used as a fish poison and to contain potentially carcinogenic glucosides and an alkaloid, hydrocotylin; it has been used in folk medicine to treat leprosy and other skin ailments, and apparently shows activity against the bacterium responsible for tuberculosis (Duke 1985; Alqahtani et al. 2017; Mabberley 2017). ?

CHAEROPHYLLUM L. CHERVIL, WILD CHERVIL

A n temperate genus of ca. 65 species including *Oreomyrrhis* (Mabberley 2017). (Greek: *chaero*, delight or rejoice, and *phyllon*, a leaf, alluding to the agreeable odor of the foliage)
REFERENCES: Chung et al. 2005; Chung 2007; Naczi 2024.

Chaerophyllum tainturieri Hook., (for L.F. Tainturier des Essarts, who sent plants from Louisiana to Sir William Hooker from 1824–1836), HAIRY-FRUIT CHERVIL. Moderately to densely pubescent, erect annual 10–80 cm tall; leaves ternate-pinnately dissected; axillary umbels sessile, terminal umbels with peduncles to 4 cm long; rays 1–3(–4), 0.5–1.5 cm long at flowering, longer in fruit to 7 cm; flowers white; fruits narrowly oblong, 5–8 mm long, slightly beaked or narrowed toward the apex. Stream bottoms, woods, roadsides, and waste areas; throughout East TX; widespread in e 2/3 of TX; throughout much of e U.S. from VA s to FL and w to KS and TX. Mid-Mar–Apr. [*C. tainaturieri* var. *dasycarpum* Hook. ex S. Watson, *C. texanum* J.M. Coult. & Rose] Native of North America. Infraspecific variation is sometimes formally recognized in this species. However, we are following Turner et al. (2003), Yatskievych (2006), and Weakley, A.S., and Southeastern Flora Team (2024) in not recognizing such variants. Yatskievych (2006) noted the "the apparent lack of correlations between the distributions of character states and their seemingly random dispersal in the species range." Below is a key to the traditional varieties.

1. Ovary and fruit pubescent (this character occurs sporadically throughout the range of the species) ____ var. **dasycarpum**
1. Ovary and fruit glabrous or nearly so ____ var. **tainturieri**

CICUTA L. WATER-HEMLOCK

A n temperate genus of 8 species (Mabberley 2017); all are extremely toxic. (Latin: *cicuta*, old name for *Conium maculatum*—poison-hemlock, a deadly herb native to the Old World; the name was transferred to this genus)
REFERENCES: Mathias & Constance 1942; Mulligan 1980; Lee & Downie 2006.

Cicuta maculata L., (spotted), COMMON WATER-HEMLOCK, SPOTTED COWBANE, COWBANE, BEAVER-POISON, SPOTTED WATER-HEMLOCK, MUSQUATROOT, MUSKRAT-WEED. Glabrous and glaucous, usually tall perennial to 2 m tall (occasionally flowering when much smaller) stem reddish purple; leaves 2–3 times pinnately compound; flowers white; fruit oblong-elliptic, 2–4 mm long, ribs prominent, pale, blunt, corky. Low or wet ground; widespread in East TX; Edwards Plateau and Rolling and High Plains;

Ammoselinum butleri [LUN]

Ammoselinum popei [LUN]

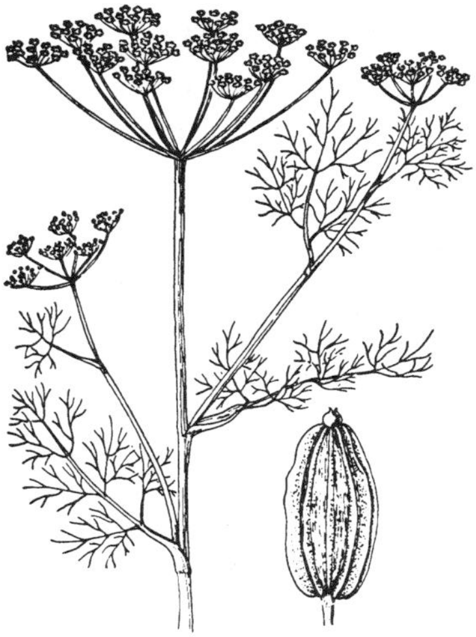

Anethum graveolens [GLE]

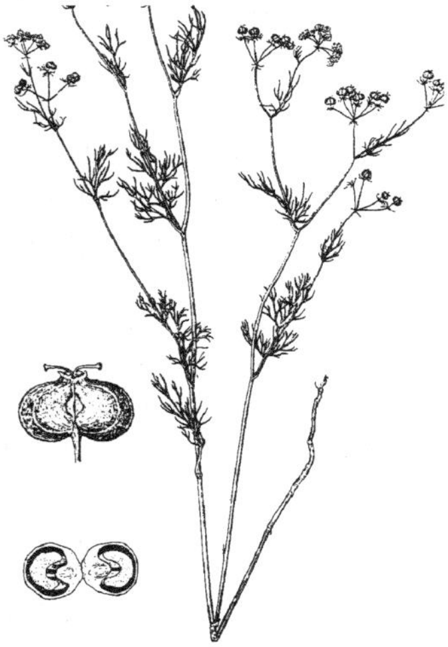

Atrema americanum [LUN]

Berula incisa [CO1]

Bowlesia incana [UCP]

throughout the U.S.; Canada and n Mexico. Late May–Jul. [*C. maculata* var. *bolanderi* (S. Watson) G.A. Mulligan, *C. mexicana* J.M. Coult. & Rose] Native of the U.S. and Canada. Virulently poisonous due to polyacetylenes (e.g., cicutoxin and cicutol), which affect the nervous system; a single bite is reported to be sufficient to kill a human; children have even been poisoned by whistles made from the stems; in extreme cases death may occur in less than an hour; some authorities consider this to be the most poisonous plant of the North Temperate zone; even non-fatal poisonings can result in post-event chronic effects such as mental deficit (Lampe & McCann 1985); fatalities are also known in browsing livestock (Stephens 1980; Turner & Szczawinski 1991; Hardin & Brownie 1993; Yatskievych 2006; Mabberley 2017). Mulligan (1980) recognized several varieties in this species based on differences in the mature fruit (seldom present in herbarium sheets). While we are not distinguishing varieties, a key is provided for those who wish to distinguish the variation found in East TX. ☠

1. Dorsal corky ribs of fruit equaling to slightly exceeding size of oil tubes in width; lateral ribs of the commissure flush against each other _____ var. **maculata**
1. Dorsal corky ribs of fruit much narrower than oil tubes; lateral ribs of the commissure separated by a groove _____ var. **bolanderi**

CONIUM L. POISON-HEMLOCK

A temperate Eurasian and s African genus of 6 species (Stratton 2024); all are very poisonous. (Greek: *coneion*, the hemlock, by which Socrates and various criminals were put to death at Athens)

Conium maculatum L., POISON-HEMLOCK, POISON-PARSLEY. Glabrous biennial to 3 m tall with distinctive unpleasant odor; root a carrot- or parsnip-like taproot; stems usually purplish-spotted; leaves pinnately decompound; peduncle 2–8 cm long; rays numerous, often 15 or more; umbellets (8–)10–20(–25), pedicel 1–5 mm long; bractlets 3–7, less commonly absent; flowers white; fruits broadly ovoid, 2–4 mm long. Stream bottoms; Bastrop, Harris, and Travis (Turner et al. 2003) cos.; Edwards Plateau, Cross Timbers and Prairies, and Trans-Pecos; nearly throughout the U.S. (less frequent in se U.S.); adjacent Canada. May–Jun. Native of Europe and Asia. All parts of the plant are extremely poisonous to humans and livestock due to coniine and other toxic alkaloids. It is often fatal if eaten due to paralysis of respiratory muscles. The leaves can be mistaken for parsley, the roots for parsnips, or the seeds for anise. Also, the similarity to white-rooted "wild carrots" (*Daucus carota*) has been the cause of many an unfortunate accident and is a strong argument against eating any wild umbellifer (Reed 2005). POISON-HEMLOCK has been recognized as poisonous since ancient times and is supposedly the source of the poison used to kill the Greek philosopher Socrates in 399 B.C.; it was apparently widely used as a means of execution by the early Greeks (Muenscher 1951; Kingsbury 1964, 1965; Hardin & Arena 1974; Schmutz & Hamilton 1979; Turner & Szczawinski 1991; Mabberley 2017). Significant skin contact with the foliage can result in symptoms of toxicity such as nausea or blurred vision. ☠

CORIANDRUM L. CORIANDER, CHINESE-PARSLEY

A nearly cosmopolitan genus of 2 species native to sw Asia (Mabberley 2017). (From the Greek name, *koriandron*, from *koris*, bug (*Cimex lectularius*, the bed bug), which supposedly can smell like CORIANDER leaves when disturbed)
REFERENCES: Mathias & Constance 1961; Yatskievych 2006; Wunderlin et al. 2020; Ball 2024b.

Coriandrum sativum L., (cultivated or sown), CILANTRO, CORIANDER, CHINESE-PARSLEY. Low glabrous annual 20–70 cm tall; basal leaves simple and lobed or pinnate, much less finely divided than stem leaves; stem leaves pinnately dissected, ultimate segments linear; peduncles 4–6 cm long; rays 3–5(–10); umbellets 10–14-flowered; bractlets 3–5, shorter to longer than pedicels; outer flowers with rather large petals, some bilobed; inner flowers with smaller petals; petals white, often tinged pink; fruits orbicular, (1.5–)2.5–3(–5) mm in diam. Escapes cultivation, waste places; w margin of East TX, especially sw; Edwards Plateau, South TX Plains, and Trans-Pecos; widely scattered in the U.S. and s Canada. Apr–May. Native of Mediterranean region. Long cultivated for the fruits, known as CORIANDER, (e.g., seeds were found in the tomb of Tutankhamun—Hepper 1990) which are used to flavor foods (in cooking, in confections, and as an ingredient of curry powder) and alcoholic beverages, and in scents and soaps (Yatskievych 2006; Mabberley 2017); the foliage of this plant provides the flavoring (CILANTRO) so widely used in Mexican salsas and other Latin American cuisines.

CRYPTOTAENIA DC. WILD CHERVIL, HONEWORT

A genus 4 species of the n temperate region and tropical African mountains (Mabberley 2017). (Greek: *crypto*, hidden, and *taenia*, band or ribbon, referring to the concealed oil tubes)
REFERENCES: Spalik & Downie 2007; Spalik et al. 2009; Nesom 2024e.

Cryptotaenia canadensis (L.) DC., (of Canada), CANADIAN HONEWORT. Largely glabrous perennial to 1 m tall; leaflets 3 or 5, thin, broadly lanceolate to ovate, sharply double-toothed, sometimes lobed; inflorescence a compound umbel, often in a paniculate-

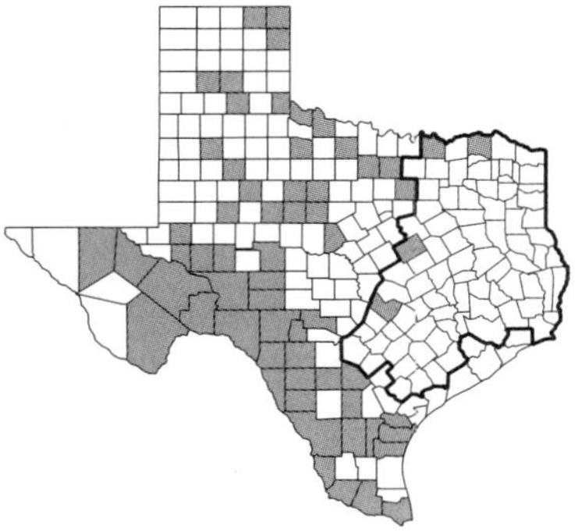

Ammoselinum popei

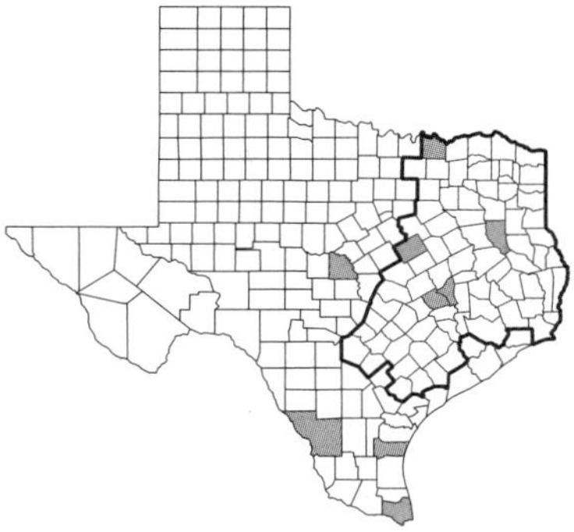

Anethum graveolens

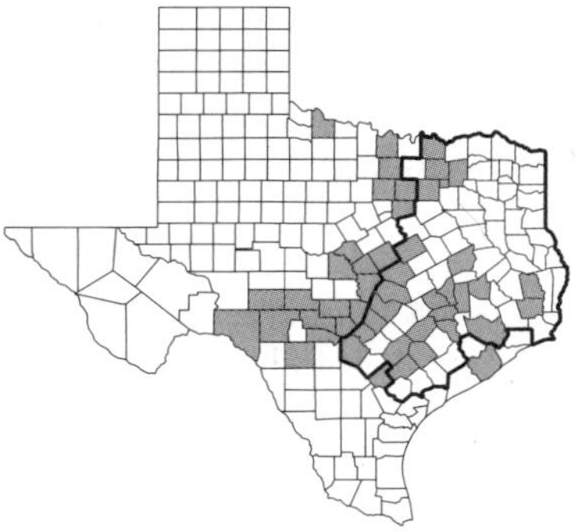

Atrema americanum

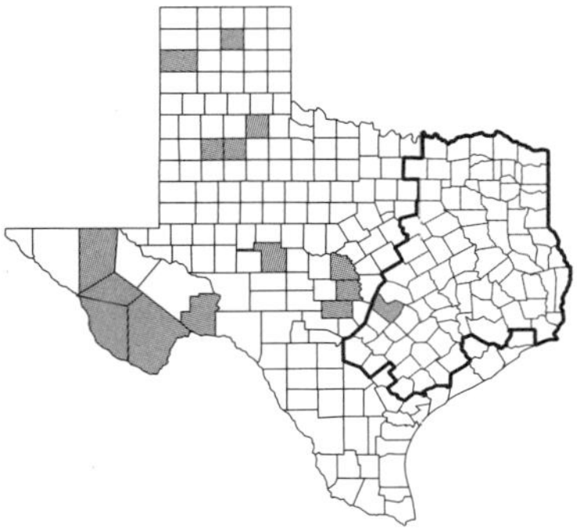

Berula erecta

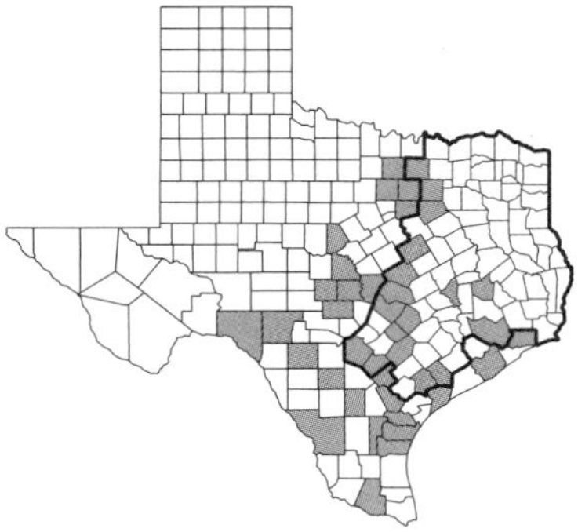

Bowlesia incana

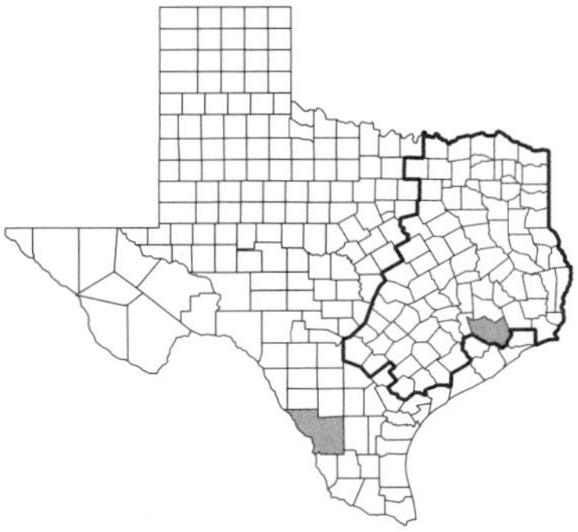

Bupleurum lancifolium

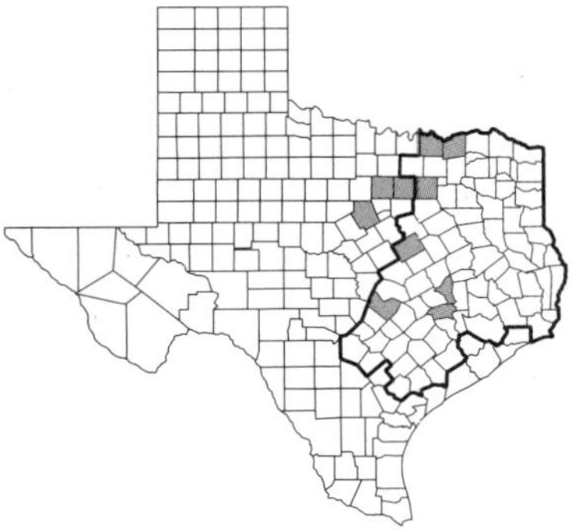

Bupleurum rotundifolium

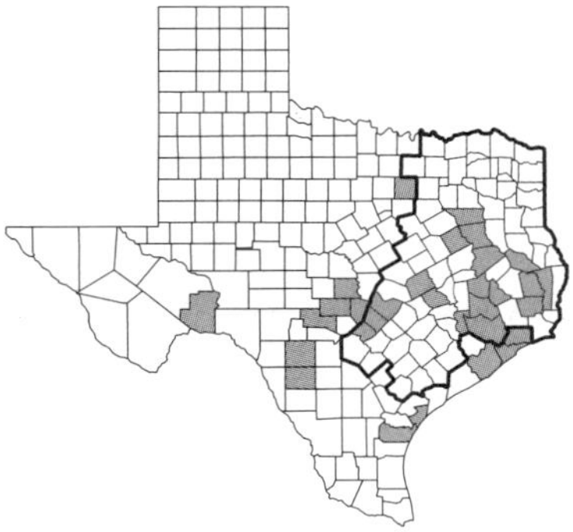

Centella asiatica

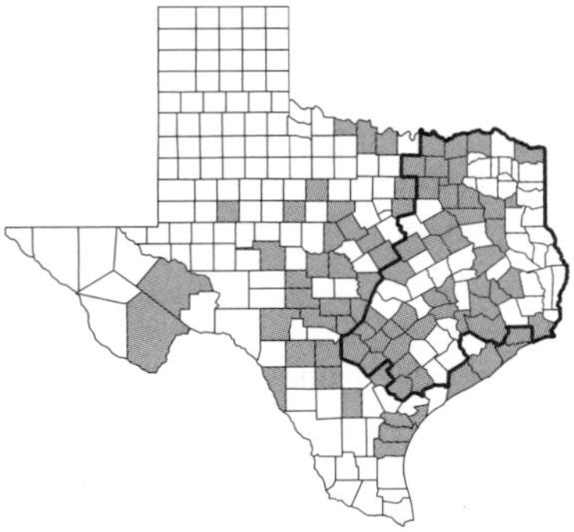

Chaerophyllum tainturieri

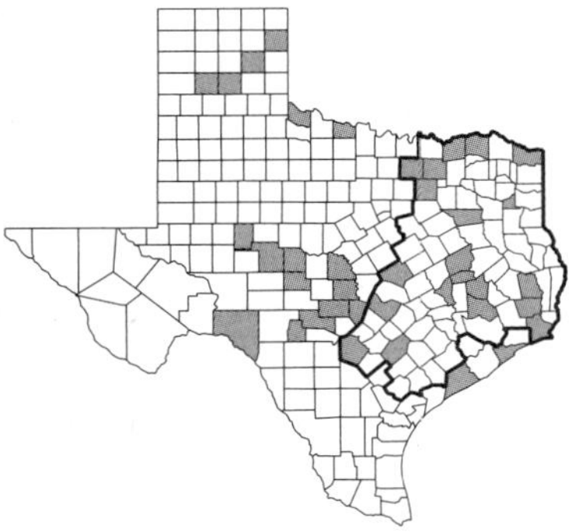

Cicuta maculata

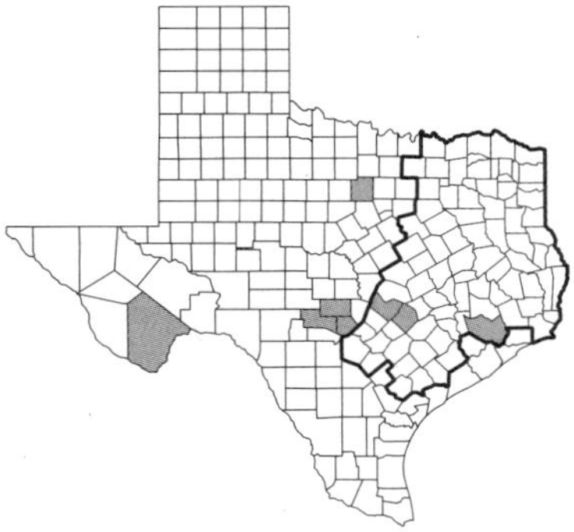

Conium maculatum

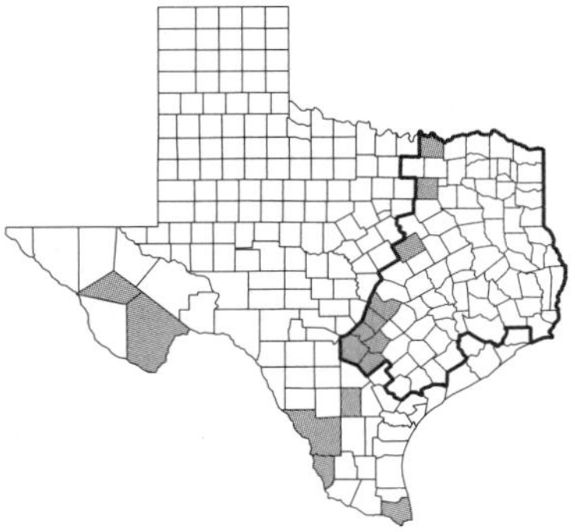

Coriandrum sativum

like array; peduncles 1.5–7.5 cm long; rays 2–7; umbellets 2–10-flowered; pedicels 2–30 mm long, strongly unequal; flowers white; fruits 4–7(–8) mm long, narrowly ellipsoidal. Woods; collected by Julien Reverchon in Dallas (Mahler 1988); recent collections in Collin, Dallas, Grayson, Harrison, Lamar, Titus (BRIT), Bowie, and Red River (Turner et al. 2003) cos.; e 1/2 of the U.S. May–Jul(–Aug). Native of U.S. and Canada.

CYCLOSPERMUM Lag. SLIM-LOBE-CELERY, MARSH-PARSLEY

A genus of 3 species native to the West Indies and Central America (Mabberley 2017); related to and previously included in the genus *Apium* which includes *A. graveolens* L. (CELERY). Molecular evidence indicates a close relationship between *Cyclospermum* and *Spermolepis* (SCALESEED) (Downie et al. 2002). The name has sometimes been spelled *Ciclospermum*; however, the accepted spelling is *Cyclospermum* (Greuter et al. 1993). (Greek: *cyclo*, circle, and *spermum*, seed)
REFERENCES: Downie et al. 2002; Wunderlin et al. 2020; Ball 2024c.

Cyclospermum leptophyllum (Pers.) Sprague ex Britton & P. Wilson, (slender-leaved), SLIM-LOBE-CELERY, MARSH-PARSLEY. Low glabrous annual 5–60 cm tall; leaves highly variable, the lowest with wide segments, the uppermost with almost thread-like segments; compound umbel sessile or short-peduncled, sometimes appearing simple; rays 1–5; umbellets 4–20-flowered; flowers minute, white; fruits ovoid, 1.4–2.3 mm long. Ditches or low areas, disturbed sites; widespread in East TX; widespread in mainly s 1/2 of TX with a few scattered locations further n; widespread in the s 1/2 of the U.S. Apr–Jun, sporadically later. [*Apium leptophyllum* (Pers.) F. Muell.] This species has in the past been considered to be native to the s U.S. (e.g., Diggs et al. 1999); however, it is now thought to be introduced into the U.S. (Ball 2024c; USDA Plants Database 2025). Native to tropical America. It is reported to be poisonous (Burlage 1968).

CYNOSCIADIUM DC. FINGER DOGSHADE

A North American monotypic genus (Yatskievych 2006). The genus, along with *Limnosciadium*, *Daucosma*, *Ptilimnium*, and *Lilaeopsis*, belongs to an informal clade known as the North American endemics clade in the tribe Oenantheae (Hardway et al. 2004). (Greek: *kyon* or *cyno*, dog, and *skiados*, an umbel)
REFERENCE: Hardway et al. 2004.

Cynosciadium digitatum DC., (finger- or hand-like), FINGER DOGSHADE. Glabrous annual, stems to 30–80(–120) cm tall; basal leaves linear-lanceolate, entire; stem leaves palmately compound or parted, with 3–5 long narrow segments, 20–150 mm long, 1–6 mm wide; rays 2–10, 1–4 cm long; umbellets 2–20-flowered, pedicels 2–12 mm long; flowers, white; fruits ovoid, 2–3 mm long. Wet places; scattered in East TX; n Gulf Prairies and Marshes; c U.S. from IL s to AL and w to OK and TX. May–Jul. Native to c and se U.S.

DAUCOSMA Engelm. & A. Gray MEADOW-PARASOL

A monotypic genus (Mabberley 2017) endemic to TX. (Possibly from *Daucus*, the genus including carrots, and Greek: *osma*, smell or odor—Hardway et al. 2004)
REFERENCES: Correll & Johnston 1970; Feist 2024a.

Daucosma laciniata Engelm. & A. Gray, (laciniate, torn), MEADOW DAUCOSMA, MEADOW-PARASOL. Annual to 1.2 m tall; crushed leaves very aromatic, pinnately compound, leaflets deeply and irregularly incised; umbels compound, peduncles to 10 cm; involucres and involucels equaling or longer than the rays and pedicels; rays 10–18; umbellets ca. 20-flowered; flowers white; fruits ovoid, 3–4 mm long, glabrous. Low disturbed areas; sw margin of East TX; adjacent Edwards Plateau; endemic to TX; Kartesz (2022), mapped it as present in 1 county (Hidalgo) in far sw NM, and Feist (2024a) also included it from Hidalgo Co., NM; however, Allred (2020) reported that the identification was in error. May–Aug.

DAUCUS L. CARROT, QUEEN-ANNE'S-LACE

Hispid-pubescent annuals or biennials; leaf blades finely cut, the ultimate divisions narrowly lanceolate; umbels compound, the outer rays becoming longer in age, the umbels then hollow in center; fruits bristly.

A genus 26 species of Europe, the Mediterranean, sw and c Asia, tropical Africa, Australia, New Zealand, and the Americas (Spooner 2024). Some authors (e.g., Banasiak et al. 2016) include several segregate genera in *Daucus* expanding the species to ca. 44. (From *daukos*, the classical Greek name)
REFERENCES: Spalik & Downie 2006; Spooner et al. 2013, 2024; Banasiak et al. 2016.

Bupleurum lancifolium [WLD]

Bupleurum rotundifolium [BB2]

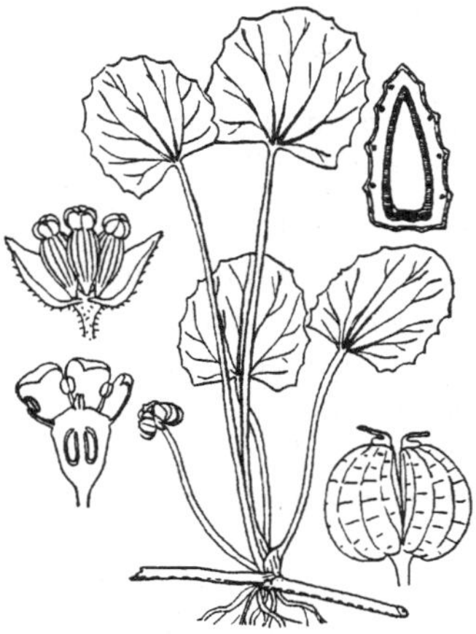

Centella asiatica [BR2]

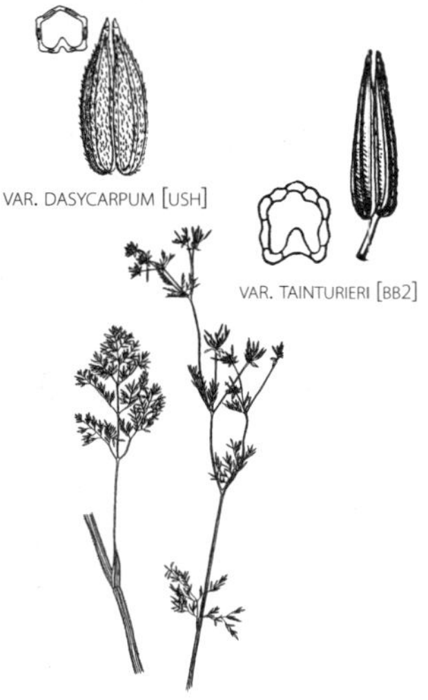

Chaerophyllum tainturieri [BT3]

Cicuta maculata [REE]

Conium maculatum [REE]

Coriandrum sativum [BL1]

Cryptotaenia canadensis [BB1]

Cyclospermum leptophyllum [CO1]

1. Petals white to yellowish, except those of central flower in umbel usually pink or purple; rays 7–30+; involucral bracts shorter than umbel in flower, divided into filiform divisions; bristles of fruit without prominent apical barbs; bractlets of umbellets with white margins, at least basally ____ **D. carota**
1. Petals white (withering yellowish), including those of central flower in umbel; rays 3–15; involucral bracts ± equal to or exceeding the umbel in flower, divided into linear or lanceolate divisions; bristles of fruit with apical barbs (easily visible with 10× hand lens); bractlets of umbellets usually without white margins ____ **D. pusillus**

Daucus carota L., (carrot), WILD CARROT, QUEEN-ANNE'S-LACE. Biennial, sometimes annual, 0.4–1.5 m tall; involucral bracts usually reflexed; fruits 3–4 mm long. Roadsides or railroads, disturbed areas; spreading rapidly in e part of TX; scattered in East TX; widely scattered in Edwards Plateau; nearly throughout the U.S. and adjacent Canada. May–Jul. Native of Europe. This plant is an aggressive invader capable of crowding out native vegetation. While the typically white roots are edible, they are not always palatable. Also, considering the very real possibility of confusing this plant with the deadly poisonous and similar appearing *Conium*, it is safest not to sample it (Reed 2005). The cultivated CARROT (*D. carota* subsp. *sativus* (Hoffm.) Schübl. & G. Martens) is by far the "most economically important member" of the family (Downie et al. 2000). The starch and sugar rich taproots of cultivated CARROTS are quite variable in terms of shape, color (including white, yellow, orange, red, and purple), and size; orange CARROTS are high in beta carotene (as the color suggests), which is converted in the human body to vitamin A (Yatskievych 2006). However, handling the wet foliage of cultivated CARROT can apparently cause skin irritation in sensitive individuals; dermatitis due to photosensitization may be involved (Tampion 1977; Duke 1985). Yatskievych (2006) noted that there is "disagreement on whether subsp. *carota* should be considered mildly poisonous" and indicated that the plants can produce polyacetylenes "that act as natural pesticides and have been documented to cause mild symptoms in livestock in Europe." Care should thus be taken with both cultivated CARROT and WILD CARROT.

Daucus pusillus Michx., (very small), RATTLESNAKE-WEED, SOUTHWESTERN CARROT, SEEDTICKS, AMERICAN WILD CARROT. Annual to ca. 0.9 m tall; fruits 3–5 mm long. Stream banks, roadsides, and waste areas; throughout East TX; throughout much of TX except the Panhandle; primarily s 1/2 of the U.S. and also w tier of states. Apr–Jun. Native of North America.

ERYNGIUM L. ERYNGO

Plants creeping to erect; essentially glabrous; leaves entire to pinnately or palmately lobed to divided, often spinose; inflorescences conspicuously head-like, globose to cylindrical, sometimes with a coma (= tuft of bracts) from apex of head; flowers white to purple; sepals well-developed, persistent in fruit; fruits variously covered with scales or tubercles.

A genus of ca. 230 species (the largest genus in the family and probably the most taxonomically complex) of tropical and temperate areas of the world (Mabberley 2017); a number are edible or cultivated as ornamentals. (From *eyringion*, the Greek name for *E. campestre* L.)

REFERENCES: Mathias & Constance 1941b; Bell 1963; Calviño et al. 2008, 2010; Calviño & Levin 2024.

1. Leaves parallel-veined, broadly linear, completely unlobed and undivided, the margin entire except for widely-spaced bristles, to 1 m long; plants monocot-like in appearance ____ **E. yuccifolium**
1. Leaves reticulate-veined, not linear, broader in shape, unlobed to usually palmately lobed or divided, to only 9 cm long, often much shorter; plants not monocot-like.
   2. Stem leaves conspicuously sharp-spinulose, palmately lobed or divided; bracts (including teeth) subtending heads > 2 mm wide (often much greater), usually spinulose.
      3. Heads 20–35 mm long, purplish to reddish; coma of 4–8 spinescent bracts 1–2 cm long projecting from apex of heads ____ **E. leavenworthii**
      3. Heads 8–15 mm long, bluish to purplish; coma bracts absent or inconspicuous, nearly entire.
         4. Plants erect, the stems solitary, branched above; basal leaves serrate or dentate, petioled; heads pedunculated; fruits 1–2.5 mm long; plants of limestone soils ____ **E. hookeri**
         4. Plants low, much-branched; basal leaves deeply palmately parted, sessile or subsessile; heads sessile or subsessile; fruits 2–2.5 mm long; plants of sandy soils ____ **E. diffusum**
   2. Stem leaves not sharp-spinulose or only slightly so, unlobed to palmately lobed or divided; bracts (including teeth if present) subtending heads very narrow, usually < 2 mm wide, entire or with 3–5 spinulose teeth.
      5. Plants erect; bracts subtending heads usually with 3–5 teeth (rarely entire); stem leaves not palmately lobed, usually serrate to laciniate or subentire; heads 5–15 mm wide, few–numerous in a cymose inflorescence terminating the stem ____ **E. integrifolium**
      5. Plants prostrate or weakly ascending; bracts subtending heads without teeth or nearly so; stem leaves entire to irregularly toothed, at least some palmately lobed; heads 2–4 mm wide, usually solitary in the leaf axils ____ **E. prostratum**

Cynosciadium digitatum [LUN]

Daucosma laciniata [LUN]

Daucus carota [BL1]

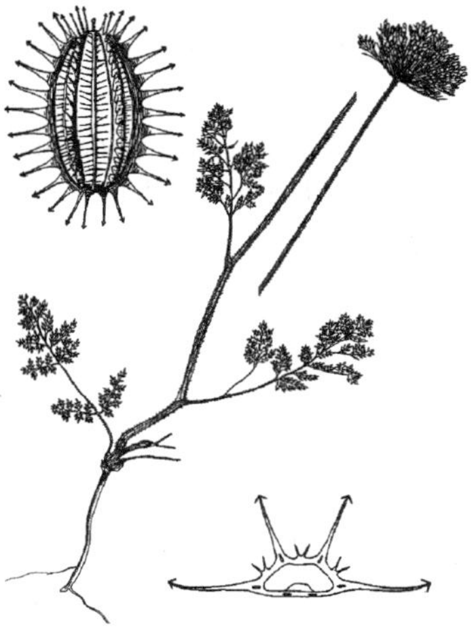

Daucus pusillus [BT3, USH]

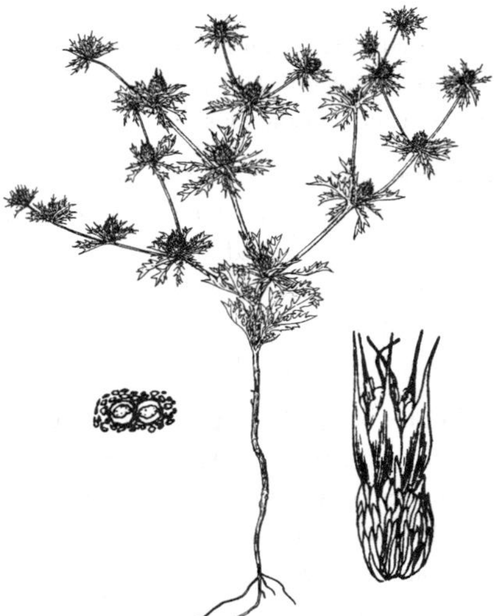

Eryngium diffusum [LUN]

Eryngium hookeri [LUN]

Eryngium integrifolium [GWO]

Eryngium leavenworthii [LUN]

Eryngium diffusum Torr., (diffuse, spreading), BUSHY ERYNGO, DIFFUSE ERYNGO, SPREADING ERYNGO. Annual or biennial 10–40 cm tall; prostrate to erect; stem leaves deeply palmately parted, the divisions spinulose-dentate or lobed; heads bluish; coma of bracts absent; fruits 2–2.5 mm long. Sandy soils; scattered in East TX in Denton, Hill, Jefferson, Travis, and Washington (Kartesz 2022) cos.; also scattered in the w 2/3 of the state; endemic to OK and TX. May–Aug. Native of OK and TX.

Eryngium hookeri Walp., (for William Jackson Hooker, 1785–1865, director of Kew Gardens), HOOKER'S ERYNGO. Erect annual 30–60 cm tall; lower stem leaves nearly sessile, lanceolate, laciniately toothed, spinulose; upper stem leaves ovate, palmately divided into 5–7 oblong, laciniate or pinnatifid lobes; heads purplish; coma of a few bracts or coma absent; fruits 1–2.5 mm long. Moist limestone soils; widespread in w and s portions of East TX; adjacent Gulf Prairies and Marshes; AR, LA, MS, OK, and TX. Jun–Nov. Native of the U.S.

Eryngium integrifolium Walter, (entire-leaved), SIMPLE-LEAF ERYNGO, BLUE-FLOWER ERYNGO. Erect perennial 30–80 cm tall; stem leaves simple, ovate to narrowly ovate or elliptic; heads hemispheric to globose, 5–15 mm in diam., bluish, subtended by 6–10 linear bracts, the bracts 5–20(–30) mm long; coma of bracts absent. Moist wooded areas; primarily e portions (Robertson Co. and to the e) of East TX; se U.S. from VA s to FL w to OK and TX. Aug–Oct. [*E. ludovicianum* Morong] Native of the U.S.

Eryngium leavenworthii Torr. & A. Gray, (for its discoverer, Melines Conklin Leavenworth, 1796–1862, s U.S. botanist, explorer, and army surgeon), LEAVENWORTH'S ERYNGO. Erect annual to 200 cm tall; vegetative tissues often purplish; upper stem leaves sessile, orbicular to obovate in outline, deeply palmately parted, the divisions pinnatifid, spiny; heads purplish to reddish; coma of 4–8 conspicuous spinescent bracts; sepals spinescent; stamens bluish; fruits 2–4 mm long, conspicuously covered with white scales. Prairies and weedy areas; mainly w margin of East TX, with a few scattered localities to the e; Edwards Plateau and Trans-Pecos; c U.S. in AR, KS, MO, OK, and TX. Jul–Sep. Native of the U.S. This strikingly colored and obviously spiny, thistle-like plant is often common and conspicuous.

Eryngium prostratum Nutt. ex DC., (prostrate, flat to the ground), CREEPING ERYNGO. Creeping, prostrate, mat-forming, or ascending perennial; stems 10–70 cm long, slender, 0.5–1 mm in diam.; basal leaves persistent, long petioled, blade 2–4(–5.5) cm long, entire to irregularly toothed, often palmately lobed; stem leaves reduced; heads very small, to 5–9 mm long, short cylindric, often pale to dark blue or purple-blue, subtended by 5–10 linear bracts to 12 mm long; coma of bracts absent. Lake margins and other moist areas; widespread in East TX except nw and sw portions; e U.S. from DE s to FL w to IL, KS, and TX. May–Sep. Native of the U.S.

Eryngium yuccifolium Michx., (with leaves like *Yucca*), RATTLESNAKE-MASTER, BUTTON SNAKE-ROOT, BUTTON ERYNGO, BRISTLE-LEAF ERYNGO. Plant glabrous, perennial from tuberous root; stems 30–180 cm tall; basal leaves persistent, blade broadly linear, to 1 m long; stem leaves like the basal but reduced upward; heads globose-ovoid, 10–25 mm in diam.; coma of bracts absent. Prairies; widespread in East TX; n Gulf Prairies and Marshes; also Gillespie Co. (Turner et al. 2003) in e Edwards Plateau; e 1/2 of the U.S. May–Aug. We are following McGregor (1986), Jones et al. (1997), and Wunderlin et al. (2020) in lumping var. *synchaetum*. [*E. yuccifolium* var. *synchaetum* A. Gray ex J.M. Coult. & Rose, *E. synchaetum* (A. Gray ex J.M. Coult. & Rose) J.M. Coult. & Rose] Native of the U.S. Some authors (Radford et al. 1968; Weakley, A.S., and Southeastern Flora Team 2024) recognize *E. yuccifolium* var. *synchaetum*, however, the difference between it and *E. yuccifoium* var. *yuccifolium* may be less distinct in their more western range (Weakley, A.S., and Southeastern Flora Team 2024). This species was extensively used medicinally by Native Americans as a diuretic (for kidney problems), as a sedative, for pain relief, and as a tonic (Moerman 1998). It was also "an important fiber plant and was used in cordage, bags, cloth, and sandals" (Yatskievych 2006). The common names refer to Native American and early settler use of poultices from the roots to treat snakebite (Tveten & Tveten 1993). It is an indicator of native prairie. The epithet, *yuccifolium*, refers to the monocot/*Yucca*-like leaves; however, *Yucca* leaves are more leathery, are in rosettes, and lack spine-like marginal structures (Yatskievych 2006).

EURYTAENIA Torr. & A. Gray TEXAS SPREADWING

A genus of 2 species (Mabberley 2017) native to TX, LA, and OK. (Greek: *eury*, wide, and *taeni*, ribbon or band)
REFERENCES: Nesom 2012a, 2015c, 2024f.

Eurytaenia texana Torr. & A. Gray, (of Texas), TEXAS SPREADWING. Largely glabrous annual 30–120 cm tall; basal leaves lobed or pinnatifid; stem leaves pinnately or ternate-pinnately dissected, the ultimate segments becoming filiform; inflorescence of loose compound umbels; rays 10–26; involucres usually of ca. 5 three-cleft bracts 5–10 mm long; involucel bractlets similar to bracts or entire; pedicels 4–8 mm long; petals white; fruits broadly elliptic to suborbicular, 4–6 mm long and 4–5 mm broad. Sandy habitats from abandoned fields, pastures, roadsides, sandy prairies, mesquite savannas, and floodplains to oak-juniper and live

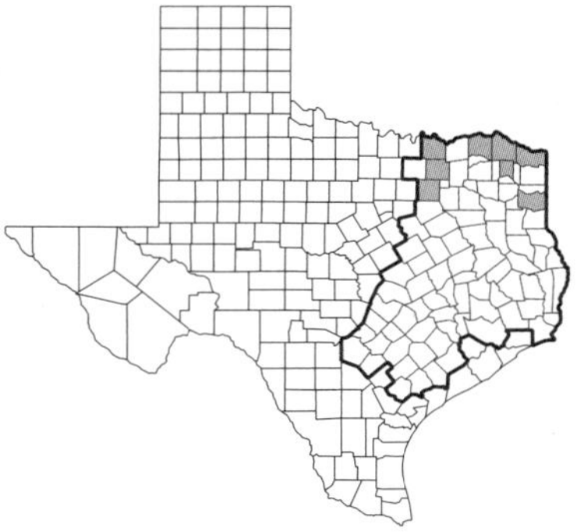
Cryptotaenia canadensis

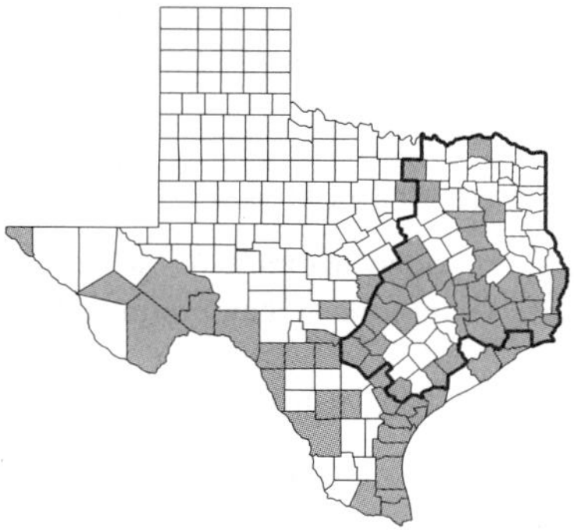
Cyclospermum leptophyllum

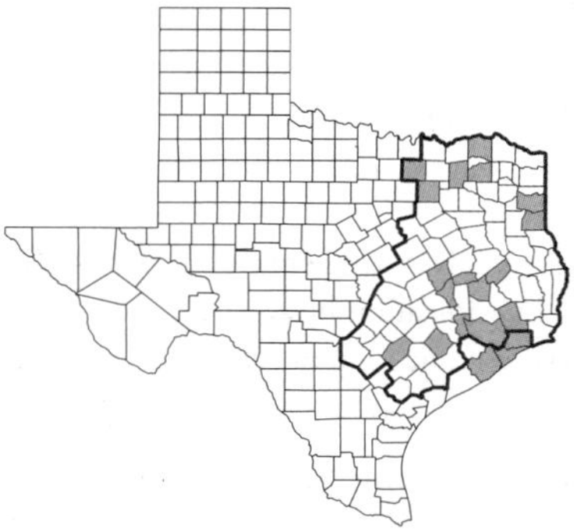
Cynosciadium digitatum

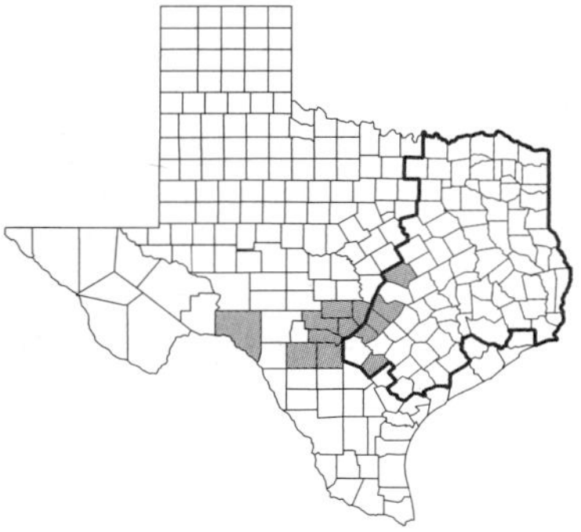
Daucosma laciniata

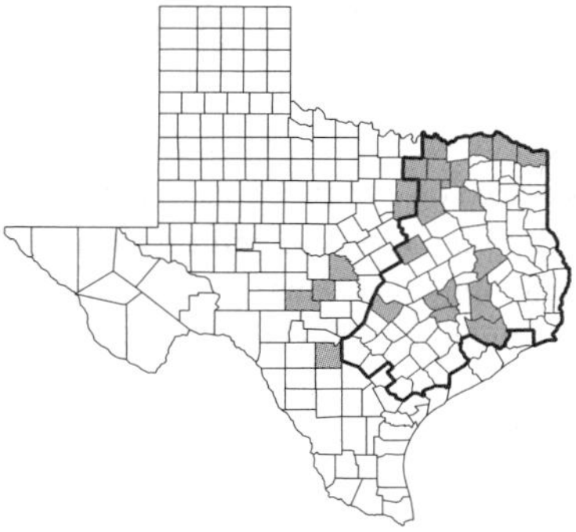
Daucus carota

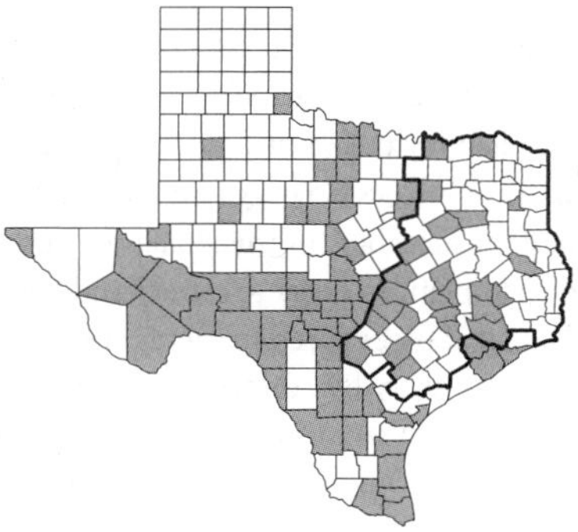
Daucus pusillus

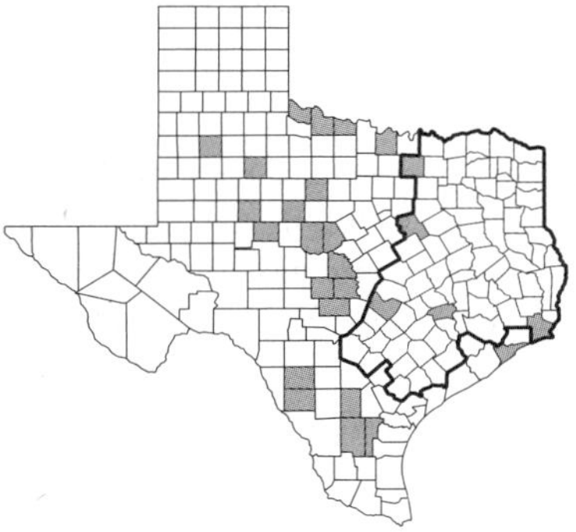
Eryngium diffusum

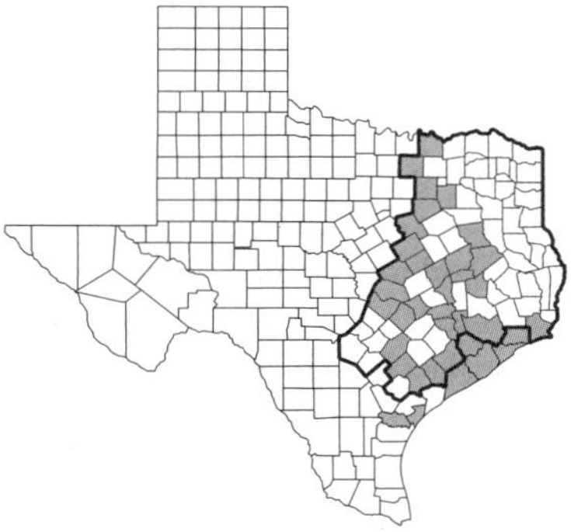
Eryngium hookeri

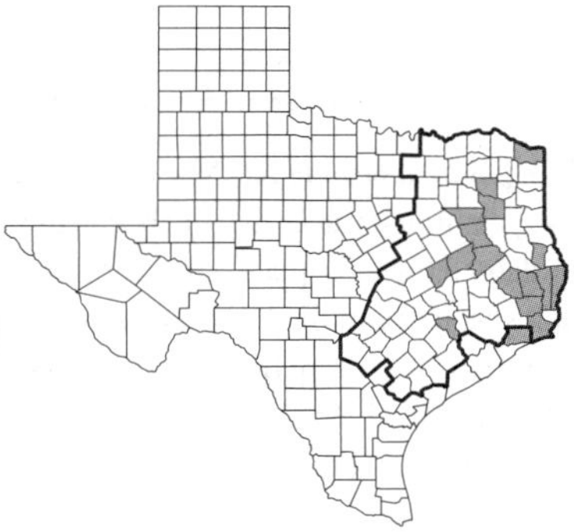
Eryngium integrifolium

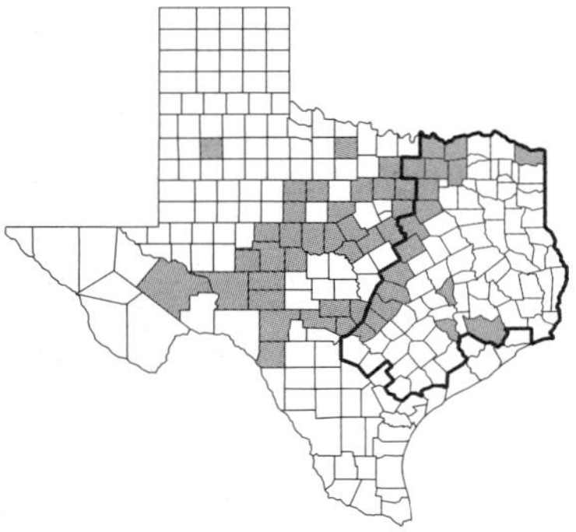
Eryngium leavenworthii

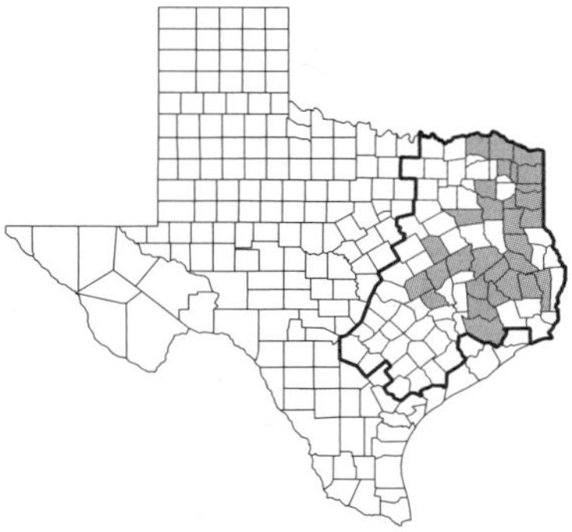
Eryngium prostratum

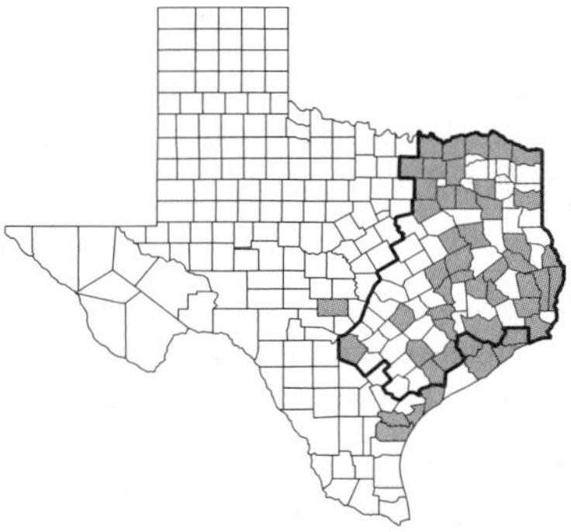
Eryngium yuccifolium

oak-post oak woodlands; scattered in East TX; Edwards Plateau, South TX Plains, and Rolling and High Plains; LA, OK, and TX. Reports of *E. texana* from New Mexico are based on misidentified specimens of *E. hinckleyi* (Nesom 2024f. (Mar–)Apr–Jul (–Aug). Native of the U.S.

FOENICULUM Mill. FENNEL

A mainly Asian genus of 5 species (Ball 2024d), widely naturalized (Mabberley 2017). It is closely related to *Anethum* (DILL) and hybridization between the two can occur. (Latin: *foenum*, hay, presumably from the abundant straw-like leaf segments; classical name for fennel)
REFERENCES: Mathias & Constance 1961; Yatskievych 2006; Ball 2024d.

Foeniculum vulgare Mill., (common), COMMON FENNEL. Glabrous strong-scented (anise/licorice-like) biennial or perennial; stems 0.9–2+ m tall, glaucous; leaves pinnately decompound, numerous ultimate thread-like segments resulting in a feathery appearance; rays 4–30; umbellets 8–25-flowered; petals yellow; fruits usually 4–5 mm long, the ribs not winged. Widely cultivated and escapes, low areas; it has been observed spreading from cultivation in Grayson Co. (G. Diggs, pers. obs.), but not actually naturalizing; Brazos and Jefferson (Turner et al. 2003) cos.; otherwise in TX known only from Cameron and Tom Green (Kartesz 2022) cos.; widely scattered in the U.S. and adjacent Canada. Spring–summer. Native of the Mediterranean region. A purplish form, BRONZE FENNEL, *F. vulgare* 'Purpureum' is often cultivated for its distinctive color. Black swallowtail butterflies (*Papilio polyxenes*) lay eggs on and their larvae feed on FENNEL as well as DILL (*Anethum*), the native PRAIRIE-PARSLEY (*Polytaenia*), and other members of the Apiaceae; larvae have been observed virtually defoliating large FENNEL plants; it is thus an excellent choice to plant in order to attract and encourage butterflies. This species is cultivated for its anise/licorice scented fruits, leaves, inflated leaf bases (used in salads and cooked), and roots. The essential oils have been used as a flavoring and in medicine since the 13th century BC. (Mabberley 2017); FENNEL is also one of the ingredients in the alcoholic drink absinthe. However, distilled oil of fennel, even in small quantities can be toxic and can cause symptoms including pulmonary edema, respiratory problems, and seizures (Duke 1985).

LILAEOPSIS Greene. GRASSWORT

A genus of 13 species of North and South America, Mexico, West Indies, Australia, and the Pacific Islands, (Levin 2024a). A mostly aquatic or wet area genus. (from *Lilaea* in the Juncaginaceae and Greek: opsis, to resemble)
REFERENCES: Affolter 1985; Hatch & Slack 2008; Bone et al. 2011; Levin 2024a.

Lilaeopsis carolinensis J.M.Coult. & Rose, (of Carolina), CAROLINA GRASSWORT. Rhizomatous, perennial herb, rooting at the nodes, glabrous; leaves usually borne directly from the horizontal rhizome, leaf blade 10–30 cm long, 2–15 mm wide, narrowly spatulate to linear, septate, apex rounded or acute, glabrous; umbels simple, axillary, 1–2 per node; peduncles 1–6(–9) cm long, much shorter than the leaves; flowers 5–15 per umbel; bracts triangular, minute; corolla white; fruit ovoid or obovoid, 1.9–2.8 mm long, 1.5–2.7 mm broad, all five ribs with abundant spongy cells. Pond margins, ditches, boggy areas; known only from one collection (Hatch & Slack 2008) in TX for Jefferson Co., also known from Caddo Par. in LA on the ne border of TX and may be found in that area of East TX; coastal U.S. from VA s to FL and w to TX; South America; Spain; and Portugal. May–Jun. [*L. attenuata* of authors, not (Hook. & Arn.) Fernald] Native of the U.S. and South America.

LIMNOSCIADIUM Mathias & Constance DOGSHADE

Slender, caulescent, branching annuals; roots fibrous; leaves petiolate, usually pinnately compound, the basal and uppermost sometimes simple; leaflets linear to lanceolate; inflorescences terminal and axillary, of compound umbels; rays 3–12, unequal; involucral bracts absent or of several linear to linear-lanceolate bracts to 6 mm long; involucels of several bractlets 1–5 mm long; pedicels 2–8 mm long; flowers small but numerous and rather showy; petals white; fruits glabrous, slightly flattened laterally, ribbed, but without wings.

A genus of 2 species of the sc United States (Mabberley 2017); sometimes treated as part of the genus *Cynosciadium*. The following treatment is modified from the one in Correll and Johnston (1970). (Greek: *limne*, a marsh, and *skiados*, an umbel)
REFERENCES: Mathias & Constance 1941a; Correll & Johnston 1970; Feist 2024b.

1. Plant erect or ascending, 10–80 cm tall; calyx lobes 0.5 mm or less long, attached well below and shorter than the stylopodium (= enlarged base of style); fruits oblong-oval, 2–4 mm long, 1–2 mm wide ____ **L. pinnatum**
1. Plant low and spreading, 5–40 cm tall or long; calyx lobes to 1.5 mm long, attached just below and as long as the stylopodium; fruits oval to orbicular, 2–3 mm long, ca. 2 mm wide ____ **L. pumilum**

Eryngium prostratum [CO1]

Eryngium yuccifolium [EN1]

Eurytaenia texana [HEA, LUN]

Foeniculum vulgare [CO1]

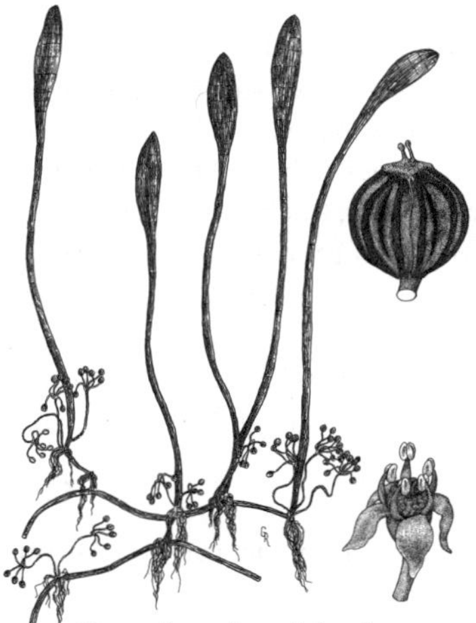
Lilaeopsis carolinensis [GWO]

Limnosciadium pinnatum [LUN]

Limnosciadium pumilum [LUN]

Limnosciadium pinnatum (DC.) Mathias & Constance, (pinnate, feather-like), ARKANSAS DOGSHADE, TANSY DOGSHADE. Stems 10–50 (–80) cm tall; stem leaves pinnately compound with 3–9(–11) linear to linear-lanceolate divisions 3–10 cm long or the lowest and uppermost leaves entire; fruits oblong-ellipsoid, 2–4 mm long. Ditches or low ground; widespread in East TX; Edwards Plateau, South TX Plains, and Cross Timbers and Prairies; sc U.S. from IL s to AL w to KS, OK, and TX. May–Jun. [*Cynosciadium pinnatum* DC.] Native of the U.S.

Limnosciadium pumilum (Engelm. & A. Gray) Mathias & Constance, (dwarf), PRAIRIE DOGSHADE. Stems 5–40; stem leaves pinnately compound with 3–7 filiform to lanceolate divisions or the lowest and uppermost leaves entire; fruits oblong-ellipsoid, 2–3 mm long. Wet areas, dry roadside ponds, adjacent to marshes, poorly-drained sandy loam, roadside ditches, in standing water; mainly s part of East TX; adjacent Gulf Prairies and Marshes and South TX Plains; AR, LA, and TX. Mar–Jun. Native of the U.S.

LOMATIUM Raf. DESERT-PARSLEY, WILD PARSLEY

A genus of 102 species of c and w North America and nw Mexico; some have edible roots (Wilson et al. 2024). Molecular evidence (Downie et al. 2002; Sun et al. 2004: Sun & Downie 2010) indicates that *Lomatium*, as traditionally viewed, is polyphyletic and closely allied to such genera as *Cymopterus*, *Tauschia*, and *Vesper* (see further discussion under *Vesper*). Until the relationships among the various genera and species are more adequately resolved and the taxonomy somewhat stabilized, we are continuing to recognize *Lomatium* in the traditional sense. (Greek, *lomation*, a little border, in reference to the winged fruit)
REFERENCES: Schlessman 1984; Downie et al. 2002; Sun et al. 2004; Sun & Downie 2010; Wilson et al. 2024.

Lomatium foeniculaceum (Nutt.) J.M. Coult. & Rose subsp. **daucifolium** (Torr. & A. Gray) W.L. Theob., (sp.: like *Foeniculum*—fennel; subsp.: leaves like *Daucus*, CARROT), CARROT-LEAF LOMATIUM, CARROT-LEAF DESERT-PARSLEY. Pubescent, dwarf, acaulescent perennial to 45 cm tall, from a thick taproot; leaves decompound; leaf segments linear or linear-lanceolate; petals yellow; ovaries glabrous or rarely with a few hairs; fruits ovate-oblong, 5–11 mm long, 3–8 mm broad, glabrous. Prairies, in calcareous clay; Collin, Dallas, Denton, Ellis, Grayson, and Travis (Kartesz 2022) cos.; scattered to the w of East TX; c US from IA and NE s to TX. Mar–Apr. [*Lomatium daucifolium* (Torr. & A. Gray) J.M. Coult. & Rose] Native of the U.S. This subsp. differs from the more northerly subsp. *foeniculaceum* (with hairy ovaries and fruit) primarily in terms of pubescence. While we are following a number of authors (e.g., McGregor 1986b; Hatch et al. 1990; Yatskievych 2006; Kartesz 2022; Weakley, A.S., and Southeastern Flora Team 2024) in recognizing subspecies within this species, the taxonomic value of such a distinction is questionable (e.g., not recognized by Turner et al. 2003) and awaiting detailed study.

OSMORHIZA Raf. SWEET-CICELY, ANISEROOT

A genus of 11 species native to the Americas and e Asia (Applequist & Lowery 2024). (Greek: *osme*, a scent, and *rhiza*, a root, in reference to the odor of the root)
REFERENCES: Constance & Shan 1948; Lowry & Jones 1984; Wen et al. 2002; Applequist & Lowery 2024.

Osmorhiza longistylis (Torr.) DC., (long-styled), ANISEROOT, LONG-STYLE SWEETROOT. Pubescent perennial 0.6–1 m tall, the leaves and particularly roots with strong anise-like odor; leaves 2–3-ternate; leaflets thin, ovate to trullate (= trowel-shaped), sharply toothed; rays of umbel usually 4–6; petals white; fruits (10–)15–20 mm long, with some hairs on the ribs, terminated by appendages 4–8 mm long. Woods, along streams; Collin (R. O'Kennon, pers. obs.; Turner et al. 2003), Dallas, and Grayson (Turner et al. 2003) cos.; widespread in the e 2/3 of the U.S. and adjacent Canada. Apr–Jun. Native of the U.S. and Canada. An extract of the root has apparently sometimes been "used as a flavoring in place of anise" (Yatskievych 2006).

OXYPOLIS Raf. HOG-FENNEL, COWBANE

A North American genus of 4 species (Mabberley 2017). (Greek: *oxys*, sharp, and *polios*, white)
REFERENCES: Tucker et al. 1983; Feist et al. 2012.

Oxypolis rigidior (L.) Raf., (rigid, stiff), COWBANE, STIFF COWBANE, WATER DROPWORT, WATER DROPWORT COWBANE. Stems 0.6–1.5 m tall; leaflets 5–11, lanceolate to linear, toothed or entire; peduncles 8–15 cm long; rays 10–many, 1–8 cm long; umbellets 10–many-flowered; fruits 4–7 mm long, 2.5–4 mm wide. Along streams, wet areas; e portions of East TX with an outlying record from Gonzales Co. (Turner et al. 2003) in the s portion of East TX; Terrell Co. (Turner et al. 2003) in the w Edwards Plateau; widespread in the e 1/2 of the U.S.; adjacent Canada. Aug–Nov. Native of the U.S. and Canada. Feist et al. (2012) examined fruit for anatomical characters and leaf morphology and found this corroborated molecular analysis showing *Oxypolis* to be polyphyletic. It was divided into three new segregates of phyllodial-leafed (= leaf without expanded tissue) taxa in the genus *Tiedemannia*, leaving four expanded-leafed taxa in *Oxypolis*. Reported to be poisonous to cattle (Burlage 1968).

Lomatium foeniculaceum subsp. daucifolium [BB2, BT3]

Osmorhiza longistylis [HO1]

Oxypolis rigidior [CO1]

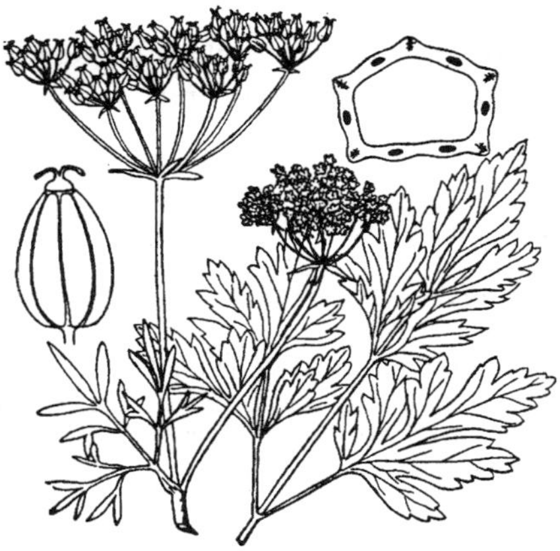

Petroselinum crispum [BB2]

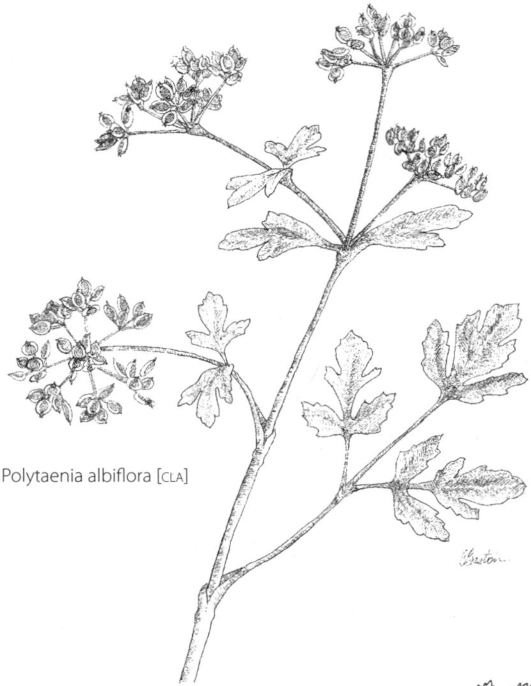

Polytaenia albiflora [CLA]

Polytaenia nuttallii [LUN]

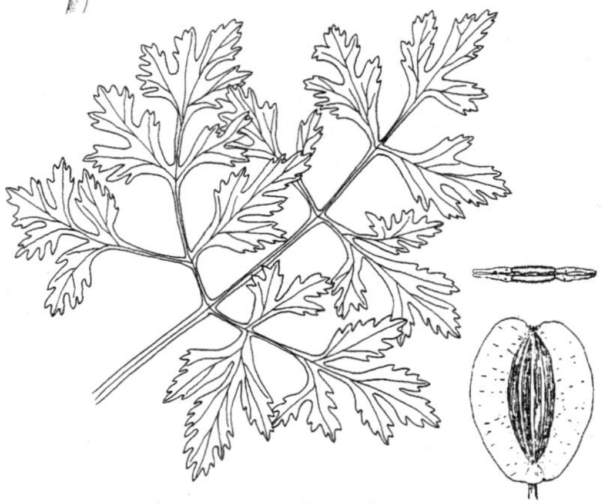

Polytaenia texana [BT3, LUN]

PETROSELINUM Hill PARSLEY

A genus of 2 species of Europe and the Mediterranean region (Mabberley 2017). (Greek: *petros*, rock, and genus name *Selinum* from *selinon*, CELERY)

REFERENCES: Mathias & Constance 1944–1945, 1961; Radford et al. 1968; Nesom 2024g.

Petroselinum crispum (Mill.) Fuss, (crisped, curled), PARSLEY, GARDEN PARSLEY, CURLY PARSLEY. Glabrous, erect, pungent-scented biennial; stems 0.2–1.3 m tall; basal and lower stem leaves ternately-pinnately or pinnately decompound; upper leaves sometimes becoming 3-lobed; petioles 10–20 cm long, blade 5–8 cm long, the ultimate segments 2–5 mm wide or wider and divided into narrow lobes (usually < 5 mm wide), flat or curled and crisped; peduncles 3–8 cm long; rays (10–)14–25(–30); umbellets 15–25-flowered, pedicels 2–5 mm long; flowers yellow or greenish yellow. Cultivated and possibly escapes in weedy areas; Bexar and Comal cos. (Kartesz 2022); widely scattered in the U.S. and Canada. Summer. [*Apium petroselinum* L.] Native of Europe. This species is used as a garnish and in various foods including salads, meat dishes, and sausages; it is high in vitamin C (Mabberley 2017). Extensive contact has been reported to cause skin inflammation (Duke 1985).

POLYTAENIA DC. PRAIRIE-PARSLEY, PRAIRIE-PARSNIP

Plants stout, 0.5–1.5 m tall, erect herbaceous perennials from taproots; leaves bipinnate or ternately-pinnate, large, crenate to incised, leaflets lobed; inflorescence terminal and axillary compound umbels, involucre absent, involucel linear to filiform or absent; flowers yellow or white; fruit broadly oval to orbicular to obovate, flattened dorsally, dorsal ribs filiform to obscure, lateral ribs corky, oil tubes distinct or obscure.

A North American genus of 3 species (Mabberley 2017). *Polytaenia* is related to a group of perennial North American genera including *Cymopterus*, *Lomatium*, *Tauschia*, *Taenidia*, *Thaspium*, *Vesper*, and *Zizia* (Downie et al. 2002). (Greek: *poly*, many, and *taenia*, a ribbon or band, which refers to the numerous oil tubes in the fruits)

REFERENCES: Downie et al. 2002; Keith 2012; Nesom 2012b, 2024h.

1. Flowers white; involucel bractlets narrowly lanceolate, mostly 1–3 mm long or absent; mature fruits (9–)11–15 mm long, (6–)7–9.5 mm broad; endemic to c TX ______ **P. albiflora**
1. Flowers yellow; involucel bractlets linear to filiform, mostly 3–5 mm long; mature fruits 5–11(–15) mm long, 4–7 mm broad.
   2. Fruits 6–9 mm long, 4–6 mm broad, lateral wings of mature fruits narrower and thicker than the body (in cross-section); dorsal oil tubes (6–)8–10, obscure; only in extreme ne TX ______ **P. nuttallii**
   2. Fruits 5–11(–15) mm long, 4–7 mm broad, lateral wings of mature fruits broader and thinner than the body (in cross-section); dorsal oil tubes 4(–6), distinct; widespread in e 1/2 of TX except Pineywoods ______ **P. texana**

Polytaenia albiflora E.L. Keith, (white-flowered), WHITE-FLOWER PRAIRIE-PARSLEY. Plant glabrous except for papillate hairs of the inflorescence; basal leaves 1–2-pinnately compound, stem leaves 1-pinnatley compound; peduncles 1–6 cm long, rays 6–15, 1–2 cm long, umbellets 8–20-flowered; flowers white; fruits oval to slightly obovate. Shade of mature forest canopy, rock crevices; Comal Co. (Keith 2012) on the extreme sw boundary (therefore not in East TX proper); also Bandera, Gillespie (?), Kendall, and Uvalde cos. (Keith 2012) on the Edwards Plateau; endemic to c TX. Apr–May.

Polytaenia nuttallii DC., (for Sir Thomas Nuttall, 1786–1859, English-American botanist), PRAIRIE-PARSLEY, TEXAS-PARSLEY, WILD DILL. Herbaceous perennial to 1 m tall; rays (6–)10–25, (10–)20–40 mm long, umbellets (6–)12–22(–42)-flowered, pedicels 2–6 mm long; flowers yellow. Blackland prairies, chalk prairies and outcrops, limestone glades, barrens, open rocky woods, rocky hillsides; in East TX found only in the extreme ne; known only from Bowie and Harrison cos. (Nesom 2012b); c U.S. from WI s to AL and w to MN, OK, and TX. Apr–Jun. Native of the U.S. There has been disagreement in the past as to whether *Polytaenia nuttallii* and *Polytaenia texana* were distinct entities (Diggs et al. 1999). This probably stems, in many cases, from misconceptions of its distribution in Texas. Correll and Johnston (1970) described the distribution of *P. nuttalli* as in the Timber Belt and the Blackland Prairies, and *P. texana* as abundant in the Blackland and Coastal Prairies. As noted above, *P. nuttallii* is known in Texas only from two counties (Bowie and Harrison), and *P. texana* is widespread in the eastern half of the state. Two species were being sought where essentially only one existed. Therefore, this mistaken inability to distinguish the two species resulted in many collectors treating *P. texana* as a synonym of *P. nuttallii*. The caveat, however, is that mature fruits are necessary to distinguish the two species with any certainty.

Polytaenia texana (J.M. Coult. & Rose) Mathias & Constance, (from Texas), TEXAS PRAIRIE-PARSLEY. Herbaceous perennial 0.5–1.5 m tall; leaves ternate–pinnately compound; peduncles 1–5 cm long, rays 8–17, umbellets (7–)12–20-flowered; flowers yellow. Blackland prairies, coastal prairies, mesquite prairies, oak-pine woods, sandy oak woods; in all but the e portion of East

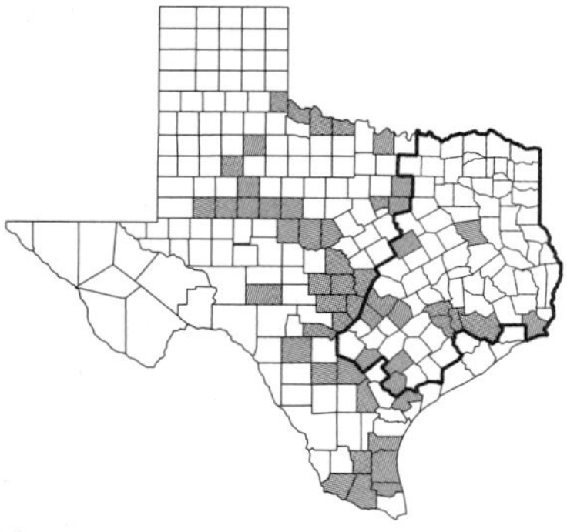
Eurytaenia texana

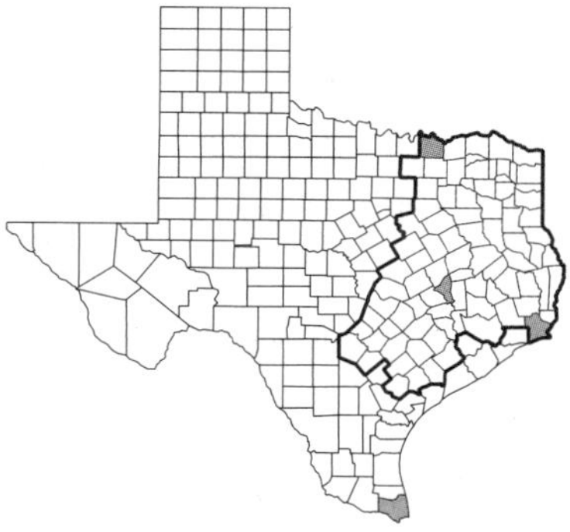
Foeniculum vulgare

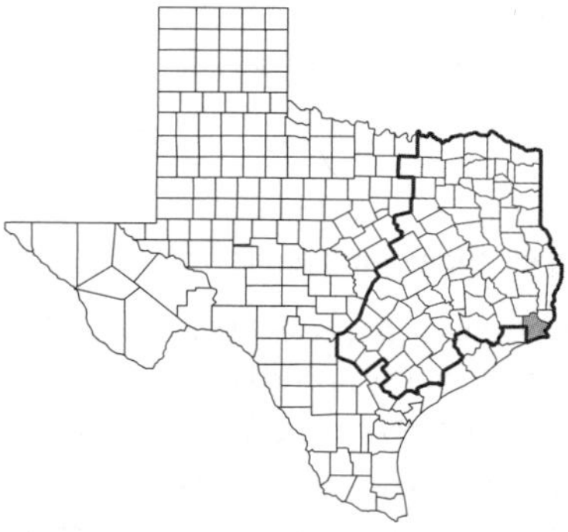
Lilaeopsis carolinienesis

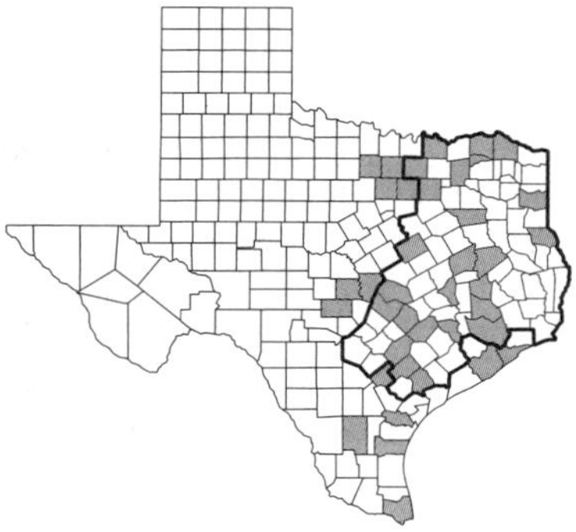
Limnosciadium pinnatum

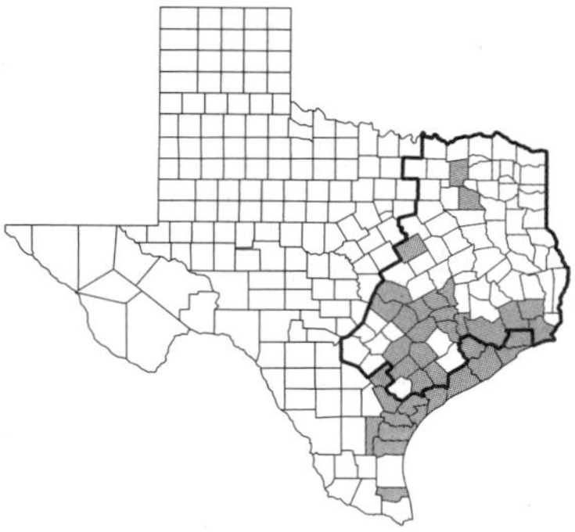
Limnosciadium pumilum

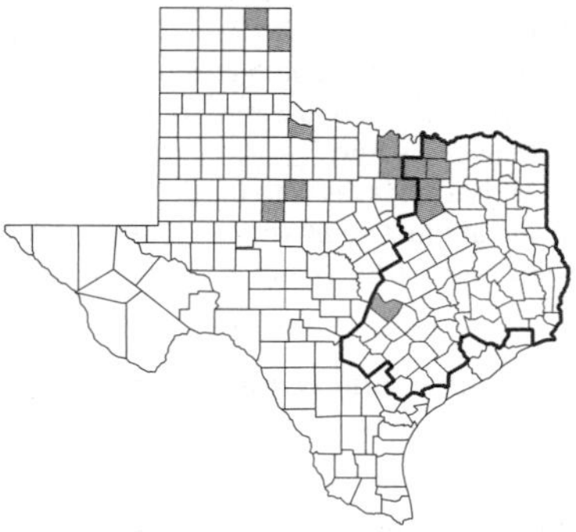
Lomatium foeniculaceum subsp. daucifolium

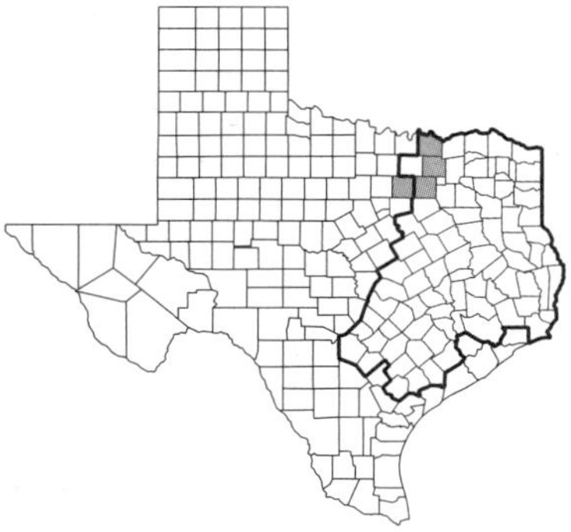
Osmorhiza longistylis

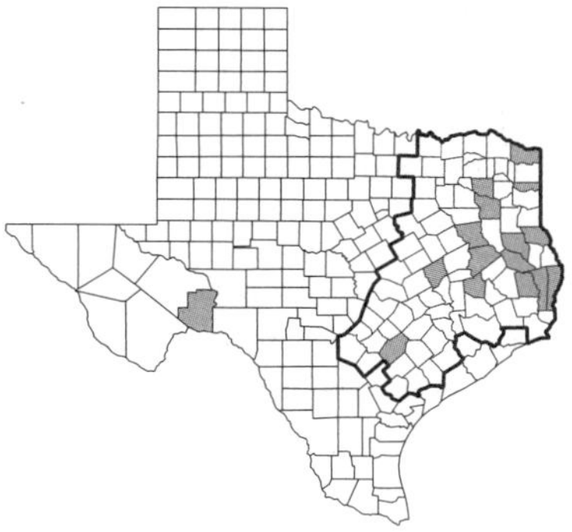
Oxypolis rigidior

Petroselinum crispum

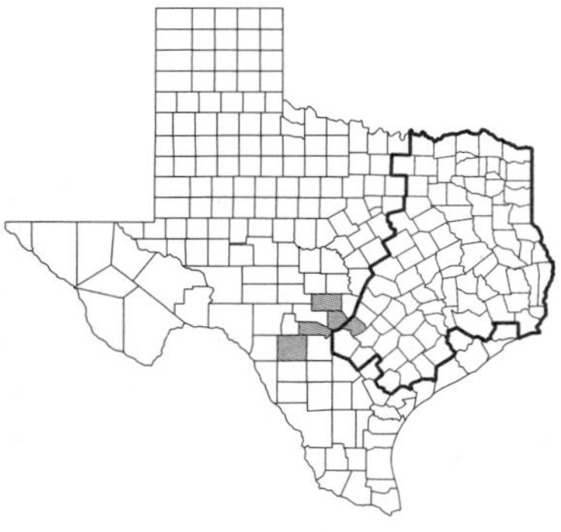
Polytaenia albiflora

Polytaenia nuttallii

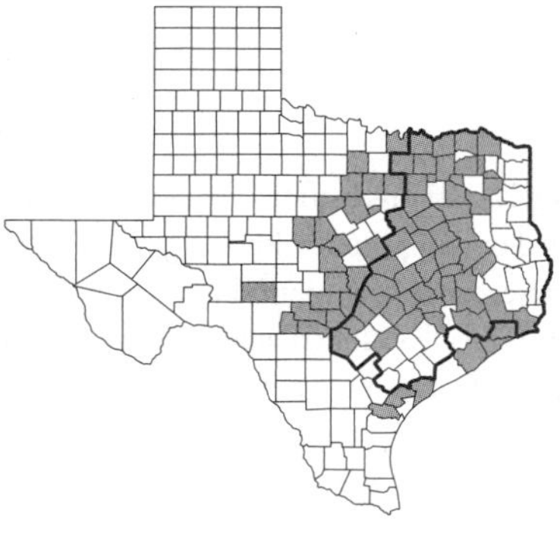
Polytaenia texana

TX; Edwards Plateau, West Cross Timbers, and Gulf Prairies and Marshes; s OK; endemic to OK and TX. Apr–Jun. [*P. nuttallii* var. *texana* J.M. Coult. & Rose]

PTILIMNIUM Raf. MOCK BISHOP'S-WEED, MOCK BISHOP

Rather low, glabrous annuals of sandy or boggy ground; leaves pinnately decompound with numerous ultimate filiform divisions; umbels compound; petals white, rarely pink.

A genus of 5 species of the U.S. and the West Indies (Feist 2024c). (Greek: *ptilon*, feather or down, and *limne*, mud, in allusion to the finely divided leaves and the habitat)

REFERENCES: Easterly 1957; Feist 2010; Feist et al. 2012; Feist 2024c.

1. Middle and upper leaf petiole bases papillate on the abaxial surface (sometimes sparsely so); mature fruit styles 0.1–0.2 mm long; calyx teeth < 0.2 mm long, deltoid ____ **P. capillaceum**
1. Middle and upper leaf petiole bases not papillate on the abaxial surface; mature fruit styles (0.3–)0.5–0.8(–1) mm; calyx teeth > 0.2 mm long, narrowly triangular.
   2. Midstem leaves with 2–4(–5) nodes along the rachis, primary leaf segments usually alternate or opposite at the nodes (not including at the apex of the petiole); involucral bracts entire, rarely with 3 segments; roots fibrous; styles (0.3–)0.4–0.6 mm long; fruit 1–1.9 mm long, dorsal ribs thick, rounded; flowering Apr–Jul, fruiting late May–early Aug ____ **P. nuttallii**
   2. Midstem leaves with (6–)7–10 nodes along the rachis, primary leaf segments whorled at the major nodes; involucral bract segments (1–)3; stem thickened and rounded at the base forming a slightly elongate corm; styles (0.3–)0.5–0.8(–1) mm long; fruit 2.2–3.5 mm long, dorsal ribs narrow, blunt; flowering Aug–Oct, fruiting Sep–Nov ____ **P. texense**

Ptilimnium capillaceum (Michx.) Raf., (with fine hairs), THREAD-LEAF MOCK BISHOP'S-WEED. Glabrous annual 10–85 cm tall, roots fibrous; peduncles 3–10 cm long; rays 4–20; involucral bracts usually 3-parted; umbellets 4–16-flowered; fruits broadly ovoid, 1.4–2.5(–3) mm long. Wet places; widespread in East TX; adjacent n Gulf Prairies and Marshes, e Cross Timbers and Prairies, and Edwards Plateau; mostly se U.S. from NJ s to FL and w to MO, OK, and TX. Late May–Aug. Native of the U.S and the West Indies.

Ptilimnium nuttallii (DC.) Britton, (for its discoverer, Sir Thomas Nuttall, 1786–1859, English-American botanist), NUTTALL'S MOCK BISHOP'S-WEED. Glabrous annual 30–70 cm tall, roots fibrous; peduncles 3–12 mm long; rays 15–30; involucral bracts usually entire; umbellets 25–30-flowered; fruits ovoid, 1–1.9 mm long. Moist prairies, sandy or silty open ground, often abundant on roadsides; widespread in East TX; East Cross Timbers and e Edwards Plateau; c U.S. from IL s to AL and w to KS, OK, and TX. Flowering Apr–Jul; fruiting May–early Aug. Native of the U.S.

Ptilimnium texense J.M. Coult. & Rose, (of Texas), TEXAS MOCK BISHOP'S-WEED. Glabrous perennial (30–)50–70 cm tall, roots from a small rounded corm; peduncles < 10 cm long; rays 8–15(–20); involucral bracts linear, entire or 3-parted, rarely more; umbellets 12–24-flowered; fruits ovate to orbicular, often maroon or purple-tinged, oil tubes dark brown. Acid soils, seeps, bogs, pine savannas; scattered in East TX; w LA and s AR; endemic to the West Gulf Coastal Plain. Flowering Aug–Apr; fruiting Sep–Nov. Native of the U.S. *Ptilimnium texense* and *P.* ×*texense* have often been synonymized under *Ptlilimnium costatum* (Elliott) Raf. Through molecular research and field and herbarium work, Feist (2010) demonstrated that *P. texense* was not a hybrid and further reinstated the taxon as a species. Her herbarium and field studies failed to locate *P. costatum* in Texas. Additionally, Weakley, A.S., and Southeastern Flora Team (2024) does not have its distribution extending to Texas. We are therefore not including *P. costatum* in the flora.

SANICULA L. BLACK-SNAKEROOT, SANICLE

Largely glabrous biennials or perennials; roots fibrous or tuberous-thickened; leaves petiolate below to subsessile above, palmately lobed or compound with 3–5 leaflets, the leaflets toothed and lanceolate to elliptic, the largest > 1 cm wide, usually much wider; inflorescences compound umbels, at first short, barely exceeding the leaves, later elongating; involucral bracts of usually 2 leaf-like, typically 3-lobed bracts; flowers relatively few per ultimate umbel (= umbellet); perfect flowers subsessile, usually three per umbellet; staminate flowers pedicellate, (0–)2–15 or more per umbellet, the umbellets sometimes with only staminate flowers present; petals inconspicuous, white to yellow-green; calyx persistent, the lobes small; styles persistent; fruits somewhat flattened laterally, bur-like, with numerous hooked bristles.

A nearly cosmopolitan genus of ca. 40 species (Ball 2024e). The fruits are animal-dispersed with the hooked bristles catching in fur, etc. This is an example of exozoochory (= external unintentional animal dispersal). (Latin: *sanare*, to heal, in reference to former medicinal use of some species)

Ptilimnium capillaceum [CO1]

Ptilimnium nuttallii [BB2]

Ptilimnium texense [CO1]

Sanicula canadensis [TOR]

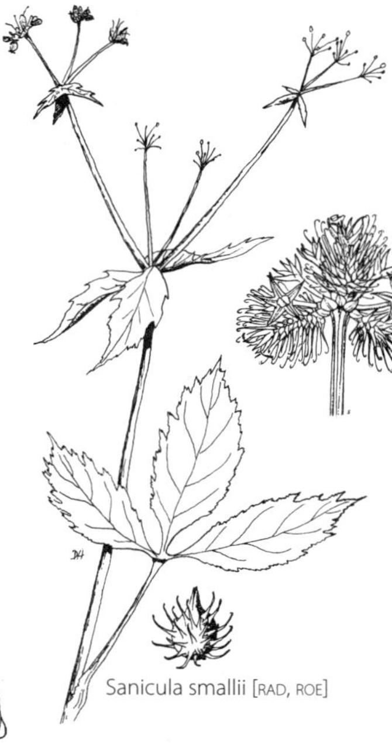

Sanicula smallii [RAD, ROE]

Sanicula odorata [TOR]

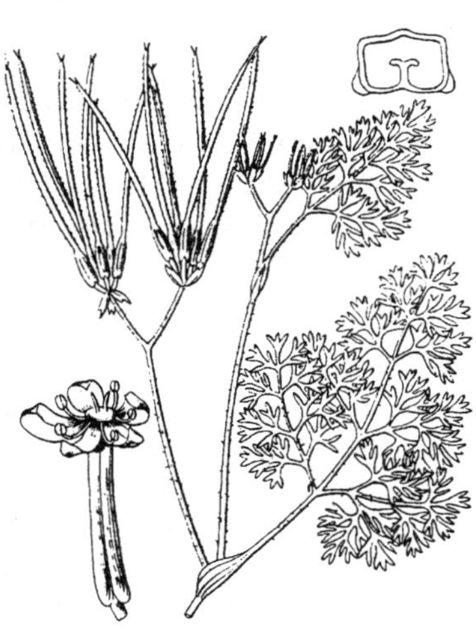

Scandix pecten-veneris [BB1]

Sium suave [CO1]

REFERENCES: Bicknell 1895; Shan & Constance 1951; Phillippe 1978; Pryer & Phillippe 1989; Ball 2024e.

1. Petals yellow-green, exceeding the ovate-lanceolate to triangular blunt-pointed calyx lobes; styles longer than bristles of fruit, ca. twice as long as persistent calyx lobes; staminate flowers 12–35 per umbellet ____ **S. odorata**
1. Petals white or greenish white, shorter than or equaling the narrowly lanceolate, sharp-pointed persistent calyx lobes; styles shorter than bristles of fruit, shorter to only slightly longer than persistent calyx lobes; staminate flowers 1–8 per umbellet.
   2. Biennial with thin fibrous roots; styles shorter than to nearly as long as the persistent calyx lobes; fruits short-stalked, the pedicels 1–1.5 mm long ____ **S. canadensis**
   2. Perennial with tuberous-thickened roots; styles slightly longer than the persistent calyx lobes; fruits sessile ____ **S. smallii**

Sanicula canadensis L., (of Canada), CANADA SANICLE, CANADA BLACK-SNAKEROOT. Stems 15–100 cm tall; umbellets with 1–3(–4) staminate flowers, inconspicuous, shorter than bisexual flowers; anthers white; fruits 3–6 mm long. Woods and thickets; throughout East TX; Edwards Plateau and Cross Timbers and Prairies; throughout e U.S.; adjacent Canada. Apr–Jun. Native of the U.S. and Canada.

Sanicula odorata (Raf.) Pryer & Phillippe, (fragrant), CLUSTER SANICLE, CLUSTERED BLACK-SNAKEROOT. Perennial, stems 30–80 cm tall; umbellets with 12–35 staminate flowers; anthers yellow; fruits 3–5 mm long. Woods, thickets; scattered in the e part of East TX and also on the w margin in Comal, Dallas, Grayson, and Travis (Turner et al. 2003) cos.; widespread in the e 1/2 of the U.S.; adjacent Canada. Apr–May. [*S. gregaria* E.P. Bicknell, *Triclinium odoratum* Raf.] Native of the U.S. and Canada. While the name *S. gregaria* has often been used for this species, Pryer and Phillippe (1989) argued convincingly that the older name *S. odorata*, is correctly applied to this species.

Sanicula smallii E.P. Bicknell, (for its discoverer, John Kunkel Small, 1869–1938, American botanist and author of numerous works including *Manual of the Southeastern Flora*), SOUTHERN SANICLE, SMALL'S BLACK-SNAKEROOT. Stems 20–75 cm tall; umbellets with 4–8 staminate flowers; anthers white; fruits 5–6 mm long. Moist woods; Angelina, Bowie, Jasper, Polk (Turner et al. 2003), and Tyler (Kartesz 2022) cos. in far e part of East TX; e U.S. from VA s to FL and w to AR and TX. Apr–May. Native of the U.S.

SCANDIX L. SHEPHERD'S-NEEDLE

A genus of 15 species of Europe and the Mediterranean region (Ball 2024f). (Greek name for chervil, another member of the carrot family)
REFERENCE: Mahler 1988.

Scandix pecten-veneris L., (comb of Venus), VENUS'-COMB, CROW-NEEDLES, SHEPHERD'S-NEEDLE, LADY'S-COMB. Hispid annual; leaves pinnately decompound; ultimate leaf divisions short and linear; petals white; fruits linear or narrowly oblong, body 6–15 mm long, 1–2 mm broad, ciliate; beak of fruit linear, 20–70 mm long, compressed laterally. Low moist areas; Dallas (Kidd Spring Park (*Couch s.n.*, 1958, SMU)), and not recorded for Dallas Co. again till 2022 (Lipscomb, pers. comm.), also Harris, Liberty, Travis, Tyler, and Waller (Kartesz 2021) cos.; Coastal Prairies and Marshes; scattered in the U.S. and s Canada. Mar–Apr. Native of Eurasia.

SIUM L. WATER-PARSNIP

A North American and Eurasian genus of ca. 9 species (Weakley, A.S., and Southeastern Flora Team 2024). Some are grown ornamentally, while the leaves or roots of others are eaten. However, some are reported to contain polyacetylenes (similar to those in the *Cicuta*, WATER-HEMLOCK) and the group should thus be considered poisonous (Yatskievych 2006). Molecular research (Spalik & Downie 2006; Spalik et al. 2009) demonstrates a close relationship of *Sium* with *Berula* and some *Sium* taxa have been transferred to *Berula*. (Greek: *sion*, the name of some wet area plant)
REFERENCES: Spalik & Downie 2006; Spalik et al. 2009; Nesom 2024i.

Sium suave Walter, (sweet), WATER-PARSNIP, HEMLOCK. Perennial; stems erect, 0.5–1.2 m tall; superficially very similar to *Cicuta*; leaves only once pinnate, rarely simple (submerged leaves can be decompound); leaflets lanceolate to linear, 1–7(–9) cm long, serrate or incised; petals white; fruits oval to orbicular, 1.5–3 mm long, the ribs prominent. Swamps, wet areas; Franklin, Gonzales, and possibly McLennan (Kartesz 2022) cos.; widespread in the n 1/2 of the U.S. and adjacent Canada, scattered in the s 1/2 of the U.S. May–Sep. [*S. cicutifolium* Schrank] Native of the U.S. and Asia. Molecular and biogeographic work (Spalik & Downie 2006) indicates that this species migrated to North America through Beringia ca. 0.6 million years ago. *Sium* species are reported to contain polyacetylenes (similar to those in *Cicuta*) and should be considered poisonous (Yatskievych 2006); there are reports of this species causing livestock poisoning (Kingsbury 1964).

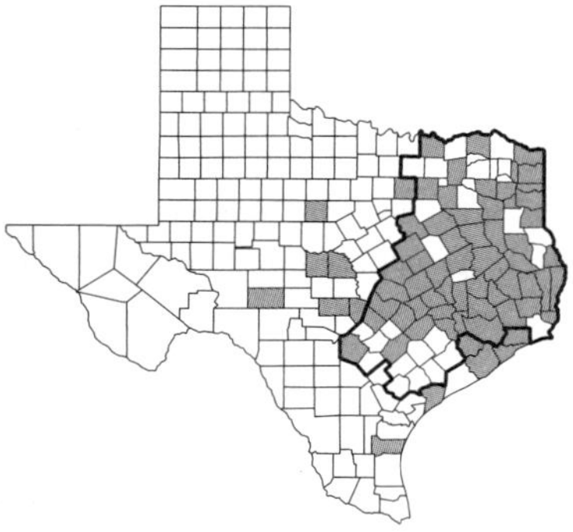
Ptilimnium capillaceum

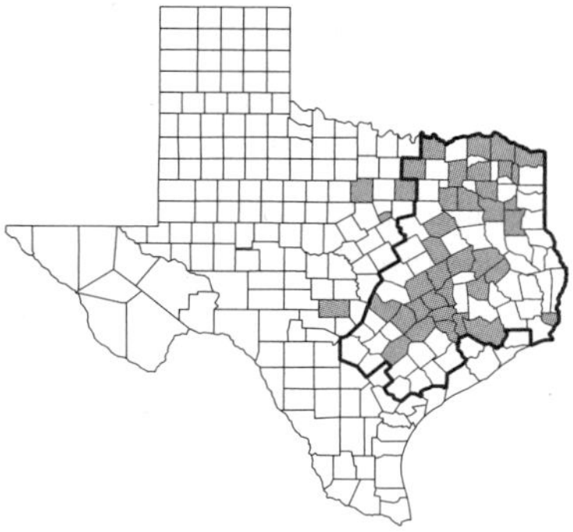
Ptilimnium nuttallii

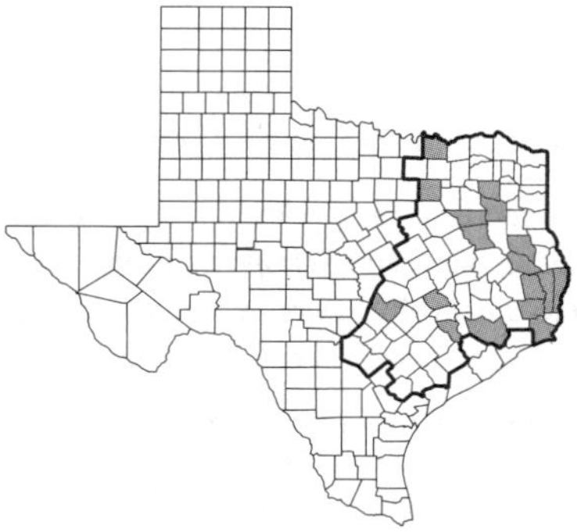
Ptilimnium texense

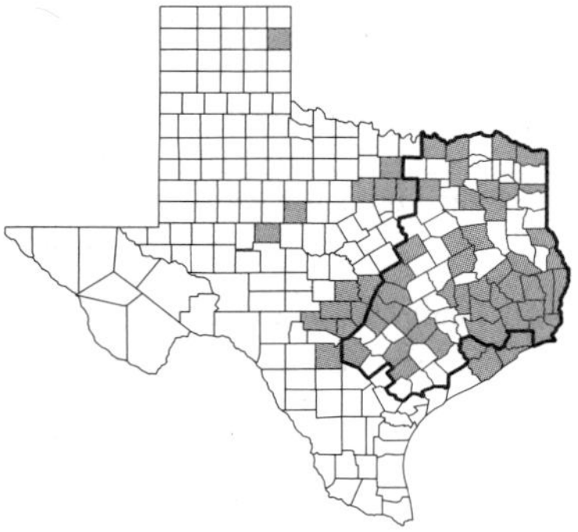
Sanicula canadensis

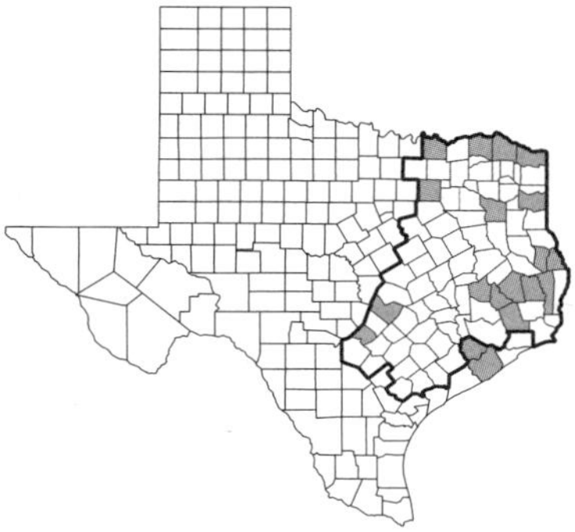
Sanicula odorata

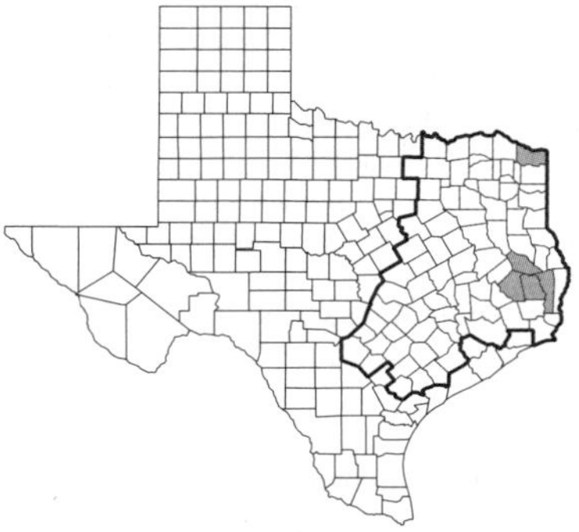
Sanicula smallii

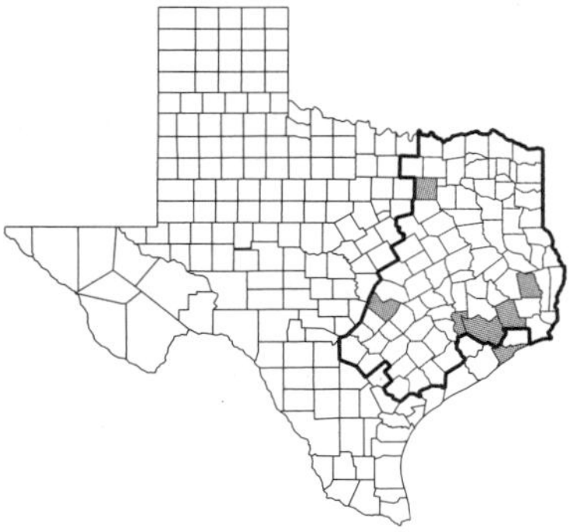
Scandix pecten-veneris

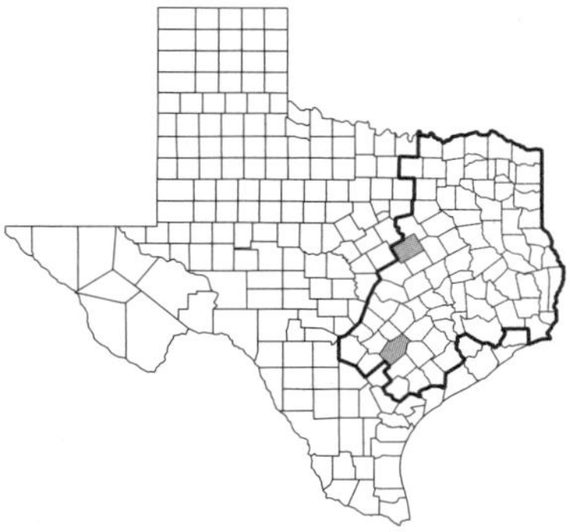
Sium suave

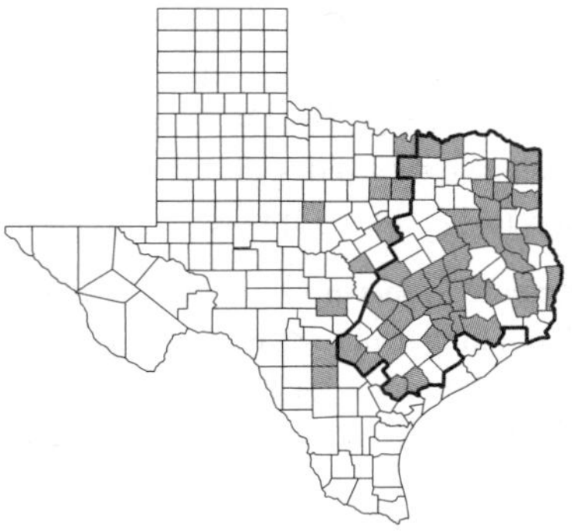
Spermolepis diffusa

Spermolepis divaricata

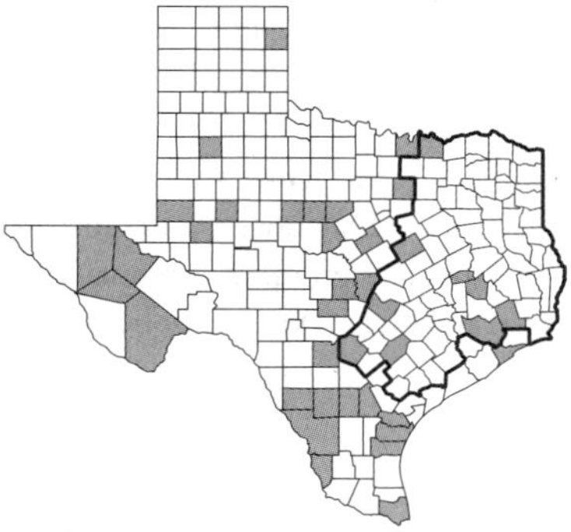
Spermolepis echinata

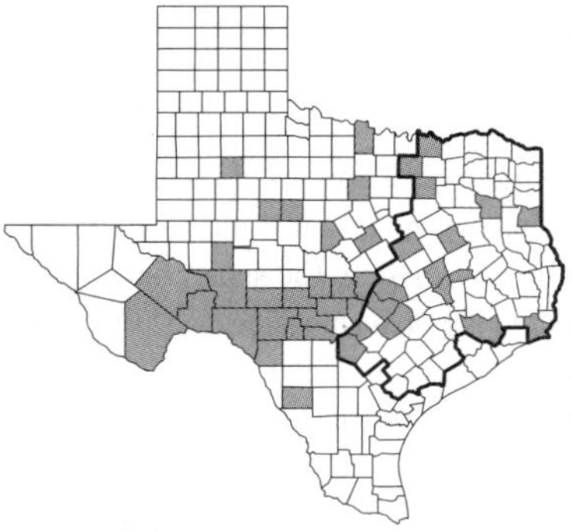
Spermolepis inermis

SPERMOLEPIS Raf. SCALESEED

Low, glabrous annuals; leaves ternately or ternate-pinnately decompound, the ultimate divisions linear to filiform; inflorescence of loose compound umbels; petals white; fruits 1–2 mm long, with or without bristles.

A genus of 11 species in North America, Argentina, and Hawaii (Nesom 2012c). Molecular evidence indicates a close relationship between *Cyclospermum* and *Spermolepis* (SCALESEED) (Downie et al. 2002). (Greek: *sperma*, seed, and *lepis*, scale, alluding to the scruffy or bristly fruits of some species)

REFERENCES: Downie et al. 2002; Nesom 2012c, 2024j.

1. Fruits with hooked short bristles ____ **S. echinata**
1. Fruits scabrous with minute hairs, essentially smooth, or tuberculate, but without bristles.
   2. Fruits essentially smooth ____ **S. laevis**
   2. Fruits scabrous with minute hairs or roughened with tubercles, but without bristles.
      3. Fruits roughened with densely arranged tubercles ____ **S. inermis**
      3. Fruits sparsely to densely scabrous with minute ascending hairs; fruiting pedicels (0–)2–32 mm long; fruiting rays 5–33 mm long; some flowers of the umbellet sessile to subsessile OR all flowers of the umbellet pedicellate.
         4. Fruiting pedicels (0–)2–9 mm long; fruiting rays 5–17 mm long; central 1–2 flowers of each umbellet sessile to subsessile ____ **S. divaricata**
         4. Fruiting pedicels (8–)14–32 mm long; fruiting rays 15–33 mm long; all flowers of each umbellet with subequal pedicels, none sessile or subsessile ____ **S. diffusa**

Spermolepis diffusa (Nutt. ex DC.) G.L. Nesom, (loosely speading), SPREADING SCALESEED. Plant 15–75 cm tall; peduncles 2–5 cm long; fruiting rays 2–4(–6), spreading, subequal, 15–33 mm long. Sandy clay, sand, roadsides, fencerows, pastures; widespread in East TX; AR, KS, LA, MO, OK, and TX. Apr–May(–Jun).

Spermolepis divaricata (Walter) Raf., (spreading, widely divergent), FORKED SCALESEED. Plant erect, 7–40 cm tall; peduncles 1.7–4(–5) cm long; fruiting rays 3–6, spreading, appearing open, unequal, 5–17 mm long. Open sandy areas, open woodlands; known in TX only from Austin and Liberty (Nesom 2012c) cos. of East TX; mainly se U.S. from NC s to FL w to TX. Mar–Apr(–May).

Spermolepis echinata (Nutt. ex DC.) A. Heller, (prickly), BEGGAR'S-LICE, BRISTLY SCALESEED. Plant erect, often spreading, to 40 cm tall; peduncles (1–)2–5(–6.5) cm long; fruiting rays 5–9(–12), (0–)1–15 mm long; central umbellet sessile to subsessile. Sand, gravel, sandy roadsides, disturbed areas; scattered in East TX and TX except for the Panhandle; s 1/3 of the U.S. from VA s to FL and w to AZ. (Mar–)Apr–May(–Jun). Native of the U.S. and n Mexico

Spermolepis inermis (Nutt. ex DC.) Mathias & Constance, (unarmed), SPREADING SCALESEED. Plant erect, 8–80 cm tall; peduncles 2–7 cm long; fruiting rays 5–11, suberect, evidently clustered, 1–13 mm long; central umbellet subsessile. Sandy or gravelly soils; scattered in East TX; Edwards Plateau, Cross Timbers and Prairies, and Trans-Pecos; c U.S. from IN s to AL and w to NE and NM. Apr–May(–Jun). Native of the U.S. and n Mexico

Spermolepis laevis G.L. Nesom, (smooth), SMOOTH SCALESEED. Plant 8–48 cm tall; peduncles 2–5 cm long; fruiting rays 3–8, suberect, evidently clustered, unequal, (0–)2–9(–13) mm long; central umbellet sessile to subsessile. Granite outcrops, limestone gravel, sandy fields; in East TX known only from Bell Co. (Nesom 2012c); also Burnet, Gillespie, Hamilton, Llano, Mason, and Tarrant (Nesom 2012c) cos.; endemic to OK and TX. Apr–May(–Jun). Native of the U.S.

TAENIDIA (Torr. & A. Gray) Drude YELLOW-PIMPERNEL

A genus of 2 species of the e U.S. (Mabberley 2017); the genus was previously considered monotypic (e.g., Guthrie 1969), but *Pseudotaenidia* (also monotypic) of the e U.S. is now included (Cronquist 1982). Further molecular research may either confirm or dispute the separation. *Taenidia* is related to a group of perennial North American genera including *Cymopterus*, *Lomatium*, *Polytaenia*, *Tauschia*, *Thaspium*, *Vesper*, and *Zizia* (Downie et al. 2002). (Greek: *tainidion*, a little band, in reference to the inconspicuous ribs)

REFERENCES: Guthrie 1969; Cronquist 1982.

Taenidia integerrima (L.) Drude, (quite entire), YELLOW-PIMPERNEL, GOLDEN ALEXANDER. Plants glabrous, glaucous perennials 30–85(–100) cm tall, with "light, pleasant, celery-like odor" (Cronquist 1982); stems branched, purplish near base; tap root tuberous-thickened; leaves usually 2–3 times ternately compound; leaflets entire; inflorescences compound umbels; terminal and axillary; rays 5–15(–20), unequal, the central shorter and bearing mostly sterile flowers; involucres and involucels absent;

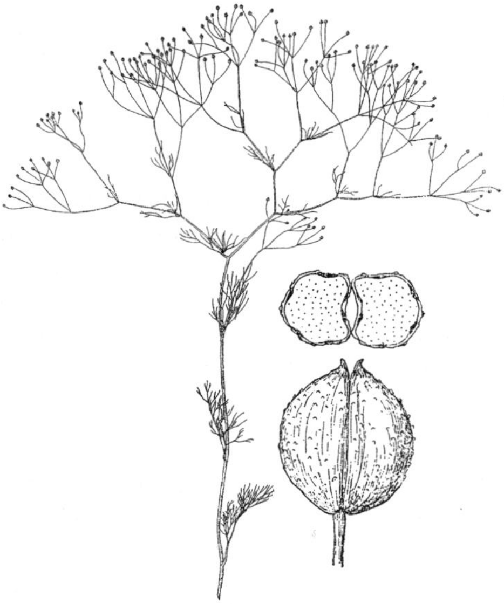

Spermolepis diffusa [LUN]

Spermolepis divaricata [WF2]

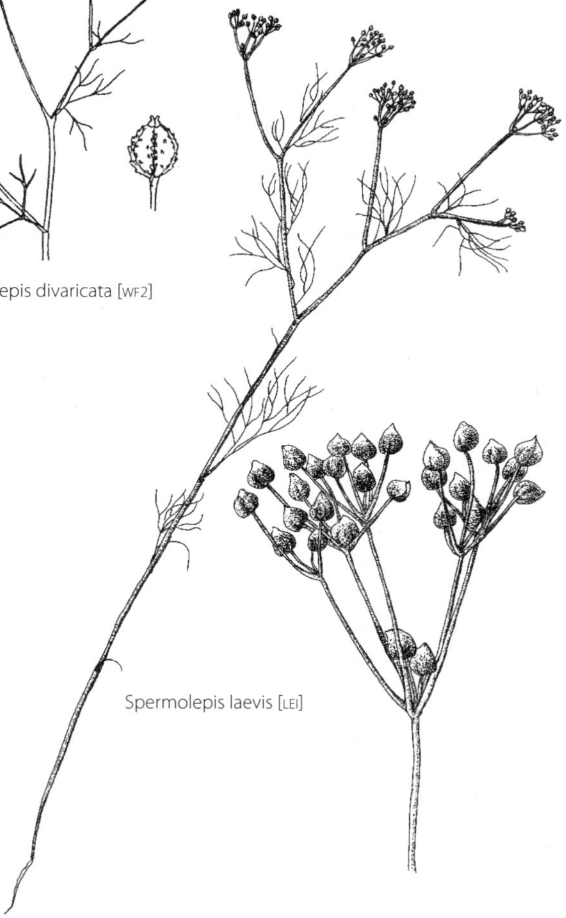

Spermolepis laevis [LEI]

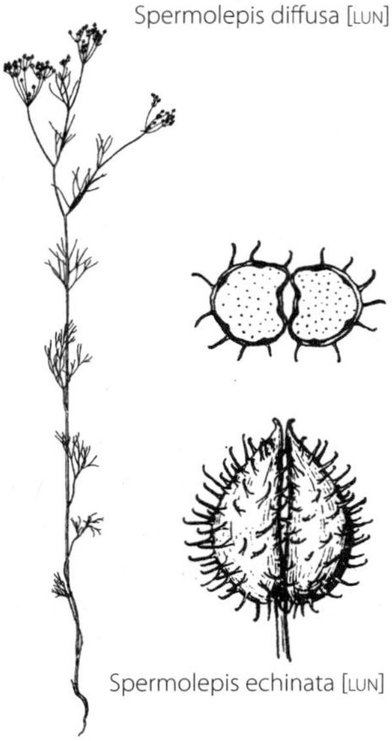

Spermolepis echinata [LUN]

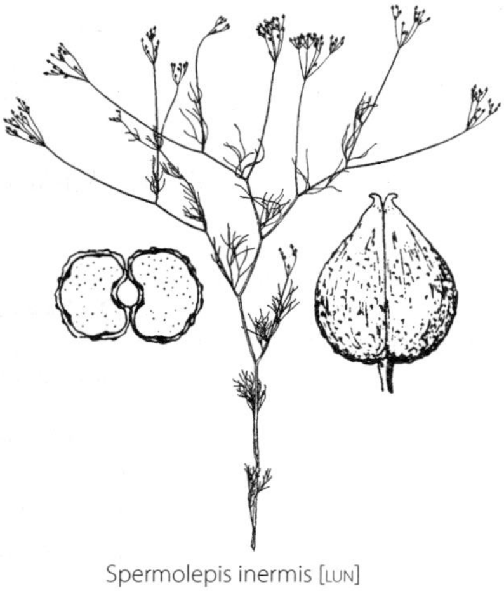

Spermolepis inermis [LUN]

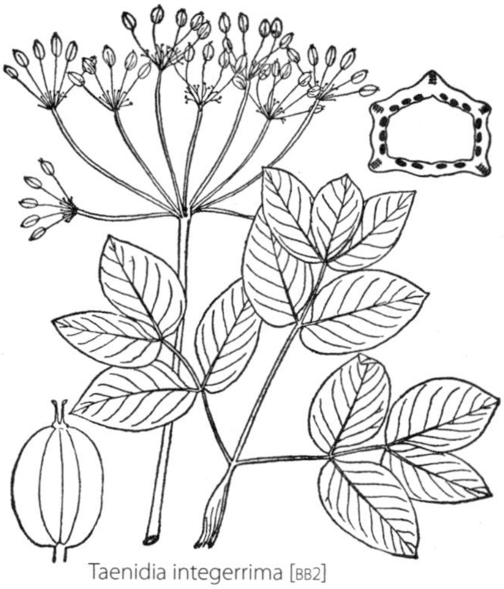

Taenidia integerrima [BB2]

Tauschia texana [LUN]

umbellets with numerous flowers, the central ones sterile and with very short pedicels; flowers yellow; sepals absent; fruits ca. 4–5 mm long, slightly compressed laterally, low ribbed, wingless, with 3 oil tubes between sets of ribs. Rocky open or low woods, bluffs; Anderson, Harrison, Red River, Sabine, and San Augustine (Kartesz 2022) cos.; widespread in the e 1/2 of the U.S. w to SD, OK, and TX; adjacent Canada. May–Jun. Native of the U.S. and Canada.

TAUSCHIA Schltdl. UMBRELLA-WORT

A genus of ca. 25 species ranging from the w U.S. to tropical South America (Wilson 2024). Molecular evidence (Downie et al. 2002; Sun et al. 2004) indicates that *Tauschia*, as traditionally viewed, is polyphyletic and closely allied to such genera as *Cymopterus* and *Lomatium* (see further discussion under *Vesper*). Until the relationships among the various genera and species are more adequately resolved and the taxonomy somewhat stabilized, we are continuing to recognize the genera in the traditional sense. The following treatment is modified from the one in Correll and Johnston (1970). (Named for Ignaz Friedrich Tausch (1793–1848), Czech professor of botany, director of the garden of the Duke of Canal de Malabaillas in Prague, and author of a flora of Bohemia in 1831 (Charter 2005)).

REFERENCES: Correll & Johnston 1970; Constance & Affolter 1997; Downie et al. 2002; Sun et al. 2004.

Tauschia texana A. Gray, (of Texas), TEXAS UMBRELLA-WORT. Acaulescent to short-caulescent glabrous perennial, 10–40 cm tall, from a taproot; leaves pinnately compound; leaflets, at least larger ones, pinnately parted or lobed; inflorescences compound umbels; peduncles 1–3(–4) dm long; fertile rays 5–8, unequal; involucres absent or of a single bract; involucels of several linear to lanceolate bractlets; pedicels 1–2 mm long; flowers yellow; calyx minute; fruits 3–4 mm long, oval, glabrous, slightly compressed laterally. Alluvial thickets, wet woods; s portions of East TX; n Gulf Prairies and Marshes; endemic to TX. Feb–Jun. [*Museniopsis texana* (A. Gray) J.M. Coult. & Rose]

THASPIUM Nutt. MEADOW-PARSNIP

Slender perennials; roots fibrous; stems usually erect; leaves basal and cauline, alternate, toothed; leaf blades simple or once or more compound; inflorescences compound umbels; involucral bracts absent or 1–3, minute; flowers all short-stalked, the pedicels 1–4 mm long; bractlets of involucel small; sepals usually minute; petals creamy white to golden yellow; fruits slightly flattened laterally (at right angles to the commissure), 3–5 mm long, with mericarps usually 5-ribbed, all or most of the ribs winged, glabrous or minutely roughened between the ribs, with oil tubes between each set of ribs.

A North American genus of 4 species (Tennessee Flora Committee 2015; Floden 2019). Sometimes only three according to some authors (Yatskievych 2006; Mabberley 2017; Wunderlin et al. 2020) who synonymize *Thaspium chapmanii* under *Thaspium barbinode*. Except for fruit differences (winged in *Thaspium*, unwinged or only slightly so in *Zizia*, these two genera are morphologically very similar and apparently related evolutionarily (Downie et al. 2000); preliminary DNA data indicate that *Thaspium* is monophyletic. According to Yatskievych (2006), "*Thaspium* and *Zizia* are sometimes cited in botany textbooks as an example of genera whose morphologies have converged over time to the point that they are often difficult to distinguish (homoplasy). They may be determined at flowering by the sessile or nearly sessile central flower of each umbellet in *Zizia* vs. the noticeably short-stalked central flowers in *Thaspium*, but this character requires practice and patience to discern with confidence. Plants of *Zizia* also have rootstocks with clusters of slightly to moderately tuberous-thickened roots, whereas the roots of *Thaspium* species are fibrous and not fleshy or thickened." *Thaspium* and *Zizia* are related to a group of perennial North American genera including *Cymopterus*, *Lomatium*, *Polytaenia*, *Taenidia*, *Tauschia*, and *Vesper* (Downie et al. 2002). Solitary bees and flies appear to be the primary pollinators (Lindsey 1984; Lindsey & Bell 1985). ("Name a play upon *Thapsia*, the name of a related genus, so called from the peninsula of Thapsus"—Fernald 1950)

REFERENCES: Lindsey 1975; Brown & Brown 1984; Cooperrider 1985; Swink & Wilhelm 1994; Downie et al. 2002; Yatskievych 2006; Tennessee Flora Committee 2015; Floden 2019; Wunderlin et al. 2020.

1. Stem leaves usually 2–3 times ternate; margins of leaves hairy, not whitened; petals creamy white to pale yellow ____ **T. chapmanii**
1. Stem leaves usually once ternately lobed or compound; margins of leaves glabrous, with a narrow whitish border; petals golden yellow ____ **T. trifoliatum**

Thaspium chapmanii (J.M. Coult. & Rose) Small, CHAPMAN'S MEADOW-PARSNIP. Plant 8–15 dm tall, much branched; basal leaves usually once to twice ternately compound; basal leaflets ovate to lanceolate, margin irregularly and coarsely serrate to lobed; stem leaves twice to three times ternately compound; rays 9–16, to 2 cm long; flowers pale yellow to creamy-white; peduncles and umbel rays scaberulous; mature fruits 4–5 mm long, minutely spinulose to scaberulous between wings. Glade edges, bluff tops, prairies, rocky banks, open woodlands; in TX known only from Houston Co. (*Orzell & Bridges 6417* BRIT); ec U.S. from OH s to FL and w to MN and TX. Apr–Jun. [*T. barbinode* var. *chapmanii* J.M. Coult. & Rose] Native of the U.S. and Canada.

Thaspium trifoliatum (L.) A. Gray var. **aureum** (L.) Britton, (sp.: three-leaved; var.: golden), PURPLE MEADOW-PARSNIP, SMOOTH MEADOW-PARSNIP. Plants nearly glabrous (stems with minute band of hairs at base of upper leaf sheaths, the hairs visible only using magnification), usually 20–65(–80) cm tall, sparingly branched; basal leaves usually simple, cordate to reniform, or once ternately lobed or compound; leaflet margins finely and evenly toothed; stem leaves varying from simple to usually once to twice ternately compound, leaflet bases often asymmetrical; rays 6–10(–18), to 3 cm long, unequal; fruits 3–5 mm long, the ribs winged, glabrous. Moist forests, upland prairies, savannas, glades, and ledges; mostly in the se portion of East TX; e U.S. from PA s to FL and w to MN, KS, and TX; adjacent Canada. Apr–May; Jul–Aug. [*T. aureum* (L.) Nutt.] Native of the U.S. and Canada.

Thaspium barbinode (Michx.) Nutt., (with bearded nodes), HAIRYJOINT MEADOW-PARSNIP. According to Brock (2024) the single collection from the state of this species from Houston Co., TX (*Orzell & Bridges 6417* TEX) is in early flowering condition and its identification is uncertain. The key to differentiating it from *T. chapmanii* from Brock (2024) is provided below.

1. Leaflets of leaves subtending terminal umbels wider than or equal to length of rays; fruit surfaces smooth to minutely papillose-scabrous _____ *T. barbinode*
1. Leaflets of leaves subtending terminal umbels narrower than length of rays; fruit surfaces minutely papillose-scabrous, rarely smooth _____ **T. chapmanii**

TIEDEMANNIA DC. FAKE DROPWORT

A genus of 2 species of se U.S. and the West Indies (Feist 2024d). Analysis of fruit anatomy, leaf morphology, and molecular techniques revealed that the genus *Oxypolis* (formerly 7 species) was polyphyletic. Feist et al. (2012) reinstated the genus *Tiedemannia* to accommodate two former *Oxypolis* species (see further discussion under *Oxypolis*). (Commemorates Friedrich Tiedemann (1781–1861), German zoologist)
REFERENCES: Feist & Downie 2008; Feist et al. 2012.

Tiedemannia filiformis (Walter) M.A. Feist & S.R. Downie, (thread-like), WATER FAKE DROPWORT. Stems 0.6–1.5 m tall; the reduced leaves (phyllodes) 20–60 cm long, linear, septa inconspicuous; inflorescence a terminal or axillary compound umbel, peduncles 3–18 cm long; involucral bracts 4–11, 0.5–2.5 cm long, linear-lanceolate; rays (5–)10–25, 1–4.5 cm long; umbellets 10–22-flowered, pedicel 2–10 mm long, bractlets 2–4, 3–10 mm long; flowers white; fruits 3–8 mm long, 1–5 mm wide. Wet areas; in TX limited to the se portion of East TX in Austin, Hardin, Houston, Jefferson, Tyler (Turner et al. 2003), and Newton (*Thomas 101876* USCH:MAIN) cos.; se U.S. from NC s to FL w to TX. Jul–Sep. [*Oxypolis filiformis* (Walter) Britton] Native of se U.S. and the West Indies.

TORILIS Adans. HEDGE-PARSLEY

Hispid annuals; leaves once pinnate or pinnately decompound; petals white; fruits in ours 3–5 mm long, bristly.

A genus of ca. 15 species of Eurasia and n Africa (Ball 2024g). (Name used by Adanson in 1763; possibly from Latin: *torus*, cushion)
REFERENCE: Ball 2024g.

1. Flowers pedicelled, in terminal, peduncled, open umbels; plants erect _____ **T. arvensis**
1. Flowers subsessile, in axillary, sessile or short-peduncled, head-like umbels; plants reclining to suberect _____ **T. nodosa**

Torilis arvensis (Huds.) Link, (pertaining to cultivated fields), HEDGE-PARSLEY. Plant erect 15–100 cm tall; all fruits bristly with microscopic barbs. Waste or disturbed ground, especially low or shady places, chiefly in limestone areas; mainly w parts of East TX; Edwards Plateau and Trans-Pecos; e U.S. from NY s to FL and w to MN, KS, and TX, also far w U.S.; adjacent Canada. May–Jun. [*T. japonica* of authors, not (Houtt.) DC.] Native of the Mediterranean region. HEDGE-PARSLEY can be a problematic weed in some areas. The mericarps of this species were used in the successful prosecution of a kidnapping/child molestation case in Fort Worth in 1995. A 2-year-old girl was pulled from the window of her apartment, kidnapped, sexually molested, and then left in a weedy area ca. 100 meters from where she was abducted. Law enforcement officials found plant fragments on the suspect's shoes and also collected assorted plant material from the crime scene. Botanists at BRIT identified the material from both the shoes and the crime scene as mericarps of *Torilis arvensis*, thus linking the suspect with the scene of the crime. This forensic botanical evidence, in combination with various other types of evidence, led to a conviction and a 99 year jail term (Lipscomb & Diggs 1998).

Torilis nodosa (L.) Gaertn., (with nodes), KNOTTED HEDGE-PARSLEY. Plant decumbent or procumbent, branched from base; stems to 15–70 cm long; at least some fruits partially tuberculate but some also partially bristly. Waste or disturbed ground, especially low or shady places, chiefly in limestone areas, can become a pest in lawns; widespread in East TX; Edwards Plateau and South TX Plains; widely scattered in se U.S., also AZ, CA, and OR. Apr–Jun. Native of Mediterranean region.

TREPOCARPUS Nutt. ex DC. WHITE-NYMPH

A monotypic genus of the se United States (Mabberley 2017). (Greek: *trep*, turn, and *karpos*, fruit)
REFERENCES: Baskin et al. 2003; Downie et al. 2008; Nesom 2024k.

Trepocarpus aethusae Nutt. ex DC., (presumably referring to resemblance to *Aethusa*—fool's parsley, another genus of Apiaceae), WHITE-NYMPH. Glabrous erect annual 3–100 cm tall; leaves pinnately decompound, the ultimate divisions linear; petals white; fruits oblong or broadly oblong-linear, 8–10 mm long. Woods and thickets, especially in low ground; widespread in East TX except for sw portion; n Gulf Prairies and Marshes and Tarrant Co. (R. O'Kennon, pers. obs.) just to the w of East TX; se U.S. from KY s to FL w to MO, OK, and TX. May–Jun. Native of se U.S. A detailed study of the life cycle of this species indicates that it is a winter annual whose seeds germinate in the fall, with bolting, flowering, and seed production the following growing season (Baskin et al. 2003).

VESPER R.L. Hartm. & G.L. Nesom VENUS-PARSLEY

A w North American genus of 6 species (Hartman & Nesom 2012). This genus, segregated from *Cymopterus*, forms a monophyletic group. The genus is characterized by its combination of thick taproots, acaulescent habit, compact inflorescence, white to cream, pink, or purple petals, and basally connate involucel bracts (Hartman & Nesom 2012). The genus is part of a monophyletic group of about 200 species of endemic perennial Apiaceae in w North America. However, most of the genera in the group are highly polyphyletic, including *Lomatium*, *Cymopterus*, and *Tauschia*. Traditional generic boundaries have been based largely on fruit characters, which now appear to be highly variable and of little use taxonomically. Based on DNA sequence data, it has been hypothesized that the complicated (and still unresolved) relationships among these species is due to their rapid evolution, which resulted from "late Tertiary and Quaternary geologic and climatological events that gave rise to widespread aridity and great habitat diversification in western NA" (Sun et al. 2004). Undoubtedly, generic concepts in the group will be greatly changed. (Name from the Latin *vesper*, evening or west, sometimes referring to the "evening star" (usually Venus) seen at sunset in the western sky. "The name alludes to the team of Sun and Downie, who have provided molecular analyses (Sun & Downie 2004, 2010; and including Downie et al. 2002) indicating that evolutionary relationships among many of the currently and historically recognized genera of western North American (subfamily) Apioideae are complex and apparently reticulate." —Hartman & Nesom 2012)
REFERENCES: Downie et al. 2002, 2010; Sun & Downie 2004, 2010; Sun et al. 2004; Hartman & Nesom 2012.

Vesper macrorhizus (Buckley) R.L. Hartm. & G.L. Nesom, (large-rooted) BIG-ROOT VENUS-PARSLEY. Glabrous, glaucous, dwarf perennial to 35 cm tall; from a thick, soft-woody, fusiform to subglobose root; leaves few, crowded at base of plant, pinnate or bipinnate; flowers pink or nearly white; anthers purple-black, prominent; fruits ovoid to ovoid-oblong, 4–9 mm long, 3–8 mm wide, the wings obvious. Rocky limestone prairies; w margin of East TX, Bastrop, Collin, Dallas, Hill, McLennan, and Travis (Kartesz 2022) cos.; Edwards Plateau, Cross Timbers and Prairies, and Rolling and High Plains; endemic to NM, OK, and TX. Mar–Apr. [*Cymopterus macrorhizus* Buckley].

ZIZIA W.D.J. Koch ALEXANDERS

A North American genus of 3 species (Levin 2024b). *Zizia* and *Thaspium* are morphologically very similar and apparently related evolutionarily (Downie et al. 2000); preliminary DNA data indicate that *Zizia* is monophyletic. *Zizia* and *Thaspium* are related to a group of perennial North American genera including *Cymopterus*, *Lomatium*, *Polytaenia*, *Taenidia*, *Tauschia*, and *Vesper* (Downie et al. 2002). Solitary bees and flies appear to be the primary pollinators (Lindsey 1984; Lindsey & Bell 1985). (Named for Johann Baptist Ziz, 1779–1829, a Rhenish (= of the Rhine region) botanist)
REFERENCES: Lindsey 1975; Cooperrider 1985; Downie et al. 2002; Levin 2024b.

Zizia aurea (L.) W.D.J. Koch, (golden), GOLDEN-ALEXANDERS, GOLDEN ZIZIA. Glabrous perennial 30–110 cm tall; leaves once or twice compound; ultimate leaflets ovate- or oblong-lanceolate, sharply toothed; petals yellow; fruits 2–4 mm long. Low woods; n and e parts of East TX; widespread primarily in the e 1/2 of the U.S.; adjacent Canada. Apr–May. Native of the U.S. and Canada.

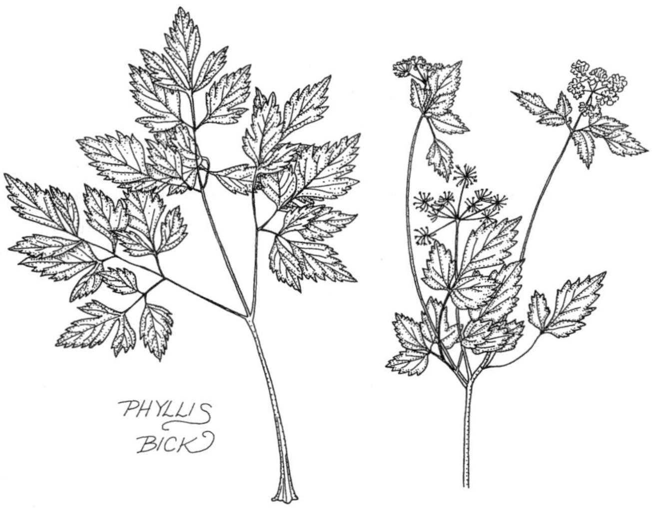

Thaspium chapmanii [YA2]

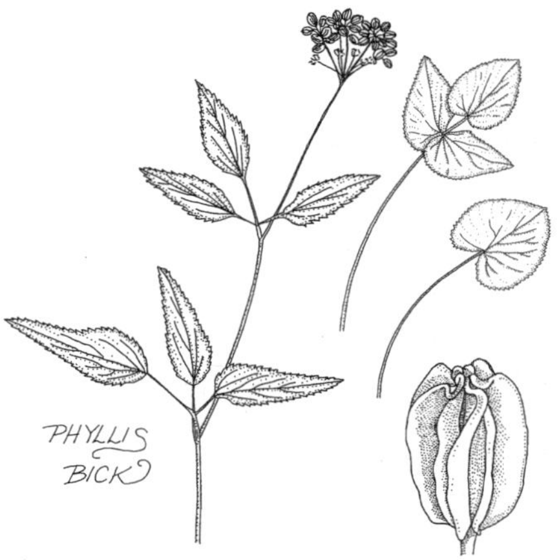

Thaspium trifoliatum var. aureum [YAT]

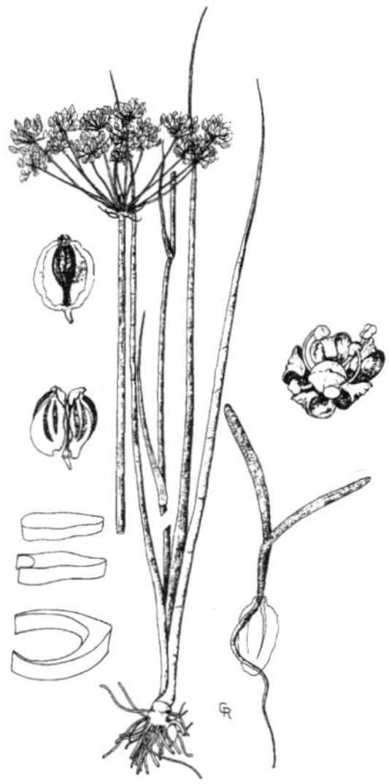

Tiedemannia filiformis [GWO]

Torilis arvensis [HEA]

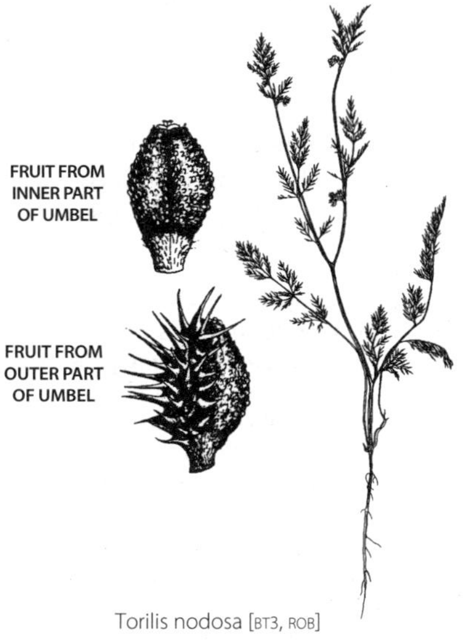

Torilis nodosa [BT3, ROB]

APOCYNACEAE Juss.

OLEANDER OR DOGBANE FAMILY

Ours perennial herbs or vines with milky juice or not so; leaves opposite, whorled, or alternate, sessile or petiolate, simple, entire or undulate, deciduous or evergreen; flowers axillary, *extra* axillary, or terminal, solitary, in cymes, or umbelliform inflorescences; sepals 5, barely united at base; corollas salverform, funnelform, campanulate, cylindric, or urceolate, with 5 lobes, the throat or summit of tube usually with a constriction, dense zone of hairs, or crown; crown (= corona) of various sorts (e.g., ring-like, inflated, flat, stamen-like—in *Asclepias* and *Funastrum* often called hoods when inflated, sometimes very conspicuous) present around base of the short stamen-tube or associated with corolla lobes; stamens 5, separate, pollen not in pollinia (former Apocynaceae s.s.); OR, in the case of the former Asclepiadaceae, the filaments united (except in *Periploca*), the anthers large, united to form an anther head enclosing the two pistils (the combined column-like or disk-like structure made up of the stamens and pistils is called the gynostegium); anthers usually with a terminal outgrowth of the connective called an anther appendage and with lateral wing-like margins (= anther wings); anther wings from adjacent anthers are positioned close together and thus form a slit-like opening or groove between them (through which the legs or other appendages of pollinators can fit); pollen grains coherent in a mass (= pollinium) (except in *Periploca*); the 2 pollinia from adjacent anthers connected by a wishbone-shaped structure (= translator); clip-like portion (= corpusculum) of translator attaches the pair of pollinia to the pollinator's appendage as it is pulled out of the groove between adjacent anthers (pollinia from other flowers can enter the groove and thus be brought into contact with the enclosed stigmatic surfaces); pistils 2, united by styles and/or stigmas only (superficially flowers can appear to have 2 separate pistils); ovaries superior; fruit of 2 (or 1 by abortion) follicles; sometimes seeds with a coma (= tuft of hairs).

A large (ca. 4,800 species in ca. 460 genera), nearly worldwide family, especially in the tropics and subtropics (Fishbein et al. 2023); vegetatively, it includes mainly lianas, trees, shrubs, and herbs; they have ubiquitous laticifer systems, glycosides, and alkaloids. The family contains many ornamental, poisonous, and medicinal species including the beautiful *Plumeria rubra* L. (FRANGIPANI), the widely planted but very poisonous *Nerium oleander* L. (OLEANDER), *Rauvolfia serpentina* (L.) Benth. ex Kurz, the source of the medicinal alkaloid reserpine, *Acokanthera* species, from which deadly arrow poisons (cardiac glycosides including ouabain with an effect like digitalin) are obtained in Africa, *Strophanthus*, an Old World genus yielding arrow poisons and the cardiac drug strophanthin, and *Catharanthus*, the source of alkaloids used in cancer treatments (see below). OLEANDER, a commonly used ornamental, contains extremely poisonous cardiac glycosides (e.g., neriin and oleandrin) resembling digitalin in action; a single leaf can kill an adult; children have been poisoned by sucking nectar or chewing leaves; poisoning may result from using twigs as skewers for food or even inhaling the smoke; even water in which flowers have been placed is toxic (Schmutz & Hamilton 1979; Morton 1982; Powell 1988). The sap of all species should be avoided. Apocynaceae are closely related to Asclepiadaceae (MILKWEED family), but were treated separately as late as the last half of the 20th century. Ample morphological and molecular phylogenetic evidence provides a sound basis for uniting the two families (Sennblad & Bremer 1996; Endress and Bruyns 2000; Stevens 2001 onwards). The Apocynaceae are most closely related to (in addition to Asclepiadaceae) the Gelsemiaceae, Loganiaceae, Gentianaceae, and Rubiaceae (Stevens 2001 onwards) (subclass Asteridae—Cronquist; order Gentianales—APG IV)

FAMILY RECOGNITION IN THE FIELD: herbs or vines with simple, entire, often opposite leaves, milky sap, sympetalous corollas usually with a distinct tube, and 2 pistils united only by their styles and/or stigmas. Also, flowers with a distinctive corona and a gynostegium (= combined structure composed of stamens and pistils).

REFERENCES: Woodson 1938; Rosatti 1989; Judd et al. 1994; Sennblad & Bremer 1996; Endress & Bruyns 2000; Fishbein et al. 2023.

1. Androecium and gynoecium fused into a gynostegium (except in *Periploca*); coronas usually associated with gynostegium, often conspicuous with inflated hoods (*Asclepias* and *Funastrum*) (5), or collar-like crown of laminar or other variously shaped segments; pollen massed into two rigid sacs (pollinia); follicles usually solitary (paired in *Periploca*).
   2. Central column of flower surrounded by 5 separate, fleshy-inflated or fleshy-thickened, erect or spreading stamen-like appendages (= hoods).
      3. Stems not twining, the plants erect to prostrate herbs; corollas without a fleshy collar-like ring around the base of the hoods; hoods usually with an internal horn or crest (horn or crest absent in a few species) ____ **Asclepias**
      3. Stems twining, at least towards the tip, the plants vines; corollas with a fleshy collar-like ring around the base of the hoods; hoods without internal horn or crest ____ **Funastrum**
   2. Central column of flower or its base with a crown of 5 segments variously separate, united, or lobed (sometimes long in *Periploca* and *Cynanchum*) OR with a low, minute collar-like cup.

4. Plants woody nearly throughout; corona with 5 filament-like segments ca. 7 mm long; follicles paired; introduced species ____ **Periploca**
4. Plants herbaceous or woody only in lower half; corona not as above; follicles usually solitary; native species.
    5. Corolla lobes maroon, olive green, or green-reticulate; leaf base cordate; corona of 5 variously united fleshy segments, often with inner appendages OR corona a low collar-like cup (can be minute), coronas < 1/3 as long as corolla lobe.
        6. Corolla lobes maroon, olive green, green, or green-reticulate; corona of 5 variously united, fleshy segments OR a low minute collar-like cup; follicles muricate (= with short stout spines) ____ **Matelea**
        6. Corolla lobes olive green to green; corona a prostrate-undulating ring, minute; follicles not muricate ____ **Gonolobus**
    5. Corolla lobes white to cream or green with white margin; leaf base cordate, sagittate, or cuneate; corona of 5 mostly distinct segments (may be lobed or divided) fleshy or thin, ca. 1/3 as long to as long as corolla lobes.
        7. Inner surface of corolla lobes pubescent ____ **Metastelma**
        7. Inner surface of corolla lobes glabrous.
            8. Leaf blades ovate to deltate, base cordate or sagittate ____ **Cynanchum**
            8. Leaf blades linear, base cuneate ____ **Pattalias**

1. Stamens distinct; anthers connivent (= converging) or not, but androecium not fused with gynoecium into a gynostegium; corolla throat or summit of tube with a constriction, crown or dense zone of hairs; follicles usually paired.
    9. Plant a trailing and climbing vine; corollas cream or yellowish white to pale yellow.
        10. Leaves thin and membranous; corolla lobes 3–4 mm long, 2–3 mm wide, spreading to reflexed ____ **Thyrsanthella**
        10. Leaves coriaceous; corolla lobes 7–10 mm long, 4–5 mm wide, spreading ____ **Trachelospemum**
    9. Plant erect or a trailing vine with ascending branches, not climbing; corollas variously colored white to green, yellow, or pink to lavender, purple or red.
        11. Leaves alternate; flowers blue to lavender-blue (rarely white), in many-flowered branched clusters usually terminating the stems; throat with a dense zone of hairs ____ **Amsonia**
        11. Leaves opposite or whorled; flowers either 1–2 in leaf axils OR if in many-flowered branched clusters then variously colored (white to green, yellow, or pink to lavender, purple or red).
            12. Plant an erect woody perennial typically to ca. 3 m tall; corollas large (limb 2.5–5 cm across) AND in many-flowered branched clusters; sap not milky; crown a collar-like lacerate ring ____ **Nerium**
            12. Plant a trailing vine or a small (< 0.6 m tall) tender perennial (grown as an annual) or suffrutescent perennial; flowers EITHER much smaller (ca. 2–4 mm across) and in clusters OR 1–2 in leaf axils; sap milky OR not so.
                13. Plants trailing vines with ascending branches; corollas blue to lavender-blue (rarely white); sap not milky ____ **Vinca**
                13. Plants erect; corollas variously colored (white to green, yellow, or pink, lavender, purple or red); sap milky.
                    14. Flowers in a many-flowered branched cluster; corollas campanulate to cylindric or urceolate, white to green or yellow, small, 2–4 mm across; plants to 1 m tall ____ **Apocynum**
                    14. Flowers usually 1–2 per leaf axil or solitary at stem tip; corollas salverform, variously colored, from white to rosy pink or red, sometimes with a differently colored "eye," large, > 20 mm across; plants < 0.6 m tall, often much less.
                        15. Introduced tender perennial widely grown in East TX as an annual; lower surface of leaf blades glabrous or with sparse hairs; corollas white to lavender, rosy pink or red, sometimes with a dark eye, with a tube 30 mm or less long; seeds without a coma of hairs ____ **Catharanthus**
                        15. Native suffrutescent perennial known only from extreme sw margin of East TX; gynostegium and corona absent; lower surface of leaf blades yellowish-white due to dense tomentum; corollas white (can be tinged with pink), with a tube 4.6 cm (not including expanded throat) or more long; seeds with a coma of hairs ____ **Mandevilla**

AMSONIA Walter BLUESTAR, SLIMPOD

Largely glabrous to sparsely pubescent, herbaceous perennials from woody rootstocks; leaves numerous, alternate, filiform to lanceolate or elliptic; flowers in compact terminal cymes; corollas salverform, light to deep blue or lavender, rarely white, with narrow lobes; follicles erect to pendulous, slender; seeds numerous, without a coma.

A genus of 17 species (Tō Ngô & Applequist 2023) native to North America and Japan; a number have alkaloids and some are cultivated as ornamentals. It is "one of the few wholly temperate genera in the family" (Bogler 2006). *Amsonia* species "are taxonomically problematic. Species of the southeastern and south-central United States are highly variable with, in some cases, no definitive species or varietal boundaries and persistent nomenclatural issues (Tō Ngô & Applequist 2023)." (Named for Charles Amson, 18th century Gloucester, Virginia physician and friend of John Clayton)

REFERENCES: Woodson 1928, 1929; McLaughlin 1982; Lemke 1987; Bogler 2006; Doffitt et al. 2014; Williams 2019; Tō Ngô & Applequist 2023.

1. Corolla tube 28–41 mm long, sharply constricted at the orifice ______ **A. longiflora** var. **salpignantha**
1. Corolla tube 6–9 mm long, not constricted at the orifice.
   2. Corolla glabrous on the outer surface; stem usually glabrous.
      3. Leaf blades heteromorphic (= main stem and branch leaf blades dissimilar); main stem leaf blades wider, narrowly elliptic, to elliptic, 3.5–6.5 cm long, 5–15 mm wide; branch leaf blades narrower, linear to very narrowly elliptic, 3.2–5.6 cm long, 2–6 mm wide ______ **A. ciliata** var. **texana**
      3. Leaf blades not heteromorphic; elliptic, narrowly oblong, lanceolate or obovate 3.9–8.4 (–9.3) cm long, (5–) 8–18(–23) mm wide ______ **A. rigida**
   2. Corolla pubescent on the outer surface; stem glabrous (or sometimes pubescent when young in *A. ludoviciana*).
      4. Leaf blades densely to moderately tomentose below; follicles usually pubescent ______ **A. ludoviciana**
      4. Leaf blades glabrous to somewhat pubescent below, never tomentose; follicles glabrous ______ **A. tabernaemontana**

Amsonia ciliata Walter var. **texana** (A. Gray) J.M. Coult., (sp.: fringed; var.: of Texas), TEXAS SLIMPOD, TEXAS AMSONIA. Stems basally ascending to erect, 15–49(–67) cm long, glabrous; leaves heteromorphic; main stem leaf blades 3.5–6.5 cm long, 5–15 mm wide, narrowly elliptic to elliptic, margins slightly to strongly revolute, usually ciliate; branch leaf blades 3.2–5.6 cm long, 2–6 mm wide; linear to very narrowly elliptic; sepals narrowly deltate, apex acuminate or acute, 1–1.5(–2) mm long, sparsely ciliate with long hairs; corolla tube (6–)7–9(–10) mm long, lobes (5–)6–8(–9) mm long, margins often minutely ciliate; follicle (7–)11–13.5 cm long. Prairies, meadows, rocky ground; Dallas (*Cory 54192* LL), Denton (*Lundell 13396* LL), and McLennan (*Mauldin s.n.*, 15 Apr 1970, SMU); Cross Timbers and Prairies and Edwards Plateau; OK and TX. Spring. [*A. texana* (A. Gray) A. Heller]

Amsonia longiflora Torr. var. **salpignantha** (Woodson) S.P. McLaughlin, (sp.: with a long flower; var; Greek: *salpingnos*, an ancient Greek trumpet-like instrument, and *anthus*, flower), TRUMPET SLIMPOD. Stems erect, 16–60 cm long, sparsely to moderately pubescent; leaves slightly heteromorphic, blades 3–6.5(–9) cm long, 1–2.8(–5) mm wide, linear to ligulate, margins occasionally somewhat revolute, at least sparsely ciliate, surfaces sparsely to moderately short-pubescent (seldom glabrate or glabrous on one or both surfaces); calyx pubescent to glabrous, sepal margins ciliate (or not); corolla cream to bluish, corolla tube 28–41 mm long, lobes 8–14 mm long; follicles 6–11(–20) cm long. Rocky hills, on limestone; on the sw margin of East TX, Bell, Bexar, Comal, and Travis (Kartesz 2022) cos.; Edwards Plateau and Trans-Pecos; NM and TX. Mar–May. [*A. salpignantha* Woodson] Variety *longiflora* is limited in TX to the Trans-Pecos.

Amsonia ludoviciana Vail ex Small, (of Louisiana), LOUISIANA BLUESTAR. Stems erect, 30–41(–78) cm tall, glabrous or sparsely pubescent; leaves not heteromorphic, main stem leaf blades, 5–7.5(–9) cm long, 10–40(–45) mm wide, lanceolate, elliptic, or ovate, margins entire, not revolute, moderately to densely ciliate, apex acuminate, surfaces densely (to moderately) tomentose abaxially, glabrous adaxially; sepals deltate, 0.8–1.1 mm long; corolla tube purplish, greenish above, 7–8 mm long, lobes blue, 6–8 mm long, outer surface of corolla pubescent, less often glabrous; follicles erect, 8.5–10 cm long, pubescent at least distally or glabrous. Low wet woods, road banks, ditches, open disturbed sites; in TX known only from Newton Co. (*Allen 22212* BRIT); GA, LA, MS, and TX. Apr–Jun. Considered rare in TX. NatureServe: G3, vulnerable. ⚠

Amsonia rigida Shuttlew. ex Small, (stiff, inflexible), STIFF BLUESTAR. Stems erect (somewhat ascending at base), 48–92 cm long, glabrous or seldom sparsely pubescent; leaves not heteromorphic, main stem leaf blades 3.9–8.4 (–9.3) cm long, (5–)8–18(–23) mm wide, elliptic, narrowly oblong, or lanceolate, margins often slightly revolute, usually irregularly so, surfaces glabrous; branch leaf blades similar to stem leaf blades; corolla tube blue, greenish above, 6–8 mm long, lobes blue to white, 5–7.5 mm long, outer surface of corolla glabrous; follicles erect, 9–13(–15) cm long, glabrous. Pond edges, steam bottoms, moist woods; in the se portion of East TX; Gulf Prairies and Marshes; se U.S from FL and GA w to TX. [*A. glaberrima* Woodson]

Amsonia tabernaemontana Walter, (for Jacob Theodore von Bergzabern, (d. 1590), Heidelberg botany professor who Latinized his name as Tabernaemontanus), EASTERN BLUESTAR. Stems erect (ascending), 27–102 cm long, glabrous to sparsely pubescent; leaves not heteromorphic, main stem leaf blades 3.3–12 cm long, 11–50 mm wide, lanceolate to narrowly lanceolate, ovate, narrowly elliptic, or elliptic, surfaces sparsely pubescent abaxially or glabrous; sepals deltate to narrowly deltate, glabrous to tomentose; corolla tube bluish (rarely to lilac), sometimes greenish above, 6–8 mm long, lobes pale blue (rarely to dark or lilac-blue or white), 6–10 mm long, outer surface of corolla with long hairs; follicles erect to pendulous 7–15 cm long, glabrous. Stream bottom thickets moist woods, wet meadows. Mar–Apr.

1. Leaf blades ovate to oblong-elliptic; inflorescence open, below to barely rising above the leaves, of 15–50 flowers; calyx glabrous; fruits 1–4 pairs per inflorescence ______ var. **tabernaemontana**
1. Leaf blades narrowly elliptic to linear-lanceolate; inflorescence congested, ascending above the leaves, from 50–80 flowers; calyx pubescent with long hairs; fruits 5–10 pairs per inflorescence ______ var. **repens**

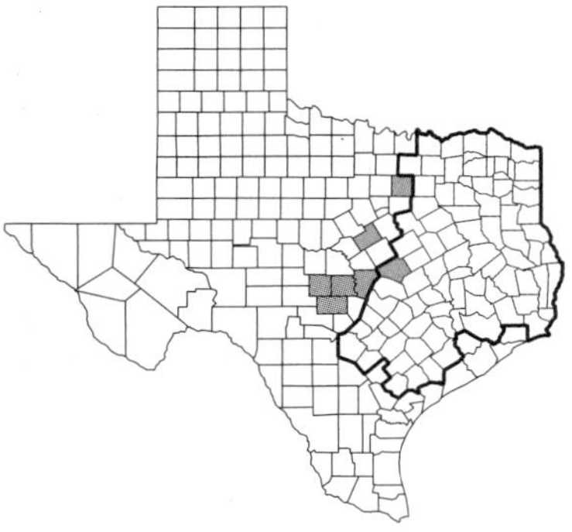

Spermolepis laevis

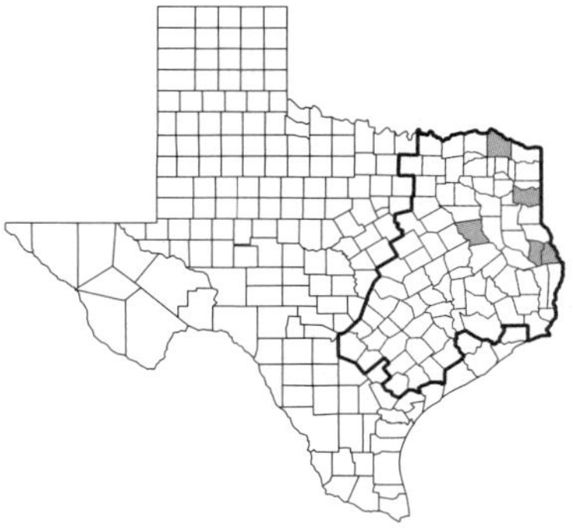

Taenidia integerrima

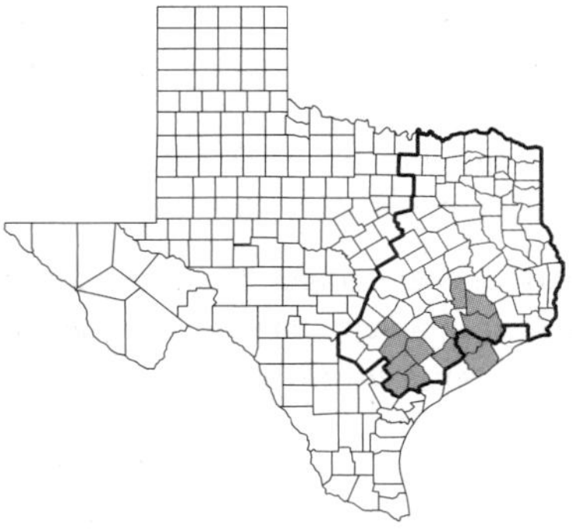

Tauschia texana

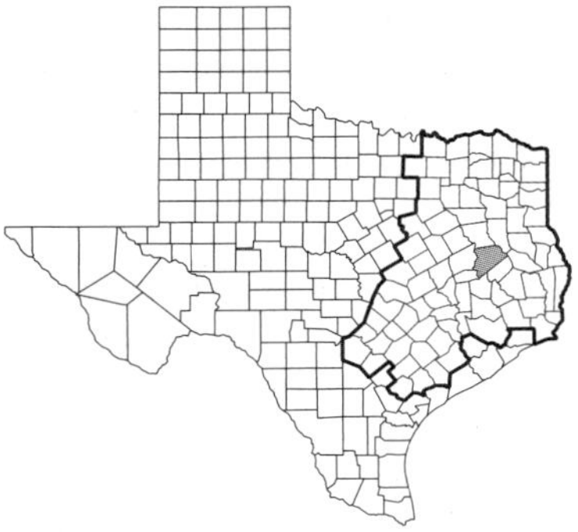

Thaspium chapmanii

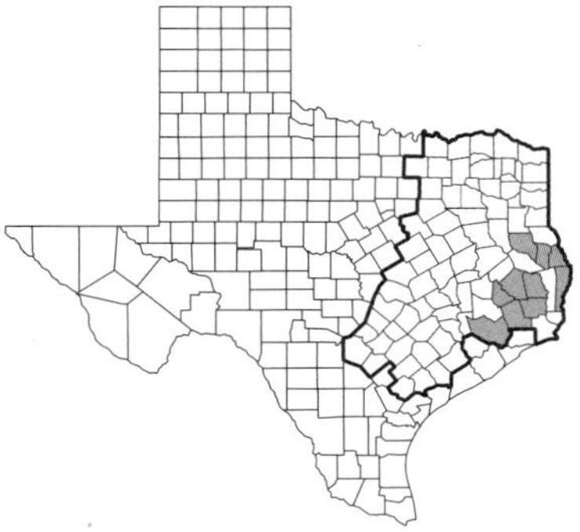

Thaspium trifoliatum var. aureum

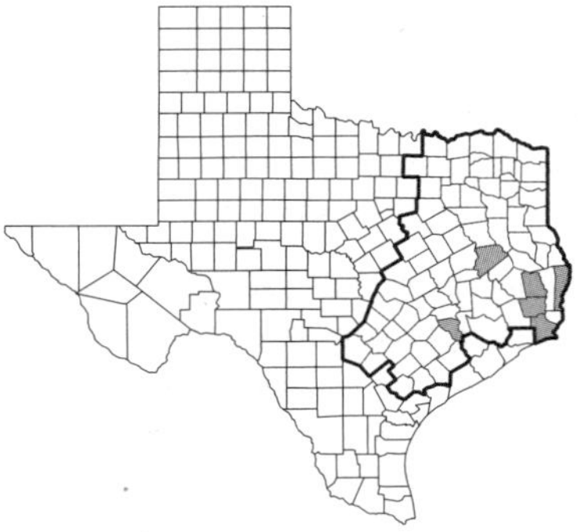

Tiedemannia filiformis

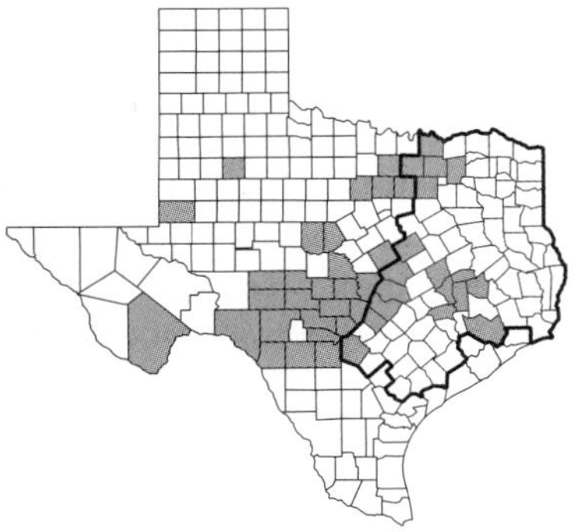

Torilis arvensis

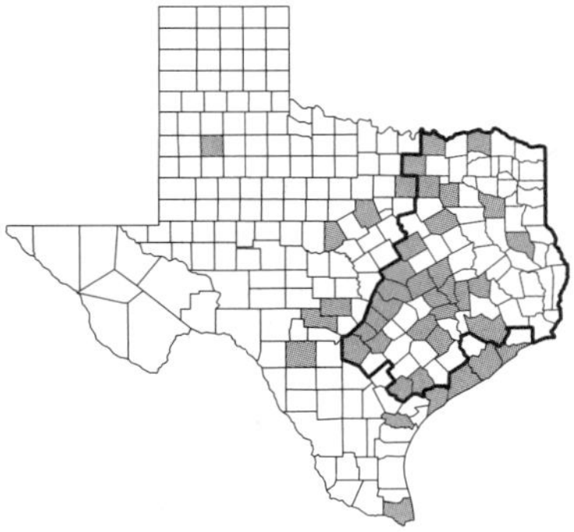

Torilis nodosa

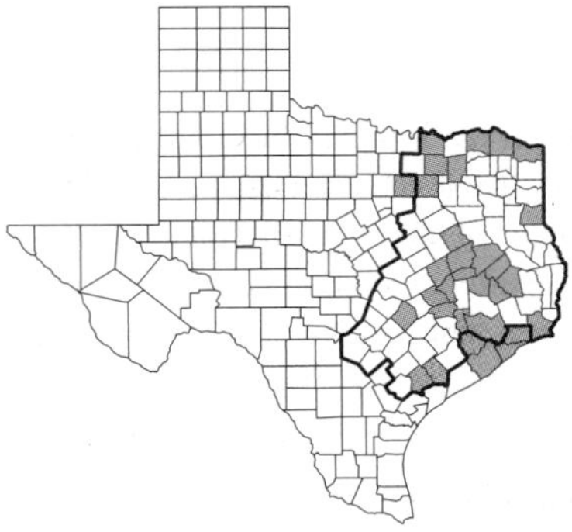

Trepocarpus aethusae

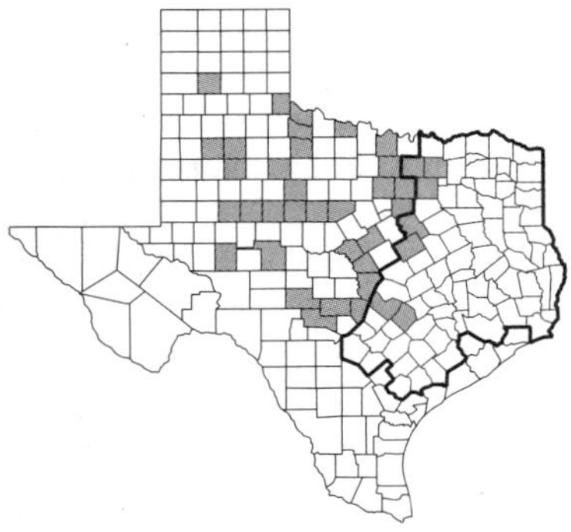

Vesper macrorhizus

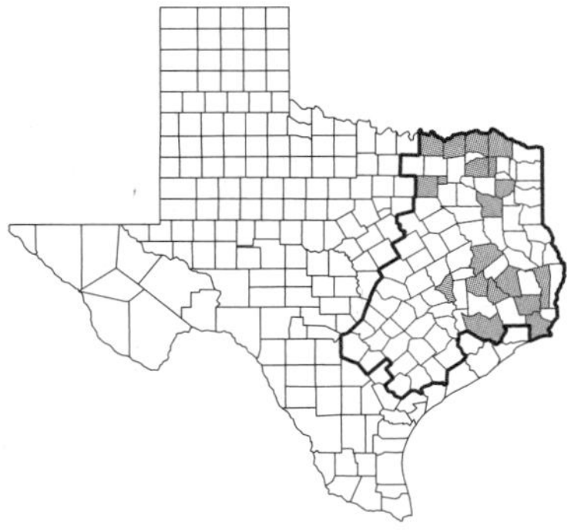

Zizia aurea

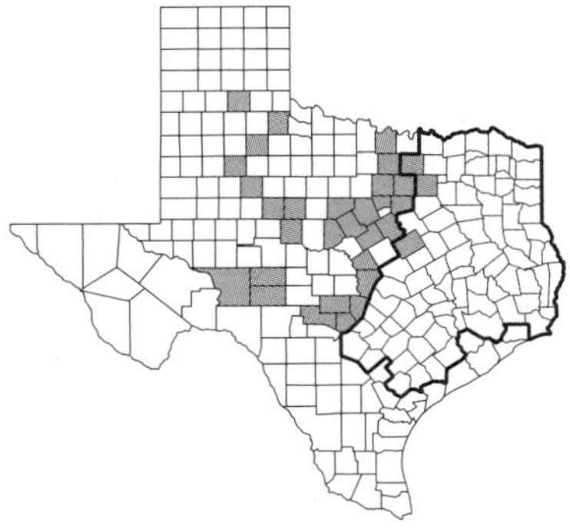

Amsonia ciliata var. texana

var. **repens** (Shinners) J.K. Williams, (creeping and rooting), CREEPING SLIMPOD. Main stem leaf blades (2.5–)3.3–7.4 cm long, (8–)11–21(–25) mm wide, often sparsely ciliate especially around apex; inflorescence congested, 50–80 flowers; sepals usually narrowly deltate, (0.6–)1.1–1.8(–2.3) mm long, pubescent with long, soft hairs, rarely glabrous; corolla tube (5.3–)6.5–7.5(–8) mm long, lobes (6.5–)7–9(–10.5) mm long; follicles erect, 11–15 cm long. Widespread in East TX; also East Cross Timbers and n Gulf Prairies and Marshes; OK and TX. [*A. repens* Shinners]

var. **tabernaemontana.** Main stem leaf blades (5.3–)8–12(–14) cm long, (17–)24–50 mm wide, margin almost always short-ciliate or minutely scabrous, sparsely pubescent mostly on midrib and veins abaxially; inflorescence somewhat open, 15–50 flowers; sepals usually deltate, 0.5–0.9(–1.1) mm long, glabrous (rarely bearing single long hairs); corolla tube (5–)6–7.5 mm long, lobes (5.5–)6.4–8.4(–9.5) mm long; follicles erect, (5–)7–12 cm long. Gregg, Harrison, and Wood (Turner et al. 2003) cos. in the ne part of East TX; se U.S from VA s to GA and w to OK and TX.

Amsonia hubrichtii Woodson, (for Leslie Hubricht, 1908–2005, American biologist and malacologist (one who studies mollusks)), HUBRICHT'S BLUESTAR. Similar to *Amsonia ciliata* var. *texana* in having a corolla glabrous on the outer surface and stems glabrous. But the leaf blades, both main stem and branch, in *A. hubrichtii* are linear and only slightly heteromorphic; whereas in *A. ciliata* var. *texana*, leaf blades are distinctly heteromorphic, with main stem leaf blades elliptic to narrowly elliptic and branch leaf blades linear to very narrowly elliptic. The main stem leaf blades in *A. hubrichtii* are longer, to 11 cm vs. to ca. 8.4 cm in *A. ciliata* var. *texana*. Low, wet woods, roadbanks, ditches; AR and OK. Included based on presence in McCurtain Co. OK (Kartesz 2022) on the ne border of East TX. Spring(–early summer).

Amsonia tabernaemontana Walter var. *illustris* (Woodson) J.K. Williams, (sp.: for Jacob Theodore von Bergzabern, (d. 1590), Heidelberg botany professor who Latinized his name as Tabernaemontanus); var.: excellent, superior). Similar to *Amsonia tabernaemontana* var. *repens* but, *A. tabernaemontana* var. *illustris* has calyx with ciliate hairs on the margins and/or the body of the calyx lobes and calyx tube glabrous vs. *A. tabernaemontana* var. *repens* has calyx tomentose with hairs on the calyx tube and calyx lobes glabrous. Also, *A. tabernaemontana* var. *illustris* has adaxial leaf blade midrib glabrous to glabrescent vs. leaf blade midrib villous in *A. tabernaemontana* var. *repens*. Gravel bars, stream beds and banks, wetlands, low woods; AL AR, IL, KS, KY, MO, OK, and TN; included based on presence mapped in Little River Co., AR (Kartesz 2022) on the far ne border of East TX. Spring. [*A. illustris* Woodson]

APOCYNUM L. DOGBANE, INDIAN-HEMP

A genus 3 species (including one hybrid) native to North America and Mexico (Lemke 2023a). Some have been used medicinally (some contain alkaloids and glycosides that affect the heart) or for their fiber (e.g., in rope, nets, sails, bags, clothing). The number of species in *Apocynum* have historically varied widely. As many as 250 species and infraspecific taxa have been named at one time. Woodson (1930) reduced the number to 7 species and 17 varieties, all in North America. Lemke (2023a) takes an even more conservative view with only 3 species (including 1 hybrid) (Greek: *apo*, far from, and *kyon or cyon*, a dog, from ancient use as dog poison and ancient name of the Old World DOGBANE)
REFERENCES: Woodson 1930; McGregor 1984b; Hartman 1986a; Bogler 2006; Lemke 2023a.

Apocynum cannabinum L., (resembling *Cannabis*–hemp), INDIAN-HEMP, HEMP DOGBANE, PRAIRIE DOGBANE, HAIRY DOGBANE, SMOOTH DOGBANE, WILLOW DOGBANE, CHOCTAWROOT. Herbaceous perennial with deep root, glabrous to pubescent; sap milky; stems erect to ascending, to 1 m tall; leaves usually opposite, the blades mostly more than 2.5 times as long as wide, variable in shape, linear-lanceolate, ovate to oval, sessile or short petiolate, narrowed to rounded or cordate; flowers in usually terminal branched inflorescences, 5-merous; corollas white, yellow, or greenish, 3–6 mm long, as long as or longer than calyx; follicles 4–22 cm long, pendulous at maturity, glabrous; seeds numerous, each terminated by a coma of long, silky hairs. Open or disturbed, often moist ground, sandy or less often gravelly or eroding clayey soils; nearly throughout TX, except South TX Plains; throughout the U.S.; Canada. Apr–Jul. [*A. cannabinum* L. var. *glaberrimum* A. DC., *A. cannabinum* var. *pubescens* (R. Br.) A. DC., *A. sibiricum* Jacq.] The common name, INDIAN-HEMP, results from the use of the plant as a source of fiber by Native Americans; it was widely used for making twine, rope, fishing lines, nets, bags, thread, etc.; early settlers also utilized the fiber (Cheatham & Johnston 1995). The root has been used as an emetic and cardiac stimulant (Mabberley 2017); however, it is poisonous to animals such as cats, dogs, and livestock due to the presence of the cardiac glycoside apocynamarin, resins, and other toxins; sickness and death have been reported in humans from medicinal use (Kingsbury 1964; Schmutz & Hamilton 1979; Fuller & McClintock 1986; Mulligan & Munro 1990). The species is also reported as sometimes being a "troublesome weed in cultivated fields, pastures, and gardens" (Bogler 2006).

ASCLEPIAS L. MILKWEED, SILKWEED

Perennial herbs; stems erect to reclining; leaves alternate, opposite, or whorled, sessile or petioled; flowers in terminal or lateral cymose umbels; corollas becoming reflexed or shallowly campanulate; hoods with or without an exserted appendage (= horn); fruit a follicle; seeds with a coma (= tuft of hairs).

A genus of ca. 130 species of the Americas, especially the United States (Fishbein 2023a); some are cultivated as ornamentals; others have been used medicinally since early times. Many if not most species are poisonous; however, all species are distasteful to livestock and severe losses usually occur only when animals are forced to eat the plants. Abundant MILKWEEDS in a pasture are often a sign of severe overgrazing. The poisonous principles include resinoids, cardiac glycosides, and alkaloids; the milky latex can cause dermatitis in susceptible individuals (Kingsbury 1964; Lewis & Elvin-Lewis 1977; Turner & Szczawinski 1991). Monarch butterflies (*Danaus plexippus*) feed as larvae on *Asclepias* species and obtain cardiac glycosides which provide them with protection from bird predators; the butterflies are poisonous only if they fed on poisonous plants as larvae (Scott 1986). The long silky hairs on the seeds were formerly used in making candle wicks (Ajilvsgi 1984). (Named from Greek *Asklepios*, god of medicine, alluding to its medicinal properties)

REFERENCES: Woodson 1947, 1953, 1954, 1962; Edwards & Wyatt 1994; White 2008; Fishbein 2023a.

1. Midstem leaf blades linear to linear-lanceolate, 10 times as long as wide or more.
  2. Leaves mostly in whorls of 3 or 4.
    3. Plants with 1–8 stems; reduced branchlets bearing pairs or clusters of short leaves in the axils of many main leaves; midstem leaf blades 0.1–0.4 cm wide, glabrous on both surfaces; anthers 1.5–2 mm long; hood margin entire ____ **A. subverticillata**
    3. Plants rarely with more than one stem; axillary branchlets absent or rare; midstem leaf blades 0.1–0.2 cm wide, adaxially puberulent to glabrate; anthers 1.2–1.5 mm long; hood margin shallowly lobed to sharply toothed ____ **A. verticillata**
  2. Leaves opposite, subopposite, or alternate.
    4. Leaves drooping; stems glabrous; corolla lobes adaxially pale green; hoods without horns ____ **A. engelmanniana**
    4. Leaves not drooping; stems puberulent or if glabrous then corolla lobes adaxially red to bright orange; hoods with or without horns, when present the horns free OR horns adnate with hood bodies and visible as the middle lobe of the hood.
      5. Hoods without horns.
        6. Flowers 34–112 per umbel; pedicels and calyces hirtellous with straight hairs; hoods 1.5–2 mm long, not exceeding the point of the anther wing; follicles 1–2 cm wide ____ **A. hirtella**
        6. Flowers 13–28(–36) flowers per umbel; pedicels and calyces puberulent with curved hairs; hoods 2–2.5 mm long, mostly reaching to the point of the anther wing; follicles 0.7–1.2 cm wide ____ **A. longifolia**
      5. Hoods with horns free or adnate with hood bodies.
        7. Leaves opposite; horns free, longer or shorter than hood bodies.
          8. Horns longer than hood bodies ____ **A. linearis**
          8. Horns shorter than hood bodies ____ **A. lanceolata**
        7. Leaves alternate; horns adnate and visible as the middle lobe of the hood ____ **A. stenophylla**
1. Midstem leaf blades narrowly lanceolate to lanceolate, ovate to oblong-orbicular, < 10 times as long as wide (sometimes narrower in *A. asperula*), often only 1.5–5 times as long as wide.
  9. Leaves opposite.
    10. Umbels terminating the stems or some in upper leaf axils, held erect (*A. incarnata*, *A. perennis*, and *A. texana* sometimes may appear lateral); peduncles to 7 cm long (to 13 cm in *A. rubra*; to 30 cm in *A. amplexicaulis*)
      11. Corollas greenish, tinged with rose or purple; umbels long-peduncled, the peduncle greatly exceeding the upper leaves; leaf blades glaucous, clasping the stem; apex mostly rounded to broadly rounded ____ **A. amplexicaulis**
      11. Corollas white, white tinged with various shades of pink, or corollas reddish purple; peduncles various; leaf blades variously pubescent to glabrate, not clasping the stem; apex rounded, obtuse to acute to acuminate.
        12. Midstem leaf blades 4–10 cm wide; corollas white or reddish purple, the lobes 6–10 mm long.
          13. Corollas white, lobes 6–8 mm long; hoods 2.5–4 mm long ____ **A. variegata**
          13. Corollas reddish purple; lobes 7–10 mm long; hoods 5–6 mm long ____ **A. purpurascens**
        12. Midstem leaf blades mostly 4 cm or less wide; corollas white or various shades of pink or reddish purple, the lobes 3–6 mm long OR 7–10 mm long.
          14. Corolla lobes 7–10 mm long; hoods 5–7 mm long, light pink to bright red.
            15. Leaf apices mostly acute to obtuse; hoods usually bright red, 5–6 mm long ____ **A. purpurascens**
            15. Leaf apices mostly acuminate; hoods usually white to light pink, hoods 6–7 mm long ____ **A. rubra**
          14. Corolla lobes 3–6 mm long; hoods 1.5–3 mm long; white to pink tinged.

16. Corollas light to bright pink, lobes (4.5–)5–6 mm long; hoods 2–2.5 mm long; larger leaf blades often abruptly narrowed or truncate basally; stems stout, 0.3–1.5 m tall; inflorescences usually paired at upper nodes ____ **A. incarnata**
16. Corollas white (sometimes tinged with pink); lobes 3–5 mm long; hoods 1.5–3 mm long; larger leaf blades gradually or abruptly narrowed basally; stems slender, 0.3–0.6 m tall; inflorescences solitary at upper nodes.
    17. Corolla lobes 3–4 mm long; hoods 1.5–2.5 mm long; follicles pendulous; seeds naked, without a coma; midstem leaf blades narrowly lanceolate to lanceolate, to 15(–20) mm wide; mostly in low swampy areas ____ **A. perennis**
    17. Corolla lobes 3.5–5 mm long; hoods 2–3 mm long; follicles erect on erect pedicels; seeds with a coma of hairs to ca. 20 mm long; midstem leaf blades lanceolate to ovate or elliptic, to 30 mm wide; mostly in rocky shaded areas ____ **A. texana**

10. Umbels terminating the stems or some in upper leaf axils, held erect (*A. incarnata*, *A. perennis*, and *A. texana* sometimes may appear lateral); peduncles to 7 cm long (to 13 cm in *A. rubra*; to 30 cm in *A. amplexicaulis*)
10. Umbels lateral, in leaf axils (but often in upper axils), held in a horizontal or oblique manner; peduncles to 4 cm long, usually less (to 8 cm long in *A. curassavica*).
    18. Calyx lobes and pedicels slightly to conspicuously white tomentose (= long tangled hairs); stems and lower surface of leaf blades slightly or persistently and conspicuously white-tomentose to glabrate in age.
        19. Stems, lower leaf surface, and particularly pedicels and calyx persistently and conspicuously tomentose; petioles 7–17 mm long; leaf blade margins parallel to slightly diverging toward the apex ____ **A. arenaria**
        19. Stems and lower leaf surfaces slightly tomentose to glabrate in age; pedicels and calyx slightly tomentose; petioles 0–4 mm long; leaf blade margins rounded ____ **A. latifolia**
    18. Calyx lobes, pedicels, stems, and lower surface of leaf blades variously pubescent; glabrate to puberulent, hirtellous, velutinous or pilose, hairs straight or curved but not white tomentose.
        20. Hoods without horns; hoods slightly shorter than gynostegium; inflorescences 22–60-flowered; stems solitary; corolla lobes green to yellowish green ____ **A. viridiflora**
        20. Hoods with horns; hoods longer or shorter than gynostegium; inflorescences 4–37-flowered; stems solitary OR 1–7; corolla lobes red or greenish or yellowish, sometimes tinged with a darker color.
            21. Hood length ≤ to gynostegium length; hoods 3–4 mm long; corolla lobes 7–9 mm long ____ **A. tomentosa**
            21. Hoods longer than gynostegium; hoods 3–10 mm long; corolla lobes 5–9 mm long.
                22. Hoods 3.5–5.5 mm long; corolla lobes red or green or green tinged with red or brown, 5–9 mm long.
                    23. Corolla lobes green sometimes tinged with red or brown; hoods much longer that gynostegium; horns somewhat obscure; inflorescences 4–8-flowered ____ **A. emoryi**
                    23. Corolla lobes red; hoods only slightly longer than gynostegium; horns distinct; inflorescences 5–22-flowered ____ **A. curassavica**
                22. Hoods 5–10 mm long; corolla lobes greenish white, yellow or green tinged with red, 7–12 mm long.
                    24. Plants with rhizomes; corolla lobes greenish white or yellow, (7–)9–12 mm long; hoods (5–) 7–10 mm long; horns emerging at the top of the hoods ____ **A. oenotheroides**
                    24. Plants without rhizomes; corolla lobes green tinged with red, 7–9 mm long; hoods 5–8 mm long; horns emerging from hoods well below the top ____ **A. obovata**

9. Leaves, except lowest, alternate.
    25. Stems pilose; corollas yellow, orange, or red; hoods with horns; plants without milky latex ____ **A. tuberosa**
    25. Stems glabrous or short pubescent; corollas green to white; hoods without horns; plants with milky latex; (previously segregated as genus *Asclepiodora*).
        26. Corolla lobes 7–12 mm long; midstem leaf blades 0.4–3.7 cm wide, narrowly oblong or ovate-lanceolate, the apex narrowly acuminate; anther head much wider than tall ____ **A. asperula**
        26. Corolla lobes 12–15 mm long; midstem leaf blades 1.5–6 cm wide, lanceolate, narrowly ovate to broadly oblong, the apex usually obtuse to acute; anther head ca. as wide as tall ____ **A. viridis**

Asclepias amplexicaulis Sm., (clasping the stem), BLUNT-LEAF MILKWEED. Plant glabrous; stems erect, 35–100+ cm tall; leaves opposite, upper leaves sessile, clasping, margins ruffled; midstem leaf blades 6–14 cm long, 3–7 cm wide, ovate to elliptic-oblong, apically rounded to broadly rounded and often mucronate; corolla lobes 8–11 mm long, greenish, tinged with rose or purple; gynostegium purple; hoods reddish purple to cream, with horns. Sandy open woods, roadsides, and old fields; throughout the e 1/2 of East TX, also in the sw portion; throughout the e 1/2 of the U.S. Apr–May.

Asclepias arenaria Torr., (of sand), SAND MILKWEED. Stems solitary, 20–100 cm tall, pubescent; leaves opposite, petioles usually 7–17 mm long; midstem leaf blades 5–12 cm long, 3–8 cm wide, oblong or obovate to ovate or oval; corolla lobes 7–8 mm long, pale green, wide-spreading; gynostegium green; hoods with horns. Sandy soils, grasslands, riparian areas, roadsides; Anderson, Bastrop, Freestone, Lee, and Leon (Kartesz 2022) cos; c U.S. from SD s to TX and w to WY and NM. Jun–Aug.

Trepocarpus aethusae [LUN]

Vesper macrorhizus [LUN]

Zizia aurea [BB1]

Amsonia cilitata var. texana [SH1]

Amsonia longiflora var. salpignantha [AMB, HEA]

Amsonia ludoviciana [GOR]

Amsonia rigida [GOR]

Amsonia tabernaemontana [HEA]

Apocynum cannabinum [REE]

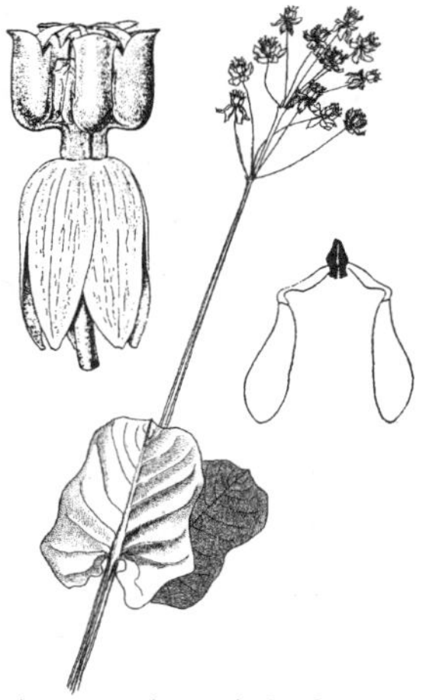

Asclepias amplexicaulis [AMB]

Asclepias asperula (Decne.) Woodson subsp. **capricornu** (Woodson) Woodson, (sp.: rough; subsp.: like a goat's horn), ANTELOPE-HORNS, TRAILING MILKWEED. Stems partly decumbent or low-spreading, 15–60 cm long, minutely pubescent or largely glabrous; petioles 2–4 mm long, midstem leaf blades 5–17 cm long, 0.4–3.7 cm wide, lanceolate to linear; corolla lobes pale yellowish green, sometimes purple-tinged; gynostegium greenish cream to dark purple; hoods wide-spreading, without horns. Rocky or sandy prairies, hills, stream banks; mostly in the w 1/3 of East TX; KS, NE, NM, OK, and TX. Apr–Jun, sporadically through summer, fall. [*A. asperula* var. *decumbens* (Nutt.) Shinners, *A. capricornu* Woodson, *Asclepiodora decumbens* (Nutt.) A. Gray] The common name ANTELOPE-HORNS is said to come from the curved fruits (Kirkpatrick 1992). Reportedly poisonous to livestock (Hartman 1986b).

Asclepias curassavica L., (from Curaçao), BLOODFLOWER. Stems 30–150 cm long; petioles 4–25 mm long, midstem leaf blades 4–18 cm long, 1–4.5 cm wide, elliptic or oval to narrowly lanceolate; corolla bright red or sometimes yellow; gynostegium yellow; hoods orange, with horns; follicles spindle-shaped, 6–10 cm long. Disturbed sites, moist or wet soil; Brazos, Harris, and Jefferson (Kartesz 2022) cos.; also extreme s TX; se U.S. from TN s to FL and w to AR and TX, mostly scattered except throughout FL. Apr–Aug. Native of the Neotropics. Often an escape.

Asclepias emoryi (Greene) Vail, (named for William H. Emory, 1811–1887, prominent American surveyor and civil engineer), EMORY'S COMET MILKWEED. Stems erect to spreading, 6–30 cm long; leaves sessile or petioles to 17 mm long, midstem leaf blades 3–7.5 cm long, 0.4–2.5 cm wide, elliptic to lanceolate or lance-ovate; corolla lobes 5–7 mm long, green, sometimes tinged with red or brown; gynostegium pale greenish to cream; hoods greenish proximally, white distally, with horns. Hills, slopes, rocky or gravelly soils; in East TX known only from Bexar and Guadalupe (Turner et al. 2003) cos.; South TX Plains and Edwards Plateau; in the U.S known only from TX; also Mexico. Mar–Sep.

Asclepias engelmanniana Woodson, (for George Engelmann, 1809–1884, German-born botanist and physician of St. Louis), ENGELMANN'S MILKWEED. Stems erect, 40–160 cm tall, glabrous or nearly so; leaves sessile, midstem leaf blades 5–19 cm long, 0.1–0.3(–0.5) cm wide; linear; corolla lobes 4–5 mm long, pale green, purple-tinged; gynostegium yellowish; hoods without horns. Sandy or rocky ground; Bexar, Collin, Comal, and Robertson (Kartesz 2022) cos.; w 2/3 of TX; c U.S. from NE s to TX and w to WY and NM. Jun–Sep.

Asclepias hirtella (Pennell) Woodson, (minutely hairy), GREEN MILKWEED. Stems 30–125 cm long, pubescent; leaves opposite or alternate, sessile or petioles to 3 mm long, midstem leaf blades 4–16 cm long, 0.2–1.5 cm wide, linear to linear-lanceolate; corolla lobes 3–5 mm long, green to greenish cream, reddish maroon at tips; gynostegium green; hoods greenish cream, without horns. Prairies, open woods, sandy soil, roadsides; mostly in the se portion of East TX; ec U.S. from WV s to GA and w to MN, KS and TX; also adjacent Canada. May–Sep. [*A. longifolia* var. *hirtella* (Pennell) B.L. Turner]

Asclepias incarnata L., (flesh-colored), SWAMP MILKWEED. Stems erect, usually 30–150 cm tall; petioles 1–15 mm long, midstem leaf blades 5–15 cm long, to ca. 4 cm wide, narrowly lanceolate or ovate-elliptic; corolla lobes (4.5–)5–6 mm long, bright pink (rarely white); gynostegium light brown to green; hoods pink to white, with horns. Wet ground. May–Oct.

1. Plants glabrous or nearly so ______ subsp. **incarnata**
1. Plants pubescent ______ subsp. **pulchra**

subsp. **incarnata.** Dallas, Gonzales, Guadalupe, Leon, and Wood (Kartesz 2022) cos.; Edwards Plateau, High Plains; mostly n 2/3 of U.S. from ME s to GA and w to MT and AZ, but throughout ME s to KY and w to ND and KS.

subsp. **pulchra** (Ehrh. ex Willd.) Woodson, (handsome). Cherokee, Hood, Leon, and Madison (Kartesz 2022) cos.; e U.S. ME s to FL and w to TN and AL, TX; also Canada [*A. incarnata* L. var. *pulchra* (Ehrh. ex Willd.) Pers.]

Asclepias lanceolata Walter, (lance-like, referring to the narrow leaves), LANCE-LEAF MILKWEED, FEW-FLOWER MILKWEED. Stems erect, 60–125 cm tall; leaves essentially sessile, midstem leaf blades 7–25 cm long, 0.2–1.2(–1.5) cm wide, linear to linear-lanceolate; inflorescences terminal but also sometimes extra-axillary at an upper node, peduncle 1–7.5 cm long; corolla lobes 9–10 mm long, red; gynostegium green, tinged yellow to red; hoods yellow to reddish orange, with horns. Marshes, wet pine barrens, riparian woods; Hardin, Harris, Jasper, Jefferson, Newton, and Orange (Kartesz 2022) cos.; mostly coastal U.S. from NJ s to FL and w to TX. May–Aug.

Asclepias latifolia (Torr.) Raf., (with broad leaves), BROAD-LEAF MILKWEED. Stems erect, 25–100 cm tall; leaves sessile or petioles to 4 mm long, midstem leaf blades 5.5–14 cm long, 3–14 cm wide, oval, oblong, or orbiculate; corolla lobes green, 7–9 mm long; gynostegium green; hoods cream to yellow, with horns. Prairies, roadsides, railways, sandy soils; in East TX known only from

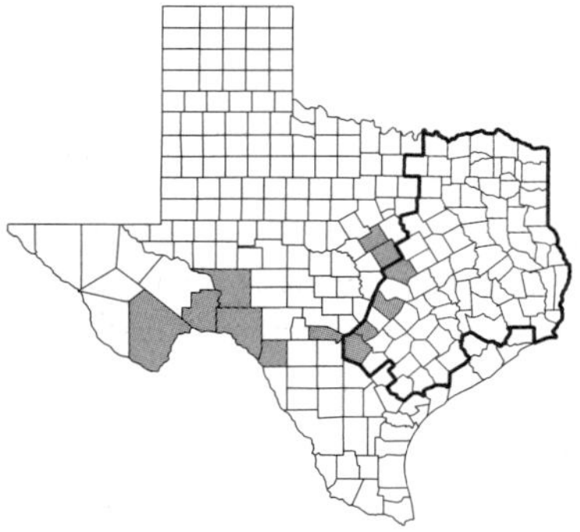
Amsonia longiflora var. salpignantha

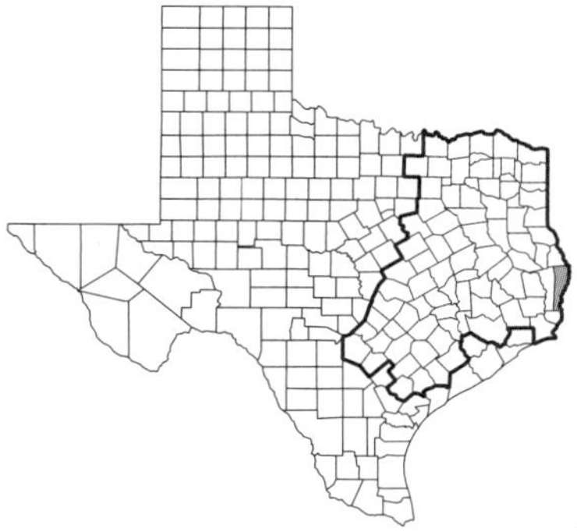
Amsonia ludoviciana

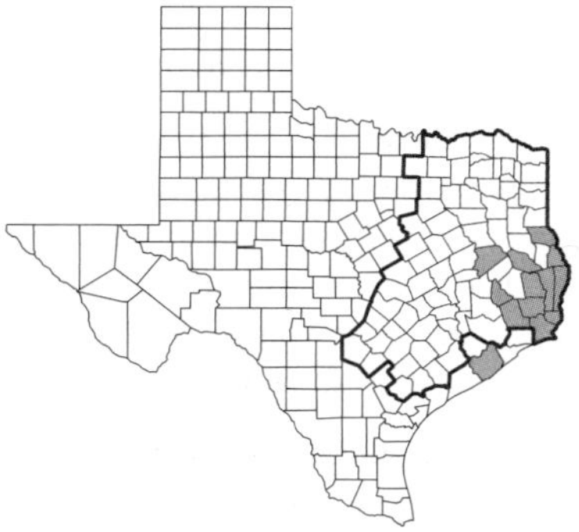
Amsonia rigida

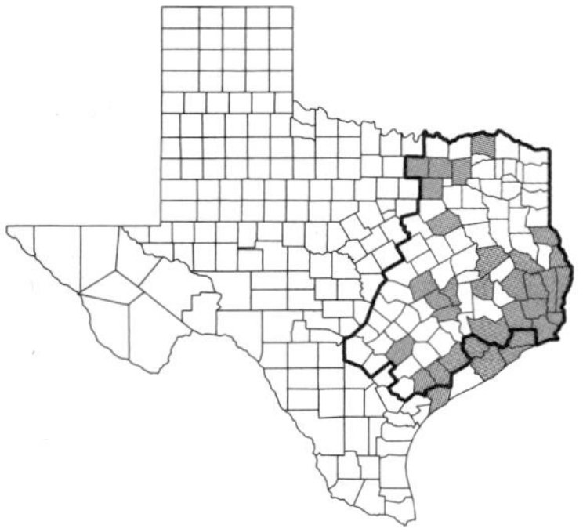
Amsonia tabernaemontana var. repens

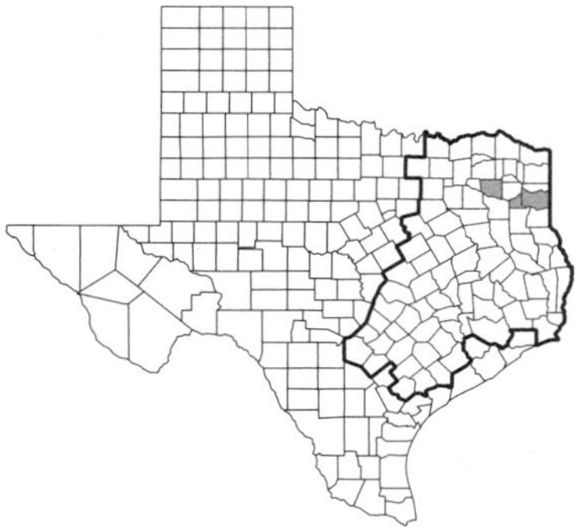
Amsonia tabernaemontana var. tabernaemontana

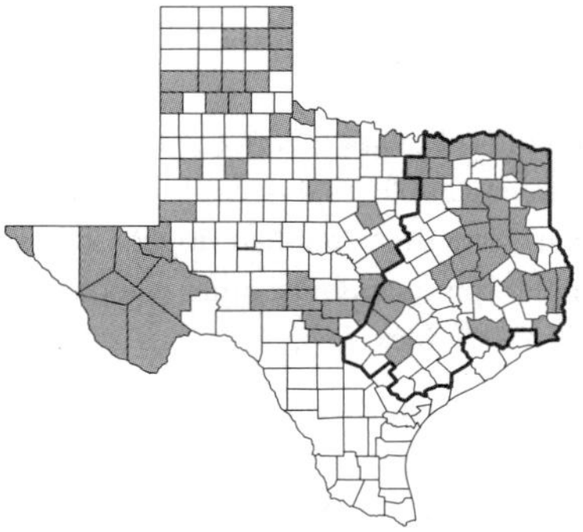
Apocynum cannabinum

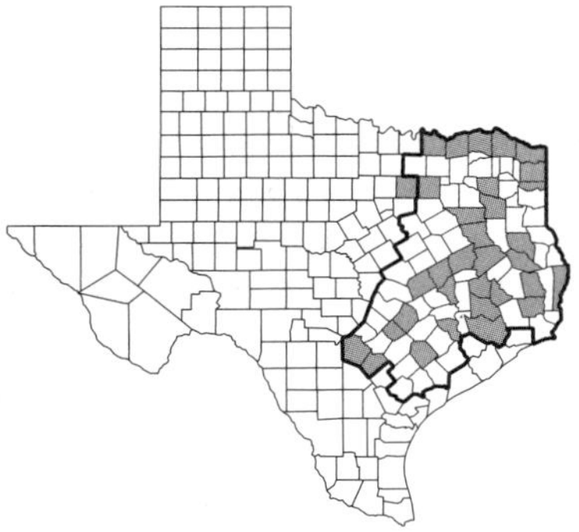
Asclepias amplexicaulis

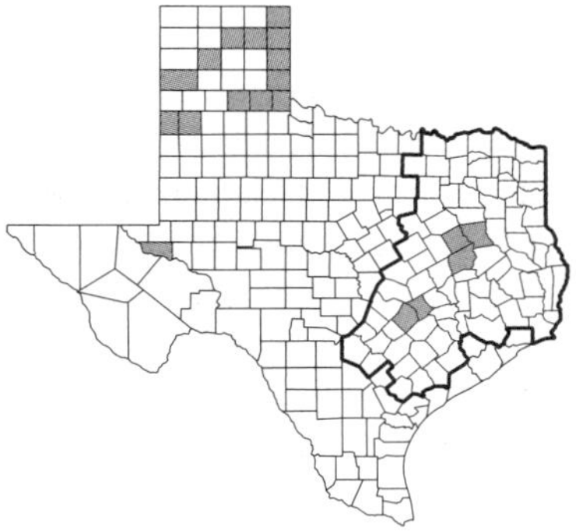
Asclepias arenaria

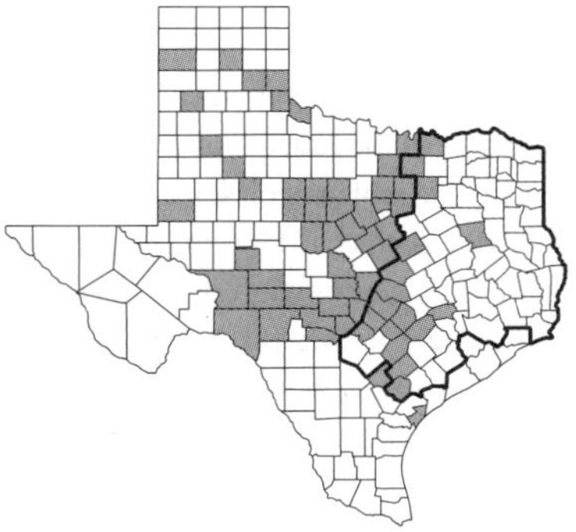
Asclepias asperula subsp. capricornu

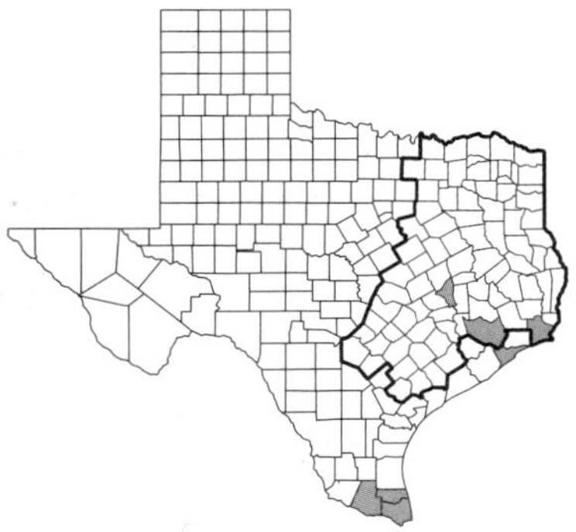
Asclepias curassavica

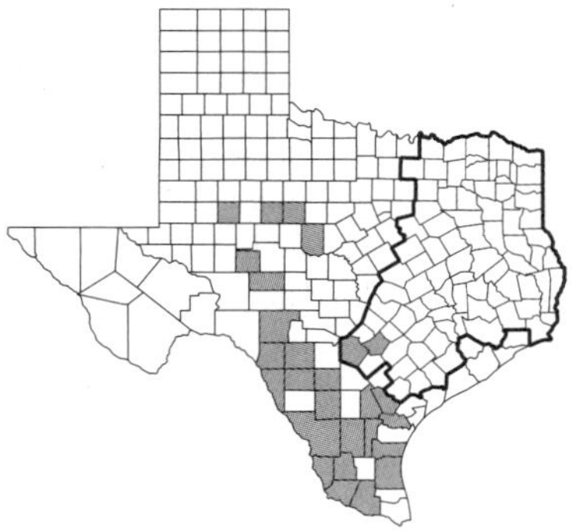
Asclepias emoryi

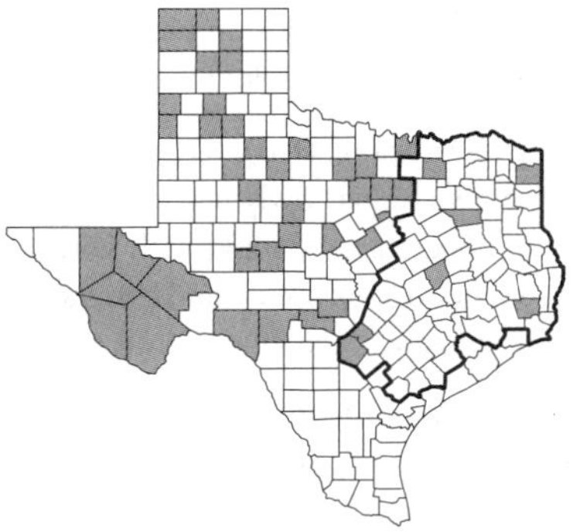
Asclepias engelmanniana

Bastrop, Bexar, Leon, and Travis (Turner et al. 2003) cos.; Edwards Plateau, Trans-Pecos and High Plains; sw U.S. quadrant from NE s to TX and w to UT and AZ, also CA. May–Aug.

Asclepias linearis Scheele, (narrow, with sides nearly parallel), SLIM MILKWEED. Stems erect, usually 20–50 cm tall, glabrous or inconspicuously pubescent; leaves sessile or subsessile; midstem leaf blades 3–8 cm long, 0.1–0.4 cm wide, linear; umbels axillary on upper part of stem; corolla lobes 3–4 mm long, greenish white (rarely brownish or purplish); gynostegium pinkish white; hoods cream, with horns longer than hoods. Dry prairies, ditches, pastures; collected at Buzzards Spring, Dallas, in October 1902 by Reverchon (Mahler 1988), also Milam Co. (Woodson 1954), scattered mostly in the s portion of East TX; Gulf Prairies and Marshes, and South TX Plains; in the U.S. known only from TX; also n Mexico. May–Oct.

Asclepias longifolia Michx., (having long leaves), LONG-LEAF MILKWEED. Stems 1–few, spreading to decumbent, unbranched, 25–70 cm long; leaves sessile or petioles to 2 mm long, midstem leaf blades 5–18 cm long, 0.2–1 cm wide, linear to linear-lanceolate; inflorescence terminal and at upper nodes; corolla lobes 3–5 mm long, white to greenish cream, purple at tips; gynostegium green or brown, longer than hoods; hoods pinkish to lavender to red-violet, often white apically. Bogs, swamps, flats, depressions; in East TX known only from Jasper and Newton cos. (Turner 2009); e U.S. from MD s to FL and w to TX. (Jan–)Feb–Sep(–Oct).

Asclepias obovata Elliott, (obovate), PINELAND MILKWEED. Stems erect to spreading, 40–70(–200) cm long; petioles 1–4 mm long, midstem leaf blades 4–9 cm long, 1–3.5 cm wide, oblong or elliptic to obovate or ovate; corolla lobes 7–9 mm long, usually green, sometimes reddish tinged; gynostegium greenish to brown; hoods dull reddish brown, lighter distally, with horns. Longleaf pine sandhills, other woodlands and savannas; mostly in the se portion of East TX; se U.S. from SC s to FL and w to OK and TX. May–Sep.

Asclepias oenotheroides Schltdl. & Cham., (resembling *Oenothera*—evening primrose), SIDE-CLUSTER MILKWEED, HIERBA DE ZIZOTES. Stems erect or spreading, 10–45 cm long, minutely scabrous-pubescent; petioles 2–20 mm long, midstem leaf blades 2–11 cm long, 1.5–6 cm wide, ovate or oblong or elliptic, from broadly to narrowly so; corolla lobes (7–)9–12 mm long, greenish white or yellow; gynostegium greenish brown; hoods green, often light tan and yellowish, with horns. Sandy or gravelly ground, especially disturbed places; widespread in the sw quadrant of East TX; throughout TX except for far East TX; AZ, CO, LA, NM, OK, and TX. Apr–Aug.

Asclepias perennis Walter, (perennial), SHORE MILKWEED, THIN-LEAF MILKWEED. Stems erect, 30–50 cm tall; plant glabrous except for upper stem, peduncles, and pedicels; petioles to 12 mm long, midstem leaf blades 5–14 cm long, 1.5(–2) cm wide, narrowly to broadly lanceolate; corolla lobes white, usually suffused with pale pink or lavender; gynostegium pale rose to brown; hoods with horns; seeds ca. 15 mm long. Damp woods, swampy areas; collected by Reverchon at Dallas, not found there recently (Mahler 1988); se portion of East TX; n Gulf Prairies and Marshes; se U.S. from SC s to FL and w to AR and TX, also IL, IN, KY, MO, and TN. Jun–Jul. According to Woodson (1954), the pendulous follicles and large naked seeds may be adaptations to dispersal by water. Woodson (1954) indicated that he could not distinguish this species from *A. texana* except by the strikingly different fruit and seed, and the geographical distribution.

Asclepias purpurascens L., (purplish), PURPLE MILKWEED. Stems erect, 50–120 cm tall; petioles 4–18 mm long, midstem leaf blades 6–20 cm long, 2–10 cm wide, ovate or oval to lanceolate or elliptic; corolla lobes 7–10 mm long, light red to reddish purple; gynostegium brown; hoods 5–6 mm long, longer than gynostegium, with horns. Thickets, moist woods, pond edges, roadsides, prairies, fields; in East TX known only from Bowie Co. (*Correll 18829*, *18853* SMU) collected on the same date (24 May 1958) in two different localities; e U.S. from NH s to GA and w to MN and TX, except for AL, FL, and SC. May–Jun.

Asclepias rubra L., (red), RED MILKWEED. Stems erect, 40–110 cm tall; petioles 1–2 mm long, midstem leaf blades 5–16 cm long, 1–4.5 cm wide, narrowly lanceolate to ovate; corolla lobes 7–9 mm long, light to dark pink; gynostegium green to greenish brown; hoods usually pink, with horns. Similar to *A. lanceolata* but *A. rubra* has hood apex acute, and corollas light to dark pink, vs. hood apex obtuse and corollas orange, yellow, and red in *A. lanceolata*. Bogs, baygalls, marshes, wet meadows, low pine barrens; mostly in the Pineywoods region of East TX; e U.S. from NY s to FL and w to TX. Jun–Aug.

Asclepias stenophylla A. Gray, (narrow-leaved), SLIM-LEAF MILKWEED, NARROW-LEAF MILKWEED. Stems erect to spreading, 15–85 cm long, minutely puberulent to glabrate; leaves sessile; midstem leaf blades usually 5–16 cm long, 0.1–0.5 cm wide, linear; corolla lobes pale greenish cream or yellow, 3–5 mm long; gynostegium pale greenish; hoods cream, often tinged with green, 3–4 mm long, with horns, horn appears as a middle lobe at summit of hood. Prairies or open ground; widely scattered in East TX; Trans-Pecos, scattered in Cross Timbers and Prairies, and High Plains; mostly in the Great Plains. Jun–Aug.

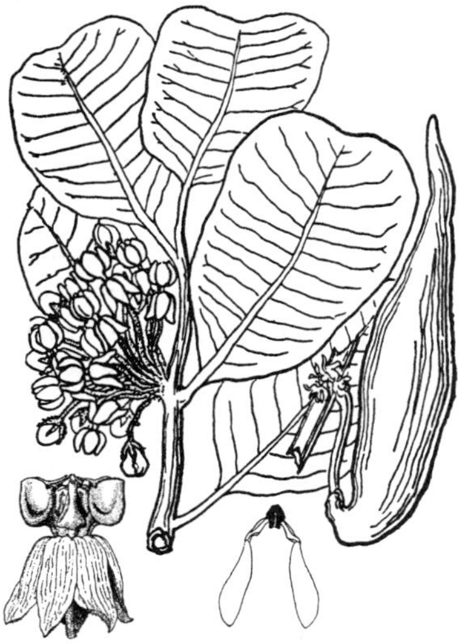

Asclepias arenaria [AMB, BB2]

Asclepias asperula subsp. capricornu [AMB, BT3]

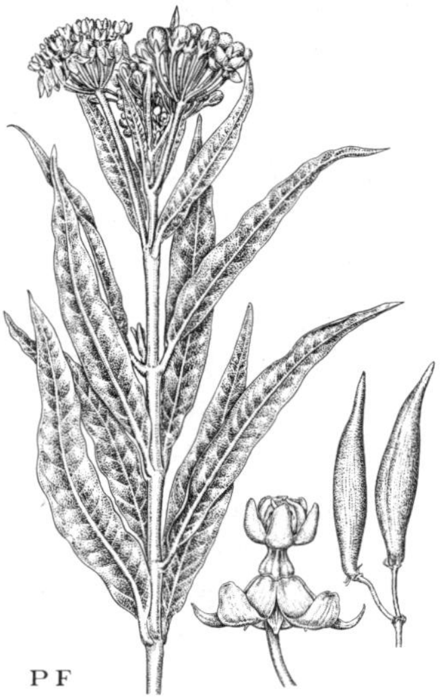

Asclepias curassavica [CO3]

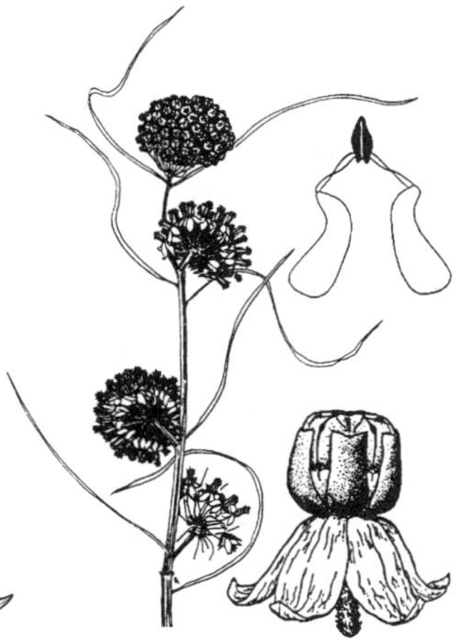

Asclepias engelmanniana [AMB, SHI]

Asclepias hirtella [YAT]

Asclepias emoryi [NAZ]

Asclepias incarnata subsp. incarnata [AMB, COR]

Asclepias subverticillata (A. Gray) Vail, (almost whorled), HORSETAIL MILKWEED. Stems erect 20–90 cm tall; leaves whorled, 3–4, midstem leaf blades 3–13 cm long, 0.1–0.4 cm wide, linear; corolla 3.5–4.5 mm long, pale green to cream, sometimes pink- or tan-tinged; gynostegium yellow to greenish; hoods cream, sometimes green- or pink-tinged, with horns, exerted portion of horn as long as or longer than hood. Sandy and rocky grounds, roadsides, riparian areas; in East TX known only from Grayson (Turner et al. 2003), Hays (*Stanfield s.n.*, 1897, NYBG), and Victoria (Woodson 1954) cos.; Trans-Pecos and High Plains; w U.S. from NE s to TX and w to WY and AZ. May–Sep.

Asclepias texana A. Heller, (of Texas), TEXAS MILKWEED. Stems erect, to 25–90 cm tall, with inconspicuous pubescence in lines; midstem leaf blades 2–7 cm long, 0.5–3 cm wide, lanceolate to ovate or elliptic, inflorescences solitary at the uppermost nodes; flowers showy; corolla lobes white, 3.5–5 mm long; gynostegium brown; hoods white, with horns; follicles erect, 9–13 cm long; seeds ca. 8 mm long, with a white coma to ca. 20 mm long. Rocky shaded areas, stream banks, oak woods; along the sw margin of East TX; mainly Edwards Plateau and Trans-Pecos; in the U.S. known only from TX; also Mexico. Jun–Aug.

Asclepias tomentosa Elliott, (densely pubescent), TUBA MILKWEED. Stems erect, 25–150 cm tall, densely puberulent; petioles 2–9 mm long, midstem leaf blades 3.5–10 cm long, 1–5 cm wide, lanceolate to ovate, oblong to elliptic, or obovate, margins ruffled, puberulent to densely puberulent; corolla lobes 7–9 mm long, green, often red- or purple-tinged; gynostegium green; hoods green with cream apex, often pink- or purple-tinged, with horns. Sandy soil, pine barrens, longleaf pine sand ridges; on the w edge of the Pineywoods region of East TX; FL, GA, NC, SC, and TX. May–Jul.

Asclepias tuberosa L. subsp. **interior** Woodson, (sp.: tuberous; subsp.: presumably referring to its range in the interior of the North American continent), BUTTERFLY-WEED, BUTTERFLY MILKWEED, ORANGE MILKWEED, CHIGGER-FLOWER, CHIGGER-WEED, PLEURISY-ROOT. Plant without milky latex; stems spreading to erect, to 90 cm long, pilose; leaves many, crowded, midstem leaf blades 2–12 cm long, 0.5–3 cm wide, narrowly elliptic, or lanceolate to oblong, sometimes linear, ± sessile and clasping or petioles to 4 mm long; corolla lobes usually orange (rarely yellowish or reddish); gynostegium usually orange to brownish, occasionally yellowish; hoods reddish orange or yellow, with horns. Sandy open woods or occasionally silty clay prairies; throughout East TX; Edwards Plateau, Cross Timbers and Prairies, High Plains, and Trans-Pecos; throughout much of the e U.S., also AZ, CO, NM, and UT; se Canada. May–Jul. [*A. tuberosa* var. *interior* (Woodson) Shinners, *A. tuberosa* L. subsp. *terminalis* Woodson] Reportedly poisonous to livestock (Hartman 1986b).

Asclepias variegata L., (variegated, irregularly colored), WHITE-FLOWER MILKWEED. Plant glabrous except for upper stem, peduncles, and pedicels; stems erect, 30–120 cm tall; petioles 10–25 mm long, midstem leaf blades 5–15 cm long, 4–9 cm wide, oval to ovate, obovate or suborbicular; corolla lobes 6–8 mm long, white; gynostegium brown; gynostegial column red-purple; hoods white, with short horns. Sandy woods; in the e 1/3 portion of East TX, found by Reverchon at Dallas, not seen there recently (Mahler 1988), possibly Gonzales Co. (Woodson 1954); e U.S. from NY s to FL and w to IL, MO, OK, and TX; also Canada. Apr–Jun.

Asclepias verticillata L., (whorled), WHORLED MILKWEED, EASTERN WHORLED MILKWEED. Stems erect, 30–90 cm tall, puberulent in lines; leaves mostly in whorls; leaves sessile or nearly so; midstem leaf blades 1.5–7 cm long, 0.1–0.2 cm wide, filiform to linear, often revolute; corolla lobes greenish white to cream, sometimes light brownish pink, ca. 3.5 mm long; gynostegium green; hoods 1.5–2 mm long, cream, with horns. Sandy open woods, rocky slopes, or prairies; throughout East TX; West Cross Timbers and Edwards Plateau, also Trans-Pecos; throughout much of e 1/2 of U.S., also MT; also adjacent Canada. May–Jul, sporadically later. Reported to be poisonous to livestock (Hartman 1986b).

Asclepias viridiflora Raf., (green-flowered), GREEN-FLOWER MILKWEED, GREEN ANTELOPE-HORNS, WAND MILKWEED. Stems erect, (10–)20–125 cm tall, glabrous or densely and finely pubescent; leaves sessile or petioles to 5 mm long; midstem leaf blades 2–13(–14) cm long, 0.8–6 cm wide, quite variable, narrowly lanceolate to oblong-obovate or suborbicular; corolla lobes 5–7 mm long, green to yellowish green; gynostegium green; hoods without horns. Sandy or calcareous areas, gravelly open ground, or prairies; throughout TX, except South TX Plains; throughout the e 2/3 of U.S.; also adjacent Canada. Jun–Aug. [*A. viridiflora* Raf. var. *lanceolata* Torr.]

Asclepias viridis Walter, (green), GREEN MILKWEED, ANTELOPE-HORNS. Plant glabrous or upper stem and young leaves minutely pubescent; stems low-spreading to erect, usually 15–70 cm long; petioles 2–6 mm long, midstem leaf blades 3–13 cm long, 1.5–6 cm wide, lanceolate, narrowly ovate to broadly oblong; corolla lobes 12–15 mm long, oval, pale green; gynostegium brown; hoods wide-spreading pale to dark purple, upper margin white, without horns. Prairies, ditch banks, pastures, and disturbed ground; throughout East TX; West Cross Timbers, Edwards Plateau, and Gulf Prairies and Marshes; ec U.S. from OH s to FL and

Asclepias lanceolata [CO1]

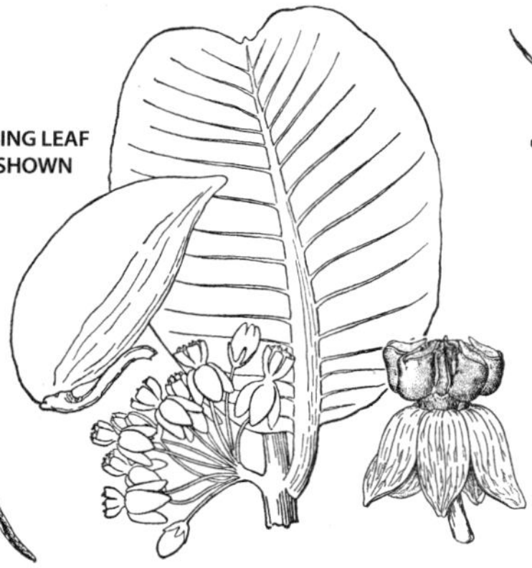

Asclepias latifolia [BB2, AMB]

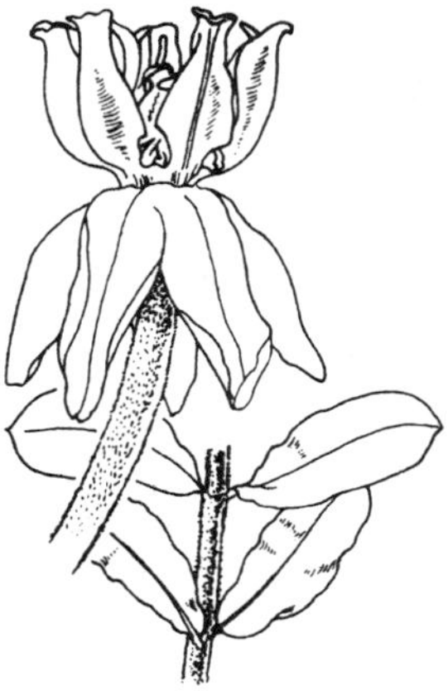

Asclepias obovata [RAD]

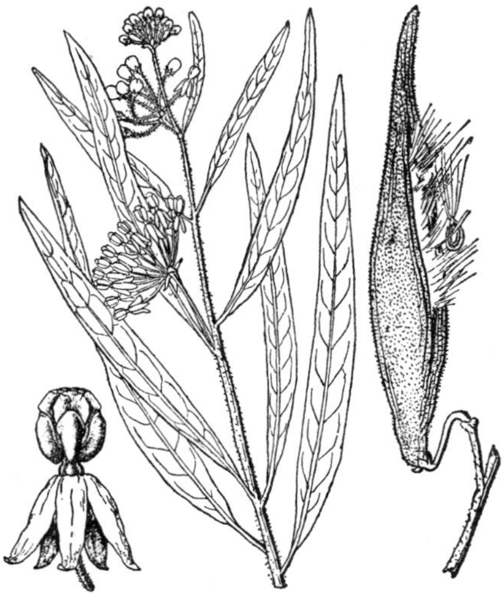

Asclepias longifolia [AMB, BB2]

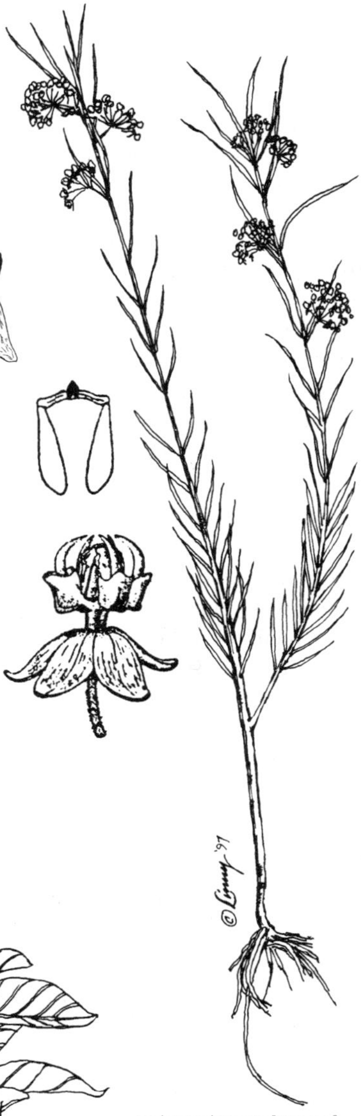

Asclepias linearis [AMB, HEA]

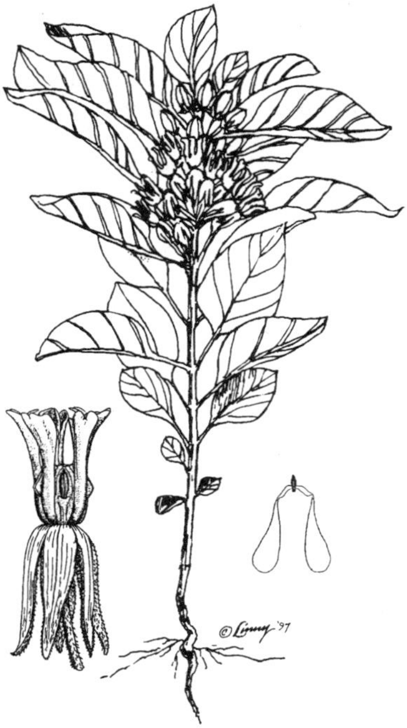

Asclepias oenotheroides [AMB, HEA]

w to NE and TX. Apr–Jun, sporadically in summer and fall. [*Asclepiodora viridis* (Walter) A. Gray] Probably the most common milkweed in East TX; sometimes extremely abundant in overgrazed pastures.

CATHARANTHUS G. Don PERIWINKLE

A tropical genus of 8 species, 7 endemic to Madagascar, 1 India and Sri Lanka; previously recognized in *Vinca*, but differing in a variety of ways including having ± sessile anthers without a terminal appendage (Mabberley 2017). (Greek: *katharos*, pure, and *anthos*, flower)
REFERENCES: Bailey & Bailey 1976; Lemke 2023b.

Catharanthus roseus (L.) G. Don, (rose-colored), MADAGASCAR PERIWINKLE, ROSE PERIWINKLE, OLD-MAID, CAYENNE-JASMINE, CAPE PERIWINKLE. Erect perennial (grown here as an annual) herb 20–60 cm tall; leaves usually opposite, blades 2.5–7 cm long, 0.8–4 cm wide, elliptic to obovate or oblanceolate, apex rounded, acute, or obtuse, mucronate, entire; inflorescence axillary, flowers 1–2 in upper leaf axils; peduncles 2–4 mm long; corollas showy, salverform, the tube 20–30 mm long, lobes 10–28 mm long, 10–25 mm wide, spreading, broadly obovate, variously colored, from white to rosy pink or red, sometimes with a differently colored "eye"; follicles solitary or paired, erect, ± straight, striately ridged, puberulent; seeds numerous, glabrous. Very commonly grown in flower beds in summer, and sometimes appearing spontaneously in landscapes; not frost-resistant; Brazos, Harris, McLennan, Travis, and Washington (Kartesz 2022) cos., also Tarrant Co. (Diggs et al. 1999) to the w of East TX; e U.S. from NC s to FL and w to KS and TX, mostly scattered except throughout much of FL. Jun–Oct. Native of Madagascar, now pantropical. [*Vinca rosea* L.] This species is widely cultivated as an ornamental and also produces at least 80 alkaloids including vinblastine, vincristine, and others useful in the treatment of certain cancers including Hodgkin's disease and leukemia; it is considered the most important plant to date in terms of cancer treatment; if misused the alkaloids are potentially toxic; they can act as an abortifacient and the plant has been reported to be poisonous when eaten by cattle (Bailey & Bailey 1976; Duke 1985; McGuffin et al. 1997; Mabberley 2017).

CYNANCHUM L. MILKWEED VINE

Twining vines to 3 m or more long; leaves opposite, petioled, basally cordate; inflorescences axillary; follicles ovoid to lanceolate in outline, smooth.

A genus of 300 species of tropical and warm areas of the world (Fishbein 2023b). Woodson (1941) adopted a very broad circumscription of *Cynanchum*, and since then many species have been assigned to various other genera (e.g., *Orthosia*, *Pattalias*, and *Metastelma*). (Greek: *cyon*, dog, and *anchein*, to strangle; an ancient name for some plant supposedly poisonous to dogs)
REFERENCES: Sundell 1981; Liede & Kunze 2002; Liede & Täuber 2002; Fishbein 2023b.

1. Latex clear; peduncle mostly shorter than petiole of subtending leaf; inflorescence 6–30-flowered; corona segments ca. as long as the corolla lobes, distally deeply divided into 2 linear segments ____ **C. laeve**
1. Latex white; peduncle about as long as or longer than petiole of subtending leaf; inflorescence 4–20-flowered; corona segments < 2/3 as long as the corolla lobes, broadly oblong with 3-lobed summit ____ **C. unifarium**

Cynanchum laeve (Michx.) Pers., (smooth), BLUEVINE, SANDVINE, SMOOTH SWALLOW-WORT, SMOOTH ANGLEPOD. Petioles to 20(–90) mm long, leaf blades 2–11 cm long, 1.5–10 cm wide, ovate or deltate, base cordate to sagittate, apex acute to acuminate; flowers in peduncled umbellate clusters or abbreviated racemes, corolla lobes white, twisted; follicles 7–15 cm long. Low moist woods or fields; scattered in the w and s portions of East TX; Edwards Plateau and Cross Timbers and Prairies; e U.S. from NY s to FL and w to MN, NE, and TX. Jun–Sep.

Cynanchum unifarium (Scheele) Woodson, (of a single rank or series), TALAYOTE. Petioles to 50 mm long, leaf blades 3–11 cm long, 1.5–9 cm wide, broadly ovate to ovate-lanceolate or triangular-lanceolate with stipule-like small leaves from leaf axils, base cordate to sagittate; corolla lobes green with white margin, in racemes; corona segments apically 3-lobed, mostly ca. as long as gynostegium, middle lobe sometimes extended as a long tooth; follicles 8–12 cm long. Sandy forests and thickets; in the sw portion of East TX; Edwards Plateau, South TX Plains, and Trans-Pecos; in the U.S. known only from TX; also Mexico. May–Oct. [*C. racemosum* (Jacq.) Jacq. var. *unifarium* (Scheele) Sundell]

FUNASTRUM E. Fourn. TWINEVINE

Twining vines; leaves opposite, petiolate, with cordate to truncate bases; inflorescences of peduncled, umbellate clusters, extra-axillary; corona composed of 2 parts, an outer fleshy collar-like ring and an inner whorl of inflated hood-like segments; follicles smooth, fusiform, not angled.

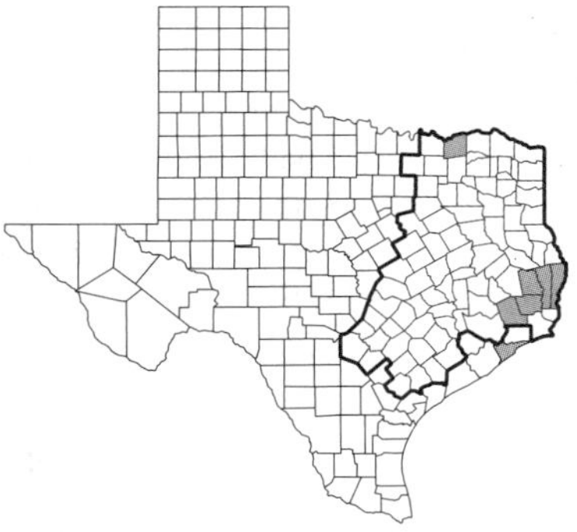
Asclepias hirtella

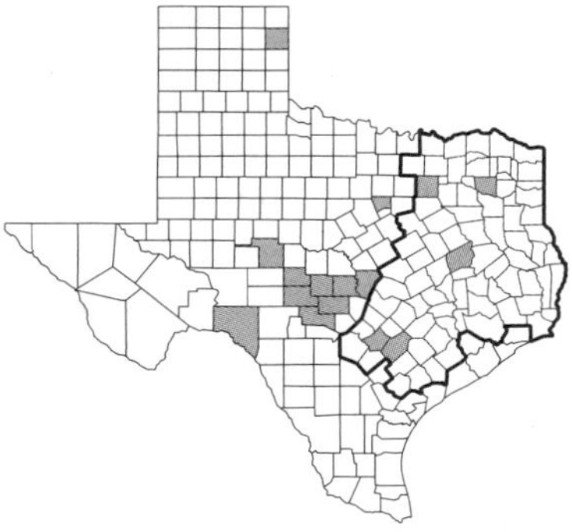
Asclepias incarnata subsp. incarnata

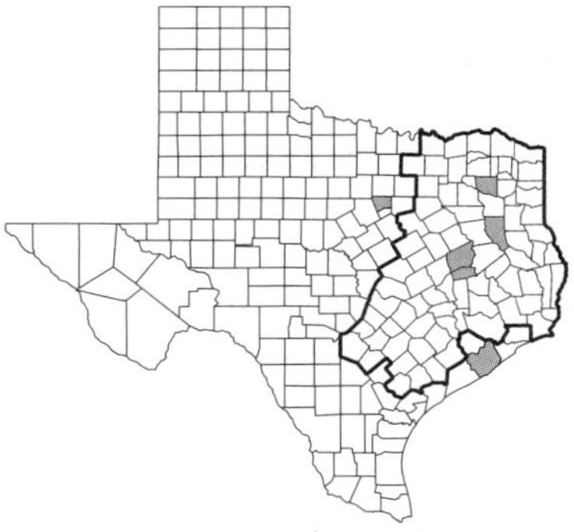
Asclepias incarnata subsp. pulchra

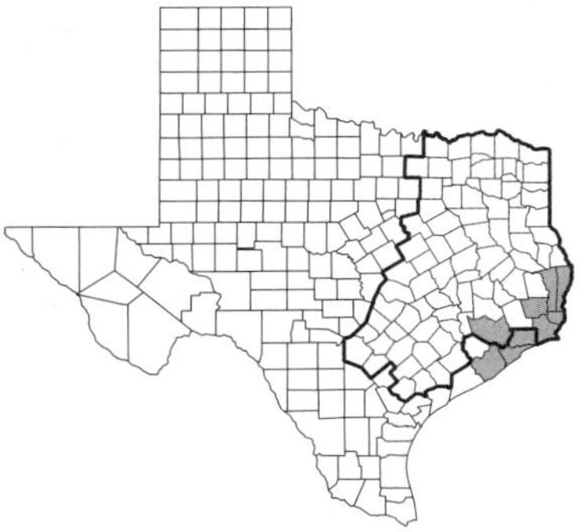
Asclepias lanceolata

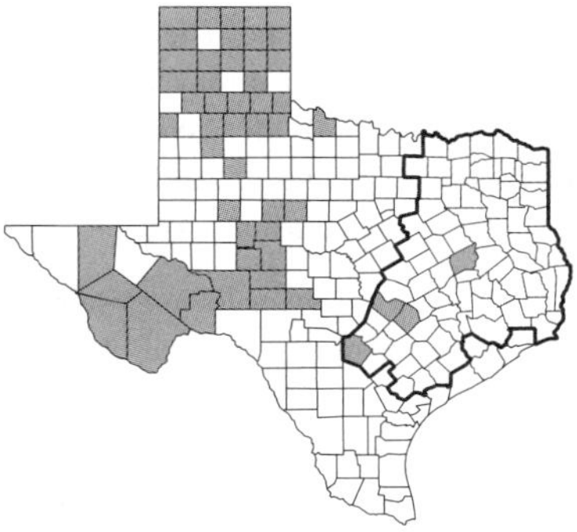
Asclepias latifolia

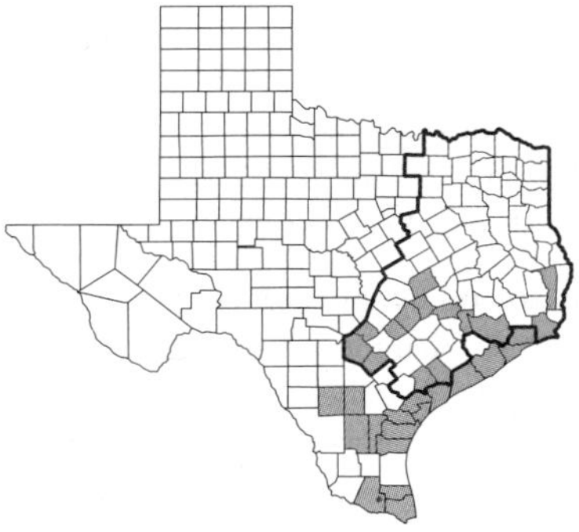
Asclepias linearis

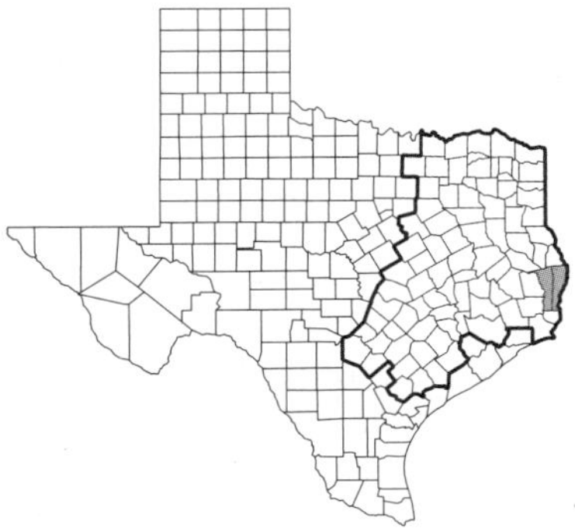
Asclepias longifolia

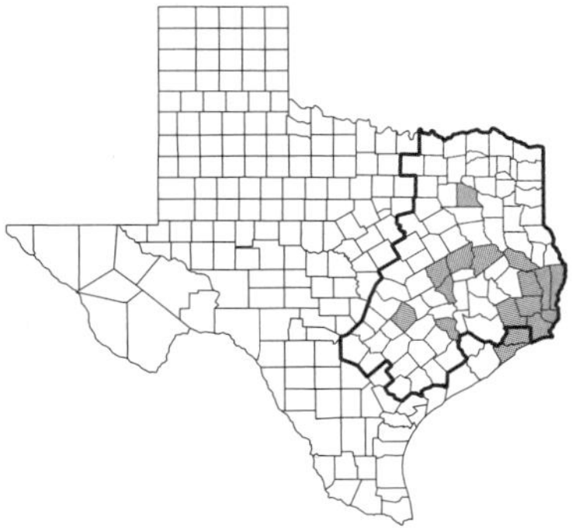
Asclepias obovata

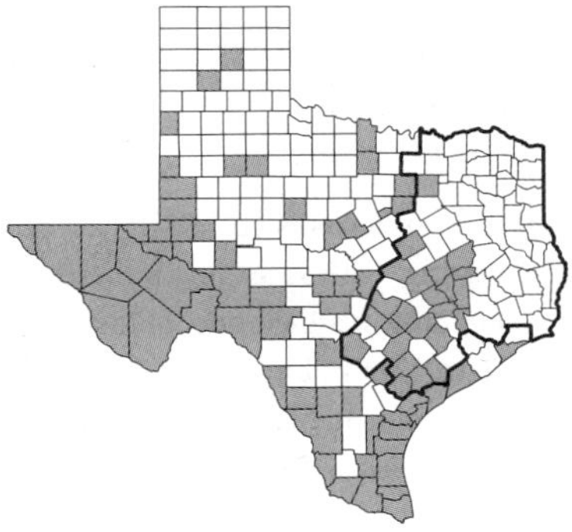
Asclepias oenotheroides

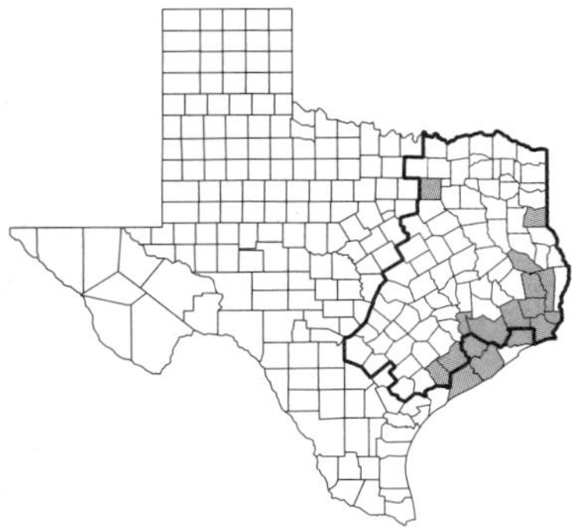
Asclepias perennis

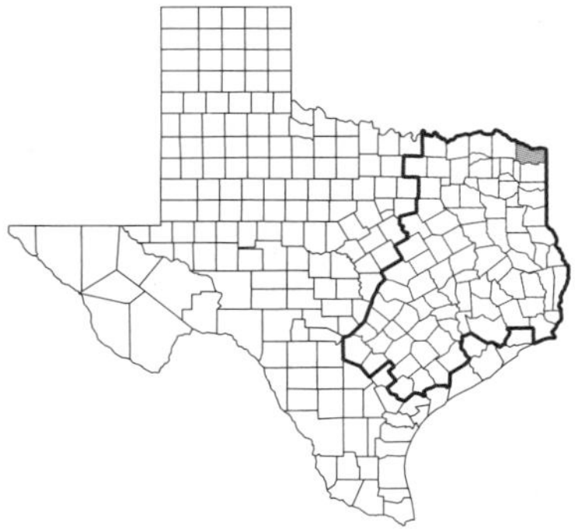
Asclepias purpurascens

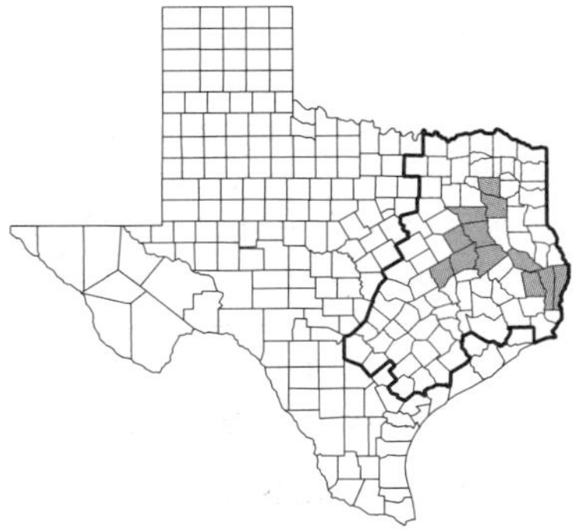
Asclepias rubra

About 18 species in the New World (Fishbein 2023c). The species here treated in *Funastrum* have traditionally been placed in *Sarcostemma* (e.g., Correll & Johnston 1970; Kartesz 1994; Jones et al. 1997), a genus in the broad sense of ca. 40 species of leafy or leafless vines with photosynthetically active stems (Liede 1996); it is native to tropical and warm areas of the world. Based on a cladistic analysis, Liede (1996) argued convincingly that *Sarcostemma* (later synonymized with *Cynanchum* (Liede & Täuber 2002)) be restricted to Old World taxa, while the ca. 18 species in the New World be recognized in *Funastrum*; the similar floral structures of the 2 groups are interpreted as parallel evolution. We are following Liede (1996), Kartesz (2022), and Fishbein (2023c) in recognizing *Funastrum*. (Greek: *funis*, rope, presumably in reference to the vine habit)
REFERENCES: Holm 1950; Jones & Jones 1991; Liede 1996; Liede & Kunze 2002; Liede & Täuber 2002; Fishbein 2023c.

1. Leaf margins distinctly crisped or ruffled; corollas brownish green or bronze; calyx lobes mostly > 3 times as long as wide ____ **F. crispum**
1. Leaf margins neither crisped nor ruffled; corollas white to pale lavender or pinkish; calyx lobes mostly < 3 times as long as wide ____ **F. cynanchoides**

Funastrum crispum (Benth.) Schltr., (crisped, curled), WAVY-LEAF TWINEVINE. Glabrous, low-climbing or tangle-forming vine; leaf blades 3–10 cm long, 0.2–3 cm wide, lanceolate to linear, acuminate, blue-green, with ruffled, occasionally purple-tinged margin; inflorescence solitary at nodes, 4–13-flowered; calyx lobes 2.5–4 mm long, narrowly lanceolate; follicles 10–12 cm long. In gravelly or silty limestone soils; Dallas (*Reverchon 2094*, 1900, MO "West Dallas"), Bexar, Comal, McLennan, Travis (Kartesz 2022), and Williamson (*Gibson 6428* TEX) cos., South TX Plains; Edwards Plateau, Cross Timbers and Prairies, High Plains, and Trans-Pecos; AZ, CA, CO, NM, OK, and TX. Apr–Aug. [*Sarcostemma crispum* Benth.]

Funastrum cynanchoides (Decne.) Schltr., (resembling *Cynanchum*—swallow-wort), CLIMBING-MILKWEED, TWINEVINE, WHITE TWINEVINE. Glabrous to sparsely puberulent twining to trailing vine; leaf blades 1.2–8 cm long, 0.7–4.5 cm wide, broadly to narrowly ovate-lanceolate to triangular-lanceolate, acute to acuminate, foliage with noticeably very unpleasant odor; inflorescence solitary or sometimes paired at nodes, 6–31-flowered; calyx lobes 2–3 mm long, ovate to ovate-linear; follicles 6–9 cm long. Sandy or rocky soils, waste places; mostly on the sw margin of East TX; mainly w 2/3 of TX; AR, AZ, NM, OK, and TX. Apr–Sep. [*Sarcostemma cynanchoides* Decne.]

GONOLOBUS Michx. ANGLEPOD

A genus of 100–150 species of the Americas; introduced in Africa (Krings 2023). (Greek: *gonia*, angle and *lobos*, pod, alluding to the follicle shape)
REFERENCE: Krings 2023.

Gonolobus suberosus (L.) R. Br. var. **granulatus** (Scheele) Krings & Q.Y. Xiang, (sp.: corky; var.: granular), ANGLEPOD. Plant high-climbing, with spreading pubescence; leaves opposite, petioles 2.8–5 cm long, blades 2.9–15(–20) cm long, 2.5–17.5 cm wide, deltate to ovate-elliptic, oblong or suborbiculate, base cordate, apex acuminate; inflorescence extra-axillary, racemiform, peduncle to 17 mm long, 2–12-flowered; calyx lobes 1.6–5.4 mm long, lanceolate; corolla tube 1–1.4 mm long; lobes 2.6–11 mm long, uniformly colored adaxially, olive-green at anthesis, corona a prostrate-undulating ring, minute, shorter than the anthers; follicles 8–15 cm long, 1.5–2.8 cm wide, smooth, sharply angled. Stream bottom woods; throughout East TX; East Cross Timbers, Lampasas Cut Plain, and Edwards Plateau; e U.S. from IN s to AL and w to IL, OK, and TX. May–Jun. [*G. granulatus* Sheele]

MANDEVILLA Lindl. ROCK-TRUMPET

A genus of ca. 170 species (Lemke 2023c). This genus has traditionally been included in *Macrosiphonia* (Woodson 1933). Henrickson (1996) separated the North American species as *Telosiphonia* from the South American species (which he continued to recognize in *Macrosiphonia*). Molecular work (Simões et al. 2004, 2006) supports the separation of *Telosiphonia* and *Macrosiphonia* into two separate clades, but includes both in the larger genus *Mandevilla*, which would be paraphyletic without these two small clades. Henrickson (1996) noted that *Telosiphonia* species (treated here as *Mandevilla*) "apparently strongly respond to variation in available water resources," with the smallest corollas often half the sizes of the largest. (for John Henry Mandeville, 1773–1861, British Diplomat in Argentina and avid gardener)
REFERENCES: Woodson 1933; Henrickson 1996; Simões et al. 2004, 2006; Lemke 2023c).

Mandevilla macrosiphon (Torr.) Pinchon, (large- or long-tubed), PLATEAU ROCK-TRUMPET, FLOR DE SAN JUAN, LONG-TUBE TRUMPET-FLOWER. Suffrutescent perennial with white latex; stems ± erect, 10–40(–50) cm tall, densely tomentose; leaves opposite (–subopposite), petioles 3–6(–9) mm long, leaf blades 20–40 mm long, 14–35 mm wide, ovate to obovate to oblong-elliptic or rarely suborbicular, entire, often revolute and undulate marginally, the upper and lower surfaces differently colored, the upper green

Asclepias perennis [AMB, CO1]

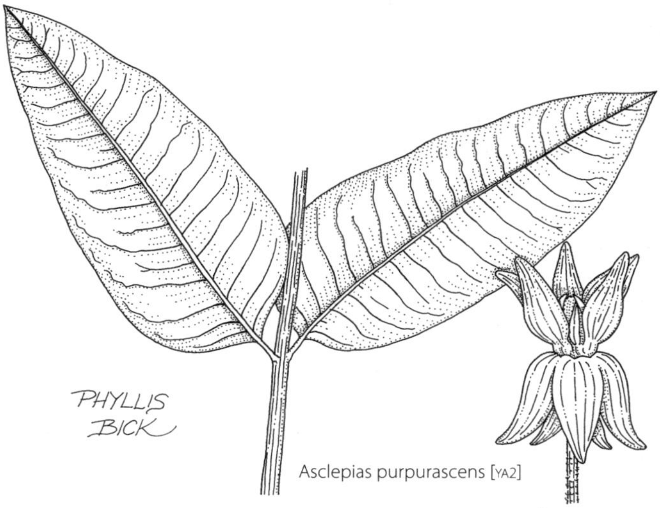

Asclepias purpurascens [YA2]

Asclepias rubra [CO1]

Asclepias stenophylla [AMB, BB2]

Asclepias subverticillata [CR2]

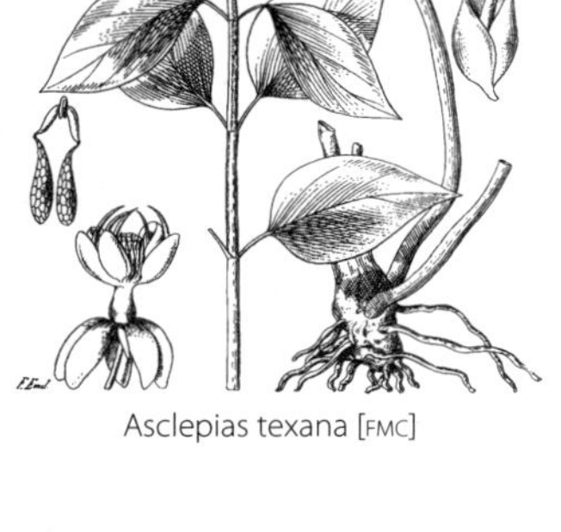

Asclepias texana [FMC]

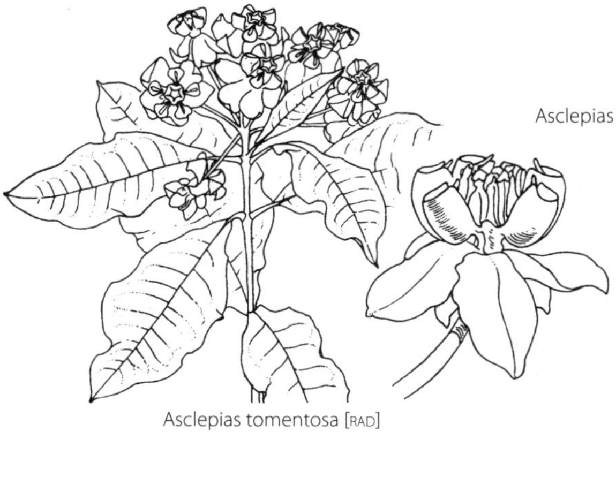

Asclepias tomentosa [RAD]

to gray-green, the lower yellowish-white due to dense tomentum, the lower surface with veins strongly raised; flowers terminal, usually solitary; sepals 5, 13–18 mm long; corollas salverform, showy, long-tubed, white, often tinged with pink, 10–14 cm long, the tube narrow (ca. 1.5 mm wide), 4.6–9.5 cm long, ca. 2–3 times as long as the throat, expanded for up to 3 cm into a somewhat wider throat, the limb equally 5-lobed, the lobes 22–40 mm long, overlapping; androecium and gynoecium not united into a gynostegium, corona absent; fruits of paired elongate follicles usually 10–15 cm long, with numerous 9–11 mm long seeds bearing an apical coma of elongate hairs. Dry, rocky, limestone or igneous-derived slopes; Comal (*Lindheimer 984*, 1849, TEX) and Travis (Kartesz 2022) cos. at sw margin of East TX; mainly Edwards Plateau and Trans-Pecos; in the U.S. known only from TX; also Mexico. May–Sep, following rains. Moth pollination is predicted (Grant 1983). [*Echites macrosiphon* Torr., *Macrosiphonia macrosiphon* (Torr.) A. Heller, *M. berlandieri* A. Gray, *Telosiphonia* (Woodson) Henrickson *macrosiphon* (Torr.) Henrickson]

MATELEA Aubl. MILKVINE, ANGLEPOD

Plants variously pubescent or pilose to glabrate, often with burned-rubber odor; stems trailing to suberect, or twining and climbing; leaves opposite, short- to rather long-petioled, blades orbicular to ovate to oblong-ovate or lanceolate, with cordate or sagittate bases; flowers extra-axillary; corollas rotate; corona composed of a low ring of 5 variously shaped laminar segments; follicles muricate (= with spine-like tubercules), angled or not.

A genus of ca. 200 species native to tropical and temperate regions of the New World (Fishbein & McDonnell 2023). (Derivation of generic name not explained by original author)

REFERENCES: Shinners 1950d; Correll 1965; Fishbein & McDonnell 2023.

1. Flowers in pedunculate, extra-axillary, several- or many-flowered umbels or umbel-like racemes or cymes; stems twining or climbing; leaf blades to 17 cm long, typically > 6 cm.
    2. Corollas greenish brown to brownish purple, without a network of fine veins on the upper surface, the lobes oblong, narrowly deltate, or narrowly lanceolate.
        3. Corolla lobes < 8 mm long, adaxially hirtellous ____ **M. hirtelliflora**
        3. Corolla lobes mostly > 8 mm long (often much longer), adaxially glabrous ____ **M. decipiens**
    2. Corollas green, with a network of fine veins on at least part of the upper surface, the lobes ovate-lanceolate or elliptic.
        4. Upper surface of corollas densely puberulent; only distal part of upper surface of corolla lobes with network of fine veins (with dark green parallel veins below middle); stems and leaves with only sparse short pubescence of curved or appressed hairs; crown (= corona of flower) with 5 short, distinct, fleshy, spreading appendages; peduncles to 12 mm long, shorter than the subtending petiole ____ **M. edwardsensis**
        4. Upper surface of corollas glabrous; entire upper surface of corollas with network of fine veins; stems and leaves with both long spreading hairs and short glandular hairs; crown without 5 distinct appendages, instead with a low minute collar-like cup surrounding the gynostegium; peduncles usually as long as or longer than the subtending petiole ____ **M. reticulata**
1. Flowers mostly in 2s, with pedicels appearing to be attached directly in leaf axils; stems trailing to suberect, not twining; leaf blades 5(–6.5) cm or less long, often much less.
    5. Corolla lobes adaxially pilose to hirsute; pedicels 3–11 mm long; corolla lobes 3–6 mm long, maroon to dark brown ____ **M. biflora**
    5. Corolla lobes adaxially glabrous; pedicels 3–5 mm long; corolla lobes 2.5–4 mm long, green to brown or maroon ____ **M. cynanchoides**

Matelea biflora (Raf.) Woodson, (two-flowered), TWO-FLOWER MILKVINE. Plant trailing or decumbent; stems 4–10, 10–40 cm long, pubescent with long eglandular hairs and short glandular hairs; petioles 0.5–2.5 mm long, leaf blades 0.8–5 cm long, 0.6–3.2 cm wide, ovate to deltate, base cordate; inflorescence 1–2-flowered, pedicels 3–11 mm long; calyx lobes spreading; corolla maroon to dark brown, not reticulate, rotate-campanulate, lobes 3–6 mm long, spreading, adaxially pilose to hirsute; follicles 4.5–8.5 cm long, densely muricate. Prairies or open ground, usually limestone areas; mostly in the w 1/3 of East TX; Blackland Prairie w to Rolling Plains, also Edwards Plateau; NM, OK, and TX. Mar–Oct.

Matelea cynanchoides (Engelm. & A. Gray) Woodson, (resembling *Cynanchum*—SWALLOW-WORT), MILKVINE. Plant with odor like that of burned rubber, usually ascending to decumbent; stems, 3–10, 10–70 cm long; petioles 0.5–1.3 cm long, leaf blades 2.2–5.4 cm long, 1–4.5 cm wide, ovate to deltate (elliptic or orbiculate); inflorescence 2-flowered, sometimes appearing terminal with more than 2 flowers; pedicels 3–5 mm long; calyx lobes 2.5–4 mm long, spreading; corolla green to brown or maroon, not reticulate, lobes spreading, glabrous adaxially; follicles 5–9.7 cm long, sparsely muricate. Loose sandy soils, open oak woods; throughout much of East TX.; se and e TX w to Edwards Plateau and Rolling Plains; AR, LA, OK, and TX. Apr–Aug.

Asclepias tuberosa subsp. interior [BT3]

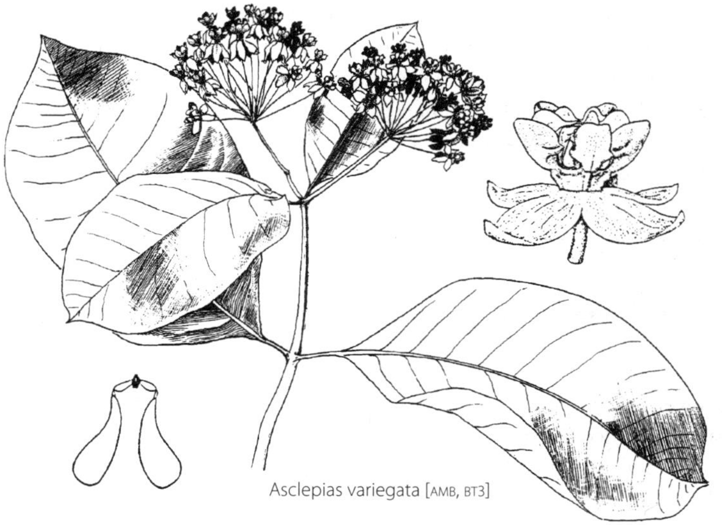

Asclepias variegata [AMB, BT3]

Asclepias verticillata [REE]

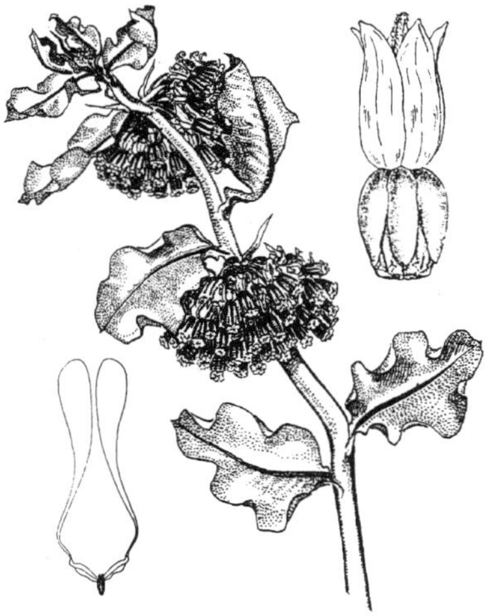

Asclepias viridiflora [AMB, BT3]

Asclepias viridis [AMB, STE]

Catharanthus roseus [BT3, NIC]

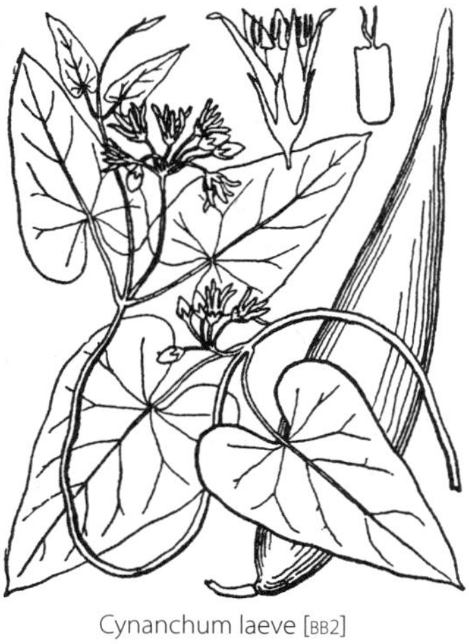

Cynanchum laeve [BB2]

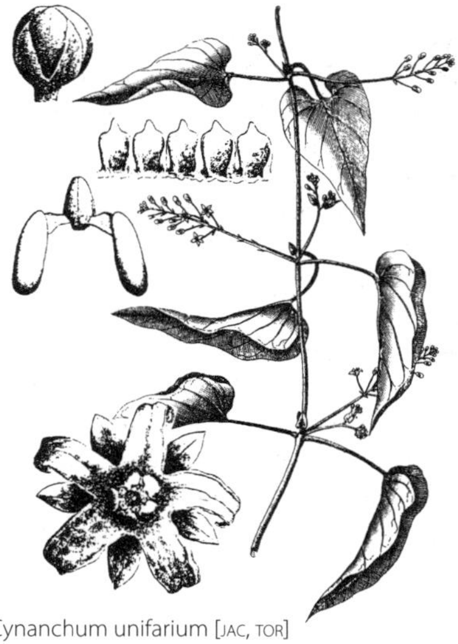

Cynanchum unifarium [JAC, TOR]

Matelea decipiens (Alexander) Woodson, (deceptive, not obvious), CLIMBING-MILKWEED. Plant climbing on other plants, with spreading pubescence and minute glandular pubescence; petioles 1.5–9.5 cm long, leaf blades 4–14 cm long, 2.5–16 cm wide, ovate to oval to orbiculate; inflorescence solitary or paired, simple or compound umbelliform, pedunculate, 10–40-flowered, pedicels 10–25 mm long; calyx lobes elliptic to lanceolate; corolla maroon to green tinged with maroon, not reticulate; corolla lobes 7–18 mm long, narrowly oblong or narrowly lanceolate; erect to spreading, adaxially glabrous; follicles somewhat angled, 8–12 cm long, moderately muricate. Sandy soils, wooded areas; mostly in the e 1/2 of East TX; however, it's distribution may be more limited (see discussion under *M. hirtelliflora*); AR, KS, LA, MO, OK, and TX. Apr–Jul. [*Gonolobus decipiens* (Alexander) L.M. Perry] According to Fishbein and McDonnell (2023), *M. decipiens* is not found e of the Mississippi River and *M. carolinensis* (Jacq.) Woodson is not found w of the Mississippi River. Therefore, we are not including *M. carolinensis* as a part of the East TX flora.

Matelea edwardsensis Correll, (of the Edwards Plateau), PLATEAU MILKVINE. Plant climbing on other plants, stems twining; petioles 1–4 cm long, blades 1.5–8 cm long, 1–6 cm wide, ovate; inflorescence umbelliform to racemiform, pedunculate, 3–12-flowered, peduncle 0.3–1.5 cm long, pedicels 4–10 mm long; calyx pubescence short; calyx lobes 2–3 mm long; corolla pale green to yellow-green, with darker basal longitudinal lines, becoming reticulate at lobe tips; corolla lobes 3–4 mm long; follicles 8–10 cm long, moderately muricate, glabrate. On limestone, open woods; mostly on the sw margin of East TX along the Edwards Plateau, but also Bell Dallas and McLennan (Kartesz 2022) cos.; Edwards Plateau; endemic to TX. Apr–May. NatureServe: G3, vulnerable.

Matelea hirtelliflora McDonnell & Fishbein, (with a pubescent flower), HAIRY-FACED SPINY POD. Plant perennial herbaceous twining vine; largest leaves with petioles 2.3–6.8 cm long, blades 7.5–17 cm long, 3.7–12 cm wide, blades ovate to oval; inflorescence an extra-axillary, dichasial cyme with 1–6 flowers, peduncles 0.5–6 cm long, pedicels 0.6–1.3 cm long; sepals adaxially glabrous, abaxially hirsute and glandular; corolla campanulate-rotate, 0.9–1.4 cm in diam., deeply 5-lobed, lobes 3.2–7.5 mm long, oblong to narrowly deltate, light to deep purple-maroon, both surfaces hirtellous; follicles 6–8 cm long, muricate; coma white. Wooded areas near water, areas under lain by sand or sandy clay soil; in the northeastern portion of East TX; possibly endemic to TX. Though Kartesz (2022) maps this species for TX only, Fishbein and McDonnell (2023) state, "There is photographic documentation of the species in extreme southwestern Arkansas (Miller County), but we have not seen specimens." We have seen an image of a specimen from Miller Co., Arkansas (*Keesling 20-1114* ANHC; TORCH 2023) on the northeastern border of TX with AR, but cannot verify its identification. Additionally, McDonnell and Fishbein (2016) state, "Though the treatment for Apocynaceae has not yet been completed for the Flora of East Texas (Diggs et al. 2006) … We have seen only one specimen of *M. decipiens* from the flora area in eastern Jasper County. All other specimens in the flora area are attributable to *M. hirtelliflora*." Apr–Jun. NatureServe: G3, vulnerable. ?

Matelea reticulata (Engelm. ex A. Gray) Woodson, (netted) NET-VEIN MILKVINE, GREEN MILKWEED VINE. Plant climbing on other plants; stems 1–5, twining, 30–200 cm long; petioles 1–6 cm long, leaf blades 2–10 cm long, 1.5–7 cm wide ovate, apex acute; inflorescence umbelliform to racemiform, pedunculate, 3–10-flowered; peduncle 1–7 cm long, pedicels 12–25 mm long; calyx lobes spreading, lanceolate, 3–3.5 mm long; corolla pale green with darker green reticulations on lobes, rotate, lobes broadly ovate to nearly orbiculate, 4–5 mm long, hirsute abaxially, glabrous adaxially; follicles 8–15 cm long, sparsely to moderately muricate, glabrate. Thickets on rocky hillsides; mostly in the sw portion of East TX; Edwards Plateau, Cross Timbers and Prairies, and Trans-Pecos; in the U.S. known only from TX; also Mexico. Mar–Sep.

Matelea baldwyniana (Sweet) Woodson, (for William Baldwyn, 1779–1819, American physician and botanist), BALDWYN'S MILKVINE. *Matelea baldwyniana* is different from all East TX *Matelea* species in having white to cream corollas; sometimes faintly reticulate. Riparian systems, oak hickory forest, silty clay loam; AL, AR, FL, GA, MO, and OK; included based on several collections from McCurtain Co., OK on the extreme ne border of East TX (see TORCH 2023).

Matelea parviflora (Torr.) Woodson, (having small flowers), SMALL-FLOWER MILKVINE. This species is known from just to the south of East TX in the South TX Plains. It can be distinguished from the 2 non-twining species in East TX (*M. cynanchoides* and *M. biflora*) by having a pedunculate inflorescence; peduncle 0.7–10 cm long vs. inflorescence non-pedunculate with pedicels attached directly in leaf axils in *M. cynanchoides*; and *M. biflora*; having more flowers (2–20) in the inflorescence vs. 1–2 in the other 2 species; and corolla faintly to strongly reticulate in *M. parviflora* vs. not reticulate in *M cynanchoides* and *M. biflora*. Plains, hillsides, arroyos, sandy and gravel soils; South TX Plains; in the U.S. known only from TX; also Mexico. Mar–Oct.

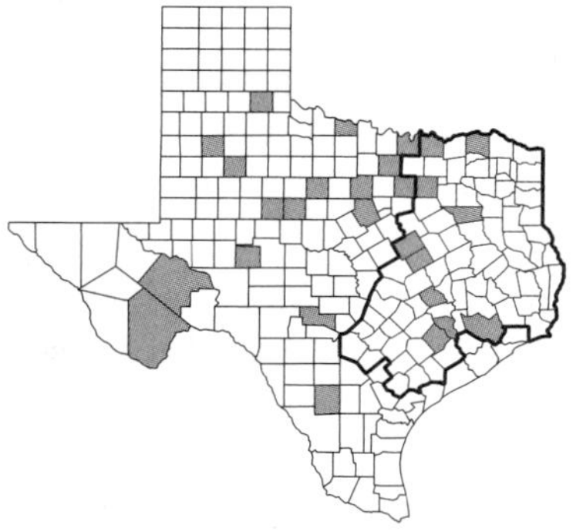

Asclepias stenophylla

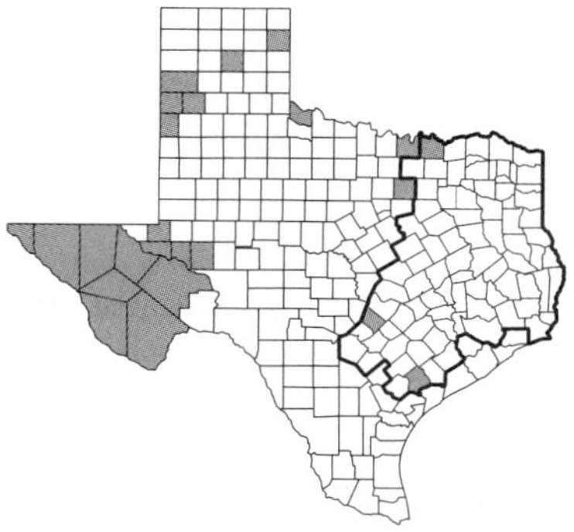

Asclepias subverticillata

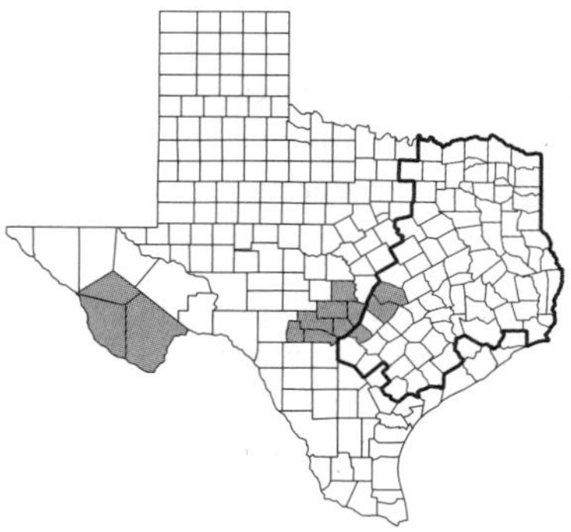

Asclepias texana

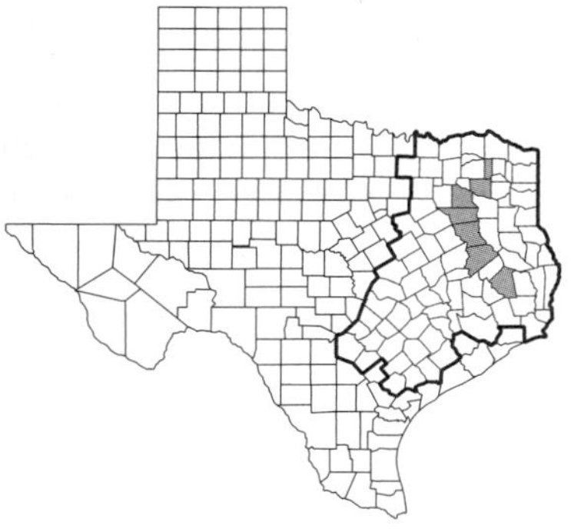

Asclepias tomentosa

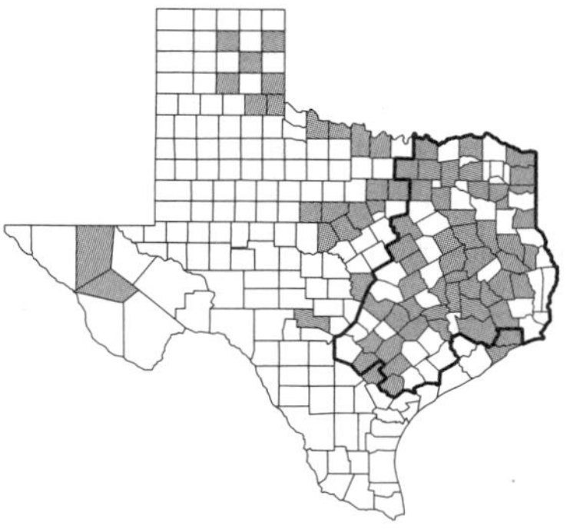

Asclepias tuberosa subsp. interior

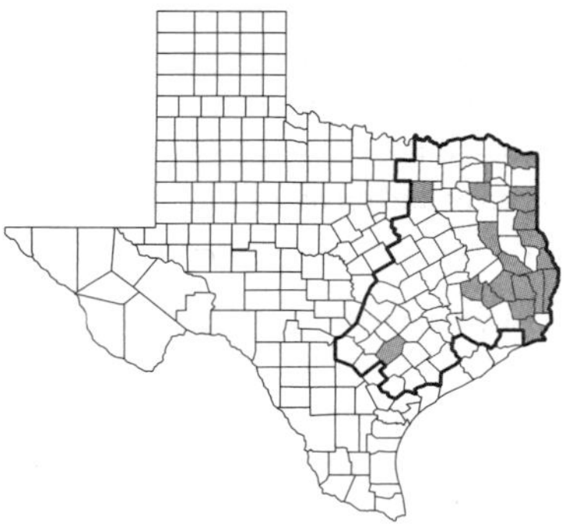

Asclepias variegata

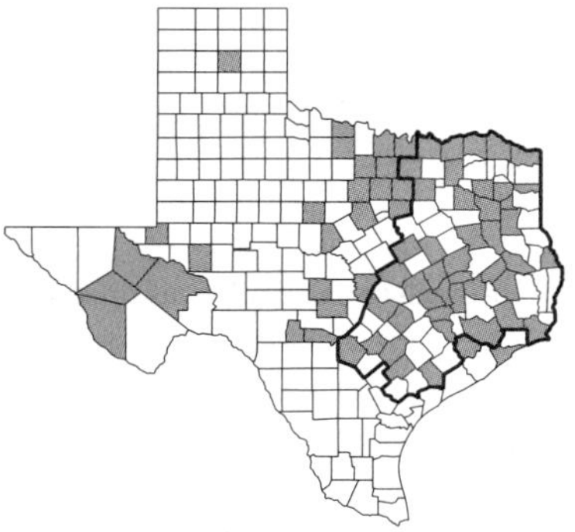

Asclepias verticillata

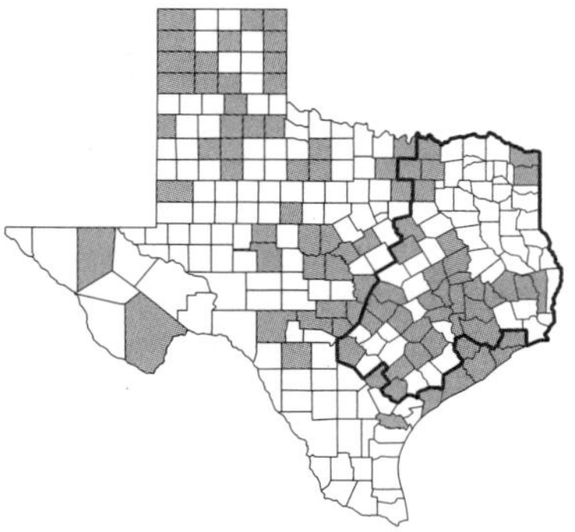

Asclepias viridiflora

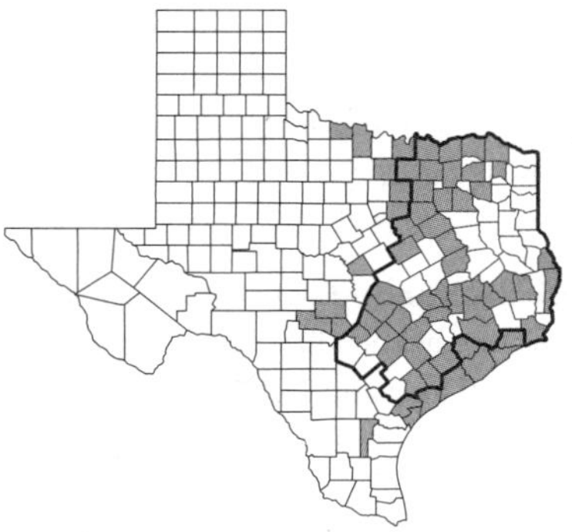

Asclepias viridis

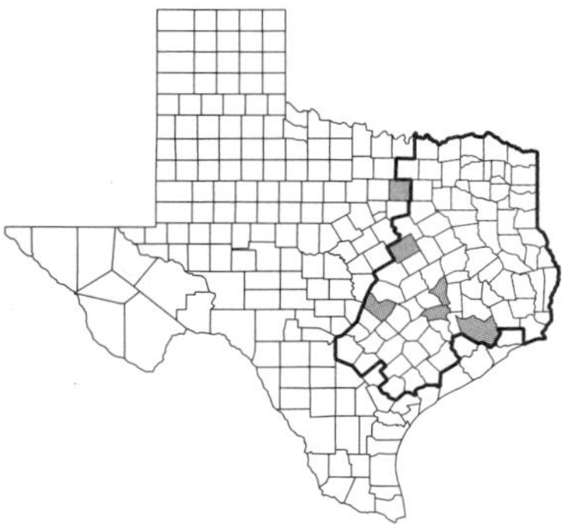

Catharanthus roseus

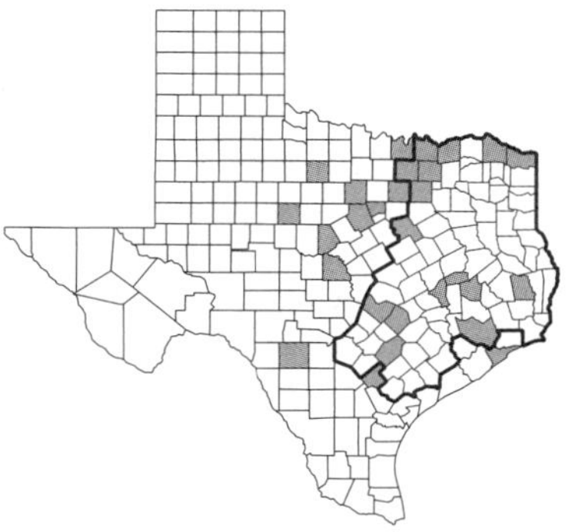

Cynanchum laeve

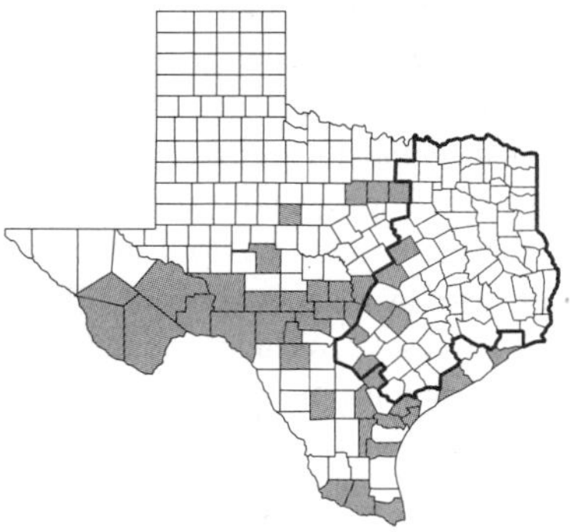

Cynanchum unifarium

METASTELMA R. Br. SWALLOW-WORT

Vines, herbaceous or somewhat woody at base, glabrous to pubescent; perennial; latex white; stems twining; leaves opposite, petioles 1.5–5 mm long; leaf blades linear, oblong, elliptic or lanceolate, base rounded; inflorescences umbelliform to racemiform, sessile to short-pedunculate, few flowered; corollas campanulate, cream to yellowish cream; corona of 1 whorl of 5 free, laminar or acicular segments; follicles dark brown.

A genus of ca. 80 species of s U.S., Mexico, South and Central America (Fishbein 2023d). Woodson (1941) included *Metastelma* in *Cynanchum* but its inclusion has been shown to render *Cynanchum* polyphyletic. (Greek: *meta*, other side, next to and *stelma*, a girdle or belt, referring to the corolla)
REFERENCE: Fishbein 2023d.

1. Adaxial surface of corolla lobe apices puberulent with short, matted, flattened hairs, lobes 1.5–2 mm long ____ **M. palmeri**
1. Adaxial surface of corolla lobe apices villous with long, erect, flattened hairs, lobes 2.5–3.5 mm long ____ **M. barbigerum**

Metastelma barbigerum Scheele, (bearded, presumably in reference to the densely long pilose inner surfaces of the corolla lobes), BEARDED SWALLOW-WORT, THICKET THREADVINE. Leaf blades linear to linear-oblong or elliptic-lanceolate, to 5 cm long, usually shorter; flowers white or creamy white, in umbellate clusters of 3–6 flowers; corolla lobes 2.5–3.5 mm long, erect with spreading to recurved tips; corona segments acicular, 1–1.5 mm long, somewhat included in the corolla; follicles 4–6 cm long. In open rocky areas and on herbs and shrubs; in the sw portion of East TX; South TX Plains, Gulf Prairies and Marshes, Edwards Plateau and Trans-Pecos; in the U.S. known only from TX; also Mexico. May–Jul. [*Cynanchum barbigerum* (Scheele) Shinners]

Metastelma palmeri S. Watson, (for its discoverer, Edward Palmer, 1831–1911, English-American surgeon and botanist, collected in w U.S. and Mexico), PALMER'S LECHVINE. Leaf blades ca. 3 cm long, linear-lanceolate, elliptic, or lanceolate, base rounded to cuneate; flowers cream to yellowish cream, in racemiform clusters of 3–8 flowers, corolla ca. 2.5 mm long, corolla lobes erect with spreading tips; corona segments laminar, 1 mm long; follicles dark brown, 4–7 cm long. Limestone, rocky or sandy soils, thorn scrub; in East TX known only from Comal Co. (*Brown 12227b* WIS); South TX Plains, and Edwards Plateau; in the U.S. known only from TX; also Mexico. Apr–Oct. [*Cynanchum maccartii* Shinners]

NERIUM L. OLEANDER, ROSE-BAY

A monotypic genus occurring from the Mediterranean to the Cape Verde Islands and Japan (Mabberley (2017). (ancient Greek name *Nerion* for OLEANDER used by Dioscorides, perhaps from Greek: *neros*, moist, alluding to the habitat and/or evergreen habit—Quattrocchi 2000)
REFERENCES: Li et al. 1995; Lemke 2023d.

Nerium oleander L., (from Italian common name), COMMON OLEANDER, LAUREL ROSA. Shrub or small tree to ca. 3 m tall, glabrous or nearly so; sap not milky; leaves evergreen, typically in whorls of 3, sometimes opposite or in 4s, leathery, oblong-lanceolate, usually 10–22 cm long and 1–3.5 cm wide, acute to acuminate, entire; inflorescence corymbose, much-branched; flowers showy, very variable in color ranging from white, yellow, or pink to lavender, purple or red, often sweetly fragrant; calyx 5-parted, the lobes 3–10 mm long; corolla funnelform, with a narrowly cylindric tube (ca. 1.5 mm in diam.), a dilated throat (ca. 7 mm in diam.), and a deeply 5-lobed spreading limb 2.5–5 cm in diam., with a corona of 5 fringed segments in the throat; stamens attached at apex of tube; anthers with an appendage; ovaries 2, distinct; follicles 2, united until just before dehiscence, 12–23 cm long; seeds numerous, comose at apex. This species is widely cultivated as an ornamental in the s parts of East TX, persists, and has been reported as being "naturalized" in TX (Toxic Plant Database 2000); scattered in the s portion of East TX; South TX Plains and Gulf Prairies and Marshes; s 1/3 of U.S. from NC s to FL and w to CA and NV. This species is only slightly frost-tolerant and often suffers topkill; it typically survives in USDA hardiness zone 8b (minimum temp. 15–20° F; e.g., Angelina, Brazos, and Travis cos.) (Cathey 1990) and possibly in zone 8a (min. temp. 10–15° F; e.g., as far n as Dallas). Spring–fall. Native of the Mediterranean region to Asia. This species has showy nectarless flowers and exhibits deceit pollination; the seeds are reported to be dispersed by water (Mabberley 2017). There are numerous cultivars including some with double flowers. The plants contain cardiac glycosides, including oleandrin (used as a rat poison), and a single leaf is a potentially fatal dose in humans (Mabberely 2017); symptoms include vomiting, diarrhea, colic, weakness, salivation, gastrointestinal irritation, cardiac abnormalities, coma, and death (sometimes sudden); while all parts of the plant are poisonous, most animals are poisoned by eating the leaves (e.g., reports of 1/4 pound of leaves or even 10–20 leaves have been cited as being fatal to a horse) (Toxic Plant Database 2000).

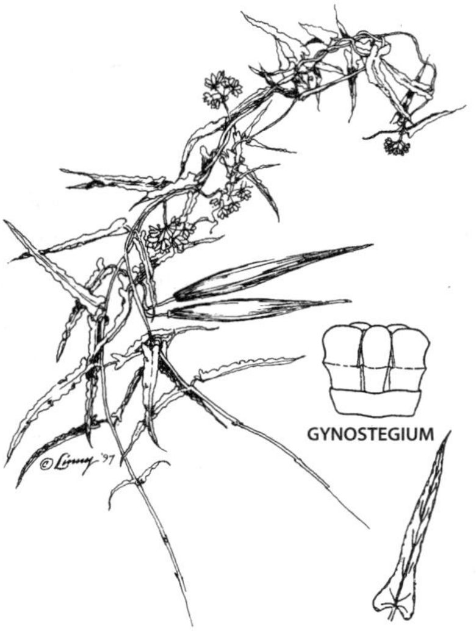

Funastrum crispum [AMB, HEA]

Funastrum cynanchoides [AMB HEA]

Gonolobus suberosus var. granulatus [BB2]

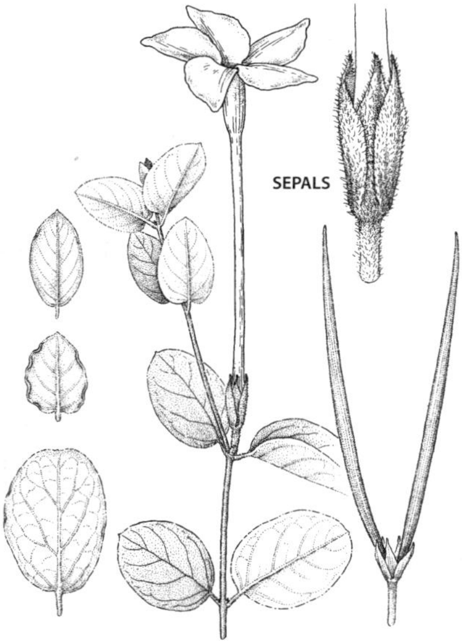

Mandevilla macrosiphon [ALS]

Matelea biflora [SHI]

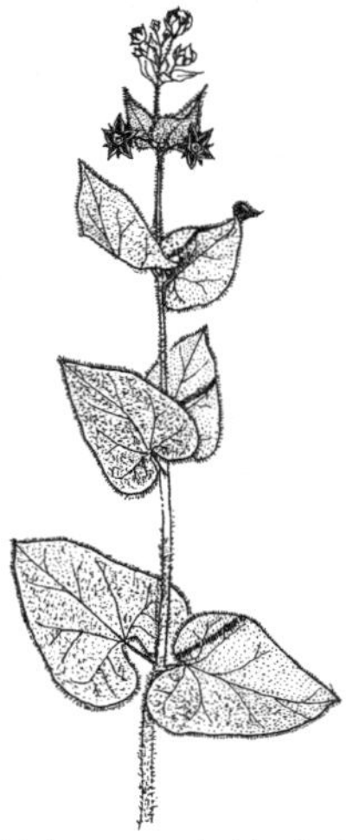

Matelea cynanchoides [BT3]

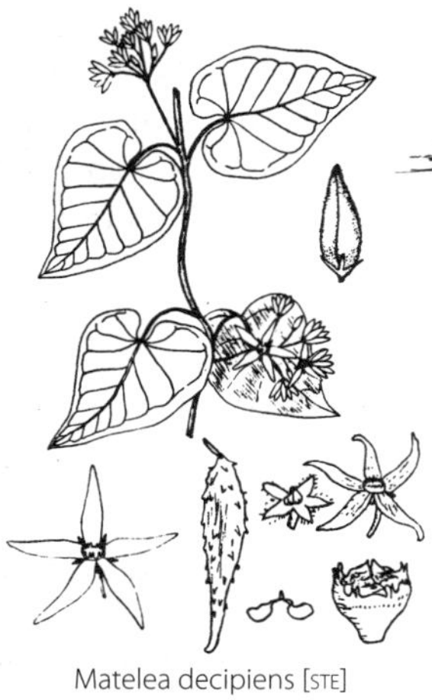

Matelea decipiens [STE]

Matelea edwardsensis [HEA]

PATTALIAS S. Watson SERPENTVINE

A genus of 2 or 3 species of the U.S., Mexico, the West Indies, and Central America (Fishbein 2023e). (Greek: *pattalias*, stake, alluding to the stout, conical beak surmounting the stigma in *P. palmeri* S. Watson, the type species)
REFERENCE: Fishbein 2023e.

Pattalias paluster (as palustre) (Pursh) Fishbein, (swamp-loving), GULF COAST SERPENTVINE. Herbaceous vine with latex white; stems twining; leaves opposite, sessile, blades 2–8 cm long, 0.1–0.6 cm wide, linear; inflorescence extra-axillary, solitary at nodes, umbelliform, peduncle 1–5 cm long, 7–20-flowered; flowers green or cream, often brown- or purple-tinged, campanulate; calyx 1–2.5 mm long, ciliate; corolla tube ca. 1 mm long, lobes spreading to reflexed, lanceolate to narrowly lanceolate, 3–4 mm long, corona of 1 whorl of 5 laminar, somewhat fleshy segments opposite stamens, connate only at very base, united to column at base of anthers, 1.5–2 mm long; follicles 4–8.5 cm long, typically solitary, variously oriented, brown, narrowly fusiform, terete, smooth, glabrous. Salt marshes, coastal dunes, flats, roadsides; in East TX known only from Harris Co. (Singhurst et al. 2018), the voucher was collected on a coastal prairie in s Harris Co. and the species may not be found growing specifically in the East TX delineated area; Gulf Prairies and Marshes; coastal U.S. from NC to TX. Apr–Oct. [*Cynanchum angustifolium* Pers., *C. palustre* (Pursh) A. Heller, *Funastrum angustifolium* (Pers.) Liede & Meve]

PERIPLOCA L. SILKVINE

A genus of 13 species native to s Europe, Asia, and tropical Africa (Endress et al. 2018); some are cultivated as ornamentals; others are used medicinally. *Periploca* is a member of the Periplocoideae, a less specialized subfamily, in contrast to the rest of the East TX members which are in subfamily Asclepiadoideae. (Greek: *peri*, around and *ploke*, twining, alluding to the climbing habit of some species)
REFERENCES: Venter & Verhoeven 1997; Endress et al. 2018.

Periploca graeca L., (of Greece), SILKVINE. Twining woody vine to 5 m long; leaves opposite, entire, petiolate; flowers in long pedunculate terminal cymes; corollas brown-purple, rotate, deeply lobed, densely villous in lines; corona a ring of 5 broad lobes, each alternating with a filament like coronal segment ca. 7 mm tall; gynostegium absent, filaments distinct; follicles linear, smooth, not angled. Cultivated and escapes; Dallas and Travis (Kartesz 2022) cos.; very scattered in KS, NM, PA, TN, and TX. May. Introduced from se Europe and w Asia.

THYRSANTHELLA Pichon CLIMBING-DOGBANE

A monotypic genus endemic to se U.S. (Lemke 2023e). It was once treated in *Trachelospermum* as the only New World species in an otherwise se Asian genus. (Greek: *thyrsus*, thyrse or panicle, and *anthos*, flower, and Latin: *ella*, diminutive, alluding to the inflorescence)
REFERENCE: Lemke 2023e.

Thyrsanthella difformis (Walter) Pichon, (of unusual or differing forms), AMERICAN STAR-JASMINE, CLIMBING-DOGBANE. Deciduous, twining vine climbing on trees; leaves opposite, petioles 1–10(–15) mm long, leaf blades 2.7–6(–12) cm long, 0.4–3.5(–7) cm wide, lanceolate to elliptic, ovate, or even suborbicular, membranous, base acute to rounded or obtuse, apex acuminate or apiculate; inflorescence axillary, thyrsiform, peduncle 2.5–5 cm long, pedicels 4–8 mm long; calyx lobes ovate-lanceolate, 3–4 mm long, pubescent or glabrous; corollas creamy white to pale yellow, corolla tube 2.5–3.5 mm long, throat 2.5–3.5 mm long with red-orange inner vertical lines, lobes spreading or reflexed, obovate, 3–4 mm long, and 2–3 mm wide; anthers not exserted; follicles 10–23 cm long, 0.1–0.3 cm wide: seeds 8–11 mm long, 1.5–2 mm wide. Along streams, forest margins, weedy areas; e 1/2 of East TX; also n Gulf Prairies and Marshes; e U.S. from NJ s to Fl and w to MO, OK, and TX. Apr–Jun. Krings (2003) discussed the nomenclatural history of this species. [*Echites difformis* Walter, *Trachelospermum difforme* (Walter) A. Gray]

TRACHELOSPERMUM Lem. CLIMBING-DOGBANE

A genus of ca. 15 species of climbing or clambering vines with milky sap native to se Asia; a number are cultivated as ornamentals (Lemke 2023f). (Greek: *trachelos*, a neck, and *sperma*, seed, referring to the narrow seeds)
REFERENCES: Krings 2003; Lemke 2023f.

Trachelospermum jasminoides (Lindl.) Lem., (resembling *Jasminum*—JASMINE), STAR-JASMINE, CONFEDERATE-JASMINE. Plant climbing and trailing; leaves opposite, evergreen, coriaceous, deep green in color, petioles 2.5–5 mm long, blade 2–6(–12) cm long, elliptic to ovate or obovate; flowers fragrant; corollas white, or with yellowish tinge near tube, tube 2.5–3.5 mm long, throat 3–4 mm long, lobes 7–10 mm long. Cultivated and spreading, climbing up sides of buildings; Grayson Co. (Austin College campus

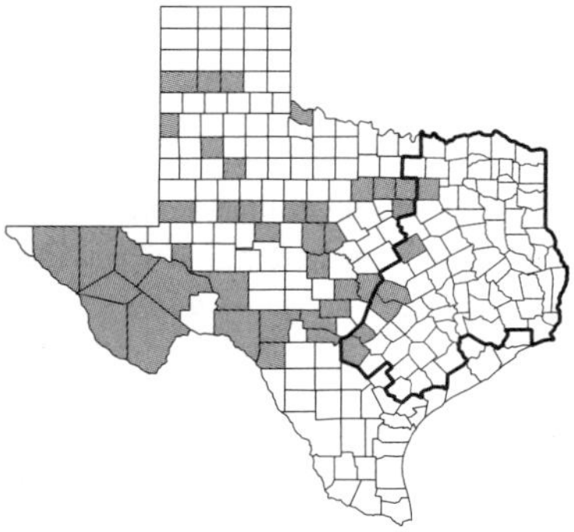
Funastrum crispum

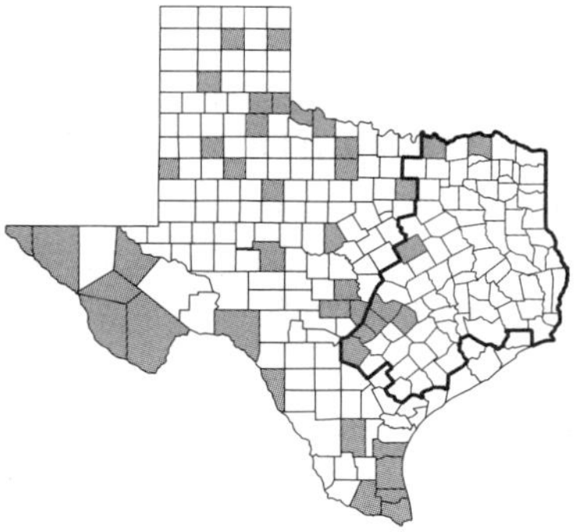
Funastrum cynanchoides

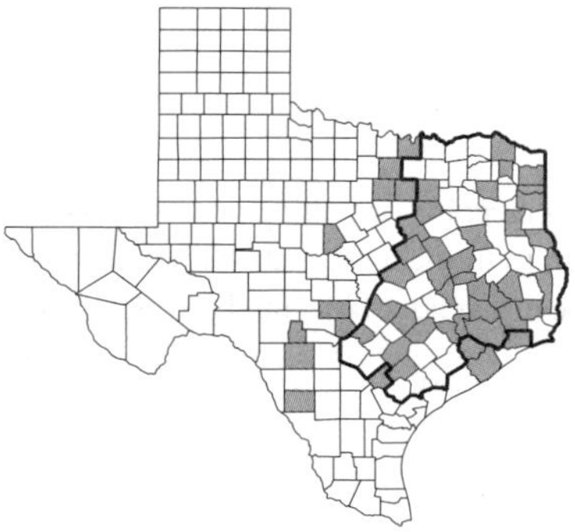
Gonolobus suberosus var. granulatus

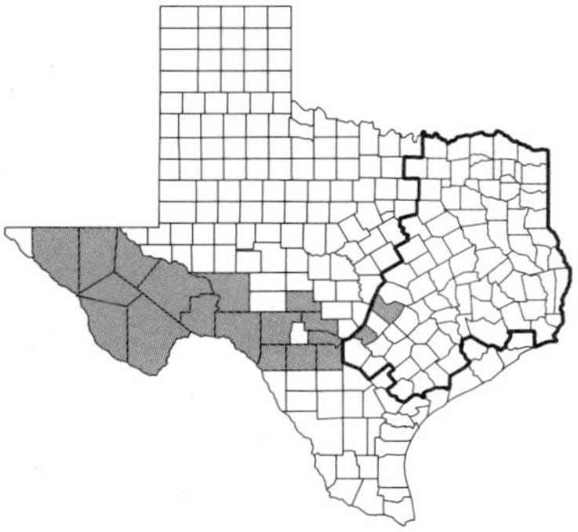
Mandevilla macrosiphon

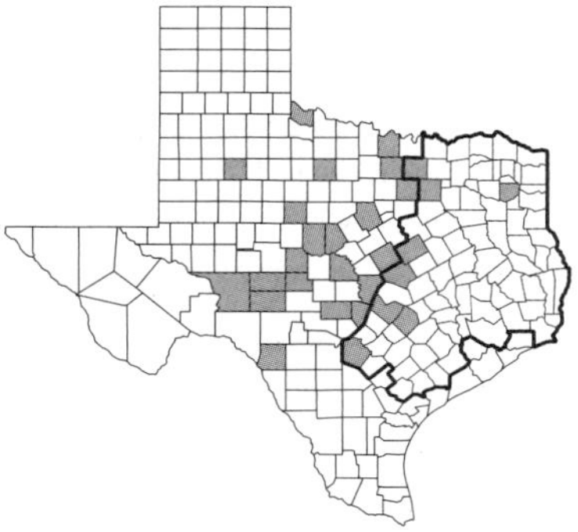
Matelea biflora

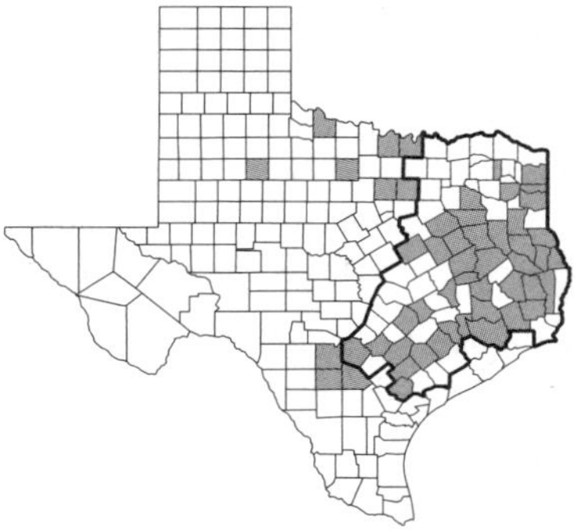
Matelea cynanchoides

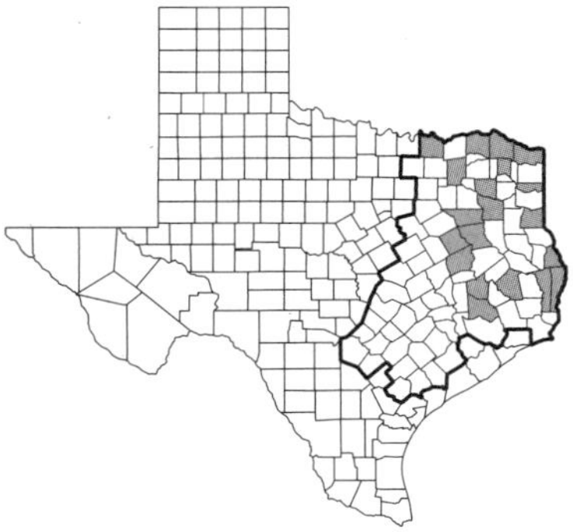
Matelea decipiens

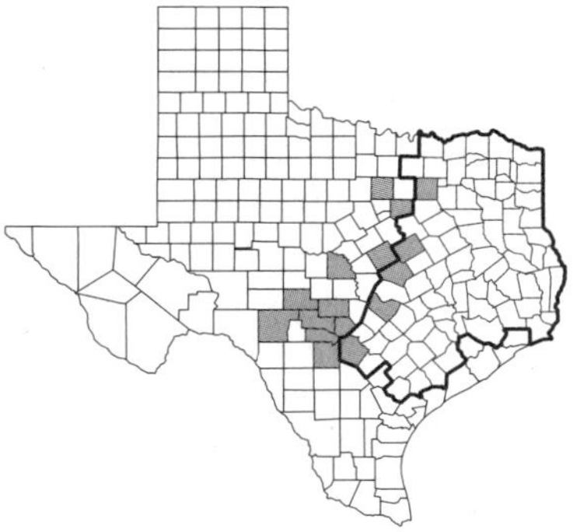
Matelea edwardsensis

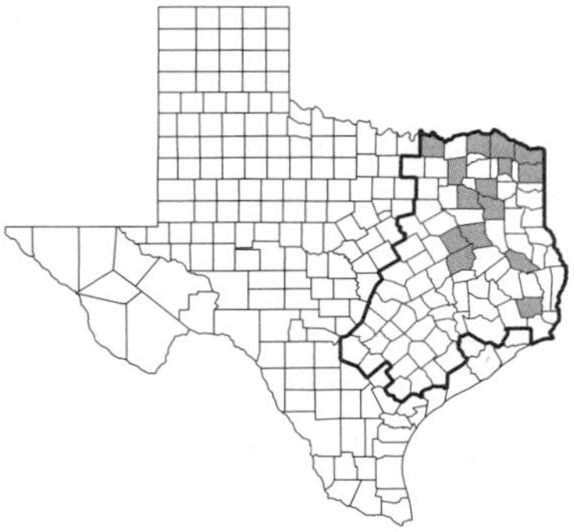
Matelea hirtelliflora

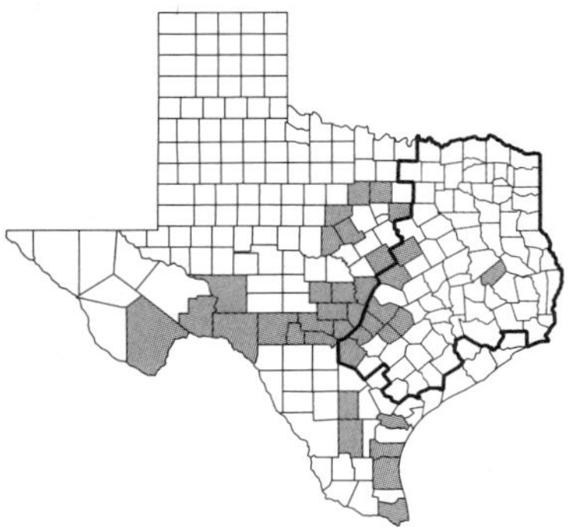
Matelea reticulata

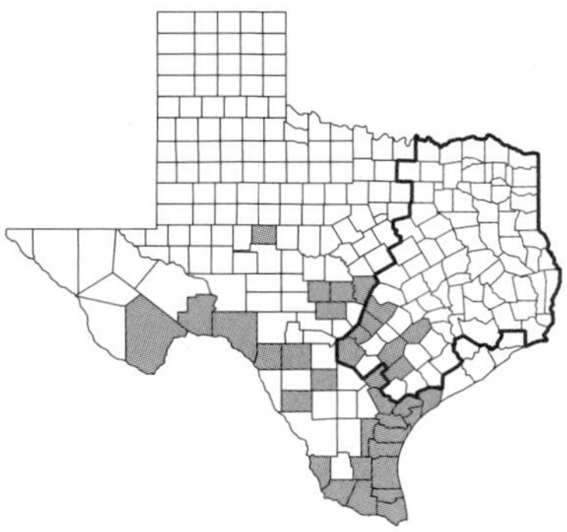
Metastelma barbigerum

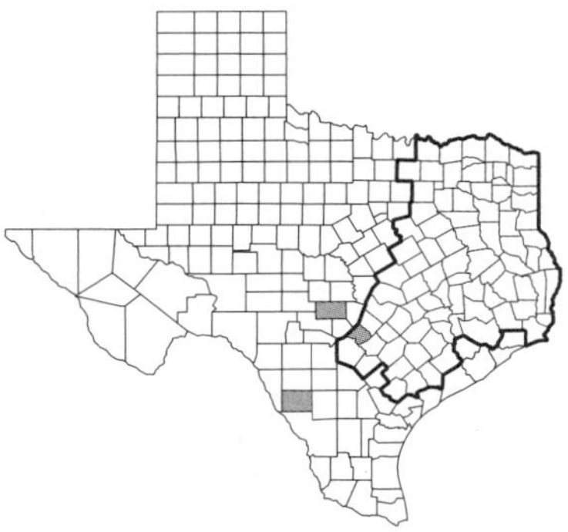
Metastelma palmeri

BRIT), also McLennan Co. (Hannick et al. 2013); widely scattered in se U.S. from SC s to FL and w to TX. May–early Jun. Native of China.

VINCA L. PERIWINKLE

Ours creeping or trailing perennials; sap not milky; leaves opposite, evergreen; flowers axillary, usually solitary; corollas salverform to funnelform with cylindrical tube and 5 equal lobes widely spreading, twisted to the left, thickened or hairy at throat, blue to lavender-blue, rarely white; paired follicles erect, glabrous, smooth; seeds without a coma.

A genus of 7 species native from Europe to w Asia including cultivated ornamentals (Lemke 2023g); some species contain alkaloids. (Name abbreviated from Latin: *vinca pervinca*, from *vincio*, to bind, alluding to the use of the foliage in wreaths; the ancient name reflected in the colloquial Italian *pervinca*, French *pervenche*, Middle English *per wynke*, and English *periwinkle*).

1. Leaves petiolate (5–15 mm long); leaf blades with ciliate margin; calyx lobes 7–15 mm long ____ **V. major**
1. Leaves short petiolate (1–2 mm long), leaf blades without ciliate margin; calyx lobes 3–4 mm long ____ **V. minor**

Vinca major L., (large), BIG-LEAF PERIWINKLE. Leaf blades to 9 cm long, to 6 cm wide, ovate to broadly ovate or lanceolate; corolla usually blue-purple, tube 4–5 mm long, throat 8–12 mm long, lobes 15–20 mm long. Cultivated, especially in limestone areas; readily spreads in woods and along streams and can overcome native herbaceous vegetation; scattered in e 1/2 of TX; Cross Timbers and Prairies and Edwards Plateau; widespread in the se U.S. and the Pacific states. Mainly Mar–Apr (but year round; observed in cultivation in Sherman flowering virtually every month of the year). Native of se Europe and sw Asia. This species has been reported to become invasive in some parts of the e and nw U.S. (Bogler 2006).

Vinca minor L., (little), COMMON PERIWINKLE. Leaf blades 2–4.5(–6) cm long, 1–1.5(–2.5) cm wide, lanceolate to elliptic; corolla usually blue-violet, tube 3–6 mm long, throat 5–7 mm long, lobes 10–15 mm long. Cultivated, wooded areas; scattered in East TX; throughout the e 1/2 of the U.S. and scattered in the w; also Canada. Native of s Europe. Mar–May. This species apparently does not produce fruits in North America. However, it can spread vegetatively in some parts of North America. It can become aggressive and potentially problematic (covering larger areas, climbing into trees, etc.) (Bogler 2006).

APODANTHACEAE Tiegh. ex Takht.

APODANTHES FAMILY

A very small (ca. 10 or 11 species in 2 genera), disjunctly distributed (North to South America, mainland Africa, Iran, and Australia, mainly tropical) family of tiny, nonphotosynthetic, endoparasitic herbs that live permanently inside trees or shrubs of the families Fabaceae or Salicaceae and that emerge outside the host tissue only to flower. They are completely parasitic, lacking chlorophyll, with the vegetative structures being made up entirely of the mycelium-like filamentous tissue inside the host and the bracts that subtend the tiny flowers. The family is extremely reduced morphologically (no roots, stems, or leaves) and is one of the world's least known families of flowering plants (Bellot & Renner 2014a; Yatskievych 2015). Its evolutionary relationships have long been unclear, both because of its morphological reduction and because of lateral gene transfer between Apodanthaceae and their hosts (Filipowicz & Renner 2010); it has previously been included in a very broadly conceived and polyphyletic Rafflesiaceae (Malpighiales) (e.g., Diggs et al. 1999), but recent molecular and morphological data strongly support its inclusion in the Cucurbitales, though its exact location in that order is still unclear (Stevens 2001 onwards; Filipowicz & Renner 2010). The Rafflesiaceae, as previously treated broadly, was composed of three or four separate unrelated lineages (Nickrent et al. 2004) and is now treated more narrowly to include only three genera of s China, Assam, Bhutan, Thailand, and w Malesia characterized by gigantic flowers (Stevens 2001 onwards; Yatskievych 2015). Pollination in Apodanthaceae is thought to be by flies and bees, and possibly also wasps (Bellot & Renner 2013; Sipes et al. 2014). Family name from *Apodanthes*, a monotypic, endoparasitic, Neotropical genus occurring from Guatemala to Brazil. (Greek: *a-*, without, *pous* or *podium*, foot, and *anthos*, flower, presumably in reference to the sessile flowers) (subclass Rosidae—Cronquist; order Cucurbitales—APG IV)

FAMILY RECOGNITION IN THE FIELD: the sole East TX species is a stem parasite with the only structures visible outside the host being the small, bud-like flowers and a few subtending, tiny, scale-like bracts.

REFERENCES: Mat Salleh 1991; Meijer 1993; Blarer et al. 2004; Nickrent et al. 2004; Filipowicz & Renner 2010; PPC 2010; Bellot & Renner 2014a, 2014b; Yatskievych 2015.

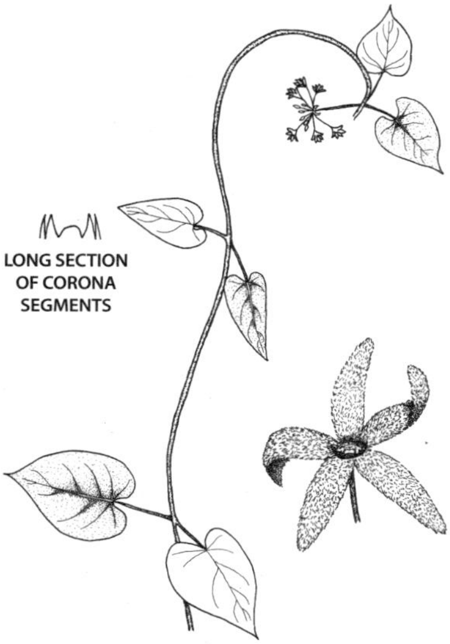

Matelea hirtelliflora [SYS2]

Matelea reticulata [BT3]

Metastelma barbigerum [HEA]

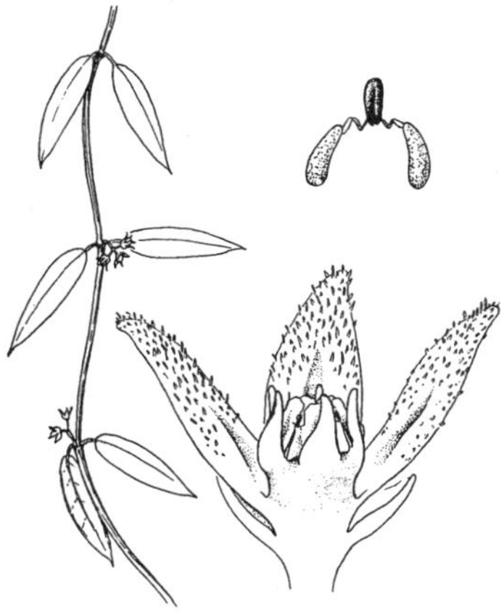

Metastelma palmeri [SID]

Nerium oleander [BR1]

Pattalias paluster [SWF]

Periploca graeca [LAM]

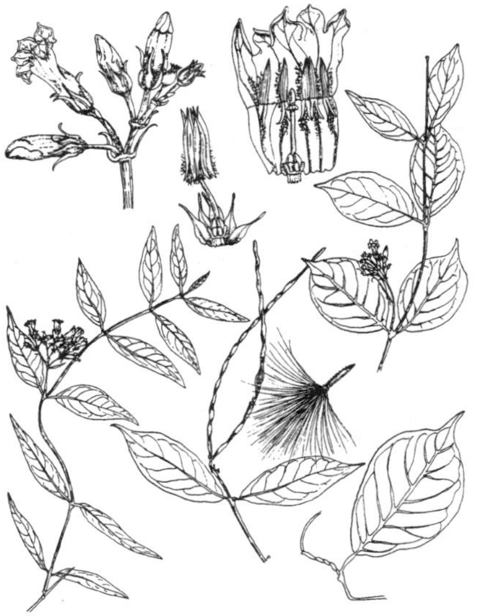

Thyrsanthella difformis [CO1]

PILOSTYLES Guill. STEMSUCKER

A genus of ca. 9–10 species (Stevens 2001 onwards) mainly of the New World (CA and TX to the tropics) but with 2 in sw Australia, 1 in Iran, and 1 in tropical Africa. They all parasitize species of Fabaceae. The genus is now treated as including *Berlinianche* (an invalid genus name) which is composed of one unusual African species (Bellot & Renner 2014a). The flowers develop within the host stems and rupture through the stem surface (Yatskievych 2015). Molecular research has recently shown that *Pilostyles* is monophyletic if *Berlinianche* is included (Bellot & Renner 2014a). According to Yatskievych (2015), though apparently dioecious, "It is not known definitely whether species of this genus are monoecious or dioecious." (Greek: *pilos*, cap, and *stylus*, column, stake, or style; presumably from the shape of the style/stigma)
REFERENCES: Rose 1909; Rutherford 1970; Kuijt et al. 1985; Bellot & Renner 2015; Arias-Agudelo et al. 2019.

Pilostyles thurberi A. Gray, (for George Thurber, 1821–1890, botanist and quartermaster with the U.S.-Mexican Boundary Survey), THURBER'S PILOSTYLES, THURBER'S STEMSUCKER. Parasite with vegetative structures ± entirely within the tissues of the host plant with only the small flowers and sometimes a few subtending scale-like bracts visible externally (complete visible portion of plant bud-like); floral bracts usually 4–7, broadly ovate, 1–1.5 mm long; flowers unisexual, solitary per short (to 1.5 mm long) peduncle, brownish to whitish or reddish, produced internally and erupting through host stems; calyx segments in 3 whorls of 3–5, similar to bracts; ringed nectary at base of sepals; petals absent; staminate flowers with many sessile anthers in 1–3 series around a column that is expanded at tip into a disk ca. 1 mm in diam.; pistillate flowers with stigmas in ring along disk margin; ovary inferior; fruit a globose capsule 1–1.5 mm in diam.; seeds numerous. Parasitic on *Dalea frutescens* (Fabaceae) in TX and also on *D. formosa* to the s and w, and also reported on various *Dalea*, *Parosela* and *Psorothamnus* (Bellot & Renner 2014a); Dewitt, Ellis (USDA Plants Database 2019), and Travis (Kartesz 2015) cos.; also Brown, Comanche, Hamilton, and Mills (HPC) cos. w of East TX, but mainly w 1/2 of TX. May–Jul. [*P. covillei* Rose]

AQUIFOLIACEAE Bercht. & J. Presl.
HOLLY FAMILY

A medium-sized (over 600 species in 1 genus, *Ilex* L.), almost cosmopolitan family of usually evergreen trees and shrubs (Mabberley 2017). Family name conserved from *Aquifolium*, a genus now treated as *Ilex* (the name *Ilex* was published earlier and thus has priority in terms of nomenclature). In the past, Aquifoliaceae has included four genera (*Ilex*, *Nemopanthus*, *Phelline*, and *Shemostemon*) (Mabberley 1987); we, however, are following Mabberley (2017) and Judd et al. (2002) in treating the Aquifoliaceae as a monogeneric family. (subclass Rosidae—Cronquist; order Aquifoliales—APG IV)
FAMILY RECOGNITION IN THE FIELD: shrubs or small trees; leaves simple, alternate; flowers usually unisexual, small, inconspicuous, axillary, often 4–5-merous; fruit a small, red, yellow, or orange berry-like drupe.
REFERENCES: Lundell 1961b; Brizicky 1964a; Judd et al. 2002.

ILEX L. HOLLY

Ours shrubs or small trees with mostly smooth, light gray bark; leaves alternate, short-petioled, simple; leaf blades nearly entire to crenulate to serrulate, or spinose, evergreen and leathery or deciduous and membranous; stipules minute, falling early; flowers 4–9-parted, small, axillary, fascicled, solitary or in sessile or pedunculate, umbel-like or cymose clusters, usually unisexual (plants dioecious, but often with some perfect flowers; however, according to Clark (2023) functional bisexual flowers have never been experimentally verified); calyces rotate or shallowly funnelform; petals rotate, oblong-elliptic, united, greenish to yellow to white; stamens (or staminodia) 4–9; pistil 1; ovary superior; fruit a small berry-like drupe, globose or nearly so, usually with 4–10 stones.

Ilex is economically important for its hard white wood, as an ornamental, and as a source of Christmas decorations. Various species are a source of stimulants (contain alkaloids including caffeine and theobromine). *Ilex paraguariensis* A. St.-Hil. (YERBA MATÉ) is used as a tea in South America. The drupes are eaten by a variety of wildlife species (Burrows & Tyrl 2001). Apparently the Roman use of holly for Saturnalia was taken over by Christians for Christmas. The fruits of *Ilex* species, while an important source of food for native birds, are considered somewhat poisonous due to the presence of saponins; they cause vomiting, diarrhea, and stupor if eaten in quantity (Hardin & Arena 1974; Lampe & McCann 1985). (Ancient Latin name of the holly-oak (*Quercus ilex* L.), rather than of the holly).
REFERENCES: Lundell 1961b; Godfrey & Wooten 1981; Manen et al. 2002; Loizeau et al. 2005; Wunderlin et al. 2020; Clark 2023; Weakley, A.S., and Southeastern Flora Team 2024.

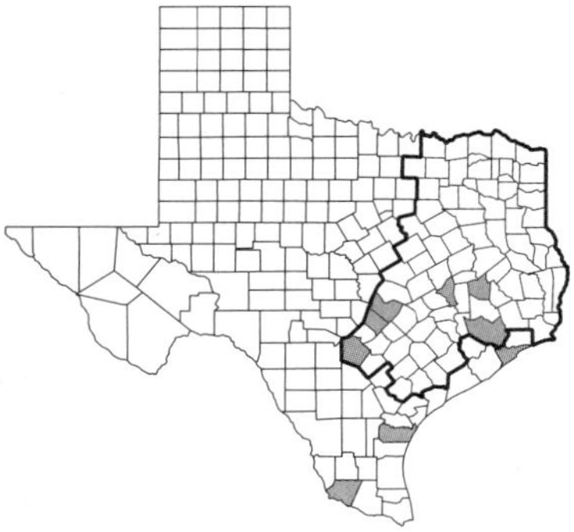

Nerium oleander

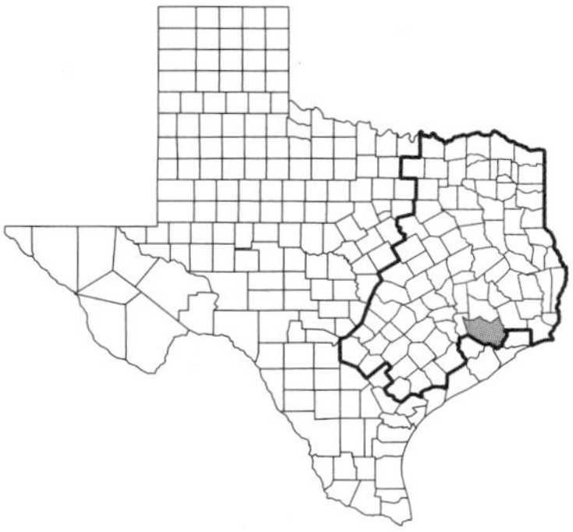

Pattalias paluster

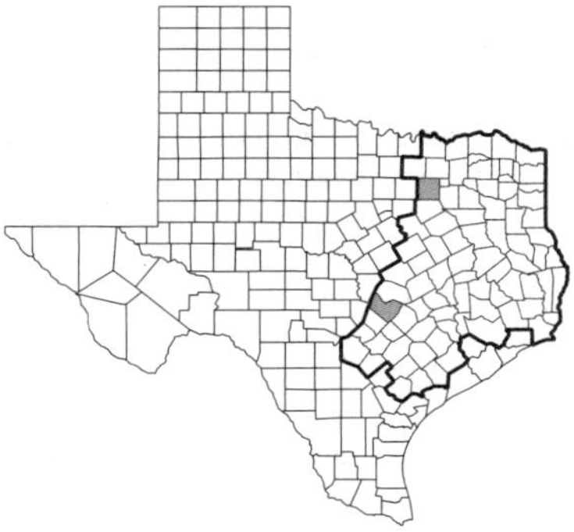

Periploca graeca

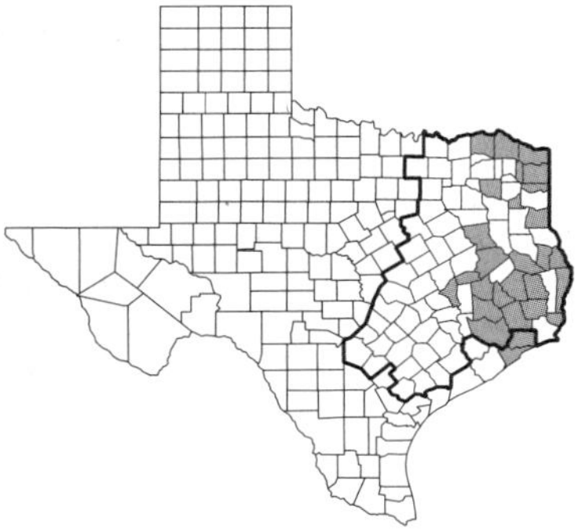

Thyrsanthella difformis

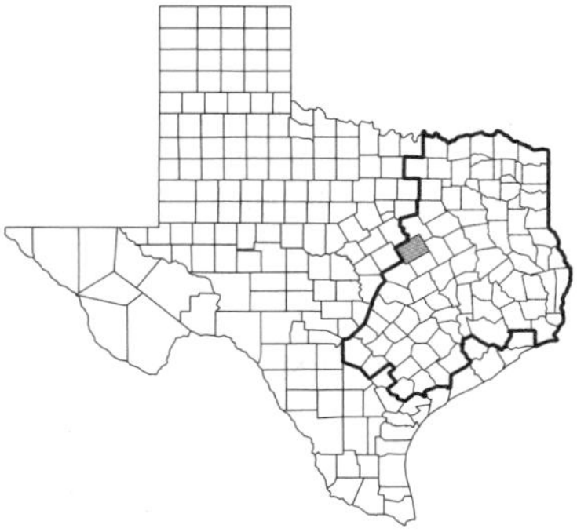

Trachelospermum jasminoides

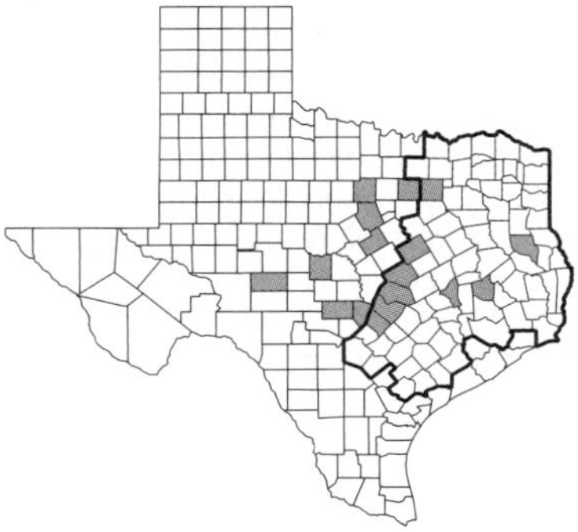

Vinca major

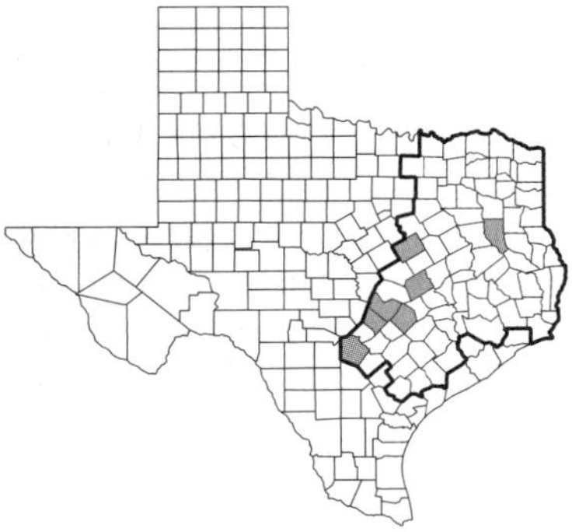

Vinca minor

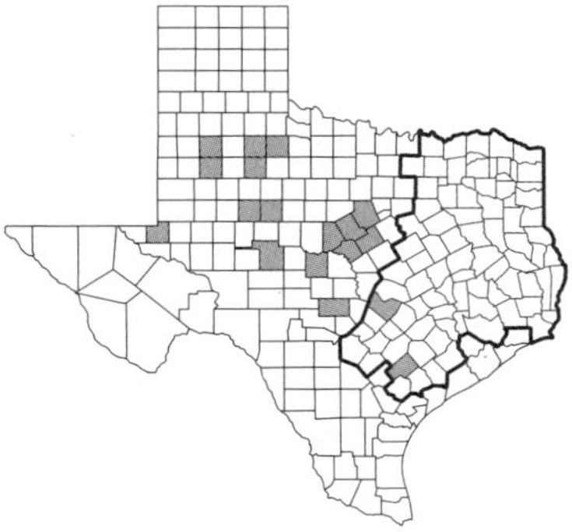

Pilostyles thurberi

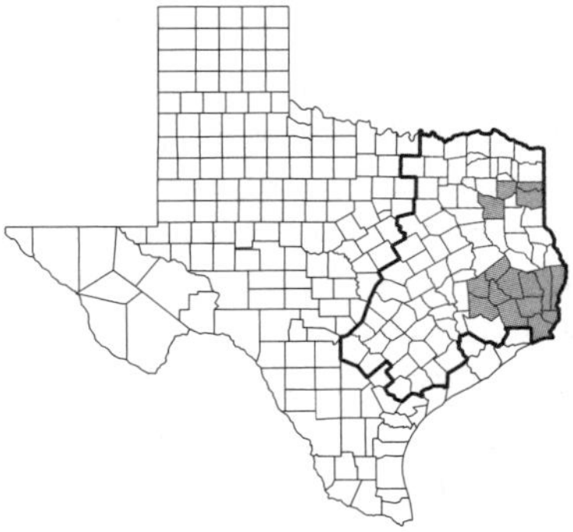

Ilex ambigua

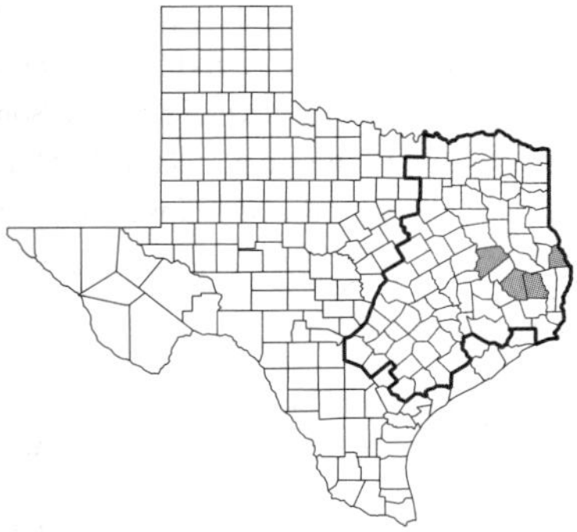

Ilex cassine

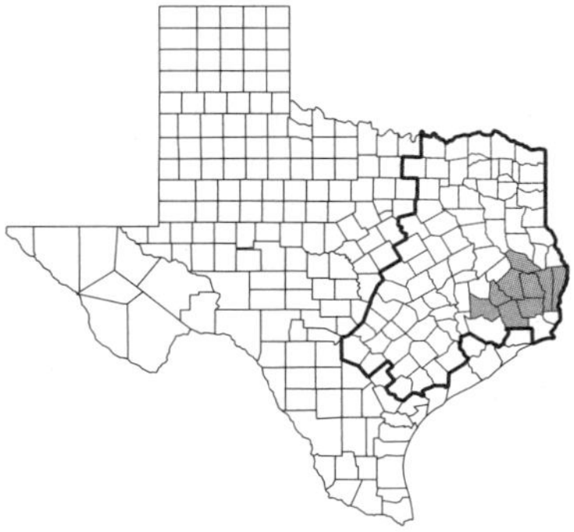

Ilex coriacea

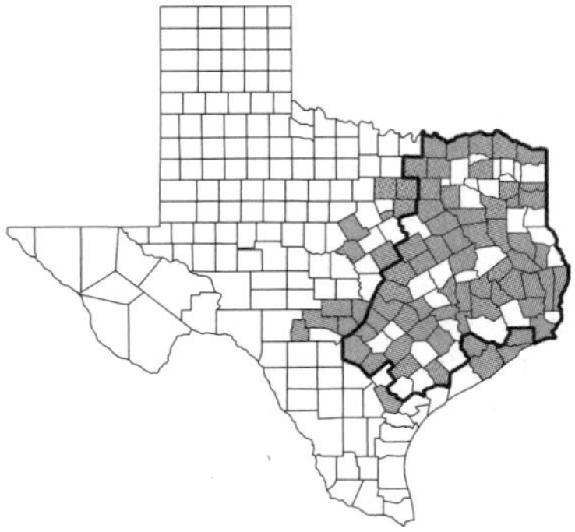

Ilex decidua

1. Leaves coriaceous, evergreen.
   2. Leaf blade margins usually distinctly crenate or crenate-serrate from base to apex; lower leaf blade surface not punctate ______ **I. vomitoria**
   2. Leaf blade margins entire, or sometimes with marginal indurate (= hardened) spines or weak teeth; lower leaf blade surface punctate or not so.
      3. Leaf blade margins entire with apical indurate (= hardened) spine or also with 2 to several marginal indurate spines; spines ≥ 2 mm long.
         4. Flowers in 1–few-flowered cymes on growth of same year; leaf blade margins entire with only an apical indurate spine or also often with several robust marginal spines, then blade often scalloped; stones ridged on the rounded back ______ **I. opaca** var. **opaca**
         4. Flowers in axillary clusters on growth of previous year; leaf blade margins entire with an apical indurate spine or also often with 2 or 4 marginal indurate spines, then blade often quadrate; stones rugose (= wrinkled) throughout ______ **I. cornuta**
      3. Leaf blade margins entire with weak apical tooth or also with few to several weak marginal teeth; teeth ≤ 2 mm long.
         5. Lower leaf blade surface punctate (minute) ______ **I. coriacea**
         5. Lower leaf blade surface not punctate ______ **I. cassine**
1. Leaves membranous, deciduous.
   6. Leaf blades mostly oblanceolate to obovate, widest above middle, base narrowly cuneate, lower surface pubescent at least on midvein.
      7. Pedicels of pistillate flowers 2–6 mm long; pedicels of staminate flowers (2–)4–8(–16) mm long; leaf blades often revolute toward base, lower surface pubescent mostly on the midvein, tomentose when abundant, margins rarely ciliate ______ **I. decidua**
      7. Pedicels of pistillate flowers (5.5–)10–30 mm long; pedicels of staminate flowers (10–)15–26 mm long; leaf blades rarely revolute, lower surface with strigose pubescence on the midvein, margins often ciliate ______ **I. longipes**
   6. Leaf blades elliptic or ovate, broadest near middle (sometimes obovate in *I. verticillata*), base rounded to broadly cuneate, lower surface glabrous to pubescent at least on midvein.
      8. Plants rare in se East TX; stones (5–)6–8 per fruit, smooth on the rounded back; staminate flowers on peduncles 2–4(–)6 mm long; pistillate flowers with entire corolla margins ______ **I. verticillata**
      8. Plants common in se East TX; stones 4–5 per fruit, with striate ridges on the rounded back; staminate flowers pedicillate (without a peduncle) or on peduncles to 2 mm long; pistillate flowers with ciliate corolla margins ______ **I. ambigua**

Ilex ambigua (Michx.) Torr., (ambiguous), CAROLINA HOLLY. Shrub or small tree rarely to 6 m high; leaves with petioles to 1 cm long, blades elliptic to ovate-elliptic, sometimes suborbiculate, apex acute to short-acuminate, lower surface glabrous to pubescent at least on midvein, margin crenate or crenate-serrulate, the lower third usually entire; staminate flowers fasciculate; pistillate flowers usually solitary, pedicels 1–4 mm long; drupe slightly globose, 5–7(–10) mm diam. In swamps, low hummocks, and along streams; Pineywoods in se and ne East TX; se U.S. from VA s to FL and w to OK and TX; also disjunct to Mexico. Mar–Apr. [*I. beadlei* Ashe ex. Kearney, *I. buswellii* Small, *I. montana* Torr. & A. Gray var. *beadlei* (Ashe ex Kearney) Fernald, *I. montana* Torr. & A. Gray var. *mollis* (A. Gray) Britton] Native of the U.S.

Ilex cassine L., (the genus *Cassine* L. in the Celastraceae was applied here as the specific epithet by Linnaeus, after a Florida Native American name for *Cassine glauca* (Rottb.) O. Kuntze or for *Ilex vomitoria* Aiton), DAHOON. Shrub or tree to 10 m tall; leaf blades to 11 cm long, 5 cm wide, size very variable, broadly to narrowly elliptic, or oblong, sometimes wider above middle, margin with a few short teeth, often entire with only the midvein extended as a small apical tooth; drupe 4–8 mm in diam.; stones 3–5, ribbed on the rounded back. Flatwoods, wet hammocks, swamps, pond margins; Polk (*Tharp s.n.*, 1942, NYBG), possibly Houston, Sabine, and Tyler (Kartesz 2022) cos., also included based on presence in TX in Wunderlin et al. (2020) and Weakley, A.S., and Southeastern Flora Team (2024); se U.S. from NC s to FL and w to TX. May–Jun, Oct–Nov. Native of the U.S. Some authors (e.g., Wunderlin et al. 2020) recognize *I. cassine* var. *myrtifolia* (Walter) Sarg., while others (e.g., Clark 2023 and Weakley, A.S., and Southeastern Flora Team 2024) recognize *I. myrtifolia* Walter. The Polk Co. record above appears to be *I. cassine* var. *cassine*. A key to the varieties from Wunderlin et al. (2020) is reproduced below.

1. Leaf blades > 1 cm wide; branchlets short-pilose or occasionally glabrous; secondary branches at an angle of < 45 degrees from main branch ______ var. **cassine**
1. Leaf blades < 1 cm wide; branchlets minutely strigose-puberulent; secondary branches at an angle > 45 degrees from main branch ______ var. **myrtifolia**

Ilex coriacea (Pursh) Chapm., (leathery), LARGE GALLBERRY, SWEET GALLBERRY. Shrub or small tree to 5 m high; leaves with petioles 3–9 mm long, blades 3–8 cm long and 1.5–3.5 cm wide, elliptic to oblong-elliptic or obovate-elliptic to oblanceolate, apex acute to obtuse or rarely slightly acuminate, base acute, margin entire (not crenate) and usually with few to several small, divergent

Trachelospermum jasminoides [HEA]

Vinca major [BEN]

Vinca minor [ORA]

Pilostyles thurberi [EMO]

Ilex ambigua [BR1]

Ilex cassine [FSM, SA4]

spinescent teeth (not prickly); flowers solitary or fasciculate; drupe globose, black rarely red; stones smooth. Flatwoods, streams, swamps, and bogs; Pineywoods in se East TX; se U.S. from VA s to FL and w to TX. Mar–Apr. Native of se U.S.

Ilex cornuta Lindl. & Paxton, (horned), HORNED HOLLY. Shrub or small tree 1–4 m tall; leaf blades 4–9 cm long, 2–4 cm wide, often quadrangular with 2 or 4 marginal spines and an apical spine, or margins entire with only apical spine and ovate, oblong-obovate, or oblong, spines conspicuously deflected above or below the plane of the leaf blade; fruiting pedicels 1.3–1.4 cm long; stones 4, rugose throughout. Escaping and possibly naturalizing; possibly scattered as an escape in TX; se U.S. from VA s to FL and w to TX and AR. Feb–Apr. Native of China. According to Weakley, A.S., and Southeastern Flora Team (2024) 'Burford' cultivars have nearly entire margins but naturalizing plants have spined margins. No county distribution map provided.

Ilex decidua Walter, (deciduous), DECIDUOUS HOLLY, POSSUMHAW, WINTERBERRY, BEARBERRY, HOLLY, MEADOW HOLLY, PRAIRIE HOLLY, WELK HOLLY. Shrub or small tree to 10 m high; leaves membranous but often thickening with age, often fascicled, petioles to 15 mm long, blades to 6(–8) cm long and 3(–4.5) cm wide, oblanceolate to spatulate or obovate to narrowly obovate, apex obtuse to rounded, often slightly emarginate, base mostly cuneate, often narrowly so, margin crenate to crenate-serrulate from near base to apex; flowers solitary or fasciculate; staminate flowers often on the ends of spur shoots; drupe globose or depressed-globose, bright red or orange, to 10 mm long; stones ribbed on the rounded back. Rock outcrops, ravines, and disturbed areas; throughout East TX; from East TX w to Edwards Plateau and Cross Timbers and Prairies; e U.S. from VA s to FL and w to MO, OK, and TX. Mar–May. [*I. curtissii* (Fernald) Small, *I. decidua* var. *curtissii* Fernald] Native of the U.S. The persistent, red, yellow or orange fruits are often conspicuous during the winter on the otherwise bare branches. Fruits poisonous (Lampe & McCann 1985).

Ilex longipes Chapm. ex Trel., (long-stalked), GEORGIA HOLLY, CHAPMAN'S HOLLY. Shrub or small tree rarely to 7 m high; leaves with petioles 3–10 mm long, blades to 6 cm long and 3 cm wide, obovate or narrowly oblanceolate, sometimes elliptic, apex short-acuminate to acute, base broadly to narrowly cuneate, margins crenate-serrulate to crenate; staminate flowers solitary or fasciculate; pistillate flowers usually solitary; stones of drupe with inconspicuous ridges on the rounded back. In woods, often on sandy banks of streams; in se East TX; se U.S. from NC s to FL and w to AR and TX. Mar–May. [*I. decidua* var. *longipes* (Chapm. ex Trel.) H.E. Ahles, *I. longipes* var. *hirsuta* Lundell] Native of the U.S.

Ilex opaca Aiton, (shady), AMERICAN HOLLY, CHRISTMAS HOLLY. Tree to 16 m high; leaves with petioles 5–12(–18) mm long, blades to 12 cm long and 6 cm wide, elliptic to obovate or oval, apex acute to acuminate and spine-tipped, base obtuse to rounded or sometimes acutish or subcuneate; margin entire often scalloped with several indurate spines; occasionally spine only at apex; spines deflected < 30 degrees from plane of leaf blade; drupe globose or ellipsoid, red, rarely yellow to 12 mm in diam.; stones ribbed on the rounded back. In moist woods, hammocks, along banks of streams, and in swamps; throughout se East TX; e U.S. from NY s to FL and w to MO, OK, and TX. Apr–Jun. Native of the U.S.

Ilex verticillata (L.) A. Gray, (whorled), DOWNY WINTERBERRY, BLACK-ALDER. Shrub to small tree to 8 m high; leaves with petioles 1–1.6 cm long, blades 8 cm long and 3.5 cm wide, usually elliptic or broadly lanceolate, sometimes obovate, apex acute to acuminate, upper surface rugose (= wrinkled), margin serrulate to crenate-serrulate; drupe bright-red to yellow. In or about swamps, pond margins, river banks, and damp thickets; in TX recorded only for Harris (*Brown 9990* BRIT) and Orange (Kartesz 2022) cos.; e U.S. from ME s to FL and w to MN, AR, and TX. Apr–Jun. [*I. bronxensis* Britton, *I. fastigiata* E.P. Bicknell, *I. verticillata* var. *cyclophylla* B.L. Rob.] Native of the U.S.

Ilex vomitoria Aiton, (emetic), YAUPON, YAUPON HOLLY, CASSINE, CASSENA, CASSIO-BERRY BUSH, CHOCOLATE DEL INDIO, EMETIC HOLLY, EVERGREEN CASSENA, EVERGREEN HOLLY, INDIAN BLACKDRINK, SOUTH-SEA-TEA, PUKE HOLLY. Shrub or tree to 8 m high; leaves with petioles usually 1–3 mm long, leaf blades usually 1–3 cm or sometimes up to 5.5 cm long (largest leaves usually found on the midstem (Stalter & Kincaid 1994) and 0.8–2.5 cm wide, elliptic to oblong or oblong-elliptic to ovate-elliptic, apex obtuse or rounded and often minutely emarginate and mucronulate, base obtuse to rounded or acutish, lower leaf surface glabrous, margin crenate or sometimes crenate-serrulate; inflorescences pedunculate or not so; drupe bright red, to ca. 6.5 mm in diam.; stones shallowly ribbed on the rounded back. Commonly cultivated, low woods, sandy areas; throughout much of se East TX, scattered to the n; e Edwards Plateau and Gulf Prairies and Marshes; se U.S., from TN s to FL and w to OK and TX; disjunct in Mexico. Mid–late Apr. Native of se U.S. This species is one of the most widely used evergreen landscape shrubs in East TX. The leaves contain caffeine and were used by Native Americans to make a ceremonial drink (known as "black drink"); during religious festivals they drank large amounts of a strong preparation which caused vomiting—thus the specific epithet (Havard 1896; Alston & Schultes 1951; Correll & Johnston 1970; Cox & Leslie 1991). Fruits poisonous (Lampe McCann 1985).

Ilex amelanchier M.A. Curtis ex Chapm., (name from the genus *Amelanchier*, which it resembles), SARVIS HOLLY. ILEX AMELANCHIER can be easily identified because no other East TX deciduous *Ilex* has a stone with a prominent narrow midsagittal

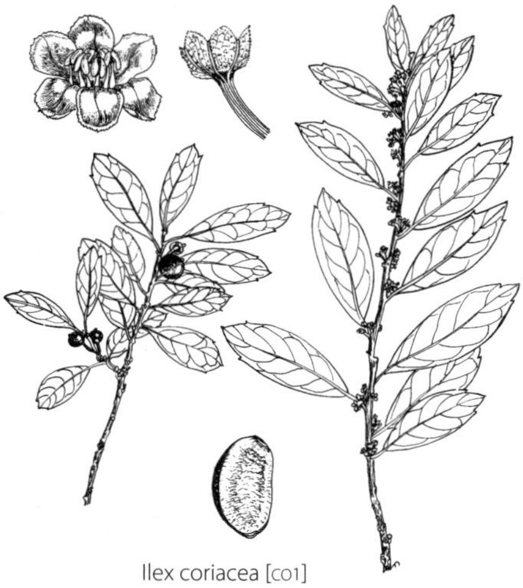

Ilex coriacea [CO1]

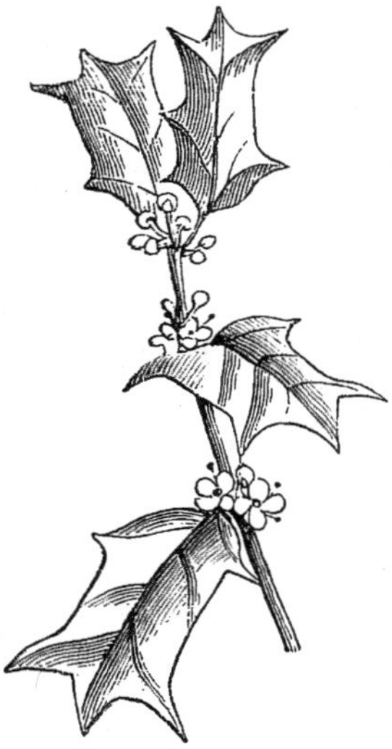

Ilex cornuta [NIC]

Ilex decidua [SA3]

Ilex longipes [G&F]

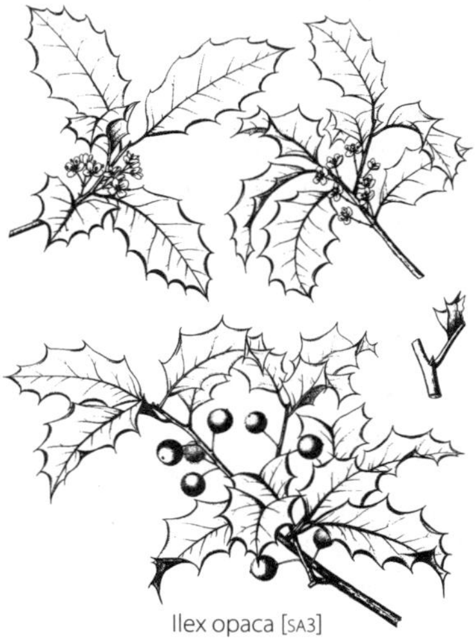

Ilex opaca [SA3]

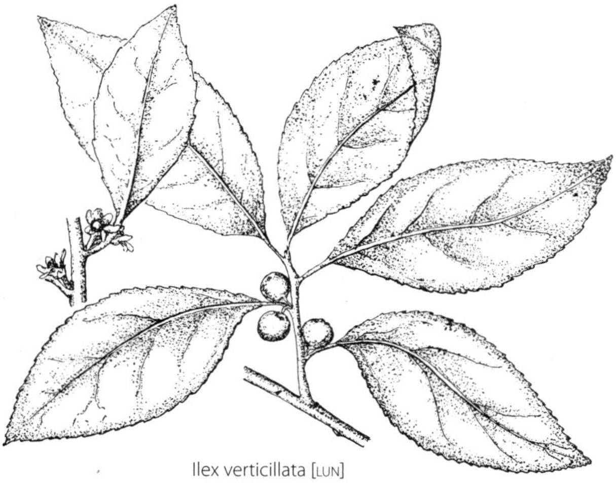

Ilex verticillata [LUN]

Ilex vomitoria [SA3]

ridge and a lower blade surface with conspicuous areoles bordered by pubescent tertiary veins. Wooded stream banks, pond edges, low woods, and sandhills; on the e border of ET in Vernon Par., LA; se U.S. from NC s to FL and w to LA; endemic to se U.S. Apr–May.

Ilex glabra (L.) A. Gray, (smooth, without hair), INKBERRY. Very similar vegetatively and in fruit to *Ilex coriacea*, except *I. glabra* has smaller leaf blades 1.5–7 cm long and 0.7–2 cm wide (almost never > 2 cm wide), margins often slightly crenate above middle. *Ilex coriacea* usually has larger leaf blades 3–8 cm long and 1.5–3.5 cm wide, margins usually entire (not crenate). In low sandy soil, usually in pine lands, and swamps; on the e border of East TX in Sabine Par., LA (*Pavlu 1135* LTU); cited in Thomas and Allen (1996); coastal U.S. from MA s to FL and w to LA. Feb–Jul. Native of the U.S.

ARALIACEAE Juss.

GINSENG OR ARALIA FAMILY

Trees, shrubs, woody vines, or perennial herbs; leaves usually alternate, simple or 2–3-compound, petiolate, deciduous or evergreen; inflorescences simple or compound umbels or panicles of umbels; flowers small, usually numerous, mostly perfect, radially symmetrical; petals 5, separate; stamens 5; ovary inferior; fruits drupes or schizocarps.

A medium-sized (1,450–1,625 species in 38–43 genera (Stevens 2001 onwards; Mabberley 2017; Kang et al. 2023), mainly tropical family with a few in temperate regions. It contains trees, shrubs, lianas, woody epiphytes, or rarely herbs; leaves are often large and compound. The family includes a number of ornamentals (e.g., Hedera, Schefflera) as well as the economically important Panax quinquefolius L. (GINSENG), which supposedly acts as a stimulant and aphrodisiac. It has long been known that Araliaceae are closely related to Apiaceae and some workers have argued that the two should be combined into a single family (Judd et al. 1994). Extensive molecular research has been done on the two families and though related in the Apiales, the most recent research and modern treatments support the view that the families are two distinct lineages, with Apiaceae more closely related to Myodocarpaceae, a small family of 2 genera and 19 species of New Caledonia, Malesia, and Queensland, Australia (Plunkett & Lowry 2001; Stevens 2001 onwards; Chandler & Plunkett 2004; Soltis et al. 2018). Nonetheless, some changes in traditional delimitation have been necessary—the most striking example is that Hydrocotyle and closed related taxa (e.g., the genus Trachymene) are now treated in the Araliaceae rather than the Apiaceae; their subfamily, the Hydrocotyloideae, is sister to the rest of the Araliaceae (Chandler & Plunkett 2004; Plunkett et al. 2004; Nicolas & Plunkett 2009). (subclass Rosidae—Cronquist; order Apiales—APG IV)

FAMILY RECOGNITION IN THE FIELD: the East TX species all have numerous small flowers in umbels (these umbels are solitary or variously arranged) and include a conspicuously *prickly shrub or small tree*, an introduced, *woody, evergreen vine* with alternate leaves exhibiting leaf dimorphism—at least some leaves are 3–5-palmately lobed, or *creeping herbs* with leaves peltate or with petiole attached to blade at base of a deep notch. In general in East TX, the similar Apiaceae (e.g., also with small flowers in umbels and inferior ovaries) are herbaceous and have schizocarp fruits while Araliaceae are usually (except the morphologically distinct *Hydrocotyle*) woody with drupes. The Araliaceae also lack the distinctive plant defense compounds, furanocoumarins, found in Apiaceae (Stevens 2001 onwards).

REFERENCES: Smith 1944; Graham 1966; Eyde & Tseng 1971; Judd et al. 1994; Plunkett & Lowry 2001; Wen et al. 2001; Frodin & Govaerts 2003; Chandler & Plunkett 2004; Lowry et al. 2001, 2004; Plunkett et al. 1996, 1997, 2004; Nicolas & Plunkett 2009; Kang et al. 2023.

1. Plant a conspicuously prickly erect shrub or small tree to 12 m or more tall; leaves compound ____ **Aralia**
1. Plant unarmed, creeping or a trailing or high-climbing vine; leaves simple.
   2. Plant a glabrous creeping herb, of damp ground or shallow water; leaf blades round to kidney-shaped, either peltate or petiole attached to blade at base of deep notch; fruit a schizocarp 4 mm or less in largest dimension ____ **Hydrocotyle**
   2. Plant a woody trailing or high-climbing vine of various habitats with 4–10 rayed stellate pubescence; leaf blades ovate to lanceolate or 3–5-palmately lobed, cuneate to shallowly cordate at base; fruit a berry-like drupe 5–8 mm in diam. ____ **Hedera**

ARALIA L. HERCULES'-CLUB

A genus of 71 species of North America, e Asia, and Malesia (Mabberely 2017). As now treated the genus includes the segregate genera *Pentapanax* and *Sciadodendron* (Lowry et al. 2004). *Aralia* is closely related to the well-known, medicinally important genus *Panax* (GINSENG) (Wen et al. 2001a). Some species are cultivated as ornamentals, others are used medicinally, and some are edible. (From *aralie*, the French Canadian name—Hyam & Pankhurst 1995)

REFERENCES: Wen & Reveal 1992; Wen 1993, 2000, 2001a, 2011; Wen et al. 1998.

Aralia spinosa L., (spiny), DEVIL'S-WALKINGSTICK, AMERICAN ANGELICA TREE, HERCULES'-CLUB, TEAR-BLANKET. Shrub or small tree to 12+ m tall, unbranched or few-branched, with the large, often ± clustered leaves giving the plant an umbrella-like or palm-like appearance; rhizomes produced, sometimes resulting in colonies; stems and leaves armed with stout prickles; leaves alternate, deciduous, 2–3-pinnately compound, to 1 m or more long, red to bronze-red with yellow in fall; leaflets elliptic to ovate, 3–10 (–13) cm long, to 7 cm wide, acute to acuminate, serrulate to serrate, petiolulate; inflorescence a large (to 6,000 or more small flowers—White 1984), terminal, compound panicle of umbels, the main branches to 60 cm long, the branches and pedicels with pubescence, when in fruit purplish red; flowers numerous, pedicellate, the pedicels 5–10 mm long; petals 5, creamy white, 2–3 mm long; stamens 5; carpels (4–)5(–6); ovary inferior; fruits berry-like drupes, 4–6 mm in diam., capped by persistent style bases, purplish black to ± black. Wooded areas and thickets, often along streams; e part of East TX, w to Anderson, Brazos, and Red River (Turner et al. 2003; Kartesz 2015) cos.; widespread in the e U.S. from ME s to FL w to IL, MO, OK, and TX. May–Jun. The species is cultivated for its palm-like growth form, the large white-flowered inflorescences, and the beautiful, red to bronze-red with yellow fall foliage. The leaves are one of the largest (to > 1 m long), if not the largest, of any tree in temperate North America. Smith (1982) noted that there is significant variation in the amount of prickles (= outgrowths of the epidermis and cortex) on the leaves, with some leaves having few or no prickles; he concluded that the prickle-less leaves were on "adult" or reproductive *branches* or *individuals* versus the heavily prickled "juvenile" or non-reproductive phase; White (1988), suggested that the amount of prickles reflects vulnerability to browsers and the allocation of resources towards defense during establishment; he noted that young tissues within 3 meters of the ground would be expected to have greatest prickle density; further, he found more prickles on sprouts after injury. The bark has been used medicinally and the bird-dispersed fruits are said to have been used to dye hair (Yatskievych 2006). The leaves and fruits are reported as potentially poisonous (Hardin & Brownie 1993) and the bark and roots can cause dermatitis in some individuals (Yatskievych 2006).

HEDERA L. IVY

A genus of 12 species (Mabberley 2017) of vines native to Europe, the Mediterranean region, and e Asia; thought to be related to *Schefflera*, a widely used ornamental. The genus is complex taxonomically, with a great deal of morphological and chromosomal variability and diploids, tetraploids, hexaploids, and octoploids known (2n = 48 to 192) (Vargas et al. 1999; Yi et al. 2004). (The ancient Latin name)

REFERENCES: McAllister & Rutherford 1990; Vargas et al. 1999; Ackerfield & Wen 2002; 2003; Jury 2004; Yi et al. 2004; Metcalfe 2005; Melzer et al. 2012; Valcárcel et al. 2014.

Hedera helix L., (Greek, *helisso*, "I twist myself around"—Paulson et al. 2010; for a twining plant), ENGLISH-IVY. Woody, evergreen, trailing or high-climbing vine; young twigs with 4–10-rayed stellate to peltate pubescence, and with often dense unbranched adventitious roots used in attachment and climbing; leaves alternate, evergreen, long lived (3–4 years), dark green, often with paler markings and paler below, sometimes tinged with purple in winter, dimorphic, the juvenile and adult differently shaped, the juvenile 3–5-palmately lobed, the adult ovate to lanceolate, only on reproductive shoots which are borne exclusively on vertical stems; flowers small, radially symmetrical, 5-merous, in terminal, solitary or racemose, globose umbels; petals yellowish-green, 3–4 mm long; ovary inferior; fruit a small, bluish-black or greenish-black, 2–5-stoned berry-like drupe 6–8 mm in diam., the pulp purple. Widely cultivated ornamental that long persists around old homesites and occasionally spreads to adjacent vegetation, wooded areas; widely scattered in East TX—e.g., Grayson (BRIT!; G. Diggs, pers. obs.), Bexar, Dallas, Harris, Travis, Walker (Turner et al. 2003), Jasper, and Newton (Kartesz 2015) cos.; also scattered to the w of East Texas—e.g., Gillespie (Turner et al. 2003) and Tarrant (R. O'Kennon, pers. obs.) cos.; widespread in the e U.S. w to Michigan, MO, and TX, also coastal states in the w U.S. and adjacent British Columbia and scattered in inland w states. Flowering in fall. Native of Europe, w Asia, and n Africa. This invasive species can form a dense ground cover, excluding other plants, and can potentially damage trees by overtopping the foliage and stealing light, thus reducing the tree's photosynthetic output. The species reproduces vegetatively (stem fragments sprout when contacting the soil) but also by seeds. The juvenile stems bear lobed shade leaves in a single plane while the adult flowering stems have unlobed sun leaves spirally circling the stem (Graham 1966; Metcalfe 2005). The leaves and fruits contain the saponic glycoside, hederagenin, which if ingested can cause vomiting, diarrhea, fever, muscular weakness, breathing difficulties, and coma; the sap can cause dermatitis with blistering and inflammation—this is apparently due to polyacetylene compounds (e.g., falcarinol) which are also present; gloves should be used when trimming the plant (Schmutz & Hamilton 1979; Hausen et al. 1987; Spoerke & Smolinske 1990; Turner & Szczawinski 1991; Metcalfe 2005; Paulsen et al. 2010; Texas Invasive Species Institute 2014). Ollerton et al. (2007) discussed pollination in this species and noted that while it was visited by a number of different insects, it is functionally quite specialized, with wasps in the genus *Vespula* being the main pollinators. The species is complex taxonomically with a number of subspecies, varieties, or forms (e.g., see Ackerfield & Wen 2003; Jury 2004); there are numerous (400+) cultivars including variegated forms. Our material is presumably subsp.

helix; however, we are not distinguishing infraspecific taxa. Further, Weakley et al. (2012) pointed out that in VA it is possible that some escaped material may be the morphologically similar *H. hibernica* (G. Kirchn.) Bean, IRISH IVY, the situation is possibly the same in TX, though we have not carried out research on TX material; it should be looked for in TX. This tetraploid species, which has sometimes been considered to be part of *H. helix* in the broad sense, can be distinguished from the diploid *H. helix* as follows (McAllister & Rutherford 1990): *H. hibernica* with leaf sinuses mostly deep and often strongly arched with the leaves somewhat funnel-shaped; stellate hairs on leaves appearing less dense due to the hair branches lying parallel on the leaf surface; sap odor strong, often pine-like and sweet; *H. helix* with leaf sinuses sometimes deep but the blade usually flat; stellate hairs with branches projecting at various angles and becoming enmeshed, not simply parallel to leaf surface; sap odor much weaker and usually unpleasant.

HYDROCOTYLE L. WATER-PENNYWORT, NICKELS-AND-DIMES

Ours glabrous creeping perennials of damp ground or shallow water, rooting at the nodes; leaves long-petiolate; leaf blade peltate or deeply divided to about the middle, sometimes ± round, shiny above; umbels simple, branched, or verticillate (= whorled); flowers inconspicuous; sepals absent; petals yellow-green or white; fruits schizocarps of 2 mericarps, flattened laterally, without secondary ribs, not reticulate.

A cosmopolitan genus of ca. 130–200 species (Stevens 2001 onwards; Mabberley 2017) often with peltate leaves; some are grown as ornamental ground covers or cultivated in water gardens or other aquatic habitats; in certain situations they can spread aggressively. This genus has traditionally been included in the Apiaceae (e.g., Correll & Johnston 1970; Hatch et al. 1990; Diggs et al. 1999). The Araliaceae and Apiaceae are clearly related and some workers in the past advocated that the two families should be combined (e.g., Judd et al. 1994). Others have suggested that if Araliaceae and Apiaceae are recognized as separate families (as done here), then the Hydrocotylaceae should be recognized as well (e.g., Thorne 1992a, 1992b). The most recent molecular research (e.g., Chandler & Plunkett 2004; Nicolas & Plunkett 2009; Kang et al. 2023) supports recognizing the two families, but transferring *Hydrocotyle* to the Araliaceae (the approach we are following in this treatment); its subfamily, the Hydrocotyloideae, is sister to the rest of the Araliaceae (Stevens 2001 onwards; Plunkett et al. 2004). (Greek: *hydr*-, water, and *kotyle*, cup, alluding to the hollows in the center of the leaves of some species)

REFERENCES: Diggs & Lipscomb 2005; Nicolas & Plunkett 2009; Konstantinova & Yembaturova 2010.

1. Leaves peltate (= petiole attached to middle of lower surface of leaf blade).
  2. Flowers in a simple umbel (= all flowers attached at the same point) ____ **H. umbellata**
  2. Flowers in a branched umbel or in whorls along an inflorescence axis (all flowers definitely not attached at the same point).
    3. Flowers in a branched umbel, the branches more than 2; species mainly coastal (e.g., Jefferson Co. near extreme se margin of East TX), otherwise known in East TX only from a surprising occurrence in Travis Co. ____ **H. bonariensis**
    3. Flowers in whorls along an unbranched inflorescence axis forming an interrupted spike or spike-like raceme or the axis with only 2 branches; species widespread in East TX ____ **H. verticillata**
1. Leaves not peltate, the petiole attached at base of a notch.
  4. Plants aquatic; stems and petioles fleshy; individual flowers and fruits with short but distinct pedicels; widely scattered in East TX ____ **H. ranunculoides**
  4. Plants terrestrial; stems and petioles thread-like; flowers and fruits sessile; reported in TX only from Dallas Co. ____ **H. sibthorpioides**

Hydrocotyle bonariensis Lam., (of or from Buenos Aires), LARGELEAF WATER-PENNYWORT, BEACH WATER-PENNYWORT, COASTAL PLAIN WATER-PENNYWORT, SALT WATER-PENNYWORT, SOMBRERILLO. Leaf blades to 12(–15) cm in diam., but usually smaller with 12–19 shallow lobes, variable in size; petioles to 37 cm long; inflorescences branched, with more than 2 branches; umbels many-flowered. Wet places; Hardin (*S.D. Jones 14366* ASU), Harris (Singhurst et al. 2018), Jefferson (*C.L. Lundell & A.A. Lundell 11271* LL), Orange (*J. Norsworthy s.n.*, 1962, VSC) cos. near se margin of East TX; also Travis Co. (*B.C. Tharp s.n.*, 15 May 1938, TEX; however, the herbarium sheet has an annotation by B.L. Turner, "doubtful locality (A Tharp collection!) probably from coast") on w margin of East TX; there are also three geographically surprising collections to the w of East TX in Parker and Tarrant cos. (BRIT; TORCH 2025); mainly Gulf Prairies and Marshes and s tip of South TX Plains; se U.S. from VA s to FL w to TX; apparently native from North to South America. [*H. bonariensis* var. *texana* J.M. Coult. & Rose] May–Nov. According to Carr (2002a), the "Texas range of this water-penny is mostly confined to somewhat saline soils of swales and wet grasslands along the Gulf of Mexico; its occurrence here [Travis Co.], so far from maritime influence, is somewhat peculiar." The species is reported to be an aggressive lawn weed in Australia (Mabberley 2017).

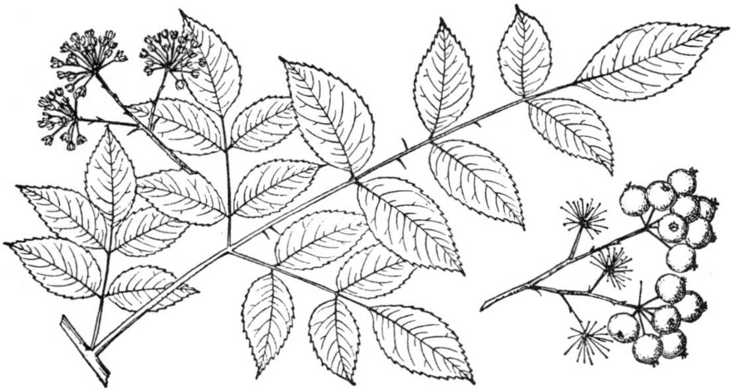

Aralia spinosa [SA2]

Hedera helix [BL1]

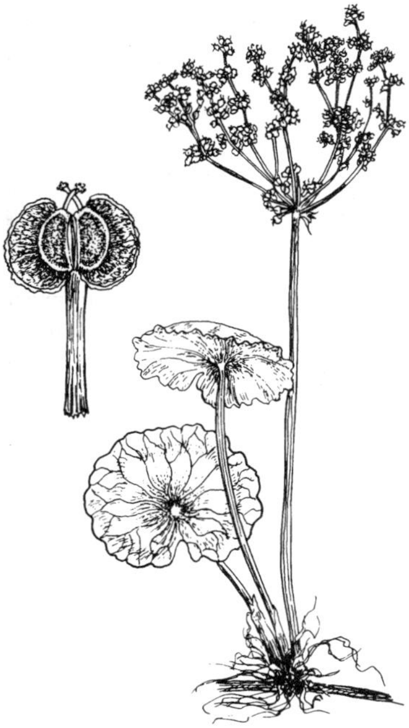

Hydrocotyle bonariensis [CO1]

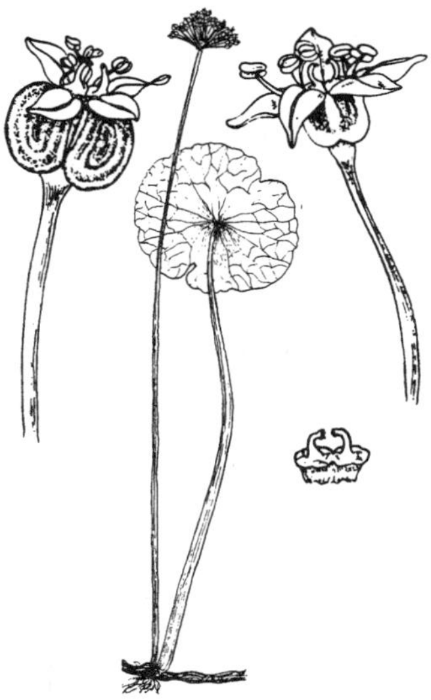

Hydrocotyle umbellata [LUN]

Hydrocotyle ranunculoides [MAS]

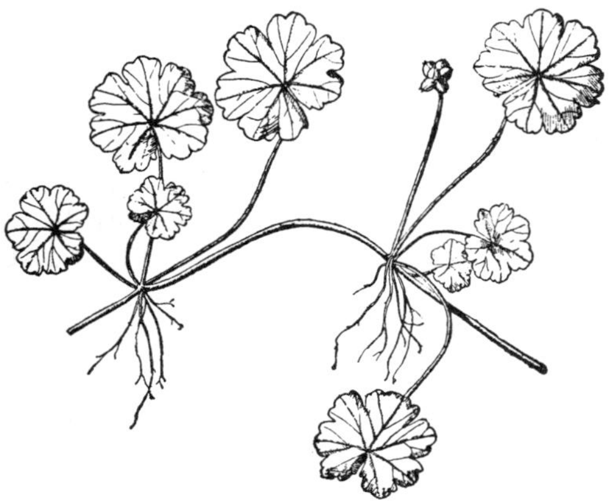

Hydrocotyle sibthorpioides [BA1]

Hydrocotyle verticillata var. triradiata [MAS]

Hydrocotyle ranunculoides L. f., (resembling *Ranunculus*—buttercup), FLOATING WATER-PENNYWORT, BUTTERCUP WATER-PENNYWORT. Leaf blades roundish reniform, to 8 cm long and wide; petioles to ca. 35 cm long; umbels simple, 5–10-flowered. Wet places; widely scattered in East TX mainly e of the Blackland Prairie, also Dallas (*E. Fritz s.n.*, 1979, SMU) and Gonzales (*B.C. Tharp s.n.*, 10 Apr 1934, TEX) cos.; also n Gulf Prairies and Marshes; widespread in the e U.S. w to IL, KS, OK, and TX, also AZ to CA n to WA. Apr–Oct.

Hydrocotyle sibthorpioides Lam., (resembling the genus *Sibthorpia*, (Plantaginaceae, previously treated in the Scrophulariaceae) which was named for Humphrey Sibthorp (1713–1797), a professor of botany at Oxford Univ.), LAWN WATER-PENNYWORT, LAWN MARSH-PENNYWORT. Delicate terrestrial herb; leaf blades reniform to suborbicular, 3–12+ mm wide, shallowly 5–7-lobed, minutely crenate; petioles 0.5–2+ cm long; inflorescences simple, with 3–10 flowers. Lawn weed; Dallas Co. (*B. Lipscomb 3502* BRIT!); widely introduced in the e U.S., also CA and HI (Diggs & Lipscomb 2005). Mar–Sep. Native of Asia. This species was first reported for TX by Diggs and Lipscomb (2005), who noted, that because the "plant is low-growing and inconspicuous, particularly when growing with taller grasses, we suspect that it is more widespread in Texas yards and has simply escaped notice." [*H. rotundifolia* Roxb.]

Hydrocotyle umbellata L., (with umbels), UMBRELLA WATER-PENNYWORT, OMBILIGO DE VENUS. Leaf blades to 7.5 cm in diam.; petioles to ca. 40 cm long; umbels simple, many-flowered. Wet places; widespread in East TX; mainly e 1/2 of TX; widespread in the e 1/2 of the U.S. w to WI, OK, and TX, also CA and OR, also Nova Scotia. May–Oct.

Hydrocotyle verticillata Thunb., (verticillate, whorled). Leaf blades to ca. 6 cm in diam., with 8–14 shallow lobes; petioles to 35 cm long; inflorescences interrupted, with 2–15-flowered whorls. Wet places; widespread in e 1/2 of TX; also Edwards Plateau and Trans-Pecos; widespread mostly in the s 1/2 of U.S. The county distribution map does not distinguish varieties. Turner et al. (2003) did not recognize varieties in this species. May–Oct.

1. Fruits sessile or subsessile; inflorescences often bifurcate (= 2-branched) ____ var. **verticillata**
1. Fruits pedicellate, the pedicels 1–10 mm long; inflorescences rarely bifurcate ____ var. **triradiata**

var. **triradiata** (A. Rich.) Fernald, (three-rayed). Inflorescences with few 4–15-flowered verticils.

var. **verticillata.** WHORLED WATER-PENNYWORT. Inflorescences with 2–7 few-flowered verticils.

ARISTOLOCHIACEAE Juss.

PIPEVINE OR BIRTHWORT FAMILY

Perennial herbs or woody vines; stems aerial, erect to procumbent or twining, or aerial stems absent (rhizomes present); leaves alternate, simple, petiolate, deciduous or evergreen, often aromatic; leaf blades often palmately veined, unlobed, often cordate to hastate or sagittate at base, with entire margins; flowers in ours solitary (in the leaf axils or at the end of a rhizome) or in racemose inflorescences from near the base of the plant, bisexual; calyces of fused sepals (usually 3), curved and strikingly pipe-like in shape or straight; corollas absent; stamens 5, 6, or 12; ovary inferior or partly inferior; fruit a capsule.

A medium-sized family of ca. 587–660 species in 7–9 or more genera (Stevens 2001 onwards; Mabberley 2017); it consists primarily of herbs, shrubs, and lianas (= woody vines) or in the case of *Hydnora* and *Prosopanche* (until recently recognized in the Hydnoraceae) echlorophyllous root parasites. While primarily tropical, it extends from the tropics to temperate areas, especially in the Americas; they are usually aromatic and contain alkaloids or aristolochic acid; a number are cultivated for their unusual flower shapes. The relationships of the family have been unclear and our understanding has changed over the past few years. Previously, some taxonomists referred to the Aristolochiaceae as "paleoherbs" (a group viewed as including Aristolochiales, Piperales, and Nymphaeales) and believed them to be an early branch off the evolutionary line leading to monocots; this view was supported by characters such as the 3-merous flowers (Zomlefer 1994) and early molecular data which placed the "paleoherbs" as the immediate sister group of the monocots (Chase et al. 1993). More recent molecular work has indicated that the family is in the Piperales (with Piperaceae, PEPPER family, and Saururaceae, LIZARD'S-TAIL family), an order in the "magnoliids" (a group of orders including the Magnoliales and Laurales) (Stevens 2006 [2001 onwards]; APG IV 2016). The family is thus interesting evolutionarily because it is in the approx. 3% of flowering plants near the base of the angiosperm lineage that are neither in the monocots or eudicots. More recently it has become clear that two small enigmatic groups, *Lactoris* Phil. (1 species) and Hydnoraceae (2 genera and 15 species) should be included in the Aristolochiaceae (Stevens 2001 onwards; Massoni et al. 2014; APG IV 2016). (subclass Magnolidae—Cronquist; order Piperales—APG IV)
FAMILY RECOGNITION IN THE FIELD: herbs or woody vines with alternate, simple, often palmately veined leaves often basally cordate to hastate or sagittate and apetalous flowers with a curved and pipe-like or straight and tubular, 3-merous calyx; fruit a capsule.

REFERENCES: Gregory 1956; Huber 1993; Barringer & Whittemore 1997; Stevens 2001 onwards; Neinhuis et al. 2005; Massoni et al. 2014; APG IV 2016.

1. Plant with a distinct, easily visible aerial (above ground) stem; flowers bilaterally symmetrical, the calyx tube usually bent or curved; stamens 5–6; ovary inferior ____ **Aristolochia**
1. Plant without aerial stem (stem a below ground rhizome); flowers radially symmetrical, the calyx tube straight; stamens 12; ovary ca. 1/3 inferior ____ **Asarum**

ARISTOLOCHIA L. PIPEVINE, BIRTHWORT, TACOPATE

Perennial herbs or woody vines; aerial stems present; leaves alternate, simple; leaf blades entire, often palmately veined to pinnately veined, often cordate to hastate or sagittate at base; flowers in ours solitary in the leaf axils or in racemose inflorescences from near the base of the plant; calyces corolla-like, with tube curved, strikingly pipe-like in shape; corollas absent; stamens 5–6, the filaments fused with the style into a column; ovary inferior; capsule dry; seeds flattened or rounded.

A genus of ca. 500 (Barringer 1997; Mabberley 2017) species of vines, scramblers, or herbs, widespread in tropical and warm areas, with less diversity in Africa and Australia. The species contain poisonous alkaloids or aristolochic acid (carcinogenic and nephrotoxic—McGuffin et al. 1997); some are used medicinally, in treating snakebites, or as an arrow poison; the genus is complex biochemically with many physiologically active compounds of different classes (Kuo et al. 2012). According to Crosswhite and Crosswhite (1985), "*Aristolochia* was the source of the active ingredient in 'snakeroot oil' sold by itinerant snakeroot doctors who staged medicine shows in the western United States in the Nineteenth Century." The leaves of a number of species are eaten by larvae of pipe-vine swallowtail butterflies such as *Battus philenor*; the aristolochic acid thus obtained is sequestered and presumably makes them unpalatable to birds (Howe 1975; Barringer & Whittemore 1997). The common name, PIPEVINE, comes from the striking, curved, pipe-like calyx. The genus has often been treated as a large variable single genus, *Aristolochia* sensu lato (s.l.) (e.g., Stevens 2001 onwards; Mabberley 2017); however, other authors have recognized a variable number of segregates (e.g., González & Stevenson 2002; Kelly & González 2003; Ohi-Toma et al. 2006; Wanke et al. 2006, 2007; Weakley et al. 2012; Zhu et al. 2019). Since it is widely agreed based on molecular evidence that *Aristolochia* s.l. is monophyletic (e.g., Neinhuis et al. 2005; Massoni et al. 2014), because different authors recognize different numbers of segregates, and because of the long usage of the binomials under *Aristolochia*, we follow the authors mentioned earlier (Stevens 2001 onwards; Mabberley 2017) in recognizing the genus in the broad sense. This is an example of a situation in which there is no clear-cut correct answer; as long as the phylogeny of the group is accurately reflected, we prefer the broader approach. Some flowers trap flies which then pollinate them; the flowers of many species in the genus have fetid odors and purple-brown coloration (Barringer 1997), presumably to attract carrion feeders. According to Sakai (2002), "All the *Aristolochia* species studied so far are pollinated by saprophagous [= feeding on decaying matter] flies of different families, which are presumably deceived by floral odor. Flowers of many species have trap-and-release mechanisms. The flowers attract and imprison pollinators during the female stage first day of flowering and release them after anther dehiscence." The inside of the perianth tube is covered with hairs, acting as an insect-trap. The hairs eventually wither, releasing the pollinators covered with pollen (iNaturalist 2024). (Greek: *aristos*, best, and *lochia*, delivery, from supposed value in aiding childbirth; apparently derived through the ancient Doctrine of Signatures from the fancied resemblance of the flower bud to a fetus; the plants were used as a source of medicine to alleviate the pain of childbirth—Pfeifer 1966)

REFERENCES: Pfeifer 1966; Barringer 1997; Murata et al. 2001; González & Stevenson 2002; Sakai 2002; Ohi-Toma et al. 2006; Weakley et al. 2012; Ohi-Toma & Murata 2016; Zhu et al. 2019.

1. High-climbing, twining, woody vines to 25 m long; flowers from leaf axils on upper part of the stem; leaf blades large, 8–15(–20) cm wide; seeds ca. 10 mm long ____ **A. tomentosa**
1. Erect to sprawling herbs to 60 cm tall; flowers borne mostly near the base of the stems; leaf blades ca. 7 cm or less wide; seeds 5 mm or less long.
   2. Leaf blade definitely narrowed at base (cuneate to attenuate) ____ **A. erecta**
   2. Leaf blade not narrowed at base (truncate to cordate, auriculate, sagittate, or hastate).
      3. Leaf blades approx. as wide as long, with palmate venation; auricles (essentially lobes) at base of leaf blade strikingly obvious, each of the two auricles ca. half as long as the central part of the blade or longer; species reported in East TX only from Travis and Williamson cos. at sw margin of area ____ **A. coryi**
      3. Leaf blades longer than wide, with pinnate or palmate-pinnate venation; auricles of leaf blade much smaller; species widespread in East TX.
         4. Leaves subsessile, conspicuously clasping; leaf blades subcoriaceous, prominently reticulate veined beneath, obtuse to subacute at apex; stems with conspicuous spreading hairs ____ **A. reticulata**
         4. Leaves with distinct petioles, not clasping; leaf blades thin, delicate, not prominently reticulate veined beneath, apically acuminate to acute; stems glabrous to with inconspicuous pubescence ____ **A. serpentaria**

Aristolochia coryi I.M. Johnst., (for V.L. Cory, 1880–1964, Texas A&M Univ. and Southern Methodist Univ., botanist and collector of TX plants), CORY'S DUTCHMAN'S-PIPE. Perennial herb; stems procumbent, to 40 cm long; leaves petiolate, the petioles 1–3 cm long, blades deltate, to 5 cm long and wide but usually much smaller, basally cordate, obviously auricled; inflorescences axillary on new growth, of solitary flowers, the peduncles 1–4(–8) cm long; perianth yellow-green to yellow-brown to brownish purple; capsules elliptic to ± globose, ca. 15–20 mm long and 12–15 mm wide. Rocky areas, crevices or slopes, usually on limestone; this species was reported from Travis Co. by Johnston and Riskind (1975) and Kartesz (2015), and Travis and Williamson cos. are documented on iNaturalist (2024); it is found primarily on "limestone strata in the Trans-Pecos and western Edwards Plateau" (Carr 2009), and now also known from Travis and Williamson cos.; in the U.S. known only from TX; also adjacent Mexico. Apr–Sep. Presumably contains aristolochic acid. ☠

Aristolochia erecta L., (erect), SWAN FLOWER. Perennial herb; stems erect to sprawling, to ca. 25 cm long; leaves with petioles only 0.1–0.8 cm long, blades linear-lanceolate, to 15 cm long and 1(–1.5) cm wide, cuneate to attenuate at base; inflorescences axillary on new growth, of solitary flowers, the peduncles 1–4 cm long; perianth brownish purple; capsules ovoid to obovoid, ca. 25 mm long. Grassy areas, hillsides, riverbanks, chaparral, sandy soils; s part of East TX as far n as Robertson Co. (Kartesz 2015), also slightly to the w on the Edwards Plateau, and the South TX Plains; TX; n Mex. Mar–Nov. [*A. longifolia* Engelm. ex A. Gray] Presumably contains aristolochic acid. ☠

Aristolochia reticulata Nutt., (netted, in reference to the leaf venation), TEXAS DUTCHMAN'S-PIPE, TEXAS TURPENTINE-ROOT. Perennial herb; stems erect to sprawling, to ca. 40 cm long, ± zigzag; leaves few, subsessile, the petioles only 0.1–0.8 cm long, blades oblong to broadly ovate, 7–12 cm long and 3–7 cm wide; inflorescences racemose, several-flowered, arising from near base of plant (often covered by fallen leaves); perianth brownish purple; capsules subglobose, ca. 10–30 mm in diam. Wooded areas, sandy soils; widespread in the e part of East TX, and w disjunct to Young Co. (Pfeifer 1966; *T. Drummond 226* NY—TORCH 2024) on w edge of nc TX; from TX and se OK to AR and LA. May–Jul. [*Endodeca reticulata* (Nutt.) Floden & Weakley, *Isotrema reticulatum* (Nutt.) X.X. Zhu, S. Liao, & J.S. Ma, *Siphisia reticulata* (Nutt.) Klotz.] The dried rhizome has been sold as a serpentary (= treatment for snakebite); it is also used as a tonic; however, the active ingredient is aristolochic acid, a gastric irritant that in large doses can cause respiratory paralysis (Barringer 1997) and is known to be carcinogenic and nephrotoxic (McGuffin et al. 1997). ☠

Aristolochia serpentaria L., (referring to medicinal use for snakebite), VIRGINIA DUTCHMAN'S-PIPE, VIRGINIA SNAKEROOT, NASH DUTCHMAN'S-PIPE, TURPENTINE-ROOT, SERPENT BIRTHWORT. Perennial herb; stems erect to decumbent, to 60 cm tall; leaf blades quite variable, ovate to narrowly lanceolate, cordate to hastate, sagittate, or truncate at base, 5–15 cm long, 1–5 cm wide, basal auricles, if present small; petioles 0.5–3.5 cm long; flowers solitary, at the ends of scaly branches arising from the lowest nodes of the stem; perianth purplish brown; capsules subglobose, 8–20 mm in diam. Moist wooded areas; widely scattered in e part of e TX; also Fort Hood (Bell or Coryell cos.—Sanchez 1997) to the w, also disjunct to the Edwards Plateau, and in s OK just n of the Red River; widespread in e US from NY s to FL and w to IA, OK, and TX. This TX distribution is just one example of the East TX-Edwards Plateau disjunction seen in numerous genera and species of East TX plants; this disjunction seems to be the result of survival of floristic elements of a much more widespread ice age flora in the comparatively moist modern areas of East TX and certain microhabitats on the Edwards Plateau (see discussion in the introduction of Diggs et al. 2006). May–Jun. [*A. serpentaria* var. *hastata* Duch., *Endodeca serpentaria* (L.) Raf., *Isotrema serpentarium* (L.) X.X. Zhu, S. Liao, & J.S. Ma] According to Barringer (1997), closed, apparently cleistogamous flowers are often present. Even though it has distinctive leaves, this is a small, easily overlooked plant of the forest floor. Also, this species exhibits a great deal of variation in leaf shape; this variability is apparently related to the fact that eastern pipe-vine swallowtail butterflies use leaf shape as a search image when looking for *Aristolochia* leaves on which to lay their eggs (Barringer 1997). The species, which contains a number of physiologically active components including aristolochine and aristolochic acid, was long used by Native Americans and early settlers for a variety of medicinal applications including snakebite; it has been rather popular recently as an herbal medicine; however, it can irritate the gastrointestinal tract and kidneys, and potentially cause coma and death due to respiratory paralysis (Duke 1985; Cheatham & Johnston 1995); aristolochic acid is carcinogenic and nephrotoxic (McGuffin et al. 1997). ☠

Aristolochia tomentosa Sims, (densely woolly, with matted hairs), WOOLLY DUTCHMAN'S-PIPE, PIPEVINE. Vine to 25 m long, densely pubescent except on old growth; leaf blades cordate, 3-ribbed from base; petioles 1–5.5(–9) cm long; flowers solitary (rarely paired), axillary; perianth green or yellowish green with brown or purplish stripes inside, with curved, funnelform tube abruptly inflated above the ovary, and flaring limb; stamens 6, united with style; anthers sessile; pistil 1; ovary appearing inferior (tightly enclosed by base of perianth tube); fruit a cylindric capsule to 8 cm long and 3 cm in diam.; seeds flat, triangular. Thickets and woods, sandy or silty soils, frequent, but not often found in flower; widespread in East TX; se and e TX w to West Cross Timbers, also disjunct to the Edwards Plateau; FL and GA n to IN w to KS, OK, and TX, also disjunct to w NC. Apr. [*Isotrema tomentosum* (Sims) H. Huber] See discussion of the TX disjunction under *A. serpentaria*. Presumably contains aristolochic acid. ☠

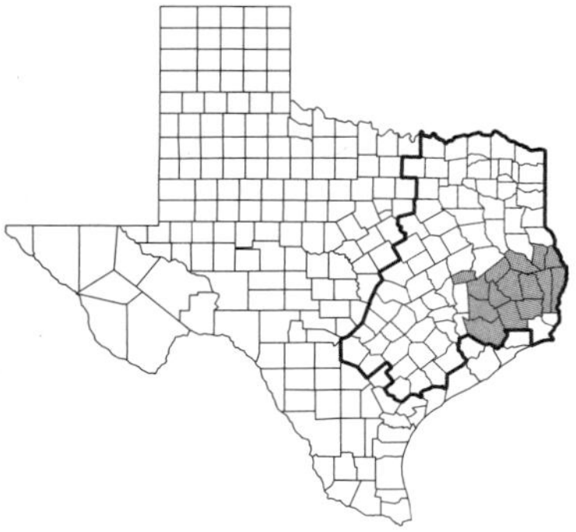
Ilex longipes

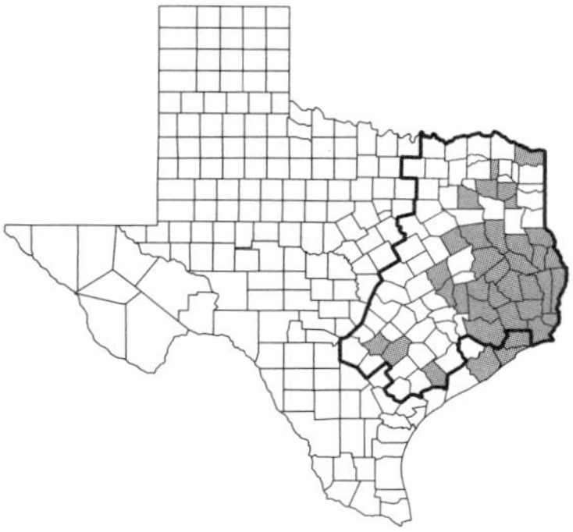
Ilex opaca

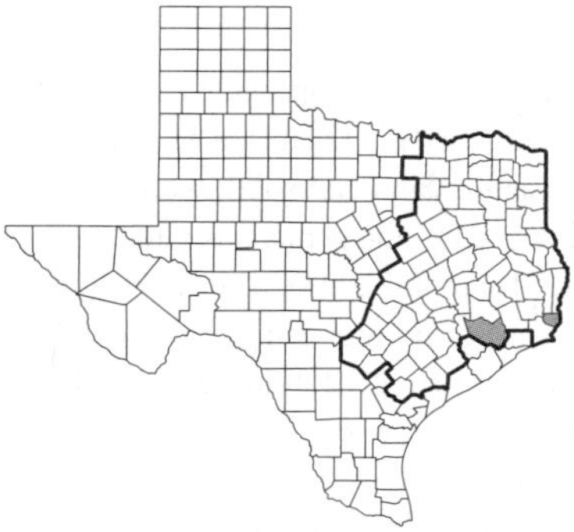
Ilex verticillata

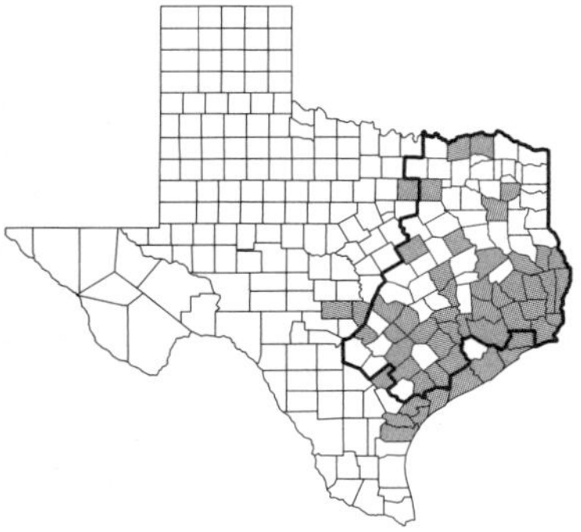
Ilex vomitoria

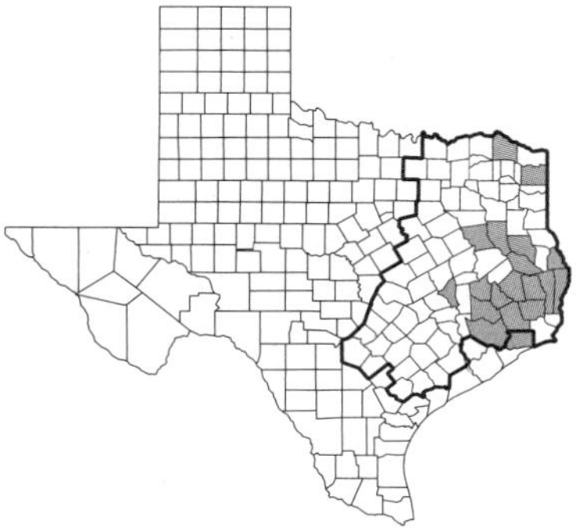
Aralia spinosa

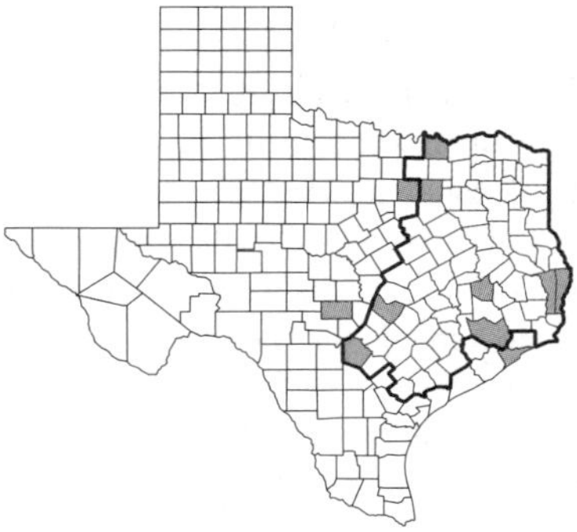
Hedera helix

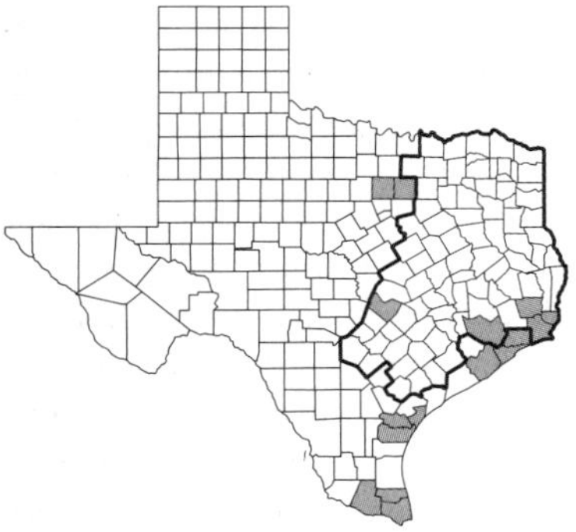
Hydrocotyle bonariensis

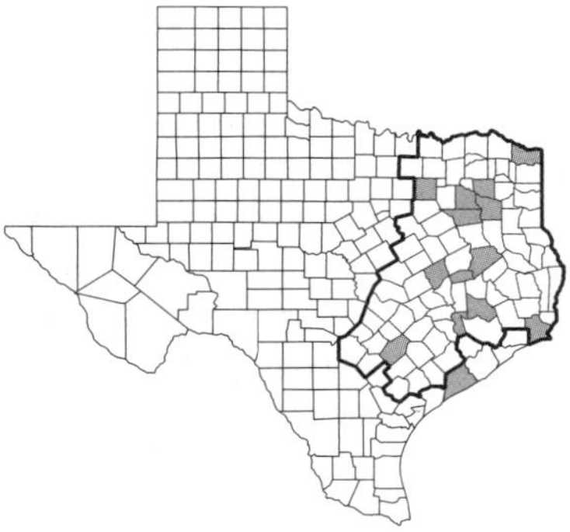
Hydrocotyle ranunculoides

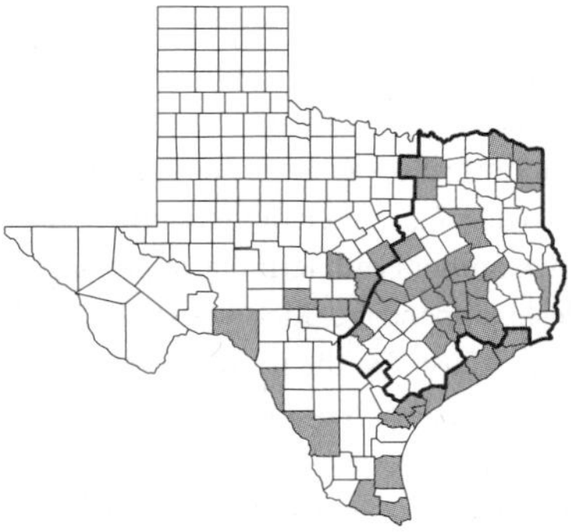
Hydrocotyle umbellata

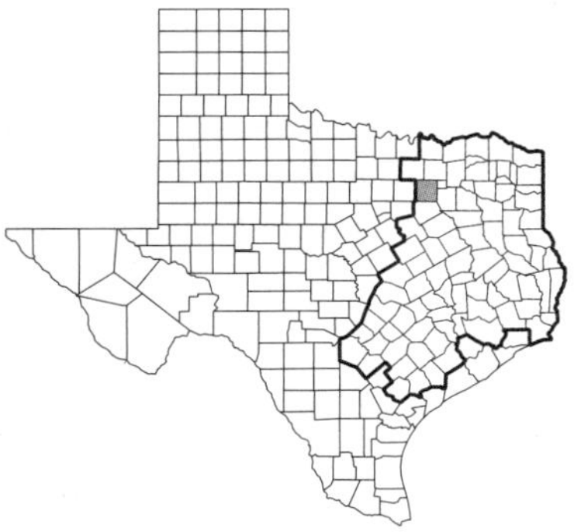
Hydrocotyle sibthorpioides

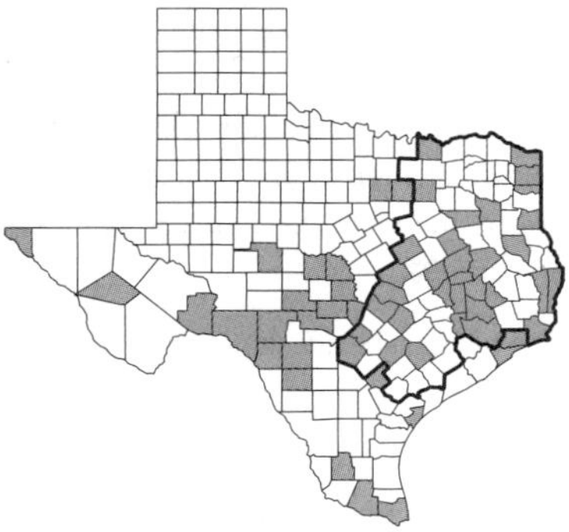
Hydrocotyle verticillata

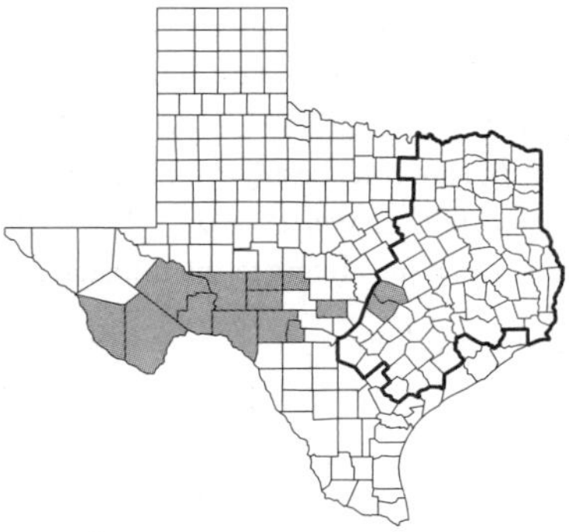
Aristolochia coryi

ASARUM L. WILD-GINGER

A n temperate, especially e Asian, genus of ca. 80–ca.100 (or even –128) species (Stevens 2001 onwards; Sinn et al. 2015; Mabberley 2017; Takahashi & Setoguchi 2017). The majority of the species are in Asia (China, Taiwan, and Japan to Vietnam and the Korean Peninsula), with up to ca. 25 in North America, and 1 in Europe (Sinn et al. 2015; Mabberley 2017; Keener & Todia 2021). The genus has often been treated in the broad sense (Stevens 2001 onwards; Sinn et al. 2015, 2018; Mabberley 2017; Diamond 2023). Alternatively, it has been divided into as many as six segregate genera including *Hexastylis* Raf. (HEART-LEAF), a North American group of ca. 18–19 species (e.g., Blomquist 1957; Whittemore & Gaddy 1997; Kelly & González 2003; Weakley et al. 2012; Kartesz 2015; Keener 2020; Keener & Todia 2021). It is generally agreed that the genus treated broadly is monophyletic (e.g., Kelly 1997, 1998; Sinn et al. 2015, 2018), and we are following Sinn et al. (2015) who, based on both plastid and nuclear molecular data, argue that "*Asarum* is best recognized as a single genus of three subgenera and six sections" with *Hexastylis* as one of these sections. Takahashi and Setoguchi (2017) supported recognizing *Asarum* in the broad sense and provided molecular evidence that *Hexastylis* is possibly polyphyletic; thus recognizing it as a segregate genus would not accurately reflect evolutionary history. There has been considerable work on the genus in recent years in North America, with four new species named from the southeastern U.S. (Sinn 2015, 2017; Keener 2020; Keener & Todia 2021). The dark brown, resiny-scented flowers are visited by insects such as flies or fungus-gnats; a number produce fungal-like appendages in the flowers upon which the fungus-gnats lay their eggs; however, the larvae cannot eat the tissue (Mabberley 2017); there is also self-pollination in some species (Kelly 2001). The fleshy seed appendages are used in myrmecochory (= ant-dispersal). Some species have been used medicinally and also as an ingredient of snuffs, and some are grown ornamentally (Mabberley 2017). Aristolochic acid, which is known to be toxic (see discussion under the genus *Aristolochia*), has been found in some species, and the plants should thus be considered toxic (Schaneberg et al. 2002). (from ancient Greek *asaron*, the name for this genus used by Dioscorides—Charters 2023).

REFERENCES: Blomquist 1957; Soltis 1984; Gaddy 1988; Barringer 1993; Kelly 1997, 1998, 2001; Whittemore & Gaddy 1997; Stevens 2001 onwards; Sinn et al. 2015, 2018; Takahashi & Setoguchi 2017; Keener 2020; Keener & Todia 2021; Diggs, Lipscomb, & George 2025.

Asarum arifolium Michx., (refers to the resemblance of leaves to those of the *Arum* plant, which have an arrowhead shape), LITTLE BROWN JUG, ARROW-LEAF GINGER, ARROWHEAD WILD GINGER. Rhizomatous, evergreen, aromatic perennial herb; aerial stems absent; leaves distinctly petiolate, the blades variable, but usually triangular-sagittate to subhastate, glabrous, variegated with a pattern due to variation in color; inflorescences of a single flower, borne on a peduncle at end of rhizome, sometimes a number of inflorescences present per rhizome (flowers often not obvious under leaves and in leaf litter, etc.); calyces corolla-like, the sepals fused for much of their length, the tube urceolate-campanulate, not curved, the lobes spreading; corollas absent; stamens 12, the anthers with terminal appendage absent or rudimentary; ovary ca. 1/3 inferior; styles 6, 2-cleft; fruit a capsule; seeds with a fleshy appendage. Rich woods; Tyler Co. (G. Watson, pers. comm.; Diggs, Lipscomb, & George 2025); disjunct to extreme e part of East TX, the nearest known occurrence is in e Louisiana; se U.S. from VA s to FL w to KY, LA, and TX. Late winter–spring. [*Hexastylis arifolia* (Michx.) Small in Britton] Geraldine Watson (see biographical info. in introduction in Vol. 1 of this flora) reported (pers. comm.) finding this species in the "Canyonlands" in the ne part of the Pineywoods; she subsequently transplanted some plants to her preserve where they continued to grow (G. Diggs, pers. obs.; Photos: BRIT). To the best of our knowledge, no herbarium specimens from the Canyonlands location are known and it is not known if the location still exists in unaltered form. We did not make an herbarium specimen from the Watson preserve because of the paucity of plant material. Because of our limited knowledge of the species, we are not recognizing intraspecific taxa. The species, which is aromatic, was used medicinally by Native Americans to treat such conditions as stomach pain, heart trouble, backache, whooping cough, and asthma (Moerman 1998). However, the nephrotoxic compound aristolochic acid is found in some members of the genus and the plants are thus best considered toxic (Schaneberg et al. 2002). The seeds are ant-dispersed (González 1972; Bierzychudek 1982; Sorrells & Warren 2011). The term "ginger" in two of the common names refers to the aromatic tissues, while the name "little brown jug" refers to the unusual flower shape.

Asarum canadense L., (CANADIAN WILD GINGER) is a related species that occurs just north of East TX in McCurtain Co. OK, directly n of Bowie Co. TX. It differs from *A. arifolium* as follows:

1. Leaf blades roughly kidney-shaped, not variegated; sepals distinct; anthers tipped with a prominent appendage; styles fused to form a column; ovary inferior ____ **A. canadense**
1. Leaf blades elongated so as not to be kidney-shaped, typically variegated with a pattern due to variation in color; sepals connate for most of their length; anthers not tipped with an appendage or appendage vestigial; styles distinct; ovary superior or 1/3-inferior ____ **A. arifolium**

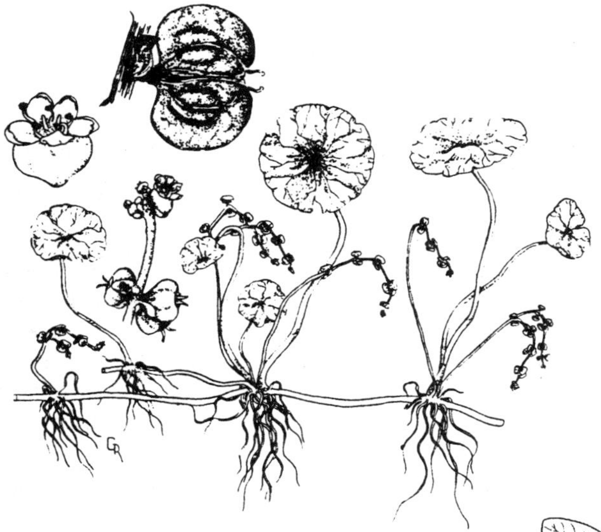

Hydrocotyle verticillata var. verticillata [GWO]

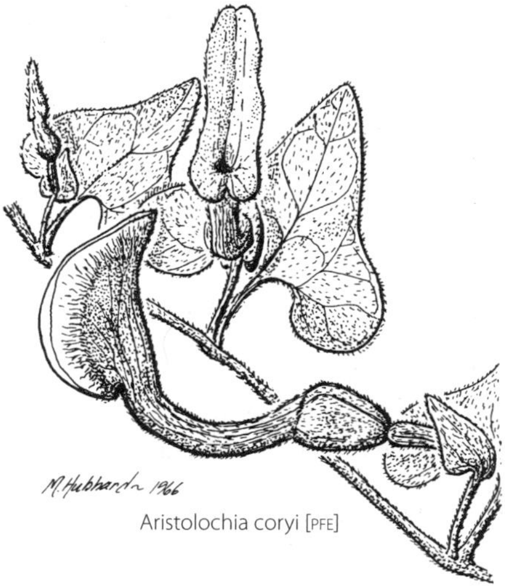

Aristolochia coryi [PFE]

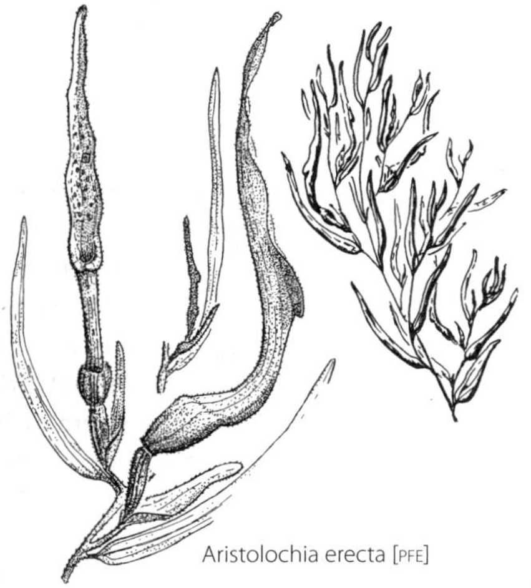

Aristolochia erecta [PFE]

Aristolochia reticulata [AMB]

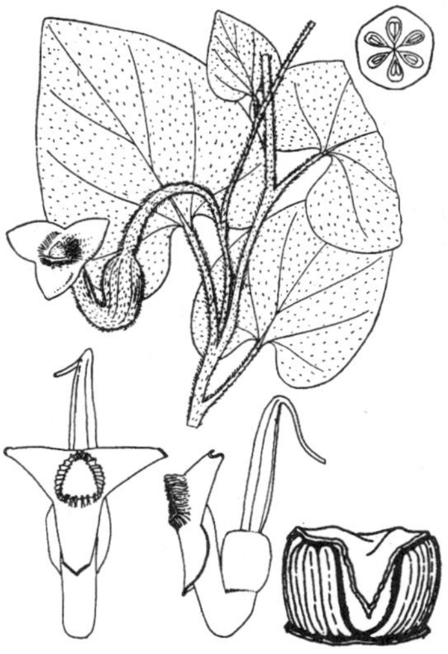

Aristolochia tomentosa [AMB, BB2]

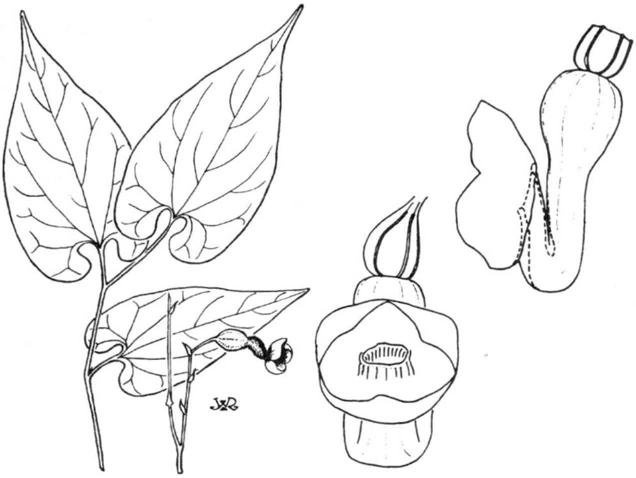

Aristolochia serpentaria [AMB, STR]

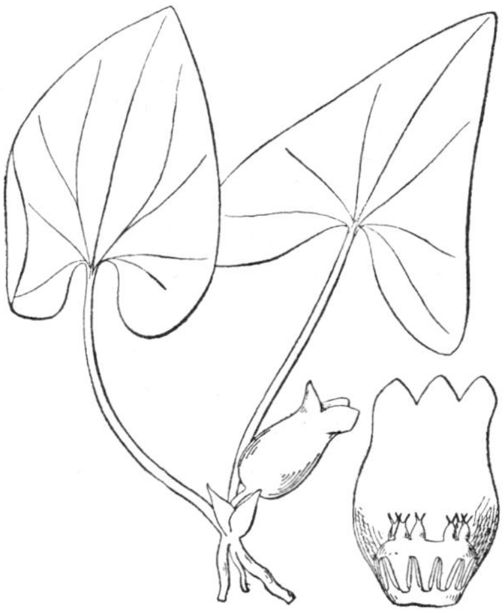

Asarum arifolium [BB2]

ASTERACEAE Bercht. & J. Presl (COMPOSITAE Giseke)
SUNFLOWER, ASTER, OR DAISY FAMILY

Contributed by **Amanda K. Neill** (ex BRIT-SMU-VDB)

Unless otherwise noted, number of taxa in generic synopses are from Flora of North America vols. 19, 20, and 21.—akn

Localities are generally from USDA Plants maps (USDA Plants Database 2016). If Turner et al. 2003 or Kartesz 2015 (online) have additional counties, they are cited within taxon locality statements.

Ours annual, biennial, or perennial herbs, or less frequently subshrubs, shrubs, or vines; leaves basal, alternate, opposite, or whorled, simple or compound, entire, toothed, or lobed, not stipulate (but small basal lobes sometimes resemble stipules); inflorescence a single involucrate capitulum (= head), or several or many involucrate heads in spikes, racemes, corymbs, or panicles, each head simulating a single flower; involucre of one or more rows of separate or united bracts (= phyllaries, so termed to avoid confusion with bracts on peduncles) imitating sepals; flowers grouped inside the involucre on a platform (= receptacle), which may be smooth or pitted, or bristly or hairy, or may have chaffy bracts (= pales) associated with each flower; flowers without typical calyx, but often with a modified calyx of bristles, awns, scales, teeth, or minute crowns (= pappus) on the summit of the inferior ovary; flowers of two basic types (with a few exceptions): ray flowers, which may be perfect or styliferous and fertile, sterile, or neuter, with corollas of a small basal tube and a broad, often showy, strap-shaped main portion (= ligule); and disk flowers, which may be perfect and fertile or functionally staminate and sterile, with corollas mainly tubular (varying from threadlike to tubular-funnelform or campanulate, with a cylindrical basal tube), with typically 4–5 equal or unequal lobes; stamens none or (4–)5, attached inside the corolla tube, anthers separate or united into a ring; pistil 1; style 1, commonly with 2 branches (in perfect or styliferous flowers); ovary inferior; fruit a single-seeded achene (also known as a cypsella), often dispersed with aid from the pappus (see Illustrated Glossary).

Asteraceae is one of the two largest families of flowering plants (Orchidaceae is the other), containing ca. 25,000 species in 1,620 genera (Stevens 2001 onwards). It is a cosmopolitan family of mainly herbs to shrubs and is of significant economic importance as a source of food plants (e.g., *Lactuca*—LETTUCE, CYNARA—ARTICHOKE), oil (e.g., *Helianthus*—SUNFLOWER, *Carthamus*—SAFFLOWER), spices and flavorings (e.g., *Artemisia*—TARRAGON and ABSINTH), herbal teas and medicines (e.g., *Matricaria* or *Chamaemelum*—CHAMOMILE, *Echinacea*—PURPLE CONEFLOWER), and ornamentals (e.g., *Aster*, *Chrysanthemum*, *Cosmos*, *Dahlia*, *Gerbera*, *Helianthus*, *Tagetes*, and *Zinnia*). Many are weeds including *Taraxacum* (DANDELIONS) and *Cirsium* (THISTLES); ☠ In some parts of the world poisonous species of *Senecio* are a major cause of livestock poisoning; wind-pollinated genera such as *Ambrosia* (RAGWEED), *Iva* (SUMPWEED), and *Artemisia* (SAGEBRUSH), which produce large quantities of allergenic pollen, are important causes of hay fever (Lewis et al. 1983). Cronquist (1981) suggested that the evolutionary success of the family may be in part due to a diversified chemical defense system, including polyacetylenes and sesquiterpene lactones. Asteraceae is the largest family in the flora of North America north of Mexico, with 418 genera and 2,413 species (Barkley et al. 2006). The 351 species of Asteraceae found in OK make it the largest family in that state (Taylor & Taylor 1994) and the 620 species (almost 13% of the TX flora) in TX likewise make it the largest TX family (Hatch et al. 1990). Asteraceae is also the largest family in the nc TX flora; the 263 species represent nearly 12% of the 2,223 species known in nc TX (Diggs et al. 1999). Asteraceae is the largest family in East TX, with 138 genera and 423 species, representing just over 12% of the 3,526 species estimated to occur in East TX, and is over 9% more speciose than the next largest East TX family, Poaceae (Diggs et al. 2006). The tribes and generic circumscriptions (and for the most part, species circumscriptions) in this treatment follow Barkley et al. (2006). (subclass Asteridae—Cronquist; order Asterales—APG IV)

FAMILY RECOGNITION IN THE FIELD: usually herbs or rarely shrubs or vines with a characteristic inflorescence: flowers in a compact *head* subtended by bracts (= *phyllaries*)—the inflorescence resembling a single flower (heads are often grouped together to form compound inflorescences); corollas sympetalous, 5-merous; stamens united by their anthers; fruit a 1-seeded achene often topped by a *pappus* of bristles, awns, or scales.

REFERENCES: Solbrig 1963; Carlquist 1976; Cronquist 1977, 1980, 1985; Gandhi & Thomas 1989; Jansen et al. 1991, 1992; Michaels et al. 1993; Bremer 1994, 1996; Nesom 1994a; Jansen & Kim 1996; Taylor 1997; Panero & Funk 2002; Anderberg et al. 2006; Barkley et al. 2006; Funk et al. 2009; Turner 2016.

Keys to Genera of Asteraceae

Modified from Barkley et al. (2006), Diggs et al. (1999), and Taylor (1997) and contributed by
Amanda K. Neill (ex BRIT-SMU-VDB)

Artificial Key to the Genera of Asteraceae

(For Technical Key to Tribes of Asteraceae see page 139)

1. Heads with flowers all similar and usually all perfect; only ray flowers present (with strap-shaped ligules) OR all corollas neither ray-like nor disk-like, all 2-lipped.
  2. Plant sap usually milky; heads with only ray flowers (with strap-shaped ligules) ____ **Use Key to Cichorieae (Lactuceae) Tribe** (page 143)
  2. Plant sap usually watery; heads with all corollas neither ray-like nor disk-like, all 2-lipped ____ **Use Key to Mutisieae Tribe** (page 148)
1. Heads with flowers all similar and perfect or not so; heads with disk flowers present (with 4–5-lobed, tubular, eligulate corollas); ray flowers absent OR present at the periphery of the heads and pistillate or neuter.
  3. Heads with ray flowers, the ligules usually evident.
    4. Ray ligules yellow to orange or red to reddish brown, sometimes with reddish brown or maroon marks.
      5. Receptacle chaffy, a bract (pale) subtending each flower ____ **Key 1**
      5. Receptacle without chaffy bracts or pales associated with each flower (but sometimes rough or bearing bristles) ____ **Key 2**
    4. Ray ligules white, yellowish cream, tawny, pink, lavender, bluish, or purple.
      6. Disk flower pappus (at least the innermost series) of capillary bristles or hairs ____ **Key 3**
      6. Disk flower pappus of awns, scales, or a minute crown, or pappus absent ____ **Key 4**
  3. Heads with only disk flowers or apparently so (ray flowers sometimes present but their corollas inconspicuous or reduced, < 0.7 mm long).
    7. Plants sometimes spiny; phyllaries with hooked or straight spines or tubercles (at least on pistillate heads when heads unisexual), sometimes united at maturity to form a bur-like fruit or phyllaries with margins variously lobed, toothed, or bristle-fringed (COCKLEBURS, RAGWEEDS, THISTLES, KNAPWEEDS, and BASKET-FLOWERS) ____ **Key 5**
    7. Plants not spiny; phyllaries without hooked or straight spines or tubercles or large teeth or bristles, not united into a bur.
      8. Disk corollas white, yellowish cream, greenish, tawny, pink, lavender, bluish, or purple (or infrequently obscured by pubescence, enclosed by receptacle pales, or corollas vestigial or absent) ____ **Key 6**
      8. Disk corollas yellow to orange or red to reddish brown or maroon, sometimes with reddish brown or maroon marks ____ **Key 7**

Key 1. Ray and disk flowers both present; ray flowers yellow to orange or red to reddish brown, sometimes with reddish brown or maroon marks; receptacle chaffy.

1. Only ray flowers fertile (= maturing achenes), their achenes much larger than those of the sterile disk flowers.
  2. Leaf blades usually ± palmately 3–5-lobed; ray achenes thick, rounded, without wings ____ **Smallanthus**
  2. Leaf blades unlobed or pinnately lobed; ray achenes ± strongly flattened, with or without wings.
    3. Achenes broadly winged.
      4. Taprooted annuals; phyllaries 8–10, in 2 series; ray flowers usually 5, in 1 series; involucres 5–12 mm wide ____ **Lindheimera**
      4. Coarse perennials; phyllaries 11–45, in 2–4 series; ray flowers 8–38, in 2–3 series; involucres 10–30 mm wide ____ **Silphium**
    3. Achenes not winged (but may be flattened).
      5. Leaves crenate, serrate, dentate, or lyrate-pinnatifid; involucres 12–30+ mm wide; ray ligules with green, red, or maroon veins on the lower surface; achenes 4.5–6 mm long ____ **Berlandiera**
      5. Leaves deeply pinnatifid; involucres 6–10 mm wide; ray ligules entirely yellow on the lower surface, without colored veins; achenes 3–4 mm long ____ **Engelmannia**
1. Disk flowers perfect and fertile; ray flowers present and fertile OR sterile.
  6. Ray flowers neuter OR pistillate and sterile; only disk flowers fertile (= maturing achenes), their achenes much larger than those of the sterile ray flowers.
    7. Leaves mostly opposite (NOT both alternate and auriculate-clasping); phyllaries apparently in ± 3 series, dimorphic, the outer (secondary) series herbaceous (and sometimes foliaceous), the inner 2 (primary) series often very unlike the outer, usually scarious-margined; achenes flattened parallel to the phyllaries or 3–4-angled and linear to spindle-shaped.
      8. Primary (inner) phyllaries united for 1/5 to 1/2 or more of their lengths; disk corolla lobes often lance-linear (sometimes triangular) ____ **Thelesperma**
      8. Phyllaries all separate to their bases or nearly so; disk corolla lobes triangular to ovate.
        9. Achenes (at least central ones) ± 4-angled, ± linear to spindle-shaped, often with tips narrowed or beaked (no achenes winged).

10. Disk flowers 10–20 per head; stamen filaments hairy near anthers (use lens); achenes usually with 1 groove on each face ______ **Cosmos** (in part)
10. Disk flowers (5–)12–150+ per head; stamen filaments not hairy; achenes with 0 or 2 grooves on each face ______ **Bidens** (in part)
9. Achenes all ± obcompressed or flattened (sometimes winged).
11. Achenes sometimes thickened or winged; pappus usually of barbellate (rarely smooth) awns or sometimes absent ______ **Bidens** (in part)
11. Achenes (some or all) usually thin-margined or ± winged; pappus usually absent a crown or of 2 bristly scales (or achene wing shoulders may be bristly) ______ **Coreopsis**
7. Leaves alternate or opposite; phyllaries in 1–several series, usually not strongly dimorphic (IF strongly dimorphic THEN leaves alternate and blades auriculate and clasping); achenes various but usually not flattened parallel to the phyllaries.
12. Receptacles hemi-spheric to high-conic or columnar (mostly 8–20+ mm high); achenes not winged.
13. Leaves lyrate-pinnate to 1–2-pinnatifid; phyllaries unequal (outer noticeably longer than inner); achenes linear-lanceolate to oblong, strongly compressed, with one margin usually toothed, fringed, or ciliate ______ **Ratibida**
13. Leaves simple to deeply lobed; phyllaries subequal or unequal; achenes obpyramidal and ± 4-angled, not strongly compressed, margins not toothed, fringed, or ciliate ______ **Rudbeckia**
12. Receptacles mostly flat to convex or conic (mostly 0–5 mm high); achenes winged or unwinged.
14. Receptacle pales completely enclosing each achene, forming a hardened, warty, fruit-like structure ______ **Sclerocarpus**
14. Receptacle pales sometimes enfolding an achene, sometimes shed with it, but not completely enclosing each achene in a fruit-like structure.
15. Stems sometimes winged with decurrent leaf bases; achenes strongly flattened and papery-winged; pappus usually of 2 persistent subulate scales or awns to 2 mm long (or sometimes pappus absent) ______ **Verbesina** (in part)
15. Stems not winged, leaf bases not decurrent; achenes various but not strongly flattened and papery-winged (achenes may be somewhat flattened, with thin margins); pappus of 1–2+ scales, these often fragile and deciduous (plus sometimes shorter scales or squamellae) or pappus sometimes absent.
16. Petiole bases dilated, fused to form disks at nodes; ray flowers styliferous (= styles present); achenes somewhat flattened and thin-margined ______ **Simsia**
16. Petiole bases not dilated, not forming nodal disks; ray flowers neuter, styles and stigmas absent; achenes ± compressed, often obpyramidal.
17. Involucres 4–40+ mm wide; stamen filaments not hairy; pappus fragile, readily falling ______ **Helianthus**
17. Involucres 7–10 mm wide; stamen filaments hairy; pappus persistent or tardily falling ______ **Viguiera**
6. Ray flowers present and fertile (rays sometimes inconspicuous); disk flowers perfect and fertile; both kinds of flowers producing mature achenes (though these may differ in shape).
18. Stems sometimes winged with decurrent leaf bases; achenes strongly flattened and papery-winged ______ **Verbesina** (in part)
18. Stems not winged, leaf bases not decurrent; achenes various but not strongly flattened and papery-winged (achenes may be somewhat flattened, with thin margins).
19. Plants usually strongly aromatic; leaves usually alternate; receptacle chaff (= pales) readily falling; pappus of 20–40 bristles, 3–5.5 mm long ______ **Chrysactinia**
19. Plants sometimes aromatic; leaves opposite; receptacle chaff persistent or falling; pappus of a few awns or bristles OR of scales or a minute crown OR pappus lacking or a combination of 2 of these.
20. Outer 4 phyllaries broadly lanceolate, foliaceous, forming square "buds" prior to anthesis, overlapping at anthesis, notably larger than inner, scarious phyllaries ______ **Tetragonotheca**
20. Outer phyllaries larger than inner or not, sometimes herbaceous, but not notably different in texture, not forming 4-sided "buds."
21. Ray ligules sessile, not borne on corolla tubes, 10–40 mm long, usually persistent and becoming papery in age.
22. Plants perennial; leaves petiolate and usually toothed; ray ligules yellow to orange; achenes 4–5 mm long; native species ______ **Heliopsis**
22. Plants annual; leaves sessile and entire; ray ligules usually red (or yellow in cultivars); achenes 6–10 mm long; horticultural exotics escaping cultivation ______ **Zinnia**
21. Ray ligules usually borne on corolla tubes, not sessile, 2–15 mm long, not persistent, not becoming papery in age.
23. Plants often decumbent to prostrate; ray ligules to 5 mm long.
24. Receptacles conic to short-columnar; phyllaries 8–15+; disk flowers 25–100(–200+); pappus of 1–3 bristle-like awns, or pappus absent ______ **Acmella**
24. Receptacles convex; phyllaries 5; disk flowers 10–20; pappus of 2(–5+) stout awns ______ **Calyptocarpus**

23. Plants erect or decumbent to prostrate; ray ligules 6–15 mm long.
  25. Leaves usually villous to sericeous; phyllaries 20–40+; phyllaries and pales spine-tipped ______ **Borrichia**
  25. Leaves usually hispid to scabrous; phyllaries 8–16+; phyllaries and pales not spine-tipped.
    26. Prostrate herbaceous perennials, often rooting at nodes; achenes strongly biconvex to plumply 3–4-angled; pappus a small fringed crown or absent ______ **Sphagneticola**
    26. Small erect shrubs; achenes mostly strongly compressed and 3–4-angled, usually ± winged; pappus a small cup-like crown plus sometimes 1–3 bristles or awns to 6 mm long ______ **Wedelia**

Key 2. Ray and disk flowers both present; ray flowers yellow to orange or red to reddish brown, sometimes with reddish brown or maroon marks; receptacle without chaffy bracts or pales (but sometimes rough or bearing bristles).

1. Disk flower pappus of scales OR a few (≤ 10) stout bristles or awns or a minute crown OR pappus lacking.
  2. Plants usually strongly aromatic (sometimes unpleasantly so); phyllaries (and usually also leaves) noticeably dotted or streaked with pellucid glands containing strong-scented oils.
    3. Leaves linear, margins entire and glandular-punctate, basally ciliate; phyllaries distinct to bases or nearly so; disk flowers 7–20 per head, corollas sometimes weakly 2-lipped ______ **Pectis**
    3. Leaves linear and entire to coarsely toothed OR wider and pinnately lobed to pinnatisect, not basally ciliate; phyllaries fused 1/3+ of their lengths (at least the inner series) or sometimes nearly separate; disk flowers (10–)20–150+ per head, corollas 5-lobed.
      4. Leaves pinnately lobed or pinnatisect, the lobes or segments usually lanceolate to ovate; secondary (subtending) phyllary-like bracts absent; disk corollas 7–12+ mm long; pappus of 2–6 elements; horticultural exotics escaping cultivation ______ **Tagetes**
      4. Leaves simple and entire to toothed or pinnately lobed, the leaves or lobes usually filiform to linear; secondary phyllary-like bracts often present; disk corollas 2–4.5 mm long; pappus of 10–12 elements; native species.
        5. Plants 40–90 cm tall; leaf bases and secondary phyllary-like bracts often with a few stiff bristles; phyllaries 10–12; ray flowers 7–12, ligules 10–15 mm long ______ **Dysodiopsis**
        5. Plants 15–30 cm tall; leaf bases and secondary phyllary-like bracts (if any) lacking stiff bristles; phyllaries 12–22; ray flowers 10–21, ligules 2–10 mm long ______ **Thymophylla**
  2. Plants sometimes aromatic; leaves sometimes with sessile or stalked glands, but these generally minute and not noticeably pellucid; phyllaries eglandular or glandular, or sometimes notably resinous.
    6. Plants taprooted; stems, leaves, and/or phyllaries resinous or gummy; leaves often gland-dotted, sometimes releasing a turpentine scent when crushed.
      7. Leaves usually crenate to dentate or serrate; involucres (6–)8–25 mm wide; disk flowers usually 100+ per head; achenes glabrous ______ **Grindelia** (in part)
      7. Leaves entire; involucres 1–5 mm wide; disk flowers 1–50 per head; achenes pubescent or bristly.
        8. Ray and disk flowers with strongly dissimilar pappus types, the ray pappus a small crown, the disk pappus of 5–8 white, narrowly spatulate, basally-fused scales ______ **Amphiachyris**
        8. Pappus not dimorphic, or only weakly so (ray pappus may differ in size from disk pappus but will not differ in pappus type) OR pappus lacking.
          9. Plants subshrubs, perennials, or annuals, to 140 cm tall; receptacles hairy; ray flowers 1–23 per head, ligules 1.5–6 mm long; disk pappus a short scaly crown 0.1–1 mm high, or rarely absent ______ **Gutierrezia**
          9. Plants shrubs, to 150(–200) cm tall; receptacles naked; ray flowers 4–9 per head, ligules 2–3 mm long; pappus absent, or a minute toothed crown < 0.1 mm high ______ **Gymnosperma**
    6. Plant root types various; herbage not resinous or gummy; leaves gland-dotted or eglandular, aromatic or not when crushed (lacking a turpentine scent).
      10. Leaves opposite, ± succulent; phyllaries usually 2; ray flowers usually 1 per head; pappus absent ______ **Flavaria**
      10. Leaves either basal or mostly alternate (at least on upper stems), seldom succulent; phyllaries 5+; ray flowers 2+ per head; pappus present or absent.
        11. Ray flowers 2–5 per head.
          12. Stems and leaves green, glabrous or sparsely hairy; disk flowers 16–22 per head, 4-lobed; pappus usually of 1–2 bristles to 1.8 mm long plus minute vestigial scales or a callous crown, or rarely pappus absent ______ **Perityle**
          12. Stems and leaves gray-green, cobwebby-villous; disk flowers 5–8(–12) per head, 5-lobed; pappus of 4–6 scales, 2.5–3+ mm long ______ **Psilostrophe**
        11. Ray flowers 6+ per head.
          13. Ray and disk flowers with strongly dissimilar achenes, the ray achenes prismatic or club-shaped, 4–5 mm long, the disk achenes compressed and often winged, 6–8 mm long; pappus absent; horticultural exotics escaping cultivation ______ **Dimorphotheca**

13. Ray and disk achenes not dimorphic as described above (they may be dimorphic in other ways) OR either ray or disk flowers sterile and not producing achenes; pappus of awns or scales; native species.
14. Stems usually winged by decurrent leaf bases; achenes 0.6–1.5 mm long (IF stems unwinged THEN achenes to only 1.3 mm long) ________ **Helenium** (in part)
14. Stems not winged; leaf bases not decurrent; achenes 1.5+ mm long.
15. Leaves usually sharply crenate to dentate or serrate (rarely entire); phyllaries in 5–6(–9) series, usually strongly recurved; achenes glabrous ________ **Grindelia** (in part)
15. Leaves entire to toothed or lobed to pinnatifid; phyllaries in 2–5 series, erect to reflexed at maturity; achenes pubescent or rarely glabrous.
16. Disk flowers functionally staminate and sterile, only ray flowers producing achenes; ray flower pappus of several series of barbellate bristles 1–3 mm long; disk pappus of 1–3 awn-like bristles ________ **Bradburia** (in part)
16. Disk flowers perfect and fertile, ray flowers fertile or sterile; ray and disk pappus similar, of scales.
17. Herbage noticeably sweet-scented, like newly-mown sweet-clover; achenes narrowly obconic, prominently 10-ribbed; pappus scales not awn-tipped ________ **Amblyolepis**
17. Herbage not noticeably sweet-scented (may be otherwise aromatic); achenes usually obpyramidal and 4-sided; pappus scales awn-tipped.
18. Phyllaries usually strongly reflexed in fruit; receptacles often bristly; ray ligules (when present) frequently in colors other than solid yellow; disk corollas usually brown-purple to red-brown, or tipped with these colors; pappus of 7–12 awn-tipped scales, 5–9 mm long ________ **Gaillardia** (in part)
18. Phyllaries mostly spreading to erect in fruit; receptacles naked; ray ligules usually uniformly some shade of yellow (rarely tipped or striped with white); disk corollas usually solid yellow (or sometimes purple-tipped); pappus of 4–7 awn-tipped scales, 1.2–4 mm long.
19. Phyllaries in 2 series of identical number, glabrous or inconspicuously short-pubescent, margins not noticeably scarious; ray flowers either 6–8 and ligules minute (to 3 mm long) in two rare species OR ray flowers 8–13 and ligules 8.5–11 mm long in one common species, withering and twisting in age ________ **Hymenoxys**
19. Phyllaries in 3 series, usually ± hairy, margins usually scarious (at least inner series); ray flowers 9–26, ligules 8.2–22 mm long, often not withering, remaining attached and strongly recurved in age ________ **Tetraneuris**

1. Disk flower pappus of 20+ capillary bristles or hairs, with or without an additional outer series of scales or awns or bristles.
20. Perennials or subshrubs, evergreen, strongly aromatic; leaves and phyllaries noticeably dotted or streaked with pellucid glands containing strong-scented oils ________ **Chrysactinia**
20. Perennials, biennials, or annuals, evergreen or deciduous, sometimes aromatic; leaves and/or phyllaries sometimes with sessile or stalked glands, but these generally minute and not noticeably pellucid.
21. Primary phyllaries in 1–2 series, equal or subequal, usually subtended by 1–20 secondary phyllary-like bracts.
22. Plants vining, to 500 cm tall; ray ligules orange to brick-red; disk corollas orange; horticultural escape in subtropical parts of TX ________ **Pseudogynoxys**
22. Plants herbaceous, to 80 cm tall; ray ligules yellow; disk corollas yellow; native or weedy plants widespread in TX.
23. Plants annual or perennial; stem leaves deeply toothed to pinnately lobed or compound, bases sometimes auriculate and clasping ________ **Packera**
23. Plants annual; stem leaves nearly entire or shallowly toothed, bases usually auriculate and clasping ________ **Senecio**
21. Primary phyllaries in 3–6+ series, usually unequal, not subtended by secondary phyllary-like bracts.
24. Outer pappus series of scales or flat-based awns, 0.25–1.4 mm long (use hand lens), surrounding inner series of longer bristles.
25. Basal leaves persistent, linear to lanceolate or oblanceolate, grass-like in appearance, 3–11-parallel veined, pubescence silky-sericeous ________ **Pityopsis**
25. Basal leaves (if present) spatulate to obovate or ovate to elliptic (not grass-like), usually 1-nerved, with pinnate or reticulate veins, pubescence scabrous to hispid-strigose or pilose to sericeous-cobwebby.
26. Plants fibrous-rooted, sometimes with short lateral rhizomes; herbage pilose to sericeous-cobwebby; basal leaves sessile; ray ligules 2–3 mm wide ________ **Chrysopsis**
26. Plants taprooted, sometimes rhizomatous; herbage scabrous to hispid-strigose or pilose; basal leaves (if present) petiolate or subsessile; ray ligules 1–2 mm wide.

27. Plants annual or perennial (IF annual, THEN achenes dimorphic and ray achenes lacking pappus); leaves scabrous to hispid or strigose, often with sparse to dense stalked glands (IF eglandular THEN leaves silvery gray-green) ________ **Heterotheca**
27. Plants annual; leaves pilose, eglandular; ray and disk achenes similar; pappus always present ________ **Bradburia** (in part)

24. Pappus solely of thin-based, capillary bristles in 1–4 series (outer series may be shorter than inner).
28. Achenes dimorphic, the ray achenes ± 3-sided, the disk achenes ± compressed.
29. Herbage eglandular; basal and lowermost stem leaves pinnately lobed or pinnatifid; receptacle with a persistent network of subulate, chaff-like scales; inner phyllaries basally narrowed and stalk-like, expanded above ________ **Xanthisma**
29. Herbage with prominent sessile or stalked glands; basal and lowermost stem leaves coarsely serrate to entire; receptacle naked; inner phyllaries not basally narrowed ________ **Rayjacksonia**
28. Ray and disk achenes similar in shape.
30. Involucres 10–15 mm high and 15–25 mm wide; ray flowers 25–45 per head ________ **Grindelia** (in part)
30. Involucres 3–10 mm high and 1.7–10 mm wide; ray flowers 2–30 per head.
31. Taprooted annuals or biennials; heads borne singly or in loose, widely-branched paniculate arrays ________ **Croptilon**
31. Rhizomatous perennials; heads usually borne in clusters in corymbose, paniculate, or racemose arrays.
32. Leaves entire, with glandular-punctate dots; heads in a ± flat-topped corymbose arrangement; achenes 2–4-nerved ________ **Euthamia**
32. Leaves usually serrate, without glandular-punctate dots; heads in pyramidal paniculate or club-shaped racemose arrangements (or sometimes in axillary clusters); achenes 8–10-nerved ________ **Solidago**

Key 3. Ray and disk flowers both present; ray flowers white, yellowish cream, tawny, pink, lavender, bluish, or purple; disk pappus (at least the innermost series) of capillary bristles or hairs.

1. Ray flowers inconspicuous, ligules < 1 mm long, barely exceeding the phyllaries ________ **Conyza** (in part)
1. Ray flowers conspicuous, ligules 1.3+ mm long, exceeding the phyllaries.
2. Corollas of disk flowers bilaterally symmetrical, creamy white to purple, sometimes maturing to crimson ________ **Chaptalia**
2. Corollas of disk flowers radially symmetrical, most often true yellow, infrequently yellowish cream or yellow-orange, sometimes becoming pink to brown with age.
3. Plants subshrubs or rush-like perennials, colonial; stems green and glaucous, green thorns often present; few leaves present at flowering; phyllaries (1–)3(–5)-nerved ________ **Chloracantha**
3. Plants herbaceous annuals or perennials, solitary or colonial; stems not becoming thorny; at least upper leaves persistent through flowering; phyllaries usually 1-nerved (sometimes 3-nerved in *Erigeron* and *Symphyotrichum*)
4. Plants 1–3 cm tall; heads solitary and nearly sessile, nestled in basal rosettes; involucres (10–)12–25+ mm wide ________ **Townsendia**
4. Plants usually 10+ cm tall; heads solitary or borne in various arrays, usually well above any basal leaves; involucres often < 12 mm wide.
5. Heads usually borne in flat-topped corymbose arrays; ray ligules white, 2–7 per head; disk flowers usually < 20 per head; disk flower pappus innermost bristles with strongly flattened tips (= clavate) ________ **Doellingeria**
5. Heads solitary or borne in various arrays; ray ligules white or various shades of pink, blue, violet, or purple, 7–100+ per head; disk flowers usually > 20 per head; disk flower pappus innermost bristles with rounded tips (or perhaps obscurely flattened).
6. Leaves narrow, stiff, evenly distributed; phyllaries strongly keeled; ray and disk achenes slightly dimorphic; achene pubescence silky strigose-sericeous ________ **Ionactis**
6. Leaves variously shaped and distributed; phyllaries flat or only swollen at bases, not strongly keeled; ray and disk achenes either monomorphic OR strongly dimorphic; achenes glabrous to strigose but usually not silky-sericeous.
7. Plants taprooted annuals; achenes strongly dimorphic, ray achenes 1.3–1.5 mm long and sparsely hairy, disk achenes 2–3 mm long and glabrous or nearly so; pappus lacking on ray achenes ________ **Psilactis**
7. Plants annual or perennial, taprooted or fibrous rooted or rhizomatous; achenes of various sizes, glabrous to variously pubescent, but generally monomorphic within a head; ray achenes with some pappus present, similar to disk achenes or differing.
8. Leaves glabrous to minutely scabrous, margins usually entire and often revolute; phyllaries glabrous and tips often squarrose or strongly recurved; ray flowers 15–30 per head, ligules usually violet-purple, rarely white; achenes 9–16-nerved ________ **Eurybia**
8. Leaves glabrous or variously pubescent, margins various; phyllaries glabrous or variously pubescent, tips appressed or erect to variously recurved; ray flowers 8–350+ per head; ligules often white, but also sometimes pinkish to bluish or purplish; achenes 2–5(–10)-nerved.

9. Heads borne singly or in loose corymbose or paniculate arrays; peduncles of buds often nodding; phyllaries with orange-resinous veins, tips usually lacking a well-differentiated green zone; ray flowers usually 44–350+ per head, ligules usually < 1 mm wide; plants flowering Mar–Jun (sporadically into Oct) **Erigeron**
9. Heads usually borne in paniculate or corymbose arrays; peduncles of buds often erect; phyllaries lacking orange-resinous veins, tips usually with a well-differentiated green zone, or sometimes foliaceous; ray flowers usually 8–50 per head, ligules usually > 1 mm wide; plants flowering Aug–Nov (rarely Mar–Apr) **Symphyotrichum**

Key 4. Ray and disk flowers both present; ray flowers white, yellowish cream, tawny, pink, lavender, bluish, or purple; disk pappus of awns, scales, or a minute crown, or pappus absent.

1. Receptacles chaffy, a bract (pale) subtending each flower.
2. Only ray flowers fertile (= maturing achenes), their achenes much larger than those of the sterile disk flowers.
3. Inner phyllaries each subtending and clasping a ray achene; mature achenes enclosed by and shed within the subtending phyllary, forming a fruit-like or bur-like structure.
4. Heads borne in sessile clusters; ray flowers 5–8 per head, ligules inconspicuous, minute; achenes shed in a bur-like structure, 7–9+ mm long, with hooked or straight spines or tubercles **Acanthospermum**
4. Heads borne singly on peduncles; ray flowers 7–13 per head, ligules obvious, showy; achenes shed in a ± conical or boot-shaped, fruit-like structure, 1.3–2.6 mm long **Melampodium**
3. Inner phyllaries sometimes subtending and clasping a ray achene; mature achenes sometimes shed with the subtending phyllary or other parts, but not completely enclosed within the phyllary and not forming a fruit-like or bur-like structure.
5. Ray flowers 5(–8) per head, ligules to 2 mm long; achenes compressed but not winged, 1.5–4+ mm long (each shed with the subtending phyllary, two adjacent disk flowers, and their pales) **Parthenium**
5. Ray flowers 14–30 per head, ligules 10+ mm long; achenes flattened and winged, 12–26 mm long (shed without accessory structures) **Silphium**
2. Disk flowers perfect and fertile; ray flowers fertile OR sterile.
6. Ray flowers neuter OR pistillate and sterile; only disk flowers fertile (= maturing achenes), their achenes much larger than those of the sterile ray flowers.
7. Leaves all opposite; phyllaries apparently in ± 3 series, dimorphic, the outer (secondary) series herbaceous (and sometimes foliaceous), the inner 2 (primary) series often very unlike the outer, usually scarious-margined.
8. Ray ligules 15–50 mm long; disk corollas 5–7 mm long; stamen filaments hairy near anthers (use lens); achenes usually with 1 groove on each face **Cosmos** (in part)
8. Ray ligules 1–15+ mm long; disk corollas 2–5 mm; stamen filaments not hairy; achenes with 0 or 2 grooves on each face **Bidens**
7. Leaves mostly alternate; phyllaries in 1–several series, usually not strongly dimorphic, herbaceous or scarious-margined.
9. Stems winged by decurrent leaf bases; achenes strongly flattened and papery-winged; pappus usually of 2 persistent subulate scales or awns to 2 mm long (or sometimes pappus absent) **Verbesina** (in part)
9. Stems unwinged, leaf bases not decurrent; achenes rounded or 4-angled in cross-section, not strongly flattened or papery-winged; pappus absent or a small crown, sometimes with a few teeth.
10. Ill-scented annuals; leaves 1–2-pinnately lobed; involucres 5–9 mm wide; ray ligules white **Anthemis**
10. Non-aromatic perennials; leaves entire to dentate or serrate; involucres 12–40 mm wide; ray ligules pink to purple or rarely white **Echinacea**
6. Ray flowers fertile; disk flowers perfect and fertile; both kinds of flowers producing mature achenes (though these may differ in shape).
11. Leaves all opposite; achenes 3–4-angled or compressed.
12. Leaves serrate to subentire; involucres 3–5 mm wide; ray ligules ca. 2 mm long **Eclipta**
12. Leaves entire; involucres 15–25 mm wide; ray ligules 10–35 mm long **Zinnia**
11. Leaves mostly alternate; achenes strongly flattened and winged.
13. Plants usually > 50 cm tall; stems winged by decurrent leaf bases; ray ligules 3–7+ mm long **Verbesina** (in part)
13. Plants often < 50 cm tall; stems unwinged, leaf bases not decurrent; ray ligules 1.5–3 mm long **Achillea**
1. Receptacles naked, not bearing chaff or pales (but sometimes bearing bristles).
14. Disk flower pappus either absent OR a minute crown < 1 mm high (crown may be of thickened rings, scales, awns, or hairs).
15. Leaves opposite, ± succulent; phyllaries usually 2; disk flowers 0–1(–2) per head **Flaveria**
15. Leaves alternate, not succulent; phyllaries 10+; disk flowers 5+ per head.
16. Heads borne in loose, leafy corymbose or paniculate arrays; achenes dimorphic, ray achenes ± 3-sided and sometimes winged, disk achenes ± compressed **Boltonia** (in part)
16. Heads borne singly; achenes not dimorphic, or only weakly so and not as described above (pappus may be dimorphic).
17. Involucres 12–20+ mm wide; ray ligules often > 15 mm long **Leucanthemum**

17. Involucres < 12 mm wide; ray ligules usually < 15 mm long.
    18. Plants aromatic (smelling of sweet hay when bruised); leaves finely and narrowly divided, usually 2–3-pinnately lobed; achenes 0.75–0.9 mm long ________ **Matricaria** (in part)
    18. Plants not especially aromatic; leaves entire to pinnatifid; achenes usually 1+ mm long.
        19. Plants often small and delicate; receptacles flat to convex, smooth (can be roughened); ray flowers 5–19 per head; disk flowers 4–22; achenes 1.5–2.2 mm long ________ **Chaetopappa** (in part)
        19. Plants delicate or not; receptacles conic, pitted; ray flowers (7–)13–75 per head; disk flowers 25–400; achenes 1–1.6 mm long.
            20. Leaves varying from entire to deeply pinnatifid, achenes columnar, 4-angled to somewhat rounded, 4–12-ribbed ________ **Aphanostephus**
            20. Leaves usually entire or sometimes toothed; achenes obovoid to oblanceoloid, compressed, 2-nerved (nerves marginal) ________ **Astranthium**

14. Disk flower pappus (at least on the central achenes) with at least some elements usually > 1 mm long, of scales, awns, hairs, a cup-like crown, or bristles, OR any combination of these.
    21. Plants dainty taprooted annuals, 3.5–30 cm tall; pappus of 5 scales alternating with 5 slender awns ________ **Chaetopappa**
    21. Plants annual or perennial, of various heights; pappus various but not of 5 scales and 5 awns alternating.
        22. Achenes dimorphic, ray achenes ± 3-sided and sometimes winged, disk achenes ± compressed; pappus of 2–3 awns (sometimes with 1–12 shorter bristles or scales) ________ **Boltonia** (in part)
        22. Achenes not dimorphic, or only weakly so and not as described above; pappus various (may be dimorphic), usually not of 2–3 awns.
            23. Ray flowers usually 44–350+ per head; disk achene pappus with a inner series of (5–)8–36 bristles, often with an outer series of scales, or shorter bristles, or a crown; achenes ± compressed or flattened ________ **Erigeron**
            23. Ray flowers < 40 per head; disk achene pappus of scales (may be awn-tipped), not of bristles; achenes ± 4-angled or club-shaped, not compressed.
                24. Small plants, < 20 cm tall; ligules inconspicuous, 2–3 mm long, whitish to pale yellow; pappus of 5–6 awn-tipped scales, 1.4–2.5 mm long ________ **Hymenoxys**
                24. More robust plants, usually > 20 cm tall; ligules usually obvious and showy, 5–30+mm long, pink to purple or white to pale yellow; pappus of 4–12 rounded or awn-tipped scales, 3–9 mm long (sometimes shorter or absent on peripheral achenes).
                    25. Leaves entire to toothed or pinnately lobed; heads borne singly; phyllaries 14–40; receptacles often bristly; achenes 1.5–4+ mm long ________ **Gaillardia** (in part)
                    25. Leaves entire; heads borne in ± corymbose arrays; phyllaries 8–14; receptacles without bristles; achenes 5–9 mm long ________ **Palafoxia** (in part)

Key 5. Disk flowers only or apparently so; phyllaries with hooked or straight spines, or tubercles (at least on pistillate heads when heads unisexual), sometimes united at maturity to form a bur-like fruit or phyllaries with margins variously lobed, toothed, or bristle-fringed; plants sometimes spiny.

1. Some or all flowers imperfect, corollas small (< 5 mm long), rudimentary, or absent; pappus absent.
    2. Stems procumbent; pistillate and functionally staminate flowers in the same head ________ **Acanthospermum**
    2. Stems generally erect; pistillate and staminate flowers in different heads (plants monoecious) (RAGWEEDS and COCKLEBURS).
        3. Staminate heads usually in racemes or spikes, their phyllaries united into a cup shape; pistillate heads forming a 1-seeded bur < 6 mm long, with phyllary tips projecting as 1–12 spines or tubercles (or rarely, membranous wings) ________ **Ambrosia**
        3. Staminate heads in terminal clusters, their phyllaries not fused; pistillate heads forming a 2-seeded bur (cocklebur) 10–30+ mm long, with phyllary tips projecting as 20+ hooked, spine-like structures ________ **Xanthium**
1. Flowers usually perfect, corollas relatively large, showy, variously colored; pappus of awns, bristles, or scales usually present (THISTLES, KNAPWEEDS, and BASKET-FLOWERS).
    4. Corollas bright yellow to yellow-orange or orange-red.
        5. Leaf margins never spiny; leaf bases long-decurrent, creating winged stems; all phyllaries spine-tipped; pappus of stiff bristles ________ **Centaurea** (in part)
        5. Leaf margins sometimes spiny; leaf bases not decurrent; stems unwinged; outer phyllaries sometimes leafy, at least the inner phyllaries spine-tipped; pappus absent or of scales ________ **Carthamus**
    4. Corollas usually pink to rosy-red or lavender to bluish purple, less often white, or rarely pale yellow.
        6. Pappus of 4–5 scales, 8–12 mm long, shed early from mature achenes ________ **Stokesia**
        6. Pappus of many awn-like scales or bristles.
            7. Leaf margins spiny.
                8. Pappus of plumose bristles (feather-like, with long side branches, at least on the upper portion) ________ **Cirsium**
                8. Pappus of awn-like scales, smooth bristles, or barbellate bristles (with short side branches).
                    9. Leaves variegated (mottled white and green above); leaf bases not decurrent; stems lacking spiny wings ________ **Silybum**

9. Leaves uniformly colored above; leaf bases long-decurrent, creating spiny-winged stems.
    10. Receptacles unpitted, bearing flattened bristles or scales; pappus bristles 11–25 mm long ____ **Carduus**
    10. Receptacles deeply pitted, lacking bristles or scales; pappus bristles 7–9 mm long ____ **Onopordum**

7. Leaf margins not spiny.
    11. Phyllary margins entire or minutely serrate, tips with a single, prominent, hooked spine; corollas of all flowers similar; the entire head shed as a spiny bur at maturity ____ *Arctium*
    11. Phyllary margins variously lobed, toothed, or bristle-fringed, not prominently spine-tipped; corollas of peripheral flowers often enlarged, simulating ray flowers; achenes generally shed individually at maturity.
        12. Involucres 10–16 mm high; peripheral (sterile) flower corollas 15–25 mm long ____ **Centaurea** (in part)
        12. Involucres 30–50 mm high; peripheral (sterile) flower corollas 35–50 mm long ____ **Plectocephalus**

Key 6. Disk flowers only or apparently so; phyllaries mostly entire; corollas white, yellowish cream, greenish, tawny, pink, lavender, bluish, or purple (or infrequently obscured by pubescence, enclosed by receptacle pales, or corollas vestigial or absent); plants not spiny.

1. Plants dioecious (or essentially so), all flowers imperfect.
    2. Shrubs or subshrubs, variable in size from 25–600 cm tall (male and female plants may differ in appearance); leaves mostly on stems; pappus of 25–50 bristles, 7–14 mm long ____ **Baccharis**
    2. Herbs, usually < 45 cm tall; leaves mostly basal; pappus of 10–20+ bristles, 2.5–8 mm long ____ **Antennaria**

1. Plants monoecious, at least some flowers in each head usually perfect.
    3. Leaves alternate, usually entire (NOT deeply divided or pinnately lobed), white or gray woolly-tomentose beneath (sometimes bicolor and green above, or with dense silvery-gray pubescence on both sides); phyllaries usually pale and notably scarious-margined; heads often with copious cottony or woolly pubescence; flowers typically small, not very showy, sometimes embedded in or obscured by pubescence.
        4. Plants rhizomatous perennials, often > 50 cm tall.
            5. Stems winged by long-decurrent leaf bases; heads in interrupted, ± cylindric, spicate arrays (not leafy); pappus of numerous bristles ____ **Pterocaulon**
            5. Stems unwinged, leaf bases not decurrent; heads in paniculate or racemose arrays (often leafy); pappus absent ____ **Artemisia** (in part)
        4. Plants taprooted annuals or short-lived fibrous-rooted perennials, usually 3–50 cm tall.
            6. Heads embedded in and ± obscured by cottony pubescence, clustered in compact, axillary or terminal arrays, subtended by leafy bracts; phyllaries 0–6; receptacles chaffy; pappus essentially absent.
                7. Pales of perfect flowers saccate, each enclosing a flower; achenes (at least the peripheral ones) strigose; known only from Gulf coastal prairie in Victoria Co. ____ **Micropsis**
                7. Pales of perfect or staminate flowers flat to concave, not enclosing flowers; achenes glabrous; widespread in East TX ____ **Diaperia**
            6. Heads often with woolly hairs but not embedded in or obscured by these, clustered in terminal, corymbose, spicate, or paniculate arrays; phyllaries more numerous, in 3–7 series; receptacles naked; pappus of bristles.
                8. Pappus of plumose bristles (use lens to see side branches) ____ **Facelis**
                8. Pappus of minutely-barbed or barbellate bristles.
                    9. Heads in spicate or narrow paniculate arrays (reduced to terminal clusters in depauperate individuals); receptacles concave in fruit; pappus of 12–28 minutely barbed bristles, deciduous as a ring; usually blooming Mar–Jun ____ **Gamochaeta**
                    9. Heads in corymbose or paniculate arrays or terminal clusters, often somewhat flat-topped; receptacles flat in fruit; pappus of 10–12 barbed bristles, these readily separating individually, in clusters, or as an easily-fragmented ring; most species blooming Aug–Dec (Apr–Jun in *P. luteoalbum*) ____ **Pseudognaphalium** (in part)
    3. Leaves variously arranged, margins various (IF white or gray woolly-tomentose beneath or with dense silvery-gray pubescence on both sides THEN deeply divided or pinnately lobed); phyllaries green or colorful or pale, sometimes scarious-margined; heads glabrous or pubescent but generally not with copious cottony or woolly pubescence; flower size various, showy or not so, usually not embedded in or obscured by pubescence.
        10. Receptacles chaffy, a bract (pale) subtending each flower.
            11. Leaves opposite; primary phyllaries subtended by 3–21+ secondary phyllary-like bracts (these may be ± herbaceous); pappus usually of 2–4 retrorsely-barbed awns 2–4 mm long ____ **Bidens**
            11. Leaves opposite or alternate; primary phyllaries not subtended by secondary phyllary-like bracts; pappus of unbarbed scales or awns or pappus absent.
                12. Plants wind-pollinated, the heads small and not at all showy nor attractive to pollinators; disk corollas minute or insignificant ____ **Iva**
                12. Plants generally adapted for pollination by insects, heads colorful or otherwise attractive; disk flowers with obvious corollas.
                    13. Pappus of ca. 40 barbellate bristles ____ **Chromolaena** (in part)
                    13. Pappus of 2–6 scales or awns or pappus absent.
                        14. Leaves entire; heads solitary or in open corymbose arrays; corollas lavender-pink to purple to white, lobes long and often contorted; pappus of 5–6 scales ____ **Marshallia**

14. Leaves dentate to deeply pinnately lobed; heads in relatively compact corymbose or paniculate arrays; corollas white to pale dingy yellow, lobes short; pappus of 2 subulate scales or awns or pappus absent.
15. Plants 100+ cm tall; stems winged with decurrent leaf bases; leaves minutely pubescent; achenes strongly flattened and papery-winged ______ **Verbesina** (in part)
15. Plants often < 100 cm tall; stems unwinged, leaf bases not decurrent; leaves usually hairy, sometimes silvery-gray; achenes compressed but not strongly flattened, unwinged ______ **Parthenium**
10. Receptacles not bearing chaff or pales (but sometimes bearing bristles).
16. Pappus of scales OR a few stout-based bristles or awns or a minute crown OR pappus lacking.
17. Pappus (at least of the central achenes) of scales OR a few stout-based bristles or awns.
18. Heads sessile; flowers 1–5+ per head.
19. Leaves mostly basal, wider than linear-filiform; several heads clustered and subtended by whorls of 2–3 leaf-like ovate-deltate bracts ca. 10 mm long; flowers (1–)4(–5+) per head, corollas pink or purple to white; pappus of 5(–6) bristle-tipped scales ______ **Elephantopus**
19. Leaves mostly on the stems, linear-filiform; heads not clustered, not subtended by ovate leaf-like bracts; flowers 3 per head, corollas white to pale yellowish; pappus of 10 scales in 2 series ______ **Thurovia**
18. Heads pedunculate; flowers 5–100+ per head.
20. All leaves usually 1–2-pinnately lobed, often tomentose and gland-dotted; phyllaries petaloid at tips; pappus of 14–20 awn-tipped scales ______ **Hymenopappus**
20. Leaves entire to toothed or pinnatifid; glabrous to sparsely villous; phyllaries not petaloid (may be colorful); pappus (at least of the central achenes) of 4–10 subulate or awn-tipped scales.
21. Leaves mostly basal, margins entire to toothed or pinnatifid; phyllaries 22–32, usually green; achenes 2 mm long ______ **Gaillardia** (in part)
21. Leaves mostly on the stems, margins entire; phyllaries 8–14, often tipped purple or white; achenes 3–8 mm long ______ **Palafoxia** (in part)
17. Pappus a minute crown OR pappus lacking.
22. Rhizomatous perennials, 20–120+ cm tall; leaves linear to lanceolate or oblong-elliptic, margins entire to toothed or pinnately lobed; heads borne in paniculate arrays; achenes 0.5–0.8 mm long ______ **Artemisia** (in part)
22. Taprooted annuals, usually 2–40 cm tall; leaves obovate to spatulate, (1–)2–3-pinnately lobed or finely dissected; heads usually borne singly; achenes 1–3+ mm long.
23. Heads sessile in leaf axils; achenes tipped by a persistent spine-like stylar sheath (painful in a manner similar to sandburs) ______ **Soliva**
23. Heads pedunculate; achenes without a persistent spine-like stylar sheath.
24. Receptacles flat to convex; peripheral flowers pistillate, central flowers functionally staminate; achenes compressed or ± flattened, obovoid to oblong, the peripheral ones larger and ± winged; pappus absent ______ **Cotula**
24. Receptacles conic to oblong-ovoid; all flowers perfect and fertile; achenes ± cylindric-obconic, ribbed; pappus a small crown ______ **Matricaria** (in part)
16. Pappus of bristles or hairs, with or without an additional, shorter, outer series of scales or bristles.
25. Leaves all or mostly opposite or whorled.
26. Perennial twining or scrambling vines; phyllaries 4; flowers usually 4 per head ______ **Mikania**
26. Non-vining annuals, perennials, or shrubs; phyllaries 8+; flowers 3–60+ per head.
27. Involucres 5–12 mm wide; achenes 10-ribbed ______ **Brickellia** (in part)
27. Involucres 1–7 mm wide; achenes (3–)4–5(–8) ribbed.
28. Pappus of 2–6+ hair-like scales or bristles to 0.5 mm long ______ **Trichocoronis** (in part)
28. Pappus of (5–)10–50 barbellate bristles or hair-like scales, much longer than 0.5 mm.
29. Phyllaries ± equal in length.
30. Receptacles conic ______ **Conoclinium**
30. Receptacles flat or convex.
31. Involucres 1–3(–5+) mm wide; flowers (4–)5(–15) per head; style bases usually puberulent ______ **Eupatorium** (in part)
31. Involucres 3–6 mm wide; flowers 10–60 per head; style bases glabrous ______ **Ageratina**
29. Phyllaries unequal (outer ones shorter).
32. Leaves whorled, with 3–7 petiolate leaves per node ______ **Eutrochium**
32. Leaves opposite, rarely whorled (IF whorled THEN sessile).
33. Flowers (4–)5(–15) per head; style bases usually puberulent and enlarged ______ **Eupatorium** (in part)
33. Flowers (6–)15–40(–75) per head; style bases glabrous, not enlarged.

34. Involucres 5–10 mm high; phyllaries in 4–6+ series; achenes (3–)5-ribbed, usually gland-dotted ________ **Chromolaena**
34. Involucres 4–5 mm high; phyllaries in 2–4 series; achenes 5(–8)-ribbed, not gland-dotted ________ **Fleischmannia**

25. Leaves all or mostly alternate.
35. Pappus of two distinctly different series, the inner row of long bristles, the outer of shorter scales or bristles ________ **Vernonia**
35. Pappus of 1–several series of bristles, usually of similar lengths.
36. Leaves deeply divided to pinnately or palmately pinnatifid, ultimate lobes linear ________ **Eupatorium** (in part)
36. Leaves entire to variously toothed or lobed but not deeply divided or pinnatifid.
37. Phyllaries usually 5, 8, 13, or 21 in 1–2 series, ± equal; pappus bristles 60–200+.
38. Plants perennial; leaves entire or rarely toothed, palmately veined (with several prominent, ± parallel longitudinal nerves converging at the base and tip); phyllaries mostly pale green to white; 5 flowers per head ________ **Arnoglossum**
38. Plants annual; leaves entire to sharply double-serrate and sometimes irregularly lobed, pinnately veined (with a single prominent midnerve and lacking additional prominent, parallel, longitudinal nerves); phyllaries mostly leaf-green; 10+ flowers per head.
39. Leaves entire to toothed or weakly lobed; corollas pink to purple or reddish purple, surpassing the phyllaries by 2–4 mm; achenes with 5 ribs or nerves ________ **Emilia**
39. Leaves sharply double-serrate to lobed; corollas whitish to greenish or yellowish cream to pale pinkish, only slightly surpassing the phyllaries; achenes with 10–12 ribs or nerves ________ **Erechtites**
37. Phyllaries 18–40+ in 2–7 series, unequal; pappus bristles usually < 40.
40. Achenes 8–11-ribbed, 2.7–9 mm long; pappus bristles barbellate to plumose.
41. Shrubs or woody-based perennials, usually much-branched; heads in paniculate or corymbose arrays; corollas cream to pale yellow or pale green to pinkish or purplish ________ **Brickellia** (in part)
41. Herbaceous perennials from a woody corm or rhizome, stems unbranched or infrequently branched; heads usually in spicate or racemose arrays, or rarely in cymose arrays; corollas pink to lavender-purple or rarely white ________ **Liatris**
40. Achenes 2–8-ribbed, < 2 mm long; pappus bristles only minutely barbed.
42. Plants usually aromatic (sometimes unpleasantly so); leaves lance-ovate or ovate to elliptic or oblong; achenes cylindric, 4–8-ribbed ________ **Pluchea**
42. Plants not aromatic; leaves lanceolate or oblanceolate to linear; achenes compressed, 1-nerved on each edge ________ **Conyza** (in part)

Key 7. Disk flowers only or apparently so; phyllaries mostly entire; corollas yellow to orange or red to reddish brown or maroon, sometimes with reddish brown or maroon marks; plants not spiny.

1. Receptacles naked; pappus of 15+ bristles or hairs.
2. Primary phyllaries in 1–2 series, equal or subequal, subtended by 2–4(–6+) secondary phyllary-like bracts ________ **Senecio**
2. Primary phyllaries in 2–6 series, unequal, not subtended by secondary phyllary-like bracts.
3. Weedy annuals; heads borne singly or in elongated paniculate or corymbose arrays; peripheral flowers pistillate and fertile, ligules absent or minute; central disk flowers perfect and fertile, corollas yellowish ________ **Conyza** (in part)
3. Perennials; heads borne in compact, terminal, corymbose arrays; all flowers perfect and fertile, corollas bright yellow.
4. Leaves mostly basal in persistent rosettes; phyllaries disposed in vertical ranks, usually yellowish; flowers 2–6 per head; achenes sparsely strigose ________ **Bigelowia**
4. Leaves mostly on the stems; phyllaries disposed spirally, green to yellowish with scarious margins; flowers 12–34 per head; achenes sericeous ________ **Isocoma**

1. Receptacles naked or bristly or paleate; pappus of 2–10 bristles, hairs or awn-tipped scales OR a few stout awns OR pappus absent.
5. Receptacles not bearing chaff or pales (but sometimes bearing bristles); primary phyllaries not subtended by secondary phyllary-like bracts.
6. Leaves opposite; phyllaries usually 2; pappus absent ________ **Flaveria**
6. Leaves alternate; phyllaries 9+; pappus present.
7. Leaves (at least 1 side) white to gray tomentose; heads small, not very showy; phyllaries white to silvery gray or yellowish ________ **Pseudognaphalium** (in part)
7. Leaves glabrous to hairy but not white-gray tomentose; heads relatively large, showy; phyllaries green.
8. Plants resinous; leaves usually crenate to serrate (rarely entire); phyllaries in 5–6(–9) series, usually strongly recurved; achenes glabrous ________ **Grindelia** (in part)
8. Plants not resinous; leaves entire to toothed or lobed; phyllaries in 2–3 series, erect to reflexed at maturity; achenes pubescent.

9. Plants perennial; stem winged by decurrent leaf bases; heads borne in paniculate arrays; achenes 1–1.2 mm long; pappus of 5–6 bristle-like scales __________ **Helenium** (in part)
9. Plants usually annual; stems unwinged, leaf bases not decurrent; heads borne singly; achenes 2–2.5 mm long; pappus of 7–10 bristle-like scales __________ **Gaillardia** (in part)

5. Receptacles chaffy, a bract (pale) subtending each flower; primary phyllaries subtended by secondary phyllary-like bracts (these may be ± herbaceous).
   10. Primary (inner) phyllaries united for 1/5 to 1/2 or more of their lengths; corollas reddish brown or yellow with reddish brown veins __________ **Thelesperma**
   10. Phyllaries all separate to their bases or nearly so; corollas solid yellow or yellow-orange to orange __________ **Bidens**

Technical Key to the Tribes

1. Plant sap usually milky; all flowers perfect, rayed or ligulate, the ligules strap-shaped (disk flowers absent) __________ **Cichorieae (Lactuceae)**
1. Plant sap usually watery, rarely milky; disk flowers with tubular, eligulate, or bilabiate corollas present; ray flowers with strap-shaped ligules present or absent, if present, these pistillate or neuter.
   2. Anther bases usually rounded or obtuse to acute (sometimes sagittate, but not tailed).
      3. Leaves usually opposite, sometimes whorled or alternate; heads of disk flowers only; corollas white to pinkish, bluish, or purple, or greenish to pale yellowish cream; style-branch appendages usually terete to club-shaped (lengths usually 2–5+ times lengths of stigmatic lines) __________ **Eupatorieae**
      3. Leaf arrangement various; ray flowers present or absent; corolla color various; style-branch appendages essentially absent, or resembling small tufts of hair, or deltate to lance-linear or filiform (lengths mostly 0–3+ times lengths of stigmatic areas or lines).
         4. Leaf margins usually 1–3 pinnately or palmately lobed or divided, or sometimes merely toothed; phyllaries usually scarious to hyaline at margins and tips; rays (if present) white to grayish white or creamy white; if only disk flowers present their corollas mostly white to yellowish cream or yellowish green; disk flower achenes with pappus usually absent or sometimes a small crown __________ **Anthemideae**
         4. Leaf margins various; phyllary margins and tips various; pappus various or sometimes absent.
            5. Ray flowers absent, heads of disk flowers only (though peripheral flowers may have enlarged or zygomorphic corollas simulating rays); corollas pink to lavender or blue to purple, or sometimes white (never yellow); style-branch appendages essentially absent __________ **Vernonieae**
            5. Ray flowers absent or present (peripheral disk flowers may have enlarged or zygomorphic corollas simulating rays); style-branch appendages usually resembling small tufts of hair, or deltate to lanceolate, or sometimes essentially absent.
               6. Leaves alternate; primary phyllaries equal to subequal, in 1–2 series, sometimes subtended by secondary phyllary-like bracts; receptacles naked; style-branch appendages usually resembling small tufts of hair or essentially absent; achenes usually columnar to prismatic, seldom compressed or flattened; pappus of many smooth to finely-barbed (not plumose) bristles __________ **Senecioneae**
               6. Leaf arrangement various; primary phyllaries subequal in 1–2 series or unequal in 3–5+ series, sometimes subtended by secondary phyllary-like bracts; receptacles chaffy, bristly, or naked; style-branch appendages usually deltate to lanceolate, seldom resembling small tufts of hair; achenes sometimes compressed or flattened; pappus various or sometimes absent.
                  7. Leaves alternate (sometimes mostly basal); secondary phyllary-like bracts absent; receptacles naked; pappus usually of bristles, sometimes of scales, rarely of awns or a crown or absent __________ **Astereae**
                  7. Leaves often wholly or partially opposite (at least on lower stems); secondary phyllary-like bracts present or absent; receptacles chaffy, bristly, or naked; pappus usually of scales, sometimes of awns or bristles, sometimes a crown or absent __________ **Heliantheae**
   2. Anther bases ± tailed.
      8. Corollas all ± 2-lipped __________ **Mutisieae** (in part)
      8. Corollas actinomorphic or zygomorphic (not all 2-lipped, usually none 2-lipped).
         9. Plants often prickly-spiny or thistle-like; phyllaries with margins often spiny, toothed, or bristle-fringed; ray flowers absent, heads of disk flowers only (peripheral flowers may have enlarged or zygomorphic corollas simulating rays) __________ **Cardueae (Cynareae)**
         9. Plants not spiny, not thistle-like; phyllary margins not spiny, toothed, or bristle-fringed; ray flowers absent or present.
            10. Ray and disk achenes with dissimilar shapes, sizes, and textures or ornamentations; pappus absent __________ **Calenduleae**
            10. All achenes within a head of similar shape, size, and texture or ornamentation; pappus various or sometimes absent.
               11. Ray flowers usually present (ligules sometimes reduced or unequally 2-lipped); achenes beaked (= with a narrowed portion above the wider achene body) __________ **Mutisieae** (in part)
               11. Ray flowers absent, heads of disk flowers only; achenes not beaked __________ **Inuleae**

Anthemideae Tribe

1. Heads with disk flowers only, or apparently so (if peripheral florets pistillate, these lacking corollas or ligules).
   2. Rhizomatous perennials, 20–120+ cm tall; leaves linear to lanceolate or oblong-elliptic, margins entire to toothed or pinnately lobed; heads borne in paniculate arrays; achenes 0.5–0.8 mm long ____ **Artemisia**
   2. Taprooted annuals, usually 2–40 cm tall; leaves obovate to spatulate, (1–)2–3-pinnately lobed or finely dissected; heads usually borne singly; achenes 1–3+ mm long.
      3. Heads sessile in leaf axils; achenes tipped by a persistent spine-like stylar sheath (painful in a manner similar to sandburs) ____ **Soliva**
      3. Heads pedunculate; achenes without a persistent spine-like stylar sheath.
         4. Receptacles flat to convex; peripheral flowers pistillate, central flowers functionally staminate; achenes compressed or ± flattened, obovoid to oblong, the peripheral ones larger and ± winged; pappus absent ____ **Cotula**
         4. Receptacles conic to oblong-ovoid; all flowers perfect and fertile; achenes ± cylindric-obconic, ribbed; pappus a small crown ____ **Matricaria** (in part)
1. Heads with both ray flowers and disk flowers, the ray flowers with evident ligules.
   5. Stem leaves toothed or lobed, but not finely dissected; involucres > 10 mm wide; ray ligules 12–20(–35+) mm long ____ **Leucanthemum**
   5. Stem leaves 1–2-pinnately lobed or finely dissected; involucres < 10 mm wide; ray ligules 1.5–15+ mm long.
      6. Plants perennial; heads borne in congested corymbose arrays; ray flowers (3–)5–8 per head; ligules 1.5–3 mm long ____ **Achillea**
      6. Plants annual; heads borne singly or in loose, few-flowered corymbose arrays; ray flowers 10+ per head, ligules 5–15+ mm long.
         7. Receptacles chaffy; ray flowers 10–15; disk flower corollas 2–2.5 mm long; achenes 1.3–2 mm long ____ **Anthemis**
         7. Receptacles naked; ray flowers (10–)14–26; disk flower corollas 1.6–1.8 mm long; achenes 0.7–0.9 mm long ____ **Matricaria** (in part)

Astereae Tribe

1. Heads with disk flowers only, or apparently so (if peripheral florets pistillate, these lacking ligules, or ligules inconspicuous and < 0.7 mm long).
   2. Shrubs or subshrubs, variable in size from 25–600 cm tall; flowers imperfect, the plants dioecious (male and female plants may differ in appearance) ____ **Baccharis**
   2. Subshrubs or herbaceous plants, from 6 to 150 cm tall; at least some flowers perfect, the plants not dioecious.
      3. Pappus of 2–10 scales or awns.
         4. Delicate plants, 6–45 cm tall; herbage eglandular; involucres 1.5–2 mm wide; flowers 3 per head, corollas white to yellowish cream; restricted to Gulf coastal plain ____ **Thurovia**
         4. Robust plants, usually 40–100 cm tall; herbage gland-dotted or resinous; involucres 8–20+ mm wide; flowers ca. 100 per head, corollas bright yellow; variously distributed in TX ____ **Grindelia** (in part)
      3. Pappus of 15–50 bristles.
         5. Weedy annuals; heads borne singly or in elongated paniculate or corymbose arrays; peripheral flowers pistillate, ligules absent or minute; central disk flowers perfect, corollas yellowish ____ **Conyza** (in part)
         5. Perennials; heads borne in compact, terminal, corymbose arrays; all flowers perfect and fertile, corollas bright yellow.
            6. Leaves mostly basal in persistent rosettes; phyllaries disposed in vertical ranks, usually yellowish; flowers 2–6 per head; achenes sparsely strigose ____ **Bigelowia**
            6. Leaves mostly on the stems; phyllaries disposed spirally, green to yellowish with scarious margins; flowers 12–34 per head; achenes sericeous ____ **Isocoma**
1. Heads with both ray flowers and disk flowers, the ray ligules usually conspicuous but sometimes concealed by phyllaries.
   7. Ray flower corollas yellow to yellow-orange.
      8. Disk flower pappus of scales OR a few stout bristles or awns or a minute crown OR pappus lacking.
         9. Involucres (6–)8–25 mm wide; disk flowers usually 100+ per head; achenes glabrous ____ **Grindelia** (in part)
         9. Involucres 1–9 mm wide; disk flowers 1–60 per head; achenes pubescent or bristly.
            10. Ray and disk flowers with strongly dissimilar pappus types, but neither lacking pappus.
               11. Ray flower pappus a small crown; disk pappus of 5–8 white, narrowly spatulate, basally fused scales ____ **Amphiachyris**
               11. Ray flower pappus of several series of barbellate bristles 1–3 mm long; disk pappus of 1–3 awn-like bristles ____ **Bradburia** (in part)
            10. Pappus not dimorphic, or only weakly so (ray pappus may differ in size from disk pappus but will not differ in pappus type) OR pappus lacking.
               12. Plants subshrubs, perennials, or annuals, to 140 cm tall; receptacles hairy; ray flowers 1–23 per head, ligules 1.5–6 mm long; disk pappus a short scaly crown 0.1–1 mm high, or rarely absent ____ **Gutierrezia**

12. Plants shrubs, to 150(–200) cm tall; receptacles naked; ray flowers 4–9 per head, ligules 2–3 mm long; pappus absent, or a minute toothed crown < 0.1 mm high ____ **Gymnosperma**
8. Disk flower pappus of 20+ capillary bristles or hairs, with or without an additional outer series of scales or awns or bristles.
13. Outer pappus series of scales or flat-based awns, 0.25–1.4 mm long (use hand lens), surrounding inner series of longer bristles.
14. Basal leaves persistent, linear to lanceolate or oblanceolate, grass-like in appearance, 3–11-parallel veined, pubescence silky-sericeous ____ **Pityopsis**
14. Basal leaves (if present) spatulate to obovate or ovate to elliptic (not grass-like), usually 1-nerved, with pinnate or reticulate veins, pubescence scabrous to hispid-strigose or pilose to sericeous-cobwebby.
15. Plants fibrous-rooted, sometimes with short lateral rhizomes; herbage pilose to sericeous-cobwebby; basal leaves sessile; ray ligules 2–3 mm wide ____ **Chrysopsis**
15. Plants taprooted, sometimes rhizomatous; herbage scabrous to hispid-strigose or pilose; basal leaves (if present) petiolate or subsessile; ray ligules 1–2 mm wide.
16. Plants annual or perennial (IF annual, THEN achenes dimorphic and ray achenes lacking pappus); leaves scabrous to hispid or strigose, often with sparse to dense stalked glands (IF eglandular THEN leaves silvery gray-green) ____ **Heterotheca**
16. Plants annual; leaves pilose, eglandular; ray and disk achenes similar; pappus always present ____ **Bradburia** (in part)
13. Pappus solely of thin-based, capillary bristles in 1–4 series (outer series may be shorter than inner).
17. Achenes dimorphic, the ray achenes ± 3-sided, the disk achenes ± compressed.
18. Herbage eglandular; basal and lowermost stem leaves pinnately lobed or pinnatifid; receptacle with a persistent network of subulate, chaff-like scales; inner phyllaries basally narrowed and stalk-like, expanded above ____ **Xanthisma**
18. Herbage with prominent sessile or stalked glands; basal and lowermost stem leaves coarsely serrate to entire; receptacle naked; inner phyllaries not basally narrowed ____ **Rayjacksonia**
17. Ray and disk achenes similar.
19. Involucres 10–15 mm high and 15–25 mm wide; ray flowers 25–45 per head ____ **Grindelia** (in part)
19. Involucres 3–10 mm high and 1.7–10 mm wide; ray flowers 2–30 per head.
20. Taprooted annuals or biennials; heads borne singly or in loose, widely-branched paniculate arrays ____ **Croptilon**
20. Rhizomatous perennials; heads usually borne in clusters in corymbose, paniculate, or racemose arrays.
21. Leaves entire, with glandular-punctate dots; heads in a ± flat-topped corymbose arrangement; achenes 2–4-nerved ____ **Euthamia**
21. Leaves usually serrate, without glandular-punctate dots; heads in pyramidal paniculate or club-shaped racemose arrangements (or sometimes in axillary clusters); achenes 8–10-nerved ____ **Solidago**
7. Ray flower corollas white, tawny, pink, rose, lavender, bluish, or purplish.
22. Disk flower pappus of scales OR a few stout bristles or awns or a minute crown OR pappus lacking.
23. Plants fibrous rooted and rhizomatous, sometimes colonial, 30–200 cm tall; achenes dimorphic (and sometimes winged), ray achenes ± 3-sided OR 2–(3)-nerved, disk achenes ± compressed OR 3–5-nerved.
24. Heads borne in loose, leafy corymbose or paniculate arrays; achenes often 2–3-winged; pappus of 2–3 awns, 0.3–1.2 mm long, plus 0–12 shorter bristles or scales ____ **Boltonia**
24. Heads borne singly; ray achenes 2–(3)-nerved, disk achenes 3–5-nerved; pappus of thickened rings or a minute, fringed crown < 0.1 mm high ____ **Chaetopappa** (in part)
23. Plants taprooted; ray and disk achenes similar, never winged.
25. Plants often small and delicate; receptacles flat to convex, smooth; ray flowers 5–19 per head; disk flowers 4–22; achenes 1.5–2.2 mm long ____ **Chaetopappa** (in part)
25. Plants delicate or not; receptacles conic, pitted; ray flowers (7–)13–75 per head; disk flowers 25–400; achenes 1–1.6 mm long.
26. Leaves varying from entire to deeply pinnatifid, achenes columnar, 4-angled to somewhat rounded, 4–12-ribbed ____ **Aphanostephus**
26. Leaves usually entire or sometimes toothed; achenes obovoid to oblanceoloid, compressed, 2-nerved (nerves marginal) ____ **Astranthium**
22. Disk flower pappus of (5–)8–90 capillary bristles or hairs, with or without an additional outer series of scales or awns or bristles.
27. Ray flowers inconspicuous, ligules < 1 mm long, barely exceeding the phyllaries ____ **Conyza** (in part)
27. Ray flowers conspicuous, ligules 1.3+ mm long, exceeding the phyllaries.
28. Plants subshrubs or rush-like perennials, colonial; stems green and glaucous, green thorns often present; few leaves present at flowering; phyllaries (1–)3(–5)-nerved ____ **Chloracantha**

28. Plants herbaceous annuals or perennials, solitary or colonial; stems not becoming thorny; at least upper leaves persistent through flowering; phyllaries usually 1-nerved (sometimes 3-nerved in *Erigeron* and *Symphyotrichum*)
  29. Plants 1–3 cm tall; heads solitary and nearly sessile, nestled in basal rosettes; involucres (10–)12–25+ mm wide ______ **Townsendia**
  29. Plants usually 10+ cm tall; heads solitary or borne in various arrays, usually well above any basal leaves; involucres often < 12 mm wide.
    30. Heads usually borne in flat-topped corymbose arrays; ray ligules white, 2–7 per head; disk flowers usually < 20 per head; disk flower pappus innermost bristles with strongly flattened tips (= clavate) ______ **Doellingeria**
    30. Heads solitary or borne in various arrays; ray ligules white or various shades of pink, blue, violet, or purple, 7–100+ per head; disk flowers usually > 20 per head; disk flower pappus innermost bristles with rounded tips (or perhaps obscurely flattened).
      31. Leaves narrow, stiff, evenly distributed; phyllaries strongly keeled; ray and disk achenes slightly dimorphic; achene pubescence silky strigose-sericeous ______ **Ionactis**
      31. Leaves variously shaped and distributed; phyllaries flat or only swollen at bases, not strongly keeled; ray and disk achenes either monomorphic OR strongly dimorphic; achenes glabrous to strigose but usually not silky-sericeous.
        32. Plants taprooted annuals; achenes strongly dimorphic, ray achenes 1.3–1.5 mm long and sparsely hairy, disk achenes 2–3 mm long and glabrous or nearly so; pappus lacking on ray achenes ______ **Psilactis**
        32. Plants annual or perennial, taprooted or fibrous rooted or rhizomatous; achenes of various sizes, glabrous to variously pubescent, but generally monomorphic within a head; rays achenes with some pappus present, similar to disk achenes or differing.
          33. Leaves glabrous to minutely scabrous, margins usually entire and often revolute; phyllaries glabrous and tips often squarrose or strongly recurved; ray flowers 15–30 per head, ligules usually violet-purple, rarely white; achenes 9–16-nerved ______ **Eurybia**
          33. Leaves glabrous or variously pubescent, margins various; phyllaries glabrous or variously pubescent, tips appressed or erect to variously recurved; ray flowers 8–350+ per head; ligules often white, but sometimes pinkish to bluish or purplish; achenes 2–5(–10)-nerved.
            34. Heads borne singly or in loose corymbose or paniculate arrays; peduncles of buds often nodding; phyllaries with orange-resinous veins, tips usually lacking a well-differentiated green zone; ray flowers usually 44–350+ per head, ligules usually < 1 mm wide; plants flowering Mar–Jun (sporadically into Oct) ______ **Erigeron**
            34. Heads usually borne in paniculate or corymbose arrays; peduncles of buds often erect; phyllaries lacking orange-resinous veins, tips usually with a well-differentiated green zone, or sometimes foliaceous; ray flowers usually 8–50 per head, ligules usually > 1 mm wide; plants flowering Aug–Nov (rarely Mar–Apr) ______ **Symphyotrichum**

Calenduleae Tribe

1. Only one genus ______ **Dimorphotheca**

Cardueae (Cynareae) Tribe

1. Corollas bright yellow to yellow-orange or orange-red.
  2. Leaf margins never spiny; leaf bases long-decurrent, creating winged stems; all phyllaries spine-tipped; pappus of stiff bristles ______ **Centaurea** (in part)
  2. Leaf margins sometimes spiny; leaf bases not decurrent; stems unwinged; outer phyllaries sometimes leafy, at least the inner phyllaries spine-tipped; pappus absent or of scales ______ **Carthamus**
1. Corollas usually pink to rosy-red or lavender to purple, less often white, or rarely pale yellow.
  3. Leaf margins spiny.
    4. Pappus of plumose bristles (feather-like, with long side branches, at least on the upper portion) ______ **Cirsium**
    4. Pappus of awn-like scales, smooth bristles, or barbellate bristles (with short side branches).
      5. Leaves variegated (mottled white and green above); leaf bases not decurrent; stems lacking spiny wings ______ **Silybum**
      5. Leaves uniformly colored above; leaf bases long-decurrent, creating spiny-winged stems.
        6. Receptacles unpitted, bearing flattened bristles or scales; pappus bristles 11–25 mm long ______ **Carduus**
        6. Receptacles deeply pitted, lacking bristles or scales; pappus bristles 7–9 mm long ______ **Onopordum**
  3. Leaf margins not spiny.
    7. Phyllary margins entire or minutely serrate, tips with a single, prominent, hooked spine; corollas of all flowers similar; the entire head shed as a spiny bur at maturity ______ *Arctium*

7. Phyllary margins variously lobed, toothed, or bristle-fringed, not prominently spine-tipped; corollas of peripheral flowers often enlarged, simulating ray flowers; achenes generally shed individually at maturity.
   8. Involucres 10–16 mm high; peripheral (sterile) flower corollas 15–25 mm long ____ **Centaurea** (in part)
   8. Involucres 30–50 mm high; peripheral (sterile) flower corollas 35–50 mm long ____ **Plectocephalus**

Cichorieae (Lactuceae) Tribe

1. Achenes (at least central ones) distinctly beaked (= with a narrowed portion above the wider achene body).
   2. Pappus with an outer series of basally flattened, awn-like scales and an inner series of plumose bristles ____ **Leontodon** (in part)
   2. Pappus of bristles or hairs (plumose or otherwise), awn-like scales lacking (minute outer crowns may be present in addition to hair-like bristles).
      3. Primary (inner) phyllaries in 3–5 series, unequal; secondary (subtending) phyllary-like bracts absent; receptacles chaffy.
         4. Leaves in a basal rosette and on lower stems; heads in loose cymose arrays or solitary; corollas 5–15 mm long, yellow to white (sometimes striped reddish or gray-green beneath); eastern 1/3 TX ____ **Hypochaeris**
         4. Leaves nearly all in a basal rosette, stem leaves few or absent; heads solitary; corollas 15–18 mm long, rose-lavender to pink to white (often yellow at the base); western 2/3 TX ____ **Pinaropappus**
      3. Primary phyllaries in 1–2 series, usually equal or subequal; secondary phyllary-like bracts often present (and sometimes intergrading with phyllaries); receptacles naked.
         5. Leaves linear, grass-like, margins entire; achenes 25–40 mm long (including beak); pappus 20–30 mm long ____ **Tragopogon**
         5. Leaves not linear or grass-like, margins entire or variously toothed or lobed; achenes < 20 mm long; pappus < 15 mm long.
            6. Plants frequently > 80 cm tall in flower; corollas yellow, bluish, purplish, or whitish (sometimes reddish-tinged); achenes elliptic to oblong, ± flattened ____ **Lactuca**
            6. Plants < 80 cm tall; corollas yellow (sometimes striped or tipped reddish beneath); achenes spindle-shaped, terete or subcylindric.
               7. Phyllaries and subtending phyllary-like bracts coarsely setose or hispid; longest pappus bristles ca. 4 mm ____ **Crepis** (in part)
               7. Phyllaries and subtending phyllary-like bracts glabrous or minutely pilose; longest pappus bristles 4–12 mm.
                  8. Plants often with branched, leafy stems, each bearing 1–several heads, or sometimes stemless; lateral leaf margins pinnately lobed (but usually not runcinately or retrorsely so); achene bodies 4–6 mm long; pappus bristles 7–12 mm long, subtended by a ring of minute, reflexed hairs (use lens) ____ **Pyrrhopappus**
                  8. Plants stemless, leaves in a basal rosette; heads all solitary; lateral leaf margins usually sharply and retrorsely runcinate; achene bodies 2–4 mm long; pappus bristles 4–7(–8) mm long, not subtended by a ring of minute hairs ____ **Taraxacum**
1. All achenes beakless.
   9. Pappus of scales (basally flattened, sometimes awn-like but not hair-like) or scales and bristles or a minute crown OR pappus lacking.
      10. Secondary (subtending) phyllary-like bracts absent; achenes 1.2–3 mm long.
          11. Lower stem leaves usually sessile; heads mostly in sessile, axillary clusters; phyllaries unequal, with gland-tipped hairs; corollas lavender-blue, rarely white or pink; achenes 3–5-angled ____ **Cichorium**
          11. Lower stem leaves petiolate; heads borne singly; phyllaries equal, glabrous; corollas yellow or yellow-orange; achenes 10–20-ribbed ____ **Krigia** (in part)
      10. Secondary phyllary-like bracts 3–12; achenes 5–12 mm long.
          12. Leaves in a basal rosette and on branching stems; heads borne singly or in loose corymbose arrays; phyllaries ± keeled and incurved at maturity, each ± enfolding a subtended peripheral achene; pappus on central achenes of 1–2 series of scales (peripheral achenes may have only a minute, fringed crown) ____ **Hedypnois**
          12. Leaves all in a basal rosette; heads borne singly on long unbranched stems; phyllaries unkeeled, reflexed at maturity; pappus with an outer series of basally flattened, awn-like scales and an inner series of plumose bristles ____ **Leontodon** (in part)
   9. Pappus solely of bristles or hairs, at least some > 1 mm long.
      13. Leaves all (or nearly all) in a basal rosette (1–2 reduced stem leaves may be present).
          14. Heads borne singly; primary phyllaries 12–16; secondary (subtending) phyllary-like bracts absent; corollas 15–25 mm long; pappus bristles of the inner series 5–8 mm long, plus an outer series of minute scales (may be obscure) ____ **Krigia** (in part)
          14. Heads borne in corymbose to paniculate arrays; primary phyllaries usually 8; secondary phyllary-like bracts 3–5+; corollas mostly 4.5–6.5 mm long; pappus bristles 2.5–3.5 mm long ____ **Youngia**
      13. Well-developed stem leaves present, at least on lower stem (leaves may be linear or wider).
          15. Corollas bright yellow (ligules sometimes striped or tipped reddish on the lower surface).

16. Stems often hollow; leaf margins usually spiny or prickly; stem leaf bases auriculate; achenes flattened ______ **Sonchus**
16. Stems not hollow; leaf margins not spiny-prickly; leaf bases not auriculate; achenes rounded in cross-section (at least the central ones).
    17. Plants annual; leaves short-hispid or with dense stalked glands; leaf margins toothed or lobed; primary phyllary bases strongly keeled and thickened; secondary (subtending) phyllary-like bracts 5–7; pappus bristles 1.5–5 mm long ______ **Crepis** (in part)
    17. Plants perennial; leaves pilose-hirsute (hairs 2–8+ mm long); leaf margins entire; primary phyllary bases not strongly keeled or thickened; secondary phyllary-like bracts 8–13+; pappus bristles 5–6.5 mm long ______ **Hieracium**

15. Corollas white, creamy yellow, greenish yellow, pink, or lavender-purple.
    18. Lower stem leaves linear, 1–8 mm wide; heads erect, borne singly; corollas pink, purple, or lavender (rarely white), 35–40 mm long; pappus bristles smooth, 10–15 mm long ______ **Lygodesmia**
    18. Lower stem leaves broader, 10–160 mm wide; heads often nodding, borne in racemose, thyrsiform, or paniculate arrays; corollas white to creamy yellow or greenish yellow, 7–17 mm long (phyllaries may be more colorful); pappus bristles minutely barbed, 5–8 mm long ______ **Prenanthes**

Eupatorieae Tribe

1. Perennial twining or scrambling vines; leaves opposite; phyllaries 4; flowers usually 4 per head ______ **Mikania**
1. Non vining annuals, perennials, or shrubs; leaves alternate, opposite, or whorled; phyllaries 8+; flowers 3–60+ per head.
   2. Achenes 8–11 ribbed (IF 8-ribbed, THEN the achenes 2.7–9 mm long).
      3. Single-stemmed or infrequently-branched herbaceous perennials from a woody corm or rhizome; leaves basal or basal and cauline at flowering; corollas usually lavender to magenta to pinkish purple (sometimes creamy white) ______ **Liatris**
      3. Shrubs or branched perennials with woody bases; leaves mostly cauline at flowering; corollas usually white to cream (sometimes greenish, purplish, or yellowish) ______ **Brickellia**
   2. Achenes (3–)4–5(–8) ribbed (IF 8-ribbed, THEN the achenes 1.8–2.8 mm long).
      4. Pappus of 2–6+ hair-like scales or bristles to 0.5 mm long ______ **Trichocoronis** (in part)
      4. Pappus of (5–)10–50 barbellate bristles or hair-like scales, much longer than 0.5 mm.
         5. Phyllaries ± equal in length.
            6. Receptacles conic ______ **Conoclinium**
            6. Receptacles flat or convex.
               7. Involucres 1–3(–5+) mm wide; flowers (4–)5(–15) per head; style bases usually puberulent (glabrous in *E. capillifolium*) ______ **Eupatorium** (in part)
               7. Involucres 3–6 mm wide; flowers 10–60 per head; style bases glabrous ______ **Ageratina**
         5. Phyllaries unequal (outer ones shorter).
            8. Leaves whorled, with 3–7 petiolate leaves per node ______ **Eutrochium**
            8. Leaves opposite or alternate, rarely whorled (IF whorled THEN sessile).
               9. Flowers (4–)5(–15) per head; style bases usually puberulent and enlarged (glabrous in *E. capillifolium*) ______ **Eupatorium** (in part)
               9. Flowers (6–)15–40(–75) per head; style bases glabrous, not enlarged.
                  10. Involucres 5–10 mm high; phyllaries in 4–6+ series; achenes (3–)5-ribbed, usually gland-dotted ______ **Chromolaena**
                  10. Involucres 4–5 mm high; phyllaries in 2–4 series; achenes 5(–8)-ribbed, not gland-dotted ______ **Fleischmannia**

Heliantheae Tribe (Including Helenieae)

1. Plants generally wind-pollinated or self-pollinated, the heads small and not at all showy nor attractive to pollinators; all flowers imperfect, pistillate and staminate flowers in same or different heads; ray flowers absent; disk corollas minute or insignificant; pappus absent.
   2. Pistillate and staminate flowers in the same head; phyllaries without spine-like structures or tubercules, the heads not bur-like ______ **Iva**
   2. Pistillate and staminate flowers in separate heads, the staminate usually uppermost; phyllaries of pistillate heads joined and enclosing achenes at maturity and forming a bur or nut-like structure with prickles, spines, or tubercles (or rarely encircled by membranous wings).
      3. Staminate heads usually in racemes or spikes, their phyllaries united and cup-shaped; pistillate heads forming a 1-seeded bur < 6 mm long, with phyllary tips projecting as 1–12 spines or tubercles (or rarely as membranous wings) ______ **Ambrosia**
      3. Staminate heads in terminal clusters, their phyllaries not fused; pistillate heads forming a 2-seeded bur (cocklebur) 10–30+ mm long, with phyllary tips projecting as 20+ hooked, spine-like structures ______ **Xanthium**

1. Plants generally adapted for pollination by insects (and sometimes hummingbirds), at least some heads colorful or otherwise attractive; usually some or all flowers perfect; ray flowers present or absent; disk flowers with obvious corollas; pappus various.
 4. Receptacles not bearing chaff or pales (but sometimes bearing bristles, or with fringed pit borders around flower attachment points), or rarely with pales associated with peripheral disk flowers only.
  5. Plants usually strongly aromatic (sometimes unpleasantly so); phyllaries (and usually also leaves) noticeably dotted or streaked with pellucid glands containing strong-scented oils.
   6. Phyllaries distinct to bases or nearly so.
    7. Evergreen perennials or subshrubs to 50+ cm tall; leaves mostly alternate; heads solitary; ray ligules 6–12 mm long; pappus of 20–40 bristles, 3–5.5 mm long ____ **Chrysactinia** (in part)
    7. Small perennials or annuals to 15 cm tall; leaves opposite; heads congested in cymose arrays; ray ligules 3–5(–7) mm long; pappus of 1–7 bristles or awns, or sometimes a minute crown, usually < 1 mm high ____ **Pectis**
   6. Phyllaries fused 1/3+ of their lengths (at least the inner series).
    8. Leaves pinnately lobed or pinnatisect, the lobes or segments usually lanceolate to ovate; secondary (subtending) phyllary-like bracts absent; disk corollas 7–12+ mm long; pappus of 2–6 elements; horticultural exotics escaping cultivation ____ **Tagetes**
    8. Leaves simple and entire to toothed or pinnately lobed, the leaves or lobes usually filiform to linear; secondary phyllary-like bracts often present; disk corollas 2–4.5 mm long; pappus of 10–12 elements; native species.
     9. Plants 40–90 cm tall; leaf bases and secondary phyllary-like bracts often with a few stiff bristles; phyllaries 10–12; ray flowers 7–12, ligules 10–15 mm long ____ **Dysodiopsis**
     9. Plants 15–30 cm tall; leaf bases and secondary phyllary-like bracts (if any) lacking stiff bristles; phyllaries 12–22; ray flowers 10–21, ligules 2–10 mm long ____ **Thymophylla**
  5. Plants sometimes aromatic; leaves sometimes with sessile or stalked glands, but these generally minute and not noticeably pellucid; phyllaries usually eglandular.
   10. Phyllaries 5–14; corollas white to cream or pink to lavender to dark purple (but usually not maroon or brownish purple).
    11. Leaves usually 1–2-pinnately lobed (but may be simple or merely lobed), often conspicuously white-hairy on lower surface; ray flowers absent; pappus of 14–20 scales ____ **Hymenopappus**
    11. Leaves entire, sparsely to moderately pubescent; ray flowers absent or 3–13; pappus of 4–10 scales (sometimes absent on peripheral achenes) ____ **Palafoxia** (in part)
   10. Phyllaries 2–60+; at least some corollas partially or completely pale yellow, golden yellow, orange, reddish brown, or maroon-purple (rarely white); IF some corollas completely purplish to pinkish THEN phyllaries usually > 14.
    12. Leaves opposite, ± succulent; phyllaries usually 2; disk flowers 0–1(–2) per head; pappus absent ____ **Flaveria**
    12. Leaves either basal or mostly alternate (at least on upper stems), seldom succulent; phyllaries 5+; disk flowers 5+ per head; some pappus usually present.
     13. Ray flowers 2–5 per head
      14. Stems and leaves green, glabrous or sparsely hairy; disk flowers 16–22 per head, 4-lobed; pappus usually of 1–2 bristles to 1.8 mm long plus minute vestigial scales or a callous crown, or rarely absent ____ **Perityle**
      14. Stems and leaves gray-green, cobwebby-villous; disk flowers 5–8(–12) per head, 5-lobed; pappus of 4–6 scales, 2.5–3+ mm long ____ **Psilostrophe**
     13. Ray flowers either 6+ per head or absent.
      15. Stems usually winged by decurrent leaf bases; achenes 0.6–1.5 mm long (IF stems unwinged THEN achenes to only 1.3 mm long) ____ **Helenium** (in part)
      15. Stems not winged; leaf bases not decurrent; achenes 1.5+ mm long.
       16. Herbage noticeably sweet-scented, like newly-mown sweet-clover; achenes narrowly obconic, prominently 10-ribbed; pappus scales not awn-tipped ____ **Amblyolepis**
       16. Herbage not noticeably sweet-scented (may be otherwise aromatic); achenes usually obpyramidal and 4-sided; pappus scales awn-tipped.
        17. Phyllaries usually strongly reflexed in fruit; receptacles often bristly; ray ligules (when present) frequently in colors other than solid yellow; disk corollas usually brown-purple to red-brown, or tipped with these colors; pappus of 7–12 awn-tipped scales, 5–9 mm long ____ **Gaillardia** (in part)
        17. Phyllaries mostly spreading to erect in fruit; receptacles naked; ray ligules usually uniformly some shade of yellow (rarely tipped or striped with white); disk corollas usually solid yellow (or sometimes purple-tipped); pappus of 4–7 awn-tipped scales, 1.2–4 mm long.

18. Phyllaries in 2 series of identical number, glabrous or inconspicuously short-pubescent, margins not noticeably scarious; ray flowers either 6–8 and ligules minute (to 3 mm long) in two rare species OR ray flowers 8–13 and ligules 8.5–11 mm long in one common species, withering and twisting in age ________ **Hymenoxys**
18. Phyllaries in 3 series, usually ± hairy, margins usually scarious (at least inner series); ray flowers 9–26, ligules 8.2–22 mm long, often not withering, remaining attached and strongly recurved in age ________ **Tetraneuris**

4. Receptacles chaffy, a bract (pale) subtending each flower.
19. Only ray flowers fertile (= maturing achenes), their achenes much larger than those of the sterile disk flowers; ray ligules often showy but sometimes insignificant or corollas absent.
20. Ray achenes thick, rounded, without wings (may be shed enclosed in a fruit-like or bur-like structure).
21. Leaf blades usually ± palmately 3–5-lobed; inner phyllaries subtending but not clasping the achenes; ray ligules yellow; achenes obovoid, shed separately from phyllaries ________ **Smallanthus**
21. Leaf blades entire to toothed or pinnately lobed; inner phyllaries each subtending and clasping a ray achene; ray ligules white to cream or pale yellow (may be minute); achenes enclosed by and shed within the subtending phyllary, forming a fruit-like or bur-like structure.
22. Heads borne in sessile clusters; ray flowers 5–8 per head, ligules inconspicuous, minute; achenes shed in a bur-like structure 7–9+ mm long, with hooked or straight spines or tubercles ________ **Acanthospermum**
22. Heads borne singly on peduncles; ray flowers 7–13 per head, ligules obvious, showy; achenes shed in a ± conical or boot-shaped, fruit-like structure, 1.3–2.6 mm long ________ **Melampodium**
20. Ray achenes ± strongly flattened, with or without wings.
23. Achenes broadly winged.
24. Taprooted annuals; phyllaries 8–10, in 2 series; ray flowers usually 5, in 1 series; involucres 5–12 mm wide ________ **Lindheimera**
24. Coarse perennials; phyllaries 11–45, in 2–4 series; ray flowers 8–38, in 2–3 series; involucres 10–30 mm wide ________ **Silphium**
23. Achenes not winged.
25. Ray ligules white to pale dingy yellow, < 2 mm long ________ **Parthenium**
25. Ray ligules lemon to bright yellow (at least on the upper surface), 10–20 mm long.
26. Leaves crenate, serrate, dentate, or lyrate-pinnatifid; involucres 12–30+ mm wide; ray ligules with green, red, or maroon veins on the lower surface; achenes 4.5–6 mm long ________ **Berlandiera**
26. Leaves deeply pinnatifid; involucres 6–10 mm wide; ray ligules entirely yellow on the lower surface, without colored veins; achenes 3–4 mm long ________ **Engelmannia**
19. Disk flowers perfect and fertile; ray flowers present and fertile OR sterile OR absent.
27. Ray flowers neuter OR pistillate and sterile OR absent; only disk flowers fertile (= maturing achenes), their achenes much larger than those of the sterile ray flowers (if present).
28. Leaves mostly opposite (NOT both alternate and auriculate-clasping); phyllaries apparently in ± 3 series, dimorphic, the outer (secondary) series herbaceous (and sometimes foliaceous), the inner 2 (primary) series often very unlike the outer, usually scarious-margined; achenes flattened parallel to the phyllaries or 3–4-angled and linear to spindle-shaped.
29. Primary (inner) phyllaries united for 1/5 to 1/2 or more of their lengths; disk corolla lobes often lance-linear (sometimes triangular) ________ **Thelesperma**
29. Phyllaries all separate to their bases or nearly so; disk corolla lobes triangular to ovate.
30. Achenes (at least central ones) ± 4-angled, ± linear to spindle-shaped, often with tips narrowed or beaked (no achenes winged).
31. Disk flowers 10–20 per head; stamen filaments hairy near anthers (use lens); achenes usually with 1 groove on each face ________ **Cosmos**
31. Disk flowers (5–)12–150+ per head; stamen filaments not hairy; achenes with 0 or 2 grooves on each face ________ **Bidens** (in part)
30. Achenes all ± obcompressed or flattened (sometimes winged).
32. Achenes sometimes thickened or winged; pappus usually of barbellate (rarely smooth) awns or sometimes absent ________ **Bidens** (in part)
32. Achenes (some or all) usually thin-margined or ± winged; pappus usually absent or of 2 bristly scales (or achene wing shoulders may be bristly) ________ **Coreopsis**
28. Leaves alternate or opposite; phyllaries in 1–several series, usually not strongly dimorphic (IF strongly dimorphic THEN leaves alternate and blades auriculate and clasping); achenes various but usually not flattened parallel to the phyllaries.
33. Receptacles spheric to high-conic or columnar (mostly 8–20+ mm high); ray flowers present; achenes not winged.
34. Pales longer than disk corollas, the awn-like tips apparent between the flowers at anthesis; ray ligules pink to purple or rarely white ________ **Echinacea**
34. Pales not awn-like, tips not emerging between flowers at anthesis; ray ligules yellow to yellow-orange, maroon, reddish brown, or a bicolor combination of these.

35. Leaves lyrate-pinnate to 1–2-pinnatifid; phyllaries unequal (outer noticeably longer than inner); achenes linear-lanceolate to oblong, strongly compressed, with one margin usually toothed, fringed, or ciliate ____ **Ratibida**
35. Leaves simple to deeply lobed; phyllaries subequal; achenes obpyramidal and ± 4-angled, not strongly compressed, margins not toothed, fringed, or ciliate ____ **Rudbeckia**
33. Receptacles mostly flat to convex or conic (mostly 0–5 mm high); ray flowers present or absent; achenes winged or unwinged.
36. Receptacle pales completely enclosing each achene, forming a hardened, warty, fruit-like structure ____ **Sclerocarpus**
36. Receptacle pales sometimes enfolding an achene, sometimes shed with it, but not completely enclosing each achene in a fruit-like structure.
37. Stems sometimes winged with decurrent leaf bases; achenes strongly flattened and papery-winged; pappus usually of 2 persistent subulate scales or awns to 2 mm long (or sometimes pappus absent) ____ **Verbesina** (in part)
37. Stems not winged, leaf bases not decurrent; achenes various but not strongly flattened and papery-winged (achenes may be somewhat flattened, with thin margins); pappus of 1–2+ scales, these often fragile and deciduous (plus sometimes shorter scales or squamellae) or of 5+ persistent scales (or sometimes pappus absent).
38. Ray flowers absent; disk corollas pale lavender, purple, pink, or white; achenes usually 5-angled and 10-ribbed; pappus of 5–6 scales ____ **Marshallia**
38. Ray flowers present, ligules yellow to yellow-orange; disk corollas yellow to orange or reddish purple; achenes ± compressed, often obpyramidal; pappus of 1–2+ scales, these often fragile and deciduous (plus sometimes shorter scales or squamellae).
39. Petiole bases dilated, fused to form disks at nodes; ray flowers styliferous (= styles present); achenes somewhat flattened and thin-margined ____ **Simsia**
39. Petiole bases not dilated, not forming nodal disks; ray flowers neuter, styles and stigmas absent; achenes ± compressed, often obpyramidal.
40. Involucres 4–40+ mm wide; stamen filaments not hairy; pappus fragile, readily falling ____ **Helianthus**
40. Involucres 7–10 mm wide; stamen filaments hairy; pappus persistent or tardily falling ____ **Viguiera**
27. Ray flowers present and fertile (rays sometimes inconspicuous); disk flowers perfect and fertile; both kinds of flowers producing mature achenes (though these may differ in shape).
41. Stems sometimes winged with decurrent leaf bases; achenes strongly flattened and papery-winged ____ **Verbesina** (in part)
41. Stems not winged, leaf bases not decurrent; achenes various but not strongly flattened and papery-winged (achenes may be somewhat flattened, with thin margins).
42. Plants usually strongly aromatic; leaves usually alternate; pales readily falling; pappus of 20–40 bristles, 3–5.5 mm long ____ **Chrysactinia** (in part)
42. Plants sometimes aromatic; leaves opposite; pales persistent or falling; phyllaries usually eglandular; pappus of a few awns or bristles OR of scales or a minute crown OR pappus lacking.
43. Outer 4 phyllaries broadly lanceolate, foliaceous, forming square "buds" prior to anthesis, overlapping at anthesis, notably larger than inner, scarious phyllaries ____ **Tetragonotheca**
43. Outer phyllaries larger than inner or not, sometimes herbaceous, but not notably different in texture, not forming 4-sided "buds."
44. Ray ligules sessile, not borne on corolla tubes, 10–40 mm long, usually persistent and becoming papery in age.
45. Plants perennial; leaves petiolate and usually toothed; ray ligules yellow to orange; achenes 4–5 mm long; native species ____ **Heliopsis**
45. Plants annual; leaves sessile and entire; ray ligules usually red (or white, yellow, or pink-purple in cultivars); achenes 6–10 mm long; horticultural exotics escaping cultivation ____ **Zinnia**
44. Ray ligules usually borne on corolla tubes, not sessile, 2–15 mm long, not persistent, not becoming papery in age.
46. Plants often decumbent to prostrate; ray ligules to 5 mm long.
47. Ray ligules 20–40, white or whitish; pappus a small crown (sometimes plus 2 teeth) ____ **Eclipta**
47. Ray ligules 3–20, yellow; pappus of a few awns or bristles, or pappus absent.
48. Receptacles conic; phyllaries 8–15+; disk flowers 25–100(–200+); pappus of 1–3 bristle-like awns, or absent ____ **Acmella**
48. Receptacles convex; phyllaries 5; disk flowers 10–20; pappus of 2(–5+) stout awns ____ **Calyptocarpus**
46. Plants erect or decumbent to prostrate; ray ligules 6–15 mm long.

49. Leaves usually villous to sericeous; phyllaries 20–40+; phyllaries and pales spine-tipped ______ **Borrichia**
49. Leaves usually hispid to scabrous; phyllaries 8–16+; phyllaries and pales not spine-tipped.
50. Prostrate herbaceous perennials, often rooting at nodes; achenes strongly biconvex to plumply 3–4-angled; pappus a small fringed crown or absent ______ **Sphagneticola**
50. Small erect shrubs; achenes mostly strongly compressed and 3–4-angled, usually ± winged; pappus a small cup-like crown plus sometimes 1–3 bristles or awns to 6 mm long ______ **Wedelia**

Inuleae Tribe (Including Plucheeae and Gnaphalieae)

1. Plants without conspicuous white or gray woolly pubescence; middle and upper stem leaves broad, 10–70 mm wide; herbage usually glandular and strongly aromatic (sometimes unpleasantly so) ______ **Pluchea**
1. Plants with conspicuous white or gray woolly pubescence (most often on leaf undersurfaces); middle and upper stem leaves narrow, 1–20 mm wide; herbage sometimes glandular but not strongly aromatic.
2. Plants often > 50 cm tall; stems winged by long-decurrent leaf bases ______ **Pterocaulon**
2. Plants usually 3–50 cm tall; stems unwinged; leaf bases not or only briefly decurrent.
3. Heads embedded in and ± obscured by cottony pubescence, clustered in compact, terminal arrays, subtended by leafy bracts; phyllaries 0–6; receptacles chaffy; pappus essentially absent.
4. Pales of perfect flowers saccate, each enclosing a flower; achenes (at least the peripheral ones) strigose; known only from Gulf coastal prairie in Victoria Co. ______ **Micropsis**
4. Pales of perfect or staminate flowers flat to concave, not enclosing flowers; achenes glabrous; widespread in East TX ______ **Diaperia**
3. Heads often with woolly hairs but not embedded in or obscured by these, clustered in terminal, corymbose, spicate, or paniculate arrays; phyllaries more numerous, in 3–7 series; receptacles naked; pappus of bristles.
5. Pappus of plumose bristles (use lens to see side branches) ______ **Facelis**
5. Pappus of smooth or barbellate bristles.
6. Leaves mostly basal, these 12–45 mm wide, obovate-spatulate to rhombic-ovate or suborbicular; heads typically few, in tight, terminal, corymbose arrays; staminate and pistillate heads usually on separate plants ______ **Antennaria**
6. Well-developed stem leaves present; most leaves < 20 mm wide (usually much less), lance-linear to narrowly spatulate or oblanceolate; heads usually numerous, in elongate spicate or paniculate arrays or terminal clusters; all heads with pistillate peripheral flowers and perfect central flowers.
7. Heads in spicate or narrow paniculate arrays (reduced to terminal clusters in depauperate individuals); receptacles concave in fruit; pappus of 12–28 minutely barbed bristles, deciduous as a ring; usually blooming Mar–Jun ______ **Gamochaeta**
7. Heads in corymbose or paniculate arrays or terminal clusters, often somewhat flat-topped; receptacles flat in fruit; pappus of 10–12 barbed bristles, these readily separating individually, in clusters, or as an easily-fragmented ring; most species blooming Aug–Dec (Apr–Jun in *P. luteoalbum*) ______ **Pseudognaphalium** (in part)

Mutisieae Tribe

1. Basal rosette leaves with scattered straight hairs (and sometimes stalked glands), margins entire to lyrate, undulately shallowly lobed, finely denticulate; peripheral flower corollas usually ray-like, central flower corollas shorter and bilaterally symmetrical; achenes beaked ______ **Chaptalia**
1. Basal rosette leaves nearly glabrous above, thinly to densely gray or white tomentose beneath, margins pinnately lobed, dentate-prickly to spinulose; all corollas similar, bilabiate and bilaterally symmetrical; achenes unbeaked ______ **Acourtia**

Senecioneae Tribe

1. Corollas yellow or orange to brick-red; heads usually with both ray and disk flowers (ray flowers absent in 1 species).
2. Plants vining, to 500 cm tall; ray ligules orange to brick-red; disk corollas orange; horticultural escape in subtropical parts of TX ______ **Pseudogynoxys**
2. Plants herbaceous, to 80 cm tall; ray ligules yellow or absent; disk corollas yellow; native or weedy plants widespread in TX.
3. Stem leaves deeply toothed to pinnately lobed or compound; outer (shortest) phyllaries not black-tipped; plants annuals or perennials; ray flowers present ______ **Packera**
3. Stem leaves nearly entire or shallowly toothed to pinnately lobed (IF pinnately lobed, THEN outer (shortest) phyllaries black-tipped and ray flowers absent); plants annuals; ray flowers absent or present ______ **Senecio**

1. Corollas white, yellowish cream, greenish, pinkish, or purple-red; heads with only disk or disk-like flowers.
   4. Plants perennial; leaves entire or rarely toothed, palmately veined (with several prominent, ± parallel longitudinal nerves converging at the base and tip); phyllaries mostly pale green to white; 5 flowers per head ________ **Arnoglossum**
   4. Plants annual; leaves entire to sharply double-serrate and sometimes irregularly lobed, pinnately veined (with a single prominent midnerve and lacking additional prominent, parallel, longitudinal nerves); phyllaries mostly leaf-green; 10+ flowers per head.
      5. Leaves entire to toothed or weakly lobed; corollas pink to purple or reddish purple, surpassing the phyllaries by 2–4 mm; achenes with 5 ribs or nerves ________ **Emilia**
      5. Leaves sharply double-serrate and lobed; corollas whitish to greenish or yellowish cream to pale pinkish, only slightly surpassing the phyllaries; achenes with 10–12 ribs or nerves ________ **Erechtites**

VERNONIEAE TRIBE

1. Involucres 25–45 mm wide; phyllary margins (at least the outer) spinose-toothed; peripheral corollas simulating rays, enlarged and zygomorphic ________ **Stokesia**
1. Involucres < 10 mm wide; phyllary margins entire; peripheral and central corollas similar in size.
   2. Heads sessile, in dense clusters subtended by whorls of 2–3 leaf-like ovate-deltate bracts ca. 10 mm long; phyllaries 8; flowers (1–)4(–5+) per head; pappus of 5(–6) bristle-like scales ________ **Elephantopus**
   2. Heads mostly pedunculate, not borne in clusters subtended by a whorl of leaf-like bracts; phyllaries 25–70+; flowers 9–55+ per head; pappus of 20+ outer scales and 20+ inner, longer bristles ________ **Vernonia**

ACANTHOSPERMUM Schrank STARBUR

A mainly South American genus of 6 species, with several becoming widespread and weedy. (Greek: *acantha*, spine, and *sperma*, seed, alluding to spiny "fruits") (tribe Heliantheae)

REFERENCES: Blake 1921; Stuessy 1970; Strother 2006a.

Acanthospermum australe (Loefl.) Kuntze, (of the south) PARAGUAY STARBUR. Annual herbs; stems procumbent, dichotomously branching, 10–60(–120) cm long, often rooting at the nodes; leaves opposite, deltate to rhombic-ovate, 13–37 mm long, 7–32 mm wide, scaberulous or appressed-pubescent to glabrate, gland-dotted, margins toothed above the middle; heads 3–5 mm wide, appearing axillary; phyllaries in 2 series, outer 4–6 herbaceous and ovoid, inner 5–8 spiny, enclosing ray achenes at maturity; receptacles chaffy; ray flowers pistillate, fertile, ligules 0.8–1 mm long, white to yellowish cream; disk flowers staminate, corollas white to yellowish cream; mature achenes enclosed within and shed with inner phyllaries as a bur, these "fruits" plumply ellipsoid, 7–9+ mm long, with hooked prickles, reminiscent of cockleburs (*Xanthium* spp.); pappus absent. Disturbed areas, sandy soils; collected only once in TX in San Jacinto Co. in 1933 (Correll & Johnston 1970); recorded from Vernon Par. in w LA (Gandhi & Thomas 1989); e and se U.S. from MA s to FL and w to TX; introduced elsewhere. Mar–Oct. [*A. xanthioides* DC., *Melampodium australe* Loefl.] Native of South America; now a weed in many countries worldwide.

ACHILLEA L. YARROW

A mainly n temperate genus of 115 species, usually aromatic; many are used medicinally or cultivated as ornamentals. (So named because its healing powers were said to have been discovered by Achilles of Greek mythology; he supposedly stopped the flow of blood from the wounds of his Myrmidon warriors, perhaps with this plant—Arriagada & Miller 1997) (tribe Anthemideae)

REFERENCES: Mulligan & Bassett 1959; Tyrl 1975; Arriagada & Miller 1997; Trock 2006a.

Achillea millefolium L., (thousand-leaved), MILFOIL, WESTERN YARROW, COMMON YARROW. Aromatic, woolly-pubescent, rhizomatous, perennial herbs 35–100 cm tall, usually unbranched up to the inflorescence; leaves alternate, sessile, lanceolate, 2–3 times pinnately compound, fern-like in appearance, the crowded, narrow leaflets finely lobed or toothed; heads small, in a dense, terminal corymbose arrangement; phyllaries 20–30 in ± 3 series; receptacles chaffy; ray flowers (3–)5–8, pistillate, fertile, corollas white, rarely pinkish to purplish, the ligules ca. 2 mm long; disk flowers bisexual, fertile, corollas grayish white or creamy white; achenes 1–2 mm long, margins broadly winged; pappus absent. Roadsides, disturbed sites, prairies, and pastures; widespread in TX but most common in c and e TX; throughout most of the U.S. and Canada; Mexico; introduced elsewhere. Apr–May. [*A. lanulosa* Nutt., *A. millefolium* var. *lanulosa* (Nutt.) Piper, *A. millefolium* subsp. *lanulosa* (Nutt.) Piper, *A. millefolium* var. *occidentalis* DC., *A. occidentalis* (DC.) Raf. ex Rydb.] While this species is sometimes split into subspecific taxa (e.g., Kartesz 1994; Jones et al. 1997), Tyrl (1975) regarded the species as so variable that infraspecific names in North America were relatively meaningless. Cronquist (1980) indicated that *A. millefolium* is a highly variable polyploid complex with both native and introduced forms, thus, several varieties are likely to be combined in our concept and distribution of this species. We are therefore not recognizing subspecies or varieties. When the foliage is crushed, it releases a spicy odor. MILFOIL was reputed to have extensive curative powers by herbals of the Middle Ages, supposedly being useful in treating such conditions as influenza, gout, and ailments of the kidneys and liver (Wills & Irwin 1961). It was also used medicinally by Native Americans and is still used in folk remedies; however, ingestion has been reported to cause irritation of mucous membranes and gastric and abdominal pain (Burlage 1968). ☠

ACMELLA Rich. EX Pers. SPOTFLOWER

A mainly tropical genus of 30 species. Previously included in and closely related to *Spilanthes*. (Greek: *akme*, a point, the highest point, bloom, in reference to the taste of the foliage—Quattrocchi 2000) (tribe Heliantheae)
REFERENCES: Jansen 1981, 1985a, 1985b; Strother 2006b.

Acmella repens (Walter) Rich., (creeping), CREEPING SPOTFLOWER. Herbaceous rhizomatous perennials, creeping, rooting at nodes; flowering branches ascending; leaves opposite, petiolate, blades ovate, rhombic to deltoid, 20–40 mm long, margins serrate, dentate to entire; heads few, solitary, on axillary peduncles 5–15 cm long; phyllaries linear, in ± 2 series; receptacles elongate, conical, 8–10 mm long in fruit, chaffy, the pales about equaling disk flowers; ray flowers few, pistillate, fertile, the ligules ca. 5 mm long, yellow; disk flowers perfect, fertile, yellow; achenes black, compressed, 1–2.5 mm long; pappus absent or rarely of 1–2(–3) bristles. Low moist areas, ditches, ponds; mainly se and e TX; Blackland Prairie in Dallas, Ellis, and Brazos cos., sw to the e edge of Edwards Plateau in Travis and Bexar cos.; se U.S. from NC s to FL and w to MO and TX; widespread in tropical America. May–Nov. [*A. oppositifolia* (Lam.) R.K. Jansen var. *repens* (Walter) R.K. Jansen, *Spilanthes americana* (Mutis ex L. f.) Hieron. var. *repens* (Walter) A.H. Moore]

ACOURTIA D. Don DESERT-PEONY

A North American genus of ca. 41 species. *Acourtia* was long treated as a section of *Perezia*, but Reveal and King (1973) resurrected the genus, transferring all of the cauline-leaved members of *Perezia* from North America into *Acourtia*; Turner (1978) subsequently transferred the scapiform species. So treated, *Perezia* is confined to South America; *Acourtia* to North America and Mexico. (According to its author, the genus is named for Mrs. A'Court of Heytesbury House, Wiltshire, England, out of respect for her "botanical taste and knowledge.") (tribe Mutiseae)
REFERENCES: Bacigalupi 1931; Reveal & King 1973; Turner 1978, 1993; Simpson 2006.

Acourtia runcinata (Lag. ex D. Don) B.L. Turner, (for its runcinate or saw-like leaves), SAWTOOTH ACOURTIA, FEATHERLEAF DESERT-PEONY. Scapose perennial herbs 5–30 cm tall with fasciculated tuberous-fusiform roots; leaves in a basal rosette; petioles 0.5–9 cm long, blades 2.5–23 cm long, 3–8 cm wide, deeply pinnately lobed, glandular-pubescent to nearly glabrate, the margins spinulose-dentate; heads 1–3 per scape, scapes about as long as the longer leaves, ultimate peduncles 1–3 cm long; phyllaries in 2–4 series, subulate to lanceolate, 3–15 mm long; receptacles naked; flowers all perfect, fertile, corollas pink or lavender, 12–22 mm long, zygomorphic, 2-lipped, the outer lip ligule-like and 3-toothed, the inner lip smaller, 2-lobed, the lobes often curled; achenes fusiform, ribbed, 4–8 mm long; pappus of many tan or white ciliate bristles to 17 mm long. Dry calcareous hillsides; sw border of East TX along Edwards Plateau from Travis Co. s to Bexar and Goliad cos.; Trans-Pecos to South TX Plains; Mexico. Mar–Aug. [*Clarionia runcinata* Lag. ex D. Don, *Perezia runcinata* (Lag. ex D. Don) Lag. ex A. Gray.]

AGERATINA Spach SNAKEROOT

Ours shrubs or perennials; leaves opposite, petiolate, simple; heads clustered in compact, corymbose arrays; phyllaries in 2(–3) series, ± equal, herbaceous; receptacles naked; ray flowers absent; disk flower corollas white or pinkish; achenes spindle-shaped, usually 5-angled; pappus of minutely barbed bristles.

A genus of ca. 250 species of the Americas; formerly included in *Eupatorium*; a few species are cultivated. (Named for the resemblance to the related genus *Ageratum*, and Latin: *-ina*, diminutive) (tribe Eupatorieae)
REFERENCES: King & Robinson 1970b, 1987; Clewell & Wooten 1971; Nesom 2006c.

1. Herbaceous perennials; petioles usually 10–30 cm long; corolla lobes short-villous; achenes glabrous _______ **A. altissima**
1. Rounded, much-branched, evergreen shrubs; petioles usually 3–10 cm long; corolla lobes glabrous; achenes hispid _______ **A. havanensis**

Ageratina altissima (L.) R.M. King & H. Rob., (tallest), WHITE SNAKEROOT, SNAKEROOT, RICHWEED, FALLPOISON. Herbaceous perennials from fibrous crowns, sometimes rhizomatous, 30–120 cm tall; stems erect or sprawling; leaves with petioles (5–)10–30(–50) mm long, blades deltoid-ovate to broadly lance-ovate, sometimes cordate, margins coarsely double-serrate; involucres 4–5 mm high; corollas white, lobes sparsely short-villous; achenes glabrous. Low woods, sandy soils, along streams; e and se TX w to Edwards Plateau; e 1/2 of North America. Sep–Nov. [*A. altissima* (L.) R.M. King & H. Rob. var. *angustata* (A. Gray) Clewell &

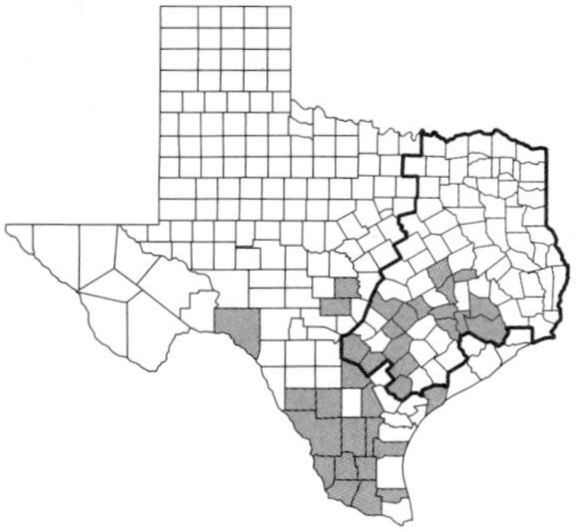
Aristolochia erecta

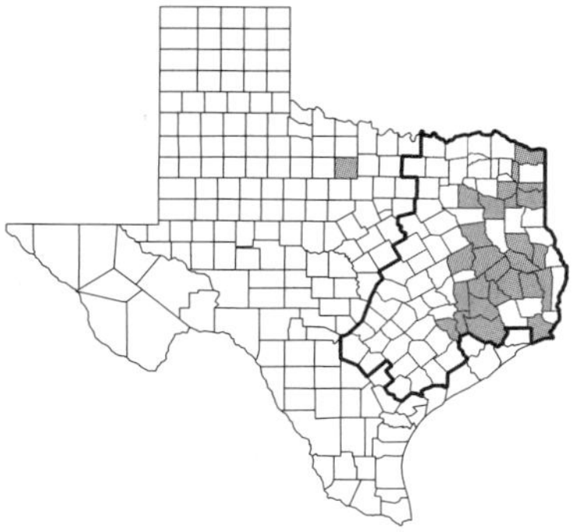
Aristolochia reticulata

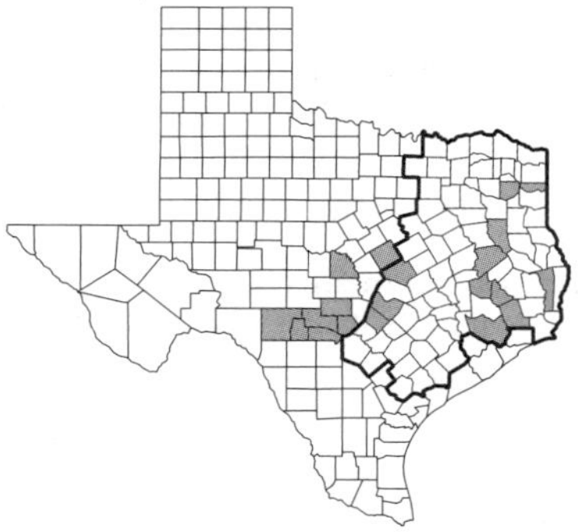
Aristolochia serpentaria

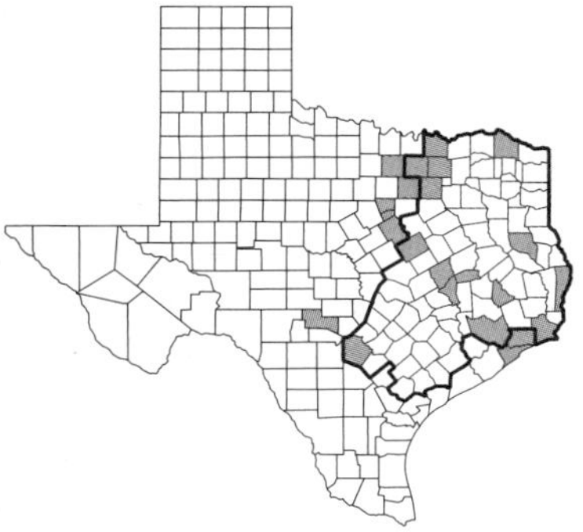
Aristolochia tomentosa

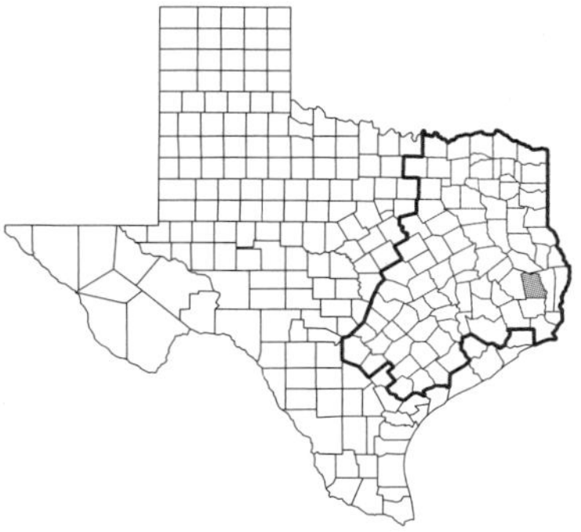
Asarum arifolium

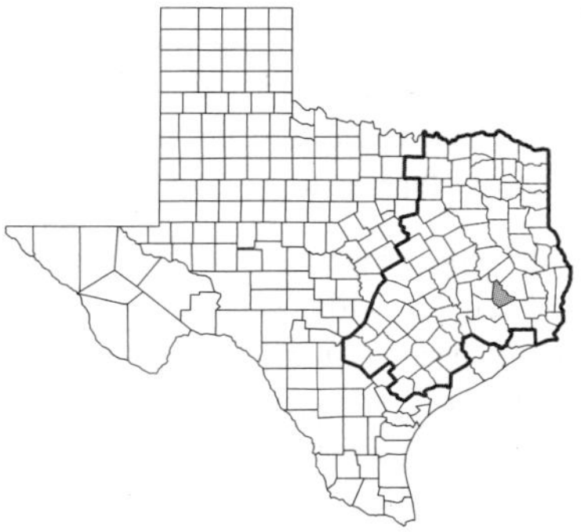
Acanthospermum australe

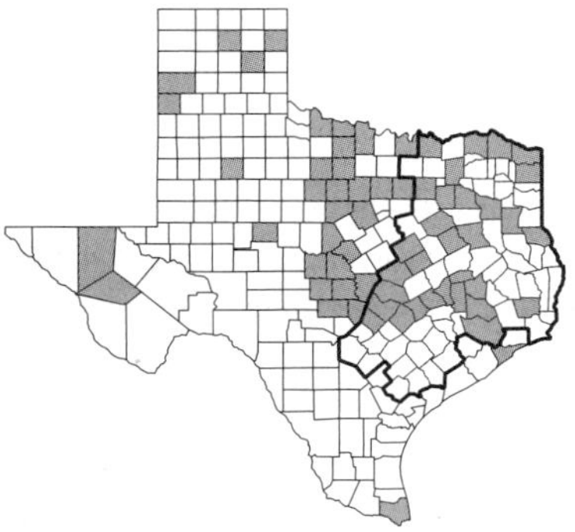
Achillea millefolium

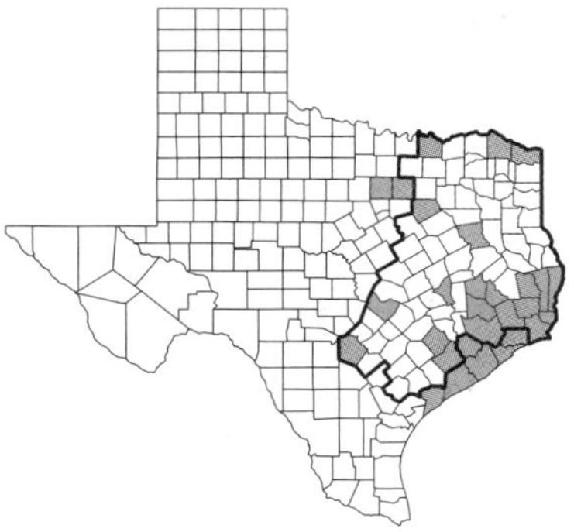
Acmella repens

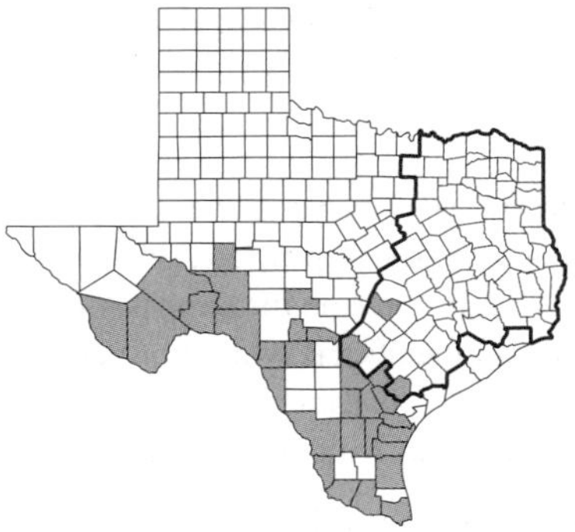
Acourtia runcinata

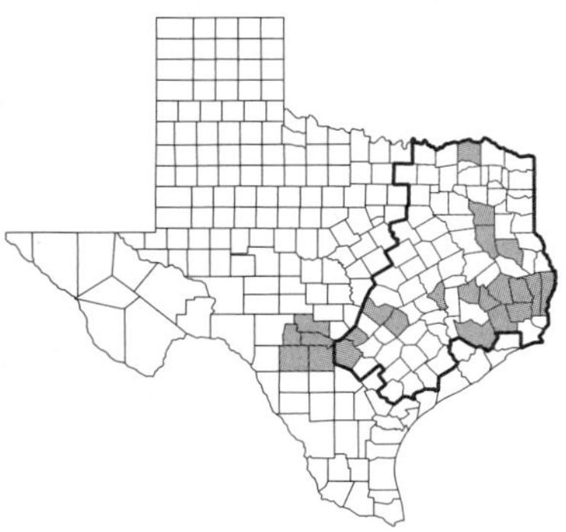
Ageratina altissima

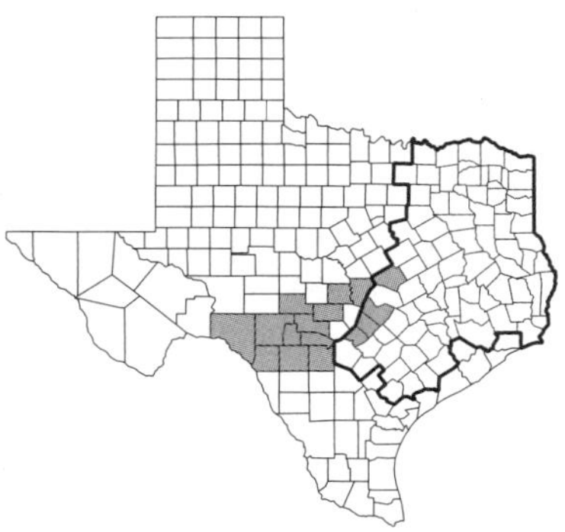
Ageratina havanensis

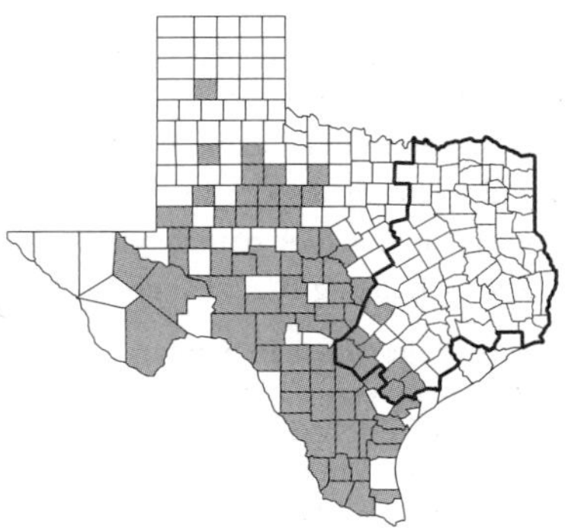
Amblyolepis setigera

Wooten, *Eupatorium rugosum* Houtt., *E. urticifolium* Reichard] All parts of WHITE SNAKEROOT contain tremetol, a complex alcohol, and glycosides; it is poisonous to livestock. During colonial times cows eating SNAKEROOT passed the poison through their milk causing "milk sickness" in humans (Hardin & Arena 1974). So many deaths resulted that in some areas the human population was reduced to less than 1/2 the original number, and whole villages were abandoned (Kingsbury 1964); in livestock the condition is known as "trembles" (Stephens 1980). Other species are not known to contain the toxin. ☠

Ageratina havanensis (Kunth) R.M. King & H. Rob., (of Havana), HAVANA SNAKEROOT, SHRUBBY BONESET. Rounded, much-branched shrubs, 30–200 cm tall; stems erect, brittle; leaves evergreen-persistent, petioles 3–10(–15) mm long, blades deltoid to broadly ovate or hastate, margins dentate; involucres 4–6 mm high; corollas pinkish white to white, lobes glabrous; achenes hispid. Rocky limestone outcrops, oak-juniper woodlands, along streams; sw border of East TX from Bell Co. s to Comal Co.; Edwards Plateau and s part of Lampasas Cut Plain; Mexico and West Indies. (Sep–)Oct–Nov. [*Eupatorium havanense* Kunth, *E. texense* (Torr. & A. Gray) Rydb.] As the epithet implies, this species was first described from Cuba, but it is also native to TX.

AMBLYOLEPIS DC. HUISACHE DAISY

A monotypic genus endemic to the sw U.S. and Mexico; closely related to *Helenium* and *Gaillardia* but distinguished by a number of characters, including the strong smell of coumarins, present even in dried specimens after years of storage. (Greek: *ambly*, blunt, and *lepis*, scale, in reference to the pappus scales—Turner 2016) (tribe Heliantheae)
REFERENCES: Bierner 1990, 2006a.

Amblyolepis setigera DC., (bristle-bearing, presumably for the bristle-like hairs on the phyllaries), HUISACHE DAISY. Small annuals, 10–60 cm tall, with sweet odor reminiscent of freshly-mown sweet-clover (*Melilotus*); stems erect or decumbent with few branches distally; stems and leaf margins with long, wide-spreading, cottony hairs; leaves alternate, sessile, entire, the lowest oblanceolate, the upper ovate to lanceolate, subclasping; heads long-peduncled and large, the involucres 7–11(–17) mm long; phyllaries in 2 series, lanceolate, the outer series green and pilose, the inner series scarious; receptacles hemispherical-ovoid, usually with short bristles, rarely with peripheral flowers subtended by pales; ray flowers 8–13(–20), pistillate, fertile, ligules yellow (sometimes striped or tipped with white), 10–22 mm long; disk flowers perfect, fertile, yellow; achenes 3–4.5 mm long, with a cage-like exterior of 8–10 stout, separate ribs; pappus of 5–6 ovate to obovate scales, 2–3(–4.1) mm long. Sandy or rocky open ground; w 1/2 of TX; along the sw border of East TX from Travis, Comal, and Bexar cos. s to Goliad and Victoria cos.; mainly High Plains, Edwards Plateau, and South TX Plains; n Mexico. Mar–Jul. [*Helenium setigerum* (DC.) Britton & Rusby]

AMBROSIA L. RAGWEED

Ours annual or perennial herbs (or rarely shrubs), 30–400+ cm tall; leaves alternate, alternate above and opposite below, or opposite, sessile or petiolate, palmately lobed or pinnatifid, sometimes aromatic; receptacles chaffy; flowers unisexual, in separate heads, these on the same plant; staminate heads in spicate or racemose arrays, their phyllaries united into a cup shape, corollas insignificant, whitish or purplish; pistillate heads 1–few-flowered, axillary (usually below the staminate heads, rarely intermixed); phyllaries often toothed, ± united, enclosing the achenes at maturity, forming a hard, indehiscent "bur" or "fruit," the bur usually with the tips of the phyllaries projecting as spines, tubercles, or wings; pistillate corollas and pappus absent.

A cosmopolitan, but especially American genus of ca. 40 species. Most of the widespread species are exceedingly variable, both in vegetative and fruiting characters. Several of the species are known to form natural hybrids, which has probably confounded specific delimitation (Strother 2006c; Turner 2016). The flowers of *Ambrosia* are small and wind-pollinated; the abundant air-borne pollen is a problematic cause of allergic reactions during the fall and is considered the leading cause of hay fever in the U.S. (Lewis & Elvin-Lewis 1977). The allergic response is initiated when pollen grain proteins (antigens) attach to receptors on antibodies (immunoglobulin E—IgE) linked to immune system cells. This results in the immune cells releasing histamines, the molecules directly responsible for the symptoms known as hay fever (Kuby 1997; Lim 1998). (Early Greek name for aromatic plants; the mythic food of the gods) (tribe Heliantheae)
REFERENCES: Rydberg 1922; Payne 1964; Lee & Dickinson 1980; Lee 1981; Strother 2006c.

1. Shrubs; fruits with membranous wings; rare in East TX ____ **A. monogyra**
1. Annual or perennial herbs; fruits with spines or tubercles; including plants common in East TX.
    2. Leaves 3–5-lobed or sometimes unlobed.
        3. Plants < 1 m tall; sap not red; stem leaves sessile or nearly so; side lobes of leaves basically a large tooth on each side, these much smaller than the middle lobe; lobes or whole leaf (if unlobed) 12 mm wide or less; staminate heads sessile ____ **A. bidentata**
        3. Plants 1–3(–5) m tall; sap often red; stem leaves petiolate; side lobes of leaves not much smaller than the middle lobe; lobes or whole leaf > 12 mm wide (often much more); staminate heads on short peduncles ____ **A. trifida**

2. Leaves pinnatifid, divided into numerous narrow segments.
   4. Leaves 1–4 times pinnatifid, parsley-like in appearance, with petioles to 4 cm long (longest on lower leaves); spines of the bur (1–)5–12+, with tips hooked, in several series on the upper 2/3 of the body of the bur; rare in East TX ____ **A. confertiflora**
   4. Leaves 1–3 times pinnatifid, but not parsley-like, subsessile or with petioles to only 3 cm long; spines or tubercles of the bur 3–7, with tips straight or blunt, in a single series near the apex (= beak) of the bur; widespread throughout East TX.
      5. Taprooted annuals; stem leaves petiolate ____ **A. artemisiifolia**
      5. Perennials from creeping rootstocks; stem leaves essentially sessile ____ **A. psilostachya**

Ambrosia artemisiifolia L., (with leaves like *Artemisia*—sagebrush, wormwood), COMMON RAGWEED, ROMAN WORMWOOD, SHORT RAGWEED, ALTAMISA, HOGBRAKE. Tap-rooted annuals; stems 30–100 cm tall; leaves opposite or uppermost sometimes alternate, pinnatifid to bipinnatifid, with petioles usually 10–30 mm long, uppermost nearly sessile; staminate heads short-stalked; "fruits" ca. 3 mm long, obovoid in outline, the spines 3–7. Disturbed areas; se and e TX, w to nc TX and Edwards Plateau; throughout most of Canada and the U.S.; occasionally introduced elsewhere. Aug–Nov. [*A. elatior* L., *A. glandulosa* Scheele, *A. monophylla* (Walter) Rydb.]

Ambrosia bidentata Michx., (two-toothed), SOUTHERN RAGWEED, LANCE-LEAF RAGWEED. Annuals; stems 30–100 cm tall; leaves alternate above, opposite below, sessile or nearly so, narrowly ovate to lanceolate to lance-linear, 3–5 cm long, usually with a broad tooth or small lobe on each side at the blade base; staminate heads sessile with oblique involucre; "fruits" ca. 5 mm long, 4-angled, the spines 4(–5). Prairies; extreme n and e TX; from Grayson and Lamar cos. to Bowie Co. in Red River Area, also Sabine Co. in Pineywoods; ec U.S. from NY s to GA and w to MN and TX. Aug–Sep.

Ambrosia confertiflora DC., (with flowers crowded together), FIELD RAGWEED, BUR-SAGE. Perennials, forming colonies; stems 30–60(–150) cm tall; leaves mostly alternate, petiolate, 1–4-pinnatifid, often with 1–several pair of small lobes on the petiolar bases below the main blade; staminate heads short-stalked; "fruits" 2–4 mm long, the spines usually (1–)5–12+, ca. 0.5–1 mm long, with hooked tips. Disturbed areas; mainly w 1/2 of TX; along the sw border of East TX in Travis, Comal, and Bexar cos., disjunct n to Dallas Co.; sw U.S. from CA to OK; Mexico; occasionally introduced elsewhere. Aug–Nov. [*A. simulans* Shinners, *Franseria caudata* Rydb., *F. confertiflora* (DC.) Rydb.]

Ambrosia monogyra (Torr. & A. Gray) Strother & B.G. Baldw., (one-whorled), BURROBUSH, SINGLE-WHORL BURROBUSH. Woody shrubs, to 150(–400) cm tall; leaves mostly alternate, sessile, filiform, unlobed or sometimes with 3(–5+) filiform lobes, finely scabrous on upper surface, staminate heads sessile or short-stalked; "fruits" 4–5 mm long, spindle- or pear-shaped, membranously-winged in a single whorl around the middle, wing segments 7–12+. Disturbed areas, alluvial washes, ravines; in East TX recorded only from Travis Co. in collections from the early 20th century (*H.H. York s.n.* and *B.C. Tharp 44189* TEX-LL), at least the latter from xeric sandpiles along the Colorado River; Edwards Plateau and Trans-Pecos; sw U.S. from TX to CA. May–Oct. [*Hymenoclea monogyra* Torr. & A. Gray] Perhaps not a persistent member of the East TX flora.

Ambrosia psilostachya DC., (naked spike), WESTERN RAGWEED, PERENNIAL RAGWEED. Perennials, forming colonies; stems 30–60 (–100+) cm tall; leaves alternate above, opposite below, sessile, pinnatifid; staminate heads short-stalked; "fruits" ca. 2.5 mm long, obovoid in outline, spines or tubercles 1–6, acute or blunt. Disturbed areas; nearly throughout TX; w 1/2 of the U.S.; Mexico. (Jun–)Aug–Nov. [*A. lindheimeriana* Scheele, *A. rugelii* Rydb.] Used as a medicinal tea by Native Americans; also apparently inhibits the growth of some other organisms through allelopathy (Cheatham & Johnston 1995).

Ambrosia trifida L., (three-parted), GIANT RAGWEED, BLOOD RAGWEED, BUFFALOWEED. Annuals; sap blood-red; stems often very tall, 100–300(–500) cm; leaves usually opposite, petiolate, palmately 3(–5)-lobed or uppermost rarely unlobed, scabrous on both sides, extremely so on upper surface; staminate heads stalked; "fruits" ca. 4 mm long, obovoid in outline, the tubercles 4–8, small or obsolete. Disturbed areas, ditches, often extremely abundant; nearly throughout TX; widespread in the U.S. and Canada. Jul–Nov. [*A. aptera* DC., *A. trifida* var. *texana* Scheele] The sap from broken stems stains the hands red.

AMPHIACHYRIS (A. DC.) Nutt. BROOMWEED

Taprooted glabrous annuals, with strong turpentine scent when crushed; stems usually much-branched in the upper portion; leaves alternate, sessile, linear to lanceolate, gland-dotted, margins entire; heads borne in paniculate or corymbose arrays; phyllaries in 1–3 series, unequal, whitish-resinous, bases hardened, margins hyaline; receptacles not chaffy (may be hairy); ray flowers pistillate, fertile, corollas yellow; disk flowers functionally staminate, corollas yellow; achenes obovate, 4–9-ribbed, hairy; pappus of ray achenes a low crown of scales.

A genus of 2 species of the c U.S. The species treated here have also been treated in *Gutierrezia* and *Xanthocephalum*. The molecular analysis by Suh and Simpson (1990) indicated *Amphiachyris* and *Thurovia* (a monotypic genus) form a sister group to *Gutierrezia*, and supported these as separate genera. As in *Gutierrezia*, plant populations tend to increase under overgrazing. (Greek: *amphi*, around, and *achyron*, chaff or husks, alluding to the pappus) (tribe Astereae)
REFERENCES: Shinners 1951c; Solbrig 1960, 1961; Ruffin 1974; Lane 1979, 1982; Suh & Simpson 1990; Nesom 2006d.

1. Plants to 40(–60) cm tall; leaves 0.2–1(–2) mm wide; heads scattered in open panicles; receptacles with hooked, swollen-based hairs; achenes 4–6-ribbed, with long, bristly pubescence ____ **A. amoena**
1. Plants to 100(–200) cm tall; leaves 0.5–6 mm wide; heads many to very numerous in crowded corymbs; receptacles glabrous; achenes 7–9-ribbed, with short, bristly pubescence ____ **A. dracunculoides**

Amphiachyris amoena (Shinners) Solbrig, (charming), ANNUAL BROOMWEED, TEXAS BROOMWEED. Plants to 40(–60) cm tall; primary stems usually (2–)5 mm in diameter; leaves linear, rarely linear-lanceolate, 0.2–1(–2) mm wide; heads scattered in open paniculate arrays; receptacles with hooked, swollen-based hairs; phyllaries 3–4 mm long; achenes 2–3 mm long, 4–6-ribbed, with long, bristly pubescence. Calcareous soils on or near limestone outcrops; endemic to TX; w border of East TX in Denton Co., Bell Co., and Travis to Comal cos.; c to nc TX and Edwards Plateau. Sep–Oct. [*Gutierrezia amoena* (Shinners) Diggs, Lipscomb, & O'Kennon, *Xanthocephalum amoenum* Shinners, *X. amoenum* var. *intermedium* Shinners]

Amphiachyris dracunculoides (DC.) Nutt., (resembling *Artemisia dracunculus* L.—taragon), COMMON BROOMWEED, PRAIRIE BROOMWEED. Plants to 100(–200) cm tall; primary stems 0.3–1(–2) mm in diameter; leaves narrowly to broadly lanceolate, 0.5–6 mm wide; heads many to very numerous in crowded corymbose arrays; receptacles glabrous; phyllaries 2–3 mm long; achenes 1.2–2.2 mm long, 7–9-ribbed, with short, bristly pubescence. Disturbed habitats, pastures, often in large populations in overgrazed areas; nearly throughout TX but less common in East TX; sc U.S. from TN and AL w to KS and NM; adventive in states further e and n (Kartesz 2015); Mexico. May–Nov. [*Gutierrezia dracunculoides* (DC.) S.F. Blake, *Xanthocephalum dracunculoides* (DC.) Shinners] This species can cause contact dermatitis in humans, eye inflamation in humans and livestock, and gastrointestinal upsets in cattle (Gates 1945; Ajilvsgi 1984). The plants were tied to sticks by early settlers and used as brooms (Ajilvsgi 1984).

ANTENNARIA Gaertn. PUSSY-TOES, EVERLASTING, LADIES'-TOBACCO

Ours low herbaceous perennials to 35(–45) cm tall, stoloniferous; leaves alternate, entire, tomentose beneath; basal leaves petiolate, elliptic to broadly obovate, 2–6 cm wide; stem leaves reduced and ± sessile; heads in congested corymbose arrays, unisexual, the heads containing all female or all male flowers (these usually on separate plants); phyllaries imbricate in 3–5 series, narrow, membranous or scarious; receptacles naked; ray flowers absent; disk flowers either functionally staminate or pistillate and fertile; corollas creamy white to pinkish; achenes minute, elliptical; pappus of white, hair-like bristles.

A genus of 45 species of small dioecious herbs of temperate and alpine-arctic regions of the Americas and Eurasia; some are cultivated as rock garden subjects. (Latin: *antenna*; the pappus of the male flowers have swollen tips resembling a butterfly's antennae) (tribe Inuleae)
REFERENCES: Bayer & Stebbins 1982; Bayer 1984, 1985, 2006; Anderberg 1991.

1. Basal leaves tomentose or glabrous on upper surface; pistillate involucres (7–)8–13 mm high; staminate corollas 3.5–5 mm long; pistillate corollas 4–7 mm long; pistillate flower pappus 5–8 mm long ____ **A. parlinii**
1. Basal leaves tomentose on upper surface; pistillate involucres 5–7 mm high; staminate corollas 2–3.5 mm long; pistillate corollas 3–4 mm long; pistillate flower pappus 3.5–5.5 mm long ____ **A. plantaginifolia**

Antennaria parlinii Fernald, (for its discoverer, John Crawford Parlin, 1863–1948, of Maine), LARGE-LEAF PUSSY-TOES, PLAIN-LEAF PUSSY-TOES, PARLIN'S PUSSY-TOES. Plants to 35(–45) cm tall; leaves obovate-spatulate, prominently 3- to 5-nerved, lower surface gray-pubescent with matted, woolly hairs; involucres of pistillate heads (7–)8–13 mm high; phyllaries with scarious or white tips; staminate flowers with prominent black-brown anthers; pistillate corollas 4–7 mm long; achenes 1–2 mm long; pistillate flower pappus 5–8 mm long. Sandy woods; East TX w to West Cross Timbers; e 1/2 of North America. Mar–Apr. Thought to be a polyploid of hybrid origin (Bayer & Stebbins 1982; Bayer 1985).

1. Stems usually eglandular; upper surface of basal leaves tomentose ____ subsp. **fallax**
1. Stems usually with purple glandular hairs (at least near summits of young flowering stems); upper surface of basal leaves glabrous ____ subsp. **parlinii**

subsp. **fallax** (Greene) Bayer & Stebbins, (deceptive), DECEITFUL PUSSY-TOES. Basal leaves usually gray-tomentose on both sides. Widespread from East TX w to West Cross Timbers. [*A. fallax* Greene]

Acanthospermum australe [BB2]

Achillea millefolium [REE]

Acmella repens [CO2]

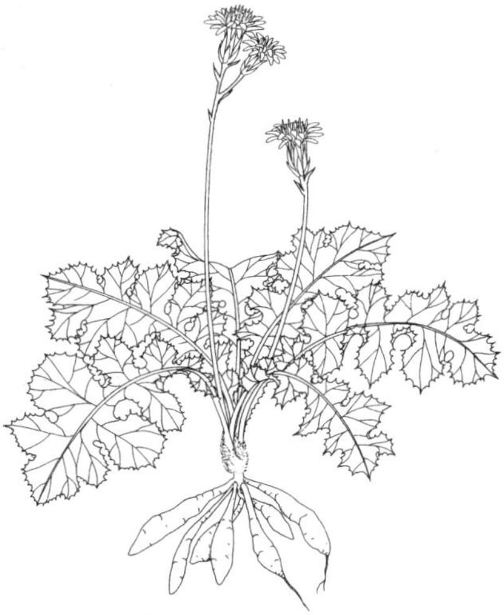

Acourtia runcinata [FNA]

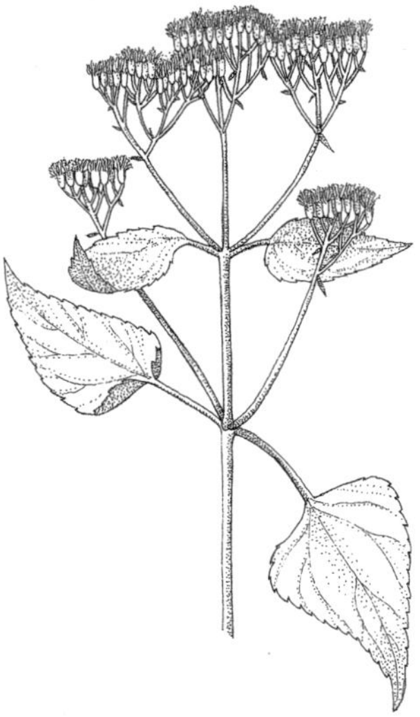

Ageratina altissima [FOV]

Ageratina havanensis [HEA]

Amblyolepis setigera [HEA]

Ambrosia artemisiifolia [REE]

subsp. **parlinii**, PARLIN'S PUSSY-TOES. Basal leaves gray-tomentose beneath, the upper surface green and glabrous. Rare in TX; recorded only in Montgomery Co. (Brown 2014). [*A. arnoglossa* Greene]

Antennaria plantaginifolia (L.) Richardson, (with leaves resembling *Plantago*—plantain), PLANTAIN-LEAVED PUSSY-TOES. Plants to 20(–25) cm tall; leaves obovate-orbiculate, 3–5(–7)-nerved, lower surface tomentose, upper surface green and glabrescent to gray-pubescent; involucres of pistillate heads 5–7 mm high; phyllaries with white tips; pistillate corollas 3–4 mm long; achenes 0.5–1.6 mm long; pistillate flower pappus 3.5–5.5 mm long. Sandy woods; rare in TX; recorded from Lamar (TEX) and Morris cos.; e 1/2 of North America. Mar–Apr. [*A. caroliniana* Rydb., *Gnaphalium plantaginifolium* L.]

ANTHEMIS L. MAYWEED, DOG-FENNEL, CHAMOMILE

An Old World genus of ca. 175 species ranging from Europe through the Mediterranean region to Iran and e Africa; some species are medicinal or are cultivated as ornamentals. (Greek name for the related *Chamaemelum nobile* (L.) All., chamomile; probably derived from *anthemon*, flower) (tribe Anthemideae)
REFERENCES: Arriagada & Miller 1997; Watson 2006a.

Anthemis cotula L., (resembling species belonging to the genus *Cotula*), MAYWEED, CHAMOMILE, DOG-FENNEL, STINKING CHAMOMILE, STINKWEED. Rankly aromatic annuals to 60(–90) cm tall; slightly and inconspicuously pubescent; leaves alternate, finely and deeply cut, twice compound or pinnately lobed; heads borne singly or in loose corymbose arrays; phyllaries appressed, scarious-margined; receptacles ovoid-conical, chaffy; ray flowers 10–15 per head, sterile, ligules white, ca. 7–12 mm long, persistent and reflexed in age; disk corollas greenish yellow; achenes ribbed, 1.3–2 mm long; pappus absent. Sandy roadsides, pastures, disturbed areas; cultivated and escaped; se and e TX w to nc TX and Edwards Plateau; throughout most of the U.S. and Canada. Apr–May(–Jul). Native of Europe; now a widespread weed in temperate regions. Reverchon (1880) stated that its introduction in Dallas Co. dates from 1875. Source of an insecticide and reported to taint cow's milk (Mabberley 2017). According to Stubbendieck et al. (1994) it is poisonous to both cattle and poultry, the sap containing caustic substances which irritate membranes. Related to *Chamaemelum nobile*, which is the source of the stomach remedy chamomile.

APHANOSTEPHUS DC. LAZY DAISY, DOZE DAISY

Annual or perennial herbs; leaves alternate, with narrowed, petiolar bases; blades linear to oblong-oblanceolate, entire to pinnately lobed; heads solitary; phyllaries ovate-lanceolate with broad, scarious margins; receptacles naked; ray flowers pistillate, fertile, with ligules white to pink, often rose-red or rose-purple beneath, spreading from mid-morning to late afternoon, erect at night; disk flowers perfect, fertile, corollas yellow; achenes terete or 4-angled and multi-ribbed; pappus absent or minute, of scales, awns, or cilia 1 mm or less long.

A genus of 4 species endemic to the U.S. and Mexico. The common names refer to the flowers remaining closed until mid-morning or later (Ajilvsgi 1984); the often rose-red or rose-purple undersides of the erect ray flowers are conspicuous (Kirkpatrick 1992). (Greek: *aphanes*, obscure or hidden, and *stephanos*, crown, presumably from the inconspicuous crown-like pappus of some species) (tribe Astereae)
REFERENCES: Shinners 1946c; Turner 1984; Elisens et al. 1992; Nesom 2006e.

1. Plants soft-pubescent to hispid-pubescent with spreading to reflexed hairs (usually < 1 mm long); heads not crowded, on long peduncles naked for 15–100 mm below the heads (to 150 mm in age); ray flowers 16–85 per head.
   2. Ray flowers with ligules 5–10 mm long; bases of disk corollas only slightly enlarged at maturity; taprooted annuals or clump-forming perennials; sw border of East TX.
      3. Ray flowers usually 20–41 per head; plants annual from slender taproot ____ **A. ramosissimus**
      3. Ray flowers 30–75 (as few as 25 in second flowering of summer and fall) per head; plants perennial from branched, woody base ____ **A. riddellii**
   2. Ray flowers with ligules 8–15 mm long; bases of disk corollas becoming swollen (to 2 times or more their original diam.) and hardened at maturity; taprooted annuals; widespread in East TX ____ **A. skirrhobasis**
1. Plants conspicuously hispid with long (0.7–2.2 mm long), coarse, jointed, translucent hairs spreading at right angles; heads ± crowded, on short peduncles naked for 3–12 mm below the heads (to 25 mm in age); ray flowers 12–18 per head ____ **A. pilosus**

Aphanostephus pilosus Buckley, (pilose, with long soft hairs), HAIRY LAZY DAISY. Small annuals; stems 6–33 cm tall; stem hairs spreading at right angles; involucres 4.5–5.5 mm high, 4–6 mm wide; ray flower ligules 5–7 mm long, white to rosy-lavender, especially upon withering; pappus to 0.7 mm long. Sandy or silty soils; low prairies, draws, and ditches; Rolling Plains and West Cross Timbers with a disjunct record from Red River Co. (Kartesz 2015), rare in East TX; endemic to n TX and OK. Mar–May(–Jun).

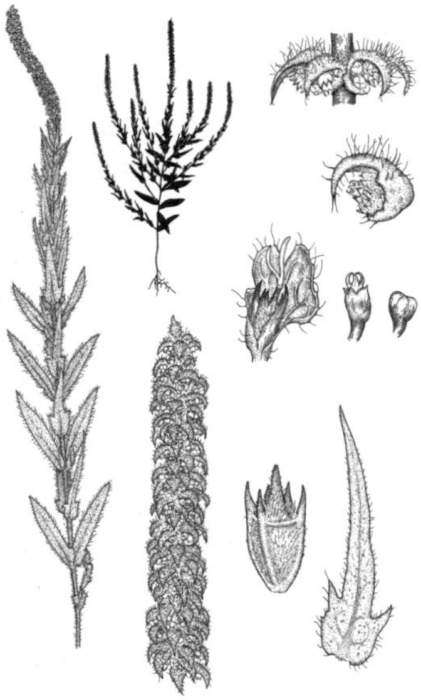

Ambrosia bidentata [MO4]

Ambrosia confertiflora [ARF]

Ambrosia monogyra [VIN]

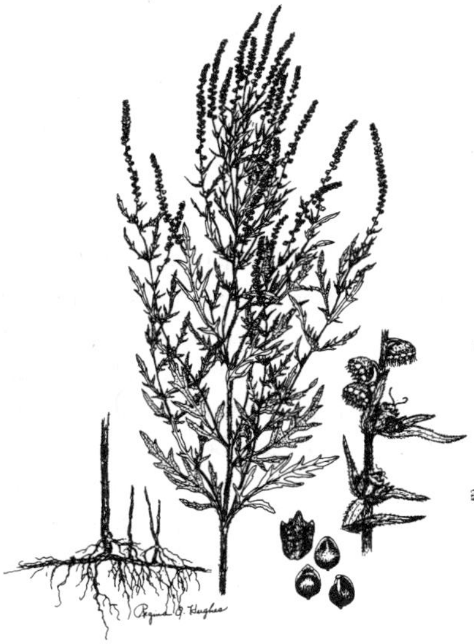

Ambrosia psilostachya [REE]

Ambrosia trifida [REE]

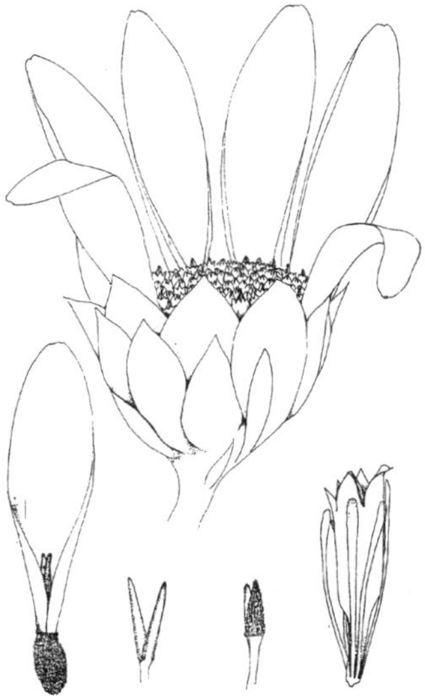

Amphiachyris amoena [SID]

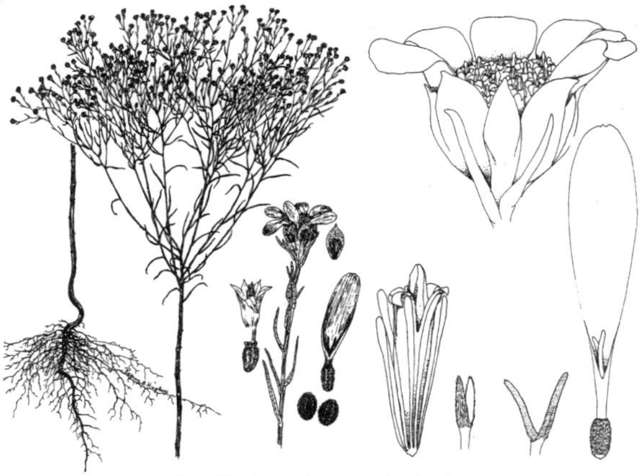

Amphiachyris dracunculoides [REE, SID]

Aphanostephus ramosissimus DC. var. **ramosissimus** (much-branched), PLAINS LAZY DAISY, ARIZONA LAZY DAISY. More or less soft-pubescent annuals, often widely branched and partly decumbent; stems 20–45 cm tall; involucres 3.3–5 mm high, 5–9 mm wide; ray flower ligules 5.5–7.5 mm long, white or becoming rose to purple, especially on lower surface; pappus to 0.6 mm long, a small crown or cup. Sandy open woods, fields, prairies, and roadsides; along the sw border of East TX from Bastrop (TEX), Bexar, and Travis (TEX) cos. s to DeWitt and Goliad cos.; High Plains s to South TX Plains; NM, OK, and TX; n Mexico. Apr–Jun(–Aug).

Aphanostephus riddellii Torr. & A. Gray, (for John Leonard Riddell, 1807–1865, botanist), RIDDELL'S LAZY DAISY. Hispid-pubescent perennials; stems woody at the base, 10–50 cm tall; stem hairs spreading or bent downward or upwards; heads usually numerous, rather large and long-peduncled; involucres 4.5–6.2 mm high, 9–14 mm wide; ray flower ligules 7–10 mm long, white; pappus to 0.3 mm long. Rocky or sandy ground, chiefly on limestone; along the sw border of East TX from Bexar, Comal, and Travis cos. s to Hays and Goliad cos.; High Plains s to South TX Plains; NM; n Mexico. Apr–Jun(–Jul–Oct).

Aphanostephus skirrhobasis (DC.) Trel. var. **skirrhobasis,** (Greek: *skirros*, hard, and *basis*, base, referring to the bases of disk flowers becoming hardened at maturity—Holloway 2005), ARKANSAS LAZY DAISY. Soft, gray-pubescent, taprooted annuals; stems to 50 cm tall; heads rather large; involucres 6–8 mm high, 7–13 mm wide; ray flowers 20–44 with ligules 8–15 mm long, white or reddish to rosy on the lower surface, the color often streaked; pappus to 0.3 mm long. Sandy open woods, fields, prairies, and roadsides; widespread in TX; AR, KS, LA, NM, OK, and TX; n Mexico. Apr–Jun, sporadically later. Sometimes cultivated as an ornamental.

A similar variety, *A. skirrhobasis* (DC.) Trel. var. *thalassius* Shinners, is restricted to Gulf Coast sand dunes but enters our area in Jefferson Co. It differs in having leaves that are thickened and densely felted with gray hairs.

ARCTIUM L. BURDOCK

A genus of 10 species native to Eurasia and n Africa; widely introduced and weedy worldwide. (Greek: *arktion*, from *arktos*, bear, perhaps alluding to the bur-like involucre) (tribe Cardueae)
REFERENCES: Duistermaat 1996; Keil 2006a.

Arctium minus Bernh., (small), LESSER BURDOCK, COMMON BURDOCK, BARDANE, BEGGAR'S BUTTON. Herbaceous biennials, 50–300 cm tall; stems erect, openly branched; basal leaves with long, hollow petioles, blades 30–60 cm long, green above, thinly gray-tomentose beneath, margins coarsely dentate to subentire (rarely deeply dissected), not spiny; heads in racemose or paniculate clusters; involucres 15–40 mm wide; phyllaries in many series, linear to lanceolate, margins often minutely serrate; tips hooked-spiny; receptacles bristly; ray flowers absent; disk flowers 30+, corollas pink to purple or white; achenes obovoid, ± compressed, 5–8 mm long, enclosed within mature involucres that are shed as spiny burs; pappus falling early, of many bristles, 1–3.5 mm long. Waste places, roadsides, fields; not yet reported in TX but included here based on records from adjacent Bryan and McCurtain cos., OK (OKL); widespread in U.S. but less frequent in s states; Canada; Eurasia; introduced elsewhere. Jul–Sep. [*Lappa minor* Hill] Native to Eurasia; an invader of pastures and hayfields, now a noxious weed on several continents.

ARNOGLOSSUM Raf. INDIAN-PLANTAIN

Ours perennial herbs, mostly glabrous, with fleshy fibrous roots; stems usually 1, erect, 3–30 cm tall; leaves mostly conspicuously large and basal, fewer and reduced upwards, alternate, petiolate or sessile; leaf blades cordate to ovate to lanceolate, with prominent ± parallel, longitudinal nerves eventually converging toward the tip (reminiscent of larger-leaved *Plantago* spp.); heads numerous in a broad corymbose arrangement at apex of plant; involucres mostly cylindrical; phyllaries 5, in 1–2 series, flattened or keeled dorsally; receptacles naked; ray flowers absent; disk flowers 5, perfect, fertile, corollas cream or greenish white; achenes cylindric, ribbed, glabrous; pappus of 100+ white capillary bristles.

A genus of 8 species native to e North America; previously recognized in *Cacalia*. (From Greek: *arnos*, lamb, and *glossa*, tongue; an ancient name for some species of *Plantago*) (tribe Senecioneae)
REFERENCES: Shinners 1950c; Robinson 1974; Pippen 1978; Anderson 2006.

1. Stems rounded; leaves membranous; phyllaries not keeled or winged on their dorsal midvein; plants flowering late summer–fall ____ **A. ovatum**
1. Stems striate-angled; leaves firm and fleshy; phyllaries keeled or winged on their dorsal midvein; plants usually spring-flowering ____ **A. plantagineum**

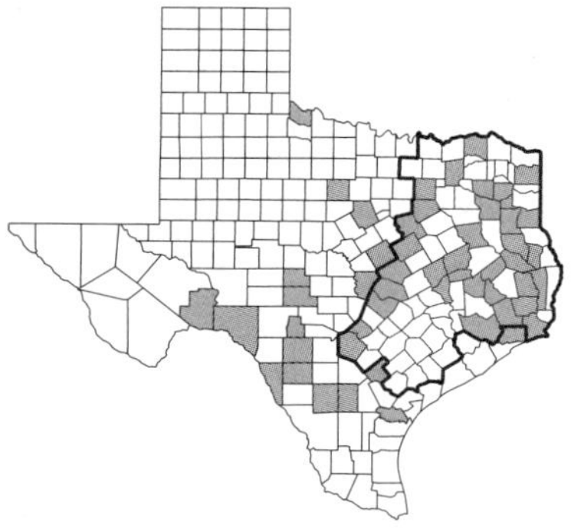

Ambrosia artemisiifolia

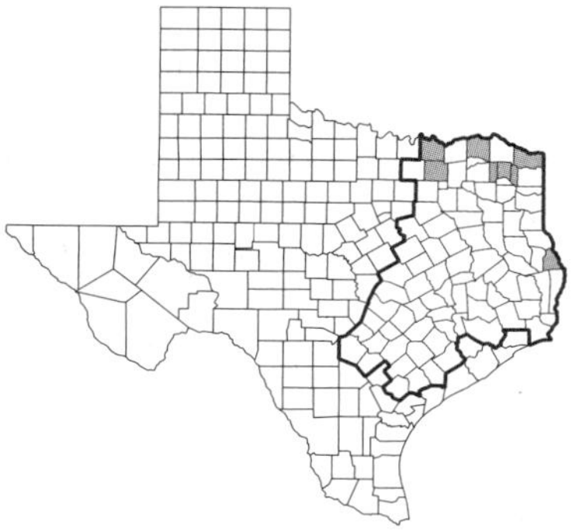

Ambrosia bidentata

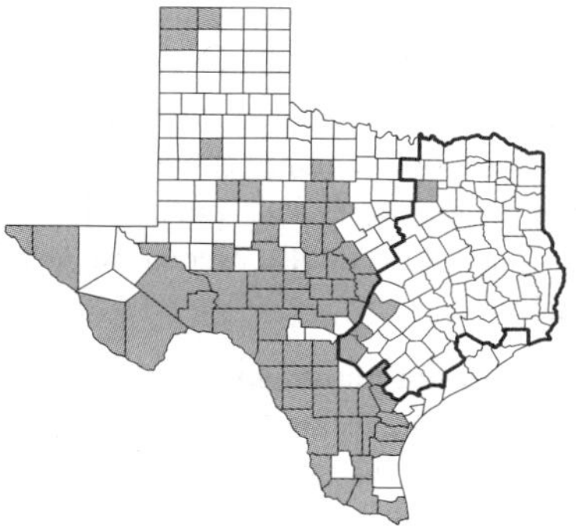

Ambrosia confertiflora

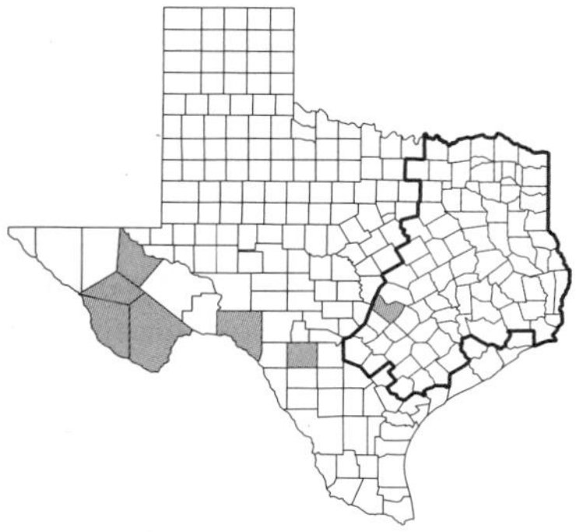

Ambrosia monogyra

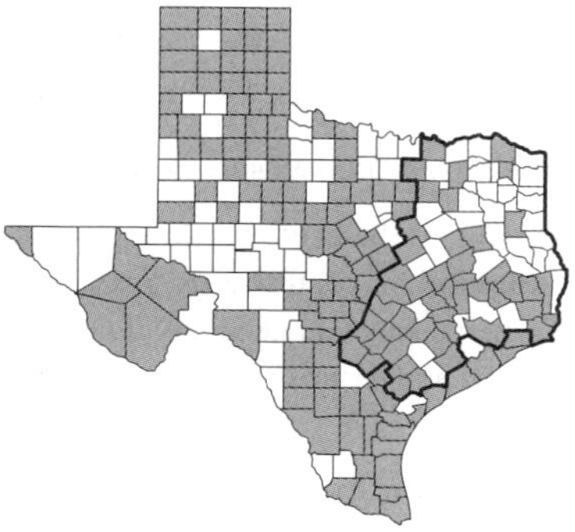

Ambrosia psilostachya

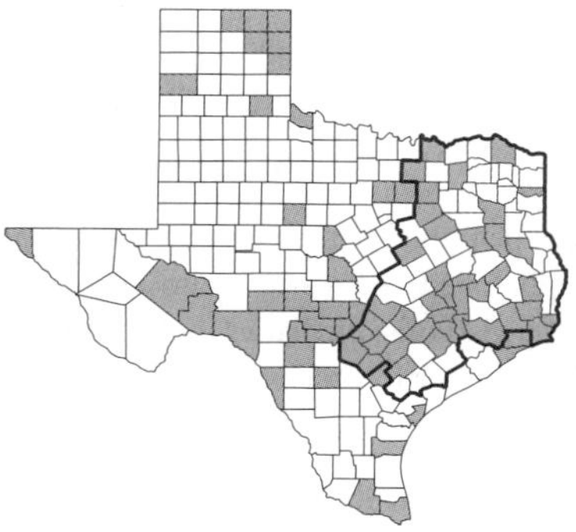

Ambrosia trifida

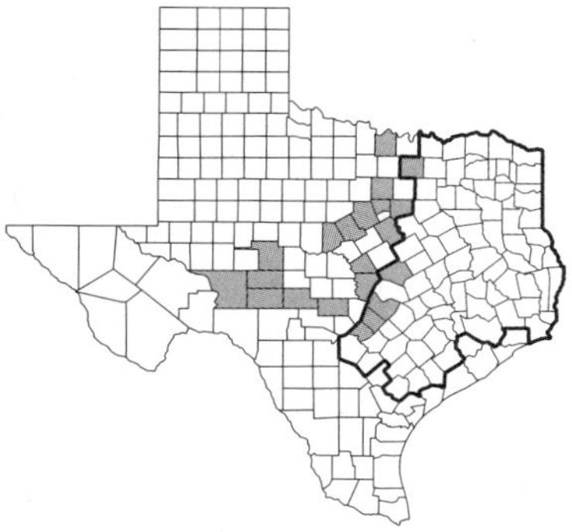

Amphiachyris amoena

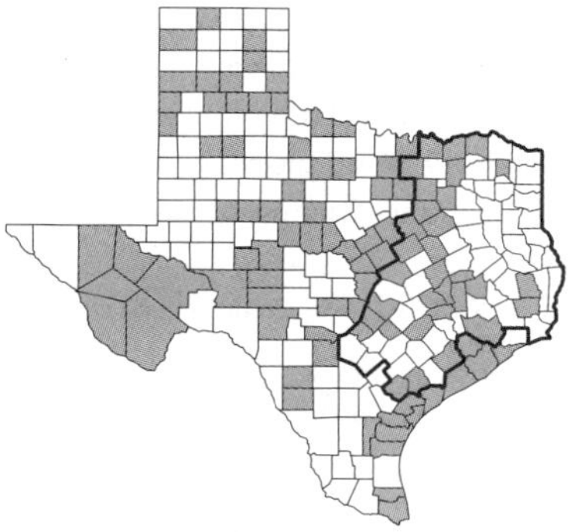

Amphiachyris dracunculoides

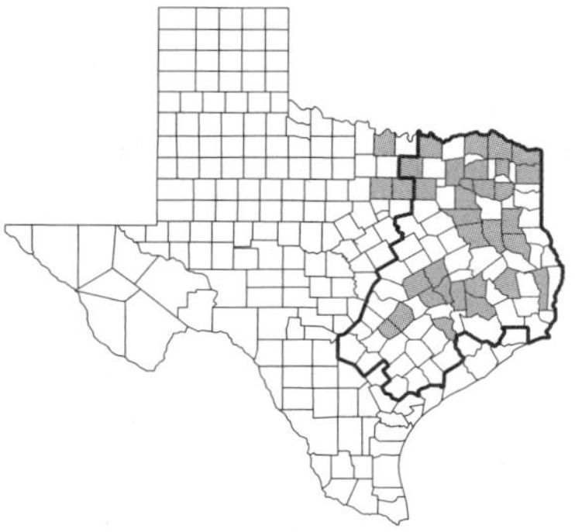

Antennaria parlinii subsp. fallax

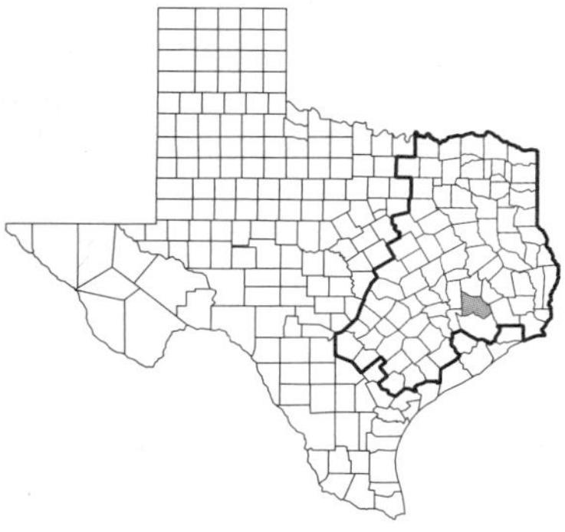

Antennaria parlinii subsp. parlinii

Antennaria plantaginifolia

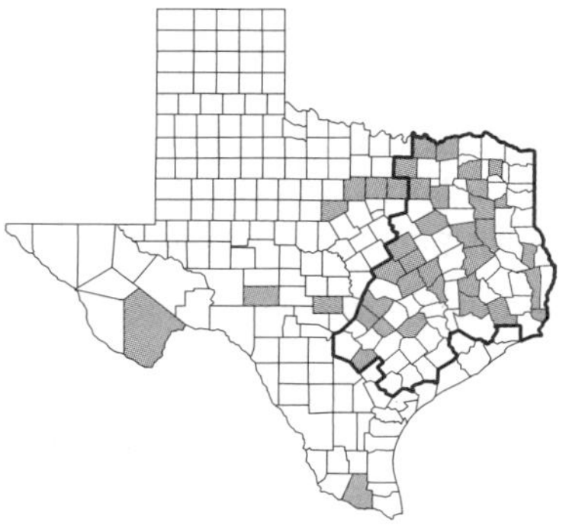

Anthemis cotula

Arnoglossum ovatum (Walter) H. Rob., (for the ovate leaf blades), LANCE-LEAF INDIAN-PLANTAIN. Plants 50–100(–300) cm tall, weakly rhizomatous; stems rounded; basal leaves to 30+ cm long, long-petioled, the blades ovate to narrowly lanceolate or lance-linear, usually entire, membranous; upper stem leaves smaller and subsessile; phyllaries lance-ovate, pale green to white, sometimes purple-tinged at tips, flattened, the midvein not keeled or winged dorsally; achenes 4–5 mm long; pappus bristles to 7.5 mm long. Sandy soils, prairies, fields, and moist open woods; se TX in Pineywoods and Gulf Prairies; se U.S. from NC w to TX. (May–)Jul–Nov. [*A. ovatum* var. *lanceolatum* (Nutt.) D.B. Ward, *Cacalia elliottii* (Harper) Shinners, *C. lanceolata* Nutt., *C. ovata* Walt.]

Arnoglossum plantagineum Raf., (resembling *Plantago*—plantain), PRAIRIE-PLANTAIN, GROOVE-STEM INDIAN-PLANTAIN, TUBEROUS INDIAN-PLANTAIN. Plants 50–100 cm tall, weakly rhizomatous; stems striate-angled, often dark purple; basal leaves 5–17 cm long, long-petioled, the blades elliptic or ovate, entire or rarely toothed, thick and fleshy; stem leaves relatively few, alternate, much smaller, the upper subsessile; phyllaries ovate, pale green to white, with a narrow, sharp, prominent keel or wing on the midvein dorsally; achenes 4–5 mm long; pappus bristles to 8(–9) mm long. Prairies, fields, roadsides, and moist open woods; se and e TX w to nc TX and sw to Travis and Comal cos.; adjacent cos. on the e border of Edwards Plateau; e 1/2 of North America. Apr–Jun, sometimes late Sep–Nov. if moist enough. [*Cacalia plantaginea* (Raf.) Shinners] The striking leaves and erect flowering stalks make this species a conspicuous spring component of many East TX prairies, fields, and roadsides.

ARTEMISIA L. SAGEBRUSH, WORMWOOD, SAGEWORT, SAGE, MUGWORT

Ours perennial herbs or slightly subshrubby; white-pubescent or nearly glabrous, often aromatic; leaves alternate, entire or deeply lobed, the segments linear to oblong, elliptical, or ovate; heads small, in paniculate arrays; involucres white-woolly or nearly glabrous; phyllaries in 4–7 series, unequal; receptacles naked; ray flowers absent; peripheral disk flowers pistillate, with minute filiform corollas; central disk flowers perfect and fertile or functionally staminate, corollas minute, pale yellowish cream; achenes ca. 1 mm or less long; pappus absent.

A genus of 350–500 species of the n temperate zone, w South America, and s Africa; usually in dry areas. *Artemisia* species are wind-pollinated and cause allergies. The genus contains many aromatic shrubs and herbs that have been employed for culinary, medicinal, and spiritual purposes. *Artemisia tridentata* Nutt. is the famous SAGEBRUSH widespread in the arid intermountain w U.S.; the culinary herb TARRAGON is the Eurasian *A. dracunculus* L.; the WORMWOOD of the Bible is thought to be *A. herba-alba* Asso; the green liqueur absinthe is made from the European *A. absinthium* L.; the Asian *A. annua* L. has been long used in traditional Chinese medicine and is now being used to treat malaria. The common name MUGWORT is from Old English *mucgwyrt*, *mucg* meaning midge, and referring to the use of several Old World species in repelling flies and midges (Shultz 2006). (Ancient name of *Artemisia vulgaris* L., from Greek goddess of the hunt, Artemis, or possibly from Artemisia, wife of Mausolus, king of Caria) (tribe Anthemideae)

REFERENCES: Hall & Clements 1923; McArthur & Welch 1984; Arriagada & Miller 1997; Shultz 2006.

1. Leaf blades nearly glabrous, usually entire; phyllaries glabrous ____ **A. dracunculus**
1. Leaf blades conspicuously grayish or white-woolly, at least on lower surfaces, entire to lobed or dissected; phyllaries gray-tomentose ____ **A. ludoviciana**

Artemisia dracunculus L., (little dragon, for its ancient reputation to cure snakebite), WILD TARRAGON. Rhizomatous perennial or subshrub, 50–120(–150) cm tall, often strongly aromatic; leaves variable, linear to oblong, mostly entire, sometimes irregularly lobed, bright green to olive or gray-green, glabrous to sparsely pubescent; heads in loose corymbose panicles; phyllaries green to tan, glabrous, with hyaline margins; achenes 0.5–0.8 mm long. Desert scrub, fields, roadsides; in East TX recorded only from Comal Co., apparently based on early 19th-century collections by Ferdinand Lindheimer, possibly from cultivated plants; Trans-Pecos and Panhandle; w 1/2 of North America; Mexico; Eurasia. Aug–Sep. [*A. aromatica* A. Nelson, *A. dracunculoides* Pursh, *A. dracunculoides* Pursh var. *dracunculina* (S. Watson) S.F. Blake, *A. glauca* Pall. ex Willd.] Widely cultivated as a culinary herb; likely introduced in parts of its range.

Artemisia ludoviciana Nutt., (of Louisiana), WHITE SAGEBRUSH. Perennial rhizomatous herb, rarely slightly subshrubby, 20–80(–120) cm tall, aromatic; leaves variable in shape, usually some lobed or dissected but varying to entire, also variable in terms of pubescence, the young leaves white-woolly above and beneath, older leaves glabrate and dark green above and white-woolly beneath; heads in compact to diffuse panicles; phyllaries densely gray-tomentose; achenes ca. 0.5 mm long. Prairies, roadsides, disturbed habitats; widespread in TX but more frequent in c and w TX; throughout North America; Mexico. Aug–Dec. [*A. albula* Wooton, *A. gnaphalodes* Nutt., *A. ludoviciana* Nutt. var. *mexicana* Willd. ex Spreng., *A. mexicana* Willd., *A. neomexicana* Greene ex Rydb., *A. redolens* A. Gray] We are following Turner (2016) in recognizing *A. ludoviciana* as a single, widespread, highly variable species without meaningful subspecific units.

Antennaria parlinii subsp. fallax [HEA]

Antennaria parlinii subsp. parlinii [FOV]

Antennaria plantaginifolia [USE]

Anthemis cotula [REE]

Aphanostephus pilosus [BT3]

ASTRANTHIUM Nutt. WESTERN DAISY

A genus of 12 species endemic to the s U.S. and Mexico. (Greek, *astron*, star, and *anthos*, flower) (tribe Astereae)
REFERENCES: Larsen 1933; DeJong 1965; Nesom 2005b, 2006f.

Astranthium ciliatum (Raf.) G.L. Nesom, (ciliate, fringed), WESTERN DAISY, COMANCHE WESTERN DAISY, BLUE DAISY. Low pubescent annuals from a taproot; stems usually 1, 10–40(–50) cm tall; leaves alternate, the lowest with narrow, petiolar bases, the upper sessile, blades oblanceolate, entire; heads solitary at the ends of long peduncles, usually several per main stem; phyllaries subequal in 2–3 series; receptacle rounded-conical, naked; ray flowers (7–)13–24, perfect, fertile, the ligules white to bluish white or pinkish white, not curling, (4–)6–10(–12) mm long; disk flowers perfect, fertile, yellow; achenes 1–1.6 mm long; pappus absent or a minute ring or crown of scales. Open woods, prairies, and roadsides, in sandy or silty clay soils; se and e TX w to Rolling Plains and Edwards Plateau; AR, KS, MO, OK, and TX; n Mexico. Mar–May. [*A. integrifolium* var. *ciliatum* (Raf.) Larsen, *A. integrifolium* (Michx.) Nutt. subsp. *ciliatum* (Raf.) DeJong, *A. integrifolium* var. *triflorum* (Raf.) Shinners, *Bellis ciliata* Raf.]

BACCHARIS L. GROUNDSELTREE

Shrubs or subshrubs, dioecious, pistillate and staminate plants with a somewhat differing appearance; leaves alternate, subulate to obovate, entire, serrate, or dentate, 1- or 3-nerved, glandular or punctate glandular; heads in paniculate or corymbose arrays; involucres hemispheric to narrowly cylindrical; phyllaries strongly graduated in several series; receptacles naked; ray flowers absent; staminate heads with corollas funnelform, white to yellowish brown; pistillate heads with corollas filiform, yellowish white to brown; achenes 5–10-ribbed, yellow to reddish; pappus of numerous white to brownish bristles, those of pistillate heads generally elongating and exceeding phyllaries at maturity.

An American genus of ca. 400 species of dioecious shrubs; some are used medicinally, while others are cultivated as ornamentals. Pistillate plants are necessary for definitive identification. ☠ The leaves and flowers of a number of species contain cardioactive glycosides and are considered dangerous, even fatal, to livestock (Hardin & Brownie 1993) (Name derived from Bacchus, Roman god of wine, allusion obscure and likely used originally for a different plant) (tribe Astereae)
REFERENCES: Mahler 1955; Mahler & Waterfall 1964; Nesom 1990d; Sundberg & Bogler 2006.

1. Leaf blades linear to narrowly lanceolate, ca. 1–2 mm wide; achenes 6–8-ribbed, 3–5 mm long, slightly to prominently glandular-scabrous; pistillate pappus of many series, light reddish brown; plants < 1 m tall ____ **B. texana**
1. Leaf blades linear to oblanceolate, elliptic, rhomboid or obovate, usually > 2 mm wide (often much more); achenes 8–10-ribbed, 1–2 mm long, glabrous; pistillate pappus in 1–2 series, whitish; plants usually 1–3(–6) m tall (but can flower at 0.5 m tall).
   2. Leaf blades elliptic to rhomboid or obovate, the larger ones 22–37 mm wide, the upper leaves gradually reduced, narrower; pistillate involucres 3–5 mm long ____ **B. halimifolia**
   2. Leaf blades very narrowly elliptic to linear or oblanceolate, 2–8(–15) mm wide; pistillate involucres 4–8 mm long.
      3. Pistillate involucres to 5 mm long; leaves linear to very narrowly elliptic, usually 2–4(–5) mm wide; widespread in East TX ____ **B. neglecta**
      3. Pistillate involucres 5–9 mm long; leaves oblanceolate, usually 4–8(–15) mm wide; uncommon in East TX ____ **B. salicina**

Baccharis halimifolia L., (with leaves like *Halimium* of the Cistaceae), EASTERN BACCHARIS, SEA-MYRTLE, CONSUMPTION-WEED, TREE-GROUNDSEL, GROUNDSELTREE, MANGLIER, SALTBUSH, SILVERLING. Shrubs 1–6 m tall; leaves petiolate, blades rhombic to oblong, gland-dotted and resinous, with 1 prominent nerve and 2 lateral nerves, bases cuneate, margins entire in lower half, the upper half entire or with few to several teeth; achenes 1–1.7 mm long; pistillate pappus 8–12 mm long. Wet fields, edges of swamps and marshes, along creeks in woods; widespread in e 1/3 of TX; e and se U.S. from MA to FL, w to OK and TX; n Mexico and West Indies; introduced elsewhere. Jul–Nov. Salt tolerant. In the w part of its distribution it appears to hybridize with *B. neglecta*, making distinction between these species difficult (Turner 2016). This species is a serious invasive pest in Tridens Prairie and similar areas in Lamar Co. Potentially toxic to livestock (Duncan et al. 1957; Burlage 1968). ☠

Baccharis neglecta Britton, (neglected, overlooked), ROOSEVELT-WEED, NEW DEAL WEED, JARA DULCE. Shrubs 1–4 m tall; leaves sessile or short petiolate, blades linear to narrowly elliptic, gland-dotted, 1-nerved with lateral nerves obscure, margins entire or finely serrate with few teeth; achenes ca. 1.2 mm long; pistillate pappus 7–12 mm long. Calcareous soils and disturbed areas; Blackland Prairie and Post Oak Savannah, widely distributed from nc to s TX and w to Trans-Pecos; OK and TX; n Mexico. Aug–Nov. Several common names for this species are in reference to its abundance in fallow fields during the Great Depression of the 1930s (Turner 2016).

Aphanostephus ramosissimus var. ramosissimus [FNA]

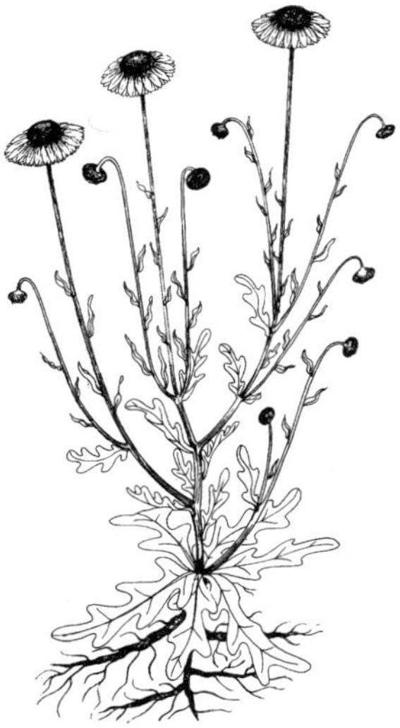

Aphanostephus riddellii [JON]

Aphanostephus skirrhobasis [BT3]

Arnoglossum ovatum [GWO]

Arnoglossum pantagineum [MO4]

Artemisia dracunculus [EN2]

Artemisia ludoviciana [HOL]

Astranthium ciliatum [SM1, STE]

Baccharis halimifolia [GLE]

Baccharis salicina Torr. & A. Gray, (resembling *Salix*—willow), SEEP-WILLOW, WATER-WILLOW, WATER-WALLY, JARA, WILLOW BACCHARIS. Shrubs 1–3 m tall; leaves nearly sessile, blades oblong to oblanceolate, gland-dotted, the wider ones 3-nerved, margins usually coarsely serrate, sometimes entire; achenes 1.2–2 mm long; pistillate pappus 8–12 mm long. Alluvial, often sandy or saline soils; uncommon in East TX, Bell, Rusk, and Walker cos. (TAMU), Brazos Co. (TAES); mainly Trans-Pecos and Plains Country, sw U.S. from CA e to KS and TX; n Mexico. Aug–Nov.

Baccharis texana (Torr. & A. Gray) A. Gray, (of Texas), PRAIRIE BACCHARIS. Rhizomatous subshrubs or shrubs 0.25–0.6 m tall; leaves sessile, leaf blades linear to narrowly lanceolate, gland-dotted, 1-nerved, margins minutely undulate; achenes 3–5 mm long; pistillate pappus 11–13 mm long. Calcareous soils, prairies; Bell, Bexar, Goliad, Grimes, and McLennan cos., also Dallas Co., but not collected there since 1900 (Mahler 1988); widely distributed from nc to s TX; OK and NM; n Mexico. Aug–Oct. [*Linosyris texana* Torr. & A. Gray]

BERLANDIERA DC. GREENEYES

Pubescent, herbaceous to suffrutescent perennials; leaves alternate, the margins toothed to lyrate-pinnatifid; heads 1–several in corymbose clusters; phyllaries in 2–3 series, broadly ovate, herbaceous; receptacles chaffy, pales each folded around a disk flower and apically widened and green; ray flowers pistillate, fertile, with ligules yellow to orange-yellow with conspicuous green or red to maroon veins on lower surface; disk flowers staminate, in ours reddish to maroon; achenes broadly obovate, thin-edged, each adhering to and shed with subtending phyllary; pappus absent or inconspicuous.

A genus of 7 species native to the s U.S. and Mexico. The disk flowers are subtended by pales, their tips giving the receptacle a green appearance, hence the common name GREENEYES. (Named for Jean Louis Berlandier, 1805–1851, French botanist who explored TX, NM, and Mexico, and one of the first botanists to make extensive collections in TX). (tribe Heliantheae)
REFERENCES: Turner & Johnston 1956; Pinkava 1967, 2006; Nesom & Turner 1998.

1. Ligules of ray flowers with veins on lower (= abaxial) surface red to maroon; at least some leaves lyrate-pinnatifid; peduncles scabrous to subscabrous; stems arising from a persistent basal rosette; rare in East TX ____ **B. lyrata**
1. Ligules of ray flowers with veins on lower surface green; leaves variously toothed, not lyrate-pinnatifid; peduncles with spreading or matted hairs; stems not arising from a persistent basal rosette; scattered or common in East TX.
   2. Middle stem leaves usually with evident petioles; peduncles with matted hairs; stems not densely leafy, the internodes often 3 cm or more long ____ **B. pumila**
   2. Middle stem leaves sessile or with inconspicuous short petioles, peduncles with ± spreading hairs; stems very densely leafy, with most internodes 3 cm or less long ____ **B. betonicifolia**

Berlandiera betonicifolia (Hook.) Small, (with leaves like *Betonica*—betony, now = *Stachys* in Lamiaceae), TEXAS GREENEYES. Plants weakly suffrutescent; stems to 100 cm tall, producing a new crop of flowering branches from summit after spring blooming period; leaves evenly distributed or crowded near summit; central stem leaves sessile or short-petioled, blades ovate, stiffly hirsute to subscabrous or loosely hairy but not velvety, margins toothed; ray flowers with ligules 10.5–17 mm long, veins on lower surface green; achenes 4.5–5.5 mm long, 3–4 mm wide. Sandy soils, open post oak woods; patchy distribution across TX; Blackland Prairie in Dallas, Fannin, Grayson, and Lamar cos.; Pineywoods in Shelby and San Augustine cos., and w border of Edwards Plateau from Travis and Bexar to Goliad and Victoria cos.; sc U.S. from NM to MO to LA. Jun–Sep. [*B. texana* DC.] We are following Nesom and Turner (1998) for nomenclature of this species. The name *B. betonicifolia* must be used since the type collection of *B. betonicifolia* (named in 1835) is of the same species as *B. texana* (named in 1836) (Nesom & Turner 1998). Plants from se TX which had previously gone under the name *B.* ×*betonicifolia* (Hook.) Small are now being recognized as *B. pumila*.

Berlandiera lyrata Benth., (lyre-shaped), LYRE-LEAF GREENEYES, BERLANDIER'S DAISY, CHOCOLATE DAISY, GREEN-EYED LYRE-LEAF. Plants herbaceous; stems to 120 cm tall; leaves crowded toward base of plant in a persistent basal rosette, blades velvety, lyrate to pinnatifid; ray flowers with ligules 10–14 mm long, sometimes the entire lower surface red to maroon; achenes 4.5–6 mm long, 2.7–3.7 mm wide. Dry rocky limestone areas, roadsides; rare in East TX; Denton Co. and Grayson Co. (Kartesz 2015), mostly High Plains and Trans-Pecos; sw U.S. from CO and AZ e to KS, OK, and TX; n Mexico. Nearly throughout the growing season. [*B. incisa* Torr. & A. Gray] The flowers have a chocolate-like aroma (Kirkpatrick 1992).

Berlandiera pumila (Michx.) Nutt., (dwarf), SOFT GREENEYES. Plants herbaceous to suffrutescent; stems to 70 cm tall; central stem leaves often with long petioles, blades velvety, oblong to ovate; ray flowers with ligules 12–20 mm long, veins on lower surface green; achenes 4.5–6 mm long, 3–4 mm wide. Roadsides, open wooded areas, often in sandy soils; throughout e and sc TX; se U.S. from SC, GA, and FL w to LA and TX. Apr–Oct. [*B. dealbata* (Torr. & A. Gray) Small, *B. tomentosa* (Pursh) Nutt.]

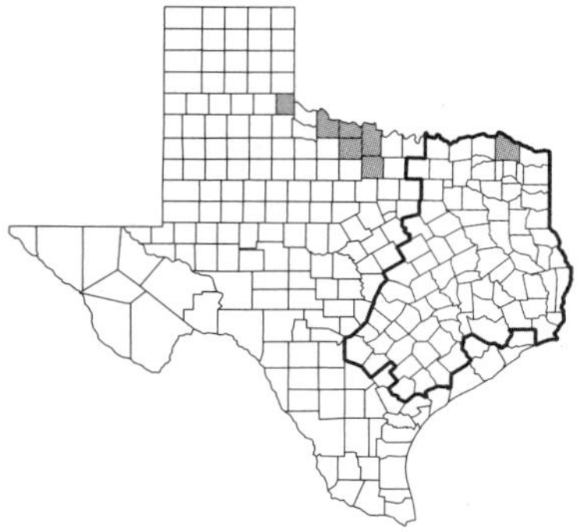

Aphanostephus pilosus

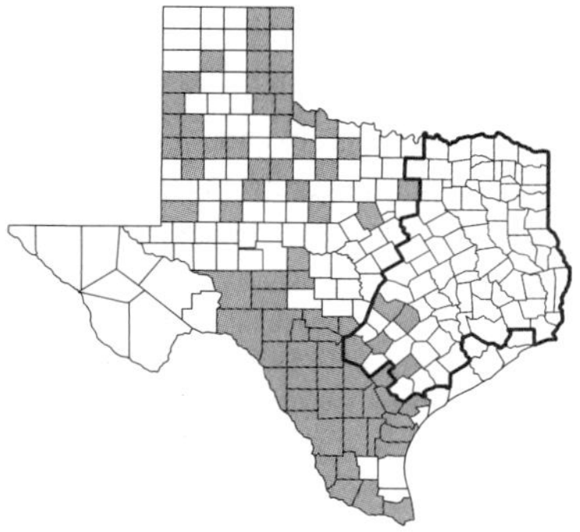

Aphanostephus ramosissimus var. ramosissimus

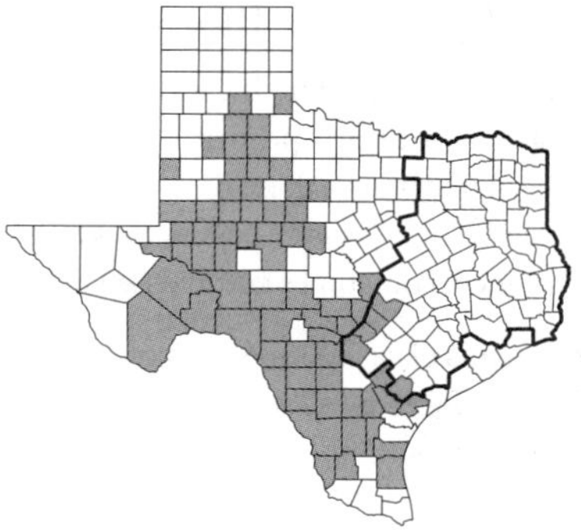

Aphanostephus riddellii

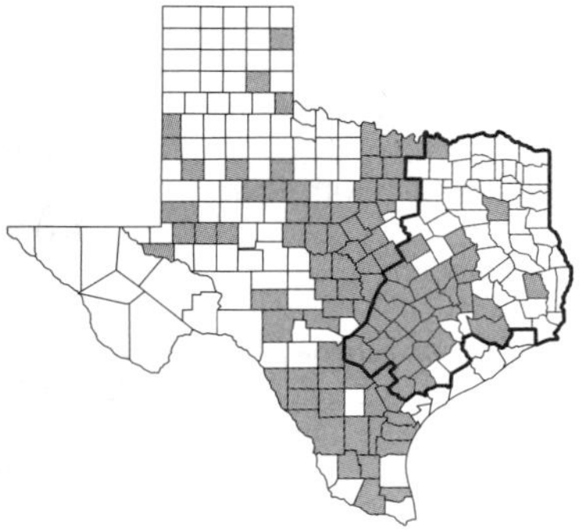

Aphanostephus skirrhobasis var. skirrhobasis

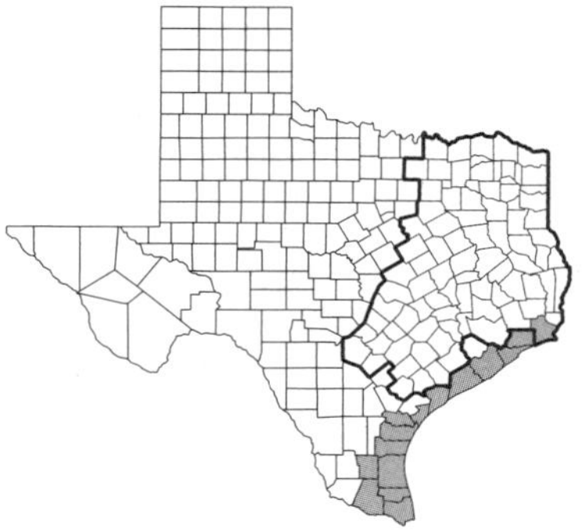

Aphanostephus skirrhobasis var. thalassius

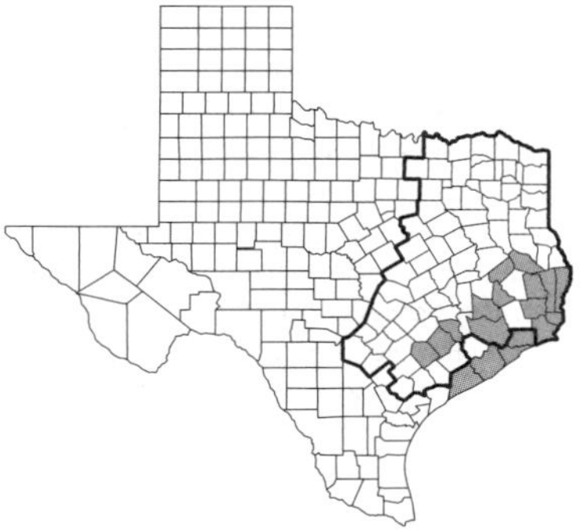

Arnoglossum ovatum

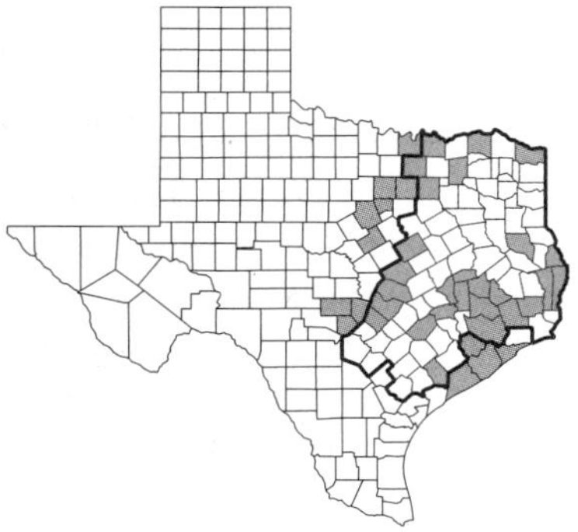

Arnoglossum plantagineum

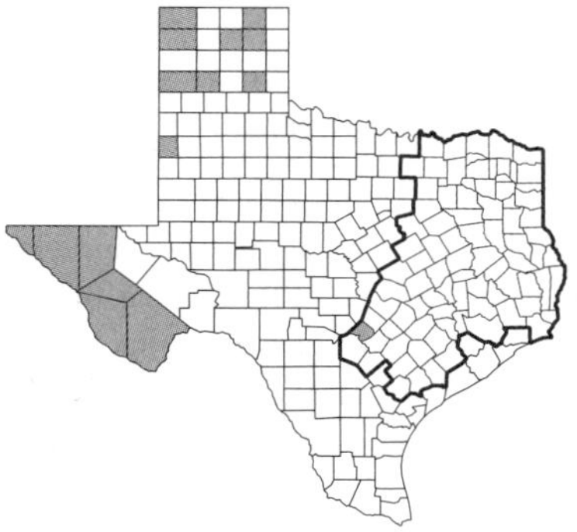

Artemisia dracunculus

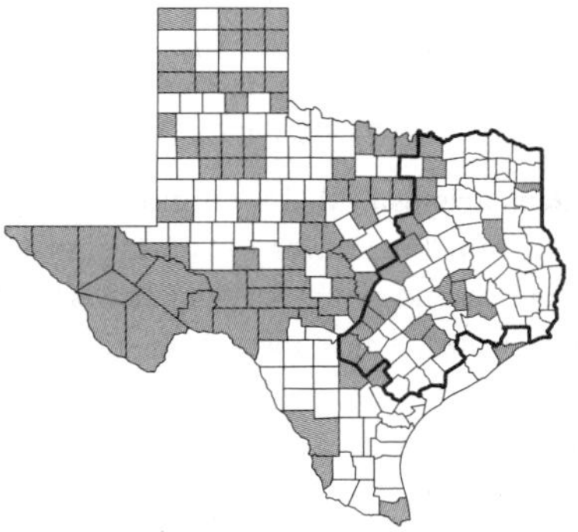

Artemisia ludoviciana

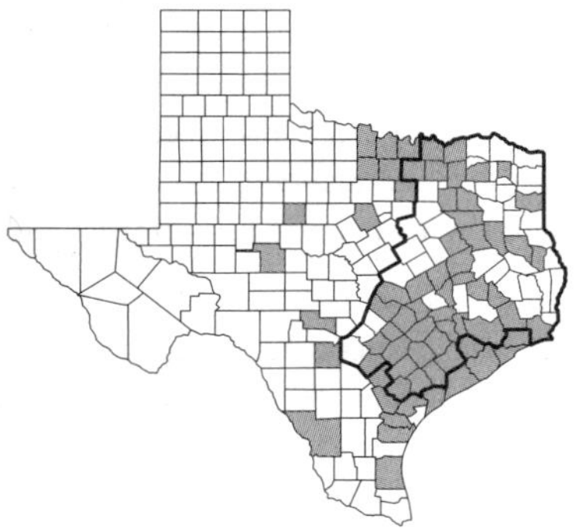

Astranthium ciliatum

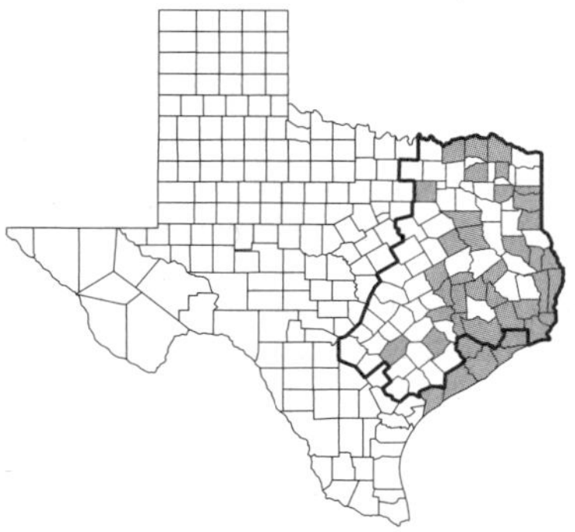

Baccharis halimifolia

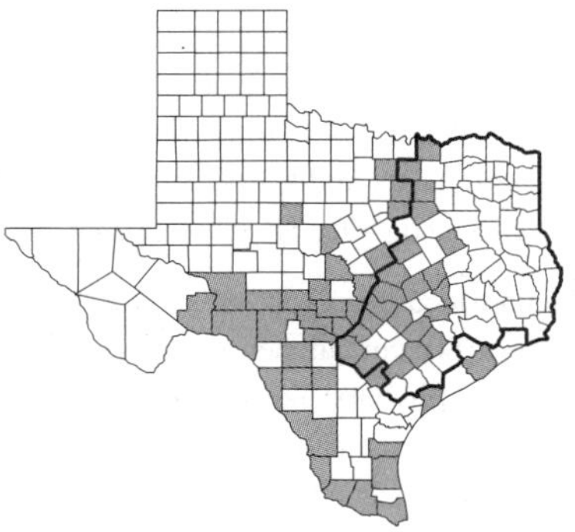

Baccharis neglecta

BIDENS L. BEGGAR-TICKS, BUR-MARIGOLD, TICKSEED-SUNFLOWER

Annual or perennial herbs; leaves opposite (upper sometimes alternate), simple, divided, or compound; heads borne singly or in corymbose arrays; primary (inner) phyllaries in 2–3 series, hyaline- or yellow-margined, subtended by a series of secondary, phyllary-like bracts, these narrower, herbaceous; receptacles chaffy; ray flowers neuter or pistillate and infertile, or rarely absent, corollas in ours usually yellow or whitish (rarely pinkish); disk flowers perfect, fertile, yellow or orange-yellow; disk achenes 3–4 angled, sometimes compressed or flattened; pappus of (1–)2–4 usually barbed awns or sometimes reduced or absent.

A genus of ca. 200 species, cosmopolitan, but especially speciose in Mexico; a few are cultivated ornamentals while some are considered weeds. Achenes with retrorsely-barbed pappus awns are dispersed by attaching to hair or clothing. (Latin: *bis*, twice, and *dens*, a tooth, in allusion to the two awns on the achenes of some species) (tribe Heliantheae)

REFERENCES: Sherff 1937; Sherff & Alexander 1955; Hall 1967; Lipscomb & Smith 1977; Ballard 1986; Tadesse et al. 1995, 2001; Strother & Weedon 2006.

1. Midstem leaves sessile, simple, the margins dentate-serrate to nearly entire ____ **B. laevis**
1. Midstem leaves petiolate, variously dissected or compound, not simple, variously margined.
   2. Central achenes equally 4-angled, linear-fusiform, widest in the middle.
      3. Midstem leaves usually 2–3 times pinnately dissected or compound with numerous segments or leaflets; ray flowers when present with ligules yellowish to whitish, 1–3 mm long ____ **B. bipinnata**
      3. Midstem leaves once-pinnately lobed with 3–7 segments or leaflets; ray flowers when present with ligules white (rarely pinkish or yellowish), 2–15+ mm long ____ **B. pilosa**
   2. Central achenes flattened or unequally 4-angled, wedge-shaped, widest in the upper 1/3 portion.
      4. Ray flowers absent, or if present, with ligules < 4 mm long.
         5. Ray flowers present or absent; secondary phyllary-like bracts with margins usually ciliate; achenes 7–10 mm long; pappus awns 2–5 mm long ____ **B. frondosa**
         5. Ray flowers absent; secondary phyllary-like bracts with margins seldom ciliate; achenes 4–6+ mm long; pappus awns usually 1–2.4 mm long ____ **B. discoidea**
      4. Ray flowers with conspicuous ligules 10+ mm long.
         6. Central achenes 3.5–5 mm long, the margins not barbed or ciliate and usually not winged ____ **B. mitis**
         6. Central achenes 5–8 mm long, the margins usually barbed or ciliate and often ± corky-winged.
            7. Secondary phyllary-like bracts usually 8–12, shorter than the primary phyllaries; central achenes 5–7 mm long; pappus awns when present usually 2–4 mm long ____ **B. aristosa**
            7. Secondary phyllary-like bracts 12–21+, usually longer than the primary phyllaries; central achenes 7–8 mm long; pappus awns or scales when present usually 0.1–0.5 mm long ____ **B. polylepis**

Bidens aristosa (Michx.) Britton, (bearded), BEARDED BEGGAR-TICKS, TICKSEED-SUNFLOWER. Annuals, 30–60(–150) cm tall; petioles 10–30 mm long; leaves 1–2 times pinnately lobed, the segments linear to lanceolate or narrowly ovate, the margins laciniate to serrate; secondary phyllary-like bracts 8–12+, linear, (4–)5–7(–12) mm long, margins entire, ciliate; primary phyllaries 7–8(–13), lance-ovate to lance-linear, 6–8(–12) mm long; ray flowers usually 8–10, ligules 10–25 mm long, golden yellow; central achenes flattened or unequally 3–4-angled, wedge-shaped, 5–7 mm long, the margins barbed and sometimes corky-winged; pappus awns absent or 2(–4), 2–4 mm long, antrorsely or retrorsely barbed. Low moist areas; e 1/4 TX; also Travis Co.; e 1/2 of North America. (Jun–)Sep–Nov.

Bidens bipinnata L., (twice-pinnate), SPANISH-NEEDLES. Annuals, 30–100(–150+) cm tall; petioles 20–50 mm long; leaves 2–3 times pinnately dissected or compound with numerous segments, the segments obovate or lanceolate, the ultimate margins entire; secondary phyllary-like bracts 7–10, linear to lanceolate, 3–5 mm long, margins ciliate; primary phyllaries 8–12, lanceolate to linear, 4–6 mm long; ray flowers absent or 3–5+, ligules 1–3 mm long, yellowish or whitish; central achenes equally 4-angled, linear-fusiform, 12–18 mm long; pappus of (2–)3–4 awns, 2–4 mm long, retrorsely barbed. In moist soils, disturbed sites; se and e TX w to nc TX and Edwards Plateau; rare in rest of TX; throughout the U.S. except northerly, mountainous states; introduced elsewhere. Aug–Oct. [*B. bipinnata* var. *biternatoides* Sherff] Possibly native to e Asia; a weed in many warm parts of the world.

Bidens discoidea (Torr. & A. Gray) Britton, (discoid, for its rayless heads), SMALL BEGGAR-TICKS, RAYLESS BEGGAR-TICKS. Annuals, 20–60(–180) cm tall; petioles 10–40(–60) mm long; leaves 3-foliolate, the leaflets lance-ovate to lanceolate, the margins serrate; secondary phyllary-like bracts usually 3–5, spatulate to linear, (3–)12–25 mm long, margins seldom ciliate; primary phyllaries 5–7, oblong to lanceolate, 4–6+ mm long; ray flowers absent; central achenes flattened, narrowly wedge-shaped, 4–6+ mm long; pappus of 2 erect awns, usually 1–2.4 mm long, antrorsely barbed or smooth. In permanently moist soils; e and se TX, infrequent; e 1/2 of North America. Aug–Oct.

Baccharis neglecta [BB2]

Baccharis salicina [RYD]

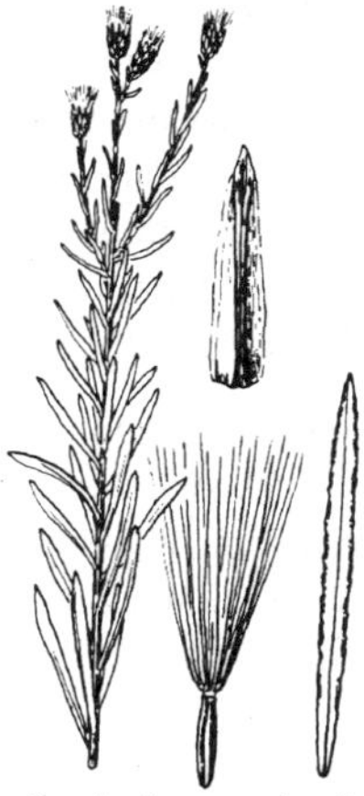

Baccharis texana [POW]

Berlandiera betonicifolia [DEL]

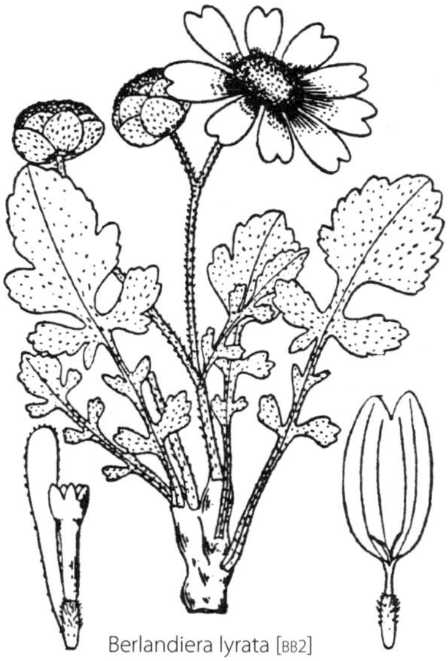

Berlandiera lyrata [BB2]

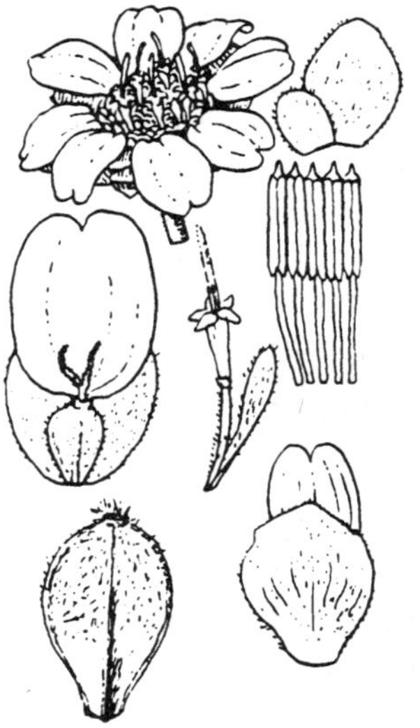

Berlandiera pumila [SM1]

Bidens aristosa [MO4]

Bidens bipinnata [REE]

Bidens frondosa L., (leafy), BEGGAR-TICKS, STICKTIGHTS, DEVIL'S BEGGAR-TICKS. Annuals, 20–60(–180) cm tall; petioles 10–40(–60) mm long; leaves once-pinnately or ternately compound, the 3–5 leaflets lanceolate to lance-ovate, the margins dentate to serrate; secondary phyllary-like bracts 5–10, linear-spatulate, sometimes leaf-like, often very long (to 60 mm), margins ciliate; primary phyllaries 6–12, oblong or ovate to lance-ovate, shorter than the outer phyllaries, 5–9 mm long; ray flowers absent or 1–3+, ligules 2–3.5 mm long, golden yellow; central achenes flattened, wedge-shaped, 7–10 mm long, the margins barbed; pappus of 2 awns, 2–5 mm long, antrorsely or retrorsely barbed. Moist areas; se and e TX w to Rolling Plains and Edwards Plateau; also Panhandle and Trans-Pecos; throughout the U.S.; Mexico; introduced elsewhere. Aug–Nov. A weed in many parts of the world.

Bidens laevis (L.) Britton, Sterns & Poggenb., (smooth), SMOOTH BEGGAR-TICKS, WILD GOLDENGLOW. Annuals or perennials, 20–60 (–120+) cm tall; leaves sessile, simple, linear to lanceolate to obovate, the margins usually coarsely dentate to serrate; heads erect at flowering, sometimes nodding in fruit; secondary phyllary-like bracts 5–9+, linear-lanceolate to oblanceolate, often leaf-like, (6–)10–12(–20+) mm long, margins usually ciliate; primary phyllaries 8–12, obovate to lance-oblong, (4–)6–8(–10+) mm long, tips often orange to purplish; ray flowers 7–8 (rarely absent), ligules (10–)15–25(–30) mm long, orange-yellow; central achenes flattened or unequally 3–4-angled, wedge-shaped, 6–10 mm long, the margins ciliate or barbed; pappus of 2–4 awns, 3–5 mm long, retrorsely barbed. In wet areas, margins of ponds and streams; scattered across TX but most frequent in East TX w to nc TX and Edwards Plateau; New England and s 1/2 of U.S. from ME w to CA; Mexico to South America; introduced elsewhere. (May–)Jun–Nov. [*B. elegans* Greene, *B. nashii* Small, *Helianthus laevis* L.] A weed in many parts of the world.

Bidens mitis (Michx.) Sherff, (mild or gentle), SMALLFRUIT BEGGAR-TICKS. Annuals or perennials, 30–100+ cm tall; petioles 5–20(–30+) mm long; leaves once-pinnately lobed or dissected, the segments lance-ovate to linear, the margins serrate or entire; secondary phyllary-like bracts 7–10, linear to spatulate, 5–10 mm long, margins usually ciliate; primary phyllaries 6–8+, lance-ovate to lanceolate, 4–5 mm long; ray flowers 8–13, ligules 12–25+ mm long, golden yellow; central achenes flattened or unequally 3–4-angled, wedge-shaped, 3.5–5 mm long; pappus absent or of 2 awns or scales, 0.5–1 mm long. In wet areas, marshes; rare in TX; recorded only from Newton (LSU) and Trinity (Kartesz 2015) cos.; se U.S. from NJ s to FL and w to MO and TX. Sep–Nov. [*Coreopsis mitis* Michx.]

Bidens pilosa L. (hairy), HAIRY BEGGAR-TICKS, BLACK-JACK, SPANISH-NEEDLES. Annuals or perennials, 30–60(–180+) cm tall; petioles 10–30(–70) mm long; leaves ovate to lanceolate or 3-foliolate or once-pinnately lobed, the segments or leaflets ovate to lanceolate; the ultimate margins serrate or nearly entire; secondary phyllary-like bracts 6–9(–13+), linear to spatulate, 3–5 mm long, margins ciliate; primary phyllaries 7–13, lanceolate to oblanceolate, 4–6 mm long; ray flowers absent or 3–8+, ligules 2–15+ mm long, white (rarely pinkish or yellowish); central achenes equally 4-angled, linear-fusiform, 7–16 mm long; pappus absent or of 2–3(–5) awns, (0.5–)2–4 mm long, retrorsely barbed. Moist areas, disturbed sites; Montgomery (Brown & Elsik 2002 (as *B. alba*)) and Harris cos. in se TX; also s TX and Trans-Pecos; scattered across North America in mostly coastal states; Mexico and Central America; introduced elsewhere. All seasons, but mostly summer–fall. [*B. alba* (L.) DC., *B. odorata* Cav.] We are following Strother and Weedon's (2006) broader circumscription of this species to include Ballard's (1986) rayless or short-rayed *B. pilosa*, and the white or pinkish-rayed *B. alba* and *B. odorata*. Native of Mexico; now a widespread weed.

Bidens polylepis S.F. Blake, (many scales), AWNLESS BEGGAR-TICKS, TICKSEED-SUNFLOWER. Annuals or biennials, 30–100+ cm tall; petioles 5–30 mm long; leaves pinnately lobed or laciniate-pinnatifid, the segments linear to lanceolate, the margins incised to dentate to serrate; secondary phyllary-like bracts 12–21+, linear, (6–)8–12(–20) mm long, margins ciliate; primary phyllaries 6–8+, lance-ovate to lanceolate, 4–7 mm long; ray flowers ca. 8, ligules 10–25+ mm long, golden yellow; central achenes flattened, wedge-shaped, 7–8 mm long, the margins barbed or ciliate and ± corky-winged; pappus absent or of 2 retrorsely-barbed awns or scales, 0.1–0.5(–1.5) mm long. Low moist areas; e 1/4 TX; e 1/2 of North America. May–Jul. [*B. polylepis* var. *retrorsa* Sherff] Similar to and sometimes included in *B. aristosa* (Michx.) Britton.

Bidens cernua L., (nodding), NODDING BEGGAR-TICKS, which would key to *B. laevis* here, can be distinguished by the sessile, simple leaves (rarely 3-foliolate), heads sometimes nodding in flower, usually nodding in fruit, 6–8 orange-yellow ray ligules 2–15 (–18) mm long (rays sometimes absent), and achenes with thickened or winged margins. While widespread in wet habitats across North America, this species is only known in TX from a single county in the Panhandle; it has been recorded near East TX in Caddo Par., LA, and several counties in sw AR, and might be expected in ne TX.

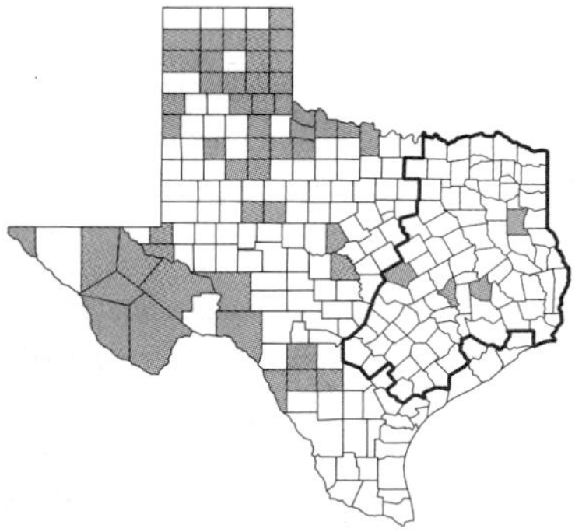

Baccharis salicina

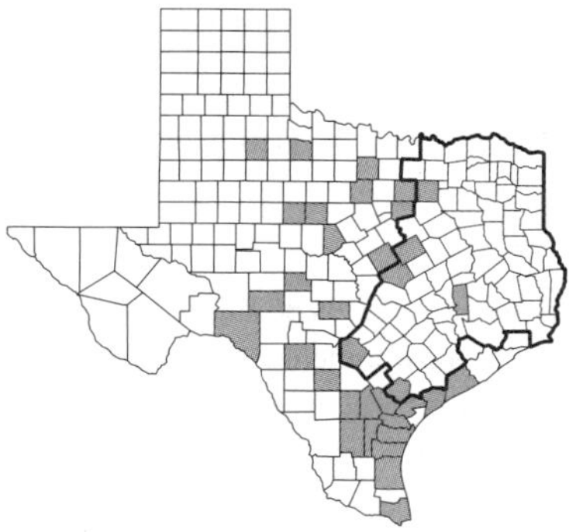

Baccharis texana

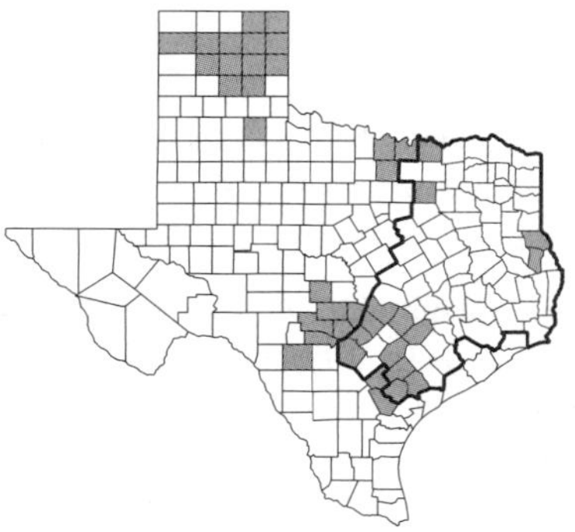

Berlandiera betonicifolia

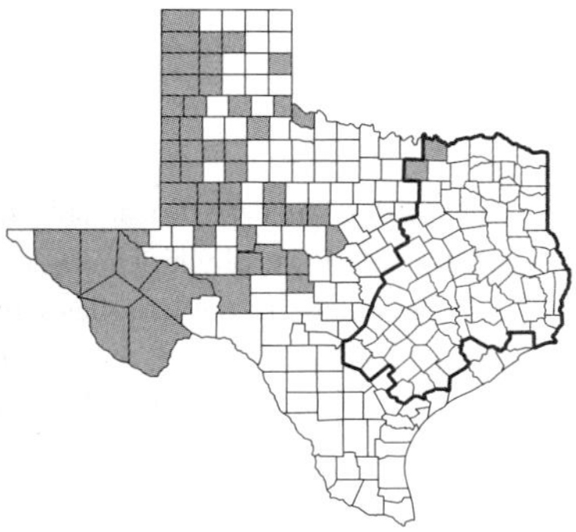

Berlandiera lyrata

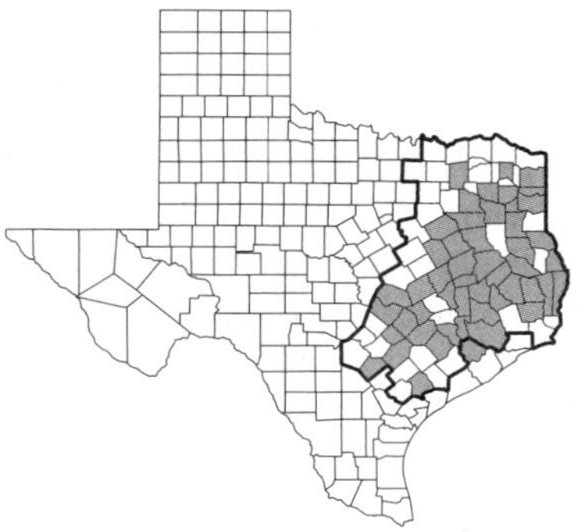

Berlandiera pumila

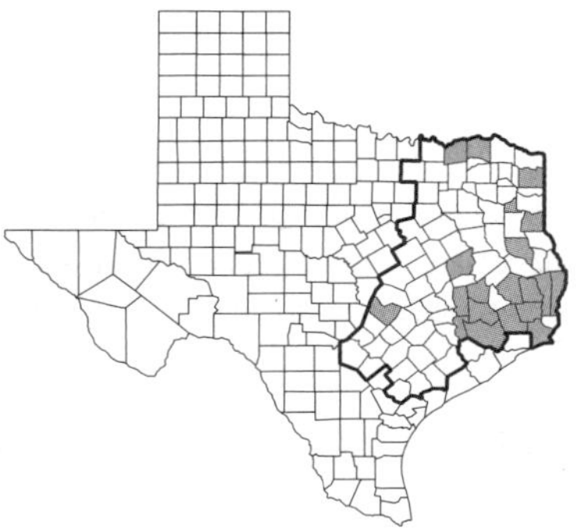

Bidens aristosa

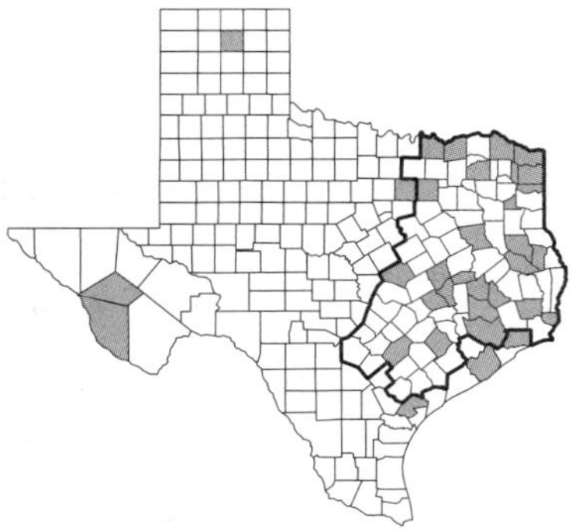

Bidens bipinnata

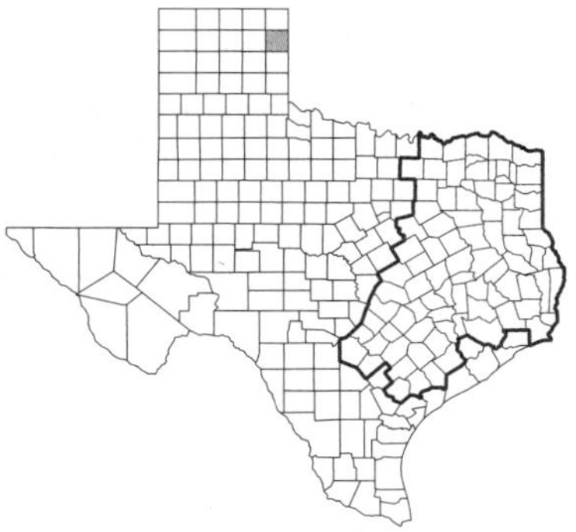

Bidens cernua

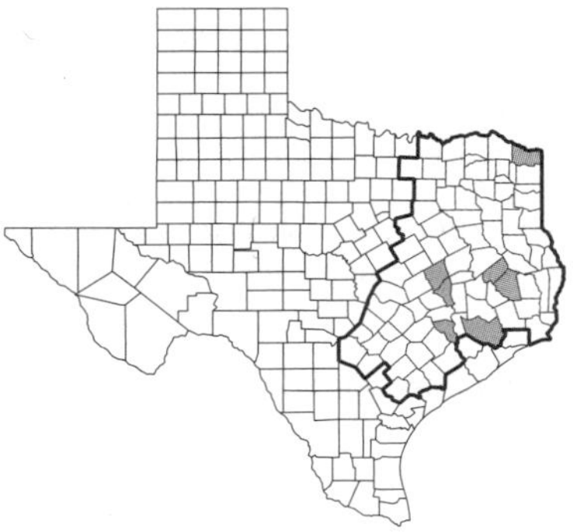

Bidens discoidea

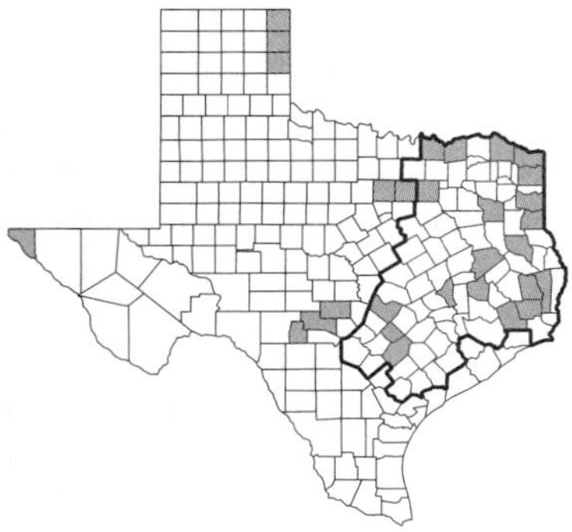

Bidens frondosa

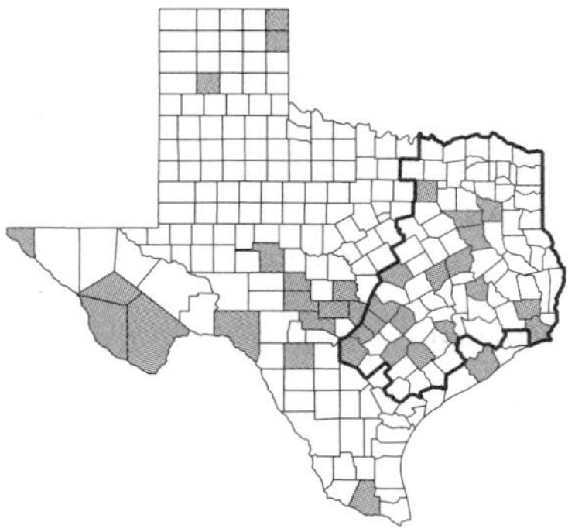

Bidens laevis

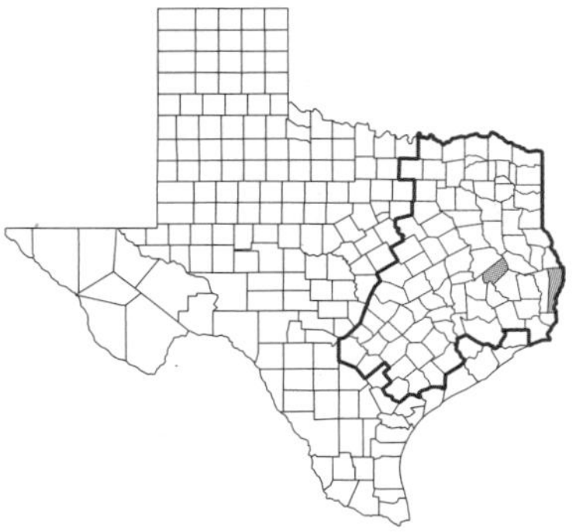

Bidens mitis

BIGELOWIA DC. RAYLESS-GOLDENROD

A genus of 2 species native to the se U.S. (Named for Dr. Jacob Bigelow, 1787–1879, Boston physician who assisted George Engelmann on the U.S.–Mexican boundary mission and collected plant specimens regularly while visiting his patients in the Boston area on horseback) (tribe Astereae)
REFERENCES: Anderson 1970, 1977; Nesom 2006g.

Bigelowia nuttallii L.C. Anderson, (for Sir Thomas Nuttall, 1786–1859, English-American botanist), SLENDER BIGELOWIA, NUTTALL'S RAYLESS-GOLDENROD. Perennials with rhizome-like caudex branches, growing in colonies or clumps; stems rigidly erect, 30–70 cm tall, unbranched except in head-bearing region; leaves basal in persistent rosettes, 1–2 mm wide, alternate, sessile, linear, entire, often resinous, bases persistent and sheathing lowest branches; heads crowded in a corymbose arrangement; involucres 4.5–9 mm long; phyllaries in 3–4 series, unequal, narrowly lanceolate to linear; receptacles naked; ray flowers absent; disk flowers 4–6 per head; corollas ca. 4.5 mm long, yellow; achenes with stiff antrorse hairs; pappus of numerous bristles, ca. 3.8 mm long. Sandy soils, sandstone outcrops; se TX; se U.S. from GA and FL w to TX. Aug–Nov. [*B. virgata* of authors, not (Nutt.) DC]. Despite previous erroneous reports, the other species in the genus, *B. nudata* (Michx.) DC., is not found in TX (G.L. Nesom, pers. comm. 2016).

BOLTONIA L'Hér. DOLL'S DAISY

Perennial fibrous-rooted herbs, stoloniferous, stiffly erect, ± glabrous; leaves alternate, sessile; blades 1-nerved, linear, subulate to lanceolate or oblanceolate; heads small, borne in diffusely-branched paniculate or corymbose arrays; phyllaries 30–55 in several series, linear to spatulate, unequal to subequal; receptacles naked; ray flowers 20–60, pistillate, fertile, corollas white (sometimes pink upon drying) to lilac; disk flowers perfect, fertile, corollas yellow; ray achenes 3-sided, 3-ribbed, the margins narrowly winged; disk achenes obovoid to wedge-shaped, laterally flattened, the margins winged or not; pappus usually of 2–3 awns plus sometimes a ring of shorter bristles or scales.

A genus of 5 species of c and e North America and e Asia; some are cultivated as ornamentals. Turner (2016) states the distinctions between the two species treated here are nebulous at best, there being a large series of intergradations between the characters that supposedly distinguish them. (Named for James Bolton, 1758–99, English botanist) (tribe Astereae)
REFERENCES: Fernald 1940b; Morgan 1966; Karaman-Castro & Urbatsch 2006.

1. Heads usually in corymbose arrays (creating a somewhat flat-topped appearance, with many heads in the same horizontal plane); upper stem leaves leaf-like, usually > 20 mm long, not reduced to bracts; ray ligules 5–13 mm long ________ **B. asteroides**
1. Heads usually in paniculate arrays (creating a more pyramidal appearance, without many heads in the same horizontal plane); upper stem leaves reduced to bracts, not leaf-like, usually < 20 mm long; ray ligules 3–6.3 mm long ________ **B. diffusa**

Boltonia asteroides (L.) L'Hér. var. **asteroides,** (*Aster*-like), WHITE BOLTONIA, LARGE-FLOWER DOLL'S DAISY. Plants 30–100(–200) cm tall; leaves entire; upper stem leaves leaf-like, only reduced to bracts on peduncles; heads in corymbose arrays; involucres 2.4–3.8 mm high, 3.7–8.7 mm wide; phyllaries in 3–5 series, subulate to linear, 0.3–1 mm wide; ray flowers 20–60, ligules 5–13 mm long; achenes 1–3 mm long, obovoid, wings 0.1–0.5 mm wide; pappus awns 0.4–1.2 mm long. Wet prairies, pond margins, woodland borders; rare in TX, known only from Lamar Co. (TEX), Brazos and Madison cos. (TAMU), and Newton Co. (Kartesz 2015); e and s coastal states from NY s to FL and w to TX. Jul–Oct. [*B. asteroides* (L.) L'Hér. var. *glastifolia* (Hill) Fernald]

Boltonia diffusa Elliott var. **diffusa,** (diffuse, spreading), SMALL-HEAD BOLTONIA, DOLL'S DAISY. Plants 30–100(–200) cm tall; leaves entire, upper stem leaves reduced to bracts, not leaf-like; heads in paniculate arrays; involucres 2.5–3 mm high, 2.5–6 mm wide; phyllaries in 4–6 series, subulate, 0.2–0.7 mm wide; ray flowers 20–40, ligules 3–6.3 mm long; achenes 1.2–2.5 mm long, obovoid, wings 0.2–0.4 mm wide; pappus awns 0.3–0.7 mm long. Disturbed areas, pond margins, roadside ditches, moist sand; e and se TX; se and sc U.S. from VA and IL s to FL and TX. Jul–Oct.

BORRICHIA Adans. SEA OX-EYE DAISY

A genus of 2 mostly coastal species of the se U.S., Mexico, and West Indies. (Named for Ole Borch (Olaus Borrichius), 1626–1690, Danish botanist) (tribe Heliantheae)
REFERENCES: Semple 1978, 2006a.

Borrichia frutescens (L.) DC., (shrubby), SEA OX-EYE DAISY, BUSHY SEASIDE TANSY. Perennial bushy herbs or subshrubs to 150 cm tall; sap sticky; leaves opposite, often leathery or subsucculent; petioles or blade bases usually with 1–2+ spine-tipped teeth

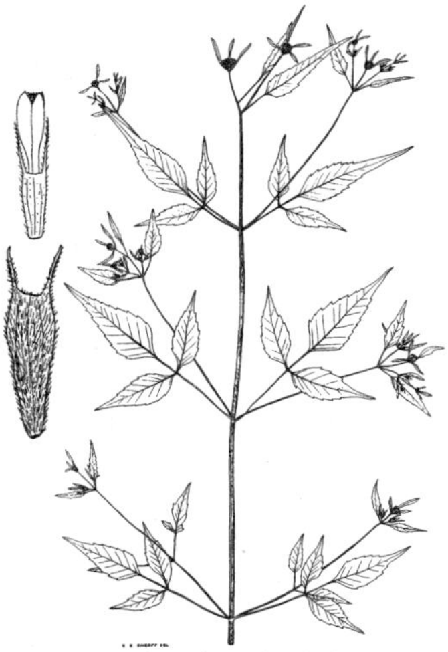

Bidens discoidea [FIE]

Bidens frondosa [REE]

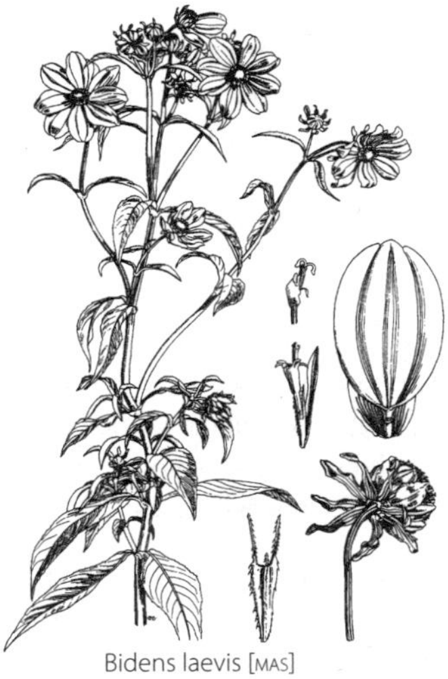

Bidens laevis [MAS]

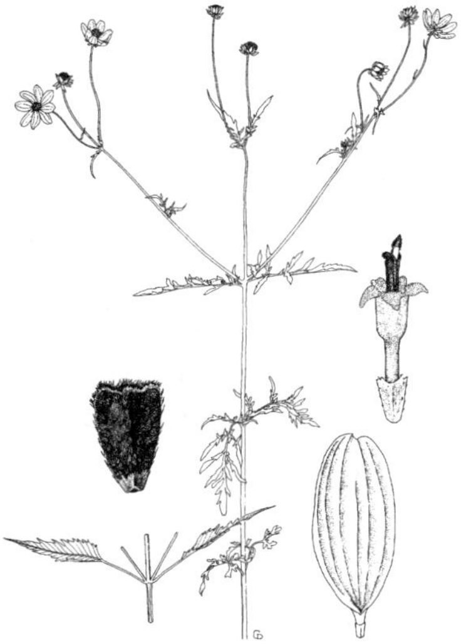

Bidens mitis [CO1, GWO]

Bidens pilosa [YAT]

Bidens polylepis [FAS]

Bigelowia nuttallii [HEA]

Boltonia asteroides [GEO]

blades elliptic to obovate or oblanceolate, usually ± pubescent, margins dentate to serrate in the upper 1/2; heads borne singly; phyllaries 20–40 in 3–4 series, spine-tipped, reflexed and swollen at maturity; receptacles chaffy, pales spine-tipped; ray flowers 15–30, pistillate, fertile, corollas yellow, ligules 6–8 mm long; disk flowers perfect, fertile, corollas yellow; achenes 2.5–4(–5) mm long, 3–4-angled; pappus a short crown of scales. Saline flats and marshes, brackish backwaters, rocky limestone; DeWitt, Goliad, Gonzales, and Harris cos.; coastal TX and lower Rio Grande Valley; se coastal U.S. from MD s to FL and w to TX; Mexico and West Indies. Nearly throughout the growing season.

BRADBURIA Torr. & A. Gray GOLDEN-ASTER

Taprooted annual or perennial herbs, 15–80 cm tall, sparsely pilose; basal leaves petiolate, stem leaves sessile, leaves alternate, hispid to pilose; heads borne singly or in loose paniculate arrays; peduncles with stalked glands; phyllaries in 3–5 series, strongly unequal; receptacles naked; ray flowers pistillate, fertile, corollas yellow; disk flowers perfect and fertile or functionally staminate, corollas yellow; achenes smooth or slightly ribbed, short-strigose; pappus in 2–3 series, the inner series of bristles, the outer series of bristles or small scales.

A genus of 2 species native to the se U.S. *Bradburia* species have sometimes been treated in *Chrysopsis* or *Heterotheca*; generic boundaries for the GOLDEN-ASTERS have long been problematic. (Named for John Bradbury (1768–1823), English naturalist and collector in Missouri Territory) (tribe Astereae)
REFERENCES: Shinners 1951b; Harms 1968, 1974; Semple 1977, 1981, 1996, 2006b; Semple et al. 1980; Semple & Chinnappa 1984; Nesom 1991b, 1991c, 1997c.

1. Midstem leaves 3–8 mm wide; ray flowers with a pappus of barbellate bristles 1–3 mm long, the outer series shorter than and grading into the inner series; disk flowers 11–25, functionally staminate (not developing fertile achenes) ____ **B. hirtella**
1. Midstem leaves 10–25 mm wide; ray flowers with a pappus in 2 distinct series, the outer series of flat scales 0.5–1.1 mm long, and the inner series of barbellate bristles 5–6 mm long; disk flowers 25–60, perfect and fertile (developing achenes) ____ **B. pilosa**

Bradburia hirtella Torr. & A. Gray, (hirtellous, somewhat hairy), TEXAS GOLDEN-ASTER, MAUCHIA. Annuals, rarely perennials; basal leaves usually withered before flowering; midstem leaves linear to linear-oblanceolate, 3–8 mm wide, hispid-pilose with stalked glands, entire or obscurely dentate; ray flowers (7–)10(–14); disk flowers 11–25, functionally staminate and infertile; ray pappus of 2–3 series of bristles, the outer slightly shorter. Open, disturbed areas, usually on sand or gravel, roadsides; e and se TX; most common in the Post Oak Savannah of sc TX; LA and TX. Apr–Oct. [*Chrysopsis texana* G.L. Nesom]

Bradburia pilosa (Nutt.) Semple, (pilose, with long soft hairs), SOFT GOLDEN-ASTER. Annuals; basal leaves oblanceolate; midstem leaves linear-elliptic, 10–25 mm wide, short-pilose, sparsely glandular, entire or dentate; ray flowers (11–)16(–24); disk flowers 25–60, perfect and fertile; ray and disk pappus similar, the outer series of minute, flat scales, and the inner series of long, barbellate bristles. Sandy woods, old fields, and roadsides; se and e TX; East Cross Timbers and South TX Plains; se U.S. from VA s to FL and w to KS and TX. Summer–fall. [*Chrysopsis pilosa* Nutt., *C. nuttallii* Britton, *Heterotheca pilosa* (Nutt.) Shinners]

BRICKELLIA Elliott BRICKELLBUSH, FALSE BONESET

Ours perennial herbs or small shrubs; stems solitary to several, erect; leaves simple; leaf blades resin-dotted, 1- or 3-nerved from the base, reticulate-veined beneath; involucres cylindrical; phyllaries in several series, the outer much shorter than the inner, linear or lanceolate to oblong, ± chartaceous; receptacles naked; ray flowers absent; achenes 10-ribbed, cylindrical; pappus of bristles.

A New World genus of 100 species native from w U.S., Mexico, and Central America to Argentina; some are cultivated as ornamentals. Several taxa were previously treated in *Kuhnia*. (Named for Dr. John Brickell, 1749–1809, of Savannah, GA, amateur botanist and helpful correspondent of Muhlenberg, Fraser, and others) (tribe Eupatorieae)
REFERENCES: Robinson 1917; Shinners 1946d, 1971; King & Robinson 1987; Turner 1989; Scott 2006.

1. Pappus bristles plumose, with conspicuous side branches many times longer than the width of the central bristle axis (use lens); heads in ± hemispherical or flat-topped corymbose arrays; widespread in TX on many soil types ____ **B. eupatorioides**
1. Pappus bristles scabrous or barbellate, not feathery, the side branches ca. as long as the width of the central bristle axis (use lens); heads usually in ± elongated racemose or paniculate arrays; generally restricted to c TX and Edwards Plateau on limestone and gravel.
   2. Woody subshrubs with sharply-ascending branches, the stems becoming white and corky with age; leaves along upper portions of primary stems mostly alternate; achenes glabrous or nearly so ____ **B. dentata**

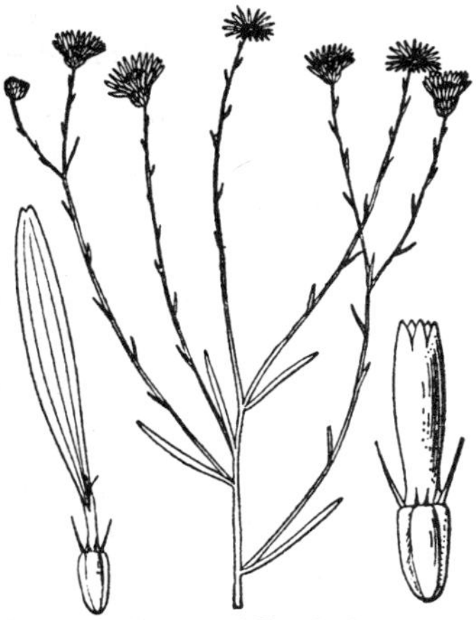

Boltonia diffusa [BB2]

Borrichia frutescens [CO1]

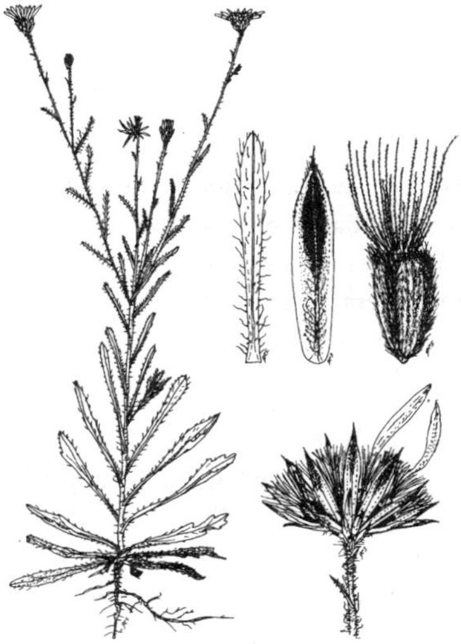

Bradburia hirtella [SEM]

Bradburia pilosa [GLE]

Brickellia cylindracea [MGH]

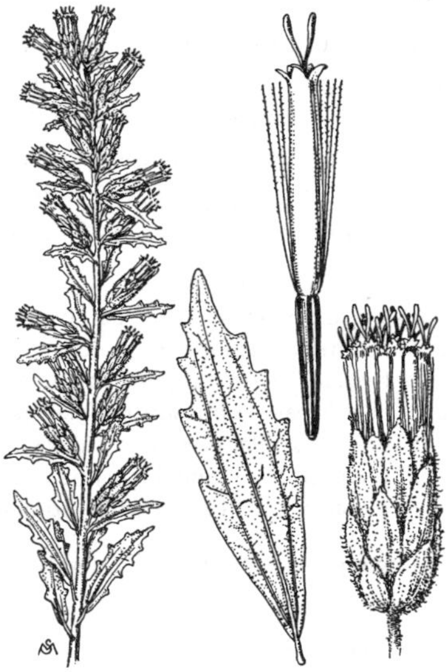

Brickellia dentata [MGH]

Brickellia eupatorioides var. corymbulosa [BB2]

Brickellia eupatorioides var. eupatorioides [GLE]

2. Sprawling or shrubby herbs, woody only at the base, the stems not corky-white; leaves along upper portions of primary stems mostly opposite; achenes pubescent ______ **B. cylindracea**

Brickellia cylindracea A. Gray & Engelm., (for the cylindrical heads), GRAVEL-BAR BRICKELLBUSH. Herbaceous perennials from a woody base, to 120 cm tall; leaves mostly opposite, sessile or with petioles to 3 mm long, blades lanceolate to obovate, densely tomentose and gland-dotted beneath, margins crenate to serrate; heads with peduncles 2–15 mm long, in elongated paniculate arrays; phyllaries often purplish-tinged; disk corollas cream to pale yellow; achenes 3.5–5 mm long, pubescent; pappus of 30–35 white to tawny bristles, usually barbellate. Various habitats, but often on rocky limestone soils; sw border of East TX from McLennan and Bell cos. s to Comal and Bexar cos.; c TX to Edwards Plateau. Jun–Nov. [*B. cylindracea* var. *laxa* A. Gray] This species exhibits notable variability in leaf size, shape, texture, and arrangement of heads (Correll & Johnston 1970; Scott 2006).

Brickellia dentata (DC.) Sch. Bip., (toothed), LEAFY BRICKELLBUSH. Small shrubs to 120 cm tall; stems branched from the base and strictly erect, corky-white and peeling in age; leaves alternate, petioles 1–2 mm long, blades lanceolate to lance-ovate, gland-dotted beneath, margins usually coarsely dentate, sometimes entire; heads nearly sessile with peduncles 1–2 mm long, in elongated paniculate arrays; phyllaries greenish or straw-colored; disk corollas pale green or creamy yellow; achenes 3.5–4 mm long, usually glabrous; pappus of 30–35 tawny bristles, usually barbellate, sometimes subplumose. Calcareous soils, gravel of limestone streambeds; sw border of East TX in Williamson, Travis, Hays, and Comal cos.; Edwards Plateau. Oct–Nov. Records from Trinity Co. in se TX likely represent labeling errors (Mahler 1981; Carr 2009; Turner 2016).

Brickellia eupatorioides (L.) Shinners, (resembling *Eupatorium*—BONESET), FALSE BONESET, PLAINS KUHNIA, PRAIRIE KUHNIA. Herbaceous perennials from a woody base, 30–150+ cm tall; leaves opposite or alternate, sessile or with petioles to 1 cm long, blades linear to lanceolate to broadly rhombic-lanceolate, resin-dotted, margins entire to dentate; heads with peduncles 5–100 mm long, in rounded or flat-topped corymbose arrays; disk flowers cream to yellow-green or pinkish to purplish; achenes 2.7–5.5 mm long; pappus of 20–28 white to tawny bristles, usually ± plumose. [*Kuhnia eupatorioides* L.] Most of the varieties in TX appear to intergrade along their overlapping ranges (Turner 2016).

1. Midstem leaves narrowly linear, < 1.5 mm wide; restricted to Edwards Plateau stream beds ______ **var. gracillima**
1. Midstem leaves linear-lanceolate to narrowly ovate, 2–40 mm wide; not restricted to Edwards Plateau.
    2. Midstem leaves alternate; middle and outer phyllaries long-acuminate, usually with conspicuous sickle-shaped or twisted, filiform tips; middle and outer phyllaries usually more than 1/2 as long as inner ones ______ **var. texana**
    2. Midstem leaves opposite; middle and outer phyllaries acute or acuminate, tips not sickle-shaped or contorted; middle and outer phyllaries usually < 1/2 as long as inner ones.
        3. Involucres 8–15 mm high; heads with 15–35 flowers; mostly c and nc TX and Panhandle ______ **var. corymbulosa**
        3. Involucres 7–11 mm high; heads with 6–15 flowers; mostly East TX ______ **var. eupatorioides**

var. **corymbulosa** (Torr. & A. Gray) Shinners, (with flowers in small corymbs), PLAINS KUHNIA. Silty-clay soils, grasslands, open post oak woodland; nw border of East TX in Fannin, Dallas, and Kaufman cos.; nc TX s to Edwards Plateau and w to Plains and Panhandle; c U.S. from MI and OH w to MT and s to NM, TX, and AR; Canada. Jul–Nov. [*Kuhnia eupatorioides* L. var. *corymbulosa* Torr. & A. Gray]

var. **eupatorioides.** FALSE BONESET. Sandy and silty soils, open pine-oak woods; e 1/3 of TX from Lamar and Red River cos. s to Harris Co. and w to Milam, Bastrop, and Gonzales cos.; e and se U.S. from NJ and PA s to FL and w to IL, MO, and TX. Aug–Nov. [*Kuhnia glutinosa* Elliott]

var. **gracillima** (A. Gray) B.L. Turner, (slenderest, for the narrow leaves). Dry stream beds; sw border of East TX in Travis and Comal cos.; Edwards Plateau. Sep–Nov. [*B. leptophylla* (Scheele) Shinners, *Kuhnia eupatorioides* var. *gracillima* A. Gray, *K. leptophylla* Scheele]

var. **texana** (Shinners) Shinners, (of Texas), PRAIRIE KUHNIA. Calcareous soils, prairies; w border of East TX from Red River Area (Grayson to Lamar cos.) s to Bexar and Fayette cos.; nc and c TX to Edwards Plateau; sc U.S. prairies from IL s to AR and w to KS and TX. Sep–Nov. [*B. eupatorioides* var. *ozarkana* (Shinners) Shinners, *Kuhnia eupatorioides* var. *ozarkana* Shinners, *K. eupatorioides* var. *texana* Shinners]

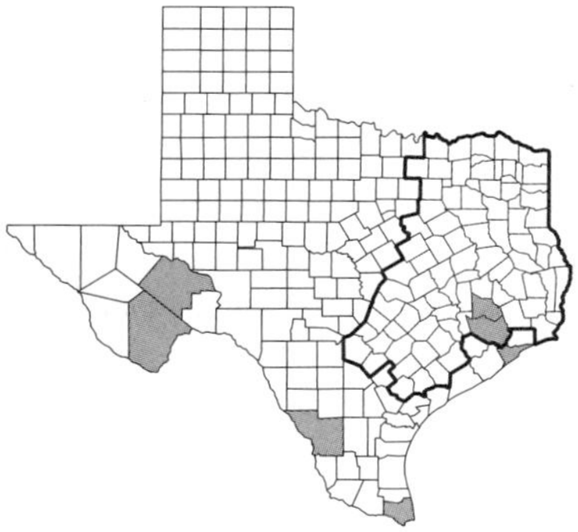
Bidens pilosa

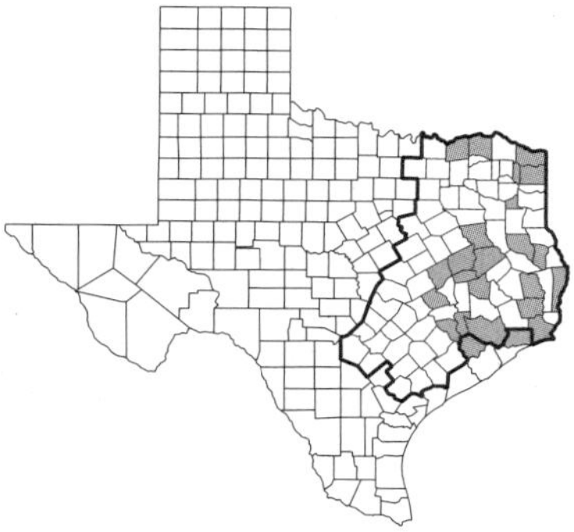
Bidens polylepis

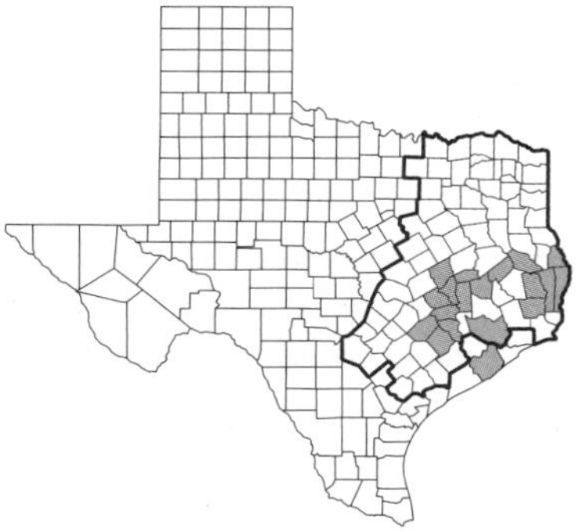
Bigelowia nuttallii

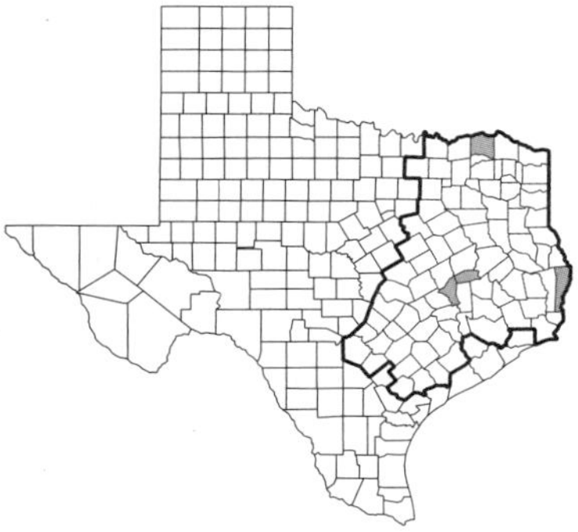
Boltonia asteroides

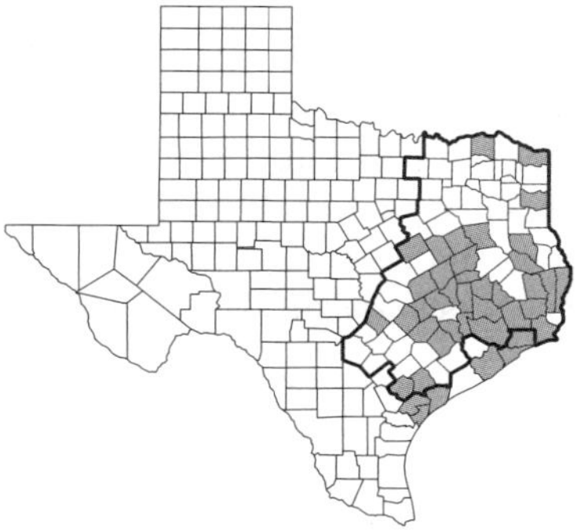
Boltonia diffusa var. diffusa

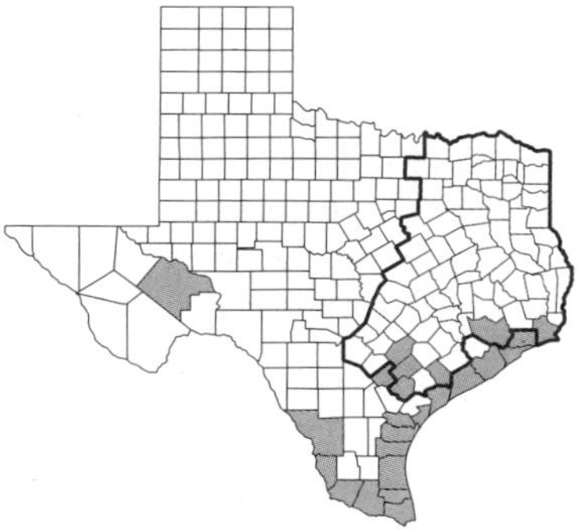
Borrichia frutescens

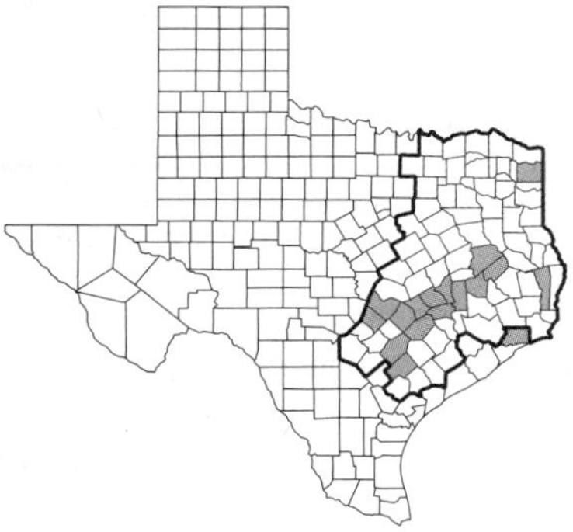
Bradburia hirtella

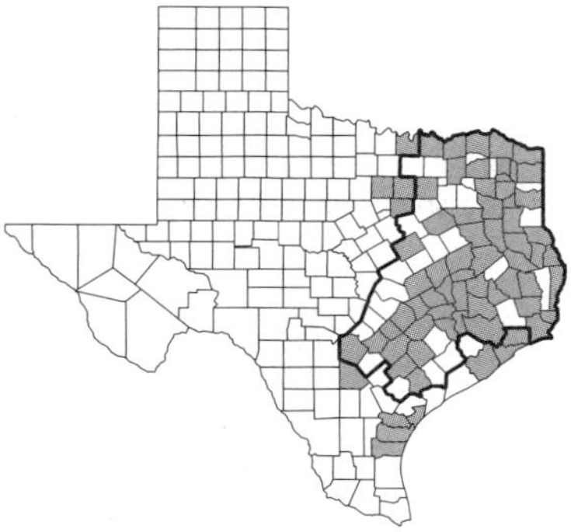
Bradburia pilosa

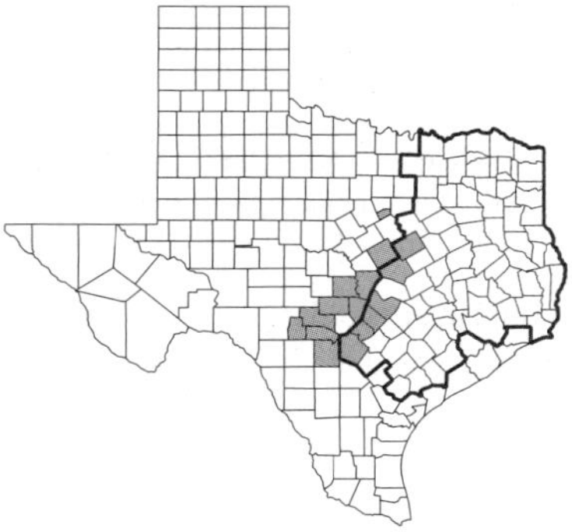
Brickellia cylindracea

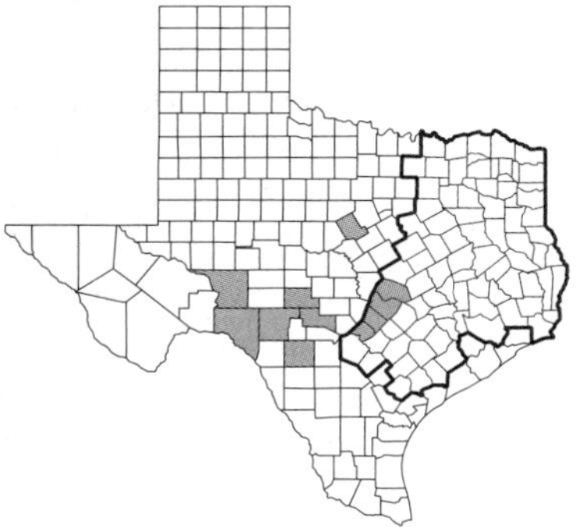
Brickellia dentata

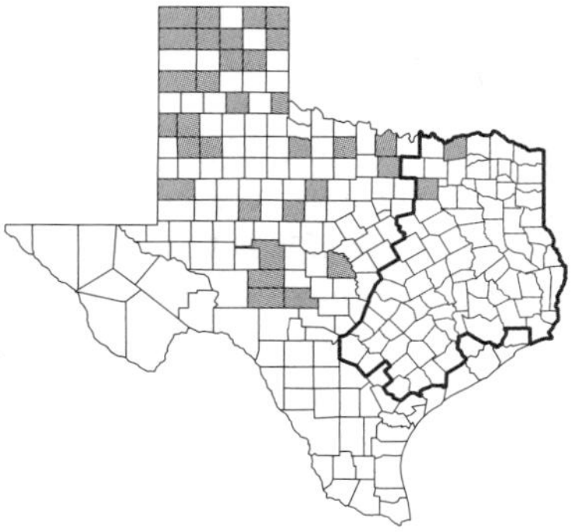
Brickellia eupatorioides var. corymbulosa

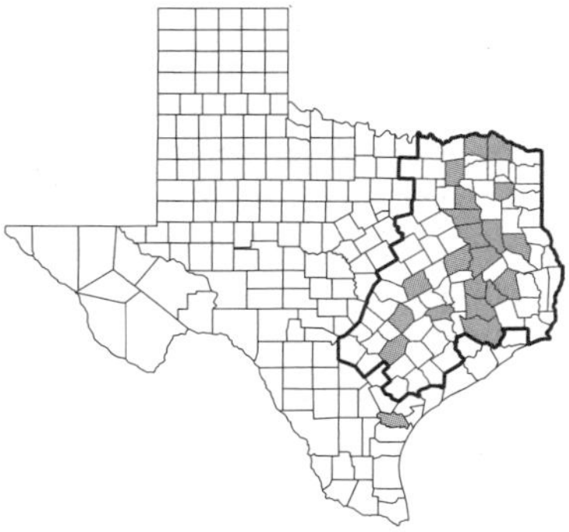
Brickellia eupatorioides var. eupatorioides

CALYPTOCARPUS Less. HORSE HERB

A genus of 2 species native from TX to Guatemala. (Greek: *kalypto*, covered or wrapped around, and *karpos*, fruit, for the achenes' location in the involucral cup) (tribe Heliantheae)
REFERENCES: McVaugh & Smith 1967; Strother 2006d.

Calyptocarpus vialis Less., (of the path or roadside), PROSTRATE LAWNFLOWER, HIERBA DEL CABALLO, HORSE HERB, STRAGGLER DAISY. Small, trailing, pubescent, perennial herbs 5–30 cm tall; leaves opposite, petiolate, blades ovate to ovate-lanceolate or deltoid, 1–3(–4) cm long, margins toothed; heads small, short-peduncled, solitary in leaf axils; phyllaries 4–5, subequal in length, obtuse, overlapping laterally; receptacles chaffy; ray flowers 3–8, pistillate, fertile, corollas yellow, ligules 2–5 mm long; disk flowers perfect, fertile; corollas yellow or yellow-orange; achenes obpyramidal, 2–4 mm long; pappus of 2 persistent awns and sometimes also 1–3 rudimentary awns. Lawn weed; common throughout most of TX except for the Plains Country; s U.S. from SC and FL w to AZ, also IL; Mexico and Central America; spreading elsewhere pantropically. Nearly throughout the growing season. This sometimes troublesome lawn weed tolerates dry shade well and is occasionally purposely planted as a groundcover.

CARDUUS L. THISTLE, PLUMELESS-THISTLE

Ours spiny annual or biennial herbs; leaves alternate, lobed or pinnatifid, with prickly margins, the basal leaves tapering to winged petioles, the stem leaves sessile, shorter, less divided, decurrent as spiny wings along stem; phyllaries many in 7–10+ series, imbricated, the tips spiny; receptacles bristly; ray flowers absent; disk corollas pink to lavender or purple; achenes ovoid, slightly compressed; pappus of numerous hair-like bristles, these not plumose, sometimes barbellate with short projections to ± as long as width of bristle axis, the bristles joined basally and deciduous as a ring.

A genus of ca. 90 species native to Eurasia, the Mediterranean, and n Africa. At least seven species have been introduced to North America and some have become persistent and noxious weeds of pastures and rangelands; populations increase under grazing pressure as more palatable plants are consumed (Desrochers et al. 1988). (The ancient Latin name) (tribe Cardueae)
REFERENCES: McCarty 1978; Desrochers et al. 1988; McGregor 1985a; Keil 2006b.

1. Peduncles 2–30 cm long, unwinged nearest the heads or throughout; heads 20–70 mm wide, nodding, often ± solitary ____ **C. nutans**
1. Peduncles 0–4(–10) cm long, usually winged throughout or only unwinged nearest the heads; heads 7–25 mm wide, erect, often in clusters of 2–5.
   2. Heads sessile or short-pedunculate; peduncles 0–2 cm long, spiny-winged throughout or unwinged nearest the heads; involucres cylindric to ellipsoid, 7–15 mm wide; pappus bristles 15–20 mm long ____ **C. pycnocephalus**
   2. Heads pedunculate; peduncles to 4(–10) cm long, spiny-winged throughout; involucres hemispheric, 10–25 mm wide; pappus bristles 11–13 mm long ____ **C. acanthoides**

Carduus acanthoides L., (spiny), SPINY PLUMELESS THISTLE. Prickly annuals or biennials, 30–200+ cm tall; basal leaves 10–30 cm long, 1–2-pinnately lobed, lobes spine-tipped; heads borne singly or 2–5 in corymbose arrays; peduncles to 4(–10) cm long, spiny-winged throughout; involucres hemispheric, 10–25 mm wide; corollas pink to lavender-purple; achenes 2.5–3 mm long; pappus 11–13 mm long. Waste ground, roadsides, pastures; in East TX recorded only from Collin Co. (*J.G. Saunders 3418* TEX); also Schleicher Co. in Edwards Plateau; across much of North America; Eurasia; introduced elsewhere. May–Jun. Native of s Europe; unknown in TX until 1992. This species is listed as a noxious weed in several states. If subspecies are recognized, ours would be subsp. *acanthoides*.

Carduus nutans L., (nodding), MUSK-THISTLE, NODDING THISTLE. Extremely prickly, stout annuals or biennials, 40–200+ cm tall; basal leaves 10–40 cm long, 1–2-pinnately lobed, lobes spine-tipped; heads borne singly or in corymbose arrays, at least terminal head conspicuously long-pedunculate, often nodding; peduncles 2–30 cm long, unwinged nearest the heads or throughout; involucres hemispheric, 20–70 mm wide; corollas pink to purple; achenes 4–5 mm long; pappus 13–25 mm long. Waste ground, roadsides, pastures; Anderson Co., Travis Co., and w edge of East TX in nc TX in Collin, Dallas, and Grayson cos.; scattered across TX but mostly in nc TX and Edwards Plateau; across most of North America; Eurasia; introduced elsewhere. May–Jul. [*C. macrocephalus* Desf., *C. macrolepis* Peterm., *C. nutans* L. subsp. *macrocephalus* (Desf.) Nyman] Native of Eurasia. This species was apparently first collected in TX in Sutton Co. on the Edwards Plateau in 1939 (along with the following species); it was supposedly introduced through a shipment of hay from California (Cory 1940). This species is listed as a noxious weed in many states; it is unpalatable to wildlife and livestock and forms dense, impenetrable stands in pastures and rangelands (Keil 2006b).

Carduus pycnocephalus L., (thick-headed), ITALIAN THISTLE, ITALIAN PLUMELESS THISTLE. Prickly annuals, 20–200 cm tall; basal leaves 10–25 cm long, 2–5-lobed, lobes spine-tipped; heads borne singly or clustered in groups of 2–5; peduncles 0–2 cm long, spiny-winged throughout or unwinged nearest the heads; involucres cylindric to ellipsoid, 7–15 mm wide; corollas pink to lavender-purple; achenes 4–6 mm long; pappus 15–20 mm long. Waste ground, roadsides, pastures; w border of East TX in Bell, Williamson, and Travis cos. and Hays Co (TEX); nc TX to Edwards Plateau; scattered across U.S. but most common in CA; Europe; introduced elsewhere. Apr–Jul. Native of Mediterranean Europe. This species was apparently first collected in TX in Sutton Co. on the Edwards Plateau in 1939 (along with the previous species); it was supposedly introduced through a shipment of hay from California (Cory 1940). This species is listed as a noxious weed in five states and has the potential of becoming a problematic pernicious weed much like *C. nutans*. If subspecies are recognized, ours would be subsp. *pycnocephalus*.

CARTHAMUS L. SAFFLOWER, DISTAFF-THISTLE

Mostly annual herbs; leaves usually alternate, spiny, upper stem leaves sessile and sometimes clasping; heads terminal, solitary or in corymbose arrays; at least inner phyllaries spiny or spine-tipped, outer phyllaries spreading, conspicuously leafy; receptacles scaly; ray flowers absent; disk corollas tubular, in ours yellow to orange; achenes oblong, ± 4-angled, with an attachment scar at base; pappus of scales or absent.

A genus of about 14 species native from the Mediterranean area to c Asia. (From Arabic or Hebrew: *quarthami*, to paint, alluding to the dye obtained from the flowers of *C. tinctorius*) (tribe Cardueae)

REFERENCES: Shinners 1958; Kessler 1987; Keil 2006c.

1. Leaves ± pinnatifid, usually woolly-pubescent, coarsely and conspicuously spine-toothed; corollas yellow; filaments bearded; achenes (at least peripheral ones) rugose, tan to brown; pappus present (at least on central achenes) ________ *C. lanatus*
1. Leaves usually simple (rarely pinnatifid), glabrous, entire or inconspicuously spine-toothed; corollas bright orange or yellow-orange; filaments glabrous; achenes ± smooth, white; pappus usually absent ________ **C. tinctorius**

Carthamus tinctorius L., (belonging to the dyers, of dyes), SAFFLOWER, FALSE SAFFRON, BASTARD SAFFRON. Glabrous taprooted herbs; stems 0.3–1 m tall; leaves alternate to subopposite above, lanceolate to broadly ovate or oblong, margins dentate; corollas yellow to red; achenes glabrous, ivory white, ca. 7 mm long and 4 mm wide; pappus usually absent (rarely of rudimentary or well-developed scales to 4 mm long). Yard weed or waif, likely from commercial birdseed; recorded from Brazos and McLennan cos. and Ellis and Hill cos. (BRIT), also Hidalgo Co. (*Karcher s.n.*, 1963, TAES); also Tarrant Co.; many states, mostly in the w U.S.; w Canada. Summer. Native of Old World, possibly from the Near East. Cultivated since ancient times as a dye plant (orange from the flowers used in dyeing food and cosmetics) and now grown for its edible oil (from the seeds). Seeds were found in the tomb of Tutankhamun, and mummy wrappings were sometimes dyed with this plant (Hepper 1990).

Carthamus lanatus L., (woolly), DISTAFF-THISTLE, SAFFRON-THISTLE. Viciously spiny glandular herbs, usually with woolly, spider-web-like pubescence; stems 0.4–1.8 m tall; leaves alternate with spine-tipped lobes; corollas yellow with red or black veins; achenes brownish, 5–6 mm long; pappus of ± rigid, flat, bristle-like, ciliate scales to 1.3 cm long as well as some outer short scales. Disturbed roadsides, rangelands, pastures; *not yet recorded in East TX but to be expected*; just w of our area in Edwards Plateau and nc TX; recorded in Coryell and Lampasas cos. (TEX), also Palo Pinto Co. (TEX); AZ, CA, ID, MA, OK, OR, SC, TX, and WY; w Canada; invasive in many countries worldwide. May-Jun. A noxious weed native to the Mediterranean region. In rangelands it is known to injure the eyes and mouths of livestock.

CENTAUREA L. STAR-THISTLE, KNAPWEED

Annuals, biennials, or perennials; leaves alternate, gradually reduced upwards, margins often toothed or pinnatifid, not prickly; heads solitary or in corymbose-paniculate arrays; phyllaries in several series, erect or spreading, usually dentate or fringed, sometimes prickly or spinose; receptacles bristly; ray flowers absent; peripheral disk flowers sterile, their corollas enlarged, often simulating rays; central disk flowers perfect, fertile, their corollas deeply (4–)5-lobed, the lobes linear, pink-purple to blue or yellow, rarely white; anthers with elongated appendages; achenes ± barrel-shaped, ± compressed, obliquely attached; pappus of numerous stiff bristles.

A large genus (ca. 500 species) of herbs and subshrubs native to the Mediterranean, Turkey, the Near East, n Eurasia, tropical Africa, and Australia; some are cultivated as ornamentals, others are noxious weeds. The taxonomic limits of *Centaurea* have been controversial; various taxonomists have attempted to separate *Centaurea* into smaller genera and recent molecular work has reinforced some of these separations (e.g., Garcia-Jacas et al. 2001; Susanna et al. 2011). The two species native to the U.S., formerly *C. americana* (AMERICAN BASKETFLOWER) and *C. rothrockii*, fall outside redefined generic boundaries and are now treated as *Plectocephalus* (Keil & Ochsmann 2006). In recognition of the traditional placement of AMERICAN BASKETFLOWER in *Centaurea*, we have included it in this key as well. *Centaurea stoebe* L. (SPOTTED KNAPWEED), a European native now naturalized and invasive in parts of the U.S. and recently recorded in TX, is suspected of being carcinogenic; a field worker in Idaho developed tumors in his fingers after pulling plants of this species and apparently getting sap in breaks in his skin (Jerry Niefoff, pers. comm.); as a precaution, gloves should be worn when handling *Centaurea* species. Some species are known to cause toxic effects in horses (Kingsbury 1964; James & Welsh 1992). (Greek: *kentaur*, centaur; the centaur Chiron in Greek mythology was said to know the medicinal value of plants) (tribe Cardueae)

REFERENCES: Moore 1972; Keil & Ochsmann 2006.

1. Corollas yellow; phyllaries with a conspicuous sharp spine at tip.
   2. Phyllary spine 5–10 mm long, often darker than phyllary body ______ **C. melitensis**
   2. Phyllary spine 10–25 mm long, straw-colored, usually lighter than phyllary body ______ **C. solstitialis**
1. Corollas blue, pink, purple, or white; phyllaries without a sharp spine at tip, margins variously toothed, lobed, or fringed.
   3. Heads large, the involucres of well-developed heads 3–5+ cm wide; pappus bristles 6–14 mm long; native species; formerly known as *Centaurea americana* ______ see **Plectocephalus**
   3. Heads small, the involucres 2.5 cm or less wide; pappus bristles 2–5 mm long; introduced species.
      4. Annuals; involucres of well-developed heads 1.2–1.6 cm wide; corollas usually blue (sometimes purple or pink, rarely white); peripheral corollas enlarged, appearing broadly ligulate, 20–25 mm long; achenes 4–5 mm long ______ **C. cyanus**
      4. Biennials or perennials; involucres of well-developed heads 1–1.3 cm wide; corollas usually pink to purple; peripheral corollas slender and only slightly enlarged or broadened, 15–25 mm long; achenes 3–3.5 mm long ______ **C. stoebe**

Centaurea cyanus L., (blue), BACHELOR'S-BUTTON, CORNFLOWER, BLUEBOTTLE. Taprooted annuals, 20–100 cm tall; lower and basal leaves linear to narrowly oblanceolate, 3–10 cm long, loosely gray or white tomentose, entire or remotely pinnately lobed or toothed; upper stem leaves decurrent, usually entire; phyllary tips white to dark brown or black, with numerous small teeth; peripheral flowers with expanded ray-like corollas; corollas blue (sometimes pink-purple or white); achenes 4–5 mm long; pappus of many stiff bristles, 2–4 mm long. Cultivated, escaping from "wildflower seed" mixes, naturalized in disturbed areas, roadsides; scattered across c and e TX; widespread in North America; introduced elsewhere. Mar–Jun. [*Leucacantha cyanus* (L.) Nieuwl. & Lunell] Native of Mediterranean region.

Centaurea melitensis L., (of Malta), STAR-THISTLE, MALTA CENTAUREA, MALTA STAR-THISTLE, TOCALOTE. Annuals or rarely biennials, 10–100 cm tall; herbage gray-tomentose; basal and lower leaves 2–15 cm long, entire to dentate or pinnately lobed; upper stem leaves linear, decurrent; phyllaries purplish in upper portion, spiny-margined and spine-tipped, the central spine to 1 cm long; corollas yellow; achenes ca. 2.5 mm long; pappus of many stiff bristles, 2.5–3 mm long. Disturbed calcareous soils, roadsides, agricultural areas; w edge of East TX from Bell Co. s to Bexar and Guadalupe cos., also Dallas and Harris (SBSC) cos.; widespread in c and w TX but especially Edwards Plateau; scattered across w North America; Mexico; introduced elsewhere. Mar–Jun. Native of Europe. Cory (1940) reported that this species was first collected in TX in Bexar Co. in 1934.

Centaurea solstitialis L., (of the summer solstice), YELLOW STAR-THISTLE, BARNABY'S STAR-THISTLE. Annuals; 10–100 cm tall; basal leaves pinnatifid, 5–15 cm long, gray-tomentose to short-bristly; upper stem leaves linear, decurrent; phyllaries spiny-margined and spine-tipped, the central spine tan or yellowish and 1–2.5 cm long; corollas yellow; achenes 2–3 mm long; pappus of many fine bristles, 2–4 mm long. Disturbed areas, roadsides, pastures; uncommon in TX; nc TX in Dallas and Ellis cos., also Edwards Plateau in Guadalupe Co. (TEX); scattered throughout North America; introduced elsewhere. May–Jun. Native of Eurasia. Reportedly causes brain lesions and a nervous syndrome, "chewing disease," in horses; the toxic principle is not known (Kingsbury 1964). First noted in Dallas Co. in 1962; this species has the potential of becoming a problematic weed in TX. It is considered noxious in many western states, where it has invaded millions of acres of rangeland (Keil & Ochsmann 2006).

Centaurea stoebe L. subsp. **micranthos** (S.G. Gmel. ex Gugler) Hayek, (sp.: obscure, possibly from Greek, *stoibe*, stuffing or padding; subsp.: small-flowered), SPOTTED KNAPWEED. Biennials or short-lived perennials, 30–150 cm tall; herbage loosely gray-tomentose; basal and lower leaves pinnately lobed or pinnatifid, 10–15 cm long; upper stem leaves entire or dissected; phyllary tips dark brown or black, with numerous small teeth; corollas pink to purple or rarely white; achenes 3–3.5 mm long; pappus of many stiff bristles, to 5 mm long. Disturbed areas, roadsides, pastures; first confirmed in TX in 2014 in Travis Co. at the Balcones Canyonlands National Wildlife Refuge by botanist Arthur C. Gibson and vouchered by wildlife biologist Chuck Sexton (BCNWR; EDDMapS 2016); also collected in Tarrant Co. (R. O'Kennon, pers. comm.); scattered throughout North America; introduced elsewhere. May–Sep. [*C. maculosa* Lam., *C. maculosa* Lam. subsp. *micranthos* S.G. Gmel. ex Gugler, *C. biebersteinii*

Brickellia eupatorioides var. texana [YAT]

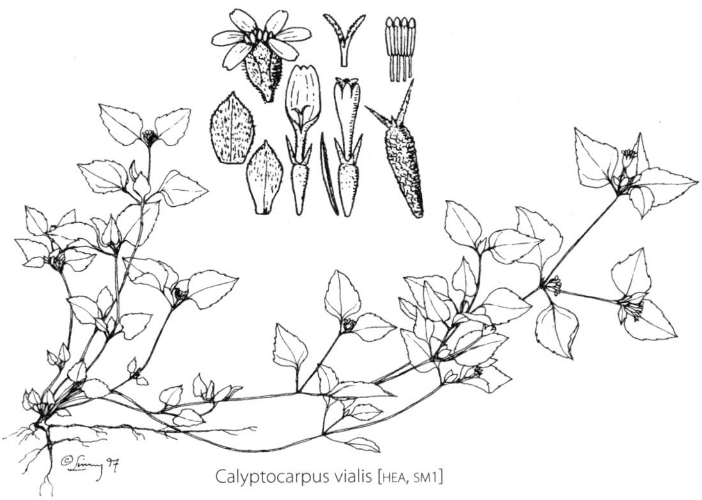

Calyptocarpus vialis [HEA, SM1]

Carduus acanthoides [BEN]

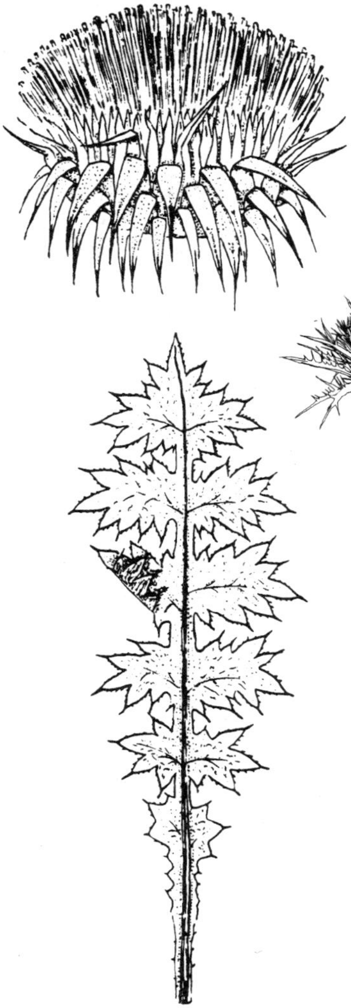

Carduus nutans [MIT]

Carthamus lanatus [HEA]

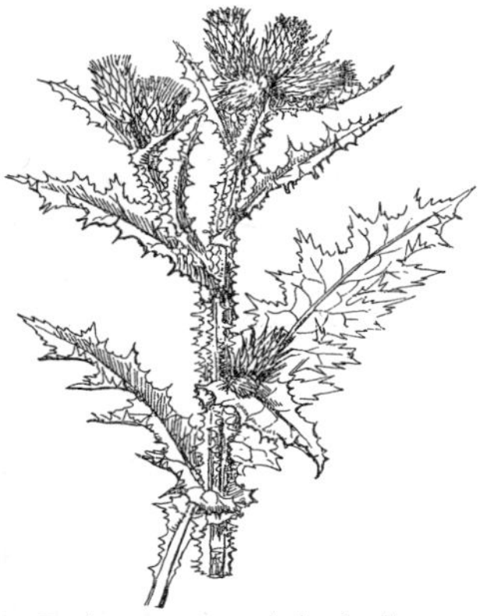

Carduus pycnocephalus [CTG]

DC.] Native of Europe. This species has the potential of becoming a problematic weed in TX; it is considered noxious in many western states.

CHAETOPAPPA DC. LEAST DAISY

Small annual or perennial herbs; stems strigose to pilose; leaves alternate, small, entire; heads usually solitary at the ends of the upper branches; phyllaries in 2–6 series, elliptic to lanceolate, unequal, the margins whitish-scarious; receptacles naked; ray flowers pistillate and fertile, white to bluish, pinkish, or lavender, the ligule tips sometimes curling under; disk flowers perfect and fertile or functionally staminate, yellow; achenes cylindric to narrowly obovoid, 2–10-nerved; pappus of rings, crowns, scales, and/or bristles.

A genus of 11 species native to s North America; including the previously-recognized monotypic genus *Leucelene*. (Greek: *chaite*, hair or bristle, and *pappos*, down, fuzz, pappus) (tribe Astereae)
REFERENCES: Shinners 1946a, 1946b; Van Horn 1973; Nesom 1988, 2006h.

1. Perennials from fibrous roots and woody rhizomes; ray achenes with 2(–3) ribs or nerves ____ **C. effusa**
1. Taprooted annuals; ray achenes with 5+ ribs or nerves.
   2. Pappus in 2 series, with short outer scales, alternating with inner bristles 1.5–3 mm long; ray achenes 5(–6)-nerved; common throughout e 1/2 of TX ____ **C. asteroides**
   2. Pappus in 1 series, of thickened rings or a minute cup-like crown; ray achenes 8–10-nerved; endemics of rocky calcareous soils on Edwards Plateau, or of deep sands of the Carrizo Formation in sc TX.
      3. Pappus of minute, thickened rings; ray achenes 10-nerved; endemic to Edwards Plateau ____ **C. bellidifolia**
      3. Pappus a white-hyaline, rough-margined, asymmetric, cup-like crown; ray achenes 8-nerved; mostly concentrated in Carrizo Formation sands in sc TX ____ **C. imberbis**

Chaetopappa asteroides Nutt. ex DC. var. **asteroides,** (resembling *Aster*), COMMON LEAST DAISY, ARKANSAS LEAST DAISY. Small taprooted annuals, 3.5–30 cm tall; midstem leaves linear to oblanceolate, 4–15 mm long; ray flowers 5–15, ligules white, turning bluish, pinkish, or lavender; ray achenes 1.6–2 mm long, 5(–6)-nerved; pappus in 2 series, the outer series of ca. 5 scales 0.1–0.8 mm long, the inner series of ca. 5 awn-like bristles to 3 mm long. Sandy or gravelly open areas; widespread in e 1/2 TX; sc U.S. from MO s to LA and w to KS and TX; Mexico. Apr–Jun, sporadically later. Nesom (1988, 2006h) recognized two varieties of this species, var. *asteroides* and var. *grandis* Shinners; ours would be the former, as the latter is restricted to s TX along the Rio Grande Valley. Turner (2016) found the characters defining those varieties to be sporadic across the range of the species.

Chaetopappa bellidifolia (A. Gray & Engelm.) Shinners, (leaves resembling *Bellis*—English daisy), DWARF WHITE ASTER, HAIRY LEAST DAISY, WHITERAY LEAST DAISY. Small taprooted annuals, 8–40 cm tall; midstem leaves oblong-spatulate to oblanceolate, 7–35 mm long; ray flowers 6–15, ligules white, turning bluish, pinkish, or lavender; ray achenes 1.5–2.2 mm long, 10-nerved; pappus of a minute, thickened ring. Calcareous soils, sandy or rocky open areas; sw border of East TX from McLennan Co. s to Bexar Co.; Edwards Plateau. Mar–May, sporadically later. [*Keerlia bellidifolia* A. Gray & Engelm.]

Chaetopappa effusa (A. Gray) Shinners, (sprawling, for the open arrangement of the heads), SPREADING LEAST DAISY. Perennials, 20–80 cm tall, with fibrous roots and woody rhizomes; midstem leaves oblong to lanceolate-oblong, 10–20 mm long, bases subclasping; ray flowers 6–9, ligules white; ray achenes 1.6–2.2 mm long, 2(–3)-nerved; pappus of thickened rings or a minute crown. Limestone hillsides and canyons, seeps, oak-juniper woodlands; sw border of East TX from Travis Co. s to Bexar Co.; Edwards Plateau. May–Oct. [*Keerlia effusa* A. Gray]

Chaetopappa imberbis (A. Gray) G.L. Nesom, (beardless, for the lack of pappus bristles), AWNLESS LEAST DAISY, CARRIZO CHAETOPAPPA. Small taprooted annuals, 4–15 cm tall; midstem leaves oblanceolate-spatulate to oblanceolate-obovate, 8–26 mm long; ray flowers 8–19, ligules white; ray achenes 1.5–1.8 mm long, 8-nerved; pappus a minute, cup-like crown. Deep sands in oak woodlands, Carrizo Formation; concentrated in sc TX from Bastrop Co. s to Bexar, Wilson, and Gonzales cos.; also Pleistocene sands on the Gulf Coast in Goliad and Victoria cos. (Carr 2009). Mar–May, sporadically later. [*C. asteroides* Nutt. ex DC. var. *imberbis* A. Gray]

CHAPTALIA Vent. SUNBONNETS

Acaulescent perennial herbs, 3–40(–100) cm tall; fibrous-rooted; leaves in a basal rosette, blades elliptic to obovate, margins entire, lobed, or denticulate, usually densely pubescent beneath; heads borne singly, erect in flowering, often nodding in bud and again at maturity; phyllaries in 2–5 series, lanceolate to lance-linear, unequal; receptacles naked; heads with 2–3 series of flowers varying by season; the peripheral series of ray-like flowers, pistillate, fertile, ligulate or unequally 2-lipped, corollas creamy white to purple, sometimes with a darker stripe beneath; the middle series (when present) of disk-like flowers, unequally 2-lipped or

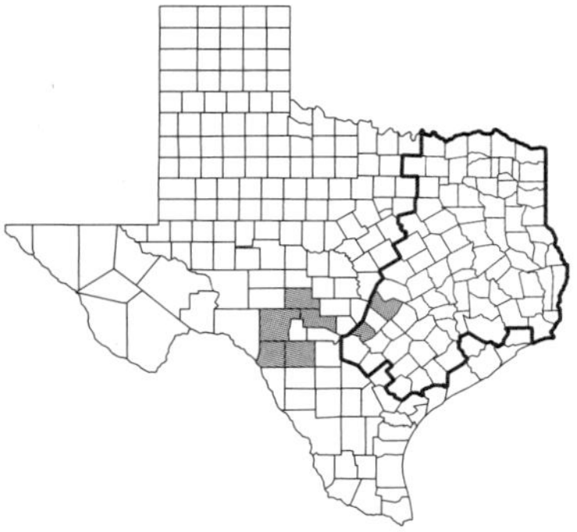

Brickellia eupatorioides var. gracillima

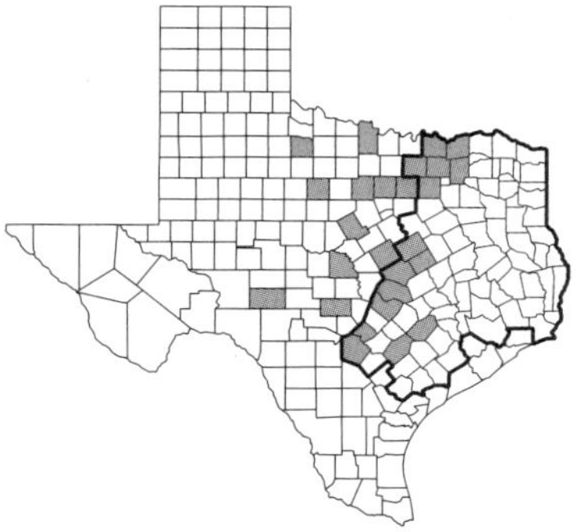

Brickellia eupatorioides var. texana

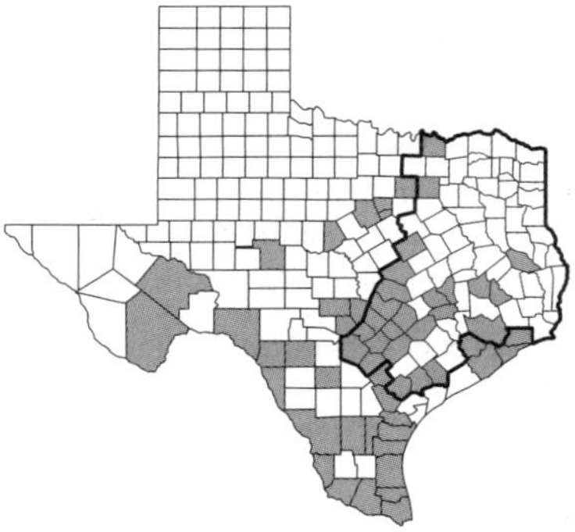

Calyptocarpus vialis

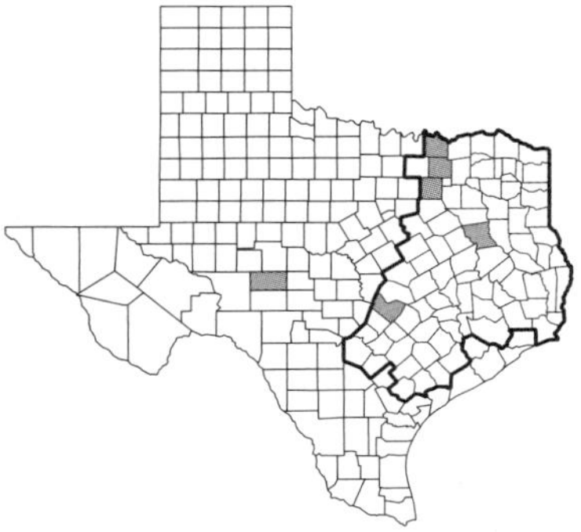

Carduus acanthoides

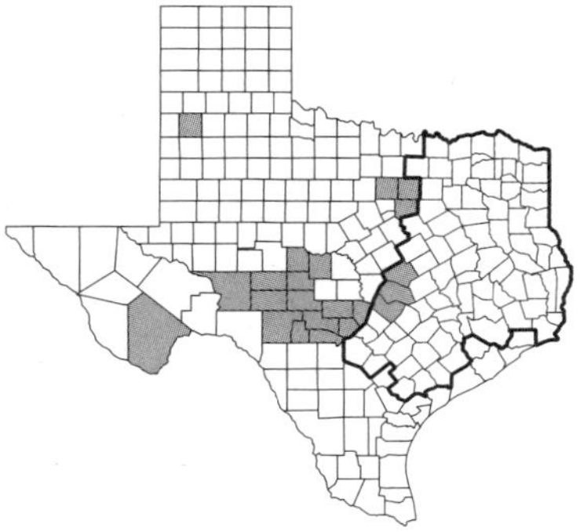

Carduus nutans

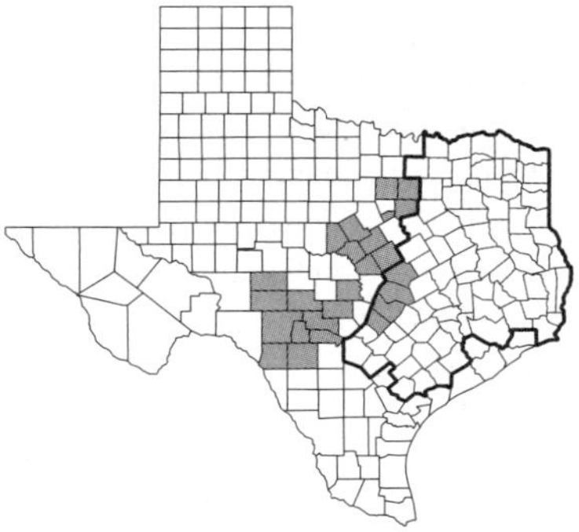

Carduus pycnocephalus

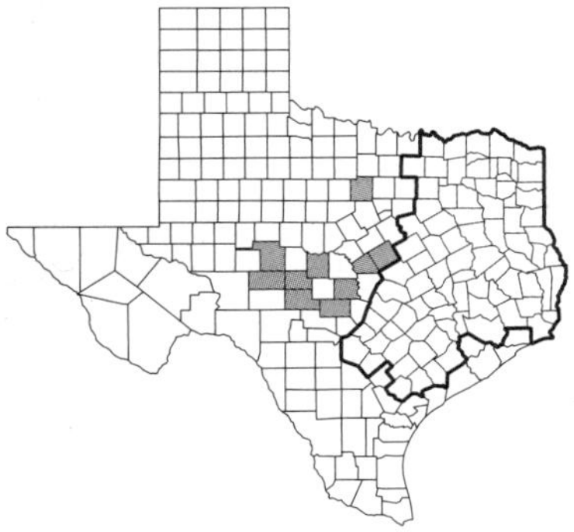

Carthamus lanatus

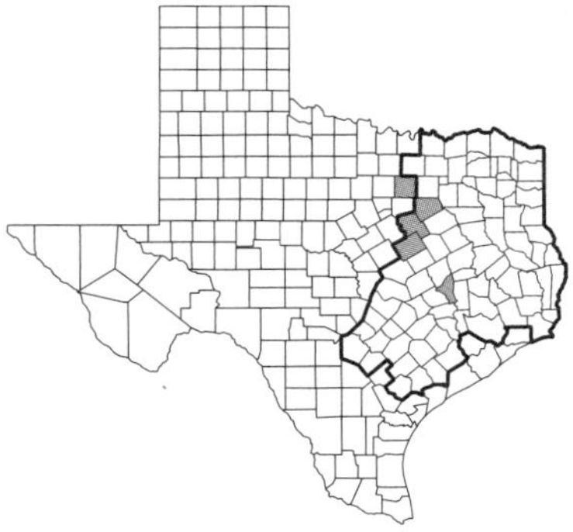

Carthamus tinctorius

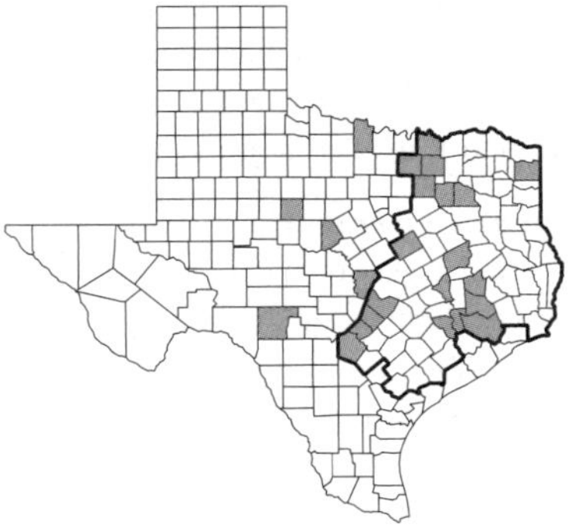

Centaurea cyanus

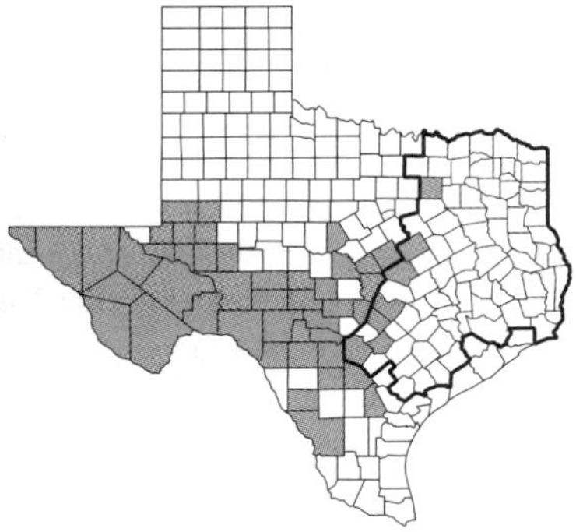

Centaurea melitensis

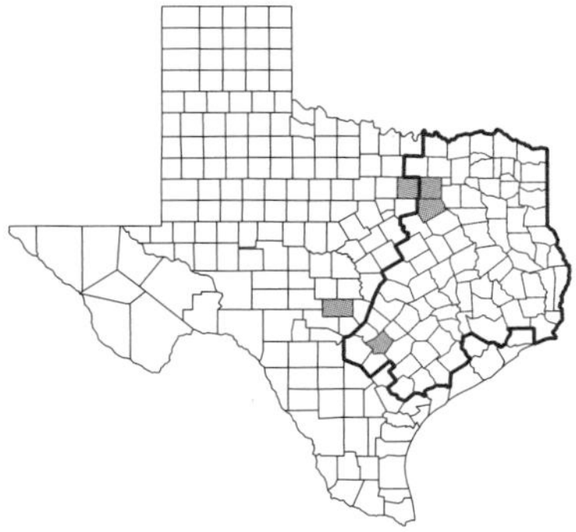

Centaurea solstitialis

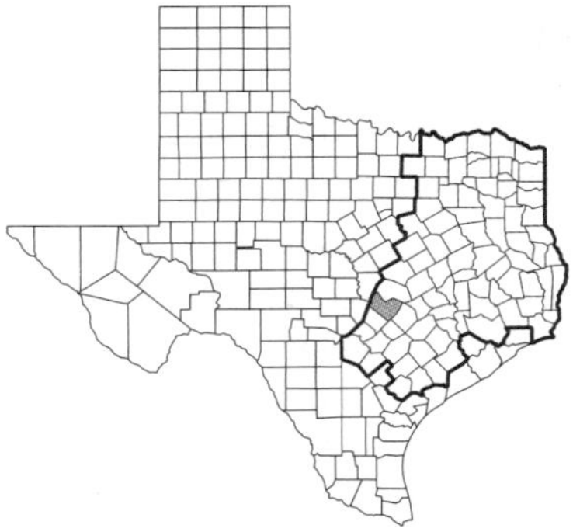

Centaurea stoebe subsp. micranthos

tubular, pistillate, fertile; the central series of disk flowers, perfect, fertile or functionally staminate, unequally 2-lipped, corollas whitish to pinkish; achenes fusiform, ribbed, beaked; pappus of 50+ bristles, pale tan to pinkish, finely-barbed.

A genus of ca. 60 species of warm areas of the Americas. The key below is based on Harms (2011). (Named for J.A.C. Chaptal, 1756–1831, agricultural chemist, who invented the wine-making process of chaptalization, which makes the wine more alcoholic by adding sugar at the same time as squeezing the grape) (tribe Mutisieae)
REFERENCES: Vuilleumier 1969b; Simpson 1978; Nesom 1995, 2006i; Harms 2011.

1. Ligules of peripheral ray-like flowers with a purple midstripe beneath; central disk flowers functionally staminate; deep se TX ______ **C. tomentosa**
1. Ligules of peripheral ray-like flowers uniformly colored on both sides; central disk flowers perfect and fertile; Edwards Plateau, s, and w TX.
   2. Ligules of peripheral ray-like flowers maturing to crimson, barely extending beyond phyllaries; phyllaries rigid, usually > 1.5 mm wide in the middle; pappus of mature achenes usually > 10.5 mm long ______ **C. carduacea**
   2. Ligules of peripheral ray-like flowers maturing to yellow (rarely crimson), extending beyond phyllaries; phyllaries flexible, usually ca. 1 mm wide in the middle; pappus of mature achenes usually < 10.5 mm long ______ **C. texana**

Chaptalia carduacea Greene, (thistle-like), SILVERPUFF, TEXAS SUNBONNETS. Leaves obovate to elliptic, sessile or short-petioled, margins undulate-lobed and finely denticulate, upper surfaces green and nearly glabrous, lower surfaces thinly tomentose; peduncles to 34 cm long, without bracts or rarely with 1–2 minute bracts; phyllaries rigid, usually > 1.5 mm wide in the middle; peripheral ray-like flowers creamy white, maturing to crimson, corollas 10–12.5 mm long; central disk flowers perfect and fertile; achenes 10–12.8 mm long, 1/2–2/3 of which is the beak; beaks filiform and papillate. Calcareous areas, oak woods, canyons; sw border of East TX from Travis Co. sw to Comal Co. and se to Goliad Co.; Edwards Plateau to s TX; endemic to TX?; also likely n Mexico. Mar–Apr, sporadically all year. [*C. leonina* Greene] Formerly included in *C. texana* and recently resurrected by Harms (2011). ?

Chaptalia texana Greene, (of Texas), SILVERPUFF, TEXAS SUNBONNETS. Leaves obovate to elliptic to sublyrate, petiolate, margins undulate-lobed and finely denticulate, upper surfaces green and nearly glabrous, lower surfaces thinly gray-tomentose; peduncles to 46 cm long, without bracts or rarely with 1–2 minute bracts; phyllaries flexible, usually ca. 1 mm wide in the middle; peripheral ray-like flowers creamy white, maturing to yellowish (rarely to crimson), corollas 11–13.2 mm long; central disk flowers perfect and fertile; achenes 11.5–13 mm long, 1/2–2/3 of which is the beak; beaks filiform and papillate. Calcareous areas, thin soils, oak woods, canyons; sw border of East TX from Bastrop and Travis cos. s to Bexar and Wilson cos.; Edwards Plateau to Trans-Pecos; also n Mexico. Mar–Apr, sporadically all year. [*C. nutans* (L.) Polak.

Chaptalia tomentosa Vent., (woolly), WOOLLY SUNBONNETS. Leaves elliptic to obovate, sessile, margins denticulate, upper surfaces green and ± glabrous, lower surfaces densely white-tomentose; peduncles to 40 cm long, without bracts; peripheral ray-like flowers creamy white with purple midstripe beneath, corollas to 12 mm long; central disk flowers functionally staminate; achenes 3.8–5.1 mm long, 1/4 of which is the beak; beaks stout and hairy. Coastal plains, pineywoods, moist sands, acidic soils; known only from deep se TX in Hardin, Jasper, Newton, and Tyler cos.; se coastal U.S. from VA s to FL and w to TX. Feb–Apr.

CHLORACANTHA G.L. Nesom, Y.B. Suh, D.R. Morgan, S.D. Sundb., & B.B. Simpson
MEXICAN DEVILWEED, SPINY-ASTER

A monotypic genus occurring from the sw U.S. to Central America. It is distinctive, but of uncertain phylogeny, and has been placed in a variety of genera including *Aster*, *Erigeron*, and *Leucosyris*; based on DNA evidence, *C. spinosa* seems more closely related to *Boltonia* and *Heterotheca* than to either *Aster* or *Erigeron*; it is thus recognized in its own genus (Nesom et al. 1991). (Greek: *chloro*, green, and *acanthos*, spine or thorn, in reference to the evergreen thorny stems) (tribe Astereae)
REFERENCES: Nesom et al. 1991; Sundberg 1991; Nesom 1994b; Sundberg & Nesom 2006.

Chloracantha spinosa (Benth.) G.L. Nesom var. **spinosa,** (spiny), MEXICAN DEVILWEED, SPINY-ASTER, DEVILWEED-ASTER. Coarse, nearly leafless rhizomatous perennial, 50–120(–150) cm tall, the plant superficially rush-like in appearance, sometimes forming dense clumps; stems glabrous or glabrate, green and glaucous, strictly erect, freely branched, alive for up to ca. 4 growing seasons; branchlets sometimes modified into thorns 10–20 mm long; usually ± leafless or leaves early deciduous, or with a few small alternate leaves to 40 mm long; heads numerous, in loose paniculate arrays; phyllaries in 4–5 series, lanceolate; involucres 4–5.5(–6) mm high; receptacles naked; ray flowers 20–33, pistillate, fertile, ligules to 5(–7) mm long, white, sometimes bluish tinged; disk flowers perfect, fertile, yellow; achenes fusiform, glabrous; pappus of 30–60 tawny, barbellate bristles. Low disturbed or weedy areas, along streams, roadsides; widespread in TX but more common in sc and w parts of the state; sw U.S. from CA e to TX; OK, and LA; Mexico. May–Oct. [*Aster spinosus* Benth., *Leucosyris spinosa* (Benth.) Greene]

Carthamus tinctorius [EN2]

Centaurea cyanus [GAR, M&F]

Centaurea melitensis [M&F]

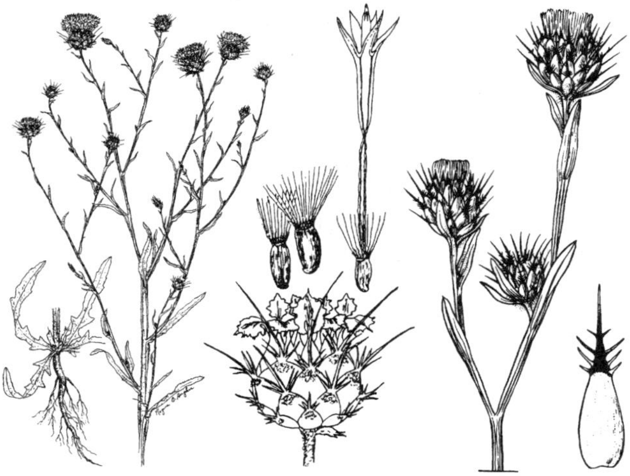

Centaurea solstitialis [M&F, REE]

Centaurea stoebe subsp. micranthos [REE]

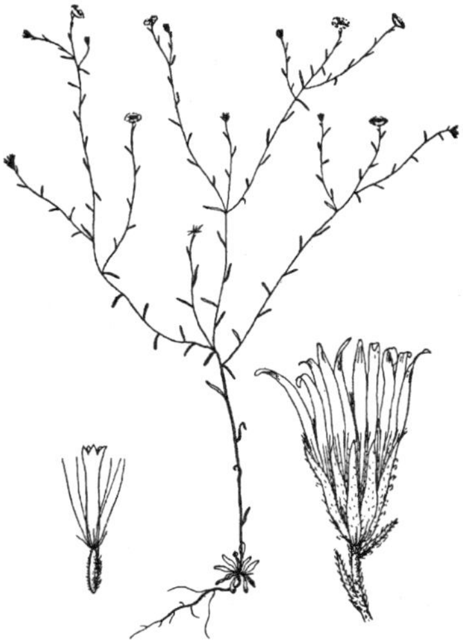

Chaetopappa asteroides [GLE]

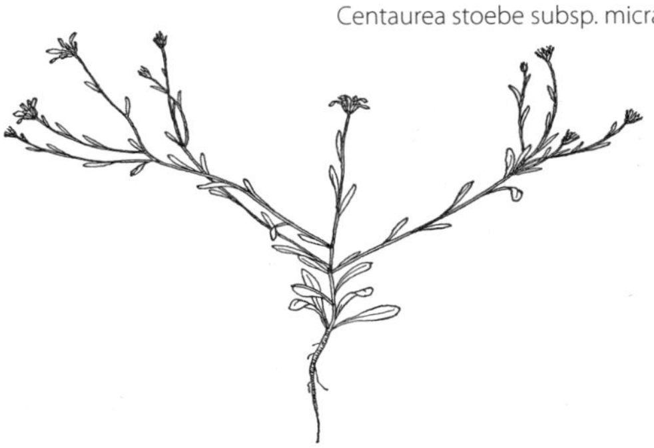

Chaetopappa bellidifolia [UCP]

CHROMOLAENA DC. THOROUGHWORT

Perennials or subshrubs; leaves opposite, blades usually 3-nerved from bases; heads borne in corymbose arrays; phyllaries in 4–6+ series, unequal, usually deciduous at maturity; receptacles naked or chaffy; ray flowers absent; disk flowers with corollas usually light blue or lavender, sometimes nearly white or pinkish purple; achenes 3–5-ribbed, usually gland-dotted; pappus of ca. 40 barbellate bristles.

A genus of ca. 165 species of the warm-temperate and tropical Americas, especially Brazil; some are cultivated; a few are widespread weeds. (Greek: *chroma*, color, and *laina*, cloak, evidently alluding to the colored phyllaries of the type species) (tribe Eupatorieae)
REFERENCES: King & Robinson 1980, 1987; Nesom 2006j.

1. Leaves sessile or with petioles to 1(–3) mm long, blades linear-lanceolate to elliptic-lanceolate, 0.4–1.2 cm wide, bases tapered; involucres 5–6 mm high; tips of inner phyllaries ± spreading ____ **C. ivifolia**
1. Leaves with petioles 5–20 mm long, blades narrowly lanceolate to deltate-lanceolate or ovate-lanceolate, 1–4 cm wide, bases ± abruptly contracted; involucres (7–)8–10 mm high; tips of inner phyllaries appressed ____ **C. odorata**

Chromolaena ivifolia (L.) R.M. King & H. Rob., (with leaves resembling *Iva*—sumpweed), IVY-LEAF THOROUGHWORT, IVY-LEAF FALSE THOROUGHWORT. Perennials or subshrubs, 50–150 cm tall; stems erect; leaves sessile or with petioles 0.5–1(–3) mm long, blades linear-lanceolate to elliptic-lanceolate, margins denticulate to nearly entire; involucres cylindric to campanulate, 5–6 mm high; inner phyllary tips ± spreading; corollas light blue to purplish or reddish. Floodplains, bottomlands, roadsides, disturbed sites; se TX in Hardin, Harris, Jefferson, and Liberty cos.; se US from FL w to TX; Mexico; West Indies; Central and South America. Aug–Nov. [*Eupatorium ivifolium* L., *Osmia ivifolia* (L.) Sch. Bip.]

Chromolaena odorata (L.) R.M. King & H. Rob., (fragrant), BLUE MISTFLOWER, FRAGRANT BONESET, JACK-IN-THE-BUSH, SIAM WEED. Perennials or subshrubs, 80–250 cm tall; stems erect or sprawling; leaves with petioles 5–20 mm long, blades narrowly lanceolate to deltate-lanceolate or ovate-lanceolate, margins coarsely dentate to nearly entire; involucres cylindric, (7–)8–10 mm high; inner phyllary tips appressed; corollas light blue to purplish or nearly white to pinkish. Thickets, pine flats, canal banks, disturbed sites; Jefferson, Harris, Jackson, and Victoria cos.; se Gulf Prairies to s TX; FL and TX; Mexico and West Indies. Oct–Dec. [*Eupatorium odoratum* L., *Osmia odorata* (L.) Sch. Bip.] Native to North America, now a widespread and troublesome weed on many continents.

CHRYSACTINIA A. Gray

A genus of 5 species occurring from the sw U.S. to Mexico. (Greek: *chrysos*, gold, and *actinos*, ray) (tribe Heliantheae)
REFERENCES: Strother 1977, 2006e.

Chrysactinia mexicana A. Gray, (of Mexico), DAMIANITA. Bushy evergreen perennial herbs or subshrubs to 50 cm tall; aromatic; leaves usually alternate, simple, linear to club- or needle-shaped, entire, semi-succulent, with conspicuous oil glands; heads borne singly; involucres 4–5 mm high; phyllaries ca. 13 in ± 2 series, lance-linear, each tipped with an oil gland; receptacles naked or rarely chaffy (pales soon deciduous); ray flowers 8(–13), pistillate, fertile, ligules bright yellow, 6–12 mm long; disk flowers perfect, fertile, yellow to orange; achenes cylindric or fusiform, 3–4 mm long; pappus of 20–40 bristles, 3–5.5 mm long. Calcareous soils, juniper woodlands; sw border of East TX in Bexar, Comal, Travis, and Williamson cos.; Edwards Plateau and Trans-Pecos; NM and TX; Mexico. Mar–Nov. This species is gaining popularity as a drought-tolerant garden plant.

CHRYSOPSIS (Nutt.) Elliott GOLDEN-ASTER

A genus of 11 species native to the se U.S., especially Florida; some are rare or threatened. *Chrysopsis* was formerly treated as including many species now segregated to *Bradburia*, *Heterotheca*, and *Pityopsis*; generic boundaries for the GOLDEN-ASTERS have long been problematic. (Greek: *chrysos*, gold, and *opsis*, appearance, for the golden inflorescences) (tribe Astereae)
REFERENCES: Shinners 1951b; Harms 1968, 1974; Semple 1977, 1981, 1996, 2006c; Semple et al. 1980; Semple & Chinnappa 1984; Nesom 1991b, 1991c, 1997c.

Chrysopsis mariana (L.) Elliott, (of Maryland), SOFT GOLDEN-ASTER. Biennial or short-lived perennial herbs, 20–90 cm tall; fibrous-rooted or with short lateral rhizomes; stems often purplish, silky-sericeous; basal leaves oblanceolate to spatulate, stem leaves alternate, sessile, lanceolate to elliptic-oblong, glabrate to long silky-sericeous, margins entire or obscurely dentate, reduced up the stem; heads in crowded subumbelliform arrays; peduncles with dense stalked glands; involucres 7–10 mm high phyllaries in 4–5 series, unequal, with dense stalked glands; receptacles naked; ray flowers pistillate, fertile, ligules yellow; disk

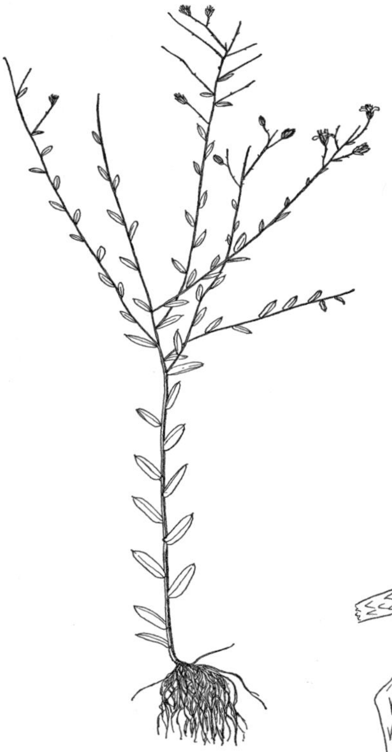

Chaetopappa effusa [UCP]

Chaetopappa imberbis [THO]

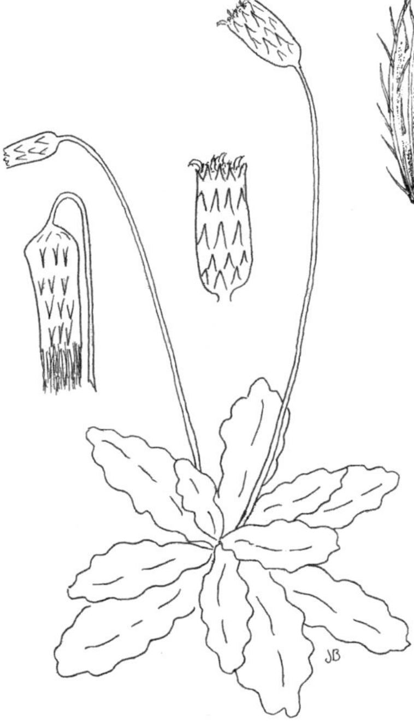

Chaptalia carduacea [BAK]

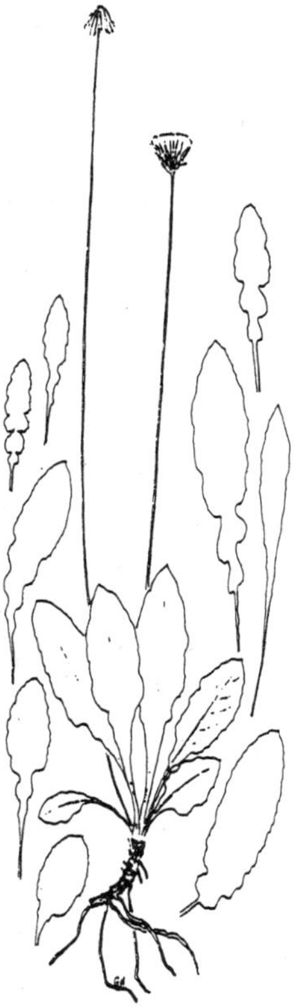

Chaptalia texana [PHY]

Chaptalia tomentosa [RAD, SM1]

flowers perfect, fertile, corollas yellow; achenes shallowly ribbed or smooth, 2–3 mm long, short-strigose; pappus of 2–3 series of bristles, the outer series 0.5–1 mm long, the inner series 4–6 mm long. Sandy and clay soils, open woods, old fields, roadsides; se TX; e and se U.S. from RI s to FL and w to OH, TN, and TX. (Apr–)Aug–Nov. [*C. mariana* (L.) Elliott var. *macradenia* Fernald, *Heterotheca mariana* (L.) Shinners]

CICHORIUM L. CHICORY

A genus of 6 species native to Europe, the Mediterranean, and Ethiopia. The Old World *C. endivia* L., ENDIVE, has long been cultivated for its edible leaves. (Arabic: *chikourych*, a salad vegetable (Holloway 2005) (tribe Cichorieae)
REFERENCES: Vuilleumier 1973; Strother 2006f.

Cichorium intybus L., (old generic name, from Greek *intubos*, for endive or chicory), COMMON CHICORY, BLUE-SAILORS, SUCCORY, WITLOOF, RAGGED-SAILORS. Glabrous perennial herbs with deep taproots, 30–100(–170) cm tall, sap milky; basal leaves toothed or pinnately lobed, oblong-lanceolate; stem leaves alternate, much reduced; heads sessile and axillary, or terminating naked branchlets, the interrupted arrays racemose; phyllaries in 2 series, the outer much shorter, all with hardened, whitish bases and loose, green tips; receptacles hispid, usually lacking pales; flowers all ligulate, corollas lavender-blue, rarely white or pink, open during the morning, later on cloudy days, ligules 15–25 mm long; achenes prismatic, 2–3 mm long; pappus a minute crown of scales to 0.2 mm long. A cool-climate species, uncommon in TX; roadsides; nc TX in Collin, Grayson, and Rockwall cos.; c TX in McLennan Co. and Bell and Travis cos. (TEX); also Harris Co. (Brown 2014); a few scattered records elsewhere throughout the state; throughout the continental U.S. and s Canada. May–Oct. Native of Eurasia, now a widespread weed. The roots, when ground and roasted, are the chicory often mixed with coffee; the blanched leaves are reported to be edible (Mabberley 2017).

CIRSIUM Mill. THISTLE, PLUMED THISTLE, TRUE THISTLE

Erect annual, biennial, or perennial herbs; stems sometimes spiny-winged, sometimes hollow; leaves sometimes in a basal rosette, stem leaves alternate, simple to deeply pinnatifid, prickly-toothed or -lobed, glabrate to often strikingly white or gray tomentose on one or both sides; heads borne singly or in racemose or paniculate to corymbose arrays; phyllaries many, in 5–20 series, imbricated, usually unequal, appressed, spine-tipped, spines deflexed or spreading to erect; receptacles covered with bristles or hair-like scales, but lacking pales; ray flowers absent; disk flower corollas purplish, pink-lavender, white, or rarely pale yellow to reddish, long-tubular, the lobes linear; achenes ovoid, ± compressed, smooth, glabrous; pappus of 3–5 series of many plumose bristles or narrow, plumose scales, united at bases and persistent, or falling in rings.

A n temperate genus of ca. 200 species; some are problematic weeds. *Cirsium* is taxonomically complex and prone to hybridization and subsequently has become nomenclaturally overwrought; overlapping morphological characters make keying difficult. Collection of the entire plant, including below-ground parts, is recommended. Keil (2006d) opines several of the New World species complexes of *Cirsium* "...are apparently evolutionary works in progress;" our understanding of this group is still in flux. (Greek: *kirsion*, thistle, from *kirsos*, a swollen vein, for which thistle was a reputed remedy) (tribe Cardueae)
REFERENCES: Howell 1959; Moore & Frankton 1969; O'Kennon & Nesom 1988; Dabydeen 1997; Kelch & Baldwin 2003; Keil 2004, 2006d.

1. Upper surface of leaf blades covered with short, yellowish, appressed, bristle-like spines or prickles; leaf blade bases decurrent along stem as spiny wings ____ **C. vulgare**
1. Upper surface of leaf blades glabrous to variously pubescent but without bristle-like spines or prickles; leaf blades decurrent along stem or not.
   2. Involucres closely subtended by an involucre-like ring of spiny-margined bracts, equaling the phyllaries in length ____ **C. horridulum**
   2. Involucres without whorl of spiny-margined bracts just below the phyllaries (peduncles may be naked, leafy-bracted, or overtopped by upper leaves, but these features not in a whorl).
      3. Outer phyllaries not or only minutely spine-tipped, any spines to 0.5 mm long; plants of wet soil (marshes and bogs) ____ **C. muticum**
      3. Outer phyllaries tipped with spines 1–12 mm long; plants of moist or dry habitats.
         4. Involucres 1.2–2 cm high.
            5. Midstem and lower leaves linear to narrowly elliptic, bases tapered, cuneate, not decurrent; corollas 15–20 mm long; pappus 12–14 mm long; mostly restricted to e 1/4 TX ____ **C. carolinianum**
            5. Midstem and lower leaves ovate to oblong or elliptic, broadly sessile, sometimes auriculate-clasping or short-decurrent; corollas 20–25 mm long; pappus 15–16 mm long; widespread in TX but less common in East TX ____ **C. texanum**
         4. Involucres usually 2.5–5 cm high.

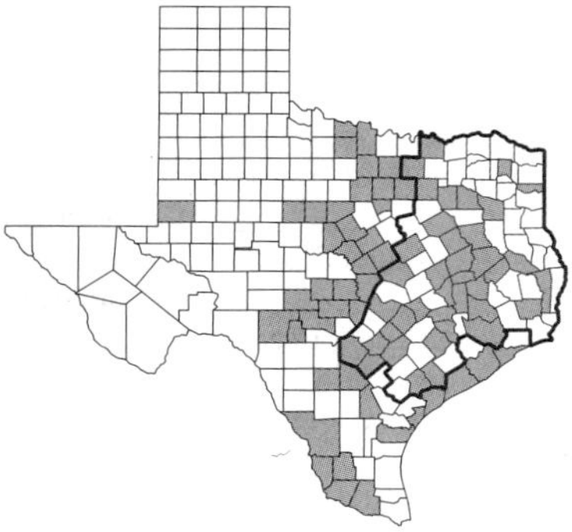

Chaetopappa asteroides

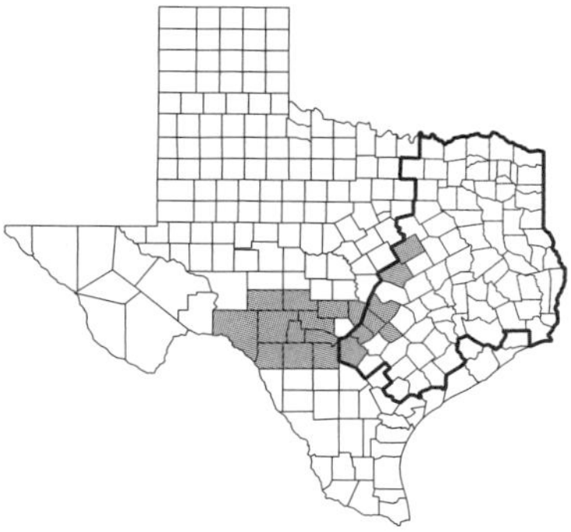

Chaetopappa bellidifolia

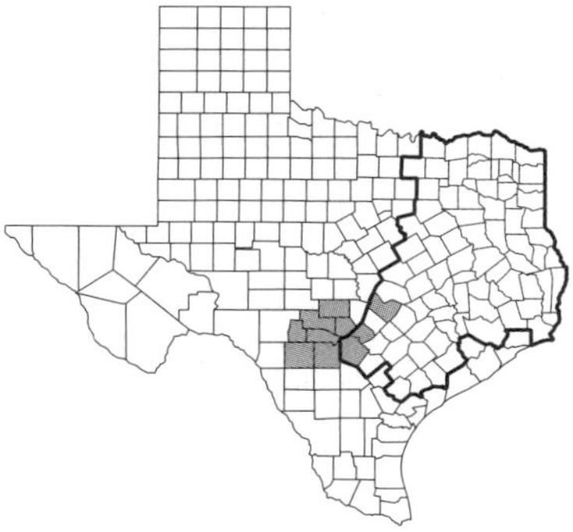

Chaetopappa effusa

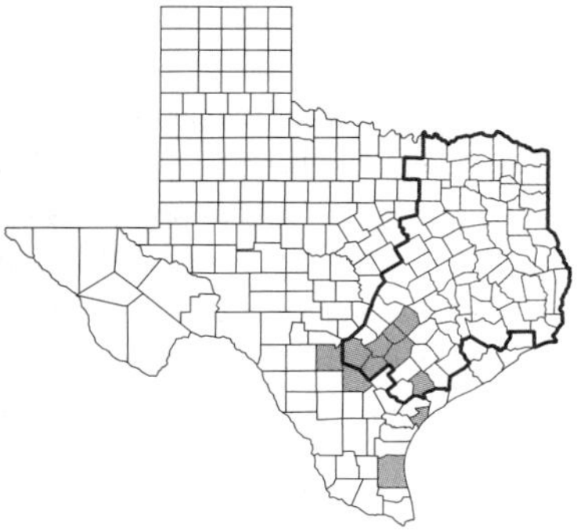

Chaetopappa imberbis

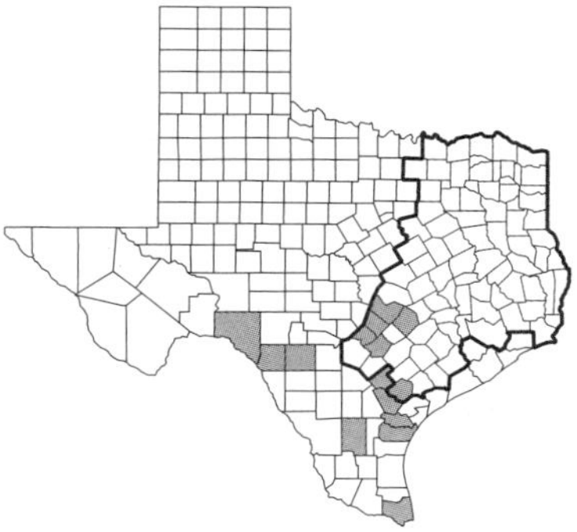

Chaptalia carduacea

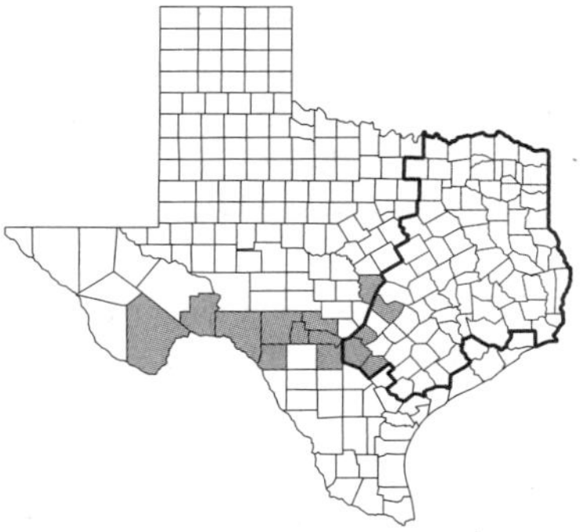

Chaptalia texana

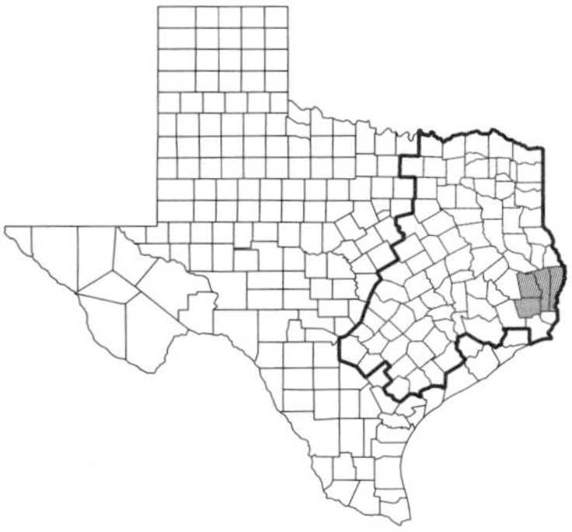

Chaptalia tomentosa

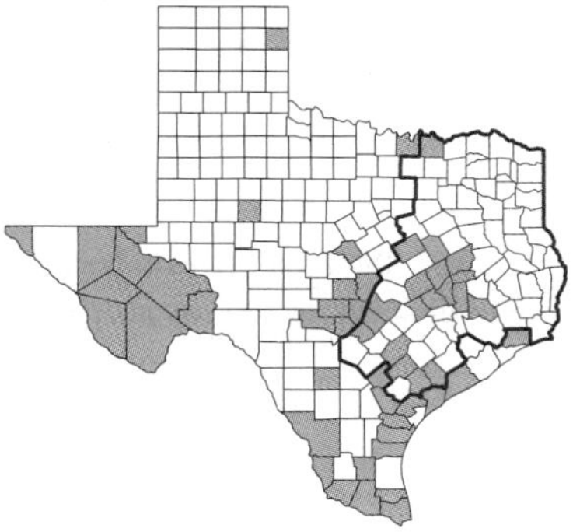

Chloracantha spinosa var. spinosa

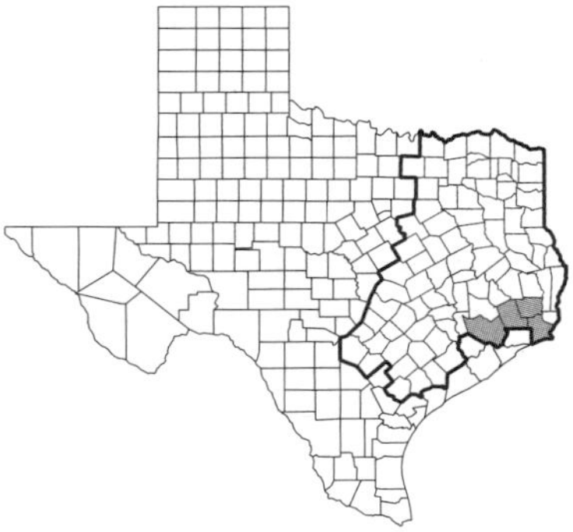

Chromolaena ivifolia

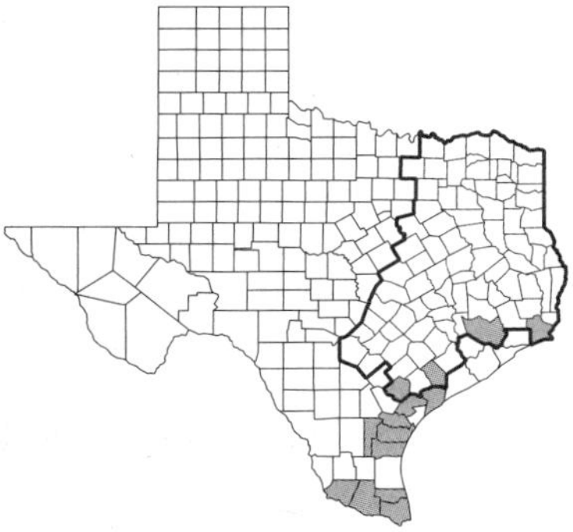

Chromolaena odorata

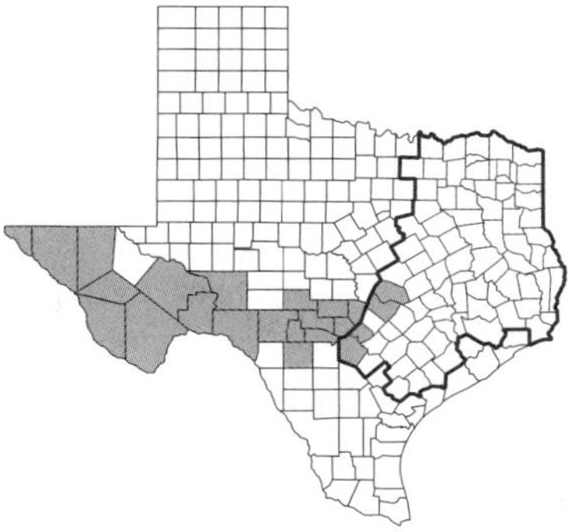

Chrysactinia mexicana

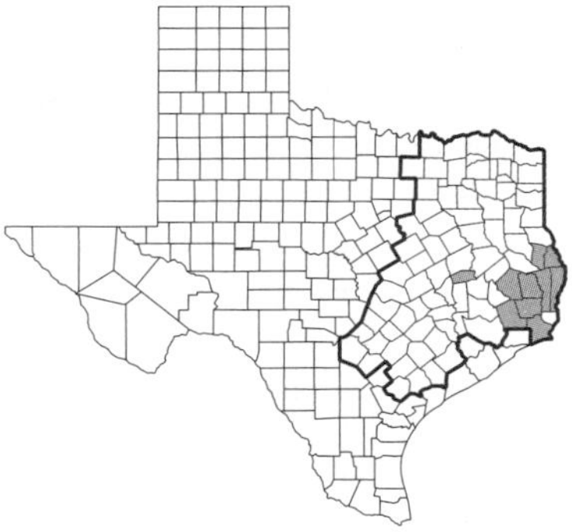

Chrysopsis mariana

6. Biennials or short-lived perennials with taproots and clusters of coarse, fibrous roots; stems thinly tomentose when young, ± glabrate or tomentum persisting on the upper stems in age; midstem leaves leaves green and glabrate on upper surfaces, bases narrowed or weakly clasping; achenes 4–6 mm long.
   7. Usually some roots with tuber-like enlargements; leaf margins revolute, upper stem leaves reduced and bract-like; peduncles essentially naked (or with much-reduced bracts); pappus 25–30 mm long ____ **C. engelmannii**
   7. Roots without tuber-like enlargements; leaf margins flat, upper stem leaves well-developed; peduncles leafy-bracted; pappus 12–24 mm long ____ **C. altissimum**
6. Perennials with crown sprouts and/or runner roots that produce adventitious buds; stems uniformly and persistently gray-tomentose; midstem leaves thinly to densely gray- or white-tomentose on upper surfaces (at least on young leaves), bases ± clasping-auriculate or decurrent into long or short stem wings; achenes 6–9 mm long.
   8. Stem leaves not decurrent, or with decurrent wings to 1 cm long; outer phyllaries tipped with spines 1.5–5 mm long ____ **C. undulatum**
   8. Stem leaves decurrent, stem wings 1–5+ cm long; outer phyllaries tipped with spines 3–12 mm long ____ **C. ochrocentrum**

Cirsium altissimum (L.) Spreng., (very tall), IOWA THISTLE, TALL THISTLE, ROADSIDE THISTLE. Biennials or short-lived perennials, to 300 (–400) cm tall, taprooted and often with a cluster of coarse fibrous roots; stems glabrate, or thinly tomentose in the upper portion; basal leaves often absent at flowering, stem leaves green and glabrate to villous above, white-tomentose beneath, undivided and spiny-toothed to coarsely toothed or shallowly pinnatifid, bases narrowed and sometimes weakly clasping, upper leaves well-developed; involucres usually 2.5–3.5 cm high; outer phyllary spines abruptly spreading, 3–4 mm long; corollas 20–35 mm long; achenes 4–5.5 mm long; pappus 12–24 mm long. Stream bottom thickets, pastures, often calcareous soils; scattered from ne to se TX; nc TX; e 1/2 of the U.S. Jun–Sep. [*C. altissimum* (L.) Hill var. *biltmoreanum* Petr., *C. iowense* (Pammel) Fernald] Hybrids with *C. discolor* are well documented (Dabydeen 1997) and have sometimes been treated as *Cirsium* ×*iowense* (Pammel) Fernald.

Cirsium carolinianum (Walter) Fernald & B.G. Schub., (of the Carolinas), SOFT THISTLE, CAROLINA THISTLE, PURPLE THISTLE, SMALLHEAD THISTLE. Biennials, to 180 cm tall, taprooted with slender, lateral fibrous roots; stems glabrous to ± tomentose; basal leaves often present at flowering, stem leaves green and glabrous or sparsely villous above, gray-tomentose beneath, unlobed and spiny-toothed to irregularly dentate or pinnatifid, bases tapered, petiolate or sessile, upper leaves reduced or bract-like; involucres 1.2–2 cm high; outer phyllary spines ascending to spreading, 1–4 mm long; corollas 15–20 mm long; achenes 3–4 mm long; pappus 12–14 mm long. Wooded areas, fields, roadsides; scattered from ne to se TX; se U.S. from OH s to GA and w to IL, MO, and TX. Jun–Sep. [*C. flaccidum* (Small) Petr.]

Cirsium engelmannii Rydb., (for George Engelmann, 1809–1884, German-born botanist and physician of St. Louis), ENGELMANN'S THISTLE, BLACKLAND THISTLE. Biennials or short-lived perennials, to 200 cm tall, taprooted and often with a cluster of coarse fibrous roots with tuber-like thickenings; stems glabrate to thinly arachnose-tomentose; basal leaves usually absent at flowering, stem leaves green and glabrate to villous above, white-tomentose beneath, usually deeply pinnatifid, lobes entire or spiny-toothed to few-toothed or sharply lobed, bases narrowed and sometimes weakly clasping, upper leaves reduced or bract-like; involucres 2.5–3.5 cm high; outer phyllary spines abruptly spreading to deflexed, 2–4 mm long; corollas 32–38 mm long; achenes 5–6 mm long; pappus 25–30 mm long. Prairies, oak savannas, in calcareous clay, rarely sandy soils; Blackland Prairie w to Grand Prairie and s to Edwards Plateau, also se TX; AR, LA, OK, and TX. May–Sep. [*C. altissimum* (L.) Hill var. *filipendulum* (Engelm.) A. Gray, *C. terrae-nigrae* Shinners, *C. virginianum* (L.) Michx. var. *filipendulum* A. Gray] Previously included by some in *C. altissimum*, and often confused with *C. discolor* (Muhl. ex Willd.) Spreng., which according to Keil (2006d) does not occur in TX.

Cirsium horridulum Michx., (prickly, horribly armed), BULL THISTLE, YELLOW THISTLE, HORRID THISTLE. Winter annuals, biennials or perennials, to 250 cm tall, taprooted with a cluster of fleshy lateral roots, often perennating by root sprouts; stems glabrous to densely tomentose; basal leaves present at flowering, stem leaves green and glabrous to loosely tomentose above and beneath shallowly to deeply pinnatifid, lobes spiny-dentate to coarsely lobed, bases often ± auriculate-clasping, upper leaves well-developed, often spinier than lower leaves; involucres 3–5 cm high, immediately subtended by a conspicuous ring of spiny-margined bracts as long as the phyllaries; outer phyllaries reddish-tinged, spines to 1 mm long; corollas pink to purple or rarely pale yellow to reddish, 30–47 mm long; achenes 4–6 mm long; pappus 25–35 mm long. Open woods, low fields, roadsides mainly e and sc TX; nc TX to c TX and s TX. Mar–Jun. Two of the varieties apparently have overlapping ranges in East TX; var. *megacanthum* was published by Keil in 2004, too recently for reassessment of many TX specimens previously identified as var. *horridulum*, therefore we are not providing TX ranges or maps for these varieties. Keil (2006d) divided the two East TX varieties of this variable species as follows:

Chloracantha spinosa [ABR]

Chromolaena ivifolia [JUJ]

Chromolaena odorata [JUJ]

Chrysactinia mexicana [FNA]

Chrysopsis mariana [GEO]

Cichorium intybus [REE]

Cirsium altissimum [BB2]

Cirsium carolinianum [STE]

1. Stems densely tomentose; main leaf spines mostly 5–10 mm long; involucres ± densely tomentose ____________ var. **horridulum**
1. Stems glabrous or sparsely tomentose; main leaf spines 10–30 mm long; involucres glabrous ____________ var. **megacanthum**

var. **horridulum.** E and s U.S. from ME s to FL and w to TX. [*C. horridulum* Michx. var. *elliottii* Torr. & A. Gray]

var. **megacanthum** (Nutt.) D.J. Keil, (large-spined). Se U.S. from FL w to OK and TX. [*C. megacanthum* Nutt.] Thomas Nuttall, the original author of this taxon, described it as "one of the most terribly armed plants in the genus" (Keil 2006d).

Cirsium muticum Michx., (muted, blunt, for the phyllaries), SWAMP THISTLE, DUNCE-NETTLE, HORSETOPS. Biennials, to 230 cm tall, taproots fleshy; stems glabrate or villous to thinly tomentose; basal leaves usually absent at flowering, stem leaves green and glabrate to thinly tomentose above, green and thinly pilose beneath, deeply pinnatifid, lobes irregularly few-toothed or lobed, bases tapered and sometimes ± clasping, petiolate or sessile, upper leaves gradually reduced, often spinier than lower leaves; involucres 1.7–3 cm high; outer phyllary spines erect, to 0.5 mm long; corollas 16–32 mm long; achenes 4.5–5.5 mm long; pappus 12–20 mm long. Marshes, bogs, wet soil in open woods; Anderson, Gonzales, and Wood cos., also Bastrop Co. (Kartesz 2015); e 1/2 of North America but more common in ne U.S. and Canada. Sep–Oct. [*C. muticum* Michx. var. *monticola* Fernald] Hybridizes with *C. discolor* (Keil 2006d).

Cirsium ochrocentrum A. Gray var. **ochrocentrum,** (yellow-centered), YELLOWSPINE THISTLE. Perennials, to 90 cm tall, with crown sprouts or runner roots; stems densely gray-tomentose; basal leaves usually present at flowering, stem leaves thinly gray-tomentose above, densely white-tomentose beneath, coarsely dentate or shallowly to deeply pinnatifid, bases usually long-decurrent as spiny stem wings, upper leaves much reduced; involucres 2.5–4.5 cm high; outer phyllary spines spreading, 3–12 mm long; corollas 25–45 mm long; achenes 6–9 mm long; pappus 20–40 mm long. Sandy or rocky prairies and roadsides, mesquite and juniper scrub, desert grasslands; w border of East TX in Ellis, Bell, Hays, and DeWitt cos.; nc TX to Edwards Plateau, Trans-Pecos, and Panhandle; w U.S. from SD s to TX and w to OR and CA; introduced in a few states; Mexico. May–Jul.

Cirsium texanum Buckley, (of Texas), TEXAS THISTLE, SOUTHERN THISTLE. Annuals or biennials, to 200 cm tall, taprooted; stems glabrate to tomentose; basal leaves often absent at flowering, stem leaves green and glabrous or thinly arachnose above, thinly grayish or whitish arachnose-tomentose beneath, undivided and spiny-toothed to irregularly dentate or shallowly to deeply pinnatifid, bases broadly ± auriculate-clasping or decurrent, upper leaves reduced or bract-like; involucres 1.5–2 cm high; outer phyllary spines spreading, 1–5 mm long; corollas 20–25 mm long; achenes 3–5 mm long; pappus 15–16 mm long. Prairies and roadsides, silty-clay and calcareous soils; nearly throughout the state but less common in deep East TX, Panhandle, and Trans-Pecos; sc U.S. from MO s to LA and w to NM; Mexico. Mar–Jun. [*C. austrinum* (Small) E.D. Schultz, *C. texanum* Buckley var. *stenolepis* Shinners] May hybrizide with *C. undulatum* (Correll & Johnston 1970). Goldfinches are reported to use the plumose pappus bristles to line their nests (Ajilvsgi 1984).

Cirsium undulatum (Nutt.) Spreng., (undulated, wavy), WAVY-LEAF THISTLE, PASTURE THISTLE, GRAY THISTLE. Perennials, to 230 cm tall, with runner roots; stems densely gray-tomentose; basal leaves sometimes present at flowering, stem leaves thinly gray-tomentose above, densely gray-tomentose beneath, coarsely dentate or shallowly to deeply lobed, bases broadly ± auriculate-clasping or short-decurrent, upper leaves reduced, often spinier than lower leaves; involucres 2.5–4.5 cm high; outer phyllary spines spreading, 1.5–5 mm long; corollas 25–50 mm long; achenes 6–7 mm long; pappus 20–38 mm long. Prairies, pastures, sagebrush deserts, roadsides and disturbed areas; Blackland Prairie from ne and nc TX to sc TX; Edwards Plateau, Trans-Pecos, and Panhandle; mostly w 1/2 of North America; introduced in a few e states; Mexico. May–Sep. [*C. megacephalum* (A. Gray) Cockerell, *C. undulatum* (Nutt.) Spreng. var. *megacephalum* (A. Gray) Fernald] May hybrizide with *C. texanum* (Correll & Johnston 1970).

Cirsium vulgare (Savi) Ten., (common), BULL THISTLE, COMMON THISTLE, SPEAR THISTLE. Biennials, to 200(–300) cm tall, taprooted; stems villous; basal leaves sometimes present at flowering, stem leaves green and covered with short and appressed bristle-like spines above, gray-tomentose beneath, coarsely spinose-dentate or coarsely 1–2-pinnatifid, bases decurrent as long spiny stem wings, upper leaves reduced, often spinier than lower leaves; involucres 3–4 cm high; outer phyllary spines spreading, 2–5 mm long; corollas 25–35 mm long; achenes 3–4.5 mm long; pappus 20–30 mm long. Roadsides, pastures, waste areas; Red River Co (R. O'Kennon, pers. obs.) and w border of East TX in Bell, Hays, Comal, and Bexar cos. (all at TEX-LL); Edwards Plateau and nc TX; first collected in TX in Gillespie Co. in 1987 (O'Kennon & Nesom 1988); throughout North America but less common ir se U.S.; Eurasia; introduced and weedy elsewhere. Jun–Sep. Native of Eurasia; a noxious weed that is distasteful to livestock anc increases in heavily-grazed pastures.

Cirsium arvense (L.) Scop., (of the fields), CANADA THISTLE, CREEPING THISTLE, FIELD THISTLE. which would possibly have keyed to C *muticum* or *C. carolinianum* here, can be distinguished by its colonial habit, small heads (1–2 cm high and wide), outer phyllar

Cirsium engelmannii [HEA]

Cirsium horridulum [BT3]

Cirsium muticum [BA1]

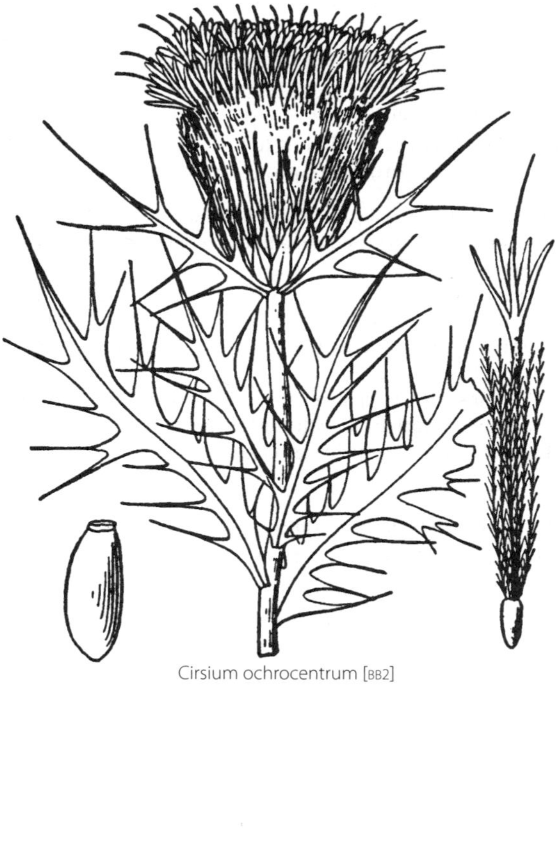

Cirsium ochrocentrum [BB2]

tips with no spines or minute spines to 1 mm long, and unusual dioecious habit (with most plants having either all staminate flowers or all pistillate flowers). Roadsides, pastures, waste areas; reported as present in TX by Keil (2006d); included here based on records from adjacent Bryan Co., OK (OKL); throughout North America except se U.S.; Eurasia; introduced and weedy in temperate regions worldwide. Jun–Oct. Despite its common name implying Canadian origin, a native of Eurasia; a highly noxious weed that is distasteful to livestock, increases in heavily-grazed pastures, and easily proliferates in cultivated fields.

CONOCLINIUM DC. THOROUGHWORT, MISTFLOWER

A genus of 4 species occurring from e North America to Mexico; formerly included in *Eupatorium*. *Conoclinium dissectum*, GREGG'S MISTFLOWER, is growing in popularity as a garden plant; its native range begins just to the sw of East TX (Edwards Plateau). (Greek: *konos*, cone, and *kline*, bed, referring to the conical receptacles) (tribe Eupatorieae)
REFERENCES: King & Robinson 1970c; Wooten & Clewell 1971; Patterson 1994; Patterson & Nesom 2006.

Conoclinium coelestinum (L.) DC., (sky-blue), BLUE MISTFLOWER, WILD AGERATUM, BLUE BONESET. Rhizomatous perennial herbs; stems erect or decumbent, 50–200 cm long, sometimes rooting at nodes; leaves opposite, petiolate, blades triangular to deltoid or ovate, gland-dotted, margins serrate to crenate; heads in tight, corymbose arrays; involucres 3–6 mm across; phyllaries in 2–3 series, ± equal; receptacles naked; ray flowers absent; disk flowers with corollas blue to violet-purple, rarely white; achenes 5-angled, 1–1.5 mm long; pappus of barbellate bristles. Moist sandy soils, low woods, ravines, stream banks; throughout e 1/2 of TX; extending into s TX; e U.S. from NY s to FL and w to NE and TX; Canada. (Jun–)Aug–Nov. [*Eupatorium coelestinum* L.] This species is gaining popularity as a native TX garden plant, although it can spread aggressively in favorable sites.

CONYZA Less. HORSEWEED

Annual herbs; stems glabrous to hispid or strigose; leaves alternate, mostly lanceolate or linear, usually strigose to hispid; heads numerous, small, borne in paniculate or corymbose arrays or ± singly (in 1 species); phyllaries in 2–4 series, unequal, appressed (usually reflexed in fruit); receptacles naked; peripheral "ray" flowers pistillate, fertile, corollas whitish, cream, or pinkish purple, the tubes filiform, the ligules mostly inconspicuous, shorter than the tubes and scarcely if at all surpassing the pappus, or sometimes ligules lacking; disk flowers perfect, fertile; corollas yellowish; achenes biconvex, 2-ribbed, 1–1.5 mm long; pappus of hair-like, minutely-barbed bristles, extending beyond phyllaries at maturity.

A genus of ca. 40 species of warm-temperate and subtropical areas; similar to *Erigeron* and previously often lumped with that genus; some species once treated in *Conyza* are now included in *Laennecia*. (Ancient name for fleabane, perhaps from Greek: *konops*, flea, or *konis*, dust, alluding to insect-repelling properties when dried and powdered) (tribe Astereae)
REFERENCES: Cronquist 1943; Shinners 1949e; Nesom 1978, 1990b; Strother 2006g.

1. Plants low and spreading, branching from near base, < 30 cm tall; receptacles 0.7–1 mm wide in fruit ____ **C. ramosissima**
1. Plants stiffly erect, often single-stemmed, branching only in the upper portion, usually > 30 cm tall; receptacles 1–5 mm wide in fruit.
   2. Phyllaries usually strigose to somewhat hispid; heads 3–5 mm wide in fruit; pistillate flowers 60–150+ per head; pappus 3–4+ mm long ____ **C. bonariensis**
   2. Phyllaries glabrous to sparsely strigose; heads 1–3 mm wide in fruit; pistillate flowers 20–45+ per head; pappus 2–3 mm long.
      3. Phyllaries usually sparsely strigose; pistillate flower ligules absent or to 0.3 mm long; usually at least a few achenes in each head with reddish nerves ____ **C. sumatrensis**
      3. Phyllaries usually glabrous; pistillate flower ligules 0.3–1+ mm long; achenes uniformly colored, without reddish nerves ____ **C. canadensis**

Conyza bonariensis (L.) Cronquist, (of Bonaria, near Buenos Aires, Argentina, where first collected), ASTHMAWEED, HAIRY FLEABANE Plants erect, 10–100(–150+) cm tall, branched in the upper portion; leaves oblanceolate to linear, entire or toothed; heads borne in paniculate or racemose arrays; involucres 3.5–5 mm high, 3–5 mm wide at maturity; pistillate flower ligules 0–0.3 mm long pappus 3–4+ mm long. Disturbed sites, roadsides; se TX; also Trans-Pecos; s and w U.S.; South America; introduced elsewhere Mostly late summer–fall. [*Erigeron bonariensis* L.] Thought to be native to South America; now a widespread weed.

Conyza canadensis (L.) Cronquist, (of Canada), HORSEWEED, HORSE-TAIL CONYZA, CANADA FLEABANE. Plants erect, 50–200(–350+ cm tall, branched in the upper portion; leaves oblanceolate to linear, entire or toothed; heads borne in paniculate arrays; involucres 3–4 mm high, receptacles 1–1.5(–3) mm wide at maturity; pistillate flower ligules 0.3–1 mm long; pappus 2–3 mm long. Disturbed sites, eroded ground, waste areas; common and widespread across TX; throughout North America; Mexico and Central America; introduced and weedy elsewhere. Mostly summer–fall. [*C. canadensis* (L.) Cronquist var. *pusilla* (Nutt. Cronquist, *Erigeron canadensis* L., *E. canadensis* var. *glabratus* A. Gray]

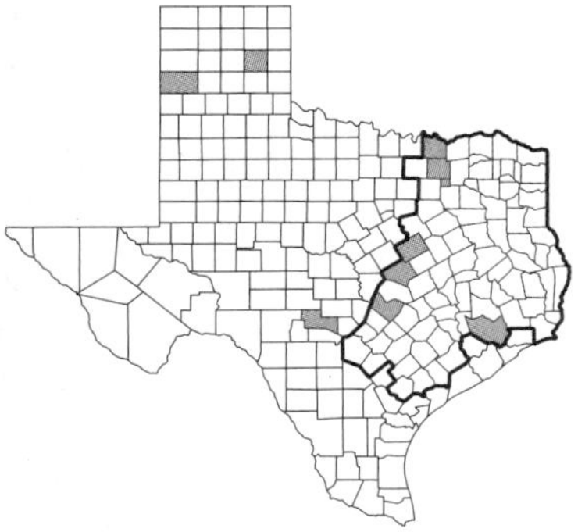

Cichorium intybus

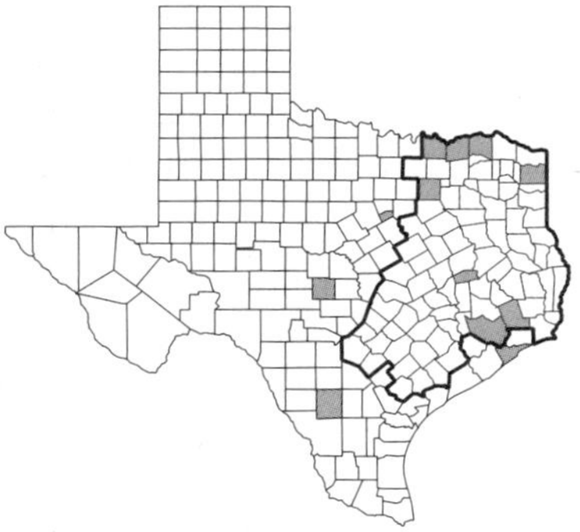

Cirsium altissimum

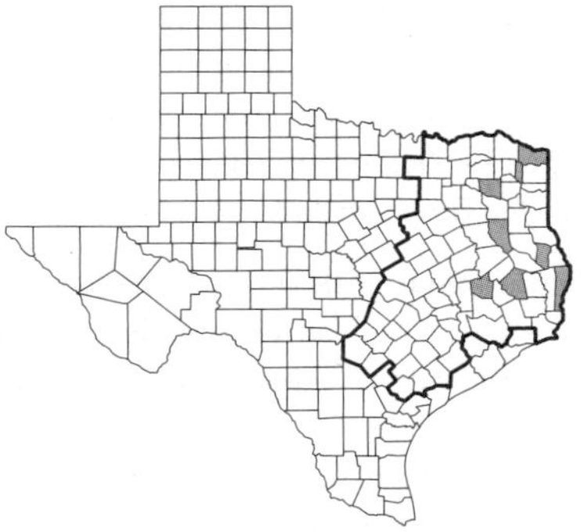

Cirsium carolinianum

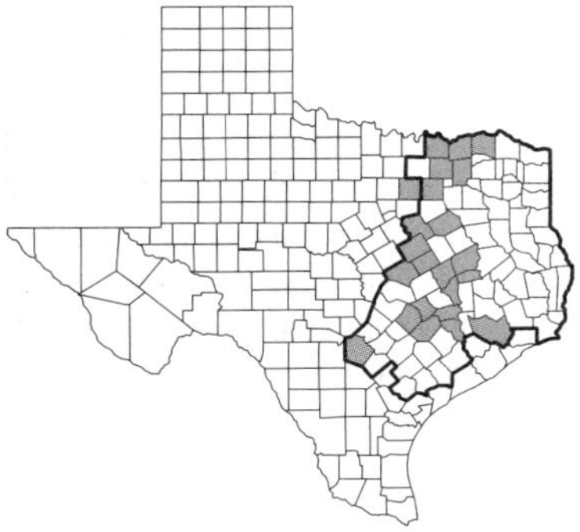

Cirsium engelmannii

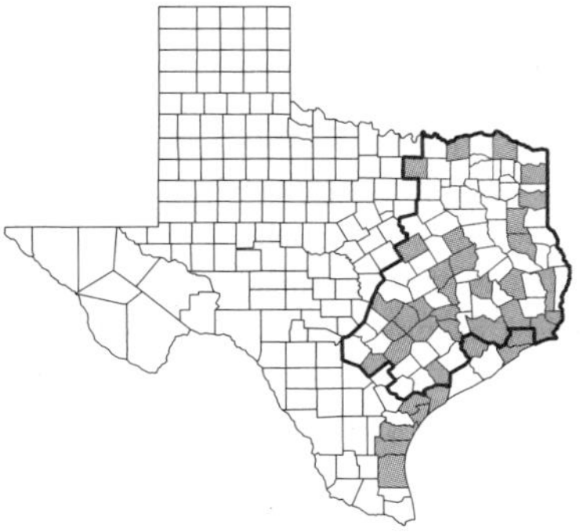

Cirsium horridulum

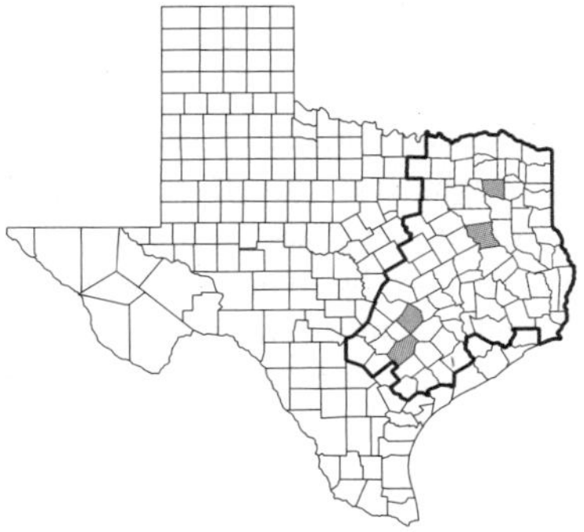

Cirsium muticum

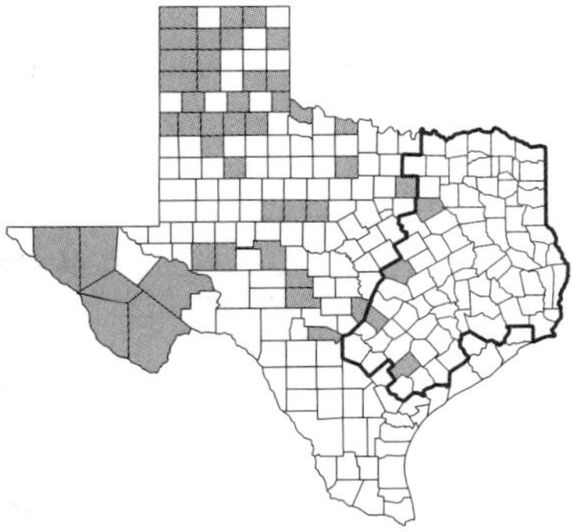

Cirsium ochrocentrum var. ochrocentrum

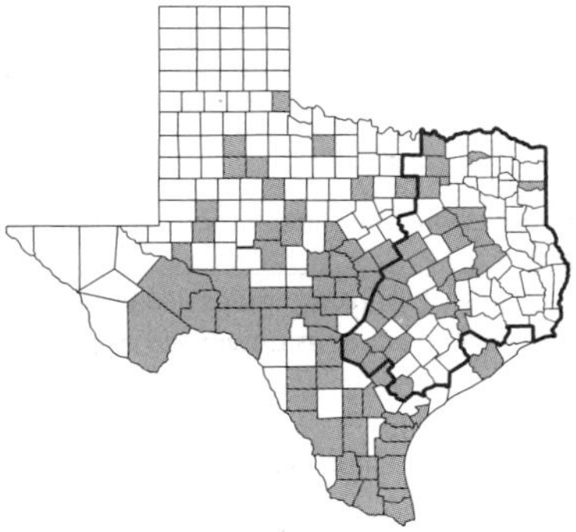

Cirsium texanum

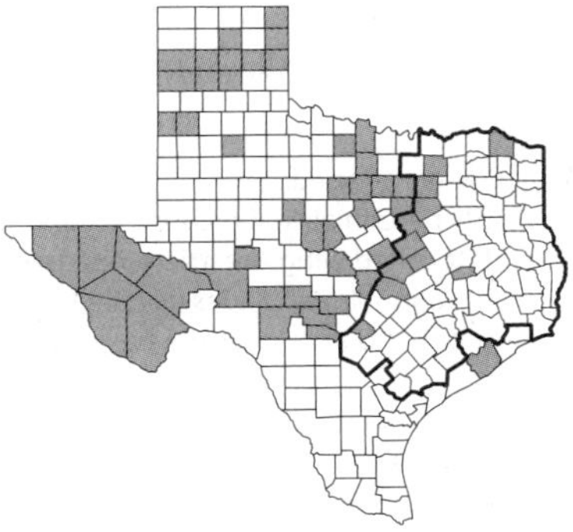

Cirsium undulatum

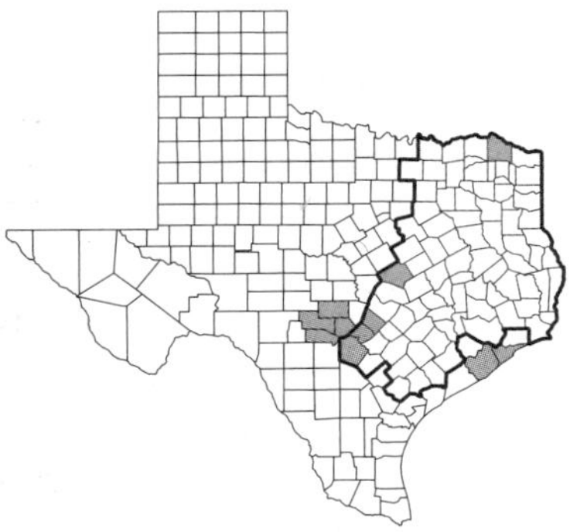

Cirsium vulgare

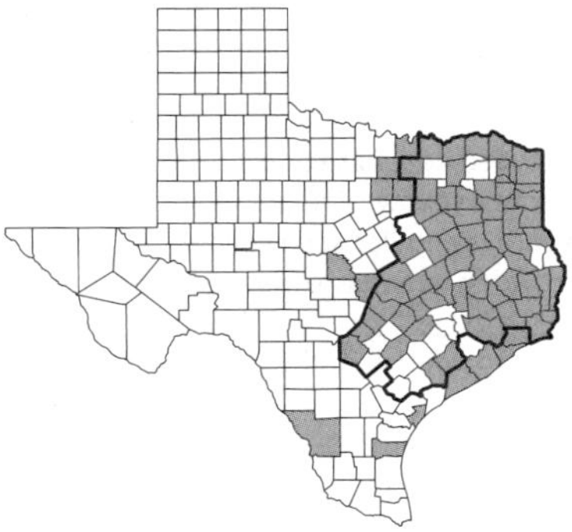

Conoclinium coelestinum

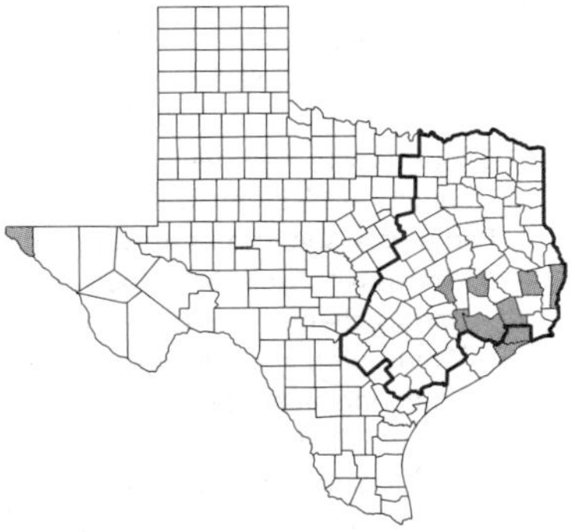

Conyza bonariensis

Conyza ramosissima Cronquist, (much-branched), DWARF HORSEWEED, LOW FLEABANE, LOW CONYZA, SPREADING FLEABANE. Plants spreading, 5–25+ cm tall, branched throughout; stems ± strigose; leaves narrowly spatulate to linear, entire; heads borne ± singly or in loose corymbose arrays; involucres 3–4 mm high, 0.7–1 mm wide at maturity; pistillate flower ligules 0.3–0.8 mm long; pappus 2–2.5 mm long. Clay soils, disturbed habitats; ne and nc TX s to Edwards Plateau; also Brazos Co.; Plains Country; c U.S. Jun–Oct. [*Erigeron divaricatus* Michx.]

Conyza sumatrensis (Retz.) E. Walker, (of Sumatra), BILBAO FLEABANE. Plants erect, 30–150+ cm tall, branched in the upper portion; leaves narrowly oblanceolate to linear, entire or toothed; heads borne in paniculate arrays; involucres 3–4 mm high, receptacles 1–2+ mm wide at maturity; pistillate flower ligules 0–0.3 mm long; pappus 2–3 mm long. Disturbed sites; in East TX recorded only from Waller Co. (iNaturalist.org observation *408969, Jason Singhurst*, TPWD State Botanist, Sep 2013 (as *C. floribunda*; iNaturalist.org 2016)); first reported for TX in 2013 in Chambers Co. (as *C. floribunda*; Singhurst & Holmes 2013); AL, CA, GA, and TX; South America; introduced elsewhere. Summer–fall. [*C. albida* Willd. ex Spreng., *C. bonariensis* (L.) Cronquist var. *leiotheca* (S.F. Blake) Cuatrec., *C. floribunda* Kunth, *Erigeron bilboanus* (Remy) Cabrera] Thought to be native to South America; now a widespread weed. The complexity of this species is discussed in Pruski and Sancho (2006).

COREOPSIS L. GOLDEN-WAVE, TICKSEED

Ours herbaceous annuals or perennials, glabrous or pubescent; leaves sometimes in a basal rosette, stem leaves usually opposite, sometimes alternate, usually petiolate, simple and entire to pinnately or pedately dissected or compound; heads pedunculate, solitary or in corymbose arrays; primary (inner) phyllaries usually 8, in 2 series, membranous, forming a cup, margins ± scarious, subtended by secondary, phyllary-like bracts, these distinct and herbaceous; receptacles chaffy; ray flowers ca. 8 (more in "double" cultivars), neuter, the ligules yellow, in some species with a reddish brown to purplish basal spot or flecks, often lobed or toothed at the tip; disk flowers perfect, fertile, corollas solid yellow or with lobes reddish brown to purplish, 4–5-lobed; achenes flattened, usually thin-margined or winged; pappus none, or of 2 bristly scales, or sometimes shoulders of achene wings ± bristly and pappus-like.

A genus of ca. 35 species, mostly of North America, also tropical New World and Old World; closely related to *Bidens*. Some are cultivated as ornamentals and for the cut-flower trade. (Greek: *korios*, bedbug, and *opsis*, appearance, from the form of the achene) (tribe Heliantheae)

REFERENCES: Sherff & Alexander 1955; Smith & Parker 1971; Smith 1974, 1976; Jansen et al. 1987; Tadesse et al. 1995, 2001; Strother 2006h.

1. Disk flowers 4-lobed.
   2. Plants perennial; leaf blades entire or sometimes with 1–2 lateral lobes; ray ligules usually solid yellow; achenes with toothed wings; restricted to deep East TX ____ **C. gladiata**
   2. Plants annual; leaf blades usually 1–2(–3)-pinnately lobed; ray ligules usually yellow with a red-brown to purplish spot or band at or near the base, rarely solid yellow; achenes unwinged or with entire wings; widespread in TX ____ **C. tinctoria**
1. Disk flowers 5-lobed.
   3. Leaf blades usually 3-foliolate; ray ligules oblong to elliptical, widest near the middle, the tips entire or with inconspicuous notches ____ **C. tripteris**
   3. Leaf blades simple or pinnately or pedately lobed; ray ligules ± cuneate, widest above the middle, the tips usually lobed or toothed.
      4. Annuals (rarely persisting); ray ligules yellow, usually with a red-brown to purplish spot, band, or flecks at or near the base.
         5. Stems with well-developed leaves nearly to the top (3/4+ of plant height); secondary phyllary-like bracts 6–9(–12+) mm long; disk corolla lobes red-brown to purple; achenes wingless ____ **C. basalis**
         5. Stems with well-developed leaves crowded near base (most on lower 1/2 of stem); secondary phyllary-like bracts 4–6(–8) mm long; disk corolla lobes yellow; achenes winged ____ **C. nuecensis**
      4. Perennials (sometimes flowering first year); ray ligules solid yellow.
         6. Stems with well-developed leaves crowded near base (most on lower 1/2 of stem); disk corollas 6–7.5 mm long; achenes (2.6–)3–4 mm long ____ **C. lanceolata**
         6. Stems with well-developed leaves not crowded near the base (>1/2 of plant height); disk corollas 2.2–5.4 mm long; achenes 2–3 mm long.
            7. Leaf blades rarely simple, usually 1(–2)-pinnately or pedately lobed with (3–)5–9+ lobes, terminal lobes (or blades) narrowly lanceolate to linear or filiform ____ **C. grandiflor**
            7. Leaf blades usually simple, rarely with 1–2(–3+) lateral lobes, blades (or terminal lobes) oblong-elliptic to ovate or lanceolate to oblanceolate.
               8. Leaf blades (or terminal lobes) oblong to elliptic or ovate; primary phyllaries 9–12 mm long; ray ligules 20–30+ mm long; disk corollas 3.4–4.4 mm long ____ **C. intermedi**

8. Leaf blades (or terminal lobes) lance-elliptic to lanceolate or oblanceolate; primary phyllaries 5–8+ mm long; ray ligules 12–15+ mm long; disk corollas 4.6–5.4 mm long ________ **C. pubescens**

Coreopsis basalis (A. Dietr.) S.F. Blake, (basal), GOLDEN-MANE COREOPSIS, GOLDEN-MANE TICKSEED. Annuals, 10–50+ cm tall; leaves opposite, blades simple or 1(–2)-pinnately lobed; secondary phyllary-like bracts 6–9(–12+) mm long; ray ligules 15–20+ mm long, yellow, usually with a basal red-brown to purple spot or band; disk corollas 5-lobed, lobes red-brown to purplish; achenes 1.2–1.8(–2) mm long, wingless. Sandy soils, calcareous prairies, disturbed areas; se and sc TX; nc TX s to Edwards Plateau and s TX; AR, LA, OK, and TX; apparently adventive in se U.S. and a few other states (Kartesz 2015). Mar–Jun. [*Calliopsis basalis* A. Dietr., *Coreopsis basalis* (A. Dietr.) S.F. Blake var. *wrightii* (A. Gray) S.F. Blake, C. wrightii (A. Gray) H.M. Parker ex E.B. Sm.]

Coreopsis gladiata Walter, (sword-like), COASTAL PLAIN TICKSEED, TEXAS TICKSEED. Perennials, 30–70 cm tall; leaves usually alternate, sometimes opposite, blades simple or sometimes with 1(–2) lateral lobes; secondary phyllary-like bracts 2–6+ mm long; ray ligules (12–)15–20+ mm long, yellow; disk corollas 4-lobed, lobes red-brown to purplish; achenes 3–4 mm long, wings toothed. Peaty bogs, pine barrens, sandy soils; se TX in Angelina, Hardin, Newton, Polk, and Tyler cos.; se U.S. from VA s to FL and w to TX. Aug–Oct. [*C. gladiata* Walter var. *linifolia* (Nutt.) Cronquist, *C. helianthoides* Beadle, *C. linifolia* Nutt.] We are following Strother (2006h) in including *C. linifolia* as a synonym.

Coreopsis grandiflora Hogg ex Sweet, (large-flowered), LARGE-FLOWER TICKSEED, SHOWY TICKSEED. Perennials, 40–60+ cm tall; leaves opposite, blades usually irregularly pinnately or pedately lobed (usually with 5–9+ lobes), rarely simple; secondary phyllary-like bracts 3.5–9+ mm long; ray ligules 12–25+ mm long, yellow; disk corollas 5-lobed, lobes yellow; achenes 2–3+ mm long, wings entire or toothed. Sandy woods, roadsides, disturbed sites; scattered in e 1/2 TX; se and sc U.S. from NC s to FL and w to NM; records in the ne and sw U.S. are apparently adventive (Kartesz 2015); Canada. Mar–Jun. [*C. grandiflora* Hogg ex Sweet var. *harveyana* (A. Gray) Sherff, *C. grandiflora* Hogg ex Sweet var. *longipes* (Hook.) Torr. & A. Gray] We are following Strother (2006h) in not recognizing varieties. Widely cultivated as an ornamental and spreading outside of its native range.

Coreopsis intermedia Sherff, (intermediate), GOLDEN-WAVE TICKSEED. Perennials, 60–90+ cm tall; leaves opposite, blades usually simple, rarely with 1(–3+) lateral lobes; secondary phyllary-like bracts 6–9 mm long; ray ligules 20–30+ mm long, yellow; disk corollas 5-lobed, lobes yellow; achenes 2.2–2.8 mm long, wings entire. Sandy soils, openings in oak woods, roadsides, disturbed sites; e 1/3 TX; AR, LA, OK, and TX. May–Jul.

Coreopsis lanceolata L., (lanceolate, lance-shaped), LANCE-LEAF TICKSEED, LANCE COREOPSIS. Perennials, 10–30(–60+) cm tall; leaves opposite, blades simple or with 1–2+ lateral lobes; secondary phyllary-like bracts 4–8(–12) mm long; ray ligules 15–30+ mm long, yellow; disk corollas 5-lobed, lobes yellow; achenes (2.6–)3–4 mm long, wings entire. Sandy soils, roadsides, disturbed areas; e 1/2 TX, mainly se TX; adventive in Trans-Pecos; Great Lakes s to FL and w to KS and TX; apparently adventive in ne and w U.S. (Kartesz 2015); Canada. Apr–Jul. [*C. lanceolata* L. var. *villosa* Michx.] Widely cultivated as an ornamental and popular in wildflower seed mixes; spreading outside of its native range, especially along highways.

Coreopsis nuecensis A. Heller, (of the Nueces River area of TX), NUECES TICKSEED, CROWN COREOPSIS, RIO GRANDE TICKSEED. Annuals, 10–30(–50+) cm tall; leaves opposite, blades simple or 1(–2)-pinnately lobed; secondary phyllary-like bracts 4–6(–8) mm long; ray ligules 12–18+ mm long, yellow, usually with basal flecks of red-brown to purple; disk corollas 5-lobed, lobes yellow; achenes 2.5–4 mm long, wings entire. Sandy soils, open oak and mesquite woods; se and sc TX; s TX; LA and TX; adventive in a few states; Mexico (probably). Mar–Jun. [*C. coronata* Hook., *C. nuecensoides* E.B. Sm.] We are following Strother (2006h) in including *C. nuecensoides* as a synonym.

Coreopsis pubescens Elliott, (pubescent), STAR TICKSEED. Perennials, 20–70(–90+) cm tall; leaves opposite, blades usually simple, rarely with 1–2(–3+) lateral lobes; secondary phyllary-like bracts 3–7 mm long; ray ligules 12–15+ mm long, yellow; disk corollas 5-lobed, lobes yellow; achenes 2.5–3 mm long, wings entire. Sandy soils, open pine-oak woods, ditches and roadsides; in TX recorded only from Newton and Jasper cos; se U.S. from WV s to FL and w to IL, KS, and TX; apparently adventive in ne U.S. (Kartesz 2015). Jul–Aug. [*C. pubescens* Elliott var. *debilis* (Sherff) E.B. Sm. We are following Strother (2006h) in not recognizing varieties.

Coreopsis tinctoria Nutt., (of dyes), PLAINS COREOPSIS, GOLDEN TICKSEED, CARDAMINE COREOPSIS, MANZANILLA SILVESTRE, CALLIOPSIS. Annuals, 10–70(–150+) cm tall; leaves usually opposite, blades usually 1–2(–3)-pinnate, rarely simple; secondary phyllary-like bracts 1–3+ mm long; ray ligules (7–)12–18+ mm long, usually yellow with a basal red-brown to purple spot or band, rarely solid yellow; disk corollas 4-lobed, lobes red-brown to purplish; achenes 1.3–3(–4+) mm long, wingless or wings entire. Low moist areas, often sandy soils; throughout TX; throughout U.S. but mostly in Great Plains, adventive farther e and w; Canada; Mexico. May–Aug (sporadically year-round). [*C. cardaminefolia* (DC.) Torr. & A. Gray, *C. tinctoria* Nutt. var. *similis* (Boynt.)

H.M. Parker] We are following Strother (2006h) in not recognizing varieties. This species has been used as a source of red and yellow dyes—hence the specific epithet (Tveten & Tveten 1993). Widely cultivated for its showy flowers and popular in wildflower seed mixes, spreading along highways and well-established outside its native range.

Coreopsis tripteris L., (3-winged), TALL TICKSEED, TALL COREOPSIS. Perennials, 90–180+ cm tall; leaves opposite, blades usually 3-foliolate, leaflets usually simple, sometimes ± pinnately lobed; secondary phyllary-like bracts 1.5–5 mm long; ray ligules 12–22+ mm long, yellow; disk corollas 5-lobed, red-brown to purplish; achenes 4–5(–6) mm long, wings entire. Bogs and seeps, moist sands, along streams; Anderson, Bowie, Henderson, Jasper, Newton, and Wood cos.; e U.S. from PA s to FL and w to WI, KS, and TX; Canada. Jul–Sep. [*C. tripteris* L. var. *deamii* Standl., *C. tripteris* L. var. *smithii* Sherff]

Coreopsis palmata Nutt., (palmate, for the leaves), STIFF TICKSEED, which would not have keyed satisfactorily to any other species here, can be distinguished by its perennial habit, palmately 3(–5+) lobed leaves that are nearly sessile, ray ligules that are solid yellow and widest near the middle, yellow-lobed disk flowers, and winged achenes that are 5–6 mm long. Prairies, open woods; not yet reported in TX but included here based on records from adjacent Choctaw and McCurtain cos., OK (OKL); c U.S. from MI s to LA and w to MN, NE, and OK. Jun–Aug.

COSMOS Cav.

Ours showy annual herbs; stems usually 1, erect; leaves opposite, blades usually 1–3-pinnately-lobed, sometimes entire, ultimate margins usually entire; heads solitary or few to numerous in terminal corymbose arrays; primary (inner) phyllaries usually 8, in 2 series, equal, subtended by secondary, phyllary-like bracts, these usually 8; receptacles chaffy; ray flowers usually 8 (more in "double" cultivars), neuter, corollas white to pink to purple or yellow to red-orange; disk flowers perfect, fertile, corollas yellow to orange (at least the lobes); achenes spindle-shaped, usually with 1 groove, beaked; pappus of 2–4(–8) barbed awns, or sometimes absent.

A genus of ca. 26 species of the tropical and subtropical Americas; most speciose in Mexico. Some have a long history of being cultivated as ornamentals; cultivars may vary considerably from these descriptions in both stature and corolla coloration. (Greek: *kosmo*, ornament) (tribe Heliantheae)
REFERENCES: Safford 1918; Sherff 1955; Melchert 1975; Kiger 2006.

1. Ray ligules white, pink, or purplish ______ **C. bipinnatus**
1. Ray ligules yellow to red-orange ______ **C. sulphureus**

Cosmos bipinnatus Cav., (bipinnate), GARDEN COSMOS. Plants 30–200 cm tall; leaves sessile or petiolate, 6–11 cm long, heads 7–15 mm across; primary phyllaries lanceolate to ovate-lanceolate; ray ligules white, pink, or purplish, 15–50 mm long, toothed at the tips; achenes 7–16 mm long; pappus absent, or of 2–3 awns 1–3 mm long. Disturbed sites, roadsides; recorded in Harris and Polk cos. (Brown et al. 2007); scattered across U.S.; Mexico; introduced elsewhere. Summer–fall. Native of Mexico and sw U.S.; widely cultivated and escaping and naturalizing in warm climates worldwide. In TX, occurring as an occasional escape or waif, often introduced in "wildflower" seed mixes, and probably not persisting in the flora.

Cosmos sulphureus Cav., (yellow-orange), YELLOW COSMOS, SULPHUR COSMOS. Plants 30–200 cm tall; leaves petiolate, 5–12(–25) cm long, heads 6–10 mm across; primary phyllaries oblong-lanceolate; ray ligules yellow to red-orange, 18–30 mm long, toothed at the tips; achenes 15–30 mm long; pappus absent, or of 2–3 awns, 1–7 mm long. Disturbed sites, roadsides; recorded in Bexar, Brazos, Dallas, Travis, and Tyler cos.; a few other counties across TX; scattered across U.S.; Mexico; introduced elsewhere Summer–fall. Native of Mexico; widely cultivated and escaping and naturalizing in warm climates worldwide. According to Safford (1918), this species was used by the Aztecs as a dye plant and was known as *xochipalli*, flower paint. In TX, occurring as an occasional escape or waif, often introduced in "wildflower" seed mixes, and probably not persisting in the flora.

COTULA L. WATER-BUTTONS

A genus of ca. 55 species of the s hemisphere, mostly native to the Old World. (Greek: *kotule*, small cup, probably in reference to the appearance of the heads) (tribe Anthemideae)
REFERENCES: Lloyd 1972; Watson 2006b.

Cotula australis (Sieber ex Spreng.) Hook. f., (southern), AUSTRALIAN WATER-BUTTONS, SOUTHERN BRASSBUTTONS. Annual herbs 2–25 cm tall, pubescent; stems branched throughout, sometimes decumbent and rooting at the nodes; leaves alternate, petiolate or sessile, blades obovate to spatulate, 2–3-pinnately lobed, ultimate lobes narrowly spatulate or linear; heads borne singly; phyllaries in 2–3 series, subequal, tips brown; receptacles naked; ray flowers pistillate, fertile, in 1–3 series, lacking corollas; disk flowers functionally staminate, corollas yellowish white to pale creamy yellow; peripheral achenes stalked, 1–1.2 mm long, winged

Cirsium texanum [HEA]

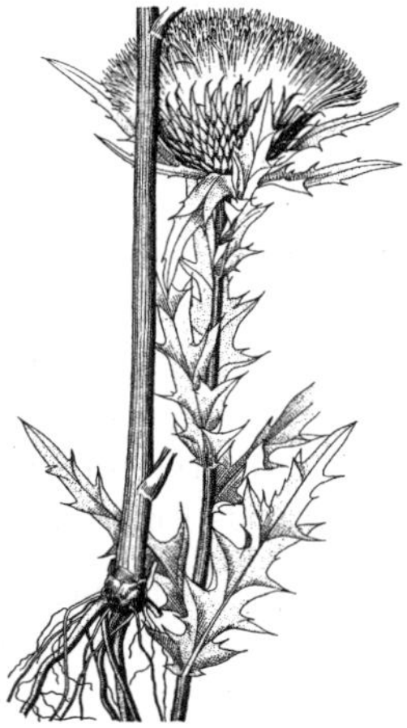

Cirsium undulatum [GLE]

Cirsium vulgare [REE]

Conoclinium coelestinum [CO1]

Conyza bonariensis [DIL]

Conyza canadensis [REE]

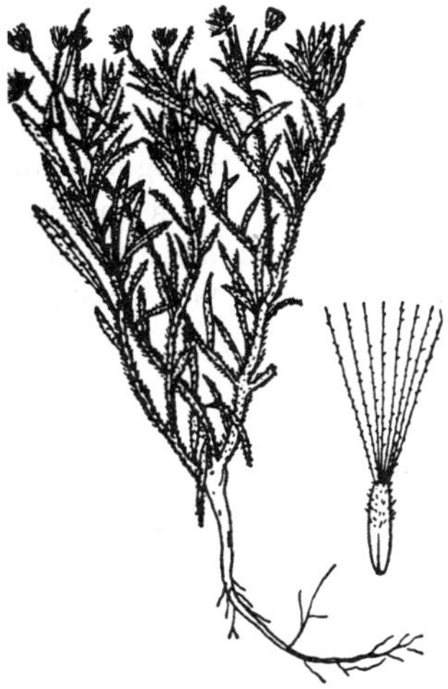

Conyza ramosissima [BB2]

Conyza sumatrensis [FSA]

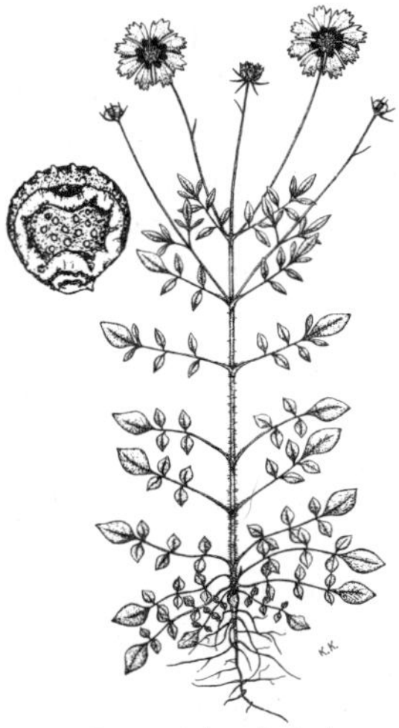

Coreopsis basalis [SID]

central achenes sessile, 0.8–1 mm long, unwinged; pappus absent. Disturbed, moist sites; rare in TX; recorded in Brazos Co. in 1970 (Correll & Johnston 1970); scattered in coastal states, mostly w and sw U.S.; introduced elsewhere. Jan–Sep. Native of Australia; widely introduced as a weed worldwide. In TX, occurring as an occasional escape and probably not persisting in the flora.

CREPIS L. HAWK'S-BEARD

Ours annual herbs, often pubescent, usually taprooted; sap milky; leaves in a basal rosette, alternate and reduced up the stem, variously lobed, toothed, or pinnatifid; heads solitary or few to numerous in terminal cymose arrays; inner (primary) phyllaries in 1–2 series, lanceolate, equal to subequal, subtended by secondary, phyllary-like bracts, these reduced and subulate to lanceolate; receptacles naked or hairy, usually not chaffy; all flowers ligulate, perfect, fertile, corollas usually yellow; achenes narrowly columnar, ribbed, the apices usually tapered or beaked; pappus of numerous hair-like, minutely barbed bristles.

A genus of ca. 200 species of the n hemisphere, s Africa, and South America; a number are weeds and a few are cultivated as ornamentals. (Greek: *crepis*, a classical Greek name of some plant that also meant "sandal" or "boot," possibly alluding to the shape of the achene) (tribe Cichorieae)

REFERENCES: Babcock & Stebbins 1938; Babcock 1947; Clausen 1949; Stebbins 1949; Vuilleumier 1973; Bogler 2006a.

1. Stems dichotomously branched; heads solitary and ± sessile in branch axils; primary phyllaries becoming basally swollen and angular at maturity, enclosing and partially fusing to achenes ____ **C. zacintha**
1. Stems not dichotomously branched; heads many, in terminal cymose arrays, not sessile; primary phyllaries strongly keeled but not enclosing nor partially fusing to achenes at maturity.
   2. Primary phyllaries coarsely hispid; achenes beaked ____ **C. setosa**
   2. Primary phyllaries glabrous, achenes with apices sometimes narrowed but not beaked ____ **C. pulchra**

Crepis pulchra L., (handsome), SHOWY HAWK'S-BEARD, SMALLFLOWER HAWK'S-BEARD. Taprooted annuals, 50–100 cm tall; stem pubescence glandular; leaves petiolate, with glandular pubescence like the stems, blades oblanceolate, often runcinate, margins deeply pinnately lobed to denticulate; heads numerous; involucres 8–12 mm high; secondary phyllary-like bracts 5–7, ovate to lanceolate; primary phyllaries 10–14, lanceolate, strongly keeled, glabrous; corollas yellow, ligules 5–12 mm long; achenes 4–6 mm long, subcylindric, 10–12 ribbed, apices tapered; pappus bristles dusky white, 4–5 mm long. Disturbed and weedy areas; scattered in nc, c, and e TX; se and c U.S. and CA and OR; introduced elsewhere. Apr–May. Native of Eurasia; now a widespread weed.

Crepis setosa Haller f., (bristly), BRISTLY HAWK'S-BEARD. Shallowly-taprooted annuals, 8–80 cm tall; stems often reddish, coarsely bristly; leaves petiolate, finely hispid, blades oblanceolate, often runcinate or lyrate, margins dentate to pinnately lobed or laciniate; heads numerous; involucres 6–10 mm high; secondary phyllary-like bracts 10–14, linear; primary phyllaries 12–16, lanceolate, strongly keeled, coarsely hispid; corollas yellow, sometimes reddish beneath, ligules 8–10 mm long; achenes 3–5 mm long including beak (beaks 1–2 mm long), fusiform, 10-ribbed; pappus bristles white, 4 mm long. Disturbed and weedy areas; recorded only from Travis (TEX) and Williamson cos.; scattered in North America in c and ne U.S. and Pacific and mountain states; introduced elsewhere. Apr–May. Native of the Mediterranean; now a widespread weed.

Crepis zacintha (L.) Babc., (likely from *Zacynthus*, Latinized from Zakynthos, an island in the Ionian Sea), STRIPED HAWK'S-BEARD. Shallowly-taprooted annuals, 20–30 cm tall; stems hispid; leaves petiolate, finely hispid, blades lyrate, margins pinnately lobed; heads borne singly in branch axils, ± sessile; involucres 5–7 mm high; secondary phyllary-like bracts 5, lanceolate; primary phyllaries 10, lanceolate, becoming basally swollen and angular at maturity (then often purplish), enclosing and partially fusing to achenes; corollas yellow, striped or tipped purplish red beneath, ligules 7 mm long; achenes 2–2.5 mm long, triangular and flattened or columnar with apices tapered; pappus bristles white, 1.5 mm long. Disturbed and weedy areas; recorded only from Henderson Co. (SBSC) and Waller Co., which in 1992 was the first record for the U.S. (Gandhi & Brown 1993); introduced elsewhere. Apr–May. Native of the Mediterranean.

CROPTILON Raf. SCRATCH DAISY

Ours usually annual, taprooted herbs; leaves alternate, sessile, 3-nerved, narrowly oblanceolate to linear, margins entire to serrate; heads in loose, paniculate arrays or solitary; phyllaries in 3–5 series, strongly unequal; receptacles naked; ray flowers pistillate, fertile, ligules bright yellow to yellow-orange, coiling at maturity; disk flowers perfect, fertile, corollas yellow; achenes columnar, 6–14+-nerved, pubescent, 2–3 mm long; pappus of persistent, barbed bristles.

A genus of 3 species of the se U.S. and n Mexico; previously treated by some in *Haplopappus*. (Greek: *kropion*, scythe, and *ptilon*, wing or feather, perhaps alluding to the appearance of the leaves) (tribe Astereae)

REFERENCES: Hall 1928; Smith 1965, 1981; Nesom 1991c, 2006k.

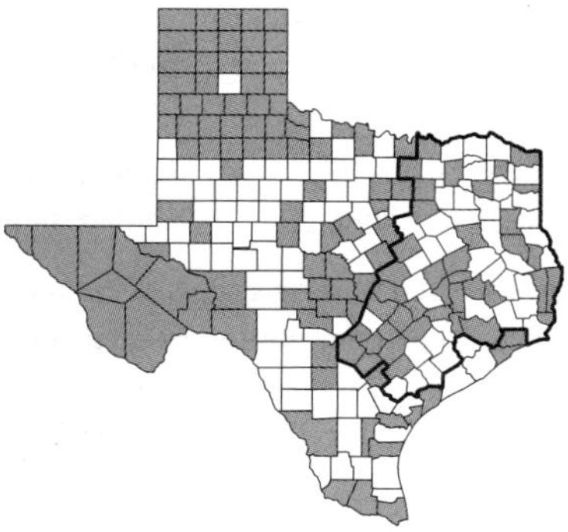

Conyza canadensis

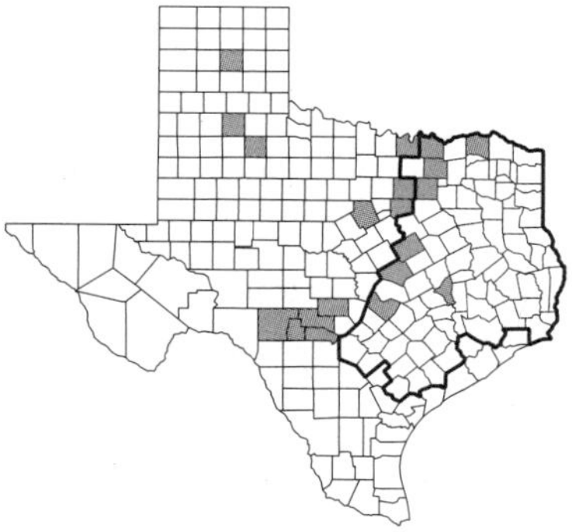

Conyza ramosissima

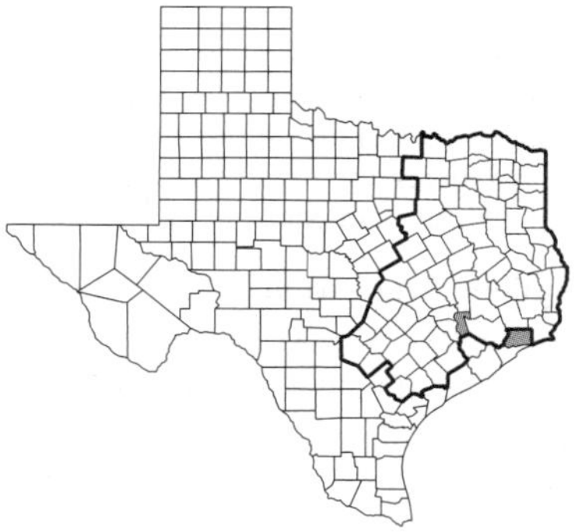

Conyza sumatrensis

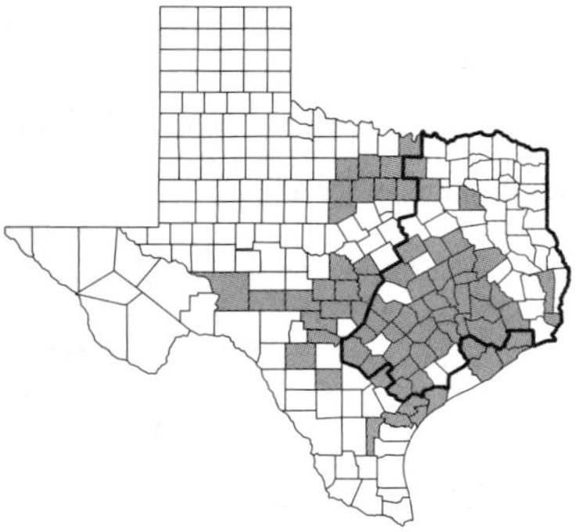

Coreopsis basalis

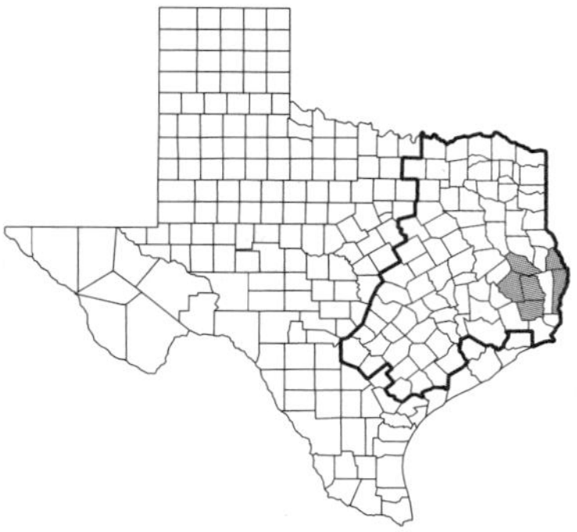

Coreopsis gladiata

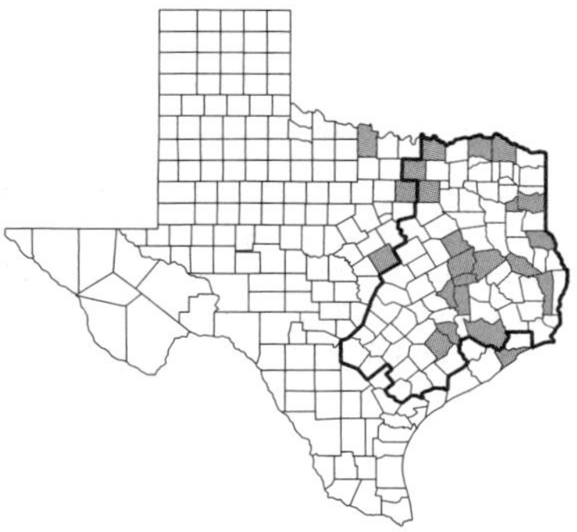

Coreopsis grandiflora

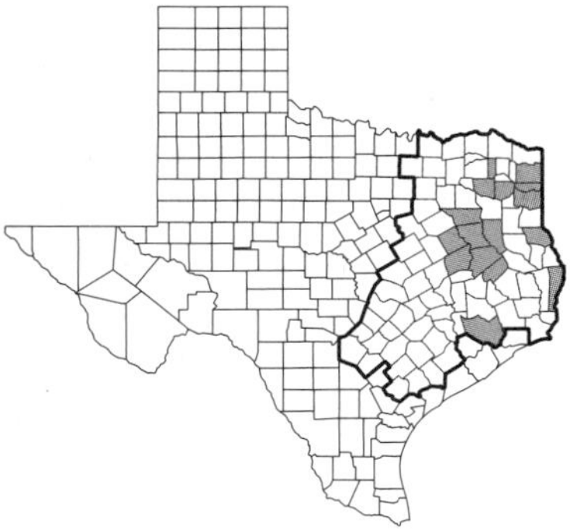

Coreopsis intermedia

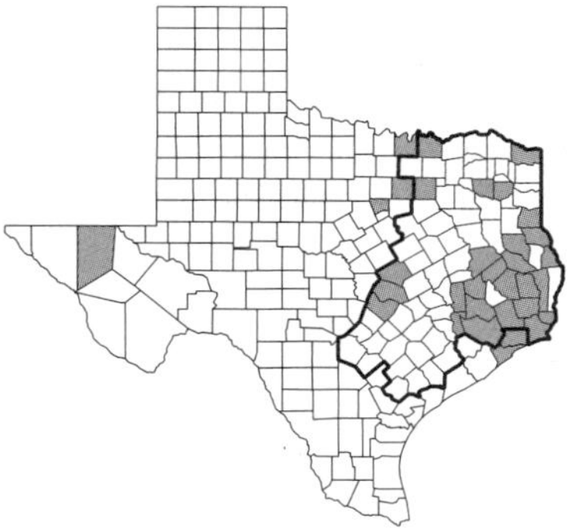

Coreopsis lanceolata

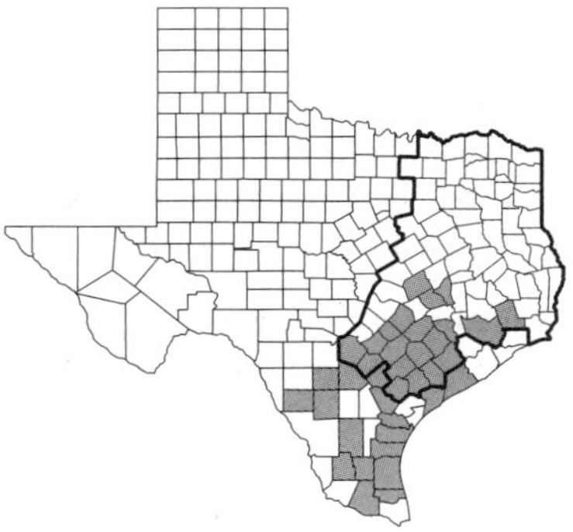

Coreopsis nuecensis

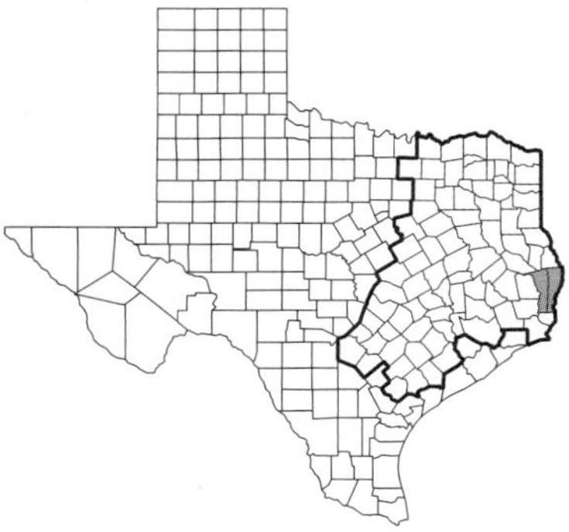

Coreopsis pubescens

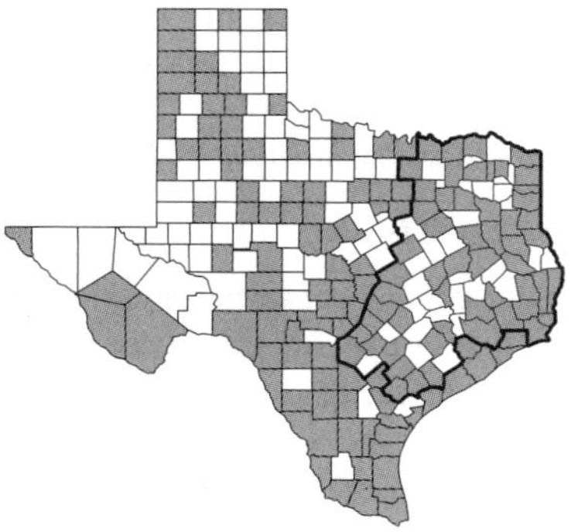

Coreopsis tinctoria

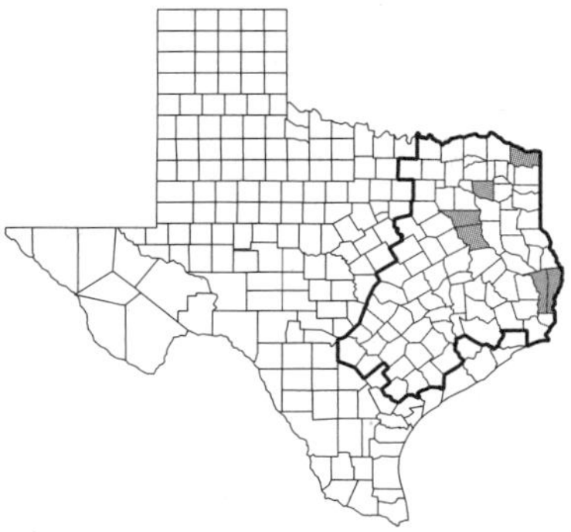

Coreopsis tripteris

1. Stem leaves spatulate to lanceolate, not much reduced up the stem, margins usually entire, sometimes serrate in the upper portion; peduncles eglandular ____ **C. rigidifolium**
1. Stem leaves lanceolate to linear, much reduced up the stem and becoming bract-like, margins usually serrate in the upper portion, rarely entire; peduncles with stalked glands.
   2. Involucres turbinate, (2–)2.5–5 mm wide; ray flowers 5–11, ligules 4–6 mm long ____ **C. divaricatum**
   2. Involucres campanulate, (3.5–)5–8(–10) mm wide; ray flowers (10–)13–21(–29), ligules 6–12 mm long ____ **C. hookerianum**

Croptilon divaricatum (Nutt.) Raf., (spreading, widely divergent), SLENDER GOLDENWEED, SCRATCH DAISY, SPREADING GOLDEN-ASTER. Erect annuals, 20–100+ cm tall; stem leaves lanceolate to linear, margins usually serrate in the upper portion or rarely entire; leaves reduced up the stem and becoming bract-like; peduncles with stalked glandular hairs; involucres turbinate, (2–)2.5–5 mm wide; ray flowers 5–11, ligules 4–6 mm long. Sandy soils, open woods, roadsides; ne to se and sc TX; se U.S. from VA s to FL and w to KS and TX. (May–)Aug–Oct(–Nov). [*Haplopappus divaricatus* (Nutt.) A. Gray]

Croptilon hookerianum (Torr. & A. Gray) House, (for William Jackson Hooker, 1785–1865, director of Kew Gardens), HOOKER'S SCRATCH DAISY, HOOKER'S GOLDEN-ASTER. Erect annuals, 20–80(–120) cm tall; stem leaves lanceolate to linear, margins usually serrate in the upper portion or rarely entire; leaves reduced up the stem and becoming bract-like; peduncles with stalked glandular hairs; involucres campanulate, (3.5–)5–8(–10) mm wide; ray flowers (10–)13–21(–29), ligules 6–12 mm long. Jul–Oct. According to Nesom (2006k), three varieties exist, more easily distinguished by geography and habitat than by morphology:

1. Peduncles with long-stalked glandular pubescence (longest hairs 0.4–0.5 mm long including gland), often slightly hispid just below heads; ray ligules 1–2.5 mm wide; nc TX to Panhandle ____ var. **validum**
1. Peduncles with short-stalked glandular pubescence (longest hairs 0.2–0.3 mm long including gland), never hispid; ray ligules (1.2–)2–3(–4) mm wide; c and s TX to Edwards Plateau.
   2. Peduncles densely stalked-glandular; ray ligules 1.2–2.2 mm wide; sandy habitats ____ var. **hookerianum**
   2. Peduncles sparsely stalked-glandular; ray ligules (1.5–)2–3(–4) mm wide; granitic habitats ____ var. *graniticum*

var. *graniticum* (E.B. Sm.) E.B. Sm., (of granite). Stems and leaves sparsely stalked-glandular, never hispid; ray ligules (1.5–)2–3 (–4) mm wide. Granite outcrops, gravelly slopes; *not yet reported in East TX but included here based on records from adjacent counties just to the w of our area;* central mineral region of Edwards Plateau: Burnett, Llano, and Mason cos., also Gillespie and Kimble cos. (TEX-LL). Jun–Oct. [*C. divaricatum* (Nutt.) Raf. var. *graniticum* (E.B. Sm.) Shinners, *Haplopappus validus* (Rydb.) Cory subsp. *graniticus* E.B. Sm.]

var. **hookerianum.** Stems and leaves densely stalked-glandular, never hispid; ray ligules 1.2–2.2 mm wide. Sandy loam, silt, stream banks, fields; sw edge of East TX from Milam Co. s to Goliad and Victoria cos.; Edwards Plateau to South TX Plains. [*C. divaricatum* var. *hookerianum* (Torr. & A. Gray) Shinners, *Haplopappus divaricatus* (Nutt.) A. Gray var. *hookerianus* (Torr. & A. Gray) Waterfall]

var. **validum** (Rydb.) E.B. Sm., (strong). Stems and leaves densely long-stalked-glandular, often slightly hispid just below heads; ray ligules 1–2.5 mm wide. Sandy gravel, sandstone outcrops, open oak woodlands; nw border of East TX in Grayson, Denton, and McLennan cos.; Blackland Prairie in nc TX w to Rolling Plains and Panhandle; AR, KS, OK, and TX. [*Haplopappus validus* (Rydb.) Cory]

Croptilon rigidifolium (E.B. Sm.) E.B. Sm., (stiff-leaved), STIFF-LEAF SCRATCH DAISY, STIFF-LEAF GOLDEN-ASTER. Erect or decumbent annuals (sometimes seemingly overwintering), 20–150 cm tall; stem leaves spatulate to lanceolate, margins usually entire or sometimes serrate in the upper portion; leaves not much reduced up the stem; peduncles without glands; involucres turbinate to nearly cylindrical, (4–)5–7 mm wide; ray flowers (5–)9–22, ligules usually 4.5–6.3 mm long. Sandy soils, creek terraces, prairies; Post Oak Savanna in Leon, Madison, and Robertson cos. s to Wilson, Victoria, and Wharton cos.; South TX Plains and Gulf Prairies; Mexico. Sep–Nov. [*C. divaricatum* (Nutt.) Raf. var. *hirtellum* (Shinners) Shinners, *Haplopappus rigidifolius* E.B. Sm.]

DIAPERIA Nutt. RABBIT-TOBACCO, EVAX, PYGMY CUDWEED

Very small annuals, 3–25 cm tall, taprooted, densely and conspicuously white- to gray-pubescent with cottony hairs; stems ascending or ± prostrate; leaves basal and alternate or uppermost crowded and apparently whorled, small (usually < 20 mm long), oblanceolate to obovate, entire; heads sessile, in dense clusters, axillary or terminal, small and inconspicuous, usually embedded in cottony pubescence; phyllaries (2–)4–6; receptacles chaffy; ray flowers absent; disk flowers of 2 series, the peripheral pistillate and fertile, the central functionally staminate or perfect and fertile, corollas pinkish or obscured by hairs; achenes minute, ellipsoid; pappus absent.

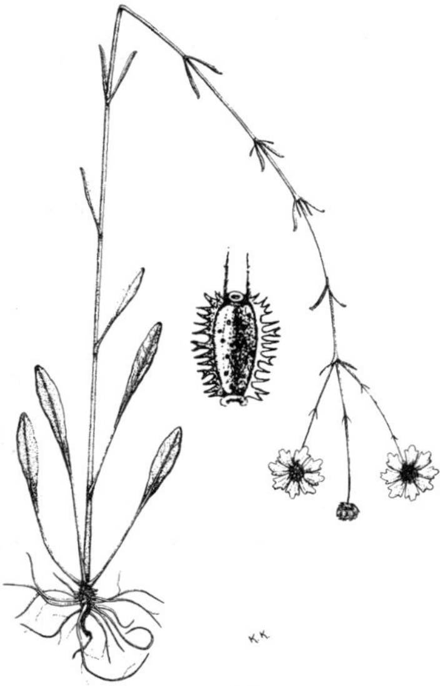
Coreopsis gladiata [SID]

Coreopsis grandiflora [YAT]

Coreopsis intermedia [SID]

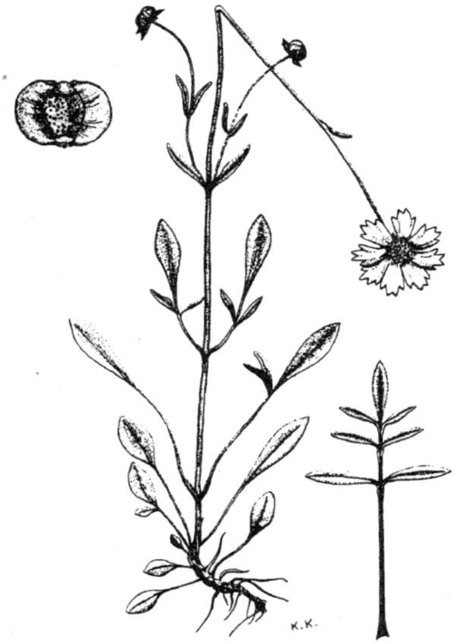
Coreopsis lanceolata [SID]

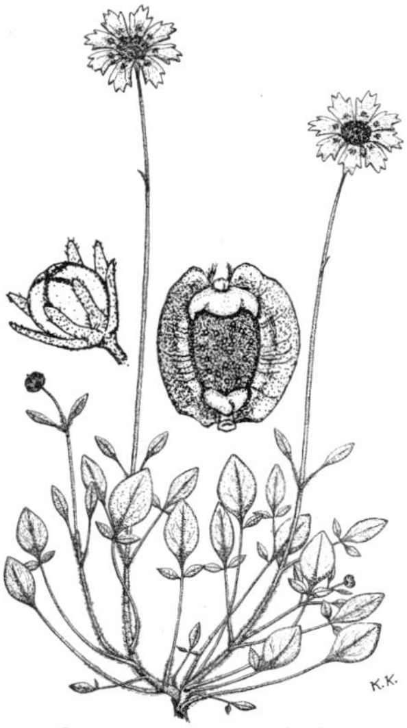
Coreopsis nuecensis [BTT]

Coreopsis pubescens [SID]

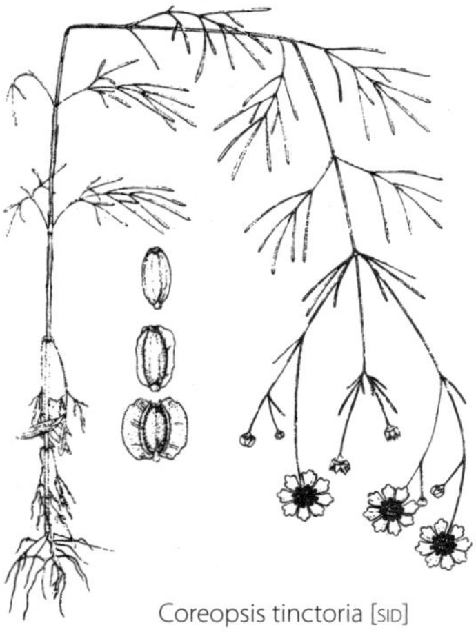
Coreopsis tinctoria [SID]

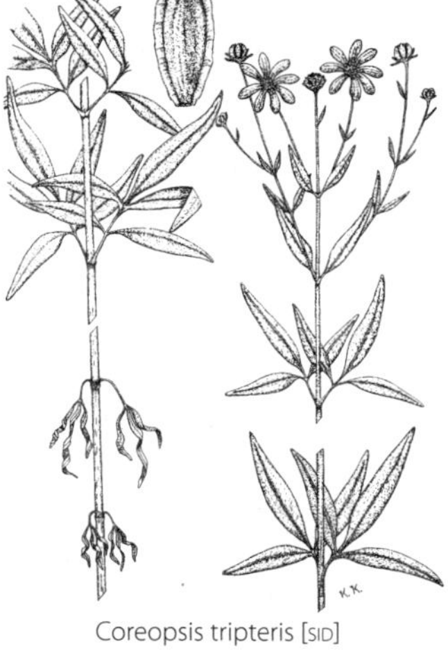
Coreopsis tripteris [SID]

Cosmos bipinnatus [BA1]

A genus of 3 species native to sc U.S. and n Mexico; formerly and variously included in the genera *Evax, Calymmandra,* and *Filago*. (Greek: *diapero*, to pass through, alluding to what was interpreted as multifurcating (as opposed to bifurcating) branching patterns in the first species described in the genus) (tribe Inuleae)
References: Shinners 1951e, 1964; Anderberg 1991; Morefield 1992, 2004, 2006a.

1 Plants unbranched, or only branched from the base; clusters of heads axillary, in racemose or spicate arrays; heads 1.5–2 mm high; central flowers of heads usually fertile (producing achenes) ____ **D. candida**
1. Plants usually branched at the base and along the stem, rarely unbranched; clusters of heads nearly all terminal at the branch tips; heads 2–4.5 mm high; central flowers of head usually infertile (functionally staminate).
 2. Leafy bracts among the heads shorter than or equal to the heads in length (bracts thus largely concealed by dense, cottony pubescence associated with the heads); heads ± spherical, 2–3.3 mm high and about as wide ____ **D. verna**
 2. Leafy bracts among the heads exceeding the heads in length (bracts thus conspicuously visible, protruding from between the crowded heads); heads ellipsoid, 3.5–4.5 mm high, the height 2–3 times the width ____ **D. prolifera**

Diaperia candida (Torr. & A. Gray) Benth., (pure white, shining), silver rabbit-tobacco, silver evax. Plants silvery gray, with dense silky hairs; stems usually unbranched, or only branched from the base; clusters of heads axillary, in racemose or spicate arrays; leafy bracts subtending the clusters of heads, and sometimes also among the heads, shorter than the heads in length; heads 1.5–2 mm high; central flowers of heads usually fertile; achenes 0.5–0.6 mm long. Dry sandy soils, open woods, roadsides, disturbed sites; e 1/2 TX; AR, LA, OK, and TX. Mar–Jun. [*Calymmandra candida* Torr. & A. Gray, *Evax candida* (Torr. & A. Gray) A. Gray, *Filago candida* (Torr. & A. Gray) Shinners]

Diaperia prolifera (Nutt. ex DC.) Nutt., (producing offshoots), big-head rabbit-tobacco. Plants grayish green to silvery white, with silky or woolly hairs; stems branched at the base and along the stem, rarely unbranched; clusters of heads nearly all terminal at the branch tips; leafy bracts among the heads exceeding the heads in length; heads 3.5–4.5 mm high; central flowers of head usually infertile; achenes 0.9–1.2 mm long.

1. Plants silvery white, with dense silky hairs; leafy bracts subtending the head clusters erect and simulating an involucre; heads borne singly or 2–3 in largest clusters ____ var. **barnebyi**
1. Plants grayish to greenish, with loose woolly hairs; leafy bracts subtending the head clusters ± spreading, not involucral in appearance; heads 4–40+ in largest clusters ____ var. **prolifera**

var. **barnebyi** Morefield, (for Rupert C. Barneby (1911–2000), of the New York Botanical Garden, who collected the type specimen), barneby's rabbit-tobacco. Thin rocky or gravelly soils, usually over limestone or gypsum, dry drainages; Bell and McLennan cos. (Kartesz 2015); scattered in w 1/2 TX; OK and NM. Apr–Jun.

var. **prolifera,** big-head rabbit-tobacco, big-head evax. Dry, silty soils, prairies, rocky slopes, often over limestone; Blackland Prairies and Post Oak Savanna; Rolling Plains and Edwards Plateau; c U.S. from MS n to MT and s to AZ and TX. Mostly May–Jun. [*Evax prolifera* Nutt. ex DC., *Filago nuttallii* Shinners, *F. prolifera* (Nutt. ex DC.) Britton]

Diaperia verna (Raf.) Morefield var. **verna,** (of Spring), spring rabbit-tobacco, many-stem evax. Plants grayish green, with woolly hairs; stems branched at the base and along the stem, rarely unbranched; clusters of heads nearly all terminal at the branch tips; leafy bracts among the heads shorter than or equal to the heads in length; heads 2–3.3 mm high; central flowers of head usually infertile; achenes 0.7–0.9 mm long. Sandy or eroding silty or rocky ground, disturbed sites, roadsides; nearly throughout TX; s U.S. from GA w to AZ; Mexico. Apr–Jun. [*Evax multicaulis* DC., *E. verna* Raf., *Filago verna* (Raf.) Shinners]

Dimorphotheca Moench CAPE MARIGOLD, AFRICAN DAISY

A genus of 7–18 species from South Africa; some are cultivated as ornamentals. (Greek: *di*-, two, *morphe*, shape, and *theca*, case or container, alluding to two forms of achenes in each head) (tribe Calenduleae)
References: Keith 2004a; Strother 2006i.

Dimorphotheca sinuata DC., (sinuous), cape marigold, african daisy. Showy annual herbs 5–30+ cm tall; leaves alternate, narrowly oblong to lanceolate or linear, margins usually sinuately fine-toothed, sometimes serrate or entire, rarely pinnatifid; heads borne singly; terminating stems on long peduncles; phyllaries in 2–3 series, lance-linear; receptacles naked; ray flowers 10–21+, ligules 15–20(–30+) mm long, mostly yellow to orange, sometimes marked with purple; disk flowers perfect, fertile, or the central flowers sometimes functionally staminate, corollas yellow to orange, lobes purplish; ray achenes 4–5 mm long, disk achenes 6–8 mm long; pappus absent. Disturbed sites; rare in TX; recorded in Walker Co. (Keith 2004a); scattered in w and sw U.S.; introduced elsewhere. Feb–May. Native of South Africa. In TX, occurring as an occasional escape or waif and probably not persisting in the flora.

Cosmos sulphureus [BA1]

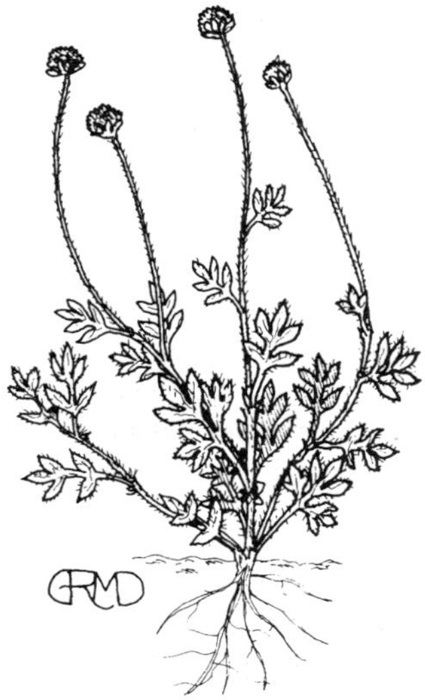

Cotula australis [FSA]

Crepis pulchra [UCP]

Crepis setosa [UCP]

Crepis zacintha [UCP]

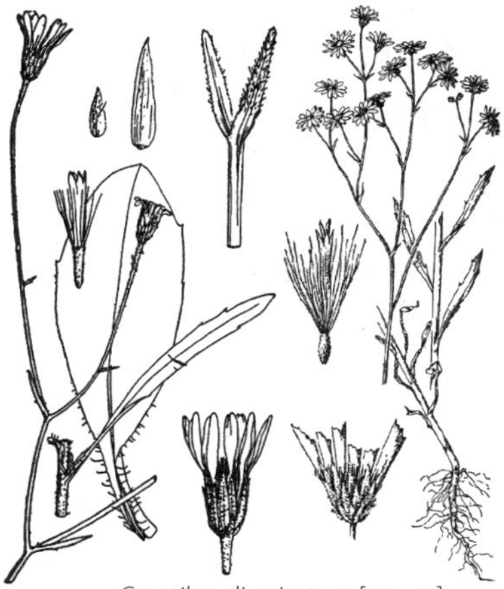

Croptilon divaricatum [GLE, HAL]

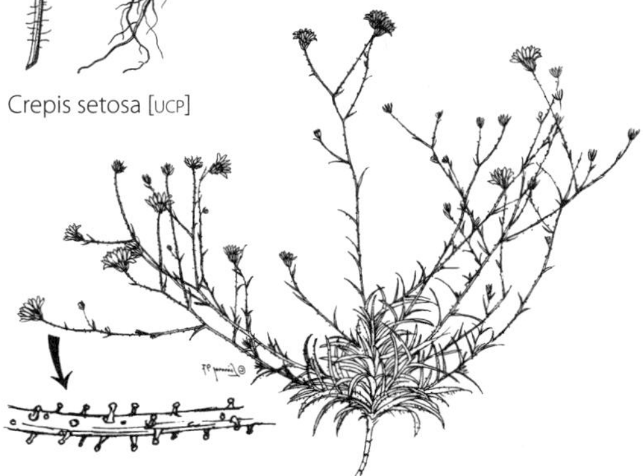

Croptilon hookerianum var. validum [HEA]

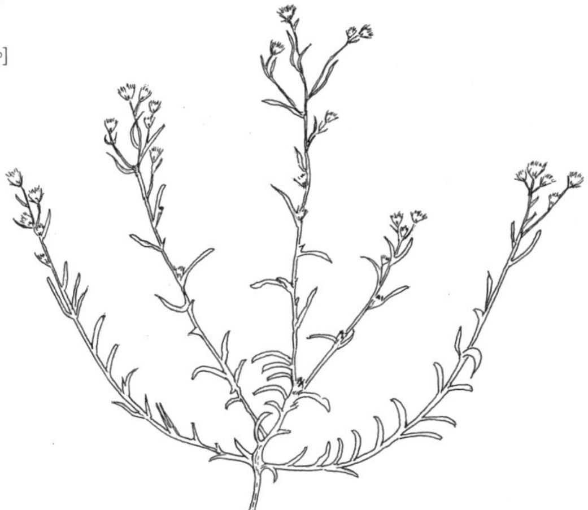

Croptilon rigidifolium [BAK]

DOELLINGERIA Nees WHITETOP ASTER

A genus of 3 species of e North America, previously included in *Aster*. (For Ignatz Doellinger (1770–1841), German botanist) (tribe Astereae)

REFERENCES: Nesom 1993b, 1994b; Semple & Chmielewski 2006.

Doellingeria sericocarpoides Small, (silky-fruited), SOUTHERN WHITETOP ASTER, SOUTHERN FLAT-TOPPED ASTER. Perennial herbs, rhizomatous; stems 50–150 cm tall; leaves alternate, sessile, blades lanceolate to ovate, margins entire, the basal and sometimes lower leaves withering before flowering, midstem and upper leaves crowded and reduced up the stem; heads in flat-topped corymbose arrays; phyllaries in 3–4 series, unequal, the tips broadly rounded; receptacles naked; ray flowers 2–7, pistillate and fertile, corollas white; disk flowers perfect and fertile, corollas pale yellow; achenes terete, 1.5–3.7 mm long, 6–8-ribbed, sparsely strigose; pappus of an outer series of short bristles and 3 inner series of minutely-barbed bristles, 4–7 mm long. Bogs, wet thickets, permanently wet areas; uncommon in East TX; e and se U.S. from NJ s to FL and w to OK and TX. Oct–Nov. [*Aster sericocarpoides* (Small) K. Schum., *A. umbellatus* Mill. var. *latifolius* A. Gray, *D. umbellata* (Mill.) Nees var. *latifolia* (A. Gray) House]

DYSODIOPSIS (A. Gray) Rydb. DOGWEED, FOETID-MARIGOLD

A monotypic genus native to the s U.S.; previously recognized in *Dyssodia*; we are following Karis and Ryding (1994) and Strother (2006j) in treating it separately. (From related genus *Dyssodia*, Greek: *dysodia*, an ill smell, and *opsis*, aspect or appearance) (tribe Heliantheae)

REFERENCES: Johnston 1956; Johnston & Turner 1962; Strother 1969, 1986, 2006j; Karis & Ryding 1994.

Dysodiopsis tagetoides (Torr. & A. Gray) Rydb., (resembling *Tagetes*—marigold), MARIGOLD DOGWEED. Annuals or short-lived perennials 0.4–0.9 m tall; glabrous; leaves mostly opposite, becoming alternate on upper stems, linear, 2–6 mm wide, with conspicuous yellowish brown or orange dot-like glands, unpleasantly aromatic, the margins with coarse, conspicuous teeth; involucres 9–12 mm high; phyllaries in 2 series, connate, with conspicuous orange glands, subtended by 5–8 bractlets; receptacles naked; ray flowers 7–12, pistillate and fertile, with ligules 10–15 mm long, lemon yellow to greenish yellow; disk flowers perfect, fertile, corollas brownish yellow; achenes obpyramidal, 3–3.5 mm long; pappus of 10–12 scales, 1–2.5 mm long, with 1(–3) awns or bristles at the tip of each scale. Calcareous soils, limestone outcrops; Blackland Prairie and West Cross Timbers from the Red River s to c TX; OK and TX. Jun–Aug. [*Dyssodia tagetoides* Torr. & A. Gray, *Hymenatherum tagetoides* (Torr. & A. Gray) A. Gray]

ECHINACEA Moench CONEFLOWER, PURPLE CONEFLOWER

Perennial herbs; basal and lower stem leaves long-petioled, alternate, blades simple, usually entire or subentire, rarely serrate or dentate, often with 3–5 prominent veins; heads large, solitary, terminal, long-peduncled; phyllaries in 2–4 series, usually spreading, recurved, or reflexed; receptacles convex to conical, chaffy, pales sharp-pointed and stiff, reddish orange to dark reddish purple, projecting beyond the disk corollas and partially surrounding achenes; ray flowers neuter, with ligules spreading to reflexed, purple to pink or rarely white; disk flowers perfect and fertile, corollas pinkish, greenish, reddish purple, or yellow; achenes 3–4-angled; pappus a minute crown, sometimes with a few teeth.

A genus of 9 species of the c and e U.S.; several species and many cultivars of *Echinacea* are popular ornamentals. *Echinacea* species were widely used medicinally by Native Americans for a variety of purposes, and are still valued today as herbal remedies that apparently act as immune system stimulants; unfortunately, some wild populations are under pressure from over-collecting (Foster 1991; Kindscher 2006, 2016). (Greek: *echinus*, spiny, for the prickly pales that protrude beyond the disk flowers) (tribe Heliantheae)

REFERENCES: McGregor 1968a, 1968b; Cox & Urbatsch 1990; Baskin et al. 1993; Binns et al. 2002; Kindscher 2006, 2016; Urbatsch et al. 2006; Keller 2014.

1. Plants fibrous-rooted; leaf blades abruptly contracted to the petiole, bases rounded to cordate, margins usually serrate to dentate, rarely entire ____ **E. purpurea**
1. Plants with ± branched taproots; leaf blades gradually tapering to base, bases cuneate or attenuate; margins usually entire or nearly so.
    2. Hairs of stems and peduncles appressed to ascending.
        3. Ray ligules 19–35 mm long; ray achenes glabrous in the upper portion; pales with tips usually straight ____ **E. atroruben**
        3. Ray ligules 30–70 mm long; ray achenes hairy in the upper portion (at least on angles); pales with tips often incurved ____ **E. paradox**
    2. Hairs of stems and peduncles spreading.
        4. Ray ligules 15–40 mm long and 5–8 mm wide ____ **E. angustifoli**

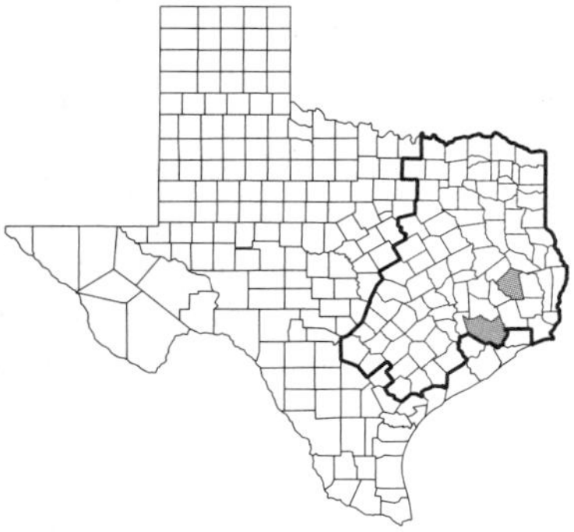
Cosmos bipinnatus

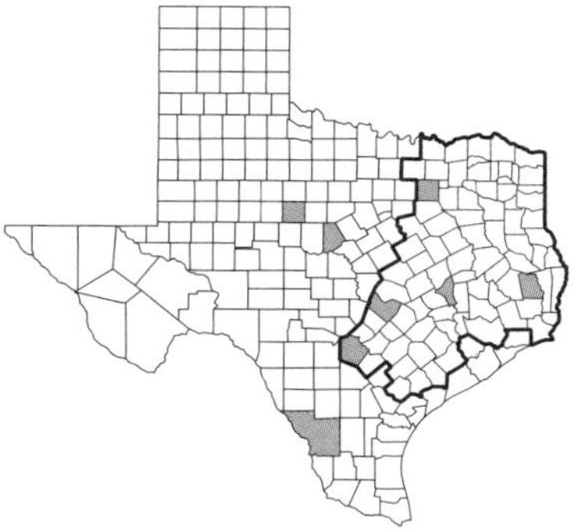
Cosmos sulphureus

Cotula australis

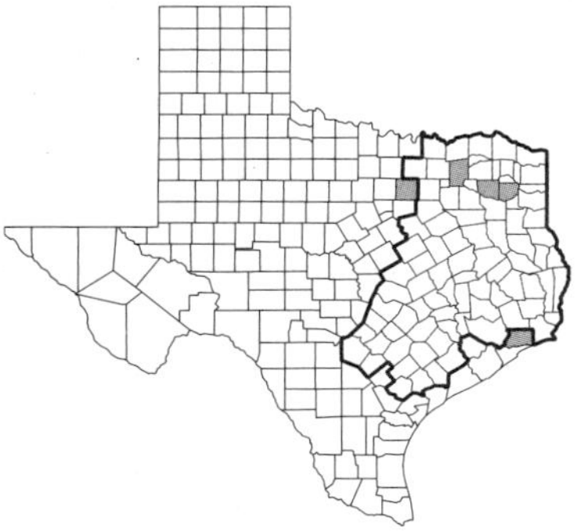
Crepis pulchra

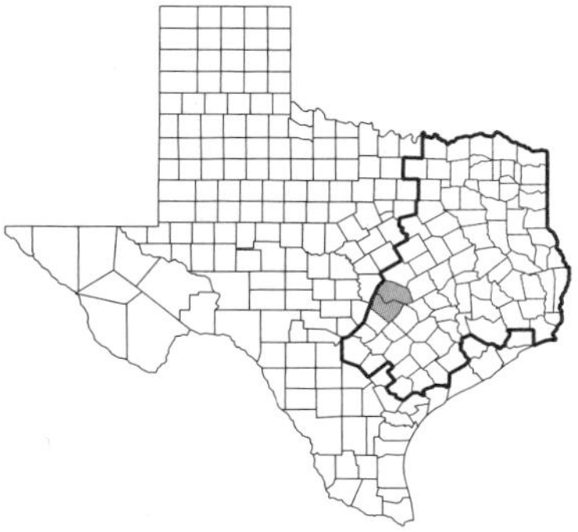
Crepis setosa

Crepis zacintha

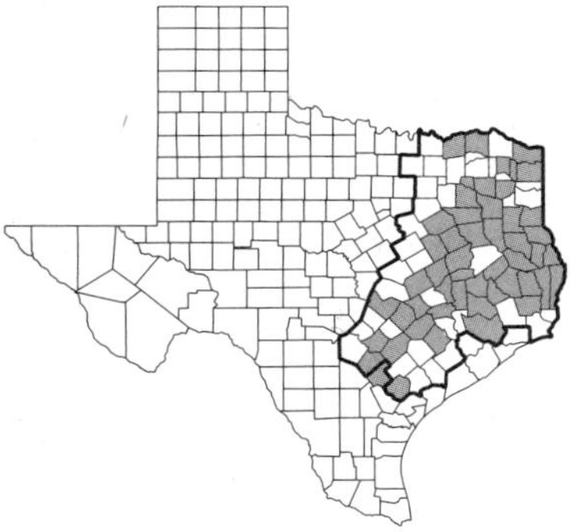
Croptilon divaricatum

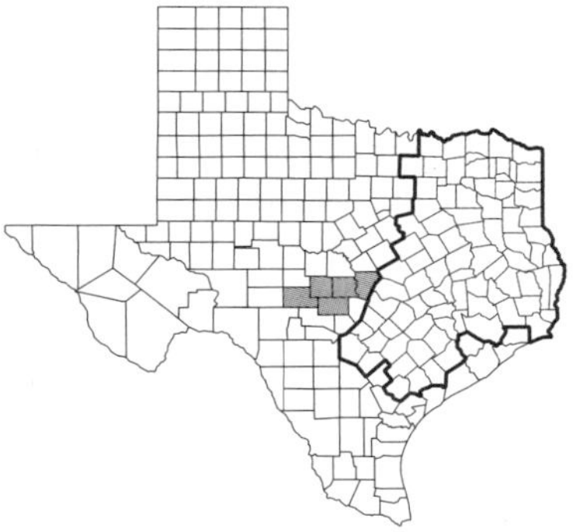
Croptilon hookerianum var. graniticum

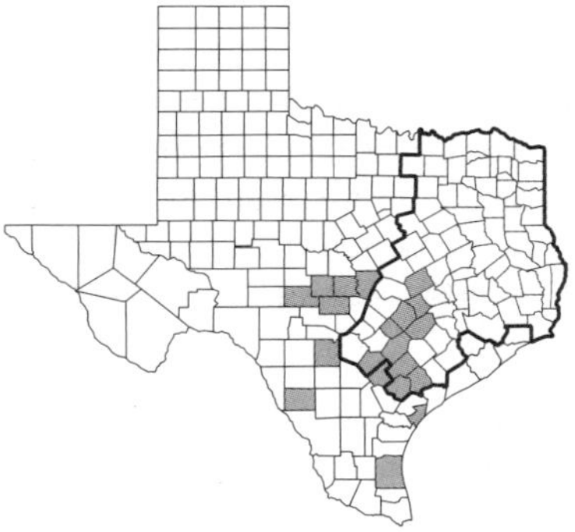
Croptilon hookerianum var. hookerianum

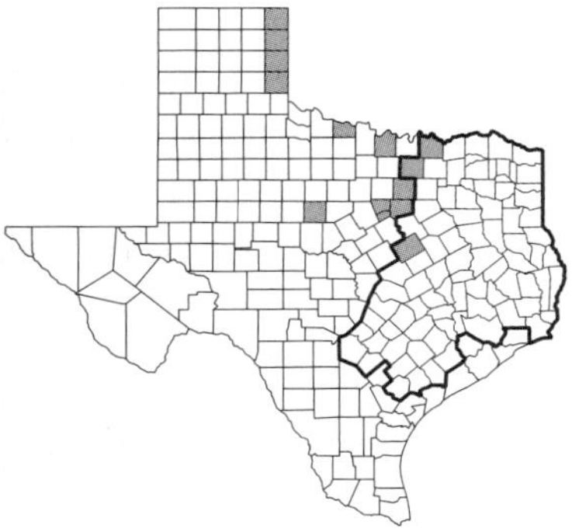
Croptilon hookerianum var. validum

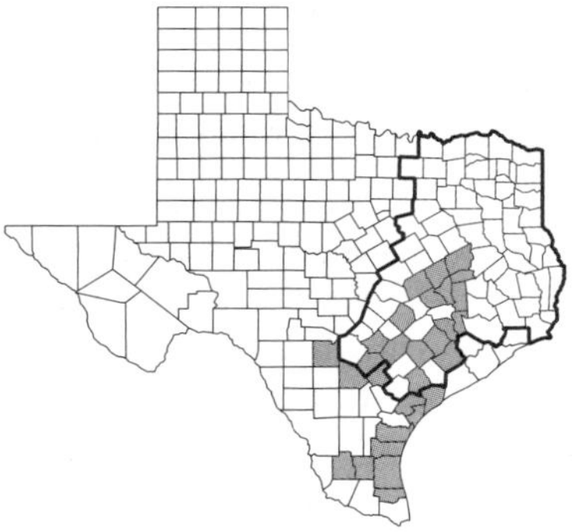
Croptilon rigidifolium

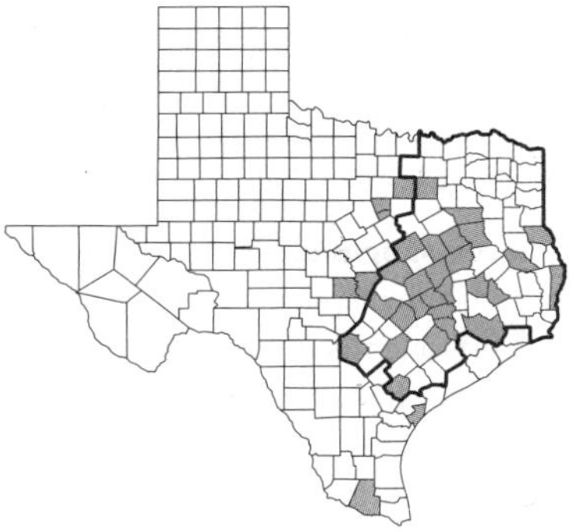
Diaperia candida

4. Ray ligules 40–90 mm long and 3–4 mm wide.
   5. Pales 8–11 mm long, tips rounded; pollen yellow ____ **E. sanguinea**
   5. Pales 9–14 mm long, tips sharp-pointed; pollen usually white, rarely lemon yellow ____ **E. pallida**

Echinacea angustifolia DC., (narrow-leaved), BLACKSAMSON, NARROW-LEAVED PURPLE CONEFLOWER. Plants to 70 cm tall, with ± branched taproots; stems green to purplish, moderately to densely rough-hairy, hairs spreading; basal leaves elliptic to lanceolate, bases cuneate to attenuate; pales with tips purple, straight, sharp-pointed; ray ligules 15–40 mm long and 5–8 mm wide, pink to purplish, reflexed; disk corolla lobes usually purple; achenes 4–5 mm long. Gravelly or rocky limestone prairies (on mixed sand and gravel farther w); scattered in East TX; mostly nc TX Blackland Prairie w to Plains Country and s to Edwards Plateau; Great Plains of c North America. May–Jun. [*E. angustifolia* DC. var. *strigosa* R.L. McGregor, *E. pallida* (Nutt.) Nutt. var. *angustifolia* (DC.) Cronquist, *E. pallida* (Nutt.) Nutt. var. *strigosa* (R.L. McGregor) Gandhi]

Echinacea atrorubens (Nutt.) Nutt., (dark red), TOPEKA PURPLE CONEFLOWER. Plants to 90 cm tall, with ± branched taproots; stems light green to tan, hairy, hairs appressed to ascending (rarely glabrous); basal leaves linear to lanceolate, rarely ovate, bases attenuate; pales with tips red to orange, usually straight, sharp-pointed; ray ligules 19–35 mm long and 2–7 mm wide, purple or rarely pink or white, reflexed; disk corolla lobes greenish to pink or purple; achenes 4–5 mm long. Limestone hillsides, dry prairies, sandstone outcrops; ne to se TX on Blackland Prairie; endemic to a narrow band of prairie from KS through OK to TX. May–Jul. [*Rudbeckia atrorubens* Nutt.]

Echinacea pallida (Nutt.) Nutt., (pale), PALE ECHINACEA; PALE PURPLE CONEFLOWER. Plants to 140 cm tall, with ± branched taproots; stems green to purplish, sparsely to densely hairy, hairs spreading; basal leaves elliptic to lanceolate, bases cuneate to attenuate; pales with tips purple, usually incurved, sharp-pointed; ray ligules 40–90 mm long and 3–4 mm wide, pale pink to reddish purple, reflexed; disk corolla lobes pink to purple; pollen usually white, rarely lemon yellow; achenes 2.5–5 mm long. Silty-clay hillsides, open woods, rocky prairies; sc to ne TX, mostly in Red River drainage; nc TX; scattered in e 1/2 of North America but concentrated in c U.S. plains and prairies; apparently introduced in several states in ne U.S. May–Jun. [*Rudbeckia pallida* Nutt.]

Echinacea paradoxa (J.B.S. Norton) Britton var. **neglecta** McGregor, (sp.: a paradox, illogical; var.: neglected), BUSH'S PURPLE CONEFLOWER. Plants to 90 cm tall, with ± branched taproots; stems yellowish green, sparsely to densely hairy, hairs appressed to ascending; basal leaves linear to lanceolate, rarely ovate, bases usually attenuate; pales with tips red to orange, often incurved, sharp-pointed; ray ligules 30–70 mm long and 3–8 mm wide, light purple to pink or white, reflexed; disk corolla lobes pinkish to yellowish; achenes 4–5.5 mm long. Rocky limestone prairies, open wooded hillsides; *may or may not occur in TX, depending on the source*; mapped as occurring in Montgomery Co. (Kartesz 2015) or Harris Co., potentially only as a historical record (Poole et al. 2010); Urbatsch et al. (2006) include TX within the range; Kindscher (2006, 2016) does not map it in TX; the var. is apparently otherwise restricted to the Arbuckle Mts. in sc OK. Late spring. [*E. atrorubens* (Nutt.) Nutt. var. *neglecta* (McGregor) Binns, B.R. Baum & Arnason] This subspecies is of conservation concern; OK populations are considered to be imperiled (Kindscher 2006). ⚠

Echinacea purpurea (L.) Moench, (purple), EASTERN PURPLE CONEFLOWER. Plants to 120 cm tall, with fibrous roots; stems usually brownish green, hairy, hairs spreading to ascending (sometimes glabrous); basal leaves ovate to narrowly lanceolate, bases usually rounded to cordate; pales with tips red-orange, straight or slightly curved, sharp-pointed; ray ligules 30–80 mm long and 7–19 mm wide, pink to purple, spreading to recurved; disk corolla lobes greenish to pink or purple; achenes 3.5–5 mm long. Silty-sandy soils, open oak woods, prairies; rare in TX outside of cultivation; recorded only from Bowie Co.; scattered in e 1/2 of North America but concentrated in the Midwest; apparently introduced in numerous states outside of that area. Jun–Jul. [*Rudbeckia purpurea* L.] This *Echinacea* species and its cultivars are horticulturally popular as ornamentals and in "wildflower" seed mixes, and are also grown commercially for herbal uses; naturalized and persisting populations continue to extend the native range (Urbatsch et al. 2006).

Echinacea sanguinea Nutt., (blood-red), WOODLAND ECHINACEA, SANGUINE PURPLE CONEFLOWER. Plants to 120 cm tall, with ± branched taproots; stems green to purplish, glabrate or sparsely to densely hairy, hairs spreading; basal leaves elliptic to lanceolate, bases attenuate; pales with tips purple, slightly curved, usually rounded; ray ligules 40–70 mm long and 3–4 mm wide, pink to reddish purple, reflexed; disk corolla lobes usually purple; pollen usually yellow; achenes 2.5–5 mm long. Sandy soils, open pine-oak woods, prairies; throughout e and se TX; also sc TX; AR, LA, OK, and TX. May–Jun. [*E. pallida* (Nutt.) Nutt. var. *sanguinea* (Nutt.) Gandhi & R.D. Thomas]

Diaperia candida [HEA]

Diaperia prolifera [BB2]

Diaperia verna [SHI]

Dimorphotheca sinuata [FNA]

Doellingeria sericocarpoides [SEM]

Dyssodiopsis tagetoides [HEA]

ECLIPTA L.

A genus of 3 or 4 species found in warm areas of the world. (Greek: *ecleipo*, to be deficient or lack, alluding to the absence of pappus) (tribe Heliantheae)
REFERENCE: Strother 2006k.

Eclipta prostrata (L.) L., (prostrate, flat to the ground), FALSE DAISY, PIEPLANT, YERBA DE TAGO. Low, prostrate to erect annuals, appressed-scabrous-pubescent, usually freely-branched and rooting at nodes; leaves opposite, sessile or short petiolate, lanceolate to narrowly elliptic, entire or slightly toothed, dark green; heads small, short-peduncled, terminal and in upper leaf axils, rather inconspicuous; phyllaries in 2–3 series, lanceolate; receptacles chaffy; ray flowers numerous, pistillate and fertile, with ligules linear, very short (1–2 mm long), white; disk flowers perfect and fertile; disk corollas minute, whitish; achenes weakly 3–4-angled, 2–2.5 mm long; pappus absent or a minute crown of scales. Ditches, shorelines, stream banks; damp disturbed areas; throughout TX; e and s U.S.; Central and South America; introduced elsewhere. Jun–Nov. [*E. alba* (L.) Hassk., *Verbesina prostrata* L.] Gandhi and Thomas (1989) explained why the correct name is *E. prostrata*. This species has been introduced into the Old World and in India is used as a blackish dye for hair and tattooing (Mabberley 2017).

ELEPHANTOPUS L. ELEPHANT'S-FOOT

Perennial herbs, often rhizomatous or stoloniferous; leaves often forming basal rosettes, stem leaves alternate, simple, margins usually toothed; heads sessile, clustered in corymbose or paniculate arrays, clusters subtended by a few foliaceous bracts; phyllaries 8, the outer 4 ovate, the inner 4 lanceolate; receptacles naked; ray flowers lacking; disk flowers usually 4, perfect, fertile, corollas pink to purple or white, unevenly 5-lobed, sometimes simulating rays; achenes 10-ribbed; pappus of 5(–6) bristle-tipped scales.

A mainly tropical and warm-temperate genus of ca. 15 species; a few are weedy. (Greek: *elephantos*, elephant, and *pous*, foot; translation of aboriginal name, perhaps in reference to the large basal rosettes of some species) (tribe Vernonieae)
REFERENCES: Gleason 1906, 1922; Clonts & McDaniel 1978; Strother 2006l.

1. Basal leaves absent at flowering time, stem leaves well-developed ____ **E. carolinianus**
1. Basal leaves present at flowering time, stem leaves smaller or absent.
   2. Inner phyllaries 6–7.5 mm long; achenes 2.5–3 mm long; pappus 3–4.5 mm long ____ **E. nudatus**
   2. Inner phyllaries 9–11+ mm long; achenes (3–)4–5 mm long; pappus 6–8 mm long ____ **E. tomentosus**

Elephantopus carolinianus Raeusch., (of Carolina), CAROLINA ELEPHANT'S-FOOT, LEAFY ELEPHANTOPUS. Stems (10–)30–80(–120) cm tall; basal leaves absent at flowering time; stem leaves broadly elliptic to ovate or lanceolate, sparsely pilose to hirsute, bracts subtending clusters of heads rounded-deltate to lance-deltate; inner phyllaries 8–10 mm long, sparsely short-pubescent; achenes 2.5–4 mm long; pappus 4–5 mm long. Low wooded areas, moist sandy soils; se and e TX w to Grand Prairie; se U.S. from NJ s to FL and w to KS, OK, and TX. Aug–Oct(–Nov). [*E. flexuosus* Raf.]

Elephantopus nudatus A. Gray, (naked, smooth), SMOOTH ELEPHANT'S-FOOT. Stems (10–)30–110+cm tall; leaves mostly basal at flowering time, oblanceolate to spatulate or sometimes elliptic, sparsely pilose to hirsute; bracts subtending clusters of heads rounded-deltate to lance-deltate; inner phyllaries 6–7.5 mm long, sparsely strigose or hispidulous; achenes 2.5–3 mm long; pappus 3–4.5 mm long. Pine forests, moist sandy soils; e and se TX in Hardin, Jefferson, Marion, Polk, and Tyler cos.; also Galveston Co.; se U.S. from DE s to FL and w to TX. Aug–Oct(–Nov).

Elephantopus tomentosus L., (hairy, tomentose), DEVIL'S GRANDMOTHER, HAIRY ELEPHANT'S-FOOT. Stems (10–)30–60+ cm tall; leaves mostly basal at flowering time, obovate to oblanceolate or spatulate, densely pilose, at least beneath; bracts subtending clusters of heads rounded-cordate; inner phyllaries 9–11+ mm long, finely short-pubescent; achenes (3–)4–5 mm long; pappus 6–8 mm long. Low wooded areas, sandy soils; e 1/4 TX; se U.S. from MD s to FL and w to OK and TX; Mexico. Jul–Oct. [*E. carolinianus* Raeusch. var. *simplex* Nutt.]

EMILIA Cass. TASSELFLOWER

An Old World tropical genus of 50+ species; a few are cultivated; several are widespread weeds in warm regions of the world. (Presumably named for someone known to Cassini, the author) (tribe Senecioneae)
REFERENCES: Nicolson 1980; Williams 1994; Spencer & Williams 2005; Barkley 2006a.

Emilia fosbergii Nicolson, (for Francis Raymond Fosberg (1908–1993), a botanist at the Smithsonian Institution), FLORIDA TASSELFLOWER. Annual taprooted herbs; stems 20–100 cm tall; leaves alternate, sessile and auriculate to winged-petiolate and

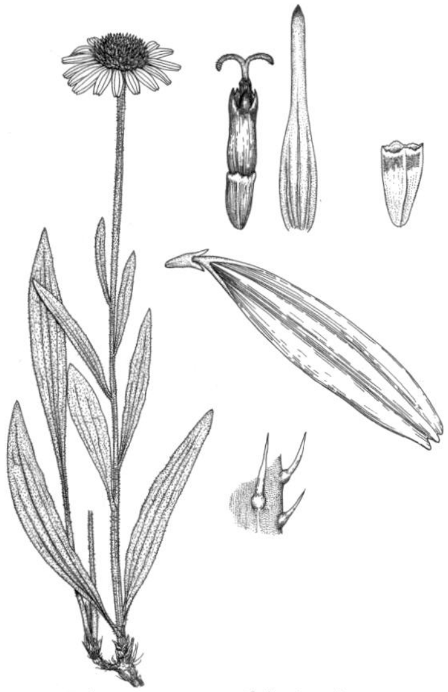

Echinacea angustifolia [MO4]

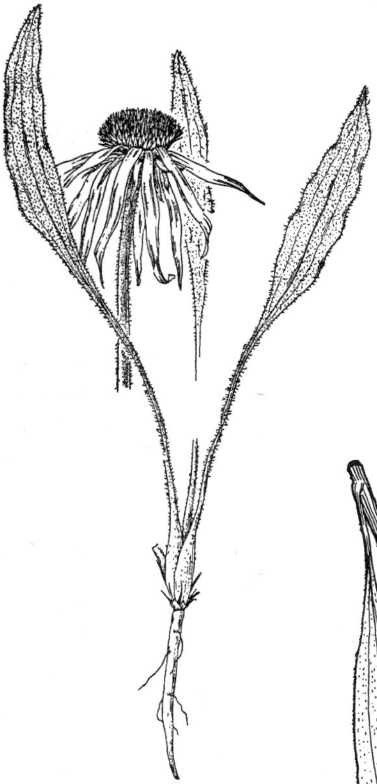

Echinacea pallida [GLE]

Echinacea atrorubens [HEA]

Echinacea paradoxa [BB2]

Echinacea purpurea [GLE]

clasping, blades oblanceolate to pandurate, mostly 5–10 cm long, margins entire, toothed or weakly lobed; heads in loose corymbose arrays; involucres 9–14 mm high; phyllaries 8 or 13, equal, reflexed in fruit; receptacles naked; ray flowers absent; disk flowers all perfect and fertile or central ones functionally staminate, corollas pinkish to reddish purple; achenes spindle-shaped, 5-ribbed; pappus of many white, minutely-barbed bristles. Disturbed areas, lawn weed, nursery escape; recorded from Harris Co. (*Brown 14320* SBSC), Travis Co. (*JK Williams s.n.* TEX), and Walker Co. (*Spencer 1* SHST); also s TX in Cameron and Hidalgo cos. (Richardson 2011); CA, FL, GA, LA, and TX; introduced elsewhere. Oct–Apr, or perhaps year-round. Native of tropical Asia; the first records of this species in TX are discussed in Williams (1994) and Spencer and Williams (2005).

ENGELMANNIA A. Gray ex Nutt. ENGELMANN'S DAISY, CUT-LEAF DAISY

A monotypic genus native to North America; sometimes cultivated as an ornamental. (Named for Dr. George Engelmann, 1809–1884, German-born botanist and physician of Saint Louis, MO) (tribe Heliantheae)
REFERENCES: Turner & Johnston 1956; Goodman & Lawson 1992; Keil 2006e.

Engelmannia peristenia (Raf.) Goodman & C.A. Lawson, (possibly from Greek: *perisso*, beyond the regular number or size, odd in number, and *tenia*, band, ribbon), ENGELMANN'S DAISY, CUT-LEAF DAISY. Hispid-pubescent perennials with woody taproots; stems spreading to erect, 20–100 cm tall; leaves alternate, blades 8–30 cm long, pinnately divided nearly to midrib, the lobes again lobed or toothed; inflorescences of several long-peduncled heads; involucres 6–10 mm high; phyllaries in 3–4 series, the outer with green, ciliate tips; receptacles chaffy; ray flowers 8–13, pistillate and fertile, ligules ca. 1 cm long, yellow, curling under during daytime, expanded from late afternoon to mid-morning; disk flowers numerous, functionally staminate, dark yellow; achenes 3–4 mm long, detaching with subtending phyllary; pappus of 2–4 minute scales. Clayey or occasionally sandy prairies and limestone outcrops, roadsides; nearly throughout TX but least frequent in Pineywoods; sc U.S. from SD sw to AZ and se to LA; n Mexico. Apr–Jul (rarely later). [*E. pinnatifida* Torr. & A. Gray] Goodman and Lawson (1992) documented that *E. peristenia* is an older name (published 1832) than the long established *E. pinnatifida* (published 1840). Farmers and ranchers refer to this as an "ice cream" plant because it is preferred by livestock; it is therefore now rarely seen except along roadsides or other areas inaccessible to grazing animals (J. Stanford, pers. comm.).

ERECHTITES Raf. BURNWEED, FIREWEED

A genus of ca. 12 species native to the Americas and Australia; some species widely distributed as weeds; some contain alkaloids. (An ancient Greek name, *erechthites*, used by Dioscorides for some species of *Senecio*—groundsel) (tribe Senecioneae)
REFERENCES: Belcher 1956; Vuilleumier 1969a; Barkley & Cronquist 1978; Barkley 2006b.

Erechtites hieraciifolius (L.) Raf. ex DC. var. **hieraciifolius,** (with leaves like *Hieracium*—hawkweed), AMERICAN BURNWEED, FIREWEED. Annual herbs, glabrescent or with sparse, jointed hairs; stem 50–200 cm tall; leaves alternate, numerous, evenly distributed up stem, blades ovate-lanceolate to oblanceolate, 3–20 cm long, subentire to sharply doubly serrate and sometimes irregularly lobed, the uppermost reduced and clasping; heads few–many in corymbose arrays; involucres cylindrical; primary phyllaries ca. 20, in 1 or 2 series, ± equal, 9–17 mm long, subtended by a few, very small bracts; receptacles naked; heads o disk flowers only, peripheral flowers pistillate, with tubular-filiform corollas, central flowers perfect and fertile or functionally staminate, corollas whitish or greenish to yellowish cream or pinkish; achenes 10–12-nerved, columnar, 2–3 mm long; pappus of 60–120 capillary bristles, deciduous. Wet areas, open woods; throughout se and e TX, a few c TX records (Bell, Gonzales Guadalupe cos.); e 1/2 of North America and sporadically on the w coast; West Indies and Mexico. Aug–Nov. [*E. hieraciifoliu* var. *intermedia* Fernald, *Senecio hieraciifolius* L.] The name *Erechtites* has been treated traditionally as grammatically feminine the International Code of Botanical Nomenclature (Art. 62.4) states that generic names ending in "*-ites*" are to be treated as masculine, therefore the correct epithet for this species is *heiraciifolius* and not *heiraciifolia* as it is sometimes written.

ERIGERON L. FLEABANE, DAISY FLEABANE, DAISY

Ours annual or perennial herbs; leaves alternate, sessile or the lower ones petiolate, oblong or oblanceolate to lanceolate o linear, margins entire or toothed; heads solitary or in loose corymbose or paniculate arrays; phyllaries numerous, in 2–5 series receptacles naked; ray flowers numerous, often 30–150, in 1(–2+) series, pistillate, fertile, ligules narrow, white or lavender-blu to pink, never yellow, often curling under at night or in age; disk flowers perfect and fertile, corollas yellow; achenes flattened usually strongly 2+-ribbed; pappus various, of hair-like bristles (at least on disk achenes), sometimes with a second series c small bristles or scales.

A cosmopolitan genus (especially of North America) of ca. 390 species, closely related to *Conyza* and *Aphanostephus;* severa are cultivated as ornamentals. *Erigeron* species are sometimes difficult to distinguish; complete basal plant parts are necessar

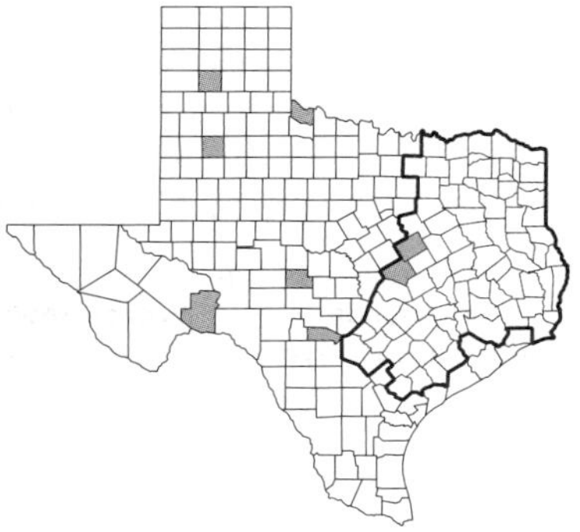

Diaperia prolifera var. barnebyi

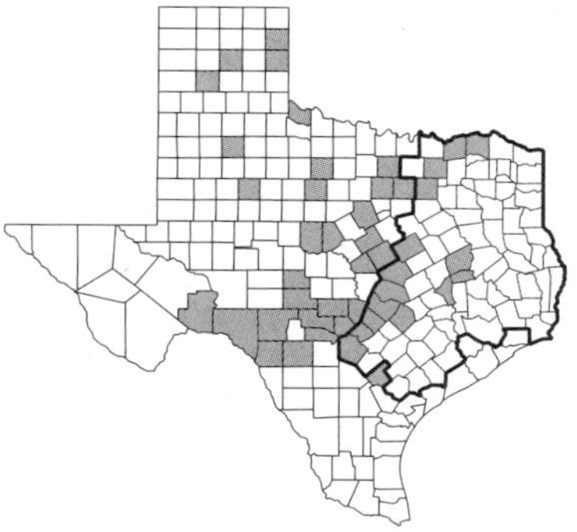

Diaperia prolifera var. prolifera

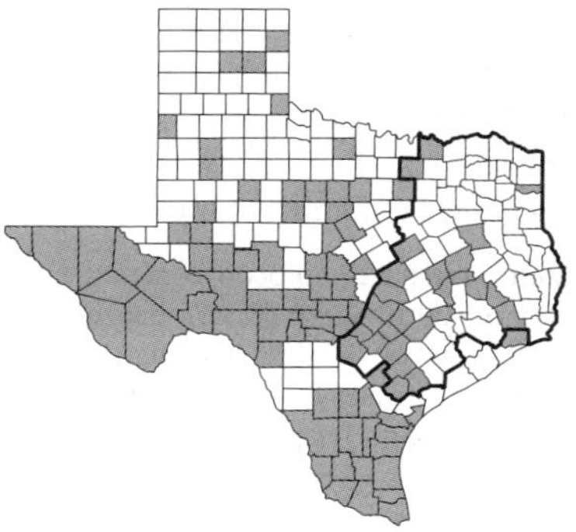

Diaperia verna var. verna

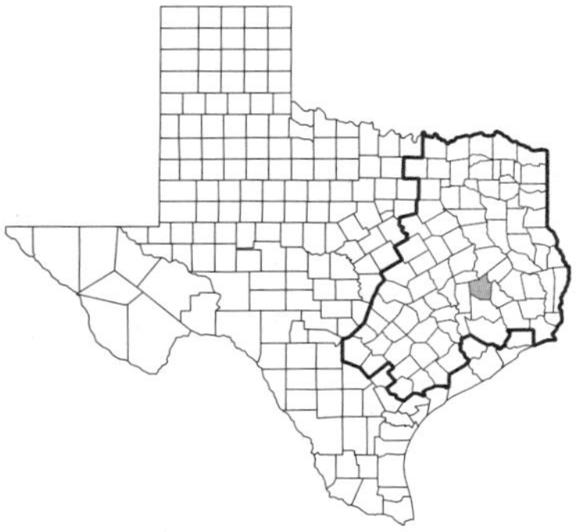

Dimorphotheca sinuata

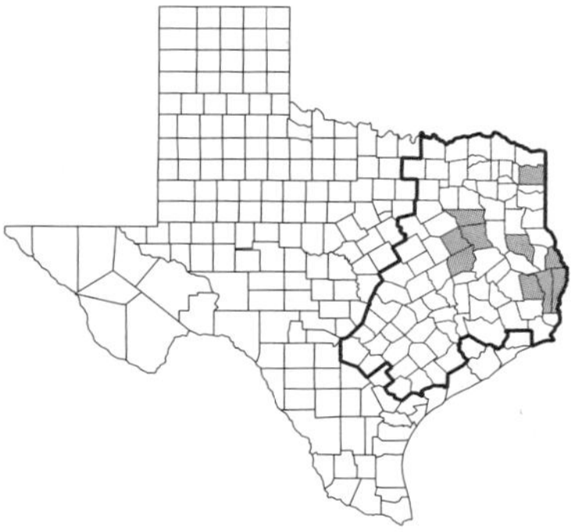

Doellingeria sericocarpoides

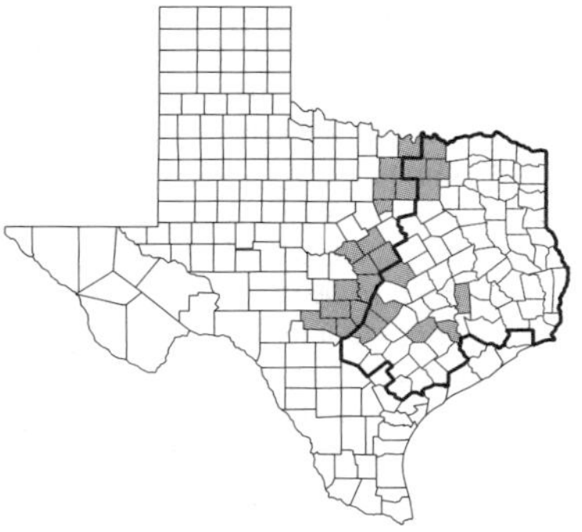

Dysodiopsis tagetoides

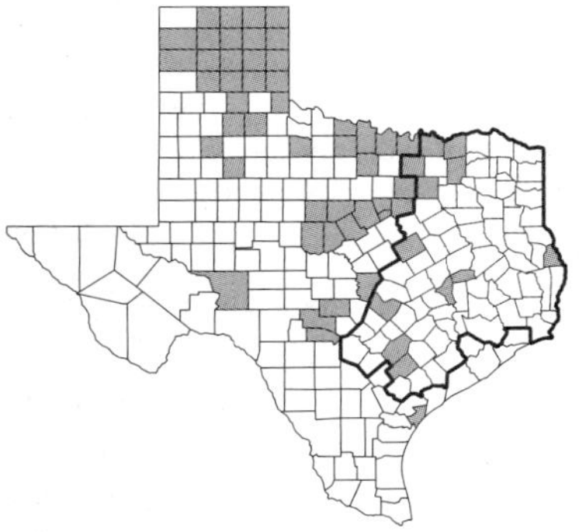

Echinacea angustifolia

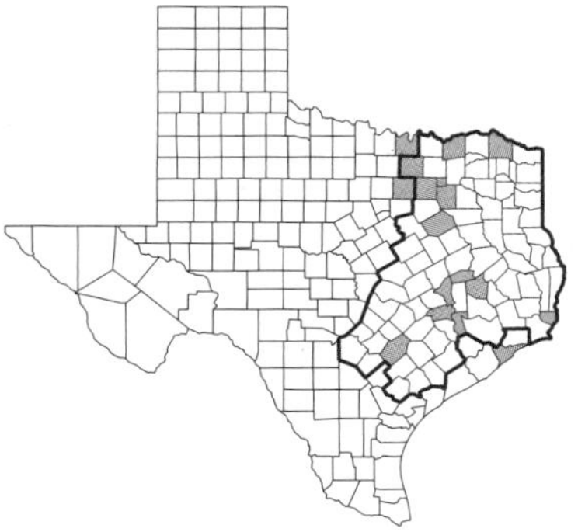

Echinacea atrorubens

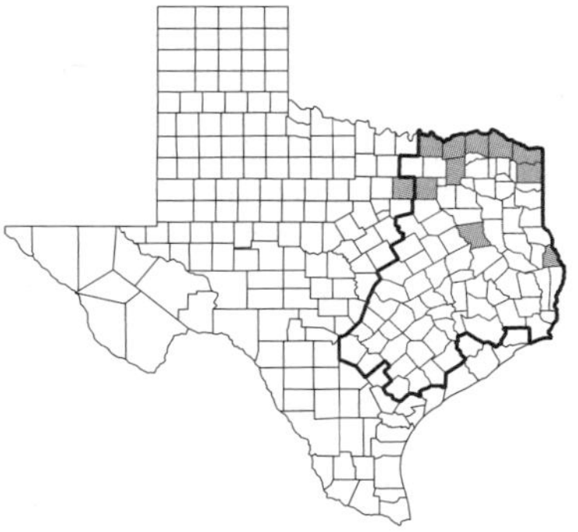

Echinacea pallida

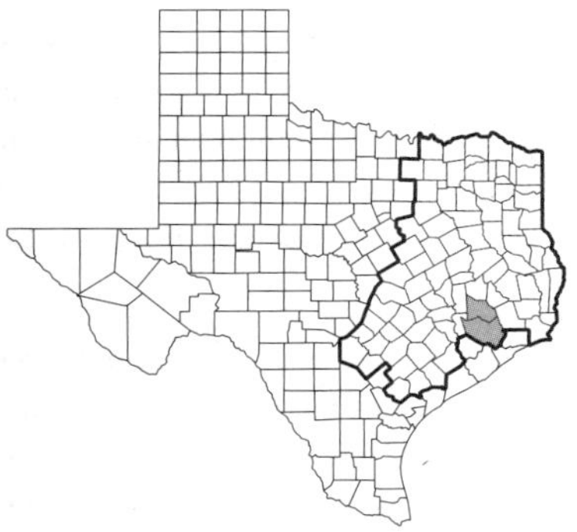

Echinacea paradoxa var. neglecta

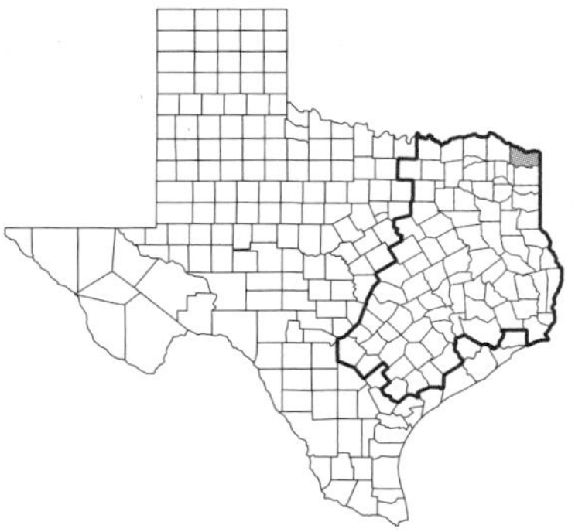

Echinacea purpurea

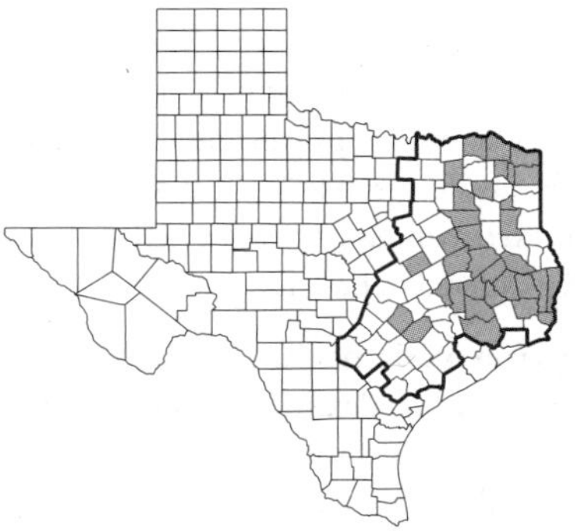

Echinacea sanguinea

to identify some species. Barkley (1986) indicated, "As with many widespread genera in the Asteraceae, the species boundaries are not always sharp, and morphological intermediates are to be expected." Polyploidy is common in *Erigeron* (Nesom 2006l). (Greek: *eri*, early, or *erio*, woolly, and *geron*, old man; perhaps alluding to the gray pappus or woolly heads of some species) (tribe Astereae)

References: Cronquist 1943, 1947; Shinners 1947a; Van Vleet 1951; Nesom 1978, 1989a, 2004a, 2006l; Nesom & Noyes 1999.

1\. Stems prostrate (= procumbent) and creeping along the ground, rooting at nodes; heads solitary, on naked peduncles mostly 10–20 cm long; most common in dune sands of Gulf Coast, uncommon inland in deep sands ____ **E. procumbens**

1\. Stems erect or variously sprawling but not prostrate/creeping nor rooting at nodes; heads few to many in corymbose or paniculate arrays, with peduncles usually < 10 cm long; plants of various habitats.

- 2\. Inner (longer) pappus series bristles absent on ray achenes but present on disk achenes.
  - 3\. Annuals, usually 60–150 cm tall; lower stems sparsely pilose-hispid (hairs spreading); basal leaves usually withering by flowering; upper leaves not much reduced compared to midstem leaves; leaf blades lanceolate to oblong or ovate ____ **E. annuus**
  - 3\. Annuals or biennials, usually 30–70 cm tall; lower stems strigose (hairs usually ascending and appressed, rarely spreading); basal leaves usually present at time of flowering; upper leaves gradually reduced compared to midstem leaves; leaf blades spatulate to oblanceolate or linear ____ **E. strigosus**
- 2\. Inner (longer) pappus series bristles present on both ray and disk achenes.
  - 4\. Plants with a distinct taproot (obviously larger/longer than lateral roots).
    - 5\. Annuals with slender taproots; lower stems herbaceous; stem leaf bases sometimes clasping to subclasping; inner series of pappus bristles 9–11; achenes 0.7–0.9 mm long; usually in sandy habitats ____ **E. geiseri**
    - 5\. Perennials with thick, woody taproots; lower stems often woody and persistent; stem leaf bases not clasping nor subclasping; inner series of pappus bristles (5–)12–17; achenes 1.1–1.4 mm long; usually in calcareous habitats ____ **E. modestus**
  - 4\. Plants with fibrous roots (primary taproot, if any, shortened in comparison to lateral roots).
    - 6\. Stem leaves with bases clasping or subclasping; involucres 4–7 mm high and 6–20 mm wide; ray ligules often white (sometimes pinkish), 5–10 mm long, sometimes coiling and recurving in age; inner series of pappus bristles 15–36.
      - 7\. Plants with slender, scale-leaved, herbaceous rhizomes; stems usually abruptly widened (dilated) below the heads; involucres 5–7 mm high; ray flowers 50–80(–100); inner series of pappus bristles (22–)28–36, outer (shorter) series sometimes absent; achenes 1.3–1.8 mm long ____ **E. pulchellus**
      - 7\. Plants fibrous-rooted, lacking rhizomes; stems not abruptly widened below the heads; involucres 4–6 mm high; ray flowers 150–250(–400); inner series of pappus bristles 15–20(–30), outer series usually present; achenes 0.6–1.1 mm long ____ **E. philadelphicus**
    - 6\. Stem leaves not clasping nor subclasping; involucres (2–)2.5–4 mm high and 5–10 mm wide; ray ligules often bluish or lavender (sometimes white), 3–5 mm long, not coiling nor recurving in age; inner series of pappus bristles 11–15 ____ **E. tenuis**

Erigeron annuus (L.) Pers., (annual), annual fleabane, eastern daisy fleabane. Taprooted or fibrous-rooted annuals, (10–)60–150 cm tall; lower stems sparsely pilose-hispid with spreading hairs; stem leaf blades lanceolate to oblong, the upper ones not much reduced; involucres 3–5 mm high and 6–12 mm wide; ray flowers 80–125, ligules white, 4–10 mm long, coiling under in age; achenes 0.8–1 mm long, 2-nerved; disk pappus with an outer minute crown and an inner series of 8–11 bristles; ray pappus lacking bristles. Prairies, disturbed areas, roadsides, various soils; mostly se TX; scattered in c and nc TX; throughout North America but most common in e 1/2; Mexico and Central America; introduced and weedy elsewhere. (Mar–)Apr–Jun(–Aug). [*Aster annuus* L., *E. annuus* (L.) Pers. var. *discoideus* (Vict. & J. Rousseau) Cronquist]

Erigeron geiseri Shinners, (for Samuel Wood Geiser, 1890–1983, professor at Southern Methodist Univ.), geiser's fleabane Slenderly-taprooted annuals, 5–40 cm tall; lower stems sparsely villous with spreading or downward-pointing hairs; stem lea blades lanceolate to lance-linear, subclasping, the upper ones gradually reduced; involucres 2.5–3.5 mm high and 5–7 mm wide; ray flowers 44–70, ligules white, drying white to pink, 3–6 mm long, not coiling nor reflexing; achenes 0.7–0.9 mm long 2-nerved; pappus of an outer series of small scales and an inner series of 9–11 bristles. Sandy prairies, roadsides; mostly nc to s TX, scattered in East TX; OK and TX. Mar–May.

Erigeron modestus A. Gray, (modest), plains fleabane. Thickly-taprooted perennials, 8–40 cm tall, often grayish-pubescen taproot and caudex woody; branches increasing in number through the season; lower stems usually woody, loosely strigose t villous with hairs sometimes spreading; stem leaf blades oblanceolate to narrowly spatulate, the upper ones gradually reduce peduncles decreasing in length through the season; involucres 3–4.5 mm high and 6–9(–12) mm wide (size decreasing throug the season); ray flowers 24–65(–170), ligules white, often with a darker midstripe beneath, drying lilac, 4–7(–11) mm long, n coiling nor reflexing; achenes 1.1–1.4 mm long, 2-nerved; pappus of an outer series of short bristles and an inner series of (5– 12–17 bristles. Gravelly or rocky limestone, sandy clay soils, oak-juniper scrubland; w border of East TX from McLennan Co

Echinacea sanguinea [THO]

Eclipta prostrata [MAS]

Elephantopus carolinianus [BB2]

Elephantopus nudatus [BB2]

Elephantopus tomentosus [BB2, GEO]

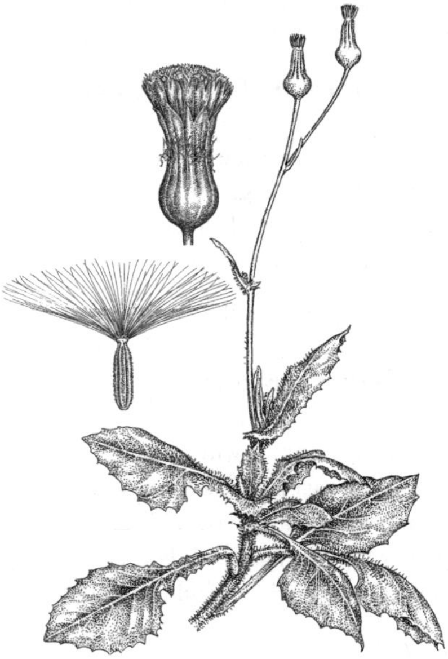
Emilia fosbergii [CO3]

Engelmannia peristenia [RBM]

Erechtites hieraciifolius [PNW]

Erigeron annuus [REE]

to Bexar Co.; mostly Edwards Plateau and Trans-Pecos, scattered in nc TX and Panhandle; AZ, KS, NM, OK, and TX; Mexico. Apr–Jun(–Oct). [*E. lobatus* A. Nelson var. *warnockii* Shinners, *E. plateauensis* Cronquist, *E. warnockii* (Shinners) Shinners]

Erigeron philadelphicus L. var. **philadelphicus,** (of the Philadelphia region), PHILADELPHIA FLEABANE. Fibrous-rooted perennials, 4–70 cm tall; lower stems hirsute-villous; stem leaf blades oblong-oblanceolate to lanceolate, clasping to auriculate-clasping, the upper ones gradually reduced; involucres 4–6 mm high and 6–15 mm wide; ray flowers 150–250(–400), ligules usually white, sometimes pinkish, 5–10 mm long, either not coiling or coiling in age; achenes 0.6–1.1 mm long, 2-nerved; pappus of an outer series of short bristles and an inner series of 15–20(–30) bristles. Low prairies and stream banks, roadsides, calcareous clays; se and e TX w to West Cross Timbers and Edwards Plateau; throughout North America; introduced elsewhere. Mar–Jun. [*E. philadelphicus* L. var. *scaturicola* (Fernald) Fernald]

Erigeron procumbens (Houst. ex Mill.) G.L. Nesom, (prostrate), CORPUS CHRISTI FLEABANE. Taprooted perennials; stems prostrate, 5–40+ cm long, rooting at nodes; lower stems ± hirsute; stem leaf blades oblanceolate to spatulate or cuneate, the more distant ones not reduced; heads solitary on long peduncles; involucres 6–8 mm high and (8–)10–16 mm wide; ray flowers 225–350, ligules usually white, sometimes pinkish, 5–7 mm long, tightly coiling under in age; achenes 0.7–1 mm long, 2-nerved; pappus with an outer series of short bristles and an inner series of 12 (ray) or 16–23 (disk) bristles. Coastal flats and dunes, salt marshes, sandy roadsides; infrequent inland in se and sc TX; mostly Gulf Coast and South TX Plains; LA, MS, and TX; Mexico. All seasons, but mostly Mar–Jun. [*Aster procumbens* Houst. ex Mill., *E. myrionactis* Small]

Erigeron pulchellus Michx. var. **pulchellus,** (beautiful), HAIRY FLEABANE, ROBIN'S PLANTAIN. Fibrous-rooted, rhizomatous perennials, 15–60 cm tall; lower stems hirsute to villous; stem leaf blades oblanceolate to obovate or suborbiculate, clasping and slightly auriculate, the upper ones reduced; involucres 5–7 mm high and 6–20 mm wide; ray flowers 50–80(–100), ligules light blue to pinkish, 6–10 mm long, coiling under in age; achenes 1.3–1.8 mm long, 2(–4)-nerved; pappus of an outer series of short bristles (sometimes absent) and an inner series of 22–28(–36) bristles. Low prairies, creek bottoms, oak-pine forest; ne to se TX; e 1/2 of North America. Mar–May.

Erigeron strigosus Muhl. ex Willd. var. **strigosus,** (strigose, with stiff bristles), COMMON EASTERN FLEABANE, PRAIRIE FLEABANE, WHITETOP. Fibrous-rooted annuals or biennials, 30–70 cm tall; lower stems strigose with hairs usually ascending; stem leaf blades spatulate to narrowly oblanceolate, the upper ones gradually reduced; involucres 4–5 mm high and 5–12 mm wide; ray flowers 50–100, ligules white, less commonly pinkish or bluish, 4–6 mm long, coiling under in age; achenes usually 0.9–1.2 mm long, 2-nerved; disk pappus with an outer minute crown of bristles or scales and an inner series of 8–15 bristles; ray pappus lacking bristles. Prairies, open woods, pastures, roadsides; widespread in e 1/2 TX; throughout North America but most common in e 1/2; introduced and weedy elsewhere. Apr–Jun, sporadically to Oct. [*E. ramosus* (Walter) Britton, Sterns, & Poggenb., *E. strigosus* Muhl. ex Willd. var. *beyrichii* (Fisch. & C.A. Mey.) Torr. & A. Gray ex A. Gray, *E. strigosus* Muhl. ex Willd. var. *traversii* (Shinners) Noyes, *E. traversii* Shinners]

Erigeron tenuis Torr. & A. Gray, (slender, thin), SLENDER FLEABANE, SLENDER-LEAF FLEABANE. Fibrous-rooted, short-lived perennials or biennials, 10–45 cm tall; lower stems strigose or rarely hirsute; stem leaf blades narrowly oblanceolate to obovate or spatulate, the upper ones abruptly reduced; involucres (2–)2.5–4 mm high and 5–10 mm wide; ray flowers 60–120, ligules pale blue to lavender or white, drying blue to purplish, 3–5 mm long, not coiling nor reflexing; achenes 1–1.2 mm long, 2(–4)-nerved; pappus of an outer series of short bristles and an inner series of 11–15 bristles. Sandy open woods, roadsides, pastures; mostly e 1/2 TX; Edwards Plateau and South TX Plains; se and sc U.S. from FL n to MO and w to KS and TX. Mar–May.

EUPATORIUM L. BONESET, THOROUGHWORT

Ours herbaceous perennials, 30–200 cm tall; stems stiffly erect, often puberulent or pubescent; lower leaves usually opposite sometimes whorled, or upper leaves sometimes alternate; heads small, borne in corymbose or paniculate arrays; phyllaries in several series; receptacles naked; ray flowers absent; disk flowers usually white, rarely pinkish; achenes brownish to black 5-ribbed; pappus of many persistent, minutely-barbed bristles.

A genus of ca. 45 species of e North America, Europe, and e Asia; some are cultivated as ornamentals; several have a history of use in folk medicine and are under investigation for potential medicinal use (Sharma et al. 1998). *Eupatorium* in the traditional broad sense (= *sensu lato*) was treated as including 1000+ species; more recent molecular analyses support its subsequent segregation into a number of smaller genera (King & Robinson 1987; Bremer et al. 1994; Schilling et al. 1999; Schmidt & Schilling 2000). The East TX species of *Eupatorium sensu lato* are now split into 6 different genera: *Ageratina, Chromolaena, Conoclinium, Eupatorium (sensu stricto), Eutrochium,* and *Fleischmannia.* Hybridization and polyploidy complicate species identification (Siripun & Schilling 2006). The common name BONESET is apparently derived from the historic use in treating dengue fever, als

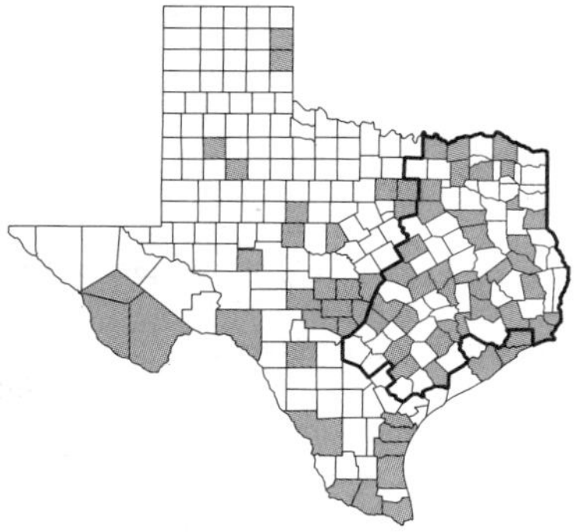

Eclipta prostrata

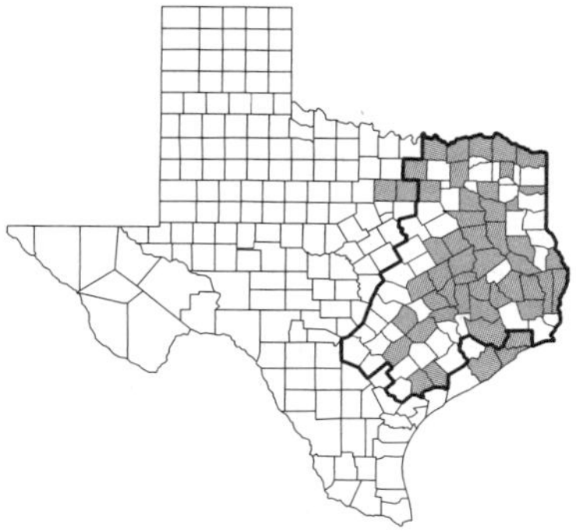

Elephantopus carolinianus

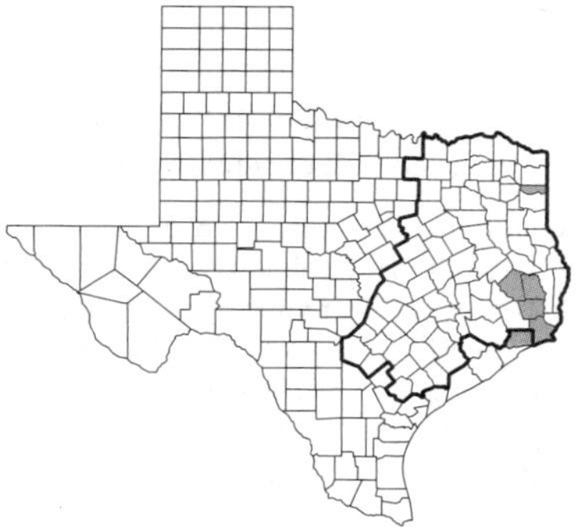

Elephantopus nudatus

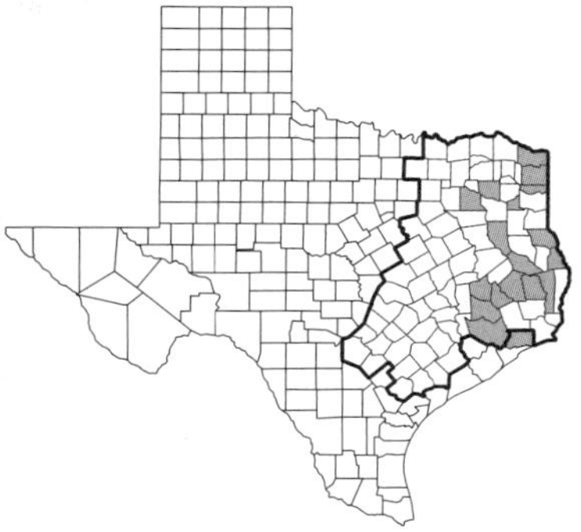

Elephantopus tomentosus

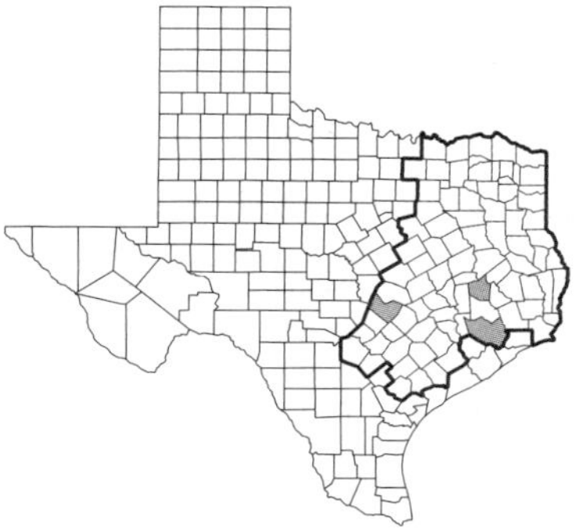

Emilia fosbergii

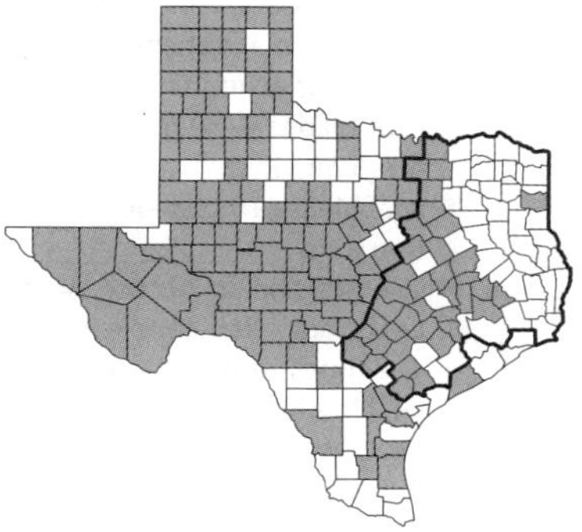

Engelmannia peristenia

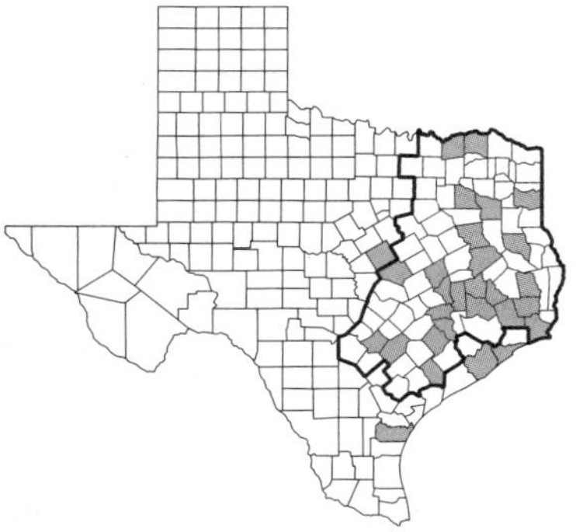

Erechtites hieraciifolius var. hieraciifolius

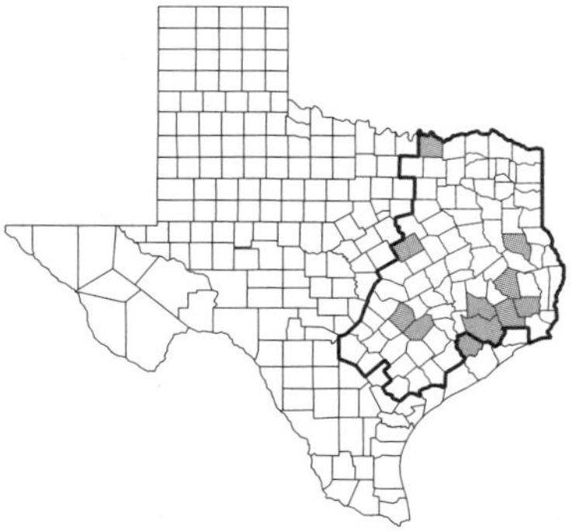

Erigeron annuus

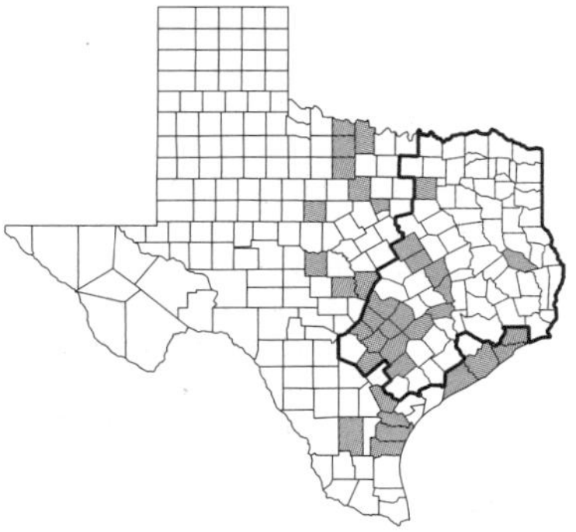

Erigeron geiseri

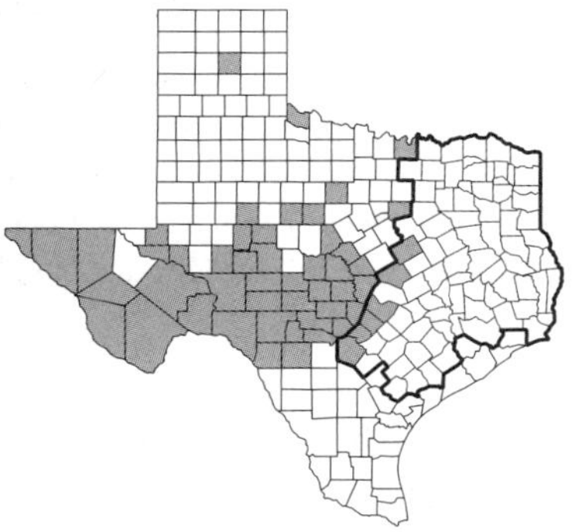

Erigeron modestus

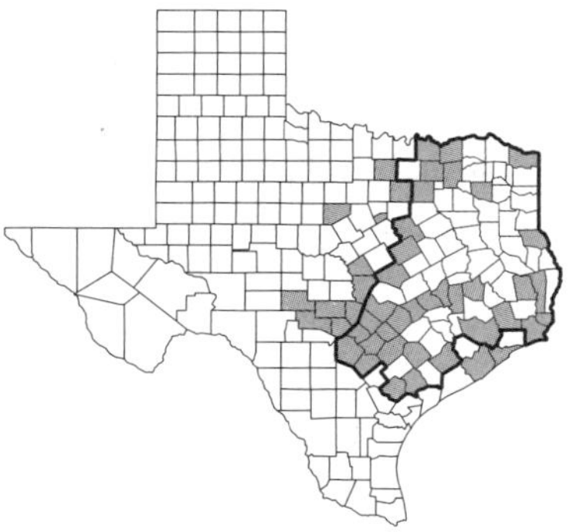

Erigeron philadelphicus var. philadelphicus

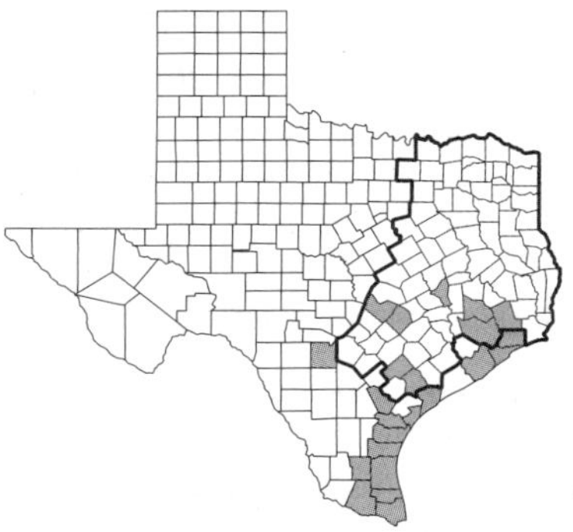

Erigeron procumbens

known as break-bone fever (Tveten & Tveten 1993). (Named for Mithridates Eupator (132–63 B.C.), King of Pontus, who is said to have used a species of the genus in medicine or as an antidote for poison) (tribe Eupatorieae)

REFERENCES: King & Robinson 1970e, 1987; Sullivan 1975; McVaugh 1982; Robinson & King 1985; Bremer et al. 1994; Schilling et al. 1999; Schmidt & Schilling 2000; Siripun & Schilling 2006.

1. Main stem leaves pinnatifid, or pinnately or palmately lobed, lobes relatively narrow.
   2. Leaves mostly palmately or 1–2-pinnately lobed; flowers 7–9 per head ____ **E. ×pinnatifidum**
   2. Leaves usually pinnately or palmately lobed or pinnatifid; flowers usually 5 per head.
      3. Leaf lobes thread-like, 0.2–0.5(–1) mm wide, margins strongly revolute; phyllaries usually glabrate or glabrous, not or only slightly gland-dotted; pappus bristles 2–2.5 mm long ____ **E. capillifolium**
      3. Leaf lobes broader, not thread-like, 0.5–2.5(–4) mm wide, margins not revolute; phyllaries usually puberulent (especially on midveins), usually gland-dotted; pappus bristles 3–3.5 mm long ____ **E. compositifolium**
1. Main stem leaves unlobed, margins entire, crenate, or serrate to serrulate.
   4. Leaves petiolate, petioles 10–30 mm long ____ **E. serotinum**
   4. Leaves sessile or subsessile, petioles absent or < 10 mm long.
      5. Leaf bases connate-perfoliate (fused in pairs around the stem); flowers 7–11 per head ____ **E. perfoliatum**
      5. Leaf bases distinct, not connate-perfoliate; flowers (4–)5 per head.
         6. Phyllary tips acuminate to attenuate, usually whitish.
            7. Leaf blades 4–10 mm wide; phyllaries 8–10, narrowly elliptic; corollas 3–3.5 mm long; pappus bristles 4.5–5 mm long ____ **E. leucolepis**
            7. Leaf blades 10–25(–40) mm wide; phyllaries 9–15, linear; corollas 4–4.5 mm long; pappus bristles 3.5–4.5 mm long ____ **E. album**
         6. Phyllary tips acute, obtuse, or rounded, sometimes with narrow, whitish (hyaline) margins.
            8. Leaf blades deltate to suborbiculate, sometimes ovate, usually widest below the middle (lengths mostly 1–2 times widths), bases broadly rounded to truncate ____ **E. rotundifolium**
            8. Leaf blades of various shapes but usually widest near or above the middle (lengths usually more than 2 times widths), bases ± cuneate.
               9. Leaves strongly to obscurely 3-nerved from bases; IF lateral veins are obscure, THEN the leaves linear, 2–5 mm wide.
                  10. Stems (at least the lower portion) glabrous to glabrate (sometimes pilose); leaves 30–50 mm long, glabrous beneath; phyllaries 0.2–0.8 mm wide; corollas 2.5–3 mm long ____ **E. lancifolium**
                  10. Stems pubescent throughout; leaves 20–60(–120) mm long, but the larger usually 50+ mm long, scabrous, puberulent, or villous beneath; phyllaries 0.5–1.5 mm wide; corollas 3–3.5 mm long.
                      11. Leaf blades 20–60 mm long and 2–5 mm wide, linear, margins entire or finely and obscurely serrate; plants of se TX ____ **E. hyssopifolium**
                      11. Leaf blades 50–120 mm long and 5–20 mm wide, lance-elliptic to oblanceolate, margins entire in the lower portion and sharply serrate in the upper portion; plants of ne TX ____ **E. altissimum**
               9. Leaves 3-nerved but lateral nerves arising higher on the midrib, not at leaf base.
                  12. Leaves recurved, upper blade surface glabrous or glabrate; stems sometimes from tuberous-thickened rhizomes; phyllaries 0.2–0.5 mm wide; achenes 1–2 mm long ____ **E. mohrii**
                  12. Leaves spreading, upper blade surface puberulent to villous; stems from short caudices or rhizomes; phyllaries 0.5–1 mm wide; achenes 1.5–3 mm long.
                      13. Stems densely branched in upper portion; leaf blades (30–)50–70 mm long, margins usually serrate; phyllaries elliptic (larger ones 2.5–3 mm long); corollas 2.5–3 mm long ____ **E. semiserratum**
                      13. Stems usually branched at or near bases; leaf blades mostly 20–35(–45) mm long, margins weakly serrate to subentire; phyllaries lanceolate (larger ones usually 4–5 mm long); corollas 3–3.5 mm long ____ **E. linearifolium**

Eupatorium album L. var. **album,** (white), WHITE EUPATORIUM, WHITE THOROUGHWORT. Plants with short caudices or stout rhizomes; leaves usually opposite (at least the lower ones), sessile, blades elliptic to oblanceolate, 3-nerved from above base, pubescent, usually gland-dotted, bases narrowly cuneate, margins coarsely serrate; phyllaries pubescent and gland-dotted, tips acuminate to attenuate; flowers (4–)5; achenes 2.5–3.5 mm long; pappus 3.5–4.5 mm long. Sandy pinelands; rare in TX; in East TX reported from Anderson Co. (Singhurst et al. 1998); also Jefferson Co.; e and se U.S. from MA s to FL and w to IN and TX. Aug–Nov. [*E. album* L. var. *glandulosum* (Michx.) DC.]

Eupatorium altissimum L., (very tall), TALL EUPATORIUM, TALL THOROUGHWORT. Plants with short caudices or stout rhizomes; leaves usually opposite, sessile or nearly so, blades lance-elliptic to oblanceolate, strongly 3-nerved from bases, puberulent or villous, gland-dotted, base ± cuneate, margins serrate in the upper portion; phyllaries pubescent, tips rounded to acute; flowers 5; achenes 2–3 mm long; pappus 3.5–4 mm long. Open pine-oak woodlands, thickets, most often on limestone soils; ne TX near Red River drainage; nc TX; e U.S. from MA s to FL and w to MN, NE, and TX; Canada. Aug–Oct. *E. altissimum* hybridizes with *E. serotinum* (Siripun & Schilling 2006).

Erigeron geiseri [HEA]

Erigeron modestus [HEA]

Erigeron philadelphicus [GLE]

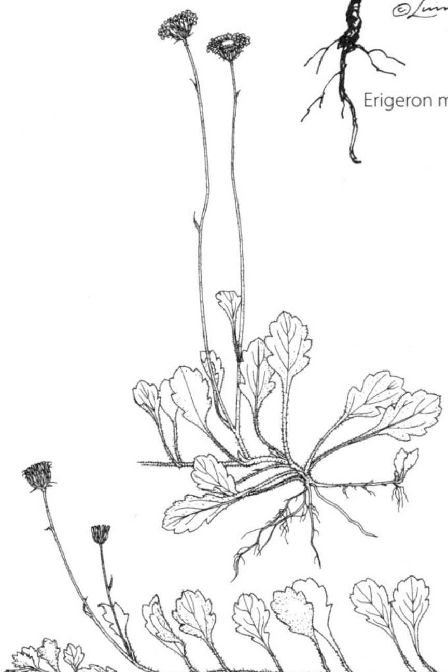

Erigeron procumbens [FNA]

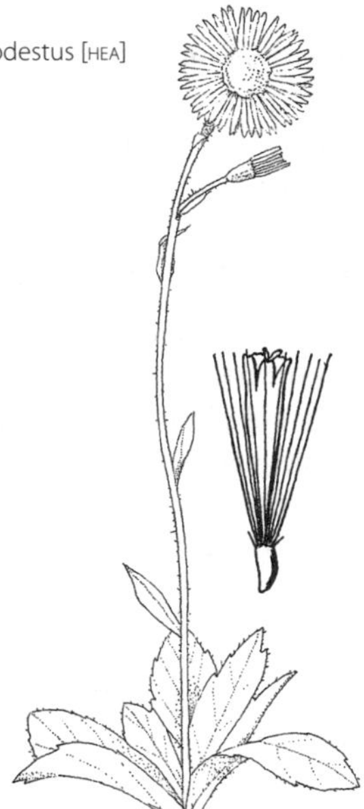

Erigeron pulchellus [BB2, FOV]

Eupatorium capillifolium (Lam.) Small, (hair-leaved), DOGFENNEL, DOGFENNEL EUPATORIUM, CYPRESSWEED. Plants with short caudices; leaves usually alternate (at least the upper ones), sessile, blades dissected and pinnately lobed, lobes linear, glabrate, gland-dotted, bases ± cuneate, margins strongly revolute; phyllaries nearly glabrous, not or only minimally gland-dotted, tips acuminate and mucronate; flowers 5; achenes 1–1.7 mm long; pappus 2–2.5 mm long. Disturbed areas, sandy soils, old fields; East TX but most common in se TX; nc TX, Edwards Plateau, and South TX Plains; se U.S. from MA s to FL and w to MO and TX. Aug–Nov. A hybrid of *E. capillifolium* and *E. linearifolium* from Hardin Co. was described by Jordan (1991).

Eupatorium compositifolium Walter, (compound-leaved), YANKEEWEED. Plants with short caudices; leaves usually alternate (at least the upper ones), sessile, blades pinnately or palmately lobed, lobes linear, puberulent, gland-dotted, bases cuneate, margins entire; phyllaries puberulent, usually gland-dotted, tips acuminate and mucronate; flowers 5; achenes 1–1.7 mm long; pappus 3–3.5 mm long. Roadsides, disturbed areas, sand dunes; widespread across East TX to Gulf Coast and South TX Plains; se U.S. from NC s to FL and w to OK and TX. Oct–Nov. May hybridize with *E. serotinum* (Jordan 1991; Turner 2015).

Eupatorium hyssopifolium L. var. **hyssopifolium,** (with leaves resembling the genus *Hyssopus*), HYSSOP-LEAVED EUPATORIUM, HYSSOP-LEAF THOROUGHWORT. Plants with short caudices or stout rhizomes; leaves usually opposite or whorled in 3s and 4s (at least the lower ones), sessile, blades linear, 3-nerved from base (sometimes weakly), scabrous, bases cuneate, margins entire or obscurely and finely serrate; phyllaries pubescent, tips obtuse to acute; flowers 5; achenes 2–3 mm long; pappus 3.5–4 mm long. Disturbed sites, roadsides; uncommon in deep se TX; e U.S. from MA s to FL and w to WI, MO, and TX. Aug–Oct. [*E. hyssopifolium* L. var. *calcaratum* Fernald & B.G. Schub., *E. lecheifolium* Greene]

Eupatorium lancifolium (Torr. & A. Gray) Small, (lance-leaved), LANCE-LEAVED EUPATORIUM, LANCELEAF THOROUGHWORT. Plants with short rhizomes; leaves usually opposite (at least the lower ones), sessile or subsessile, blades lanceolate to lance-linear, 3-nerved from base, glabrous above, sparsely pilose beneath, gland-dotted, bases narrowly cuneate, margins serrate; phyllaries pilose, gland-dotted, tips rounded; flowers 5; achenes 2.5–2.8 mm long; pappus 3.5–4 mm long. Dry clay soils, rolling terrain, pine and oak woods; se and sc TX; AL, AR, LA, MS, and TX. Aug–Sep. [*E. parviflorum* Elliott var. *lancifolium* Torr. & A. Gray, *E. semiserratum* DC. var. *lancifolium* Torr. & A. Gray]

Eupatorium leucolepis (DC.) Torr. & A. Gray var. **leucolepis,** (white-scaled), JUSTICEWEED, WHITE-BRACTED THOROUGHWORT. Plants with short caudices or stout rhizomes; leaves usually opposite (at least the lower ones), sessile, blades lance-oblong to linear-oblong, pinnately nerved, strongly folded along the midribs, scabrous above, villous beneath, bases rounded to cuneate, margins entire or serrate; phyllaries puberulent, gland-dotted, tips acuminate to attenuate; flowers 5; achenes 2–3 mm long; pappus 4.5–5 mm long. Pine barrens, pond margins, moist sandy soils; deep se TX in Angelina, Hardin, Jasper, Newton, Orange, and Tyler cos.; e and se U.S. from NY s to FL and w to KY and TX. Aug–Oct.

Eupatorium linearifolium Walter, (linear-leaved), WAXY EUPATORIUM, WAXY THOROUGHWORT. Plants with short caudices; leaves usually opposite (at least the lower ones), sessile or subsessile, blades oblong to lance-oblong, 3-nerved from above bases, finely puberulent, gland-dotted, bases cuneate, margins entire or serrate in the upper portion; phyllaries puberulent, gland-dotted, tips rounded to acute; flowers 5; achenes 2.5–3 mm long; pappus 3–5 mm long. Dry sandy soils, pine and oak woods, old fields; se and sc TX to Gulf Coast and South TX Plains; se U.S. from VA s to FL and w to TX. Aug–Nov. [*E. cuneifolium* Willd., *E. tortifolium* Chapm.] A hybrid of *E. linearifolium* and *E. capillifolium* from Hardin Co. was described by Jordan (1991).

Eupatorium mohrii Greene, (for Charles Theodor Mohr (1924–1901), German-born pharmacist and botanist of the se U.S.) MOHR'S EUPATORIUM, MOHR'S THOROUGHWORT. Plants with tuberous rhizomes; leaves usually opposite (at least the lower ones), sessile or subsessile, blades oblanceolate, 3-nerved from above base, nearly glabrous, gland-dotted, bases cuneate, margins serrate in the upper portion; phyllaries puberulent, gland-dotted, tips rounded; flowers 5; achenes 1–2 mm long; pappus 2.5–3 mm long. Moist sandy soils, pond margins; deep se TX; se U.S. from VA s to FL and w to TX. Aug–Sep. [*E. recurvans* Small] May hybridize with *E. serotinum* and *E. rotundifolium* (Siripun & Schilling 2006).

Eupatorium perfoliatum L., (perfoliate, with leaf bases surrounding the stem), BONESET, COMMON THOROUGHWORT, AGUEWEED. Plants with short caudices; leaves usually opposite (sometimes whorled), sessile, blades oblong, pinnately nerved, pilose, gland-dotted beneath, bases connate-perfoliate, margins serrate; phyllaries villous or puberulent, gland-dotted, tips acute to acuminate; flowers 7–11+; achenes 1.5–2 mm long; pappus 3–3.5 mm long. Moist sandy soils, marshes, roadsides; e 1/3 TX; also Panhandle; e 1/2 of North America. Aug–Oct. [*E. chapmanii* Small, *E. perfoliatum* L. var. *cuneatum* Engelm.] Hybridizes with *E. serotinum* and potentially other species of *Eupatorium* (Siripun & Schilling 2006). This species contains the glucoside eupatorin and has been used in the traditional medicine of Native Americans; recent experiments indicate possible anti-inflammatory effects and activity against *Plasmodium falciparum*, the parasite that causes malaria (Duke 1985; Hensel et al. 2011).

Erigeron strigosus var. strigosus [BB2]

Erigeron tenuis [GLE]

Eupatorium album [BB2]

Eupatorium altissimum [BB2]

Eupatorium capillifolium [REE]

Eupatorium compositifolium [TRP]

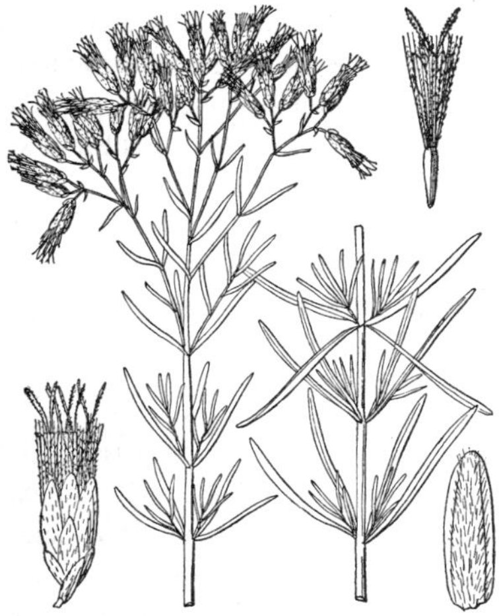
Eupatorium hyssopifolium [BB2]

Eupatorium lancifolium [THO]

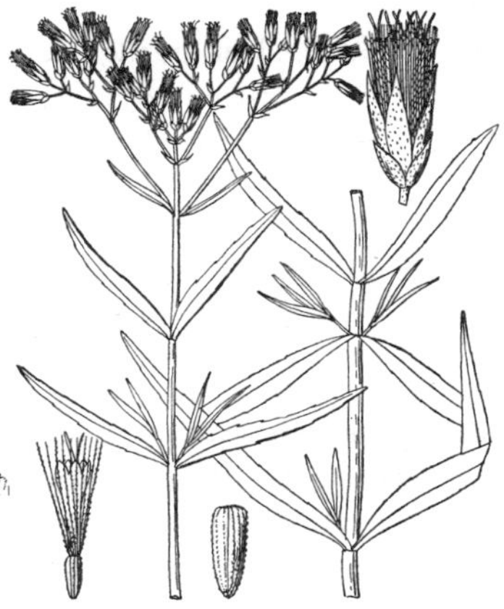
Eupatorium leucolepis [BB2]

Eupatorium ×pinnatifidum Elliott [*capillifolium* or *compositifolium* × *perfoliatum* or *serotinum*], (pinnatifid), THOROUGHWORT. This is the single formally-recognized hybrid name in the genus for our area, which apparently represents a series of recurrent hybrids according to Siripun and Schilling (2006), and tends to be intermediate to the parent species. Roadsides, open pine woods, pond banks; deep se TX; se U.S. from VA s to FL and w to TX. Aug–Oct. [*E. eugenei* Small, *E. pectinatum* Small]

Eupatorium rotundifolium L., (round-leaved), FALSE HOREHOUND, ROUND-LEAVED EUPATORIUM, ROUNDLEAF THOROUGHWORT. Plants with short rhizomes; leaves usually opposite (at least the lower ones), sessile or subsessile, blades deltate to orbiculate, sometimes ovate, 3-nerved, puberulent to villous, gland-dotted, bases rounded to truncate, margins crenate to serrate; phyllaries puberulent, gland-dotted, tips acute; flowers 5; achenes 2–3 mm long; pappus 3.5–4 mm long. Jun–Oct. Three varieties are recognized; 1 is common in East TX and 2 are to be expected. Siripun and Schilling (2006) separated the 3 intergrading varieties as follows:

1. Leaf blades 3-nerved from the base, margins crenate ______ var. **rotundifolium**
1. Leaf blades 3-nerved from above base, margins serrate.
   2. Leaf blades 30–70 mm long and 30–60 mm wide, broadest near middle ______ var. *ovatum*
   2. Leaf blades 20–50 mm long and 15–30 mm wide, broadest near base ______ var. *scabridum*

var. *ovatum* (Bigelow) Torr., (oval). Moist sandy soils, roadsides, low ground; *to be expected in East TX based on records from adjacent parishes in LA* (Beauregard, Caddo, Calcasieu, Sabine, and Vernon parishes); e and se U.S. from ME s to FL and w to AR and LA. [*E. ovatum* Bigelow, *E. pubescens* Muhl. ex Willd.] Potentially of hybrid origin.

var. **rotundifolium,** (round-leaved). Moist sandy soils, roadsides, pond margins; East TX; nc TX; e and se U.S. from NY s to FL and w to OK and TX. [*E. verbenifolium* Reichard]

var. *scabridum* (Elliott) A. Gray, (rough). Disturbed sites; *to be expected in East TX based on records from adjacent parishes in LA* (Beauregard, DeSoto, and Sabine parishes); se U.S. from SC s to FL and w to MO and LA. [*E. scabridum* Elliott] Potentially of hybrid origin.

Eupatorium semiserratum DC., (semi-serrate), SAW-LEAF EUPATORIUM, SMALLFLOWER THOROUGHWORT. Plants with short rhizomes; leaves usually opposite (at least the lower ones), sessile or subsessile, blades elliptic to lance-elliptic, 3-nerved from above bases, puberulent to villous, densely gland-dotted, bases narrowly cuneate, margins usually serrate; phyllaries puberulent, gland-dotted, tips rounded to acute; flowers 5; achenes 1.5–2 mm long; pappus 2.5–3 mm long. Moist sandy soils, bogs, swamp margins; se to sc TX; se U.S. from MD s to FL and w to MO and TX. Jul–Oct(–Nov). [*E. cuneifolium* Willd. var. *semiserratum* (DC.) Fernald & Grisc.] May hybridize with *E. rotundifolium* and *E. hyssopifolium* (Siripun & Schilling 2006).

Eupatorium serotinum Michx., (late-flowering), LATE EUPATORIUM, LATE-FLOWERING THOROUGHWORT, FALL BONESET. Plants with short caudices; leaves usually opposite, long-petiolate, blades lanceolate, 3-nerved, puberulent, gland-dotted, bases rounded to slightly oblique, margins entire or serrate; phyllaries puberulent, gland-dotted, tips slightly rounded to acute; flowers 9–15; achenes 1–1.5 mm long; pappus 2–2.5 mm long. Open areas, disturbed sites, stream banks; se and e TX w to Rolling Plains and Edwards Plateau; e 1/2 of North America. Aug–Oct. Hybridizes with *E. perfoliatum* (Siripun & Schilling 2006).

Eupatorium pilosum Walter, (pilose), ROUGH BONESET, which would have keyed to *E. album* or *E. rotundifolium* here, can be distinguished by its leaves that are alternate and entire (on the upper stems), 3-nerved from above the bases, blade lengths that are 2–2.5 times the widths, rounded-cuneate bases, and lance-oblong phyllaries (larger ones 5–7 mm long). To be expected in East TX based on records from adjacent parishes in LA (Calcasieu and DeSoto parishes); e and se U.S. from MA s to FL and w to LA. Jul–Sep. [*Eupatorium rotundifolium* L. var. *saundersii* (Porter ex Britton) Cronquist, *E. verbenifolium* Michx.]

EURYBIA (Cass.) Cass. ASTER

A North American genus of 23 species; formerly included in *Aster.* (Greek: *eurys*, wide, and *baios*, few, perhaps alluding to the relatively few, broad ray ligules) (tribe Astereae)

REFERENCES: Jones 1980; Nesom 1994b; Brouillet 2006a.

Eurybia hemispherica (Alexander) G.L. Nesom, (for the hemispheric heads), SOUTHERN PRAIRIE ASTER, SINGLE-STEMMED BOG ASTER. Perennial herbs, 20–100 cm tall, forming clonal clumps from creeping rhizomes; stems erect or ascending, usually unbranched, often reddish; basal leaves withering by flowering time; stem leaves alternate, simple, sessile and subclasping, blades lance-ovate or linear-lanceolate to linear, gradually reduced up the stem, margins usually entire; heads in elongate spicate or racemose arrays; involucres hemispheric, 8–12(–15) mm high; phyllaries in 4–6 series, unequal, the outer strongly reflexed; receptacles naked; ray flowers 15–30, pistillate, fertile, the ligules violet-purple (rarely white); disk flowers perfect, fertile, corolla pale yellow; achenes slightly flattened, ribbed, 2.6–3.7 mm long; pappus of tawny bristles 6–7.5 mm long, equaling the di

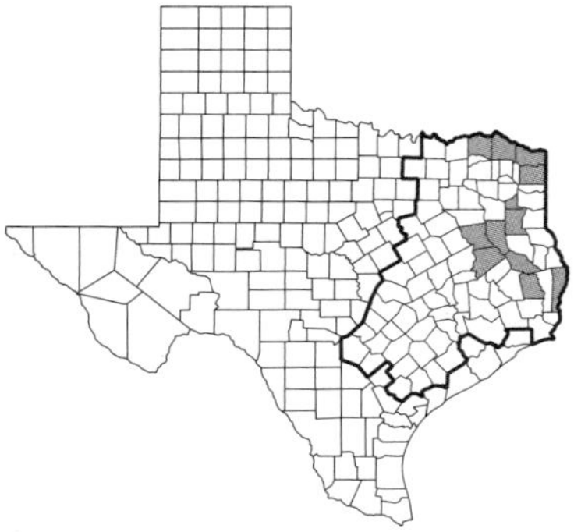
Erigeron pulchellus var. pulchellus

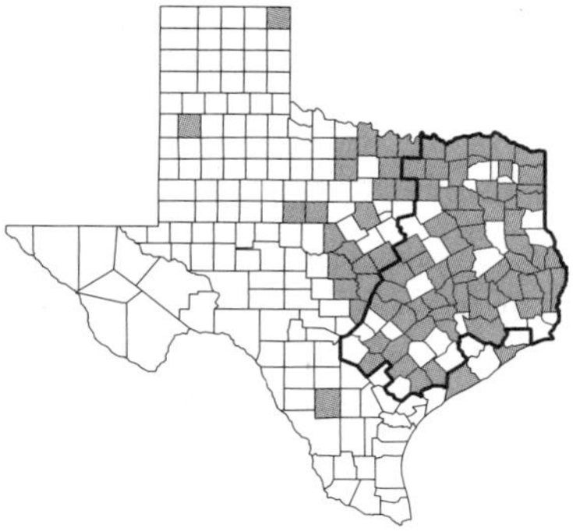
Erigeron strigosus var. strigosus

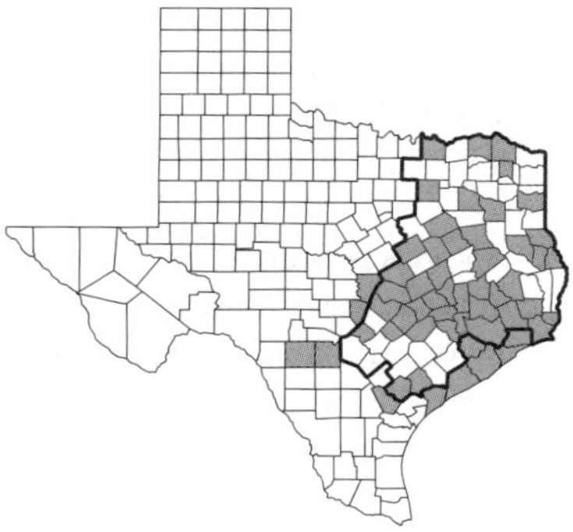
Erigeron tenuis

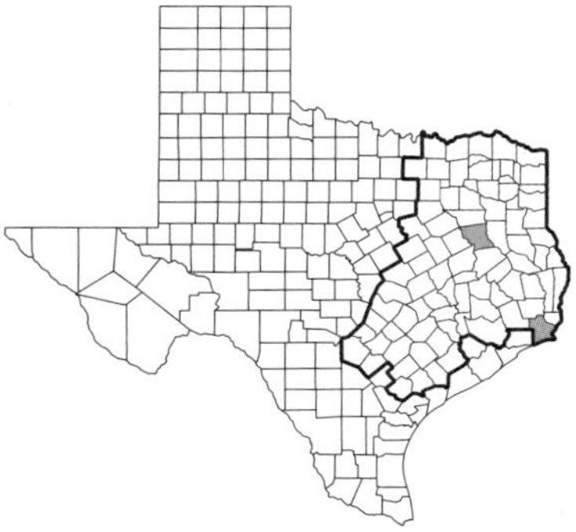
Eupatorium album var. album

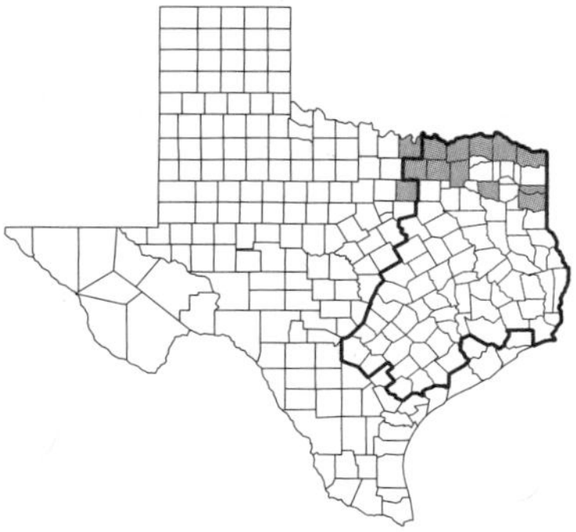
Eupatorium altissimum

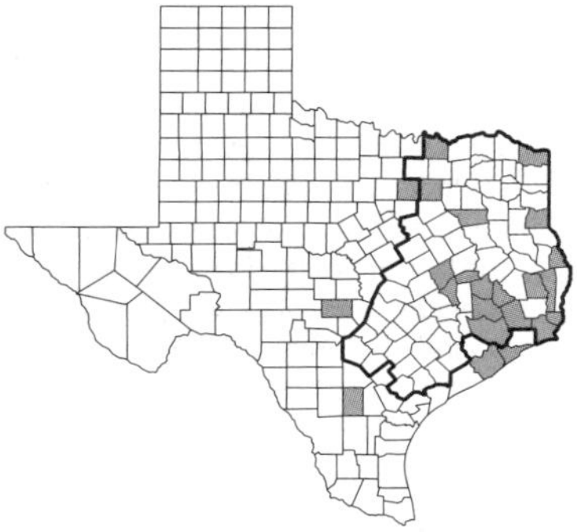
Eupatorium capillifolium

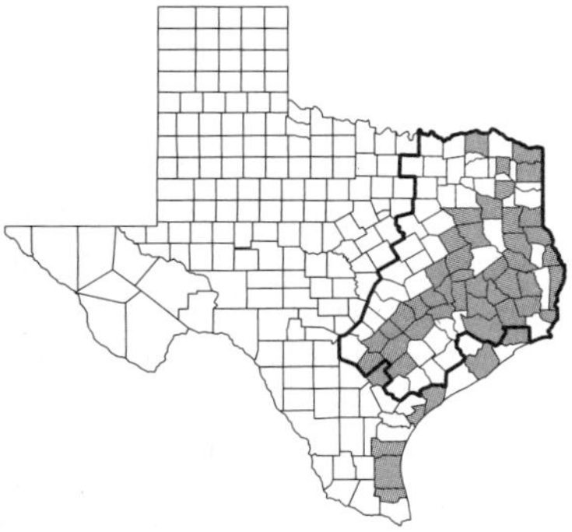
Eupatorium compositifolium

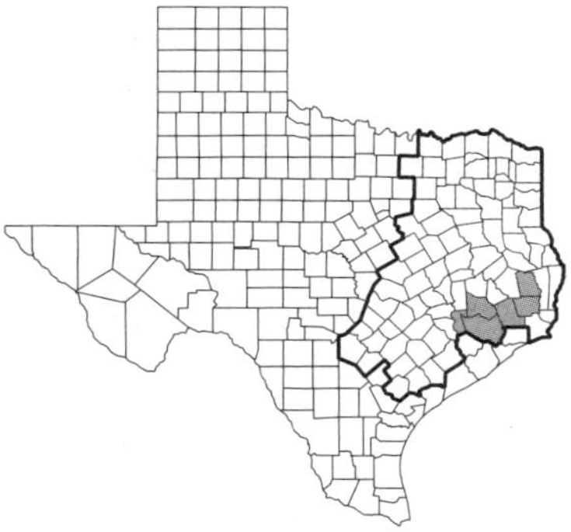
Eupatorium hyssopifolium var. hyssopifolium

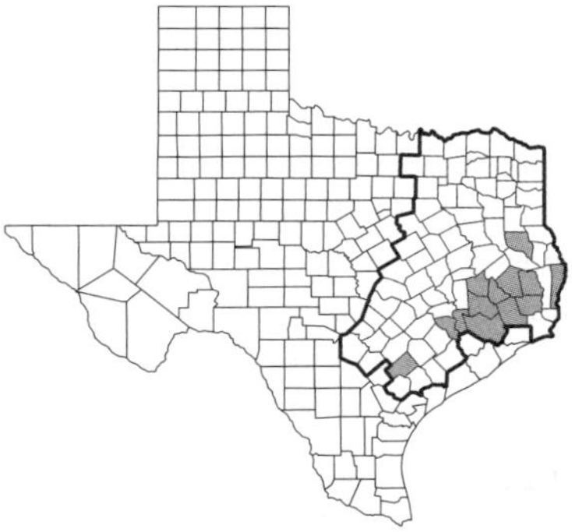
Eupatorium lancifolium

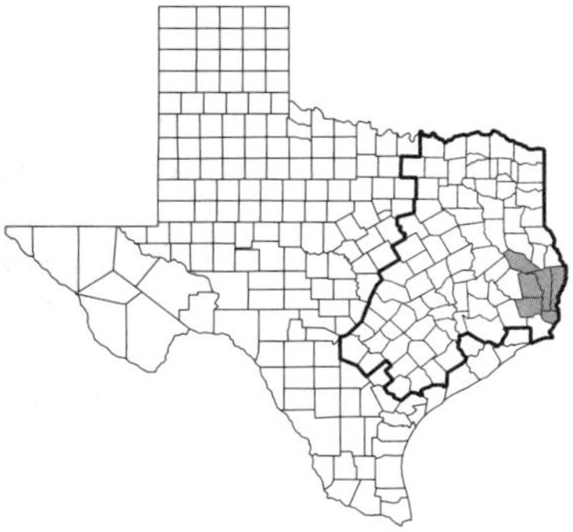
Eupatorium leucolepis var. leucolepis

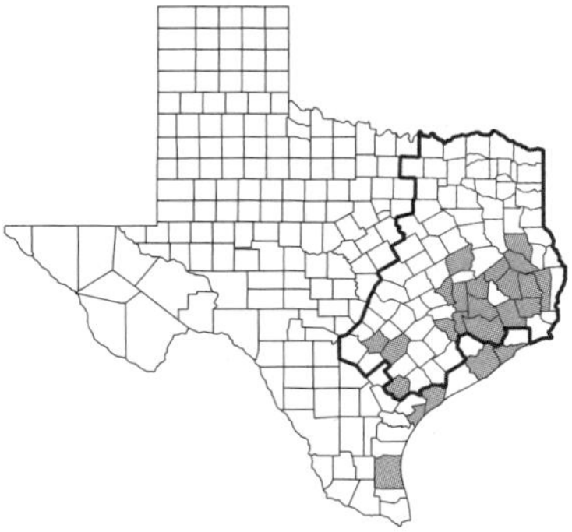
Eupatorium linearifolium

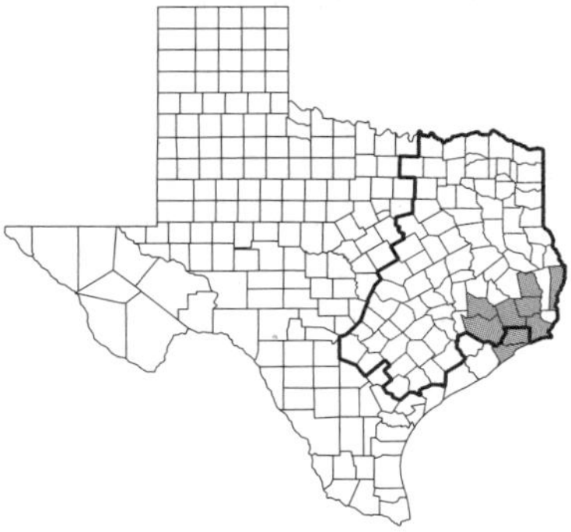
Eupatorium mohrii

corollas. Sandy-loamy soils, open woods, prairies, bottomlands; e 1/3 TX; se U.S. from KY s to FL and w to KS and TX. Sep–Oct. [*Aster hemisphericus* Alexander, *A. paludosus* Aiton subsp. *hemisphericus* (Alexander) Cronquist, *Heleastrum hemisphaericum* (Alexander) Shinners]

EUTHAMIA Nutt. ex Cass. FLAT-TOPPED GOLDENROD, GOLDENTOP

Our herbaceous perennials; fibrous-rooted with creeping rhizomes; stems stiffly erect; basal and lower stem leaves sometimes early deciduous; leaves alternate, sessile, linear to lanceolate, ± uniform in shape along the stems, sparsely to densely gland-dotted, margins entire; heads clustered to produce flat-topped or slightly rounded corymbose arrays; phyllaries in 3–5 series, glabrous; receptacles naked; ray flowers pistillate and fertile, the corollas yellow; disk flowers perfect and fertile, usually fewer than the ray flowers, the corollas yellow; achenes oblong to ellipsoid, 2–4-nerved; strigose; pappus of 20–30 white, barbed bristles.

A North American genus of 5 species; similar to and formerly lumped into *Solidago* as a section or subgenus. *Euthamia* is capable of "tremendous" morphological variation (Haines 2006). (Greek: *eu*, well, and *thama*, crowded, alluding to the dense inflorescences) (tribe Astereae)

REFERENCES: Sieren 1981; Taylor & Taylor 1983; Semple 1992; Haines 2006.

1. Involucres 3–4.7(–5.3) mm high; rare in TX, known only from Gulf Coast prairies ____ **E. caroliniana**
1. Involucres 4–6.2 mm high; common plants of wider distribution in TX (including Gulf Coast prairies).
   2. Leaf blades linear to lanceolate, lengths 12–49 times widths, gradually reduced up the stem, prominently and abundantly gland-dotted (but not blistered or pustulate); flowering section of plant (from the top of the plant to the bottom of the lowest head-bearing branches) usually > 35% of plant height ____ **E. gymnospermoides**
   2. Leaf blades lanceolate to narrowly lanceolate, lengths 8–18 times widths, abruptly reduced up the stem, sparingly and obscurely gland-dotted (sometimes blistered or pustulate); flowering section of plant (from the top of the plant to the bottom of the lowest head-bearing branches) usually < 35% of plant height ____ **E. leptocephala**

Euthamia caroliniana (L.) Greene ex Porter & Britton, (of the Carolinas), COASTAL PLAIN GOLDENTOP, SLENDER FLAT-TOPPED GOLDENROD. Plant 25–100 cm tall; stems ± glabrous; leaves linear, 1–3 mm wide, lengths 7–42.2 times widths, abruptly reduced up the stem, abundantly gland-dotted; involucres 3–4.7(–5.3) mm high; disk corollas 3.8–4.8 mm long. Moist sandy soils; in East TX recorded only from Harris Co. (*J. Singhurst & W. Newman 17900* (BAYLU); Singhurst et al. 2014); first reported in TX from Chambers Co. (Singhurst et al. 2009); e U.S. from NY s to FL and w to MI, IL, and TX; Canada. Oct–Nov. [*E. microcephala* Greene, *E. remota* Greene, *Solidago lanceolata* L. var. *minor* Michx., *S. tenuifolia* Pursh var. *pycnocephala* Fernald]

Euthamia gymnospermoides Greene, (resembling the genus *Gymnosperma*), GREAT PLAINS GOLDENTOP, VISCID EUTHAMIA, FLAT-TOPPED GOLDENROD. Plant 40–150 cm tall; stems glabrous or with scabrous lines; leaves linear to lanceolate, 1.4–4(–8) mm wide, lengths 12–49 times widths, gradually reduced up the stem, abundantly gland-dotted; involucres (4–)4.5–6.2 mm high; disk corollas (3–)3.3–4.8 mm long. Open woods, prairies, sandy soils; se and e TX, nc TX; Plains Country and Gulf Coast; c U.S. from MI s to LA and w to ND, CO, and TX; Canada. Oct–Nov. [*E. camporum* Greene, *E. pulverulenta* Greene, *Solidago gymnospermoides* (Greene) Fernald, *S. graminifolia* (L.) Salisb. var. *gymnospermoides* (Greene) Croat, *S. texensis* Friesner]

Euthamia leptocephala (Torr. & A. Gray) Greene ex Porter & Britton, (narrow-headed), MISSISSIPPI VALLEY GOLDENTOP, BUSHY FLAT-TOPPED GOLDENROD. Plant 30–100 cm tall; stems glabrous; leaves lanceolate to narrowly lanceolate, 3–6(–9) mm wide, lengths 8–18 times widths, abruptly reduced up the stem, obscurely or sparingly gland-dotted, sometimes with tiny blisters (= pustulate); involucres 4–6 mm high; disk corollas 3.3–4.4 mm long. Open woods, prairies, moist sandy soils; e and sc TX; se U.S. from IL s to GA and w to OK and TX. Sep–Nov. [*Solidago leptocephala* Torr. & A. Gray]

Euthamia graminifolia (L.) Nutt., (grass-leaved), COMMON GOLDENTOP, FLAT-TOPPED GOLDENROD, which would have keyed to *E. caroliniana* here, can be distinguished by its wider leaves (usually 3–12 mm wide vs. 1–3 mm wide) that are obscurely or sparingly gland-dotted. Excluded from TX by Haines (2006); potentially to be found in ne TX based on a record from adjacent Choctaw Co. in OK (Kartesz 2015); most common in the ne U.S. with scattered records w to WA and s to MS; Canada; introduced in Europe and Asia. Sep–Oct. [*E. nuttallii* Greene, *Solidago graminifolia* (L.) Salisb.]

Eupatorium linearifolium [GLE]

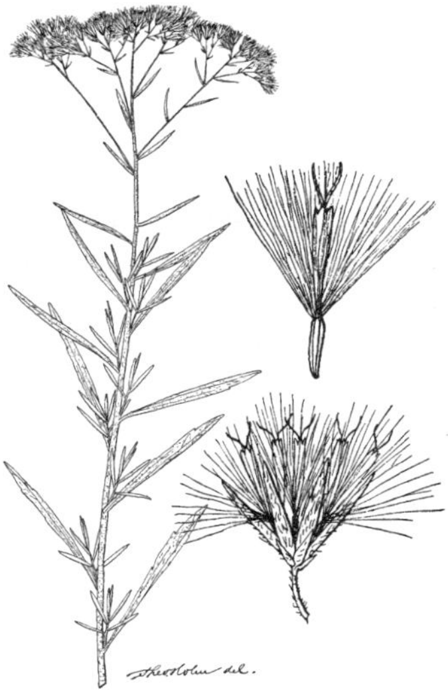
Eupatorium mohrii [USH]

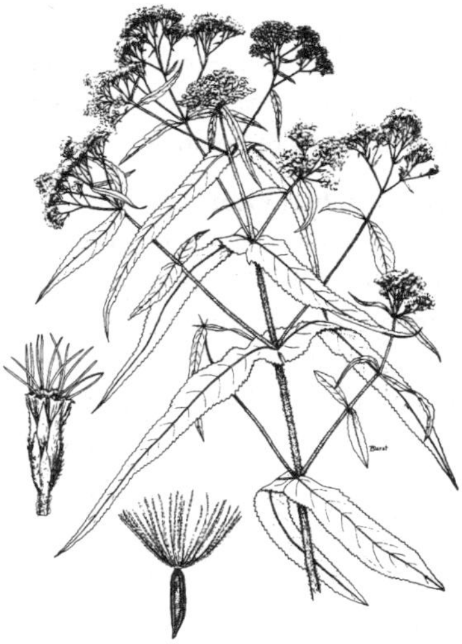
Eupatorium perfoliatum [GWO]

Eupatorium xpinnatifidum [USH]

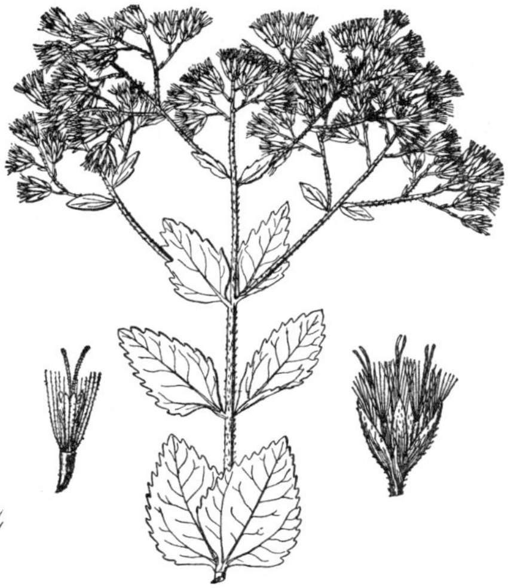
Eupatorium rotundifolium [BB2]

Eupatorium semiserratum [GWO]

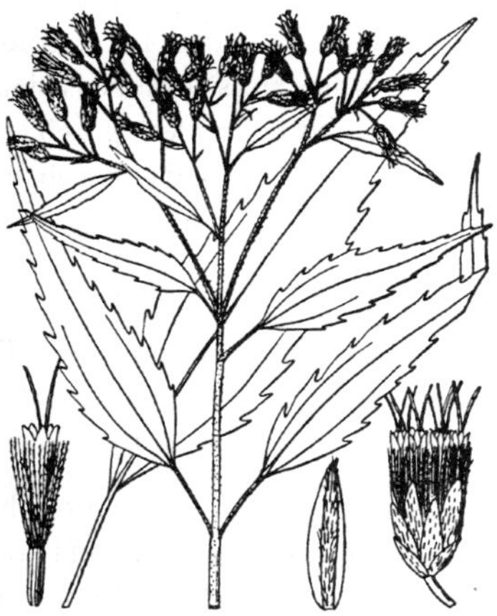
Eupatorium serotinum [BB2]

Eurybia hemispherica [YAT]

Euthamia caroliniana [BB2]

EUTROCHIUM Raf. JOEPYEWEED

A genus of 5 species of North America; formerly included in *Eupatorium*. (Greek: *eu*, well or truly, and *trocho*, wheel-like, alluding to the whorled leaves) (tribe Eupatorieae)
REFERENCES: King & Robinson 1987; Lamont 1995, 2006.

Eutrochium fistulosum (Barratt) E.E. Lamont, (hollow, for the stems), HOLLOW JOEPYEWEED, TRUMPETWEED. Herbaceous perennial, 60–350+ cm tall; stems erect, unbranched, usually purplish throughout, glaucous, hollow (at least in the lower stem); leaves whorled in 4s–6s(–7s), petiolate, blades narrowly to broadly lanceolate, margins finely serrate; heads in rounded, compound corymbose arrays; involucres 6.5–9 mm high, often purplish; phyllaries in 5–6 series, unequal; ray flowers absent; disk flowers (4–)5–7, corollas pale pinkish or purplish; achenes 5-angled, 3–4.5 mm long; pappus of numerous barbed bristles. Low woods, acidic sandy soils, along streams, bogs; se TX; e U.S. from NY s to FL and w to MI and TX. Jul–Aug. [*Eupatoriadelphus fistulosus* (Barratt) R.M. King & H. Rob., *Eupatorium fistulosum* Barratt]

Eutrochium purpureum (L.) E.E. Lamont, (purple), SWEETSCENTED JOEPYEWEED, SWEET JOEPYEWEED, can be distinguished by the usually solid (not hollow) stems that are purplish only at the nodes. Not yet reported in TX but included here based on records from adjacent counties in Oklahoma: McCurtain Co. (OKL) and Choctaw Co. (Kartesz 2015); e U.S. from NH s to FL and w to MN and OK. Late summer–early fall. [*Eupatoriadelphus purpureus* (L.) R.M. King & H. Rob., *Eupatorium purpureum* L.]

FACELIS Cass.

A South American genus of 3 or 4 species. (Greek: *phakelos*, a bundle) (tribe Inuleae)
REFERENCES: Anderberg 1991; Nesom 2006m.

Facelis retusa (Lam.) Sch. Bip., (retuse, notched slightly at a rounded apex, referring to the leaves), TRAMPWEED. Annual herb from a taproot; stems 3–20(–30) cm tall, erect or with decumbent branches, white-gray woolly; leaves alternate, simple, entire, linear-spatulate, 7–20(–30) mm long, 1.5–4 mm wide, ± sessile; heads in a terminal cluster; involucres 8–11 mm high; phyllaries mostly scarious or partly greenish, the inner often purplish-pigmented near tip; ray flowers absent; disk flowers of 2 types: the central ca. 3–5 perfect and fertile, with corollas 5-toothed, white; the peripheral 10–25 pistillate, with corollas white, filiform, truncate or obscurely toothed; achenes white-hairy, 1.6 mm long; pappus of numerous whitish, strongly plumose bristles 8–11 mm long, surpassing the corollas. Lawn weed, roadsides, sandy disturbed areas; East TX w to Edwards Plateau, se U.S. from VA s to FL and w to KS and TX; South America; introduced elsewhere. Apr–May. [*Gnaphalium retusum* Lam.] Native of South America.

FLAVERIA Juss. YELLOWTOPS

A mostly Mexican genus of 21 species; a few are widespread weeds. (Latin: *flavus*, pure yellow) (tribe Heliantheae)
REFERENCES: Powell 1978; Yarborough & Powell 2006a.

Flaveria trinervia (Spreng.) C. Mohr, (3-nerved), CLUSTERED YELLOWTOPS. Annual glabrate herbs, to 200+ cm tall; stems much-branched, erect or decumbent; leaves opposite, petiolate or sessile, blades ± succulent, 3-nerved, lanceolate to elliptic or oblanceolate, to 150 mm long, margins serrate to spinulose; heads in compact, sessile, axillary clusters; involucres 3.8–4.8 mm high; phyllaries usually 2, oblong, clasping and falling with mature achenes; ray flowers absent or 1 (usually on peripheral heads in a cluster), pistillate, fertile, ligules yellow or whitish, inconspicuous, 0.5–1 mm long; disk flowers 0–1(–2), perfect, fertile, corollas yellow; achenes 10-ribbed, glabrous, 2–2.6 mm long (ray achenes longer); pappus absent. Moist saline areas, disturbed habitats; in East TX known only from Harris Co. from a 1922 collection (*G.L. Fisher s.n.* TEX-LL) and possibly not persisting, as it is not included in Brown (2014); scattered in w 1/2 TX; scattered in mostly s and coastal U.S.; Mexico; introduced and weedy in many parts of the world. Mar–Dec. [*Oedera trinervia* Spreng.]

FLEISCHMANNIA Sch. Bip. THOROUGHWORT

A genus of ca. 80 species of the Americas; formerly included in *Eupatorium*. (Named for Gottfried F. Fleischmann (1777–1850), professor at Erlangen, Germany) (tribe Eupatorieae)
REFERENCES: King & Robinson 1970a, 1987; Wooten & Clewell 1971; Nesom 2006n.

Fleischmannia incarnata (Walter) R.M. King & H. Rob., (flesh-colored), PINK EUPATORIUM, PINK THOROUGHWORT. Perennial herb from fibrous-rooted crown; stems sprawling or scandent, to 200 cm long; leaves opposite, petiolate, blades deltoid, gland-dotted beneath, margins crenate to serrate; heads in loose corymbose arrays; involucres 4–5 mm high; phyllaries in 2–4 series, unequal; ray flowers absent; disk flowers with corollas usually pink-purple or whitish with pink to lilac lobes, rarely all white; achenes

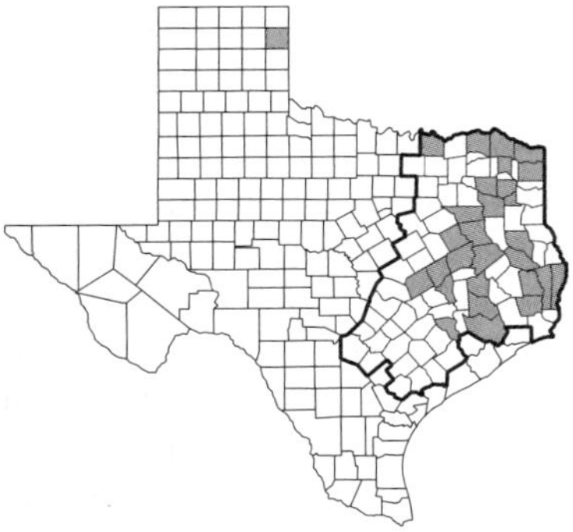

Eupatorium perfoliatum

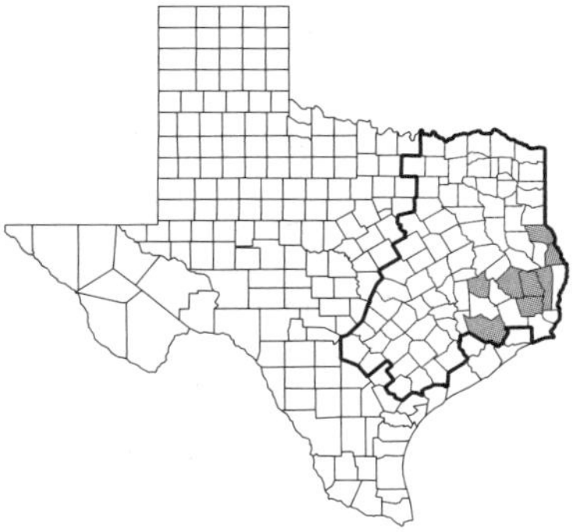

Eupatorium ×pinnatifidum

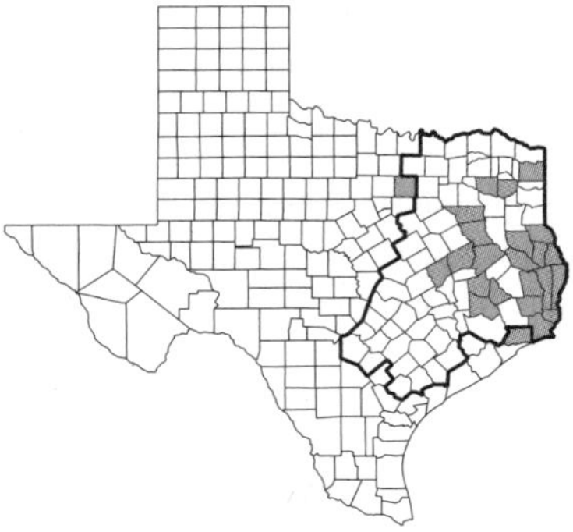

Eupatorium rotundifolium var. rotundifolium

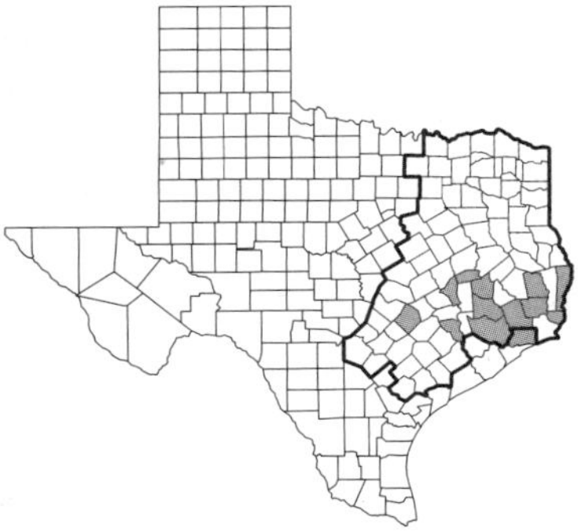

Eupatorium semiserratum

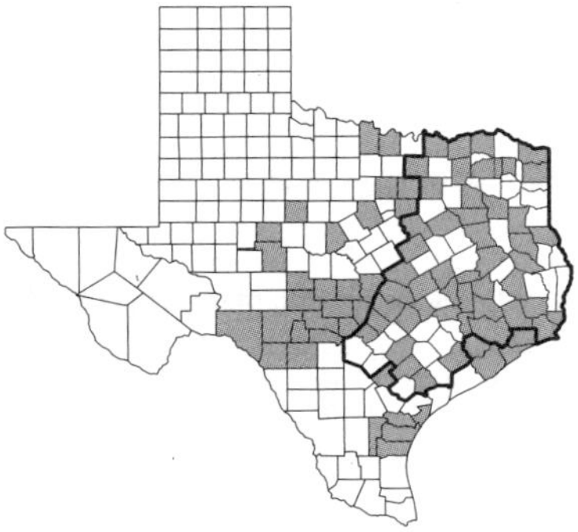

Eupatorium serotinum

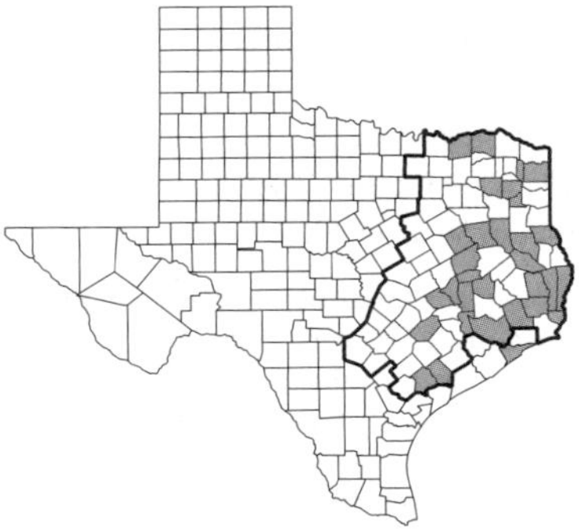

Eurybia hemispherica

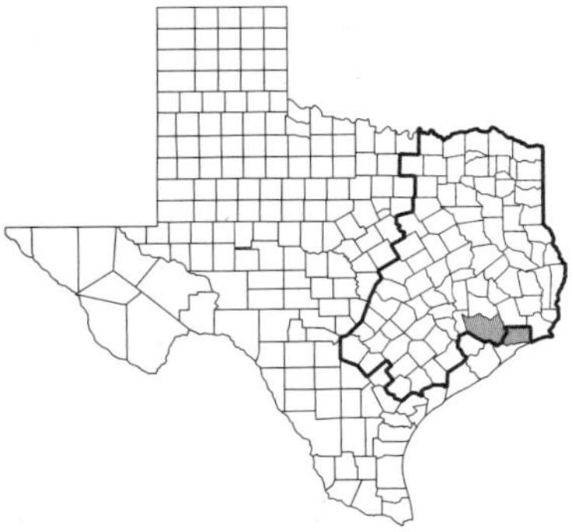

Euthamia caroliniana

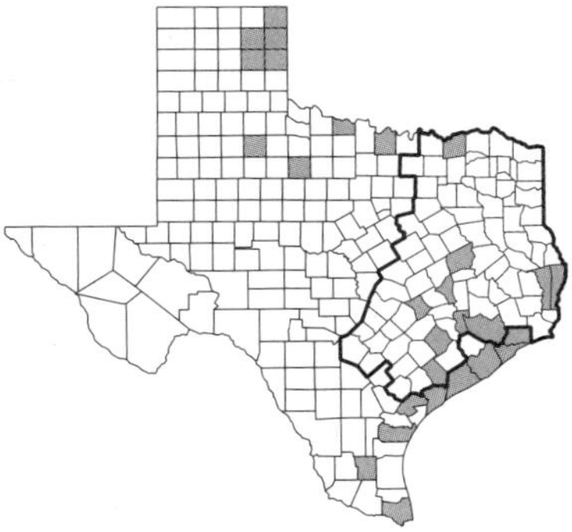

Euthamia gymnospermoides

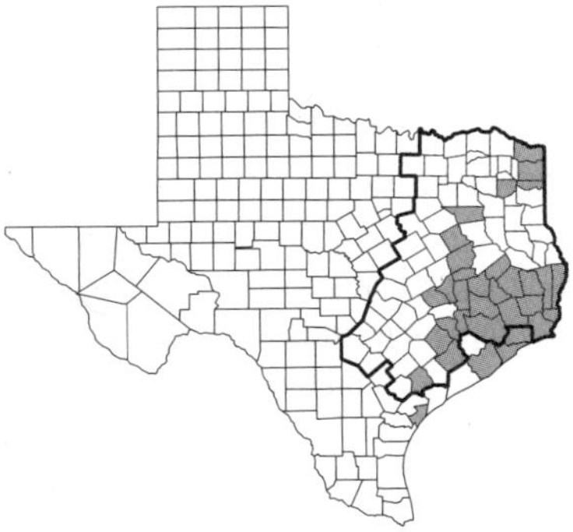

Euthamia leptocephala

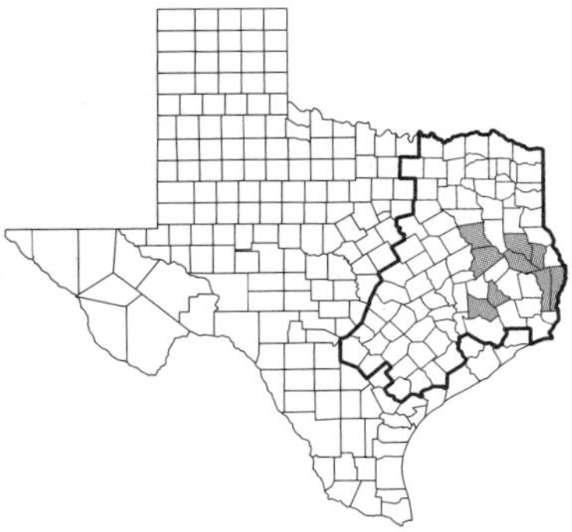

Eutrochium fistulosum

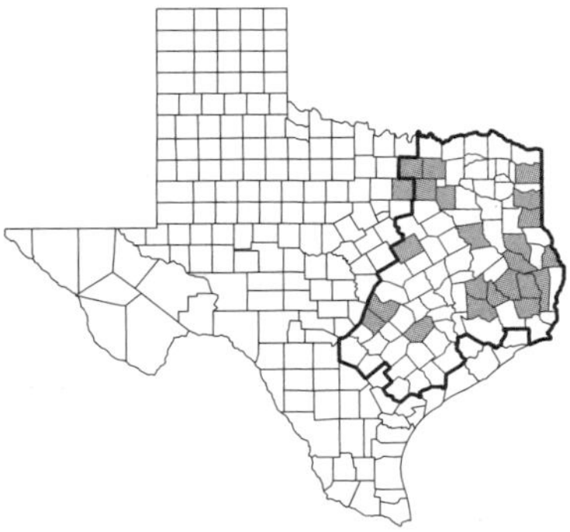

Facelis retusa

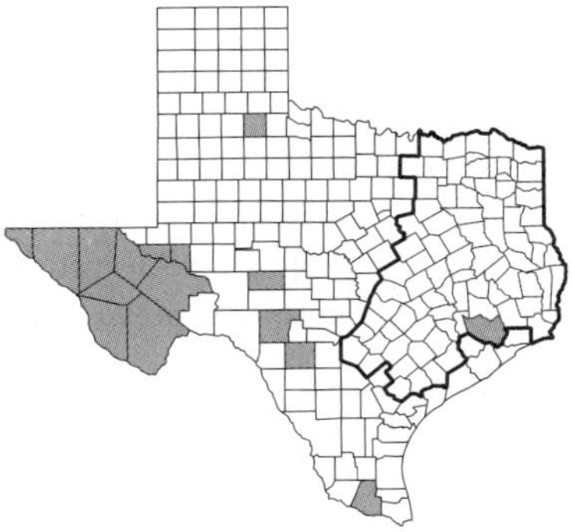

Flaveria trinervia

5-angled, 1.8–2.8 mm long; pappus of barbellate bristles. Bottomlands, moist woodlands, ditches; East TX and Cross Timbers to South TX Plains; se and sc U.S. from OH s to FL and w to MO and TX. (Sep–)Oct–Dec. [*Eupatorium incarnatum* Walter]

GAILLARDIA Foug. INDIAN-BLANKET, BLANKET-FLOWER

Ours low annual or perennial herbs, taprooted or rhizomatous; leaves alternate, sometimes also forming a basal rosette, reduced upward, petiolate to ± clasping, blades linear to elliptic or oblong to spatulate, glabrous to scaberulous or villous, sometimes gland-dotted, margins entire or toothed to lobed or pinnatifid; heads large, showy, usually borne singly on long peduncles; phyllaries in 2–3+ series, subequal, herbaceous or with hardened bases, narrow, loose, reflexed in fruit; receptacles convex to hemispheric, often with bristles intermixed with disk flowers; ray flowers absent or 6–15, neuter or rarely pistillate and fertile or sterile, ligules widened and 3-toothed apically, yellow to dark orange or red to brownish purple or pinkish (rarely cream or white), sometimes bicolored; disk flowers perfect, fertile, corollas yellow to reddish purple or reddish brown, hairy, sometimes enlarged and ray-like peripherally in eligulate heads; achenes obpyramidal to club-shaped, ± 4-angled, ± hairy; pappus of 7–12 lanceolate to obovate, usually awn-tipped scales.

A genus of ca. 17 species native to North America, Mexico, and temperate South America; includes several cultivated ornamentals. (Named for M. Gaillard de Merentonneau (or Charentonneau), 18th century French magistrate and patron of botany) (tribe Heliantheae)

REFERENCES: Biddulph 1944; Turner & Whalen 1975; Heywood & Levin 1984; Strother 2006m; Marlowe & Hufford 2007; Turner & Watson 2007.

1. Leaves all ± basal, blades glabrous or very sparsely villous, lacking glands; ray flowers absent or with ligules inconspicuous, 10–15 mm long ____ **G. suavis**
1. Leaves conspicuously extending up the stem, alternate, blades scaberulous to villous, usually gland-dotted; ray flowers usually present (sometimes absent), ligules often showy, 13–30+ mm long.
   2. Receptacles lacking bristles, or with bristles < 1 mm long; lobes of disk corollas attenuate and conical, bearing jointed hairs to 0.3 mm long ____ **G. aestivalis**
   2. Receptacles usually with bristles 0.3–3 mm long; lobes of disk corollas ovate to deltate, sometimes attenuate, bearing jointed hairs 0.3+ mm long.
      3. Ray corolla ligules usually uniformly red to purple, rarely yellow; disk corolla lobes 0.5–1 mm long; achenes dimorphic, the peripheral ones longer than the central ones and with shorter, awnless pappus scales ____ **G. amblyodon**
      3. Ray corolla ligules usually bicolored (brownish purple to reddish, tipped with yellow or orange); disk corolla lobes 1–3+ mm long; achenes all alike ____ **G. pulchella**

Gaillardia aestivalis (Walter) H. Rock, (summer), PRAIRIE GAILLARDIA, EASTERN GAILLARDIA, LANCELEAF BLANKET-FLOWER, YELLOW INDIAN-BLANKET. Annuals or Perennials (sometimes flowering first year), 10–60+ cm tall, sometimes rhizomatous; leaf blades usually closely scaberulous, margins remotely toothed or entire; receptacle bristles absent or to 0.5+ mm long; ray flowers with ligules 13–25 mm long, pinkish to maroon-purple or yellow (to rarely cream or white), or ray flowers sometimes absent; disk corolla lobes attenuate and terete, 1.5–3 mm long; achenes 1.5–2 mm long; pappus scales awned, 5–7 mm long. Sandy open woods sandy and calcareous prairies, disturbed areas; mostly e 1/2 TX; also c TX and Panhandle; se and sc U.S. from NC s to FL and w to KS and TX. May–Oct. [*G. aestivalis* (Walter) H. Rock var. *austrotexana* B.L. Turner, *G. aestivalis* (Walter) H. Rock var. *flavovirens* (C. Mohr) Cronquist, *G. aestivalis* (Walter) H. Rock var. *winkleri* (Cory) B.L. Turner, *G. fastigiata* Greene, *G. lanceolata* Michx., *G. lutea* Greene] We are following Strother (2006m) in not recognizing the numerous and often intergrading varieties that are sometimes recognized in this species.

Gaillardia amblyodon J. Gay, (blunt-tipped), MAROON BLANKET-FLOWER, TEXAS GAILLARDIA. Annuals, 20–45+ cm tall; leaf blade scaberulous and/or ± villous, margins usually entire or obscurely toothed, rarely lobed; receptacle bristles 2.5–3 mm long; ra[y] flowers with ligules 15–25 mm long, mostly bright red to dark maroon-purple, rarely yellow; disk corolla lobes ovate-deltate 0.5–1 mm long; achenes 2–4+ mm long, dimorphic, the peripheral ones longer and with awnless pappus scales, the central one shorter and with pappus scales awned, 5–6 mm long. Open areas, deep sandy soils; se and sc TX; South TX Plains. Apr–Ju[n]. Strother (2006m) notes this species may not prove to be distinct from *G. pulchella*.

Gaillardia pulchella Foug., (handsome), INDIAN-BLANKET, FIRE-WHEEL, ROSE-RING GAILLARDIA. State wildflower of OK (Tyrl et a[l.] 1994). Annuals (sometimes persisting), 5–35(–60+) cm tall; leaf blades strigillose to ± villous, margins usually entire, some[-] times toothed or lobed; receptacle bristles 1.5–3 mm long; ray flowers 8–14, rarely absent, ligules 13–30+ mm long, usual[ly] bicolored (brownish purple to reddish, tipped with yellow or orange), rarely yellow, reddish, or purplish throughout, or ligul[es] sometimes absent; disk corolla lobes deltate to ovate, often attenuate, 1–3+ mm long; achenes 2–2.5 mm long; pappus scal[es] awned, 4–7 mm long. Prairies, disturbed areas, sandy or calcareous soils; throughout TX; across much of North America b

Euthamia gymnospermoides [STE]

Euthamia leptocephala [BB2]

Eutrochium fistulosum [STR]

Facelis retusa [MAR]

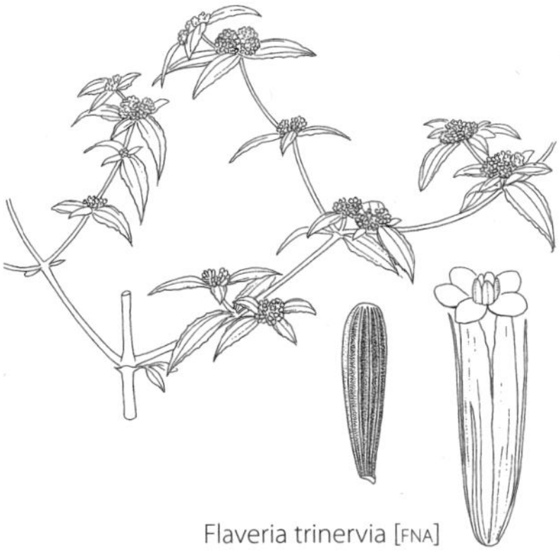

Flaveria trinervia [FNA]

Fleischmannia incarnata [BB2]

Gaillardia aestivalis [STE]

Gaillardia amblyodon [ADD]

Gaillardia pulchella [JON]

concentrated in se U.S., s plains, and desert sw; populations beyond these areas are likely adventive (Kartesz 2015); Mexico. Apr–Aug, sporadically in other seasons. [*G. drummondii* (Hook.) DC., *G. neomexicana* A. Nelson, *G. pulchella* Foug. var. *australis* B.L. Turner & M.A. Whalen, *G. pulchella* Foug. var. *picta* (D. Don) A. Gray] We are following Strother (2006m) in not recognizing the numerous and often intergrading varieties that are sometimes recognized in this species. *G. pulchella* is showy and often abundant on roadsides, and is one of the most common wildflowers in TX. This species continues to spread as it is often included in wildflower seed mixes and heavily utilized in highway beautification programs; cultivars of this species (and its hybrids) are popular garden plants.

Gaillardia suavis (A. Gray & Engelm.) Britton & Rusby, (sweet), RAYLESS GAILLARDIA, FRAGRANT GAILLARDIA, GLOBE-FLOWER, PERFUME-BALLS. Perennials (sometimes flowering first year), 20–80+ cm tall; leaf blades glabrous or very sparsely villous, margins pinnatifid to toothed or entire; receptacle bristles absent or to 0.5+ mm long; ray flowers absent, or with ligules 10–15 mm long, maroon-purple to red-orange; disk corolla lobes narrowly triangular, 1–1.2 mm long; achenes 2 mm long; pappus scales awned, 6–9 mm long. Sandy or rocky prairies, juniper scrub, roadsides; mostly w 2/3 of TX, also Jefferson Co.; KS, OK, and TX; Mexico. Apr–Jun, sporadically later. [*Agassizia suavis* A. Gray & Engelm., *G. odorata* Lindh. ex A. Gray]

GAMOCHAETA Weddell CUDWEED, EVERLASTING

Ours annual or biennial herbs, whitish to grayish with felty or woolly pubescence; leaves alternate, sessile, blades linear to oblanceolate or spatulate, often white- or gray-tomentose beneath, margins entire or sometimes sinuate; heads small, nearly sessile, clustered in small groups in the upper axils, in spicate or sometimes paniculate arrays; phyllaries in 3–7 series, brownish to tan or sometimes purplish-tinged, tips chartaceous to scarious; ray flowers absent; disk flowers of two kinds, those at the periphery pistillate and fertile, those in the center fewer and perfect, corollas all yellowish cream or with purplish lobes; achenes small, oblong, pubescent; pappus of minutely barbed bristles, deciduous as a ring. (note: involucres often buried in arachnose indument must be dissected out to determine indument of involucre base itself)

A genus of ca. 50 species native to the Americas; most are South American, with 12 in North America; several are weedy and naturalized in other parts of the world. *Gamochaeta* was previously included in *Gnaphalium*. (Greek: *gamos*, union, and *chaete*, bristle, in allusion to the basally connate pappus bristles that fall together as a ring) (tribe Inuleae)
REFERENCES: Nesom 1990a, 2006o, 2007b; Anderberg 1991.

1. Leaves bicolor (with upper surfaces greenish and glabrous or sparsely arachnose, and lower surfaces white or gray with dense pubescence); achenes 0.5–0.7 mm long.
   2. Upper leaf surfaces glabrous or glabrate; involucres 2.5–3 mm high, bases glabrous; outer phyllary tips rounded to obtuse; perfect flowers 2–3 ____ **G. coarctat**
   2. Upper leaf surfaces sparsely arachnose (use 10x lens); involucres 3–4.5 mm high, bases sparsely arachnose; outer phyllary tips acute to acuminate; perfect flowers 3–6.
      3. Stem leaves oblanceolate to spatulate; involucres 4–4.5 mm high; inner phyllary tips acute (not apiculate); perfect flowers 3–4; achenes 0.6–0.7 mm long ____ **G. purpure**
      3. Stem leaves oblanceolate to oblanceolate-oblong or oblanceolate-obovate; involucres 3–3.5 mm high; inner phyllary tips truncate-rounded and apiculate; perfect flowers 4–5(–6); achenes 0.5–0.6 mm long ____ **G. argyrine**
1. Leaves concolor (= of uniform color) to weakly bicolor (with both surfaces green to grayish green), any pubescence not thick enough to obscure leaf surface; achenes 0.4–0.5 mm long.
   4. Lower stem leaves 4–16 mm wide; bracts among heads spatulate to oblanceolate, at least the lower ones extending beyond the clusters of heads ____ **G. pensylvanic**
   4. Lower stem leaves 2–9 mm wide; bracts among heads linear to oblanceolate, extending beyond clusters of heads or not.
      5. Involucres 2.5–3 mm high, usually purplish, bases sparsely arachnose; phyllaries in 3–4(–5) series, the outer ones 1/2–2/3 as long as the inner, tips narrowly to broadly acute ____ **G. antillan**
      5. Involucres 3–3.5 mm high, brownish (not purplish), bases sparsely arachnose or glabrous; phyllaries in 5–7 series, the outer ones 1/3–1/2 as long as the inner, tips acute-acuminate ____ **G. calvice**

Gamochaeta antillana (Urb.) Anderb., (of the Antilles), DELICATE EVERLASTING, ANTILLES CUDWEED. Taprooted annuals, 6–40 c tall; basal leaves withering before flowering; stem leaves 2–3(–5) mm wide, concolor (with both sides green to grayish gree and loosely tomentose; heads initially in uninterrupted, cylindric, spicate arrays, becoming interrupted in late flowering, arra leafy-bracted throughout; involucres 2.5–3 mm high, bases sparsely arachnose; phyllaries often purple-tinged. Sandy roadside disturbed areas, pond banks; throughout e 1/2 TX; s TX; se U.S. from VA s to FL and w to OK and TX; South America; intr duced elsewhere. (Feb–)Mar–May. [*G. subfalcata* (Cabrera) Cabrera, *Gnaphalium antillanum* Urban, *Gnaphalium subfalcatu* Cabrera] *G. antillana* and *G. calviceps* have been previously combined in concept and often misidentified as *G. falcata*, a Sou American species that has not been recorded from North America (Nesom 2006o).

Gamochaeta argyrinea G.L. Nesom, (silvery), SILVERY CUDWEED, SILVERY EVERLASTING. Fibrous-rooted annuals (rarely taprooted), 12–40 cm tall; basal leaves present through flowering; stem leaves 5–12(–18) mm wide, bicolor with upper surfaces greenish and sparsely arachnose and lower surfaces white-woolly, heads initially in uninterrupted, cylindric, spicate arrays, sometimes becoming interrupted in late flowering; involucres 3–3.5 mm high, bases sparsely arachnose; phyllaries tawny or purple-tinged. Sandy or clayey roadsides, disturbed areas, lawns; throughout e 1/2 TX; e and se U.S. from PA s to FL and w to KS and TX; also CA; West Indies. Mar–Jun(–Oct). *Gamochaeta argyrinea* specimens have often been misidentified as *G. purpurea*.

Gamochaeta calviceps (Fernald) Cabrera, (bald-headed), NARROWLEAF PURPLE EVERLASTING, SILKY CUDWEED. Taprooted or fibrous-rooted annuals, 8–45(–55) cm tall; basal leaves usually withering before flowering; stem leaves 2–9 mm wide, commonly folded along midveins, concolor (with both sides green to grayish green) or weakly bicolor with appressed woolly hairs; heads initially in continuous or interrupted spicate arrays, becoming paniculate in late flowering; involucres 3–3.5 mm high, bases glabrous or nearly so; phyllaries brownish (not purple). Sandy or clayey roadsides, disturbed areas, fields; scattered in e and se TX; nc TX and Edwards Plateau; se U.S. from VA s to FL and w to OK and TX; also CA; South America; introduced elsewhere. Apr–Jun. [*Gnaphalium calviceps* Fernald] *G. antillana* and *G. calviceps* have been previously combined in concept and often misidentified as *G. falcata*, a South American species that has not been recorded from North America (Nesom 2006o).

Gamochaeta coarctata (Willd.) Kerguélen, (crowded), ELEGANT CUDWEED, GRAY EVERLASTING. Fibrous-rooted annuals or biennials, 15–35(–50) cm tall; basal leaves present in rosettes through flowering; stem leaves 6–15(–22) mm wide, bicolor with upper surfaces greenish and nearly glabrous and lower surfaces white-woolly, heads initially in uninterrupted, dense, spicate arrays, sometimes becoming interrupted and branched in late flowering; involucres 2.5–3 mm high, bases glabrous; phyllaries rosy or purple-tinged. Shady moist sites, sidewalk cracks, ditches; scattered in e and se TX; nc TX; se U.S. from VA s to FL and w to TX; also CA; Mexico; Central and South America; and West Indies; introduced elsewhere. Apr–Jun. [*G. spicata* Cabrera, *Gnaphalium coarctatum* Willd., *Gnaphalium spicatum* Lam.] Native of South America, now a widespread weed.

Gamochaeta pensylvanica (Willd.) Cabrera, (of Pennsylvania), PENNSYLVANIA CUDWEED, PENNSYLVANIA EVERLASTING. Taprooted annuals, 10–50 cm tall; basal leaves usually present at flowering; stem leaves 4–16 mm wide, concolor (with both sides green to grayish green) or weakly bicolor and loosely tomentose; heads in continuous or interrupted spicate arrays; involucres 3–3.5 mm high, bases sparsely arachnose; phyllaries often purple-tinged. Shady sites, disturbed areas, moist sandy soils; throughout most of TX with the exception of Plains Country and Trans-Pecos; e and se U.S. from PA and OH s to FL and w to OK and TX; also CA; Mexico; Central and South America; introduced elsewhere. Mar–Jun. [*Gnaphalium pensylvanicum* Willd., *Gnaphalium peregrinum* Fernald, *Gnaphalium spathulatum* Lam.] Despite the epithet, a native of South America, now a widespread weed.

Gamochaeta purpurea (L.) Cabrera, (purple), PURPLE CUDWEED, SPOON-LEAF EVERLASTING. Fibrous-rooted or taprooted annuals, 10–40(–50) cm tall; basal leaves usually withering before flowering; stem leaves 5–14 mm wide, bicolor with upper surfaces greenish and glabrate to sparsely arachnose and lower surfaces white-woolly, heads initially in uninterrupted spicate arrays, becoming interrupted in late flowering; involucres 4–4.5 mm high, bases sparsely arachnose; phyllaries purple-tinged in bud and whitish or silvery at maturity. Sandy roadsides, open woods, disturbed sites; scattered in e 1/2 TX; e and se U.S. from ME s to FL and w to NE and TX; also AZ; Canada; Mexico; Central and South America; West Indies; introduced and weedy elsewhere. Mar–Jun(–Jul). [*Gnaphalium purpureum* L.]

GRINDELIA Willd. GUMWEED, TARWEED, ROSINWEED, GUMPLANT

Ours annual, biennial, or perennial taprooted herbs, usually gland-dotted and/or resinous; leaves alternate, simple, the lower petiolate and the upper sessile and usually clasping, margins usually serrate to spiny-toothed, sometimes entire; heads borne singly or in corymbose to paniculate arrays; phyllaries in several series, subequal or unequal, filiform to linear or lanceolate, glabrous and resinous, the tips of outer phyllaries often recurved; ray flowers 12–45 (or sometimes absent), pistillate, fertile, corollas yellow to orange-yellow; disk flowers perfect, the peripheral fertile, the central sometimes sterile, corollas yellow; achenes glabrous, dimorphic in some species, ± compressed, sometimes 3–4-angled; pappus of scales, awns, or bristles in 1 series, or in *G. ciliata*, an outer series of bristles and an inner series of bristle-like awns or scales.

A genus of ca. 30 species of w North America, Mexico, and South America; some are cultivated as ornamentals and others are used as medicinal herbs. We are following Nesom et al. (1993) and Strother and Wetter (2006) in including *G. ciliata* here, which had previously been separated in the monotypic genus *Prionopsis*; other authors continue to segregate it based on its persistent, 2-seriate pappus (Lane & Hartman 1996; Bartoli & Tortosa 2012). (Named for professor David Hieronymus Grindel, 1776–1836, Latvian botanist) (tribe Astereae)

References: Steyermark 1934; Suh & Simpson 1990; Nesom 1990c, 1992a; Nesom et al. 1993; Strother & Wetter 2006; Bartoli & Tortosa 2012.

1. Stem leaves with margins serrate to dentate or denticulate, toothed from base to tip or sometimes teeth crowded toward the tips, teeth fine and sharp; or leaves sometimes entire.
   2. Pappus bristles 30+ per achene, united in a ring at base, ± persistent but upon full maturity falling as a unit or in groups ______ **G. ciliata**
   2. Pappus bristles, awns, or scales only 2 per achene, not united at base, falling off at the slightest touch______ **G. lanceolata**
1. Stem leaves with margins usually crenate to serrate, usually toothed from base to tip, teeth obtuse or blunt and usually resin-tipped; or rarely some or all leaves entire.
   3. Biennials or short lived perennials; upper stems glabrous; pappus 2.5–4.5 mm long, shorter than disk corollas ______ **G. squarrosa**
   3. Annuals; upper stems villous to hirtellous, sometimes with stalked glands; pappus 3.5–6 mm long, equaling or surpassing disk corollas.
      4. Leaf blades with stalked glands (glands sometimes sessile, rarely in pits); achene faces rugose and transversely fissured ______ **G. pusilla**
      4. Leaf blades with pitted glands (glands sometimes sessile, rarely stalked); achene faces smooth or slightly rugose (little if at all transversely fissured).
         5. Plants 30–130 cm tall; stem leaves 15–60(–90) mm long, marginal teeth 8–14 per cm; involucres 8–12 mm high and 10–20 mm wide; pappus 5–6 mm long ______ **G. adenodonta**
         5. Plants 15–40 cm tall; stem leaves 10–30(–45+) mm long, marginal teeth 6–8 per cm; involucres 6–9 mm high and 8–1 mm wide; pappus 3.5–5 mm long ______ **G. microcephala**

Grindelia adenodonta (Steyerm.) G.L. Nesom, (sticky-toothed), gland-tooth gumweed, prairie gumweed. Annuals, 30–130 cm tall; upper stems ± hirtellous to villous; stem leaves triangular to oblong or obovate, usually scabrous to hirtellous, usually with pitted glands, margins crenate, teeth blunt and resin-tipped; heads usually subtended by leaf-like bracts; mature achenes 3–4.5 mm long, faces of peripheral achenes rugose (not transversely fissured), faces of central achenes striate; pappus of 2 bristles or awns, 5–6 mm long, equaling or surpassing disk corollas. Clay soils, prairies, along streams; Blackland and Gulf Coast prairies of se and s TX; nc TX. Jun–Sep. [*G. microcephala* DC. var. *adenodonta* Steyerm.]

Grindelia ciliata (Nutt.) Spreng., (for its many-bristled pappus), saw-leaf daisy, goldenweed, spanish gold, giant gumweed. Annuals or biennials, 20–150+ cm tall; upper stems glabrous, usually unbranched except near top; stem leaves oblong to obovate, glabrous, scarcely or not gland-dotted, margins dentate, teeth fine and sharp; mature achenes 2–4 mm long, faces smooth or striate; pappus of an outer series of 25–40 barbellate bristles 3–7+ mm long, and inner series of 8–15+ awns or bristle-like scales 7–10 mm long, the longer ones surpassing disk corollas. Disturbed areas, prairies, roadsides; widespread in TX; mostly c U.S. from MI s to LA and w to CA; also MD. (Jun–)Aug–Oct. [*G. papposa* G.L. Nesom & Suh, *Haplopappus ciliatus* (Nutt.) DC., and previously recognized in the monotypic genus *Prionopsis* (as *P. ciliata* (Nutt.) Nutt.)] Avoided by grazing livestock.

Grindelia lanceolata Nutt., (lance-shaped), gulf gumweed, narrowleaf gumweed. Biennials or perennials, 30–150 cm tall; upper stems glabrous; stem leaves triangular, ovate, or oblong to lanceolate or linear, usually glabrate, scarcely or not gland-dotted, margins serrate to dentate, teeth fine and sharp; mature achenes 2–6 mm long, faces smooth or striate; pappus of 2 awns, 4–8 mm long, equaling or slightly surpassing disk corollas. Calcareous and sandy soils, rocky prairies; se and e TX w to Rolling Plains and Edwards Plateau; scattered in ne U.S., mostly in sc U.S. from IL s to AL and w to CO and NM. Jun–Oct. [*G. littoralis* Steyerm., *G. texana* Scheele] If varieties are recognized following Bartoli and Tortosa (2012), ours would be var. *texana* (Scheele) Shinners.

Grindelia microcephala DC., (small-headed), little-head gumweed. Annuals, 15–40 cm tall; upper stems villous and/or with stalked glands; stem leaves oblong or spatulate, usually hirtellous, usually with pitted glands, margins crenulate, teeth blunt and resin-tipped; heads usually subtended by leaf-like bracts; mature achenes 3.5–4 mm long, faces rugose (not or only slightly transversely fissured); pappus of 2–3 bristles or awns, 3.5–5 mm long, equaling or slightly surpassing disk corollas. Bottomlands, thickets, near streams; Bexar and Comal cos. (Bartoli & Tortosa 2012); South TX Plains. (Nov–)Mar–Jun(–Aug). [*G. inuloides* Willd. var. *microcephala* (DC.) A. Gray] *G. adenodonta* and *G. pusilla*, previously considered varieties of *G. microcephala*, were elevated to species by Nesom et al. (1993).

Grindelia pusilla (Steyerm.) G.L. Nesom, (small), little gumweed. Annuals, 30–70 cm tall; upper stems minutely villous, with stalked glands; stem leaves spatulate or oblong to lanceolate, usually hirtellous, usually with stalked glands, margins crenate, teeth blunt and resin-tipped; heads usually subtended by leaf-like bracts; mature achenes 2.5–4 mm long, faces rugose (transversely fissured); pappus of 2–3 awns, 4.5–5 mm long, equaling or surpassing disk corollas. Disturbed areas, dry open sites; Bexar Co., and an apparent disjunct record from Tarrant Co. (*S.D. Lusk s.n.* BRIT); Bartoli & Tortosa 2012); South TX Plains; Mexico. Mar–Jun. [*G. microcephala* DC. var. *pusilla* Steyerm.]

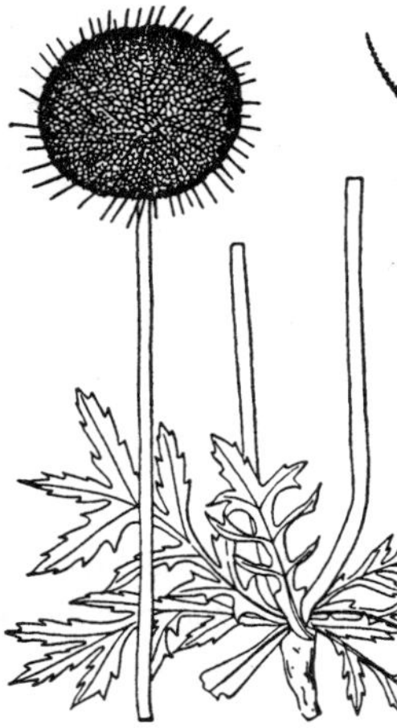
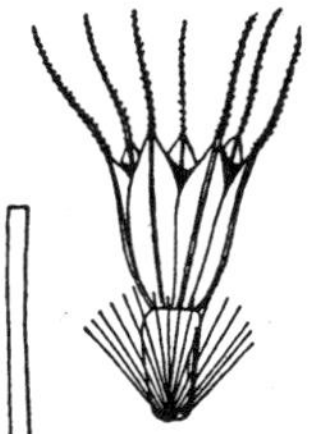

Gaillardia suavis [BB2]

Gamochaeta argyrinea [THO]

Gamochaeta antillana [NZB]

Gamochaeta calviceps [FNA]

Gamochaeta pensylvanica [HEA]

Gamochaeta coarctata [FNA]

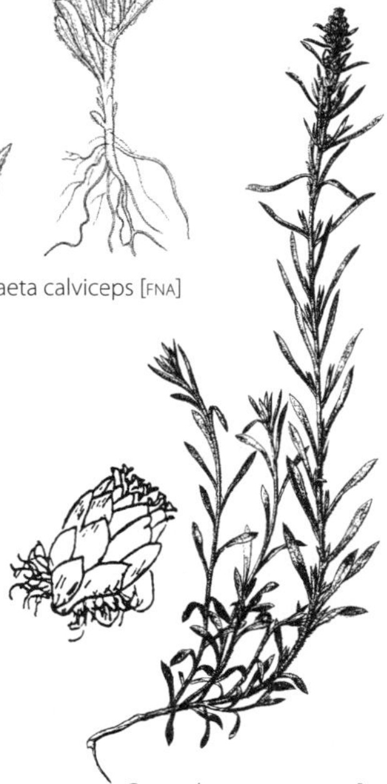

Gamochaeta purpurea [MAR, SM1]

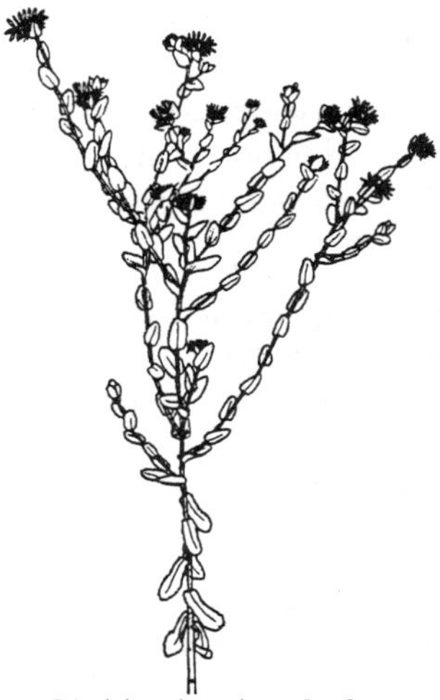

Grindelia adenodanta [AMB]

Grindelia squarrosa (Pursh) Dunal, (with parts spreading or recurved), CURLY-CUP GUMWEED, RAYLESS GUMWEED. Biennials or short-lived perennials, (10–)40–100 cm tall; upper stems glabrous; stem leaves ovate, oblong, spatulate, or oblanceolate to lanceolate or linear, glabrous, strongly gland-dotted, margins usually crenate to serrate, teeth blunt and resin-tipped; ray flowers sometimes lacking (rayless plants were formerly treated as *G. nuda*); mature achenes 1.5–4.5 mm long, faces smooth, striate, or furrowed; pappus of 2–3(–8) scales or awns, 2.5–4.5 mm long, shorter than disk corollas. Open or disturbed areas, roadsides, along streams; w 3/4 TX; nearly throughout North America, but mostly in the w 1/2, and perhaps adventive in e U.S.; Mexico; introduced in Asia. (Jul–)Aug–Oct. [*G. nuda* Alph. Wood, *G. nuda* Alph. Wood var. *aphanactis* (Rydb.) G.L. Nesom, *G. squarrosa* (Pursh) Dunal var. *nuda* (Alph. Wood) A. Gray] Used medicinally by Native Americans and in folk remedies; however, it contains the carcinogen, safrole (Duke 1985). ☠

GUTIERREZIA Lag. BROOMWEED, SNAKEWEED

Ours herbaceous annuals or perennial subshrubs, often resinous or glutinous; stems often much-branched; leaves alternate, sessile or petiolate and decurrent, linear to lanceolate or spatulate, glabrous or minutely pubescent, gland-dotted, margins entire, basal and lower stem leaves usually absent at flowering; heads borne in loose or dense paniculate or corymbose arrays; phyllaries in 2–4 series, unequal, usually green-tipped; receptacles naked; ray flowers pistillate, fertile, corollas yellow; disk flowers perfect, fertile or functionally staminate, corollas yellow; achenes club-shaped or cylindrical, 5–8-nerved, hairy; pappus of a low crown, or of 5–10 irregular, often rough-margined, whitish scales, or pappus absent.

A genus of 28 species of w North America and South America. The species treated here have been variously placed in *Amphiachyris* and *Xanthocephalum*. The molecular analysis by Suh and Simpson (1990) indicated *Amphiachyris* (with 2 species) and *Thurovia* (a monotypic genus) form a sister group to *Gutierrezia*, and supported these as separate genera. ☠ Several species are toxic to grazing livestock due to saponins or unknown toxins (Kingsbury 1964; James & Welsh 1992); plant populations increase under overgrazing. (Possibly named for Pedro Gutierrez, a 19th century Spanish nobleman, botanist, and apothecary at the Madrid Botanical Garden) (tribe Astereae)

REFERENCES: Shinners 1951c; Solbrig 1960, 1961; Ruffin 1974; Lane 1980, 1982, 1985; Suh & Simpson 1990; Nesom 1991d, 2006p.

1. Plants annual herbs; stems glabrous, typically branching from ca. the middle upwards; heads in loose arrays; involucres 2–3 mm wide; ray flowers 5–23; achenes loosely strigose; pappus a minute crown < 1 mm high, or sometimes absent ____ **G. texana**
1. Plants perennial subshrubs; stems minutely hispid, usually branching from the base upwards; heads in dense, flat-topped corymbose arrays; involucres 1.5–2(–3) mm wide; ray flowers (2–)3–8; achenes densely silky-strigose; pappus of small scales ____ **G. sarothrae**

Gutierrezia sarothrae (Pursh) Britton & Rusby, (thought by Pursh to resemble *Hypericum sarothra* Michx.), BROOM SNAKEWEED, PERENNIAL BROOMWEED, MATCHWEED, MATCHBRUSH, KINDLING-WEED, TURPENTINE-WEED, YERBA DE VÍBORA, ESCOBA DE LA VÍBORA. Woody-based subshrubs, 10–60(–100) cm tall; stems minutely hispid; leaves linear to lanceolate; involucres 1.5–2(–3) mm wide; ray flowers usually 3–8, corollas 3–5.5 mm long; achenes 0.8–1.6(–2.2) mm long, densely silky-strigose; pappus of small scales. Grasslands, open or disturbed sites, rocky calcareous slopes; Austin, Dallas, Grimes, Harris, and Travis cos.; mostly w 1/2 TX; w 1/2 of U.S.; adventive in NY; Mexico. Jul–Nov. [*Xanthocephalum sarothrae* (Pursh) Shinners, *X. tenue* (Greene) Shinners] This species is an indicator of overgrazing; a number of common names refer to the highly flammable nature of the dried stems which were used historically in starting fires (Ajilvsgi 1984). This species causes toxicity to livestock, as in *G. microcephala,* due to saponins or unknown toxins (Kingsbury 1964; James & Welsh 1992). It is reported to contain a protein with antitumor activity (Mabberley 2017). ☠

Gutierrezia texana (DC.) Torr. & A. Gray var. **texana,** (of Texas), TEXAS BROOMWEED, TEXAS SNAKEWEED, KINDLING-WEED. Annual herbs, 10–80(–100) cm tall; stems glabrous; leaves linear; involucres 2–3 mm wide; ray flowers usually 5–23, corollas 3–6 mm long; achenes 1.3–1.8 mm long, loosely strigose; pappus a minute crown < 1 mm long, or sometimes absent. Disturbed habitats, rocky prairies, roadsides; se and sc TX; nc TX, Rolling Plains, Edwards Plateau, and s TX; AR, LA, OK, and TX; apparently adventive in several states farther e (Kartesz 2015); Mexico. Jul–Nov. [*Xanthocephalum texanum* (DC.) Shinners, *Hemiachyris texana* DC.]

Gutierrezia microcephala (DC.) A. Gray, (small-headed), SMALL-HEAD SNAKEWEED, THREAD-LEAF SNAKEWEED, which would have keyed to *G. sarothrae* here, can be distinguished by the small heads (1–1.5 mm wide) and reduced number of ray flowers (1–2) and disk flowers (1–2). Grasslands, chaparral, usually over limestone or gypsum; USDA Plants Database (2016) shows this species occurring in McLennan Co. (a significantly disjunct locality) based on Hannick et al. (2013), but Kartesz (2015) indicates this is

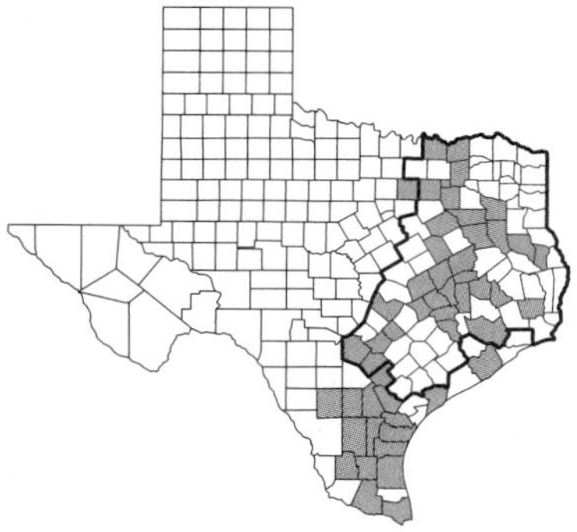
Fleischmannia incarnata

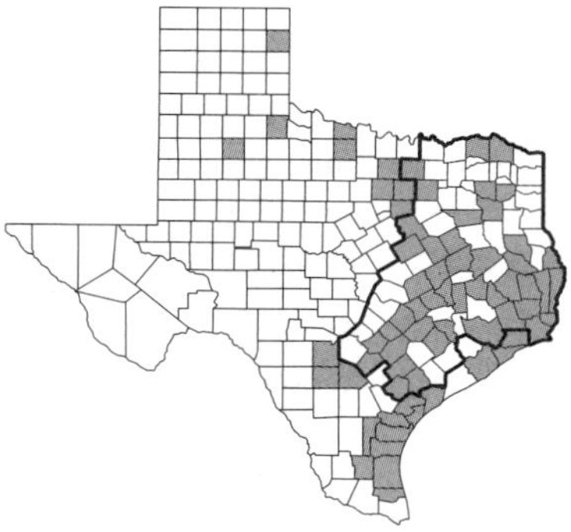
Gaillardia aestivalis

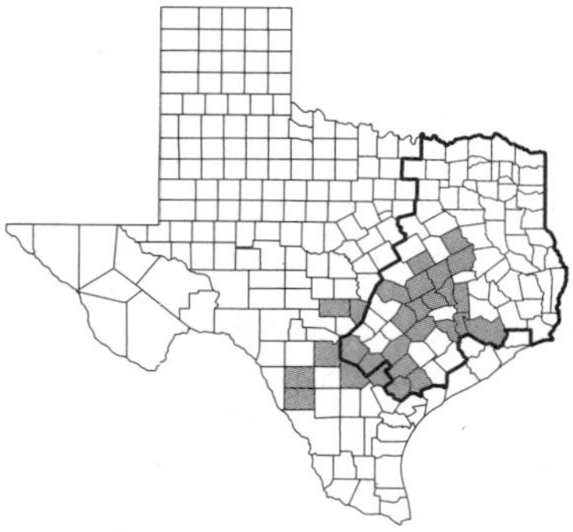
Gaillardia amblyodon

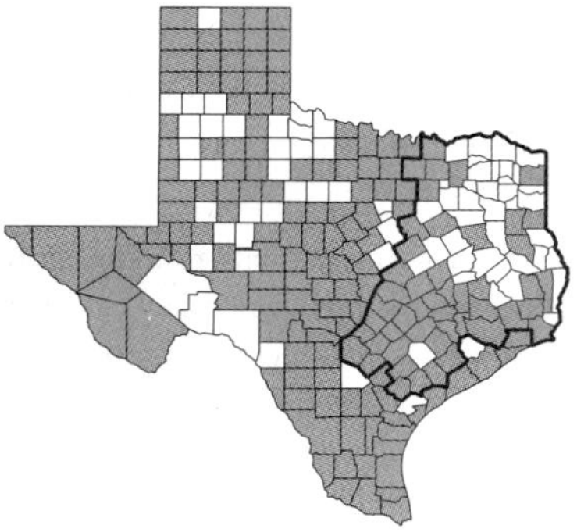
Gaillardia pulchella

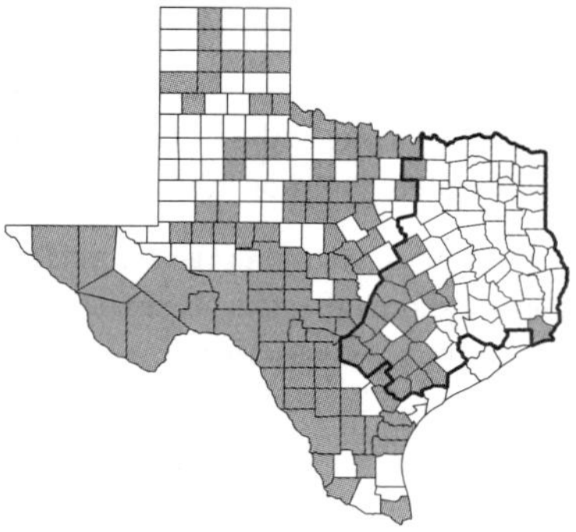
Gaillardia suavis

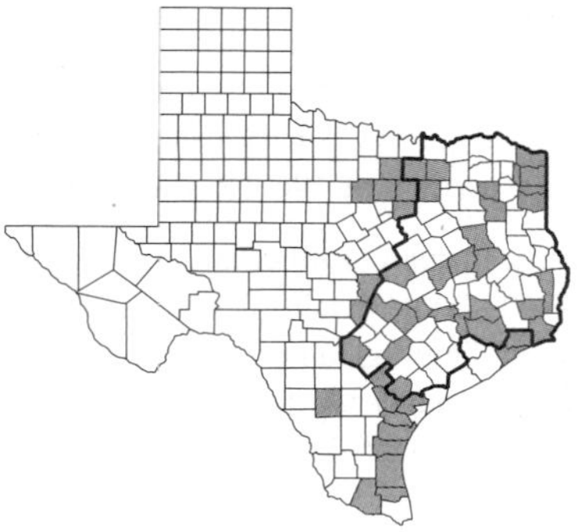
Gamochaeta antillana

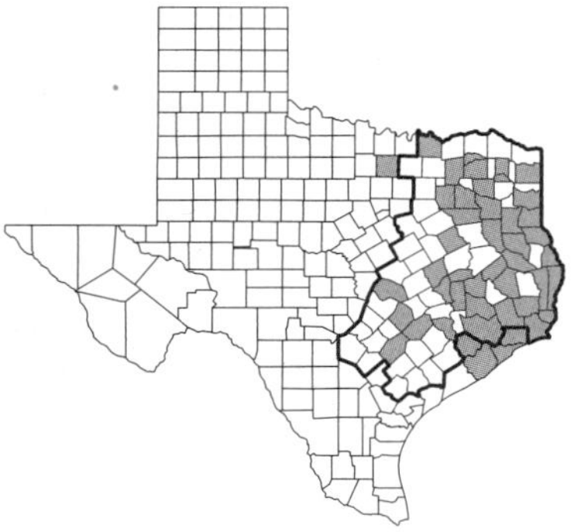
Gamochaeta argyrinea

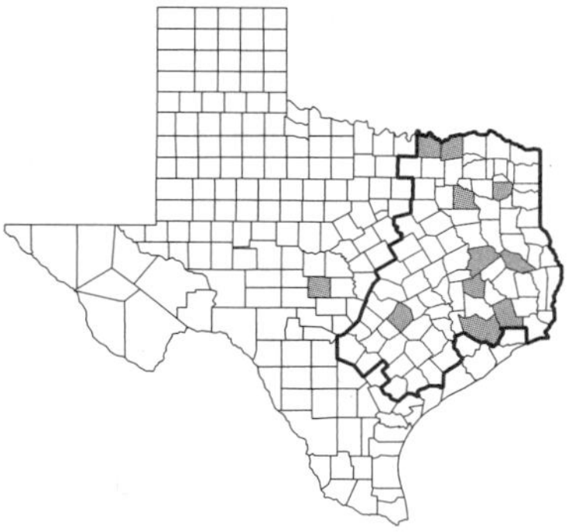
Gamochaeta calviceps

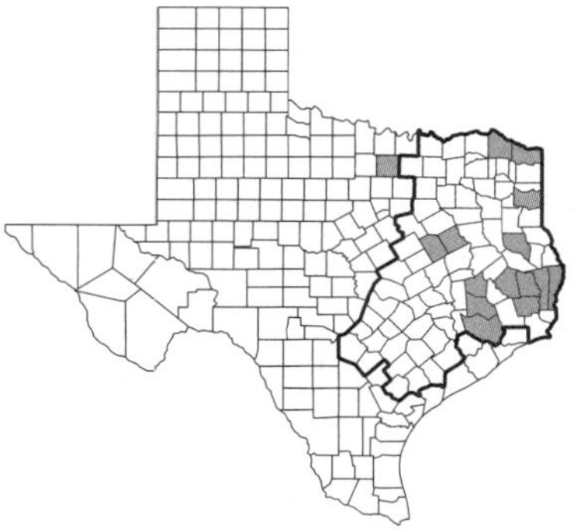
Gamochaeta coarctata

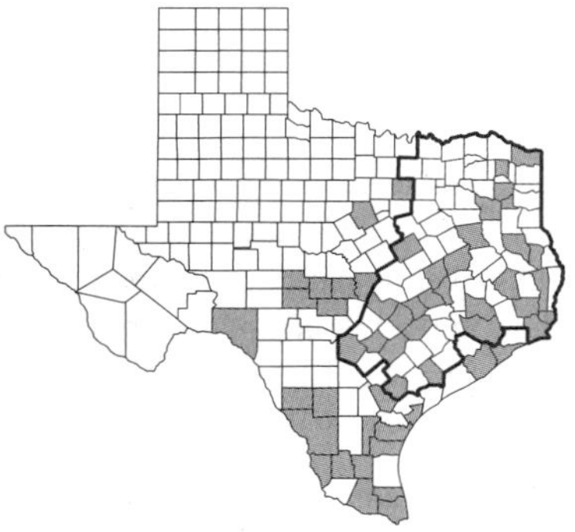
Gamochaeta pensylvanica

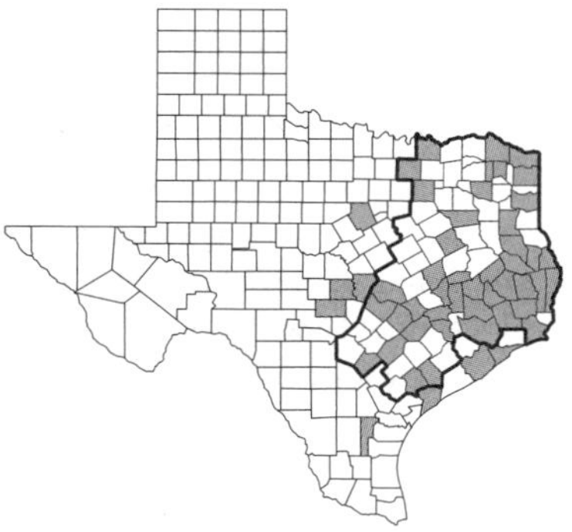
Gamochaeta purpurea

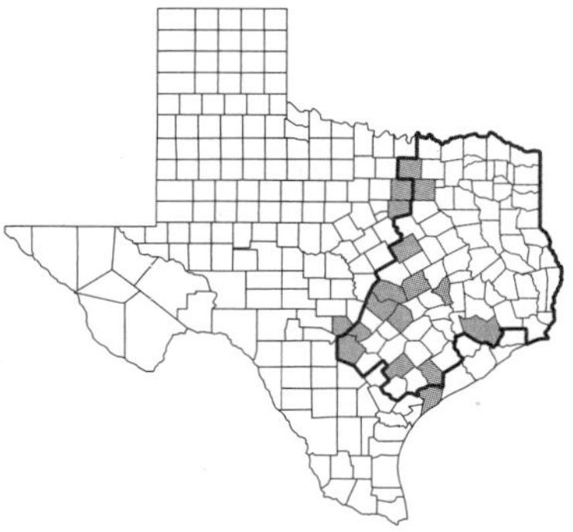
Grindelia adenodonta

a false occurrence record, presumably based on a specimen of *G. sarothrae*; included here based on a record just to the sw of East TX in Medina Co.; s TX, Edwards Plateau, and Trans-Pecos; sw U.S. from TX and NM n to CO and w to CA; Mexico. Jul–Dec. [*G. sarothrae* (Pursh) Britton & Rusby var. *microcephala* (DC.) L.D. Benson, *Xanthocephalum microcephalum* (DC.) Shinners] This species causes poisoning and abortion in sheep and cattle, apparently due to the presence of saponins; this has resulted in significant economic losses in w TX (Kingsbury 1964); symptoms include loss of appetite, drooping of the head, and hematuria (Burlage 1968). This species is an indicator of overgrazed range conditions (Powell 1988). ☠

Gutierrezia sphaerocephala A. Gray, (spherical-headed), ROUND-HEAD BROOMWEED, ROUND-HEAD SNAKEWEED, which would have keyed to *G. texana* here, can be distinguished by the papillate-scabrous stems, slightly larger involucres (2.5–5 mm wide), larger number of disk flowers (18–37 vs. 7–13 in *G. texana*), achenes that are densely silky-strigose on ribs, and larger pappus of scales (0.2–1 mm long). Disturbed habitats, grasslands, oak-pine-juniper woodlands; included here based on a record just to the sw of East TX in Medina Co.; s TX, Edwards Plateau, and Trans-Pecos to Panhandle; AZ and NM; Mexico. May–Dec. [*G. eriocarpa* A. Gray, *Xanthocephalum sphaerocephalum* (A. Gray) Shinners]

GYMNOSPERMA Less.

A monotypic genus of sw North America, Mexico, and Central America; closely related to *Amphiachyris*, *Gutierrezia*, and *Thurovia*. (Greek: *gymnos*, naked, and *sperma*, seed, referring to the absence of pappus) (tribe Astereae)
REFERENCES: Solbrig 1960; Ruffin 1974; Lane 1982; Suh & Simpson 1990; Nesom 1991d, 2006q.

Gymnosperma glutinosum (Spreng.) Less., (glutinous, sticky) GUMHEAD, TATALENCHO, STICKY SELLOA. Heavily resinous shrubs, 50–100(–200) cm tall, glabrous; stems erect; leaves alternate, sessile, bases decurrent, linear to narrowly lanceolate or oblanceolate to elliptic, gland-dotted, margins entire; heads borne in terminal, compact corymbose arrays; involucres cylindrical, 1.5–2 mm wide; phyllaries in 2–4 series, unequal, mostly whitish and hardened; receptacles naked; ray flowers 4–9, pistillate, fertile, corollas yellow, 2–3 mm long; disk flowers perfect, fertile or functionally staminate, corollas orange-yellow; achenes columnar or spindle-shaped, 1–1.4 mm long, white-strigillose; pappus absent, or a minute crown < 0.1 mm high. Rocky slopes, streambeds, creosote and juniper scrublands; Bexar, Travis, and Wilson cos.; mostly Trans-Pecos, Edwards Plateau, and s TX; AZ, NM, and TX; Mexico and Guatemala. May–Nov. [*Selloa glutinosa* Spreng., *Xanthocephalum glutinosum* (Spreng.) Shinners] This species has a history of herbal usage in Mexico and is under investigation for potential antibacterial and antifungal qualities. It is infrequently grazed by livestock.

HEDYPNOIS Mill.

A genus of 2 species native from Macaronesia to Iran. (Greek name used by Pliny for a kind of wild endive) (tribe Lactuceae)
REFERENCES: Sell 1976; Strother 2006n.

Hedypnois cretica (L.) Dom. Cours., (of Crete), CRETANWEED. Taprooted annual, ± hairy; stems spreading and ascending, 10–30 cm long, sparingly branched; leaves alternate; basal leaves pinnately lobed, usually with winged petioles; stem leaves few, reduced, sessile; heads solitary at the branch tips; involucres ca. 10 mm high; inner phyllaries scabrous-hispid, after flowering becoming conspicuously incurved and boat-shaped (with a very rounded keel); outer phyllaries < 1/2 as long as inner; flowers perfect, mostly fertile; corollas all ligulate, yellow, often tipped with reddish purple; achenes ca. 5–7.5 mm long, ± cylindrical, beakless, the peripheral achenes incurved; pappus of peripheral achenes a crown to 1 mm long, that of central achenes of short outer scales plus several bristle-like scales 2–5 mm long. Sandy or rocky soils, roadsides; scattered primarily se to sc TX; sw U.S. from TX to CA; introduced elsewhere. Mar–May. Native of Mediterranean region, Canary Islands.

HELENIUM L. SNEEZEWEED, BITTERWEED

Annual or perennial herbs, glabrous or pubescent; stems often winged by decurrent leaf bases; leaves alternate, entire to deeply and finely lobed, resin-dotted; heads terminal, solitary or in corymbose or paniculate arrays; phyllaries usually in 2 series, gland-dotted, herbaceous, usually reflexed in fruit; disks globose or hemispheric to ovoid, usually radiate; receptacles naked; ray flowers pistillate and fertile or neuter, with ligules widened and 3-toothed apically, yellow to reddish brown or purplish, often reflexed; disk flowers perfect, fertile, corollas yellow to reddish brown or purplish; achenes obpyramidal, 4–5-angled; pappus of 5–12 scales, variously entire, lacerate, and/or elongated into an awn-like tip.

An American genus of ca. 32 species, including some cultivated ornamentals. ☠ Most species are poisonous and unpalatable to grazing animals (Barkley 1986); some contain sesquiterpene lactones (e.g., helenalin) or glycosides (Blackwell 1990; Hardin & Brownie 1993). (Greek: *helenion*, the name of a similar plant that, according to Gerard (1597), "took the name *Helenium* of Helena [of Troy], wife of Menelaus, who had her hands full of it when Paris stole her away into Phrygia.") (tribe Heliantheae)

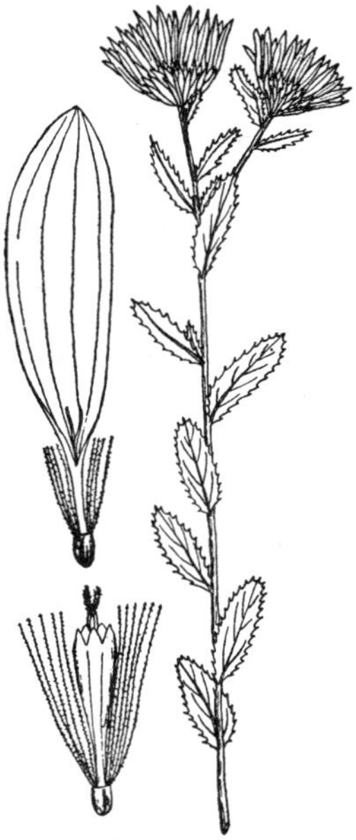
Grindelia ciliata [BB2, STE]

Grindelia lanceolata [BB2]

Grindelia microcephala [AMB]

Grindelia pusilla [AMB]

Grindelia squarrosa [KIN]

Gutierrezia texana [SID]

Gutierrezia sarothrae [BB2, SID]

Gymnosperma glutinosum [FNA]

References: Rock 1957; Bierner 1972, 1989, 2006b; Stanford & Turner 1988.

1. Leaves or leaf segments linear to thread-like, usually < 2 mm wide, the bases not extending down the stem, the stems thus unwinged ____ **H. amarum**
1. Leaves of various shapes but usually > 2 mm wide, not thread-like, the bases decurrent (= leaf tissue extending down the stem), the stems thus winged.
   2. Annuals with a taproot; pappus of entire scales, neither lacerate nor awn-tipped.
      3. Upper stem leaves linear, reduced, entire; heads peduncled, held well above the foliage; disks globose, as wide as high or wider; ligules 5.5–11.5 mm long ____ **H. elegans**
      3. Upper stem leaves ovate to oblong to obovate, little reduced, often with at least a few teeth; heads short-peduncled, barely held above the foliage; disks ovoid, higher than wide; ligules 2.6–4.8 mm long ____ **H. microcephalum**
   2. Perennials with fibrous roots from subrhizomatous base; pappus scales entire and awn-tipped, OR deeply lacerate and lacking awned tips.
      4. Ray flowers pistillate and fertile; plants of c and w TX, usually along and w of a line drawn from Dallas through Austin to San Antonio ____ **H. autumnale**
      4. Ray flowers neuter (lacking stamens or stigmas) and sterile, or absent; plants of East TX, usually well e of a line drawn from Dallas through Austin to San Antonio.
         5. Disk corolla lobes usually dark reddish purple or maroon; pappus scales entire and awn-tipped; heads (1–)5–50 per plant ____ **H. flexuosum**
         5. Disk corolla lobes usually yellow; pappus scales deeply lacerate and lacking awned tips; heads 1(–3) per plant ____ **H. drummondii**

Helenium amarum (Raf.) H. Rock, (bitter). Weedy annuals 10–100 cm tall; stems unwinged; leaves all linear or filiform except for lowest which are sometimes pinnatifid (these sometimes withered before flowering time); heads 5–9 mm high; ray flowers 8–10, pistillate, ligules yellow, 4.5–14 mm long; disk globose; disk corollas yellow with lobes yellow to yellow-brown or reddish brown to purple; achenes 0.7–1.3 mm long; pappus of 6–8 entire, awn-tipped scales. Open woods, fields, pastures, and disturbed areas. The two varieties are both found in East TX and may intergrade.

1. Lower and basal leaves typically withered at flowering time; basal leaves entire to toothed to occasionally pinnatifid; disk corollas yellow with lobes yellow to yellow-brown ____ var. **amarum**
1. Lower and basal leaves often present at flowering time; basal leaves pinnatifid; disk corollas yellow with lobes reddish brown to purple ____ var. **badium**

var. **amarum,** BITTERWEED, SNEEZEWEED. Most common in e 1/2 TX, w to Rolling Plains and Edwards Plateau; e 1/2 of the U.S., adventive in CA and some eastern states. May–Nov, winter and fall when moisture available. [*H. tenuifolium* Nutt.] The foliage is extremely bitter, largely avoided by livestock, and thus plants increase under conditions of overgrazing; conspicuously dominant on abused pastures. All parts of the plant contain the glycoside dugaldin and cause toxic symptoms in animals; cows that have eaten even small amounts of the plant produce extremely bitter milk; honey made from the flowers is also reportedly bitter (Correll & Johnston 1970; Stephens 1980; Ajilvsgi 1984).

var. **badium** (A. Gray ex S. Watson) Waterfall, (reddish brown), BASIN SNEEZEWEED. Similar to variety *amarum* except for the characters in the key. W border of East TX from Dallas and Grayson cos. sw to Bastrop, Comal, and Travis cos., also Bell and Brazos cos.; mostly w 1/2 of TX; NM, OK and TX; Mexico. Apr–Jul, sporadically to Oct. [*H. badium* (A. Gray) Greene, *H. tenuifolium* Nutt. var. *badium* A. Gray ex S. Watson]

Helenium autumnale L., (of autumn), COMMON SNEEZEWEED, TALL SNEEZEWEED, STAGGERWORT, SWAMP-SUNFLOWER. Perennials 50–130 cm tall; stems strongly winged; leaves obovate to lanceolate, margins dentate or entire, lower leaves usually withered before flowering time; heads 8–20 mm high; ray flowers 8–21, pistillate, ligules yellow, 10–23 mm long; disk globose; disk corollas yellow with lobes yellow to yellow-brown; achenes 1–2 mm long; pappus of 5–7 entire, awn-tipped scales. Low, moist, calcareous soils, ditches, streambanks; sw border of East TX in Bell, Gonzales, Hays, and Travis cos.; Rolling Plains and Edwards Plateau; across North America. Aug–Oct. [*H. latifolium* Mill., *H. parviflorum* Nutt., *H. autumnale* L. var. *canaliculatum* (Lam.) Torr. & A. Gray] Poisonous to cattle and sheep; it is reported to cause "spewing sickness" with symptoms including weakness, depression, vomiting, excessive salivation, frothing, irregular pulse, and kidney and liver damage (Burlage 1968).

Helenium drummondii H. Rock, (for Thomas Drummond, 1780–1835, Scottish botanist and collector in North America), FRINGED SNEEZEWEED, DRUMMOND'S SNEEZEWEED. Perennials 20–60 cm tall; stems moderately to strongly winged; leaves lanceolate to lance-linear, margins entire; disks 10–18 mm high; ray flowers 13–30, neuter, ligules yellow, 14–25 mm long; disk corollas yellow; achenes 1.2–1.5 mm long; pappus of 5–12 deeply-lacerate scales, without awned tips. Low woods, ditches, bogs; deep se TX; LA and TX. Mar–May. [*H. fimbriatum* of authors, not (Michx.) A. Gray]

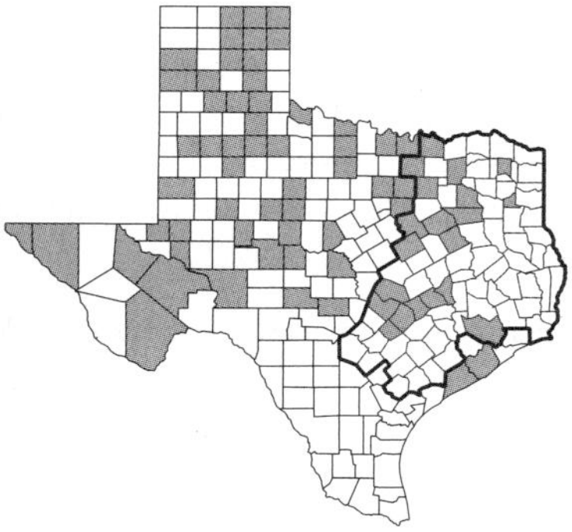
Grindelia ciliata

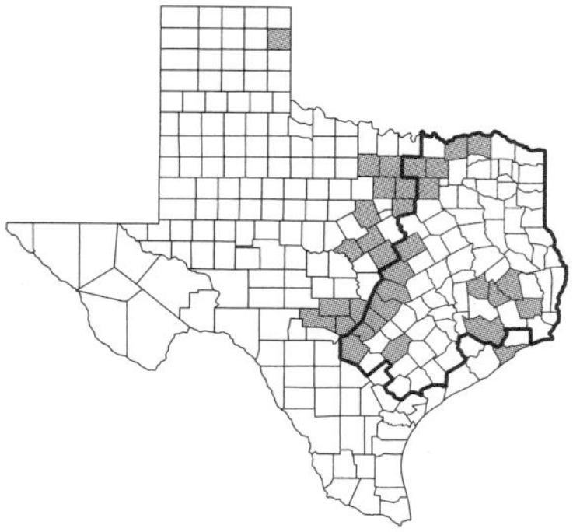
Grindelia lanceolata

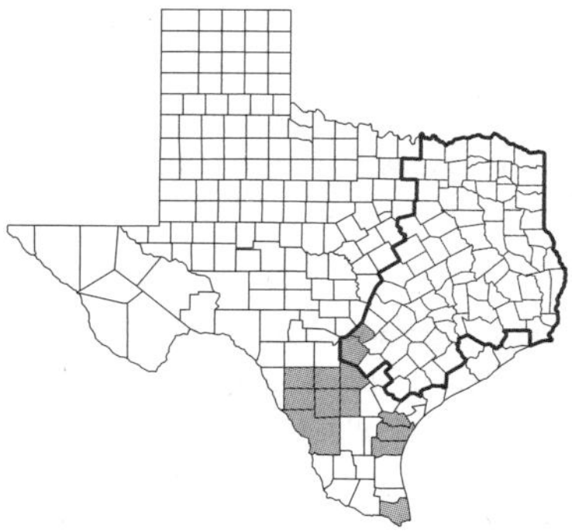
Grindelia microcephala

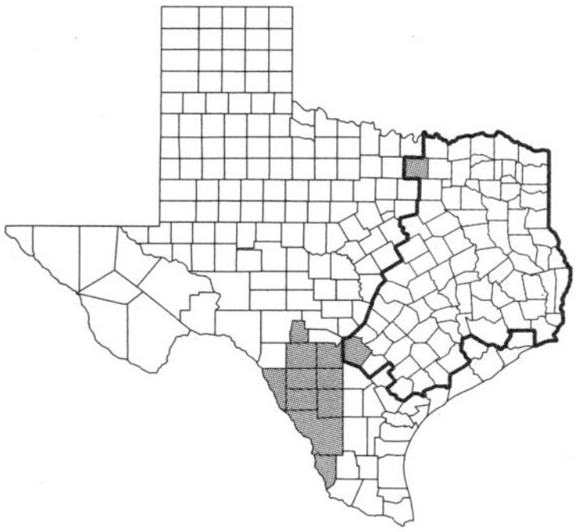
Grindelia pusilla

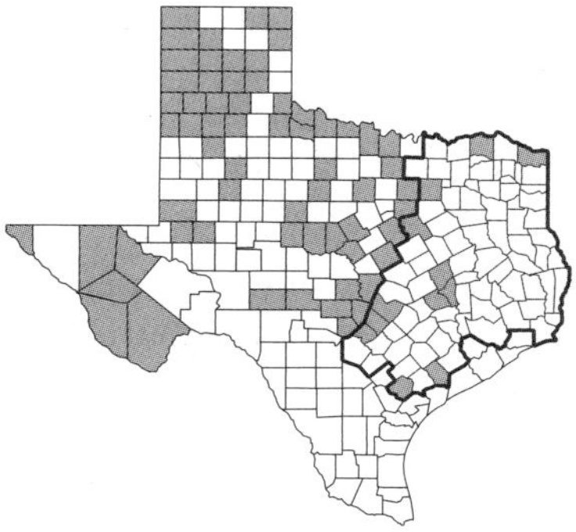
Grindelia squarrosa

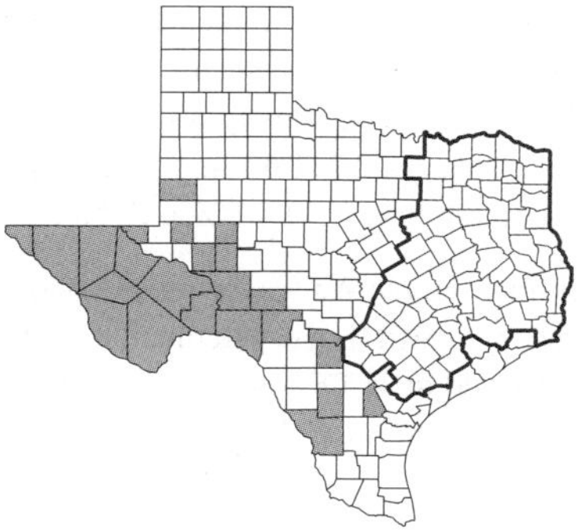
Gutierrezia microcephala

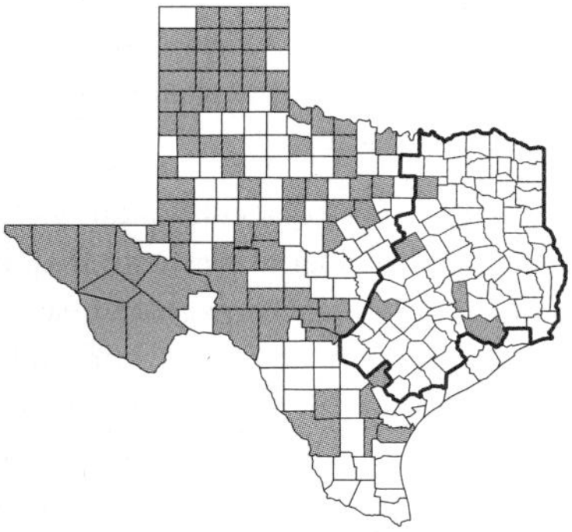
Gutierrezia sarothrae

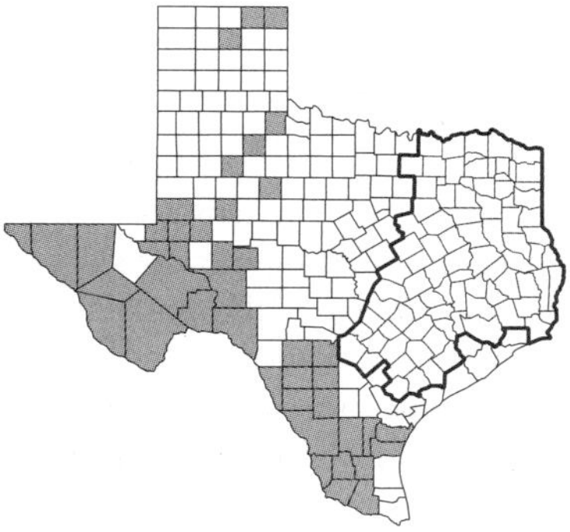
Gutierrezia sphaerocephala

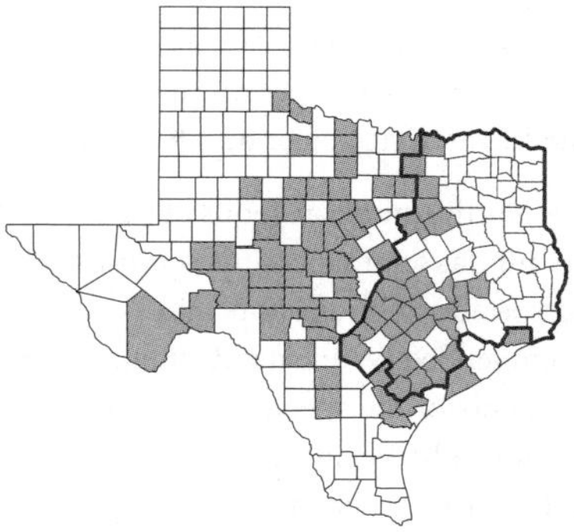
Gutierrezia texana var. texana

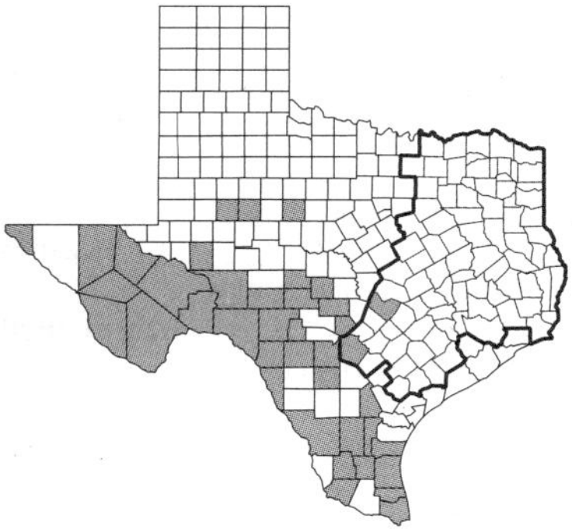
Gymnosperma glutinosum

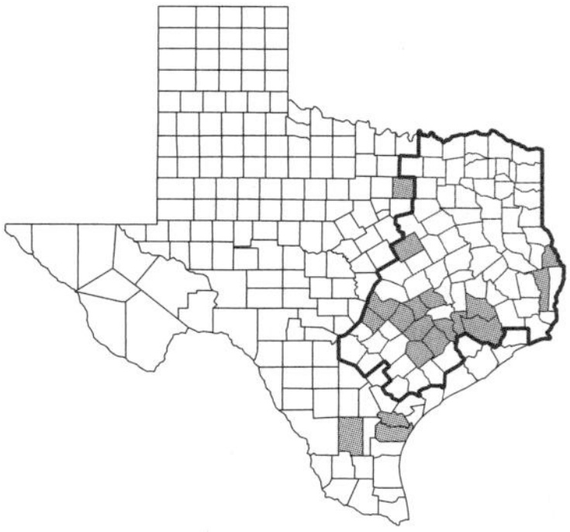
Hedypnois cretica

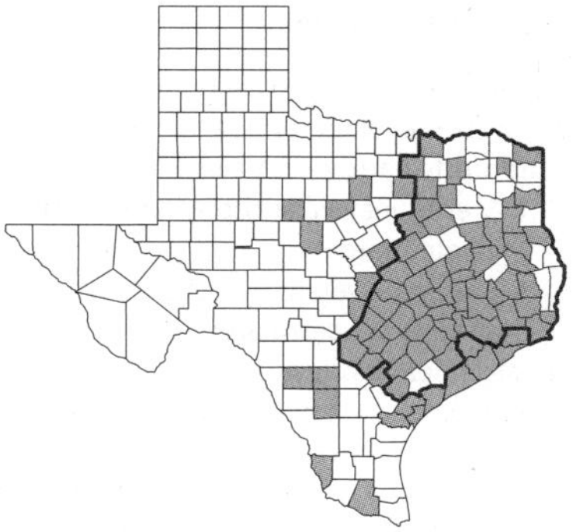
Helenium amarum var. amarum

Helenium elegans DC. var. **elegans,** (elegant). PRETTY SNEEZEWEED. Annuals 20–120 cm tall; stems moderately to strongly winged; leaves lanceolate to narrowly elliptic to linear, margins usually entire; disks 5–8 mm high; ray flowers 10–14, pistillate, ligules yellow throughout or reddish brown at the base and yellow at the tips, 5.5–11.5 mm long; disk corollas yellow with lobes yellow to yellow-brown or reddish brown; achenes 0.6–0.9 mm long; pappus of 6–7 entire scales, without awned tips. Gravelly stream bottoms or seepy slopes on limestone; w and sw border of East TX, from Grayson and Dallas cos. s to Bexar Co. and e to Gonzales Co.; mostly Edwards Plateau and Cross Timbers; AR, OK, and TX. May–Aug.

Helenium flexuosum Raf., (flexuous), PURPLE-HEAD SNEEZEWEED. Perennials 30–100 cm tall; stems strongly winged; leaves oblanceolate to lance-linear, margins toothed or entire; disks globose to ovoid, 8–17 mm high; ray flowers 8–13 and neuter or sometimes 0, ligules yellow to reddish brown to maroon, 10–20 mm long; disk corollas purple or yellow with lobes purple; achenes 1–1.2 mm long; pappus of 5–6 entire, awn-tipped scales. Ditches, fields, along streams; e 1/4 TX; e 1/2 of North America. May–Jul, sporadically all year. [*H. nudiflorum* Nutt.]

Helenium microcephalum DC. var. **microcephalum,** (small-headed), SMALL-HEAD SNEEZEWEED, SMALL SNEEZEWEED. Annuals 30–120 cm tall; stems moderately to strongly winged; leaves narrowly elliptic to narrowly oblong-elliptic, margins serrate to undulate, lower leaves usually withered before flowering time; disks 4–8 mm high; ray flowers 7–13, pistillate, ligules usually yellow throughout, sometimes reddish at the bases, 2.6–4.8 mm long; disk corollas yellow to yellow-green with lobes reddish brown; achenes 0.7–1.4 mm long; pappus of 6 entire scales, without awned tips. Moist sandy or calcareous soils, dried-up ponds, ditches; w 3/4 TX; AZ, CO, NM, OK, and TX; Mexico. Jun–Nov.

Helenium vernale Walt., (spring), SAVANNAH SNEEZEWEED, SPRING SNEEZEWEED, which would have keyed to *H. drummondii* here, can be distinguished by the perennial habit, weakly-winged stems, and pappus of 8 entire or lacerate scales without awned tips. To be expected in se TX based on records from adjacent parishes in LA (Calcasieu, Cameron, and Vernon parishes); se U.S. from NC s to FL and w to LA. Mar–May.

HELIANTHUS L. SUNFLOWER

Herbaceous annuals or perennials, to 3+ m tall; leaves sessile or petioled, alternate, opposite, or opposite below and alternate above, simple; heads relatively large, sessile or peduncled, borne singly or in various arrays; phyllaries in 2–3+ series, subequal to unequal, herbaceous; receptacles with persistent chaff (= pales); ray flowers neuter (not producing mature achenes), the corollas yellow; disk flowers perfect, fertile, the corolla lobes reddish purple or yellow; achenes obpyramidal, somewhat compressed laterally, often striped or mottled; pappus readily falling, of 2(–3) awn-like scales, plus 0–8 shorter scales.

A North American genus of 52 species, including the widely-cultivated SUNFLOWER, *H. annuus*. Species identification is often difficult; taxonomic confusion in *Helianthus* can be blamed on notorious developmental and ecological plasticity, frequent interspecific hybridization, and polyploidy (Schilling 2006a). Collection of the entire plant, including below-ground parts, is recommended. While many putative or proven hybrids exist, there are little data on their distribution; notes on hybrids follow treatments of the species involved in the crosses. Some species of *Helianthus* could be confused with *Silphium*, however, in *Silphium* only the ray flowers produce mature achenes, which have thin, papery edges, while in *Helianthus* only the disk flowers produce mature achenes, and these are not thin-edged. (Greek: *helios*, the sun, and *anthos*, flower, presumably in reference to the turning of the inflorescences towards the sun) (tribe Heliantheae)
REFERENCES: Heiser 1954; R. Jackson 1963; Heiser et al. 1969; Rogers et al. 1982; Sims & Price 1985; Chandler et al. 1986; Rieseberg 1991; Schilling 2006a.

1. Annuals or short-lived perennials from taproots.
   2. Larger leaf blades 10–40 cm wide; phyllaries ovate to lance-ovate, the larger ones usually 5–8 mm wide, the tips abruptly narrowed; pales 9–11 mm long.
      3. Stems, leaves, and phyllaries usually green and hispid; leaf margins usually serrate; phyllaries usually with ciliate margins ____ **H. annuus**
      3. Stems, leaves, and phyllaries white-silvery with dense cottony or silky pubescence; leaf margins usually entire, sometimes serrulate; phyllaries usually not marginally ciliate ____ **H. argophyllus**
   2. Larger leaf blades usually < 12 cm wide; phyllaries usually lanceolate to lance-ovate, the larger ones usually < 5 mm wide, the tips gradually narrowed; pales < 8 mm long.
      4. Leaves mostly opposite, petioles 5–9 cm long, blades sometimes constricted near the middle ____ **H. praecox**
      4. Leaves mostly alternate, petioles 1–7 cm long, blades usually not constricted near the middle.
         5. Peduncles 20–50 cm long; pales 7.5–8 mm long; achenes 2.5–3.2 mm long, glabrous or sparsely hairy; plants of e, sc, and s TX ____ **H. debilis**

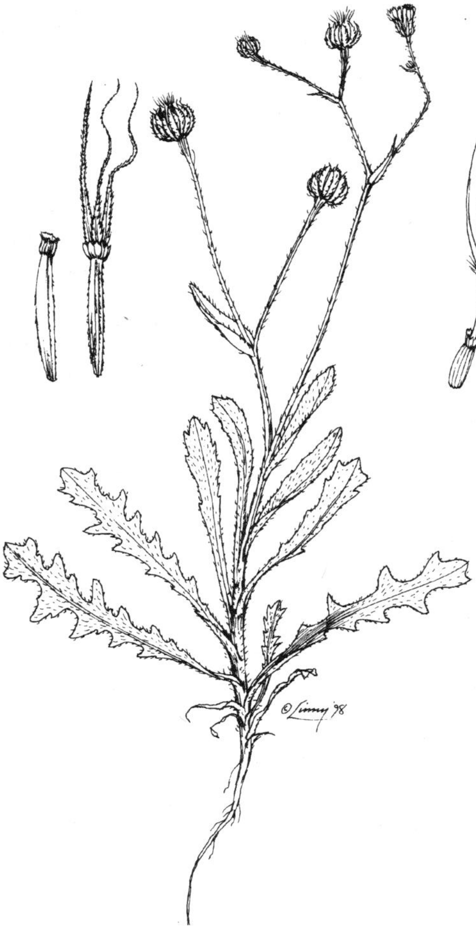

Hedypnois cretica [HEA]

Helenium amarum var. amarum [REE]

Helenium autumnale [GWO]

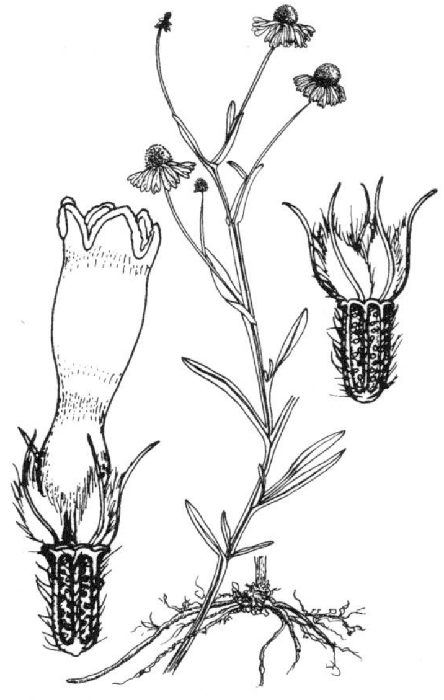

Helenium drummondii [CO1]

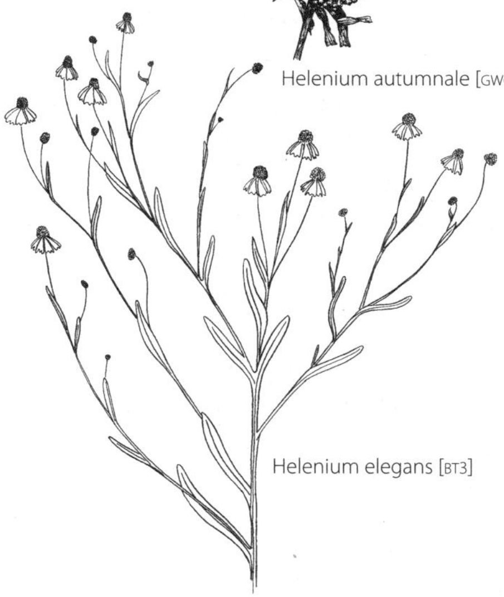

Helenium elegans [BT3]

5. Peduncles 4–15(–40) cm long; pales 4.5–7.5 mm long; achenes 3–4.5 mm long, ± villous; plants of nc, nw, and w TX ____ **H. petiolaris**

1. Perennials from crown buds or rhizomes.
6. Plants 40–70 cm tall, with slender rhizomes or creeping roots, forming colonies; stems glabrous; stems and leaves blue-green glaucous; leaves almost all opposite, sessile, margins usually ciliate and undulate; ray ligules < 10 mm long ____ **H. ciliaris**
6. Plants usually much taller, and without the above combination of characters.
7. Stems glabrous or glabrate, at least below the inflorescence arrays, sometimes also glaucous.
8. Leaves sessile or subsessile, blades linear to lance-linear, 0.2–1.2 cm wide; disk corolla lobes reddish; longest pappus scales 3–3.6 mm long ____ **H. salicifolius**
8. Leaves with petioles 1+ cm long, blades lanceolate to lance-ovate or ovate, 1.2–10 cm wide; disk corolla lobes yellow; longest pappus scales 1–2.5 mm long.
9. Pales 7–8 mm long; phyllaries 10–14 mm long, tips attenuate; ray ligules 23–40 mm long; achenes 3–4 mm long ____ **H. grosseserratus**
9. Pales 5.3–6.5 mm long; phyllaries 5.5–10 mm long, tips acute to acuminate; ray ligules 12–20(–30) mm long, achenes 4–5.5 mm long ____ **H. strumosus**
7. Stems pubescent ± throughout (scabrous, strigose, pilose, hispid, or villous), rarely glaucous.
10. Lower stem leaves or basal leaves (if present) with petioles 2–10 cm long.
11. Leaves mostly basal, greatly reduced up the stem, the few stem leaves opposite; petioles unwinged; pales 5–7 mm long; phyllaries not gland-dotted; achenes 3–4(–5) mm long ____ **H. occidentalis**
11. Leaves evenly distributed up the stem, not much reduced in size, upper leaves alternate; petioles often winged; pales 8–9 mm long; phyllaries gland-dotted; achenes 5–7 mm long ____ **H. tuberosus**
10. Lower stem leaves or basal leaves (if present) sessile or short-petiolate, any petioles < 2 cm long.
12. Leaves narrowly lanceolate to linear, margins entire or subentire, revolute (sometimes strongly so); pales 5.5–6.5 mm long; achenes 2–3 mm long; longest pappus scales < 2.5 mm long.
13. Plants 50–150+ cm tall, rhizomes absent or poorly-developed; leaves sessile or subsessile, blades 0.15–0.5(–1) cm wide; phyllaries 4–9 mm long; ray ligules 10–20 mm long ____ **H. angustifolius**
13. Plants 150–260 cm tall, rhizomes well-developed; leaves with petioles 0.5–1 cm long, blades 0.7–4 cm wide; phyllaries 6–17 mm long; ray ligules 16–40 mm long ____ **H. simulans**
12. Leaves broadly lanceolate to lance-ovate or oblong-lanceolate, margins entire or serrulate to serrate, flat (not revolute); pales 7–11 mm long; achenes 3–4.5 mm long; longest pappus scales > 2.5 mm long.
14. Leaves all or mostly opposite, sessile, velvety hispid-tomentose, gray-green, bases mostly cordate-clasping; achenes villous ____ **H. mollis**
14. Leaves opposite or alternate, sessile or short-petiolate, scabrous to hispid (but not velvety), gray-green or green, bases cuneate to rounded or truncate (not cordate-clasping); achenes glabrate to puberulent.
15. Lower leaf surfaces sparsely, if at all, gland-dotted; phyllaries ovate; disk corolla lobes usually reddish, sometimes yellow; achenes 5–6 mm long; longest pappus scales 4.5–5 mm long ____ **H. pauciflorus**
15. Lower leaf surfaces gland-dotted; phyllaries lanceolate; disk corolla lobes yellow; achenes 3–4.5 mm long; longest pappus scales 2.8–4.1 mm long.
16. Stems hirsute; leaves 3-nerved from the base, usually not folded lengthwise; heads 1–7 per main stem, in cymose or paniculate arrays; phyllaries not gland-dotted; ray ligules 15–20 mm long ____ **H. hirsutus**
16. Stems scabrous-hispid; leaves 1-nerved from the base, often folded lengthwise; heads usually 3–15 per main stem, in racemose or spicate arrays; phyllaries gland-dotted; ray ligules (15–)25–40 mm long ____ **H. maximiliani**

Helianthus angustifolius L., (narrow-leaved), SWAMP SUNFLOWER. Perennials from crown buds, rhizomes absent or poorly-developed; stems usually ± hairy; leaves opposite or alternate, sessile or subsessile, blades narrowly lanceolate to linear, 3-nerved from near the base, strigose to hispid or hirsute beneath, bases cuneate, margins entire and usually strongly revolute; involucres 10–20 mm wide; phyllaries loose, lanceolate, usually gland-dotted, scabrous or rarely glabrous, tips acute to slightly acuminate; ray flowers 10–20, ligules 10–20 mm long; disk corolla lobes purple-brown, base sometime yellow; achenes 2–3 mm long, glabrate. Open to shaded, usually moist places; widespread in e 1/3 TX; e and se U.S. from NY s to FL and w to IL, OK, and TX. (May–)Jul–Oct. [*H. angustifolius* L. var. *planifolius* Fernald]

Helianthus annuus L., (annual), COMMON SUNFLOWER, MIRASOL. Annuals, extremely variable in size; stems usually hispid; leaves mostly alternate, petiolate, blades ovate to lance-ovate, usually hispid beneath, bases cuneate to cordate, margins serrate; involucres 15–40(–200+) mm wide; phyllaries ovate to lance-ovate, usually gland-dotted, hirsute to hispid or rarely glabrous, margins usually ciliate, tips long-acuminate; ray flowers (13–)17–30 (to 100+ in cultivars), ligules 25–50 mm long (in colors other than yellow in some planted or escaped cultivars); disk corolla lobes usually reddish, sometimes yellow; achenes (3–)4–5 mm long (to 15 mm long in cultivars), glabrate. Open areas, fencerows, disturbed areas; throughout TX; throughout North America but introduced and adventive in most e states; Mexico; introduced nearly worldwide. Apr–Oct. [*H. aridus* Rydb., *H. annuus* subsp. *lenticularis* (Douglas ex Lindl.) Cockerell, *H. annuus* subsp. *texanus* Heiser, *H. jaegeri* Heiser] The cultivated SUNFLOWER, widely grown for the seeds, for oil, and as an ornamental, is derived from *H. annuus*; it is one of the world's most important oil crops (Rieseberg & Seiler 1990; Mabberley 2017). Achenes of *H. annuus* were gathered as food by Native Americans of the w U.S. (Heiser 1951), and the species was being cultivated in parts of North America when Europeans arrived (Heiser 1993); it is one of the few agricultural crops to originate in North America north of Mexico (Nabhan 1979). *Helianthus annuus* forms hybrids with other annual *Helianthus* species. Because it easily invades annual crop fields, it is considered a noxious weed in a few states.

Helianthus argophyllus Torr. & A. Gray, (silvery-leaved), SILVERLEAF SUNFLOWER. Annuals; stems usually tomentose or cottony, appearing silvery-white; leaves mostly alternate, petiolate, blades ovate to lance-ovate, often silvery, cottony or silky pubescent beneath, bases truncate to subcordate, margins entire or serrulate; involucres 20–30 mm wide; phyllaries ovate to lance-ovate, gland-dotted, usually densely white-villous, tips long-attenuate; ray flowers 15–20, ligules 20–30 mm long; disk corolla lobes reddish; achenes 4–6 mm long, glabrate. Sandy soils, open areas; sc TX in Dewitt, Goliad, Gonzales, Guadalupe, Jackson, and Wilson cos.; also introduced and spreading in Harris Co. (Brown 2014); Gulf Coast and s TX; introduced and adventive in NC and FL; introduced in South America, Africa, and Australia. Jun–Oct. Hybridizes with *H. annuus*.

Helianthus ciliaris DC., (ciliate, fringed with hairs), BLUE-WEED SUNFLOWER, BLUE-WEED, TEXAS BLUE-WEED. Perennials from rhizomes or creeping roots, often forming extensive colonies; stems glabrous or glabrate, glaucous; leaves mostly opposite, sessile, blades linear to lanceolate, 1- or 3-nerved, often bluish green, glabrous to hispid, bases ± cuneate, margins entire or toothed/lobed, usually ciliate and undulate; involucres 12–25 mm wide; phyllaries ovate to lance-ovate, not gland-dotted, glabrate to strigose, margins ciliate, tips obtuse to acute; ray flowers 10–18, ligules 8–9 mm long; disk corolla lobes reddish; achenes 3–3.5 mm long, glabrous. Sandy or sandy clay soils, cultivated fields, drainage areas; Denton Co.; w 2/3 of TX, sw U.S. from KS s to TX and w to WA and CA; Mexico. (Mar–)Jun–Oct. Spreads aggressively in cultivated fields and considered a noxious weed in some states.

Helianthus debilis Nutt., (weak, frail), WEAK SUNFLOWER, BEACH SUNFLOWER. Annuals or perennials from taproots; stems glabrate to hispid; leaves mostly alternate, petiolate, blades deltate-ovate or lance-ovate to ovate, glabrate to hispid beneath, bases cordate to truncate or broadly cuneate, margins usually serrate; disks 16–20 mm wide; phyllaries lanceolate, not gland-dotted, glabrous to hispid, tips acute to long-attenuate; ray flowers 11–20, ligules 15–30 mm long; disk corolla lobes usually reddish, sometimes yellow; achenes 2.5–3.2 mm long, glabrous or sparsely hairy. Schilling (2006a) described the two intergrading subspecies in East TX, either of which may hybridize with *H. annuus*:

1. Leaf blades 4–9 cm long; peduncles to 50 cm long, not notably slender; disks 16–18(–20) mm wide ________ subsp. **cucumerifolius**
1. Leaf blades (6–)8–14 cm long; peduncles to 40 cm long, relatively slender; disks 14–15(–16) mm wide ________ subsp. **silvestris**

subsp. **cucumerifolius** (Torr. & A. Gray) Heiser, (with leaves like *Cucumis*—cucumber), CUCUMBER-LEAF SUNFLOWER. Open areas, sandy soils; McLennan Co. and sc TX; South TX Plains and Gulf Coast; also scattered records in se TX; native to se U.S. from FL to TX; adventive in e coast states from ME to GA, and also MI. Apr–Aug. [*H. debilis* Nutt. var. *cucumerifolius* (Torr. & A. Gray) A. Gray, *H. cucumerifolius* Torr. & A. Gray]

subsp. **silvestris** Heiser, (of the forest), FOREST SUNFLOWER. Sandy soils, pine and oak woods; e, se, and sc TX;. Jun–Nov.

Helianthus grosseserratus M. Martens, (large-toothed), SAWTOOTH SUNFLOWER. Rhizomatous perennials; stems glabrous below and glabrate to strigillose above, glaucous; leaves mostly alternate (often opposite on lower stem), petiolate, blades lanceolate to lance-ovate, 3-nerved from above the base, puberulent to tomentose beneath, bases cuneate, margins usually coarsely to shallowly serrate; involucres 15–25 mm wide; phyllaries loose, spreading, lance-linear, not gland-dotted, glabrous or puberulent, margins ± ciliate, tips attenuate; ray flowers 14–20, ligules 23–40 mm long; disk corolla lobes yellow; achenes 3–4 mm long, glabrate. Wooded stream bottoms, persistently wet areas; nc and ne TX, sc TX to Edwards Plateau; Panhandle; native to c U.S. from OH w to ND and s to TX; adventive in e and se U.S. and also WA; Canada. (Jul–)Aug–Oct(–Nov). [*H. instabilis* E.E. Watson] Hybrids with *H. maximiliani* are known as *H. ×intermedius* R.W. Long; hybrids with *H. mollis* are known as *H. ×brevifolius* E.E. Watson.

Helianthus hirsutus Raf., (hairy), HAIRY SUNFLOWER, STIFF-HAIR SUNFLOWER. Rhizomatous perennials; stems hirsute; leaves mostly opposite, short-petiolate, blades lanceolate to ovate, 3-nerved from the base, hirsute beneath, bases truncate or broadly rounded to cuneate, margins subentire to serrate, flat (not revolute); involucres 10–25 mm wide; phyllaries loose, spreading, lanceolate, not gland-dotted, margins ciliate, tips acute to short-acuminate; ray flowers 10–15, ligules 15–20 mm long; disk corolla lobes yellow; achenes 4–4.5 mm long, glabrate or puberulent. Sandy open woods, forest margins, roadsides; se and e TX; nc TX to Edwards Plateau; e 1/2 of the U.S.; Canada. Jun–Oct. [*H. hirsutus* Raf. var. *trachyphyllus* Torr. & A. Gray, *H. stenophyllus* (Torr. & A. Gray) E.E. Watson]

Helianthus maximiliani Schrad., (for its discoverer, Prince Maximilian von Wied-Neuwied (1782–1867), German botanist and traveler in Brazil and North America), MAXIMILIAN SUNFLOWER, MICHAELMAS DAISY. Rhizomatous perennials, sometimes forming extensive colonies; stems scabrous to minutely hispid; leaves mostly alternate, sessile or short-petiolate, blades lanceolate, 1-nerved and often folded lengthwise (sickle-shaped), light green to gray-green, scabrous to hispid, bases cuneate, margins usually entire or sometimes serrulate; involucres 13–28 mm wide; phyllaries lanceolate to linear-lanceolate, gland-dotted, gray-pubescent, margins ciliate, tips acute to attenuate; ray flowers 10–25, ligules (15–)25–40 mm long; disk corolla lobes yellow; achenes 3–4 mm long, glabrate. Low moist areas, roadsides, prairies; widespread in TX; native to c U.S. Great Plains; adventive farther e and w; Canada and Mexico. Sep–Nov. [*H. dalyi* Britton] Gaining popularity in horticulture despite its tendency to spread aggressively if given any encouragement. Hybrids with *H. grosseserratus* are known as *H. ×intermedius* R.W. Long.

Helianthus mollis Lam., (softly hairy), ASHY SUNFLOWER, HAIRY SUNFLOWER, DOWNY SUNFLOWER. Rhizomatous perennials; stems hirsute to villous; leaves mostly opposite (sometimes alternate among heads), sessile, blades lanceolate to broadly ovate, 3-nerved from above the base, grayish green, hispid to tomentose beneath, bases cordate-clasping, margins entire or serrulate; involucres 12–25 mm wide; phyllaries lanceolate, densely gland-dotted, densely hispid to tomentose, tips usually acute or sometimes acuminate; ray flowers 17–22, ligules 25–30 mm long; disk corolla lobes yellow; achenes 3.5–4 mm long, villous on the upper portion. Sandy soils, roadsides, prairies; e, se, and sc TX to nc TX; native to e and c U.S. from OH s to AL and w to WI, SD, and TX; adventive farther e; Canada. (Jun–)Jul–Sep(–Oct). [*H. mollis* Lam. var. *cordatus* S. Watson] Hybrids with *H. grosseserratus* are known as *H. ×brevifolius* E.E. Watson.

Helianthus occidentalis Riddell subsp. **plantagineus** (Torr. & A. Gray) Shinners, (sp.: western; subsp.: resembling *Plantago*—plantain), SHINNERS'S SUNFLOWER, WESTERN SUNFLOWER, FEW-LEAF SUNFLOWER. Rhizomatous perennials, sometimes stoloniferous as well; stems pilose to appressed-hairy, often reddish; leaves mostly basal, opposite, petiolate, blades oblong-lanceolate or elliptic to ovate, strigose or glabrous, bases cuneate, margins usually serrate; involucres 9–14 mm wide; phyllaries lanceolate, not gland-dotted, glabrate to sparsely pilose, margins usually ciliate, tips acute to short-acuminate; ray flowers 8–14, ligules 18–22 mm long; disk corolla lobes yellow; achenes 3–4 (–5) mm long, glabrate or sparsely villous. Dry open areas, fields; mostly sc TX, e to Harris Co. (Brown 2014), also Newton and Wood cos.; s TX; AR, LA, and TX. Jul–Oct. [*H. occidentalis* Riddell var. *plantagineus* Torr. & A. Gray] Potentially of conservation concern for its somewhat restricted range (Poole et al. 2010). ⚠

Helianthus pauciflorus Nutt. subsp. **pauciflorus,** (few-flowered), STIFF SUNFLOWER. Rhizomatous perennials; stems scabrous-hispid; leaves mostly basal, leaves opposite on lower stem, alternate above, sessile or short-petiolate, blades oblong-lanceolate to lance-ovate, sparsely hispid beneath, bases cuneate, margins serrate to subentire; involucres 15–23 mm wide; phyllaries ovate, glabrate to hispid, margins ciliate, tips acute; ray flowers 10–20, ligules 20–35 mm long; disk corolla lobes usually reddish, sometimes yellow; achenes 5–6 mm long, glabrate. Prairies, open woods, sandy soils; Hunt and Lamar cos. and Harris Co. (Brown 2014); nc TX and disjunct records on se TX Gulf Coast; native to c U.S. from MI s to AR and w to MT, SD, and TX; introduced and adventive in a few e states; Canada. Aug–Oct. [*H. laetiflorus* Pers. var. *rigidus* (Cass.) Fernald; *H. rigidus* (Cass.) Desf.] Hybrids with *H. tuberosus* are known as *H. ×laetiflorus* Pers., which is frequently cultivated.

Helianthus petiolaris Nutt. subsp. **petiolaris,** (with a petiole or leaf stalk), PLAINS SUNFLOWER, PRAIRIE SUNFLOWER. Annuals; stems hispid to strigose, rarely hirsute or glabrate; leaves mostly alternate, petiolate, blades lanceolate to deltate-ovate, often bluish green, strigose beneath, bases subcordate or truncate to cuneate, margins entire or serrate; involucres 10–24 mm wide; phyllaries lance-linear to lance-ovate, glabrate or minutely hispid to sparsely hirsute, margins sometimes ciliate, tips short-attenuate; ray flowers 10–30, ligules 15–20 mm long; disk corolla lobes usually reddish, rarely yellow; achenes 3–4.5 mm long, ± villous. Sandy or rocky prairies; Dallas and McLennan cos. and Denton Co. (Kartesz 2015); s and w to w TX; native to c U.S. Great Plains; adventive further e and w; Canada. May–Oct. [*H. couplandii* B. Boivin] Known to hybridize with *H. annuus* and *H. debilis*.

Helianthus praecox Engelm. & A. Gray, (early or precocious, for the flowering time), TEXAS SUNFLOWER. Annuals; stems hispid to hirsute; leaves mostly opposite, petiolate, blades deltate to ovate, sometimes constricted near middles, hirsute to hispid beneath, bases cordate to truncate or cuneate, margins serrate to serrulate; involucres 13–18 mm wide; phyllaries lanceolate, minutely hispid to hirsute, tips short-attenuate or short-acuminate; ray flowers 11–15, ligules 16–26 mm long; disk corolla lobes reddish; achenes 2.5–3.3 mm long, ± villous. Jun–Nov. Schilling (2006a) described the two subspecies potentially in East TX:

1. Branches horizontal; leaves sometimes constricted near middle; phyllary tips short-acuminate; middle teeth of pales puberulent; ray flowers 14–15; plants of the se TX Gulf Coast ____ subsp. **praecox**
1. Branches erect; leaves usually not constricted near middles; phyllary tips short-attenuate; middle teeth of pales white-villous-bearded; ray flowers 11–13(–14); plants of sc TX Gulf Coast and s TX ____ subsp. *runyonii*

Helenium flexuosum [GEO]

Helenium microcephalum [HEA]

Helianthus angustifolius [GWO]

Helianthus annuus [BT3]

Helianthus argopyllus [NIC]

Helianthus ciliaris [CO1]

Helianthus debilis subsp. cucumerifolius [MTB]

subsp. **praecox.** Deep sands, coastal dunes; Jefferson Co. (Kartesz 2015); se TX coast in Brazoria, Chambers, and Galveston cos. [*H. debilis* Nutt. subsp. *praecox* (Engelm. & A. Gray) Heiser] Potentially of conservation concern due to its restricted range (Schilling 2006a).

subsp. *runyonii* (Heiser) Heiser, (for Robert Runyon (1881–1968), photographer, botanist, and mayor of Brownsville, TX), RUNYON'S SUNFLOWER. Coastal prairies, sandy plains; *not yet recorded in East TX, but potentially in sc TX based on records from several cos. just sw of our area in Atascosa, Bee, and Refugio cos.*; s TX. [*H. debilis* Nutt. subsp. *runyonii* Heiser, *H. praecox* Engelm. & A. Gray var. *runyonii* (Heiser) B.L. Turner]

Helianthus salicifolius A. Dietr., (with leaves like *Salix*—willow), WILLOW-LEAF SUNFLOWER. Rhizomatous perennials; stems glabrous, glaucous; leaves alternate, sessile or subsessile, blades linear to lance-linear, glabrate beneath, bases ± attenuate, margins serrulate to subentire, flat (not revolute); involucres 10–18 mm wide; phyllaries linear to lance-linear, not gland-dotted, glabrous or glabrate, margins ciliate, tips long-attenuate; ray flowers 10–20, ligules 28–35 mm long; disk corolla lobes reddish; achenes 4–6 mm long, glabrous. Calcareous soils, prairies; Dallas, Grayson, and Hill cos.; nc TX; KS, MO, OK, and TX; introduced and adventive in some ne states. Aug–Sep. [*H. filiformis* Small] Sometimes cultivated and occasionally escaping.

Helianthus simulans E.E. Watson, (resembling), MUCK SUNFLOWER. Rhizomatous perennials; stems strigose to hirsute; leaves opposite or alternate, short-petiolate, blades linear-lanceolate to lanceolate, 3-nerved from above the base, minutely hispid-tomentose beneath, bases cuneate, margins entire or subentire, usually slightly revolute; involucres 13–18 mm wide; phyllaries loose, linear to lanceolate, gland-dotted, glabrate to sparsely strigose, tips acuminate; ray flowers 12–23, ligules 16–40 mm long; disk corolla lobes yellow or reddish; achenes 2.5–3 mm long, glabrous. Wet soils, ditches, roadsides; although Schilling (2006a) concedes only that this species "may be present" in TX, it was first reported in 1971 in Robertson Co. (Lonard & Waller 1971) and there are numerous recently published records from se TX: Harris Co. (Harcombe et al. 2007), Liberty Co. (Brown 2014), Polk Co. (Brown et al. 2006a), San Jacinto Co. (Kartesz 2015), and Walker Co. (Nesom & Brown 1998); native to AR, LA, and MS; adventive in se U.S. This species is becoming popular as a garden plant and this may increase the occurrence of hybrids with *H. angustifolius*.

Helianthus strumosus L., (with swellings, alluding to the rhizomes), ROUGH SUNFLOWER, PALE-LEAF WOODLAND SUNFLOWER. Rhizomatous perennials; stems glabrous or glabrate; leaves mostly opposite or the upper ones alternate, short-petiolate, blades lanceolate to lance-ovate or ovate, 3-nerved from above the base, glabrous or minutely tomentose beneath, bases subcordate to cuneate, margins entire or serrate; involucres 8–20 mm wide; phyllaries squarrose or erect, sometimes loose, lanceolate, usually not gland-dotted, glabrous to minutely hispid, tips acute to acuminate; ray flowers 10–20, ligules 12–20(–30) mm long; disk corolla lobes yellow; achenes 4–5.5 mm long, glabrate. Woods, roadsides, prairies; ne TX in Gregg, Harrison, Panola, and Upshur cos. (Kartesz 2015); e 1/2 of North America. Jul–Oct. [*H. montanus* E.E. Watson, *H. saxicola* Small] A variable species that may hybridize with *H. hirsutus* and *H. tuberosus*.

Helianthus tuberosus L., (tuberous), JERUSALEM-ARTICHOKE, GIRASOLE. Rhizomatous (producing tubers late in the season) perennials, producing tubers late in the season; stems scabrous-hispid to hirsute, sometimes glaucous; leaves mostly alternate, petiolate, petioles often ± winged, blades lanceolate to ovate, 3-nerved from near the base, puberulent to slightly hirsute-tomentose beneath, bases cuneate, margins entire or serrate, flat (not revolute); involucres 8–12 mm wide; phyllaries spreading or reflexed, lanceolate, gland-dotted, puberulent to minutely hispid, margins ciliate, tips acuminate; ray flowers 10–20, ligules 25–40 mm long; disk corolla lobes yellow; achenes 5–7 mm long, glabrous or hairy in the upper portion. Wooded areas, roadsides, fields; Dallas and Grayson cos. and Gregg Co. (Kartesz 2015); native to c U.S. Midwest and Great Plains; adventive farther e and w; Canada; cultivated and adventive in Europe. Aug–Oct. [*H. tomentosus* Michx., *H. tuberosus* L. var. *subcanescens* A. Gray] Cultivated for the edible, starchy-sweet tubers, which were used as a food by Native Americans (Mabberley 2017). Hybrids with *H. pauciflorus* are known as *H.* ×*laetiflorus* Pers.

Helianthus divaricatus L., (diverging), WOODLAND SUNFLOWER, which would key to *H. salicifolius* here, can be distinguished by its lanceolate to lance-ovate leaves that are wider (1–5 cm wide), yellow disk corolla lobes, and shorter pappus awns (2.2–2.5 mm long). To be expected in ne TX based on records from adjacent DeSoto Par. (VSC) and Caddo Par., LA; e U.S. from ME s to FL and w to WI, IA, and OK; Canada. Jul–Sep. [*H. divaricatus* var. *angustifolius* Kuntze]

Helianthus floridanus A. Gray ex Chapm., (of Florida), FLORIDA SUNFLOWER, which would key to *H. simulans* here, can be distinguished by its shorter leaves (4–15 cm long vs. 9–22 cm long) with undulate margins, and obtuse outer phyllary tips. To be expected in East TX based on records from adjacent Sabine Par., LA; se U.S. from NC s to FL and w to LA. Aug–Oct.

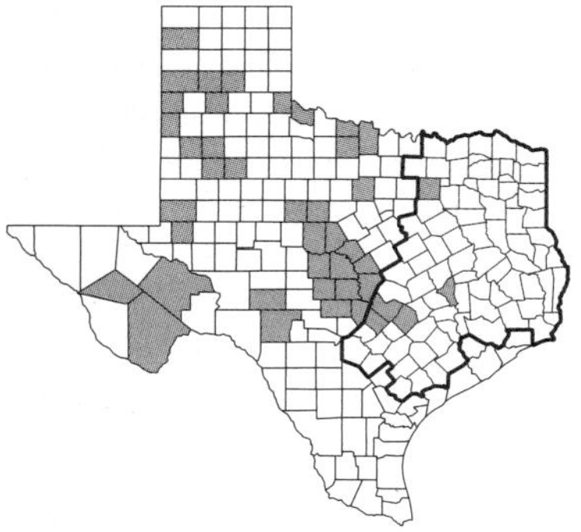
Helenium amarum var. badium

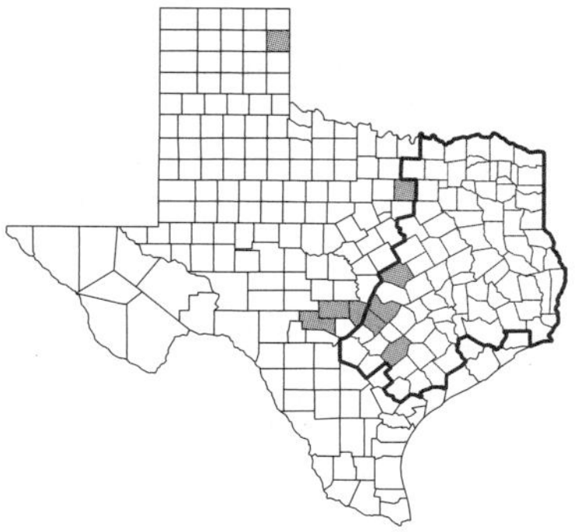
Helenium autumnale

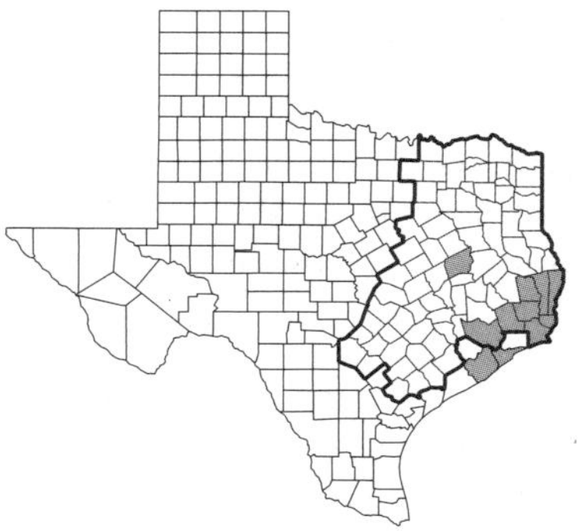
Helenium drummondii

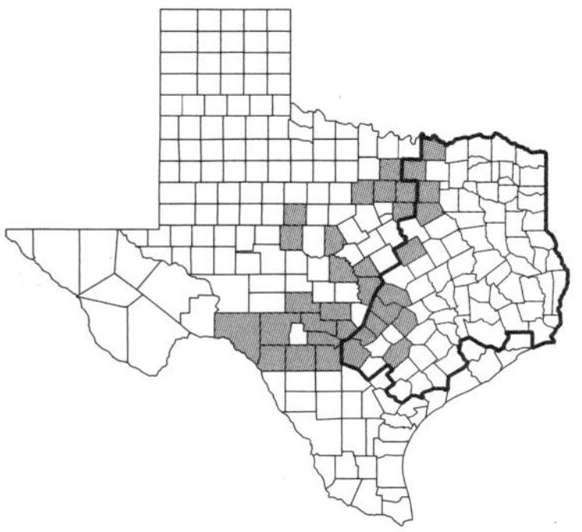
Helenium elegans var. elegans

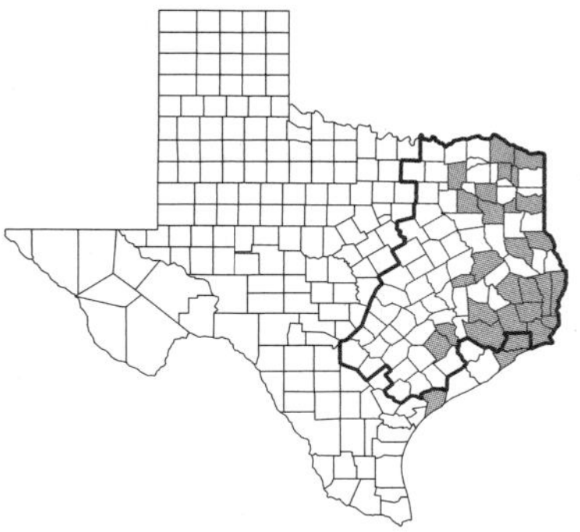
Helenium flexuosum

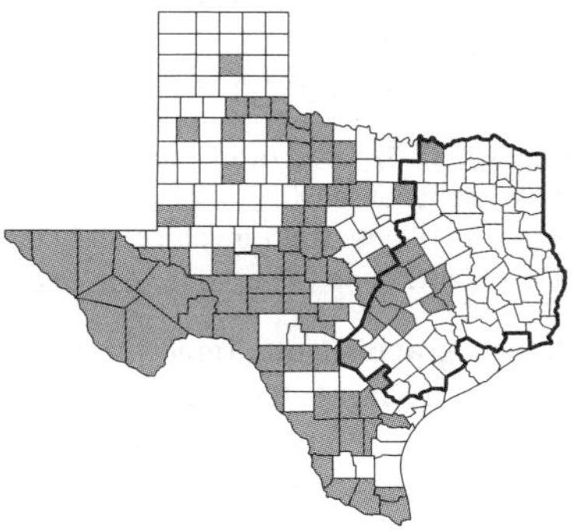
Helenium microcephalum var. microcephalum

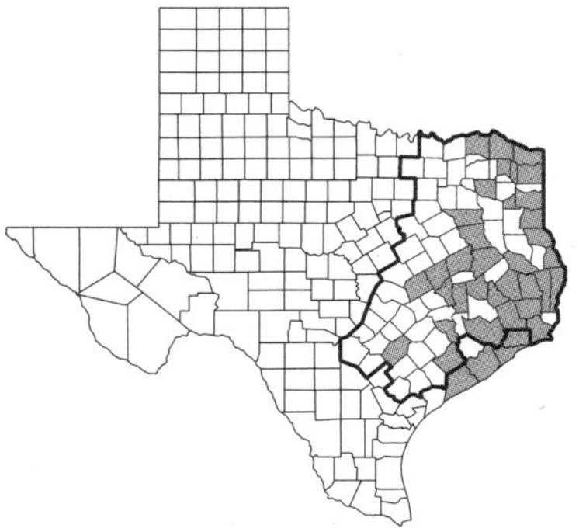
Helianthus angustifolius

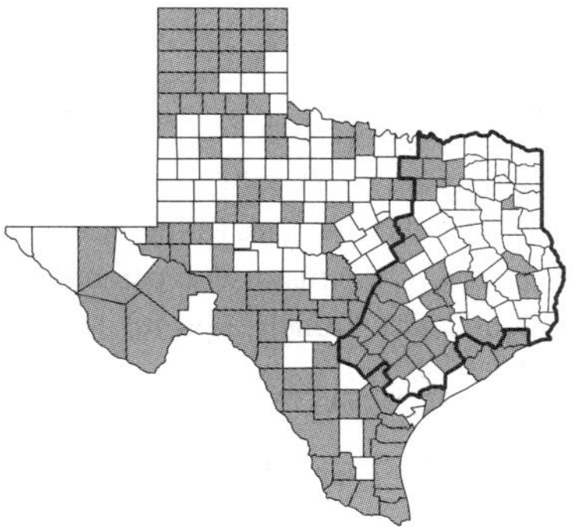
Helianthus annuus

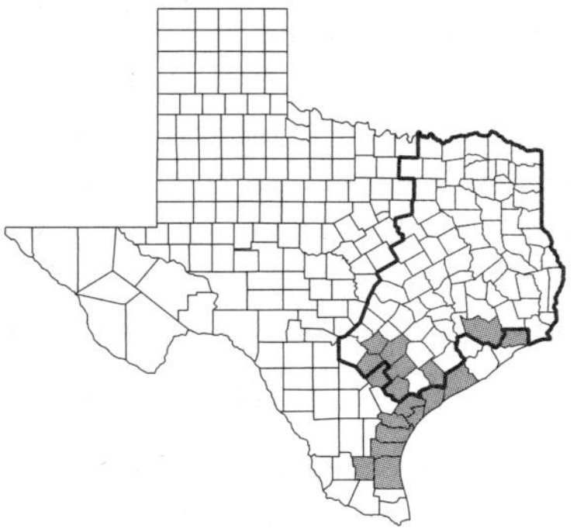
Helianthus argophyllus

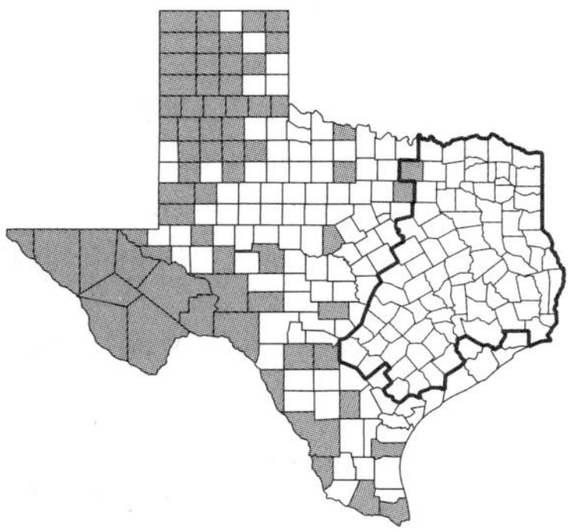
Helianthus ciliaris

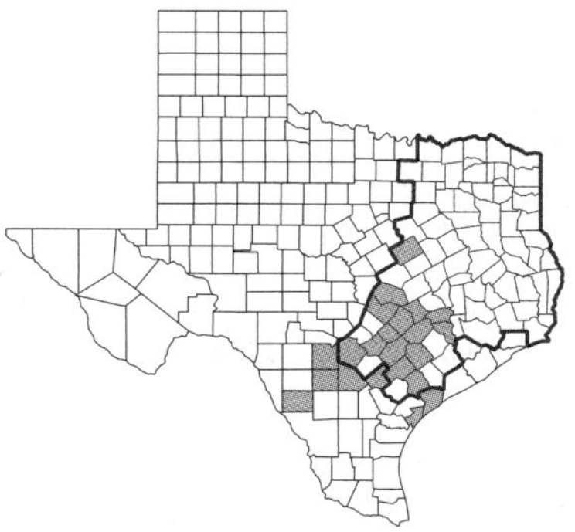
Helianthus debilis subsp. cucumerifolius

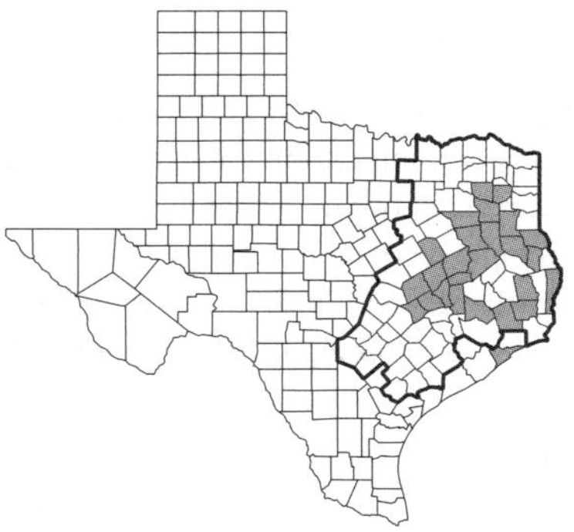
Helianthus debilis subsp. silvestris

Helianthus silphioides Nutt., (resembling *Silphium*—rosinweed or compass-plant), ROSINWEED SUNFLOWER, OZARK SUNFLOWER, which would key to *H. occidentalis* or *H. pauciflorus* here, can be distinguished from the former by its leaves that are not gland-dotted beneath and longer pales (9–10 mm long vs. 5–7 mm long); it can be distinguished from the latter by its shorter achenes (3–4 mm long vs. 5–6 mm long) and shorter pappus awns (2.5–2.7 mm long vs. 4.5–5 mm long). To be expected in ne TX based on records from adjacent McCurtain Co., OK (OKL); se U.S. from KY s to AL and w to IL, OK, and LA. Aug–Oct. [*H. kentuckiensis* McFarland & W.A. Anderson]

HELIOPSIS Pers. OXEYE, SUNFLOWER EVERLASTING

Ours perennial herbs; leaves opposite, petiolate, ± 3-nerved; heads radiate, borne singly; phyllaries in 2–3 series, subequal, the outer often longer and more foliaceous; ray flowers pistillate, fertile; corollas yellow to orange, the ligules sessile and persistent on the achenes, becoming papery; disk flowers perfect, fertile, corollas yellow to brownish purple; achenes 3–4-angled; pappus absent, or a crown plus 1–3 tooth-like scales.

A genus of ca. 18 species of North America, Mexico, and Central and South America. (Greek: *helios*, sun, and *-opsis*, likeness) (tribe Heliantheae)

REFERENCES: Fisher 1957, 1958; Turner 1988a; Smith, A. 2006a.

1. Heads 1–3(–5) per stem, involucres 7–15 mm in diameter; ray flowers 6–8(–13), ligules 10–22 mm long and 4–8 mm wide; plants of se TX ____ **H. gracilis**
1. Heads 1–15+ per stem, involucres 12–25 mm in diameter; ray flowers 10–18, ligules 20–40 mm long and 6–13 mm wide; plants of ne TX ____ **H. helianthoides**

Heliopsis gracilis Nutt., (graceful, slender), SMOOTH OXEYE, PINEWOODS OXEYE. Plants 30–40(–80) cm tall; stems glabrous, from creeping rhizomes 4–10+ cm long and 1–2 mm wide; leaf blades glabrous or nearly so, ovate-lanceolate to lanceolate, margins regularly dentate to nearly entire; heads 1–3(–5) per stem, involucres 7–15 mm in diam.; ray flowers 6–8(–13), corollas golden yellow, ligules 10–22 mm long and 4–8 mm wide; disk flowers pale brown-yellow; achenes 4–5 mm long; pappus a crown plus 1–3 tooth-like scales. Sandy slopes, open pine woods; Harris Co. (Brown 2014); Chambers Co.; se U.S. from SC s to FL and w to TX. Spring-summer. [*H. helianthoides* (L.) Sweet var. *gracilis* (Nutt.) Gandhi & R.D. Thomas, *H. laevis* Pers. var. *minor* Hook.]

Heliopsis helianthoides (L.) Sweet var. **scabra** (Dunal) Fernald, (resembling a sunflower, *Helianthus*), FALSE SUNFLOWER, ROUGH OXEYE. Plants (40–)80–150 cm tall; stems glabrous or hairy, from creeping rhizomes to 4 cm long and 2–6 mm wide; leaf blades ± scabrous, deltate to narrowly ovate-lanceolate, margins coarsely dentate; heads 1–15+ per stem, involucres 12–25 mm in diam.; ray flowers 10–18, corollas golden yellow, ligules 20–40 mm long and 6–13 mm wide; disk flowers yellow to brownish yellow; achenes 4–5 mm long; pappus absent, or of 2–4 minute, tooth-like scales. Sandy soils, open woods, prairies, roadbanks; ne TX in Bowie, Harrison, Morris, Red River, Smith, and Titus cos.; apparently reports from Harris Co. were in error (Brown 2014); c U.S. from MI s to LA and w to ND, CO, and NM; Canada. Apr–Jul. [*H. scabra* Dunal]

HETEROTHECA Cass. GOLDENASTER, GOLD-ASTER, CAMPHORWEED

Annual or perennial herbs, often aromatic, strigose to hispid or pilose; leaves alternate, simple; heads radiate, solitary or in corymbose or paniculate arrays; phyllaries in 3–5 series, usually strongly unequal, 1-nerved, lanceolate to linear-lanceolate; ray flowers pistillate, fertile, ligules yellow, curling under in age; disk flowers perfect, fertile, corollas yellow; achenes sometimes dimorphic; pappus of an outer series of inconspicuous, short, narrow scales, and an inner series of long, barbellate bristles, or pappus absent from ray flowers.

A w North American genus of ca. 30 species; some taxa have been treated under *Chrysopsis* and *Pityopsis*. *Heterotheca* is taxonomically complex, fraught with hybridization, polyploidy, and environmentally-induced plasticity, and subsequently has been often-revised; our understanding of this group is still in flux. Here we attempt to reconcile differing opinions of circumscription by Semple (2006d) and Nesom (2006b, 2007a). (Greek: *heteros*, different, and *thece*, container, alluding to the dissimilar ray and disk achenes—this translates well to the modern difficulty in circumscription) (tribe Astereae)

REFERENCES: Shinners 1951b; Wagenknecht 1960; Harms 1965, 1968; Vernon 1965; Semple et al. 1980; Nesom 1991c, 2006b, 2007a; Semple 1996, 2006d.

1. Plants annual or biennial from taproots; usually with one or a few stems from base; blades of upper and middle stem leaves not much narrowed at base, sessile and slightly clasping; pappus of ray flowers absent ____ **H. subaxillaris**
1. Plants perennial from tough woody bases, often rhizomatous; stems usually several–many; blades of upper and middle stem leaves tapered to slender, petiole-like bases, not clasping; pappus of ray flowers present.

Helianthus grosseserratus [GLE, MTB]

Helianthus hirsutus [MO4]

Helianthus maximiliani [STE]

Helianthus mollis [MTB]

Helianthus occidentalis [YAT]

Helianthus pauciflorus [HOL]

Helianthus petiolaris [MTB]

Helianthus praecox subsp. praecox [THO]

2. Leaf blades conspicuously silvery-gray in appearance, softly and densely pubescent, the hairs usually without swollen bases; leaves lacking resin glands ____ **H. canescens**
2. Leaf blades distinctly green, bristly pubescent, the hairs usually with swollen bases (use lens); foliage with abundant resin glands (use lens) ____ **H. stenophylla**

Heterotheca canescens (DC.) Shinners, (grayish white), HOARY GOLDENASTER, GRAY GOLD-ASTER. Perennials, 15–40(–65) cm tall, taprooted, frequently rhizomatous; stems eglandular; leaves rather numerous and crowded, ascending, petiolate below and sessile above, linear-oblanceolate, densely strigose-canescent (appearing silvery-gray), eglandular, margins entire and ciliate; uppermost leaves not much reduced; phyllaries moderately strigose, eglandular; ray flower ligules 5–9(–10.5) mm long; achenes 1.2–3.1 mm long, 6–10-ribbed; pappus monomorphic. Sandy or gravelly prairies, rock outcrops, roadsides; w and sw border of East TX, from Dallas Co. s to Bexar Co. and e to DeWitt Co.; mostly w 2/3 TX; sc U.S. in KS, NM, OK, and TX. Jul–Oct. [*Chrysopsis berlandieri* Greene, *C. canescens* (DC.) Torr. & A. Gray, *C. villosa* (Pursh) Nutt. ex DC. var. *canescens* (DC.) A. Gray]

Heterotheca stenophylla (A. Gray) Shinners, (narrow-leaved), STIFF-LEAF GOLDENASTER, NARROW-LEAF GOLD-ASTER. Perennials, 18–46(–65) cm tall, taprooted, sometimes rhizomatous; stems glandular; leaves rather numerous and crowded, ascending, petiolate below and sessile above, linear-oblanceolate to narrowly oblanceolate, sparsely hispid-strigose (appearing bright green to dark green), with glands, margins entire and ciliate; uppermost leaves not much reduced; phyllaries usually sparsely strigose, eglandular or with glands; ray flower ligules (5–)7–11.5(–16.5) mm long; achenes usually 1.5–2.5 mm long, 6–10-ribbed; pappus monomorphic. Sandy prairies, rocky hills; w edge of East TX in McLennan, Travis, and Hays cos.; mostly c TX and Panhandle; c U.S. from IA and SD s to TX and NM. Jun–Oct. [*Chrysopsis hispida* (Hook.) DC. var. *stenophylla* A. Gray, *C. scabrifolia* A. Nelson, *C. stenophylla* (A. Gray) Greene, *C. villosa* (Pursh) Nutt. ex DC. var. *stenophylla* (A. Gray) A. Gray] We are treating this taxon at the species level only following Nesom (2006b, 2007a); according to Semple (2006d) this would be *H. stenophylla* (A. Gray) Shinners var. *stenophylla*.

Heterotheca subaxillaris (Lam.) Britton & Rusby, (subaxillary), CAMPHORWEED, GOLDEN-ASTER, CAMPHOR DAISY. Aromatic annuals or biennials, 10–200 cm tall, taprooted; stems glandular; leaves ovate-lanceolate to elliptic, hispid-strigose, glandular, margins coarsely serrate to entire; lower leaves petiolate, petiole bases auriculate-clasping, usually withering by flowering; upper leaves reduced, sessile, becoming cordate and subclasping; phyllaries sometimes sparsely strigose-ciliate, glandular; ray flower ligules 3–7(–9) mm long; achenes dimorphic, 2–3-ribbed, ray achenes triangular in cross-section, 1.5–2.5 mm long, disk achenes flattened, 2–4 mm long; pappus of ray flowers absent. Sandy or rocky ground, open woods, fields, roadsides; common throughout TX; s 1/2 of the U.S; Mexico. Apr–Dec(–Feb). [*H. latifolia* Buckley, *H. subaxillaris* var. *latifolia* (Buckley) Gandhi & R.D. Thomas] If subspecies are recognized, ours would be subsp. *latifolia*. The foliage has a strong camphor-like odor and the plants are avoided by livestock (Ajilvsgi 1984).

HIERACIUM L. HAWKWEED

Ours perennial herbs, strikingly pilose in the field; sap milky; leaves alternate, in basal rosettes and on lower stem, much reduced or absent on upper stem; heads ligulate, in corymbose or paniculate arrays; phyllaries in 2+ series, subequal to unequal, lanceolate to linear, reflexed in fruit; disk flowers absent; ray flowers perfect, fertile; corollas yellow, ligules mostly 4–5-cleft at the tips; achenes cylindrical, usually 10-ribbed; pappus of many minutely-barbed, straw-colored bristles.

A taxonomically complicated temperate and tropical montane genus of 250+ species (with possibly ca. 1000 asexually-reproducing microspecies), most speciose in Eurasia. Our species are sometimes treated in the genus *Pilosella*. (Greek: *herax*, a hawk; the ancients, as recorded by Pliny and others, supposing that hawks ate these or similar plants to improve their eyesight) (tribe Cichorieae)

REFERENCES: Fernald 1943, 1950; Vuilleumier 1973; Beaman 1990; Strother 2006o.

1. Midstem hairs mostly < 6 mm long; midstem leaves with lengths 2–4+ times widths; corollas 8–9+ mm long ____ **H. gronovii**
1. Midstem hairs 6–15+ mm long; midstem leaves with lengths 4–7+ times widths; corollas ca. 7 mm long ____ **H. longipilum**

Hieracium gronovii L., (for Jan Friedriech Gronovius 1690–1762, Dutch botanist at Leiden), GRONOVIUS' HAWKWEED, QUEEN-DEVIL. Plants 30–45(–80+) cm tall; lower stems pilose-hirsute and sometimes also stellate-pubescent; leaves oblanceolate to obovate or elliptic, pilose with long tawny hairs; heads 7–10 mm high; phyllaries glabrous or stellate-pubescent, sometimes with stalked glands; corollas 8–9+ mm long; achenes 3.5–4.5+ mm long; pappus bristles ca. 5 mm long. Sandy soils, openings in pine-oak woods, bogs; e 1/3 TX; e U.S. from MA s to FL and w to MI, KS, and TX; Canada. May–Jun(–fall).

Hieracium longipilum Torr., (long-haired), HAIRY HAWKWEED. Plants 30–75(–200) cm tall; lower stems pilose-hirsute and sometimes also stellate-pubescent; leaves oblanceolate, pilose with long tawny hairs; heads 6–8(–10) mm high; phyllaries stellate-

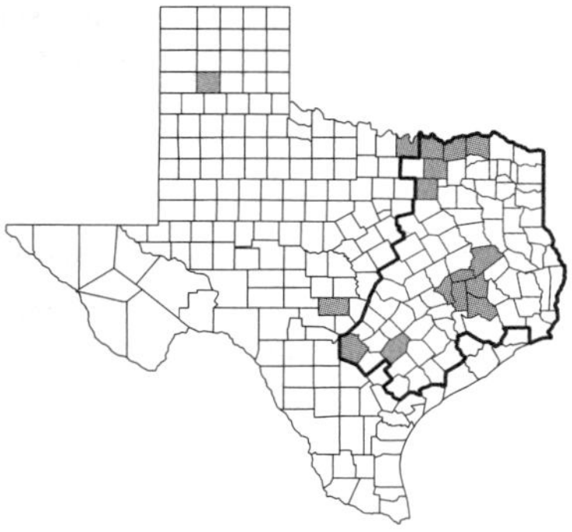

Helianthus grosseserratus

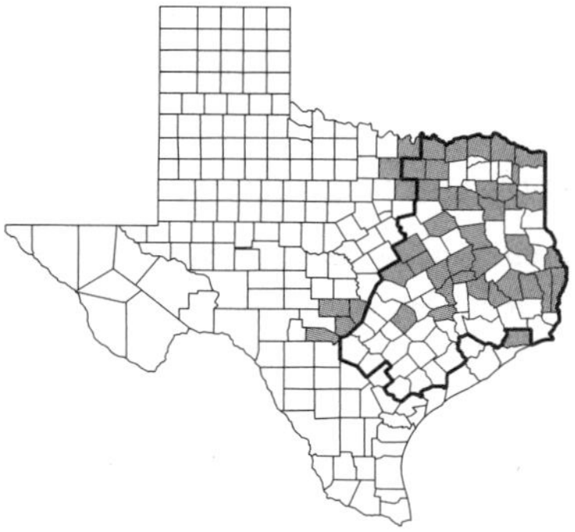

Helianthus hirsutus

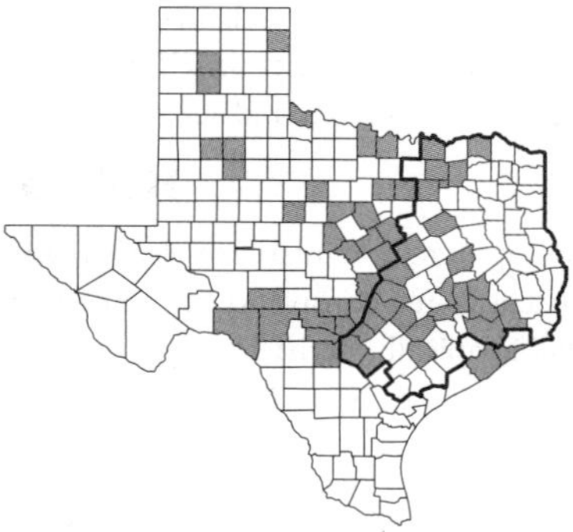

Helianthus maximiliani

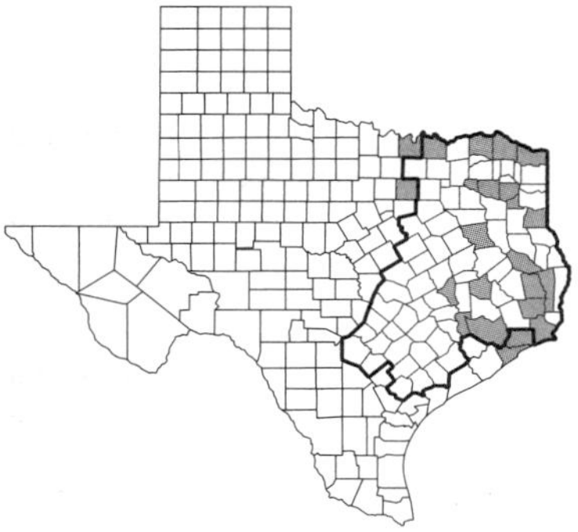

Helianthus mollis

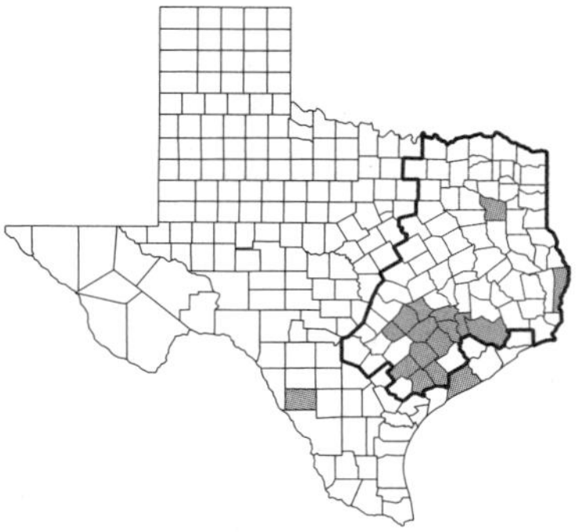

Helianthus occidentalis subsp. plantagineus

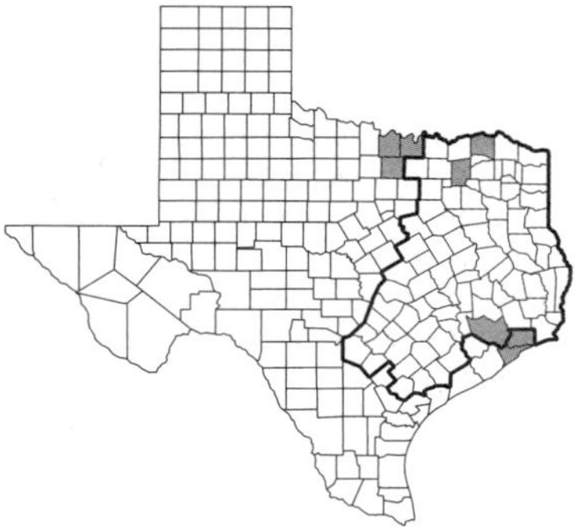

Helianthus pauciflorus subsp. pauciflorus

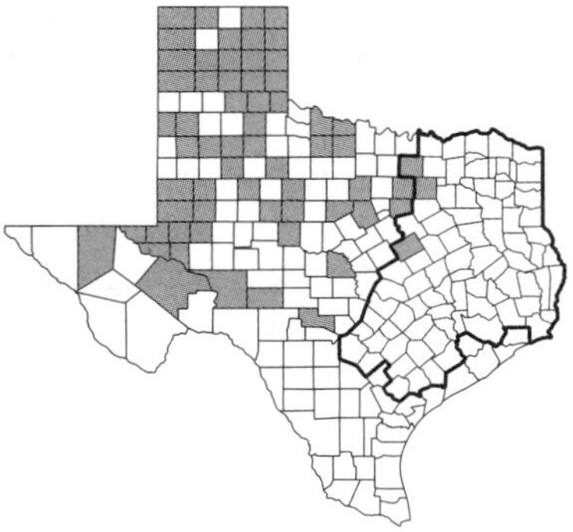

Helianthus petiolaris subsp. petiolaris

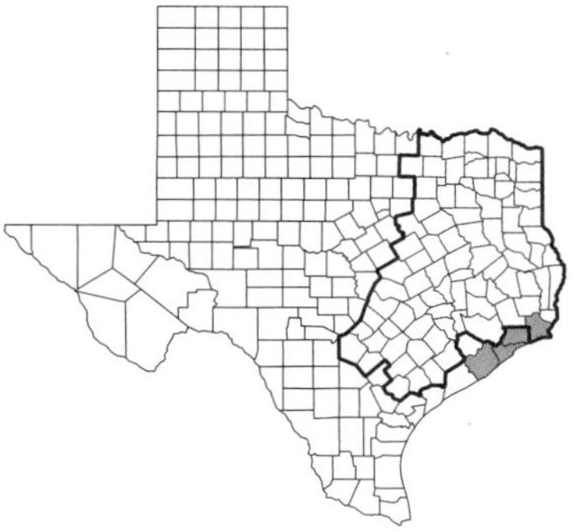

Helianthus praecox subsp. praecox

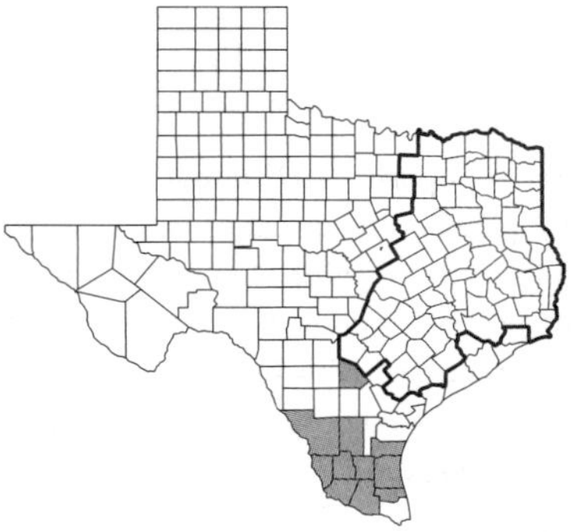

Helianthus praecox subsp. runyonii

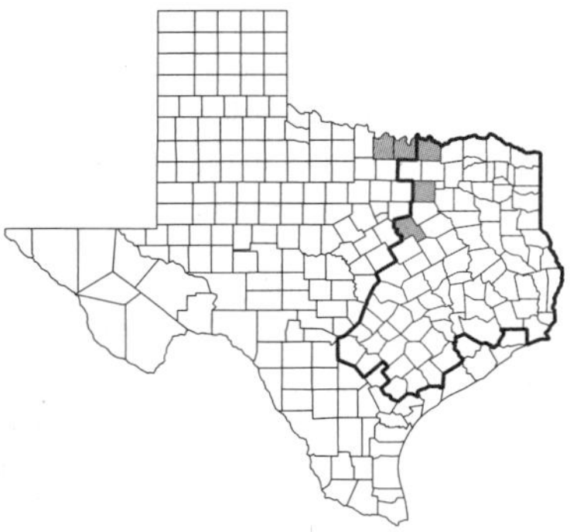

Helianthus salicifolius

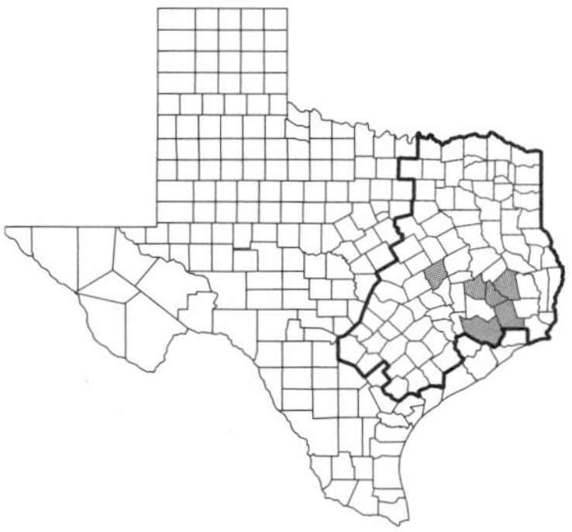

Helianthus simulans

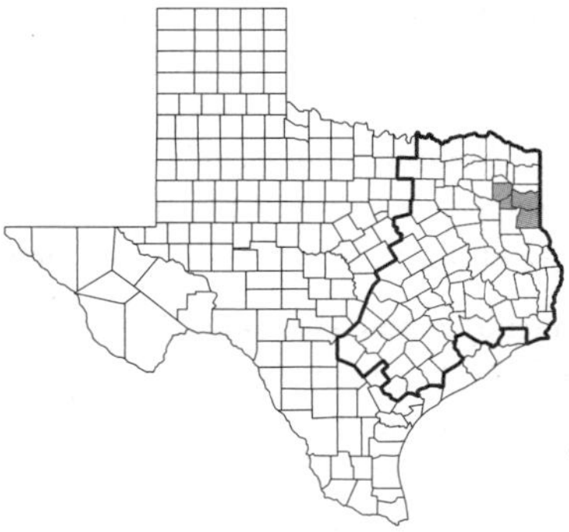

Helianthus strumosus

pubescent, with stalked glands; corollas ca. 7 mm long; achenes 3–4+ mm long; pappus bristles 5.5–6.5 mm long. Sandy soils, prairies, fields, roadsides; in TX recorded only from San Jacinto Co. and Lamar Co. (*W.R. Carr 13133* TEX); ec U.S. from MI s to LA and w to MN, NE, and TX; Canada. Sep–Oct.

HYMENOPAPPUS L'Hér. WOOLLY-WHITE, OLD-PLAINSMAN

Ours taprooted biennial herbs; leaves basal (rosette) and cauline, alternate, petiolate, the blades often pinnatifid, usually glandular-punctate, often conspicuously white-hairy on lower surface; heads in corymbose arrays; phyllaries in 2 or 3 series, subequal, petaloid apically, herbaceous basally; ray flowers (in our species) absent; disk flowers perfect, fertile; corollas whitish or purplish, the lobes becoming reflexed; anther column long-exserted; achenes obpyramidal, 4(–5)-angled, hirtellous to villous; pappus of 14–20 ovate to spatulate scales.

A s North American (including Mexico) genus of 12 species; some were used medicinally by Native Americans. (Greek: *hymen*, membrane, and *pappus*, down, fuzz, pappus) (tribe Heliantheae)

REFERENCES: Turner 1956; Strother 2006p.

1. Basal leaves twice-pinnate, with ultimate segments to 1.5 mm wide, elongate and linear to filiform.
   2. Stems equally leafy throughout, the leaves not much reduced upwards; corollas 3–4 mm long (to tips of corolla lobes), the throats ± funnelform ____ **H. carrizoanus**
   2. Stems mostly leafy below, the leaves much reduced upwards; corollas 2.5–3 mm long (to tips of corolla lobes), the throats ± campanulate ____ **H. tenuifolius**
1. Basal leaves entire to once- or twice-pinnate, with ultimate segments mostly 2–30 mm wide, variously shaped, but wider than linear to filiform.
   3. Basal leaves simple to coarsely once-pinnate (rarely bipinnate); corollas whitish to pinkish or purplish (rarely completely white); usually on sandy soils ____ **H. artemisiifolius**
   3. Basal leaves finely once- to twice-pinnate; corollas white to cream; usually on clay or limestone soils ____ **H. scabiosaeus**

Hymenopappus artemisiifolius DC. var. **artemisiifolius**, (with leaves like *Artemisia*—sagebrush or wormwood), RAGWEED WOOLLY-WHITE, WOOLLY-WHITE, WILD CAULIFLOWER. Plants 40–100 cm tall; leaves entire to once-pinnate, densely white-tomentose beneath; phyllaries 5–12 mm long, 2–7 mm wide, white or red-tinged; corollas whitish to pinkish or purplish, 3.5–5 mm long; achenes 3.5–4 mm long; pappus scales 0.3–1(–1.5) mm long. Sandy soils, open woods, fields, disturbed areas; common across East TX, w to nc TX and s to Edwards Plateau and South TX Plains; AR, LA, and TX. Mar–May. [*H. scabiosaeus* L'Hér. var. *artemisiifolius* (DC.) Gandhi & R.D. Thomas]

Hymenopappus carrizoanus B.L. Turner, (of the Carrizo Sands), CARRIZO SANDS WOOLLY-WHITE. Plants 45–150 cm tall; leaves twice-pinnate, the divisions linear, nearly glabrous beneath; phyllaries (3.5–)5–6 mm long, (1.5–)2–5 mm wide, gray-green; corollas white (drying pinkish), 3–4 mm long; achenes ca. 4 mm long; pappus scales 1–1.5 mm long. Deep sands, grasslands, open post oak woodlands; in East TX restricted to a line of counties corresponding to the Carrizo Sands formation: Anderson, Bastrop, Bexar, Caldwell, Guadalupe, Leon, and Robertson; several more TX counties to the sw. Apr–Jun. Lacking state or federal status, but a species of conservation concern because of its restricted range and habitat loss, as discussed in Poole et al. (2007). ⚠

Hymenopappus scabiosaeus L'Hér. var. **corymbosus** (Torr. & A. Gray) B.L. Turner, (sp.: from Latin: *scabies*, itch; var.: with flowers in corymbs), OLD-PLAINSMAN. Plants 40–150 cm tall; leaves mostly twice-pinnate, densely white-tomentose beneath; phyllaries 5–9 mm long, 2–4 mm wide, white to yellowish; corollas white to cream, 3–5.5 mm long; achenes 3–5 mm long; pappus scales 0.1–1 mm long. Calcareous soils and clays, prairies, open woods, roadsides; widespread from nc to sc TX, less common far e and w TX; sc U.S. from NE s to TX; Mexico. Mar–Jul. [*H. corymbosus* Torr. & A. Gray] The other variety of this species, var. *scabiosaeus*, is not recorded from TX but may be expected based on records from adjacent counties in OK (Choctaw and McCurtain cos., OK (OKL)). It differs from var. *corymbosus* in having peduncles subtended by membranous bracts 5–14 mm long and larger phyllaries (7–15 mm long and 4–8 mm wide).

Hymenopappus tenuifolius Pursh, (slender-leaved), CHALKHILL WOOLLY-WHITE, WOOLLY-WHITE, OLD-PLAINSMAN. Plants 40–150 cm tall; leaves twice-pinnate, the divisions linear, densely white-tomentose beneath; phyllaries 5–8 mm long, 2–4 mm wide, white to yellowish; corollas white to cream, 2.5–3 mm long; achenes 3.5–4.5 mm long; pappus scales 1–1.5(–2) mm long. Gravelly prairies, limestone outcrops, sandy soils; w margin of East TX from Lamar s to Travis and Hays cos.; mostly c TX and Panhandle; SD s to TX and w to WY and NM. Apr–Aug.

Helianthus salicifolius [MTB]

Helianthus simulans [MTB]

Helianthus strumosus [MO4]

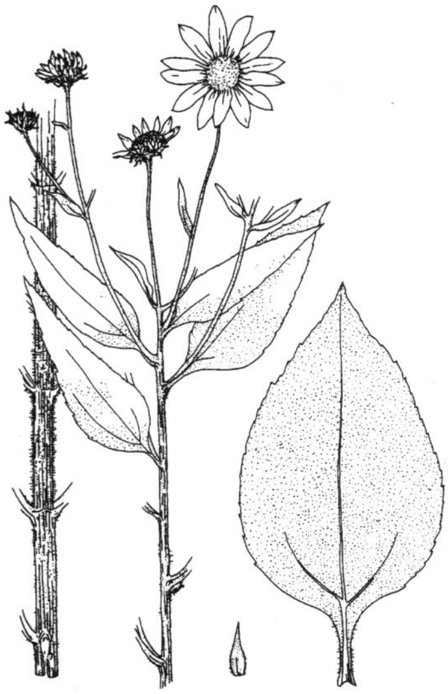

Helianthus tuberosus [MTB]

Heliopsis gracilis [LEI]

Heliopsis helianthoides var. scabra [FNA]

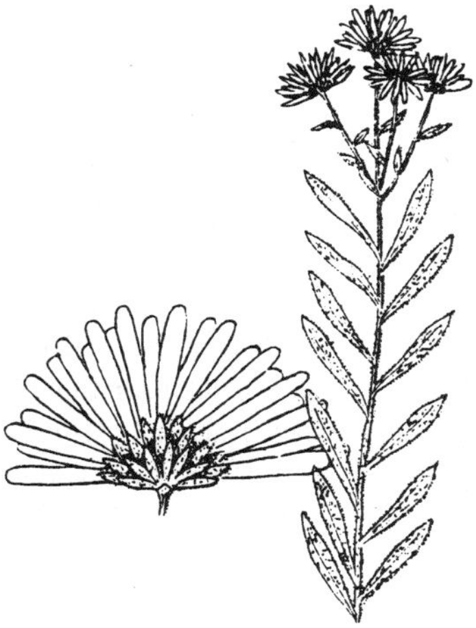

Heterotheca canescens [STE]

Heterotheca stenophylla [UWA]

Heterotheca subaxillaris [BB2]

HYMENOXYS Cass. PRAIRIE DAWN, BITTERWEED, RUBBERWEED

Ours annual herbs; leaves alternate, ± gland-dotted; heads solitary or borne in loose paniculate arrays; phyllaries in 2 series, unequal or subequal; ray flowers pistillate, fertile, corollas usually yellow; disk flowers perfect, fertile; corollas yellow to yellow-brown; achenes obpyramidal; pappus of scales, usually awn-tipped.

A genus of ca. 26 species depending on circumscription, native from s North America to Argentina; sometimes treated broadly to include *Tetraneuris*. A molecular study by Bierner and Jansen (1998) supported the recognition of *Tetraneuris* as a separate genus. Some species contain sesquiterpene lactones toxic to sheep and goats; other species were formerly used as chewing gum by Native Americans. ⚠ Two of the species treated here are of conservation concern. (Greek: *hymeno*, membrane, and *oxys*, sharp, alluding to the awned pappus scales) (tribe Heliantheae)

REFERENCES: Mahler 1983; Bierner & Jansen 1998; Bierner 2006c; Mink et al. 2012; Singhurst et al. 2014.

1. Robust plants 10–80 cm tall; ray flowers 8–13, conspicuous and extending well beyond the phyllaries, corollas 8.5–11 mm long ____ **H. odorata**
1. Delicate plants 5–10(–15) cm tall; ray flowers 6–8, inconspicuous and usually not much surpassing the phyllaries, corollas 2–3 mm long.
   2. Basal rosettes 1.3–2.4 cm in diameter, withering at or before time of flowering; disk flowers 17–32; rare endemic of ne TX ____ **H. perpygmaea**
   2. Basal rosettes 4–7+ cm in diameter, persistent, usually present at time of flowering; disk flowers 40–75+; rare endemic of se TX ____ **H. texana**

Hymenoxys odorata DC., (fragrant), POISON BITTERWEED, WESTERN BITTERWEED, BITTER RUBBERWEED. Robust taprooted annuals, 10–80 cm tall, inconspicuously pubescent, becoming bushy-branched when not crowded, forming rounded mounds; foliage aromatic when crushed; leaves simple or pinnately divided into 3–19 linear-filiform segments, minutely gland-dotted; heads terminal, solitary or corymbose; ray flowers 8–13, the ligules conspicuous, golden yellow, 8.5–11 mm long, extending well beyond the phyllaries, widened and toothed apically; disk flowers 50–150+, corollas yellow-orange; achenes 1.7–2.5 mm long; pappus of 5–6 obovate, awn-tipped scales 1.6–2.3 mm long. Disturbed sites, roadsides, ditches, becoming abundant in overgrazed areas; Hill Co., also possibly Burleson Co. (Turner et al. 2003); mostly w 1/2 TX; sw U.S. from KS s to TX and w to CA; Mexico. Mar–Jul(–Sep). Poisonous to livestock, especially to sheep under starvation conditions; the sesquiterpenes cause gastrointestinal irritation resulting in a wasting condition after short periods of grazing; POISON BITTERWEED causes considerable loss of livestock and has been one of the main causes of the decline of sheep ranching in the sw U.S. (Kingsbury 1964; James & Welsh 1992).

Hymenoxys perpygmaea W.C. Holmes, Singhurst, & Mink, (very small), PYGMY PRAIRIE DAWN. Delicate, ephemeral, taprooted annuals, 5–6.9 cm tall, glabrous; basal rosette leaves elliptic to obovate, withering at or before time of flowering, the rosettes 1.3–2.4 cm in diameter; stem leaves simple, margin entire, sometimes irregularly toothed, minutely gland-dotted, margins entire; heads terminal, solitary, 1–4 per plant; ray flowers ca. 7, the ligules inconspicuous, yellowish, not much surpassing the phyllaries, ca. 2.8 mm long, widened and toothed apically; disk flowers 17–32, corollas yellow; achenes ca. 2 mm long; pappus of ovate, awn-tipped scales ca. 2.5 mm long. Saline prairie barrens in the presence of mima mounds (pimple mounds); an extremely rare endemic apparently restricted to Lamar Co. Apr–May. This species, described in 2012, currently lacks state or federal protection, but should be considered critically imperiled due to its extremely restricted range and highly specialized habitat (Mink et al. 2012). ⚠

Hymenoxys texana (J.M. Coult. & Rose) Cockerell, (of Texas), PRAIRIE DAWN, TEXAS PRAIRIE DAWN. Delicate taprooted annuals, 5–10(–15) cm tall, glabrous or inconspicuously pubescent; leaves in a persistent basal rosette 4–7+ cm in diameter, blades obovate or spatulate, present at time of flowering; stem leaves simple or 3–5-lobed, minutely gland-dotted, margins entire or pinnately toothed; heads terminal, solitary or corymbose; ray flowers 6–8, the ligules inconspicuous, whitish or pale yellow, not much surpassing the phyllaries, 2–3 mm long, widened and toothed apically; disk flowers 40–75+, corollas pale yellow to deep yellow; achenes 1.5–1.8 mm long; pappus of 5 obovate, awn-tipped scales 1.4–1.8 mm long. Saline prairie barrens, often in the presence of mima mounds (pimple mounds); a rare endemic restricted to the TX Gulf Coastal Plain, known from only 3 counties: Harris and Trinity cos.; Fort Bend Co. to the se of East TX in the Gulf Prairies and Marshes. Mar–Apr. [*Actinella texana* J.M. Coult. & Rose] This species is listed as endangered at the state and federal level due to its restricted range and highly specialized habitat, both threatened by urban development around Houston (Poole et al. 2007; Singhurst et al. 2014). ⚠

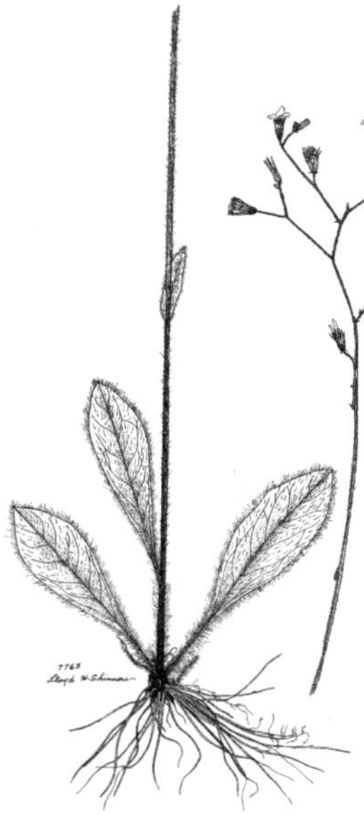

Hieracium gronovii [BT3]

Hieracium longipilum [YAT]

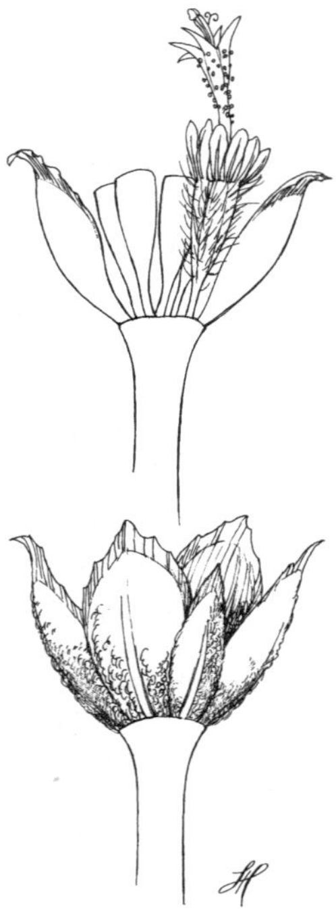

Hymenopappus carrizoanus [HEA]

Hymenopappus artemisiifolius [HEA]

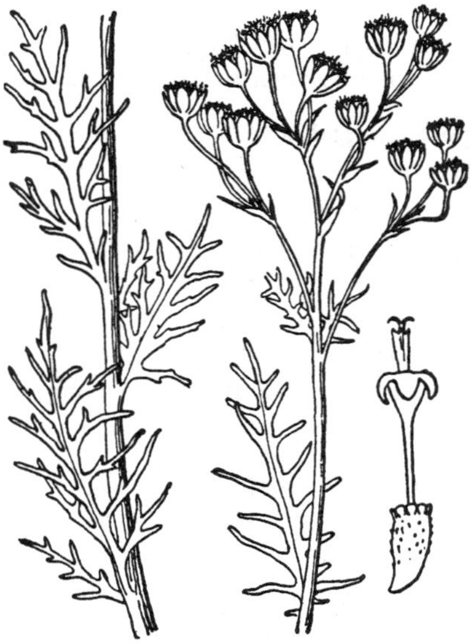

Hymenopappus scabiosus var. corymbosus [BB2]

HYPOCHAERIS L. CAT'S-EAR

Annual, biennial, or perennial herbs with milky sap; leaves mostly basal or flowering stems with a few ± well-developed leaves, at least toward the base; leaves alternate, entire to dentate or pinnatifid; heads borne singly or in loose, few-flowered arrays; phyllaries in 3–4 series, unequal, linear-lanceolate; flowers all ligulate, corollas yellow or in one species white; achenes cylindrical or spindle-shaped, ribbed, all alike and beaked, or of two kinds, the peripheral achenes beakless and the central ones beaked; pappus of 1 or 2 series of white or tan bristles, the single or inner series plumose, the outer, when present, shorter and barbellate.

A genus of ca. 60+ species native to Europe, Asia, n Africa, and especially South America; including some weeds and cultivated ornamentals. The generic name has sometimes been spelled *Hypochoeris* (e.g., Shinners 1966; Tomb 1974; DeFilipps 1976), probably based on the spelling by Linnaeus (1754) in *Genera Plantarum*. However, the accepted spelling is *Hypochaeris* (Greuter et al. 1993), based on Linnaeus' (1753) original spelling in *Species Plantarum*. (According to Quattrocchi (2000), Greek: *hypochoiris*, used by Theophrastus for some plant, and also mentioned by Pliny as *hypochoeris*, perhaps derived from *hypo*, beneath or under, and *choiros*, pig, the animals being fond of its roots) (tribe Cichorieae)

REFERENCES: Shinners 1966; Vuilleumier 1973; Tomb 1974; Cronquist 1980; Diggs et al. 1997b; Bogler 2006b.

1. Leaves all or mostly basal (at most with reduced leafy bracts abruptly much smaller than the numerous basal rosette leaves); pappus bristles in 2 series: an outer series of short, merely barbellate bristles and an inner series of much longer, plumose bristles.
   2. Plants annual, essentially glabrous; heads not very showy, the ligules about equaling the phyllaries; peripheral achenes usually beakless, only the inner achenes beaked ____ **H. glabra**
   2. Plants perennial, with hispid leaves; heads showy, the ligules surpassing the phyllaries; all achenes usually with well-developed beaks ____ **H. radicata**
1. Flowering stems with a few ± well-developed leaves, at least toward the base; pappus bristles in 1 series, of long plumose bristles.
   3. Heads relatively broad, the involucres campanulate; phyllaries hirsute; corollas yellow; achenes 4–5-nerved ____ **H. chillensis**
   3. Heads narrow, the involucres cylindric; phyllaries glabrous or sparsely tomentose; corollas white; achenes 10-nerved ____ **H. microcephala**

Hypochaeris chillensis (Kunth) Britton, (of Chile), BRAZILIAN CAT'S-EAR. Taprooted biennial or perennial; stems 30–70 cm tall, glabrous to pilose; leaves basal and on lower stems, blades elliptic to oblanceolate, coarsely and sharply toothed to pinnatifid; heads 10–20 mm high; corollas yellow, 5–7 mm long, equaling phyllaries at flowering; achenes all beaked, 4–5-nerved; pappus bristles in 1 series, 6–8 mm long. Waste areas, sandy soils, roadsides, lawns; first reported for TX in Anderson Co. by Tomb (1974), also reported from McLennan Co. by Hannick et al. (2013); se U.S. from NC w to TX; South America. Apr. [*H. brasiliensis* (Less.) Griseb. var. *tweedyi* (Hook. & Arn.) Baker, *H. tweedyi* Hook. & Arn.] Native of South America.

Hypochaeris glabra L., (smooth, hairless), SMOOTH CAT'S-EAR. Taprooted annual or winter annual; stems 10–50 cm tall, glabrous, leafless or minutely bracteate; leaves basal, blades oblanceolate to oblong, nearly entire to toothed or pinnatifid; heads 8–16 mm high; corollas yellow or rarely whitish, 5–8 mm long, equaling phyllaries at flowering; achenes of two types, the peripheral achenes stout and beakless, the central achenes with well-developed beaks as maturity, 10-nerved; pappus bristles in 2 series, the longest 9–10 mm. Sandy soils, disturbed areas, roadsides; scattered in e and se TX; first reported for TX in Williamson Co. by Diggs et al. (1997b); reported in Harris, Leon, and Waller cos. by Brown and Elsik (2002) and in Brazos and Burleson cos. by M. Reed (pers. comm. 2009); 8 more counties in East TX have been vouchered at TEX-LL since 1992; scattered across North America, mostly se U.S. and w coast; Europe; introduced elsewhere. Feb–Jun. Native of Europe; apparently on its way to becoming a common weed in East TX.

Hypochaeris microcephala (Sch. Bip.) Cabrera var. **albiflora** (Kuntze) Cabrera, (sp.: small-headed; var.: white-flowered), SMALL-HEAD CAT'S-EAR. Taprooted perennial; stems 10–60 cm tall, glabrous to sparsely hairy; leaves basal and on lower stems, blades oblanceolate, entire to sharply toothed or deeply pinnatifid; heads 8–15(–18) mm high; corollas white, 5–7 mm long, not surpassing phyllaries at flowering; achenes all beaked, 10-nerved; pappus bristles in 1 series, 7–8 mm long. Grazed woodlands, ditches, sandy soils; mostly se TX, w to Williamson Co.; se U.S. from GA w to TX; South America. Apr–Jun. [*H. albiflora* (Kuntze) Azevdo-Gon. & Matzenb., *H. brasiliensis* (Less.) Griseb. var. *albiflora* Kuntze] Native of South America.

Hypochaeris radicata L., (having conspicuous roots), HAIRY CAT'S-EAR. Taprooted perennial; stems 10–60 cm tall, glabrous to coarsely hirsute, leafless or sparsely bracteate; leaves basal, ± hirsute, blades oblanceolate, lyrate, coarsely toothed to pinnatifid; heads 10–25 mm high; corollas yellow, sometimes striped grayish green or reddish beneath, 10–15 mm long, surpassing phyllaries at flowering; achenes all beaked, 10–12-nerved; pappus bristles in 2 series, the longest 10–12 mm. Oak-pine forest, dry

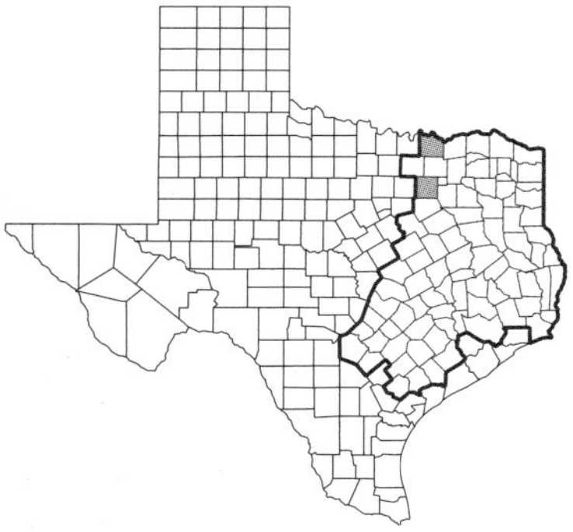
Helianthus tuberosus

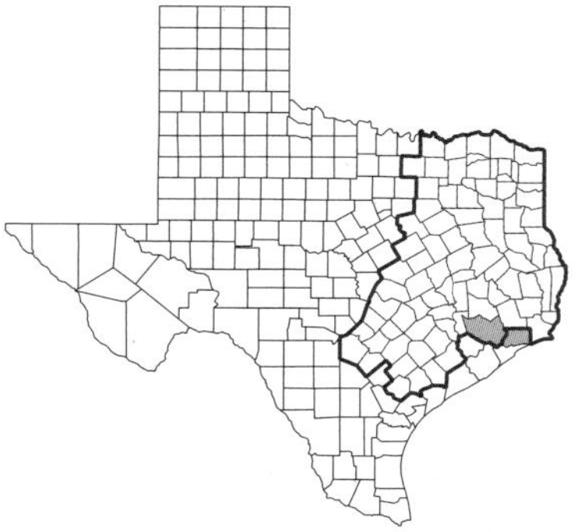
Heliopsis gracilis

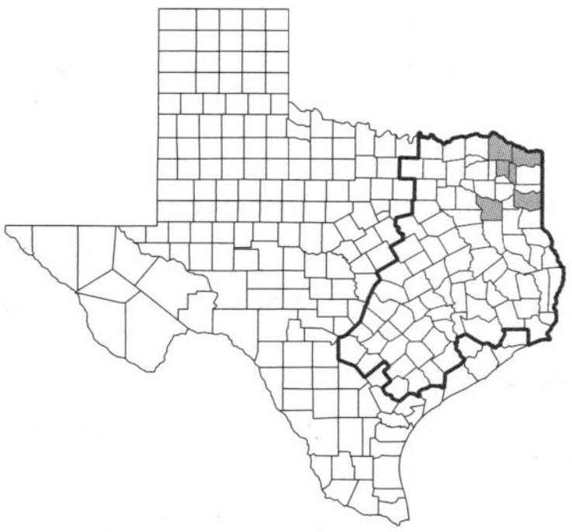
Heliopsis helianthoides var. scabra

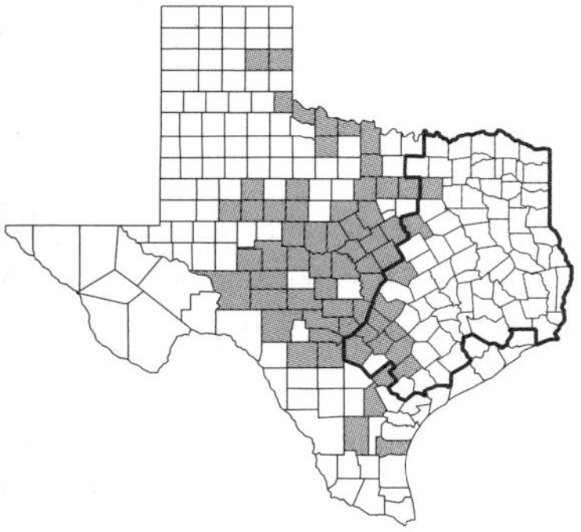
Heterotheca canescens

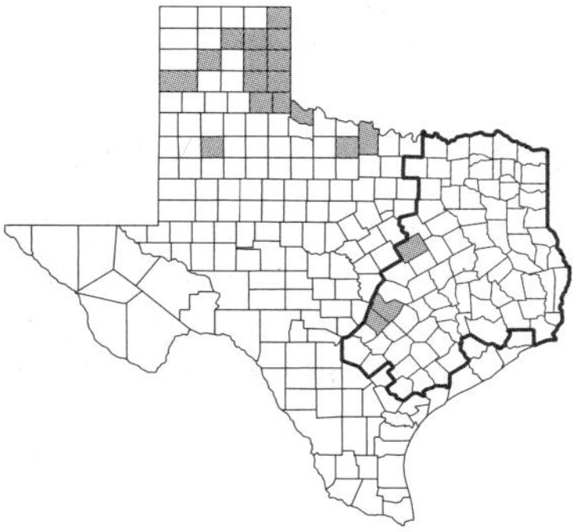
Heterotheca stenophylla

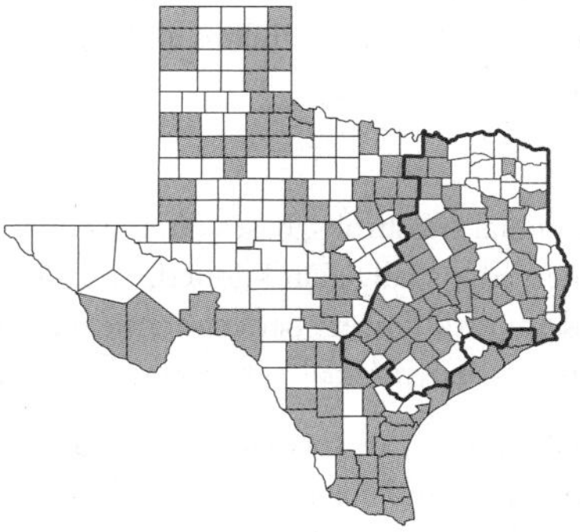
Heterotheca subaxillaris

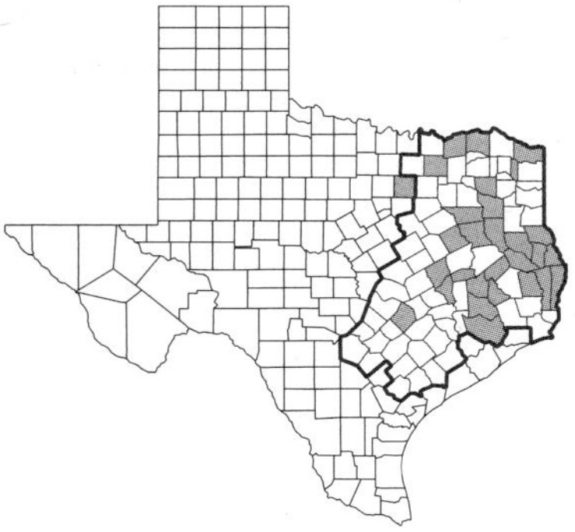
Hieracium gronovii

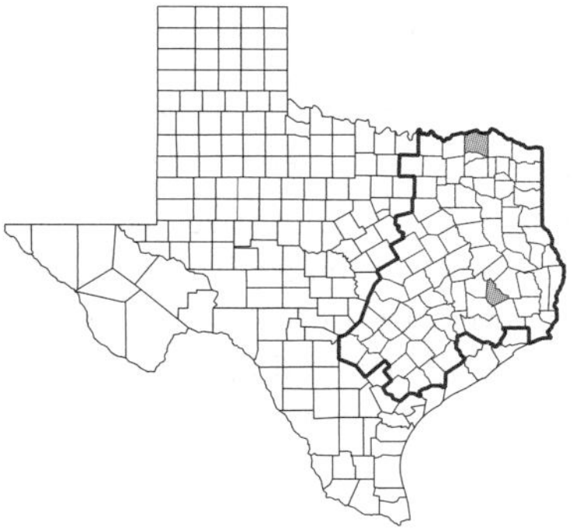
Hieracium longipilum

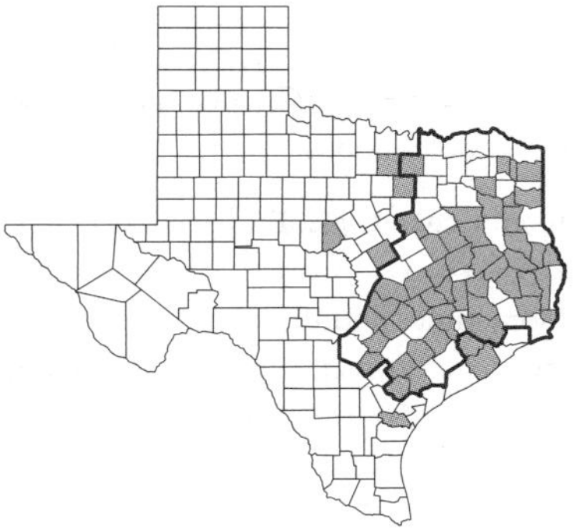
Hymenopappus artemisiifolius var. artemisiifolius

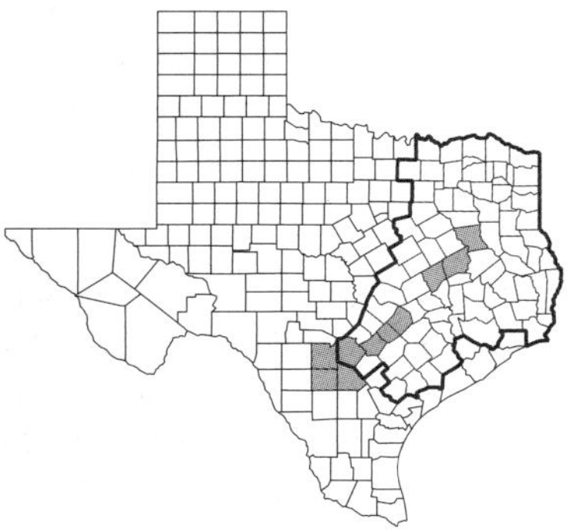
Hymenopappus carrizoanus

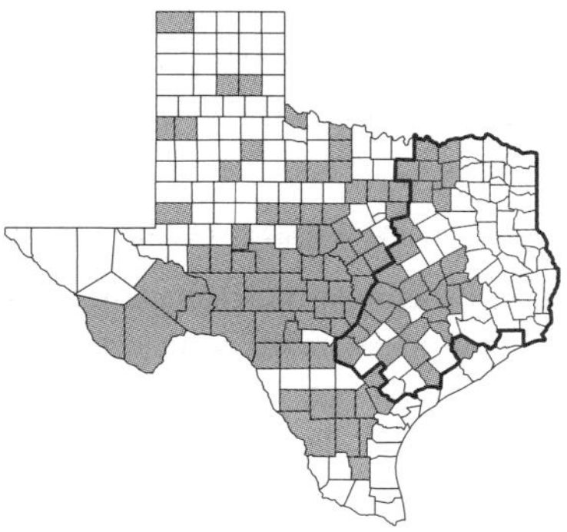
Hymenopappus scabiosaeus var. corymbosus

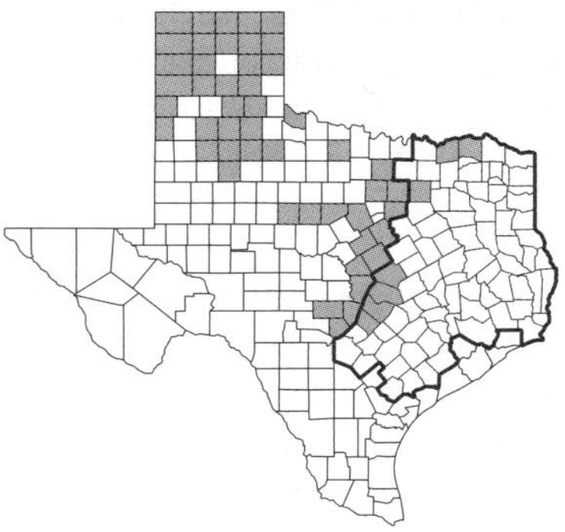
Hymenopappus tenuifolius

fields, ditches, lawns; recorded from Brazos, Harrison, and Newton cos.; scattered across most of North America; Europe and Asia; widely introduced elsewhere. May–Jun. Native of Europe.

IONACTIS Greene ANKLE-ASTER

A genus of 5 species native to North America. The genus has often been included in *Aster*; Nesom and Leary (1992) resurrected the genus. (Greek: *ion*, violet, and *aktis*, rays, alluding to the color of the ligules) (tribe Astereae)

REFERENCES: Nesom & Leary 1992; Nesom 1992b, 1994b, 2006r.

Ionactis linariifolia (L.) Greene, (leaves like flax, *Linum*), FLAX-LEAF ANKLE-ASTER, FLAX-LEAF WHITETOP, NARROW-LEAF IONACTIS. Herbaceous or slightly woody perennial, 10–50(–70) cm tall; leaves alternate, sessile, blades linear to narrowly oblong or lanceolate, 12–40 mm long, glabrous, entire; heads borne singly or in loose corymbose arrays; involucres 6–9 mm high; phyllaries appressed, linear-lanceolate, unequal; ray flowers pistillate, fertile, corollas violet to bluish or rarely white; disk flowers perfect, fertile, corollas yellow; achenes flattened, (2.5–)3.5–4 mm long; pappus of 2 series, the outer of short bristles, the inner of barbellate bristles. Sandy habitats, seeps or other moist sites, longleaf pine woods; restricted to deep se TX in Jasper and Newton cos., and Sabine Co. (Phillips et al. 2007); e U.S. from ME s to FL and w to WI, OK, and TX; Canada. Sep–Nov. [*Aster linariifolius* L.]

ISOCOMA Nutt. GOLDENBUSH, JIMMYWEED

Perennial subshrubs, bases often woody; stems usually erect, few-branched, usually glandular; leaves alternate, sessile, blades 1-nerved, usually gland-dotted; heads borne in compact, corymbose arrays; phyllaries in (3–)4–6 series, unequal; ray flowers absent; disk flowers perfect, fertile; corollas yellow with dark orange veins, goblet-shaped, the tubes elongating at maturity; achenes obpyramidal, sericeous, 5–11-ribbed; pappus of many thick, unequal, barbellate bristles.

A genus of 16 species native to the sw U.S. and n Mexico; it has been included by many authors in *Haplopappus*. (Greek: *isos*, equal, and *kome*, hair of the head, "so called from its equal flowers," according to the original generic description) (tribe Astereae)

REFERENCES: Nesom 1991a, 2006s; Lane & Hartman 1996.

1. Heads 2.5–4 mm wide; phyllary tips green to greenish yellow to yellowish, slightly gland-dotted if at all; flowers 12–15 per head; leaves resinous ____ **I. coronopifolia**
1. Heads 4.5–5.5(–7) mm wide; phyllary tips usually green, usually strongly gland-dotted; flowers (18–)22–34 per head; leaves not resinous ____ **I. drummondii**

Isocoma coronopifolia (A. Gray) Greene, (presumably, with leaves like *Coronopus* (now *Lepidium*)—SWINECRESS—Turner 2016), COMMON JIMMYWEED, COMMON GOLDENBUSH. Plants glabrous; leaves resinous, blades oblanceolate, margins usually pinnatifid (or sometimes entire), segments spine-tipped; involucres 5–6(–7) mm tall and 2.5–4 mm wide; phyllary tips green to greenish yellow to yellowish, slightly gland-dotted if at all; flowers 12–15 per head. Alkaline soils, sandy flats, shrublands; Bexar and Wilson cos.; South TX Plains and Rio Grande Valley; Mexico. May–Nov. [*I. coronopifolia* (A. Gray) Greene var. *pedicellata* (Greene) G.L. Nesom, *I. pedicellata* Greene]

Isocoma drummondii (Torr. & A. Gray) Greene, (for Thomas Drummond, 1780–1835, Scottish botanist and collector in North America), DRUMMOND'S JIMMYWEED, DRUMMOND'S GOLDENBUSH. Plants glabrous; leaves not resinous, blades narrowly oblong to oblanceolate-oblong, margins usually entire (sometimes toothed in the upper portion); involucres (5.5–)6.5–8 mm tall and 4.5–5.5(–7) mm wide; phyllary tips usually green, usually strongly gland-dotted; flowers (18–)22–34 per head. Sandy soils, brushy prairies, mesquite scrub, coastal dunes; DeWitt Co.; s coastal TX; Mexico. Mar–Jul(–Sep). [*Haplopappus drummondii* (Torr. & A. Gray) S.F. Blake, *I. megalantha* Shinners]

IVA L. SUMPWEED, MARSH-ELDER

Annual or perennial herbs, subshrubs, or shrubs; leaves opposite or sometimes alternate above, entire to toothed, often aromatic and gland-dotted; inflorescences spicate or racemose in paniculate arrangements; heads small, inconspicuous (plants wind pollinated), sessile or subsessile in the axils of bracts; phyllaries in 1–3+ series, distinct or united; ray flowers lacking or ± vestigial; heads with flowers of two kinds, the peripheral disk or disk-like flowers pistillate and fertile, the central disk flowers functionally staminate; corollas inconspicuous; achenes obovate to pear-shaped, 2–3 mm long, usually gland-dotted; pappus none.

A genus of ca. 11 species native from North America to the West Indies; pollen from some species causes hay fever. (Latin name from the mint *Ajuga iva* Schreb., with similar odor) (tribe Heliantheae)

Hymenopappus tenuifolius [RYD]

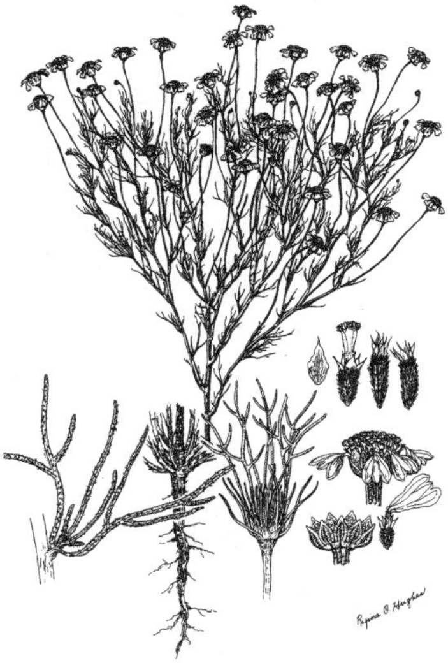

Hymenoxys odorata [REE]

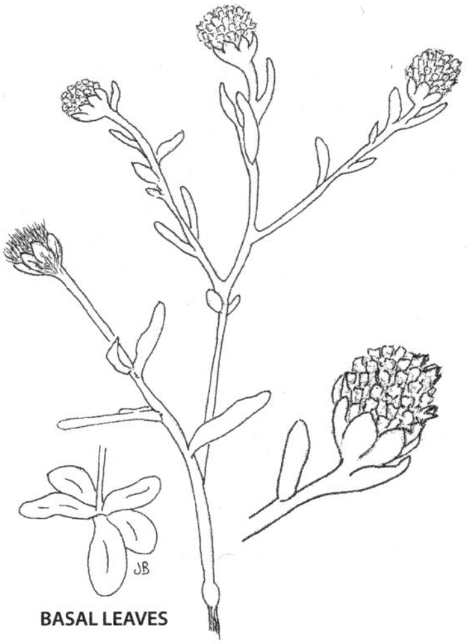

Hymenoxys perpygmaea [BAK]

Hymenoxys texana [SID, TPW]

Hypochaeris chillensis [MAR]

Hypochaeris glabra [BEN]

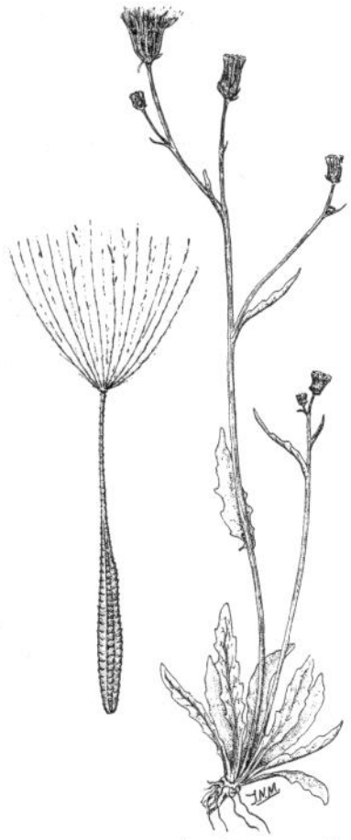

Hypochaeris microcephala var. albiflora [JON]

Hypochaeris radicata [GEO]

References: Rydberg 1922; Jackson 1960; Bolick 1985; Strother 2006q; Turner 2009a, 2009b.

1. Midstem leaves sessile or nearly so, blades linear to lance-linear, usually 1–4 mm wide, outer phyllaries united ______ **I. asperifolia**
1. Midstem leaves with petioles 5+ mm long, blades variously shaped but broader than lance-linear, usually 5–45 mm wide, outer phyllaries distinct.
   2. Plants shrubby or subshrubby perennials (lower stems woody); peduncles 1–3+ mm long; restricted to moist, brackish sites of the Gulf Coast ______ **I. frutescens**
   2. Plants herbaceous annuals or perennials; peduncles 0–1 mm long; of various habitats in e and c TX.
      3. Plants herbaceous annuals, usually < 100 cm tall; heads about as wide as high; pales (chaff) of the receptacle linear; central functionally staminate flowers 8–12+; widespread in e and c TX ______ **I. annua**
      3. Plants herbaceous perennials, to 200+ cm tall; heads wider than high; pales of the receptacle oblanceolate-ovoid; central functionally staminate flowers 5–10; known only from Travis Co. ______ **I. corbinii**

Iva annua L., (annual), ANNUAL MARSH-ELDER, SHARP-BRACT SUMPWEED, PELOCOTE. Annual herbs, (10–)50–100(–150+) cm tall; leaves petiolate, blades deltate to lanceolate or ovate to elliptic, usually 30–100 mm long and 8–45 mm wide, margins ± toothed; heads sessile; outer 3–5 phyllaries distinct, herbaceous. Low moist areas, calcareous or sandy soils, disturbed sites, sometimes forming extensive stands; widespread, but mainly e and c TX; se and sc U.S., becoming adventive in ne states; Mexico. Jul–Nov. [*I. annua* L. var. *caudata* (Small) R.C. Jacks, *I. caudata* Small, *I. ciliata* Willd.] The wind-borne pollen is a cause of hay fever (Jackson 1960); the achenes were formerly eaten by Native Americans who domesticated this species (Heiser 1990; Mabberley 2017).

Iva asperifolia Less., (rough-leaved). Annual or perennial herbs, 20–60(–100+) cm tall; leaves sessile or nearly so, blades lance-linear to linear, usually 12–25 mm long and 1–4 mm wide, margins entire; heads sessile or nearly so; outer 3–5 phyllaries united, ± herbaceous. Turner (2009a) moved the formerly recognized *I. angustifolia* into synonymy under *I. asperifolia*, and described the two intergrading varieties in East TX:

1. Plants annual, usually 75 cm tall or less; widespread in e 1/2 of TX ______ var. **angustifolia**
1. Plants perennial, to 150 cm tall; restricted to Gulf Coast and South TX Plains ______ var. **latior**

var. **angustifolia** (Nutt. ex DC.) B.L. Turner, (narrow-leaved), NARROW-LEAF SUMPWEED, NARROW-LEAF MARSH-ELDER. Annual herbs, to 75 cm tall. Low moist areas, prairies, and post oak woodlands; widespread in e 1/2 of TX; AR, KS, LA OK, and TX. Jul–Oct. [*I. angustifolia* Nutt. ex DC. var. *angustifolia*]

var. **latior** (Shinners) B.L. Turner, (broad), TEXAS SUMPWEED, TEXAS MARSH-ELDER, PELOCOTE. Perennial herbs, to 150 cm tall. Sandy soils, dunes; sw border of East TX in Bexar, Goliad, and Gonzales cos. (Turner 2009a); mostly c and s TX Gulf Coast; Mexico. All seasons. [*I. angustifolia* Nutt. ex DC. var. *latior* Shinners, *I. texensis* R.C. Jacks.]

Iva corbinii B.L. Turner, (for Robert Corbin, naturalist of Austin, TX, and first collector), CORBIN'S SUMPWEED. Perennial tap-rooted herbs, 200+ cm tall; leaves petiolate, blades ovate to deltoid, margins entire or toothed; heads sessile or nearly so; outer 3–5 phyllaries distinct, herbaceous. Moist silty-clay; first described in 2009 from small islands/sandbars in the Colorado River in Travis Co. (Turner 2009b). Jul–Aug? ⚠? ?

Iva frutescens L., (shrubby), MARSH-ELDER, JESUIT'S BARK. Perennial shrubs or subshrubs, 50–350 cm tall; leaves petiolate, blades ovate to elliptic to lanceolate, usually 30–60 mm long and 5–30 mm wide, margins usually toothed; heads subsessile; outer 5 phyllaries distinct, herbaceous. Moist brackish areas, saline marshes, beaches; Harris, Jackson, Jefferson, and Orange cos.; TX Gulf Coast; Atlantic and Gulf coasts from NJ s to FL and w to TX. Jul–Oct. [*I. frutescens* L. subsp. *oraria* (Bartlett) R.C. Jacks.]

KRIGIA Schreb. DWARF-DANDELION

Small, ± glabrous annual or perennial herbs; sap milky; leaves basal or alternate; leaf blades oblanceolate or oblong-lanceolate to linear, entire or unevenly toothed or lobed; heads generally small, usually solitary and terminal or few together; phyllaries nearly equal; flowers all ligulate; corollas yellow or yellow-orange, open during morning hours in sunny weather; achenes columnar, fusiform, or obconic, ribbed, not beaked; pappus none or single or double, of scales or an inner ring of bristles and an outer ring of scales.

A North American genus of 7 species. The 5 species treated here all differ in chromosome number (Kim & Turner 1992). (Named for David Krig or Krieg, died 1713, a German or Hungarian physician, who was among the first to collect plants in Maryland) (tribe Cichorieae)

References: Shinners 1947b; Vuilleumier 1973; Kim & Mabry 1991; Kim & Turner 1992; Kim et al. 1992a, 1992b; Chambers 2004; Chambers & O'Kennon 2006.

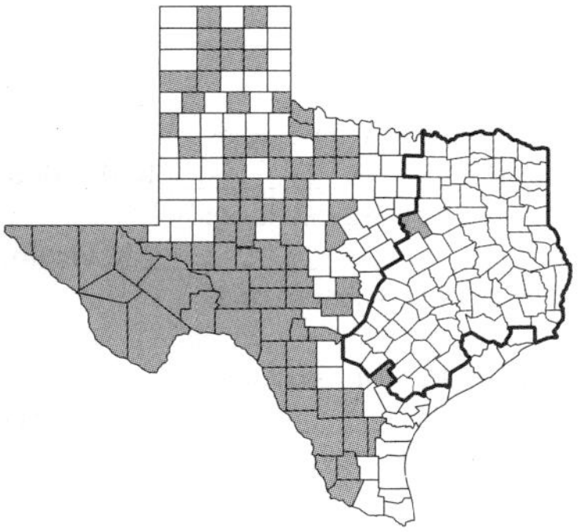

Hymenoxys odorata

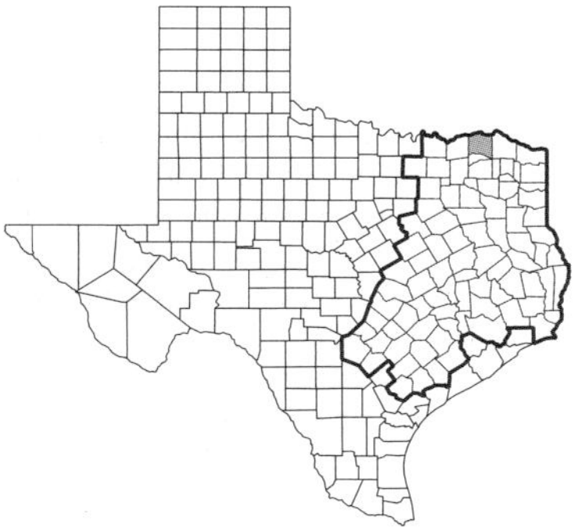

Hymenoxys perpygmaea

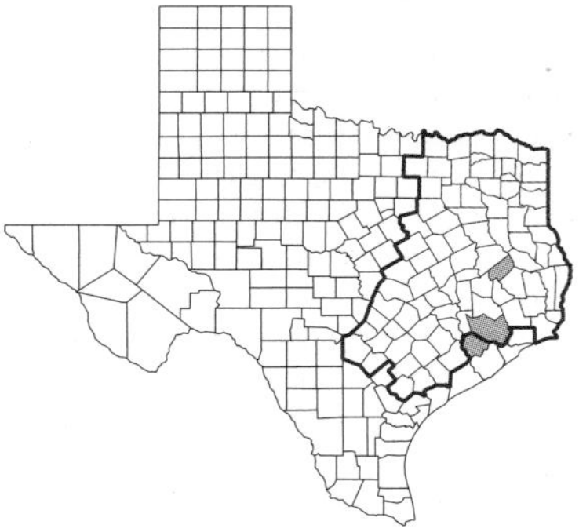

Hymenoxys texana

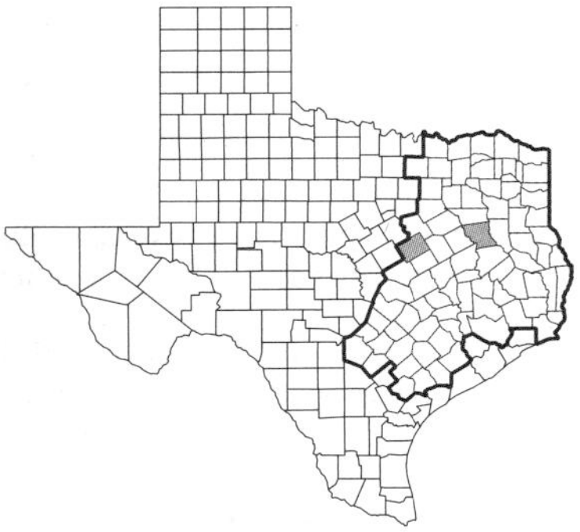

Hypochaeris chillensis

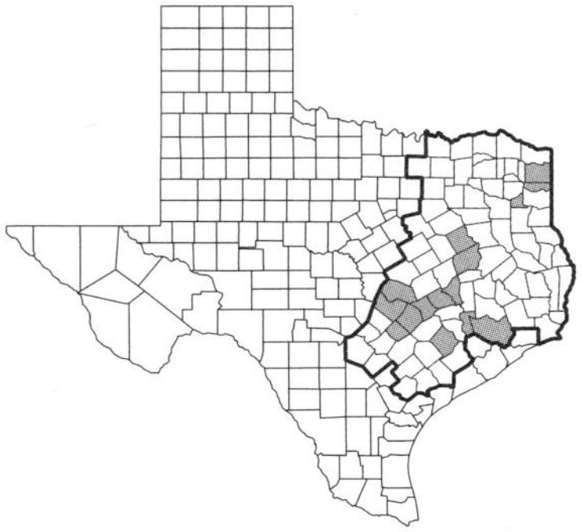

Hypochaeris glabra

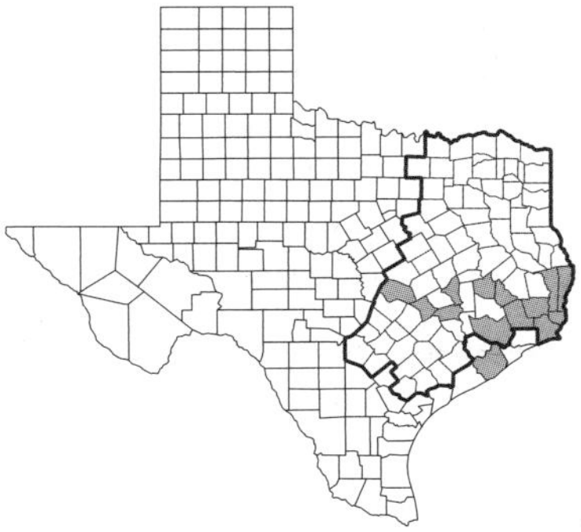

Hypochaeris microcephala var. albiflora

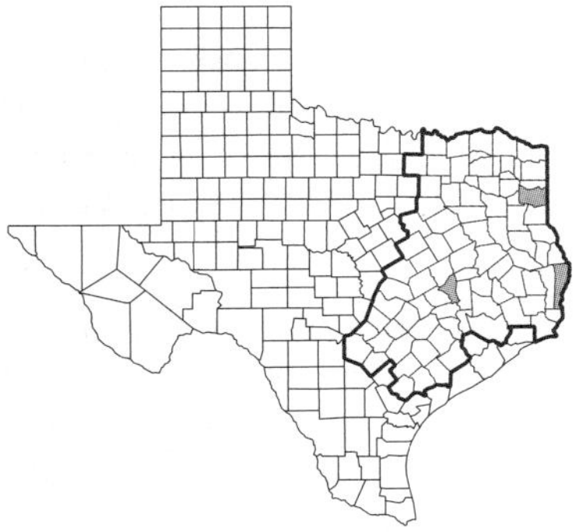

Hypochaeris radicata

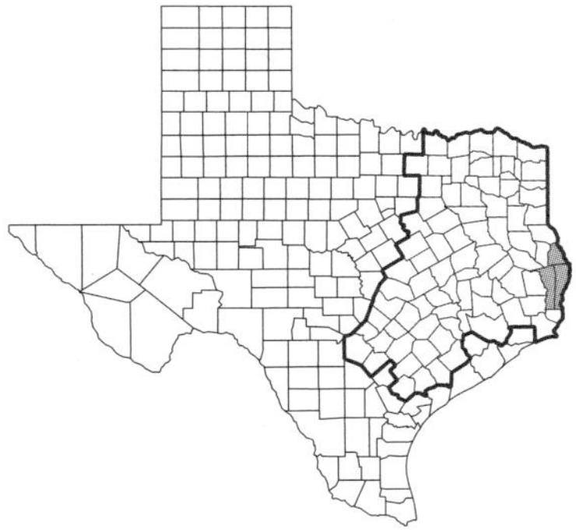

Ionactis linariifolia

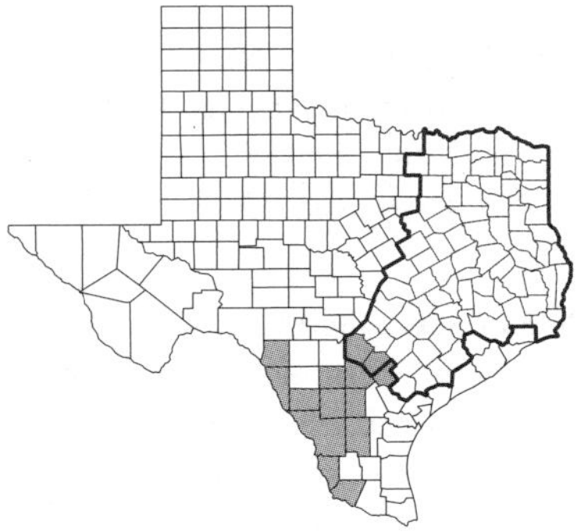

Isocoma coronopifolia

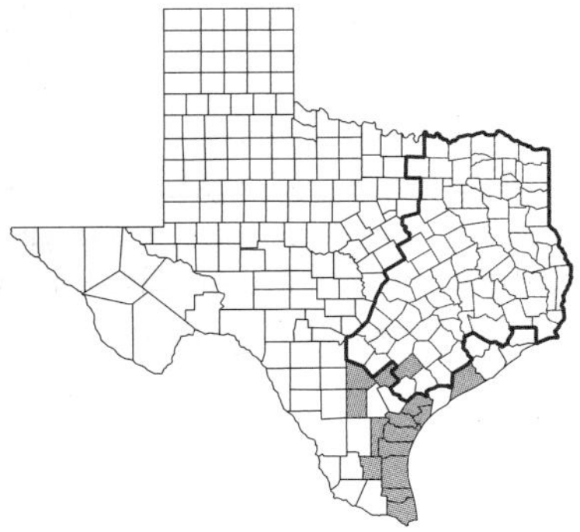

Isocoma drummondii

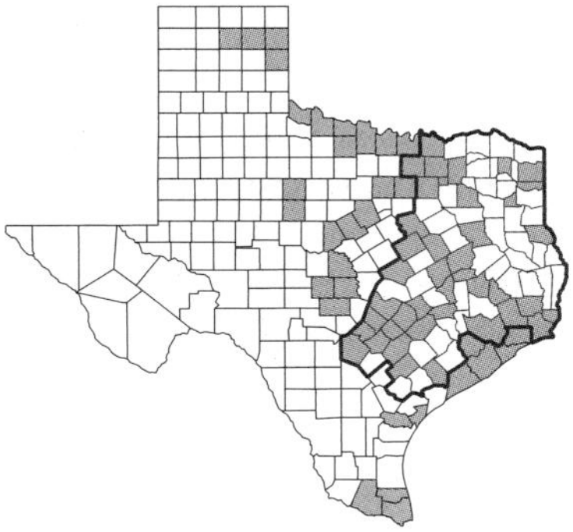

Iva annua

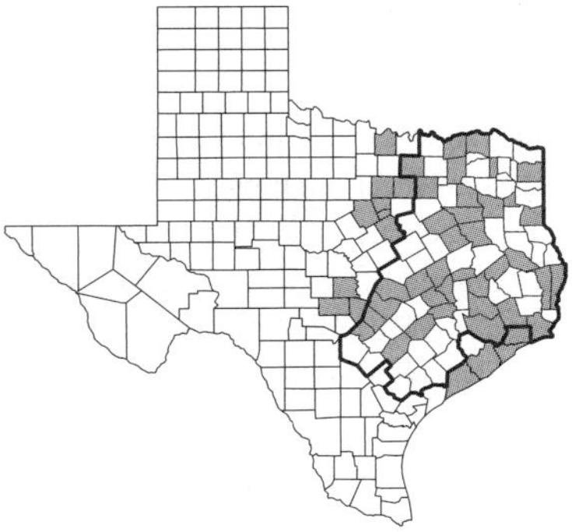

Iva asperifolia var. angustifolia

1. Phyllaries 4–8 times as long as wide, linear-lanceolate to oblong-lanceolate, numerous (8–16), becoming shrivelled or reflexed in age, never keeled; inner pappus of bristles 4–8 mm long, outer of as many or fewer short scales < 1/4 as long; stems leafless.
   2. Plants perennials with a small tuber a few cm below ground, developing slender, whitish rhizomes; involucres 10–15 mm high; corollas 15–17 mm long; pappus with numerous bristles and numerous short scales ____________ **K. dandelion**
   2. Plants annuals with a taproot, lacking tubers; involucres 4–6.5 mm high (to 9 mm in age); corollas 5–12 mm long; pappus with 5 bristles and 5 scales ____________ **K. virginica**
1. Phyllaries 1.5–3 times as long as wide, lanceolate to ovate, few (4–10), remaining erect in age; pappus absent or of short scales or bristles up to 2 mm long; stems leafless or leafy.
   3. Stems leafless, unbranched (but robust plants or stems buried by loose soil may develop short, basal internodes); pappus of 5 well-developed short scales or both 5 well-developed short scales and short bristles ____________ **K. occidentalis**
   3. Stems leafy, branched (but beginning to flower with leaves crowded at base and branches scarcely developed); pappus absent or of much reduced scales and bristles.
      4. Pappus of much reduced scales and bristles present on at least some achenes (use hand lens or scope) ____________ **K. wrightii**
      4. Pappus absent on all achenes ____________ **K. cespitosa**

Krigia cespitosa (Raf.) K.L. Chambers, (growing in tufts). Variable taprooted annual; stems leafy, branched; phyllary midveins evident but not prominent nor forming curved keels; achenes 1.4–1.7 mm long; pappus absent. Shinners (1947b) treated the following 2 taxa as separate species, Kim and Turner (1992) treated them as formae of *K. cespitosa*, and Chambers (2004) elevated them to varieties; they may occasionally hybridize (Chambers & O'Kennon 2006).

1. Corollas 2–6 mm long; involucres 2–4.5 mm high in flower, 3.5–5 mm high in age; heads 3–6 mm in diam. ________ var. **cespitosa**
1. Corollas 6.5–11 mm long; involucres 4.5–7 mm high in flower, 5.5–8.5 mm high in age; heads 6–9 mm in diam. ________ var. **gracilis**

var. **cespitosa**, WEEDY DWARF-DANDELION. Plant 35–40 cm tall; phyllaries 5–8. Stream banks, damp woods, roadsides, and disturbed areas, especially on sandy soils; widespread in e 1/2 of TX; se and sc U.S. from VA s to FL and w to NE and TX; Mexico. Mar–May. [*K. oppositifolia* Raf., *Serenia cespitosa* (Raf.) Kuntze]

var. **gracilis** (DC.) K.L. Chambers, (graceful, slender). Plant 8–40 cm tall; phyllaries 8–10. Low spots in prairies, or borders of woods, disturbed areas, chiefly calcareous clay, less often sandy soils; Post Oak Savannah and Blackland Prairie w to West Cross Timbers and Edwards Plateau; LA, OK, and TX. Apr–May. [*K. gracilis* (DC.) Shinners]

Krigia dandelion (L.) Nutt., (from its resemblance to *Taraxacum*—dandelion), TUBER DWARF-DANDELION, POTATO-DANDELION. Scapiform tuberous perennial 10–45 cm tall; leaves all basal; heads large and showy; phyllaries ca. 15, reflexed in age; corollas 15–17 mm long; achenes 2.5 mm long; pappus of numerous bristles 5–8 mm long and 10 small scales. Sandy open woods, roadsides, and lawns; widespread in e 1/3 of TX; se and c U.S. from NJ s to FL and w to IA and TX. Apr–May. [*Cynthia dandelion* (L.) DC.]

Krigia occidentalis Nutt., (western), WESTERN DWARF-DANDELION. Scapose taprooted annual 4–16 cm tall; leaves mostly basal; phyllaries 4–7, their midveins becoming prominent and forming curved keels in age; corollas 5–9 mm long; achenes 1.2–1.8 mm long; pappus usually of 5 bristles 1.2–2 mm long and 5 small scales, sometimes absent. Sandy open woods, fields, and roadsides; widespread in e 1/2 of TX; sc U.S. from MO and KS s to TX and LA. Feb–May. [*Cymbia occidentalis* (Nutt.) Standl.]

Krigia virginica (L.) Willd., (of Virginia), CAROLINA DWARF-DANDELION. Taprooted annual 4–30 cm tall; ± scapose; phyllaries ca. 10, reflexed in age; corollas 5–12 mm long; achenes 1.5–2.3 mm long; pappus of 5 bristles 4–6 mm long and 5 small scales. Sandy open woods, fields, and roadsides; scattered throughout e 1/2 of TX; e U.S. from ME s to FL and w to WI and TX. Mar–Apr, rarely repeating as late as Oct. [*Hyoseris virginica* L.]

Krigia wrightii (A. Gray) K.L. Chambers ex K.J. Kim, (for Charles Wright, 1811–1885, TX collector). Taprooted annual 4–25 cm tall; stems leafy, branched; phyllaries 5–9, their midveins becoming prominent and forming curved keels in age; corollas 4–7 mm long; achenes 1.3–1.6 mm long; pappus of minute scales and bristles. Roadsides, weedy areas, low spots, often in sandy soils; scattered throughout e 1/2 of TX; AR, LA, OK, and TX. [*Apogon wrightii* A. Gray] Extremely similar to the dwarf form of *K. cespitosa* and the taller individuals of *K. occidentalis*; chromosomal and molecular data (Kim & Turner 1992) indicate the species is more distinctive than morphology would suggest.

Ionactis linariifolia [STR]

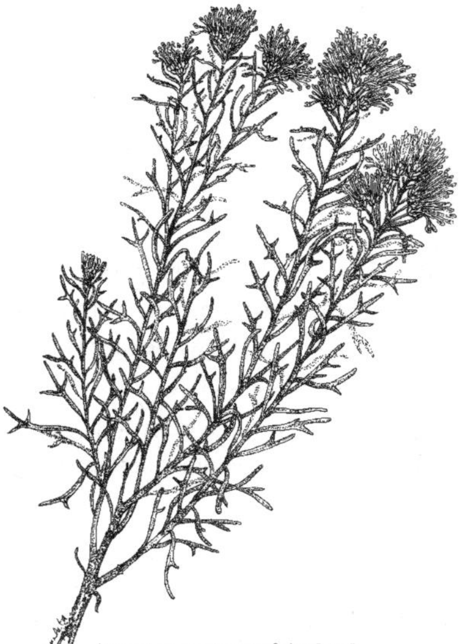

Isocoma coronopifolia [THO]

Iscoma drummondii [HAL]

Iva annua [LAM]

Iva asperifolia var. angustifolia [BT3, UKS]

Iva corbinii [THO]

Iva frutescens [GLE]

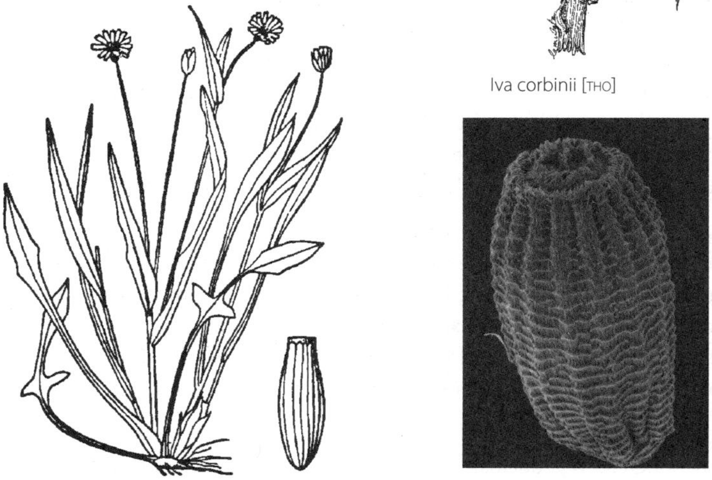

Krigia cespitosa var. cespitosa [BB2, BTT]

LACTUCA L. LETTUCE

Coarse annual or biennial taprooted herbs; sap milky or brownish; stems usually 1, erect; leaves alternate; heads small, in terminal paniculate arrays (these at first dense and spicate); involucres narrow; phyllaries in ± 2 series, subequal to equal; flowers all ligulate; corollas yellow, bluish, or whitish, sometimes reddish-tinged, open during the morning, withering greenish or bluish; achenes elliptic to oblong, often somewhat flattened, ribbed, beaks stout or filiform; pappus of numerous hair-like, smooth or barbellate bristles.

A genus of ca. 75 species; cosmopolitan but especially n temperate. Young (pre-flowering) plants of *Lactuca sativa* L., the cultivated LETTUCE or GARDEN LETTUCE, lack the bitter sesquiterpene lactones omnipresent in wild species. (The ancient name of the lettuce, *L. sativa*; from Latin: *lac*, milk, in allusion to the white sap) (tribe Cichorieae)

REFERENCES: Stebbins 1937; Vuilleumier 1973; Strother 2006r.

1. Corollas usually bluish or white; beak of achenes stout, 0.1–0.5(–1) mm long ____ **L. floridana**
1. Corollas usually yellow, sometimes bluish or purplish, sometimes reddish-tinged; beak of achenes filiform (thread-like), > 1 mm long.
    2. Faces of achene 1(–3)-nerved.
        3. Involucres usually 10–12 mm high, achene bodies 2.5–3.5 mm long (not including beak) ____ **L. canadensis**
        3. Involucres usually 12–20 mm high; achene bodies 4.5–6 mm long (not including beak).
            4. Lower stem leaves usually undivided, blades spatulate to lance-linear; corollas bluish to purplish ____ **L. graminifolia**
            4. Lower stem leaves usually pinnatifid or deeply runcinate, blades ovate or oblanceolate to spatulate; corollas usually yellow, sometimes red- or orange-tinted, rarely bluish.
                5. Leaf margins entire or denticulate, seldom prickly; flowers 12–24+ per head ____ **L. hirsuta**
                5. Leaf margins often toothed and prickly; flowers 20–50+ per head ____ **L. ludoviciana**
    2. Faces of achene (3–)5–9-nerved.
        6. Lower stem leaves with margins entire to denticulate but not prickly; upper stem leaves lanceolate to linear or filiform, phyllaries usually erect in fruit; achenes without bristles at the base of the beak ____ **L. saligna**
        6. Lower stem leaves with margins denticulate-toothed and prickly; upper stem leaves oblong to obovate or lanceolate, phyllaries usually reflexed in fruit; achenes with conspicuous bristles at the base of the beak ____ **L. serriola**

Lactuca canadensis L., (of Canada), WILD LETTUCE. Biennial; stems (15–)40–200(–450+) cm tall; leaf blades deeply pinnatifid on the lower stem, oblong to obovate, upper stem leaves undivided, lanceolate to lance-linear, narrowed to nonclasping base or sagittate and clasping, margins entire to denticulate, midribs sometimes sparsely pilose; involucres 10–12+ mm high; phyllaries usually reflexed in fruit; corollas yellow or sometimes bluish; achenes ± flattened, somewhat winged, bodies 2.5–3.5 mm long, the beak filiform, 1–3 mm long; pappus 5–6 mm long. Sandy soils, roadsides, thickets; se and ne TX w to nc TX; across North America but more common in e 1/2; Mexico and Central America; Eurasia. May–Sep. [*L. canadensis* L. var. *latifolia* Kuntze, *L. canadensis* L. var. *longifolia* (Michx.) Farw., *L. canadensis* L. var. *obovata* Wiegand, *L. sagittifolia* Elliott] In some cases *L. canadensis* and *L. ludoviciana* are difficult to distinguish; according to Correll and Johnston (1970), in TX the two intergrade and introgressive hybridization has possibly occurred.

Lactuca floridana (L.) Gaertn., (of Florida), WOODLAND LETTUCE, FLORIDA LETTUCE, BLUE-FLOWERED LETTUCE. Annual or biennial; stems 25–50(–200+) cm tall; leaf blades often deeply runcinate-pinnatifid on the lower stem, the terminal portion large, triangular, the lateral lobes large, 1 or 2(–3) per side, upper stem leaves undivided, oblong to elliptic, margins entire to denticulate, midribs sometimes sparsely pilose; involucres (8–)10–12+ mm high; phyllaries usually reflexed in fruit; corollas light blue or sometimes whitish; achenes somewhat flattened or rounded, unwinged, bodies 4–5 mm long, the beak stout, 0.1–0.5(–1) mm long; pappus 4–5 mm long. Disturbed moist areas, margins of thickets and woods; East TX w to West Cross Timbers and Edwards Plateau; e 1/2 of the U.S. Jun–Sep. [*L. floridana* var. *villosa* (Jacq.) Cronquist]

Lactuca graminifolia Michx., (grass-leaved), GRASSLEAF LETTUCE, NARROW-LEAF LETTUCE. Biennial; stems 25–90(–150+) cm tall; leaf blades usually undivided, rarely dissected, spatulate to lance-linear, reduced up the stem, margins entire to denticulate, midribs sometimes with stiff bristles; involucres 10–20+ mm high; phyllaries usually reflexed in fruit; corollas bluish to purplish; achenes ± flattened, somewhat winged, bodies 5–6 mm long, the beak filiform, 2–4 mm long; pappus 5–9 mm long. Sandy ridges, pine forests; reported from deep sands in Angelina Co. (Brown et al. 2007); Trans-Pecos; mostly coastal/southern states from SC s to FL and w to AZ; Mexico and Central America. Jun–Sep.

Lactuca hirsuta Muhl. ex Nutt., (hairy), HAIRY LETTUCE, REDDISH-FLOWERED LETTUCE. Biennial; stems 15–80(–120) cm tall; leaf blades often deeply runcinate-pinnatifid on the lower stem, sometimes hairy, upper stem leaves undivided, ovate, margins denticulate, midribs usually pilose-spiny; involucres 12–18+ mm high; phyllaries usually reflexed in fruit; corollas usually yellow,

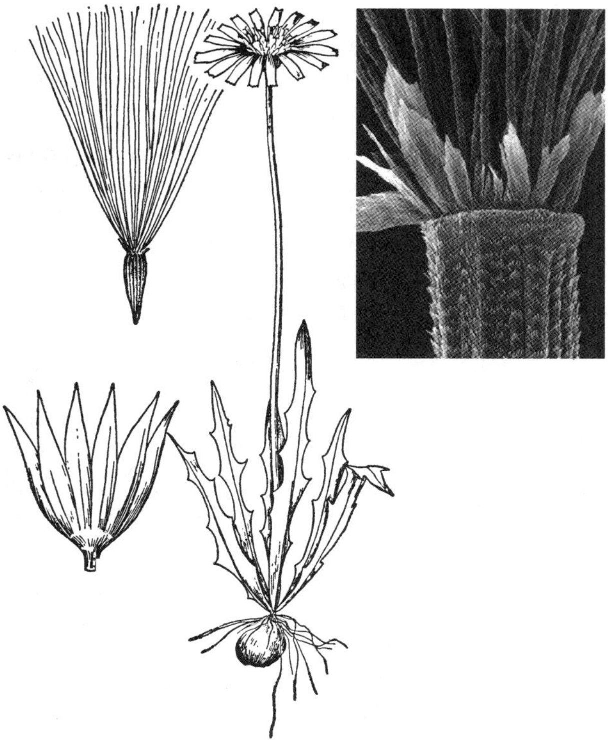

Krigia dandelion [BTT, GLE]

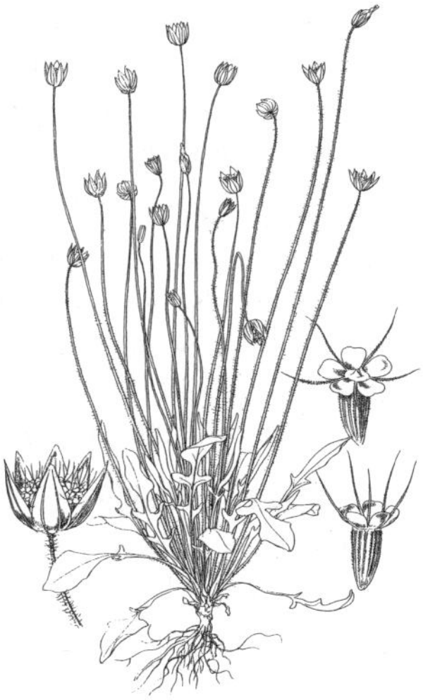

Krigia occidentalis [IPL]

Krigia occidentalis [BTT]

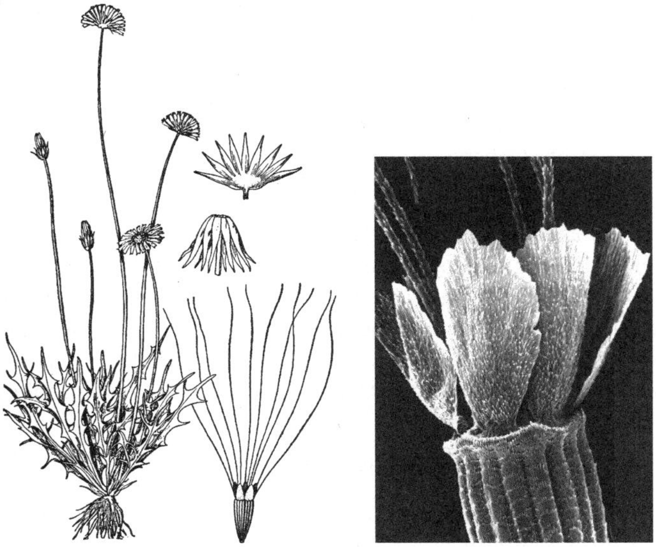

Krigia virginica [BTT, GLE]

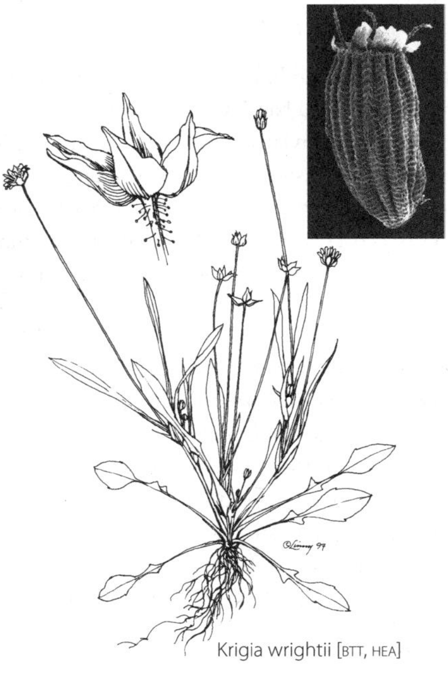

Krigia wrightii [BTT, HEA]

sometimes red- or orange-tinted, sometimes drying bluish; achenes ± flattened, winged, bodies 4.5–5+ mm long, the beak ± filiform, 2.5–3.5 mm long; pappus 6–8(–10+) mm long. Sandy soils, openings in woods; e and se TX; e U.S. from ME s to GA and w to MI, OK, and TX. (Apr–)Jun–Sep. [*L. hirsuta* Muhl. ex Nutt. var. *albiflora* (Torr. & A. Gray) Shinners, *L. hirsuta* Muhl. ex Nutt. var. *sanguinea* (Bigelow) Fernald]

Lactuca ludoviciana (Nutt.) Riddell, (of Louisiana), WESTERN WILD LETTUCE. Biennial; stems 15–150 cm tall; leaf blades pinnately lobed on the lower stem, upper leaves undivided, auriculate-clasping, obovate to spatulate, margins denticulate-ciliate or toothed and prickly, midribs usually pilose-spiny; involucres 12–15+ mm high; phyllaries usually reflexed in fruit; corollas usually yellow or rarely bluish; achenes ± flattened, somewhat winged, bodies 4.5–5+ mm long, the beak ± filiform, 2.5–4.5 mm long; pappus 5–7(–11) mm long. Usually in calcareous soils, prairies, openings in woods; scattered across e and c TX w to Edwards Plateau and Panhandle; c and w North America. (Apr–)Jun–Sep. [*L. campestris* Greene]

Lactuca saligna L., (resembling *Salix*—Willow), WILLOW-LEAF LETTUCE. Annual; stems 15–70(–100+) cm tall; leaf blades pinnatifid on lower stem, sagittate or auriculate, upper leaves undivided, linear to filiform, margins entire to denticulate, midribs sometimes prickly; involucres 6–9(–13+) mm high; phyllaries usually erect in fruit; corollas yellow, sometimes bluish or pinkish beneath; achenes ± flattened, unwinged, bodies 2.5–3.5 mm long, the beak ± filiform, (2–)5–6 mm long; pappus 5–6 mm long. Disturbed sites; Denton and Nacogdoches cos.; first reported in TX in Tarrant Co. by O'Kennon et al. (1998); ec and w U.S.; introduced elsewhere. Aug–Sep. Native of Eurasia; it is not clear whether it will become an established member of the TX flora.

Lactuca serriola L., (possibly from Latin: *serrula*, a small saw, referring to the toothed leaves, or a corruption of scariola, an old name for wild lettuce), PRICKLY LETTUCE. Annual; stems 30–70(–100+) cm tall; leaf blades often deeply pinnatifid on lower stem, sagittate or auricled and clasping, upper leaves undivided, clasping, oblong to obovate or lanceolate, margins denticulate and usually prickly, midribs usually prickly-spiny; involucres 9–10(–12) mm high; phyllaries usually reflexed in fruit; corollas yellow; achenes somewhat flattened, unwinged, bodies 2.5–3.5 mm long, the beak filiform, 2.5–4 mm long, beak bases hispid; pappus (3–)4–5 mm long. Disturbed sites, roadsides; scattered in se and ne TX; mainly w 1/2 TX; common across North America; introduced and weedy nearly worldwide. May–Sep. Native of Eurasia; name often but not originally written *scariola*. The widely cultivated GARDEN LETTUCE, *L. sativa*, is thought to be derived from *L. serriola* (Moore et al. 1976). Cattle have been poisoned by eating large quantities of the young growth (Kingsbury 1964).

LEONTODON L. HAWKBIT

A genus of ca. 50 species native to Europe, w Asia, and n Africa; some are now widely distributed as weeds. (Greek: *leon*, lion, and *odons*, tooth, alluding to the deeply toothed leaves) (tribe Cichorieae)
REFERENCE: Bogler 2006c.

Leontodon hispidus L., (bristly), BRISTLY HAWKBIT. Herbaceous perennials with milky sap; stems 10–60 cm tall, usually unbranched, from a basal rosette; leaves oblanceolate, usually hispid or hirsute, margins coarsely dentate to deeply lobed; heads borne singly; phyllaries subequal, linear-lanceolate; disk flowers absent; ray flowers 30–50+, the ligules bright yellow, the peripheral ones sometimes tinged orange or reddish, 12–15 mm long; achenes cylindric, ± beaked, 6–12 mm long; pappus brown, the outer series of bristle-like scales, the inner series of plumose bristles. Lawns, roadsides; recorded from Harrison (*R.J. Fleetwood* 6 TEX; Turner 2016) and Panola (Turner et al. 2003) cos.; infrequently recorded in c and ne North America; Europe; introduced elsewhere. Mar–Sep. [*L. hastilis* L.]

LEUCANTHEMUM Mill. OX-EYE DAISY

A genus of ca. 30 species of Europe and n Asia, including a number of cultivated ornamentals; previously treated in *Chrysanthemum*. The widely cultivated CHRYSANTHEMUMS or MUMS are now treated in the segregate genus *Dendranthema*. Another segregate of *Chrysanthemum*, *Tanacetum*, includes the European *T. cinerariifolium* (Trevir.) Sch. Bip. [*Pyrethrum cinerariifolium* Trevir.] (PYRETHRUM, DALMATIAN INSECT-FLOWER), the flower heads of which yield the monoterpene pyrethrum, used since ancient times as an insecticide and now widely used by organic gardeners (Fuller & McClintock 1986). The leaves and flowers of some species can cause contact dermatitis in sensitive individuals (Spoerke & Smolinske 1990). (Greek: *leucos*, white, and *anthemon*, flower) (tribe Anthemideae)
REFERENCES: Arriagada & Miller 1997; Strother 2006s.

Leucanthemum vulgare Lam., (common), OX-EYE DAISY, WHITE DAISY, WHITEWEED, MOON DAISY, DOG DAISY, MARGUERITE. Perennial; stems glabrous, 0.3–1 m tall; leaves mostly basal, on slender petioles, the blades spatulate-obovate; stem leaves alternate, the

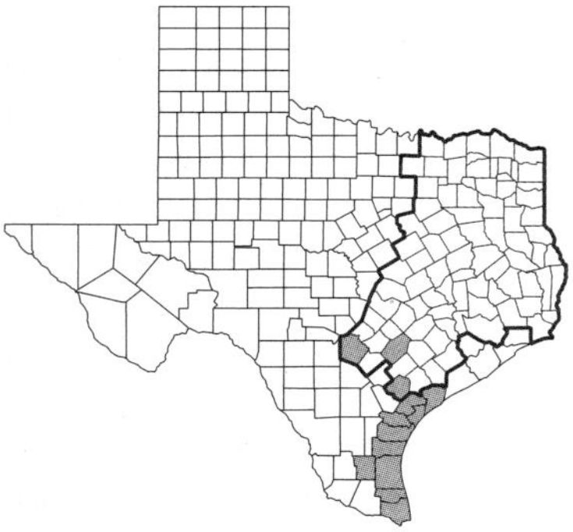

Iva asperifolia var. latior

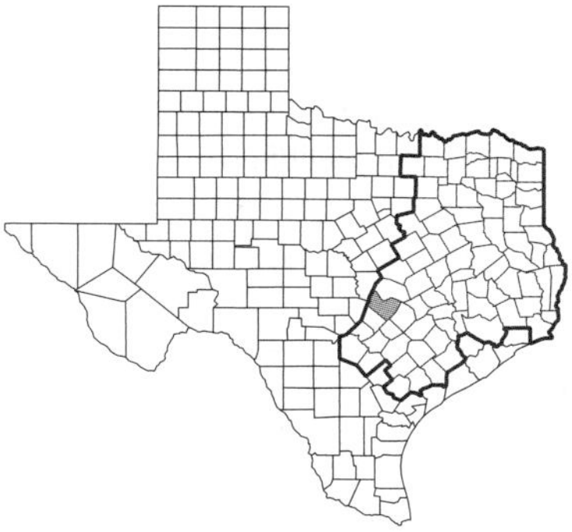

Iva corbinii

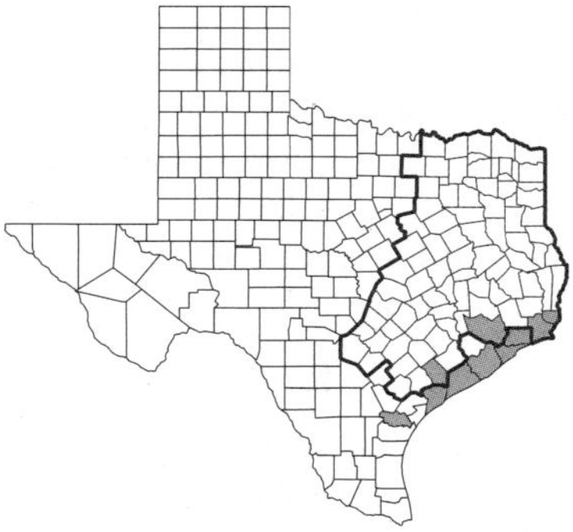

Iva frutescens

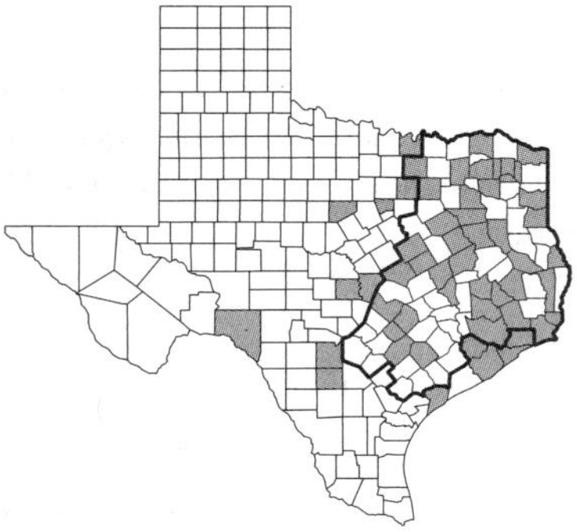

Krigia cespitosa var. cespitosa

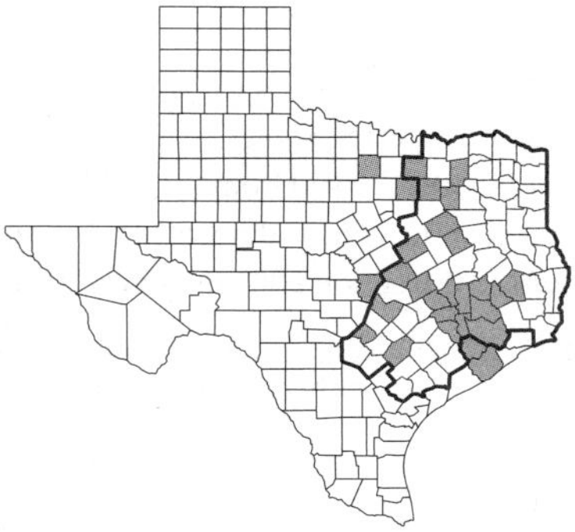

Krigia cespitosa var. gracilis

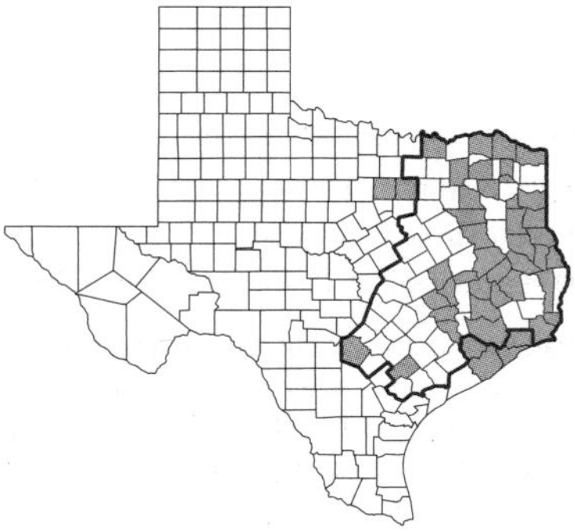

Krigia dandelion

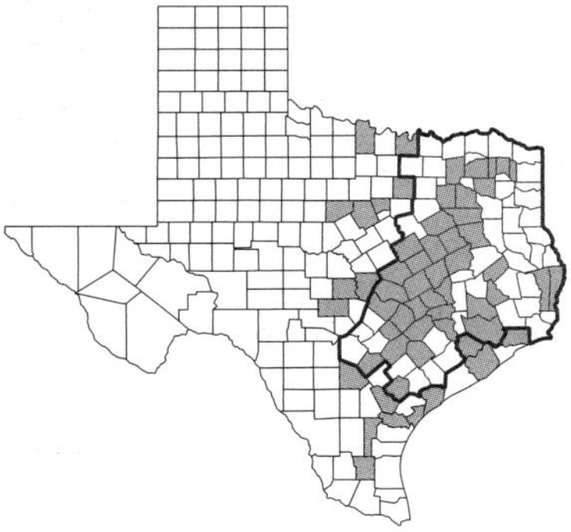

Krigia occidentalis

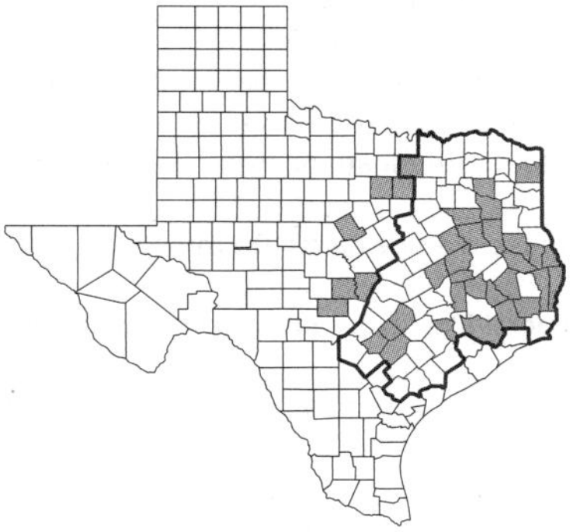

Krigia virginica

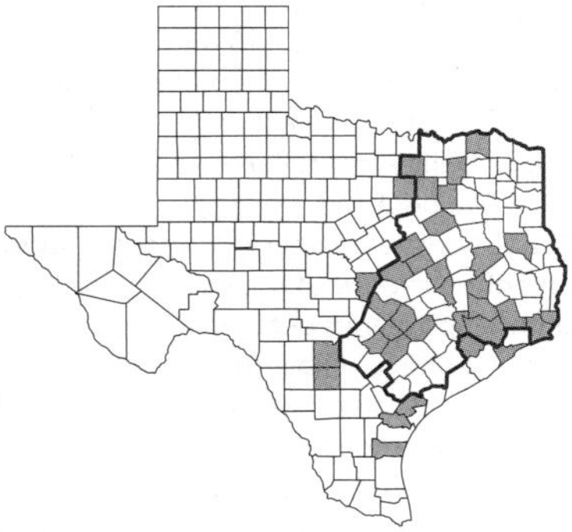

Krigia wrightii

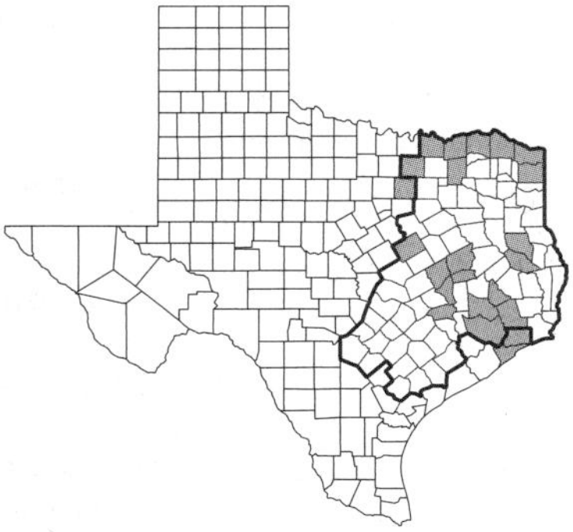

Lactuca canadensis

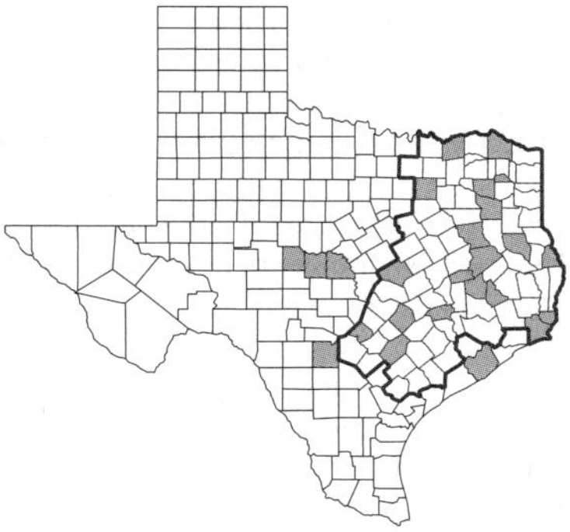

Lactuca floridana

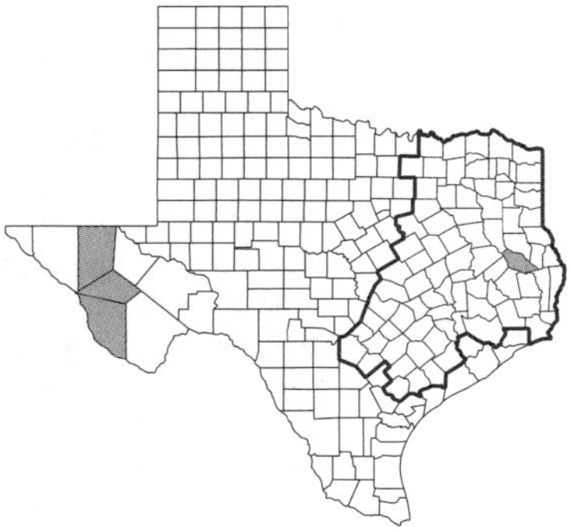

Lactuca graminifolia

blades serrate to pectinate to pinnately lobed, attenuate basally; heads solitary on long peduncles; receptacles naked; phyllaries in 3–4 series, unequal, margins and apices scarious; ray flowers pistillate, fertile, the ligules 3-toothed at apex, white, 12–20 (–30+) mm long, 2–10 mm wide; disk flowers perfect, fertile, the corollas yellow; achenes 1.5–2 mm long; pappus a very short crown. Open grassy areas, spreading in native prairie from "wildflower seed" plantings; nc TX in Grayson and Denton cos.; se-c TX in Robertson and Brazos cos. s to Harris Co.; escaping more commonly in the e U.S.; recorded throughout North America. May–Jul(–fall). Native of Europe and Asia. [*Chrysanthemum leucanthemum* L., *C. leucanthemum* var. *pinnatifidum* Lecoq & Lemotte, *Leucanthemum leucanthemum* Rydb.]

LIATRIS Gaertn. ex Schreb. BLAZINGSTAR, GAYFEATHER, BUTTON-SNAKEROOT

Herbaceous perennials from swollen, underground corms; stems erect; stem leaves alternate, simple, ± sessile, the blades entire, essentially linear, usually punctate with resinous dots; heads usually in spicate or racemose arrays; phyllaries in (2–)3–7 series, usually unequal, sometimes toothed or ciliate, sometimes resinous-dotted, sometimes petaloid, green to purplish, pink, or white; ray flowers absent; disk flowers perfect, fertile, corollas pink to purplish lavender or rarely white; achenes ± cylindrical, 8–11-ribbed, usually gland-dotted; pappus of 12–40 barbellate or plumose bristles, usually ± equaling corollas.

An e and c North American (including ne Mexico) genus of 37 species including some cultivated as ornamentals. Hybridization has been recorded between numerous species, sometimes causing morphological characters to intergrade where ranges overlap. The common name BUTTON SNAKEROOT is derived from the folk-medicinal use of the corms in treating snake bites (Ajilvsgi 1984; Kindscher 1992). (Derivation of generic name unknown) (tribe Eupatorieae)

REFERENCES: Gaiser 1946; Shinners 1959; Nesom & O'Kennon 2001; Nesom 2005a, 2006t.

1. Pappus bristles merely barbellate, the side hairs < 0.5 mm long, inconspicuous even with a lens.
  2. Lower stem leaves usually 1-nerved (sometimes 3-nerved); heads in loose spicate or racemose arrays, the central axis easily visible; flowers 10+ per head.
    3. Phyllaries usually reflexed, strongly puckered at the delimitation of the broadly hyaline, lacerate margin, the tips rounded or flattened ____ **L. aspera**
    3. Phyllaries erect to spreading or reflexed but not puckered, margins not (or only weakly) hyaline, entire or ciliate but not lacerate, the tips acuminate to rounded.
      4. Involucres (6–)8–12 mm high; corolla tubes pilose inside ____ **L. squarrulosa**
      4. Involucres 10–13 mm high; corolla tubes glabrous inside ____ **L. tenuis**
  2. Lower stem leaves 3–5 nerved; heads in dense spicate arrays, the central axis obscured; flowers < 10 per head.
    5. Phyllary tips erect and ± appressed, acuminate to acute; flowers (2–)3–4(–5) per head ____ **L. acidota** (in part)
    5. Phyllary tips reflexed to spreading, acute to acuminate or sometimes enlarged and ± petaloid; flowers (4–) 5–8 per head ____ **L. pycnostachya**
1. Pappus bristles feathery, plumose or subplumose, the side hairs > 0.5 mm long and usually conspicuous with a lens and often even with the naked eye.
  6. Inner phyllaries petaloid (with prolonged petal-like tips), conspicuously longer than the corollas or pappus, the tips whitish, yellowish, or pink ____ **L. elegans**
  6. Inner phyllaries not usually petaloid, the tips green or purplish and not exceeding the corollas or pappus.
    7. Stems and leaves with stalked glands ____ **L. glandulosa**
    7. Stems and leaves without stalked glands (but may be gland-dotted).
      8. Heads in open, branched arrays; peduncles 20+ mm long ____ **L. cymosa**
      8. Heads in unbranched racemose or spicate arrays; peduncles either absent or < 10 mm long.
        9. Lower stem leaves 1-nerved.
          10. Heads in loose spicate arrays, the central axis easily visible; flowers 8–14(–16) per head ____ **L. bracteata**
          10. Heads in dense spicate arrays, the central axis obscured; flowers 3–8 per head.
            11. Phyllaries in 2–3 series, weakly graduate in length, the outermost 1/2+ the length of the inner; lateral veins on phyllaries usually continuing to the tip; involucres 3.5–5 mm wide; flowering mostly July–August ____ **L. aestivalis**
            11. Phyllaries in 3–6 series, strongly graduate in length, the outermost 1/3 the length of the inner or less; lateral veins on phyllaries usually not evident on the upper 1/3; involucres (4–)5–7 mm wide; flowering mostly September–October ____ **L. punctata**
        9. Lower stem leaves 3–5 nerved.
          12. Involucres < 10 mm high; flowers < 10 per head ____ **L. acidota** (in part)
          12. Involucres > 10 mm high; flowers 10+ per head.
            13. Phyllaries ± equal (outer usually as long as inner); achenes 4–5.5 mm long ____ **L. squarrosa**
            13. Phyllaries ± unequal (outer shorter than inner); achenes 5.5–6.5 mm long ____ **L. hirsuta**

Liatris acidota (Engelm. & A. Gray) Kuntze, (possibly from Latin: *acidus*, sour, sharp, for the phyllary tips), GULF COAST GAYFEATHER, SHARP GAYFEATHER. Stems 20–90(–130) cm tall; stems and leaves glabrous; lower stem leaves 3–5-nerved; heads in dense

Lactuca canadensis [BB2]

Lactuca floridana [BB2, GLE]

Lactuca graminifolia [FIE]

Lactuca hirsuta [YAT]

Lactuca saligna [STR]

Lactuca ludoviciana [BB2, GLE]

Lactuca serriola [REE]

spicate arrays, the central axis usually obscured; involucres 6–7(–10) mm high; phyllaries glabrous, often purple, margins hyaline, the tips erect to appressed, acuminate to acute; flowers (2–)3–4(–5); achenes 4–4.5 mm long; pappus bristles barbellate to subplumose. Damp acidic soils, coastal prairies, oak-pine savannas; se TX; LA and TX. Jun–Oct.

Liatris aestivalis G.L. Nesom & R. O'Kennon, (summer), SUMMER GAYFEATHER. Stems 20–65 cm tall; glabrous; leaves glabrous, margins sometimes sparsely ciliate, lower stem leaves 1-nerved; heads in dense, cylindrical spicate arrays, the central axis obscured; involucres 9–12 mm high; phyllaries glabrous, often dark purple, margins not hyaline, the tips usually acute to acute-attenuate; flowers 3–4(–5); achenes 4.5–6 mm long; pappus bristles plumose. Shallow soils over limestone outcrops, sandy slopes; concentrated in nc TX; also Anderson, Hays, and Travis cos.; OK. Jul–Aug(–Sep).

Liatris aspera Michx., (rough), TALL GAYFEATHER, ROUGH GAYFEATHER. Stems 30–180 cm tall, hispidulous-puberulent, especially in the inflorescence; leaves glabrous or hispidulous-puberulent, often gland-dotted, lower stem leaves 1-nerved; heads in loose spicate arrays, the central axis usually easily visible; involucres (9–)10–16 mm high; phyllaries glabrous, usually reflexed, strongly puckered at the delimitation of the broadly hyaline lacerate margin, the tips rounded or nearly truncate; flowers (14–)18–24 (–30); achenes usually 4.5–6 mm long; pappus bristles barbellate. Sandy soils, prairies, old fields; e and se TX; e and nc U.S. (Jul–)Aug–Oct. [*L. aspera* Michx. var. *salutans* (Lunell) Shinners]

Liatris bracteata Gaiser, (with bracts), SOUTH TEXAS GAYFEATHER, COASTAL GAYFEATHER. Stems 25–75 cm tall, glabrous; leaves glabrous, gland-dotted, lower stem leaves 1-nerved; heads in loose spicate arrays, the central axis usually easily visible; involucres 11–12 (–15) mm high; phyllaries glabrous, erect to spreading, margins not hyaline, the upper margins sharply inrolled, forming a point at the tip; flowers 8–14(–16); achenes 6–9 mm long; pappus bristles plumose. Coastal prairies; Colorado and Harris cos., and Waller Co. (Poole et al. 2007); Gulf Coast of c TX. Sep–Nov. Lacking state or federal status, but a species of conservation concern because of its restricted habitat and range, as discussed in Poole et al. (2007)

Liatris cymosa (H. Ness) K. Schum., (for the branched inflorescence), AGGIELAND GAYFEATHER, BRANCHED GAYFEATHER. Stems 20–75 cm tall, strigose-puberulent; leaves glabrous, may be ciliate-margined, lower stem leaves 1-nerved; heads in open, branched arrays; involucres 14–17 mm high; phyllaries glabrous, sparsely short-pilose, margins ciliate, not hyaline, the tips rounded, sometimes with a small terminal point; flowers 20–25; achenes 6–7 mm long; pappus bristles plumose. Post oak woodlands, clay slopes, fencerows; endemic to a small region in ec TX in Brazos, Burleson, Grimes, Lee, Madison, Walker, and Washington cos., and San Jacinto Co. (Kartesz 2015). Jul–Oct. Lacking state or federal status, but a species of conservation concern because of its restricted habitat and range, as discussed in Poole et al. (2007)

Liatris elegans (Walter) Michx., (elegant), PINK-SCALE GAYFEATHER, HANDSOME BLAZINGSTAR. Stems 30–120 cm tall, puberulent to hirsute; leaves glabrous or sparsely pubescent, gland-dotted, lower stem leaves 1-nerved; heads in dense spicate arrays, the central axis sometimes visible; involucres 12–20 mm high; phyllaries strigose, erect to spreading, margins hyaline, the inner phyllaries elongated and petaloid, white to yellow or pinkish to light purple; flowers 4–5, corollas white or purple; achenes usually 3.5–5 mm long; pappus bristles plumose. Nesom 2006t recognized 3 varieties in East TX that may intergrade morphologically:

1. Petaloid phyllary tips usually yellowish to creamy white ______ var. **bridgesii**
1. Petaloid phyllary tips usually pink, lavender-blue, or magenta-purple.
    2. Corms broadly club-shaped in outline, 2–4 times as long as wide ______ var. **carizzana**
    2. Corms mostly spherical in outline, ca. as long as wide ______ var. **elegans**

var. **bridgesii** Mayfield (named for Edwin L. Bridges (1957–), noted plant collector of the se U.S.). Corms depressed-globose; petaloid phyllary tips usually yellowish to creamy white, rarely pale lavender. Post oak savannas, fields, sandy roadsides; ec TX from Wood Co. sw to Williamson and Bastrop cos. and se to Angelina Co. Jul–Nov.

var. **carizzana** Gaiser, (of the Carrizo Sands). Corms elongate and tapering, or sometimes ± globose; petaloid phyllary tips usually pink, lavender-blue, or magenta-purple. Deep sandy soils, dunes, live oak savannas, roadsides; sc TX from Bastrop and Fayette cos. sw to Bexar Co. and se to Goliad and Victoria cos.; s coastal TX plains. Jul–Oct.

var. **elegans.** Corms depressed-globose; petaloid phyllary tips usually pink, lavender-blue, or magenta-purple. Sandy clay soils, oak-pine savannas; Widespread in East TX; se U.S. from SC to TX. Aug–Oct. [*Liatris elegans* (Walter) Michx. var. *flabellata* (Small) Gaiser] A hybrid of *L. elegans* var. *elegans* and *L. tenuis* from Angelina and Newton cos. (*L.* ×*orzellii* G.L. Nesom) was described in White and Nesom (2012).

Liatris glandulosa G.L. Nesom & R. O'Kennon, (for the glandular pubescence), STICKY GAYFEATHER. Stems 30–60 cm tall; stems and leaves with stalked glands; lower stem leaves 1-nerved; heads in dense, cylindrical spicate arrays, the central axis obscured;

Leontodon hispidus [CPJ]

Leucanthemum vulgare [REE]

Liatris acidota [RBM]

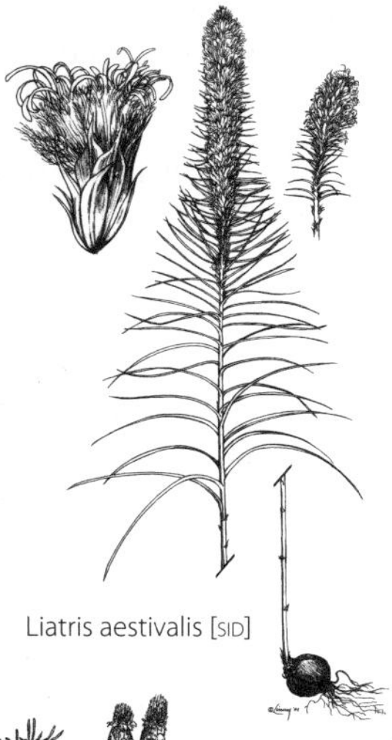

Liatris aestivalis [SID]

Liatris aspera [GLE]

Liatris bracteata [THO]

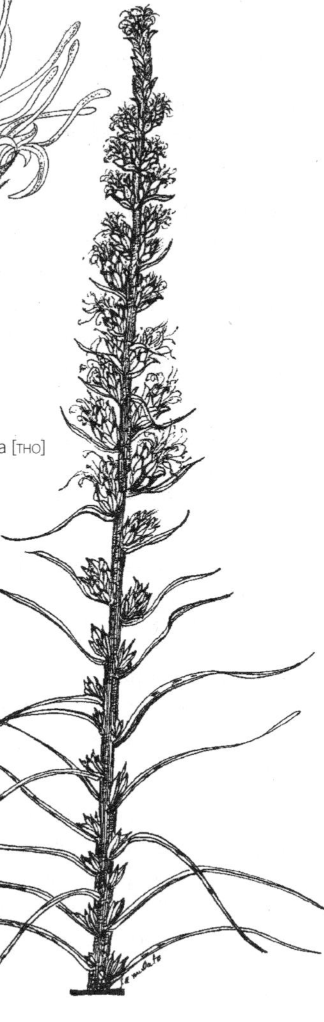

Liatris cymosa [TOR]

Liatris elegans var. elegans [FNA]

involucres 8–12 mm high; phyllaries densely glandular, green or rarely purplish, margins not hyaline, the tips acute to acuminate, sometimes with a small terminal point; flowers 3–4(–5); achenes 5–6 mm long; pappus bristles plumose. Limestone outcrops, shallow gravelly soils; Dallas, McLennan, and Travis cos.; Bosque Co. Aug–Oct.

Liatris hirsuta Rydb., (hairy), HAIRY GAYFEATHER. Stems 20–70 cm tall; stems and leaves pilose-hirsute; lower stem leaves 3–5-nerved; heads in loose racemose or spicate arrays, the central axis easily visible; involucres 11–17 mm high; phyllaries spreading to reflexed, usually sparsely hirsute, margins not hyaline, the tips acute-acuminate; flowers 15–30; achenes 5.5–6.5 mm long; pappus bristles plumose. Prairies, pine-oak woods, rocky slopes, roadsides; Harrison, Polk, and Smith cos. (Kartesz 2015); c U.S. from IA and NE s to TX and GA. Jul–Sep. [*L. squarrosa* (L.) Michx. var. *hirsuta* (Rydb.) Gaiser]

Liatris punctata Hook. (gland-dotted), PLAINS GAYFEATHER, DOTTED GAYFEATHER. Stems 15–85 cm tall, glabrous; leaves glabrous or sparsely pilose-hirsute, gland-dotted, margins sometimes ciliate, lower stem leaves 1-nerved; heads in dense spicate arrays, the central axis usually obscured; involucres 7–14 mm high; phyllaries erect to slightly spreading, glabrous, margins not hyaline, sometimes ciliate, the tips acute-acuminate to rounded-truncate, often with a small terminal point; flowers 3–8; achenes usually 6–8.5 mm long; pappus bristles plumose. Nesom 2006t recognized 2 varieties in East TX, which may intergrade morphologically:

1. Corms spherical; involucres 7–9(–12) mm high ____ var. **mucronata**
1. Corms elongate and branched, or large and obovate, not spherical; involucres 10–14 mm high ____ var. **punctata**

var. **mucronata** (DC.) B.L. Turner, (mucronate, pointed). Corms spherical; involucres 7–9(–12) mm high; phyllaries in (4–)5–6 series; flowers (3–)4–5(–6). Prairies, mesquite grasslands, silty clay soils; e to nc and sc TX; sc U.S. from MO s to LA and w to KS and TX. Jul–Nov. [*Liatris mucronata* DC.]

var. **punctata.** Corms elongate and branched, or large and obovate, not spherical; involucres 10–14 mm high; phyllaries in 4 (–5) series; flowers 4–8. Sagebrush prairies, gravelly slopes, calcareous soils; se TX; mostly w 1/2 TX; plains of c North America. Jul–Nov.

Liatris pycnostachya Michx., (thick-spiked), KANSAS GAYFEATHER, PRAIRIE GAYFEATHER, HAIRY BUTTON-SNAKEROOT. Stems (40–)60–120(–180) cm tall; stems and leaves glabrous to densely pilose-puberulent; leaves weakly gland-dotted, lower stem leaves 3–5-nerved; heads in dense spicate arrays, the central axis obscured; involucres 7–9(–10) mm high; phyllaries spreading to reflexed, glabrous to pilose, margins hyaline, sometimes ciliate, the tips acute-acuminate or sometimes petaloid; flowers (4–)5–8; achenes usually 3.8–4.5 mm long; pappus bristles barbellate. At one time, a poultice of the roots was used in treating snakebite, leading to the common name (Tveten & Tveten 1993). Nesom 2006t recognized 2 varieties in East TX, which may intergrade morphologically:

1. Corms globose to elongate; lower stems moderately to densely pilose-puberulent; leaves moderately to densely pilose-puberulent to nearly glabrous; typically flowering from July–Sept(–Nov) ____ var. **lasiophylla**
1. Corms globose; lower stems usually glabrous or sparsely pilose; leaves glabrous; typically flowering June–Aug ____ var. **pycnostachya**

var. **lasiophylla** Shinners, (downy- or woolly-leaved). Corms globose to elongate; lower stems moderately to densely pilose-puberulent; leaves moderately to densely pilose-puberulent to nearly glabrous. Sandy woods, bogs, pine savannas; scattered from ne to se TX; LA, MS, and TX. Jul–Nov. [*L. serotina* (Greene) K. Schum.]

var. **pycnostachya.** Corms globose; lower stems usually glabrous or sparsely pilose; leaves glabrous. Prairies, rocky ridges, sandy woods, roadsides; se TX; c U.S. from WI s to MS and w to ND and TX. Jun–Jul. [*L. bebbiana* Rydb.] Frequently grown as an ornamental. A hybrid of *L. pycnostachya* and *L. squarrosa* var. *glabrata* (*L.* ×*johnsonii* M. White) from Lamar Co. was described in White and Nesom (2012).

Liatris squarrosa (L.) Michx., (with parts spreading or recurved), LOOSESCALE GAYFEATHER, COLICROOT. Stems 30–80 cm tall; stems and leaves glabrous to pilose, puberulent-strigose, or villous; leaves weakly gland-dotted, lower stem leaves 3–5-nerved; heads borne singly or in loose racemose or spicate arrays, the central axis easily visible; involucres 12–20(–25) mm high; phyllaries erect to spreading-reflexed at tips, glabrous or pubescent, margins sometimes ciliate, not hyaline, the tips acute to acuminate; flowers (10–)23–45(–60); achenes 4–5.5 mm long; pappus bristles plumose. Nesom 2006t recognized 2 varieties in East TX, which may intergrade morphologically:

1. Stems, leaves, and outer phyllaries glabrous; widespread in East TX ____ var. **glabrata**
1. Stems usually puberulent-strigose or villous; leaves pilose to nearly glabrous; phyllaries glabrous or pubescent, the margins usually ciliate; in East TX usually restricted to counties near the Red River ____ var. **squarrosa**

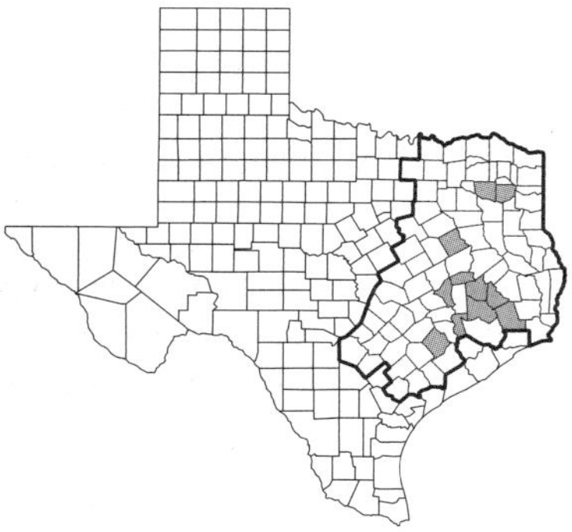
Lactuca hirsuta

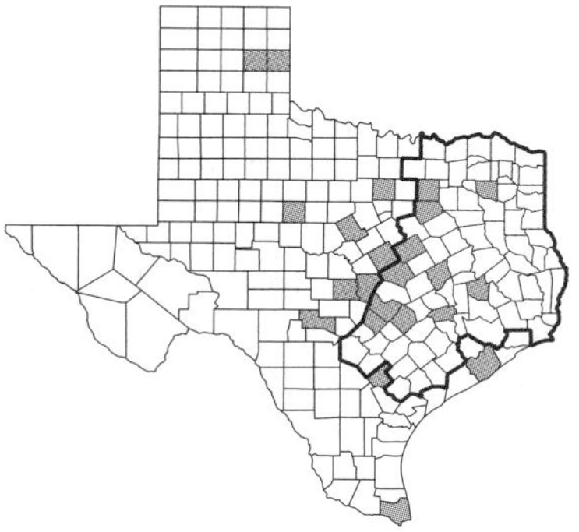
Lactuca ludoviciana

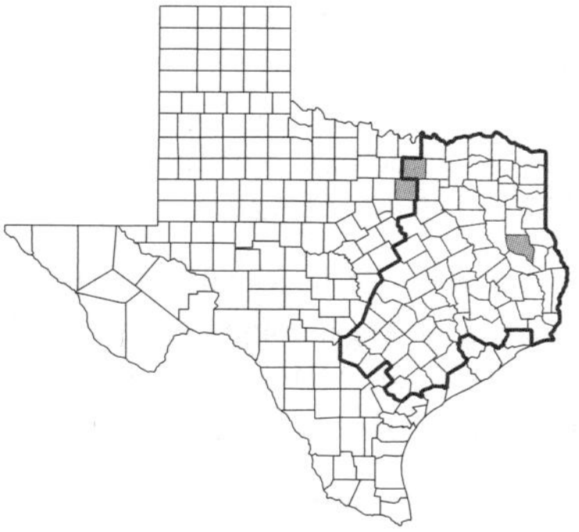
Lactuca saligna

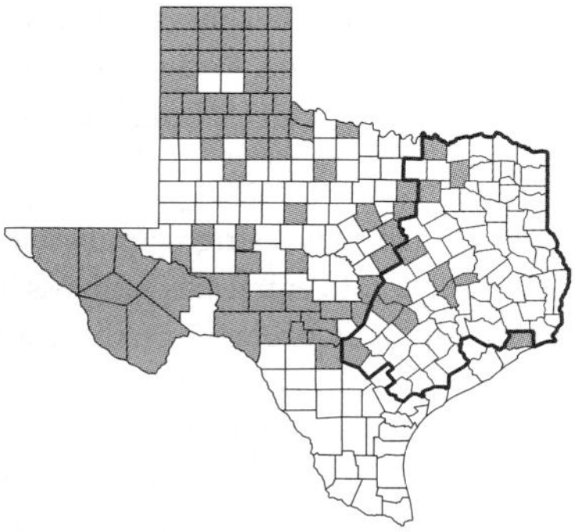
Lactuca serriola

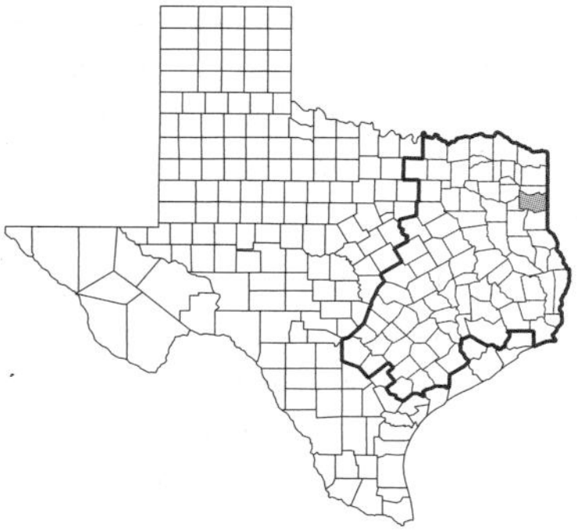
Leontodon hispidus

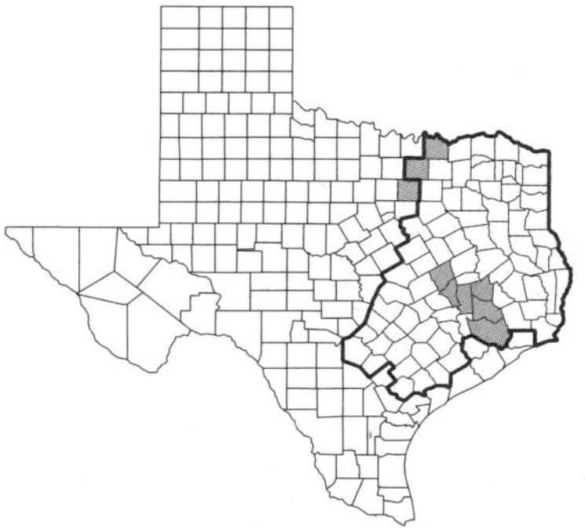
Leucanthemum vulgare

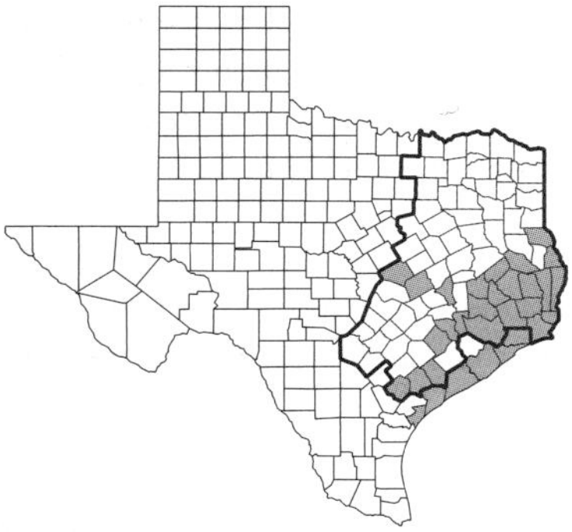
Liatris acidota

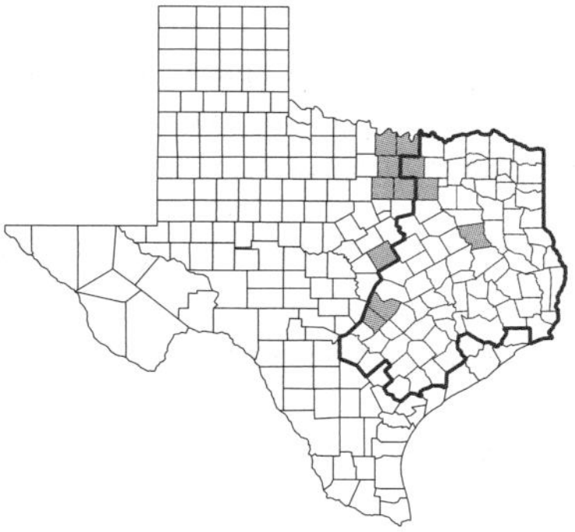
Liatris aestivalis

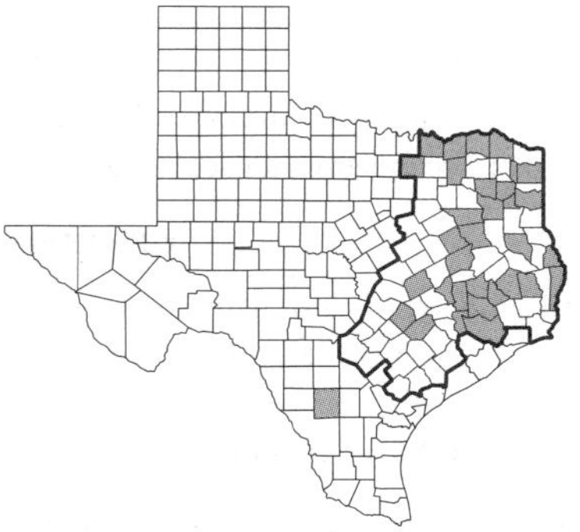
Liatris aspera

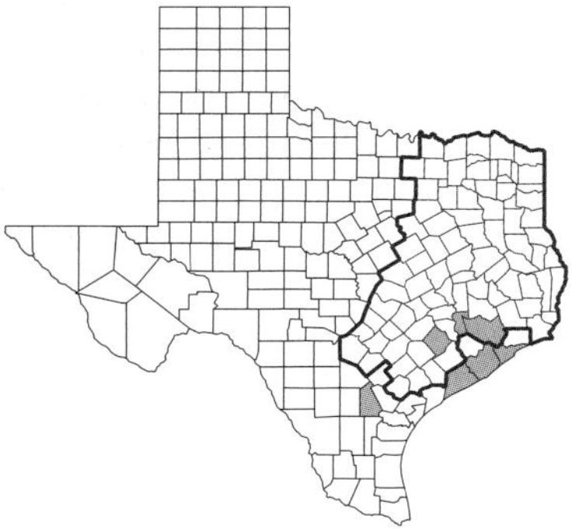
Liatris bracteata

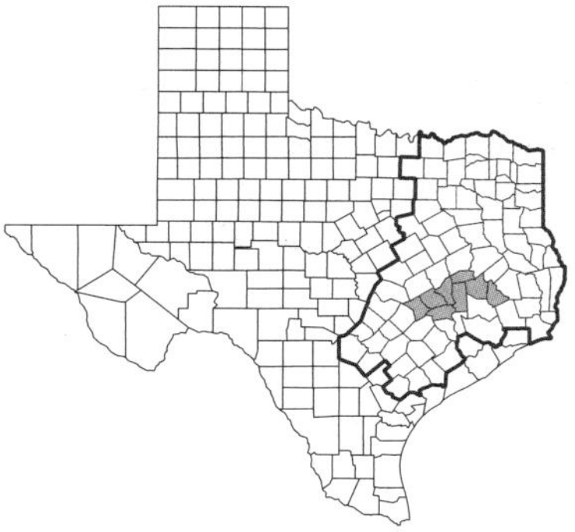
Liatris cymosa

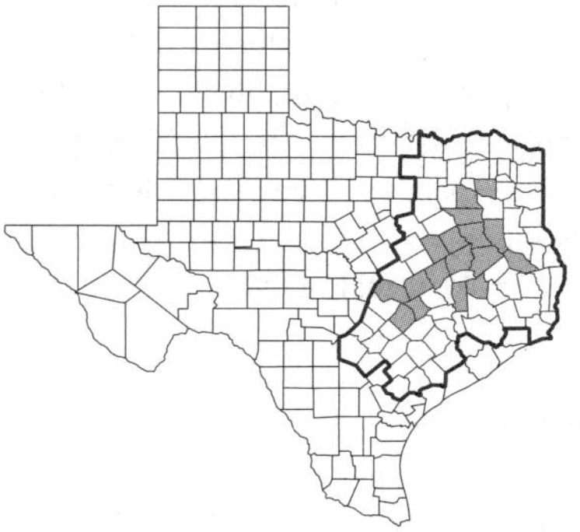
Liatris elegans var. bridgesii

var. **glabrata** (Rydb.) Gaiser, (somewhat glabrous, becoming glabrous), SMOOTH GAYFEATHER. Stems, leaves, and outer phyllaries glabrous. Prairies, sandy and gravelly soils, roadsides; near the Red River in ne TX in Denton, Fannin, Grayson, Hunt, Lamar, and Red River cos.; Cross Timbers and Panhandle; c U.S. from MN s to AR and w to SD, CO, and TX. Jul–Aug. [*Liatris glabrata* Rydb.] A hybrid of *L. squarrosa* var. *glabrata* and *L. pycnostachya* from Lamar Co. (*L.* ×*johnsonii* M. White) was described in White and Nesom (2012).

var. **squarrosa.** Stems usually puberulent-strigose or villous, the hairs soft, upturned, or appressed; leaves pilose to nearly glabrous; phyllaries glabrous or pubescent, the margins usually ciliate. Sandy-loam soils, oak-pine woods, fencerows; mainly se and e TX; e U.S. from MD s to FL and w to MI, MO, and TX. Jun–Sep. [*L. glabrata* Rydb. var. *alabamensis* (Alexander) Shinners, *L. squarrosa* (L.) Michx. var. *gracilenta* Gaiser]

Liatris squarrulosa Michx., (with parts slightly spreading or recurved at the ends), SOUTHERN GAYFEATHER, APPALACHIAN BLAZING-STAR. Stems 25–80(–130) cm tall; stems puberulent; leaves glabrous to puberulent, weakly gland-dotted, if at all, lower stem leaves 1-nerved; heads in loose to dense racemose or spicate arrays, the central axis usually visible; involucres (6–)8–12 mm high; phyllaries erect to spreading-reflexed at tips, glabrous or pubescent, margins sometimes weakly hyaline, the tips acute to obtuse to rounded; flowers 11–26(–28); achenes usually 3.5–5.5 mm long; pappus bristles barbellate. Rocky ridges, oak-hickory-pine woods, prairies, fencerows; recorded from Harris, Sabine, and Walker cos. (Kartesz 2015); e U.S. from VA s to GA and w to IL, OK, and TX. Aug–Nov.

Liatris tenuis Shinners, (slender), SHINNERS' GAYFEATHER, SLENDER GAYFEATHER. Stems 30–55 cm tall; stems strigose-puberulent; leaves sparsely pilose, gland-dotted; lower stem leaves 1(–3)-nerved; heads in loose spicate arrays, the central axis easily visible; involucres 10–13 mm high; phyllaries nearly glabrous or sparsely pilose, margins not hyaline, the tips loose and spreading, acute to acuminate, innermost sometimes obtuse; flowers 10–12; achenes 4.2–4.5 mm long; pappus bristles barbellate or plumose near the bases. Sandy soils of fire-maintained upland longleaf pine savannas; deep se TX in Angelina, Jasper, Newton, Sabine, San Augustine, and Tyler cos.; LA and TX. Jun–Sep. Lacking state or federal status, but a species of conservation concern because of its restricted range and habitat loss, as discussed in Poole et al. (2007). A hybrid of *L. tenuis* and *L. elegans* var. *elegans* from Angelina and Newton cos. (*L.* ×*orzellii* G.L. Nesom) was described in White and Nesom (2012). ⚠

Liatris spicata (L.) Willd. var. *spicata*, (spiked), which would not have keyed to any other species satisfactorily here, can be distinguished by the 3–5-nerved lower leaves, heads in dense spicate arrays, involucres 7–11 mm high, phyllaries erect to appressed with tips rounded to obtuse, flowers (4–)5–8(–14) per head, and barbellate pappus bristles. To be expected in se TX based on records from adjacent parishes in LA (Beauregard, Calcasieu, and Vernon parishes); e 1/2 of North America. Jul–Sep.

LINDHEIMERA A. Gray & Engelm. TEXAS-STAR, STAR DAISY

A monotypic genus native to the sw U.S. and n Mexico. (Named for Ferdinand Jacob Lindheimer, 1801–1879, German-born collector of TX plants and correspondent of Asa Gray and George Engelmann, resided in New Braunfels—Geiser 1948) (tribe Heliantheae)

REFERENCES: Turner & Johnston 1956; Turner & Woodruff 1993; Strother 2006t.

Lindheimera texana A. Gray & Engelm., (of Texas), TEXAS-STAR, YELLOW TEXAS-STAR, LINDHEIMER'S DAISY. Hispid-pubescent taprooted annual 10–30(–65) cm tall; leaves alternate and subopposite or the uppermost opposite, blades oblanceolate to rhombic- or ovate-lanceolate, entire or coarsely toothed, sessile or short-petioled; heads usually solitary or few and corymbose, on peduncles 1–3(–6) cm long; involucres 10–15 mm tall, the phyllaries in 2 series, the outer series narrower, longer, and more acute than the conspicuous inner; ray flowers usually 5, pistillate and fertile, the ligules ca. 1 cm long, yellow or orange-yellow, 2-toothed at apex; disk flowers few, functionally staminate, dull yellow; achenes 4.5–6 mm long, flattened, each shed with its subtending phyllary; pappus absent, shoulders of achene wings projecting and sometimes interpreted as 2 stout awns. Clayey or occasionally sandy prairies, roadsides; common from nc to sc TX; Post Oak Savannah and Blackland Prairie w to Rolling Plains and Edwards Plateau; LA, OK, and TX; n Mexico. Mar–May.

LYGODESMIA D. Don SKELETON-WEED, SKELETON-PLANT

A genus of ca. 5 species of the Americas, especially w North America. (Greek: *lygos*, a plant twig, and *desme*, a bundle, from clustered stick-like or twiggy stems) (tribe Cichorieae)

REFERENCES: Vuilleumier 1973; Tomb 1980; Bogler 2006d.

Lygodesmia texana (Torr. & A. Gray) Greene, (of Texas), TEXAS SKELETON-PLANT, RUSH-PINK, ROSE-RUSH, RUSHWEED, PURPLE-DANDELION, FLOWERING-STRAW, MILK-PINK. Glabrous and glaucous perennial with deep, thick, woody root producing in age numerous vertical

Liatris glandulosa [SID]

Liatris hirsuta [YAT]

Liatris punctata var. punctata [YAT]

Liatris pycnostachya [CO1, GLE]

Liatris squarrulosa [BT3]

Liatris squarrosa [BB2]

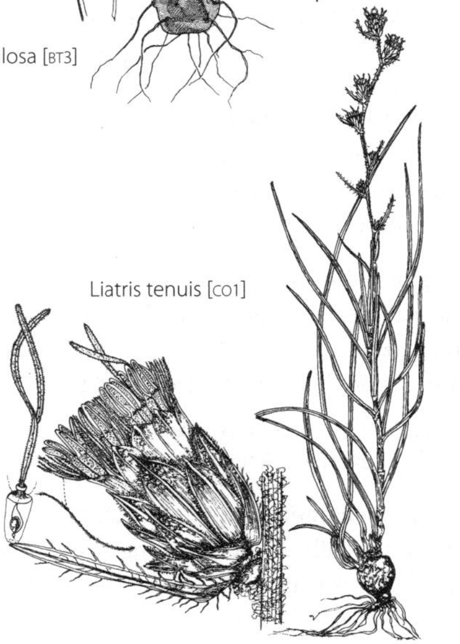
Liatris tenuis [CO1]

to oblique rhizomes, terminating in above-ground stems, forming clumps; sap milky; stems 20–65 cm tall, with few reduced leaves or bracts, openly and stiffly branched toward summit, blades mostly basal, linear to lanceolate, entire or usually deeply pinnately lobed; heads large; involucres subcylindrical, 18–25 mm high; phyllaries in one row, linear, equal in length, margins scarious; flowers all ligulate, with heavy sweet scent; corollas purple to pale lavender (rarely white), open during morning hours; achenes 13–17 mm long; pappus of hair-like bristles, 10–15 mm long. Limestone outcrops and rocky calcareous soils; widespread in TX except the extreme e part of the state; NM, OK, and TX; n Mexico. Apr–Nov. [*L. aphylla* (Nutt.) DC. var. *texana* Torr. & A. Gray]

MARSHALLIA Schreb. BARBARA'S BUTTONS

Herbaceous perennials, usually 20–80 cm tall; leaves in a basal rosette, alternate on the stems, entire, glabrous; heads 20–32 mm wide; phyllaries in ± 2 series, subequal, herbaceous; ray flowers absent; disk flowers sweet-scented, corollas pale lavender, purple, pink, or white, usually hairy, lobes long and often contorted; achenes 5-angled, 10-ribbed; pappus of 5(–6) scales.

A genus of seven species native to the e U.S. Watson and Estes (1990) indicated that morphological variation is continuous among the species and that there are few abrupt boundaries. (Named for Dr. Moses Marshall, 1785–1813, nephew of the better-known early American pioneer Humphrey Marshall) (tribe Heliantheae)
REFERENCES: Channell 1957; Watson & Estes 1990; Watson 2006c.

1. Corollas usually white, sometimes pale lavender; phyllaries 9–12 mm long; flowering primarily Apr–Jun ____ **M. caespitosa**
1. Corollas usually pale lavender to purple, rarely white; phyllaries 4–9.5 mm long; flowering primarily Jul–Oct ____ **M. graminifolia**

Marshallia caespitosa Nutt. ex DC., (growing in tufts), PUFFBALLS, BARBARA'S BUTTONS. Aerial stems unbranched, usually clustered; basal leaves linear, linear-elliptic, or linear-oblanceolate; heads borne singly; phyllaries 9–12 mm long; corollas usually white, sometimes pale lavender. Clay prairies, limestone outcrops, sandy soils; East TX w to nc TX, e Rolling Plains, Edwards Plateau, and sc TX; sc U.S. from MO s to LA and w to KS and TX. Apr–Jun. [*M. caespitosa* Nutt. ex DC. var. *signata* Beadle & F.E. Boynt.] We are following Watson (2006c) in not recognizing varieties.

Marshallia graminifolia Walter (Small), (grass-leaved), GRASS-LEAVED BARBARA'S BUTTONS. Aerial stems simple, or branched in the lower portion; basal leaves oblanceolate; heads borne singly or in open, long-pedunculate, corymbose arrays; phyllaries 4–9.5 mm long; corollas usually pale lavender to purple, rarely white. Wet soils, pine savannas, pitcher plant bogs; se TX; se U.S. from NC s to FL and w to TX. Jul–Oct. [*M. graminifolia* (Walter) Small subsp. *tenuifolia* (Raf.) L. Watson, *M. tenuifolia* Raf.] If varieties are recognized, ours would be var. *cynanthera* (Elliott) Beadle & F.E. Boynt.

Marshallia trinervia (Walter) Trel., (3-nerved), BROAD-LEAVED BARBARA'S BUTTONS, which would key to *M. caespitosa* here (but would more likely be mistaken for *Monarda fistulosa*—WILD BERGAMOT or LAVENDER BEEBALM, in the Lamiaceae), can be distinguished by its ovate leaves, which are not reduced up the stem, and its pink corollas. To be expected in East TX based on records from adjacent Vernon Par., LA; se U.S. from NC s to GA and w to LA. May–Jun.

MATRICARIA L. MAYWEED, GERMAN CHAMOMILE

Taprooted annuals; often aromatic; to 80 cm tall (but usually much smaller); leaves alternate, usually 2–3-pinnately lobed, the bases sheathing or clasping, ultimate margins entire; phyllaries in several series, subequal, margins and tips scarious; ray flowers pistillate and fertile or absent, corollas white (when present); disk flowers perfect and fertile, corollas yellow to greenish yellow; achenes ribbed, glabrous, usually asymmetric; pappus of a small crown, or auricles (on ray achenes), or absent.

A genus of 7 species native to North America, Eurasia, and n Africa; several have a long history of cultivation for herbal and medicinal uses; a few are weedy. (Greek: *matrix*, womb, and *aria*, pertaining to, alluding to reputed medicinal properties) (tribe Anthemidae)
REFERENCES: Rydberg 1916; Brouillet 2006b.

1. Heads with both ray and disk flowers, ray corollas white; plants smelling of sweet hay when bruised ____ **M. chamomilla**
1. Heads with only disk flowers; plants smelling of pineapple when bruised ____ **M. discoidea**

Matricaria chamomilla L., (chamomile), GERMAN CHAMOMILE. Plants smelling of sweet hay when bruised; heads borne singly; peduncles usually 20–50 mm long; ray flowers present, corollas white; disk corollas yellow to yellowish green, 1.6–1.8 mm long, lobes 5; achenes 0.75–0.9 mm long; pappus usually absent, sometimes a small crown, or toothed auricles on ray achenes. Disturbed sites, roadsides, persisting in abandoned gardens; Gonzales Co.; also s TX in Nueces, Starr, and Webb cos.; scattered across much of North America; Eurasia. Apr–Jul. [*Chamomilla recutita* (L.) Rauschert, *M. courrantiana* DC., *M. recutita* L.]

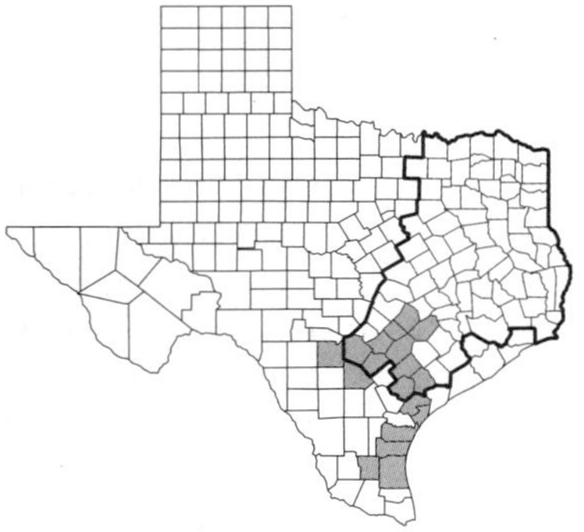

Liatris elegans var. carizzana

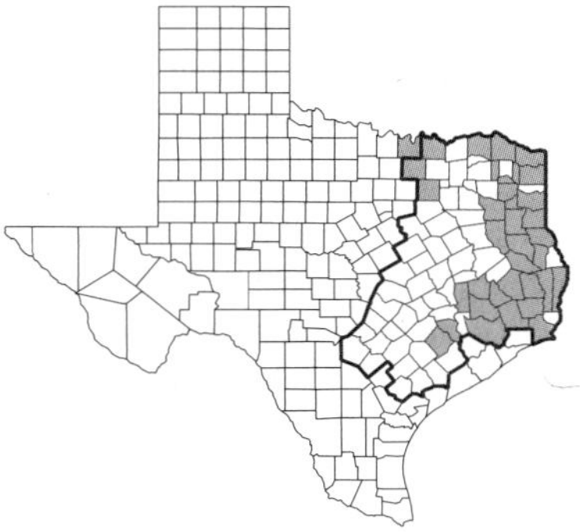

Liatris elegans var. elegans

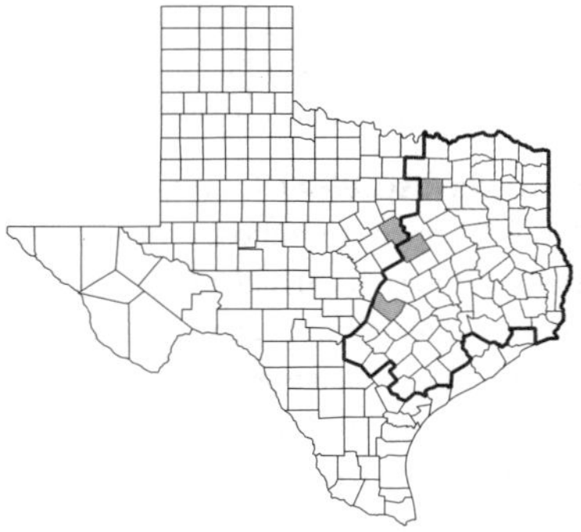

Liatris glandulosa

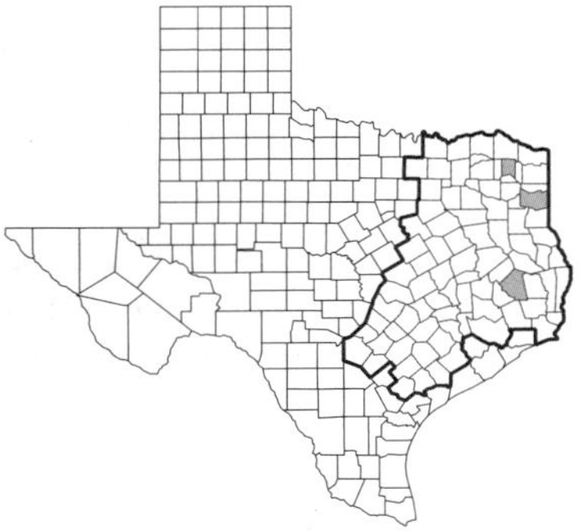

Liatris hirsuta

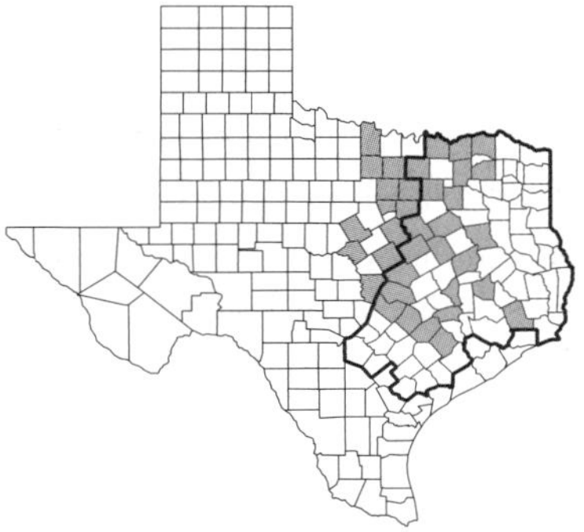

Liatris punctata var. mucronata

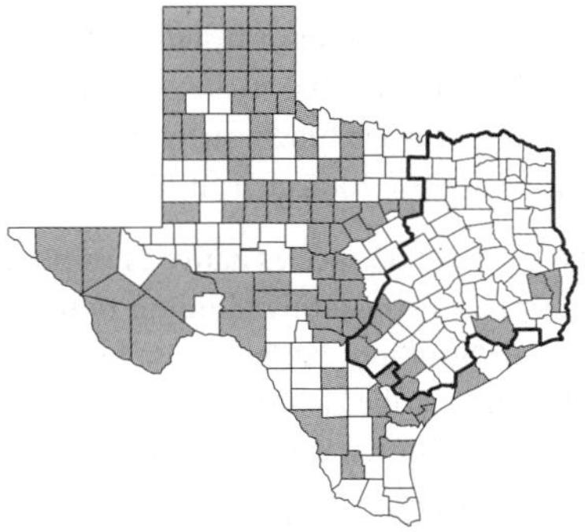

Liatris punctata var. punctata

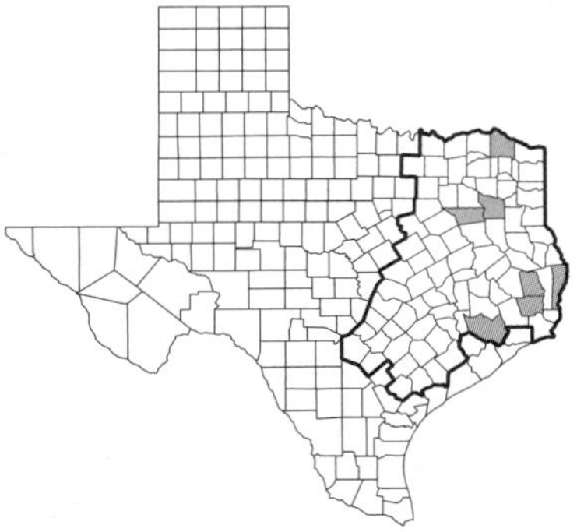

Liatris pycnostachya var. lasiophylla

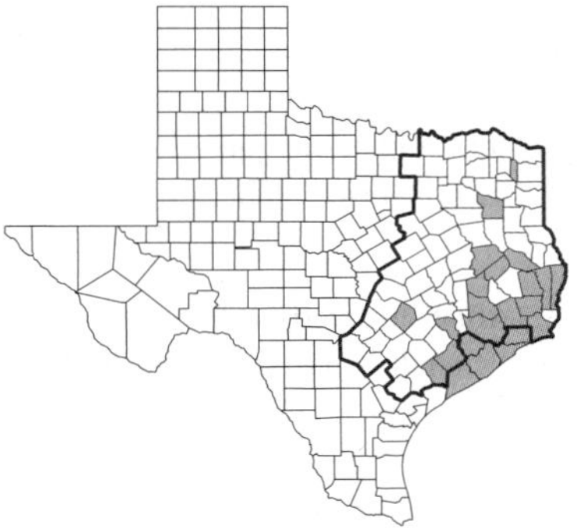

Liatris pycnostachya var. pycnostachya

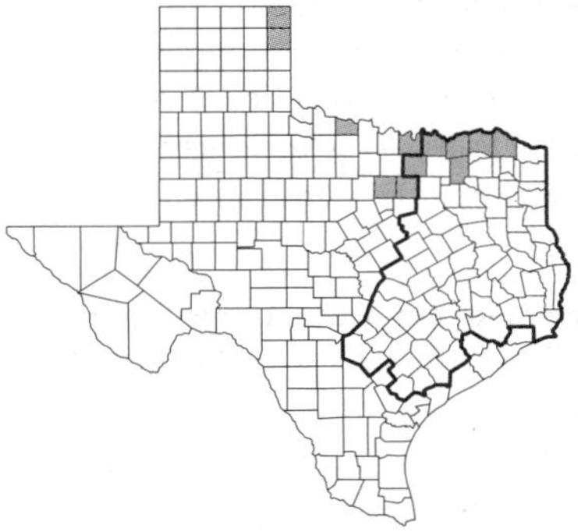

Liatris squarrosa var. glabrata

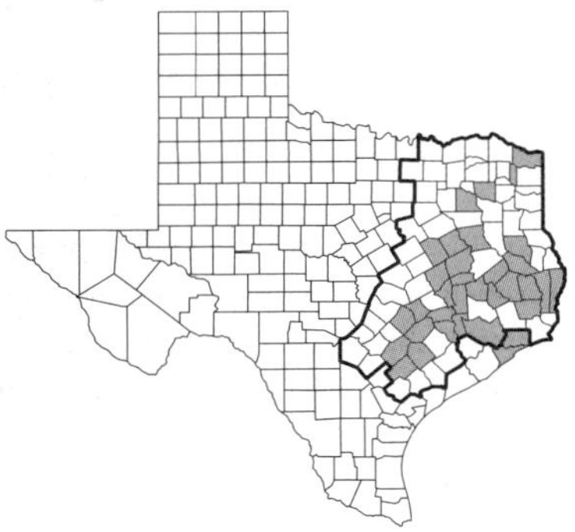

Liatris squarrosa var. squarrosa

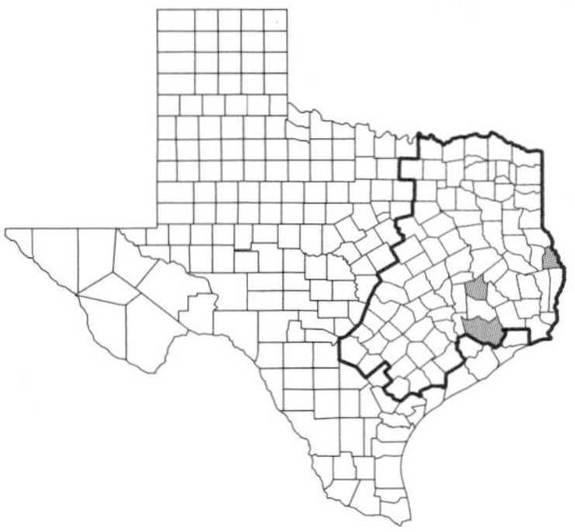

Liatris squarrulosa

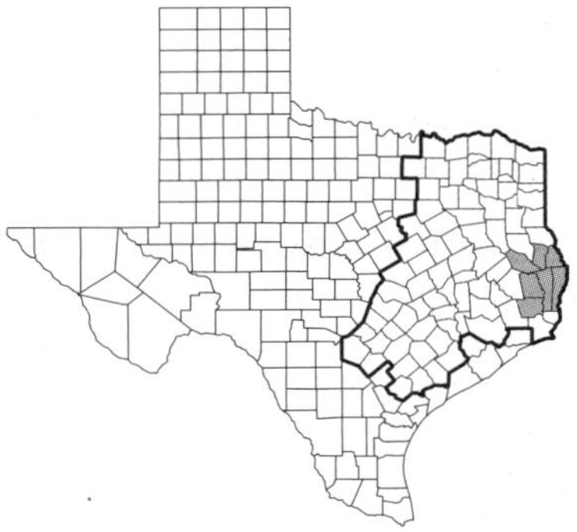

Liatris tenuis

Native of Eurasia. A popular herbal remedy of long historical use, particularly as a tea made from the dried heads, it has been found to have antimicrobial and anti-inflammatory properties (McKay & Blumberg 2006). *M. chamomilla* closely resembles *Anthemis nobilis*, ROMAN CHAMOMILE, in appearance, and has similar uses, but is not as sweetly-scented.

Matricaria discoidea DC., (disc-shaped), DISC MAYWEED, PINEAPPLE WEED. Plants smelling of pineapple when bruised; heads borne singly or in open corymbose arrays; peduncles usually 2–25 mm long; ray flowers absent; disk corollas yellowish green, 1.1–1.3 mm long, lobes 4(–5); achenes 1.15–1.5 mm long; pappus a small crown. Disturbed sites, roadsides, persisting in abandoned gardens; collected in Brazos Co. in 1959 (*F.W. Gould 8640* TAES); also recorded in Brown Co. (iNaturalist.org observation *2918215, Donna Bohnsack, Apr 2016* (iNaturalist.org 2016)); across most of North America except deep se U.S. states; introduced in Eurasia and Australia. Spring–summer. [*Chamomilla discoidea* (DC.) J. Gay ex A. Braun, *C. suaveolens* (Pursh) Rydb., *Santolina suaveolens* Pursh]

MELAMPODIUM L. BLACK-FOOT DAISY, ROCK DAISY

Ours herbaceous perennials or weak subshrubs; stems erect or prostrate; leaves opposite; heads borne singly; phyllaries in 2 series, the outer 5 often ± fused and forming a cupule; each inner phyllary subtending and clasping a ray flower ovary, creating a fruit-like structure that encloses the achene and is shed with it upon maturity; ray flowers pistillate and fertile, ligules white (in ours); disk flowers functionally staminate, yellow to orange; achenes each partially enclosed within and shed with the subtending phyllary as a conical, fruit-like structure; pappus absent.

A genus of 36 species native to tropical and warm areas of the Americas, especially Mexico. (Greek: *melas*, black, and *podos*, foot or base, apparently alluding to the fruit-like structure enclosing the mature achene, which turns black upon maturity; however, Strother (2006u) maintains this is erroneous, and the name derives from *Melampus*, a soothsayer renowned in Greek mythology) (tribe Heliantheae)

REFERENCES: Turner & King 1962; Stuessy 1971, 1972, 1979; Stuessy & Crisci 1984; Strother 2006u.

1. Outer 5 phyllaries united 1/3 their lengths or less; ray ligules 2–8+ mm long and 1–3+ mm wide __________ **M. cinereum**
1. Outer 5 phyllaries united 1/2 their lengths or more; ray ligules 7–13 mm long and 2.5–8 mm wide __________ **M. leucanthum**

Melampodium cinereum DC., (ashy, for the soft grayish hairs on the leaves), SOUTH TEXAS ROCK DAISY, HOARY BLACK-FOOT. Herbaceous or weakly subshrubby perennial; stems 10–20+ cm tall; leaves sessile, blades linear-oblong to linear, 10–55 mm long, usually pinnately lobed; outer 5 phyllaries united 1/3 their lengths or less; ray flowers 7–13, ligules 2–8+ mm long, 1–3+ mm wide; fruits enclosing achenes 1.3–2.2 mm long. Open sites, gravel, juniper scrublands, grasslands; on sw edge of East TX in Bexar Co.; Edwards Plateau and South TX Plains; Mexico. Feb–Oct(–Nov). If varieties are recognized, ours would be var. *cinereum*.

Melampodium leucanthum Torr. & A. Gray, (white-flowered), BLACK-FOOT DAISY, ROCK DAISY, PLAINS BLACK-FOOT. Herbaceous or weakly subshrubby perennial; stems 12–40(–60) cm tall; leaves sessile, blades lanceolate to linear-oblong or linear, 20–35(–45) mm long, entire to pinnately lobed; outer 5 phyllaries united 1/2 their lengths or more; ray flowers 8–13, ligules 7–13 mm long, 2.5–8 mm wide; fruits enclosing achenes 1.5–2.6 mm long. Calcareous soils, juniper scrublands, grasslands, roadcuts; w edge of East TX from McLennan Co. sw to Bexar Co.; also Washington Co.; Lampasas Cut Plain and East Cross Timbers w to High Plains and Trans-Pecos; Mexico. Apr–Sep. Increasingly popular as a native xeriscaping ornamental.

MICROPSIS DC. STRAIGHTJACKETS

A genus of 5 species of the sc U.S. and s South America; resembling and related to *Diaperia* (*Evax*). (Greek: the genus *Micropus*, and *opsis*, resembling, for its similarity to that genus) (tribe Inuleae)

REFERENCES: Beauverd 1913; Morefield 2004, 2006b.

Micropsis dasycarpa (Griseb.) Beauverd, (thick-fruited), BIG-HEAD STRAITJACKETS. Very small annuals, to 10 cm tall, taprooted; leaves alternate, spatulate, silvery or greenish gray; heads small, 4–5 mm high and inconspicuous, congested in dense terminal clusters, subtended by leafy bracts, usually embedded in white cottony pubescence; ray flowers absent; disk flowers of 2 series, most pistillate and fertile, a few perfect and fertile and enclosed by pales (chaff); achenes minute; pappus usually absent. Known only from one site in the U.S., in drying mud at the edge of a Gulf Coast Prairie pool in Victoria Co. (*R.C. Barneby 18201* TEX), where it is either a recent introduction or possibly an overlooked native species; se South America. Mar–May. [*Filago dasycarpa* Griseb.] ?

Lindheimera texana [HEA]

Lygodesmia texana [SBM]

Marshallia caespitosa [BT3]

Marshallia graminifolia [RAD, SM1]

Matricaria chamomilla [BEN]

MIKANIA Willd. HEMPWEED, CLIMBING HEMPWEED

Perennial vines twining to 3+ m; leaves opposite, petiolate, ovate to cordate or deltoid; heads arranged in spicate or corymbose panicles; phyllaries 4, ± equal; ray flowers absent; disk flowers 4, corollas white to cream or pinkish to purplish; achenes usually 5-ribbed; pappus of 30–60 bristles.

A mainly tropical genus of ca. 450 species of climbers; our two species are the only Asteraceae in the East TX flora that are twining vines. (Named for Joseph Gottfried Mikan, 1743–1814, professor at the University of Prague) (tribe Eupatorieae)
REFERENCES: Holmes 1981, 1990, 2006.

1. Stems 6-angled; heads 7–10 mm high; phyllaries 6–8 mm long, elliptic-ovate, acute to slightly rounded at tips; corollas white, the lobes linear; achenes 3–4 mm long; rare species of deep se TX ____ **M. cordifolia**
1. Stems rounded or slightly 6-angled; heads 6–7 mm high; phyllaries 5–6 mm long, linear-lanceolate, acuminate; corollas pinkish or purplish, sometimes white, the lobes triangular; achenes 1.8–2.2 mm long; common across e and se TX ____ **M. scandens**

Mikania cordifolia (L. f.) Willd., (heart-leaved), FLORIDA KEYS HEMPVINE, GUACO. Perennial twining vine; stems 6-angled; leaf blades deltoid to ovate, 5–10 cm long, acute to acuminate, bases cordate, margins subentire to undulate-dentate; heads 7–10 mm high; phyllaries 6–8 mm long, elliptic to narrowly ovate, acute to slightly rounded at tips; disk corollas white, the lobes linear; achenes 3–4 mm long, sparsely gland-dotted; pappus white, 4–5 mm long. Low woodlands, seeps, along streams; in TX recorded only from Newton and Tyler cos.; se U.S. from GA s to FL and w to TX; West Indies; Mexico; Central and South America. Sep–Dec. [*Cacalia cordifolia* L. f.]

Mikania scandens (L.) Willd., (scandent, climbing), CLIMBING HEMPWEED, HEMPVINE, CLIMBING-BONESET. Perennial twining vine; stems slightly 6-angled or rounded; leaf blades deltoid to ovate, 3–15 cm long, generally acuminate, bases cordate to hastate, margins subentire to undulate, crenate, or dentate; heads 6–7 mm high; phyllaries 5–6 mm long, linear-lanceolate, acuminate; disk corollas pinkish to purplish, sometimes white, the lobes triangular; achenes 1.8–2.2 mm long, densely gland-dotted; pappus pinkish or purplish, 4–4.5 mm long. Low woodlands, swamps, thickets along streams; e and se TX w to East Cross Timbers and Edwards Plateau; also far s TX; e and se U.S. from ME s to FL and w to MI and TX; West Indies and Mexico. Jun–Dec. [*Eupatorium scandens* L., *M. scandens* (L.) Willd. var. *pubescens* (Nutt.) Torr. & A. Gray]

ONOPORDUM L. COTTON-THISTLE, SCOTCH-THISTLE

A genus of ca. 60 species of prickly herbs native to Europe, the Mediterranean region, and w Asia. (Ancient Greek name of the plant, from *onos*, donkey, and *porde*, flatulence, Pliny stating that it produced flatulence when eaten by donkeys) (tribe Cardueae)
REFERENCES: Dress 1966; Keil 2006f.

Onopordum acanthium L., (prickly), SCOTCH-THISTLE. Coarse biennial 0.5–3(–4) m tall, often gray from sparse to dense cottony pubescence; stems with wings 5–15 mm wide, these with spine-tipped lobes; leaves alternate, lobed or toothed, conspicuously spiny-margined, sessile, decurrent; heads terminal, solitary or 2–3 in ± loose clusters, large (2.5+ cm in diam.); phyllaries in 8–10 series, linear-subulate, ascending-spreading, tapering from base to spine-tip; ray flowers absent; disk corollas purplish or pinkish, sometimes white, slender, 22–25 mm long; achenes 4–5 mm long; pappus of numerous barbellate bristles 7–9 mm long. Roadsides and disturbed areas; first seen in TX at the Fort Worth stockyards (Tarrant Co.) in 1938; in East TX recorded only in Collin Co., Dallas Co. (Kartesz 2015), and Bell Co. (*L.L. Hansen-Sanchez 3562* BRIT); also a few cos. in nc TX and Edwards Plateau; scattered across North America and naturalized in many countries. May–Jul. Native of Eurasia; the national emblem of Scotland. A noxious weed and a potential problem species for TX.

PACKERA (L.) Á. Löve & D. Löve RAGWORT, GROUNDSEL, BUTTERWEED

Ours annual or perennial herbs, glabrous to woolly pubescent; leaves alternate, toothed to pinnately compound; inflorescences terminal, the heads cymosely or umbellately arranged; primary (inner) phyllaries in a single series, linear, with hyaline margins, reflexing at maturity, subtended by a secondary series of short, phyllary-like bracts; ray flowers in ours 8–13, pistillate, fertile, corollas yellow; disk flowers perfect, fertile, corollas yellow; achenes cylindrical, usually prominently 5–10-ribbed; pappus of numerous, finely barbed, white bristles, easily detached from the achene.

A genus of 64 species centered in temperate North America, but with 16 species in Mexico and 1 or 2 in Siberia; it is well known for imprecise species boundaries and intergrading characters (Barkley et al. 1996; Trock 2006b) and was formerly included in the genus *Senecio*. As with *Senecio*, some *Packera* species contain pyrrolizidine alkaloids (e.g., senecionine)

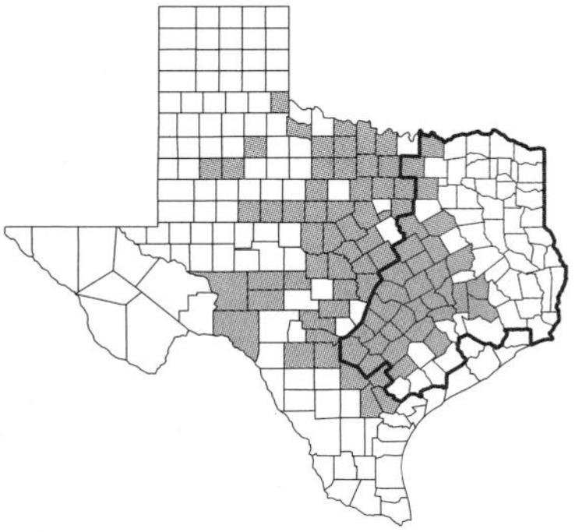
Lindheimera texana

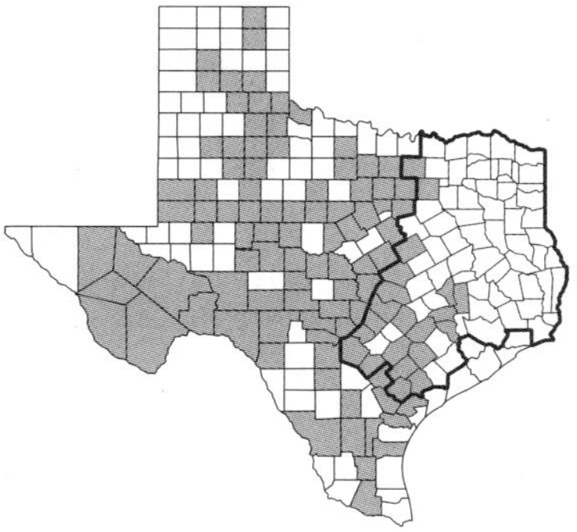
Lygodesmia texana

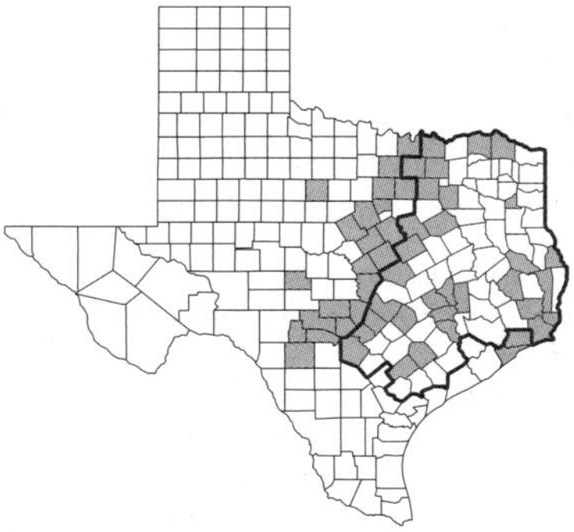
Marshallia caespitosa

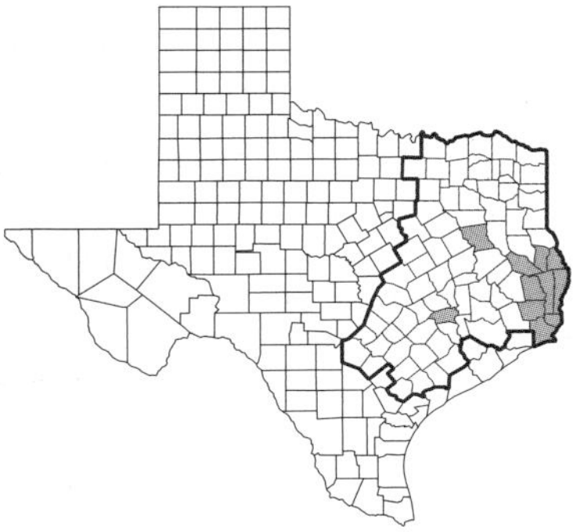
Marshallia graminifolia

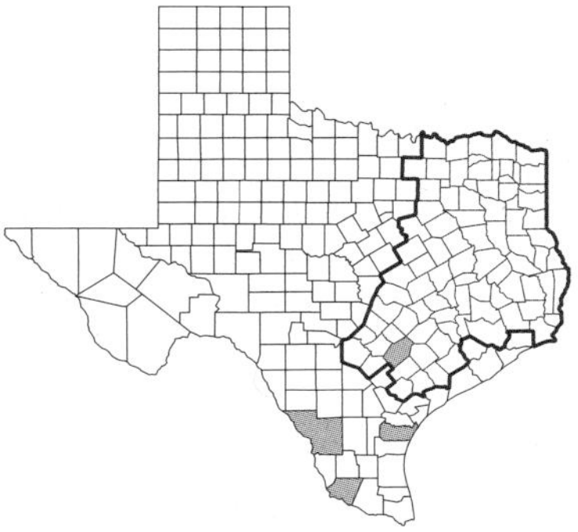
Matricaria chamomilla

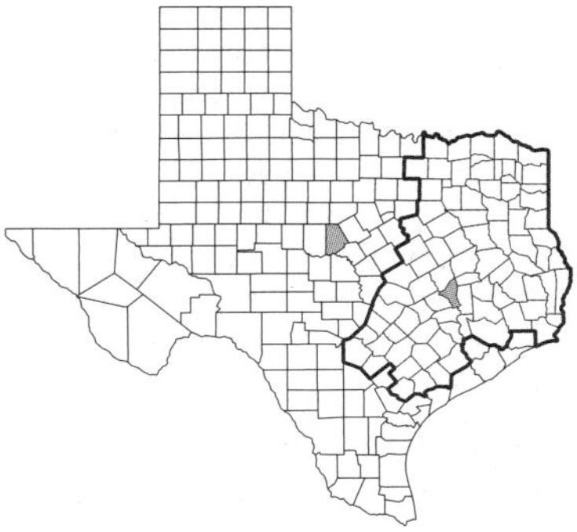
Matricaria discoidea

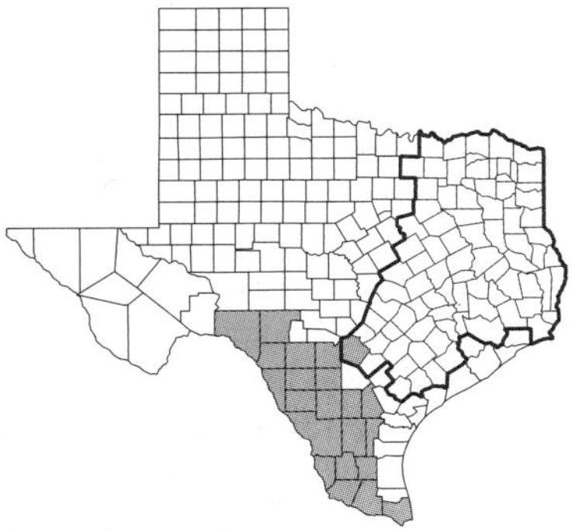
Melampodium cinereum

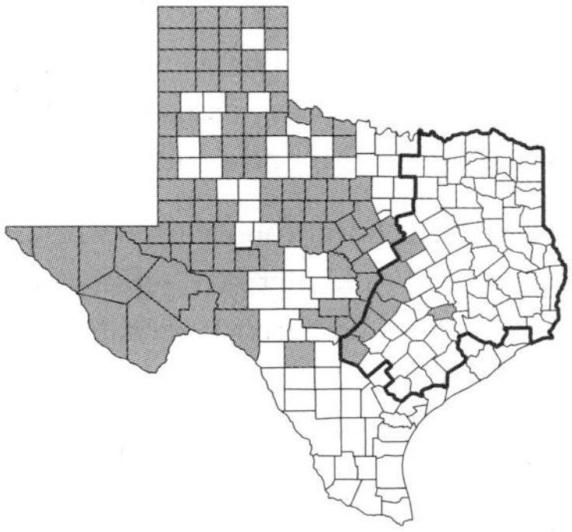
Melampodium leucanthum

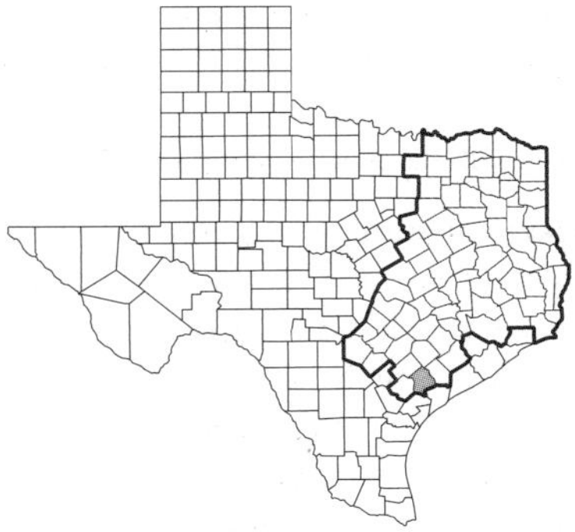
Micropsis dasycarpa

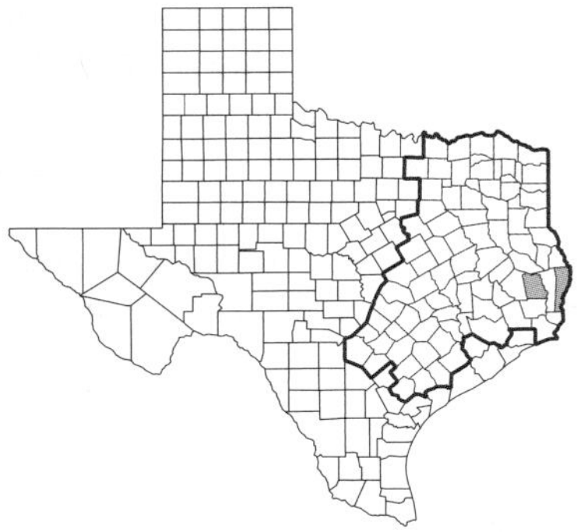
Mikania cordifolia

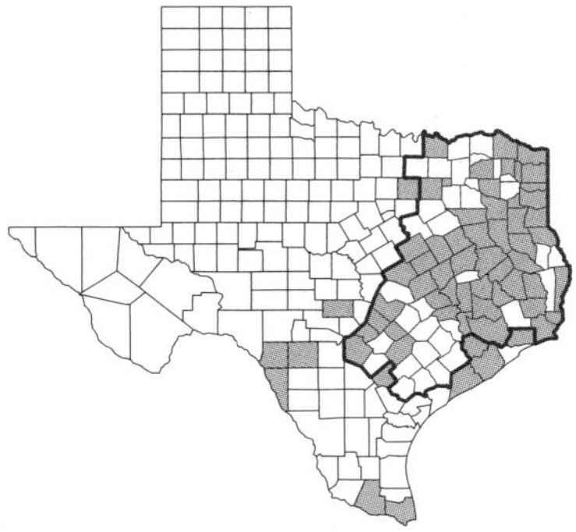
Mikania scandens

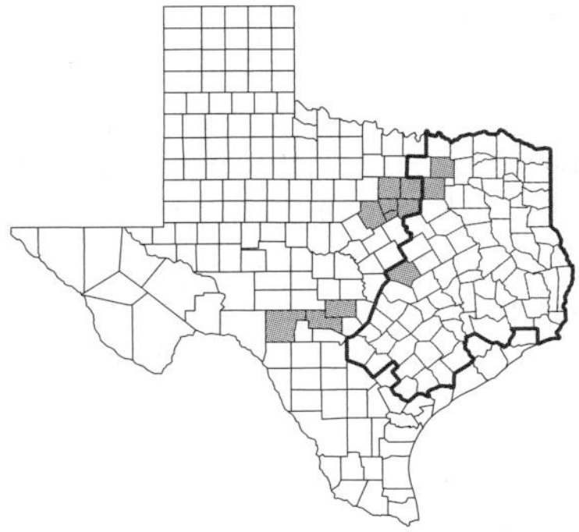
Onopordum acanthium

which can cause severe liver damage and even death in livestock (Morton 1982); because similar effects are to be expected in people, the plants should not be used as herbal teas or otherwise ingested. (Named for John G. Packer, botanist and friend of Áskell and Doris Löve, who named the genus) (tribe Senecioneae)

REFERENCES: Greenman 1915–1918; Barkley 1962, 1978, 1980, 1981, 1985a, 1985b, 1986, 1988; Vuilleumier 1969a; Freeman & Barkley 1995; Barkley et al. 1996; Trock 2006b.

1. Basal leaves pinnately compound or very deeply lobed; basal leaves and stem leaves usually similar; large leaves ± well-distributed along the stem, only gradually reduced upwards; annuals and biennials, without rhizomes or stolons.
   2. Plants from a cluster of fibrous roots; stems usually hollow; ray ligules 7–9 mm long; disk corolla tubes 2.5–3.5 mm long; achenes sparsely hirtellous on ribs or glabrous ____ **P. glabella**
   2. Plants from a delicate, distinct taproot; stems usually solid; ray ligules 3–7 mm long; disk corolla tubes 1.5–2.5 mm long; achenes hirtellous on ribs ____ **P. tampicana**
1. Basal leaves nearly entire to shallowly toothed or deeply lobed; basal leaves and stem leaves quite different in appearance; largest leaves crowded near base of plant, progressively reduced up the stem; perennials, sometimes with rhizomes or stolons.
   3. Plants glabrous except in leaf axils; basal leaf blades orbiculate, ovate or obovate, < 1.5 times as long as wide ____ **P. obovata**
   3. Plants with sparse or dense woolly or cobwebby pubescence on the lower stem internodes, leaves, and nodes of the inflorescences; basal leaf blades narrowly elliptic to elliptic-ovate or oblanceolate, sublyrate, or lanceolate, > 1.5 times as long as wide.
      4. Stems 1–3 in clusters; basal leaves rosette-forming, sparsely hairy beneath; ray flowers 8–10 per head, ligules 9–10 mm long; achenes 1.5–2.5 mm long, hirtellous or glabrous; plants mostly of c and w TX ____ **P. plattensis**
      4. Stems usually 1; basal leaves usually not rosette-forming, usually with cobwebby or thick cottony pubescence beneath; ray flowers 10–13 per head, ligules 6–8+ mm long; achenes 1–1.5 mm long, hispid; plants of deep e and ne TX ____ **P. tomentosa**

Packera glabella (Poir.) C. Jeffrey, (nearly hairless), BUTTERWEED, YELLOWTOP, CRESS-LEAF GROUNDSEL. Annuals or biennials, 20–70+ cm tall, fibrous-rooted; stems usually hollow; basal and lower leaves short-petiolate, blades deeply pinnatifid or once-pinnate, terminal leaf segments sub-orbicular to ovate or obovate, margins crenate to irregularly undulate, upper leaves reduced and weakly clasping; primary phyllaries 5–7 mm long; ray ligules 7–9 mm long; achenes 1–1.5 mm long; pappus 3–4 mm long. Partially-shaded areas, disturbed sites, along streams; e and se TX, w to nc and sc TX; c and se U.S. from OH s to FL and w to NE and TX. Mar–May. [*Senecio glabellus* Poir., *S. carolinianus* Spreng., *S. densiflorus* M. Martens, *S. lobatus* Pers.] This species is superficially similar to *P. tampicana*, but tends to be larger. Liver damage and death have been reported in livestock that have consumed this species; symptoms may not appear for 1 or 2 months after grazing (Morton 1982). ☠

Packera obovata (Muhl. ex Willd.) W.A. Weber & Á. Löve, (obovate), GOLDEN GROUNDSEL, ROUND-LEAF GROUNDSEL, ROUND-LEAF RAGWORT. Perennials, 20–75+ cm tall, rhizomatous and stoloniferous (forming colonies in moist sites); basal and lower leaves petiolate, simple, blades orbiculate or ovate to obovate, often purplish beneath, margins crenate to serrate, upper leaves reduced and clasping, sublyrate or pinnatisect; primary phyllaries 3–6 mm long; ray ligules 7–10 mm long; achenes 1–1.5 mm long; pappus 3–6 mm long. Stream bottoms, woodland meadows, rocky hillsides; e and se TX, w to nc TX and Edwards Plateau; also Trans-Pecos; e and sc U.S. from NH s to FL and w to MI, KS, and NM; Canada; Mexico. Feb–Apr. [*Senecio obovatus* Muhl. ex Willd., *S. aureus* L. var. *obovatus* (Muhl. ex Willd.) Torr. & A. Gray, *S. elliottii* Torr. & A. Gray, *S. elongatus* Pursh, *S. rotundus* (Britton) Small] This is one of the earliest-flowering native wildflowers in East TX.

Packera plattensis (Nutt.) W.A. Weber & Á. Löve, (of the Platte River region), PRAIRIE GROUNDSEL, PRAIRIE RAGWORT. Perennials or biennials, 20–60+ cm tall, rhizomatous and fibrous-rooted, sometimes stoloniferous; basal and lower leaves petiolate, blades suborbiculate or elliptic-ovate to obovate or sublyrate, often with cobwebby hairs, margins nearly entire to crenate or serrate-dentate to pinnately lobed, upper leaves reduced and petiolate, sublyrate or pinnatisect; primary phyllaries 5–6+ mm long; ray ligules 9–10 mm long; achenes 1.5–2.5 mm long; pappus 6.5–7.5 mm long. Prairies and open woods, roadsides, calcareous soils; uncommon in East TX, mostly nc TX to sc TX, w to Rolling Plains and s to Edwards Plateau; also Trans-Pecos; mostly c 1/3 of North America, scattered farther e. Feb–May. [*Senecio plattensis* Nutt., *S. pseudotomentosus* Mack. & Bush] Reported to be poisonous (Kingsbury 1964). ☠

Packera tampicana (DC.) C. Jeffrey, (of Tampico, Mexico), YELLOWTOP, GREAT PLAINS RAGWORT. Annuals, 20–50+ cm tall, taprooted; basal and lower leaves petiolate, often variously anthocyanic, blades deeply pinnatifid or once-pinnate, terminal leaf segments reniform to orbiculate, margins crenate or dentate to irregularly incised or lobed, upper leaves reduced and auriculate-clasping; primary phyllaries 3–7 mm long; ray ligules 3–7 mm long; achenes 1–1.5 mm long; pappus 3–5 mm long. Weedy in low prairies, disturbed areas, roadsides, typically in moist, open places; populations in some areas can number in the tens of thousands;

Melampodium leucanthum [BB2]

Melampodium cinereum [ELB]

Matricaria discoidea [WLJ]

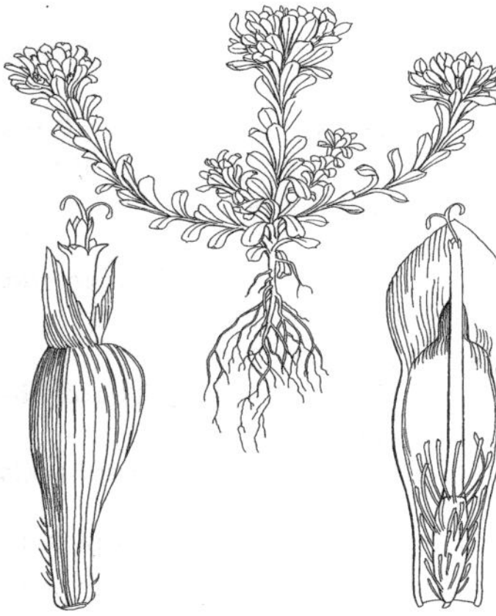

Mikania scandens [BT3]

Mikania cordifolia [SID]

Micropsis dasycarpa [FNA]

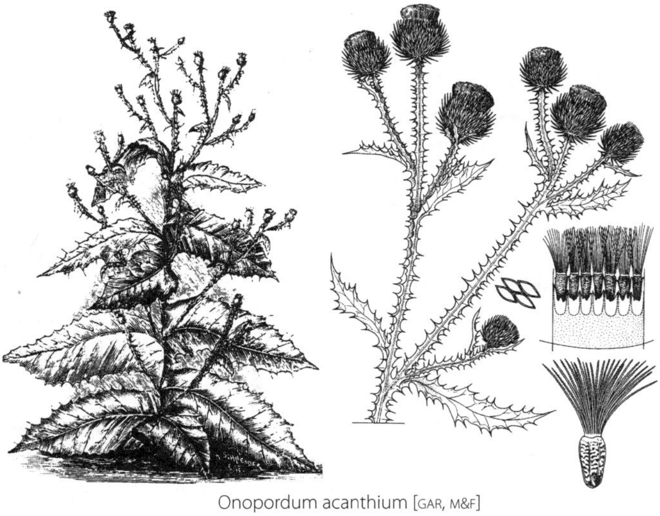

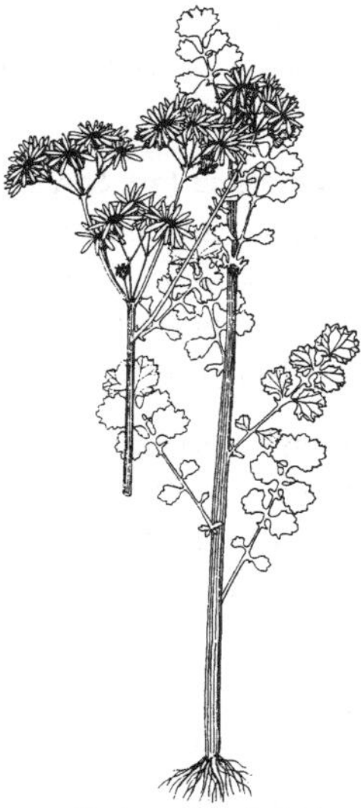

Onopordum acanthium [GAR, M&F]

Packera glabella [STE]

widespread in TX, more abundant in e 1/2; AR, KS, LA, OK, and TX; Mexico. Feb–May. [*Senecio imparipinnatus* Klatt., *S. greggii* Rydb., *S. tampicanus* DC.]

Packera tomentosa (Michx.) C. Jeffrey, (woolly pubescence), WOOLLY GROUNDSEL, WOOLLY RAGWORT. Perennials, 30–60+ cm tall, taprooted, sometimes stoloniferous; basal and lower leaves petiolate, blades lanceolate to narrowly ovate or elliptic, with cobwebby or cottony pubescence beneath, margins nearly entire to crenate or serrate-dentate, upper leaves reduced and petiolate or sessile and weakly clasping, dentate to pinnately lobed; primary phyllaries 5–8 mm long; ray ligules 6–8+ mm long; achenes 1–1.5 mm long; pappus 5–7 mm long. Meadows, roadsides, sandy or shallow soils; deep e and ne TX; se U.S. from NJ s to FL and w to OK and TX. Apr–May. [*Senecio alabamensis* Britton ex Small, *S. tomentosus* Michx.]

PALAFOXIA Lag.

Ours annual or perennial herbs, sometimes glandular; leaves alternate or lowest ones opposite, sessile or petiolate, blades linear to broadly lanceolate, sparsely to moderately pubescent; heads in corymbose arrays; phyllaries in 2–3 series, subequal or unequal, tips often purplish or whitish, reflexed at maturity; receptacles naked; ray flowers pistillate and fertile or absent, corollas pinkish to purplish or white, ligules prominently 3-toothed; disk flowers perfect and fertile, radially symmetrical or irregularly lobed and sometimes simulating rays, corollas pinkish to purplish or white; achenes obpyramidal, 4-angled, pubescent; pappus of 4–10 scales, minute to awn-like, sometimes varying within a head.

A genus of 12 species native to the s U.S. and Mexico; some are cultivated as ornamentals. (Named for General José de Palafox, 1780–1847, a Spanish duke) (tribe Heliantheae)

REFERENCES: Baltzer 1944; Cory 1946; Shinners 1952; Turner & Morris 1976; Strother 2006v.

1. Stems without stalked glands; heads with disk flowers only; all achenes with similar pappus.
  2. Phyllaries 3–5 mm long; corollas 5–6 mm long; pappus scales of central achenes mostly 0.3–1 mm long ____ **P. callosa**
  2. Phyllaries 5–10 mm long; corollas 7–10 mm long; pappus scales of central achenes (1.5–)3–8 mm long.
    3. Leaf blades narrowly lanceolate to linear, 2–6(–12) mm wide; phyllaries usually with stalked glands ____ **P. rosea**
    3. Leaf blades mostly lanceolate, (5–)10–20 mm wide; phyllaries usually lacking stalked glands ____ **P. texana**
1. Stems with stalked glands (at least in the upper portion); heads with conspicuous ray flowers; ray and disk achenes with pappus differing in size.
  4. Leaf blades narrowly lanceolate to linear, 2–6 mm wide; phyllaries 6–8 mm long; achenes 5–6 mm long ____ **P. reverchonii**
  4. Leaf blades broadly to narrowly lanceolate, 3–25 mm wide; phyllaries 7–16+ mm long; achenes 6–9 mm long.
    5. Stems usually with stalked glands ± throughout; phyllaries mostly 3–5 mm wide; ray corollas 10–20 mm long, ligules 9–14 mm long; common in East TX ____ **P. hookeriana**
    5. Stems with stalked glands only in the upper portion and on the peduncles, or lacking stalked glands entirely; phyllaries mostly 1.5–3 mm wide; ray corollas 15–25 mm long, ligules 5–8(–12) mm long; rare in East TX ____ **P. sphacelata**

Palafoxia callosa (Nutt.) Torr. & A. Gray, (thick-skinned), SMALL PALAFOXIA. Annuals, 20–60 cm tall; stems lacking stalked glands; leaf blades linear; phyllaries 3–5 mm long; ray flowers absent; disk corollas 5–6 mm long; achenes 3–5 mm long; pappus scales of central achenes 0.3–1 mm long. Calcareous and rocky soils, disturbed habitats; e and sc TX w to nc TX, Rolling Plains, and Edwards Plateau; AR, LA, MO, MS, OK, and TX; Mexico. (May–)Jul–Oct. [*P. callosa* (Nutt.) Torr. & A. Gray var. *bella* (Cory) Shinners]

Palafoxia hookeriana Torr. & A. Gray, (for Sir William Jackson Hooker, 1785–1865, director of Kew Gardens), SHOWY PALAFOXIA, HOOKER'S PALAFOXIA, SAND PALAFOXIA. Annuals, 25–180 cm tall; stems usually with stalked glands throughout; leaf blades broadly to narrowly lanceolate; phyllaries 7–16+ mm long; ray corollas 10–20 mm long, ligules 9–14 mm long; disk corollas 10–12 mm long; achenes 6–9 mm long; pappus scales of central achenes 5–8 mm long. Sandy soils, often in forested areas; mainly se to sc TX; South TX Plains; endemic to TX. Sep–Nov. [*P. hookeriana* Torr. & A. Gray var. *minor* Shinners]

Palafoxia reverchonii (Bush) Cory, (for Julien Reverchon, 1837–1905, a French-American immigrant to Dallas and important botanical collector of early TX), REVERCHON'S PALAFOXIA. Annuals, 10–90 cm tall; stems usually with stalked glands in the upper portion; leaf blades narrowly lanceolate; phyllaries 6–8 mm long; ray corollas 15–25 mm long, ligules 6–12 mm long; disk corollas 5–8 mm long; achenes 5–6 mm long; pappus scales of central achenes 3–6 mm long. Sandy soils, wooded areas; e and se TX; endemic to TX. Sep–Oct.

Palafoxia rosea (Bush) Cory, (rose-colored), ROSY PALAFOXIA, SLENDER PALAFOXIA. Annuals, 10–50 cm tall; stems lacking stalked glands except in branches of inflorescence; leaf blades linear-lanceolate; phyllaries 5–10 mm long; ray flowers absent; disk corollas 7–10 mm long; achenes 5–8 mm long; pappus scales of central achenes (1.5–)3–8 mm long. Sandy soils, disturbed habitats;

Packera obovata [MO4]

Packera plattensis [MO4]

Packera tampicana [HEA]

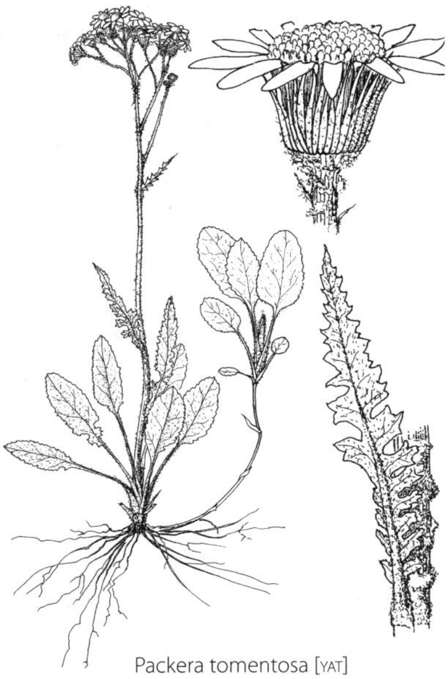

Packera tomentosa [YAT]

Palafoxia callosa [BT3]

Palafoxia hookeriana [NIC]

mainly se and sc TX; c TX, Trans-Pecos, and Panhandle; CO, KS, NM, OK, TX, and WY. Jun–Nov. [*P. rosea* (Bush) Cory var. *macrolepis* (Rydb.) B.L. Turner & Morris]

Palafoxia sphacelata (Nutt. ex Torr.) Cory, (dead, withered), RAYED PALAFOXIA, WESTERN PALAFOXIA, OTHAKE. Annuals, 10–90 cm tall; stems usually with stalked glands in the upper portion; leaf blades broadly to narrowly lanceolate; phyllaries 9–12+ mm long; ray corollas 15–25 mm long, ligules 5–12 mm long; disk corollas 10–14 mm long; achenes 6–9 mm long; pappus scales of central achenes 7–9 mm long. Sandy soils, grasslands; rare in East TX, with a single disjunct record in Grayson Co. in Red River drainage (Turner & Morris 1976); mainly Trans-Pecos and Panhandle e to Rolling Plains; CO, KS, NM, OK, SD, and TX; Mexico. Jul–Oct. [*Othake sphacelata* (Nutt. ex Torr.) Rydb.]

Palafoxia texana DC., (of Texas), TEXAS PALAFOXIA. Annuals or perennials, 20–80 cm tall; stems lacking stalked glands; leaf blades ovate-lanceolate to linear-lanceolate; phyllaries 5–8 mm long; ray flowers absent; disk corollas 7–10 mm long; achenes 4–6 mm long; pappus scales of central achenes 2–6 mm long. Calcareous and gravelly soils, limestone hills, shrublands; at extreme sw margin of East TX in Bexar and Jackson cos. and Goliad Co. (TEX); South TX Plains; FL, LA, and TX; Mexico. Feb–Nov. [*P. texana* DC. var. *ambigua* (Shinners) B.L. Turner & Morris]

PARTHENIUM L. FEVERFEW

Ours annual, biennial, or perennial herbs or subshrubs; leaves sometimes in a basal rosette, alternate on stems, petiolate or sessile, blades coarsely toothed or 1–2-pinnately lobed, usually hairy and gland-dotted (at least beneath); heads small, borne in corymbose or paniculate arrays; involucres 3–8(–12+) mm wide; phyllaries in 2 series; receptacles with pales (chaff) enfolding some or all disk flowers; ray flowers 5(–8), pistillate, fertile, corollas whitish to pale dingy yellow, ligules minute (to 2 mm long) or sometimes absent; disk flowers functionally staminate, not producing mature achenes, corollas whitish to pale dingy yellow, often obscured by pubescence; achenes black, oblanceolate, each shed with a subtending phyllary and 2 contiguous disk flowers and their subtending pales (chaff); pappus absent (achene shoulders may bear 1–4 pappus-like short awns or scales which sometimes can be observed projecting from the head adjacent to the ligules at anthesis).

A genus of ca. 16 species native to the New World. The genus also contains the GUAYULE RUBBER PLANT (*P. argentatum* A. Gray), a source of natural rubber (Rollins 1950) native to w TX and Mexico; during World War II, because of rubber shortages, a program to grow that species was carried out by the U.S. Forest Service (McGinnies in Foster et al. 1983). (Greek: *parthenos*, virgin (allusion unclear), or *parthenion*, an ancient name of some plant) (tribe Heliantheae)
REFERENCES: Rollins 1950; Mears 1975; Strother 2006w.

1. Subshrubs; leaves usually 1.5–2.5 cm long; outer phyllaries 1.5–2 mm long ____ **P. incanum**
1. Annuals, biennials, or herbaceous perennials; leaves usually 3+ cm long; outer phyllaries 2–5 mm long.
   2. Perennials; leaf blade margins usually crenate to serrate, sometimes coarsely toothed or somewhat lobed (but then only toward bases); pappus-like achene shoulder projections absent or if present then narrowly awn-like or thread-like ____ **P. integrifolium**
   2. Annuals or biennials; leaf blades mostly 1–2-pinnately lobed; pappus-like achene shoulder projections usually 2, deltate to ovate.
      3. Biennials (sometimes flowering first year or persisting); leaf blades usually 1-pinnately lobed, strigillose beneath, usually also with erect hairs 1–2 mm long ____ **P. confertum**
      3. Annuals (rarely persisting); leaf blades usually 2-pinnately lobed, scaberulous beneath, seldom with additional erect hairs 1–2 mm long ____ **P. hysterophorus**

Parthenium confertum A. Gray, (crowded, for its closely-packed heads), GRAY'S FEVERFEW. Biennials, usually 10–30 cm tall; most leaves 1-pinnately lobed, strigillose, usually also with erect hairs, and gland-dotted; outer phyllaries 2.5–3+ mm long; ray ligules absent or 0.1–0.5 mm long; achenes 2–3 mm long, pappus-like achene shoulder projections deltate to ovate. Sandy plains, mesquite grasslands, shallow soils; sw border of East TX in Bexar, Hays, and Travis cos.; Edwards Plateau, Trans-Pecos, and s TX; AZ, NM, and TX; Mexico. Apr–Nov. [*P. confertum* A. Gray var. *divaricatum* Rollins, *P. confertum* A. Gray var. *lyratum* (A. Gray) Rollins, *P. confertum* A. Gray var. *microcephalum* Rollins] We are following Strother 2006w in not recognizing varieties of this highly variable species.

Parthenium hysterophorus L., (old generic name, from Greek: *hystéra*, womb or uterus, and *phoros*, bearing), FALSE RAGWEED, RAGWEED PARTHENIUM, SANTA-MARIA, FEVERFEW, CICUTILLA. Annuals, usually 30–120+ cm tall; most leaves 2-pinnatifid, scaberulous and gland-dotted; outer phyllaries 2–4 mm long; ray ligules 0.3–1 mm long; achenes 1.5–2(–3.5) mm long, pappus-like achene shoulder projections deltate to ovate. In and near towns and farms, disturbed areas; widespread in TX but less common in the far e and w parts of the state; scattered in e and c U.S. but more common in the se; Mexico; introduced and weedy in tropical and

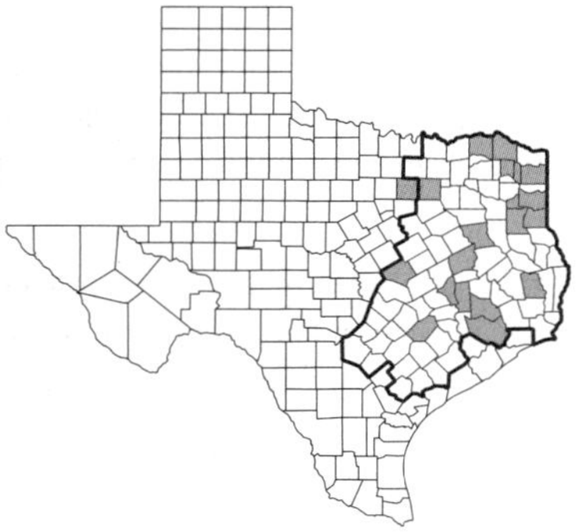

Packera glabella

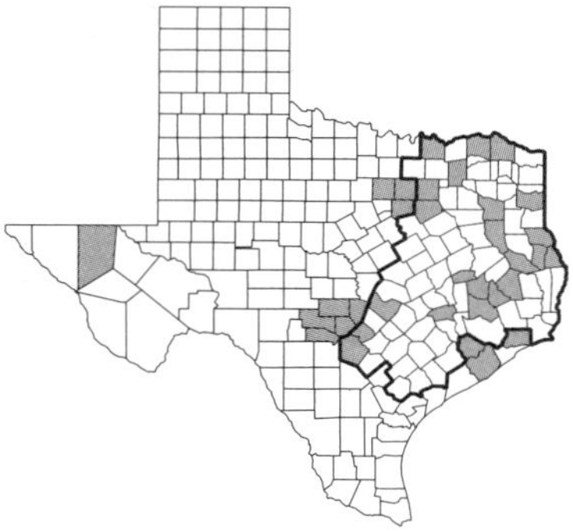

Packera obovata

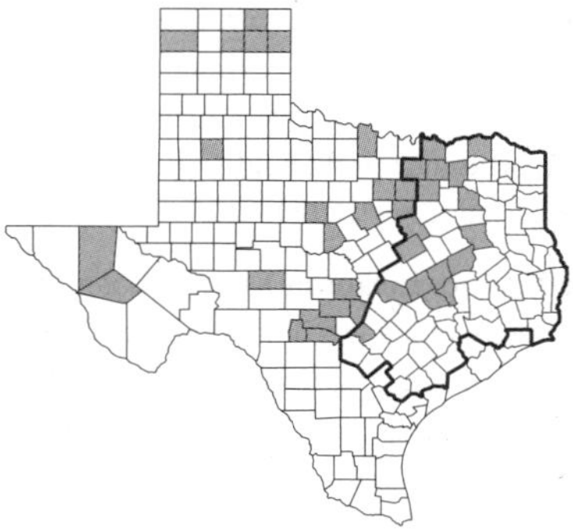

Packera plattensis

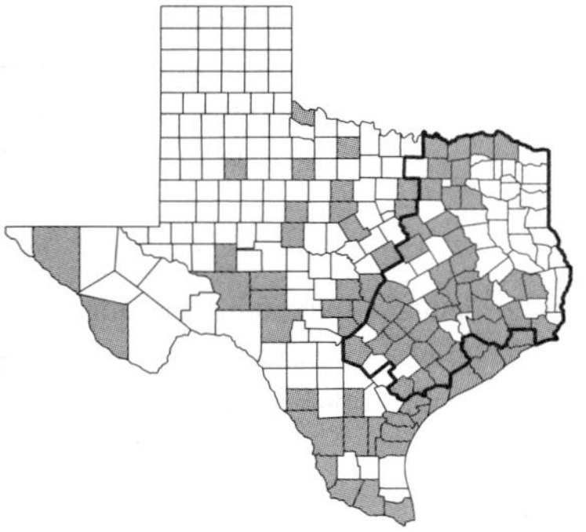

Packera tampicana

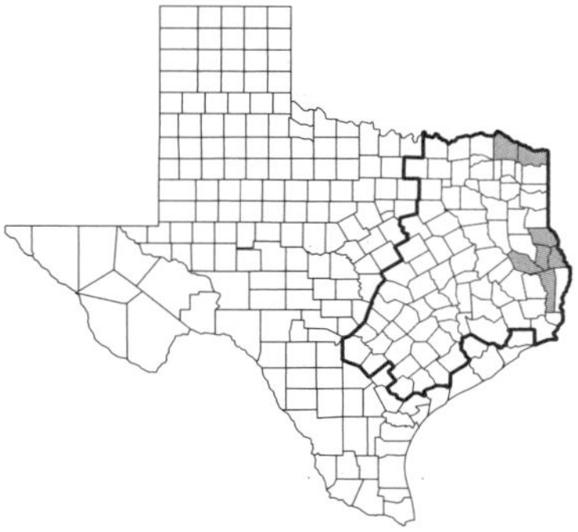

Packera tomentosa

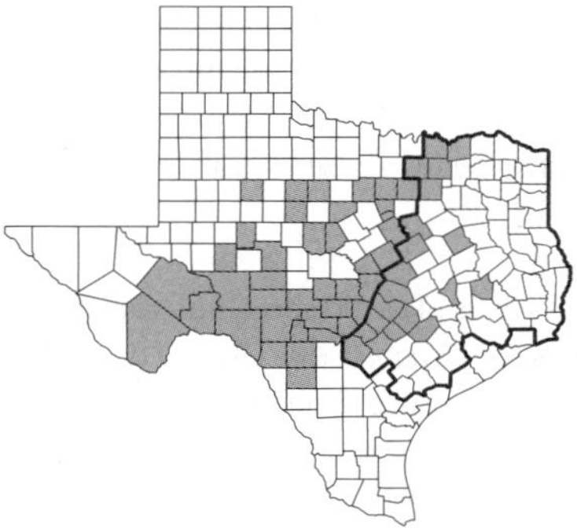

Palafoxia callosa

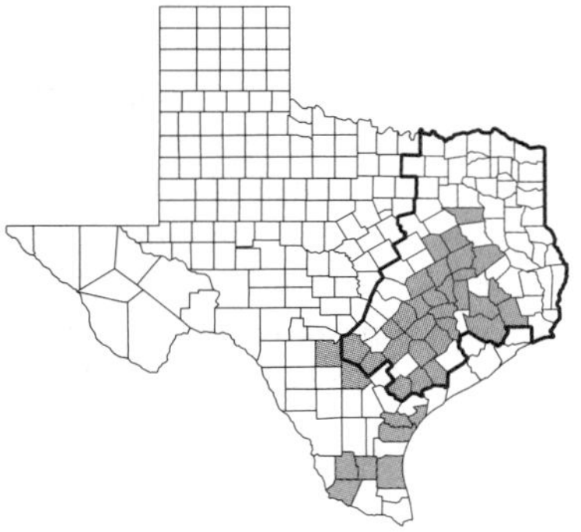

Palafoxia hookeriana

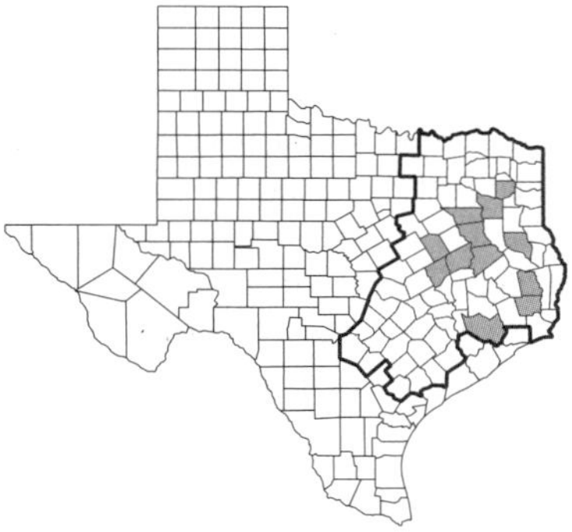

Palafoxia reverchonii

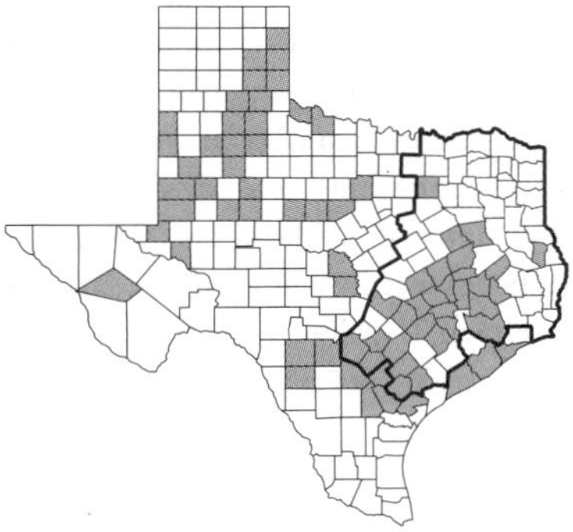

Palafoxia rosea

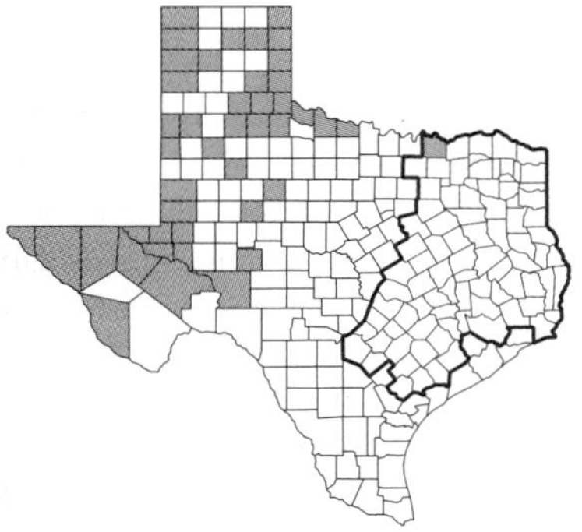

Palafoxia sphacelata

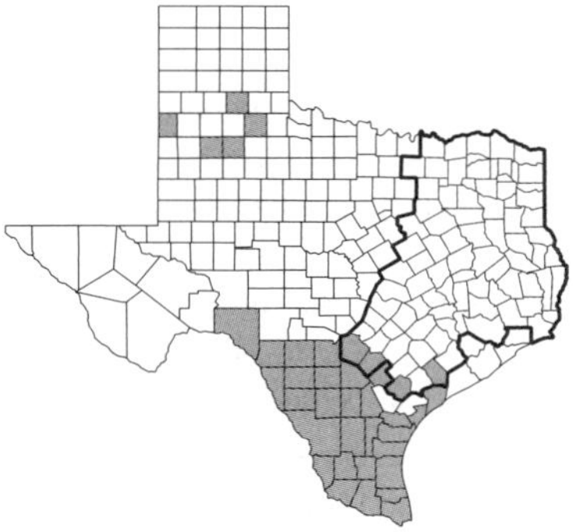

Palafoxia texana

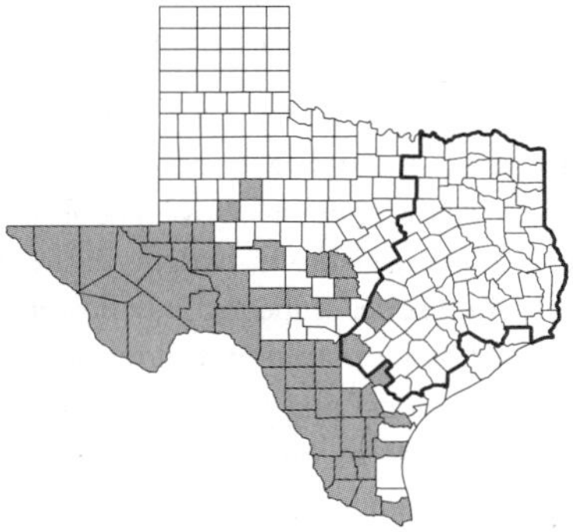

Parthenium confertum

warm temperate zones worldwide. Summer–fall (sporadically in all seasons). Hairs and sessile capitate glands on the leaf surfaces contain sesquiterpene lactones (parthenin and ambrosin) which can cause serious contact dermatitis in some individuals; this Central American native was introduced in India (first noted in 1956) presumably as a contaminant with cereal grains, where it has become a problematic alien invader, crowding out the native flora and causing dermatitis in agricultural workers (Lampe 1986; Mabberley 2017).

Parthenium incanum Kunth, (hoary-gray), RAYWEED, HOARY PARTHENIUM, MARIOLA. Subshrubs, 30–100+ cm tall; leaves pinnately or lyrately lobed, gray-white tomentose and gland-dotted; outer phyllaries 1.5–2 mm long; ray ligules 1–1.5 mm long; achenes 1.5–2 mm long, pappus-like achene shoulder projections ± subulate. Openings in desert scrub, often on limestone soils; recorded in Travis Co., disjunct from other TX records; mostly Trans-Pecos, also High Plains and s TX; AZ, NM, NV, TX, and UT; Mexico. May–Oct.

Parthenium integrifolium L., (entire-leaved), WILD QUININE. Perennials, usually 30–60 cm tall; leaves coarsely toothed or somewhat lobed (then mostly toward bases), hispid to scabrous and gland-dotted; outer phyllaries 3–5 mm long; ray ligules 1–2+ mm long; achenes 3–4+ mm long, pappus-like achene shoulder projections absent or awn-like to thread-like. Calcareous glades and barrens, sandy prairies, disturbed areas; ne TX in Bowie, Grayson, Harrison, and Titus cos.; e and c U.S. from MA s to GA and w to MN, KS, and TX. May–Jul. [*P. auriculatum* Britton, *P. hispidum* Raf., *P. hispidum* Raf. var. *auriculatum* (Britton ex Britton) Rollins, *P. integrifolium* L. var. *hispidum* (Raf.) Mears]

PECTIS L.

A genus of ca. 90 species of tropical and warm areas of the New World and the Galápagos Islands. Like many members of the genus, *Pectis papposa* A. Gray, of the sw U.S. and Mexico, is strongly-scented; it was used by Native Americans for flavoring foods and as a perfume (Bradley & Haagen-Smit 1949). (Greek: *pectis*, to comb, when the marginal leaf glands become broken, the leaf has a somewhat serrated effect marginally—Bradley & Haagen-Smit 1949) (tribe Heliantheae)
REFERENCES: Fernald 1897; Bradley & Haagen-Smit 1949; Keil 1977, 2006g.

Pectis angustifolia Torr. var. **fastigiata** (A. Gray) D.J. Keil, (sp.: narrow-leaved; var.: having branches close together and erect), TEXAS CHINCHWEED. Spicy-scented, fibrous-rooted, much-branched perennial or some individuals annual, 5–15 cm tall, sometimes woody at base; leaves opposite, linear, blade 10–40 mm long, 1–2 mm wide, glabrous, basally ciliate with bristles 1–2 mm long, often revolute, marginally glandular-punctate; heads congested at ends of branches in cymose arrays; peduncles 3–30 mm long; phyllaries in one series, subequal, 2.5–4.5 mm long, keeled, broadest near apex, with a conspicuous subterminal oil gland 0.5–1 mm long and 1 or 2 pairs of smaller submarginal glands; ray flowers 8, pistillate, fertile, inserted on phyllary bases, ligules 3–5(–7) mm long, yellow; disk flowers 7–20, perfect, fertile, corollas 2.5–3.5 mm long, yellow; achenes 2.5–3.5 mm long, the ray achenes usually falling attached to a subtending phyllary; pappus usually < 1 mm long, of 1–7 short bristles or awns, forming a low crown. Usually on limestone, dry uplands; w border of East TX in Bell, Hays, Travis, and Williamson cos.; Edwards Plateau and Lampasas Cut Plain; endemic to TX. Jul–Nov. Two other varieties of this species occur in s and w TX and Mexico. [*P. fastigiata* A. Gray, *P. texana* Cory]

PERITYLE Benth. ROCK DAISY

A genus of ca. 66 species native to sw North America, with 1 in Chile and Peru. (Greek: *peri*, around, and *tyle*, knot, knob, or callus, from the thickened margin around the fruit) (tribe Heliantheae)
REFERENCES: Powell 1973; Yarborough & Powell 2006b.

Perityle lindheimeri (A. Gray) Shinners var. **lindheimeri,** (for Ferdinand Jacob Lindheimer, 1801–1879, German-born TX collector), LINDHEIMER'S ROCK DAISY, BLUFF DAISY. Subshrub or perennial from a woody base; plants (10–)18–45(–60) cm tall; leaves opposite below, alternate above, petioles 4–10 mm long, blades broadly ovate to ovate-lanceolate, 2–5 cm long, minutely punctate, essentially glabrous, margins entire or serrate-lobed; heads in corymbose arrays; involucres ca. 4–6 mm high; phyllaries usually in 2 series, subequal; ray flowers 3–5, pistillate and fertile, the ligules 2.5–3 mm long, 1.5–2 mm wide, yellow; disk corollas 2.3–3.5 mm long, yellow; achenes 2–2.8 mm long, linear to narrowly obconical; pappus usually of 1(–2) bristles, 0.5–1.8 mm long, plus vestigial scales or a minute crown, rarely absent. Crevices in Cretaceous limestone bluffs, often beside streams; on the sw border of East TX in Bexar, Comal, Hays, and Travis cos.; Edwards Plateau; endemic to TX. Apr–Sep. [*Laphamia halimifolia* A. Gray subsp. *lindheimeri* (A. Gray) Niles, *Laphamia rotundata* Rydb., *P. rotundata* (Rydb.) Shinners]

Palafoxia reverchonii [RHO]

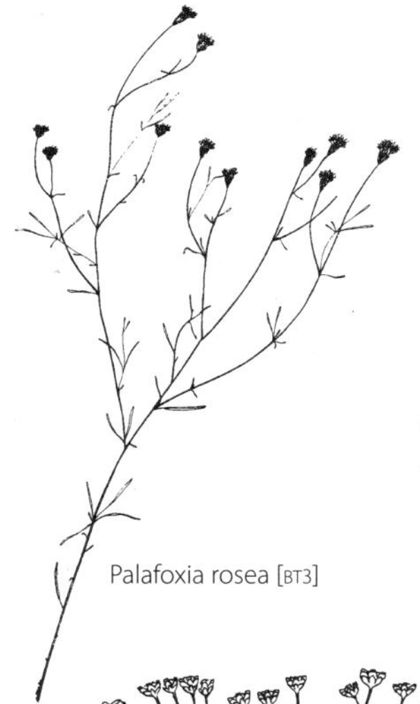
Palafoxia rosea [BT3]

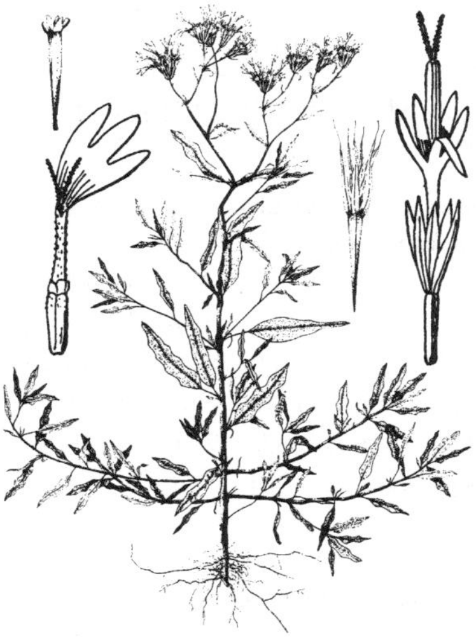
Palafoxia sphacelata [BB2, RHO]

Palafoxia texana [RHO]

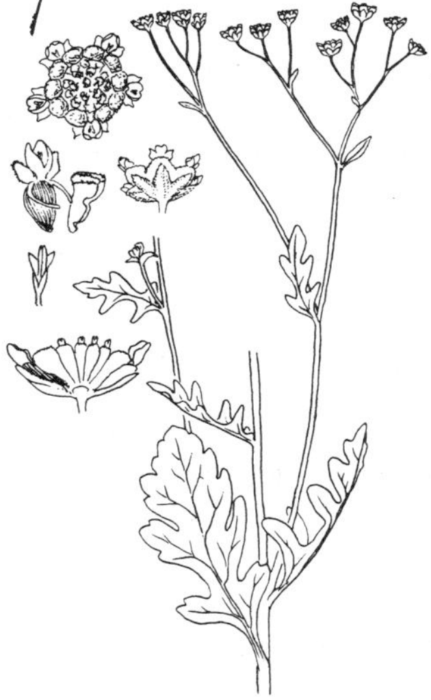
Parthenium confertum [IV2]

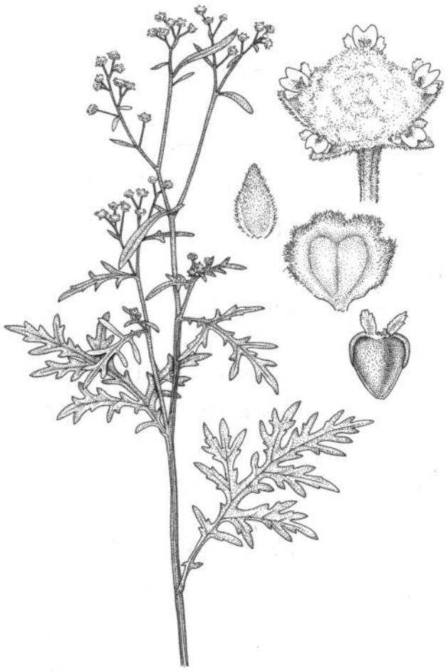
Parthenium hysterophorus [MO4]

Parthenium incanum [HUM]

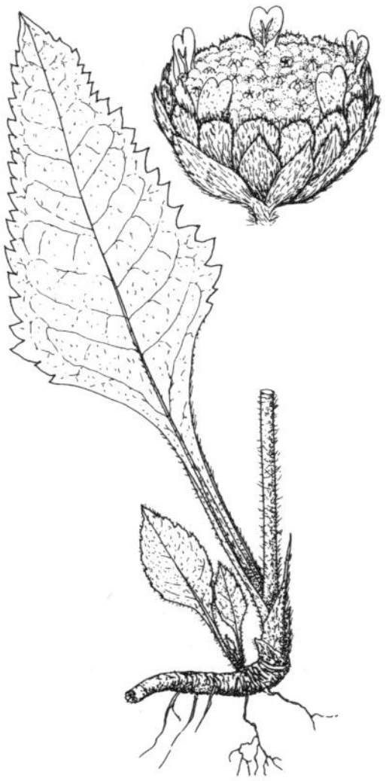
Parthenium integrifolium [YAT]

Pectis angustifolia var. fastigiata [RHO]

PINAROPAPPUS Less. PINK-DANDELION, ROCK-LETTUCE

A genus of ca. 8 species native to s North America and Central America. (Greek: *pinaro*, dirty, and *pappus*, down, fuzz, pappus, in reference to the tawny pappus) (tribe Cichorieae)
REFERENCES: McVaugh 1972; Bogler 2006e.

Pinaropappus roseus (Less.) Less., (rose-colored), SMALL ROCK-LETTUCE. Glabrous, small (10–40 cm tall) perennial with a woody, branching root; leaves mostly basal, alternate, blades oblong-lanceolate, becoming linear and bracteate on upper stem, entire or coarsely toothed or lobed; heads large (15–25 mm high including corollas), terminal, solitary; phyllaries semi-scarious, in several overlapping rows, usually with pinkish margins and prominent brown tips; flowers all ligulate; corollas pink to deep rose-lavender beneath, paler to white above, yellowish basally; achenes fusiform, 5–6 mm long, beaked; pappus of numerous tawny, hair-like bristles. Limestone outcrops and calcareous soils; w edge of East TX from Denton Co. s to Bell Co., and Edwards Plateau from Travis Co. to Bexar and DeWitt cos.; mostly c and sw TX; AZ, NM, OK, and TX; Mexico. Apr–May. [*Achyrophorus roseus* Less.]

PITYOPSIS Nutt. GRASS-LEAF GOLDEN-ASTER

Perennial rhizomatous herbs, forming clumps; stems erect, ours usually pubescent and glandular; leaves overwintering in a basal rosette, alternate on the stem, sessile, often grass-like, blades 3–11-parallel-nerved, usually with long silvery-silky pubescence, margins entire; heads in corymbose or paniculate arrays; phyllaries in 3–5 series, unequal, lanceolate, margins scarious; ray flowers pistillate and fertile, corollas yellow; disk flowers perfect and fertile, corollas yellow; achenes spindle-shaped, 8–10-ribbed, strigose; pappus in (3–)4 series, the outermost series of linear scales, the inner series of longer, minutely-barbed, tawny bristles.

A genus of 7 species of the e U.S., Central America, and West Indies. Previously included in *Chrysopsis* or *Heterotheca*. (Greek: *pitys*, pine or fir, and *opsis*, appearance or resembling, apparently for the pine-needle-like appearance of the narrow leaves) (tribe Astereae)
REFERENCES: Shinners 1951b; Semple 1977, 2006e; Semple et al. 1980; Semple & Bowers 1985, 1987.

1. Stem leaves (above the basal rosette) numerous, usually 20+; heads numerous, usually 10–100+; phyllaries eglandular or with sparse stalked glands; outer pappus series scales 0.4–0.9 mm long ____ **P. graminifolia**
1. Stem leaves (above the basal rosette) few, usually 2–6; heads few, usually 1–5; phyllaries with dense stalked glands; outer pappus series scales 0.8–1.4 mm long ____ **P. oligantha**

Pityopsis graminifolia (Michx.) Nutt., (grass-leaved), NARROW-LEAVED SILK-GRASS, GRASS-LEAVED GOLDEN-ASTER. Stems 20–80 cm tall, 1–5+ per plant; stems and leaves with long, appressed, silvery-white, silky pubescence; leaves linear to lanceolate, reduced in size up the stem; phyllaries often pilose, sometimes with sparse stalked glands; achenes 2.5–4.5 mm long. Pine-oak woods, roadsides, waste areas, often in sandy soils; e 1/2 TX, mainly se and e TX; se U.S. from DE s to FL and w to OH, OK, and TX; Mexico; Central America; and West Indies. (Aug–)Sep–Nov. Two of the varieties apparently have overlapping ranges in East TX; the infraspecific range maps provided by Semple and Bowers (1985), Kartesz (2015), and USDA Plants Database (2016) are incomplete or irreconcilable, therefore we are not providing ranges or maps for these varieties. Semple (2006e) divided the two East TX varieties of this highly variable species as follows:

1. Involucres 8–12 mm high; ray flowers 10–16, ligules 7–15 mm long ____ var. **latifolia**
1. Involucres 4.5–8 mm high; ray flowers 8–12, ligules 4–7 mm long ____ var. **tenuifolia**

var. **latifolia** (Fernald) Semple & F.D. Bowers, (broad-leaved). [*Chrysopsis argentea* (Pers.) Elliott, *C. correllii* Fernald, *C. graminifolia* (Michx.) Elliott var. *latifolia* Fernald, *Heterotheca nervosa* (Willd.) Shinners]

var. **tenuifolia** (Torr.) Semple & F.D. Bowers, (narrow-leaved). [*Chrysopsis graminifolia* (Michx.) Elliott var. *microcephala* (Small) Cronquist, *Heterotheca microcephala* (Small) Shinners, *P. graminifolia* (Michx.) Nutt. var. *microcephala* (Small) Semple]

Pityopsis oligantha (Chapm. ex Torr. & A. Gray) Small, (few-flowered), FEW-HEADED GRASS-LEAVED GOLDEN-ASTER, LARGE-FLOWERED GOLDEN-ASTER, COASTAL PLAIN SILK-GRASS. Stems 20–50 cm tall, usually 1 per plant, with dense stalked glands throughout, and with long, appressed, silvery-white, silky pubescence on lower 1/3; leaves linear to broadly oblanceolate, reduced in size up the stem, with pubescence similar to the lower stem; phyllaries sericeous, with dense stalked glands; achenes 4–5 mm long. Longleaf pine savannas, sandy saline soils; in TX known only from Jasper (*W.C. Holmes & J.R. Singhurst 11100* BAYLU; Holmes & Singhurst 2012) and Tyler (TAES) cos., presumably rare in the state; AL, FL, GA, LA, MS, and TX. Sep–Oct. [*Chrysopsis oligantha* Chapm.

ex Torr. & A. Gray, *Heterotheca oligantha* (Chapm. ex Torr. & A. Gray) V.L. Harms]. Likely of conservation concern due to its limited range. ⚠

PLECTOCEPHALUS D. Don. BASKET-FLOWER

A genus of 4 species native to North America; formerly included in *Centaurea*. (Greek: *plektos*, woven, and *kephale*, head, alluding to interwoven fringes of phyllaries) (tribe Cardueae)
REFERENCES: Roalson & Allred 1998; Keil 2006h.

Plectocephalus americanus (Nutt.) D. Don, (of America), BASKET-FLOWER, AMERICAN BASKET-FLOWER, POWDERPUFF THISTLE, THORNLESS-THISTLE, CARDO DEL VALLE, AMERICAN KNAPWEED, STAR-THISTLE. Annual; stems 0.5–2 m tall; stems usually 1; leaves sessile, ± entire, oblanceolate to narrowly obovate, glabrous or sparsely scabrous and gland-dotted; heads solitary at the ends of branches, large, (3–)4–8(–10) cm across including the corollas, quite showy, 1–few per plant; phyllaries appressed, imbricate, of 2 distinct parts, the lower part light green, entire, the upper part straw-colored (sometimes pinkish or light brown) and pectinately dissected into narrow straw-colored lobes or teeth; ray flowers absent; disk corollas pink to occasionally deep purple-red (rarely white); peripheral flowers neuter with corollas 3.5–5 cm long, ± simulating rays; central flowers perfect, fertile, with corollas 2–2.5 cm long; achenes grayish brown to black, 4–5 mm long, with a basal attachment scar; pappus of stiff bristles, 6–14 mm long, minutely barbed. Prairies, disturbed areas, roadsides; widespread in TX; c U.S. from WI s to TX and w to AZ; Mexico. May–Aug. [*Centaurea americana* Nutt.] Sometimes cultivated; occasionally escapes outside its native range. The basket-like appearance of the overlapping phyllary teeth gives this species its common name. The stamens are reported to be sensitive to touch; when touched by insects they suddenly contract and push pollen out onto the pollinator (Kirkpatrick 1992).

PLUCHEA Cass. MARSH-FLEABANE, STINKWEED, CAMPHORWEED

Annual or perennial unpleasantly aromatic herbs, usually glandular; stems erect, 0.5–2+ m tall, ± pubescent; leaves alternate, simple, ovate to elliptic or lanceolate, glabrous to puberulent or tomentose, sometimes with resin-globules, margins usually serrate to dentate or crenate; heads numerous in corymbose panicles; phyllaries in 3–6 series, sometimes pink or purplish; ray flowers absent; disk flowers pink to rose-purple or creamy white to yellowish, of two types; peripheral disk flowers pistillate and fertile, the corolla lobes 3–4, filiform; central disk flowers functionally staminate, the corollas (4–)5-lobed; achenes cylindrical, ribbed, < 1 mm long; pappus of fine barbellate bristles in a single series.

A genus of ca. 50 species of warm areas of the world; many species are found in seasonally wet habitats. The foliage is very aromatic, for better or for worse. (Named for Abbé Nöel-Antoine Pluche, 1688–1761, French naturalist) (tribe Inuleae)
REFERENCES: Godfrey 1952; Robinson & Cuatrecasas 1973; Gillis 1977; Nesom 1989b, 2006u; Keeley & Jansen 1991.

1. Middle and upper leaves sessile, usually broad at the base and ± auriculate-clasping (although sometimes basally cuneate or subclasping).
   2. Involucres 4–6 mm tall; phyllaries and corollas rose-pink to purplish; phyllaries with appressed, spider-web-like hairs, sometimes also with glandular hairs ____ **P. baccharis**
   2. Involucres 5–8 mm tall; phyllaries and corollas usually creamy-white, sometimes greenish to yellowish or pale pink; phyllaries thinly pubescent with spider-web-like hairs and sparsely dotted with resin globules ____ **P. foetida**
1. Middle and upper leaves petiolate, tapering to a narrow base, not auriculate-clasping.
   3. Stems usually with appressed, spider-web-like hairs; heads usually arranged in a convex, elongate panicle, with the central uppermost axis maturing first, but the lateral branches never equaling or exceeding it; involucres 4–6 mm tall and 3–4 mm wide; inner phyllaries glabrous (resin globules may be present) ____ **P. camphorata**
   3. Stems pubescent or not, but lacking spider-web-like hairs; heads usually arranged in a flat-topped cyme, with the younger lateral branches elongating and equaling or exceeding the more central ones; involucres 5–6 mm tall and 4–6(–7) mm wide; inner phyllaries clearly puberulent, at least in the upper portion (resin globules may also be present) ____ **P. odorata**

Pluchea baccharis (Mill.) Pruski, (some species in the genus were originally described under *Baccharis*), ROSY CAMPHORWEED. Perennial to 60 cm tall; leaves sessile, the blades ovate to oblong-elliptic, 2–7 cm long, puberulent to sparsely villous, glandular, bases cuneate to clasping, margins toothed; involucres 4–6 mm tall and 5–9 mm wide; phyllaries rose-pink to purplish, with appressed, spider-web-like hairs, sometimes with glandular hairs; corollas rose-pink to purplish. Low drainage areas, ditches, pond margins; Hardin, Harris, Jefferson, Newton, and Tyler cos.; deep se TX; coastal plains from NC s to FL and w to TX. Jul–Sep. [*Conyza baccharis* Mill., *P. rosea* Godfrey]

Pluchea camphorata (L.) DC., (camphor), CAMPHORWEED, PLOUGHMAN'S-WORT. Annual or perennial to 200+ cm tall; leaves with a camphor-like odor when crushed, petiolate, the blades elliptic to oblong-elliptic, 6–15 cm long, puberulent, glandular, margins dentate-serrate or entire; heads usually arranged in a convex panicle; involucres 4–6 mm tall and 3–4 mm wide; phyllaries usually cream, sometimes purplish, nearly glabrous to sparsely puberulent, sparsely dotted with resin globules; corollas rosy to purplish. Low drainage areas, stream channels; e and c TX w to Edwards Plateau; e and c U.S. from NJ s to FL and w to WI, KS, and TX. Aug–Oct. [*Erigeron camphoratus* L.]

Pluchea foetida (L.) DC., (fetid, bad-smelling), STINKING PLUCHEA, STINKING MARSH-FLEABANE. Annual or perennial to 100 cm tall; stems often purplish; leaves with an unpleasant body-odor scent when crushed, sessile, the blades oblong-elliptic to lance-ovate, 3–10(–13) cm long, sparsely dotted with resin globules, bases clasping, margins denticulate; heads often arranged in a flat-topped cyme; involucres 5–8 mm tall and 6–9(–12) mm wide; phyllaries usually creamy white, sometimes greenish or yellowish, thinly pubescent with spider-web-like hairs, sparsely dotted with resin globules; corollas creamy white to pale yellowish or pale pink. Wet areas, bogs, swampy woods; e and sc TX; mainly coastal plains from NJ s to FL and w to TX. Jul–Nov. [*Baccharis foetida* L., *P. tenuifolia* Small] If varieties are recognized, TX plants would be var. *foetida*.

Pluchea odorata (L.) Cass. var. **odorata,** (fragrant), CANELA, PURPLE PLUCHEA, SWEETSCENT. Annual or perennial to 80(–200) cm tall; leaves usually petiolate, the blades lance-ovate to ovate-elliptic, 4–15 cm long, nearly glabrous to densely tomentose, margins shallowly serrate; heads usually arranged in a flat-topped cyme; involucres 5–6 mm tall and 4–6(–7) mm wide; phyllaries usually cream, sometimes purplish, nearly glabrous to puberulent, sparsely dotted with resin globules; corollas pink to rosy-purple. Low drainage areas, springs, marshes, estuaries; nearly throughout TX but most common in e 1/2; s 1/2 of the U.S. from MD w to CA; introduced elsewhere, often in saline habitats. Aug–Nov. [*Conyza odorata* L., *P. purpurascens* (Sw.) DC.]

PRENANTHES L. RATTLESNAKE-ROOT

Perennial taprooted herbs, often producing offshoots from slender rhizomes; stems erect, often purplish, producing milky sap; leaves in a basal rosette, alternate on the stem, entire or toothed to pinnately cleft or palmately lobed, the lowermost sometimes appearing compound; heads often nodding at flowering, in racemose or paniculate arrays; primary phyllaries in 1 series, equal, yellow-green to purplish, often hairy, subtended by a secondary series of small, unequal bractlets; ray flowers perfect and fertile, corollas creamy white to yellowish or pinkish, the ligules toothed apically; disk flowers absent; achenes spindle-shaped or columnar, 5–12-ribbed, not beaked; pappus of many minutely barbed, whitish to tawny bristles.

A genus of 26–30 species of North America and Eurasia, with 1 species in South Africa. Sometimes treated in the genus *Nabalus*. (Greek: *prenes*, drooping, and *anthos*, flower, alluding to the nodding heads of some species) (tribe Cichorieae)
REFERENCES: Milstead 1964; Singhurst et al. 2004; Bogler 2006f; Singhurst & Holmes 2010.

1. Lower leaves present at flowering; heads (4–)5(–6)-flowered; phyllaries (4–)5(–6), glabrous or sparsely hairy; pappus 5–6 mm long ____ **P. altissima**
1. Lower leaves usually withered by flowering; heads 8–19-flowered; phyllaries 6–10, coarsely hispid or bristly, at least along midveins; pappus 7–8 mm long.
   2. Lower stem leaf blades spatulate in overall outline; heads ± ascending; phyllaries yellow-green to tan; achenes 5–6 mm long ____ **P. aspera**
   2. Lower stem leaf blades ± elliptic in overall outline; heads nodding; phyllaries usually purple or lavender; achenes 8–10 mm long ____ **P. barbata**

Prenanthes altissima L., (tallest), TALL RATTLESNAKE-ROOT. Stems 40–250 cm tall; lower leaves petiolate, petioles winged, blades ovate or triangular, bases truncate or hastate to cordate, margins entire or shallowly dentate, often deeply 3-lobed, reduced in size and lobing up the stem; phyllaries pale green, glabrous or sometimes sparsely hairy; corollas pale yellow to greenish yellow; achenes 4–5 mm long. Open hardwood floodplains, creek banks and terraces; in TX recorded only from Jasper and Newton cos.; e U.S. from ME s to GA and w to MI, MO, and TX. Oct–Nov. [*Nabalus altissimus* (L.) Hook., *P. altissima* L. var. *cinnamomea* Fernald]

Prenanthes aspera Michx., (rough), ROUGH RATTLESNAKE-ROOT. Stems 35–170 cm tall; lower leaves petiolate, blades spatulate, bases attenuate, margins entire or weakly dentate, reduced in size up the stem and there sessile and clasping; phyllaries yellow-green to tan, coarsely setose; corollas pale yellow to creamy white; achenes 5–6 mm long. Calcareous pimple-mound prairies, tallgrass prairies, open woodlands; in TX known only from Bowie Co. (*J.R. Singhurst & W.C. Holmes s.n.*, *J.R. Singhurst 18337* BAYLU; Singhurst & Holmes 2010), apparently rare in the state; c U.S. from OH s to MS and w to SD and TX. Sep–Oct. [*Nabalus asperus* (Michx.) Torr. & A. Gray]

Prenanthes barbata (Torr. & A. Gray) Milstead, (barbed), BARBED RATTLESNAKE-ROOT. Stems 50–150 cm tall; lower leaves petiolate, petioles winged, blades ± elliptic, bases narrowed, margins coarsely or irregularly dentate or serrate, unlobed, reduced in size

Perityle lindheimeri var. lindheimeri [GR2, SID]

Pityopsis graminifolia [BB2]

Pityopsis oligantha [UWA]

Pinaropappus roseus [BT3]

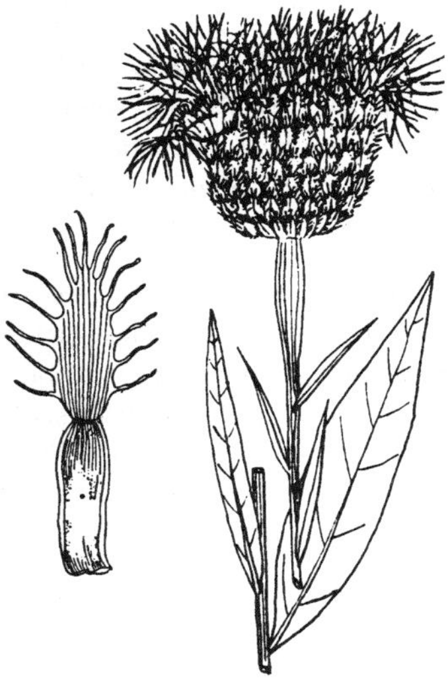

Plectocephalus americanus [GLE, STE]

Pluchea baccharis [JEM]

Pluchea camphorata [GLE]

up the stem and there sessile and cuneate or clasping; phyllaries usually purple or lavender, coarsely hispid or bristly along midveins; corollas white to yellowish white; achenes 8–10 mm long. Oak-hickory terraces, moist ravines, seepage slopes; mostly se TX, also ne TX; se U.S. from KY s to GA and w to OK and TX. Sep–Oct. [*P. serpentaria* Pursh var. *barbata* Torr. & A. Gray] While recorded in numerous counties in East TX, this species is becoming rarer due to habitat loss through road expansions and clear-cutting on private lands (Poole et al. 2007). ⚠

Prenanthes carrii Singhurst, O'Kennon & W.C. Holmes, (for William "Bill" R. Carr, modern prolific collector of TX plants, formerly of Texas Parks and Wildlife Department and Texas Nature Conservancy), CARR'S RATTLESNAKE-ROOT, CANYON RATTLESNAKE-ROOT, which would not have keyed to any other species satisfactorily here, can be distinguished by its long-petiolate, sagittate lower and midstem leaves, and its presence in the Edwards Plateau (as opposed to East TX). Endemic to rich woodlands and canyons of Bandera, Gillespie, Kerr, and Real cos. in sw Edwards Plateau, just w of the sw border of East TX (Singhurst et al. 2004). Aug–Nov. Only recently recognized as a distinct species, this endemic has no federal or state protection but is considered imperiled due to its restricted range (Poole et al. 2007, 2010). ⚠

PSEUDOGNAPHALIUM Kirp. RABBIT-TOBACCO, CUDWEED

Ours annual herbs, whitish to grayish with cottony or woolly pubescence (at least when young); stems erect or ascending; leaves sometimes in a basal rosette (usually withering before flowering), alternate on the stem, sessile, sometimes short-decurrent, blades linear to narrowly lanceolate or oblanceolate to subspatulate; inflorescences usually with numerous heads clustered in corymbose arrays; phyllaries in several series, unequal, scarious; receptacles naked; ray flowers absent; disk flowers all fertile but only the central ones perfect, corollas usually yellowish; achenes compressed-oblong or cylindric; pappus of 10–12 bristles, these readily separating individually, in clusters, or as an easily-fragmented ring.

A genus of ca. 100 species in the New World, Africa, and Asia, mostly of temperate regions; formerly lumped into *Gnaphalium*. (Greek: *pseudo*, deceptively similar, and the genus name *Gnaphalium*, from an ancient Greek name of some downy plant, derived from *gnaphalon*, lock of wool) (tribe Inuleae)

REFERENCES: Hillard & Burtt 1981; Anderberg 1991; Nesom 2001a, 2001b, 2001c, 2004b, 2006v.

1. Leaves at maturity ± concolor (both surfaces usually gray to grayish green, tomentose), bases usually decurrent, creating wings extending 1–2 mm down the stem; corollas red-tipped; achenes unribbed, conspicuously dotted with whitish, globular hairs ____ **P. luteoalbum**
1. Leaves at maturity strongly to weakly bicolor (upper surfaces green, lower surfaces gray to white and tomentose), bases not decurrent; corolla lobes yellow, lacking red tips; achenes longitudinally ribbed (may additionally be roughened with tiny bumps).
   2. Leaves with bases subclasping, margins strongly revolute; involucres 4.5–5 mm high; achenes ridged and roughened with tiny bumps ____ **P. austrotexanum**
   2. Leaves with bases not clasping, margins flat; involucres 5–7 mm high; achenes ridged, smooth between ridges.
      3. Plants thickly-fibrous-rooted; stems greenish, with stalked glands; perfect flowers 9–15 per head ____ **P. helleri**
      3. Plants tap-rooted; stems white-woolly tomentose, eglandular; perfect flowers 4–8(–11) per head ____ **P. obtusifolium**

Pseudognaphalium austrotexanum G.L. Nesom, (of south Texas), SOUTH TEXAS RABBIT-TOBACCO. Taprooted annuals, to 70 cm tall; stems densely white-woolly, not glandular; leaves bicolor (green above, densely stipitate glandular, and white-tomentose beneath), bases subclasping, not decurrent, margins strongly revolute; involucres 4.5–5 mm high; phyllaries silvery-white, glabrous; achenes ridged and roughened with tiny bumps. Pastures, grasslands, open disturbed sites, often in sandy soils; Brazos and Harris cos.; South TX Plains and Gulf Prairies; Mexico. Oct–Dec.

Pseudognaphalium helleri (Britton) Anderb., (for A.A. Heller (1867–1944), early collector in TX), HELLER'S CUDWEED, HELLER'S RABBIT-TOBACCO. Thickly-fibrous-rooted annuals, to 100 cm tall, sometimes aromatic; stems greenish, with stalked glands; leaves bicolor (green above and white to gray beneath, lightly tomentose), with stalked glands, bases not clasping, not decurrent, margins flat; involucres 6–7 mm high; phyllaries whitish, ± tomentose; achenes ridged, smooth between ridges. Pine-oak forests, in sandy or clay soils; ne TX in Bowie, Cass, Harrison, and Titus cos., and Marion Co. (Kartesz 2015); se U.S. from NC s to FL and w to OK and TX. Sep–Nov. [*Gnaphalium helleri* Britton, *G. obtusifolium* L. var. *helleri* (Britton) S.F. Blake]

Pseudognaphalium luteoalbum (L.) Hilliard & B.L. Burtt, (yellowish white), JERSEY CUDWEED, RED-TIP RABBIT-TOBACCO. Tap- or fibrous-rooted annuals, to 40 cm tall; stems loosely white-woolly, not glandular; leaves concolor or weakly bicolor (usually gray-tomentose on both sides), not glandular, bases subclasping, usually decurrent 1–2 mm down the stem, margins weakly revolute; involucres 3–4 mm high; phyllaries silvery to yellowish, glabrous; corollas red-tipped; achenes unridged, conspicuously dotted with whitish, globular hairs. Roadsides, ditches, disturbed sites; recorded from Marion (TEX), Hardin (MacRoberts et al.

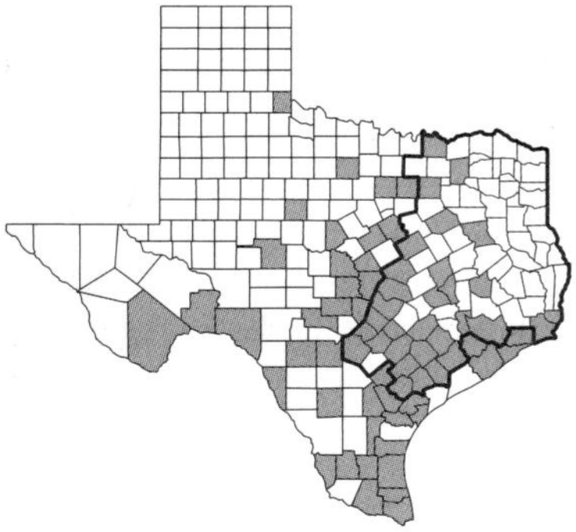

Parthenium hysterophorus

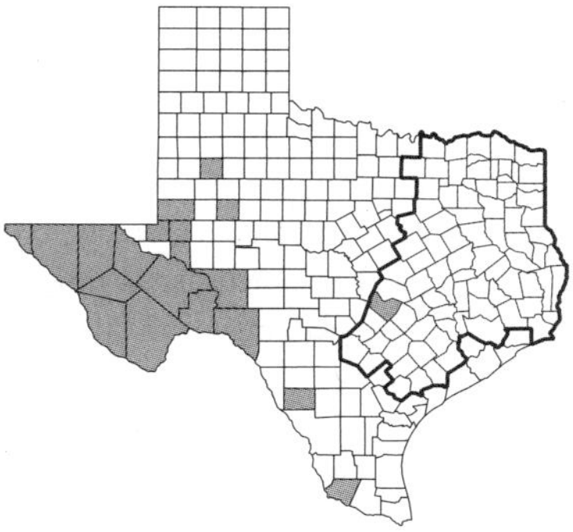

Parthenium incanum

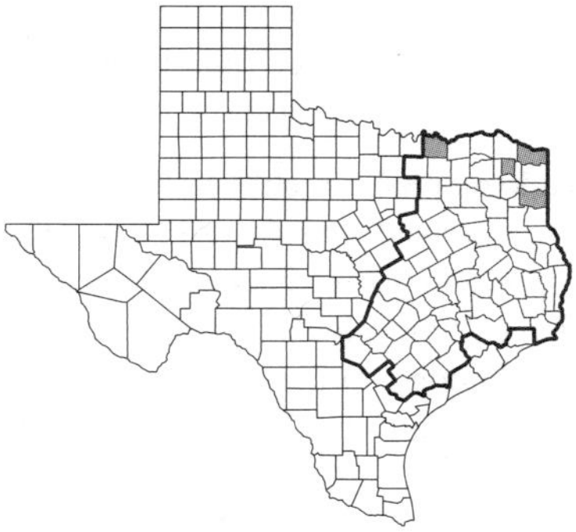

Parthenium integrifolium

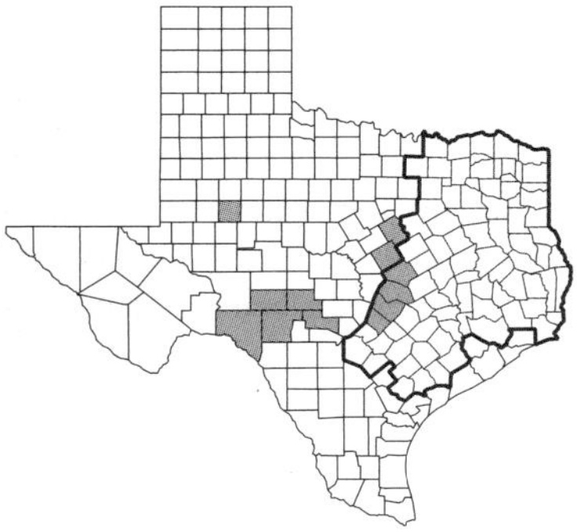

Pectis angustifolia var. fastigiata

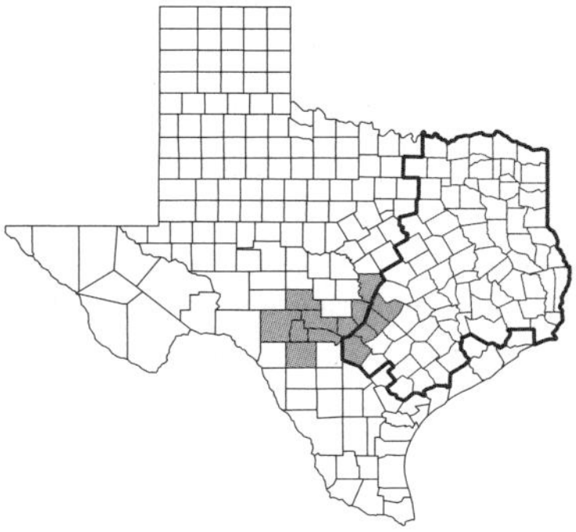

Perityle lindheimeri var. lindheimeri

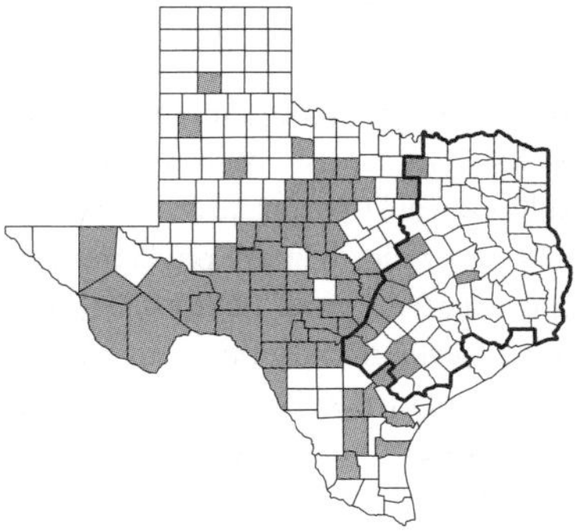

Pinaropappus roseus

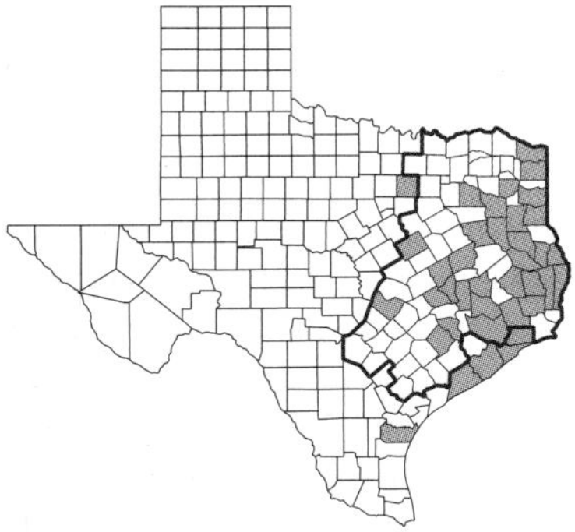

Pityopsis graminifolia

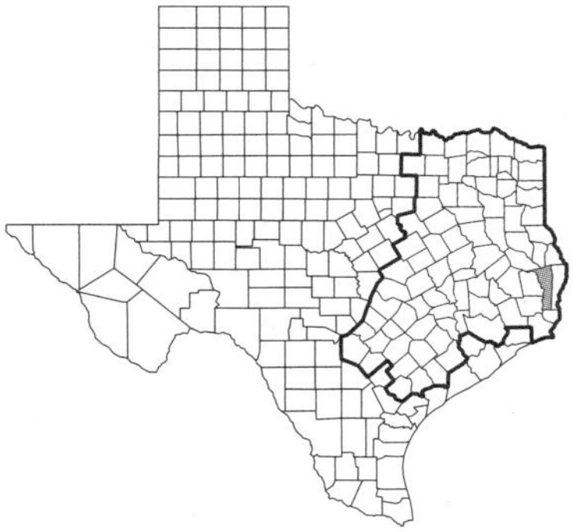

Pityopsis oligantha

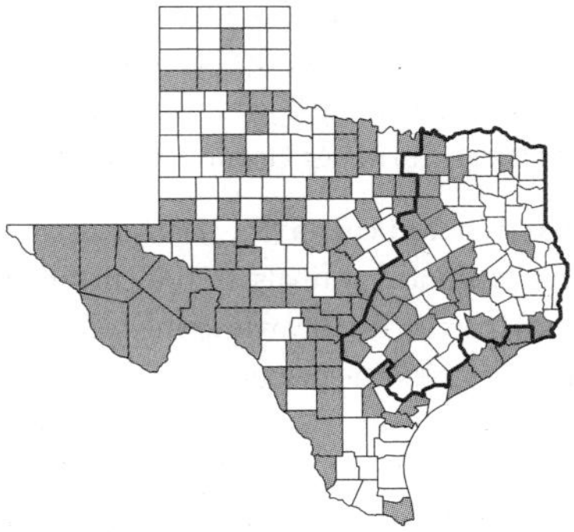

Plectocephalus americanus

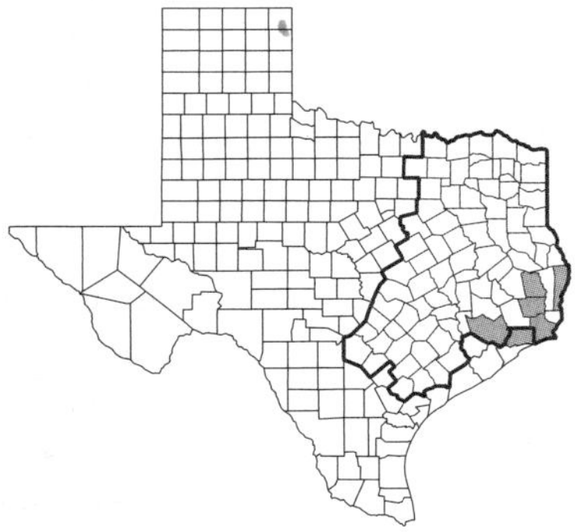

Pluchea baccharis

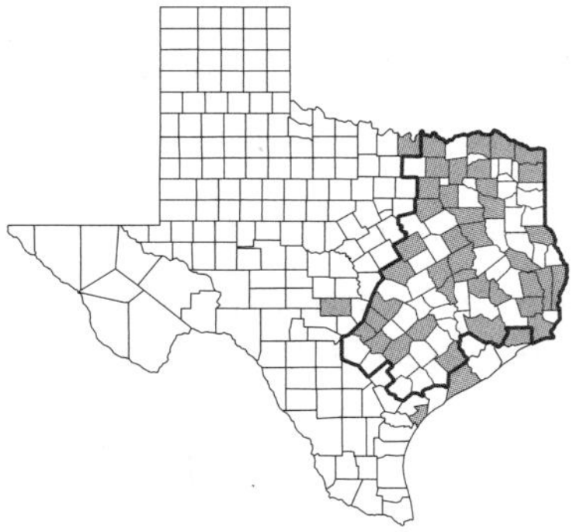

Pluchea camphorata

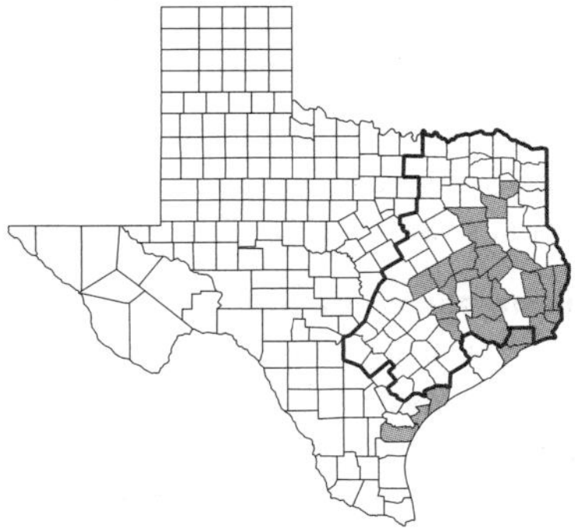

Pluchea foetida

2010), and Liberty (Brown et al. 2007) cos.; scattered in nc TX, Edwards Plateau, and Trans-Pecos; w U.S. from WA s to CA and e to UT and TX; Mexico; Eurasia; introduced and weedy in warm areas throughout the world. Apr–Jun. [*Gnaphalium luteoalbum* L., *Laphangium luteoalbum* (L.) Tzvelev] Native of Eurasia.

Pseudognaphalium obtusifolium (L.) Hilliard & B.L. Burtt, (blunt-leaved), EASTERN RABBIT-TOBACCO, FRAGRANT CUDWEED, CAT'S-FOOT. Taprooted annuals, to 100 cm tall, sometimes aromatic; stems sparsely to densely white-woolly, usually not glandular; leaves bicolor (green above and white-tomentose beneath), sometimes slightly glandular, bases not clasping, not decurrent, margins flat; involucres 5–7 mm high; phyllaries whitish, glabrous or the bases tomentose; achenes ridged, smooth between ridges. Prairies, open woods, roadsides, often in sandy soils; e, se, and sc TX; nc TX, Edwards Plateau, and s TX; e 1/2 of North America. Aug–Oct. [*Gnaphalium obtusifolium* L.]

PSEUDOGYNOXYS (Greenm.) Cabrera FLAMEVINE

A genus of ca. 12 species native to the New World tropics; some are cultivated as ornamentals. (Greek: *pseudo*, deceptively similar, and the genus name *Gynoxys*, for its resemblance to that genus) (tribe Senecioneae)
REFERENCES: Robinson & Cuatrecasas 1977; Barkley 2006c.

Pseudogynoxys chenopodioides Kunth, (*Chenopodium*-like, for the leaves' resemblance to some GOOSEFOOT species), MEXICAN FLAMEVINE, ORANGE GLOW VINE. Vines to 5+ m long; stems twining; leaves alternate, petiolate, blades ovate to lanceolate, bases broadly cuneate to truncate or rounded, margins dentate; heads showy, borne in corymbose arrays or singly; involucres usually 12–20 mm wide; primary phyllaries in 1–2 series, subequal, lance-linear, subtended by small bractlets; ray flowers 6–15+, pistillate, fertile, ligules 12–20(–25+) mm long, orange to brick-red; disk flowers perfect, fertile, corollas orange; achenes (2–)4 mm long, cylindric; pappus of many white, finely-barbed bristles, 3–8+ mm long. Disturbed sites, waste grounds, abandoned gardens; recorded in Bexar and Harris cos.; also s TX; FL and TX; Mexico; Central and South America; and West Indies. Year-round (mostly winter). [*Senecio confusus* Britten] Native of tropical America; possibly escaping from cultivation and naturalizing in subtropical TX; its potential to spread and become a permanent member of the flora is not known.

PSILACTIS A. Gray TANSY-ASTER

A genus of 6 species native to sw U.S., Mexico, and South America; strongly resembling and related to *Symphyotrichum* but easily distinguished by the heteromorphic achenes and strongly glandular phyllaries. (Greek: *psilos*, bare, and *actis*, ray, referring to the absence of ray flower pappus) (tribe Astereae)
REFERENCES: Turner & Horne 1964; Morgan 1993, 2006.

Psilactis heterocarpa (R.L. Hartm. & M.A. Lane) D. Morgan, (differently-fruited, for the two types of achenes), SAN PATRICIO TANSY-ASTER. Taprooted annuals, to 100 cm tall; stems appressed-hairy, with stalked glands in head-bearing region; leaves sometimes in a basal rosette (usually withering before flowering), alternate on the stem, sessile, blades obovate or lanceolate to linear, margins entire; heads borne in loose corymbose arrays; involucres 4–6 mm high; phyllaries in 2–3 series, scarious-margined, with stalked glands; ray flowers 15–25, pistillate, fertile, corollas white to bluish or purple; disk flowers perfect, fertile, corollas yellow; achenes of ray flowers 1.3–1.5 mm long, lacking pappus; achenes of disk flowers 2–3 mm long; disk pappus of many barbellate bristles, 3.2–4 mm long. Grasslands, weedy pastures, open mesquite woodlands; endemic to the Gulf Coastal Plain of sc TX in Goliad, Jackson, and Victoria cos.; also Karnes, Kleberg, Nueces, Refugio, and San Patricio cos. Sep–Nov. [*Machaeranthera heterocarpa* R.L. Hartm. & M.A. Lane] This species is of conservation concern due to its restricted range (Poole et al. 2007)

PSILOSTROPHE DC. PAPER-FLOWER

A genus of 7 species native to sw U.S. and Mexico. (Greek: *psilos*, bare, and *strophe*, turn, perhaps alluding to the naked receptacles) (tribe Helliantheae)
REFERENCES: Brown 1978; Turner et al. 1988; Strother 2006x.

Psilostrophe villosa Rydb., (hairy), WOOLLY PAPER-FLOWER. Biennials or perennials (rarely flowering the first year), 10–45(–60+) cm tall, from a woody taproot; stems and leaves gray-green, cobwebby-villous; leaves alternate, blades oblanceolate, margins entire; heads showy, borne in ± crowded corymbose arrays; involucres 4–5(–7) mm high; phyllaries in 1–2 series, erect in fruit, the bases indurate; receptacles naked; ray flowers (2–)3(–5), pistillate, fertile, ligules golden-yellow, usually 3–4 mm long, tips toothed or lobed; disk flowers 5–8(–12), perfect, fertile, corollas golden-yellow; achenes cylindric, usually glabrous, sometimes gland-dotted; pappus of 4–6 lanceolate to lance-subulate scales, 2.5–3+ mm long, entire. Grassy plains, desert flats, gypsum and limestone soils; Fayette Co., also Brazos, Caldwell, Goliad, and Montgomery cos. (Kartesz 2015); mostly Plains Country, Edwards Plateau, and s TX; KS, NM, OK, and TX; likely also Mexico. Mar–Oct. [*P. tagetina* (Nutt.) Greene var. *cerifera* (A.

Pluchea foetida [ROE]

Pluchea odorata [CO1]

Prenanthes altissima [VIC]

Prenanthes aspera [YAT]

Prenanthes barbata [GEN]

Pseudognaphalium austrotexanum [SID]

Pseudognaphalium helleri [BB2]

Nelson) B.L. Turner] Though apparently the plants are not distasteful to livestock, they are poisonous and have caused deaths in grazing sheep (Kingsbury 1964). ☠

PTEROCAULON Elliott BLACKROOT

A genus of 17 species largely confined to South America, with a few found in North America and Australasia. (Greek: *pteron*, wing, and *kaulos*, stem, alluding to stems winged by decurrent leaf bases) (tribe Inuleae)
REFERENCES: D'Arcy 1975; Cabrera & Ragonese 1978; Nesom 2006w.

Pterocaulon virgatum (L.) DC., (winged), WAND BLACKROOT. Perennial herbs, 40–150 cm tall; stems simple, erect, winged by decurrent leaf bases; leaves alternate, sessile, blades linear to narrowly elliptic or linear-lanceolate, strongly bicolor (green above and whitish-tomentose beneath), usually glandular, margins entire or minutely denticulate, revolute; heads sessile, borne in cylindric, interrupted, spicate arrays at ends of branches; involucres 4–5 mm high; phyllaries in 4–6 series, unequal, narrowly lanceolate; ray flowers absent; disk flowers peripherally pistillate and fertile, centrally functionally staminate, corollas yellowish cream; achenes cylindric, ribbed, 1–1.4 mm long; pappus of many pale, finely-barbed bristles. Moist woods, ditches, marshes, sandy or silty-clay soils; e, se, and sc TX; s TX; LA and TX; Mexico; Central and South America; and West Indies. Jul–Oct. [*Gnaphalium virgatum* L.]

PYRRHOPAPPUS DC. FALSE DANDELION, NATIVE-DANDELION

Annuals or perennials; sap milky; leaves basal, usually in a rosette, alternate and reduced upwards; leaf blades oblanceolate or oblong- to elliptic-lanceolate, entire or usually toothed or lobed; heads solitary or few, large and showy; the inner (primary) phyllaries linear-lanceolate, subtended by a secondary series of short, phyllary-like bracts; flowers all ligulate; corollas yellow, open during morning hours; achenes fusiform, beaked; pappus of abundant hair-like bristles 7–12 mm long.

A North American genus of 4 or 5 species; the roots of some were eaten by Native Americans. (Greek: *pyrros*, flame-colored or red, and *pappus*, down, fuzz, pappus, referring to the reddish pappus found in some species) (tribe Cichorieae)
REFERENCES: Northington 1971, 1974; Vuilleumier 1973; Barber & Estes 1978; Turner & Kim 1990; Strother 2006y.

1. Perennials with small tuberous swelling 5–20 mm thick located 2–15 cm below ground at end of roots; stems above rosette with 0–2 leaves or leafy bracts, usually pubescent or short-pilose ____ **P. grandiflorus**
1. Annuals with tapering taproot; stems above rosette with 1–9 leaves, pubescent or glabrous.
   2. Stems pubescent with curly hairs (rarely glabrous), stem leaves 1–5, the uppermost with blades deeply several-lobed; secondary phyllary-like bracts usually 1/3 as long as primary phyllaries (or less); mostly calcareous soils of c TX ____ **P. pauciflorus**
   2. Stems glabrous, stem leaves 3–9, the uppermost with blades entire or merely toothed or with one pair of basal lobes; secondary phyllary-like bracts mostly 1/3–2/3 as long as primary phyllaries; mostly sandy soils of East TX ____ **P. carolinianus**

Pyrrhopappus carolinianus (Walter) DC., (of Carolina), CAROLINA FALSE DANDELION. Annual or biennial from a taproot; stems usually glabrous, (5–)30–100 cm tall; basal leaves in a rosette, often early deciduous; stem leaves reduced upwards, entire or with 1–2 basal lobes; primary phyllaries 15–25 mm long; achenes 12–16 mm long (including beak). Sandy soils, woodlands and fields; throughout e 1/3 of TX; se U.S. from PA s to FL and w to NE and TX. Apr–Jul, sporadically to Sep. [*P. carolinianus* (Walter) DC. var. *georgianus* (Shinners) H.E. Ahles, *P. georgianus* Shinners]

Pyrrhopappus grandiflorus (Nutt.) Nutt., (large-flowered), TUBER FALSE DANDELION. Perennial with a tuber giving rise to 1 or several, simple or branched, erect or suberect rhizomes, these developing near ground level; basal leaves in a rosette; flowering stems 10–30(–40) cm tall; stems usually rather densely pubescent with curly hairs, rarely glabrous; primary phyllaries 15–20 mm long; achenes 10–13 mm long (including beak). Rocky, clayey, or sandy prairies; on the w borders of East TX; Grand Prairie s and w to Panhandle and w TX; KS, OK, and TX. Apr–May. [*P. scaposus* DC.]

Pyrrhopappus pauciflorus (D. Don) DC., (few-flowered), MANY-STEM FALSE DANDELION, TEXAS DANDELION, PATA DE LEON. Annual 20–60 cm tall; basal leaves in a rosette; stems usually branching from the base, usually rather densely pubescent with curly hairs, rarely glabrous; primary phyllaries 16–20 mm long; achenes 11–14 mm long (including beak). Calcareous soils, prairies, roadsides; Blackland Prairie and Grand Prairie; se and e TX w to nc TX and Edwards Plateau; AR, LA, OK, and TX; n Mexico. (Feb–)Mar–Jun(–Aug). Hybridizes with *P. carolinianus* along highways where sand and limestone gravel have been brought in for fill. Petersen et al. (1990) indicated that infraspecific taxa are not warranted. [*P. geiseri* Shinners, *P. multicaulis* DC., *P. multicaulis* var. *geiseri* (Shinners) North.]

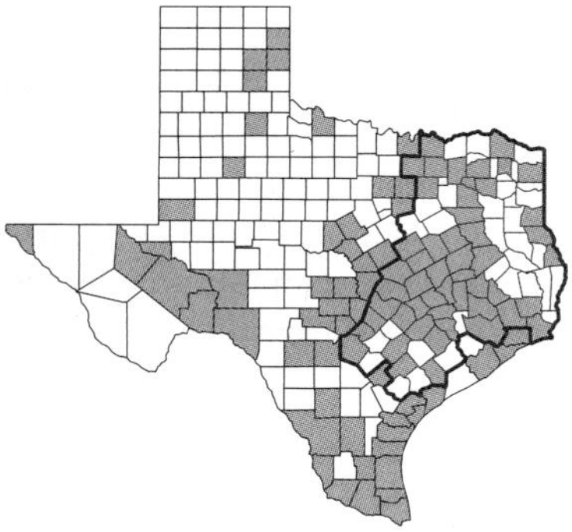

Pluchea odorata var. odorata

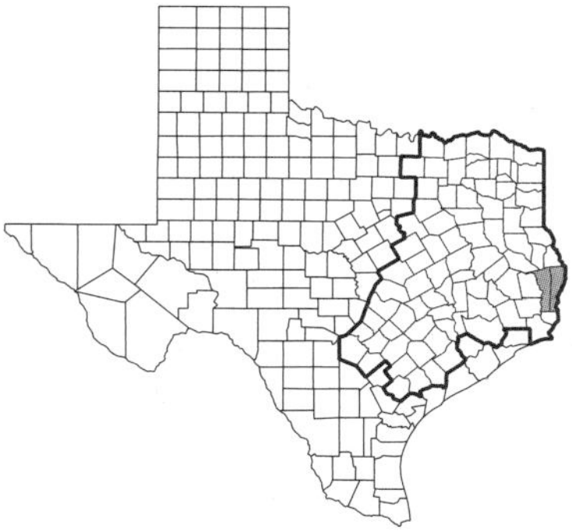

Prenanthes altissima

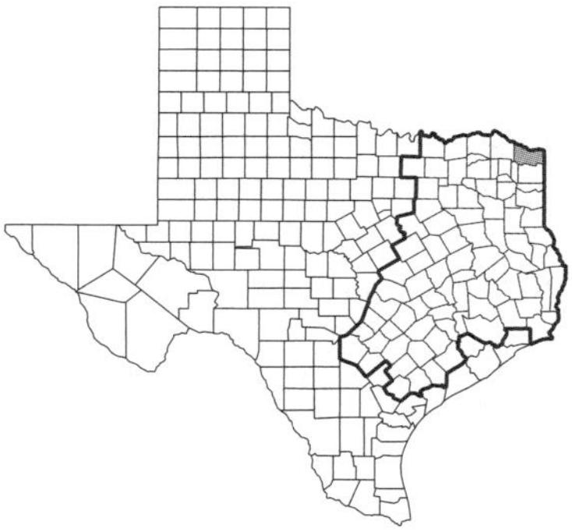

Prenanthes aspera

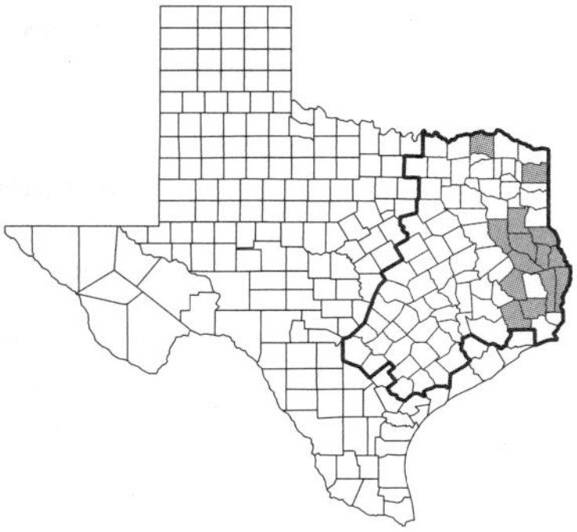

Prenanthes barbata

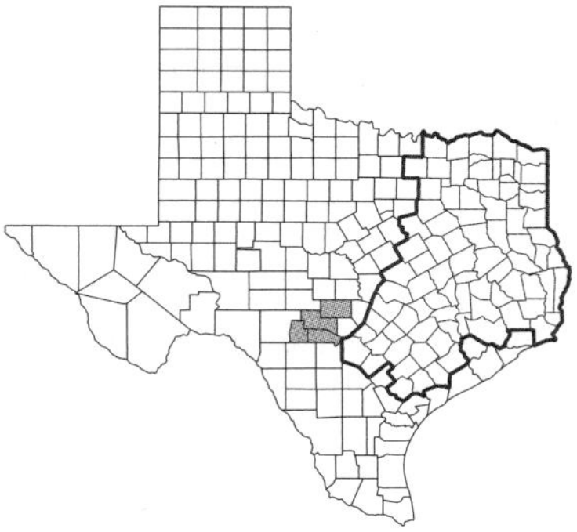

Prenanthes carrii

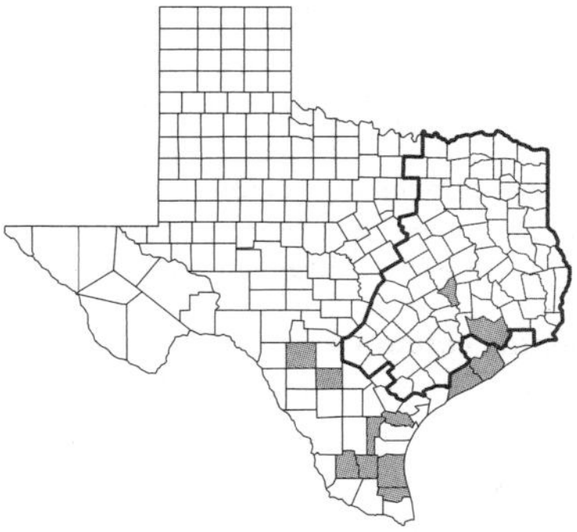

Pseudognaphalium austrotexanum

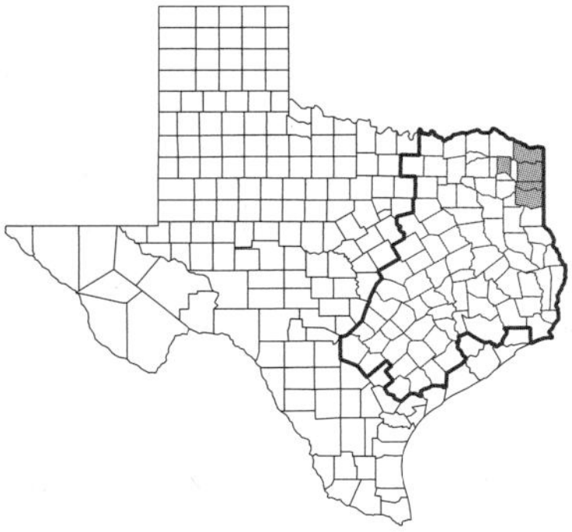

Pseudognaphalium helleri

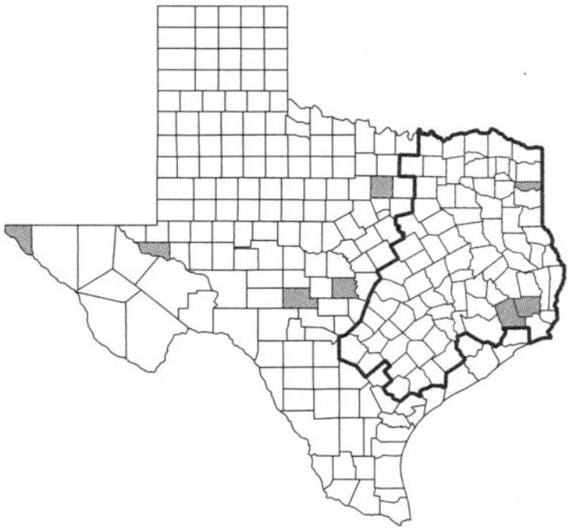

Pseudognaphalium luteoalbum

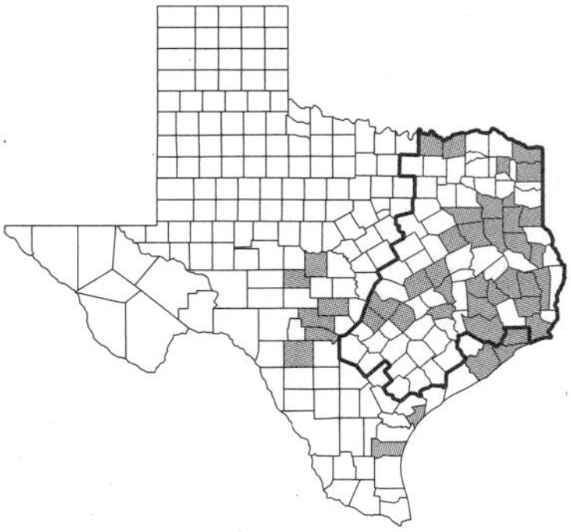

Pseudognaphalium obtusifolium

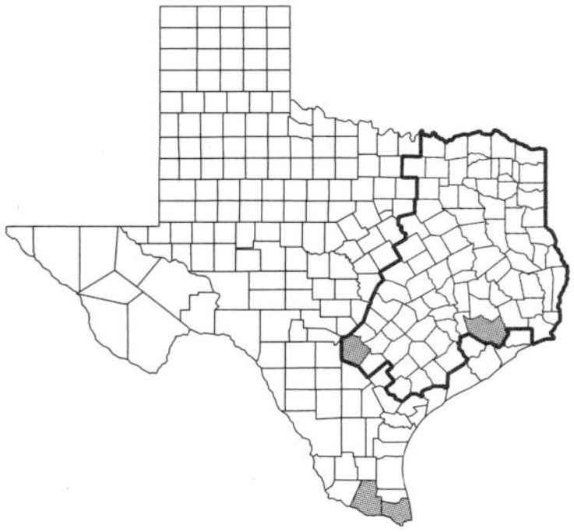

Pseudogynoxys chenopodioides

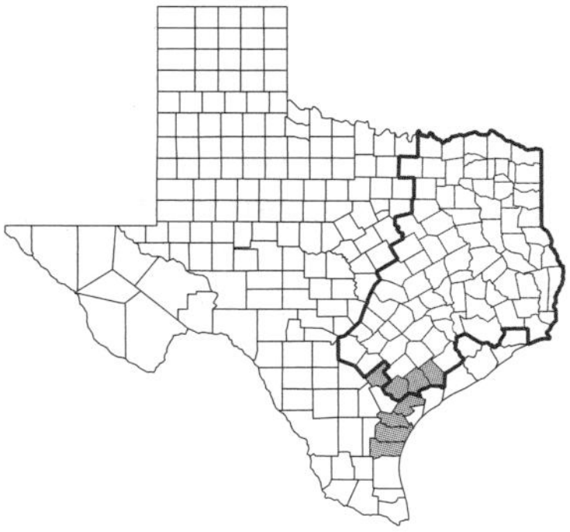

Psilactis heterocarpa

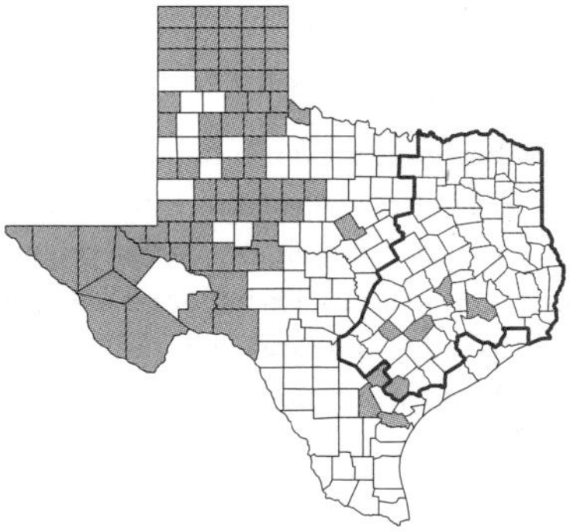

Psilostrophe villosa

RATIBIDA Raf. MEXICAN-HAT, PRAIRIE CONEFLOWER

Herbaceous perennials; stems strigose-hirsute, gland-dotted; leaves in a basal rosette and alternate up the stem, petiolate, pinnately lobed to 1–2-pinnatifid, hirsute or strigose, usually gland-dotted; heads solitary or borne in loose corymbose arrays, long-peduncled (in ours), peduncles often with prominent, tan ribs; phyllaries in 2 series, unequal, usually reflexed in fruit; receptacles (cones) subspheric to columnar, pales (chaff) folded lengthwise, strigose and gland-dotted; ray flowers 4–15, neuter, corollas yellow, maroon, or both; disk flowers perfect, fertile, corollas yellowish green, sometimes purplish; achenes oblanceolate to broadly ovate, black, strongly compressed, margin(s) often ciliate, fringed, or comb-like; pappus of 1–2 tooth-like projections, or absent.

A genus of 7 species of North America and Mexico; some are cultivated as ornamentals. (According to Fernald (1950), the derivation of the name is "... like most work of its author [Rafinesque], not clear") (tribe Heliantheae)
REFERENCES: S. Jackson 1963; Richards 1968; Cox & Urbatsch 1990, 2006; Urbatsch & Jansen 1995.

1. Plants fibrous-rooted; cone (convex part of head, including pales and disk flowers) ellipsoid to globular or ovoid, 10–25 mm high; pappus absent, or of 1–2 tooth-like projections ____ **P. pinnata**
1. Plants taprooted; cone (convex part of head, including pales and disk flowers) columnar, 10–70 mm high; pappus of 1–2 tooth-like projections, or sometimes absent.
   2. Disk flower style branch tips rounded; disk corollas 1–2.5 mm long; achenes with one margin glabrous and the opposing margin glabrous or ciliate ____ **P. columnifera**
   2. Disk flower style branch tips subulate; disk corollas 2.5–3 mm long; achenes with one margin narrowly winged to fringed or comb-like and the opposing margin fringed or comb-like ____ **P. peduncularis**

Ratibida columnifera (Nutt.) Wooton & Standl., (bearing columns), MEXICAN-HAT, REDSPIKE MEXICAN-HAT, THIMBLE-FLOWER, PRAIRIE CONEFLOWER, UPRIGHT PRAIRIE CONEFLOWER. Taprooted perennials, to 105+ cm tall; leaves 1–2-pinnatifid; cones columnar, 10–50 mm high; ray ligules 4–12, yellow or maroon, sometimes bicolor, 7–35 mm long; achenes 1.2–3 mm long, with one margin glabrous, and the opposing margin glabrous or ciliate; pappus of 1–2 tooth-like projections, or sometimes absent. Prairies, roadsides, waste areas; throughout most of TX; throughout North America but most common in c plains; Mexico. (Mar–)May–Oct (–Nov). [*R. columnaris* (Sims) D. Don, *Rudbeckia columnifera* Nutt.] Often included in wildflower seed mixes, contributing to an increasing geographic range; populations e of the Mississippi River may be adventive (Kartesz 2015).

Ratibida peduncularis (Torr. & A. Gray) Barnhart, (peduncled), NAKED MEXICAN-HAT, NAKED PRAIRIE CONEFLOWER. Taprooted perennials, to 115+ cm tall; leaves 2-pinnatifid to lyrate-pinnate; cones columnar, 15–70 mm high; ray ligules 4–14, yellow or maroon, often bicolor, 5–15 mm long; achenes 2.3–5 mm long, with one margin narrowly winged to fringed or comb-like, and the opposing margin fringed or comb-like; pappus of 1–2 tooth-like projections. Sandy sites, coastal areas; scattered from se to sc TX, mostly Gulf Coast and South TX Plains; LA and TX; Mexico. (Mar–)Apr–Nov. The two varieties hybridize with each other, as well as with *R. columnifera* (S. Jackson 1963; Richards 1968) and they have overlapping ranges in East TX; no infraspecific range maps are provided by Richards (1968) or USDA Plants Database (2016), therefore we are not providing ranges or maps for these varieties. Cox and Urbatsch (2006) divided the two East TX varieties as follows:

1. Leaves 2-pinnatifid; ray corollas mostly yellow; achenes with both margins fringed or comb-like ____ var. **peduncularis**
1. Leaves lyrate-pinnate; ray corollas mostly maroon (sometimes with some yellow); achenes with only one margin fringed or comb-like ____ var. **picta**

var. **peduncularis,** (peduncled). [*Lepachys peduncularis* Torr. & A. Gray]

var. **picta** (A. Gray) Sharp, (painted, colored), [*Lepachys peduncularis* Torr. & A. Gray var. *picta* A. Gray]

Ratibida pinnata (Vent.) Barnhart, (pinnate), PINNATE PRAIRIE CONEFLOWER, GRAYHEAD PRAIRIE CONEFLOWER. Fibrous-rooted perennials, to 125+ cm tall, arising from stout rhizomes or woody crowns; leaves pinnatifid to pinnate; cones ellipsoid to globular or ovoid, 10–25 mm high; ray ligules yellow, 16–60 mm long; achenes 2–4 mm long, with margins usually glabrous, or ciliate on one margin; pappus absent, or of 1–2 tooth-like projections. Tallgrass prairies, woodland openings, limestone outcrops; in TX recorded only from Bowie Co., where it was first collected in 2009 (*J. Singhurst 17500* BAYLU; Singhurst et al. 2010); e 1/2 of North America but most common in the c prairies. May–Oct. [*Rudbeckia pinnata* Vent.]

RAYJACKSONIA R.L. Hartm. & M.A. Lane CAMPHOR-DAISY, TANSY-ASTER

Annuals or perennials, 20–100 cm tall, taprooted; herbage glabrous, glandular; leaves alternate, sessile to subpetiolate, margins coarsely serrate or rarely entire; heads in paniculate or corymbose arrays; involucres hemispheric; phyllaries in 4–5 series, subequal or unequal, tips herbaceous; ray flowers pistillate, fertile, corollas yellow; disk flowers perfect, fertile, corollas yellow;

Pseudognaphalium luteoalbum [CRO]

Pseudognaphalium obtusifolium [DAR]

Pseudogynoxys chenopodioides [FNA]

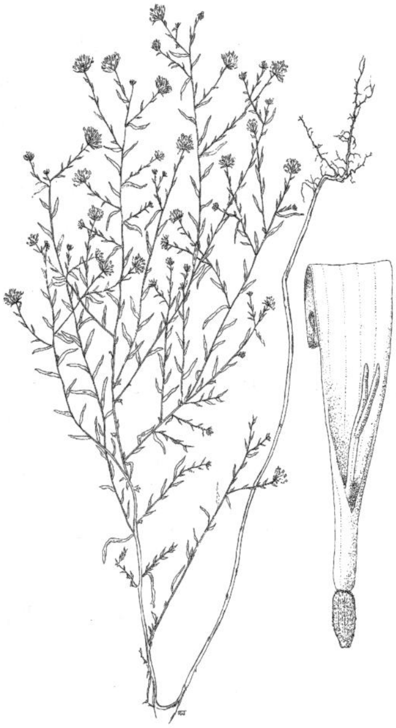

Psilactis heterocarpa [BTT]

Psilostrophe villosa [BB2]

Pterocaulon virgatum [FNA]

achenes dimorphic, strigose-sericeous, ribbed (sometimes absent), ray achenes 3-angled, disk achenes compressed; pappus of 30–40 brownish, unequal, barbellate bristles.

A genus of 3 species of North America and Mexico; it was previously included in *Haplopappus* and the taxa have been treated in a variety of genera including *Machaeranthera*. (Named for Dr. Raymond C. Jackson, American botanist and plant geneticist) (tribe Astereae)

References: Hartman 1990; Kawaguchi 1996; Lane & Hartman 1996; Morgan 1997; Nesom 2006x; Nesom et al. 2013.

1. Midstem leaf blades linear to linear-oblanceolate, 1–3(–4) mm wide; involucres 4–7 mm high and 10–15 mm wide; rays 14–19, corollas 6.5–9.5 mm long ____ **R. aurea**
1. Midstem leaf blades oblanceolate to oblong or oblong-lanceolate, (1–)4–15(–25) mm wide; involucres (4.5–)6–15 mm high and 9–30 mm wide; rays (13–)17–38, corollas 6–19 mm long.
   2. Heads on short peduncles, not surpassed by subtending leaves; phyllaries strongly unequal, 0.9–1 mm wide, tips broadly spreading to recurved ____ **R. annua**
   2. Heads essentially sessile, often surpassed by subtending leaves; phyllaries subequal, 1.3–1.7 mm wide, tips erect to spreading ____ **R. phyllocephala**

Rayjacksonia annua (Rydb.) R.L. Hartm. & M.A. Lane, (annual), viscid camphor-daisy, viscid tansy-aster. Annuals; leaves oblanceolate to oblong-lanceolate; heads short-peduncled, not surpassed by subtending leaves; involucres usually 6–9 mm high; phyllaries strongly unequal, 0.9–1 mm wide, tips broadly spreading to recurved; ray corollas 6–11(–14) mm long. Roadsides, prairies, alkaline flats, sandy soils; recorded in Gonzales Co. (*B.C. Tharp s.n.* TEX); otherwise far to the w in Plains Country and Trans-Pecos; c U.S. from NE s to TX and w to WY and NM. Aug–Oct. [*Haplopappus annuus* (Rydb.) Cory, *Machaeranthera annua* (Rydb.) Shinners, *Sideranthus annuus* Rydb.]

Rayjacksonia aurea (A. Gray) R.L. Hartm. & M.A. Lane, (golden), houston camphor-daisy, houston tansy-aster. Annuals; leaves linear to linear-oblanceolate; heads short-peduncled, not surpassed by subtending leaves; involucres 4–7 mm high; phyllaries strongly unequal, ca.1 mm wide, tips erect to slightly spreading; ray corollas 6.5–9.5 mm long. Saline prairie barrens, often in the presence of mima mounds (pimple mounds), sandy soils; endemic to the TX Gulf Coastal Plain, known from only 3 counties:, Fort Bend, Galveston, and Harris cos. Oct–Nov(–Dec). [*Haplopappus aureus* A. Gray, *Machaeranthera aurea* (A. Gray) Shinners, *Sideranthus aureus* (A. Gray) Small] This species is not listed at the state or federal level, but is of conservation concern due to its restricted range and highly specialized habitat, both threatened by urban development around Houston (Poole et al. 2007). ⚠

Rayjacksonia phyllocephala (DC.) R.L. Hartm. & M.A. Lane, (leafy-headed), gulf coast camphor-daisy. Annuals or perennials; leaves oblong to oblanceolate; heads essentially sessile, often surpassed by subtending leaves; involucres usually 10–15 mm high; phyllaries subequal, 1.3–1.7 mm wide, tips erect to spreading; ray corollas 8.5–19 mm long. Saline flats, sand dunes, canal banks, rarely inland as a waif; Gonzales, Harris, Jackson, and Victoria cos., also Austin and Brazos cos. (Nesom et al. 2013); TX Gulf Coast; FL, LA, MS, and TX; Mexico. Sep–Dec (sporadically all year). [*Haplopappus phyllocephalus* DC., *Machaeranthera phyllocephala* (DC.) Shinners var. *megacephala* (Nash) Shinners, *Sideranthus megacephalus* (Nash) Small] If varieties are recognized, ours would be var. *phyllocephala* (Nesom et al. 2013).

RUDBECKIA L. CONEFLOWER, BLACK-EYED-SUSAN, BROWN-EYED-SUSAN

Annual, biennial, or perennial herbs; leaves alternate, sometimes also basal, the blades simple to deeply lobed; heads terminal, large, solitary or corymbose; phyllaries in usually 1–3 series, usually subequal, herbaceous, loose; receptacles or "cones" subspheric or conic to columnar, often elongating after flowering; receptacle chaff (= pales) concave, each ± clasping a flower, the awn-like tips not apparent between flowers at anthesis; ray flowers neuter, with ligules yellow to yellow-orange, sometimes with a splotch or zone of maroon proximally; disk flowers perfect, fertile, with corollas yellowish green to maroon or brownish purple; achenes black, 4-angled; pappus a crown, or of a few short scales, or absent.

A North American genus of 23 species; several are cultivated as ornamentals. (Named for the Swedish professors Rudbeck: Olaf, the father (1630–1702) and Olaf, the son (1660–1740), predecessors of Linnaeus at Uppsala) (tribe Heliantheae)

References: Perdue 1957; Cox & Urbatsch 1990, 1994; Urbatsch & Jansen 1995; Urbatsch & Cox 2006; Campbell & Seymour 2013.

1. Lower stem and midstem leaves 3–9-lobed or pinnatifid (basal leaves, if present, may be entire).
   2. Lower stem and midstem leaves 5–9-lobed or pinnatifid; disk corollas yellow to yellowish green with yellow lobes; achenes 4.2–6 mm long ____ **R. laciniata**
   2. Lower stem and midstem leaves 3–5-lobed; disk corollas mostly brownish purple (may be yellowish green basally); achenes 1.9–3.5 mm long.

3. Tips of pales cuspidate, awn-like, glabrous; ray ligules 8–17 mm long ____ **R. triloba**
3. Tips of pales acute, hirsute; ray ligules 20–40 mm long ____ **R. subtomentosa**

1. Lower stem and midstem leaves unlobed (leaves may be entire or variously toothed).
   4. Stems and leaves blue-green, glaucous and glabrous; midstem leaves sessile, the bases auriculate and clasping.
      5. Taprooted annuals, usually 25–60 cm tall; ray ligules 12–30 mm long; achenes 1.8–2.5 mm long; pappus absent ____ **R. amplexicaulis**
      5. Robust, fibrous-rooted perennials, to 250 cm tall; ray ligules 30–80 mm long; achenes 6.5–7 mm long; pappus of 4–6 scales to 1.5 mm long ____ **R. maxima**
   4. Stems and leaves green (not blue-green), usually not glaucous, often with some pubescence present; midstem leaves sessile or petiolate but not clasping (bases may be auriculate).
      6. Pales 6–8 mm long, not surpassing achenes; crown-like pappus > 1 mm long.
         7. Midstem leaves sessile, blades hairy, bases auriculate or rounded; heads borne in arrays of up to 15; achenes 3.5–5 mm long ____ **R. scabrifolia**
         7. Midstem leaves petiolate or sessile, blades glabrous to sparsely hairy, bases attenuate or cuneate; heads borne singly or in arrays of 2–5; achenes 5–7.5 mm long ____ **R. texana**
      6. Pales 2.5–6.5 mm long, surpassing achenes; crown-like pappus < 0.6 mm long, or absent.
         8. Annuals, biennials, or short-lived perennials, taprooted or fibrous-rooted (not rhizomatous); tips of pales hairy; style branches of disk flowers elongate, awl-like; pappus absent ____ **R. hirta**
         8. Perennials, often rhizomatous or stoloniferous, crowns sometimes woody; tips of pales glabrous or sparsely strigose; style branches of disk flowers short and blunt; pappus present, crown-like (often minute).
            9. Plants to 80 cm tall; basal leaf blades linear to narrowly spatulate, 0.5–2 cm wide ____ **R. missouriensis**
            9. Plants to 120 cm tall; basal leaf blades lanceolate to elliptic or ovate, 2–11 cm wide.
               10. Plants with a thick woody root or crown; leaf blades gland-dotted beneath (with microscopic resin globules that appear golden in light with lens); pales 5–6.5 mm long, the tips acute to cuspidate (awn-tipped), pubescent ____ **R. grandiflora**
               10. Plants stoloniferous; leaf blades without microscopic resin globules; pales 2.5–4 mm long, the tips obtuse to acute, glabrous except for ciliate margin ____ **R. fulgida**

Rudbeckia amplexicaulis Vahl, (stem-clasping), CLASPING CONEFLOWER, CLASPING-LEAF CONEFLOWER. Annuals, 25–60(–120) cm tall, taprooted; herbage glabrous and glaucous; basal leaves seldom persisting to flowering, midstem leaves sessile, blades elliptic to oblong or lanceolate to oblanceolate, unlobed, bases auriculate-clasping, margins toothed or entire; receptacles ovoid to conic; ray flowers 6–10+, ligules solid yellow or sometimes partly orange or maroon basally, 12–30 mm long; disk corollas greenish yellow with purplish lobes; achenes 1.8–2.5 mm long; pappus absent. Disturbed habitats, moist areas, floodplains; e 2/3 of TX; sc U.S. from MO se to FL and w to KS and TX; records outside this range are probably adventive. Primarily Apr–May. Previously treated as the monotypic species [*Dracopis amplexicaulis* (Vahl) Cass.]. This species is often included in wildflower mixes and has become weedy in some agricultural croplands.

Rudbeckia fulgida Aiton var. **palustris** (Eggert ex C.L. Boynt. & Beadle) Perdue, (sp.: fulgid, shining; var.: marsh-loving), PRAIRIE CONEFLOWER, MARSH CONEFLOWER, ORANGE CONEFLOWER. Perennials, to 120 cm tall, stoloniferous, rosettes forming at stolon tips; basal leaves usually present at flowering, midstem leaves petiolate or sessile, blades lanceolate to ovate, reduced up the stem, unlobed, sparsely to moderately hairy, bases cuneate to attenuate, margins entire to serrate or sometimes lacerate; receptacles ovoid to hemispheric; ray flowers 8–14, ligules solid yellow to yellow-orange, 15–25 mm long; disk corollas yellowish green with brownish purple lobes; achenes 2.2–4 mm long; pappus to 0.2 mm long. Calcareous soils, moist and wet sites; Upshur, Hardin, and San Jacinto cos.; Edwards Plateau; AR, MO, OK, and TX. Jul–Nov. [*R. coryi* Shinners, *R. palustris* Eggert ex C.L. Boynt. & Beadle]

Rudbeckia grandiflora (D. Don) J.F. Gmel. ex DC., (large-flowered), ROUGH CONEFLOWER, LARGE-FLOWER CONEFLOWER. Perennials, to 120 cm tall, fibrous-rooted and often woody-crowned; basal leaves usually persisting to flowering, midstem leaves petiolate to nearly sessile, blades elliptic to lanceolate or ovate, often folded longitudinally, strigose, unlobed, bases rounded to cuneate, margins entire or remotely serrate; receptacles ovoid to hemispheric; ray flowers 12–25, ligules solid yellow, 20–50 mm long; disk corollas maroon, with greenish yellow bases and lobes; achenes 2–3 mm long; pappus to 0.5 mm long. Sandy open woods, moist to dry prairies; mostly e 1/3 TX; also nc TX; sc U.S. from MO s to LA and w to KS and TX; scattered records further e are apparently adventive (Kartesz 2015), perhaps due to the inclusion of this species in wildflower seed mixes. (May–)Jun–Jul. Several popular garden cultivars have been developed from this species. The two varieties apparently have overlapping ranges in East TX; no infraspecific range maps are provided by Turner et al. (2003) or USDA Plants Database (2016), therefore we are not providing ranges or maps for these varieties. Perdue (1957), Cox and Urbatsch (1990, 1994), Urbatsch and Jansen (1995), Urbatsch and Cox (2006), and Campbell and Seymour (2013) divided the two varieties as follows:

1. Lower stems glabrous or sparsely hairy, upper stems hairy, hairs ascending, mostly < 0.5 mm long; leaves scabrous to puberulent ____ var. **alismifolia**

1. Stems hairy throughout, hairs spreading to ascending, ca. 1 mm long; leaves hirsute ____ var. **grandiflora**

var. **alismifolia** (Torr. & A. Gray) Cronquist, (with leaves like *Alisma*—WATER-PLANTAIN). [*R. alismifolia* Torr. & A. Gray]

var. **grandiflora,** [*Centrocarpha grandiflora* Sweet]

Rudbeckia hirta L., (hairy), BLACK-EYED-SUSAN, BROWN-EYED-SUSAN. Annuals, biennials, or short-lived perennials, to 100 cm tall, taprooted or fibrous-rooted; herbage scabrous to hispid or sericeous; basal leaves often withering by flowering, midstem leaves petiolate or sessile, blades elliptic to broadly linear or spatulate to oblanceolate, unlobed, bases attenuate to cuneate, margins entire or serrate; receptacles hemispheric to ovoid; ray flowers 8–16, ligules solid yellow or sometimes partly orange or maroon, 15–45 mm long; disk corollas yellowish green with brownish purple lobes; achenes 1.5–2.7 mm long; pappus absent. Prairies, roadsides, old fields; mostly e 1/2 TX; also Edwards Plateau and s TX; across much of North America but mostly in e 2/3; scattered records further w are apparently adventive (Kartesz 2015). (Mar–)May–Jul(–Sep). Some strains of this species are cultivated and/or used in wildflower seeds mixes; its weedy tendencies make it popular for large-scale "wildscaping" and erosion control. *Rudbeckia hirta* is reported to cause poisoning in cattle, sheep, and hogs (Pammel 1911; Kingsbury 1964). ☠ Two varieties apparently have overlapping ranges in East TX; no infraspecific range maps are provided by Turner et al. (2003) or USDA Plants Database (2016), therefore we are not providing ranges or maps for these varieties. Cox and Urbatsch (2006) divided the two East TX varieties as follows:

1. Stems branched mostly at or near midstem height, leafy mostly toward bases (leaves reduced up the stem); peduncles at least 1/2 plant height ____ var. **angustifolia**
1. Stems branched mostly above midstem height, leafy ± throughout; peduncles to 1/3 plant height ____ var. **pulcherrima**

var. **angustifolia** (T.V. Moore) Perdue, (narrow-leaved). [*R. divergens* T.V. Moore, *R. floridana* T.V. Moore var. *angustifolia* T.V. Moore]

var. **pulcherrima** Farw., (very handsome), [*R. bicolor* Nutt., *R. longipes* T.V. Moore, *R. sericea* T.V. Moore, *R. serotina* Nutt.]

Rudbeckia laciniata L. var. **laciniata,** (cut or slashed), CUT-LEAF CONEFLOWER, GREEN-HEAD CONEFLOWER. Perennials, 50–300 cm tall, fibrous-rooted and rhizomatous, plants colonial; basal leaves often withering before flowering, midstem leaves petiolate or sessile, blades broadly ovate to lanceolate, 5–9-lobed or pinnatifid, glabrous or sparsely hairy, bases cuneate to attenuate or cordate, margins entire or dentate; receptacles globose to ovoid; ray flowers 8–12, ligules solid yellow, 20–45 mm long; disk corollas yellow to yellowish green with yellow lobes; achenes 4.2–6 mm long; pappus 0.1–0.7 mm long. Wet sites, along streams, edges of woods; in TX recorded only from Jasper Co.; mostly e 1/2 of North America but less frequent in s states; some w North American records are apparently adventive (Kartesz 2015). Jun–Jul. This species is grown as an ornamental and occasionally escapes cultivation. Cultivated varieties often have many times the number of rays and no visible disk.

Rudbeckia maxima Nutt., (largest), GIANT CONEFLOWER, GREAT CONEFLOWER, CABBAGE CONEFLOWER. Robust perennials, to 250 cm tall, fibrous-rooted and rhizomatous, forming large colonies; herbage glabrous and heavily glaucous; basal leaves usually persisting to flowering, midstem leaves sessile, blades ovate to pandurate, unlobed, bases auriculate-clasping, margins crenate; receptacles ovoid to conic, conspicuously elongating in fruit; ray flowers 10–20, ligules solid yellow, 30–80 mm long; disk corollas maroon (at least the lobes); achenes 6.5–7 mm long; pappus to 1.5 mm long. Low ground, moist ditches, sandy or silty soils; East TX to East Cross Timbers; AR, LA, OK, and TX; adventive in SC. May–Jun. This species is cultivated as an ornamental and is sometimes a problematic pasture weed.

Rudbeckia missouriensis Engelm. ex C.L. Boynt. & Beadle, (of Missouri), MISSOURI ORANGE CONEFLOWER, OZARK BLACK-EYED-SUSAN. Perennials, to 80 cm tall, rosettes forming at stem bases; basal leaves usually present at flowering, midstem leaves petiolate or sessile, blades linear to spatulate, unlobed, hirsute, bases cuneate to attenuate, margins entire or remotely serrulate; receptacles mostly hemispheric; ray flowers 9–15, ligules solid yellow to yellow-orange, 10–25 mm long; disk corollas greenish yellow with brownish purple lobes; achenes 1.5–2.7 mm long; pappus to 0.1 mm long. Dry or rocky prairies, limestone glades; Polk, Tyler, and Walker cos., also Montgomery Co. (Brown 2014); sc U.S. from IL s to LA and w to OK and TX. Jun–Aug. [*R. fulgida* Aiton var. *missouriensis* (Engelm. ex C.L. Boynt. & Beadle) Cronquist]

Rudbeckia scabrifolia L.E. Brown, (rough-leaved), ROUGH-LEAF CONEFLOWER, BOG CONEFLOWER. Perennials, to 200 cm tall, fibrous-rooted and rhizomatous; basal leaves usually persisting to flowering, midstem leaves sessile, blades ± pandurate, hairy, unlobed, bases auriculate (not clasping) or rounded, margins entire or coarsely toothed; receptacles ovoid to ellipsoid; ray flowers 10–15, ligules solid yellow, 20–40 mm long; disk corollas greenish yellow with maroon lobes; achenes 3.5–5 mm long; pappus to 2.5 mm long. Acid seeps, longleaf pine forests, hillside bogs; Angelina, Jasper, Newton, Sabine, and Shelby cos.; LA and TX. Jun–Aug.

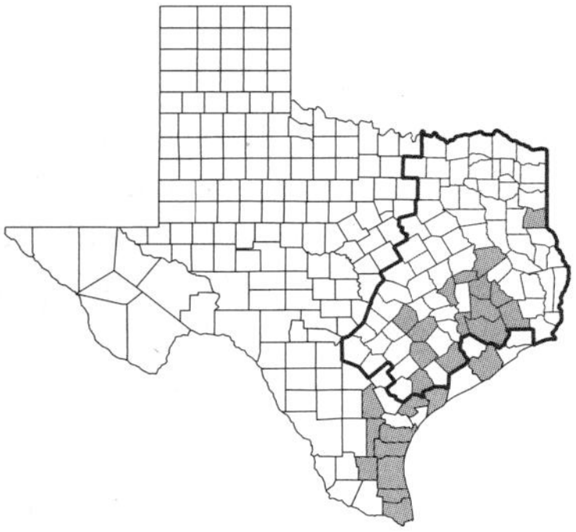

Pterocaulon virgatum

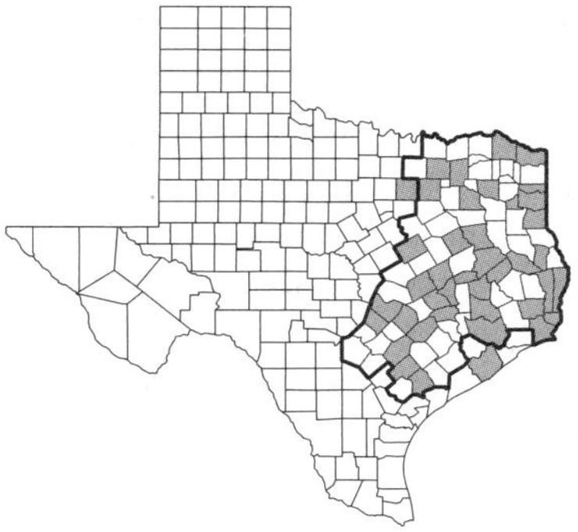

Pyrrhopappus carolinianus

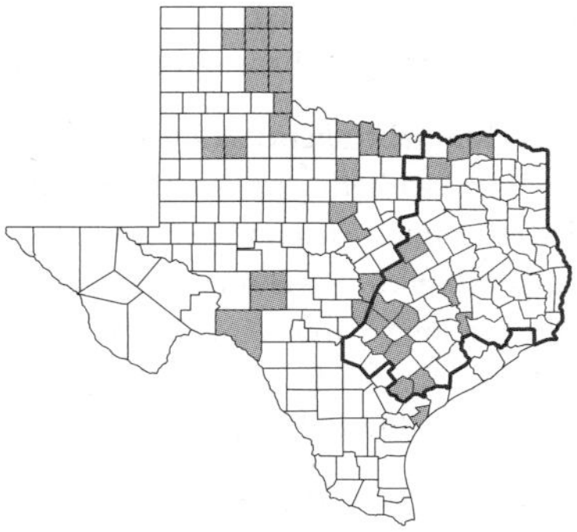

Pyrrhopappus grandiflorus

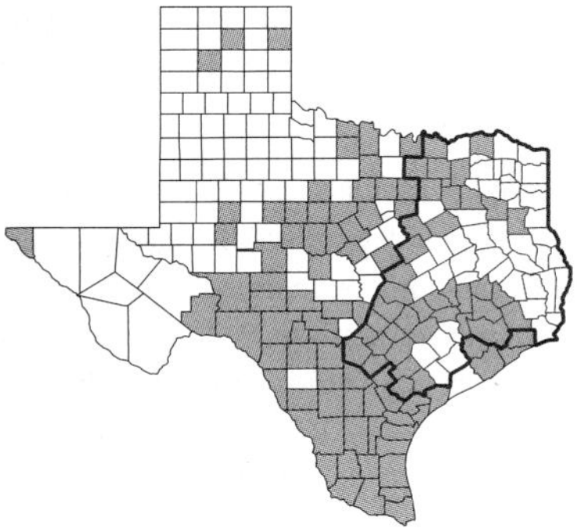

Pyrrhopappus pauciflorus

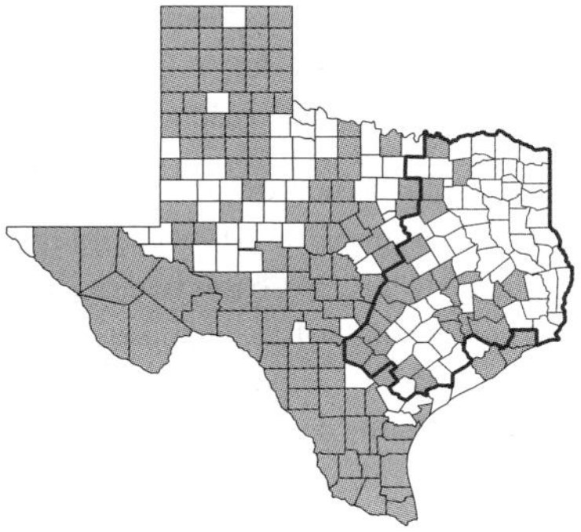

Ratibida columnifera

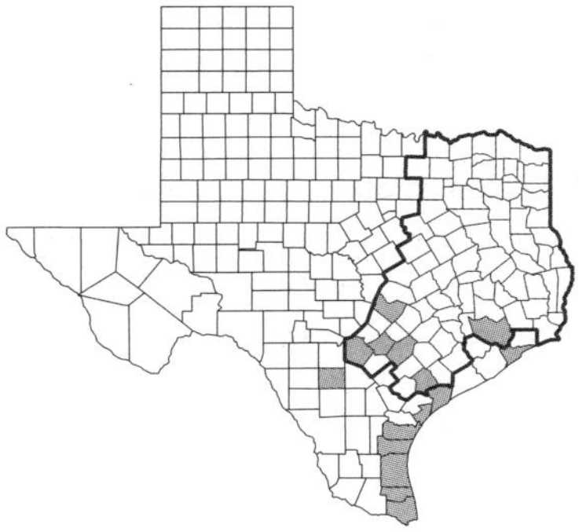

Ratibida peduncularis

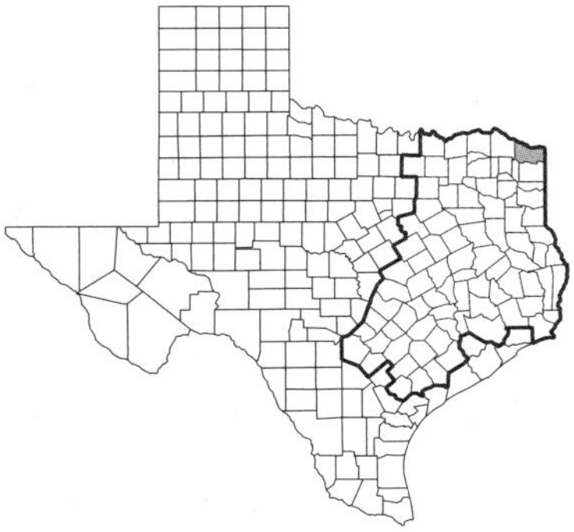

Ratibida pinnata

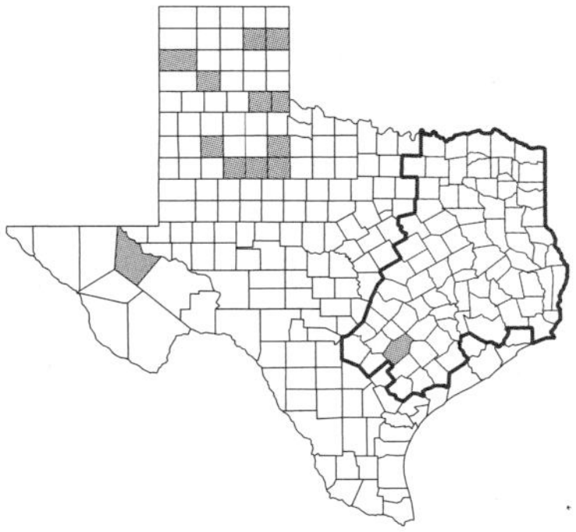

Rayjacksonia annua

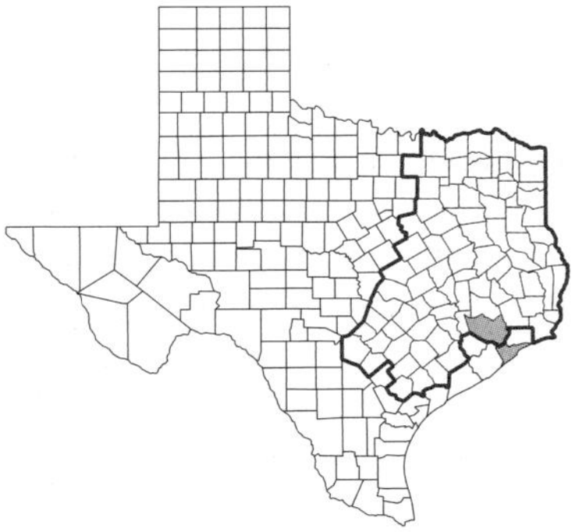

Rayjacksonia aurea

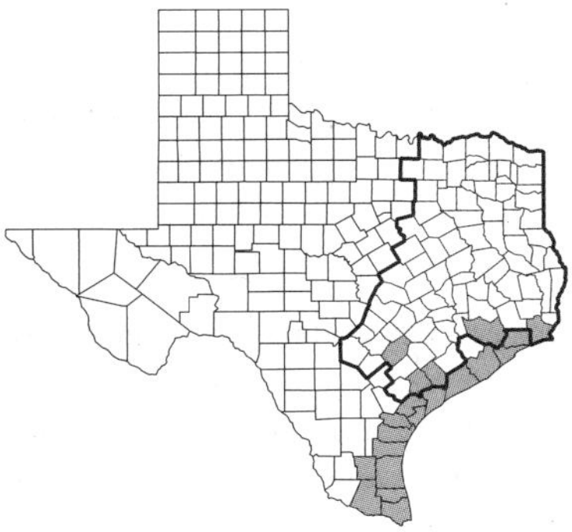

Rayjacksonia phyllocephala

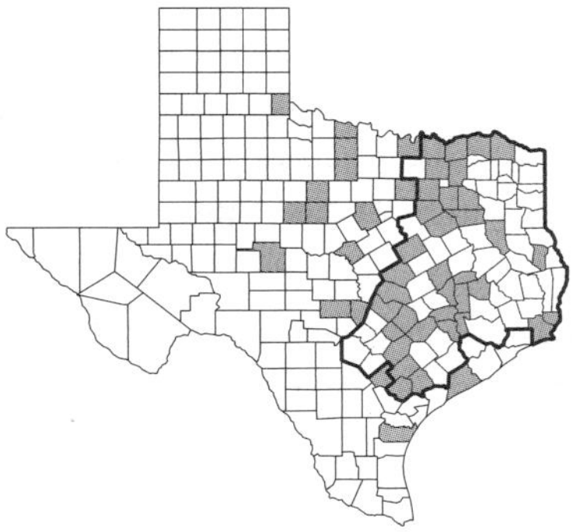

Rudbeckia amplexicaulis

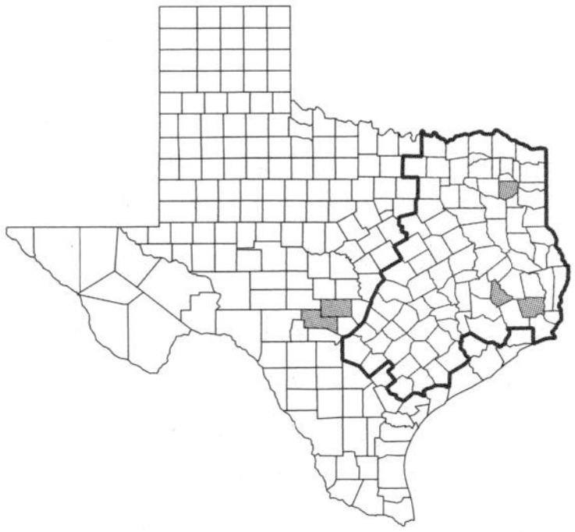

Rudbeckia fulgida var. palustris

This species is not listed at the state or federal level, but is of conservation concern due to its restricted range and specialized habitat; habitat disturbance and fire suppression threaten some populations of this fire-dependent species (Poole et al. 2007). ⚠

Rudbeckia subtomentosa Pursh, (slightly tomentose), SWEET CONEFLOWER. Perennials, to 200 cm tall, rhizomes stout; basal leaves often withering by flowering, midstem leaves petiolate, blades ovate to elliptic, 3–5-lobed, densely hirsute and gland-dotted, bases truncate to cuneate or rounded, margins denticulate to serrate; receptacles conic to hemispheric; ray flowers 10–16, ligules solid yellow to yellow-orange, 20–40 mm long; disk corollas mostly brownish purple but yellowish green basally; achenes 2–3.5 mm long; pappus to 0.2 mm long. Moist prairies, stream banks, openings in woods; Angelina, Hardin, and Tyler cos.; c U.S. from MI s to MS and w to WI, KS, and TX; scattered records further e are apparently adventive (Kartesz 2015). Jun–Jul. Sometimes cultivated as an ornamental.

Rudbeckia texana (Perdue) P. Cox & Urbatsch, (of Texas), TEXAS CONEFLOWER. Perennials, to 150 cm tall, fibrous-rooted and rhizomatous; basal leaves usually persisting to flowering, midstem leaves petiolate or sessile, blades elliptic to lanceolate, glabrous or sparsely hairy, unlobed, bases attenuate to cuneate, margins entire or toothed to serrate; receptacles ovoid to ellipsoid; ray flowers 10–16, ligules solid yellow, 20–50 mm long; disk corollas greenish yellow with maroon lobes; achenes 5–7.5 mm long; pappus to 1.5 mm long. Coastal prairies, ditches and bayous, swales; se TX w to Montgomery and Waller cos. (Brown 2014); LA and TX. May–Jul (sporadically later). [*R. nitida* Nutt. var. *texana* Perdue]

Rudbeckia triloba L. var. **triloba,** (three-lobed), YELLOW DAISY, BROWN-EYED-SUSAN. Short-lived perennials, to 150 cm tall, rhizomatous; basal leaves often withering by flowering, lower leaves 3(–5)-lobed, midstem leaves petiolate or sessile, blades ovate to elliptic, reduced and fewer-lobed up the stem, hirsute to strigose, bases truncate or rounded to cordate, margins serrate; receptacles conic to subhemispheric; ray flowers 8–15, ligules solid yellow or sometimes partly orange or maroon, 8–17 mm long; disk corollas mostly brownish purple but yellowish green basally; achenes 1.9–2.8 mm long; pappus to 0.2 mm long. Creek bottoms, moist thickets, roadsides; Hardin, Lamar, and San Augustine cos., and Nacogdoches Co. (Kartesz 2015); mostly e 1/2 of the U.S.; introduced in a few w states and Canada. May–Oct.

SCLEROCARPUS Jacq. BONE-BRACT

A genus of 12 species native to the s U.S., Mexico, and Central America, with one species in the Old World tropics. (Greek: *skleros*, hard, and *carpos*, fruit, alluding to the hardened pales enfolding disk achenes) (tribe Heliantheae)
REFERENCES: Feddema 1972; Turner 1988b; Harriman 2006.

Sclerocarpus uniserialis (Benth.) Hemsl. var. **uniserialis,** (for the phyllaries, which appear to be in a single series), MEXICAN BONE-BRACT. Annual or perennial taprooted herbs or subshrubs, 60–200 cm tall; stems sprawling to erect, often densely hairy; leaves usually opposite below and alternate above, usually petiolate, blades 3-nerved, 2–5 cm long, ovate to rhombic or lanceolate, glabrous to scabrous, margins coarsely toothed; heads borne singly; involucres 12–20 mm wide; phyllaries in 1(–2) series, green, loose, oblong to linear; pales (receptacle chaff) each enfolding a disk flower, forming a hardened "fruit" around the achene; ray flowers 5–8, neuter, ligules yellow to orange, rotundly elliptic, 5–15 mm long, readily falling; disk flowers perfect, fertile, corollas yellow with reddish veins; achenes each enclosed within a hardened, brown, warty "fruit" formed by the subtending pale; pappus absent. Disturbed sites, caliche, limestone, sandy soils; sc TX and Gulf Coast; s TX; Mexico and Central America. Jul–Dec. [*Gymnopsis uniserialis* Hook.] This species apparently intergrades in s TX with var. *austrotexanus* B.L. Turner, which has purplish brown disk corollas.

SENECIO L. RAGWORT, SQUAW-WEED, GROUNDSEL

Ours annual herbs, glabrate to usually pubescent; leaves alternate, the blades toothed to pinnately lobed; inflorescences terminal, the heads cymosely arranged; primary (inner) phyllaries in a single series, linear, with hyaline margins, subtended by a secondary series of short, phyllary-like bracts; ray flowers usually pistillate, fertile, with yellow ligules, or ray flowers absent; disk flowers perfect, fertile, the corollas yellow; achenes cylindrical; pappus of numerous white capillary bristles.

When broadly considered, *Senecio* is a huge, cosmopolitan genus of 1,250–3,000 species depending on circumscription. Species vary in growth form from trees to shrubs, vines, herbs, or desert succulents, and include the GIANT GROUNDSELS of the mountains of e Africa. We are following Freeman and Barkley (1995), Barkley et al. (1996), and Barkley (2006d) in treating the aureoid group of Senecios as *Packera*. Even when this and numerous other segregates are removed, *Senecio* is still one of the largest genera of seed plants. ☠ A number of species are toxic and can be problematic if eaten by livestock, especially horses; acute illness and death can result from hepatotoxic pyrrolizidine alkaloids that cause liver damage. Milk from grazing animals and honey from *Senecio* nectar reportedly contain the alkaloids (Lewis & Elvin-Lewis 1977; Fuller & McClintock 1986; Blackwell

Rayjacksonia phyllocephala [FNA]

Rudbeckia amplexicaulis [MO4]

Rudbeckia fulgida var. palustris [STE]

Rubeckia laciniata [BA1, CO1]

Rubeckia grandiflora var. grandiflora [YAT]

Rudbeckia hirta var. pulcherrima [BT3]

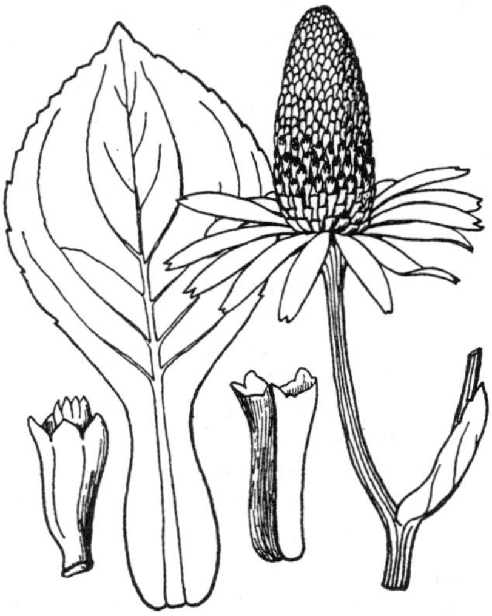

Rudbeckia maxima [BB2]

1990). Human deaths have been reported from ingestion of *Senecio* herbal teas as the result of liver damage—there is no known cure (Lampe & McCann 1985; Tveten & Tveten 1993). Some *Senecio* species were in the past used in poultices for wounds and abscesses; the common name GROUNDSEL apparently is derived through a series of changes from the Anglo-Saxon *grundeswelge*, pus-absorber (Tveten & Tveten 1993). (Latin name of a plant, from *senex*, an old man, alluding to the hoariness of many species, or to the white hairs of the pappus) (tribe Senecioneae).

REFERENCES: Greenman 1915–1918; Vuilleumier 1969a; Barkley 1978, 1985a, 1985b, 1986, 2006d; Barkley et al. 1996.

1. Ray flowers yellow, prominent; primary phyllaries and subtending phyllary-like bracts green-tipped; leaves shallowly toothed to subentire; native species ____ **S. ampullaceus**
1. Ray flowers usually absent, but if present the ligules greatly reduced and inconspicuous; phyllary-like bracts subtending primary phyllaries prominently black-tipped; leaves usually shallowly pinnately lobed; introduced weedy species ____ **S. vulgaris**

Senecio ampullaceus Hook., (flask-shaped), TEXAS GROUNDSEL. Loosely woolly-pubescent to nearly glabrate annual (10–)20–80 cm tall; single-stemmed; lowermost leaves narrowed to a subpetiolar base, auriculate-clasping at very base; middle and upper leaves often truncate-clasping; involucres 7–11 mm high; secondary phyllary-like bracts and primary phyllaries green-tipped; ray flowers ca. 8, ligules 8–15 mm long. Sandy open woods, fields, roadsides; widespread in e 1/2 of TX; once thought to be endemic to TX but now reported from AR, LA, MO, and OK. Mar–May.

Senecio vulgaris L., (common), COMMON GROUNDSEL. Annual 10–30(–60) cm tall, woolly-pubescent to glabrate; lower leaves petiolate; middle and upper leaves sessile, auriculate, ± clasping, undulate to pinnately lobed, 2–10 cm long, 0.5–2(–4) cm wide; involucres 6–10 mm high; secondary phyllary-like bracts prominently black-tipped; ray flowers usually absent. Disturbed areas; nc TX (Dallas, Denton, Grayson cos.), c TX (McLennan Co.), sc TX (Bastrop and DeWitt cos.), and se TX (Brazos and Harris cos.); apparently first collected in TX in Dimmit Co. in 1944 (Cory 1947, 1948a); probably more common in TX than the records imply and undercollected; throughout North America; a common weed the world over. Feb–May. Native of Europe. This species is among those in the genus that contain toxic pyrrolizidine alkaloids; animal deaths have been reported; the toxins are not lost upon drying and are still present in hay (Burlage 1968; Lampe & McCann 1985; Fuller & McClintock 1986).

SILPHIUM L. ROSINWEED, COMPASSPLANT

Herbaceous, coarse, sunflower-like perennials; stems erect; leaves alternate to opposite, petiolate or sessile, simple, the blades entire or toothed to deeply lobed or pinnatifid, stiff, leathery, often conspicuously scabrous; heads large and showy, solitary or in corymbose or spicate-racemose arrays; phyllaries in 2–4 series, the outer series foliaceous, the inner series smaller and thinner; ray flowers pistillate, fertile, with ligules yellow or white (in one species); disk flowers functionally staminate, corollas yellow or white; achenes flattened, thin-edged, ± winged; pappus absent, or of 2 awns projecting from achene shoulders.

An e and c North American genus of 12 species. The common name ROSINWEED refers to the sticky secretions on the stems and leaves of some species (Ajilvsgi 1984). *Silphium* is a taxonomically difficult genus that formerly had 23 recognized species (Perry 1937); differing assessments of phenotypic variability and hybridization have led to frequent species concept adjustments and nomenclatural changes. Complete specimens with mature inflorescences and lower leaves are often needed for definitive identification. Some species of this genus could be confused with *Helianthus*, however, in *Silphium* only the ray flowers produce mature achenes, which have thin, papery edges, while in *Helianthus* only the disk flowers produce mature achenes, and these are not thin-edged. (Greek: *silphion*, the ancient name of some resinous plant, applied by Linnaeus to this wholly American genus) (tribe Heliantheae)

REFERENCES: Perry 1937; Settle & Fisher 1970; Clevinger 2006.

1. Plants taprooted; leaves (at least the lower ones) deeply pinnatifid or 1–2-pinnately lobed.
   2. Ray flowers 14–30, ray and disk corollas white; stems 1 m or less tall; achenes 12–26 mm long ____ **S. albiflorum**
   2. Ray flowers 27–38, ray and disk corollas yellow; stems usually 1–3 m tall; achenes 10–18 mm long ____ **S. laciniatum**
1. Plants fibrous-rooted; leaves entire to toothed.
   3. Leaves both strictly opposite throughout and clasping the stem, leaves sessile, bases cordate to rounded ____ **S. integrifolium**
   3. Leaves alternate or opposite, sessile or petiolate (if strictly opposite and sessile then not clasping the stem), bases attenuate to rounded.
      4. Outer faces of phyllaries usually glabrous (rarely scabrous) ____ **S. asteriscus** (in part)
      4. Outer faces of phyllaries glandular, hirsute, hispid, or scabrous.
         5. Basal leaves present at flowering ____ **S. radula** (in part)
         5. Basal leaves withering before flowering.
            6. Ray flowers 20–30+ ____ **S. radula** (in part)
            6. Ray flowers 12–20 ____ **S. asteriscus** (in part)

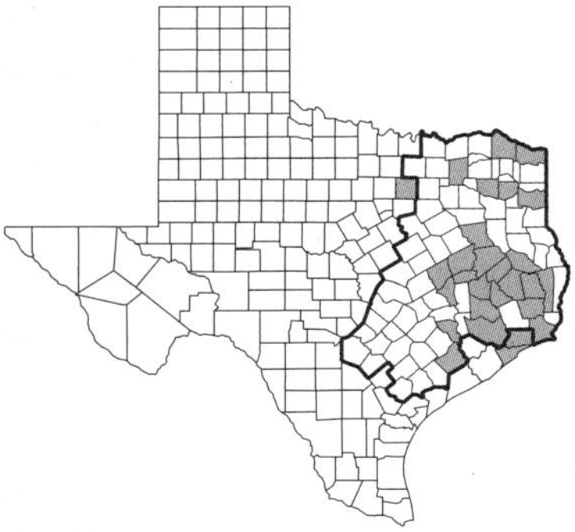

Rudbeckia grandiflora

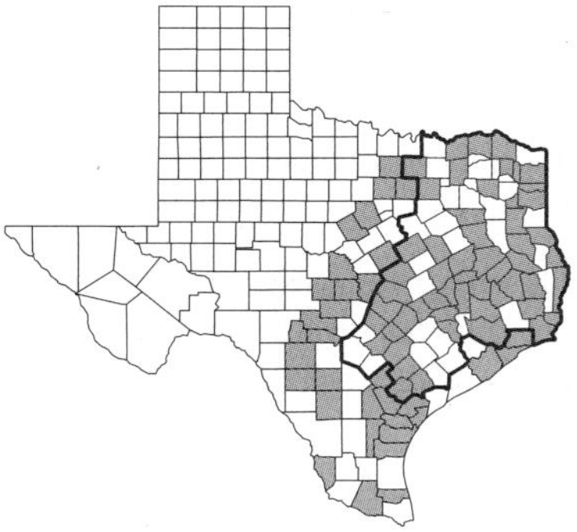

Rudbeckia hirta

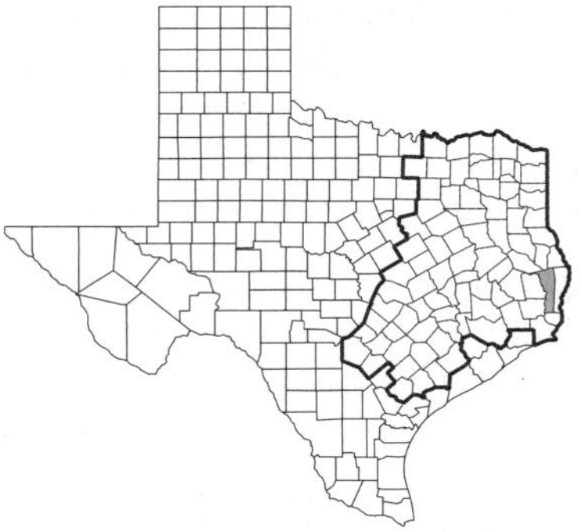

Rudbeckia laciniata var. laciniata

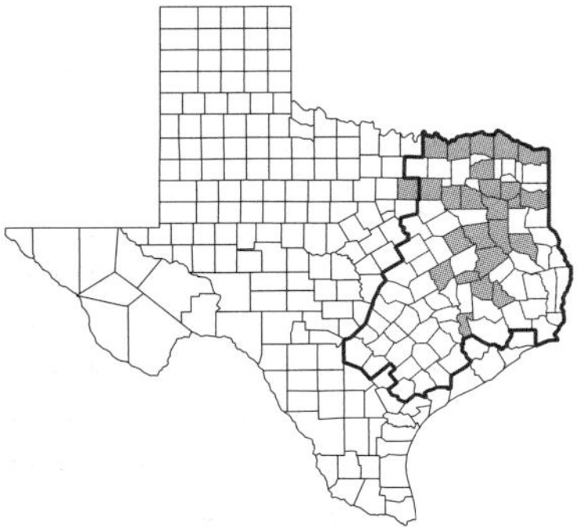

Rudbeckia maxima

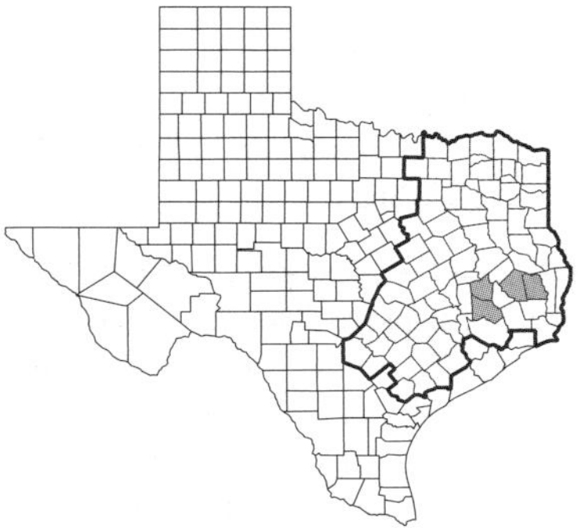

Rudbeckia missouriensis

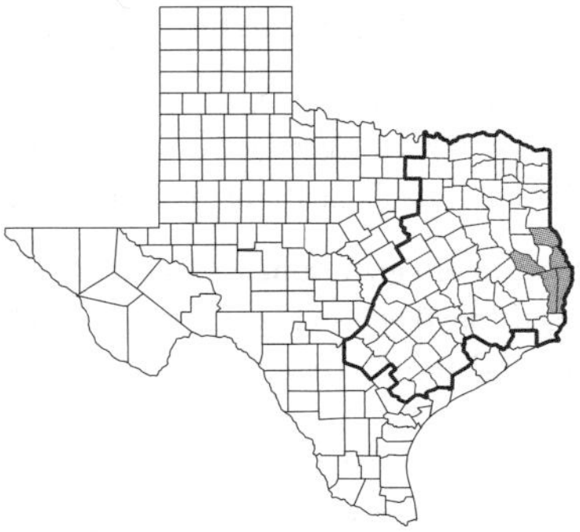

Rudbeckia scabrifolia

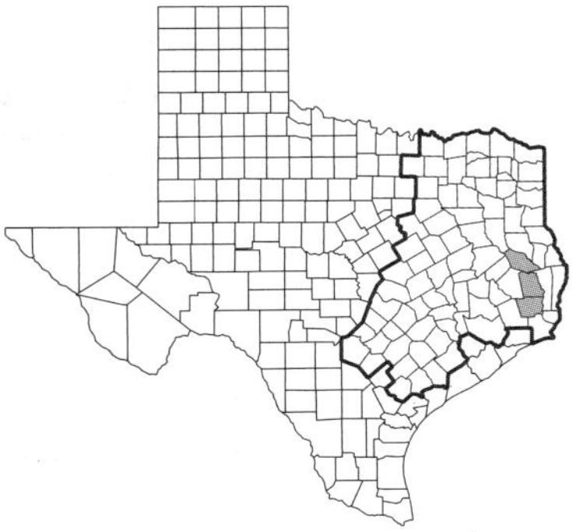

Rudbeckia subtomentosa

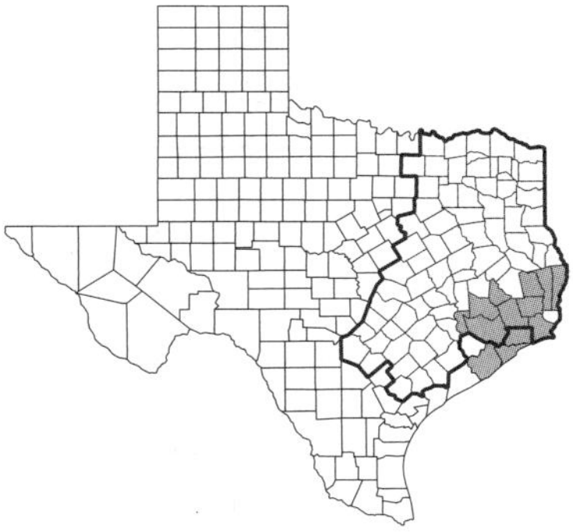

Rudbeckia texana

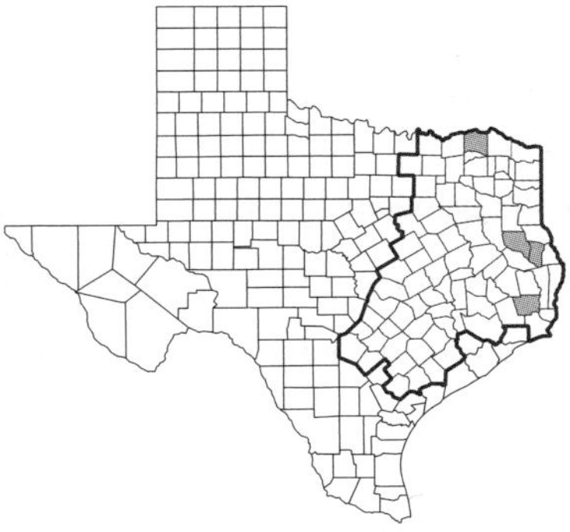

Rudbeckia triloba var. triloba

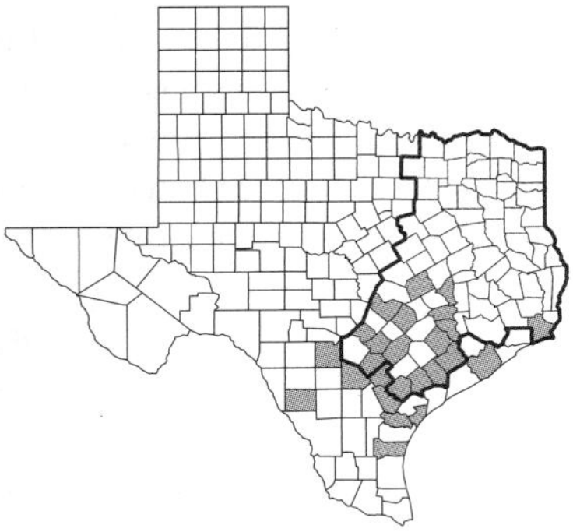

Sclerocarpus uniserialis var. uniserialis

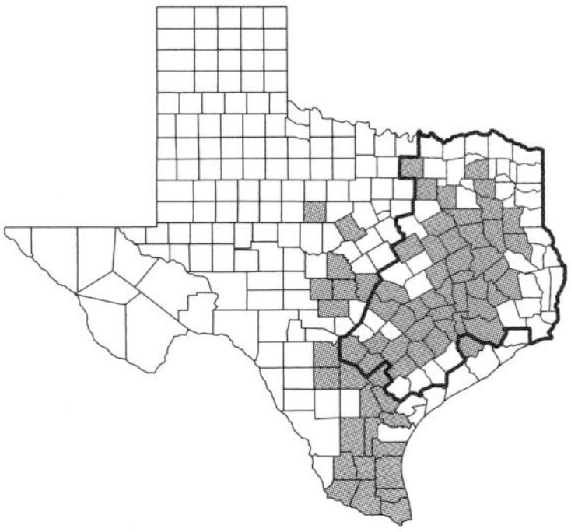

Senecio ampullaceus

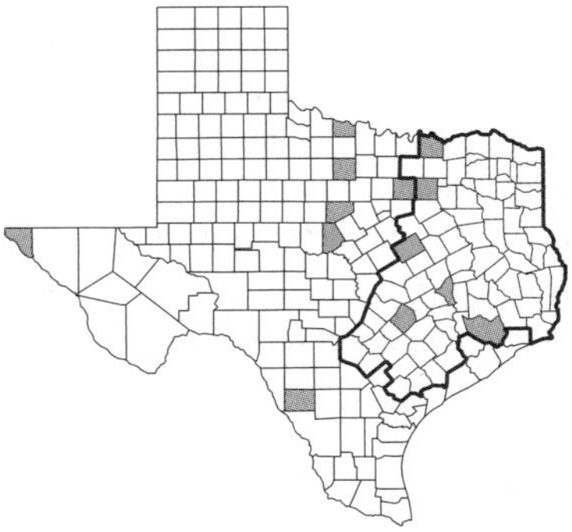

Senecio vulgaris

Silphium albiflorum A. Gray, (white-flowered), WHITE ROSINWEED, WHITE COMPASSPLANT. Plants 20–75 cm tall, taprooted; basal leaves present at flowering, stem leaves usually alternate, petiolate or sessile, scabrous to hirsute, 1–2-pinnately lobed, margins unevenly toothed or entire, bases attenuate to cuneate or truncate; ray and disk corollas white; achenes 12–26 mm long; pappus 2–5 mm long. Calcareous soils, open fields and prairies; w border of East TX from Dallas to McLennan to Travis cos.; Blackland Prairie and Grand Prairie from nc TX s to Edwards Plateau and w to Rolling Plains; endemic to TX. May–Jul.

Silphium asteriscus L., (Greek: *asteriskos*, little star), STARRY ROSINWEED. Plants 30–200 cm tall, fibrous-rooted; basal leaves present or absent at flowering, stem leaves alternate or opposite, petiolate or sessile, hirsute or hispid to scabrous, margins dentate to serrate or entire, bases attenuate to rounded; ray and disk corollas yellow; achenes 6–15 mm long; pappus absent or to 5 mm long. Prairies, open forests, roadsides. Jun–Sep.

1. Basal leaves withering before flowering; pales not glandular ________ var. **asteriscus**
1. Basal leaves present at flowering; pales with stalked glands ________ var. **simpsonii**

var. **asteriscus.** E and c TX; Edwards Plateau; se U.S. from VA s to GA and w to OK and TX. [*S. asperrimum* Hook. (misapplied to *S. radula* Nutt.), *S. asteriscus* L. var. *scabrum* Nutt.]

var. **simpsonii** (Greene) Clevinger, (for Joseph H. Simpson (1841–1918), botanist of Florida and collector of the type). Harris Co. (*L.E. Brown 33879* SBSC; Brown 2014); Angelina, Hardin, and Liberty (TAES) cos.; Galveston Co. in Gulf Prairies and Marshes; sparsely distributed from GA and FL w to MS and disjunct to TX. [*S. simpsonii* Greene]

Silphium integrifolium Michx., (entire-leaved), WHOLE-LEAF ROSINWEED. Plants 40–200 cm tall, fibrous-rooted; basal leaves withering before flowering, stem leaves opposite, sessile, hispid to scabrous or glabrous, margins finely serrate or entire, bases rounded to cordate, often clasping stems; ray and disk corollas yellow; achenes 9–14 mm long; pappus 1–4 mm long. Calcareous soils, prairies, roadsides; reported from a calcareous pimple-mound prairie in Bowie Co. (Singhurst & Holmes 2010); c U.S. from MI s to AL and w to WY, CO, and NM; Canada. May–Sep. According to Clevinger (2006), only one variety (var. *laeve*) occurs in TX; according to Kartesz (2015), both varieties occur in TX but are only recorded from Bowie Co.; due to chronic nomenclatural changes resulting in previously recognized varieties of *S. integrifolium* being removed to other species, no complete range maps are available for the two varieties as currently circumscribed, therefore we are not providing ranges or maps for these varieties. Clevinger (2006) divided the two potential TX varieties as follows:

1. Outer faces of phyllaries usually hispid, pilose, or scabrous, rarely glabrous; ray flowers 12–22 ________ var. **integrifolium**
1. Outer faces of phyllaries glabrous; ray flowers 20–36+ ________ var. **laeve**

var. **integrifolium.** [*S. integrifolium* var. *deamii* L.M. Perry]

var. **laeve** Torr. & A. Gray, (smooth). [*S. speciosum* Nutt.]

Silphium laciniatum L., (laciniate, torn), COMPASSPLANT. Plants usually 100–300 cm tall, taprooted; basal leaves present at flowering, stem leaves alternate, petiolate or sessile, hispid to hirsute or scabrous, 1–2-pinnately lobed, margins unevenly toothed or entire, bases attenuate to truncate; ray and disk corollas yellow; achenes 10–18 mm long; pappus 1–3 mm long. Calcareous or sandy soils, prairies, open or disturbed sites; scattered in e and c TX; Cross Timbers and Edwards Plateau; e and c U.S. from NY s to AL and w to MN, SD, and NM; Canada. Jun–Aug. [*S. laciniatum* L. var. *robinsonii* L.M. Perry] The common name refers to the supposedly north-south orientation of the leaves.

Silphium radula Nutt., (rough), ROUGH-STEM ROSINWEED, ROUGH-LEAF ROSINWEED. Plants 30–250 cm tall, fibrous-rooted; basal leaves present or absent at flowering, stem leaves alternate or opposite, petiolate or sessile, hispid to hirsute or scabrous, margins coarsely dentate or entire, bases attenuate to rounded; ray and disk corollas yellow; achenes 10–17 mm long; pappus 2–5 mm long. Calcareous or sandy soils, prairies, fencerows. Jun–Jul. The common name refers to the sticky resinous material secreted along the stems and leaves, which was used medicinally by Native Americans and also as a chewing gum (Ajilvsgi 1984).

1. Basal leaves present at flowering; ray flowers 12–18 ________ var. **gracile**
1. Basal leaves withering before flowering; ray flowers 20–30+ ________ var. **radula**

var. **gracile** (A. Gray) Clevinger, (graceful), SLENDER ROSINWEED. Se and sc TX; LA and TX. [*S. gracile* A. Gray]

var. **radula,** ROUGH-STEM ROSINWEED, SIMPSON ROSINWEED. E and sc TX; West Cross Timbers; AR, LA, KS, MO, OK, and TX. [*S. reverchonii* Bush, *S. simpsonii* Greene var. *wrightii* L.M. Perry]

Rudbeckia missouriensis [YAT]

Rudbeckia scabrifolia [PHY]

Rudbeckia subtomentosa [YAT]

Rudbeckia texana [GOR]

Rudbeckia triloba var. triloba [MO4]

Sclerocarpus uniserialis [HO5]

Senecio ampullaceus [CUR]

Senecio vulgaris [GLE]

Silphium albiflorum [CUR]

SILYBUM Adans. MILK-THISTLE

A genus of 2 species native to the Mediterranean region, both now widely introduced elsewhere, especially in areas used for cattle grazing (Turner 2016). (Greek: *silybon*, name for similar thistle-like plants) (tribe Cardueae)
REFERENCE: Keil 2006i.

Silybum marianum (L.) Gaertn., (with white-mottled leaves; the spots were supposed to have resulted from drops of Mary's milk falling on the leaves), BLESSED MILK-THISTLE, OUR-LADY'S-THISTLE, HOLY-THISTLE. Coarse prickly annual or biennial 0.5–1.5+ m tall; stems loosely woolly-pubescent, becoming glabrous; leaves alternate, mottled green and white, the lower pinnately lobed, the middle and upper toothed; stem leaves auricled-clasping, all prickly or spiny-margined, not decurrent; heads large (2.5–6 cm in diam.), solitary, terminal; phyllaries to 4 cm long, with spinescent margins, constricted below middle, with broad bases and green, sub-leafy apical portions tapering to a spiny tip to 8 mm long; ray flowers absent; disk corollas pink to rosy purple; achenes ellipsoid, 6–8 mm long; pappus of 2 series, the outer of numerous white, awn-like scales 15–20 mm long, deciduous as a ring, the inner of minute smooth bristles. Pastures and roadsides, waste areas; scattered from nc TX to Edwards Plateau; mostly along the w edge of East TX; first collected in TX in Sutton Co. in 1938 (supposedly introduced through a shipment of hay from California) (Cory 1940); scattered across North America and naturalized in many countries. May–Jun. [*Carduus marianus* L.] Native of Mediterranean region. The fruit is used medicinally and considered since the time of Dioscorides to protect the liver; it contains flavonoids effective as an antidote for mushroom (*Amanita*) poisoning—the mode of action is to displace the toxin from cell membrane receptors (Mabberley 2017); its clinical use in Europe is widespread (Leung & Foster 1996).

SIMSIA Pers. BUSH-SUNFLOWER

A New World genus of ca. 20 species extending from the sw U.S. to Argentina with 1 taxon endemic to Jamaica (Schilling & Spooner 1988; Spooner 1990). (Named for John Sims, 1749–1831, of England, editor of *Curtis's Botanical Magazine*) (tribe Heliantheae)
REFERENCES: Schilling & Spooner 1988; Spooner 1990, 2006.

Simsia calva (Engelm. & A. Gray) A. Gray, (naked), AWNLESS BUSH-SUNFLOWER. Perennial, herbaceous from woody base, coarsely pubescent; roots fusiform or woody; stems 30–150 cm long; herbage dark green; leaves opposite, petiolate, the petioles sometimes auricled, the auricles fused to form a nodal disk, blades deltoid, sometimes 3-lobed, serrate; stipules toothed or serrate, often united; heads usually solitary at the end of long peduncles; phyllaries in several series, often reflexed; ray flowers 8–21, pistillate, sterile, the ligules yellow or orange-yellow, often spotted or striped beneath with purple or red; disk flowers perfect, fertile; corollas yellow or orange-yellow, sometimes with purplish lines, gibbous basally; achenes flattened, emarginate, 3.5–5.7 mm long; pappus of 2 short scales, often absent on mature achenes. Calcareous soils, roadsides, brushlands; along the w and sw border of East TX from Bell Co. s to Bexar and Goliad cos.; also Austin Co.; widespread in Edwards Plateau, South TX Plains, and Trans-Pecos; NM and TX; Mexico. May–Nov. [*Barrattia calva* A. Gray & Engelm.]

SMALLANTHUS Mack. ex Small LEAFCUP

A genus of ca. 20 species of tropical and warm areas of the Americas; previously recognized in the genus *Polymnia*. (Named for John Kunkel Small, 1869–1938, American botanist and author of numerous works including *Manual of the Southeastern Flora*) (tribe Heliantheae)
REFERENCES: Wells 1965; Robinson 1978; Turner 1988c; Strother 2006z.

Smallanthus uvedalia (L.) Mack. ex Small, (presumably from Latin *uvida*, humid or moist, for its tendency to occur in wet habitats), BEAR'S-FOOT, HAIRY LEAFCUP. Herbaceous perennial 1–3 m tall; stems erect, purple-spotted; leaves opposite, at least lower, petioles to 12 cm, blades often huge, to ca. 70 cm long and 40 cm wide, sessile or with broad conspicuous wings to base of petiole; leaf blades ovate to deltoid, palmately 3- to 5-lobed and veined, margins irregularly dentate; heads clustered in loose leafy cymes; phyllaries in 2 series, the outer series numbering 4–6, herbaceous, 10–20 mm long, to ca. 10–12 mm wide, ovate or ovate-lanceolate; ray flowers 7–13, pistillate, fertile, the ligules yellow, 12–20(–30) mm long; disk flowers numerous, staminate, yellow; achenes obovoid, 5–6 mm long and ca. 4 mm wide; pappus absent. Wooded bottomlands; scattered across East TX, to nc TX and Edwards Plateau; e and se U.S. from NY s to FL and w to MI, KS, and TX; Mexico. Jul–Oct. [*Osteospermum uvedalia* L., *Polymnia uvedalia* (L.) L., *Polymnia uvedalia* (L.) L. var. *densipilis* S.F. Blake]

SOLIDAGO L. GOLDENROD

Herbaceous perennials from woody caudices or rhizomes; leaves alternate, sessile or short-petioled, margins often serrate, basal leaves sometimes withering before flowering time; inflorescence terminal, usually branched in cylindrical, flat-topped, or pyramidal arrays; heads numerous, relatively small, sometimes secund (in a one-sided arrangement); phyllaries imbricate, in 3–5 series, usually unequal, margins scarious; ray flowers usually 2–15(–24), pistillate, fertile, the ligules yellow in ours (sometimes minute or inconspicuous); disk flowers perfect, fertile, the corollas yellow; achenes cylindrical, usually 8–10-ribbed; pappus of numerous, white, minutely-barbed bristles.

A mainly North American genus of ca. 100 species, with a few taxa in Mexico, South America, and Eurasia. Because of marked phenotypic plasticity, polyploidy, and hybridization, *Solidago* is a taxonomically difficult genus. Complete specimens with mature inflorescence arrays and lower leaves are often needed for definitive identification. Despite their bad reputation as allergens, GOLDENRODS are insect-pollinated and thus do not produce significant windblown pollen; most fall allergies in TX are probably due to wind-pollinated species of *Ambrosia* (RAGWEEDS) that flower inconspicuously at the same time as the much showier GOLDENRODS. The related genus *Euthamia*, sometimes included in *Solidago*, is here treated separately. Two East TX species (*S. nitida* and *S. rigida*) are sometimes segregated in the genus *Oligoneuron*; we are following Semple and Cook (2006) in treating these traditionally and including them in *Solidago*; appropriate synonymy is given below. (Latin: *solidus*, whole, and *ago*, resembling or becoming, alluding to medicinal properties) (tribe Astereae)

REFERENCES: Taylor & Taylor 1983, 1984; Heard & Semple 1988; Gandhi & Thomas 1989; Nesom 1990e, 1993a, 2008; Semple 1992, 2013, 2016; Semple & Cook 2006; Semple et al. 2012, 2016.

1. Inflorescence arrays flat-topped or slightly rounded (= corymbose); petiole bases of basal leaves persistent (remaining into the next year); phyllaries striate with 3–7 prominent nerves (segregate genus *Oligoneuron*).
   2. Leaf blades, stems, and peduncles moderately to densely short-hairy; midstem leaves ovate to rhombic or sometimes lanceolate, 15–17 mm wide; ray flowers 6–13 per head; achenes 0.8–1.7 mm long; pappus 3–4 mm long ____ **S. rigida**
   2. Leaf blades, stems, and peduncles glabrous to sparsely hairy, midstem leaves linear to oblanceolate, 1–7 mm wide; ray flowers 1–4 per head; achenes 2–3 mm long; pappus 4–5 mm long ____ **S. nitida**
1. Inflorescence arrays elongate, not flat-topped, either racemose, paniculate, cylindric, or pyramid-shaped and broadest near base or midsection; petiole bases not persisting on the plant; phyllaries not striate (*Solidago* in a narrower sense).
   3. Midstem leaves ovate, blade bases or petiole bases usually cordate and auriculate-clasping ____ **S. auriculata**
   3. Midstem leaves of various shapes, blade bases or petiole bases not cordate and not auriculate-clasping (petiole bases may be winged and sheathing stem).
      4. Heads arranged in discrete, relatively small, racemose or paniculate axillary clusters along the arching and glaucous main stem, usually interrupted by bare internodes and interspersed with leaves far surpassing in length the clusters of heads, the leaves continuing to the summit of the stem ____ **S. caesia**
      4. Plants not as described above.
         5. Heads arranged in thyrsiform, racemose, or paniculate arrays, not nodding or arched at the summit, lateral array branches (if any) erect, not arching or recurved.
            6. Basal and lowest stem leaves usually present at flowering; petioles of lower stem leaves nearly completely sheathing stem; plants of sandy moist soils, pine flatwoods, and coastal brackish marshes, mostly of se and coastal TX ____ **S. virgata**
            6. Basal and lowest stem leaves often absent at flowering; petioles or bases of lower stem leaves not sheathing stem; plants of generally drier soils throughout East TX (including se and coast).
               7. Peduncles 2–15 mm long; phyllaries with stalked glands; rays (5–)7–9; achenes 3–4 mm long ____ **S. petiolaris**
               7. Peduncles 1.5–3 mm long; phyllaries eglandular; rays (2–)3–7(–9); achenes 1.6–2.5 mm long ____ **S. speciosa**
         5. Heads arranged in paniculate arrays, often pyramidal or rhombic, with summits usually nodding or arching, and/or lateral array branches arching or recurved.
            8. Basal and lowest stem leaves usually present at flowering and relatively large; middle to upper stem leaves much reduced.
               9. Petioles of lower stem leaves nearly completely sheathing stem; plants of wet, sandy-saline soils, dunes, and brackish marshes of coastal and adjacent counties of se and sc TX ____ **S. sempervirens**
               9. Petioles or bases of lower stem leaves not sheathing stem; plants of generally drier soils throughout East TX (including se and coast).
                  10. Lower stem leaves ovate to lanceolate.
                      11. Stems square in cross-section, at least lower parts of the stem 4-angled, sometimes winged; upper leaf surfaces strongly scabrous; disk corollas 2.8–3 mm long; pappus 2–3 mm long ____ **S. salicina**
                      11. Stems rounded in cross-section; upper leaf surfaces glabrous to slightly scabrous or strigose; disk corollas 3.5–5 mm long; pappus 3–4 mm long.

12. Plants without slender stoloniferous rhizomes; ray ligules 4–5.5 mm long; achenes 1.5–2 mm long ____ **S. arguta**
12. Plants with slender stoloniferous rhizomes in addition to deeper rhizomes; ray ligules ca. 2 mm long; achenes 2–3 mm long ____ **S. ludoviciana**
10. Lower stem leaves lanceolate to linear-lanceolate.
13. Stems and leaves usually glabrous ____ **S. missouriensis**
13. Stems and leaves sparsely to densely hairy.
14. Plants with short-branched caudices (root crowns); stems thickly short-pubescent with grayish hairs; leaves densely puberulent; peduncles 2–3.5 mm long ____ **S. nemoralis**
14. Plants sometimes with creeping rhizomes; stems scabrous to loosely puberulent; leaves scabrous; peduncles 0.5–2 mm long ____ **S. radula**
8. Basal and lowest stem leaves usually withering by flowering or relatively small; middle to upper stem leaves gradually reduced.
15. Lowermost stem leaves usually the largest on the stem; lower and midstem leaves reticulate-nerved with multiple, prominent lateral nerves, not 3-nerved.
16. Stems developing from elongate, creeping rhizomes; lowermost stem leaves sessile ____ **S. rugosa**
16. Stems developing from branched caudices or short rhizomes; lowermost stem leaves petiolate.
17. Lower stem leaves scabrous to hirsute; peduncles 1.7–2 mm long; achenes 1–1.6 mm long ____ **S. ulmifolia**
17. Lower stem leaves glabrous; peduncles 2–8 mm long; achenes 1.4–2.3 mm long.
18. Stems puberulent in lines; lower stem leaves entire, minutely translucently gland-dotted (view with back lighting), usually anise-scented when crushed; pappus 2.4–3 mm long ____ **S. odora**
18. Stems glabrous; lower stem leaves serrate, not gland-dotted, not anise-scented; pappus 1.5–2 mm long ____ **S. delicatula**
15. Lowermost stem leaves usually not the largest on the stem; lower and midstem leaves usually 3-nerved (lateral nerves sometimes obscure in *S. tortifolia*).
19. Midstem leaves linear to linear-lanceolate, often twisted; ray flowers mostly 2–8 ____ **S. tortifolia**
19. Midstem leaves lanceolate to oblanceolate, not twisted; ray flowers mostly 8–15.
20. Upper stems usually glabrous and sometimes glaucous; leaves glabrous above ____ **S. gigantea**
20. Upper stems short-hairy to villous-tomentose; leaves strigose to pilose above.
21. Upper stems short-hairy; leaves scabrous above and more densely strigillose beneath, especially on nerves; inflorescences in broad or narrow pyramidal arrays, usually not more than 2 times as long as wide ____ **S. altissima**
21. Upper stems densely villous-tomentose; leaves evenly moderately to densely pilose on both sides; inflorescences in narrow pyramidal arrays, 2–4 times as long as wide ____ **S. juliae**

Solidago altissima L. subsp. **altissima,** (tallest), LATE GOLDENROD. Plants to 200 cm tall; rhizomes creeping, short or long; stems short-hairy (at least among the flowering branches); lower leaves withering by flowering; midstem leaves sessile, blades oblanceolate to lanceolate, scabrous above, strigillose beneath, especially on nerves, margins minutely serrate to entire; inflorescence arrays paniculate, pyramidal, branches divergent, ascending or recurved, heads secund; achenes 0.5–1.5 mm long, strigillose; pappus 2.5–3.5 mm long. Disturbed areas, fields, roadsides; throughout e 1/2 TX; Edwards Plateau and s TX; e 1/2 of North America; Mexico; introduced worldwide. Aug–Oct. [*S. altissima* L. var. *pluricephala* M.C. Johnst., *S. canadensis* L. var. *scabra* (Muhl. ex Willd.) Torr. & A. Gray] Semple et al. (2015) and Semple (2016) indicate the correct name for the infraspecific taxon in TX may be var. *pluricephala* M.C. Johnst.

Solidago arguta Aiton var. **boottii** (Hook.) Palmer & Steyerm., (sp.: sharp-toothed; var.: named for Francis Boott (1792–1863), American physician and botanist), BOOTT'S GOLDENROD, CUT-LEAF GOLDENROD. Plants to 120 cm tall; caudices branching; stems strigose among the flowering branches; basal leaves often present at flowering, bases tapered to thin, winged petioles, margins sharply serrate; midstem leaves sessile, blades lanceolate, strigose to strigillose; inflorescence arrays paniculate, open, leafy-bracted, branches recurved and hairy, heads secund; phyllaries eglandular; achenes 1.5–2 mm long, strigillose; pappus 3–3.5 mm long. Dry woods; Anderson Co. (Kartesz 2015) and Hardin, Harrison, Marion, and Titus cos.; se U.S. from SC and GA n to IL and w to KS and TX. Sep–Oct. [*S. boottii* Hook., *S. arguta* Aiton subsp. *boottii* (Hook.) G. Morton]

Solidago auriculata Shuttlw. ex S.F. Blake, (auriculate, for the clasping leaf bases), EARED GOLDENROD, CLASPING-LEAF GOLDENROD. Plants to 150 cm tall; caudices or rhizomes short, stout; stems velvety or finely hirsute; basal leaves often present at flowering; basal and midstem leaves with petioles winged, bases flared and auriculate-clasping, blades broadly ovate and cordate, glabrate or scabrous to strigose, margins serrate; inflorescence arrays paniculate, branches few, short or arching, heads secund; phyllaries short-strigose, eglandular; achenes 2–2.5 mm long, distinctively ribbed, short-strigose; pappus ca. 2 mm long, shorter than

Silphium asteriscus [BB2, LAM]

Silphium integrifolium [GLE]

Silphium radula var. gracile [HEA]

Silphium laciniatum [GLE, STE]

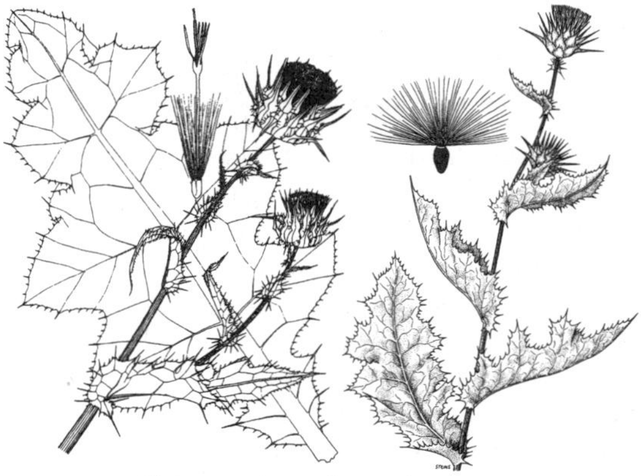

Silybum marianum [M&F, PNW]

achenes. Rocky wooded slopes, streambeds, mesic woods; far e and se TX; se U.S. from SC s to FL and w to OK and TX. Sep–Oct. [*S. amplexicaulis* Torr. & A. Gray ex Chapm., *S. notabilis* Mack. ex Small]

Solidago caesia L. var. **caesia,** (blue-gray), BLUESTEM GOLDENROD, WREATH GOLDENROD, WOODLAND GOLDENROD. Plants to 80(–100) cm tall; rhizomes woody; stems strongly arching, glaucous, sometimes appearing blue to purple, glabrous, or slightly strigose among the flowering branches; basal leaves withering by flowering; midstem leaves sessile, blades narrowly lanceolate, sparsely hairy above, margins serrate or entire; inflorescence arrays racemose, interrupted, leafy-bracted, heads in short, secund axillary clusters; phyllaries without stalked glands; achenes 1–2(–2.5) mm long, strigose; pappus 2–3(–3.7) mm long. Shady mesic areas, oak-hickory forests; deep e and se TX; e and se U.S. from ME to FL and w to MI, MO, and TX; Canada. Sep–Oct. [*S. axillaris* Pursh, *S. caesia* L. var. *axillaris* (Pursh) A. Gray]

Solidago delicatula Small, (somewhat delicate), ELM-LEAF GOLDENROD, SMOOTH ELM-LEAF GOLDENROD. Plants to 120 cm tall; caudices compact, branching, woody; stems glabrous; basal leaves withering by flowering, bases tapered to short petioles; midstem leaves subpetiolate or sessile, blades elliptic-lanceolate, glabrous, margins serrate; inflorescence arrays paniculate, arching, branches elongate and arching, heads secund; phyllaries glabrous, eglandular; achenes 1.5–2 mm long, 6–9-ribbed, sparsely strigose; pappus 1.5–2 mm long. Sandy soils, roadsides, openings in oak woods, drier upland sites; ne and e TX; nc and sc TX; AR, KS, LA, OK, and TX. Aug–Oct(–Nov). [*S. microphylla* (A. Gray) Engelm. ex Small, *S. ulmifolia* Muhl. ex Willd. var. *microphylla* A. Gray] According to Semple and Cook (2006), this taxon was found to be sufficiently distinct from *S. ulmifolia,* and has therefore been returned to species status.

Solidago gigantea Aiton, (gigantic), GIANT GOLDENROD, SMOOTH GOLDENROD. Plants to 200 cm tall; rhizomes creeping, short or long; stems glabrous or sparsely strigose among the flowering branches, sometimes glaucous; lower leaves withering by flowering; midstem leaves sessile, blades lanceolate, glabrous or pilose on nerves beneath, margins sharply serrate; inflorescence arrays paniculate, broadly pyramidal, branches divergent, recurved, heads secund; achenes 1.3–1.5 mm long, sparsely strigillose; pappus 2–2.5 mm long. Low wet areas, roadsides, margins of ponds and streams, disturbed sites; widespread in TX; w 3/4 of North America; Mexico. Aug–Oct. [*S. gigantea* Aiton var. *pitcheri* (Nutt.) Shinners, *S. gigantea* Aiton var. *serotina* (Kuntze) Cronquist, *S. shinnersii* (Beaudry) Beaudry]

Solidago juliae G.L. Nesom, (named for Julia Wells Nesom, wife of the species author Guy L. Nesom), JULIA'S GOLDENROD. Plants to 250 cm tall; rhizomes short; stems densely and evenly villous-tomentose, hairs white; lower leaves withering by flowering; midstem leaves sessile, blades lanceolate to narrowly lanceolate, moderately to densely short-pilose, margins shallowly crenate to serrate or entire; inflorescence arrays paniculate, pyramidal, branches spreading and slightly arching, heads secund; phyllaries glabrous; achenes 1.4–1.6 mm long, sparsely strigillose; pappus ca. 3 mm long. Wet calcareous soils, stream and lake edges, grasslands; Hays and Travis cos.; Edwards Plateau and Trans-Pecos; Mexico. Feb–Apr and Aug–Oct. [*S. canadensis* L. var. *canescens* A. Gray]

Solidago ludoviciana (A. Gray) Small, (of Louisiana), LOUISIANA GOLDENROD. Plants to 90(–150) cm tall; rhizomes short or slender; stems glabrous, or short-strigose among the flowering branches; basal rosette leaves often present at flowering, bases tapered to long, winged petioles, margins serrate; midstem leaves sessile, blades linear-elliptic to elliptic-oblanceolate, glabrous or scabrous-strigose; inflorescence arrays paniculate, open, branches ascending or arching, heads sometimes secund; phyllaries eglandular; achenes 2–3 mm long, 4–7-ribbed, short-strigose; pappus 3–4 mm long. Roadsides, edges of woods, usually sandy soils; throughout e 1/3 TX; nc TX; AR, LA, OK, and TX. Sep–Oct. [*S. boottii* Hook. var. *ludoviciana* A. Gray, *S. strigosa* Small] Nesom (2009a) determined that *S. dispersa* Small was the correct name for this taxon, but as there seems to be a lack of consensus, we are following Semple and Cook (2006) and Semple (2016) in treating it as *S. ludoviciana.*

Solidago missouriensis Nutt., (of Missouri), MISSOURI GOLDENROD. Plants to 80 cm tall; rhizomes short to long; stems glabrous or sometimes sparsely strigose among the flowering branches; basal rosette leaves often present at flowering, bases tapered to long, winged petioles, blades entire to serrulate; midstem leaves sessile, blades lanceolate to linear, glabrous; inflorescence arrays paniculate, pyramidal or rhombic, branches glabrous, with secund heads spreading and arching, sometimes ascending, with non-secund heads; phyllaries glabrous, eglandular; achenes 1–2 mm long, glabrous or sparsely strigose; pappus 2.5–3 mm long. Upland prairies, pastures, roadsides; ne TX in Bowie, Dallas, and Lamar cos., and an apparent disjunct population in se TX in Austin, Harris, and Waller cos. (possibly adventive); nc TX and Rolling Plains; c and w North America; adventive in a few ne states; Mexico. Aug–Oct. [*S. glaberrima* M. Martens, *S. glaberrima* M. Martens var. *moritura* (E.S. Steele) E.J. Palmer & Steyerm., *S. glaucophylla* Rydb., *S. missouriensis* Nutt. var. *fasciculata* Holz.]

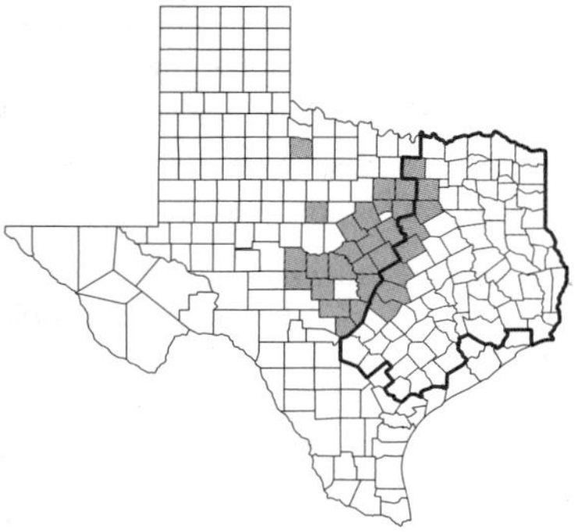

Silphium albiflorum

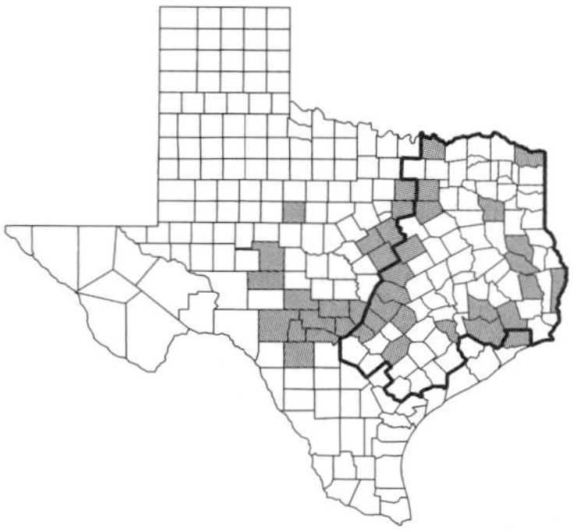

Silphium asteriscus var. asteriscus

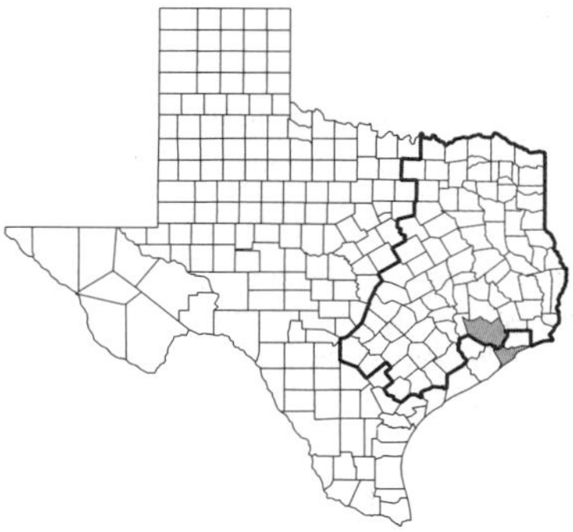

Silphium asteriscus var. simpsonii

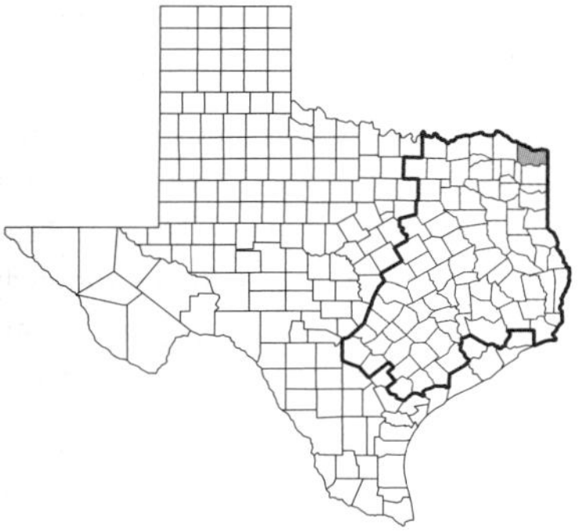

Silphium integrifolium

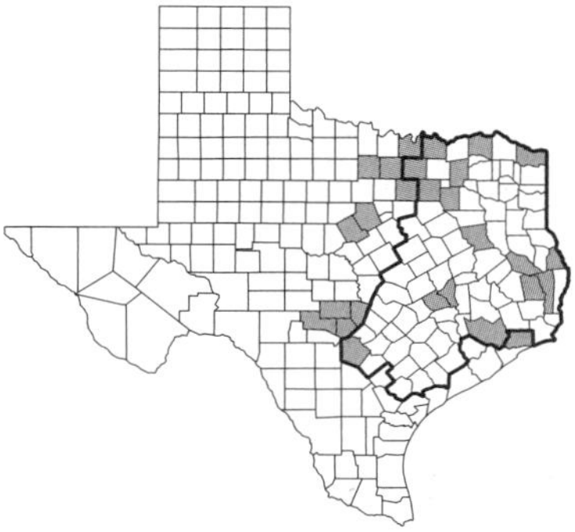

Silphium laciniatum

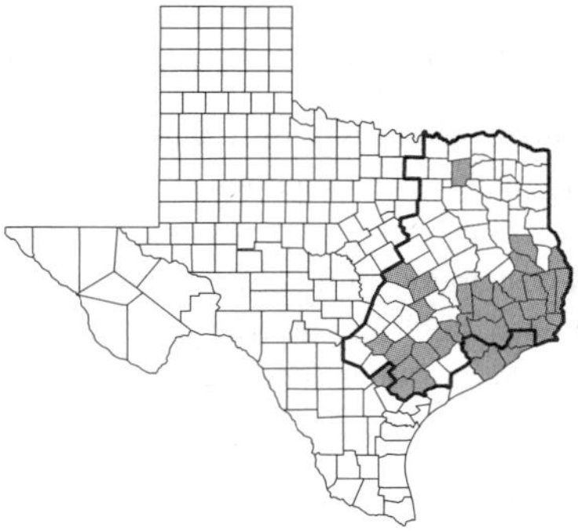

Silphium radula var. gracile

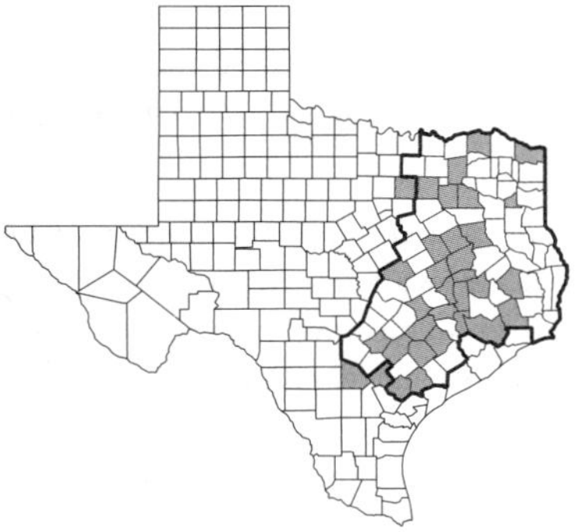

Silphium radula var. radula

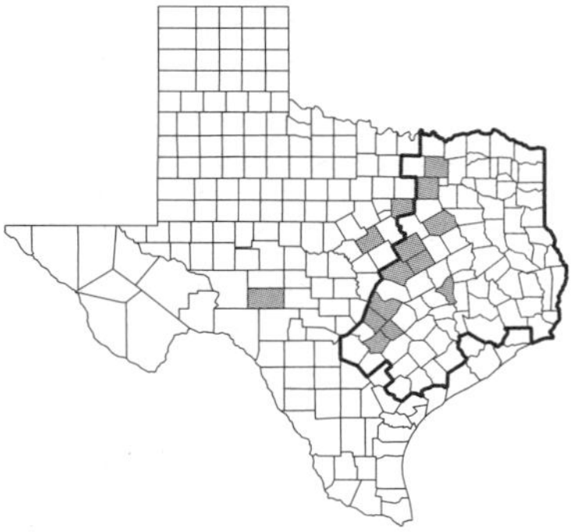

Silybum marianum

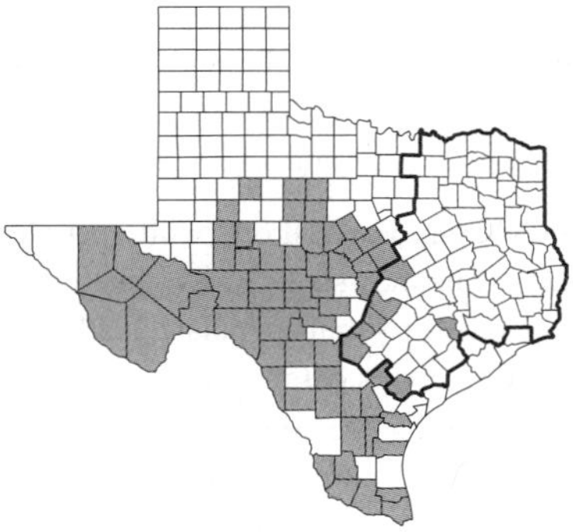

Simsia calva

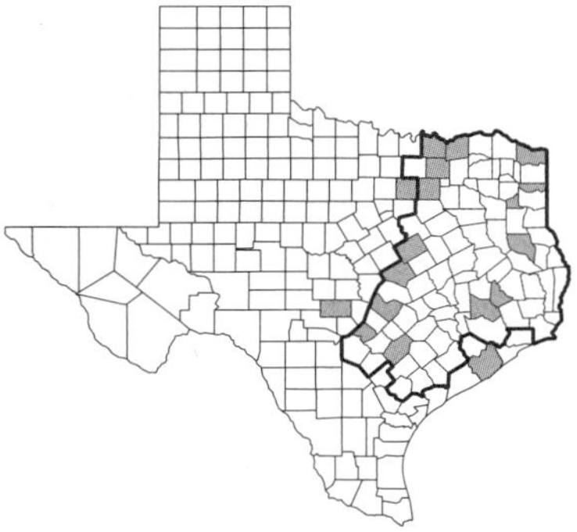

Smallanthus uvedalia

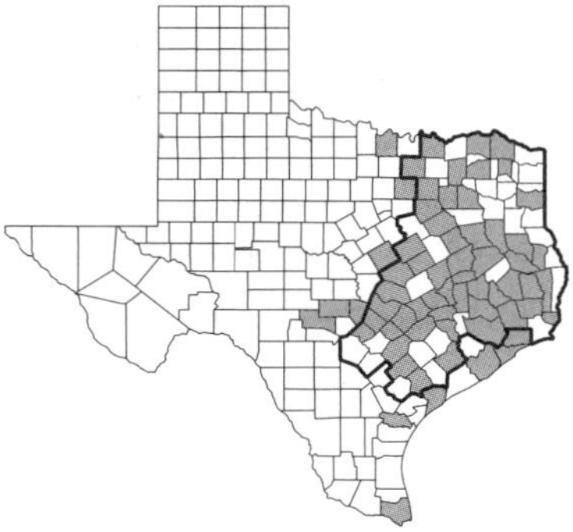

Solidago altissima subsp. altissima

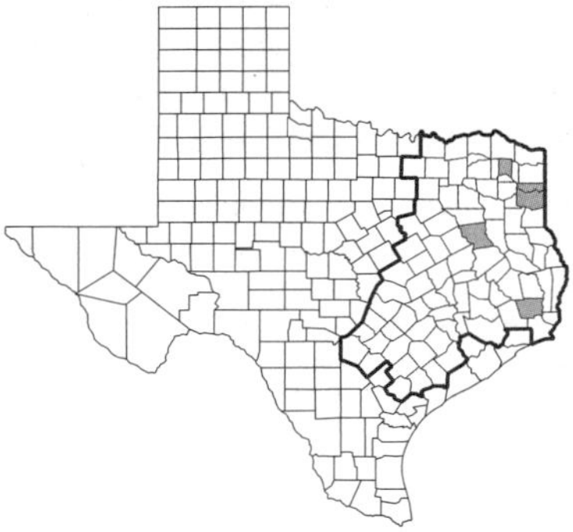

Solidago arguta var. bootii

Solidago nemoralis Aiton, (of the woods), OLD-FIELD GOLDENROD, GRAY GOLDENROD. Plants to 100 cm tall; caudices short-branched; stems short-canescent; basal rosette leaves often present at flowering, bases tapered to long, winged petioles; midstem leaves sessile, blades spatulate to linear-oblanceolate, densely puberulent, crenate to entire; inflorescence arrays paniculate, wand-like, elongate and apically recurved, heads secund; phyllaries without stalked glands; achenes 0.5–2 mm long, strigose; pappus 2–4 mm long. Wooded upland sites, sandy and gravelly soils, roadsides. Sep–Oct. Two subspecies are found in East TX:

1. Basal leaves usually entire, often linear-oblanceolate; involucres usually 4.6–5.8 mm high; pappus bristles usually exceeding ray flower tubes; plants of se, sc, and c TX ____ subsp. **decemflora**
1. Basal leaves usually crenate, oblanceolate to obovate; involucres usually 2.6–4.2 mm high; pappus bristles usually not or barely exceeding ray flower tubes; plants of n, ne, and far e TX ____ subsp. **nemoralis**

subsp. **decemflora** (DC.) Brammall ex Semple, (ten-flowered). Grayson Co., c TX from McLennan to Bexar cos., and sc TX; e Rolling Plains and Edwards Plateau; c U.S. from MI s to TX and w to MT and NM; Canada. [*S. decemflora* DC., *S. nemoralis* Aiton var. *decemflora* (DC.) Fernald, *S. nemoralis* Aiton var. *longipetiolata* (Mack. & Bush) E.J. Palmer & Steyerm.]

subsp. **nemoralis.** Red River drainage and Gregg and Newton cos.; nc TX; e 1/2 of the U.S.; Canada. [*S. nemoralis* Aiton var. *haleana* Fernald, *S. nemoralis* Aiton var. *arenicola* Burgess]

Solidago nitida Torr. & A. Gray, (shining), SHINY GOLDENROD, FLAT-TOPPED GOLDENROD. Plants to 100 cm tall; caudices branching; stems glabrous or scabrous-puberulent among the flowering branches; petiole bases of rosette leaves persistent; basal rosette leaves often present at flowering, long-petiolate; midstem leaves sessile, blades linear-lanceolate to linear, glabrous or scabrous, shiny, margins entire or sparsely serrulate; inflorescence arrays corymbose, compact, flat-topped or sometimes somewhat rounded; phyllaries glabrous, eglandular, 3–7 striately-nerved; achenes 2–3 mm long, glabrous, 7–10-nerved; pappus 4–5 mm long. Prairies, open woods, disturbed areas; e and se TX; nc TX and s TX; AR, LA, MS, OK, and TX. (Jun–)Aug–Oct. Sometimes placed in the segregate genus *Oligoneuron* [as *O. nitidum* (Torr. & A. Gray) Small]

Solidago odora Aiton subsp. **odora,** (odorous, fragrant), FRAGRANT GOLDENROD, SWEET GOLDENROD, ANISE-SCENTED GOLDENROD. Plants to 120 cm tall; caudices short, stout; stems puberulent (at least among the flowering branches), usually in vertical lines; basal leaves withering by flowering, bases tapered to broadly-winged petioles; midstem leaves sessile, blades linear-lanceolate to narrowly ovate, glabrous, finely translucent gland-dotted, usually anise-scented when crushed, margins entire; inflorescence arrays paniculate, pyramidal, branches ascending to spreading and recurved, heads secund; phyllaries glabrous, eglandular; achenes 1.4–2.3 mm long, glabrate to strigose; pappus 2.4–3 mm long. Wet sandy soils, roadsides, oak and pine woods; mainly se and e TX; c and sc TX; e and se U.S. from NH s to FL and w to OH, AR, and TX. Sep–Oct.

Solidago petiolaris Aiton, (with a petiole), DOWNY RAGGED GOLDENROD. Plants to 150 cm tall; caudices stout, sometimes with long, slender rhizomes; stems finely pubescent (at least among the flowering branches); basal leaves withering by flowering; midstem leaves sessile to short-petiolate, blades usually lanceolate-elliptic or ovate, glabrous or scabrous, sometimes resinous or shiny beneath, margins entire or few-toothed; inflorescence arrays paniculate (rarely racemose), usually elongate and leafy-bracted, branches stiffly ascending, not recurved, heads not secund; phyllaries glabrous or strigose, with sparse to dense stalked glands; achenes 3–4 mm long, glabrous or glabrate; pappus ca. 4 mm long. Sandy soils, prairies, open woods; throughout most of TX with the exception of far s and w TX; se and sc U.S. from NC s to FL and w to KS, CO, and NM; Mexico. Sep–Oct. Nesom (1990e, 2008) discussed and mapped three varieties of *S. petiolaris*; Semple and Cook (2006) did not treat any subspecific taxa; Semple (2016) resurrected these varieties, based on research done after the 2006 FNA treatment was prepared. Two of the varieties have overlapping ranges in East TX; the infraspecific range maps provided by Nesom (2008), Kartesz (2015), Semple (2016), and USDA Plants Database (2016) are incomplete or irreconcilable, therefore we are not providing ranges or maps for these varieties. Nesom (2008) divided the two East TX varieties as follows:

1. Phyllaries densely glandular (may appear viscid), essentially glabrous ____ var. **angusta**
1. Phyllaries scarcely glandular, usually finely strigose or sometimes glabrous ____ var. **wardii**

var. **angusta** (Torr. & A. Gray) A. Gray, (narrow). [*S. angusta* Torr. & A. Gray, *S. lindheimeriana* Scheele]

var. **wardii** (Britton) Fernald, (named for Lester F. Ward (1841–1913), American botanist and collector), WARD'S GOLDENROD. [*S. wardii* Britton]

Solidago radula Nutt., (rough), ROUGH GOLDENROD, WESTERN ROUGH GOLDENROD. Plants to 90 cm tall; with caudices and sometimes creeping rhizomes as well; stems scabrous to puberulent; basal leaves usually withering by flowering, bases tapered to long, winged petioles; midstem leaves sessile or subsessile, blades elliptic to oblanceolate, distinctly scabrous, margins serrate or crenate;

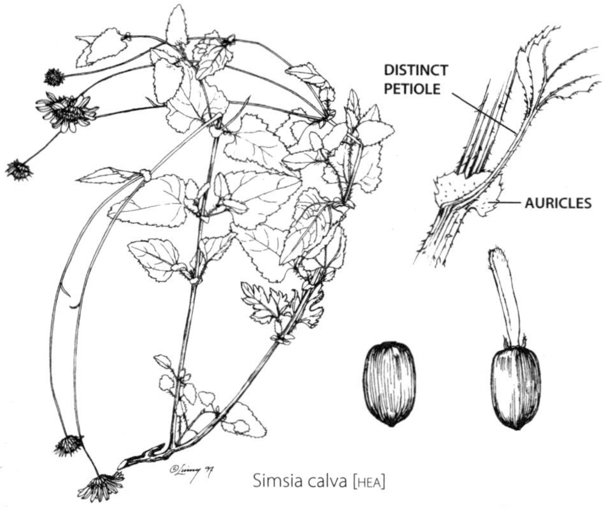

Simsia calva [HEA]

Smallanthus uvedalia [REE]

Solidago altissima subsp. altissima [BB2]

Solidago arguta var. bootii [STR]

Solidago auriculata [SEM]

Solidago caesia [BA1, BB2]

Solidago delicatula [HEA]

Solidago gigantea [GLE]

inflorescence arrays paniculate, pyramidal, branches recurved, heads secund; phyllaries without stalked glands; achenes 1.5–2.5 mm long, short-strigose; pappus 3 mm long. Dry rocky bluffs, upland xeric sites, calcareous soils; e and c TX; Rolling Plains and Edwards Plateau; mostly c U.S. from IL s to TX, scattered in se U.S. Sep–Nov. [*S. pendula* Small, *S. radula* Nutt. var. *rotundifolia* (DC.) A. Gray, *S. scaberrima* Torr. & A. Gray]

Solidago rigida L., (rigid), STIFF GOLDENROD, RIGID GOLDENROD. Plants to 150 cm tall; caudices branching, woody; stems glabrous or ± hispid; petiole bases of rosette leaves persistent; basal rosette leaves usually present at flowering, long-petiolate; midstem leaves sessile, blades ovate to lanceolate or rhombic, glabrous or strigose to hispid, margins entire or serrate; inflorescence arrays in flat-topped or corymbose, compact or branches long and spreading, usually somewhat rounded; phyllaries glabrous or strigillose, eglandular, 3–5 striately-nerved; achenes 0.8–1.7 mm long, glabrous or strigillose, ribbed; pappus 3–4 mm long. Prairies, open woods, dry open areas; e and se TX; Panhandle, nc TX, and South TX Plains; w 3/4 of the U.S.; Canada. Sep–Oct(–Nov). Sometimes placed in the segregate genus *Oligoneuron* [as *O. rigidum* (L.) Small]. Three subspecies have overlapping ranges in East TX (Heard & Semple 1988; Semple & Cook 2006), and the infraspecific range maps provided by Heard and Semple (1988), Kartesz (2015), Semple (2016), and USDA Plants Database (2016) are incomplete or irreconcilable, therefore we are not providing ranges or maps for these subspecies. Semple and Cook (2006) divide the subspecies as follows:

1. Outer phyllaries glabrous; leaves and stems glabrous or ± hispid ____ subsp. **glabrata**
1. Outer phyllaries strigillose; leaves and stems moderately to densely strigose or hispid.
    2. Inner phyllaries conspicuously strigillose, often linear; plants usually short (40–70(–90) cm tall); inflorescence arrays compact; leaves and stems finely and densely hispid (> 50 hairs/sq. mm); achenes usually strigillose on one end ____ subsp. **humilis**
    2. Inner series phyllaries glabrate to sparsely strigillose, oblong to spatulate; plants more robust (60–150 cm tall); inflorescence arrays loose, open; leaves and stems coarsely hispid (< 50 hairs/sq. mm); achenes glabrous ____ subsp. **rigida**

subsp. **glabrata** (E.L. Braun) Heard & Semple, (glabrate). [*Oligoneuron rigidum* (L.) Small var. *glabratum* (E.L. Braun) G.L. Nesom, *S. rigida* L. var. *glabrata* E.L. Braun, *S. rigida* L. var. *laevicaulis* Shinners]

subsp. **humilis** (Porter) Heard & Semple, (dwarf). [*Oligoneuron canescens* Rydb., *O. rigidum* (L.) Small var. *humile* (Porter) G.L. Nesom, *S. rigida* L. var. *humilis* Porter]

subsp. **rigida.** [*Oligoneuron grandiflorum* (Raf.) Small, *S. grandiflora* Raf., *S. rigida* L.]

Solidago rugosa Mill. subsp. **aspera** (Aiton) Cronquist, (sp.: rugose, wrinkled; subsp.: rough), ROUGH-LEAF GOLDENROD, WRINKLE-LEAF GOLDENROD. Plants to 200 cm tall; rhizomes long-creeping; stems glabrous or strigose to densely hispid; basal leaves withering by flowering, midstem leaves sessile, blades lanceolate to elliptic or ovate, hispid-strigose beneath (at least on main nerves), margins serrate to subentire; inflorescence arrays paniculate, pyramidal, branches divergent and recurved, leafy-bracted, heads secund; phyllaries eglandular; achenes 0.9–1.5 mm long, strigillose; pappus 1.8–2.5 mm long. Roadside ditches, along streams, moist sites in oak-hickory forests; mostly e 1/3 TX; sc TX; e and se U.S. from ME s to FL and w to WI, IA, and TX; Canada. Sep–Nov. This subspecies is further divided into two varieties with overlapping ranges in East TX (Semple & Cook 2006), and the range maps provided by Kartesz (2015), Semple (2016), and USDA Plants Database (2016) are incomplete or irreconcilable, therefore we are not providing ranges or maps for these varieties. Semple and Cook (2006) divide the TX varieties as follows:

1. Upper stem leaf blades lanceolate to elliptic, not much reduced up the stem ____ var. **aspera**
1. Upper stem leaf blades ovate, much reduced up the stem ____ var. **celtidifolia**

var. **aspera** (Aiton) Fernald [*S. aspera* Aiton]

var. **celtidifolia** (Small) Fernald, (with leaves resembling *Celtis*), CELTIS-LEAF GOLDENROD. [*S. celtidifolia* Small]

Solidago salicina Elliott, (resembling *Salix*—WILLOW), SOUTHERN ROUGH-LEAF GOLDENROD, WILLOW GOLDENROD. Plants to 150 cm tall; caudices short, rhizomes creeping and elongate; stems angular in cross-section, glabrous, or sparsely hairy among the flowering branches; basal leaves often present at flowering, bases tapered to long, winged petioles, margins serrate; midstem leaves sessile, blades lanceolate, scabrous above; inflorescence arrays paniculate, pyramidal, open, branches ascending to recurved, heads secund; phyllaries eglandular; achenes 0.6–1.5 mm long, strigillose; pappus 2–3 mm long. Swamp margins, wet meadows; e and se TX; nc TX; se U.S. from VA s to FL and w to OK and TX. Sep–Oct. This taxon was included in *S. patula* [as *S. patula* Muhl. subsp. *strictula* (Torr. & A. Gray) Semple] in Semple and Cook (2006), but Semple et al. (2012) concluded the subspecies was sufficiently distinct to be returned to species status.

Solidago sempervirens L. var. **mexicana** (L.) Fernald, (sp.: evergreen; var.: of Mexico), SEASIDE GOLDENROD, MEXICO GOLDENROD. Plants to 200 cm tall; caudices short, stout; stems glabrous throughout, or hairy among the flowering branches; basal rosette

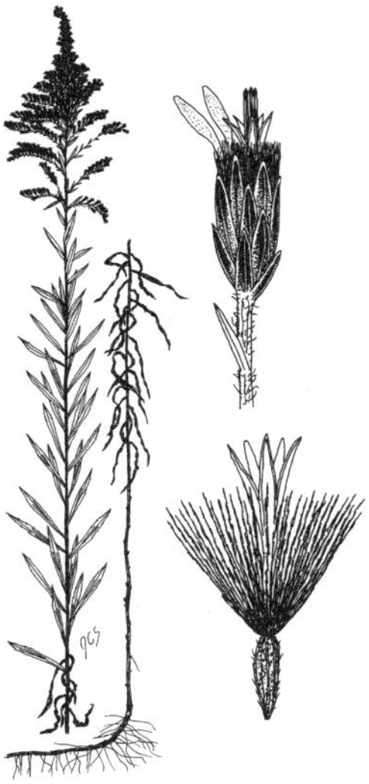

Solidago juliae [SEM]

Solidago ludoviciana [HEA]

Solidago missouriensis [BB2]

Solidago nemoralis subsp. nemoralis [BB2]

Solidago nemoralis subsp. decemflora [HEA]

leaves present at flowering, petioles winged and sheathing stems or nearly so; midstem leaves sessile, blades lanceolate, thick or fleshy, bases sometimes subclasping, margins entire; inflorescence arrays paniculate and narrowly elongate, branches spreading-recurved, heads secund; phyllaries eglandular; achenes 1–1.5 mm long, strigose; pappus 3.8–4 mm long. Wet sandy-saline soils, dunes, brackish marshes; coastal and adjacent cos. of se and sc TX; s TX; Atlantic and Gulf coastal states from MA s to FL and w to TX; Mexico; Central America; and West Indies. Sep–Dec. [*Aster mexicanus* (L.) Kuntze, *S. mexicana* L.] Semple et al. (2016) suggest this taxon be returned to species status as *S. mexicana* L., based on a multivariate study of the *S. sempervirens* complex.

Solidago speciosa Nutt. var. **rigidiuscula** Torr. & A. Gray, (sp.: showy; var.: somewhat rigid), PRAIRIE GOLDENROD, NOBLE GOLDENROD, SHOWY GOLDENROD. Plants to 200 cm tall; caudices stout, woody; stems glabrous, or strigillose among the flowering branches; basal leaves sometimes withering by flowering; midstem leaves sessile, blades lanceolate to ovate-elliptic, somewhat scabrous, glabrate to sparsely strigillose, serrate to crenate or entire; inflorescence arrays paniculate or thyrsiform, compact and elongate, branches stiffly ascending, heads not secund; phyllaries glabrous, eglandular; achenes 1.6–2.5 mm long, glabrous; pappus 3–4.5 mm long. Sandy and gravelly soils, prairies, open woods; e and se TX; nc TX, c TX, and Plains Country; c U.S. from MI s to GA and w to ND and TX; Canada. Sep–Oct. [*S. rigidiuscula* (Torr. & A. Gray) Porter, *S. venulosa* Greene] Semple (2016) suggests this taxon be returned to species status as *S. rigidiuscula* (Torr. & A. Gray) Porter.

Solidago tortifolia Elliott, (contorted leaves), TWIST-LEAF GOLDENROD. Plants to 130 cm tall; rhizomes creeping, elongate; stems strigose-villous; lower leaves withering by flowering, often coiling; midstem leaves sessile, blades linear to linear-lanceolate, glabrous or finely strigillose, margins serrulate to entire; inflorescence arrays paniculate, broadly pyramidal, branches recurved and elongate, heads secund; achenes 1 mm long, sparsely strigillose; pappus 2–3 mm long. Dry sandy soils, pine woods; ne and se TX; sc TX; se U.S. from VA s to FL and w to TX. Aug–Nov.

Solidago ulmifolia Muhl. ex Willd. var. **ulmifolia,** (with leaves like *Ulmus*—ELM), ELM-LEAF GOLDENROD. Plants to 120 cm tall; caudices branching, woody; stems glabrous or nearly so; basal leaves withering by flowering, tapering to short, winged petioles; midstem leaves subsessile to sessile, blades lanceolate, hirsute to scabrous above, hirsute on main nerves beneath, margins entire; inflorescence arrays paniculate, open, branches widely divergent, elongate or short, heads secund; phyllaries eglandular; achenes 1–1.6 mm long, finely hairy; pappus ca. 2.5 mm long. Openings in woodlands, dry upland sites; e and se TX; nc and sc TX; e 1/2 of North America. Aug–Oct. [*S. helleri* Small]

Solidago virgata Michx., (wand-like), WAND GOLDENROD, WILLOW-LEAF GOLDENROD. Plants to 200 cm tall; caudices short, rhizomes long, stolon-like; stems glabrous; basal rosette leaves present at flowering, nearly sessile or petioles winged and sheathing stems; midstem leaves sessile, blades lanceolate-oblong to linear, glabrous, margins entire; inflorescence arrays thyrsiform and narrowly elongate, not branching, heads not secund; phyllaries glabrous, eglandular; achenes 1.5–2.5 mm long, strigose; pappus 3 mm long. Sandy moist soils, pine flatwoods, coastal brackish marshes; se TX and Harrison Co.; s TX; se U.S. from NC s to FL and w to TX; Mexico and Central America. Sep–Nov. [*S. angustifolia* Elliott, *S. stricta* Aiton var. *angustifolia* (Elliott) A. Gray] Semple (2013) explains the separation of *S. stricta* from *S. virgata*, and why the latter is the correct name for this taxon in TX.

Solidago hispida Muhl. ex Willd., (hairy, hispid), HAIRY GOLDENROD, which would key to *S. speciosa* here, can be distinguished from that species by the usually moderately to densely hispid-villous stems and leaves. Not yet reported in TX but included here based on records from adjacent McCurtain Co., OK (OKL); e U.S. from ME s to GA and w to SD and OK. Aug–Oct. [*S. bicolor* L. var. *concolor* Torr. & A. Gray, *S. bicolor* L. var. *hispida* (Muhl. ex Willd.) Britton, Sterns, & Poggenb., *S. hirsuta* Nutt., *S. lanata* Hook.]

SOLIVA Ruiz & Pav. BURRWEED, PIQUANTE, STICKERWEED, STICKERS

Low-growing taprooted annual herbs, much branched, mat-forming or ascending; stems short (usually < 15 cm tall) or nearly stemless; leaves alternate, often crowded basally and at nodes, blades 1–3-pinnatifid, reminiscent of ferns or parsley; heads greenish, bur-like, sessile in leaf axils; involucres 2–8+ mm wide; ray flowers seemingly absent; peripheral disk flowers numerous, pistillate, without corollas, styles sheathed by tissue, becoming hardened and spine-like; central disk flowers 4–8+, functionally staminate with minute 4-toothed whitish yellow or green-translucent corollas; achenes flattened, with wings or ± wing-like lateral projections, tipped by the persistent spine-like stylar sheath, conspicuously hairy to glabrous apically; pappus absent.

A South American genus of 4–8 species; some were formerly included in the genus *Gymnostyles*. Easily recognized by the spine-like stylar sheaths which make *Soliva* species painfully noxious lawn weeds, similar in effect to *Cenchrus* (SANDBURS). (Named for Dr. Salvador Soliva, 18th century physician to the Spanish court and botanist at the Royal Botanic Garden, Madrid, where he studied medicinally useful plants—Arriagada & Miller 1997) (tribe Anthemideae)

REFERENCES: Cabrera 1949; Tutin 1976; Ray 1987; Arriagada & Miller 1997; Watson 2006d.

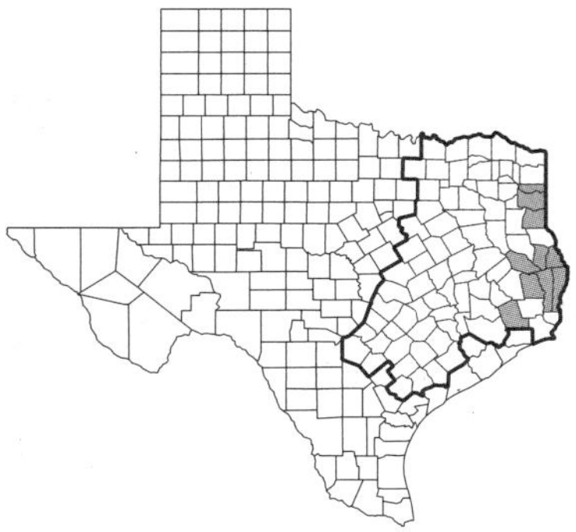

Solidago auriculata

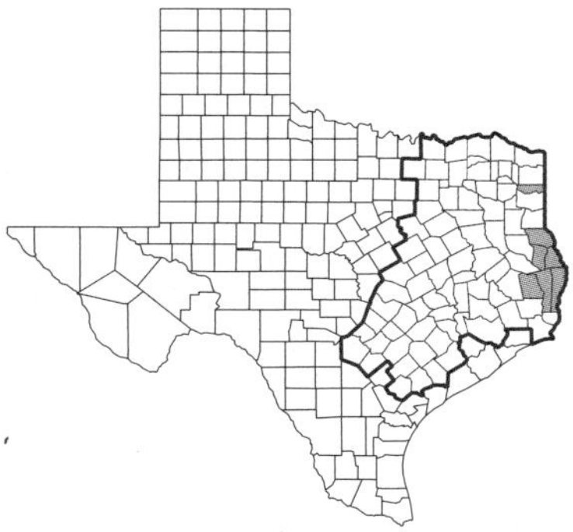

Solidago caesia var. caesia

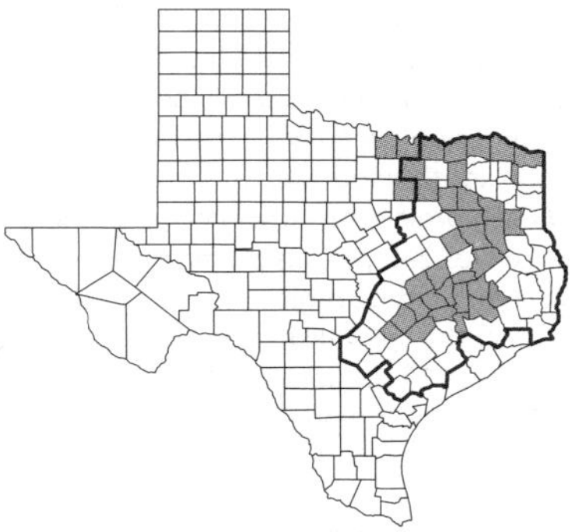

Solidago delicatula

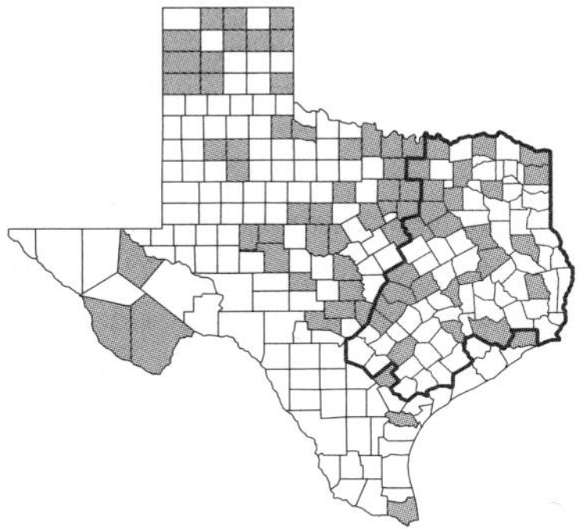

Solidago gigantea

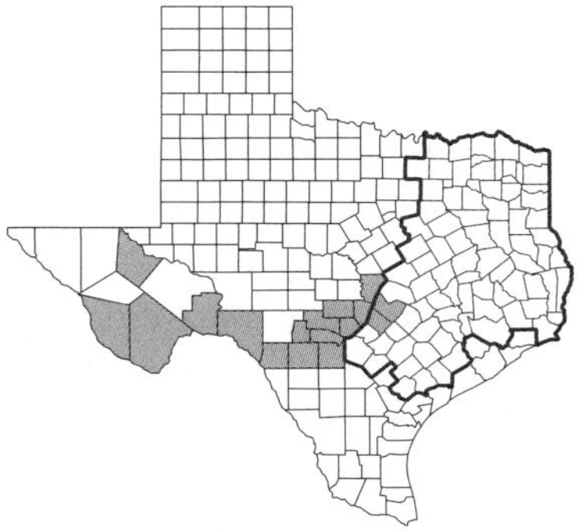

Solidago juliae

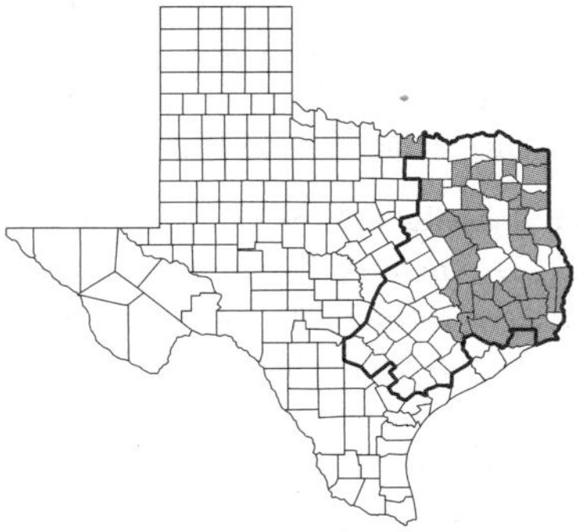

Solidago ludoviciana

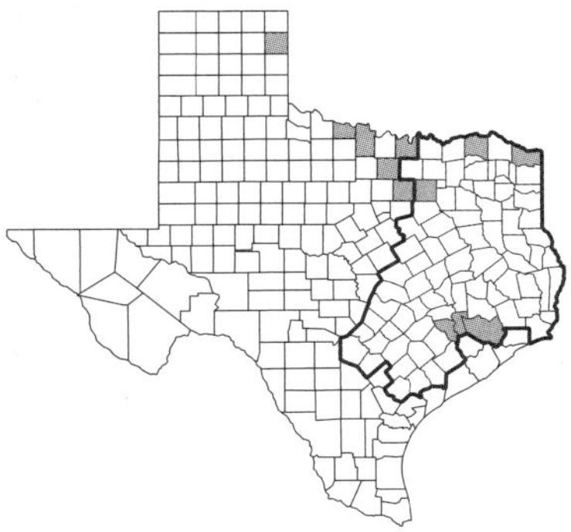

Solidago missouriensis

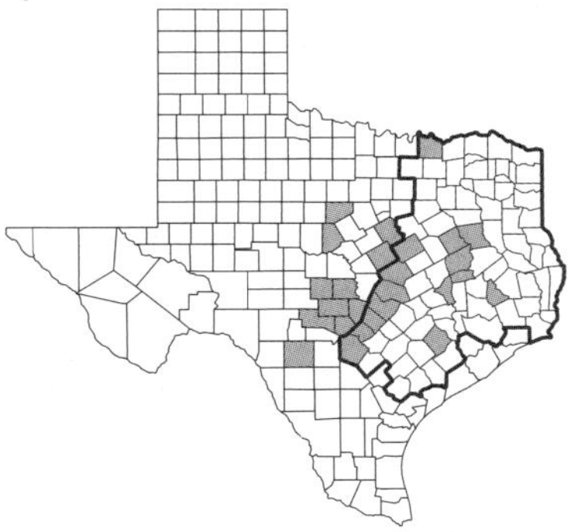

Solidago nemoralis subsp. decemflora

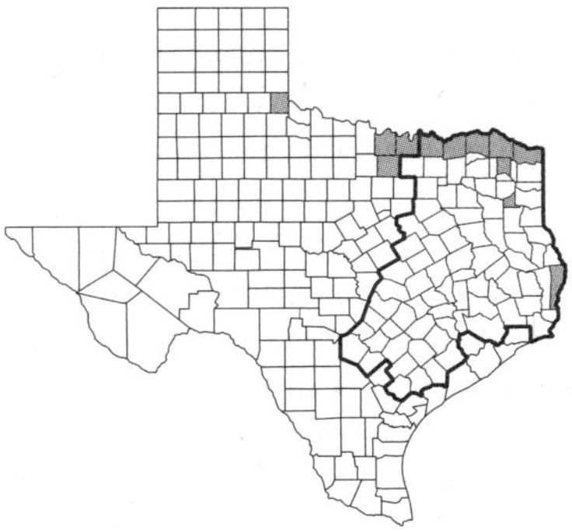

Solidago nemoralis subsp. nemoralis

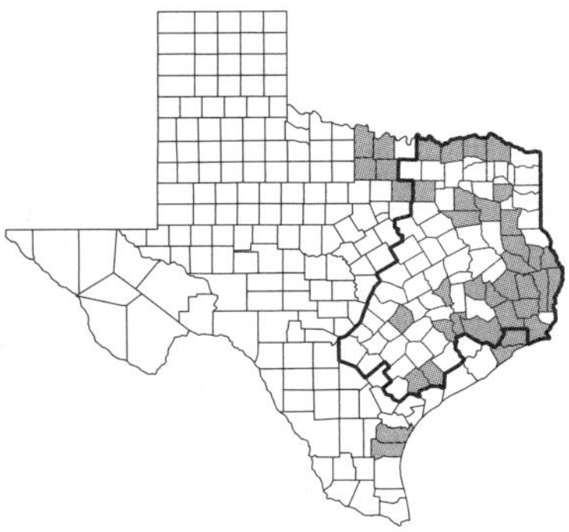

Solidago nitida

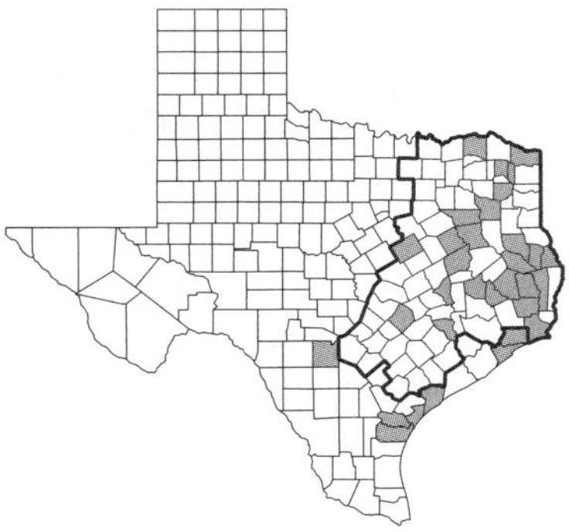

Solidago odora subsp. odora

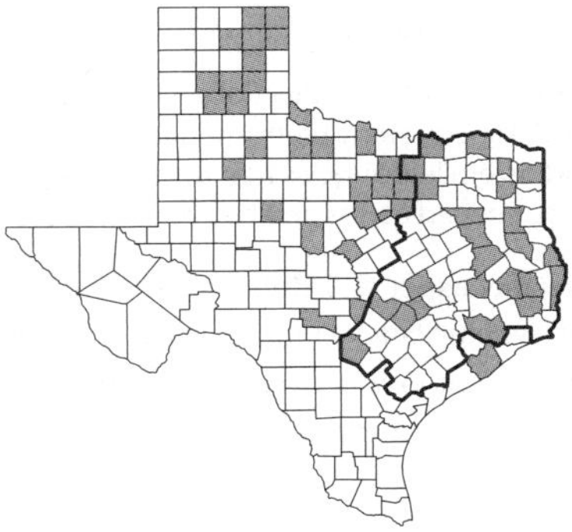

Solidago petiolaris

1. Achenes obovate to oblanceolate, glabrous or short-hairy; lateral projections of achenes wing-like, broad, thin, smooth, without raised lines, margins sometimes conspicuously indented toward lower half, forming 2 basal lobes, each wing shoulder usually projecting as a spine-like tooth on each side of the spine-like style ____ **S. sessilis**
1. Achenes oblanceolate to oblong-cuneate, usually with longer hairs at the tip; lateral projections of achenes thick, only slightly wing-like, with conspicuous raised lines running across them, margins not indented, the upper portion of each projection either not or only slightly spinose.
   2. Leaves 2–3-pinnati-palmately lobed; heads mostly clustered in leaf axils at ground level, rarely scattered along stems; lateral projections of achenes ± smooth in distal 1/3, without any apical awns, projections, or shoulders near the spine-like style ____ **S. anthemifolia**
   2. Leaves 1(–2)-pinnati-palmately lobed; heads mostly scattered along stems; lateral projections of achenes with raised lines nearly to apex, with 2 divergent acute awns, projections, or shoulders at apex of achene, 1 on each side of the spine-like style ____ **S. stolonifera**

Soliva anthemifolia (Juss.) Sweet, (with leaves resembling *Anthemis*—CHAMOMILE), BUTTON BURRWEED, SMOOTH-STICKERS. Stems sometimes rooting at nodes; leaves 3–8(–15) cm long, 2–3-pinnati-palmately lobed; heads mostly clustered in leaf axils at ground level, rarely scattered along stems; pistillate flowers (20–)50–100+; achenes oblanceolate to oblong-cuneate, 1.5–2+ mm long, winged, wings transversely ribbed in the lower 2/3, wing shoulders not spinose; spinose stylar sheath 1.5–3 mm long. Disturbed sites, lawns, roadsides; se and sc TX; s TX; AR, FL, LA, MS, and TX; South America; introduced elsewhere. Mar–May. [*Gymnostyles anthemifolia* Juss., *S. mutisii* Kunth] Native of South America.

Soliva sessilis Ruiz & Pav., (for the sessile heads), LAWN BURRWEED, STICKERS, JO-JO WEED. Stems often rooting at nodes; leaves 1–2(–3+) cm long, 2(–3)-pinnati-palmately lobed; heads mostly scattered along stems; pistillate flowers 5–8(–17+); achenes obovate to oblanceolate, usually 2.5–3+ mm long, usually winged, wings entire or sinuate to incised, each wing shoulder usually projecting as a spine-like tooth; spinose stylar sheath 1–2+ mm long. Disturbed sites, lawns, roadsides; e and sc TX; nc TX; se U.S. from VA s to FL and w to OK and TX, also w coast from WA s to CA; Canada; South America; introduced elsewhere. Mar–May. [*S. daucifolia* Nutt., *S. pterosperma* (Juss.) Less] Native of South America.

Soliva stolonifera (Brot.) Sweet, (bearing stolons), CARPET BURRWEED, TRAILING-STICKERS. Stems often rooting at nodes; leaves 1–2(–3) cm long, 1(–2)-pinnati-palmately lobed; heads mostly scattered along stems; pistillate flowers (20–)40–60+; achenes oblanceolate to oblong-cuneate, 1.8–2.2 mm long, winged, wings transversely ribbed in the lower 9/10, each wing shoulder ± spinose; spinose stylar sheath 0.8–1.5 mm long. Disturbed sites, lawns; Bastrop, Brazos, and Leon cos.; also Nueces Co. in s TX; se U.S. from SC s to FL and w to AR and TX; South America; introduced elsewhere. Mar–Apr. [*Gymnostyles stolonifera* (Brot.) Tutin, *S. nasturtiifolia* (Juss.) DC.] Native of South America.

SONCHUS L. SOW-THISTLE

Ours annual or biennial taprooted herbs, glabrous or glandular-pubescent, often ± glaucous, with milky sap; stems usually hollow; midstem leaves sessile, blades spatulate to obovate or oblong to lanceolate, bases auricled-clasping, margins coarsely or spiny-toothed or runcinate to pinnately lobed; heads small, in corymbose arrays; phyllaries in 3–5+ series, unequal, narrow, green; receptacles naked; flowers all ligulate, corollas yellow; achenes compressed, 2–3.5+ mm long, unbeaked, ribbed, glabrous; pappus of many white, hair-like bristles, 5–9 mm long.

A genus of ca. 60 species native from Eurasia to Australasia and tropical Africa; several have become widespread weeds. (Greek: *sonchos*, ancient name used for a prickly plant) (tribe Cichorieae)
REFERENCES: Boulos 1973, 1976; Vuilleumier 1973; Hyatt 2006.

1. Midstem leaf auricles (lobe-like extensions on both sides at base) rounded, sometimes made irregularly so by prominent, spiny teeth; mature achenes smooth between the several ribs (not wrinkled-roughened) ____ **S. asper**
1. Midstem leaf auricles triangular, acutely pointed, often also sharp-toothed; mature achenes transversely wrinkled-roughened across and between the several ribs ____ **S. oleraceus**

Sonchus asper (L.) Hill, (rough), PRICKLY SOW-THISTLE, SPINY SOW-THISTLE, ACHICORIA DULCE. Stems 10–120(–200+) cm tall; midstem leaf base auricules rounded, often recurved or curled; phyllaries usually with stalked glands; corolla tubes longer than the ligules; achenes smooth between ribs. Gardens and lawns, roadsides, disturbed areas; nearly throughout TX; widespread in North America; Europe; introduced and weedy nearly worldwide. Mar–Jun (sporadically all year). [*S. oleraceus* L. var. *asper* L.] Native of Eurasia.

Sonchus oleraceus L., (of the vegetable garden, a potherb used in cooking), COMMON SOW-THISTLE. Stems 10–140(–200) cm tall; midstem leaf base auricules deltate to lanceolate, usually straight; phyllaries usually glabrous, sometimes with stalked glands; corolla tubes ca. equaling the ligules; achenes transversely wrinkled or roughened across and between ribs. Gardens and lawns,

Solidago nitida [BT3]

Solidago odora [GLE]

Solidago petiolaris [FNA]

Solidago rigida [HOL]

Solidago radula [BB2]

Solidago rugosa subsp. aspera var. aspera [SWF]

roadsides, disturbed areas; widespread in TX; widespread in North America; Europe; introduced and weedy nearly worldwide. Mar–Jun (sporadically all year). Native of Eurasia.

Sonchus tenerrimus L., (very slender), SLENDER SOW-THISTLE, which would probably have keyed to *S. oleraceous* here because they share transversely wrinkled-roughened achenes, can be distinguished by its leaves that have lobes that are constricted at the bases, terminal lobes that are ± equal to the lateral lobes (as opposed to terminal lobes usually larger than the laterals), and corolla tubes that are shorter than the ligules. Disturbed sites; excluded from TX by Hyatt (2006) but included as a Gregg Co. record in Kartesz (2015), based on a pers. comm. from the same author; AL, CA, NY, PA, and TX?; possibly Canada; s Europe; introduced in w Asia and n Africa. Mar–Jun. Native of Eurasia.

SPHAGNETICOLA O. Hoffm. CREEPING-OXEYE

A genus of ca. 5 species of tropical areas, mostly of the Americas; previously recognized in the genus *Wedelia*. (Evidently from Latin: *sphagnum*, a moss, and *cola*, dwelling in, perhaps usually alluding to wet habitats) (tribe Heliantheae)
REFERENCES: Strother 2006aa; Brown et al. 2007.

Sphagneticola trilobata (L.) Pruski, (three-lobed, for the leaves), CREEPING-OXEYE, TRAILING DAISY, WEDELIA. Tender herbaceous perennial; stems prostrate, 30–200 cm long, branching and rooting at nodes, mat-forming; leaves opposite, sessile or short-petioled, the blades 50–180 mm long, usually 3-nerved, often 3-lobed, somewhat succulent, margins entire or irregularly toothed; heads borne singly; phyllaries in 2–3 series, the outer series larger, ± foliaceous; ray flowers 4–10+, pistillate, fertile, the ligules yellow to gold-orange, 6–15 mm long; disk flowers perfect, fertile, corollas yellow to orange; achenes 3–5 mm long; pappus absent, or a minute fringed crown to 0.6 mm long. Wet soils, disturbed areas, coastal dunes; rare in TX; recorded as escaping from cultivation in Cherokee Co. (Brown et al. 2007) and Harris Co. (Brown 2014); se U.S. from FL w to TX; introduced elsewhere. Flowering nearly year-round. [*Wedelia trilobata* (L.) Hitchc.] Native of tropical America; cultivated and escaping; now a widespread weed in tropical and subtropical regions around the world.

STOKESIA L'Hér. STOKES'S ASTER

A monotypic genus of the se U.S. (Named for Jonathan Stokes (1755–1831), English physician and botanist) (tribe Vernonieae)
REFERENCES: Strother 2006bb.

Stokesia laevis (Hill) Greene, (smooth), STOKES'S ASTER. Herbaceous perennial, 20–50 cm tall; basal leaves winged-petiolate, blades simple, ovate to lanceolate or lance-linear, 8–15 cm long, stem leaves alternate, ± sessile, glabrous or glabrate, resin-gland-dotted, bases ± clasping, margins entire or spinose-toothed; heads showy, borne singly or in loose corymbose arrays; involucres 25–45 mm wide; phyllaries in 5–7 series, tomentulose, gland-dotted, the outer ones with lower margins spiny-toothed and the upper portions foliaceous; ray flowers absent; disk flower corollas usually blue to purplish blue (rarely white or lilac), peripheral corollas zygomorphic and simulating rays, 15–25+ mm long, central corollas ± actinomorphic, 12–15+ mm long; achenes columnar, 5–8 mm long, 3–4-angled, glabrous; pappus falling early, of 4–5 scales, 8–12 mm long. Openings in woodlands, bogs, near springs. In TX known only from Jasper Co. (Loos et al. 2019) in far se East TX; se U.S. from NC s to FL and w to TX. Jun–Sep. [*Carthamus laevis* Hill] This species has gained popularity as a garden ornamental; several cultivars have been developed that extend the corolla color range to deep purples, pinks, and yellows.

SYMPHYOTRICHUM Nees. ASTER

Ours mostly herbaceous perennials (2 species annual); stems often clumped or spreading colonially by rhizomes; leaves alternate, simple, entire or toothed, the basal and stem leaves often strongly differing in shape; uppermost leaves of stem branchlets sometimes greatly reduced; heads usually numerous, usually in paniculate or corymbose arrays; phyllaries in several series, often strongly graduated, the bases often whitish and hardened, the margins often hyaline, the tips usually with a green zone; receptacles naked; ray flowers pistillate and fertile, corollas white to pink, blue, violet, or purplish, never yellow, ligules often coiling under at maturity; disk flowers perfect and fertile, corollas cream to yellow, often becoming pinkish to purplish brown at maturity; achenes somewhat flattened, several-nerved; pappus of hair-like, white to brownish, barbellate bristles.

A genus of ca. 90 species mostly native to North America; also found in Central and South America, the West Indies, and Eurasia; a number are cultivated; many species hybridize and intergrade; definitive identification is often difficult. *Symphyotrichum* was traditionally included in the very large genus *Aster*; modern morphological and molecular studies provided strong support for their division (Brouillet et al. 2006). Nesom (1994b) treated *Aster* as an essentially Old World genus of ca. 180 species and segregated the New World taxa into 13 genera: *Almutaster*, *Ampelaster*, *Canadanthus*, *Chloracantha*, *Doellingeria*,

Solidago salicina [HEA]

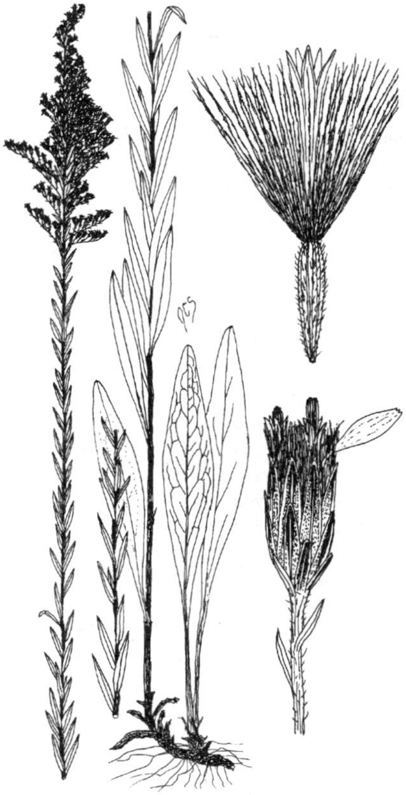

Solidago sempervirens var. mexicana [SEM]

Solidago speciosa var. rigidiuscula [BB2]

Solidago ulmifolia [YAT]

Solidago tortifolia [GLE]

Eucephalus, *Eurybia*, *Ionactis*, *Oclemena*, *Oreostemma*, *Psilactis*, *Sericocarpus*, and *Symphyotrichum*. (Greek: *symphysis*, junction, and *trichos*, hair, apparently alluding to what the author perceived as a basal fusion of pappus bristles) (tribe Astereae)

References: Shinners 1953a; Jones, A. 1978a, 1978b, 1980, 1983, 1984, 1987, 1992; Semple & Brouillet 1980a, 1980b; Jones & Young 1983; Jones, R. 1983; Semple & Chmielewski 1987; Nesom 1994b, 1997a, 2005c, 2006a; Semple et al. 1996; Yatskievych 2004; Brouillet et al. 2006.

1. Plants perennial and superficially reed-like, forming colonies from creeping rhizomes; usually some branchlets developed into thorns or spines; usually without well-developed leaves ____ see **Chloracantha**
1. Plants annual or perennial, of various habits but lacking thorns/spines; leaves present.
  2. Plants annual from a slender taproot.
    3. Heads in a dense, pyramidal paniculate array; phyllaries with narrowly-lanceolate central green zone, usually extending the entire length of the phyllary, bases hardened for a short distance or not at all; restricted to salt marshes and tidal flats of the Gulf Coast ____ **S. subulatum**
    3. Heads in open, diffuse paniculate arrays; phyllaries with a lanceolate to elliptic green zone in the upper portion, bases conspicuously hardened; across much of TX in many habitats, usually not in wet-saline sites ____ **S. divaricatum**
  2. Plants perennial from rhizomes or branched, woody roots.
    4. Uppermost leaves crowding heads, not much reduced in size in comparison to midstem leaves, longer and broader than phyllaries, partly hiding phyllaries or grading into them.
      5. Outer phyllaries 2–3 mm wide, glabrous to slightly pubescent, margins with a fringe of long hairs; achenes 5–8-nerved; ray flowers usually 13–15; sandy soils throughout East TX ____ **S. pratense**
      5. Outer phyllaries 1.3–2 mm wide, moderately to densely canescent-sericeous, margins not distinctly fringed; achenes 7–10-nerved; ray flowers usually 15–30; calcareous soils of sc TX and Edwards Plateau ____ **S. sericeum**
    4. Uppermost leaves subtending heads not crowding or hiding phyllaries, usually reduced in size in comparison to midstem leaves, often smaller than phyllaries.
      6. Basal and lower stem leaves cordate or truncate at the base, with long petiolar bases (sometimes winged) quite distinct from the much wider well-delimited blade portion.
        7. Midstem leaves usually cordate (sometimes truncate) where narrowed to petiolar base; upper stem leaves obovate to elliptic or lanceolate, not linear; middle and sometimes even upper stem leaves sharply serrate ____ **S. drummondii**
        7. Midstem leaves usually rounded or truncate or even attenuate where narrowed to petiolar base; upper stem leaves linear; middle and upper stem leaves usually entire ____ **S. oolentangiense** (in part)
      6. Basal and lower stem leaves sessile or gradually narrowed to base, not distinctly differentiated into petiolar and blade portions (if differentiated, the petiole very short).
        8. Midstem and upper leaves sessile and conspicuously to weakly auriculate-clasping; ray ligules usually some shade of lavender, blue, or purple, seldom white or pinkish.
          9. Lower stems usually densely and uniformly hirsute, with stiff, spreading hairs; phyllaries all nearly the same length, the tips long-acuminate or caudate to attenuate, strongly recurved, sometimes foliaceous; restricted to saturated soils of bogs and seeps in ne TX ____ **S. puniceum**
          9. Lower stems glabrous or pubescent with the hairs softer and appressed; phyllaries strongly unequal, the outer shorter than the inner, the tips obtuse, acute, or acuminate but not long-acuminate, appressed or only slightly spreading; of various (generally drier) habitats in East TX.
            10. Stems generally glabrous and glaucous; leaves glabrous; phyllaries without stalked glands; achenes glabrous or glabrate, 4–5-nerved ____ **S. laeve**
            10. Stems sparsely to densely pubescent; leaves scabrous to hirsute; phyllaries sometimes with stalked glands; achenes pubescent, faintly 7–10-nerved ____ **S. patens**
        8. Midstem and upper leaves sessile or petiolate, not auriculate-clasping (may be weakly clasping); ray ligules of various colors or white.
          11. Phyllaries, peduncles, and often also the leaves and upper stems glandular (use lens); basal and lower stem leaves sessile ____ **S. oblongifolium**
          11. Phyllaries, peduncles, leaves, and stems varying from glabrous to hairy, but not glandular; basal and lower stem leaves sessile to short-petiolate.
            12. Phyllaries tipped with white or clear spines; disk corollas yellow, turning brown with age.
              13. Heads numerous and densely crowded, often secund on branches; upper stem leaves as small as 10 mm long and 1.5 mm wide; involucres 2.5–4.5(–5) mm high; achenes 1.2–2 mm long; pappus 3–4 mm long ____ **S. ericoides**
              13. Heads fewer and not densely clustered, usually not secund; upper stem leaves usually longer than 25 mm and wider than 2 mm; involucres (4.5–)5–8 mm high; achenes 2–2.5 mm long; pappus 4.5–6 mm long ____ **S. falcatum**
            12. Phyllaries not tipped with white or clear spines (inrolled phyllaries may appear spine-like and green); disk corollas various shades of cream to yellow, usually turning pinkish, reddish, or purplish brown with age.
              14. Glabrous, fleshy plants of coastal salt marshes and tidal flats; stems flexuous, wiry; leaves narrow, linear; midstem leaves < 6 mm wide; restricted to the Gulf Coast ____ **S. tenuifolium**

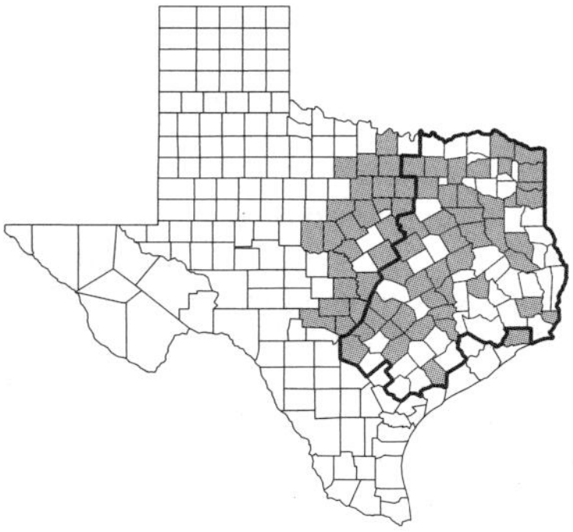
Solidago radula

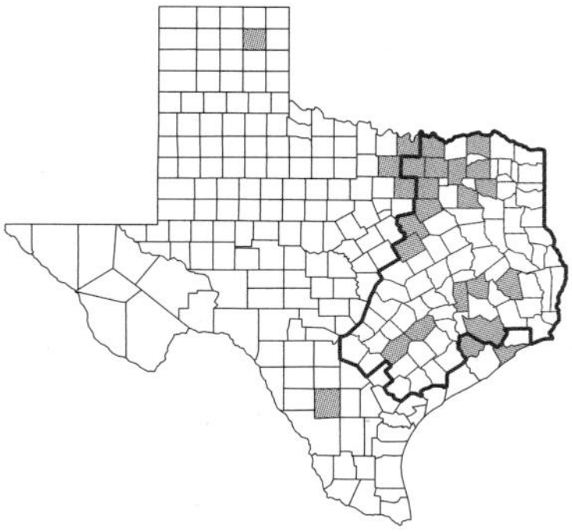
Solidago rigida

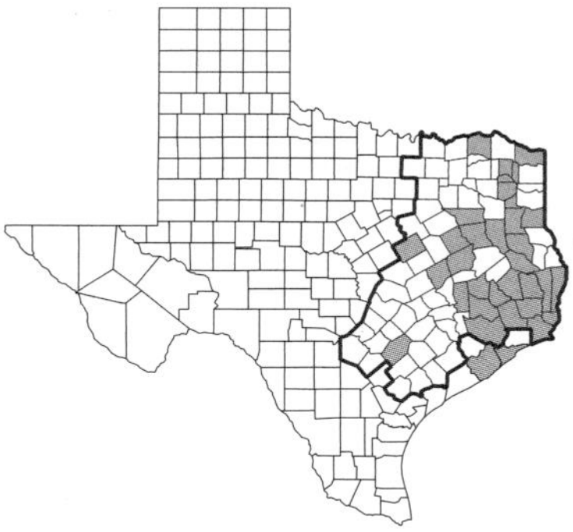
Solidago rugosa subsp. aspera

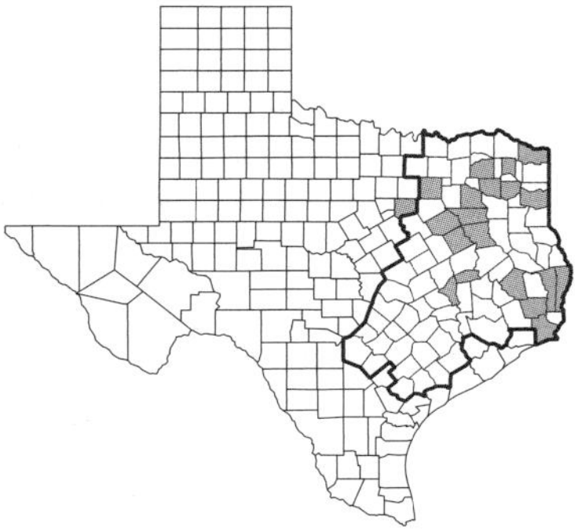
Solidago salicina

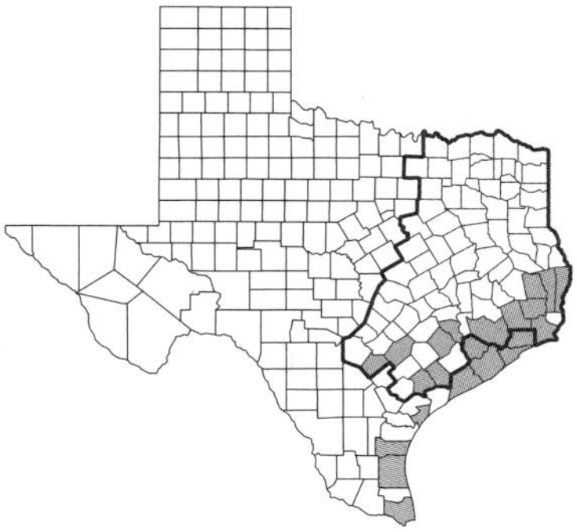
Solidago sempervirens var. mexicana

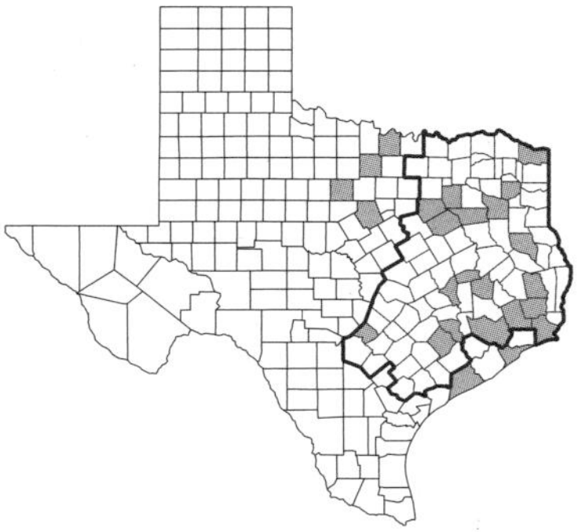
Solidago speciosa var. rigidiuscula

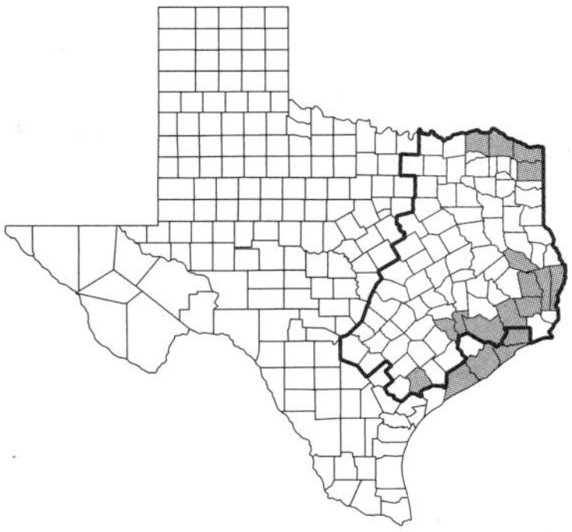
Solidago tortifolia

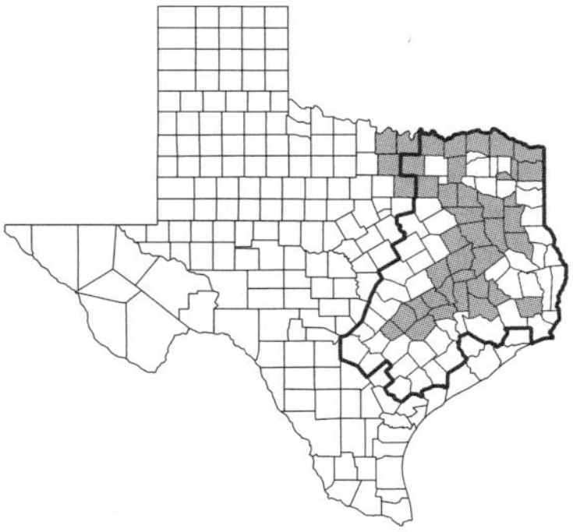
Solidago ulmifolia var. ulmifolia

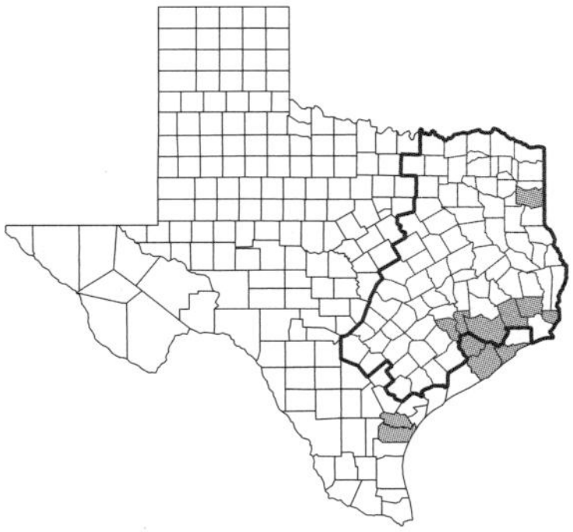
Solidago virgata

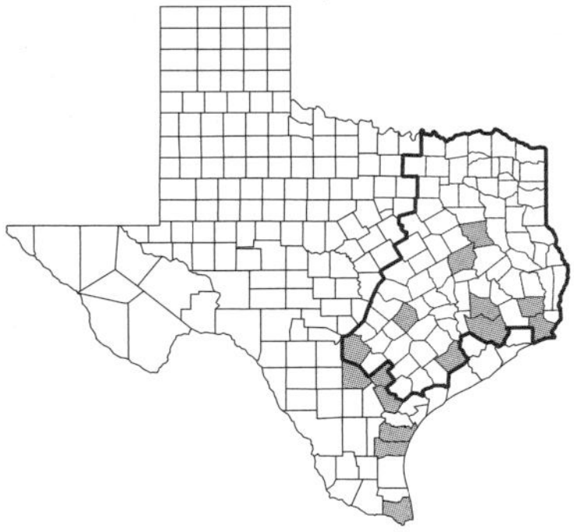
Soliva anthemifolia

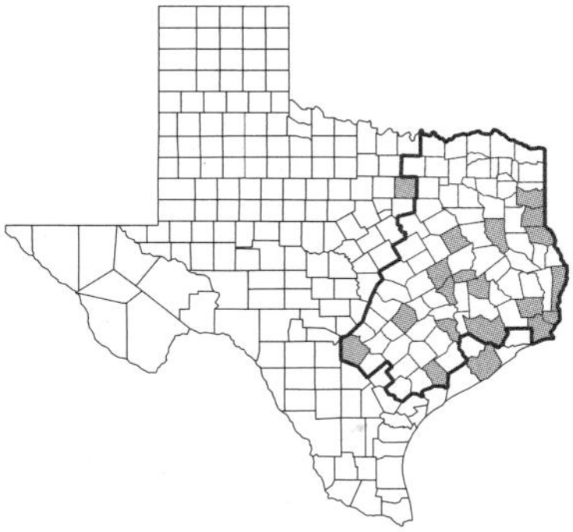
Soliva sessilis

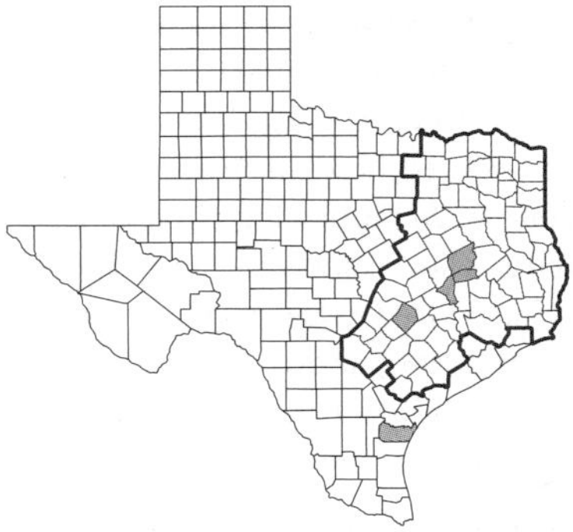
Soliva stolonifera

14. Plants variable in habit, habitat, and leaf morphology, but generally not of brackish coastal habitats.
    15. Lobes of the mature disk corolla comprising (45–)50–75% or more of the flaring portion above the tube (= limb), flaring or recurved.
        16. Leaf blades glabrous beneath or pubescent only on veins; ray flowers 8–15(–23); lobes of the mature disk corolla comprising 50–75% of the flaring portion above the tube (= limb) ______ **S. lateriflorum**
        16. Leaf blades sparsely to densely pubescent beneath, on and between veins; ray flowers (10–)15–26; lobes of the mature disk corolla comprising 45–55(–65)% of the flaring portion above the tube (= limb) ______ **S. ontarionis**
    15. Lobes of the mature disk corolla comprising 15–45% of the flaring portion above the tube (= limb), erect to ascending or sometimes recurved.
        17. Midstem leaves usually < 10 mm wide (often linear and < 3 mm wide).
            18. Upper stems glabrate to strigose; peduncular bracts numerous, crowded, < 4 mm long; involucres usually 4.5–6.3 mm high; phyllaries of the inner series twice as wide as those of the outer series; ray flowers usually 15–33, the ligules some shade of blue, pink, or lavender, less frequently white; pappus ca. 4 mm long ______ **S. dumosum**
            18. Upper stems glabrous to glabrate; peduncular bracts fewer, not crowded, > 4 mm long; involucres usually 3.5–4.5 mm high; phyllaries of the inner series < twice as wide as those of the outer series; ray flowers usually 16–20, the ligules usually white, rarely pink; pappus 2.5–3.5 mm long ______ **S. racemosum**
        17. Midstem leaves usually > 10 mm wide, not linear.
            19. Phyllaries not greatly differentiated in size, the outer smallest ones 1/2 or more as long as the longest inner ones, OR if phyllaries more strongly graduated, ONLY THEN should the following characters be evident: peduncles ± pilose; peduncular bracts 1–3(–5) and 4+ mm wide; ray ligules usually < 8 mm long.
                20. Leaf margins flat (may appear revolute in dried material); reticulate brownish veins usually conspicuous on lower leaf surface (use lens), the enclosed areolae ± isodiametric; peduncular bracts 5–12; ray flowers usually 20–35, the ligules usually bluish white or lavender (rarely white); achenes brown or purple at maturity ______ **S. praealtum**
                20. Leaf margins ± revolute; reticulate brownish veins not conspicuous on the lower leaf surface, the inconspicuous areolae not isodiametric; peduncular bracts 1–3(–5); ray flowers usually 17–47, the ligules usually white; achenes gray or tan at maturity ______ **S. lanceolatum**
            19. Phyllaries more differentiated in size, the outer smallest ones 1/3 or less as long as the longest inner ones, OR if phyllaries less strongly graduated, ONLY THEN should the following characters be evident: peduncles glabrous; peduncular bracts more than 5 and > 2 mm wide; ray ligules usually > 8 mm long.
                21. Plants long-rhizomatous and colonial; stems glabrous; leaves glabrous, margins ± revolute; peduncular bracts 5–10, not crowded, leaf-like; ray flowers usually 27–36; pappus 4.5–5.8 mm long ______ **S. eulae**
                21. Plants cespitose from a branched, woody caudex, or short-rhizomatous; upper stems usually with hairs in lines; leaves pilose-strigose or hirsute beneath, margins flat; peduncular bracts numerous, densely crowded, scale-like; ray flowers usually 13–20; pappus 3–4 mm long ______ **S. oolentangiense** (in part)

Symphyotrichum divaricatum (Nutt.) G.L. Nesom, (spreading, for the growth pattern), WIREWEED, BLACKWEED, SLIM ASTER. Taprooted annual, (10–)60–100(–200+) cm tall; leaves glabrous; basal and lower stem leaves withering by flowering; midstem leaves petiolate or sessile, blades narrowly lanceolate to subulate, margins entire or rarely with a few teeth; uppermost leaves much reduced; heads in open paniculate arrays; involucres usually 6–6.9 mm high; phyllaries unequal, bases hardened, margins hyaline, often purple-tinged; ray flowers usually 17–30, the ligules (3.5–)4.5–7 mm long, lavender to pale blue or white; disk corollas yellow, sometimes tinged with purple with age; achenes usually 1.7–2.5 mm long, 5-nerved; pappus white, (3–)3.5–4.6 mm long. Weed in lawns, ditches, disturbed moist areas, roadsides; abundant throughout TX; sc U.S. from VA s to AL and w to KS and NM; Mexico. (Aug–)Sep–Dec. [*Aster divaricatus* (Nutt.) Torr. & A. Gray [nom. illeg.], *A. neomexicanus* Wooton & Standl., *A. subulatus* var. *ligulatus* Shinners, *S. subulatum* (Michx.) G.L. Nesom var. *ligulatum* S.D. Sundberg] We are following Nesom (2005c) and Turner (2016) in treating this taxon at the specific, rather than varietal level.

Symphyotrichum drummondii (Lindl.) G.L. Nesom, (for its discoverer, Thomas Drummond, 1780–1835, Scottish botanist and collector in North America), TEXAS ASTER, DRUMMOND'S ASTER. Perennial with short rhizomes, 30–120 cm tall; basal and lower stem leaves withering by flowering; midstem leaves ± petiolate, petioles winged, blades ovate to oblong or linear-lanceolate, bases

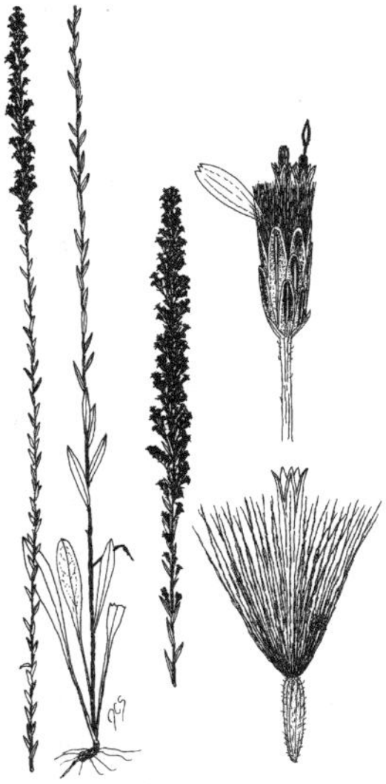

Solidago virgata [SEM]

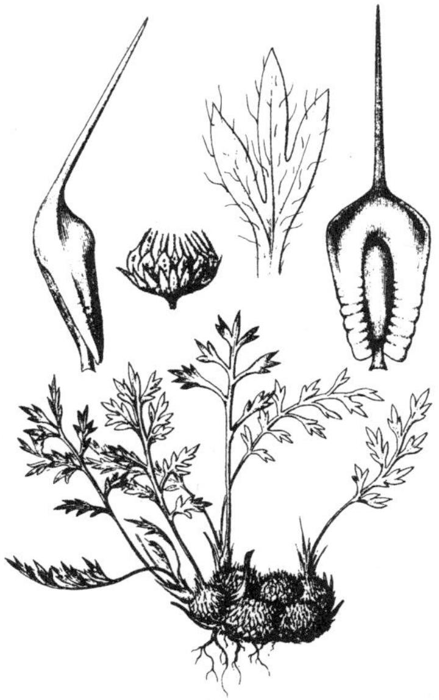

Soliva anthemifolia [MAR]

Soliva sessilis [FNA]

Soliva stolonifera [HEA]

Sonchus asper [REE]

Sonchus oleraceus [REE]

Sphagneticola trilobata [DHA]

Stokesia laevis [ALI]

cordate to cuneate, margins crenate to serrate; uppermost leaves reduced; heads in open paniculate arrays; involucres usually 4.5–6 mm high; phyllaries unequal, bases hardened, margins hyaline or sometimes bodies green to margins, tips often purple; ray flowers usually 10–15, the ligules (5–)7–10(–12) mm long, usually blue-lavender to deep purple (sometimes white); disk corollas cream to yellow, turning reddish purple with age; achenes (1.5–)2–3 mm long, 5–6-nerved; pappus cream to pinkish, 3–4 mm long. Two varieties exist; 1 is common in East TX and 1 is to be expected. Brouillet et al. (2006) separated the 2 varieties as follows:

1. Plants 40–120 cm tall; lower stems sparsely hirsute; leaves firm; involucres campanulate, usually 4.5–7 mm high; ligules blue-lavender to deep purple (sometimes white); achenes glabrous ____ var. *drummondii*
1. Plants 30–80 cm tall; lower stems glabrate to densely hirsute; leaves membranous, involucres hemispheric, usually 3.8–5.2 mm high; ligules bluish white; achenes finely pubescent ____ var. **texanum**

var. *drummondii*, DRUMMOND'S ASTER. Open woods, stream banks, roadside ditches; to be expected in East TX based on records from adjacent counties/parishes in other states: Caddo Par., LA (LSU), and Marshall, Bryan, Choctaw, and McCurtain cos., OK (OKL); Midwest and sc U.S. from PA s to AL and w to MN and TX. Sep–Oct. [*Aster drummondii* Lindl., *A. sagittifolius* Wedemeyer ex Willd. var. *drummondii* (Lindl.) Shinners]

var. **texanum** (Burgess) G.L. Nesom, (of Texas), TEXAS ASTER. Open woods and prairies, bottomlands; e and c TX to Edwards Plateau; sc U.S. from KY s to AL and w to KS and TX; Mexico. Oct–Nov, rarely Mar–Apr. [*Aster drummondii* Lindl. var. *texanus* (Burgess) A.G. Jones, *A. texanus* Burgess, *A. texanus* Burgess var. *parviceps* Shinners, *S. drummondii* (Lindl.) G.L. Nesom var. *parviceps* (Shinners) G.L. Nesom]

Symphyotrichum dumosum (L.) G.L. Nesom, (bushy), BUSHY ASTER, RICE-BUTTON ASTER. Rhizomatous perennial, (20–)30–100 cm tall; leaves scabrous, tipped with a small white spine; basal and lower stem leaves withering by flowering; midstem leaves sometimes withering by flowering, sessile or subpetiolate with petioles widely winged and clasping, blades linear-oblanceolate, bases cuneate, margins crenate-serrate; uppermost leaves minute, bract-like; heads in open paniculate arrays; involucres usually 4.5–6.3 mm high; phyllaries strongly unequal, bases hardened, margins hyaline, tips sometimes reddish; ray flowers 15–33, the ligules (4–)5–7(–8) mm long, pale blue, pink, or lavender to white; disk corollas cream to pale yellow, turning pink with age; achenes 1.5–2.5 mm long, prominently 3–4-nerved; pappus whitish, 4 mm long. Low areas, roadsides, marshes, prairies, moist woods and thickets; mostly e 1/4 of TX; e 1/2 of the U.S.; Canada. Sep–Nov. [*Aster dumosus* L., *A. coridifolius* Michx., *A. dumosus* var. *coridifolius* (Michx.) Torr. & A. Gray] We are following Brouillet et al. (2006) in not distinguishing varieties.

Symphyotrichum ericoides (L.) G.L. Nesom var. **ericoides,** (resembling *Erica*—HEATH), HEATH ASTER, WHITE PRAIRIE ASTER, WREATH ASTER. Rhizomatous, colonial perennial, usually much-branched, 20–80(–100) cm tall; leaves sessile, usually entire, ± white spine-tipped; basal and lower stem leaves withering by flowering; midstem leaves usually withering by flowering, linear to lanceolate to oblong, strigose or hirsute, bases cuneate, margins coarsely ciliate; uppermost leaves much reduced; heads in crowded paniculate arrays; involucres 2.5–4.5(–5) mm high; phyllaries unequal, bases hardened, margins hyaline, spine-tipped; ray flowers usually 10–18, the ligules 6–12(–20) mm long, white, rarely pink or bluish; disk corollas yellow, turning brown with age; achenes 1.2–2 mm long, faintly 7–9-nerved; pappus whitish, 3–4 mm long. Disturbed or open areas; widespread in TX; e and c North America; Mexico. Sep–Oct(–Nov). [*Aster* ericoides L., *A. exiguus* (Fernald) Rydb., *A. multiflorus* Aiton, *A. polycephalus* Rydb., *S. ericoides* (L.) G.L. Nesom var. *prostratum* (Kuntze) G.L. Nesom]

Symphyotrichum eulae (Shinners) G.L. Nesom, (for Eula Whitehouse, 1892–1974, of Southern Methodist Univ., TX collector, artist, and author of *Texas Flowers in Natural Color*), EULA'S ASTER. Rhizomatous, colonial perennial, 5–150 cm tall; leaves scabrous, margins ± revolute; basal and lower stem leaves withering by flowering; midstem leaves petiolate or nearly sessile and clasping, blades elliptic to lanceolate, bases attenuate to cuneate, margins crenate-serrate; uppermost leaves reduced, sessile, entire; heads in open paniculate arrays; involucres usually 5–6.5 mm high; phyllaries unequal, bases hardened, margins hyaline; ray flowers usually 27–36, the ligules 8–11 mm long, usually white to bluish lavender, rarely pinkish; disk corollas yellow, turning reddish brown with age; achenes 1.4–2.9 mm long, 4–5-nerved; pappus whitish, 4.5–5.8 mm long. Shady low areas, disturbed sites, moist woods and thickets; East TX range according to Kartesz (2015) from nc TX in Dallas, Hunt, and Wood cos. to sc TX in Bexar and Travis cos.; TEX-LL holds specimens from Burleson, Walker, Waller, and Washington cos.; the only *Symphyotrichum* species endemic to TX. Oct–Nov. [*Aster eulae* Shinners] This species was formerly considered a hybrid of *S. lanceolatum* and *S. praealtum* but we are following Brouillet et al. (2006) in not treating it as such here.

Symphyotrichum falcatum (Lindl.) G.L. Nesom var. **commutatum** (Torr. & A. Gray) G.L. Nesom, (sickle-shaped, for the leaves), WHITE PRAIRIE ASTER. Rhizomatous, colonial perennial, 10–80 cm tall, sometimes cespitose, usually densely hairy; leaves entire,

Symphyotrichum divaricatum [BB2]

Symphyotrichum drummondii var. texanum [HEA]

Symphyotrichum dumosum [HEA]

Symphyotrichum ericoides [DEL]

Symphyotrichum eulae [GOR]

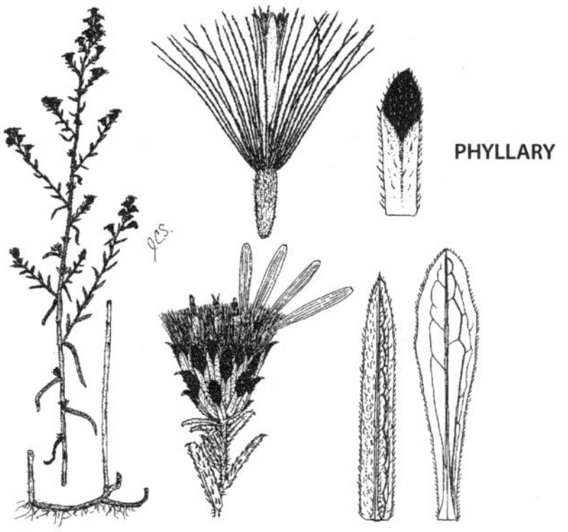

Symphyotrichum falcatum var. commutatum [UWA]

Symphyotrichum laeve var. purpuratum [BB2]

sessile, ± spine-tipped; basal leaves withering by flowering; midstem leaves strigose, blades linear-oblong to linear-lanceolate, bases cuneate; uppermost leaves somewhat reduced; heads in diffuse paniculate or racemose arrays; involucres (4.5–)5–8 mm high; phyllaries unequal, bases hardened, margins hyaline, tips strongly reflexed; ray flowers usually 20–35, the ligules (8–)18–30 mm long, usually white, sometimes blue or pink; disk corollas yellow, turning brown with age; achenes 2–2.5 mm long, faint-nerved; pappus whitish, 4.5–6 mm long. Dry soils, prairies, roadsides, stream banks; recorded in Hardin and Walker cos. in se TX, and Bastrop and Williamson cos. on the edge of Edwards Plateau; mostly w TX and Panhandle; c U.S. from WI s to TX and w to MT and AZ; Canada; Mexico. Aug–Nov. [*Aster commutatus* (Torr. & A. Gray) A. Gray, *A. crassulus* Rydb., *A. falcatus* Lindl. subsp. *commutatus* (Torr. & A. Gray) A.G. Jones, *A. multiflorus* Aiton var. *commutatus* Torr. & A. Gray, *S. falcatum* (Lindl.) G.L. Nesom var. *crassulum* (Rydb.) G.L. Nesom]

Symphyotrichum laeve (L.) Á. Löve & D. Löve, (smooth, for the stems and foliage), SMOOTH ASTER, SMOOTH BLUE ASTER. Cespitose, short-rhizomatous perennial, (15–)20–70(–120) cm tall; leaves glabrous, glaucous; basal and lower stem leaves withering by flowering; midstem leaves sessile or nearly so, blades ovate-oblong or lance-ovate to linear, or sometimes ± panduriform, bases attenuate to rounded or auriculate-clasping, margins entire or shallowly serrate-crenate; uppermost leaves reduced; heads in broad or flattened paniculate arrays; involucres usually 5–7 mm high; phyllaries ± unequal, bases hardened, margins hyaline; ray flowers usually 13–23, the ligules (6–)7.2–11.3(–14.6) mm long, pale to dark blue or purple, seldom white; disk corollas yellow, turning purplish red with age; achenes 2–3.5 mm long, 4–5-nerved; pappus tawny to red-tinged, 5–7 mm long. Dry prairies, open deciduous woods. Sep–Oct. We are following Brouillet et al. (2006) in separating the 2 potential East TX varieties as follows:

1. Leaves lanceolate to ovate, conspicuously auriculate-clasping, lengths < 5 times widths ____ var. *laeve*
1. Leaves linear-lanceolate to linear, slightly auriculate-clasping, lengths often > 5 times widths ____ var. **purpuratum**

var. *laeve*. *To be expected in ne TX based on records from adjacent counties in other states*: Le Flore and McCurtain cos., OK (A. Jones 1992); e 1/2 of North America; cultivated and introduced/escaping elsewhere. [*Aster laevis* L.]

var. **purpuratum** (Nees) G.L. Nesom, (purple). Jasper, Polk, and Tyler cos. in se TX; se U.S. from GA w to AR and TX. [*Aster attenuatus* Lindl., *A. laevis* L. var. *purpuratus* (Nees) A.G. Jones, *A. purpuratus* Nees]

Symphyotrichum lanceolatum (Willd.) G.L. Nesom, (lance-shaped), PANICLED ASTER, TALL WHITE ASTER, LANCE-LEAVED ASTER. Rhizomatous, colonial perennial, 30–150(–200) cm tall; leaves glabrous or sparsely scabrous; basal and lower stem leaves withering by flowering; midstem leaves sessile or nearly so, withering by flowering, blades lance-ovate to linear-lanceolate, bases cuneate, margins serrate; uppermost leaves slightly reduced, entire; heads in leafy paniculate arrays; involucres 4–7.2 mm high; phyllaries ± unequal, bases hardened, margins hyaline; ray flowers 17–47, the ligules 3.2–10.1 mm long, usually white to pinkish to pale blue-violet; disk corollas yellow, turning purple with age; achenes 1.5–2 mm long, 4–5-nerved; pappus white to tawny, 3.7–6.4 mm long. Low thickets, moist prairies, edges of deciduous woods, roadside ditches. Aug–Nov. We are following Semple and Chmielewski (1987) in separating the 2 East TX subspecies as follows:

1. Outer phyllaries at least 2/3 length of inner ones; heads usually subtended by large leafy bracts ____ subsp. **hesperium**
1. Outer phyllaries 1/3–2/3 length of inner ones; heads not usually subtended by large leafy bracts ____ subsp. **lanceolatum**

subsp. **hesperium** (A. Gray) G.L. Nesom, (of the west), SISKIYOU ASTER. Harris, Jasper, Liberty, and Van Zandt cos.; sparsely distributed across TX; w 2/3 of the U.S; Canada; Mexico. [*Aster hesperius* A. Gray, *A. lanceolatus* Willd. subsp. *hesperius* (A. Gray) Semple & Chmielewski, *Symphyotrichum hesperium* (A. Gray) Á. Löve & D. Löve]

subsp. **lanceolatum,** PANICLED ASTER, TALL WHITE ASTER, LANCE-LEAVED ASTER. Scattered across e 1/2 TX; e 2/3 of the U.S.; Canada. [*Aster lanceolatus* var. *simplex* (Willd.) A.G. Jones, *A. simplex* Willd., *Symphyotrichum simplex* (A. Gray) Á. Löve & D. Löve] This subspecies is sometimes further divided into 4 varieties (Brouillet et al. 2006).

Symphyotrichum lateriflorum (L.) Á. Löve & D. Löve, (with one-sided inflorescences), CALIFORNIA ASTER, CALICO ASTER, SIDE-FLOWER ASTER, STARVED ASTER, WHITE WOODLAND ASTER. Cespitose perennial, 20–120(–150) cm tall; new basal rosettes often forming at flowering; leaves glabrate to scabrous; lower stem leaves sometimes persisting into flowering; midstem leaves sessile or subpetiolate with petioles winged and sheathing, blades ovate to elliptic or lanceolate, bases attenuate to cuneate, margins crenate-serrate; uppermost leaves gradually reduced; heads in pyramidal or spicate arrays; involucres usually 4–6 mm high; phyllaries unequal, bases hardened, margins hyaline or sometimes reddish; ray flowers 8–15(–23), the ligules (3–)4–5(–8) mm long, white, rarely pinkish or purplish; disk corollas cream to light yellow, turning pink or reddish purple with age; achenes (1.3–)1.8–2.2 mm long, 3–5-nerved; pappus white to pinkish, 3–4 mm long. Well-drained uplands, deciduous woods, meadows,

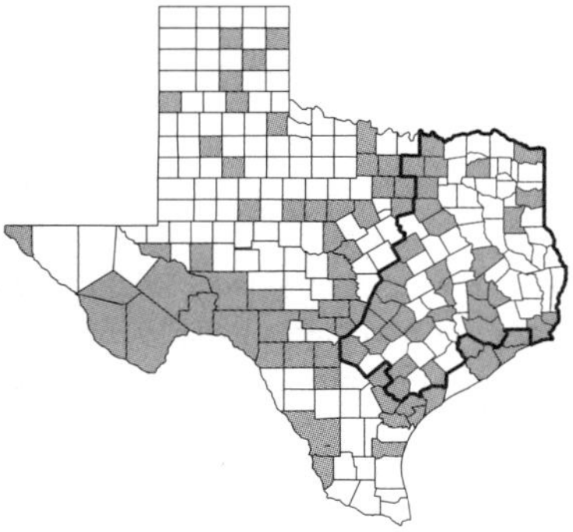

Sonchus asper

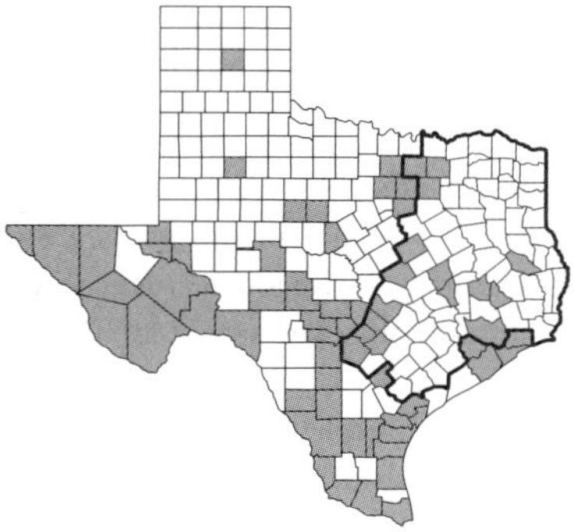

Sonchus oleraceus

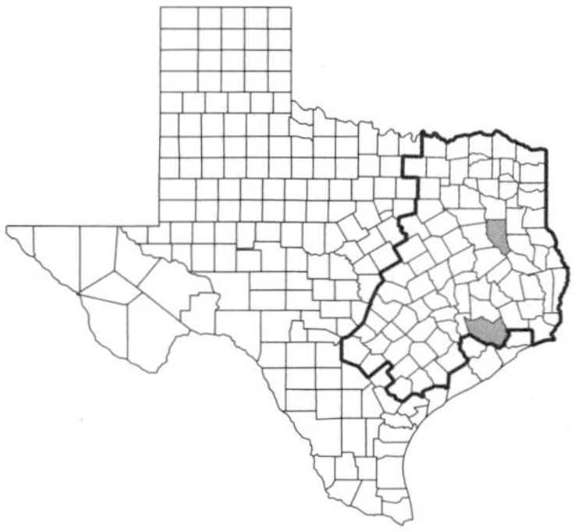

Sphagneticola trilobata

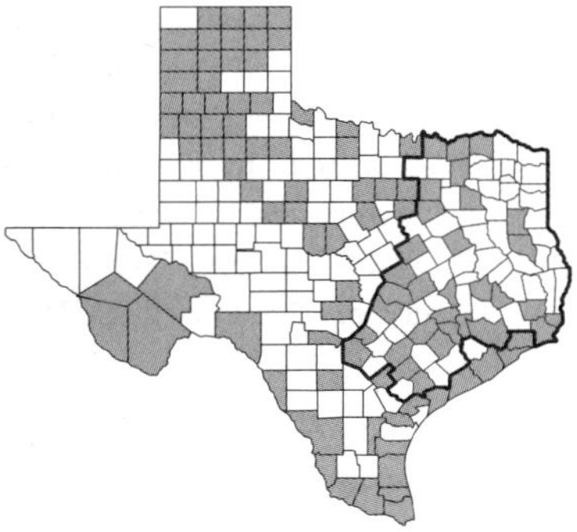

Symphyotrichum divaricatum

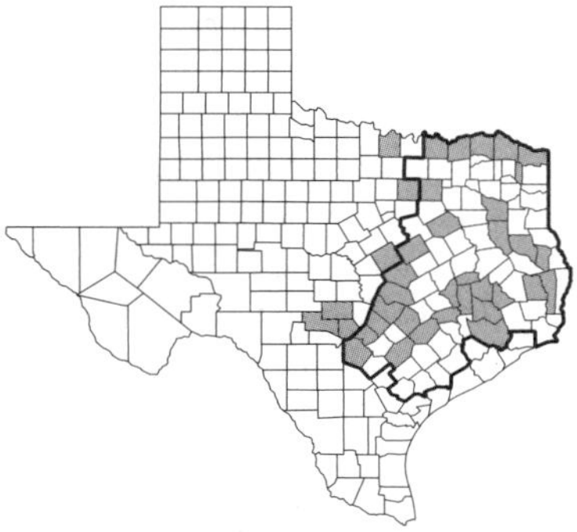

Symphyotrichum drummondii var. texanum

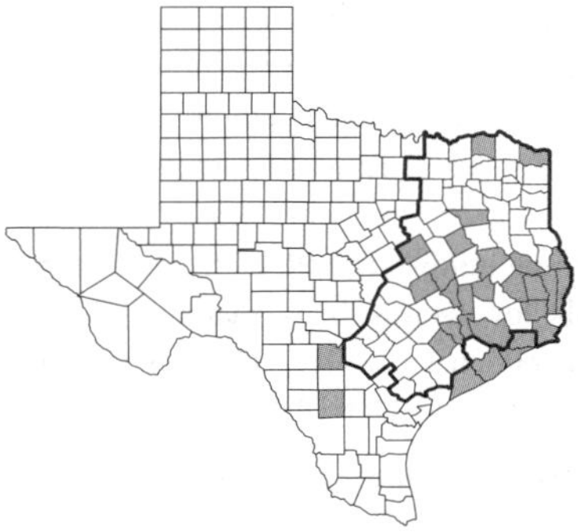

Symphyotrichum dumosum

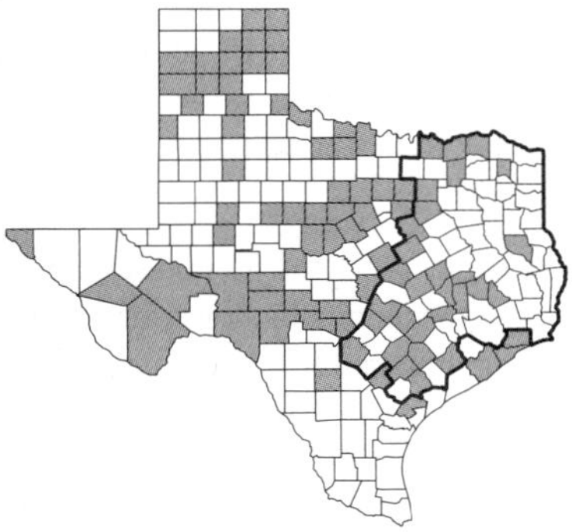

Symphyotrichum ericoides var. ericoides

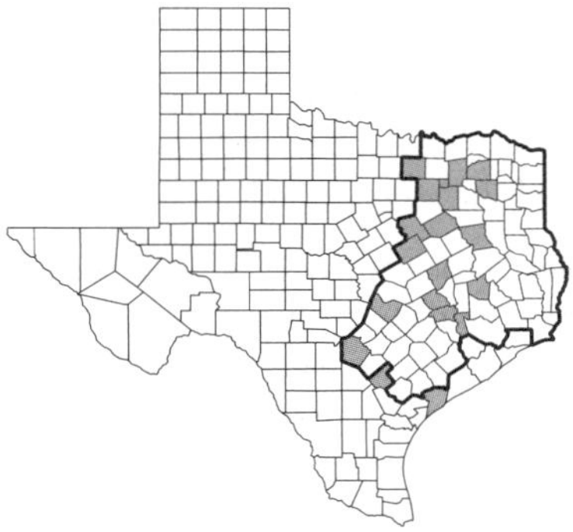

Symphyotrichum eulae

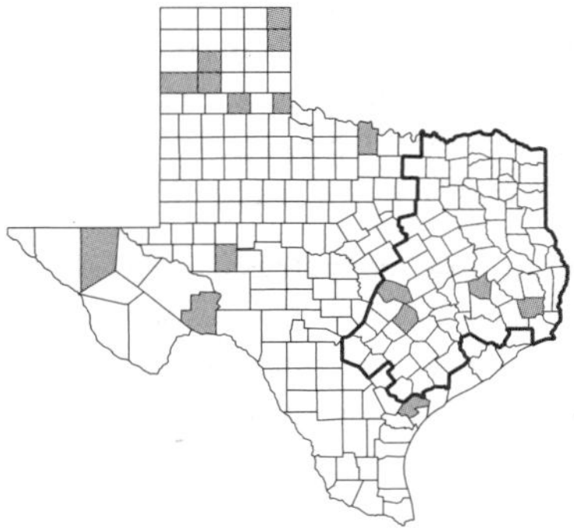

Symphyotrichum falcatum var. commutatum

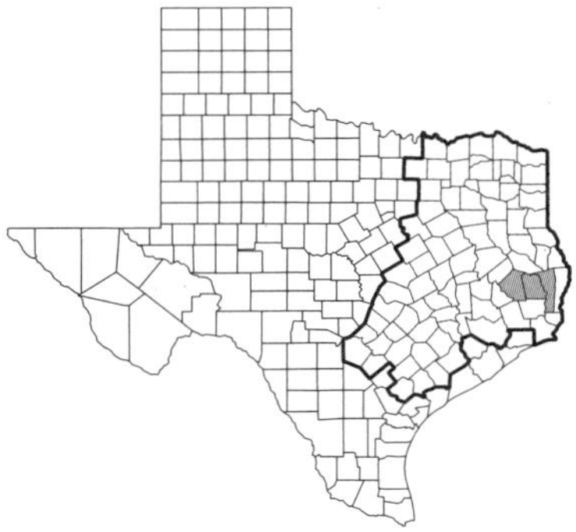

Symphyotrichum laeve var. purpuratum

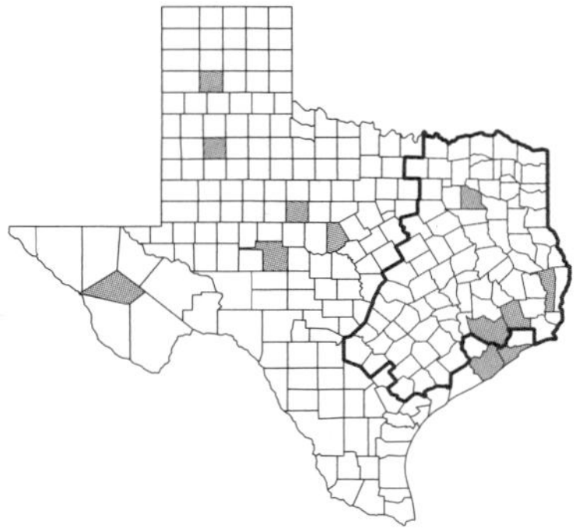

Symphyotrichum lanceolatum subsp. hesperium

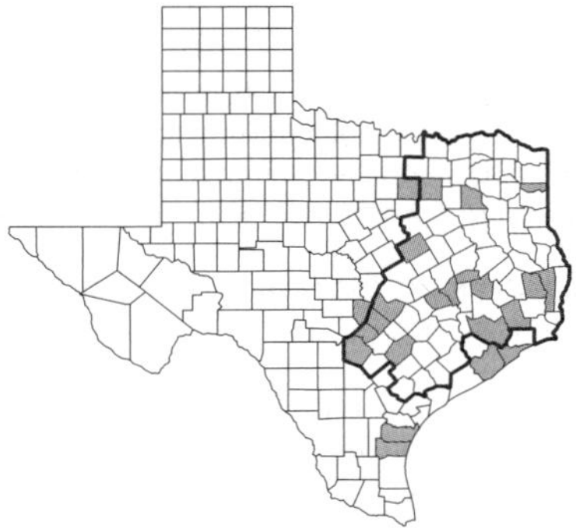

Symphyotrichum lanceolatum subsp. lanceolatum

roadsides; e 1/3 TX; e 1/2 of North America. Sep–Nov. [*Aster lateriflorus* (L.) Britton var. *flagellaris* Shinners, *A. lateriflorus* (L.) Britton var. *indutus* Shinners, *S. lateriflorum* (L.) Á. Löve & D. Löve var. *hirsuticaule* (Lindl. ex DC.) G.L. Nesom] We are following Brouillet et al. (2006) in not recognizing varieties.

Symphyotrichum oblongifolium (Nutt.) G.L. Nesom, (oblong-leaved), AROMATIC ASTER, OBLONG-LEAF ASTER. Rhizomatous, colonial perennial, 10–80(–100) cm tall; leaves entire, usually hirsute, rarely glabrous; basal and lower stem leaves sometimes persisting into flowering; midstem leaves sessile, blades oblong to linear-lanceolate, bases cuneate or rounded to slightly clasping; uppermost leaves much reduced; heads in diffuse paniculate arrays; involucres usually 7–9 mm high; phyllaries ± unequal, bases hardened, margins hyaline; ray flowers usually 25–35, the ligules 9–13(–15) mm long, rose-purple to lavender or deep purple; disk corollas yellow, turning brown or reddish purple with age; achenes 2–2.5 mm long, 7–10-nerved; pappus tawny, 3.5–6 mm long. Calcareous soils, prairies, outcrops, roadsides; w border of East TX from Denton to Travis to Bexar cos.; mostly nc and c TX; c and e U.S.; Mexico. Sep–Nov. [*Aster oblongifolius* Nutt.]

Symphyotrichum ontarionis (Wiegand) G.L. Nesom var. **ontarionis,** (of Ontario), ONTARIO ASTER, BOTTOMLAND ASTER. Rhizomatous, colonial perennial, 20–120 cm tall; leaves usually strigose, margins crenate-serrate or entire; basal and lower stem leaves usually withering by flowering; midstem leaves sessile or petiolate with petioles narrowly winged and ± clasping, blades ovate or lance-ovate to elliptic-lanceolate or oblanceolate, bases attenuate or cuneate; uppermost leaves progressively reduced; heads in open paniculate arrays; involucres 3–5.5 mm high; phyllaries unequal, bases hardened, margins hyaline; ray flowers (10–)15–26, the ligules 3.5–5.5(–8) mm long, white, rarely pinkish to lavender or blue; disk corollas cream or light yellow, turning reddish purple with age; achenes 1.2–1.8(–2) mm long, 3–5-nerved; pappus whitish, 3–3.5 mm long. Damp alluvial soils, low woods, bogs, marshes; Gonzales Co. (*F.A. Barkley & B.C. Tharp 13968* TEX), Jasper Co. (*D.S. Correll 28634* LL), Nacogdoches Co. (*B.C. Tharp & W.J. Brown 53-130* TEX), Red River Co. (*D.S. Correll & H.B. Correll 24801* LL); e and c North America. Aug–Oct. [*Aster ontarionis* Wiegand, *A. missouriensis* Britton var. *thyrsoides* (A. Gray) Wiegand]

Symphyotrichum oolentangiense (Riddell) G.L. Nesom, (of Oolentangy [Olentangy] River in Ohio), AZURE ASTER, SKY-BLUE ASTER. Cespitose perennial, 20–150 cm tall; leaves scabrous; basal and lower stem leaves usually withering by flowering; midstem leaves sessile or short-petiolate, petioles slightly winged and sheathing, blades ovate to lanceolate, bases rounded or attenuate, margins crenate-serrate to nearly entire; uppermost leaves reduced; heads in diffuse paniculate arrays; involucres 4.5–8 mm high; phyllaries strongly unequal, bases hardened, margins hyaline; ray flowers usually 13–20, the ligules (5–)8–12(–14) mm long, blue to violet-purple, rarely pinkish or white; disk corollas light yellow, turning deep purple with age; achenes 1.8–2 mm long, 4–5-nerved; pappus cream or pinkish, 3–4 mm long. Dry sandy areas, prairies, open deciduous woods; scattered in e 1/4 of TX; e and c U.S.; Canada; Mexico. Sep–Nov. [*Aster azureus* Lindl., *A. oolentangiensis* Riddell]

Symphyotrichum patens (Aiton) G.L. Nesom, (spreading), SPREADING ASTER, LATE PURPLE ASTER, SKY-DROP ASTER. Cespitose perennial, 10–100(–120) cm tall; leaves scabrous, margins entire to finely serrate; basal leaves usually withering by flowering; midstem leaves sessile, blades ovate to lanceolate, bases strongly cordate to auriculate and clasping; uppermost leaves much reduced, appressed; heads in paniculate arrays; involucres 5.5–12 mm high; phyllaries strigillose to canescent, strongly unequal, bases hardened, margins hyaline, tips sometimes glandular-pubescent; ray flowers 12–24+, the ligules 10–18(–20) mm long, light lavender-violet, rarely pinkish or white; disk corollas creamy-white or pale yellow, turning purple with age; achenes 2–3.5 mm long, faintly 7–10-nerved; pappus tawny or pinkish, 4.5–6.5 mm long. Disturbed or open, often sandy areas, open woods, roadsides. Sep–Nov. We are following Nesom (2006a) in distinguishing only 2 varieties, 1 common in East TX, and 1 potentially present in ne TX:

1. Involucres 8–12 mm high, broadly turbinate; phyllaries of middle series ovate-lanceolate, 1.2–1.7 mm wide, obtuse, not squarrose, eglandular or sparsely glandular ____ var. *patentissimum*
1. Involucres usually < 8 mm high, campanulate or slenderly turbinate; phyllaries of middle series linear-lanceolate, usually < 1.2 mm wide, acute to acuminate, often at least somewhat squarrose, usually distinctly glandular ____ var. **patens**

var. **patens.** SKY-DROP ASTER. Widespread in e 1/3 TX, w to West Cross Timbers and Edwards Plateau; e and sc U.S. from NH s to FL and w to KS and TX. [*Aster patens* Aiton, *S. patens* (Aiton) G.L. Nesom var. *gracile* (Hook.) G.L. Nesom]

var. *patentissimum* (Lindl. ex DC.) G.L. Nesom, (much-spreading). *To be expected in ne TX based on records from adjacent counties/parishes in other states*: Little River Co., AR (UARK) and McCurtain Co., OK (OKL); c U.S. from IL s to MS and w to KS and OK. [*Aster patens* Aiton var. *patentissimus* (Lindl. ex DC.) Torr. & A. Gray]

Symphyotrichum praealtum (Poir.) G.L. Nesom, (very tall), WILLOW-LEAF ASTER, WILL ASTER, TALL ASTER. Rhizomatous, colonial perennial, (10–)50–150(–220) cm tall; upper stems usually with pubescence in lines; leaves scabrous, margins often revolute;

Syphyotrichum lanceolatum subsp. hesperium [UWA]

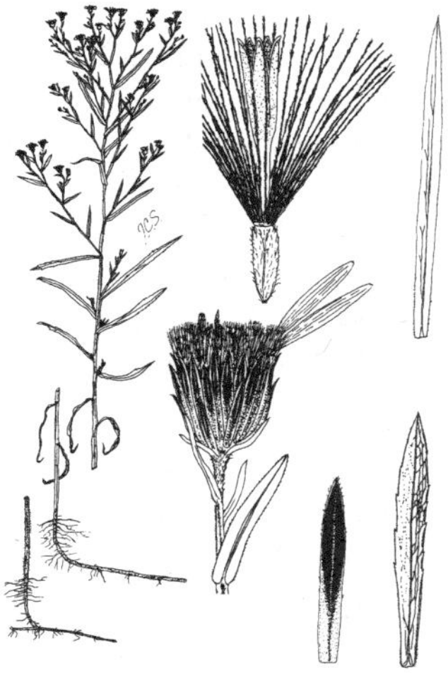

Symphyotrichum lanceolatum subsp. lanceolatum [UWA]

Symphyotrichum lateriflorum [STE]

Symphyotrichum oblongifolium [GLE]

Symphyotrichum ontarionis [HOL]

Symphyotrichum oolentangiense [STE]

Symphyotrichum patens var. patens [STE]

Symphyotrichum praealtum [YAT]

basal leaves usually withering by flowering; midstem leaves sessile or nearly so, blades linear-lanceolate to elliptic or oblanceolate, bases cuneate or rounded; uppermost leaves reduced, linear to subulate; heads in dense and leafy paniculate arrays; involucres usually 5–7 mm high; phyllaries unequal, bases hardened, margins scarious or hyaline, tips sometimes reddish purple; ray flowers usually 20–35, the ligules 5–10(–12) mm long, light lavender-violet, bluish, or pinkish, rarely white; disk corollas cream or pale yellow, turning pinkish purple with age; achenes 1.5–2 mm long, 4–5-nerved; pappus white, 4–6.5 mm long. Low disturbed areas, open woods, moist banks, roadsides; scattered across East TX; w to Rolling Plains and s to Edwards Plateau and s TX; mostly e 1/2 of North America; Mexico; cultivated and escaping elsewhere. (Sep–)Oct–Nov. [*Aster coerulescens* DC., *A. praealtus* var. *coerulescens* (DC.) A.G. Jones, *A. praealtus* var. *texicola* Wiegand, *A. salicifolius* Aiton] We are following Brouillet et al. (2006) in not recognizing varieties.

Symphyotrichum pratense (Raf.) G.L. Nesom, (of the meadows), SILKY ASTER. Cespitose perennial, 40–60 cm tall; leaves sessile, glabrous or moderately strigose or scabrous, margins usually entire; basal and lower stem leaves withering by flowering; midstem leaves with blades oblong-lanceolate, bases cuneate or rounded and subclasping; uppermost leaves not much reduced; heads in open paniculate arrays; involucres 6–11 mm high; phyllaries leaf-like (at least the outer series), subequal, glabrous to sparsely pubescent, margins fringed with long hairs, bases ± hardened, tips sometimes purplish; ray flowers usually 13–15, the ligules 8–13 mm long, rose-purple to blue-violet; disk corollas pink, turning purple with age; achenes ca. 3 mm long, 5–8-nerved; pappus tan, 3.5–6 mm long. Open woods, fields, roadsides, usually sandy soils; widespread in e 1/2 TX; se U.S. from VA s to FL and w to OK and TX. Sep–Nov. [*Aster phyllolepis* Torr. & A. Gray, *A. pratensis* Raf.] Differences between this species and the closely-related *S. sericeum* (Vent.) G.L. Nesom are discussed in R. Jones et al. (2008).

Symphyotrichum puniceum (L.) Á. Löve & D. Löve var. **scabricaule** (Shinners) G.L. Nesom, (sp.: purple; var.: rough-stemmed), PURPLESTEM ASTER, ROUGH-STEMMED ASTER, SWAMP ASTER. Cespitose perennial, 100–250(–300) cm tall, sometimes rhizomatous; lower stems densely and uniformly hirsute; leaves scabrous, margins entire or remotely toothed, abaxial midveins pilose; basal and lower stem leaves withering by flowering; midstem leaves sessile and auriculate or subpetiolate and attenuate to cuneate with petioles widely winged and clasping, blades lanceolate or oblanceolate to oblong; uppermost leaves gradually reduced; heads in open paniculate or corymbose ± leafy arrays; involucres usually 8–12 mm high; phyllaries sometimes leaf-like and spreading (at least the outer series), subequal, bases hardened, tips sometimes purplish; ray flowers usually 20–50, the ligules (7–)12–18 (–21) mm long, blue to lavender or purple, rarely white or pink; disk corollas cream or yellow, turning pink or purple with age; achenes (2–)2.5–3.5(–4) mm long, prominently 3–4-nerved; pappus white, 3.8–6 mm long. Saturated soils, oak and pine woodlands, seeps and bogs, pond margins; ne TX in Anderson, Cherokee (TEX-LL), Franklin, Henderson, Hopkins, Kaufman, Smith, and Van Zandt cos.; AL, LA, MS, and TX. Sep–Nov. [*Aster puniceus* L. var. *scabricaulis* (Shinners) A.G. Jones, *A. scabricaulis* Shinners] The conservation status of this taxon, which is currently considered imperiled at the state, national, and global level, is discussed in Nesom (1997b) and Poole et al. (2007). ⚠

Symphyotrichum racemosum (Elliott) G.L. Nesom, (clustering), SMOOTH WHITE OLDFIELD ASTER, OLDFIELD ASTER, SMALL WHITE ASTER. Cespitose or colonial perennial, 30–90(–100) cm tall, glabrous or nearly so; new basal rosettes often forming at flowering; leaves with margins often ± revolute; basal and lower stem leaves withering by flowering; midstem leaves sessile or subpetiolate with petioles winged, bases clasping or attenuate, blades elliptic-lanceolate to linear-lanceolate, notably unequal in size, margins serrulate or entire; uppermost leaves abruptly reduced; heads in diffuse paniculate arrays; involucres usually 3.5–4.5 mm high; phyllaries unequal, bases hardened, tips sometimes purplish; ray flowers usually 16–20, the ligules 5–8 mm long, usually white, rarely pink; disk corollas cream or pale yellow, turning pink or red with age; achenes 1–1.8 mm long, faintly 4–5-nerved; pappus white, 2.5–3.5 mm long. Moist alluvial soils, open woodlands, marshes, wet prairies; e and sc TX from Anderson Co. s to Hardin Co. and w to Gonzales Co; e U.S. from ME s to Fl and w to WI, OK, and TX. Aug–Oct. [*Aster brachypholis* Small, *A. racemosus* Elliott]

Symphyotrichum sericeum (Vent.) G.L. Nesom, (silky), SILKY ASTER, SILVER ASTER. Cespitose perennial, (20–)30–70 cm tall, with thickened, woody stem bases; leaves sessile, silky-pilose, margins usually entire; basal and lower stem leaves withering by flowering; midstem leaves with blades oblanceolate to linear-lanceolate or elliptic, bases cuneate or rounded and subclasping; uppermost leaves not much reduced; heads in open paniculate arrays; involucres usually 7.5–10 mm high; phyllaries ± leaf-like (at least the outer series), unequal, densely silky-pilose, bases ± hardened, tips sometimes purplish; ray flowers usually 15–30, the ligules 8.5–11 mm long, rose-purple to violet, rarely white; disk corollas pink, turning purple with age; achenes 2–3 mm long, 7–10-nerved; pappus white or tawny, 6–7 mm long. Limestone slopes, fields, rocky calcareous soils, outcrops; Bastrop, Bexar, Comal, Hays, and Travis cos.; also Washington Co. (TEX-LL) and reported from Anderson Co. (Singhurst et al. 2003); an early Lindheimer collection from Harris Co. at BRIT may be a labeling error (R. Jones et al. 2008); Edwards Plateau and Lampasas Cut

Plain; MI w to ND and s to AR and TX; Canada. Aug–Nov. [*Aster sericeus* Vent., *A. sericeus* var. *microphyllus* DC.] Differences between this species and the closely related *S. pratense* (Raf.) G.L. Nesom are discussed in R. Jones et al. (2008).

Symphyotrichum subulatum (Michx.) G.L. Nesom, (awl-shaped, presumably for the tips of the phyllaries), EASTERN SALTMARSH ASTER. Tap-rooted annual, 30–120(–150) cm tall; leaves glabrous; basal and lower stem leaves withering by flowering; midstem leaves petiolate or sessile, blades narrowly lanceolate to subulate, margins entire or rarely with a few teeth; uppermost leaves much reduced; heads in a dense, pyramidal paniculate array; involucres usually 6–7 mm high; phyllaries unequal, bases hardened for a short distance or not at all, margins hyaline, often purple-tinged; ray flowers usually 16–30, the ligules (1.3–)1.5–2.6(–3.1) mm long, white (drying white or lavender); disk corollas yellow, sometimes tinged with purple with age; achenes usually 1.5–2.5 mm long, 5-nerved; pappus white, (3.5–)4–5.5 mm long. Salt marshes, brackish shores, salt-polluted areas; Orange Co., Newton Co. (*C. Allen & S. Allen 22496* TEX-LL); also Chambers and Galveston cos.; mostly coastal regions in the e and se U.S. from ME to TX; Canada; introduced elsewhere. Sep–Oct. [*Aster subulatus* Michx., *A. subulatus* Michx. var. *euroauster* Fernald & Grisc., *A. subulatus* Michx. var. *obtusifolius* Fernald, *S. subulatum* (Michx.) G.L. Nesom var. *subulatum*] We are following Nesom (2005b) and Turner (2016) in treating this taxon at the specific, rather than varietal level.

Symphyotrichum tenuifolium (L.) G.L. Nesom var. **tenuifolium,** (narrow-leaved), PERENNIAL SALTMARSH ASTER. Rhizomatous, colonial perennial, 40–60(–100) cm tall; stems branching sparely, wiry, often purplish or brown, glabrous or sometimes with lines of hairs on upper stems; new basal rosettes often forming at flowering; leaves entire, glabrous; basal and lower stem leaves withering by flowering; midstem leaves sessile, blades lanceolate to subulate, bases cuneate, somewhat reduced upwards; heads in diffuse paniculate arrays; involucres usually 6–9.5 mm high; phyllaries subequal, bases hardened, margins hyaline, tips often tinged with purple; ray flowers usually 17–25, the ligules (4.5–)5–8.5(–9.5) mm long, white or pink; disk corollas yellow, turning purplish with age; achenes 2.8–4(–4.5) mm long, 5–6-nerved; pappus tawny to white, 5–6.1 mm long. Coastal salt marshes, tidal flats, brackish shores; Harris Co. (*D.J. Rosen 6184* TEX and *L.E. Brown 15785* SBSC) and Jackson Co.; e and c coastal TX; e U.S. coast from ME to FL to TX. Sep–Nov. [*Aster tenuifolius* L.]

Symphyotrichum anomalum (Engelm.) G.L. Nesom, (anomalus), MANYRAY ASTER, which would key to *S. drummondii* here, can be distinguished by the entire midstem leaves, long-acuminate, reflexed phyllary tips, and ray petals that are more numerous (18–45) with a longer ligule (usually 9–15 mm or longer). Not yet reported in TX but included here based on records from adjacent counties in other states: McCurtain Co., OK (OKL) and Little River Co., AR (UARK); c U.S. from IL s to AR and w to OK. Aug–Oct. [*Aster anomalus* Engelm. ex Torr. & A. Gray]

Symphyotrichum cordifolium (L.) G.L. Nesom, (heart-shaped leaves), HEARTLEAF ASTER, BLUE WOOD ASTER, which would key to *S. drummondii* here, can be distinguished by the usually unwinged petioles, and leaves with less pubescence beneath, often restricted to the veins. Not yet reported in TX but included here based on records from adjacent counties in other states: Bryan Co., OK (OKL) and several cos. in sw AR; e 1/2 of North America. Aug–Oct. [*Aster cordifolius* L., *S. lowrieanum* (Porter) G.L. Nesom, *S. sagittifolium* (Wedem. ex Willd.) G.L. Nesom]

Symphyotrichum pilosum (Willd.) G.L. Nesom var. *pilosum*, (hairy), FROST ASTER, OLDFIELD ASTER, HAIRY WHITE ASTER, which would key to *S. dumosum* or *S. racemosum* here, and is often confused with *S. ericoides*, can be distinguished by the basal rosette of oblanceolate leaves usually present at flowering, and phyllaries with an inrolled margin that creates a green spinulose tip. To be expected in ne TX based on records from adjacent counties/parishes in other states: McCurtain Co., OK. (OKL), Little River and Miller cos., AR (UARK), Caddo Par., LA (LSU); e 1/2 of North America; introduced and sometimes weedy elsewhere. Aug–Oct. [*Aster pilosus* Willd., *A. ericoides* L. var. *pilosus* (Willd.) Porter]

Symphyotrichum squamatum (Spreng.) G.L. Nesom, (scaly), TROPICAL ASTER, which would key to *S. divaricatum* here, can be distinguished by the broadly lanceolate green zones of the phyllaries, which do not extend to the phyllary tip, phyllary tips that are appressed rather than loose, ray flowers with erect ligules often inrolled lengthwise, and fewer disk florets (usually 7–14). This South American species of saline habitats is known in TX only from the city of Galveston (Galveston Co.; Nesom 2005c) and might be expected in other coastal TX counties; also in Cameron Par., LA; AL, LA, and NC; introduced elsewhere. Jul–Nov. [*Aster subulatus* Michx. var. *australis* (A. Gray) Shinners, *A. subulatus* Michx. var. *sandwicensis* (A. Gray) A.G. Jones, *S. subulatum* (Michx.) G.L. Nesom var. *squamatum* (Spreng.) S.D. Sundb.]

Symphyotrichum turbinellum (Lindl.) G.L. Nesom, (turban-shaped), PRAIRIE ASTER, SMOOTH VIOLET PRAIRIE ASTER, which would key to *S. eulae* or *S. oolentangiense* here, can be distinguished by the large turbinate heads with involucres 7–12 mm high, phyllaries in 6–9 series, rounded-keeled inner phyllaries, 8–15 uncrowded peduncular bracts, and 14–20 ray flowers that are 12–20 mm long. Not yet reported in TX but included here based on records from adjacent counties in other states: McCurtain Co., OK (OKL) and several cos. in sw AR; central U.S. from IL s to LA and w to KS and OK. Aug–Oct. [*Aster turbinellus* Lindl.]

TAGETES L. MARIGOLD

A genus of 40+ species of the tropical and warm-temperate Americas, especially Mexico; several species are cultivated for ornamental, culinary, medicinal, and ceremonial purposes; a few species have naturalized in other regions and are considered invasive. The common name MARIGOLD is derived from Mary's-gold, a name first applied to a similar plant native to Europe, *Calendula officinalis*. (Name thought to be derived from a common name for a French or African plant so called to honor the early Etruscan deity, *Tages*) (tribe Heliantheae)
REFERENCES: Soule 1993; Keith 2004b; Strother 2006cc.

Tagetes erecta L., (erect), AZTEC MARIGOLD, AFRICAN MARIGOLD, FRENCH MARIGOLD. Highly aromatic annuals to 120+ cm tall; leaves opposite, the blades 30–120(–250+) mm long, usually pinnately lobed or pinnatisect with 9–25 segments; heads borne singly; involucres 10–22 mm high; phyllaries in 1–2 series, fused for most of their lengths, usually with oil glands; ray flowers few to 100+ in "double" cultivars, pistillate, sometimes fertile, the ligules usually some combination of yellow, gold, orange, or red-brown, occasionally other colors in cultivars; disk flowers perfect, fertile, corollas greenish yellow to orange, sometimes tipped with red-brown; achenes 6–11 mm long; pappus of 2–4 linear-oblong scales, 2–6+ mm long, plus 0–2 narrower scales, 6–12+ mm long. Disturbed sites; rare in TX; recorded in Walker Co. (Keith 2004b) and Harris Co. (Brown 2014); scattered across the U.S.; Mexico; introduced elsewhere. Spring–Fall. [*T. patula* L.] This is one of several *Tagetes* species in which the roots produce thiophenes, shown to be an effective control for nematodes in crops (Strother 2006x). Native of Mexico; widely cultivated and sometimes escaping; possibly not persisting in the TX flora.

TARAXACUM F.H. Wigg. DANDELION

Small perennials, sometimes behaving as winter annuals, with coarse taproot; sap milky; leaves all in a basal rosette, the blades oblong-lanceolate, coarsely toothed to deeply pinnately lobed; scapes hollow, ± pubescent, to 40(–60) cm tall; heads solitary, terminal; phyllaries green, the outer series short and ± curved-spreading or reflexed, the inner 2 series long, erect, reflexed at maturity; flowers all ligulate, frequently all pistillate and parthenocarpic; corollas sulphur-yellow; achenes oblanceoloid, with slender beak 1–2.5 times as long as the body; pappus of numerous minutely barbed bristles, 4–7(–8) mm long; mature achenes with their umbrella-like pappi form a conspicuous, easily disrupted, ball-like arrangement.

A genus of at least 60 species of the n temperate zone and temperate South America, including some cosmopolitan weeds. Asexual reproduction, hybridization, and polyploidy have made the taxonomy of this group exceedingly complex. Our two species are difficult to distinguish without mature achenes, and in each case the nomenclature is confusing and not yet satisfactorily resolved (Brouillet 2006c). *Taraxacum* species have been used medicinally, as edible greens, and to produce wine; they are particularly rich sources of vitamin C (Small & Catling 1999). The name DANDELION comes from the French: *dent de lion*, tooth of the lion, apparently in reference to the leaf shape (Tveten & Tveten 1993). (Persian: *talkh chakok*, bitter herb, through medieval Latin *tarasacon*) (tribe Cichorieae)
REFERENCES: Shinners 1949a; Vuilleumier 1973; Taylor 1987; Brouillet 2006c.

1. Ligules 12–16 mm long; mature achenes red to purplish red or brownish red; leaf blades typically consistently lacerate, deeply lobed for their whole length; the terminal lobe often with straight or concave sides ________ **T. erythrospermum**
1. Ligules 15–22 mm long; mature achenes greenish tan or brownish to straw-colored; leaf blades occasionally lacerate, usually shallowly to deeply lobed or toothed; the terminal lobe often rounded with convex sides ________ **T. officinale**

Taraxacum erythrospermum Andrz. ex Besser, (red-seeded), RED-SEED DANDELION. Leaf blades consistently lacerate, with terminal lobe usually triangular, with straight or concave, rarely convex, sides; corollas 12–16 mm long; mature achenes red to purplish red or brownish red, 15-ribbed. In lawns, disturbed areas, and under trees in towns; widespread in TX but frequently ignored by collectors; a widespread weed in North America and throughout temperate regions of the world. Dec–Jun, sporadically to Nov, often drying up and disappearing during summer heat. Native of Eurasia. [*T. laevigatum* (Willd.) DC., *T. officinale* var. *erythrospermum* (Andrz.) Bab.]

Taraxacum officinale F.H. Wigg., (medicinal), COMMON DANDELION. Leaf blades shallowly to deeply lobed or toothed, with terminal lobe rounded-triangular with convex sides; corollas 15–22 mm long; mature achenes greenish tan or brownish to straw-colored, 4–12-ribbed. In lawns, disturbed areas, and under trees in towns; widespread in TX but frequently ignored by collectors; a widespread weed in North America and throughout temperate regions of the world. Dec–Jun, sporadically to Nov, often drying up and disappearing during summer heat. Native of Eurasia. [*Leontodon taraxacum* L.]

Symphyotrichum pratense [HEA]

Symphyotrichum puniceum var. scabricaule [HEA]

Symphyotrichum racemosum [YAT]

Symphyotrichum subulatum [GLE]

Symphyotrichum sericeum [STE]

TETRAGONOTHECA L. NERVERAY, SQUAREHEAD

Herbaceous perennials; stems sometimes from a basal rosette; stem leaves opposite, distinct or connate-perfoliate, the blades deltoid or lanceolate to ovate; heads large, 2–4+ cm broad (including ligules), borne singly or in loose corymbose arrays; phyllaries in ± 2 series; outer phyllaries 4, broadly lanceolate, foliaceous, conspicuous, inner series more numerous, smaller, scarious, each subtending a ray flower; ray flowers pistillate, fertile, with ligules conspicuous, yellow, often with some reddish or brownish venation; disk flowers numerous, perfect, fertile, the corollas yellow, sometimes with darker stripes or reddish lobes; achenes 4(–5)-angled, finely ribbed; pappus of scales or absent.

A genus of 4 species of the se U.S. and n Mexico; easily recognized by the 4 large outer phyllaries that give the buds a square appearance. (Greek: *tetragonos*, four-angled, and *theca*, a case, from the shape of the involucre) (tribe Heliantheae)
REFERENCES: Turner & Dawson 1980; Urbatsch & Fischer 1989; Strother 2006dd.

1. Leaves mostly basal, reduced upwards, narrowed into petiole-like bases and not connate-perfoliate; ray flowers 16–21+, the ligules 20–30+ mm long; sandy soils on sw border of East TX from Bexar to Goliad cos. ______ **T. repanda**
1. Leaves generally well-developed up the stems, connate-perfoliate; ray flowers < 15, the ligules 8–18 mm long; various habitats in e and c TX and Edwards Plateau.
   2. Leaf blades usually toothed, the upper stem leaves often broadest near the base; achenes 4–6 mm long; pappus of 16–30 obovate-spatulate scales, 0.5–2 mm long; sandy habitats in e and c TX ______ **T. ludoviciana**
   2. Leaf blades usually lyrate to pinnatifid, the upper stem leaves usually not broadest near the base; achenes 2.5–4 mm long; pappus absent, or of 1–10+ narrow or fringe-like scales to 0.5(–1.5) mm long; calcareous habitats in Edwards Plateau ______ **T. texana**

Tetragonotheca ludoviciana (Torr. & A. Gray) A. Gray ex H.M. Hall, (of Louisiana), SAWTOOTH NERVERAY, LOUISIANA NERVERAY. Plant to 100 cm tall; leaves mostly cauline, connate-perfoliate, the blades rounded-deltoid to ovate, 8–12 cm long, prominently and coarsely toothed; ray flowers 12–13, the ligules 8–18 mm long; achenes 4–6 mm long; pappus of 16–30 scales, 0.5–2 mm long. Sandy soils, oak woodlands; mainly se and e TX; also nc and sc TX; AR, LA, and TX. Apr–Aug. [*Halea ludoviciana* Torr. & A. Gray]

Tetragonotheca repanda (Buckley) Small, (having an undulating margin, for the leaves), SHOWY NERVERAY. Plant to 50 cm tall; leaves mostly basal, reduced upwards, narrowed into petiole-like bases and not connate-perfoliate, the blades rounded-deltoid to ovate or lanceolate, 5–12 cm long, repand or irregularly toothed; ray flowers 16–21, the ligules 20–30+ mm long; achenes 4–5.5 mm long; pappus of 20–30 scales, 1–2 mm long. Deep sandy soils; sw border of East TX in Bexar and Goliad cos.; South TX Plains; possibly Mexico. Mar–Nov. [*Halea repanda* Buckley]

Tetragonotheca texana Engelm. & A. Gray ex A. Gray, (of Texas), PLATEAU NERVERAY, SQUARE-BUD DAISY. Plant to 50 cm tall; leaves mostly cauline, connate-perfoliate, the blades ovate to lanceolate, 3–7 cm long, lyrate to pinnatifid or sometimes nearly entire; ray flowers 8–13, the ligules 10–18 mm long; achenes 2.5–4 mm long; pappus absent, or of 1–10+ scales to 0.5(–1.5) mm long. Calcareous soils, uplands, dry limestone ridges; w border of East TX in Comal, Hays, and Travis cos.; Edwards Plateau, Trans-Pecos, and far s TX; Mexico. Apr–Jun.

TETRANEURIS Greene FOUR-NERVE DAISY

Ours small annuals or perennials, 16–50+ cm tall, ± soft-pilose or woolly-pubescent to glabrate; stems erect or decumbent; leaves basal and/or cauline, alternate, simple, the blades spatulate to oblanceolate or linear, with petiolar base, entire or few-lobed, usually gland-dotted; heads terminal, rather large, long-peduncled, solitary or in corymbose arrays; phyllaries in 3 series, the outer herbaceous and ± hairy, the inner with scarious margins; ray flowers pistillate, fertile; ligules golden yellow or orange-yellow, with 4 prominent veins, usually 3–lobed at tip, often reflexed and persistent in age and becoming almost whitish; disk corollas perfect, fertile, orange-yellow; achenes obpyramidal, hairy; pappus of 4–7 awn-tipped scales.

A genus of 9 species, mostly confined to w North America and Mexico; it is a segregate of *Hymenoxys* and has sometimes been treated in that genus (e.g., Karis & Ryding 1994; Mabberley 2017). (Greek: *tetra*, four, and *neuron*, cord or nerve, alluding to the venation of the ray ligules) (tribe Heliantheae)
REFERENCES: Seeligmann & Alston 1967; Bierner & Jansen 1998; Bierner & Turner 2003, 2006.

1. Taprooted annuals with branched, conspicuously leafy stems; peduncles 8–14(–19) cm long ______ **T. linearifolia**
1. Woody-crowned perennials with leaves all basal or nearly so; peduncles 12–50 cm long.
   2. Plants not colonial; stems erect; leaves sparsely to densely hairy (but not woolly); involucres 7–12 mm wide; pappus 1.6–2.3 mm long; widespread in w 1/2 TX ______ **T. scaposa**

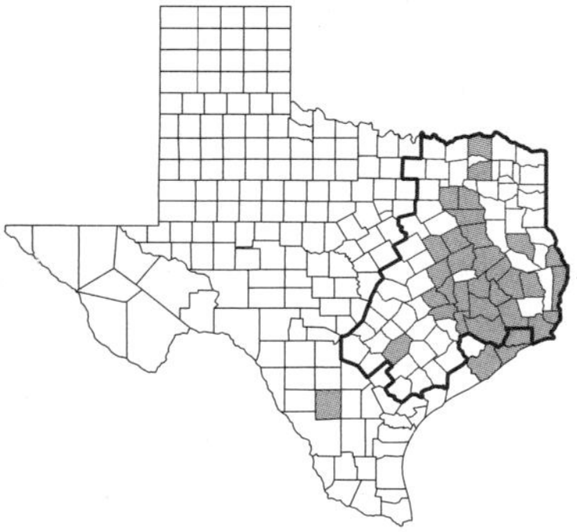

Symphyotrichum lateriflorum

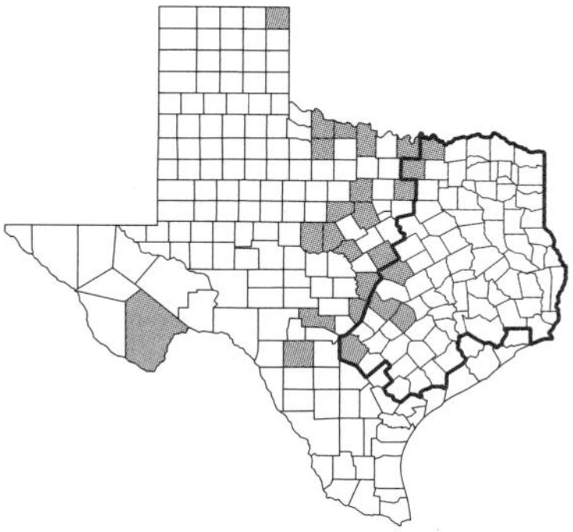

Symphyotrichum oblongifolium

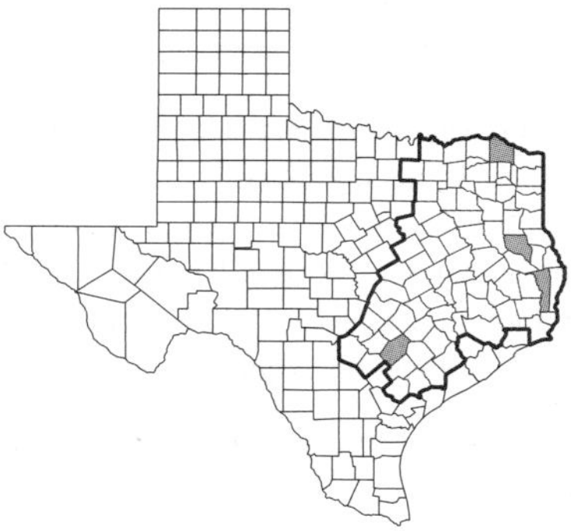

Symphyotrichum ontarionis

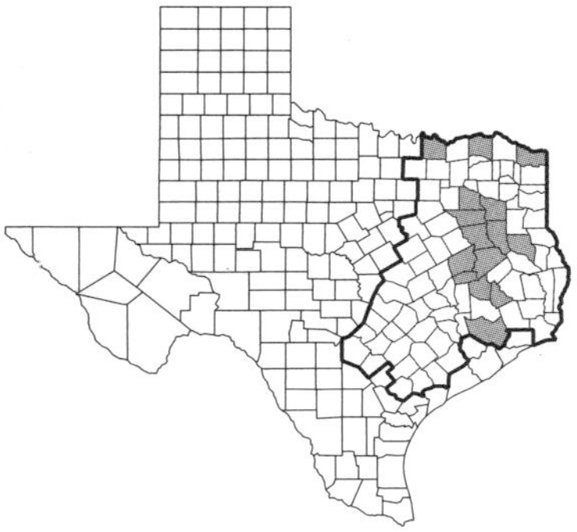

Symphyotrichum oolentangiense

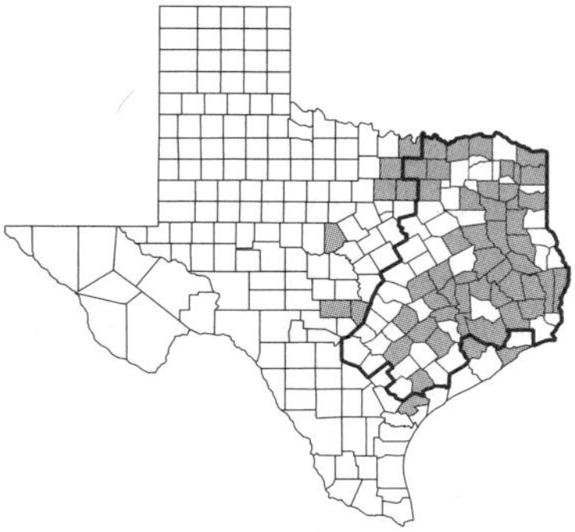

Symphyotrichum patens var. patens

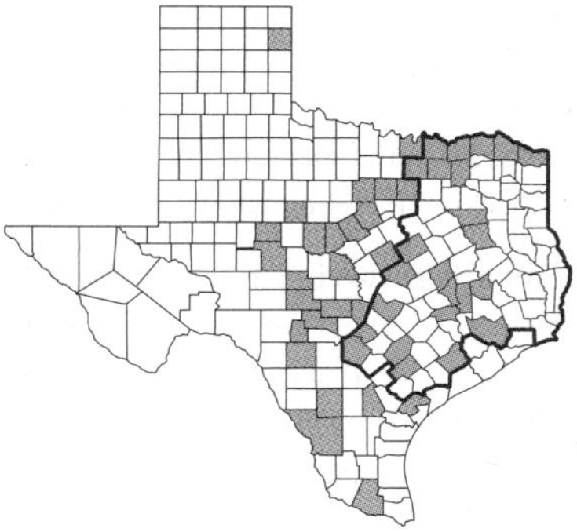

Symphyotrichum praealtum

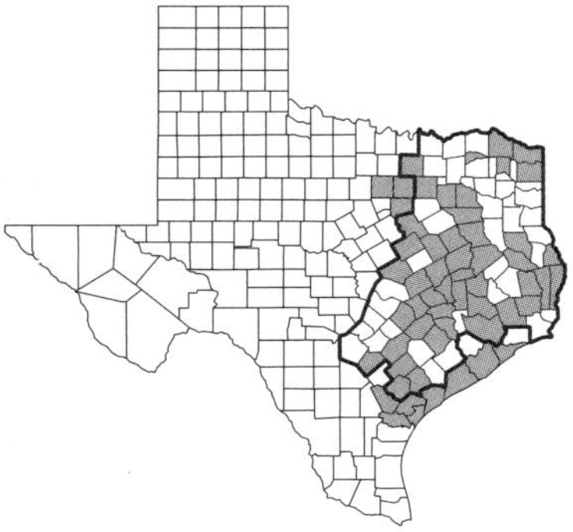

Symphyotrichum pratense

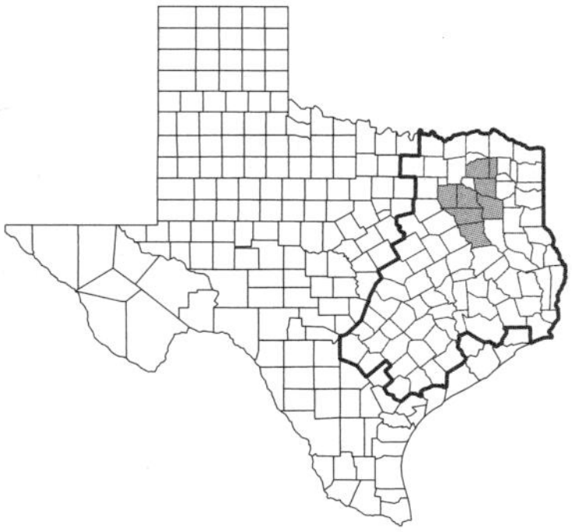

Symphyotrichum puniceum var. scabricaule

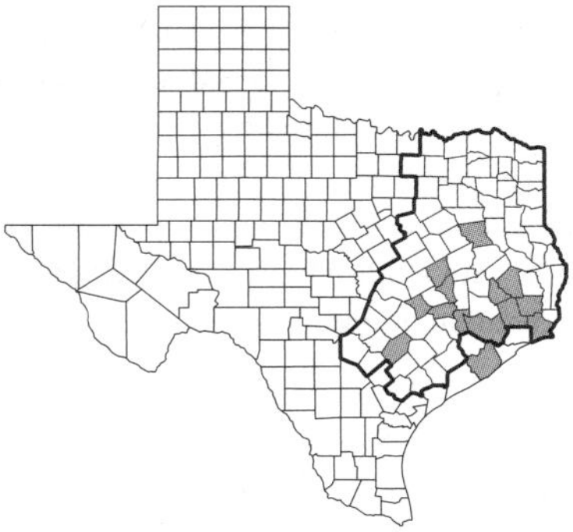

Symphyotrichum racemosum

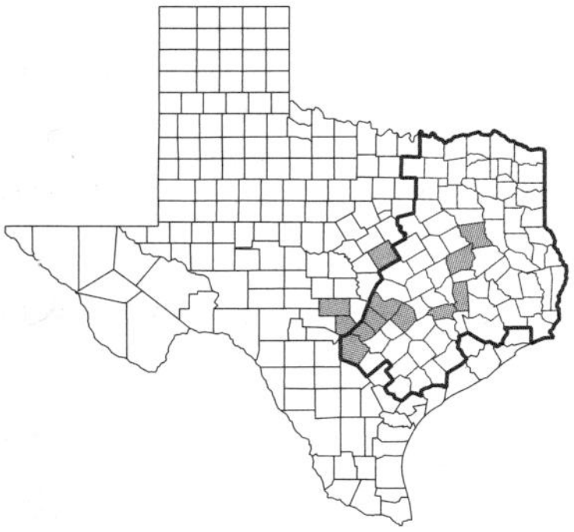

Symphyotrichum sericeum

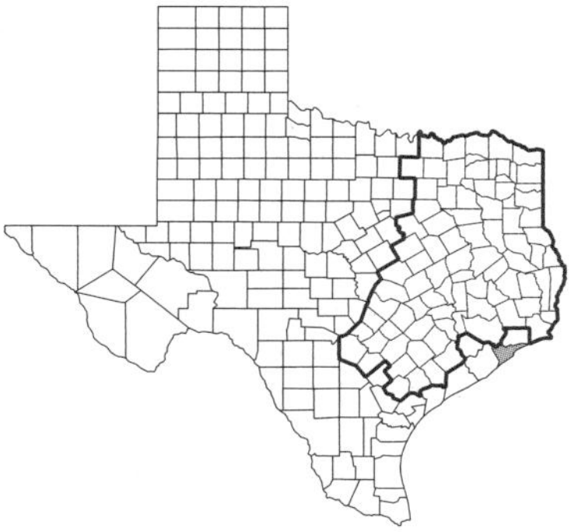

Symphyotrichum squamatum

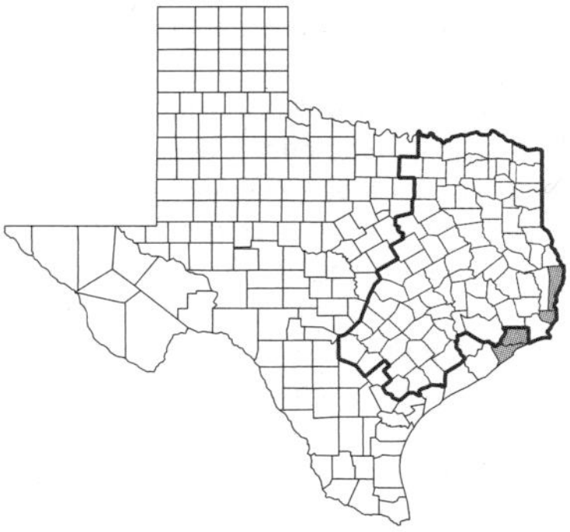

Symphyotrichum subulatum

2. Plants forming colonies by short lateral stolons; stems erect or decumbent; leaves moderately to densely silky or woolly; involucres 12–20 mm wide; pappus 3–4 mm long; restricted to sc TX _____ **T. turneri**

Tetraneuris linearifolia (Hook.) Greene var. **linearifolia,** (linear-leaved), FINE-LEAF FOUR-NERVE DAISY. Taprooted annuals to 40+ cm high; stems branched in the upper portion; leaves mostly on the stems, blades ± hairy, sparsely to densely gland-dotted, margins entire or with 2–6 teeth or lobes; involucres usually 8–11 mm wide; ray flowers 9–20; achenes 1.5–2.1(–2.5) mm long; pappus scales 1.2–1.9(–2.5) mm long. Prairies, roadsides, in calcareous clay or rocky soils; nearly throughout TX but most common in c, s, and sw TX; KS, NM, OK, and TX; Mexico. Mar–May, re-branching and producing additional smaller heads to Jul. [*Hymenoxys linearifolia* Hook.]

Tetraneuris scaposa (DC.) Greene var. **scaposa,** (with scapes), PLAINS YELLOW DAISY, YELLOW PAPER-FLOWER, FOUR-NERVE DAISY. Woody-crowned perennials; stems sometimes branched near the bases; leaves nearly all basal, blades ± hairy, usually densely gland-dotted, margins entire or with 2 teeth or lobes; involucres 7–12 mm wide; ray flowers 12–26; achenes 2–3 mm long; pappus scales 1.6–2.3 mm long. Gravelly or rocky prairies on limestone, roadsides, pastures; w border of East TX from McLennan to Travis to Bexar cos., also Brazos Co.; c and w TX and Panhandle; CO, KS, NE, NM, OK, and TX; Mexico. Mar–May, sporadically through summer, often repeating in Sep–Oct. [*Hymenoxys glabra* (Nutt.) Shinners, *H. scaposa* (DC.) K.L. Parker, *H. scaposa* (DC.) K.F. Parker var. *villosa* Shinners, *T. glabra* (Nutt.) Greene, *T. linearis* (Nutt.) Greene] The leaves are dotted with thin granules of a resin-like substance which gives them a bitter taste (Ajilvsgi 1984). The common name YELLOW PAPER-FLOWER is probably derived from the tendency of the rays to remain on the heads long after maturity, eventually turning almost white.

Tetraneuris turneri (K.F. Parker) K.F. Parker, (for Billie Lee Turner, 1925–2020, long-time professor of botany at Univ. of TX, Austin), TURNER'S FOUR-NERVE DAISY. Woody-crowned perennials; stems sometimes branched near the bases; leaves nearly all basal, blades moderately to densely woolly, usually densely gland-dotted, margins entire or with 2–4 teeth or lobes; involucres 12–20 mm wide; ray flowers 12–24; achenes 2.6–3.4 mm long; pappus scales 3–4 mm long. Gravelly or rocky calcareous soils, roadsides, pastures; Goliad Co.; s TX in Bee, Jim Wells, Karnes, Live Oak, and San Patricio cos.; Mexico. May–Jun. [*Hymenoxys turneri* K.F. Parker]

THELESPERMA Less. GREENTHREAD

Perennial or annual herbs, usually glabrous; leaves mostly opposite, usually compound or deeply lobed, the segments narrow; heads terminal on naked peduncles, borne singly or in loose corymbose arrays; primary (inner) phyllaries in 2 series, scarious-margined, basally fused for 1/5 to 1/2 of their lengths or more, subtended by a secondary series of phyllary-like bracts, these herbaceous, often spreading; ray flowers usually 8 (sometimes 0), neuter, with ligules yellow to golden yellow or bicolored (yellow and reddish brown to purple), toothed at the tips, or ray flowers absent; disk flowers bisexual and fertile, the corollas yellow with reddish brown veins, or dark reddish brown, sometimes unequally lobed; achenes linear to linear-oblong; pappus of 2 prominent or minute awns, or pappus absent.

A genus of ca. 12 species native to w North America, Mexico, and s South America; similar in appearance to and sometimes confused with *Coreopsis*; the inner phyllaries of *Thelesperma* are basally united for 1/5 to 1/2 of their lengths or more, while those of *Coreopsis* are separate or nearly so. (Greek: *thele*, nipple, and *sperma*, seed, in reference to the papillose achenes of some species) (tribe Heliantheae)

REFERENCES: Shinners 1950a; Melchert 1990; Greer 1997; Strother 2006ee; Turner 2007.

1. Throats of disk corollas equal to or longer than lobes; pappus usually absent, or rarely of two awns < 0.5 mm long _____ **T. simplicifolium**
1. Throats of disk corollas shorter than lobes; pappus usually of two awns or scales (0.5–)1–3 mm long, or rarely absent.
   2. Perennials (sometimes flowering first year); secondary phyllary-like bracts 3–5+, ovate to oblong, 1–3 mm long; ray ligules 4–8(–12+) mm long _____ **T. ambiguum**
   2. Annuals or short-lived perennials; secondary phyllary-like bracts 5–9, linear to narrowly triangular, (2–)4–8+ mm long; ray ligules 10–20+ mm long.
      3. Ray ligules largely reddish brown to purple (usually tipped with yellow) _____ **T. burridgeanum**
      3. Ray ligules yellow or golden-yellow throughout (may be reddish tinged at base).
         4. Robust plants 30–70(–120+) cm tall; internodes 35–100+ mm long; disk corollas yellow, often with reddish brown nerves; plants of deep sandy soils in e and sc TX _____ **T. flavodiscum**
         4. Non-robust plants 10–40(–70+) cm tall; internodes mostly 10–35(–50+) mm long; disk corollas reddish brown or yellow with reddish brown nerves; plants of calcareous clays throughout TX _____ **T. filifolium**

Thelesperma ambiguum A. Gray, (ambiguous), COLORADO GREENTHREAD. Perennials, (10–)30–50 cm tall, leaves simple, trifid, or usually 1–2-pinnatifid, the ultimate segments linear; secondary phyllary-like bracts 3–5+, ovate to oblong, 1–3 mm long; ray

Symphyotrichum tenuifolium [GWO]

Tagetes erecta [YAT]

Taraxacum erythrospermum [BB1]

Taraxacum officinale [REE]

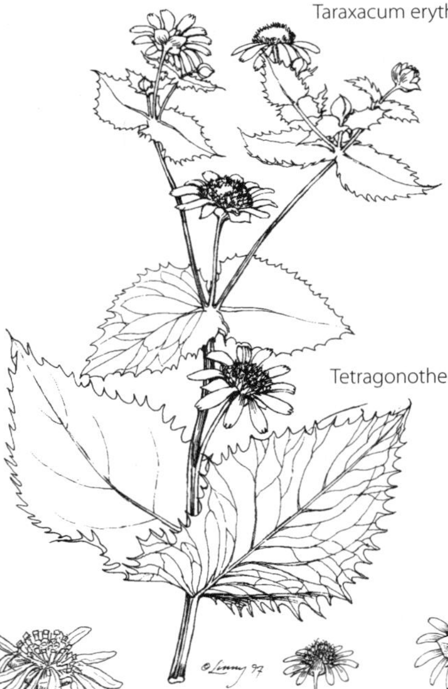

Tetragonotheca ludoviciana [HEA]

Tetragonotheca repanda [GEN]

Tetragonotheca texana [HEA]

flowers usually 8, rarely 0, ligules 4–8(–12+) mm long, yellow; disk corollas reddish brown, throats shorter than lobes; achenes 4–5+ mm long; pappus of 2 awns, 1.5–2+ mm long, or rarely absent. Open limestone areas, sands and clays, disturbed sites; sw border of East TX in Bexar and Wilson cos.; mainly s TX and Edwards Plateau; NM and TX; Mexico. Apr–Jun. [*T. fraternum* Shinners, *T. megapotamicum* (Spreng.) Kuntze var. *ambiguum* (A. Gray) Shinners]

Thelesperma burridgeanum (Regel, Korn. & Rach.) S.F. Blake, (for a 19th century horticulturist named Burridge, who brought this species into cultivation in Europe), BURRIDGE'S GREENTHREAD, COSMIDIUM. Annuals, 30–40(–70+) cm tall, leaves usually pinnatifid, the ultimate segments usually linear to filiform; secondary phyllary-like bracts 5–6(–9), linear to narrowly triangular, 2–4 mm long; ray flowers 8, ligules 7–9(–12+) mm long, mostly reddish brown to purplish, often with yellow tips; disk corollas reddish brown to purplish, throats shorter than lobes; achenes 3.5–4 mm long; pappus of 2 awns, 0.5–1+ mm long. Deep sandy soils; sw border of East TX in Bexar and Wilson cos.; interior South TX Plains. Mar–Jun. [*Cosmidium burridgeanum* Regel]

Thelesperma filifolium (Hook.) A. Gray, (thread-leaved), GREENTHREAD, THREAD-LEAF THELESPERMA. Annuals or short-lived perennials blooming the first year, 10–40(–70+) cm tall, leaves pinnatifid, the ultimate segments mostly linear to filiform; secondary phyllary-like bracts 7–8+, linear to narrowly triangular, (2–)4–8+ mm long; ray flowers 8, ligules 12–20+ mm long, yellow to golden-yellow, sometimes reddish-tinged at the base; disk corollas reddish brown or yellow with reddish brown nerves, throats shorter than lobes; achenes 3.5–4+ mm long; pappus of 2 awns, 0.5–1(–2+) mm long. Calcareous clays, prairies, roadsides, disturbed areas; widespread in TX but less common in East TX; c U.S. from MS n to ND and w to AZ; Mexico. Apr–Jun, sporadically to Sep. [*Coreopsis filifolia* Hook., *T. filifolium* (Hook.) A. Gray var. *intermedium* (Rydb.) Shinners, *T. intermedium* Rydb.] If varieties are recognized, ours would be var. *filifolium*.

Thelesperma flavodiscum (Shinners) B.L. Turner, (for the yellow disk flowers), EAST TEXAS GREENTHREAD. Annual, 30–70(–120+) cm tall, leaves pinnatifid, the ultimate segments mostly linear to filiform; secondary phyllary-like bracts 7–9, linear to narrowly triangular, 4–6(–8) mm long; ray flowers 8, ligules 12–20+ mm long, yellow; disk corollas yellow, often with reddish brown nerves, throats shorter than lobes; achenes 4–5 mm long; pappus of 2 awns, 0.5–1.5+ mm long. Deep sandy soils, post oak savannas; Cass Co. sw to Wilson Co. and e to Newton Co.; South TX Plains; AR, LA, and TX. May–Jul. [*T. filifolium* (Hook.) A. Gray var. *flavodiscum* Shinners] Turner (2007) discusses the distinction between this species and *T. filifolium*.

Thelesperma simplicifolium A. Gray, (simple-leaved), SLENDER GREENTHREAD. Perennial, (20–)30–70+ cm tall, usually glaucous; leaves simple and thread-like or 3–5-parted, often few, the plant thus sometimes with a naked appearance; secondary phyllary-like bracts 6–8+, linear to subulate, 1–3+ mm long; ray flowers usually 8, rarely 0, ligules 9–15(–20+) mm long, yellow; disk corollas yellow with red-brown nerves, throats equal to or longer than lobes; achenes 3–4 mm long; pappus usually absent, or rarely of 2 minute tooth-like awns < 0.5 mm long. Limestone hillsides, oak-juniper woodlands, desert scrub; w border of East TX from McLennan s to Bexar and Wilson cos.; mostly c TX, Edwards Plateau, and South TX Plains; NM and TX; Mexico. May–Sep. [*T. curvicarpum* T.E. Melchert]

THUROVIA Rose THREE-FLOWER SNAKEWEED

A monotypic genus endemic to TX; it was formerly treated in the genus *Gutierrezia*. (Named for collector Fredrick William Thurow (1852–1952), originally from Germany, later of Hockley, near Houston) (tribe Astereae)
REFERENCES: Lane 1985; Suh & Simpson 1990; Nesom 2006y.

Thurovia triflora Rose, (three-flowered), THREE-FLOWER SNAKEWEED, THREE-FLOWER BROOMWEED. Taprooted annuals or short-lived perennials, 6–45 cm tall; stems erect, glabrous, branching mostly in the upper portion; leaves alternate, sessile, usually 2–10 mm long, linear-filiform, glabrous, ascending-appressed and reduced up the stem, margins entire; heads numerous, small, sessile, in leafy-bracted spicate arrays; involucres narrowly cylindric, 1.5–2 mm wide; phyllaries 5–9, glabrous; ray flowers absent; disk flowers 3, corollas white to yellowish cream, to 2.8 mm long; achenes 0.9–1.7 mm long, densely white-sericeous; pappus of 10 silvery-white, ovate-lanceolate scales in 2 series, equaling corollas. Saline prairie barrens and slicks, coastal flats; a rare endemic restricted to the TX Gulf Coastal Plain, known from Harris and Jackson cos. and Waller Co. (historical, possibly extirpated (Poole et al. 2007)); also Aransas, Brazoria, Calhoun, Fort Bend, Galveston, Matagorda, Refugio, and San Patricio cos. Sep–Nov. [*Gutierrezia triflora* (Rose) M.A. Lane] This species is not protected at the state or federal level, but is of conservation concern due to its restricted range, small population numbers, and specialized habitat; it is threatened by urban development and the expanding oil and gas industry around Houston (Poole et al. 2007).

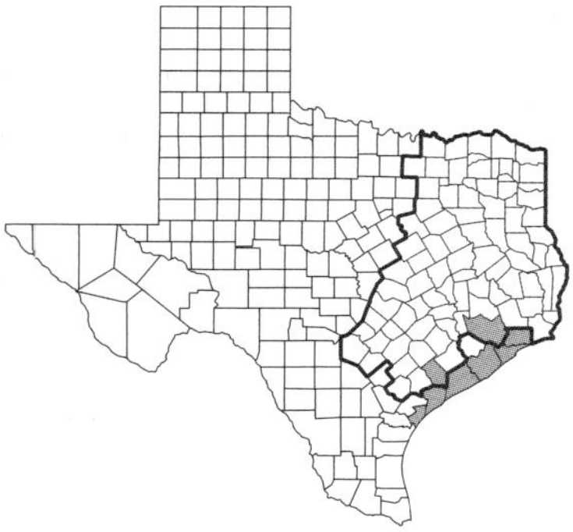
Symphyotrichum tenuifolium var. tenuifolium

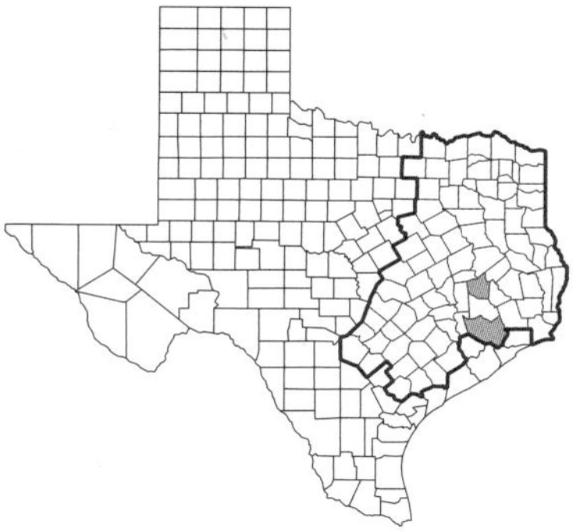
Tagetes erecta

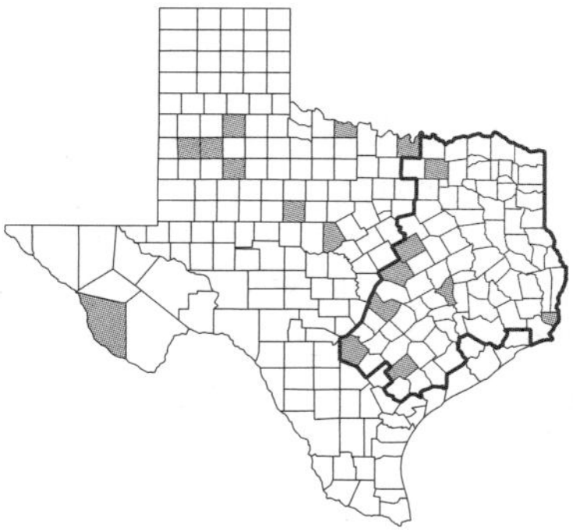
Taraxacum erythrospermum

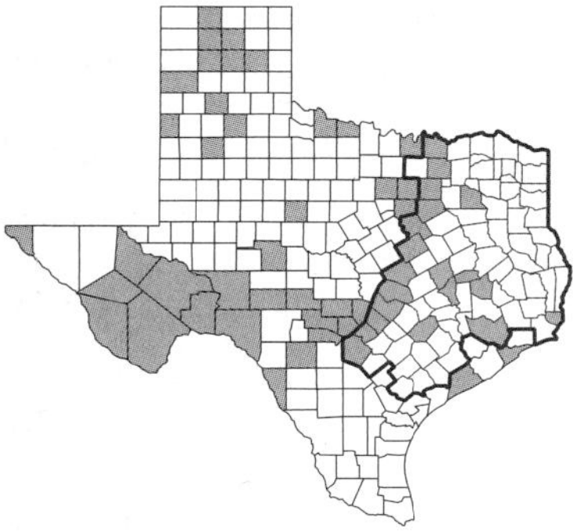
Taraxacum officinale

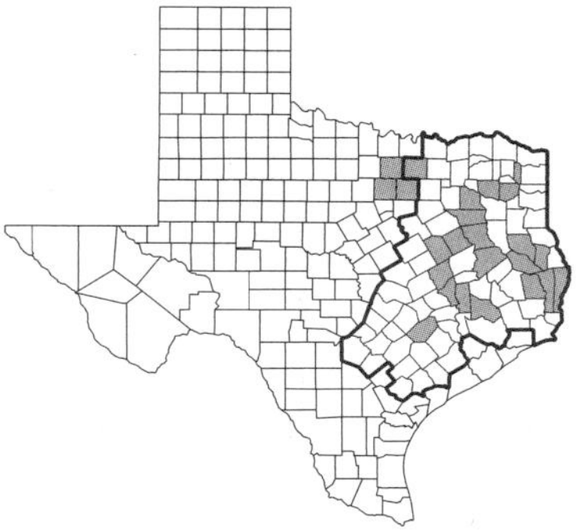
Tetragonotheca ludoviciana

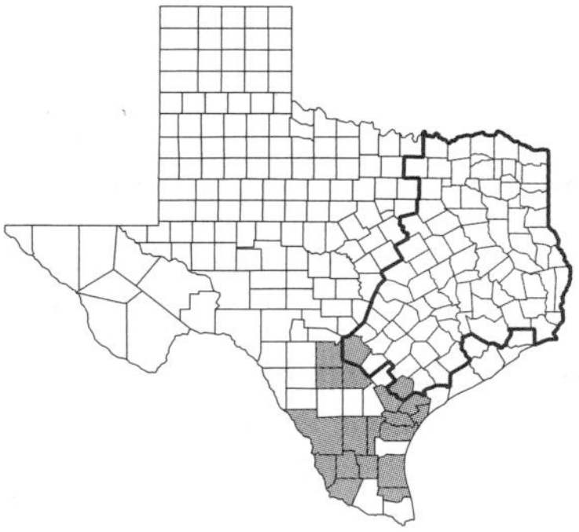
Tetragonotheca repanda

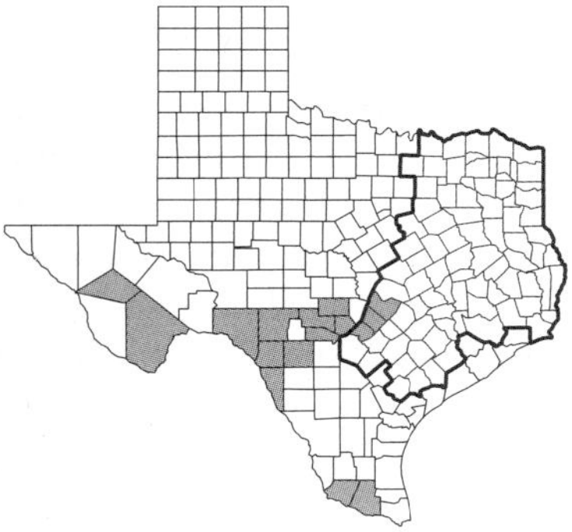
Tetragonotheca texana

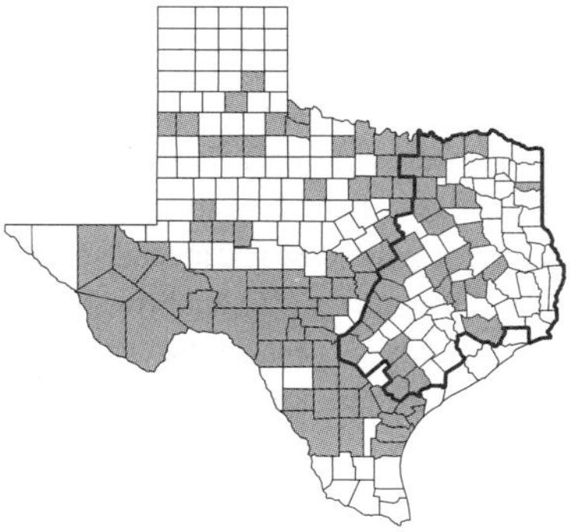
Tetraneuris linearifolia var. linearifolia

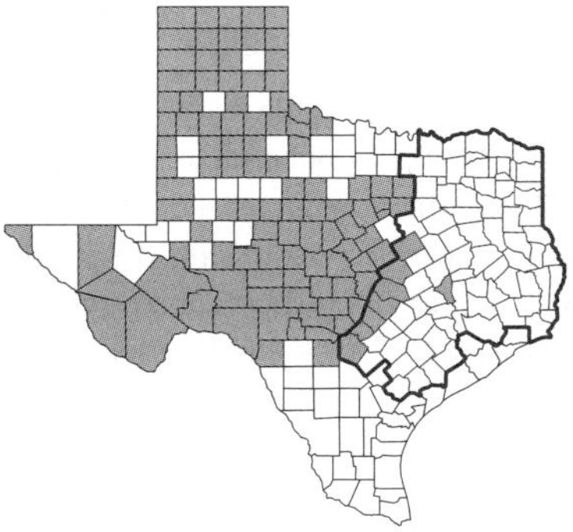
Tetraneuris scaposa var. scaposa

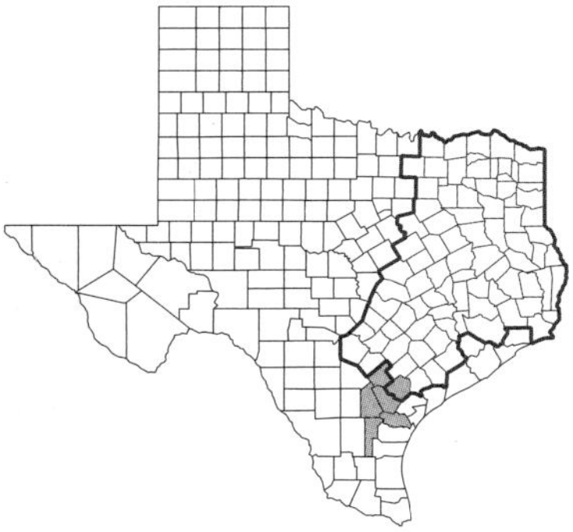
Tetraneuris turneri

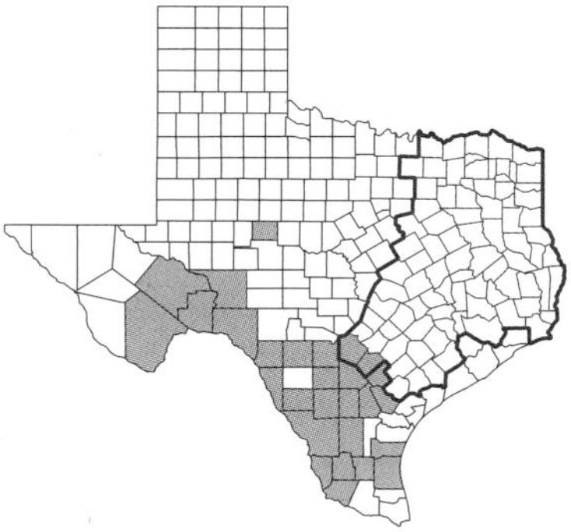
Thelesperma ambiguum

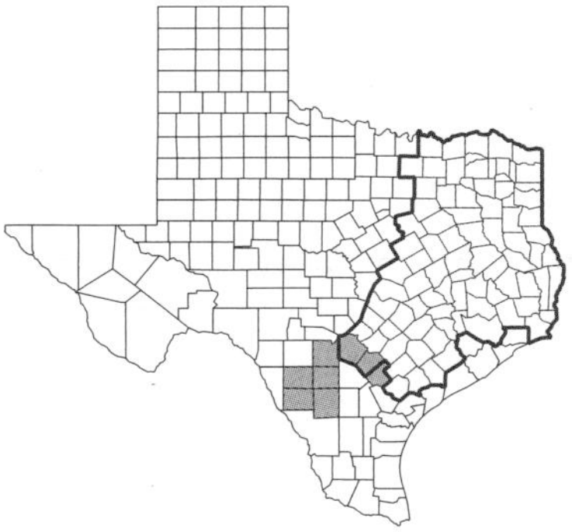
Thelesperma burridgeanum

THYMOPHYLLA Lag. DOGWEED, PRICKLYLEAF

Ours low annual or perennial herbs or subshrubs, aromatic, to 30 cm tall; stems frequently much-branched, sometimes forming dense clumps; leaves opposite or alternate, petiolate or sessile, often pinnately lobed, often with scattered oil glands; heads mostly borne singly; involucres 4–7 mm high; primary phyllaries in 2 series, fused for a portion of their lengths, usually with oil glands, sometimes subtended by a series of small bractlets; ray flowers (8–)10–21, pistillate, fertile, corollas yellow to yellowish orange; disk flowers perfect, fertile, corollas yellow to yellowish orange; achenes linear to obconic, glabrous or pubescent; pappus of awn-tipped and sometimes rough-margined scales, to 3.5 mm long.

A genus of ca. 13 species native to the sw U.S. and Mexico; a few are weedy or are cultivated as ornamentals. *Thymophylla* species were formerly included in the genus *Dyssodia*. (Greek: *thymon*, thyme, and *phyllon*, leaf) (tribe Heliantheae)
REFERENCES: Johnston 1956; Strother 1969, 1986, 1989, 2006ff; Gandhi & Thomas 1984; Nesom 2009b; Turner 2009c.

1. Perennial herbs or subshrubs (sometimes flowering first year); leaves mostly opposite; pappus of alternating awn-tipped scales and scales without awns.
   2. Leaf lobes mostly 9–11; phyllaries usually glabrous or glabrescent, rarely hairy, margins of outer phyllaries fused ca. 1/2 their lengths ____ **T. pentachaeta**
   2. Leaf lobes mostly 5–7; phyllaries densely puberulent, margins of outer phyllaries fused ca. 2/3 their lengths ____ **T. puberula**
1. Annuals (rarely persisting); leaves mostly alternate, at least the upper ones; pappus of awn-tipped scales.
   3. Leaf blades rigid, pinnatisect, with 7–15 subulate lobes; pappus scales each terminating in 3–5 awns ____ **T. tenuiloba**
   3. Leaf blades relatively lax, seldom lobed (linear to filiform, usually entire, rarely with a few pairs of tooth-like lobes); pappus scales usually terminating in a single awn (or rarely 2) ____ **T. wrightii**

Thymophylla pentachaeta (DC.) Small, (five-bristled), COMMON DOGWEED, PARRALENA, FIVE-NEEDLE PRICKLYLEAF. Perennials or subshrubs, ± grayish or green, glabrous or puberulent; leaves mostly opposite, leaf lobes (3–)9–11, linear to filiform, stiff and sharp-tipped; peduncles 6–10 cm long; involucres 4–6 mm high; phyllaries usually glabrescent or glabrous, margins of outer phyllaries fused ca. 1/2 their lengths; achenes 2–3 mm long; pappus of 10 alternating awnless and awned scales. Sandy, often calcareous soils, scrublands, disturbed sites; sw margin of East TX from Williamson to Hays to Bexar cos., DeWitt and Goliad cos., and Colorado Co. (BRIT-SMU); Rolling Plains, Edwards Plateau, and s TX; Mexico; Argentina. Mar–Jul, Sep–Nov. [*Dyssodia pentachaeta* (DC.) B.L. Rob., *T. pentachaeta* (DC.) Small var. *pentachaeta*] The foliage has an unpleasant scent if handled (Ajilvsgi 1984).

Thymophylla puberula Rydb., (finely pubescent), DOGWEED, PARRALENA, FIVE-NEEDLE PRICKLYLEAF. Perennials or subshrubs, ± grayish green, glabrous or puberulent; leaves mostly opposite, leaf lobes 3–7(–10), linear to filiform, stiff and sharp-tipped; peduncles 4–10 cm long; involucres 4.5–5.5 mm high; phyllaries densely puberulent, margins of outer phyllaries fused ca. 2/3 their lengths; achenes 2–3 mm long; pappus of 10 alternating awnless and awned scales. Calcareous bluffs and slopes, scrublands; sw margin of East TX in Travis and Hays cos. and Williamson Co. (TEX); Edwards Plateau; Mexico. Mar–Jul, Sep–Nov. [*Dyssodia pentachaeta* (DC.) B.L. Rob. var. *puberula* (Rydb.) Strother, *T. pentachaeta* (DC.) Small var. *puberula* (Rydb.) Strother] This taxon was treated as *T. pentachaeta* var. *puberula* in Strother (2006ff), but Turner (2009c) concluded the variety was sufficiently distinct to be returned to species status.

Thymophylla tenuiloba (DC.) Small var. **tenuiloba,** (slender-lobed), BRISTLE-LEAF DYSSODIA, BRISTLE-LEAF PRICKLYLEAF, TINY-TIM. Annuals (rarely persisting), green, glabrous or sparsely hirtellous; leaves mostly alternate, rigid, leaf lobes 7–15, subulate; peduncles 3–8 cm long; involucres 5–7 mm high; phyllaries glabrous or sparsely hirtellous, margins of outer phyllaries fused more than 4/5 their lengths; achenes 2–3.5 mm long; pappus of 10–12 similar scales, each bearing 3–5 awns. Sandy, often calcareous soils, roadsides, disturbed areas; se and c TX; Plains Country to w and s TX; NM and TX; recently adventive in CA and se U.S.; introduced in West Indies; Asia; and Africa. Apr–Oct. [*Dyssodia tenuiloba* (DC.) B.L. Rob.] DAHLBERG DAISY and GOLDEN FLEECE are names used for cultivars of this variety in the horticultural trade.

Thymophylla wrightii (A. Gray) Small, (for Charles Wright, 1811–1885, prolific collector of TX and elsewhere), WRIGHT'S PRICKLYLEAF. Annuals (rarely persisting), green, glabrous or sparsely hirtellous; leaves mostly alternate, relative lax, usually entire (rarely with a few pairs of tooth-like lobes), linear to filiform; peduncles 3–8 cm long; involucres 5–7 mm high; phyllaries glabrous or sparsely hirtellous, margins of outer phyllaries fused more than 4/5 their lengths; achenes 2–3.5 mm long; pappus of 10–12 similar scales, each bearing a single awn (or rarely 2). Sandy soils, coastal grasslands, roadsides; endemic to sc TX; Bastrop, DeWitt, Goliad, Gonzales, Guadalupe, and Lavaca cos., also Wilson Co. (TEX); s to TX coast. Apr–Oct. [*Dyssodia wrightii* (A. Gray) B.L. Rob., *D. tenuiloba* (DC.) B.L. Rob. subsp. *wrightii* (A. Gray) Strother, *T. tenuiloba* (DC.) Small var. *wrightii* (A. Gray) Strother] This taxon was treated as *T. tenuiloba* var. *wrightii* in Strother (2006ff), but Nesom (2009b) concluded the variety was sufficiently distinct to be returned to species status.

Tetraneuris linearifolia var. linearifolia [IPL]

Tetraneuris scaposa var. scaposa [FNA]

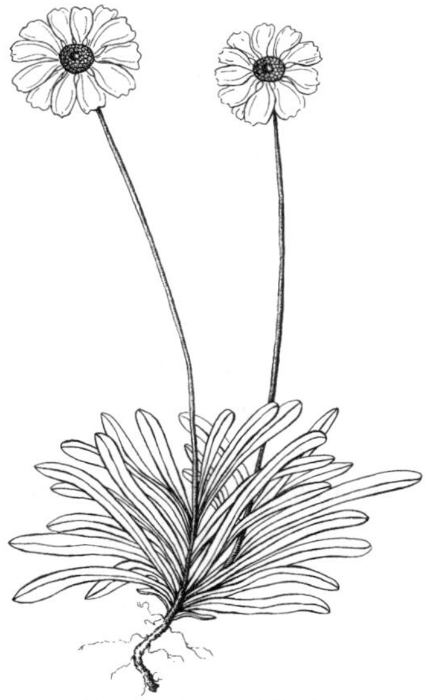

Tetraneuris turneri [HAR]

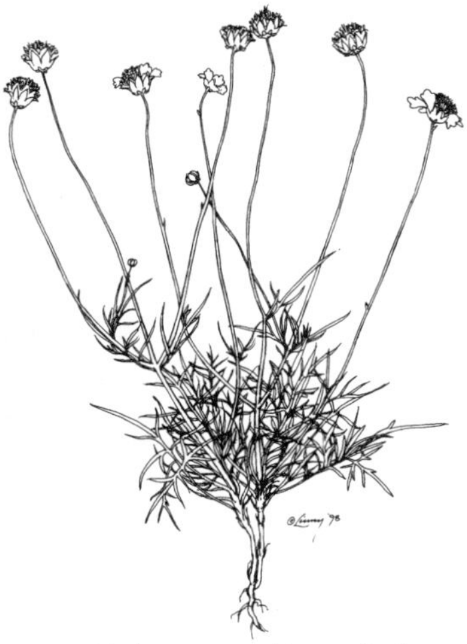

Thelesperma ambiguum [HEB]

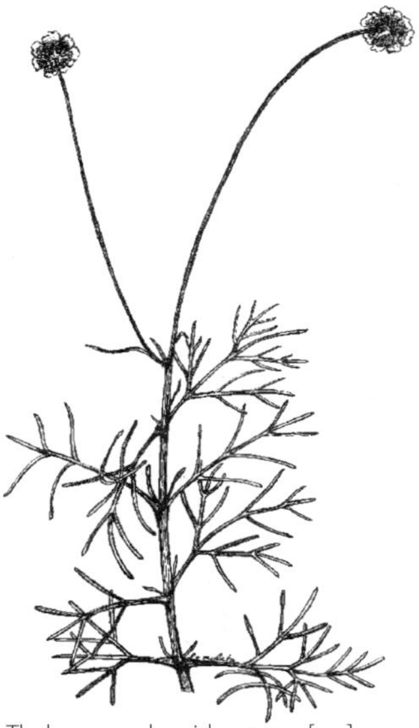

Thelesperma burridgeanum [THO]

Thelesperma filifolium [BB2]

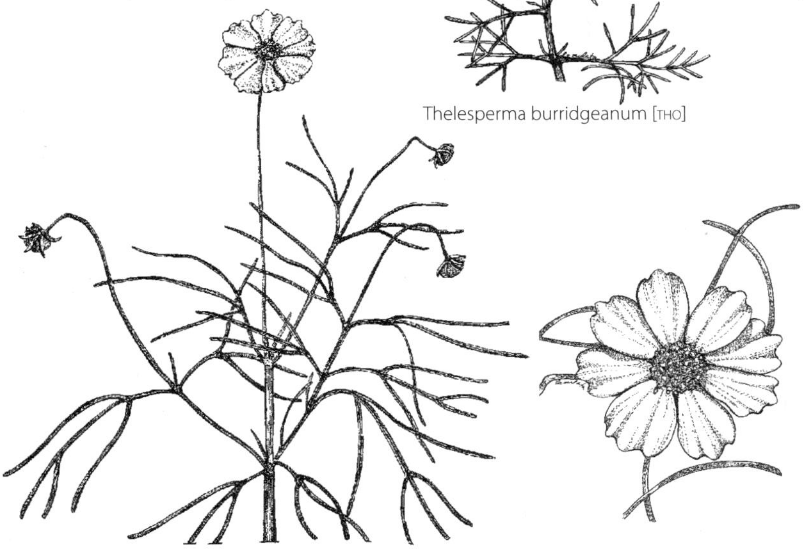

Thelesperma favodiscum [THO]

Thelesperma simplicifolium [HEA]

TOWNSENDIA Hook. EASTER DAISY

A genus of 27 species native to w North America and Mexico; some are cultivated as ornamentals. (Named for David Townsend, 1787–1858, amateur U.S. botanist from Pennsylvania) (tribe Astereae)
REFERENCES: Larsen 1927; Beaman 1957; Strother 2006gg.

Townsendia exscapa (Richardson) Porter, (without scapes), STEMLESS TOWNSENDIA, EASTER DAISY. Tufted or matted, appressed-pubescent, dwarf perennial, the woody stems barely out of the ground; leaves ± in rosettes, blades spatulate to linear-oblanceolate, 1.2–4(–7.5) cm long, 2–6 mm wide, entire; heads large in relation to plant size, 1–2.5 cm across, sessile or short-pedunculate; phyllaries in 4–6 series, linear to narrowly lanceolate; ray flowers 20–40, pistillate and fertile, with ligules white to pinkish or rosy-lavender, often with a darker stripe beneath, showy, 12–18+ mm long, curling under at night or in age; disk flowers perfect and fertile, corollas yellow, often pink- or purple-tipped or -tinged; achenes 2(–3)-ribbed, 4–6 mm long; pappus of minutely-barbed bristles 6–11+ mm long. Eroding limestone slopes; rare in East TX; mostly Plains Country and Trans-Pecos, with disjunct occurrence in Cross Timbers in Bell Co. and adjacent cos. to the w; widespread in w 1/2 of North America excepting coastal states; Mexico. Jan–Apr. [*Aster exscapus* Richardson, *T. sericea* Hook.]

TRAGOPOGON L. SALSIFY, GOAT'S-BEARD

Biennial herbs with stout taproot; sap milky; stems 15–100+ cm tall; leaves alternate, sessile and clasping, to 30 cm long, grass-like, tapered from base to a long, narrow tip, entire; heads solitary, terminal, large, on swollen peduncles; phyllaries in one row, equaling or longer than corollas, ca. 2.4–4 cm long in flower; flowers all ligulate, open mainly in morning hours; achenes 25–40 cm long including elongate beak; pappus of very conspicuous plumose bristles, 20–30 mm long; mature achenes with their pappi in a ball-like arrangement resembling the infructescence of a giant dandelion.

A temperate Eurasian and Mediterranean genus of ca. 110 species of taprooted herbs with monocot-like leaves. While both species occurring in East TX are diploids, hybridization and polyploidy are well-documented in the genus (Soltis et al. 1995; Soltis 2006). (Greek: *tragos*, goat, and *pogon*, beard, in reference to the conspicuous pappus) (tribe Cichorieae)
REFERENCES: Shinners 1949a; Ownbey 1950; Vuilleumier 1973; Roose & Gottlieb 1976; Soltis & Soltis 1989, 1991; Novak et al. 1991; Soltis et al. 1995; Soltis 2006.

1. Corollas lemon-yellow; phyllaries typically ca. 13 per head (sometimes less on depauperate plants) ____ **T. dubius**
1. Corollas purple; phyllaries typically ca. 8 per head ____ **T. porrifolius**

Tragopogon dubius Scop., (doubtful), GOAT'S-BEARD, WESTERN SALSIFY, NOON-FLOWER. Plant 0.3–1 m tall; phyllaries 13–17; corollas lemon-yellow. Disturbed sites, roadsides; mostly w 1/3 of TX, uncommon further e; Grayson and Collin (TEX) cos. in nc TX and Bell (TEX) and McLennan (Hannick et al. 2013) cos. in c TX; a widespread weed in North America and other temperate regions. Apr–Jun. Native of Eurasia. [*T. major* Jacq.]

Tragopogon porrifolius L., (with leaves like leek, in the Porrum group of the genus *Allium*), SALSIFY, VEGETABLE-OYSTER SALSIFY, OYSTERPLANT. Plant 0.4–1 m tall; phyllaries 8–9(–12); corollas purple. Disturbed sites, roadsides; scattered in w 1/3 of TX; in East TX known only from Dallas Co.; reported by Reverchon as "naturalized in gardens" in Dallas in 1903 (Mahler 1988); a single plant found in vacant lot in University Park, Dallas, in 1946; a 1977 collection from freeway waste area in downtown Dallas; also observed in Harry S. Moss Park, Dallas (R. May, pers. comm.); a widespread weed in North America and other temperate regions. Apr–Jun. Native of Eurasia. Cultivated for its edible root (Mabberley 2017).

TRICHOCORONIS A. Gray BUGHEAL

A genus of 2 species native to TX and Mexico. (Greek: *trichos*, hair, and *koronos*, crown, alluding to the pappus) (tribe Eupatorieae)
REFERENCES: King & Robinson 1970d; Nesom 2006z.

Trichocoronis wrightii (Torr. & A. Gray) A. Gray var. **wrightii**, (for Charles Wright, 1811–1885, prolific collector of TX and elsewhere), LIMESTONE BUGHEAL, WRIGHT'S HAIRY-CROWN. Perennial subaquatic herbs, fibrous-rooted, sometimes colonial; stems decumbent to erect, glabrate to pilose, sometimes rooting at nodes; leaves mostly opposite, sessile, blades oblong to elliptic or lanceolate, 10–25 mm long, margins ± dentate (at least the upper portion); involucres usually 3–4 mm wide, borne singly or in loose corymbose arrays; phyllaries in 2–3 series, lanceolate, ± equal; ray flowers absent; disk flowers 75–125, corollas whitish to pinkish or purplish, ca. 1 mm long; achenes spindle shaped, 4–5-ribbed, ca. 1 mm long; pappus of 2–6+ bristle-like scales, to 0.5 mm long. Edges of ponds, ditches, wet depressions in clay and sand; Colorado, Gonzales, Lavaca, and Victoria cos.; s to TX coast; adventive in CA; Mexico. Feb–Jul(–Oct). [*T. riparia* (Greene) Greene]

Thurovia triflora [USH]

Thymophylla puberula [THO]

Thymophylla wrightii [THO]

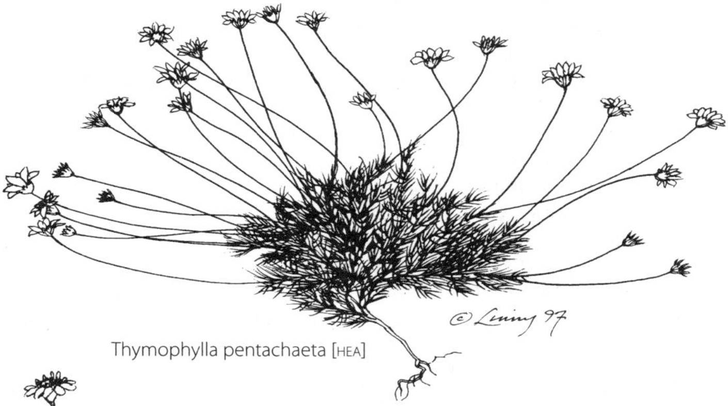

Thymophylla pentachaeta [HEA]

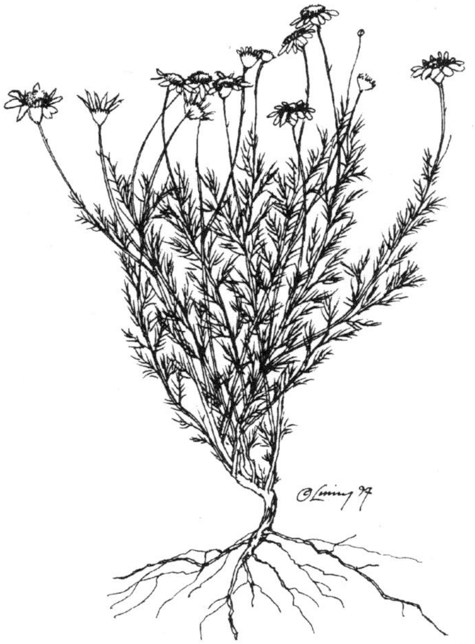

Thymophylla tenuiloba var. tenuiloba [HEA]

Townsendia exscapa [HO1]

VERBESINA L. CROWNBEARD, FLAT-SEED SUNFLOWER, WINGSTEM

Ours herbaceous annuals or perennials, ± pubescent; leaves simple, alternate or opposite; margins usually toothed; petioles usually winged when present, the wings in some species extending down along the stem; phyllaries in 2 to several series, subequal; ray flowers pistillate, fertile or infertile, the ligules yellow or whitish, or ray flowers absent; disk flowers perfect, fertile, the corollas yellow or whitish, usually the same color as ray ligules; achenes flattened, usually winged; pappus of 2 awns or scales, or pappus rarely absent.

A genus of 200+ species of trees, shrubs, and herbs of warm areas of the Americas; some are cultivated as ornamentals. (Name said to be modified from *Verbena*, reason unknown) (tribe Heliantheae)

REFERENCES: Robinson & Greenman 1899; Coleman 1966; Olsen 1979, 1985; Strother 2006hh.

1. Corollas white to dingy pale yellow or grayish yellow (but not bright yellow); ray ligules absent or < 8 mm long.
   2. Involucres 8–10 mm in diameter; ray flowers absent; disk flowers 40–60+ ____ **V. walteri**
   2. Involucres 3–7 mm in diameter; ray flowers (1–)2–12+; disk flowers 8–25+.
      3. Involucres 3–5 mm in diameter; ray flowers (1–)2–3(–7), the ligules 3–4(–7+) mm long; disk flowers 8–12 (–15) ____ **V. virginica**
      3. Involucres 5–7 mm in diameter; ray flowers (9–)10–12+, the ligules 4–6 mm long; disk flowers 20–25+ ____ **V. microptera**
1. Corollas bright yellow; ray ligules 8–30+ mm long.
   4. Leaves all or mostly opposite (sometimes alternate on uppermost stems); phyllaries 24–30+, oblong to orbiculate or ovate; achenes ca. 7 mm long; pappus of disk flowers 0–0.5 mm long ____ **V. lindheimeri**
   4. Leaves all or mostly alternate (sometimes opposite on lowermost stems); phyllaries 8–21+, spatulate or lance-ovate to lanceolate or linear; achenes 3.5–5+ mm long; pappus of disk flowers 0.5–2 mm long.
      5. Plants annual from a taproot; stem internodes unwinged; leaf blades usually gray-canescent beneath; ligules of ray flowers prominently toothed at the tips, the lobes mostly 2–4 mm long; disk flowers 80–150+ ____ **V. encelioides**
      5. Plants perennial from fibrous or fleshy roots or rhizomes, stem internodes, at least on lower stems, winged (with leaf-like tissue extending from the leaves down along the stem); leaf blades green and not gray-canescent beneath; ligules of ray flowers only slightly toothed or rounded at the tips, the lobes (if any) mostly < 1 mm long; disk flowers 40–80+.
         6. Heads several to numerous, (3–)8–25(–50+) per array; phyllaries 8–12+, spreading or reflexed; ray flowers (2–)6–8+; pappus 1.5–2 mm long ____ **V. alternifolia**
         6. Heads few to several, 2–5(–10+) per array; phyllaries 16–21+, erect; ray flowers 8–13+; pappus 0.5–1.5 mm long ____ **V. helianthoides**

Verbesina alternifolia (L.) Britton ex Kearney, (alternate-leaved), WINGSTEM. Rhizomatous perennials, 30–100(–200+) cm tall; lower stem internodes winged; leaves alternate (lower leaves sometimes opposite), sessile, blades lance-elliptic or lanceolate to lance-linear, margins coarsely toothed or nearly entire; heads several to numerous in corymbose or paniculate arrays; ray flowers (2–)6–8+, the ligules 15–25+ mm long, yellow; achenes 4.5–5 mm long; pappus 1.5–2 mm long. Woodlands, along streams; Collin, Dallas, Fannin, and Grayson cos.; e 1/2 of North America. Sep.

Verbesina encelioides (Cav.) Benth. & Hook. f. ex A. Gray, (resembling *Encelia*, another genus of Asteraceae), COWPEN DAISY, GOLDEN CROWNBEARD, FEVERWEED, BUTTER DAISY. Annuals, 10–50(–120+) cm tall; stem internodes unwinged; leaves alternate (lower leaves sometimes opposite), petioles usually winged and sometimes auriculate at base, blades deltate-ovate or rhombic to lanceolate, ashy-white to green above, gray-canescent beneath, margins coarsely toothed and lobed to nearly entire; heads solitary or 3+ in loose corymbose arrays, large and conspicuous; ray flowers (8–)12–15+, the ligules 8–10(–20+) mm long, yellow, with prominent terminal teeth; achenes 3.5–5+ mm long; pappus of disk flowers 0.5–1(–2) mm long. Disturbed areas, livestock pens, often in sandy soils; nearly throughout TX; widespread in U.S.; Mexico; West Indies; and South America; introduced and weedy elsewhere. Apr–Nov. [*V. encelioides* (Cav.) Benth. & Hook. f. ex A. Gray var. *exauriculata* B.L. Rob. & Greenm., *Ximenesia encelioides* Cav.] The foliage has an unpleasant odor if crushed or touched; Native Americans reportedly used the plant in treating skin diseases (Ajilvsgi 1984).

Verbesina helianthoides Michx., (resembling *Helianthus*—SUNFLOWER), GRAVELWEED CROWNBEARD. Rhizomatous perennials, 60–120+ cm tall; stem internodes winged; leaves alternate (lower leaves sometimes opposite), sessile, blades lance-ovate to lanceolate, margins ± serrate; heads few to several in corymbose arrays, large and conspicuous; ray flowers 8–13+, the ligules 20–25(–30+) mm long, yellow; achenes 5 mm long; pappus of disk flowers 0.5–1.5 mm long. Disturbed areas; moist sandy woods; East TX and Red River drainage, w to nc TX; ec U.S. from OH s to GA and w to KS and TX. May–Aug.

Verbesina lindheimeri B.L. Rob. & Greenm., (for Ferdinand Jacob Lindheimer, 1809–1879, German-born TX collector), LINDHEIMER'S CROWNBEARD. Perennials, 30–70+ cm tall; stem internodes unwinged; leaves opposite, sometimes alternate upward,

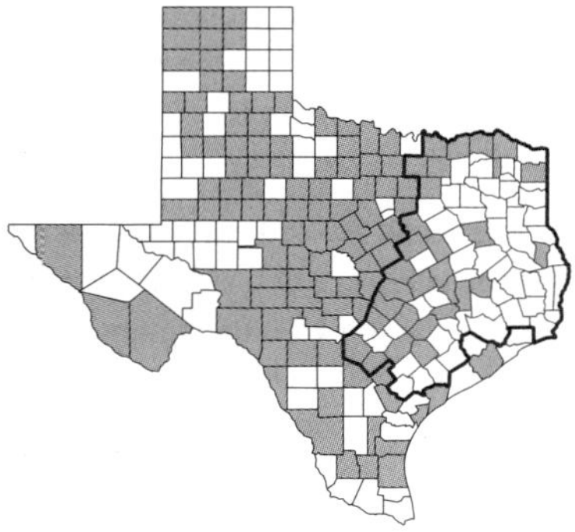

Thelesperma filifolium

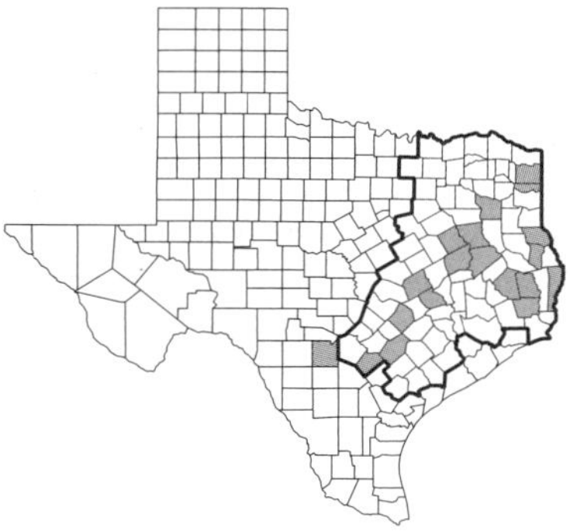

Thelesperma flavodiscum

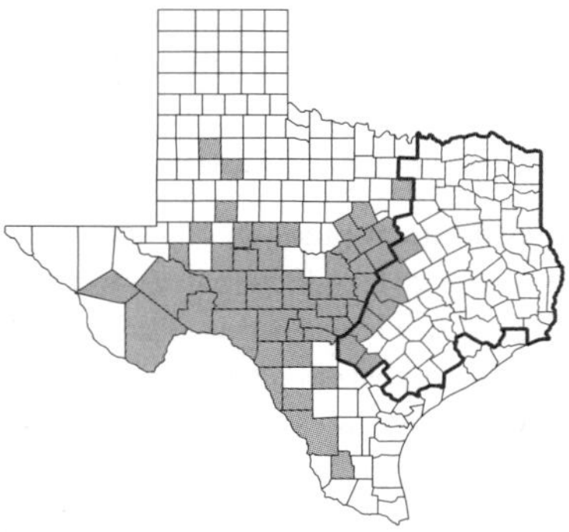

Thelesperma simplicifolium

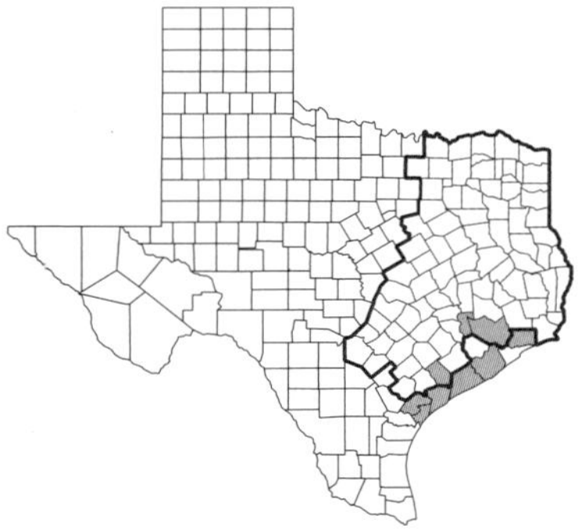

Thurovia triflora

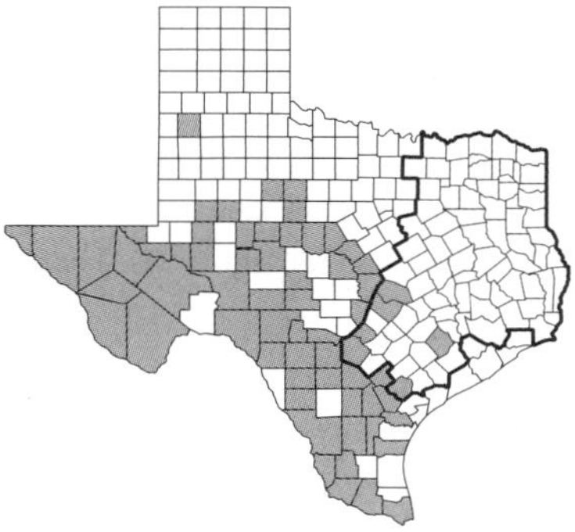

Thymophylla pentachaeta

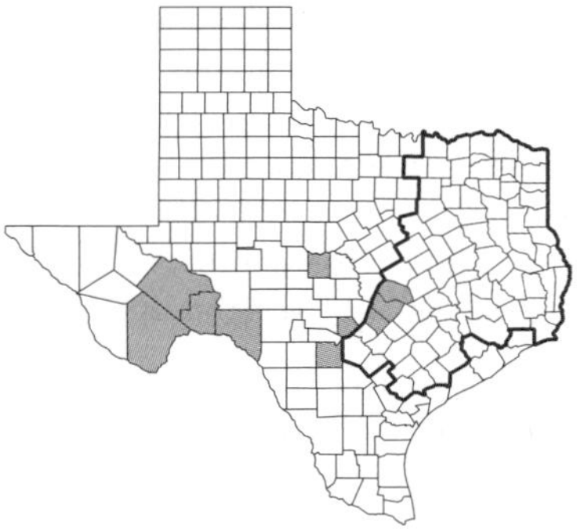

Thymophylla puberula

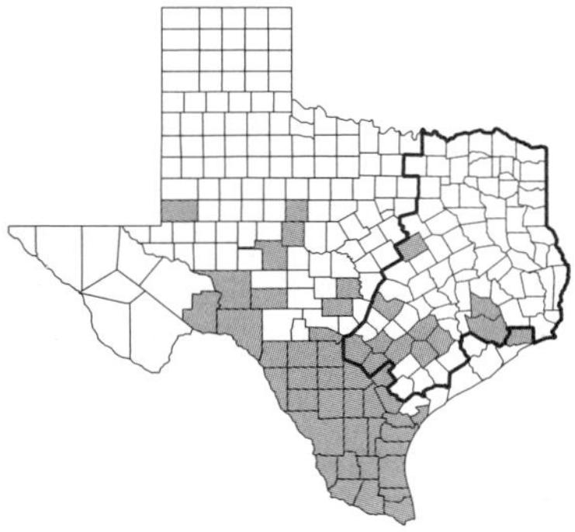

Thymophylla tenuiloba var. tenuiloba

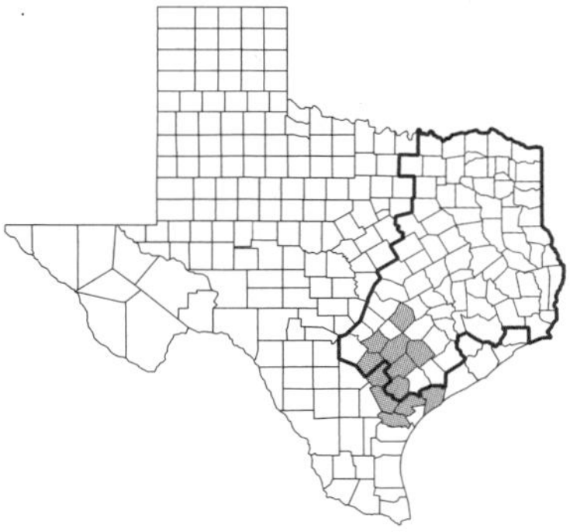

Thymophylla wrightii

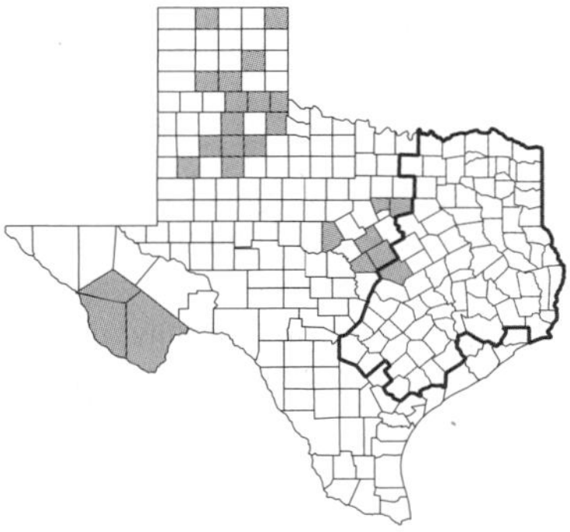

Townsendia exscapa

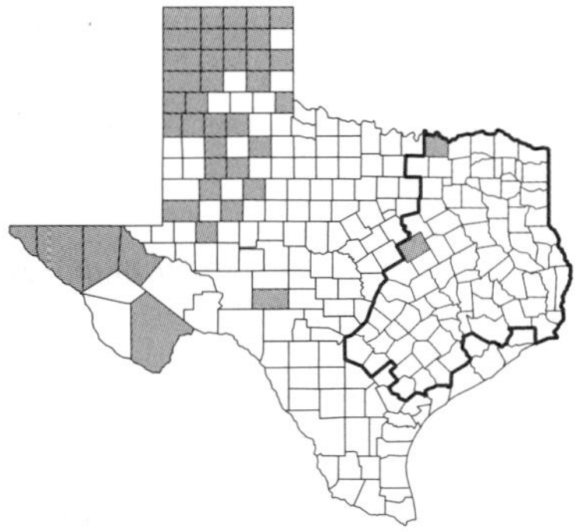

Tragopogon dubius

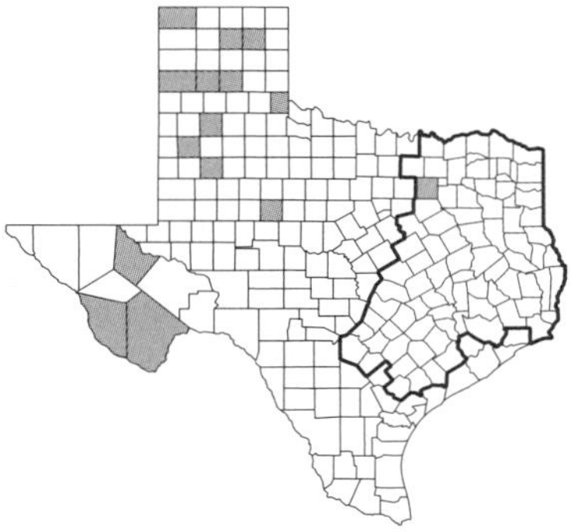

Tragopogon porrifolius

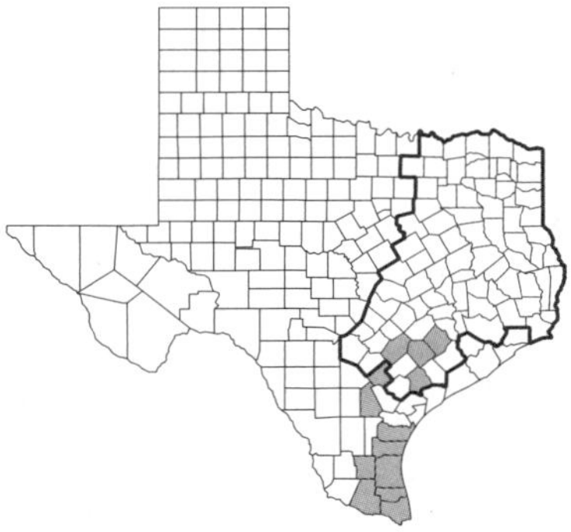

Trichocoronis wrightii var. wrightii

sessile, blades deltate-ovate or ovate to lance-ovate, margins toothed to entire; heads solitary or few in corymbose arrays, large and conspicuous; ray flowers ca. 13, the ligules 25 mm long, yellow; achenes 7 mm long; pappus 0–0.5 mm long. Juniper-oak woodlands, shaded slopes, on limestone; sw border of East TX from Bell to Comal cos., also apparently Jackson Co.; generally considered endemic to Lampasas Cut Plain and e part of Edwards Plateau. May–Sep(–Nov).

Verbesina microptera DC., (small-winged), TEXAS CROWNBEARD, MEXICAN CROWNBEARD. Perennials, 50–250 cm tall; stem internodes winged; leaves alternate (lower leaves sometimes opposite), petioles winged, blades deltate-ovate or deltate to lanceolate, margins coarsely toothed to nearly entire; heads numerous in paniculate arrays; ray flowers (9–)10–12+, the ligules 4–6 mm long, white to pale dingy yellow; achenes 4–5 mm long; pappus 1.5–2+ mm long. Disturbed sites, silty-clay soils; sw border of East TX in Wilson Co. (*W. Carr 20230* TEX) and Goliad Co; South TX Plains and Gulf Prairies; Mexico. Aug–Nov. [*V. texana* Buckley]

Verbesina virginica L., (of Virginia), FROSTWEED, ICEPLANT, WHITE CROWNBEARD, VIRGINIA CROWNBEARD, RICHWEED, SQUAW-WEED, INDIAN-TOBACCO. Perennials, (50–)100–250+ cm tall; lower stem internodes winged; leaves alternate (lower leaves sometimes opposite), petioles winged, blades lance-ovate to lance-linear, margins coarsely toothed to nearly entire; heads numerous in paniculate or corymbose arrays; ray flowers (1–)2–3(–7), the ligules 3–4(–7+) mm long, white to pale dingy yellow; achenes 3.5–5+ mm long; pappus 1.5–3+ mm long. Moist disturbed sites, bottomlands, woodland borders; se and e TX w to West Cross Timbers and Edwards Plateau; e and c U.S. from MD s to FL and w to IA and TX. Aug–Nov. [*V. laciniata* Walter, *V. virginica* var. *laciniata* A. Gray] When exposed to the first freezing temperatures of the year, the stems split and exude a sap that freezes into conspicuous ice formations—hence several of the common names (Ajilvsgi 1984). The roots were used by Native Americans to relieve cramps, chills, and fevers (Burlage 1968).

Verbesina walteri Shinners, (for Thomas Walter (c. 1740–1789), British-born American botanist and original describer of *Athanasia paniculata*, later moved to *Verbesina*), CAROLINA CROWNBEARD, WALTER'S CROWNBEARD. Perennials, 100–400 cm tall; lower stem internodes winged; leaves alternate (lower leaves sometimes opposite), petioles winged, blades lance-elliptic to lance-linear, margins coarsely toothed to nearly entire; heads several to numerous in paniculate or corymbose arrays; ray flowers absent; disk flowers white to pale dingy yellow; achenes 3–4.5+ mm long; pappus 1–2 mm long, sometimes with shorter scales between the awns. Bottomlands, floodplains, woodland borders; in TX reported only from Jasper and Newton cos. (Singhurst & Holmes 2001); se U.S. from NC s to GA and w to OK and TX. Aug–Sep. [*Athanasia paniculata* Walter]

VERNONIA Schreb. IRONWEED

Ours clump-forming or rhizomatous perennial herbs; leaves evenly distributed along the stem, alternate, simple, often pubescent and gland-dotted beneath, margins toothed; heads many, rather small, in corymbose or paniculate arrays; phyllaries in 4–7 series, unequal, herbaceous but rather stiff or hard, sometimes purple-tinged, strongly imbricated; ray flowers absent; disk corollas purple to lavender-pink, usually more than 12 per head; achenes ± columnar, 8–10-ribbed; pappus of 2 series of numerous scales or bristles, the inner series much longer than the outer.

A genus of ca. 20 species of e and c North America, Mexico, and South America; some species are cultivated as ornamentals and some are used medicinally. Interspecific hybridization is well known in *Vernonia* (Urbatsch 1972; Jones & Faust 1978; Strother 2006ii). Hybrids tend to be intermediate to the parental species and more difficult to key (Turner 2016); three recognized hybrids of East TX are listed following the species in this treatment. The rich purple corollas make *Vernonia* species a showy component of the late summer–fall flora of East TX. The common name alludes to the strength of the erect stems. (Named for William Vernon, 1666–1711, English botanist who collected in Maryland) (tribe Vernonieae)

REFERENCES: Gleason 1922; Faust 1972; Urbatsch 1972; Faust & Jones 1973; King & Jones 1975; Chapman & Jones 1978; Robinson 1999; Strother 2006ii.

1. Phyllaries densely silky-tomentose, appearing white or grayish; leaf blades densely white-woolly beneath; stems densely gray-woolly ____ **V. lindheimeri**
1. Phyllaries glabrous or slightly pubescent, not whitish or grayish; leaf blades glabrous or variously pubescent but not densely white-woolly beneath; stems glabrous to pubescent, but not gray-woolly.
   2. Leaf blades with numerous small but conspicuous pits on the lower surface (best seen with a 10x lens; can appear as black dots under low magnification).
      3. Heads campanulate, involucres 5–7(–8+) mm high and 4–6 mm wide, in congested, corymbose arrays; achenes 3.5–4 mm long; pappus brownish gray or purplish; rare in ne TX (Red River Co.) ____ **V. fasciculata**
      3. Heads obconic to hemispheric, involucres 4.5–6 mm high and 5–7 mm wide, in open scorpioid-paniculate arrays; achenes 2–3 mm long; pappus whitish to straw-colored (rarely purplish); common throughout e and c TX ____ **V. texana**

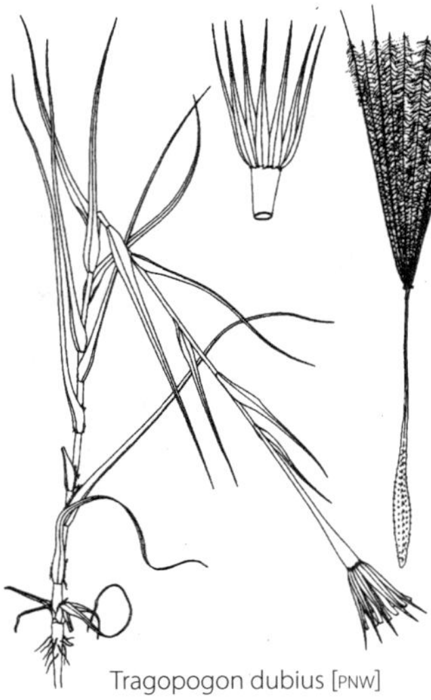
Tragopogon dubius [PNW]

Tragopogon porrifolius [PNW]

Trichocoronis wrightii [CO1]

Verbesina alternifolia [GLE]

Verbesina lindheimeri [HEA]

Verbesina encelioides [BB1]

Verbesina helianthoides [BB1]

2. Leaf blades gland-dotted or not, but without conspicuous pits on the lower surface.
    4. Leaf blades usually scaberulous on the lower surface (with appressed, awl-shaped hairs) and not or only sparsely gland-dotted beneath ____ **V. gigantea**
    4. Leaf blades usually puberulent to tomentose on the lower surface (with erect or curled hairs) and conspicuously gland-dotted beneath.
        5. Heads broadly campanulate to hemispheric, 4–6(–8+) mm high and 4–7+ mm wide; phyllaries often gland-dotted; flowers (15–)20–25(–35+); achenes 2.5–3 mm long; pappus brownish gray or purplish ____ **V. baldwinii**
        5. Heads broadly campanulate to urceolate, (6–)7–10+ mm high and 5–9+ mm wide; phyllaries seldom gland-dotted; flowers 30–55; achenes 3.5–4 mm long; pappus whitish to straw-colored ____ **V. missurica**

Vernonia baldwinii Torr., (for its discoverer, William Baldwin, 1779–1819, botanist and physician of Pennsylvania), BALDWIN'S IRONWEED, WESTERN IRONWEED. Plants 60–100(–150) cm tall; midstem leaves with blades elliptic to lance-ovate or lanceolate, usually 8–15 cm long and 2–4.5 cm wide, gland-dotted and puberulent to tomentose beneath; phyllaries usually puberulent, often gland-dotted at tips; achenes 2.5–3 mm long; pappus brownish gray or purplish, outer scales 0.2–1 mm long, inner bristles 5–7 mm long. Open woods, low ground, prairies; e and c TX and Panhandle; c U.S. from MN s to LA and w to CO and NM. Jun–Oct. [*V. baldwinii* Torr. subsp. *interior* (Small) Faust, *V. interior* Small] The bitter foliage apparently prevents herbivory by cattle and the plant is thus sometimes common in overgrazed pastures (Ajilvsgi 1984).

Vernonia fasiculata Michx., (bundled or clustered), PRAIRIE IRONWEED. Plants 30–120+ cm tall; midstem leaves with blades ± lanceolate, usually 5–12 cm long and 0.5–1.8 cm wide, gland-dotted and scaberulous beneath; phyllaries nearly glabrous; achenes 3.5–4 mm long; pappus brownish gray or purplish, outer scales or bristles 0.5–3+ mm long, inner bristles or scales 5–7+ mm long. Low prairies, bottomlands, ditches; in TX known only from Red River Co. (*M. White s.n.* BAYLU; White 2012); c U.S. from MI s to KY and w to MT, CO, and TX; Canada. Jul–Sep. [*V. fasiculata* Michx. subsp. *corymbosa* S.B. Jones]

Vernonia gigantea (Walter) Trel., (gigantic), TALL IRONWEED, GIANT IRONWEED. Plants 80–200(–300) cm tall; midstem leaves with blades ± lanceolate, 12–25+ cm long and 2–6+ cm wide, scaberulous to nearly glabrous beneath, not or only slightly gland-dotted; phyllaries glabrous or nearly so; achenes 2.5–3.5 mm long; pappus usually purplish, sometimes straw-colored, outer scales 0.5–1 mm long, inner bristles 4.5–6+ mm long. Weedy areas, floodplains, low moist sites; se TX in Hardin Co. (Brown 2014), Harris Co. (TEX), Jasper, Jefferson, Orange (TAES) and Montgomery and Polk cos.; e 1/2 of the U.S.; Canada. Jun–Nov. [*V. altissima* Nutt., *V. gigantea* subsp. *ovalifolia* (Torr. & A. Gray) Urbatsch, *V. ovalifolia* Torr. & A. Gray]

Vernonia ×guadalupensis A. Heller [*baldwinii* × *lindheimeri*], (for the Guadalupe River area of the Edwards Plateau). This hybrid is known from several counties on the w border of East TX, from Bell Co. (Hansen 2010) and Williamson and Travis cos., to Hays Co. (TEX); c TX and Edwards Plateau.

Vernonia ×illinoensis Gleason [*gigantea* × *missurica*], (of Illinois). In TX, this hybrid is known only from Jefferson Co.; ec U.S. from MI s to FL and w to OK and TX.

Vernonia lindheimeri A. Gray & Engelm., (for Ferdinand Jacob Lindheimer, 1809–1879, German-born TX collector), WOOLLY IRONWEED, LINDHEIMER'S IRONWEED. Plants 20–50(–80+) cm tall; stems densely gray-woolly; midstem leaves with blades linear, 5–8 cm long and 2–4 mm wide, densely white-woolly beneath; phyllaries silky-tomentose, appearing white or gray; achenes 3–4 mm long; pappus straw-colored to purplish, outer scales 0.6–1.1 mm long, inner bristles 6–7+ mm long. Limestone outcrops, calcareous prairies, creekbeds; w border of East TX from Dallas Co. s to Bexar Co.; c TX, Edwards Plateau, and Trans-Pecos; AR and TX; Mexico. Jun–Sep.

Vernonia missurica Raf., (of Missouri), MISSOURI IRONWEED. Plants 60–120(–200+) cm tall; midstem leaves with blades elliptic to lance-ovate or lanceolate, usually 6–16 cm long and 1.8–4.8 cm wide, usually puberulent to tomentose beneath, rarely glabrate, gland-dotted; phyllaries nearly glabrous; achenes 3.5–4 mm long; pappus straw-colored to whitish, outer scales 0.5–1.1 mm long, inner bristles 6–8+ mm long. Bogs in open woods, sandy prairies, marsh edges; e 1/4 TX; ec U.S. from MI s to FL and w to NE and TX. Jul–Oct.

Vernonia texana (A. Gray) Small, (of Texas), TEXAS IRONWEED. Plants 40–80(–100+) cm tall; midstem leaves with blades lance-ovate to lanceolate, usually 5–12+ cm long and 1.2–2.5+ cm wide, glabrous or nearly so beneath and gland-pitted; phyllaries nearly glabrous; achenes 2–3 mm long; pappus straw-colored to whitish (rarely purplish), outer scales or bristles 0.3–1+ mm long, inner bristles or scales 6–7+ mm long. Sandy woods, pine savannas, prairies; widespread from East TX to sc TX; AR, LA, MS, OK, and TX. Jun–Sep. [*V. angustifolia* Michx. var. *texana* A. Gray]

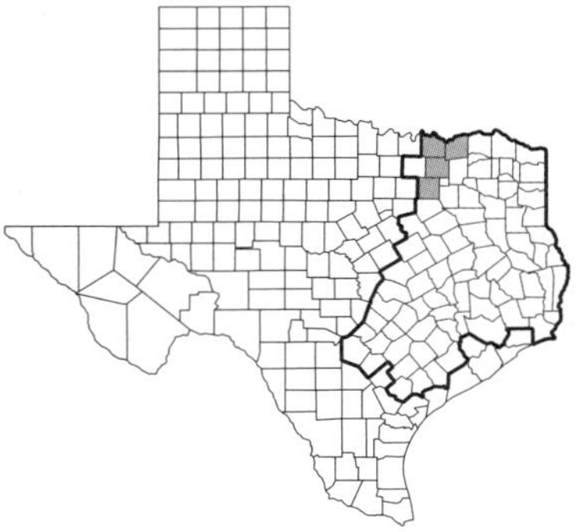

Verbesina alternifolia

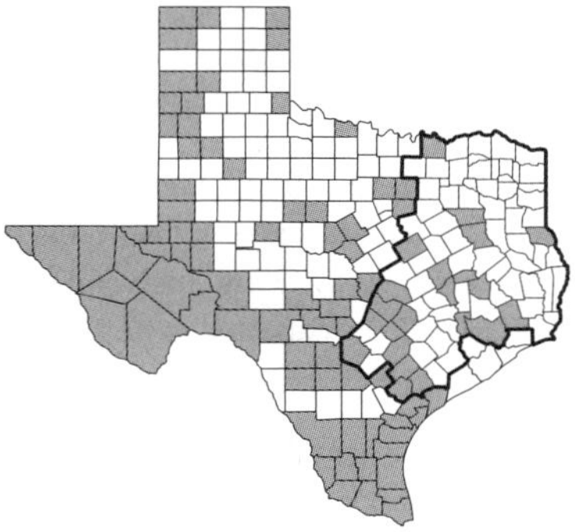

Verbesina encelioides

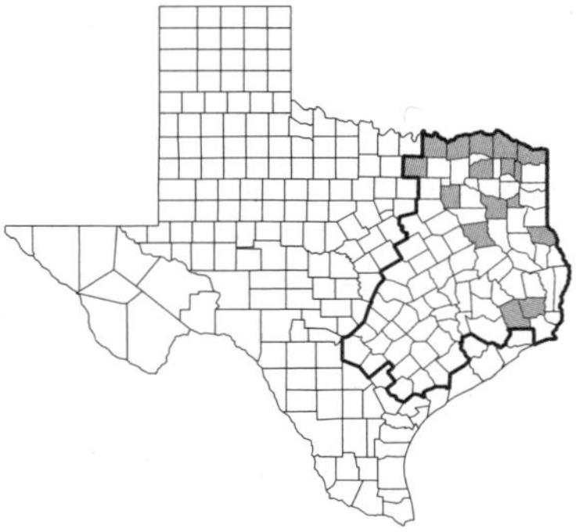

Verbesina helianthoides

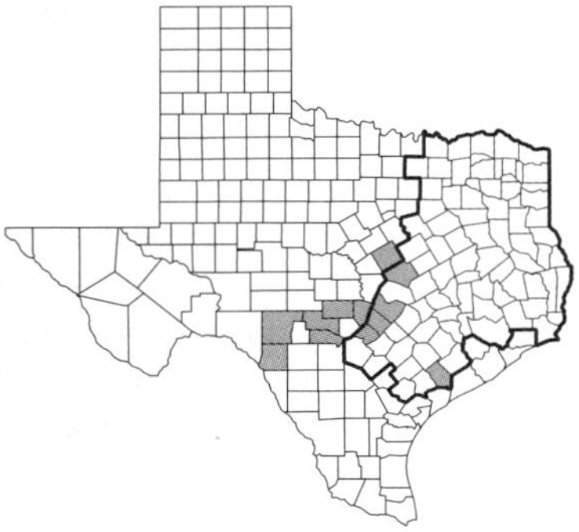

Verbesina lindheimeri

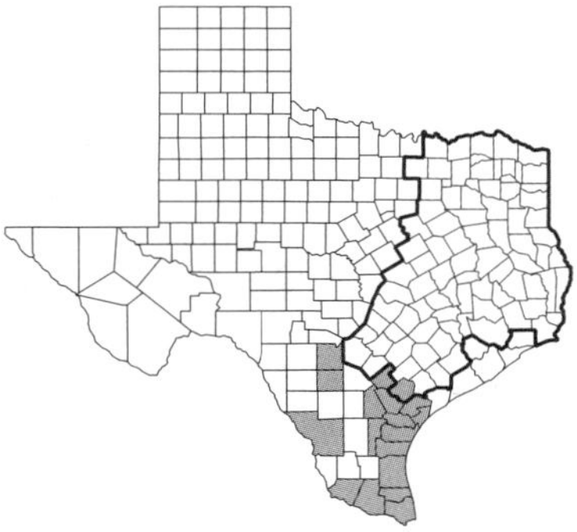

Verbesina microptera

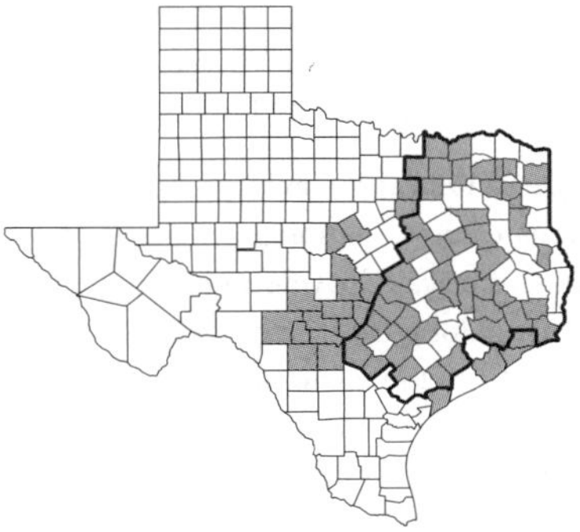

Verbesina virginica

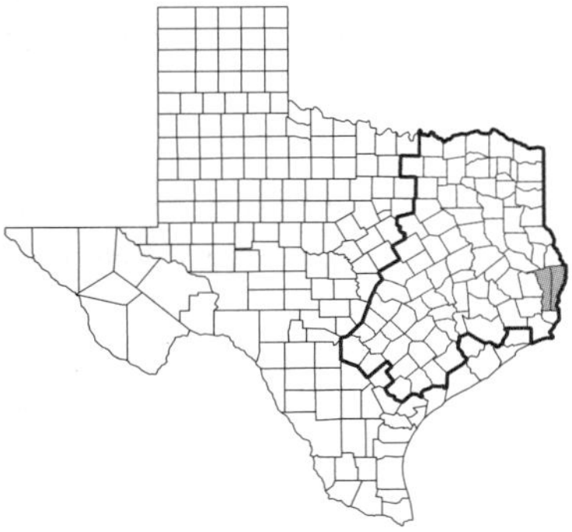

Verbesina walteri

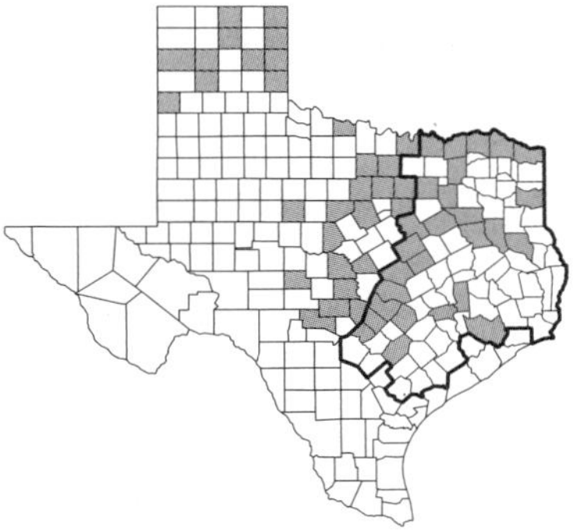

Vernonia baldwinii

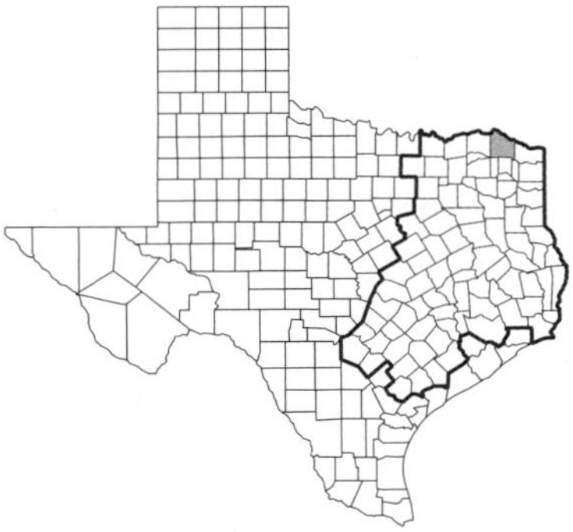

Vernonia fasciculata

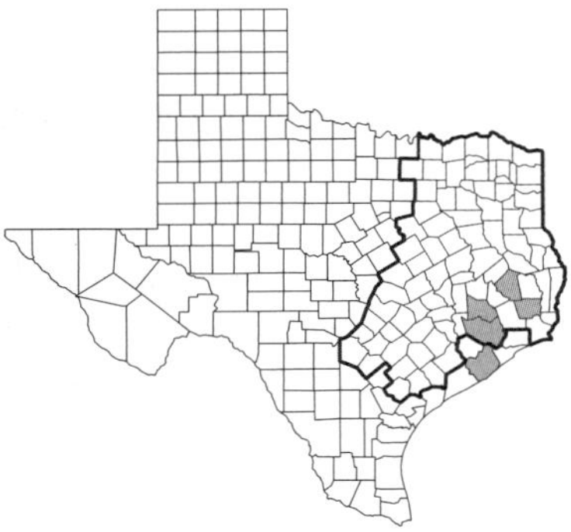

Vernonia gigantea

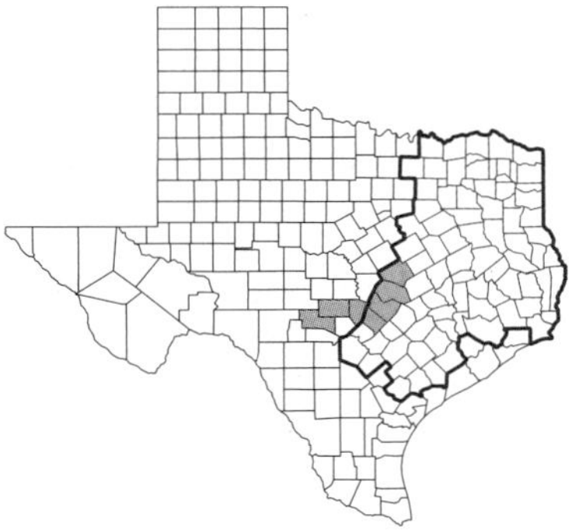

Vernonia ×guadalupensis

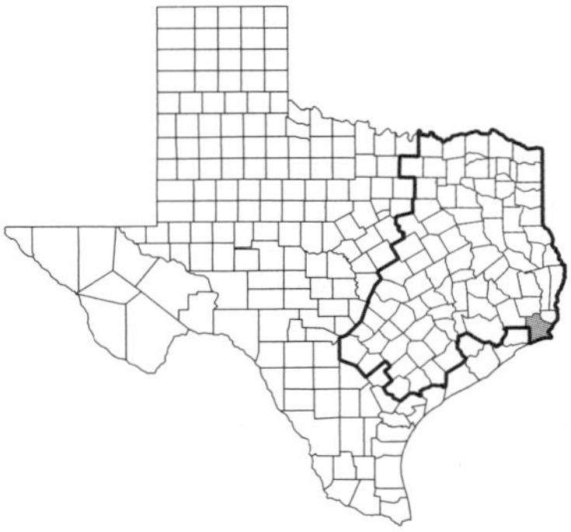

Vernonia ×illinoensis

Vernonia ×*vulturina* Shinners [*baldwinii* × *marginata*], (for Buzzards Spring in Dallas, TX). This hybrid is known only from an apparently large plant collected by Reverchon ca. 1880 at Buzzards Spring, in the White Rock Creek drainage, Dallas (since filled and now a residential area); Shinners (1954) noted the population was likely extirpated.

Vernonia lettermannii Engelm. ex A. Gray, (for George Washington Letterman (1840–1913), amateur botanist of the c U.S. and collector of one of the type specimens), which would have keyed to *V. fasciculata* or *V. texana* here, can be distinguished by the filiform leaves, narrowly campanulate heads that are 7–9(–10+) mm high and 4–5 mm wide, and flowers numbering 10–12+ (as opposed to 12–25+ in *V. fasciculata* and *V. texana*). To be expected in ne TX based on records from adjacent counties in OK (Bryan and McCurtain cos. (OKL)); AR. Jul–Aug.

Vernonia ×*peralta* Daniels [*baldwinii* × *missurica*], (highest or tallest). This hybrid may be expected in East TX based on a record from adjacent Vernon Par., LA.; c U.S. from IA s to LA and w to OK (Kartesz 2015).

VIGUIERA Kunth GOLDEN-EYE

A genus of ca. 150 species native to warm and tropical areas of the Americas; a few are cultivated as ornamentals. (Named for L.G.A. Viguier, 1790–1867, French physician and botanist) (tribe Heliantheae)
REFERENCES: Blake 1918; Robinson 1977; Schilling 2006b.

Viguiera dentata (Cav.) Spreng., (toothed), SUNFLOWER GOLDEN-EYE. Herbaceous perennial 0.5–1.5 m tall; leaves opposite below, alternate above, petiolate, blades ovate or rhombic-ovate to lanceolate, cuneate to truncate, margins serrate; phyllaries in 3–4 series, imbricated, ovate and indurate basally, with herbaceous linear tips; ray flowers 10–13, pistillate, neuter, the ligules yellow, 7–15 mm long, 3–7.5 mm wide; disk flowers perfect, fertile, the corollas yellow; achenes black or mottled, appressed-pubescent, 3.5–3.8 mm long; pappus awns 2, 2.2–2.8 mm long, squamellae (additional small pappus scales) 2–4, squarish, fimbriate. Dry calcareous soils, roadsides; w and sw edge of East TX from Hill Co. s to Bexar, Fayette, and Washington cos., also Austin, DeWitt, and Hardin (TAES) cos.; Edwards Plateau to Trans-Pecos; AZ, NM, and TX; Mexico. Oct–Nov. [*Helianthus dentatus* Cav.]

WEDELIA Jacq.

A genus of ca. 25 species of tropical and warm areas of the world; 1 species is found in the U.S.; a number of taxa have sometimes been treated in *Zexmenia*. (Named for Georg Wolfgang Wedel, 1645–1721, professor of botany at Jena, Germany) (tribe Heliantheae)
REFERENCES: Turner 1988d, 1992; Strother 1991, 2006jj.

Wedelia acapulcensis Kunth var. **hispida** (Kunth) Strother, (sp.: of Acapulco; var.: with stiff hairs) ORANGE ZEXMENIA, HAIRY WEDELIA, ORANGE DAISY. Small shrub 0.5–1.5 m tall; herbage strigose-hispid; leaves opposite, sessile, blades narrowly ovate, usually 5 cm or more long, cuneate basally, coarsely toothed to slightly 3-lobed; heads long-peduncled, solitary or in a cyme of 3; phyllaries in 2–3 series; ray flowers pistillate, fertile, the ligules conspicuously yellow-orange; disk flowers perfect, fertile, the corollas similar in color to ligules of ray flowers; achenes 2- or 3-angled, usually winged; pappus of (0–)1–3 awns, 4.5–6 mm long, subtended by a crown of minute hyaline scales. Calcareous soils, rocky slopes; sw edge of East TX from Travis Co. s to Bexar and DeWitt cos.; sw part of nc TX s through Edwards Plateau to s TX. Apr–Nov. [*W. hispida* Kunth, *W. texana* (A. Gray) B.L. Turner, *Zexmenia hispida* (Kunth) A. Gray] This species is gaining popularity in TX as a long-blooming, drought-tolerant ornamental.

XANTHISMA DC. SLEEPY DAISY

A genus of 17 species native to w North America and Mexico; some were previously included in the genera *Haplopappus*, *Machaeranthera*, or *Sideranthus*. The common name refers to the late morning expansion of the ray flowers. (Greek: *xanthos*, yellow, and *ismos*, condition or quality, referring to the bright yellow corollas) (tribe Astereae)
REFERENCES: Semple 1974; Hartman 2006.

Xanthisma texanum DC., (of Texas), TEXAS SLEEPY DAISY. Annuals (or rarely biennials), 10–75 cm tall; stems mostly glabrous; leaves alternate, sessile or subsessile, basal leaves pinnatifid to coarsely serrate, rarely persisting, stem leaves reduced, blades lanceolate to spatulate, glabrous, margins entire to toothed or ciliate; heads mostly solitary and terminal on main stem and branches; receptacles naked; involucres 11–20 mm wide; phyllaries in 3–4 series, basally narrowed and stalk-like, expanded above, glabrous, margins often ciliate; ray flowers pistillate, fertile, ligules yellow, expanded from late morning to late afternoon, becoming erect and with tips nearly touching during the night; disk flowers perfect, fertile, corollas yellow; achenes 1.6–1.8 mm long, the ray achenes ± 3-sided, the disk achenes compressed; pappus of many basally-flattened bristles, 5–6.5 mm long. May–Jul. Burlage (1968) reported the plant to contain a saponin and to be poisonous.

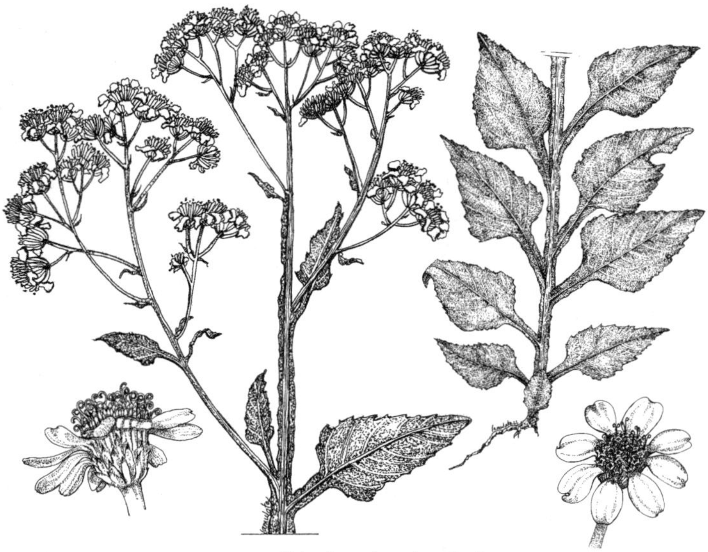

Vebesina microptera [THO]

Verbesina virginica [MO4]

Verbesina walteri [RAD]

Vernonia baldwinii [REE]

Vernonia fasciculata [GEO]

Vernonia gigantea [GLE]

Vernonia lindheimeri [BT3]

1. Phyllaries abruptly expanded above stalk-like base, mostly ovate, the portion above the widest part longer than wide; relatively widespread in TX ____ var. **drummondii**
1. Phyllaries somewhat expanded above stalk-like base, mostly orbiculate to elliptic, the portion above the widest part shorter than wide; restricted to s TX.
    2. Phyllary tips cuspidate, flared near tip ____ var. **orientale**
    2. Phyllary tips rounded, with little or no flare near tip ____ var. **texanum**

var. **drummondii** (Torr. & A. Gray) A. Gray, (for Thomas Drummond, 1780–1835, Scottish botanist and collector in North America), DRUMMOND'S SLEEPY-DAISY. Calcareous soils, prairies, pastures, roadsides; se to c TX; nc TX and Panhandle; NM, OK and TX; apparently introduced in AZ. [*X. texanum* DC. subsp. *drummondii* (Torr. & A. Gray) Semple]

var. **orientale** Semple, (eastern). Sandy prairies, open woods, disturbed sites; Goliad Co. and Wilson Co. (TEX-LL); s TX and s Gulf Coast; probably Mexico.

var. **texanum.** Carrizo Formation, deep sands, pastures, roadsides; Bexar Co. and Wilson Co. (TEX-LL); s TX. [*X. texanum* DC. subsp. *texanum*]

XANTHIUM L. COCKLEBUR, BURWEED

Coarse, monoecious annuals with taproots; leaves alternate; heads small, axillary, nearly sessile, unisexual; ray flowers absent; staminate heads with 1–3 series of separate phyllaries; disk flower corollas minute, whitish; anthers 5, separate; pistillate heads with 6–12+ series of phyllaries united into a spiny bur or "fruit" with hooked prickles; bur completely enclosing 2 flowers that lack corollas; achenes 2 per bur; pappus absent.

A cosmopolitan (now) genus of ca. 3 species with conspicuous, accrescent involucres covered with hooked prickles effecting dispersal by mammals. According to the VELCRO® Industries website (www.velcro.com), in the early 1940s, a Swiss inventor, George de Mestral, after a walk noticed "cockleburrs" [presumably *Xanthium*] on his dog and his pants. He examined the hooked prickles under a microscope and derived the idea for the well-known two-sided fastener—one side with stiff, cocklebur-like "hooks" and the other side with soft "loops" like the cloth of his pants. The word velcro comes from the French words *velours*, velvet, and *croché*, hooked. (Greek name of some plant used to dye the hair, from *xanthos*, yellow) (tribe Heliantheae)
REFERENCES: Millspaugh & Sherff 1919; Rydberg 1922; Löve & Dansereau 1959; Karis 1995; Strother 2006kk.

1. Nodes with a conspicuous, 1–3-pronged, yellowish spine; bur ca. 1 cm long (including prickles); leaf blades tapering or wedge-shaped at base, silvery-pubescent beneath; rare in East TX ____ **X. spinosum**
1. Nodes without spines (the only spiny or prickly part of the plant is the bur or "fruit"); bur usually (1–)2–3 cm long (including prickles); leaf blades truncate or cordate at base, green and scabrous beneath; widespread in East TX ____ **X. strumarium**

Xanthium spinosum L., (spiny), SPINY COCKLEBUR, CLOTBUR. Stems 0.3–1(+) m tall; nodes with spines; leaf blades to 12 cm long, entire or with a few teeth or lobes, slightly pubescent or glabrate on upper surface, densely silvery-pubescent on lower surface, tapering to a short petiole; bur 1–1.2 cm long, finely pubescent, with numerous, hooked prickles ca. 2 mm long. Disturbed areas, seasonally wet waste places; rare in East TX, with the only record from Orange Co. (Turner et al. 2003); mainly Edwards Plateau and Trans-Pecos; scattered throughout North America; Central and South America; introduced elsewhere. Jul–Oct. [*X. ambrosioides* Hook. & Arn.] Region of origin unclear, probably the Americas. The plant can cause mechanical injuries; ingestion by pigs can result in toxic symptoms including prostration and convulsions (Burlage 1968).

Xanthium strumarium L., (of swellings or tumors), COCKLEBUR, COMMON COCKLEBUR, ABROJO. Stems 0.4–2 m tall; nodes lacking spines; leaf blades to 15 cm long, ovate to deltoid, suborbicular, or reniform, serrate or with shallow lobes, scabrous; bur (1–)2–3+ cm long, ca. 1.5 cm in diam., terminated by 2 prominent spines, covered with stiff, hooked prickles ca. 5 mm long; bur and bases of spines densely pubescent. Disturbed moist areas or sandy soils; throughout TX; scattered throughout North America; Central and South America; introduced elsewhere. Fruiting Jul–Nov. [*X. italicum* Moretti, *X. strumarium* L. var. *canadense* (Mill.) Torr. & A. Gray] Region of origin unclear, probably the Americas. Recognition of a dozen or more taxa (treated as species, subspecies, varieties, and/or forms) has been proposed for plants treated together here as *Xanthium strumarium* (Strother 2006kk). Poisonous and potentially fatal to pigs and other livestock; the poisonous principle is hydroquinone or a diterpenoid glycoside and occurs in the seeds and seedlings (Sperry et al. 1955; Kingsbury 1964, 1965; Stephens 1980; Hardin & Brownie 1993). The burs can cause mechanical injury when eaten by livestock and are sometimes referred to as "porcupine eggs" (Barkley 1986).

YOUNGIA Cass.

An Asian genus of ca. 30 species. (Named for "*deux Anglais célèbres, l'un comme poète, l'autre comme physicien,*" both named Young; the poet may have been Edward Young (also dramatist), 1683–1765; the physician may have been Thomas Young (also physicist and Egyptologist), 1773–1829 (Spurr 2006)). (tribe Cichorieae)
REFERENCES: Babcock & Stebbins 1939; Vuilleumier 1973; Spurr 2006.

Youngia japonica (L.) DC., (of Japan), JAPANESE-HAWKWEED. Annual herb from taproot; sap milky; stems 20–50(–90+ cm tall; basal leaves pinnatifid, crowded; stem leaves alternate, few and reduced; heads numerous, small, 6–7 mm long including pappus, in corymbose panicles; involucres 3.5–5.7 mm high, glabrous; phyllaries usually 8, linear-lanceolate, subequal, subtended by a series of small phyllary-like bracts; flowers all ligulate, ligules yellow or yellow-orange, sometimes purplish beneath; achenes 1.5–2.5 mm long, fusiform, ribbed; pappus of numerous hair-like bristles 2.5–3.5 mm long, slightly surpassing phyllaries. Flowerbeds, gardens, weedy metropolitan areas; Dallas and Travis cos.; Brazos Co. se to Harris and Jefferson cos. (see recent county records for Liberty and Montgomery cos. in Brown et al. (2009) and Brown (2014); scattered across the ne and se U.S. from NY s to FL and w to TX; also CA; introduced elsewhere. All seasons, but mostly Oct–Nov. Native of Asia; a widespread subtropical weed.

ZINNIA L. ZINNIA

A genus of ca. 17 species from the sw U.S. to Argentina, especially Mexico; herbs and low shrubs with opposite or whorled leaves and alkaloids; some are cultivated as ornamentals. (For Johann Gottfried Zinn, 1727–1759, German professor of botany at Göttingen known for his botanical work in Mexico) (tribe Heliantheae)
REFERENCES: Torres 1963; Smith 2006b.

Zinnia violacea Cav., (elegant), ELEGANT ZINNIA, GARDEN ZINNIA, YOUTH-AND-OLD-AGE. Annual herb to ca. 1 m tall; stems unbranched or sparingly branched; leaves opposite, ovate to oblong, 6–10 cm long, 2–6 cm wide, scaberulous to glabrate; heads terminating stems on peduncles 8+ cm long; involucres 10–15 mm high; phyllaries obovate; ray flowers 8–21 (more in "double" cultivars), pistillate, fertile; ligules to 35 mm long, spatulate to obovate, usually red (or white, yellow, or pink-purple in cultivars); disk flowers perfect, fertile, the corollas yellow; pales red to purple; achenes 6–10 mm long, ray achenes 3-angled, disk achenes compressed; pappus absent. Disturbed sites; rare in TX; recorded in Walker Co. (Keith 2004a) and Harris Co. (Kartesz 2015); scattered across the se U.S. and CA; Mexico; Central and South America; and West Indies; introduced elsewhere. Summer–fall. [*Z. elegans* Jacq.] Native of Mexico. Originally cultivated by the Aztecs, now the most widely cultivated *Zinnia*; escaping and naturalized in humid subtropical regions around the world. In TX, occurring as an occasional escape or waif and probably not persisting in the flora.

BALSAMINACEAE A. Rich.

TOUCH-ME-NOT FAMILY

A medium-sized (ca. 1,000 species in 2 genera), mainly Old World tropical family with a few in temperate regions, with none in Central or South America or Oceania (Stevens 2001 onwards; Mabberley 2017); they are generally perennial herbs, with a few annuals or rarely subshrubs (Fischer 2004; Mabberley 2017). *Hydrocera*, native to Indomalesia, has only a single species. Family name conserved from *Balsamina*, a genus now treated as *Impatiens* (the name *Impatiens* was published earlier and thus has priority in terms of nomenclature). Early molecular research suggested a relationship of the family with the Ericales (Fischer 2004), and more recent work confirmed the link; the family is now widely accepted to be part of the Ericales (APG IV). (Possibly from Greek: *balsamon*, a fragrant gum, or from ancient Arabic: *balassam*, for some species of *Impatiens*) (subclass Rosidae—Cronquist; order Ericales—APG IV)
FAMILY RECOGNITION IN THE FIELD: herbs with translucent watery stems, alternate simple leaves, showy bilaterally symmetrical flowers with a conspicuous *spur*, and *explosively dehiscing* capsules.
REFERENCES: Rydberg 1910; Wood 1975; Fischer 2004.

IMPATIENS Riv. ex L.

TOUCH-ME-NOT, JEWELWEED, BALSAM, SNAPWEED, BUSY-LIZZY

Ours erect, annual herbs; stems watery, translucent, sometimes with swollen nodes; leaves alternate, simple; flowers bisexual; sepals 3, 2 small, 1 large and conspicuous, petal-like, saccate, and with a long narrow nectar spur; petals apparently 3, the 2 laterals 2-lobed (representing fused petals); anthers 5, united around stigma; pistil 1; ovary superior; fruit an explosively dehiscent, 5-valved capsule.

A genus of ca. 1,000 species of tropical and n temperate areas, especially Africa, Madagascar, India, the Himalaya region, Indo-China, and adjacent se Asia, with only a few species from temperate areas; many are found in montane rainforests and the genus is remarkable for its high degree of endemism (Stevens 2001 onwards; Fischer 2004; Mabberley 2017). Classification within the genus is in serious need of revision (Janssens et al. 2006). Many *Impatiens* species are used as ornamentals (e.g., *I. walleriana* Hook. f.—BUSY-LIZZY, SULTAN'S-FLOWER). Several Middle Eastern and African species were used as a source of dyes (red, yellow, or black) (Baumgardt 1982). Fischer (2004) and Mabberley (2017) noted that the common names TOUCH-ME-NOT and BUSY-LIZZY are from the explosively dehiscent capsule type fruit (Hayashi et al. 2009). Tension develops in the fleshy fruit wall and when ripe, the slightest touch or other disturbance can cause the valves of the fruit to suddenly roll up inwards starting at their base, thus forcibly ejecting the seeds. The mechanism is energy efficient and is similar to a catapult in which "stored elastic energy is rapidly released and converted into kinetic energy of the seeds" (Deegan 2012). Floral diversity is extreme; *Impatiens* has been called the "dicot counterpart of orchids" and the "evolutionary trend towards specialized pollination has implied abundant speciation and often accounts for high rates of endemism" (Fischer 2004). (Latin: *impatiens*, impatient, alluding to the explosive release of the seeds when the capsule is touched)

REFERENCES: Rust 1977; Janssens et al. 2006; Hayashi et al. 2009; Deegan 2012.

1. Flowers of various colors, with extremely narrow opening; capsules densely pubescent; stems usually pubescent; petioles 0.5–2 cm long, often with conspicuous glands ____ **I. balsamina**
1. Flowers orange with crimson spots, with obvious, open, funnel-like shape; capsules glabrous; stems glabrous; petioles 0.5–10 cm long, without glands ____ **I. capensis**

Impatiens balsamina L., (balsamic, similar to balsam), GARDEN BALSAM, ROSE BALSAM, BALSAM. Stems to ca. 0.75 m tall; leaf blades elliptic-lanceolate to lanceolate, 5–10(–15) cm long, 1–3 cm wide, serrate; petioles 0.5–2 cm long; flowers axillary, solitary or in groups of 2–3, large and showy, ca. 25–50 mm across, of various colors including white to pink, red, or yellow, often spotted, often double-flowered (= with extra petals); capsules 1–3 cm long. Widely cultivated; escaped to a creek bottom in Dallas Co. (*Perrine s.n.*, 1973, SMU; Diggs et al. 1999; Kartesz 2015); widely scattered in the e U.S. w to WI, MO, LA, and TX. Nov. Native of Asia. This species is extensively used as a showy ornamental, has long been used in Asian medicine to treat rheumatism, fractures, and fingernail inflammation (Wang et al. 2009), and the flowers have been used to dye fingernails (Mabberley 2017).

Impatiens capensis Meerb., (mistakenly thought to be from the Cape of Good Hope), SPOTTED TOUCH-ME-NOT, JEWELWEED, LADY'S-EARRINGS, ORANGE BALSAM. Plant glabrous; stems to 1.5 m tall, succulent; leaves ovate to ovate-elliptic or elliptic, to 10(–13) cm long and to 8 cm wide (often much smaller), crenate-serrate with mucronate teeth; petioles 0.5–10 cm long, those on flowering branches often rather short; flowers usually 2–4 on axillary inflorescences, pendulous on pedicels to ca. 20 mm long; flowers of 2 kinds, some large and showy, others small and cleistogamous; larger flowers 20–30 mm long; large sepal colored like the petals, spurred, the spur usually ca. 1/3 to 1/2 of the sepal length and bent backward parallel to the body of the sepal; stamens 5, pistil 1, ovary superior; capsules ca. 2 cm long. Stream bottom woods; Nacogdoches, San Augustine, Shelby (Turner et al. 2003), Cherokee, and Sabine (Kartesz 2015) cos. in the e part of East TX and in the Red River drainage w to Lamar and Fannin (BRIT) cos. (TORCH 2024), also Bastrop Co. (Kartesz 2015) further w; widespread in the U.S. and adjacent Canada except for the sw U.S. May–Jul. [*I. biflora* Walter] This species is cultivated and naturalized in Great Britain (Mabberley 2017). The sap has long been reported to prevent POISON-IVY dermatitis when rubbed on the skin (McGregor 1986c); it was used by Native Americans for this and other skin problems and its saponins have recently been shown to be effective against POISON-IVY (Motz et al. 2015). Despite the specific epithet, this species is native to North America including East TX. Austin (1975) considered this species adapted for pollination by ruby-throated hummingbirds (*Archilochus colubris*) but it can also attract various bees and butterflies. Droplets of rain or dew can bead up on the leaves and "sparkle"; this is possibly the origin of the common name JEWELWEED.

BASELLACEAE Raf.

MADEIRA-VINE OR BASELLA FAMILY

A very small (ca. 19–20 species in 4 genera) family of mainly tropical and warm areas, apparently native to the Americas and Africa, with the center of diversity in nw South America, introduced into India and e Asia (Stevens 2001 onwards; Vincent 2003; Eriksson 2007; Holtum et al. 2018). They are herbaceous climbing vines containing betalain pigments (lacking anthocyanins); a number are cultivated as ornamentals and several are edible. *Ullucus tuberosus* Caldas, ULLUCO, OLLUCO, is a widely grown and economically important tuber crop in the Andes and was probably important in pre-Columbian times, maybe as far back as 4,200 b.p. (Sperling 1987). The family is thought to be related to such families as the Portulacaceae (Applequist & Wallace 2001), Talinaceae, and Cactaceae (Stevens 2001 onwards). Family name from *Basella*, MALABAR-NIGHTSHADE or MALABAR-SPINACH, a genus of 5 species of Madagascar, e Africa, and one pantropical (Mabberley 2017). *Basella alba* L., MALABAR SPINACH, a sometimes

Vernonia missurica [BB2]

Vernonia texana [BT3]

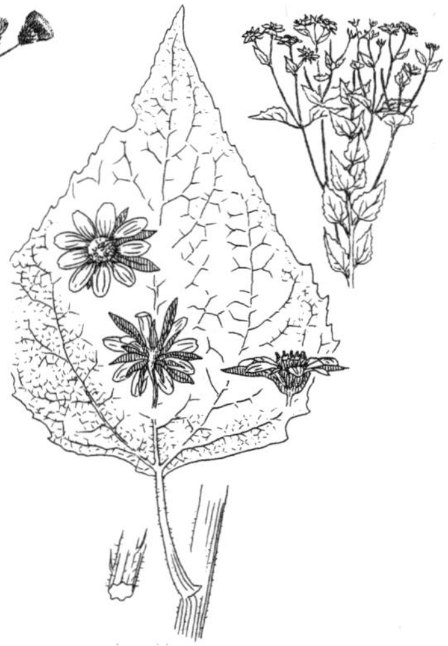

Viguiera dentata [IVE]

Wedelia acapulcensis var. hispida [USH]

Xanthisma texanum var. drummondii [BT3, RBM]

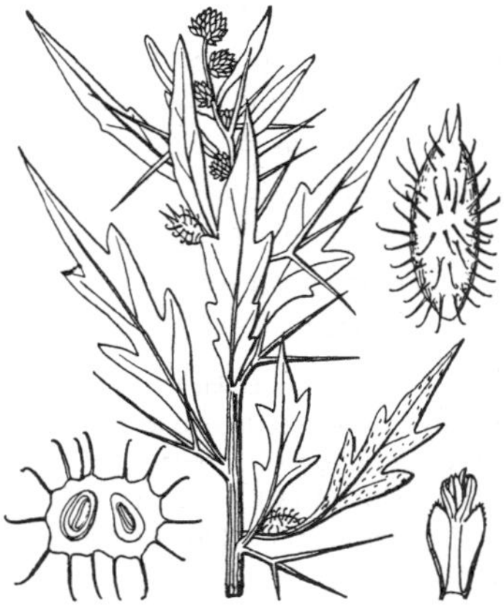

Xanthium spinosum [BB2]

Xanthium strumarium [REE]

Youngia japonica [BB2]

Zinnia violacea [BA1]

colorful (due to betalain pigments) rapidly growing vine, is widely cultivated in the tropics and subtropics and is used as a leaf vegetable and the source for a red pigment (Vincent 2003; Chaurasiya et al. 2021); it is sometimes grown in TX both for food and as an ornamental. (Latinized version of the Malabar name for the plant) (subclass Caryophyllidae—Cronquist; order Caryophyllales—APG IV)

FAMILY RECOGNITION IN THE FIELD: unarmed, non-tendriled, glabrous, herbaceous vines with alternate, simple, entire, succulent leaves, often with either axillary aerial tubers or tubers at stem base, and small cream-white flowers on axillary spike-like racemes or panicles.

REFERENCES: Wilson 1932; Bogle 1969; Sperling 1987; Sperling & Bittrich 1993; Behnke & Mabry 1994; Nowicke 1996; Applequist & Wallace 2001; Vincent 2003; Eriksson 2007.

ANREDERA Juss. MADEIRA-VINE

Twinning, glabrous, herbaceous, succulent, mucilaginous vines, without tendrils or armature (= spines or thorns); perennial from fleshy fibrous or tuber-like root; leaves alternate, simple, petiolate, the blades entire, thin-fleshy, orbicular to ovate, elliptic, or cordate, apically acute to obtuse; stipules absent; inflorescences indeterminate, axillary or terminal, racemes or panicles of racemes, with numerous, small, bisexual (but often functionally unisexual) flowers; pedicels subtended by a bract and with 2 minute bracteoles persistent at tip; flowers fragrant; sepals 2 (these have sometimes been considered bracteoles), distinct or basally connate, slightly shorter than the petals, adnate with the petals basally, winged in fruit or not so; petals 5, connate at base; stamens 5; pistils 3-carpellate; ovule 1; styles 3, basally connate 1/4–3/4 their length, the stigmas bifid or not so; ovary superior, with basal placentation; fruits utricles, partially enclosed in dry persistent perianth.

A genus of 12 species of warm areas of the Americas; a number are cultivated as ornamentals and some have been used medicinally (Vincent 2003). Facultative CAM metabolism (an adaptation for arid environments) has recently been found in a species of *Anredera*; both this adaptation and C_4 photosynthesis have independently evolved numerous times in the Caryophyllales (Holtum et al. 2018). Both of the species growing in TX have been used medicinally (Vincent 2003). (Derivation "for Anreder, about whom nothing else is known"—Vincent 2003, or from the Spanish word *enredadera*, which refers to a twining or climbing plant—Bihrmann's Caudiciforms 2025)

REFERENCES: Bogle 1969; Sperling & Bittrich 1993; GISD 2006; CABI 2019.

1. Stems often with axillary tubers; perianth drying dark brown; larger leaf blades, especially those proximal to the inflorescences, cordate to truncate to rounded at base; pedicel bracts basally connate, persisting at apex of pedicel; flowers drying dark brown ____ **A. cordifolia**
1. Stems without axillary tubers, but tubers present at base of stem and underground; perianth drying pale; larger leaf blades tapering towards base, attenuate to cuneate at base; pedicel bracts distinct, deciduous; flowers drying pale ____ **A. vesicaria**

Anredera cordifolia (Ten.) Steenis, (with heart-shaped leaves), MADEIRA-VINE, MIGNONETTE-VINE. Twining herbaceous vine to ca. 6 m long, often producing axillary, wart-like, aerial tubers; perennial from a tuber-like root, hardy in East TX; petioles 6–12 mm long, blades to ca. 11 cm long, broadly ovate to subcordate or cordate; inflorescences spike-like racemes or panicle of racemes 4–35+ cm long; flowers drying dark brown; pedicels 1.5–2.5 mm long; sepals unwinged, cream-white; petals cream-white, to ca. 2.3 mm long; styles 3, the stigmas clavate to capitate, undivided; fruit not seen (fruit and seed-set are reported to be rare, with tubers the main means of propagation—Vincent 2003). Cultivated as an ornamental and escaping, disturbed areas; Dallas (Kartesz 2015), Williamson (*J.W. Walker 68* TEX), and Travis (*W.R. Carr 15847* TEX; *G. Nesom & G. Nesom 1* NY, U, CR) cos. in East TX; also Tarrant (*L. Seth s.n.*, 8 Oct 1995, BRIT, NY) and Gillespie (Kartesz 2015) cos. to the w of East TX; we are not aware of other TX localities; also CA, FL, and LA; native to tropical South America but cultivated and naturalized in tropical and subtropical areas worldwide. Late summer–fall. [*A. baselloides* (Kunth) Baill., *Boussingaultia cordifolia* Ten., *B. gracilis* Miers] In tropical areas (e.g., ne Australia, Hawaii) this species is an aggressive invader, can smother vegetation and even damage trees, and is considered notoriously difficult to control because of its vegetative reproduction (the easily detached aerial tubers and tuberous roots) and rapid growth (stems elongating up to 1 m per week or 6 m per season) (GISD 2006; CABI 2019); we do not have evidence of it being problematic in East TX, but given its reputation, care should be taken because of its invasive potential.

Anredera vesicaria (Lam.) C.F. Gaertn., (bladder-like), TEXAS MADEIRA-VINE. SACASILE, SUELDA CON SUELDA. Twining herbaceous vine to ca. 8 m long, without axillary tubers, but tubers present at base of stems and underground; petioles 3–18 mm long, blades to 16 cm long, ovate to elliptic; inflorescences racemes or panicles of racemes, to 35+ cm long; flowers bisexual or functionally unisexual, drying pale; pedicels 1–3 mm long; sepals keeled, becoming winged in fruit, cream-white, to 2 mm long; petals cream-white, 1.5–2.3 mm long; styles 3, the stigmas often bifid (= divided into two parts); utricles obovoid, 1–1.4 mm long. Late summer–fall. Disturbed areas, thickets, roadsides, fencerows; this species is known in East TX only from Wilson Co. (*G.L. & J. Nesom 7215* BAYLU, NY, TEX) in sw corner of the area; mainly s TX n to Wilson and Karnes cos. (Kartesz 2015); in the U.S. known only from FL (introduced—Wunderlin & Hansen 2008) and TX (native); native to s North America, Mexico, Central America, the Caribbean, and n South America, widely escaped in other tropical and subtropical parts of the world. May–Dec. [*A. leptostachys* (Moq.) Steenis, *Basella vesicaria* Lam., *Boussingaultia leptostachys* Moq.] This species, with showy inflorescences and fragrant flowers, is cultivated as an ornamental. Eriksson (2007) noted, following Sperling (1987), that this species is treated as having two morphologically distinct forms that are apparently ± functionally staminate and ± functionally pistillate forms of the same species (the pistillate not producing pollen, the staminate only rarely producing fruit). Eriksson further noted that, "Both forms produce tubers at the base of the stem and below ground, and it is the apparent means of reproduction of the staminate plants."

BATACEAE Mart. ex Perleb

SALTWORT FAMILY

A family of 1 genus and 2 halophytic species of warm-temperate tropical Americas and Australia (Thorne 2010). Early taxonomists placed the family in the Chenopodiales because of similarities to halophytic members of the Chenopodiaceae; however, because of the absence of betalains, dissimilarity of sieve-element plastid types, and different chromosome numbers, this relationship was ruled out. Family name from *Batis*. (subclass Dilleniidae—Cronquist; order Brassicales—APG IV
FAMILY RECOGNITION IN THE FIELD: dioecious, halophytic subshrub of coastal environs with minute reduced flowers in axillary catkins (= a dense spike), leaves opposite, fleshy.
REFERENCES: Godfrey & Wooten 1981; Thorne 2010.

BATIS P. Browne SALTWORT

A genus of 2 species. (possibly from Greek via Latin for another coastal plant, or possibly from Greek: *batos*, bramble)

Batis maritima L., (confined to the coast), SALTWORT. Subshrub usually less than 1 meter tall, often with prostrate stems, main stems 2–4 cm in diam. at base, rooting at the nodes; leaves opposite, simple, blade 5–20 mm long, 2–3 mm wide, clavate, apex rounded to acuminate, base with a loose flange, upper surface flattened, fleshy; spikes 6 mm long, 4 mm wide, flowers inconspicuous; staminate inflorescence several-flowered, flowers enclosed in a sac and spitting into 2 lobes (sometimes described as 2 sepals), tepals spatulate or tapered with a long claw, stamens 4 or 5; pistillate inflorescence 4–12 flowered, flowers connate at maturity, perianth lacking; fruit a fleshy spike of cohering berry-like pistils. Salt marshes, salt flats, tidal shores; in East TX known only from Jefferson Co. (Kartesz 2022); Turner et al. 2003 also cites Gonzalez Co.; otherwise coastal in TX; AL, CA, FL, GA, LA, NC, SC and TX; Mexico; West Indies; Central and South America; Galapagos Islands; and Hawaii (probably introduced). Apr–Sep. [*B. californica* Torr.] Native of the US.

BERBERIDACEAE Juss.

BARBERRY FAMILY

Ours perennial herbs or shrubs; leaves alternate or apparently opposite, simple or compound, entire, toothed, or lobed; flowers solitary or in racemes or panicles; perianth parts in 3s, all nearly alike or distinctly differentiated into sepals and petals (1 to several rows of each), the outer usually falling as or soon after the flowers open, the innermost sometimes with glands or modified into prominent staminode-like nectaries; perianth sometimes with basal bracts; stamens as many as the petals or twice as many; anthers opening by apical valves or by a longitudinal slit; pistil 1; ovary superior; fruit in ours a berry.

A medium-sized family of ca. 650–700 species in 13–19 genera (Whetstone et al. 1997a; Stevens 2001 onwards; Hsieh et al. 2022) mostly in e Asia and e North America, but also South America, with some general north temperate, and scattered in Africa. They often have alkaloids; tissues are often colored yellow due to the isoquinoline berberine; many are valued as ornamentals including *Alloberberis*, *Berberis*, *Mahonia*, and *Nandina* and many have been used medicinally (Peng et al. 2006). Although the family has sometimes been split with *Podophyllum* and related genera recognized as the Podophyllaceae, molecular studies (e.g., Kim & Jansen 1996 [1997]; Wang et al. 2007; Hoot et al. 2015; Sun et al. 2018) show the Berberidaceae *sensu lato* is a monophyletic group related to (sister group of) the Ranunculaceae, and then to the Menispermaceae. (Latinized from *berberys*, Arabic name of the fruit) (subclass Magnoliidae—Cronquist; order Ranunculales—APG IV)
FAMILY RECOGNITION IN THE FIELD: evergreen shrubs or perennial herbs with compound or deeply palmately lobed leaves; *flower parts in 3s* is a good field character for this otherwise obviously dicot family.

References: Ernst 1964a; Nowicke & Skvarla 1981; Loconte & Estes 1989; Loconte 1993; Kim & Jansen 1996 [1997], 1998; Whetstone et al. 1997a; Kim et al. 2004a; Wang et al. 2007, 2009; Ying et al. 2011; Hoot et al. 2015; Sun et al. 2018.

1. Evergreen shrubs; leaves numerous per plant, once compound with 3–13 leaflets or 2–3 times pinnately compound, not peltate; flowers numerous.
    2. Leaves once compound with 3–13 leaflets; leaflets with spiny lobe-like teeth ____ **Alloberberis**
    2. Leaves 2–3 times pinnately compound; leaflets without spiny teeth ____ **Nandina**
1. Herbs; leaves 1 or 2 per plant, simple, deeply palmately lobed, peltate; flower 1 per plant ____ **Podophyllum**

Alloberberis C.C. Yu & K.F. Chung BARBERRY

Evergreen shrubs; wood yellow; stems spineless, dimorphic with elongate primary shoots and short spurs; leaves once compound, with 3–11(–13) leaflets; leaflets with spiny lobe-like teeth resembling those of holly, very rigid; inflorescences terminal, loose, relatively few-flowered racemes; flowers small, yellow; perianth of 9 ± petal-like sepals, and 6 narrower, longitudinally cupped petals; anthers opening by apical valves; fruits usually ± globose, yellowish tan to orangish brown or reddish.

A genus of 11 species (Yu & Chung 2017) of sw U.S. and adjacent Mexico of shrubs with non-spiny stems, compound leaves, and alkaloids (but note that the leaves have spiny lobe-like teeth); it has sometimes been recognized as *Mahonia* sect. Horridae. Among many taxonomists over the past few decades there has been a tendency to recognize a very broad *Berberis*, including the compound-leaved species formerly known as *Mahonia* (Whittemore 1997d; Stevens 2001 onwards; Kim et al. 2004b; Harms 2007; Adhikari et al. 2015). For example, Adhikari et al. (2015) argued based on molecular evidence that "*Berberis* s.str. (simple-leaved *Berberis*) is nested within the compound-leaved *Berberis*, formerly known as *Mahonia*, and that these two genera should be treated as a single genus." However, Yu and Chung (2017) argue, based on more detailed molecular evidence, that *Berberis sensu lato* should be divided into four strongly supported monophyletic clades—the genera *Alloberberis* (= *Mahonia* sect. Horridae, 11 spp.), *Berberis* (ca. 500 species), *Mahonia* (ca. 100 species), and the monotypic *Moranothamnus* (= *B. claireae*, a unique spineless Baja California endemic species) (Yu & Chung 2017). This is a situation in which either a monophyletic *Berberis sensu lato* or the four monophyletic segregate genera accurately reflect evolutionary history; in other words there is no definitive correct answer. However, we are following Yu and Chung (2017), Sun et al. (2018), Hsieh et al. (2022), and Weakley, A.S., and Southeastern Flora Team (2025) in recognizing *Alloberberis* at the generic level (as well as *Berberis, Mahonia*, and *Moranothamnus*). We base this decision not only on the morphological and ecological distinctness of these four clades, but primarily on the molecular evidence indicating that the four segregate genera separated very long ago in geologic history, probably on the order of 35 million years (Yu & Chung 2017); in other words "their deep stem divergences [that are] equivalent to other genera of Berberidaceae" (Yu & Chung 2017). Recognizing the related genus *Mahonia* (which also has compound leaves but is a separate evolutionary lineage and is characterized by having monomorphic branches) also allows many names known under that genus to be used in their traditional sense. A number of *Mahonia* species are cultivated as ornamentals (e.g., *M. aquifolium* (Pursh) Nutt.—oregon-grape, holly-leaf oregon-grape) and the berries of some species are edible. Some members of the related genus *Berberis* serve as alternate hosts for cereal rusts, caused by the fungus *Puccinia graminis* Pers., black stem rust of wheat (Peterson 2018; Barnes et al. 2020); as a result, the sale or transport of certain species is illegal in the U.S. and Canada. Whittemore (1997d) noted that *Mahonia trifoliolata* (=*Alloberberis trifoliolata*) was susceptible to infection by black stem rust of wheat. However, according to rust scholar J. Hennen (pers. comm.), it is not susceptible; he indicated that the species of *Berberis* here treated in the genus *Alloberberis* are not infected by *P. graminis*; rather they are susceptible to rusts of the genus *Cumminsiella*, e.g., *C. texana* (Holw. & Long.) Arthur, an autoecious (= requires only 1 host) rust (J. Hennen, pers. comm.; Baxter 1957). Species of both *Alloberberis* and *Mahonia* have been used medicinally since ancient times in a number of cultures (e.g., Native American, Chinese) and are still used in some today (Moerman 1998; He & Mu 2015; Srivastava et al. 2015). Isoquinoline alkaloids such as berberine are present in some species of both *Alloberberis* and *Mahonia*, and a number, including those in East TX, are considered toxic (Powell 1988). (According to Yu & Chung (2017)—the authors of the genus, "We propose the new name *Alloberberis*, another or different *Berberis*, suggested by Dr. Peter H. Raven, to emphasize its distinctness").
References: Ahrendt 1961; Durand 1972; McCain & Hennen 1982; Breckenridge 1983; Kim & Jansen 1994; Whittemore 1997d; Kim et al. 2004b; Harms 2007; Adhikari et al. 2015; Yu & Chung 2017; Chen et al. 2020; Hsieh et al. 2022.

1. Leaves pinnately compound with 5–13 leaflets; terminal leaflet usually stalked (rachis segment to 15 mm long) ____ **A. swaseyi**
1. Leaves palmately trifoliolate; leaflets all sessile ____ **A. trifoliolata**

Alloberberis swaseyi (Buckley) C.C. Yu & K.F. Chung, (for the botanist Dr. Swasey of Mobile, AL), texas barberry. Shrub to 2.4 m tall, usually taller than wide, loosely intricately branched, the stems with numerous lateral branches; leaves once pinnately compound with (5–)7–11(–13) leaflets; leaflets with (3–)4–8(–9) pairs of spiny lobe-like teeth, grayish-green, the lateral leaflets sessile, the terminal leaflet stalked, lanceolate, usually longer and narrower than the lanceolate to elliptic or even suborbicular

Impatiens balsamina [BL1]

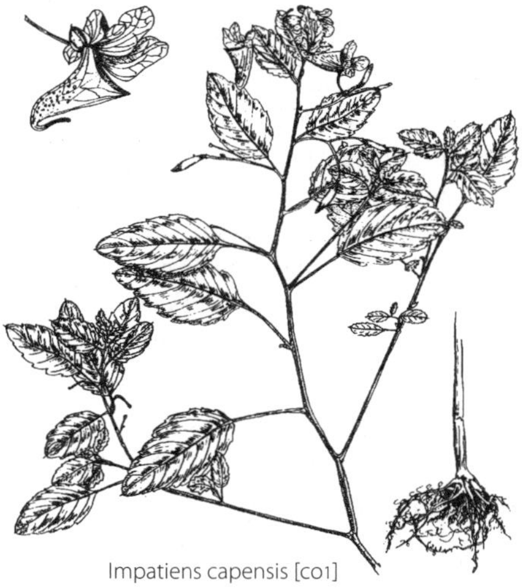

Impatiens capensis [CO1]

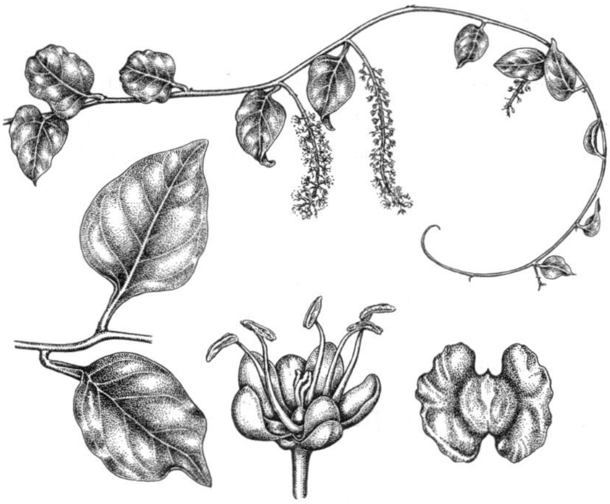

Anredera vesicaria [CO3]

Anredera cordifolia [HEA]

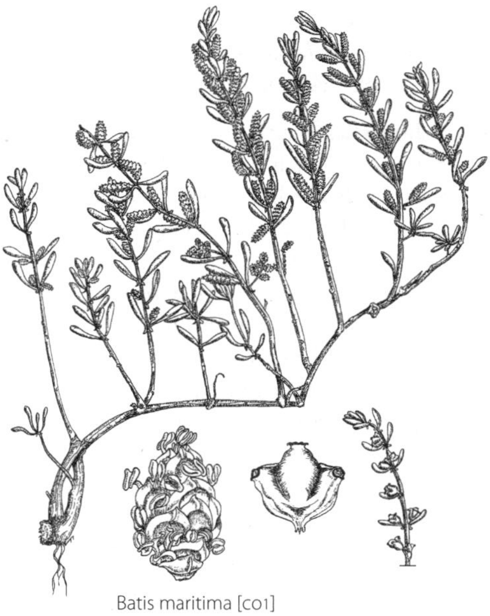

Batis maritima [CO1]

lateral leaflets; inflorescences appearing with new leaves, usually short-racemose, with (3–)4–5(–6) flowers; flowers yellow; filaments with a pair of tooth-like lateral appendages; fruits 8–13 mm in diam., somewhat hollow and inflated, style lost during fruit development. Primarily bottomland habitats, rarely in steeper areas (then with other woody vegetation); species narrowly endemic to Comal (*J.R. Ott s.n.*, 16 Mar 2007, TEX), Hays (*R.T. Harms 53* TEX), and Travis (*B.C. Tharp 1437* TEX) (Harms 2007) cos. at extreme sw margin of East TX, a few counties in the e Edwards Plateau, and Frio Co. (Turner et al. 2003) in the n South TX Plains; endemic to TX. (Mar–)Apr–early May. [*Berberis swaseyi* Buckley; *Mahonia swaseyi* (Buckley) Fedde] Harms (2007) described numerous ways in which *A. swaseyi* and *A. trifoliolata* differ. He also documented the existence of hybrids between the two (based on morphology and phenology) from Blanco, Comal (e.g., *R.T. Harms 61* TEX), Hays (e.g., *R.T. Harms 56* TEX), and Travis (Harms 2007) cos. and presented evidence suggesting introgression. However, molecular research (e.g., Yu & Chung 2017) indicate that *A. swaseyi* and *A. trifoliolata* are not each others' closest relatives within the genus. The acidic fruits were used to make jelly by early settlers and are eaten by birds; the flowers are visited by bees (Durand 1972). The species is not susceptible to infection by *Puccinia graminis* (see discussion under *B. trifoliolata*). The alkaloid berberine is present (Durand 1972) in the roots so care should be taken if they are used.

Alloberberis trifoliolata (Moric.) C.C. Yu & K.F. Chung, (with three leaflets), ALGERITAS, AGARITO, CURRANT-OF-TEXAS. Densely leafy shrub to 2 m tall, usually wider than tall; stems straight to arching, multiple from a swollen woody below ground base, with relatively few lateral branches; leaves with 3 leaflets; leaflets narrowly oblong-lanceolate, yellow-green to blue-green, glaucous, with 1–3 pairs of spiny lobe-like teeth, stiff, glabrous; inflorescences of compact, axillary, umbel-like racemes; flowers yellow; filaments without tooth-like lateral appendages; fruits 6–10 mm in diam., juicy, not noticeably hollow, with persistent style. Rocky limestone soils in a variety of habitats; Bell (BRIT), Bastrop, Bexar, Comal, Hays, McLennan, and Travis (Turner et al. 2003) cos. on sw margin of East TX; widespread in the w 2/3 of TX except the Panhandle; AZ, NM, and TX; also ne Mexico. Early Feb–early Apr. [*Berberis trifoliolata* Moric.; *Mahonia trifoliolata* (Moric.) Fedde] See discussion of changing views of which genus to recognize this species in under the generic synopsis. Jones et al. (1997) spelled the epithet *trifoliata*. According to Correll and Johnston (1970), an excellent jelly can be made from the fruits. The wood and roots were used as a source of yellow dye by early settlers (Kirkpatrick 1992). This species is sometimes cultivated in native plant gardens. The stamens are reported to be touch sensitive; when touched by a pollinating insect, they spring out throwing pollen onto the pollinator (Wills & Irwin 1961; Kirkpatrick 1992). The roots contain berberine and associated alkaloids and were used in folk medicine; however, in high concentrations they can cause poisoning (Powell 1988).

NANDINA Thunb. SACRED-BAMBOO, HEAVENLY-BAMBOO

A monotypic genus native from India to China and Japan; the single species is a much cultivated ornamental shrub, especially in Japan (Mabberly 2017). This plant is very distinctive and in the past has sometimes been put into a separate family, the Nandinaceae (e.g., Takhtajan 1997); however, it is now generally recognized in subfamily Nandinoideae with several related taxa (including the genus *Caulophyllum*) (Song et al. 2022; Stevens 2001 onwards). It is widely known to be toxic. (Chinese name meaning "plant from the south" or from the Japanese name, *nanten*)

REFERENCES: Whetstone et al. 1997b; Kim et al. 2004a; Wang et al. 2007; Song et al. 2022.

Nandina domestica Thunb., (domesticated, frequently used as a house plant), SACRED-BAMBOO, HEAVENLY-BAMBOO. Erect, evergreen shrub to ca. 2 m tall, few branched, the branchlets reddish when young; wood yellow; leaves 2–3 pinnately compound, long petiolate, without stipules; petioles and petiolules swollen basally; leaf blade 30–50 cm long, the leaflets 9–81, lanceolate to elliptic-lanceolate, to ca. 10 cm long, sessile or nearly so, entire, glabrous, often with reddish or reddish purple coloration; inflorescence a panicle to 20+ cm long, often with hundreds of flowers; flowers bisexual, 3-merous, fragrant, 5–8 mm long; sepals numerous, spirally arranged, intergrading somewhat with the petals; petals 6, usually larger than the sepals, whitish or cream-colored; stamens 6; anthers opening by a longitudinal slit; fruits 6–8(–9) mm in diam., red; seeds 1–3, but frequently 2. Widely cultivated and escapes into sandy woods; widespread in c and e TX; scattered elsewhere in the state; MD s to FL w to OK and TX; also CA. May–Jun. Native from India to e Asia. This species has cyanogenic glycosides, alkaloids, and a serotonin antagonist, and is toxic to a variety of species ranging from humans to ruminants and house pets; the brightly colored fruits have been reported to be toxic and potentially fatal to some birds (e.g., cedar waxwings) (Abrol et al. 1966; Burrows & Tyrl 2001; Woldemeskel & Styer 2010; Forrester 2018). The species can be invasive in the se U.S. (Mabberly 2017), including parts of East TX.

PODOPHYLLUM L. MAY-APPLE, MANDRAKE

A genus variously treated as having from 1 to as many as ca. 14 (Mabberley 2017) species (including species sometimes recognized in the genera *Diphylleia*, *Dysosma*, and *Sinopodophyllum*), and native to e North America and the Himalayas to e Asia. All of these species have sometimes been segregated into the Podophyllaceae and are now usually treated as a subfamily, the Podophylloideae, in a monophyletic Berberidaceae (e.g., Stevens 2001 onwards; Ye et al. 2018, 2022). We are treating *Podophyllum* as containing only a single species, *P. peltatum*, native to e North America. Molecular data suggests that *Podophyllum* is the sister group to *Dysosma* (7 species from e Asia) (Ye et al. 2022). Ye et al. (2022) further found that *Sinopodophyllum hexandrum* [sometimes treated as *Podophyllum hexandrum*] is sister to the rest of the Podophylloideae. *Podophyllum* and *Dysosma* are an example of a disjunct e Asia–e North America distribution (Ye et al. 2022). For more information on this biogeographic pattern, see the discussion under *Campsis* (Bignoniaceae). (Greek: *podos*, foot, and *phyllon*, a leaf, probably referring to the stout petioles of the radical leaf)

REFERENCES: Swanson & Sohmer 1976; O'Rourke George 1997; Moraes et al. 2000; Shaw 2000, 2002; Liu et al. 2002; Ye et al. 2018, 2022.

Podophyllum peltatum L., (peltate, shield-shaped), MAY-APPLE, WILD JALAP, AMERICAN-MANDRAKE. Rhizomatous perennial herb to 50 cm tall; leaves 2 (sometimes 1), appearing opposite at summit of stem, long-petioled, blades large, to 30 cm or more in diam., typically peltate, suborbicular, with 5–9 segments, inconspicuously pubescent beneath; flower solitary, nodding, over-topped by the leaves; perianth white (rarely rose), of 12–15 broad tepals, the 6 outer shorter (= sepals) than the inner, falling early, the inner (= petals) 2–3.5 cm long; stamens twice as many as the petals; anthers opening by a longitudinal slit; fruit an ovoid, many-seeded, fleshy berry 3–5 cm long, yellow to purplish at maturity. Open woods and thickets, sandy soils; widespread in e part of East TX w to Hunt Co. in South Sulphur River drainage (Mahler 1988), locally w in Red River drainage to Grayson Co. (e.g., Buckner Preserve; BRIT), and to Kaufman Co. (BRIT) in Trinity River drainage; e U.S. and adjacent Canada w to MN, NE, KS, OK, and TX. Late Mar–mid-Apr. The unripe fruit is toxic but reported to be edible when ripe and is sometimes used to make preserves; all other parts of the plant are toxic, apparently including the seeds of the ripe fruit; there are over 15 biologically active compounds including the poisonous podophyllin, a bitter resin containing lignins and flavonols; podophyllin has strong cathartic and antineoplastic properties. Children have been poisoned by the unripe fruit and poisoning has resulted from the misuse of medicinal preparations. Eating large quantities of the plant or repeated applications of the resin to the skin can be fatal; workers handling the rootstocks sometimes develop dermatitis and conjunctivitis. The plant has long been used medicinally by Native Americans for warts; recently it has been used against venereal warts and testicular cancer (Muenscher 1951; Kingsbury 1964; Morton 1977; Mabberley 1987; Turner & Szczawinski 1991; Leung & Foster 1996; Moraes et al. 2000). A rust fungus (*Puccinia podophylli* Schwein.), commonly known as MAY-APPLE RUST, sometimes causes yellow to orange leaf lesions (J. Hennen, pers. comm.). ☠

BETULACEAE Gray

BIRCH OR HAZELNUT FAMILY

Ours monoecious shrubs to small or large trees; leaves alternate, simple, deciduous, short-petioled, blades pinnately straight-veined, serrulate or doubly serrate; stipules deciduous; perianth small or absent; staminate flowers in spreading or drooping catkins (= aments); pistillate flowers in spike-like or cone-like inflorescences; pistil 1; ovary inferior; fruit a 1-seeded samara (2-winged) or nutlet (unwinged).

A small (ca. 125–160+ species in 6 genera) (Furlow 1997; Stevens 2001 onwards; Mabberley 2017; Yang et al. 2018), mainly n temperate family extending s to tropical mountains (e.g., in the Andes and Sumatra) and n to the Arctic tundra (e.g., *Betula nana* L., *Alnus* spp.). They are wind-pollinated trees and shrubs usually with ectotrophic mycorrhizae in the roots; species are variously important economically for timber, as a source of edible nuts, as ornamentals, and as aids in soil nitrification and stabilization; the two additional genera not occurring in Texas are the north temperate *Corylus* (source of hazelnuts and filberts) and the Asian *Ostryopsis*. The family was previously often split into two families, Betulaceae (*Alnus* and *Betula*) and Corylaceae (*Carpinus, Ostrya, Corylus,* and *Ostryopsis*); most modern authorities now recognize these groups at the subfamilial level (Furlow 1997; Kato et al. 1999; Grimm & Renner 2013; Zhu et al. 2018). *Corylus americana* Walter, AMERICAN HAZEL or HAZELNUT, is known from Johnson and McCurtain (Kartesz 2015) cos. OK near the Red River (state boundary with TX); it should therefore be looked for in TX. This species is a shrub usually 3 m or less tall and can be distinguished from TX Betulaceae by it relatively large (1 cm or more in diam.) nuts ("hazelnuts") surrounded by 2 leaf-like bracts that are distinct nearly to base and slightly longer than to 2 times the length of the nuts. (subclass Hamamelidae—Cronquist; order Fagales—APG IV)

FAMILY RECOGNITION IN THE FIELD: shrubs or trees with alternate, simple, usually doubly sharp-toothed leaves, early deciduous stipules, and male flowers in long *catkins*; fruits 1-seeded, in cone-like or spike-like, obviously *bracteate* infructescences.
REFERENCES: Hardin 1971; Little 1971; Furlow 1990, 1997; Bousquet et al. 1992; Kubitzki 1993a; Chen et al. 1999; Kato et al. 1999; Milliken 2010; Grimm & Renner 2013; Zhu et al. 2018.

1. Leaves truncate (= as if cut or squared off) or broadly cuneate (= wedge-shaped) at base; fruits tiny winged samaras, the wings narrower than the body, subtended by a small (3–8 mm long) bract and grouped with bracts into a cone-like structure.
   2. Trees with shaggy, conspicuously exfoliating bark; lower surface of leaves often grayish white, densely hairy when young; pistillate inflorescences solitary, bracts 3-lobed, 6–8 mm long, papery and eventually deciduous (the cone-like inflorescences thus shattering); buds not stalked ____ **Betula**
   2. Shrubs or small trees with ± smooth bark; lower surface of leaves green, glabrous or nearly so even when young; pistillate inflorescences in groups of 2–3, bracts not lobed, 3–4 mm long, becoming woody and persistent (the cone-like inflorescences not "shattering"); buds stalked ____ **Alnus**
1. Leaves rounded to subcordate at base; fruits wingless nutlets (but wing-like bract can be present), enclosed in or subtended by conspicuous sac-like or 3-lobed, wing-like bracts 8 mm or more long (well-developed bracts usually at least 15 mm long), these ± densely clustered together in a spike-like inflorescence.
   3. Fruit subtended (not enclosed) by a 3-lobed bract; bark smooth, gray, the trunk appearing like rippling muscles; lower surface of mature leaves glabrous except for some hairs in the axils of the veins; none of the main side veins near the base of the leaf forked ____ **Carpinus**
   3. Fruit enclosed by a pouch- or sac-like unlobed bract; bark rough, shreddy, brownish, the trunk not appearing rippled; lower surface of mature leaves pubescent; some of the main side veins near the base of the leaf forked ____ **Ostrya**

ALNUS Mill. ALDER

A genus of ca. 35 species (Stevens 2001 onwards; Mabberley 2017) ranging from the n temperate zone s to Assam, se Asia, and the Andes of South America and n to the Arctic tundra. ALDERS have a symbiotic relationship with species of the actinomycete *Frankia* (filamentous gram positive bacteria) resulting in root nodule formation and fixation of atmospheric nitrogen (Furlow 1997); as a result they can be important ecologically. Some species also produce useful timber. (Latin name for alder)
REFERENCES: Furlow 1997; Navarro et al. 2003; Chen & Li 2004; Ren et al. 2010.

Alnus serrulata (Aiton) Willd., (somewhat serrate or saw-toothed), HAZEL ALDER, COMMON ALDER, SMOOTH ALDER. Shrub or small tree 2–5+ m tall, usually with multiple stems; bark grayish brown or blackish gray to reddish brown; leaf blades 1.6–12 cm long, elliptic to obovate, usually widest beyond middle, cuneate to broadly cuneate basally, usually rounded to acute at apex, usually doubly serrulate and occasionally somewhat undulate; flowers appearing well before the leaves; staminate catkins to 3–8.5(–10) cm long; staminate flowers 3 per bract; pistillate inflorescences cone-like, ca. 15–20 mm long, the bracts becoming woody and persistent; samaras ca. 2–3.5 mm long, with wings narrower than body. Along streams and wet areas; mainly e part of East TX, disjunct w to Milam and Bastrop (Kartesz 2015) cos. near the w margin of East TX; e U.S. and adjacent Canada w to IL, KS, OK, and TX. Mar–Apr. [*Betula serrulata* Aiton, *A. noveboracensis* Britton, *A. rubra* Bong., *A. rugosa* (Du Roi) Spreng. var. *serrulata* (Aiton) H. Winkl.] Native Americans used preparations of this species medicinally in a variety of ways (e.g., for pain of childbirth, emetic and purgative, for coughs, fevers, indigestion, heart trouble, toothaches, high blood pressure, etc.) (Moerman 1998).

BETULA L. BIRCH

A genus of ca. 35–60 species of n temperate and boreal zones of the n hemisphere (Stevens 2001 onwards; Wang et al. 2016; Mabberley 2017); related to *Alnus* but the female cone-like inflorescences shatter when ripe; some provide timber for furniture and plywood; the bark of *B. papyrifera* Marshall (PAPER BIRCH) is impervious to water and was used by Native Americans for canoes, baskets, cups, and wigwam covers. The wind-borne pollen is a major cause of hay fever in Europe (Schenk et al. 2008). Dwarf birches can be an important component of tundra vegetation, making up a substantial part of the biomass (Chapin & Körner 1995). With the possible exception of *Salix* (Salicaceae), *Betula* is reported to have the most northern distribution of any arborescent angiosperm genus in the Northern Hemisphere, extending above the Arctic Circle (Hora 1981; Li et al. 2005). Species relationships within the genus are complex, due in part to hybridization, introgression, and polyploidy, and a number of molecular studies have been done to elucidate them. (Latin name for birch)
REFERENCES: Jarvinen et al. 2004; Li et al. 2005; Schenk et al. 2008; Ashburner & McAllister 2013; Wang et al. 2016.

Betula nigra L., (black), RIVER BIRCH, RED BIRCH. Tree to 30 m tall; bark soft, shaggy, pinkish to tan, freely shedding; leaf blades 4–10 cm long, roughly triangular to ovate, truncate to broadly cuneate at base, acute at apex, conspicuously doubly serrate at least

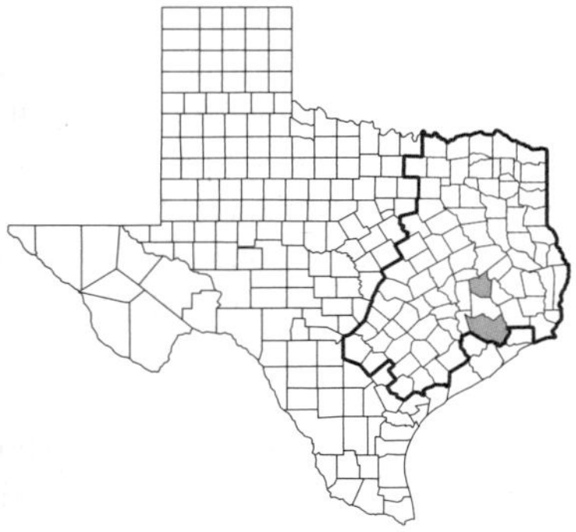

Zinna violacea

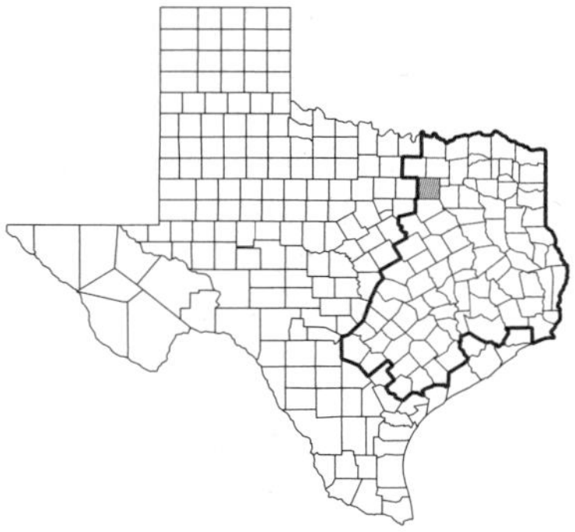

Impatiens balsamina

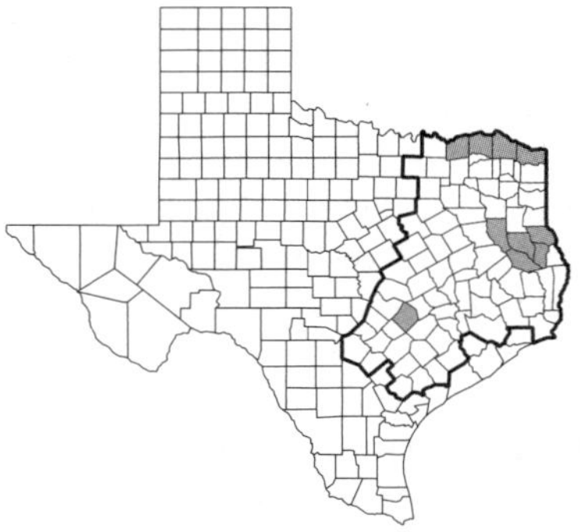

Impatiens capensis

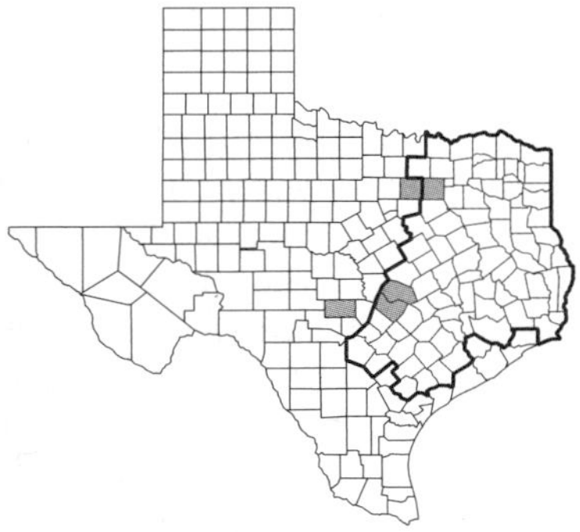

Anredera cordifolia

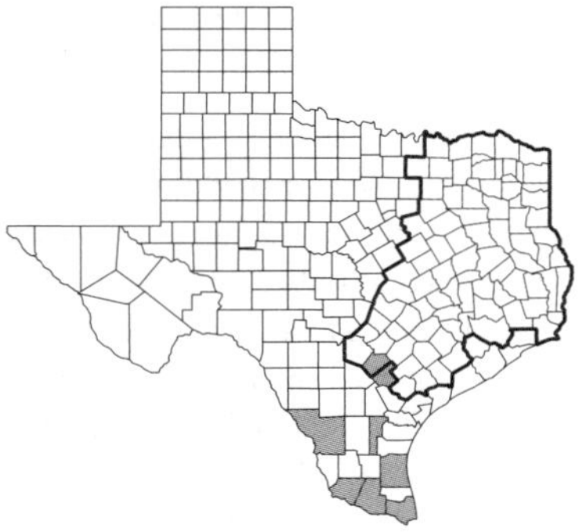

Anredera vesicaria

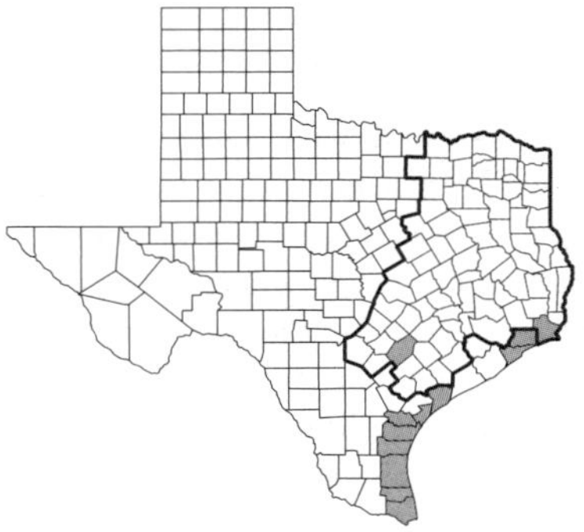

Batis maritima

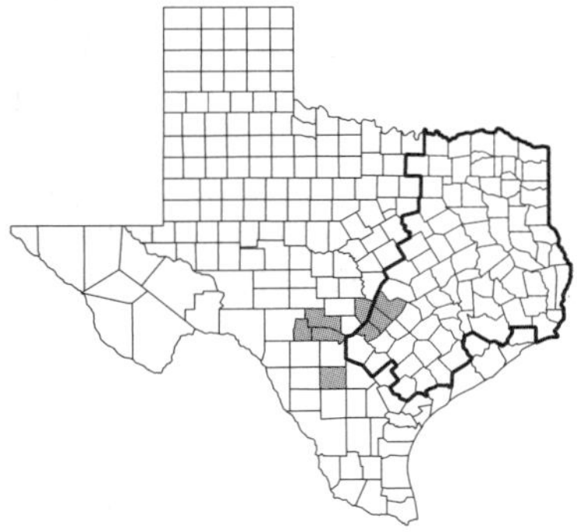

Alloberberis swaseyi

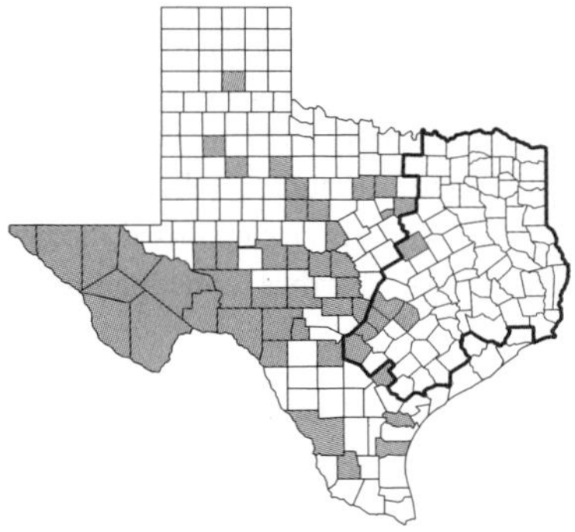

Alloberberis trifoliolata

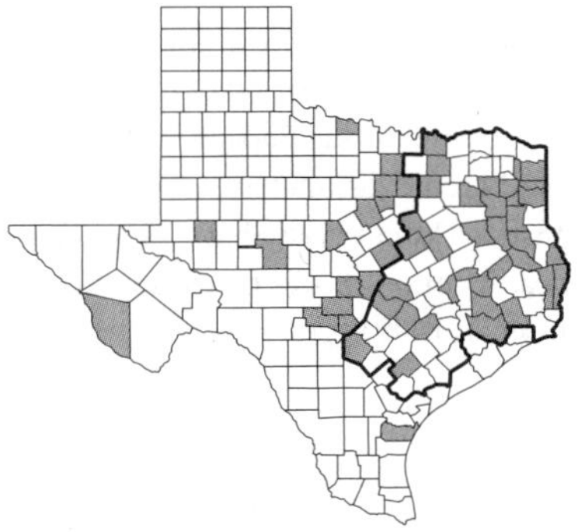

Nandina domestica

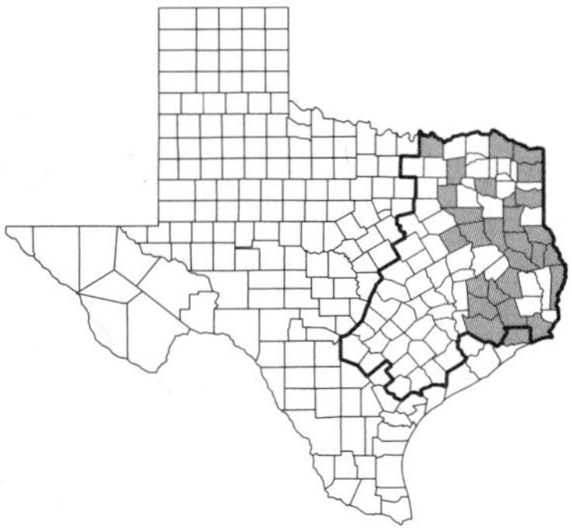

Podophyllum peltatum

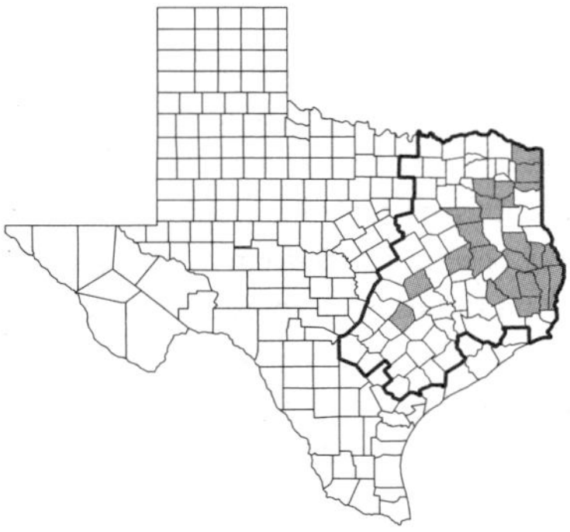

Alnus serrulata

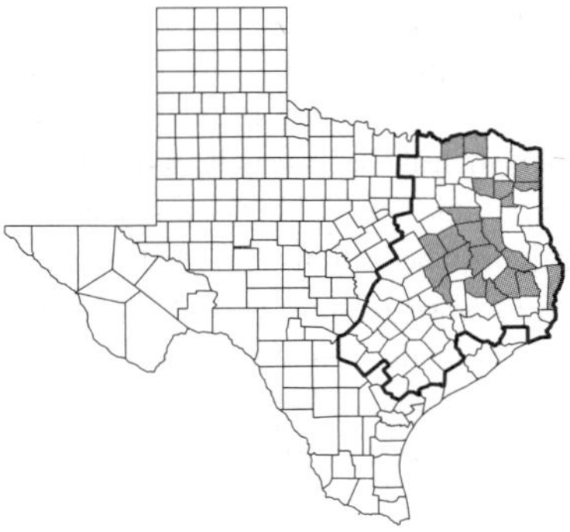

Betula nigra

beyond middle; staminate inflorescences to ca. 9 cm long; staminate bracts ovate to suborbicular; staminate flowers 3 per bract, with calyx; pistillate inflorescences up to 25–35 mm long in fruit, cone-like, but shattering when ripe, not persistent; pistillate bracts divided from apex into 3 oblong-linear lobes, papery and eventually deciduous; samaras ca. 3–5.5 mm long, with wings narrower than the body. Along streams and in low woods, often where periodically flooded; mainly e and se parts of East TX w to Fannin Co. (Little 1971) in Red River drainage, also w to Henderson, Limestone (Kartesz 2015) and Robertson (USDA Plants Database 2019) cos.; widespread in the e U.S. w to MN, KS, OK, and TX. Mar–Apr; fruits released in early summer. [*B. rubra* F. Michx.] This species was used medicinally by Native Americans (e.g., to treat dysentery, colds etc.) (Moerman 1998).

CARPINUS L. HORNBEAM, IRONWOOD

A genus of ca. 25–41 species (Furlow 1997; Mabberley 2017) of the n temperate zone, especially e Asia; also s to Central America, with only 1 species native to the New World; some provide good timber for turnery and tools; the very hard wood is used for such things as mallet heads, tool handles, and levers. According to Peattie (1948), the common name is derived from horn, for toughness, and beam, an old word for tree, related to the German, *baum*, tree. (Latin: *carpinus*, hornbeam, possibly derived from Latin: *carpentum*, a Roman horse-drawn vehicle with wheels made from its hard wood—Furlow 1997)
REFERENCES: Furlow 1987a, 1987b; Li et al. 2018.

Carpinus caroliniana Walter, (of Carolina), AMERICAN HORNBEAM, BLUE-BEECH, WATER-BEECH, MUSCLETREE, LECHILLO, LEANTREE. Small tree with very hard wood to ca. 10 m tall; bark smooth, neither shedding nor peeling; leaf blades 3–10(–12) cm long, usually elliptic to ovate, sharply serrate, inconspicuously doubly so; staminate inflorescences 2–6 cm long; staminate flowers 1 per bract, without calyx; staminate bracts acute, not awned; pistillate inflorescences 1–2.5 cm long, becoming 2.5–12 cm in fruit; pistillate bracts with a long terminal lobe and 2 small lateral lobes near base; fruit a small nutlet that is shed with its subtending bract, the unit functioning as "autorotating" wind-dispersed small "helicopters" (see Lentink et al. 2009 for physics of movement of *Carpinus* fruit/bract); nutlets ribbed, 4–6 mm long. Rich or low woods and along streams; e TX w in Red River drainage to Lamar Co., also Henderson Co., also Hopkins Co. (Little 1971); widespread across the e U.S. and adjacent Canada w to MN, OK, and TX; also Mexico to n Central America. Mar–May. [*C. americana* Michx.] This species is easily distinguished by its smooth gray bark and the rippled, muscle-like appearance of the stems. *Carpinus caroliniana* has an interesting geographic distribution—occurring mainly in the e U.S. but with disjunct populations in the mountains of Mexico, Guatemala, and Honduras (Miranda & Sharp 1950; Milliken 2010). This pattern is the result of a middle to late Miocene (Miocene epoch—24.6–5.1 mya) extension (during a period of climatic cooling) of deciduous forest and associated fauna (particularly amphibians) into Mexico. Subsequently, during Pliocene (Pliocene epoch—5.1–2 mya) and later times as the climate warmed and dried, these deciduous forests became disjunct, surviving in Mexico and Central America only in isolated pockets of appropriate microclimate in the highlands (Miranda & Sharp 1950; Graham 1993). This species was used medicinally by Native Americans to treat a variety of health conditions (e.g., cloudy urine, consumption, diarrhea, constipation, as an astringent, and to facilitate childbirth) (Moerman 1998).

OSTRYA Scop. HOP-HORNBEAM, IRONWOOD

A genus of ca. 5–9 species (Furlow 1997; Mabberley 2017) ranging from the n temperate zone to Central America; some provide timber. The hard wood has been used for such things as sleigh runners, airplane propellers, and mallet heads (Furlow 1997). (Greek: *ostrys*, shell, alluding to the inflated floral bracts)
REFERENCES: Fehrenbach 1984; Reveal 1990a.

Ostrya virginiana (Mill.) K. Koch, (of Virginia), IRONWOOD, EASTERN HOP-HORNBEAM, AMERICAN HOP-HORNBEAM, WOOLLY AMERICAN HOP-HORNBEAM, LEVERWOOD. Small tree with very hard wood to ca. 20 m tall; bark roughened, shredding into ragged strips, shaggy, brownish; leaf blades usually 6–10 cm long, usually elliptic to ovate, sharply doubly serrate, the lower surface pubescent; staminate catkins 2–5 cm long; staminate bracts abruptly narrowed to a spine tip (= awned); staminate flowers 1 per bract, without calyx; pistillate catkins 0.8–2 cm long in flower, becoming 3.5–6 cm in fruit; pistillate bracts pouch-like; nutlets 4–6.5 mm long, each enclosed within a bract. Lowland or upland woods; widespread in e part of East TX w to Bastrop Co. (Kartesz 2015), with a disjunct collection in the East Cross Timbers at Handley, Tarrant Co., in 1910, but not found there recently (Mahler 1988); widespread across the e and c U.S. w to ND, WY, OK, and TX; also Mexico to El Salvador and Honduras (see discussion of similar distribution under *Carpinus caroliniana*) (Milliken 2010). Late Feb–Mar. [*Carpinus virginiana* Mill.] The species was used medicinally by Native Americans in a variety of ways (e.g., treat toothache, for coughs, kidney trouble, etc. (Moerman 1998).

Alloberberis swaseyi [VIN]

Alloberberis trifolioata [LYN, POW]

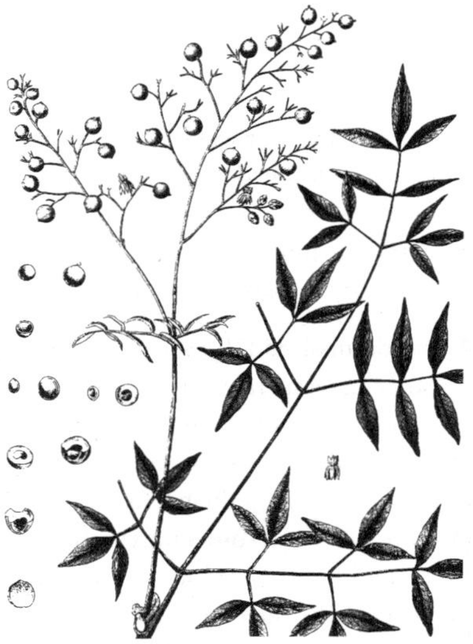

Nandina domestica [LAM]

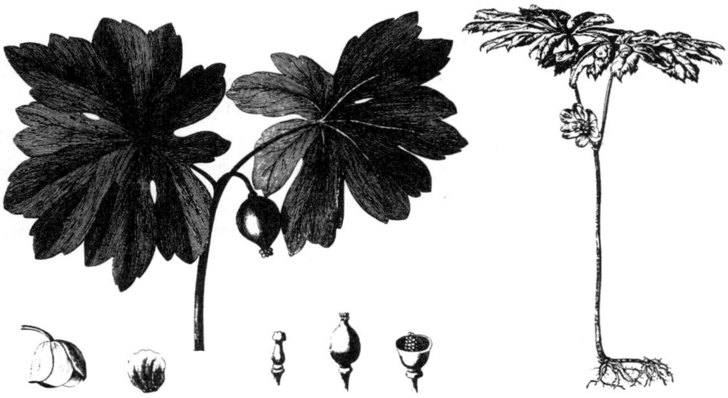

Podophullum peltatum [LAM, NIC]

Alnus serrulata [CO1]

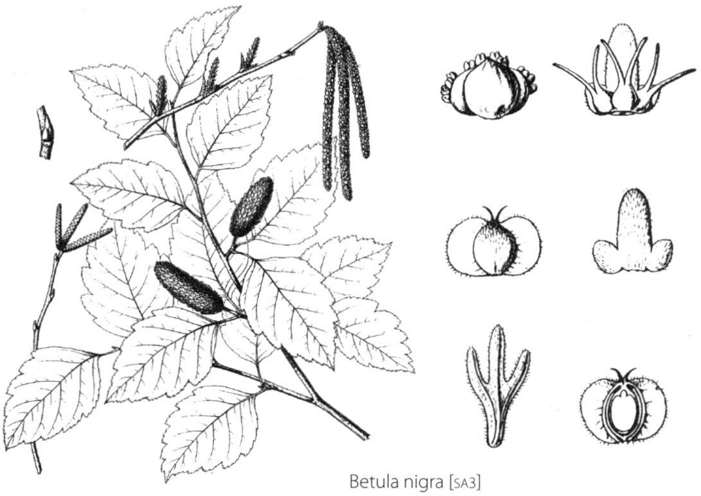

Betula nigra [SA3]

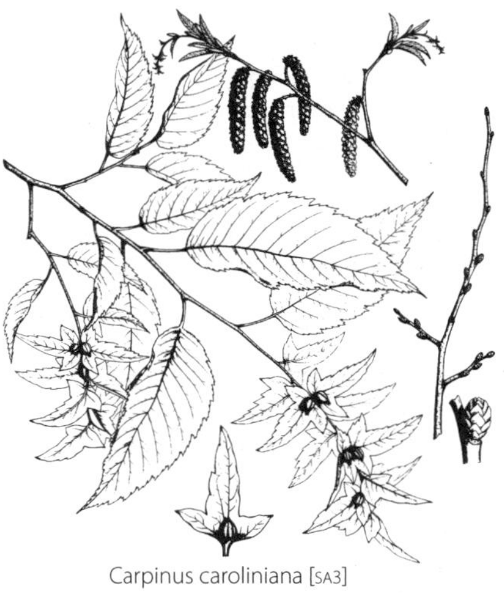

Carpinus caroliniana [SA3]

BIGNONIACEAE Juss.

CATALPA OR TRUMPETVINE FAMILY

Ours shrubs, trees, or woody vines; leaves usually opposite or in whorls of 3, or sometimes alternate, simple or pinnately compound, entire, toothed, or shallowly lobed, without stipules; flowers rather large and showy, sympetalous, in terminal or axillary inflorescences; calyces short, splitting into 2-lipped or unequally 4- to 5-toothed; corollas campanulate to funnelform, radially symmetrical to bilabiate, 5-lobed, showy; stamens 2 or 4, inserted on the corolla; staminodes sometimes present; ovaries superior; pistil 2-carpellate; style 1; fruit an elongate 2-valved capsule; seeds large, flattened, numerous, winged.

A medium-sized (790 species in 110 genera) (Stevens 2001 onwards), mainly tropical family centered in n South America with some in temperate regions; nearly all species are woody, ranging from numerous lianas to trees and shrubs, or rarely herbs; in the tropics it is one of the most important families of lianas. The Bignoniaceae includes many showy tropical ornamentals such as *Crescentia* (CALABASH-TREE), *Jacaranda*, *Kigelia* (SAUSAGETREE), *Spathodea* (AFRICAN-TULIPTREE, FLAMETREE), and *Tabebuia* (POUI) as well as some taxa valuable for timber including *Tabebuia*. *Paulownia*, previously treated in the Bignoniaceae (e.g., Correll & Johnston 1970) and sometimes in the Scrophulariaceae (e.g., Radford et al. 1968), is now based on molecular data treated in its own family, the Paulowniaceae, and is widely recognized (e.g., Olmstead et al. 2009; Refulio-Rodriquez & Olmstead 2014) to be more closely related to a number of families in the Lamiales including the Orobanchaceae, Phrymaceae, and Lamiaceae, than to the Bignoniaceae. As noted by Yatskievych (2006) it is most easily distinguished "from the superficially similar *Catalpa* by its bluish purple corollas and ovoid fruits." There are some fascinating plant-animal relationships in the family with two examples being: 1) *Kigelia africana* (Lam.) Benth., SAUSAGE TREE, is bat-pollinated and has huge (typically 11–22 pounds), sausage-shaped fruits that hang beneath the foliage and are eaten by a variety of animals from monkeys to elephants; 2) *Crescentia alata* Kunth, CALABASH, of Central America, has large, spherical, cannonball-like fruits that fall beneath the tree and rot; however, it is thought to be anachronistic, with its fruits once being eaten by long extinct dispersal agents such as Pleistocene gomphotheres, members of the elephant clade (Janzen & Martin 1982). (subclass Asteridae—Cronquist; order Lamiales—APG IV)

FAMILY RECOGNITION IN THE FIELD: shrubs, trees, or woody vines with large showy flowers usually in racemes, panicles, or axillary clusters; corollas sympetalous, tubular, bilaterally symmetrical, with 2 or 4 separate, epipetalous stamens; capsules usually large, elongate, woody, with winged seeds.

REFERENCES: Shinners 1961; Spangler & Olmstead 1999; Manning 2000; Fischer et al. 2004; Olmstead et al. 2009; Olmstead 2013.

1. Corollas yellow to orange-red or red-orange; leaves mostly compound with (1–)2–13 leaflets; high-climbing or trailing vines or in one species small shrubs.
   2. Plants small shrubs to 2(–3) m tall; leaflets 5–13; corollas yellow ____ **Tecoma**
   2. Plants high-climbing vines; leaflets (1–)2 or in one species 7–13 (if 7–13 then corollas orange to orange-red); corollas yellow or orange or red-orange or orange-red.
      3. Tendrils present; leaflets (1–)2, entire or toothed, evergreen or nearly so; flowers in axillary or terminal inflorescences.
         4. Flowers yellow, in a terminal inflorescence; tendrils hooked, claw-like; petioles 2.2–7 cm long; fruits 18–136 cm long ____ **Dolichandra**
         4. Flowers red-orange, in axillary inflorescences; tendrils not hooked, not claw-like; petioles 1–2 cm long; fruits 10–17 cm long ____ **Bignonia**
      3. Tendrils absent; leaflets 7–13, sharply toothed, deciduous; flowers in a terminal inflorescence ____ **Campsis**
1. Corollas white to light pink to lavender with orange or purple markings inside; leaves simple; shrubs or trees.
   5. Leaves subsessile, mostly alternate, the blades linear-lanceolate to linear, glabrous, unlobed; fertile stamens 4, one staminode also present ____ **Chilopsis**
   5. Leaves petioled, mostly opposite or in whorls of 3, the blades ovate, usually with pubescence beneath, the larger sometimes lobed; fertile stamens 2, three minute staminodes typically present ____ **Catalpa**

BIGNONIA L. CROSSVINE, QUARTERVINE

A genus of 28 species of wet to dry tropical forests and warm temperate forests from the se U.S. s to Argentina (Lohmann & Taylor 2014). In the past, *Bignonia* was often considered to be a monotypic genus (e.g., Correll & Johnston 1970; Diggs et al. 1999) but has recently been circumscribed much more broadly based on both molecular and morphological characters to include species previously treated in a number of tropical genera (Lohmann & Taylor 2014). (Named for the Abbè Jean-Paul Bignon, 1662–1743, court-librarian at Paris and a friend of Tournefort)

REFERENCE: Lohmann & Taylor 2014.

Bignonia capreolata L., (winding or twining), CROSSVINE, QUARTERVINE. High-climbing, woody, evergreen vine; leaves opposite, typically 2-foliate, terminated by a three-branched, tendril, the tendril not hooked, not claw-like; leaflets 3–18 cm long, lanceolate to elliptic, ovate, or oblong, cordate at base, acute to acuminate apically, entire; petioles 1–2 cm long; flowers in axillary clusters, fragrant; corollas red-orange, lighter inside, 4–5 cm long; stamens 4; staminode absent; capsules linear, flattened, 10–17 cm long. Moist woods, mainly in se and e TX, cultivated and spreads or escapes further w (e.g., Dallas and Grayson cos.) and disjunct to the Edwards Plateau (Bandera and Gillespie cos.—Kartesz 2015); mainly se U.S. from MD s to FL w to IL, MO, OK, and TX. See discussion of this disjunct TX distribution under *Aristolochia serpentaria* (Aristolochiaceae). Mar–Jul. [*Anisostichus capreolata* (L.) Bureau, *Doxantha capreolata* (L.) Miers] The common name, CROSSVINE, is said to come from the cross-shaped pattern seen in a cut stem (Metcalfe & Chalk 1950). The species has been used medicinally and recently was shown to contain the alkaloid resperine (Clark & Lund 2012).

CAMPSIS Lour. TRUMPET-CREEPER

A genus of 2 species of adventitious root-climbers; both are cultivated as ornamentals; one occurs in e North America and the other (*C. grandiflora* (Thunb.) K. Schum.) in e Asia (Wen & Jansen 1995; Mabberley 2017). The striking disjunct e Asia–e North America distribution pattern exhibited by this and other genera has complex origins and is of interest to plant geographers. In the geologic past, dispersal between the Eurasian and North American continents was possible across both the Bering and North Atlantic land bridges, and the combined area is considered a single "Holarctic" biogeographic region. The fossil record shows that many plants had extensive distributions across the Northern Hemisphere—for example, temperate forests with tropical elements occurred very broadly and reached their maximum extension in the mid-Tertiary (the Tertiary extended from 65 million years ago to 1.8 mya). This widespread flora has been referred to as the Arcto-Tertiary flora, the Tertiaro-mesophytic flora, the boreotropical flora, or a mixed mesophytic forest. Geohistorical events from the mid-Tertiary to the present have included alterations in the shapes of the northern land masses, fluctuations in sea levels, plate tectonic movements, mountain building, glaciation, and other changes in the climate. As a result, there have been great changes in both the composition and the distribution of the flora, and the ranges of many plants have been greatly restricted (e.g., eliminated from Europe and w North America). A significant number of plants that were once much more widespread now survive in only two areas, e North America and e Asia. While this disjunct distribution pattern has complex and multiple origins, with similar present day distributions differing in time and manner of origin, the consensus is that the e Asia–e North America pattern is in general a relict of the maximum development of Northern Hemisphere temperate forests (with tropical elements) in the Tertiary, with a more or less continuous distribution that spanned the Northern Hemisphere. Subsequently, there was greater survival in e Asia, and secondarily in e North America, and higher rates of extinction in Europe and w North America. According to Wen (1999), ca. 65 genera of seed plants have this disjunct distribution. The genus *Campsis* is an example (other East TX examples include, but are not limited to *Aletris*, *Ampelopsis*, *Apios*, *Brachyelytrum*, *Carya*, *Catalpa*, *Diarrhena*, *Halesia*, *Hamamelis*, *Lindera*, *Lyonia*, *Menispermum*, *Nyssa*, *Parthenocissus*, *Penthorum*, *Phryma*, *Podophyllum*, *Sassafras*, *Saururus*, *Stewartia*, *Tipularia*, *Trachelospermum*, *Triosteum*, and *Zizania*) (Li 1952; Little 1970; Graham 1972; Boufford & Spongberg 1983; Hamilton 1983; Hsü 1983; Wu 1983; Ying 1983; Tiffney 1985a, 1985b; Cox & Moore 1993; Graham 1993a, 1999a; Schnabel & Wendel 1998; Xiang et al. 1998a; Wen 1999, 2001a; Dilcher 2000; Donoghue et al. 2001; Manos & Donoghue 2001; Xiang & Soltis 2001; Liu et al. 2002; Milne & Abbott 2002; Donoghue & Moore 2003; Donoghue & Smith 2004; Manchester et al. 2009; Manos & Meireles 2015; Lu et al. 2018; Melton et al. 2020; Lin et al. 2022). (Greek: *campsis*, curvature or bend, from the curved stamens)
REFERENCE: Wen & Jansen 1995.

Campsis radicans (L.) Bureau, (rooting), COMMON TRUMPET-CREEPER, COWITCH VINE, TRUMPET-HONEYSUCKLE, HELL-VINE. High-climbing, woody, deciduous vine (climbing to 10 m or more), with aerial rootlets; leaves opposite, odd-pinnate with 7–13 leaflets; tendrils absent; leaflets 2–8 cm long, ovate to lanceolate, toothed; petioles 2–8 cm long; flowers in terminal corymbs; calyx teeth about 1/5 as long as the tube; corollas 2–2.5 cm long, with 5 short recurved lobes, orange to orange-red with orange-red to scarlet lobes; stamens 4; staminodes absent; capsules cylindric-oblong, 8–12 cm long. Stream banks, disturbed ground, along fences, also cultivated; e 1/2 of TX; widespread in the e 1/2 of the U.S. w to ND, OK, and TX, scattered further w. Jun–Aug. [*Bignonia radicans* L., *Tecoma radicans* (L.) Juss. ex DC.] This species is sometimes planted as an ornamental but can get out of control and become problematic. The challenge "is usually not how to grow them but how to restrain them, in large part because they sucker profusely from underground runners and freely self-seed, often forming impenetrable colonies that can choke out other plants" (North Carolina Cooperative Extension 2024). The showy flowers are visited by and presumably pollinated by ruby-throated hummingbirds (*Archilochus colubris*) (James 1948). Dermatitis has been reported in some individuals from handling the leaves or flowers (Muenscher 1951; Burlage 1968).

CATALPA Scop. INDIAN CIGARTREE, CIGARTREE

Deciduous trees, leaves simple, opposite or usually in whorls of 3, distinctly petiolate; flowers in panicles; calyces bilabiately split to near base; corollas broadly campanulate, strongly bilabiate with 3 lower lobes longer than the upper two, white, with purple and yellow patterning; fertile stamens 2, 3 minute staminodes also present; fruit an elongate, cylindric, cigar-like capsule with 2 locules; seeds flat, with wings.

A monophyletic genus of 8 species of e North America (and the Greater Antilles, from Jamaica across Cuba and Hispaniola to the Bahamas) and e Asia; this e Asia-e North America disjunct distribution pattern is discussed under the genus *Campsis* (Bignoniaceae). *Catalpa* is most closely related to *Chilopsis* and hybrids are known between the two. A number of species are cultivated as ornamentals and used for timber; the CHINESE CATALPA, *C. ovata*, has been cultivated for nearly 2,000 years and has been widely planted; hybrids are known between this species and both *C. bignonioides* and *C. speciosa* (Olsen & Kirkbride 2017). The Catalpa Sphinx (a hawk moth), *Ceratomia catalpae*, is a specialist herbivore of *Catalpa*; the caterpillars have overcome the plants chemical defense and sequester an iridoid glycoside, catacol, from the leaves to use in their own defense (Lampert et al. 2011). During outbreaks the caterpillars can defoliate *Catalpa* trees; they are used as a fishing bait in some parts of the se U.S. *Catalpa* species have extrafloral nectaries consisting of many closely crowded glands on the undersurface of the leaves in the axils of major veins; the nectar produced attracts ants that defend the plant against herbivores such as the Catalpa Sphinx. Research on *C. bignonioides* has shown that additional extrafloral nectar production is stimulated (a type of "induced defense") when leaves are damaged; the result is that the increased nectar in turn attracts a greater number of *Forelius pruinosus* (ants who serves as "bodyguards" against the leaf-eating herbivores) giving added protection to the damaged leaves and the whole plant (Stephenson 1982; Ness 2003). This treatment draws heavily from the revision of the genus by Olsen and Kirkbride (2017). (The Native American name)

REFERENCES: Paclt 1952; Weniger 1996; Li et al. 2006; Li 2008; Kirkbride & Olsen 2011; Coder 2016; Olsen & Kirkbride 2017.

1. Corolla tubes 1.5–2.5 cm long, 8–12 mm wide at throat; stamens 16–22 mm long; styles 12–20 mm long; capsules 9–12 mm in diam., with septum (separating the locules) 4–5 mm wide; seeds with body (not counting wings) 4.5–5.5 mm long; inflorescences typically with 60–80 flowers; fresh foliage with a fetid odor when crushed; leaves abruptly acuminate; bark of old trunks plated ____ **C. bignonioides**
1. Corolla tubes (2.3–)2.5–4 cm long, 13–19 mm wide at throat; stamens 22–26 mm long; styles 20–28 mm long; capsules 10–17 mm in diam., with septum 5–8 mm wide; seeds with the body 5.5–7.5 mm long; inflorescences typically with 28–40(–75) flowers; fresh foliage odorless or nearly so when crushed; leaves long-acuminate; bark of old trunks vertically furrowed ____ **C. speciosa**

Catalpa bignonioides Walter, (resembling *Bignonia*), SOUTHERN CATALPA, COMMON CATALPA, CATAWBA, INDIAN-BEAN CATALPA, CATALPA. Deciduous tree to 21.5(–30) m tall; old trunks with bark plated; leaves opposite or in whorls of 3, those of a whorl unequal in size; petioles 10–18 cm long, blades (10–)15–31 cm long, ovate to broadly so, entire or rarely weakly 3-lobed, cordate truncate, obtuse or rarely cuneate basally, apically long acuminate; inflorescences showy thyrses usually with 60–80 flowers; flowers with pedicels 3–16 mm long, calyces 8–12 mm long, spitting into 2(–3) lobes; corollas white with a few purple dots near the base of tube outside, the lower lobes with rows of purple dots along veins inside tube, with 2 bright yellow spots at mouth of tube on inside between lower lobes, turning dark red with age, the tube 1.5–2.5 cm long; stamens 16–22 mm long, white; staminodes 3–6 mm long; capsules 24–56 cm long, 9–12 mm wide; seeds with body narrowly oblong, 4.5–5.5 mm long. Apparently native in East TX and also widely cultivated and escapes; mainly se part of East TX, also Travis Co. (USDA Plants Database 2019) further west; native in the se U.S. in AL, FL, GA, LA, and MS, and according to some authorities the s edge of AR and the eastern most edge of TX (Weniger 1996; Olsen & Kirkbride 2017); now widespread in the e 1/2 of the U.S. and scattered to the w. Late May, usually flowering after *C. speciosa*. [*Bignonia catalpa* L.] The species is cultivated in temperate regions worldwide (Olsen & Kirkbride 2017).

Catalpa speciosa Teas, (showy, good-looking), NORTHERN CATALPA, CATAWBA-TREE, CIGAR-TREE, HARDY CATALPA. Deciduous tree to 24 m tall; old trunks with bark vertically furrowed; leaves opposite or in whorls of 3, those of a whorl unequal in size; petioles 10–16 cm, blades 15–40 cm long, ovate to broadly so, entire or rarely 2–3-lobed, basally truncate to cordate or obtuse basally, acuminate apically; inflorescence a showy thyrse, usually with 28–40 flowers (rarely more); flowers with pedicels 5–20 mm long; calyces 8–13 mm long, splitting into 2(–3) lobes; corollas white with a few purple dots near base of tube outside, the lower lobes with rows of purple dots along veins and inside tube, with 2 bright yellow spots at mouth of tube on inside between lower lobes, turning dark red with age, the tube (2.3–)2.5–4 cm long; stamens 22–26 mm long, white; staminoides 4–9 mm long; capsules 2.2–5.5 cm long, 10–17 mm wide; seeds with body narrowly oblong, 5.5–7.5 mm long. Native in the ne part of TX according to Olsen and Kirkbride (2017), though Weniger (1996) disagrees, and also widely cultivated and escaped throughout, roadsides, along streams; mainly e 1/2 of TX; native of middle Mississippi and lower Ohio River valleys sw to ne TX (Olsen & Kirkbride 2017); widespread in e 2/3 of the U.S. May. [*C. bignonioides* Walter var. *speciosa* (Teas) Meechan] The species is widely cultivated in the e U.S. and in other temperate areas of the world (Olsen & Kirkbride 2017). Pollination is by bumblebees and carpenter bees and at night by moths (Stephenson & Thomas 1977). The authority of this species is often incorrectly given as (Warder) Warder ex Engelm.; however, Kirkbride and Olsen (2011) clarified the nomenclature.

CHILOPSIS D. Don DESERT-WILLOW, MIMBRE

A monotypic genus of the Chihuahuan and Sonoran deserts of the sw United States and Mexico; the branches are used to make baskets. The genus is most closely related to *Catalpa* (Li et al. 2006; Li 2008; Olsen & Kirkbride 2017) and hybrids are known (e.g., Li et al. 2006). (Greek: *chil*, lip, and *opsi*, appearance, from flower shape resembling lips)
REFERENCES: Fosberg 1936; Henrickson 1985; Gentry 1992; Li et al. 2006; Li 2008; Olsen & Kirkbride 2017.

Chilopsis linearis (Cav.) Sweet, (narrow, with sides nearly parallel), DESERT-WILLOW, FLOWERING-WILLOW, WILLOW-LEAF CATALPA, DESERT-CATALPA, FLOR DE MIMBRE, BOW-WILLOW. Shrub or small tree 2–10 m tall; leaves simple, mostly alternate, sometimes partially opposite or whorled, subsessile, blades linear-lanceolate to linear, entire, glabrous; inflorescences terminal racemes or racemose panicles to 30 cm long; flowers sweetly fragrant, sometimes sterile (J. Stanford, pers. comm.); corollas broadly tubular-campanulate, ca. 2–5 cm long; variable in color, usually white to light pink or lavender with purple lower lip and yellow lines in throat to rarely both lips white or purplish red with purple stripes; stamens 4, a staminode also present; fruits capsules, linear, terete, to 30 cm long, usually ca. 6 mm thick. Cultivated and long persists, planted along highways, escapes; also long persisting (ca. 100 years) in Hamilton Co. (J. Stanford, pers. comm.) scattered in mainly the w part of East TX and to the w, native in w TX; native of Mexico and sw U.S. from CA and NV e to wc TX, introduced elsewhere in the U.S. May–Sep. [*Bignonia linearis* Cav.] The species is sometimes divided into 2 subspecies, one of which is divided into 2 varieties (Henrickson 1985). It is often planted in TX in landscapes emphasizing native plants.

DOLICHANDRA Cham. CLAW-VINE, DOLICHANDRA

A genus of 9 species of Neotropical lianas occurring from Mexico and the Caribbean s to Argentina. As currently recognized it includes the species previously treated in *Macfadyena*, *Melloa*, and *Parabignonia* (Fonseca et al. 2017). (Greek: *dolichos*, long, and *andro*, male, presumably from the exerted anthers of the type species; the derivation not explained in the type description by Chamisso)
REFERENCES: Lohmann & Taylor 2014; Fonseca & Lohmann 2015a, 2015b; Fonseca et al. 2017.

Dolichandra unguis-cati (L.) L.G. Lohmann, (from Latin *unguis-cati*, cat's claws, in reference to the unique, hooked, claw-like, 3-pronged tendrils adapted for climbing), UÑA-DE-GATO, CAT-CLAW-VINE. High-climbing, woody, evergreen vine; leaves typically 2-foliated, with a terminal trifid, uncinate tendril 12–40 mm long; petioles 2.2–7 cm long; petiolules sulcate, 1–35 mm long; leaflets 3.2–26 cm long, 1–13.5 cm, wide, extremely variable, ovate to elliptic, apically acute to acuminate, basally cuneate to truncate, symmetric or slightly asymmetric, marginally entire or serrate; inflorescences a thyrse or leafy thyrse; floral pedicels 7–42 mm long; floral bracts deciduous, rarely present, lanceolate to elliptic, 1–10 mm long, membranaceous, green; bracteoles deciduous, rarely present, linear to lanceolate; flowers fragrant, with abundant nectar; calyces campanulate, inflated, with truncate to sinuous margin, 6–30 mm long, 7–22 mm wide, membranaceous, green; corollas yellow, with lobes dark yellow, tubular-campanulate, with open mouth, with upper two lobes diverging from axis at almost 90° and the lower three lobes spreading, membranaceous, with nectar guides present, the overall corollas 37–100 mm long, 12–25 mm wide, the tubes 29–71 mm long, the lobes 10–29 mm long, 12–25 mm wide; stamens 4; staminode present; anthers included; styles 13–28 mm long; stigmas included; fruits capsules, flattened and linear, narrowed at both ends, extremely variable in length, 18–136 cm long, 1–2.2 cm wide, slightly woody, smooth, brown, glabrous; seeds numerous, winged, oblong, 27–68 mm long. Weedy areas, disturbed places, forests; Bexar (Kartesz 2015; *J. Garland, Sp. Obs. #21380*, 26 Jul 2015—Texasinvasives.org), Harris (Kartesz 2015; *J. Aplaca Sp. Obs. #22578*, 15 Dec 2015—Texasinvasives.org), Hays (Kartesz 2015; *L. Jones, Sp. Obs. #24952*, 23 Sep 2018—Texasinvasives.org), and Travis (Kartesz 2015; *A. Rodriguez, Sp. Obs. #18728*, 24 Feb 2014—Texasinvasives.org) cos. in East TX; also Fort Bend (USDA Plants Database 2019), Goliad (*J. Garland Sp. Obs. #22471*, 20 Nov 2015—Texasinvasives.org), Karnes (*J. Garland Sp. Obs. #22475*, 20 Nov 2015—Texasinvasives.org), and Nueces (*M.G. Lopez Sp. Obs. #1959*, 6 Jan 2008—Texasinvasives.org) cos. near e TX, and a number of counties in s TX (Kartesz 2015); FL, GA, LA, SC, and TX; native from Mexico and the Caribbean s to Argentina. Spring–? [*Batocydia unguis-cati* (L.) Mart. ex Britton, *Bignonia unguis-cati* L., *Doxantha unguis-cati* (L.) Miers, *Macfadyena unguis-cati* (L.) A.H. Gentry] This species is widely used as an ornamental and has been used medicinally in a number of ways (Fonseca et al. 2017). It can be a problematic invader, for example in Australia and South Africa where biological control efforts are being attempted using a variety of insect species (Fonseca et al. 2017); it is an invasive in TX (Texasinvasives.org 2015) and is considered "High Risk of becoming weedy or invasive in the United States (APHIS 2019). The floral morphology suggests pollination by large to medium sized bees (Fonseca et al. 2017). ⊘

TECOMA Juss. TRUMPET-FLOWER

A genus of 13–14 species of trees and shrubs with alkaloids and extrafloral nectaries of tropical and s Africa and tropical America (especially the Andes) n to AZ and TX (Stevens 2001 onwards; Mabberley 2017). A number are cultivated as ornamentals and several used medicinally. (Name from *tecomaxochitl*, a Mexican name for these plants—Hyam & Pankhurst 1995)
REFERENCES: Shinners 1961; Pelton 1964; Gentry 1992; Turner 2013.

Tecoma stans (L.) Juss. ex Kunth, (self-supporting, upright, erect, standing, in reference to the upright inflorescence), YELLOW TRUMPET-BUSH, YELLOW-BELLS, YELLOW-ELDER, ESPERANZA, TRONADORA, MIÑONA, GINGER-THOMAS. Perennial deciduous shrub to ca. 2(–3) m, capable of stump-sprouting following damage, plants sometimes dying to the ground in winter; leaves opposite, with petioles to 4 cm long, odd pinnately compound, the leaflets usually 5–13; leaflet blades to 10 cm long, 0.8–4 cm wide, ovate-lanceolate to elliptic or linear-lanceolate, bright green, serrate to incised, sessile, glabrous or pubescent; inflorescences terminal racemes or panicles; flowers fragrant; calyces 5-toothed, 3–6 mm long; corollas 3.5–6.5 cm long, bright yellow, funnelform-campanulate, abruptly contracted near base, with 5 subequal lobes, the limb 3–5 cm wide; stamens 4, a sterile 5th stamen much reduced; anthers ca. 4 mm long; fruits capsules, linear, subterete, ca. 6–8 mm wide, 13–22 cm long; seeds ca. 6–8 mm long, flat, oval to narrowly elliptic, with 2 large, thin, membranous, transparent wings; Bexar, Goliad, Hays, Polk, Victoria, and Wilson cos. in s half of e TX (USDA Plants Database 2019); the e TX locations all presumably persisting from cultivation or naturalized, also far s TX, native in rocky areas in the mts. of the Trans-Pecos; species with an extremely broad natural range—from TX w to AZ and s through Mexico and Central America to South America as far as n Argentina, as well as in s FL (apparently introduced), through much of the Caribbean. Apr–Nov. [*Bignonia stans* L., *Stenolobium stans* (L.) Seem.] Plants native to the sw US (including w TX) and adjacent n Mexico are often recognized as *Tecoma stans* var. *angustatum* Rehder [*Stenolobium incisum* Rose & Standl. ex Wooton & Standl.] YELLOW TRUMPET-FLOWER. It is shorter, more drought-tolerant, and more cold-tolerant than some of the tropical varieties sold in nurseries. Plants escaped or naturalized in the s U.S. (including e TX) are var. *stans*, which is apparently native to the south of the U.S. Turner (2013) noted that individuals with leaflets "markedly pubescent" (vs. glabrous to sparsely pubescent in typical var. *stans*), which sometimes have been recognized as var. *velutina* DC., occur throughout the range of var. *stans*, often at higher elevations. He further noted that the different pubescence forms often occur together and "numerous intermediates can be found where they coexist, nothing but pubescence separating the two. In short, they are not geomorphologically coherent, nor do they form meaningful populational aggregates, hence their treatment as but forms." We are following Turner (2013) in treating this variant as forma *velutina* (DC.) B.L. Turner. *Tecoma stans* is widely used as an ornamental and is locally used as a diuretic (Mabberley 2017). The species is reportedly pollinated by hummingbirds, but also visited by bees and by other insects as well (Pelton 1964). The calyx has extra-floral nectaries which attract ants (Pelton 1964). The species is unpalatable to cattle. It is the official flower of the United States Virgin Islands and the floral emblem of The Bahamas. It has been widely introduced in many parts of the world (e.g., Asia, Africa, Australia) and can be a problematic invader (e.g., Queensland, Australia; Kenya). Cultivars range in corolla color from yellow, to with orange markings in the throat of the corolla, to reddish orange. Turner (2013) separated the taxa as follows:

1. Leaflets mostly linear-lanceolate, 0.8–1.5 cm wide, glabrous, having margins sharply serrate, their apices sharply acute; north-central Mexico and closely adjacent U.S.________ var. *angustatum*
1. Leaflets more nearly ovate to lanceolate, 1–4 cm wide, to some degree pubescent (rarely not), margins not sharply serrate, their apices not sharply acute.
   2. Leaflets nearly glabrous to sparsely pubescent________ var. **stans**
   2. Leaflets markedly pubescent ________ forma *velutina*

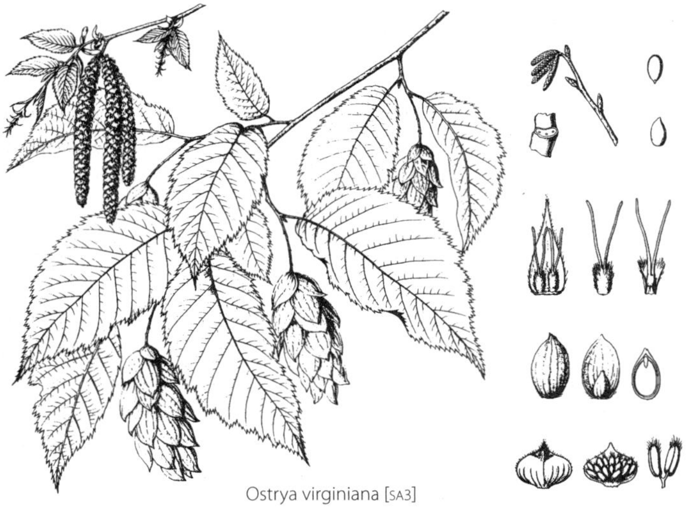
Ostrya virginiana [SA3]

Bignonia capreolata [RAD]

Campsis radicans [REE]

Catalpa bignonioides [SA3]

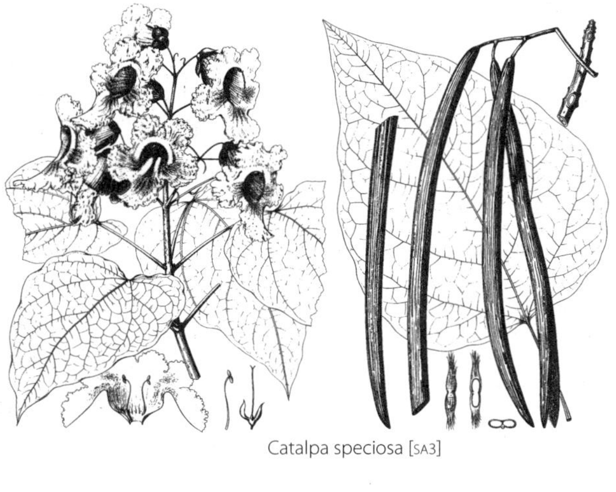
Catalpa speciosa [SA3]

Chilopsis linearis [SA3]

BORAGINACEAE Juss.

FORGET-ME-NOT OR BORAGE FAMILY

Ours pubescent, bristly-hispid annual or perennial herbs; leaves alternate, simple, entire, usually sessile; flowers in terminal or axillary cymes, or more often in uncurling, 1-sided, spike-like or raceme-like inflorescences (= scorpioid cymes); axis of fruiting inflorescence often elongating; sepals 5, barely united at base; corollas 5-toothed or -lobed, often base of lobes with inflated appendages (= fauces) which partially or completely obstruct the corolla throat, tubular, funnelform, or salverform; stamens 5, attached near base of corolla or higher in tube; pistil 2-carpellate; ovary superior, 4-lobed; style and stigma 2- or 4-lobed; fruit usually breaking into 4 one-seeded, achene-like mericarps ("nutlets"), sometimes reduced to fewer by abortion.

A large (1,600–1,700 species in ca. 90 genera) (Luebert et al. 2016) family of mainly temperate and subtropical annual and perennial herbs to shrubs, including a number of ornamentals such as *Mertensia* (VIRGINIA BLUEBELLS) and *Myosotis* (FORGET-ME-NOT); some, including *Borago officinalis* L. (BORAGE) and *Symphytum officinale* L. (COMFREY), have been used for flavorings or medicinally. Family name from *Borago*, a genus of three species native to the Mediterranean region, Europe, and Asia; cultivated as ornamentals. The roots of many species of Boraginaceae, particularly in the genera *Alkana*, *Anchusa*, *Arnebia*, *Cynoglossum*, *Echium*, and *Lithospermum*, yield red to purple naphthaquinone dyes commonly known as alkannins. These dyes were used by the early Romans and by the Japanese as coloring agents for wood, medicine, cosmetics, and wines (Al-Shehbaz 1991). (Possibly from Latin: *burra*, a hairy garment, alluding to the leaves that are usually densely hairy). Most members of the family are characterized by a 1-sided inflorescence that uncoils as it matures; alkaloids are often present. As with many families, the Boraginaceae have been circumscribed variously in the past, alternating between various broad and narrow interpretations. In its broad sense, it has at times included the Heliotropiaceae, Namaceae, Cordiaceae, Ehretiaceae, and others. We are following Stevens (2001 onwards) and Luebert et al. (2016) and treating it in its strictest sense, which does not include any of the above families. However, these families have been included in the key to the genera below. (subclass Asteridae—Cronquist; order Boraginales—APG IV)

FAMILY RECOGNITION IN THE FIELD: herbs with alternate, coarse, rough-hairy leaves and round stems; flowers usually in *scorpioid* cymes (= 1-sided, uncurling inflorescences), with sympetalous, radially symmetrical corollas and 5 stamens; fruits of 4 mericarps.

REFERENCES: Johnston 1954a, 1967; Al-Shehbaz 1991; Chacón et al. 2016; Luebert et al. 2016; Weakley, A.S., and Southeastern Flora Team 2024.

1. Plants trees or shrublets; fruit a fleshy drupe or 4 mericarps ____ Ehretiaceae
1. Plants herbaceous to suffrutescent; fruit a capsule, 1–4 mericarps, or a dry drupe.
  2. Plants herbaceous; fruit a capsule ____ Namaceae
  2. Plants herbaceous to suffrutescent; fruit 1–4 mericarps or a dry drupe covered with a sac-like calyx.
    3. Plants suffrutescent; fruit a dry drupe ____ Cordiaceae
    3. Plants mostly herbaceous annuals or perennials; various habits; fruit 1–4 mericarps.
      4. Ovary entire or slightly lobed; stigmas conical ____ Heliotropiaceae
      4. Ovary distinctly lobed; stigmas not conical.
        5. Flowers in naked, spike-like or raceme-like cymes; bracts present in the lower portions of some species.
          6. Calyces with ± appressed hairs; corollas usually pale to deep blue (rarely white); basal leaf blades to 30 cm long.
            7. Plants perennial; stems spreading hirsute; limb (= open face of corolla) 8–12 mm wide ____ **Andersonglossum**
            7. Plants biennial; stems strigose (= stiff, appressed hairs); limb (= open face of corolla) 5–7 mm wide ____ **Cynoglossum**
          6. Calyces with widely spreading or reflexed hairs; corollas white (rarely slightly bluish) or yellow-orange; basal leaf blades < 10 cm long.
            8. Flowers sessile; calyx hairs all straight and sharp-pointed; axis of spike with both wide-spreading (usually conspicuous with a hand lens) and appressed hairs.
              9. Corollas white with greenish yellow eye, < 3 mm long; calyx lobes remaining close together and thus appearing fused even at maturity, only slightly spreading apart apically; mericarps with very minute warts and a groove along one side ____ **Cryptantha**
              9. Corollas yellow-orange, 4–5 mm long; calyx lobes often spreading and clearly separated at maturity; mericarps with toothed ridges and a conspicuously warty surface, but without a groove ____ **Amsinckia**
            8. Flowers pedicellate; calyx hairs (at least some) hooked; axis of racemes (except near base) with uniformly incurved, subappressed hairs ____ **Myosotis**
        5. Flowers in bracteate spike-like or raceme-like cymes (bracts sometimes small or absent in *Cryptantha*).
          10. Corollas tubular, white to greenish white, large, 8–20 mm long; leaf blades with 4–6 secondary veins

conspicuously visible on the lower surface ____ **Lithospermum** (*Onosmodium*)

10. Corollas salverform or funnelform, white to blue, 1.5–7 mm long OR yellow to orange, 4.5–48 mm long; leaf blades with midrib only visible or occasionally 4–6 secondary veins slightly visible on the lower surface.
    11. Plants perennial; corollas yellow to yellow-orange, 11–48 mm long OR corollas 4.5–6 mm long ____ **Lithospermum**
    11. Plants annual or biennial; corollas white to pale blue (sometimes with yellow or greenish yellow eye), ≤ 7.5 mm long.
        12. Plants with midstem leaf blades with 4–6 secondary veins visible on the lower surface; corollas white to pale blue, ca. 2 mm long; plants biennial ____ **Hackelia**
        12. Plants with midstem leaf blades with only midvein conspicuous on lower leaf blade surface; corollas white to pale blue (usually with yellow or greenish yellow eye), 2–7.5 mm long; plants annual.
            13. Stems with wide-spreading hairs.
                14. Corollas 4–5.5 mm long, white with a yellow eye; fruits (= mericarps) not obscured by the calyx at maturity, stout bristles absent ____ **Lithospermum matamorense**
                14. Corollas 3 mm or less long, white to pale blue (usually with a yellow or greenish yellow eye); fruits not obscured by the calyx, easily visible at maturity, stout bristles present OR fruit completely obscured by the calyx at maturity and stout bristles absent.
                    15. Calyces rather coarsely pubescent with slender ± appressed hairs only; calyx not significantly obscuring the fruit from view; fruits with stout bristles sometimes partly or almost completely united into a wing or cup-like structure; corollas with a yellow eye ____ **Lappula**
                    15. Calyces prickly-hispid with spiny, conspicuously spreading hairs and pubescent with slender ones as well; calyx completely obscuring the fruit from view; fruits essentially smooth, stout bristles absent; corolla with a greenish yellow eye ____ **Cryptantha**
            13. Stems with appressed hairs only ____ **Buglossoides**

AMSINCKIA Lehm. FIDDLENECK, TARWEED

A genus of 15 species of w North America and w temperate South America, 2 naturalized in Europe (Mabberley 2017). (Named for Wilhelm Amsinck Burchard-Motz, a burgomaster (mayor) of Hamburg, who early in the 19th century gave important support to the botanical garden of that city)

REFERENCES: MacBride 1917; Ray & Chisake 1957a, 1957b, 1957c; Al-Shehbaz 1991.

Amsinckia menziesii (Lehm.) A. Nelson & J.F. Macbr., (for Archibald Menzies, 1754–1842, British naval surgeon and botanist who accompanied Captain George Vancouver on his expedition to explore North America's northwestern Pacific Coast, 1791–1795), SMALL-FLOWER FIDDLENECK. Bristly annual 30–60 cm tall; stems simple or freely branched with spreading pubescence, usually decumbent; inflorescences spike-like, becoming greatly elongated, with few (in lower portion of the inflorescence) or no bracts; fruiting calyces 6–8 mm long, the lobes narrowly to broadly elongate; corollas yellow-orange, narrowly funnelform, 4–5 mm long, the lobes small; mericarps 4, triangular, 2.5–3 mm long, the surface with toothed ridges and a conspicuously warty surface. Grasslands and dry areas; Bell, Bexar, Brazos, and Lee cos. (Kartesz 2022); Crockett Co. (Kartesz 2022) in w Edwards Plateau; mostly w U.S.; Canada. Mar–May. [*A. micrantha* Suksd.] Native of nw U.S., found as a waif elsewhere and in adjacent Canada.

ANDERSONGLOSSUM J.I. Cohen WILD COMFREY

A genus of 3 species in North America, (named for William Russell Anderson 1942–2013)

REFERENCE: Cohen 2015.

Andersonglossum virginianum (L.) J.I. Cohen, (of Virginia), BLUE COMFREY, BLUE HOUND'S-TONGUE, WILD COMFREY. Erect, pubescent perennial, 30–80 cm tall; stems spreading hirsute; basal leaf blades several, to ca. 30 cm long, large, blade elliptic-oblong, petioles to ca. 7 cm long, stem leaf blades smaller, sessile and some ± clasping; inflorescences branched, long pedunculate with appressed hairs; calyx lobes ≤ width of individual mericarps; corollas blue to whitish, ca. 8–12 mm wide, corolla lobes often somewhat erose, large fauces (= inflated appendages) at base of lobes; mericarps 5.5–7 mm long, covered with bristles with grapple-hook-like barbs and sticking to hair or clothing. Sandy woods, low ground; mainly e East TX; e U.S. from ME s to FL and w to MN, OK, and TX; Canada. Mar–Apr. [*Cynoglossum virginianum* L.] Native of e U.S. Cohen (2015) using DNA sequencing data and morphological characters determined that New World native species of *Cynoglossum* were distinct from those native to the Old World and placed them in the genus *Andersonglossum*. According to Burlage (1968), reputed to be poisonous. ☠

BUGLOSSOIDES Moench

A temperate Eurasian genus of 7 species (Wunderlin et al. 2018); it is sometimes lumped into *Lithospermum* (e.g., Mabberley 1997, 2008). (From the old genus name, *Buglossum*, and the Greek, *-oides*, resembling)

Buglossoides arvensis (L.) I.M. Johnst., (pertaining to cultivated fields), CORN GROMWELL. Erect annual, 20–70 cm tall; stem leaf blades lanceolate to linear; flowers in leafy bracted raceme-like cymes which at maturity are loosely flowered; corollas white (rarely yellowish or purplish), 5–7 mm long; mericarps light brown, tuberculate, wrinkled, or pitted, ca. 3 mm long. Ditch banks, roadsides, disturbed sites; throughout East TX; Cross Timbers and Prairies, Edwards Plateau, and Trans-Pecos; throughout e U.S. and in the nw; Canada; South America; Europe; and Australia. Mar–May. Native of Europe. Sometimes recognized in the genus *Lithospermum* [as *L. arvense* L.] (e.g., Jones et al. 1997). Recently (Prasad et al. 2019), *B. arvensis* seeds have been found to be high in omega-3 fatty acids and may be a potential source on par with fish oil.

CRYPTANTHA Lehm. ex G. Don CAT'S-EYES

A genus of 160 species of w North America (Mabberley 2017). (Greek: *krypto*, to hide, and *anthos*, flower, from cleistogamous flowers of some species)
REFERENCES: Johnston 1925b; Payson 1927.

Cryptantha texana (A. DC.) Greene, (of Texas), TEXAS CRYPTANTHA. Prickly-hispid annual with erect to decumbent stems to 40 cm long; leaf blades conduplicate, spatulate to elongate oblanceolate; inflorescences to ca. 15 cm long, without bracts; calyces with conspicuous sharp hairs, lobes linear; corollas white with greenish yellow eye, < 3 mm long; mericarp usually 1, gray, with a groove along one side and covered with minute numerous warts. Loose sandy soils; in the sw 1/3 of East TX; South TX Plains, Edwards Plateau, and Rolling and High Plains; endemic to TX. Apr–May. Kartesz (2022) shows *C. texana* as present in Choctaw Co., OK on the n border of East TX, most likely based on a voucher (*Lundell 14864* GH) from that county (voucher image seen via the internet). However, the voucher in question is definitely not that taxon. It is most likely *Myosotis* sp.; Kartesz (2022) also shows a voucher (*Hoagland*, collection number not accessible, OKL) for Pittsburg Co., OK. This is a misidentified specimen of *Myosotis verna* (pers. comm., A. Buthod, OKL).

CYNOGLOSSUM L. HOUND'S-TONGUE, BEGGAR'S-LICE

A temperate and warm area, especially Old World genus of ca. 200 species (Chacón et al. 2016); according to Chacón et al. (2016), a whole range of segregate genera have been proposed and their phylogenetic relationships are not at all resolved; some have alkaloids; a few are cultivated as ornamentals and the European *C. officinale* L. (HOUND'S TONGUE) was formerly used medicinally. (Greek: *kyon* or *cyno*, dog, and *glossa*, tongue, alluding to the strap-shaped, lumpy leaves)
REFERENCES: Correll & Johnston 1970; Wunderlin et. al 2018.

Cynoglossum zeylanicum (Vahl ex Hornem.) Thunb. ex Lehm., (of Ceylon), CEYLON HOUND'S TONGUE. Plant biennial, erect, 3–9 dm tall, with strigose (= stiff, appressed) hairs; stems without spreading or lax hairs; basal leaf blades to 26 cm long; long-petioled, narrowly elliptic, persistent at flowering time; stem leaf blades to 10 cm long, 2.5 cm wide, narrowly elliptic, sessile, softly hairy, secondary veins often prominent, the upper leaf blades with a narrowly cuneate base; racemes forked; calyx lobes ovate, in fruit often 3 mm long and broadly oblong, calyx lobes ≥ width of individual mericarps; corollas blue, 4–5 mm long, 5–7 mm wide, pedicellate at least the lower, pedicels often 3–6 mm long; mericarps 2.5–4 mm long with scattered barbed bristles. Disturbed sites; in TX recorded only from Sabine Co. (*McVaugh 8443*, TEX); AR, FL, LA, MS, OK, and TX. May. [*C. furcatum* Wall.] Native of India, Sri Lanka, and Malaysia, introduced in various parts of the world.

HACKELIA Opiz STICKSEED

A north temperate genus of about 45 species; also Australia, Central, and South America (Mabberley 2017). (Named for Josef Hackel, 1783–1869, often credited as a Czech botanist; however, according to Quattrocchi (2000), he was specifically a Bohemian botanist)
REFERENCES: Correll & Johnston 1970; Yatskievych 2006; Weakley et. al 2012.

Hackelia virginiana (L.) I.M. Johnst., (of Virginia), BEGGAR'S LICE, STICKSEED. Biennial plant 3–15 dm tall; stems much branched above, pubescent with short fine reflexed hairs, spreading to ascending above midstem; leaf blades with 4–6 evident arcuate secondary veins; basal leaf blades 10–30 cm long, 3–15 cm wide, broadly ovate to cordate, often withered at flowering time, petiole to 15 cm long, stem leaf blades 4–25 cm long, narrowly elliptic to narrowly ovate, acuminate at both ends, sessile, progressively reduced above becoming narrower; inflorescences bracteate, terminal and sometimes in upper axils, often forked,

Dolichandra unguis-cati [SA3]

Tecoma stans [APG]

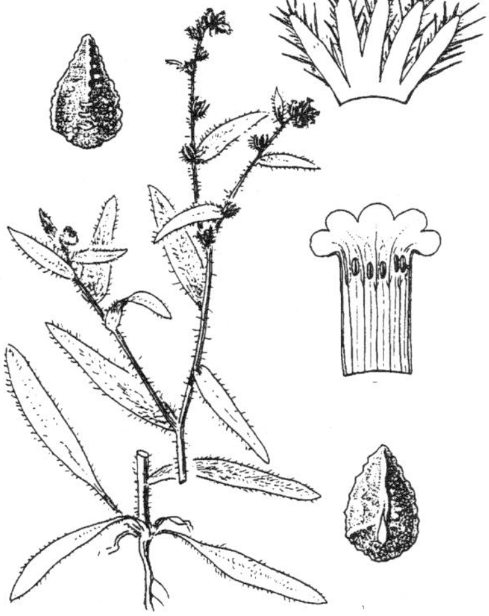

Amsinckia menziesii [PNW]

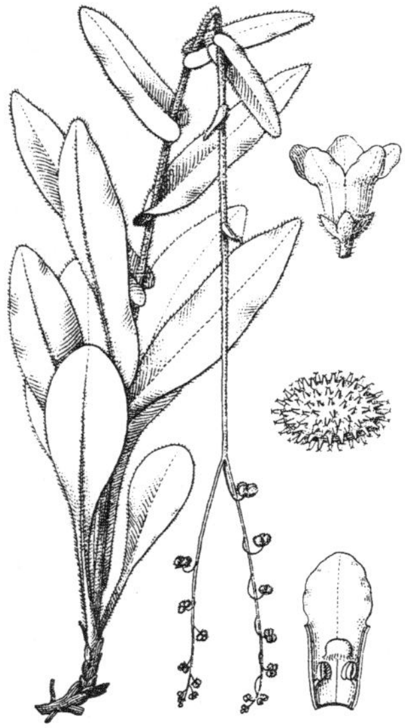

Andersonglossum virginianum [EN1]

Buglossoides arvensis [REE]

numerous, and paniculate; flowers of inflorescences at first clustered but axis elongating into a scorpiod raceme-like cyme; flowers bracteate at base of raceme-like cyme, becoming minutely so or ebracteate towards the tip, pedicels 1–8 mm long, becoming recurved in fruit, calyx 1.5–2.5 mm long, corolla white to pale blue, salverform, 1.7–2.5 mm long, 2 mm wide, throat partially constricted by inflated appendages (= fauces) at the base of each corolla lobe; fruits dividing into 4 mericarps, 2–3 mm long, ± angular-ovoid, back surface with numerous stout, barbed (shaped like a grappling hook) bristles. Mesic to dry forests, old fields, and other disturbed, shaded habitats, open woods; Bowie, Dallas, Red River, and Smith (Kartesz 2022) cos.; otherwise only known in TX from Hemphill, Tarrant, and Wise (Kartesz 2022) cos.; e U.S. except for the se; Canada. Jun–Sep. [*Lappula virginiana* (L.) Greene, *Myosotis virginiana* L.] Native of the U.S.

LAPPULA Moench STICKSEED

A genus of 50–60 species of Eurasia and w North America (Weakley, A.S., and Southeastern Flora Team 2024). (Latin: *lappa*, a bur, presumably in reference to the fruit)
REFERENCE: Johnston 1924.

Lappula occidentalis (S. Watson) Greene, (western), FLAT-SPINE SHEEPBUR. Annuals 10–50 cm tall, hispid-villous with spreading or ascending hairs; stems usually single, sometimes several from the base; leaf blades narrowly oblanceolate or spatulate, secondary veins not conspicuous; basal leaf blades 3–5 cm long, (2–)3–5(–8) mm wide, forming a rosette; stem leaf blades usually numerous reducing in size above; inflorescence a heavily bracted, raceme-like cyme, single or several on ends of stems and branches, becoming loosely flowered, 1–3 dm long at maturity; calyx 2–3 mm long at anthesis, linear or lanceolate, usually ascending, fruiting calyx 3–5 mm long; corollas funnelform, white to pale blue, usually with yellow eye, usually about 3 mm long, the tube about 1.2 mm long, throat partially constricted by inflated appendages (= fauces) at the base of each corolla lobe; fruits divided into 4 mericarps, smooth but with margins of barbed bristles, the bristles separate or united into wings or cup-like structures. Sandy or gravelly open ground; Bexar, Gonzales, Hays, and Travis (Kartesz 2022) cos.; throughout most of the w 1/2 of TX; w 1/2 of U.S.; Canada; Mexico; South America; Asia; and Russia. Late Mar–May. Native of w North America, Europe, and Asia.

1. Bristles of fruits partly or almost completely joined into spiny-margined wings or cup-like structures ______ var. **cupulata**
1. Bristles of fruits separate or nearly so ______ var. **occidentalis**

var. **cupulata** (A. Gray) L.C. Higgins, (cup-like), HAIRY STICKSEED, CUPSEED. Bexar, Hays, and Travis, (Kartesz 2022) cos.; throughout most of w 1/2 of TX; throughout w U.S.; w Canada and Mexico. [*L. texana* (Scheele) Britton]

var. **occidentalis,** (western) FLAT-SPINE STICKSEED. Bexar, Gonzales, and Travis (Kartesz 2022) cos.; throughout most of w 1/2 of TX; throughout w U.S.; w Canada and Mexico. [*L. redowskii* var. *occidentalis* (S. Watson) Rydb.]

LITHOSPERMUM L. PUCCOON, GROMWELL

This genus now includes, among others, the genus *Onosmodium* (Cohen & Davis 2009). Therefore, as treated here, *Lithospermum* in our area can now be divided into two groups with somewhat different characters. *Lithospermum* as treated in a strict sense in the past was a low perennial or annual with rather showy, salverform or funnelform, yellow to orange-yellow flowers (or flowers small and white with yellow eye, one species) in leafy-bracted terminal inflorescences; mericarps smooth or pitted, sometimes with a constriction above the base. *Onosmodium*, treated here as *Lithospermum*, in its former strict sense was a conspicuously hairy, leafy perennial from a taproot; leaves alternate, entire, 4–6 conspicuously visible secondary veins; flowers usually numerous in uncurling, 1-sided, terminal or branched raceme-like cymes or occasionally axillary; corollas tubular, slightly enlarged just below junction of tube and 5 erect lobes, greenish white to whitish or yellowish white, 9–20 mm long; fruit usually of (by abortion) 1(2–4) mericarp(s) 5 mm or less long, conspicuously white to whitish brown.

A genus of ca. 60 species (Cohen 2018) mostly in temperate regions of the world except Australia; some are cultivated as ornamentals or used medicinally; the roots of some U.S. species were used by Native Americans and early settlers as a source of a red dye; it is still used today in dying weaver's wool (Ajilvsgi 1984). This genus has been treated more strictly in the past with as few as 45 species. It now includes among others the genus *Onosmodium*. (Greek: *lithos*, stone, and *sperma*, seed, from the hard mericarps)
REFERENCES: MacKenzie 1906; Johnston 1952, 1954a, 1954b; Govoni 1973; Turner 1995a; Cohen & Davis 2009; Weakley et al. 2012; Luebert et al. 2016; Cohen 2018.

1. Corollas tubular (lobes not reflexed to form a flattened limb (= open face of corolla), white to greenish white, 9–20 mm long; styles long exerted; lower leaf blade surface with 4–6 conspicuous secondary veins.

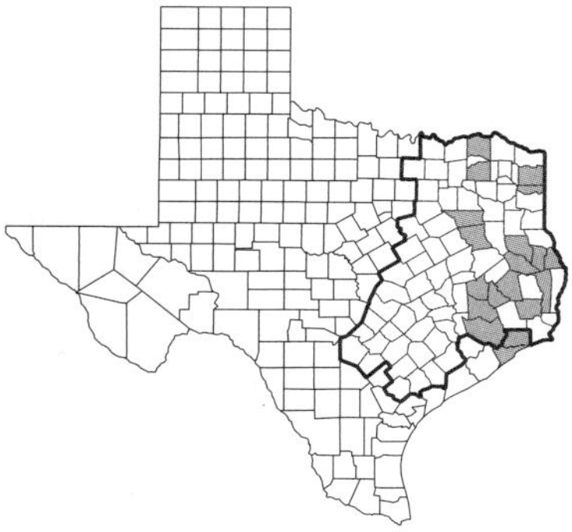

Carpinus caroliniana

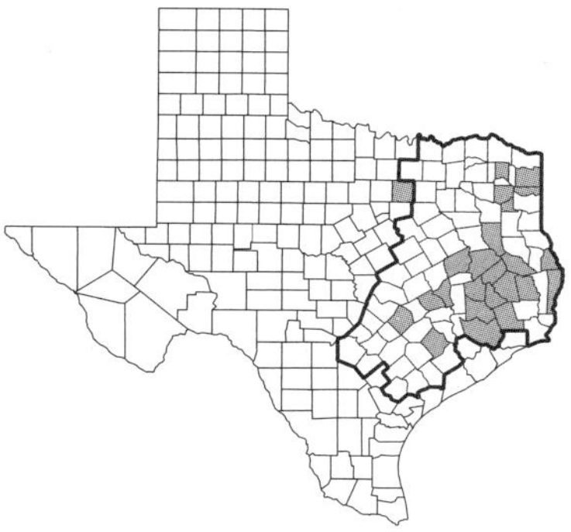

Ostrya virginiana

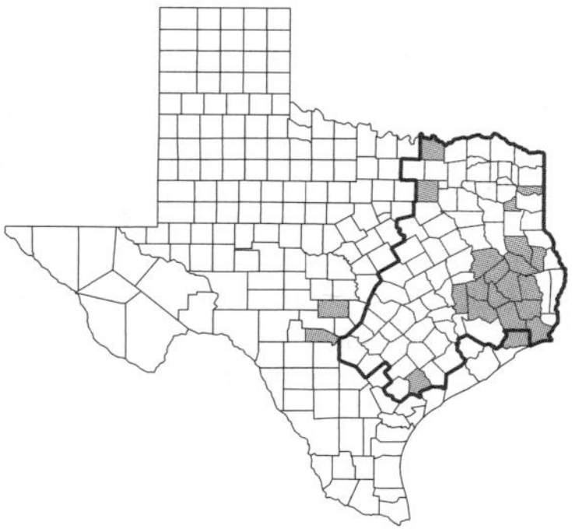

Bignonia capreolata

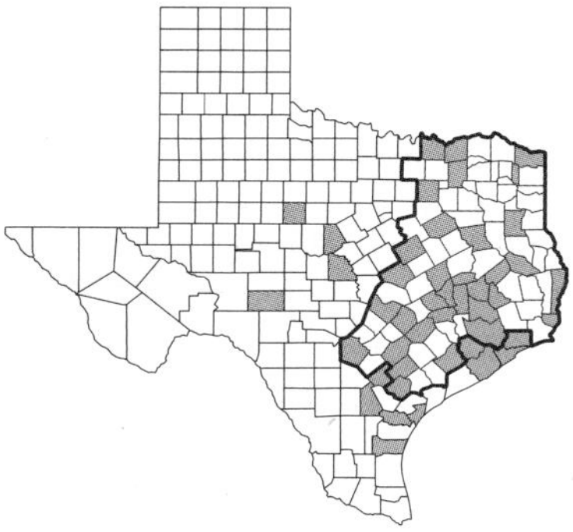

Campsis radicans

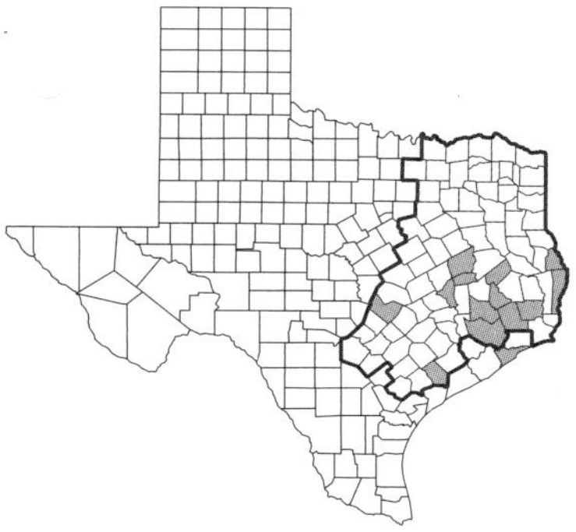

Catalpa bignonioides

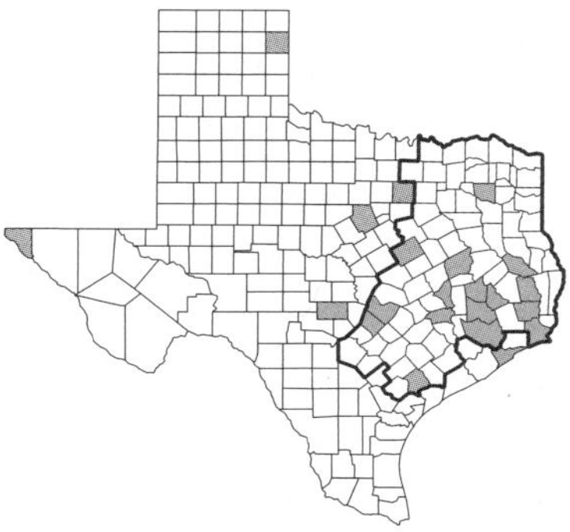

Catalpa speciosa

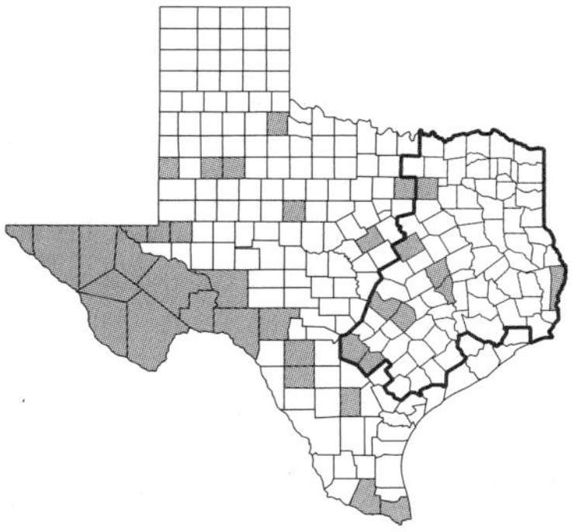

Chilopsis linearis

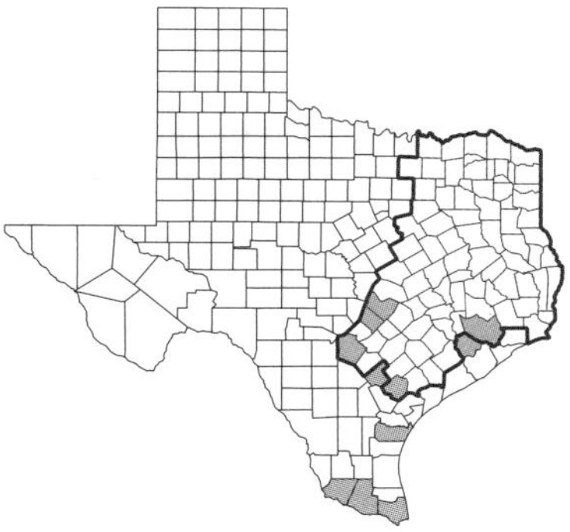

Dolichandra unguis-cati

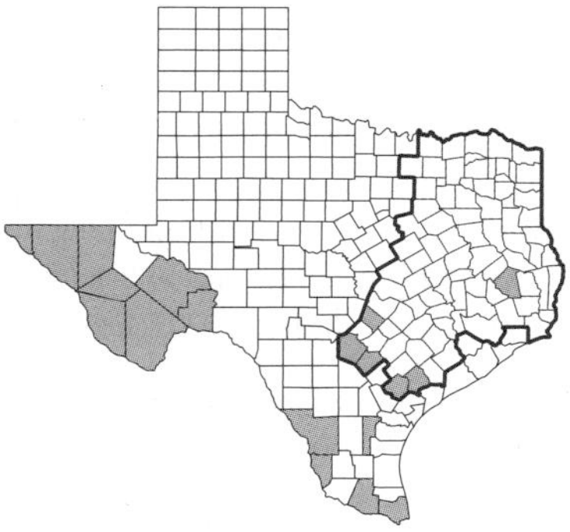

Tecoma stans

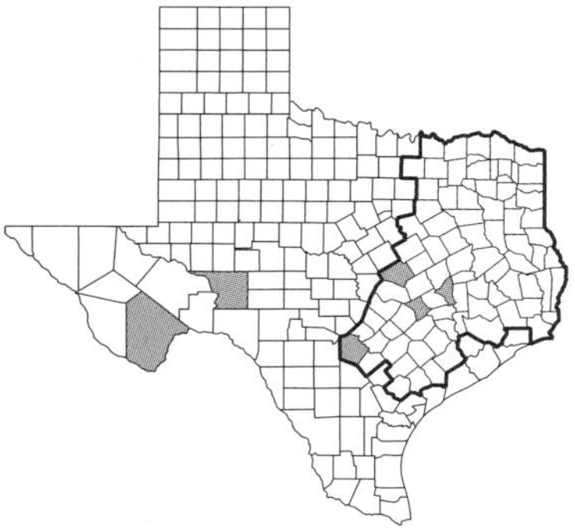

Amsinckia menziesii

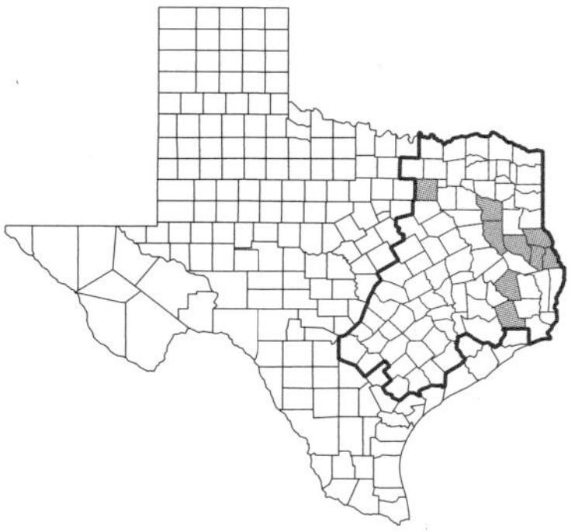

Andersonglossum virginianum

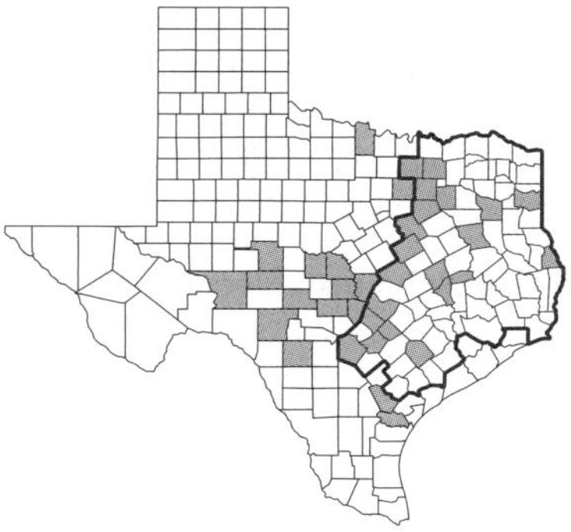

Buglossoides arvensis

2. Pedicels of flowers (both at flowering time and later) short, 0.5–5(–7) mm long; mericarps 2–5 mm long; mericarp base rounded or nearly so or constricted into a collar.
   3. Larger corollas 11–20 mm long; mericarp base rounded or nearly so; fruiting pedicels 0.5–5(–7) mm long; mericarps 2–5 mm long ____ **L. occidentale**
   3. Larger corollas 6–10 mm long; mericarp base constricted into a short collar; fruiting pedicels to 4 mm long; mericarps 2–3.5 mm (in East TX known only from Dallas Co.) ____ **L. parviflorum**
2. Pedicels of flowers (both at flowering time and later) elongate, usually 5–15 mm long, the lower ones often 10–15 mm long; mericarps 3–3.6 mm long; mericarp base rounded (confined to 12 cos. in c TX) ____ **L. helleri**

1. Corollas salverform or funnelform, lobes reflexed to form a somewhat flattened limb (= open face of corolla), variously colored but not green to greenish white, 4–48 mm long; styles included or essentially so; lower leaf blade surface with only the midvein conspicuous or lower leaf blade surface with 2–4 somewhat weak to conspicuous secondary veins.
   4. Plants annual; corollas white with yellow eye; corolla tubes ≤ 2 mm long; corolla lobe margins entire ____ **L. matamorense**
   4. Plants perennial; corollas yellowish white to yellow to orange-yellow, corolla tubes > 2 mm long; corolla lobe margins erose and or ruffled OR corolla lobe margins entire or nearly so.
      5. Corolla lobe margins erose and or ruffled ____ **L. incisum**
      5. Corolla lobe margins entire or nearly so.
         6. Leaf blades at base of stem not clustered, scale-like or imperfect, 5–15 mm long ____ **L. caroliniense**
         6. Leaf blades at base of stem clustered and much longer than 15 mm.
            7. Corolla tubes 10–18 mm long; stem hairs appressed; lower leaf blade surface with only midvein evident ____ **L. mirabile**
            7. Corolla tubes 3–4 mm long; stem hairs spreading; lower leaf blade surface with 4–6 secondary veins evident ____ **L. tuberosum**

Lithospermum caroliniense (J.F. Gmel.) MacMill. (of Carolina), PUCCOON, CAROLINA GROMWELL, CAROLINA PUCCOON. Erect perennial 3–10 dm tall; stems few to numerous from the base; basal rosette not formed, leaf blades toward the base scale-like, 5–15 mm long, ovate to lanceolate, upper stem leaf blades 3–7 times as long as wide, to ca. 10 mm wide; inflorescence a heavily bracted spike-like raceme; corollas orange-yellow, the limb (= open face of corolla) to ca. 22 mm across; mericarps 3–3.5 mm long, smooth or pitted. Sandy woods; common in East TX; Cross Timbers and Prairies and Panhandle; se U.S. from VA s to FL and w to OK and TX. Apr–May. Native of the U.S.

Lithospermum helleri (Small) J.I. Cohen, (for its discoverer, Amos Arthur Heller, 1867–1944), HELLER'S MARBLESEED. Erect perennial herb 30–60 cm tall; stems pubescent with both reflexed and spreading coarse hairs 1–2 mm long; basal leaf blades present at time of flowering, basal leaf blades spatulate, long attenuate, midstem leave blades mostly 4–12 cm long, 2–5 cm wide, narrowly to broadly ovate and narrowly to broadly elliptic; inflorescence a raceme-like cyme; calyx lobes 6–10 mm long; larger corollas creamy white, 9–13 mm long; mericarps 3–3.6 mm long. Juniper-oak woodlands, on crumbly, rather bare limestone soils, typically on slopes; narrowly endemic to a cluster of 12 counties in c TX; East TX cos. include Bexar, Comal, Hays, Travis, and Williamson (Kartesz 2022). Endemic to TX. Mar–May. [*Onosmodium helleri* Small]

Lithospermum incisum Lehm., (incised, cut), NARROW-LEAF GROMWELL, NARROW-LEAF PUCCOON. Erect perennial to ca. 40 cm tall, basal leaf blades sometimes dried up or absent by flowering time; stem leaf blades usually 3–6 cm long, 3–6 mm wide, upper stem leaf blades 8–20 times as long as wide; spring inflorescences usually with apical, sterile clustered flowers, not elongating; corollas showy, lemon- or deep-yellow, lobes decidedly erose and/or ruffled, limb (= open face of corolla) 9–20 mm across; late spring and summer inflorescences with minute, very fertile, cleistogamous flowers with permanently closed corollas 1–2.5 mm long; mericarps 2.5–3 mm long, smooth or pitted. Prairies, open woods, and roadsides; throughout TX except far East TX; throughout c and w U.S. from WI s to TX and w to MT and NV; Canada and Mexico. Late Mar–Apr. Native of the U.S. and Canada. After the production of conspicuous flowers in the spring, plants of this species typically produce large numbers of small cleistogamous flowers that are extremely fertile (Johnston 1952). The roots were the source of a red dye for Native Americans and settlers (Tveten & Tveten 1993).

Lithospermum matamorense DC., (of Matamoros, Mexico), ROUGH GROMWELL. Decumbent to erect annual, stems to 40 cm long; leaf blades spatulate, obtuse, rounded, or retuse apically, gradually reduced up the stem; largest basal leaf blades to 9 cm long, 2 cm wide, often present at flowering time, secondary veins sometimes present; bracts of inflorescence large and very similar to upper leaf blades and, therefore, flowers appearing to be in the axils of leaves rather than subtended by bracts of the inflorescence; corolla 4–5.5 mm long, white with yellow eye, limb (= open face of corolla) 4–7 mm across, corolla tube 1.5–2 mm long; mericarps usually only one maturing, 2.5–3 mm long, pitted. Thickets, open woods, floodplains, and weedy areas; in 7 East TX cos. along the se border, also Harris Co.; mainly Edwards Plateau to s TX; in the U.S. known only from TX; Mexico. Mar–Apr. Native of TX and Mexico.

Cryptantha texana [HEA]

Cynoglossum zeylanicum [COL]

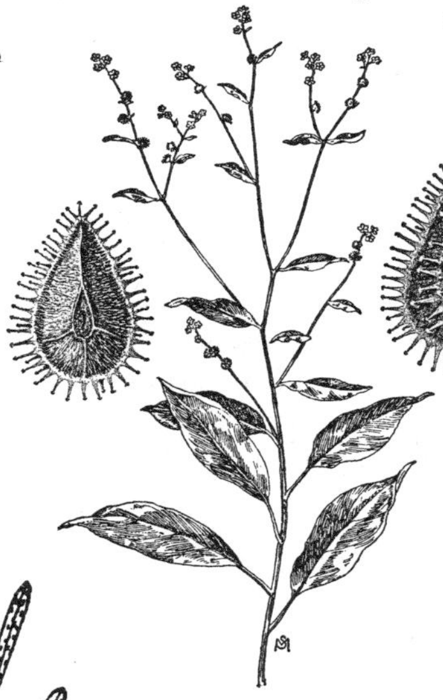

Hackelia virginiana [GEO, PAM]

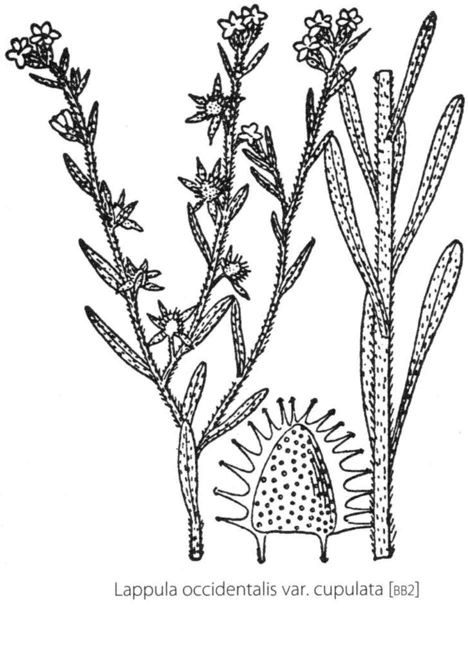

Lappula occidentalis var. cupulata [BB2]

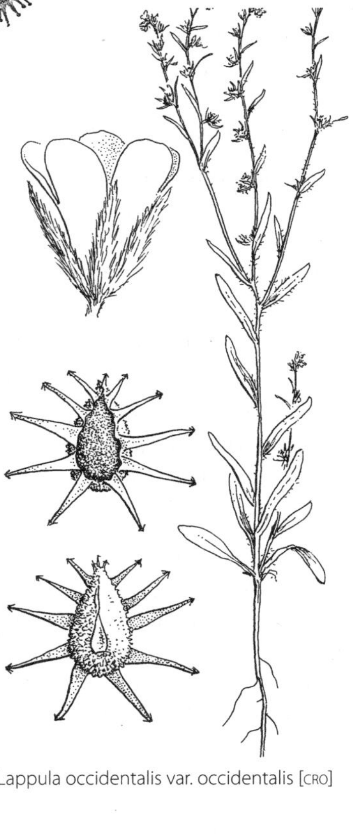

Lappula occidentalis var. occidentalis [CRO]

Lithospermum mirabile Small, (wonderful), ROUGH GROMWELL. Perennial, stems erect, 10–40 cm tall, one to several from the base, pubescence appressed; basal leaf blades clustered, 4–7 cm long, 0.7–1.3 cm wide, oblanceolate; stem leaf blades numerous, gradually reducing up the stem, midstem leaf blades 20–40 mm long, 3–5 mm wide, usually linear-oblong, sessile; inflorescence terminal on stems and branches, bracteate; flowers of inflorescences at first clustered but axis elongating into raceme-like cymes; early season flowers chasmogamous, late season flowers cleistogamous; chasmogamous fruiting calyx 6–8 mm long, pedicels 1–4 mm long, erect; corollas yellow, salverform, margin of corolla lobes entire or sometimes inconspicuously erose, tube 10–18 mm long, throat partially constricted by inflated appendages (= fauces) at the base of each corolla lobe; both cleistogamous, and chasmogamous flowers fertile; mericarps 4, surface ridged and/or coarsely and unevenly pitted; mericarps of chasmogamous and cleistogamous flowers similar. Woods, limestone hills, thorn scrub, usually associated with limestone; on se margin of East TX; Edward Plateau, Trans-Pecos, and South TX Plains; NM and TX; also Mexico. Mar–Jun (includes chasmogamous and cleistogamous flowering) [*L. parksii* I.M. Johnst.]

Lithospermum occidentale (Mack.) Weakley, Witsell, & D. Estes, (western). Erect perennial 6–12 dm tall, branching above; stems with ascending to spreading midstem hairs 1–4 mm long; well-developed stem leaf blades 8–13 cm long, 0.8–2 cm wide, elliptic to narrowly elliptic to narrowly oblanceolate, or lanceolate with 4–6 prominent secondary veins; calyx lobes 6–12 mm long; inflorescence a spike-like raceme; larger corollas 11–20 mm long, whitish or greenish white; mericarps mostly 3–5 mm long. River banks, limestone soil, blackland prairie, sandy hills, grassy banks, and open woodlands; scattered in n and s East TX; Edwards Plateau and South TX Plains; c U.S. from WI s to TX and w to MT and NM; also Canada. Apr–May. [*L. onosmodium* J.I. Cohen, *Onosmodium bejariense* DC., *O. bejariense* var. *occidentale* (Mack.) B.L. Turner, *O. molle* Michx. var. *occidentale* (Mack.) I.M. Johnst., *O. molle* subsp. *occidentale* (Mack.) Cochrane] This taxon, as witnessed by the above synonymy, has been treated at three different ranks (species, subspecies, and variety) with many names and variously included or excluded from the *Onosmodium molle* complex. Native of the U.S., Canada, and Mexico.

Lithospermum parviflorum Weakley, Witsell, & D. Estes, (small-flowered), SHAGGY MARBLESEED. Erect perennial 3–12 dm tall; midstem hairs ascending to spreading, to 2.2 mm long; midstem leaf blades 8 cm long, 1.4–2.5 cm wide, narrowly ovate to narrowly elliptic to lanceolate, sessile with 4–6 prominent secondary veins; inflorescence a raceme-like cyme; calyx lobes 4–6 mm long; corolla 6–10 mm long; mericarps 2–3.5 mm long, base constricted into a short collar. Usually in rocky or gravelly calcareous soils in prairies, on banks and in openings of glades, and sandy roadsides; in East TX recorded only in Dallas Co. in 1880 (*Reverchon 649* herbarium not known) (Johnston 1967); in Cameron Par., LA on the se margin of East TX and Little River Co., AR on the ne margin of East TX; e U.S. from NY s to AL and w to WI, AR, and TX; also e Canada. Mar–May. [*Onosmodium bejariense* DC. var. *hispidissimum* (Mack.) B.L. Turner, *O. hispidissimum* Mack., *O. molle* Michx. var. *hispidissimum* (Mack.) Cronquist, *O. molle* subsp. *hispidissimum* (Mack.) B. Boivin] Native of the U.S. and Canada.

Lithospermum tuberosum Rugel ex A. DC., (swollen), TUBEROUS GROMWELL. Perennial; stems one to several from the base to 7 dm tall with spreading hairs; leaf blade to 10 cm long and 2.5 cm wide, oblanceolate to elliptic or ovate-elliptic; inflorescence laxly few-flowered, bracts similar to stem leaf blades but smaller; calyx shorter than the corolla tube or at most equal in length; corollas small, yellow to yellowish white, 4.5–6 mm long, tube 3–4 mm long and lobes 1.5–2 mm long, sometimes slightly ruffled and ascending, longer than wide; mericarps often abundantly punctuate. Low, often dry woods, mixed hardwood forest, mesic deciduous forest, rocky woods; scattered in se East TX; widespread in se U.S. May–Jun. Native of the U.S.

Lithospermum canescens (Michx.) Lehm., (grayish white), HOARY PUCOON. Similar to *L. caroliniense* with calyces 7–12 mm long; corollas 14–25 mm long; mericarps 4–3 mm long. *L. canescens* has calyces are 3–6 mm long; corollas 12–14 mm long; mericarps 2–3 mm long. Dry woodlands, barrens, and limestone outcrops; found on the ne border of East TX in Choctaw and McCurtain cos. of OK and in Little River Co. of AR; e U.S. from PA s to SC and w to ND and OK; Canada. Apr–May. [*Batschia canescens* Michx.]

Lithospermum croceum Fernald, (saffron-colored). Similar to *L. caroliniense* with leaf blade pubescence with slender bases; mature calyx lobes flat; plants with 15–25 well-developed leaves below the inflorescence. *Lithospermum croceum* has leaf blade pubescence with pustular bases; mature calyx lobes strongly keeled; plants with (30–)35–45 well-developed leaves below the inflorescence. Upland woods and prairies; found on the ne and se border of East TX in Choctaw and McCurtain cos. of OK, Little River and Miller cos. of AR, and Vernon and Beauregard parishes of LA; e U.S. from NY and PA w to MN, SD, OK and LA; Canada. Apr–May. [*L. caroliniense* (J.F. Gmel.) MacMill. var. *croceum* (Fernald) Cronquist] Some authors (Kaul 1986; Yatskievych 2006) feel that the characters used to define *L. caroliniense* and *L. croceum* are too variable to allow consistent recognition as distinct taxa. However, we are following Weakley, A.S., and Southeastern Flora Team (2024) in treating them as separate species.

Lithospermum caroliniense [GLE]

Lithospermum helleri [HEA]

Lithospermum incisum [HO1]

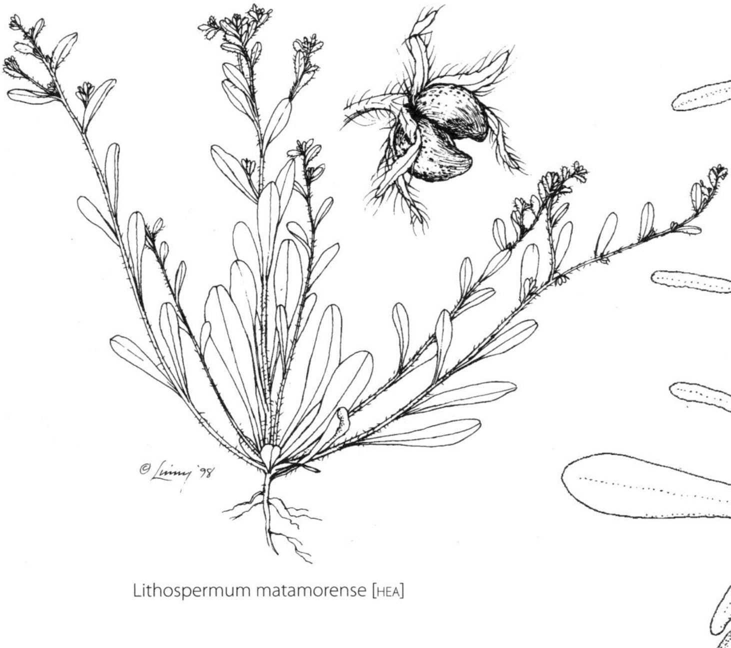

Lithospermum matamorense [HEA]

Lithospermum mirabile [DEN]

Lithospermum subsetosum (Mack. & Bush) Weakley, Witsell, & D. Estes, (slightly bristly-hairy), SMOOTH MARBLESEED. Similar to *L. helleri* with stems with both spreading and reflexed hairs 1–2 mm long; mericarps 3–3.6 mm long. *Lithospermum subsetosum* has lower half of stem nearly glabrous, upper half of stem with appressed-ascending bristly hairs ca. 1 mm long; mericarps 2–3 mm long. Glades, savannas, bottomland forests, and mesic to dry upland forests; found on the ne border of ET in McCurtain Co., OK; mostly in AR and MO, also AL, OK, and TN. May–Jul. [*Onosmodium bejariense* DC. var. *subsetosum* (Mack. & Bush) B.L. Turner, *O. molle* Michx. subsp. *subsetosum* (Mack. & Bush) Cochrane, *O. subsetosum* Mack. & Bush]

Lithospermum virginianum L. (of Virginia), WILD JOB'S TEARS, VIRGINIA MARBLESEED. L. *virginianum* with corollas yellow orange; mericarps 2–2.8 mm long can be distinguished from *L. occidentale* with corollas whitish to greenish; mericarps 3–5 mm long and *L. helleri* with corollas creamy white; mericarps 3–3.6 mm long. Glades and woodlands over calcareous rock; on the e border of East TX in Vernon Par., LA; e U.S. from NY s to FL and w to LA; more concentrated in the se U.S. Apr–Sep. [*Onosmodium virginianum* (L.) A. DC., *O. virginianum* var. *hirsutum* Mack.]

MYOSOTIS L. SCORPION-GRASS, FORGET-ME-NOT

Ours small hairy annuals; inflorescences raceme-like cymes, becoming loosely flowered; calyces hairy with stiff hairs, some hooked; corollas inconspicuous, usually white (or slightly bluish), 2–3.5 mm long; mericarps 4, ovoid to ellipsoid, smooth, light brown, and glassy in appearance.

A genus of ca. 90 species of temperate areas and tropical mountains (Mabberley 2017); a number are cultivated as ornamentals and some have corollas changing from pink to blue in color. (Greek: *myos*, of a mouse, and *oto*, ear, from shape of the leaves or the soft leaves in some species)

REFERENCES: Correll & Johnston 1970; Weakley et. al 2012; Wunderlin et. al 2018.

1. Fruiting pedicels closely ascending or erect in their basal part, rather abruptly outcurved near tip; fruiting calyces deciduous, 3–10 mm long, calyces dense with abruptly hooked hairs; racemes elongating rapidly, the calyces soon becoming > 10 mm apart ____ **M. macrosperma**
1. Fruiting pedicels obliquely spreading from base, straight or only slightly curved; fruiting calyces persistent, 3–5.5 mm long, covered at the base with a few abruptly hooked hairs; racemes elongating gradually, the calyces becoming < 10 mm apart ____ **M. verna**

Myosotis macrosperma Engelm., (large-seeded), SPRING FORGET-ME-NOT. Erect or lax annual 1.5–7 dm tall, loosely branching; basal leaf blades forming a short-lived rosette, larger stem leaf blades 3–7.5 cm long; inflorescences raceme-like cymes at the ends of branches and in axils of lower leaf blades and sometimes from the base, usually with bracts in lower portion; calyx lobes unequal; corollas white or sometimes slightly bluish, limb (= open face of corolla) 2 mm wide. Sandy or silty woods, stream banks, fencerows, and roadsides; throughout East TX; Cross Timbers and Prairies and Edwards Plateau; e U.S. from PA s to FL and w to IL, OK, and TX; Canada. Apr–early May. [*M. verna* Nutt. var. *macrosperma* (Engelm.) Chapm., *M. virginica* L. var. *macrosperma* (Engelm.) Fernald] Native of the U.S.

Myosotis verna Nutt., (of Spring), EARLY SCORPION-GRASS, SOUTHERN FORGET-ME-NOT. Erect annual 0.5–4 dm tall, simple or branched; basal leaf blades forming a short-lived spring rosette, stem leaf blades 1–5 cm long, oblong or narrowly elliptic, lowermost petiolate, others sessile; inflorescences in raceme-like cymes, 3–32 cm long, usually with bracts in lower portion; calyces unequal; corollas white, limb (= open face of corolla) 1–2 mm wide. Sandy open woods, roadsides, and old fields; scattered in East TX; Cross Timbers and Prairies and e Edwards Plateau; throughout e U.S., in the nw, and CA; Canada; reported to be indigenous in Argentina (Johnston 1925a). Apr–early May. [*M. virginica* of authors, not (L.) Britton, Sterns, & Poggenb.] Native of the U.S.

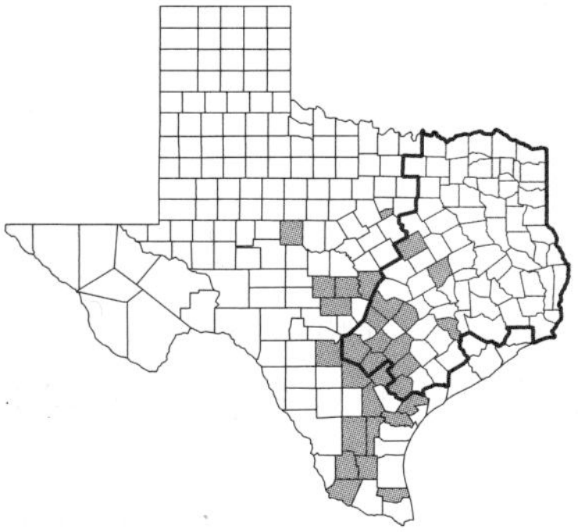

Cryptantha texana

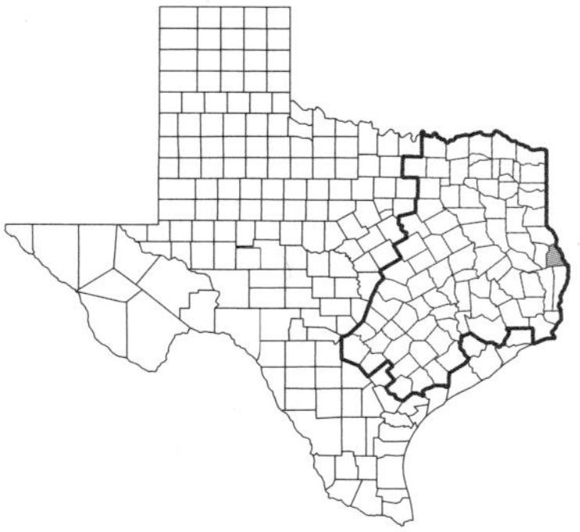

Cynoglossum zeylanicum

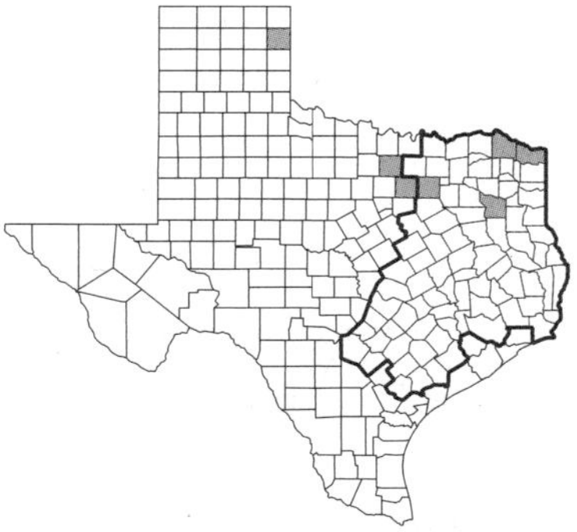

Hackelia virginiana

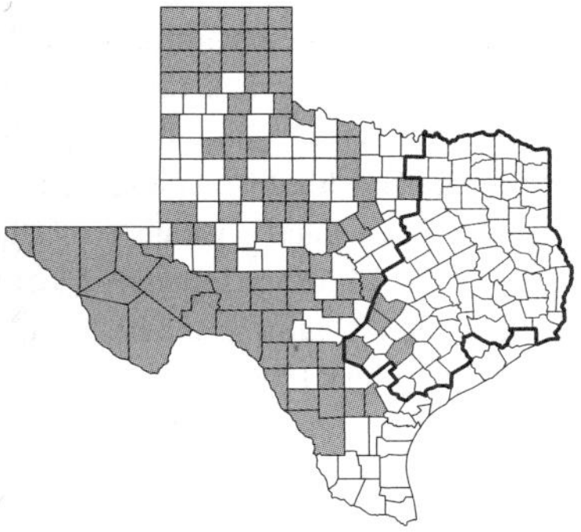

Lappula occidentalis

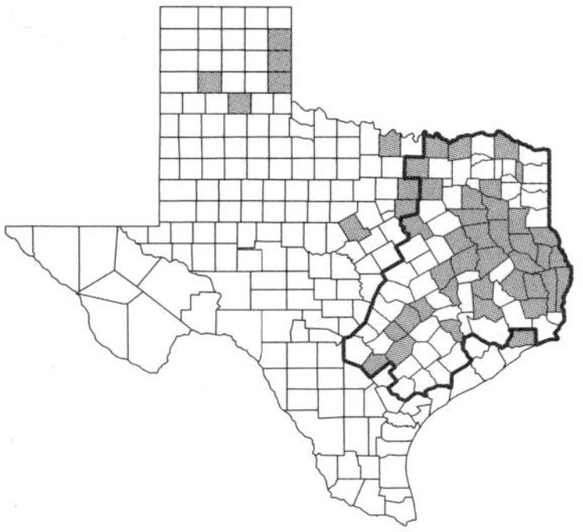

Lithospermum caroliniense

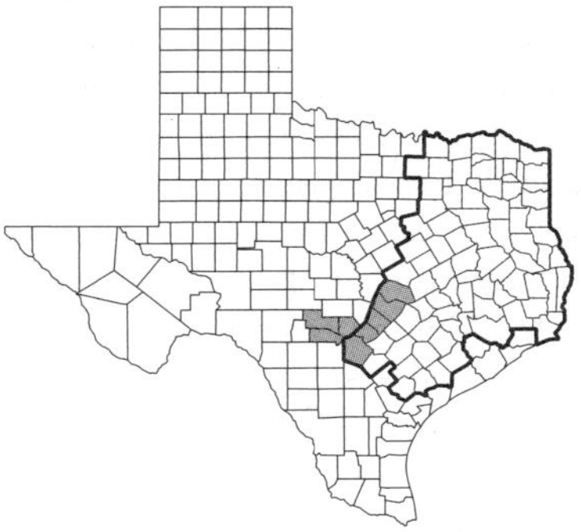

Lithospermum helleri

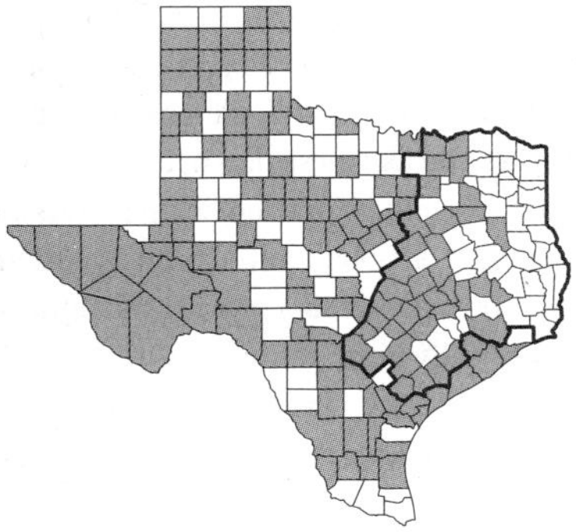

Lithospermum incisum

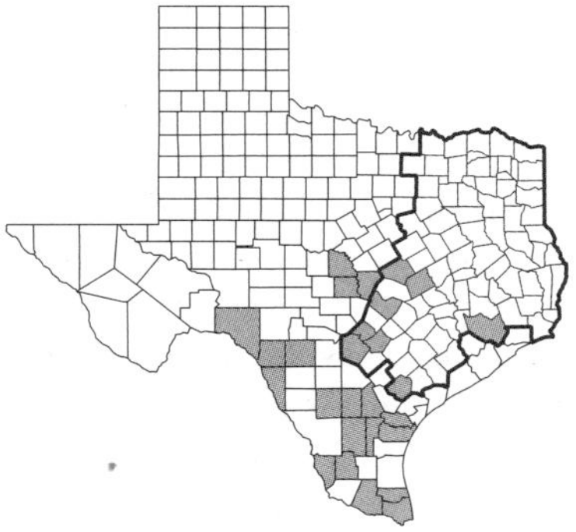

Lithospermum matamorense

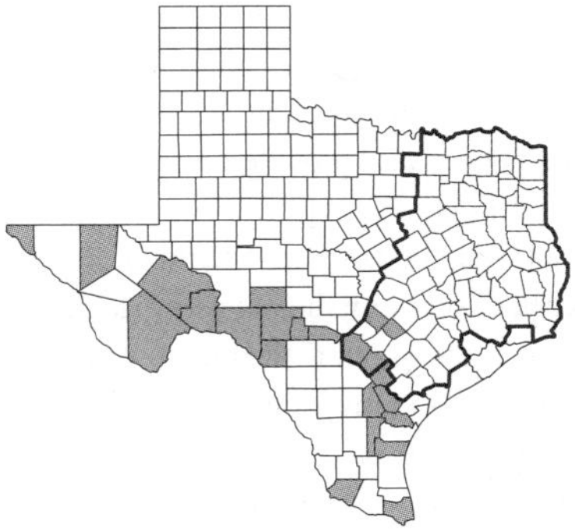

Lithospermum mirabile

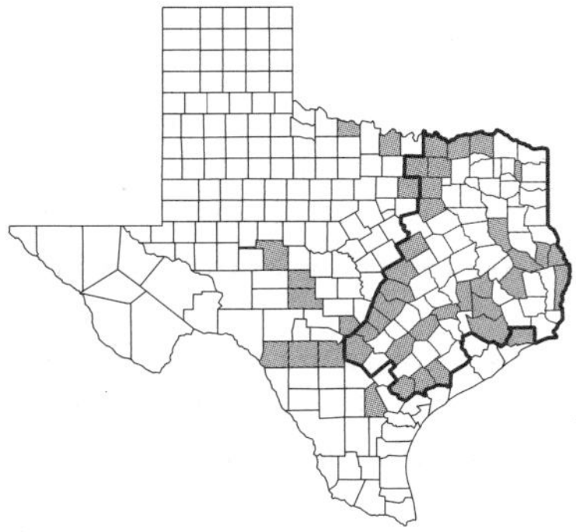

Lithospermum occidentale

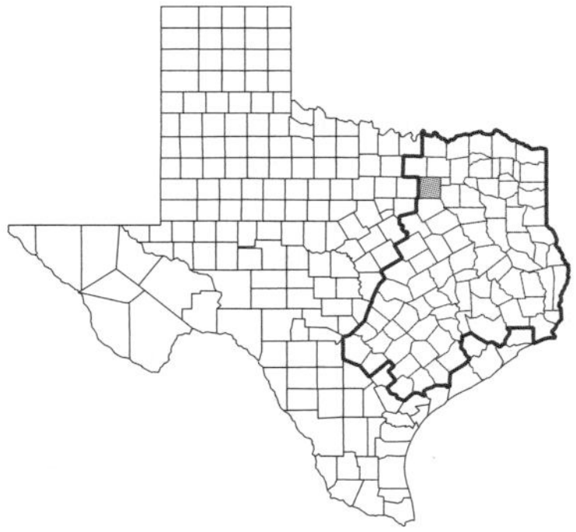

Lithospermum parviflorum

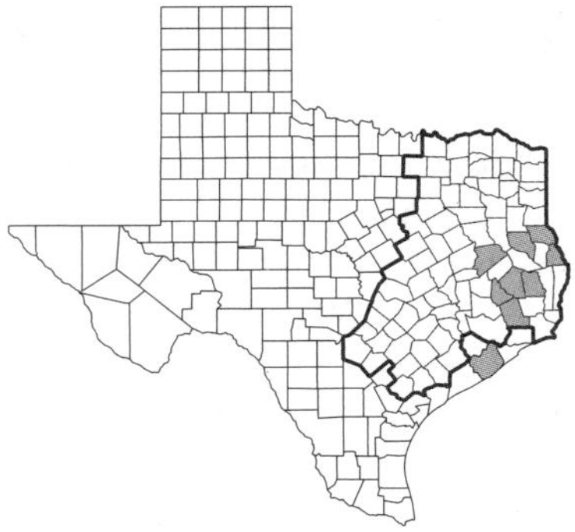

Lithospermum tuberosum

BRASSICACEAE Burnett (CRUCIFERAE)

MUSTARD OR CABBAGE FAMILY

Modified from Al-Shehbaz (2010b) and Diggs et al. (1999),
and contributed by
Amanda K. Neill (ex BRIT-SMU-VDB)

Unless otherwise noted, number of taxa in generic synopses are from Al-Shehbaz 2010b.—akn

Localities are generally from USDA Plants maps (USDA Plants Database 2021). If Turner et al. 2003 or Kartesz 2015 (online) have additional counties, they are cited within taxon locality statements.

Ours annual, biennial, or perennial herbs, less frequently rhizomatous or woody-crowned perennials, or rarely submerged aquatics; sap watery, often peppery-pungent or bitter; trichomes (when present) simple or branching or peltate-stellate, sometimes glandular; leaves basal or alternate (rarely opposite or whorled in some *Cardamine* species), not stipulate, blades simple or compound, entire or toothed or lobed; inflorescences of terminal or axillary racemes, usually corymbose, sometimes paniculate or umbellate, or rarely flowers solitary on pedicels from basal rosette, bracts sometimes present; flowers perfect, usually radially symmetrical; sepals 4, distinct and separate; petals 4 (infrequently absent or vestigial), distinct and separate, usually longer than the sepals, typically equal in a cross-like arrangement alternate with the sepals (or rarely unequal, with 2 larger and 2 smaller petals), typically with a narrowed basal *claw* widening abruptly into an apical *blade*; stamens distinct and separate, typically 6 and tetradynamous (the outer 2 shorter than the inner 4) or infrequently stamens 2 or 4(–5); pistil 1, ovary superior, 2-locular (rarely 1-locular), sessile on the pedicel or sometimes borne on a stipe (= gynophore); style 1, often persisting as a beak; stigma 1, often rounded, sometimes 2-lobed; fruits typically dry, dehiscent capsules, of various terete or flattened shapes, termed *siliques* if length ≥ 3 times width, or *silicles* if length < 3 times width, with fruit halves (= *valves*) splitting from the bottom, leaving behind a longitudinal partition termed the *replum* or septum, or sometimes fruits transversely 2-segmented and 1 or both segments indehiscent or sterile, or rarely fruits loment-like and breaking into multiple 1-seeded segments; ovules 1–many per ovary; seeds sometimes winged, sometimes mucilaginous when wetted.

A medium-sized, economically-important family containing ca. 3,400 species in 351 genera (Koch et al. 2020), mostly herbaceous, native to every continent (except Antarctica) and especially diverse in n temperate regions; some species are crops of economic importance, others are widespread and noxious weeds. ☠ Bitter, sulfur-containing compounds (glucosinolates) are often present and provide chemical defense for the plants; tissue damage (e.g., by herbivores) breaks these down, releasing toxic compounds such as isothiocyanates (mustard oils), thiocyanates, and nitriles, which can cause illness or even death in livestock (Blackwell 1990; Young-Matthews 2012). However, some glucosinolates have been identified as potent anti-cancer agents and may have other medicinal uses (Halkier & Gershenzon 2006; Verkerk et al. 2009); these compounds also provide the pungent, sharp, bitter, or "hot" flavors of some members of this family. *Brassica* is the most economically important genus in the family, and includes many long-cultivated vegetable or "cole" crops (e.g., CABBAGE, KALE, BROCCOLI, TURNIP, BOK CHOY, and MUSTARD GREENS), and oilseed crops (e.g., MUSTARD, CANOLA, RAPESEED). Other genera used as food or flavoring include *Armoracia* (HORSERADISH), *Eruca* (ARUGULA), *Eutrema* (WASABI), *Nasturtium* (WATERCRESS), and *Raphanus* (RADISH). A number of genera are grown as garden ornamentals or used in floral design, such as *Lobularia* (SWEET ALYSSUM), *Lunaria* (HONESTY or MONEYPLANT), and *Matthiola* (STOCK). *Arabidopsis thaliana* (MOUSE-EAR CRESS) was the first plant to have its entire genome sequenced and is a model organism in evolutionary biology research—which has facilitated subsequent, extensive molecular studies throughout the family and beyond. Tribal and generic reassignments have been frequent (e.g., Al-Shehbaz 2010b, 2012; Koch et al. 2020), and some are still in flux; tribal affiliations cited in treatments follow Al-Shehbaz (2012). Generic delimitation is difficult and relies primarily on fruit characters, but specimens with both flowers and mature fruit are often needed for species determination. In the treatments, petal length measurements include both blade and claw (if present). Ovule number may be higher than number of developed seeds. The old family name and common appellation CRUCIFER derive from the distinctive, cross-like arrangement of the 4 petals. (subclass Dilleniidae—Cronquist; order Brassicales—APG III)

FAMILY RECOGNITION IN THE FIELD: herbs with alternate and/or basal leaves, flowers with 4 petals in a cruciform or cross arrangement (flower color variable, but often yellow or white), stamens usually 4 long + 2 short, and characteristic capsular fruits (silicles or siliques) with a longitudinal partition between the 2 locules, borne in racemes. Similar and closely-related to Capparaceae and Cleomaceae, but those families more often have somewhat bilaterally symmetrical flowers (not cross-like), inflorescence bracts (often absent in Brassicaceae), stamens rarely 4+2, and fruits lacking a longitudinal partition.

REFERENCES: Rollins 1993; Koch et al. 2003, 2020; Al-Shehbaz et al. 2006; Bailey et al. 2006; Dixon 2007; Al-Shehbaz 2010b, 2012; Warwick et al. 2010; Franzke et al. 2011.

Keys to Genera of Brassicaceae

1. All trichomes simple (unbranching) OR plants glabrous and trichomes absent (use a hand lens to inspect stems, leaves, rachis, and pedicels).
   2. Salt-loving, ± fleshy-succulent herbs of coastal beaches and dunes; petals white to lavender, fruits fleshy, transversely 2-segmented, segments dissimilar and single-seeded, becoming corky and dry when mature, the lower segment retained on the pedicel ______ **Cakile**
   2. Plants without the above combination of characters.
      3. Fresh-water aquatic perennials; glabrous; rhizomatous and rooting at nodes; leaves of submerged lower stems finely 1–4-pinnatisect (divisions filiform); leaves of emergent stems ± sessile, margins entire or dentate; petals white, 4–8 mm long; fruits silicles (infrequently developing) ______ **Rorippa** (in part)
      3. Plants without the above combination of characters.
         4. Ovaries and young fruits linear; mature fruits siliques (length ≥ 3 times width) ______ **Key A**
         4. Ovaries and young fruits not linear; mature fruits silicles (length < 3 times width) ______ **Key B**
1. At least some trichomes forked, branching, stellate, or peltate (scale-like).
   5. Ovaries and young fruits linear; mature fruits siliques (length ≥ 3 times width) ______ **Key C**
   5. Ovaries and young fruits not linear; mature fruits silicles (length < 3 times width) ______ **Key D**

Key A

All trichomes simple (unbranching) OR plants glabrous and trichomes absent; ovaries and young fruits linear; mature fruits siliques (length ≥ 3 times width)

1. Plants scapose (with pedicels or racemes emerging directly from the basal rosette), apparently stemless; stem leaves absent; petals yellow, 5–10 mm long.
   2. Basal leaves simple or lyrate-pinnatisect with 1–2 pairs of leaflets below the larger and broadly oblong terminal leaflet; flowers usually on solitary pedicels emerging directly from the basal rosette; petal apex notched (emarginate); ovules 4–14 per ovary; seeds winged; rare plants of glades and barrens ______ **Leavenworthia**
   2. Basal leaves sinuate or lyrate-pinnatifid with 2–6 lobes per side; flowers in racemes; petal apex rounded or truncate; ovules 20–36 per ovary; seeds unwinged; ill-smelling weeds of disturbed areas ______ **Diplotaxis** (in part)
1. Plants not scapose; stems and stem leaves usually present; petals various colors (including yellow) and various sizes, OR petals vestigial/absent.
   3. Stem leaves (at least the uppermost) sessile, with bases sagittate to auriculate or cordate (and frequently clasping the stem) OR if stem leaves petiolate then the petiole bases auriculate-clasping.
      4. Aquatic perennials, rhizomatous and rooting at nodes; all leaves petiolate with petiole bases auriculate; lower leaves simple (or with a few small lateral leaflets and a large terminal leaflet); upper stem leaves pinnately compound with 3–9(–13) leaflets; petals white or pinkish, ≤ 6 mm long ______ **Nasturtium**
      4. Plants without the above combination of characters; annuals, biennials, or perennials of dry or wet habitats, infrequently rhizomatous; upper stem leaves short-petiolate to sessile, blades entire or toothed to lobed or pinnatisect (if pinnatisect, then upper leaves typically with fewer divisions than basal leaves).
         5. Plants glabrous and glaucous; basal leaves ± sessile and entire; stem leaves sessile and entire, elliptic, bases deeply cordate-clasping; petals white to greenish-white to pale yellow, 7–12 mm long; fruits linear, 5–14 cm long, terete or 4-angled, keeled; seeds unwinged ______ **Conringia**
         5. Plants without the above combination of characters; basal leaves usually petiolate, their margins sinuate to dentate or lyrate-pinnatifid or pinnatisect (rarely entire).
            6. Developing racemes ± corymbose, with open flowers often overtopping or only moderately exceeded in height by young buds; sepals yellow to green; petals yellow to pale yellow or white OR petals absent/vestigial at anthesis; fruits terete or 4-angled or slightly compressed; seeds unwinged.
               7. Sepals ≤ 2.6 mm long; petals ≤ 3 mm long OR absent/vestigial; fruits ellipsoid to oblong or linear; seeds arranged in two rows per locule ______ **Rorippa** (in part)
               7. Sepals 3+ mm long; petals 5+ mm long; fruits linear; seeds arranged in 1 row per locule.
                  8. Upper stems striate-angular; stem leaf margins entire or dentate to lyrate-pinnatifid; corolla blades spatulate or oblanceolate, gradually attenuated into an obscure claw; fruits not transversely segmented, style 1–3.5 mm long ______ **Barbarea**
                  8. Upper stems not striate; stem leaf margins entire or sinuate; corolla blades obovate, narrowing abruptly into a distinct claw; fruits transversely 2-segmented, with the lower segment multi-seeded and linear, and the terminal segment 0–2-seeded and 3–22 mm long ______ **Brassica** (in part)
            6. Developing racemes elongating early, with young buds often exceeding open flowers in height; sepals green to whitish or pinkish to purple or maroon; petals white or pink to lavender or purple to magenta; fruits terete or flattened; seeds unwinged or winged.
               9. Flowers radially symmetrical; sepals pubescent, green; petals white, 3–5 mm long; fruits flattened; seeds winged ______ **Borodinia** (in part)

9. Flowers radially or bilaterally symmetrical; sepals glabrous or pubescent, green to whitish or pinkish to purple or maroon; petals often similarly-colored (as sepals), 7–27 mm long; fruits terete or flattened; seeds unwinged or winged.
   10. Upper stem leaves toothed; petals lavender or pinkish to nearly white, blade 1.5–3 mm wide; stigma rounded; fruits terete, 1.5–4 cm long; seeds unwinged __________ **Iodanthus**
   10. Upper stem leaves entire or sinuate to dentate; petals lavender to purple or magenta, blade 5–11 mm wide; stigma 2-lobed; fruits flattened, 4–14.5 cm long; seeds winged __________ **Streptanthus** (in part)

3. Stem leaves petiolate or sessile, bases of blades (or petioles) not clasping, neither sagittate nor auriculate nor cordate.
   11. Plants pubescent with stiff recurved or retrorse trichomes; most leaves deeply lobed or pinnatifid, transitioning to leaf-like bracts subtending at least the lower pedicels of the raceme __________ **Erucastrum**
   11. Plant pubescence and leaf shape various; racemes without bracts.
      12. Plants with stalked-glandular pubescence; petals pink to lavender or purple; fruits loment-like, breaking apart into 8–12 corky 1-seeded segments, leaving behind the persistent marginal replum __________ **Chorispora**
      12. Plants lacking glandular trichomes; petals various colors; fruits various (if breaking into 1-seeded segments these usually ≤ 8 and not leaving behind a persistent marginal replum).
         13. Plants in flower. [*Specimens with flowers and fruits should ideally be keyed through both halves of Key A couplet 13*]
            14. Petals absent/vestigial at anthesis.
               15. Stem leaves simple, the margins lobed or sinuate to dentate; sepals yellow; stamens 6; ovules >150 per ovary __________ **Rorippa** (in part)
               15. Stem leaves pinnately compound; sepals green to pinkish or whitish; stamens 4 (rarely 5 or 6); ovules 14–40 per ovary __________ **Cardamine** (in part)
            14. Petals present.
               16. Petals 12–26 mm long (including claw), the blades with some veins darker than the rest of the blade (veins usually purplish or brownish).
                  17. Plants usually glabrous; upper stem leaves linear-lanceolate, their margins entire; flowers radially or bilaterally symmetrical; calyx campanulate; petal blades 1–4 mm wide; stamens in 3 pairs of unequal length __________ **Streptanthus** (in part)
                  17. Plants glabrous to hispid; upper stem leaves broader, their margins often lobed or toothed; flowers radially symmetrical; calyx not campanulate; petal blades 3–9 mm wide; stamens strongly tetradynamous.
                     18. Petals white to pale yellow or pink to purple; ovaries transversely 2-segmented into unequal parts, with the basal segment sterile and stipe-like, and the terminal segment ovule-bearing; style ± terete __________ **Raphanus** (in part)
                     18. Petals white to creamy yellow; ovaries transversely 2-segmented into nearly equal lengths, with the basal segment developing ovules, and the terminal segment sterile (beak-like) and ± flattened __________ **Eruca** (in part)
               16. Plants without the above combination of characters; petal length variable, blades and veins of similar color intensity, OR if veins are darker, then petals < 12 mm long.
                  19. Petals bright yellow to pale yellow.
                     20. Uppermost stem leaves deeply pinnatifid or pinnately compound with several ± evenly-spaced lobes/leaflets per side OR if few-lobed then upper portion of leaf narrow with an acute apex; style or beak short and blunt.
                        21. Stems usually prostrate or decumbent; larger leaf blades with obtuse lobes obtuse; petals 1–2 mm long; stigma ± rounded; ovules in 2 rows per locule ___ **Rorippa** (in part)
                        21. Stems usually erect; larger leaf blades with acute or linear lobes; petals 2.5–10 mm long; stigma ± 2-lobed; ovules in 1 row per locule __________ **Sisymbrium** (in part)
                     20. Uppermost stem leaves undivided or shallowly lobed OR deeply lobed/divided only in the lower portion of the blade (= lyrate-pinnatifid) with the upper portion usually relatively broad, apex often obtuse; style or beak short and blunt or longer and attenuate.
                        22. Stems often ± scapose, leaves crowded at base of plant; few simultaneously-open flowers per raceme; ovary terminal segment bluntly cylindrical (style-like); ovules in 2 rows per locule __________ **Diplotaxis** (in part)
                        22. Stems often branching and leafy well above base; several simultaneously-open flowers per raceme; ovary terminal segment linear or conic to spear-shaped (style- or beak-like); ovules in 1 row per locule.
                           23. Sepals usually erect or ascending; ovaries/fruits with lower segment valves each prominently midveined; terminal segment seedless, linear or conic __________ **Brassica** (in part)

23. Sepals usually spreading; ovaries/fruits with lower segment valves each prominently 3–7-veined; terminal segment seedless or 1-seeded, conic or spear-shaped ______ **Sinapis** (in part)

19. Petals white to pinkish or lavender.

24. Stem leaves petiolate, their blades all palmately or pinnately compound or deeply pinnatifid or pinnatisect, with 1–15 leaflets or lobes per side.

25. Plants annual or biennial with leaves pinnately compound or pinnatisect (with 2–10 leaflets or lobes per side) OR rhizomatous perennials with palmately 3-foliolate leaves (often whorled on stem); petals white to pale pink, 1.5–20 mm long ______ **Cardamine** (in part)

25. Plants annual; leaves deeply pinnatisect or pinnatifid, with 2–15 lobes or leaflets per side; petals white, 2–3 mm long ______ **Planodes** (in part)

24. Stem leaves petiolate or ± sessile, at least the uppermost blades undivided, their margins entire to sinuate-repand or toothed.

26. Basal leaves long-petiolate and lyrate-pinnatifid; uppermost stem leaves petiolate, with blades linear-lanceolate and margins entire; petals white to pinkish or lavender, 4–8 mm long; stigma ± 2-lobed ______ **Streptanthus** (in part)

26. Basal leaves petiolate or sessile, simple; uppermost stem leaves sessile; all leaf margins entire or repand to dentate; petals white (rarely pale pink), 2–16 mm long; stigma rounded.

27. Rhizomatous perennials; rhizomal leaves usually reniform to cordate or ovate; stem leaves sessile, blade bases ± truncate; petals white (rarely pale pink), obovate, 6–16 mm long; sepals glabrous ______ **Cardamine** (in part)

27. Annuals or biennials; basal leaves spatulate to ovate-elliptic or oblanceolate; stem leaves short-petiolate or sessile, blade bases cuneate; petals white, spatulate, 2–5 mm long; sepals glabrous or pubescent.

28. Plants annual; stems to 50 cm tall; sepals glabrous or their tips sparsely pubescent; maturing ovaries and their pedicels directed upwards or spreading outwards (above the horizontal) from the rachis ______ **Arabidopsis** (in part)

28. Plants biennial; stems to 100+ cm tall; sepals pubescent; maturing ovaries and their pedicels directed horizontally or downwards from the rachis ______ **Borodinia** (in part)

13. Plants in fruit. [*Specimens with flowers and fruits should ideally be keyed through both halves of Key A couplet 13*]

29. Fruits transversely 2-segmented with the lower valvular segment very short (stipe-like) and seedless, and the terminal segment usually 2–8-seeded (seeds in a single row), spongy or corky at maturity, 2.5–15 mm wide, indehiscent or separating into 1-seeded segments ______ **Raphanus** (in part)

29. Fruits unsegmented OR transversely 2-segmented with the lower segment multi-seeded and dehiscing by valves (not stipe-like).

30. Fruits transversely 2-segmented, the terminal segment seedless or 1-seeded, often beak- or style-like, 1–30 mm long.

31. Seeds in 2 rows per locule.

32. Stems often ± scapose, leaves crowded at base of plant; fruits 1.5–2.5 mm wide, with terminal segment bluntly cylindrical, midveined, 1–3 mm long, much shorter than lower segment ______ **Diplotaxis** (in part)

32. Stems often branching and leafy well above base; fruits 2.5–5 mm wide, with terminal segment ± flattened and spear-shaped, 5-veined, 4–11 mm long, equaling or slightly shorter than lower segment ______ **Eruca** (in part)

31. Seeds in 1 row per locule.

33. Fruits with lower segment valves prominently midveined; terminal segment seedless, linear or conic, shorter than lower segment ______ **Brassica** (in part)

33. Fruits with lower segment valves each prominently 3–7-veined; terminal segment seedless or 1-seeded, conic or spear-shaped, longer or shorter than lower segment ______ **Sinapis** (in part)

30. Fruits unsegmented (obscurely 2-segmented taxa can also key here); style or beak 0–12 mm long.

34. Fruits elastically dehiscent, the valves spirally coiling at separation ______ **Cardamine** (in part)

34. Fruit valves neither dehiscing elastically nor coiled at separation.

35. Fruits ± terete (rounded in cross section); seeds unwinged.

36. Fruiting pedicels 0.5–5.3 mm long; fruits oblong to linear, valves obscurely veined; seeds in 2 rows per locule ______ **Rorippa** (in part)

36. Fruiting pedicels 1.5–37 mm long; fruits linear to subulate, valves each prominently midveined; seeds in 1 or 2 rows per locule.

37. Uppermost stem leaves compound or deeply pinnatifid with several ± evenly-spaced lobes/leaflets per side OR if few-lobed then upper portion of leaf narrow with an acute apex; fruit valves each distinctly 3-veined __ **Sisymbrium** (in part)
37. Uppermost stem leaves undivided or toothed to shallowly lobed OR deeply lobed/divided only in the lower portion of the blade (=lyrate-pinnatifid) with the upper portion usually relatively broad, apex often obtuse; fruit valves each prominently 1-veined.
38. Fruits 0.5–0.8 mm wide; seeds 0.3–0.5 mm long, in 1 row per locule ______ **Arabidopsis** (in part)
38. Fruits 1.5–2.5 mm wide; seeds 0.9–1.3 mm long, in 2 rows per locule ________ **Diplotaxis** (in part)
35. Fruits flattened parallel to the replum; seeds winged or narrowly margined.
39. Stem leaves petiolate, blades deeply pinnatifid or pinnatisect, with 4–15 lobes or leaflets per side; fruits < 3.5 cm long ________ **Planodes** (in part)
39. Stem leaves petiolate or ± sessile, at least the uppermost blades undivided, their margins entire to sinuate-repand or toothed; fruits > 3.5 cm long.
40. Fruits horizontal to pendent, usually curving downward ________ **Borodinia** (in part)
40. Fruits erect or ascending-spreading, straight or slightly curved upward _____ **Streptanthus** (in part)

Key B

All trichomes simple (unbranching) OR plants glabrous and trichomes absent; ovaries and young fruits not linear; mature fruits silicles (length < 3 times width)

1. Small, glabrous annuals; basal and stem leaves pinnatisect, as are the leaf-like bracts subtending each pedicel (racemes bracteate); petals yellow and 8–13 mm long; styles 3–12 mm long; rare plants of limestone glades and barrens (known from San Augustine and Fannin cos.) ________ **Selenia**
1. Plants without the above combination of characters; racemes without bracts.
2. Some stem leaves sessile (often the uppermost), with bases sagittate to auriculate or cordate (and frequently clasping the stem).
3. Plants glabrous, sometimes glaucous.
4. Petals white; fruits strongly flattened perpendicular to the replum, winged and deeply notched at apex; style obsolete or to 0.3 mm long.
5. Plants to 55(–80) cm tall, unpleasantly-scented; sepals 1–1.5 mm wide; fruits (6–)9–20 mm long with wings 3.5–5 mm wide at apex; ovules 6–16 per ovary; seeds reddish brown, patterned with concentric ridges ________ **Thlaspi**
5. Plants to 28(–40) cm tall; not ill-scented; sepals 0.5–1 mm wide; fruits 3–6.5(–8) mm long with wings 1–2 mm wide at apex; ovules 4–8 per ovary; seeds yellowish brown, ± smooth ________ **Noccaea**
4. Petals yellow to pale yellow or white; fruits terete or somewhat flattened, neither winged nor deeply notched at apex; style 0.2–2 mm long.
6. Stem leaves entire or dentate; petals 3–5 mm long; fruits obpyriform to clavate-obcordiform (club-shaped or inverted heart- or pear-shaped), indehiscent and woody at maturity; ovules 2 per ovary; style flattened, triangular, to 2 mm long ________ **Myagrum**
6. Stem leaves lyrate-pinnatisect; petals 1.5–3 mm long; fruits oblong-elliptic, dehiscent at maturity; ovules 20–90 per ovary; style blunt, to 1.2 mm long ________ **Rorippa** (in part)
3. Plants pubescent, at least on lower stems.
7. Petals white OR absent/vestigial at anthesis; stamens 2 or 6; fruits flattened perpendicular to the replum, notched at apex; ovules 2(–4) per ovary ________ **Lepidium** (in part)
7. Petals yellow to pale yellow or white; stamens 6; fruits terete or flattened parallel to the replum, apex not notched; ovules 8–90 per ovary.
8. Stem leaves entire or dentate; stamens in 3 pairs of unequal length; fruits pyriform to obovoid, flattened parallel to the replum; style 1–3.5 mm long ________ **Camelina** (in part)
8. Stem leaves lyrate-pinnatisect; stamens tetradynamous (2 short + 4 long); fruits oblong-elliptic, ± terete; style 0.2–1.2 mm long ________ **Rorippa** (in part)
2. Stem leaves petiolate or sessile, blade bases not clasping, neither sagittate nor auriculate nor cordate.
9. Petals pale yellow and 6–11 mm long; stamens 6; fruits transversely 2-segmented, with the lower segment cylindrical (persistent, may appear stipe-like) and the terminal segment globose or ovoid (falling at maturity) ________ **Rapistrum**
9. Petals white and up to 3 mm long OR petals vestigial/absent at anthesis; stamens 2; fruits flattened perpendicular to the replum or inflated (didymous), but not transversely segmented ________ **Lepidium** (in part)

Key C

At least some trichomes forked, branching, stellate, or peltate; ovaries and young fruits linear; mature fruits siliques (length ≥ 3 times width)

1. Stem leaves 1–3-pinnately compound ______ **Descurainia** (in part)
1. Stem leaves simple, the margins entire to sinuate-repand or toothed to pinnatifid.
   2. Petals white and < 6 mm long OR petals vestigial/absent at anthesis; fruits glabrous or pubescent, 0.8–10 cm long.
      3. Plants biennial; stems to 125 cm tall; fruits 4–10 cm long; seeds winged ______ **Borodinia** (in part)
      3. Plants annual; stems to 50 cm tall; fruits ≤ 2 cm long; seeds winged or unwinged.
         4. Branching trichomes forked; stem leaves present; sepals glabrous or their tips sparsely pubescent; petals present at anthesis; fruits linear, 0.5–0.8 mm wide, terete; seeds in 1 row per locule, unwinged ______ **Arabidopsis** (in part)
         4. Branching trichomes 2–4-rayed; stem leaves present or absent (when plants scapose); sepals often pubescent; petals sometimes absent at anthesis; fruits elliptic-obovate or linear to lanceolate, 0.9–4 mm wide, flattened parallel to the replum; seeds in 2 rows per locule, sometimes winged ______ **Draba** (in part)
   2. Petals white or yellow to orange or pink-purple to brownish, 6–30 mm long; fruits pubescent, 2.5–15 cm long.
      5. All trichomes sessile and branching/stellate with appressed rays; petals yellow to greenish yellow or orange; stigma rounded or slightly 2-lobed ______ **Erysimum**
      5. Branching trichomes stalked with the rays variously directed (simple and/or glandular trichomes may also be present); petals pink to purple or yellow to brownish or white; stigma prominently 2-lobed, sometimes with 2 hornlike projections ______ **Matthiola**

Key D

At least some trichomes forked, branching, stellate, or peltate; ovaries and young fruits not linear; mature fruits silicles (length < 3 times width)

1. Plants scapose (with racemes emerging directly from the basal rosette), apparently stemless; stem leaves absent ______ **Draba** (in part)
1. Plants not scapose; stems and stem leaves present.
   2. Stem leaves (at least the uppermost) sessile, with bases sagittate to auriculate or cordate (and frequently clasping the stem).
      3. Petals white to pinkish (or rarely, petals vestigial/absent at anthesis); fruits obdeltoid to obcordiform (inverted triangular or heart-shaped), strongly flattened perpendicular to the replum; style 0.2–0.7 mm long ______ **Capsella**
      3. Petals yellow to pale yellow or white (sometimes drying purple); fruits globose to orbicular or pyriform to obovoid, terete or somewhat flattened (either parallel or perpendicular to the replum); style 1–3.5 mm long.
         4. Petals 1–2 mm wide; stamens in 3 pairs of unequal length; fruiting pedicels glabrate; fruits 2–5 mm wide, slightly flattened parallel to the replum ______ **Camelina** (in part)
         4. Petals 4–9 mm wide; stamens tetradynamous (2 short + 4 long); fruiting pedicels pubescent; fruits 4–9 mm wide, terete or ± flattened perpendicular to the replum ______ **Paysonia**
   2. Stem leaves petiolate or sessile, blade bases not clasping, neither sagittate nor auriculate nor cordate.
      5. Stem leaves 1–3-pinnately compound ______ **Descurainia** (in part)
      5. Stem leaves simple, margins entire to sinuate-repand or toothed.
         6. Branching trichomes sessile, with 2 opposing appressed rays; petals white or lavender to purple; ovules usually 2 per ovary (1 seed typically produced in each locule) ______ **Lobularia**
         6. Branching trichomes not as above, either sessile or stalked, 2–7-rayed (rays distinct or sometimes fused into scales); petals white or yellow to orange OR petals vestigial/absent at anthesis; seeds 4+ per ovary.
            7. Basal leaves with margins entire or dentate; sepals 0.7–2.5 mm long; petals white and < 5.5 mm long or petals vestigial/absent at anthesis; fruits flattened parallel to the replum; style < 0.5 mm long ______ **Draba** (in part)
            7. Basal leaves with margins entire to repand or toothed to pinnatifid; sepals 2.5–10 mm long; petals yellow to orange or rarely white, 4–14 mm long; fruits terete or only slightly compressed; style 1.5–5.5 mm long ______ **Physaria**

ARABIDOPSIS Heynh. ROCKCRESS

A genus of 10 species of circumboreal and north-temperate regions; 1 is native to North America and 9 are native to Europe. The circumscription of the genus has long been in controversy and at one time it included more than 50 species; molecular research has resulted in a major revision of generic concepts (O'Kane & AL-Shehbaz 2003; Al-Shehbaz 2010c). (From the genus *Arabis*, and Greek: *-opsis*, aspect, based on resemblance to that genus) (tribe Camelineae)

REFERENCES: O'Kane & Al-Shehbaz 1997, 2003; Al-Shehbaz 2010c.

Arabidopsis thaliana (L.) Heynh., (for Johann Thal, 1542–1583, German physician and botanist, who first described the species), MOUSE-EAR CRESS, THALE CRESS. Petite herbaceous annuals, to 30(–50) cm tall; glabrous or pubescent with simple trichomes (sometimes mixed with forked ones); stems simple or few from the base, erect or spreading; basal leaves forming a rosette, short-petiolate, blades spatulate or obovate to elliptic, margins entire or repand to dentate; stem leaves smaller, nearly sessile, margins usually entire, bases cuneate; racemes few- to several-flowered, not elongated in fruit; petals white, spatulate, 2–3.5(–4) mm long, blade gradually narrowed into claw; stamens slightly tetradynamous; fruiting pedicels slender, spreading; siliques dehiscent, linear, terete, (0.8–)1–1.5(–1.8) cm long, 0.5–0.8 mm wide, valves each distinctly midveined; style to 0.5 mm long; ovules 40–80 per ovary; seeds unwinged, in 1 row per locule. Sandy areas, roadsides, railroads, university grounds; recorded in Marion, Smith, and Upshur cos.; also El Paso Co.; scattered across the U.S. but most frequent in e 1/3; Canada; Europe; introduced elsewhere. Feb–May. [*Arabis thaliana* L., *Sisymbrium thalianum* (L.) J. Gay & Monn.] This species is currently the most widely-used model organism in plant biology; its diminutive size, rapid growth cycle (4–6 weeks), and small genome (completely sequenced in 2000) have made it the ideal subject for intensive research in developmental and evolutionary genetics. Native to Europe; now introduced nearly worldwide.

BARBAREA W.T. Aiton. YELLOW ROCKET, WINTERCRESS

A genus of ca. 25 species of North America, Europe, Asia, n Africa, and Australia; some species are edible, and some are widespread weeds. (Possibly named for Saint Barbara, 3rd century Christian saint of Phoenecia) (tribe Cardamineae)
REFERENCES: Fernald 1909; Al-Shehbaz 2010d.

Barbarea vulgaris W.T. Aiton, (common), YELLOW ROCKET, WINTERCRESS, BITTERCRESS. Somewhat showy herbaceous biennials or perennials; glabrous throughout or leaf margins ciliate; stems erect, striate-angled, branching, to 90(–120) cm tall; basal leaves petiolate, blades usually lyrate-pinnatifid with 1–3(–5) lobes per side, the terminal lobe largest and ovate to orbicular; upper stem leaves sessile, subentire to coarsely dentate or lyrate-lobed, bases glabrous and conspicuously auriculate; racemes elongated in fruit; sepals glabrous, margins scarious; petals yellow, spatulate to oblanceolate, (5–)6–9(–10) mm long, claw obscurely differentiated from blade; stamens tetradynamous; fruiting pedicels thickened, diverging or ascending; siliques dehiscent, short-stipitate, linear, terete or 4-angled or compressed parallel to the replum, (0.7–)1.5–3 cm long, 1.2–2 mm wide, valves each with a prominent midvein and marginal veins; style 1–3.5 mm long; ovules 18–28 per ovary; seeds unwinged, in 1 row per locule. Riverbanks, moist roadsides, disturbed areas. Not yet reported in TX but included here based on a record from adjacent Little River Co., AR, and likely in far ne TX; across the U.S. excepting warmest regions; Canada; Eurasia; and n Africa; introduced elsewhere. Apr–Jul. [*B. arcuata* (Opiz ex C. Presl) Rchb., *B. vulgaris* var. *arcuata* (Opiz) Fr., *Erysimum barbarea* L., *E. arcuatum* Opiz ex C. Presl] Native to Eurasia; sometimes cultivated as a potherb or ornamental; now widespread and weedy in temperate areas; it may be overlooked in our flora as it superficially resembles species of *Brassica*, *Raphanus*, *Rapistrum*, and *Sinapis*, among others.

BORODINIA N. Busch SICKLEPOD ROCKCRESS, FALSE ROCKCRESS

A genus of 8 species; 7 of these are native to e North America and 1 is native to Siberia and far e Russia. Our only representative was long treated in the "highly unnatural" large genus *Arabis*, then segregated to *Boechera* (along with most North American former *Arabis* species) (Al-Shehbaz 2003); recent molecular molecular research has supported fragmentation of the polyphyletic *Boechera*, with most w North American species retained in that genus, a few segregated to 7 other genera, and all e North American species placed in *Borodinia* (Alexander et al. 2013). (Named for Ivan Parfenievich Borodin, 1847–1930, founding president of the Russian Botanical Society) (tribe Boechereae)
REFERENCES: Rollins 1993; Mulligan 1995; Al-Shehbaz 2003; Kiefer et al. 2009; Al-Shehbaz & Windham 2010; Alexander et al. 2013.

Borodinia canadensis (L.) P.J. Alexander & Windham, (of Canada), SICKLEPOD ROCKCRESS, CANADA FALSE ROCKCRESS. Herbaceous biennials to 100(–125) cm tall; stems usually solitary, erect, usually unbranched, at least the lower portion pubescent; trichomes simple (or mixed with 2- or 3-rayed ones); basal leaves forming a winter rosette (usually withered by flowering), short-petiolate, blades obovate to oblanceolate, margins dentate; stem leaves evenly spaced, sessile or nearly so, blades lanceolate to elliptic or oblanceolate, sparsely pubescent, margins dentate or entire, bases cuneate or the lowermost minutely auriculate; racemes lax (sometimes paniculate-branching), elongated in fruit; sepals green, pubescent; petals white, spatulate to oblanceolate, 3–5 mm long, claw undifferentiated from blade; stamens tetradynamous; fruiting pedicels slender, usually curved, horizontal to descending; siliques dehiscent, linear, strongly flattened parallel to the replum, glabrous, often arcuate downward, 4–10 cm long, 2.5–4 mm wide, valves keeled; style to 1.1 mm long; ovules 40–62 per ovary; seeds with marginal wings 0.4–1 mm wide, in 1 row per locule. Lakeshores and streambanks, rocky bluffs, open woods; mainly ne and nc TX, s to Walker Co.; w to Cross Timbers; e U.S.; Canada. May–Jul. [*Arabis canadensis* L., *A. falcata* Michx., *A. ovata* (Pursh) Poir., *Boechera canadensis* (L.) Al-Shehbaz]

Lithospermum occidentale [HEA]

Lithospermum parviflorum [GL2, HOL]

Lithospermum tuberosum [GLE]

Myosotis macrosperma [STR]

Myosotis verna [GLE]

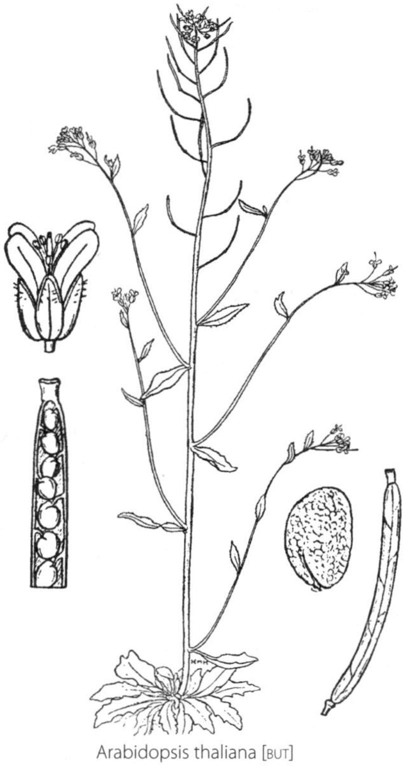

Arabidopsis thaliana [BUT]

BRASSICA L. CABBAGE, MUSTARD, TURNIP

Ours herbaceous annuals or biennials, often rather large and showy; glabrous or pubescent with simple trichomes, sometimes glaucous; stems erect, sometimes branching in the upper portion; basal leaves sometimes forming a winter rosette, petiolate, blades sinuate-lobed to lyrate-pinnatifid or pinnatisect with 1–6 lobes per side; stem leaves sessile or short-petiolate, margins entire or sinuate to dentate, bases sometimes auriculate or clasping; racemes considerably elongated in fruit (not paniculately branched); sepals often yellowish, usually erect or ascending; petals bright yellow to pale yellow, ovate to obovate, claw differentiated from blade; stamens tetradynamous; siliques linear, glabrous, transversely 2-segmented; lower fruit segment dehiscent, multi-seeded, terete or 4-angled or flattened parallel to replum, valves each prominently midveined; terminal fruit segment (beak) indehiscent, usually seedless (or 1–2-seeded), conical or linear, 1–22 mm long; ovules 4–60 per ovary; seeds unwinged, in 1 row per locule.

A genus of 35 species native to Eurasia and n Africa; closely related to the genus *Sinapis*, whose species were once included here. *Brassica* is the most economically important genus in the family, and includes many vegetable or "cole" crops, such as CABBAGE, KALE, and BROCCOLI (all selected forms of *B. oleracea*), TURNIP and BOK CHOY (*B. rapa*), and MUSTARD GREENS (*B. juncea*), as well as sources of condiment MUSTARD and CANOLA oil; human domestication of *Brassica* began at least 8000 years ago (Warwick 2010a). Several taxa are the products of hybridization, and extant taxa may hybridize with each other; this genetic flexibility is reflected in "the greatest diversity of products used by man derived from a single genus" (Dixon 2007); this has also contributed to the difficulty in reliably separating species. Mustard oil glycosides in *Brassica* exert protection against some cancers (Verkerk et al. 2009); these compounds can be poisonous to livestock when consumed in large quantities (Young-Matthews 2012). (The Latin name for CABBAGE) (tribe Brassiceae)

REFERENCES: Lemke & Worthington 1991; Dixon 2007; Warwick 2010a.

1. Stem leaves sessile, their bases auriculate or clasping ____ **B. rapa**
1. Stem leaves sessile or short-petiolate, their blade bases narrowed or cuneate (not auriculate or clasping).
   2. Lower stems glabrous; fruiting pedicels (5–)10–15(–20) mm long, fruits and pedicels not appressed to rachis; mature fruits often >3 cm long, torulose (constricted between seeds) ____ **B. juncea**
   2. Lower stems sparsely to densely hirsute; fruiting pedicels (2–)3–5(–6) mm long, fruits and pedicels ± appressed to rachis; mature fruits <2.8 cm long, not torulose ____ **B. nigra**

Brassica juncea (L.) Czern., (for its resemblance to the rush, *Juncus*), MUSTARD GREENS, CHINESE MUSTARD, LEAF MUSTARD, BROWN MUSTARD. Plants annual, ± glaucous; stems ± glabrous, to 100 cm tall; basal leaf blades with 1–3 lobes per side (basal leaves shed early); stem leaves usually short-petiolate, rarely sessile, blade bases tapered or cuneate (not auriculate); petals pale yellow, (7–)9–13 mm long; fruiting pedicels spreading to diverging-ascending; fruits subcylindrical or somewhat flattened, torulose, (2–)3–5(–6) cm long, 2–5 mm wide; lower segment with 6–20 seeds per locule; terminal segment conic, (4–)5–10(–15) mm long. Roadsides, weedy areas, sometimes as an escape or waif from cultivation; recorded in Brazos, Harris, Lamar, Liberty, Robertson, Travis, and Walker cos.; a few cos. to the w and s; scattered across most of the U.S.; Canada; Mexico; Central and South America; West Indies; Eurasia; introduced elsewhere. Apr–May (–Sep). [*B. japonica* (Thunb.) Siebold ex Miq., *B. juncea* var. *crispifolia* L.H. Bailey, *B. juncea* var. *japonica* (Thunb.) L.H. Bailey, *Sinapis juncea* L.] The species is likely derived from hybridization of *B. nigra* and *B. rapa*; Warwick (2010a) does not distinguish subspecific taxa. Native to se Europe and w Asia; long-cultivated for greens and oil-and mustard-producing seeds.

Brassica nigra (L.) W.D.J. Koch, (black), BLACK MUSTARD. Plants annual, lower parts not glaucous (often glaucous above); stems sparsely to densely hirsute-hispid (at least on lower stem), to 200 cm tall; basal leaf blades with 1–3 lobes per side; stem leaves sessile or nearly so, blade bases tapered (not auriculate); petals yellow, 7–11(–13) mm long; fruiting pedicels erect; fruits appressed to rachis, ± 4-angled, not torulose, (1–)2.5–2.7 cm long, (1.5–)2–3(–4) mm wide; lower segment with 2–8 seeds per locule; terminal segment linear, (1–)2–5(–6) mm long. Fields, roadsides, disturbed areas; recorded from Brazos, Limestone, and Travis cos. (Kartesz 2015) and Madison and McLennan cos.; a few cos. further w and s; recorded throughout the U.S.; Canada; Mexico; Central and South America; Eurasia; and n Africa; introduced elsewhere. Feb–May (—sporadically year-round). [*B. sinapioides* Roth, *Erysimum glabrum* C. Presl., *Raphanus sinapis-officinalis* Crantz, *Sinapis erysimoides* Roxb., *S. nigra* L.] Native to Mediterranean Europe and the Middle East; now a cosmopolitan weed. BLACK MUSTARD has long been cultivated for medicinal uses; the seeds are a source of condiment mustard.

Brassica rapa L., (the Latin name for turnip), FIELD MUSTARD, TURNIP-MUSTARD, WILD RAPE, BIRDSEED RAPE. Plants annual or biennial, green to slightly glaucous; stems glabrous or sparsely hairy, to 100 cm tall; basal leaf blades with 2–4(–6) lobes per side; stem leaves sessile (at least on upper stem), blade bases auriculate to clasping; petals yellow to deep yellow, 6–11(–13) mm long; fruiting pedicels spreading to ascending; fruits terete, torulose, (2–)3–8(–11) cm long, 2–4(–5) mm wide; lower segment with 8–15 seeds per locule; terminal segment conical, 8–22 mm long. Cultivated, in modified varieties, as TURNIP, CHINESE CABBAGE, BOK CHOY or PAK CHOI, NAPA CABBAGE, RAPINI, and RAPE or CANOLA; reseeds in gardens, around birdfeeders, abandoned fields, weedy roadsides; a waif or escape from cultivation, sometimes persisting, especially on sandy soils; scattered across e and c TX; a few cos. further s and w; recorded throughout the U.S.; Canada; Mexico; Central and South America; West Indies; Eurasia; introduced elsewhere. Feb–early May. [*B. campestris* L., *B. campestris* var. *oleifera* DC., *B. chinensis* L., *B. perfoliata* Crantz, *B. rapa* var. *sylvestris* (Lam.) Briggs, *Raphanus campestris* (L.) Crantz, *Sinapis rapa* (L.) Brot.] According to Warwick (2010a), *B. rapa* subsp. *oleifera* is the most likely identity of naturalized weeds of this species in the U.S. Derived from plants native to Eurasia; long-cultivated for edible greens, taproots, forage, and oil-producing seeds; selectively bred to encourage numerous morphologically-diverse varieties; now widely cultivated; weedy and naturalizing in temperate regions worldwide. FIELD MUSTARD can be toxic to livestock if consumed in large quantities over time (Young-Mathews 2012).

Brassica napus L., (turnip-like), CANOLA, OILSEED RAPE, RAPESEED, RUTABAGA, SWEDISH TURNIP, which would here have keyed to *B. rapa* (and overlaps that species in many morphological characters), can be distinguished by its flowering racemes in which the buds sometimes overtop open flowers, its slightly larger petals that are 10–16 mm long and golden yellow to creamy or pale yellow, and its lower fruit segments that have 12–20(–30) seeds per locule; these taxa are difficult to distinguish without flowers. Roadsides, railroads, abandoned fields; a waif or escape from cultivation; not reported from East TX but included here based on records from adjacent Caddo Par., LA, and Little River Co., AR; recorded across most of the U.S., but Warwick (2010a) notes some reports of *B. napus* in the se states may be misidentifications of *B. rapa*, with which it has been known to hybridize; Canada; Mexico; Eurasia; introduced elsewhere. Feb–May. [*B. napus* var. *oleifera* (Moench) Delile, *B. oleifera* Moench, *B. rugosa* (Roxb.) L.H. Bailey, *Raphanus napus* (L.) Crantz., *Sinapis napus* (L.) Brot.] The species is derived from hybridization of *B. oleracea* and *B. rapa*; according to Warwick (2010a), *B. napus* subsp. *napus* (CANOLA, OILSEED RAPE, RAPESEED) is more likely to escape from cultivation. Native to Mediterranean Europe; primarily of importance as an oilcrop.

Brassica oleracea L., (of the vegetable garden, an edible potherb), collectively known as COLE CROPS, which would here have keyed to *B. rapa,* can be distinguished by its fleshy, glaucous leaves, its much larger petals that are (15–)18–25(–30) mm long and lemon yellow to white, its smooth (not torulose) fruits, and its shorter terminal fruit segment that is (3–)4–10 mm long. Cultivated, in modified varieties, as CABBAGE, KALE, COLLARDS, CAULIFLOWER, BROCCOLI, BRUSSELS SPROUTS, and KOHLRABI; reseeds in gardens, abandoned fields, weedy areas; a waif or escape from cultivation, rarely persisting; included here based on records from adjacent DeSoto and Sabine parishes, LA; Lemke & Worthington (1991) stated there were several sporadically-adventive collections of *B. oleracea* from the e part of TX; also reported from Gillespie Co. (Kartesz 2015); scattered across the U.S. and perhaps only naturalizing in CA; Mexico; Canada; Eurasia; introduced elsewhere. Mar–Jun (–sporadically later). [*B. capitata* (L.) Pers., *B. maritima* Tardent, *B. oleracea* var. *sylvestris* L., *B. oleracea* var. *acephala* DC.] Derived from plants native to coastal Mediterranean Europe; a vegetable crop selectively bred to encourage numerous morphologically-diverse varieties; now cultivated nearly worldwide.

CAKILE Mill. SEA-ROCKET

Salt-loving, herbaceous annuals (rarely perennials) of coastal beaches and dunes (entering the flora region in Jefferson Co.); ± fleshy-succulent, glabrous or sparsely pubescent with simple trichomes; stems prostrate to erect, branching basally; basal leaves (when present) not forming a rosette; stem leaves usually petiolate, margins entire or crenate to dentate or pinnatifid; racemes considerably elongated in fruit; petals white to lavender (rarely absent), claw and blade sometimes differentiated; stamens tetradynamous; fruiting pedicels slender or stout; silicles or siliques stipitate, indehiscent, oblong to lanceoloid, fleshy, becoming corky and dry when mature, transversely 2-segmented, sometimes constricted at the joint; lower fruit segment persistent, terete to conical or obpyramidal (may be mistaken for a stipe or extension of the pedicel), sometimes with horn-like projections, usually 1-seeded; terminal fruit segment deciduous at maturity, terete or 4-angled or 8-ribbed, usually 1-seeded, beaked; ovules 2–4 per ovary; seeds unwinged.

A genus of 7 species native to the coasts of North and Central America, the Caribbean, Europe, and n Africa; one species is found in the deserts of the Middle East. Dispersal by ocean currents has resulted in broad "native" ranges for some species; additional introductions have come from ship ballast and beach fortifications. Overlapping ranges, hybridization, and morphological variation contribute to the difficulty of this genus, as does a long history of nomenclatural confusion; mature fruits are very helpful in identification. (From an old Arabic name for this plant: *qaqulleh*) (tribe Brassiceae)
REFERENCES: Rodman 1974, 2010.

Only one species, *C. constricta*, is recorded in our flora region, in Jefferson Co.; four additional taxa are found on the coasts of se TX and sw LA. Given their ease of dispersal and identical habitat preference, any of these may potentially be encountered.

1. Leaf blade margins sinuately lobed or deeply pinnatifid (at least on lower stem); petals 3–6 mm wide, usually lavender, rarely white; lower fruit segment with 2 opposite lateral horns at apex (below joint) ________ *C. maritima* subsp. *maritima*
1. Leaf blade margins various; petals 1.2–4.5 mm wide, white or lavender; lower fruit segment without lateral horns.
   2. Plants usually sprawling; leaves not markedly fleshy; racemes often 30+ cm long; petals usually white, rarely lavender, 3–4.5 mm wide ________ *C. lanceolata* subsp. *pseudoconstricta*
   2. Plants erect to prostrate, rarely sprawling; leaves usually fleshy; racemes 10–20 cm long; petals white or lavender, < 3 mm wide.
      3. Fruiting rachis geniculate (bent at sharp angles from one node to the next) and thus appearing zig-zag; petals 1.2–1.9 mm wide ________ *C. geniculata*
      3. Fruiting rachis not geniculate; petals 1.3–3 mm wide.
         4. Fruits terete or 4-angled, 3–4 mm wide; beak conic, apex acute ________ **C. constricta**
         4. Fruits 8-ribbed or 4-angled, 5–9 mm wide; beak flattened, apex usually blunt or retuse (rounded, with a notch) ________ *C. edentula* subsp. *harperi*

Cakile constricta Rodman, (constricted, for the fruit segment joint), GULF COAST SEA-ROCKET. Stems erect, to 30 cm long; leaf margins crenate to dentate or sinuate (not pinnatifid); fruiting rachis usually 10–20 cm long, not geniculate; petals white or lavender, 5.3–8 mm long, claw distinct; fruits 14–22 mm long, 3–4 mm wide, terete or 4-angled, constricted at joint; terminal segment fusiform, apex acute, beak conic. Sandy coastal shores; se margin of East TX in Jefferson Co. (iNaturalist.org observation *14891727, Eric Keith, Jun 2018* (iNaturalist.org 2021)); Brazoria and Galveston cos. (Kartesz 2015); se U.S. from FL to TX. Spring–Fall.

Cakile edentula (Bigelow) Hook. subsp. *harperi* (Small) Rodman, (sp.: without teeth; subsp.: for Roland McMillan Harper, 1878–1966, se U.S. botanist and geologist), HARPER'S SEA-ROCKET, AMERICAN SEA-ROCKET. Stems usually erect, to 80 cm long; leaf margins sinuate (not pinnatifid); fruiting rachis straight, congested, 10–20 cm long; petals white to pale lavender, 4.6–9.7 mm long, claw not distinct; fruits 20–29 mm long, 5–9 mm wide, 8-ribbed; terminal segment cylindrical to fusiform, apex usually retuse, beak sometimes flattened or blunt. Sandy beaches; to be expected in e coastal TX based on records from adjacent Cameron Par., LA (Kartesz 2015); se U.S. from NC to LA. Mar–Oct. [*C. harperi* Small]

Cakile geniculata (B.L. Rob.) Millsp., (bent at sharp angles), GULF SEA-ROCKET. Stems erect, to 1 m tall; leaf margins sinuate to dentate or pinnately lobed; fruiting rachis geniculate (bent at sharp angles), 10–20 cm long, pedicels widely spaced; petals white to pale lavender, 4–6.1 mm long, claw distinct; fruits 20–27 mm long, 3–5 mm wide, 8-ribbed; terminal segment slenderly conical, apex acute or blunt (often curved). Sandy beaches; coastal TX and LA; Mexico. Mar–Aug(–year-round). [*C. lanceolata* (Willd.) O.E. Schulz var. *geniculata* (B.L. Rob.) Shinners, *C. maritima* Scop. var. *geniculata* B.L. Rob.]

Cakile lanceolata (Willd.) O.E. Schulz subsp. *pseudoconstricta* Rodman, (sp.: lance-shaped; subsp.: similar to *C. constricta*), COASTAL SEA-ROCKET. Stems usually prostrate, often 50+ cm long; leaf margins deeply pinnatifid; fruiting rachis straight, 30+ cm long; petals usually white or rarely lavender, 4.9–9.4 mm long, claw distinct; fruits 15–24 mm long, 3–4 mm wide, weakly 4-angled or terete, constricted at joint; terminal segment slenderly conical, apex acute. Sandy beaches; coastal TX; AL, FL, and LA; Mexico. Year-round.

Cakile maritima Scop. subsp. *maritima*, (of the seaside) EUROPEAN SEA-ROCKET. Stems erect to prostrate, to 80 cm long; leaf margins sinuately lobed to deeply pinnatifid on lower stem, often becoming subentire on upper stem; fruiting rachis straight, to 40 cm long; petals usually lavender or rarely white, 8–14 mm long, claw distinct; fruits 12–27 mm long, 4–8 mm wide, hastate; lower segment conical, with 2 opposite lateral horns below joint; terminal segment usually 4-angled, conic or mitre-shaped, apex acute or blunt, usually compressed. Sandy beaches; Aransas, Brazoria, and Matagorda cos. (verified research-grade observations at iNaturalist.org); scattered in U.S., naturalized from coastal WA to CA; Mexico; Canada; Europe; introduced elsewhere. Spring–Fall(–year-round). [*Bunias cakile* L, *C. cakile* (L.) H. Karst.] Native of Europe.

CAMELINA Crantz FALSEFLAX

Ours annual herbs; stems unbranched or branched in the upper portion; lower stems moderately to densely hirsute, the simple trichomes often mixed with fewer, branched ones; upper stems nearly glabrous; basal leaves petiolate (sometimes withered by flowering), blade margins entire or toothed; stem leaves sessile, blades lanceolate to oblong, pubescent, margins entire or finely toothed to ciliate, bases sagittate or auriculate; racemes elongated in fruit; petals pale yellow (often drying white) to creamy yellow or white, oblanceolate, claw and blade somewhat differentiated; stamens in 3 pairs of unequal length; fruiting pedicels slender, ascending; silicles short-stipitate, dehiscent, obpyriform or obovoid, slightly flattened parallel to the replum, valves obscurely veined; style 1–3.5 mm long; ovules 8–25 per ovary; seeds unwinged or narrowly margined, usually in 2 series per locule.

Borodinia canadensis [DEL]

Brassica juncea [ROB]

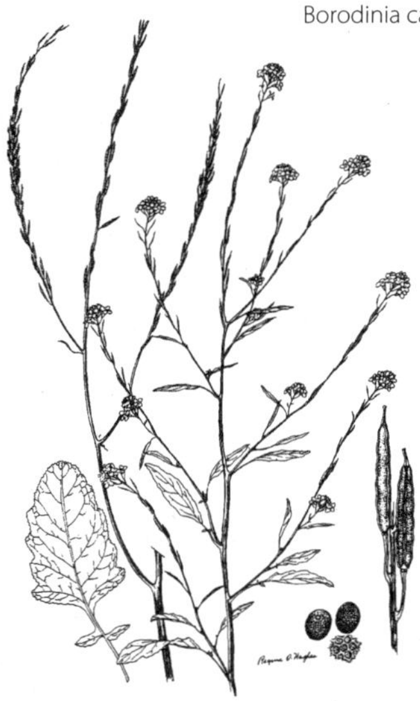

Brassica nigra [REE]

Brassica rapa [SMI]

Camelina microcarpa [DEL]

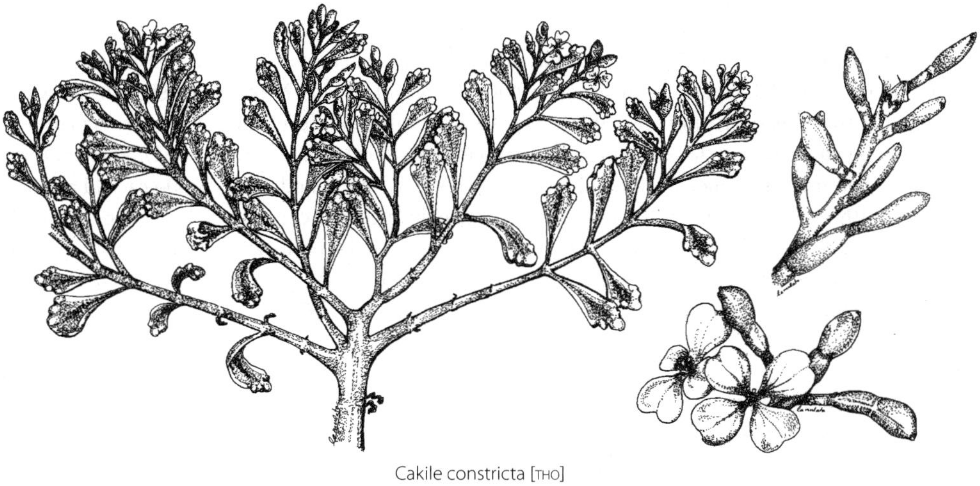

Cakile constricta [THO]

A genus of 8 species native to Europe and the Mediterranean to c Asia, including several widely naturalized throughout the temperate world as weeds (Francis & Warwick 2009). *Camelina sativa* or GOLD-OF-PLEASURE has been cultivated as an oilseed crop since the Bronze Age (Zohary & Hopf 2000). (Greek: *chamai*, dwarf or on the ground, and *linon*, flax, from allelopathic suppression of the growth of flax plants) (tribe Camelineae)

REFERENCES: McGregor 1984, 1985; Al-Shehbaz & Beilstein 2010.

1. Basal leaves usually withered by anthesis; unbranched trichomes of the lower stem to 2.5 mm long; petals (2.5–) 3–4(–6) mm long; fruits obpyriform to narrowly so, 3.5–5(–7) mm long ____ **C. microcarpa**
1. Basal leaves still green at anthesis and into fruiting; unbranched trichomes of the lower stem to 3.5 mm long; petals (5–)6–8(–9) mm long; fruits obpyriform to obovoid, 5–7 mm long ____ **C. rumelica**

Camelina microcarpa Andrz. ex DC., (small-fruited), LITTLEPOD FALSE FLAX, SMALLSEED FALSE FLAX. Erect annuals, stems 20–80 (–100) cm tall; unbranched trichomes of the lower stem to 2.5 mm long; basal leaves withering early; petals pale yellow (often drying white), (2.5–)3–4(–6) mm long; fruits obpyriform or narrowly so, 3.5–5(–7) mm long, 2–4(–5) mm wide. Fields, roadsides, railroads, and disturbed areas; recorded across nc TX, also in Harris (Brown 2014), Walker, and Gregg cos., and McLennan (Hannick et al. 2013) and Travis cos.; scattered to w TX and the Panhandle; adjacent border parishes in LA; OK; throughout North America; Europe; Asia; introduced elsewhere. Late Mar–May. [*C. sativa* (L.) Crantz subsp. *microcarpa* (Andrz. ex DC.) Hegi & Em. Schmid] Native to Europe, now widely introduced as a weed of fields and waste areas.

Camelina rumelica Velen., (of Rumelia, part of the Ottoman Empire), GRACEFUL FALSE FLAX. Erect annuals, stems 15–40(–60) cm tall; unbranched trichomes of the lower stem to 3.5 mm long; basal leaves persistent past anthesis; petals creamy yellow to white, (5–)6–8(–9) mm long; fruits obpyriform to obovoid, 5–7 mm long, 3.5–5 mm wide. Along railroads, disturbed areas; not naturalizing in TX according to Al-Shehbaz and Beilstein (2010), perhaps an infrequent waif and rarer than previously reported due to mis-identification or hybridization; recorded from Denton Co. (*L. Shinners 14098* SMU!); possibly also adjacent Wise Co.; Taylor Co. (Kartesz 2015) and the Panhandle; CO, KS, MN, NV, OK, OR, and TX; Canada; Eurasia; introduced elsewhere. Late Mar–Apr. Native to Europe; a weed of fields and waste areas.

CAPSELLA Medik. SHEPHERD'S PURSE

A genus currently understood to be of 5 species (Hurka et al. 2012), native to temperate and subtropical Eurasia and n Africa. (Diminutive of Latin: *capsa*, a box) (tribe Camelineae)

REFERENCES: Hurka & Neuffer 1997; Slotte et al. 2006; Al-Shehbaz 2010e.

Capsella bursa-pastoris (L.) Medik., (shepherd's purse, the fruits resembling the scrotum of a sheep used as a coinpurse), SHEPHERD'S PURSE, PICKPOCKET. Erect annual herbs; stems to 60(–70) cm tall, sparsely to densely pubescent, trichomes stellate, often mixed with longer simple hairs on the lower stem; basal leaves in a rosette, petiolate, blade margins pinnatifid to dentate; stem leaves mostly sessile, blades narrowly oblong to lanceolate or linear, entire or sinuate to dentate, bases sagittate to auriculate-clasping; racemes elongated in fruit; petals white or cream, obovate, (1.5–)2–4(–5) mm long, claw differentiated from blade; stamens tetradynamous; fruiting pedicels slender, spreading; silicles dehiscent, obdeltoid to obcordiform (inverted triangular or heart-shaped), strongly flattened perpendicular to the replum, keeled, (3–)4–9(–10) mm long, (2–)3–7(–9) mm wide, valves prominently veined; style to 0.7 mm long; ovules usually 20–40 per ovary; seeds unwinged, in 1 row per locule. Roadsides, fields, lawns, and disturbed sites; Reverchon (1880) indicated that it was seen for the first time in Dallas Co. in 1865; widespread in TX but more common in e 2/3 of state; throughout most of the U.S. and Canada; Mexico; Eurasia; n Africa; widely introduced elsewhere. Feb–May. [*Bursa bursa-pastoris* (L.) Shafer, *Lepidium bursa-pastoris* (L.) Willd., *Thlaspi bursa-pastoris* L.] Native to Eurasia, now cosmopolitan; estimated to be one of the most prolific weeds on Earth (Coquillat 1951). Apparently used by humans as a source of food and medicine since neolithic times—cached seeds were found at an 8,000-year-old archaeological site in Turkey (Mabberley 1997); still used in traditional medicine and cultivated as a commercial food crop in Asia (Flora of China Editorial Committee 2003). The seeds of SHEPHERD'S PURSE appear to be protocarnivorous; the wetted seed coat releases mucilage that appears to attract and eventually kill mosquito larvae, nematodes, protozoans, and bacteria; following this process, germination and growth of the plants is enhanced (Roberts et al. 2018).

CARDAMINE L. BITTERCRESS, TOOTHWORT

Ours herbaceous cool-season annuals, biennials, and spring-ephemeral perennials; glabrous or pubescent with simple trichomes; stems erect to decumbent; basal leaves petiolate, sometimes in a rosette, sometimes withered by flowering, blades simple or compound; stem leaves petiolate or sessile, blades simple or pinnatisect to compound, bases not auriculate; racemes elongated in fruit; sepals falling early; petals white to pale pink (rarely absent), claw sometimes indistinct; stamens 6 (rarely 4),

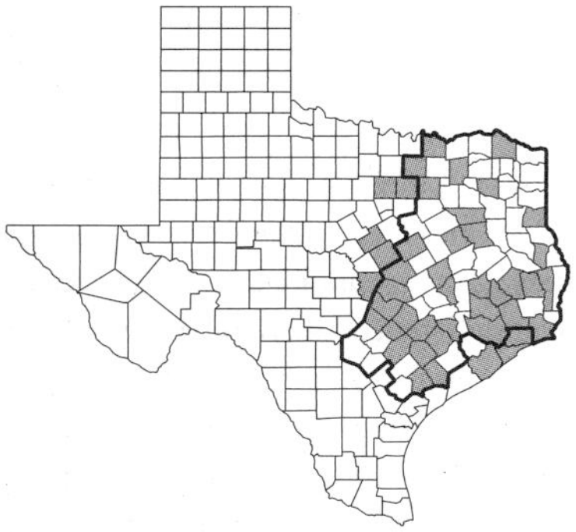
Myosotis macrosperma

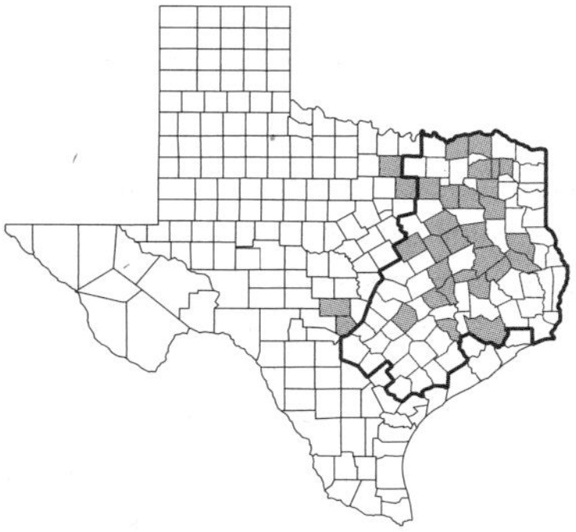
Myosotis verna

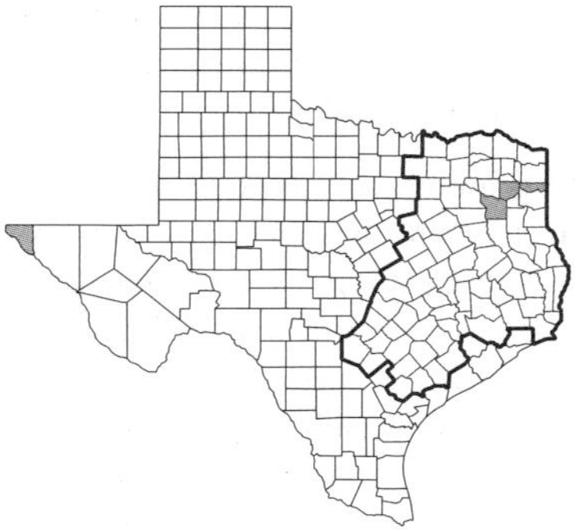
Arabidopsis thaliana

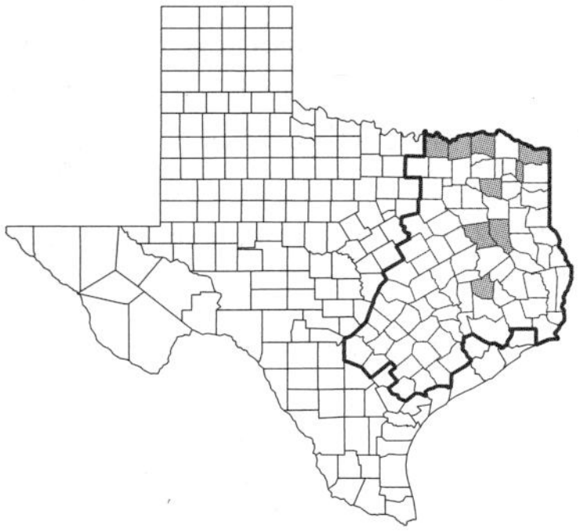
Borodinia canadensis

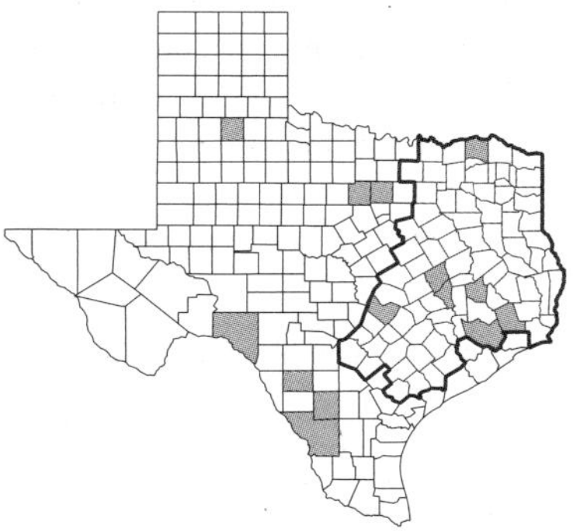
Brassica juncea

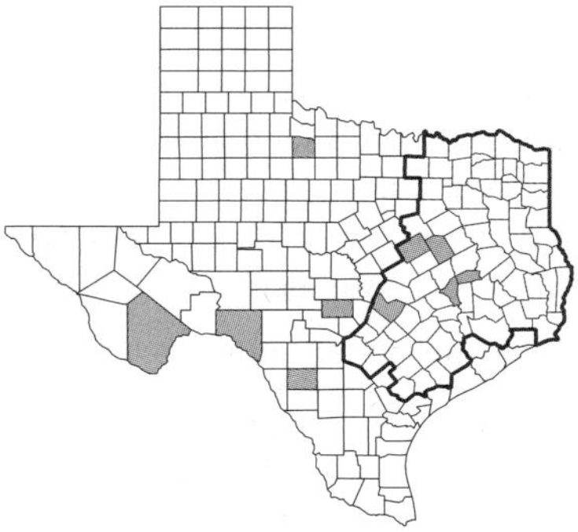
Brassica nigra

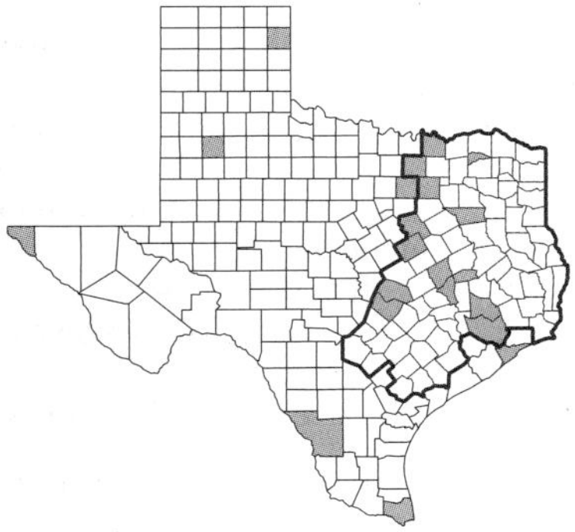
Brassica rapa

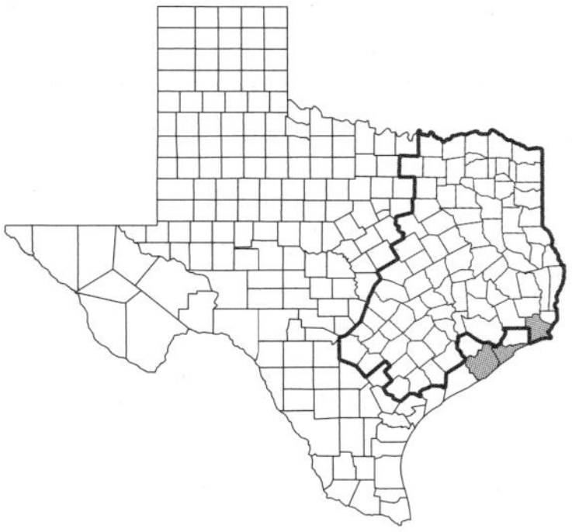
Cakile constricta

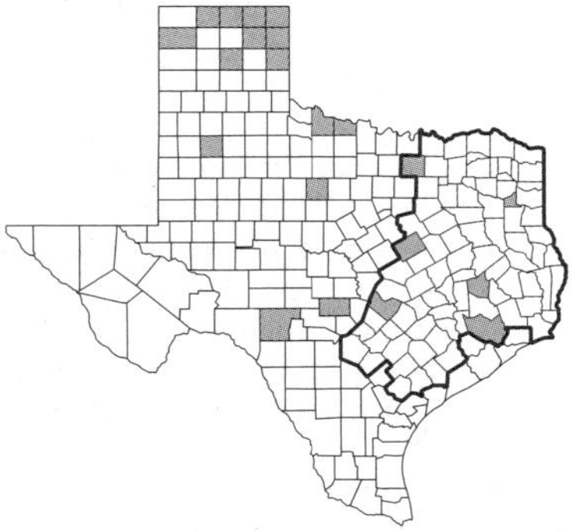
Camelina microcarpa

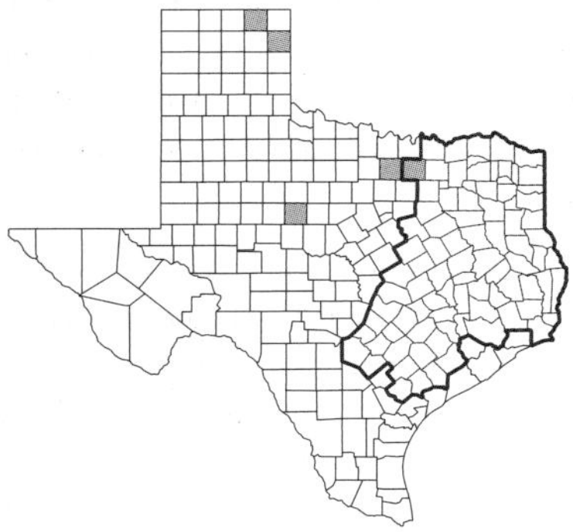
Camelina rumelica

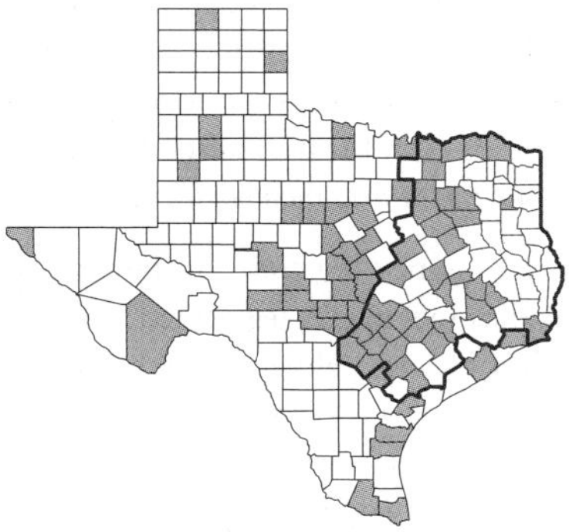
Capsella bursa-pastoris

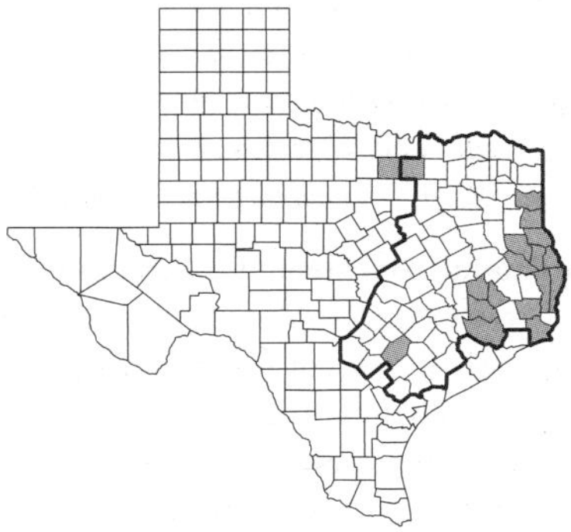
Cardamine bulbosa

filaments equal or unequal in length; fruiting pedicels slender or stout; siliques elastically dehiscent (valves spirally coiling), linear or linear-lanceolate, compressed parallel to the replum, valves unveined; style blunt or slender; ovules 10–80 per ovary; seeds narrowly margined or unwinged, in 1 row per locule.

A genus of ca. 200 species worldwide, native to every continent (excluding Antarctica); it includes narrow endemics, cultivated ornamentals, and some invasive weeds; some species were previously placed in *Dentaria*. Considerable morphological variability and plasticity are common in *Cardamine*; some species readily hybridize; others form morphologically-similar species complexes that are difficult to distinguish, especially when geographic ranges and habitat preferences overlap. Much confusion has reigned over species' names, origins, and defining characters: among the short-lived, non-rhizomatous *Cardamine* species formerly or currently treated as present in East TX, the epithet *debilis* was previously used to describe 2 species by separate authors; the epithets *debilis*, *flexuosa*, *hirsuta*, and *pensylvanica* have been interchanged across and within species by multiple authors; European *C. flexuosa* has often been treated as a single entity in North America, but recent molecular studies have revealed some of these plants to be the invasive Asian *C. occulta* (Lihova et al. 2006; Marhold et al. 2016). Pubescence is an important character to separate similar taxa; complete herbarium specimens are required. Since floras and keys available for the U.S. do not include all of the species now known to be present, a high percentage of misidentifications are expected in TX herbaria, and complete range maps are not available for some species as currently circumscribed. (A Greek name, *kardamon*, used by Dioscorides for some cress with medicinal uses, perhaps in treating heart problems; presumably from *cardia* or *kardia*, heart) (tribe Cardamineae)

REFERENCES: Brown & Marcus 1998; Lihova et al. 2006; Carlsen et al. 2009, Post et al. 2009, 2011; Al-Shehbaz et al. 2010a; Marhold et al. 2016; Šlenker et al. 2018.

1. Plants perennial, arising from rhizomes or tubers; all leaves simple or 3-foliolate; petals white or pale pink, 6–20 mm long; style 2+ mm long in fruit.
   2. Stem leaves 4–10, simple ____ **C. bulbosa**
   2. Stem leaves 2 or 3, 3-foliolate ____ **C. concatenata**
1. Plants annual or biennial, arising from weak taproots or fibrous roots; some leaves pinnately compound with 5+ leaflets; petals white, 1.5–5 mm long (or rarely absent/vestigial); style < 2 mm long in fruit.
   3. Basal leaves in a rosette, persistent to anthesis.
      4. Stems flexuous, usually pubescent; basal rosette loose; upper leaf surface markedly pubescent; stamens 6 ____ **C. flexuosa**
      4. Stems not flexuous, usually glabrous; basal rosette compact; petiole bases sparsely hirsute; stamens 4 (rarely 5 or 6) ____ **C. hirsuta**
   3. Basal leaves not forming a rosette, often withered by anthesis.
      5. Terminal leaflet of midstem leaves narrowly oblong or linear, 1–3 mm wide ____ **C. parviflora**
      5. Terminal leaflet of midstem leaves reniform to suborbicular or obovate to oblanceolate, > 3 mm wide.
         6. Stems erect, to 55(–70) cm, glabrous or pubescent basally; lateral leaflets of stem leaves often slightly decurrent along rachis ____ **C. pensylvanica**
         6. Stems erect to decumbent, to 25(–29) cm, sparsely hirsute basally; lateral leaflets of stem leaves usually distinct, not decurrent along rachis ____ **C. occulta**

Cardamine bulbosa (Schreb. ex Muhl.) Britton, Sterns & Poggenb., (bulbous), BULBOUS BITTERCRESS, SPRING CRESS. Plants perennial; rhizomes fleshy, tuberous and bulb-like; stems erect, to 60 cm tall, unbranched, glabrous or sparsely pubescent in the upper portion; basal leaves simple, blades reniform to cordate or ovate, margins usually repand to entire; stem leaves usually 4–10, short-petiolate becoming sessile on upper stem, simple, blade margins entire or repand to dentate, bases not auriculate; petals white or rarely pale pink, obovate, (6–)7–12(–16) mm long; stamens 6; fruits 2–3.5(–4) cm long, 1.4–1.7 mm wide. Low woods, stream banks, swamps and bogs; deep e and se TX from Harrison Co. s to Harris Co. and the LA border; with apparent disjuncts in Gonzales and Wise cos.; e 1/2 U.S; Canada. Feb–Apr. [*Arabis bulbosa* Schreb. ex Muhl., *A. rhomboidea* Pers., *C. rhomboidea* (Pers.) DC., *C. rhomboidea* var. *hirsuta* O.E. Schulz, *Dentaria rhomboidea* (Pers.) Greene]

Cardamine concatenata (Michx.) O. Schwarz., (joined together, for the segmented rhizomes), CUTLEAF TOOTHWORT, CRINKLEROOT, PEPPER ROOT, CROW'S-TOES. Plants perennial; rhizomes fragile, segmented; stems erect, to 40(–55) cm tall, unbranched, glabrous or sparsely pubescent in the upper portion; basal leaves palmately 3-foliolate, leaflet margins dentate to lobed or incised to laciniate; stem leaves (2 or) 3, petiolate, 3-foliolate, usually whorled or opposite, blades similar in morphology to basal leaves, terminal leaflet lanceolate to oblanceolate or linear; petals white to pale pink, oblanceolate, (8–)10–20 mm long, stamens 6; fruits (2–)2.5–3.8(–4.8) cm long, 1.5–3 mm wide. Low woods, creek bottoms, rocky banks; rare in TX and apparently restricted to deep e TX in Sabine and San Augustine cos., Newton Co. (*C. Allen et al. 21362* LSU! BRIT), and Shelby Co. (iNaturalist.org observation *14080474*, *Matt Buckingham, Mar 2015* (iNaturalist.org 2021)); e 1/2 of the U.S; Canada. Feb–Mar. [*C. laciniata*

(Muhl. ex Willd.) Alph. Wood, *Dentaria concatenata* Michx., *D. laciniata* Muhl. ex Willd.] NatureServe: G5, secure; S1, critically imperiled due to the restricted range in TX.

Cardamine flexuosa With., (flexuous), WOODLAND BITTERCRESS, ROADSIDE BITTERCRESS. Plants annual or biennial; not rhizomatous; stems erect or ascending, to 29(–40) cm long, branched, flexuous, pubescent; basal leaves persistent, in a rosette, similar to stem leaves but leaflets broader; midstem leaves petiolate, blades adaxially markedly pubescent, pinnately 7–14(–16)-foliolate, terminal leaflet elliptical or broadly ovate with margins repand or 1–7(–9)-lobed (sinuses deep or shallow); petals white, spatulate to obovate or oblanceolate, 2.7–4.6 mm long, less than twice as wide as sepals; stamens 6; fruits 2.7–4.6 cm long, 1.1–1.8 mm wide. Roadsides and lawns, damp meadows, open disturbed areas; range unknown but likely in e 1/2 of TX (see discussion below); mainly e and se 1/2 of the U.S. and w coast; Canada; Europe; introduced elsewhere. Feb–Apr. [*C. flexuosa* var. *debilis* (O.E. Schulz) T.Y. Cheo & R.C. Fang, *C. pusilla* Schur, *C. sylvatica* Link] For discussions on this species see Lihova et al. (2006), Post et al. (2011), Marhold et al. (2016), and Šlenker et al. (2018). With a confounding overlap in application of epithets, "*C. flexuosa*" has been used in U.S. floras (e.g., Al-Shehbaz et al. 2010a) to describe plants that were in part *C. flexuosa* and in part *C. occulta*; "*C. debilis*" has been given as a synonym (or alternate) name for these in some publications, but this name is not a synonym of *C. flexuosa* as currently circumscribed. *C. debilis* D. Don is an illegitimate synonym of *C. occulta* Hornem., which has been referred to in some floras as "Asian *C. flexuosa*," with the synonym of *C. flexuosa* subsp. *debilis* O.E. Schulz (but NOT *C. flexuosa* var. *debilis* (O.E. Schulz) T.Y. Cheo & R.C. Fang, which is a synonym of *C. flexuosa* as circumscribed here). *C. debilis* Banks ex DC. is from New Zealand and is not found in North America. Native to Europe; a weedy invader of many parts of the world. We are unable to separate reported localities of this taxon from those reported for *C. flexuosa*, so the maps provided are identical for both taxa.

Cardamine hirsuta L., (hairy), HAIRY BITTERCRESS, FLICKWEED. Plants annual; not rhizomatous; stems erect to decumbent, to 35 (–45) cm long, unbranched or branched, not flexuous, usually glabrous; basal leaves persistent, in a compact rosette, pinnately (5–)8–15(–22)-foliolate, terminal leaflet reniform or orbicular with margins entire or repand to dentate or lobed; stem leaves few, petiolate, petiole bases sparsely hirsute, blades similar to basal leaves in leaflet number but smaller or lobes/leaflets fewer and narrower; petals white (rarely absent), spatulate, 2.5–4.5(–5) mm long, usually less than twice as wide as sepals; stamens 4 (rarely 5 or 6); fruits (0.9–)1.5–2.5(–2.8) cm long, (0.8–)1–1.4 mm wide. Roadsides and lawns, weedy fields, creekbanks; mainly e 1/3 of TX sw to DeWitt Co.; e and se U.S. and w coast; Canada; Europe; introduced elsewhere. Feb–Apr. [*C. hirsuta* subsp. *hirsuta* L., *C. multicaulis* Hoppe ex Schur, *C. tenella* E.D. Clarke, *C. umbrosa* Andrz. ex DC.] Native to Europe; now a weedy invader of drier regions on every continent (Marhold et al. 2016).

Cardamine occulta Hornem., (hidden), HIDDEN BITTERCRESS, ASIAN BITTERCRESS. Plants annual or biennial; not rhizomatous; stems erect to decumbent to 25(–29) cm long, branched, flexuous, glabrous or pubescent; basal leaves not in a rosette; midstem leaves petiolate, blades adaxially ± glabrous, pinnately 5–13(–15)-foliolate, terminal leaflet reniform to ovate, obovate or suborbicular with margins (1–)3–5(–7)-lobed (sinuses deep and sharply angled); petals white, spatulate to obovate or oblanceolate, 3–4.7 mm long, more than twice as wide as sepals; stamens 6; fruits 1.3–2.5 cm long 0.6–0.9 mm wide. Greenhouses and irrigated flowerbeds, moist urban and disturbed areas, rice fields; range unknown but likely in e 1/2 of TX (see discussion below); mainly e 1/2 of the U.S. and w coast; Canada; Mexico; West Indies; e Asia; introduced elsewhere. Feb–Apr. [*C. debilis* D. Don [nom. illeg.], *C. flexuosa* subsp. *debilis* O.E. Schulz, *C. flexuosa* var. *occulta* (Hornem.) O.E. Schulz, *C. hamiltonii* G. Don] For discussions on this species see Lihova et al. (2006), Post et al. (2011), Marhold et al. (2016), and Šlenker et al. (2018). *C. occulta* has been referred to in some floras (e.g., Al-Shehbaz et al. 2010a), as "Asian *C. flexuosa*," with the synonym of *C. flexuosa* subsp. *debilis* O.E. Schulz (but NOT *C. flexuosa* var. *debilis* (O.E. Schulz) T.Y. Cheo & R.C. Fang, which is a synonym of *C. flexuosa* as it is now circumscribed). With a confounding overlap in application of epithets, "*C. debilis*" has been used in U.S. floras to describe plants that were in part *C. flexuosa* and in part *C. occulta*, but *C. debilis* D. Don is an illegitimate synonym of *C. occulta*, and *C. debilis* Banks ex DC. is from New Zealand and is not found in North America. Native to e Asia, a potentially increasing weed of urban and disturbed areas, often introduced in potting soil of nursery plants. We are unable to separate reported localities of this taxon from those reported for *C. flexuosa*, so the maps provided are identical for both taxa.

Cardamine parviflora L., (small-flowered), SMALL-FLOWER BITTERCRESS, SAND BITTERCRESS. Plants annual; not rhizomatous; stems erect, to 30(–40) cm tall, often branched in the upper portion, somewhat flexuous, glabrous or pubescent; basal leaves often withering early, usually not in a rosette; stem leaves petiolate, pinnately (5–)9–15(–17)-foliolate, terminal leaflet linear or narrowly oblong with margins entire or toothed; petals white, oblanceolate, (1.5–)1.8–2.5(–3) mm long, usually less than twice as wide as sepals; stamens 6; fruits (0.5–)1.2–2.5 cm long, 0.6–0.9 mm wide. Roadsides, stream banks, moist sandy soils; mainly e 1/3 of TX sw to Gonzales Co.; Edwards Plateau; e 1/2 of the U.S.; Canada; Mexico; Eurasia. Feb–Apr. [*C. arenicola* Britton, *C.*

parviflora var. *arenicola* (Britton) O.E. Schulz] We are following recent authors in not recognizing infraspecific taxa; if varieties are recognized, ours would be var. *arenicola*.

Cardamine pensylvanica Muhl. ex Willd., (of Pennsylvania), PENNSYLVANIA BITTERCRESS, QUAKER BITTERCRESS. Plants annual or biennial; not rhizomatous; stems erect, to 55(–70) cm tall, unbranched or branched in the upper portion, not flexuous, glabrous except for hispid lower portion; basal leaves withering early, not in a rosette; midstem leaves petiolate, pinnately (5–)7–13(–19)-foliolate, terminal leaflet suborbicular to obovate or oblanceolate with margins entire or repand to lobed; petals white, narrowly spatulate to oblanceolate, (2–)3.5–4 mm long, less than twice as wide as sepals; stamens 6; fruits (1.4–)1.7–2.7(–3.2) cm long, 0.8–1.1 mm wide. Urban lawns, roadside ditches, stream and lake margins; scattered in e and se TX; widespread throughout the U.S.; Canada. Dec–Apr. [*C. flexuosa* With. subsp. *pensylvanica* (Muhl. ex Willd.) O.E. Schulz, *C. hirsuta* L. var. *pensylvanica* (Muhl. ex Willd.) P.W. Graff, *C. multifolia* Rydb., *Dracamine pensylvanica* (Muhl. ex Willd.) Nieuw.] A native yet weedy species; highly morphologically variable.

CHORISPORA R.Br. ex DC. CROSSFLOWER

A genus of 11 species, native to Europe, Asia, and n Africa. (Greek: *choris*, asunder or separate, and *spora*, seed, from the fruit breaking apart into one-seeded segments) (tribe Chorisporeae)
REFERENCES: Al-Shehbaz 2010f; German et al. 2011.

Chorispora tenella (Pall.) DC., (tender, soft), CROSSFLOWER, BLUE MUSTARD, MUSK MUSTARD. Herbaceous annuals, to 40(–56) cm tall; musky-scented, roughly stipitate-glandular, trichomes unbranched; stems erect or decumbent, often branching near the base; basal leaves not forming a rosette, often withered by flowering, petiolate, blade margins sinuate-dentate to pinnatifid, bases cuneate to attenuate; stem leaves similar, decreasing in size up the stem, the uppermost nearly sessile; racemes slightly or considerably elongated in fruit; sepals purplish, glandular-pubescent; petals pink to lavender to purple, 8–10(–12) mm long, claw strongly differentiated from and longer than blade (blade-claw transition often more intensely colored); stamens strongly tetradynamous; fruiting pedicels stout, spreading; fruits linear, terete, curved-ascending, (1.4–)1.8–2.5(–3) cm long, 1.5–2 mm wide, glandular-pubescent, loment-like, with 8–12 constrictions on each side, breaking and splitting into single-seeded corky segments, leaving behind the persistent marginal replum; beak 6–22 mm long; ovules 5–30 per ovary; seeds unwinged, in 1 row per locule. Roadsides, fields, disturbed areas; reported as new for TX by Lipscomb (1984); recorded from Grayson Co. (*B. Faber 77* BRIT) and McLennan and Dallas cos.; in nc TX in Wichita Co. (*A. Nelson s.n.* TAC) and Tarrant Co.; also Erath Co. (iNaturalist.org observation *22309016*, *Sam Kieschnick*, *Apr 2019* (iNaturalist.org 2021)), El Paso Co., and a few cos. in the Panhandle; OK; widespread in w 1/2 of the U.S., with scattered records from the Midwest to New England and s to TN and LA; w Canada; Eurasia; n Africa; introduced elsewhere. Mar–May. [*Chorispermum tenellum* (Pall.) R.Br., *Raphanus tenellus* Pall.] A weedy native of Eurasia.

CONRINGIA Heist. ex Fabr. HARE'S-EAR MUSTARD

A genus of 6 species, native to sc Europe, the e Mediterranean region, and c Asia. (Named for Hermann Conring, 1606–1681, professor at Helmstedt, Germany) (tribe Conringieae)
REFERENCES: Al-Shehbaz 1985; Rollins & Al-Shehbaz 1986; Warwick 2010b.

Conringia orientalis (L.) D. Presl, (eastern), HARE'S-EAR MUSTARD, RABBIT EARS. Herbaceous annuals to 70 cm tall, glaucous, glabrous; stems erect, often unbranched; basal leaves not forming a rosette, nearly sessile, blades slightly fleshy, oblanceolate to obovate, margins ± entire; stem leaves sessile, often not much reduced in size (or only on uppermost stems), oblong to elliptic or lanceolate, margins entire, bases deeply cordate-clasping, apices rounded to obtuse; racemes considerably elongated in fruit; petals white to greenish white or pale yellow, 7–12 mm long, claw differentiated from and usually longer than the blade; stamens slightly tetradynamous; fruiting pedicels stout, ascending; siliques dehiscent, linear, strongly 4-angled and keeled or nearly terete, ± torulose, (5–)8–14 cm long, 2–2.5 mm wide, valves each with prominent midvein; style 0.5–4 mm long; ovules many; seeds unwinged, in 1 row per locule. Railroads, roadsides, abandoned fields, disturbed areas; gravelly or sandy soils; recorded from Grayson, Tarrant, Brazos, Fayette, and Comal cos., as well as Denton, Dallas, Navarro, and Upshur cos. (Kartesz 2015); also recorded in Tarrant Co. (Mahler 1988) and Taylor and Gillespie cos.; widespread in the U.S. and Canada; Mexico; Europe; introduced elsewhere. Mar–early May. [*Brassica orientalis* L., *Erysimum orientale* (L.) Crantz] A weedy native of Eurasia.

DESCURAINIA Webb & Berthel. TANSY-MUSTARD, FLIXWEED

Ours herbaceous annuals; usually pubescent at least on lower stems, trichomes mostly branching, sometimes glandular; stems erect, sometimes branching; basal leaves in a rosette, sometimes withered by flowering, petiolate, 1–3-pinnately compound;

Camelina rumelica [HEA]

Capsella bursa-pastoris [REE]

Cardamine bulbosa [CO1]

Cardamine concatenata [BB2]

Cardamine flexuosa [BB2, WU1]

Cardamine hirsuta [BB2]

Cardamine occulta [SBS]

Cardamine parviflora [CO1]

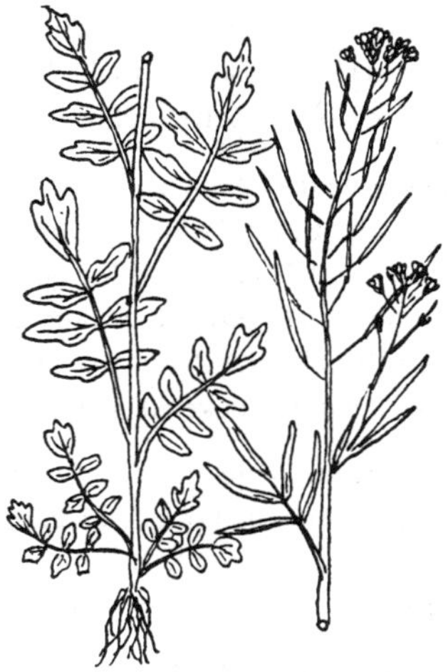

Cardamine pensylvanica [RYD]

stem leaves pinnatifid or pinnatisect, similar to basal, sessile or short-petiolate, blades decreasing in size up the stem, bases not auriculate or clasping; racemes considerably elongated in fruit; sepals yellowish to rose or purplish; petals small, yellow to whitish, slightly shorter to slightly longer than the sepals, claw sometimes indistinct; stamens tetradynamous; fruiting pedicels slender; fruits siliques or silicles, dehiscent, linear or clavate, terete, valves each distinctly midveined; style minute; ovules 16–48 per ovary, seeds unwinged, in 1 or 2 rows per locule.

A genus of ca. 45 species, primarily of temperate and cool areas including North America, South America, Eurasia, the Canary Islands, and n Africa. *Descurainia* is notoriously difficult, with extensive morphological variation and hybridization resulting in a "taxonomic nightmare" according to Goodson and Al-Shehbaz (2010); taxonomically reliable characters are few, and limited to specimens with mature fruits, resulting in a significant percentage of misidentifications in North American herbaria. (Named for François Descurain, 1658–1740, French botanist and apothecary) (tribe Descurainieae)
REFERENCES: Goodson & Al-Shehbaz 2010; Goodson et al. 2011.

1. Plants eglandular; stem leaves glabrate to sparsely pubescent; fruits ≤ 1 mm wide, usually narrowly linear, torulose (swollen and constricted at intervals); seeds in 1 row per locule ____ **D. sophia**
1. Plants glandular or eglandular; stem leaves densely pubescent; fruits > 1 mm wide, usually clavate (wider toward the apex), not torulose; seeds usually in 2 rows per locule ____ **D. pinnata**

Descurainia pinnata (Walter) Britton, (pinnate, feathery), WESTERN TANSY-MUSTARD, GREEN TANSY-MUSTARD. Plants to 57(–92) cm tall; glandular or eglandular; stem leaves densely pubescent; sepals yellowish to rose or purplish, pubescent; petals 1–3 mm long; fruits usually clavate, not torulose, 4–13(–17) mm long, 1.2–2.2 mm wide; seeds usually in 2 rows per locule. Roadsides, open stream banks, waste ground, often on sandy or limestone soils; nearly throughout TX, but absent from deep e TX; mostly w, c, and se U.S.; Canada; n Mexico. Mar–Apr. Reportedly poisonous to livestock, with symptoms similar to those of selenium poisoning; animals can become blind, wander aimlessly, and lose the ability to swallow; the condition is sometimes referred to as blind staggers or paralyzed tongue (Kingsbury 1964). ☠ This species displays tremendous variability in every conceivable character and is "one of the most complex assemblages in North American Brassicaceae" (Goodson & Al-Shehbaz 2010). Up to 11 subspecies have been recognized, with some later being transferred to other species; the remaining four recognized subspecies fall into "relatively distinct" groups (Goodson & Al-Shehbaz 2010). Due to chronic nomenclatural changes and highly variable characters within a given population, earlier keys are unreliable, and complete range maps are not available for the subspecies as currently circumscribed. Therefore, we are not providing TX ranges or maps for the three subspecies found in TX:

1. Plants gray-pubescent, often eglandular; sepals rose to purplish ____ subsp. **ochroleuca**
1. Plants not gray-pubescent, often glandular; sepals yellowish or rose.
   2. Fruiting pedicels spreading to horizontal or descending, 4–14(–17) mm long; sepals rose (at least the tips); petals 1–1.8 mm long ____ subsp. **pinnata**
   2. Fruiting pedicels spreading to ascending, (7–)10–18(–23) mm long; sepals yellowish; petals (1.7–)2–3 mm long ____ subsp. **brachycarpa**

subsp. **brachycarpa** (Richardson) Detling, (short-fruited). Plants usually glandular, usually not gray-pubescent; stems branching in upper portion; fruiting pedicels spreading to ascending, (7–)10–18(–23) mm long; sepals yellowish; petals (1.7–)2–3 mm long. TX; mainly c and w U.S.; Canada. [*D. brachycarpa* (Richardson) O.E. Schulz, *D. intermedia* (Rydb.) Daniels, *D. pinnata* var. *brachycarpa* (Richardson) Fernald, *D. pinnata* subsp. *intermedia* (Rydb.) Detling, *Sisymbrium brachycarpum* Richardson, *Sophia brachycarpa* (Richardson) Rydb.]

subsp. **ochroleuca** (Wooton) Detling, (yellow and white or light yellow). Plants usually eglandular, distinctly gray-pubescent; stems branching near the base; fruiting pedicels spreading to ascending, (4–)6–12 mm long; sepals rose to purplish; petals 1.5–2 mm long. TX; AZ and NM; n Mexico. [*Descurainia andrenarum* (Cockerell) Cory, *D. halictorum* (Cockerell) O.E. Schulz var. *andrenarum* (Cockerell) O.E. Schulz, *D. menziesii* (DC.) O.E. Schulz var. *ochroleuca* (Wooton) O.E. Schulz, *D. pinnata* subsp. *halictorum* (Cockerell) Detling, *D. pinnata* var. *ochroleuca* (Wooton) Shinners, *Sisymbrium ochroleucum* (Wooton) K. Schum., *Sophia andrenarum* Cockerell, *S. ochroleuca* Wooton]

subsp. **pinnata.** Plants usually glandular, usually not gray-pubescent; stems branching in upper portion; fruiting pedicels spreading to horizontal or descending, 4–14(–17) mm long; sepals rose (at least the tips); petals 1–1.8 mm long. TX; se U.S. from NC to FL and w to TX. [*Descurainia canescens* (Nutt.) Prantl, *D. multifida* (Pursh) O.E. Schulz, *Erysimum pinnatum* Walter, *Sisymbrium canescens* Nutt., *Sophia californica* (Torr. & A. Gray) Rydb., *S. millefolia* Rydb., *S. pinnata* (Walter) Howell]

Descurainia sophia (L.) Webb ex Prantl, (from the old genus name in Greek: wisdom, alluding to reputed medicinal properties), FLIXWEED TANSY-MUSTARD, HERB-SOPHIA. Plants to 70(–100) cm tall; eglandular; basal leaves 2–3-pinate; stem leaves often glabrous; sepals yellowish, glabrate to sparsely pubescent; petals 2–3 mm long; fruits narrowly linear, torulose, (12–)15–27(–30) mm long, 0.5–0.8(–1) mm wide; seeds in 1 row per locule. Roadsides, pastures, croplands, waste ground; Denton and Grayson cos., historically recorded in Travis Co., perhaps based on a misidentification (Carr 2008); recorded in Erath Co. (Cotton & Nelson 2018), mostly w 1/2 of TX, especially the Panhandle; mostly w 1/2 of the U.S.; Canada; Mexico; Eurasia; introduced elsewhere. Mar–May. [*Hesperis sophia* (L.) Kuntze, *Sisymbrium sophia* L., *S. parviflorum* Lam., *Sophia parviflora* (Lam.) Standl.] Native to Europe, now an invasive weed of dry lands on many continents, and considered noxious in some regions. No county distribution map provided.

Diplotaxis DC. WALLROCKET

A genus of 25–30 species, native to Europe, Asia, and the Mediterranean. (Greek: *diplo-*, double, and *taxis*, row or arrangement, alluding to the two rows of seeds in each locule) (tribe Brassiceae)

References: Eschmann-Grupe et al. 2003; Martinez-Laborde 2010.

Diplotaxis muralis (L.) DC., (of walls), annual wallrocket, stinking wallrocket, sandrocket. Herbaceous annuals or short-lived, tap-rooted perennials, to 60 cm tall; ill-scented; often scapose (with racemes emerging directly from the basal rosette) or stems sometimes branching, moderately pubescent to hispid in the lower portion, trichomes simple; basal rosette leaves petiolate, blades rather thick and fleshy, margins sinuate to lyrate-pinnatifid, with 2–4(–6) lobes per side; stem leaves (when present) sessile or short-petiolate, margins entire or dentate, bases not auriculate or clasping; racemes often few-flowered, considerably elongated in fruit; petals yellow (sometimes aging to rose-orange before falling), 5–8(–10) mm long, obovate, with a short claw-like base; stamens tetradynamous; fruiting pedicels slender, ascending or spreading; siliques dehiscent, sometimes stipitate, linear, terete or slightly flattened parallel to the replum, (1.5–)2–4 cm long, 1.5–2.5 mm wide, transversely 2-segmented; lower segment many-seeded, valves each midveined; upper segment seedless, beak-like or resembling a style, 1–3 mm long; ovules 20–36 per ovary; seeds unwinged, in 2 rows per locule. Roadsides, railroads, around buildings, waste ground; first collected in TX in 1947 (Cory 1948), recorded from Grayson, Collin, Dallas, and Ellis cos., naturalized and apparently increasing around the DFW metroplex, and see numerous recent "research-grade" iNaturalist.org observations documenting further spread along the I-35 corridor, also se-c TX in Washington (Kartesz 2015) and Madison cos., and sc TX in Bexar (*R. Gutierrez 1082* ASU) and Travis cos.; mainly nc TX; scattered across the U.S.; Canada; Mexico; introduced elsewhere. Mar–Sep. [*Sinapis muralis* (L.) R. Br., *Sisymbrium murale* L.] Native to Europe; now introduced in many parts of the world.

Draba L. WHITLOW-GRASS

Ours small herbaceous annuals; pubescent, trichomes branched, sometimes mixed with simple ones; often scapose or nearly so (stemless, the inflorescences arising from a basal rosette) or in some species stems present, erect to decumbent, sometimes branching; basal leaves sometimes in a rosette, obscurely short-petioled, blades simple, elliptic to spatulate or obovate to oblanceolate, margins entire or dentate; stem leaves (when present) similar to basal, sessile, bases cuneate; racemes corymbose and elongated in fruit or subumbellate and not elongating; sepals green or pinkish; petals small (or absent in late-season cleistogamous flowers), white, longer than the sepals, obovate to spatulate, sometimes emarginate, claw distinct or obscure; stamens tetradynamous; fruiting pedicels slender, straight; fruits silicles or siliques; dehiscent, linear to lanceolate or oblong to obovate, flattened parallel to the replum, valve midveins distinct or obscure; style minute; ovules 8–88 per ovary, seeds unwinged or winged, in 2 rows per locule.

As treated here, the largest genus in Brassicaceae, with ca. 400 species; native to all continents except Australia and Antarctica, but most diverse in subarctic regions, w North America, and the Himalayas and Andes; some are cultivated as rock-garden ornamentals. *Draba* is a difficult and controversial genus; frequent hybridization and polyploidy events have greatly complicated our understanding of its evolutionary history (Jordon-Thaden et al. 2010). Our taxa have been sometimes treated under segregate genera *Abdra* and *Tomostima* (Al-Shehbaz 2012; Jordon-Thaden et al. 2013); current respected taxonomic resources disagree regarding generic circumscription, therefore we treat *Draba* in the traditional and familiar sense, following Al-Shehbaz et al. (2010b), Kartesz (2015), and GBIF.org. Species determination may require complete specimens, with flowers and mature fruit, with the caution that similar species may be intermingled at the same locality. (Greek: *drabe*, acrid—applied by Dioscorides to describe the taste of such plants) (tribe Arabideae)

References: Fernald 1934; Al-Shehbaz et al. 2010b, 2012; Jordon-Thaden et al. 2010, 2013; Hohmann et al. 2015.

1. Stems and leaves scabrous-pubescent, with sessile 4-branched (cruciform) trichomes (use hand lens or dissecting scope); mature fruits (2–)2.5–5(–6) mm long, glabrous; ovules 8–16(–20) per ovary; seeds winged (=*Abdra*) **D. brachycarpa**
1. Stems (at least lower) and leaves pubescent to hirsute, with stalked 2–4-branched trichomes; mature fruits ≥5 mm long, glabrous or pubescent; ovules ≥20 per ovary; seeds unwinged (=*Tomostima*).

2. Stem leaves 3–8(–14), often scattered up the stem, blade margins dentate; fruits obovate to obovate-oblong, 5–8.2(–9.5) mm long, (2.4–)2.8–3.7(–4) mm wide, pubescent with all trichomes simple ______ **D. platycarpa**
2. Stem leaves absent or 1–6, crowded on lower 1/3 of stem, blade margins entire or dentate; fruits oblong to linear or lanceolate, (5–)7–16(–20) mm long, 1.2–2.7(–3) mm wide, glabrous or pubescent with simple and/or branched trichomes.
   3. Some leaf blades coarsely dentate; racemes densely pubescent, 10–50(–70)-flowered, corymbose, elongated in fruit ______ **D. cuneifolia**
   3. All leaf blades entire; racemes usually glabrous (rarely sparsely pubescent), (3–)5–12(–16)-flowered, subumbellate, not elongating in fruit ______ **D. reptans**

Draba brachycarpa Nutt. ex Torr. & A. Gray, (short-fruited), SHORT-POD DRABA, SHORT-POD WHITLOW-GRASS. Plants scabrous-pubescent, trichomes sessile and 4-branched (cruciform); stems present (plants not scapose), usually branching, to 19(–22) cm long, often much shorter; stem leaves 4–6(–11), often scattered up the stem, blades lanceolate to oblong or linear, margins usually entire; racemes pubescent as stem, (20–)25–65(–74)-flowered, corymbose, elongated in fruit; petals (when present) spatulate, 2–3 mm long; fruits linear to elliptic (rarely ovate-elliptic), (2–)2.5–5(–6) mm long, 0.9–1.4(–1.9) mm wide, glabrous; ovules 8–16(–20) per ovary; seeds winged. Open woods, roadsides, cedar glades, often on sandy soils; e TX w to Rolling Plains and Edwards Plateau; mostly se and sc U.S. Feb–Mar. [*Abdra brachycarpa* (Nutt. ex Torr. & A. Gray) Greene]

Draba cuneifolia Nutt. ex Torr. & A. Gray, (wedge-leaved), WEDGE-LEAF DRABA, WEDGE-LEAF WHITLOW-GRASS. Plants pubescent to hirsute, trichomes stalked and 2–4 rayed, sometimes mixed with simple ones; stems obscure (scapose) or unbranched, to 27(–37) cm long; basal leaf blades oblanceolate to spatulate or broadly obovate; stem leaves absent or 1–6 on lower 1/3 of stem, similar to basal, margins coarsely dentate in upper portion; racemes pubescent as stem, 10–50(–70)-flowered, corymbose, elongated in fruit; petals (when present) spatulate, (2–)2.5–4.5(–5) mm long; fruits oblong to linear or lanceolate, (3–)6–12(–16) mm long, 1.7–2.7(–3) mm wide, usually puberulent (rarely glabrous); ovules (12–)24–66(–72) per ovary; seeds unwinged. Rocky slopes, pastures, roadsides, sandy and gravelly soils. Late Feb–Apr. [*Tomostima cuneifolia* (Nutt.) Al-Shehbaz, M.A. Koch & Jordon-Thaden] Al-Shehbaz et al. (2010b) tentatively recognized 3 varieties; 2 may be found in East TX:

1. Stems hirsute, with at least some simple trichomes ______ var. **cuneifolia**
1. Stems pubescent, with 3- or 4-rayed trichomes (stalked) ______ var. **integrifolia**

var. **cuneifolia.** (wedge-leaved). Stems hirsute, with at least some simple trichomes; racemes often composing upper 1/2–1/3 of scape; fruits to 12(–16) mm long, with simple trichomes (and sometimes also 2-branched ones). In East TX from nc to sc TX and Edwards Plateau, with a few deep e TX outliers; mainly western 3/4 TX; sw and sc U.S., rare in se states (Kartesz 2015); Mexico. [*D. cuneifolia* var. *foliosa* Mohlenbr. & J.W. Voigt, *D. cuneifolia* var. *helleri* (Small) O.E. Schulz, *D. cuneifolia* var. *leiocarpa* O.E. Schulz, *D. helleri* Small]

var. **integrifolia** S. Watson, (entire-leaved). Stems pubescent, with stalked 3- or 4-rayed trichomes; racemes composing upper 3/4 of scape; fruits to 12 mm long, with stalked 3- or 4-rayed trichomes (and sometimes also 2-branched ones). Perhaps not a permanent member of our flora; in East TX only recorded from Navarro Co.; possibly as a waif (Hatch et al. 1990; Kartesz 2015); also Val Verde Co.; AZ, CA, NM, NV, and TX; Mexico. [*D. integrifolia* (S. Watson) Greene]

Draba platycarpa Torr. & A. Gray, (broad-fruited), BROAD-POD DRABA, BROAD-POD WHITLOW-GRASS. Plants pubescent to hirsute, trichomes stalked and 2–4 rayed, sometimes mixed with simple ones; stems obscure (scapose) or often unbranched, to 32(–40) cm long; basal leaf blades obovate to spatulate or oblanceolate; stem leaves 3–8(–14), often scattered up the stem, similar to basal, margins dentate; racemes pubescent as stem, 12–45(–60)-flowered, corymbose, elongated in fruit; petals (when present) oblanceolate, (1.7–)2.5–3.5(–4) mm long; fruits obovate to obovate-oblong, 5–8.2(–9.5) mm long, (2.4–)2.8–3.7(–4) mm wide, pubescent, trichomes simple; ovules (28–)40–70(–82) per ovary; seeds unwinged. Rocky slopes, prairie barrens, disturbed sites, sandy and gravelly soils; e Blackland Prairie w to Rolling Plains and Edwards Plateau; AR, AZ, LA, OK, TX and Pacific nw; probably Mexico. Mar–Apr (main period 1 or 2 weeks later than *D. cuneifolia*). [*D. cuneifolia* Nutt. ex Torr. & A. Gray var. *platycarpa* (Torr. & A. Gray) S. Watson, *D. viperensis* H. St. John, *Tomostima platycarpa* (Torr. & A. Gray) Al-Shehbaz, M.A. Koch & Jordon-Thaden]

Draba reptans (Lam.) Fernald, (creeping), CAROLINA DRABA, CAROLINA WHITLOW-GRASS. Plants sparsely to densely pubescent (upper stems may be glabrous), trichomes stalked and 2–4 rayed, sometimes mixed with simple ones; stems obscure (scapose) or un branched, to 12(–16) cm long; basal leaves elliptic to spatulate or obovate to suborbicular; stem leaves usually absent or rarely 1–3 on lower 1/3 of stem, similar to basal, margins entire; racemes glabrous (rarely sparsely pubescent), (3–)5–12(–16)-flowered, subumbellate, not elongating in fruit; petals (usually present) spatulate, 2–4.5 mm long; fruits linear or linear-oblong, (5–)7–16(–20) mm long, 1.2–2.3 mm wide, glabrous or pubescent, trichomes simple (and also sometimes 2-branched); ovules

Chorispora tenella [EN2, PNW]

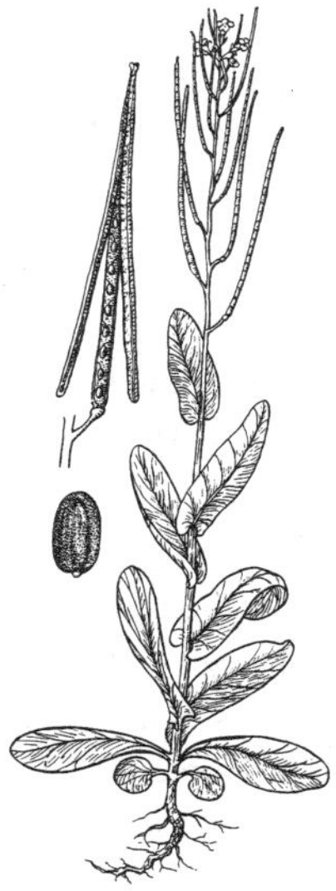

Conringia orientalis [ROB]

Descurainia pinnata [ARF]

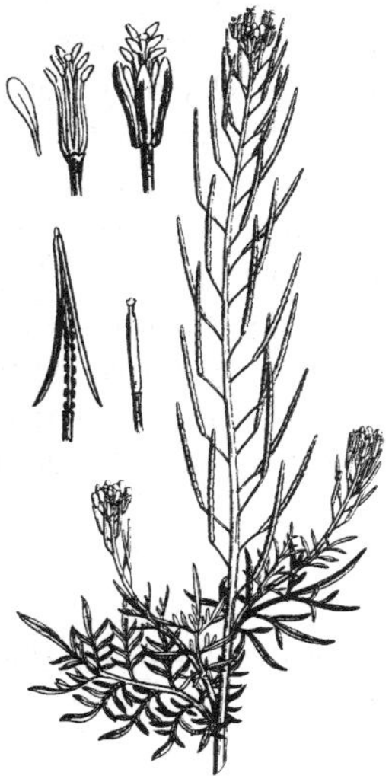

Descurainia sophia [SMI]

Diplotaxis muralis [BB2]

Draba cuneifolia [BT3]

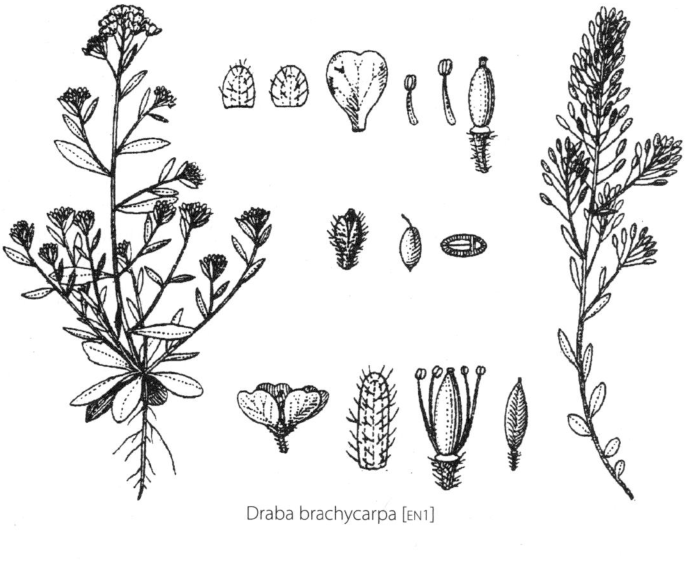

Draba brachycarpa [EN1]

32–88 per ovary; seeds unwinged. Open woods, sandy prairies, rocky hillsides, roadsides; recorded in Dallas, McLennan, Travis, Gonzales, and Brazos cos.; Blackland Prairies from nc to sc TX, w to High Plains; mainly c and w U.S., rarer in e U.S.; Canada. Feb–Mar. [*Arabis reptans* Lam., *D. caroliniana* Walter; *D. micrantha* Nutt., *D. reptans* var. *micrantha* (Nutt.) Fernald, *D. umbellata* Muhl., *Tomostima caroliniana* (Walter) Nieuwl., *T. hispidula* (Michx.) Raf., *T. micrantha* (Nutt.) Lunell, *T. reptans* (Lam.) Al-Shehbaz, M.A. Koch & Jordon-Thaden]

ERUCA Mill. ARUGULA, ROCKET SALAD

A genus of 3 or 4 species, native to Europe, the Mediterranean, and n Africa; taxa sometimes treated as subspecies of a single species, *E. vesicaria.* (Possibly from Latin: *urere*, to burn, or Archaic: *roc*, harsh or rough—probably alluding to its pungently peppery flavor) (tribe Brassiceae)
REFERENCES: Warwick 2010c; Pignone & Gómez-Campo 2011.

Eruca vesicaria (L.) Cav. subsp. **sativa** (Mill.) Thell., (sp.: bladder-like, perhaps for the calyx; subsp.: cultivated), ARUGULA, ROCKET SALAD, GARDEN ROQUETTE. Herbaceous annuals, to 80(–100) cm tall; glabrous to hispid with trichomes simple and often retrorse; stems erect, usually branching from the base; basal rosette leaves (often withered by fruiting) petiolate, blade widely oblanceolate, sinuately lobed to lyrate-pinnatifid or pinnatisect, with 3–9 lobes per side, the terminal lobe largest, lobe margins entire or dentate; stem leaves similar to basal, subsessile; racemes corymbose, greatly elongated in fruit; sepals green or rose to purplish, glabrous or with a tuft of trichomes near the apex; petals white to creamy yellow with some veins purplish or brownish, (12–)15–20(–26) mm long, claw differentiated from blade; stamens strongly tetradynamous; fruiting pedicels stout, erect to ascending to subappressed to the rachis; siliques dehiscent, linear to oblong, terete or slightly flattened parallel to the replum, (1.1–)1.5–3.5(–4) cm long, (2.5–)3–5 mm wide, transversely 2-segmented; lower segment many-seeded, valves each prominently midveined; upper segment seedless, beak-like, flattened to a point, 4–11 mm long; ovules 10–50 per ovary; seeds unwinged, in 2 rows per locule. Roadsides, old pastures, disturbed sites; recorded on sw border of East TX in Bexar Co. (Kartesz 2015); Edwards Plateau and Trans-Pecos; scattered across the U.S.; Canada; Mexico; Eurasia; introduced elsewhere. Feb–May(–Sep). [*Brassica eruca* L., *B. vesicaria* L., *E. sativa* Mill.] Native to Mediterranean Europe and w Asia; now a crop escape or weed in many temperate parts of the world. This species has been used as a food and medicine since ancient Greek and Roman times and is mentioned in the Bible and Talmud; it was once valued as an aphrodisiac and remedy for impotence (Pignone & Gómez-Campo 2011); it retains wide popularity as a raw or cooked leafy vegetable, condiment, and source of cooking oil, and is also used as animal feed (Warwick 2010c).

ERUCASTRUM C. Presl DOG MUSTARD

A genus of 19 species, native to Europe, Asia, and Africa; some are weedy. (Latin: *Eruca*, a similar Mediterranean genus, and *-astrum*, resemblance) (tribe Brassiceae)
REFERENCES: Luken et al. 1993; Warwick 2010d.

Erucastrum gallicum (Willd.) O.E. Schulz, (of Gaul or France), COMMON DOG MUSTARD, ROCKETWEED, HAIRY ROCKET. Herbaceous annuals, to 65(–80) cm tall; sparsely to densely pubescent, trichomes simple, stiff and recurved or retrorse; stems erect or ascending, simple or branching; basal rosette leaves petiolate, blade oblanceolate, deeply lobed to pinnatifid, with 3–10 lobes per side, the terminal lobe largest; stem leaves similar to basal, sessile or short-petiolate, decreasing in size up the stem, the uppermost becoming bract-like on the raceme; racemes greatly elongated in fruit, at least the lower portion bracteate; sepals sparsely hispid toward the apex; petals pale yellow to creamy white, 4–8 mm long, gradually narrowed into a short claw; stamens tetradynamous; fruiting pedicels slender, ascending or spreading; siliques dehiscent, linear, subterete or 4-angled, slightly torulose, 1–4.5 cm long, 1–2(–2.7) mm wide, valves each prominently midveined, fruits transversely 2-segmented; lower segment many-seeded; upper segment seedless, style-like, 1.5–4 mm long; ovules 10–60 per ovary; seeds unwinged, in 1 row per locule. Roadsides, railroads, disturbed sites; Austin, Denton, Grayson, Hill, Red River, and Travis cos., Robertson Co. (Kartesz 2015); also Edwards Plateau in Gillespie, Terrell, and Val Verde cos.; first collected in TX in 1926 (Luken et al. 1993); scattered across the U.S. but most common in the Midwest; Canada; Europe; introduced elsewhere. Mar–Apr. [*E. pollichii* K.F. Schimp. & Spenn., *Sisymbrium gallicum* Willd.] Native to sw and c Europe; now naturalized in other parts of Europe, e Asia, and North America.

ERYSIMUM L. WALLFLOWER

Ours herbaceous annuals or biennials; pubescent with sessile branched or stellate hairs; stems erect, sometimes branching; basal leaves sometimes forming a rosette, petiolate, often withered by flowering/fruiting; stem leaves petiolate or sessile on upper stem, blade margins entire or sinuate to repand or dentate, bases not auriculate; racemes densely-flowered, considerably

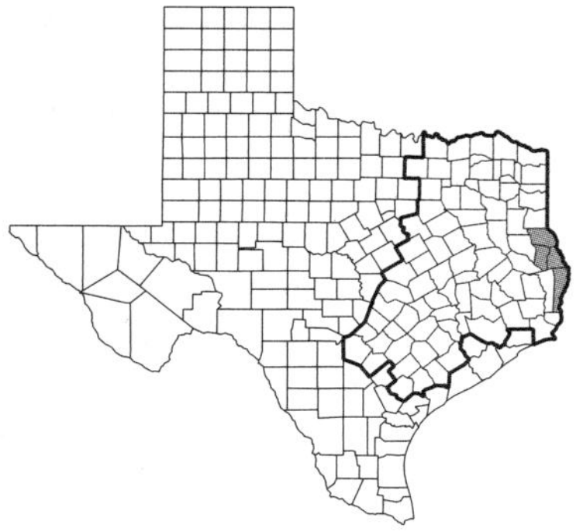

Cardamine concatenata

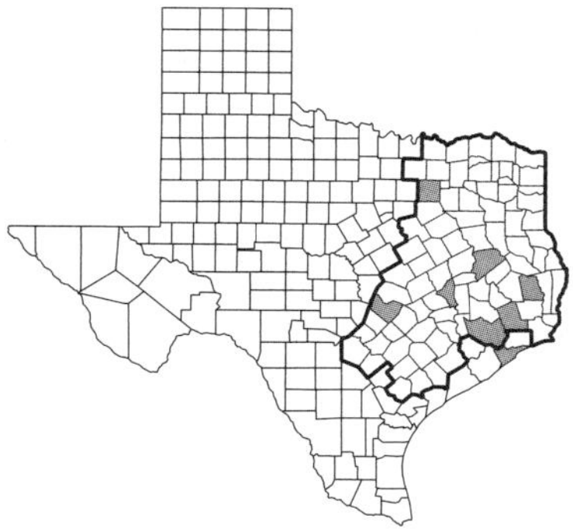

Cardamine flexuosa

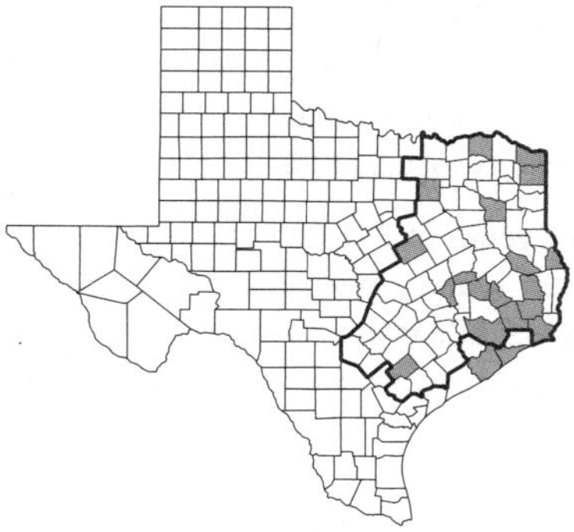

Cardamine hirsuta

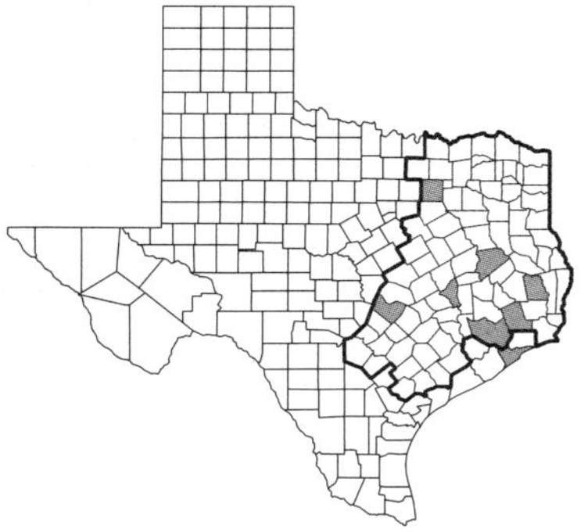

Cardamine occulta

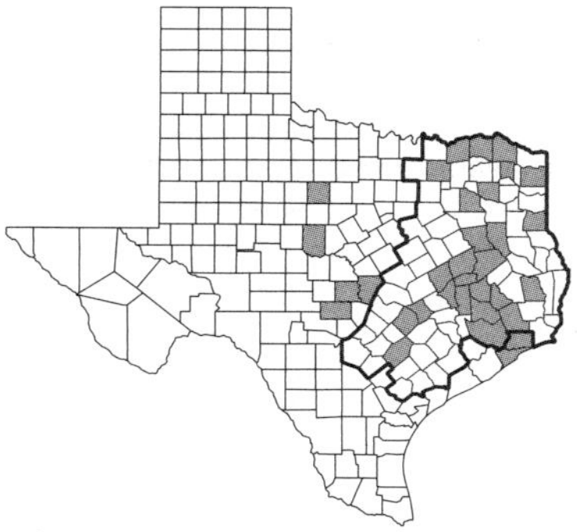

Cardamine parviflora

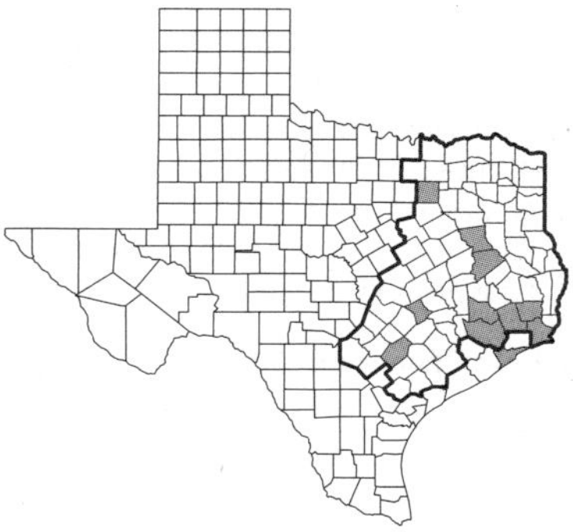

Cardamine pensylvanica

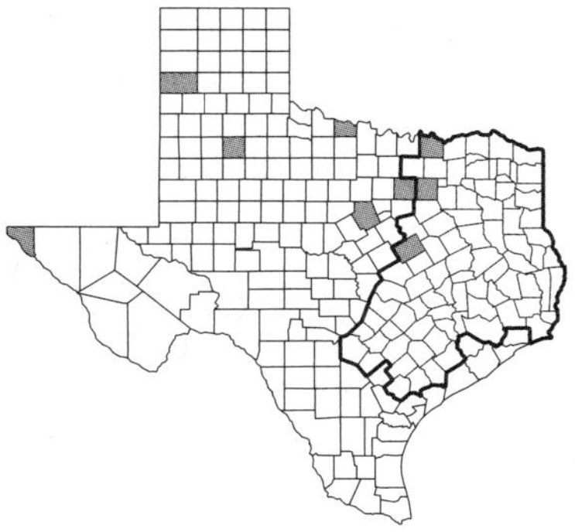

Chorispora tenella

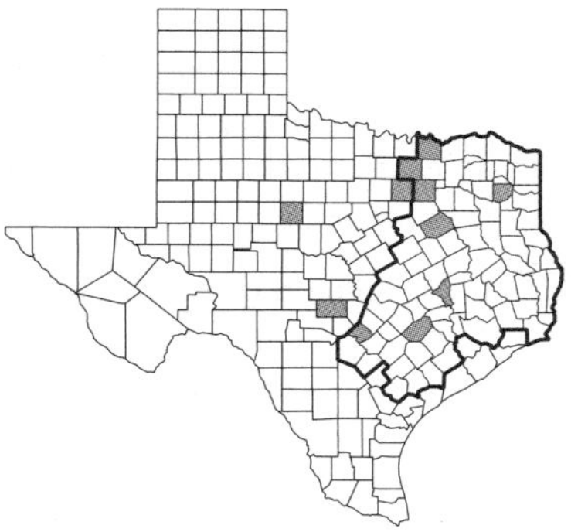

Conringia orientalis

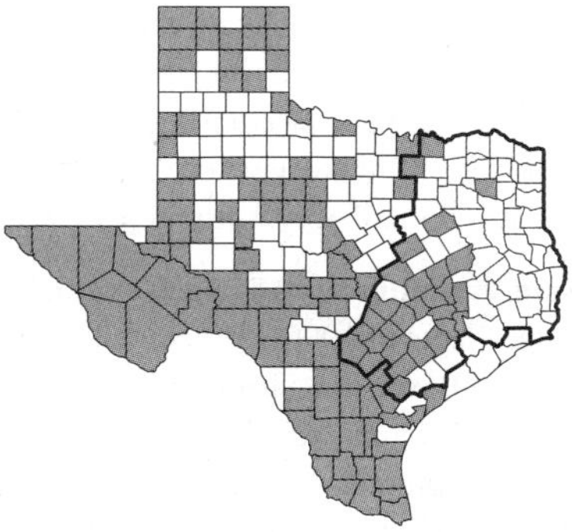

Descurainia pinnata

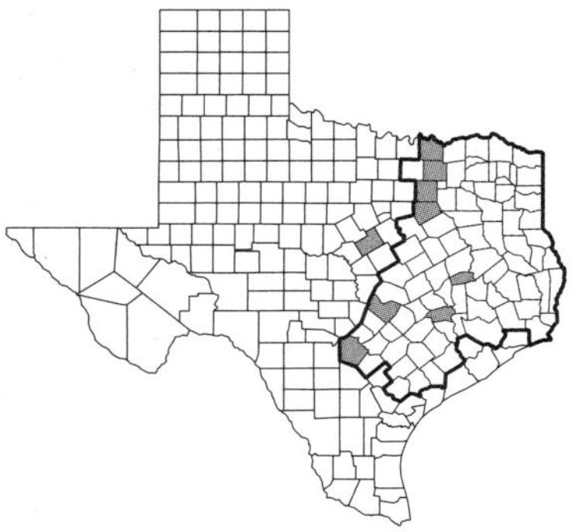

Diplotaxis muralis

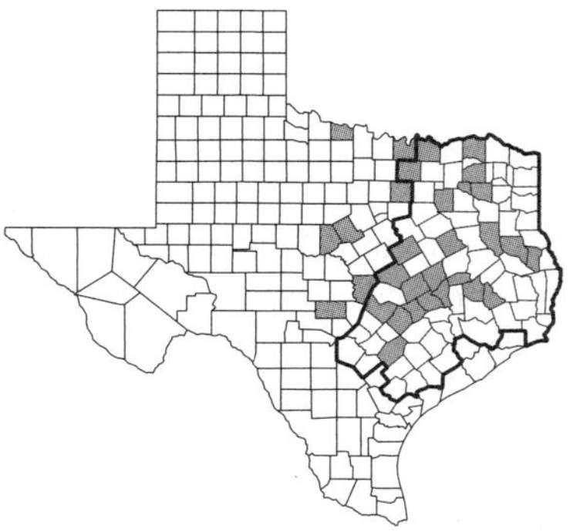

Draba brachycarpa

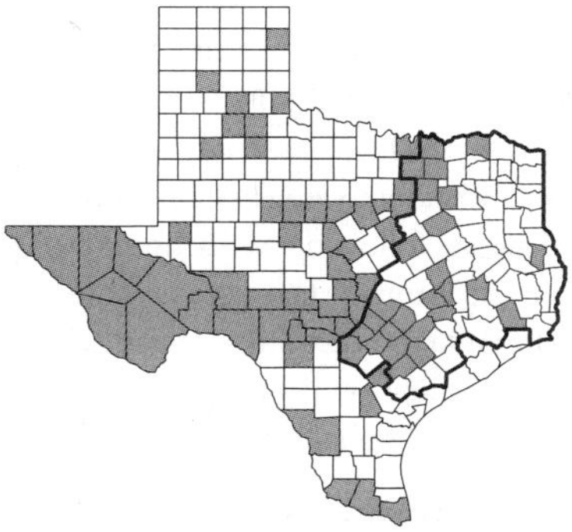

Draba cuneifolia var. cuneifolia

elongated in fruit; petals yellow to greenish yellow or orange, claw and blade differentiated, claw often equaling or exceeding sepals in length; stamens tetradynamous; siliques dehiscent, pubescent, narrowly linear, 4-angled or flattened parallel to the replum, valves each prominently midveined; style stout or slender, to 4 mm long; ovules 40–120 per ovary, seeds unwinged or partially winged, in 1 or 2 rows per locule.

A genus of 150–275 species, widespread in the n hemisphere, mainly in Asia and Europe, with 17 species native to North America, and a few native to n Africa, Macaronesia, Baja California, and Central America; some are cultivated for their showy flowers; others are cosmopolitan weeds. *Erysimum* is a difficult genus in need of comprehensive study (Al-Shehbaz 2010g); definitive identification requires the presence of basal leaves, flowers, mature fruits, and seeds—a combination nearly impossible to find on a single plant or specimen. Species native to North America may hybridize, further complicating species boundaries. (Greek: *eryso*, to ward off or cure, from supposed medicinal properties of some species) (tribe Erysimeae)
REFERENCES: Rossbach 1958; Turner 2006b; Al-Shehbaz 2010g; Koch et al. 2020.

1. Petals yellow to greenish yellow (sometimes basally splotched with white), 6–8 mm long; fruiting pedicels 2–4(–6) mm long, stout (as wide as fruit) ____ **E. repandum**
1. Petals yellow to orange, 12–25(–30) mm long; fruiting pedicels 4–17(–25) mm long, stout or slender, but generally narrower than fruit.
   2. Basal leaves to 1.5(–2.4) mm wide; petals yellow (rarely orange-yellow); mature fruits 4-angled, prominently 4-striped, trichomes between midvein and replum 2-rayed; seeds usually unwinged ____ **E. asperum**
   2. Basal leaves 3–15(–30) mm wide; petals orange to golden yellow; mature fruits flattened parallel to the replum, rarely 4-angled, not striped longitudinally, trichomes between midvein and replum 2–5-rayed; seeds usually at least partially winged ____ **E. capitatum**

Erysimum asperum (Nutt.) DC., (rough), WESTERN WALLFLOWER, PRAIRIE ROCKET. Plants to 65(–80) cm tall; rough-textured with trichomes of leaves 2- or 3-rayed; petals yellow, obovate to suborbicular, 13–22 mm long; fruiting pedicels slender (narrower than fruit), horizontal to diverging; fruits (3–)5–12(–14) cm long, 1.2–2.7 mm wide, 4-angled and longitudinally 4-striped, trichomes between midvein and replum 2-rayed. Prairies, sandy hillsides, limestone bluffs; in Bell, Bexar, Brazos, Ellis, Grimes, Limestone, and Washington cos.; mainly Panhandle and a few other scattered cos.; Great Basin and w Plains states; Canada. Apr–May. [*Cheiranthus asper* Nutt., *Cheirinia aspera* (Nutt.) Rydb., *E. asperum* (Nutt.) DC. var. *asperum*] Turner (2006) felt the *Erysimum asperum-capitatum* complex was best treated as a single species, *E. asperum*—following numerous previous authors—under which he treated this taxon as *Erysimum asperum* subsp. *asperum*; we are following Al-Shehbaz (2010g) and Koch et al. (2020) in treating *E. asperum* as a species distinct from *E. capitatum*. Given the rampant taxonomic confusion over time, herbarium annotations from any given institution present a discordant picture of taxon distribution, and the few available printed range maps are contradictory; with these caveats, we are basing our range maps on Kartesz (2015).

Erysimum capitatum (Douglas ex Hook.) Greene var. **capitatum,** (headed), SAND-DUNE WALLFLOWER, WESTERN WALLFLOWER, PRAIRIE ROCKET. Plants to 100(–120) cm tall; trichomes of leaves 3(–7)-rayed; petals orange to golden yellow, obovate to suborbicular, 12–25(–30) mm long; fruiting pedicels stout or slender but narrower than fruit, ascending to diverging; fruits 3.5–11(–15) cm long, 1.3–3.3 mm wide, flattened parallel to the replum or rarely 4-angled, not striped longitudinally, trichomes between midvein and replum 2–5-rayed. Rocky outcrops, roadcuts, dry streambeds, open wooded slopes; reported in East TX by Kartesz (2015) in Bell, Brazos, Ellis, Grimes, and Limestone cos., also in Washington Co. (*E. Brockett 105* TEX!), and Bexar Co. (*E.D. Schulz s.n.* TEX!); mostly in the Trans-Pecos; s Plains, Great Basin, and w coast, also scattered across the midwest (perhaps as sporadic introductions); Mexico. Apr–Jun. [*Cheiranthus capitatus* Douglas ex Hook., *C. elatus* (Nutt.) Greene, *E. arkansanum* Nutt., *E. asperum* subsp. *capitatum* (Douglas ex Hook.) B.L. Turner, *E. asperum* var. *elatum* (Nutt.) Torrey, *E. capitatum* var. *angustatum* (Greene) Rossbach, *E. elatum* Nutt.] Turner (2006) felt the *Erysimum asperum-capitatum* complex was best treated as a single species, *E. asperum*—following numerous previous authors—under which he treated this taxon as *Erysimum asperum* subsp. *capitatum* (Douglas) B.L. Turner, but noting the correct name may be *E. asperum* var. *elatum* (Nutt. ex. Torr. & A. Gray) Torr.; we are following Al-Shehbaz (2010g) and Koch et al. (2020) in treating *E. capitatum* as a species distinct from *E. asperum*. Given the rampant taxonomic confusion over time, herbarium annotations from any given institution present a discordant picture of taxon distribution, and the few available printed range maps are contradictory; with these caveats, we are basing our range maps on Kartesz (2015), who does not distinguish varieties within *E. capitatum*. According to habitats and ranges described by Al-Shehbaz (2010g), plants in East TX are probably *E. capitatum* var. *capitatum*.

Erysimum repandum L., (with wavy margins), SPREADING WALLFLOWER, BUSHY WALLFLOWER. Plants to 45(–70) cm tall; trichomes of leaves 2-rayed, mixed with fewer 3-rayed ones; petals yellow to greenish yellow (sometimes basally splotched with white), narrowly oblanceolate to spatulate, 6–8 mm long; fruiting pedicels stout, as wide as fruit, diverging; fruits (2–)3–8(–10) cm long, 1.5–2 mm wide, 4-angled, not striped longitudinally, trichomes between midvein and replum 2-rayed, mixed with fewer 3-rayed ones. Disturbed sites, roadsides, pastures, railroads; Collin, Dallas, Denton, Fannin, and Grayson cos., also Harris Co. (Kartesz 2015) and Brazos and McLennan cos.; mainly nc TX and Panhandle, scattered s to Edwards Plateau and se TX; scattered across most of the U.S.; Canada; Eurasia; n Africa; introduced elsewhere. Apr–May. [*Cheirinia repanda* (L.) Link, *E. ramosissimum* Crantz, *E. rigidum* DC.] Native to Europe, now a widespread weed.

IODANTHUS (Torr. & A. Gray) Steud. PURPLE ROCKET

A monotypic genus of the e and c U.S.; three endemic Mexican species previously included in *Iodanthus* have been moved to the genus *Chaunanthus* (Price & Al-Shehbaz 2001). (Greek: *iodes*, violet-colored, and *anthos*, flower) (tribe Cardamineae)
REFERENCES: Rollins 1942; Al-Shehbaz 2010h.

Iodanthus pinnatifidus (Michx.) Steud., (pinnately-cut), PURPLE ROCKET, VIOLET ROCKET. Herbaceous perennials to 80(–100) cm tall; glabrous or sparsely pubescent with simple trichomes; stems erect, striate, branching in the upper portion; basal leaves not forming a rosette, withered by flowering, petiolate, denticulate or lobed or sometimes lyrate-pinnatifid; lower stem leaves short-petiolate (petiole winged) becoming sessile on upper stem, blades ovate to lanceolate, margins toothed (rarely subentire), bases cuneate or the uppermost ± auriculate; racemes lax, considerably elongated in fruit; flowers sometimes ± bilaterally symmetrical; sepals often pinkish or purplish; petals lavender or pink to nearly white, (7–)12–14 mm long, claw differentiated from blade; stamens tetradynamous; fruiting pedicels spreading to ascending, slender (but nearly as thick as fruit); siliques dehiscent, sometimes short-stipitate, linear, terete, (1.5–)2–3.5(–4) cm long, 1–1.5 mm wide, valves each prominently midveined; style (1–)2–4 mm long; ovules 22–36 per ovary; seeds unwinged, in 1 row per locule. Low rich woods, streamsides, shady thickets; uncommon, from ne to sc TX, recorded in Grimes and Victoria cos., also in Jackson Co. (*G.L. Webster 3147* BRIT), Nacogdoches Co. (*R. Kral 68611* (TROY)), Panola Co. (*K.R. Norton et al. 622* BRIT), and Wharton Co. (iNaturalist.org observation *21929156, Suzanne Simpson, Apr 2019* (iNaturalist.org 2021)); also Brazoria and Fort Bend cos.; e and c U.S., from NJ s to AL and w to MN, OK, and TX. Apr–Jun. [*Arabis hesperidoides* (Torr. & A. Gray) A. Gray, *Cheiranthus hesperidoides* Torr. & A. Gray, *Hesperis pinnatifida* Michx., *I. hesperidoides* (Torr. & A. Gray) A. Gray, *Thelypodium pinnatifidum* (Michx.) S. Watson]

LEAVENWORTHIA Torr. GLADECRESS

A genus of 8 species, native to the se U.S.; typically restricted to thin limestone soils, glades, and outcrops; most are narrow endemics. (Named for Melines Conklin Leavenworth, 1796–1862, se U.S. botanist, explorer, and army surgeon) (tribe Cardamineae)
REFERENCES: Mahler 1987; Poole et al. 2007; Al-Shehbaz & Beck 2010.

Leavenworthia texana Mahler, (of Texas), TEXAS GOLDEN GLADECRESS, TEXAS GLADECRESS. Small herbaceous annuals, to 20 cm tall; glabrous; usually scapose (stemless); basal leaves in a rosette, petiolate, blades simple or lyrate-pinnatisect with 1–2 small leaflets per side, terminal leaflet (or simple leaf blades) broadly oblong to reniform, margins slightly lobed or shallowly dentate; flowers usually solitary on pedicels emerging directly from the basal rosette (rarely in few-flowered racemes); petals bright yellow, central veins and claws darker yellow (entire perianth often aging to rose-orange before falling), narrowly obovate, 7.3–9 mm long, apex shallowly emarginate, claw differentiated from and much shorter than blade; stamens strongly tetradynamous; fruiting pedicels slender, 3–7 cm long (solitary flowers) or 2–4 cm long (in racemes); siliques dehiscent, oblong, flattened parallel to the replum, torulose, 1.5–2.3 cm long, valves obscurely veined; style 1.7–3 mm long; ovules 4–14 per ovary; seeds in 1 row per locule, with wings 0.2–0.4 mm wide. Vernally moist, thin soils on ironstone or greensand barrens and glades of the Weches Formation; endemic to East TX; natural populations known only from Sabine and San Augustine cos., experimentally introduced in Nacogdoches Co. but perhaps not persisting there (Poole et al. 2007). Feb–May. [*L. aurea* Torr. var. *texana* (Mahler) Rollins] This species was first described by Mahler (1987), then reduced to a subsp. of *L. aurea* by Rollins (1993); it is now reaffirmed as a distinct species (Al-Shehbaz & Beck 2010). TEXAS GOLDEN GLADECRESS is known from only a few sites in Sabine and San Augustine cos.; these populations are threatened by road and pipeline construction, gravel mining, grazing, and herbicides (Poole et al. 2007). Listed as a federally-protected endangered species in 2013. NatureServe: G1, critically imperiled; S1, critically imperiled.

LEPIDIUM L. PEPPERWEED, PEPPER-GRASS, PEPPERWORT

Ours herbaceous annuals and biennials (rarely perennials); glabrous or pubescent with simple trichomes; stems erect to ascending or decumbent, branching or unbranched; basal leaves petiolate, sometimes in a rosette, sometimes withered by flowering, blade margins entire to dentate or lyrate to pinnatifid/pinnatisect; stem leaves petiolate or sessile, blades and margins diverse, bases sometimes sagittate or auriculate; racemes elongated in fruit; sepals usually falling early; petals white, small (sometimes rudimentary or absent), claw often indistinct; stamens 2 and equal in length, or 6 and tetradynamous; fruiting pedicels slender or stout, ascending to horizontal; fruits typically silicles and dehiscent (or rarely, didymous schizocarps), broadly oblong to ovate or orbicular to obovate, compressed perpendicular to the replum or somewhat inflated, sometimes keeled, usually apically winged, valves usually unveined (rarely prominently veined); style obsolete or to 0.7 mm long; ovules 2(–4) per ovary; seeds unwinged or winged.

A large and cosmopolitan genus of ca. 220 species, mainly of the temperate and subtropical zones, native to every continent (except Antarctica). Several are invasive weeds; some are rare endemics; a few are cultivated; many have been used in folk medicines and harvested for their edible, peppery sprouts, leaves, tuberous roots, fruits, and seeds (Tull 2013). Some species were previously placed in the now-subsumed genera *Cardaria*, *Coronopus*, and *Stroganowia* (Al-Shehbaz et al. 2002). *Lepidium* is not a difficult genus, according to Al-Shehbaz and Gaskin (2010), but prior North American treatments overemphasized pubescence (despite the variability of that character), resulting in a plethora of unwarranted infraspecific divisions. Species determination requires complete specimens, with basal leaves, flowers, and mature fruit. *Lepidium* seeds produce copious mucilage upon wetting, facilitating long-distance transport by migratory birds (Mummenhoff et al. 2001) and human activities, likely contributing to their invasive success. Beyond the species treated and noted here, there are nearly a dozen exotic species that may eventually appear in our region, either as unproductive waifs or as naturalizing adventives; none of these has been verified in or near East TX at the time of this writing, so we merely list them for future consideration: *L. africanum* (Burm. f.). DC., *L. appelianum* Al-Shehbaz, *L. bonariense* L., *L. chalepense* L., *L. coronopus* (L.) Al-Shehbaz, *L. draba* L., *L. graminifolium* L., *L. heterophyllum* Benth., *L. latifolium* L., *L. perfoliatum* L., and *L. sativum* L. (Greek: *lepidion*, little scale, alluding to the shape of the fruit) (tribe Lepidieae)
REFERENCES: Hitchcock 1936; Al-Shehbaz 1986; Mummenhoff et al. 2001; Al-Shehbaz et al. 2002; Al-Shehbaz & Gaskin 2010.

1. Stem leaves (at least uppermost) sessile, blade bases auriculate to sagittate.
  2. Petals present, >1 mm long; stamens 6; fruits 4–6.5 mm long; style distinct (0.2 mm or longer) ____ **L. campestre**
  2. Petals absent, or rudimentary and <1 mm long; stamens 2; fruits 2.2–3.5 mm long; style obsolete or to 0.1 mm long ____ **L. oblongum** (in part)
1. Stem leaves usually petiolate, or if sessile, blade bases attenuate or cuneate (not auriculate or sagittate).
  2. Fruits schizocarpic, didymous (with 2 distinct rounded lobes), 1.3–1.7 mm long, unwinged, apically notched; valves thick, rough-textured, prominently veined ____ **L. didymum**
  2. Fruits dehiscent silicles, broadly oblong to ovate or orbicular to obovate, ± compressed perpendicular to the replum, 1.5–4 mm long, apically winged and notched; valves thin, smooth, not prominently veined.
    3. Stems often several from the base, additionally branching, decumbent to ascending or erect; rachis of inflorescence hirsute; upper stem leaves moderately to deeply lobed or pinnatifid ____ **L. oblongum** (in part)
    3. Stems simple from the base, branching in the upper portion, erect to ascending; rachis of inflorescence pubescent to puberulent or glabrous; upper stem leaves entire or shallowly dentate-serrate (rarely deeply pinnatifid).
      4. Fresh plants fetid; basal leaf blades usually 2- to 3-pinnatisect (sometimes 1-pinnatisect); stem leaves sessile; fruits elliptic ____ **L. ruderale**
      4. Fresh plants not fetid; basal leaf blades dentate to serrate or lyrate to pinnatifid; stem leaves short-petiolate; fruits elliptic-obovate to obovate-orbicular.
        5. Plants often densely hirsute; basal leaves pinnatifid; pubescence of inflorescence rachis of mixed curved and straight trichomes; style 0.05–0.1 mm long ____ **L. austrinum**
        5. Plants puberulent or glabrous; basal leaves dentate to serrate or lyrate to pinnatifid; pubescence of inflorescence rachis either straight or curved trichomes OR rachis glabrous; style 0.1–0.2 mm long.
          6. Pubescence of inflorescence rachis of straight trichomes; petals absent or rudimentary and 0.3–0.9 mm long; fruits obovate to obovate-suborbicular, widest above the middle ____ **L. densiflorum**
          6. Pubescence of inflorescence rachis of curved trichomes OR rachis glabrous; petals usually present, 1–2(–2.5) mm long; fruits orbicular, widest at the middle ____ **L. virginicum**

Lepidium austrinum Small (southern), SOUTHERN PEPPERWORT, SOUTHERN PEPPERWEED. Plants annual or biennial; often densely hirsute; stems erect, to 67(–94) cm, often simple from base, branching in the upper portion; basal leaves in a rosette (later withering), blade margins pinnatifid; stem leaves short-petiolate, blades oblanceolate to nearly linear, margins entire or dentate, bases attenuate to cuneate (not auriculate); petals white (sometimes absent), oblanceolate, 0.4–1.6 mm long; stamens 2; fruits elliptic-obovate to orbicular, 2.4–3.2 mm long, 1.8–2.5 mm wide, apically winged, valves unveined; style 0.05–0.1 mm long, included in apical notch. Disturbed areas, roadsides, fields; Blackland Prairie w to Rolling Plains and s to s TX; MS, LA, KS, OK, and NM, with adventive records in OH and SC (Kartesz 2015); Mexico. Late Mar–May. [*L. austrinum* var. *orbiculare* Thell., *L. lasiocarpum* Nutt. var. *orbiculare* (Thell.) C.L. Hitchc.]

Lepidium campestre (L.) W.T. Aiton, (of fields), FIELD PEPPERWEED, CREAM-ANTHER FIELD PEPPERWORT. Plants annual; densely hirsute; stems erect, to 50(–63) cm, simple from base, unbranching or branching in the upper portion; basal leaves in a rosette, blade margins entire or lyrate to pinnatifid; stem leaves sessile, blades oblong to deltate-lanceolate, margins subentire or dentate, bases

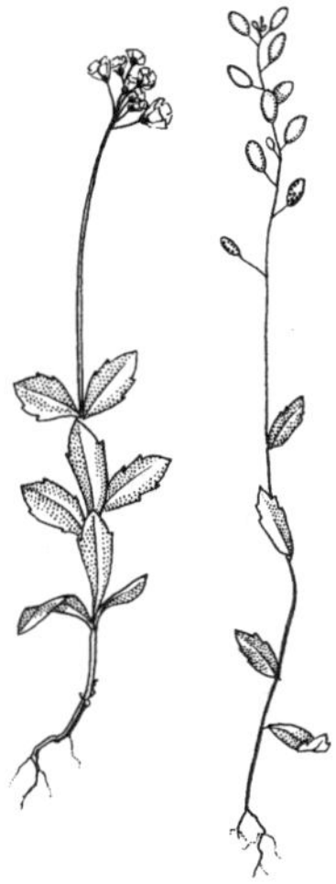
Draba platycarpa [BT3]

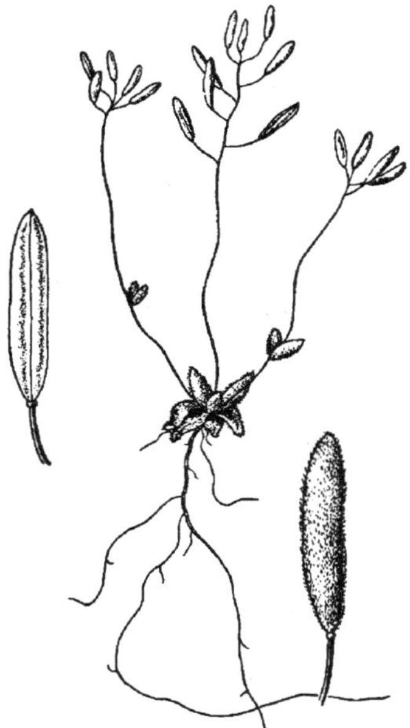
Draba reptans [GLE]

Eurca vesicaria subsp. sativa [HOL]

Erucastrum gallicum [YAT]

Erysimum asperum [BB1]

Erysimum capitatum [ABR]

Erysimum repandum [BB2]

Iodanthus pinnatifidus [BB2]

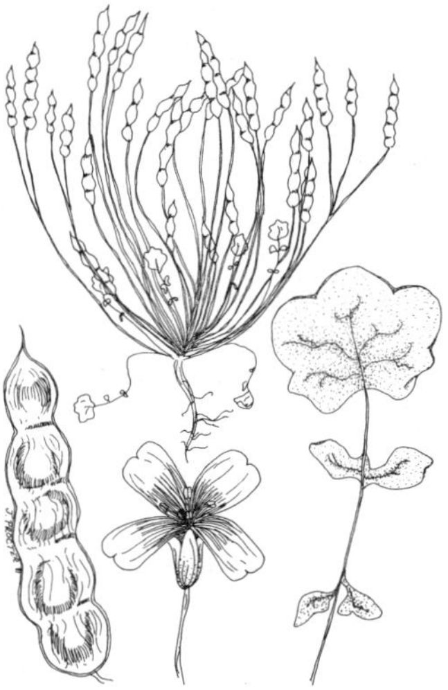
Leavenworthia texana [SID]

sagittate or auriculate; petals white, spatulate, (1.5–)1.8–2.5(–3) mm long; stamens 6; fruits oblong to obovate or ovate, (4–)5–6 (–6.5) mm long, (3–)4–5 mm wide, apically curved upward and broadly winged, valves unveined; style 0.2–0.5(–0.7) mm long, included in apical notch or slightly exserted. Disturbed areas, roadsides, piney woods; in TX reported from Montgomery (*L.E. Brown 27994, 27 Apr 2003—fide* Brown et al. 2007) and Denton (*Parks & Cory 28720*, 1938, TAES) cos.; reported in bordering OK cos. (Kartesz 2015); widespread in U.S. but less common in the Gulf states; Canada; Eurasia; introduced elsewhere. Apr–May. [*Neolepia campestris* (L.) W.A. Weber, *Thlaspi campestre* L.] Native to Eurasia; now a weed on every continent.

Lepidium densiflorum Shrad., (densely-flowered), COMMON PEPPERWEED, PRAIRIE PEPPERWEED, GREEN-FLOWER PEPPER-GRASS, MINER'S PEPPERWORT. Plants annual or biennial; puberulent or glabrous; stems erect, to 50(–65) cm, simple from base, branching in the upper portion; basal leaves in a rosette (withering early), blade margins coarsely serrate or pinnatifid; stem leaves short-petiolate, blades narrowly oblanceolate or linear, margins usually entire or serrate to dentate (or rarely pinnatifid), bases attenuate to cuneate (not auriculate); petals absent or, if rudimentary, then white, filiform, 0.3–0.9 mm long; stamens 2; fruits obovate to suborbicular, (2–)2.5–3(–3.5) mm long, 1.5–2.5(–3) mm wide, apically winged, valves unveined; style 0.1–0.2 mm long, included in apical notch. Disturbed areas, roadsides, fields, often on sandy soils; scattered throughout TX; widespread in U.S. but less common in the se states; apparently considered native w of the Mississippi River, and adventive e of it (Kartesz 2015); Canada; Mexico; introduced elsewhere. Feb–Jun. [*L. densiflorum* var. *elongatum* (Rydb.) Thell., *L. densiflorum* var. *macrocarpum* G.A. Mulligan, *L. densiflorum* var. *pubicarpum* (A. Nelson) Thell., *L. densiflorum* var. *ramosum* (A. Nelson) Thell., *L. elongatum* Rydb., *L. neglectum* Thell., *L. pubicarpum* A. Nelson, *L. ramosum* A. Nelson] We follow Al-Shehbaz and Gaskin (2010) in not recognizing varieties; they report this species hybridizes with *L. virginicum*.

Lepidium didymum L., (in pairs, twins), LESSER PEPPERWORT, SWINE WARTCRESS, LESSER SWINECRESS. Plants annual; rank-scented, glabrous or pilose; stems erect to decumbent, to 50(–70) cm, few to several from base, branching in the upper portion; basal leaves not in a rosette (withering early), blades 1- or 2-pinnatisect; stem leaves short-petiolate or subsessile, blades similar to basal, smaller and less divided on upper stem, bases not auriculate; racemes mostly axillary (lateral); petals white, elliptic to linear, 0.4–0.5 mm long; stamens 2; fruits schizocarpic, didymous (2-lobed, plump), 1.3–1.7 mm long, 2–2.5 mm wide, not winged apically, valves rugose and each strongly veined; style obsolete, included in apical notch. Roadsides, urban lawns, disturbed areas, often in sandy soils; scattered mainly from nc to sc and se TX, but probably undercollected; recorded in most of the U.S. excluding the central plains states; Mexico; Central and South America; Canada; introduced elsewhere. Apr–May. [*Carara didyma* (L.) Britton, *Coronopus didymus* (L.) Sm., *Senebiera didyma* (L.) Pers., *S. incisa* Willd., *S. pinnatifida* DC.] Formerly recognized in the subsumed genus *Coronopus*. Native to South America; now a weed on every continent.

Lepidium oblongum Small, (oblong), VEINY PEPPERWEED, VEINY PEPPERWORT. Plants annual; hirsute; stems erect to decumbent, to 24(–32) cm, often several from base, branching in the upper portion; basal leaves not in a rosette, blades 1- or 2-pinnatifid; stem leaves usually sessile or rarely short-petiolate, blades obovate to oblanceolate, margins dentate to laciniate or pinnatifid, bases cuneate or at least the uppermost usually auriculate; petals absent or if rudimentary then white, linear-oblanceolate, 0.7–1 mm long; stamens 2; fruits orbicular to obovate or elliptic, 2.2–3.5 mm long, 2–3 mm wide, apically winged, valves unveined; style to 0.1 mm long, included in apical notch. Prairies, floodplains, disturbed areas; Denton, McLennan, and Travis cos.; mainly Rolling Plains and Edwards Plateau w to w TX; sc and sw U.S. and w coast; Mexico and Central America. Mar–Apr. [*L. greenei* Thell., *L. oblongum* var. *insulare* C.L. Hitchc.]

Lepidium ruderale L., (of rubbish or waste ground), ROADSIDE PEPPERWEED, STINKING PEPPERWEED, NARROWLEAF PEPPERWORT. Plants annual or biennial; rank-scented, puberulent; stems erect to ascending, to 35(–55) cm long, simple from base, branching in the upper portion; basal leaves in a rosette, blades usually 2- to 3-pinnatisect (sometimes 1-pinnatisect); stem leaves sessile, blades linear, margins entire or with a few remote teeth, bases cuneate (not auriculate); petals absent or if rudimentary then white, linear, 0.2–0.5 mm long; stamens 2; fruits elliptic, (1.5–)1.8–2.5(–3) mm long, 1.5–2(–2.3) mm wide, apically winged, valves unveined; style to 0.1 mm long, included in apical notch. Roadsides, disturbed areas, pastures; Anderson, Bastrop, Brazos, Burleson, Dallas, and Goliad cos.; also Chambers Co.; very scattered in the U.S., mainly New England and Midwest; Canada; South America; Eurasia; introduced elsewhere. Apr–Jul. [*L. texanum* Buckley, *L. virginicum* L. subsp. *texanum* (Buckley) Thell.] Native to Eurasia.

Lepidium virginicum L., (of Virginia). Plants annual; puberulent; stems erect, to 55(–70) cm, simple from base, branching in the upper portion; basal leaves not in a rosette (withering early), blade margins pinnatifid to lyrate or dentate; stem leaves short-petiolate, blades oblanceolate or linear, margins serrate or entire, bases attenuate to cuneate (not auriculate); petals white (rarely rudimentary), spatulate to oblanceolate, 1–2(–2.5) mm long; stamens 2; fruits orbicular, 2.5–3.5(–4) mm in diam., apically winged, valves unveined; style 0.1–0.2 mm long, included in apical notch. Roadsides, fields, disturbed areas. Mar–May.

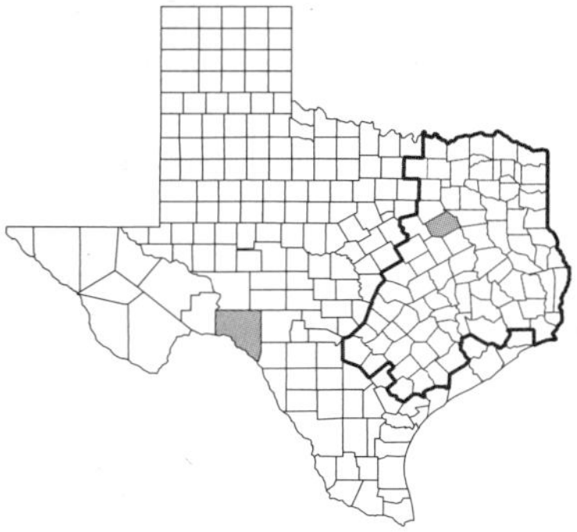

Draba cuneifolia var. integrifolia

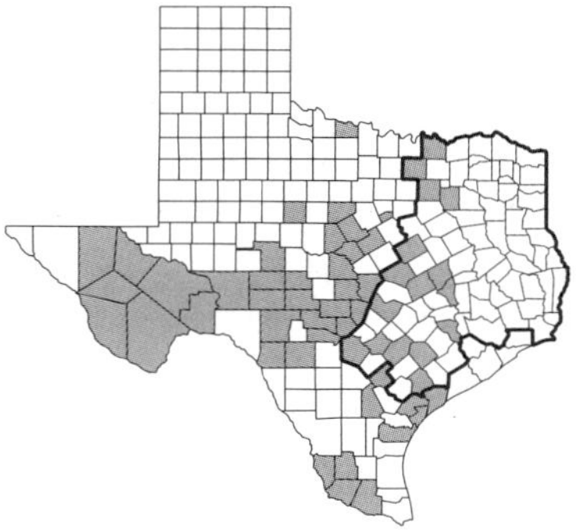

Draba platycarpa

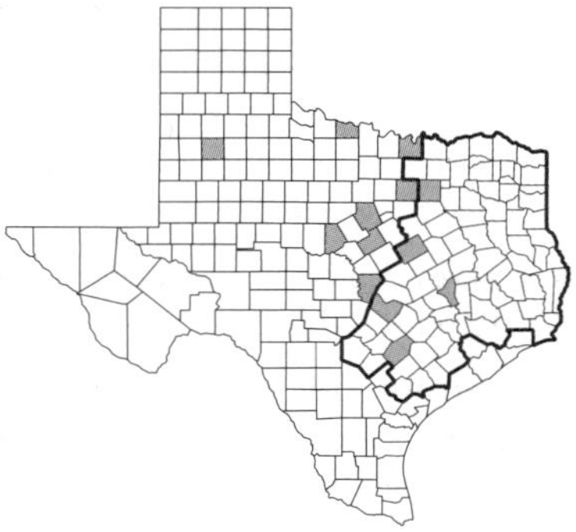

Draba reptans

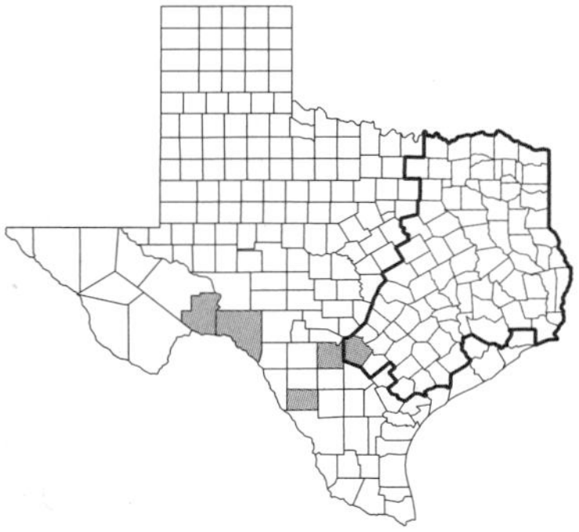

Eruca vesicaria

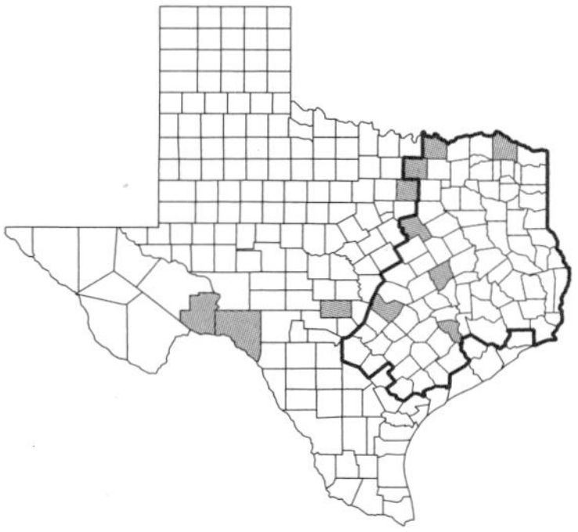

Erucastrum gallicum

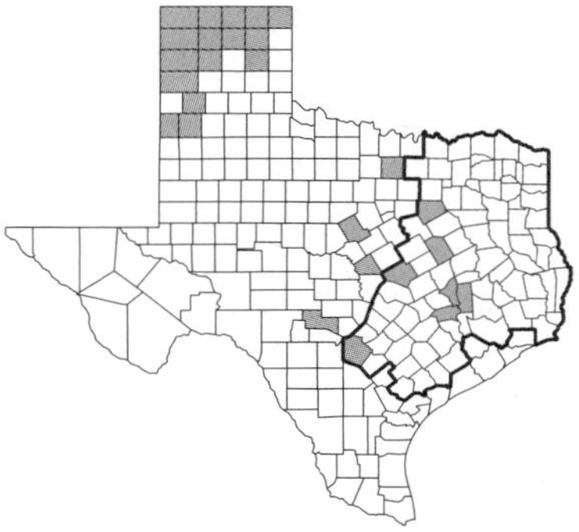

Erysimum asperum

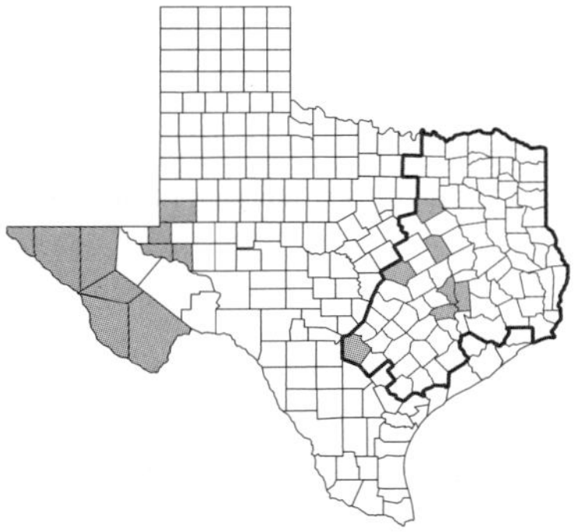

Erysimum capitatum

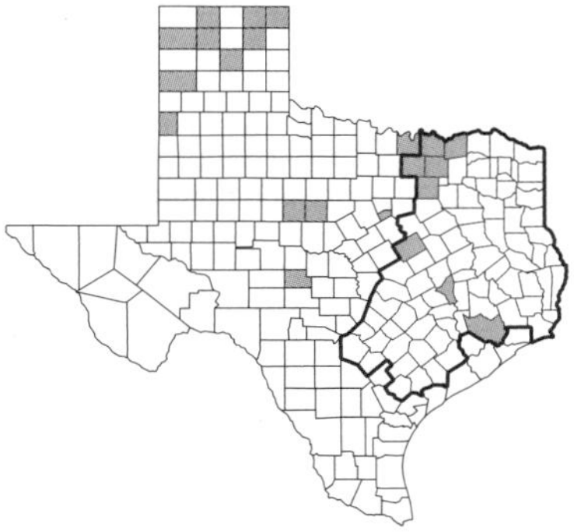

Erysimum repandum

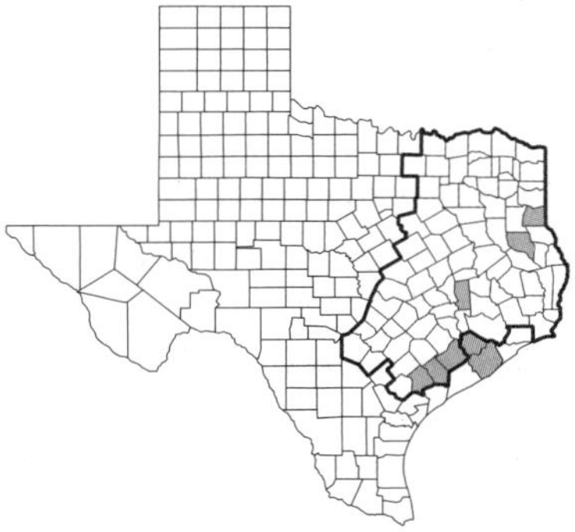

Iodanthus pinnatifidus

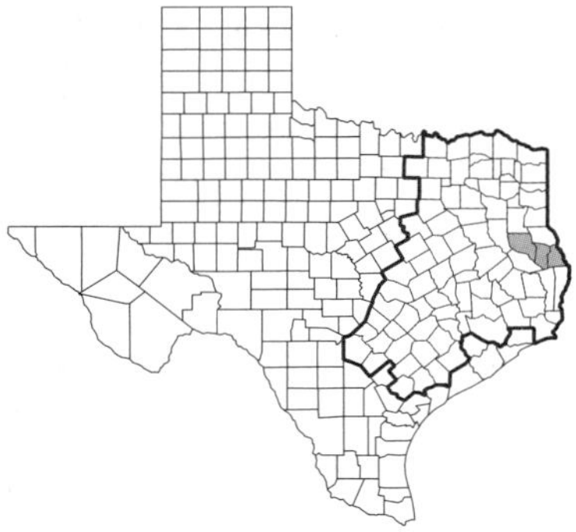

Leavenworthia texana

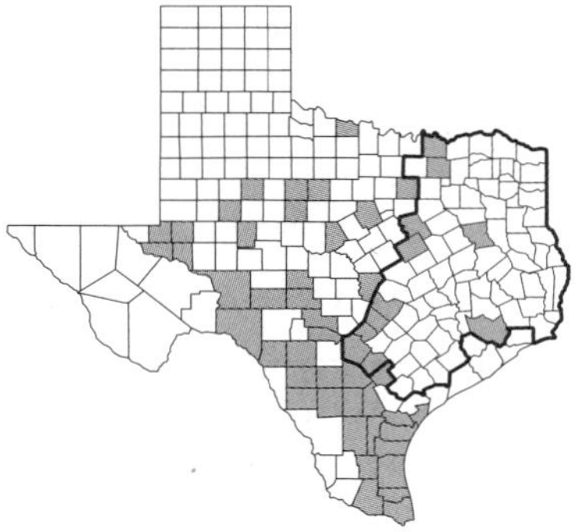

Lepidium austrinum

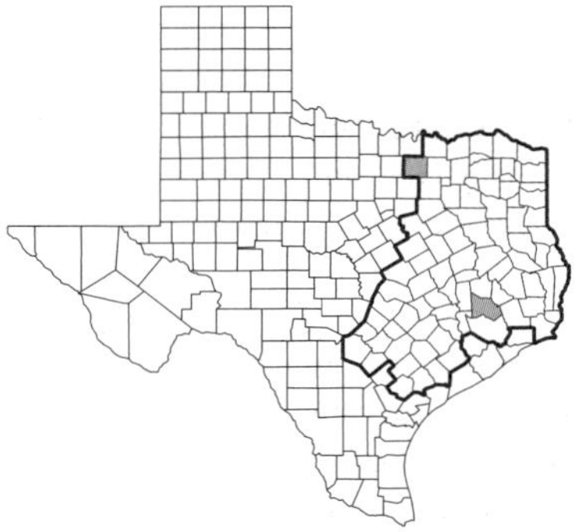

Lepidium campestre

This is a native yet weedy species that apparently hybridizes with *L. densiflorum* (Al-Shehbaz & Gaskin 2010), complicating identification in TX, where their ranges overlap. Several infraspecific taxa of *L. virginicum* recognized by previous authors (e.g., Hitchcock 1936; Rollins 1993) were unfounded, and of the two currently recognized subsp. found in TX, characteristics of the cotyledons are an unfortunate distinction; our key is based on fruiting characters only, with the caveat that Diggs et al. (1999) did not find consistent differences after examination of nc TX material. Range maps are based on Kartesz (2015).

1. Fruit valves glabrous; fruiting pedicels terete, 0.15–0.2 mm wide ______ subsp. **virginicum**
1. Fruit valves glabrous or puberulent; fruiting pedicels flattened near the apex, (0.2–)0.3–0.4 mm wide ______ subsp. **menziesii**

subsp. **menziesii** (DC.) Thell., (named for Archibald Menzies, 1754–1842, Scottish surgeon, botanist, and appointed naturalist on the global circumnavigation of HMS *Discovery*), MENZIES' PEPPERWEED, INTERMEDIATE PEPPER-GRASS, HAIRY PEPPERWORT. Fruiting pedicels flattened near the apex, (0.2–)0.3–0.4 mm wide, fruit valves glabrous or puberulent. Scattered, mainly c and sw TX; OK and w 1/2 of the U.S., Canada; Mexico. [*L. glaucum* Greene, *L. hirsutum* Rydb., *L. intermedium* var. *pubescens* Greene, *L. medium* Greene, *L. menziesii* DC., *L. virginicum* var. *californicum* Jepson, *L. virginicum* var. *medium* (Greene) C.L. Hitchc., *L. virginicum* var. *menziesii* (DC.) C.L. Hitchc., *L. virginicum* var. *pubescens* (Greene) Thell., *L. virginicum* var. *robinsonii* (Thell.) C.L. Hitchc., *Nasturtium majus* (Darracq) Kuntze]

subsp. **virginicum,** VIRGINIA PEPPERWEED, VIRGINIA PEPPER-GRASS, POORMAN'S-PEPPER, LENTEJILLA. Fruiting pedicels terete, 0.15–0.2 mm wide; fruit valves glabrous. Throughout TX; common in e 1/2 of the U.S., adventive in some w states; Canada; Mexico; an introduced weed on all other continents. [*L. danielsii* C.L. Hitchc., *L. praecox* (Raf.) DC., *L. virginicum* var. *linearifolium* Farw., *Nasturtium virginicum* (L.) Kuntze]

Lepidium alyssoides A. Gray, (resembling *Alyssum*), MESA PEPPERWORT. This species which would not have keyed satisfactorily to any other species here, can be distinguished by its subshrubby perennial habit, glabrous or puberulent, pinnately-lobed basal leaves, linear stem leaves that are entire and basally attenuate, large white petals 2–3 mm long, and style 0.2–0.6 mm long, exserted beyond the apical notch. Sandstone outcrops, gravelly roadsides, sagebrush flats; not reported in East TX but included here based on a single, historic record from adjacent Wise Co. (*E.J. Palmer 30493, 6 Jun 1926* (MO)); a few Panhandle and Edwards Plateau cos.; mainly Trans-Pecos; AZ, CO, NE, NM, UT, and WY; Mexico. May–Jul. [*L. alyssoides* A. Gray var. *angustifolium* (C.L. Hitchc.) Rollins, *L. montanum* Nutt. var. *alyssoides* (A. Gray) M.E. Jones] Al-Shehbaz and Gaskin (2010) do not recognize subspecific varieties.

LOBULARIA Desv. SWEET ALYSSUM

An Old World genus of 4 species native to s Europe, sw Asia, and Mediterranean Africa. (Latin: *lobulus*, a little lobe, in reference to the fruit) (tribe Anastaticeae)
REFERENCE: Borgen 2010.

Lobularia maritima (L.) Desv., (of the coast), SWEET ALYSSUM, SEA ALYSSUM. Herbaceous perennials to 25(–40) cm tall; trichomes sessile, with 2 opposite-pointing rays; stems erect or spreading, much-branched; basal leaves absent; stem leaves sessile, linear or lanceolate-oblanceolate, margins entire, bases attenuate; racemes densely-flowered, elongated in fruit; flowers sweetly fragrant, petals white to lavender or purple, broadly obovate, 2.3–2.8(–3.1) mm long, blade abruptly contracted into claw; stamens slightly tetradynamous; fruiting pedicels slender, ascending; silicles dehiscent, ± flattened parallel to the replum, elliptic-suborbicular, (1.9–)2.3–2.7(–4.2) mm long, (1.2–)1.6–2(–2.9) mm wide, valves each midveined; style 0.4–0.6 mm long; ovules 2 per ovary (1 in each locule); seeds unwinged. Infrequently and perhaps only ephemerally escaping from cultivation in TX, weedy areas; recorded in Walker Co. (Kartesz 2015) and Harris and Travis cos.; also Coryell and El Paso cos. (Kartesz 2015); scattered across U.S., more likely to naturalize in coastal areas; Canada; Mexico; Europe; introduced elsewhere. Spring–summer. [*Alyssum maritimum* (L.) Lam., *Clypeola maritima* L., *Koniga maritima* (L.) R. Br.] Native to s Europe; a popular garden ornamental and drought-tolerant edging plant with many cultivars; now introduced in many parts of the world.

MATTHIOLA W.T. Aiton STOCK

Ours herbaceous annuals, biennials, or perennials; often gray-pubescent, hairs usually branching and sometimes glandular; stems erect to decumbent, sometimes branching; basal leaves sometimes forming a vegetative rosette, petiolate, margins entire or repand to toothed or pinnatisect; stem leaves petiolate or sessile, similar to basal, bases attenuate to cuneate; racemes elongated in fruit; sepals connivent; petals pink to purple or brownish to yellow or white, much longer than the sepals, claw and blade differentiated; stamens strongly tetradynamous; fruiting pedicels slender or stout, diverging or ascending; siliques dehiscent, terete or flattened parallel to replum, linear, pubescent, valves each prominently midveined; style obsolete or to 5 mm long, stigma prominently 2-lobed or with 2 hornlike appendages; ovules 15–60 per ovary; seeds in 1 row per locule, wing 0.1–0.5 mm wide.

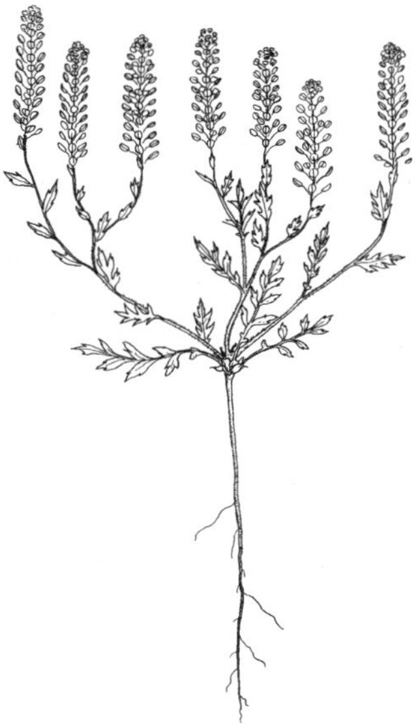

Lepidium austrinum [BT3]

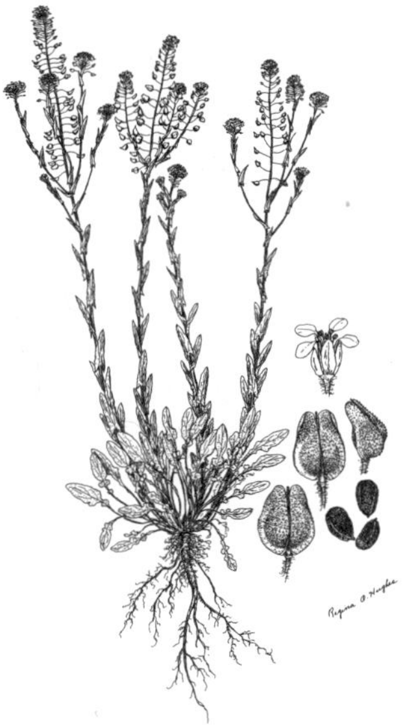

Lepidium campestre [REE]

Lipidium densiflorum [ABR]

Lepidium didymus [BB1]

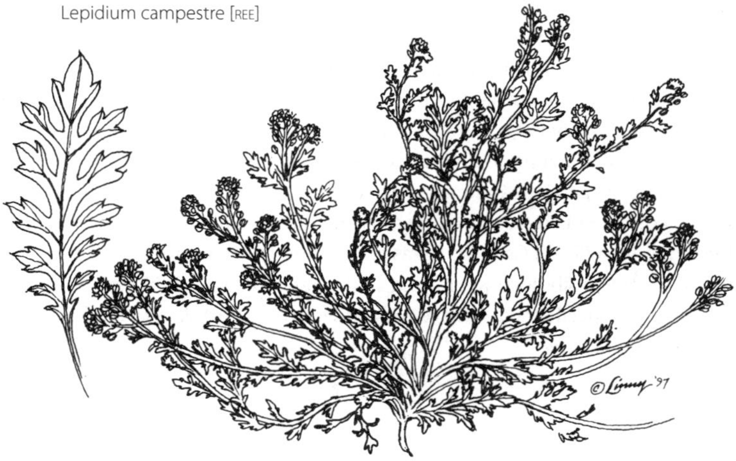

Lepidium oblongum [HEA]

Lepidium ruderale [SMI]

Lepidium virginicum [REE]

Lobularia maritima [BA1, SM1]

An Old World genus of ca. 50 species native to Eurasia and Africa; some are cultivated as ornamentals. (Named for Pietro Andrea Mattioli, 1501–1577, Italian botanist, artist, and physician) (tribe Anchonieae)
REFERENCE: Al-Shehbaz 2010i.

1. Stem leaf blade margins usually entire or repand (or rarely sinuate); petals 7–15 mm wide; fruiting pedicels 6–25 mm long; fruits 3–6 mm wide; stigma lacking hornlike appendages ____ **M. incana**
1. Stem leaf blade margins usually sinuate or dentate to pinnatifid, rarely entire; petals 2–5 mm wide; fruiting pedicels 0.5–3 mm long; fruits 1–2 mm wide; stigma with 2 hornlike appendages ____ **M. longipetala**

Matthiola incana (L.) W.T. Aiton, (very gray, hoary), HOARY STOCK, TENWEEKS STOCK, GILLYFLOWER. Plants biennial or perennial, rarely annual; usually densely tomentose; stems erect, to 60(–90) cm long; basal leaves often in vegetative rosettes; stem leaves short-petiolate or sessile, blades linear-oblanceolate to oblong or lanceolate, margins usually entire or repand (rarely sinuate); petals purple or violet to pink or white, obovate to ovate, 20–30 mm long, 7–15 mm wide; fruiting pedicels (6–)10–20(–25) mm long, narrower than fruit; siliques flattened, (4–)6–12(–15) cm long, (3–)4–6 mm wide; stigma without apical horns; seeds 2.5–3.2 mm long. Weedy roadsides, abandoned gardens, sandy coastal sites; rarely and perhaps only ephemerally escaping from cultivation in TX; recorded only in Harris Co. (Kartesz 2015); in the U.S. apparently only naturalizing in CA; South America; Europe; introduced elsewhere. Mar–Apr. [*Cheiranthus incanus* L.] Al-Shehbaz (2010i) does not address subspecific taxa. Native to s Europe; a widely-cultivated garden ornamental and fragrant cut flower; now introduced in many temperate parts of the world.

Matthiola longipetala (Vent.) DC., (long-petaled), NIGHT-SCENTED STOCK, EVENING STOCK, PERFUME PLANT. Plants annual; pubescent, sometimes glandular; stems erect or ascending to decumbent, to 50(–60) cm long; basal leaves usually not forming a vegetative rosette; stem leaves petiolate becoming sessile on upper stems, blades linear-lanceolate to oblanceolate, margins usually pinnatisect to sinuate or dentate, rarely entire; petals purple to pink or yellow to brownish or rarely white, oblong to linear-lanceolate, (15–)18–23(–27) mm long, 2–4(–5) mm wide; fruiting pedicels (0.5–)1–2(–3) mm long, nearly as thick as fruit; siliques terete, (2.5–)4–8.5(–10) cm long, 1–2 mm wide; stigma apex forked into 2 conspicuous horns that are 2–8(–12) mm long; seeds 1–2 mm long. Weedy areas, abandoned gardens, rocky waste places; occasionally escaping cultivation in TX and infrequently persisting; recorded in Bell Co.; also Edwards and Sutton cos. (Kartesz 2015), and El Paso and Kimble cos. (UTEP); sporadically naturalizing in sw U.S.; Canada; Eurasia; introduced elsewhere. Mar–Apr. [*Cheiranthus bicornis* Sibth. & Sm., *C. longipetalus* Vent., *M. bicornis* (Sibthorp & Sm.) DC., *M. longipetala* subsp. *bicornis* (Sibthorp & Sm.) P.W. Ball] Al-Shehbaz (2010i) does not address subspecific taxa. Native to Mediterranean Eurasia and n Africa; a popular garden plant now introduced in other temperate regions.

MYAGRUM L. BIRD'S-EYE CRESS

A genus of a single species, native to s Europe and the Middle East. (Greek: *muagron* or *muagros*, a name used by Dioscorides for another species of mustard—perhaps CAMELINA) (tribe Isatideae)
REFERENCES: Al-Shehbaz 2010k; USDA APHIS 2017.

Myagrum perfoliatum L., (perfoliate, with leaves surrounding the stem), BIRD'S-EYE CRESS, MUSKWEED, MITRE CRESS. Herbaceous annuals to 50(–60) cm tall, glabrous, sometimes glaucous; stems erect, rigid, branching in upper portion; basal leaves forming a rosette, sometimes withered by anthesis, short-petiolate, blade margins pinnatifid or runcinate to dentate or sinuate, midveins often strikingly pale; stem leaves sessile, blades gradually reduced in size up the stem, oblong to lanceolate, margins entire or denticulate, bases clasping and auriculate or sagittate; racemes considerably elongated in fruit; petals light yellow, oblong to oblanceolate, 3–5 mm long, claw undifferentiated from blade; stamens tetradynamous; fruiting pedicels stout, widest at the apex, hollow, appressed to rachis; silicles indehiscent, obpyriform to clavate-obcordiform (club-shaped, much wider above the middle), slightly flattened perpendicular to the replum, woody at maturity, 4.5–7.5(–8.5) mm long, 2.5–4.5(–5) mm wide, unsegmented; basal locule 1(–2)-seeded (may appear stipe-like), upper 2 locules seedless, hollow; style flattened, triangular, (0.8–)1.2–2 mm long; ovules 2 per ovary; seeds unwinged. Pastures and roadsides, fields, often in black clay; naturalized and apparently increasing across ne and nc TX (e.g., numerous, recent, research-grade observations at iNaturalist.org), also in Walker Co. (Brown et al. 2007) and Goliad Co. (Lehman 2009); thought to have been introduced into TX in Delta Co. with vetch seed from Oregon in 1949 (Cory 1950); possibly only present as a waif in OK, KS, and VA (Al-Shehbaz 2010k; USDA APHIS 2017); Canada; s Europe and the Middle East; introduced elsewhere. Apr–May. A native of Europe and the Middle East; an invasive weed that reproduces copiously on croplands and other favorable sites.

NASTURTIUM W.T. Aiton WATERCRESS, YELLOWCRESS

A genus of 5 species of North America, Europe, Asia, and n Africa; previously treated under the genera *Rorippa* and *Cardamine*; 3 are narrow endemics and 2 are widespread weeds; all are restricted to freshwater aquatic or wetland habitats. (Greek: *nasus*, nose, and *tortus*, twisted, alluding to the plants' pungent qualities) (tribe Cardamineae)
REFERENCES: Green 1962; Al-Shehbaz & Price 1998; Al-Shehbaz 2010l.

Nasturtium officinale W.T. Aiton, (medicinal), WATERCRESS, WHITE WATERCRESS, TWO-ROWED WATERCRESS. Aquatic or semiaquatic, herbaceous, rhizomatous perennials, resembling *Cardamine*; glabrous or sparsely pubescent with simple trichomes; stems 10–100+ cm long, unbranched, erect to prostrate (or floating), rooting at nodes in mud or water; basal leaves of young plants in a loose rosette, petiolate, simple, entire or few-lobed; stem leaves petiolate, petiole bases auriculate, blades simple on deeply-submerged stems, or on emerged stems pinnately compound with 1–4(–6) leaflets per side, the terminal leaflet largest and ovate to rhombic-orbicular or lanceolate; racemes elongated in fruit; petals white to pinkish, spatulate or obovate, 2.8–4.5(–6) mm long, claw undifferentiated from blade; stamens tetradynamous; fruiting pedicels slender, diverging or descending; siliques dehiscent, linear to oblong, terete or ± flattened parallel to the replum, slightly torulose, (0.6–)1–1.8(–2.5) cm long, (1.8)2–2.5(3) mm wide, valves obscurely veined; style a bluntly conical beak 0.5–1(–1.5) mm long; ovules 28–60 per ovary; seeds unwinged, in 2 rows per locule. In shallow water of streams or springs, sometimes stranded in wet mud; mainly along the w border of East TX from Grayson Co. s to Bexar Co., also recorded from Polk, Walker, and Washington cos.; Edwards Plateau, Trans-Pecos, c TX, and a few cos. in the Panhandle; recorded throughout the U.S.; Canada; Mexico; Central and South America; Europe and the Mediterranean; introduced elsewhere. Mar–Jun (sporadically later). [*Cardamine nasturtium* (Moench) Kuntze, *N. aquaticum* Wahlenb., *N. siifolium* Rchb., *Rorippa nasturtium-aquaticum* (L.) Hayek, *Sisymbrium nasturtium-aquaticum* L.] Native to Europe; introduced to North America as a food plant (Les 2017) and still cultivated; raw or cooked, the young leaves of WATERCRESS are enjoyed for their peppery-sharp flavor; now widespread and weedy in many parts of the world.

NOCCAEA Moench PENNYCRESS

A controversial genus of ca. 128 species, when treated in the broad sense; native to Europe, Asia, n Africa, Patagonia, and w North America; previously included in the large and polyphyletic genus *Thlaspi*. (Named for Dominico Nocca, 1758–1841, Italian clergyman and botanist) (tribe Coluteocarpeae)
REFERENCES: Koch & Al-Shehbaz 2004; Al-Shehbaz 2010j, 2014; Ali et al. 2016; Özüdoğru et al. 2019; Neill & Kieschnick 2021.

Noccaea perfoliata (L.) Al-Shehbaz, (perfoliate), CLASPLEAF PENNYCRESS, PERFOLIATE PENNYCRESS, THOROUGHWORT PENNYCRESS. Diminutive, herbaceous annuals; glabrous and glaucous; stems to 28(–40) cm long, often purple-tinged, unbranched or branching; basal leaves in a loose rosette (some withered by fruiting), petiolate, blades elliptic to ovate, margins entire or remotely and obtusely dentate; stem leaves sessile, ovate-cordate, margins entire to repand or with a few obtuse teeth, bases cordate-clasping, sometimes auriculate; racemes several-flowered, considerably elongated in fruit; sepals green with white margins, apices often pinkish or purplish; petals white, 2–3.5(–4.7) mm long, spatulate to oblanceolate, claw obscure; stamens slightly tetradynamous; fruiting pedicels slender, spreading or horizontal; silicles dehiscent, strongly flattened perpendicular to the replum, obcordate, 3–6.5(–8) mm long, (2.5–)3–6(–7) mm wide, valves keeled, marginal wings 1–2 mm wide (widest apically), apical notch 1–1.5 mm deep; style obsolete or to 0.3 mm long; ovules 4–8 per ovary; seeds unwinged, in 1 one row per locule, seed coat smooth. Roadsides, waste ground, fields; an adventive weed reported as new to TX by Neill and Kieschnick (2021), apparently first recorded in TX in 2019 in Grayson Co. (iNaturalist.org observation *20000793*, *Sam Kieschnick, Jan 2019* (iNaturalist.org 2021)), and 2020 in Collin Co. (iNaturalist.org observation *39559646*, *S. Kieschnick, May 2020* (iNaturalist.org 2021)) and Dallas Co. (iNaturalist.org observation *39684417*, *Annika Lindqvist, Mar 2020* (iNaturalist.org 2021)); ME and MA s to NC and LA, and w to KS, OK, and TX, also IA (iDigBio.org) and WA; Canada; Europe, e Asia, n Africa; introduced elsewhere. Jan–Apr. [*Microthlaspi perfoliatum* (L.) F.K. Mey., *Thlaspi perfoliatum* L.] CLASPLEAF PENNYCRESS was treated as *Microthlaspi perfoliatum* in Al-Shehbaz (2010j) and Ali et al. (2016); the recent generic synopsis of *Noccaea* (Al-Shehbaz 2014) subsumed *Microthlaspi*; phylogenetic studies (e.g., Özüdoğru et al. 2019) continue to seek resolution on the correct generic placement of this species. Native to Europe, e Asia, n Africa.

PAYSONIA O'Kane & Al-Shehbaz EARED BLADDERPOD, TWINPOD, LESQUERELLA

Ours herbaceous, low-growing annuals; often densely pubescent with mixed simple and stalked-branching trichomes, or sometimes glabrous; stems erect to decumbent, often branched basally; basal leaves forming a rosette, petiolate, margins entire or sinuate-dentate to lyrate or 1–2-pinnatifid; stem leaves sessile, blade margins entire or sinuate-dentate to incised or lobed, bases auriculate (at least on uppermost stem); racemes slightly to strongly elongated in fruit; sepals falling early, pubescent; petals yellow or light yellow, broadly obovate, claw and blade slightly differentiated; stamens tetradynamous; fruiting pedicels slender, pubescent; silicles dehiscent, globose to subglobose or sometimes slightly flattened perpendicular to the replum, valves unveined, glabrous or pubescent; style slender, 1–2 mm long; ovules 8–40 per ovary; seeds narrowly winged, ± in 2 rows per locule.

A genus of 8 species native to the se and sc U.S. and n Mexico; some are narrow endemics of conservation concern. These species, along with those now belonging to *Physaria*, were formerly included in the large and polyphyletic genus *Lesquerella*; species segregated to *Paysonia* have sessile upper stem leaves with auriculate bases, and seeds with narrow wings. (For Edwin B. Payson, 1893–1927, American botanist who first monographed *Lesquerella*) (tribe Physarieae)

REFERENCES: Payson 1921; Rollins 1955; Rollins & Shaw 1973; O'Kane & Al-Shehbaz 2002; O'Kane 2010a; Borja 2013.

1. Petals light yellow (often drying purplish), 5–9 mm long; anther filaments not dilated at the base; fruiting pedicels recurved; fruit valves densely pubescent ____ **P. lasiocarpa**
1. Petals yellow, 7–12 mm long; anther filaments dilated at the base; fruiting pedicels ascending to diverging but not recurved; fruit valves glabrous.
   2. Stems 5–20 cm tall; lower stems usually hirsute, with longest hairs unbranched; petals 4–5 mm wide; fruiting racemes dense, not much elongated ____ **P. auriculata**
   2. Stems 20–70 cm tall; lower stems usually pubescent, with longest hairs branching (5-rayed); petals 6–9 mm wide; fruiting racemes lax and elongated ____ **P. grandiflora**

Paysonia auriculata (Engelm. & A. Gray) O'Kane & Al-Shehbaz, (eared, for the auriculate leaf bases), EAR-LEAF BLADDERPOD, PLAINS EARED BLADDERPOD. Plants usually hirsute, longer simple trichomes mixed with shorter branching ones; stems to 20 cm tall; basal leaf blade margins lyrate to sinuate-dentate or entire; upper stem leaf blade margins entire or dentate, bases auriculate; petals yellow, 7–10 mm long; fruiting racemes dense and not much elongated; fruiting pedicels ascending to diverging; fruits globose or subglobose, 4–6(–8) mm long, 4–6 mm wide, valves glabrous. Prairies, pastures, disturbed soils; infrequently recorded in TX; ne and ec TX in Kaufman, McLennan, Navarro, and Upshur cos,. and se-c TX in Austin Co.; KS, OK, and TX. Mar–May. [*Alyssum auriculatum* (Engelm. ex A. Gray) Kuntze, *Lesquerella auriculata* (Engelm. & A. Gray) S. Watson, *Vesicaria auriculata* Engelm. & A. Gray]

Paysonia grandiflora (Hook.) O'Kane & Al-Shehbaz, (large-flowered), BIG-FLOWER BLADDERPOD. Plants usually pubescent with 5-rayed trichomes, or sometimes glabrous; stems to 70 cm tall; basal leaf blade margins irregularly dentate to 2-pinnatifid; upper stem leaf blade margins dentate, bases often cuneate, becoming auriculate (at least the uppermost); petals yellow, 8–12 mm long; fruiting racemes loose and greatly elongated; fruiting pedicels ascending to diverging; fruits globose or subglobose, 4–6 mm long, 4–5 mm wide, valves glabrous. Loose sandy soils, roadsides, grasslands, scrubland; se TX from Harris Co. (Brown 2014) and Brazos Co., w to Travis Co.; Edwards Plateau and South TX Plains; endemic to TX or possibly to be found in ne Mexico (O'Kane 2010a). Mar–May. [*Alyssum grandiflorum* (Hook.) Kuntze, *Lesquerella grandiflora* (Hook.) S. Watson, *Vesicaria brevistyla* Torr. & A. Gray, *V. grandiflora* Hook. var. *pinnatifida* A. Gray] ?

Paysonia lasiocarpa (Hook. ex A. Gray) O'Kane & Al-Shehbaz subsp. **lasiocarpa**, (woolly-fruited), ROUGHPOD BLADDERPOD. Plants usually pubescent, longer simple trichomes mixed with shorter branching ones; stems to 50 cm tall; basal leaf blade margins sinuate-dentate to lobed or incised; upper stem leaf blade margins sinuate-dentate to incised, bases auriculate (at least the uppermost); petals light yellow (often drying purplish), 5–9 mm long; fruiting racemes lax and elongated; fruiting pedicels recurved; fruits subglobose or slightly flattened perpendicular to the replum, 5–9 mm long, 4–9 mm wide, valves densely pubescent. Roadsides, brushland, sandy and gravelly soils; on the sw border of East TX in Gonzales Co. and historically recorded in Travis Co. (Carr 2008); South TX Plains, also Burleson Co. (*Janne 156* TAES); n Mexico. (Jan–)Feb–Apr. [*Lesquerella lasiocarpa* (Hook. ex A. Gray) S. Watson var. *lasiocarpa*]

PHYSARIA (Nutt. ex Torr. & A. Gray) A. Gray BLADDERPOD, TWINPOD, LESQUERELLA

Ours herbaceous, taprooted annuals and biennials, or somewhat woody perennials arising from a caudex; scurfy- pubescent (often densely so), hairs stellate-branching, sessile or short stalked, the rays sometimes basally fused (stellate-peltate); stems erect or decumbent to prostrate, unbranched or branched in the upper portion; basal leaves usually forming a rosette, petiolate, blade margins entire or repand to variously toothed or pinnatifid; stem leaves sessile or petiolate, blade margins entire or sinuate to repand or toothed, bases cuneate or attenuate (not auriculate); racemes corymbose (rarely paniculate or subumbellate), sometimes elongated in fruit; petals yellow to orange or rarely white (sometimes drying purplish), cuneate to spatulate or ovate to orbicular or obdeltate, claw often short, sometimes differentiated from blade; stamens tetradynamous; fruiting pedicels usually slender; silicles dehiscent, sometimes stipitate, often ± globose (sometimes ellipsoid or obovoid to obpyriform), terete or only slightly compressed, valves glabrous (or rarely papillose, or pubescent inside), not retaining seeds after dehiscence; style slender,

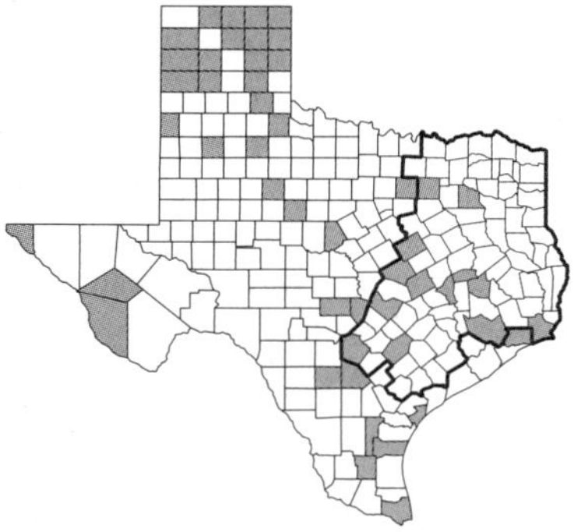

Lepidium densiflorum

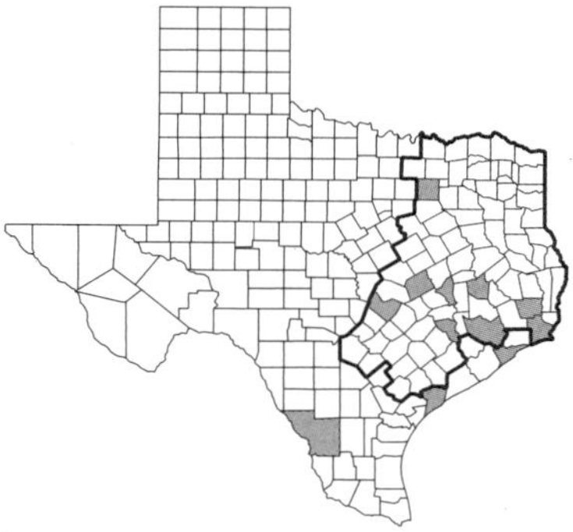

Lepidium didymum

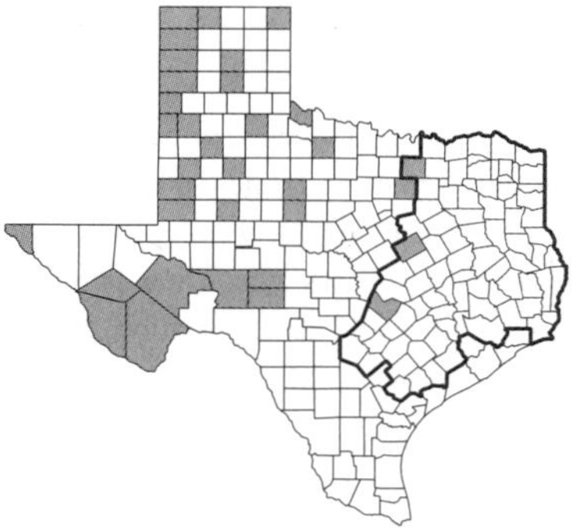

Lepidium oblongum

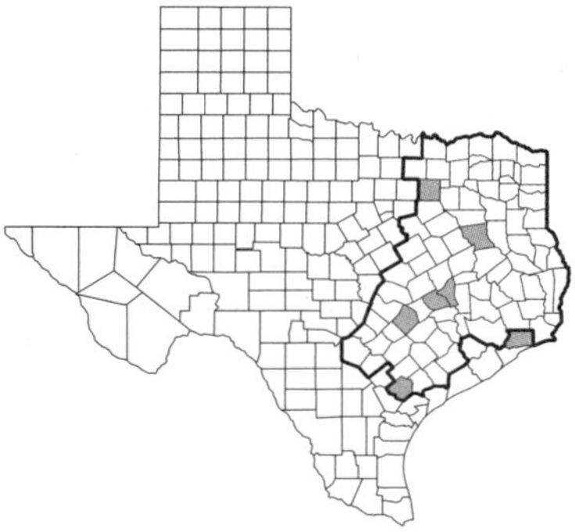

Lepidium ruderale

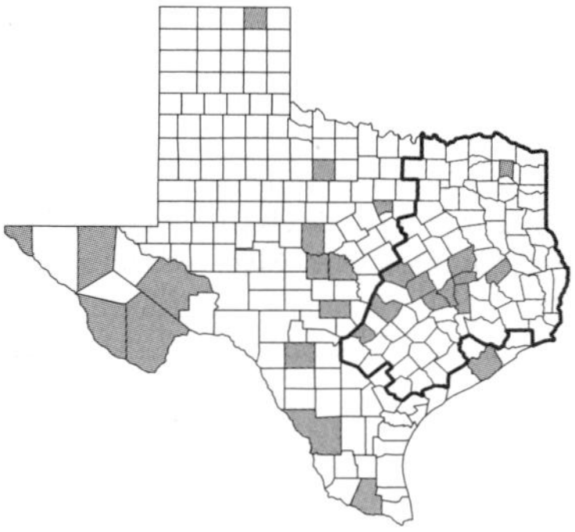

Lepidium virginicum subsp. menziesii

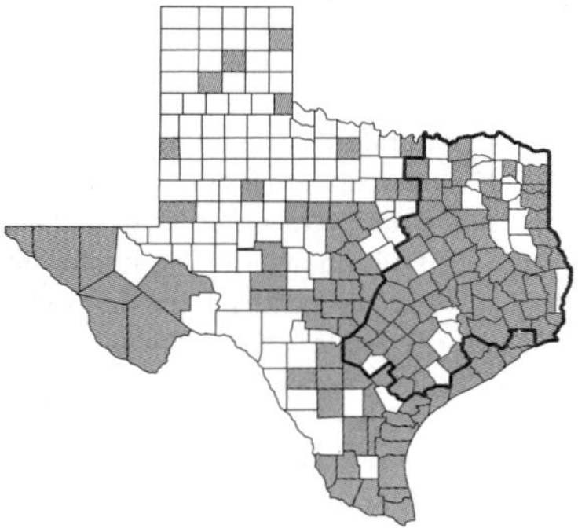

Lepidium virginicum subsp. virginicum

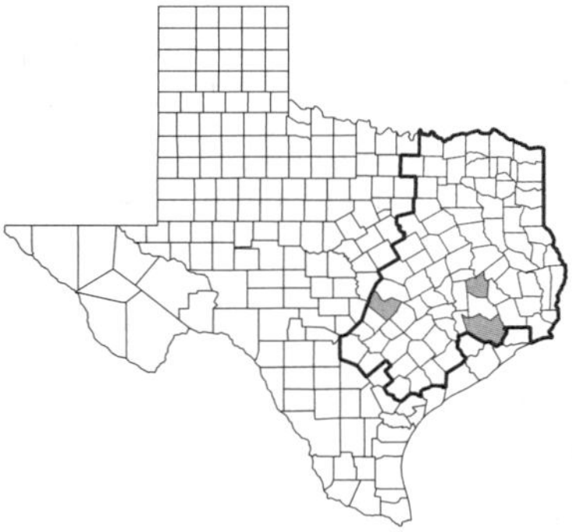

Lobularia maritima

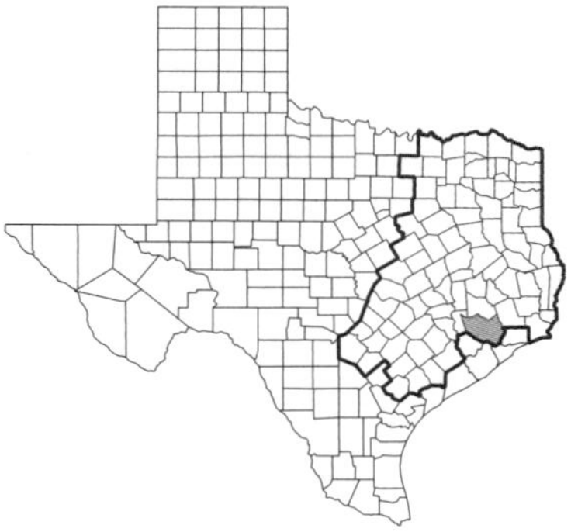

Matthiola incana

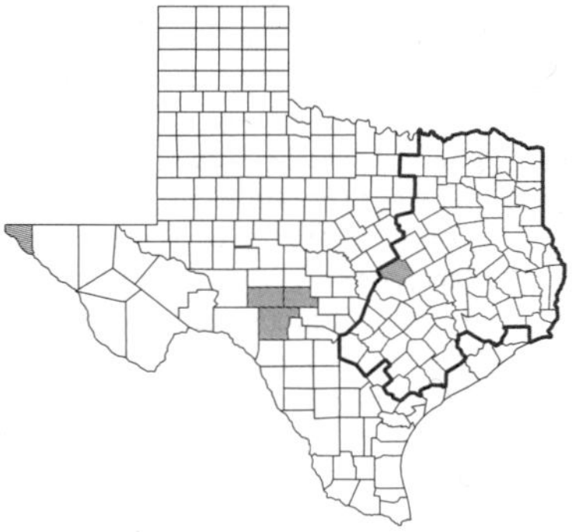

Matthiola longipetala

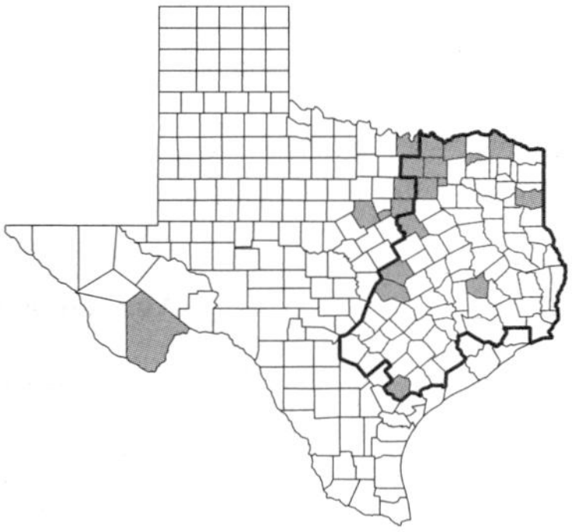

Myagrum perfoliatum

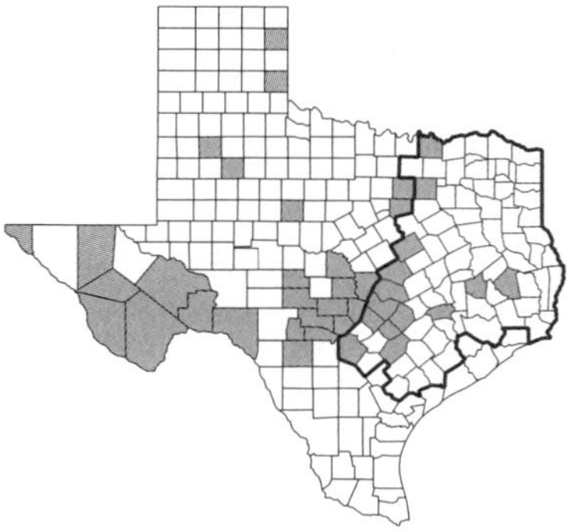

Nasturtium officinale

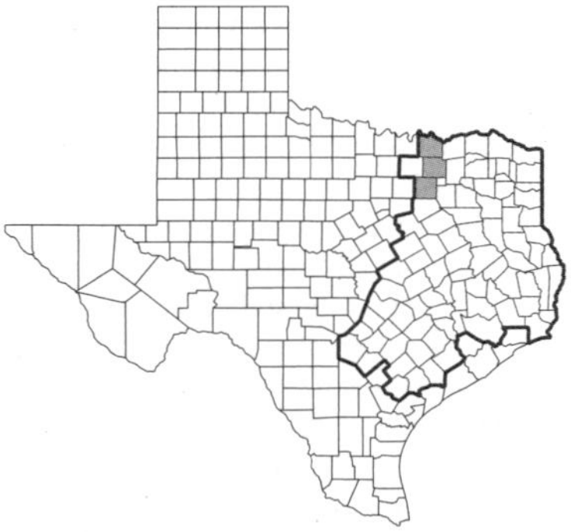

Noccaea perfoliata

usually persisting as a beak 1.5–5.5 mm long; ovules (4–)8–32 per ovary; seeds usually unwinged (sometimes narrowly margined), in 2 rows per locule.

A genus of 106 species, primarily of North America and n Mexico, with some in s South America, and one panboreal species; many are narrow endemics of conservation concern. Our species, along with those now belonging to *Paysonia*, were formerly included in the large and polyphyletic genus *Lesquerella*; species moved to *Physaria* have upper stem leaves with cuneate or attenuate blade bases (these may be sessile or petiolate), and the seeds tend to be unwinged. Root structures and mature fruits may be required for determination. (Greek: *physa*, bladder, alluding to the inflated fruits) (tribe Physarieae)

REFERENCES: Payson 1921; Rollins 1981; Rollins & Shaw 1973; Nixon et al. 1983; Al-Shehbaz & O'Kane 2002; O'Kane 2010b.

1. Plants annual or biennial (or rarely short-lived perennials), herbaceous, arising from a slender taproot.
  2. Racemes paniculate; petals white (may be veined with yellow); plants of ironstone or greensand glades in pine-oak forests of deep East TX; known only from San Augustine Co. **P. pallida**
  2. Racemes corymbose or subumbellate; petals yellow to orange (sometimes drying purplish); plants of other parts of TX (unlikely in San Augustine Co).
    3. Plants sparsely pubescent; fruiting pedicels 5–10(–15) mm long, uniformly recurved (bent downward) and the fruits therefore pendent **P recurvata**
    3. Plants often densely pubescent; fruiting pedicels 5–20(–25) mm long, straight or variously curved but usually not uniformly recurved.
      4. Fruits stipitate (connected to the pedicel by a narrow stipe), stipe ≥ 0.5 mm long.
        5. Fruiting racemes loose and elongated; fruiting pedicels straight or slightly curved **P. gracilis**
        5. Fruiting racemes dense; fruiting pedicels often sigmoid (curved into an S-shape).
          6. Fruits with a stipe 0.5–1 mm long **P. gordonii** (in part)
          6. Fruits with a stipe 1–2 mm long **P. lindheimeri**
      4. Fruits ± sessile on the pedicel.
        7. Ovules 2 in each locule; plants of ne TX; known only from Red River Co. **P. angustifolia**
        7. Ovules 4+ in each locule; plants of other parts of TX (unlikely in Red River Co.).
          8. Stellate trichome rays unbranching (use a hand lens or dissecting scope); racemes usually dense (elongating in fruit); fruit valves smooth and glabrous inside and out; plants wide-ranging from nc to se TX and w to the Edwards Plateau and sc TX **P. densiflora** (in part)
          8. Stellate trichome rays branching or forked; racemes loose; fruit valves papillose outside and pubescent inside; plants restricted to the Edwards Plateau and sc TX **P. sessilis**
1. Plants perennial (or rarely biennial), often arising from a simple or branching caudex or rootstock; lower stems and caudex sometimes woody.
  9. Stellate hairs with a deep notch between two of the rays on one side (best viewed with a dissecting scope).
    10. Racemes subumbellate, dense; petals 8–14 mm long; fruiting pedicels 10–15 mm long; fruits short-stipitate (connected to the pedicel by a narrow stipe) **P. engelmannii**
    10. Racemes corymbose, dense but elongating in fruit; petals (4.5–)7–10(–11) mm long; fruiting pedicels 7–10 mm long; fruits ± sessile on the pedicel **P. densiflora** (in part)
  9. Stellate hairs lacking a deep notch between two of the rays on one side.
    11. Stem leaves sometimes overlapping, erect and often appressed; stellate trichome rays unbranching, distinct or basally fused; racemes loose; fruits sessile on the pedicel **P. argyraea**
    11. Stem leaves not overlapping, neither erect nor appressed; stellate trichome rays branching or forked, distinct to the base; racemes dense; fruits short-stipitate (connected to the pedicel by a narrow stipe) **P. gordonii** (in part)

An alternate key by Monique D. Reed based on Rollins and Shaw (1973) and O'Kane (2010) is provided below.

1. Plants perennial from a simple or branched caudex (*P. gordonii* and *P. argyraea*, short-lived perennials sometimes without a caudex, will key under both halves of this couplet, as will *P. densiflora* which is annual or biennial but which may have a caudex).
  2. Stellate hairs with a deep, U-shaped notch between two of the rays on one side, this notch extending nearer the center of the hair than the other, narrower, V-shaped notches and the U-shaped notch often occupying the center of the hair. (Use high magnification).
    3. Racemes subumbellate and usually dense even in fruit (occasionally elongating); petals 8–14 mm long; fruits short-stipitate; stellate hairs with a central umbo (which may partly obscure the notch) **P. engelmannii**
    3. Racemes corymbose, dense but elongating in fruit; petals (4.5–)7–10(–11) mm long; fruits ± sessile; stellate hairs without an umbo **P. densiflora**
  2. Plants without the above combination of characters; hairs without a deep U-shaped notch on one side.
    4. Rays of stellate hairs usually simple, not forked, free to the base or only slightly fused; fruits sessile; mature fruiting pedicels generally all or nearly all sigmoid **P. argyraea** (in part)
    4. Rays of stellate hairs commonly branched above the base, usually at least one such on each hair; mature fruiting pedicels various – sigmoid (usually at least some) or straight, divaricate to ascending, often not all alike on the same plant **P. gordonii**

Matthiola incana [WU1]

Matthiola longipetala [HEA]

Mayagrum perfoliatum [GLE, LAM]

Nasturtium officinale [REE]

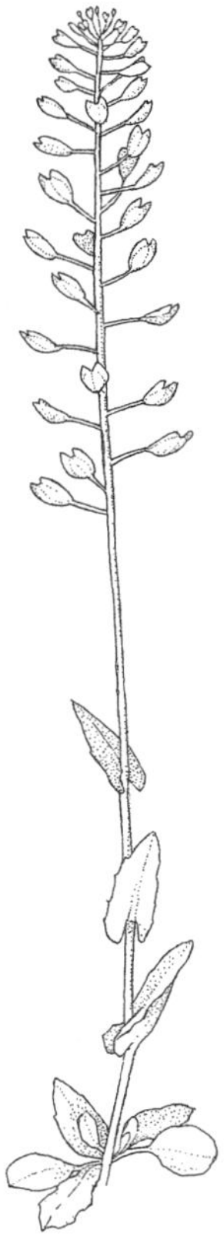

Noccaea perfoliata [FO2]

Paysonia auriculata [HEA]

Paysonia grandiflora [AMB]

Paysonia lasiocarpa [AM1]

1. Plants annual, biennial, or short-lived perennials from (usually slender) taproots.
  5. Flowers white; inflorescences paniculate; plants known only from San Augustine Co. ____ **P. pallida**
  5. Flowers yellow; inflorescences racemes; plants with various distributions but unlikely in San Augustine Co.
    6. Stellate hairs minute, usually < 0.3 mm wide (use high magnification).
      7. Ovules only 2 per locule/4 per ovary; most cauline leaves linear ____ **P. angustifolia**
      7. Ovules more than 4 per ovary; cauline leaves mostly narrowly oblanceolate or narrowly elliptic to obovate or elliptic.
        8. Lowest leaves gradually narrowed to a short petiole; fruiting pedicels strongly to slightly curved, divaricate to slightly recurved; plants of Central Texas eastward, generally absent from the South TX Plains and Coastal Prairies ____ **P. gracilis**
        8. Lowest leaves abruptly constricted to the petiole; fruiting pedicels sigmoid to horizontal or recurved but upright at the tip; plants primarily of the Coastal Prairies and South TX Plains ____ **P. lindheimeri**
    6. Stellate hairs larger, 0.3–0.6 mm wide or wider.
      9. Stellate hairs with a U-shaped notch on one side, the sinus between 2 of the rays rounded and extending to include the center of the hair, the other sinuses shallower ____ **P. densiflora**
      9. Stellate hairs without a U-shaped notch on one side that reaches to the center of the hair.
        10. Fruit minutely papillose externally and stellate-pubescent inside ____ **P. sessilis**
        10. Fruit smooth externally and glabrous inside.
          11. Rays of stellate hairs usually simple, only rarely forked, free to the base or very slightly fused__ **P. argyraea** (in part)
          11. Rays of stellate hairs free to their bases, generally at least one ray on each hair forked above the base.
            12. Fruiting pedicels filiform, those of mature fruits uniformly recurved ____ **P. recurvata**
            12. Fruiting pedicels not filiform, divaricate to ascending, straight to sigmoid, often not all alike on the plant ____ **P. gordonii**

Physaria angustifolia (Nutt. ex Torr. & A. Gray) O'Kane & Al-Shehbaz, (narrow-leaved), NARROWLEAF BLADDERPOD, THREADLEAF BLADDERPOD. Plants annual, taprooted; ± densely pubescent, trichomes stellate or peltate, rays forked; stems erect, to 40 cm; basal leaf blade margins entire to repand or toothed to pinnatifid; stem leaves short-petiolate or sessile, blades linear or narrowly obovate, margins entire or repand to shallowly toothed; racemes usually loose; petals yellow, obovate to obdeltate, 6–10 mm long, apex often emarginate; fruiting pedicels diverging or horizontal; fruits ± globose, 4–6 mm long, ± sessile; ovules 4 per ovary. Shallow calcareous soils, sometimes in disturbed sites; in TX known only from Red River Co.; otherwise restricted to a few se OK cos. Apr–May. [*Lesquerella angustifolia* (Nutt.) S. Watson, *L. longifolia* Cory, *Vesicaria angustifolia* Nutt.] Of conservation concern, due to its restricted range. NatureServe: G3, vulnerable; S1, critically imperiled. ⚠

Physaria argyraea (A. Gray) O'Kane & Al-Shehbaz, (silvery), SILVER BLADDERPOD. Plants perennial, from a woody caudex; ± densely pubescent, trichomes stellate or stellate-peltate, rays unbranched; stems erect to prostrate, to 40(–70) cm; basal leaves usually shed early, blade margins entire or pinnatifid; stem leaves petiolate, blades elliptic to obovate or rhombic, margins entire or sinuate to dentate; racemes loose; petals orange to yellow (rarely drying purplish), spatulate or obovate to obdeltate, 4.5–8.5(–11) mm long; fruiting pedicels erect to spreading, usually sigmoid or sometimes recurved; fruits subglobose to ellipsoid, 4–8 mm long, sessile; ovules 16–32 per ovary. Sandy granitic or calcareous soils; mainly sw portion of East TX, scattered from Navarro and McLennan cos. s to to Bexar and Goliad cos., also Brazos Co.; Edwards Plateau, Trans-Pecos, and South TX Plains; n Mexico. Feb–May(–Jun). [*Alyssum argyraeum* (A. Gray) Kuntze, *Lesquerella argyraea* (A. Gray) S. Watson var. *argyraea*, *Vesicaria argyraea* A. Gray] In the U.S. known only from TX; a second subsp. is endemic to Mexico.

Physaria densiflora (A. Gray) O'Kane & Al-Shehbaz, (densely-flowered), DENSE-FLOWER BLADDERPOD, LOW BLADDERPOD. Plants annual or biennial, from a simple or branched caudex; densely pubescent, trichomes stellate, rays unbranched, with a U-shaped notch between two of the rays on one side; stems erect or decumbent, to 40 cm; basal leaf blade margins lyrate-pinnatifid; stem leaves short-petiolate or sessile, blades narrowly obovate to elliptic, margins entire or repand to dentate; racemes dense, elongated in fruit; petals yellow to orange-yellow, obovate to obdeltate, (4.5–)7–10(–11) mm long, apex often emarginate; fruiting pedicels spreading to diverging or recurved; fruits globose to broadly obovate, 4–6 mm long, ± sessile; ovules 8–16 per ovary. Sandy or gravelly granitic or calcareous soils, rocky prairies, limestone outcrops; c to se TX from McLennan Co. to Brazos and Montgomery cos., in sc TX in Gonzales, DeWitt, and Victoria cos., also Travis Co.; mainly c TX, Edwards Plateau, and Rolling Plains; endemic to TX. Mar–May. [*Lesquerella densiflora* (A. Gray) S. Watson, *Vesicaria densiflora* A. Gray]

Physaria engelmannii (A. Gray) O'Kane & Al-Shehbaz, (for George Engelmann, 1809–1884, German-born botanist and physician of St. Louis, MO), ENGELMANN'S BLADDERPOD. Plants perennial, from a woody caudex; ± densely pubescent, trichomes stellate or peltate, rays unbranched or forked, with a deep U-shaped notch between two of the rays on one side; stems erect, to 40(–60) cm; basal leaf blade margins entire or sinuate to toothed; stem leaves petiolate or sessile, blades oblanceolate to linear, margins

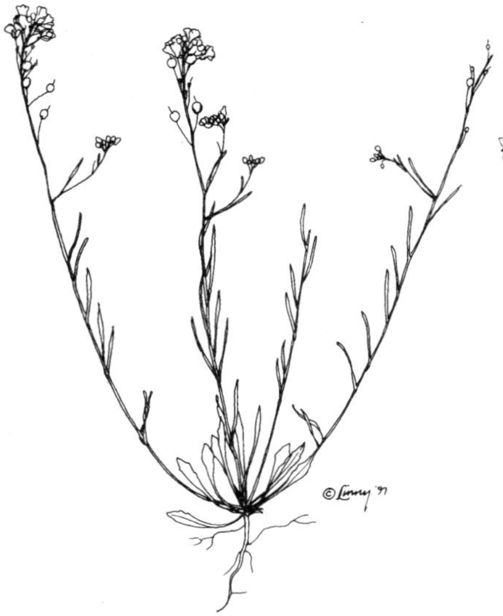

Physaria angustifolia [HEA]

Physaria argyraea [CP1]

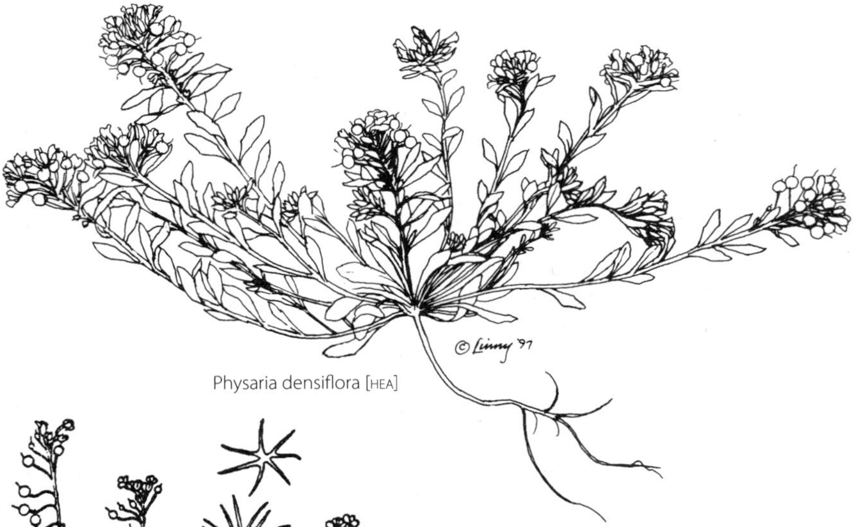

Physaria densiflora [HEA]

Physaria engelmannii [GR1]

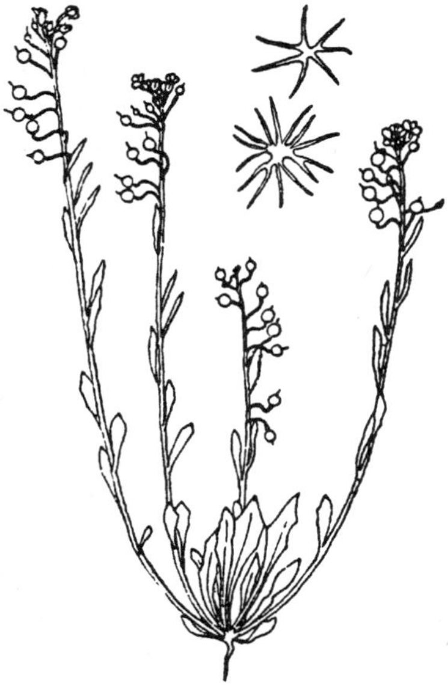

Physaria gordonii [AMB]

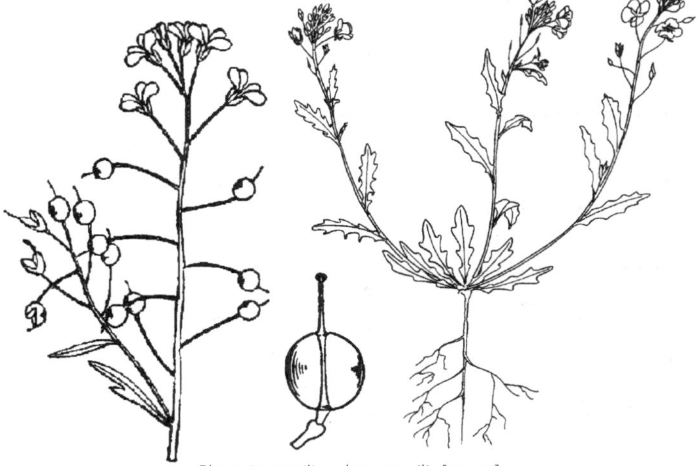

Physaria gracilis subsp. gracilis [BB2, SHI]

entire; racemes dense (subumbellate); petals bright yellow, obovate to elliptic, 8–14 mm long; fruiting pedicels ascending; fruits globose to ellipsoid, 5–8 mm long, short-stipitate; ovules 8–20 per ovary. Limestone outcrops, rocky calcareous prairies, thin caliche soils; along the w border of East TX from Grayson Co. s to Comal and Guadalupe cos.; mostly in a narrow band from nc TX to the e border of the Edwards Plateau; OK and TX. Apr–May. [*Alyssum engelmannii* (A. Gray) Kuntze, *Lesquerella engelmannii* (A. Gray) S. Watson, *Vesicaria engelmannii* A. Gray, *V. engelmannii* var. *elatior* A. Gray, *V. pulchella* Kunth & Bouché]

Physaria gordonii (A. Gray) O'Kane & Al-Shehbaz, (for James Gordon, d. ca. 1781, correspondent of Linnaeus and London nurseryman), GORDON'S BLADDERPOD, POPWEED, BEAD-POP. Plants annual, biennial, or perennial (short-lived), taprooted; usually densely pubescent, trichomes stellate, rays free to the base, at least 1 ray forked above the base; stems erect to prostrate, to 35(–45) cm; basal leaf blade margins entire or dentate to lyrate-pinnatifid; stem leaves petiolate or sessile, blades linear to oblanceolate, often falcate, margins entire or repand to dentate; racemes dense; petals yellow to orange, cuneate or obovate to obdeltate, 5–8(–10) mm long; fruiting pedicels ascending to spreading, sometimes sigmoid; fruits globose, (3–)4–8 mm long, short-stipitate; ovules 8–26 per ovary. Sandy and gravelly soils, brushland, pastures, roadsides; mostly w border of East TX in Hill Co. (Kartesz 2015), Bexar, Ellis, and Travis cos., also in Washington Co.; mainly w 2/3 of TX; AZ, KS, NM, OK, and TX. Mar–Jun. [*Alyssum gordonii* (A. Gray) Kuntze, *Lesquerella gordonii* (A. Gray) S. Watson, *L. gordonii* subsp. *densifolia* (Rollins) O'Kane & Al-Shehbaz, *P. gordonii* var. *densifolia* (Rollins) B.L. Turner, *Vesicaria gordonii* A. Gray] When stepped upon, the dry fruits pop, giving rise to some of the common names (Kirkpatrick 1992).

Physaria gracilis (Hook.) O'Kane & Al-Shehbaz, (graceful or slender). Plants annual or biennial, taprooted; pubescent, trichomes stellate, rays branching or forked; stems erect or decumbent; basal leaf blade margins lyrate-pinnatifid to dentate or repand; stem leaves petiolate or sessile, blades oblanceolate to oblong, margins dentate to repand; racemes loose, elongated in fruit; petals yellow to orange, broadly obovate, 6–11 mm long; fruiting pedicels spreading to diverging or horizontal to recurved; fruits globose to obovoid or obpyriform, 3–9 mm long, stipitate; ovules 8–28 per ovary. Calcareous soils, prairies, fields, roadsides. Mar–May. [*Alyssum gracile* (Hook.) Kuntze, *Lesquerella gracilis* (Hook.) S. Watson, *Vesicaria gracilis* Hook.] O'Kane (2010b) recognizes two subspecies, both found in East TX:

1. Fruits globose or subglobose, 3–6 mm long, basally rounded ____ subsp. **gracilis**
1. Fruits obpyriform (reverse pear-shaped) to narrowly obovoid, (5–)6–9 mm long, basally truncate ____ subsp. **nuttallii**

subsp. **gracilis**, SPREADING BLADDERPOD, SLENDER BLADDERPOD, LAX BLADDERPOD, CLOTH-OF-GOLD. Stems to 70 cm; midstem leaf blade margins usually entire to dentate, rarely repand; fruits globose or subglobose, 3–6 mm long, basally rounded. From ne TX s to Harris and Brazoria cos. and w to the Edwards Plateau (Kartesz 2015); AL, LA, MS, OK, and TX; introduced in IA, IL, MO, and TN. [*Lesquerella polyantha* (Schltr.) Small; *Vesicaria polyantha* Schltr.]

subsp. **nuttallii** (Torr. & A. Gray) O'Kane & Al-Shehbaz, (for Sir Thomas Nuttall, 1786–1859, English-American botanist), NUTTALL'S BLADDERPOD, PEAR-FRUIT BLADDERPOD. Stems usually < 30 cm; midstem leaf blade margins frequently repand or occasionally dentate; fruits obpyriform to narrowly obovoid, (5–)6–9 mm long, basally truncate. In ne TX in Dallas, Grayson, Kaufman, Lamar, and Red River cos. (Kartesz 2015); KS, OK, and TX. [*Alyssum nuttallii* (Torr. & A. Gray) Kuntze, *A. repandum* (Nutt.) Kuntze, *Lesquerella gracilis* (Hook.) S. Watson subsp. *nuttallii* (Torr. & A. Gray) Rollins & E.A. Shaw, *L. gracilis* var. *repanda* (Nutt.) Payson, *L. nuttallii* (Torrey & A. Gray) S. Watson, *Physaria gracilis* var. *repanda* (Nutt.) B.L. Turner, *Vesicaria nuttallii* Torr. & A. Gray]

Physaria lindheimeri (A. Gray) O'Kane & Al-Shehbaz, (for Ferdinand Jacob Lindheimer, 1809–1879, German-born TX collector), LINDHEIMER'S BLADDERPOD. Plants annual or biennial, taprooted; densely pubescent, trichomes stellate, rays branching or forked; stems erect to decumbent, to 80 cm; basal leaf blade margins pinnatisect to repand; stem leaves petiolate or sessile, blades elliptic, margins entire or deeply dentate; racemes dense; petals yellow to orange-yellow (sometimes drying purplish), suborbicular or broadly ovate, 4.5–7(–9) mm long; fruiting pedicels horizontal to recurved; fruits globose to ellipsoid, (4–)5–8 mm long, stipitate; ovules 8–20 per ovary. Sandy or clay soils; coastal prairies, thickets, field margins; sc TX in Lavaca (Kartesz 2015), Goliad, Jackson, and Victoria cos., also Brazos and Travis cos. (Kartesz 2015); South TX Plains; Mexico. Dec–Apr. [*Alyssum lindheimeri* (A. Gray) Kuntze, *Lesquerella gracilis* (Hook.) S. Watson var. *pilosa* Lundell, *L. lindheimeri* (A. Gray) S. Watson, *Vesicaria lindheimeri* A. Gray]

Physaria pallida (Torr. & A. Gray) O'Kane & Al-Shehbaz, (pale), WHITE BLADDERPOD. Plants annual, taprooted; sparsely pubescent, trichomes stellate, minute, rays branching or forked; stems erect, to 60 cm; basal leaf blade margins entire to sinuate-dentate or sometimes lobed; stem leaves short-petiolate or sessile, blades similar to basal leaves; racemes paniculate; petals white (may be veined with yellow), broadly ovate, to 12 mm long; fruiting pedicels diverging-ascending or slightly recurved; fruits globose or

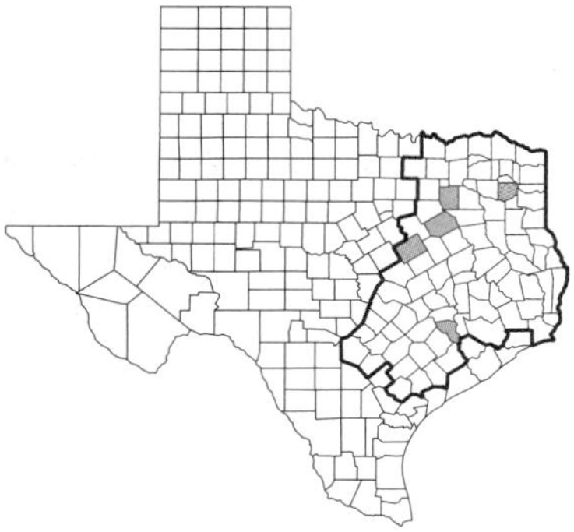

Paysonia auriculata

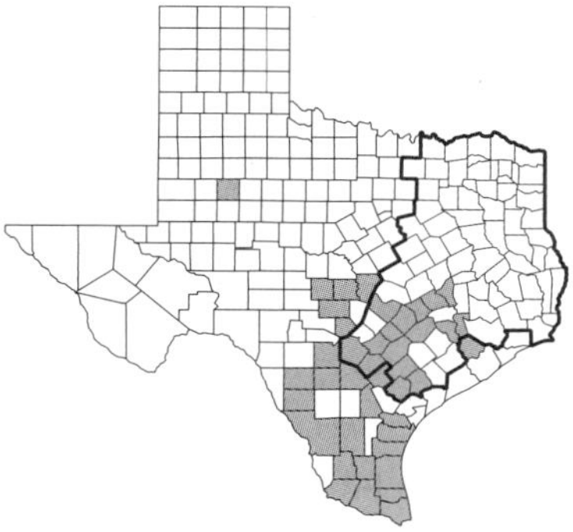

Paysonia grandiflora

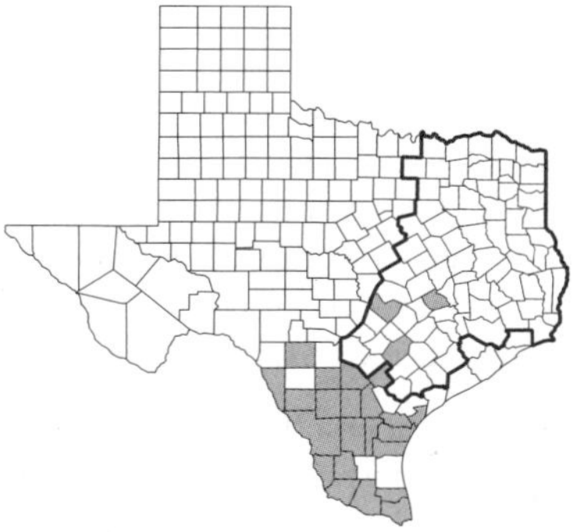

Paysonia lasiocarpa

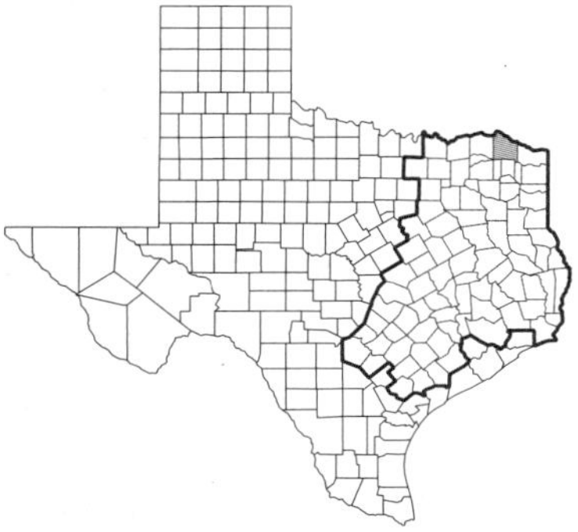

Physaria angustifolia

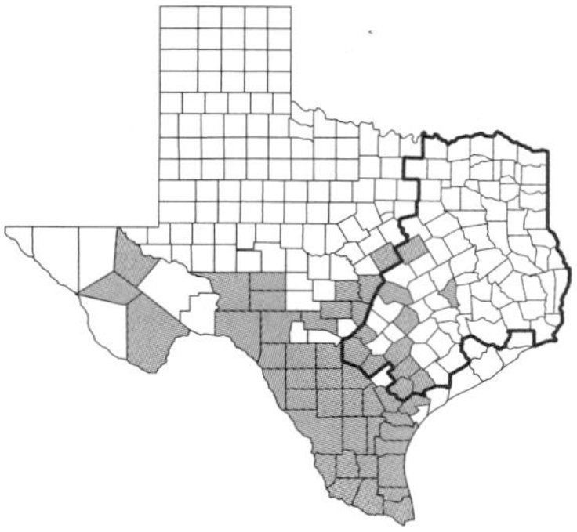

Physaria argyraea

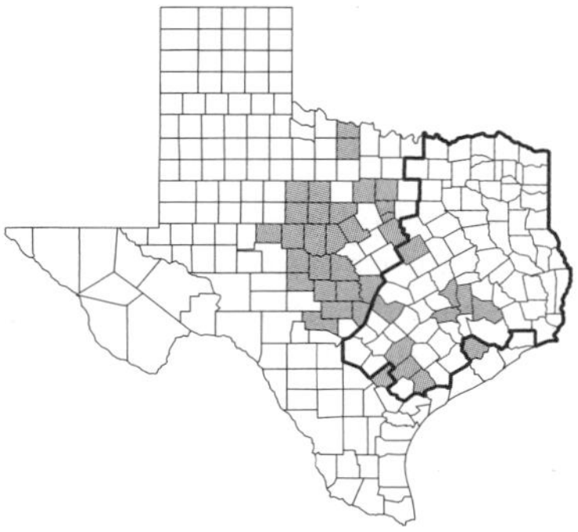

Physaria densiflora

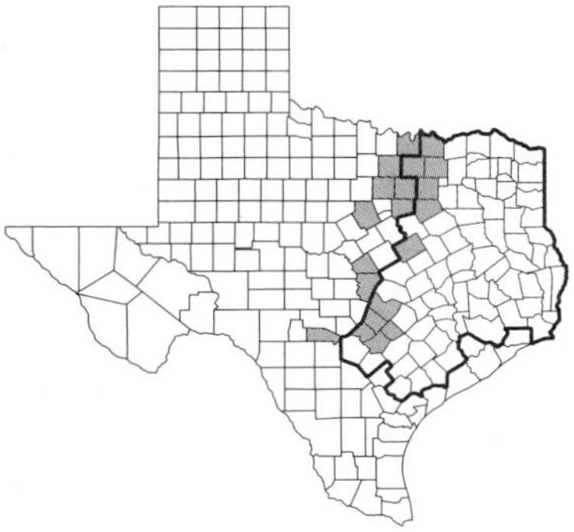

Physaria engelmannii

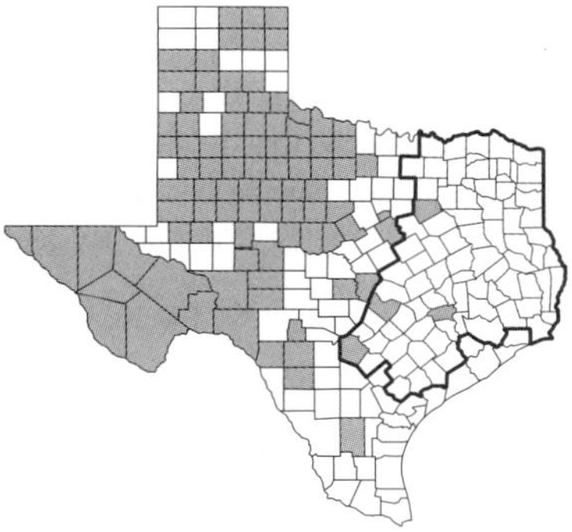

Physaria gordonii

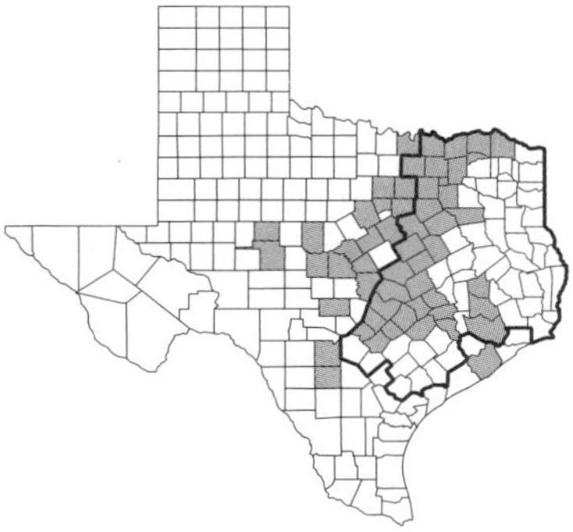

Physaria gracilis subsp. gracilis

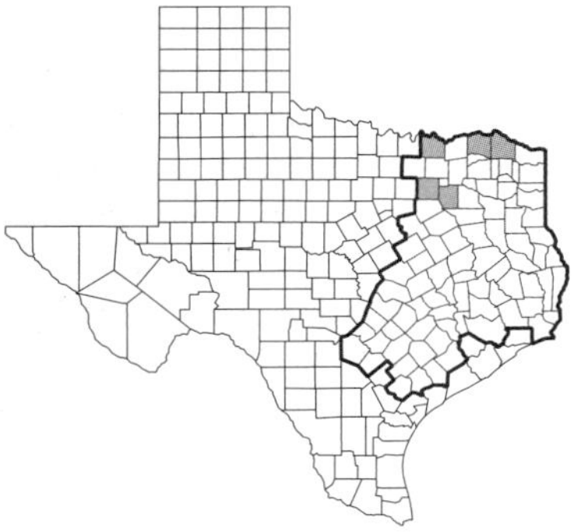

Physaria gracilis subsp. nuttallii

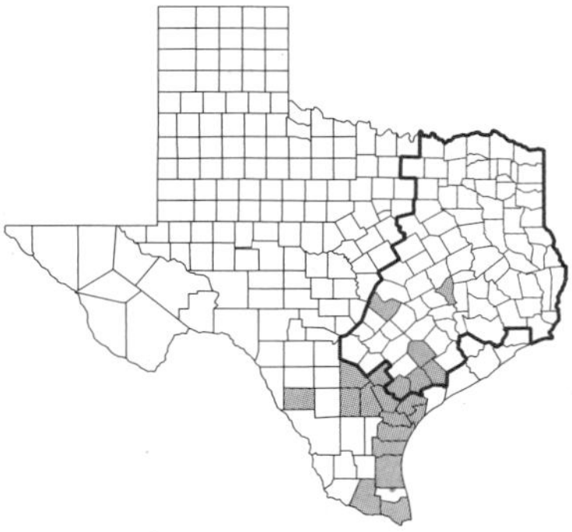

Physaria lindheimeri

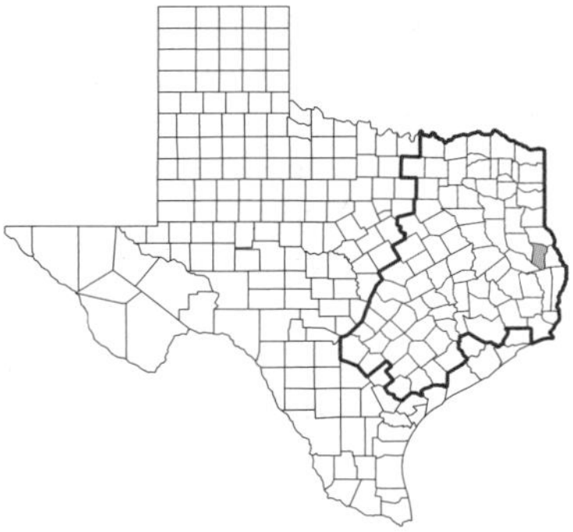

Physaria pallida

subglobose, 3–10 mm long, short-stipitate; ovules 8–12 per ovary. Vernally moist, sandy soils on ironstone or greensand glades of the Weches Formation, in pine-oak forests; endemic to East TX; known only from San Augustine Co., (Poole et al. 2007). Apr–May. [*Alyssum pallidum* (Torr. & A. Gray) Kuntze, *Lesquerella pallida* (Torr. & A. Gray) S. Watson, *Vesicaria grandiflora* Hook. var. *pallida* Torr. & A. Gray, *V. pallida* (Torr. & A. Gray) Nutt.] WHITE BLADDERPOD was rediscovered in 1983 when it was collected for the first time in 150 years (Nixon et al. 1983, George & Nixon 1990); it is now known from only a few sites in San Augustine Co.; these populations are threatened by road construction and maintenance, gravel mining, herbicides, and encroachment of woody plants (Poole et al. 2007). Listed as a federally-protected endangered species in 1987. NatureServe: G1, critically imperiled; S1, critically imperiled.

Physaria recurvata (Engelm. ex A. Gray) O'Kane & Al-Shehbaz, (recurved), PLATEAU BLADDERPOD, SLENDER BLADDERPOD, GASLIGHT BLADDERPOD. Plants annual or biennial, taprooted; sparsely pubescent, trichomes stellate, rays forked; stems erect or decumbent, to 50 cm; basal leaf blade margins entire or lyrate-pinnatifid; stem leaves petiolate or sessile, blades rhombic or obovate to elliptic, margins entire or sinuate to toothed; racemes loose; petals yellow to orange-yellow, obovate to cuneate, 4–7 mm long, apex sometimes retuse; fruiting pedicels recurved in age; fruits globose or subglobose, (2–)3–5(–7) mm long, ± sessile; ovules 4–20 per ovary. Limestone outcrops, gravelly calcareous prairies, open rocky areas; along the w border of East TX from Ellis Co. (Kartesz 2015) s to Bexar Co., also Waller Co.; mostly nc TX s to Edwards Plateau; endemic to TX. Mar–Apr. [*Lesquerella recurvata* (Engelm. ex A. Gray) S. Watson, *Vesicaria recurvata* Engelm. ex A. Gray]

Physaria sessilis (S. Watson) O'Kane & Al-Shehbaz, (sessile), SESSILE BLADDERPOD. Plants annual or biennial, taprooted; densely pubescent, trichomes stellate, rays branching or forked; stems erect, to 60 cm; basal leaf blade margins entire or dentate to sinuate or lyrate-pinnatifid; stem leaves short-petiolate or sessile, blades narrowly elliptic to linear, margins entire or repand to dentate; racemes loose; petals yellow, obovate or deltate, 5–10 mm long, margins undulate; fruiting pedicels diverging-ascending to widely spreading; fruits globose or subglobose, 3–6 mm long, sessile; ovules 8–18 per ovary. Shallow calcareous soils, sandy roadsides, gravelly limestone hills, mesquite brushland; on the sw border of East TX in Bexar, Comal, Hays, and Victoria cos.; mainly Edwards Plateau and South TX Plains; endemic to TX. Apr–Jun. [*Lesquerella gracilis* (Hooker) S. Watson var. *sessilis* S. Watson, *L. sessilis* (S. Watson) Small]

Physaria fendleri (A. Gray) O'Kane & Al-Shehbaz, (for August Fendler, 1813–1883, one of the first botanists to collect in New Mexico and Venezuela), FENDLER'S BLADDERPOD, POPWEED, which would here have keyed to *P. argyraea* and has a similar perennial habit and silvery-scurfy pubescence, can be distinguished by its stellate-peltate trichomes with unbranched rays that are basally fused about 1/2 their length, and showy orange-yellow petals that are 8–12 mm long. Limestone outcrops, gravelly calcareous soils, sandy roadsides; not reported in East TX but included here based on records from adjacent Coryell, Kendall, and Bandera cos.; c and sw TX to Trans-Pecos and the Panhandle; AZ, CO, KS, NM, OK, TX, and UT; Mexico. Mar–May(–sporadically later).

PLANODES Greene WINGED ROCKCRESS

A genus of 2 species, native to the se and c U.S. and Mexico; previously shuffled between *Sibara*, *Arabis*, and *Cardamine*. (Greek: *planis*, wanderer, and *–odes*, resemblance, for its movement from another genus) (tribe Cardamineae)
REFERENCES: Rollins 1947; Al-Shehbaz 2010a, 2010m.

Planodes virginicum (L.) Greene, (of Virginia), VIRGINIA WINGED ROCKCRESS, COMMON ROCKCRESS, VIRGINIA SIBARA. Weedy herbaceous annuals, resembling *Cardamine*; stems simple or many from the base, often branching, erect to decumbent, to 35(–55) cm long; lower stem and leaves usually pubescent with simple trichomes; basal leaves in a loose rosette, petiolate, blades pinnatifid to pinnatisect, with 4–15 ± evenly-spaced lobes per side, the terminal lobe often similar and not much larger; stem leaves short-petiolate, similar to basal but smaller; racemes considerably elongated in fruit; sepals often pink-tinged; petals white, oblanceolate, 2–3 mm long, claw undifferentiated from blade; stamens slightly tetradynamous; fruiting pedicels slender, glabrous; siliques dehiscent, linear, flattened parallel to the replum, slightly torulose, (1–)1.5–2.5(–3.2) cm long, 1–1.5 mm wide, valves each obscurely midveined; style blunt, 0.2–0.7 mm long; ovules 20–44 per ovary; seeds narrowly winged, in 1 row per locule. Floodplains, roadside ditches, lawns and gardens; mainly e 1/2 of TX; a few counties further w; se-sc U.S. from PA s to FL and w to NE, OK, and TX; introduced in CA; n Mexico. Feb–Apr. [*Arabis ludoviciana* (Hook.) C.A. Mey., *A. virginica* (L.) Poir., *Cardamine ludoviciana* Hook., *C. virginica* L., *Sibara virginica* (L.) Rollins]

RAPHANUS L. RADISH

Ours herbaceous annuals or biennials; roots slender or fleshy; often scabrous to hispid with simple trichomes, or sometimes glabrous; stems erect, often unbranching; basal leaves petiolate, not forming a rosette, blades lyrate or pinnatifid to pinnatisect or sometimes undivided with margins crenate to dentate; stem leaves subsessile, smaller, often undivided, margins toothed, bases

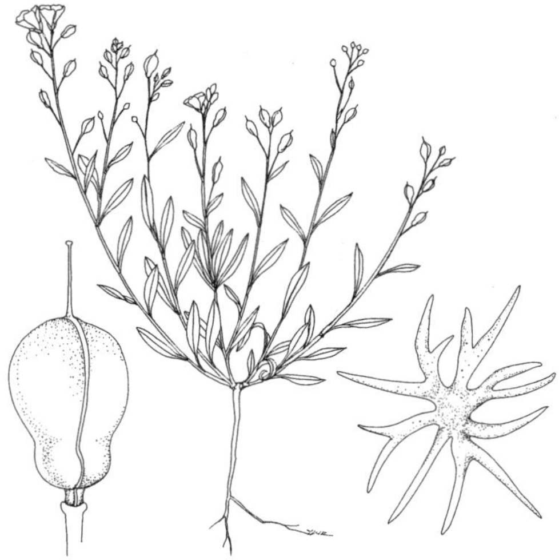

Physaria gracilis subsp. nuttallii [FNA]

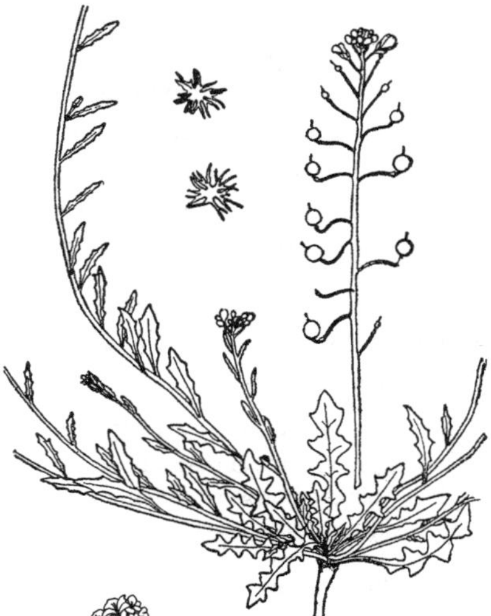

Physaria lindheimeri [AMB]

Physaria pallida [SID]

Physaria sessilis [CP1]

Physaria recurvata [AMB]

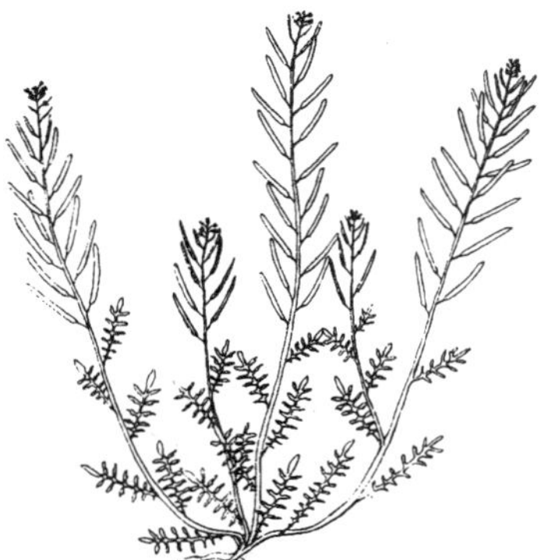

Planodes virginicum [BB2]

Raphanus raphanistrum [NCS]

not clasping; racemes sometimes paniculately branched, greatly elongated in fruit; sepals sometimes tinged pink or purple; petals creamy white to yellow or pink to purple, broadly obovate, with veins darker than blade (often purplish or brownish), claw differentiated from blade and often longer than sepals; stamens strongly tetradynamous; siliques indehiscent, terete or angular, transversely 2-segmented; lower fruit segment valvular but seedless, short, nearly as wide as pedicel (thus appearing stipe-like); terminal segment wider, 2–8-seeded, spongy or corky at maturity, either indehiscent and falling at maturity or loment-like and separating into 1-seeded segments; style persisting as an attenuated beak 10–50 mm long; 2–22 ovules per ovary; seeds unwinged, in a single row per fruit.

A genus of 2 or 3 species, native to Eurasia. The GARDEN RADISH is an important, long-cultivated root crop; both introduced *Raphanus* species can be weedy and are known to hybridize. (Greek: *raphanos*, radish) (tribe Brassiceae)
REFERENCE: Warwick 2010e.

1. Petals yellow to creamy white; fruits (2.5–)3–8(–11) mm wide, strongly constricted between seeds, breaking into corky single-seeded units at maturity; seeds usually 4–8 per fruit ________ **R. raphanistrum**
1. Petals usually pink to purple or sometimes white; fruits (5–)7–13(–15) mm wide, rarely constricted between seeds, spongy, usually not breaking into single-seeded units at maturity; seeds usually 2 or 3 per fruit ________ **R. sativus**

Raphanus raphanistrum L., (resembling a radish), WILD RADISH, JOINTED CHARLOCK. Plants annual, to 80 cm tall; slender-rooted; sparsely to densely pubescent; basal leaf blade margins lyrate or pinnatifid (or sometimes undivided), lobes 1–4 per side, the terminal lobe largest; sepals sparsely pubescent; petals yellow to creamy white, with dark brown or purple veins, 15–25 mm long; fruiting pedicels diverging or ascending; fruits cylindrical or narrowly lanceolate; lower segment 1–1.5 mm long; terminal segment cylindrical or lanceolate, (1.5–)2–11(–14) cm long, (2.5–)3–8(–11) mm wide, strongly ribbed and constricted between seeds, usually breaking into corky 1-seeded segments at maturity; style conical, 10–50 mm long; seeds usually 4–8 per fruit. Fields and roadsides, disturbed areas; recorded from Harris, San Jacinto, and Walker cos. (Kartesz 2015); also Fort Bend Co.; scattered throughout the U.S.; Canada; Mexico; Central and South America; Europe and Asia; introduced elsewhere. Dec–Jun. [*Raphanistrum innocuum* Moench, *Raphanus sylvestris* Lam., *Rapistrum arvense* All., *R. raphanistrum* (L.) Crantz] According to Warwick (2010e), *R. raphanistrum* subsp. *raphanistrum* is the only subspecies represented in the U.S. A native of Eurasia and the Mediterranean region, now a cosmopolitan weed.

Raphanus sativus L., (cultivated), GARDEN RADISH, RABANO, DAIKON. Plants annual or biennial, to 130 cm tall; fleshy-rooted in cultivated forms; sparsely scabrous or hispid or sometimes glabrous; basal leaf blade margins lyrate or pinnatifid (or sometimes undivided), lobes 1–12 per side, the terminal lobe largest; sepals glabrous or sparsely pubescent; petals pink to purple or sometimes white, with darker veins, 15–25 mm long; fruiting pedicels spreading to ascending; fruits usually fusiform or lanceolate (or sometimes ovoid or cylindrical when few-seeded); lower segment 1–3.5 mm long; terminal segment fusiform or lanceolate, (1–)3–15(–25) cm long, (5–)7–13(–15) mm wide, not strongly ribbed, smooth or slightly constricted between seeds, spongy, falling at maturity (not usually breaking into 1-seeded segments); style conical or linear, 10–40 mm long; seeds usually 2 or 3 per fruit. Cultivated; reseeds in gardens, abandoned fields, weedy roadsides; a waif or escape from cultivation, not usually persisting; recorded from Bexar, Brazos, Harris, and Walker cos. (Kartesz 2015); also Gillespie, Terrell, and Val Verde cos.; scattered throughout the U.S.; Canada; Mexico; South America; Europe; and Asia; introduced elsewhere. Mar–May. [*Raphanistrum gayanum* Fisch. & C.A. Mey, *Raphanus chinensis* Mill., *R. sativus* var. *longipinnatus* L.H. Bailey, *R. sativus* var. *radicula* Pers.] Domesticated in pre-Roman times for its peppery-flavored fleshy root, with an ancestral origin in s Asia or the Mediterranean; the unripe fruits are also flavorful; cultivated but sometimes escaping and weedy in temperate regions worldwide.

RAPISTRUM Crantz BASTARD-CABBAGE, TURNIPWEED

An Old World genus of 2 species, native to s Europe, the Mediterranean region, and w Asia, now worldwide weeds. (Latin: *rapa*, turnip or wild rape, and *-astrum*, resemblance) (tribe Brassiceae)
REFERENCES: Lemke & Worthington 1991; Warwick 2010f.

Rapistrum rugosum (L.) All., (rugose, wrinkled), ANNUAL BASTARD-CABBAGE, ANNUAL TURNIPWEED. Herbaceous winter annuals to 1(–1.5) m tall, resembling species of *Brassica*; lower parts hispid with simple trichomes; stems erect, unbranched or branched basally; basal leaves often numerous (and appearing rosette-like in young plants), petiolate, blades usually lyrate-pinnatifid with 1–5 lobes per side, the terminal lobe largest and suborbicular to ovate, or blades rarely undivided and margins dentate; stem leaves subsessile to short-petiolate, blades simple or sinuately lobed, margins subentire or dentate, bases not auriculate; racemes greatly elongated in fruit; petals pale to bright yellow, obovate, 6–11 mm long, claw differentiated from blade; stamens tetradynamous; fruiting pedicels slender to stout, erect (appressed to rachis); silicles transversely 2-segmented, constricted at the joint; lower fruit segment dehiscent, valvular, narrow (appearing stipe-like), ellipsoid, 0.7–3 mm long, 0.5–1.5 mm wide, seedless or

1(–3)-seeded; upper fruit segment indehiscent, falling at maturity, globose or ovoid, usually rugose or ribbed, 1.5–3.5 mm long, 1–2.8 mm wide, usually 1-seeded; style slender, 1–3(–5) mm long; seeds unwinged, in 1 row per fruit. Fields and roadsides, disturbed areas; apparently increasing in TX; frequent from nc to sc and se TX; also s TX and Edwards Plateau; scattered across U.S.; Canada; Mexico; Europe; introduced elsewhere. (Nov–)Mar–Jun. [*Myagrum rugosum* L., *R. hispanicum* (L.) Crantz] We are following Warwick (2010f) in not recognizing subspecies. A weedy native of Eurasia and the Mediterranean region, now invasive in many parts of the world; a noxious weed now found in near-monoculture on many formerly-wildflowered springtime roadsides of TX.

RORIPPA Scop. YELLOWCRESS, MARSHCRESS

Ours herbaceous annuals, biennials, or perennials, often of moist habitats (or sometimes aquatic); glabrous or puberulent with simple trichomes; stems usually branching, erect or decumbent to prostrate; basal leaves petiolate (sometimes absent or withered early), sometimes forming a rosette, blade margins dentate-crenate or lobed to lyrate-pinnatifid or pinnatisect; stem leaves sessile or short-petiolate, blade margins similar to basal or not (sometimes entire), bases sometimes auriculate or clasping; racemes elongated in fruit; sepals often yellowish; petals yellow to pale yellow or white (sometimes absent), spatulate to obovate, claw sometimes indistinct; stamens usually 6 and tetradynamous (rarely 4 and equal); fruiting pedicels slender or stout; fruits siliques or silicles, dehiscent, linear or oblong to ellipsoid, terete or slightly flattened parallel to the replum, valves obscurely veined; style often blunt (sometimes slender); ovules 20–242 per ovary; seeds unwinged, in 2 rows per locule.

A moisture-loving genus of ca. 85 species, of every continent but Antarctica; North America hosts both native and introduced species; some are weedy. Members have been previously and variously included in the genera *Armoracia*, *Cardamine*, *Nasturtium*, *Neobeckia*, and *Sisymbrium*; the rare TX native LAKECRESS (*Rorippa aquatica*) should not be confused with the introduced European edible WATERCRESS (*Nasturtium officinale*, formerly known as *Rorippa nasturtium-aquaticum*). (Old Saxon: *rorippen*, used for plants of this genus) (tribe Cardamineae)
REFERENCES: Rollins 1961; Stuckey 1972; Al-Shehbaz & Bates 1987; Les 1994; Al-Shehbaz 2010n.

1. Plants aquatic; lower stem (submerged) leaf blades 1–4-pinnatisect, their divisions filiform; petals white, 4–8 mm long; fruits silicles (when developed), style 2–4 mm long ____ **R. aquatica**
1. Plants aquatic or terrestrial; lower stem leaf blades variously toothed or lobed to pinnatifid or pinnatisect, but any divisions not filiform; petals pale to bright yellow and 1–3 mm long OR petals absent; fruits silicles or siliques, style ≤ 1.5 mm long.
   2. Plants glabrous; petals absent at anthesis (petals completely lacking); fruiting pedicels 0.5–2(–4) mm long ____ **R. sessiliflora**
   2. Plants glabrous or pubescent; petals present at anthesis; fruiting pedicels usually ≥ 2 mm long.
      3. Stems erect; petals (1.5–)1.8–2.5(–3) mm long; fruits oblong-ovoid to ellipsoid, (2.5–)4–10 mm long ____ **R. palustris** subsp. **palustris**
      3. Stems usually prostrate or decumbent, petals 1–2 mm long; fruits linear to oblong-linear, 8–14(–21) mm long ____ **R. teres**

Rorippa aquatica (Eaton) E.J. Palmer & Steyerm., (aquatic), LAKECRESS. Plants perennial, aquatic; rhizomatous and rooting at lower nodes; glabrous; stems erect, to 85(–110) cm, lower stems submerged, flowering branches emergent; basal leaves absent; submerged stem leaves short-petiolate, blades finely 1–4-pinnatisect, divisions filiform; emergent stem leaves sessile or short-petiolate, blades lanceolate to oblong, usually undivided, margins entire or dentate, bases not auriculate; petals white, spatulate to obovate, 4–8 mm long; fruiting pedicels (5–)7–15 mm long; silicles (infrequently developed) oblong to ellipsoid, straight, 4–7 mm long, 2.5–3 mm wide; style 2–4 mm long. Springs, streams, lake margins; rare in TX, recorded in Liberty Co. (Brown & Marcus 1998), Madison Co. (*A.K. Neill 1402* (TAMU!) and https://www.inaturalist.org/observations/*38952586* (iNaturalist.org 2021)), Red River Co. (iNaturalist.org observation *13238701, Bob O'Kennon, May 2018* (iNaturalist.org 2021)), and Tyler Co. (Al-Shehbaz & Bates 1987), with additional literature reports from Freestone and Wood cos. (Gabel and Les 2001); uncommon across e 1/2 of the U.S., mainly Mississippi Basin; Canada. Apr–Jun. [*Armoracia lacustris* (A. Gray) Al-Shehbaz & V. Bates, *Cochlearia armoracia* L. var. *aquatica* Eaton, *C. aquatica* (Eaton) Eaton, *Nasturtium lacustre* A. Gray, *N. natans* DC. var. *americanum* A. Gray, *Neobeckia aquatica* (Eaton) Greene, *Radicula aquatica* (Eaton) B.L. Robinson, *Rorippa americana* (A. Gray) Britton] LAKECRESS is possibly dispersed by waterfowl (Weakley, A.S., and Southeastern Flora Team 2020), as detached pieces of stems and submerged leaves readily produce new plantlets; within some populations all plants may be clonal and perhaps self-incompatible, explaining poor fruit development (Brown & Marcus 1998, Al-Shehbaz 2010n). LAKECRESS is native and should not be confused with the introduced European edible WATERCRESS (*Nasturtium officinale*, formerly known as *Rorippa nasturtium-aquaticum*). LAKECRESS lacks federal protection but is considered uncommon, with 20–80 extant populations estimated in North America. NatureServe: G4, apparently secure; S1, critically imperiled.

Rorippa palustris (L.) Besser subsp. **palustris,** (marsh-dwelling), MARSH YELLOWCRESS, BOG YELLOWCRESS, BOG MARSHCRESS. Plants annual or rarely perennial; glabrous or sometimes sparsely pubescent; stems erect, to 100(–140) cm; basal leaves in a rosette, blades lyrate-pinnatisect; stem leaves petiolate or subsessile, blades lyrate- pinnatisect, bases auriculate or clasping; petals yellow or pale yellow, spatulate, (1.5–)1.8–2.5(–3) mm long; fruiting pedicels (2.5–)3–10(–14) mm long; silicles (rarely siliques) oblong-ovoid to ellipsoid, (2.5–)4–10 mm long, (1.5–)1.7–3(–3.5) mm wide; style 0.2–1.2 mm long. Margins of lakes and streams, roadside ditches, thickets and bogs; scattered mainly in e 1/2 of TX; throughout the U.S.; Canada; n Mexico; Eurasia; introduced elsewhere. Apr–Sep. [*Nasturtium semipinnatifidum* Hook., *N. terrestre* (With.) W.T. Aiton var. *occidentale* S. Watson, *Radicula pacifica* (T.J. Howell) Greene, *Rorippa islandica* (Oeder) Borbás subsp. *fernaldiana* (Butters & Abbe) Hultén, *R. palustris* var. *fernaldiana* (Butters & Abbe) Stuckey, *R. palustris* var. *occidentalis* (S. Watson) Rollins, *Sisymbrium palustre* (L.) Leyss.]

Rorippa sessiliflora (Nutt.) Hitchc., (sessile-flowered), STALKLESS YELLOWCRESS. Plants annual, glabrous; stems erect, to 45(–63) cm; basal leaves not forming a rosette, margins dentate to crenate or lobed to lyrate-pinnatifid; stem leaves short-petiolate or sessile, margins dentate to sinuate or lobed to lyrate, bases sometimes auriculate; sepals greenish yellow, 1.2–2 mm long; petals absent; fruiting pedicels 0.5–2(–4) mm long; siliques oblong to oblong-linear, (4–)6–9(–12) mm long, (1.4–)1.8–2.8(–3.5) mm wide; style 0.1–0.5 mm long. Shores of streams and lakes, floodplains, roadsides; scattered across e 2/3 of TX; e U.S. from NJ s to FL and w to MN, KS, and TX. Apr–May, less commonly Aug–Sep. [*Nasturtium limosum* Nutt., *N. sessiliflorum* Nutt., *Radicula limosa* (Nutt.) Greene, *R. sessiliflora* (Nutt.) Greene]

Rorippa teres (Michx.) Stuckey, (terete, circular in cross-section), SOUTHERN YELLOWCRESS, SOUTHERN MARSH YELLOWCRESS, TANSY-LEAF YELLOWCRESS. Plants annual or rarely biennial; puberulent or glabrous; stems usually prostrate or decumbent, to 40 cm; basal leaves in a rosette, pinnatifid; stem leaves short-petiolate, margins lyrate-pinnatifid to pinnatisect, bases sometimes auriculate; petals yellow, spatulate, 1–2 mm long; fruiting pedicels (1.5–)2.3–4.7(–5.3) mm long; siliques linear to oblong-linear, 8–14(–21) mm long, 1–2.5 mm wide; style 0.2–1.1 mm long. Mudflats, stream banks, moist sandy soils; scattered in TX, mainly s 1/2 of state but ranging into ne TX in Anderson, Dallas, and Wood cos.; se U.S. from NC s to FL and w to AR and TX; Mexico; Central America. Dec–May. [*Cardamine teres* Michx., *Nasturtium micropetalum* Fisch. & C.A. Mey., *N. obtusum* Nutt., *N. tanacetifolium* (DC.) Hook. & Arn., *R. obtusa* (Nutt.) Britton, *R. teres* var. *rollinsii* Stuckey, *R. walteri* (Elliott) Mohr, *Sisymbrium teres* (Michx.) Torr. & A. Gray, *S. walteri* Elliott]

Rorippa sylvestris (L.) Besser, (of the forest, wild), CREEPING YELLOWCRESS, YELLOW FIELDCRESS, KEEK, which would not have keyed satisfactorily to any other species here, can be distinguished by its rhizomatous perennial habit, deeply pinnatisect upper stem leaves with bases usually not auriculate, fruiting pedicels 3–12 mm long, yellow petals (2.2–)2.8–5.5(–6) mm long, and linear siliques 10–25 mm long. Ditches, lakeshores, sandy beaches, wet roadsides; not yet reported in TX but likely present in East TX; included here based on a record from adjacent Beauregard Par., LA (NLU); scattered across the U.S., mostly Midwest and New England; Canada; Eurasia; introduced elsewhere. May–Jun. [*Cardamine sylvestris* (L.) Kuntze, *Nasturtium sylvestre* (L.) W.T. Aiton, *Radicula sylvestris* (L.) Druce, *Sisymbrium sylvestre* L.] Native to Eurasia, now a weed on several continents, and considered noxious in some regions.

SELENIA Nutt. SELENIA

A genus of 5 species, native to the c and sw U.S. and ne Mexico. (Greek: *selene*, moon, referring to a resemblance to *Lunaria*) (tribe Cardamineae)

REFERENCES: Al-Shehbaz 2010o; Holmes et al. 2012; Singhurst & Holmes 2014.

Selenia aurea Nutt., (golden), GOLDEN SELENIA. Small but striking herbaceous annuals to 27(–35) cm tall; glabrous, stems and leaves sometimes purple-tinged; stems erect to ascending, simple or branched from the base; basal leaves often rosette-like, short-petiolate, blades somewhat fleshy, pinnatisect, with 3–18 ± evenly-spaced lobes per side (these often with secondary lobes or teeth that may twist and project beyond the plane, giving the leaf a 3-dimensional toothiness), the terminal lobe largest; stem leaves similar to basal, pinnatisect, decreasing in size up the stem, the uppermost bract-like and subtending each flower; racemes several-flowered, bracteate, elongated in fruit; sepals yellowish; petals yellow, spatulate, (8–)10–13 mm long, claw differentiated from blade; stamens tetradynamous; fruiting pedicels slender, spreading to ascending; silicles short-stipitate, dehiscent, oblong to elliptical or rarely suborbicular, usually flattened parallel to replum, (5–)10–25(–30) mm long, (3.5–)5–8(–11) mm wide, valves obscurely net-veined; style slender, 3–9(–12) mm long; ovules 8–20 per ovary; seeds winged, in 2 rows per locule. Vernal pools on limestone glades, sandstone barrens, rocky prairies; rare in TX and only collected a few times, in San Augustine Co. ca. 1832 (*M.C. Leavenworth s.n.* PH!, NY!) and in Fannin Co. in 2001 (*Singhurst & White 10332* BAYLU!), as detailed in Holmes et al. (2012) and Singhurst and Holmes (2014); AR, KS, MO, OK, and TX. Mar–May. [*S. aptera* (S. Watson) Small, *S. aurea* var. *aptera* S. Watson] As this species forms large colonies in its preferred habitats in other states, it is lacking any federal protection;

Raphanus sativus [LAM]

Rapistrum rugosum [ALL]

Rorippa aquatica [CO1]

Rorippa palustris subsp. palustris [GLE]

Rorippa sessiliflora [GR1]

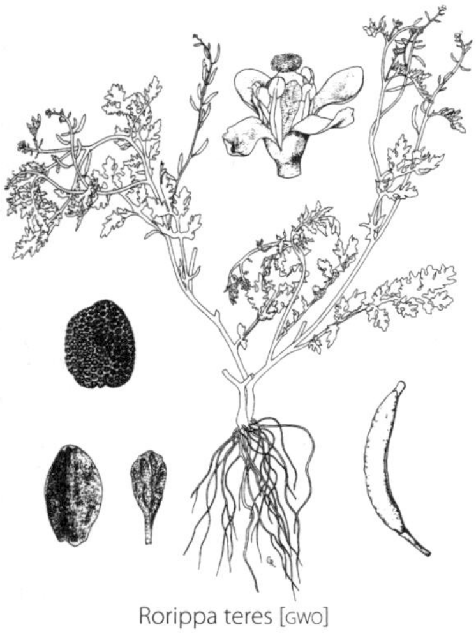

Rorippa teres [GWO]

Selenia aurea [BB2]

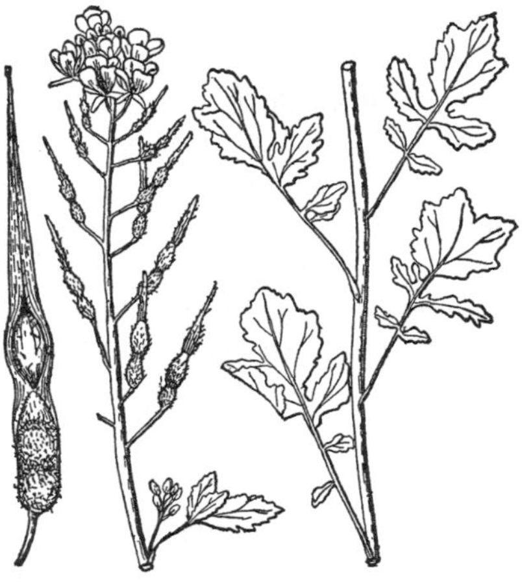

Sinapis alba [BB2]

such habitats in ne TX are rare and often mined for gravel, or otherwise disrupted. *S. aurea* should be considered imperiled in TX and worthy of protection. ⚠

SINAPIS L. MUSTARD, FIELD MUSTARD, WHITE MUSTARD

Ours rather showy herbaceous annuals, resembling species of *Brassica*; often hirsute-hispid, trichomes simple; stems usually 20–100 cm tall, erect, sometimes branching; basal leaves petiolate, usually not forming a rosette, blades lyrate to pinnatifid or pinnatisect with 1–4 lobes per side (or blades sometimes undivided); stem leaves usually short-petiolate, margins similar to basal or shallowly lobed or merely toothed; bases not auriculate; racemes considerably elongated in fruit; sepals often yellowish, usually spreading; petals bright yellow to pale yellow, obovate, claw differentiated from blade; stamens tetradynamous; fruiting pedicels usually stout; siliques linear or lanceolate, transversely 2-segmented; lower fruit segment dehiscent, terete or slightly flattened parallel to the replum, torulose, 2–8(–12) seeds per locule, valves each with 3–5(–7) prominent veins; terminal fruit segment indehiscent, conical or spear-shaped (beak-like), seedless or 1-seeded; seeds unwinged, in 1 row per locule.

A genus of 5 species generally considered native to Europe and the Mediterranean region. Closely related to and formerly included in the genus *Brassica*; as with *Brassica nigra*, an edible mustard is made from the seeds of *S. alba*. ☠ Some species are poisonous to livestock due to the presence of mustard oil glycosides (Stephens 1980). (Greek: *sinapi*, mustard) (tribe Brassiceae)
REFERENCES: Lemke & Worthington 1991; Warwick 2010g.

1. Petals pale yellow; fruiting pedicels diverging; fruits lanceolate; lower segment hispid, usually 3.5–5 mm wide; terminal segment flattened, equal in length or longer than valves, seedless; seeds yellow to pale brown ______ **S. alba** subsp. **alba**
1. Petals bright yellow; fruiting pedicels ascending or suberect; fruits linear; lower segment glabrous or pubescent, usually 2.5–3.5 mm wide; terminal segment terete, much shorter than valves, seedless or 1-seeded; seeds reddish brown to black ______ **S. arvensis**

Sinapis alba L., (white), WHITE MUSTARD, YELLOW MUSTARD. Stems usually hispid, rarely glabrous; basal leaf blade margins lyrate to pinnatifid or pinnatisect; stem leaves similar or less-divided; petals pale yellow, (7–)8–12(–14) mm long; fruiting pedicels spreading; fruits lanceolate, (1.5–)2–4.2(–5) cm long, (2–)3–5.5(–6.5) mm wide; lower segment hispid, equal to or shorter than terminal segment, 2–5 seeds per locule; terminal segment spear-shaped, flattened, seedless. Roadsides, weedy areas, sometimes as an escape or waif from cultivation; we have located no records from TX and are unable to provide a range map for the state; Turner et al. (2003) did not include this taxon for TX, and Aplaca (2010) was apparently unable to locate a voucher for the state; however, Warwick (2010g) includes TX in the range for this taxon, and Kartesz (2015) and USDA Plants Database (2021) indicate it is present in TX based on Hatch et al. (1990); Lemke and Worthington (1991) describe the range of this taxon in TX (as *Brassica hirta*) as "roadsides and railroad rights-of-way in widely scattered parts of the state;" scattered across most of the U.S.; Canada; Mexico; Central and South America; Eurasia; introduced elsewhere. Mar–May (probably). [*Brassica alba* (L.) Rabenh., *B. hirta* Moench, *Raphanus albus* (L.) Crantz, *Rorippa coloradensis* Stuckey, *S. foliosa* Willd.] Native to Eurasia and long-cultivated there for its oil-and mustard-producing seeds and for greens; now cultivated in temperate regions worldwide and occasionally reported as a weedy escape. According to Warwick (2010g), *S. alba* subsp. *alba* is the only subsp. naturalized in the New World. We are unable to provide a TX range map for this taxon (see discussion above).

Sinapis arvensis L., (of cultivated fields), FIELD MUSTARD, CHARLOCK. Stems usually hirsute to hispid or sometimes glabrous; basal leaf blade margins lyrate to pinnatifid or sometimes undivided; stem leaves often undivided; petals bright yellow, (8–)9–12(–17) mm long; fruiting pedicels ascending or suberect; fruits linear, (1.5–)2–4.5(–5.7) cm long, (1.5–)2.5–3.5(–4) mm wide; lower segment glabrous or pubescent, longer than terminal segment, (2–)4–8(–12) seeds per locule; terminal segment ± conical, terete, seedless or 1-seeded. Roadsides, cultivated fields, disturbed areas; widespread in East TX; nc to s TX, Edwards Plateau, a few counties further w; recorded throughout the U.S.; Canada; Mexico; Eurasia; introduced elsewhere. Feb–May. [*Brassica arvensis* (L.) Rabenh., *B. kaber* (DC.) L.C. Wheeler, *B. sinapis* Vis., *B. sinapistrum* Boiss., *Sinapis arvensis* var. *arvensis* L., *S. kaber* DC.] We are following Warwick (2010g) in not recognizing infraspecific taxa, which are deemed too variable to be meaningful. Generally considered native to Eurasia, but perhaps originally circumboreal in native distribution, or dispersed to North America in pre-Columbian times, based on archaeological evidence (as *Brassica kaber*, in Jacobson et al. 1988); now a weed worldwide, especially of cultivated fields. Young leaves are palatable, but poisoning and death have been reported in livestock from eating large quantities of the plant or seeds (Mulligan & Bailey 1975). ☠

SISYMBRIUM L. HEDGE-MUSTARD, TUMBLE-MUSTARD, ROCKET

Ours herbaceous annuals; glabrous or pubescent with simple trichomes; stems erect, often branching; basal leaves sometimes in a rosette, petiolate, blade margins runcinate to lyrate-pinnatifid or pinnatisect, with 1–8 lobes per side, terminal lobe often largest; stem leaves petiolate or subsessile, blades similar to basal or fewer-lobed or -toothed, becoming smaller and narrower

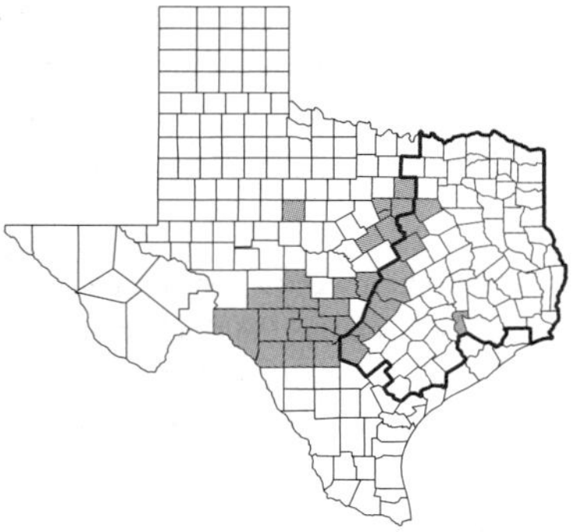

Physaria recurvata

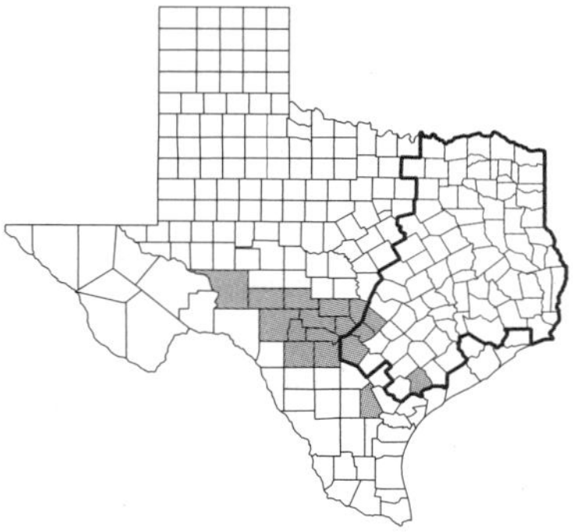

Physaria sessilis

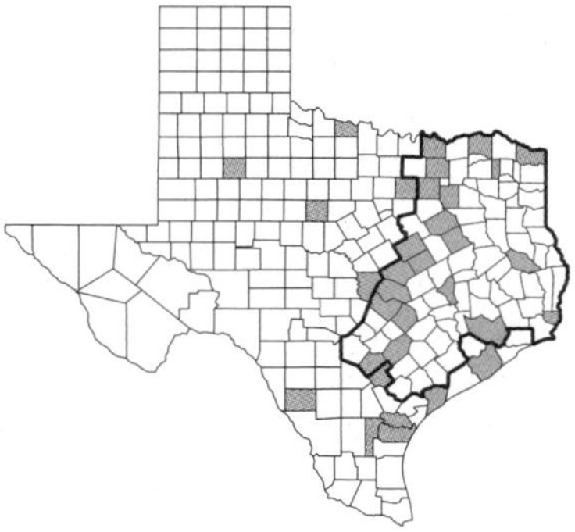

Planodes virginicum

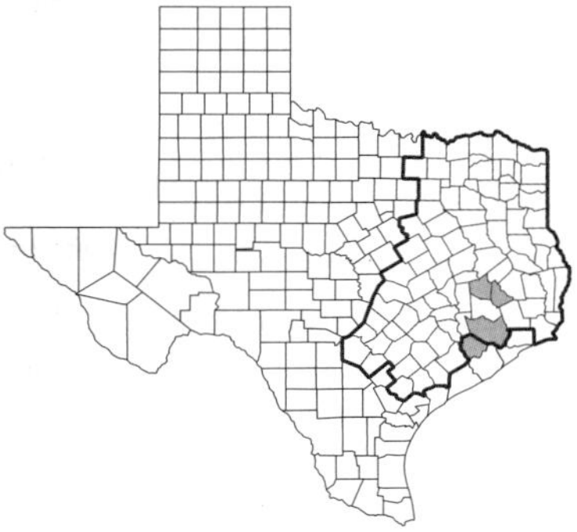

Raphanus raphanistrum

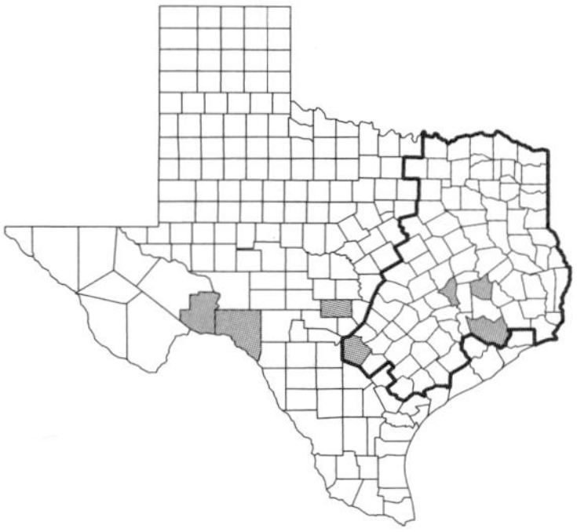

Raphanus sativus

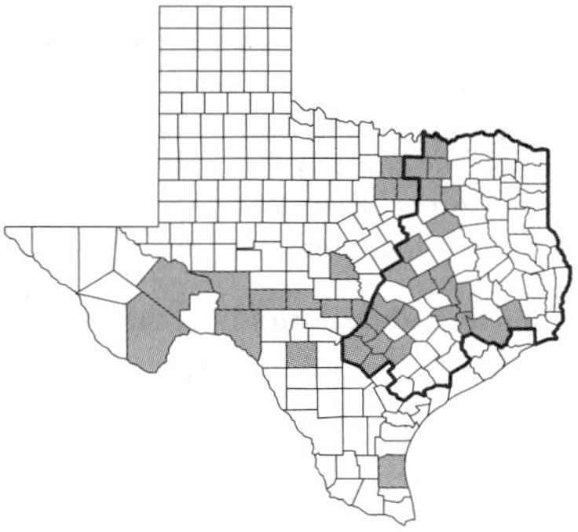

Rapistrum rugosum

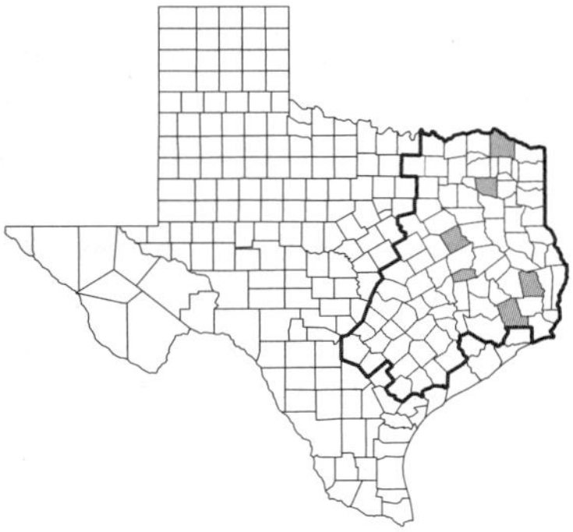

Rorippa aquatica

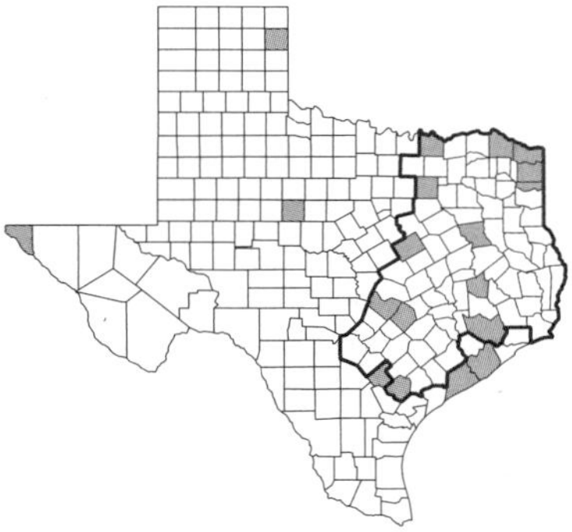

Rorippa palustris subsp. palustris

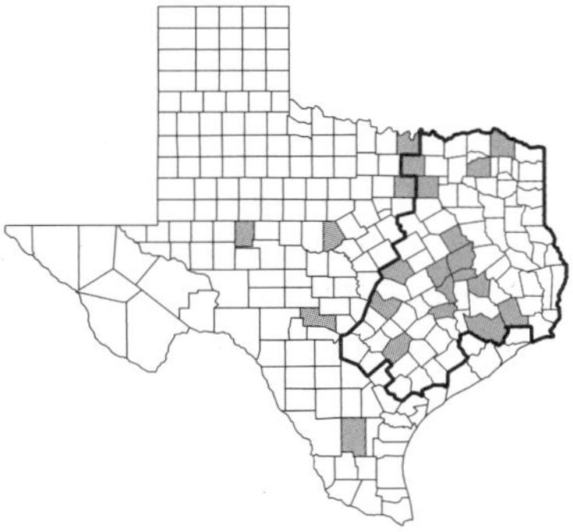

Rorippa sessiliflora

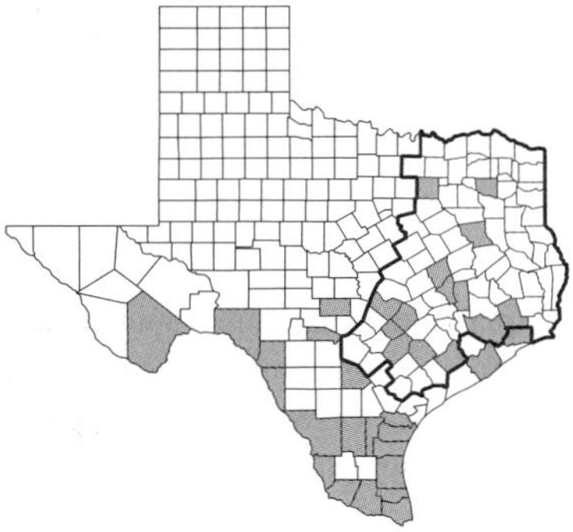

Rorippa teres

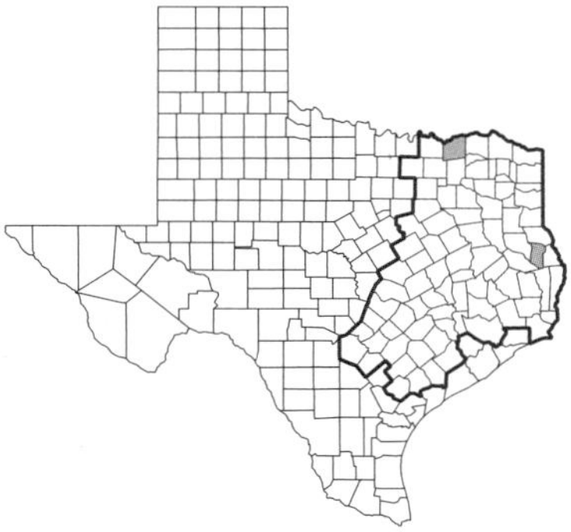

Selenia aurea

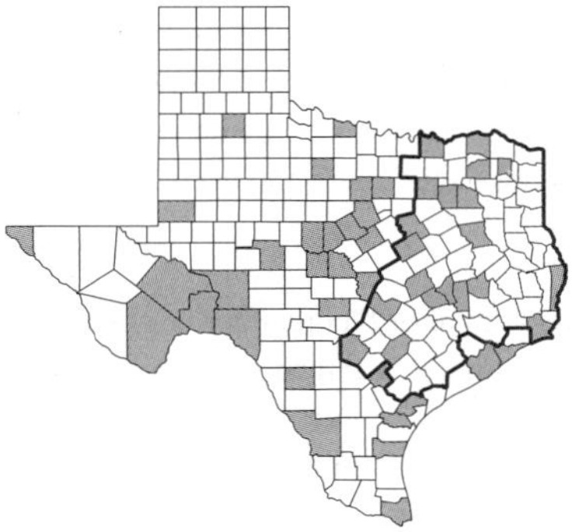

Sinapis arvensis

or entire on upper stem, bases narrowed or sometimes basal lobes hastate (not auriculate); racemes often elongated in fruit; sepals often yellowish, erect to spreading; petals yellow, spatulate or oblong-oblanceolate, claw and blade differentiated; stamens tetradynamous; fruiting pedicels slender or stout, spreading or ascending to erect; siliques dehiscent, linear or subulate, terete or slightly torulose, valves each prominently 3-veined; style 0.2–2 mm long, stigma 2-lobed; ovules 10–120 per ovary; seeds unwinged, in 1 row per locule.

A genus of ca. 40 species native to Europe, Asia, and Africa, including some now-cosmopolitan weeds. *Sisymbrium* species have been used medicinally and as edible greens; the ground seed has a mustard flavor and has been used to thicken beverages and soups (Moerman 1998). Called TUMBLE-MUSTARDS because entire, mature plants of some species break off at the base and roll in the wind like tumbleweeds, spreading the seeds. (Latinized from an ancient Greek name for some plant of this family) (tribe Sisymbrieae)

REFERENCES: Payson 1922; Al-Shehbaz 2010p.

1. Upper stem leaf blades pinnatisect with all lobes linear or filiform; fruiting pedicels ca. as wide as fruit; petals 5–10 mm long; mature fruits 4.5–12 cm long ____ **S. altissimum**
1. Upper stem leaf blades subentire to dentate or lobed (lobes not linear); fruiting pedicels narrower than fruit; petals 2.5–4 mm long; mature fruits 0.7–5 cm long.
   2. Fruiting pedicels spreading or ascending, 5–20 mm long; lower stems glabrous or sparsely pubescent; mature fruits > 2 cm long ____ **S. irio**
   2. Fruiting pedicels closely appressed to rachis, 1.5–4 mm long; lower stem usually sparsely to densely hirsute with retrorse trichomes; mature fruits < 2 cm long ____ **S. officinale**

Sisymbrium altissimum L., (tallest), TALL TUMBLE-MUSTARD. Plants to 160 cm tall; stems sparsely to densely hirsute basally, glabrate in the upper portion; basal leaves in a rosette, blades with 3–8 lobes per side; upper stem leaf blades pinnatisect with all lobes linear or filiform; petals yellow, (5–)6–8(–10) mm long; fruiting pedicels spreading, nearly as wide as fruit; fruits narrowly linear, (4.5–)6–9(–12) cm long, 1–2 mm wide. Railroads, roadsides, pastures, waste ground; Grayson, Hardin, and McLennan cos. (Kartesz 2015) and Dallas and Polk cos.; scattered in n 2/3 of TX; a widespread weed throughout most of the U.S. and Canada; Eurasia and n Africa; introduced elsewhere. Feb–Jun. [*Hesperis altissima* (L.) Kuntze, *Norta altissima* (L.) Britton] Native to Europe and w Asia.

Sisymbrium irio L., (Latin for a mustard-like plant), ROCKET-MUSTARD, LONDON ROCKET. Plants to 75 cm tall; stems glabrous or sparsely pubescent at least basally; basal leaves not forming a rosette, blades with 1–8 lobes per side; upper stem leaf blades few-lobed or entire; petals yellow, (2.5–)3.5–4 mm long; fruiting pedicels spreading or ascending, narrower than fruit; fruits narrowly linear (2.5–)3–4(–5) cm long, 0.9–1.1 mm wide. Fields, roadsides, waste ground, disturbed sites; Harris Co. (Brown 2014), McLennan Co. (Kartesz 2015), and Brazos and Dallas cos., also edge of Edwards Plateau in Bexar, Guadalupe, Hays, and Travis cos.; scattered in TX but more common in w and s TX; sporadic across the U.S., most frequent in sw states; Mexico; Canada; Eurasia; and n Africa; introduced elsewhere. (Nov–)Feb–May. [*Descurainia irio* (L.) Webb & Berthel., *Erysimum irio* (L.) Farw., *Norta irio* (L.) Britton] Native to Europe and w and c Asia, now a weed in many temperate regions of the world.

Sisymbrium officinale (L.) Scop., (medicinal), HEDGE-MUSTARD, TANSY-MUSTARD. Plants to 110 cm tall; stems usually sparsely to densely hirsute (trichomes retrorse), rarely glabrate in the upper portion; basal leaves usually in a rosette, blades with 2–5 lobes per side; upper stem leaf blades few-lobed to dentate or subentire; petals yellow, 2.5–4 mm long; fruiting pedicels erect, appressed to rachis, stout; fruits subulate-linear (0.7–)1–1.4(–1.8) cm long, 1–1.5 mm wide. Fields, thickets, waste ground, construction sites; Anderson, Denton, Grayson, McLennan, Nacogdoches, Navarro, Robertson, Travis, Van Zandt, and Walker cos.; also nc and sw TX; throughout the U.S. but most frequent in the Midwest, ne states, and the w coast; Canada; Eurasia; and n Africa; introduced elsewhere. Apr–Jun. [*Erysimum officinale* L., *S. officinale* (L.) Scop. var. *leiocarpum* DC.] Native to Eurasia, now a weed in many regions of the world.

STREPTANTHUS Nutt. JEWELFLOWER, TWISTFLOWER

Ours annual or biennial herbs; glabrous throughout or sparsely pubescent with simple trichomes, sometimes glaucous; stems to 130 cm tall, unbranched or branched in the upper portion; basal leaves sometimes in a rosette, sometimes withered by flowering, usually petiolate, blades entire or dentate to sinuate or hastate- or lyrate-pinnatifid; stem leaves sessile or short-petiolate, on lower stems similar to basal, on upper stems becoming entire, bases sometimes auriculate-clasping, blades decreasing in size and sometimes continued as bracts on the raceme; racemes usually elongating in fruit, sometimes bracteate; flowers sometimes bilaterally symmetrical; calyx usually campanulate, sepals purplish (often similarly-colored as petals); petals white, lavender, pink, purplish, or magenta, veins often darker than blades, claw and blade usually differentiated; stamens tetradynamous or in

3 pairs of unequal length; fruiting pedicels slender or stout, spreading or ascending; siliques short-stipitate, dehiscent, linear, flattened parallel to the replum; valve midveins often prominent; style 0.3–3.5 mm long; ovules 28–120 per ovary; seeds winged, in 1 row per locule. Given measurements of floral parts and fruits apply to early-season flowers of the lower racemes.

A genus of ca. 35 species of s and w U.S. and n Mexico. Calluses on the leaf margins of some species mimic pierid butterfly eggs, reducing egg deposition and thus herbivory by larvae (Buck et al. 1993). Some species are endemic to serpentine soils and are known to hyperaccumulate nickel (Kruckeberg & Reeves 1995). (Greek: *streptos*, twisted, and *anthos*, flower, from the curved and wavy-margined petals of some species) (tribe Thelypodieae)
REFERENCE: Al-Shehbaz 2010q.

1. Stem leaves sessile, blade bases auriculate-clasping.
   2. Plants of e 1/3 TX; sepals 4–9 mm long; seeds 2–2.5 mm long ____ **S. maculatus**
   2. Plants of the Edwards Plateau and Trans-Pecos; sepals 8–12 mm long; seeds 3–5 mm long.
      3. Racemes usually lacking bracts, at least in the upper portion; petals 16–27 mm long; stamens in 3 pairs of unequal length; fruits 4.5–6 mm wide ____ **S. platycarpus**
      3. Racemes with bracts subtending each pedicel; petals 14–19 mm long; stamens tetradynamous; fruits 2.5–4 mm wide ____ **S. bracteatus**
1. Stem leaves petiolate or sessile, blade bases not auriculate-clasping.
   4. Midstem leaf blade margins lyrate-pinnatifid; sepals 3–4.5 mm long; fruits 3.5–5 mm wide ____ **S. petiolaris**
   4. Midstem leaf blade margins entire; sepals 7–11 mm long; fruits 1.5–2 mm wide ____ **S. hyacinthoides**

Streptanthus bracteatus A. Gray, (with bracts), BRACTED JEWELFLOWER, BRACTED TWISTFLOWER. Plants annual or biennial; glaucous, usually glabrous; basal leaves not in a rosette, long-petiolate, blades lyrately lobed to irregularly dentate; stem leaves sessile, blades oblong to ovate, margins shallowly dentate or entire, bases auriculate-clasping; racemes with bracts subtending each pedicel; calyx campanulate, sepals purplish; petals lavender to lilac-purple, 14–19 mm long; stamens tetradynamous; fruits 8–14.5 cm long, 2.5–4 mm wide. Oak-juniper glades, limestone bluffs, moist gravelly clay-loams; endemic to the Balcones Canyonlands of the Edwards Plateau; sw margin of East TX in Bexar, Comal, Hays, and Travis cos. (Fowler 2014); further w in Bandera, Medina, Real, and Uvalde cos. (Poole et al. 2007) and Blanco Co. (Kartesz 2015); endemic to TX. Apr–Jun. [*Erysimum bracteatum* (A. Gray) Kuntze] Under consideration for federal endangered species status. Threatened by white-tail deer herbivory and urban development; some populations have been impacted or extirpated by housing developments (Poole et al. 2007, Fowler 2014). NatureServe: G1, critically imperiled; S1, critically imperiled.

Streptanthus hyacinthoides Hook., (resembling *Hyacinthus*—HYACINTH), SMOOTH JEWELFLOWER, SMOOTH TWISTFLOWER. Plants annual; usually glabrous; basal leaves withering early; stem leaves short-petiolate or sessile, blades usually linear to lanceolate, margins entire, bases not auriculate; racemes ebracteate; calyx campanulate, sepals purplish; petals purple to lavender or whitish, 13–20 mm long, veins and blade bases deep purple; stamens in 3 unequal pairs; fruits 3.7–11.2 cm long, 1.5–2 mm wide. Sandy prairies, dry sandhills, open woods, roadsides; mostly ne and ec TX; Bastrop, Freestone, Hardin, Van Zandt (Kartesz 2015) cos., Austin (*Tharp s.n.*, 1940, TEX), Franklin (*Ajilvsgi 6981* BRIT), Limestone (*Holmes 7685* BAYLU), and Marion (*R. Kral 68604* (APSC)!) cos., also Anderson, Leon, Smith, and Wood cos. nc TX; AR, KS, LA, OK, and, TX. May–Jun(–sporadically to Aug). [*Erysimum hyacinthoides* (Hook.) Kuntze, *Euklisia hyacinthoides* (Hook.) Small, *Icianthus atratus* Greene, *I. glabrifolius* (Buckley) Greene, *I. hyacinthoides* (Hook.) Greene, *S. glabrifolius* Buckley]

Streptanthus maculatus Nutt., (spotted), CLASPING JEWELFLOWER, CLASPING TWISTFLOWER. Plants annual; glaucous, usually glabrous; basal leaves in a rosette (soon withering), petiolate, blades entire; stem leaves sessile, blades broadly ovate to oblong or lanceolate, margins entire, bases auriculate-clasping; racemes ebracteate; calyx campanulate, sepals purplish; petals lavender to magenta, 11–21 mm long, veins and blade bases deeper magenta; stamens tetradynamous; fruits 6–11.4 cm long, 2–3 mm wide. Sandy pine and oak woods, creekbeds, rocky bluffs, roadbanks; ne and ec TX in Anderson, Cherokee, Nacogdoches, San Augustine, and Smith cos. (Kartesz 2015), Henderson Co. (iNaturalist.org observation *25059081*, *Bill Freiheit, May 2019* (iNaturalist.org 2021)), and Angelina, Gregg, Sabine, and Wood cos.; AR, OK, and TX. Apr–May. [*Erysimum maculatum* (Nutt.) Kuntze, *S. maculatus* subsp. *obtusifolius* (Hook.) Rollins, *S. obtusifolius* Hook.] We are following Al-Shehbaz (2010q) in not recognizing subspecies.

Streptanthus petiolaris A. Gray, (with a petiole), BRAZOS JEWELFLOWER, BRAZOS ROCKCRESS. Plants annual; pilose in the lower portion; basal leaves in a semi-rosette, long-petiolate, blades lyrate-pinnatifid; stem leaves petiolate, blades lyrate-pinnatifid, hastate, becoming linear-lanceolate on upper stem, margins becoming entire, bases not auriculate; racemes ebracteate; calyx subcampanulate, sepals lavender; petals white to pink or lavender, 4–8 mm long, veins sometimes darker than blades; stamens tetradynamous; fruits 4–9 cm long, 3.5–5 mm wide. Rocky limestone prairies, juniper-oak woods, canyons, roadsides; mostly c and sc TX; endemic to TX. Mar–May. [*Arabis petiolaris* (A. Gray) A. Gray, *Erysimum petiolare* (A. Gray) Kuntze, *S. brazoensis* Buckley] This species has often been treated under *Arabis*, but it has a chromosome number universal in *Streptanthus* and lacks the branched trichomes found in *Arabis* and its segregate genera (Al-Shehbaz 2003, 2010q). Once considered for federal listing as a threatened or endangered species, but found to be more common than previously assumed (Carr 2008).

Streptanthus platycarpus A. Gray, (wide-fruited), BROAD-POD JEWELFLOWER, SPARSE-FLOWER JEWELFLOWER. Plants annual; glaucous, usually glabrous; basal leaves in a semi-rosette (soon withering), petiolate, margins sinuate to lyrate or pinnatifid; stem leaves sessile, blades ovate, margins sinuate-dentate becoming entire distally, bases auriculate-clasping; racemes bracteate in the lower portion or ebracteate; calyx campanulate, sepals purplish to maroon; petals lavender to purplish, 16–27 mm long, veins darker than blades; stamens in 3 unequal pairs; fruits 4–9.5 cm long, 4.5–6 mm wide. Cliff ledges, rocky creek beds, roadsides, often on limestone; once recorded in Travis Co. (*M.S. Young s.n.*, 17 Apr 1917, TEX-LL) and perhaps only a rare waif on the sw border of East Texas, since all populations recorded in the current century are further w (Carr 2008); mostly w Edwards Plateau and Trans-Pecos; NM and TX; ne Mexico. Mar–Apr. [*Erysimum platycarpum* (A. Gray) Kuntze, *S. sparsiflorus* Rollins.] The formerly-recognized *S. sparsiflorus* was found by Al-Shehbaz and Mayer (2008) to be indistinguishable in most aspects from *S. platycarpus* and was reduced to synonymy.

THLASPI L. PENNYCRESS

A genus of 4 species, native to Eurasia and n Africa; including one worldwide weed. *Thlaspi* once included ca. 75 species, but molecular and morphological studies have resulted in the removal of most to other genera. (Greek: *thlao-*, to compress, alluding to the flattened silicle) (tribe Thlaspideae)

REFERENCES: Al-Shehbaz 2010r; Esmailbegi et al. 2018.

Thlaspi arvense L., (of cultivated fields), FIELD PENNYCRESS, FRENCHWEED, STINKWEED, FANWEED. Unpleasantly-scented, herbaceous annuals; glabrous, sometimes glaucous; stems erect, to 55(–80) cm; sometimes branching; basal leaves not forming a rosette, petiolate, blades oblanceolate to spatulate or obovate, margins entire or repand-sinuate to dentate; stem leaves sessile, blades oblong, margins similar to basal, apices rounded, bases auriculate or sagittate; racemes considerably elongated in fruit; petals white, 2.4–5 mm long, narrowed to claw-like base; stamens slightly tetradynamous; fruiting pedicels slender, diverging; silicles dehiscent, obovate to suborbicular, strongly flattened perpendicular to the replum, (6–)9–20 mm long, (5–)7–20 mm wide, valves keeled, marginal wings 3.5–5 mm wide (widest apically), apical notch ca. 5 mm deep; style 0.1–0.3 mm long; ovules 6–16 per ovary; seeds unwinged, in 1 row per locule, seed coat concentrically striate. Roadsides, railroads, fields, disturbed sites; scattered across East TX in Brazos, Denton, Grayson, Henderson, Robertson, and Travis cos., also Hardin Co. (Kartesz 2015), Red River (*R.H. Smith 9* BRIT), and Smith (*Moore 629* GH) cos.; a few other cos. farther w; throughout the U.S.; Canada; Eurasia; introduced elsewhere. Mar–early May. [*Teruncius arvensis* (L.) Lunell] Native to Eurasia; an invader of agricultural cropland, now a cosmopolitan weed. Formerly used medicinally (Mabberley 1997); the plant can be toxic to animals; reported to cause poisoning and death of cattle when seeds are consumed in large quantities (Warwick et al. 2002).

CABOMBACEAE Rich. ex A. Rich.

WATER-SHIELD FAMILY

Aquatic, rhizomatous, perennial herbs rooting in mud; stems floating; vegetative tissues with conspicuous air chambers; leaves floating or both floating and submerged, without stipules; flowers axillary, solitary, on thick peduncles, usually borne at or above water surface, radially symmetrical, the flower parts all free; sepals and petals each usually 3(–4), persistent in fruit; stamens 3–38(–51), with flattened filaments; pistils (1–)2–18, separate, usually with 1–3(–5) ovules or seeds per unicarpellate pistil; ovaries superior; fruits indehiscent, leathery; seeds 1–3.

A very small (6 species in 2 genera—Wiersema 1997c; Stevens 2001 onwards; Mabberley 2017) mainly pantropical and warm temperate family of aquatics widely scattered worldwide. It has often been lumped with the Nymphaeaceae (e.g., Correll & Johnson 1970) and is closely related to that family, but differs in a variety of ways including having all the flower parts free, the stems floating, and lacking alkaloids (Stevens 2001 onwards). Further, the two families have had separate evolutionary lineages for tens of millions of years (Wikström et al. 2001; Yoo et al. 2005). *Cabomba* and *Brasenia* have flower parts in 3s like those of monocotyledons and the family has often been considered to be among the "paleoherbs," believed to be an early evolutionary lineage of flowering plants. Indeed, recent molecular evidence (e.g., Soltis et al. 2000; Stevens 2001 onwards; Qiu et al. 2010) places the Cabombaceae (in the Nymphaeales along with the related Nymphaeaceae) on a branch near the very base of the angiosperms (the second most basal branch) (see Appendix 4). Thus the Cabombaceae is extremely interesting evolutionarily

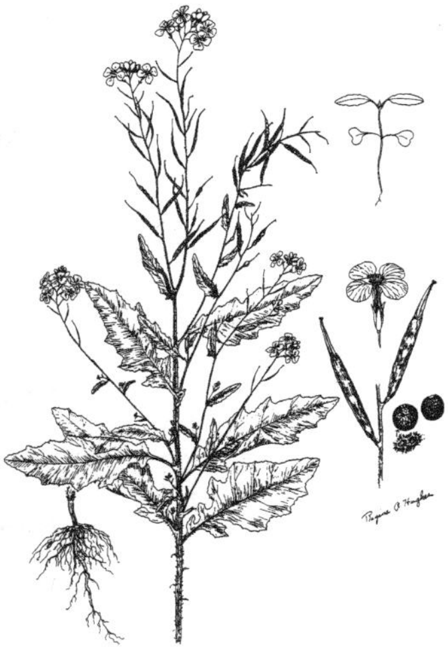
Sinapis arvensis [REE]

Sisymbrium altissimum [REE]

Sisymbrium irio [GLE]

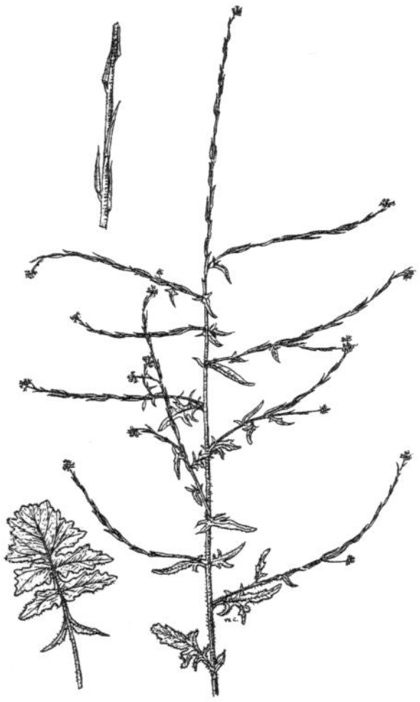
Sisymbrium officinale [ROB]

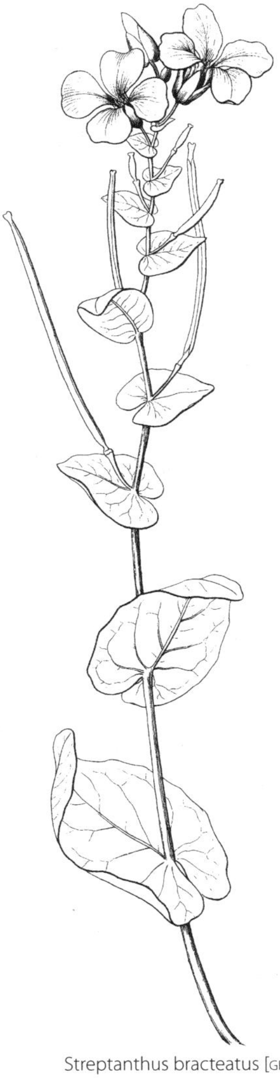
Streptanthus bracteatus [GR1]

Streptanthus maculatus [ANS]

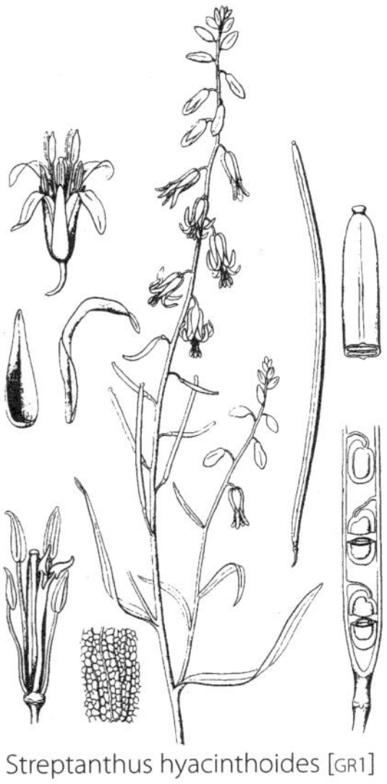
Streptanthus hyacinthoides [GR1]

since it is one of the most ancient surviving lineages of flowering plants. The family has a long fossil history with fossils of *Pluricarpellatia*, probably Cabombaceae or close to it, known from 115 million year old rocks in northeast Brazil (Mohr et al. 2008). (subclass Magnoliidae—Cronquist; order Nymphaeales—APG IV)

FAMILY RECOGNITION IN THE FIELD: aquatics with peltate floating leaves and either with conspicuous mucilage on underwater surfaces or with much dissected submerged leaves; perianth usually 3-merous; carpels free.

REFERENCES: Wood 1959; Ørgaard 1991; Osborn et al. 1991; Williamson & Schneider 1993a; Wiersema 1997c; Schneider et al. 2003; Aoki et al. 2004; Yoo et al. 2005; Löhne et al. 2008.

1. Plants with only undivided, alternate, peltate, floating leaves, the blades 3.5–11(–13.5) cm long; lower surfaces of leaf blades and petioles coated with a heavy, very conspicuous layer of jelly-like mucilage; stamens 18–38(–51); perianth purplish, 10–20 mm long ____ **Brasenia**
1. Plants with much dissected, opposite (rarely whorled), submerged leaves and also usually small (blades ca. 1–3 cm long), alternate, peltate, floating leaves; lower surfaces of leaf blades and petioles without conspicuous mucilage layer, the tissues only barely mucilaginous; stamens 3–6; perianth usually whitish (rarely purplish or yellowish), 4–12 mm long ____ **Cabomba**

BRASENIA Schreb. WATER-SHIELD, PURPLE WEN-DOCK

A monotypic genus known nearly worldwide (America, Africa, India, temperate e Asia, and Australia) except for Europe (where it is known from fossils) (Wiersema 1997c; Stevens 2001 onwards). Stems with vascular bundles in two pairs and no vascular cambium (Stevens 2001 onwards; Mabberley 2017). Submerged parts are conspicuously covered with mucilaginous jelly; this is possibly an anti-herbivore defense mechanism (Keddy 2010; Thompson et al. 2014) or is useful allelopathically in competition with other plants. (Named for Christopher Brasen, 1734–1774, Moravian missionary and plant collector in Greenland and Labrador—Pringle 1995; Wiersema 1997c)

REFERENCES: Elakovich & Wooten 1987; Osborn & Schneider 1988; Osborn et al. 1991; Pringle 1995; Taylor & Williams 2009; Thompson et al. 2014; Acuña-Castillo et al. 2021.

Brasenia schreberi J.F. Gmel., (for Johann Christian Daniel von Schreber, 1739–1810, German botanist), PURPLE WEN-DOCK, WATER-SHIELD, SCHREBER'S WATERSHIELD. Rooted aquatic; jelly-like mucilage layer thick and obvious; stems to 2 m long; leaves alternate, long-petiolate, blades elliptic to broadly oval, the ends rounded, 3.5–11(–13.5) cm long, 2–6.5 cm wide, centrally peltate, entire or slightly crenate; flowers on peduncles to 15 cm long, emergent when open; perianth 12–20 mm long, the 3 petals slightly longer and narrower than the 3 sepals; sepals not petaloid; petals without proximal auricles; pistils 2–18; fruits fusiform, 5–10 mm long, with 1–2 seeds. Ponds, lakes, slow streams; Pineywoods and Post Oak Savannah, w in Red River drainage to Lamar Co. (*W.R. Carr 13993* TEX); widespread throughout the U.S. and adjacent Canada, except the sw U.S.; the species scattered worldwide except Europe. May–Jul. [*B. peltata* Pursh] This species has a very wide distribution: "Scattered, tropical and (warm) temperate, but also S.E. Australia and S.E. Siberia" (Stevens 2001 onwards). Cultivated as an ornamental, as a vegetable in China, the young leaves eaten in Japan, and the roots were consumed by Native Americans (Mabberley 2017; Les 2017). This species is wind-pollinated (Osborn & Schneider 1988; Osborn et al. 1991; Wiersema 1997c; Taylor & Williams 2009). *Brasenia* can in some situations become dominant, excluding other aquatic plants; it has anti-bacterial, anti-algal, and phytotoxic properties which contribute to such dominance; it is possibly useful as an allelopathic way of controlling aquatic weeds (Elakovich & Wooten 1987). The mucilage is thought to reduce leaf herbivory (Thompson et al. 2014).

CABOMBA Aubl. FANWORT

A genus of 5 species (Wiersema 1997c; Stevens 2001 onwards) of mostly tropical to warm areas of the Americas. Pollination is by insects, in particular flies (Wiersema 1997a; Taylor & Williams 2009) (Probably an aboriginal name, possibly from Guiana)

REFERENCES: Fassett 1953; Schneider & Jeter 1982; Ørgaard 1991; Teixeira de Lima et al. 2014; Invasive.org 2024.

Cabomba caroliniana A. Gray, (of Carolina), CAROLINA FANWORT, FANWORT. Rooted aquatic; stems to 2 m or more long; submersed leaves opposite or rarely whorled, the blades palmately 2–3 times dissected into linear-filiform segments; floating leaves alternate, few, petiolate, the blades peltate, entire, linear-elliptic, often notched at one end, 1–3 cm long; flowers solitary on axillary peduncles; sepals 3; petals 3; perianth 4–12(–18) mm long, the petals slightly longer than the petaloid sepals; sepals and petals similarly colored; petals proximally with yellow, nectar-bearing auricles; pistils (1–)2–4; fruits ovoid, 1–3–seeded, 4–7 mm long. Ponds, lakes, quiet streams; widespread in Pineywoods, also Colorado Co. (*B.H. Warnock et al. 46440* TEX) in Post Oak Savannah and Comal (Turner et al. 2003), Hays (*D.S. Correll & E.C. Ogden 25368* TEX), and Travis (*R.W. Strandtmann s.n., 1940*, TEX) cos. near w edge of East TX; also n Gulf Prairies and Marshes and Edwards Plateau; widespread in the e U.S. and adjacent Ontario w to MI, OK, and TX, and WA, OR, and CA; s to South America. Apr–Jul. [*C. caroliniana* var. *pulcherrima* R.M.

Harper, *C. pulcherrima* (R.M. Harper) Fassett] While native to the se U.S., this species is considered an aggressive and difficult to eradicate weed in the ne U.S. and in other parts of the world where introduced (Crow & Hellquist 2000a). It "can form extremely dense stands which can clog drainage systems and interfere with recreational activities such as swimming and boating. It is still sold in the aquarium trade. It is believed the introduction of *Cabomba caroliniana* outside of its native range is a result of improper disposal of aquarium contents. The plants are able to root from vegetative parts and thus are easily spread. In the north it appears that the seeds do not readily germinate. "However, viable shoots have been observed in January beneath ice" (Invasive.org 2024). *Cabomba* species have nectaries on the petals and are pollinated by flies (Schneider & Jeter 1982; Stevens 2001 onwards; Taylor & Williams 2009). ⦸

CACTACEAE Juss.

CACTUS FAMILY

Fleshy or soft-woody, usually armed (with spines), succulent, green-stemmed pincushions to shrubs or small trees; leaves in ours absent, or present as fleshy points on new growth; stems (primary photosynthetic organ, also modified for water storage) segmented or unsegmented, sometimes with tubercles or ribs, with specialized, axillary, cushion-like bud areas (technically short shoots = areoles) usually bearing spines (technically derived from reduced leaves) and sometimes barbed hairs or bristles (= glochids); flowers solitary per areole (but can be crowded together), sessile, closed at night in most species (one exception being *Echinocereus coccineus*), often showy, radially symmetrical; sepal-like perianth parts (= outer tepals) 5 or more, grading into the many, thin-textured petal-like structures (= inner tepals), all joined at base to form a cup or tube (= hypanthium) on summit of the thick, pedicel-like, inferior ovary; stamens many; fruit berry-like, dry or fleshy.

A medium-large (ca. 1,800 species in 125–130 genera—Parfitt & Gibson 2003) family of stem-succulent xerophytes native almost exclusively to the New World (North to South America, West Indies, and the Galápagos Islands)—only *Rhipsalis* Gaertn., a tropical, epiphytic, bird-dispersed, "mistletoe cactus" has reached the Old World (Africa, Madagascar, and Sri Lanka), apparently without assistance from humans (Parfitt & Gibson 2003). Indeed, members of the family are among the most visually striking, morphologically distinctive, and characteristic plants of warm arid regions of the Western Hemisphere. Various cacti have been introduced in many parts of the world for human or animal food, as ornamentals, or for the cochineal dye industry (see discussion under *Opuntia*), and cacti are now widely naturalized and some have become problematic invaders. In addition to the typical growth forms, the family includes a number of epiphytes. Xerophytic adaptations include a thick cuticle, large volume to surface ratio, widespread shallow root system (to take in any available rain), crassulacean acid metabolism (CAM photosynthesis—allows night absorption and storage of CO_2 thereby reducing water loss through transpiration during the day—Crosswhite 1984), and spines (condense dew and protect from herbivores). Numerous species are utilized as ornamentals, both as house plants and in landscapes, and some are endangered due to overcollecting (e.g., "cactus rustling"); cacti protected by the Convention of International Trade in Endangered Species (CITES) (Barthlott & Hunt 1993). Many species have alkaloids (e.g., *Lophophora williamsii* (Lem. ex Salm-Dyck) J.M. Coult.—PEYOTE, which contains the hallucinogen, mescaline, and is used in religious ceremonies by Native Americans). Cactaceae have a long history of human use—Anderson (2001) indicated that they have been used for 12,000 years; some were very important to Native Americans for food, etc. Because of their extreme morphological specializations, the taxonomic placement of the Cactaceae was problematic in the past; however, biochemical and molecular markers now indicate that the family is a monophyletic group whose relationship is with the Caryophyllales. A study by Downie and Palmer (1994) suggested the family might be derived from Portulacaceae. On the basis of molecular data, Hershkovitz and Zimmer (1997) also indicated that the family is derived from within the Portulacaceae and that the molecular divergence "between pereskioid cacti and the genus *Talinum* (Portulacaceae) is less than that between many Portulacaceae genera." Molecular research (e.g., Applequist & Wallace 2001) confirms the relationship between a "portulacaceous cohort" and the Cactaceae. The ancestral condition within the family is clearly observable in the small subfamily Pereskioideae, composed of shrubs and trees with only slightly succulent stems and well-developed, broad, seasonally deciduous leaves. This subfamily is an excellent example of an evolutionary transitional form—it nicely bridges the gap between "normal"-appearing plants and the specialized morphology typical of most Cactaceae. Molecular research (Nyffeler 2002; Edwards et al. 2005) confirms that *Pereskia* is basal within the family and is in fact paraphyletic, with the rest of the family arising from within the genus. The molecular research also suggests that the family originated in South America (where many basal family members reside), possibly in the mid-late Tertiary Period (20–25 million years ago) associated with the climate instability resulting from the formation of the central Andes Mountains. Family name conserved from *Cactus*, a generic name that was deemed illegitimate and rejected in favor of *Mammillaria* (Farr et al. 1979). (Greek: *kaktos* or Latin: *cactus*, a spiny or prickly plant) (subclass Caryophyllidae—Cronquist; order Caryophyllales—APG IV)

FAMILY RECOGNITION IN THE FIELD: *leafless* stem *succulents* with watery sap, areoles (= pad-like axillary buds unique to the family) typically bearing *spines*, and flowers showy, with numerous perianth parts (= tepals) and numerous stamens.
REFERENCES: Britton & Rose 1919–1923, 1937, 1963; Schulz & Runyon 1930; Marshall & Bock 1941; Benson 1969, 1982; Weniger 1970, 1984; Grant & Grant 1979a; Hunt & Taylor 1990; Barthlott & Hunt 1993; Behnke & Mabry 1994; R.S. Wallace 1995; Hershkovitz & Zimmer 1997; Hunt 1999; Anderson 2001; Applequist & Wallace 2001; Butterworth et al. 2002; Gorelick 2002; Landrum 2002; Nyffeler 2002; Parfitt & Gibson 2003; Powell & Weedin 2004; Edwards et al. 2005; Powell & Worthington 2018; Weakley, A.S., and Southeastern Flora Team 2022.

1. Stems jointed, cylindrical OR flattened; glochids (= small barbed hairs) present at areoles.
   2. Stems segments cylindrical; spines with entire epidermis separating into a thin, paper-like sheath (sheath as long as spine) (CHOLLAS) ____ **Cylindropuntia**
   2. Stem segments flattened; spines without an epidermal sheath (PRICKLY-PEARS) ____ **Opuntia**
1. Stems not jointed, variously shaped (typically ribbed or tuberculate) but never flattened; glochids absent.
   3. Stems vertically ribbed (ribs sometimes with tubercles) (sometimes spirally ribbed in *Hamatocactus*).
      4. Central spine (of an areole) strongly hooked at end, 12–38 mm long; inner tepals yellow, red basally ____ **Hamatocactus**
      4. Central spine absent OR present, if present to 80 mm, appressed to projecting, but not hooked; flowers variously red to red and yellow, pink to light purple, purplish red or reddish basally and then orange and pink to whitish apically.
         5. Stems usually multiple, cylindric, 15 cm or less in diam., usually clearly rising above soil surface; spines, if present, not cross-ribbed, variable in number, but often more than 8 per areole, including central and radial spines; flowers red to red and yellow, pink to light purple, or purplish red ____ **Echinocereus**
         5. Stems usually single, hemispheric, to 30 cm in diam., the exposed portion often ± flush with soil surface; spines stout, cross-ribbed (= annulate), 6–8 per areole, including central and radial spines; flowers reddish basally and then orange and pink to whitish apically ____ **Echinocactus**
   3. Stems not ribbed, but tuberculate (= with nipple-like protrusions).
      6. Flowers variously colored, located on new growth near the stem apex; plants NEITHER with stems ± flat or slightly convex and usually solitary NOR with radial spines hair-like, very different from central spines AND stems usually branching; tubercles usually with a longitudinal groove on their upper side; including species in several East TX cos.
         7. Inner tepals yellow (clear yellow to satiny golden yellow, rarely greenish yellow) usually red basally; outer tepals not fimbriate (i.e., entire); seeds smooth, dark reddish brown, sometimes minutely punctate; central spines 0–4; areolar glands sometimes conspicuous ____ **Coryphantha**
         7. Inner tepals greenish yellow to bronze; outer tepals fimbriate; seeds pitted, black; central spines usually none (sometimes 1(–2); areolar glands absent ____ **Pelecyphora**
      6. Flowers pink, white, yellowish, cream, or some mixture, located on old growth of preceding seasons below/somewhat distant from the stem apex (flowers thus in a ring around the stem); plants EITHER with stems ± flat or slightly convex and usually solitary OR with radial spines hair-like, very different from central spines AND stems usually branching; tubercles without a longitudinal groove on their upper side; species known in East TX only from Bexar Co. at extreme sw corner of area ____ **Mammillaria**

CORYPHANTHA (Engelm.) Lem. PINCUSHION CACTUS, NIPPLE CACTUS, BEEHIVE CACTUS, CORY CACTUS

A genus of ca. 45 species (Sánchez et al. 2022) of North America. Some are cultivated as ornamentals. This genus has been variously treated in the past. A number of authorities (e.g., Hunt 1978, 1999; Taylor 1978; Barthlott & Hunt 1993; Diggs et al. 1999; Kartesz 2005, 2022) have separated some species of *Coryphantha* as the segregate genus *Escobaria*. Hunt (1978), for example, believed "… that *Escobaria* (with *Neobesseya*) represent an evolutionary lineage independent of *Coryphantha*, that is to say that they are distinct phyletic groups at an approximately analogous stage." Zimmerman and Parfitt (2003d) treated *Escobaria* as a synonym of *Coryphantha*, as did Powell et al. (2008); however, Powell and Worthington (2018) separated the two genera. The results of a study by Sánchez et al. (2022) supported the monophyly of *Coryphantha*, and found *Pelecyphora* and *Escobaria* as sister to *Coryphantha*. Additionally, they found that *Escobaria* and *Pelecyphora* form a monophyletic group. Since the name *Pelecyphora* has priority, they transferred all *Escobaria* species to *Pelecyphora*.
REFERENCES: Britton & Rose 1919–1923, 1937, 1963; Castetter et al. 1975; Hunt 1978; Taylor 1978; Fischer 1980; Zimmerman 1985; Zimmerman & Parfitt 2003d.

Coryphantha sulcata (Engelm.) Britton & Rose, (sulcate, furrowed), PINEAPPLE CACTUS, GROOVED NIPPLE CACTUS, FINGER CACTUS. Plants unbranched or usually branched and forming clumps; spines at first yellow and pink or pink, later overlain by gray or white, with darker tips; central spines (0–)1–3(–4) per areole, spreading, curving or straight, the longer 9–12(–15) mm long; radial spines 8–15(–18) per areole; flowers opening late morning, closing late afternoon; outer tepals entire; inner tepals yellow (clear yellow to satiny golden yellow, rarely greenish yellow) apically, usually red basally; fruits usually green, but can become

Streptanthus petiolaris [HEA]

Streptanthus platycarpus [ELB]

Thlaspi arvense [REE]

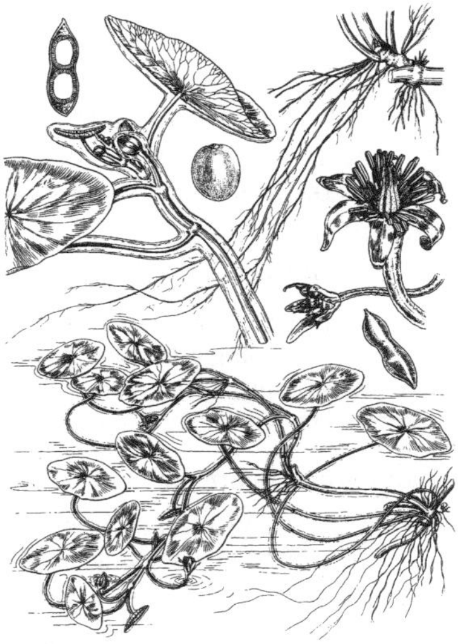

Brasenia schreberi [CO1]

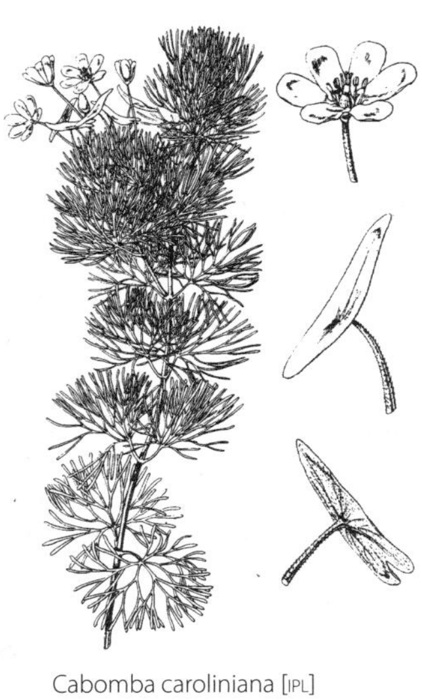

Cabomba caroliniana [IPL]

Coryphantha sulcata [HEA]

dull red, 1.5–2.5 cm long; seeds reddish brown, smooth. Prairies, sandy or gravelly soils; Denton (BRIT), Austin, Bastrop, Bexar, Brazos, Comal, Fayette, Hays, Lee, and Travis (Benson 1982; Turner et al. 2003) cos; w part of East TX and nc TX s to sc TX; in the U.S. known only from TX; probably also Mexico. Apr–Jun. Zimmerman and Parfitt (2003d) indicated that "From Austin, Texas, eastward, *C. sulcata* has converged on the morphology of the sympatric *Pelecyphora. missouriensis*; the vegetative resemblance between these two species sometimes is nearly perfect, and numerous misidentifications occur in the literature and herbaria. The porrect (= ± perpendicular to the surface of the stem segment) central spines of adult *C. sulcata*, when present, often curve downward, unlike the straight spines of *P. missouriensis*." According to Powell and Weedin (2004), "*Coryphantha sulcata* is traditionally endemic to Texas, but certain taxa from northeast Mexico are obviously related and nearly identical. . . ." Likewise, Zimmerman and Parfitt (2003d) noted that the species "probably will prove conspecific with taxa in Mexico." [*Mammillaria sulcata* Engelm.] ?

CYLINDROPUNTIA (Engelm.) F.M. Knuth CHOLLA

Erect, much branched shrubs or small trees; stems jointed, cylindrical; areoles usually with spines and small, brown, retrorsely barbed hairs (= glochids); young stems with a small, single, fleshy, vestigial, early deciduous leaf below each areole; fruits fleshy.

A genus of ca. 35 species of sw and sc United States, Mexico, and the West Indies (Pinkava 2003a). This genus has traditionally been included within *Opuntia* (e.g., Benson 1969, 1982; Correll & Johnson 1970). However, we are following Pinkava (2003a) who says in the Cactaceae treatment for *Flora of North America* that new "research findings in morphology, anatomy, palynology, seed characteristics, host-herbivore relations, and chloroplast and nuclear DNA make untenable the maintenance of the single genus *Opuntia*." Further, *Cylindropuntia* is quite distinct morphologically and no hybrids are known between the segregate genera formerly included within *Opuntia*. In addition to the obvious spines, glochids (= minute barbed hairs or bristles) are also present and can cause dermatitis; they can be extremely irritating; gloves should be worn when handling the plants (Spoerke & Smolinske 1990). (Latin: *cylindrus*, cylinder-shaped, and *Opuntia*, the related genus from which this genus was segregated)

REFERENCES: Champie 2003; Pinkava 1999, 2003a.

1. Stem segments (= joints) usually whorled or subwhorled, 10–40 cm long, usually 20–40 mm in diam.; inner tepals dark pink to red-purple; fruits 24–45 mm long; tubercles distinct ____ **C. imbricata**
1. Stem segments usually alternate, 2.5–8 cm long, 3–5 mm in diam.; inner tepals yellow to greenish yellow, the tepals sometimes tipped with red; fruits (9–)12–15(–27) mm long; tubercles barely discernable ____ **C. leptocaulis**

Cylindropuntia imbricata (Haw.) F.M. Knuth var. **imbricata,** (overlapping), TREE CHOLLA, COYONOSTYLE. Small tree 1–3(–5) m tall; tubercles prominent; spines (5–)8–15(–30) per areole (rarely spineless or nearly so); inner tepals dark pink to red-purple, 15–35 mm long, usually opening in late morning; fruits yellow, tuberculate, sometimes persistent until following season. Open areas; Bexar and Victoria (Benson 1982) cos. at extreme s margin of East TX, the herbarium specimens from Bexar and Victoria cos. cited in Benson (1982) are from locations outside of East TX proper; mainly w 1/2 of TX; AZ, CO, KS, NM, OK, TX, and UT; also adjacent Mexico. May–Jul. This species, which is cultivated and sometimes persists, is one of the dominant chollas of the Chihuahuan Desert (Pinkava 2003a). Extrafloral nectaries are present at the upper margins of the areoles, producing droplets of liquid, and sometimes attracting patrolling ants (Powell & Weedin 2004), which presumably provide protection against herbivores in exchange for the nectar. [*Opuntia imbricata* (Haw.) DC.]

Cylindropuntia leptocaulis (DC.) F.M. Knuth, (thin-stemmed), DESERT CHRISTMAS CACTUS, PENCIL CACTUS, PENCIL CHOLLA, CHRISTMAS CHOLLA, TASAJILLO, TESAJO, RAT-TAIL CACTUS, SLENDER-STEM CACTUS. Bush or erect shrub 0.5–0.7(–1.8) m tall; tubercules almost absent; spines 1(–3) per areole (rarely spineless), usually (1.4–)2.5–5 cm long; inner tepals yellow to greenish yellow, 5–8 mm long, sometimes tipped with red, usually opening in the afternoon; fruits yellow to bright red, smooth, sometimes persisting until the next season. Clay or alluvial soils; mostly in the sw 1/3 of East TX; East TX w through w 2/3 of TX; AZ, NM, OK, and TX; also adjacent Mexico. May–Sep. [*Opuntia leptocaulis* DC.]

ECHINOCACTUS Link & Otto BARREL CACTUS

A genus of 6 species (Zimmerman & Parfitt 2003b) of ribbed cacti often with long spines native to the sw U.S. and Mexico; some are cultivated as ornamentals. While now treated as a small genus, *Echinocactus* formerly included most cacti with ribbed stems (Barthlott & Hunt 1993). (Greek: *echinos*, hedgehog, and *Cactus*, an old genus name, from Greek: *cactus*, the name for another spiny plant)

REFERENCES: Maddox 1986; Zimmerman & Parfitt 2003b.

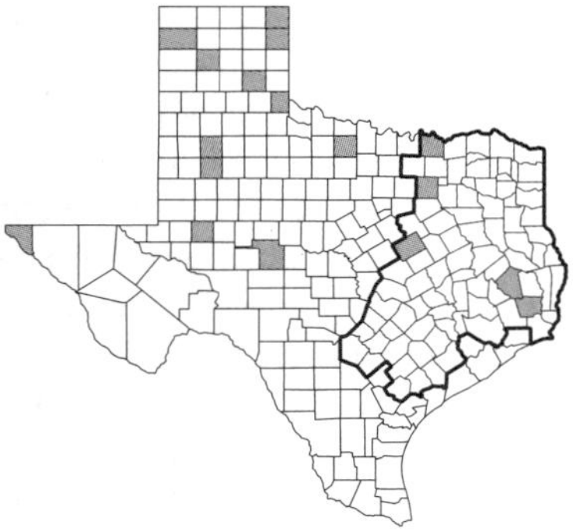

Sisymbrium altissimum

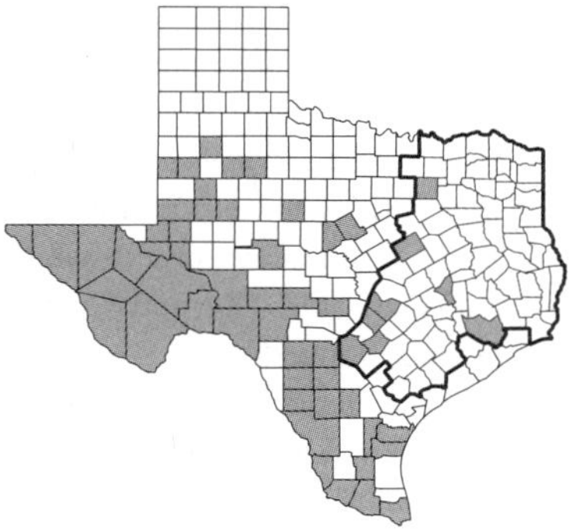

Sisymbrium irio

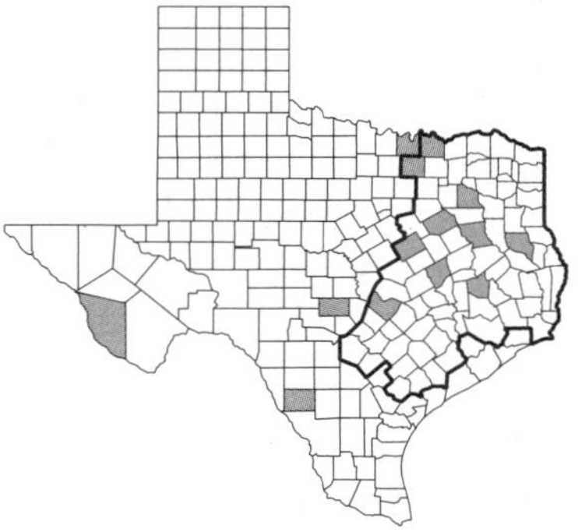

Sisymbrium officinale

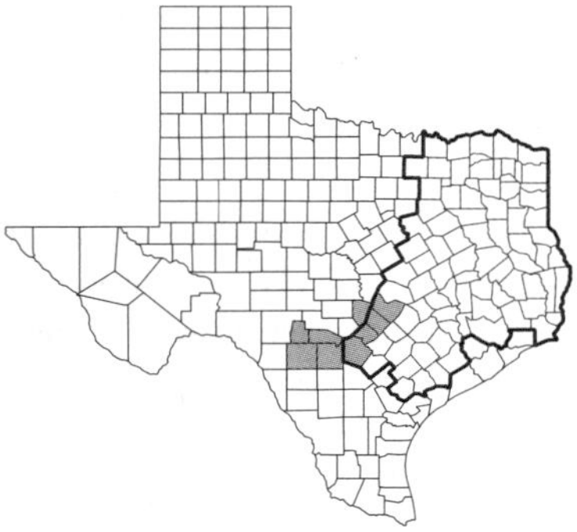

Streptanthus bracteatus

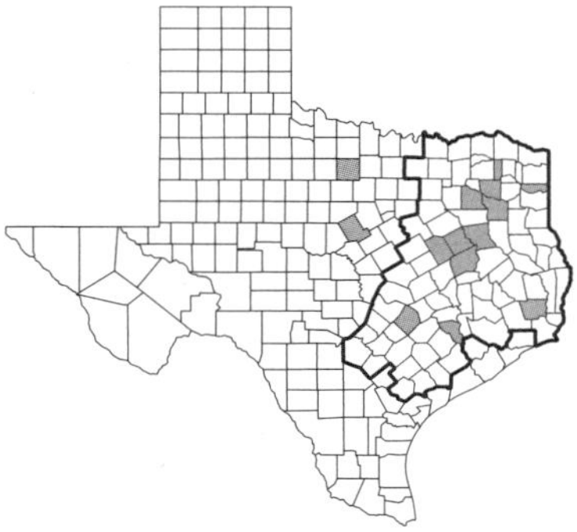

Streptanthus hyacinthoides

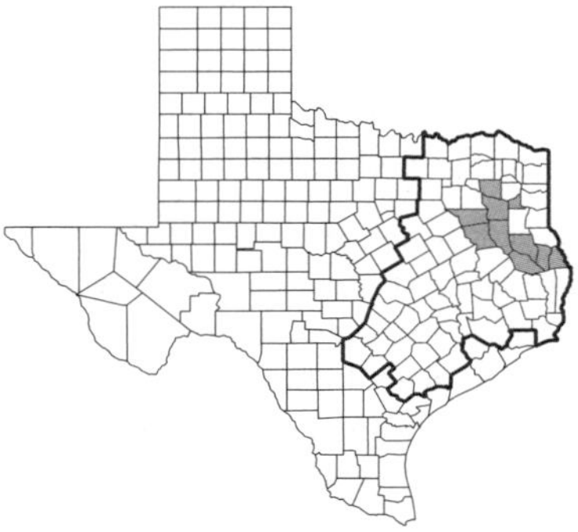

Streptanthus maculatus

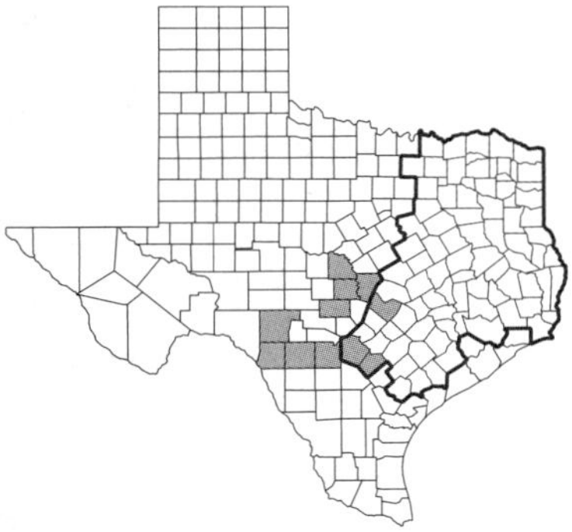

Streptanthus petiolaris

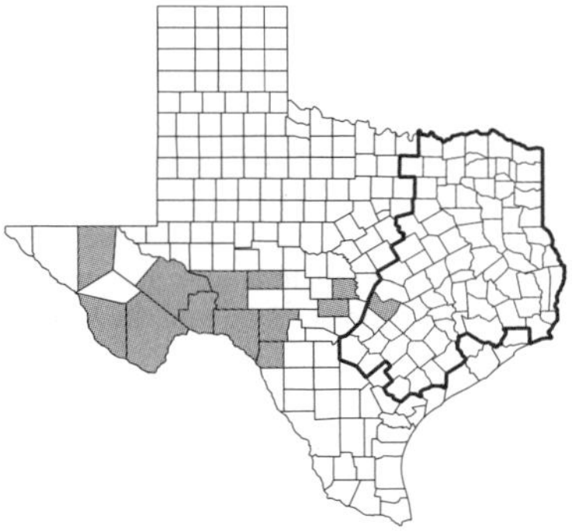

Streptanthus platycarpus

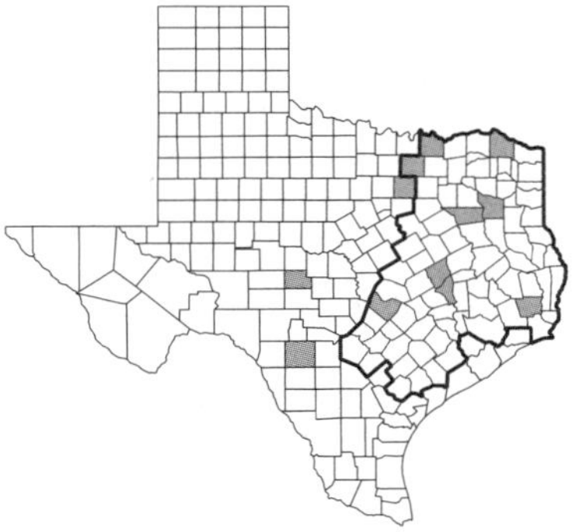

Thlaspi arvense

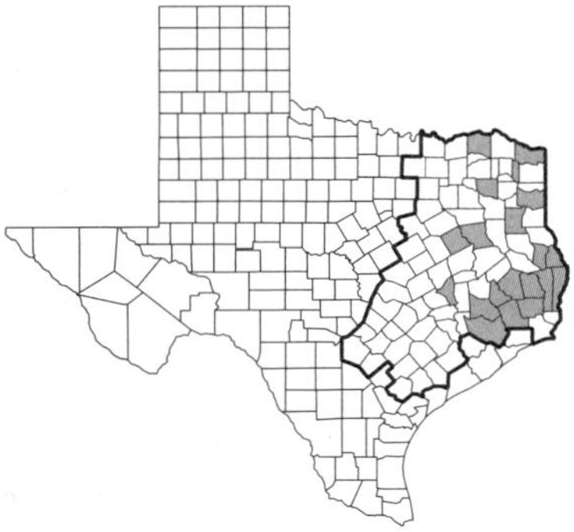

Brasenia schreberi

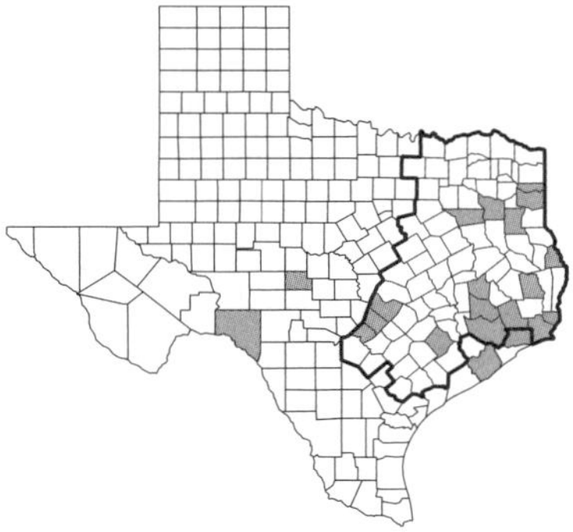

Cabomba caroliniana

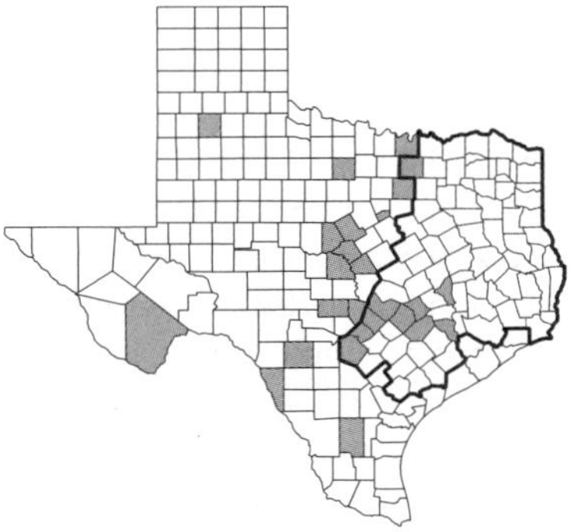

Coryphantha sulcata

Echinocactus texensis Hopffer, (of Texas), HORSECRIPPLER, DEVIL'S-HEAD CACTUS, DEVIL'S-PINCUSHION, DEVIL'S-CLAW, MANCA CABALLO. Stems solitary, green, hemispheroidal, 12.5–20 cm long, up to 30 cm in diam., deeply seated in the soil, the exposed portion often ± flush with soil surface (can expand to several centimeters above soil level under ample moisture conditions); ribs 13–27; spines stout, cross-ribbed; central spine 1 per areole, curved downward; radial spines 5–7 per areole; inner tepals basally reddish, then orange and pink to whitish apically, with midribs purplish to violet; filaments red; anthers yellow; fruits red, fleshy becoming dry, splitting irregularly with deciduous scales. Sandy or limestone soils; Bexar, Comal, Dewitt, Hays, Travis, and Williamson (Benson 1982; Turner et al. 2003) cos.; mainly w 2/3 of TX; OK, NM, and TX; also Mexico. Apr–May. [*Homalocephala texensis* (Hopffer) Britton & Rose] Grazing animals can be badly injured by the strong spines and older plants are reported capable of puncturing pickup tires; as a result, this species has been eliminated from many areas (Wills & Irwin 1961; Kirkpatrick 1992). Powell and Weedin (2004) noted that the stems "are as stiff as rubber tires and can be stepped on by heavy animals without harming the plant." Zimmerman and Parfitt (2003b) indicated that a "dense cover of ephemeral herbs or shallow blanket of snow can hide this species completely from view."

ECHINOCEREUS Engelm. HEDGEHOG CACTUS, STRAWBERRY HEDGEHOG CACTUS

Stems to only ca. 40 cm long, usually branching near ground level (resulting in clumps), unsegmented, cylindroid, with 5–19 ribs, usually with tubercles projecting from the ribs; areoles without glochids; spines smooth (not cross-ribbed); fruits on old growth of preceding year above areoles, fruits fleshy at maturity, bearing slender spines.

A genus of ca. 49 species of the w U.S. and Mexico (Zimmerman & Parfitt 2003a); all are cultivated (Mabberley 2017). The name STRAWBERRY HEDGEHOG CACTUS, which is used for several species, "refers to the strawberry-like flavor of the ripe fruits, which are avidly harvested by human foragers" (Zimmerman & Parfitt 2003a) (Greek: *echinos*, hedgehog, and genus *Cereus*, a genus of columnar cactus, from Latin: *cereus*, a wax taper (candle), from the upright, candle-like habit of some species)

REFERENCES: Taylor 1985; Miller 1988; Ferguson 1989; Hoffman 1992; Cota & Philbrick 1994; Blum et al. 1998; Baker & Routson 2003; Zimmerman & Parfitt 2003a.

1. Inner tepals predominantly light yellow, proximally orange-red to purple; radial spines 7–10 per areole; species known in East TX only from Bexar Co. ____ **E. papillosus**
1. Inner tepals predominantly pink to light purple, magenta (= purplish red), red or red and yellow; radial spines 4–36 per areole; including species widely known in East TX.
   2. Inner tepals predominantly red or red and yellow (no blue mixture); radial spines usually 4–7 per areole; species known only on the far sw border of East TX ____ **E. coccineus**
   2. Inner tepals predominantly pink to light purple to magenta (can have contrasting proximal colors); radial spines 8–36 per areole; including species widely known in East TX.
      3. Areoles usually 14–25 mm apart, with 8–10(–13) radial spines per areole, these usually 9–19 mm long, not pectinately arranged in a row on each side of an elongate areole; stems usually (3.2–)3.8–4.5 cm in diam., often sprawling, sometimes prostrate; inner tepals pink or magenta, darkest proximally; central spines (12–) 20–44+ mm long ____ **E. enneacanthus**
      3. Areoles usually 1–6 mm apart, with 12–36 radial spines per areole, these usually 2–8 mm long, pectinately arranged in a row on each side of an elongate areole; stems usually 4–10 cm in diam., erect; inner tepals pink to magenta, usually white, crimson, green, or multicolored proximally; central spines (sometimes absent) 0–6(–15) mm long ____ **E. reichenbachii**

Echinocereus coccineus Engelm., (scarlet), CLARET-CUP CACTUS, TEXAS CLARET-CUP CACTUS, RED-FLOWER HEDGEHOG, SCARLET HEDGEHOG CACTUS. Stems few, 15–20 cm long, 6.2–10 cm in diam., with 5–9 ribs; central spines usually 0(–1); radial spines usually 4–7 per areole, nearly straight, often grayish, 38 mm or less long; inner tepals predominantly red or red and yellow (no blue mixture), 2.5–3.8 cm long, reportedly remaining open for 3–4 days, even at night (Powell & Weedin 2004); fruits red. Rocky limestone soils; Comal, Hays, (Powell et al. 2008) and Williamson (Ferguson 1989; Turner et al. 2003) cos. on w margin of East TX; also s part of Cross Timbers and Prairies, Edwards Plateau, and Trans-Pecos; AZ, CO, NM, and TX; also Mexico. Late Mar–Jun. [*E. coccineus* var. *paucispinus* (Engelm.) D.J. Ferguson, *E. triglochidiatus* Engelm. var. *paucispinus* (Engelm.) W.T. Marshall] While it has been variously treated taxonomically in the past (e.g., Ferguson 1989; Blum et al. 1998; Powell & Weedin 2004), we are following Zimmerman and Parfitt (2003a) in recognizing this taxon at the specific level and in not recognizing any infraspecific taxa. Some TX populations of this species have in the past been treated as *E. triglochidiatus* Engelm. var. *paucispinus* (Engelm.) W.T. Marshall (Correll & Johnston 1970) or recognized as *E. coccineus* var. *paucispinus* D.J. Ferguson (Diggs et al. 1999; Turner et al. 2003). Turner et al. (2003) also mapped some individuals of this species (including from Williamson Co.) as *E. triglochidiatus*. However, Zimmerman and Parfitt (2003a) and Powell and Weedin (2004) noted that *E. coccineus* is a tetraploid species, while the related *E. triglochidiatus* is a diploid endemic to AZ, CA, CO, NM, NV, and UT. Zimmerman and Parfitt (2003a) also discussed spine

Cylindropuntia imbricata [BEL]

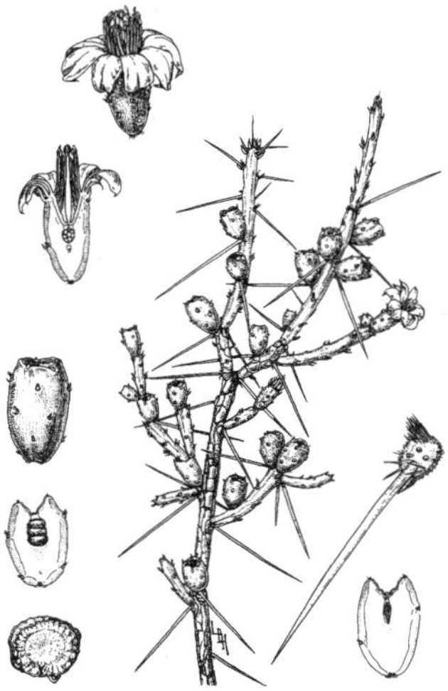

Cylindropuntia leptocaulis [BEL]

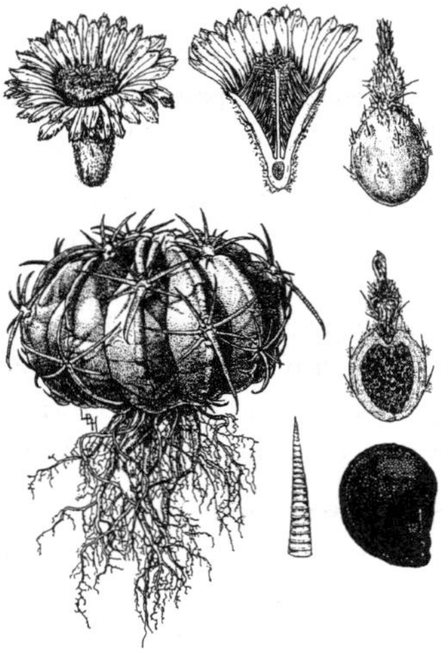

Echinocactus texensis [LUN]

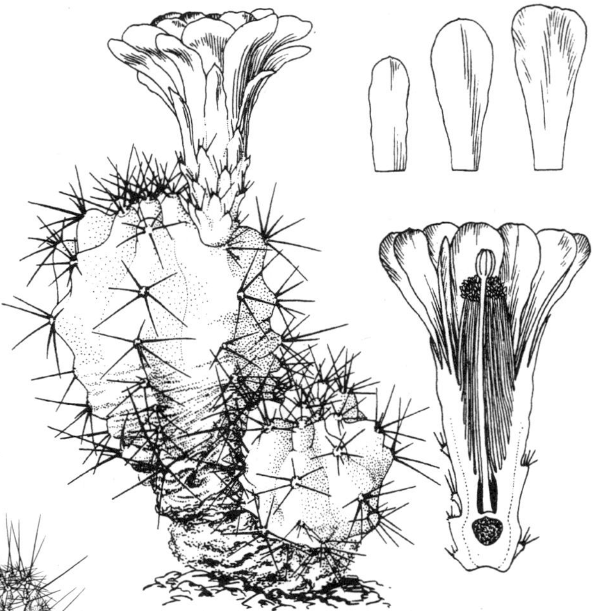

Echinocereus coccineus [TAN]

Echinocereus enneacanthus [EMO]

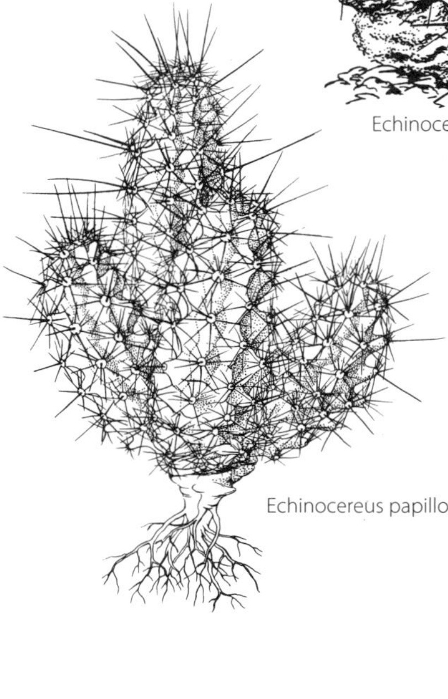

Echinocereus papillosus [TAN]

variation in this species, indicating that it is "sparsely-spined" in parts of TX, but more "densely-spined" further west. Indeed, the description above reflects the fewer-spined easternmost forms of the species. There are large nectar chambers and the flowers "appear adapted to pollination by hummingbirds" (Powell & Weedin 2004). They (Powell & Weedin 2004) also observed bee pollinators. Hoffman (1992) found that in NM pollination was by halictid bees.

Echinocereus enneacanthus Engelm. var. **brevispinus** (W.O. Moore) L.D. Benson, (sp.: nine-spined; var.: short-spined), STRAWBERRY CACTUS, PITAYA, ALICOCHE. Stems in clumps, 8–40 cm long, usually (3.2–)3.8–4.5 cm in diam., often sprawling, cylindric, with 6–10 ribs; central spines 1–4; radial spines 8–10(–13) per areole, nearly straight, often rather light in color and tipped with brown, usually ca. 9–19 mm long; inner tepals pink or magenta, darkest proximally; fruits yellow-green to reddish brown or dull reddish. Rocky limestone soils; Bexar (Correll & Johnston 1970) and Hays (Turner et al. 2003) cos. on sw margin of East TX; mainly South TX Plains, Edwards Plateau, and Trans-Pecos; in the U.S. known only from TX; also Mexico. Apr–Jul. Variety *enneacanthus* is limited to the Trans-Pecos and Mexico.

Echinocereus papillosus Linke ex Haage, (with pimple-like or nipple-like structures), YELLOW-FLOWERED ALICOCHE, ALLICOCHE CACTUS. Stems in clumps, 4–20 cm long, usually 2–3(–5) cm in diam., erect, often becoming decumbent in age, cylindric, with 6–10 ribs; central spines usually 1; radial spines 7–10 per areole, straight, to 25 mm long; inner tepals predominantly light yellow, proximally orange-red to purple; fruits usually green. Open prairie, thorn scrub, limestone soils; Bexar Co. (Turner et al. 2003) at extreme sw margin of East TX; mainly South TX Plains; in the U.S. known only from TX; also Mexico. Early spring. [*E. papillosus* var. *angusticeps* (Clover) W.T. Marshall] This species may not be a component of East TX proper. Much of Bexar Co. is either South TX Plains or Edwards Plateau (see front inside flap). NatureServe: G3, vulnerable; S3, vulnerable.

Echinocereus reichenbachii (Terscheck ex Walp.) J.N. Haage, (presumably for Heinrich Gottlieb Ludwig Reichenbach, 1793–1879, German naturalist), LACE CACTUS, WHITE LACE CACTUS, BROWN LACE CACTUS, HEDGEHOG CACTUS. Stems solitary or in clumps, usually 7–15(–30) cm long, 4–10 cm in diam., erect, cylindric, with (10–)12–19 ribs; central spines usually absent (rarely 1–3+, if present, smaller than the radials); radial spines 12–21(–36) per areole, appressed and pectinately (= comb-like) arranged in a row on both sides of the elongate areole, ± obscuring the stem, straight or slightly curved, usually 2–8 mm long, often straw-colored to pale gray, distally pink to reddish; inner tepals pink to light purple to magenta, usually white, crimson, green, or multicolored proximally, opening about midday and closing before night; fruits green with conspicuous, deciduous, woolly hairs from the areoles. Gravelly, rocky, or sandy soils, especially limestone; widespread in sw part of East TX n to McLennan Co. (Benson 1982); widespread in w 2/3 of TX; CO, KS, NM, OK, and TX; also Mexico. Apr–Jun. [*E. reichenbachii* subsp. *caespitosus* (Engelm.) W. Blum & Mich. Lange]

HAMATOCACTUS Britton & Rose TWISTED-RIB CACTUS

A monotypic genus of the sw U.S. and Mexico. The single species of *Hamatocactus* has sometimes been treated in various other genera including *Ferocactus*, *Glandulicactus*, or *Thelocactus*, but it is taxonomically isolated and now considered worthy of generic recognition (Zimmerman & Parfitt 2003c). (Latin: *hamatus*, hooked, referring to the hooked central spine of the areoles, and *Cactus*, an old genus name, from Greek: *kaktos* or Latin: *cactus*, the name for another spiny plant)
REFERENCE: Zimmerman & Parfitt 2003c.

Hamatocactus bicolor (Terán & Berland.) I.M. Johnst. (two-colored, probably in reference to the flowers that are yellow except red basally), TWISTED-RIB CACTUS, HEDGEHOG CACTUS, FISHHOOK CACTUS. Miniature barrel cactus with unsegmented stems solitary to sometimes several to numerous, spheric, ovoid, or cylindroid, 3.8–10(–20) cm long, 3.8–5(–12) cm in diam., vertically or spirally ribbed; ribs 13; spines ± smooth, not cross-ribbed, dense but not obscuring stem; main central spine strongly hooked, 12–38 mm long; 1–3 smaller, straight, central spines also present; radial spines usually 12–15 per areole, ± straight; inner tepals clear yellow, red basally, the largest 20–25 mm long; fruits red, globular, ca. 8–13 mm in diam.; seeds black. Black or clay soils, grasslands, mesquite thickets; Bexar, Comal, Hays, Travis (Turner et al. 2003), and Victoria (Benson 1982) cos. at sw margin of East TX; also Brown (Stanford 1971) and San Saba (Turner et al. 2003) cos. in Prairies and Cross Timbers, South TX Plains, and Edwards Plateau; in the U.S. known only from TX; also Mexico. Early spring–fall. [*Echinocactus setispinus* Engelm., *Ferocactus setispinus* (Engelm.) L.D. Benson, *Thelocactus setispinus* (Engelm.) E.F. Anderson] Benson (1982) treated this species in the genus *Ferocactus*, while in TX (e.g., Diggs et al. 1999) it went under the name *Thelocactus setispinus*.

MAMMILLARIA Haw. PINCUSHION CACTUS, FISHHOOK CACTUS, NIPPLE CACTUS, GLOBE CACTUS, CHILITA

Stems solitary or branching basally and forming clumps, unsegmented, with tubercles, but without ribs, with or without milky latex; tubercles without an upper longitudinal groove; areoles of two types, the vegetative areoles, bearing spine clusters, at tips of tubercles, the reproductive areoles in upper tubercle axils; aerolar glands absent; flowers located on old growth of preceding seasons below/somewhat distant from the stem apex (flowers thus in a ring around the stem); fruits fleshy, red, smooth, obovoid to clavate (= club-shaped); seeds reddish brown or black, pitted.

A genus of 164 species ranging from the sw U.S. to Mexico, Central America, and the West Indies (Zimmerman & Parfitt 2003e); as treated here it is the most species-rich genus in the family (Harpke et al. 2006). However, molecular research (e.g., Butterworth & Wallace 2004) indicated that the genus as presently viewed is polyphyletic and its circumscription will need to be modified in the future as additional information becomes available. Breslin et al. (2021), using molecular techniques, proposed segregating a good number of *Mammillaria* species and transferred them to the genus *Cochemiea* (K. Brandegee) Walton (formerly a genus of 5 species), and formally made 25 new combinations, including some varieties (ca. 17 taxa had already been named in *Cochemiea*). As convincing as the study seems to be, we are continuing to treat *Mammillaria* in a traditional manner until further corroborating studies are published. There are no newly circumscribed species of *Cochemiea* in East TX. The species are low, often tuft-forming and some produce a milky latex. A number are cultivated as ornamentals and there is a society (Mammillaria Society) with "the aim of promoting and furthering the study of the cactus genus *Mammillaria*." (Latin: *mamilla*, nipple, in reference to the shape of the tubercles, which in some species produce a "milky" white latex—Zimmerman & Parfitt 2003e)

REFERENCES: Craig 1945; Hunt 1971; Pilbeam 1980; Zimmerman & Parfitt 2003e; Butterworth & Wallace 2004; Harpke et al. 2006.

1. Radial spines hair-like, very unlike central spines, winding and twisting, 30–60 per areole; central spines (5–)8–12 per areole; stems usually branched, forming clumps clearly above level of soil, without milky latex ________ **M. prolifera**
1. Radial spines rigid, ± similar to central spines, ± straight, (6–)13–22(–26) per areole; central spines (0–)1(–4) per areole; stems usually unbranched, hardly protruding above level of soil, with milky latex ________ **M. heyderi**

Mammillaria heyderi Muehlenpf., (for Herr Heyder, 1808–1884, a noted cultivator of cacti in Berlin), FLATTENED MAMMILLARIA, HEYDER'S NIPPLE CACTUS, HEYDER'S PINCUSHION CACTUS, NIPPLE CACTUS, LITTLE NIPPLE CACTUS, LITTLE-CHILIS, PANCAKE CACTUS, BIZNAGA DE CHILITOS. Plant low-growing, spiny pincushion; sap milky; stems typically solitary to rarely clustered, turbinate to subglobose, but the above ground portion ± flat to shallowly convex, (4–)7.5–10(–15) cm in diam.; tubercles prominent, ± conical from a subpyramidal base, 9–12(–20) mm long; areoles apical on the tubercles; spines dense but not obscuring the stem; central spines (0–)1(–4) per areole, the longer ones 3–9 mm long, straight; radial spines usually (6–)13–22(–26) per areole, the longer ones 6–16 mm long; inner tepals pink, white, cream, or some mixture, usually with pink or greenish-brown midstripes; fruits red, 12–40 mm long, 5–9 mm broad; seeds reddish brown. Limestone soils in grasslands or dry areas, typically in partial shade; in East TX known only from Bexar Co. (Turner et al. 2003); also Brown (Stanford 1971) and San Saba (Turner et al. 2003) cos. in Cross Timbers and Prairies s to s TX and w to Trans-Pecos; AZ, NM, OK, and TX; also Mexico. Spring. [*M. applanata* Engelm., *M. gummifera* Engelm. var. *applanata* (Engelm.) L.D. Benson, *M. gummifera* var. *hemisphaerica* (Engelm.) L.D. Benson, *M. heyderi* var. *applanata* (Engelm.) Engelm., *M. heyderi* var. *hemisphaerica* Engelm.] The plants in East TX represent var. *heyderi*; the only other currently recognized variety, var. *bullingtoniana* Castetter, P. Pierce, & K.H. Schwer., is known only from AZ, NM, and far w TX (Powell & Worthington 2018), and Mexico. Zimmerman and Parfitt (2003e) noted "plants with fewer radial spines, often segregated as var. *hemisphaerica*, are frequent in populations of var. *heyderi* near the Texas coast." The fruits are reported to be edible (Kirkpatrick 1992).

Mammillaria prolifera (Mill.) Haw. var. **texana** (Engelm.) Borg, (sp.: producing offsets or bunched growth; var.: of Texas), TEXAS NIPPLE CACTUS, HAIR-COVERED CACTUS, GRAPE CACTUS. Stems branched, 2–9 cm long, 1.2–2.5(–5) cm in diam., forming clumps; sap not milky; tubercles mammiform, 4.5–10 mm long; spines dense, somewhat obscuring the stem; central spines (5–)8–12 per areole, often reddish with darker tips, straight, relatively rigid, (4–)6–9 mm long; radial spines 30–60 per areole, white to pale yellow, hair-like, spreading irregularly, winding and twisting, ca. as long as the centrals; inner tepals predominantly whitish to yellowish, sometimes with some pinkish in the midstripe; fruits red, 8–12(–20) mm long, 4–5 mm broad; seeds black. Grasslands; in East TX known only from Bexar Co. (Correll & Johnston 1970; Turner et al. 2003) at extreme sw corner of East TX; mainly Edwards Plateau and South TX Plains); in the U.S. known only from TX; also n Mexico. Spring. Zimmerman and Parfitt (2003e) noted that the unique hair-like radial spines "provide an instant means of identifying this species …" Plants of var. *texana* are tetraploid; two other varieties, occurring in Mexico and the West Indies, are variously diploid, tetraploid, or hexaploid (Zimmerman & Parfitt 2003e).

OPUNTIA Mill. PRICKLY-PEAR

Mat- or clump-forming perennials to shrubs or small trees; stems jointed, flattened; areoles usually with spines and small, brown, retrorsely barbed hairs (= glochids); young stems with a small, single, fleshy, vestigial, early deciduous leaf below each areole; fruits dry or fleshy, sometimes sweet and edible.

A genus of ca. 150 species native from Massachusetts and British Columbia s to Straits of Magellan and the Galapagos Islands (Pinkava 2003b); stems are flattened; some are tree-like; a number have edible fruits (PRICKLY-PEARS) and the flattened stems of others are eaten as a vegetable. *Opuntia* species have long been used as host plants in the historically important cochineal dye industry (e.g., Kennedy 1841). A red dye, which was extremely valuable prior to modern synthetic dyes, was obtained from the dried bodies of the homopteran cochineal insects (*Dactylopius coccus*) that feed on the cacti. Such insects can be found on native *Opuntia* in East TX today. They can be easily spotted by the white secretions in which they hide—removing and crushing an insect allows easy observation of the bright red color. Such dyes are still used in Mexico in the production of high quality "natural" weavings. In some parts of TX during periods of drought, flame-throwers are used to burn the spines off *Opuntia* species so they can be used as an emergency food for livestock. Some species of *Opuntia* have been introduced into the Old World and have become problematic invaders in certain areas (e.g., Australia). In addition to the obvious spines, glochids (= minute barbed hairs or bristles) are also present and can cause dermatitis; they can be extremely irritating; gloves should be worn when handling the plants (Spoerke & Smolinske 1990). The PRICKLEY-PEAR was designated the state plant of Texas by the 74th state legislature; all members of *Opuntia* (with flat stems) are considered the state plant, while those of the related genus *Cylindropuntia* (with cylindrical stems; sometimes included in *Opuntia*) are not (Jones et al. 1997). Many species of *Opuntia* are known to hybridize in addition to demonstrating phenotypic plasticity, and can closely mimic other species, even those of distant kinship. This often makes accurate identification of species difficult. Key to species of *Opuntia* adapted from Majure et al. (2017) and Majure (2022). (Greek name for a spiny plant (not Cactaceae) that grew near Greek town of Opus (Opuntis))

REFERENCES: Grant & Grant 1979b, 1979c, 1982; Grant & Hurd 1979; Grant et al. 1979; Parfitt & Pinkava 1988; Leuenberger 1991, 1993; Pinkava 2003a, 2003b, 2003c; Powell & Weedin 2004; Powell et al. 2008; Majure 2012, 2022; Majure et al. 2017; Powell & Worthington 2018.

1\. Plants forming low, spreading shrubs.

  2\. Inner tepals yellow with red or maroon bases (sometimes all yellow in *O. mackensenii*); stem segments 3.8–25 cm long.

    3\. Erect-spreading shrubs, chains of stem segments radiating from center of plant, the flat surface generally held perpendicular to the ground surface; stem segments remaining turgid throughout the year; stigma usually green or green to cream; fruit barrel-shaped (= widest near the middle) or clavate (= widest toward tip or club-shaped).

      4\. Stigma green to cream; fruit clavate **O. mackensenii**

      4\. Stigma green; fruit barrel-shaped **O. phaeacantha**

    3\. Spreading shrubs, plants clump forming or with chains of stem segments spreading from the central axis of the plant, the flat surface generally parallel to the ground surface; stem segments becoming cross-wrinkled during fall and winter months, fruit clavate.

      5\. Central spines mostly 1–2 (when 2, both spines generally in the same plane, i.e., both reflexed or both erect); spines monomorphic (central spines only produced) **O. cespitosa**

      5\. Central spines mostly 1–3 per areole (when 2–3, the lower spines generally deflexed and the upper spine porrect (= ± perpendicular to stem segment surface) on upper stem segments; spines monomorphic (central spines only) or dimorphic (with smaller hair-like radial spines and larger central spines) **O. macrorhiza**

  2\. Inner tepals entirely yellow, rarely tinged pink basally; stem segments 4.5–8.4 cm long **O. nemoralis**

1\. Plants forming erect or ascending shrubs, small trees OR tall trees.

  6\. Plants tall trees, 3–6 m; main trunk 0.6–1.2 m tall; larger terminal stem segments(20–)40–60 cm long, 20–30+ cm wide; fruits 5–10 cm long; escaped cultivar rare in East TX **O. ficus-indica**

  6\. Plants erect or ascending shrubs or small trees 0.3–3 m tall, main trunk, if present, very short; larger terminal joints 3.8–34(–40) cm long, except 45–90(–120) cm long in locally rare cultivar, *O. engelmannii* var. *linguiformis*); fruits 2.1–7 cm long; including native East TX taxa, some widespread in the area.

    7\. Inner tepals yellow with red or reddish maroon bases **O. phaeacantha**

    7\. Inner tepals yellow or greenish yellow.

      8\. Glochid pattern generally of an adaxial crescent in the areole, i.e., glochids forming a dense fascicle in the upper portion of the areole, not widely separated throughout; central spines yellow; stem segments typically tuberculate with scalloped margins **O. stricta**

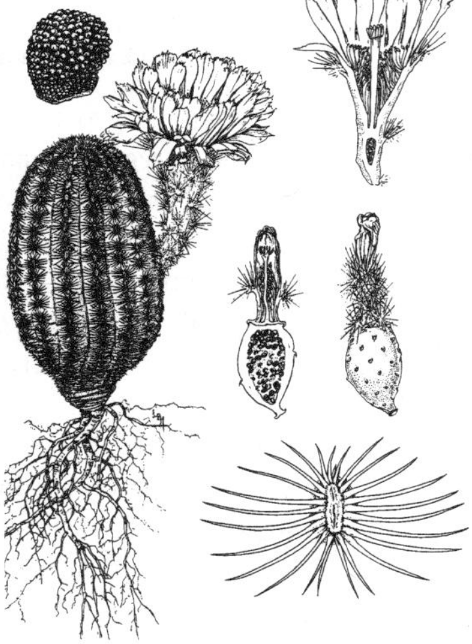

Echinocereus reichenbachii [BEL, GAR]

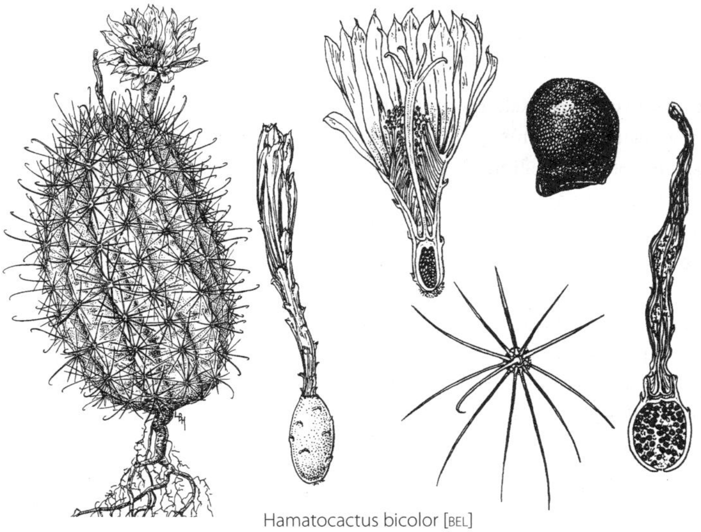

Hamatocactus bicolor [BEL]

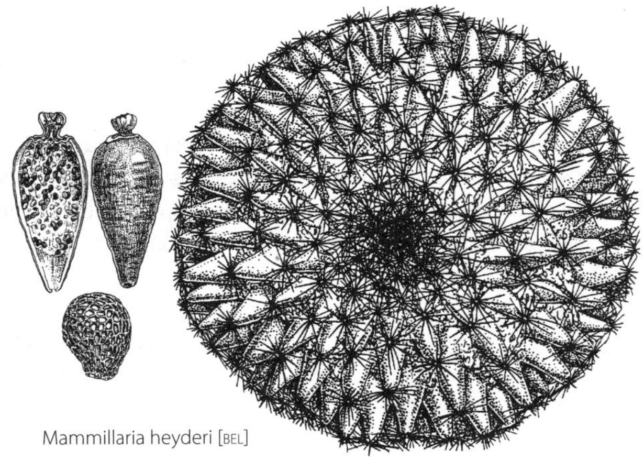

Mammillaria heyderi [BEL]

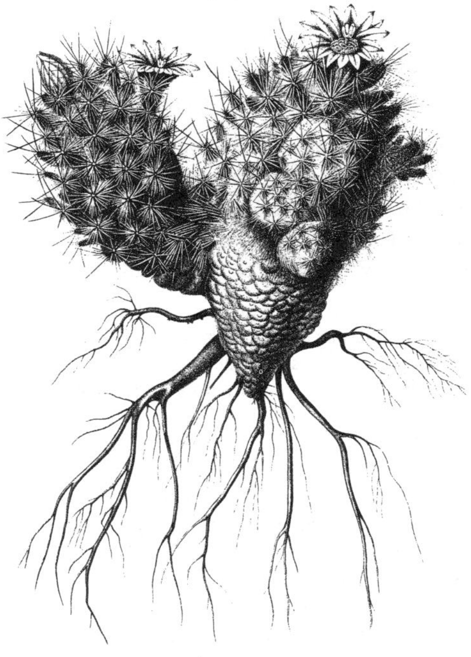

Mammillaria prolifera var. texana [EMO]

Opuntia cespitosa [USW]

8. Glochid pattern generally of a pincushion type, with the glochids widely separated; central spines white with dark (red-brown) bases, yellow, or yellow with dark (red-brown) bases; stem segments not strongly tuberculate, the margins smooth not scalloped ____ **O. engelmannii**

Opuntia cespitosa Raf., (growing in tufts), WESTERN PRICKLY-PEAR. Sprawling shrub, to 0.3 m tall, with chains of up to 2–6 stem segments, the segments generally produced with the flat (broad) surface parallel to the ground surface; stem segments mostly obovate, rotund, or elliptical in outline, margins not scalloped, with 4–6 (generally 5) areoles per diagonal row, strongly glaucous-green (gray-green) when developing, aging dark green or light gray-green, cross-wrinkling during the winter months, 3.8–18.7 cm long, 3.2–11.3 cm wide; spines robust or delicate, smooth to the touch, 1–2(–3) per areole (most commonly 1), 1.5–4.3 cm long, 0.78–0.87 mm in diam., dark reddish brown at the base during development but maturing bony-white, and finally dark gray in age, typically spreading in one plane from the areoles (i.e., in line with one another) with primarily 1 spine, or occasionally 2 of roughly the same length or 1 long and 1 short and slightly deflexed, rarely 3 spines produced from the areoles with the central spine typically not porrect (= ± perpendicular to the surface of the stem segment); in age the mid-stem segment and especially the basal stem segment spines tend to deflex; glochids dark red, crimson red, or dark amber, aging light to dark brown; inner tepals yellow, basally tinged dark red, crimson, orange-red, or reddish-pink; fruits dark red, or orange-red, 2.7–4.5 cm long. Sandy soils or blackland prairies, juniper glades, limestone rock outcrops; in East TX known only from Lamar Co. (Majure et al. 2017); e U.S. from NY s to GA and w to MI, MO, AR, and TX, except not known from FL, LA, NC, or SC. May–July. [*O. humifusa* of authors, not (Raf.) Raf., *O. rafinesqueii* Engelm. var. *microsperma* Engelm. & J.M. Bigelow, *O. rafinesqueii* var. *minor* Engelm.] *O. humifusa* (Raf.) Raf. has a long history of confused descriptions and distributions that were often widespread (see Majure et al. 2017). Majure et al. (2017) convincingly demonstrated that what was once thought to be a widespread *O. humifusa* was actually several species in the *humifusa* complex. According to Majure et al. (2017), *O. humifusa* is not in TX, but rather *O. cespitosa* and *O. nemoralis* Griffiths of the *humifusa* complex are.

Opuntia engelmannii Salm-Dyck ex Engelm., (for George Engelmann, 1809–1884, German-born botanist and physician of St. Louis who did extensive work on the Cactaceae), ENGELMANN'S PRICKLY-PEAR, TEXAS PRICKLY-PEAR, CACTUS-APPLE. Large, suberect (sometimes with a small trunk) to sprawling shrub 1–3 m high; stem segments green to blue-green; spines mostly white or mostly yellow, usually 1–6 per areole (sometimes absent), usually on all but lower areoles, 1.2–4(–5+) cm long; inner tepals usually yellow; fruits purple to dark red, 3.1–7 cm long. Spring.

1. Larger terminal stem segments usually obovate to orbicular, 15–35 cm long.
   2. Central spines usually white with red-brown bases ____ var. **engelmannii**
   2. Central spines usually yellow often with red-brown bases ____ var. **lindheimeri**
1. Larger terminal stem segments elongate, lanceolate to linear-lanceolate, 45–90(–120) cm long ____ var. **linguiformis**

var. **engelmannii,** ENGELMANN'S PRICKLY-PEAR. Plants 0.9–2 m tall; stem segments 20–35 cm long, 15–30 cm wide, pale green, usually obovate, broadly elliptic or circular, wool tan, aging gray to blackish; spines over most of the stem segment, white with red-brown bases. Rocky, dry habitats, sandy soils, woodlands, grasslands; Bexar, McLennan, and Travis (Kartesz 2022) cos.; w 2/3 of TX; AZ, CA, NM, NV, TX, and UT; also Mexico. [*O. phaeacantha* Engelm. var. *discata* (Griffiths) L.D. Benson & Walk.]

var. **lindheimeri** (Engelm.) B.D. Parfitt & Pinkava, (for Ferdinand Jacob Lindheimer, 1801–1879, German-born TX collector), TEXAS PRICKLY-PEAR, NOPAL PRICKLY-PEAR. Sprawling shrub 0.5–1.2(–3) m tall, commonly broader than tall, often with a short trunk; stem segments 15–27(–30) cm long, 12–22(–30) cm wide, green to blue-green, broadly obovate to subcircular. Sandy, gravelly, or alluvial soils; grasslands; widespread in s and w parts of East TX n to Dallas Co. (Turner et al. 2003) and e to Smith (Benson 1982) and Jefferson (Kartesz 2022) cos.; South TX Plains and Trans-Pecos; AR, AZ, LA, NM, and TX; also Mexico. [*Opuntia lindheimeri* Engelm.] This variety is reported to hybridize with *O. strica*, forming the hexaploid *O. ×alata* Griffiths (Pinkava 2003b). According to Crosswhite (1980), during pioneer days the fruits of var. *lindheimeri* were eaten fresh, made into preserves or syrup, or boiled and fermented into "colonche"; the young stem segments were also eaten; however, eating excessive amounts of older stem segments can cause oxalic acid poisoning. ☠

var. **linguiformis** (Griffiths) B.D. Parfitt & Pinkava, (tongue-shaped), COW'S-TONGUE, COW-TONGUE PRICKLY-PEAR, LENGUA DE VACA. Plant erect 1–2 m tall, often with spreading branches; stem segments lanceolate to narrowly lanceolate, becoming very elongated; spines 1–3, entirely yellow, or absent. Cultivated as an ornamental because of the tongue-like stem segments, escapes; Bexar (Benson 1982; Turner et al. 2003) and Madison (Neill & Wilson 2001) cos.; also Brown (Stanford 1971) and Comanche (Turner et al. 2003) cos. in Cross Timbers and Prairies, Edwards Plateau, and Trans-Pecos; AZ and TX; also Mexico. Spring. [*O. lindheimeri* var. *linguiformis* (Griffiths) L.D. Benson] According to Benson (1982), "… this cultivated and escaped variety is known only from plants descended from those collected near San Antonio, Texas, by Griffiths in 1906." Powell and Weedin (2004) consider this taxon a "mere cultivar" of var. *lindheimeri* and note that it supposedly "originated by a single-gene mutation."

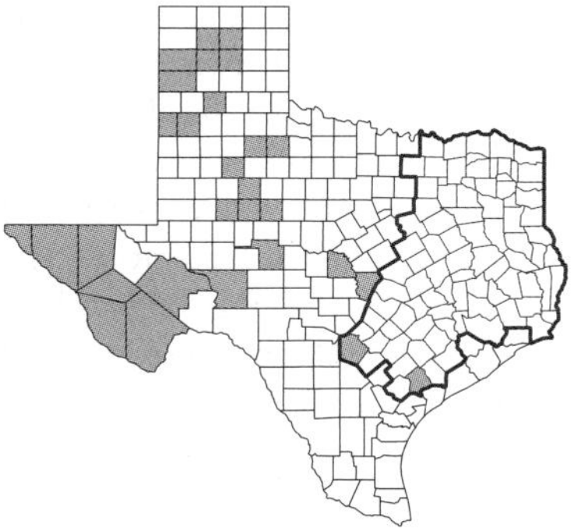
Cylindropuntia imbricata

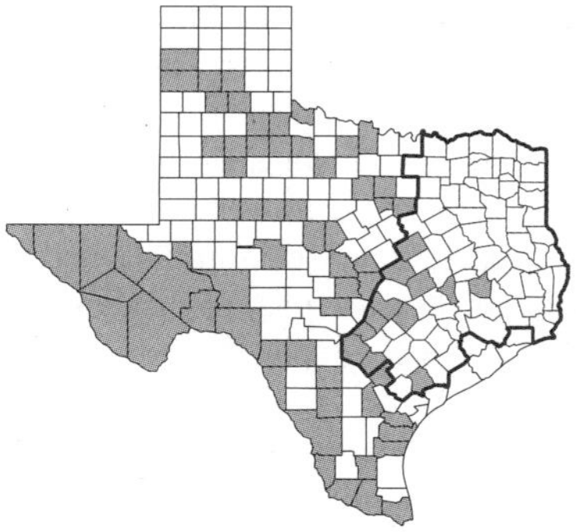
Cylindropuntia leptocaulis

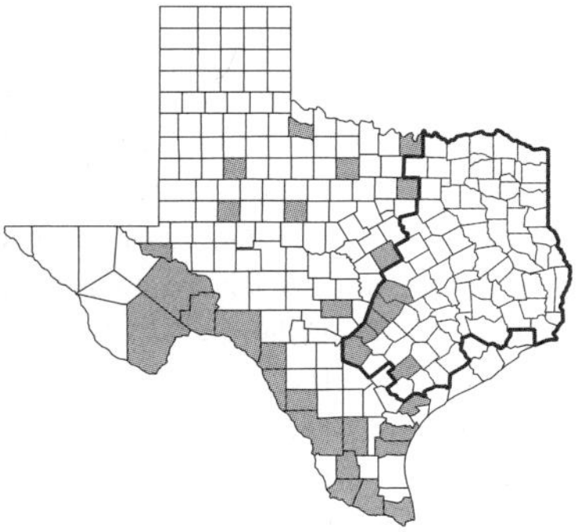
Echinocactus texensis

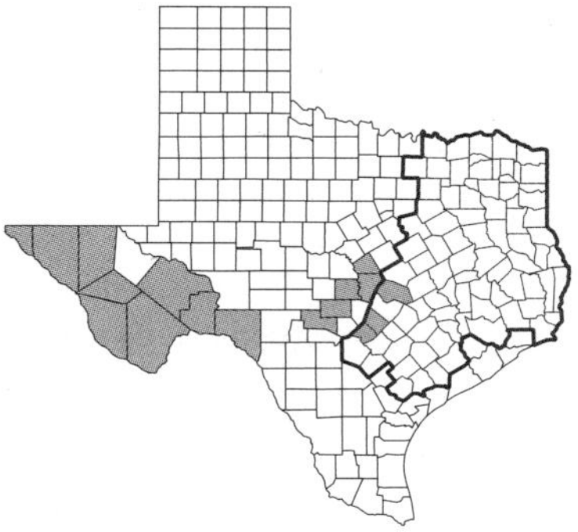
Echinocereus coccineus

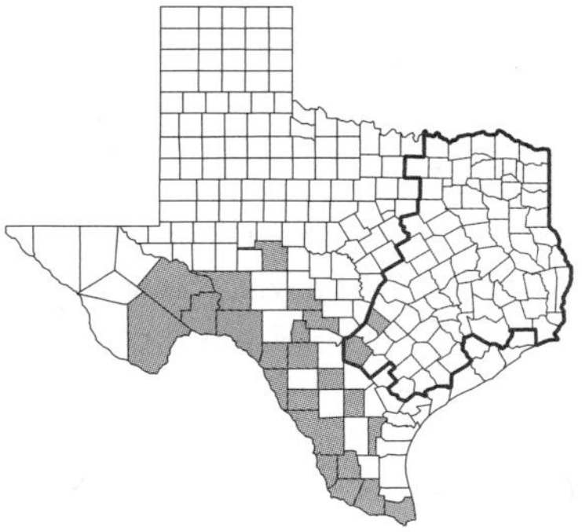
Echinocereus enneacanthus var. brevispinus

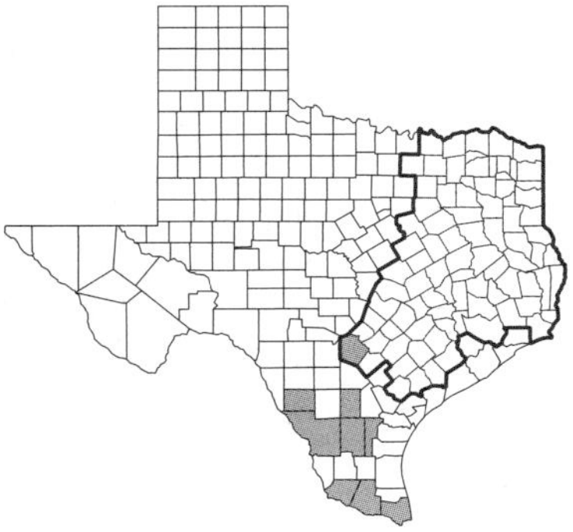
Echinocereus papillosus

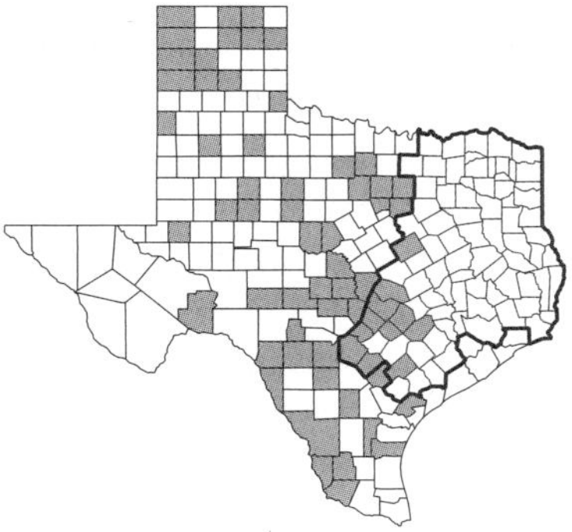
Echinocereus reichenbachii

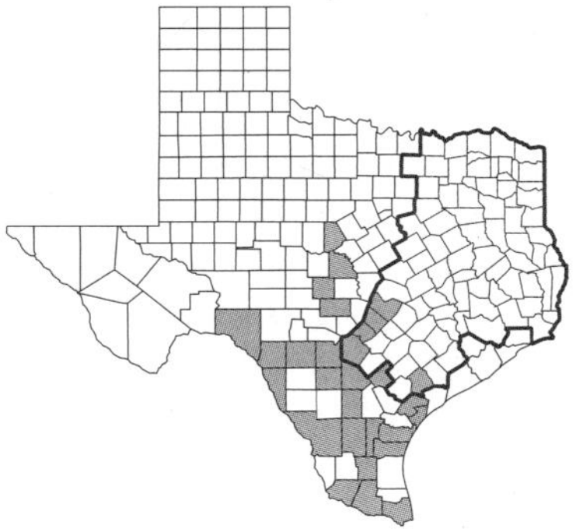
Hamatocactus bicolor

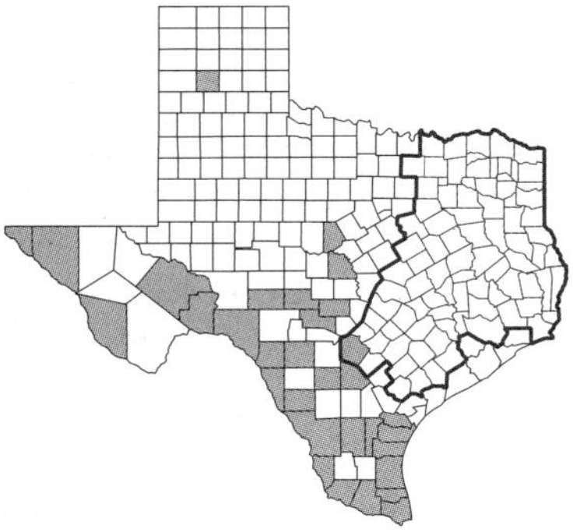
Mammillaria heyderi

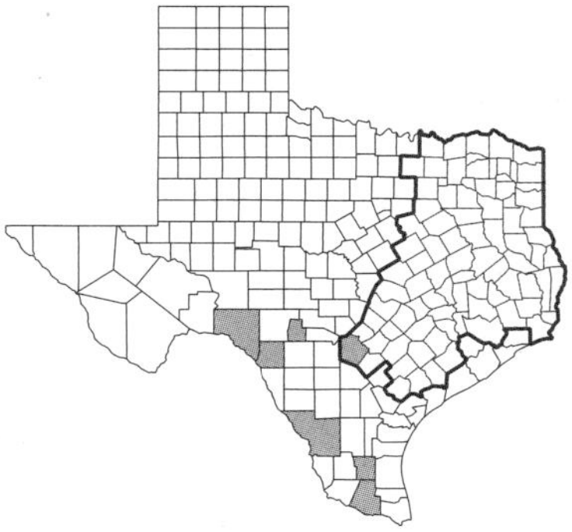
Mammillaria prolifera var. texana

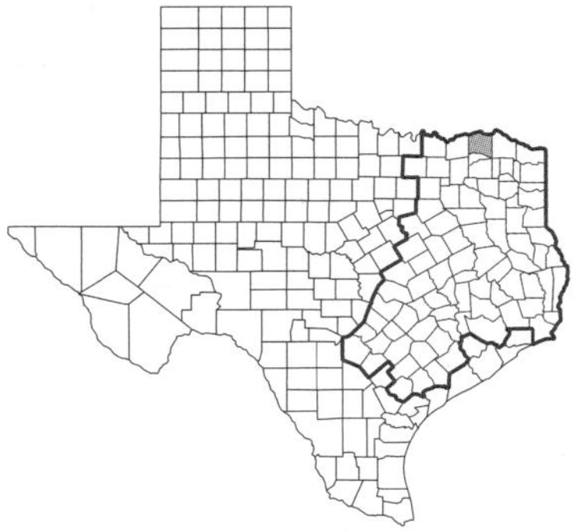
Opuntia cespitosa

Opuntia ficus-indica (L.) Mill., (Indian fig, "a reference to the fig-like fruit that was used by North American Indians—Powell & Weedin 2004), INDIAN-FIG. Tree 3–6 m tall with a main trunk 0.6–1.2 m tall; larger terminal stem segments green, broadly to narrowly obovate or oblong to elliptic; spines sometimes absent to abundant, usually 1–6 per areole, flattened basally, the longer ones 1–2.5(–4) cm long; glochids along adaxial margin of areole, small, less than 2 mm long, yellowish, aging brown; inner tepals yellow or orange-yellow throughout; fruits yellow, orange, red, or purplish, 5–10 cm long, usually spineless. Cultivated and escapes; Comal (Benson 1982) and Walker cos. (Turner et al. 2003); also to the w of East TX in Brown (Stanford 1971) and Lampasas (Benson 1982) cos. and the Trans-Pecos; AZ, CA, and TX, but probably native of Mexico (although original habitat unknown). Spring. Spineless forms have been valued for their fruits since prehistoric times and were probably spread by trading; this species is widely cultivated and has become naturalized in many warm parts of the world and is a pest in some areas (Benson 1982). It has been used as cattle feed, to culture cochineal scale insects (for their red dye), for the fruits ("tunas"), and for the young stem segments (= "nopalitos"), cut into strips and boiled (Pinkava 2003b); nopalitos can sometimes be found fresh or canned in TX supermarkets; a candy-like jelly, known as "queso de tuna," is made from the fruits in Mexico. Crook and Mottram (1997) indicated that it has been cultivated in Mexico for about 8,000 years.

Opuntia mackensenii Rose, (for Bernard Mackensen, 1863–1914, Texas botanist, original collector of the species and author of *The Trees and Shrubs of San Antonio and Vicinity*), MACKENSEN'S PRICKLY-PEAR. Low shrubs, 2–4.5 dm high, branches sprawling, flat stem segment surfaces generally held perpendicular to the ground; roots sometimes with tuberous thickenings; stem segments green to pale green, narrowly to broadly obovate, circular, or obovate-rhombic, 10–20 cm long, to 20 cm wide; spiniferous areoles distributed over most of the stem segment; mid-stem segment areoles 2–3.5 mm apart, oblong, 4–6 mm long, 2.5–4 mm wide, woolly pubescence gray; central spines 2–5(–6) per areole, usually white, ashy-white, or gray, sometimes black-speckled, basally brownish to brownish-black, tip reddish-brown, 2 spines often dominating in each areole, 1–2 deflexed, largest 4–5.5 cm long, 0.7–1 mm wide, proximally flattened; radial spines sometimes present, 3–6 mm long, slender, terete, white, tips yellow; glochids dense in upper part of areoles, 2–3 mm long, reddish, reddish-brown, brown or yellowish; flowers 7–8 cm across; inner tepals all yellow or basally reddish; fruits red, obovoid 3–6 cm long, spineless. Rocky limestone substrates, grassy habitats, woodlands; Hays (*Majure 1970* MISSA), Travis (*Majure 1969* MISSA), and Williamson (*Majure 1967* MISSA) cos.; Edwards Plateau, Trans-Pecos, Rolling and High Plains; expected in Mexico. Endemic to TX? May. [*O. edwardsii* V.E. Grant & K.A. Grant] Most likely this species is only found on the Edwards Plateau just to the w of East TX. Pinkava (2003b) synonymized *O. mackensenii* under *O. tortispina* Engelm. & J.M. Bigelow; however, Powell and Worthington (2018) treat *O. tortispina* as a separate species. They note, however, "Most recent workers have not recognized *O. mackensenii*."

Opuntia macrorhiza Engelm., (large-rooted), PLAINS PRICKLY-PEAR, WESTERN PRICKLY-PEAR, GRASSLAND PRICKLY-PEAR, CHAIN PRICKLY-PEAR, TUBEROUS-ROOT PRICKLY-PEAR. Low, clump-forming perennial 7.5–12.5(–16) cm high; stem segments glaucous, dark green to bluish green, 5–11 cm long, orbicular to obovate, rooting where they touch the ground; spines (0–)1–4(–6) per areole, mostly in upper areoles, the longer spines 3.8–6 cm long, usually deflexed, others if present, porrect (= ± perpendicular to stem segment); glochids to 5 mm long, in a dense tuft; inner tepals yellow, usually with base red-tinged; fruits green to yellowish to dull red or reddish purple, 2.5–3.8 cm long. Sandy or rocky soils; scattered in East TX; widely scattered in TX; mostly c and sw U.S.; also Mexico. Apr–Jun. [*O. compressa* (Salisb.) J.F. Macbr. var. *macrorhiza* (Engelm.) L.D. Benson]

Opuntia nemoralis Griffiths, (growing in woods and groves), OUACHITA MOUNTAINS PRICKLY-PEAR. Plants forming small, spreading shrubs, oftentimes these forming masses (piles) of stem segments resulting in large patches, mounds, or clones with stem segments ascending to 30 cm tall in the summer; stem segments small, gray-green, glaucous, 4.5–8.4 cm long, 2.8–5.8 cm wide, oblong, elliptical, or obovate, the terminal stem segments easily detaching, becoming strongly cross-wrinkled during the winter; spines 1–6 per areole (typically 2), white or yellowish during development, aging bright white when mature, gray in age, strongly retrorsely barbed when developing and into maturity, 1.4–3 cm long, 0.5–0.8 mm in diameter; glochids bright yellow when young turning a dull brown in age; inner tepals 7–8, yellow (or rarely tinged pink basally), obovate with a mucronate tip, 2.7–3.5 cm long; fruits clavate, 2.3–4.2 cm long, dark red to pink, or occasionally light green at maturity. Saline barrens; in East TX known only from Gregg Co. (Majure et al. 2017); Gulf Prairies and Marshes; AR, LA, and TX. Apr–May. [*O. macateei* Britton & Rose] *Opuntia nemoralis* is part of the *O. humifusa* complex. See discussion under *O. cespitosa*. *O. nemoralis* is not listed on NatureServe, but Majure et al. (2017) consider it vulnerable to habitat loss and fragmentation. ⚠

Opuntia phaeacantha Engelm., (dark-thorned), BROWN-SPINED PRICKLY-PEAR, ENGELMANN'S PRICKLY-PEAR, PURPLE-FRUIT PRICKLY-PEAR. Large, prostrate or sprawling to suberect shrub 30–90+ cm high; stem segments green to dark green, with some purplish pigmentation in cold weather, 10–25+ cm long; spines (0–)2–7 per areole (over most of stem segment to only on terminal third or less), central spines usually 2 per areole, whitish to tan throughout or often reddish-brown or reddish-black proximally, tips sometimes red-brown, lowermost spine deflexed, 2–3 cm long, second spine deflexed or projecting, 3.5–5(–6.3) cm long; 3–5

Opuntia engelmannii var lindheimeri [LUN]

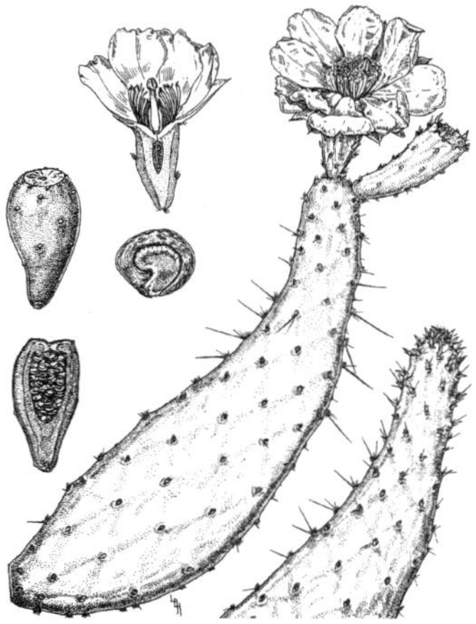
Opuntia engelmannii var. linguiformis [BEL]

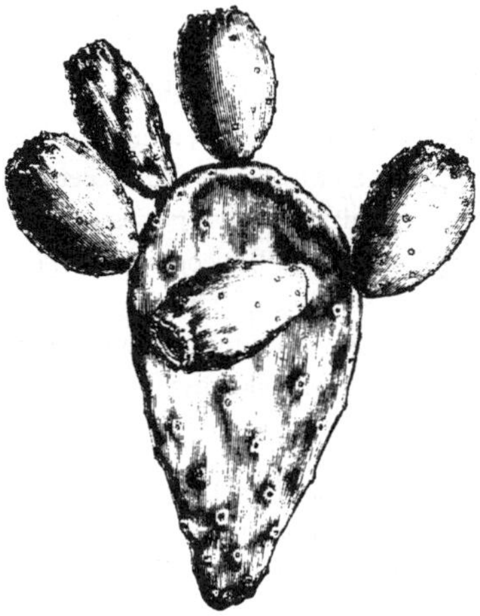
Opuntia ficus-indica [NIC]

Opuntia mackensenii [CLA]

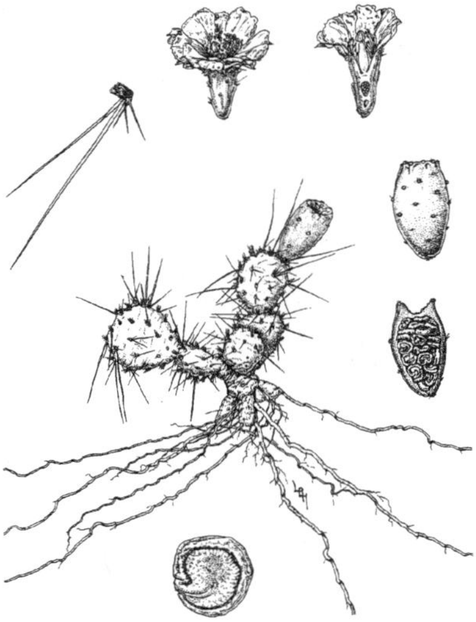
Opuntia macrorhiza [BEL]

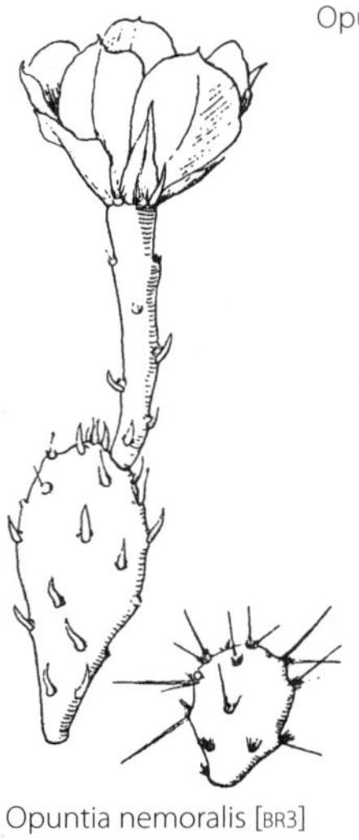
Opuntia nemoralis [BR3]

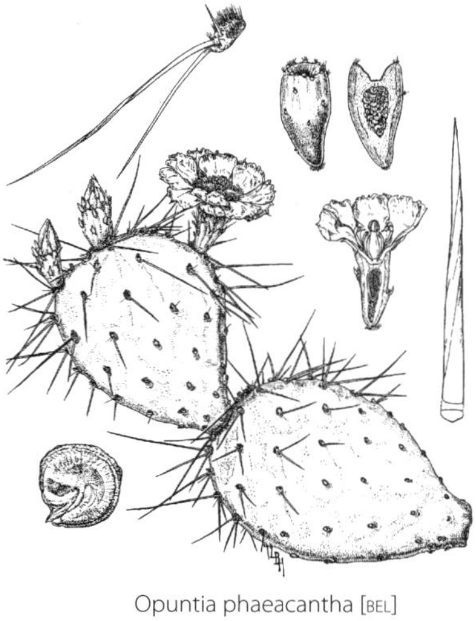
Opuntia phaeacantha [BEL]

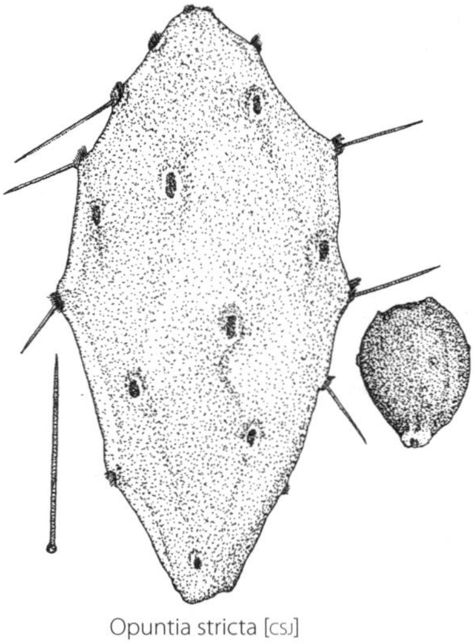
Opuntia stricta [CSJ]

central spines often present in some areoles; radial spines usually 2, at the lower areole periphery, 4.5–8 mm long; glochids tufted, 2–5 mm long; inner tepals yellow, basally pale to prominently reddish; fruits purplish or reddish, 3–6.2 cm long. Hillsides, rocky, gravelly, or sandy soils; Denton, Ellis, McLennan, Williamson, Travis (Turner et al. 2003), Comal, and Hays (Benson 1982) cos.; widespread mainly in w 2/3 of TX; sw U.S. from KS s to TX w to CA. May–Jun. [*O. phaeacantha* var. *camanchica* (Engelm. & J.M. Bigelow) L.D. Benson; *O. phaeacantha* var. *major* Engelm.] A number of infraspecific taxa have in the past been recognized in this species (e.g., Correll & Johnston 1970; Benson 1982; Diggs et al. 1999). More recently, Powell and Worthington (2018) recognized *O. phaeacantha* var. *camanchica* at the species level, but noted, "Our current interpretation of *O. camanchica* in the Trans-Pecos is that it includes several populational hexaploid morphotypes, all in need of further study, but probably warranting separate taxonomic recognition." Their description of its distribution appears to exclude East TX. However, we are following the treatment by Pinkava (2003b) in not formally recognizing variation below the level of species. This species is known to hybridize with several others including *O. engelmannii* and *O. ficus-indica* (Pinkava 2003b).

Opuntia stricta (Haw.) Haw., (upright, erect), PEST PRICKLY-PEAR. Erect to sprawling perennial usually 0.6–2 m tall; stem segments with areoles somewhat raised (tuberculate) resulting in the stem segments appearing slightly scalloped marginally; spines 0–11 per areole, yellow, 1.2–4(–6.2) cm long; inner tepals light yellow throughout; fruits purplish, 3.8–6 cm long. Sandy areas; Anderson (Fleming et al. 2002), Hardin (Brown et al. 2006b), Harris (Kartesz 2022), and Lee (Turner et al. 2003) cos; otherwise in TX limited to near the coast; AL, FL, GA, and SC. Spring–summer. This species is closely tied to coastal habitats and the surprising inland records need confirmation. This species is known to hybridize with *O. engelmannii* forming the hexaploid hybrid, *O. ×alata*, along the se TX coast and in LA (Pinkava 2003b). [*O. stricta* var. *dillenii* (Ker Gawl.) L.D. Benson] Authors have varied in their treatment of *O. stricta*. Some separating *O. stricta* into vars. *stricta* and *dillenii*. Most have not recognized varieties; however, Majure (2022) made the distinction. He does not map var. *dillenii* anywhere in TX, and only questionably maps var. *stricta* in TX (possibly in East TX). His key to differentiate the two varieties is presented below. We, however, are following Pinkava (2003b) in not recognizing varieties.

1. Stem segments spiny throughout, the spines 0–11 per areole, generally curved and spreading from the areole margins, the radial spines usually strongly flattened at the base, then with 1–4 central spines, which are dark yellow and commonly with dark brown or black bands towards the base and middle of the spine ______ var. *dillenii*
1. Stem segments usually few-spined, spines mostly 0–3 per areole, generally straight, produced erect from the areole or only slightly spreading, usually rounded, slightly twisted, or only slightly flattened at base, yellow ______ var. *stricta*

PELECYPHORA Ehrenb.

A genus of 20 species (Sánchez et al. 2022) of mostly w U.S., Mexico, and s Canada. *Pelecyphora*, formerly recognized as *Escobaria*, has often been placed in *Coryphantha* (e.g., Benson 1982; Zimmerman & Parfitt 2004d; Powell et al. 2008). However, we are following Powell and Worthington (2018) and Kartesz (2022), who recognize *Escobaria*, and Sánchez (2022) and Weakley, A.S., and Southeastern Flora Team (2022) who recognize *Pelecyphora* by treating *Pelecyphora* (*Escobaria*) separately from *Coryphantha*. (Greek: *pelekys*, ax, and *phoreus*, bearer, referring to the shape of the tubercles)
REFERENCES: Sánchez et al. 2022; Weakley, A.S., and Southeastern Flora Team 2022.

Pelecyphora missouriensis (Sweet) D. Aquino & Dan. Sánchez, (of Missouri), PLAINS NIPPLE CACTUS, MISSOURI FOXTAIL CACTUS. Plants usually branched; areolar glands absent; spines white to gray or tan, becoming dark gray or yellowish brown, pubescent, the tips often differently colored; central spines usually none (sometimes 1(–2)), if present; radial spines (6–)12–15(–20) per areole, (4–)10–20 mm long; outer tepals fringed (rarely entire); inner tepals greenish yellow to bronze; fruits red, globose to ellipsoid (previous year's fruits can be present at spring flowering time); seeds black, pitted. Grayson (BRIT), Bexar, Dallas, Denton, Kaufman, Lee, Travis, Walker, Washington, and Williamson (Benson 1982; Turner et al. 2003) cos.; High Plains e to w East TX and s to Edwards Plateau; wc U.S. from ND s to TX and w to MT and AZ. Apr–May. [*Coryphantha missouriensis* (Sweet) Britton & Rose, *C. missouriensis* var. *caespitosa* (Engelm.) L.D. Benson, *C. missouriensis* var. *robustior* (Engelm.) L.D. Benson, *Escobaria missouriensis* (Sweet) D.R. Hunt, *E. missouriensis* var. *robustior* (Engelm.) D.R. Hunt, *E. missouriensis* var. *similis* (Engelm.) N.P. Taylor, *Mammillaria missouriensis* Sweet, *M. similis* Engelm., *Neobesseya missouriensis* (Sweet) Britton & Rose] This species has long been confused taxonomically, having been treated in four genera (*Coryphantha*, *Escobaria*, *Mammillaria*, and *Neobesseya*), divided into a number of varieties, and partially confused with *C. sulcata* (e.g., Kartesz 1994; Jones et al. 1997). However, in an exhaustive molecular study, Sánchez et al. (2022) recovered the genus *Pelecyphora* (a genus of northeastern Mexico) in a clade with *Escobaria* with almost all species of *Coryphantha* included in the study falling in a separate clade. *Pelecyphora* has priority over *Escobaria* and they transferred all species of *Escobaria* to *Pelecyphora*. Zimmerman and Parfitt (2003d) state that *P. missouriensis* is sometimes misidentified as *Coryphantha sulcata* (see *C. sulcata* description for discussion).

CALYCANTHACEAE Lindl.

SWEET SHRUB, SPICEBUSH, OR STRAWBERRY-SHRUB FAMILY

A family of 3–4 genera and 10–13 species of aromatic shrubs and trees of e Asia, temperate e and far w North America, and tropical ne Australia (Stevens 2001 onwards; Li et al. 2004; Mabberley 2017). The disjunct occurrence in four widely scattered areas is probably a relict of a once more widespread distribution, a conclusion supported "by the occurrence of Cretaceous Calycanthaceae fossils in Brazil" (Zhou et al. 2006). As treated here, following the Angiosperm Phylogeny Group (APG IV 2016), the family includes the Idiospermaceae (one species endemic to Australian rainforests). However, Li et al. (2004) have pointed out that "sequence divergence [of *Idiospermum*] in both nuclear and chloroplast regions is much higher than that among other genera of Calycanthaceae ... and *Idiospermum* is morphologically distinctive from other genera." Species are variously used as ornamentals, medicinally, and as spices and for scents. The family shares a number of unspecialized flower characters with the Magnoliaceae—beetle-pollinated flowers with numerous spirally arranged perianth parts and numerous stamens and carpels. This is not surprising because it is one of the "magnoliids" (a group of four related orders—Canellales, Laurales, Magnoliales, Piperales) which are among the approx. 3% of flowering plants near the base of the angiosperm lineage that are neither in the monocots or eudicots (see Appendices 3 & 4). (subclass Magnoliidae—Cronquist; order Laurales—APG IV)

FAMILY RECOGNITION IN THE FIELD: the only East TX species is an aromatic shrub with opposite, entire, exstipulate leaves and flowers solitary at the branch tips having numerous reddish brown to maroon tepals and with numerous achenes inside of a persistent expanded fruit-like hypanthium.

REFERENCES: Pollard 1908; Wood 1958, Nicely 1965; Endress 1993a; Bygrave 1996; Johnson 1997; Li & Li 2000; Li et al. 2004; Zhou et al. 2006; Li & Bartholomew 2008.

CALYCANTHUS L. SWEET SHRUB, ALLSPICE, SPICEBUSH, OR STRAWBERRY-BUSH

A genus of 3 species of the e and sw U.S. and China (Li et al. 2004), including *C. chinensis* [*Sinocalycanthus chinensis*], CHINESE SWEET SHRUB, previously recognized as a monotypic genus. However, molecular evidence suggests that *C. chinensis* should be included in *Calycanthus* (Zhou et al. 2006). The sw U.S. species, *C. occidentalis*, and the e U.S. species, *C. floridus*, are most closely related to each other, and these in turn are related to the Asian *C. chinensis*. Molecular data "lend support to the vicariant hypothesis that this well-known floristic disjunction represents the fragmentation of a more or less continuous Mixed Mesophytic forest community that existed [broadly across the Northern Hemisphere] during the Tertiary" (Xiang et al. 1998a). The species are cultivated as ornamentals. ☠ Calycanthine, an alkaloid similar to strychnine, is present primarily in the fruit/seeds; it is toxic to both livestock and humans via neurological effects (Burrows & Tyrl 2001; Xu & Cheng 2015). The inner tepals, stamens, and staminodes of *Calycanthus* are reported to have white food bodies at their tips (Nicely 1965; Johnson 1997), presumably to attract beetle pollinators. (Greek: *kalyx*, covering, cup, calyx, and *anthos*, flower, from the lack of clear differentiation of the tepals into sepals and petals—Hyam & Pankhurst 1995)

REFERENCES: Boufford & Spongberg 1981; Hardin 1984; Ferry & Ferry 1987; Xiang et al. 1998a; Brown & Elsik 2002.

Calycanthus floridus L. var. **glaucus** (Willd.) Torr. & A. Gray, (sp.: free flowering, producing abundant flowers; var.: whitened with a coating or bloom), CAROLINA ALLSPICE, SWEET SHRUB, EASTERN SWEET SHRUB, SWEET-BETSY, STRAWBERRY SHRUB, BUBBY-BUSH. Aromatic (when foliage crushed) shrub 1–3.5 m tall, the odor of the leaves and bark sometimes referred to as camphor-like; twigs, petioles, and lower leaf surfaces glabrous or with only scattered pubescence; buds naked, the lateral buds partially hidden by petiole bases; leaves deciduous, opposite, simple, petiolate, without stipules, blades elliptic to oblong or ovate, usually entire, to 15 cm long, acute to acuminate; flowers solitary, terminal on branches, radially symmetrical, perfect, with a cup-shaped hypanthium, with strawberry or pineapple scent; perianth of numerous free tepals, these reddish brown to maroon, 2–4 cm long; stamens numerous, on the rim of the hypanthium, the inner reduced to staminodes, the fertile with the filament extending slightly beyond the anthers; pistils 10–35 per flower; fruits achenes developing inside the expanded persistent hypanthium (= pseudocarp or false fruit), the fruit-like hypanthium apically constricted, brownish, fibrous, to ca. 6 cm long. Woods; Newton Co. (*Brown & Elsik 24437* BRIT, SBSC; *Lewis, Lewis, & Elsik 6169* SBSC, TEX—fide Brown & Elsik 2002); otherwise unknown in TX; mainly se U.S. from PA, MD, and VA s to FL w to TN and AL and possibly native as far w as LA and TX. Late spring. At least 15 individuals of this species were reported from woods near a residence by Brown and Elsik (2002). The owners reported that the "plants were here when they established a residence on the site" and Brown and Elsik (2002) concluded that they "are uncertain of their native status here; however, they are well established and naturalized shrubs and should be considered a member of the Texas flora." Likewise, Thomas and Allen (1996) were uncertain regarding the native status of a collection of this species from West Feliciana Parish, LA. [*C. glaucus* Willd., *C. fertilis* Walter, *C. floridus* var. *laevigatus* (Willd.) Torr. & A. Gray, *C. floridus* var. *oblongifolius* (Nutt.) Boufford & Spongberg, *C. nanus* Loisel.] This species is cultivated as an ornamental, and the bark has been dried and used as a substitute for cinnamon (Mabberley 2017). The species has been used medicinally, including by Native Americans (Johnson 1997; Mabberley 2017), but is also poisonous (especially the seeds) due to the presence of the alkaloid calycanthine (Wood 1958; Nicely 1965; Burrows & Tyrl 2001). Pollination is apparently by beetles (Nicely 1965). The similar var. *floridus* also occurs in the se U.S.; it differs in having pubescent twigs, petioles, and lower leaf surfaces. The two varieties are reported to intergrade in some areas and "determination of variety is not always possible" (Johnson 1997).

CAMPANULACEAE Juss., Nom. Cons.

BELLFLOWER, BLUEBELL, OR HAREBELL FAMILY

Ours annual, biennial, or perennial herbs; leaves alternate, simple, entire or toothed, sessile or the lower petioled; stipules absent; flowers bisexual, axillary or terminal, solitary or in racemes or spike-like inflorescences; calyces usually with 5 lobes; corollas sympetalous at least at base, radially symmetrical and 5-lobed or bilaterally symmetrical and 2-lipped or (in cleistogamous forms of *Triodanis*) vestigial or absent; stamens 5; filaments free or united above into a tube; anthers free or united into a ring; pistil 1, of 2–5 carpels; ovary inferior or partly inferior, with placentation axile; fruit a capsule; seeds numerous.

A medium-large (2,250–2,380 species in 84–85 genera—Stevens 2001 onwards; Mabberley 2017) cosmopolitan family of herbs or rarely shrubs with a network of laticifers; it includes a variety of ornamentals in genera such as *Campanula*, *Downingia*, and *Lobelia*; a number of species contain toxic alkaloids. The family is composed of 5 monophyletic subfamilies (Gustafsson & Bremer 1995; Stevens 2001 onwards), with the two largest being very distinctive—the Campanuloideae (non-resupinate radially symmetrical flowers, free anthers), and the Lobelioideae (with resupinate, bilaterally symmetrical, 2-lipped flowers, united anthers), which is sometimes recognized as a distinct family, the Lobeliaceae (e.g., Lammers 1992). Cronquist (1988), however, included the Lobelioideae and treated the Campanulaceae in the broad sense as have most recent authors (e.g., Stevens 2001 onwards; Yatskievych 2006; Lammers 2007; APG IV 2016; Crowl et al. 2016; Morin in press). Molecular analyses (e.g., Cosner et al. 1994; Crowl et al. 2016) support the monophyly of the lineage, which is consistent with the single family approach. Stevens (2001 onwards) noted that "generic limits need a great deal of attention (an understatement), *Campanula*, *Lobelia* and *Wahlenbergia* and their relatives (i.e., most of the family) all being particular problems." As traditionally treated, the genus *Campanula* is "seriously polyphyletic" and is now being subdivided (e.g., Morin 2020) and *Lobelia* also is non-monophyletic (Crowl et al. 2016) and "patently artificial" (Lammers 2011). Thus, changes in how genera are treated in the future is to be expected. *Sphenoclea zeylanica* Gaertn. (PIEFRUIT, CHICKENSPIKE), which we consider to be in the Sphenocleaceae, was previously placed in the Campanulaceae; it is not closely related to Campanulaceae, instead being in the Solanales and related to families including the Convolvulaceae, Hydroleaceae, and Solanaceae (Stevens 2001 onwards). Species of *Campanula* typically have a complex pollination mechanism with protandrous flowers (= anthers maturing before pistils) and pollen actually presented on the style (a type of secondary pollen presentation). The 5 anthers, which in bud are arranged in a tight cylinder (though not fused together) around the immature style, release their pollen to the inside (within the cylinder) where it encounters a dense brush of pollen-collecting hairs on the outside of the style, including the unexpanded and still unreceptive stigma lobes. As the lower part of the style expands, the pollen is swept out of the cylinder and exposed on the style, where pollinating insects (attracted by nectar) can have access. Eventually, after all or most of the pollen is removed, the stigma lobes open exposing the now receptive female surfaces. Since the male and female parts mature at different times, the system in general ensures outcrossing. However, self-pollination could possibly happen if the unfolding stigma lobes come into contact with pollen grains remaining on the style (Shetler 1979; Nyman 1993a, 1993b). (Diminutive of Latin: *campana*, bell, in reference to the shape of the flowers) (subclass Asteridae—Cronquist; order Asterales—APG IV)

FAMILY RECOGNITION IN THE FIELD: herbs with alternate simple leaves and sometimes milky sap, inferior ovaries maturing into capsules with many seeds, showy, bilaterally (2–lipped) or radially symmetrical corollas, and sometimes fused anthers.

REFERENCES: McVaugh 1943, 1961; Shetler 1979; Diggs 1982; Rosatti 1986; Shetler & Morin 1986; Lammers 1992, 2007; Nyman 1993a, 1993b; Cosner et al. 1994, 2004; Gustafsson & Bremer 1995; Savolainen et al. 2000; Eddie et al. 2003; Lundberg & Bremer 2003; Shulkina et al. 2003; Roquet et al. 2009; Mansion et al. 2012; Crowl et al. 2016; Xu & Hong 2021; Morin (in press).

1. Corollas bilaterally symmetrical and conspicuously 2-lipped (the dorsal lip with 2 lobes separated by a deep cleft, the ventral lip with 3 lobes); anthers permanently united into a ring surrounding the style; filaments free at base but united above into a tube; ovary of 2 carpels (Subfamily Lobelioideae) ______ **Lobelia**
1. Corollas radially symmetrical (can be vestigial or absent in cleistogamous flowers); anthers separate; filaments separate; ovary of 3(–5) carpels (Subfamily Campanuloideae).

2. Corollas 20–30 mm in diam., rotate (= saucer shaped, with a ± flat face), usually deep to pale blue, rarely white, with white or pale blue raised ring at base; lower leaves with petioles 10–50 mm long; styles very distinctive, declined in basal half, curved downward in apical half; plants currently unknown in TX, but present just across the Red River in se OK _______ *Campanulastrum*
2. Corollas < 20 mm in diam., campanulate to funnelform or rotate, blue to purple, violet, or white, without a distinct white or pale blue raised ring at base; lower leaves with petioles 7 mm or less long; styles not as above; plants widespread in TX.
   3. Flowers mostly sessile in the axils of sessile, often clasping bracts/leaves, typically both chasmogamous and cleistogamous (= permanently closed, with vestigial corolla; usually at lower nodes) flowers present; stems usually unbranched _______ **Triodanis**
   3. Flowers mostly long-pedicellate (= long-stalked), all chasmogamous, the bracts/leaves not clasping; stems branched.
      4. Plants annuals with stem branched, but not in clumps; capsules opening above the middle by lateral pores; corollas (5–)9–13(–17) mm long; leaves 15 mm or less long; habitats alluvial sandy or gravelly areas; species known in East TX only from extreme w margin of area (Travis Co.) _______ **Poolea**
      4. Plants annuals or short-lived perennials with stems usually several to many in clumps; capsules opening by apical pores (= within the calyx lobes); corollas 5–8(–12) mm long; leaves 10–40 mm long; habitats typically disturbed areas, roadsides, old fields, banks of streams; species known in TX only from extreme e margin of East TX (se Pineywoods) _______ **Wahlenbergia**

CAMPANULASTRUM Small AMERICAN BELLFLOWER, TALL BELLFLOWER

A monotypic genus of e and c North America. It has often been included within a broadly conceived *Campanula* (e.g., Yatskievych 2006; Weakley et al. 2012), but recent research shows that it is part of a clade with *Poolea reverchonii*, *Legousia*, and *Triodanis*. Its chromosome number of $2n = 58$ is unique for counts of *Campanula sensu lato* (in the broad sense) (Morin in press). Based on this evidence, it is inappropriate to keep *Campanulastrum americanum* in *Campanula*. (From *Campanula*, because of the resemblance to that genus, and Latin, *astrum*, in reference to the star-shaped corolla)

REFERENCES: Shetler 1963; Baskin & Baskin 1984; Nowicke et al. 1992; Kalisz & Wardle 1994; Mansion et al. 2012; Barnard-Kubow et al. 2015; Morin. in press

Campanulastrum americanum (L.) Small, (of America), AMERICAN BELLFLOWER, TALL BELLFLOWER. Plants annual or biennial (20–) 50–210 cm tall; stems erect, branched or unbranched; leaves alternate; the blade margins crenate to serrate or biserrate, the lower leaves petiolate with petioles 10–50 mm long, ovate to broadly lanceolate, the upper sessile, narrowly lanceolate; inflorescences usually spikes or spike-like racemes, rarely branched, (4–)10–40 cm long; pedicels to 4 mm long; flowers all chasmogamous; calyx lobes linear-lanceolate, 5–12 mm long; corollas rotate, 10–30 mm in diam., usually deep to pale blue, rarely white, with white or pale blue raised ring at base; styles somewhat s-shaped, declined in basal half, curved upward in apical half; capsules dehiscing by a lateral pore near apex. Open mesic woods, forest margins, floodplains, disturbed areas; currently unknown from TX, but present in McCurtain, Choctaw, and Bryan cos. OK directly across the Red River from the ne part of East TX (Kartesz 2015); widespread in the e 1/2 of the US w to SD, NE, KS, OK, AK, and LA (but primarily from the Appalachian Mts. to just w of the Mississippi River (Morin in press). (May–)Jul–Sep. [*Campanula americana* L., *C. americana* var. *illinoensis* (Fresen.) Farw., *C. illinoensis* Fresen., *Campanulastrum americanum* var. *illinoense* (Fresen.) Mohlenbr.] Morin (in press) has concluded that the type specimen of *Campanula illinoensis* Fresen. "does not differ from most *Campanulastrum americanum*" and as a result does not recognize infraspecific taxa based on this name. This species is sometimes cultivated in gardens and can be extremely showy, with "hundreds of flowers open simultaneously" (Morin in press). Barnard-Kubow et al. (2015) presented evidence on the possible genetic effects of multiple glacial refugia on this species.

LOBELIA L. LOBELIA

Perennial, biennial, or annual herbs often with milky juice; inflorescences terminal racemes, these sometimes spike-like because of the relatively short pedicels, often 1-sided; flowers resupinate (= inverted as a result of twisting of flower stalk); calyces 5-cleft, radially symmetrical; corollas bilaterally symmetrical, 2-lipped, the ventral lip 3-lobed, the corolla split nearly to base between 2 lobes of the dorsal lip, the tube fenestrate (= with lateral openings) or not so, red, blue, purple, to white; stamens 5; filaments united above into a tube; anthers united into a tube/ring, some or all the anthers with apical tufts of hairs; 2-lobed stigma exerted through tip of anther tube; ovary 2-carpellate, inferior or partly inferior; capsules opening by apical pores.

A genus of ca. 415–435 species (Stevens 2001 onwards; Lammers in press) of tropical and warm areas, especially the Americas, Australia, and Africa; also a few in temperate regions (Mabberley 2017); it is the second largest genus in the family, following only *Campanula*. According to Lammers (in press), "In light of the molecular data available so far, the genus is patently artificial as currently circumscribed and will require dismemberment once sufficient data have accumulated." Further, Stevens (2001 onwards) noted that "*Lobelia* is wildly paraphyletic, all other genera of the subfamily being embedded within it." Therefore generic circumscription and name changes should be expected in the future. ☠ Many have toxic alkaloids including lobelamine and lobeline (Krochmal et al. 1972; Hardin & Arena 1974) and are strongly poisonous; according to some sources (e.g., Heywood 1993), even smelling *Lobelia tupa* L. of Chile may cause sickness; some have been used medicinally; lobeline has an effect on the central nervous system similar to nicotine (Blackwell 1990); a number are cultivated as ornamentals. Some *Lobelia* species can be hybridized (e.g., *L. cardinalis* × *L. siphilitica* = *L.* ×*speciosa* Sweet) (e.g., Bowden 1964a, 1982; Thompson & Lammers 1997); hybrids are rarely found in nature (McVaugh 1943) and are unknown from East TX. Yatskievych (2006) noted that the flowers of *Lobelia* and its relatives "are functionally similar to those of the Asteraceae in that the stigma elongates through the center of the ring of fused anthers, with a brush of hairs near the style tip pushing the pollen out of the flower, and the stigma lobes do not spread and become receptive until after the pollen has been released." Such a mechanism increases the probability of outcrossing. (Named for Matthias de l'Obel, 1538–1616, Flemish herbalist/botanist and physician to James I of England)

REFERENCES: McVaugh 1936, 1940; Turner 1950; Bowden 1959a, 1959b; 1960a, 1960b, 1961, 1964a, 1982; Krochmal et al. 1972; McGregor 1985d; Lammers 1993, 2006, 2011, (in press); Thompson & Lammers 1997; Molano-Flores 2002; Antonelli 2007; Spaulding & Barger 2016.

1. Corollas bright red; flowers (including calyx) 30–50 mm long; filaments including tube portion (but not anther tube) (15–)19–33 mm long ____ **L. cardinalis**
1. Corollas various shades of blue, purple, lavender or violet to white; flowers 9–33 mm long; filaments including tube portion 2–15 mm long.
    2. Flowers 14–33 mm long; filaments including tube portion 6–15 mm long.
        3. Corollas white-striped in throat; flowers 23–33 mm long; corolla tube often 5–10 mm wide; filaments including tube portion (10–)12–15 mm long; calyx lobes often with conspicuous basal auricles 2–5 mm long; bracteoles at or above middle of pedicel ____ **L. siphilitica**
        3. Corollas not white-striped in throat (but can have a central white "eye" or spot where bottom lip of corolla joins the corolla tube); flowers 14–24 mm long; corolla tube 3–4 mm wide; filaments including tube portion 6–11 mm long; calyx lobes with small basal auricles 1.5 mm or less long (except auricles large in *L. reverchonii*); bracteoles at or below middle of pedicel, often near its base.
            4. Stems usually with obvious pubescence; corolla tube fenestrate (= with 2 lateral slit-like openings as well as the dorsal split).
                5. Stems usually puberulent; pubescence on the base of the calyx and pedicels rather short; inflorescences with flowers usually closely spaced; sepals narrowly triangular with small triangular or rounded auricles; filament tube 6–8 mm long ____ **L. puberula**
                5. Stems usually long hirsute/pilose; pubescence on the base of the calyx and pedicel densely shaggy; inflorescences with flowers widely spaced; sepals ovate to ovate-lanceolate with large, deeply cordate auricles; filament tube 8–11 mm long ____ **L. reverchonii**
            4. Stems glabrous or sparsely hirtellous; corolla tube not or only rarely fenestrate.
                6. Basal leaves 0; cauline leaves 3–20; sepals narrowly triangular, the auricles round, 0.8–1.2 mm long; pedicels 2-bracteolate at or below middle; corollas blue ____ **L. flaccidifolia**
                6. Basal leaves 3–10; cauline leaves 3–5; sepals triangular or deltate, the auricles very small; pedicels 2-bracteolate at base; corollas pale purple or white ____ **L. floridana**
    2. Flowers 7–15 mm long; filaments including tube portion 2–4 mm long.
        7. Calyx lobes without auricles (= earlobe-like appendages) at base; seeds smooth; species known in East TX only from Travis Co. on extreme w margin of area ____ **L. berlandieri**
        7. Calyx lobes with auricles at base; seeds rough; species known almost entirely from e 1/2 of East TX.
            8. Calyx lobes glabrous or nearly so (ciliate only near tips), with thread-like or very short and triangular auricles; leaves at middle of stem narrowed to base, not clasping ____ **L. spicata**
            8. Calyx lobes ciliate, with usually conspicuous, flat, membranous, lanceolate auricles 3–5 mm long; leaves at middle of stem subsessile with broad base or clasping the stem ____ **L. appendiculata**

Lobelia appendiculata A. DC., (with an appendage), EARFLOWER, PALE LOBELIA. Slender annual 0.15–0.9(–1.1) m tall; stems glabrous or with a few basal hairs; basal leaves tardily deciduous, obovate; cauline leaves oblong-lanceolate to ovate, obtuse, subentire or toothed; middle stem leaves usually clasping or subsessile with broad base, ± glabrous; flowers in loose, ± secund (= 1-sided) spike-like raceme (pedicels short) to 32 cm long; pedicels 4–11 mm long, 2 bracteolate at base; calyx lobes linear-lanceolate, 4–8 mm long, usually densely bristly-ciliate, sometimes only near tips, with conspicuous basal auricles that are flat, ciliate, and lanceolate; corollas light violet or light lavender-blue to rarely white; filament tube 2–4 mm long; anther tube 1.9–2.5 mm long, bluish gray. Sandy open woods, fields, prairies, and roadsides; widespread in e 1/2 of East TX and also Travis Co. (Turner et al. 2003) further w; also n Gulf Prairies and Marshes and San Saba Co. (Turner et al. 2003) in Cross Timbers and Prairies; KY s to AL w to KS, OK, and TX. Apr–Jun. Assumed to contain alkaloids and to be poisonous. ☠

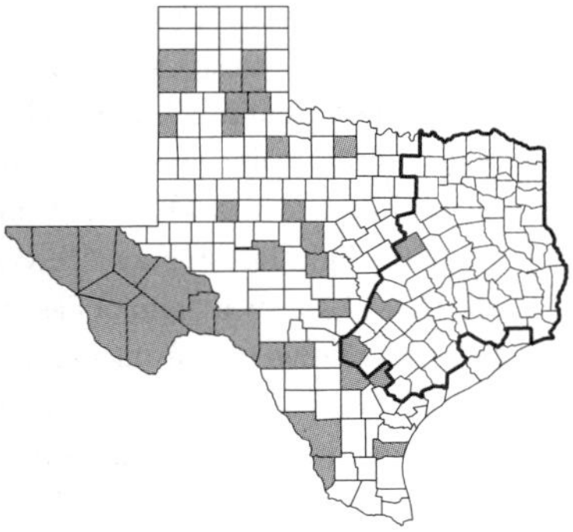

Opuntia engelmannii var. engelmannii

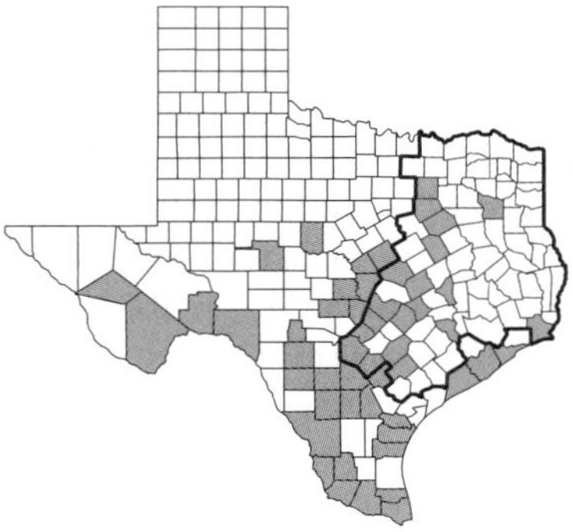

Opuntia engelmannii var. lindheimeri

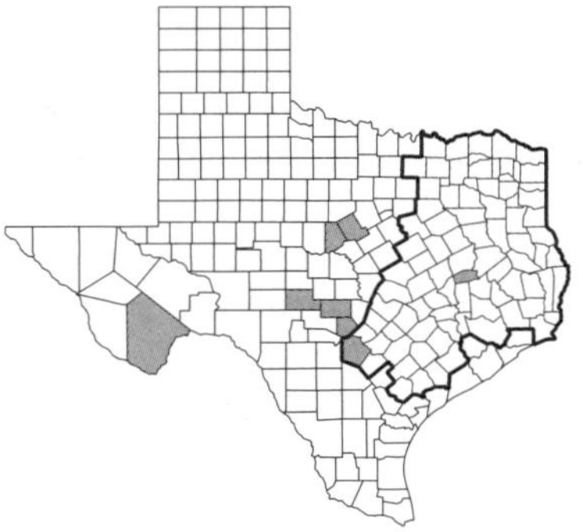

Opuntia engelmannii var. linguiformis

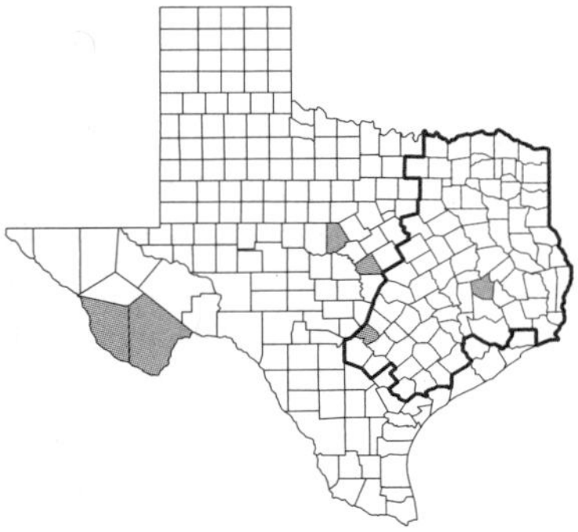

Opuntia ficus-indica

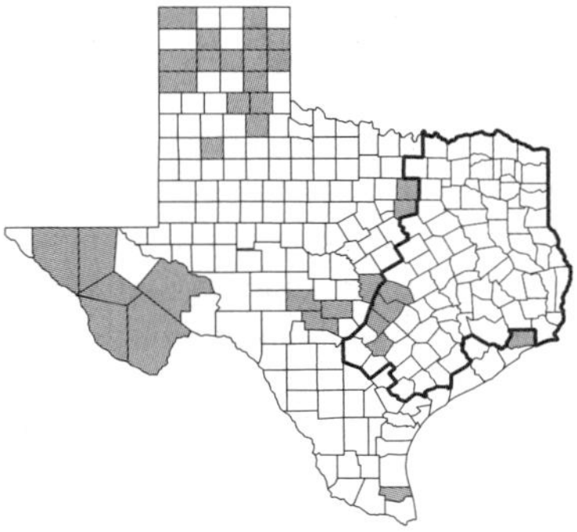

Opuntia mackensenii

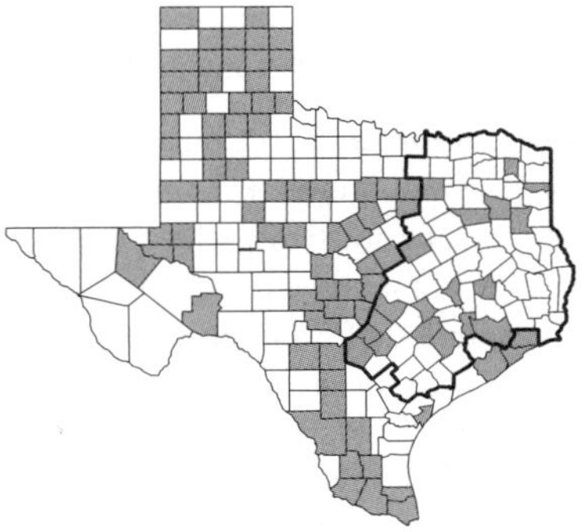

Opuntia macrorhiza

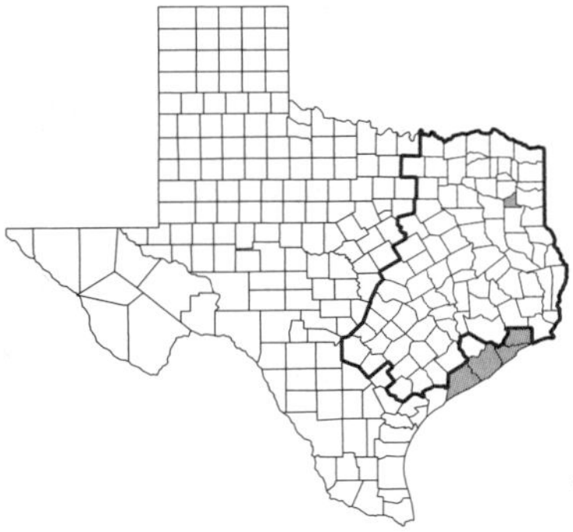

Opuntia nemoralis

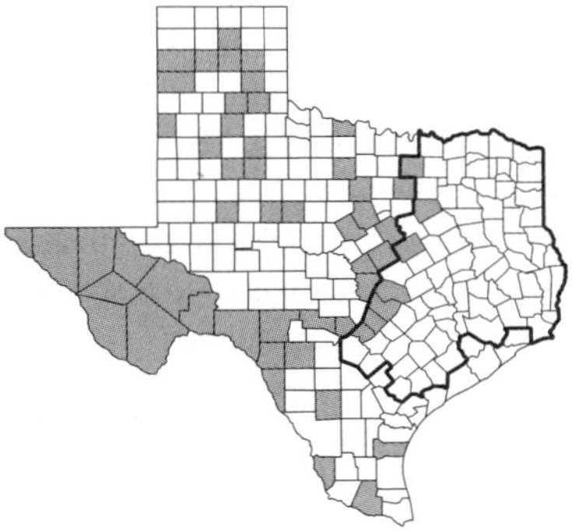

Opuntia phaeacantha

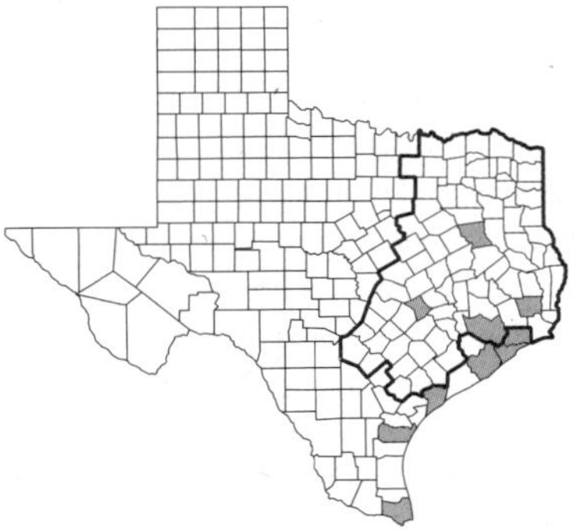

Opuntia stricta

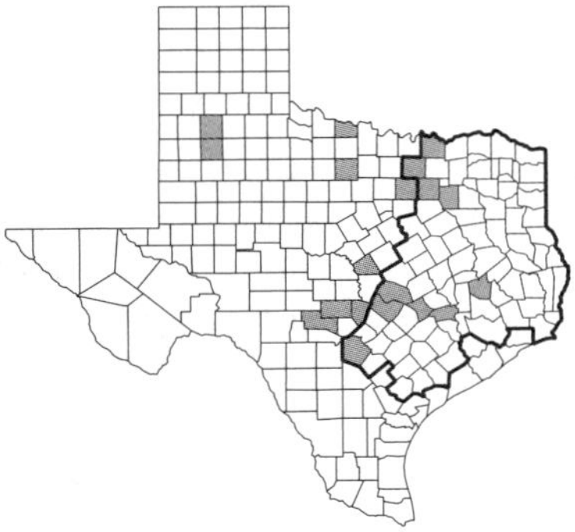

Pelecyphora missouriensis

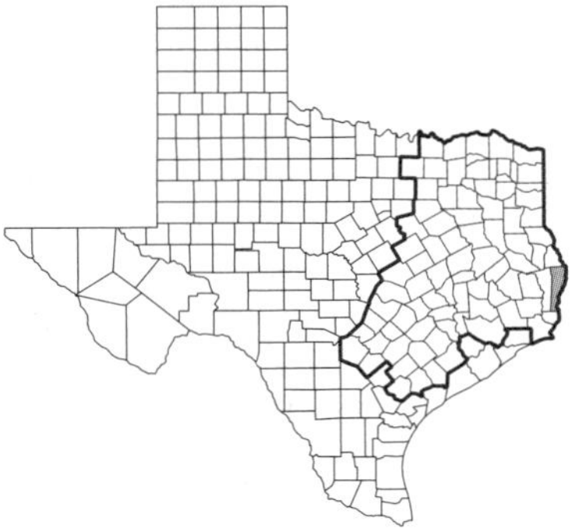

Calycanthus floridus var. glaucus

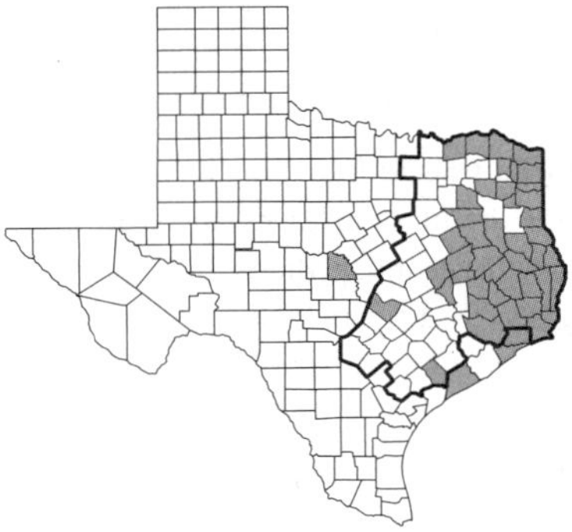

Lobelia appendiculata

Lobelia berlandieri A. DC. subsp. **brachypoda** (A. Gray) Lammers (sp.: for Jean Louis Berlandier, 1805–1851, French botanist who explored TX, NM, and Mexico, and one of the first botanists to make extensive collections in TX, subsp.: short-footed), BERLANDIER'S LOBELIA. Annual to ca. 0.6 m tall; stems glabrous or with sparse hairs near base; leaves mostly cauline, 2.5–5 cm long, elliptic to lanceolate or ovate, obtuse apically, subentire or toothed; raceme to ca. 25 cm long; pedicels relatively straight, 4–10 mm long; calyx lobes linear-subulate, ciliate, without basal auricles; corollas pale to bright purplish-blue with a white eye (rarely all white); corolla tube not fenestrate; filament tube 2.5–4 mm long; anther tube 1–1.7 mm long, light bluish gray. Alluvium along the Colorado River; in East TX known only from Travis Co. (e.g., *B.C. Tharp s.n.*, 1935, TEX-LL) at sw margin of East TX; mainly Edwards Plateau and Trans-Pecos, also s TX; also adjacent Mexico. In the US this species is known only from TX. May–Sep. [*L. berlandieri* var. *brachypoda* (A. Gray) McVaugh] Turner et al. (2003) recognized subsp. *brachypoda* at the species level. Lammers (2006) argued that the great number of morphologically intermediate plants in the limited zone of contact between otherwise geographically separated populations is "exactly the sort for which I prefer to use subspecific rank." We are following him in treating the taxon as a subspecies. Subspecies *berlandieri*, which is distinguished by its glabrous calyx lobes, leaves mostly in a basal rosette, and curved pedicels usually 12–25 mm long or rarely longer, is known in TX only well to the s of East TX, along the Gulf Coast. Lammers (in press) noted that "plants of intermediate morphology are sometimes found in their zone of sympatry." Assumed to contain alkaloids and to be poisonous.

Lobelia cardinalis L., (cardinal, referring to the red robes worn by Catholic cardinals (Coffey 1993), in reference to the flower color), CARDINAL-FLOWER. Perennial to 1.3(–2) m tall, with basal offshoots; raceme terminal, to 50 cm long; calyx lobes foliaceous, to 14 mm long and 6 mm wide; corollas bright crimson red (rarely white or pink); corolla tube fenestrate; filament tube 14–37 mm long; anther tube usually 4–5.5 mm long, bluish gray; capsules 8–10 mm long. Low moist areas, stream banks; nearly throughout TX; e US and adjacent Canada, s half of c and w US, and s through Mexico and Central America to Colombia. Mainly Sep–Oct. [*L. cardinalis* subsp. *graminea* (Lam.) McVaugh, *L. cardinalis* var. *phyllostachya* (Engelm.) McVaugh, *L. fulgens* Humb. & Bonpl. ex Willd., *L. splendens* Humb. & Bonpl. ex Willd.] A BRIT/SMU sheet annotated by McVaugh as *L. cardinalis* var. *phyllostachya* is known from Dallas Co. Bowden (1960b) indicated that plants from TX are difficult to classify and apparently represent hybrid swarms between *L. cardinalis* subsp. *cardinalis* and *L. cardinalis* subsp. *graminea* var. *phyllostachya*. As can be seen from these examples, infraspecific taxa are often recognized in this species (e.g., Kartesz 1994; Jones et al. 1997). However, because of apparently unstable character combinations, we are following Brooks (1986b), Thompson and Lammers (1997), Yatskievych (2006), and Lammers (in press) in not recognizing infraspecific taxa. Thompson and Lammers (1997) concluded that all populations of the *L. cardinalis* complex (sometimes divided into four species) occurring from s Canada to n Colombia should be recognized as a single species without infraspecific taxa. They noted "that the complex comprises a single species, *L. cardinalis*, and that this species cannot be divided into infraspecific taxa." Yatskievych (2006) commented that about the most that can be said regarding variation within the *L. cardinalis* complex is "that there is a trend for populations in the southwestern U.S. to have narrower leaves." *Lobelia cardinalis* is most closely related to *L. siphilitica*, with which it is known to rarely hybridize (Bowden 1982; Thomas & Lammers 1997; Yatskievych 2006; Spaulding & Barger 2016); the hybrids are intermediate morphologically, have magenta (purplish-red) flowers (sometimes described as "royal purple"—Lammers in press), and have been named *L.* ×*speciosa*, SHOWY LOBELIA. The corolla "is smaller and stouter than the red corolla of *L. cardinalis*, but larger and less occluded than the blue corolla of *L. siphilitica* (Lammers in press). No collections of the hybrid are known from East TX. The flowers of this species are visited by and presumably pollinated by ruby-throated hummingbirds (*Archilochus colubris*) (James 1948). The entire plant is poisonous to humans and livestock, sometimes fatally so, due to a mixture of pyridine alkaloids including lobeline; the sap can also irritate the skin; it was at one time used to treat conditions including syphilis and worm infestations but because of severe illness or even death resulting from overdoses, its use has long been discontinued (Krochmal et al. 1972; Turner & Szczawinski 1991).

Lobelia flaccidifolia Small, (weak-leaved), FOLD-EAR LOBELIA. Annual 0.3–1 m tall; stems glabrous or sparsely hirtellous; basal leaves absent; stem leaves lanceolate to oblong or oblanceolate, acute to rounded apically, subentire to toothed; racemes to 30 cm long, not secund; pedicels 3–11 mm long, 2-bracteolate at or below middle; calyx lobes narrowly triangular, with small auricles usually 0.8–1.5 mm long; corollas bluish to bluish lavender or nearly white, with a white "eye," 14–20 mm long; corolla tube not fenestrate or rarely so; filament tube 5–8 mm long; anther tube 2–3 mm long, bluish-gray. Moist pinelands, savannahs, prairies, and other wet places; known in TX only from se corner of East TX from Harris, Montgomery, and Trinity cos. e to the LA border, also Galveston Co. to the s (Turner et al. 2003; Kartesz 2015); AL, FL, GA, LA, and TX. Apr–Aug. Assumed to contain alkaloids and to be poisonous.

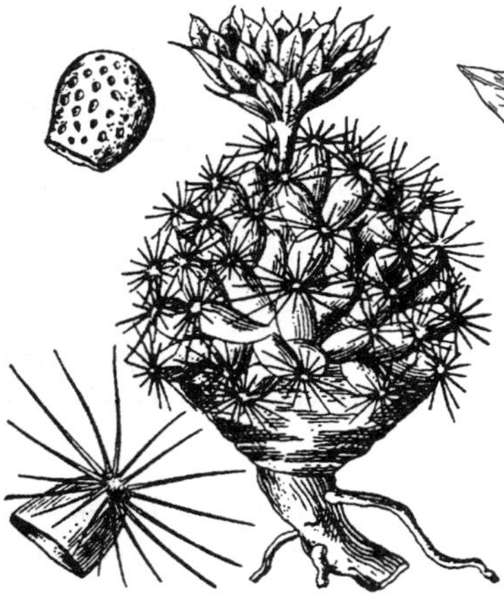

Pelecyphora missouriensis [BB2]

Calycanthus floridus var. glaucus [GL2]

Lobelia appendiculata [HEA]

Lobelia cardinalis [GLE]

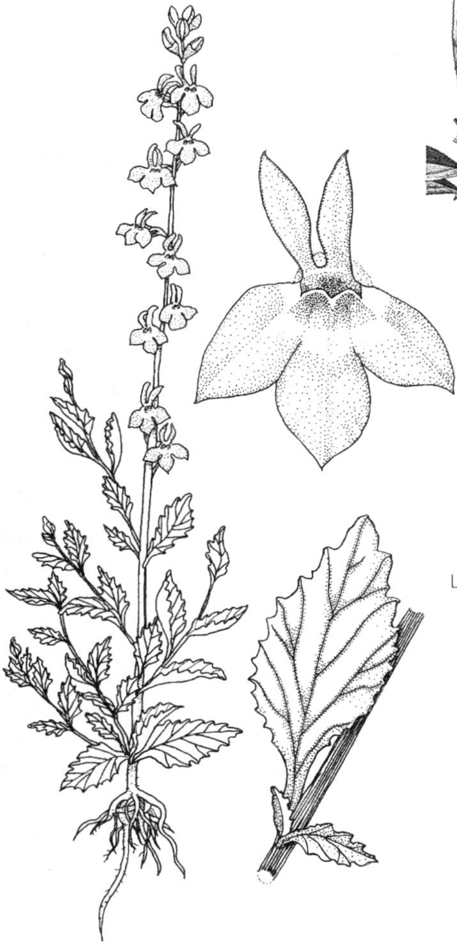

Lobelia berlandieri subsp. brachypoda [WEL]

Lobelia flaccidifolia [CO1]

Lobelia floridana Chapm., (of Florida), FLORIDA LOBELIA. Perennial (from basal offsets), glabrous; stems to 150 cm tall; leaves basal and cauline, the basal 3–10, persistent, the blades oblanceolate to narrowly elliptic, the cauline 3–5, the blades lanceolate; racemes not secund or secund; pedicels 2–6 mm long; calyx lobes triangular or deltate, with very small auricles; corollas pale purple or white, 12–18 mm long; corolla tube not fenestrate or only rarely so; filament tube 6–11 mm long; anther tube 2.5–3.5 mm long. Ponds, wet places, moist to wet pinelands; cited for vegetational area 2 by Hatch et al. 1990 and by Correll and Johnston (1970) as occurring "in savannahs in s.e. Tex." Lammers (in press) cites the species as occurring in TX. While we cannot find a confirmed TX herbarium specimen, we are including this species in the East TX flora based on Lammers' citation. Because of the lack of confirmed locations, no map is included. Further, it has been reported from Calcasieu Parish, LA, adjacent to the TX border (Kartesz 2015) across the Sabine River from Orange and Newton cos., and thus is reasonably expected in East TX near the state margin; NC s to FL w to LA and TX. May–Aug. [*L. paludosa* Nutt. var. *floridana* (Chapm.) A. Gray] Assumed to contain alkaloids and to be poisonous.

Lobelia puberula Michx., (somewhat pubescent), DOWNY LOBELIA, PURPLE-DEWDROP. Perennial; stems to 1.6(–2.7) m tall, densely short hirsute or puberulent throughout; leaves lanceolate to oblanceolate or oblong, obtuse to acute apically, nearly entire to toothed; raceme to 50 cm long, with up to 75 flowers, often 1-sided; calyx lobes narrowly triangular, with small basal auricles; corollas blue to purple with a white eye (rarely white), 14–22 mm long; corolla tube fenestrate; anther tube 3–3.5 mm long, bluish gray; capsules 4–8 mm long. Wooded areas to prairies or fields, usually in wet places; widespread in East TX; also n Gulf Prairies and Marshes; widespread in the e US from NJ s to FL and w to IL, MO, OK, and TX. Aug–Dec. [*L. puberula* var. *mineolana* E. Wimm., *L. puberula* var. *simulans* Fernald] While varieties are sometimes recognized (e.g., Kartesz 1994; Jones et al. 1997), we are following most modern authorities (e.g., Lammers in press) in not distinguishing infraspecific taxa in this variable species. Yatskievych (2006) noted that the various varieties and forms were "described based on minor components of morphological variation." *Lobelia reverchonii*, here treated as a distinct species (see below), has sometimes been treated as a variety or subspecies of *L. puberula* (e.g., McVaugh 1961). Lobeline and three other alkaloids are present in quantities similar to those in *L. cardinalis* (Krochmal et al. 1972).

Lobelia reverchonii B.L. Turner, (for Julien Reverchon, 1837–1905, a French-American immigrant to Dallas and important botanical collector of early TX), REVERCHON'S LOBELIA. Perennial from basal offsets; stems to 1.4 m tall, usually long hirsute/pilose; leaves elliptic, oblanceolate or lanceolate to oblanceolate or oblong, obtuse to acute apically, nearly entire to serrulate; raceme with 3–20 flowers, sometimes 1-sided; calyx lobes ovate, the auricles deeply cordate, large; corollas blue to purplish blue with a white eye (rarely white), 18–22 mm long; corolla tube fenestrate; anther tube 3.5–4 mm long; capsules 4–8 mm long. Bogs, wet places; Smith Co. (*J. Reverchon 3206* BRIT, GH [Bush designated this collection as the "type"]), also Henderson, Montgomery, Robertson, and Wood cos. e to the LA border; e TX and LA. Jul–Oct. [*L. puberula* var. *pauciflora* Bush, not *L. pauciflora* Kunth 1819, *L. puberula* subsp. *pauciflora* (Bush) Bowden, *L. vaughiana* E. Wimm.] This species is somewhat similar to *L. puberula* and was originally recognized as *Lobelia puberula* var. *pauciflora* Bush in 1906. It was first recognized as a distinct species by Turner in 1950 (mistakenly as "*reverchoni*"). Turner (1950) remarked that when he found *L. reverchonii* growing side by side with *L. puberula* in Texas, that: "Their dissimilarity was very strong, even to one unfamiliar with the species. In the area concerned, the two species seemed to be well marked and showed no signs of integration." McVaugh (1961), however, submerged it into *L. puberula*. Turner et al. (2003) treated it as a separate species as did Spaulding and Barger (2016), Kartesz (2024), and Lammers (in press). We are convinced by the evidence that it a distinct species and are recognizing it as such—Turner's familiarity with the taxon in the field is particularly convincing, as is its morphological distinctiveness. It will be interesting to see what insights possible future molecular studies will provide. Assumed to contain alkaloids and to be poisonous.

Lobelia siphilitica L. (syphilitic, from supposed medicinal value), BIG BLUE LOBELIA, BLUE CARDINAL-FLOWER, GREAT LOBELIA, LOUISIANA LOBELIA. Nearly glabrous perennial with basal offshoots; stems to 1.5 m tall; basal leaves absent; stem leaves usually 1.5 cm or less wide; raceme to 30(–50) cm long, often with fewer than 20 flowers, not 1-sided; calyx lobes foliaceous, to 14 mm long, with broad basal auricles 2–5 mm long; corollas blue, white striped in throat; corolla tube fenestrate; filament tube 12–15 mm long; anther tube 4–5 mm long, bluish gray; capsules 8–10 mm long. Low moist areas, prairies, woodlands; collected by Reverchon in Dallas Co. (*J. Reverchon s.n.*, 1875—cited in McVaugh 1961), also a Grayson Co. population (no longer extant) was observed for a number of years by Sally Crosthwaite (pers. comm.), and McVaugh (1961) cited a J. Boll collection (*s.n.*, 1871) of undetermined county; USDA Plants Database (2025) also cites the two counties just mentioned; otherwise apparently unknown in TX; e 2/3 of the U.S. (including nearby areas in OK) and adjacent Ont. and Man. in Canada. Aug–Oct. [*L. siphilitica* var. *ludoviciana* A. DC.] This species was used by Native Americans and early settlers as a treatment for syphilis (Larsen 1940); according to Steyermark (1963), it was also formerly used medicinally as an emetic and to relieve spasms of the air passages in cases of laryngitis and asthma. However, it is poisonous when taken in overdoses (Steyermark 1963); lobeline and two other alkaloids are present

Lobelia floridana [GWO]

Lobelia puberula [CO1]

Lobelia reverchonii [THO]

Lobelia siphilitica [STE]

Lobelia spicata [GLE]

(Krochmal et al. 1972). Despite the flower color differences, this species is closely related to *L. cardinalis* and morphologically intermediate hybrids between the two are known; see *L. cardinalis* for discussion. Variety *ludoviciana* is sometimes recognized (e.g., Correll & Johnston 1970; Diggs et al. 1999), distinguished by having narrower leaves and fewer flowers per inflorescence. Spaulding and Barger (2016) noted that "after examining numerous collections throughout the range of the species, we find that the characters do not hold up." Likewise, Lammers (in press) does not distinguish varieties; we are following these recent authors. Our description reflects the western material of the species.

Lobelia spicata Lam., (spicate, with spikes), PALE-SPIKE LOBELIA, HIGHBELIA, SPIKED LOBELIA. Perennial 0.2–1.1 m tall; stems densely short pubescent to nearly glabrous; leaves lanceolate to oblanceolate to obovate or oblong; inflorescence a spike-like raceme (pedicels short); calyx lobes subulate or linear-lanceolate, with thread-like or very short and triangular auricles; corollas light blue to white; corolla tube not fenestrate; filament tube 2–4 mm long; anther tube 1.7–2 mm long, pale-bluish-gray. Grassy areas or open woods; se corner of East TX from Trinity s to Walker, Montgomery, and Harris cos. and e to the LA border; also Galveston Co. further s; e and c US and adjacent Canada. May–Aug. [*L. hirtella* (A. Gray) Greene, *L. leptostachys* A. DC., *L. spicata* var. *campanulata* McVaugh, *L. spicata* var. *hirtella* A. Gray, *L. spicata* var. *leptostachys* (A. DC.) Mack. & Bush, *L. spicata* var. *parviflora* A. Gray, *L. spicata* var. *scaposa* McVaugh] Five infraspecific taxa are sometimes recognized in this species (e.g., McVaugh 1936; USDA Plants Database 2025). However, only one of these is known from TX. McVaugh (1936, 1943) did not map any TX locations for *L. spicata,* but the variety coming closest to se TX was his var. *scaposa* McVaugh, which he cited for LA. In 1961, McVaugh indicated that the TX material of this species collected in 1945 is "essentially a slender form of var. *leptostachys* (A. DC.) Mack. & Bush." USDA Plants Database (2025) treats TX plants as *L. spicata* var. *scaposa* McVaugh. Because there is intergradation and apparent confusion among the varieties of this species, and because only one entity is known for TX, we are not recognizing infraspecific taxa in *L. spicata*. This seems the most prudent course, particularly in light of McVaugh's (1961) comment that in the w outlying part of the species' range "… it is not always possible to assign the plants to any of the five principal varieties which may be distinguished farther north and east." Further, Lammers (in press) noted that "with few gaps in the patterns of variation and correlation with geography that is modest at best, recognition of infraspecific taxa is ill-advised." Reported to be poisonous (Burlage 1968). Assumed to contain alkaloids and to be poisonous.

Lobelia inflata L. (Latin, *inflatus*, to blow into, puff up, in reference to the inflated fruiting capsules) INDIAN-TOBACCO, PUKE WEED. Annual to 100 cm tall; stems pubescent at least basally; basal leaves absent; cauline leaves elliptic to ovate, oblong or obovate; inflorescence branched or unbranched, not secund, with 3–30 flowers; pedicels 1–7 mm long, the bracteoles near the base; calyx lobes subulate or linear, 2–8 mm long, without auricles; corollas pale bluish white or white, 3–8 mm long; filament tube 2–3 mm long; anther tube 1.2–1.8 mm long; fruiting capsules strongly inflated, 6–12 mm long. While not known from TX, this species should be looked for in the ne corner of the state. It is known from McCurtain Co., OK just across the Red River from Red River and Bowie cos. in the ne corner of the state; e 1/2 of the US w to MN, NB, KS, OK, and LA. Jul–Nov. It can be recognized by being an annual with short corollas (3–8 mm long) and inflated capsules 6–12 mm long. Native Americans used to smoke the dried leaves as a tobacco substitute, hence one of the common names. Poisonous, the species is "the commercial source of the pyridine alkaloid lobeline, which has been used pharmaceutically as a respiratory stimulant and emetic as well as in combating nicotine addiction." (Lammers in press).

POOLEA Morin BASIN BELL-FLOWER

A monotypic genus endemic to central TX. This genus was long treated in a broadly conceived *Campanula* (e.g., Correll & Johnston 1970) but "recent phylogenetic studies conclude that *Campanula* is polyphyletic" (Morin 2020). Modern research shows that *Poolea* is part of a clade with *Campanulastrum*, *Legousia*, and *Triodanis* (Morin in press). Morin (2020), in preparation of the treatment of Campanuloideae for Flora of North America, has made a number of name changes in order to make classification in the group better fit evolutionary reality–*Poolea* is just one example. (Named in honor of Jackie M. Poole, botanist with the Texas Parks and Wildlife Department for more than thirty years, an effective and passionate advocate for plant conservation, especially of the Texas flora—Morin 2020).

REFERENCES: Gray 1886; Correll & Johnston 1970; Walters & Wyatt 1982; Poole et al. 2008; Wendling et al. 2011; Morin 2020, ined.

Poolea reverchonii (A. Gray) Morin, (for Julien Reverchon, 1837–1905, a French-American immigrant to Dallas and important botanical collector of early TX), BASIN BELLFLOWER, TEXAS BELLFLOWER. Erect, annual, taprooted herb to ca. 35 cm tall; stem much-branched usually starting at base, the branching sympodial; leaves to ca. 15 mm long, spatulate (basal) to lanceolate (cauline) or the upper ones almost filiform; inflorescence loosely and much-branched, the internodes long; bracts narrow; pedicels 0.5–2(–4) cm long; flowers radially symmetrical, all chasmogamous, long-stalked; calyx with 5 narrowly triangular to linear-subulate lobes, these usually < 1/2 the length of the corolla; corollas (5–)9–13(–17) mm long, radially symmetrical, narrowly funnelform, the lobes slightly shorter than the tube, light blue; stamens 5, free; capsules 3.5–6.5 mm long, ellipsoid or obovoid, opening by lateral pores above the middle (opening from the top of the pore downward). Alluvial sandy or gravelly areas; *Poolea reverchonii* is currently known from only approximately 15 populations; it is essentially narrowly endemic to the Llano Uplift (typically on granitic sands or gravel) in Kendall (*Parks & Cory 12932* TAES); Burnet, Gillespie, Llano, and Mason cos. in e Edwards Plateau (USDA Plants 2019) to the w of East TX (Walters & Wyatt 1982) but also known from Travis Co. (*B.C. Tharp s.n.*, Jul 1938, CM; *B.C. Tharp s.n.*, 03 May 1938, TEX-LL; *B.H. Warnock 20558* TEX-LL) at extreme w edge of East TX; endemic to TX. The species is of conservation concern. May–Jul. [*Campanula reverchonii* A. Gray] This species is inconspicuous and easily overlooked in the field. Morin (in press) noted that, "The unusual opening of the capsule pore, from top to bottom, is otherwise only seen in *Triodanis coloradoensis*, which also occurs on the Llano Uplift."

TRIODANIS Raf. VENUS'-LOOKING-GLASS

Annuals with 1 or several, usually unbranched, densely leafy stems to ca. 1 m tall but usually much shorter; leaves cauline, often clasping, usually < 3(–7) cm long, the upper gradually smaller, the blades crenate or serrate marginally, callose-toothed or teeth mucronate, the lower leaves sessile or with short petioles, the upper sessile; flowers in axils of bracts/leaves, 1–3 per axile, sessile or on pedicels 0–2 mm long, overall giving the inflorescence a spike-like appearance; flowers both cleistogamous and chasmogamous, the early and typically lower flowers cleistogamous (obligately self-pollinated) usually with calyx lobes 3–4, these sometimes unequal, the corolla greatly reduced/vestigial, not opening, the stamens and styles also reduced, maturing fruit very early; chasmogamous flowers with calyx lobes usually 5, equal, the corollas radially symmetrical, rotate to campanulate or funnelform, rather deeply 5-lobed, blue to blue-purple to purple or rarely white, white basally, the stamens usually 5, free; capsules subcylindrical, opening by lateral pores (like window shades), the pores positioned from near the middle of the capsule to its apex, or near base, or by lateral slits; seeds lenticular (quadrangular in 1 species) in cross-section.

A genus of 7 species, all native to North America (Morin in press) and also all native to East Texas (1 extending to South America), including two TX endemics. Given the two TX endemics and the fact that all of the species occur in the state, one could conclude that TX is the center of diversity of the genus. In the past, a Mediterranean species, *Legousia falcata* (Ten.) Fritsch ex Janchen [= *Triodanis falcata* (Ten.) McVaugh], was considered to be in *Triodanis* (McVaugh 1945; Mabberley 1997). However, molecular studies indicate that *L. falcata* should not be included in *Triodanis* because it is as closely related, if not more related, to a number of other genera in the Campanulaceae (e.g., *Poolea*, *Campanulastrum*, *Legousia*, *Physoplesis*) (Cellinese et al. 2009; Haberle et al. 2009; Wendling et al. 2011; Crowl et al. 2014; Mavrodiev 2015; Jones et al. 2017). *Legousia falcata* also differs in chromosome number with $2n = 26$, compared to $2n = 20$ in other *Legousia* and $2n = 56$ in *Triodanis* (Gadella 1966). In the future, *L. falcata* may need to be put in a separate monotypic genus, but for now it seems clear that it does not belong in *Triodanis*. (Greek: *tri*, three, and *odontos*, a tooth, possibly in reference to the three calyx lobes of some cleistogamous flowers)
REFERENCES: McVaugh 1945, 1948; Fernald 1946b; Gadella 1966; Bradley 1975; Ward 1978; Gara & Muenchow 1990; Cellinese et al. 2009; Haberle et al. 2009; Wendling et al. 2011; Goodwillie & Stewart 2013; Crowl et al. 2014; Mavrodiev 2015; Jones et al. 2017; Morin (in press).

1. Plants branched at nodes near to or distal to the middle of plant; leaves (not bracts subtending flowers) with pubescence on the upper surface; calyx lobes 5, all in a given flower ± alike, usually 8–15 mm long; capsules of both cleistogamous and chasmogamous flowers usually with 3 locules and thus 3 pores; pores of capsule opening first at apex with covering of pore curling toward capsule base; species known in East TX only from extreme sw margin of area ____ **T. coloradoensis**
1. Plants unbranched or branched from base; leaves glabrous on upper surface (or with a few bristles near tips); calyx lobes 3–5, often 3 in cleistogamous flowers or sometimes 1 or 2 much smaller than others, usually < 10 mm long (often much less); capsules of cleistogamous flowers usually with 1 or 2 locules and thus usually 1 or 2 pores (chasmogamous flowers have 3 locules and 3 pores); pores of capsule opening first at base with covering of pore curling toward capsule apex; including species widespread in East TX.
    2. Floral leaves (= bracts subtending flowers) narrow, oblong-lanceolate or lanceolate to nearly linear, 3.5–10 times as long as wide, not clasping; capsules of cleistogamous flowers often curved, subulate, usually 8–12(–20) mm long, dehiscent by fractures/longitudinal slits near apex or rarely by a single (–2) pore near apex; capsules of chasmogamous flowers 15–20(–25) mm long ____ **T. leptocarpa**
    2. Floral leaves broader, rhombic-lanceolate to semi-orbicular to reniform, 0.5–3 times as long as wide, not clasping to strongly clasping; capsules of cleistogamous flowers straight, not subulate, usually < 8(–12) mm long, dehiscent by 2(–3) pores near capsule apex or middle; capsules of chasmogamous flowers usually 12 mm or less long.
        3. Pores of capsules near apex.
            4. Floral leaves longer than wide, mostly < 1 cm wide; inflorescences with terminal flower on each main branch chasmogamous, most others cleistogamous; seeds 0.4–0.65 mm long; widespread in East TX ____ **T. biflora**

4. Floral leaves wider than long to about as long as wide, mostly > 1 cm wide; inflorescences with flowers on distal nodes chasmogamous, those on proximal 1/4–2/3 of nodes cleistogamous; seeds 0.7–1 mm long; rare in East TX, known mainly from far n portion of area ____ **T. lamprosperma**

3. Pores of capsules near middle.

5. Pores linear or very narrowly oblong, 0.2–0.5 mm wide, the cartilage (which rolls out resulting in pore) before dehiscence with only very narrow scarious margins; seeds minutely low-tuberculate ____ **T. holzingeri**

5. Pores oblong-ovate or broader, 0.5–1.5 mm wide, the cartilage with broad scarious margins (relative to cartilage width); seeds smooth, roughened by minute points, or reticulate (= with network of ridges).

6. Stems and underside of leaves glabrous, scabrous, or main veins pilose; seeds smooth or roughened by minute points ____ **T. perfoliata**

6. Stems and underside of leaves densely hirsute or pilose; seeds reticulate, the pattern resembling a basket-weave ____ **T. texana**

Triodanis biflora (Ruiz & Pav.) Greene, (two-flowered), SMALL VENUS'-LOOKING-GLASS. Stems 10–45(–80) cm tall; leaves not clasping, the blades longer than wide, apically acute to rarely obtuse, marginally crenate, the upper and middle leaves with blades elliptic to ovate; floral bracts longer than wide, somewhat clasping, mostly < 1 cm wide; cleistogamous flowers at nearly all nodes, each stem usually with 1 open terminal flower; corollas of chasmogamous flowers 5–9(–11) mm long; capsules usually 7(–9) mm or less long; pores of capsules oval to nearly round, usually 0.5 mm or less from apex of capsule; seeds 0.4–0.65 mm long, smooth, glossy. Disturbed areas, weedy places, ditches, edges of wet areas, forest margins; widespread mainly in the e 1/2 of TX; also Trans-Pecos; widespread primarily in the s 2/3 of the US; also Mexico and South America. Apr–Jun. [*Campanula biflora* Ruiz & Pav., *Legousia biflora* (Ruiz & Pav.) Britton, *Specularia biflora* (Ruiz & Pav.) Fisch. & C.A. Mey., *T. perfoliata* (L.) Nieuwl. subsp. *biflora* (Ruiz & Pav.) Lammers, *T. perfoliata* var. *biflora* (Ruiz & Pav.) T.R. Bradley] This taxon has been variously treated as a distinct species (e.g., Greene 1894; Nieuwland 1914; Goodwillie & Stewart 2013; Weakley, A.S., and Southeastern Flora Team 2025; Morin in press), a subspecies of *T. perfoliata* (e.g., Lammers 2006), and a variety of *T. perfoliata* (e.g., Bradley 1975; Diggs 1982; Diggs et al. 1999). Bradley (1975) presented evidence of the presence of hybrids, morphological intermediacy of the hybrids, and the occurrence of backcrosses that supported recognition at the varietal level. Ward (1978), however, noted that "previous workers have not reported difficulty in distinguishing the populations which constitute the two recognized species, and observation of sympatric situations during the present study revealed few or no intermediates. It is clear that isolating mechanisms are operating under field conditions." More recently, Goodwillie and Steward (2013) provided more evidence that in spite of the two sometimes occurring together *T. biflora* persisted as a distinct phenotypic entity and that hybridization seemed limited. Despite some hybridization, the two species are rather easily distinguished morphologically and they remain distinct even when growing together—this evidence supports recognizing *T. biflora* as a separate species as we are doing here. This follows the recent trend among writers of floras who believe species status is justified (e.g., Yatskievych 2006; Weakley et al. 2012; Morin (in press); Weakley, A.S., and Southeastern Flora Team 2025). As noted by Morin (in press), *T. biflora* is "generally more slender than the other short-capsuled species in the genus, with longer internodes and narrower bracts. It has been found with or near *T. coloradoensis*, *T. holzingeri*, *T. lamprosperma*, and *T. leptocarpa*. It co-occurs most frequently with *T. perfoliata*, which tends to grow in more shaded microhabitats."

Triodanis coloradoensis (Buckley) McVaugh, (of the Colorado River area), COLORADO VENUS'-LOOKING-GLASS, TEXAS VENUS'-LOOKING-GLASS, LINDHEIMER'S VENUS'-LOOKING-GLASS. Plants branched at nodes near to or distal to the middle of plant (unlike other species which are unbranched or branched from base); stems 25–100 cm tall; leaves not clasping; upper and middle leaves with blades to 7 cm long and 15 mm wide, oblanceolate to elliptic, sessile; lower leaves broader, short petiolate; floral bracts sessile, lanceolate, attenuate; flowers all chasmogamous or a few lower ones cleistogamous; corollas of chasmogamous flowers 9–18 mm long; capsules from cleistogamous flowers 11–18 mm long, those from chasmogamous flowers to 23 mm long; pores of capsules 1.5–2.5 mm from apex of capsule, 2–4 mm long, 0.8–1.7 mm wide, opening first at apex with covering of pore curling toward capsule base (unlike other species with pores opening first at base); seeds 0.8–1 mm long, smooth, lustrous. Dry hillsides, bluffs, rocky ledges, woods, gravel bars, floodplains; Bell, Bexar, Comal, Guadalupe, Hays, Travis, and Williamson (McVaugh 1945, 1961) cos. on w margin of East TX; mainly Edwards Plateau and adjacent areas; endemic to TX. Apr–early Jun. [*Campanula coloradoense* Buckley, *Legousia coloradoensis* (Buckley) A. Heller, *Specularia coloradoensis* Buckley ex Small, *S. lindheimeri* Vatke] Morin (in press) noted that: "This species differs from others in the genus in having elongated branches from distal nodes, axillary branches in the inflorescence, nearly all flowers chasmogamous, hypanthium somewhat constricted at the apex, and a distinct area where sepals and hypanthium join, which are character states shared with *Legousia* and not otherwise found in *Triodanis*." She also pointed out that *T. coloradoensis* has pore capsules that open from apex to base rather than from base to apex—this is a condition found in only one other member of the Campanulaceae, *Poolea reverchonii*, another species endemic to or near the Edwards Plateau. In a molecular analysis, Mavrodiev (2015) found *T. coloradoensis* sister to *Poolea reverchonii*, which together were sister to the *Legousia* clade. Morin noted that these "morphological and genetic differences would support of

Poolea reverchonii [THO]

Triodanis biflora [GLE]

Triodanis holzingeri [HEA]

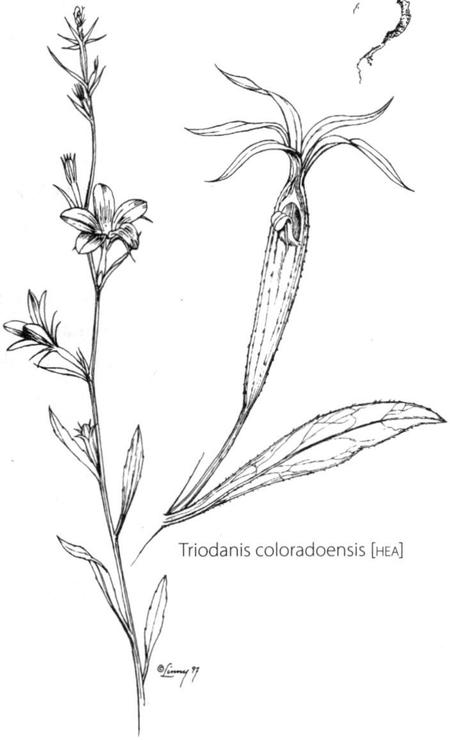

Triodanis coloradoensis [HEA]

Triodanis lamprosperma [HEA]

separation of *T. coloradoensis* from *Triodanis*." It is thus possible that in the future *T. coloradoensis* will be recognized as a distinct monotypic genus. However, such a consideration is premature at this time. Morin (in press) also noted that the "long, narrow capsules and sepals, and long, spreading bracts give *T. coloradoensis* a rangy, spiky look."

Triodanis holzingeri McVaugh, (for John Mitchell Holzinger, 1853–1929, German-born bryologist, who first noted its distinctions). Stems to 20–60(–90) cm tall; leaves not clasping, the blades ovate to elliptic or obovate, marginally crenate; floral bracts broader than leaves, ovate, cordate-clasping; corollas of chasmogamous flowers 7–9(–11) mm long; pores of capsules linear, 0.2–0.4(–0.5) mm wide, the narrowest of any East TX species, (0.5–)1–2.7 mm from apex of capsule; seeds 0.3–0.7 mm long, minutely tuberculate. Sandy open woods, open ground, washes, limestone outcrops, and grasslands; Bastrop, Bell, Bexar, Grayson, and Travis cos. on w margin of East TX; scattered in w 2/3 of TX; mainly Great Plains (AR, CO, KS, MO, NE, OK, TN, TX, and WY), disjunct to sw TX and mountains in sc AZ. May–Jun. [*Specularia holzingeri* (McVaugh) Fernald]

Triodanis lamprosperma McVaugh, (with shining seeds). Stems usually 20–60(–85) cm tall; leaves not clasping, the blades broadly elliptic to obovate or ovate, sometimes broadly so, apically acute, basally attenuate, marginally crenate to nearly entire; floral bracts often wider than long, to 25 mm wide, cordate clasping at base; corollas of chasmogamous flowers usually 9–12 mm long; pores of capsules broadly elliptic, 0.7–1.5 mm wide, immediately below apex of capsule; seeds 0.7–1 mm long, smooth, lustrous. Stream bottom woods, sandy or rocky areas; Fannin (*McVaugh 7132* SMU, McVaugh 1949), Dallas, Grayson, Harrison, Hunt, Lamar, and Walker (Turner et al. 2003) cos.; rare, mainly n part of East TX; also Brazoria Co. (Turner et al. 2003) in the Gulf Prairies and Marshes; AK, KS, LA, MO, OK, and TX. May–Jun. [*Specularia lamprosperma* (McVaugh) Fernald] Morin (in press) noted that: "*Triodanis lamprosperma* occurs primarily in the Ozark-Ouachita Interior Highlands. In all the *Triodanis* with clasping bracts, some seeds accumulate in the subtending bract and are dispersed from there. This is particularly true in *T. lamprosperma*, and may provide an additional method of seed dispersal beyond being shaken from the capsule, perhaps through water splashing."

Triodanis leptocarpa (Nutt.) Nieuwl., (slender-fruited), SLIMPOD VENUS'-LOOKING-GLASS, SLENDER VENUS'-LOOKING-GLASS. Stems 10–50(–75) cm tall; leaves not clasping, the blades narrowly elliptic to lanceolate or oblanceolate, apically acute to obtuse, basally attenuate, marginally crenate; floral bracts narrower than leaves, lanceolate to linear, sessile, not clasping, apically acute to acuminate (vs. acute to obtuse in other TX species except *T. coloradoensis*); corollas of chasmogamous flowers mostly 7–10 mm long; capsules dimorphic: capsules of chasmogamous flowers ± straight, linear, 15–20(–25) mm long (usually longer and relatively narrower than in other species), dehiscent by fractures/longitudinal slits near apex or rarely by 1(–2) pore near apex; capsules of cleistogamous flowers curved, subulate; seeds 0.7–1 mm long, smooth, lustrous. Rocky limestone outcrops or less often sandy soils, grasslands, wooded areas, roadsides, disturbed areas; scattered in w part of East TX; also Cross Timbers and Prairies and Edwards Plateau; central 2/3 of the U.S from IN s to AR, OK, and TX, and w to CO, ND, and WY. May–Jun. [*Specularia leptocarpa* (Nutt.) A. Gray] Morin (in press) noted that: "*Triodanis leptocarpa* is unique in Campanulaceae in having capsules that usually dehisce by lateral slits splitting through the apex of the capsule. It is the only species in the genus with parietal placentation, a character state found in few other Campanuloideae. In fruit the capsules and sepals become woody and, because the sepals reflex, may snag on passing animals, resulting in dispersal."

Triodanis perfoliata (L.) Nieuwl., CLASPING VENUS'-LOOKING-GLASS. HEN-AND-CHICKENS. Stems usually 10–80(–100) cm tall; leaves except the lower one clasping, the blades about as wide or wider than long, apically broadly acute to rounded, marginally crenate to serrate; each stem typically with more than 1 open flower; corollas of chasmogamous flowers usually 8–14 mm long; floral bracts cordate-clasping, reniform to suborbicular; pores of capsules elliptic to rounded, 0.5–1.5 mm wide, 1–1.5(–2.5) mm from apex of capsule. Widespread mainly in the e 1/2 of TX, also Trans-Pecos; widespread across the U.S and adjacent Canada. Apr–Jul. [*Campanula perfoliata* L., *Legousia perfoliata* (L.) Britton, *Specularia perfoliata* (L.) A. DC.]

Triodanis texana McVaugh, (of Texas). Stems usually 25–50(–75) tall; leaves ovate to elliptic or obovate, apically rounded to obtuse (rarely acute), marginally crenate; corollas of chasmogamous flowers 7–14 mm long; pores of capsules oval, 1.5–2.5 mm from apex of capsule; seeds 0.4–0.6 mm long, reticulate, the pattern resembling a basket-weave. Sandy open woods and open ground; mainly w portion of East TX, but also Hardin Co. (Turner et al. 2003) in e part of the area; also Denton (BRIT) and Erath (McVaugh 1945, 1961) cos. in the Cross Timbers and Prairies and in the e Edwards Plateau; endemic to TX. Apr–May. According to Shetler and Morin (1986), this species has seed morphology unique within the genus (quadrangular in cross-section, different surface texture). [*Specularia texana* (McVaugh) Fernald] Morin (in press) pointed out that "*Triodanis texana* occurs in pinelands, plains, and woods, on the deep, loose sand of the Carrizo Formation from eastern Texas west to north-central Texas. It may grow together with *T. holzingeri* and *T. perfoliata*. The shorter calyx lobes, sculptured seeds, and nearly glabrous bracts distinguish *T. texana* from these other species."

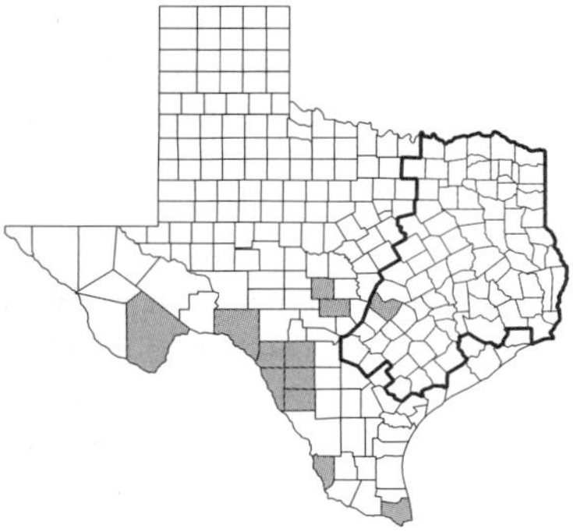

Lobelia berlandieri subsp. brachypoda

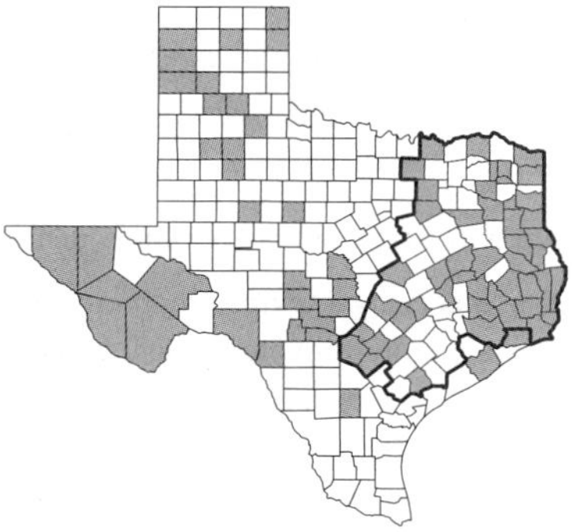

Lobelia cardinalis

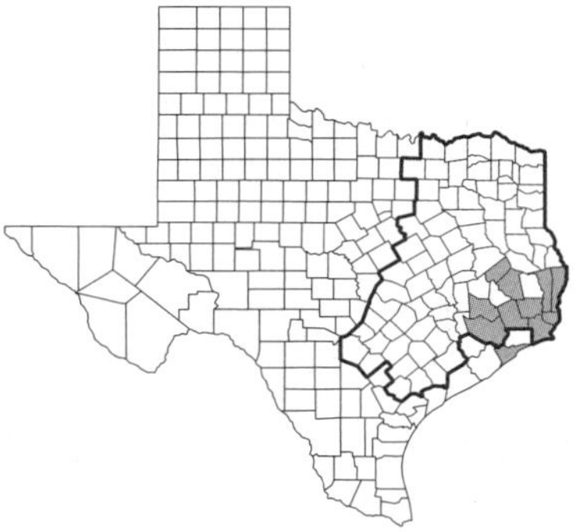

Lobelia flaccidifolia

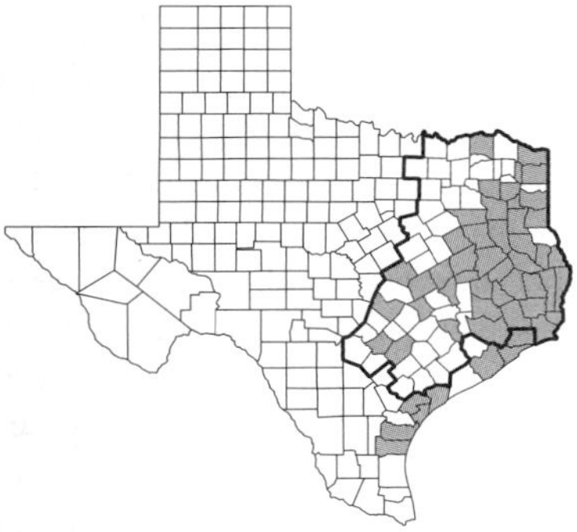

Lobelia puberula

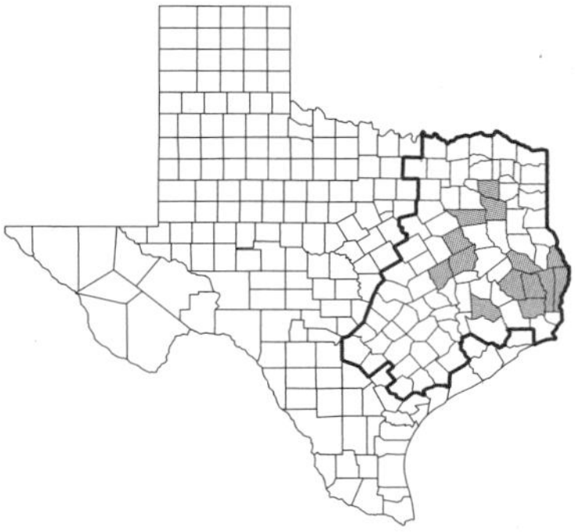

Lobelia reverchonii

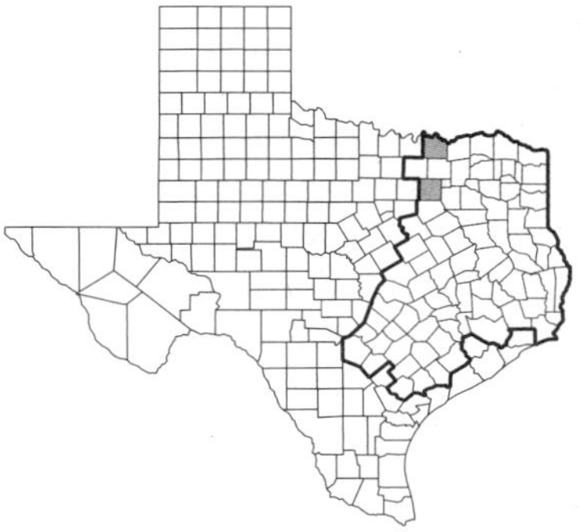

Lobelia siphilitica

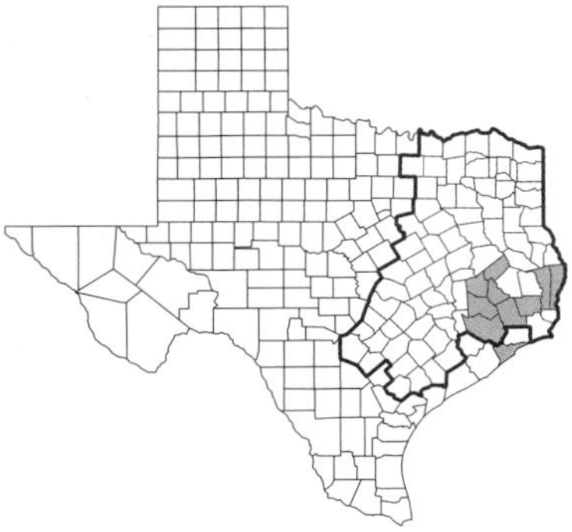

Lobelia spicata

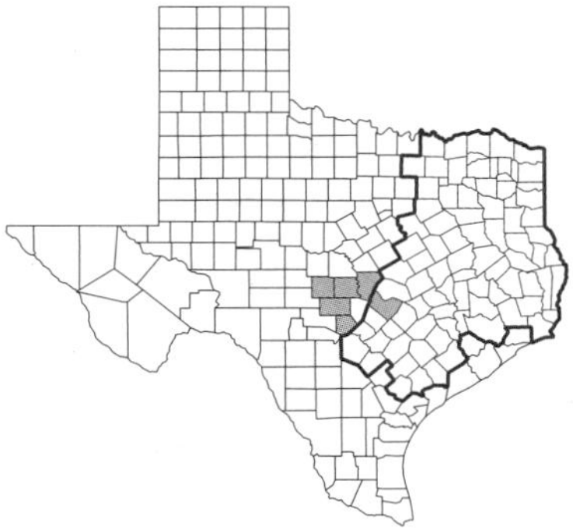

Poolea reverchonii

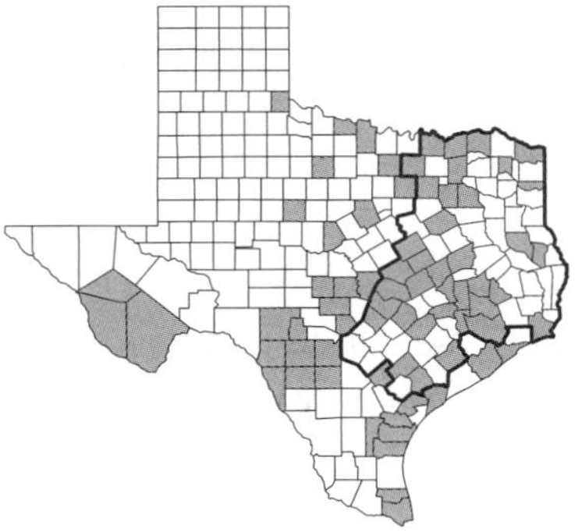

Triodanis biflora

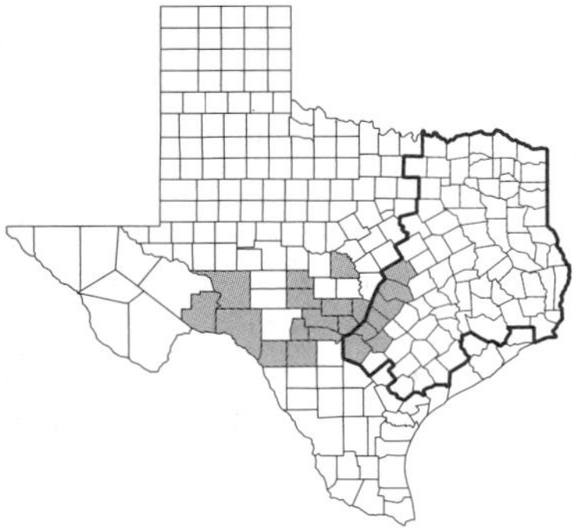

Triodanis coloradoensis

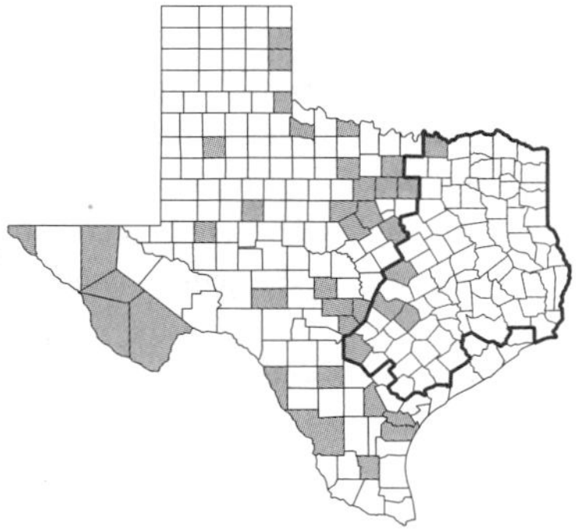

Triodanis holzingeri

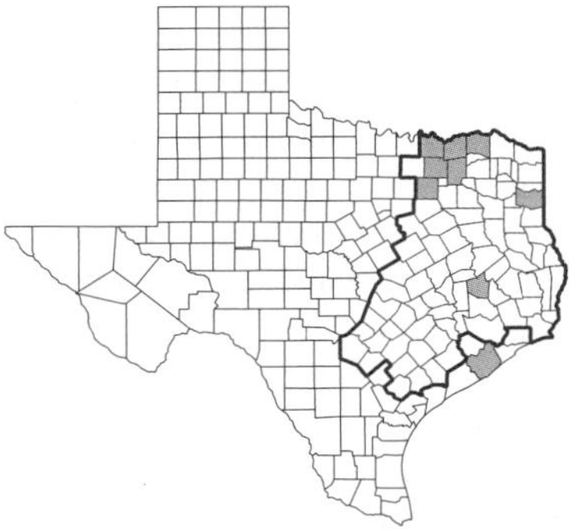

Triodanis lamprosperma

Individuals intermediate between a number of the taxa above are known. These include those that combine the characters of *T. perfoliata* and *T. biflora*; *T. perfoliata* and *T. texana*; and *T. perfoliata* and *T. holzingeri* (McVaugh 1945). McVaugh (1945) listed two East TX collections (Bexar Co., Brazos Co.) for the first of the pairs, two (Brazos Co., Erath Co.) of the second of these pairs, and two (Bell Co., DeWitt Co.) for the last pair mentioned.

WAHLENBERGIA Schrad. ex Roth ROCKBELL

A subcosmopolitan genus of ca. 260 species, especially S temperate, particularly diverse in tropical Africa and Madagascar and Australia, including pachycauls (= plants with proportionately thick trunks) at high altitudes on St. Helena (Mabberley 1997, 2017). Its primary distribution in the Southern Hemisphere "is generally considered … a development parallel to that of *Campanula* in the Northern Hemisphere" (Rosatti 1986). It is somewhat similar to *Campanula* but has capsules with apical valves (Mabberley 2017). Stevens (2001 onwards) noted that generic limits in *Wahlenbergia* need attention. A number are cultivated as ornamentals. (Named for Georg Wahlenberg, 1780–1851, Swedish professor of botany at Uppsala)
REFERENCES: McVaugh 1949; Shinners 1957; Correll & Johnston 1970; Ward 1978; Rosatti 1986; Smith 1992; Cupido et al. 2014; Hong & Wang 2015; Morin (in press).

Wahlenbergia marginata (Thunb.) A. DC., (margined), SOUTHERN ROCKBELL. Annual or short lived perennial herb usually 10–40 cm tall; stems usually several to many in clumps, erect to ascending, freely branched, with leaves mostly on lower half or less (plant almost scapose); leaves linear to lanceolate near stem base, the upper reduced to bracts, the petioles 0–7 mm long, the blades 10–40 mm long, entire to serrate, sometimes undulate; inflorescences of solitary, usually terminal flowers, these rarely axillary, the flowers together appearing like a very open inflorescence; flowers radially symmetrical, all chasmogamous, long-pedicellate; calyx with 5 linear lobes 1.5–3 mm long; corollas 5–8(–12) mm long, the tube funnelform to campanulate, the lobes slightly shorter than to nearly equal the tube, spreading to rotate, blue to violet; stamens 5, free; stigmatic lobes 3; hypanthium obovoid to ellipsoid; capsules 3–6(–8) mm long, opening by apical pores. Disturbed areas, roadsides, old fields, banks of streams; Bastrop (*W.R. Carr 33843* TEX) Co., Newton (*J. Singhurst 11213* BAYLU), Hardin, Tyler (Turner et al. 2003), Harris, Montgomery, and Polk (Kartesz 2015) cos. in the s Pineywoods, also Jim Wells Co. (*W.R. Carr 25860* BRIT, TEX) further south; NC s to FL w to TX, also CA and additionally introduced in HI. Spring–fall. Native of e Asia to Malaysia, Australia, and New Zealand (Rosatti 1986). [*Campanula marginata* Thunb., *W. gracilis* (G. Forst.) A. DC.] This species was first collected in the U.S. in Florida in 1937 (Ward 1978; see also McVaugh 1949); Shinners (1957) reported it from LA and noted its "steady movement westward."

CANNABACEAE Martinov

HEMP FAMILY

Shrubs, trees, or herbs; leaves opposite or alternate, simple or palmately compound; petiolate, stipulate; plants dioecious, monoecious, or polygamomonoecious (= plants with bisexual flowers and staminate and pistillate flowers); flowers mostly unisexual, actinomorphic; staminate flowers in axillary, cymose panicles or fascicles, sepals 4–5, free or connate; petals absent, stamens 4–5; pistillate flowers solitary or in axillary, cymose spikes or fascicles, sepals 4–5, free or connate, petals absent, stamens small and sterile or absent; carpels 2, 1-locular, stigmas 1–2; fruit an achene or drupe.

The Cannabaceae is a small family (ca. 180 species in 10 genera, nearly worldwide—Yang et al. 2013) of trees, shrubs, and herbs. Traditionally, the family was treated in the Urticales (subclass Hamamelidae) (Cronquist 1988) along with families Moraceae, Ulmaceae, and Urticaceae. However, it is now placed in the Rosales. In earlier works (e.g., Correll & Johnston 1970; Diggs et al. 1999; Turner et al. 2003), the genus *Celtis* was treated in the Ulmaceae. *Celtis* is here treated in the Cannabaceae. Though some intergeneric relationships were not well resolved, Yang et al. (2013) found the Cannabaceae as here treated to be monophyletic. The beer flavoring HOPS comes from *Humulus*. The family is sometimes referred to as the Cannabidaceae; however, the name Cannabaceae is conserved. Family name from the genus *Cannabis*. (subclass Hamamelidae—Cronquist; order Rosales—APG IV)
FAMILY RECOGNITION IN THE FIELD: introduced erect annual herb with PALMATELY compound leaves with 3–9(–11) coarsely toothed leaflets and small, unisexual flowers; fruits achenes OR native trees with alternate, simple, pinnately veined, entire to toothed leaves with *asymmetric bases*; flowers small, unisexual or bisexual, greenish, inconspicuous; fruits small, rounded drupes.
REFERENCES: Miller 1970; Kubitzki 1993b; Judd et al. 1994; Small 1997; Yang et al. 2013.

1. Plants trees or shrubs; leaves simple and unlobed ____ **Celtis**
1. Plants erect herbs with palmatley compound leaves ____ **Cannabis**

CANNABIS L. HEMP

A monotypic genus of c Asia; long in cultivation. (From the Greek name, *kannabis*, thought by some to come from the Persian name, *kanab*, or the Arabic, *kinnab*)
REFERENCES: Emboden 1974; Small & Cronquist 1976; Small et al. 1976; Small 2015.

Cannabis sativa L., (cultivated), HEMP, MARIJUANA. Erect annual herb to 4 m tall; stems angled, minutely pubescent; leaves palmately compound, the lower opposite, the upper alternate; leaflets 3–9(–11), narrowly lanceolate, coarsely and sharply toothed, scabrous above, pubescent beneath; stipules lance-linear; flowers unisexual, staminate and pistillate flowers on separate plants (sometimes on same plant in cultivars); staminate inflorescence a narrow, terminal nearly leafless panicle, flowers on pedicels 0.5–3 mm long, the perianth greenish, 5-parted; pistillate inflorescence mostly in axillary, small, spike-like clusters, flowers sessile, enclosed by sepal-like bracts; achenes 2–5 mm long. Found in the city of Dallas in Jun 1876 by Reverchon (Mahler 1988), Walker Co. (Turner et al. 2003); probably scattered in other localities in TX; probably present in most or all U.S. states. Jul–Oct. Native of c Asia. [*C. indica* Lam., *C. sativa* subsp. *indica* (Lam.) E. Small & Cronquist] While numerous infraspecific taxa are sometimes recognized (e.g., Small & Cronquist 1976; Jones et al. 1997) and hotly debated, according to Small (1997), "the chemical and morphologic distinctions by which *Cannabis* has been split into taxa are often not readily discernible, appear to be environmentally modifiable, and vary in a continuous fashion. For most purposes it will suffice to apply the name *Cannabis sativa* to all plants encountered in North America." We are taking a conservative approach in not recognizing taxa below the specific level. Cultivated for use as a psychoactive drug and as a source of the fiber known as hemp, which has numerous uses including rope, paper, cloth, and nets; due to demand for rope, *Cannabis* was a major economic crop in the American colonies; during World War II, because fiber supplies were low, the U.S. government encouraged and subsidized the cultivation of *Cannabis* for hemp. It is one of the world's oldest cultivated plants having been domesticated for ca. 8500 years; the psychoactive ingredients are tetrahydrocannabinol (THC) and related phenolic resins; THC is controlled in the U.S. by the Comprehensive Drug Abuse Prevention and Control Act of 1970 and classified in the Controlled Substances Act and is used illegally as a recreational drug, though several U.S. states have legalized recreational use; medicinally, *Cannabis* is used in the treatment of glaucoma and in combating the nausea experienced during cancer chemotherapy; and in recent years numerous other non-intoxicating and mildly intoxicating cannabinoids (e.g., CBD (cannabidiol)) have been isolated and are used in various forms for recreation and various medicinal purposes; there are adverse effects from concentrated doses or heavy usage of THC and the pollen can be a cause of hay fever; a single inflorescence has been estimated capable of producing more than 500 million pollen grains (Tippo & Stern 1977; Stephens 1980; Fuller & McClintock 1986; Blackwell 1990; Turner & Szczawinski 1991; Kubitzki 1993b; Small 1997, 2015; Mabberley 2017).

CELTIS L. HACKBERRY, SUGARBERRY, NETTLE-TREE

Trees or shrubs, to 30 m tall; crowns spreading; bark usually gray, smooth or often fissured and conspicuously warty; branches without or with thorns, slender, glabrous or pubescent; leaves stipulate, stipules falling early; leaf blade deltate to ovate to oblong-lanceolate, base oblique or cuneate to rounded, margins entire or serrate-dentate; venation 3(–5)-pinnate; staminate inflorescences cymes or fascicles; pistillate solitary or few-flowered clusters; flowers usually unisexual, staminate and pistillate on same plants, along with a few bisexual flowers, pedicellate on branches of current year, appearing in mid or late spring; staminate flower filaments incurved in bud, exserted after anthesis; gynoecium minute, rudimentary; pistillate flower calices slightly to deeply 4(–5)-lobed; stamens 4–5, often nonfunctional, filaments usually shorter than in staminate flowers, rarely absent; ovaries sessile, ovoid, 1-locular, styles short, sessile, divided into 2 divergent, elongate, reflexed lobes, lobes entire or 2-cleft; fruits fleshy drupes, ovoid or globose; outer mesocarp thick, firm, inner mesocarp thin, fleshy; stones thick walled, ripening in autumn, persisting after leaves fall.

A genus of ca. 60 species (Mabberley 2017), mostly of the tropics, but also in temperate areas; the fruits are often edible. HACKBERRIES nearly always exhibit galls or other disease manifestations on the leaves or twigs. The fruits are popular as a food for winter birds such as cedar waxwings (Martin et al. 1951). Sherman-Broyles et al. (1997b) indicated that the group is taxonomically complex and in need of revision. Whittemore (2013) cautions that identification solely from vegetative leafy twigs, especially on herbarium specimens, is often not reliable. Leaves from different parts of the plant and at different stages of development can look very different. Leaves subtending a fruit or flower are best used in identification. A conspicuous structure known as witches' broom (a dense mass of deformed twigs) is sometimes formed on HACKBERRIES by an infection involving both an insect and a non-rust fungus (J. Hennen, pers. comm.). (Latin: *celtis*, a kind of lotus; name used by Pliny for *Celtis australis* L., the lotus of the Ancients, with sweet fruits)
REFERENCES: Sherman-Broyles et al. 1997b; Whittemore 2013; Weakley, A.S., and Southeastern Flora Team 2022.

1. Branches armed with thorns 5–25 mm long ______ **C. pallida**
1. Branches unarmed.
   2. Leaf lower surface densely white-pubescent ______ **C. lindheimeri**
   2. Leaf lower surface glabrous or with inconspicuous hairiness mainly on the veins.
      3. Leaf blades 2–4.5(–7) cm long; upper leaf surface usually harshly scabrous; leaf apex obtuse to acute (rarely nearly acuminate); fruit (5–)8–10 mm in diam. ______ **C. reticulata**
      3. Leaf blades 3.5–10 cm long; upper leaf surface smooth or slightly scabrous; leaf apex acute to acuminate; fruit 5–8 mm in diam.
         4. Plants shrubs or small trees, to 7 m tall, with ascending trunks and horizontal or arching upper branches; bark often nearly smooth (sometimes with corky warts near the base or around wounds); leaf lower surface glaucous or pale green, usually distinctly lighter than the upper surface; flower stalks usually hairy (puberulent = with fine short hairs); twigs usually moderately to densely hairy (sometimes glabrous); anthers small and indehiscent, pollen sterile ______ **C. pumila**
         4. Plants usually trees, to 30 m tall, typically single-trunked, with erect upper branches; bark with corky warts on trunks and major branches; leaf lower surface usually bright green, little if at all paler than the upper surface; flower stalks glabrous (rarely with a few hairs); twigs glabrous or sparsely hairy; anthers well-developed, pollen viable ______ **C. laevigata**

Celtis laevigata Willd. (smooth), SUGARBERRY, SUGAR HACKBERRY, PALO BLANCO. Tree 8–30 m tall; bark of trunk and large branches with prominent corky warts, twigs glabrous or sparsely pubescent; leaves alternate, petioles (of leaves subtending fruits) 6–10 mm long, blade 4.5–10 cm long, 2–4 cm wide, lanceolate or narrowly ovate, thin and flexible, both surfaces usually bright green to yellowish green, lower surface usually sparsely pubescent, sometimes glabrous, margin entire or sometimes with 1–2 teeth on one side (leaves of vigorous shoots sometimes more toothed); flower stalks 7–16 mm long; fruits globose, 5–8 mm in diam. Rich bottomlands, floodplains, rocky slopes; throughout most of East TX; mostly in the e 2/3 of TX; e U.s. from VA s to FL and w to KS and TX; also Mexico. May–Oct. [*C. laevigata* var. *texana* (Scheele) Sarg.]

Celtis lindheimeri Engelm. ex K. Koch, (for Ferdinand Jacob Lindheimer, 1801–1879, German-born TX collector), LINDHEIMER'S HACKBERRY. Tree to 12 m tall; trunk to 15 dm in diam.; crown widely spreading, much branched; bark with corky warts; branches spreading to pendulous, smooth, young branches and twigs villous-pubescent; leaves alternate, petiole 3–4(–11) mm long, blade 4–9 cm long, 2–5 cm wide, ovate to ovate-lanceolate or oblong-lanceolate, leathery, surfaces abaxially white-tomentose, adaxially dark green, scabrous, base rounded to cordate, margins entire or with a few serrations, apex obtuse to acute or shortly acuminate; inflorescences erect, dense clusters, 2–9-flowered, at base of leaves; fruits light brown, globose, 7–9 mm in diam., smooth; seeds ovoid, prominently 4-ribbed, reticulate. Ravines and brushlands; Bexar (*Van Auken BUL1215* BRIT), Comal (*McGregor 36735* NY), Gonzales (*Loucks s.n.*, 1975, LL), and Travis (*Sexton 788* BCNWR) cos. in the sw portion of East TX; South TX Plains and Edwards Plateau; also Mexico. Mar–May. NatureServe: G3, vulnerable. ⚠

Celtis pallida Torr., (pale), DESERT HACKBERRY, SPINEY HACKBERRY. Shrub 3–6 m tall; crown rounded; bark gray, smooth, branches spreading, flexuous, whitish gray, with thorns, puberulent; thorns single or in pairs, mostly 5–25 mm long; leaf blades 2–3 cm long, 1.5–2 cm wide, ovate to ovate-oblong, thickish, surfaces scabrous, margins entire or crenate-dentate, apex rounded to acute; inflorescences in 3–5-flowered cymes, longer than petiole, flowers mostly staminate on proximal branches, terminal flower bisexual; fruits usually orange, ovoid, 6–7 mm long; pedicel 1–2 mm long. Thickets, brushland; Bexar, Comal, Dewitt, Goliad, Jackson, Travis, Victoria, and Wilson (Turner et al. 2003) cos; South TX Plains, Edwards Plateau, and Trans-Pecos; AZ, FL, NM, and TX; also Mexico; Central and South America. Mar–May. [*C. spinosa* Spreng. var. *pallida* (Torr.) M.C. Johnst.]

Celtis pumila Pursh, (low), DWARF HACKBERRY. Apomictic (= producing seeds without fertilization) shrub or small spreading tree, 3–7 m tall; bark with corky warts at base and around wounds, otherwise smooth; twigs usually moderately to densely pubescent, glabrous with age; leaves alternate, those subtending fruits with petiole 3–8 mm long, blade 3.5–8.5 cm long, 1.8–5.5 cm wide, lanceolate-triangular to triangular-ovate, lower surface pale green or glaucous, much lighter than the upper surface, base obliquely truncate or asymmetrically shallowly cordate, margins entire or with 1–13 small teeth on each side; leaves of vigorous shoots and juveniles somewhat different; fruits globose, 5–8 mm in diam., dark brownish purple when mature. Glades, ledges, edges of dry upland forest, savannas; mostly in the e 1/3 of East TX; Cross Timbers and Prairies; e U.S. from PA s to FL and w to KS and TX; also Canada. Apr–May. [*C. tenuifolia* Nutt.]

Celtis reticulata Torr. (netted), NETLEAF HACKBERRY, PALO BLANCO. Shrub or tree 3–8(–16) m tall; bark gray with corky ridges; twigs villous when young; leaf petiole 3–8 mm long, blade 2–4.5(–7) cm long, 1.5–3.5 cm wide, ovate, thick, rigid, surfaces pubescent, lower surface yellow-green, upper surface gray-green, usually scabrous, base cordate or occasionally oblique, margins entire or serrate above the middle, apex obtuse to acute, rarely nearly acuminate; inflorescence 1–4-flowered; fruits reddish or reddish black when mature, globose. Dry hills, ravine banks, rocky outcrops; widespread in East TX except for se portion;

Triodanis leptocarpa [BB2]

Triodanis perfoliata [STE]

Triodanis texana [HEA]

Wahlenbergia marginata [SUL]

Cannabis sativa [REE]

Celtis laevigata [SA3]

Celtis lindheimeri [LUN]

throughout TX; w U.S. from KS, AR and LA w to WA and CA; widespread in sw U.S.; also Mexico. Late winter–spring. [*C. laevigata* Willd. var. *reticulata* (Torr.) L.D. Benson]

CAPRIFOLIACEAE Juss.

HONEYSUCKLE FAMILY

Herbs, shrubs, or woody vines; leaves opposite, sessile or petioled, simple, entire, toothed, or lobed; flowers in involucrate heads (resembling Asteraceae) in *Dipsacus and Scabiosa*, OR axillary or terminal, solitary, in pairs, or in spikes, whorls, umbels, or cymes; calyx lobes absent or minute to large, 3–5 (pappus-like of 5 bristles in *Scabiosa* and cup-shaped in *Dipsacus*); corollas 2-lipped or radially symmetrical and 4–5-lobed, funnelform, campanulate, or tubular (superficially somewhat like the ray flowers of Asteraceae in *Scabiosa*); stamens 3, 4, or 5; pistil 1; style and stigma 1; ovary inferior; fruit in ours an achene, drupe, or a berry.

A medium-sized (ca. 900 species in 39 genera) (Mabberley 2017), nearly cosmopolitan family of shrubs or small trees, lianas or herbs; it includes a number of ornamentals such as *Abelia*, *Lonicera*, *Symphoricarpos*, and *Weigela*. We are following Stevens (2001 onwards) in including *Sambucus*, *Viburnum*, and relatives in the Viburnaceae. Stevens (2001 onwards) further includes the Valerianaceae, Linnaeaceae, and Dipsacaceae with the remaining Caprifoliaceae to form a more inclusive monophyletic Caprifoliaceae. Family name from *Caprifolium*, a genus now treated as *Lonicera*. (Latin: *capra*, goat, and *folius*, leaved, from the foliage being used as goat fodder) (subclass Asteridae—Cronquist; order Dipsacales—APG IV)
FAMILY RECOGNITION IN THE FIELD: herbs, shrubs, or vines with similarities to the Rubiaceae (e.g., opposite leaves, inferior ovaries); leaves, entire or divided; lacking stipules, flower parts mostly in 5s. (Rubiaceae have stipules and flower parts often in 4s).
REFERENCES: Ferguson 1966a; Brooks 1986c; Donoghue et al. 1992, 2001; Judd et al. 1994; Bell et al. 2001; Yatskievych 2006; Weakley et al. 2012.

1. Plants herbs.
   2. Flowers in terminal involucrate heads resembling the Asteraceae; corollas lavender to lilac, rose, dark purple, or white.
      3. Plants spiny, thistle-like; upper leaves entire to toothed or scalloped; involucral bracts much longer than flower heads; heads 3–10 cm long; corollas 4-lobed ____ **Dipsacus**
      3. Plants not spiny, not at all thistle-like; upper leaves pinnately parted; involucral bracts shorter than flower heads; heads ca. 1.5 cm long; corollas 5-lobed ____ **Scabiosa**
   2. Flowers axillary, solitary or 2–4 per axil OR flowers terminal in head-like, cymose inflorescences; corollas greenish to yellowish white or yellow OR white or white with bluish tinge (rarely pale pinkish).
      4. Flowers axillary, solitary or 2–4 per axil; plants perennial herbs; stems unbranched; corollas greenish to yellowish white, yellow, or rarely orangish, 12–17 mm long ____ **Triosteum**
      4. Flowers terminal in head-like cymose inflorescences; plants annual herbs; stems dichotomously branched multiple times; corollas white or white with bluish tinge (rarely pale pinkish), shorter than 12 mm long, typically much shorter ____ **Valerianella**
1. Plants vines or shrubs.
   5. Plants shrubs; flowers in crowded axillary spikes, rather inconspicuous OR flowers solitary or few in axillary cymes; corollas either 3–5 mm long OR ca. 20 mm long with conspicuously expanded throat.
      6. Flowers in small crowded axillary spikes (not whorled), rather inconspicuous; corollas 3–5 mm long ____ **Symphoricarpos**
      6. Flowers solitary or few in axillary or terminal cymes; corollas larger, ca. 20 mm long with expanded throat ____ **Abelia**
   5. Plants vines or shrubs; flowers axillary or both axillary and terminal, mostly in 2s, heads, or in loose spikes forming whorls; corollas showy, 12–50 mm long, corolla tube not conspicuously expanded ____ **Lonicera**

ABELIA R. Br. ABELIA

A genus of 40 species of shrubs of s and e Asia (Mabberley 2017); some cultivated; *Abelia* ×*grandiflora* (Rovelli ex André) Rehder [*A. chinensis* R. Br. × *A. uniflora* R. Br.] is widely cultivated in North America and elsewhere. Kartesz (2015) placed *Abelia* in the Linnaeaceae; we, however, are including it in the Caprifoliaceae as suggested by Bell et al. (2001) and others. (Commemorates Abel Clarke (1780–1826), British physician and naturalist)
REFERENCES: Bailey 1900; Jones 1940; Nesom & Brown 1998; Qiner & Landren 2011; Wunderlin et al. 2020.

Abelia ×**grandiflora** (Rovelli ex André) Rehder, (large-flowered), LARGEFLOWER ABELIA. Shrub 2(–3) m tall, branchlets red, pubescent when young, glabrous in age, bark light brown, shedding; leaves opposite, petiole 1–3 mm long, slightly winged, blades 2–6 cm long, 0.5–2.5 cm wide, ovate to narrowly ovate, apex acute, mucronate, base cuneate, upper surface glabrous, lower surface glabrous or sparsely pubescent on veins or in vein axils, margin irregularly crenate-serrate; flowers solitary or few in axillary or terminal cymes, pedicel 2–4 mm long, bracts ca. 1 mm long, 4, lanceolate, hypanthium 3–4 mm long, elliptic, outer surface glabrous or sparsely pubescent, calyx lobes 2–5, 6–8 mm long, elliptic-oblanceolate, reddish, corolla lobes 5, bilabiate, white tinged with pink, corolla ca. 20 mm long, campanulate, gibbous at the base, lower lip hirsute on inner surface, stamens 5, epipetalous, ovary inferior, 3-carpellate and 3-loculate; style 1; fruit 8–10 mm long, an achene, oblong, sparsely pilose or glabrous, with persistent sepals at the apex. Disturbed sites; Dallas (Kartesz 2022) and Walker (Nesom & Brown 1998) cos.; escaped from cultivation; AL, AZ, FL, GA, MS, NC, and TX; Mexico; Europe; Africa; Asia; and Australia. Summer–fall. Native of Asia.

DIPSACUS L. TEASEL

A genus of 20 species of Eurasia, the Mediterranean, and tropical African mountains; some are cultivated as ornamentals (Mabberley 2017). Kartesz (2015) placed *Dipsacus* in the Dipsacaceae, we, however, are including it in the Caprifoliaceae as suggested by Bell et al. (2001) and others. The leaves "are basally connate, forming rainwater-collecting troughs, which may prevent insects climbing up to the flowers" (Mabberley 2017). The rigid, down-curved bracted heads of *D. sativus* (L.) Honck. (FULLERS'S TEASEL) were used for centuries in producing textiles (particularly those of wool) by serving as a natural comb; they were used to clean and raise the nap on cloth—hence the name TEASEL. Ryder (1996) states that it is still used today and "no man-made substitute has been found for its gentle action on the finest cloths." In nature the rigid bracts act as fruit catapults, causing the seeds to be thrown when animals brush against the heads (Dallimore 1912; Mabberley 2017). (Greek: *dipsakos*, the classical name for teasel, apparently from *dipsa*, thirsty or dry, in reference to the leaf bases that in some species can hold water—(McClintock 1993; Mabberley 2017)

REFERENCES: Ferguson & Brizicky 1965; Ryder 1996; Singhurst & Holmes 2001; Caputo et al. 2004.

Dipsacus fullonum L., (pertaining to fullers, a worker who cleans and raises the nap on fabrics in the process known as fulling; the species is unfortunately named since this function is actually better served by the closely related, cultivated *D. sativus*), COMMON TEASEL, WILD TEASEL. Biennial herb to 2(–3) m tall; stems prickly; leaves opposite; basal leaves lanceolate to oblanceolate, to 30(–40) cm long, sessile to short-petiolate, marginally scalloped to toothed and often prickly, prickly along midvein beneath, usually dead or absent at time of flowering; stem leaves sessile or sometimes inconspicuously fused basally, marginally similar to basal, the uppermost narrower and entire; flowers numerous in pedunculate, involucrate, dense heads, perfect, each flower subtended by a persistent receptacular bract ending in a spine-like awn and by a calyx-like involucel of fused bractlets; involucral bracts unequal, to 15 cm long, at least some usually longer than the heads, narrowly linear; heads ovoid to subcylindric, to 10 cm long; calyces cup-shaped; corollas with a slender tube and 4 short unequal lobes, 10–15 mm long, the tube white, the lobes usually lavender, rarely white; stamens 4, exerted, usually lavender; pistil 1; ovary inferior, with a single ovule; fruits achenes, 3–5 mm long, crowned by persistent calyx, shed with involucel. Disturbed areas; Collin Co. (*Singhurst 9379* BAYLU, TEX); reported for TX (Singhurst & Holmes 2001) and apparently known in the state only from a single county; widespread in the U.S., but more commonly in the n 1/2 of the country. Jun–Oct. [*D. sylvestris* Huds.] Native of Europe and Asia. The dried heads are used in floral arrangements (Mabberley 2017). This species has persistent receptacular bracts which are not as rigid as those of *D. sativus* nor are they recurved as in *D. sativus* and are thus not as useful for fulling; it is a wild weedy relative of the cultivated *D. sativus* (Yatskievych 2006). It is considered a noxious weed in some states (e.g., MO) (Yatskievych 2006).

LONICERA L. HONEYSUCKLE

Vines or shrubs; leaves short-petioled or sessile, entire; flowers axillary or terminal, rather large and showy; fruit a several-seeded berry.

A genus of 180 species of the n hemisphere s to Mexico, the Philippines, and China (Mabberley 2017); many are cultivated as ornamentals and a number have become serious invasive weeds; some have very fragrant flowers. (Named for Adam Lonitzer, 1528–1586, German naturalist and herbalist)

REFERENCES: Rehder 1903; Luken & Thieret 1995, 1996; Miyake & Yahara 1998; Schierenbeck 2004.

1. Uppermost leaves united around stem; flowers terminal, in heads or in whorls forming spikes; twining, trailing or climbing vines.
    2. Corollas 20–50 mm long, red to yellow to orange.
        3. Upper surface of leaf olive to dark green; corollas 30–50 mm long, nearly actinomorphic (corolla lobes nearly equal) __________ **L. sempervirens**
        3. Upper surface of leaf bright green; corollas 20–35 mm long, strongly zygomorphic (lobes of corolla divided into an upper and lower lip) __________ **L. flava**
    2. Corollas 12–18 mm long, white to yellowish white __________ **L. albiflora**
1. Uppermost leaves separate; flowers axillary or both axillary and terminal, mostly in 2s; vines or shrubs.
    4. Twining, trailing, or climbing vines; corolla tube > 15 mm long __________ **L. japonica**
    4. Shrubs; corolla tube < 15 mm long.

5. Corollas nearly actinomorphic (upper two lobes somewhat fused) ______ **L. tatarica**
5. Corollas strongly zygomorphic.
    6. Leaves acuminate apically; branches hollow with brown pith ______ **L. maackii**
    6. Leaves rounded or acute apically (but short apiculate); branches with solid white pith ______ **L. fragrantissima**

Lonicera albiflora Torr. & A. Gray, (white-flowered), WHITE HONEYSUCKLE, BUSHY HONEYSUCKLE. Evergreen, low, shrubby vine with arched to twining branches; young stems usually glabrous (rarely with sparse pubescence); leaves sessile or subsessile, blades (1.5–)2–4(–6.5) cm long, variable in shape, elliptic to ovate, suborbicular, rhombic, or obovate, ± coriaceous, upper surface of leaf blades gray-green or yellow-green, the lower surface variously glaucous; typically only the terminal pair of leaves perfoliate; corollas white to yellowish white, < 2 cm long; fruits reddish orange, 5–15 mm in diam. Limestone outcrops; w margin of East TX in Blackland Prairie on Austin Chalk; Bell, Comal, Dallas, Hays, Travis, Williamson (Turner et al. 2003), and McLennan (Hannick et al. 2013) cos.; Edwards Plateau and Trans-Pecos; sw U.S. from OK and TX w to AZ. Apr–May. [*L. albiflora* var. *dumosa* (A. Gray) Rehder] Native of sw U.S. and Mexico. The fruits are reported to contain a substance that induces vomiting (Powell 1988).

Lonicera flava Sims, (yellow), YELLOW HONEYSUCKLE. Glabrous, deciduous twinning vine; young stems glabrous; leaves sessile to petioles 6 mm long, blades mostly 3–9 cm long, 2–6.5 cm wide, elliptic to broadly elliptic, or rounded, upper surface of leaf blades glabrous, bright green, lower surface usually pale green, those of the upper 1 or few pairs perfoliate, slightly constricted where bases join; flowers in 1or 2 whorls of 6; corollas bright yellow to orange, sometimes reddish-tinged, 20–35 mm long, strongly zygomorphic, divided into 2 recurved lips of ± equal length, the upper lip (3–)4-lobed, the lower lip with one lobe; fruits orangish red to red, 5–10 mm in diam. Rocky soils in woods, slopes, bluffs, ledges, and stream margins; Kartesz (2022) does not show *L. flava* in TX, but was added to the Texas flora in 2021 (Singhurst et al. 2021) with records from Cass (*Singhurst 5570* BAYLU), Morris (*Carr 11029* TEX-LL), and Red River (*Holmes et al.14537* BAYLU) cos.; concentrated in AR, OK, and MO, also se U.S.; Apr–Jun. [*L. flavida* Cockerell ex Rehder, *L. flava* var. *flavescens* Gleason] Native of the U.S.

Lonicera fragrantissima Lindl. & Paxton, (very fragrant), WINTER HONEYSUCKLE, SWEET-BREATH-OF-SPRING. Semi-evergreen shrub to ca. 3 m tall; leaves short petioled, blades broadly oval to obovate, to ca. 8 cm long, strongly apiculate; flowers fragrant, several pairs in leaf axils; corollas ca. 16 mm long, creamy white, sometimes tinged with pink, the tube ca. 5–6 mm long; fruits orange to red, 8–10 in diam. Widely cultivated and escapes, forest margins; Collin (Turner et al. 2003), Fannin (EDDMapS 2024), McLennan (Hannick et al. 2013) cos., also Tarrant Co. (R. O'Kennon, pers. obs.) on the w border of East TX; e U.S. from NY s to AL and w to OK and TX, also UT; Europe and China. Jan–Apr. [*Xylosteon fragrantissima* (Lindl. & Paxton) Small] Native of China.

Lonicera japonica Thunb., (of Japan), JAPANESE HONEYSUCKLE. Trailing or twining vine, evergreen; young stems usually densely pubescent to almost glabrous; leaves short-petioled, blades to 8 cm long, oblong-ovate or elliptic, those on new spring or summer growth (=juvenile leaves) often pinnatifid; flowers very fragrant, in axillary pairs; corollas 25–40 mm long, creamy white or white tinged with purple, becoming yellow with age, eventually changing to brownish, the tube pubescent outside, the upper lip with at least 2 of its lobes united more than half way; fruits black, 5–8 mm in diam. Cultivated, escaped, and locally established in disturbed areas, woods, and thickets; widespread in e 1/2 of TX, with a few localities further w; pervasive in e U.S. from ME s to FL and w to NE and TX, also scattered in sw U.S. and n to WA; Canada; Mexico; South America; Asia; Australia; and Europe. This species can be invasive overtaking the habitat of native species. Mar–Jul. [*Nintooa japonica* (Thunb.) Sweet] Native of Asia.

Lonicera maackii (Rupr.) Maxim., (for Richard Maack, 1825–1886, Russian naturalist), AMUR HONEYSUCKLE, BUSH HONEYSUCKLE, TREE HONEYSUCKLE, MAACK'S HONEYSUCKLE. Shrub to ca. 6 m tall, deciduous but leafing out very early and holding leaves late; leaves short-petioled, blades to ca. 8 cm long, ovate to lance-ovate, long-acuminate; flowers fragrant, in axillary pairs; corollas white becoming yellow, ca. 20 mm long; fruits dark red, 4–7 mm in diam. Widely cultivated and escapes; calcareous slopes and forest margins; Brazos, Dallas (Turner et al. 2003), Grayson (*Valek 21* BRIT), McLennan (Hannick et al. 2013), and Rains (*Baker 68* BRIT) cos.; also Tarrant Co. (R. O'Kennon, pers. obs.) on w border of East TX; e U.S. from ME s to AL and w to NE and TX, also OR and WA; Canada; Asia; and Europe. Apr–May. Native of c and ne China, the Amur River and Ussuri River valleys, Korea, and Japan (Luken & Thieret 1996). This species was apparently first introduced to North America in the late 1800s; it was first referred to as a weed in the U.S. in Chicago in 1924 and is now considered aggressively invasive and problematic in some areas of the e and midwestern U.S. (Luken et al. 1995; Luken & Thieret 1995, 1996; Yatskievych 2006). By leafing out earlier in the spring than its competitors, it gains an advantage, but also reduces the available light for spring ephemerals (Yatskievych 2006). The fruits are reported to be mildly toxic to humans, but also quite unpalatable (Luken & Thieret 1995).

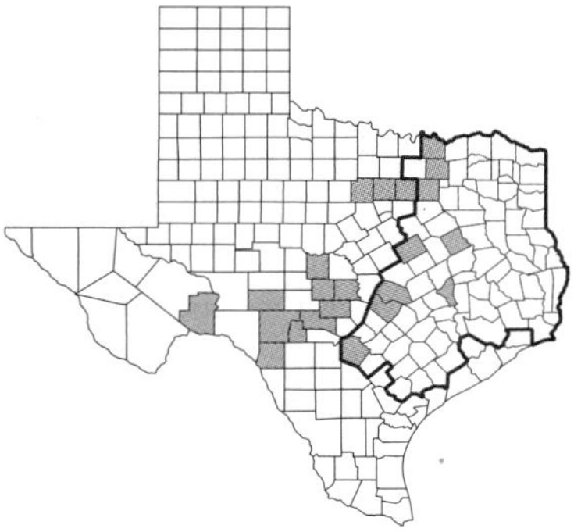

Triodanis leptocarpa

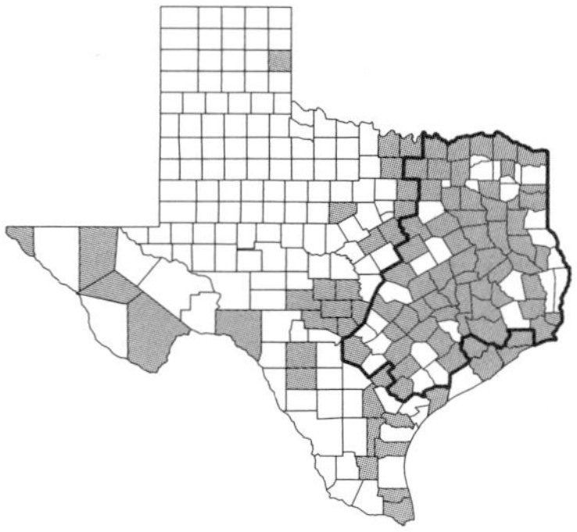

Triodanis perfoliata

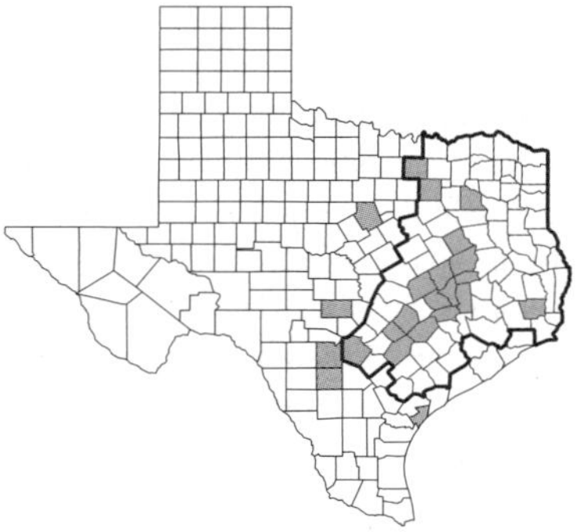

Triodanis texana

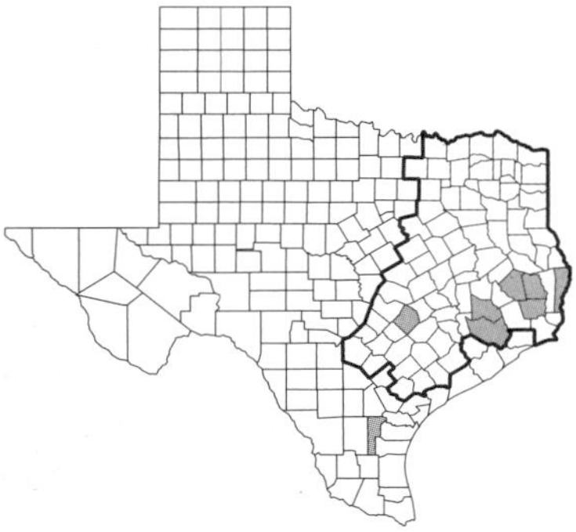

Wahlenbergia marginata

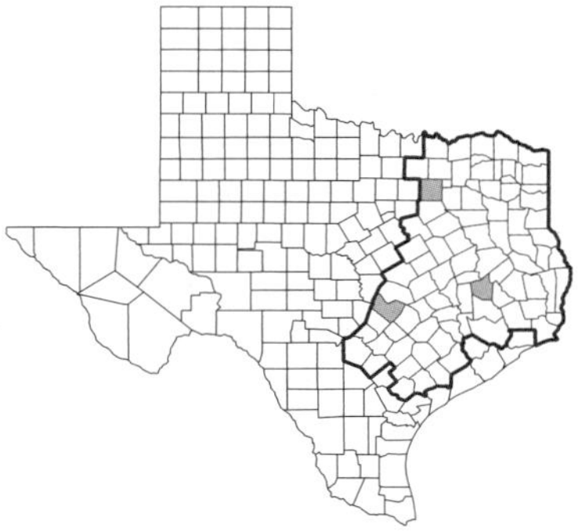

Cannabis sativa

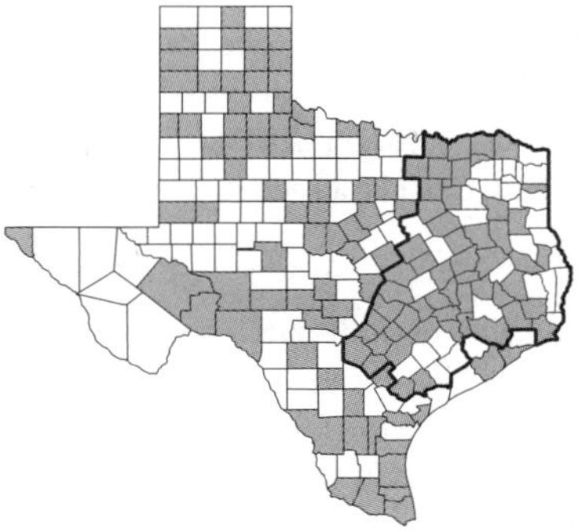

Celtis laevigata

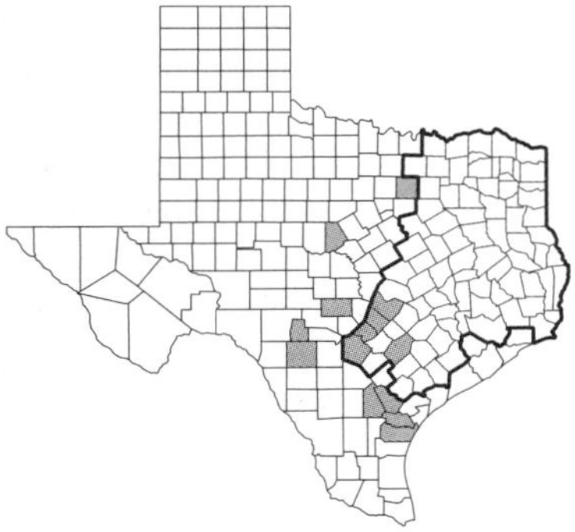

Celtis lindheimeri

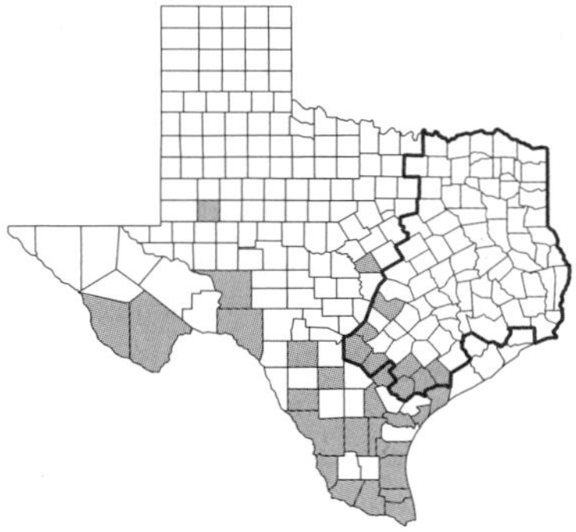

Celtis pallida

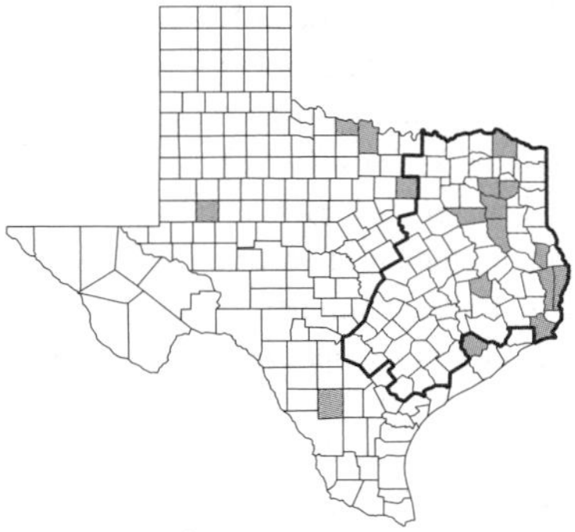

Celtis pumila

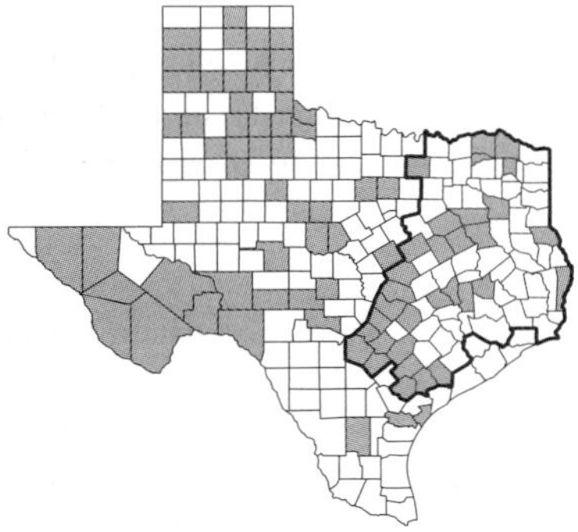

Celtis reticulata

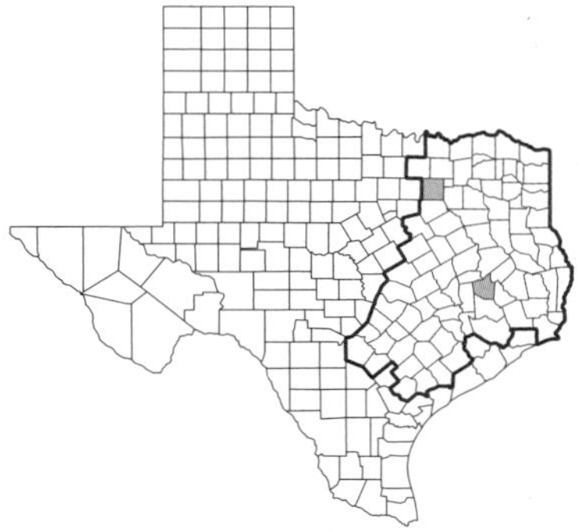

Abelia ×grandiflora

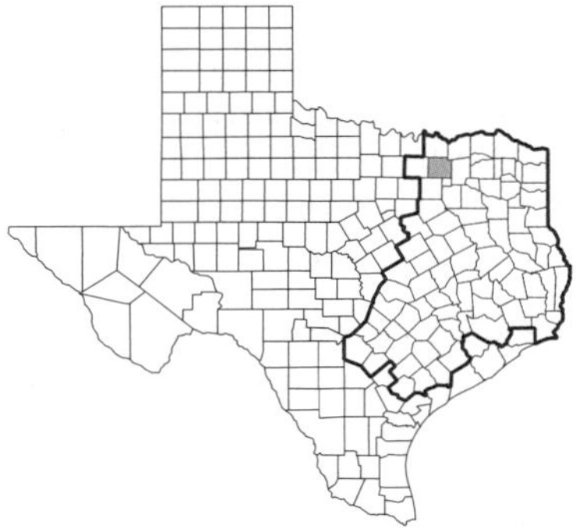

Dipsacus fullonum

Lonicera sempervirens L., (evergreen), CORAL HONEYSUCKLE, TRUMPET HONEYSUCKLE, EVERGREEN HONEYSUCKLE. Glabrous, evergreen, twining vine to ca. 5 m long; young stems glabrous or nearly so; leaves sessile or subsessile, blades 3–6(–7) cm long, 2–4 cm wide, ovate to elliptic or obovate, subcoriaceous, upper surface of leaf blades olive-green or dark green, lower surface lighter than the upper, often glaucous; terminal 1–2(–several) pairs of leaves perfoliate; flowers in whorls of 1–4; corollas 30–50 mm long, usually red to orangish red outside, orange inside, tubular, shallowly lobed, the lobes nearly equal; stamens and style exerted; fruits red or reddish orange, 6–10 mm in diam. Stream banks or hillside woods, also cultivated; East TX w to Cross Timbers and Prairies, Edwards Plateau, and n Gulf Prairies and Marshes; e U.S. from ME s to FL and w to IA and TX; also Europe. Apr–Jun. [*L. sempervirens* var. *hirsutula* Rehder, *L. sempervirens* var. *minor* Aiton, *Phenianthus sempervirens* (L.) Raf.] Native of the U.S. The scentless flowers of this species are visited by and presumably pollinated by ruby-throated hummingbirds (*Archilochus colubris*) (James 1948).

Lonicera tatarica L., (probably from Tatary, an old name applied to a portion of Asia), TWINSISTERS, TATARIAN HONEYSUCKLE. Deciduous, glabrous shrub to 3 m tall; leaf petiole 4–8 mm long, leaf blades ovate to oblong or oval, glabrous, base rounded to subcordate or cordate, apex acute to obtuse; flowers paired in leaf axils; corollas 14–20 mm long, nearly actinomorphic, pink or white, sometimes red, lobes ca. as long as tube; fruits red to orange or yellow, 5–6 mm in diam., globose. Cultivated but easily escapes and is a noxious weed; disturbed forests, forest edges, and near old homesites; reported in only two counties in TX, Dallas Co. based on EDDMapS (2021) and McLennan Co. (Hannick et al. 2013); mostly in the n 2/3 of the U.S., but also in TX and NM; Canada; all across Europe; Russia; and China. May–Jun. Native of Eurasia.

SCABIOSA L. PINCUSHIONS

A genus of ca. 30 species native to temperate Eurasia and the Mediterranean region (Mabberley 2017); herbs usually with an umbrella-like epicalyx (= involucel of fused bracteoles) functioning in wind dispersal of the fruit. Some species are cultivated (e.g., *S. columbaria* L.). Though sometimes included in the Dipsacaceae (e.g., Diggs et al. 1999); we, however, are including it in the Caprifoliaceae as suggested by Bell et al. (2001), Kartesz (2022) and others. (Latin: *scabies*, itch; the rough leaves were used medicinally to treat skin complaints)

REFERENCES: Jasiewicz 1976; Mayer 1995; Mayer & Ehrendorfer 1999.

Scabiosa atropurpurea L., (dark purple), PINCUSHIONS, SWEET SCABIOUS. Annual unarmed herb 20–60 cm tall; leaves opposite; basal leaves dentate; upper leaves pinnately parted, the lobes entire to dentate; flowers in peduncled involucrate heads; involucral bracts narrowly lanceolate, herbaceous, distinct; the marginal flowers usually larger, superficially somewhat like the ray flowers of Asteraceae, each flower subtended by an umbrella-like epicalyx (= involucel of fused bracteoles); corollas white to rose, lilac, or dark purple, bilaterally symmetrical; calyces pappus-like, of 5 bristles ± 5 mm long; corollas 5-lobed, the lobes unequal; stamens 4, distinct, epipetalous; ovary inferior; achenes with 5 persistent conspicuous calyx bristles. Garden escape; Collin, Dallas, Fannin, Grayson, Hopkins, Hunt (Turner et al. 2003), Denton (*Voss s.n.*, 2005, TCSW), Ellis (*Frary 135* BRIT), and Rockwall (*Holmes 10156* BAYLU) cos.; also Johnson, Tarrant, and Wise cos. to the w of East TX; scattered in the U.S. in AR, CA, KS, MA, ME, MO, OR, TX, and WA. Jun–Sep. Native of Europe. This species is rapidly spreading and becoming a seriously problematic invasive weed in fields, roadsides, and prairies; it is apparently being spread by highway mowing equipment in parts of Collin Co. where it forms virtual monocultures on some roadsides. It is listed in the *Invasive Plant Atlas of the United States* (www.invasiveplantatlas.org).

SYMPHORICARPOS Duhamel CORAL-BERRY, SNOWBERRY

A genus of 16 species of deciduous shrubs of North America, including Mexico (15 species) and China (1) (Mabberley 2017); some are cultivated as ornamentals. Birds are reported to eat the fruits during the winter, but often not until other foods are unavailable (Yatskievych 2006). The fruits of some species are reportedly toxic if ingested in quantity (Lampe & McCann 1985). (Greek: *symphorein*, to bear together, and *carpos*, fruit; from the clustered fruits)

REFERENCE: Jones 1940.

Symphoricarpos orbiculatus Moench, (orbicular, round), INDIAN-CURRANT, CORAL-BERRY, BUCKBRUSH. Short, rhizomatous (forming colonies), inconspicuously pubescent, erect to ascending shrub to ca. 2 m tall; leaves very short-petioled, blades usually < 4 cm long, rarely to 6 cm long, ovate to elliptic, oblong or suborbicular, entire or with a few blunt teeth; flowers in small dense axillary spikes, rather inconspicuous; corollas funnelform, 3–5 mm long, greenish white, sometimes partly brown-red; drupes coral-red to pink or purplish, 3–7 mm long, long persisting, usually with 2 stones. Frequently cultivated, woods and thickets; East TX w to West Cross Timbers and Edwards Plateau; also n Gulf Prairies and Marshes; e U.S. from NY s to FL and w to NE and TX, also AZ, CO, and UT; Europe. This species is increasing under disturbance. Jun–Aug. Native of the U.S. The leaves are reported to have been used medicinally (Mabberley 2017) and the species to be poisonous (Burlage 1968).

Celtis pallida [EMO]

Celtis pumila [SA4, TOR]

Celtis reticulata [SA3]

Lonicera albiflora [VIN]

Dipsacus fullonum [GEO, VIN]

Abelia ×grandiflora [BA1]

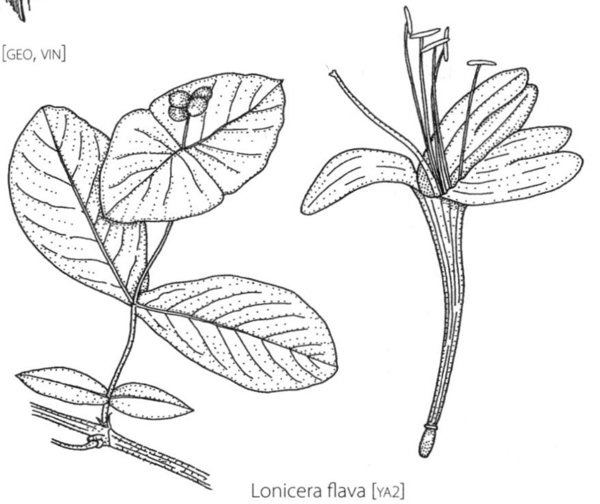

Lonicera flava [YA2]

TRIOSTEUM L. FEVERWORT, HORSE-GENTIAN

Coarse perennial herbs; stems erect, unbranched, pubescent; leaves simple, opposite, entire, sessile, or with broadly winged petiole, perfoliate or connected by a ridge of tissue around the stem, apically acute to acuminate; flowers axillary, solitary or ± 2–3 per axil, each flower subtended by paired bracts; corollas slightly bilateral, tubular, with 5 lobes, with glandular pubescence; fruit a drupe with 3 stones.

A genus 6 species of perennial herbs of e Asia (3) and e North America (3) (Gould & Donoghue 2000); this disjunct distribution pattern is discussed under the genera *Campsis* (Bignoniaceae) and *Carya* (Juglandaceae), see also the introduction to volume one of the *Illustrated Flora of East Texas*, page 205. (Greek: *tri*, three, and *osteon*, bone, alluding to 3 bony nutlets)
REFERENCES: Bicknell 1901; Wiegand 1923; Wilkinson 1949; Lewis & Frantz 1973; Gould & Donoghue 2000; Yatskievych 2006.

1. Leaves not perfoliate (but joined by a ridge of tissue around the stem); corollas usually yellowish white, yellowish or greenish (rarely orangish or reddish); leaf blades 2–5.5 cm wide; stems long-hirsute, the hairs 1–4 mm long, sometimes also with shorter minutely gland-tipped hairs; styles included or only slightly exerted (not more than 2 mm beyond corolla lobes); flowers usually 1 per leaf axil ____ **T. angustifolium**
1. Leaves (at least those of midstem) perfoliate; corollas usually dull red to purplish red (rarely greenish yellow); leaf blades 4–9 cm wide; stems densely short pubescent, the minutely gland-tipped hairs 0.5 mm or less long, and also with longer straight hairs usually 1.5 mm or less long; styles slightly to distinctly exerted (to 5 mm); flowers (1–)2–3 (–4) per leaf axil ____ **T. perfoliatum**

Triosteum angustifolium L., (narrow-leaved), YELLOW-FLOWERED HORSE-GENTIAN. Perennial herb to ca. 80 cm tall; leaves not perfoliate (bases joined by a ridge of tissue around the stem), blades 15(–19) cm long, 5.5 cm or less wide, narrowly obovate to lanceolate, the largest sometimes somewhat fiddle-shaped, pubescent; flowers usually solitary in the leaf axils (2 per node); calyx lobes 7–13 cm long; corollas tubular, yellowish white, or greenish, rarely orangish or reddish, 12–15 mm long; fruit a drupe 5–7(–10) mm in diam., pale orange maturing red, usually with 3 stones. Wooded areas, thickets; Bowie, Smith (Kartesz 2022), Grayson (BRIT), and Lamar (Turner et al. 2003) cos. in Red River drainage, also Cass, Cherokee and Smith (Kartesz 2022) cos., otherwise not known in TX; e U.S. from PA s to GA and west to KS and TX. Mar–May. Native of the U.S.

Triosteum perfoliatum L., (perfoliate, with a leaf surrounding the stem), COMMON HORSE-GENTIAN, FEVERWORT, TINKER'S-WEED, WILD COFFEE. Perennial herb to ca. 1.2 m tall; leaves (midstem pairs) perfoliate, blades to 20 cm long and 4–9 cm wide, obovate to oblanceolate or elliptic, the largest sometimes somewhat fiddle-shaped, pubescent; flowers (1–)2–3(–4) per leaf axil, (2–)4–6 (–8) per node); calyx lobes 10–18 mm long; corollas 12–17 mm long, dull red to purplish red, sometimes with base and inner surface with cream-colored spots, occasionally greenish yellow; fruit 7–10 mm in diam., pale orange to orange. Moist woods, clearings, disturbed sites; in TX known only from Anderson Co. (Holmes et al. 1999); e U.S. from NY s to AL and w to MN and TX; pervasive in mw U.S. May–Jun. Native of the U.S. This species was formerly used as a coffee substitute (Mabberley 2017).

VALERIANELLA Mill. LAMB'S-LETTUCE, CORNSALAD

Ours small annual herbs, dichotomously branched; often with a distinctive "wet dog" odor when dry; leaves, except lowest, sessile, often clasping, simple, opposite, entire or with few coarse, basal teeth or lobes; flowers small, terminal, in compact, head-like, dichotomous, leafy-bracted cymose inflorescences; calyx lobes absent; corollas usually white, bluish, or rarely pale pinkish, funnelform, subequally 5-lobed, the tube saccate with a nectary; stamens 3; pistil 3-carpellate, inferior, of 1dorsal fertile locule (often somewhat flattened) and 2 ventral sterile (= empty) locules (the sterile locules sometimes inflated); fruits dry, 1-seeded, indehiscent, usually < 3 mm long.

A genus of 64 species of the n temperate zone s to n Africa (Mabberley 2017); some are used as potherbs; many North American taxa exhibit fruit polymorphisms in which a single population of a species may have up to three different genetically controlled fruit morphs. Kartesz (2015) placed *Valerianella* in the Valerianaceae, we, however, are including it in the Caprifoliaceae as suggested by Bell et al. (2001) and others. For *Valerianella* nomenclature, we are following Eggers (1969). Mature fruits examined with a hand lens are necessary for positive identification of most of our taxa. Donna Eggers-Ware has annotated a number of BRIT specimens without mature fruits as indistinguishable between *V. radiata* and *V. woodsiana*. (Diminutive of *Valeriana*, a medieval name, perhaps from Latin: *valere*, to be strong or healthy, possibly alluding to medicinal properties of some species, or said by Linnaeus to honor Publius Aurelius Licinius Valerianus, Roman emperor from 253–260 B.C. and patron and friend of botanists)
REFERENCES: Dyal 1938; Eggers 1969; Eggers-Ware 1983; Hidalgo et al. 2004.

1. First stem dichotomy of wide spreading plants (with multiple cymules) within the lower 2/5 of flowering plant; fruits with a corky hump on the fertile locule; flowers often with bluish tinge ____ **V. locusta**

Lonicera fragrantissima [CUR]

Lonicera japonica [REE]

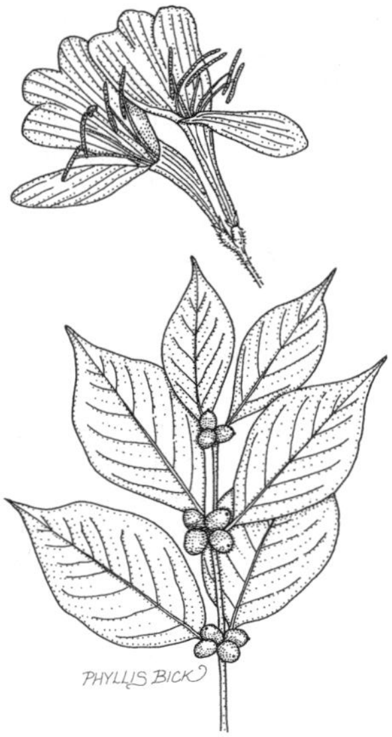

Lonicera maackii [YA2]

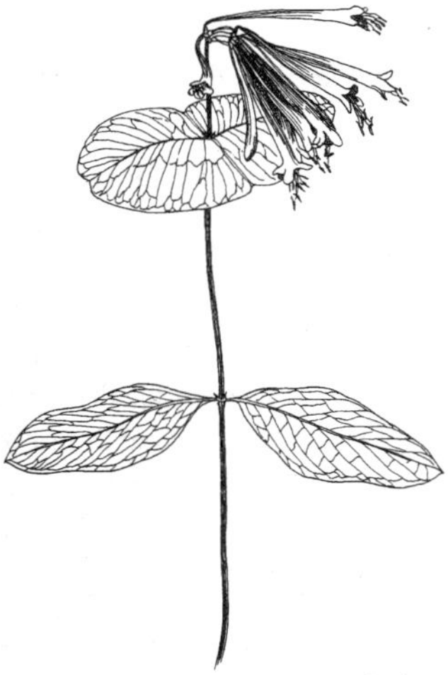

Lonicera sempervirens [BT3]

Lonicera tatarica [GLE]

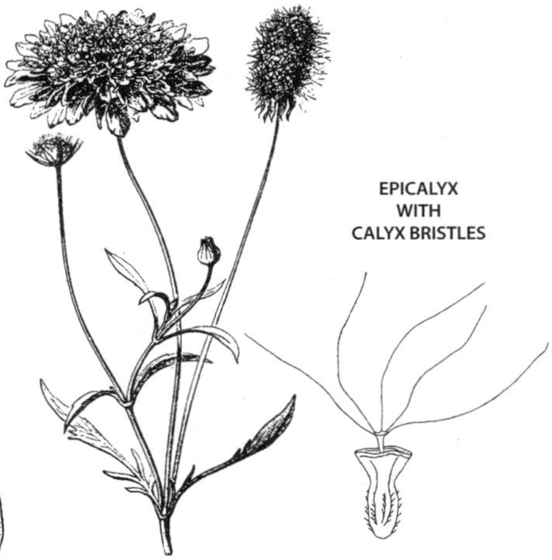

Scabiosa atropurpurea [BL1, RJG]

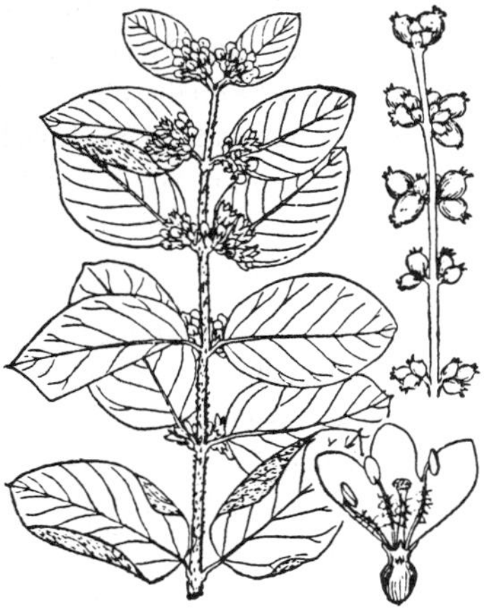

Symphoricarpos orbiculatus [BB2]

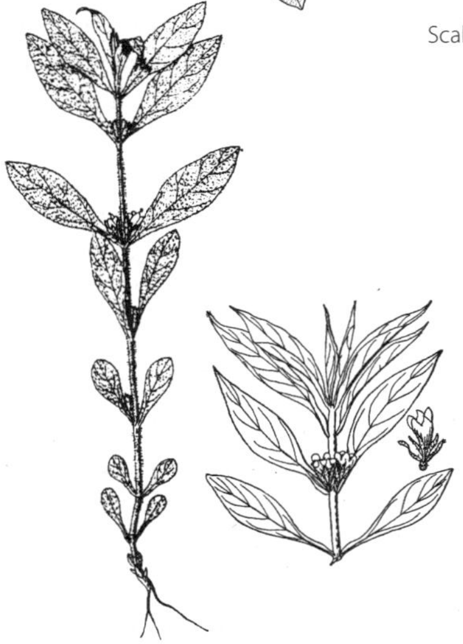

Triosteum angustifolium [BB2, STE]

Triosteum perfoliatum [GEO]

1. Plants without the combination of characters as above.
  2. Floral bracts with margins conspicuously glandular-denticulate ____ **V. florifera**
  2. Floral bracts not as above (some plants with bracts weakly glandular-dentate toward the tip).
    3. Flowers showy, the corollas 2.5–4.5 mm across; fruits with conspicuous hairs having a minute hook at their ends OR fruits glabrous or rarely with straight hairs.
      4. Mature fruits broadly ovate to nearly round in outline when viewed dorsally (looking down on the fertile locule), with conspicuous hairs having a minute hook at their ends; stems, leaves, and bracts essentially glabrous except for tufts of hairs on each side of leaf base ____ **V. amarella**
      4. Mature fruits narrowly elliptic or lanceolate in outline when viewed dorsally, glabrous or rarely with straight hairs; stems pubescent on their angles, leaves and bracts pubescent on their margins ____ **V. stenocarpa**
    3. Flowers not showy, the corollas 0.6–2 mm across; fruits glabrous or with straight hairs.
      5. Combined width of the 2 sterile locules distinctly wider than the single fertile locule (fertile locule conspicuously exceeded laterally by the widely divergent sterile locules); mature fruits wide, broadly elliptic to nearly round in outline ____ **V. woodsiana**
      5. Combined width of the 2 sterile locules narrower to barely wider than fertile locule (fertile locule narrower to barely exceeded laterally by the sterile locules which are not divergent or only slightly so); mature fruits narrower, narrowly lanceolate to elliptic or ovate in outline ____ **V. radiata**

Valerianella amarella (Lindh. ex Engelm.) Krok, (bitter, in reference to the taste of the leaves), HAIRY CORNSALAD. Plant (6–)9–20 (–25) cm tall; stem glabrous; rather showy, usually many flowers open at once, over-topping the small bracts; corollas 3–4.5 mm across, white; fruits with the 2 sterile locules much smaller than the large fertile locule, the groove between sterile locules narrow, very shallow or inconspicuous. Limestone outcrops; Blackland Prairie—on Austin Chalk on w edge of East TX; Cross Timbers and Prairies and Edwards Plateau, also Montgomery, San Jacinto and Walker (Nesom & Brown 1998) cos; OK and TX; endemic to OK and TX. Late Mar–May. [*Fedia amarella* Lindh. ex Engelm.] Hundreds or even thousands of plants of *V. amarella* can make an extremely showy display when in flower on nearly bare limestone outcrops.

Valerianella florifera Shinners, (flower-bearing), TEXAS CORNSALAD. Plant (10–)15–25(–30) cm tall, glabrous; leaves entire, glabrous or the lower slightly scabrous-ciliate, especially near the apex; lower leaves oblong or oblong-spatulate, sessile; upper leaves oblong to oblong-lanceolate or narrowly deltoid-oblong, sessile; bracts lanceolate or elliptic-lanceolate, acute; floral bracts, glandular-denticulate to lacerate, glabrous; inflorescence rather open, cymules few but many-flowered; corolla 4–5 mm across, white; stamens and style exerted; fruit wider than deep, fertile locule narrowly elliptic-oblong when viewed dorsally, sterile locules divergent, their combined width somewhat wider to slightly more than 2 times as wide as the fertile locule, surfaces glabrous or short pubescent. In sandy, moist soil in post oak woods; mostly in the sw portion of East TX, Blackland Prairie and Post Oak Savannah; endemic to TX. Mar–May. This species, which occurs on sandy soils, has relatively large corollas as in *V. amarella* and *V. stenocarpa*, but differs from all other East TX species in having margins of floral bracts conspicuously glandular-denticulate (vs. non-glandular or essentially so).

Valerianella locusta (L.) Laterr., (possibly meaning growing in an enclosed space), EUROPEAN CORNSALAD, LAMB'S-LETTUCE. Plant (10–)15–25(–40) cm tall, erect to low and spreading, dichotomously branching within lower 2/5 of flowering plant; stems with spreading or retrorse (sometimes minute) cilia on angles; leaves of short main stem few (if any) narrowly oblong with blunt tips and ciliate margins; leaves or bracts subtending dichotomies narrowly oblong to ovate-oblong to lanceolate with blunt tips and margins that are entire or toothed at base (rarely with teeth scattered along margin) and ciliolate; outer bracts of flowering cymules with ciliolate margins to nearly glabrous, often scarious; corolla white, usually with bluish tinge on the lobes, < 1.5 mm across; fruits (1.7–)2–2.5(–4) mm long with corky hump on the fertile locule (entire fruit ca. 0.8 mm wide when viewed from the side). Dry sandy soil, disturbed areas, moist bottomland; in East TX known only from Bowie Co. (Brown et al. 2007); e U.S. from NY s to AL and w to IL and TX, Pacific nw, also MT, ID, and UT; South America and Europe. Apr–May [*Valerianella olitoria* (L.) Pollich] Native of Europe. Sometimes used as a potherb and sold in French and Swiss markets (Couplan 1998).

Valerianella radiata (L.) Dufr., (having rays), BEAKED CORNSALAD, NARROW-CELL CORNSALAD. Plant (10–)15–60(–80) cm tall; stems often pubescent along angles; upper leaves often lobed-toothed toward base; bracts oblong, often abruptly angled distally to triangular tip, or oblanceolate; corollas < 2 mm across, white (rarely pale pinkish). Damp ground, disturbed areas; se and e TX, Cross Timbers and Prairies; e U.S. from MD s to FL and w to KS and TX. Mar–May. [*Fedia radiata* (L.) Michx.] Native of the U.S. Eggers-Ware (1983) indicates that the 3 subspecific taxa of *V. radiata* found in TX are distinguishable only by fruit shape. Because they sometimes occur in mixed colonies and because of known genetically controlled fruit polymorphisms in other members of the genus, we are following Eggers-Ware in recognizing these taxa at the forma level. The county level map only shows distribution of the species.

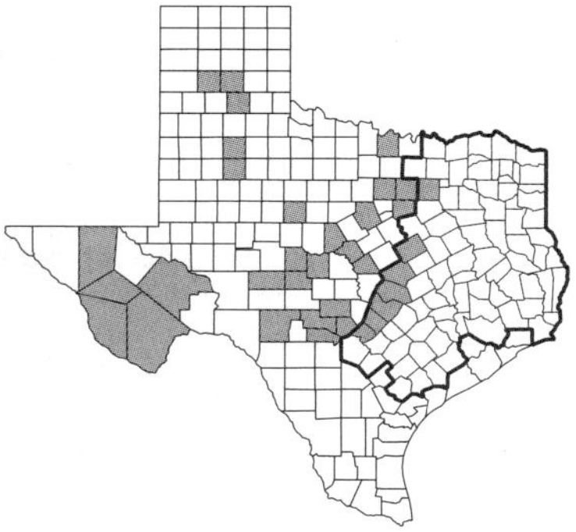

Lonicera albiflora

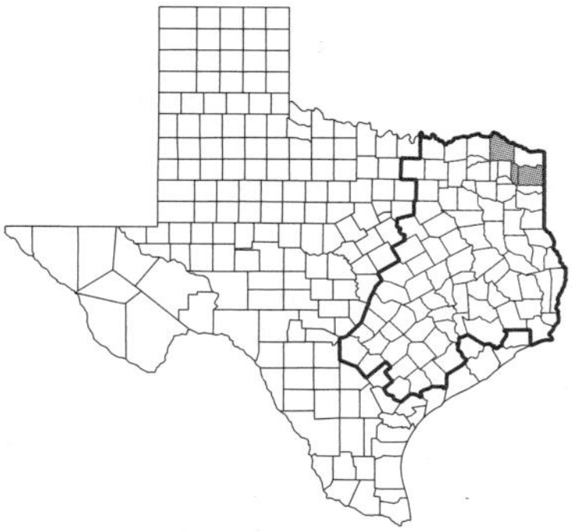

Lonicera flava

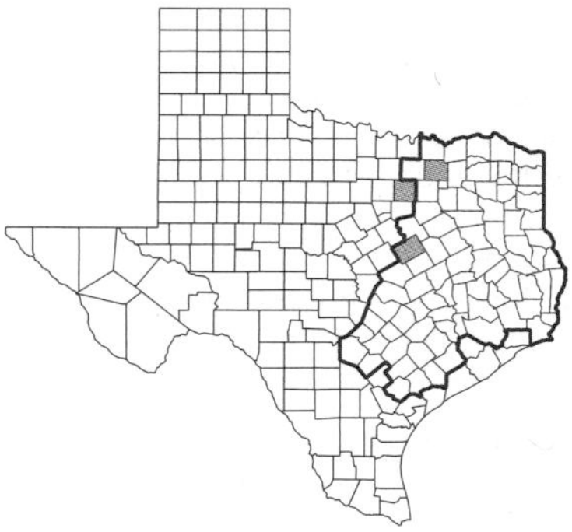

Lonicera fragrantissima

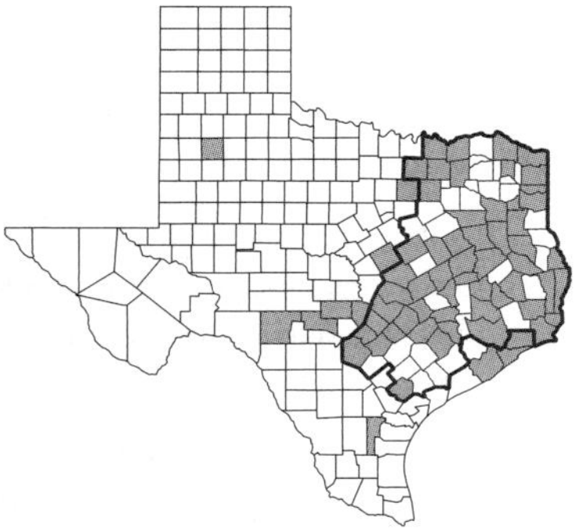

Lonicera japonica

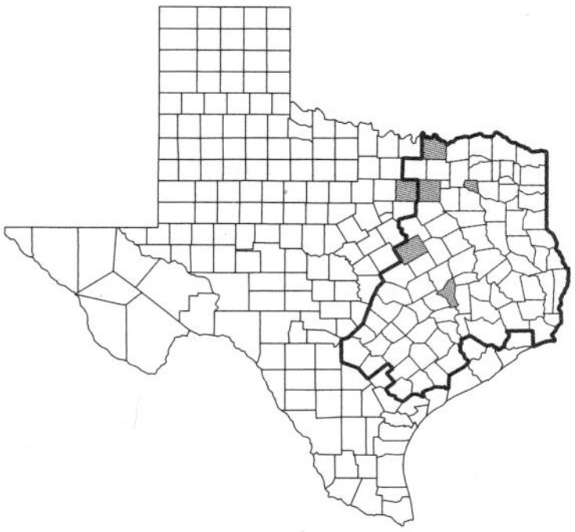

Lonicera maackii

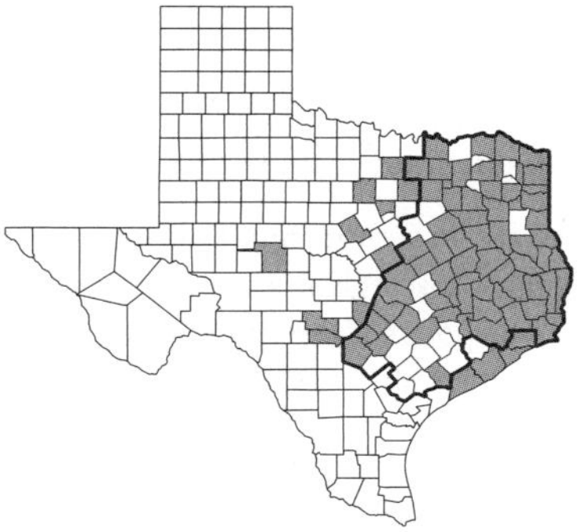

Lonicera sempervirens

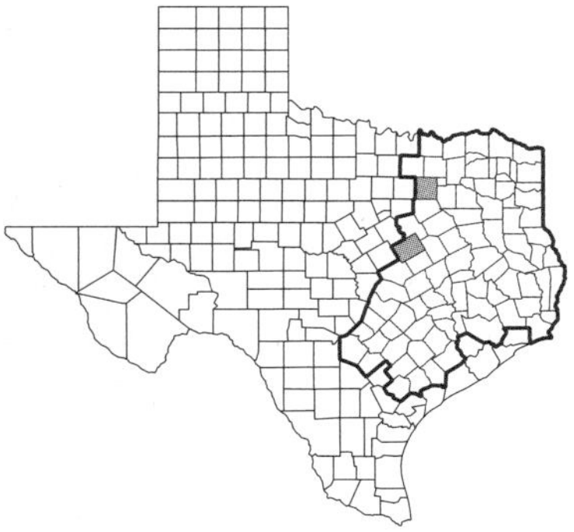

Lonicera tatarica

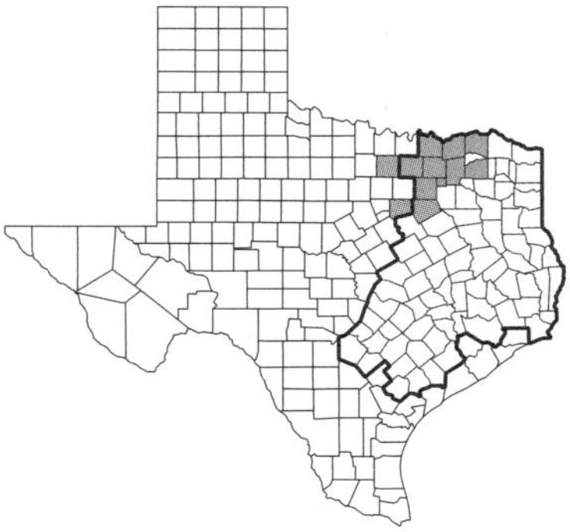

Scabiosa atropurpurea

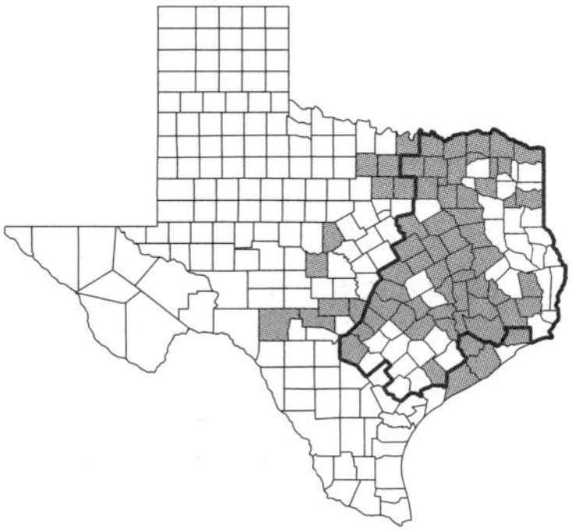

Symphoricarpos orbiculatus

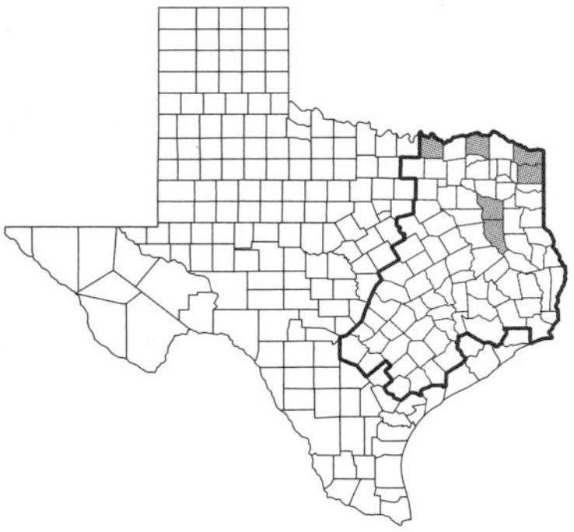

Triosteum angustifolium

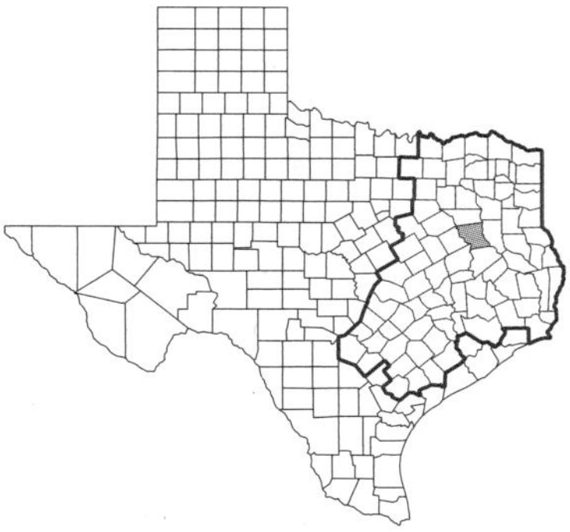

Triosteum perfoliatum

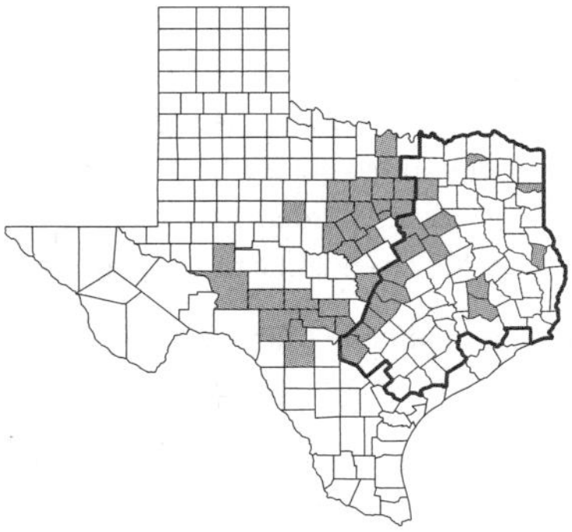

Valerianella amarella

1. Single fertile locule somewhat humped across its width when viewed in cross-section; mature fruits narrowly lanceolate to narrowly elliptic in outline, usually more than twice as long as wide ______ **V. radiata f. parviflora**
1. Single fertile locule nearly flat across its width when viewed in cross-section; mature fruits elliptic to ovate in outline, ca. twice as long as wide or less.
   2. The 2 sterile locules slightly spreading apart from each other, with a deep valley-like groove between them; combined width of sterile locules slightly narrower to distinctly wider than fertile locule, but < 1.5 times as wide ______ **V. radiata f. radiata**
   2. The 2 sterile locules not spreading apart from each other, with only a shallow line-like groove between them; combined width of sterile locules distinctly narrower than fertile locule (fertile locule clearly visible when fruit viewed ventrally—from sterile locule side) ______ **V. radiata f. fernaldii**

f. **fernaldii** (Dyal) Egg. Ware, (for Merritt Lyndon Fernald, 1893–1950, author of *Gray's Manual of Botany*, 8th ed.). Kaufman Co.; otherwise in TX known only from Gonzales and Travis (Eggers 1969) cos. [*V. radiata* var. *fernaldii* Dyal]

f. **parviflora** (Dyal) Egg. Ware, (small-flowered). Throughout much of East TX; Cross Timbers and Prairies. [*V. stenocarpa* (Engelm.) Krok var. *parviflora* Dyal] This taxon has a hump-backed fruit (not as dramatic as *V. locusta*) similar to that of *V. stenocarpa*; however, based on a variety of other floral and vegetative characters (e.g., flower size), it is more appropriate to recognize it as a form of *V. radiata* (Eggers 1969).

f. **radiata.** Throughout much of East TX; Cross Timbers and Prairies. [*V. radiata* (L.) Dufr. var. *radiata*]

Valerianella stenocarpa (Engelm.) Krok, (narrow-fruited), NARROW-CELL CORNSALAD. Plant 10–50 cm tall; stems often pubescent on angles; bracts glabrous or the middle and outer ± hispid-pubescent on margins, mostly linear- to oblong-lanceolate and rather narrowly pointed; corollas 2.5–4 mm across, white. Margins of thickets or woods, roadsides, fields, and fencerows; s and w East TX; Cross Timbers and Prairies and Edwards Plateau; endemic to TX. Apr–May. [*Fedia stenocarpa* Engelm.]

Valerianella woodsiana (Torr. & A. Gray) Walp., (for Joseph Woods, 1776–1864, British botanist and student of this genus), WOODS' CORNSALAD. Plant 15–50 cm tall; stems often pubescent along the angles; corollas white. Damp ground, especially sandy soils; Denton and Navarro cos.; se and e TX w to nc TX; e U.S. from SC s to AL and w to KS, and TX. Apr–May. [*Fedia woodsiana* Torr. & A. Gray] Native of the U.S. Though Kartesz (2022) treated *V. woodsiana* as a synonym of *V. radiata* we are following Eggers-Ware (1983) and recognizing *V. woodsiana*. Eggers-Ware indicated that *V. woodsiana* can be reliably distinguished from *V. radiata* only by fruit characters and suggested the need for further study to determine whether it is a separate species or another fruit morph of *V. radiata*. Yatskievych (2013) treated it as a synonym of *V. radiata*. No county level map provided.

Valerianella longiflora (Torr. & A. Gray) Walp., (with long flowers), LONG-TUBE CORNSALAD This taxon can distinguished from all other East Texas *Valerianella* taxa in having a salverform corolla; corolla tube pink to red-violet, 7–11 mm long. Open scrub woodland, along roadside embankments and fence rows and in cedar glades; included based on presence in Choctaw and McCurtain cos., OK (Kartesz 2022); endemic to AR and OK. Apr–May. [*Fedia longiflora* Torr. & A. Gray, *Siphonella longiflora* (Torr. & A. Gray) Small]

Valerianella longiflora (Torr. & A. Gray) Walp. f. *inexpansa* Egg. Ware, (not expanded). This taxon can be distinguished from f. *longiflora* in having sterile locules < 2 times as wide as fertile locule. Endemic to AR and OK.

Valerianella longiflora f. *longiflora*. This taxon can be distinguished from f. *inexpansa* in having sterile locules 2–3(–4) times as wide as the fertile fruit.

Valerianella nuttallii (Torr. & A. Gray) Walp., (for Sir Thomas Nuttall, 1786–1859, English-American botanist), NUTTALL'S CORNSALAD. This taxon can be distinguished from all other East TX *Valerianella* taxa in having a salverform corolla and differing from *V. longiflora* in having a white corolla with corolla tube 2.5–5.0 mm long. The two taxa have very similar fruit morphologies. Along woodland borders in open fields and rocky open hillsides; included based on presence in Bryan and McCurtain cos., OK (Kartesz 2022); endemic to OK and AR. Apr–May. [*Fedia nuttallii* Torr. & A. Gray]

Valerianella nuttallii (Torr. A. Gray) Walp. f. *inexpansa* Egg. Ware (not expanded). This forma can be distinguished from f. *nuttallii* in having sterile locules < 2 times as wide as fertile locule.

Valerianella nuttallii (Torr. A. Gray) Walp. f. *nuttallii*. This forma can be distinguished from f. *inexpansa* in having sterile locules 2–3(–4) times as wide as the fertile fruit. Rocky open banks and along woodland borders.

Valerianella palmeri Dyal, (named for the eminent botanist Ernest Jesse Palmer of Webb City, Missouri), PALMER'S CORNSALAD. This species is most similar to *V. amarella* in the width of the corolla but *V. palmeri* has a flattened glabrous fruit and *V. amarella* has

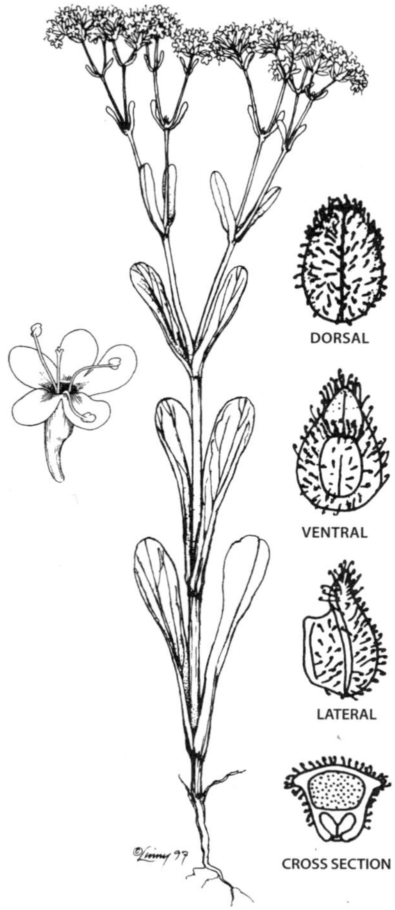

Valerianella amarella [EGG, HEA, RHO]

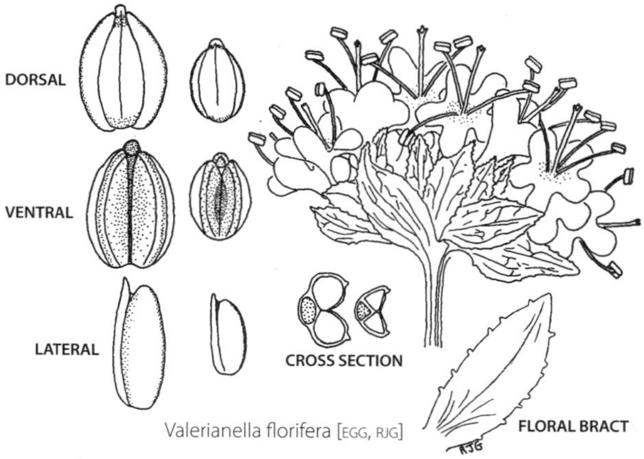

Valerianella florifera [EGG, RJG]

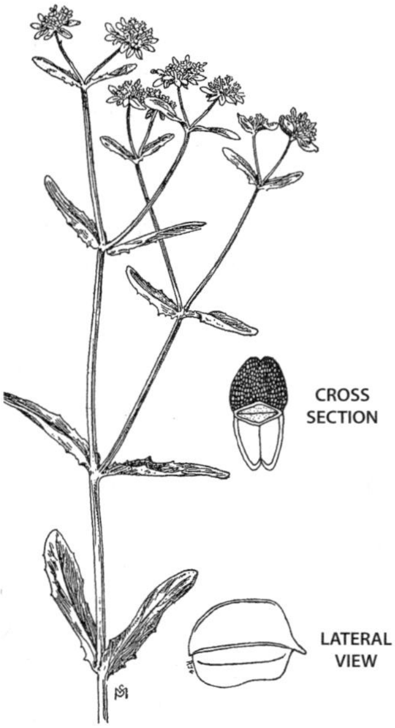

Valerianella locusta [EGG, GEO, RJG]

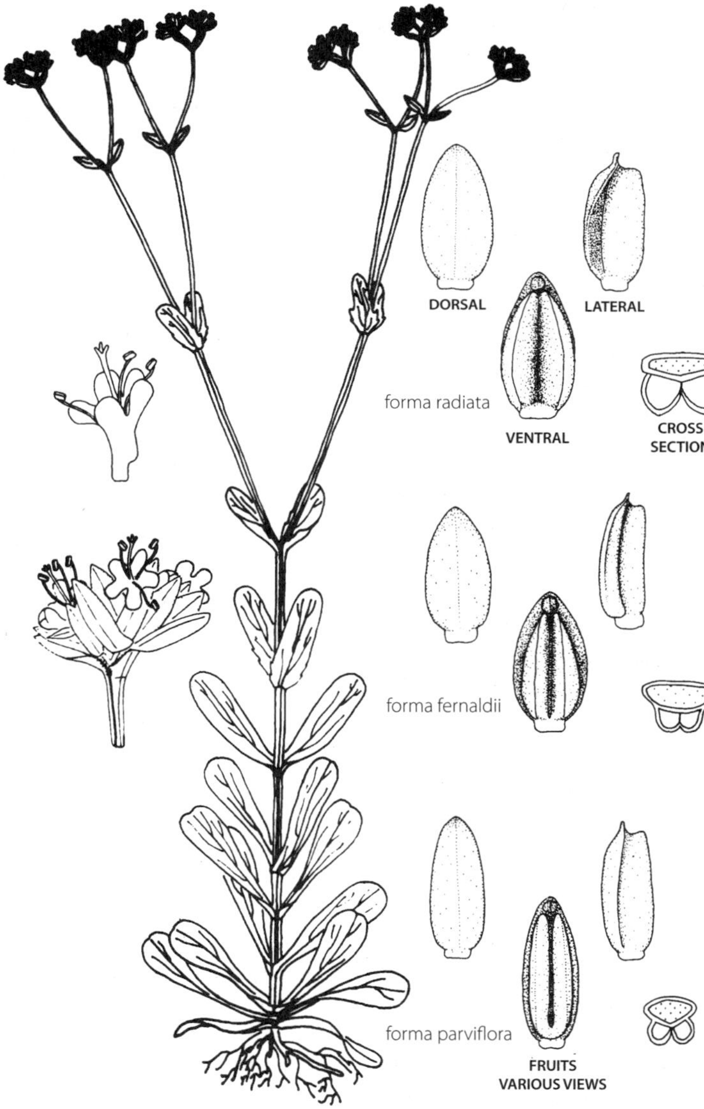

Valerianella radiata [EGG, SYS]

a fruit about as wide as deep (0.7–1.0 mm) and is pubescent with hooked hairs. Included based on presence in McCurtain Co., OK (Kartesz 2022); endemic to AR and OK. Apr–May.

CARICACEAE Dumort.

PAPAYA FAMILY

A small family of 6 genera and ca. 34 species (Carvalho 2013 onwards) of Mexico to South America, with one genus and 2 species in Africa; they are mostly small to medium sized, stout-stemmed trees or rarely herbs or vines; they are characterized by palmately-lobed or -compound leaves, latex, and large many-seeded berries. The family is most closely related to the Moringaceae and more distantly to other Brassicales (Stevens 2001 onwards; Holmes 2010; Carvalho 2013 onwards). The Caricaceae is thought to have originated in Africa and reached the New World by long-distance dispersal at ca. 35 million years ago (Carvalho 2013 onwards) (subclass Dilleniidae—Cronquist; order Brassicales—APG IV)

FAMILY RECOGNITION IN THE FIELD: The sole TX member of the family is an introduced tree with huge, very long-petiolate, shallowly to deeply palmately 5–9-lobed leaves confined to the top of the short usually unbranched trunk, with the tissues containing copious white latex.

REFERENCES: Badillo 1971, 2000; Holmes 2010; Carvalho & Renner 2012, 2013, 2014; Carvalho 2013 onwards; Carvalho et al. 2014.

CARICA L. PAPAYA

A monotypic genus native of Central America (Mexico to Costa Rica, the only place where the wild forms occur) (Carvalho 2013 onwards). It has long been thought that the closest relatives of *C. papaya* were the HIGHLAND PAPAYAS of South America (now treated in the genus *Vasconcellea*), and the genus has often been treated as having ca. 20 species (e.g., Holmes 2010). However, it is now considered to be monotypic (*C. papaya*) (e.g., Badillo 2000; Mabberley 2017). Recent research, including molecular studies, has shown that *Carica* is sister to a clade of four species endemic to Mexico and Guatemala, three herbs in the genus *Jarilla*, and a thin-stemmed tree with stinging hairs in *Horovitzia* (Carvalho & Renner 2012, 2013; Carvalho et al. 2014)—all four are morphologically quite different than *C. papaya*. (Name "Alluding to imagined resemblance of leaves or fruits to those of a fig, *Ficus carica*, erroneously thought to be from Caria in southwestern Asia Minor"—Holmes 2010)

REFERENCE: Mink et al. 2017.

Carica papaya L., (the Native American (Carib) name via Spanish), PAPAYA, LECHOSO, PAPAW, PAWPAW. Tree to 6+ m tall, usually unbranched, with soft wood and milky sap, fast-growing, usually dioecious; leaves alternate, clustered at apex of plant, evergreen, simple, with petioles 30–70(–105) cm long, blades 20–60 cm in diam., ± round in contour, shallowly to deeply palmately 5–9-lobed, the lobes lanceolate to ovate, with margins entire or usually pinnately subdivided into lobes; staminate inflorescences elongate, with numerous flowers; pistillate inflorescences with 1–several flowers; flowers unisexual or occasionally bisexual, the staminate and pistillate usually on different plants, 5-merous, sessile or subsessile; calyces 5-toothed; staminate flowers with corollas white, usually funnelform, 5-lobed, 2–3.5 cm long; stamens 10; ovary vestigial or absent; pistillate flowers with petals white to pale yellow, distinct or connate basally, 3–5 cm long; ovary (1–)5 carpellate; fruits large berries, hanging from trunk near summit, green to orange, ellipsoidal to oblong, slightly roundly 5-angled, with flesh yellow to orange or reddish; seeds numerous, brown to black. Escaped in disturbed areas, waste places, woodland openings; included based on a Lee County (*Dickschat 1* BAYLU—fide Mink et al. 2017) collection in the sw part of East TX; also naturalized in FL; native of Central America (Mexico to Costa Rica, the only place where the wild forms occur) (Carvalho 2013 onwards), but now cultivated throughout the tropics worldwide. Mink et al. (2017) indicated that two populations were found in Lee Co., with each population including a reproductively mature individual and a cluster of immature additional plants. They also noted that it is surprising that reproductively mature specimens of this frost-sensitive species can be found this far north. They consider these populations to be waifs and believe that "Permanent establishment of *Carica papaya* in the Texas flora seems unlikely based on frost-sensitivity. However, under particular cycles of unseasonably warm weather, periodic waifs, including reproductively mature specimens, could be encountered." Holmes (2010) also noted that "The species is cultivated in extreme southern Texas, where occasional seedlings arise from seeds discarded in waste places or dumps but are never persistent." This species is cultivated worldwide in the tropics for its large, sweet, melon-like fruits; the immature fruits, leaves, and flowers are also used in some places as a cooked vegetable. While most commercial fruits are smaller, some can become quite large (up to 9 kg or 20 lbs.). All parts of the plant, except the ripe fruit, have a milky latex containing the proteolytic enzyme papain, which is used to tenderize meat; even meat wrapped in the leaves becomes tender. Papaya is used in a variety of other ways including "in chewing gums, brewing, drugs for digestive ailments, treatment of gangrenous wounds, use in the textile industry, and use in production of soaps and shampoos" (Holmes

2010); it is also used in a variety of ways in traditional medicine (Duke 1985). The seeds have sometimes been used to treat intestinal parasites (Okeniyi et al. 2007).

CARYOPHYLLACEAE Juss.

PINK OR CARNATION FAMILY

Ours annual, biennial, or perennial herbs; stems often with swollen nodes; leaves opposite (–whorled), simple, entire; stipules scarious (= thin and papery) or absent; flowers solitary or inflorescences cymose, panicle-like, or capitate; sepals or tepals (4–)5; petals 0–5 (or more in cultivated forms), often notched ("pinked") at apex, frequently differentiated into claw and blade, sometimes with a crown of appendages (projections or scales) 2 per petal at junction of claw and blade; stamens (1–)5–10; pistil 1; ovary superior; placentation usually free-central or basal; fruit a capsule dehiscing apically by valves or teeth or a utricle.

A large (ca. 3,000 species in ca. 83–89 genera—Rabeler & Hartman 2005a), cosmopolitan, but especially temperate and warm n hemisphere family of mostly herbs or rarely shrubs or small trees. The number of genera may be considerably greater if a number of genera including *Silene* are subdivided as advocated by some Old World authors (Rabeler & Hartman 2006). The family includes ornamentals such as *Dianthus* (CARNATION, SWEET-WILLIAM), *Gypsophila* (BABY'S-BREATH), and *Silene* (CATCHFLY, CAMPION). It is unusual in its order, the Caryophyllales, in having anthocyanin rather than betalain pigments (Cronquist & Thorne 1994); however, molecular analyses link it with other members of the Caryophyllales (Giannasi et al. 1992; Downie & Palmer 1994), particularly Amaranthaceae and Chenopodiaceae, and indicate that the family is monophyletic (Downie et al. 1997; Rabeler & Hartman 2005a). The common name PINK is an old name probably referring to the notched or "pinked" petals (as in pinking shears). In the past, the family has been divided into 3–5 subfamilies (e.g., Bittrich 1993; Rabeler & Hartman 2005a). The subfamilies were found to be polyphyletic, and research that is more recent favors the family as best divided into several tribes (Harbaugh et al. 2010; Greenberg & Donoghue 2011). Family name conserved from *Caryophyllus* Mill., a genus now treated as *Dianthus* (the name *Dianthus* was published earlier and thus has priority in terms of nomenclature). In pre-Linnaean times, some authorities (e.g., Tournefort) referred to all PINKS as belonging to the genus *Caryophyllus*; Linnaeus, however, used the name for *Syzygium aromaticum* (L.) Merr. & L.M. Perry, CLOVES or CLOVE TREE, of the Myrtaceae; *Caryophyllus* L. has been rejected in favor of *Syzygium* (Farr et al. 1979). Linnaeus did use "caryophyllus" as an epithet for one of the familiar carnations also known as the CLOVE PINK (Bailey 1938). (Greek: *caryon*, nut, and *phyllon*, leaf, possibly in reference to the capsular or utricular fruit being subtended by bracts—R. Rabeler, pers. comm.) (subclass Caryophyllidae—Cronquist; order Caryophyllales—APG IV)

FAMILY RECOGNITION IN THE FIELD: herbs with opposite, simple, entire leaves, *swollen nodes*, and separate, often *notched petals* frequently with claw and blade; fruit often a toothed or valved capsule.

REFERENCES: Larsen 1986; Rabeler & Thieret 1988; Giannasi et al. 1992; Bittrich 1993; Behnke & Mabry 1994; Downie & Palmer 1994; Rabeler 2004; Rabeler & Hartman 2005a; Harbaugh et al. 2010; Wunderlin et al. 2017.

1. Flowers or several-flowered head-like clusters subtended by 2–3 pairs of conspicuous, broadly ovate, scarious (= dry, not green) bracts, ± equaling sepals; 2 introduced species ____ **Petrorhagia**
1. Flowers or inflorescences without bracts or bracts very different from the above (e.g., in *Dianthus* bracts linear or lance-linear, not scarious); widespread native and introduced species.
   2. Stipules present.
      3. Lower leaves usually in distinct whorls of 4 or apparently so, elliptic, ovate, or obovate; style 1 AND fruit a many-seeded capsule ____ **Polycarpon**
      3. Lower leaves not in whorls of 4, instead either opposite OR somewhat needle-like and crowded; styles EITHER 3–5 OR fruit a 1-seeded utricle.
         4. Petals absent or essentially so; stipules bristle-like to awl-like, lanceolate or ovate, 2–6 times as long as wide; sepals green to yellow, tan, brown, or red-brown, sometimes white apically.
            5. Sepals acuminate to acute or obtuse, with minute awn-point at apex, entire, green to yellow, tan, brown, or red-brown, sometimes white apically; fruit a single-seeded utricle; stipules awl-like to lanceolate or ovate; styles 1–2 ____ **Paronychia**
            5. Sepals rather abruptly narrowed to an awn-like tip about half their total length, the outer 3 with a distinct bristle-tooth on each side from about the middle, green; fruit a several-seeded capsule; stipules bristle-like; styles 3 ____ **Loeflingia**
         4. Petals present, white to pink or rosy; stipules triangular, not over 2 times as long as wide; sepals green to silvery.
            6. Leaves appearing whorled, seemingly many per node and crowded together due to the presence of 2 clusters of axillary leaves per node; styles and valves of fruits 5 ____ **Spergula**
            6. Leaves appearing opposite (axillary clusters usually absent); styles and valves of fruits 3 ____ **Spergularia**

2. Stipules absent.
  7. Sepals separate or united only at base; petals not differentiated into claw and blade OR petals absent.
    8. Petals absent; fruit a one-seeded utricle; styles 2 ____ **Scleranthus**
    8. Petals present or absent; fruit a few-many-seeded capsule; styles 3–5.
      9. Petals split more than half way to base (corollas apparently of 10 petals) ____ **Stellaria**
      9. Petals entire or divided half way to base or less or petals absent.
        10. Petals absent.
          11. Leaf blades 2–4 mm long; stems 4 cm or less long; capsules opening by 3 valves ____ **Geocarpon**
          11. Leaf blades 3–60 mm long; stems 2–70 cm long; capsules opening by 4–6 valves or 10 teeth.
            12. Leaf blades very narrow, < 1 mm wide, glabrous; capsules opening apically by 4–5 valves; calyces without a red band at base; styles equal in number to capsule valves ____ **Sagina**
            12. Leaf blades usually > 1 mm wide, glabrous or pubescent; capsules opening apically by 6–10 valves or teeth; calyces with a red band at base or not so; styles half as many as capsule valves or teeth.
              13. Capsules opening apically by 6 valves (opening to ca. middle of capsule or more, not curved; petal blades (if present) split > half way to base (corollas thus appearing to be of 10 petals) or petals absent; styles usually 3; plants ± glabrous or with inconspicuous pubescence, the pubescence if present not glandular ____ **Stellaria**
              13. Capsules opening apically usually by 10 teeth (opening only at very apex), usually curved; petal blades split < half way to base; styles usually 5; plants with obvious pubescence, the pubescence often glandular, particularly in the inflorescences ____ **Cerastium**
        10. Petals present (sometimes absent in *Arenaria*).
          14. Styles 3–4; petals 0.6–15 mm long; capsules opening apically by 3 or 6 valves or variable 6–8.
            15. Capsules opening apically by 6 valves or variable 6–8; petals 0.6–6 mm long, equal to or shorter than the sepals, entire at apex, acute or rounded or split ca. 1/3 to base.
              16. Styles 3; capsules opening apically by 6 valves; petals (when present) shorter than sepals, entire at apex, acute to rounded ____ **Arenaria**
              16. Styles 3–4; capsules opening by 6–8 valves; petals equal to sepals, split ca. 1/3 to base ____ **Cerastium**
            15. Capsules opening apically by 3 valves; petals 5–15 mm long, longer than the sepals (1.3–3 times as long), broadly notched or entire at apex.
              17. Plants annual; sepals obscurely veined; leaf blades 2–4 mm wide ____ **Geocarpon**
              17. Plants annual or perennial; sepals prominently 3- or 5-ribbed; leaf blades 0.5–3 mm wide ____ **Sabulina**
          14. Styles 5 (occasionally 4); petals ca. 1–7 mm long; capsules opening apically by 10 teeth or 4–5 valves.
            18. Petals toothed or notched at apex; leaf blades ± pubescent, oblong or lanceolate to elliptic or obovate; capsules 5–17 mm long, usually curved, opening apically by 10 teeth ____ **Cerastium**
            18. Petals entire; leaf blades glabrous, narrowly linear; capsules 2–3.5 mm long, not curved, opening apically by 4–5 valves ____ **Sagina**
  7. Sepals united 1/3 or more (may have separate bracts outside), often forming a campanulate, funnelform, cylindric or inflated calyx tube; petals often, but not always, clawed; petals sometimes absent.
    19. Plant diminutive, the stems 4 cm or less long and the leaf blades 4 mm or less long; petals absent; flowers sessile ____ **Geocarpon**
    19. Plants not diminutive, the stems and leaf blades larger; petals present, variously colored; flowers sessile to long-stalked.
      20. Flowers in dense, often head-like cymes; calyces each with 2–several bracts at base; petals white to variously colored, often minutely dotted with white ____ **Dianthus**
      20. Flowers in very open to congested cymes or flowers solitary and axillary; calyces without bracts at base; petals white to variously colored but if colored not minutely dotted with white.
        21. Flowers solitary at the tips of branches or subaxillary, on elongate pedicels; calyx lobes (12–)15–45 mm long, longer than calyx tube; styles (4–)5; petals usually purplish red varying to pink or rarely white, (15–)24–36(–40) mm long including claw, without a crown of projections or scales on upper (inner) side at junction of blade and claw ____ **Agrostemma**
        21. Flowers in very open to congested cymes; calyx lobes 0.5–5 mm long (longer in *Silene conoidea*), shorter or ca. as long as calyx tube; styles 2–3(–4); petals white to pinkish, lavender-pink, purplish, or scarlet, 5–44 mm long including claw, with OR without a crown.
          22. Stem leaves not clasping; calyces 10–30-nerved, neither angled nor winged; petals usually with a crown of projections or scales on upper (inner) side at junction of blade and claw (crown sometimes quite small or absent); plants annual or perennial.
            23. Styles 3(–4); inflorescences open; petals white to pink, rose-lavender, red-violet or scarlet; capsules opening apically by 3 or 6 teeth; calyx variable, in some species as little as 5–9 mm long, while in others as long as 20–30 mm ____ **Silene**

23. Styles usually 2; inflorescences usually congested; flowers white to pink (can dry dull purple); capsules opening apically by 4(–6) teeth; calyx 15–25 mm long ____ **Saponaria**

22. Stem leaves clasping or connate; calyces 5-nerved, strongly 5-angled or winged; petals without a crown; plants annual ____ **Gypsophila**

AGROSTEMMA L. CORN-COCKLE

A genus of 2 species of temperate Eurasia, but now introduced worldwide (Thieret 2005b). (Greek: *agros*, field, and *stemma*, a crown or garland, Linnaeus apparently believing it was suitable for such a use)

REFERENCES: Firbank 1988; Thieret 2005b; Weakley et al. 2012.

Agrostemma githago L., (an old generic name for *Agrostemma*), COMMON CORN-COCKLE. Erect taprooted annual 0.3–1 m tall, stems simple or few-branched, densely pubescent with long, appressed to spreading, non-glandular hairs; leaf blades linear to lanceolate, 4–12(–15) cm long, 1–10 mm wide, entire, sessile, connate at base; stipules absent; flowers usually solitary at branch tips, or axillary, on elongate peduncles to 20 cm long; calyx tube 10-ribbed, 10–18 mm long; calyx lobes (12–)15–45 mm long, longer than the tube, linear or lance-linear, often exceeding the petals; petals (15–)24–36(–40) mm long including claw, usually purplish red varying to pink (rarely white), with small black dots along veins near base; stamens 10; capsules 16–22 mm long, opening by (4–)5 teeth, many-seeded. Disturbed areas; Camp (*Jones & Jones 2809* HPC), Denton (*LaLonde s.n.*, 1924, TCSW), Henderson (*Stanfield s.n.*, 1900, NY), Titus (*Amerson 397* SMU), Upshur (*Jones & Jones 4583* SAT, *Shinners 14395* SMU), Walker (Kartesz 2022), and Grayson (*Rex Elmore*? (handwriting undecipherable) collected for *Gentry 1883* BRIT) cos.; otherwise known in TX only from Parker Co. (*O'Kennon s.n.*, 1998, BRIT) to the w of East TX; nearly throughout the U.S. and s Canada. Mar–Jul. Native of Europe. [*Lychnis githago* (L.) Scop.] According to Thieret (2005b), this species was formerly a common weed of grain fields, but is now becoming increasingly rare due to the mechanical screening of grain and modern herbicides. The seeds are poisonous to humans, livestock, and birds due to saponins (Thieret 2005b; Mabberley 2017). The species is sometimes cultivated as an ornamental.

ARENARIA L. SANDWORT, CHICKWEED

Our species small annuals or perennials; leaves sessile or short petiolate, often connate; stipules absent; flowers in terminal, open, leafy cymes or solitary, axillary; petals white, entire, shorter or slightly longer than the sepals; styles 3; fruit a capsule opening by 6 valves.

A genus of ca. 210 species (Hartman et al. 2005b), well represented in the n temperate area, but also more widespread (e.g., Central and South America); it includes some cultivated ornamentals. (Latin: from *arena*, sand, in which many of the species grow)

REFERENCES: Maguire 1947, 1951; McNeill 1962, 1980; Shinners 1962c; Wofford 1981; Hartman et al. 2005b.

1. Plants perennial, often with slender rhizomes; stems prostrate to trailing; flowers solitary in leaf axils ____ **A. lanuginosa**
1. Plants annual, without rhizomes; stems erect to ascending or sprawling; flowers in terminal, open, leafy, 3–numerous-flowered inflorescences.
   2. Sepals glabrous, 1-veined; leaf blades 5–15(–20) mm long; pedicels 2–15 mm long in flower, up to 40 mm in fruit; on limestone on w margin of East TX ____ **A. benthamii**
   2. Sepals minutely scabrous-pubescent or stipitate-glandular, 3-veined; leaf blades 2–7 mm long; pedicels 1–3 mm long in flower, up to 12 mm in fruit; in sand scattered in East TX ____ **A. serpyllifolia**

Arenaria benthamii Fenzl ex Torr. & A. Gray, (for George Bentham, 1800–1884, English taxonomist and president of Linnaean Society), HILLY SANDWORT. Annual to 50 cm tall; leaf blades oblong-lanceolate or oblanceolate to elliptic-lanceolate, to 5–15(–20) mm long, usually smaller; petals 1.5–3 mm long; capsules 3–4 mm long. Limestone outcrops, usually in shade of shrubs or small trees; w margin of East TX; mainly Edwards Plateau, otherwise scattered in w 1/2 of TX; LA, NM, OK, and TX; adjacent Mexico. Late Mar–May.

Arenaria lanuginosa (Michx.) Rohrb., (downy or woolly), SPREADING SANDWORT. Perennial, often rhizomatous, stems prostrate to trailing; leaf blades linear-lanceolate to narrowly elliptic or oblanceolate, 3–35 mm long; petals shorter than sepals or absent; capsules 3–6 mm long. Sandy or limestone areas; in East TX known only from Newton Co. (*Allen & Early 21744* BRIT); se U.S. from VA s to FL w to TX. Spring–summer. Hartman et al. (2005b) recognize two varieties in this species, var. *lanuginosa* (occurring in the se US) and var. *saxosa* (A. Gray) Zarrucchi, R.L. Hartm., & Rabeler (occurring in the sw U.S. including the Trans-Pecos of TX).

Arenaria serpyllifolia L., (with leaves like *Thymus serpyllum*—wild thyme), THYME-LEAF SANDWORT. Annual to 20(–40) cm tall; leaf blades elliptic or ovate, to 7 mm long; petals 0.6–2.7 mm long; capsules 3–3.5 mm long, ovoid to cylindric-ovoid, broader basally; $2n = 40$. Sandy roadsides, waste areas, or stream bottoms; widely scattered in East TX; also adjacent Cross Timbers and Prairies; nearly throughout the U.S., but concentrated in the e 1/2; s Canada. Late Mar–May. Native of Eurasia and n Africa. The plants occurring in TX are var. *serpyllifolia*; var. *tenuior* Mert. & W.D.J. Koch, which has capsules cylindric or nearly so (not significantly broader at base), is known from the e U.S. (including LA) and differs in chromosome number ($2n = 20$) (Hartman et al. 2005b).

CERASTIUM L. MOUSE-EAR CHICKWEED, CHICKWEED

Small annuals or perennials, with obvious pubescence, the hairs glandular or eglandular; stipules absent; sepals separate; petals white, notched (the blades notched < 1/2 to base), or petals sometimes absent; stamens usually 10; styles usually 5; capsules cylindrical or nearly so (except ovoid in *C. parvum*), usually curved, exceeding the calyces in age, scarious (= thin and papery), with 10 prominent teeth when opened (6–8 valves in *C. parvum*), the whole suggesting a corolla.

An almost cosmopolitan, but especially n temperate genus of ca. 100 species (Bittrich 1993; Morton 2005a; Mabberley 2017) or as many as 214 (Stevens 2001 onwards) including some cultivated annuals and many weeds. (Greek: *cerastes*, horned, alluding to the shape of the slender and often curved fruit)

REFERENCES: Volponi 1983; Good 1984; Landry et al. 1988, 1989; Turner 1995d [1996]; Rabeler & Reznicek 1997; Morton 2005a; Sharples & Tripp 2019.

1. Capsule 1–2 times as long as the sepals; petals if present shorter than to 1.5 times as long as sepals (sometimes withering very early); sepal pubescence of dense, glandular or non-glandular hairs (sparsely so to glabrous in *C. parvum*), these long (some ca. 1/3 or more as long as sepal width).
   2. Flowers solitary in axils of distal nodes; capsules ovoid; bracts absent ____ **C. parvum**
   2. Flowers in bracteate cymes (= clusters); capsules cylindric.
      3. Sepals with long, spreading hairs up to and projecting beyond the apex; upper bracts of inflorescence entirely herbaceous, without scarious (= thin and papery) margins; inflorescence or its main divisions remaining compact OR open and diffuse.
         4. Inflorescence or its main divisions remaining compact, the pedicels mostly shorter than the sepals or capsules, 0.5–3(–5) mm long; stems with mostly glandular hairs, the hairs neither long nor shiny-silvery; petal and filament bases not ciliate ____ **C. glomeratum**
         4. Inflorescence or its main divisions open and diffuse, the pedicels once to twice as long as the sepals or capsules, 5–12 mm long; stems usually with long, shiny-silvery, mostly non-glandular hairs; petals and filament bases sparsely ciliate ____ **C. brachypetalum**
      3. Sepals with subappressed to spreading hairs becoming shorter and stopping just below apex; upper bracts of inflorescence with scarious margins; inflorescence open and diffuse in age.
         5. Petals ca. as long as sepal, deeply notched (to 1/2 way to base); sepals pilose, seldom glandular-hairy, 5–7 mm long; stamens usually 10; capsules $\geq$ 9 mm long; plants perennial, often with non-flowering basal shoots ____ **C. fontanum**
         5. Petals $\leq$ sepal length, shallowly notched (to 1/4 way to base); sepals glandular-hairy, 3–5 mm long; stamens 5; capsules $\leq$ 9 mm long; plants annual.
            6. Lower bracts usually foliaceous (= green); sepals with narrow scarious margins; petals equal to sepals; capsule twice as long as sepals; inflorescence 3–15-flowered (rarely more) ____ **C. pumilum**
            6. Lower bracts with wide scarious margins; sepals with wide scarious margins; petals shorter than sepals; capsules 1.5 to 2 times as long as sepals; inflorescence 3–30-flowered ____ **C. semidecandrum**
1. Capsule 2–3 times as long as the sepals; petals often 1.5–2 times as long as the sepals; sepal pubescence of sparse, glandular hairs, these short (much < 1/3 as long as sepal width).
   7. Leaf blades usually 30 mm or less long; pedicels often reflexed at base, usually equaling or shorter than the capsules in fruit, 0.5–1.25 times the length of the flowering calyces, to 3 times the calyx length at maturity (ca. 2–13 mm long); widespread in w 1/2 of East TX ____ **C. brachypodum**
   7. Leaf blades often > 30 mm long; pedicels sharply curved or hooked just below the calyx at maturity, usually much longer than the capsules in fruit, 1–3 times the length of the flowering calyces, to 5 times or more the calyx length at maturity (longer pedicels 10–35 mm long); only in 2 cos. of East TX ____ **C. nutans**

Cerastium brachypetalum Pers., (short-petaled), GRAY MOUSE-EAR CHICKWEED, GRAY CHICKWEED. Annual, stems erect, usually covered with long, shiny-silvery, mostly non-glandular hairs (giving a grayish appearance); leaf blades spatulate, elliptic, or ovate; tip of pedicel often bent in fruit; stamens 10. Roadside ditches; a Red River Co. collection (*Rabeler 1333* MICH?) made in April of 1998 is the first report for TX, also Madison (Kartesz 2022), and Waller (*Brown 31917* BRIT) cos., also Parker Co. (*Nesom et al. DB-21* BRIT) just w of East TX; the species should be expected, but is currently not known, from elsewhere in TX; e U.S. from VA s to FL and w to KS and TX, also OR and WA. Apr. Native of Europe.

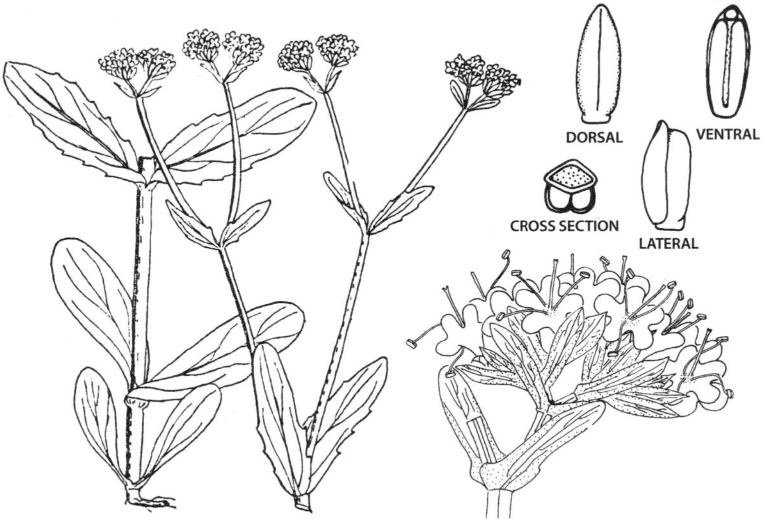

Valerianella stenocarpa [EGG]

Agrostemma githago [GLE]

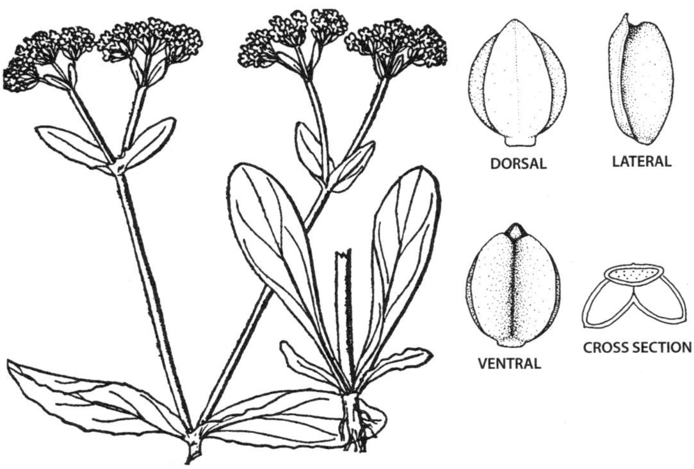

Valerianella woodsiana [BB2, SYS]

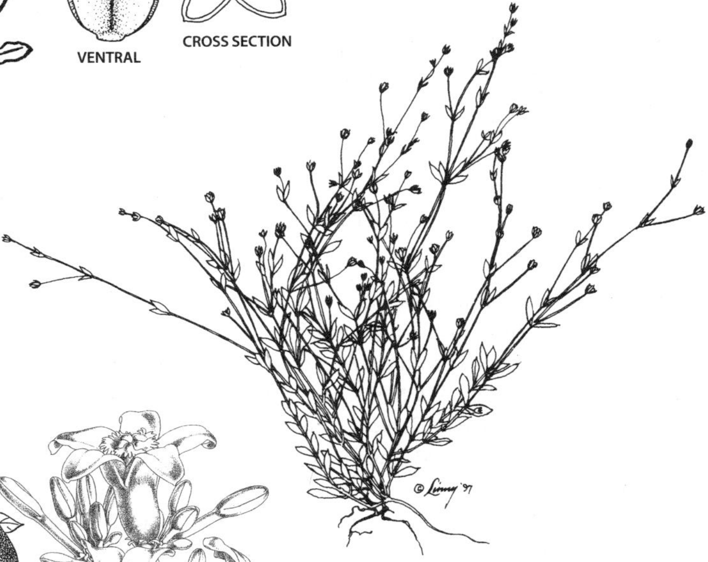

Arenaria benthamii [HEA]

Carica papaya [JLS, SA2]

Cerastium brachypodum (Engelm. ex A. Gray) B.L. Rob., (short-stalked), SHORT-STALK MOUSE-EAR CHICKWEED, SHORT-STALK CHICKWEED. Annual, stems erect, glandular-pubescent with hairs spreading at right angles; leaf blades glabrous to rather densely viscid-pubescent; pedicels usually 10 mm or less long, reflexed near base before and after flowering, erect or ascending in flower and again in mature fruit, not sharply curved or hooked just below calyx; sepals 3–4.5 mm long; petals 3–4 mm long; stamens 10; capsule 6–12 mm long. Prairies, disturbed sites; widespread in w 1/2 of East TX; mainly East TX and adjacent Cross Timbers and Prairies and Edwards Plateau; throughout much of the U.S.; s Canada. Mar–early Apr. [*C. nutans* Raf. var. *brachypodum* Engelm. ex A. Gray] Native of the U.S.

Cerastium fontanum Baumg. subsp. **vulgare** (Hartm.) Greuter & Burdet, (sp.: pertaining to springs or fountains; subsp.: common), COMMON MOUSE-EAR CHICKWEED, COMMON MOUSE-EAR. Apparently annual in East TX, though perennial farther n (Larsen 1986; R. Rabeler, pers. comm.), with eglandular or glandular pubescence, stems erect from decumbent base; leaf blades oblong-elliptic to lanceolate or oblanceolate; pedicels curved distally; petals ca. as long as sepals, deeply notched; stamens 10, occasionally 5. Sandy roadsides, disturbed sites; scattered in East TX, also Brewster, Erath, and San Saba (Kartesz 2022) cos.; Tarrant Co. (R. O'Kennon, pers. obs.); throughout the U.S. and Canada. Apr–May. [*C. holosteoides* of authors, not Fr., *C. triviale* Link, *C. vulgare* Hartm., *C. vulgatum* L. var. *hirsutum* Fr.] Native of Europe; introduced nearly worldwide. This species long went under the name *C. vulgatum*; however, that name is ambiguous and has been officially rejected (Turland & Jackson 1997; Rabeler & Hartman 2006).

Cerastium glomeratum Thuill., (glomerate, clustered), STICKY MOUSE-EAR CHICKWEED. Glandular-pilose annual, erect or ascending; leaf blades orbicular ovate to obovate; inflorescence usually dense; pedicels sometimes slightly curved; petals usually shorter than sepals; stamens 10. Roadsides, disturbed sites; widespread in e 1/2 of TX; widespread in the U.S. and Canada. Mar–early Apr. [*C. viscosum* of authors, not L., an ambiguous and officially rejected name—Turland & Jackson 1997; Rabeler & Hartman 2006] Native of Europe.

Cerastium nutans Raf., (nodding), NODDING MOUSE-EAR CHICKWEED, POWDERHORN CHICKWEED, NODDING CHICKWEED. Erect annual, stems finely glandular-pubescent; pedicels 5–20(–35) mm long, sharply curved or hooked just below calyx; stamens 10. Moist or rich woods, open areas; in East TX known only from Burleson and Waller (Kartesz 2022) cos., also Tarrant Co. (Kartesz 2022); widespread in the U.S. except the sw; s Canada. Apr–Jun. Native of the U.S.

Cerastium parvum (Pederson) M.T. Sharples & E.A. Tripp, (small), SMALL STARWORT. Mat-forming annual or perennial, much branched, stems creeping, rooting at the lower nodes, glabrous or sparsely pubescent on 2 sides; leaves sessile or nearly so, blades elliptic to narrowly obovate or spatulate, basally cuneate; flowers solitary, axillary; sepals 2.5–3 mm long; petals equaling sepals, rarely absent; stamens 4–5; capsules longer than sepals; seeds brown, tuberculate, taller than wide; 2n = 34 (Morton 2005b). Open disturbed sandy sites, wet ditches, wet areas in lawns; Liberty (Trinity River National Wildlife Refuge, *Brown & Marcus 20104, 20216* SBSC), Harris, Montgomery, and Newton (Kartesz 2022) cos.; AL, LA, FL and TX. Spring–fall. [*Stellaria parva* Pedersen] Transferred from *S. parva* in 2019 (Sharples & Tripp 2019). Native of South America. This species was reported as new to North America (in LA) by Landry et al. (1988) and further details about its occurrence were given by Landry et al. (1989). It is easily distinguished from all other *Cerastium* species in TX by it solitary, axillary flowers (versus 3–many-flowered cymes).

Cerastium pumilum Curtis, (dwarf), DWARF MOUSE-EAR CHICKWEED, CURTIS' MOUSE-EAR CHICKWEED, STICKY MOUSE-EAR CHICKWEED. Annual resembling a small plant of *C. fontanum*, with glandular hairs on the inflorescence, bracts, and sepals; stems erect or ascending; pedicels curved below the sepals; petals shallowly notched; stamens 5. Roadside ditches and open disturbed areas; a Kaufman Co. collection (*Reznicek 10336*, BRCH, MICH) was the first report for TX (Rabeler & Reznicek 1997); in April of 1998 the species was found in Cooke, Denton, Fannin, Grayson, Lamar, and Red River cos. (e.g., *Rabeler & Diggs 1321* MICH?), also Henderson, Polk, San Jacinto, and Titus (Kartesz 2022) cos., also Cooke Co. (Kartesz 2022) to the nw of East TX; this is an easily overlooked species; it is to be expected in other parts of East TX and possibly elsewhere in the state; widespread in e 1/2 of the U.S. and adjacent Canada, also ID, OR and WA. Apr–May. Native of Europe and sw Asia. [*C. glutinosum* Fr.] Morton (2005a) noted that some forms of this species could be confused with *C. glomeratum*, but that the two could be distinguished by stamen number (10 in *C. glomeratum*).

Cerastium semidecandrum L., (having 5 instead of 10 anthers), FIVE-STAMEN MOUSE-EAR CHICKWEED. Annual with slender taproot, stems erect or ascending, branching at base to 20 cm long, viscid; leaf blades oblanceolate, ovate to elliptic-oblong; pedicels curved distally, 1 to 3 times as long as capsule; petals shorter than sepals; stamens 5. Weedy in parks, vacant lots, roadsides; in TX known only from Hopkins Co. (White & Singhurst 2012); scattered in the e 1/2 of the U.S., ID, OR, and WA; Canada. Spring. Native of Eurasia, Africa.

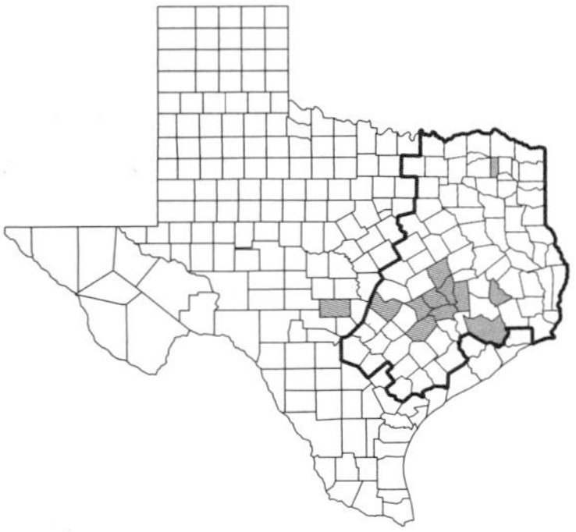

Valerianella florifera

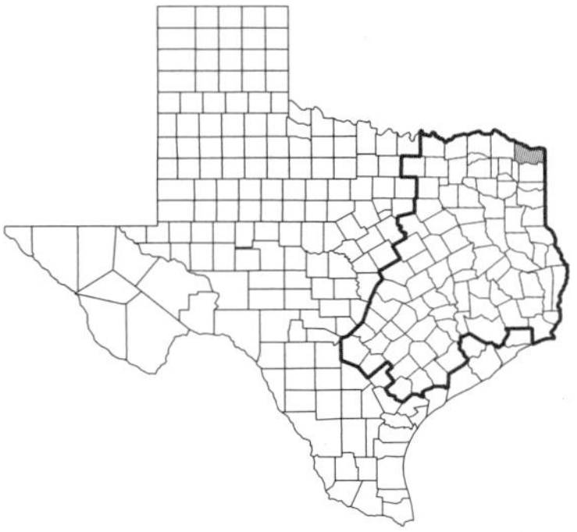

Valerianella locusta

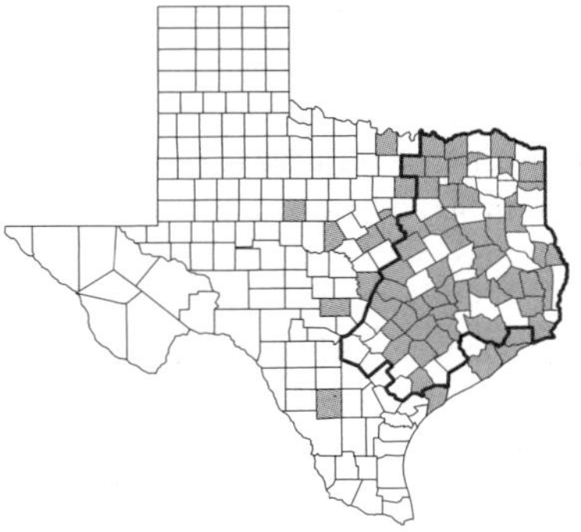

Valerianella radiata

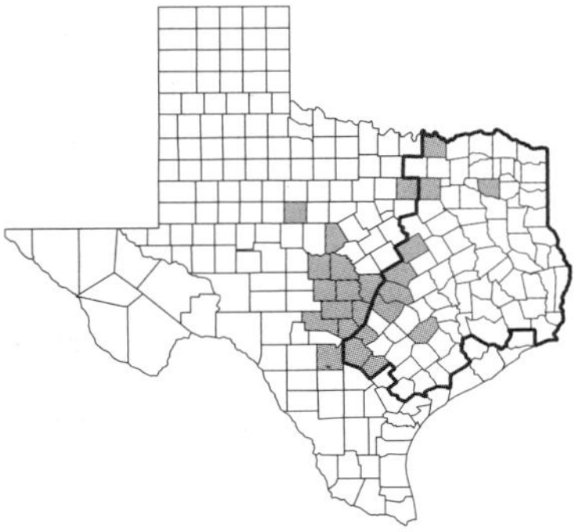

Valerianella stenocarpa

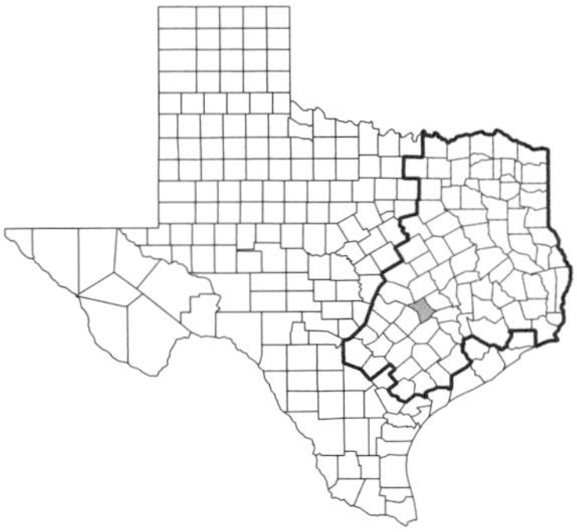

Carica papaya

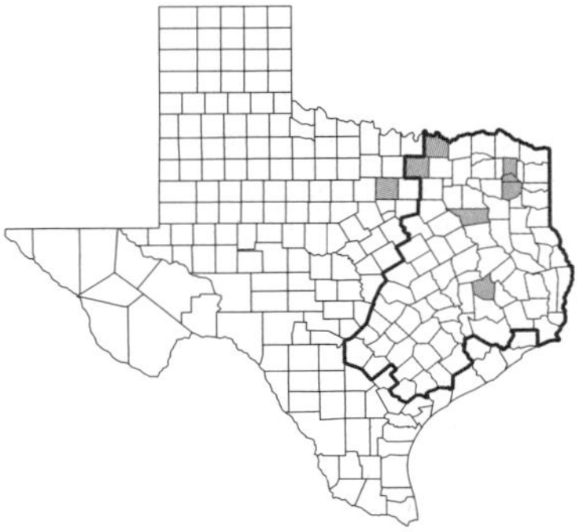

Agrostemma githago

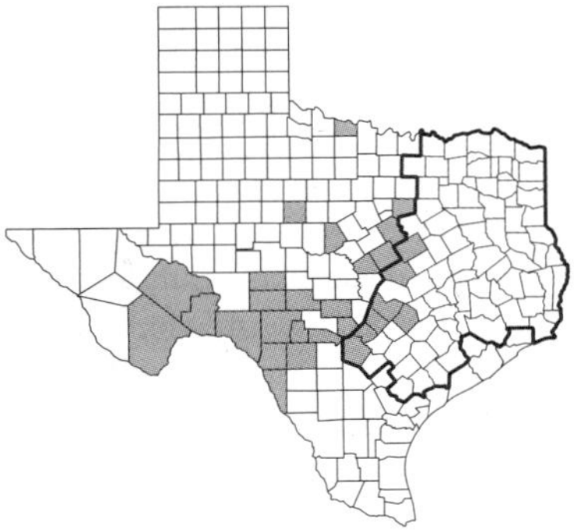

Arenaria benthamii

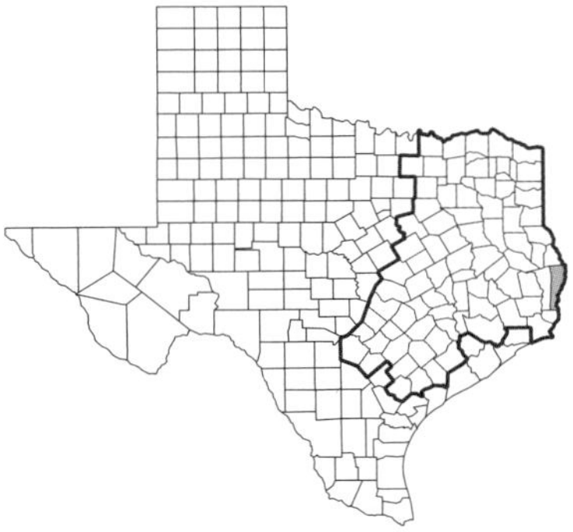

Arenaria lanuginosa

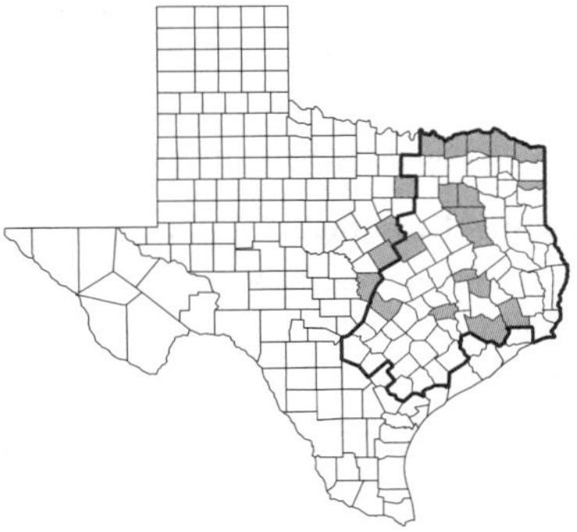

Arenaria serpyllifolia

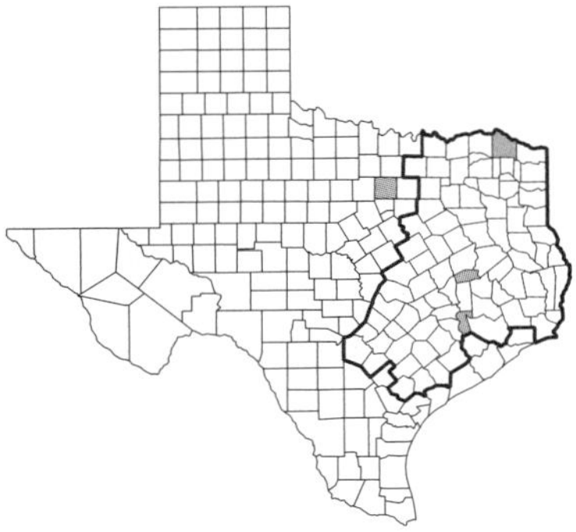

Cerastium brachypetalum

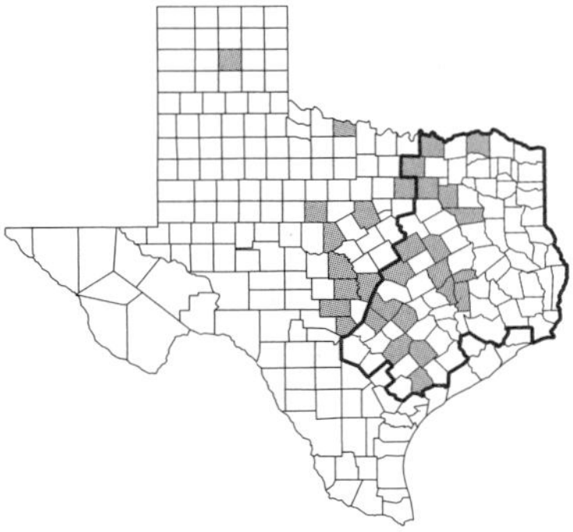

Cerastium brachypodum

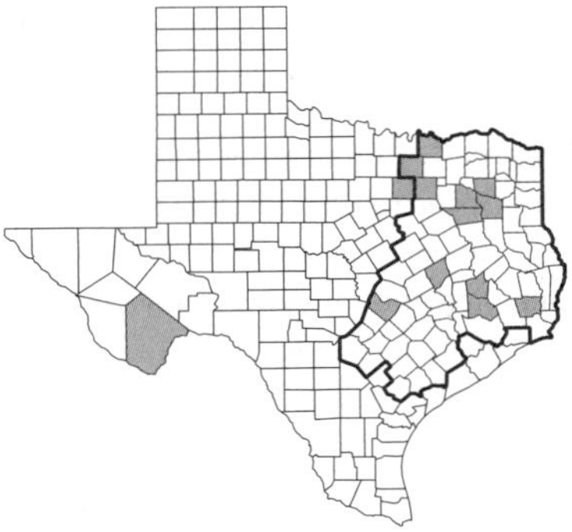

Cerastium fontanum subsp. vulgare

Holosteum umbellatum L., (genus: all bone, from Greek: *holos*, whole and *osteon*, bone; sp.: having an umbel), JAGGED CHICKWEED. *Holosteum umbellatum* will most likely key out to *Cerastium*. Plant with a distinct basal rosette, inflorescence an umbel. There are no other species of Caryophyllaceae in East TX that have an umbellate inflorescence. Also differs from any other species of *Arenaria* in having clawed petals. Recorded for Marshall Co., OK, n of Grayson Co. in East TX; e U.S. from NY s to GA and w to SD and s to TX (Trans-Pecos), also nw U.S.; Canada. Mar–May. Native of Europe, Asia and Africa.

DIANTHUS L. PINK, CARNATION

Annual to perennial herbs; leaves connate basally, forming a short sheath; stipules absent; flowers in ours subsessile, in dense terminal cymes; sepals connate below into a cylindrical tube, the tube longer than the teeth, subtended by 2–several, linear or lance-linear bracts; petals 5, long clawed, without a crown; styles 2; capsules opening by 4 teeth.

A Eurasian and n North American to s African genus of ca. 320 species (Rabeler & Hartman 2005b) with many cultivated ornamentals including CARNATIONS (*D. caryophyllus* L., *D. ×allwoodii* Hort. Allwood); CARNATION has religious significance extending back beyond the Dark Ages (Baumgardt 1982). The genus is the second largest in the family, exceeded only by *Silene*, but is in need of thorough study (Rabeler & Hartman 2005b). It diversified faster than any other plant (2.2–7.6 species per million years (Mabberley 2017). The common name PINK is an old name probably referring to the notched or "pinked" petals (as in pinking shears). (Greek: *dios*, divine, Jove, Jupiter, or Zeus, chief of the Greek gods, and *anthos*, flower; Jove's flower or divine flower, from beauty or fragrance of flowers)
REFERENCE: Rabeler & Hartman 2005b.

1. Leaf blades 2–8 mm wide, linear; stems usually pubescent to almost glabrous; plants annual or biennial; cymes dense, but not head-like, usually of 3–6 flowers ____ **D. armeria**
1. Leaf blades 10–18 mm wide, lanceolate to oblanceolate; stems glabrous; plants perennial; cymes usually head-like, usually of 5–20 flowers ____ **D. barbatus**

Dianthus armeria L., (the Latin name for a kind of *Dianthus*), DEPTFORD PINK. Plant usually 20–80 cm tall, usually dichotomously branched above; leaves grass-like, olive-green; calyx tube 10–15 mm long, lobes 3–6 mm long, pubescent, closely subtended by elongate bracts nearly or ca. equaling the calyx; petals pink to reddish or purplish, dotted with white, drying purplish, long clawed, the blades 4–5 mm long. Disturbed areas; Grayson (Diggs et al. 1999), Titus (*Ajilvsgi 7286* SMU), and Upshur (*Jones & Jones 4579* VDB) cos.; also Tarrant Co. (Diggs et al. 1999); in TX apparently only in nc and ne parts of state; widespread in the U.S. and s Canada. May–Aug. Native of Eurasia. This species was first reported for TX by Lipscomb (1984). It is occasionally cultivated (Rabeler & Hartman 2005b).

Dianthus barbatus L., (barbed, bearded), SWEET-WILLIAM. Plant usually 20–60 cm tall, glabrous; calyx tube 10–12 mm long, lobes 2–7 mm long, glabrous (the lobes can be ciliate), closely subtended by elongate bracts ca. equaling or slightly longer than the calyx; bracteoles 4 (or 6), 0.75–1.25 times as long as calyx, apex long aristate in distal 1/3(–1/2); petals whitish to pink, purplish, or dark red, often spotted, with or without dark center. Cultivated and possibly escapes; a Hunt Co. collection at TAES which is questionably identified as *D. barbatus* (R. Rabeler, pers. comm.), Erath (Kartesz 2022), Henderson (*Holmes 9438* BAYLU), and Smith (*Ricketts 114* BRIT) cos.; widespread in the e 1/2 of the U.S. and s Canada, scattered in the w U.S. Jun–Aug. Native of Eurasia.

GEOCARPON Mack. EARTH FRUIT, TINY-TIM

Small to very small annual or winter annual herbs; leaf blades 1-veined; flowers open cymes or solitary in axils, pedicellate or not so; petals present or absent; stamens 5 or 10; styles 3; fruit an ovoid or broadly ellipsoid capsule with 3 valves.

A genus of 7 annual or perennial herbaceous species centered in e North America (Schilling et al. 2022). The monotypic genus, *Geocarpon*, was originally placed in the Aizoaceae and subsequently treated in that family by many authors (e.g., Fernald 1950). However, a variety of characters (e.g., exstipulate leaves, connate sepals, free central placentation, valvular fruit, anthocyanin pigments) support its inclusion in the Caryophyllaceae (Palmer & Steyermark 1950; Bogle et al. 1971; Nienaber 2005). Dillenberger and Kadereit (2014) transferred *Geocarpon minimum* Mack. to *Mononeuria minima* (Mack.) Dillenb. & Kadereit. Schilling et al. (2022) further transferred *M. minima* back to *G. minimum* (formerly a monotypic genus). In their study, Schilling et al. (2022), transferred other species to *Geocarpon* (a total of 7 species), thus *Geocarpon* is no longer a monotypic genus. Additionally, they transferred some taxa in *Mononeuria* to *Sabulina* (see *Sabulina* for further discussion). (Greek: *ge*, earth, and *carpos*, fruit)
REFERENCES: Shinners 1949d; Palmer & Steyermark 1950; Steyermark et al. 1959; Bogle et al. 1971; Keith et al. 2004; Nienaber 2005; Schilling et al. 2022)

Arenaria lanuginosa [MAR]

Arenaria serpyllifolia [GEO]

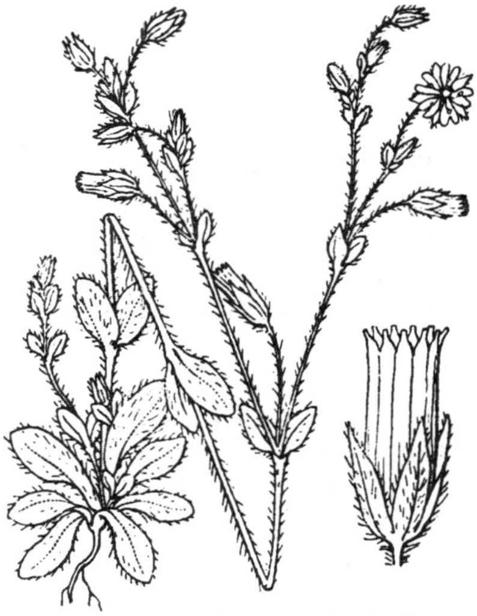

Cerastium brachypetalum [COS]

Cerastium brachypodum [BB2]

Cerastium fontanum subsp. vulgare [REE]

Cerastium glomeratum [BB2, GLE]

Cerastium nutans [GR1]

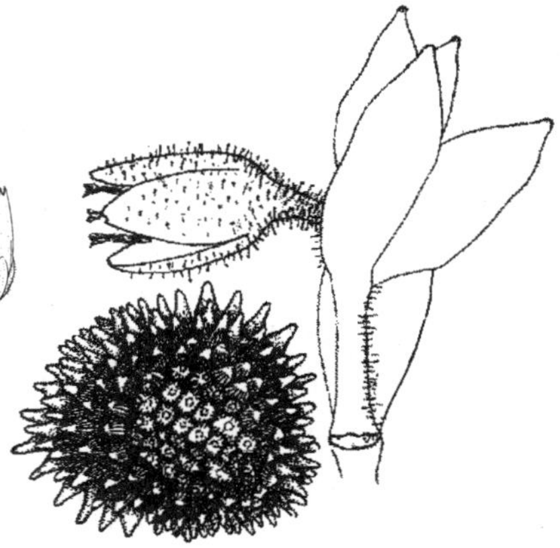

Cerastium parvum [BTK]

1. Diminutive herb; stems often reddish, 1–4 cm long; leaf blades 1–4 mm long; flowers solitary in axils; sepals 3–3.5 mm long; petals absent ____ **G. minimum**
1. Larger herb; stems to 20 cm long; leaf blades to 35 mm long; flowers in 7–12 flowered cymes; sepals 3–6 mm long; petals to 15 mm long ____ **G. nuttallii**

Geocarpon minimum Mack., (least, smallest), EARTH FRUIT, TINY-TIM. Diminutive winter annual, glabrous, stems simple or few-branched, 1–4 cm long, often reddish or purplish; leaves sessile, blades 1–4 mm long, narrowly oblong or ovate-oblong, entire, apically acute, connate at base forming a sheath; stipules absent; flowers 3–4 mm long, axillary, single at alternate nodes, sessile, with a cup-shaped hypanthium; sepals 5, greenish or reddish with white margins, fused at the base (actually the flowers are perigynous and what is apparently the fused portion of the "calyx" is actually a hypanthium—therefore, in order to avoid confusion, in the key to genera, *Geocarpon* can be reached in two ways); petals absent; stamens 5; staminodes 5, scale-like; styles 3; capsules opening by 3 apical valves; seeds usually ca. 30. Barren or shallow, mineral soils (saline or alkaline prairies or barrens, sandstone glades); in a saline prairie just above the floodplain of the Neches River in Anderson Co. (*Singhurst et al. 12921* BAYLU, TEX), where only ca. 200 plants are known (Keith et al. 2004), Harrison and Panola cos. (Kartesz 2022); endemic to sc US (AR, LA, MO, OK, and TX). Late winter–early spring. [*Mononeuria minima* (Mack.) Dillenb. & Kadereit] Listed as federally threatened, and is of conservation concern (Nienaber 2005). *Geocarpon minimum* was reported new to OK (Osage Co) in 2023 (Buthod & Miller 2023). NatureServe: G2, imperiled.

Geocarpon nuttallii (Torr. & A. Gray) E.E. Schill., (for Sir Thomas Nuttall, 1786–1859, English-American botanist), DRUMMOND'S STITCHWORT. Simple or sparingly branched annual to ca. 20 cm tall, stems and inflorescences mostly glandular-pubescent; leaf blades to 35 mm long, 2–4(–7) mm wide, linear-oblong to oblanceolate or narrowly ovate; pedicels 10–25 mm long, reflexing in fruit; flowers in dichotomous cymes, showy; petals to 15 mm long. Sandy or sandy clay soils, roadsides, fencerows, or waste areas; widespread in East TX; also n Gulf Prairies and Marshes; endemic to AR, LA, OK, and TX. Mar–May. [*Arenaria drummondii* Shinners, *Minuartia drummondii* (Shinners) McNeill, *Mononeuria nuttallii* (Torr. & A. Gray) Dillenb. & Kadereit]

GYPSOPHILA L. BABY'S-BREATH

A genus of ca. 150 species (Weakley, A.S., and Southeastern Flora Team 2024) native to Eurasia and the Mediterranean, but widely introduced elsewhere. It has been included in *Vaccaria* by some (e.g., Thieret & Rabeler 2005a). (Greek: *gypsos*, gypsum; *philein*, to love, in reference to the habitat of some species)
REFERENCES: Pringle 2005; Thieret & Rabeler 2005a.

Gypsophila vaccaria (L.) Sm. (from Latin: vacca, cow), COWCOCKLE, COWHERB. Erect, glabrous, usually wide-branched annual to 60(–100) cm tall; leaf blades ovate- to broadly oblong-lanceolate, the cauline ones sessile and auricled-clasping; stipules absent; flowers in terminal, open cymes, long-pedicelled; calyces tubular in flower, inflated in fruit, ovoid, sharply 5-angled or winged, the tube 9–17 mm long, lobes 1.5–3 mm long; petals long-clawed, lavender-pink or pink, showy; crown absent; stamens 10; capsules dehiscent by 4–(6) teeth. Roadsides and disturbed areas, at margins of grain fields; also cultivated as an ornamental; mostly on the nw margin of East TX; sparsely scattered further w in TX; widespread in the U.S. and s Canada. Late Apr–early Jun. [*Saponaria hispanica* Mill., *S. vaccaria* L., *Vaccaria hispanica* (Mill.) Rauschert, *V. pyramidata* Medik.] *Gypsophila paniculata* L. is the BABY'S-BREATH of the nursery trade. Recent changes in agriculture (better grain handling techniques, herbicides) have affected the abundance and distribution of this species (Rabeler & Hartman 2006). It was at one time "a common weed of grain fields…[but] is now increasingly rare or has been extirpated in many localities…" (Thieret & Rabeler 2005a). The seeds contain saponins and are poisonous if ingested (Thieret & Rabeler 2005a) Native of Eurasia and the Mediterranean.

Gypsophila elegans M. Beib., (graceful), SHOWY BABY'S-BREATH, ANNUAL BABY'S-BREATH. Somewhat similar to *G. vaccaria*, but *G. elegans* is generally less coarse, flowers larger, calyx much smaller (2.5–5 mm long), not winged, and calyx lobes about as long as calyx tube. Roadsides, open, sandy, or disturbed sites; Erath (*Kunselman 23* TAC) and Tarrant cos. (*Quayle 1403* BRIT) on the nw margin of East TX; widely scattered in several U.S. states, also Canada. Summer–early fall. Native of Russia, eastern Turkey and Iran. *Gypsophila elegans* is often included in mixtures of "wildflower" seeds used for roadside planting and this may account for its scattered distribution in the U.S. (Pringle 2005).

LOEFLINGIA L. PYGMYLEAF

A North American, Mediterranean, and sw Asian genus of 7 species (Bittrich 1993; Hartman & Rabeler 2005c; Mabberley 2017). (Named for P. Loefling, 1729–1756, Swedish botanist and explorer)
REFERENCES: Barneby & Twisselmann 1970; Hartman & Rabeler 2005c.

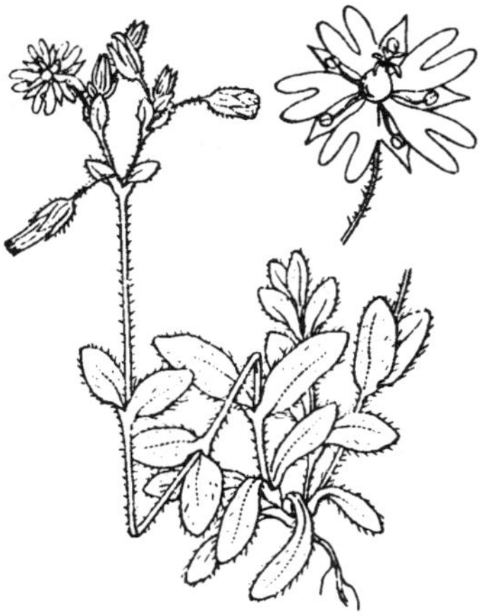

Cerastium pumilum [COS]

Cerastium semidecandrum [BB2]

Dianthus armeria [GLE]

Geocarpon minimum [STE]

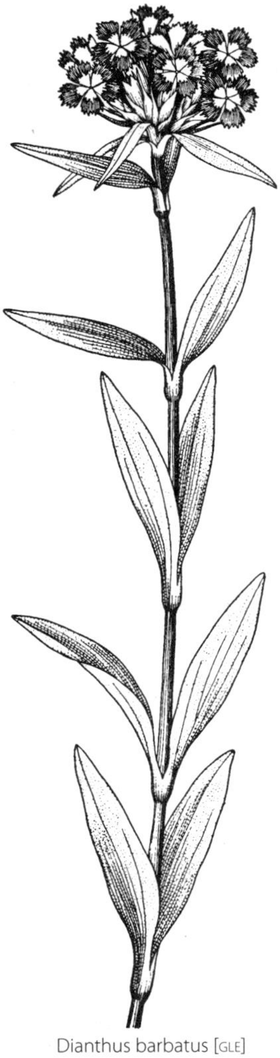

Dianthus barbatus [GLE]

Geocarpon nuttallii [HEA]

Gypsophila vaccaria [GLE]

Loeflingia squarrosa Nutt., (with recurved tips), SPREADING LOEFLINGIA. Small, minutely viscid-pubescent, erect annual to ca. 15 cm tall, stems much branched, the plants often globose in shape, about as broad as tall; leaves crowded, almost needle-like, usually 4–6 mm long; stipules bristle-like; flowers inconspicuous, axillary, sessile, usually solitary or few together in fascicles; sepals 5, resembling the leaves, rather abruptly narrowed to an awn-like tip, the outer 3 with a distinct but weak bristle ca. 1.8 mm long on each side from about the middle; petals absent or minute; styles 3; capsules slender, 3-valved. Loose, dry sand; widespread in East TX; widespread in e 1/2 of TX; scattered to the w; mostly sw U.S., e to AR and LA. Late Apr–May. [*L. squarrosa* subsp. *texana* (Hook.) Barneby & Twisselm., *L. texana* Hook.] This species is inconspicuous and more rarely collected than its widespread occurrence would suggest.

PARONYCHIA Mill. NAILWORT, WHITLOW-WORT

Annuals, biennials, or perennials; stipules usually conspicuous, awl-like to lanceolate to ovate, scarious (= thin and papery); flowers few or many, in terminal cymes, with a cup-shaped hypanthium; sepals 5, cupped or folded at apex, acuminate to acute or obtuse, with minute awn-point, entire; petals absent or essentially so; stamens usually 5; styles 1–2; fruit a 1-seeded utricle.

A cosmopolitan genus of ca. 110 species (Bittrich 1993; Hartman et al. 2005a; Mabberley 2017) including some cultivated ornamentals. This is the largest genus of Caryophyllaceae in TX (13 species—fide Turner 1983). The scaly appearance of *Paronychia* species (due to the stipules) caused the plant to be used historically in the treatment of whitlow, a condition which caused the fingernails to look scaly—this is an example of the Doctrine of Signatures, an ancient belief that a plant that resembled a portion of the human body (a sign or signature) gave clues to its use; i.e., was useful in treating an illness of the body structure it resembled (Bidlack & Jansky 2014). (Greek: *para*, close to, and *onyx*, nail, alluding to the original use of the plant to treat whitlow or felon, an inflammation of the finger, especially beneath the nail)

REFERENCES: Core 1941; Chaudhri 1968; Turner 1983, 1995b; Oxelman et al. 2002; Hartman et al. 2005a.

1. Sepals with a long (0.4–2.1+ mm) awn attached at or below apex.
  2. Sepals with white awns.
    3. Awns 0.5–0.6 mm long; leaf blades 1–5(–7) mm wide; hood prominent ____ **P. drummondii**
    3. Awns 0.6–1.3 mm long; leaf blades 0.3–1 mm wide; hood obscure ____ **P. setacea**
  2. Sepal awns green to brown or yellow.
    4. Stems herbaceous; plants annual, biennial, or perennial usually from a slender taproot.
      5. Stems glabrous to minutely puberulent; sepals of distalmost flowers extending beyond the subtending bracts ____ **P. lindheimeri**
      5. Stems hispid to finely pubescent, sepals of distalmost flowers exceeded by the subtending bracts ____ **P. setacea**
    4. Stems woody from a tough, dense caudex; plants strongly perennial and long-lived with old stems apparent.
      6. Sepals (not including the awn) 1.3–1.8 mm long, apex broadly rounded, with no apparent veins ____ **P. jamesii**
      6. Sepals (not including the awn) 2.0–2.9 mm long, lanceolate to oblong-lanceolate, with the midvein and 2 lateral veins prominent ____ **P. virginica**
1. Sepals with a short (< 0.3 mm long) mucro or cusp ____ **P. fastigiata**

Paronychia drummondii Torr. & A. Gray, (for its discoverer, Thomas Drummond, 1780–1835, Scottish botanist and collector in North America), DRUMMOND'S NAILWORT. Annual or biennial to 25(–35) cm tall, branching above into wide-spreading branches, often wider than tall; leaf blades flat, oblanceolate to linear-oblong, to 30 mm long, 1–5(–7) mm wide; sepals red-brown, brownish, or greenish, with prominent white margins and hood, awn 0.5–0.6 mm long; staminodes filiform. Loose sand; widespread in East TX; also in adjacent areas just to the south; endemic to LA, OK, and TX. May–Sep. [*P. drummondii* subsp. *parviflora* Chaudhri]

Paronychia fastigiata (Raf.) Fernald, (with crowded erect branches), CLUSTER-STEM NAILWORT, FORKED CHICKWEED, HAIRY FORKED NAILWORT. Annual to 30 cm tall, much-branched; leaf blades flat, oblanceolate to elliptic or obovate, to 15(–25) mm long, 0.5–7(–10) mm wide; sepals greenish to brownish, with narrow or obscure margins and hood, slightly mucronate, mucro < 0.3 mm long; staminodes absent. Sandy soils, dry woods or openings; Dallas, Cass, Franklin, Nacogdoches, Robertson, Smith, and Wood (Kartesz 2022) cos; in TX known only from scattered localities in n portion of East TX; throughout the e U.S.; adjacent s Canada. Jun-Aug. [*P. fastigiata* var. *paleacea* Fernald] Some authors (e.g., Chaudhri 1968; Weakley, A.S., and Southeastern Flora Team 2024) recognize 2 varieties in our area (*P. fastigiata* var. *fastigiata* and *P. fastigiata* var. *paleacea* Fernald); we, however, are following Hartman et al. (2005a) in treating *P. fastigiata* var. *paleacea* as a synonym of var. *fastigiata*. Below is a key from Weakley, A.S., and Southeastern Flora Team (2024) for those who wish to distinguish the varieties.

1. Stipular bracts subtending the flowers exceeding the flowers (calyx) ____ **P. fastigiata** var. **paleacea**
1. Stipular bracts subtending the flowers somewhat shorter than the flowers (calyx) ____ **P. fastigiata** var. **fastigiata**

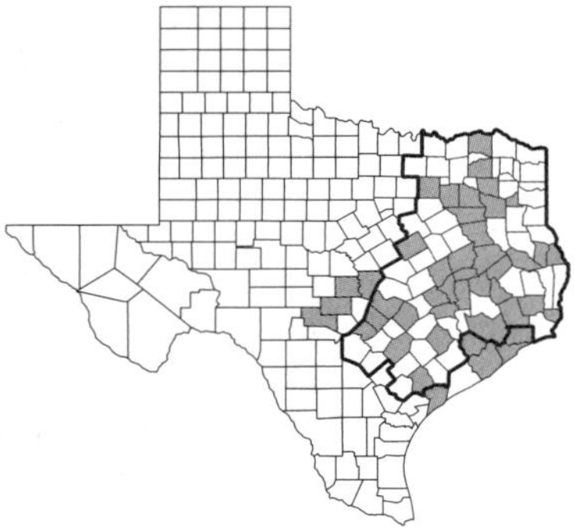

Cerastium glomeratum

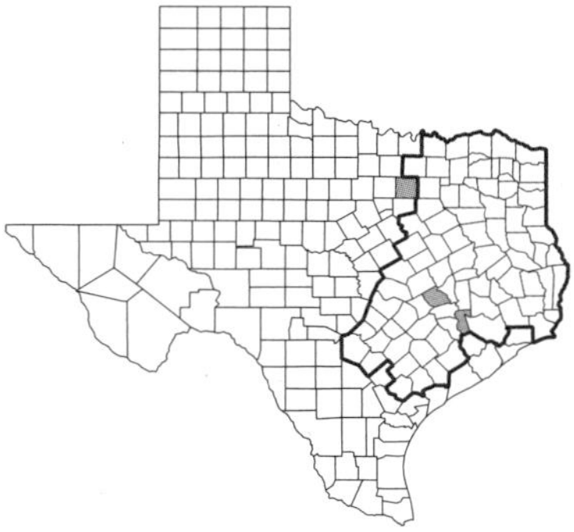

Cerastium nutans

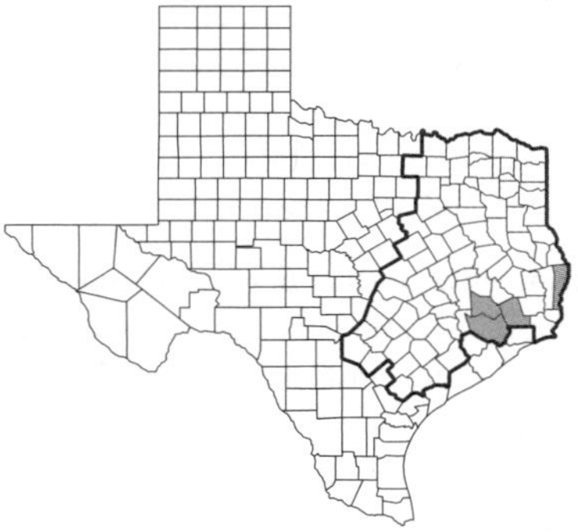

Cerastium parvum

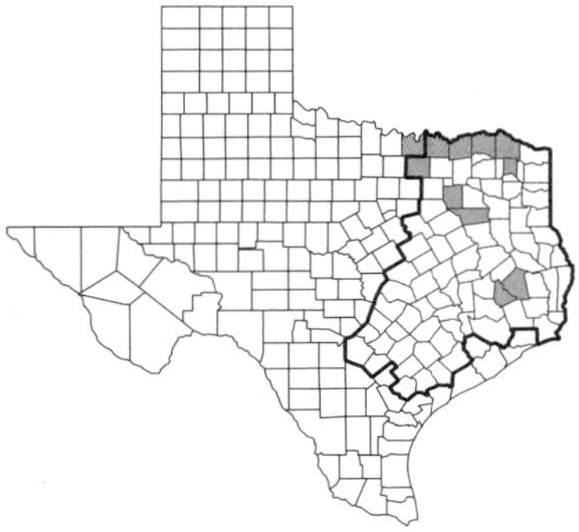

Cerastium pumilum

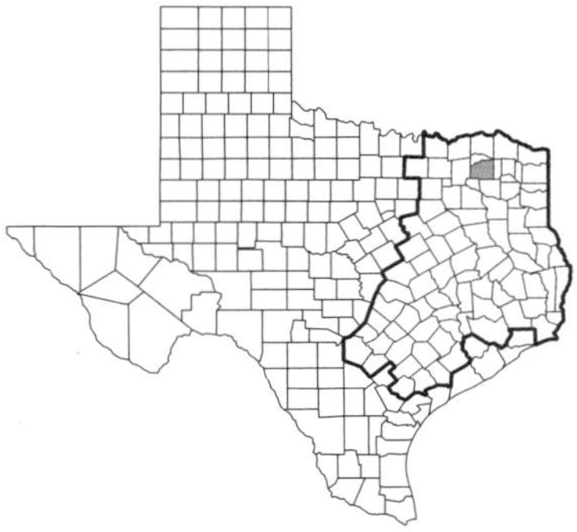

Cerastium semidecandrum

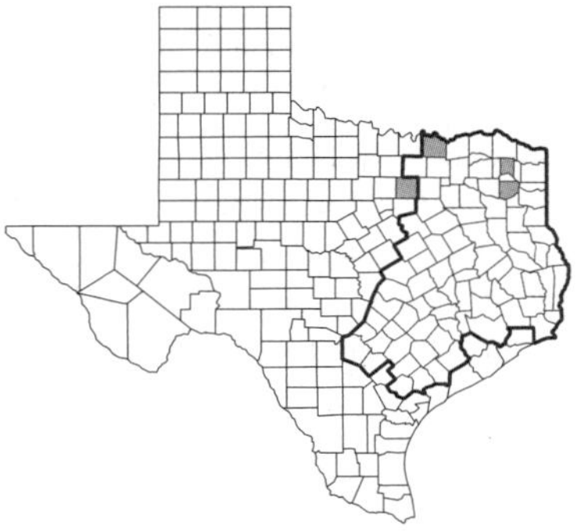

Dianthus armeria

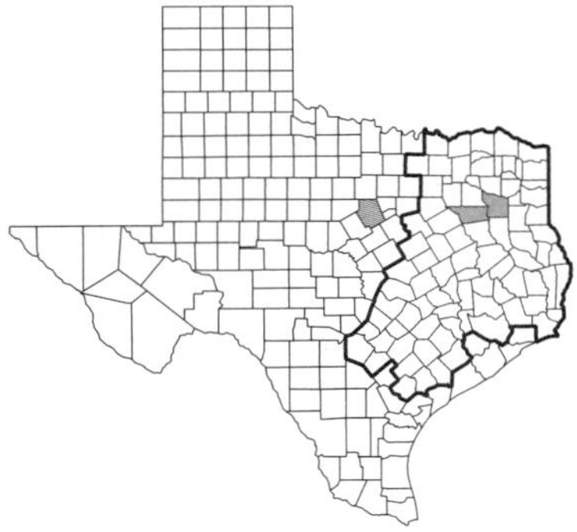

Dianthus barbatus

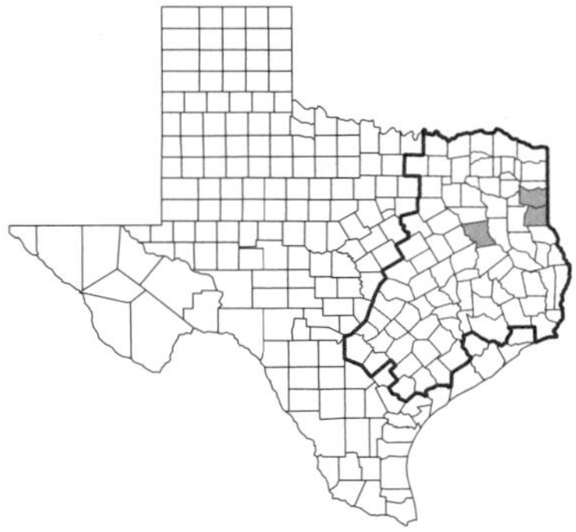

Geocarpon minimum

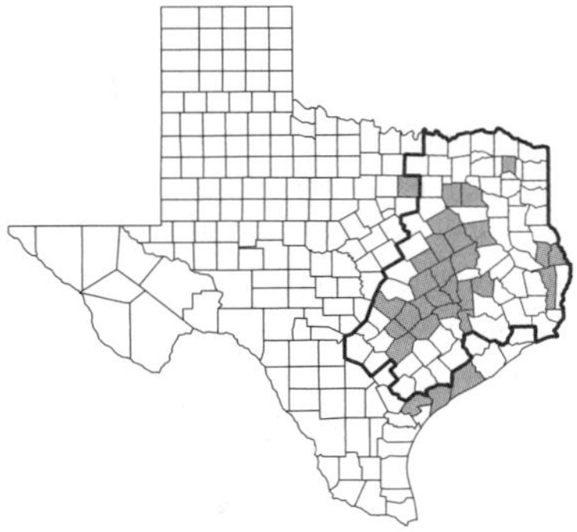

Geocarpon nuttallii

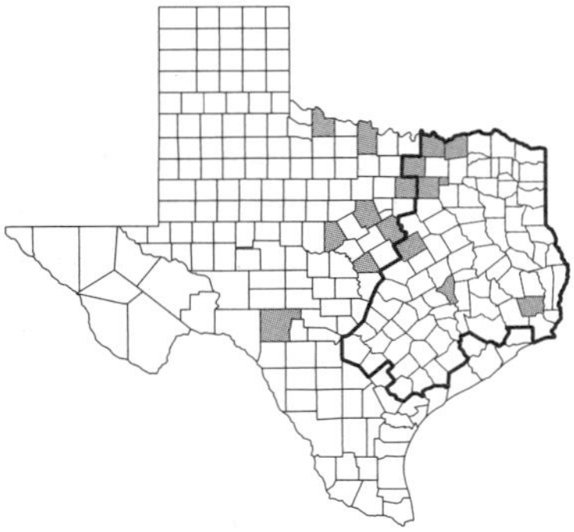

Gypsophila vaccaria

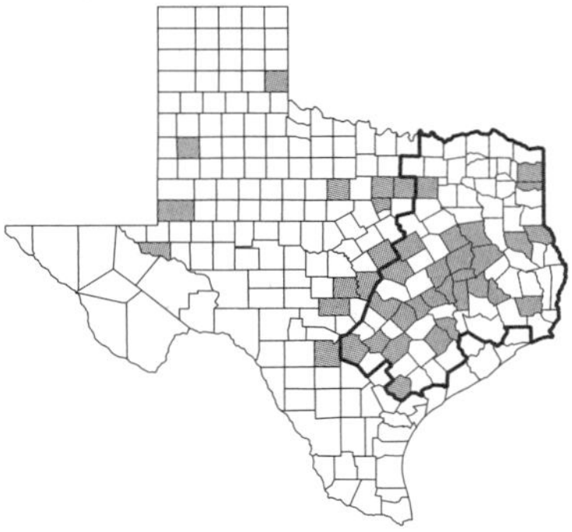

Loeflingia squarrosa

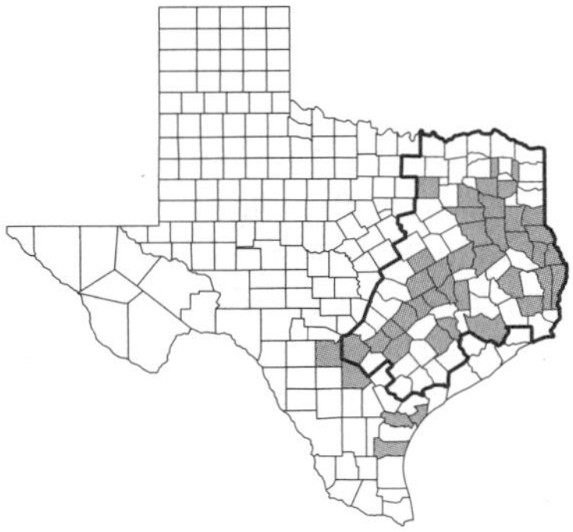

Paronychia drummondii

Paronychia jamesii Torr. & A. Gray, (for Edwin James, M.D., 1797–1861, American botanical explorer of the Rocky Mt. area), JAMES' NAILWORT. Perennial branching from woody base, to ca. 35 cm tall, stems scabrous, minutely pubescent or glabrous with age; leaves gray-green; sepals green to red-brown, 1.3–1.8 mm long excluding the awn. Rocky, sandy, open ground; Bastrop, Dewitt, and Hays (Kartesz 2022) cos. on the w margin of East TX; w 1/2 of TX; mostly in the Plains states from NE s to TX and w to WY and AZ; n Mexico. Jun–Sep. [*P. jamesii* Torr. & A. Gray var. *praelongifolia* Correll]

Paronychia lindheimeri Engelm. ex A. Gray, (for Ferdinand Jacob Lindheimer, 1801–1879, German-born TX collector), LINDHEIMER'S NAILWORT, FORKED NAILWORT. Annual to ca. 30 cm tall, much branched from near base; leaf blades to 15 mm long, 0.4–0.8 mm wide; sepals reddish brown, with slight hood, awn 0.4–0.6 mm long; staminodes filiform. Rocky limestone, sandy, or gravelly areas; mostly in sw portion of East TX; mainly e Edwards Plateau; considered endemic to TX by Hartman et al. (2005a). Jul–Nov. [*P. chorizanthoides* Small, *P. lindheimeri* var. *longibracteata* Chaudhri] Stanford (1976) recognized *P. chorizanthoides* (endemic to c TX) based on the following distinctions:

1. Flowers mostly clustered; calyces (including awns) over 2 mm long; plants pubescent or puberulent; endemic to central Texas ____ *P. chorizanthoides*
1. Flowers mostly solitary and separate; calyces (including awns) usually 2 mm long or less; [plants glabrous or at most minutely scabrous]; Edwards Plateau with disjunct stations in East Texas and Trans-Pecos ____ **P. lindheimeri**

Paronychia setacea Torr. & A. Gray, (bristle-like), BRISTLE NAILWORT. Annual to weak perennial from a taproot, to 20 cm tall, much-branched; leaf blades to 20 mm long, 0.3–1 mm wide; sepals green to tan, obscurely hooded, awn 0.6–1.3 mm long; staminodes subulate. Limestone areas, rocky, sandy, or silty soils, grasslands; scattered in the s 1/2 of East TX; Edwards Plateau, South TX Plains.; endemic to TX (Hartman et al. 2005a). Spring–fall. [*P. lundellorum* B.L. Turner, *P. setacea* var. *longibracteata* Chaudhri] Turner (1983) and Turner et al. (2003) recognized *P. lundellorum*, a TX endemic from the South TX Plains, as a distinct species; we are following Hartman et al. (2005a), who included that taxon within *P. setacea* noting that they could not distinguish it morphologically; however, they did note that further study is warranted. Turner (1983) noted that *P. lundellorum* "is most closely related to *P. setacea* but can be distinguished by its larger, more pedicellate flowers and perennial [versus annual/weak perennial] habit." Poole et al. (2007) recognized *P. lundellorum* and included it in their *Rare Plants of Texas* with a rating of G1S1.

Paronychia virginica Spreng., (of Virginia), PARKS' NAILWORT, BROOM NAILWORT, YELLOW NAILWORT, APPALACHIAN NAILWORT, VIRGINIA WHITLOW WORT, APPALACHIAN WHITLOW WORT. Perennial branching from woody base, to 45 cm tall, often becoming yellowish with age; leaf blades 10–30 mm long, 0.4–1 mm wide; sepals yellowish, greenish yellow to brownish, slightly hooded, awn 0.4–1.1 mm long; staminodes filiform. Limestone outcrops; in East TX mainly on w margin of the area and scattered to the east; also adjacent e Cross Timbers and Prairies and e Edwards Plateau; AL, AR, GA, MD, MO, OK, TX, VA, and WV. Aug–Oct. [*P. parksii* Cory, *P. virginica* Spreng. var. *scoparia* (Small) Cory] *Paronychia virginica* is found in two disjunct populations: one eastern and one western. Some authors (e.g., Weakley, A.S., and Southeastern Flora Team 2024) treat the two populations as separate taxa (variety or species): the western population as *P. scoparia* Small, and the eastern population as *P. virginica*. We are following Hartman et al. (2005a) and treating it as one entity.

PETRORHAGIA (Ser.) Link CHILDING-PINK

Plants annual, stems simple or branched; glabrous to stipitate glandular or scabrous; leaf blades linear to linear-lanceolate or oblanceolate, 3-veined; inflorescences capitate (sometimes appearing almost solitary when few-flowered and in early development); bracts enclosing flowers, scarious; petals pink to purplish; seeds 1–1.8 mm long; stamens 10; styles 2; capsules opening by 4 teeth.

An Old World genus of 33 species (R. Rabeler, pers. comm.; Rabeler & Hartman 2005c; Mabberley 2017) including *Kohlrauschia* and *Petrorhagia sensu stricto* (as recognized by Bittrich 1993) ranging from the Canary Islands and Mediterranean region to Kashmir. Some are cultivated as ornamentals. *Petrorhagia* was in the past sometimes included in a broadly circumscribed *Dianthus* (e.g., Steyermark 1963). (Greek: *petros*, rock, and *rhagas*, chink or fissure, from the principal habitat of some species being in rock crevices—Rabeler & Hartman 2005c)

REFERENCES: Ball & Heywood 1964; Shinners 1969; Rabeler 1985; Rabeler & Hartman 2005c.

1. Leaf sheaths ± as long as wide, usually 1–2 mm long; petals with apex truncate or emarginated, dark-colored veins absent ____ **P. prolifera**
1. Leaf sheaths 2–3 times as long as wide, (3–)4–9 mm long; petals with apex bifid or obcordate, darkly-colored veins near base ____ **P. dubia**

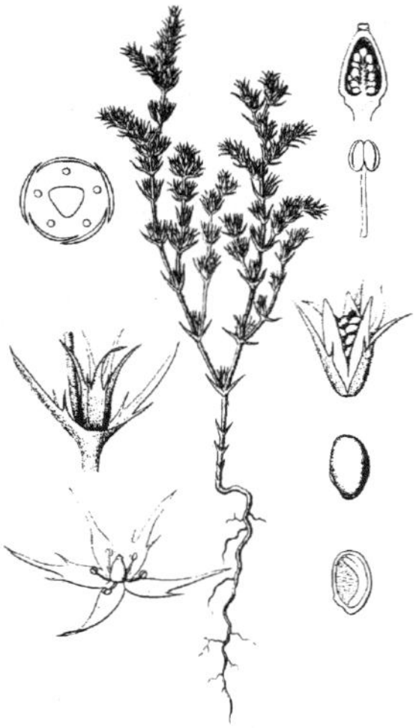
Loeflingia squarrosa [GR1]

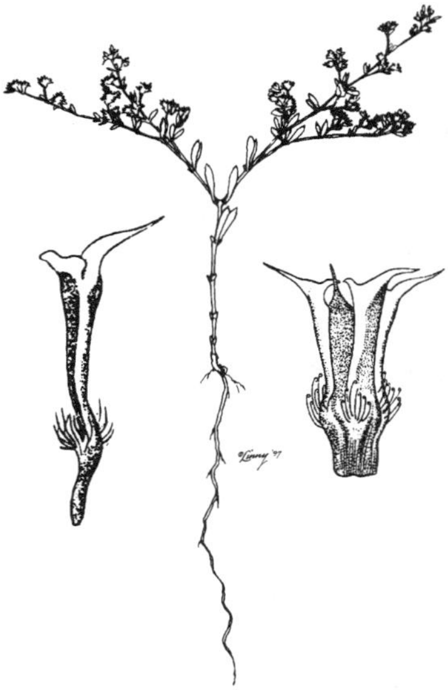
Paronychia drummondii [CHA, HEA]

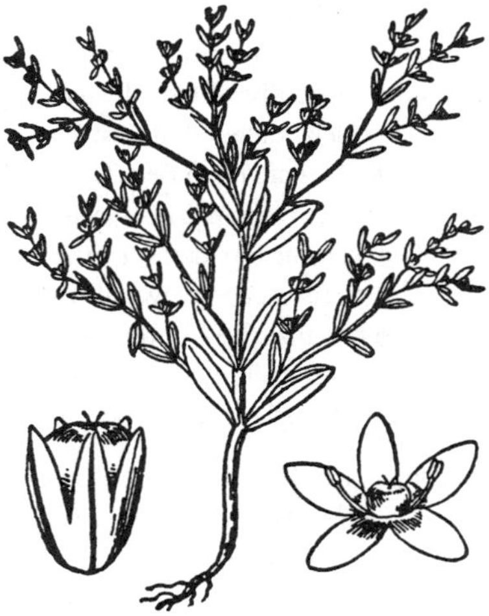
Paronychia fastigiata [BB2]

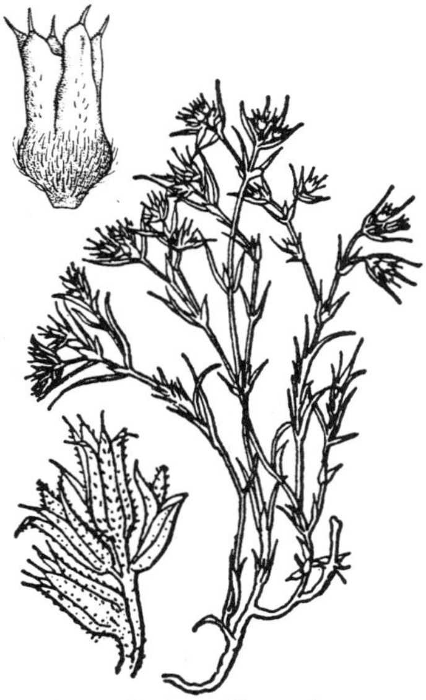
Paronychia jamesii [BB2, CHA]

Paronychia virginica [BB2]

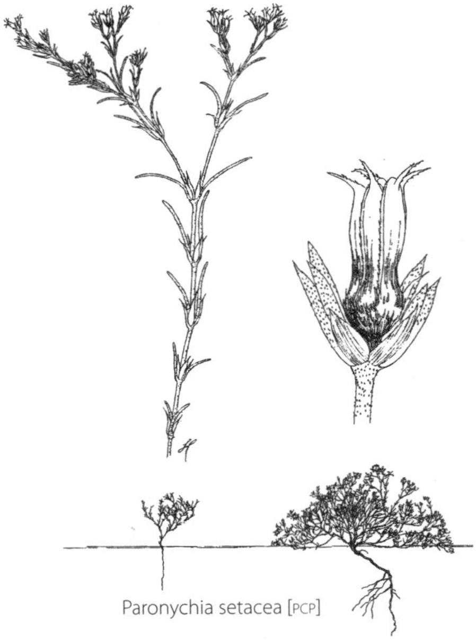
Paronychia setacea [PCP]

Petrorhagia dubia [HEA]

Petrorhagia dubia (Raf.) G. López & Romo., (doubtful), CHILDING-PINK. Erect annual to ca. 90 cm tall; stem internodes glabrous or densely stipitate glandular-pubescent; leaf blades mostly linear to occasionally oblanceolate, to ca. 5(–6) cm long; flowers in several-flowered head-like clusters; sepals 8–15 mm long; petals pink or purplish with darkly colored veins. Roadsides and fields; widely scattered in East TX; also Cooke Co. (BRIT) just to the w of East TX; AR, CA, OK, OR, and TX. Apr–Jun. [*Dianthus dubius* Raf., *P. velutina* (Guss.) P.W. Ball & Heywood, *P. prolifera* of TX auth., not (L.) P.W. Ball & Heywood] Native of the Mediterranean region. This species was first reported for TX by Shinners (1969) who suggested it was possibly introduced with *Lolium perenne* (RYE GRASS) by the Texas Highway Department. Rabeler (1985) indicated that introduction with *Trifolium incarnatum* (CRIMSON CLOVER) was also a possibility. Rabeler and Hartman (2005c) noted that the glabrous internodes of most TX plants "suggests that these populations were derived from seed that came from Italy or Sicily, the only area in the native range where plants with glabrous stems are known." Texas plants were confused in the past (e.g., Correll & Johnston 1970) with *P. prolifera* (L.) P.W. Ball & Heywood.

Petrorhagia prolifera (L.) P.W. Ball & Heywood, (becoming more numerous), CHILDING-PINK. Annual 20–30(–60) cm tall; stem internodes glabrous or midstem slightly scabrous; leaf blades linear to linear-lanceolate, 1–3 cm long; sepals 7–12 mm long; flowers in several head-like clusters; sepals 7–12 mm long; petals pink to slightly purplish (rarely white) without darkly colored veins. Roadsides and fields; in TX known only from Anderson and Bowie cos. (Kartesz 2022); e U.S. from NY s to GA and w to MO and TX, also CA and ID; sw Canada. Spring–summer. [*Dianthus prolifera* L.] Native of Europe and North Africa.

POLYCARPON Loefl. MANYSEED, POLYCARP

A widespread genus of 9–15 species (Thieret & Rabeler 2005c). Primarily, a genus of Europe with several species in South America. (Greek: *poly*, many, and *carpus*, fruit, in reference to the numerous capsules)
REFERENCES: Radford et al. 1968; Thieret & Rabeler 2005c.

Polycarpon tetraphyllum (L.) L., (four-leaved), FOUR-LEAF MANYSEED. Plant small, prostrate to decumbent or erect, glabrous, much-branched, usually annual, stems to ca. 15 cm long; leaves mostly in whorls of four or opposite, subsessile or on petioles to 1.2 mm long, blades 2–12 mm long, apically obtuse to rounded, entire; stipules scarious, ovate, 1.8–3 mm long; flowers numerous, in dense cymes; bracts scarious; pedicels 0.5–3 mm long; sepals 5, separate, 1.5–2.5 mm long, the margins whitish; petals 5, white, 0.5–0.8 mm long, oblanceolate; stamens usually 3–5; style 1, obscurely 3-branched; capsules 3-valved, many-seeded. Sandy or silty soils, openings; widespread in s 1/2 of East TX; also adjacent e Cross Timbers and Prairies and e Edwards Plateau; scattered in se and w U.S.; B.C. in sw Canada. Mar–Jul. Native of Europe and the Mediterranean. This species has been very widely spread (considered cosmopolitan), "at least in part from ships' ballast" (Thieret & Rabeler 2005c). [*Mollugo tetraphylla* L.]

SABULINA Rchb. MOCK SANDWORT

Annuals or perennials; stipules absent; inflorescences open cymes; petals white, longer than sepals, usually shallowly notched at apex; stamens 10; styles 3; fruit a capsule opening by 3 valves.

A genus of ca. 65 species widely distributed in the Northern Hemisphere, with 2 species in South America (Weakley, A.S., and Southeastern Flora Team 2024). Dillenberger and Kadereit (2014) segregated several taxa as *Sanbulina* from *Minuartia*, formerly ca. 175 species, which was known to be polyphyletic. The result being there are no species of *Minuartia* in the U.S. and worldwide there are only ca. 54 species (Dillenberger & Kadereit 2014) (from *sabulo*, growing in sandy places)
REFERENCES: Maguire 1951; McNeill 1962, 1980; Wofford 1981; Rabeler 1992; Rabeler et al. 2005; Dillenberger & Kadereit 2014; Schilling et al. 2022.

1. Perennials with stiff, almost needle-like leaf blades, with axillary tufts of smaller leaf blades; inflorescences glabrous ____ **S. michauxii**
1. Annuals with flat, soft leaf blades, without axillary tufts of smaller leaf blades; inflorescences usually glandular-pubescent (use lens).
   2. Sepals (3-or)5-ribbed; leaf blades 0.5–1.5(–1.8) mm wide, 20 mm or less long; plants usually 5–30 cm tall; seeds usually dull reddish brown (sometimes black), tubercles low and rounded ____ **S. patula**
   2. Sepals 3-ribbed; leaf blades usually 1.5–3 mm wide, 10–50 mm long; plants often larger 10–55 cm tall; seeds black, shiny, tuberculate, tubercles somewhat pointed ____ **S. muscorum**

Sabulina michauxii (Fenzl) Dillenb. & Kadereit, (for Andrè Michaux, 1746–1803, French botanist and explorer of North America), ROCK SANDWORT, MICHAUX'S MOCK SANDWORT. Densely tufted perennial to ca. 40 cm tall, glabrous; stem leaves rather crowded in basal part of plant, the blades almost needle-like, often boat-shaped, to 15(–30) mm long, ca. 0.5(–1.8) mm wide; axillary tufts of smaller leaves present; inflorescences 5–30-flowered, glabrous; pedicels 0.3–4(–6) cm long; sepals 3-ribbed,

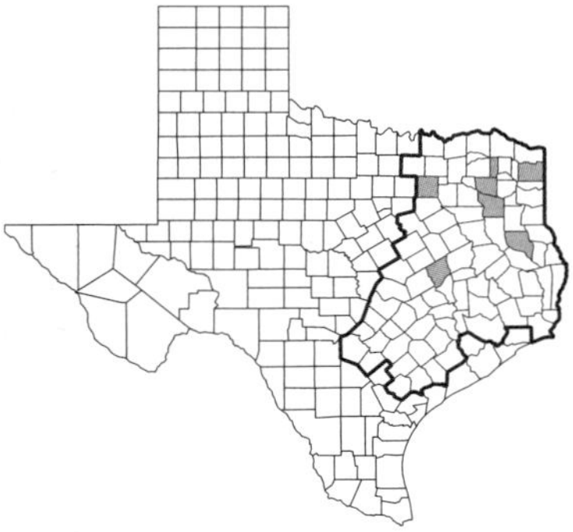
Paronychia fastigiata

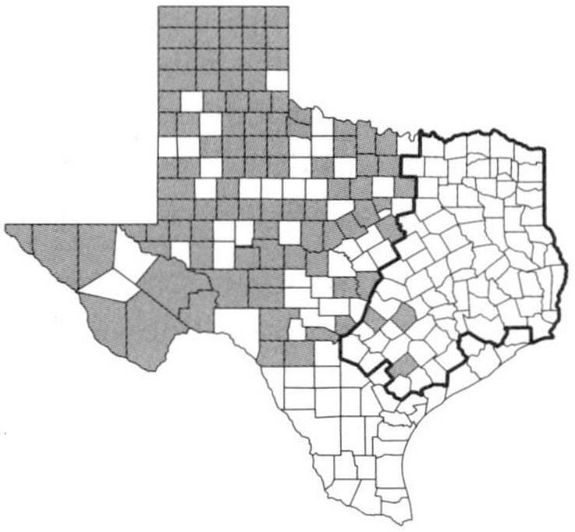
Paronychia jamesii

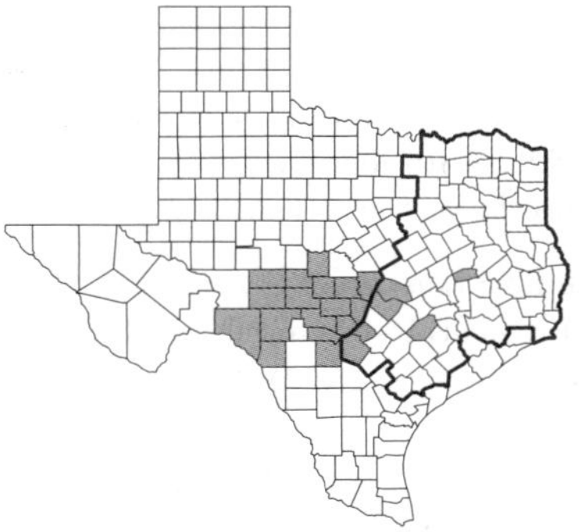
Paronychia lindheimeri

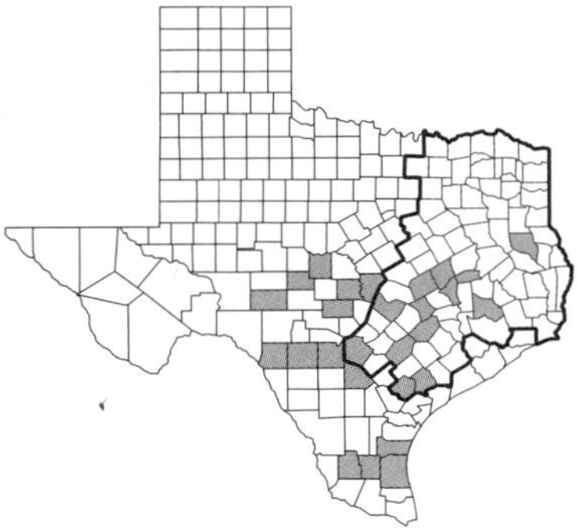
Paronychia setacea

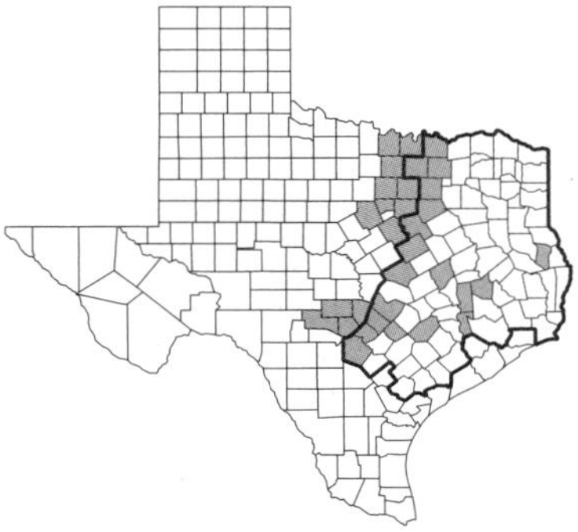
Paronychia virginica

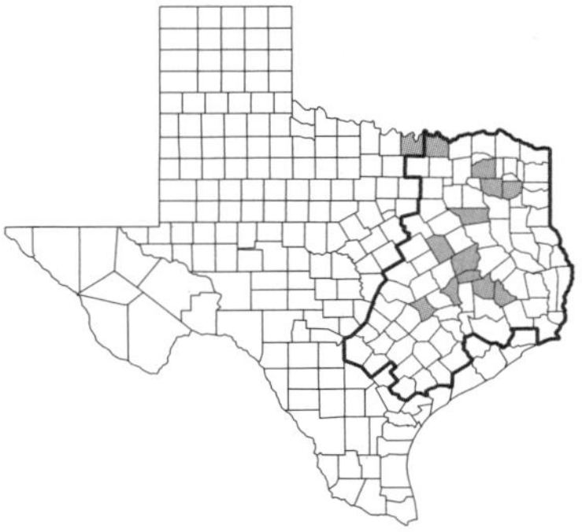
Petrorhagia dubia

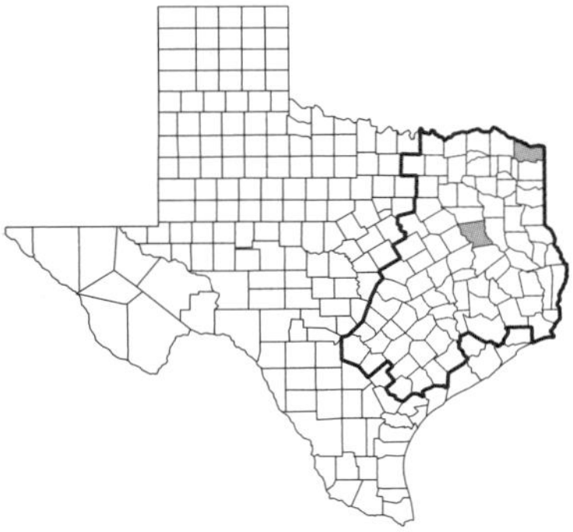
Petrorhagia prolifera

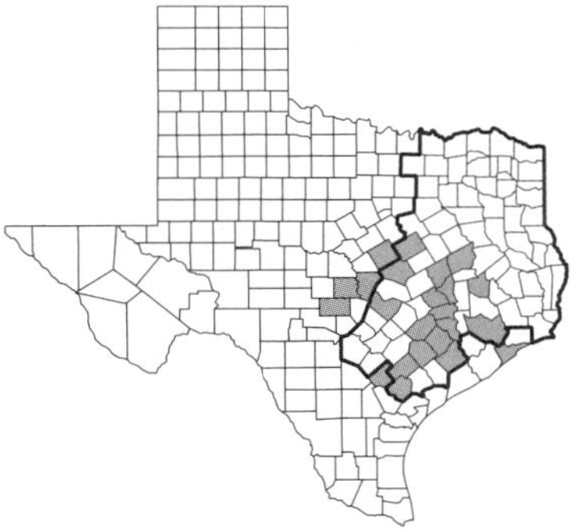
Polycarpon tetraphyllum

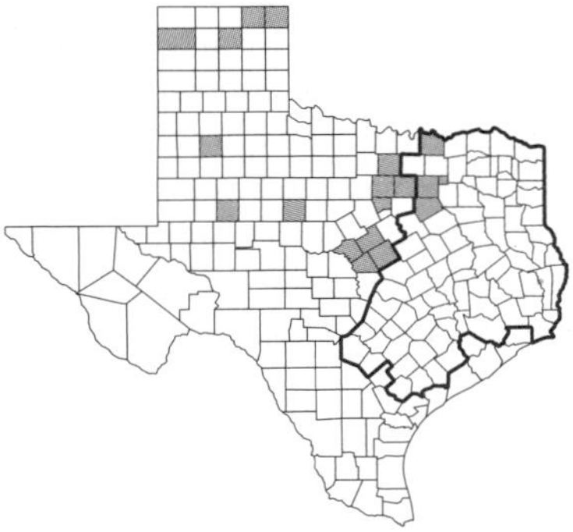
Sabulina michauxii

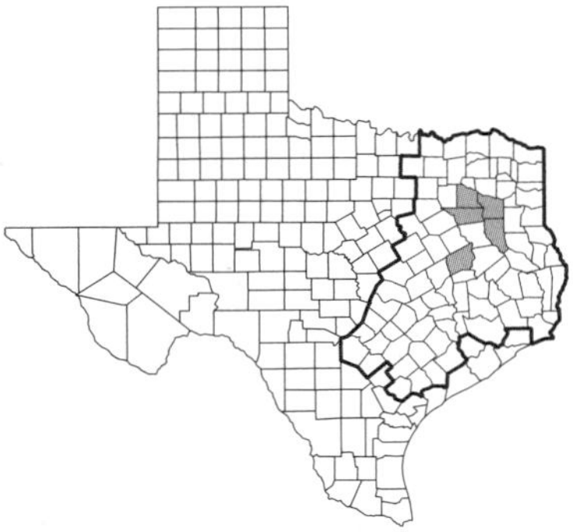
Sabulina muscorum

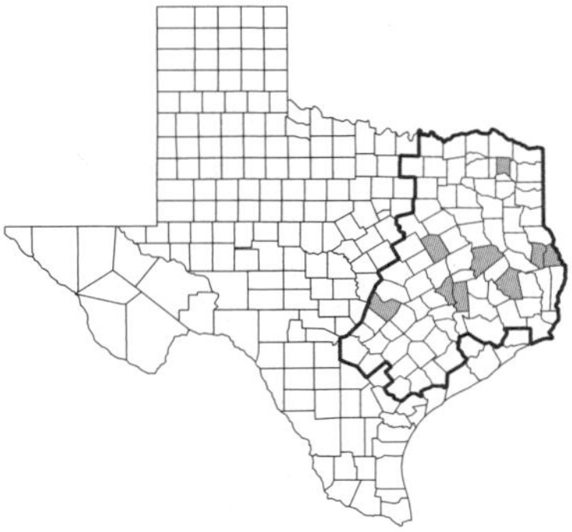
Sabulina patula

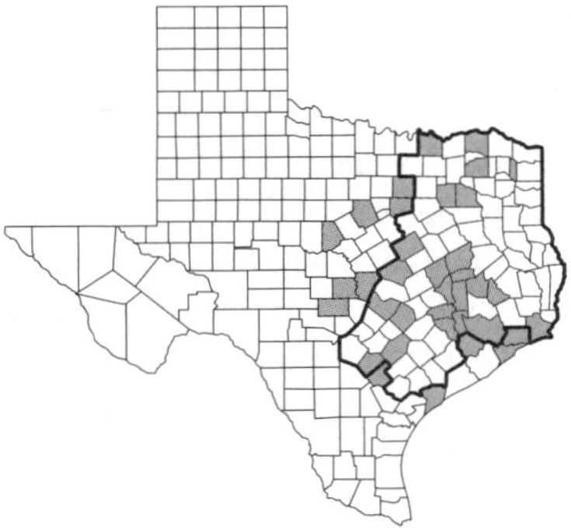
Sagina decumbens

lanceolate to ovate; petals usually 5–8 mm long. Limestone outcrops, rocky areas; Dallas (*Reverchon* in 1872, but not found there recently), Ellis, Grayson, and Rockwall (Kartesz 2022) cos. on nw margin of East TX; mainly Cross Timbers and Prairies and scattered w to the Panhandle; scattered in the e 1/2 of U.S except the se, also OR and WA; Canada. Apr–Jun. [*Arenaria stricta* Michx., *A. stricta* var. *texana* B.L. Rob., *A. stricta* subsp. *texana* (B.L. Rob. ex Britton) Maguire, *A. texana* (B.L. Rob.) Britton, *Minuartia michauxii* (Fenzl) Farw., *M. michauxii* var. *texana* (B.L. Rob. ex Britton) Mattf.]

Sabulina muscorum (Fassett) E.E. Schill., (mossy), DIXIE MOCK SANDWORT. Branched annual similar to but often larger than *S. patula*; stems 10–55 cm tall; leaf blades usually 10–50 mm long, linear-lanceolate to narrowly oblanceolate; inflorescences usually glandular-pubescent, pedicels to 5.5 cm long; sepals prominently 3-ribbed, 3–4 mm long, to 5 mm in fruit. Roadsides, stream bottoms, often in sandy soils; Cherokee, Henderson, Leon, Smith, and Van Zandt (BRIT) cos.; other locations in the state possible; endemic to the sc U.S. in AL, AR, KY, LA, MO, OK, TN, and TX. Apr–May. [*Arenaria patula* var. *robusta* (Steyerm.) Maguire, *A. muriculata* McNeill, *Minuartia muriculata* (Maguire) McNeill, *M. patula* var. *robusta* (Steyerm.) McNeill, *Mononeuria muscorum* (Fassett) Dillenb. & Kadereit, *Stellaria muscorum* Fassett] In their recircumscription of *Minartia*, Dillenberger and Kadereit (2014) transferred several taxa to *Mononeuria* including *Minuartia muscorum* (Fassett) Rabeler. However, Schilling et al. (2022) found *Mononeuria* not to be monophyletic and transferred *Mononeuria muscorum* (Fassett) Dillenb. & Kadereit to *S. muscorum*.

Sabulina patula (Michx.) Small ex. Rydb. (spreading), PITCHER'S SANDWORT, PITCHER'S STITCHWORT. Branched annual usually 5–30 cm tall; leaf blades to 20 mm long, 0.5–1.5(–1.8) mm wide; inflorescences sparingly and minutely glandular-pubescent; pedicels to 3 cm long; sepals (3-or) 5-ribbed, 4–5.5 mm long, not enlarging in fruit. Damp sandy ground, thin rocky soils; scattered in East TX; otherwise unknown in the state; e U.S. from OH s to GA and w to KS and TX. Mar. [*Arenaria patula* Michx., *A. pitcheri* Nutt., *Minuartia patula* (Michx.) Mattf., *Mononeuria patula* (Michx.) Dillenb. & Kadereit]

SAGINA L. PEARLWORT

A n temperate and tropical mountain genus of ca. 30 species (Mabberley 2017) of usually tufted herbs. (Latin: *sagina*, fodder, feasting, or fattening; previously applied to *Spergula* which was used as early forage)
REFERENCES: Crow 1978, 1979, 2005.

Sagina decumbens (Elliott) Torr. & A. Gray, (trailing with tips upright), TRAILING PEARLWORT. Inconspicuous, small, to ca. 10(–17) cm tall, erect or partly decumbent annual, largely glabrous or with some glandular hairs above; leaves sessile; basal rosette leaves soon deciduous, blades narrowly linear, 3–15(–23) mm long, very narrow, < 1 mm wide, sometimes with a purple tinge, glabrous; stipules absent; flowers axillary and terminal, long pedicelled (pedicels 3–20 mm long); sepals 4–5, separate, 1.4–2.5 mm long; petals usually 5 or absent in 4-merous flowers, entire, white, usually slightly shorter than the sepals to sometimes slightly longer; stamens usually 5 or 10, sometimes 8; styles 4–5; capsules with 4–5 valves, 2–3.5 mm long. Sandy open woods, roadsides, disturbed sites; widespread in East TX; widespread in e 1/2 of TX; widespread in the e 1/2 of the U.S. and adjacent Canada. Late Mar–early May. [*S. decumbens* var. *smithii* (A. Gray) S. Watson & J.M. Coult.]

SAPONARIA L. BOUNCING-BET, SOAPWORT

A temperate Eurasian genus of ca. 40 species (Bittrich 1993; Thieret & Rabeler 2005b; Mabberley 2017). (Latin: *sapo*, soap, referring to the mucilaginous juice of *S. officinalis* forming a lather with water)
REFERENCE: Thieret & Rabeler 2005b.

Saponaria officinalis L., (medicinal), BOUNCING-BET, SOAPWORT, FULLER'S-HERB. Erect glabrous perennial to 30–80(–150) cm tall, rhizomatous, the rhizomes long-creeping; leaf blades 3–10(–15) cm long, 1–5 cm wide, prominently 3–5-nerved; stipules absent; flowers in usually congested cymes, showy; calyx tube 12–20 mm long; calyx lobes ca. 2–5 mm long; petals 5 (can be 10 in cultivated forms), white or pink (can dry dull purple), with 2 appendages (crown or coronal scales) per petal; petal claw prominent, 10–20 mm long; petal blade 8–18 mm long, notched apically; stamens 10; styles usually 2; capsules ca. 15–20 mm long, opening apically by 4(–6) teeth. Open areas and waste places; Anderson (Kartesz 2022), Cherokee (*Ajilvsgi 4986* BRIT), and Tyler (*Tharp 51-1466* SMU) cos.; also Tarrant, just w of East TX, and Edwards, Taylor (Kartesz 2022), and Gillespie (*O'Kennon 10730c* BRIT) cos. in the Edwards Plateau; widespread nearly throughout the U.S. and s Canada. Jun–Sep. Native of Europe. Double-flowered horticultural forms exist in the wild and can be locally abundant. This species was formerly used as a soap substitute, when the rootstocks were beaten in water (Baumgardt 1982); the leaves are also reported to have been used as a source of soap (Thieret & Rabeler 2005b); all parts of the plant contain saponins; it is toxic and potentially lethal to livestock but rarely eaten because it is distasteful (Kingsbury 1964; Burlage 1968; Stephens 1980).

SCLERANTHUS L. Knawel, GERMAN KNOTWEED

A genus of 10 species of temperate Europe, Asia, and Africa (Bittrich 1993). (Greek: *scleros*, hard, and *anthos*, flower, in reference to the hardened calyx tube)

REFERENCES: Thieret & Rabeler 2005d; Hill et al. 2016; Wunderlin et al. 2017.

Scleranthus annuus L., (annual), ANNUAL KNAWEL. Annual or perennial herb, stems to 20 cm long, branched at base; leaf blades 5–20 mm long and ca. 1 mm wide, entire, connate at base with membranous often ciliate flange; inflorescence a compact axillary or terminal cyme of 2–3 flowers with urceolate hypanthia, equaled or exceeded by bracts similar to leaves; petals absent; sepals 5, 1.5–4 mm long, margins scarious, apices mostly acute; utricle enclosed by fused hypanthium and crowned by sepals which form the indehiscent "fruit." Sandy fields, roadsides weedy areas, parks; in East TX known only from Van Zandt Co. (Hill et al. 2016); otherwise in TX known only from Tarrant Co. (*Sasan 52* BRIT) just to the w of East TX; scattered in e 1/2 of U.S., w U.S., also Canada. Winter–fall. Native of Europe, Asia, and Africa.

SILENE L. CATCHFLY, CAMPION

Erect annuals or perennials; stipules absent; inflorescences simple or branched, open cymes; sepals 5, fused into a tube for much of their length, with free lobes; petals usually clawed, usually with a crown (of appendages); stamens 10; styles 3(–4); fruit in ours a 3 or 6-toothed capsule.

A mainly n hemisphere genus of nearly 700 species (Bittrich 1993; Morton 2005c; Mabberley 2017) of mostly herbs including a number cultivated as ornamentals and several important weeds. It is the largest genus in the family. (Name adopted by Linnaeus from earlier authors; probably from mythical Greek *Silenus*, intoxicated foster father of Bacchus being described as covered with foam, from sticky secretions of many species)

REFERENCES: Hitchcock & Maguire 1947; Maguire 1950; McNeill 1978; Oxelman & Linden 1995; Oxelman et al. 1997, 2000; Singhurst et al. 2002; Burleigh & Holtsford 2003; Morton 2005c.

1. Petals scarlet.
  2. Plants glabrous or nearly so; petals unlobed; stems 0.5–1.1 m tall; upper stem leaf blades mostly to 8 mm or less wide, can be much narrower; species known in TX only from se part of East TX ______ **S. subciliata**
  2. Plants pubescent; petals apically deeply 2-lobed, lobes acute; stems usually 0.2–0.8 m tall; upper stem leaf blades mostly to 15 mm wide; species known in TX only from Bowie and Cass cos. in the ne part of East TX ______ **S. virginica**
1. Petals white to pink, rose-lavender or red-violet.
  3. Calyces prominently veined with 25–30 parallel veins, inflated at maturity, 20–30 mm long ______ **S. conoidea**
  3. Calyces with 10 veins, these prominent or obscure, at maturity 5–12 mm long.
    4. Middle stem leaf blades opposite; stem leaf blades to ca. 15 mm wide; petals usually 5–14 mm long (including claw), < 1.5 times as long as calyx, unlobed or 2-lobed; plants annual.
      5. Calyces glabrous; upper stems glabrous or nearly so, often with some viscid (= sticky), dark-colored bands encircling upper internodes; plants to 0.8(–1.2) m tall ______ **S. antirrhina**
      5. Calyces with conspicuous hairs to ca. 2 mm long; upper stems usually with obvious hairs (these can be short or long or glandular), without sticky, dark-colored bands; plants 0.15–0.45 m tall ______ **S. gallica**
    4. Middle stem leaf blades in whorls of 4; stem leaf blades to ca. 40 mm wide; petals usually 14–22 mm long (including claw), ca. 2 times as long as calyx, fimbriately 4–12-lobed (almost fringed in appearance); plants perennial ______ **S. stellata**

Silene antirrhina L., (possibly for likeness to *Antirrhinum*–snapdragon, whose flowers are said to resemble a snouted dragon; from Greek: *Anti*, against, opposed to, like, and *Rhis*, nose or snout), SLEEPY CATCHFLY. Annual to 0.8(–1.2) m tall (usually much smaller), stems glabrous or nearly so or with fine hairs near base, often with sticky, dark-colored bands encircling upper internodes; leaf blades 1–7(–9) cm long, linear- or oblong-lanceolate; flowers small, open only in afternoon and evening; calyces 5–9 mm long; petals mostly white to partly or wholly rose-lavender or red-violet, occasionally absent, slightly longer than calyx; crown (of appendages) minute or absent. Open woods, prairies, disturbed sites; throughout East TX; throughout much of TX; throughout the U.S.; s Canada; Mexico; and South America. Late Apr–early Jun. [*S. antirrhina* var. *confinis* Fernald, *S. antirrhina* var. *depauperata* Rydb., *S. antirrhina* var. *divaricata* B.L. Rob., *S. antirrhina* var. *laevigata* Engelm. & A. Gray, *S. antirrhina* var. *subglaber* Engelm. & A. Gray, *S. antirrhina* var. *vaccarifolia* Rydb.] A number of varieties and forms have been recognized in the past in this species; however, we are not recognizing infraspecific taxa following Morton (2005c), who noted that "none appear to be worthy of recognition." Reported to be poisonous (Burlage 1968).

Silene conoidea L., (cone-like), LARGE SAND CATCHFLY, CONOID CATCHFLY. Annual to 1 m tall; leaf blades to 12 cm long, oblanceolate to lanceolate; flowers showy; calyces inflated in fruit, prominently veined with 25–30 parallel veins (unique among our species);

petals deep pink, lobes 8–12 mm long; crown 2–4 mm long; known in TX only from Washington Co. (Reed 2004); mainly w U.S. and sw Canada, also MO and TX. Early summer. Native of Eurasia. The only known occurrence (Reed 2004) is from a vegetable garden at Peaceable Kingdom Farm in Washington Co. Perhaps introduced by organic fertilizer.

Silene gallica L., (of France), SMALL-FLOWERED CATCHFLY, FORKED CATCHFLY. Annual to 45 cm tall; stem leaf blades oblanceolate to lanceolate, to 5(–7) cm long; flowers small; calyces to 10 mm long; petals white or pink, sometimes with a dark spot; crown 1–1.5 mm long. Roadsides, disturbed areas; widespread in East TX; also Trans-Pecos; scattered in several U.S. states and adjacent Canada. Apr–May. [*S. anglica* L., *S. quinquevulnera* L.] Native of Europe and Asia.

Silene stellata (L.) W.T. Aiton, (starry), WHORLED SILENE, STARRY CAMPION, WIDOW'S-FRILL. Perennial to 0.3–0.8(–1.2) m tall; midstem leaves in whorls of 4, blades lanceolate to elliptic or ovate, to 10 cm long, 4 cm wide; calyces 7–11 mm long; lobes broadly triangular, 2–3 mm long; petals large (ca. 2 times as long as calyx), white, showy, with wide, fimbriate outer margin; crown absent. Woods and thickets, sandy or clayey soils; n and ne parts of East TX w to Dallas and Grayson cos., also Coryell and Mills (Kartesz 2022) cos. in Edwards Plateau; otherwise unknown in the state; nearly throughout the e 1/2 of the U.S.; Canada. Late May–early Jul. Native of the U.S.

Silene subciliata B.L. Rob. (somewhat ciliate, fringed), PRAIRIE FIRE PINK, SCARLET CATCHFLY. Perennial with thick taproot to 1.1 m tall, glabrous or nearly so; leaf blades linear to narrowly oblanceolate, to 16 cm long, 12 mm wide; calyces 17–22 mm long; petals very showy, scarlet, 2–2.5 times as long as calyx, unlobed (entire or obscurely toothed); crown 4–5 mm long. Sandy soils, open woods, stream banks; limited in TX to 9 counties, Hardin, Jasper, Jefferson, Liberty, Newton, Polk, Sabine, Shelby, and Tyler (Poole et al. 2007; Kartesz 2022) cos., on se margin of East TX; endemic to w Gulf Coastal Plain of sw LA and se TX. This rare species is of conservation concern in TX and is in the Center for Plant Conservation's National Collection of Endangered Plants (Morton 2005c). NatureServe: G3, vulnerable; S3, vulnerable. ⚠

Silene virginica L., (of Virginia), FIRE PINK. Perennial usually 0.2–0.8 m tall, stems glandular-pubescent, at least above; leaf blades oblanceolate to lanceolate; calyces 16–22 mm long; petals very showy, scarlet, ca. 2 times as long as calyx, apex deeply 2-lobed, lobes acute; crown ca. 3 mm long. Base of mesic hardwood slope; known in TX only from Bowie (Kartesz 2022) and Cass (*Singhurst 10294* BAYLU) cos. in the extreme ne corner of the state; widespread in the e 1/2 of the U.S. and adjacent Canada. Spring. The 2001 collection by Singhurst was the first report of this species from TX (Singhurst et al. 2002). Although not officially recognized as such since this species is known from only two counties in the state, we consider it to be of conservation concern in TX. ⚠

Silene noctiflora L., (flowering at night), NIGHT-FLOWERING CATCHFLY. *Silene noctiflora* will most likely key out to *S. gallica*. *Silene noctiflora* differs in that the habit is taller, 60+ cm tall; has longer leaf blades, 1–11 cm long; and in having longer calyces, over 20 mm long. *Silene gallica* is a plant to 45 cm tall; leaf blades are 5(–7) cm long; and calyces ca. 10 mm long. Disturbed sites; *S. noctiflora* is known from McCurtain Co., AR and Caddo Par., LA across from the ne portion of East TX; found in all U.S. states except AL, AZ, NV, SC, and TX; Canada. Summer. Native of Europe and Asia.

SPERGULA L. SPURREY

A Eurasian genus of 5 species (Hartman & Rabeler 2005a; Mabberley 2017); somewhat similar to *Spergularia*. (Latin: *spargo*, scatter or sow, in reference to the discharge of seeds—Hartman & Rabeler 2005a)
REFERENCE: Hartman & Rabeler 2005a.

Spergula arvensis L., (of fields), CORN SPURREY, STICKWORT, STARWORT. Annual with stems erect to spreading, usually branched, to 50 cm long, glabrous or glandular-pubescent; leaves appearing whorled (technically opposite), each node with two axillary clusters of 8–15 leaves, sessile, fused basally, hair-like to linear, usually appearing terete and somewhat succulent, often revolute marginally and thus channeled beneath, 1.5–3(–5) cm long; stipules 4 per node, broadly triangular, whitish; inflorescences open terminal clusters; flowers with a short, saucer-shaped hypanthium; sepals separate (appear fused basally because of presence of hypanthium), 3.5–5 mm long; petals 5, white, nearly as long to as long as sepals; stamens usually 10; styles 5; capsules opening by 5 valves; seeds subglobose. Sandy soils, roadsides, disturbed areas; in East TX known only from Walker Co. (Kartesz 2022); also Dimmit, Lubbock, and Tom Green (Kartesz 2022) cos. around the state; widespread in the e 1/2 of the U.S. and s Canada, scattered in the w U.S. Mar–Jul. [*S. arvensis* var. *sativa* (Boenn.) Mert. & W.D.J. Koch] Native of Eurasia.

Petrorhagia prolifera [YA2]

Polycarpon tetraphyllum [LAM]

Sabulina michauxii [BB2]

Sabulina patula [BB2]

Sagina decumbens [BB2]

Sabulina muscorum [HEA]

Saponaria officinalis [GLE]

Scleranthus annuus [GEO]

SPERGULARIA (Pers.) J. Presl & C. Presl SAND-SPURREY

Annuals; leaves opposite, linear to filiform, somewhat fleshy, in ours usually without axillary clusters of leaves; stipules scarious; flowers usually numerous in cymose inflorescences, 5-merous; sepals connate at very base; petals white to pink or rosy; stamens 1–5; styles 3; capsules 3-valved.

A nearly cosmopolitan genus of ca. 60 species (Hartman & Rabeler 2005b; Mabberley 2017) including a number of halophytes. (Derived from the name of the related genus, *Spergula*, and Latin: *-aria*, pertaining to)
REFERENCES: Rossbach 1940; Hartman & Rabeler 2005b.

1. Plants glabrous; sepals 0.8–2 mm long; capsules 1.4–2.6 mm long; leaf blades linear to thread-like ____ **S. platensis**
1. Plants glandular-pubescent (at least branches of inflorescence); sepals 2.5–5 mm long; capsules 2.8–6.4 mm long and leaf blades linear OR capsules 1.2–2.5 mm long and leaf blades orbiculate to reniform.
   2. Leaf blades orbicular; stems prostrate and spreading; stipules fimbriately divided ____ *Drymaria cordata*
   2. Leaf blades linear; stems erect or ascending; stipules broadly triangular.
      3. Stipules a little longer than wide; inflorescences simple or 1–3+ times compound; seeds dull, not silver tinged ____ **S. marina**
      3. Stipules shorter than wide; inflorescences generally 4–8+ times compound; seeds shiny, silver tinged ____ **S. echinosperma**

Spergularia echinosperma Čelak., (with bristly seeds), BRISTLE-SEED SAND-SPURREY. Erect to ascending fleshy annual similar to *S. marina,* to 15 cm tall, glandular-pubescent; leaf blades linear, 0.5–3.5 cm long; stipules shorter than wide; flowers in leafy bracted cymes; petals white to pink or rosy, ca. as long as or longer than sepals; stamens 1–4(–5). Flats, sandy river banks, dunes; Anderson, Brazos, Gonzales, Harris, Jefferson, Liberty, Robertson, and Shelby (Kartesz 2022) cos.; w Edwards Plateau, Gulf Prairies and Marshes, and Trans-Pecos; AL, GA, LA, and TX.

Spergularia marina (L.) Griseb., (growing in salty places), SALT-MARSH SAND-SPURREY, MARSH SAND-SPURREY. Erect or diffuse fleshy annual ca. 5–20(–30) cm tall, glandular-pubescent (at least inflorescence); leaf blades linear, 0.8–2.5(–4) cm long, to 1.5 mm wide; stipules membranous, triangular, 1.2–3.5 mm long, a little longer than wide; flowers in leafy bracted cymes, the inflorescences often making up much of the plant; petals white to pink or rosy, ca. as long as or slightly shorter than sepals; stamens (1–)2–3(–5). In brackish or saline soils; mostly in the se portion of East TX; also Tarrant Co. (Mahler 1988) in the Cross Timbers and Prairies, Trans-Pecos, and Gulf Prairies and Marshes; widespread in the U.S. but mostly coastal from ME to TX and in the w 1/3 of the U.S.; s Canada. Apr–Sep. [*S. salina* J. Presl & C. Presl] Native of Europe and Asia, and possibly w North America (Rabeler & Hartman 2006). This species is expected to occur along highway right-of-ways where significant salt accumulates.

Spergularia platensis (Cambess.) Fenzl, (from the district of the Río de la Plata in South America), LA PLATA SAND-SPURREY. Glabrous, low, delicate annual often forming mats; leaf blades thread-like to linear, 0.5–3 cm long; flowers in loose terminal cymes, these generally 4–7+ times compound; stipules triangular, 1.5–3.5 mm long, ca. as long as wide or a little longer; petals white, small or rarely none, slightly shorter than sepals; stamens 3–5. Collected at Dallas by Reverchon in April 1880 (*Reverchon* 72 SMU) in wet sandy soil, later reported by him as "common," but not found there since, also Gonzales Co. (Kartesz 2022); otherwise in TX known only from Aransas Co. (*Whitehouse 18270* SMU) in the Gulf Prairies and Marshes; known in the U.S. only from CA and TX, also Baja CA in Mexico. Native of s South America (Argentina).

Drymaria cordata (L.) Willd. ex Schult., (heart-shaped, in reference to the leaves), DRYMARY, WEST INDIAN CHICKWEED. *Drymaria* will mostly key out to *Spergularia.* However, the habit and other characters are very different. Stems prostrate, glabrate to glandular pubescent; leaf blades orbicular to reniform; stipules divided in to 3–5 subulate to filiform segments; capsules 1.2–2.5 mm long. Lawns gardens and disturbed areas; found on the East TX border with LA in Beauregard and Vernon parishes; e U.S. from NC s to FL and w to LA. Flowering year-round. Native of Mexico, Central America, and tropical America.

STELLARIA L. CHICKWEED, STARWORT, STITCHWORT

Ours low usually annual herbs (one species can be perennial); stipules absent; flowers solitary or in terminal cymes of usually 3–7, 5-merous; petals 5, the petal blades split more than half way to base (corollas thus appearing to be of 10 petals) or petals absent; stamens 1–10; styles usually 3; capsules dehiscent by 6 valves.

A cosmopolitan, but especially n temperate genus of ca. 120 species (Morton 2005b; Mabberley 2017) including some cultivated ornamentals. The key to species is modified from Rabeler and Reznicek (1997) with additional characters from Morton (1972). We are following Kent (1997) for nomenclature of *S. pallida.* (Latin: *stella*, a star, in allusion to the star-shaped flowers)
REFERENCES: Whitehead & Sinha 1967; Morton 1972, 2004, 2005b; Rabeler 1988; Kent 1997; Rabeler & Reznicek 1997; Brown & Marcus 1998.

Silene antirrhina [BB2]

Silene conoidea [MOS]

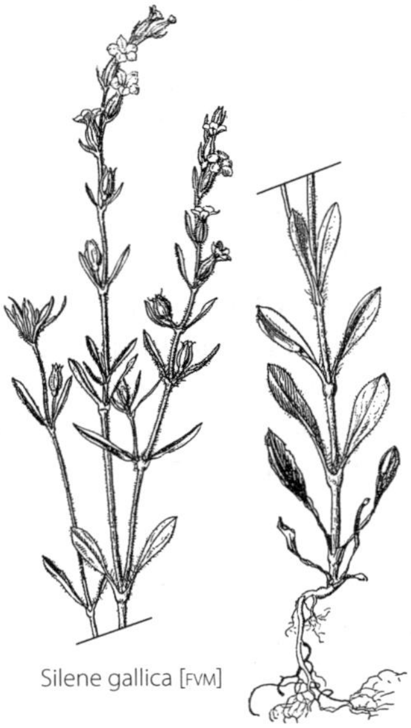

Silene gallica [FVM]

Silene stellata [GLE]

Silene subciliata [THO]

Silene virginica [BA1]

1. Leaf blades cordate to truncate (rarely abruptly rounded) at base; seeds covered with prominent stalked glands ______ **S. prostrata**
1. Leaf blades cuneate to round at base; seeds tuberculate, the tubercles broader than tall.
   2. Flowers usually open, with white petals; plants green, rarely yellowish green; sepals 4.5–5(–6) mm long, uniformly green; seeds dark brown, (0.8–)0.9–1.3 mm in diam., the surface covered with wavy, blunt papillae; capsules 3–5 mm long; stamens 3–5(–8); anthers red-violet ______ **S. media**
   2. Flowers usually cleistogamous, without petals; plants yellowish green; sepals 3–4 mm long, often with a red basal band; seeds yellowish brown, 0.5–0.9 mm in diam., the surface covered with acute papillae; capsules 2–4(–5) mm long; stamens usually 1–3; anthers gray-violet ______ **S. pallida**

Stellaria media (L.) Vill., (intermediate, the middle), COMMON CHICKWEED, TENPETAL, CHICKWEED STARWORT. Annual, possibly overwintering in protected places, stems prostrate to ascending, usually pubescent in lines; lower leaves petioled, upper sessile, blades oblong-lanceolate to ovate-lanceolate, widely tapered to abruptly narrowed at base; sepals 4.5–5(–6) mm long; petals white (sometimes absent), slightly shorter than the sepals, 2-lobed nearly to base and thus appearing as 10 petals; capsules usually pointing downward, equal or exceeding the sepals by 1–2 mm; $2n = 40, 42, 44$ (Morton 1972). Widespread weed of stream bottoms, lawns, disturbed sites; widespread in East TX; widespread in e 1/2 of TX; scattered to the w, throughout the U.S. and Canada. Feb–Apr. Native of Europe. This now cosmopolitan weed is considered by some to be "one of the most widely distributed weeds in the world" (Rabeler & Hartman 2006). It is usually autogamous (= self-pollinating). The young herbage of this species was reportedly gathered as a spinach substitute in its native range (Steyermark 1963), but it also has been reported to cause digestive disorders in sheep (Burlage 1968).

Stellaria pallida (Dumort.) Crép., (pale), LESSER CHICKWEED. Annual similar to *S. media*; differing as described in the key; flowers usually cleistogamous, ± closed at flowering, somewhat flask-shaped; petals usually absent or apparently so (if present, then minute); $2n = 22$ (Morton 1972). Disturbed areas, typically in sand; a Hopkins Co. collection (*Reznicek 10361* BRCH, MICH) was the first report for this species from TX (Rabeler & Reznicek 1997), also Dallas, Denton, Grayson, Hopkins, Lavaca (Kartesz 2022), and Grimes (*Shinners 9668* SMU) cos.; e U.S. from NY s to FL and w to NE and TX, also AZ, CA, NM, and WA. Because of its similarity to *S. media*, this is an easily overlooked species; it is to be expected in other nearby counties and possibly elsewhere in the state, especially early in the spring. Spring. Native of Europe. [*S. media* subsp. *pallida* (Dumort.) Asch. & Graebn., *S. apetala* of authors, not Ucria ex Roem.] This species was first reported for North America (Morton 1972) from a collection at Kitty Hawk, NC in 1969; however, Rabeler and Hartman (2006) indicate that, though not reported until much later, it was collected in MO in 1957. According to Morton (2005b), *S. pallida* is "automatically self-pollinated and often cleistogamous."

Stellaria prostrata Baldwin, (prostrate, flat to the ground), MEXICAN CHICKWEED, LLOVISNA. Annual, much-branched, stems decumbent, glandular-pubescent; lower leaves petiolate, upper sessile, blades ovate to deltate, basally cordate, truncate, or rarely abruptly rounded; sepals 4–5 mm long, in fruit to 8 mm long; petals white, shorter or equaling the sepals, deeply 2-lobed; capsules ± equaling the sepals; $2n = 26, 52$ (Morton 2005b). Low wooded areas; mostly in the sw portion of East TX; mainly se 1/4 of TX and Childress Co. (Kartesz 2022) in the Rolling Plains, also Trans-Pecos; FL, LA, NM, and TX. Feb–May. [*S. cuspidata* Willd. ex D.F.K. Schltdl. subsp. *prostrata* (Baldwin) J.K. Morton]

CELASTRACEAE R. Br.

BITTERSWEET OR STAFFTREE FAMILY

Ours shrubs or small trees; leaves alternate, opposite, or fascicled, deciduous or persistent, stipules present (but falling early) or absent; inflorescences terminal or axillary; flowers radially symmetrical, 4 or 5-merous, perfect or unisexual; ovary superior; fruit a capsule or drupe.

A medium-large (ca.1,200 species in 98 genera (Simmons 2004)), mainly tropical and subtropical to temperate family of mostly trees, shrubs, or woody vines; laticifers are usually present. The African *Catha edulis* (Vahl) Forssk. ex Endl. (KHAT, KAT, QAT, CAFTA) is cultivated for the leaves which are chewed by Moslems as a daily amphetamine-like stimulant in countries such as Ethiopia, Somalia, and Yemen; in the U.S. it is considered a controlled substance and was of concern when the U.S. deployed troops to Somalia in 1993 (Baker 1993; Simmons 2004). *Celastrus scandens* L., CLIMBING BITTERSWEET, AMERICAN BITTERSWEET, is a high-climbing, twining, woody vine cultivated for its colorful fruits; it is widespread in the e 1/2 of the U.S. but is known in TX only from the far w part of the state. Molecular research suggests the family is related to *Lepuropetalon* and *Parnassia* (Parnassiaceae, previously in the Saxifragaceae) (Soltis & Soltis 1997; Simpson 2004; Zhang & Simmons 2006; Simmons et al. 2023). (subclass Rosidae—Cronquist; order Celastrales—APG IV)

FAMILY RECOGNITION IN THE FIELD: shrubs or small trees with simple leaves, small flowers with a *disk* beneath or ± surrounding the ovary, and leathery colorful capsules containing seeds usually with brightly colored arils or brightly colored 2-seeded drupes.

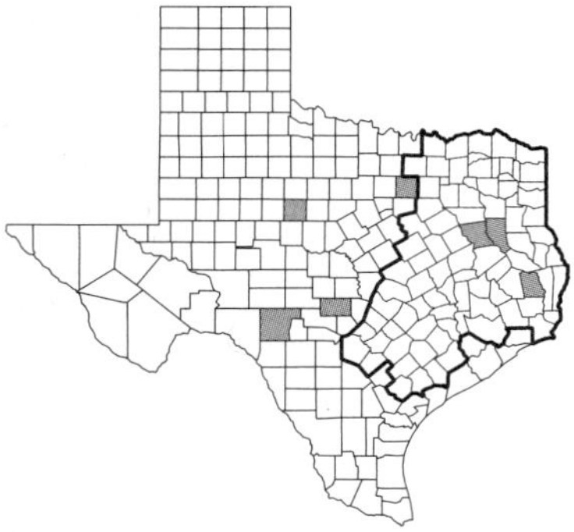
Saponaria officinalis

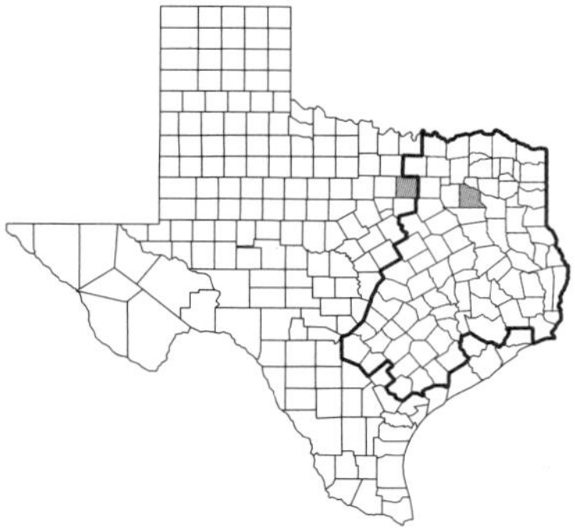
Scleranthus annuus

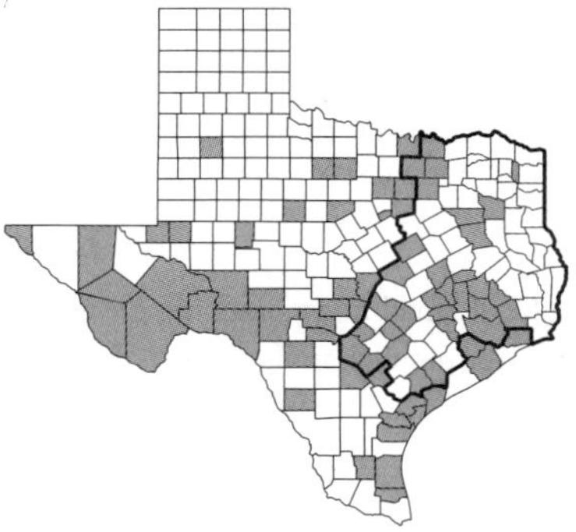
Silene antirrhina

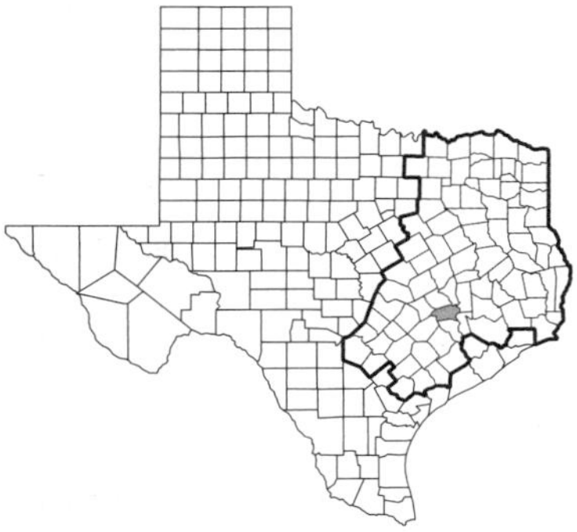
Silene conoidea

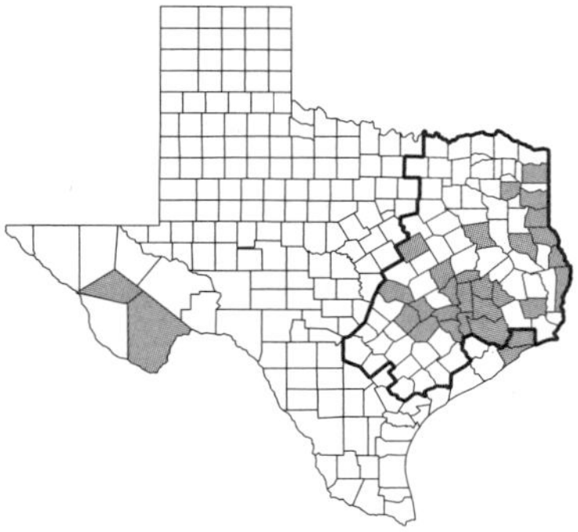
Silene gallica

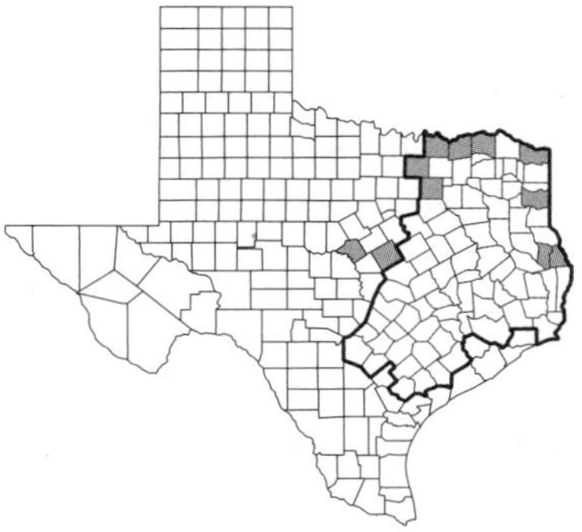
Silene stellata

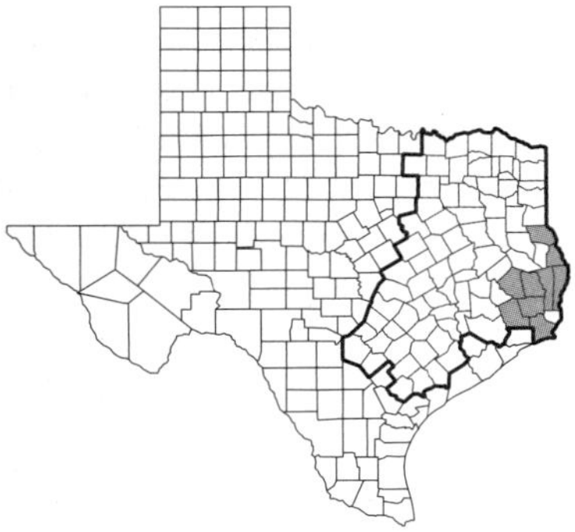
Silene subciliata

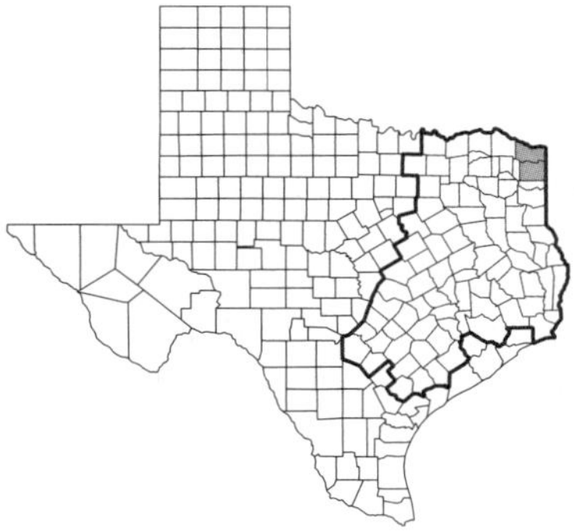
Silene virginica

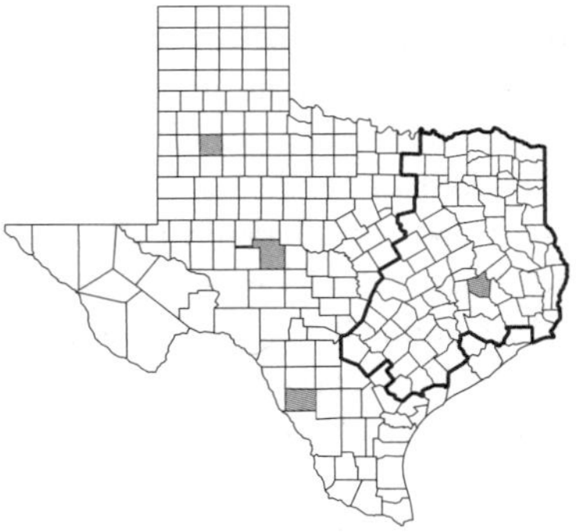
Spergula arvensis

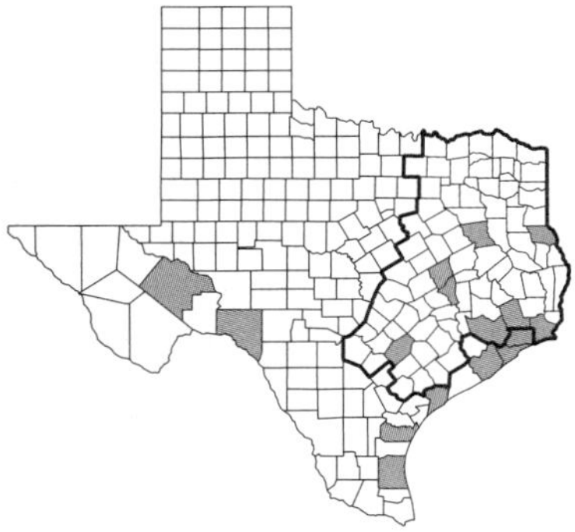
Spergularia echinosperma

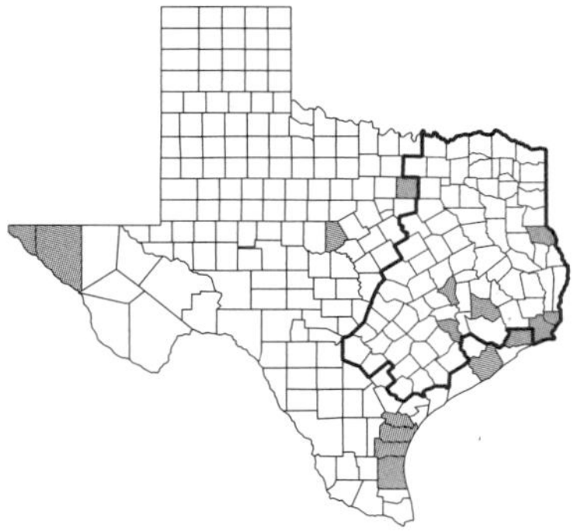
Spergularia marina

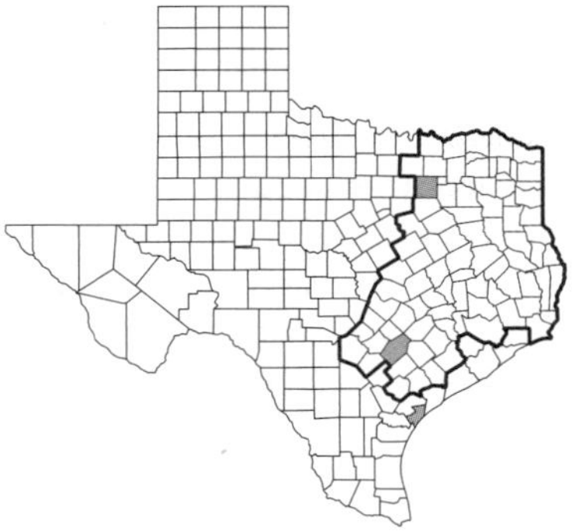
Spergularia platensis

References: Brizicky 1964a; Lundell 1969b; Simmons 2004; Zhang & Simmons 2006; Ma 2016; Ma & Levin 2016; Simmons et al. 2023.

1. Leaves opposite; stipules present (but falling early); fruit a capsule; flowers perfect, 4 or 5-merous ____ **Euonymus**
1. Leaves alternate or fascicled; stipules absent; fruit a drupe; flowers unisexual, 4-merous ____ **Schaefferia**

Euonymus L. SPINDLETREE, STRAWBERRY-BUSH

Shrubs or small trees; branchlets 4-angled or terete; leaves simple, opposite, deciduous; stipules minute, falling early; inflorescences terminal or axillary; pedicels jointed; flowers small, radially symmetrical, 4 or 5-merous, perfect; petals distinct; stamens inserted on margin of a prominent disk; ovary superior; fruit a loculicidal capsule; seeds arillate.

A genus of ca. 140 species (Ma & Levin 2016) of evergreen or deciduous trees or shrubs native to n temperate areas, especially Asia; also found in Australia; many are cultivated as ornamentals for their bright autumn foliage, evergreen leaves, or showy capsules with scarlet to orange, bird-dispersed, arillate seeds; the seeds of many species have cardiotoxic glycosides and the leaves, bark, and seeds of some species are poisonous (Hardin & Arena 1974). Simmons (2004) noted that the genus may be paraphyletic, with several other genera nested within it. The genus is sometimes spelled *Evonymus*. (Greek: *eu*, good, and *onoma*, name, but used ironically, the plants having had the bad reputation of poisoning cattle)
References: Lundell 1941; Ma & Levin 2016.

1. Leaves sessile or nearly so (petioles < 5 mm long); flowers 5-merous; fruits warty; species mostly in se part of East TX (also Bastrop Co. to the w) ____ **E. americanus**
1. Leaves petiolate (petioles usually 5–20 mm long); flowers 4-merous; fruits smooth; species scattered in w part of East TX ____ **E. atropurpureus**

Euonymus americanus L., (of America), STRAWBERRY-BUSH, BURSTING-HEART, BROOK EUONYMUS, AMERICAN BURNING-BUSH. Erect shrub to ca. 2 m tall or branches sometimes arching; leaves lanceolate to elliptic or ovate, acute, finely toothed, sessile or nearly so (petioles < 5 mm long), yellowish in autumn; inflorescences of solitary flowers or few-flowered cymes; flowers 5-merous, greenish purple; fruit pinkish to reddish, 3–5-lobed, warty, ca. 15 mm in diam.; seeds with orange to scarlet aril. Moist woods; mostly in se part of East TX, also Bastrop Co. (Turner et al. 2003) to the w; e U.S. from PA s to FL and w to MO, OK, and TX. May–Jun.

Euonymus atropurpureus Jacq., (dark purple), EASTERN WAHOO. Erect shrub or small tree to 8 m tall; bark smooth, gray or green; leaves elliptic, oval, ovate, or obovate, acuminate, finely and sharply toothed, with distinct petioles 5–20 mm long, reddish in autumn; inflorescences pedunculate, of usually 7–20 flowers; flowers perfect, 4-merous, greenish purple to dark-red; fruit pinkish to reddish or purple, usually deeply 4-lobed, smooth, ca. 15 mm in diam.; seeds with scarlet aril. Limestone soils, stream bottom woods; Bell, Burleson, Dallas, Delta, Grayson, Kaufman, and Lamar (Turner et al. 2003) cos.; throughout e U.S. from PA sw to TN and w to MN and OK; scattered from NC s to FL and w to TX. Late Apr–early Jun. This species is the source of wahoo root bark, formerly used medicinally, and a medicinal resin, called euonymin; euonymin and the fruits are apparently toxic (Duke 1985); Burlage (1968) reported the plant to be a violent purgative, dangerous to livestock and children, with symptoms ranging from vomiting to mental symptoms and loss of consciousness.

Schaefferia Jacq. DESERT-YAUPON

A genus of 5–15 species of sc and se U.S., Mexico, West Indies, Central and South America (Ma 2016). (for Jacob Christian Schaeffer, 1718–1790, German botanist, zoologist, theologian and clergyman)
References: Lundell 1969b; Ma 2016.

Schaefferia cuneifolia A. Gray, (with wedge-shaped leaves), DESERT-YAUPON. Dioecious shrub 1–2 m tall; stems ± spinescent; leaves alternate or fascicled, sessile, blade 5–23 mm long, 2–15 mm wide, obovate or oblanceolate, apex rounded sometimes shallowly emarginate, entire; inflorescence axillary, fascicled or flowers solitary, unisexual, sessile or pedicel to 2 mm long; sepals ca. 0.6 mm long; petals ca. 3 mm long, greenish white to white; stamens 4, shorter than petals; fruit a drupe, 2-locular, depressed-globose, orange to bright red; seeds 2; aril absent. Rocky hillsides, canyons; in East TX known only from Bexar Co. (Turner et al. 2003); South TX Plains, Edwards Plateau, and Trans-Pecos; n Mexico. Feb–Jul.

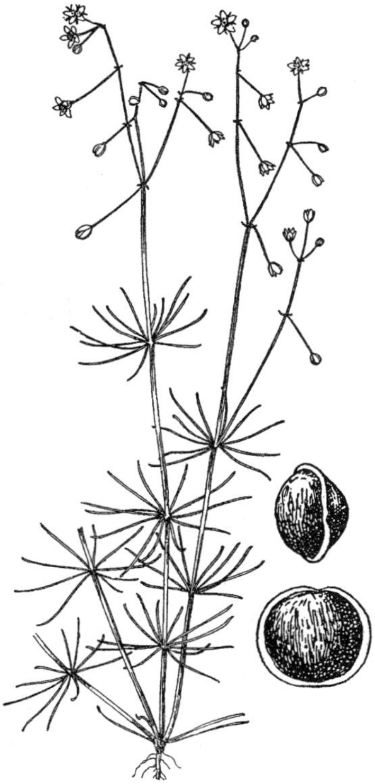

Spergula arvensis [NCS]

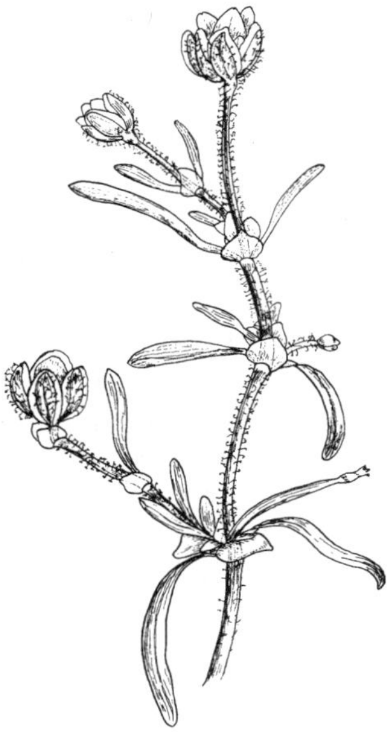

Spergularia echinosperma [CO1]

Spergularia marina [MOS]

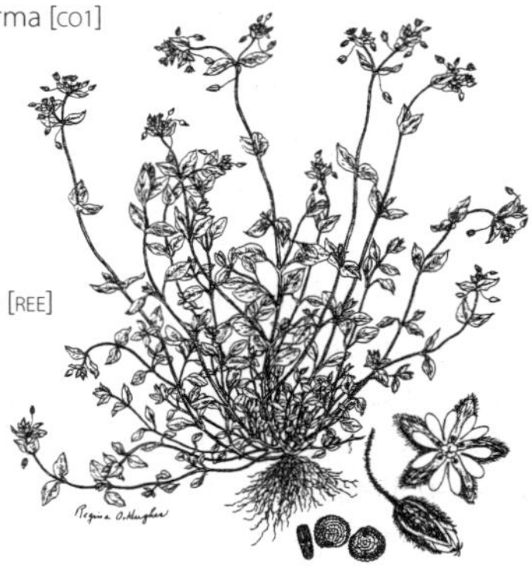

Stellaria media [REE]

Spergularia platensis [LUN]

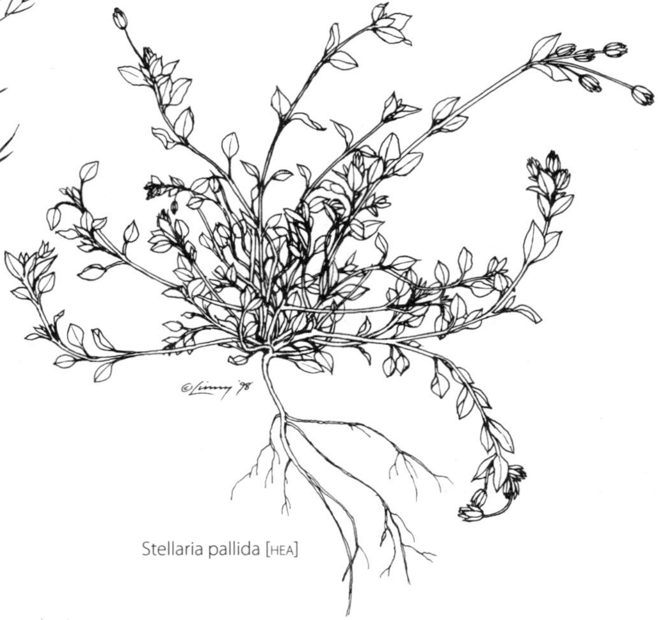

Stellaria pallida [HEA]

CERATOPHYLLACEAE Gray

HORNWORT FAMILY

A very small cosmopolitan family of 1 genus and 1–6 species, the exact number depending on taxonomic approach (Les 1997; Stevens 2001 onwards; Mabberley 2017). Mabberley (2017) states, "up to 30 recognized by some authors." *Ceratophyllum* is an extremely reduced (e.g., no roots, without vessels) and highly specialized aquatic; the resulting lack of taxonomically useful morphological characters has made its classification challenging. Early molecular analyses indicated that *Ceratophyllum* was the sister group of all other angiosperms, suggesting the genus was a vestige of an ancient angiosperm lineage that diverged early from the line leading to other living angiosperms (Chase et al. 1993; Qiu et al. 1993); a more recent molecular analysis placed *Ceratophyllum* as the sister group to the monocots (Soltis et al. 1997). Still more recently (e.g., Moore et al. 2007; Soltis et al. 2011), Ceratophyllaceae appears to be the sister group of the eudicots. Others (Moore et al. 2011) suggested that the Ceratophyllaceae are sister to Chloranthaceae. Soltis et al. (2018) stated, "Although the prevailing view is to place Ceratophyllaceae with the eudicots (followed here), we urge caution." The positon of the Ceratophyllaceae is still in question. Its unsealed carpels, among other distinctive characters, makes the genus anomalous among flowering plants (Qui et al. 1993). (subclass Magnoliidae—Cronquist; order Ceratophyllales—APG IV)

FAMILY RECOGNITION IN THE FIELD: rootless aquatics floating below water's surface, with whorled leaves *forked* into nearly thread-like segments.

REFERENCES: Wood 1959; Les 1988a, 1993, 1997; Endress 1994; Moore et al. 2007.

CERATOPHYLLUM L. HORNWORT, COON-TAIL

Perennial, bushy-branched, rootless, aquatic herbs, floating completely submerged; leaves whorled, sessile, 1–4 times dichotomously forked into nearly thread-like segments; flowers solitary in 1 axil of a whorl, sessile, without perianth but with involucre of 8–12 minute bracts, unisexual, both sexes on the same plant; stamens 4–10; pistil 1; style persistent; ovary superior; fruits achenes, with spines.

In some parts of the world, bilharzia (i.e., schistosomiasis)-carrying snails and malaria-carrying mosquito larvae are sheltered by these floating aquatics; they can also choke waterways and disturb operations of hydroelectric plants; on the positive side, they oxygenate the water and provide cover for small fishes. The flowers are water-pollinated (= hydrophilous) with the anthers breaking off and floating to the surface of the water; the pollen grains are then released at the surface and slowly sink to the female flowers—pollination thus occurs underwater; technically this type of pollination is known as hypohydrophily (= pollination below the water surface) in contrast to epihydrophily (pollination at the water surface). *Ceratophyllum* is one of only two dicot genera for which hydrophily has been documented; the other is *Callitriche* in the Plantaginaceae (Cox 1988; Philbrick 1991, 1993; Philbrick & Osborn 1994). When sterile, HORNWORTS are often confused with other aquatics, such as *Myriophyllum*, *Najas*, or the algae *Chara*; however, HORNWORTS can be distinguished by their forked leaves. The common name derives "from old translucent horny leaves" (Mabberley 2017). (Greek: *ceras*, a horn, and *phyllon*, leaf, from stiff leaf divisions)

REFERENCES: Lowden 1978; Godfrey & Wooten 1981; Les 1986a, 1986b, 1988b, 1988c, 1989; Chase et al. 1993; Qiu et al. 1993; Schneider & Carlquist 1996.

1. Leaves usually forked 1 or 2(–3) times, usually with 2–4 ultimate segments, the segments with 4–5 conspicuous, widely spaced teeth on 1 side (the teeth visible to the naked eye); achenes without spines except at very base ______ **C. demersum**
1. Leaves usually forked 2–4 times, usually with 3–8 ultimate segments, the segments entire or obscurely serrate (the teeth not visible to the naked eye); achenes with spines along lateral margins and at base ______ **C. echinatum**

Ceratophyllum demersum L., (under water). Leaves 5–12 per whorl, 1–3 cm long, ultimate segments to ca. 0.5 mm wide, to ca. 1 mm at the teeth; involucral bracts 1–2 mm long; style to ca. 10 mm long; achenes 4–6 mm long. Streams, lakes, and ponds; widespread in East TX; widespread in the e 1/3 of TX, scattered further w, mostly Edwards Plateau and Trans-Pecos; throughout the U.S.; Canada. By far the most common of the 2 species in East TX. May–Sep. Native of the U.S. This species reproduces primarily asexually; it grows prolifically and in some situations can be a serious weed (Les 1997).

Ceratophyllum echinatum A. Gray, (prickly). Similar to *C. demersum*; style to ca. 6 mm long; achenes 5–7 mm long. Streams, lakes, and ponds; Bowie (Turner et al. 2003), and Harris (*Brown 8856* BRIT-NLU) cos.; widely scattered in the e 1/2 of the U.S, also CA, OR, and WA; adjacent Canada. Summer. Jones et al. (1997) treated this taxon as *C. muricatum* Cham. subsp. *australe* (Griseb.) Les. However, Les (1997) indicated that *C. muricatum* is known in North America only from coastal se United States, while *C. echinatum* occurs widely in East TX. Native of the U.S.

CHENOPODIACEAE Vent.

GOOSEFOOT OR PIGWEED FAMILY

Ours annual, biennial, or perennial herbs (some shrubs or subshrubs) with stems often ± succulent or slightly succulent, often with a taproot, glabrous or with pubescence, sometimes glandular or mealy (due to collapsed inflated hairs); leaves alternate (rarely opposite), simple, the blades entire or rather coarsely or irregularly and bluntly toothed or lobed, often ± succulent or slightly succulent; stipules absent; flowers 1–many, glomerulate in spikes or panicles of spikes or axillary, perfect or unisexual, very small; bracts, if present, green; perianth segments 0–5, green, sometimes with white to yellowish or pink margins; petals absent; stamens 0–5, the filaments free; pistil 1; ovary usually superior; fruit a 1-seeded indehiscent or irregularly rupturing utricle; seed often lens-shaped and circular, with curved or coiled embryo (often not easily observable).

A medium-large (1,500 species in ca. 100 genera—Welsh et al. 2003) cosmopolitan family, especially of desert or dry areas; most are herbaceous with some shrubs or rarely small trees; many are halophytes (= capable of living in areas of high salt concentrations) or xerophytes. The family includes food plants such as *Beta vulgaris* L. (BEET, SUGAR BEET, SWISS CHARD), *Chenopodium quinoa* Willd. (the Andean crop QUINOA), and *Spinacia oleracea* L. (SPINACH) as well as a number of agricultural weeds. They exhibit the unusual, reddish, nitrogen-containing pigments known as betalains (characteristic of most Caryophyllidae/Caryophyllales—Cronquist & Thorne 1994; Stevens 2001 onwards) which derive their name from the genus *Beta*. Chenopodiaceae typically have CAM (Crassulacean Acid Metabolism) (Welsh et al. 2003), and many species have Kranz anatomy and the associated C4 photosynthetic pathway, an integrated set of anatomical and physiological adaptations for hot, dry conditions; these adaptations have apparently evolved a number of times in the family (Judd & Ferguson 1999). Molecular evidence indicates Chenopodiaceae is most closely related to Amaranthaceae (Downie & Palmer 1994); in fact, authors vary in their treatment of the Amaranthaceae and Chenopodiaceae. While including the Chenopodiaceae in the Amaranthaceae has been considered less so in the past (treated separately in Correll & Johnston 1970; Robertson & Clemants 2003; Turner et al. 2003), recently it has gained more proponents (e.g., Stevens 2001 onwards; Mabberley 2017; Meyers et al. 2020; Lesica 2022). However, we are maintaining the traditional separation and are following recent authors (Yatskievych 2006; Hernández-Ledezma et al. 2015; Powell & Worthington 2018; Weakley, A.S., and Southeastern Flora Team 2022) in treating the Chenopodiaceae separate from the Amaranthaceae, rather than as one family. Kühn (1993) cited references indicating that many weedy Chenopodiaceae produce allergy-causing pollen that can give rise to asthma and rhinitis, and toxicity from high levels of nitrates or oxalates has been reported for some species (Kingsbury 1964). While the small inconspicuous flowers of most are wind-pollinated, some have nectaries and may be partially pollinated by insects (Judd & Ferguson 1999). (subclass Caryophyllidae—Cronquist; order Caryophyllales—APG IV)

FAMILY RECOGNITION IN THE FIELD: herbs sometimes with succulent or slightly succulent leaves; leaves alternate, simple; flowers very small, inconspicuous, apetalous, greenish, not subtended by scarious bracts, often in spikes; filaments separate. Amaranthaceae have similar flowers (e.g., very small, lacking petals), but have scarious (= dry, papery), often colored bracts below each flower and have united filaments.

REFERENCES: Standley 1916; Reed 1969a; Kühn 1993; Wilken 1993a; Behnke & Mabry 1994; Downie & Palmer 1994; Downie et al. 1997; Gale et al. 2001; Kadereit et al. 2003; Welsh et al. 2003.

1. Leaves reduced to minute scales; stems appearing jointed, succulent; flowers embedded in cavities in the jointed succulent stem; species mainly of salt marshes ____ **Salicornia**
1. Leaves with obvious blades; stems not appearing jointed, not succulent or only slightly so; flowers not embedded in cavities in stem.
   2. Leaf blades and bracts with a spine/bristle tip, very narrow (2 mm or less wide); stem leaves sessile ____ **Salsola**
   2. Leaf blades and bracts not spine/bristle-tipped, usually > 2 mm wide (often much greater); stem leaves various, but often petioled or with narrowed, petiolar base.
      3. Root greatly thickened, often pinkish or reddish in color (can be white or yellow); fruits strongly adherent in clusters; plants glabrous; introduced cultivar (BEETS) that possibly escapes ____ **Beta**
      3. Roots not as above; fruits not adherent (but can be in clusters); plants often with conspicuous silky, woolly, mealy, or glandular pubescence OR glabrous; native and introduced species widespread in East TX.
         4. Pistillate flowers subtended by 2 bracts/bracteoles, these fused 1/2 or more of their length, enlarged and enclosing the fruit; flowers all unisexual, the staminate and pistillate often in different inflorescences or in the axils of different leaves or in different plants.
            5. Leaves and stems glabrous, green; many leaves in a basal rosette; lower leaves with conspicuous petioles 1–5 cm long; stigmas 4–5; introduced cultivated species rarely escaped in TX ____ **Spinacia**
            5. Leaves and stems usually (at least when young) covered with scales giving the plant a grayish, scurfy appearance, sometimes glabrate; few or no leaves in a basal rosette; lower leaves often sessile or nearly so OR with petioles; stigmas 2; including native species widespread in East TX ____ **Atriplex**

4. Flowers not subtended by 2 fused bracts/bracteoles; fruit subtended only by perianth segments, enclosing fruit or not; flowers generally with both sexes (some can be unisexual).
   6. Larger leaf blades typically arrowhead-shaped with 2 basal lobes; perianth with 1 perianth segment, linear-oblanceolate or linear-obovate; stamen 1 per flower ______ **Blitum**
   6. Leaf blades various, but usually not arrowhead-shaped; perianth with 4–5 perianth segments, variously shaped; stamens usually (1–)4–5 per flower (sometimes absent in 2 species of *Dysphania*).
      7. Plants mostly glabrous, glaucous; leaf blades fleshy, elliptic or circular in cross-section, sessile, linear to linear-lanceolate, entire, often tightly ascending ______ **Suaeda**
      7. Plants usually with silky, woolly, glandular or mealy pubescence; leaf blades not fleshy or only slightly so, various, but not as above.
         8. Plants with silky or woolly pubescence (sometimes glabrate except around flowers); fruiting perianth segments with a horizontal wing running across their backs.
            9. Fruiting perianth segments each with a separate horizontal wing across the back; leaf blades entire; plants not tumbleweed-like ______ **Bassia**
            9. Fruiting perianth segments not with a separate wing each, rather the perianth segments with an irregularly lobed but continuous horizontal wing going all the way around; leaf blades sinuately toothed or pinnately lobed; plants with roundish, tumbleweed-like appearance ______ **Dysphania atriplicifolia**
         8. Plants with glandular or mealy-appearing pubescence (sometimes glabrate); fruiting perianth segments rounded on back or keeled lengthwise, without a horizontal wing.
            10. Leaves and inflorescences glandular (the glands stalked or sessile) and often pubescent with non-glandular hairs; leaves coarsely toothed or pinnately lobed; plants usually with a strong odor ______ **Dysphania**
            10. Leaves and inflorescences glabrous or mealy but neither glandular nor pubescent with non-glandular hairs; leaf blades dentate, or lobed, also some entire; plants usually without a strong odor (sometimes with an odor).
               11. Mature leaves and stems mostly glabrous to slightly mealy; prominent midvein visible inside perianth segments ______ **Chenopodiastrum**
               11. Mature leaves and stems usually densely mealy; prominent midvein not visible inside perianth segments ______ **Chenopodium**

ATRIPLEX L. SALTBUSH, SPEARSCALE, ORACHE

Annual or perennial, monoecious or dioecious herbs or sometimes ± woody; surfaces usually gray-scurfy, at least when young; leaves alternate or alternate above and opposite below; leaf blades entire or undulate to dentate; inflorescence axillary or terminal, in clusters or spikes or panicles; staminate flowers ebracteate, with 5-parted perianth and usually 5 stamens; pistillate flowers subtended by 2 bracteoles, without perianth or perianth of minute scales; fruits achenes, enclosed by enlarged bracteoles.

A worldwide genus of ca. 300 species, particularly of subarctic, temperate, and subtropical areas (Mabberley 2017). Many are halophytic, and a number are used as salt-tolerant forage. A number are edible and some were used by Native Americans. The Asian *A. hortensis* L. (ORACHE) is widely cultivated for its leaves, which are used like spinach. The flowers are minute and wind-pollinated (Judd & Ferguson 1999). (Ancient Latin name)

REFERENCES: Hall & Clements 1923; Brown 1956; Bassett et al. 1983; Freeman et al. 1984; McGregor 1986a; Welsh 1995, 2003; Flores Olvera 2003.

1. Leaf blades green on both upper and lower surfaces, glabrous or nearly so, 25–120 mm long, sometimes, but not always, hastate (= arrowhead shaped with the two basal lobes turned outward near base) ______ **A. patula**
1. Leaf blades white to gray-scurfy, particularly on lower surfaces, 2–40(–50) mm long, not hastate.
   2. Leaves opposite, the blades 2–5 mm long ______ **A. matamorensis**
   2. Leaves alternate, the blades 5–50 mm long.
      3. Fruiting bracteoles with 4 conspicuous lengthwise wings (2 per bracteole); leaf blades with margins entire; plants usually 0.8–2 m tall; male and female flowers usually on separate plants ______ **A. canescens**
      3. Fruiting bracteoles various, but without 4 conspicuous wings; leaves with margins various, sinuate to dentate OR entire; plants to 1 m tall; male and female flowers on the same plant.
         4. Staminate and pistillate flowers in mixed axillary clusters; fruiting bracteole orbicular with 11 or more teeth per bracteole ______ **A. elegans**
         4. Staminate flowers in terminal or axillary spikes; pistillate flowers in axillary fascicles; fruiting bracteoles wedge-shaped with ca. 7 or fewer teeth per bracteole.
            5. Fruiting bracteoles 4.5–7 mm long, usually longer than wide ______ **A. mucronata**
            5. Fruiting bracteoles 2.5–4.5 mm long, bracteole usually wider than long ______ **A. pentandra**

Atriplex canescens (Pursh) Nutt., (gray-pubescent), FOUR-WING SALTBUSH, CHAMIZA. Dioecious (rarely monoecious) perennial shrub; stems erect, usually 0.8–2 m tall; leaves alternate, sessile or nearly so, blades 10–40(–50) mm long, linear-spatulate to

Stellaria prostrata [BR2]

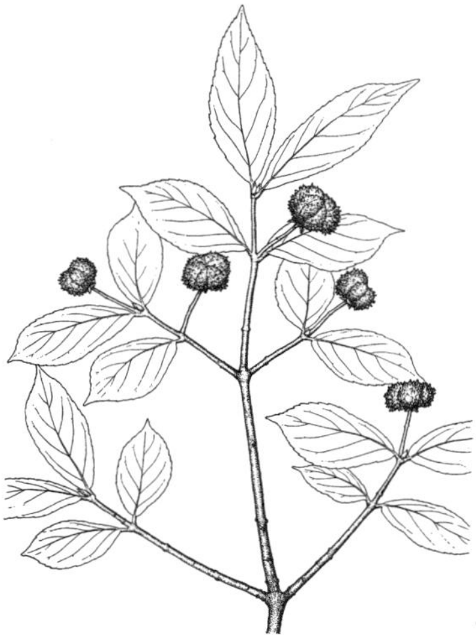

Euonymus americanus [LUN]

Euonymus atropurpureus [SA3]

Ceratophyllum demersum [REE]

Ceratophyllum echinatum [HEA]

Schaefferia cuniefolia [FNA]

Atriplex canescens [POW]

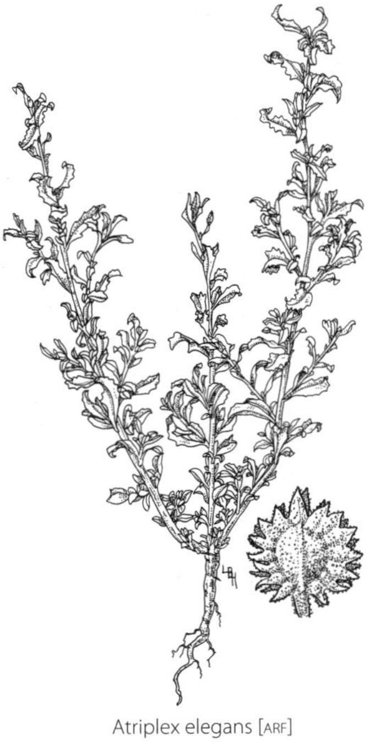

Atriplex elegans [ARF]

narrowly oblong or oblanceolate, entire, thickish, the upper and lower surfaces both gray-scurfy, becoming glabrous; staminate and pistillate flowers in glomerules in dense spikes arranged in terminal panicles; fruiting bracteoles stipitate, 8–12 mm long, with prominent longitudinal wings (2 wings per bracteole); wings usually 8–12 mm wide, toothed to entire. Saline or non-saline or alkaline soils; Bell, Brazos, Dallas, Goliad, and Gonzales (Kartesz 2022) cos.; widely scattered in w 2/3 of TX, throughout the Trans-Pecos; widespread in w 1/2 of the U.S. and adjacent Canada. Apr–Oct. [*A. nuttallii* S. Watson] Native Americans ground the seeds into a "baking powder" to use in making bread (Powell 1988). Individuals of this species are sexually labile, i.e., able to change from one sex to the other. Sex change (generally from female to male) is apparently associated with stress such as unusual cold or drought and, "... appears to confer a survival advantage to the individual" (Freeman et al. 1984). Reported to cause bloat in sheep (Burlage 1968). Four varieties are recognized in this species (Welsh 2003), but only var. *canescens* occurs in TX.

Atriplex elegans (Moq.) D. Dietr., (graceful), WHITE-SCALE SALTBUSH. Monoecious usually annual plant 5–45 cm tall; leaves subsessile to short petiolate, blades 5–30 mm long, margin entire to dentate, upper surface green, glabrate, lower surface densely scurfy; staminate and pistillate flowers mixed in small axillary clusters; fruiting bracteoles orbicular with 11 or more teeth per bracteole from base to apex. Saline soils; included based on specimen citation in Reed (1969a) for Gonzales Co.; also Trans-Pecos; sw U.S. from TX w to CA. Spring–fall. [*A. elegans* var. *thornberi* M.E. Jones, *A. thornberi* (M.E. Jones) Standl.]

Atriplex matamorensis A. Nelson, (of Matamoros, Mexico), QUELITE CENIZO. Dioecious perennial herb; stems 10–40 cm tall; leaves mostly opposite, sessile, blades quite small, almost scale-like, 2–5 mm long, oblong or oblong-lanceolate; pistillate flowers solitary or in small axillary clusters; fruiting bracteoles sessile, suborbiculate, 2–3 mm long, toothed, not winged. Saline soils; Gonzales and Victoria (Turner et al. 2003) cos. in s part of East TX; otherwise mainly coastal in TX; in the U.S. known only from TX; also adjacent Mexico. Summer–fall. [*A. oppositifolia* DC.]

Atriplex mucronata Raf., (with a sharp point), SEABEACH ORACH. Monoecious annual 1–6 dm tall; leaves alternate, or lower leaves opposite to subopposite, leaf blades often densely scurfy below and grayish green to glabrate above, staminate flowers in terminal or axillary spikes; pistillate flowers in axillary fascicles; fruiting bracteoles wedge-shaped, 4.5–7 mm long, usually longer than wide. Sandy shores, salt marshes; Jefferson Co. (Kartesz 2024); also Gulf Prairies and Marshes; coastal U.S. from NY s to FL and w to TX. Summer–fall. Native of the U.S.

Atriplex patula L., (spreading), SPEAR ORACH, SPEARSCALE. Mostly monoecious annual herb; stems erect, usually 30–90 cm tall; leaves alternate (except near base), petiolate, blades 25–70(–120) mm long, lanceolate to oblong to hastate-ovate, sometimes hastate near base (= arrowhead shaped with the two basal lobes turned outward), toothed or entire, green on both surfaces, glabrous or nearly so; fruiting bracteoles sessile, usually 2–7 mm long, ovate-rhombic, rhombic to rhombic-triangular, entire or with a few teeth, not winged. Weedy, non-saline habitats; in East TX cited only from Jefferson and Van Zandt (Turner et al. 2003) cos.; otherwise not known in the state; Kartesz (2022) and Weakley, A.S., and Southeastern Flora Team (2022) do not have *A. patula* mapped for East TX; widely scattered mainly in the n 1/2 of the U.S.; Canada. Summer–fall. [*A. hastata* L. subsp. *patula* (L.) E. Pons] Native of Eurasia. Welsh (2003) indicated that this species was probably introduced to North America in the early to mid-eighteenth century.

Atriplex pentandra (Jacq.) Standl., (with five stamens), SEASHORE ORACH, SEABEACH ORACH, SEABEACH ATRIPLEX. Monoecious annual or perennial herb; stems sprawling to erect, 30–100 cm tall; leaves alternate, sessile or short petiolate, blades 10–30 mm long, variously toothed to undulate or entire; staminate flowers in short spikes or panicles; pistillate flowers in axillary fascicles; fruiting bracteoles 2.5–4.5 mm long, usually at least as wide as long, marginally deeply toothed. Sandy seashores, coastal salt marshes; Bexar, Dewitt, Gonzales, Harris and Jefferson (Kartesz 2024) cos. in s part of East TX; mainly South TX Plains, Gulf Prairies and Marshes; coastal U.S. from NC to TX. Summer–fall. [*A. texana* S. Watson, *A. wardii* Standl.] We are treating this species in an inclusive sense, including *A. texana* and *A. wardii*, which were recognized as distinct species by Correll and Johnston (1970); Welsh (2003) indicated that the features used to distinguish them "fail singly and in combination."

BASSIA All. SMOTHERWEED

A genus of ca. 20 species of herbs and dwarf shrubs of Europe, Asia, Africa, and North America (Kadereit & Freitag 2011; Weakley, A.S., and Southeastern Flora Team 2022). Kadereit and Freitag (2011) recircumscribed *Bassia* and many taxa once in *Kochia* were transferred to *Bassia*. (Named for Ferdinando Bassi, 1710–1774, Italian botanist)

REFERENCES: Blackwell et al. 1978; Mosyakin 2003b; Kadereit & Freitag 2011.

Bassia scoparia (L.) A.J. Scott, (broom-like), MEXICAN FIREWEED, BELVEDERE SUMMER-CYPRESS, SUMMER-CYPRESS, MOCK CYPRESS, MEXICAN FIREBUSH. Annual, pubescent to glabrate; stems erect, branched, 0.3–1.5(–2) m tall, often with reddish or purplish streaking;

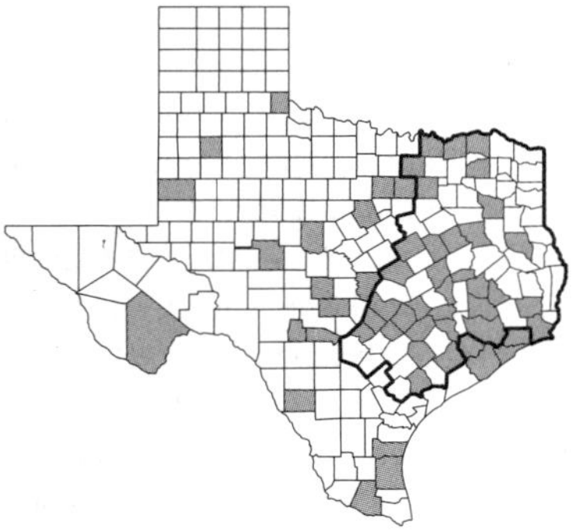
Stellaria media

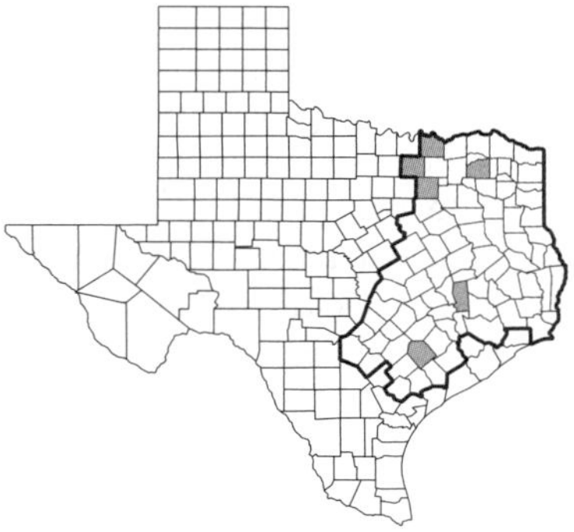
Stellaria pallida

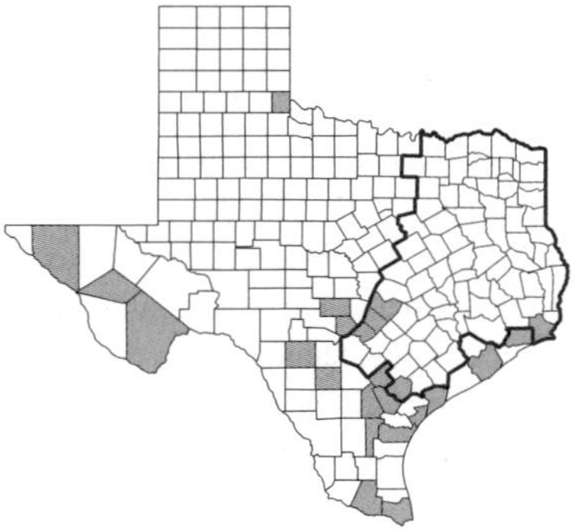
Stellaria prostrata

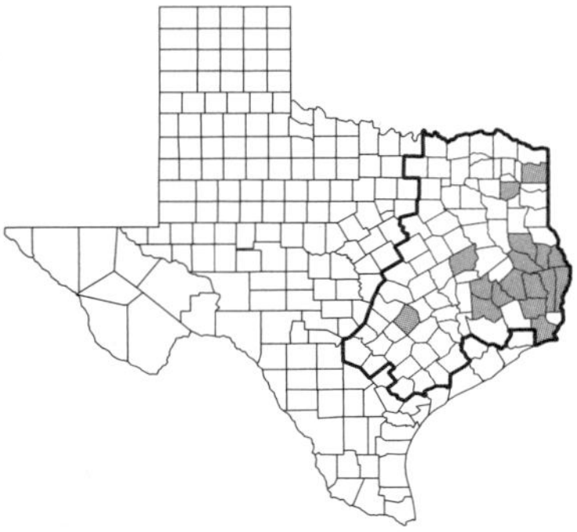
Euonymus americanus

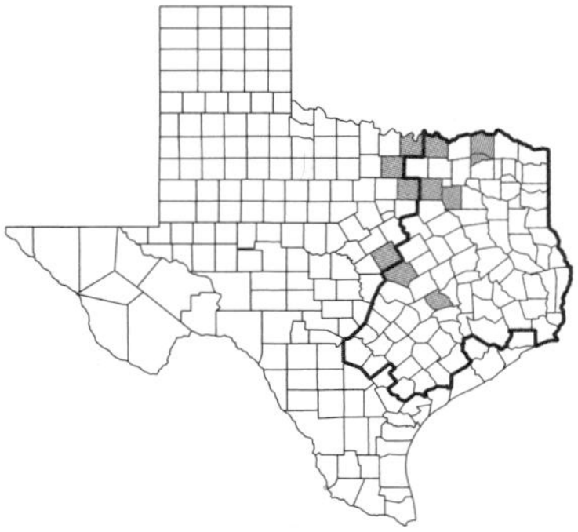
Euonymus atropurpureus

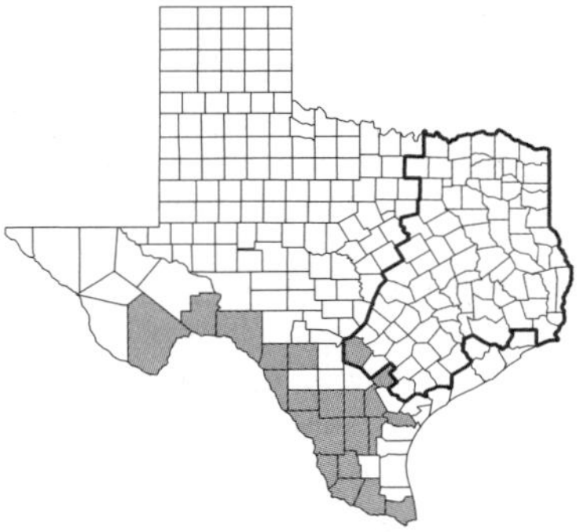
Schaefferia cuneifolia

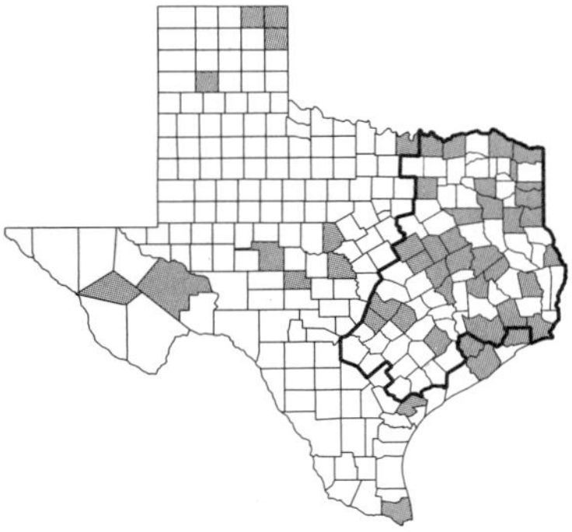
Ceratophyllum demersum

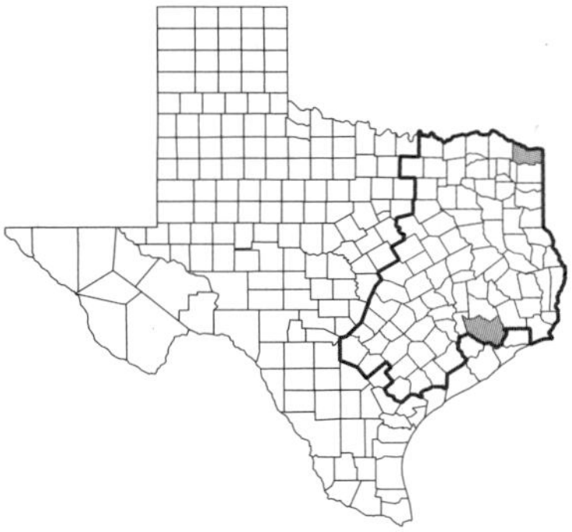
Ceratophyllum echinatum

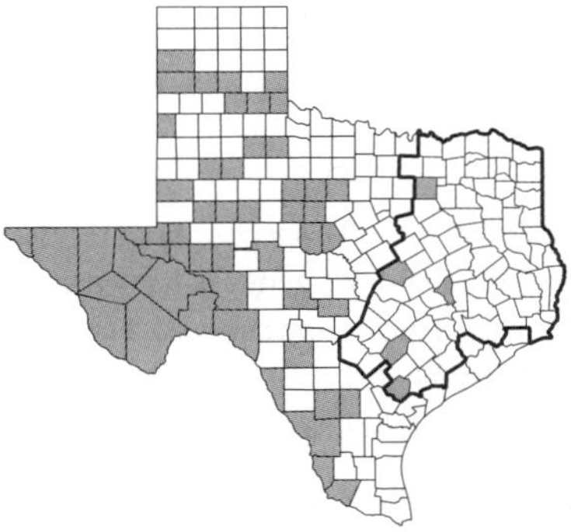
Atriplex canescens

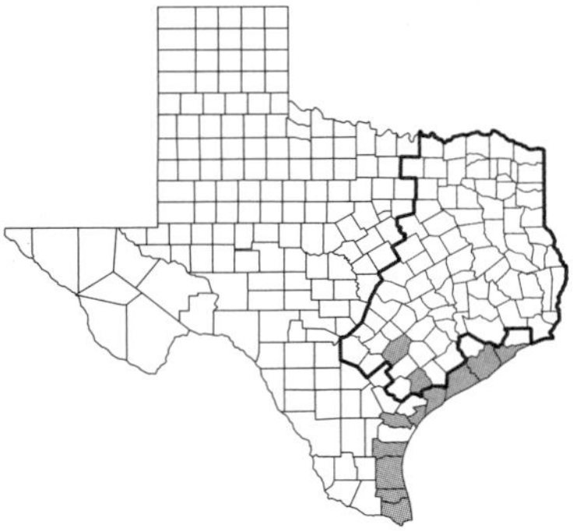
Atriplex matamorensis

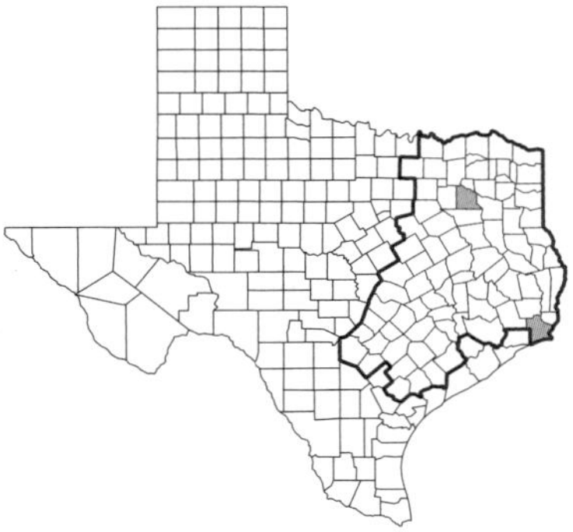
Atriplex patula

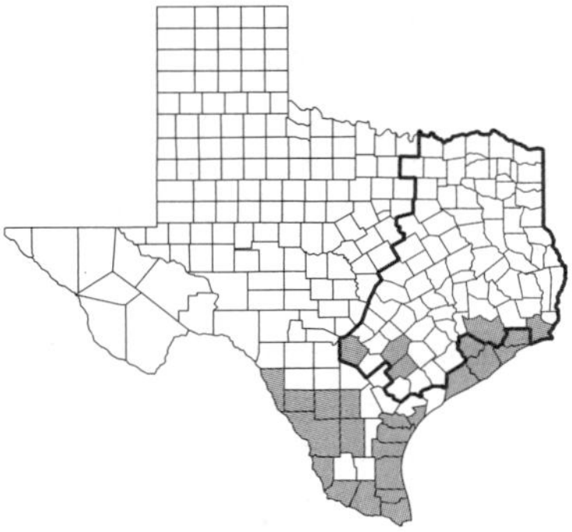
Atriplex pentandra

leaves 5–75 mm long, 1–8(–15) mm wide, linear to lanceolate, basally cuneate, sessile (sometimes narrowed into a pseudopetiole), usually 3–5-veined; mature inflorescences usually sparsely spicate, leafy, 1–2(–5) flowers in axils of bracts, some plants with flowers nearly throughout; flowers mostly perfect, some pistillate; perianth segments, each with a separate horizontal wing; stamens 5 (sometime absent); styles 2(–3); fruits utricles, covered by perianth. Waste places; Brazos, Dallas, Denton, Grayson, and McLennan (Kartesz 2022) cos.; scattered in TX (except throughout the Panhandle) but mainly in w 1/3 of the state; throughout w 1/2 of the U.S. and scattered mostly in the e 1/2; Canada. Jun–Aug. Native of e Europe and sw Asia. [*Chenopodium scoparia* L. (According to Mosyakin (2015), "it should be noted that Linnaeus used the species epithet as a noun, and thus the correction of this name to '*C. scoparium*' is incorrect,"), *Kochia scoparia* (L.) Shrad.] This species is reported to be "extremely variable" with a number of subspecies or varieties recognized in its native range (Mosyakin 2003b). A cultivated form is sometimes grown as an ornamental for its purplish red autumn foliage. The species is considered a weed, but is high in protein and readily grazed by livestock; ingestion may, however, cause photosensitization in cattle, sheep, and horses; reported to contain saponins (Burlage 1968; McGregor 1986a).

BETA L. BEET

A genus of ca. 6 species of Europe and the Mediterranean (Mabberley 2017). "The taxonomy of the genus is complicated by a long history of cultivation in which selective breeding has caused a bewildering array of diverse morphologies" (Schulz 2003). (Celtic: *bett*, red, presumably in reference to the betalain pigments)

REFERENCES: Ford-Lloyd & Williams 1975; Schulz 2003.

Beta vulgaris L., (common), BEET, SUGAR BEET, SWISS CHARD, MANGEL-WURZEL. Glabrous annual or biennial; roots dark red, pink, yellow, or white, thickened/swollen; stems to 0.6–1.2 m tall, green or often red; basal leaves in a rosette, long-petioled, reduced up the stem; flowers in glomerules, these axillary or in terminal spike-like inflorescences, perfect; perianth segments 3–5; stamens 5; ovary sunken in a disk (partly inferior); fruits achenes, strongly adherent in clusters. Widely cultivated; included based on reference that it is occasionally escaped in TX (Correll & Johnston 1970); widely scattered in the U.S. Spring–fall. Native of s Europe and adjacent areas. Cultivated since the time of the Assyrians for its edible roots, for the roots as a source of sugar (up to 20% by weight), and for its edible leaves; the roots often have high concentrations of the red pigments known as betalains (Mabberley 2017), which derive their name from this genus; ca. 30% of the world's sugar supplies are derived from BEET; plants sometimes considered a different subspecies (subsp. *cicla* (L.) Koch), SWISS CHARD, SPINACH, BEET, are cultivated for the edible leaves (Hyam & Pankhurst 1995; Mabberley 2017). No county level distribution map is provided.

BLITUM L. POVERTY-WEED

A genus of ca. 12 species (Fuentes-Bazan et al. 2012) of annual and perennial herbs of cosmopolitan distribution; some were formerly eaten. The genus is related to *Chenopodium*, with which it shares bladder (= inflated) hairs (which collapse and give the foliage a mealy appearance). Flowers are wind-pollinated (Judd & Ferguson 1999). (Latin: *blitum* or *blitus*, an old name for strawberry-blight)

REFERENCES: Holmgren 2003; Fuentes-Bazan et al. 2012.

Blitum nuttallianum Schult., (for Sir Thomas Nuttall, 1786–1859, English-American botanist), NUTTALL'S MONOLEPIS, POVERTY-WEED. Glabrous or sparsely mealy annual with prostrate to erect stems 5–35(–50) cm long; foliage pale green; lower leaves long-petioled, blades slightly fleshy, 1–3(–4) cm long, rhombic or lanceolate, with 1 or few coarse teeth or often a pair of lobes toward base, giving the leaf an arrowhead shape; flowers axillary, sessile, crowded, in upper part of stem and branches, mostly perfect or a few pistillate; the single bract-like perianth segment 0.7–1.2 mm long; stamen 1; styles 2; utricle 1–1.5 mm long. Stream bottoms, roadsides, disturbed sites; w portion of East TX as far e as Van Zandt Co. (BRIT); widespread in TX primarily in the w 2/3 of the state; widespread in w 1/2 of the U.S.; Canada. Mar–May. [*Monolepis nuttalliana* (Schult.) Greene] Native of c and w U.S.

CHENOPODIASTRUM S. Fuentes-B., Uotila, & Borsch

Weedy annual herbs, glabrous to sparsely mealy-coated foliage (due to collapsed inflated hairs); leaves alternate; flowers crowded in small bunches (= glomerules), in terminal or lateral panicles or spikes, usually perfect; perianth glabrous to white-mealy outside; petals absent; perianth with 5 segments; stamens 5; fruit a small achene with thin pericarp that is firmly attached; because of the thinness of the pericarp, the seed and fruit are essentially the same size; seed horizontal; embryo partly or completely encircling the endosperm. (Greek: *chen*, a goose, and *pous*, foot, in allusion to the supposed resemblance of the leaves of some species to the webbed feet of geese)

Atriplex matamorensis [VIN]

Atriplex mucronata [BB2, GL2]

Atriplex patula [ABR]

Bassia scoparia [REE]

Blitum nuttallianum [BB2]

Atriplex pentandra [CO3]

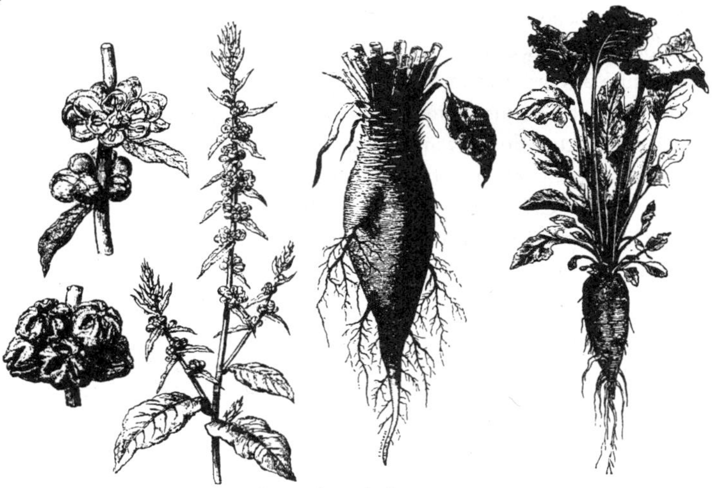

Beta vulgaris [BL2]

A genus of 6–7 n temperate species (Mabberley 2017).
REFERENCES: Clemants & Mosyakin 2003b; Fuentes-Bazan et al. 2012.

1. Stems 0.6–1.5 m long; leaf blades 3.5–15 cm long, margins with 1–5 teeth on each side; inflorescences bearing both flowers and fruits simultaneously ____ **C. simplex**
1. Stems 0.1–0.6(–1) m long; leaf blades 0.8–4(–8) cm long, margins irregularly dentate, usually with more than 5 teeth on each side; inflorescences either bearing flowers or fruits but not simultaneously ____ **C. murale**

Chenopodiastrum murale (L.) S. Fuentes-B., Uotila, & Borsch, (of walls), NETTLE-LEAF GOOSEFOOT, SOWBANE. Annual; stems erect or decumbent, 0.1–0.6(–1) m long; leaf blades thin, triangular, ovate to rhombic-ovate, coarsely sinuate-dentate, glabrous to sometimes sparsely mealy when young; perianth segments keeled; pericarp with very minute raised areas or ± smooth; seeds 1–1.5 mm in diam., dull, black. Disturbed habitats. Collected "in the streets of Dallas" in Jun 1874 and Jun 1875 by Reverchon; not found there since (Mahler 1988), also Brazos, Grayson, Hardin, Lee, and Robertson (Turner et al. 2003) cos.; widely scattered in TX, mainly in s and w part of the state; scattered in the U.S. except absent from the Great Plains; s Canada. May–Nov. [*Chenopodium murale* L.] Native of the Old World.

Chenopodiastrum simplex (Torr.) S. Fuentes-B., Uotila, & Borsch, (simple, unbranched), MAPLE-LEAF GOOSEFOOT, BIG-SEED GOOSEFOOT. Annual; stems mostly glabrous, erect, 0.3–1.5 m tall; leaf blades with 1–5 large symmetrical teeth on each side; panicles open, bractless; perianth segments slightly keeled; pericarp smooth or nearly so; seeds 1.3–1.9 mm in diam., umbonate-lenticular, tapering from raised center to margin, shiny, black. Disturbed woodlands; Collin, Dallas, Travis (Reed 1969a), Anderson, Bastrop, and Bell (Turner et al. 2003) cos.; otherwise in TX known only from a few scattered localities to the w; widespread mainly in the n 2/3 of the U.S.; s Canada. Jun–Aug. [*Chenopodium gigantospermum* Aellen, *C. simplex* (Torr.) Raf.]

CHENOPODIUM L. LAMB'S-QUARTERS, GOOSEFOOT, PIGWEED

Weedy annual (rarely biennial or perennial) herbs, often with mealy-coated foliage (due to collapsed inflated hairs); leaves alternate; flowers crowded in small bunches (= glomerules), axillary or in terminal paniculate spikes, usually perfect; perianth segments (4–)5, usually white-mealy outside; petals absent; stamens usually 5; fruit a small utricle or achene with thin pericarp that is easily separable or firmly attached; because of the thinness of the pericarp, the seed and fruit are essentially the same size; seed horizontal; embryo partly or completely encircling the endosperm.

A genus of 100+ species (Clemants & Mosyakin 2003b) of mostly weedy herbs of temperate areas; some are cultivated as "grains," as ornamentals, or for medicinal uses. *Chenopodium quinoa* Willd. (QUINOA), of Andean South America, was an important pre-Colombian "grain" and is currently making a resurgence; various QUINOA products (e.g., cereals, pastas) are at present widely available in health food stores; see Wilson (1990) for information on the origin of QUINOA. The inconspicuous flowers are primarily wind-pollinated (Judd & Ferguson 1999). Based on Wilson (1983, 1984, 1987) and Mosyakin and Clemants (2002), it now seems that "*Chenopodium* species with bladder hairs ('mealy chenopods') are probably more closely related to *Atriplex* and its satellite genera than to 'glandular chenopods" (Clemants & Mosyakin 2003b); as a result, we are following Clemants and Mosyakin (2003b) in treating the "glandular chenopods" in the genus *Dysphania*. Additionally, *Dysphania atriplicifolia* was transferred from the genus *Cycloloma* by Uotila et al. (2021). Though not glandular, but instead with woolly pubescence, it was demonstrated to be nested within *Dysphania*. *Chenopodium* species are often difficult to distinguish because of vegetative variability and the small size of the taxonomically important reproductive characters (e.g., fruits, perianth segments). Mature fruits are extremely helpful in identification. Because of the difficulties involved, several species can be reached more than one way in the key. (Greek: *chen*, a goose, and *pous*, foot, in allusion to the supposed resemblance of the leaves of some species to the webbed feet of geese)
REFERENCES: Aellen & Just 1943; Wahl 1954; Reed 1969a; Crawford 1975; Scott 1978; Bassett & Compton 1983; Wilson 1983, 1984, 1987; Crawford & Wilson 1986; Dorn 1988; Clemants & Mosyakin 2003b; Benet-Pierce & Simpson 2014, 2017; Uotila et al. 2021.

1. Mature leaf blades mostly linear, lanceolate to oblong (sometimes ovate), usually < 15(–18) mm wide (sometimes wider in *C. atrovirens*), ca. 3 times as long as wide or longer.
    2. Mature leaf blades with 3 or more veins from near base (lateral veins sometimes very faint), 4–18(–22) mm wide, typically wider toward the middle or base, not of the same width for the whole length of the blade, the sides usually not parallel.
        3. Lower surface of leaf blades glabrous or very sparsely mealy, green in color; perianth segments barely curling over the margins of the mature fruit ____ **C. standleyanum**
        3. Lower surface of leaf blades densely to sparsely mealy, usually conspicuously whitened; perianth segments spreading or enclosing the mature fruit.
            4. Perianth segments enclosing the mature fruit ____ **C. desiccatum**

4. Perianth segments partially or completely spreading from the mature fruit.
   5. Leaf blades 2–3 times as long as broad (plants rare in East TX) ____ **C. atrovirens**
   5. Leaf blades > 3 times long as wide ____ **C. pratericola**
2. Leaf blades 1-veined, usually 1–3(–4) mm wide, of about the same width for the whole length of the blade, the sides nearly parallel ____ **C. pallescens**
1. Mature leaf blades ovate, deltoid to rhombic, mostly wider than 15 mm (sometimes less in *C. album*, *C. incanum*, and *C. lenticulare*), 1–3 times as long as wide.
   6. Inflorescences usually bearing mature fruits and flowers simultaneously (maturing mixed).
      7. Plant mostly branched near base or distally; seed with an acute equator when viewed from the side, but no distinct marginal rim when viewed from the top; lower leaf blade surface densely mealy to glabrous; pericarp papillate to smooth, nonadherent; plants scattered or rare in East TX.
         8. Leaves sparsely mealy; plant usually simple proximally and branched distally; bracts absent or present; perianth segments usually partly covering mature fruit.
            9. Pericarp smooth; bracts absent; plants scattered in East TX ____ **C. standleyanum**
            9. Pericarps papillate to smooth; bracts small, entire; known only from Travis Co. in East TX ____ **C. fremontii**
         8. Leaves densely mealy; plant usually much branched from base; bracts absent; pericarp smooth; perianth segments tightly covering fruit; known only from Austin Co. in East TX ____ **C. incanum**
      7. Plant mostly branched from base; seed with a distinct equatorial margin when viewed from the side, with a distinct margin rim at least 0.2 mm wide when viewed from the top; lower leaf blade surface mealy; pericarp papillate, adherent above; plants known only from Travis Co. ____ **C. lenticulare**
   6. Inflorescences bearing flowers or fruits but usually not both simultaneously (*C. standleyanum* can also be keyed here).
      10. Pericarp with a "honeycombed" appearance (under a lens or scope) due to pericarp having alveolae ____ **C. berlandieri**
      10. Pericarp "honeycombed" or not (*C. berlandieri* can also be reached this way).
         11. Lower surface of leaf blades densely white-mealy.
            12. Pericarp "honeycombed" (with alveolae; quite clear under dissecting scope); perianth segments broadly keeled, the keel often half as wide as perianth segment ____ **C. berlandieri**
            12. Pericarp smooth to papillate, but not "honeycombed"; perianth segments not broadly keeled, the keel usually ca. 1/3 as wide as perianth segment ____ **C. album**
         11. Lower surface of leaf blades not densely white-mealy.
            13. Perianth segments broadly keeled, completely enclosing the mature fruit; glomerules of flowers usually in dense, erect inflorescences; pericarp "honeycombed" (with alveolae), rather adherent to the seed ____ **C. berlandieri**
            13. Perianth segments rounded, not or only slightly keeled, barely curling over the margins of the mature fruit; glomerules of flowers in open, sometimes nodding inflorescences; pericarp smooth, not "honeycombed," easily separable from seed ____ **C. standleyanum**

Chenopodium album L., (white), LAMB'S-QUARTERS, PIGWEED. Plant annual, pale-green; stems erect to 0.1–3 m tall, solitary; lateral branches well-developed, ascending and compacted or spreading and flexuous; nodes and infructescences with or without dark pigmentation; leaf blades coarsely toothed to entire, moderately to densely mealy, very variable in size, typically < ca. 5.5 cm long but occasionally to as much as 12 cm long, 0.5–3.8(–8) cm wide; flowers in large glomerules in axillary or terminal, dense, paniculate spikes; perianth segments keeled or keel poorly developed, slightly to densely mealy, enclosing the mature fruit; pericarp not "honeycombed"; seeds 0.9–1.6 mm in diam., shiny, black. Around old homesteads and abandoned farms, disturbed areas; scattered in East TX; widespread in TX; widespread in the U.S. and Canada. Apr–Sep. Apparently mostly native in Europe and Asia. [*C. album* var. *missouriense* (Aellen) Bassett & Crompton, *C. giganteum* D. Don, *C. missouriense* Aellen]. We are following Clemants and Mosyakin (2003b) for our concept of this species. They include *C. missouriense* (recognized as a distinct species by Reed 1969a; Diggs et al. 1999; Turner et al. 2003, and as a variety of *C. album* by Weakley, A.S., and Southeastern Flora Team 2022) in *C. album*, indicating that it appears to be a "native form of *C. album* that flowers in mid-September regardless of when it germinated" (Clemants & Mosyakin 2003b). For those wishing to distinguish the varieties, a key, modified from Weakley, A.S., and Southeastern Flora Team (2022), is provided below. As is the case in the Great Plains (Crawford & Wilson 1986), this species is probably less abundant in East TX than generally assumed because individuals of the similar *C. berlandieri* are often identified as *C. album*. However, the "honeycombed" pericarp of *C. berlandieri* clearly distinguishes it from *C. album*. It was used medicinally by Native Americans and by pioneers as a potherb; however, the plants may contain toxic quantities of nitrate and oxalic acid; poisoning and death in cattle, horses, and pigs have been reported (Burlage 1968; Schmutz & Hamilton 1979; Mulligan & Munro 1990).

1. Plant sparsely branched, with an erect habit; inflorescences in paniculate spikes; lower leaves lobed or toothed; upper leaves triangular and toothed ____ var. **album**
1. Plant densely branched from base; inflorescences in paniculate arching branches; lower leaves with a few lobes or teeth; upper leaves lanceolate and entire ____ var. *missouriense*

Chenopodium atrovirens Rydb., (dark green), PINYON GOOSEFOOT. Plant 0.7–6.5 dm tall; stems erect to ascending; leaf blades 1.5–3.5 cm long, 2–3 times long as wide, ovate, broadly oblong, or occasionally triangular, entire or with basal lobes; prominently 3-veined; perianth segments spreading from fruit at maturity; pericarp smooth; seeds 0.9–1.3 mm in diam. Open, dry, sandy areas, disturbed areas; included based on citation for Caldwell Co. (Reed 1969a) in far sw East TX; found in only 6 counties w of East TX; throughout w 1/3 of U.S., also MD, SC, and WI; w Canada. Summer–fall. [*C. aridum* A. Nelson, *C. wolfii* Simonk.] Native of w U.S. Perhaps only a waif in East TX.

Chenopodium berlandieri Moq., (for Jean Louis Berlandier, 1805–1851, French botanist who explored TX, NM, and Mexico, and one of the first botanists to make extensive collections in TX), PIT-SEED GOOSEFOOT. Plant annual; stems erect, 0.1–1.5 m tall; leaf blades often densely mealy when young, later glabrate, irregularly sinuate-dentate; inflorescences typically dense, erect, although sometimes more diffuse; perianth segments often broadly and prominently keeled, densely mealy, completely enclosing the fruit at maturity; pericarp "honeycombed"; seeds 1–2 mm in diam., shiny, brown to black. Dry disturbed habitats. Jul–Sep. Infraspecific taxa in this species do not appear clear. The keys to varieties in Reed (1968a) and Correll and Johnston (1970) versus Clemants and Mosyakin (2003b) differ greatly. The following key to varieties follows Clemants and Mosyakin (2003b). The county distribution map does not distinguish varieties.

1. Leaves not 3-lobed; inflorescences becoming bractless ____ var. **boscianum**
1. Leaves 3-lobed; inflorescences with or without bracts.
   2. Inflorescences bractless ____ var. **berlandieri**
   2. Inflorescences with leafy bracts.
      3. Style bases without yellow area ____ var. **sinuatum**
      3. Style bases with yellow area ____ var. **zschackei**

var. **berlandieri.** Scattered in East TX; scattered in the rest of the state; NM and TX; Mexico.

var. **boscianum** (Moq.) Wahl, (for its discoverer, Louis Augustin Guillaume Bosc, 1759–1828, French naturalist). Leaf blades (1–)2–4 cm long, rhombic-triangular or ovate-lanceolate; perianth segments slightly keeled; seeds 1–1.3 mm in diam. In East TX known only from McLennan Co. (Hannick et al. 2013); scattered in South TX Plains; mostly coastal U.S. from NY to TX.

var. **sinuatum** (Murr) Wahl, (having a wavy margin). Brazos, Gregg, and Upshur (Reed 1969a) cos.; scattered in TX except for c TX; mostly sw U.S. from OK s to TX and w to CA.

var. **zschackei** (Murr) Murr ex Graebn., (for Georg Zschacke, 1867–1937, German botanist). Widespread in TX; nearly throughout the U.S. (except far se); s Canada.

Chenopodium desiccatum A. Nelson, (dried up). Plant erect or more often spreading, usually branched from base, 1–1.4(–6) dm tall; leaf blades 1.5–2.5 cm long, 4–6 mm wide, linear, oblong-elliptic or ovate-lanceolate, densely mealy below, usually 3-veined; inflorescence in densely packed glomerules in axillary and terminal panicles, maturing mixed, perianth segments enclosing fruit at maturity; pericarp nonadherent, smooth; seed 0.8–1.1 mm in diam., basally widely conical, apically ± flat, seed coat warty except smooth on the apex of the conical basal side. Originally found in open, undisturbed soils, prairies, and stabilized dunes, but it has spread to disturbed open areas; scattered in East TX; scattered mostly w of East TX; throughout most of the w 1/2 of U.S., also scattered in Midwestern U.S.; w Canada. Jun–Sep. [*C. leptophyllum* (Moq.) Nutt. ex S. Watson subsp. *desiccatum* (A. Nelson) Aellen] This taxon has a very convoluted taxonomic history with many erroneous synonyms attributed to it (See Benet-Pierce & Simpson 2014). Benet-Pierce and Simpson (2014) believe *C. desiccatum* is the most misidentified *Chenopodium* in the U.S. Native of the U.S., perhaps exotic in some areas. Because of the difficulty in identifying *C. desiccatum* and the possibility of misidentifications, no county distribution map is provided.

Chenopodium fremontii S. Watson, (for John Charles Fremont, 1813–1890, American botanist, explorer, and leader of numerous expeditions in the West), FREMONT'S GOOSEFOOT. Plant erect to spreading, simple proximally, branched distally, 1–8 dm tall; leaf blades usually broadly triangular, sometimes ovate to elliptic, margins entire or with a pair of basal teeth or lobes, mealy below; inflorescence in terminal and axillary interrupted spikes, maturing mixed; perianth segments usually completely covering fruit at maturity; pericarp nonadherent, papillate to smooth; seeds 1–1.3 mm diam.; seed coat dark reddish brown to black (sometimes with light streaks), ± smooth. Moist shaded areas, riparian habitats; included based on citation for Travis Co. (Kartesz 2022), presumably a waif; mostly Trans-Pecos and High Plains; throughout w 1/2 of U.S., also SC; w Canada. Jun–Oct. Native of w U.S.

Chenopodium incanum (S. Watson) A. Heller, (gray, hoary-white), MEALY GOOSEFOOT. Plant erect to spreading; stems usually branched from base, 0.6–1.5(–3) dm tall; leaf blades 1–1.5 cm long, broadly triangular to broadly ovate, distal leaves narrowly

Chenopodiastrum murale [SMI]

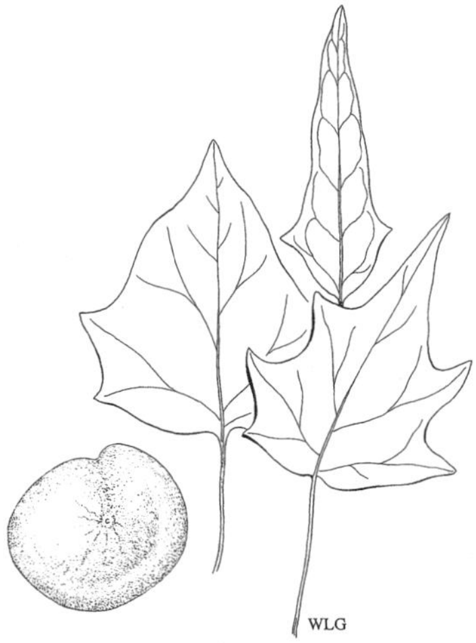

Chenopodiastrum simplex [GL2]

Chenopodium album [REE]

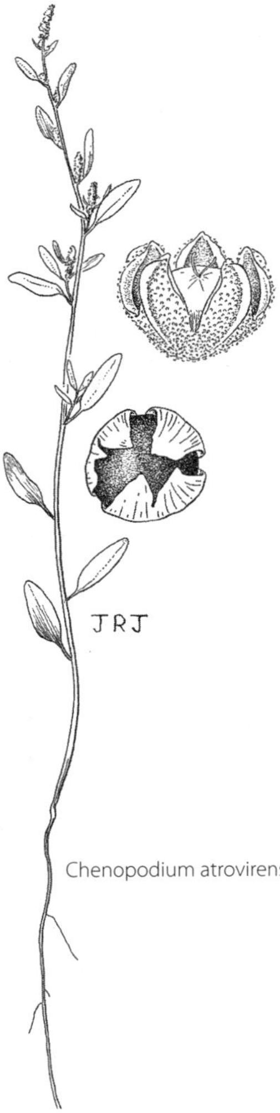

Chenopodium atrovirens [CRO]

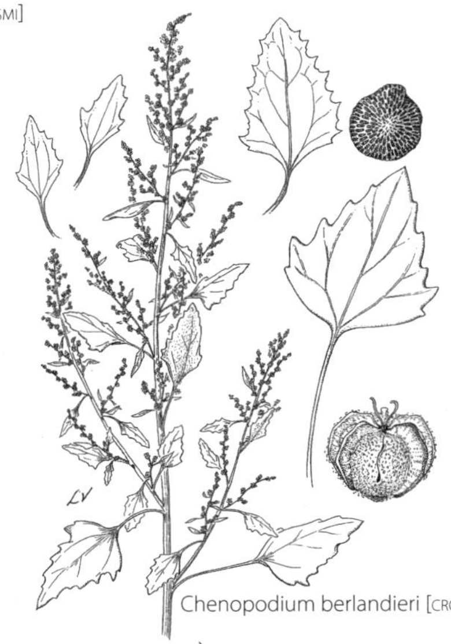

Chenopodium berlandieri [CRO]

Chenopodium desiccatum [JEP]

Chenopodium fremontii [JEP]

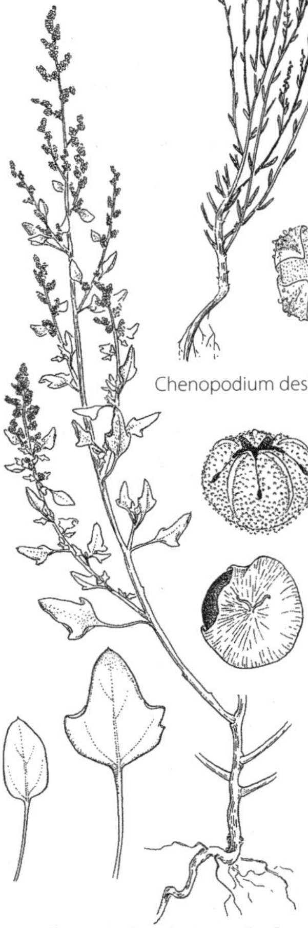

Chenopodium incanum [CR2]

triangular to narrowly ovate, margins usually with two basal teeth or lobes, teeth or lobes acute, apex acute, lower surface densely mealy; glomerules maturing mixed; perianth segments completely covering fruit at maturity; pericarp nonadherent, smooth. Seeds round, 0.9–1.1 mm in diam., margins acute when viewed from the side, without a distinct equatorial rim when viewed from the top; seed coat dark black, wrinkled. Sandy and limestone soils, waste places; included based on citation in Reed (1969a) from Austin Co. (*Parks & Cory 20920*, 1936, TAES); perhaps a waif; scattered in w 1/2 of TX; w U.S. from ND s to TX and w to MT and AZ, 1 or 2 records in IL, ME, MO, OH, and SC (Kartesz 2022); Canada. Jun–Aug. Native of the U.S.

Chenopodium lenticulare Aellen, (from Latin: lenticularis, presumably from the biconvex, lentil-shaped seed). TRANS-PECOS GOOSEFOOT. Plant erect to 7 dm tall, branching from the base; leaf blades deltoid to rhombic-ovate 0.8–2.6 cm long; fruit ca. 1.5 mm in diam.; pericarp semi-adherent below, adherent above, papillate; seed smooth, equatorial margin distinct when viewed from the side, when viewed from the top, rim at least 0.2 mm wide. Thin soil, rocky ledges, ravines; included based on type specimen (*Young 708*, 1918, G) collected in Travis Co.; Trans-Pecos; NM, TX. Jul–Oct. Native of sw U.S. Long synonymized under *C. neomexicanum* Standl., Benet-Pierce and Simpson (2017) reinstated the taxon. They did so mostly using seed and fruit characters.

Chenopodium pallescens Standl., (rather pale), LIGHT GOOSEFOOT. Annual; stems erect, 0.3–0.6 m tall; leaf blades linear, 1–3.5 cm long, 1–3 mm wide, 1-nerved, glabrous to slightly mealy below; flowers in large glomerules, widely spaced; perianth segments keeled, slightly mealy, completely enclosing the fruits; pericarp finely tuberculate; seeds 1.4–1.6 mm in diam., shiny, black. Open floodplains; Walker Co. (Turner et al. 2003); otherwise known in TX from a few scattered localities in the n and w parts of the state (e.g., Parker Co. (Mineral Wells State Park, *Lipscomb 2393* SMU)); c U.S. from IN s to LA and w to NE, and NM. Aug–Oct.

Chenopodium pratericola Rydb., (living in meadows or prairies), THICK-LEAF GOOSEFOOT, NARROW-LEAF GOOSEFOOT. Annual; stems 0.2–0.9 m long; leaf blades densely to sparingly mealy beneath; petioles ca. half as long as the blades; perianth segments white-margined, keeled, partially exposing the fruit (the apex); pericarp smooth; seeds 0.9–1.3 mm in diam., shiny, black. Stream banks, disturbed sites, sandy soils; Bastrop, Dallas, Denton, Ellis (BRIT), Gonzales, Grayson (Reed 1969a), Bexar, Brazos, Caldwell, and Travis (Kartesz 2022) cos.; scattered in TX to the w of East TX; widespread in the w U.S., scattered in e U.S.; Canada. Late May–Oct. [*C. desiccatum* A. Nelson var. *leptophylloides* (Murr) Wahl] Native of the U.S. except introduced in the e part of its range.

Chenopodium standleyanum Aellen, (for Paul Carpenter Standley, 1884–1963, student of floras of sw U.S., Mexico, and Central America), STANDLEY'S GOOSEFOOT. Annual; stems erect or arched, usually 0.6 m or less tall; (rarely taller); leaf blades 2–4.5(–8) cm long, usually 5–15 mm wide; sparsely mealy; inflorescence a slender, often nodding terminal panicle; perianth segments not or only slightly keeled, slightly mealy; pericarp smooth; seeds 0.9–1.3 mm in diam., shiny, black. Disturbed wooded habitats, floodplains; scattered in East TX; also Cross Timbers and Prairies and Edwards Plateau; widespread in the e 1/2 of the U.S.

DYSPHANIA R. Br. GOOSEFOOT, PIGWEED

Weedy annual herbs, some strongly aromatic, with woolly pubescence or glandular foliage, the glands stalked, sometimes with sessile glands also present; leaves alternate; flowers crowded in small axillary clusters (= glomerules), or in lateral or terminal spike-like or branched inflorescences, usually perfect; subtended by reduced leaves or leaf-like bracts; perianth of 4–5 segments, usually glandular or woolly outside; petals absent; stamens 1–5; fruit a small achene with thin pericarp that is easily separable or firmly attached; because of the thinness of the pericarp, the seed and fruit are essentially the same size; embryo partly or completely encircling the endosperm.

A cosmopolitan genus of ca. 50 species. The taxa have often been included in *Chenopodium* (e.g., Correll & Johnston 1970; Diggs et al. 1999; Turner et al. 2003). However, we are following Mosyakin and Clemants (2002), Clemants and Mosyakin (2003a), and Uotila et al. (2021) who treat the genus in an expanded fashion to include all "glandular" taxa previously treated in *Chenopodium*. Based on Wilson (1983, 1984, 1987) and Mosyakin and Clemants (2002), it now seems that "*Chenopodium* species with bladder hairs ('mealy chenopods') are probably more closely related to *Atriplex* and its satellite genera than to 'glandular chenopods' [= *Dysphania*]" (Clemants & Mosyakin 2003a). Thus, recognition of the "glandular chenopods" at the generic level seems the most reasonable course of action. Additionally, *D. atriplicifolia* was transferred from the genus *Cycloloma* by Uotila et al. (2021). Though not glandular, but instead with woolly pubescence, it was demonstrated to be nested within *Dysphania*. The genus name *Teloxys* has been used by some for this group, but *Dysphania* has nomenclatural priority. The key to species is modified from Weakley, A.S., and Southeastern Flora Team (2022) (Greek: *dysphanis*, obscure, in reference to the inconspicuous flowers)

REFERENCES: Scott 1978; Wilson 1983, 1984, 1987; Weber 1985; Mosyakin & Clemants 2002, 2008; Clemants & Mosyakin 2003a; Mosyakin 2003a; Uotila et al. 2021.

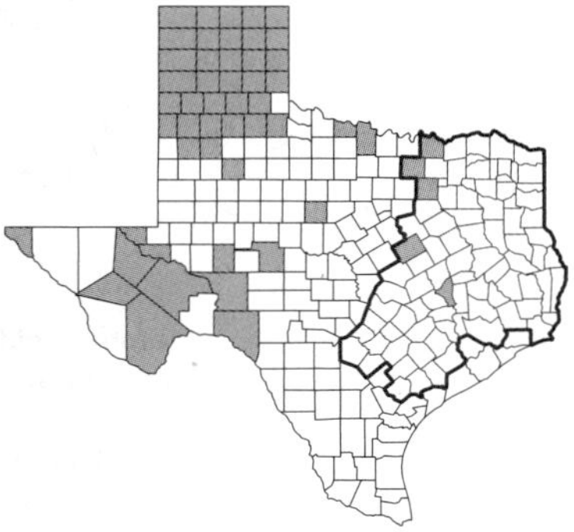

Bassia scoparia

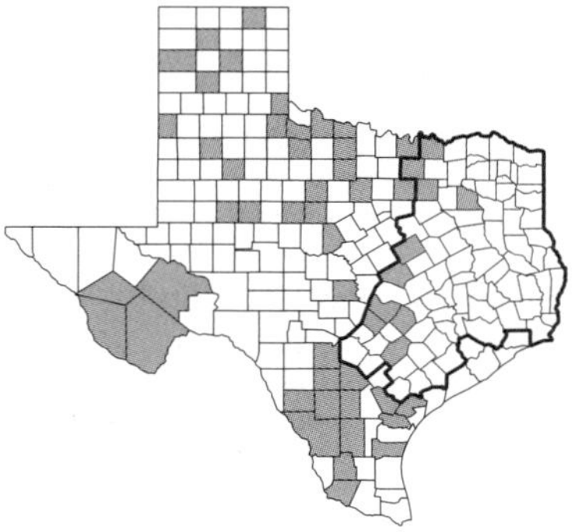

Blitum nuttallianum

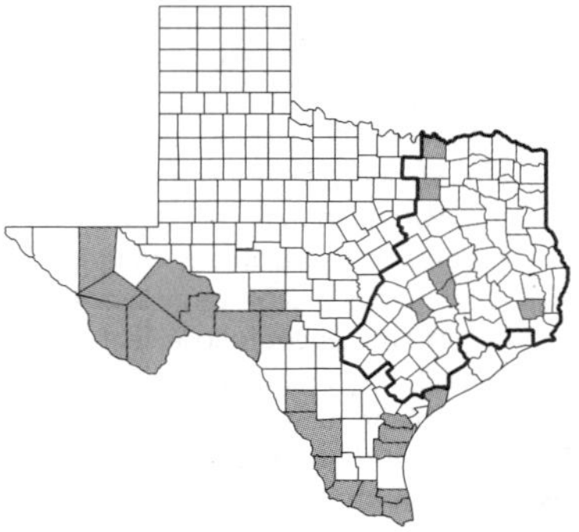

Chenopodiastrum murale

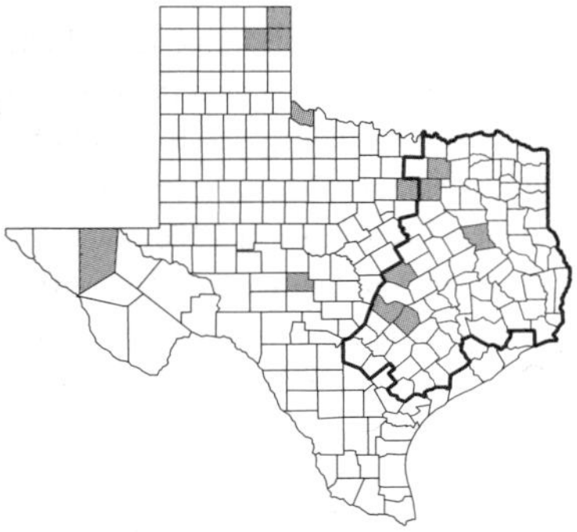

Chenopodiastrum simplex

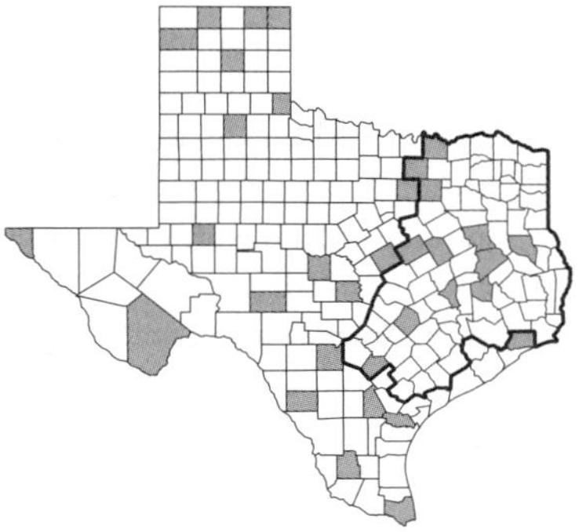

Chenopodium album

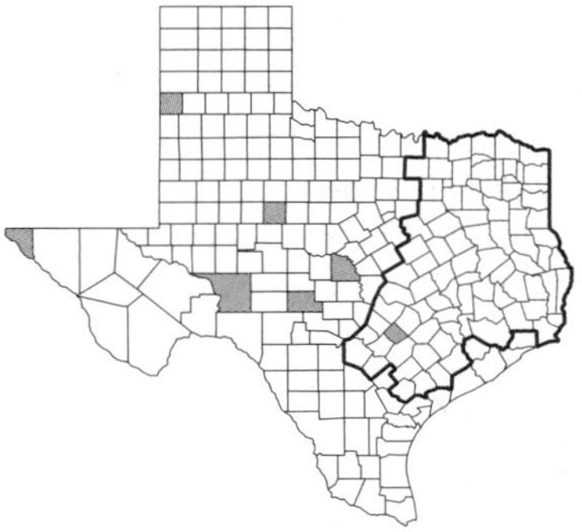

Chenopodium atrovirens

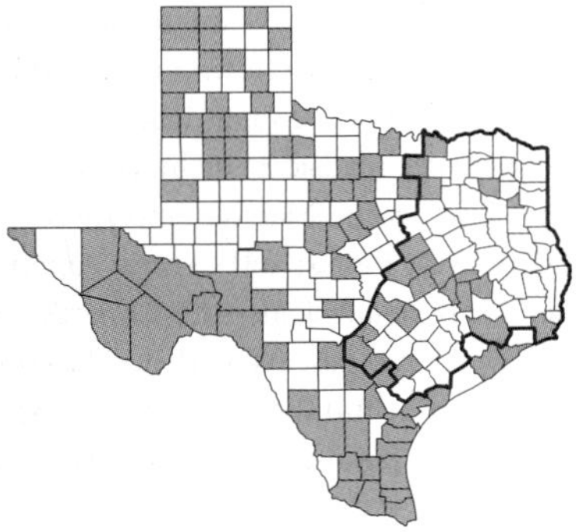

Chenopodium berlandieri

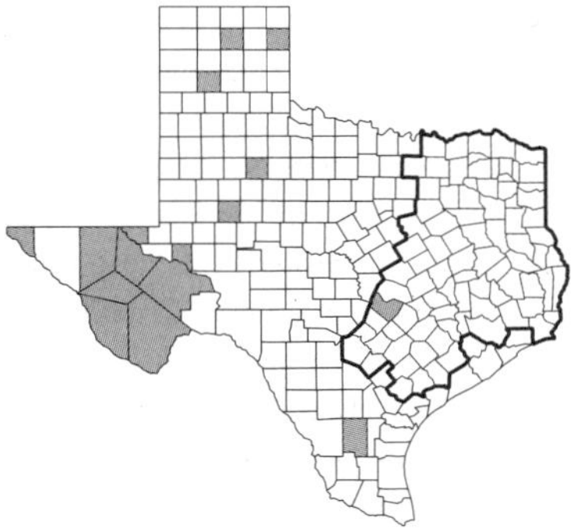

Chenopodium fremontii

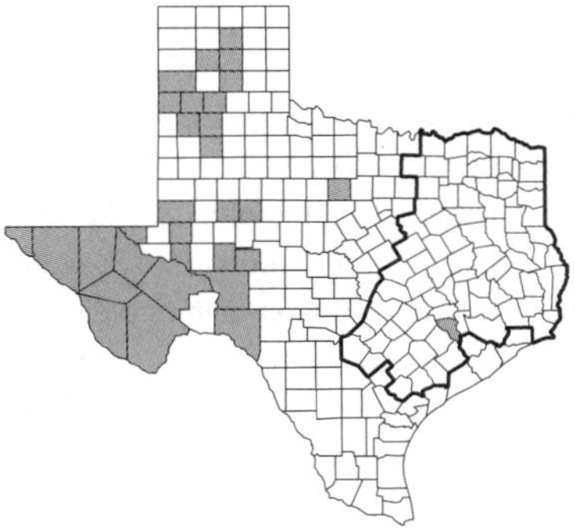

Chenopodium incanum

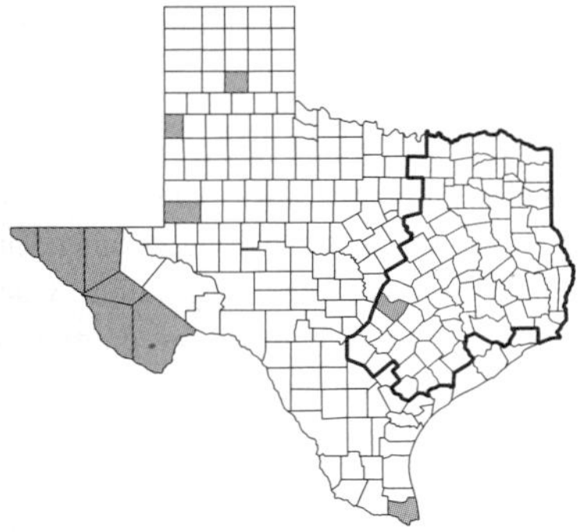

Chenopodium lenticulare

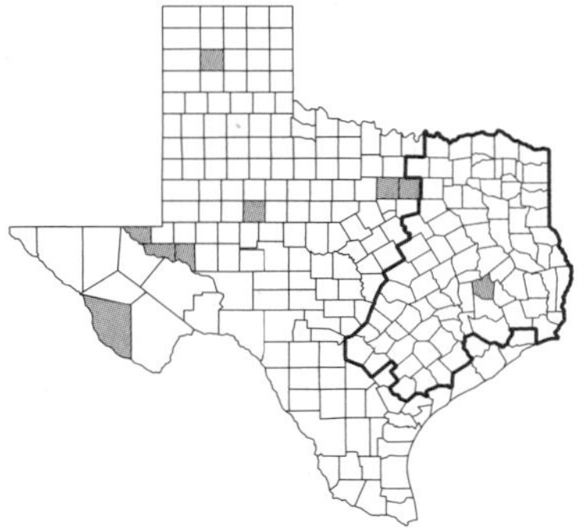

Chenopodium pallescens

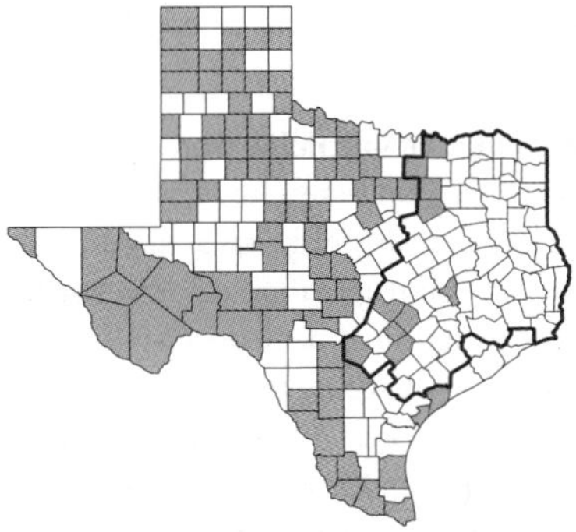

Chenopodium pratericola

1. Flowers usually in a slender, dense inflorescence (sometimes somewhat open) of lateral cymes ____ **D. botrys**
1. Flowers either solitary and spaced along the inflorescence axis, or in dense glomerules arranged secondarily into spikes and panicles.
   2. Stems with woolly pubescence; flowers solitary, sessile, and spaced along the inflorescence axis; perianth segments strongly keeled at maturity with a continuous, horizontal, scarious wing ____ **D. atriplicifolia**
   2. Stems with glandular hairs; flowers in dense glomerules arranged secondarily into spikes and panicles; perianth segments not horizontally winged.
      3. Leaf blades 2–8 cm long; seeds usually horizontal; stems 3–15 dm tall.
         4. Inflorescences foliose; leaf-like bracts 0.3–3 cm long ____ **D. ambrosioides**
         4. Inflorescences leafless; if foliose, then leaf-like bracts much smaller, to 2.2 mm long ____ **D. anthelmintica**
      3. Leaf blades 0.5–2.7 cm long; seeds vertical; stems 1–5.5 dm tall.
         5. Perianth segments rounded; stems 1–4.5 dm ____ **D. pumilio**
         5. Perianth segments keeled; stems 3–5.5 dm ____ **D. carinata**

Dysphania ambrosioides (L.) Mosyakin & Clemants, (resembling *Ambrosia*—ragweed), MEXICAN-TEA, WORMSEED, EPAZOTE, SPANISH-TEA, WORMSEED LAMB'S-QUARTERS, STINKWEED. Plant annual, strongly aromatic; stems erect or ascending, usually 0.3–1+ m tall; leaf blades densely yellow-glandular, rarely glabrous, nearly pinnatifid to coarsely toothed to entire; inflorescences of lateral spikes 3–7 cm long with leaf-like bracts 0.3–3 cm long at most nodes; perianth segments completely enclosing the fruit; seeds horizontal, rarely vertical. Disturbed habitats, sandy soils of woodlands; widespread in East TX; widespread in e 2/3 of TX; throughout the e and w U.S., scattered in c U.S.; se Canada. Aug–Nov. [*Chenopodium ambrosioides* L., *Teloxys ambrosioides* (L.) W.A. Weber] This species was once considered introduced into North America (e.g., Correll & Johnston 1970; Diggs et al. 1999). However, Clemants and Mosyakin (2003a) have clarified its distribution. It is now considered native to both South and North America, with populations in the s part of North America native and those further n introduced. The oils are allelopathic (= inhibit the germination or growth of other plants) (Clemants 1992). The odor of this aromatic species has been variously described as ill-scented, unpleasant, or "reminiscent to that of kerosene" (Yatskievych 2006), but to some it is not unpleasant (e.g., its use as a condiment or tea). There is disagreement in the validity of presence or absence of leaf-like bracts in the inflorescences of *D. ambrosioides* and *D. anthelmintica* as a discriminating character. Determinations, therefore, should be made with caution.

Dysphania anthelmintica (L.) Mosyakin & Clemants, (expelling intestinal worms), WORMWEED, EPAZOTE. Plant annual, malodorous; stems erect to ascending, 0.4–1 m long, ± glandular-pubescent, finely pubescent to almost glabrous; leaf blades 5–7 cm long, 1.5–3 cm wide, narrowly ovate to lanceolate, dentate with large widely spaced teeth, gland dotted; inflorescences terminal and lateral spikes or panicles, 3–8 cm long; bracts absent or leaf-like to 2.2 mm long, linear; perianth segments fused about 1/2 their length, rounded abaxially, glabrous; seeds horizontal, rarely vertical. Sand dunes, pinelands, roadsides; included based on map (Clemants & Mosyakin 2003a) for extreme se and extreme ne East TX; upper Gulf Prairies and Marshes; mostly coastal U.S. from NJ s to FL and w to TX, scattered in lower midwestern U.S., also CA; Mexico; West Indies; and Central America. Jun–Oct. [*Chenopodium ambrosioides* var. *anthelmintica* (L.) A. Gray, *C. anthelminticum* L., *D. ambrosioides* (L.) Mosyakin & Clemants var. *anthelmintica* (L.) Stace] *Dysphania anthelmintica* is variously considered a weed or medicinal herb and has been used as a flavoring/spice/condiment in Mexican cooking, as a tea, and as an antiflatulent. An antihelminthic (= agent that expels worms) oil (ascaridol) is obtained from the seeds with the active ingredient being a terpene; the plant has been variously used medicinally including in the treatment of malaria; the oil is toxic and overdoses have been fatal to humans and animals (Burlage 1968; Kingsbury 1964; Leung & Foster 1996; Yatskievych 2006). There is disagreement in the validity of presence or absence of leaf-like bracts in the inflorescences of *D. ambrosioides* and *D. anthelmintica* as a discriminating character. Determinations based on this character should be made with caution. Therefore, no county distribution map is provided. Native of the U.S., may be introduced inland. ☠

Dysphania atriplicifolia (Spreng.) G. Kadereit, Sukhor, & Uotila, (with leaves resembling *Atriplex*—saltbush), WINGED-PIGWEED, TUMBLE-RINGWING, PLAINS TUMBLEWEED. Bushy annual 15–100 cm tall and broad, with habit of a tumbleweed, with thin, woolly pubescence or nearly glabrous at maturity; leaves sessile or short-petioled, blades 2–8 cm long, oblong-ovate, oblong, or lanceolate, irregularly sinuately toothed or pinnately lobed; flowers sessile, solitary or few in axils of short leaf-like bracts, rather widely spaced in terminal paniculate spikes, perfect, some pistillate; perianth with a conspicuous, irregularly lobed but continuous horizontal wing going all the way around; stamens 5; styles 2–3. Loose sandy soils, roadsides and disturbed sites; scattered in the nw portion of East TX; widespread in TX except for South TX Plains and Gulf Prairies and Marshes; widespread in c and sw U.S.; Canada and n Mexico. Jun–Oct. [*Cycloloma atriplicifolium* (Spreng.) J.M. Coult.] *Dysphania atriplicifolia* was once in the monotypic genus *Cycloloma*, but Uotila et al. (2021) demonstrated that it was nested within *Dysphania*. The seeds were eaten and used medicinally for conditions including fever, rheumatism, and headache by Native Americans (Burlage 1968; Mabberley 2017).

Chenopodium pallescens [HEA]

Chenopodium lenticulare [LEI]

Chenopodium pratericola [HEA]

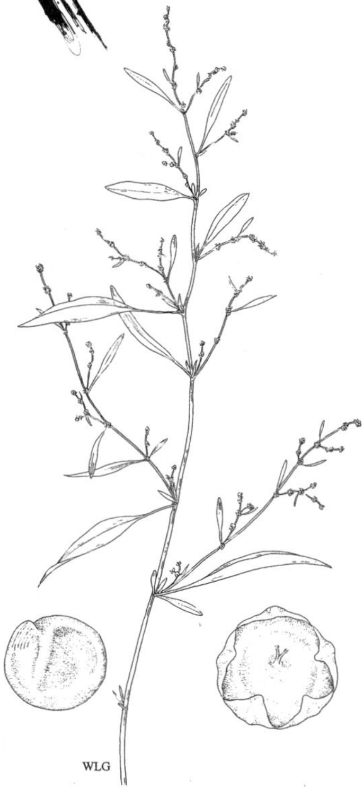

Chenopodium standleyanum [GL2]

Dysphania ambrosoides [FAW]

Dysphania anthelmintica [MAR]

Dysphania botrys (L.) Mosyakin & Clemants, (a bunch of grapes), JERUSALEM-OAK, FEATHER-GERANIUM. Annual, aromatic with unpleasant odor; stems erect, usually 0.2–0.6(–1) m tall, densely glandular-viscid throughout; leaf blades 1.3–4 cm long, 0.6–2.7 cm wide, lyrate-sinuate, pinnatifid, or sometimes entire; inflorescences cymose in axils of leaves, but without leaf-like bracts; perianth segments densely glandular-pubescent, abaxially rounded, not keeled, imperfectly enclosing the fruit; fruits 0.6–0.8 mm in diam. Disturbed habitats; Dallas (*Reverchon 819*, 1879, SMU), Denton (Kartesz 2022), Harris (Turner et al. 2003), and McLennan (Hannick et al. 2013) cos.; Brewster Co. in Trans-Pecos; e U.S. from ME s to NC and w to ND and TX, widespread in w 1/3 of the U.S. and ne U.S.; se Canada and British Columbia. Jul–Oct. Native to Eurasia. [*Chenopodium botrys* L.; *Teloxys botrys* (L.) W.A. Weber]

Dysphania carinata (R. Br.) Mosyakin & Clemants, (with a keel) KEELED WORMSEED, KEELED GOOSEFOOT. Plant annual, aromatic; stems erect to ascending, branched, 3–5.5 dm tall, pilose with glandular hairs or sessile glands; leaf blades 0.9–1.6 cm long, 0.8–1.4 cm wide, ovate to broadly ovate, usually sinuate-lobed, glandular; glomerules axillary, 1.5–2.5 mm in diam., subtended by leaf-like bracts 3–5 mm long; perianth segments distinct nearly to base, distinct portions ca. 0.9 mm long, covered with septate hairs, becoming leathery and white in fruit; seeds ovoid, ca. 0.6 mm long. Disturbed areas; in East TX known only from Polk Co. (Kartesz 2022), also Hidalgo Co. in extreme s TX; AL, AR, and TX; probably a waif. Sep–Oct. [*Chenopodium carinatum* R. Br.,] Native of Australia.

Dysphania pumilio (R. Br.) Mosyakin & Clemants, (dwarf), RIDGED GOOSEFOOT. Annual, both glandular and pubescent, aromatic with unpleasant odor; stems prostrate to suberect, ca. 1–4.5 dm long; leaf blades 0.5–2.7 cm long, ovate or oblong to lanceolate, sinuate-dentate, with lower surface conspicuously glandular; petioles ca. 1/2 as long as blades; inflorescences lateral cymes or glomerules with leaf-like bracts 3–5 mm long; seeds 0.5–0.6 mm in diam. Waste places; Polk, Travis (Turner et al. 2003), and Upshur (*Reverchon 2522*, 1901, SMU) cos.; otherwise in TX known only from Bandera (Edwards Plateau) and Brewster (Trans-Pecos) cos. further w; scattered in the e 1/2 of the U.S. and in the Pacific states. Nov. [*Chenopodium pumilio* R. Br.; *Teloxys pumilio* (R. Br.) W.A. Weber] Native of Australia. The name *Chenopodium carinatum* R. Br., has sometimes been misapplied to this species (Clemants & Mosyakin 2003a); however, that is a separate species, native to Australia, that has been introduced into AL, AR, and coastal TX and is now treated as *Dysphania carinata* (R. Br.) Mosyakin & Clemants.

SALICORNIA L. GLASSWORT, SALTWORT, SAMPHIRE

Annual or perennial with jointed, fleshy stems; leaves opposite, scale-like; flowers in terminal and/or lateral spikes, bisexual and unisexual, embedded in the fleshy stem on opposite sides, 3 flowers per side; perianth segments fused nearly to apex; stamens 1–2; style 1, stigmas 2–3; fruit an indehiscent utricle; seed 1.

A genus of 40–50 cosmopolitan, morphologically reduced, succulent species of coastal areas or other saline habitats (Weakley, A.S., and Southeastern Flora Team 2022); indeed, salt is "necessary for subsequent growth and development" (Ayala & O'Leary 1995; Judd & Ferguson 1999). Some are eaten like spinach or pickled (Mabberley 2017). The genus, in the past, has had some taxa segregated as *Sarcoconia* (see Scott 1977). Since then, authors have varied in their treat of *Salicornia/Sarcoconia.* Ball (2003b), Kadereit et al. (2007), Alonso and Crespo (2008), Wunderlin et al. (2017), and Kartesz (2022) segregate the taxa. Piirainen et al. (2017) and Weakley, A.S., and Southeastern Flora Team (2022) treat all the taxa in *Salicornia.* We are following Piirainen et al. (2017) and Weakley, A.S., and Southeastern Flora Team (2022) and treating all East TX taxa in *Salicornia.* The common name, glasswort, is the result of the plants at one time being used as "a source of sodium used in the production of glass" (Judd & Ferguson 1999). The flowers are inconspicuous and wind-pollinated. (Latin: *sal*, salt, and *cornu*, horn, referring to the physical appearance of the plant (horn-like branches) and the often saline habitat)
REFERENCES: Scott 1977; Ball 2003a, b; Alonso & Crespo 2008; Piirainen et al. 2017; Wunderlin et al. 2017; Weakley, A.S., and Southeastern Flora Team 2022.

1. Perennial from a horizontal rhizome; central flower of a group of 3 inserted at ca. the same level on the stem as the lateral 2 flowers ____ **S. ambigua**
1. Annual from a taproot; central flower of a group of 3 inserted above the lateral 2 flowers ____ **S bigelovii**

Salicornia ambigua Michx., (doubtful), WOOD GLASSWORT, PERENNIAL GLASSWORT, GUINEA-BEAD, WILD CORAL. Similar to *S. bigelovii* except, erect, procumbent, or prostrate perennial shrub, fleshy, woody only at base; inflorescences terminal and lateral, 1–6 cm long; level of flowers in 3-flowered clusters nearly even on the stem. Primarily salt marshes; Harris and Jefferson (Kartesz 2022) cos.; Gulf Prairies and Marshes; coastal U.S. from NH s to FL and w to TX; West Indies and South America. Summer–fall. [*S. virginica* of authors, not L., *Sarcocornia ambigua* (Michx.) M.Á. Alonso & M.B. Crespo] Native of the U.S. Dried specimens of *Salicornia/Sarcocornia* are often misidentified. Several factors contribute to this situation. Wide phenotypic plasticity, the small nature and insertion of the flowers in the fleshy stems, reduced leaves, and the fact that drying often distorts these characteristics

Dysphania atriplicifolia [BB2]

Dysphania botrys [BB2]

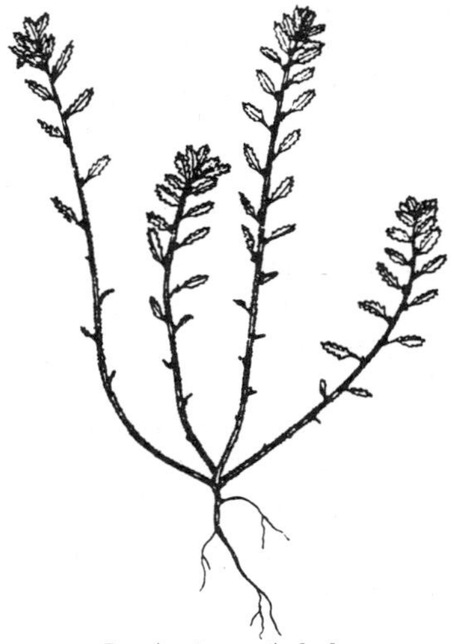
Dysphania pumilo [STE]

Dysphania carinata [VMU]

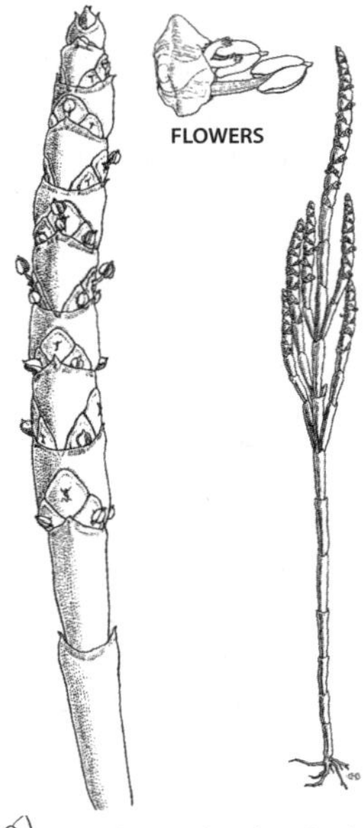

Salicornia bigelovii [MAS]

Salicornia ambigua [GWO]

make definite determinations of dried specimens difficult. This leads to descriptions that are not accurate or describe some other species and often misapplied names. Not to mention, the convoluted taxonomic history (see Alonso & Crespo 2008). For these reasons, herbarium specimens should be viewed and identified with caution. There are only 2 species of *Salicornia* known for Texas. The best method for identifying dried specimens may be only that *S. ambigua* is perennial and basally woody and *S. bigelovii* is annual and fleshy. Unfortunately, basal parts of plants are not often collected. *Salicornia virginica* Michx. is in the U.S. but coastal from ME s to GA and on the w U.S. coastal tier of states.

Salicornia bigelovii Torr., (for its discoverer, John M. Bigelow, 1804–1878, Boston physician and collector who assisted George Engelmann on the U.S.-Mexico Boundary mission). DWARF SALTWORT. Glabrous fleshy annual; stems erect, green or reddish (due to betalin pigments), unbranched or usually branched, 10–60 cm tall, apparently jointed, the joints 2–3 mm in diam., succulent; leaves reduced, scale-like, opposite, sessile, apically acute, mucronate; inflorescences cylindric terminal spikes 2–12 cm long, apparently jointed, each fertile segment (joint) consisting of 2 opposite 3-flowered clusters embedded in cavities in internode, the middle flower of each cluster slightly higher on the stem than the 2 laterals; bracts almost obscuring clusters; flowers bisexual and unisexual; perianth segments usually 3, fleshy, persistent in fruit; stamens 1–2; fruit a utricle. Primarily salt marshes; Gonzales, Hardin, Harris, Jefferson, and Orange (Kartesz 2022) cos.; otherwise in TX known only along the coast; Atlantic and Gulf coasts and coastal s CA. Late summer–early fall. The seeds of this species have a high percentage of oil (26–33%) and it has been investigated as a possible source of vegetable oil (Glenn et al. 1991).

SALSOLA L. RUSSIAN-THISTLE, SALTWORT

A cosmopolitan genus of ca. 130 species (Mosyakin 2003c) primarily of coastal or other saline habitats. Some authors (Stevens 2001 onwards; Akhani et al. 2007; Allred et al. 2020; Weakley, A.S., and Southeastern Flora Team 2022) treat *Salsola* in the strict sense with only ca. 25 species in the genus. However, we are following Mabberley (2017), Wunderlin et al. (2017), and Powell and Worthington (2018) in treating *Salsola* in the traditional broad sense. At one time, all species of *Salsola* found in the U.S. were transferred to the genus *Kali* Mill. (Akhani et al. 2007, 2014), but reverted to *Salsola* in 2017 (Wilson 2017). Both wind and insect pollination are known in the genus, with the yellow stamens of some species thought to attract pollinators; some species are known to cause hay fever (Judd & Ferguson 1999). Nomenclature for *Salsola* follows Mosyakin (1996, 2003c). (Latin: *salsus*, salty, from either the habitat of or the taste of some species)
REFERENCES: Beatley 1973; Mosyakin 1996, 2003c; Pyankov et al. 2001; Akhani et al. 2007, 2014; Wilson 2017.

Salsola tragus L., (Greek: *tragos*, a goat), RUSSIAN-THISTLE, PRICKLY RUSSIAN-THISTLE, TUMBLEWEED. Bushy, glabrous annual 1.5–10 dm or more tall and broad, with habit of a tumbleweed; stems usually with reddish or purple streaks; leaves sessile, filiform, 1.5–3.5 (–7) cm long, only 1–2 mm wide, spine/bristle-tipped, the spine < 1.5 mm long; flowers in interrupted terminal spikes when mature, enclosed or subtended by a pair of stiff, spine-pointed bracts 3–8 mm long, perfect; perianth segments 5; stamens 3–5; stigmas 2–3; fruits utricles, with prominently membranous, winged (3 smaller and 2 larger) apex, when mature, 3–6 mm wide including wing. Roadsides, disturbed sites; in East TX known only from Grayson (*Gentry 50-185* BRIT) and Wilson (Kartesz 2022) cos.; widespread primarily in w 1/2 of TX; widespread throughout the U.S. except for extreme se; Canada. Jun–Oct. [*Kali tragus* (L.) Scop., *S. australis* R. Br., *S. kali* L. var. *tenuifolia* Tausch, *S. kali* L. subsp. *tragus* (L.) Celak., *S. pestifer* A. Nelson, *S. tragus* subsp. *iberica* Sennen & Pau] Native of e Europe and Asia that is now a drought resistant agricultural pest in parts of North America; possibly introduced into the U.S. in South Dakota in 1873 or 1874 in flax seed imported from Russia (Mosyakin 1996; Tellman 1997); this species rapidly became a noxious weed in the w U.S. According to Tellman (1997), "The newly built railroad was an ideal vehicle for spreading tumbleweed throughout the west and the tumbleweed's early distribution pattern shows it moving outward along railways and roadways." Molecular studies (Ryan & Ayers 2000) suggest there are two morphologically similar cryptic populations that differ genetically; these require further study. The bushy, roughly globe-shaped plants break off at ground level in late fall; the whole plant, when blown by the wind, rolls or tumbles across open areas and disperses the seeds (Kirkpatrick 1992). Typically, one plant can produce as many as 250,000 seeds (Young 1991). Reported to contain the alkaloids, salsolidine and salsoline, and to be poisonous to livestock (Burlage 1968).

SPINACIA L. SPINACH

A genus of 3 species (Schultz 2003) of Eurasia and n Africa (Mabberley 2017). (Latin: *spina*, spine, referring to spine-tipped bracts around the fruits or Persian: *ispanakh*, spinach)
REFERENCE: Schultz 2003.

Spinacia oleracea L., (of the vegetable garden, a potherb used in cooking), SPINACH. Glabrous, monoecious annual or biennial; stems erect, 30–45 cm tall; basal leaves in a rosette, long-petioled, reduced up the stem, blades oval to triangular-ovate to hastate,

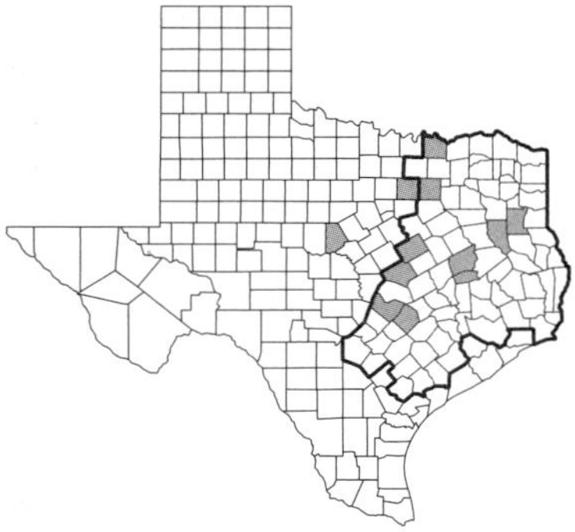

Chenopodium standleyanum

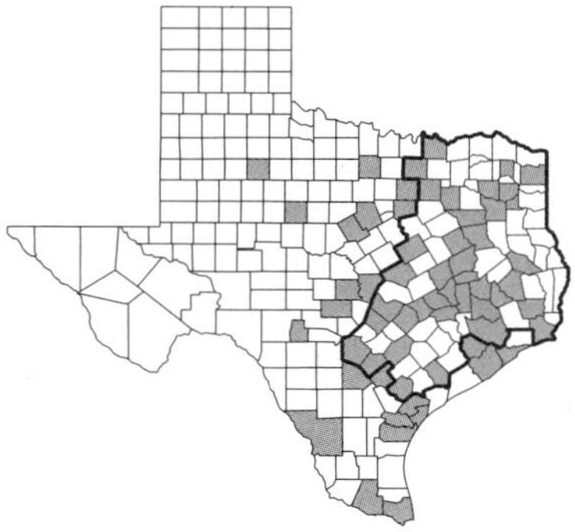

Dysphania ambrosioides

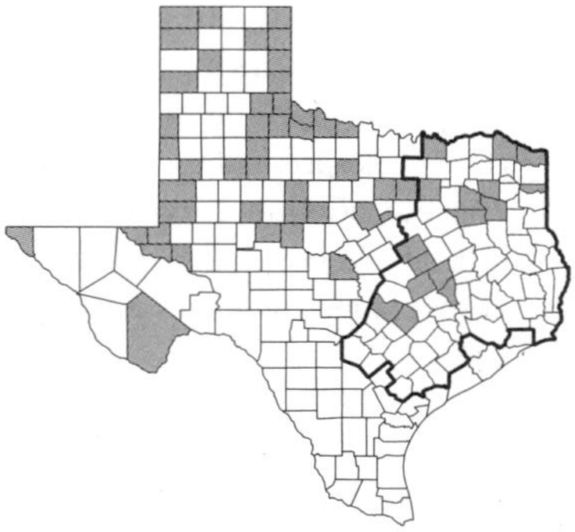

Dysphania atriplicifolia

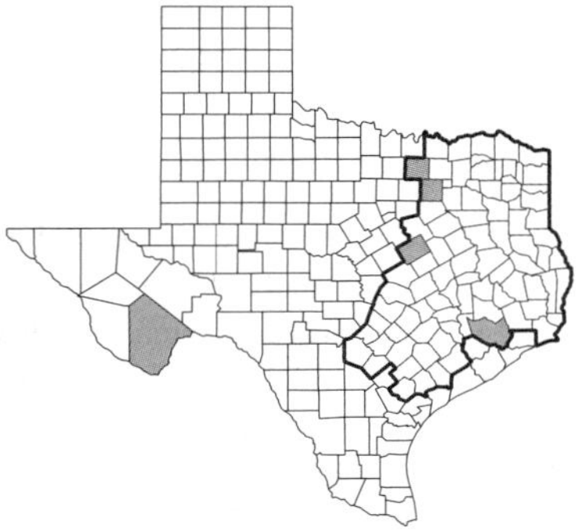

Dysphania botrys

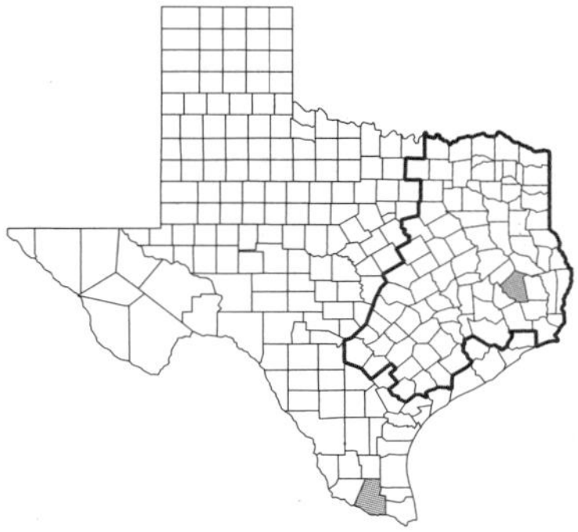

Dysphania carinata

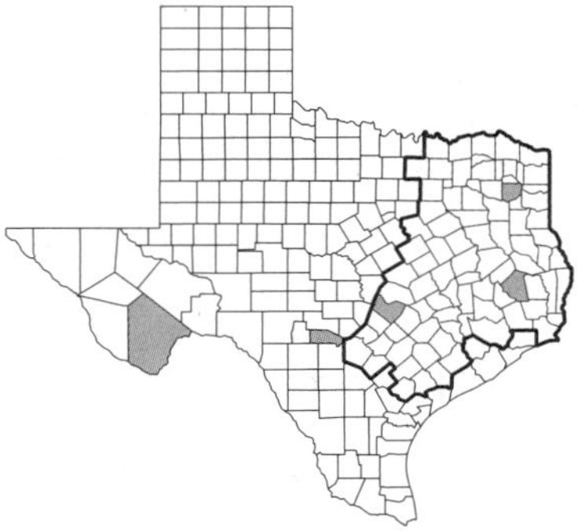

Dysphania pumilio

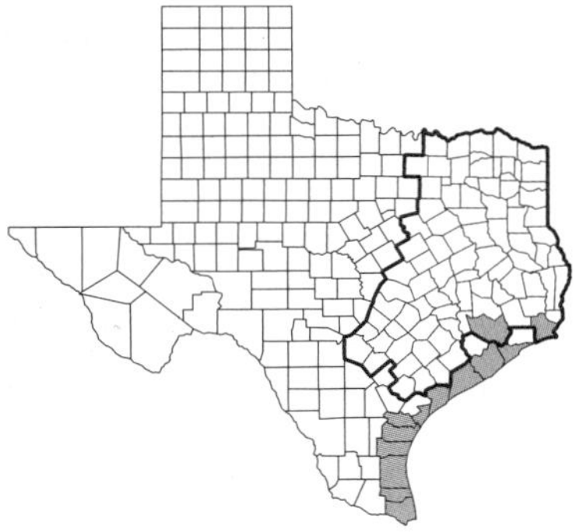

Salicornia ambigua

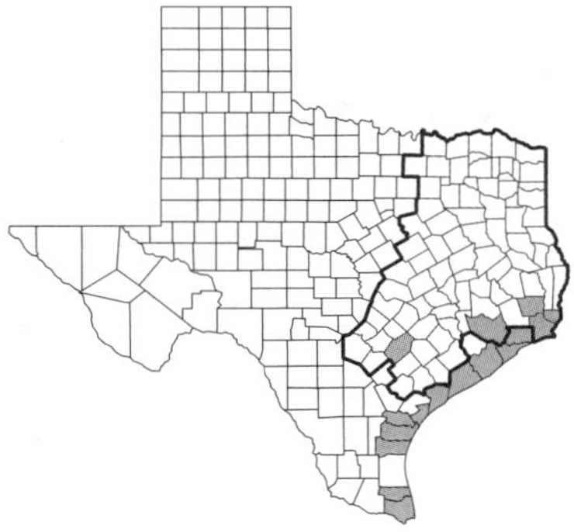

Salicornia bigelovii

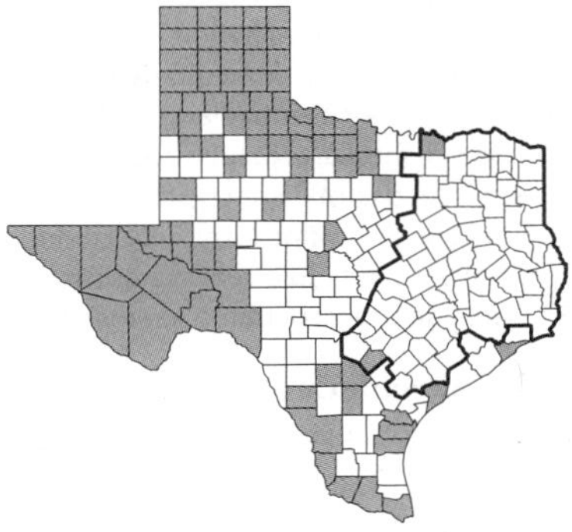

Salsola tragus

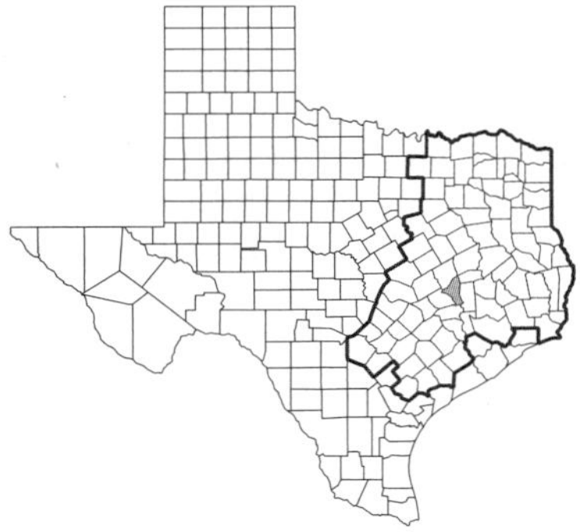

Spinacia oleracea

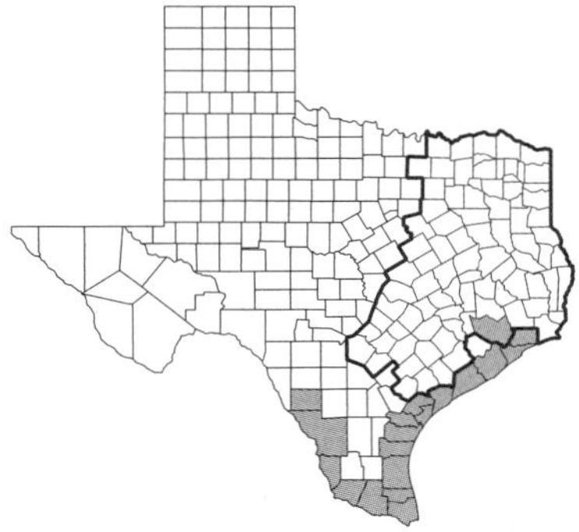

Suaeda conferta

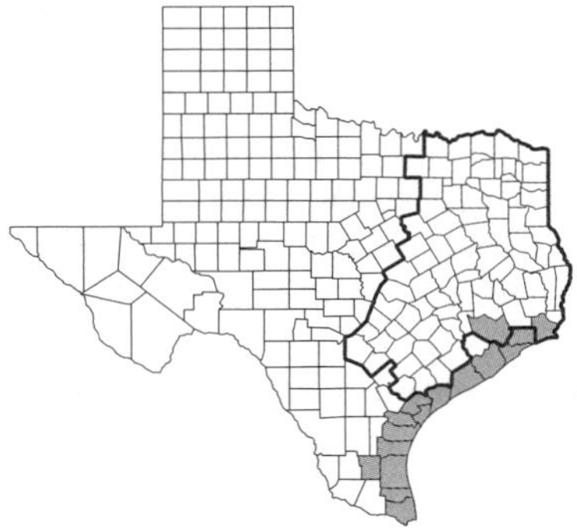

Suaeda linearis

entire or sinuate-dentate; flowers usually unisexual; staminate flowers with 4–5 perianth segments and 4–5 stamens; pistillate flowers without perianth segments, subtended by 2 usually spine-tipped bracts that grow together and enclose the fruit; stigmas 4–5; fruiting bracts 2–4 mm long. Cultivated for the edible leaves; in TX only recorded for Brazos Co. (Kartesz 2022); rarely escapes in TX; escaped in the U.S., mainly Pacific Coast states. Spring–summer. Native of sw Asia. This species has been in cultivation since ancient times and is valued as a "source of vitamins, calcium, iron, and antioxidant carotinoids…" (Shultz 2003). However, it should not be eaten in excessive amounts; the tissue contains high levels of oxalic acid that combines with calcium and can cause kidney problems or calcium deficiencies; occasional consumption rarely causes ill effects (Schmutz & Hamilton 1979).

SUAEDA Forssk. ex J.F. Gmel. SEA-BLITE, SEEPWEED

Mostly glaucous and glabrous herbs, subshrubs, or shrubs; annual or perennial; glabrous or pubescent, often glaucous; stems prostrate to erect; leaves alternate, sessile to short-petioled, fleshy, glaucous, flat to round; flowers clustered (3–8) in compound spikes or sometimes racemes, each cluster subtended by a bract; petals absent; perianths subtended by 2–4 membranous bracteoles to 1.5 mm long; flowers small, 0.7–4 mm in diam., bisexual, pistillate or staminate flowers sometimes present; perianth actinomorphic, ± zygomorphic, or irregular (i.e., perianth segments different sizes); perianth segments 5, enclosing fruit; succulent; stamens 5; stigmas usually 2–3; fruit a utricle; seeds sometimes dimorphic. Because of the succulent nature of the plants, characters become distorted upon drying making positive identifications from herbarium specimens difficult.

A cosmopolitan genus of ca. 110 species typically in saline or alkaline wetlands (Mabberley 2017); they typically exhibit salt tolerance and salt can even be necessary for growth (Judd & Ferguson 1999). Some have been cultivated as vegetables, while the seeds of others were ground and eaten by Native Americans (Ferren & Schenk 2003). "Although *Suaeda* is often considered to be wind pollinated, insect pollination appears to be common…and the flowers produce nectar" (Judd & Ferguson 1999). (Arabic: *suaed*, black, derived from the Arabic name for *Suaeda vera* (Forssk.) ex J.F. Gmel.—Ferren & Schenk 2003)
REFERENCES: Hopkins & Blackwell 1977; Bassett & Crompton 1978; Pedrol 1992; Fisher et al. 1997; Ferren & Schenk 2003.

1. Annual or perennial herbs, glabrous; perianth irregular or ± zygomorphic; perianth segments basally horizontally winged and/or vertically keeled, and/or distally horned or hooded; seeds dimorphic.
    2. Herbaceous annual; perianth segments with a basal horizontal wing and distal horn (one often larger than others); glomerules 3–5(–7)-flowered; seeds 0.8–1.7 mm in diam. ____ **S. calceoliformis**
    2. Shrub, subshrub, or facultative annual; perianth segments vertically keeled from base to tip; glomerules 1–3-flowered; seeds lenticular, 1–1.8 mm in diam. or flat, 1.5–2.6 mm in diam. ____ **S. linearis**
1. Shrubs, subshrubs, or facultative annuals, glabrous or pubescent; perianth actinomorphic; perianth segments often hooded; seeds not distinctly dimorphic (may vary in size and color).
    3. Stems usually pubescent with crinkled hairs at least distally; perianth segment apex rounded; segments fused basally or to middle; leaf scars ± smooth ____ **S. nigra**
    3. Stems glabrous; perianth segment apex obtuse; segments only basally fused; leaf scars knobby ____ **S. conferta**

Suaeda calceoliformis (Hook.) Moq., (from Greek: *calceolus*, slipper, the flower? resembling the form or shape of a slipper), HORNED SEA-BLITE, PURSH'S SEEPWEED. Annual herb; stems prostrate to erect, to 8(–10) dm tall; leaf blades usually 5–40 mm long, to 15 mm wide; linear to linear-lanceolate, apex acute to apiculate; leaves and bracts subtending spikes with a distinct lower thin margin; spikes 1–6 cm long; glomerules 3–5(–7)-flowered; bracts similar to leaves, sometimes wider; perianth 1–4 mm in diam., segments 5 (1–3 larger than others), horizontally winged proximally, keeled abaxially and/or horned or hooded distally; seeds dimorphic; lenticular, 0.8–1.7 mm in diam., black; or flat, 1–1.5 mm in diam., brown. Saline or alkaline habitats; tentatively included based on mapped distribution in Ferren and Schenk (2003); even though we have seen no East TX specimens, and the species is not mapped for East TX by Kartesz (2022) or Weakley, A.S., and Southeastern Flora Team (2022), the distribution map in the *Flora of North America* treatment (Ferren & Schenk 2003) shows this species in all but the Pineywoods of East TX; widespread in w 1/2 of the U.S. Summer–fall. This species has been reported to have been cooked and eaten as a vegetable (Steyermark 1963). In TX (e.g., Correll & Johnston 1970) this species has sometimes mistakenly (McNeill et al. 1977) gone under the name [*S. depressa* (Pursh) S. Watson]. No county level distribution map is provided.

Suaeda conferta (Small) I.M. Johnst., (closely crowded together), BEACH SEEPWEED. Perennial shrub forming dense mats, 2–10 cm tall, 2–50 cm wide; stems glabrous, prostrate to decumbent, leaf scars knobby; leaf blades 2–10 mm long, 1–2 mm wide, apex rounded; glomerules 1–5-flowered, usually distal in densely crowed compound spikes; perianth segments basally fused, apex obtuse; seed coat black. Sandy beaches and dunes, salt flats; in East TX known only from Harris Co. (Kartesz 2022); Gulf Prairies and Marshes and along the s Rio Grande; in the U.S. known only from TX; Mexico and West Indies.

Suaeda linearis (Elliott) Moq., (of uniform width), ANNUAL SEEPWEED. In our area usually a perennial herb; stems erect, 2–9 dm tall, woody at base; leaf blades 7–50 mm long, 0.8 to 1.8 mm wide, narrowly linear, apex acute; spikes 1–12 cm long; glomerules 1–3-flowered; perianth 1.2–2.5 mm in diam., ± irregular or zygomorphic; perianth segments keeled (at least some) but not horned distally, apex acute; seeds dimorphic, lenticular, 1–1.8 mm in diam., black, finely reticulate; or flat, 1.5–2.6 mm in diam., brown. Salt marshes, sandy beaches, coastal wetlands; Harris and Jefferson (Kartesz 2022) cos; coastal U.S. from ME to TX; Mexico and West Indies. Aug–Dec. Native of the U.S.

Suaeda nigra (Raf.) J.F. Macbr., (black), SHRUBBY SEEPWEED, QUELITE SALADO. Shrubs, subshrubs, or facultative annuals, ± erect, 2–15 dm tall, usually pubescent with crinkled hairs at least distally, leaf blades linear to narrowly lanceolate, (5–)10–30 mm long, 1–2 mm wide, apex obtuse to acute or acuminate, glabrous to pubescent; glomerules usually mostly distal, 1–12-flowered, arranged in compound spikes, bract usually shorter than leaves, perianth segments fused basally or to middle. Salt marshes, sandy beaches, coastal wetlands; in East TX known only from Gonzales Co. (*Tharp s.n.*, 1930, TEX; *Tharp s.n.*, 1936, TEX); Coastal Prairies and Marshes and Trans-Pecos; throughout much of w 1/3 of the U.S.; w Canada. Summer–fall. [*S. moquinii* (Torr.) Greene, *S. nigrescens* var. *glabra* I.M. Johnst., *S. suffrutescens* S. Watson] Native of the U.S. and Canada. Presumably a waif in East TX

CISTACEAE Juss.

SUN-ROSE OR ROCK-ROSE FAMILY

Ours low perennial herbs; leaves alternate, opposite or whorled, sessile to subsessile, simple, linear to elliptic or oblanceolate, entire; flowers small, terminal or in upper leaf axils, solitary, racemose, or in panicles with compact to loose clusters; sepals 5 dimorphic, 2 outer more or less linear and three inner more or less ovate, united at base; petals 3 or 5, opening in sunshine, lasting for one day only (or flowers cleistogamous); stamens 3–many; pistils 2–3-carpellate; ovary superior; fruit a capsule.

A small (170–180 species in 8 genera—Strother 2015), mainly n temperate and warm area, especially Mediterranean and e U.S. family of shrubs, subshrubs, or herbs; some are important as ornamentals, for example in rock gardens. Family name from *Cistus*, ROCK-ROSE, a mostly Mediterranean genus of 20 species of shrubs. The leaves of some *Cistus* species exude an aromatic resin, ladanum or labdanum, long used in incenses and in making perfumes and soaps and formerly in medicines (Page 2007). Old Testament biblical references to myrrh (e.g., in the story of Joseph and his brothers in Genesis) are generally considered mistranslations of the Hebrew word "loth" (= ladanum) (Page 2007). Ladanum is completely different from laudanum, which is a tincture of opium and alcohol. The flowers usually do not produce nectar, with pollen serving as the reward for pollinators. (Greek: *kistos*, classical name for these plants) (subclass Dilleniidae—Cronquist; order Malvales—APG IV)
FAMILY RECOGNITION IN THE FIELD: perennial, sometimes stellate-pubescent herbs with simple entire leaves; petals convolute (= twisted), quickly falling; ovary many-seeded, superior; stamens sometimes numerous.
REFERENCES: Brizicky 1964b; Wilbur 1969; Correll & Johnston 1970; Arrington & Kubitzki 2003; Strother 2015.

1. Petals 5 (except petals absent in late-appearing cleistogamous flowers), yellow, 4–18 mm long; plants pubescent with stellate hairs _____ **Crocanthemum**
1. Petals 3, usually reddish to purplish (can also vary to yellowish white or greenish), 1.5–3 mm long; plants pubescent with simple hairs to nearly glabrous _____ **Lechea**

CROCANTHEMUM Spach RUSH-ROSE, ROCK-ROSE, FROSTWEED

Ours perennial herbs with stellate pubescence on stems, leaves, and in inflorescence; leaves alternate and entire; flowers dimorphic; chasmogamous flowers fugacious (= falling or withering early); sepals dimorphic, 2 outer mostly linear and 3 inner mostly ovate; petals 5, yellow, conspicuous, obovate to obdeltoid with 15–36(–50) stamens; cleistogamous flowers often present (rare in *C. carolinianum*); stamens 3–8; without petals; fruit a capsule.

A genus of 21 species (Sorrie 2015) native to North America, Central America, and southern South America. Molecular research (Arrington & Manos, unpubl., cited in Arrington & Kubitzki 2003; Guzmán & Vargas 2009) suggests that *Helianthemum* as traditionally viewed is polyphyletic and that the New World species should be recognized at the generic level as *Crocanthemum* (21 species). We are thus treating North American *Helianthemum* in the genus *Crocanthemum*. (Greek: *krokos*, saffron, and *anthos*, flower (Quattrocchi 2000), in reference to the flower color)
REFERENCES: Wilbur & Daoud 1964; Daoud & Wilbur 1965; Sorrie 2015.

1. Stems with sparse, stellate hairs up to 2.5 mm long; lower surface of stem leaves clearly visible (not obscured by dense stellate hairs); leaves not revolute _____ **C. carolinianum**

1. Stems with stellate hairs similar to leaves; lower surface of stem leaves obscured by dense stellate hairs; leaves revolute.
   2. Stems leaves usually 4.5–9 mm wide; petals 6–12 mm long; inner sepals of cleistogamous flowers 3–4.2 mm long; flowers in loose racemose cymes terminating main stem and principle branches ____ **C. georgianum**
   2. Stem leaves usually < 5 mm wide; petals 4–6 mm long; inner sepals of cleistogamous flowers < 2 mm long; flowers solitary (chasmogamous) or glomerulate (cleistogamous) terminating main stem and principle branches ____ **C. rosmarinifolium**

Crocanthemum carolinianum (Walter) Spach, (of Carolina), CAROLINA FROSTWEED. Perennial herb with one to several ascending to erect branches; stems 10–38 cm tall with tufted, stellate hairs to 2.5 mm long; stem leaves usually narrower than basal; basal leaves often present, spatulate to obovate or elliptic, lower surface sparsely stellate-pubescent; stem leaves few and remote, similar to basal leaves, margins entire to slightly denticulate, not revolute; inflorescence of chasmogamous flowers in loose scorpioid cymes of 1–3 flowers, petals, 8–18 mm long, broadly spatulate to obdeltoid; Sandy fields, roadsides and open woods; se and e TX; Pineywoods, Post Oak Savannah, and Blackland Prairie; se U.S. from NC s to FL and w to AR and TX; Mar–May. [*Helianthemum carolinianum* (Walter) Michx.]

Crocanthemum georgianum (Chapm.) Barnhart, (of Georgia), GEORGIA SUN-ROSE, GEORGIA ROCK-ROSE, HOARY SUN-ROSE. Perennial herb with one to several ascending to erect branches; stems 10–40 cm tall with stellate hairs; basal leaves sometimes present, blades 10–28 mm long and 4.5–11 mm wide, spatulate-oblanceolate, lower surface densely stellate-pubescent; stem leaves 10–28 (–35) mm long, usually 4.5–9 mm wide, oblanceolate to narrowly elliptic, lower surface densely stellate-pubescent, margins revolute; inflorescence of chasmogamous and cleistogamous flowers in loose racemose cymes of 2–7 flowers, terminal on main stem and principle branches; chasmogamous flower petals 6–12 mm long, obovate to obdeltoid. Sandy fields, disturbed open areas, and open woods; throughout East TX; e 2/3 of TX; se U.S. from NC s to FL and w to OK and TX; Apr–Jun. [*Helianthemum georgianum* Chapm.] Native of the U.S.

Crocanthemum rosmarinifolium (Pursh) Janch., (with leaves like *Rosmarinus officinalis* L.—rosemary), ROSEMARY SUN-ROSE. Perennial herb with few to many ascending to erect branches; stems 13–50 cm tall; basal leaves sometimes present, blades obovate, lower surface densely stellate-pubescent; stem leaves 5–14 times as long as wide, linear to narrowly oblanceolate, lower surface densely stellate-pubescent, margins revolute; inflorescences both chasmogamous and cleistogamous in upper axils of main stems and principle branches; chasmogamous flowers solitary, rising from and above dense, glomerulate, cleistogamous flowers. Sandy open woods; throughout East TX; Gulf Prairies and Marshes and e Edwards Plateau; se U.S. from NC s to FL and w to OK and TX. Mid-May–Jun (cleistogamous flowers Jun–Jul). [*Helianthemum rosmarinifolium* Pursh] Native of the U.S.

LECHEA L. PINWEED

Ours perennial herbs lacking stellate pubescence; basal stems produced late in the growing season (overwintering), flowering stems produced in the spring; leaves alternate, opposite, almost opposite, or whorled; inflorescences densely or loosely paniculate or racemose; flowers small; sepals dimorphic, 2 mostly linear outer sepals and 3 mostly ovate to obovate inner sepals; petals 3, usually reddish to purplish (can be yellowish white or greenish), fugacious (= falling or withering early); stamens (3–)6–15(–25); style absent; fruit a capsule; seeds 2–6 per fruit.

A genus of 18 species (Lemke 2015) native to North and Central America and the West Indies; most species are in the se U.S. (Arrington & Kubitzki 2003). (Named for Johan Leche, 1704–1764, Swedish botanist)

REFERENCES: Hodgdon 1938; Wilbur & Daoud 1961; Wilbur 1966, 1969; Barringer 2004; Lemke 2015; Wunderlin et al. 2017.

1. Flowering stems with spreading villous hairs ____ **L. mucronata**
1. Flowering stems with appressed hairs or nearly glabrous.
   2. Pedicels 3–5 mm long; seeds (4–)6 ____ **L. san-sabeana**
   2. Pedicels 0.5–2.3 mm long; seeds 2–3(–5).
      3. Outer sepals equal to or longer than inner sepals.
         4. Main flowering stem leaves 4–7 mm wide; capsules ellipsoid to ellipsoid-ovoid ____ **L. minor**
         4. Main flowering stem leaves 0.5–1.5 mm wide; capsules broadly ovoid ____ **L. tenuifolia**
      3. Outer sepals shorter than inner sepals ____ **L. pulchella** var. **ramosissima**

Lechea minor L., (small), THYME-LEAF PINWEED. Flowering stems erect to 50 cm tall, sparsely pubescent; leaves whorled or opposite, blade of flowering stems 8–15 mm long, 4–7 mm wide, elliptic to lanceolate, margin entire; inflorescence in cylindric panicles; outer sepals longer than inner sepals; pedicels 0.5–2 mm; capsules ± equaling calyx. Dry, sandy, open ground, lowlands; recorded only in Hardin Co. (Brown et al. 2007); e TX and LA; e U.S. from NH s to FL and w to WI and s to AL, LA, and TX; possibly extirpated from Canada. Late Jul–Aug. Native of the U.S.

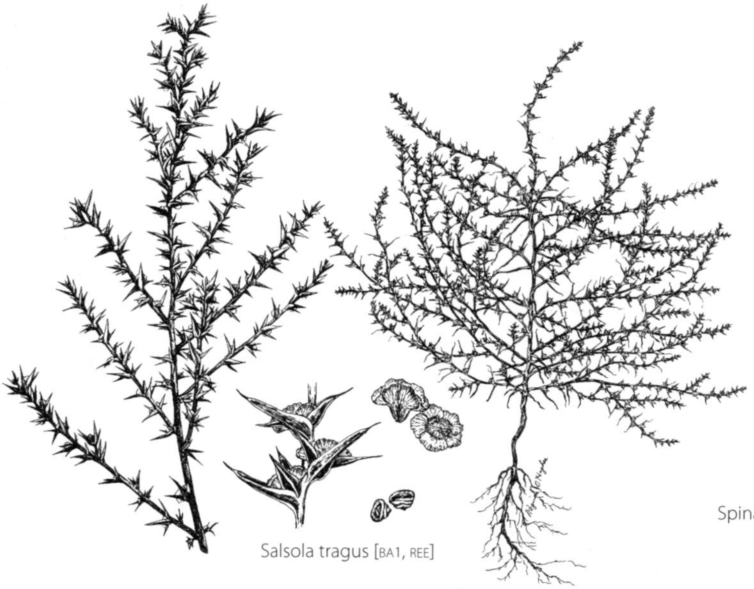
Salsola tragus [BA1, REE]

Spinacia oleracea [HEA]

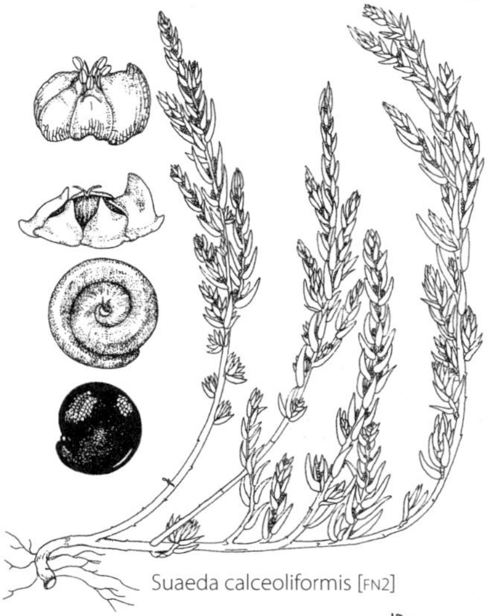
Suaeda calceoliformis [FN2]

Suadea conferta [COZ]

Suaeda linearis [WF2]

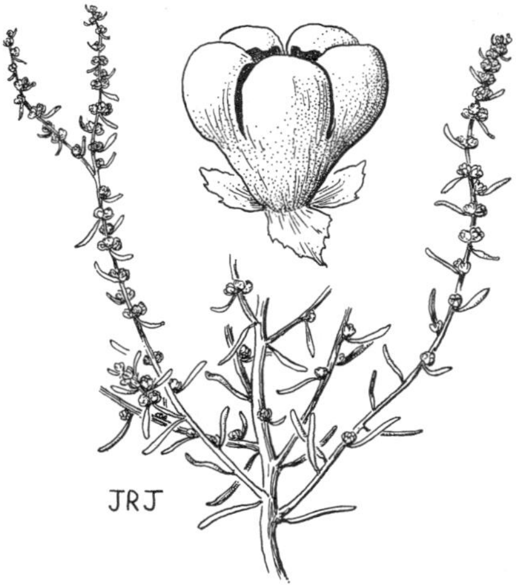

Suadea nigra [PN2]

Crocanthemum carolinianum [WFG]

Lechea mucronata Raf., (mucronate, pointed), HAIRY PINWEED, PINE PINWEED. Flowering stems usually 30–90 cm tall with spreading, villous pubescence; leaves opposite or whorled, blade of flowering stems 10–30 mm long, 3–4 mm wide, elliptic to ovate, mucronate, margins entire; inflorescence in subcylindric panicles; flowers few–several in a compact cluster, outer sepals shorter than inner, pedicels 0.8–1.5 mm long; capsules subglobose ± equaling calyx. Sandy woods, old fields, and roadsides; throughout East TX; South TX Plains and Edwards Plateau; widespread in e 1/2 of U.S.; also Canada. Jun–Jul. [*L. villosa* Elliott] Native of the U.S. and Canada.

Lechea pulchella Raf. var. **ramosissima** (Hodgdon) Sorrie & Weakley, (sp.: handsome; var.: much-branched). Flowering stems erect 25–80 cm tall, sparsely pubescent to almost glabrous; leaves opposite or almost opposite, blades of flowering stems 15–30 mm long, 1.5–3 mm wide, linear to narrowly elliptic, apex acute, hardened, margin entire; inflorescence in open elliptic, subglobose, or subcorymbiform panicles; flowers solitary or in few-flowered clusters, outer sepals shorter than inner sepals, pedicels 1–2.3 mm long, capsules subglobose, longer than calyx. Upland woods, pastures, damp flatwoods, savannahs, pinewoods; recorded for TX only in Orange Co. (Turner et al. 2003); se U.S. from VA s to FL and w. to TX. Jul–early Aug. [*L. leggettii* Britton & Hollick var. *ramosissima* Hodgdon] Native of se U.S.

Lechea san-sabeana (Buckley) Hodgdon, (of San Saba Co., TX), SAN SABA PINWEED, DRUMMOND'S PINWEED. Similar to *L. tenuifolia*; flowering stems 15–35 cm tall, densely to sparsely pubescent; leaves alternate to almost opposite, blade of flowering stems 6–18 mm long, 0.6–1.3 mm wide, linear, entire; inflorescence in a loose open racemose panicle, often drooping terminally; flowers, secund (= along one side of axis), loosely distributed, outer sepals longer than inner sepals; pedicels 3–5 mm long; capsules depressed-globose, ± equal to or longer than calyx. Sandy open woods; scattered in East TX; se and e TX w to West Cross Timbers, Edwards Plateau, and South TX Plains; nw LA. Near endemic to TX; outside of TX known only from LA in Caddo and De Soto parishes (Reid et al. 2008). Apr–Jun. [*Linum san-sabeanum* Buckley] Native of s U.S.

Lechea tenuifolia Michx., (slender-leaved), NARROW-LEAF PINWEED. Flowering stems erect 12–40 cm tall, sparsely pubescent; leaves whorled or opposite, blade of flowering stems 7–20 mm long, 0.5–1.5 mm wide, linear to narrowly oblanceolate, margin entire; inflorescence in broad panicles, flowers, sometimes secund (= along one side of axis), outer sepals equal to or longer than inner sepals; pedicels 0.5–1.5 mm long; capsules broadly ovoid, shorter than or ± equal to calyx. Sandy woods; throughout East TX; se and e TX w to West Cross Timbers and Edwards Plateau; e 1/2 of the U.S. May–Jul. [*L. tenuifolia* var. *occidentalis* Hodgdon] Native of the U.S.

CLEOMACEAE Bercht. & J. Presl

SPIDER-FLOWER FAMILY

Ours annuals with alternate, palmately compound leaves and often an odor; leaflets entire or nearly so; stipules minute or lacking; flowers in terminal, bracted racemes; sepals 4; petals 4, ± unequal, clawed; stamens 6 to many; pistil 1; ovary superior; fruit a capsule.

A family of ca. 150 species in 17 genera (Tucker & Vanderpool 2010); 300 species in 10 genera (Judd et al. 2016); ca. 300 species in 1 genus (*Cleome*) (Mabberley 2017); ca. 175 species in ca. 18 genera (Weakley, A.S., and Southeastern Flora Team 2022); nearly worldwide; tropical and temperate regions; mainly herbs producing mustard oil glucosides (glucosinolates). Traditionally, the Cleomaceae have had few genera (ca. 10 or less), with most of the species in the genus *Cleome*. It has long been thought that *Cleome* was polyphyletic, and recently it has had many segregates separated from it. However, generic limits are far from worked out as evident by the different numbers stated for the Cleomaceae above. In fact, Soltis et al. (2018) state, "It now appears that Cleomaceae consist of just *Cleome*." Cleomaceae, Capparaceae, and Brassicaceae are all closely related. In the past, some have included Cleomaceae in the Capparaceae (e.g., Mabberley 1997; Diggs et al. 1999; Kers 2003). The more current thinking is that the two families should be separate (e.g., Stevens 2001 onwards; Tucker & Vanderpool 2010; Judd et al. 2016; Soltis et al. 2018). For our treatment of the Cleomaceae, we are following the generic limits of Tucker and Vanderpool (2010) with some modification. (subclass Dilleniidae—Cronquist; order Brassicales—APG IV)

FAMILY RECOGNITION IN THE FIELD: annuals (many woody in other parts of the world) with palmately compound leaves (3–7 leaflets) and often with an odor; similar to Brassicaceae (e.g., flowers in racemes, 4 sepals, 4 petals) but with somewhat bilaterally symmetrical flowers (not cross-like as in Brassicaceae), stamens never 4 + 2 (the condition in the Brassicaceae), inflorescence bracts often present (these absent in Brassicaceae), and fruits lacking a transverse partition (this present in Brassicaceae).

REFERENCES: Iltis 1957, 1958 [1959]; Ernst 1963a; Judd et al. 1994; Hall et al. 2002; Kers 2003; Tucker & Vanderpool 2010; Judd et al. 2016; Soltis et al. 2018; Weakley, A.S., and Southeastern Flora Team 2022.

Crocanthemum georgianum [HEA]

Crocanthemum rosmarinifolium [HEA]

Lechea minor [HOL]

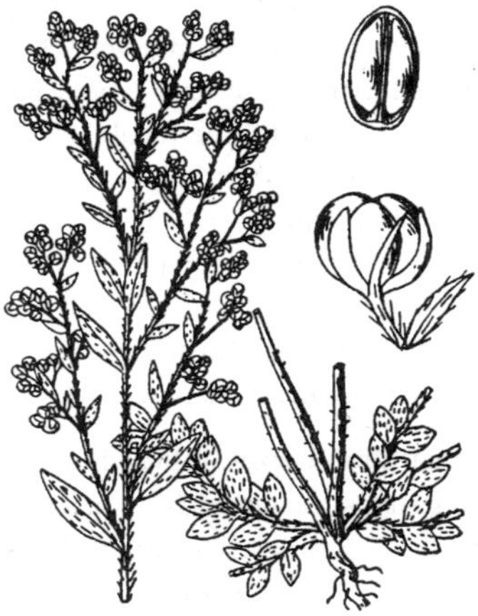
Lechea mucronata [CO1]

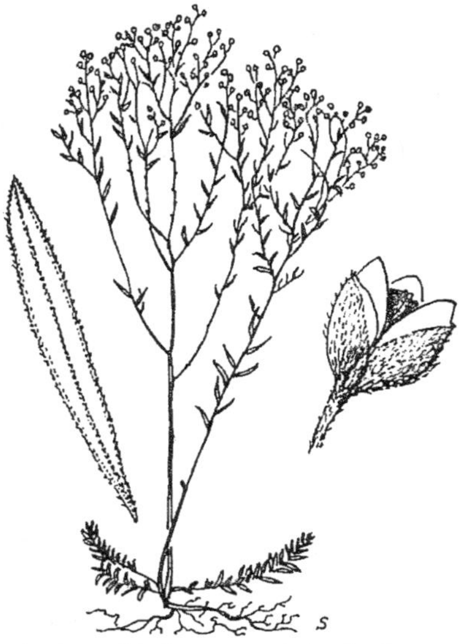
Lechea pulchella var. ramosissima [WF2]

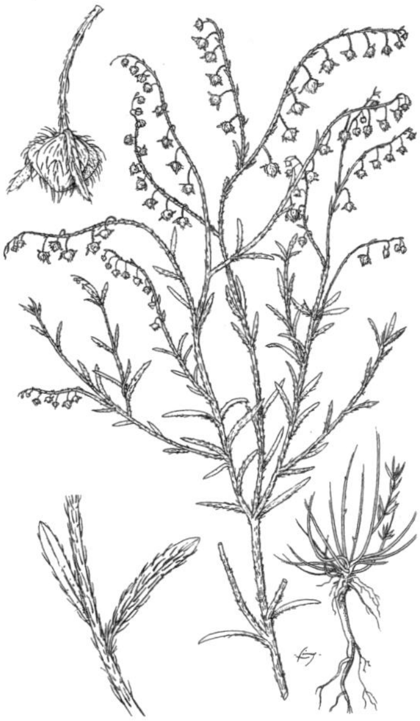
Lechea san-sabeana [FN2]

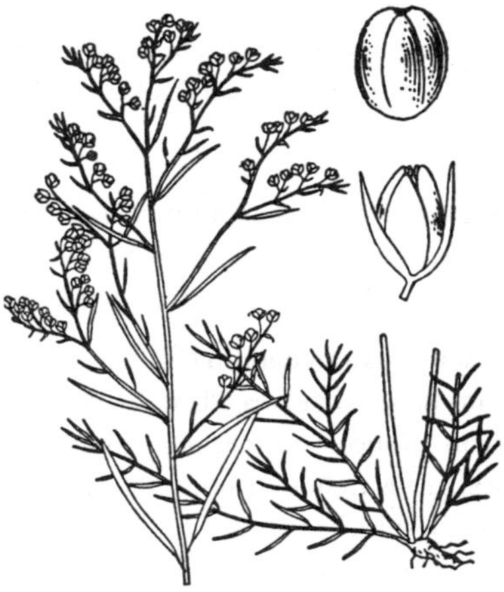
Lechea tenuifolia [BB2]

1. Stems glabrous; petals deep yellow; mature capsules ca. as wide as long **Cleomella**
1. Stems glandular-pubescent, at least in upper part (use lens); petals pink to purple, white or light yellow, sometimes with pinkish or purple tinge; mature capsules much longer than wide.
   2. Stamens usually 6–28; petals sometimes entire, usually notched or irregularly lacerate at the apex; gynophore (stipe of the pistil, above the calyx) (0–)2–14 mm long in fruit; leaflets 3 **Polanisia**
   2. Stamens 6 (except 14–25 in *Corynandra*); petals obtuse or acute at the apex; gynophore (stipe of the pistil, above the calyx) (0–)1–80 mm long; leaflets 3–7.
      3. Plants with nodal spines (and sometimes with prickles on petioles and leaf veins); leaflets 5 or 7; gynophore 30–80 mm long in fruit **Tarenaya**
      3. Plants lacking nodal spines and lacking prickles on petioles and leaf veins; leaflets 3 or 5; gynophore (0–)10–14 mm long in fruit.
         4. Petals white to pinkish or purple with a distinct claw; filaments fused to lower half of gynophore (evident from scars near the midpoint of the gynophore of fruiting specimens); gynophore 10–14 mm long in fruit **Gynandropsis**
         4. Petals light yellow with a short claw or merely attenuate; gynophore obsolete; filaments inserted on discoid or conical androphore **Corynandra**

CLEOMELLA DC. RHOMBOPOD

A genus of 10 species of sc and sw U.S. and Mexico (Weakley, A.S., and Southeastern Flora Team 2022). (Diminutive of *Cleome*)

REFERENCES: Payson 1922b; Iltis 1959; Vanderpool 2010; Weakley, A.S., and Southeastern Flora Team 2022.

Cleomella angustifolia Torr., (narrow-leaved), NARROW-LEAF RHOMBOPOD. Plant glabrous, erect, (0.3–)0.4–1.5(–2) m tall; leaves 3-foliolate; leaflets 2–8 mm wide, linear to elliptic; petals 4–6 mm long; capsules 5–9 mm wide, rhomboid-obdeltoid (like a distorted triangle); seeds 3–6 per capsule. Draws, ditch bottoms, roadsides, sandy soils; Anderson (Brown et al. 2007) Brazos, Gonzales, Harris (Turner et al. 2003), Dallas (*Reverchon 202* MISSA), and Gregg (*York s.n.*, 1931, GH) cos.; also Brazoria Co. (Turner et al. 2003) in Gulf Prairies and Marshes and West Cross Timbers w to High Plains; CO, KS, NE, OK, and TX. Jul–Oct. Native of the U.S.

CORYNANDRA Schrad. ex Spreng. TICKWEED

A genus of ca. 10 species of Asia and Africa. (Greek: *coryn*, club, and *andra*, male (i.e., stamen), reference unclear)

REFERENCE: Tucker 2010a.

Corynandra viscosa (L.) Cochrane & Iltis, (sticky), TICKWEED, YELLOW CLEOME. Annual herb, densely viscid throughout, 0.3–1 m tall; stems usually sparsely branched; leaves alternate, stipules absent, petiole 1.5–4.5(–8) cm long; leaflets 3 or 5, 2–6 cm long, 0.5–3.5 cm wide, ovate to oblanceolate-elliptic, margin entire and glandular ciliate; racemes 5–10 cm long, 10–15 cm in fruit; sepals 5–10 mm long, petals light yellow, 7–14 mm long, 3–4 mm wide; stamens 14–25, 5–9 mm long, dimorphic, 4–10 shorter; filaments inserted on a discoid or conical androphore; gynophore obsolete; capsules, partly dehiscent, 3–10 cm long, 2–4 mm wide; seeds light brown. Disturbed roadsides, railroads; Harris Co. (Brown et al. 2011); very widely scattered in AL, FL, GA, LA, MD, PA, SC, TX, and VA; however, Tucker (2010) states, "Records from New Jersey and Pennsylvania are based on non-persistent ballast plants from the late 1800s." Summer. [*Arivela viscosa* (L.) Raf., *Cleome viscosa* L., *Polanisia viscosa* (L.) DC.] Native of the tropics from Africa to northern Australia, and widely introduced elsewhere. The seeds are used as a cumin substitute in Sri Lanka (Mabberley 2017)

GYNANDROPSIS DC. SPIDERWISP

A genus of 2 species (Tucker 2010); annual or short-lived perennial herbs; pantropical. (From the genus *Gynandra* (Orchidaceae), and Greek: *opsis*, resemblance)

REFERENCES: Tucker 2010b; Weakley, A.S., and Southeastern Flora Team 2022.

Gynandropsis gynandra (L.) Briq., (female and male, presumably in reference to the stamens being borne on the gynophore), SPIDERWISP, CAT-WHISKERS. Plant 0.4–1.5 m tall, ± glandular-pubescent, unarmed; leaflets 3 or 5, 2.5–5 cm long, rhombic to obovate to oblanceolate; racemes with few to many flowers, to 40 cm long in fruit; petals white to pinkish or purple, not covering the stamens in bud; filaments 8–30 mm long; gynophore with stamen scars visible even when capsules present, 10–14 mm long; capsules linear-cylindric, to 4 mm thick. Weedy roadsides; Bexar, Colorado, Grayson (Turner et al. 2003), Goliad, Harris, Jackson, and Travis (Kartesz 2022) cos.; also Cameron, Karnes, Kennedy, and Willacy (Turner et al. 2003) cos. to the s of East TX; se U.S. from NC s to FL and w to OK and TX. May–Oct. [*Cleome gynandra* L.] Native of Africa. This species is reported to be used as a potherb (Mabberley 2017).

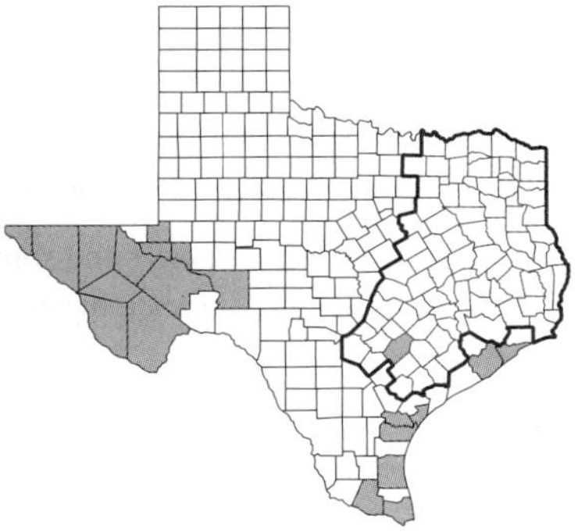

Suaeda nigra

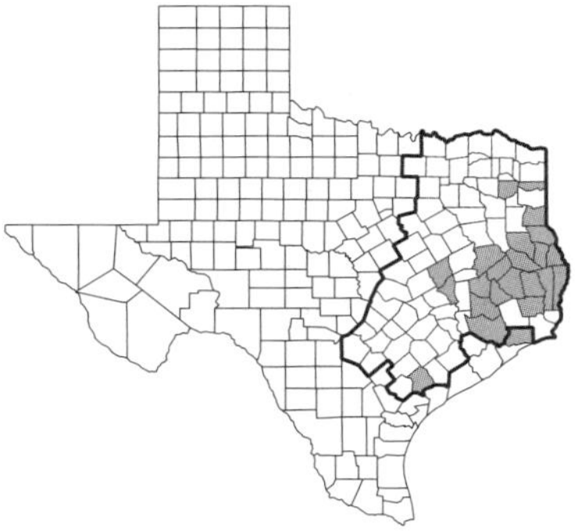

Crocanthemum carolinianum

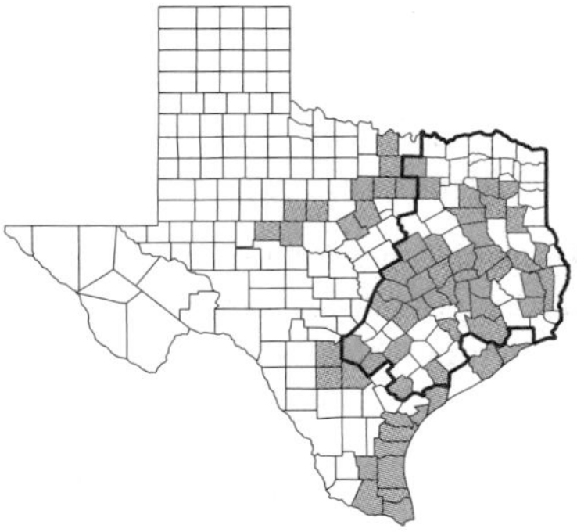

Crocanthemum georgianum

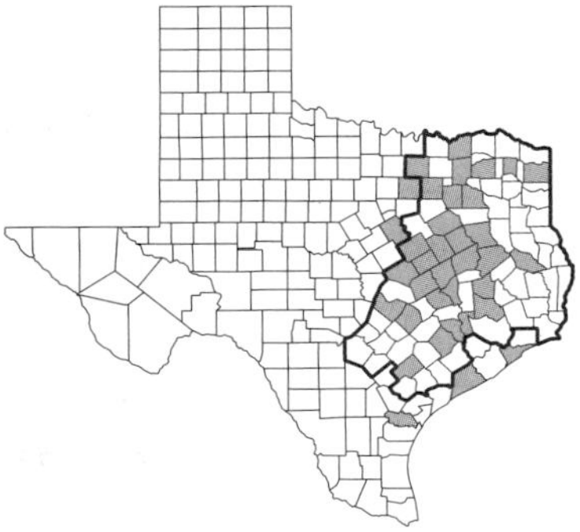

Crocanthemum rosmarinifolium

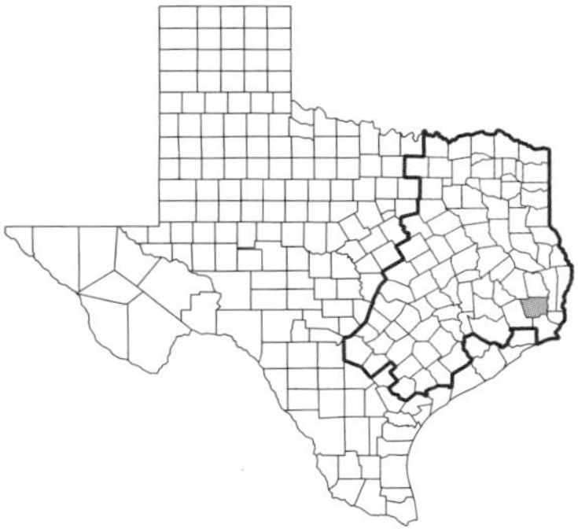

Lechea minor

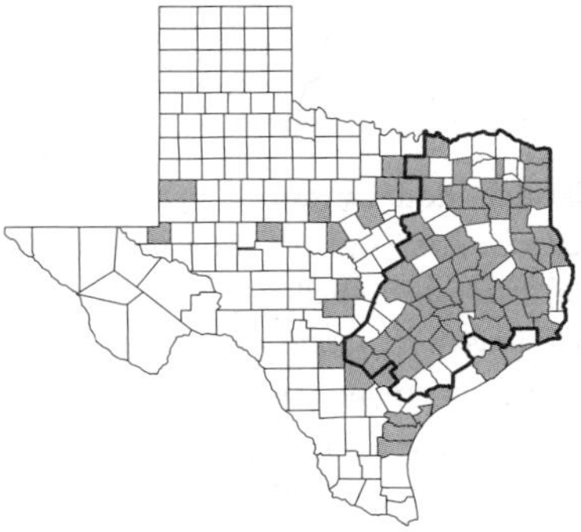

Lechea mucronata

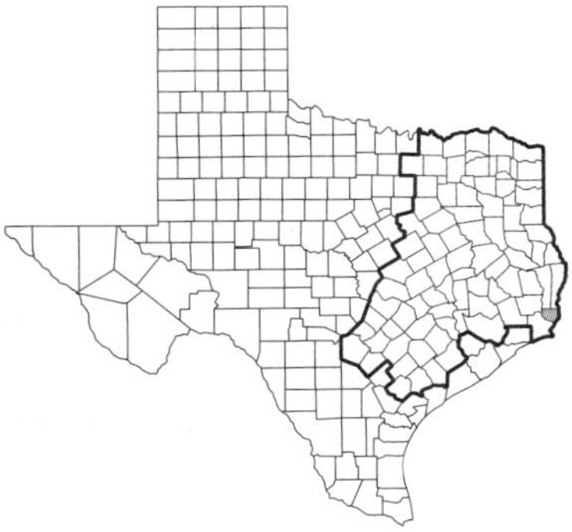

Lechea pulchella var. ramosissima

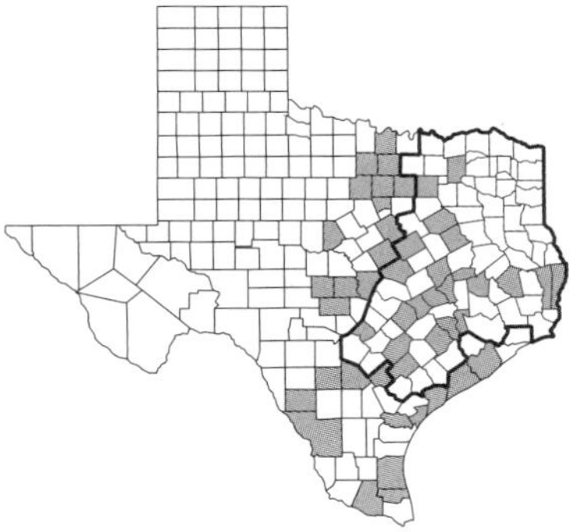

Lechea san-sabeana

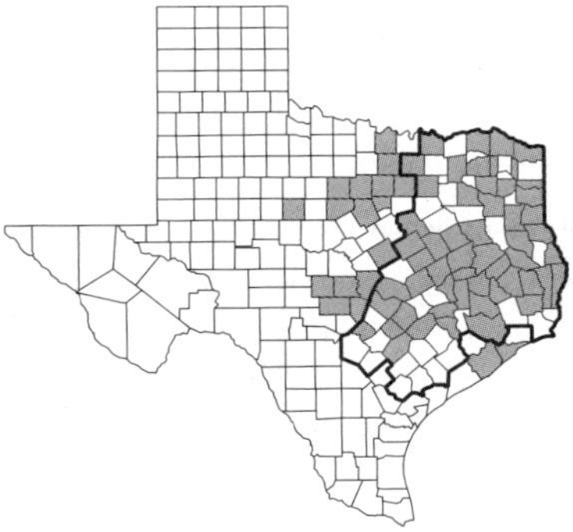

Lechea tenuifolia

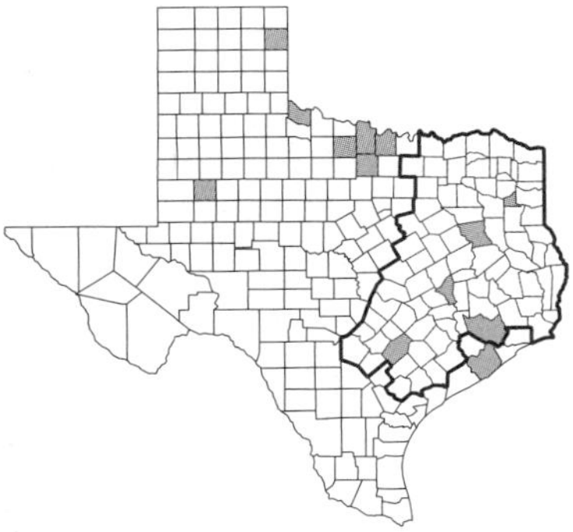

Cleomella angustifolia

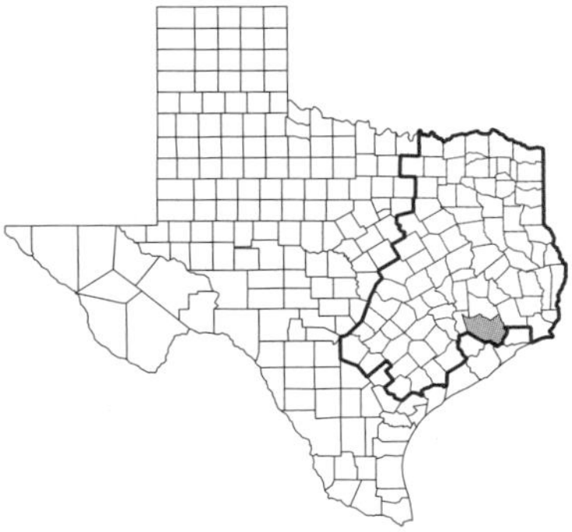

Coryandra viscosa

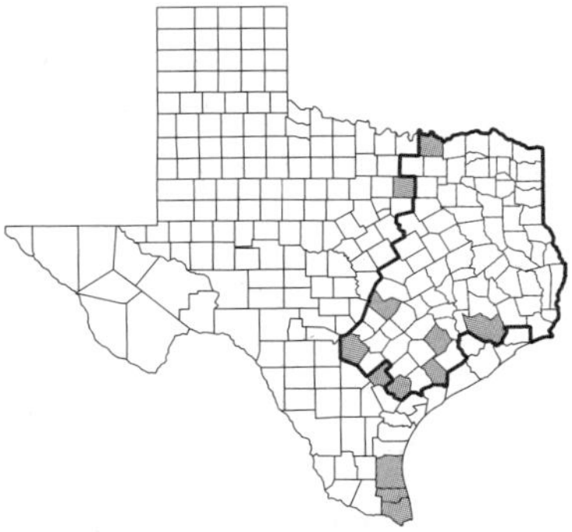

Gynandropsis gynandra

POLANISIA Raf. CLAMMYWEED

Herbaceous glandular-pubescent annuals with strong odor; leaves palmately 3-foliolate, petiolate; flowers with prominent nectary gland or not; petals (in ours) with a long claw, of 2 different lengths; stamens 6–28, exposed in bud (not covered by petals), of unequal lengths, with staggered maturation; capsules elongate, erect, dehiscing apically by valves; seeds numerous.

A North American genus of 5 species (Tucker 2010); thought to be related to xerophytic African *Cleome* species and sometimes lumped into that genus. Some are cultivated as ornamentals. (Greek: *polys*, many, and *anisos*, unequal, in reference to the characters by which the stamens differ from those of *Cleome*)
REFERENCES: Iltis 1958, 1966; Tucker 2010c; Weakley, A.S., and Southeastern Flora Team 2022.

1. Leaflets 5–20 mm wide, oblanceolate to obovate or broadly elliptic; petals slightly unequal, sometimes entire, usually notched; stamens 7–30 mm long; capsules 5–9 mm wide ____ **P. dodecandra**
1. Leaflets much narrower, only 1–2 mm wide, linear to narrowly elliptic; petals very unequal (upper 2 much larger than lower), with deeply ragged margin; stamens 3–13 mm long; capsules 2–4(–5) mm wide.
   2. Larger 2 petals 7–11 mm long; stipe (=gynophore) 7–14 mm long in fruit (beyond the 12–22 mm pedicel); stamens 10–13 mm long; style in fruit 2.5–4.5 mm long ____ **P. erosa**
   2. Larger 2 petals 3–5 mm long; stipe in fruit 2–4 mm long (beyond the 12–22 mm pedicel); stamens 3–5 mm long; style in fruit 1–2 mm long ____ **P. jamesii**

Polanisia dodecandra (L.) DC., (with twelve stamens), CLAMMYWEED. Plant 20–60(–90+) cm tall, glandular-pubescent and sticky; leaflets 5–20 mm wide; petals 3–16 mm long, white or sometimes with purplish tinge; stamens 6–28(–40); nectary gland solid, truncate to shallowly concave, not conspicuous in fruit, yellowish, the apex orange to orange-red, drying olive green; style deciduous in fruit; capsules 4–7 cm long; seeds many. Sandy stream banks, roadsides, disturbed sites. May–Oct. Native of North America and Mexico. Touching the foliage transfers a strong-smelling substance that makes the hands sticky or clammy.

1. Larger petals 3.4–6.5(–8) mm long; longest stamens 4–10(–14) mm long ____ subsp. **dodecandra**
1. Larger petals (7–)8–13(–16) mm long; longest stamens (9–)12–30 mm long ____ subsp. **trachysperma**

var. **dodecandra.** In TX known only from Harris Co. (Kartesz 2022); disjunct from the main population; e U.S. from NH s to VA and TN and west to MN to n AR; also Canada.

var. **trachysperma** (Torr. & A. Gray) Iltis, (rough-seeded). Mostly in the e and s portions of East TX, also Sabine (*MacRoberts & MacRoberts 8672* TEX) and San Augustine (*MacRoberts & MacRoberts 3263* TEX) cos.; throughout much of TX, except South TX Plains; mostly w 2/3 of U.S.; also Mexico and Canada. [*P. trachyspema* Torr. & A. Gray]

Polanisia erosa (Nutt.) Iltis, (jagged as if gnawed, in reference to petal margin), LARGE CLAMMYWEED, LARGE CRESTPETAL. Plant 20–60 cm tall, glandular throughout; leaflets 9–35 mm long, 1–2 mm wide, linear to narrowly elliptic; upper (larger) petals 7–11 mm long, white to pale-yellow, sometimes with pink tinge, the claw purplish pink; stamens 6–15; nectary gland tubular, apex truncate, usually 3.5–5.5 mm long, 0.9 mm in diam., yellow, drying pink-purple, persistent in fruit at base of gynophore; style persistent in fruit; capsules 2–6 cm long, with 12–40 seeds. Loose sand, prairies, fields; nearly throughout East TX; also Parker, Tarrant (BRIT) and Comanche (Turner et al. 2003) cos. in the Cross Timbers and Prairies s to e Edwards Plateau and n Gulf Prairies and Marshes; AR, LA, OK, and TX. Apr–Oct. Native of the U.S.

Polanisia jamesii (Torr. & A. Gray) Iltis, (for Dr. Edwin James, 1797–1861, American botanical explorer of the Texas Panhandle and the Rocky Mt. area), CRISTATELLA, JAMES' CLAMMYWEED. Plant 10–40 cm tall, glandular throughout; leaflets linear, 1–2 mm wide; inflorescences few-flowered; petals white, the upper (larger) ones 3–5 mm long, the lower (smaller) ones only 1.5–3 mm long; stamens 6–9; nectary gland tubular, 2–3.5 mm long, yellow, drying pink-purple; style persistent in fruit; capsules 8–30 mm long, with 18–36 seeds. Loose sand; Kaufman (*Tyler s.n.*, 1904, LL), Nacogdoches (*Whitehouse 11911* NY), and Shelby (*Thomas 129205* NLU) cos.; Rolling and High Plains, Cross Timbers and Prairies, and Trans-Pecos; mostly CO, KS, NE, NM, OK, SD, and TX. Late May–Oct. Sometimes segregated into the genus *Cristatella* [as *C. jamesii* Torr. & A. Gray]. Native of the U.S.

TARENAYA Raf. pink-queen

A genus of 33 species (Iltis & Tucker 2010) of annual herbs of South America. (Origin of generic epithet obscure)
REFERENCES: Iltis 1959; Iltis & Tucker 2010.

Tarenaya hassleriana (Chodat) Iltis, (for Emile Hassler, 1861?–1937, Swiss botanical collector and physician who settled in Paraguay), SPIDERPLANT, SPIDER-FLOWER, SPINY SPIDER-FLOWER, PINK-QUEEN. Plant robust, 1–2 m tall, armed with stipular spines 1–3 mm long, glandular-pubescent, strongly scented; leaflets usually 5 or 7, 2–6(–12) cm long, oblanceolate-elliptic to nearly

Cleomella angustifolia [IPL]

Corynandra viscosa [BR2]

Gynandropsis gynandra [COZ]

Polanisia dodecandra [GR1]

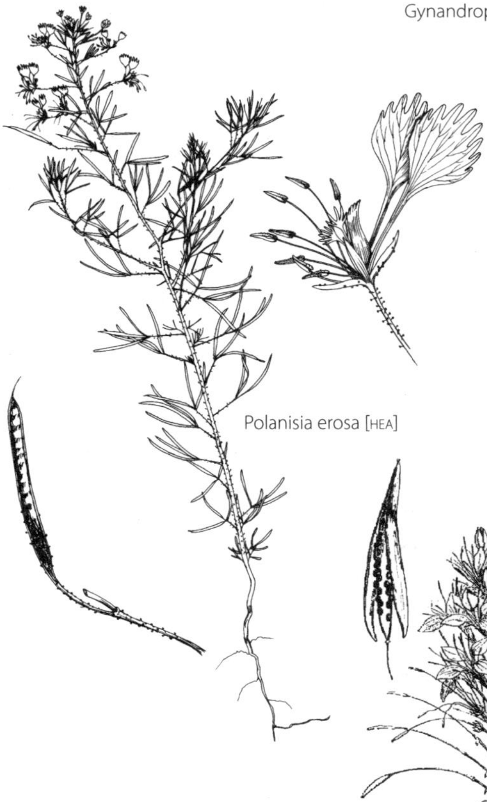
Polanisia erosa [HEA]

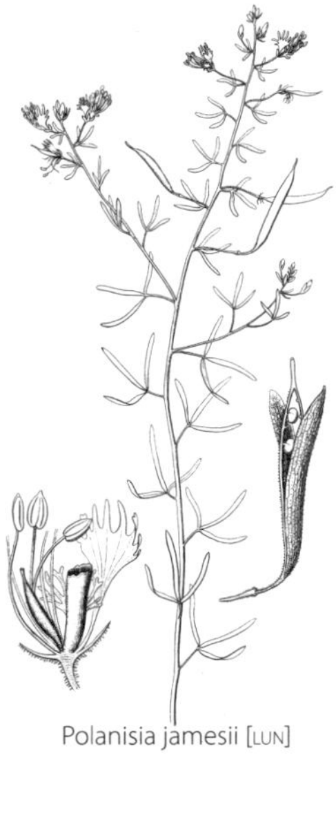
Polanisia jamesii [LUN]

Tarenaya hassleriana [BL2]

elliptic; racemes dense, 5–30 cm long, to 80 cm long in fruit; petals showy, pink to purple (rarely white), fading on second day, covering the stamens in bud; filaments sometimes 30–50 mm long; anthers ca. 1 cm long; capsules linear-cylindric, to 4 mm thick, divergent to deflexed on elongate (30–80 mm long) stipe (= gynophore). Cultivated and apparently escapes; Hardin (*Amerson 1290* SMU), Panola (*Janni KJETX2* BRIT), and Tyler (Kartesz 2022) cos.; widespread in e U.S. Apr–Oct. Native of South America. [*Cleome hassleriana* Chodat, *C. spinosa* of authors, not Jacq.]

CLETHRACEAE Klotzsch

CLETHRA, SWEET-PEPPER-BUSH OR LILY-OF-THE-VALLEY-TREE FAMILY

A small family of 2 genera and ca. 75–83 species primarily of tropical montane America as far s as Argentina and se Asia with some to temperate North America and Asia and 1 species on Madeira (Stevens 2001 onwards; Tucker & Jones 2009; Mabberley 2017); the species are mostly tropical and subtropical, evergreen or deciduous shrubs or small trees (Heywood 1993). A number are grown as fragrant-flowered ornamentals. The family was until recently often recognized as monogeneric (e.g., Heywood 1993), but is now considered to include *Purdiaea*, a genus of ca. 12 species from Cuba and Belize to Peru and characterized by indehiscent fruits, which was previously considered to be in the related Cyrillaceae (Anderberg & Zang 2002; Schneider & Bayer 2004). In the past some authorities (e.g., Mabberley 1997) included Clethraceae in Cyrillaceae; however, molecular evidence suggests the family is sister to Cyrillaceae/Ericaceae (Anderberg et al. 2002; Anderberg & Zang 2002; Rose et al. 2018) and APG IV recognizes it as a separate family. (subclass Dilleniidae—Cronquist; order Ericales—APG IV)

FAMILY RECOGNITION IN THE FIELD: the sole TX member of the family is a shrub with alternate, simple, toothed leaves, white flowers in erect racemes, and 3-valved dehiscent capsules drying brownish and persisting through the winter.

REFERENCES: Thomas 1961; Anderberg et al. 2002; Anderberg & Zang 2002; Schneider & Bayer 2004; Gonzales-Villarreal 2009; Tucker & Jones 2009; Rose et al. 2018.

CLETHRA L. SWEET-PEPPER-BUSH

A genus of ca. 65 species of tropical America, temperate North America (1 species), se Asia and temperate Asia, and Madeira (1 species) (Stevens 2001 onwards; Tucker & Jones 2009); some authorities have considered the genus to have as many as 120 species (Heywood 1993). *Clethra* species are found mostly growing in acid soils (as is also the case with Ericaceae) (Gonzalez-Villarreal 2009). Some are cultivated as ornamentals while the timber from some tropical American species is used for furniture and construction (Schneider & Bayer 2004). (Greek: *klethra*, the classical name for alder, *Alnus*, because of the resemblance of some species)

REFERENCES: Sleumer 1967; Fior et al. 2003.

Clethra alnifolia L., (with leaves like those of the genus *Alnus* in the Betulaceae), COASTAL SWEET-PEPPER-BUSH, SUMMER-SWEET, COASTAL WHITE-ALDER. Shrub to 3 m tall; leaves alternate, simple, deciduous, with petioles to 3.5(–6) cm long, blades 5–9(–12) cm long, obovate to oblong, widest beyond middle; inflorescences racemes, erect, solitary or in clusters of up to 4, to 15(–20) cm long, the axes sparsely stellate-pubescent; flowers bisexual, radially symmetrical, 5-merous, fragrant, each with a single bract at base of pedicel; bracts usually shorter than the flowers; pedicels 2–4(–5) mm long, stellate-pubescent; sepals triangular ovate, 3.5–5 mm long, stellate-pubescent; petals connate proximally, white, rarely pinkish, 5–8 mm long; nectary disk absent; stamens 10; ovary superior; carpels 3; capsules 3-valved, subglobose, 2.5–3(–4) mm long, drying light brownish and persisting through winter; seeds numerous, flattened, slightly winged. Swamps, wet woods, around lakes; mainly extreme se part of East TX in Angelina, Jasper, Jefferson, Hardin, and Tyler cos. (Big Thicket area), and Wood Co. further n (Turner et al. 2003; USDA Plants Database 2019); ME s to FL w to TX. Jul–Sep. [*C. alnifolia* var. *glabella* Michx., *C. alnifolia* var. *michauxii* Zabel, *C. alnifolia* var. *paniculata* (Aiton) G. Nicholson, *C. paniculata* Aiton] The species has been called "poor man's soap due to the flowers forming a lather when crushed (Correll & Johnson 1970). It is cultivated as an ornamental for its fragrant flowers (Schneider & Bayer 2004).

CONVOLVULACEAE Juss.

MORNING-GLORY FAMILY

Ours annual or perennial, herbaceous, some with milky sap; twining and climbing vines or stems trailing to erect; or parasitic, without chlorophyll and thread-like with twining stems in *Cuscuta*; leaves alternate, simple or palmately compound, entire or lobed or with few, coarse teeth; or scale-like in *Cuscuta*; exstipulate; flowers perfect, axillary, solitary or few in small cymes (the cymes head-like in *Jacquemontia*); calyx lobes or sepals 5; or (3–)4–5 in *Cuscuta*; corollas sympetalous, funnelform to campanulate, salverform, or rotate, 5-toothed, -angled, or -lobed; (3–)4–5-lobed in *Cuscuta*; stamens 5; or (3–)4–5, equal to the number of lobes in *Cuscuta*, distinct, attached at base or on tube of corolla; pistil usually 2-carpellate; ovary superior, styles 1–2; fruit a capsule with usually 1–4 seeds.

A medium-large family of 56 genera and 1,600–1,700 species (Austin 2023). It is a cosmopolitan family of mostly herbaceous climbers (always turning to the right), lianas, herbs, shrubs, or rarely trees; chlorophyll-less, thread-like twining stems in *Cuscuta*; leaves simple entire or lobed, usually palmately veined; scale-like in *Cuscuta*; stipules absent; a number are showy ornamentals; *Ipomoea batatas* (L.) Lam. (SWEET-POTATO) is widely cultivated; some are problematic weeds (e.g., *Convolvulus* and *Ipomoea*) (Mabberley 2017). (subclass Asteridae—Cronquist; order Solanales—APG IV)

FAMILY RECOGNITION IN THE FIELD: *twining*, *climbing*, or *trailing* herbs (a few species erect to decumbent) with alternate, sometimes cordate leaves; sap sometimes milky; corollas often showy, sympetalous, radially symmetrical, variable in shape but often funnelform to campanulate; ovary superior OR parasitic herbs without chlorophyll, stem yellowish, orangish, or greenish, leaves reduced to scales, flowers small (*Cuscuta*).

REFERENCES: Wilson 1960c; Austin 1986a, 1990, 1998, 2023; Stefanović et al. 2003; Felger et al. 2012; Wunderlin et al. 2019.

1. Parasitic vines, yellowish to orange or greenish; leaves absent (only scales present) ____ **Cuscuta**
1. Plants green, chlorophyllous; leaves present.
   2. Corollas deeply lobed, white to greenish white, 1.5–7 mm broad; leaf blades reniform (= kidney-shaped) or not; petioles longer than blade OR leaves essentially sessile; capsules deeply lobed or ovoid.
      3. Corollas white, salverform, 5–6.5 mm long; sepals erect; leaf blades elliptical to lanceolate, essentially sessile; capsules ovoid; in East TX known only from Gonzales Co. ____ **Cressa**
      3. Corollas white to greenish white, shallowly funnelform, 1.5–4 mm long; sepals spreading; leaf blades reniform (= kidney shaped) to orbiculate, petioles much longer than blade; capsules deeply lobed; common in East TX ____ **Dichondra**
   2. Corollas not deeply lobed, variously colored including white, 7–70 mm or more broad; leaf blades usually not kidney-shaped, on petioles shorter to longer than the blades; capsules without 2 distinct lobes.
      4. Stigmas or stigma lobes 2 or 4 per flower, variously shaped (filiform to linear, oblong or elliptic-flattened) but neither peltate, capitate, globose, nor subglobose; corollas either 20(–30) mm or less long OR if longer, (in *Calystegia*) then calyx enclosed by two large bracts.
         5. Flowers numerous in a densely crowded, leafy-bracted, head-like inflorescence; sepals with conspicuous, long (ca. 2–3 mm), tawny hairs easily visible to the naked eye; corollas blue ____ **Jacquemontia**
         5. Flowers solitary or in 2s or 3s, if more then not in densely crowded, leafy-bracted, head-like inflorescences; sepals glabrous to pubescent, but without long hairs conspicuous to the naked eye; corolla color various including blue.
            6. Erect to ascending or sometimes prostrate herbs with stems to 50 cm long; flowers sessile or on stalks (peduncle + pedicel), much shorter than blade of subtending leaf; sepals 3–5 mm long OR if stalks as long as or longer than subtending leaf blade, then sepals 2–3 mm long; corollas rotate to shallowly campanulate; stigmas 4 per flower ____ **Evolvulus**
            6. Trailing, twining, climbing and sometimes prostrate herbs with elongate stems that can reach 1–3 m in length; flowers on stalks (peduncle + pedicel) 3/4 as long or longer than subtending leaf blade; sepals 4–16+ mm long (3–5 mm in *Convolvulus arvensis*); corollas funnelform to campanulate; stigmas 2 per flower.
               7. Leaf blade entire, base cordate, attenuate, or rounded ____ **Stylisma**
               7. Leaf blade lobed, base hastate to sagittate.
                  8. Calyces ebracteate; corollas 1.2–3 cm long, usually with some color, sometimes all white ____ **Convolvulus**
                  8. Calyces immediately subtended by 2 large floral bracts; corollas 4–7 cm long, white, sometimes tinged with pink ____ **Calystegia**
      4. Stigmas 2, peltate to capitate or stigma 1, globose and unlobed or with 2–3 globose or subglobose lobes; corollas mostly ≥ 2 cm long (includes most Ipomoea species), if < 2 cm long then corolla all white (including throat) (shades of red to purple in *Stylisma aquatica*) or leaf with 5(–7) palmatisect (= incised ± to petiole tip) lobes or leaflets.
         9. Style 1, stigma 1, globose and unlobed or with 2–3 globose or subglobose lobes.
            10. Leaves with 5–7 palmate lobes that are toothed to pinnatifid; sepals 18–25 mm long and greatly enlarged and reflexed in fruit ____ **Distimake**
            10. Leaves entire or variously lobed; leaves can be palmately lobed to palmatisect, but lobes not toothed or pinnatifid; sepals usually < 18 mm long (sometimes longer in *I. hederacea*, and *I. lindheimeri*); if longer then not greatly enlarged and reflexed in fruit ____ **Ipomoea**
         9. Styles 2 (connate at base or nearly to tips); stigmas 2, peltate to capitate ____ **Stylisma**

CALYSTEGIA R. Br. FALSE BINDWEED

Trailing or twining perennials; leaves petiolate; leaf blades lobed; basally sagittate to cordate, apex acute; flowers axillary, solitary or in cymes of 2–4, bracteate (bracts longer than sepals); corollas campanulate; ovary 1-locular; style 1; stigmas 2; capsules 1–4-seeded.

A genus of ca. 30 species of North America, South America, Eurasia, Africa, and Australia (Brummitt 2023). It includes cultivated ornamentals and pernicious weeds. The genus is sometimes included in *Convolvulus*, but current authors vary in their treatment. Williams et al. (2014), Wood et al. (2015), and Weakley, A.S., and Southeastern Flora Team (2022) include *Calystegia* in *Convolvulus*. Though molecular research (e.g., Stefanović et al. 2002; Carine et al. 2004; Williams et al. 2014) indicates that *Convolvulus* is paraphyletic with respect to *Calystegia*, we are following, Wunderlin et al. (2019), Allred et al. (2020), and Brummitt (2023) in treating *Calystegia* separately from *Convolvulus*. (Greek: *kalyx*, cup, cover, calyx, and *stege*, cover, alluding to the two bracts that subtend the flowers and conceal the calyx)

REFERENCES: Brummitt 1965, 1981, 2023; Lewis & Oliver 1965; Austin 1986a, 1992; Austin et al. 1997; Spaulding 2013.

1. Leaf sinus quadrate (= nearly square); blade tissues not beginning for 2–5(–10) mm from petiole attachment (only vascular tissue near attachment); margins of bracts immediately subtending flower overlapping by > 1/2 their length ____ **C. silvatica**
1. Leaf sinus V- or U-shaped; blade tissues beginning at the point of petiole attachment; margins of bracts immediately subtending flower not or slightly overlapping only at the base.
   2. Leaf blades broadly triangular; basal lobes spreading, ± 2-pointed ____ **C. sepium** subsp. **angulata**
   2. Leaf blades linear to narrowly triangular; basal lobes slightly to abruptly spreading, 1-pointed or rounded ____ **C. sepium** subsp. **binghamiae**

Calystegia sepium (L.) R. Br. subsp. **angulata** Brummitt, (sp.: growing along hedges, subsp.: referring to the angles on the leaf lobes), HEDGE BINDWEED, TRAILING BINDWEED, ANGLED FALSE BINDWEED. Perennial herb with trailing or twinning stems, 0.5–3 m long, glabrous; leaf blades broadly triangular, glabrous, 2–15 cm long, 1–9 cm wide, basally cordate-sagittate to hastate, auricles obtuse to acute, 2–3-dentate, apically acute to acuminate; flowers usually solitary or paired, peduncles 3–13 cm long; bracts surrounding the calyx foliaceous mostly 14–26 mm long, 10–18 mm wide; corolla mostly 4–7 cm long, funnelform, white, rarely tinged with rose or pink; capsules 10–13 mm wide. Thickets, fence rows, openings, and glades; in TX known only from Jefferson Co. (Austin et al. 1997); scattered in ne U.S. and WI s to TX and w to WA and AZ; also Canada. Jun–Aug. [*C. sepium* var. *angulata* (Brummitt) N.H. Holmgren, *C. sepium* var. *repens* (L.) A. Gray, *Convolvulus repens* L., *C. sepium* L. var. *repens* (L.) A. Gray] Though the TX record for this taxon is from Jefferson Co. in sand dunes (at the extreme se corner of East TX, it is more likely to be found along the ne border with OK and LA where it is found in those states. Native of the U.S. and Canada.

Calystegia sepium (L.) R. Br. subsp. **binghamiae** (Greene) Brummitt, (sp.: growing along hedges; subsp.: for Caroline Bingham, one of the earliest American women botanists), MARSH BINDWEED, MARSH FALSE BINDWEED. Perennial herb with twining stems to 2 m long, pubescent; leaf blades 3–4 cm long, 1–2 cm wide, sagittate, leaf apex acute to acuminate; bracts 14–24 mm long, longer than the sepals, narrowly ovate, ribbed; flowers usually solitary or 2–4 in a cyme in leaf axils; corolla 4–6.5 cm long, white rarely tinged pink on the limb and upper tube; fruits obovoid to globose, 10 mm long, 8–10 mm wide; seeds ovoid, 4–5 mm long, gray to black, smooth. Fields, roadsides, and waste places; Harris, Jefferson, and Orange cos. (Austin et al. 1997); Trans-Pecos and Gulf Prairies and Marshes (Austin et al. 1997); CA, FL, GA, LA, NV, NM, NC, SC, and TX. May–Jun. [*C. sepium* (L.) R. Br. subsp. *limnophila* (Greene) Brummitt, *Convolvulus binghamiae* Greene, *C. nashii* House] No county distribution map provided. Native of the U.S. ⊘

Calystegia silvatica (Kit.) Griseb. subsp. **fraterniflora** (Mack. & Bush) Brummitt, (sp.: pertaining to the woods; subsp.: with brotherly flowers, these sometimes in pairs), TWIN-FLOWERED BINDWEED, SHORT-STALK FALSE BINDWEED. Herbaceous perennial 5–30 dm long, scrambling, trailing, or twining; leaf blade 2–8 cm long and 2–5 cm wide, ovate to ovate-lanceolate, base deeply cordate, the sinus quadrate (= square), blade tissue not beginning for 2–5(–10) mm from petiole attachment (only vascular tissue near attachment), apex usually acuminate to acute, glabrous; flowers solitary or often paired in leaf axils; peduncles 3.5–7 cm long, usually shorter than the petioles; bracts 16–28 mm long, strongly overlapping, inflated basally, ovate, rounded to blunt at tip, glabrous to sparsely pubescent; corolla 4–7 cm long, white, rarely pink tinged; capsules 10–15 mm long, ovoid, surrounded by the enlarged bracts; seeds 5–6 mm long. Borders of bottomland, mesic upland forests, banks of streams, margins of ponds, disturbed areas; in East TX known only from Anderson Co. (Turner et al. 2003), also Carson, Chambers, and Ochiltree (Turner et al. 2003) cos. just to the s of se East TX and in the Panhandle; e U.S. from NY s to SC and w to IA and NM; also Canada. May–Sep. [*C. sepium* var. *fraterniflora* (Mack. & Bush) Shinners, *Convolvulus sepium* L. var. *fraterniflorus* Mack. & Bush] Native of the U.S. and Canada.

Clethra alnifolia [G&F]

Calystegia sepium subsp. angulata [YAT]

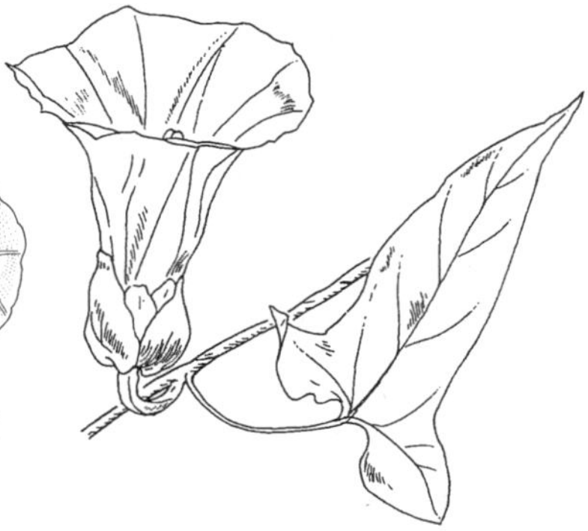

Calystegia sepium subsp. binghamiae [WF2]

Calystegia silvatica subsp. fraterniflora [BB2]

Convolvulus arvensis [NIC, REE]

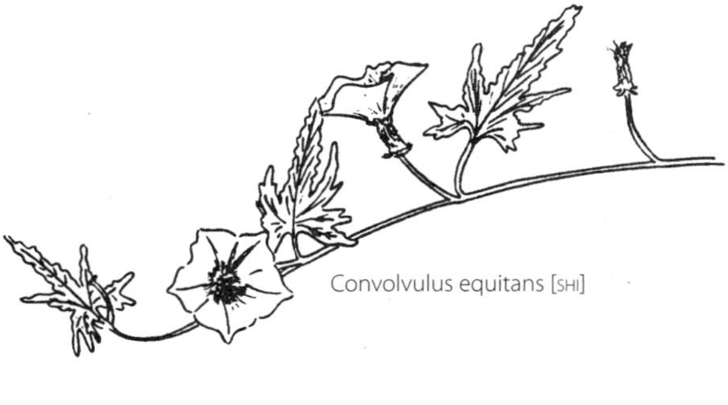

Convolvulus equitans [SHI]

Cressa truxillensis [IV2]

Calystegia macounii (Greene) Brummitt, (for John Macoun, who collected the type specimen in 1895), MACOUN'S FALSE BINDWEED. Semi-erect or sparsely twining, pubescent, perennial herb; leaf blades 2–6 cm long, 1–5 cm wide, basally cordate to subsagittate, marginally entire; petioles 0.5–40 mm long; flowers axillary, solitary, on peduncles usually 3–5 cm long; floral bracts 2, enclosing calyces, foliaceous, strongly inflated, 20–25 mm long, 10–15 mm wide; sepals 10–12 mm long, 5–7 mm wide; corollas funnelform, 4–5 cm long, white; style 1. Along a stream bank in limestone area, typically sandy soils near waterways in open grasslands or openings in woodlands; included based on presence in TX in Cooke Co. (*Shinners 13256* SMU) on the nw margin of East TX; otherwise unknown in TX; also McCurtain and Red River cos., AR, across the border from ne East TX; mostly c U.S. from ND s to TX and w to MT and AZ. May–Jun. [*Convolvulus macounii* Greene, *Convolvulus interior* House] Native of the U.S. This taxon will key out with *Calystegia sepium* subsp. *binghamiae* with narrower leaf blades (3:1 length to width ration), but with broader leaf blades (mostly 2:1 length to width ratio).

CONVOLVULUS L. BINDWEED

Trailing to decumbent or twining perennials; leaves petiolate, blades entire, toothed, or lobed, basally truncate to cordate, hastate, or sagittate; flowers axillary, solitary or in cymes of 2 or 3; corollas campanulate; ovary 2-locular; style 1; stigmas 2, linear, ± flattened, slightly acute; capsules 1–4-seeded.

A cosmopolitan, but especially temperate genus of 190 species (Austin 2023a). Some have alkaloids and a number are cultivated as ornamentals, while others can be pernicious weeds, especially *Convolvulus arvensis*. (Latin: *convolvere*, to entwine, in reference to the twining habit of some species)

REFERENCES: Lewis & Oliver 1965; Austin 1986a, 1992, 2023; Austin et al. 1997; Williams et al. 2014.

1. Calyces 3–5 mm long, inconspicuously pubescent or glabrous; corollas merely 5-angled; plants sparsely pubescent; leaf blades entire except for basal lobes, ovate to ovate-lanceolate to elliptic, often almost as wide as long ____ **C. arvensis**
1. Calyces 6–12 mm long, densely pubescent; corollas with margin having 5 sharp points or projections; plants densely gray-pubescent; leaf blades from entire to toothed or deeply lobed, variously shaped, usually much longer than wide, sometimes with the main part very long and narrow ____ **C. equitans**

Convolvulus arvensis L., (of fields), POSSESSION-VINE, BINDWEED, FIELD BINDWEED, COMMON BINDWEED, CORNBIND. Perennial from deep creeping root, forming extensive beds; stems to 1 m or more long, decumbent or trailing; flowers 1–3 in axils; calyx elliptic-orbicular, corolla 1.2–2.5 cm long, limb white to pink, or with pinkish or reddish broad bands, throat yellow, campanulate. Roadsides, disturbed sites, a problematic weed in gardens and difficult to eradicate because of the deep root; scattered in East TX, becoming more prominent to w TX; e U.S. from ME s to FL and w to OH and TX, throughout w U.S.; also Canada. May–Jul, less freely later. [*C. ambigens* House, *C. incanus* of authors, not Vahl] Native of Eurasia.

Convolvulus equitans Benth., (riding a horse, like legs of a rider around a horse), GRAY BINDWEED, TEXAS BINDWEED. Perennial from taproot, not forming extensive beds; stems to 2 m long, trailing to decumbent; leaves very variable; calyx lobes auricled basally to not auricled; corolla 1.5–3 cm long, limb white to pink, throat usually red. Prairies, disturbed areas; mostly scattered in East TX, except throughout sw portion; throughout w TX; mostly sw U.S. from NE s to TX and w to NV and AZ, also AL. May–Oct. [*C. hermannioides* A. Gray]

CRESSA L. ALKALI-WEED

A genus of 4 species of warm temperate to tropical regions (Austin 2000). (Greek, based on *kris* or *kriti*, "from Crete," a Cretan woman)

REFERENCE: Austin 2000.

Cressa truxillensis Kunth, (in reference to the city of Trujillo in Peru where the type was collected), SPREADING ALKALI-WEED. Erect to decumbent suffrutescent herb from a stout root, 8–25 cm tall, usually branched, appressed pubescent or hirsute; leaf blades 3–10 mm long, 1–4 mm wide, elliptic, lanceolate or ovate-lanceolate, apex usually acute, base cuneate, sometimes obtuse, pubescence like the stems; flowers mostly solitary, axillary, usually concentrated in the upper leaf axils, appearing almost spicate, peduncles 2–6 mm long, bracteoles ovate to ovate-lanceolate, acute, unequal, 2–3 mm long, pedicel absent; calyx about as long as corolla tube, sepals ± equal or the inner slightly longer, elliptic; corolla white, salverform, 5–6.5 mm long, tube 3–3.5 mm long, limb 5-lobed, about as long as tube, stamens exserted; fruit a capsule, ovoid, 5–6 mm long, brown; seeds usually 3–4 mm long, brown. Dry, rocky soil, waste land, disturbed areas, and grasslands; in East TX known only from Gonzales Co. (Kartesz 2022); Gulf Prairies and Marshes, South TX Plains, Edwards Plateau, High Plains, and Trans-Pecos; mostly w U.S. from TX and OK w to OR and CA; also Mexico and South America. Apr–Jul; less frequently Aug–Oct. [*C. cretica* of authors, not L., *C. depressa*

Goodd., *C. truxillensis* Kunth var. *minima* (A. Heller) Munz, *C. truxillensis* var. *vallicola* (A. Heller) Munz] Native of the U.S.; introduced in HI.

CUSCUTA L. DODDER, LOVEVINE

Ours glabrous annual herbs, ± without chlorophyll, with thread-like, orange, yellow, to whitish, irregularly twining stems; growing as parasites on other plants and attached by haustoria invading the host tissue; losing their chlorophyll upon contact with host plant as a seedling; essentially leafless, the leaves reduced to functionless scales; flowers small, usually numerous, in loose or dense cymose clusters, pedicelled or sessile; calyces of 3–5 lobes or sometimes of separate or nearly separate sepals, the lobes or sepals often overlapping laterally; corollas white, yellowish, or greenish, 3–5-lobed (equaling the number of calyx lobes), with a ring of small, flat, fringed or fimbriated scales inside at base opposite the stamens; stamens 3–5 (equaling the number of lobes), attached above the scales; ovaries superior, 2-celled; stigmas linear to capitate to globose; fruit a capsule.

This genus of ca. 200 species (Costea & Nesom 2023) includes a number of problematic parasitic weeds which can seriously reduce the yield of some crops. It is cosmopolitan in distribution, but about 75% are native to North and South America (Costea et al. 2015). Other common names are ANGEL'S-HAIR, TANGLEGUT, WITCHES'-SHOELACES, DEVIL'S-GUT, STRANGLEVINE, and SCALD. In the Sinhalese language of Sri Lanka, the name of *Cuscuta* means "plant without beginning or end" (Austin 1979a). According to Austin (1979a), the name DODDER, is an ancient one, having come from its Frissian (Middle English) [Frisian (Low German tongue closely related to Anglo-Saxon)] origin without change. The word dodder is said to signify a 'tangle of threads' in reference to the intertwined stems of the plant." The species are often difficult to identify, especially in spring when just beginning to flower. The genus is often put in its own family, the Cuscutaceae. We, however, are following Neyland (2001) and Stefanović et al. (2002) who found that *Cuscuta* is nested within the Convolvulaceae. Additionally, it is treated in the Convolvulaceae by Stevens (2001 onwards). Measurements of flowers are from rehydrated material, and flower length is measured from the base of the calyx to the tip of straightened corolla lobes. (Arabic: *kushkut* or *kusat*, a tangled wisp of hair—Austin 1979a)
REFERENCES: Yuncker 1920, 1932, 1961; Austin 1979a, 1979b, 1980, 1986b; Neyland 2001; Costea et al. 2006a, 2006b, 2006c, 2015; Costea & Nesom 2023.

1. Styles entirely separate and distinct from one another; capsules not separating near the base in a straight line of cleavage.
  2. Calyx divided to base or nearly so; flowers 5-lobed, subtended by 2–11 enlarged bracts similar to sepals; corolla lobes spreading to reflexed; capsule apex covered by persistent corolla.
    3. Flowers on pedicels 2–5 mm long in loose clusters; bracts subtending flowers not crowded ____ **C. cuspidata**
    3. Flowers essentially sessile in dense clusters, often in thick spiral ropes around stem of the host plant; bracts subtending flowers crowded.
      4. Bracts 2–several ovate-triangular to ovate-orbicular, closely appressed to the flowers, tips erect.
        5. Calyx lobes ovate, apex acute to cuspidate ____ **C. squamata**
        5. Calyx lobes orbiculate, apex rounded ____ **C. compacta**
      4. Bracts numerous, narrowly triangular or lanceolate, tips recurved ____ **C. glomerata**
  2. Calyx at least partly fused; flowers (3–)4–5-lobed; subtended by (0–)1(–3) minute bracts not similar to sepals; capsule apex covered by persistent corolla or not.
    6. Styles 0.4–1.1 mm long, subulate or uniformly stout.
      7. Flowers almost always 5-lobed; stamens shorter than corolla lobes; corolla lobes ovate to ovate-oblong, erect to spreading, apex obtuse to rounded ____ **C. obtusiflora**
      7. Flowers (3–)4-lobed; stamens almost as long as corolla lobes; corolla lobes triangular, erect, apex acute ____ **C. polygonorum**
    6. Styles (0.6–)0.7–2.5 mm long, mostly uniform in width and filiform.
      8. Flowers (3–)4–5-lobed, calyx length < to > corolla tube length; corolla lobe apex inflexed or straight.
        9. Flowers (3–)4(–5)-lobed, 2–3 mm long; calyx length < corolla tube length; calyx lobes oblong-ovate, apex obtuse; corolla lobes ovate, spreading, apex obtuse, straight ____ **C. cephalanthi**
        9. Flowers (3–)4–5-lobed, 1.7–5.3 mm long; calyx length < to > corolla tube length; calyx lobes triangular-ovate to lanceolate, apex acute to attenuate; corolla lobes triangular-ovate, suberect to erect, apex inflexed.
          10. Flowers (3–)4–5-lobed, 1.7–2.6(–3) mm long, calyx length ≥ corolla tube length; calyx lobes triangular-ovate, apex acute ____ **C. coryli**
          10. Flowers 5-lobed, 3–5.3 mm long; calyx length < to > corolla tube length; calyx lobes triangular-ovate to lanceolate, apex acute to attenuate ____ **C. indecora**
      8. Flowers (4–)5-lobed; calyx length ≤ corolla tube length; corolla lobe apex inflexed (straight in *C. gronovii*).
        11. Calyx broadly obconical; calyx sinuses broadly obtuse; calyx lobes broadly ovate-rhombic, mostly broader than long, widely overlapping ____ **C. pentagona**
        11. Calyx narrowly to broadly campanulate to cylindric or cup-shaped; calyx sinuses obtuse to acute; calyx lobes not shaped as above, as broad as long or narrower; overlapping or not.

12. Capsules depressed-globose or globose, usually about as broad as long, not at all ovoid, not thickened at style bases.
    13. Calyx lobes evidently overlapping at base, persistent corolla enveloping capsule base by 1/3 or less ____ **C. campestris**
    13. Calyx lobes not overlapping at base; persistent corolla enveloping 1/2–2/3 of the capsule base ____ **C. glabrior**
12. Capsules globose-ovoid to subpyriform (= almost pear-shaped), mostly longer than wide, ± thickened at the style bases ____ **C. gronovii**

1. Styles entirely separated, only distally separated, or wholly fused; capsules easily separating near the base in a straight line of cleavage.
   14. Styles entirely separated; calyx lobes triangular-ovate, apex acute, not overlapping ____ **C. leptantha**
   14. Styles completely fused or distally separated; calyx lobes ovate to orbicular, apex rounded or acute to obtuse, overlapping.
       15. Styles often united only part of their length, but if united easily separated; stigmas flattened or depressed ____ **C. exaltata**
       15. Styles entirely united; stigmas oval or conic ____ **C. japonica**

Cuscuta campestris Yunck., (of the fields or plains), FIELD DODDER. Inflorescences in dense glomerules; flowers 1.9–3.6 mm long, (4–)5-lobed, pedicels 0.3–2.5(–3.5) mm long; calyx cup-shaped, ca. as long as corolla tube, lobes ovate-triangular, apex obtuse to rounded, overlapping but rarely forming obscure angles at the sinuses; corolla lobes triangular-lanceolate, spreading, apex acute to acuminate, inflexed; stamens exceeding the calyx sinuses but shorter than the corolla lobes; styles 0.8–1.6 mm long, filiform; capsule depressed-globose, not thickened around the style bases, corolla persistent around 1/3 or less of the base. Growing in a wide variety of climates and soil types, parasitizing a multitude of hosts; scattered in East TX; Edward Plateau and Panhandle; throughout most of the U.S. also Canada; South America; Europe; Asia; Africa; and Australia. Jun–Oct. [*C. pentagona* of authors, not Engelm., *C. pentagona* Engelm. var. *calycina* Engelm.] The second most common species in North America, after *Cuscuta gronovii* (Yuncker 1932). ⃠

Cuscuta cephalanthi Engelm., (of *Cephalanthus*—buttonbush), BUTTONBUSH DODDER. Inflorescences in dense paniculate cymes; flowers 2–3 mm long, (3–)4(–5)-lobed, sessile to subsessile; calyx shorter than corolla tube, lobes oblong-ovate, apex obtuse; corolla cylindric-campanulate, lobes ovate, spreading, apex obtuse, straight; stamens equaling corolla lobes; styles filiform or narrowly terete, (0.6–)1–2 mm long; capsule globose to depressed-globose, apex covered by persistent corolla. Pond, marsh, and river margins, meadows, hardwood forests; found in Dallas Co. by Reverchon (Yuncker 1961), also Franklin Co. (Turner et al. 2003) of East TX; Concho and Schleicher cos. (Kartesz 2022) on the Edwards Plateau. Found on a wide range of woody and herbaceous hosts, but often on *Cephalanthus occidentalis*; widespread in the Midwest and widely scattered in much of the U.S.; Canada. Jul–Oct. [*Grammica cephalanthi* (Engelm.) W.A. Weber] Native of the U.S. and Canada. A noxious weed throughout the Midwest. ⃠

Cuscuta compacta Juss. ex Choisy, (compact, dense), COMPACT DODDER. Inflorescences often forming thick, dense clusters completely covering stems of host plant; flowers 4–6 mm long, 5-lobed, sessile; subtended by 2–4(–6) bracts, calyx shorter than the corolla tube, lobes distinct, orbiculate, broadly overlapping; corolla cylindric, lobes ovate, spreading to reflexed, apex obtuse; stamens shorter than corolla lobes; styles 1–2 mm long; capsule globose-conical, apex covered by persistent corolla. Stream bottoms; found on a wide range of herbaceous and woody hosts. Concentrated mostly in the se portion of East TX; e U.S. from NH s to FL and west to IA and TX. Jul–Oct. [*C. compacta* var. *efimbriata* Yunck.] Native of the U.S. and introduced in Canada. A noxious weed throughout the southeast. ⃠

Cuscuta coryli Engelm., (of *Corylus*—hazel or filbert; often found on members of that genus), HAZEL DODDER. Inflorescences dense to loose in paniculate clusters, pedicels 0.5–3 mm long; flowers 1.7–2.6(–3) long, (3–)4–5-lobed, usually not papillate; calyx length ≥ corolla tube length, lobes triangular-ovate, apex acute; corolla lobes triangular-ovate, suberect or erect, apex acute, inflexed; stamens shorter than corolla lobes; styles 0.7–1.8 mm long; capsules globose to depressed-globose, thickened around style bases, corolla persistent at apex or not so. Found on a wide variety of herbaceous and woody hosts, but often on *Corylus*; in TX known only from Robertson and Williamson cos. (Yuncker 1961); e U.S. from NY s to GA and w to ND and s to TX, also AZ, MT, and NM; Canada. Aug–Sep. [*Grammica coryli* (Engelm.) Hadač & Chrtek] Native of the U.S. and Canada. Related and similar to *C. indecora*; differs in the rudimentary bifid scales; *C. indecora* has fringed scales. ⃠

Cuscuta cuspidata Engelm., (cuspidate, with a sharp, stiff point), CUSP DODDER, CUSPIDATE DODDER. Inflorescences bracteate, bracts ovate, to orbicular, apices obtuse to bluntly cuspidate, pedicels distinctive with 1 or more bracts just below the calyx; flowers 3.5–4.2 mm long, pedicel 2–5 mm long; calyx lobes divided to base or slightly united, apex acute, sometimes cuspidate, shorter that the corolla tube; corolla lobes oblong to ovate-triangular, reflexed, apex obtuse to acute, sometimes cuspidate; stamens not

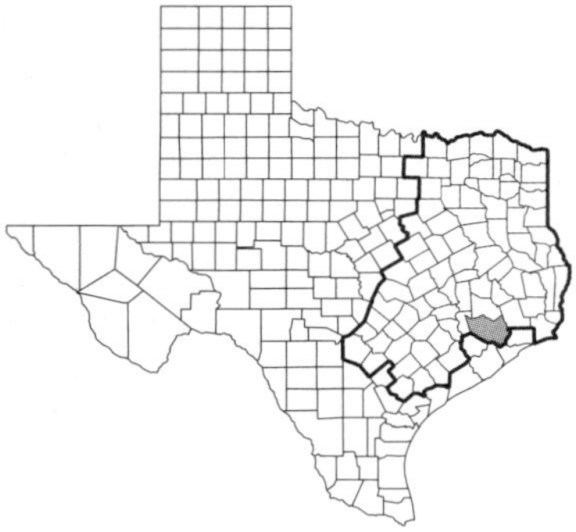

Polanisia dodecandra var. dodecandra

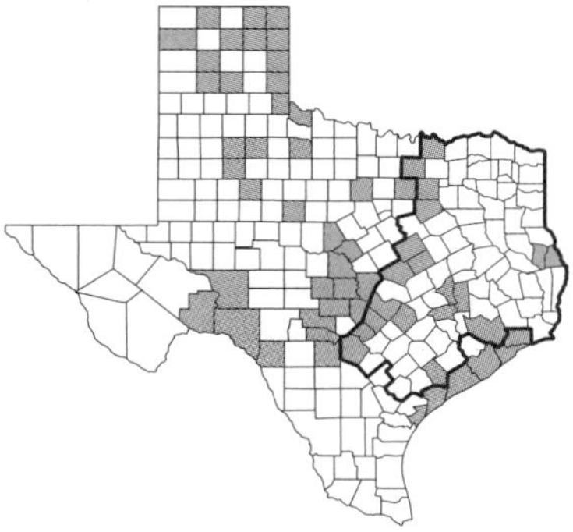

Polanisia dodecandra var. trachysperma

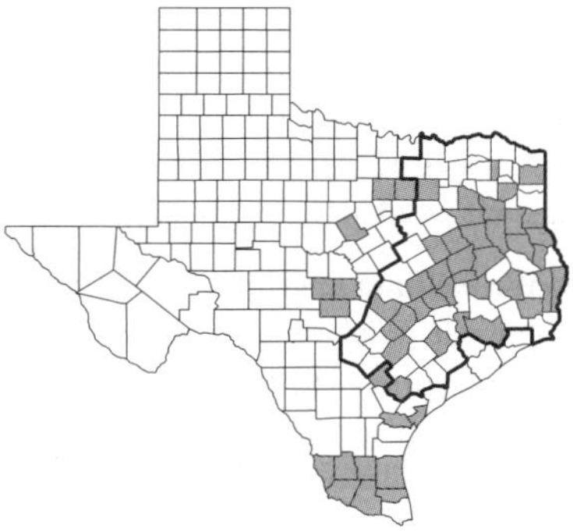

Polanisia erosa

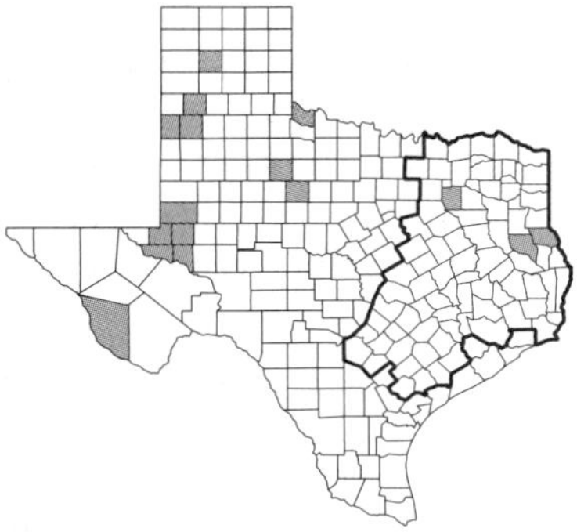

Polanisia jamesii

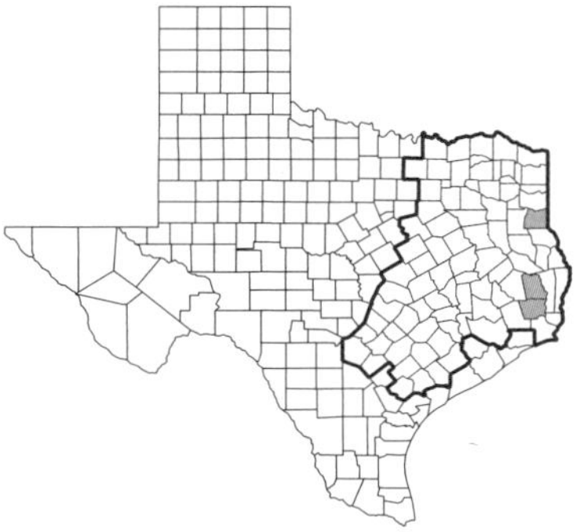

Tarenaya hassleriana

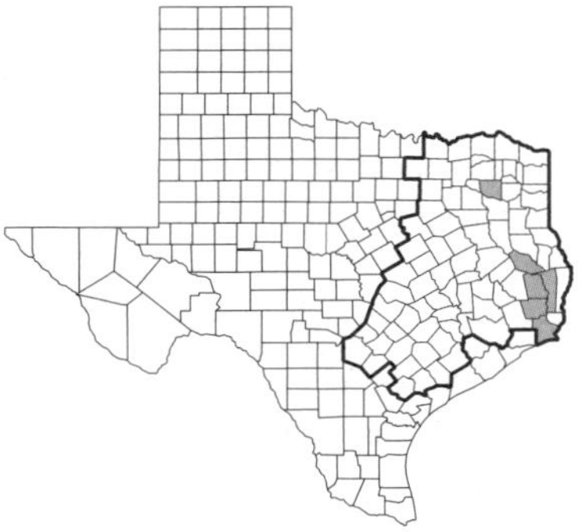

Clethra alnifolia

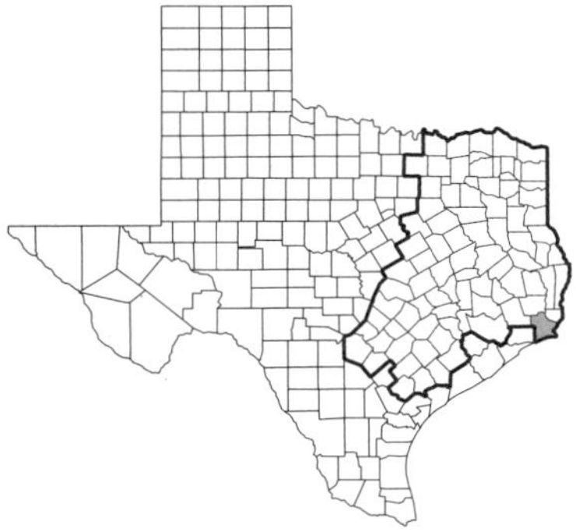

Calystegia sepium subsp. angulata

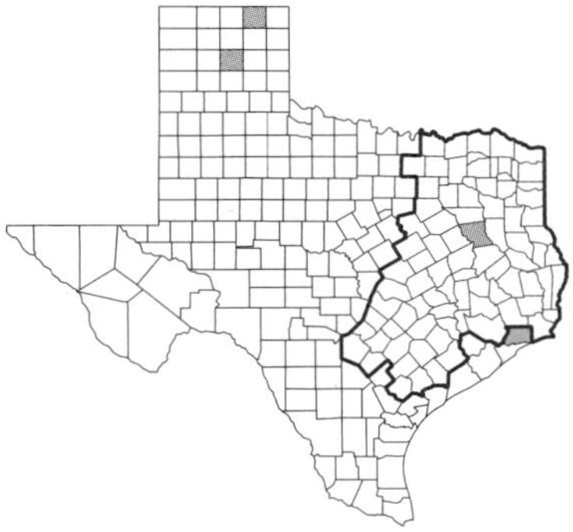

Calystegia silvatica subsp. fraterniflora

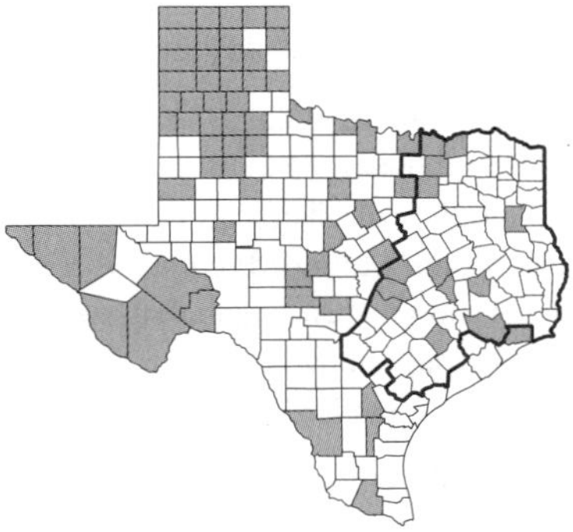

Convolvulus arvensis

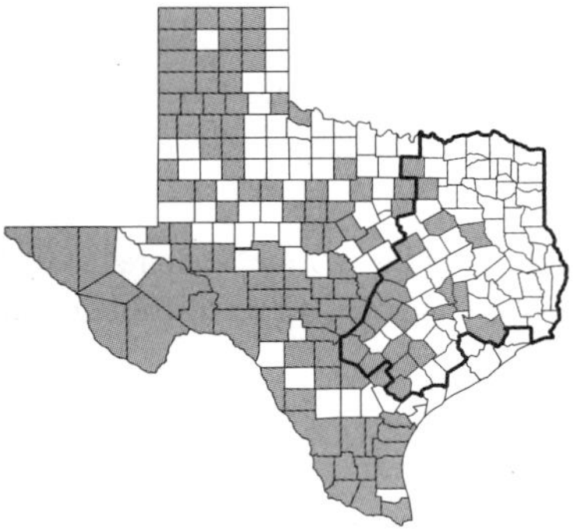

Convolvulus equitans

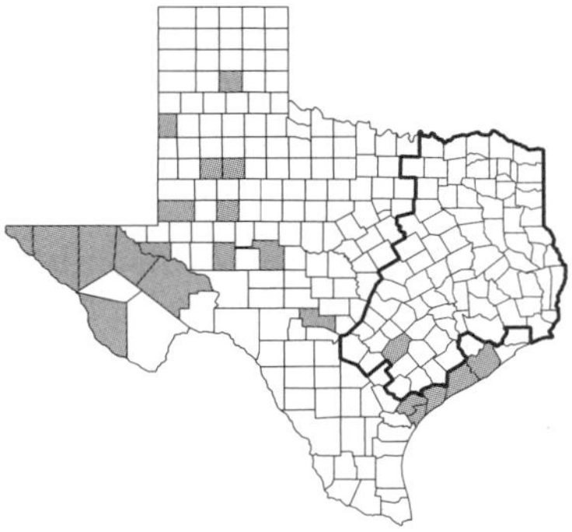

Cressa truxillensis

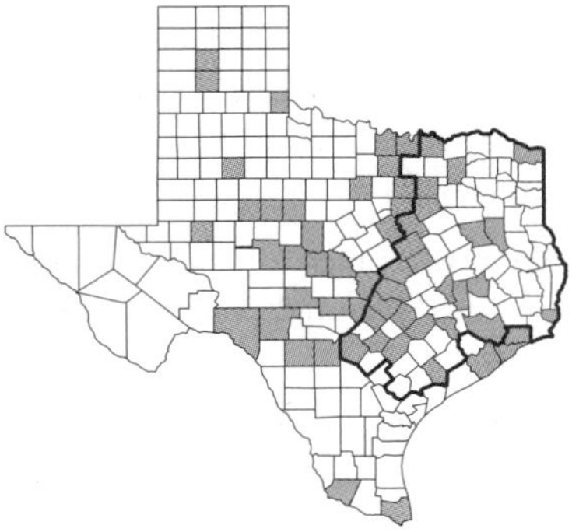

Cuscuta campestris

as long as corolla lobes, but well-exserted above the flower; style long, 2–2.6 mm; capsule globose to depressed-globose with thickened and raised collar around small interstylar aperture, apex covered by persistent corolla. Chiefly low open ground and according to Yuncker (1961) seems to prefer members of the Asteraceae including *Ambrosia*, *Baccharis*, *Helianthus*, *Iva*, and *Liatris*; widespread in East TX; Panhandle and s. TX; c U.S. from OH s to MS and w to ND and NM. Aug–Oct. Native of the U.S. A noxious weed in most U.S. states where present. ⊘

Cuscuta exaltata Engelm., (exalted, very tall), TREE DODDER. Inflorescence a spicate panicle; flowers 4–5 mm long, 5-lobed, sessile or subsessile; calyx ca. as long as corolla tube, lobes ovate-orbicular, overlapping, apex rounded; corolla tube cylindric, lobes ovate to suborbicular, erect to spreading, apex rounded; stamens at base of corolla lobe sinuses; styles completely or partially fused; capsules globose-ovoid to conic, corolla persistent at apex, enlarging to 10 mm long, circumscissile. Limestone areas; typically parasitizing woody plants, often oaks; mostly on the w margin of East TX, also Brazos Co.; Edwards Plateau; FL and TX. Sep–Oct. Native of the U.S.

Cuscuta glabrior (Engelm.) Yunck., (somewhat smooth or slightly hairy), CLUSTERED DODDER. Inflorescences loose to compact, glomerulate or corymbiform; flowers 2–3 mm long, 5-lobed, subsessile or pedicels to 5 mm long; calyx cup-shaped, ca. as long as corolla tube, lobes not overlapping, ovate-triangular, apex obtuse to subacute; corolla campanulate-globose, lobes triangular to sublanceolate, spreading to reflexed, apex acute to acuminate, inflexed; stamens exserted, shorter than corolla lobes; styles 0.9–1.6 mm long, evenly filiform; capsule depressed-globose, corolla persistent around 1/2–2/3 of the base. Parasitic on a number of herbaceous genera: *Asclepias, Convolvulus, Coreopsis, Croton, Dalea, Gaura, Liatris, Ruellia,* and *Solanum*; scattered in East TX and throughout TX; LA, NM, OK, TX, UT; also Mexico. Apr–Oct. [*C. pentagona* Engelm. var. *glabrior* (Engelm.) Gandhi, R.D. Thomas, & S.L. Hatch, *C. pentagona* var. *pubescens* (Engelm.) Yunck.] Native of the U.S. ⊘

Cuscuta glomerata Choisy, (with glomerules, clustered into ± rounded heads), CLUSTER DODDER, GLOMERATE DODDER. Stems disappearing leaving thick, yellow, rope-like spiral masses of flowers around the host stems and thus easily recognized by growth form; flowers 5–6 mm long, sessile, subtended by numerous imbricate, lanceolate, acute bracts with recurved tips; calyx of distinct sepals similar to bracts but not recurved, ca. as long as corolla tube; corolla lobes spreading to recurved, oblong to lanceolate, apex acute or obtuse; stamens shorter than corolla lobes; styles 2–3 mm long, filiform and much longer than the capsule; capsule globose with a raised column around the styles, apex covered by persistent corolla. Stream bottoms; scattered in East TX; Edwards Plateau and Panhandle; c U.S. from OH s to MS and w to ND and TX. Jul–Sep. Native of the U.S. ⊘

Cuscuta gronovii Willd., (for Jan Frederick Gronovius, 1690–1762, teacher of Linnaeus and author of "Flora Virginica"), GRONOVIUS' DODDER, COMMON DODDER. Inflorescences loose or dense panicles with cymose clusters; flowers 2–4 mm long, (4–)5-lobed, pedicellate; calyx cup-shaped to narrow-campanulate, 1/2 to as long as the corolla tube, lobes ovate to suborbicular or oblong, apex rounded or obtuse, ± basally overlapping; corolla lobes ovate, spreading or reflexed, apex rounded to obtuse; stamens shorter than to equaling corolla lobes; styles filiform, sometimes, slightly subulate, (0.6–)1.2–2.2 mm long; capsule globose-ovoid, subpyriform, corolla persistent around the base or on the apex. Stream bottoms, low ground. Aug–Nov. Native of the U.S. and Canada. The most common and widespread dodder in North America (Yuncker 1965). ⊘

1. Corolla tube broadly campanulate, 1–1.5 mm long; lobes 1/2 to equaling tube length ____ var. **latiflora**
1. Corolla tube campanulate to narrowly campanulate,1.5–2.5(–3) mm long; lobes (1/4–)1/3 length of the tube.
   2. Corolla tube campanulate.1.5–2.5 mm long; capsule loosely surrounded by persistent corolla ____ var. **gronovii**
   2. Corolla tube narrowly campanulate, 2–2.5(–3) mm long; capsule capped by persistent corolla ____ var. **calyptrata**

var. **calyptrata** Engelm., (bearing a lid or covering). This variety is the most narrowly distributed entity of the *C. gronovii* complex. Possibly Houston Co. (Yuncker 1920) and Rockwall Co. (*Cory 52529* SMU) also w LA (Yuncker 1932). Aug–Oct. [*C. calyptrata* (Engelm.) Small] Native of the s U.S. Engelmann states in his original diagnosis that it is cultivated in several botanical gardens in Germany. ⚠

var. **gronovii.** Scattered in much of East TX; Edwards Plateau; rampant throughout the e 1/2 of the U.S. and scattered to the w; also Canada, West Indies (rare) (Yuncker 1965), and Europe (Feinbrun 1972). Jun–Nov. No county level distribution map provided.

var. **latiflora** Engelm. From NY s to NC and w to IA and TX. No county level data available. Jul–Oct. According to Costea and Nesom (2023), intermediates with var. *gonovii* are relatively common.

Cuscuta indecora Choisy, (not ornamental), LARGE-SEED DODDER. Inflorescences loose or dense in paniculate-cymose clusters; flowers 3–5.3 mm long, 5-lobed, pedicels 0.5–6 mm long, papillate; calyx 1/2 to somewhat longer than the corolla tube, cup-shaped, lobes triangular-ovate to lanceolate, apex acute to attenuate, ± overlapping at the base; corolla 2.5–4(–5) mm long, tube

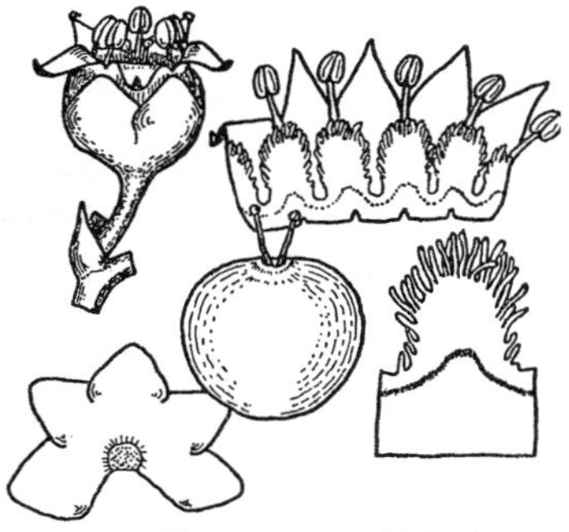

Cuscuta campestris [LUN]

Cuscuta cephalanthi [BB2]

Cuscuta compacta [GLE]

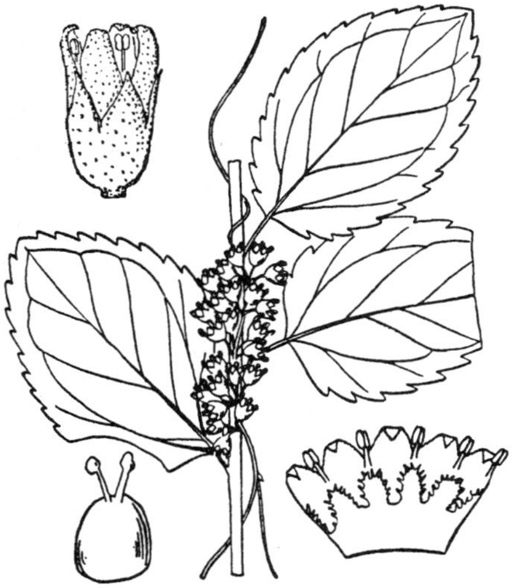

Cuscuta coryli [BB2, GLE]

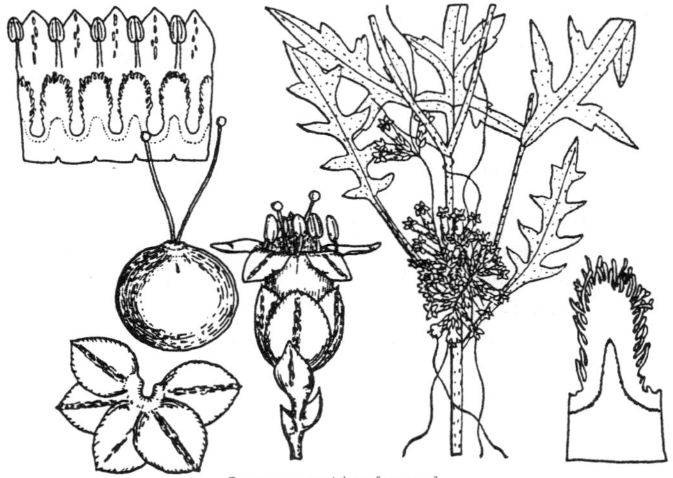

Cuscuta cuspidata [BB2, LUN]

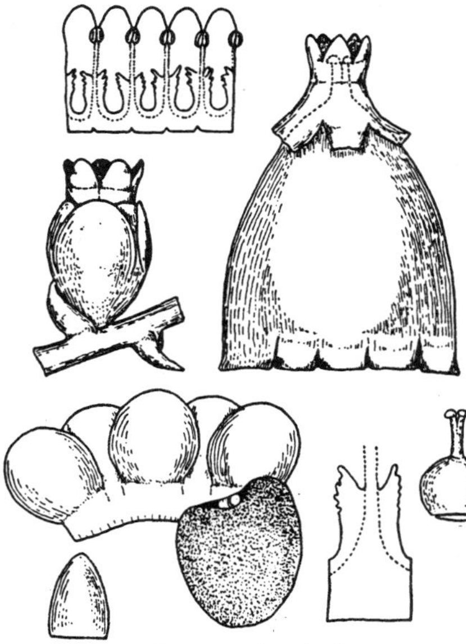

Cuscuta exaltata [LUN]

Cuscuta glabrior [LUN]

campanulate to cylindric, lobes triangular-ovate, suberect to erect, apex acute, inflexed; stamens shorter than corolla lobes; styles ± unequal, 1–2.5 mm long; capsule globose to depressed-globose, thickened around the style bases; corolla persistent around the base or on the apex. Prairies, roadsides, and stream bottoms; throughout East TX; Edwards Plateau, s TX, and Trans-Pecos; in almost every state in the U.S. except New England, NY, and PA; Canada. Jun–Oct.

1. Calyx shorter than corolla tube; calyx lobes triangular, bases overlapping, divided by 1/3–1/2 their length ____ var. **indecora**
1. Calyx longer than corolla tube; calyx lobes lanceolate, bases not overlapping, divided by 2/3 their length.
   2. Calyx lobe apex acute; flowers in loose clusters; parasitic on a wide range of species ____ var. **longisepala**
   2. Calyx lobe apex acute-attenuate; flowers in dense clusters; parasitic primarily on *Iva annua* ____ var. **attenuata**

var. **attenuata** (Waterfall) Costea, (tapered). Mudflats, floodplains, and disturbed areas; scattered in East TX; s TX; KS, OK, and TX. Late Aug–Oct. [*C. attenuata* Waterfall] NatureServe: G2, imperiled.

var. **indecora.** SHOWY DODDER, PRETTY DODDER, LARGE-SEED DODDER. Widespread in TX; found in most U.S. states; more common is sw U.S.; also Canada; Mexico; West Indies; and South America. ⃠

var. **longisepala** Yunck., (long-sepaled), LONG-SEPAL DODDER. Scattered in East TX; s TX; also Mexico and South America. Summer–fall. ⃠

Cuscuta japonica Choisy, (from Japan), JAPANESE DODDER. Inflorescences in elongate thyrsoid cymes; flowers 3–7 mm long, 5-lobed, subsessile to mostly sessile; calyx ca. 1/2 as long as the corolla tube, lobes orbicular or ovate, overlapping at the base, corolla lobes ovate to ovate-triangular, erect to spreading, apex rounded to obtuse; stamens at base of corolla lobe sinuses; styles fused the entire length, stigma conic to oval to globose; capsule ovoid to helmet-shaped, corolla persistent, covering the entire capsule, circumscissile near the base. Disturbed sites, often parasitizing *Pueraria montana* var. *lobata*; known in TX only from Harris Co. (Camilli 2003); AL, CA, FL, SC, and TX.; Eastern Asia. Sep–Nov. ⃠

Cuscuta leptantha Engelm., (bearing thin, frail flowers), SLENDER DODDER. Inflorescences in loose umbellate clusters; flowers 3–4.5 mm long, mostly 4-lobed, pedicels 0.7–7 mm long; calyx campanulate, papillate, ca. 1/2 as long as the corolla tube, lobes triangular-ovate; corolla lobes lanceolate, spreading to reflexed, apex acute; stamens exserted; stamens not as long as corolla lobes; styles entirely separated; capsule globose, corolla persistent on the apex, circumscissile. Often parasitizing the genus *Euphorbia*; in East TX known only from Bexar Co. (Turner et al. 2003); Karnes Co. (Turner et al. 2003) just s of East TX; s TX, and Trans-Pecos; NM and TX; Central Mexico and Baja California. Mar–May; Sep–Dec. Native of the U.S. and Mexico. ⃠

Cuscuta obtusiflora Kunth var. **glandulosa** Engelm., (sp.: obtuse- or blunt-flowered; var.: glandular), RED DODDER. Flowers 1.8–2.5 mm long, 5-lobed, sessile to subsessile; calyx ca. as long as the corolla tube, lobes ovate, barely overlapping, apex obtuse; corolla lobes ovate to ovate-oblong, erect to spreading, apex obtuse to rounded, straight; stamens shorter than corolla lobes; styles uniformly stout 0.4–1.1 mm long; capsules depressed-globose, corolla persistent at base. Sandy or rocky ground, often growing on *Polygonum* species (Yuncker 1961); scattered in East TX; Edwards Plateau; widely scattered in e U.S., also CA (possibly extirpated). May–Sep. [*C. glandulosa* (Engelm.) Small] Variety *glandulosa* is similar morphologically to *Cuscuta polygonorum*; it differs in its 5-merous flowers, obtuse corolla lobes, and densely fimbriate infrastaminal scales. Both taxa are usually parasitic on *Persicaria*. Variety *glandulosa* differs from *C. gronovii* var. *latiflora* in its depressed-globose capsules that are not thickened or raised around the relatively large interstylar aperture. ⃠

Cuscuta pentagona Engelm., (five-angled). Inflorescences in dense glomerules of 3–15 flowers; pedicels 0.4–3 mm long; flowers 1.4–2.1(–2.5) mm long, (4–)5-lobed; calyx ca. as long as corolla tube, broadly obconical, lobes overlapping and forming prominent angles at the sinuses, apex rounded; corolla lobes triangular-lanceolate, spreading, apex acute to acuminate, inflexed; stamens shorter than corolla lobes; styles 0.7–1.1 mm long; capsules depressed-globose, not thickened around style bases, corolla persistent at the base. Widespread in TX; widespread in the e 1/2 of the U.S.; parasitizing a variety of hosts including cultivated Fabaceae (e.g., CLOVER and ALFALFA) (Austin 1986b). May–Oct. Native of the U.S., Puerto Rico, and Canada; introduced in Hawaii. ⃠

Cuscuta polygonorum Engelm., (many-sided), SMARTWEED DODDER. Inflorescences in dense glomerules, pedicels absent to 1 mm; flowers 2–2.7 mm long, (3–)4-lobed; calyx ca. equaling corolla tube, lobes triangular-ovate, apex obtuse to rounded; corolla lobes triangular, erect, apex acute, inflexed; stamens ≤ corolla lobes; style subulate 0.4–0.9 mm long; capsule depressed-globose, corolla persistent at base. Bottomland on smartweed; Brazos Co. (*Cheatham, Marshall, & Jones s.n.*, 1996, SBSC); even though it had been included in earlier TX floristic works (e.g., Correll & Johnston 1970), according to Brown and Marcus (1998), the first confirmed TX location of this species was the aforementioned 1996 collection, also McLennan and Hayes cos. and Childress

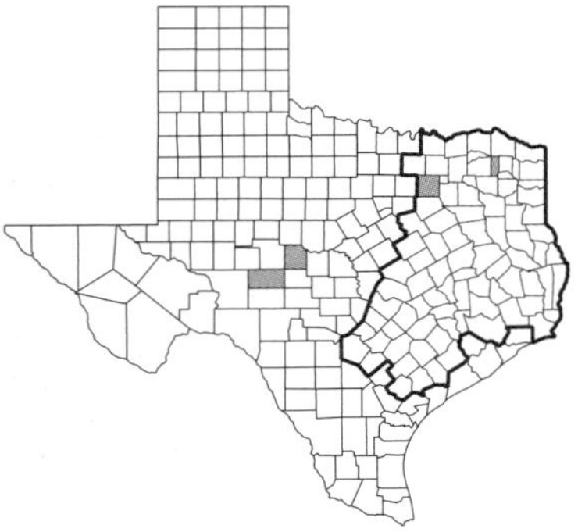

Cuscuta cephalanthi

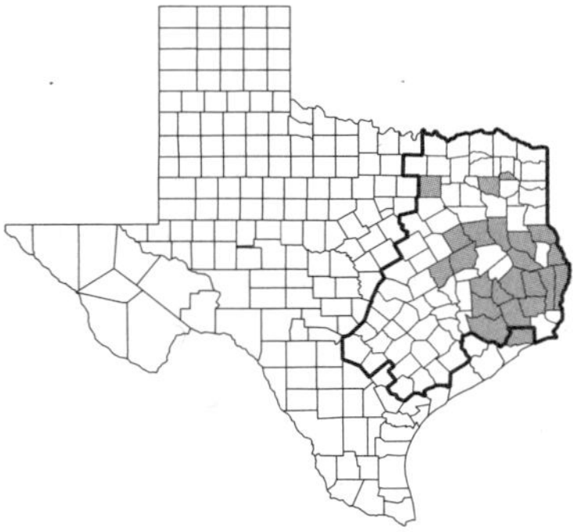

Cuscuta compacta

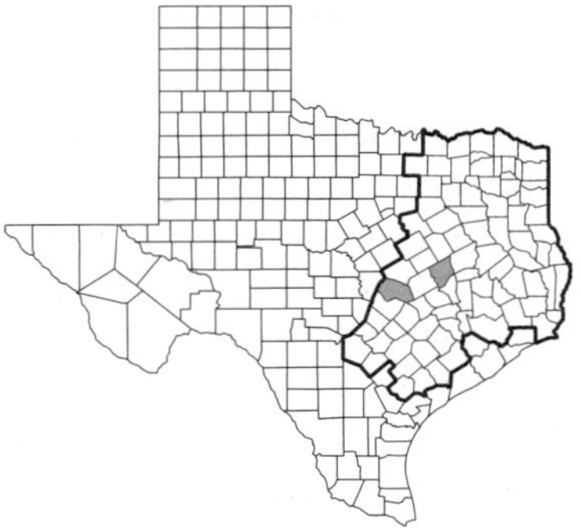

Cuscuta coryli

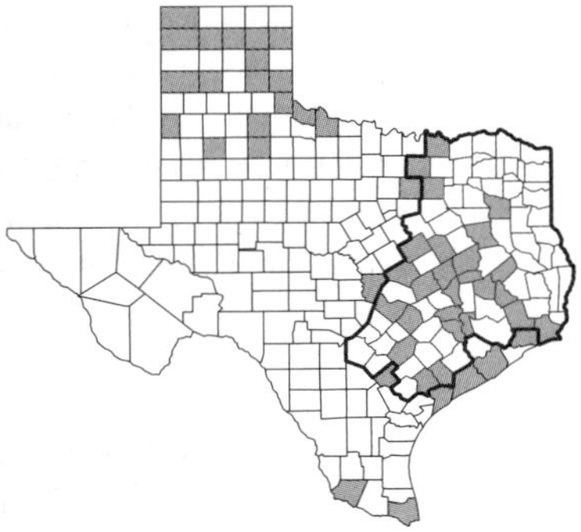

Cuscuta cuspidata

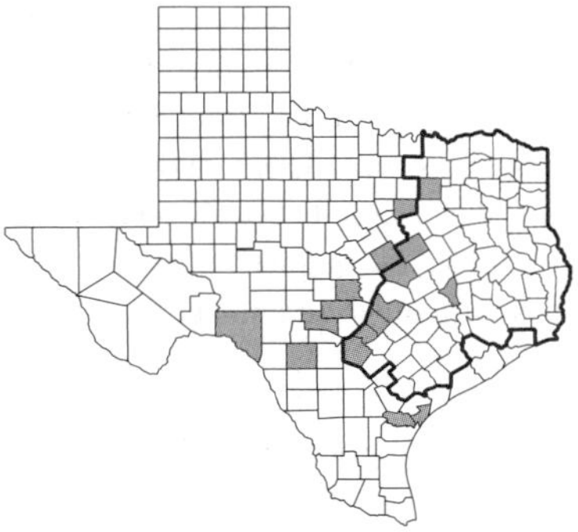

Cuscuta exaltata

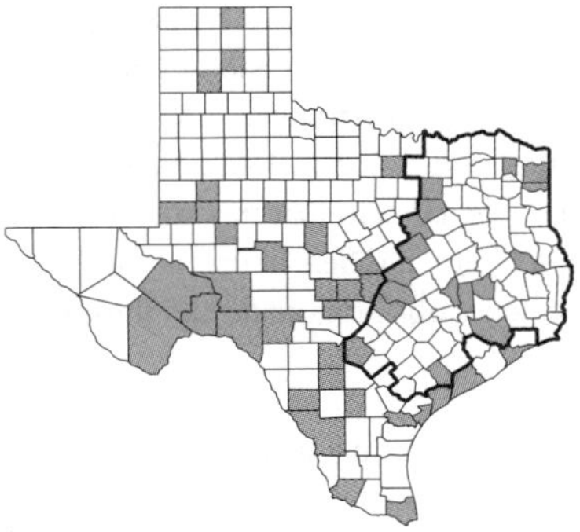

Cuscuta glabrior

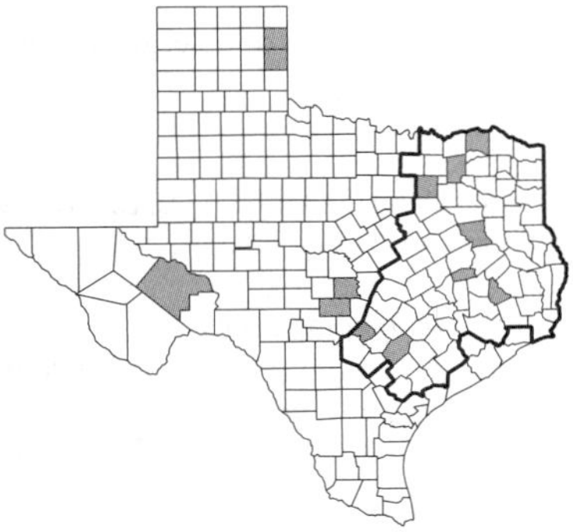

Cuscuta glomerata

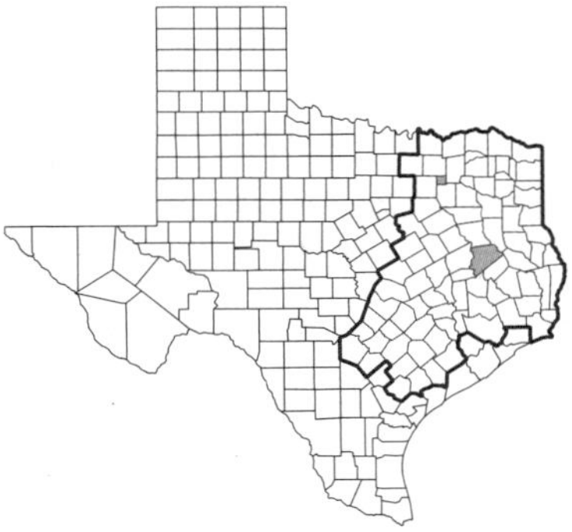

Cuscuta gronovii var. calyptrata

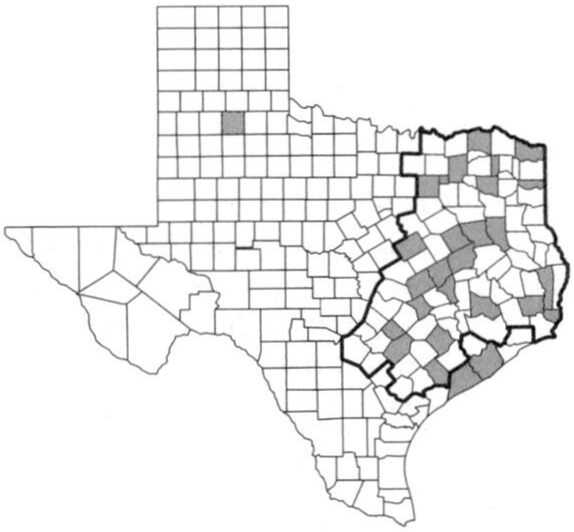

Cuscuta gronovii var. gronovii

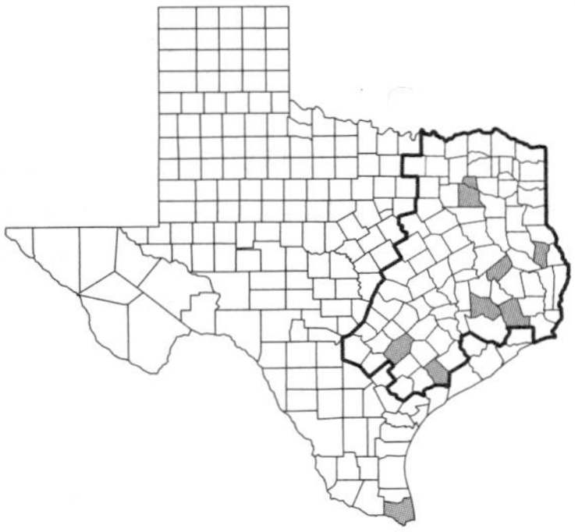

Cuscuta indecora var. attenuata

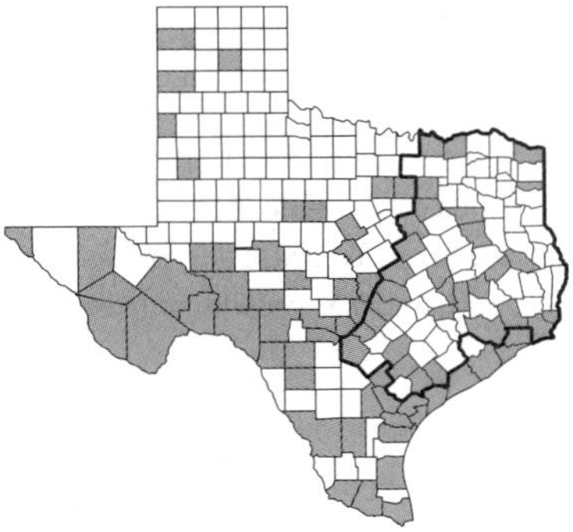

Cuscuta indecora var. indecora

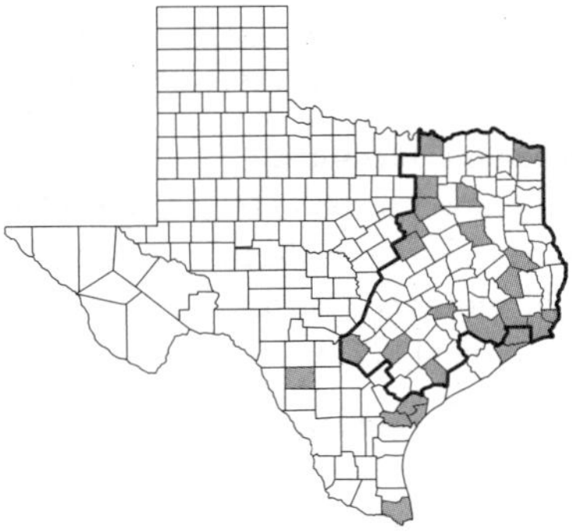

Cuscuta indecora var. longisepala

Co. (Turner et al. 2003) in Panhandle; e U.S. from NY s to VA and w to ND and TX; also Canada. Jul–Oct. Native of the U.S. and Canada. ⊘

Cuscuta squamata Engelm., (furnished with scales), SCALE-FLOWER DODDER. Inflorescences not often forming thick spiral ropes around the stem but rather isolated glomerules; flowers sessile, 5–6 mm long, subtended by (2–)4 or 5(–10) bracts, ovate-orbicular to ovate-triangular, closely appressed; calyx lobes ovate, broadly overlapping, apex acute to cuspidate, calyx shorter than corolla tube; corolla tube cylindric, lobes ovate-lanceolate, to oblong-ovate, spreading or reflexed, apex acute; stamens barely exserted, shorter than corolla lobes; styles longer than the ovary, 2.5–3.3 mm long; capsule globose and slightly conical, apex covered by persistent corolla. Disturbed roadsides; in East TX only in Hardin Co. (Correll & Correll 1972), also Trans-Pecos and Panhandle; NM and TX. Jul–Oct. ⊘

Cuscuta umbellata Kunth, (umbels, referring to the inflorescence), FLAT-GLOBE DODDER. Inflorescences in dense or loose compound cymes; flowers 2–3 mm long, 5-lobed; calyx cup-shaped, equaling the corolla tube, calyx lobes triangular-ovate, not overlapping, apex obtuse to acute; corolla lobes oblong to lanceolate, reflexed, apex obtuse to acute, straight, stamens shorter than corolla lobes; styles filiform, 0.8–1.7 mm long; stigmas capitate; capsules depressed-globose; surrounded by withered corolla, circumscissile. Sand dunes, open woods, dry wash bed in desert, sandy, gravelly slopes, sandy flats; included base on presence in Cameron Par., LA on the se border of East TX; s TX, Trans-Pecos; AZ, CO, FL, KS, LA, NM, OK, TX; Mexico and South America. Summer–fall. Native of the US, Puerto Rico, and the Virgin Islands. This species is similar to *C. exaltata*, *C. japonica*, and *C. leptantha* in that they all have capsules with corollas that easily separate near the base in a straight line of cleavage (circumscissile). It differs in that it is the only one with styles completely separate and calyx length equal to corolla tube length and calyx lobes obtuse to acute and the corolla lobes are acute to acuminate. ⊘

DICHONDRA J.R. Forst. & G. Forst. PONY-FOOT

Small, mat-forming, sparsely pubescent, creeping or trailing perennials rooting at the nodes; leaves long-petioled, blades orbicular-ovate to reniform (= kidney-shaped), entire or notched at apex; flowers small, 1–2 per axil, long-pedicelled but pedicels shorter than petioles; corolla lobes longer than the tube, light green or greenish white; styles 2, 1 from each of the 2 nearly separate carpels; capsules with 2 or 4 seeds.

A genus of 15 species (Austin 2023b) native to tropical and warm areas of the world. (Greek: *di*, two or double, and *chondros*, a grain, from the deeply lobed fruit)

REFERENCES: Tharp & Johnston 1961; Johnston 1963a; Austin et al. 1998.

1. Pedicels straight; calyx lobes 2–3 times as long as wide; corollas nearly as long as calyces; widespread in East TX __ **D. carolinensis**
1. Pedicels abruptly recurved near summit; calyx lobes 1.5–2 times as long as wide; corollas equal to 1/3 longer than calyces; mostly in sw portion of East TX or scattered.
   2. Corolla as long as or slightly longer than calyx; calyx 1.5–2 mm long in flower (to 2.5 mm in fruit); creeping stems mostly < 1 mm thick __ **D. micrantha**
   2. Corolla nearly 1/3 longer than calyx; calyx 2.5–3.2 mm long in flower (to 3.8 mm in fruit); creeping stems 1–2 mm thick __ **D. recurvata**

Dichondra carolinensis Michx., (of Carolina). Plant with stems to 12 cm long; pedicels during flowering 1/3–2/3 as long as petioles; calyx lobes oblong or oblanceolate, separate nearly to base, slightly exceeding the corolla; corolla 1.5–3 mm long. Damp sandy or silty ground, also a common lawn weed; throughout East TX; West Cross Timbers, Edwards Plateau, and South TX Plains; scattered in the se U.S. from VA s to FL and w to OK and TX; also Mexico. Mar–May [*Dichondra repens* J.R. Forst. & G. Forst. var. *carolinensis* (Michx.) Poir.] Native of the U.S.

Dichondra micrantha Urb., (bearing minute flowers), ASIAN PONY'S-FOOT. Plant 1–6 cm high, pubescent or glabrous, pedicels during flowering 1/4 to 1/2 as long as petioles, corolla ca. 2 mm long, greenish white. Common in prairies, pastures, woods, and waste places; Austin, Bexar, Brazos, and Travis (Turner et al. 2003) cos. of East TX; also South TX Plains, Edwards Plateau, and Gulf Prairies and Marshes; AL, AZ, CA, FL, GA, NM, and TX, possibly introduced elsewhere in the U.S. Mar–May. [*D. repens* of authors, not J.R. Forst. & G. Forst.]

Dichondra recurvata Tharp & M.C. Johnst., (curved backward). Small mat-forming pubescent perennial with creeping stems to 17 cm long; pedicels during flowering 1/10 to 1/2 as long as petiole; leaf blades orbicular to orbicular-reniform; flowers 1–2 per axil; flowering calyces 2.5–3.2 mm long, lobes oblong or oblanceolate, separate nearly to the base; corollas 3–4 mm long. Roadsides, brushy creek banks, lake shores; in the sw portion of East TX; Edwards Plateau; endemic to c TX. Mar–Jun.

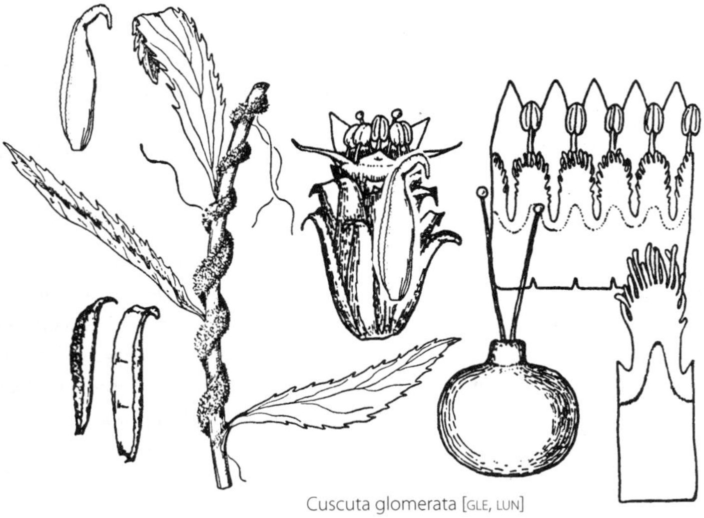

Cuscuta glomerata [GLE, LUN]

Cuscuta gronovii [BB2]

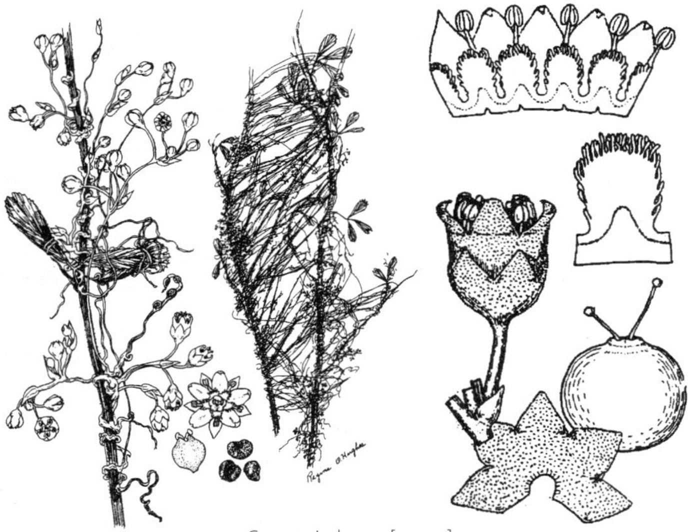

Cuscuta indecora [MTB, REE]

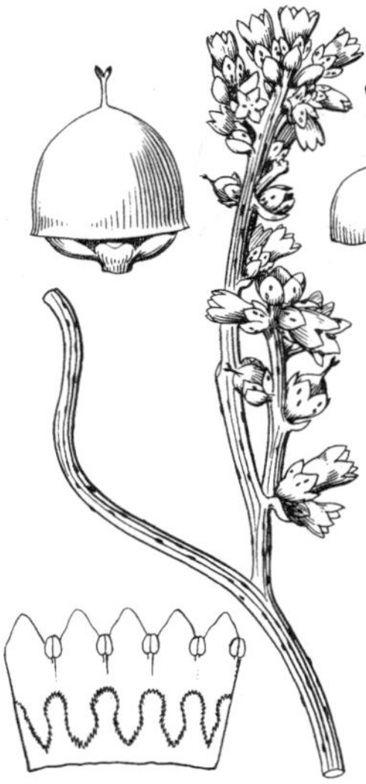

Cuscuta japonica [WU1]

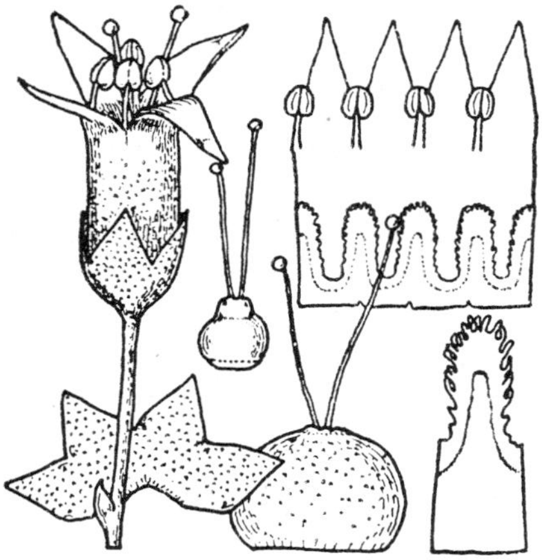

Cuscuta leptantha [LUN]

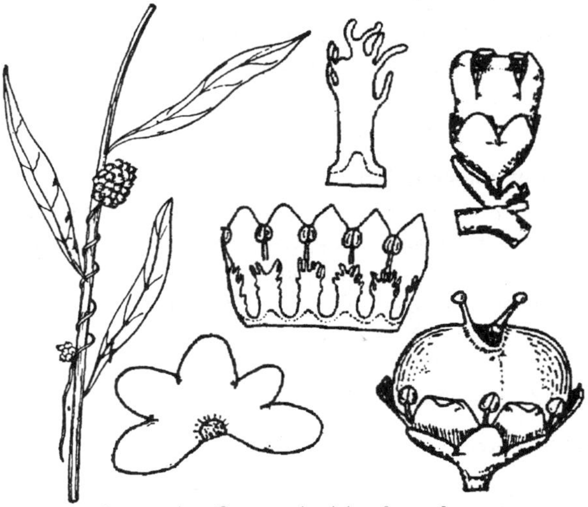

Cuscuta obtusiflora var. glandulosa [GLE, LUN]

DISTIMAKE Raf. WOOD-ROSE

A primarily tropical African and American genus of ca. 35 species (Simões & Staples 2017), including some that are serious weeds in plantations; others are cultivated as ornamentals or used medicinally. *Distimake* was formerly included in *Merremia*. Simões and Staples (2017) realigned the tribe Merremieae which includes *Merremia* and found it to be polyphyletic and assigned some of the segregates to the Rafinesque genus *Distimake*. (Rafinesque named this genus in *Flora Telluriana* 1838 [title page 1836] 4:82. His etymology (2 ac.stig.) is unclear, but it may refer to the doubly globose stigma.)
REFERENCES: Simões & Staples 2017; Austin 2023e.

Distimake dissectus (Jacq.) A.R. Simões & Staples, (dissected, deeply cut), ALAMO-VINE, CORREHUELA DE LAS DOCE. Low-climbing to trailing perennial vine, stems to 3 m long; leaf blades palmately deeply 5 or 7-lobed, 4–10 cm long, ca. as wide, the lobes toothed to pinnatifid, obtuse; peduncles 1–2-flowered; sepals oblong to oblong-ovate, mucronate, 18–25 mm long; corollas white with purple-red center, 3.5–5 cm long; fruit a 4-valved capsule, depressed globose, 1–2 cm broad; sepals greatly enlarged and reflexed in fruit. Open and disturbed areas, banks of streams; mostly in the sw portion of East TX, scattered elsewhere; South TX Plains and Edwards Plateau; FL, GA, and TX; also Mexico, Central and South America. May–Nov. Sometimes lumped into the genus *Ipomoea*. [*Ipomoea sinuata* Ortega, *Merremia dissecta* (Jacq.) Hallier f.] Native of the U.S., Puerto Rico, and the Virgin Islands.

EVOLVULUS L. DWARF MORNING-GLORY

Ours small pubescent perennials with several decumbent, erect to ascending or prostrate stems from a small woody base; never twining; leaves entire, small, 3 cm or less long, 1.5–11 mm wide; flowers solitary or in a 1–2-flowered axillary cyme; corollas 5–12 mm across, rotate or shallowly campanulate; styles 2, each with 2 branches, the 4 stigmas linear or filiform; capsules 1–4-seeded.

A genus of ca. 100 species (Austin 2023c) of warm and tropical areas of the Americas; 2 species extend to the Old World (Mabberley 2017). (Latin: *evolvere*, to untwist or unroll, alluding to the non-climbing habit compared with other members of the family)
REFERENCES: Van Ooststroom 1934; Perry 1935; Austin 2023c.

1. Peduncle + pedicel as long as or longer than subtending leaf; sepals 2–3 mm long ____ **E. alsinoides**
1. Peduncles (if present) + pedicel (if present) much shorter (< 7 mm) than subtending leaf; sepals 3–5 mm long.
   2. Leaves densely pubescent on both surfaces; sepals lanceolate to narrowly lanceolate, 4–5 mm long, spreading pilose ____ **E. nuttallianus**
   2. Leaves glabrous above or nearly so; sepals oblong lanceolate, 3–5 mm long, appressed pilose ____ **E. sericeus**

Evolvulus alsinoides (L.) L., (like the genus *Alsine*, Greek and Latin name possibly given to a chickweed), SLENDER DWARF MORNING-GLORY. Perennial herb, stems prostrate or ascending, 20–50 cm long, loosely appressed pilose, some hairs spreading; leaf blades 8–25 mm long and 3.5–11 mm wide, oblong or elliptic to lanceolate, sparsely to densely pilose an both surfaces; inflorescence 1 or 2 flowers on filiform peduncles usually longer than leaves; flowers on stalks (peduncle + pedicel) 2–4(–5) cm long, sepals lanceolate, ca. 2–3 mm long; corolla blue to pale blue to white, (5–)7–10 mm wide. Rocky or sandy ground; in East TX known only from Travis Co.; South TX Plains, Gulf Prairies and Marshes, Edwards Plateau, and Trans-Pecos; AL, AZ, FL, NM, and TX. Mar–Sep. [*E. alsinoides* var. *acapulcensis* van Ooststr., *E. alsinoides* var. *angustifolius* Torr., *E. alsinoides* var. *grisebachianus* Meisn., *E. alsinoides* var. *hirticaulis* Torr., *E. alsinoides* var. *linifolius* (L.) Baker, *E. linifolius* (L.) L.] Austin (2023c) treated all North America varieties at the species level. Native of the U.S., Puerto Rico, and the Virgin Islands.

Evolvulus nuttallianus Schult., (for Sir Thomas Nuttall, 1786–1859, English-American botanist), HAIRY EVOLVULUS, NUTTALL'S EVOLVULUS, SHAGGY EVOLVULUS, DWARF MORNING-GLORY. Suffrutescent perennial herb, stems several erect to ascending, 7–15(–35) cm long, densely spreading-pilose; leaves short petioled or sessile, leaf blades 8–20 mm long, 1.5–5 mm wide, linear-oblong, narrowly lanceolate to narrowly oblanceolate, densely pilose on both surfaces; flowers solitary in axils all along stem, peduncles short or absent; flowers on stalks (peduncle, if present, + pedicel) sessile to 4–5 mm long, reflexed in fruit, sepals lanceolate to narrowly lanceolate, 4–5 mm long; corollas rotate to broadly campanulate, 8–12 mm wide, blueish pink to blue. Sandy or rocky limestone soils; mostly on the w margin of East TX; from the Cross Timbers and Edwards Plateau w; from IL s to TX and w to MT and AZ. Apr–Jul(–Sep). [*Evolvulus pilosus* Roxb.] Native of the U.S.

Evolvulus sericeus Sw., (silky), SILKY EVOLVULUS. Perennial herb, stems woody at base, 10–30 cm long, prostrate to ascending, silky with appressed to spreading hairs, often grayish white or yellowish; leaves in two vertical rows, petiolate or sessile, leaf blades 4–25 mm long, 2–10 mm wide, lanceolate to elliptic, densely hairy below, glabrous or nearly so above; flowers solitary, axillary, stalks (peduncle + pedicel, if present) 0–2(–4) mm long, sepals oblong-lanceolate 3–5 mm long, silky; corolla usually

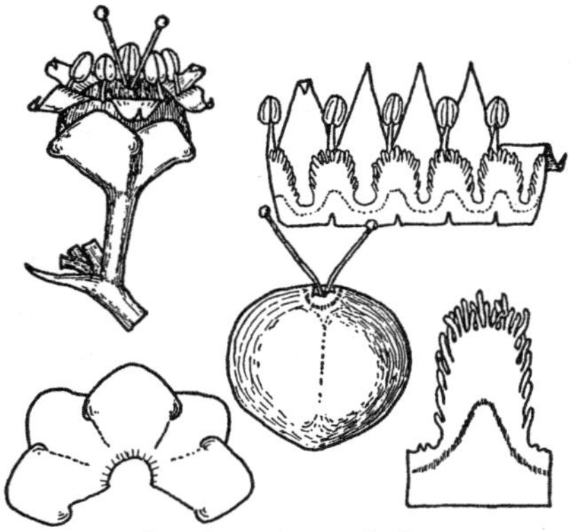

Cuscuta pentagona [LUN]

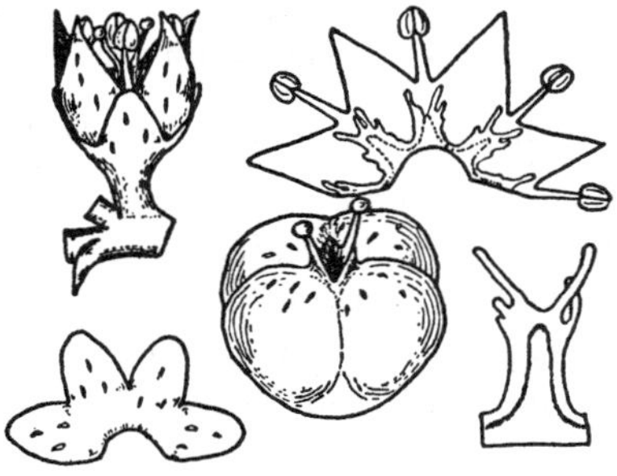

Cuscuta polygonorum [LUN]

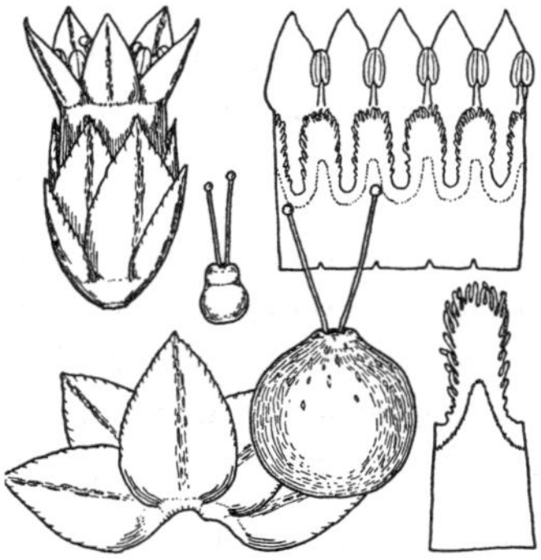

Cuscuta squamata [LUN]

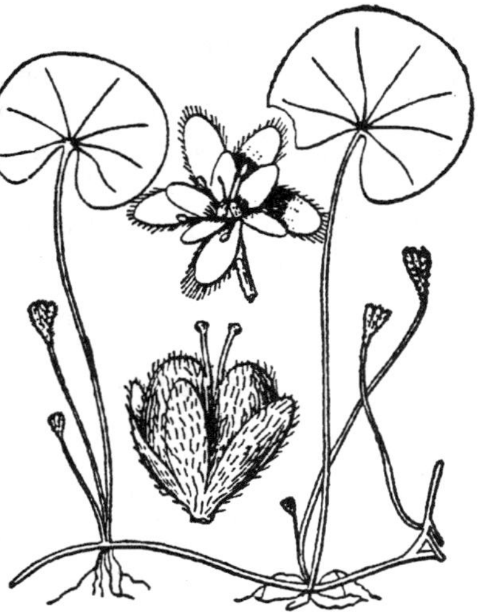

Dichondra carolinensis [BB2]

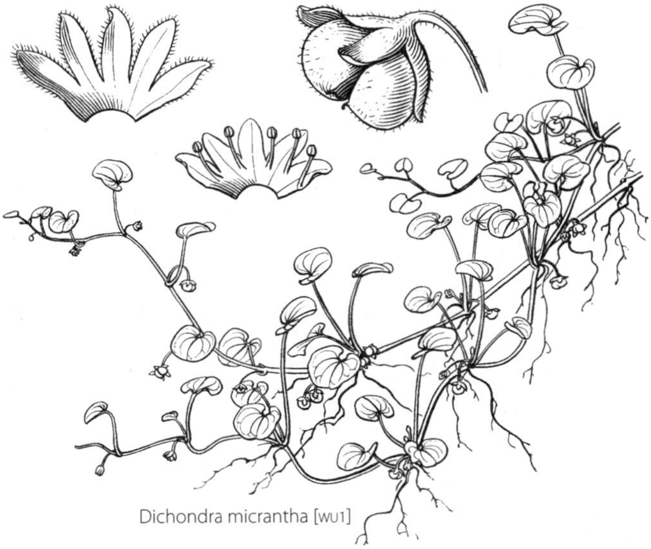

Dichondra micrantha [WU1]

Dichondra recurvata [HEA]

Distimake dissectus [BR2]

Evolvulus alsinoides [LAM]

white but also lavender to purple, rotate to broadly funnelform, 7–12 mm wide. Sandy, mixed sandy and gravelly, or rocky limestone soils, on hillsides, roadsides, and in open woods; in the sw 1/2 of East TX; Edwards Plateau and South TX Plains; s U.S. from GA and FL w to AZ. Apr–Oct. Native of the U.S., Puerto Rico and the Virgin Islands.

IPOMOEA L. MORNING-GLORY, MOONFLOWER

Trailing or twining (except erect shrub in *Ipomoea carnea* subsp. *fistulosa*) annual or perennial vines or rarely a shrubby perennial, mostly flowering in summer and fall; leaves petioled; leaf blades variable in shape, often ± cordate at base, entire to lobed or some compound; flowers axillary, solitary or in 2- to ca. 7-flowered cymose inflorescences; corollas funnelform, funnelform-campanulate, or salverform, usually large and showy, often open for less than 8 hours; style 1; stigma solitary, globose or with 2–3 globose or subglobose lobes; capsules usually with 1–4 seeds.

A huge genus of 600+ species (Austin 2023d) of climbing vines or shrubs native to tropical and warm temperate regions. Many species are cultivated as ornamentals for their flowers (e.g., HEAVENLY BLUE); some escape and become troublesome weeds; 9 others are toxic and hallucinogenic due to ergoline alkaloids; a Mexican species was used in religious ceremonies by the Aztecs (Schmutz & Hamilton 1979). The genus includes the SWEET-POTATO or YAM as discussed below. According to Strausbaugh and Core (1978), the common name, MORNING-GLORY, is derived from the tendency of the flowers of some species to open at night or in diffuse light (frequently in the morning). The rust fungus *Coleosporium ipomoeae* Burrill sometimes forms conspicuous yellow lesions on the leaves of *Ipomoea* species in TX; it is a heteroecious rust (= uses more than one host to complete its life cycle) that also infects *Pinus taeda* (LOBLOLLY PINE), but it can survive using only MORNING-GLORIES (J. Hennen, pers. comm.) (Greek: *ips*, a worm, and *homois*, resembling, from the twining habit)

REFERENCES: O'Donell 1959; Jones & Deonier 1965; Austin 1976, 1978, 1984, 1988, 2023d; Woolfe 1992; Bohac et al. 1995; Austin & Huáman 1996; Austin & Bianchini 1998; Duncan & Rausher 2013; Wood et al. 2020.

1. Erect woody shrub to 2 m+ tall ____ **I. carnea** subsp. **fistulosa**
1. Trailing, twining, prostrate, or climbing vine, or erect to decumbent and herbaceous.
   2. Corolla salverform, the long narrow tube cylindrical (with sides more-or-less parallel) for most of its length, the limb abruptly flaring at the summit of the tube.
      3. Corolla 9–14 cm long, white ____ **I. alba**
      3. Corolla 2–3 cm long, limb bright red to orange.
         4. Leaf blade pinnately divided into 11–31 (or more) linear segments ____ **I. quamoclit**
         4. Leaf blade broadly ovate, entire to dentate with 3–5 teeth ____ **I. coccinea**
   2. Corolla funnelform to campanulate, the short to long tube expanding in diameter upwards from below the middle, the limb gradually to abruptly flaring at summit of the tube.
      5. Pedicels and peduncles with reflexed hairs (hairs sometimes appressed in *I. indica*); gynoecium 3-parted.
         6. Sepals soft-pilose on the outer surface with slender hairs.
            7. Corolla limb lavender to pink, corolla 7–9 cm long; throat usually white; inflorescence solitary; leaves palmately 3–5(–7)-lobed ____ **I. lindheimeri**
            7. Corolla limb deep blue, corolla 5–7 cm long, throat pinkish to white; inflorescence pedunculate clusters; leaves ovate or 3-lobed ____ **I. indica**
         6. Sepals hispid-pilose on the outer surface, with swollen-based hairs.
            8. Sepals with slightly narrowed green tips shorter than to slightly longer than the body of the sepal; not strongly spreading ____ **I. purpurea**
            8. Sepals abruptly narrowed, the long narrow tips much longer than body of the sepal, tips strongly spreading or recurved ____ **I. hederacea**
      5. Pedicels and peduncles usually glabrous, if pubescent, then hairs not reflexed; gynoecium 2-parted.
         9. Stems trailing, rooting at the nodes; leaf apex emarginate or bilobed, truncate, or obtuse.
            10. Corolla limb lilac to pink or reddish, throat darker; larger leaves not lobed (though notched at the apex on at least the larger, better-developed leaves) ____ **I. pes-caprae** subsp. **brasiliensis**
            10. Corolla limb white, throat yellowish, sometimes purple; larger leaves 3–5-lobed ____ **I. imperati**
         9. Stems trailing but also other vine-like habits, sometimes erect to decumbent; not rooting at the nodes (except sometimes in *I. batatas*); leaf apex acute to acuminate.
            11. Leaves deeply, palmately dissected (or variously interpreted as compound, petiolules not always distinct) with 5–9(–11) lobes/leaflets.
                12. Leaf lobes/leaflets 5(–7), broadly lanceolate to lanceolate; some axillary buds developing stipule-like leaves ____ **I. cairica**
                12. Leaf lobes or leaflets 5–9(–11), linear to lanceolate; axillary buds not developing stipule-like leaves.
                    13. Leaf lobes 7–9(–11), linear to linear-lanceolate, 1.5–3 mm wide ____ **I. edwardsensis**
                    13. Leaf lobes/leaflets 5(–7), lanceolate to lance-linear, 10–15(–20) mm wide ____ **I. wrightii**
            11. Leaves entire to lobed.
                14. Leaf base deeply sagittate ____ **I. sagittata**

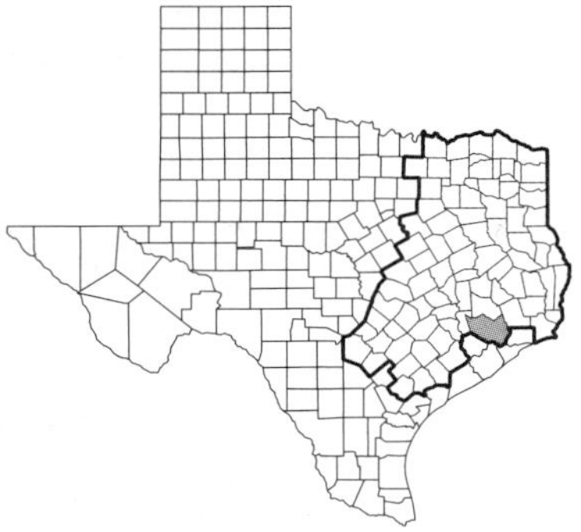

Cuscuta Japonica

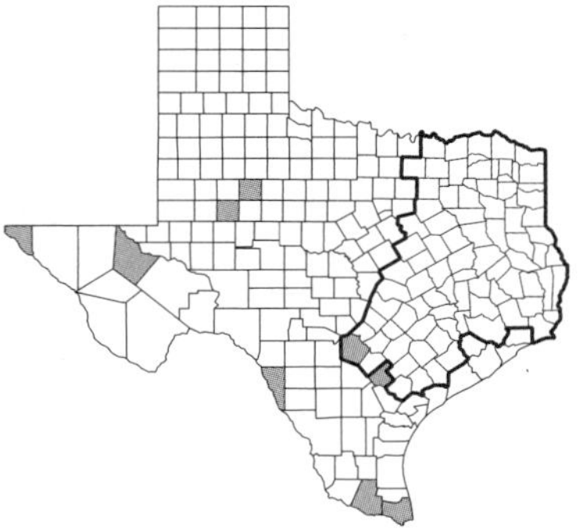

Cuscuta leptantha

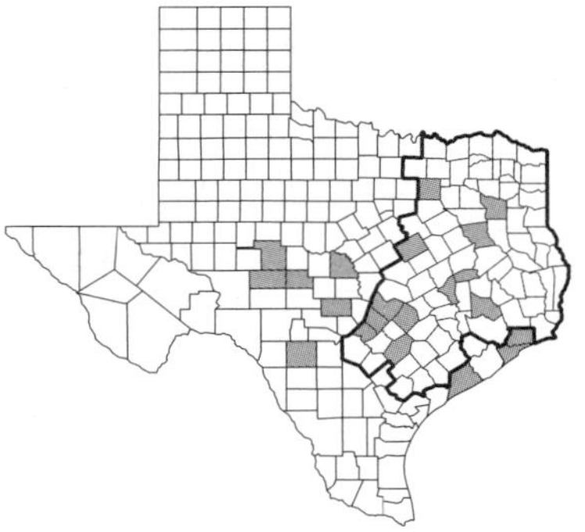

Cuscuta obtusiflora var. glandulosa

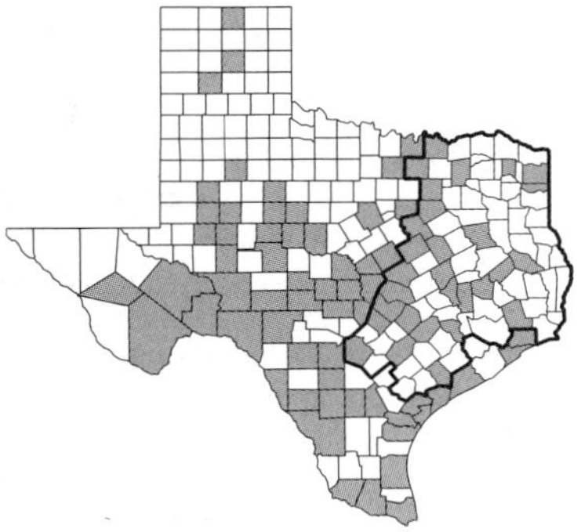

Cuscuta pentagona

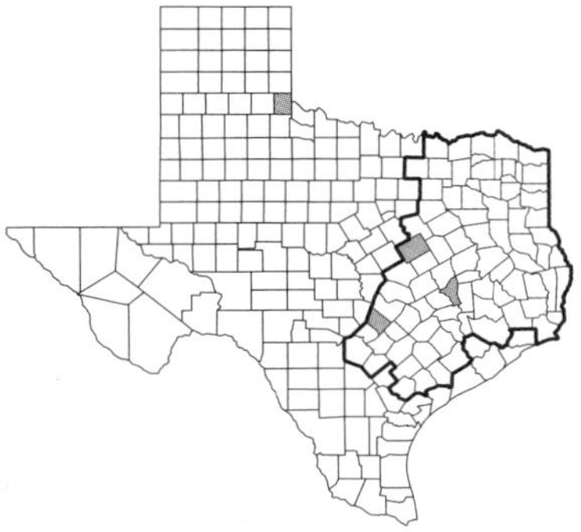

Cuscuta polygonorum

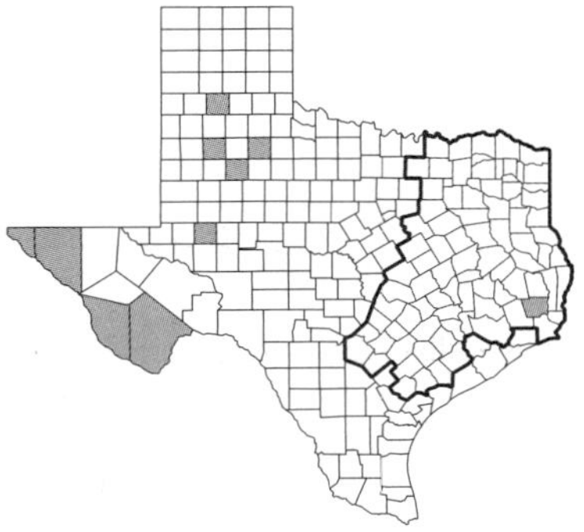

Cuscuta squamata

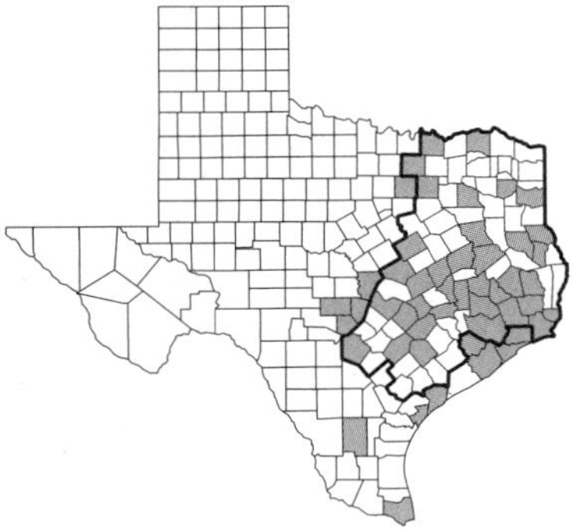

Dichondra carolinensis

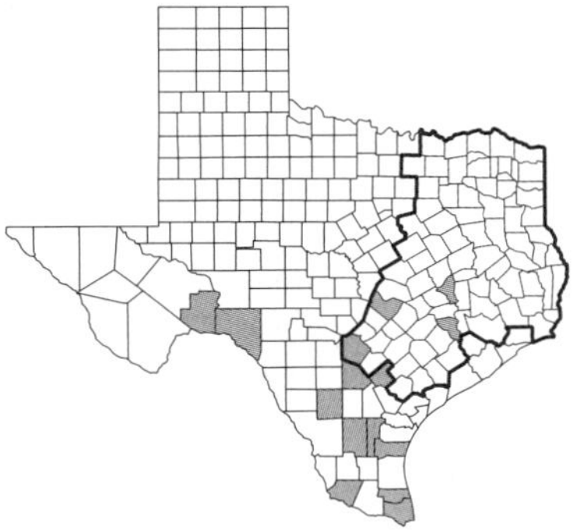

Dichondra micrantha

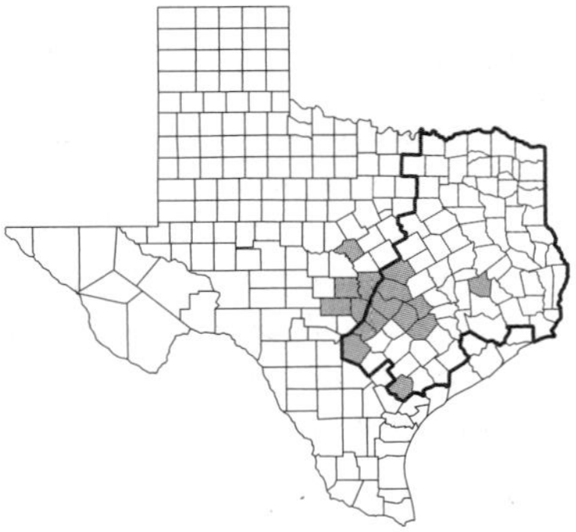

Dichondra recurvata

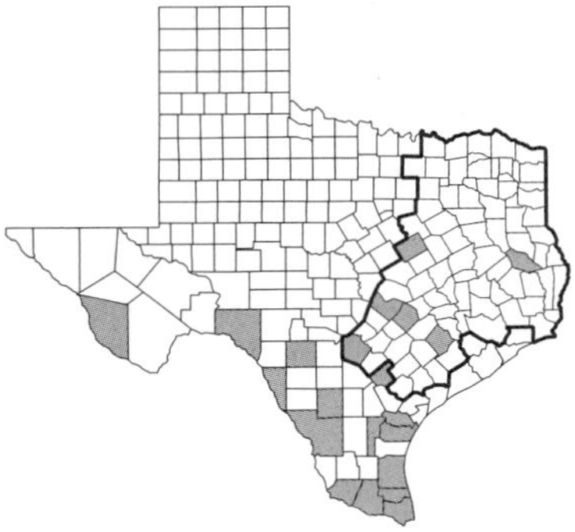

Distimake dissectus

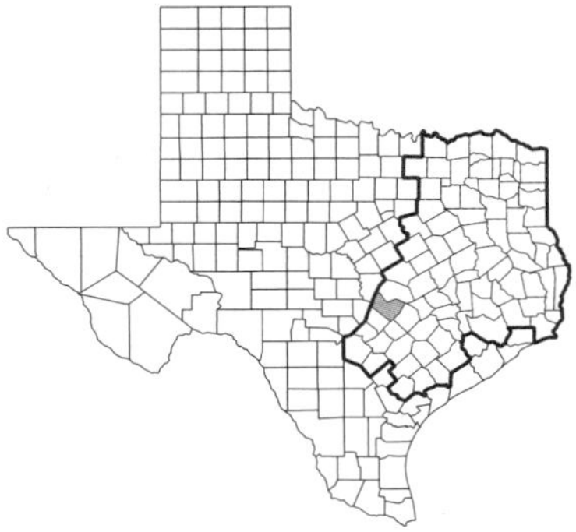

Evolvulus alsinoides

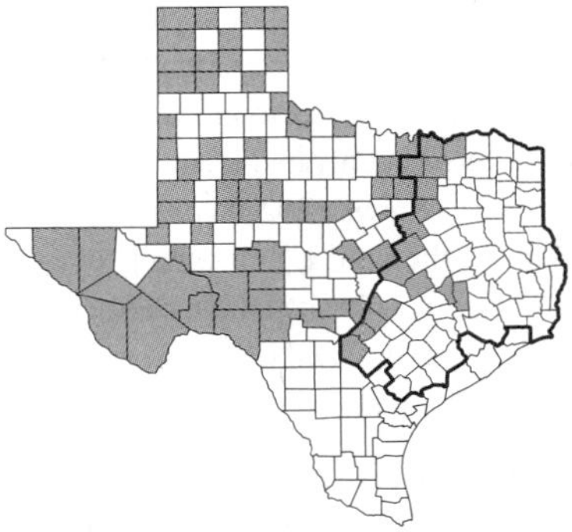

Evolvulus nuttallianus

14. Leaf base cuneate to cordate.
    15. Corolla 1.5–2(–2.5) cm long, entirely white (including throat), limb sometimes pink tinged ____ **I. lacunosa**
    15. Corollas 2–9 cm long, limb pink to lilac to purple, white, or blue, but with contrasting throat color (darker or lighter).
        16. All sepal apices rounded to obtuse.
            17. Inner sepals suborbicular, truncate, 4–5 mm long; corolla 2–3 cm long, limb pale lilac to pink or white; in East TX known only from Bexar Co. ____ **I. amnicola**
            17. Inner sepals oblong-ovate to broadly elliptic, 5–20 mm long; corollas 4–9 cm long, limb white to pink or purple-red.
                18. Corolla limb usually white; leaves ovate to 3-lobed; throughout the e 1/2 of East TX ____ **I. pandurata**
                18. Corolla limb pink to lilac or corolla completely purple-red; leaves linear; in East TX known only from Bowie Co. ____ **I. leptophylla**
        16. All sepal apices long acuminate to acute, occasionally obtuse, some apically caudate to mucronate.
            19. Sepals entirely green, ± equal to unequal, apex obtuse to acute to long-acuminate; corolla limb light pink, to lilac to purple (sometimes white in *I. batatas*), throat darker.
                20. Blades often broadly deltate, 5–7-lobed; leaves mostly 5–12 cm wide; outer sepals 4–9 mm long; in TX known from only Harrison and McLennan cos. ____ **I. batatas**
                20. Blades 3–5(–7)-lobed; leaves 1–9 cm wide; outer sepals 8–14 mm long; throughout most of East TX ____ **I. cordatotriloba**
            19. Sepals with a light margin, ± equal, apex acute; corolla limb blue, throat white to yellow ____ **I. tricolor**

Ipomoea alba L., (white), TROPICAL WHITE MORNING-GLORY. Vigorous climbing or trailing plant, stems to 10 m long; leaf blade 5–15 cm long, 4–14 cm wide, broadly ovate to triangular, sometimes 3–5-lobed to about one third, apex acuminate-aristate, base cordate, auricles sometimes with broad teeth; inflorescence of 1–3-flowers in pedunculate, axillary cymes; outer sepals lanceolate, long-awned, originating slightly below the apex abaxially; corolla 9–14 cm long, salverform, with a narrow cylindrical whitish-green tube 7–12 cm long and a spreading, white limb 4–5 cm wide. Disturbed damp ground, along streams, old home sites, pond edges; in East TX known only from Harris (*Traverse 887* TEX) and Jefferson (*Cory 50869* HUH) cos., also Cameron, Hidalgo, and Kleberg cos. in s TX, Galveston Co. (Kartesz 2022); FL, LA, SC, and TX. Mar–Nov. A weedy species found throughout the tropics.

Ipomoea amnicola Morong, (dwelling by a river), RED-CENTER MORNING-GLORY. Somewhat succulent twining or trailing perennial; leaf blade ovate, sometimes broadly so, apex acute to acuminate, base cordate with rounded auricles; inflorescence of lax to rather dense, many-flowered, pedunculate, simple or compound cymes; sepals convex, outer usually ± orbiculate, inner suborbicular, truncate, rounded; corolla 2–3 cm long, limb white to pink tinged, or pale lilac, throat darker. Roadside fallow fields, sandy alluvium, creek beds, pond margins, roadsides; in East TX known only from Bexar Co. (Turner et al. 2003); s TX (where it is weedy (Shinners 1965)); MO and TX (in MO last known record from 1921 (Shinners 1965), not seen in recent years (Yatskievych 2006)); also Mexico and South America. Aug–Nov.

Ipomoea batatas (L.) Lam., (vernacular name for this species from Chibchan-speakers of Colombia; from it is derived the English word potato—Austin 1988), SWEET-POTATO, YAM. Trailing or twining perennial herb (but cultivated as an annual) rooting from the stem and developing storage roots; stems extending to cover several meters; leaf blades rather large, 5–12 cm wide, deltate to broadly ovate or shallowly to deeply 5–7-lobed, cordate; inflorescence long-pedunculate, axillary, flowers solitary or few-flowered cymes, occasionally dense; sepals unequal, margins often but not always ciliate, apex acute to obtuse, mucronate-caudate, outer shorter than inner; corollas 3–5(–7) cm long, funnelform to funnelform-campanulate, pink to purplish, sometimes white in cultivated forms, throat darker. Widely cultivated and perhaps persisting; in TX known only from Harrison and McLennan (Kartesz 2022) cos.; scattered in se U.S. from NC s to FL and w to AR and TX; throughout tropical and subtropical regions of the world. Summer. [*Convolvulus batatas* L.] Probably native of tropical America, now widely grown in tropical and warm areas for its edible root; cultivated in more than 100 countries (Woolfe 1992) and is especially important in Japan (Mabberley 2017) and China; in terms of production, SWEET POTATO is the world's seventh largest food crop (Woolfe 1992; Bohac et al. 1995).

Ipomoea cairica (L.) Sweet, (of or from Cairo), MILE-A-MINUTE-VINE. Twining perennial herb to 3 m+ long; leaves palmately divided into 5(–7) lobes/leaflets; lobes/leaflets broadly lanceolate to ovate, acute; petioles with stipule-like outgrowths at base, 1–5 cm long; flowers usually solitary, sometimes in shortly pedunculate, 2–3-flowered axillary cymes; outer sepals ovate to broadly ovate; corolla 4.5–7 cm long, funnelform, limb lavender to pink or white; throat darker. Disturbed sites, open ground,

Evolvulus nuttallianus [STE]

Evolvulus sericeus [RUI]

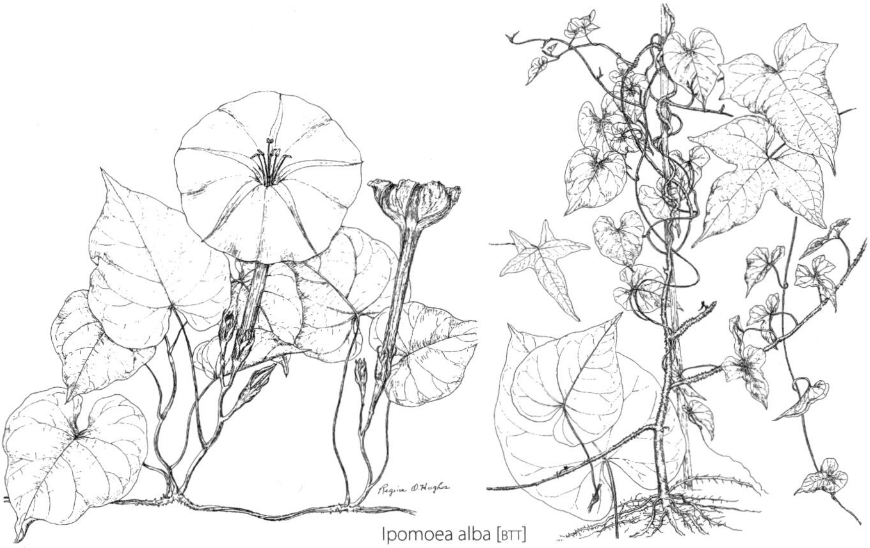

Ipomoea alba [BTT]

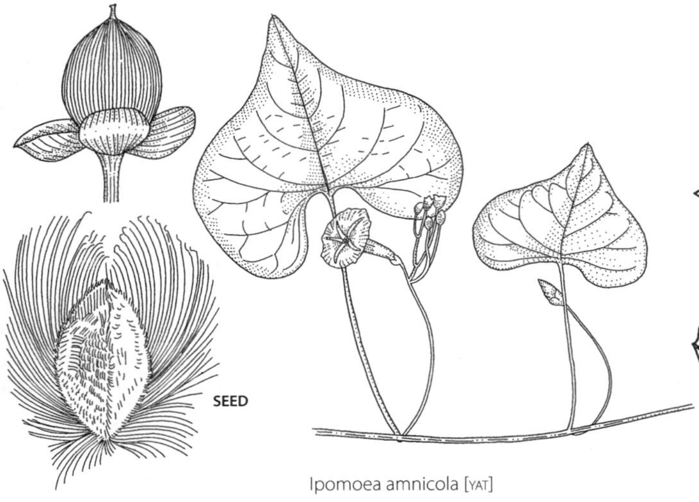

Ipomoea amnicola [YAT]

Ipomoea batatas [BR2]

along rivers, roadsides; in TX known only from Montgomery and Walker cos. (Nesom & Brown 1998); AL, CA, FL, LA, MS, and TX; throughout the tropics and subtropics. Mar–Oct. [*I. cairica* var. *hederacea*, Hallier f., *I. cairica* var. *lineariloba* (Hillebr.) O. Deg. & Ooststr.] Native of Africa.

Ipomoea carnea Jacq. subsp. **fistulosa** (Mart. ex Choisy) D.F. Austin, (sp.: flesh-colored; subsp.: hollow), BUSH MORNING-GLORY. Erect woody shrub to 2 m or more tall; leaf blade large, to 25 cm long, 17 cm wide, lanceolate, ± orbicular, or ovate, base cordate to subtruncate with rounded auricles, apex acuminate to long-acuminate; inflorescence of long-pedunculate axillary, somewhat compact cymes; sepals ± equal, ovate to suborbicular, rounded; corolla 4–9 cm long, funnelform, limb usually lavender or red-purple, sometimes white, throat darker, shallowly lobed. Persisting from cultivation; in East TX known only from Bexar, Brazos, and Hays cos. (Turner et al. 2003); s TX and Edwards Plateau; FL, GA, SC, and TX; Central and South America. Sep–Nov. [*I. fistulosa* Mart. ex Choisy] Native of South America.

Ipomoea coccinea L., (scarlet), SCARLET-CREEPER, SCARLET MORNING-GLORY, STARGLORY, RED MORNING-GLORY. Low-climbing annual vine; leaf blades broadly ovate, entire to dentate with 3–5 teeth; sepals obtuse to rounded with a narrow, tapered awn ca. 2–6 mm long, originating just below the apex abaxially; corollas 2–3 cm long, salverform, limb orange-red or red, throat yellow, the limb 1.7–1.9 cm wide. Cultivated and escapes; roads, fields, stream banks; Collin, Harris, Jefferson, Newton, Travis, (Kartesz 2022), Hardin (Brown et al. 2006b), and McLennan (Diggs et al. 1999) cos.; Gulf Prairies and Marshes, South TX Plains, and Edwards Plateau; e U.S. from NY s to GA and w to IA, and TX. Jul–Nov. Native of se U.S. Austin (1975) considered this species adapted for pollination by ruby-throated hummingbirds (*Archilochus colubris*).

Ipomoea cordatotriloba Dennst., (cordate or heart-shaped and 3-lobed), TIEVINE. Slender twining and low-climbing perennial herb (but flowering the first year), stems to 3–5 m long; leaf blade 3–5(–7)-lobed, 1–9 cm wide, the central lobe narrowed at base (very rarely unlobed), base narrowly cordate with rounded, entire or dentate auricles, apex shortly acuminate; inflorescence of axillary, pedunculate, umbelliform cymes, usually with 1–5(–9) flowers, and more lax than in *Ipomoea batatas*; sepals ± equal, apex acuminate to long-acuminate, curved outwardly, margin often ciliate with stiff spreading hairs, or glabrous, sepals 8–14 mm long, ± equal; corolla 3–4 cm long, gradually widened from base, limb pink to lilac, throat darker. Stream bottoms, fencerows, disturbed sites; throughout most of East TX; throughout TX except Panhandle; se U.S. from NC s to FL and w to TX; also Mexico. Jun–Sep. [*I. trichocarpa* Elliott, *Convolvulus carolinus* L.] Native of se U.S. According to Austin (1976), the following two varieties occupy different habitats in TX; var. *torreyana* is found mainly in prairie and plains areas, while var. *cordatotriloba* is found mainly in deciduous forest; some intermediates are known where the two habitat types meet.

1. Sepals hispid-pilose, at least at apex or on margins; leaf blades pilose, hirsute, or tomentose ______ var. **cordatotriloba**
1. Sepals glabrous; leaf blades glabrous ______ var. **torreyana**

var. **cordatotriloba,** SHARP-POD MORNING-GLORY, WILD MORNING-GLORY, PURPLE MORNING-GLORY, TIEVINE. Throughout East TX; se and e TX w to nc TX, Edwards Plateau; from NC s to FL and w to TX. [*Ipomoea trichocarpa* Elliott]

var. **torreyana** (A. Gray) D.F. Austin, (for John Torrey, 1796–1873, coauthor with Asa Gray of *Flora of North America*), COTTON MORNING-GLORY. Mostly in the w and s portion of East TX; South TX Plains, Edwards Plateau, and Trans-Pecos; TX; also Mexico. [*Ipomoea trichocarpa* Elliott var. *torreyana* (A. Gray) Shinners]

Ipomoea edwardsensis (O'Kennon & G.L. Nesom) O'Kennon, (alludes to the location of the plants on the Edwards Plateau). Slender annual herb from a filiform taproot, at first erect, becoming prostrate or climbing; stems to 3–60 cm long; leaf blades 2–3.5 cm wide, deeply palmately divided, 7–9(–11) lobes, lobes linear to linear-lanceolate, unequal; flowers solitary (rarely paired), axillary, peduncles slender, 3–22 mm long; corolla 8–11 mm long, tubular to long-campanulate, limb white, lobed, throat white. Limestone ledges and crevices, thin soil of weather pits in bedrock of limestone and granite; in East TX known only from Guadalupe (O'Kennon 2019), Bexar, and Travis cos., also Burnett, Gillespie, Llano, Uvalde (O'Kennon & Nesom 2002), and Starr (Richardson & King 2011) cos.; endemic to TX. (Aug–)Sep–Nov. [*I. costellata* var. *edwardsensis* O'Kennon & G.L. Nesom]

Ipomoea hederacea Jacq., (resembling *Hedera*—English ivy), IVY-LEAF MORNING-GLORY. Annual twining vines; stems retrorsely pubescent; leaf blades usually 3-lobed to rarely unlobed and broadly ovate to suborbiculate with auriculate base, or 5-lobed; sepals 12–24 mm long, abruptly narrowed from a broad base, apex long-acuminate, recurved, much longer than ovate sepal body, at least the body hispid-pilose; corolla funnelform, limb mostly light to deep blue, throat white or yellowish (varying to all purple or in cultivated forms other colors), sometimes drying purplish pink, 2.5–4.5 cm long. Disturbed areas, roadsides; scattered throughout East TX; Edwards Plateau, South TX Plains, and Trans-Pecos; e U.S. from ME s to FL and w to ND and TX, also AZ, CA, and NM; also Canada and Mexico, adventive in Europe. Jul–Oct. [*I. barbigera* Sweet, *I. desertorum* House]

Ipomoea cairica [NSW]

Ipomoea carnea subsp. fistulosa [MAR]

Ipomoea coccinea [BB2]

Ipomoea cordatotriloba var. cordatotriloba [BB1]

Ipomoea edwardsensis [SID]

Ipomoea hederacea [REE]

Ipomoea imperati [FLH]

Ipomoea imperati (Vahl) Griseb., (named after Ferrante Imperato (1550–1625), an apothecary of Naples), BEACH MORNING-GLORY. Perennial herb, stems trailing, rooting at the nodes, glabrous, up to 5 m long; leaf blades 1.5–8 cm long, 1.5–5 cm wide, rather small and variable, slightly succulent, linear, lanceolate or characteristically shortly oblong (± rectangular) or 3–5-lobed with the terminal larger than the laterals, apex obtuse or retuse, base truncate or very shallowly cordate, margin entire, undulate; flowers solitary (rarely 2–3), axillary; sepals unequal, glabrous, oblong-oblanceolate, acute or obtuse; corolla 3.5–5 cm long, funnelform, limb white, throat yellow, sometimes purple. Coastal sand dunes; Jefferson Co. (*Hutchinson 7871* COLO); coastal from Jefferson to Cameron cos.; coastal U.S. from NC to TX, also HI; Mexico; South America; Central America; Caribbean islands; a worldwide pantropical coastal plant. Spring–fall. [*Convolvulus imperati* Vahl, *I. stolonifera* (Cirillo) J.F. Gmel.] Native of the U.S. Found only on coastal sands just to the southeast of East TX proper.

Ipomoea indica (Burm.) Merr., (of India), OCEAN-BLUE MORNING-GLORY. Twining perennial herb; leaf blades 3–14 cm long, 3–14 cm wide, ovate or commonly, shallowly 3-lobed, both forms sometimes on the same plant, apex acuminate, base cordate with rounded auricles; inflorescence of axillary, pedunculate clusters, sometimes reduced to single flowers; sepals ± equal, 13–20 mm long, 3–5 mm wide, narrowly ovate, long-acuminate, pubescent, often somewhat spreading at maturity; corolla 5–7 cm long, limb deep blue, throat pinkish to white funnelform. Open upland woodland, disturbed areas; in East TX known only from Harris, Travis, and Walker cos. (Turner et al. 2003); also Webb Co.; se U.S. from NC s to FL and w to TX, also CA, PA, and HI; throughout tropical and subtropical countries of the world. Apr–Nov. [*I. acuminata* (Vahl) Roem. & Schult., *I. cathartica* Poir., *I. congesta* R. Br., *I. indica* var. *acuminata* (Vahl) Fosberg] ⊘

Ipomoea lacunosa L., (with air spaces, holes or pits, from the loosely reticulate venation of the leaves), PITTED MORNING-GLORY, SMALL WHITE MORNING-GLORY, WHITE MORNING-GLORY. Slender twining annual herb; leaf blade usually ovate, acuminate, base cordate with rounded auricles, sometimes 3 or 5-lobed with shortly acuminate lateral lobes; inflorescence of 1–3-flowered cymes; peduncles 0.6–6.5 cm long, sepals ± equal, 10–14 mm long, 2–4 mm wide, somewhat larger in fruit, ovate-lanceolate to lanceolate, acuminate to a long fine aristate tip, ciliate on margins, often also pilose; corolla 1.5–2(–2.5) cm long, funnelform, limb white, occasionally pale pink, lobes mucronate, throat white. Disturbed areas, river banks, fields, roadsides; throughout East TX, except sw portion; e U.S. from PA s to FL and w to IA and TX; possibly adventive in Europe. Aug–Oct. Native of the U.S. ⊘

Ipomoea leptophylla Torr., (thin-leaved), BUSH MORNING-GLORY. Perennial herb; stems erect to decumbent, bushy-branched, to ca. 1.2 m tall; leaf blades lanceolate to linear, unlobed; 3.5–10 cm long, 0.2–0.8 cm wide, petioles 1–7 mm; inflorescence of few-flowered axillary cymes; sepals unequal, outer 5–8 mm long, ovate, obtuse, inner similar but to 12 mm long, broadly elliptic and more rounded; corolla 5–9 cm long, funnelform to funnelform-campanulate, limb pink to deep pink to lilac, throat darker or corolla completely purple-red,. Dry, sandy fields, prairie hillsides, open grassland, roadsides; in East TX known only from Bowie (*Taylor & Taylor 22540* BRIT) Co.; Cross Timbers and Prairies, Panhandle, Edwards Plateau, and Trans-Pecos; c U.S. from SD s to TX and w to MT and NM; Mexico. May–Sep. Native of the U.S. ⊘

Ipomoea lindheimeri A. Gray, (for Ferdinand Jacob Lindheimer, 1801–1879, German-born Texas collector), LINDHEIMER'S MORNING-GLORY. Trailing or twining perennial herb; leaves petiolate, palmately 3–5(–7)-lobed to just over half way, base cordate, lobes usually ovate to broadly ovate, acute, narrowed at both ends, both surfaces hirsute; inflorescence of solitary, axillary flowers; sepals pilose, ± equal, lanceolate, long-acuminate, 17–28 mm long, 4–6 mm wide; corolla 7–9 cm long, narrowly funnelform, limb lavender to pink, throat white. Limestone based soils, granite outcrops; in East TX known only from Comal, Hays, and Travis (Kartesz 2022) cos.; Edwards Plateau, Trans-Pecos; AZ, NM, and TX; also Mexico. Mar–Oct. Native of the U.S. ⊘

Ipomoea pandurata (L.) G. Mey., (fiddle-shaped), BIG-ROOT MORNING-GLORY, WILD POTATO-VINE, MAN-OF-THE-EARTH, WILD POTATO, WILD SWEET-POTATO, INDIAN-POTATO. Trailing or twining perennial herb, rootstock an enlarged, woody tuber; leaf blade 2–10 cm long, 2–8 cm wide, ovate-deltoid, sometimes weakly to strongly 3-lobed, cordate with rounded auricles, shortly acuminate, usually not fiddle-shaped (= pandurate), despite its name; inflorescence of usually short, few-flowered axillary cymes; sepals unequal to almost equal; corolla 4–8 cm long, limb white, throat dark pink to maroon-red, campanulate to funnelform, undulate or lobed, the midpetaline bands terminating in small teeth. Long-leaf pine sandhills, dry forests, woodlands, roadsides; widespread in the e 1/2 of East TX and scattered to the w; Edwards Plateau and Gulf Prairies and Marshes; e U.S. from ME s to FL and w to NE and TX; Canada. Jun–Sep. [*Convolvulus panduratus* L., *I. pandurata* var. *rubescens* Choisy] Native of the U.S. and Canada. The starchy, enormously thickened roots of old plants can weigh up to 25 pounds (Wills & Irwin 1961); they were reportedly used by Native Americans as food; they are also reported to be poisonous, to have a milky resinous juice, and to contain the glucoside ipomoein (Burlage 1968). ⊘ ☠

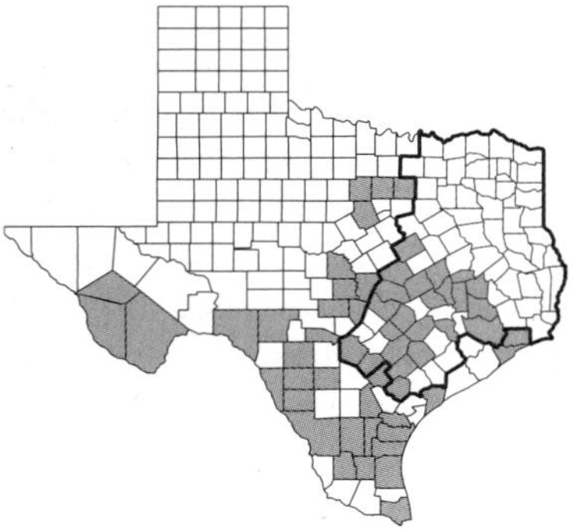

Evolvulus sericeus

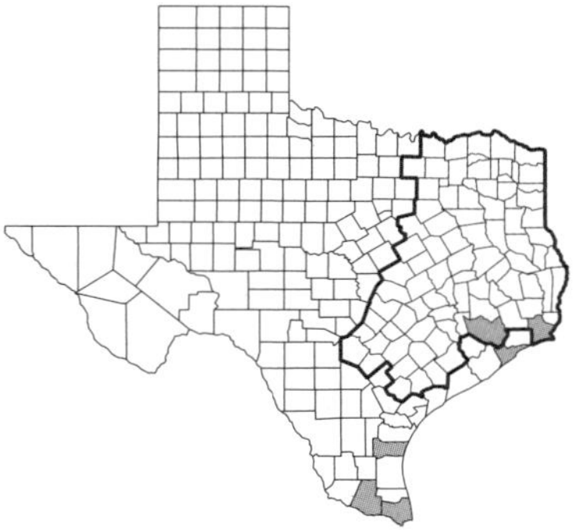

Ipomoea alba

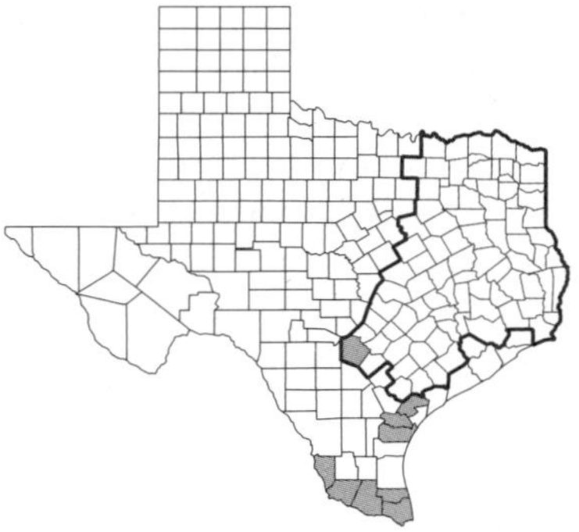

Ipomoea amnicola

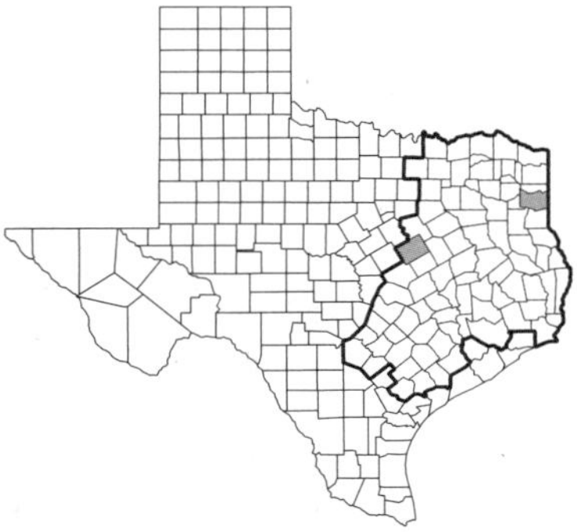

Ipomoea batatas

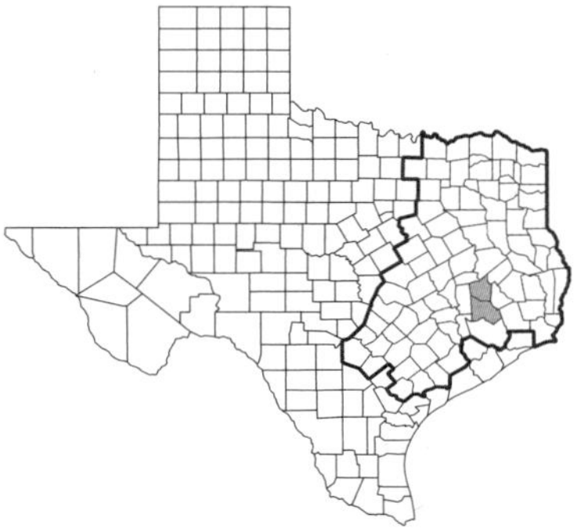

Ipomoea cairica

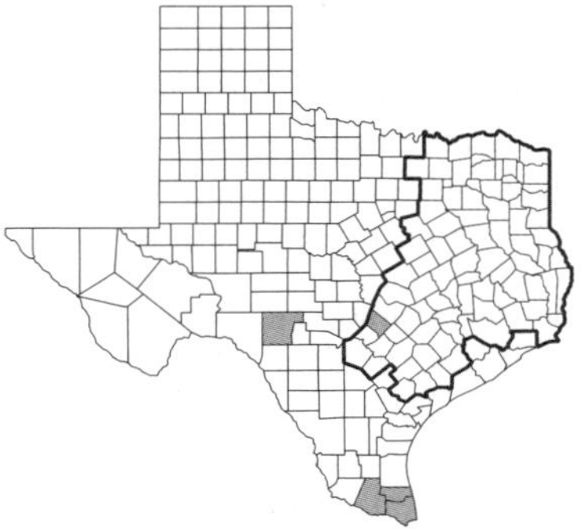

Ipomoea carnea subsp. fistulosa

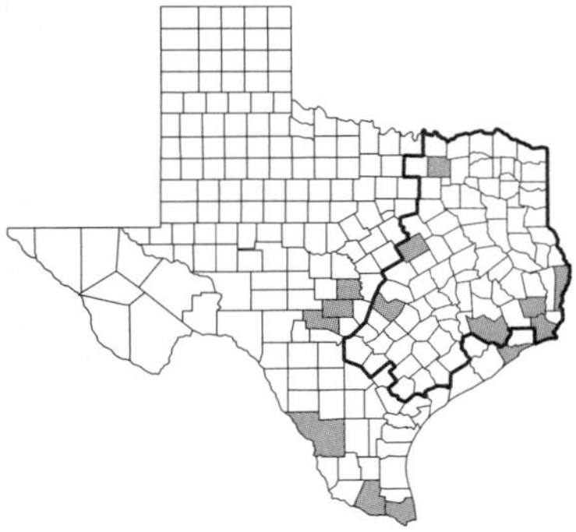

Ipomoea coccinea

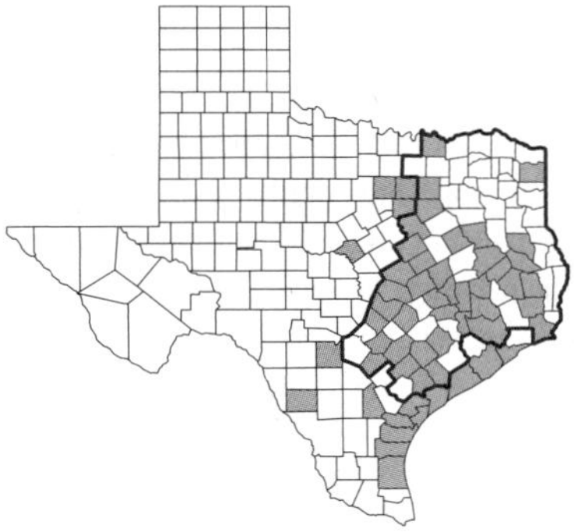

Ipomoea cordatotriloba var. cordatotriloba

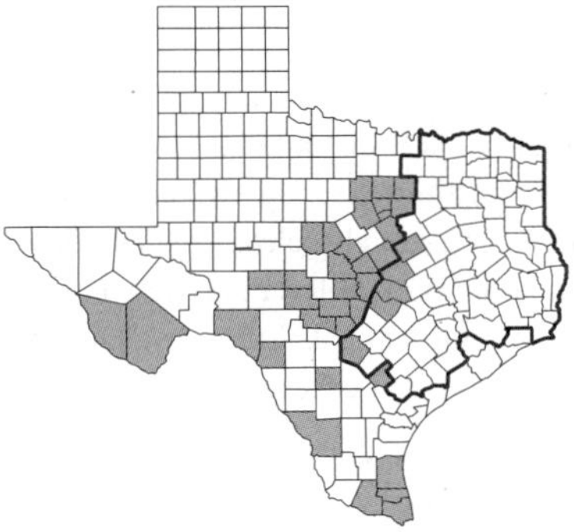

Ipomoea cordatotriloba var. torreyana

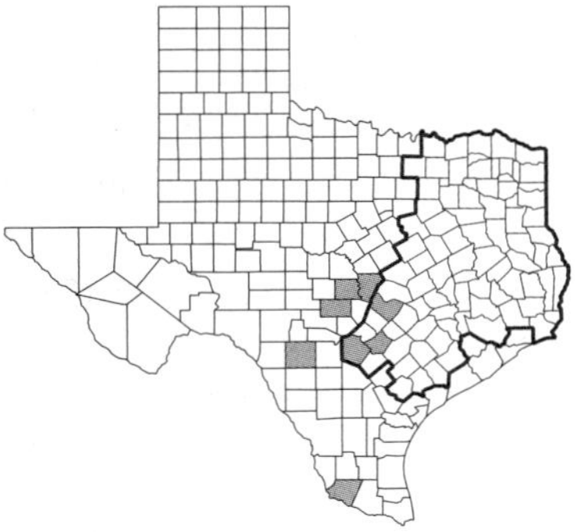

Ipomoea edwardsensis

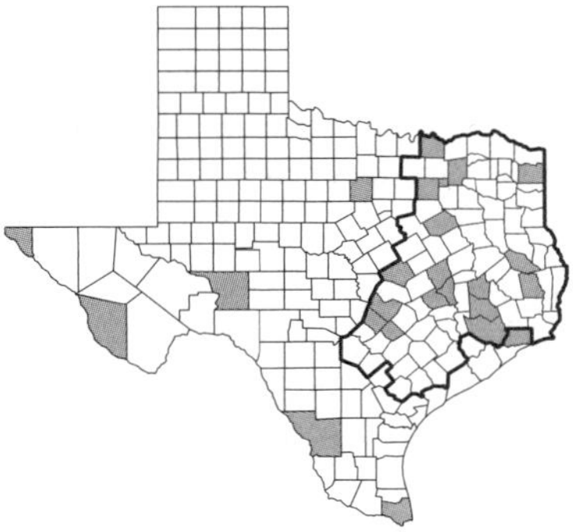

Ipomoea hederacea

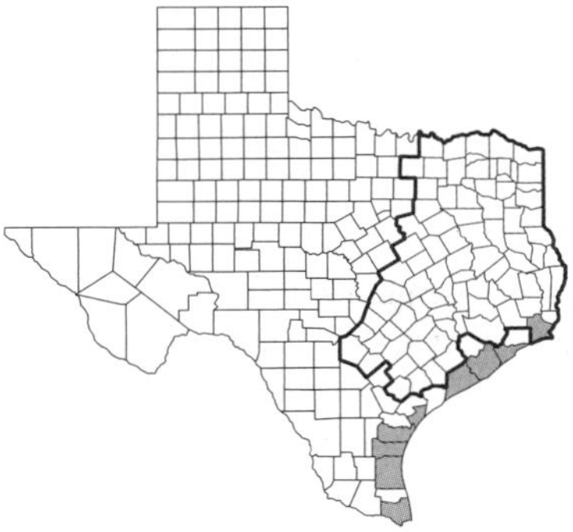

Ipomoea imperati

Ipomoea pes-caprae (L.) R. Br. subsp. **brasiliensis** (L.) Ooststr., (sp.: *pes*, Latin for foot, *caprae*, Latin for goat, referring to the shape of the leaf resembling that of a goat; subsp.: of Brazilian origin), GOATFOOT MORNING-GLORY, BEACH MORNING-GLORY. Vigorous trailing perennial to 30 m long; leaf blade 3.5–9 cm long, 3–10 cm wide, coriaceous and somewhat succulent, ovate to reniform or suborbicular, apex emarginate to shallowly bilobed (rarely rounded), base truncate to weakly cordate; inflorescence of axillary cymes; sepals slightly unequal, pale green, coriaceous, outer ovate to elliptic to broadly ovate, inner suborbicular, corolla 4–5 (–7) cm long, funnelform, limb lilac to pink or reddish, throat darker; pedicel often persistent on fallen capsule, so aiding dispersal in the sea. Mostly coastal beaches and sand dunes; in East TX known only from Orange Co. (*Alderson s.n.*, 1974, BRIT); coastal cos. from Chambers to Cameron cos.; coastal U.S. from NC to TX, also PA; a pantropical taxon usually of coastal areas, a result of its dispersal by sea-drift seeds. May–Dec. [*I. brasiliensis* (L.) Sweet, *I. pes-caprae* var. *emarginata* Hallier f.] Native of the U.S. ⊘

Ipomoea purpurea (L.) Roth, (purple), MEXICAN MORNING-GLORY, WOOLLY MORNING-GLORY, COMMON MORNING-GLORY. Twining annual herb; leaf blade, ovate (rarely 3-lobed), shortly acuminate, cordate with rounded auricles; flowers solitary or in 2–5-flowered, pedunculate, axillary cymes; sepals ± equal, 10–17 mm long, 2–4 mm wide, lanceolate to oblong-lanceolate, acute, hispid-pilose, more densely so in lower half; corolla 2.5–5(–7) cm long, limb blue-purple (variable in cultivated forms), throat white to pink, lighter in color compared to the limb, funnelform. Stream banks, disturbed or waste areas; scattered in East TX; Edwards Plateau and Trans-Pecos; in almost every state in the U.S.; also Mexico and Canada; widely distributed throughout the tropics as an escape from cultivation or as a weed. Jul–Oct. [*Convolvulus purpureus* L., *I. hirsutula* J. Jacq., *I. purpurea* var. *diversifolia* (Lindl.) O'Donell] Native of tropical America. Reported to be poisonous (Burlage 1968). ⊘

Ipomoea quamoclit L., (native Mexican name), CYPRESSVINE. Low-climbing, twining, annual vine; leaf blades pinnately divided into very narrow, linear, almost thread-like segments about as wide as the midrib (0.2–1.5 mm wide); inflorescence of 1(–5)-flowered axillary, pedunculate cymes; sepals ± equal, oblong-elliptic, obtuse to rounded and very shortly mucronate, 4–8 mm long, 2–3 mm wide, corollas salverform, deep red (white in some cultivated forms), 2–3 cm long, limb 1.8–2 cm wide, deeply lobed, lobes acute; stamens exserted, 2.5–3 cm long. Cultivated and spreads; Cherokee (Turner et al. 2003), Colorado (Kartesz 2022), Grayson (Diggs et al. 1999), Hardin (*Amerson 1273* SMU), Harris (Harcomb et al. 2007), Jasper (*Holmes 11108* NLU) McLennan (Hannick et al. 2013), and Newton (*Allen et al. 20354* BRIT) cos.; e U.S. from NY s to FL and w to IL and TX, also CA; Mexico and Canada; widely cultivated and naturalized throughout the tropics. Jul–Nov. Native of tropical America. ⊘

Ipomoea sagittata Poir., (sagittate, referring to the shape of the leaves), GLADE MORNING-GLORY. A perennial trailing or twining herb; leaf blade deltoid-sagittate, the central lobe lanceolate, acuminate to a mucronate point, 4–10 cm long, 0.2–1.5 cm wide (excluding auricles), the two auricles similar in shape but slightly shorter than the main part of blade, acute; inflorescence of solitary (very rarely paired), axillary flowers; sepals unequal, oblong-elliptic, rounded, mucronate, corolla 6–9 cm long, funnelform, limb pink to lavender, throat darker. River banks, low wet areas, marshes; several cos. in se East TX, also Gonzales and Denton (Turner et al. 2003) cos.; Gulf Prairies and Marshes and Edwards Plateau; mostly coastal U.S. from NC s to TX; coastal from Mexico to Belize and Guatemala, several islands in the Caribbean. Jun–Nov. Native of the U.S.

Ipomoea tricolor Cav., (three-colored), GRANNYVINE. Twining annual herb, stems robust and often thick (4–5 mm broad); leaves petiolate, blade 6–10 cm long, 2.5–13 cm wide, ovate, cordate with rather angular, nearly rounded auricles, apex acuminate; inflorescence of pedunculate, few-flowered axillary cymes; sepals ± equal, lance-ovate to oblong-lanceolate, acute, dark green with white margin; corolla 5–9 cm long, funnelform, limb blue, sometimes white, throat white to yellow. Roadsides, waste grounds, and disturbed areas; in East TX known only from Dallas Co. (Turner et al. 2003), perhaps based on (*Shinners 9945* FSU), which includes the annotation "self-sown in garden, blackland clay"; El Paso (*Shipes 68* UTEP), and Taylor (*Moore & Mahler 4960* BRIT) cos.; e U.S. from SC s to FL and w to MO, AR, and TX; throughout the New World, often cultivated and escapes elsewhere. Jul–Nov. Native of Mexico and Central America. Many cultivars are available, HEAVENLY BLUE being a popular one.

Ipomoea wrightii A. Gray, (for Charles Wright, 1811–1885, TX collector), WRIGHT'S MORNING-GLORY. Twining annual herb; leaves, divided into 5(–7) separate sessile leaflets; leaflets 3–7 cm long, 1–1.5(–2) cm wide, elliptic, lanceolate to narrowly lanceolate, acuminate at both ends; flowers solitary (rarely paired), axillary, peduncles slender, flexuose, sometimes coiling, sepals ± equal, 4–6 mm long, ca. 3 mm wide, obtuse to rounded; corolla 1.8–3 cm long, funnelform, limb pink to lilac, throat darker. Coastal plain, swamp edges, along rivers, disturbed areas; mostly in the sw portion of East TX; South TX Plains and Gulf Prairies and Marshes; se U.S. from SC s to FL w to OK and TX; from Mexico s to South America. Jun–Oct. The history of the specific epithet of this entity is very convoluted. We are following Correll and Johnston (1970), Wunderlin et al. (2019), Weakley, A.S., and Southeastern Flora Team (2022), and Austin (2023d) in using the name *I. wrightii*. ⊘

Ipomoea indica [POP]

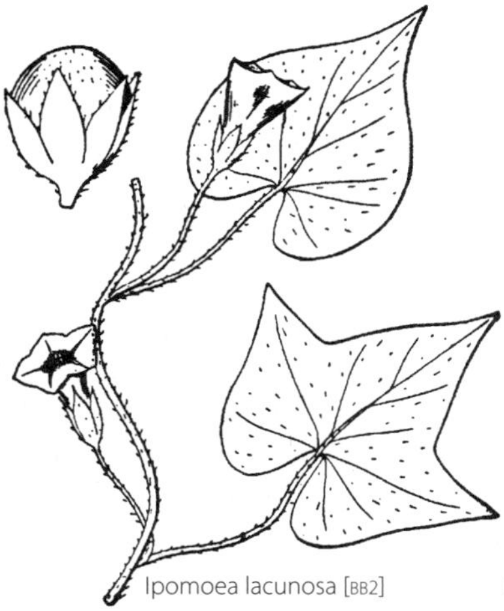

Ipomoea lacunosa [BB2]

Ipomoea leptophylla [USA]

Ipomoea lindheimeri [IV2]

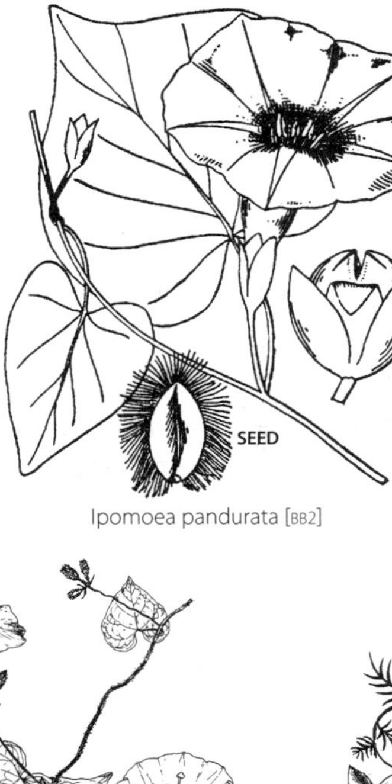

Ipomoea pandurata [BB2]

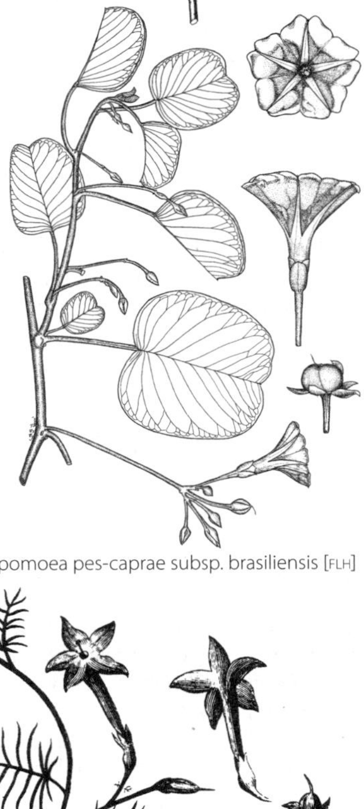

Ipomoea pes-caprae subsp. brasiliensis [FLH]

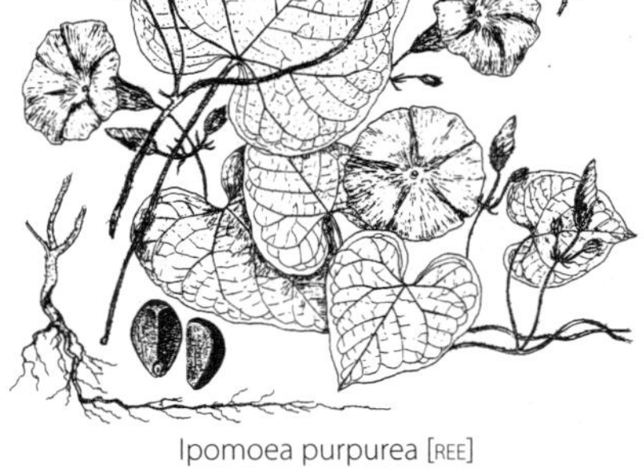

Ipomoea purpurea [REE]

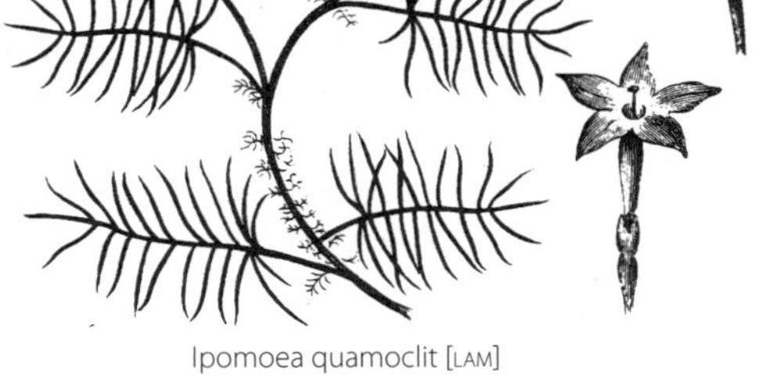

Ipomoea quamoclit [LAM]

Ipomoea ×leucantha Jacq., (bearing white flowers), WHITESTAR. Twining annual herb similar to *I. cordatotriloba*; leaf blade, 3–8 cm long, 2–7 cm wide, ovate, entire or shallowly to deeply 3–5-lobed, cordate, the auricles sometimes with a large tooth, apex shortly acuminate; flowers solitary or 2–3 in axillary cymes; sepals ± equal, 10–14 mm long, lance-ovate to lanceolate, acuminate and apiculate, glabrous; corolla 1.5–2.5 cm long, limb pink to lavender or white, throat darker. This taxon will most likely key out with *Ipomoea batatas*, but *I. ×leucantha* has shorter outer sepals and shorter corollas. Disturbed ground, wet areas, roadsides, fields, canal banks, and railroad tracks; in TX known only from Val Verde and Cameron cos. (Turner et al. 2003); AL, AZ, FL, LA, MO, SC, TX, and VA. Aug–Dec. We have included this taxon as a note because it is found in four counties along the TX/LA border. Native of the U.S. ⦸

Ipomoea shumardiana (Torr.) Shinners, (for Benjamin Franklin Shumard, 1820–1869, state geologist of TX in 1860), NARROW-LEAF MORNING-GLORY. Perennial vine similar to *I. pandurata* but leaf blades entire, never lobed, often narrower, 3–8 cm long, base truncate; sepals unequal, outer 8–9 mm long, ovate, obtuse, mucronate; inner 11–12 mm long, broadly oblong-ovate, apiculate; corollas funnelform, white (rarely pink) with purple-red throat, 5–8 cm long. Possibly a hybrid between *I. leptophylla* and *I. pandurata* (Correll & Johnston 1970). Gravelly roadside prairie; rare; Montague and Wise cos. (BRIT), also Cooke Co. (Correll & Johnston 1970; KS, OK, and TX. Jun–Aug. [*Convolvulus shumardianus* Torr., *I. carletoni* Holz.] Included based on presence in the above three TX cos. at the nw corner of East TX and Bryan Co., OK on the border of East TX.

JACQUEMONTIA Choisy CLUSTERVINE

A genus ca. 100 species (Robertson 2023) of tropical and warm areas, especially the Americas; some are cultivated as ornamentals. (Named for Victor Jacquemont, 1801–1832, French botanical explorer)
REFERENCE: Robertson 2023.

Jacquemontia tamnifolia (L.) Griseb., (with leaves resembling *Tamnus* of Dioscoreaceae), HAIRY CLUSTERVINE. Annual, stems twining or trailing, 1–4 m long, but can flower while still small and erect; leaves long-petioled, blades 3–12 cm long, 2–9 cm wide, ovate-elliptic to ovate, the larger ones basally cordate, acute to acuminate at apex; flowers many, densely arranged in a leafy-bracted, pedunculate, head-like inflorescence; peduncles shorter to longer than leaves; bracts and sepals with conspicuous long hairs; bracts lanceolate to elliptic; sepals 8–10 mm long, lance-linear; corollas 12–16 mm long, blue, style 1, stigmas 2, ± flattened; capsules 4-seeded. Widespread in the se quadrant of East TX; mostly se U.S. from VA s to FL and w to OK and TX. Jul–Oct. [*Ipomoea tamnifolia* L., *Thyella tamnifolia* (L.) Raf.] The most widespread species of *Jacquemontia*. ⦸

STYLISMA Raf. BONAMIA

Perennials with procumbent or trailing stems, rarely twining; leaves short-petioled or sessile, entire; inflorescences axillary, peduncled; flowers solitary or several; sepals 5; corolla funnelform to campanulate, shallowly 5-lobed or -angled, white or lavender, maroon, pink, purple, or red; style branches 2; stigmas peltate to capitate, minute; capsule 1(–4)-seeded

A genus of 6 species native to s and e U.S. (Mabberley 2017) (Greek: *stylus*, a pillar, stake, or column—Quattrocchi 2000, apparently in reference to the style)
REFERENCES: Fernald & Schubert 1949; Shinners 1962a; Myint 1966; Lewis 1971; Allen 2013, 2023.

1. Corollas lavender, maroon, pink, purple, or red; stamen filaments glabrous or glabrate ____ **S. aquatica**
1. Corollas white; stamens usually pubescent, at least basally, rarely glabrous.
   2. Leaf blades 10–30 mm wide; inflorescences usually 3–7-flowered, flowers rarely solitary.
      3. Sepals glabrous ____ **S. humistrata**
      3. Sepals villous ____ **S. villosa**
   2. Leaf blades 0.1–3 mm wide; inflorescences usually solitary, sometimes 2–5-flowered ____ **S. pickeringii**

Stylisma aquatica (Walter) Raf., (relating to water), WATER DAWNFLOWER. Stems procumbent or trailing, 0.5–1.5 m long, blades 2–3.5 cm long and 0.3–1 cm wide, oblong-elliptic or oblong-lanceolate, pubescent on both surfaces; inflorescence an axillary (1–)2–3-flowered cyme; sepals mostly 5–8 mm long and 3–5 mm wide, abaxially glabrous, margins ciliate; corolla 1–1.5 cm long, lavender, maroon, pink, purple, or red, style branches divided nearly to the base; filaments glabrous or nearly so; stigmas capitate. Damp sandy soils, low pinelands, and pond margins; Harris, Jasper, Liberty, Newton, Orange, and Waller (Turner et al. 2003) cos.; se U.S. from NC s to FL and w to TX. Jun–Aug. [*Bonamia aquatica* (Walter) A. Gray, *B. michauxii* (Fernald & B.G. Schub.) K.A. Wilson, *Breweria aquatica* (Walter) A. Gray, *B. michauxii* (Fernald & B.G. Schub.] Native of the U.S.

Stylisma humistrata (Walter) Chapm., (low layer in reference to a low-growing habit), SOUTHERN DAWNFLOWER. Stems procumbent or trailing, usually 1–2.5 m long, pubescent; leaf blades 2.5–8 cm long, 1.2–3 cm wide, narrowly oblong to oblong or oblong-lanceolate; inflorescence an axillary cyme, usually 3–7-flowered, rarely solitary; sepals 6–9 mm long and 3–5 mm wide,

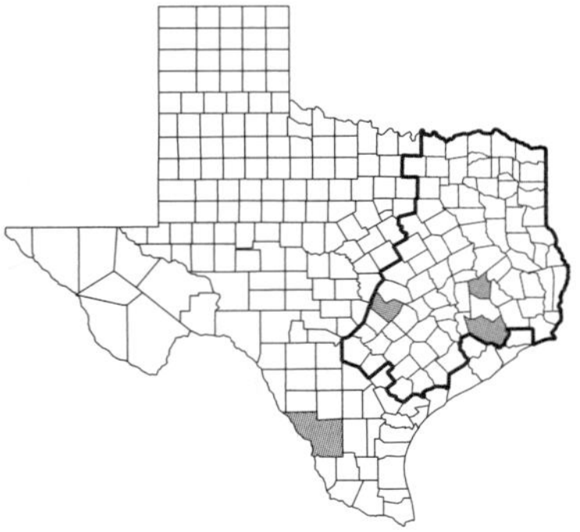

Ipomoea indica

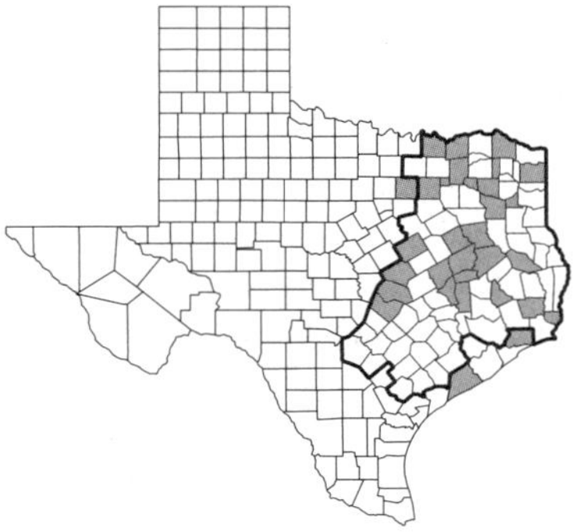

Ipomoea lacunosa

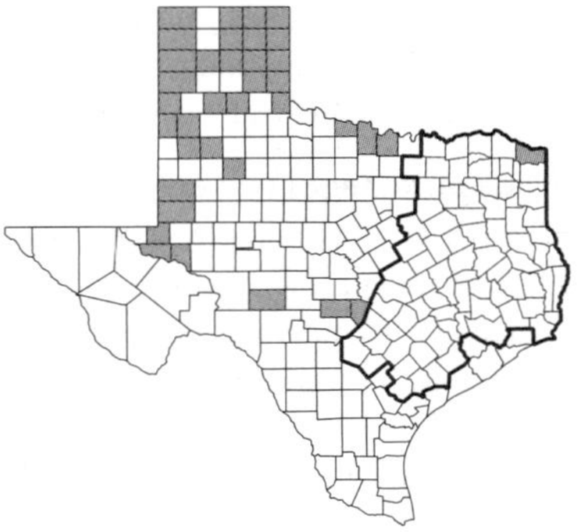

Ipomoea leptophylla

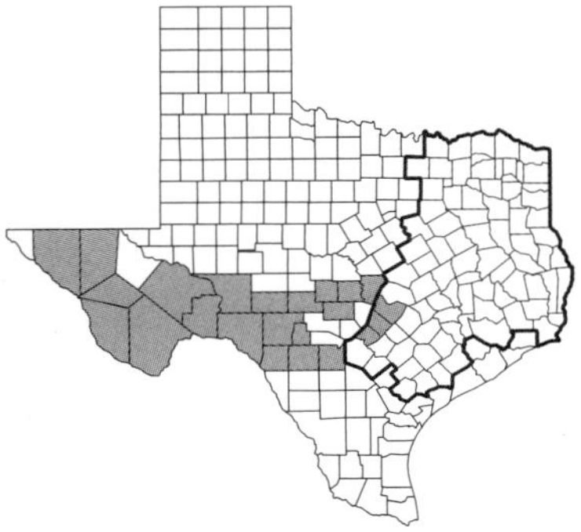

Ipomoea lindheimeri

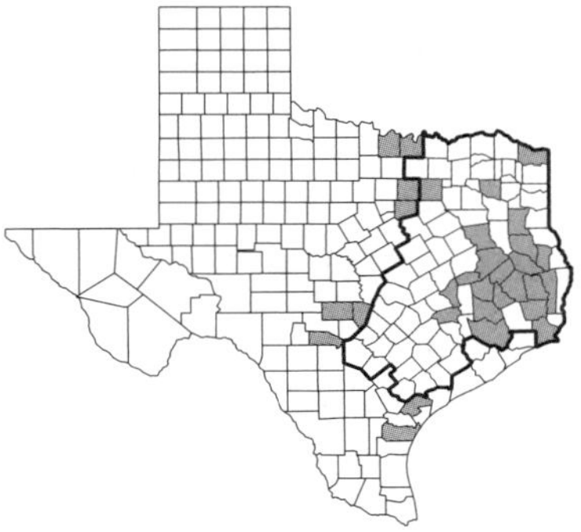

Ipomoea pandurata

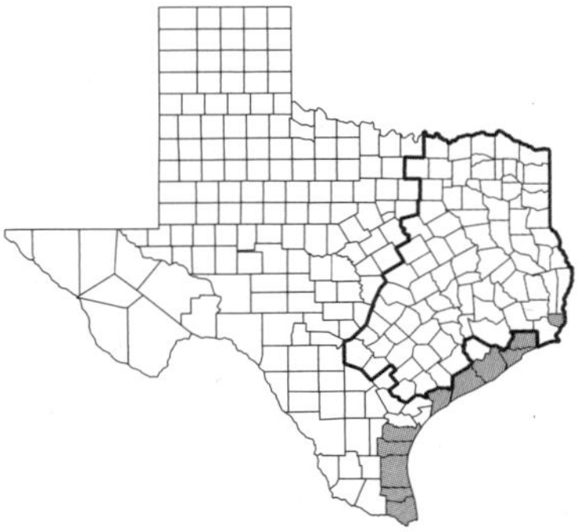

Ipomoea pes-caprae subsp. brasiliensis

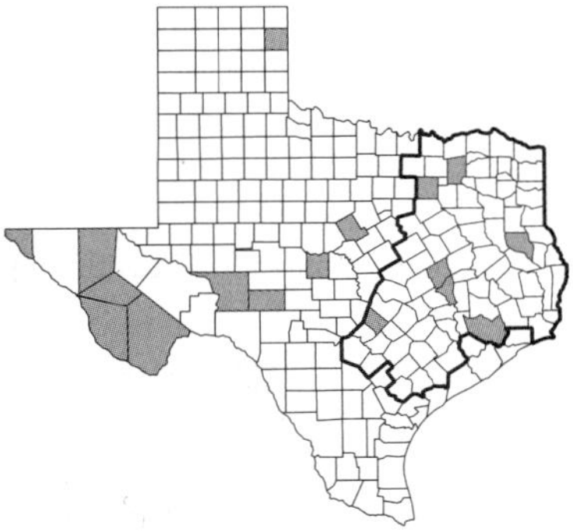

Ipomoea purpurea

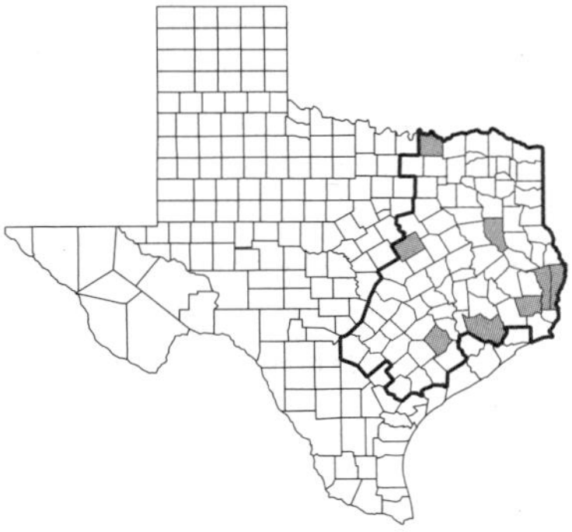

Ipomoea quamoclit

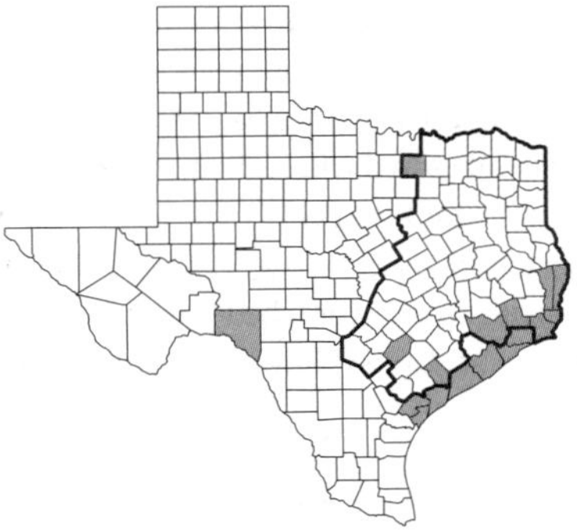

Ipomoea sagittata

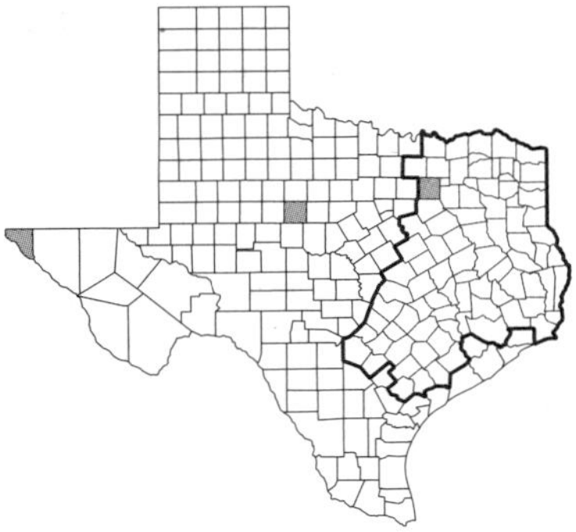

Ipomoea tricolor

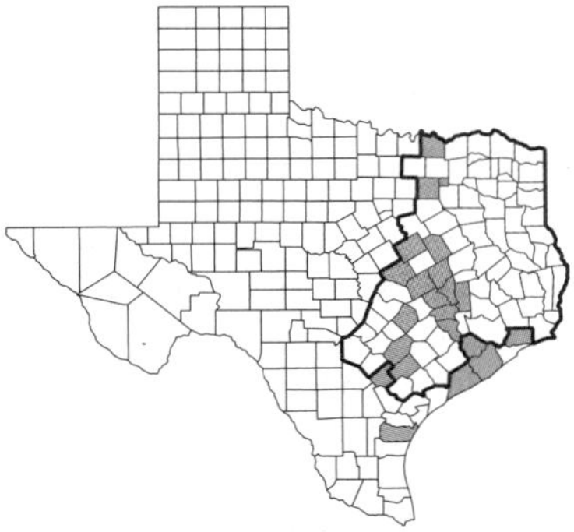

Ipomoea wrightii

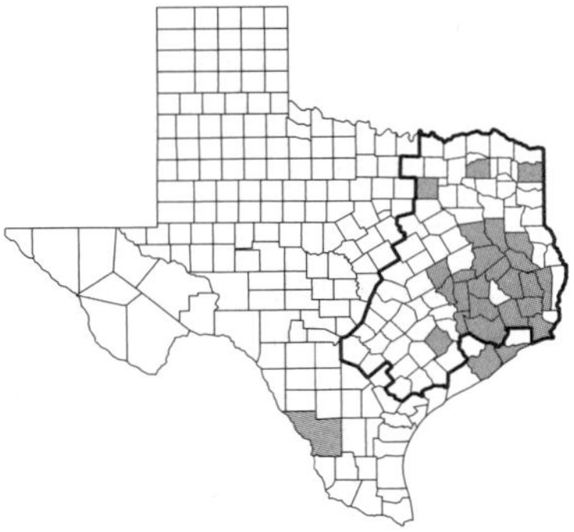

Jacquemontia tamnifolia

ovate-lanceolate, abaxially glabrous, margin ciliate; corolla 2–2.3 cm long, white, stamens basally villous. Sandy soils in damp woods, sand hills or dry sandy ridges; mostly widespread in the e 1/2 to East TX; se U.S. from VA s to FL and w to TX. May–Sep. [*Bonamia humistrata* (Walter) A. Gray, *Breweria humistrata* (Walter) A. Gray, *Convolvulus humistratus* Walter]

Stylisma pickeringii (Torr. ex M.A. Curtis) A. Gray var. **pattersonii** (Fernald & B.G. Schub.) Myint, (sp.: for Charles Pickering, 1805–1878, botanist and physician on Wilkes expedition in the Columbia River area; var.: for Harry Norton Patterson, 1853–1919, American printer, botanist, and explorer), BIG-POD BONAMIA. Stems procumbent or trailing, sparsely pubescent to 2 m long; leaf blades 25–70 mm long, 0.1–3 mm wide, entire, narrowly linear to linear; inflorescence 2–5-flowered, rarely solitary; sepals 4–6 mm long, 3–5 mm wide, white to tawny pubescent abaxially; corollas 10–18 mm long, white; filaments glabrous or nearly so; styles 2, fused at least 1/2 or more of their length. Sandy open woods; throughout East TX; c U.S. from IL s to LA and w to KS and TX, also AL and MS. May–Sep. [*Breweria pickeringii* (Torr. ex M.A. Curtis) A. Gray var. *pattersonii* Fernald & B.G. Schub.]

Stylisma villosa (Nash) House, (hairy), HAIRY DAWNFLOWER. Stems procumbent to trailing, 0.5–2 m long, usually pubescent; leaf blades 3–7 cm long and 1–2 cm wide, narrowly oblong-elliptic, or oblong-lanceolate; inflorescence 3–7-flowered; sepals 7–11 mm long and 3–5 mm wide, ovate-lanceolate, abaxially densely villous; corolla 1.5–2.5 cm long, white; stamens glandular-villous near base. Dry sandy soils in pinewoods, high pinelands, and hammocks; Austin, Harris, Montgomery, and Waller cos. in East TX; also Calhoun and Nueces cos.; AL, FL, GA, LA, NC, and TX. Apr–Aug. [*Bonamia villosa* (Nash) K.A. Wilson, *Breweria villosa* Nash] Native of the U.S.

CORDIACEAE R. Br. ex Dumort.

CORDIA FAMILY

A family of 2 genera (*Cordia* and *Varronia*) with ca. 350–400 species of trees, shrubs, or rarely lianas, distributed worldwide, especially in the tropics and subtropics (Luebert et al. 2016). A number of species of *Cordia* are important for timber and some are used as dyes, medicinally, or for their edible fruit (Mabberley 2017). The family has often been included in a broadly conceived Boraginaceae (e.g., Correll & Johnston 1970; APG IV 2016); however, based on the most recent molecular data and morphological studies (e.g., Gottschling et al. 2014, 2016; Weigend et al. 2014; Luebert et al. 2016) the botanical consensus has moved towards dividing the excessively broadly conceived Boraginaceae (Stevens 2001 onwards; Luebert et al. 2016). The Cordiaceae is now segregated as a monophyletic group related to families including the Coldeniaceae, Ehretiaceae, Heliotropiaceae, Namaceae, Hydrophyllaceae, and more distantly to Boraginaceae (in the strict sense), all these being monophyletic families in the Boraginales (Luebert et al. 2016). (subclass Asteridae—Cronquist; order Boraginales (as part of a very broad Boraginaceae)—APG IV)

FAMILY RECOGNITION IN THE FIELD: our only species is a small shrubby perennial limited to the extreme sw part of East TX with simple, alternate, toothed or lobed leaves and delicate, white funnelform flowers in a dense head.

REFERENCES: Gottschling et al. 2001, 2004, 2014, 2016; Weigend et al. 2014; Luebert et al. 2016; Hasenstab-Lehman 2017.

VARRONIA P. Browne, VARRONIA

About 100 species widespread in the New World tropics and warm temperate regions from Arizona and Texas through Argentina (Miller & Gottschling 2007). The genus has often been treated as part of *Cordia*, but recent molecular research shows that it is a distinct monophyletic lineage (Miller & Gottschling 2007) and we are following most modern authors (e.g., Stevens 2001 onwards) in recognizing it at the generic level. (for Marcus Terentius Varro, 116–27 BC, a Roman author, critic, and grammarian "who not only was responsible for transferring several hundred Greek words into the Latin language, ... but also wrote on agriculture"—Oswald & Nelson 2000)

REFERENCES: Standley 1924; Correll & Johnston 1970; Borhidi et al. 1988; Miller & Gottschling 2007; Campos-Ríos & Chiang-Cabrera 2012.

Varronia podocephala (Torr.) Borhidi ex G. Campos & F. Chiang, (Greek: *pous, podos*, foot, and *kephale*, head, perhaps in reference to the small dense head of flowers terminating an elongate peduncle; not explained in the type description by Torrey (1859)), TEXAS MANJACK, WHITE CORDIA. Small, erect, sub-shrubby perennial from a woody root; stems to ca. 60 cm tall, usually few, unbranched or with a few branches 1–2 mm thick; leaves alternate, simple, short petiolate, blades 2–9 cm long, to ca. 1.7(–2.5) cm wide, narrowly lanceolate to oblanceolate or sometimes ± ovate, apically acute to obtuse, basally acute to acuminate, pinnately veined, scabrous on both surfaces, marginally coarsely serrate (often with ca. 4–6 teeth) or sinuately toothed or lobed; inflorescence a dense globose head 5–10 mm thick before flowering, on a naked peduncle longer than the leaves, to 18 cm long, terminal or axillary; flowers 5-merous, numerous, but only a few open on a given day; calyces with lobes triangular; corollas

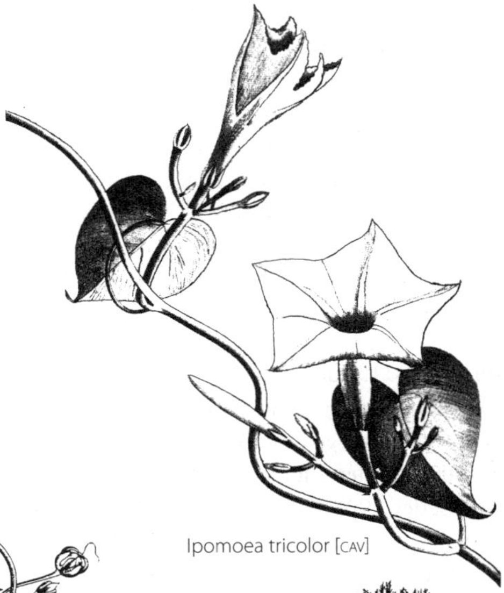

Ipomoea tricolor [CAV]

Ipomoea sagittata [GWO]

Ipomoea wrightii [WIG]

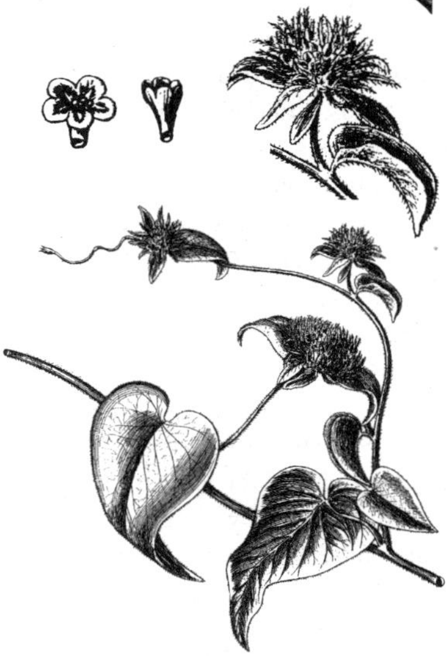

Jaquemontia tamnifolia [DIL]

Stylisma aquatica [BB2]

Stylisma humistrata [FO2]

Stylisma pickeringii var. pattersonii [BB2]

funnelform, white, delicate, 11–18 mm long, the lobes slightly notched at tip, the tube broadly flared, longer than the lobes; stamens 5, included; style twice dichotomous, resulting in 4 branches each with a stigma; ovary superior, 4-locular; drupe inside the enlarged sac-like calyx. Open brush and woodlands, canyon bottoms, open rocky slopes, roadsides; Bexar (Torrey 1859—in the type description described as "near San Antonio, Texas"; *Mrs. Blake s.n.*, June 18, "San Antonio"—Torch 2024; Turner et al. 2003), Goliad (*W.C. Holmes 8547* TEX), and Dewitt (iNaturalist 2024) cos. at far s margin of East TX; mainly s TX; in the US known only from TX; also ne Mexico (Augascalientes, Coahuila, Nuevo León, San Luis Potosí, Veracruz—Campos-Ríos & Chiang-Cabrera 2012). Apr–Jun(–Oct). [*Cordia podocephala* Torr., *Lithocardium podocephalum* (Torr.) Kuntze]

CORNACEAE Bercht. & J. Presl

DOGWOOD FAMILY

A small (80–115 species in 2 genera—Stevens 2001 onwards; Mabberley 2017) mainly n hemisphere family of primarily temperate and subtropical regions; most are trees and shrubs or rarely rhizomatous herbs; several are cultivated including the ornamentally important genus *Cornus* (DOGWOOD). Circumscription of the family has been unsettled. As Xiang et al. (1998) noted, various authorities have included from 1 to as many as 17 genera including those in the related Nyssaceae (e.g., Burckhalter 1992; Mabberley 1997; Kubitzki 2004a). Recent molecular analyses (Xiang et al. 2002; Fan & Xiang 2003; Xiang et al. 2011) suggest that Cornaceae, Nyssaceae, Hydrangaceae, and Loasaceae, as well as several other groups are related families in the Cornales, and in addition support *Cornus* and *Alangium* as most closely related; however, they do not support the lumping of Nyssaceae into Cornaceae. The Cornaceae is here recognized as a distinct family of 2 genera, *Cornus* and the Old World *Alangium*, following APG IV (2016). However, Murrell and Poindexter (2016) exclude *Alangium*, and treat the family has having only one genus. (subclass Rosidae—Cronquist; order Cornales—APG IV)

FAMILY RECOGNITION IN THE FIELD: shrubs or small trees with opposite, simple, entire leaves with *arcuate veins* (= pinnate veins arching toward leaf tip) and small 4-merous flowers in open terminal inflorescences or in some cases in heads subtended by showy bracts; inferior ovary developing into a drupe fruit. The "*Cornus* test" discussed below is also distinctive.

REFERENCES: Coulter & Evans 1890; Rickett 1945a; Ferguson 1966c; Xiang et al. 1998b, 2002, 2008, 2011; Fan & Xiang 2003; Kubitzki 2004a; Xiang & Thomas 2008; Feng et al. 2009; Murrell & Poindexter 2016; Fu et al. 2019; Thomas et al. 2021; Du et al. 2023a.

CORNUS L. DOGWOOD, CORNEL

Ours shrubs or small trees; leaves opposite (alternate in one species rare in East TX), simple, entire, with prominent pinnate veins arching toward the leaf tip, deciduous; stipules absent; flowers very small, usually perfect, radially symmetrical, in open cymose inflorescences without conspicuous bracts or in conspicuously bracteate heads terminal on branches or branchlets; calyx lobes or sepals 4; petals 4, cream to white or yellow-green, distinct; stamens 4; filaments longer than the anthers; pistil 1; ovary inferior; fruit a small red, blue, or white drupe.

A genus of ca. 60–65 species of trees, shrubs, and rhizomatous herbs native mainly to n temperate and subtropical areas with 1 species in South America and 1 in Africa (Stevens 2001 onwards; Kubitzki 2004a; Xiang et al. 2006; Murrell & Poindexter 2016). We are treating this monophyletic lineage as a single genus and not recognizing segregates. Some authors (e.g., Weakley, A.S., and Southeastern Flora Team 2024), however, subdivide the genus into a number of segregates including *Swida* and treat *Cornus* more narrowly as genus of about 6 species of Eurasia, tropical Africa, and nw North America. While it can be argued that the age of the segregate genera warrants their separate recognition, we prefer to treat the genus in the broad traditional sense following numerous recent authors (e.g., Stevens 2001 onwards; Murrell & Poindexter 2016; Mabberley 2017; Yu et al. 2017; Lindelof et al. 2020; Thomas et al. 2021; Du et al. 2023a). A number of *Cornus* species are widely cultivated for the showy, bract-surrounded inflorescences that resemble individual flowers, and many have been used medicinally. Fruits can also be showy, ranging from red to blue or white. An interesting field character which can be used to identify *Cornus* species is the "*Cornus* test"—if a leaf is very *gently* torn into 2 pieces (apical part separated from basal), the primary veins remain connected by delicate whitish threads which represent the slinky-like, unraveled, spiral thickenings of the vascular tissue (Zomlefer 1994). Pollination is generalized with visitors including beetles, bees, butterflies, and flies and seed dispersal is by a variety of birds (Eyde 1988). (Latin: *cornu*, a horn, alluding to the hardness of the wood, the European *C. sanguinea* L. having long been used for skewers by butchers, whence SKEWERWOOD in English provinces and DAGWOOD from the old English: *dagge*, a dagger or sharp pointed object)

REFERENCES: Wilson 1964 [1965]; Ferguson 1966a, 1966b; Eyde 1987, 1988; Murrell 1993; Xiang et al. 1993, 1996 [1997], 2006; Yu et al. 2017; Lindelof et al. 2020; Du et al. 2023a.

1. Leaves alternate (but internodes often short, sometimes giving the leaves a whorled appearance; petioles 20–50 mm long ____ **C. alternifolia**
1. Leaves opposite; petioles 2–25 mm long.
   2. Flowers in small compact inflorescences (heads) subtended by 4 large (to 5 cm long and 3 cm wide), white (rarely pink), petal-like bracts; petals yellow-green; fruits usually red ____ **C. florida**
   2. Flowers in open inflorescences, with no large bracts; petals cream to white; fruits white to blue or bluish.
      3. Leaf blades with abaxial (= beneath) surfaces with hairs curved upward from surface, these dense; fruits white; petioles 8–25 mm long; species widespread and common throughout the e 1/2 of the state, but less so in the far e portion of East TX ____ **C. drummondii**
      3. Leaf blades with abaxial surfaces nearly glabrous or with appressed hairs, these sparse to moderately dense; fruits either white OR variously blue; petioles 2–16 mm long; species restricted to the e portion of East TX.
         4. Fruits usually blue or at least bluish, but sometimes turning white; stems clustered from a single rootstock; plants without rhizomes; branchlets deep red, or green to bronze if shaded; 2nd year branches with lenticels not protruding or extruding tissue; inflorescences flat-topped, convex, or pyramidal; petioles 5–16 mm long ____ **C. foemina**
         4. Fruits usually white, occasionally pale blue; stems solitary; plants rhizomatous, often forming thickets or colonies; branchlets pinkish brown, turning green-maroon, and later graying; 2nd year branches with lenticels protruding or extruding tissue; inflorescences usually elongate, sometimes convex or pyramidal; petioles 2–8 mm long ____ **C. racemosa**

Cornus alternifolia L. f., (with alternate leaves), ALTERNATE-LEAVED DOGWOOD, PAGODA DOGWOOD, GREEN OSIER, PIGEON-BERRY. Shrub or small tree usually 2–6(–8) m tall, the trunk often branched basally; branches often in irregular whorls giving the appearance of tiers; twigs/branchlets usually green to yellowish green, sometimes reddish brown; leaves alternate but the internodes often short giving the appearance of the leaves being whorled, blades 4–12 cm long, with (4–)5–6 secondary veins per side, the lower surface yellow-green, appressed-hairy, the upper surface dark green, glabrate, turning yellow to red in fall; inflorescences flat-topped to hemispheric, 3–15 cm in diam., open, without large showy bracts; pedicels reddish in fruit; petals 2.5–4 mm long; drupes blue, globose, 5–8 mm in diam.; stone with a deep pit at apex (unique among our species). Rich hardwood-pine forest; known in TX only from Jasper Co. (*Keith 1104* BAYLU, SHST; Keith et al. 2017); w 1/2 of the U.S. and adjacent Canada w to MN, AR, and TX. Apr–Jun. [*Bothrocaryum alternifolium* (L. f.) Pojark., *C. alterna* Marshall, *Swida alternifolia* (L. f.) Small] Yatskievych (2006) noted that in some areas this species is widespread in understories in the ne U.S. and is "commonly cultivated for its interesting growth form, attractive foliage, and conspicuous inflorescences."

Cornus drummondii C.A. Mey., (for its discoverer, Thomas Drummond, 1780–1835, Scottish botanist and collector in North America), ROUGH-LEAF DOGWOOD. Shrub or bushy small tree to ca. 5 m tall; twigs/branchlets yellow-green abaxially, pink-maroon adaxially, sometimes reddish; leaf blades (4–)6–12 cm long, with 3–4(–5) secondary veins per side, the lower surface with dense hairs, these curved upwards from surface, pale green, the upper surface gray-green with hairs curved upwards or appressed; inflorescences flat-topped to convex, 3–8 cm in diam., the branches and pedicels pink or red; calyx lobes shorter than the tube; petals 3.5–5.5 mm long; drupes white, globose, 4–7.5 mm in diam. Stream banks, hillsides, woodlands, and thickets; e 1/2 of TX w to Clay Co. along the Red River and to Kerr, Medina, and Menard (Turner et al. 2003) cos. on the Edwards Plateau; also Gray and Hemphill (Turner et al. 2003) cos. in the n Panhandle; widespread in the e and c U.S. (but lacking from the e coastal states) w to SD, OK, and TX, also Ontario in Canada. May. [*C. asperifolia* Michx. var. *drummondii* (C.A. Mey.) J.M. Coult. & W.H. Evans, *C. priceae* Small, *Swida drummondii* (C.A. Mey.) Soják] During the winter, the naked often reddish twigs can be conspicuous.

Cornus florida L., (free flowering, producing abundant flowers), FLOWERING DOGWOOD, EASTERN DOGWOOD. Small tree with wide-spreading branches, to ca. 12 m tall; twigs/branchlets usually greenish but sometimes maroon or reddish; leaf blades with (4–)5–6(–7) secondary veins per side, the lower surface whitish, appressed-hairy, with tufts of erect hairs present in axils of secondary veins, the upper surface dark green, appressed-hairy, often scarlet in fall; inflorescence a dense cluster or head of ca. 20–30 sessile flowers subtended by 4 large bracts, overall resembling a large 4-petaled flower, the 4 bracts usually white but pink mutants are rarely seen; drupes red or sometimes blackening with age, ovoid to ellipsoid, 7–14(–18) mm long, spreading from each other. Wooded areas; common mainly in sandy woods in East TX to the e of the Blackland Prairie; also Cooke, Grayson, Fannin, Lamar, and Red River cos. in Red River drainage (Turner et al. 2003), Limestone Co. (J. Stanford, pers. comm.), and Dallas Co. (Dogwood Canyon Audubon Center in Cedar Hill); widespread in e U.S. w to e KS, OK, and TX; also disjunct to e Mexico. Late Mar–Apr. [*Benthamidia florida* (L.) Spach] The disjunct populations in the Sierra Madre Oriental of Mexico have variously been treated as a subspecies, a variety, or not distinguished taxonomically at all from *C. florida*. However, these Mexican populations may warrant recognition at the specific level as *C. urbiniana* Rose (Du et al. 2023b; Weakley, A.S., and Southeastern Flora Team 2024). This native understory species is widely cultivated as an ornamental, and is one of the showiest understory trees flowering in early spring in the e U.S.; the typically scarlet fall foliage is also very striking. Dogwood anthracnose, a fungal disease caused by a member of the genus *Discula*, *D. destructiva* Redlin, is currently damaging FLOWERING DOGWOODS in the e U.S. resulting in the death of millions of trees (Mielke & Daughtrey 1990; Daughtrey 1993; Murrell & Poindexter 2016; Weakley, A.S., and Southeastern Flora Team 2024). The wood is extremely shock resistant and was therefore widely used historically for applications including golf club heads, chisel handles, and the shuttles used in the textile industry (Peattie 1948). The fruits are bird-dispersed and have a high lipid content (up to 35% dry weight) which makes them particularly attractive to migratory birds which have high energy demands (Stiles 1984).

Cornus foemina Mill., (female, feminine), STIFF DOGWOOD, GRAY DOGWOOD. Shrubs to 8 m tall; twigs/branchlets deep red, or green to bronze if shaded, becoming grayish with age; leaf blades (3–)5–11 cm long, with 3–4(–5) secondary veins per side, with both surfaces sparsely appressed-hairy, glabrate by late summer, the lower surface usually pale green, the upper surface dark green, with midvein turning red or maroon; inflorescences flat-toped to convex or pyramidal, 2–8 cm in diam., the branches and pedicels greenish yellow, becoming red in fruit; drupes usually blue or at least bluish, but sometimes turning white, globose or somewhat ellipsoid. Swamps, wet woodlands, and stream banks; mainly far e part of East TX; other than East TX, the species is known in the state only from Wichita Co. in the ne Rolling Plains (USDA Plants Database 2019); widespread in the se US and further w to IL, MO, OK, and TX. Apr–Jun. [*C. stricta* Lam., *Swida foemina* (Mill.) Rydb., *S. stricta* (Lam.) Small] The Wichita Co. location is far to the west of other Texas occurrences of this species. As discussed by Kral (1966), MacRoberts & MacRoberts (1997), Diggs & O'Kennon (2003), and others, Texas has numerous unexpected plant occurrences which represent relicts of past climatic conditions. Many of these are probably "Ice Age holdover[s]" (MacRoberts & MacRoberts 1997) or distributional relicts from a colder and wetter period of the Pleistocene when vegetational areas were shifted considerably to the west. The decision whether to recognize this plant as *C. foemina* Mill. or *C. stricta* Lam. has varied by author. We are following Murrell and Poindexter (2016), who studied the genus throughout North America, in treating it as *Cornus foemina* (which is the oldest name dating from 1768). Lindelof et al. (2020) and Du et al. (2023a), both treating large numbers of *Cornus* species, also recognized it as *C. foemina.* Treating this species as *Swida foemina,* Weakley (2015) argued that the species' description is "very obscure," that "no type is available," that *C. foemina* is best treated as a nomen dubium, and suggests that the later name, *Swida stricta*, should instead be used. Weakley, A.S., and Southeastern Flora Team (2024), while treating the species in the genus *Cornus as C. foemina*, noted that "*C. foemina* should arguably be rejected as a nomen dubium."

Cornus racemosa Lam., (with flowers in racemes), GRAY DOGWOOD. Shrubs to 5 m tall; twigs/branchlets pinkish brown, turning green-maroon, and later graying; leaves opposite, blades usually 3.5–8(–10) cm long, with 3–4(–5) secondary veins per side, with both surfaces appressed pubescent, the lower surface pale green, the upper surface green, turning maroon in full sun, in fall dull purple with yellowish to pinkish highlights; inflorescences usually elongate, sometimes convex or pyramidal, the branches yellowish green, becoming bright red; drupes usually white, occasionally pale blue, globose or slightly ellipsoid, 4–8 mm in diam. Wooded, moist slopes, thickets, and streambanks; scattered in the e portion of East TX in Brazos (*P.A. Fryxell 3192*, "1980-07-06", NYBG, TX—det. B.L. Turner, photo on TORCH 2024); Cass (*D.S. Correll 30097*, 14 Sep 1964, NYBG—det. Z.W. Murrell, photo on TORCH 2024), Harris, and Newton (USDA Plants Database 2019) cos.; widespread in the e and c U.S. and adjacent Canada w to ND, NE, and TX. Apr–Jun. [*C. foemina* Mill. subsp. *racemosa* (Lam.) J.S. Wilson, *Swida racemosa* (Lam.) Moldenke] Turner et al. (2003) included this species in *C. foemina* and Yatskievych (2006) treated it as a subspecies of *C. foemina*; however, we are following most recent authors in recognizing it as a separate species (e.g., Murrell & Poindexter 2016; Weakley, A.S., and Southeastern Flora Team 2024).

Cornus obliqua Raf., (Latin *obliquus*, oblique, or lopsided, suggested by Bebeau (2015) to be in reference to the stone pit in the drupe sometimes being lopsided), SILKY DOGWOOD, PALE DOGWOOD. This species, which has its flowers in open inflorescences with no large bracts and blue fruits, differs from all East TX species (except the large-bracted *C. florida* and the alternate-leaves *C. alternifolia*) in its leaf blades having usually (4–)5–9 veins per side (versus 3–4(–5)). It is included as a note since it is found in two OK counties (Marshall and McCurtain) just across the Red River from East TX (USDA Plants Database 2019) and should be looked for in the state. The species is widespread in the ne and c U.S. w to ND, NE, and OK.

CRASSULACEAE J. St-Hil.

STONECROP, ORPINE, OR SUCCULENT FAMILY

Ours annual or perennial, often succulent herbs; leaves alternate, opposite or whorled, simple, entire, (or serrate in *Kalanchoë*), slightly to very thick and fleshy; stipules absent; flowers axillary or terminal, solitary or in 1-sided raceme-like or corymbiform cymes; sepals, petals, and pistils 4–5(–7); stamens 8–10(–14); ovaries superior; fruit a follicle.

Stylisma villosa [DEN]

Varronia podocephala [GOR]

Cornus alternifolia [SA3]

Cornus drummondii [GLE]

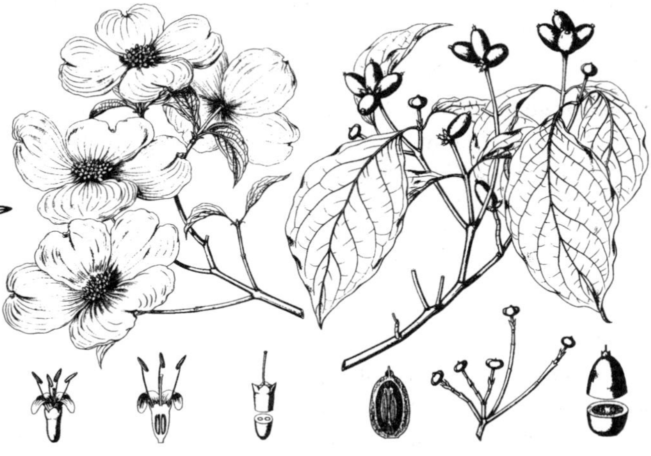

Cornus florida [SA3]

A medium-large (ca. 1,400 species in ca. 35 genera) (Moran 2009b) nearly cosmopolitan, especially s African family of usually succulent shrubs, herbs, and treelets with crassulacean acid metabolism (CAM photosynthesis—allows night absorption and storage of CO2 thereby reducing water loss through transpiration during the day). CAM photosynthesis accounts for their ability to survive in exposed, rocky habitats subject to long periods of drought along with surfaces covered in papillae, hairs, bristles, or wax. Vegetative reproduction common. Flower parts are usually in fives with some having the carpels fused basally. A number are cultivated as ornamentals including species of *Crassula* L. (e.g., JADE PLANT), *Echeveria* DC., *Kalanchoë* Adans. (MOTHER-OF-THOUSANDS), and *Sedum* L. (STONECROPS). Crassulaceae show similarities to Saxifragaceae and the two families are thought by some to be related (e.g., Heywood 1993). (subclass Rosidae—Cronquist; order Saxifragales—APG IV)

FAMILY RECOGNITION IN THE FIELD: succulent herbs with radially symmetrical flowers with separate petals and usually *separate* (at least above) carpels of the same number as the sepals. The somewhat similar Saxifragaceae are not succulent and typically have fused carpels fewer in number than sepals.

REFERENCES: Britton & Rose 1905; Spongberg 1978; Heywood 1993; Moran 2009b; Wunderlin & Hansen 2015.

1. Leaves serrate ____ **Kalanchoë**
1. Leaves entire.
   2. Flowers minute, much < 7 mm across, one per node, white to greenish white (rarely pinkish); leaves opposite; stamens as many as calyx segments ____ **Crassula**
   2. Flowers ca. 7–12 mm across, in terminal raceme-like cymes, yellow, white, pink, or lavender; leaves alternate; stamens twice as many as calyx segments ____ **Sedum**

CRASSULA L. PYGMYWEED

Ours decumbent or erect, aquatic or terrestrial annuals; leaves opposite, connate; flowers minute; flower parts in fours; petals ± twice as long as sepals; fruit a follicle, seeds 6–17 per follicle.

A nearly cosmopolitan, but especially tropical and s African genus of ca. 195 species of succulent herbs, shrubs, and treelets, including many cultivated ornamentals; many have xeromorphic features (Mabberley 2017). (Diminutive of Latin: *crassus*, thick, alluding to the succulent texture)

REFERENCES: Cody 1954; Tölkin 1977; Bywater & Wickens 1984; Moran 2009a.

1. Seeds papillate ____ **C. drummondii**
1. Seeds rugose.
   2. Follicles erect, oblong; sepals ovate to oblong, 0.5–1.5 mm long ____ **C. aquatica**
   2. Follicles ascending, obliquely lanceolate; sepals triangular-ovate to lanceolate, 0.4–0.7 mm long ____ **C. longipes**

Crassula aquatica (L.) Schöndland, (growing in or near water), WATER PIGMYWEED. Decumbent to erect, branched, aquatic (sometimes stranded) annual to 14 cm tall; stems reddish with age; leaf blades 2–6 mm long, oblanceolate to linear; inflorescence lax, flowers one per node; pedicels 0.5–20 mm long; flower parts in fours; sepals 0.5–1.5 mm long, ovate to oblong; petals 1–2 mm long, ovate to oblong; follicles erect, 6–17 seeds oblong, old follicles spreading, flat; seeds not papillate, dull, minutely rugose. Damp bare spots, pond margins, pathways, mudflats, sandy soils; in East TX scattered to the w of the Pineywoods; e Edwards Plateau; scattered in w 1/2 of U.S.; also Canada and Mexico; n Eurasia. Mar–Apr. [*Tillaea aquatica* L., *Tillaeastrum aquaticum* (L.) Britton] Native of the U.S. and Canada.

Crassula drummondii (Torr. & A. Gray) Fedde, (for Thomas Drummond, 1780–1835, Scottish botanist and collector in North America), SMALL-LEAF PYGMYWEED. Decumbent or erect annual to 10 cm tall; stems greenish red in age; leaf blades 1.5–3 mm long, narrowly lanceolate; inflorescence lax, flowers one per node; pedicels 0.5–3 mm long; flower parts in fours; sepals 0.4–0.6 mm long, triangular-ovate; petals 1.3–1.5 mm long, ovate; follicles ascending, 9–15 seeds, oblong; old follicles spreading, flat; seeds prickly-papillate, dull. Margins of ponds and streams; scattered in East TX; e Edwards Plateau and Gulf Prairies and Marshes; CO and SC; South America. Apr–May [*Tillaea drummondii* Torr. & A. Gray, *Tillaeastrum drummondii* (Torr. & A. Gray) Britton] Native of the U.S.

Crassula longipes (Rose) M. Bywater & Wickens, (long-stalked), SMALL-SEED PYGMYWEED. Erect or spreading (in water), branched annual to 25 cm long if aquatic, 5 cm long if terrestrial; stems yellow in age; leaf blades 2–5 mm long, lanceolate to oblanceolate, apex acute to obtuse; inflorescence lax; flowers one per node, flower parts in fours; pedicels 1–8 mm long; sepals 0.4–0.7 mm long, triangular-ovate, apex obtuse; petals 1.1–1.7 mm long, lanceolate; follicles ascending, 12–14 seeds; obliquely lanceoloid; old follicles spreading, flat; seeds oblong with rounded ends; not papillate, dull, rugose. Aquatic or in wet mud or sand; Bastop, Jackson, Jasper, Travis, Victoria, and Wharton (Kartesz 2022) cos.; Gulf Prairies and Marshes; AR, LA, and MS;

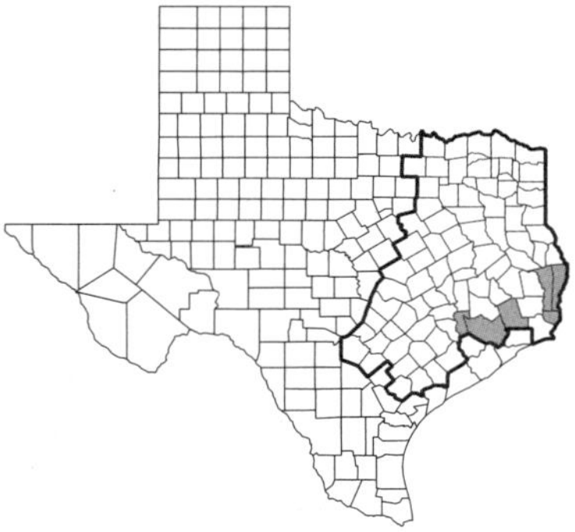
Stylisma aquatica

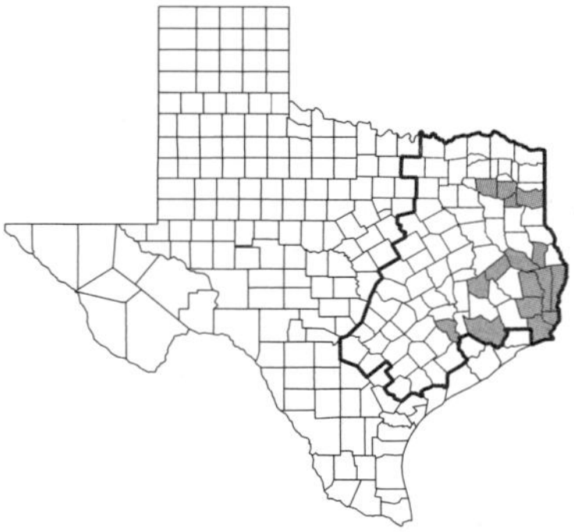
Stylisma humistrata

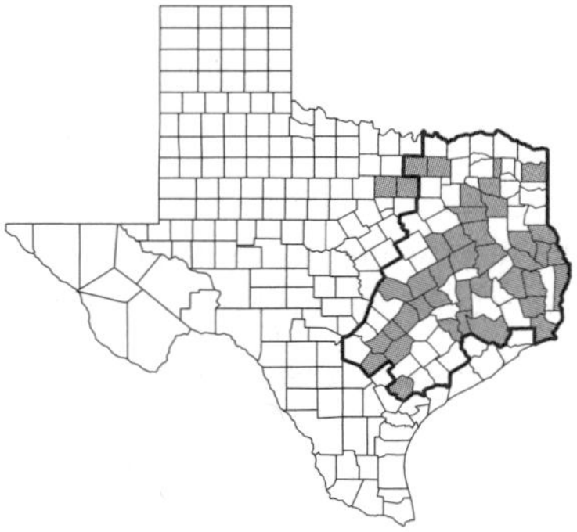
Stylisma pickeringii var. pattersonii

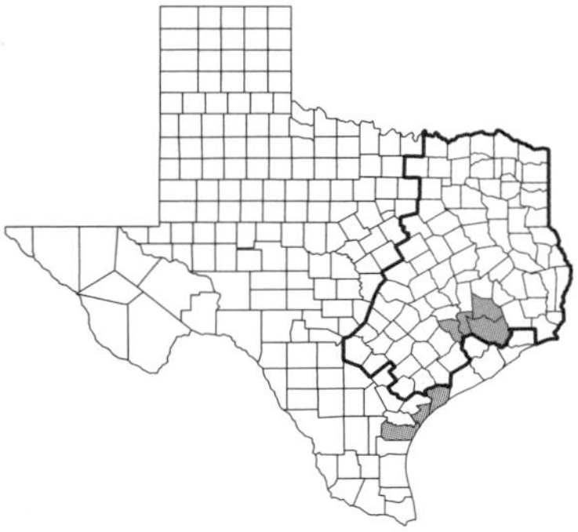
Stylisma villosa

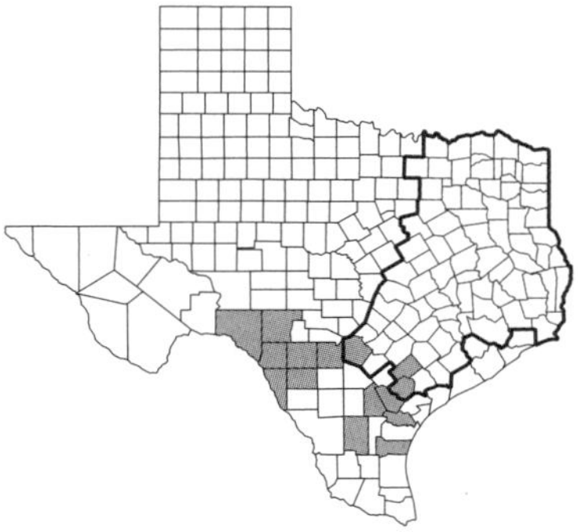
Varronia podocephala

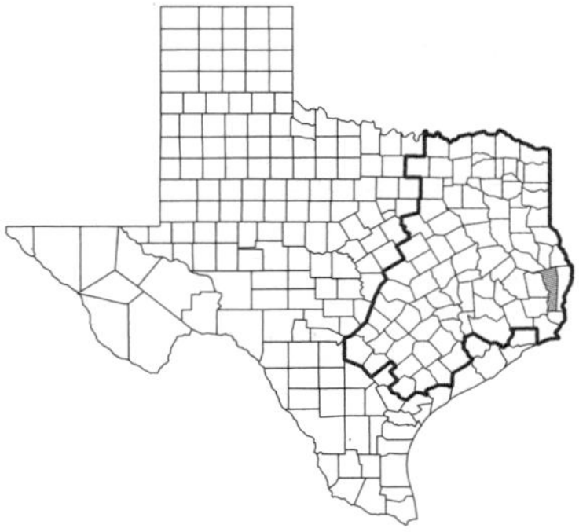
Cornus alternifolia

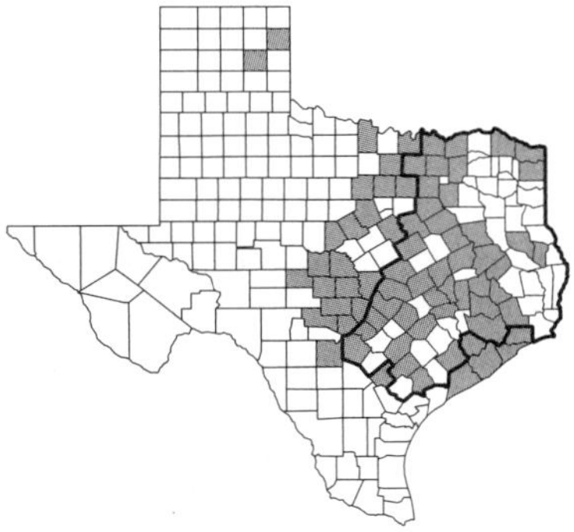
Cornus drummondii

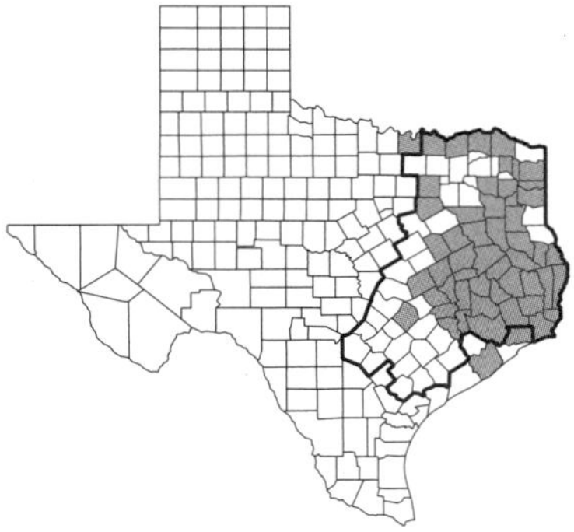
Cornus florida

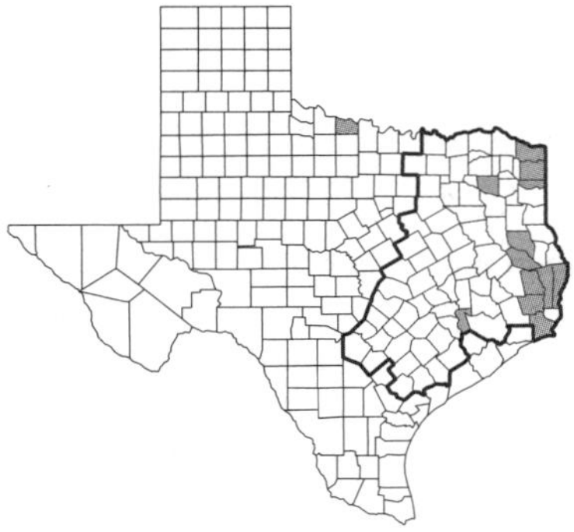
Cornus foemina

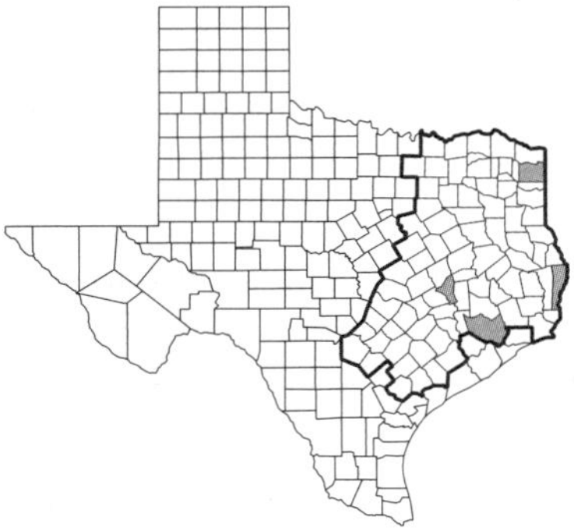
Cornus racemosa

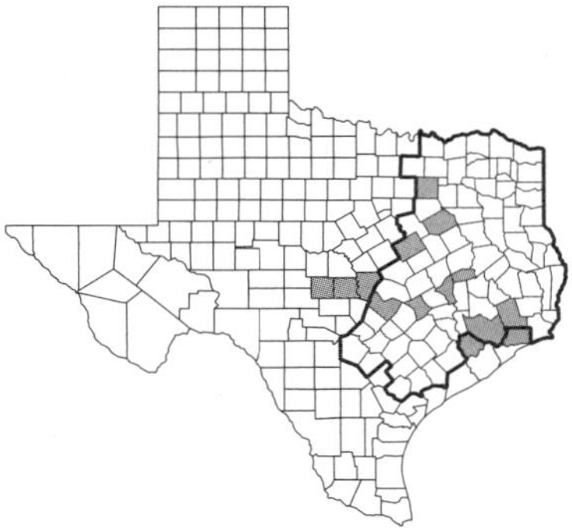
Crassula aquatica

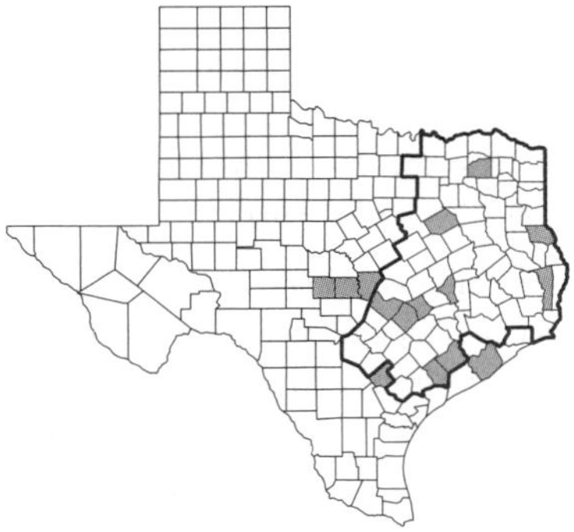
Crassula drummondii

also Mexico and South America. Mar–Apr. [*Tillaea longipes* (Rose) J. Meyrán, *Tillaeastrum drummondii* sensu (Torr. & A. Gray) Britton p.p., *T. longipes* Rose] Native of the U.S.

KALANCHOË Adans. WIDOW'S THRILL

A genus of ca. 125 species of North America, West Indies, South America, Africa, Asia, and Madagascar (Wunderlin at al. 2015). (Apparently derived from the Chinese name, Kalan Chauhuy, of one of the species) (Chernetskyy 2011)
REFERENCE: Wunderlin & Hansen 2015.

Kalanchoë delagoensis Eckl. & Zeyh., (species of Delagoa Bay, Mozambique), MOTHER-OF-THOUSANDS. Monocarpic (= flowering once and dying) herb to 2 m tall; stems erect, simple, glabrous; leaves gray-green in whorls of 3–4 or opposite, simple, sessile, subcylindric, grooved on the upper surface, 3–15 mm long, 3–6 mm wide, apex with 3–9 conic teeth with spoon-shaped bulbils; inflorescence in corymbiform cymes; flowers pendent; calyx 8–16 mm long, tube 3–6 mm long, lobes triangular 5–10 mm long; petals 2.5–4 cm long, orange to scarlet. Disturbed sites; in East TX only recorded for Harris Co. (Turner et al. 2003); Gulf Prairies and Marshes; FL and TX; West Indies; Africa; and the Pacific Islands. Winter [*Bryophyllum delagoense* (Eckl. & Zeyh.) Schinz] Native of Africa.

SEDUM L. STONECROP, ORPINE

Ours glabrous annuals and perennials; leaves alternate, crowded, terete or nearly so; sepals, petals, and pistils 4–5(–7); stamens 8–10(–14); fruit a dry, many-seeded follicle with tapered beak. (Classical Latin name for several succulent plants, from *sedo*, to sit, alluding to the manner in which many species affix themselves to rocks or walls)

A genus of ca. 420–470 species (the largest genus in the family) of n temperate areas, tropical mountains, Madagascar, and Mexico (Ohba 2009); it includes many cultivated, ornamental, succulent herbs and shrublets.
REFERENCES: Clausen 1975; Nesom & Turner 1995; White et al. 1998.

1. Petals yellow; flowers ca. 7 mm across; follicles widely divergent in fruit ____ **S. nuttallii**
1. Petals white to pink or lavender; flowers 8–12 mm across; follicles ascending in fruit ____ **S. pulchellum**

Sedum nuttallii Torr. & E. James ex Eaton, (for Sir Thomas Nuttall, 1786–1859, English-American botanist), YELLOW STONECROP. Annual, not bearing rosettes; glabrous pale green annual to 13(–20) cm tall; leaves linear-oblong, nearly as thick as wide; inflorescences with 2–5 branches, the flowers remote. Loose sandy or eroding limestone slopes, shallow soils, or on sandstone outcrops; w 1/2 of East TX; West Cross Timbers and Edwards Plateau; AR, CO, LA, MO, OK, and TX. May–Jun. [*S. nuttallianum* of authors, not Raf.] Native of the U.S.

Sedum pulchellum Michx., (beautiful), TEXAS STONECROP, ROCK-MOSS, WIDOW'S-CROSS. Annual or perennial by formation of side shoots, sometimes bearing rosettes; stems ascending, 10–30 cm tall; leaves linear, terete; inflorescences with 2–7 branches, the flowers closely spaced. Moist rocky areas; Bell Co. (Fort Hood—Sanchez 1997), Fannin, Grayson, Lamar, Limestone (White et al. 1998), Sabine, San Augustine, and Travis (Kartesz 2022) cos.; Edwards Plateau; c U.S. from OH s to GA and w to KS and TX; also West Indies; Africa; and the Pacific Islands. Winter. Native of Africa.

CUCURBITACEAE Juss.

SQUASH, GOURD OR CUCUMBER FAMILY

Herbaceous annual or perennial vines; monoecious or dioecious; stems with tendrils, trailing or climbing; leaves usually palmately lobed or compound; flowers unisexual, radially symmetrical; corollas 5-lobed, yellow to greenish yellow, yellow-orange, or white; stamens 3 or 5, free or variously united; ovary inferior, usually 3-carpellate; fruit in ours a berry (sometimes with a firm rind and then called a pepo).

A medium-sized family (ca. 990 species in 97 genera—Mabberley 2017) of mainly tropical and warm area (few temperate) vines containing numerous economically important food plants including SQUASHES, PUMPKINS, AND MELONS (see following treatments); the fruits and/or flowers of a number of species are edible; other species provide ornamental and useful gourds. (subclass Dilleniidae—Cronquist; order Cucurbitales—APG IV)
FAMILY RECOGNITION IN THE FIELD: herbaceous vines with tendrils and alternate, palmately-veined, usually lobed or compound, often conspicuous leaves; flowers usually yellow to white, unisexual; ovary inferior.
REFERENCES: Jeffrey 1975, 1990; Bates et al. 1990; Lira et al. 1997 [1998]; Nesom 2011b, 2015b.

Cornus foemina [BB2]

Cornus racemosa [STR]

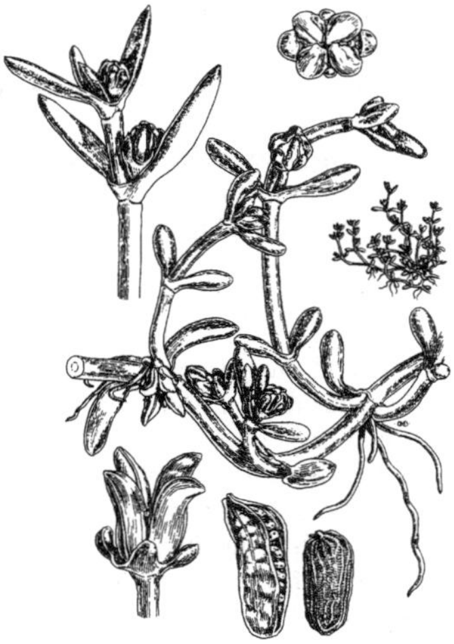

Crassula aquatica [MAS]

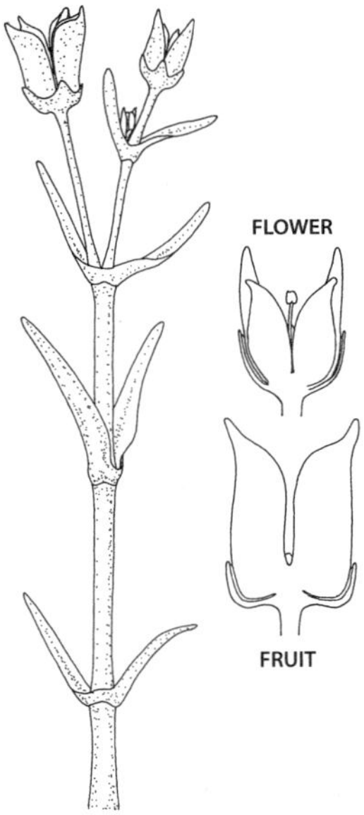

Crassula drummondii [GLE]

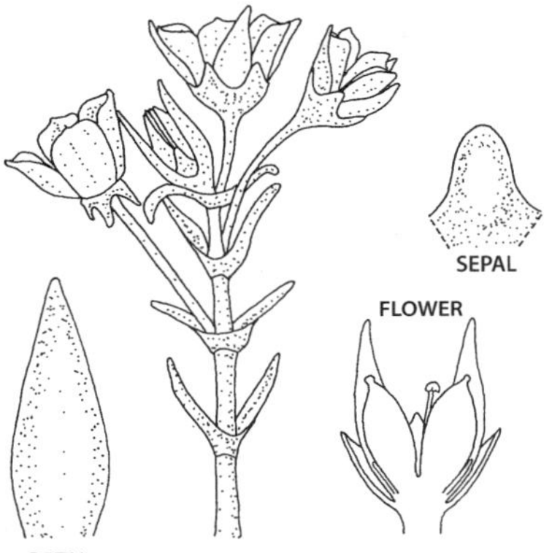

Crassula longipes [KEB]

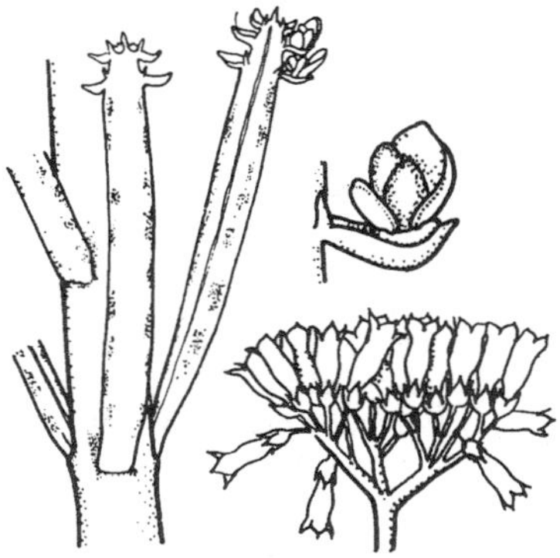

Kalanchoë delagoensis [NSW]

Sedum nuttallii [BB2]

Sedum pulchellum [BB2]

1. Corollas ≥ 5 cm long or across; fruits ≥ 5 cm long or broad (sometimes shorter in *Coccinia*), smooth.
   2. Corollas yellow to golden-yellow; stems not viscid (= sticky); tendrils 2–7-branched; leaves without small, disk--shaped glands on upper petiole or base of blade (either side of midrib) ______ **Cucurbita**
   2. Corollas white to cream; stems viscid (= sticky) or not so; tendrils unbranched or 2-branched; leaves with small disk-shaped glands on upper petiole or base of blade (either side of midrib).
      3. Plants annual, monoecious; stems sticky; tendrils 2-branched; leaf with small disk-shaped gland on each side of the base of the leaf blade at junction with petiole; fruits variously shaped (bottle, dumbbell, club, crook-necked) ______ **Lagenaria**
      3. Plants perennial, dioecious; stems not sticky; tendrils unbranched; leaf often with 3–7 small disk-shaped glands at base of blade on each side of midrib; fruits broadly cylindric to ellipsoid-cylindric ______ **Coccinia**
1. Corollas ≤ 5 cm long or across (often much less); fruits ≤ 5 cm long or broad (except fruit much larger in *Cucumis melo*, MUSKMELON, HONEYDEW, etc., and *Citrullus lanatus*, WATERMELON), smooth or prickly.
   4. Stems and petioles glabrous.
      5. Flowers greenish yellow; fruits unarmed, orange to red when ripe; leaves nearly unlobed to deeply 3–5-lobed but not divided into separate leaflets ______ **Ibervillea**
      5. Flowers white; fruits conspicuously armed with long slender spines; leaves divided into 3–7 distinct leaflets OR fruits smooth; leaves merely lobed or almost entire.
         6. Plants dioecious, perennial; fruits smooth; staminate flowers solitary; tendrils unbranched ______ **Coccinia**
         6. Plants monoecious, annual; fruits armed; staminate flowers in racemes or narrow panicles; tendrils unbranched or 2-branched ______ **Cyclanthera**
   4. Stems and/or petioles variously pubescent (use lens).
      7. Fruits smooth, reddish, 10–16 mm long, with 3–6 seeds; corollas greenish white ______ **Cayaponia**
      7. Fruits prickly or smooth, not reddish, small to very large 1–60+ cm long), with seeds 1–very numerous; corollas yellow to greenish yellow to white (if white to greenish white, then fruits 1-seeded).
         8. Corollas about 2–4 cm across, yellow; tendrils 2–3-branched; fruits 12 to 60 cm or more in length, smooth; seeds obovoid, flat, often 7–15 mm long; leaves very deeply lobed, the large central lobe itself ± pinnately lobed ______ **Citrullus**
         8. Corollas 2.5 cm or less across, white to yellow or greenish yellow; tendrils usually unbranched (if branched then fruits prickly); fruits ≤ 5 cm long (except longer in *Cucumis melo*), smooth or prickly; seeds variously shaped, 10 mm or less long (to 12 mm in *Cucumis melo*); leaves usually angled or shallowly lobed, or if deeply lobed, then the central lobe itself not pinnately lobed.
            9. Tendrils branched; fruits ovoid, 1–1.5 cm long, with a single seed, usually with prickly bristles; corollas white to greenish white ______ **Sicyos**
            9. Tendrils unbranched; fruits various shapes, 1.5–20+ cm long, many-seeded, smooth or variously warty, scaly, netted, or prickly; corollas yellow or greenish yellow.
               10. Corollas < 8 mm across; leaves mostly 5-lobed (the basal 2 lobes sometimes small); fruits green often with lighter patterning (like a baby watermelon), turning purplish black when ripe, smooth, 1.3–2 cm long ______ **Melothria**
               10. Corollas > 10 mm across; leaves either unlobed (but angled) or 3–5-lobed; fruits never purplish black, either prickly or ca. 5–20+ cm long (if ≤ 5 cm long, then stems hispid) ______ **Cucumis**

CAYAPONIA Silva Manso

A mainly tropical and subtropical New World genus of ca. 60 species (Nesom 2015b). (Caiapó, a reference to the river and/or native tribe of Amazonian Brazil)

REFERENCE: Nesom 2015b.

Cayaponia quinqueloba (Raf.) Shinners, (five-lobed). Monoecious, rhizomatous, climbing perennial; leaves long-petiolate, blades usually 3–5-lobed, 4–11(–15) cm long; tendrils unbranched or 2-branched; pistillate flowers and fruits on short stalks to 4 mm long or nearly sessile; corollas 6–10 mm long, white to cream; stamens 3; ovary 3-carpellate with 1 or 2 ovules in each cell; fruits reddish, ovoid to ellipsoid, somewhat fleshy, 10–16 mm long; seeds 3–12. River bottoms and along streams; Hardin, Harris, Jasper, Liberty, Newton, Polk, Trinity, and Van Zandt (Kartesz 2022) cos.; otherwise in TX known only from Atascosa Co. in n South TX Plains; se U.S. Jun–Aug. Native of the U.S.

CITRULLUS Shrad. WATERMELON

Annual, monoecious vines; tendrils 2–3-branched; leaves deeply 3–5-lobed, lobes pinnately lobed to shallowly lobed; staminate and pistillate flowers solitary from different axils; fruit variously colored, globose to subglobose to ellipsoid.

A genus of 6 species native to tropical and s Africa and probably also Asia (Mabberley 2017). (Diminutive of *Citrus*, from the supposed resemblance of the fruits to those of that genus of Rutaceae)

REFERENCES: Bailey 1930b; Hara 1969a; Nesom 2015b.

1. Leaf blades ovate to lanceolate-ovate or ovate-triangular in outline, mostly 8–20 cm long; fruits globose to oblong-ellipsoid, 12–35+ cm in diam.; rind tough, not durable (not persisting intact after maturity); flesh juicy, red to orange to yellow, sweet ____ **C. lanatus**
1. Leaf blades ovate in outline, 3–8 cm long; fruit globose to globose-ovoid, 15–25 cm in diam.; rind tough, durable (persisting intact after maturity); flesh dry, whitish, bitter ____ **C. amarus**

Citrullus amarus Schrad., (bitter), CITRON, PRESERVING MELON. Similar to *C. lanatus*, but differentiated by characters in the key. Sandy fields, pinelands, cotton fields, roadsides, railroad banks; Henderson and Washington (Kartesz 2022) cos.; scattered to w TX; AL, AZ, CA, FL, NM, NV, and TX. Jun–Oct(–Nov). [*C. caffer* Schrad., *C. lanatus* (Thumb.) Matsum. & Nakai var. *caffer* (Schrad.) Mansf. ex Fursa, *C. lanatus* var. *citroides* (L.H. Bailey) Mansf.] Native of Africa. The rind and flesh are candied or pickled and used in preserves and jellies (Nesom 2015b).

Citrullus lanatus (Thunb.) Matsum. & Nakai, (woolly), WATERMELON, SANDIA. Monoecious annual with tendrils 2–3-branched; stems long-trailing, prostrate, blades mostly 8–20 cm long, once or twice deeply lobed or dissected; ovary densely woolly; fruits green or mottled or striped, 12–35+ cm in diam., rind tough, not durable, the flesh sweet, succulent, edible, usually red but can be orange, yellow, or white; seeds numerous, ovoid or oblong-ovoid, usually black, rarely other colors. Commonly cultivated and found as a waif around picnic areas, trash heaps, roadsides, and waste areas; scattered in East TX; scattered mainly in e 1/2 of TX, also Trans-Pecos; sw and e 1/2 of the U.S. May–Nov. Native of Africa. [*C. citrullus* (L.) H. Karst [nom. illeg.], *C. vulgaris* Schrad.] This species has long been in cultivation; it was known in Egypt in the Bronze Age and probably much earlier (Zohary 1982); seeds were found in the tomb of Tutankhamun (Hepper 1990). Seedless fruits are triploid plants produced as hybrids between normal diploids and artificially produced tetraploids. Triploids have sterile pollen and because pollination is required to induce fruit set and enlargement, they must be interplanted with pollinizer diploids (Nesom 2015b).

COCCINIA White & Arn. IVY GOURD

A genus of ca. 30 species of s and se Asia, Africa, introduced in Pacific Islands (Nesom 2015b). (Latin: *coccineus*, scarlet, alluding to bright red mature fruit of *Coccinia grandis*)
REFERENCES: Nesom 2015b; Wunderlin & Hansen 2015.

Coccinia grandis (L.) Voigt, (large), IVY GOURD. Leaves unlobed, angular or 5-lobed; flowers connate ca. 1/2 their length, white, often with green veins; fruits 2.5 to 6 cm long, usually green with white streaks, bright red at maturity; seeds 30–50(–120). Trash dumps, fence rows, cypress swamps; in East TX known only from Goliad (*Holmes 8541* BAYLU), and Washington (*Holmes 7313* BAYLU) cos.; also Cameron Co. (*Cory s.n.*, 1940, GH) in s TX, with the note: yard of Ideal Cafe; FL, MO, and TX. May–Nov. [*Bryonia grandis* L., *C. cordifolia* of authors, not (L.) Cogn.] Native of Asia. The shoot tips and immature fruits are used in Asian and Indian cooking; long-range dispersal is often the result of introduction by humans (Nesom 2015b). Sometimes misidentified as *C. cordifolia* (L.) Cogn. Native of the Old World.

CUCUMIS L. MELON

Annual, trailing or climbing, pubescent vines; monoecious, dioecious, or andromonoecious (= staminate and bisexual flowers on the same plant); tendrils unbranched; leaf blades entire or somewhat lobed; flowers usually short-stalked and hidden by foliage; corollas yellow, bell-shaped to rotate, deeply 5-parted; fruits fleshy, prickly or smooth; seeds numerous.

Mostly an Old World tropical genus of ca. 55 species, mainly of tropical and s Africa and Asia (Nesom 2015b); cultivated since early times. It includes *C. sativus* L. (CUCUMBER). (Name from classical Latin word for cucumber)
REFERENCES: Kirkbride 1993; Nesom 2015b.

1. Leaf blades usually with 3–5 prominent lobes; ovaries and fruits prickly ____ **C. anguria** var. **longaculeatus**
1. Leaf blades usually nearly entire to slightly lobed, but usually not distinctly lobed; ovaries and fruits not prickly ____ **C. melo**

Cucumis anguria L. var. **longaculeatus** J.H. Kirkbr., (sp.: Greek name for cucumber; var.: with long prickles), BUR GHERKIN, WEST INDIAN GHERKIN. Plant dioecious; leaf blades 3–12(–15) cm long, 3–5-lobed, with sinuses rounded and lobes obtuse, sometimes nearly unlobed; fruits oval or oblong, furrowed ca. 5 cm long, on crooked peduncles, prickly, prickles 4–10(–15) mm long. Occasionally cultivated, escapes; in East TX known only from McLennan Co. (Kartesz 2022), also Cameron Co. in s TX (Kartesz 2022); scattered in the U.S., mostly FL. Jul–Oct. Native of Africa.

Cucumis melo L., (MELON). Plant monoecious or andromonoecious (= both staminate and bisexual flowers on the same plant); leaf blades 3–5-lobed or not so, 2–14(–26) cm long, margin entire to weakly serrate; staminate corolla tube 0.8–2 mm long; pistillate corolla tube 1–1.6(–3) mm long; fruit cylindric, ellipsoid to subglobose. As with many species of Cucurbitaceae in

North America, *C. melo* has long been cultivated resulting in many cultivated forms and gone under numerous names. Nesom (2011b) presented a taxonomic structure for wild and domesticated Cucurbitaceae. He suggested *C. melo* in Texas was best treated as comprising two subspecies, *C. melo* subsp. *melo* and *C. melo* subsp. *agrestis* (Naudin) Pangalo. *Cucumis melo* subsp. *melo* is the cultivated species with many known varieties such as cantaloupe, muskmelon, honeydew and others. Various varieties may be found as escapes but rarely persisting.

1. Ovaries with spreading hairs; stems glabrous or sparsely villous; fruits 10–20+ cm in diam. ____ subsp. **melo**
1. Ovaries with appressed silky hairs; stems with retrorsely hispid hairs; fruits 2.5–5 cm in diam. ____ subsp. **agrestis**

subsp. **agrestis** (Naudin) Pangalo, (growing in the fields). Plant monoecious or andromonoecious; fruit bitter; broadly ellipsoid to subglobose, rind smooth, flesh whitish. Often invading cultivated peanut fields or cotton fields, roadsides, fencerows; scattered in East TX; Edwards Plateau and South TX Plains; widely scattered in se U.S. and CA. Jul–Oct(–Nov). [*C. melo* var. *agrestis* Naudin] Native of Asia. According to Decker-Walters et al. (2002), the entity in se U.S. (including TX) is best called *C. melo* var. *texanus* Naudin. They found it to be morphologically and molecularly distinct from other varieties in the U.S. and hypothesized it was introduced intentionally or unintentionally by humans early on during post-Columbian times and before Naudin's description of the plant in 1859. However, they present other possible hypotheses that point toward a pre-Columbian introduction by other means. This idea is contrary to a long-held idea that the wild populations of the entity were escaped forms of cultivated *C. melo* var. *chito* (Morren) Naudin, *C. melo* var. *dudaim* (L.) Naudin, or less commonly, *C. melo* var. *agrestis* Naudin.

subsp. **melo,** CANTELOUPE, MUSKMELON, HOENYDEW. Plants andromonoecious. Leaf blades to 13 cm wide, rounded at apex; fruits mostly subglobose or ellipsoid with musky fragrance, the flesh usually yellowish or orangish to green; seeds slender, white. Commonly cultivated and found as a waif around picnic areas, trash heaps, roadsides, and waste areas; scattered in East TX; mainly e 1/2 of TX; throughout LA and widely scattered in the e 1/2 of the U.S and in far sw U.S. Jul–Oct. Native of Asia.

Cucumis sativus L., (cultivated), is the cultivated CUCUMBER. It has ± prickly fruit, corolla tube of staminate flowers 3.5–5(–6) mm long, corolla tube of pistillate flowers 3.5–6.5 mm long, and leaves angled or shallowly 3-lobed, with a pointed middle lobe and with ± acute sinuses. Included base on record in De Soto Par. in LA (Kartesz 2022) on the e border of East TX; widely scattered in mostly several e U.S. states. Native to s Asia.

CUCURBITA L. GOURD, SQUASH

Annuals or perennials with long running stems; leaf blades large, scabrous, to 40 cm long; flowers yellow to orange, solitary in axils; staminate flowers with long pedicels; fruits smooth, fleshy, indehiscent, with a hard rind; seeds usually numerous.

A genus of 14–22 species (Nesom 2015b) of tropical and warm parts of the Americas, including a number of widely cultivated food plants. The genus was important in pre-Columbian Mesoamerica as part of the maize/beans/squash agricultural system. (Classical Latin name for a type of gourd)

REFERENCES: Bailey 1929, 1930b; Whitaker & Bemis 1964, 1975; Rhodes et al. 1968; Bemis et al. 1970, 1978; Heiser 1979; Decker & Wilson 1986, 1987; Andres 1987; Decker 1988; Decker-Walters 1990; Nee 1990; Nesom 2015b.

1. Plants perennial; leaf blades usually longer than wide, triangular-ovate, unlobed or 2-lobed, extremely ill-smelling, especially when bruised ____ **C. foetidissima**
1. Plants annual; leaf blades usually nearly as wide as long OR wider than long, broadly ovate, kidney-shaped, triangular, or suborbiculate, angled to deeply (3–)5–7-lobed; base cordate; not noticeably ill-smelling.
   2. Fruiting peduncle abruptly expanding at point of attachment; tendrils 3–5-branched 1.5–8 cm above base; leaf blade 5–25 cm long ____ **C. moschata**
   2. Fruiting peduncle only slightly expanding at point of attachment; tendrils 2–7-branched 2–3 cm above base; leaf blade 4–15(–18) cm long ____ **C. melopepo**

Cucurbita foetidissima Kunth, (very fetid, very bad-smelling), BUFFALO GOURD, STINKING GOURD, FOETID GOURD, CALABAZILLA, CHILICOTE, CALABACILLA LOCA, CHILICOYOTE. Ill-smelling perennial; stems with leaves arranged in 2 ranks, widely running, to 6 m or more long, tendrils 3–7-branched 3–6 cm above base; leaf blades 12–30(–40) cm long, irregularly and finely toothed, unlobed or shallowly 2-lobed, thickish, often gray-green, rough to the touch; corollas to 10 cm long; fruits globose, green with lighter stripes, lemon-yellow at maturity, to 10 cm broad. Disturbed areas; mostly on the w border of East TX; Edwards Plateau, Trans-Pecos, and the Panhandle; mostly c U.S. from NE s to TX and w to CA, also the Midwest. Apr–Jul. Native of sw U.S. and n Mexico. The seeds were eaten by Native Americans; the plant was also used as soap (contains a saponin) and medicinally (Heiser 1979). During frontier days, the leaves and roots were used as a purgative (Crosswhite 1980). This species has been investigated as a source of starch and oil (Heiser 1993).

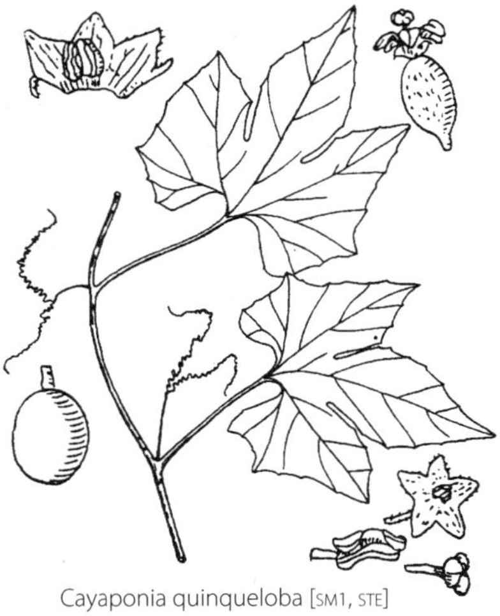
Cayaponia quinqueloba [SM1, STE]

Citrullus lanatus [NIC]

Coccinia grandis [HO5]

Citrullus amarus [ELB]

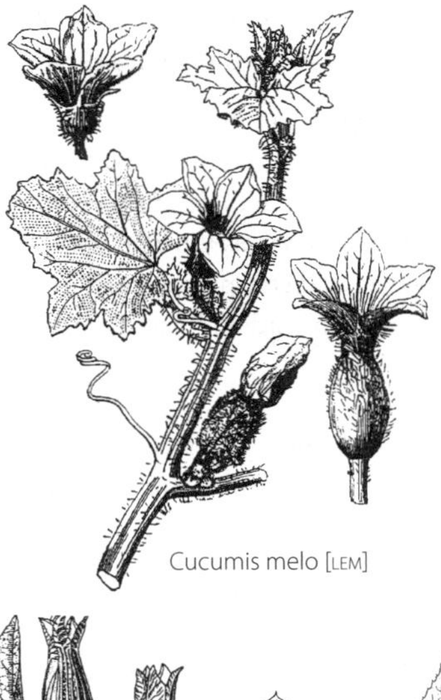
Cucumis melo [LEM]

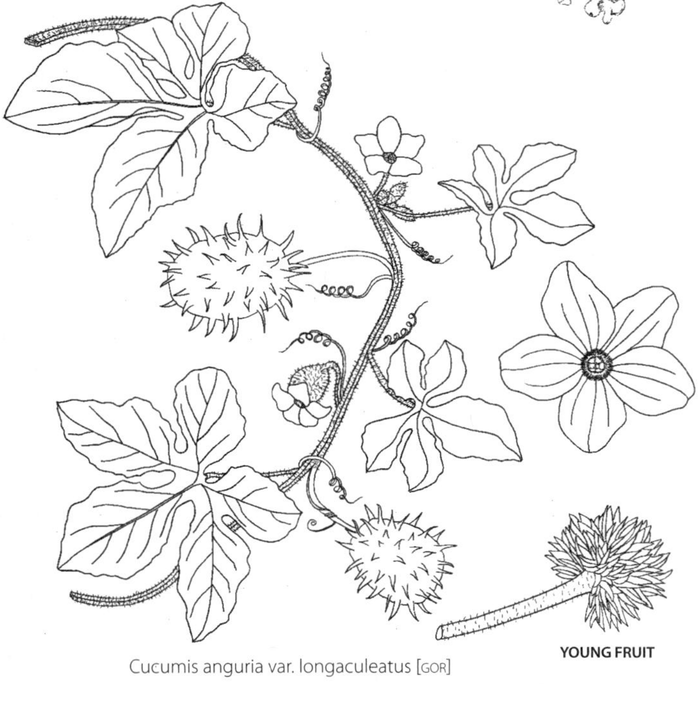

Cucumis anguria var. longaculeatus [GOR]

Cucurbita foetidissima [BB2, MUN]

Cucurbita melopepo L., (GOURD). Creeping or climbing annual to 3 m, sometimes bushy in domesticated forms, tendrils 2–7-branched; leaf blades broadly ovate-cordate to triangular-cordate or suborbicular, (3–)5–7-lobed; hypanthium 8–12 mm long; corolla 4–10 cm long.

1. Pepos solid or striped or multiple colors, including yellow and orange; rind smooth, ribbed, or warty; flesh not bitter ____ subsp. **melopepo**
1. Pepos usually green-and-white-striped, sometimes yellow; rind smooth; flesh almost always bitter ____ subsp. **texana**

subsp. **melopepo,** PATTYPAN, CROOK-NECK, ACORN, SUMMER, SPAGHETTI SQUASH, GOURDS. Various cultivated forms and escaping; fruits solid or striped, multicolors; rind smooth, ribbed or warty; flesh not bitter. Trash heaps, vacant lots, roadsides, disturbed sites; possibly in East TX as an escape; otherwise in TX known only from Hamilton Co. (Kartesz 2022) just w of East TX; AR, CA, GA, IL, KY, LA, MO, MS, TX, and WV. Jun–Oct. [*C. ovifera* L., *C. pepo* var. *ovifera* (L.) Alef.] Nesom (2015b) further divided subsp. *melopepo* into var. *melopepo*. No county level distribution map is provided.

subsp. **texana** (Scheele) G.L. Nesom, (of Texas), TEXAS GOURD. Leaves, at least some, distinctly lobed, relatively thin, green, the margins sharply serrate; corollas to 7 cm long; fruits obovoid to subglobose, to 9 cm long and 6 cm broad, bitter. Along tributaries, lake shores, low woods. Jul–Oct. Nesom (2015b) further divided subsp. *texana* into vars. *ozarkana* G.L. Nesom (confined to the central Mississippi valley and the Ozark Plateau), and var. *texana* G.L. Nesom (confined to TX?). No county level distribution map provided. A key to both varieties is provided below.

1. Pepos usually green-and-white-striped, sometimes yellow ____ var. **texana**
1. Pepos usually solid ivory, sometimes green-and-white-stripped ____ var. *ozarkana*

var. **texana** G.L. Nesom, [*C. pepo* L. var. *texana* (Scheele) D.S. Decker] Though Nesom (2015b) treated this entity as endemic to TX, Kartesz (2022) and Weakley, A.S., and Southeastern Flora Team (2024) also mapped it for AL, AR, LA, MS, NM, and OK. No county level distribution map provided. ?

Cucurbita moschata Duchesne, (musky), WINTER CROOK-NECK SQUASH, BUTTERNUT SQUASH. Creeping or climbing annual to 10+ m long, tendrils 3–5-branched; leaf blade broadly ovate to suborbiculate to reniform, broader than long, shallowly 3–5(–7)-lobed; hypanthium 5–8 mm long; corolla 5–7 cm long. Cultivated and escaping; oak-pine woods, abandoned cultivated fields, roadsides; in TX known only from Dallas and Smith (Nesom 2011a) cos.; AL, AZ, CA, FL, GA, IL, LA, NM, OH, PA, and TX; often only recorded for 1 county in these states (Kartesz 2022). May–Oct. Native of South America.

Cucurbita pepo L., (fruit like a pumpkin or gourd), PUMPKINS and ZUCCHINI. Brought into cultivation by Native Americans. Vegetatively similar to *C. melopepo* except stems hispid with persistent swollen-based hairs and understory of hispid to hirsute pubescence, whereas *C. melopepo* is mostly densely hirsute (swollen-based hairs sometimes present); leaf blades in *C. pepo* 20–30 cm long, whereas leaf blades 4–15(–18) cm long in *C. melopepo*; fruit very variable (in the different cultivated forms), but usually larger and conspicuously differently shaped and colored from *C. melopepo*. Included as note based on a record in Bryan Co., AR on the n border of East TX; mostly e U.S. from NY s to SC and w to MI and LA. Jun–Oct.

CYCLANTHERA Schrad.

Annual monoecious climbers; tendrils branched or unbranched; leaves 3-foliolate; staminate and pistillate flowers usually both from the same axils; staminate flowers numerous, in racemes or panicles; pistillate flowers solitary; fruits with spines; dehiscence explosive.

A mainly Neotropical genus of ca. 35 species (Nesom 2015b). (Greek: *cyclos*, circle, and *anthera*, anther, possibly in reference to the united stamens)

REFERENCE: Nesom 2014.

1. Floriferous portion of staminate inflorescence 0.3–2 cm long, (1–)8–27-flowered, usually without lateral branches; fruiting peduncles 10–30 mm long ____ **C. naudiniana**
1. Floriferous portion of staminate inflorescence 3–7 cm long, (50–)65–90(–130)-flowered, usually with lateral branches; fruiting peduncles 4–6 mm long ____ **C. dissecta**

Cyclanthera dissecta (Torr. & A. Gray) Arn., (dissected, deeply cut), CUT-LEAF CYCLANTHERA, BUR-CUCUMBER. Plant with slender stems to 3 m or more long; tendrils 2-branched, rarely unbranched (shortest tendrils); leaves 3-foliolate, lateral pair of leaflets deeply divided, terminal leaflet 3–4.5 cm long, coarsely serrate to shallowly or deeply lobed; staminate corolla 3.5–4.9 mm in diam.; pistillate corollas 5-parted; fruiting peduncles 4–6 mm long, fruit ovoid, armed with spines 3–4 mm long. Riparian woods

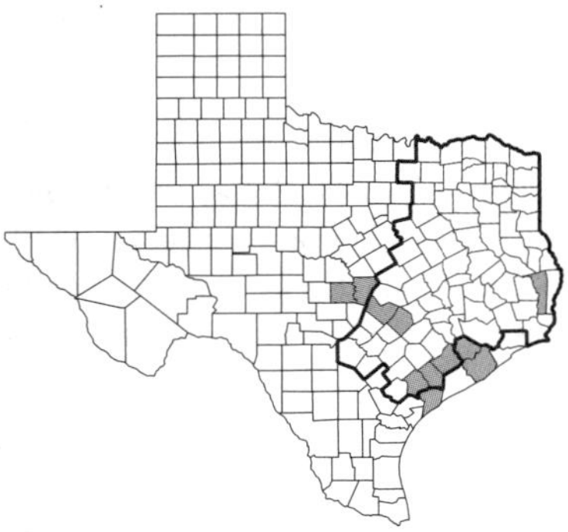

Crassula longipes

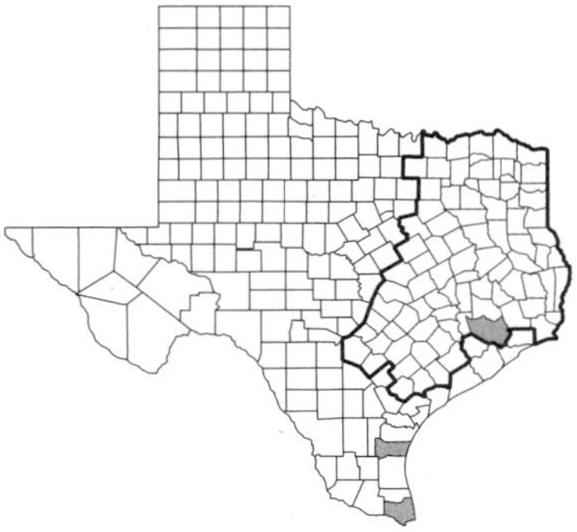

Kalanchoë delagoensis

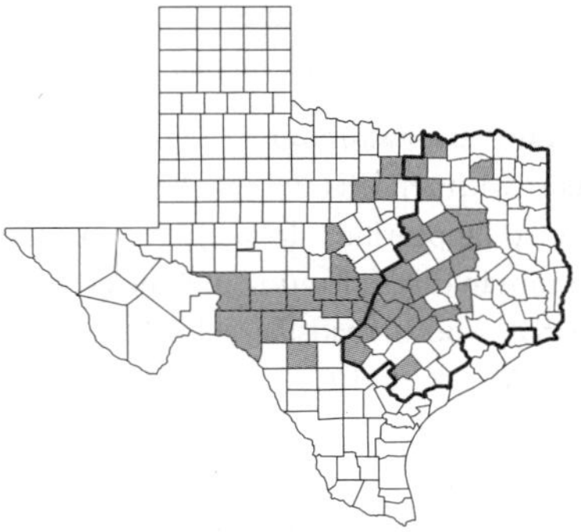

Sedum nuttallii

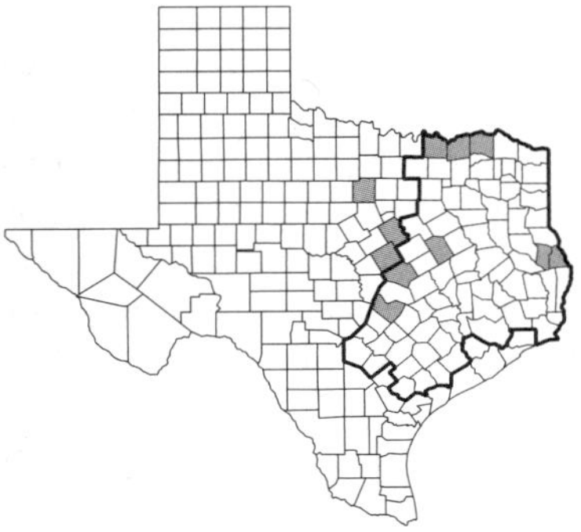

Sedum pulchellum

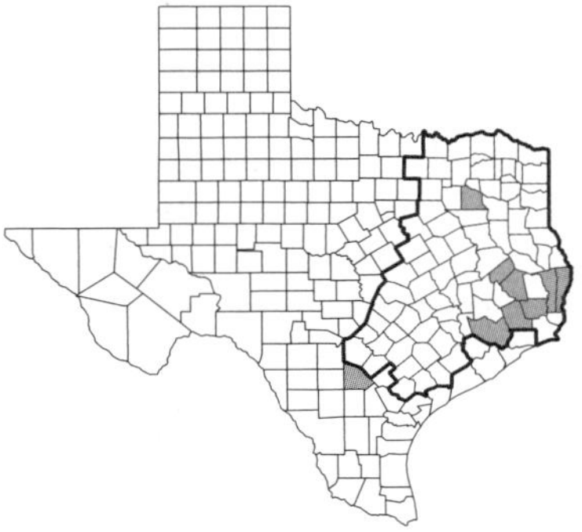

Cayaponia quinqueloba

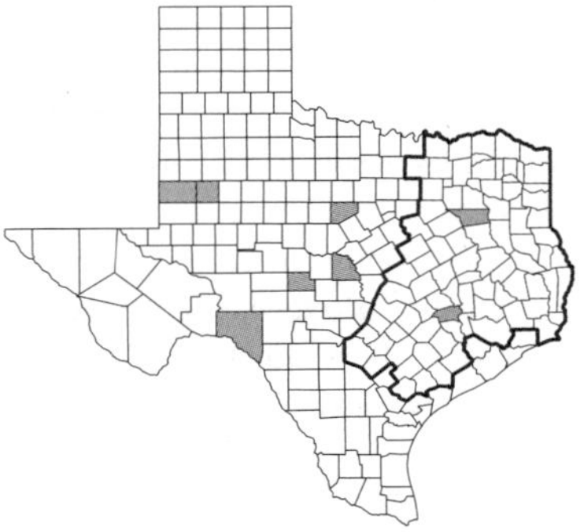

Citrullus amarus

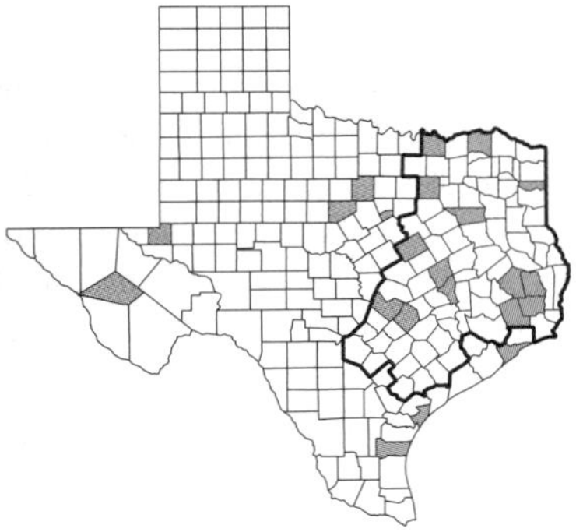

Citrullus lanatus

Coccinia grandis

Cucumis anguria var. longaculeatus

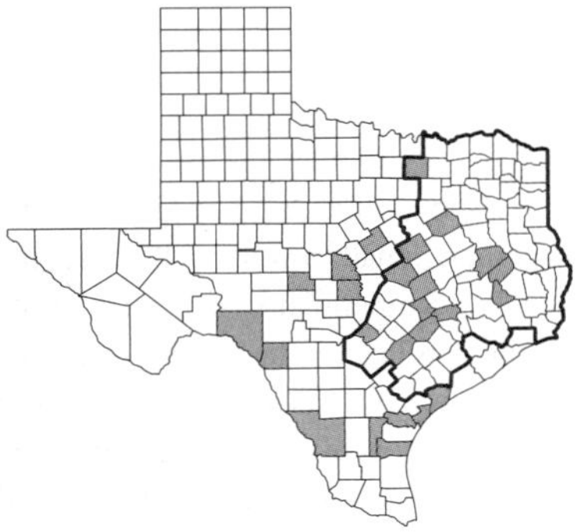

Cucumis melo subsp. agrestis

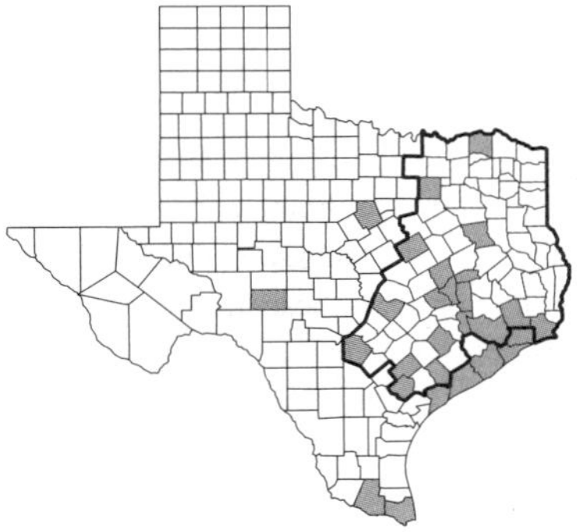

Cucumis melo subsp. melo

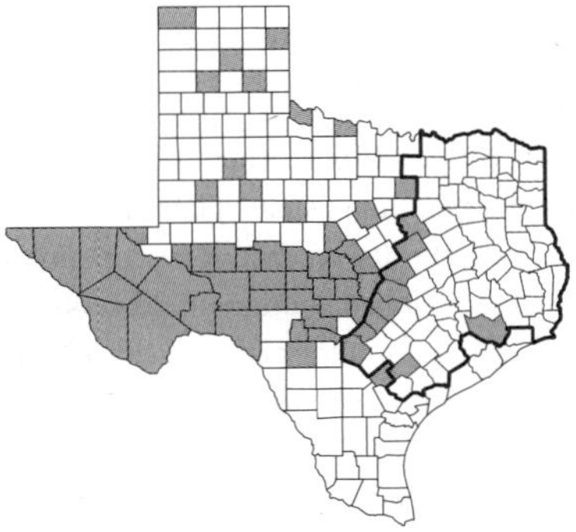

Cucurbita foetidissima

and thickets, bottomlands; Austin, Brazos, Grimes, and Victoria (Nesom 2014) cos. of East TX.; also Brazoria Co. (Nesom 2014). Endemic to coastal plain of se TX. Sep–Nov.

Cyclanthera naudiniana Cogn., (for Charles Victor Naudin, 1815–1899, French naturalist and botanist), WOODLAND CYLANTHERA. Stems with tendrils unbranched, less commonly 2-branched; leaves 3-foliolate, lateral pair of leaflets deeply lobed to nearly divided (appearing 5-foliate), terminal leaflet 3–7 cm long, coarsely serrate to shallowly or deeply lobed; staminate corolla 4.2–6.3 mm in diam.; fruiting peduncles 10–30 mm long, fruit ovoid, armed with spines 3–5 mm long. Rocky slopes, streamsides, riparian woods, roadsides; along the sw edge of East TX, from McLennan se to Bexar cos.; se edge of Edwards Plateau; CO, KS, NE, NM, OK, and TX. Endemic to sc U.S. May–Oct.

IBERVILLEA Greene GLOBEBERRY

A genus of 6 species native to the sw U.S. and Mexico (Nesom 2015b). (Derivation of generic name not explained by original author; however, possibly named for Iberville, a parish in LA)
REFERENCES: Kearns 1994; Nesom 2015b.

Ibervillea lindheimeri (A. Gray) Greene, (for Ferdinand Jacob Lindheimer, 1801–1879, German-born TX collector), BALSAM GOURD, LINDHEIMER'S GLOBEBERRY, BALSAM-APPLE. Perennial dioecious climber to 3 m long with turnip-like taproot; leaves petioled, blades mostly deeply to shallowly 3-lobed (rarely unlobed or 5-lobed), to 12 cm wide, rather fleshy, with swollen-based hairs; corollas greenish yellow; staminate flowers 5–8 per raceme, tubular, glandular-puberulent, 6–8 mm long; stamens 3; pistillate flowers solitary, funnelform, ca. 10 mm long; ovary 3-carpellate; berry globose, ca. 3 cm in diam., reddish orange when ripe, when immature green with lighter patterning (resembling a small watermelon); seeds numerous. Prairies, woodlands, thickets; widespread in the w 2/3 of East TX; widespread in TX except for the Panhandle; NM, OK, and TX; n Mexico. Apr–Sep. Native of the U.S. and Mexico.

LAGENARIA Ser. BOTTLE GOURD, GOURD

A genus of 6 species of tropical Africa and Madagascar (Nesom 2015b) with 1 species widespread in rest of tropics. (Greek: *lagenos*, flask, referring to the shape and use of the fruits)
REFERENCES: Heiser 1979; Nesom 2015b.

Lagenaria siceraria (Molina) Standl., (intoxicating, perhaps derived from use of a *Lagenaria* species in making an intoxicating drink), BOTTLE GOURD, WHITE-FLOWER BOTTLE GOURD, CALABASH GOURD. Viscid-pubescent, monoecious or rarely dioecious, climbing or trailing annual; tendrils branched; leaf blades broadly reniform or ovate, shallowly palmately 3–5-lobed or unlobed, 15–30 cm wide, cordate basally; flowers 1–2 from leaf axils; corollas white to cream, 5–10 cm across; fruits to 50 cm long or more, variously shaped, without spines or prickles but sometimes with knobs or ridges; seeds numerous. Cultivated for the fruit; the tough pericarp is used in making containers (cups, dippers, bowls, etc.) and decorations; escapes to weedy areas; Dallas, Hardin, and Jasper (Kartesz 2022) cos.; widely scattered in se U.S.; also AZ and CA; (Kartesz 2022). Aug–Oct. Native of the Old World tropics. [*L. vulgaris* Ser.] This species has long been in cultivation; BOTTLE GOURDS have been found in Egyptian tombs dating to ca. 3500–3000 BC and in New World sites dating to ca. 7000 BC (Zohary 1982).

MELOTHRIA L. MELONETTE

A genus of 12 species native to the New World; introduced in Asia (Nesom 2015b). (Altered from *melothron*, an ancient Greek name for some fruiting vine)
REFERENCE: Nesom 2015b.

Melothria pendula L., (pendulous, hanging), DROOPING MELONETTE, CREEPING-CUCUMBER, MELONCITO. Slender, usually monoecious, climbing perennial; stems glabrous or nearly so; tendrils unbranched; leaf blades 5–10 cm long, orbicular, 5-angled or -lobed, surfaces scabrous; cordate at base; petioles pubescent, the hairs sometimes stiff; flowers axillary, small; corollas yellowish green, 5-angled or -lobed; staminate flowers in racemose or corymbose inflorescences, the corollas campanulate; stamens 3; anthers free to barely united; pistillate flowers usually solitary or few in a cluster, peduncled, the corollas rotate; ovary 3-carpellate; fruits smooth, green, often with lighter patterning (like a baby watermelon), turning purplish black, ovoid, 1–2.5 cm long, peduncles 1.5–4.5 cm long; seeds numerous, white. Wooded areas, exposed sites; throughout most of East TX; Rolling Plains, Edwards Plateau, and South TX Plains; c U.S. from MD s to FL and w to KS and TX. Often divided into several varieties in the U.S. However, we are following Kartesz (2022) and Nesom (2015b) in not recognizing varieties. May–Nov. Native of North and tropical America. The fruits are reported to act as a drastic purgative (Burlage 1968).

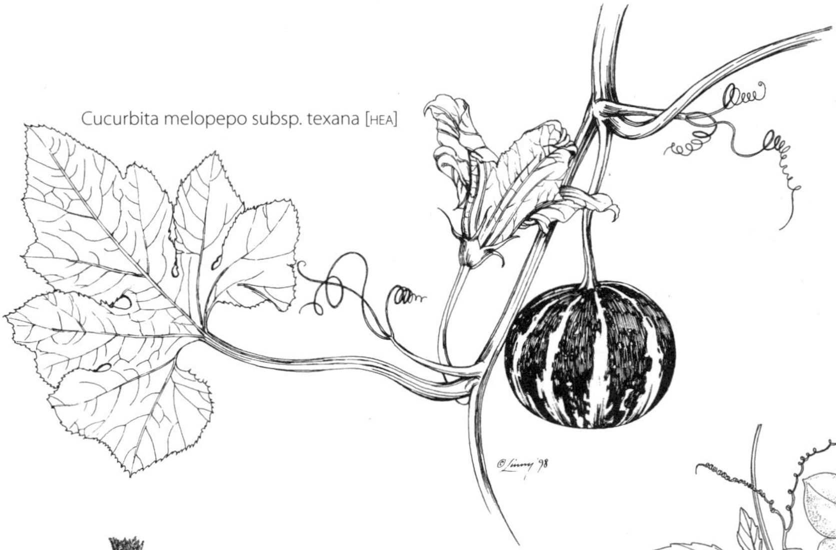

Cucurbita melopepo subsp. texana [HEA]

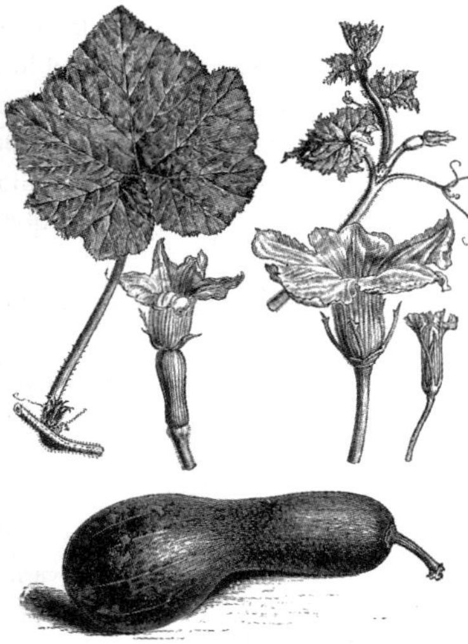

Cucurbita moschata [NIC]

Cyclanthera dissecta [BB2]

Cyclanthera naudiniana [FNA]

Ibervillea lindheimeri [EN1]

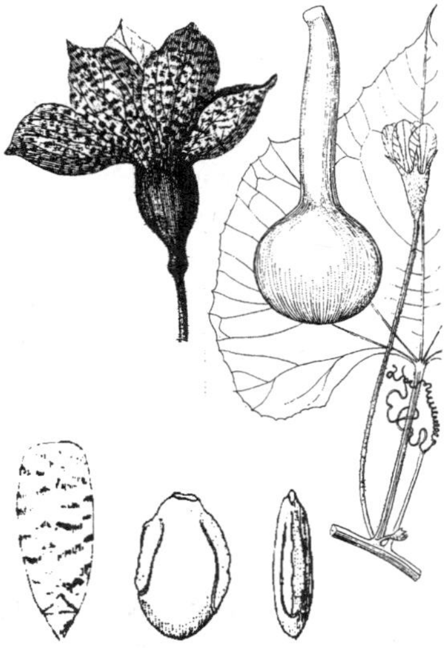

Lagenaria siceraria [LAM]

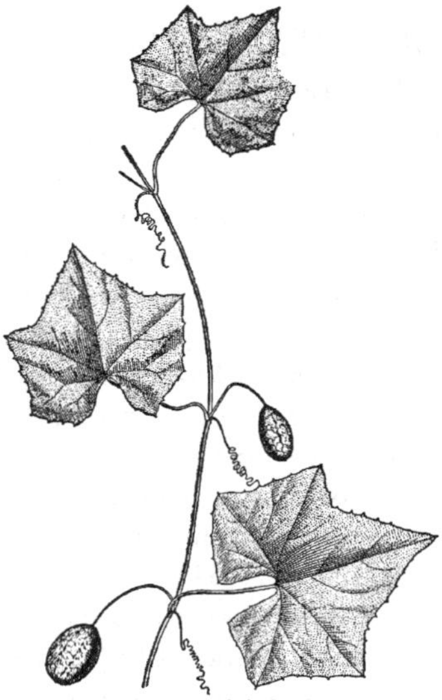

Melothria pendula [LUN]

SICYOS L. BUR-CUCUMBER

A genus of ca. 50 species of the New World, Pacific Islands, and Australia; introduced in Europe and e Asia (Nesom 2015b). (Greek: *sicyus*, cucumber or gourd)

Sicyos angulatus L., (angled), ONE-SEED BUR-CUCUMBER, WALL BUR-CUCUMBER. Glandular-pubescent, monoecious, climbing annual; tendrils 2–5-branched; leaves petioled, the petioles to 8 cm long; leaf blades 4–12(–18) cm long, suborbicular in outline, 5-angled or -lobed; staminate flowers in axillary racemes or corymbs, petals 4–5 mm long; pistillate flowers usually from the same axils in long-peduncled head-like clusters, petals 1–2 mm long; corollas white to greenish white; fruits yellowish, 1–1.5 cm long, ovoid, villous with prickly bristles, with a single seed. Wooded areas, stream banks, fallow fields; from McLennan s to Comal cos.; also Angelina Co. (Kartesz 2022); Edwards Plateau; throughout e 1/2 of the U.S. May–Sep. Sometimes cultivated as a climbing screen in landscapes (Mabberley 2017). *Sicyos edulis* Jacq. is the squash known as CHAYOTE.

CYRILLACEAE Lindl.

LEATHERWOOD OR BUCKWHEAT TREE FAMILY

A very small family of 2 genera and 2 species of wet and acidic habitats from the se U.S. s to South America (Stevens 2001 onwards; Mabberley 2017) with spirally arranged, simple, entire leaves and raceme inflorescences. *Cliftonia* is native to the se U.S. as far w as LA, while *Cyrilla* occurs from the se U.S. to the West Indies, Central and South America (Kubitzki 2004b). *Purdiaea*, previously included in the family, is now, based on both morphological and molecular evidence, included in the Clethraceae (Stevens 2001 onwards; Anderberg & Zhang 2002). Cyrillaceae are now recognized to be the sister group to the Ericaceae, with Ericaceae and Cyrillaceae sister to the Clethraceae (Anderberg & Zhang 2002; Anderberg et al. 2002; Schonenberger et al. 2005). Both genera have shallow root systems and develop numerous adventitious roots; this allows for vegetative reproduction by root sprouting and the development of extensive clonal populations (Thomas 1961; Kubitzki 2004b). (subclass Dilleniidae—Cronquist; order Ericales—APG IV)
FAMILY RECOGNITION IN THE FIELD: the sole TX member of the family is a wet area shrub or small tree with alternate, simple, entire leaves, white flowers in numerous finger-like racemes, and indehiscent few-seeded fruits.
REFERENCES: Thomas 1960, 1961; Lundell 1961a; Anderberg et al. 2002; Anderberg & Zhang 2002; Zhang & Anderberg 2002; Kubitzki 2004b; Lemke 2009; Yesilyurt 2010.

CYRILLA Garden TITI, LEATHERWOOD

A monotypic genus ranging from the se U.S. to South America (Kubitzki 2004b; Mabberley 2017). While there is considerable variation and up to 11 species have sometimes been recognized, most modern treatments recognize only one variable species (e.g., Thomas 1960; Correll & Johnson 1970; Kubitzki 2004b; Lemke 2009; Yesilyurt 2010). Fossils of *Cyrilla* are known from Europe (Worobiec 2003). (Named for Domenico Cyrillo, 1739–1799, Italian physician and professor of botany at Naples)
REFERENCES: Lundell 1961a; Dute et al. 2004.

Cyrilla racemiflora L., (flowers in racemes), TITI, BLACK TITI, LEATHERWOOD, SWAMP CYRILLA. Shrub or small tree to 6(–10) m tall, usually branching near base, often forming dense thickets by vegetative reproduction through adventitious shoots arising from the roots, glabrous; leaves alternate, often clustered towards branch tips, variable in size, to 10 cm long and to 2.5(–3) cm wide, simple, oblanceolate to elliptic, usually widest beyond middle, entire, glabrous, sometimes leathery, dark green and shiny above, paler beneath, turning reddish or yellowish in fall, tardily deciduous to evergreen or nearly so, petiolate, the petioles to 1 cm long; stipules absent; inflorescences narrow finger-like racemes, these often numerous and showy, produced on previous year's growth just below new growth, to 18 cm long, the axes angled, with flowers nearly to base; pedicels 1.5–3 mm long, with a bract at base and 2 bracteoles above the middle; flowers numerous, fragrant, perfect, radially symmetrical; sepals usually 5, sometimes fused at very base; petals usually 5, white to creamy white, ca. 2.5–3(–3.5) mm long; nectary disk present; stamens 5; ovary superior; fruits ovoid to broadly so, 2–3 mm long, dry, indehiscent, yellowish, 1–2(–4)-seeded or seeds often not developing. Wet areas of stream bottoms, swamps; Angelina, Hardin, Jasper, Liberty, Newton, Polk, and Tyler (Turner et al. 2003), Jefferson (*Clark 3930* SMU) and Trinity (Kartesz 2015) cos. in the Big Thicket area of the se corner of East TX; mostly se U.S. from DE (escaped) and VA s to FL w to se TX; also West Indies and s Mexico s to South America. May–Jul. [*C. arida* Small, *C. parvifolia* Raf., *C. racemiflora* var. *subglobosa* Fernald] This species is an important bee-tree whose flowers are used in commercial honey production (Mabberley 2017); native pollinators are also apparently bees (e.g., halicids, bumblebees, etc.) (Dute et al. 2004). The common name is pronounced as rhyming with bye-bye. Reproduction is thought to be largely vegetative, as "shallow, nearly horizontal roots produce ascending shoots giving rise to extensive clones" (Lemke 2009); the clones "may cover an acre or more" (Thomas 1961).

DROSERACEAE Salisb.

SUNDEW FAMILY

A small (ca. 100–200 species in 3 genera), subcosmopolitan, monophyletic family of insectivorous herbs. The number of species is unclear with some authorities suggesting as few as ca. 100 species (Weakley et al. 2012; Mabberley 2017), others ca. 150–170 (Rivadavia et al. 2003; Mellichamp 2015), and still others ca. 200 (Stevens 2001 onwards). The family includes *Dionaea muscipula* J. Ellis, (epithet = fly-catching), (VENUS'-FLYTRAP), from the se U.S. *Aldrovanda* and *Dionaea* are monotypic with *Drosera* having most of the species in the family. *Drosophyllum*, previously included in the Droseraceae, is now recognized as being a distinct evolutionary lineage and is recognized in the Drosophyllaceae (Kubitzki 2003b; Ravidavia et al. 2003). In *Drosera* the glandular hairs are motile, and entrap and digest insects using secreted proteolytic enzymes and ribonucleases; in *Dionaea* the 2 halves of the leaves swing together trapping the victim ("snap trap system"); *Aldrovanda*, occurring in Africa, Asia, Australia, and Europe, is a rootless aquatic with insect-trapping leaves like *Dionaea*; both have multicellular trigger hairs. While markedly different morphologically, all three genera have similar glands for digesting prey (Ravidavia et al. 2003). Most species are found in nitrogen poor habitats such as bogs or wet sandy areas. As with most carnivorous plants, nutrients (especially nitrogen), rather than calories, are gained through carnivory. Molecular research indicates that the family is in the Caryophyllales and has close relationships to two other carnivorous families, the Nepenthaceae and Drosophyllaceae (Williams et al. 1994; Soltis et al. 2000; Kubitzki 2003a; Mellichamp 2015). (subclass Rosidae—Cronquist; order Caryophyllales—APG IV)

FAMILY RECOGNITION IN THE FIELD: small wet area herbs usually having rosettes of leaves with tentacle-like hairs secreting glistening drops of sticky fluid which trap insects; flowers on a scape.

REFERENCES: Diels 1906; Wood 1960; Williams et al. 1994; Kubitzki 2003a; Weakley et al. 2012; Hernández-Ledesma et al. 2015; Mellichamp 2015.

DROSERA L. SUNDEW

Small, annual or perennial, insectivorous herbs; leaves in a basal rosette or cauline and alternate in some, fiddle-head-like in bud, filiform and bladeless to petiolate and nearly spatulate or suborbicular, with conspicuous, usually red (pale green in *D. tracyi*), gland-tipped, tentacle-like, motile hairs which secrete a clear sticky fluid used in trapping insects; stipules absent or present, adnate or nearly free; inflorescence a 1-sided scorpioid cyme, nodding apically before maturity, scapose; flowers bisexual, radially symmetrical, usually 5-merous, hypogynous, open for only a single morning, capable of self-pollination (Wood 1960); petals white to pink, roseate, or purple rose; ovary superior; fruit a 3–5-valved capsule; seeds numerous, pitted to longitudinally ridged, or papillose.

A subcosmopolitan (but especially s hemisphere—e.g., Australia > 80 species, Africa > 30 species, South America ca. 30 species) genus of ca. 100–200 species (see discussion under family synopsis) of insectivorous herbs typically found in wet areas. It is the second largest genus of carnivorous plants (after *Utricularia*, Lentibulariaceae) (Schlauer 1996) and has an "active flypaper" trapping mechanism (Rivadavia et al. 2003). Mobile, tentacle-like, glandular hairs emit a sticky, glistening, enzyme-containing liquid which attracts, ensnares, and digests insects. After capture, the hairs bend inward, pressing the insect against the surface of the leaf, the blade of which may also bend inward to enfold the insect (the leaf itself does not bend in *D. filiformis*); both hair and blade movements involve cell elongation (Wood 1960). A number of species have been used medicinally (Langer et al. 1995; Austin 2002). (Greek: *droseros*, dewy; the glands of the leaves exude drops of a clear glutinous fluid, glittering like dew drops and giving rise to both the generic and common names)

REFERENCES: Wynne 1944; Shinners 1962d; Wood 1966; Murry & Urbatsch 1979; Pietropaolo & Pietropaolo 1986; Seine & Barthlott 1994; Langer et al. 1995; Schlauer 1996; Sorrie 1998; Austin 2002; Rivadavia et al. 2003; Correa & dos Santos Silva 2005; Rice 2011.

1. Leaves filiform (= extremely long and thread-like), without an expanded blade; plant with bulbose corm-like structure at base ____ **D. tracyi**
1. Leaves not filiform, with petiole and definitely expanded blade; plant without bulbose corm-like structure.
   2. Scapes (= flowering stem) with gland-tipped hairs except near base; basal rosettes tiny, 0.8–3.5 cm in diam.; stipules absent or vestigial; seeds black, with pits ____ **D. brevifolia**
   2. Scapes glabrous; basal rosettes usually 3–12 cm in diam., rarely less; stipules present; seeds reddish brown to brown or black, without pits (either papillose or with longitudinal ridges).
      3. Petals pink, 1–3 mm wide; petioles usually shorter than blades to at most 2 times as long as blades, with at least a few long hairs; leafy stem absent (all leaves in basal rosette); scapes straight basally or nearly so; seeds with distinct longitudinal ridges (ridges can be somewhat papillose) ____ **D. capillaris**

3. Petals white (can have pinkish tinge), 3–5 mm wide; petioles usually 3–4 times as long as blades, glabrous; leafy stem 1–10(–20) cm long often present (on mature plants); scapes arched outward basally; seeds densely and ± evenly papillose, without longitudinal ridges ____ **D. intermedia**

Drosera brevifolia Pursh, (short-leaved), ANNUAL SUNDEW, DWARF SUNDEW, SHORT-LEAF SUNDEW. Leaves in a small basal rosette (3.5 cm or less in diam.), blades suborbicular (with petiole longer than blade) to nearly spatulate with enlarged terminal portion, to 7(–10) mm long; stipules absent or vestigial; scape to 8(–12) cm tall, glandular-pubescent; flowers up to 1–6(–8) per scape; pedicels glandular pubescent; sepals ovate, 2.5–4 mm long, glandular-pubescent; petals pink or roseate to whitish, 2.5–8 mm long; capsules 3.5–4 mm long; seeds black, with pits in 10–12 rows. Deep sand in woods, savannahs, or in open bogs; widespread in e part of East TX w to Henderson (BRIT), Bastrop (TEX), and Wilson (Turner et al. 2003) cos.; also n Gulf Prairies and Marshes; se U.S. from VA s to FL w to KS, OK, and TX; also Latin America s to Brazil. Feb–Jun. [*D. annua* E.L. Reed, *D. leucantha* Shinners]

Drosera capillaris Poir., (resembling hair, very slender), PINK SUNDEW. Leaves in a basal rosette, these varying from small to sometimes robust (–12 cm in diam.), blades spatulate to obovate, 5–10 mm long, 3–5 mm wide, usually shorter than petiole; stipules nearly free, fimbriate; scapes to 4–15(–20+) cm tall, with 2–20 flowers; sepals glabrous; petals pink (can fade to white), 6–9 mm long; seeds brown, with longitudinal ridges. Bogs, savannas; widespread in East TX w to Bastrop (TEX), Milam (Turner et al. 2003), and Henderson (LL) cos.; mainly coastal plain; se U.S. from VA s to FL w to WV, TN, AR, and TX; also Latin America s to South America. May–Aug.

Drosera intermedia Hayne, (intermediate), SPOON-LEAF SUNDEW, NARROW-LEAF SUNDEW. Leaves often along leafy stem to 20 cm long, or in a rosette, blades oblong-spatulate to obovate, gradually tapering to petiole; stipules adnate basally, with bristle-like segments; scapes with up to 20 flowers; sepals glabrous or with sparse pubescence; petals usually white, 5–8 mm long; seeds reddish brown to black, densely papillose. Bogs, wet areas, often in standing water; Jasper, Newton, and Tyler (Turner et al. 2003) cos. in the Big Thicket; widespread in the e 1/2 of the U.S. w to MN and TX, also adjacent Canada. Jun–Sep. [*D. americana* Willd.] Mellichamp (2015) notes that this is the only member of the genus having a distinct stem, scattered cauline leaves, and an arching scape. He also notes that the plants can reproduce vegetatively, from "winter hibernacula" that can form at the stem tips. In addition, he says that, "Its habit of growing in the wettest places, often in standing water, is characteristic."

Drosera tracyi (Diels) Macfarl., (probably for the outstanding botanist-agriculturist, Samuel M. Tracy (1847–1920), who studied the floras of LA and MS, and for whom the Tracy Herbarium at TX A&M Univ. is named), TRACY'S SUNDEW, THREAD-LEAF SUNDEW, DEW-THREADS, GULF COAST THREAD-LEAF SUNDEW. Base of leaves expanded and forming a corm-like structure 1–2 cm in diam.; glandular hairs light green; leaves filiform, to 40(–50) cm long, mostly > 1 mm wide; scape 25–60 cm tall, glabrous, with up to 24 flowers (but typically 8–12); petals 12–17(–20) mm long, purplish rose; seeds with pits. Seepage bogs, wet savannahs; the only TX record we are aware of is from individuals spreading from an introduced population in the Big Thicket in Tyler Co. (*Orzell et al. 7226* TEX); se U.S. on the Gulf Coastal Plain from FL and GA w to LA and introduced in e TX. Mid-Apr–early Jun. [*D. filiformis* Raf. var. *tracyi* Diels] While sometimes treated as a variety of the more easterly *D. filiformis* Raf. (e.g., Radford et al. 1968), Sorrie (1998) provided very convincing morphological data supporting their recognition as two distinct species. He further postulated that the two are a species pair whose ancestor survived the last glaciation in a Gulf Coastal Plain refugium, subsequently spread north along the Atlantic Coast as climate ameliorated, and later, when isolated, speciated to produce the two distinct species known today (*D. filiformis*, primarily along the Atlantic Coastal Plain, and *D. tracyi*, only on the Gulf Coastal Plain).

EBENACEAE Gürke

PERSIMMON OR EBONY FAMILY

The Ebenaceae is a medium-sized (ca. 575–855 species in 3–4 genera) family of the tropics and warm areas worldwide with a few temperate species (Stevens 2001 onwards; Mabberley 2017). Most are small to medium-sized understory trees; some have edible fruits (e.g., *Diospyros*—PERSIMMONS, JAPANESE PERSIMMONS, DATE-PLUMS, VELVET-APPLES), while others are valuable for timber including EBONY, a black, hard, heavy wood obtained from species of *Diospyros*. The dark color of the wood of a number of species is due to derivatives and oxidative decomposition products of naphthoquinones and in some species colored and uncolored parts of the wood result in a "variegated, banded, or marbled aspect" (Wallnöfer 2004) giving rise to such names as MARBLEWOOD or ZEBRAWOOD. Based on molecular research, the small South American family Lissocarpaceae (1 genus, 4–6 species) is closely related to and now included in the Ebenaceae (Berry et al. 2001; Stevens 2001 onwards; Duangjai et al. 2006). Family name conserved from *Ebenum*, a genus now treated as *Diospyros* (the name *Diospyros* was published earlier and thus has priority in terms of nomenclature) (Derivation either from Greek: *ebenos*, name used by Hippocrates for a leguminous plant or possibly

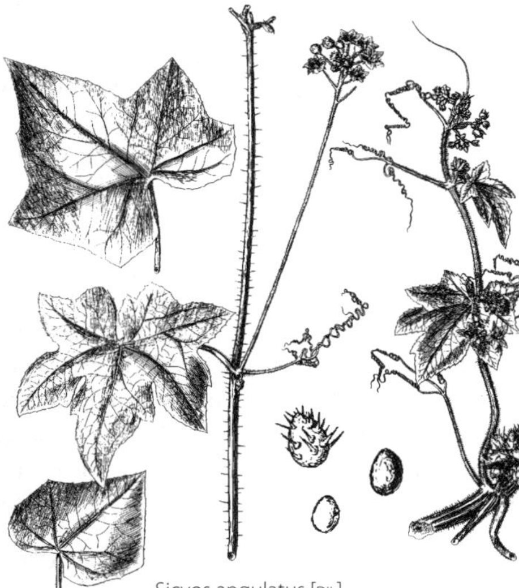
Sicyos angulatus [DIL]

Cyrilla racemiflora [ALI]

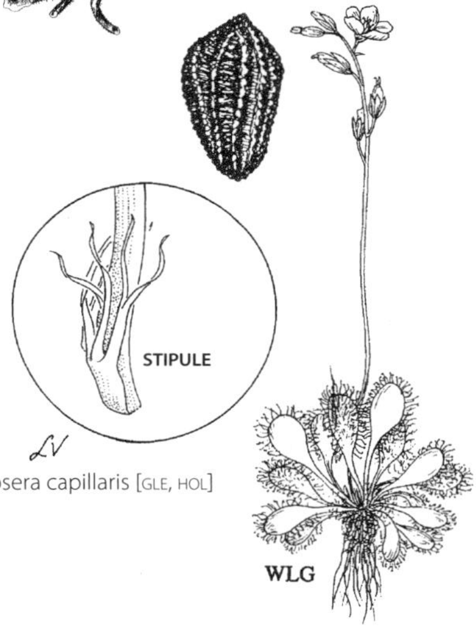

Drosera capillaris [GLE, HOL]

Drosera intermedia [GWO]

Drosera brevifolia [GWO]

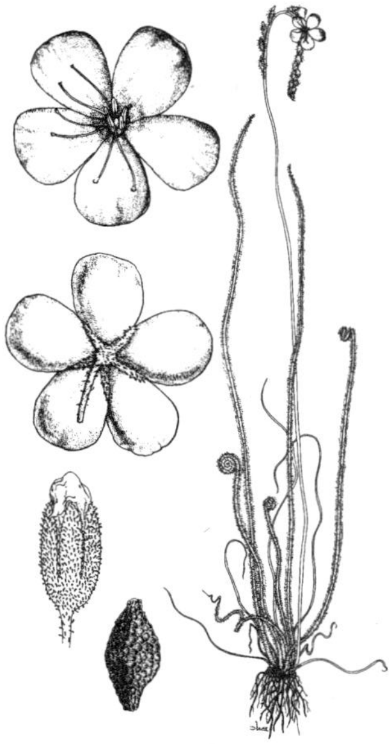
Drosera tracyi [GWO]

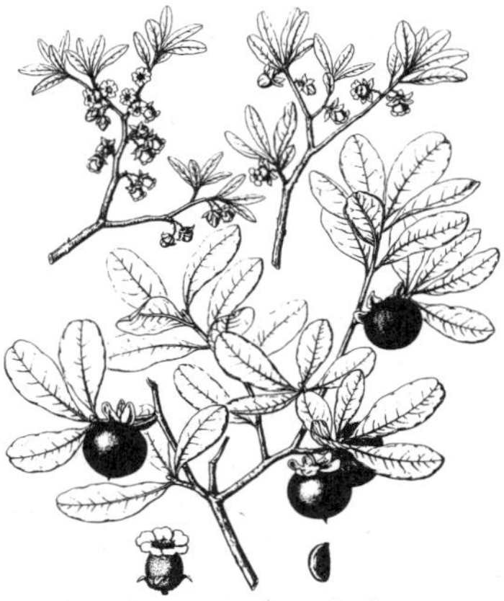
Diospyros texana [SA3]

the ebony tree, or from Latin: *ebenus*, ebony or black; the heartwood of some species is strikingly black in color) (subclass Dilleniidae—Cronquist; order Ericales—APG IV)

FAMILY RECOGNITION IN THE FIELD: shrubs or small trees with alternate, simple, entire, exstipulate leaves; milky juice absent; flowers small, axillary, usually of a single sex, with connate petals and superior ovaries; fruit a berry with several large seeds.

REFERENCES: Skallerup 1953; Wood & Channell 1960; Spongberg 1977; Morton et al. 1996 [1997]; Berry et al. 2001; Wallnöfer 2001, 2004; Anderberg et al. 2002; Briand 2005; Duangjai et al. 2006; Eckenwalder 2009.

DIOSPYROS L. PERSIMMON, EBONY

Shrubs or small trees, usually dioecious (but sometimes not completely so); leaves alternate, nearly sessile to short-petioled; leaf blades simple, oblong or elliptic, entire; stipules absent; flowers rather small, axillary, solitary or few together, short-pedicelled, drooping, mostly unisexual, the pistillate larger, with or without 8 empty filaments (staminodes, the staminate flowers smaller, with usually 16 stamens; calyces subrotate, deeply 4-lobed, enlarging in age, persistent; corollas urceolate to campanulate or salverform, with 4, short, auricled, laterally overlapping, convolute lobes, creamy or yellowish white or greenish white, the lobes sometimes more yellowish than the tube, somewhat fleshy; stamens attached to corolla near its base, with wide, pilose filaments; pistil 1; ovary superior; style divided half way or more into 4 branches; fruit a berry with 3–8 large seeds.

A mainly tropical and subtropical genus of ca. 550–800+ species including a number important for fruit and timber (Stevens 2001 onwards; Wallnöfer 2004; Mabberley 2017); it occurs from the e U.S. to South America as well as in the Asia-Pacific region (center of greatest diversity) and Africa; Eckenwalder (2009) noted that the "total species count for the genus is uncertain because new species are being described and the distinctiveness of some previously described species is questionable." The common name is apparently derived from a Native American word for the tree, *pasimenan* (Peattie 1948) or perhaps *putchamins*, possibly from roots meaning "choke-fruit" in reference to the unripe fruit (Briand 2005). *Diospyros ebenum* J. König, native of India and Sri Lanka, is the EBONY of commerce; its wood is black and extremely hard; it is used in inlays, for piano keys, and in other uses where its color and density are important. EBONY is so dense that it is one of relatively few woods that will not float in water. *Diospyros kaki* Thumb., (JAPANESE PERSIMMON, KAKI), of e Asia, produces large fruits that are widely available in U.S. supermarkets (but considered by aficionados of the native persimmon as being less flavorful); a number of other species are also cultivated for their fruits. (Greek: *dios*, of Zeus or Jove, and *pyros*, grain, alluding to the edible fruits)

REFERENCES: Hiern 1873; Skallerup 1953; Duangjai et al. 2009.

1. Leaf blades 2–5 cm long, subsessile, apically ± rounded, tomentose beneath; ovaries and fruits pubescent; fruits black at maturity; bark ± smooth, peeling in thin layers, grayish ____ **D. texana**
1. Leaf blades 6–15 cm long, with distinct petioles (ca. 7–10 mm long), apically acute to acuminate, glabrous or sparsely pubescent beneath when young; ovaries and fruits glabrous; fruits yellowish to orangish or pinkish at maturity; bark deeply divided into small blocks or plates, dark brown or blackish ____ **D. virginiana**

Diospyros texana Scheele, (of Texas), TEXAS PERSIMMON, MEXICAN PERSIMMON, BLACK PERSIMMON, CHAPOTE PERSIMMON. Shrub or small tree to ca. 3(–16) m tall, sometimes forming thickets; bark thin, peeling or flaking, with trunks "resembling those of crape-myrtle" (Correll & Johnston 1970); leaves permanently pubescent, evergreen s of the San Antonio area, but deciduous to the n (Simpson 1988); flowers appearing with expanding leaves in spring; fruits 1.5–2.5 cm in diam., sweet and edible when ripe; seeds ca. 8 mm long. Rocky woods, slopes, open areas, and along streams; mainly sw portion of East TX n to McLennan Co. (USDA Plants Database 2019); widespread mainly in s 1/2 of TX; in the U.S. found only in TX; also n Mexico. Feb–Mar. [*Brayodendron texanum* (Scheele) Small] In addition to being eaten by both humans and wildlife, the black juice of the fruits was used during frontier days to dye buckskin or leather black (Crosswhite 1980). This species is "one of Texas' premier small trees and is a good tree to use in landscape situations where there is little space" (Simpson 1988).

Diospyros virginiana L., (of Virginia), COMMON PERSIMMON, EASTERN PERSIMMON, POSSUMWOOD. Tree to ca. 30+ m tall, sometimes root suckering and forming colonies; heartwood hard and dark brown; leaves glabrous or nearly so at maturity, deciduous, turning yellow, orangish, or reddish purple in fall; flowers appearing after the leaves; fruits 2–5(–7) cm in diam., ripening in fall and some often long persistent on the tree (numerous fruits have been observed in January), sweet and edible when ripe; seeds flattened, 11–20 mm long, reddish brown. Woods, old fields, and clearings, on soils ranging from sand to the blackland clay; widespread in East TX; also n Gulf Prairies and Marshes, e Cross Timbers and Prairies, e Edwards Plateau, and Hemphill Co. (Turner et al. 2003) in Rolling Plains; widespread in the e 1/2 of the U.S. w to NE, OK, and TX, apparently naturalized in several locations further w. Mar–Jun. [*D. virginiana* var. *pubescens* Nutt., *D. virginiana* var. *platycarpa* Sarg.] The tough hard wood has been used for such things as billiard balls, golf-club heads, and textile weaving shuttles (Cox & Leslie 1991). The fruits are delicious and are variously eaten raw or in puddings, pies, syrup, jellies, or ice cream (Yatskievych 2006). However, the unripe fruits

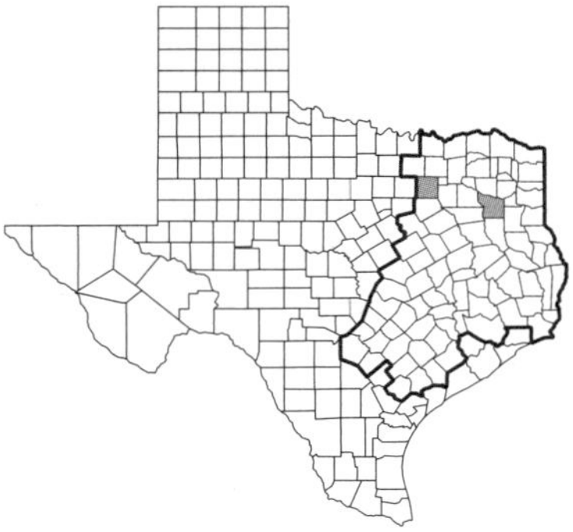
Cucurbita moschata

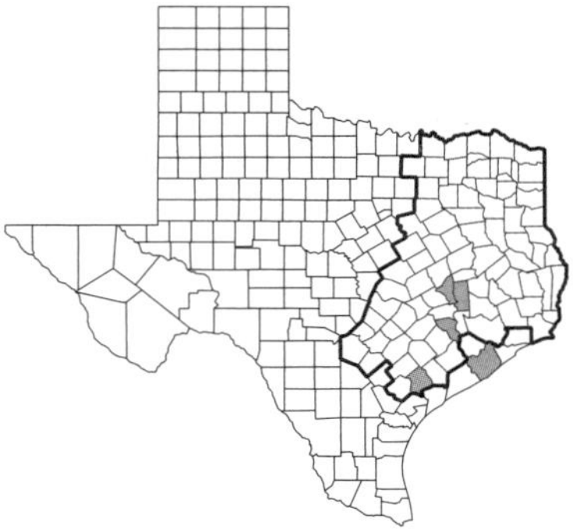
Cyclanthera dissecta

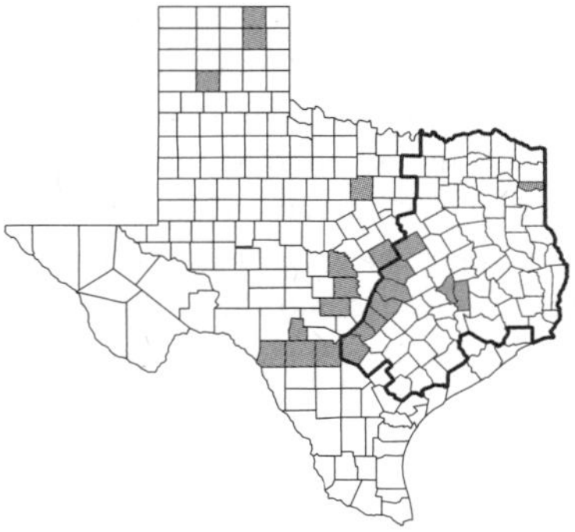
Cyclanthera naudiniana

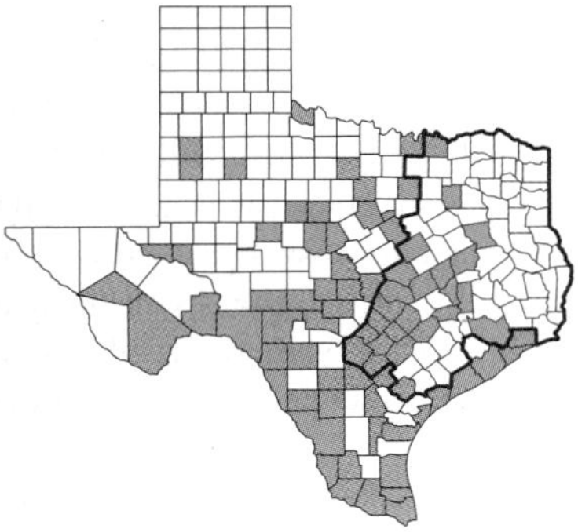
Ibervillea lindheimeri

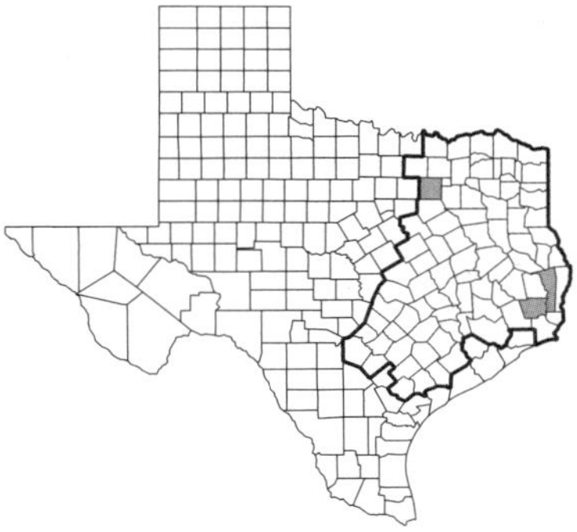
Lagenaria siceraria

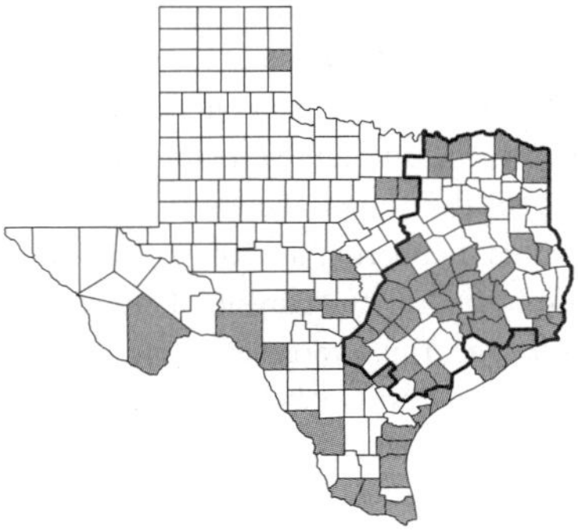
Melothria pendula

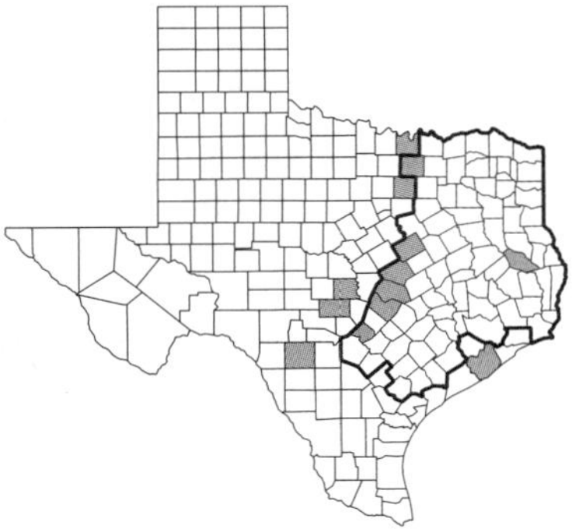
Sicyos angulatus

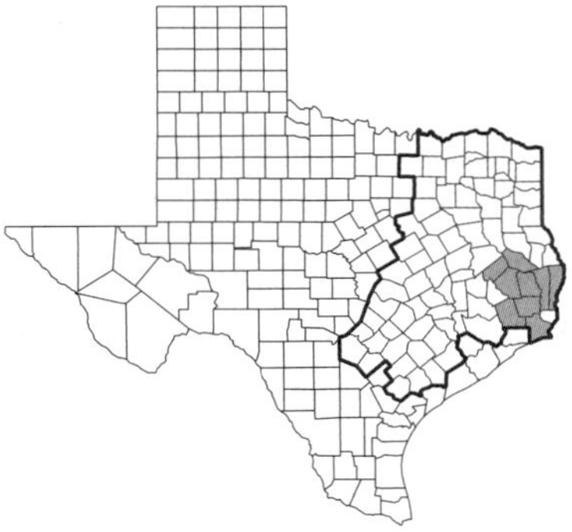
Cyrilla racemiflora

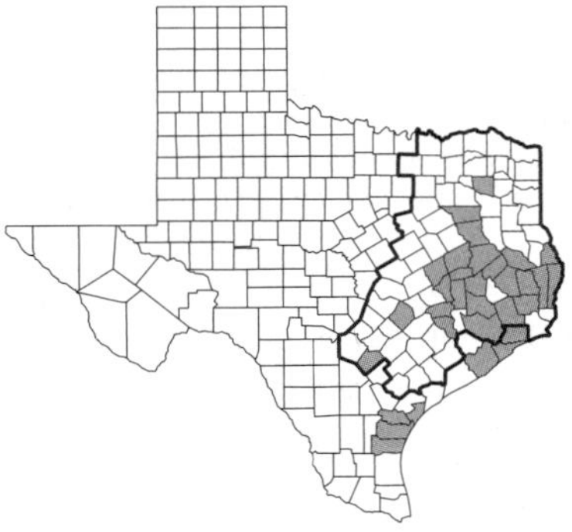
Drosera brevifolia

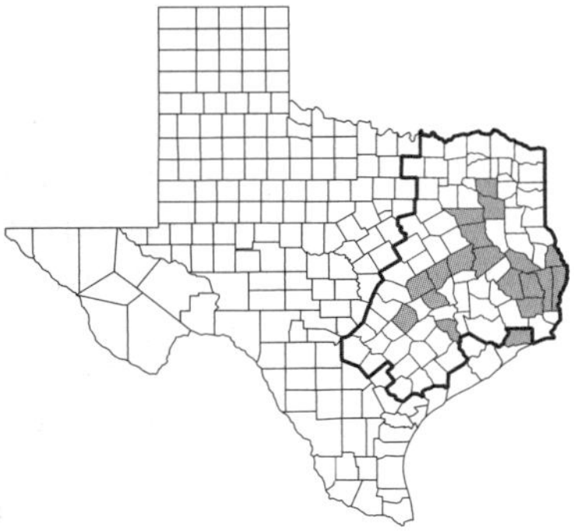
Drosera capillaris

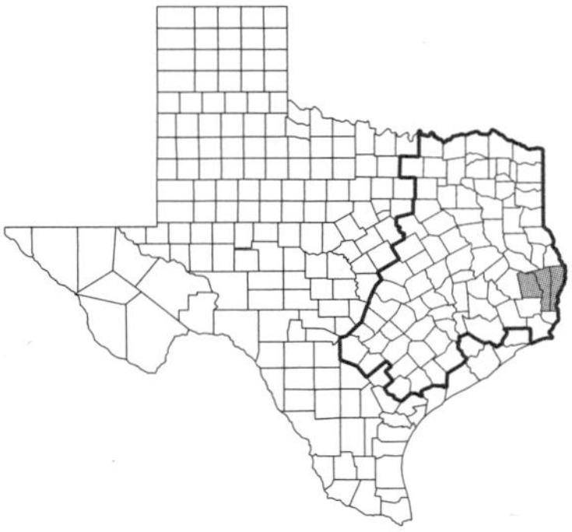
Drosera intermedia

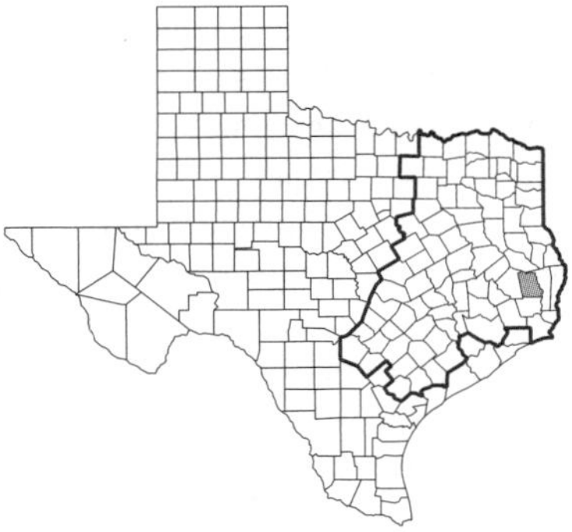
Drosera tracyi

are notoriously astringent due to the presence of tannins (Wallnöfer 2004) and provide one of the best examples of the dramatic change in flavor sometimes accompanying the process of fruit ripening. Despite anecdotes to the contrary, a frost is not necessary for ripening. In fact, some ripe fruits can often be found long before the first frost. While many fall to the ground, occasional fruits can still be found on the trees as late as January and can have a desiccated, almost raisin-like or prune-like appearance and taste. The fruits are an important wildlife food and seeds can be found in the scat of a variety of animals including coyotes and raccoons; they were reportedly used as both medicine and an important food by Native Americans—eaten raw, dried like prunes, fermented into a beer, or dried and made into flour and bread (Hedrick in Heiser 1993; Briand 2005; Yatskievych 2006). The flat seeds were sometimes used as buttons and a coffee substitute during the Civil War (Mabberley 2017).

EHRETIACEAE Mart.

EHRETIA FAMILY

A small (ca. 150 species in 7–9 genera (Stevens 2001 onwards; Leubert et al. 2016), widespread, mainly tropical and subtropical family of usually shrubs with some trees and herbs. The family has often been included in a broadly conceived Boraginaceae (e.g., Correll & Johnston 1970; Diggs et al. 1999; APG IV 2016), and indeed, the Boraginaceae *sensu lato* (including the Hydrophyllaceae) is generally agreed to be monophyletic, and it is not unreasonable to recognize it (Christenhasz et al. 2015). However, based on the most recent molecular data and morphological studies (e.g., Gottschling et al. 2014, 2016; Leubert et al. 2016) the botanical consensus has moved towards dividing the excessively broadly conceived Boraginaceae (Stevens 2001 onwards; Leubert et al. 2016). The Ehretiaceae is thus here segregated as a monophyletic group most closely related to the non-photosynthetic parasitic Lennoaceae (which has sometimes been included within it), and then to families including the Cordiaceae, Heliotropiaceae, Namaceae, Hydrophyllaceae, and Boraginaceae (in the strict sense), all monophyletic families in the Boraginales (Gottschling et al. 2001; Leubert et al. 2016). (subclass Asteridae—Cronquist; order Boraginales (as part of a very broad Boraginaceae)—APG IV)

FAMILY RECOGNITION IN THE FIELD: our only two species are a mat-forming subshrub with usually solitary pinkish, rose, or rarely white flowers in the leaf axils, and a tree with alternate, simple, entire leaves, showy displays of white flowers, and yellow to orange drupes 5–8 mm in diam.

REFERENCES: Gottschling et al. 2008, 2014, 2016; Christenhasz et al. 2015; Leubert et al. 2016; Hasenstab-Lehman 2017.

1. Small tree to 10(–15) m tall, often with multiple trunks; inflorescences terminal, cymose; corollas white, showy; leaf blades 23–60 mm long; fruits yellow to orange ____ **Ehretia**
1. Mat-forming subshrub usually < 30 cm tall; inflorescences usually of solitary flowers in the leaf axils; corollas pinkish, rose, or rarely white; leaf blades to 11.5(–13) mm long; fruits shiny light brown to dark brown ____ **Tiquilia**

EHRETIA P. Browne, EHRETIA

About 40 species of the Old and New World tropics with centers of diversity in Africa and East Asia and a few species extending into warm temperate regions (Mabberley 2017). Some species are used for timber and others medicinally. (for Georg Dionysius Ehret, 1708–1770, German born botanical illustrator/painter)

REFERENCES: Correll & Johnston 1970; Miller 1989; Gottschling & Hilger 2001.

Ehretia anacua (Terán & Berland.) I.M. Johnst., (from Spanish, *anacahuite*, which in turn is derived from the Nahuatl (Aztec) words *a¯matl*, paper, and *quatitl*, tree, possibly in reference to the bark—Riess 1958), SUGARBERRY, ANACUA, KNOCKAWAY, SANDPAPER TREE. Tree to 10(–15) m tall, often with multiple trunks, older trees with gnarled appearance; bark thick, furrowed, separating into thin reddish or gray scales; twigs puberulent; leaves alternate, simple, evergreen or partially deciduous (in n part of range); petioles to 6(–11) mm long, puberulent, blades usually 23–60 mm long, rarely longer, entire or with a few short teeth, elliptic or sometimes ovate, frequently somewhat revolute, apically obtuse and frequently with the tip apiculate to rarely acute or slightly acuminate, basally obtuse to rounded, rarely slightly cordate, the upper leaf surface dark green, scabrous, with a sandpaper feel (hence one of the common names), the lower surface paler than the upper, puberulent to somewhat scabrous; inflorescences terminal, cymose, 2.5–7 cm long; flowers bisexual, 5-merous, fragrant; sepals narrowly triangular, sometimes slightly connate basally, 2.5–3.5 mm long; corollas white, 6–7.8 mm long, with pubescence externally, 5-merous, the lobes obovate, 2.8–3.4 mm long, the tube 2.8–4.2 mm long; stamens 5; style 1, 2-branched; fruits drupaceous, yellow to orange when mature, glabrous, globose, 5–8 mm in diam., the endocarp dividing into 2, 2-locular pyrenes. Thickets, brushy areas, fencerows, forests, along streams; Austin, Harris (LL), Bexar, Comal, De Witt, Goliad, Gonzales, Hays, Jackson, Travis (TEX), Victoria (MO—*fide* Miller 1989), Brazos, and Fayette (USDA Plants Database 2019) cos.; c and s TX, in the US known only from TX; also ne Mexico (Coahuila, Nuevo León, Tamaulipas). Jun–Oct. [*E. ciliata* Miers, *E. elliptica* DC., *E. exasperata* Miers, *Gaza anacua* Terán &

Berland.] The fruits are eaten by mammals, birds, and humans (e.g., Kickapoo Indians—Latorre & Latorre 1977) and the flowers attract bees; also, the leaves are reported to be sole food source of the Texas tortoise beetle (*Coptocycla texana*) (Taber & Fleenor 2005). The hard wood has been used for fence posts and tool handles. The species is valued as a shade tree and ornamental for its very showy flowering display; Correll and Johnston (1970) noted it is planted as far north as Dallas where it can die back in "excessively cold winters." The common name KNOCKAWAY is apparently a corruption of Spanish, *anacahuite*.

TIQUILIA Pers. CRINKLEMAT

A genus of 28 species of New World deserts (Richardson 1977; Mabberley 2017). The genus was previously treated as part of *Coldenia* L. *sensu* A. Gray. However, now *Coldenia* is considered to be a monotypic genus of southern Asia, and *Tiquilia* is treated as containing the xeric species of North and South America. ("Ruiz and Pavon (1799) described *Lithospermum dichotomum*, which they collected in Peru, and subsequently, Persoon (1805) established the genus *Tiquilia* to accommodate this taxon, the generic name being derived from the vernacular name, reported by Ruiz and Pavon (1799) as "Tiquil-tiquil." It is possible that the cited vernacular name is simply a corruption of the Quechua word for flower, "t'ika" (pers. comm., Helen Barler. April, 1975; Lira 1973").—Richardson 1977.
REFERENCES: Correll & Johnston 1970; Richardson 1977.

Tiquilia canescens (A. DC.) A.T. Richardson, ("covered with short gray or white hairs, from Latin *canescens*, present participle of *canescere*, "to become gray or white"), WOOLLY CRINKLEMAT, WOODY CRINKLEMAT, DOG'S-EAR, OREJA-DE-PERRO, RAT'S-EAR COLDENIA, WOODY TIQUILIA, GRAY TIQUILIA, GRAY COLDENIA, SHRUBBY COLDENIA. Low spreading suffrutescent subshrub, typically forming mats to 60 cm in diam.; older stems gnarled, usually 10–25 cm tall; leaves alternate, simple, often clustered at nodes, on short petioles 2–7 mm long, the blades to 11.5(–13) mm long and 8.5 mm wide, ovate to elliptic, thickened, obtuse to broadly acute apically and basally, the surfaces densely covered with white hairs; flowers pinkish, rose, or rarely white, sometimes with a yellow throat, sessile, usually solitary in leaf axils, the calyces long attenuate, the corollas funnel-shaped, 3.5–6(–7.5) mm long; fruits spherical, 2–3 mm in diam., splitting into 4 bony nutlets, shiny light brown to dark brown. Dry slopes, "A very pronounced calciophile, growing on limestone, caliche or rocky calcareous soils."—Correll & Johnston 1970; in East TX known only from Bexar Co. at extreme sw part of area (*J. Singhurst 5312*, 20 Jun 1996, BAYLU; Kartesz 2015); mainly s and w TX, also sw U.S. in AZ, CA, NM, NV, and UT. Mar–Aug. [*Coldenia canescens* DC., *Coldenia canescens* DC. var. *subnuda* I.M. Johnst.] The plants in TX are var. *canescens*. They can be distinguished by corollas 3.5–6(–7.5) mm long versus 7.0–9.0 mm long in var. *pulchella* (I.M. Johnston) A.T. Richardson, which is known only from sw Arizona and se California.

ELAEAGNACEAE Juss.

OLEASTER FAMILY

A small (ca. 45 species in 3 genera—Shultz & Varga 2021) family mainly native to temperate and warm areas of the n hemisphere with some in tropical se Asia and e Australia; most are deciduous shrubs to small trees with silvery or rusty scales or hairs; they are typically armed and usually have nitrogen-fixing root nodules (which contain actinomycete fungi with nitrogenase activity) which can convert atmospheric gaseous nitrogen into a form that can be utilized by plants (Visser et al. 1991); a number are used as ornamentals including species of *Elaeagnus* and *Hippophae*. The family is related to Barbeyaceae, Cannabaceae, Cecropiaceae, Celtidaceae, Moraceae, Rhamnaceae, Rosaceae, Ulmaceae, and Urticaceae in the Rosales (Bartish & Swenson 2004). (subclass Rosidae—Cronquist; order Rosales—APG IV)
FAMILY RECOGNITION IN THE FIELD: alternate-leaved, often armed, introduced shrubs or small trees with very distinctive, peltate or stellate, silvery scales conspicuously covering the leaves and young twigs; flowers 4-merous.
REFERENCES: Graham 1964; Bartish & Swenson 2004; Qin & Gilbert 2007; Shultz & Varga 2021; Weakley, A.S., and Southeastern Flora Team 2022.

ELAEAGNUS L. SILVER-BERRY, OLEASTER

Shrubs or small trees, often armed; leaves alternate, simple, petiolate, the blades evergreen or deciduous, entire; stipules absent; flowers 1–3 in the leaf axils, usually bisexual (can sometimes be unisexual), with a ± tubular or funnelform hypanthium (derived from the receptacle and constricted above the ovary); petals absent; sepals 4, spreading, similar to the hypanthium (and together with it appearing somewhat corolla-like); stamens 4, on short filaments attached to hypanthium; carpel 1; style 1; ovary superior, but basal portion of hypanthium persistent and making ovary seem inferior, with a single ovule; fruits drupe-like (actually achenes), enclosed by the fleshy receptacle base.

A genus variously treated as having 20–45 (Bartish & Swenson 2004) or ca. 40 (Mabberley 2017) species native to Europe, North America (1 species, *E. commutata* Bernh. ex Rydb.), and Asia to ne Australia including cultivated ornamental shrubs, some species are important for their edible fruits. Seed dispersal is by birds. A number are considered invasive. (Greek: *elaia*, olive, and *agnos*, the name of the CHASTE-TREE, *Vitex agnus-castus*, from *hagnos*, pure; the name was originally applied to a willow with massed white fruit)

1. Flowering in the fall (Oct–Nov) and fruiting in the spring (Mar–Apr); leaves evergreen; lower leaf surface with silver and sometimes reddish brown umbonate scales; branches usually thorny ____ **E. pungens**
1. Flowering in the spring and fruiting in the fall; leaves deciduous; lower leaf surface with silver umbonate scales OR with silver and some reddish brown umbonate scales; branches thorny or not so.
   2. Leaf blades 3–8 times as long as wide; lower leaf surface with silver umbonate scales ____ **E. angustifolia**
   2. Leaf blades 1.5–3 times as long as wide; lower leaf surface with silver and some reddish brown umbonate scales ____ **E. multiflora**

Elaeagnus angustifolia L., (narrow-leaved), OLEASTER, RUSSIAN-OLIVE. Shrubs or small, much-branched trees to ca. 8 m tall, usually thorny; leaves deciduous, blade 3–8 cm long, to 2 cm wide, lanceolate, the surfaces with conspicuous covering of silvery umbonate scales and/or stellate hairs, the lower surface covered even at maturity, the upper losing covering and becoming green; flowers fragrant, 1 cm or less long; hypanthium silvery scaly externally; sepals silvery scaly externally, yellowish on inner surface; fruits (actually an achene enclosed in expanded hypanthium base) 10–15 mm long, silvery scaly, yellowish, dryish but sweet. Widely cultivated and presumably escaping; Dallas (Turner et al. 2003), Harris (Singhurst et al. 2018), and Travis (Kartesz 2022) cos.; Panhandle and Trans-Pecos; widespread in the U.S. (except the se) and adjacent Canada. Spring. Native of Eurasia. This species has been planted as an ornamental and a windbreak. This species is considered an exotic pest that can become invasive in some parts of the U.S. (Tellman 1997; Yatskievych 2006). This nitrogen-fixing species may change the amount of nitrogen available in some habitats and allow the invasion of other exotics (Simons & Seastedt 1999).

Elaeagnus multiflora Thunb., (many-flowered), CHERRY SILVER-BERRY. Shrubs or small, much-branched trees, 1–2.5(–3)m tall; leaves deciduous, blades 1–2.5(–5) cm long, 1–1.5(–3) cm wide, broadly lanceolate, lower surface densely covered with silvery, umbonate scales and some reddish brown umbonate scales; outer surface of sepals also covered with reddish brown umbonate scales; hypanthium 5–10 mm long; calyx lobes broadly ovate, 4–5.5 mm long; fruit oblong, 10–15(–25) mm long. Mesic woods, disturbed areas, low woods; in TX known only from Harris Co. (Kartesz 2022) in far se East TX; e U.S. from NY s to GA and w to ID and MS, also MO and TX. Native of e and se Asia.

Elaeagnus pungens Thunb., (piercing, sharp-pointed), THORNY ELAEAGNUS, THORNY-OLIVE, SILVERTHORN. Spreading shrub or tree to ca. 4 m tall, usually with thorn-like newly emerged branches; leaves evergreen, blade 4–10 cm long, 1.5–2.5 cm wide, elliptic to nearly ovate, the lower surface densely covered with silvery, umbonate scales and sometimes dotted with a few reddish brown umbonate scales giving the surface a striking dirty silver appearance, the margins undulate; flowers fragrant, with a tubular hypanthium 7–10 mm long; sepal lobes 4, petal-like, shorter than the hypanthium; petals absent; fruits (actually an achene enclosed in expanded hypanthium base) 8–12 mm long, covered with scales and tipped with the persistent distal portion of hypanthium, initially brown, eventually red. Widely cultivated and escapes; Brazos Dallas, (Turner et al. 2003), Harris, Jasper, Jefferson, Newton, and Travis (Kartesz 2022) cos.; also Tarrant Co. (Turner et al. 2003) just w of East TX; we are not aware of other TX localities; se U.S. from VA s to FL w to OK and TX. Late fall. Native of Japan and China.

ELATINACEAE Dumort.

WATERWORT FAMILY

Ours small annual herbs of wet areas; stems often rooting at lower nodes; leaves simple, opposite, entire or minutely toothed, sessile or nearly so, with paired, membranous, scale-like stipules between them; flowers 1–few in the leaf axils, small, inconspicuous, radially symmetrical, 2–5-merous; sepals and petals both present; stamens 1 or 2 times as many as the petals; pistil 1; ovary superior, 2–5-celled (= locules); fruit a subglobose capsule; seeds pitted.

A small (2 genera variously treated as having 34—Mabberley 2017 to 40–50 species—Yatskievych 2006) subcosmopolitan but especially tropical family including a number of aquatics. While the relationships of this morphologically reduced family were in the past unclear, molecular research (Davis & Chase 2004; Xi et al. 2012) has shown that it is closely related to (sister group) the Malpighiaceae (BARBADOS-CHERRY family) in order Malpighiales (which includes such families as Euphorbiaceae, Hypericaceae, Salicaceae, and Violaceae). (subclass Dilleniidae—Cronquist; order Malpighiales—APG IV)

FAMILY RECOGNITION IN THE FIELD: small wet area herbs with opposite, simple, stipulate leaves and inconspicuous flowers.

REFERENCES: Tucker 1986; Davis & Chase 2004; Xi et al. 2012.

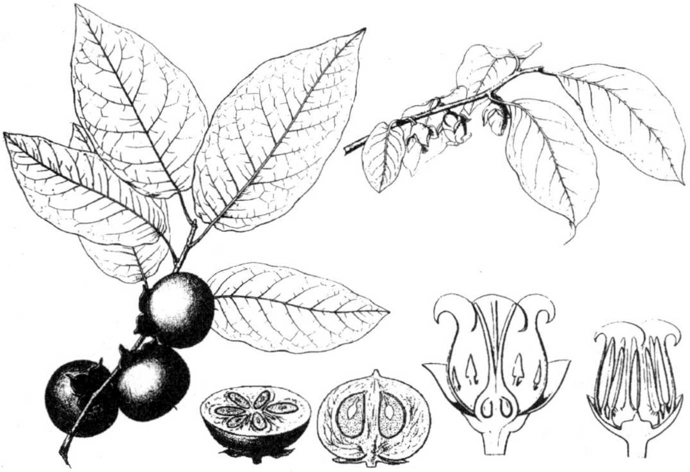

Diospyros virginiana [SA3]

Ehretia anacua [VIN]

Tiquilia canescens [CRO]

Elaeagnus angustifolia [ROS]

Elaeagnus pungens [GWO]

Elaeagnus mulitflora [BA1]

Bergia texana [MAS]

1. Plants erect to ascending, 10–40 cm tall; stems with short glandular pubescence; sepals and petals usually 5 each; sepals with thick green midrib and scarious margins, with a small sharp tooth at tip **Bergia**
1. Plants creeping and mat-forming (some stems can be ascending to erect), 10 cm or less tall; stems glabrous; sepals and petals 2–3 each; sepals without distinct midrib, without scarious margin, obtuse **Elatine**

BERGIA L. BERGIA

A genus of 20–25 (Yatskievych 2006; Tucker 2016; Mabberley 2017) species native to warm areas of the world, particularly the Old World tropics. (Named for Peter Bergius, 1730–1790, Swedish botanist and student of Linnaeus)

Bergia texana (Hook.) Seub., (of Texas), TEXAS BERGIA. Annual, usually branched at base, the whole plant ± glandular-puberulent; leaves to 30 mm long, elliptic to oblong, base attenuate, apex acute, serrulate with minute, gland-tipped teeth; flowers (–1)3–5 in the leaf axils, on short pedicels (1–)3–8 mm long; sepals 5, 2–4 mm long; petals 5, white, not exceeding the sepals; stamens 5 or 10; fruits to ca. 3 cm wide. Muddy pond margins, along creeks; Dallas, Denton, Falls, Grayson, Travis (Turner et al. 2003), and Brazos (*Parks s.n.*, 1948, TAES) cos.; widely scattered in TX; widely scattered mainly in the w 2/3 of the U.S. May–Oct.

ELATINE L. WATERWORT

Low, creeping, mat-forming, almost moss-like annuals, often rooting at the nodes; flowers 1–2 per node.

A genus of ca. 25 species (Razifard et al. 2016) with an interrupted cosmopolitan distribution (on all continents except Antarctica); some are used as aquarium plants. Yatskievych (2006) noted that *Elatine* plants growing in water can have significantly different growth forms and appearance than terrestrial individuals. Submerged plants tend to be darker green, have longer internodes, and have longer and narrower leaves. (Greek: *elatino*, fir-like, from a European species that suggests such a plant in miniature)

REFERENCES: Duncan 1964; Kaul 1986; Haynes 1998; Holmgren & Holmgren 2004; Razifard et al. 2016.

1. Leaves without a notch at tip, usually obovate to narrowly oblong-ovate (rarely linear-spatulate); largest seeds with 9–15 pits in each row **E. brachysperma**
1. Leaves often with a slight notch at tip, linear to narrowly spatulate; largest seeds with 16–25 pits in each row **E. triandra**

Elatine brachysperma A. Gray, (short-seeded), SHORT-SEED WATERWORT. Forming small mats to ca. 5 cm across, the branches ascending; leaves to 6 mm long, 2 mm wide; sepals 2 or with a third reduced; petals 3, greenish white, often pinkish. On mud or in shallow water; scattered in East TX; also scattered in c and s portions of TX; widely scattered in the U.S. Mar–Oct. [*E. triandra* Schkuhr var. *brachysperma* (A. Gray) Fassett] While we are following Tucker (1986) and Kartesz (2022) in treating this taxon at the species level, the group needs taxonomic study (see discussion under *E. triandra*) and some recent authors (e.g., Yatskievych 2006) treat *E. brachysperma* as a variety of *E. triandra*.

Elatine triandra Schkuhr, (with three anthers or stamens), AMERICAN WATERWORT. Forming mats, the branches can be ascending to erect; leaves 3–6(–12) mm long, 1–3 mm wide; perianth greenish or pinkish, usually with 3 sepals and 3 petals. On mud or in shallow water; found at Dallas by Reverchon in Mar 1874 (*Reverchon s.n.*, 1874, SMU) apparently not collected there since, also Bastrop Co. (Kartesz 2024); widely scattered in the U.S.; Canada. Mar–Oct. The taxonomy of this group is confused and needs work, with authors variously recognizing one, two, or three taxa. We apparently have two *Elatine* taxa in TX, and are following previous TX authors (e.g., Correll & Johnston 1970; Hatch et al. 1990; Diggs et al. 1999; Turner et al. 2003) in using the name *E. triandra* for this taxon. Others treat the complex as having three varieties or species (e.g., Yatskievych 2006), the two treated here plus *E. americana* (Pursh) Arn. [*E. triandra* var. *americana* (Pursh) Fassett], which occurs to the n and e of TX. Still other authorities (e.g., Kaul 1986) include *E. americana* in their concept of *E. triandra*, but recognize *E. brachysperma* as a separate species. Finally, some authors (e.g., Duncan 1964) recognize only a single species in the complex, including both *E. americana* and *E. brachysperma* within *E. triandra*. Future research will hopefully clarify and stabilize species concepts and nomenclature.

ERICACEAE Juss.

HEATH OR BLUEBERRY FAMILY

Ours shrubs, trees, or mycoparasitic herbs; leaves simple, alternate (or reduced to scale-like bracts in *Hypopitys* and *Monotropa*); petals united (separate in *Hypopitys* and *Monotropa*); stamens twice as many as corolla lobes; anthers awned or awnless, opening by terminal pores (in *Hypopitys* and *Monotropa* by slits); pollen grains in tetrads (singly in *Hypopitys* and *Monotropa*); ovary superior or inferior; fruit a capsule or berry.

A large (4,215–4,426 species in 126 genera (Stevens 2001 onwards; Schwery et al. 2014), cosmopolitan family of usually shrubs with some herbs and trees; they are usually evergreen (though sometimes deciduous) and often xeromorphic (= dry-adapted). The family is particularly prominent in Arctic and temperate areas, while the majority of tropical taxa are montane (where a number are epiphytes or lianas). Most species are found in nutrient-poor, acidic habitats and have a relationship with mycorrhizal fungi. The family includes many important ornamentals (e.g., AZALEAS and RHODODENDRONS in the genus *Rhododendron*, HEATHS in the genus *Erica*, *Arctostaphylos* (BEARBERRY and MANZANITA), and *Kalmia* (MOUNTAIN-LAUREL), as well as edible *Vaccinium* species (BLUEBERRIES, CRANBERRIES, and LINGONBERRIES) and *Gaylussacia* species (HUCKLEBERRIES). The leaves of other species (e.g., *Kalmia*—LAMBKILL, SHEEPKILL, LEUCOTHOË, and *Lyonia*) are poisonous; *Kalmia* species can be the source of poison honey, when visited by bees (Hardin & Brownie 1993). The family has sometimes been split with groups such as the mycosaprophytic/mycoparasitic *Monotropa* and its relatives raised to family rank (e.g., Kartesz 1994; Jones et al. 1997) (technically non-photosynthetic myco-heterotrophic epiparasitic (see generic synopsis of *Monotropa* for an explanation of this type of metabolism)). However, molecular studies and cladistic analyses (Kron 1996; Kron et al. 2002a; Freudenstein et al. 2016) support the inclusion of the Monotropaceae, Pyrolaceae, Epacridaceae, and Empetraceae to form a monophyletic Ericaceae. We are therefore treating the Ericaceae in the broad inclusive sense (following Stevens 2001 onwards and APG IV 2016). The family is sister to Cyrillaceae and these together to Clethraceae (Anderberg et al. 2002). Family name from *Erica*, HEATH or HEATHER, a genus of ca. 665 species of shrubs and small trees native primarily to s Africa (650 species there with 520 endemic to the Cape region), and also in tropical African mountains, the Mediterranean, Macaronesia, and Europe; *Erica* species cover large areas of moorlands in Europe. (Greek: *ereike*, heather, from *ereiko*, to break; an infusion from the leaves was supposed to break bladder stones) (subclass Dilleniidae—Cronquist; order Ericales—APG IV)

FAMILY RECOGNITION IN THE FIELD: ours usually shrubs or trees (2 species of reduced herbs without chlorophyll) with alternate, simple, often rather leathery, evergreen or deciduous leaves and usually *urn-shaped*, cylindric, campanulate or trumpet-shaped corollas; anthers usually opening by terminal pores.

REFERENCES: Small 1914a, 1914b; Wood 1961; Judd & Kron 1993; Kron & Chase 1993; Luteyn 1995; Stevens 1995; Kron 1996; Luteyn et al. 1996; Reveal 2000; Anderberg et al. 2002; Kron et al. 2002a; Stevens et al. 2004; Tucker 2009a; Schwery et al. 2014; Freudenstein et al. 2016; Kriebel et al. 2023.

1. Plants without green photosynthetic leaves; very small herbs lacking chlorophyll (plant white to yellowish, pinkish, orangish, or reddish), only 5–30 cm tall; leaves reduced to scale-like bracts.
   2. Flowers usually 3–10 (rarely reduced to only 1) per inflorescence; inflorescences (as well as rest of plant) at flowering time ranging in color from light yellow (not white) to orangish or reddish, drying brownish; plants often downy or finely pubescent; stigma often subtended by a ring of stiff hairs ____ **Hypopitys**
   2. Flower solitary per inflorescence, never more; inflorescences (as well as rest of plant) at flowering time usually waxy white or rarely tinged with pink, orange, or red or combinations of these, drying black; plants glabrous; stigma not subtended by a ring of stiff hairs ____ **Monotropa**
1. Plants with green photosynthetic leaves with well-developed blades; shrubs or trees.
   3. Corollas large and conspicuous, 20–58 mm long, trumpet-shaped, with a long narrow tube and a widely flaring limb; stamens conspicuously exserted beyond the corollas ____ **Rhododendron**
   3. Corollas much smaller, < 15 mm long, often much less, usually cylindrical to urn-shaped, usually constricted apically; stamens not exserted (except in 1 small-flowered species of *Vaccinium*).
      4. Flowers in panicles; fruit a berry with a distinctive roughened-tuberculate surface; ovary superior; plant a small tree or rarely arborescent shrub, (2–)4–6 m or more tall; leaves evergreen; species known only on sw margin of East TX ____ **Arbutus**
      4. Flowers in racemes or umbel-like clusters, rarely sub-paniculate; fruit a dry capsule OR a berry with a smooth surface; ovary superior and developing into a capsule or inferior and developing into a berry; plants varying from shrubs < 2 m tall to small trees up to 8 m tall; leaves essentially evergreen OR deciduous; including species primarily of e portion of East TX.
         5. Fruit fleshy (blueberry-like or huckleberry-like), indehiscent, crowned by the small persistent calyx teeth; ovary inferior.
            6. Fruits baccate (= berry-like and soft); seeds 2–40, testa thin; leaves not glandular punctate; includes species widely spread in East TX ____ **Vaccinium**
            6. Fruits drupaceous; with 10 seed-like pyrenes (each consisting of a seed surrounded by a bony testa); leaf blades usually glandular punctate on lower surface; species known in TX only from a single location in Newton Co. ____ **Gaylussacia**
         5. Fruit a dry capsule, dehiscent; ovary superior.
            7. Flowers in secund (= flowers on one side) racemes, the flowers all hanging; pedicels with 2 bracteoles just below calyx; capsules not thickened on the sutures; anthers with 4 apical awns; leaves deciduous ____ **Eubotrys**

7. Flowers in various inflorescences, but not in secund racemes; pedicels with bracteoles at base; capsules thickened on the sutures or not so; anthers with either 2 or no apical awns (but there can be a pair of spurs below anther/filament junction; leaves evergreen or deciduous.
   8. Inflorescences with 8–30 flowers; corollas white, 5–8 mm long; filaments straight; anthers with 2 apical awns; capsules not thickened on the sutures, oblate (= flattened at the poles, wider than long), (4.5–)5–6 mm wide; leaves evergreen ____ **Leucothoë**
   8. Inflorescences with (2–)5–8(–12) flowers; corollas white to pink or red, 2–14 mm long; filaments S-shaped; anthers without apical awns, but can have a pair of spurs below anther/filament junction; capsules thickened on the sutures, the thickenings appearing like ribs, the capsules ovoid to subglobose or globose, usually 3–5 mm wide; leaves evergreen or deciduous ____ **Lyonia**

ARBUTUS L. MADRONE, MADROÑO

A genus of 12 species, with 3 in Europe, the Middle East, and North Africa, 1 on the Canary Islands, and 8 in the New World (British Columbia s to Nicaragua) (Sörensen 1995, 2009; Gonzalez-Elizondo et al. 2012) with the greatest diversity in Mexico. Hybridization between species of *Arbutus* is reported to be common (Gonzalez-Elizondo et al. 2012). Molecular work (Hileman et al. 2001) questions the monophyly of *Arbutus* suggesting that a number of segregate genera (*Arctostaphylos, Arctous, Comarostaphylis, Ornithostaphylos,* and *Xylococcus*) are derived from within a paraphyletic *Arbutus*. If the molecular phylogeny presented by Hileman et al. (2001) is correct, since *A. unedo* (s Europe and Ireland) was named by Linnaeus and is the oldest name in the genus, the genus name *Arbutus* will need to be retained for the Old World species, while a different name will need to be used for the w North American to Central America species. However, until more information is known, we are following the recent treatment for North America (Sörensen 2009) in treating *Arbutus* in the traditional manner; if strict monophyly is followed (see Stuessy & Hörandl 2014 for comments on the topic), numerous name changes will almost certainly be necessary in the future. The typically orangish-red to red fruits with unprotected seeds are probably bird dispersed (Parker & Stickrod 2022). (Latin name for strawberry tree, probably originally applied to *Arbutus unedo* L., strawberry tree)
REFERENCES: Sörensen 1995, 2009; Hileman et al. 2001; Grant 2003; Gonzalez-Elizondo et al. 2012; Parker & Stickrod 2022; De Santis et al. 2024.

Arbutus xalapensis Kunth, (for its type locality near Jalapa or Xalapa, in the Mexican state of Veracruz), TEXAS MADRONE, MADROÑO, LADY'S-LEG, NAKED-INDIAN, AMAZAQUITL, XOXOCOTE. Ours a small tree or rarely arborescent shrub, (2–)4–6 m or more tall; bark pinkish to brick red, peeling off in large smooth flakes, the naked branches distinctive; leaves alternate, evergreen, coriaceous, the blades to 6(–7.5) cm long and 30(–45) mm wide, ovate to oval or oblong-elliptic, entire to serrate, glabrate with age, basally tapered to rarely slightly cordate, the petioles 1.2–2.5 cm long; inflorescence a panicle, often showy, the axes hairy with glandular and eglandular hairs; pedicels 1.6–3.2 mm long, with pubescence like axes; flowers perfect; calyx lobes 5, tan to pinkish white; corollas ovoid-urceolate, white, often pink-tinged, ca. 5–6(–7) mm long; stamens included, 10; anthers ca. 1.5 mm long, with a pair of slender reflexed spurs on the back, opening by terminal pores; filaments slender with swollen bases; ovary superior; stigma capitate; fruit a ± spherical berry, yellowish red to red or blackish red, 5.5–7.5(–9) mm in diam., fleshy, with ovules 2–3(–5) per locule, the fruit surface roughened-tuberculate; seeds 1–5, distinct. Wooded rocky hills, slopes, canyons, plains; Travis Co. (TEX), three occurrences are known n of the Colorado River in Travis Co. (Balcones Canyonlands Nat. Wildlife Refuge, C. Sexton, pers. comm.), also Comal, Hays, and Williamson (TEX) cos.; sw margin of East TX and adjacent Edwards Plateau and disjunct to the Trans-Pecos; also Hockley Co. (LL) in the High Plains; TX and se NM; also Mexico to Central America. Feb–Apr, the fruit ripening about 6 weeks later (Sörensen 2009). [*A. texana* Buckley, *A. xalapensis* var. *texana* (Buckley) A. Gray] According to Sörensen (1995), the TX plants with small glabrous leaves are part of a north-south cline involving a gradual change from plants with large pubescent leaves found in s Mexico and n Central America. The TX plants have often been recognized as a distinct species or variety, but we are following Sörensen (1995, 2009) in lumping them into *A. xalapensis* despite their morphological distinctiveness. The fruits are edible (Correll & Johnston 1970), though bland, and they can be made into jellies and jams; in n Mexico the Tarahumara make an alcoholic beverage from the fermented fruits (Aneyefortexas 2011). In the Trans-Pecos "many madrones … carry deep scratch marks from the claws of bears climbing to get the fruit" (Aneyefortexas 2011). TEXAS MADRONE is difficult to transplant or grow, and in the wild the "seedlings are palatable to deer so few seedlings survive to grow to maturity" (Aneyefortexas 2011). ASHE JUNIPER (*Juniperus ashei*) "has been reported to serve as a nurse tree for juvenile *A. xalapensis* plants" (Davis & Van Auken 2019).

EUBOTRYS Nutt. DECIDUOUS FETTER-BUSH

A genus of 2 low altitude species of the e U.S. (NY s to FL w to TX) (Stevens et al. 2004; Tucker 2009b); the two species have traditionally been treated in *Leucothoë* (a genus of 5 species native to the Himalayas, Japan, and North America). However, that genus, as treated more broadly in the past, is not monophyletic (Bush et al. 2010). Based on both morphological and molecular

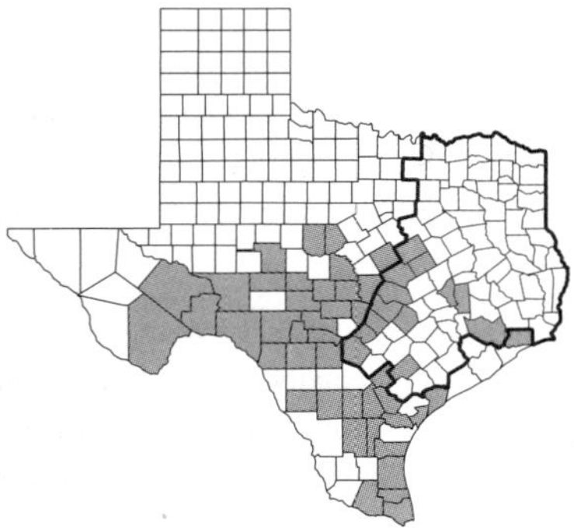

Diospyros texana

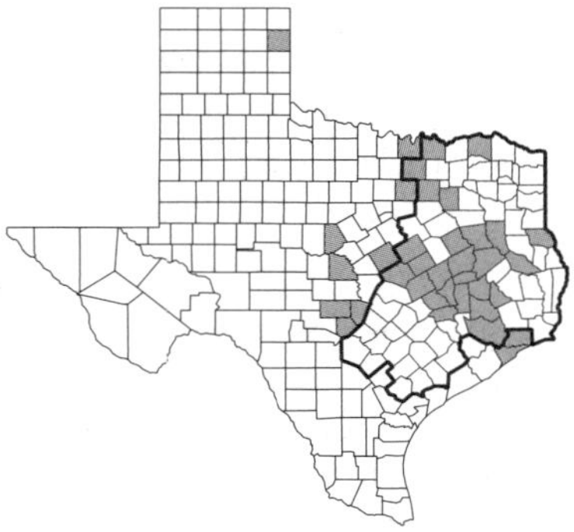

Diospyros virginiana

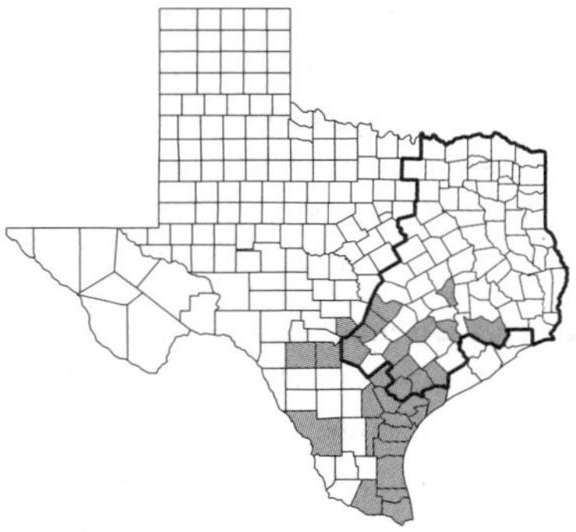

Ehretia anacua

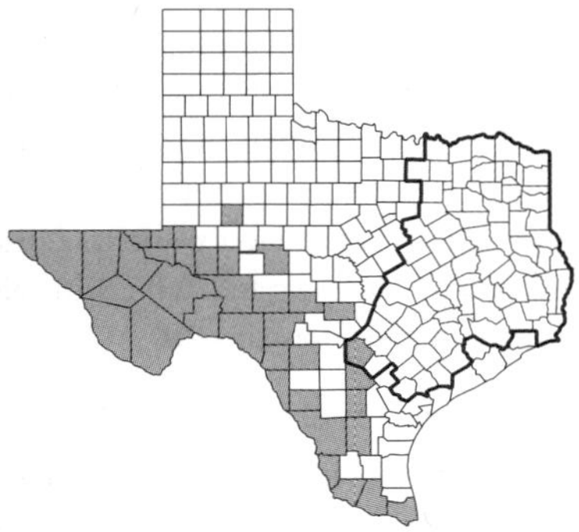

Tiquilia canescens

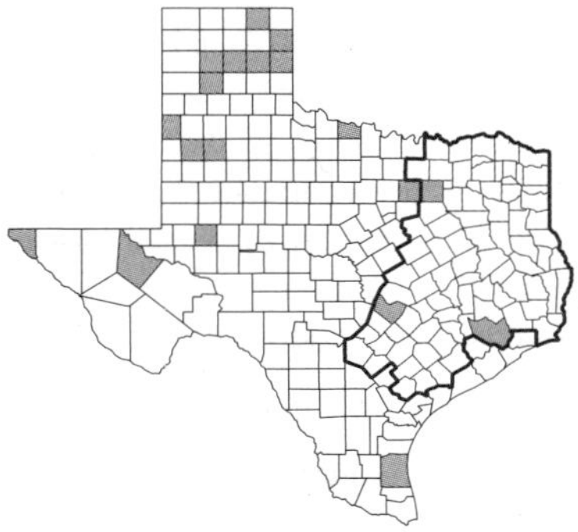

Elaeagnus angustifolia

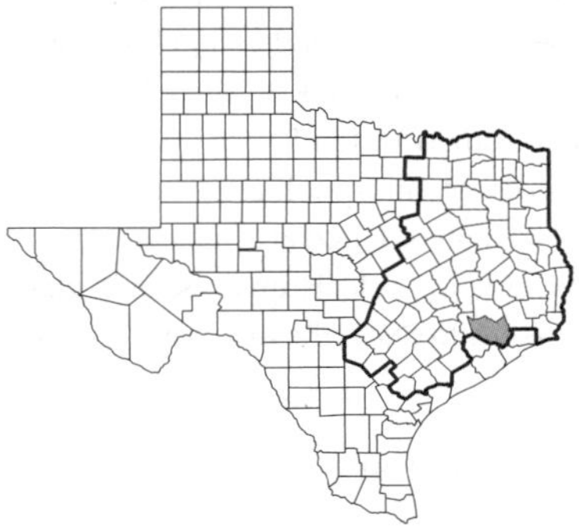

Elaeagnus multiflora

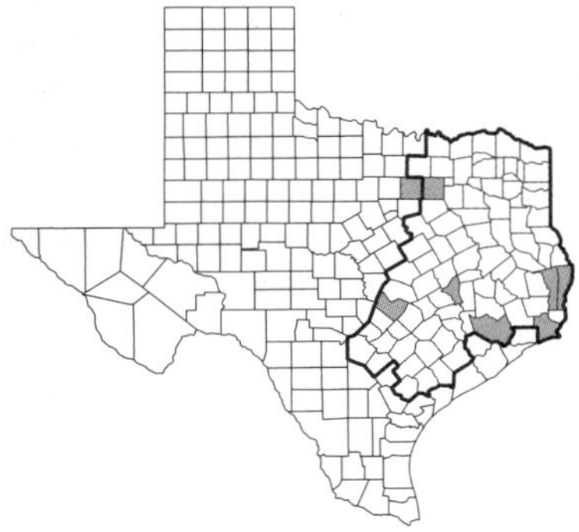

Elaeagnus pungens

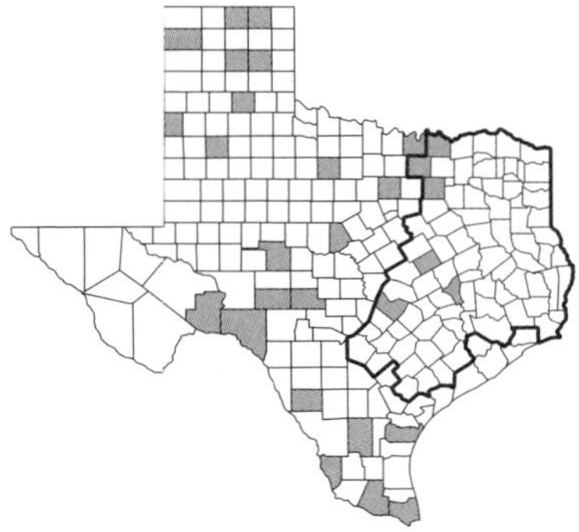

Bergia texana

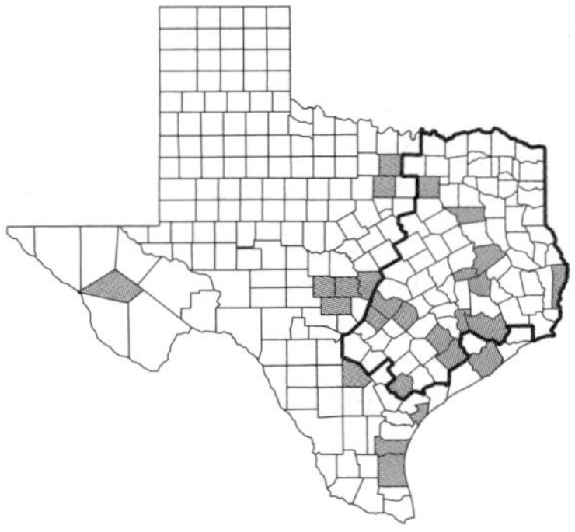

Elatine brachysperma

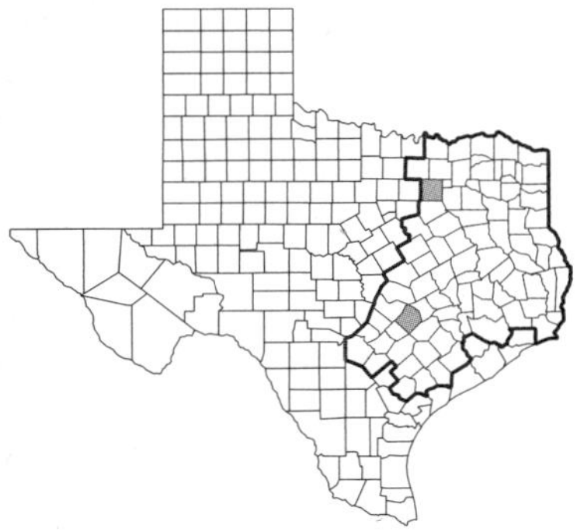

Elatine triandra

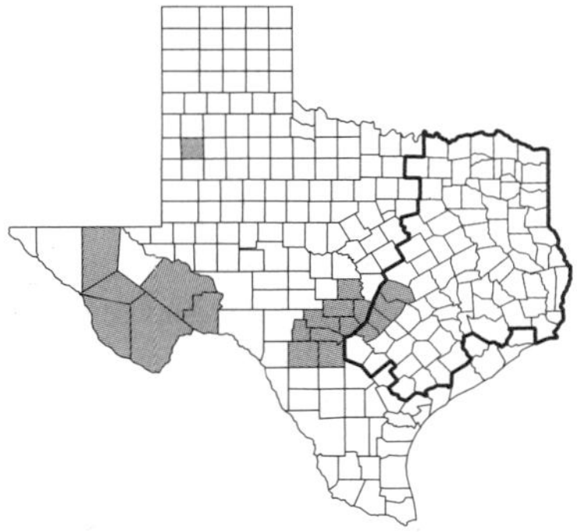

Arbutus xalapensis

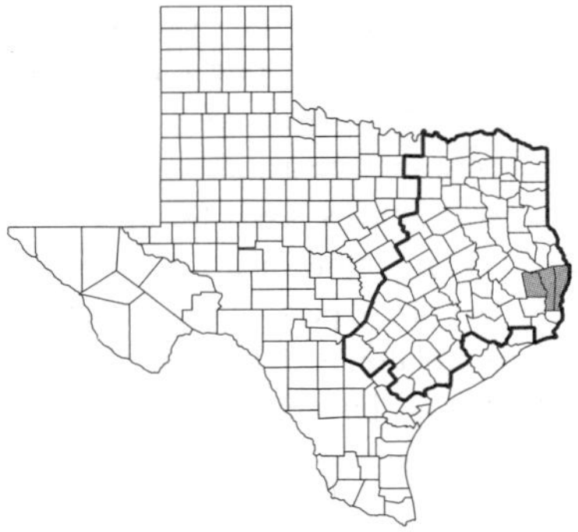

Eubotrys racemosa

data most recent authors recognize *Eubotrys* as warranting generic recognition (e.g., Powell & Kron 2001; Kron et al. 2002a; Stevens et al. 2004; Bush et al. 2009, 2010; Tucker 2009b; Judd et al. 2012); it has been shown to be more closely related to *Chamaedaphne* (LEATHER-LEAF) and *Gaultheria* (WINTERGREEN) than to *Leucothoë*. (Greek: *eu*, good, fine, and *botrys*, cluster or bunch of grapes, in reference to the capsule fruits in a tight raceme)
REFERENCES: Luteyn et al. 1996; Powell & Kron 2001; Stevens et al. 2004; Waselkov & Judd 2008; Tucker 2009b; Bush et al. 2010; Judd et al. 2012.

Eubotrys racemosa (L.) Nutt., (with a raceme type inflorescence), SWAMP DOG-HOBBLE, FETTER-BUSH, COASTAL FETTER-BUSH, DECIDUOUS FETTER-BUSH, CALFKILL. Deciduous shrub to 4 m tall; leaves alternate, short petiolate, the blades thin, usually 2–8 cm long, elliptic to ovate or obovate, serrulate, acute to acuminate apically, sometimes becoming reddish in autumn; inflorescences racemes terminating the previous year's growth, variable in length, 0.5–21 cm, these secund, bracteate, the bracts ca. 2–27 mm long, quickly deciduous; pedicels 1–3.5 mm long, with 2 bracteoles near apex; flowers 5-merous, opening before or as the leaves emerge; sepals connate basally, persistent; corollas urceolate-cylindrical, 6–8 mm long, white, sometimes with a pink tinge; stamens 8(–10); anthers with 4 awns; ovary superior; capsules depressed globose, unlobed, 1.5–3 mm long; seeds 5–10. Moist thickets, swampy woods, shorelines; se part of East TX in Jasper (*Nixon 14916*, GH—fide Judd et al. 2012), Newton (*Allen 22152* BRIT, Turner et al. 2003; Allen & Lewis 2023), and Tyler (*Correll 34224* GH—fide Judd et al. 2012) cos.; mainly e U.S. coastal states from MA s to FL w to TX. Mar–Jun. [*Andromeda racemosa* L., *Leucothoë racemosa* (L.) A. Gray] While the species shows considerable morphological variation, it is "more or less continuous" and segregating either different species or varieties is "unwarranted" (Judd et al. 2012). Marcus (2018) noted that "ingestion of the leaves, flowers, and nectar of this plant has caused poisoning in cattle. Humans should generally avoid ingesting any plant that is toxic to animals." Burrows and Tyrl (2001) also discuss its toxicity; the toxins are ericaceous diterpenoids, variously called andromedotoxins or grayanotoxins.

GAYLUSSACIA Kunth HUCKLEBERRY

A genus of 69 species of e North America and South America, with the "vast majority of species in the genus occurring in the Cerrado and Atlantic Rain Forest regions of Brazil" (Becker et al. 2024). Recent molecular research indicates that the genus evolved in temperate North America and from there moved into the tropics (Becker et al. 2024). It is a monophyletic lineage in the large monophyletic Vaccineae (the tribe that includes *Vaccinium* and its relatives). However, it is also within a non-monophyletic *Vaccinium*—that genus needs further research to clarify evolutionary lineages and the appropriate nomenclatural changes (Becker et al. 2024). The fruits are edible, notably those of *G. baccata* of e North America. The genus is somewhat similar to *Vaccinium* in having edible fleshy fruits, but can be distinguished by having fruits with 10 bony pyrenes (each with 1 seed) and usually the presence of resinous glands on the undersides of the leaves; vs. fruits with 2–40 tiny seeds and leaves without resinous glands beneath. Mabberley (2017) noted that *G. brachycera* of the e U.S. is known from an approx. 13,000 year old clone occupying 40 hectares in PA. (for Louis Joseph Gay-Lussac, 1778–1850, French chemist—Sorrie et al. 2009)
REFERENCES: Small 1897; Camp 1935, 1941; Luteyn et al. 1996; Floyd 2002; Sorrie et al. 2009; Becker et al. 2024.

Gaylussacia dumosa (Andrews) A. Gray, (from Latin *dumosus*: bushy, abounding in bushes), DWARF HUCKLEBERRY. Small shrubs usually 10–40 cm tall, forming colonies by long woody rhizomes; branches ascending from rhizomes, little branched above; leaves alternate, deciduous or semi-evergreen, with petioles 0.5–1.5 mm long, the blades 1.5–4 mm long, oblanceolate to obovate, with scattered stipulate-glandular hairs and sessile glands, subcoriaceous, cuneate basally, apically obtuse to subacute, mucronate, with margins entire or nearly so; inflorescences erect or arching, with 4–8(–10) flowers; bracts leaf-like, persistent, as long as or longer than the pedicels; pedicels with 1–2 bracteoles; sepals 5; petals 5; corollas white to pink, sometimes reddish, campanulate, 3–5 mm long, the lobes 1–1.5 mm long; stamens 10, shorter than the corolla, the anthers much longer than the filaments, without awns, dehiscent through terminal pores; ovaries inferior; fruits 5–8(–10) mm in diam., drupaceous, fleshy, black, shiny, stipitate-glandular, crowned by the tiny persistent calyx teeth; seeds 10, the testa bony (pyrenes). Dry to wet, acidic forests to savannahs, seeps; the only TX location is from Newton Co. near the LA border (along White Oak Creek off of county road 3038 S of county road 3060 W of TX 87 ca. 5 mi N of Bleakwood (*C. Allen et al.*, 22 Mar 2007, *19694* BRIT, BRIT373032)—fide Peter Fritsch, pers. comm.; also the pyrenes, which definitively distinguish *Gaylussacia* from *Vaccinium*, are visible in the dried fruits of the specimen); far e U.S. and adjacent Canada s to FL and w to LA, with a disjunct location in East TX. Mar–Aug. [*G. dumosa* var. *hirtella* (W.T. Aiton) A. Gray, *G. dumosa* var. *humilis* (P. Watson) Zabel, *G. hirtella* (W.T. Aiton) Torr. & A. Gray, *Lasiococcus dumosus* (Andrews) Small, *Vaccinium dumosum* Andrews, *V. dumosum* var. *humile* P. Watson, *V. hirtellum* W.T. Aiton]

HYPOPITYS Hill AMERICAN PINESAP

A genus of a single, widespread (North America, Central America, Asia, Europe), variable species; it was traditionally included in *Monotropa*, but molecular research supports its separation into its own genus with up to perhaps seven species (e.g., Neyland

& Hennigan 2004; Braukmann & Stefanovic 2012; Broe 2014; Freudenstein & Broe 2024; see further discussion under *Monotropa*). As noted in Weakley, A.S. and Southeastern Flora Team (2025), "These [molecular] studies also demonstrate that several cryptic to semi-cryptic species should be recognized; plants in North America are not conspecific with those in Eurasia, and at least six species appear to be present in e. North America (Broe 2014; Klooster & Culley 2010; Haines 2011)." While more species thus may be recognized in the future, until there is nomenclatural stability regarding the cryptic species, we are continuing to recognize a single variable species. *Hypopitys* is a non-photosynthetic myco-heterotrophic epiparasite (see genus synopsis of *Monotropa* for an explanation of this term). It can also be referred to as a holomycotroph since it obtains all its carbon from a fungus (Freudenstein & Broe 2024). The correct authority for the genus *Hypopitys* has a confusing history. However, a proposal by Liu et al. (2017) has recommended conserving the name as *Hypopitys* Hill. (from Greek: *hypo*, under, and *pitys*, the pine)

REFERENCES: Björkman 1960; Correll 1965; Riskind & Amerson 1977; Wallace 1977, 1995; Cullings & Hileman 1997; Kron & Johnson 1997; Cullings 2000; Bidartondo & Bruns 2001; Neyland 2004; Neyland & Hennigan 2004; Stevens et al. 2004; Wallace 2009; Klooster & Culley 2010; Haines 2011; Braukmann & Stefanovic 2012; Stoyanov et al. 2012; Broe 2014; Freudenstein & Broe 2024.

Hypopitys monotropa Crantz, (from Greek: *monos*, one, and *tropos*, turn, from the summit of the flowering stem nodding, turned to one side), AMERICAN PINESAP, DUTCHMAN'S-PIPE, FALSE BEECHDROPS, YELLOW-BIRD'S-NEST. Plants obligately myco-heterotrophic epiparasitic (= indirectly parasitize plants via fungi) fleshy herbs lacking chlorophyll and non-photosynthetic, the entire plant light yellow to orangish or reddish (but not white), drying brownish, 5–32 cm tall; stems unbranched, minutely hairy; leaves reduced to scale-like bracts 4–12 mm long; inflorescence usually a 3–10-flowered terminal raceme; pedicels slender, 2–20 mm long, to 30 mm in fruit; flowers 4–5-merous; sepals ± unlike petals, sometimes absent; petals oblong, 8–17(–20) mm long, 4–8 mm wide, separate; intrastaminal nectaries present at base of the ovary; stamens 8–10, dehiscent by slits; pollen grains shed singly (usually in tetrads in the Ericaceae); styles 1–2 mm wide; ovaries superior; capsules 4–5 segmented, 6–10 mm long not including persistent style, the segments often falling after seed dispersal; seeds numerous. Humus in woods; Lamar (*Correll & Correll 27488*, May 1963, LL; Correll 1965), Titus (*Riskind & Amerson 1678,* 1973, LL; Riskind & Amerson 1977), and Shelby (*D.S. Correll & Helen B. Correll 31124* LL) cos. in East TX; also disjunct to La Salle Co. in the South TX Plains and Culberson Co. in the Trans-Pecos (*Thomas F. Patterson 613* LL); widespread in most of the U.S. and adjacent Canada; also Mexico; Central America; (Guatemala); Asia; and Europe. Apr–Jul. [*H. americana* (DC.) Nutt., *H. fimbriata* (A. Gray) Howell, *H. lanuginosa* (Michx.) Raf., *Monotropa hypotitys* L., *M. hypopitys* subsp. *lanuginosa* (Michx.) H. Hara, *M. latisquama* (Rydb.) Hultén] Though bumblebees visit the flowers (Wallace 1977), self-pollination appears likely as pollen is shed directly onto the stigma (Yatskievych 2006). It is now known that there is "extreme specificity" in the fungi with which the epiparasite interacts (Bidartondo & Bruns 2001, 2002); in the case of *H. monotropa* it is with the basidiomycete fungal genus *Tricholoma* (Tricholmataceae) (Bidartondo & Bruns 2001). This is an extremely widespread variable species that has often been subdivided into a variety of segregates; as noted in the generic discussion above, recent research suggests that more species may be recognized in the future. However, a number of recent works (e.g., Wallace 1995, 2009; Weakley et al. 2012; Freudenstein & Broe 2024) treat all the variation within one variable species, and until the phylogeny and nomenclature of the group are clarified, we are continuing to recognize a single species. While this species has long gone under the name *Monotropa hypopitys*, for nomenclatural reasons, if placed in the genus *Hypopitys*, the correct name becomes *H. monotropa*. It should be pointed out that some authorities (e.g., Broe 2014; Weakley, A.S., and Southeastern Flora Team 2025) indicate that some plants occurring in the se U.S., including some in e TX should be separated as *Hypopitys lanuginosa* (APPALACHIAN PINESAP, HAIRY PINESAP) with the authority variously given as (Michx.) Raf., (Michx.) Nutt., or (Michx.) Small.

LEUCOTHOË D. Don HOBBLEBUSH, SWEET-BELLS, DOG-LAUREL, DOG-HOBBLE

A genus of 5 evergreen species of the e and w US and e Asia; they are unusual in that inflorescences form in autumn with meiosis also occurring in the autumn (but with flowering in spring). The genus is an example of a disjunct e Asia–e North America distribution (see discussion under *Campsis* in the Bignoniaceae). *Leucothoë* has been treated more broadly (e.g., Luteyn et al. 1996) to include the deciduous species now recognized in *Eubotrys* (2 species of the se US) and *Eubotryoides* (1 sp. of Japan). According to Judd et al. (2013), "Members of the *Leucothoë* clade, which are evergreen, are sister to the deciduous *Eubotryoides grayana*, sharing the apomorphies of winged seeds with a distinctly protuberant hilum." It is more distantly related to a lineage including *Chamaedaphne*, *Eubotrys*, and *Gaultheria* (Bush et al. 2009). Some are known to be toxic (Breen 2025). *Leucothoë fontanesiana*, native to the se US, has been widely used as an ornamental due to the combination of its evergreen foliage and striking clusters of white flowers; it is closely related to *L. axillaris* but occurs further inland and at higher elevations, and has long acuminate leaf blades, and usually longer petioles, narrower calyx lobes, and inflorescences with more flowers

(Judd et al. 2013). (from Greek named for Leucothoë, one of the many loves of Apollo, the daughter of King Orchamus of Babylonia—Quattrocchi 2000)

REFERENCES: Ingram 1961, 1979; Melvin 1981; Luteyn et al. 1996; Tucker 2009c; Bush et al. 2009, 2010; Judd et al. 2013.

Leucothoë axillaris (Lam.) D. Don, (positioned in the leaf axils), COASTAL DOG-HOBBLE, SWAMP DOG-LAUREL, DOG-HOBBLE. Shrubs 0.5–1(–2) m tall; branches spreading, arching; leaves deciduous, coriaceous, the petioles (2–)5–10 mm long, the blades to ca. 8 cm long or rarely longer, oblong to lanceolate, oblanceolate, or elliptic, marginally nearly entire to spinulose-serrulate, apically obtuse to acute to abruptly short acuminate, sometimes purplish or reddish in winter; inflorescences axillary, fascicled or solitary, sessile, 2–5.5 cm long, usually with 8–30 flowers, produced on previous year's wood; bracts more or less deciduous, ovate, usually 1.5–2.5 mm long; bracteoles 2, basal or in lower 1/3 of pedicel, partly covering pedicel; pedicels usually 1–7 mm long, rarely longer; calyx lobes whitish, ovate-triangular to ovate or widely so, 1.3–2.7 mm long, 1.2–2.3 mm wide; petals 5, connate nearly their entire lengths, white, the corolla cylindric to urceolate, 5–8 mm long, the lobes much shorter than tube; stamens usually 8, the filaments with spreading hairs, the anthers with 2 apical awns, dehiscent through terminal pores; ovary superior, glabrous; capsules not thickened on the sutures, oblate (= flattened at the poles, wider than long), 2.5–3.5 mm long, (4.5–)5–6 mm wide; seeds numerous. Flood plains, other wet areas; known in TX only from Newton Co. (*C. Allen 21371.1* BRIT—fide Allen & Lewis 2023). Mar–May. [*Andromeda axillaris* Lam., *Leucothoë axillaris* var. *ambigens* Fernald, *L. catesbaei* (Walter) A. Gray, *L. platyphylla* Small] This species, which is somewhat similar to *Eubotrys racemosa*, has in the past been attributed to TX (e.g., National Park Service 1993). Until recently, we had not been able to find any authenticated record of its occurrence in the state (e.g., Turner 2003; Kartesz 2015; USDA Plants Database 2025); se VA s to FL w to LA and the easternmost part of TX. It was previously known just two parishes east of the TX-LA border in Natchitoches Parish (Judd et al. 2013). However, the Newton Co. collection by Charles Allen cited above confirms that the species is a member of the East TX flora. The leaves and nectar of this species are reported to be toxic; if ingested they may cause salivation and nasal discharge, sweating, tingling sensation, headache, depression, weakness, abdominal pain, nausea and vomiting, diarrhea, and paralysis (Breen 2025). Burrows and Tyrl (2001) also discuss its toxicity; the toxins are ericaceous diterpenoids, variously called andromedotoxins or grayanotoxins; they can be toxic to humans and a variety of other animals. This species is sometimes used as an ornamental not only because of the inflorescences but also the purplish or reddish winter foliage (Breen 2025).

LYONIA Nutt. STAGGER-BUSH, MALEBERRY, FETTER-BUSH

Shrubs to ca. 4 m tall, usually rhizomatous; leaves alternate, deciduous or evergreen, entire or serrulate; inflorescences axillary, umbel-like clusters to racemes of fascicles or short racemes, to almost panicle-like; flowers usually 5-merous, pedicellate, the pedicels with 2 bracteoles at or near base; sepals usually slightly connate; corollas subglobose-urceolate to ± cylindrical, the lobes much shorter than the tube; white to pink or rarely reddish; stamens usually 10, with anthers opening by apical pores, lacking apical awns, the filaments S-shaped, in ours with a pair of short spurs below anther/filament junction; ovary superior; capsules with thickened sutures/ribs; seeds numerous.

A genus of 36 species native to e and se Asia, e North America including Mexico, and especially the Greater Antilles (Judd 1995, 2009); some are cultivated as ornamental shrubs. The genus is an example of a disjunct e Asia–e North America distribution (see discussion under *Campsis* in the Bignoniaceae). The S-shaped filaments act as springs, ensuring that pollinators are dusted with pollen. Judd (2009) noted that in addition to *L. mariana*, which is known to cause livestock poisoning, the leaves of other species possibly also contain andromedotoxin and may be poisonous. (Named for John Lyon, 1765–1814, Scottish-born early American botanist and explorer of the s Appalachians)

REFERENCES: Judd 1981, 1995, 2001, 2009; Kron & Judd 1997 [1998]; Brown & Elsik 2002; Benning 2015.

1. Corollas 2–4.5 mm long, subglobose-urceolate; leaves with margins minutely serrulate; calyx lobes 0.5–1.5 mm long; inflorescences racemes of fascicles or short racemes, to almost paniculate ____ **L. ligustrina**
1. Corollas 5–14 mm long, ± cylindrical; leaves with margins entire; calyx lobes 2–9.5 mm long; inflorescences umbel-like clusters.
   2. Corollas usually pink (sometimes red or white), 2–5 mm wide; leaves evergreen, leathery, with a vein just inside the leaf margin; filaments roughened, but without long hairs near base ____ **L. lucida**
   2. Corollas usually white (rarely pink), 4.5–9 mm wide; leaves deciduous, relatively thin, without a vein just inside the leaf margin; filaments with long hairs near base ____ **L. mariana**

Lyonia ligustrina (L.) DC. var. **foliosiflora** (Michx.) Fernald, (sp.: like *Ligustrum*, Oleaceae, PRIVIT; var.: flowers with many leaves, in reference to the leaf-like bracts in the inflorescence), HE-HUCKLEBERRY, MALEBERRY, SEEDY BUCKBERRY. Shrubs usually 1–4 m tall, often in colonies; leaves deciduous or semi-evergreen, the blades to 9 cm long and 4.5 cm wide, elliptic to obovate or ovate, serrulate; flowers in racemes of fascicles or short racemes, to almost paniculate, near end of previous year's growth, at least the

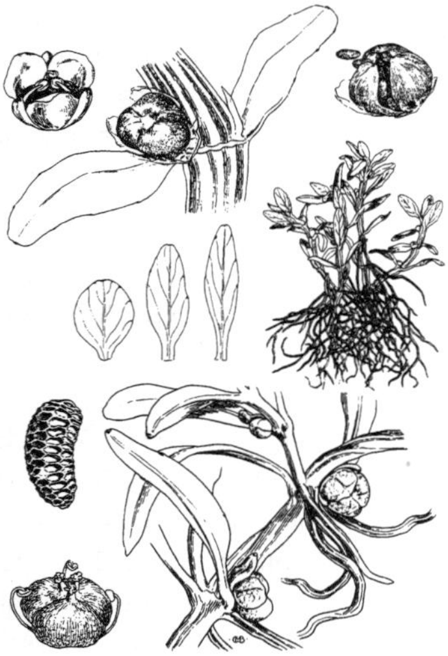

Elatine brachysperma [MAS]

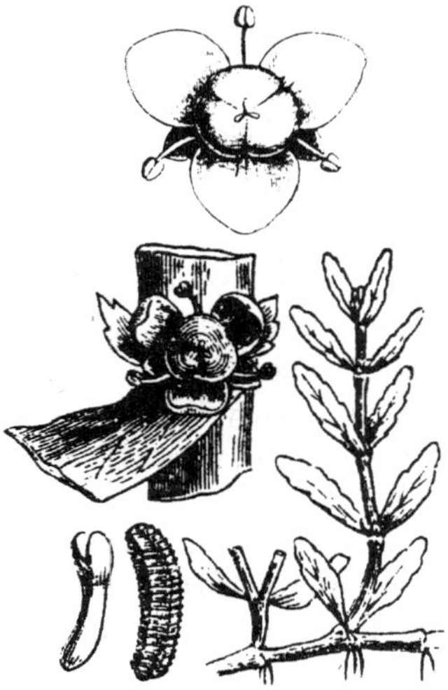

Elatine triandra [EN2, KAR]

Arbutus xalapensis [SA3]

Eubotrys racemosa [BB1]

Gaylussacia dumosa [STR]

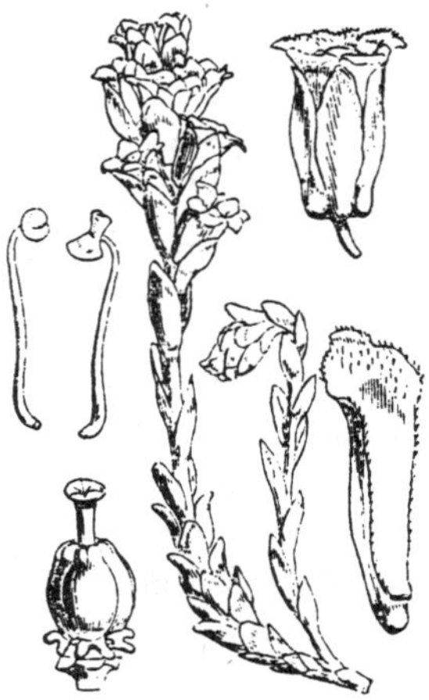

Hypopitys monotropa [BEN]

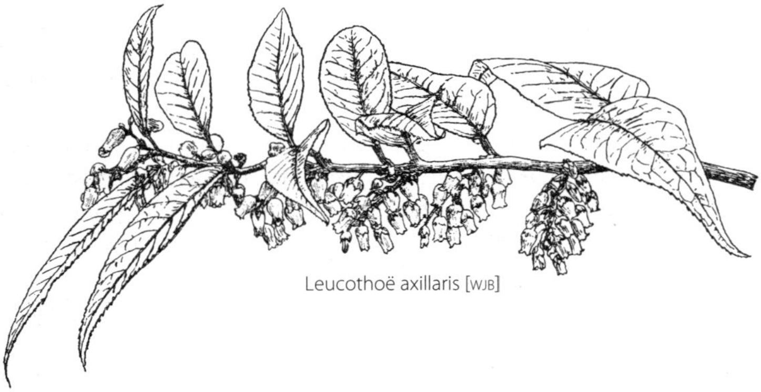

Leucothoë axillaris [WJB]

Lyonia ligustrina var. foliosiflora [CO1]

lower inflorescences with conspicuously leaf-like bracts 0.5–3 cm long, the other bracts similar to bracteoles, 1–2 mm long; pedicels 2.5–10 mm long; corollas 2–4.5 mm long, white; capsules 2–3 mm long, globose to subglobose. Bogs, seeps, wooded areas along streams; primarily e part of Pineywoods; VA s to FL w to OK and TX. Apr–Jun. [*Andromeda paniculata* L. var. *foliosiflora* Michx.] Variety *ligustrina* (e U.S. from ME s to AL and GA) can be distinguished by the inflorescences having no or only a few leaf-like bracts (bracts usually similar to the very small bracteoles). This is a variable species most closely related to the eastern Asian members of the genus (Judd 2009). Burrows and Tyrl (2001) noted that this species is of toxicological interest.

Lyonia lucida (Lam.) K. Koch, (shining or glossy), FETTER-BUSH, STAGGER-BUSH. Shrub to 2.5+ m tall, often in colonies; leaves evergreen, the blades to ca. 8.5 cm long and 4.5 cm wide, coriaceous, elliptic to obovate or ± ovate, entire, typically revolute, coriaceous; flowers in umbel-like fascicles from distal portions of branches of previous year; bracts 1 per flower, to 4 mm long; calyx lobes 2–7.5 mm long; corollas 5–9 mm long, pink, rarely reddish or white, ± cylindrical, the base swollen; capsules 3–5 mm long, ovoid to ovoid-globose. Along streams, bogs, swampy areas; Newton Co. (*Brown & Elsik 24436* BRIT, SBSC—fide Brown & Elsik 2002); se U.S. from VA s to FL w to TX; also w Cuba. Early spring. This species was discovered in TX in 2002 (Brown & Elsik 2002), even though it is native and common in the adjacent Sabine Parish of LA. They noted that there were six populations with well over a hundred total plants along Screw Pin Branch in Newton Co.; it is their opinion that the TX populations are native. Evolutionarily this species is most closely related to *L. mariana* (Judd 2009). It is sometimes used as a cultivated ornamental. Despite a flower form that suggests bumblebee pollination, recent research (Benning 2015) surprisingly shows that "*L. lucida* flowers are visited mainly by nocturnal moths, who are capable of carrying large pollen loads, and nocturnal pollination is the primary driver of fruit set." Since this species is closely related to *L. mariana*, it is possibly toxic (see discussion under *L. mariana*).

Lyonia mariana (L.) D. Don, (of Maryland), STAGGER-BUSH. Shrub usually < 1 m tall (rarely to 2 m); leaves deciduous, the blades to 9(–11) cm long and 5 cm wide, elliptic-oblong to elliptic-lanceolate or narrowly obovate, ovate or obovate, entire; flowers in axillary, umbel-like fascicles along distal portions of branches of previous year; bracts 1 per flower, to 4.5 mm long; calyx lobes 3–9.5 mm long; corollas 7–14 mm long, usually white (can be pinkish), the base not swollen; capsules 4–6(–7) mm long, ovoid. Usually moist, sandy, wooded areas; scattered in East TX w to Henderson and Leon cos. and disjunct sw to Gonzales Co. (e.g., *Tharp 7099* TEX); e U.S. from NY s to FL w to MO, OK, and TX. Mar–Jun. [*Andromeda mariana* L.] The Gonzales Co. collections are associated with the Ottine Swamp, a boggy oasis with numerous other disjunct eastern species (e.g., dwarf palmetto). This isolated wetland is probably a remnant of an Ice Age swamp that has been able to survive to the present due to its unique hydrological circumstances (Diggs et al. 2006). The leaves of this species have a number of toxic diterpenoid compounds (e.g., andromedotoxin) and are poisonous to livestock and humans (Hardin & Brownie 1993; Burrows & Tyrl 2001). The narcotic effect on livestock gave rise to the common name STAGGER-BUSH (Judd 1981). It is sometimes used as an ornamental.

MONOTROPA L. INDIAN-PIPE

A widespread mainly n temperate genus (Asia, North America, Central America, and n South America) of a single species with disjunct populations in the tropics (Neyland & Hennigan 2004). However, a number of the populations are genetically distinct and initial molecular studies indicate that "some level of interspecific or intraspecific taxonomic distinction eventually may be justified" (Neyland & Hennigan 2004). *Monotropa* and the related genus *Hypopitys* both have a complex way of getting energy/nutrition. In the past it was often incorrectly referred to as saprophytic (= obtaining nutrients from decaying organic matter via fungi), a term which though sometimes still used, perpetuates erroneous information (Leake 2005). In reality, they exhibit myco-heterotrophic epiparasitism. That is, they are *heterotrophic* organisms (= obtain their nutrition from another species, in this case a fungus—hence the term "*myco*"), but are *epiparasites* (= obtaining energy from an organism that in turn is getting energy from yet a third organism). Specifically, this interaction involves three different species where the non-photosynthetic, heterotrophic *Monotropa* parasitizes a soilborne fungus which has a mycorrhizal relationship with a photosynthetic plant; the *epiparasite* thus *indirectly parasitizes* the photosynthetic plant via the shared fungus (and ultimately obtains nutrition from the photosynthetic plant). The myco-heterotrophic plant can therefore be considered a "mycorrhizal cheater" (Bidartondo 2005) since it takes advantage of the mutualistic mycorrhizal relationship between the fungus and the photosynthetic plant. Research has shown that the epiparasite is "nourished exclusively by carbon passing through a common mycorrhizal network linked to adjacent autotrophic trees" (Leake 2005) and the relationship is thus clearly not saprophytic. It is now known that there is "extreme specificity" in the fungi with which the epiparasite interacts (Bidartondo & Bruns 2001, 2002); in the case of *M. uniflora* it is with the basidiomycete fungal genera *Russula* and *Lactaris* (Russulaceae) (Bidartondo & Bruns 2001; Kong et al. 2015). While Kartesz (1994) and Jones et al. (1997) placed *Monotropa* in the Monotropaceae, we are following Stevens (2001 onwards) and Wallace (2009) in treating the Monotropoideae as a subfamily of the Ericaceae. Based on recent molecular research (Neyland & Hennigan 2004; Braukmann & Stefanovic 2012), the two species traditionally treated in *Monotropa*, *M. uniflora* and *M.*

hypopitys, do not appear to be closest relatives. *Monotropa uniflora* is apparently more closely related to *Monotropastrum* (an Asian genus of 2 species), while *M. hypopitys* [= *Hypopitys monotropa*] seems more closely related to *Allotropa* (1 species of the w U.S.). Likewise, molecular research by Kron and Johnson (1997) and Cullings (2000), while differing in details, suggests that *Monotropa*, as traditionally conceived, is polyphyletic. Therefore, we are following Stevens et al. (2004) in recognizing a monotypic genus *Monotropa* and a monotypic *Hypopitys* (for *M. hypopitys*, which for nomenclatural reasons becomes *H. monotropa*). (Greek: *monos*, one, and *tropos*, turn, from the summit of the flowering stem nodding, turned to one side)

References: Correll 1965; Stevens 1995; Wallace 1975, 1993, 1995; Kron & Johnson 1997; Cullings 2000; Bidartondo & Bruns 2001; Neyland & Hennigan 2004; Wallace 2009; Braukmann & Stefanovic 2012; Broe 2014; Freudenstein & Broe 2024.

Monotropa uniflora L., (one-flowered), INDIAN-PIPE, GHOST-PIPE. Plant obligately myco-heterotrophic epiparasitic (= indirectly parasitize plants via fungi) fleshy herbs lacking chlorophyll and non-photosynthetic, the entire plant usually waxy white or rarely tinged pink, orange, or red or combinations of these, drying black, 5–30 cm tall; stems unbranched, glabrous; leaves reduced to scale-like bracts 5–12(–15) mm long; inflorescence a single flower, usually nodding in flower, becoming erect after pollination and erect in fruit; flowers (3–)5(–6)-merous; sepals shorter than petals or sometimes apparently absent; corollas cylindric or slightly flared distally, of separate petals, 7–12 mm long; intrastaminal nectaries present at base of the ovary; stamens 8 or 10, dehiscent by slits; pollen grains shed singly (usually in tetrads in the Ericaceae); styles 2–5 mm wide; ovaries superior; capsules 7–12 mm long not including persistent style; seeds tiny, numerous. Forest floor humus, under pines and hardwoods; scattered in the Pineywoods, especially the s part, and in the Red River drainage w to Fannin Co. (BRIT); widespread throughout the U.S. except the sw; also Canada; Mexico; Central America; South America (Colombia); and Asia. Apr–Jul. [*M. brittonii* Small]

RHODODENDRON L. RHODODENDRON, AZALEA

Ours deciduous shrubs, rhizomatous or not so; leaves entire or rarely serrulate, ciliate, petiolate; flowers in umbellate or corymb-like racemes, 5-merous, bisexual; calyces very small relative to corollas, the sepals slightly connate; corollas showy, trumpet-shaped, with a long narrow tube that expands into a widely flaring limb, usually slightly bilaterally symmetrical, white to pink, without blotch on upper lobe; stamens 5, conspicuously exserted, without appendages, the anthers with apical pores; ovaries superior; fruit a septicidal capsule; pollen agglutinate.

A mainly n hemisphere genus of ca. 1000 species (Judd & Kron 2009; Mabberley 2017) but extending from 80°N to 20°S latitudes (e.g., Greenland to Australia) (Brown et al. 2006). It is the largest genus of woody plants in the Northern Hemisphere (Xia et al. 2022). It includes mainly evergreen and deciduous trees and shrubs, but also some tropical epiphytes. It is especially diverse in China (650 species), where it is the largest genus (Mabberley 2017); secondary centers of diversity are New Guinea and e North America. The leaves are in terminal rosettes and flowers are variously colored and often extremely showy. The taxonomy of the genus is complex but with the advent of molecular work more consensus has been reached with most authors today recognizing the genus in the broad sense as treated here; all East TX species are in sect. *Pentanthera*, which comprises 15–16 species of temperate regions of the Northern Hemisphere. The genus has been extensively cultivated (over 500 species cultivated in Great Britain alone) with untold numbers of hybrids and cultivars variously known as AZALEAS and RHODODENDRONS. As noted by Yatskievych (2006), the "group now informally called azaleas (once treated as the separate genus *Azalea* L.) tend to be species that have deciduous foliage, have mostly 5 stamens, and have funnelform corollas, whereas rhododendrons tend to have evergreen leaves, 10 stamens, and more bell-shaped flowers." A number are poisonous due to the presence of diterpenes variously known as grayanotoxins, andromedotoxin, or rhodotoxin; they occur in leaves, twigs, flowers, and nectar and can cause toxicity in both humans and animals, including dogs and cats, apparently by affecting sodium channels in nerve cells; in ruminants they can cause staggers and death. Even eating the honey made from such nectar can cause low blood pressure, dizziness, cardiac rhythm disorders, nausea and/or vomiting, sweating, and impaired consciousness, the condition being known as "mad honey poisoning." (McGee 1973; Jansen et al. 2012; Mabberley 2017). While they grow extremely well and are among the most widely planted and prized ornamentals throughout the e portions of East TX, because of their preference for acidic soils, most species are extremely difficult to grow (without complete soil replacement) in the basic calcareous soils in the w part of East TX. Pollination by bumblebees has been observed (King & Buchmann 1995). The key to species is modified from Luteyn et al. (1996). (Greek: *rhodon*, rose, and *dendron*, tree)

References: Sleumer 1980; Kron & Judd 1990; Irving & Hebda 1993; Kron 1993; Chamberlain et al. 1996; Luteyn et al. 1996; Kurashige et al. 2001; Goetsch et al. 2005; Judd & Kron 2009; Weakley et al. 2012; Shrestha et al. 2018; Xia et al. 2022; Becker et al. 2024; Shen et al. 2024.

1. Flowers appearing before or with the leaves (at least some leaves still folded or some bud scales still present); corolla lobes without glandular hairs extending to near the apex externally; stamens much longer than corollas.

2. Corollas with narrow tubes that abruptly expand into lobes, the lobes shorter than the tube; pedicels usually without glandular hairs, canescent-pilose (= whitish or grayish-white due to non-glandular hairs), rarely with a few glandular hairs; leaf margins inconspicuously ciliate, the cilia appressed to the leaf margins; fruits densely hairy with non-glandular hairs ____ **R. canescens**
2. Corollas with broader tubes that gradually expand into lobes, the lobes as long as the tube or longer; pedicels usually with glandular as well as eglandular hairs; leaf margins conspicuously ciliate, the cilia ascending away from the leaf margins; fruits with scattered glandular hairs and only sparse non-glandular hairs ____ **R. prinophyllum**

1. Flowers appearing after the leaves have expanded (essentially all of the leaves have unfolded and bud scales absent); corolla lobes with glandular hairs extending to near the apex externally; stamens only slightly longer than corollas (when measured to tips of lobes), but still conspicuously exserted ____ **R. viscosum**

Rhododendron canescens (Michx.) Sweet, (covered with gray or white hairs), HOARY AZALEA, MOUNTAIN AZALEA, PIEDMONT AZALEA. Deciduous shrub(–small trees) to ca. 3+ m tall, usually non-rhizomatous; twigs, petioles, pedicels, calyx, and capsules usually ± canescent-pilose due to non-glandular hairs, scattered glandular hairs sometimes present; leaf blades 2.5–11(–13) cm long, oblanceolate to obovate or ovate, the margins entire or rarely minutely serrulate; inflorescences with 6–19 flowers; flowers opening before or with the leaves, typically with musky-sweet odor; pedicels 4–17 mm long; corollas pink, the tube often deeper pink to white with pink tube, 23–46 mm long, with scattered glandular hairs and non-glandular hairs; fruits 12–33 mm long, densely hairy. Bog margins, seeps, along streams, wooded flats; se part of East TX (e.g., Angelina, Bowie, Hardin, Jasper, Newton, Orange, Sabine, San Augustine, and Tyler cos.—Kron 1993); mainly se U.S. from PA s to FL w to OK and TX. Mar–May. [*Azalea canescens* Michx., *A. candida* Small] Probably poisonous (see generic synopsis). ☠

Rhododendron prinophyllum (Small) Millais, (presumably with the leaves of *prinus*, Latin for a type of evergreen oak; the derivation not specified in the original description), EARLY AZALEA, TEXAS HONEYSUCKLE, EARLY AZALEA, ROSE-SHELL AZALEA. Deciduous shrub to ca. 3+ m tall, usually non-rhizomatous; twigs with sparse to dense non-glandular pubescence; leaves usually clustered near branch tips, blades 3–9 cm long, oblanceolate to obovate or ovate, entire; inflorescences with 4–13 flowers; flowers opening before or with the leaves, spicy-clove fragrant; pedicels 5–26 mm long; corollas usually bright pink, the tube sometimes darker, 25–45 mm long, with scattered glandular hairs and non-glandular hairs; fruits 10–28 mm long, hairy. Bogs, swampy areas, along streams; included based on the citation of Hardin and Harris (USDA Plants Database 2019, 2025) cos., and also the recent treatment of the genus for North America (Judd & Kron 2009) noting the species as occurring in TX; the species is also known from McCurtain Co., OK, just across the state border (*Perino & Perino 942* NCSC—fide Kron 1993); ME s to AL w to IL, MO, OK, and TX, primarily in the Appalachians and Ozarks. Feb–May. [*Azalea prinophylla* Small] This species is related to and sometimes included in *R. canescens* (e.g., Turner et al. 2003); however, we are following almost all recent authors including Kron (1993), Luteyn et al. (1996), Yatskievych (2006), Judd and Kron (2009), Weakley et al. (2012), and Weakley (2015), in recognizing it as a distinct species. Probably poisonous (see generic synopsis). ☠

Rhododendron viscosum (L.) Torr., (sticky), CLAMMY AZALEA, SWAMP AZALEA. Morphologically variable deciduous shrub(–small trees) from < 1 to ca. 5+ m tall, rhizomatous or non-rhizomatous; twigs usually with non-glandular pubescence; leaf blades 2–7(–9) cm long, oblanceolate to obovate or ovate, entire or minutely serrulate; flowers opening after the leaves, with sweet fragrance; pedicels 5–27 mm long, usually with glandular pubescence; corollas white, sometimes pink-tinged or pinkish, 20–58 mm long, often conspicuously glandular pubescent with the glandular hairs on the tube and continuing to the lobes, typically sticky to the touch; fruits 7–20 mm long, usually with glandular hairs. Bogs, seeps, near streams, low woods; widespread in the Pineywoods, particularly to the s, also disjunct to Bastrop Co. to the w (Kartesz 2019); e U.S. from ME and NY s to FL w to OK and TX. Apr–Jul. [*Azalea viscosa* L., *A. oblongifolia* Small, *A. serrulata* Small, *R. coryi* Shinners, *R. oblongifolium* (Small) Millais, *R. serrulatum* (Small) Millais, *R. viscosum* var. *serrulatum* (Small) H.E. Ahles] This is a widespread, extremely variable species in which "numerous minor varieties and forms have been described" (Kron 1993) and "correlated or geographically coherent sets of character states cannot be discerned" (Judd & Kron 2009). We are following Kron (1993), Luteyn et al. (1996), and Judd and Kron (2009) in including *R. coryi* and *R. oblongifolium* (recognized as distinct species by Correll and Johnston (1970)) in *R. viscosum*. Probably poisonous (see generic synopsis). ☠

VACCINIUM L. BLUEBERRY

Ours shrubs or small trees, deciduous or evergreen, often forming colonies by stolons or rhizomes; petioles short or absent; leaf blades entire or finely toothed; flowers in mostly axillary racemes or clusters of 2–10 or sometimes solitary, 5-merous; corollas urn-shaped (urceolate) to cylindrical or bell-shaped, greenish white to white, white with pink-tinge, or pink; stamens usually 10, without spurs, anthers tapered into tubes with terminal pores, awned or awnless; ovaries inferior, 4–5-locular; fruits berries.

A genus in the recent past variously treated as having 140 species (Stevens et al. 2004) to 450 species (Mabberley 1997), but now considered to have ca. 500 (Vander Kloet 2009; Mabberley 2017; Becker et al. 2024). It is part of a monophyletic Vaccinieae (*Vaccinium* and its closest relatives), but its taxonomy is unsettled, its limits are unclear (Kron et al. 2002b), and the genus has been described as "wildly paraphyletic" (Stevens et al. 2004) or "highly polyphyletic" (Becker et al. 2024); if strict monophyly is followed (see Stuessy & Hörandl 2014 for comments on the topic), numerous name changes will almost certainly be necessary in the future (Kadereit et al. 2016). The following quote (Becker et al. 2024) helps clarify the situation:

At the extremes, options for a ranked reclassification range from narrowing *Vaccinium* to the single species *V. uliginosum* (the type of the genus) and raising the sections of *Vaccinium* to genus level, to expanding *Vaccinium* to encompass the entirety of Vaccinieae and lowering the rank of genera to sections within *Vaccinium*.

It seems clear that significant future research is needed to elucidate the appropriate classification and nomenclature of *Vaccinium*. Species range from arctic and n temperate areas to Central and South America, se Asia, the Pacific including Hawaii, Africa, and Madagascar, and often grow in acidic soils. It includes deciduous or evergreen shrubs, small trees, and lianas, almost all of which have edible fruits including BLUEBERRY, BUCKBERRY, HUCKLE-BERRY, BLUETS, BILBERRY, WHORTLE-BERRY, BLUEBERRY, CRANBERRY, COWBERRY, LINGBERRY, LINGONBERRY, and FOXBERRY; however, at least several Latin American species have toxic fruits. The CRANBERRY is *V. macrocarpon* Aiton, native to e North America, and a number of different BLUEBERRY species (e.g., *V. angustifolia*—SWEET LOWBUSH BLUEBERRY, *V. ashei*—RABBITEYE BLUEBERRY, and *V. corymbosum*, the major BLUEBERRY of commerce) are cultivated. The CRANBERRY is also used medicinally for urinary problems. Edible BLUEBERRIES are widely grown commercially in East TX. As noted by Fritsch et al. (2022 [2024]), "Blueberries represent one of only a handful of major crop plants originating in North America." The fruits are important to a variety of wildlife and also were widely used by Native Americans. A number of species of *Vaccinium* exhibit the "buzz," vibrator," or "sonication" pollination syndrome (Buchmann 1983; Cane et al. 1985; Cardinal et al. 2018; Vallejo-Marín 2018). In addition to nectar, pollen is used as food by the bees. The bees shake the anthers by vibrating/shivering their thoracic flight muscles at a certain frequency; this audible "buzz" sets up a resonance and the otherwise inaccessible pollen is released from the terminal pores of the anthers and collected by the insect. However, at least some pollen is inadvertently transferred to other flowers thereby allowing the plants to be reproductively successful. As is often the case, the use of common names can be confusing; *Vaccinium* species are widely called BLUEBERRIES. However, as noted by Vander Kloet (2009), in e eastern North America, species of *Gaylussacia* are called HUCKLEBERRIES, while in w North America some species of *Vaccinium* are called HUCKLEBERRIES. (Latin name for blueberry, probably from *bacca*, berry)

REFERENCES: Camp 1945; Ward 1974; Uttal 1987; Vander Kloet 1988, 1992, 2009; Lyrene 1995; Kron et al. 2002b; Weakley et al. 2012; Weakley 2015; Schlautman et al. 2017; Crowl et al. 2022; Becker et al. 2024; Franck & Salman 2024; Fritsch et al. 2022 [2024]; Fritsch et al. 2024b; Fritsch et al. 2025.

1. Corollas with opening widely flared, conspicuously the widest part of the corolla; stamens obviously exserted beyond the corolla ____ **V. stamineum**
1. Corollas with opening not widely flared, not conspicuously the widest part of the corolla; stamens not exserted beyond the corolla.
   2. Corollas white, usually 2.5–5.5 mm long; fruits dry and mealy, ± inedible; anthers with 2 yellow awns at junction of filament and anther; young twigs not verrucose; leaves often rounded at tip (but there can be a minute abrupt tip) or sometimes rather sharp-pointed; plants shrubs or small trees to 8 m tall, widespread in East TX ____ **V. arboreum**
   2. Corollas white, partially pink, or pink, 5–12 mm long; fruits juicy, edible (though not always good tasting); anthers awnless; young twigs verrucose (= covered with wart-like protrusions); leaves mostly sharp-pointed or narrowed to a blunt point; plants shrubs 1–4 m tall, mainly in the Pineywoods (*V. corymbosum* "complex").
      3. Leaves at maturity all 3.5 cm or less long, the margins usually serrulate, the minute teeth initially tipped by a quickly deciduous stipitate gland; corollas narrow, ca. 2 times as long as wide or more; styles included within corollas ____ **V. elliottii**
      3. Leaves at maturity mostly 3–8 cm long, the margins various, from entire to serrate; corollas < 2 times as long as wide; styles exserted from corollas.
         4. Young leaves stipitate-glandular beneath; corollas 6–12 mm long ____ **V. virgatum**
         4. Young leaves with pubescence beneath, but not glandular, 5–8 mm long ____ **V. fuscatum**

Vaccinium arboreum Marshall, (tree-like), FARKLE-BERRY, SPARKLE-BERRY. Shrub or small tree to 8 m tall; leaves obovate to oblong-elliptic, rounded at tip, entire or with some minute teeth, to ca. 7 cm long and 3.5 cm wide, coriaceous, essentially evergreen; flowers in loose leafy-bracted racemes, the bracts much smaller than the leaves; corollas usually 2.5–5.5 mm long, white, urceolate to slightly campanulate, the lobes reflexed, shorter than the tube; ovary inferior; berries 7–9 mm in diam., black or reddish black, often dryish and mealy to moist. Sandy soils, open woods, thickets, fields; widespread in East TX except w edge, also w in Red River drainage through Lamar to Fannin and Grayson (BRIT) cos.; the Grayson population (Preston Bend Park) is isolated on an outcrop of Trinity sand ca. 30 miles west of any other known populations (G. Diggs, pers. obs.); also n Gulf Prairies and Marshes; e U.S. from VA s to FL w to IL, KS, OK, and TX. Mar–May. [*V. arboreum* var. *glaucescens* (Greene) Sarg.]

Vaccinium elliottii Chapm., (for Stephen Elliott, 1771–1830, South Carolina legislator, banker, educator, botanist, and author of one of the most important works in early American botany, *A Sketch of the Botany of South-Carolina and Georgia*; apparently the first to recognize it as a separate species), ELLIOTT'S BLUEBERRY, MAYBERRY. Shrub 1–4 m tall, often with a bushy-branched crown; leaves 0.7–3(–3.5) cm long, rich green and lustrous above, non-glandular pubescent beneath, usually serrulate, resembling those of BOXWOOD or PRIVIT, tardily deciduous, not turning red in fall; flowers in fascicles of 2–6; corollas cylindric-campanulate, 4–8 mm long, white or usually pink tinged; style not exserted; berry black, with poor flavor, 5–8 mm in diam. Wooded floodplains, ravines; se part of East TX; VA s to FL w to AR and TX. Feb–Apr. This species is known to hybridize with *V. fuscatum* (Uttal 1987). Uttal (1987) argues that this taxon warrants specific recognition and notes that it is the "most distinctive" of the entities sometimes put into a very broadly defined *V. corymbosum* (e.g., Vander Kloet 1988). Fritsch et al. (2022 [2024]) noted that, "In places where *V. amoenum/V. virgatum* and other "highbush" blueberries occur, ... plants with the morphology of *V. elliottii* sensu Camp appear to be distinct. One character in particular appears to easily separate *V. elliottii* from the remaining species of [sect.] *Cyanococcus*, i.e., a style that is included within the corolla (versus exserted ...)." Franck and Salman (2025) recently typified *V. elliottii* and noted that the species occurs "south to eastern Texas."

Vaccinium fuscatum Aiton, (brownish, the leaves darkening upon drying), BLACK HIGHBUSH BLUEBERRY, HAIRY HIGHBUSH BLUEBERRY, SOUTHERN HIGHBUSH BLUEBERRY. Shrub 1–3+ m tall; twigs pubescent but non-glandular; leaves 3–8 cm long, elliptic-ovate, entire, pubescent beneath, but non-glandular, dark green, darkening upon drying, usually deciduous; corollas cylindric-urceolate, often narrowed to the tip, 5–8 mm long, greenish white to pink; berries 5–8 mm long, black or nearly so, insipid. Hillside seeps, edges of bogs, usually in wet habitats; Cass, Jasper, Nacogdoches, Newton, Polk, Shelby, Trinity, and Tyler (USDA Plants Database 2019) cos. primarily se part of East TX; widespread in the e U.S. ME s to FL w to OK and TX; adjacent Canada, s Ont. Late Feb–Apr. [*V. arkansanum* Ashe, *V. atrococcum* (A. Heller) A. Gray] Uttal (1987) noted that in a population with *V. corymbosum,* he did not observe hybridization. Fritsch et al. (2022 [2024]) pointed out that "A third species often growing with *Vaccinium amoenum/V. virgatum* and *V. elliottii* in areas where no other species of [sect. *Cyanococcus* are thought to occur (e.g., east Texas)" corresponds to some version of *V. atrococcum/V. arkansanum/V. fuscatum*. The treatment in Correll and Johnston (1970), adapted from Camp (1945), considered reports of *V. amoenum* and *V. corymbosum* to be *V. arkansanum*. We are using the name *V. fuscatum* for this species. In general, it is found in wetter habitats than *V. virgatum*.

Vaccinium stamineum L., (with prominent stamens), DEERBERRY, SQUAW-HUCKLEBERRY, SOUTHERN GOOSEBERRY. Shrub to 3+ m tall, forming clumps or colonies; twigs not verrucose; leaves 35–90 mm long, obovate to oblong, entire or nearly so, acute, often glaucous or whitish beneath, deciduous; inflorescences racemes; corollas 4–8 mm long, open-campanulate (in contrast to all our other species), the lobes divided about 1/2 way to base, greenish white to white, the veins sometimes purplish; anthers conspicuously exserted (in contrast to all our other species), awned; berries usually 10–15 mm in diam., greenish, yellow, pink, purple, or black, generally tart or bitter (though variable in taste with some individual plants yielding good fruits—Yatskievych 2006). Well drained soils; pinelands, mixed woods, wooded bluffs, savannahs; widespread in the s Pineywoods, less so further n; ME and s Ontario s to FL w to IL, KS, OK, and TX; also disjunct to n Mexico. Mar–May. [*V. caesium* Greene] This is a very variable species that has sometimes been divided into a number of segregates (e.g., *V. caesium* in Correll & Johnston 1970); however, most modern authorities agree that they do not warrant formal recognition (e.g., Vander Kloet 2009). Ward (1974) noted that it is a simple matter to find "specimens that partly or wholly bridge these [morphological] gaps and obscure the significance of these differences." This species is known to be pollinated by bees using the "buzz," vibrator," or "sonication" pollination syndrome (Cane et al. 1985; see generic synopsis for discussion).

Vaccinium virgatum Aiton, (wand-like, twiggy in growth), SMALL-FLOWER BLUEBERRY, SOUTHERN BLACK BLUEBERRY. Shrubs ranging from less than 1 m to more than 1 m tall, typically in colonies; leaves deciduous, 3–6.5 cm long, oval, shiny, serrate, conspicuously glandular beneath, the glands blackish, deciduous; corollas 6–12 mm long, usually pink-tinged to deep pink; styles strongly exserted; fruits black, 6–10 mm in diam., with poor flavor. Along streams, forest margins, savannas; Angelina, Bowie, Camp, Chambers (in n Gulf Prairies and Marshes), Cherokee, Gregg, Hardin, Harrison, Jasper, Liberty, Montgomery, Morris, Nacogdoches, Newton, Panola, Polk, Red River, Rusk, Sabine, Shelby, Smith, Titus, Trinity, Tyler, Upshur, and Wood cos. (Fritsch et al. 2025) cos. in e part of East TX; NC s to FL w to AR and TX. Mar–May. [*V. amoenum* Aiton, *V. corymbosum* var. *amoenum* (Aiton) A. Gray] According to Fritsch et al. (2022), "the abaxially stipitate-glandular-leaved plants corresponding to the Coastal Plain *Vaccinium amoenum/V. virgatum* ... appear to be distinct from the rest of ... [the] 'highbush blueberry' concept. These plants grow together with other typically non-stipitate-glandular species and are easily distinguished in the field by this character, along with a combination of usually serrate leaf blade margins and (when in flower) strongly exerted styles." In East TX, this species is similar to *V. fuscatum* which, however, has the young leaves non-glandular beneath and is usually found in wetter habitats.

Lyonia lucida [RAD]

Lyonia mariana [ROE]

Monotropa uniflora [STR]

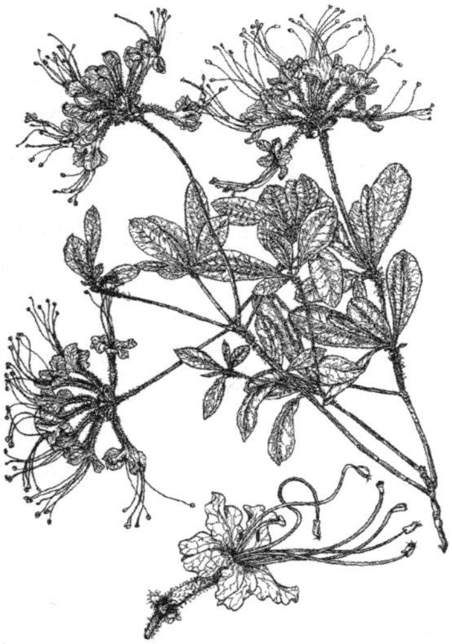

Rhododendron canescens [CO1]

Rhododendron prinophyllum [STE]

Rhododendron viscosum [BB2]

Vaccinium arboreum [SA3]

Vaccinium elliotii [BB2]

Vaccinium corymbosum L., (with corymbs—flat-topped flower clusters in which flowers are attached at different points along the inflorescence axis), HIGHBUSH BLUEBERRY. *Vaccinium corymbosum* is the major BLUEBERRY of commerce; more than 50 cultivars have been introduced (Yatskievych 2006) and naturalized populations often become established where it has been planted (e.g., Britain, Japan, New Zealand) (Vander Kloet 2009).

Many authorities (e.g., Turner et al. 2003; Vander Kloet 2009; Kartesz 2015; USDA Plants Database 2025) map *V. corymbosum* as occurring broadly in East TX. This is because they variously lump other related species into *V. corymbosum* which itself does not occur in East TX. We are recognizing just 3 species in sect. *Cyanococcus* as occurring in the state (i.e., *V. elliottii*, *V. fuscatum*, and *V. virgatum*). We are following Weakley et al. (2015) in considering *V. corymbosum* to be a species of the e US and adjacent Canada that comes no closer to TX than AL and the FL panhandle. *Vaccinium corymbosum* indeed is closely related to *V. elliottii*, *V. fuscatum*, and *V. virgatum*. Some authorities include all of these, and a number of other taxa, within a very broadly defined and heterogeneous *V. corymbosum* (e.g., Vander Kloet 1988, 2009; Luteyn et al. 1996), while others recognize some or part of the segregates (e.g., Godfrey & Wooten 1981; Uttal 1987; Bruederle & Vorsa 1994; Lyrene 1995; Turner et al. 2003; Weakley et al. 2012; Weakley 2015; USDA Plants Database 2019; Weakley, A.S., and Southeastern Flora Team 2025). There is a tremendous amount of morphological variation present in this broad "*V. corymbosum* complex," but the appropriate level at which to recognize the variation is not obvious. While hybridization is known between the taxa in some cases, especially in areas of habitat disturbance, in other instances hybridization has not been observed in mixed populations (e.g., Uttal 1987). Ward (1974) noted that the "vast bulk of individuals encountered in the field may be assigned ... to a relatively few, discrete, and wholly recognizable species." This morphological distinctiveness, in combination with significant reproductive isolation, would argue for recognition of some of the segregates at the species level. In order to highlight the differences between these usually recognizable entities, we are treating *V. elliottii*, *V. fuscatum*, and *V. virgatum* as distinct species. Of the three, the small-leaved *V. elliottii*, with small leaves and included styles is the most distinct. As noted by Lyrene (1995), if one lumps all members of the complex into a very broad *V. corymbosum*, the result is having easily recognized entities (e.g., *V. elliottii* with very small leaves) in the same overly broad "species" as entities with leaf surface areas "about 20 times larger." Occasional hybridization should not necessitate lumping species—if such an approach was followed slavishly, tremendous numbers of very well-known and distinctive species in many plant families would be eliminated since hybridization rarely or even occasionally occurs between clearly defined plant species. Lyrene (1995) gives the example of the very different *Pinus palustris* (LONGLEAF PINE) and *Pinus taeda* (LOBLOLLY PINE) which are well-known to hybridize. At the end of the day, we agree with Weakley (2015) who says, "Failure to recognize multiple entities within the highbush blueberries results in the taxonomic homogenization of the diversity of the group and obscures important phytogeographic patterns." Because past mapped locations of *V. corymbosum* in TX are unreliable, we are unsure of the identity of many county records of sect. *Cyanococcus*. Therefore our range maps for the three TX species of sect. *Cyanococcus* almost certainly do not reflect the full extent of their range within East TX.

Vaccinium darrowii Camp, (for G.M. Darrow, who first recognized this was a distinct species and brought it into cultivation—Vander Kloet 1988), SMALL-FLOWER BLUEBERRY, DARROW'S EVERGREEN BLUEBERRY. According to Vander Kloet (1988), this species was erroneously listed for TX by Correll and Johnston (1970) based on a specimen (BRIT/SMU "307153") that is actually *V. virgatum*. It is only known from AL, GA, FL, se LA, and MS, and is apparently not a component of the TX flora. Crowl et al. (2022) in a recent study of the diploid *Vaccinium* species, include a map indicating that *V. darrowii* comes nowhere near TX (the closest location is far se LA).

Vaccinium pallidum Aiton, (from the Latin *pallidus*, meaning pale), HILLSIDE BLUEBERRY, DRYLAND BLUEBERRY. While a range map in Crowl et al. (2022) indicates the species occurring in the extreme ne part of East Texas, this is in error (P. Fritsch, pers. comm.). However, it is known from just across the Red River in McCurtain Co., OK and Little River Co., AR (Kartesz 2015).

EUPHORBIACEAE Juss.

SPURGE FAMILY

Ours annual or perennial herbs, shrubs, or small trees with watery or milky juice; leaves alternate or less often opposite, simple (sometimes so deeply lobed as to appear compound, occasionally some leaves divided into palmate leaflets), entire, toothed, or lobed, with or without stipules; flowers axillary or terminal, solitary, in racemes, spikes, or heads, various in structure, unisexual or bisexual; perianth absent, or present, and of 1 type of part only, or of both sepals and petals; stamens 1–many; pistil 1, usually 3-celled and forming a 3-angled or 3-lobed ovary (a few species with these parts reduced to 2 or 1); ovary superior; fruit a capsule (sometimes considered a capsular-schizocarp), usually 3-seeded, or in 1 species of *Croton* (*C. michauxii*), an achene. Almost all East TX species of Euphorbiaceae are native to the U.S.

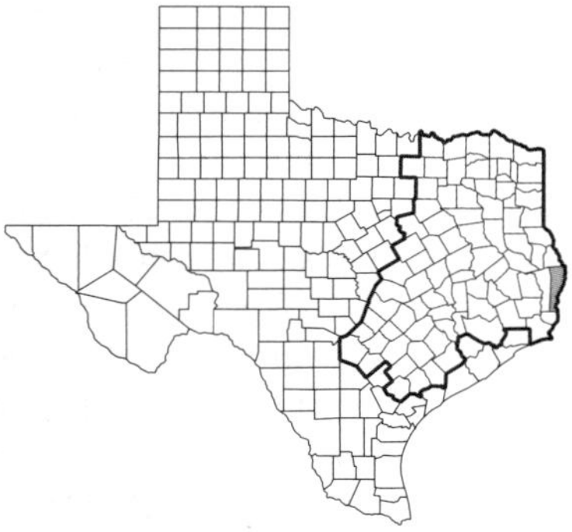
Gaylussacia dumosa

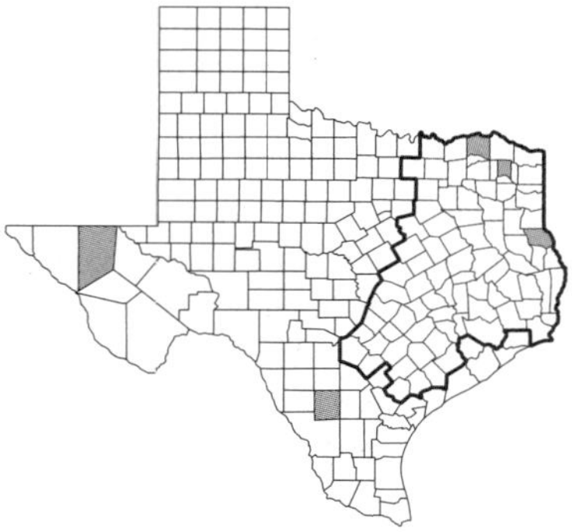
Hypopitys monotropa

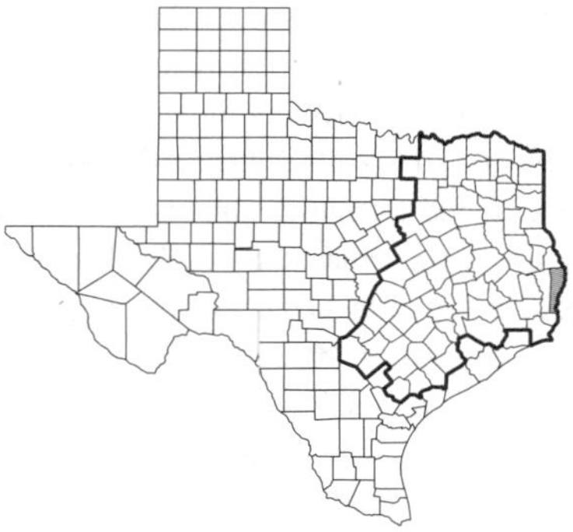
Leucothoë axillaris

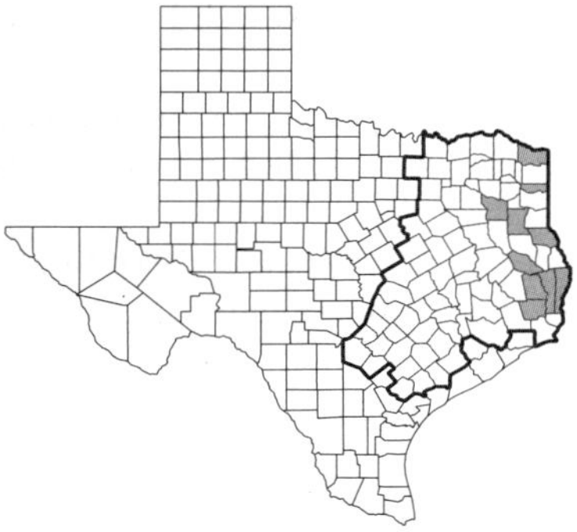
Lyonia ligustrina var. foliosiflora

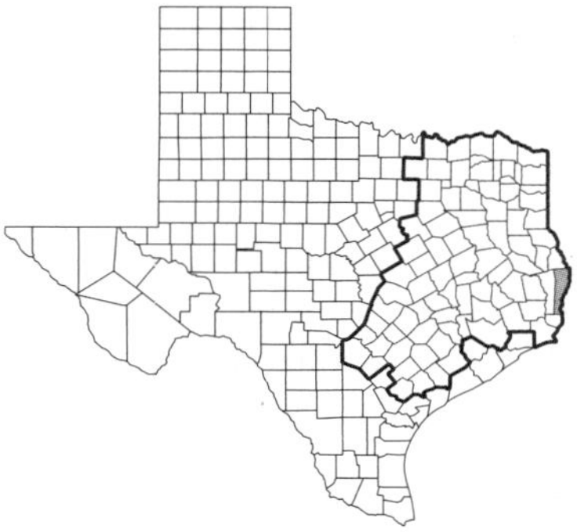
Lyonia lucida

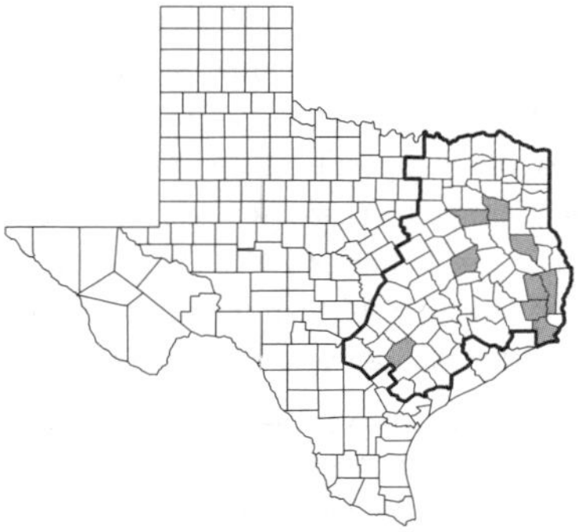
Lyonia mariana

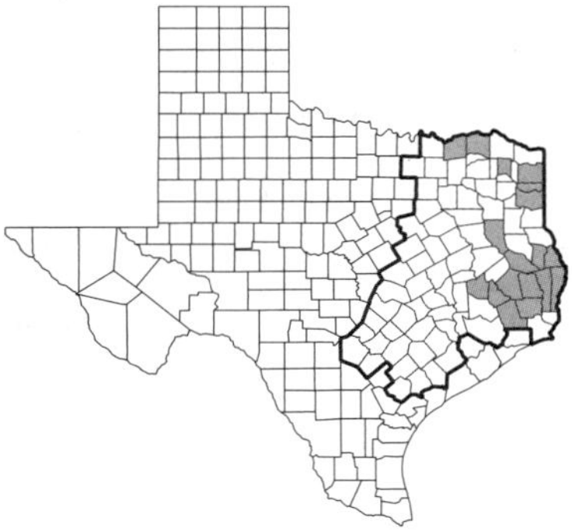
Monotropa uniflora

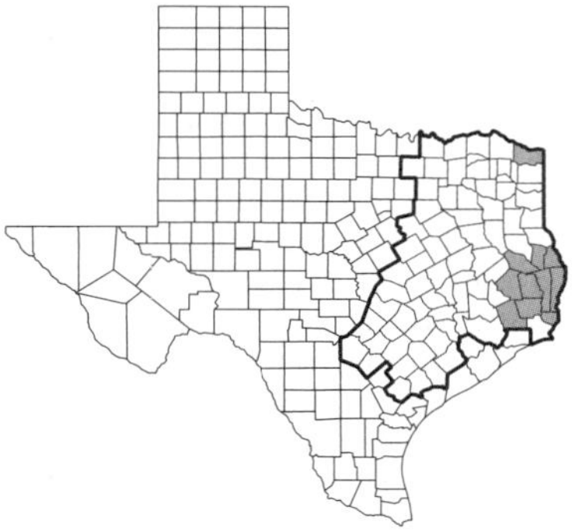
Rhododendron canescens

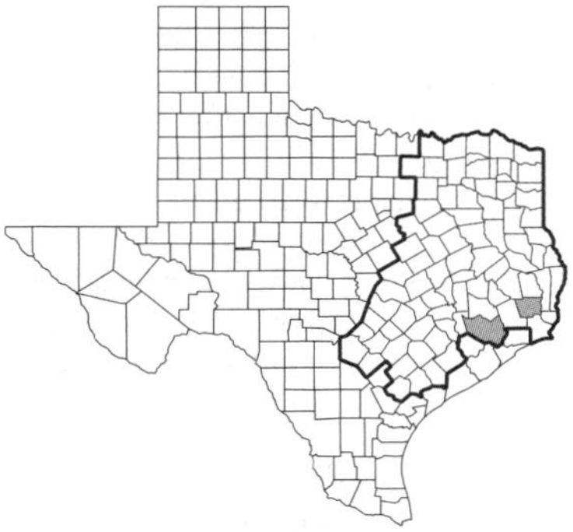
Rhododendron prinophyllum

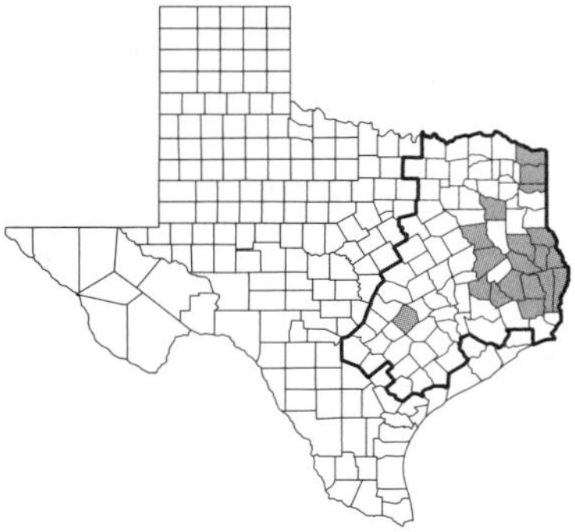
Rhododendron viscosum

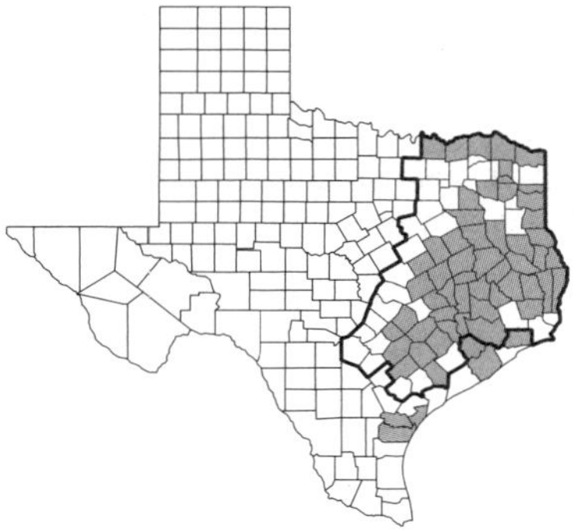
Vaccinium arboreum

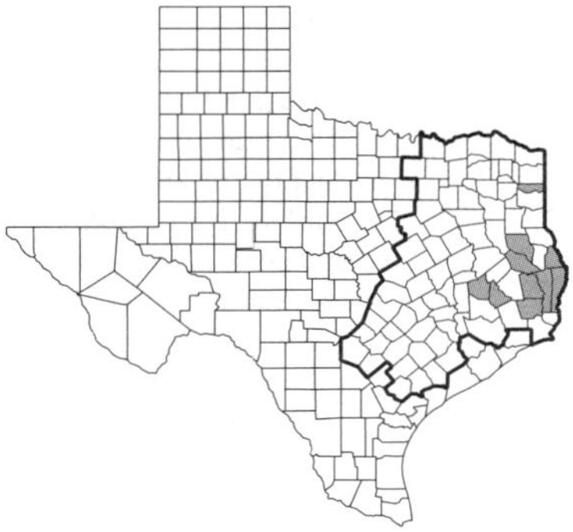
Vaccinium elliottii

A huge (ca. 6,500 species in 220 genera) (Levin & Gillespie 2016), cosmopolitan, but especially tropical family; vegetatively they vary from herbs to shrubs, lianas, large trees, or succulents; often there are specialized cells or tubes with milky or colored latex. A number are economically important including *Aleurites* species (TUNG-OIL), *Euphorbia pulcherrima* Willd. ex Klotzsch (POINSETTIA), *Hevea brasiliensis* (Willd. ex Juss.) Müll. Arg. (RUBBER, PARÁ RUBBER), *Manihot esculenta* Crantz (MANIOC, CASSAVA, or TAPIOCA), and *Ricinus communis* L. (CASTOR-BEAN); many other species are used as ornamentals. The latex of *Hevea brasiliensis*, a native of the Amazon Basin, is obtained ("rubber tapping") by making sloping incisions in the bark and collecting the white liquid in cups attached below the incisions; upon drying in the air or when coagulated by acid, the latex takes on its well-known elastic properties. A number of species are poisonous due to the presence of alkaloids and cyanogenic glycosides; others have diterpenoids that can cause irritant dermatitis and may act as co-carcinogens (= promote the action of "sub"-carcinogenic doses of known carcinogens) (Kinghorn 1979; Lampe 1986). In dry areas of the Old World, a number of *Euphorbia* species are xerophytically adapted and convergent vegetatively with Cactaceae. The family, as treated here, reflects molecular phylogenetic studies (Wurdack & Davis 2009; Xi et al. 2012) wherein the former subfamily Phyllanthoideae has been segregated as the Phyllanthaceae. The common name SPURGE is apparently derived from the Latin word *purgare*, to purge or cleanse; a number of members of the family are regarded as having cathartic or purgative properties (Tveten & Tveten 1993). (subclass Rosidae—Cronquist; order Malpighiales—APG IV)

FAMILY RECOGNITION IN THE FIELD: ours mostly herbs (a few small shrubs and 2 trees) often with *milky sap* and leaves usually alternate or less often opposite (all opposite in the former genus *Chamaesyce*, now treated in *Euphorbia*); ovary 3-celled, superior, typically developing into a capsule with 3 lobes or sections; flowers often unisexual, sometimes highly reduced.

REFERENCES: Wheeler 1943; Punt 1962; Webster 1967, 1994a, 1994b; Jensen et al. 1994; Kapil & Bhatnagar 1994; Nowicke 1994; Seigler 1994; Steinmann & Felger 1997; Govaerts et al. 2000; Radcliffe-Smith 2001; Steinmann & Porter 2002; Wurdack et al. 2004; Wurdack & Davis 2009; Xi et al. 2012; Levin & Gillespie 2016.

1. Trees to ca. 20 m tall; leaf blades cordate, sometimes with 3 shallow deltate lobes OR rhombic-ovate; apex of petiole with 2 disc-shaped glands on the upper surface.
   2. Trees to 13 m tall; leaf blades rhombic-ovate, 3–7(–10) cm long; apex of petiole with 2 yellowish green disc-shaped glands on the upper surface; invasive introduced species ____ **Triadica**
   2. Trees to 10(–20) m tall; leaf blades cordate 10–25 cm long, sometimes with 3 shallow deltate lobes; apex of petiole with 2 reddish disc-shaped glands on the upper surface; introduced species rare in East TX ____ **Vernicia**
1. Herbs, shrubs to 4 m tall, or low trees to 7 m tall (usually less); plants without the combination of characters above.
   3. Leaf blades palmately lobed, 6–60 cm wide.
      4. Plants with or without needle-like stinging hairs; leaf blades with petioles attached basally, 6–30 cm wide; lobes 3–13.
         5. Plants with conspicuous, needle-like stinging hairs; leaf blades 6–15 cm wide, not deeply divided; lobes 3–5, deltate, apex acute, coarsely toothed or lobed again; plants to 0.8(–1) m tall ____ **Cnidoscolus**
         5. Plants without needle-like stinging hairs; leaf blades 12–30 cm wide, deeply divided, almost to petiole apex; lobes 3–13, oblanceolate, sometimes lobed distally, apex acuminate; plants 1–7 m tall ____ **Manihot**
      4. Plants without needle-like stinging hairs; leaf blades peltate, 10–60 cm wide; lobes usually 7–9 ____ **Ricinus**
   3. Leaf blades entire to toothed, or crenate, not palmately lobed, usually ≤ 6 cm wide.
      6. Flowers with green perianth or sepals (sometimes minute), borne in various sorts of inflorescences (solitary, spikes, cymose clusters, etc., but not in cup-like involucres; sap usually not milky (milky only in *Stillingia*).
         7. Plants with stellate hairs (stellate hairs also in *Bernardia*) or lepidote scales on at least some parts, these often giving the foliage a distinctive silvery to yellowish or brownish color (the hairs and scales, while visible to the naked eye, are striking with a hand lens) ____ **Croton**
         7. Plants with simple (unbranched), malpighiaceous (attached at middle) or glandular hairs or glabrous; foliage green.
            8. Flowers or inflorescences axillary or at least appearing so (uppermost sometimes subtended only by much reduced leaves); sap not milky.
               9. Leaf blades often but not always rather triangular (sometimes narrowly so) in shape (leaf blades cordate in *T. cordata*), truncate to cordate or hastate basally; plants with stinging hairs mixed with soft spreading hairs ____ **Tragia**
               9. Leaf blades variously shaped but not triangular, usually not truncate, cordate, or hastate basally (except cordate in *Acalypha ostryifolia*); plants without stinging hairs.
                  10. Leaf blades entire or finely serrate; female flowers not surrounded by conspicuous folded bract; stipules absent or present; stems with broad-based glandular or malpighian (= attached at center) hairs.
                      11. Leaf blades 6–15 cm long; stipules present; stems with numerous broad-based glandular hairs ____ **Caperonia**
                      11. Leaf blades 0.5–7.5 cm long; stipules absent; stems with malpighian hairs (*A. humilis* sometimes glabrous) ____ **Argythamnia**

10. Leaf blades mostly toothed or crenate to subentire; female flowers surrounded by a conspicuous, toothed or lobed, folded bract or not so; stipules present; stem hairs not malpighian.
    12. Plants mostly herbs to 0.7 m tall; female flowers surrounded by a conspicuous, toothed or lobed, folded bract; lower leaf surface pubescence of simple hairs ____ **Acalypha**
    12. Plants shrubs, 1–3 m tall; female flowers not surrounded by folded bract; lower leaf surface pubescence stellate ____ **Bernardia**

8. Inflorescences terminating the stems (may be over-topped by leaves or leafy bracts), sometimes axillary inflorescences also present; sap milky or not so.
    13. Sap milky; inflorescences all terminal; female flowers exposed ____ **Stillingia**
    13. Sap not milky; inflorescences terminal, also some axillary present; female flowers ± concealed by a conspicuous, toothed or lobed, folded bract or not so.
        14. Plants herbs, to 0.7 m tall; female flowers ± concealed by a conspicuous, toothed or lobed, folded bract; leaf blades mostly serrate to crenate; some axillary inflorescences present; laminar glands absent ____ **Acalypha**
        14. Plants shrubs, 1–3 m tall; female flowers without folded bracts; leaf blades entire; all inflorescences terminal; laminar glands sometimes present ____ **Ditrysinia**

6. Flowers lacking a perianth; many single minute staminate flowers (without perianth) and a single larger central pistillate flower without perianth) together congested within small cup-like involucres (= cyathia) which simulate flowers, the cyathia with prominent glands and often petal-like appendages on the rim; sap milky ____ **Euphorbia**

ACALYPHA L. COPPERLEAF, THREE-SEED MERCURY

Ours annual or perennial dioecious and/or monoecious herbs or subshrubs; sap not milky; leaves alternate, petioled, blades entire or toothed, often turning reddish in fall; stipules minute; inflorescences spicate; staminate and pistillate flowers on the same or separate plants; staminate flowers: subtended by a small bract; calyces 4-parted; petals absent; stamens 4–8, united basally; pistillate flowers: each surrounded by a conspicuous folded bract that is variously toothed or lobed, sometimes allomorphic flowers (= lack bracts and are borne on elongate pedicels) present; calyces 3-parted; petals absent; ovary 3-carpellate; style branches often highly divided; seeds usually 3 per fruit (except 1 in *A. monococca*), carunculate.

A genus of ca. 450 species of tropical and warm areas; most are monoecious or dioecious shrubs with some trees and herbs and most have much-branched styles/stigmas (Levin 2016a). (Greek: *acelephe*, nettle, alluding to the similarity of the leaves to those of stinging nettle, *Urtica dioica* L.)

REFERENCES: Weatherby 1927; Miller 1964, 1969, 1970; Levin 1998, 1999, 2016a.

1. Leaf blades deeply divided, small, 5–12 mm long; staminate and pistillate flowers on different plants; species of sw portion of East TX ____ **A. radians**
1. Leaf blades variously toothed but not divided, usually > 12 mm long (usually much greater); staminate and pistillate flowers on the same plant (except sometimes on different plants in *A. monostachya*); including species widespread in East TX.
    2. Pistillate bract lobes linear, muricate (= rough with small sharp points) ____ **A. ostryifolia**
    2. Pistillate bract lobes variously deltate, lanceolate, rounded, attenuate etc., not linear, variously pubescent but not muricate.
        3. Leaf blades orbicular or reniform (= kidney-shaped); staminate and pistillate flowers on different plants or on same plant; plants sprawling low subshrubs or perennial herbs; in East TX known only from Bexar and Travis cos. ____ **A. monostachya**
        3. Leaf blades linear-lanceolate to ovate or rhombic; staminate and pistillate flowers on the same plant, either on the same or on different spikes; plants erect, annual (except the perennial *A. phleoides*); including species widespread in East TX.
            4. Base of leaf blades slightly cordate; leaf blades ovate; staminate and pistillate flowers borne on separate spikes ____ **A. ostryifolia**
            4. Base of leaf blades not cordate; leaf blades mostly linear to lanceolate to ovate or rhombic; staminate and pistillate flowers borne on the same spike.
                5. Spikes terminal, very conspicuous; plants perennial; species limited to sw margin of East TX ____ **A. phleoides**
                5. Spikes axillary; plants annual; including species widespread in East TX.
                    6. Capsules 1-seeded; leaf blades linear to linear-lanceolate; lower surface of pistillate bracts with sessile to stipitate red glands ____ **A. monococca**
                    6. Capsules 3-seeded; leaf blades linear to lanceolate, elliptic, ovate or broadly rhombic; lower surface of pistillate bracts with sessile to stipitate red glands or not so.
                        7. Leaf blades linear to lanceolate or narrowly ovate, the margins subentire to crenate; petiole usually < 1/4 as long as blade; lower surface of pistillate bracts with sessile to stipitate red glands (nonglandular hairs also present) ____ **A. gracilens**

7. Leaf blades broadly lanceolate, narrowly to broadly rhombic, or ovate to elliptic, the margins crenate to serrate; petiole 1/4 the length of blade to as long as blade; pistillate bracts without sessile to stipitate red glands (stipitate yellowish glands sometimes present).
   8. Pistillate bracts (these located at based of spike) (5–)7–9(–11) lobed; leaf blades ovate or elliptic to broadly rhombic; petioles usually ca. 1/2 to as long as leaf blades; species mainly of s portion of East TX.
      9. Pistillate bract 6–15 mm long, lobes (5–)7–9(–11), 1/3–2/3 of the bract length; styles branched; species uncommon and scattered primarily in se portion of East TX _______ **A. rhomboidea**
      9. Pistillate bract 4–5 mm long, lobes 7–9, 1/5 bract length; styles not branched; species mainly of far s TX; in East TX known only from Wilson Co. at extreme s margin _______ **A. poiretii**
   8. Pistillate bracts (9–)10–14(–16) lobed; leaf blades narrowly rhombic to broadly lanceolate; petioles usually 1/4–1/2 as long as leaf blades; species widespread primarily in n 2/3 of East TX _______ **A. virginica**

Acalypha gracilens A. Gray, (graceful), SLENDER COPPERLEAF, SLENDER THREE-SEED MERCURY. Erect annual 10–60 cm tall, simple or freely branched, usually at base, usually with dense recurved hairs; leaf blades 4–20 mm wide, linear-lanceolate or narrowly ovate; inflorescences bisexual; pistillate bracts with abaxial sessile or stipitate red glands, also pubescent with non-glandular hairs; styles 1–2 mm long, white or pinkish. This species is similar to *A. monococca* but usually occurs on deeper sandy soils under more mesic conditions, typically in openings in wooded areas; widespread in East TX; also scattered in Cross Timbers and Prairies to the w, Edwards Plateau, and in the n Gulf Prairies and Marshes; e U.S. from ME s to FL and w to WI and TX. Jun–Oct. [*A. gracilens* A. Gray var. *delzii* Lill. W. Mill., *A. virginica* L. var. *gracilens* (A. Gray) Riddell] Levin's (1998) detailed study indicated that while this species is quite variable, the variation appears as relatively smooth clines throughout its distribution and infraspecific taxa (e.g., var. *delzii*) are not warranted.

Acalypha monococca (Engelm. ex A. Gray) Lill. W. Mill. & Gandhi, (one-berried), SLENDER ONE-SEED COPPERLEAF. Erect annual similar to *A. gracilens*; leaf blades 3–12 mm wide, linear to linear-lanceolate, subentire to slightly crenate; inflorescences bisexual; pistillate bracts with abaxial often sparse sessile to stipitate red glands (this character is shared only with *A. gracilens*), also with non-glandular pubescence; capsules with only 1 seed (this is the only East TX species of *Acalypha* with 1 seed). This species is similar to *A. gracilens* but typically occurs in drier areas with more sun, little woody vegetation, and shallower, rocky or sandy soils; widespread in East TX; also scattered in Cross Timbers and Prairies to the w, Edwards Plateau, and in the n Gulf Prairies and Marshes; c U.S. from IL s to LA and w to KS and TX. Jun–Oct. [*A. gracilens* subsp. *monococca* (Engelm. ex A. Gray) G.L. Webster, *A. gracilens* var. *monococca* Engelm. ex A. Gray] We are following Levin (1998, 2016a) in recognizing this taxon at the specific level; he notes that *A. gracilens* and *A. monococca* are consistently different morphologically, differ ecologically, and that there is no evidence of hybridization between them.

Acalypha monostachya Cav., (one-spiked), ROUND COPPERLEAF. Dioecious or monoecious (staminate plants rare) subshrub or perennial herb, low sprawling; stems 15–40 cm tall; leaf blades 8–25 mm wide, orbicular or reniform, crenate, the petioles usually as long or longer than the blades; inflorescences often reddish; in monoecious plants inflorescences bisexual or staminate and terminal, pistillate inflorescences usually axillary; in dioecious plants staminate inflorescences terminal, pistillate inflorescences terminal and axillary; pistillate bract lobes ca. 1/4 of the bract length; styles bright to dark red. Dry areas, rocky or gravelly substrates; Bexar and Travis (Turner et al. 2003) cos. on sw margin of East TX; mainly to the s and w of East TX in South TX Plains, Edwards Plateau, and Trans-Pecos; in the U.S. known only from TX; also Mexico. Summer–fall. [*A. hederacea* Torr.] TX populations have in the past (e.g., Miller 1970) sometimes been separated, as *A. hederacea* (generally described as usually dioecious), from the more widespread *A. monostachya* (generally described as monoecious) which also extends well into Mexico. Levin (1999), in a detailed study, found that the sexuality of the *hederacea/monstachya* complex varied geographically, with more northern plants mostly dioecious and the more southern plants, particularly those in Mexico being mostly monoecious. Levin (1999), however, comments, "monoecious plants are common throughout the range of the complex."

Acalypha ostryifolia Riddell ex J.M. Coult., (with leaves resembling *Ostrya*—HOP-HORNBEAM), HOP-HORNBEAM COPPERLEAF, PINELAND THREE-SEED MERCURY. Erect monoecious annual to 50(–70) cm tall, usually freely branched; leaves long-petioled, the blades 1.5–5.3 cm wide, ovate, slightly cordate basally, nearly glabrous, margin serrate; pistillate spikes terminal, sometimes on short lateral branches and appearing axillary; staminate spikes axillary; capsules spiny. Stream banks or bottoms, disturbed areas, roadsides, sandy soils; scattered primarily in w 1/2 of East TX; widely scattered in TX; e U.S. from PA s to FL and w to NE and TX, also AZ. Late May–Oct.

Acalypha phleoides Cav., (resembling *Phleum*—TIMOTHY or CAT-TAIL GRASS), LINDHEIMER'S COPPERLEAF, SHRUBBY COPPERLEAF. Perennial herb, monoecious; stems several to many from a woody root, branched, spreading to erect, to 50 cm long, pilose with long,

Vaccinium fuscatum [BB2]

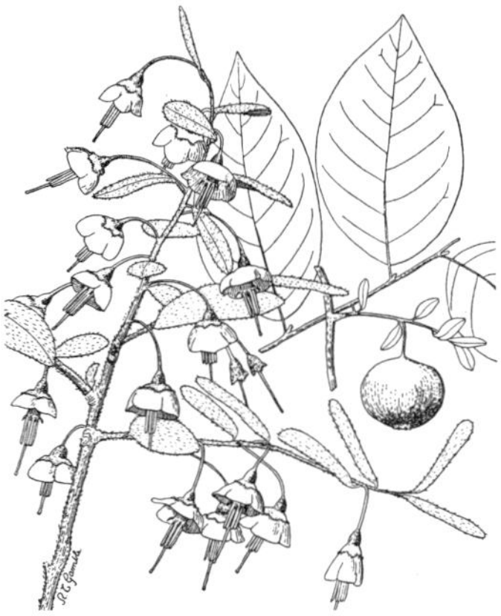
Vaccinium stamineum [BA1]

Vaccinium virgatum [BB2]

Acalypha monococca [STE]

Acalypha gracilens [STR]

Acalypha phleoides [HEA]

Acalypha monostachya [CAV]

Acalypha ostryifolia [BB2]

wide-spreading hairs or short-pubescent with curled hairs or glabrous; leaf blades rhombic-ovate to ovate, proximal suborbicular, sharply and evenly toothed; inflorescences bisexual; pistillate portion of inflorescence 4–7 cm long, 1.6–2 cm wide, staminate portion of inflorescence 0.5–3.5 cm long; pistillate bracts usually 5–7-lobed, sparsely pubescent, sparsely stipitate glandular; styles conspicuously reddish; allomorphic pistillate flowers sometimes present, bracts absent. Rocky slopes, chiefly limestone; sw portion of East TX; scattered to the w primarily in the Edwards Plateau and Trans-Pecos; AZ, NM, and TX. May–Oct. [*A. lindheimeri* Müll. Arg.] We are following Levin (2016a) for nomenclature of this species.

Acalypha poiretii Spreng., (for Jean Louis Marie Poiret, 1755–1834, French clergyman, botanist and explorer. Sprengel's name was a replacement for Poiret's name, *Acalypha macrostachyos*), POIRET'S COPPERLEAF. Annual; stems erect, to 40 cm tall, with dense pubescence of recurved hairs; leaf blades 7–35 mm wide, ovate to elliptic, crenate; petioles usually ca. 1/2 to as long as leaf blades; inflorescences bisexual; pistillate bracts shallowly divided into 7–9 lobes, pubescent and stipitate glandular abaxially; allomorphic pistillate flowers common, terminal on staminate portion of inflorescence. Dry areas, sandy or gravelly substrates; include based on a Wilson Co. collection (*Parks RX3133* TEX, not dated) at s margin of East TX; also extreme s TX; AL, FL, and TX; also Mexico. Summer–fall.

Acalypha radians Torr., (radiating outward), ROUND-CROTON, YERBA DE LA RABIA, CARDINAL'S-FEATHER. Perennial herb, dioecious; stems ± decumbent, 20–40 cm tall, densely pubescent with short stiff and long-spreading hairs; leaf blades 8–20 mm wide, reniform to orbicular; petioles 4–16 mm long, usually as long as or longer than blades; pistillate bracts 7–10 mm long, usually with 7–13 lobes; styles bright red. Dry sandy or gravelly areas; s portion of East TX n to Travis (BRIT) and Bastrop (Turner et al. 2003) cos.; widespread in s TX; in the U.S. known only from TX; also Mexico. Summer–fall.

Acalypha rhomboidea Raf., (rhomboidal, ± diamond-shaped with unequal sides), RHOMBOID COPPERLEAF, COMMON THREE-SEED MERCURY. Erect annual 10–60 cm tall; stems simple or freely branched, densely pubescent above with recurved hairs, sparsely pubescent below; leaf blades 0.8–5 mm wide, ovate to broadly rhombic; petioles usually 1/2 to as long as blades; inflorescences bisexual, axillary; pistillate bracts abaxially sparsely pubescent and stipitate glandular, deeply divided 1/3–2/3 of the bract length. Sandy soils; Dallas, Marion, Sabine (BRIT), Brazos, Harris, Madison, Montgomery, San Jacinto, and Walker (Turner et al. 2003) cos.; also Tarrant Co. (Mahler 1988) just w of East TX and n Gulf Prairies and Marshes; almost throughout the e 1/2 of the U.S.; also Canada. Jul–Oct. [*A. virginica* L. var. *rhomboidea* (Raf.) Cooperr.] While this species is cited for vegetational areas 1 through 5 (Fig. 2) by Hatch et al. (1990), according to M. Mayfield (pers. comm.), it occurs mainly in se TX; however, the identification of the Dallas and Marion county specimens were confirmed by G. Levin. This taxon is sometimes treated as a variety of *A. virginica* (e.g., Kartesz 1994; Jones et al. 1997). However, we are following Levin (2016a) in recognizing it at the specific level. According to M. Mayfield (pers. comm.), *A. rhomboidea* and *A. virginica* grow together without intergradation and are "very distinct species."

Acalypha virginica L., (of Virginia), VIRGINIA COPPERLEAF, VIRGINIA THREE-SEED MERCURY. Erect annual 10–50 cm tall; stems simple or sparsely branched, sparsely to densely pubescent with recurved and long, soft, spreading hairs; petioles usually 1/4–1/2 as long as blades; inflorescences bisexual; pistillate bract abaxially hirsute and sometimes stipitate glandular, lobes 1/4–1/2 of the bract length; styles white. Sandy open woods or low areas; n 2/3 of East TX; also adjacent Cross Timbers and Prairies; c U.S. from NY s to GA and w to SD and TX. Jul–Oct.

ARGYTHAMNIA P. Browne SILVERBUSH

Perennials, herbaceous from a woody crown, with pubescence of hairs attached at the middle (= malpighian); sap not milky; foliage sometimes darkening upon drying; leaves alternate; leaf blades usually entire; inflorescences racemose, axillary; flowers unisexual; staminate flowers: sepals 5; petals 5; glands 5, minute, opposite sepals; stamens 7–10, the filaments coherent; pistillate flowers: sepals 5; petals 5, vestigial or absent; glands 5, minute, opposite sepals, ovary 3-carpellate; ovules 1 per cell; styles 3, bifid; capsules 3-seeded; seeds ecarunculate.

A genus of ca. 80 species including subgenus *Ditaxis* (Ramírez-Amezcua 2016) ranging from the U.S. to Argentina. The species, have in the past, been treated in this broadly circumscribed manner (e.g., Correll & Johnston 1970; Ingram 1980; Hatch et al. 1990; Kartesz 1994). However, Webster (1994b) considered *Ditaxis* a distinct genus, as did Takahashi et al. (1995), Steinmann and Felger (1997), and Mabberley (2008). However, more recent authors (Ramírez-Amezcua 2016; Powell & Worthington 2018; Kartesz 2022; Weakley, A.S., and Southeastern Flora Team 2022) have reverted to treating the group in the more inclusive sense. (Greek: *argyros*, silver-white and *thamnos*, shrub, alluding to trunk and branches covered with whitish bark)

REFERENCES: Shinners 1956; Ingram 1967, 1980; Waterfall 1971 [1972]; Webster 1994b; Takahashi et al. 1995; Ramírez-Amezcua 2016; Powell & Worthington 2018; Kartesz 2022; Weakley, A.S., and Southeastern Flora Team 2022.

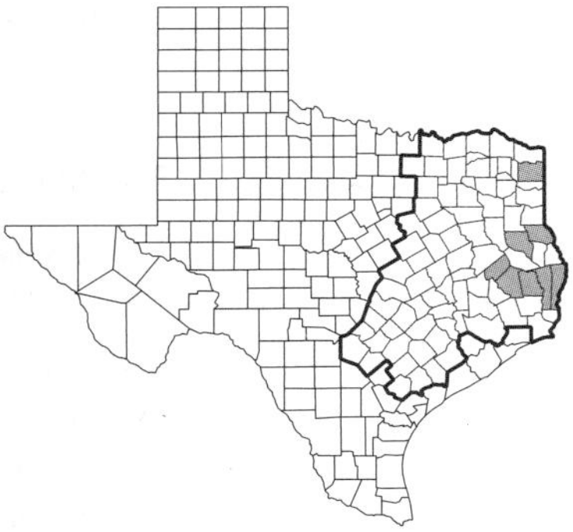

Vaccinium fuscatum

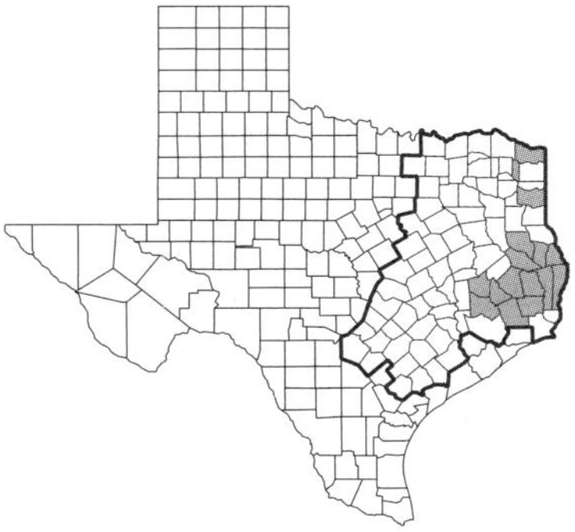

Vaccinium stamineum

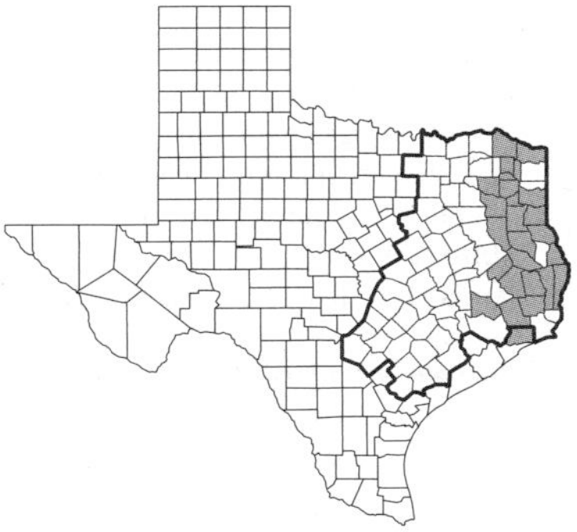

Vaccinium virgatum

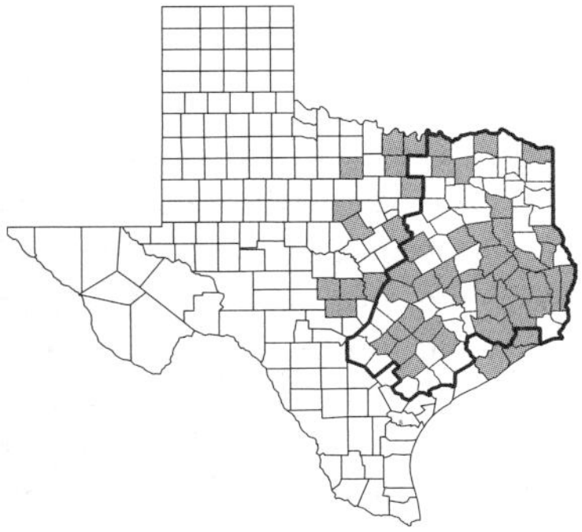

Acalypha gracilens

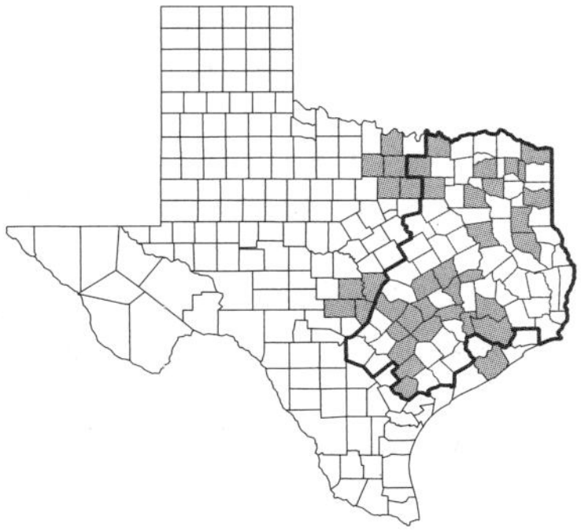

Acalypha monococca

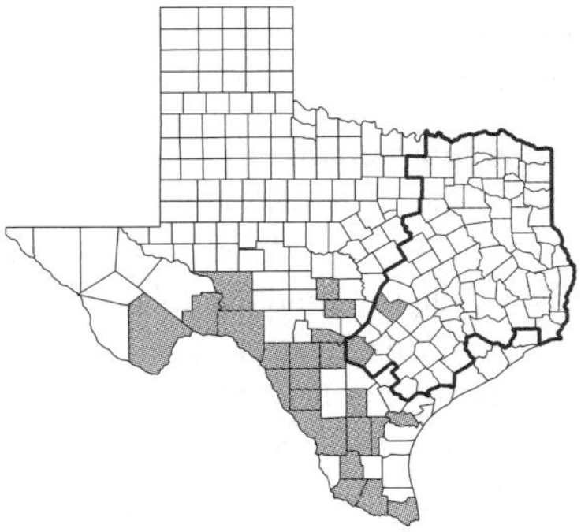

Acalypha monostachya

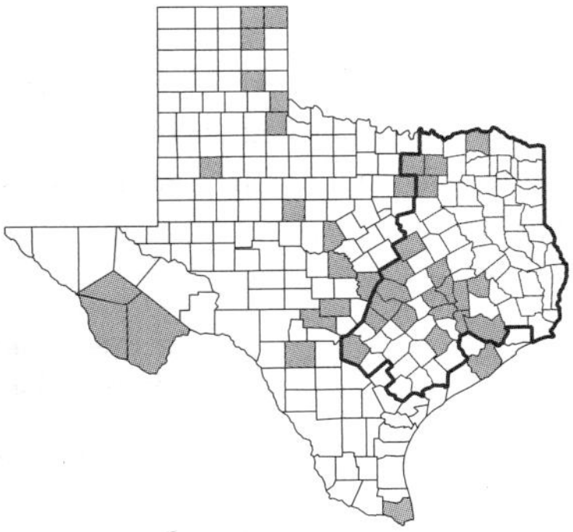

Acalypha ostryifolia

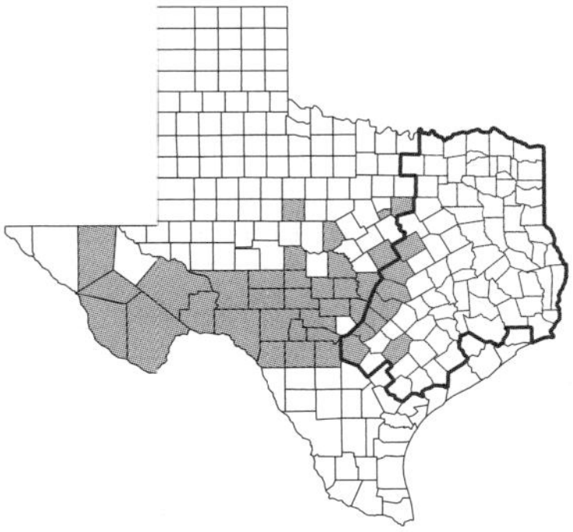

Acalypha phleoides

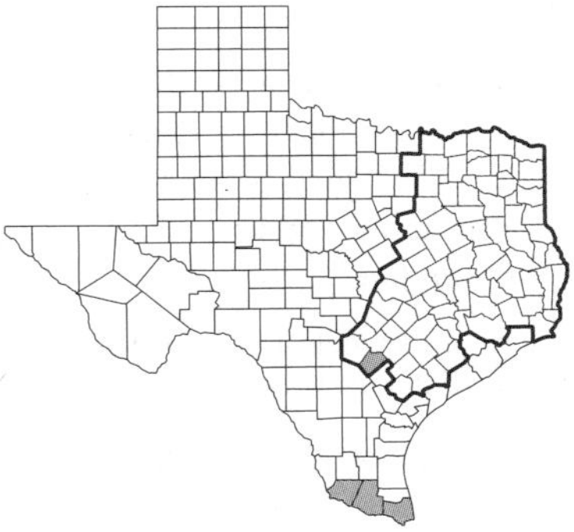

Acalypha poiretii

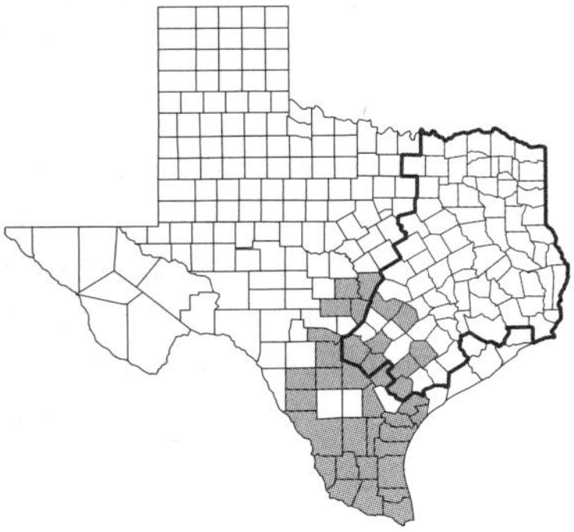

Acalypha radians

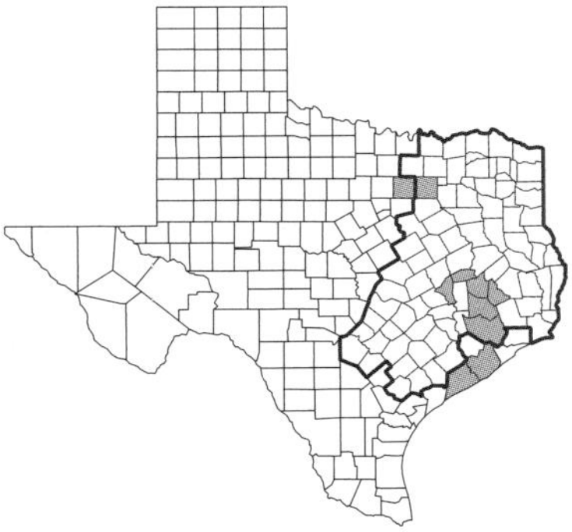

Acalypha rhomboidea

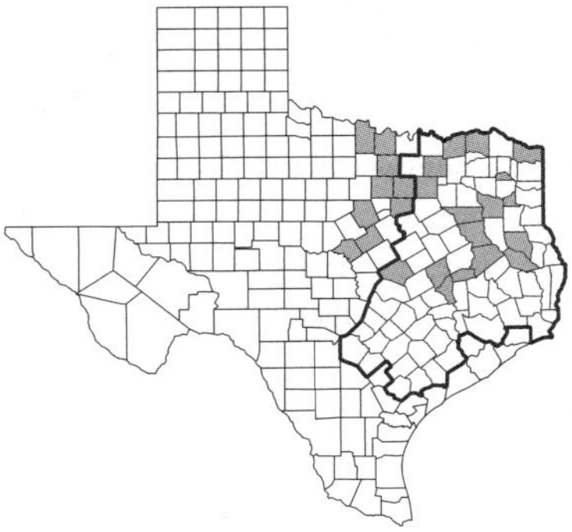

Acalypha virginica

1. Inflorescences shorter than leaves; plants with much-branched, erect to trailing or spreading stems ______ **A. humilis**
1. Inflorescences longer than leaves; plants with several, unbranched, erect to ascending stems.
   2. Glands of pistillate and staminate flowers ovate or oblong; species of sw margin of East TX.
      3. Petals of staminate flowers cuneate-obovate to cuneate-elliptic, 3.3–4 mm long, 1.6–3 mm wide; stamens 10 per flower; inflorescences unisexual ______ **A. aphoroides**
      3. Petals of staminate flowers elliptic, 2–4 mm long, 1.2–1.5 mm wide; stamens 7–8 per flower; inflorescences bisexual ______ **A. simulans**
   2. Glands of pistillate flowers linear; glands of staminate flowers linear to obovate; species more common ______ **A. mercurialina**

Argythamnia aphoroides Müll. Arg., (resembling *Aphora*, a generic synonym used by Nuttall), HILL COUNTRY WILD MERCURY, SHRUBBY DITAXIS, HILL COUNTRY SILVERBUSH. Perennial monoecious or dioecious herb to 5 dm tall, densely villous; stems usually erect; leaf blades elliptic to ovate, sessile, the main veins prominent on lower surface, palmate; staminate flowers: sepals 4–5.5 mm long, petals 3.3–4 mm long, glands oblong 0.4–0.5 mm long, stamens 10; pistillate flowers: petals absent or rudimentary, glands oblong, 0.5–0.7 mm long, styles erect, bifid. Sandy or rocky limestone soils, oak woodlands; Bexar, Comal, Hays, and Travis (Kartesz 2022) cos. in sw corner of East TX; also adjacent Edwards Plateau and Cross Timbers and Prairies; narrowly endemic to a small area of TX. Endemic to TX. Mar–May. [*Ditaxis aphoroides* (Müll. Arg.) Pax] This species will most likely only be found on the Edwards Plateau portion of the above East TX cos. NatureServe: G2, imperiled.

Argythamnia humilis (Engelm. & A. Gray) Müll. Arg., (low-growing, dwarf), LOW WILD MERCURY, LOW DITAXIS, LOW SILVERBUSH. Perennial usually monoecious herb to 3 dm tall; stems erect to prostrate; leaves sparsely to densely villous, petiole 2–10 mm long, blade obovate to linear, with veins prominent on lower surface, pubescent to glabrous; staminate flowers: petals 1.5–2.6 mm long, elliptic, glands ovate to linear, stamens (7–)10; pistillate flowers: petals 5, 0.3–1.7 mm long, glands linear to oblong, 0.3–1.5 mm long, styles erect and bifid. Rocky or disturbed calcareous clay soils; throughout much of TX, but absent from the e 1/2 of East TX; CO, KS, NM, OK, and TX; also Mexico. Apr–Oct. [*A. humilis* var. *laevis* (A. Gray ex Torr.) Shinners, *A. humilis* var. *leiosperma* Waterf., *A. laevis* (A. Gray ex Torr.) Müll. Arg., *Ditaxis humilis* (Engelm. & A. Gray) Pax, *D. humilis* var. *leiosperma* (Waterf.) Radcl.-Sm. & Govaerts, *D. laevis* (A. Gray ex Torr.) A. Heller] Varieties have sometimes been recognized in this species (e.g., Correll & Johnston 1970). Variety *leiosperma* Waterf., was named in the genus *Argythamnia* by Waterfall (1971 [1972]) apparently based on a single seed character. Both it and the more w var. *laevis* seem better lumped with var. *humilis* (Ramírez-Amezcua 2016).

Argythamnia mercurialina (Nutt.) Müll. Arg., (Mercury, messenger of the gods), TALL WILD MERCURY, TALL DITAXIS, TALL SILVERBUSH. Perennial monoecious or dioecious herb to 8 dm tall, plant dark green, villous; stems ascending, unbranched; staminate flowers: sepals 2.3–3 mm long, petals 2–2.5 mm long, elliptic to spatulate, glands linear to obovate, 0.3–0.5 mm long, stamens usually 8; pistillate flowers: petals usually absent, glands linear, ca.1.5 mm long, styles spreading, bifid. Sandy or rocky limestone soils, prairies. Apr–Jun.

1. Glands of staminate and pistillate flowers glabrous ______ var. **mecurialina**
1. Glands of staminate and pistillate flowers pubescent ______ var. **pilossisima**

var. **mercurialina.** Mostly in the w portion of East TX; Edwards Plateau, Cross Timbers and Prairies, Rolling Plains, and High Plains; AR, CO, KS, NM, OK, and TX [*Ditaxis mercurialina* (Nutt.) J.M. Coult.]

var. **pilossisima** (Benth.) Shinners, (very hairy). Mostly in the s portion of East TX, but also Brazos, Leon, and Waller (Kartesz 2022) cos.; deep s TX. According to Ramírez-Amezcua (2016) this variety it endemic to TX; however, according to Richardson and King (2011) it is found in Starr Co. in deep s TX which borders Mexico and it might be expected to occur in Mexico. [*A. pilossisima* (Benth.) Müll. Arg., *Ditaxis pilossisima* (Benth.) A. Heller]

Argythamnia simulans J.W. Ingram, (resembling), PLATEAU SILVERBUSH. Perennial herb or subshrub to 6 dm tall, usually monoecious, rarely dioecious; stems erect to ascending; leaves elliptic to ovate or obovate, sessile; male and female flowers usually in the same racemes; staminate flowers: sepals 4–5.5 mm long, petals 2–4 mm long, elliptic, glands ovate, 0.5–0.8 mm long, stamens 7–8; pistillate flowers: petals frequently absent or less than 0.5 mm long, glands ovate to oblong, 0.3–0.4 mm long, styles spreading, bifid. Rock or sandy limestone soils; Bexar Co. n to Bell Co. (BRIT) on sw margin of East TX; Edwards Plateau and n South TX Plains; in the U.S. known only from TX; also Mexico. Mar–Jul. [*Ditaxis simulans* (J.W. Ingram) Radcl.-Sm. & Govaerts]

Acalypha poiretii [SNB]

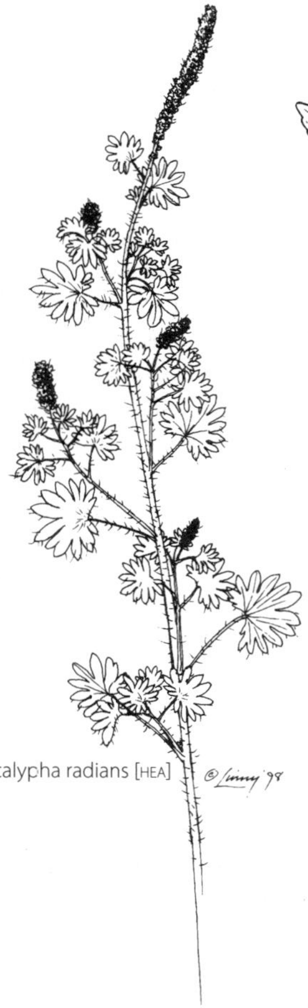

Acalypha radians [HEA]

Acalypha rhomboidea [STR]

Acalypha virginica [LUN]

Argythamnia aphoroides [HEA]

Argythamnia humilis [BB1, BT3]

Argythamnia mercurialina [EN1]

BERNARDIA Houst. ex Mill. MYRTLECROTON

A genus of ca. 70 species of warm areas of the New World (Mayfield 2016). (named for Bernard de Jussieu, French botanist, 1699–1777)
REFERENCES: Cervantes & Ramirez 2002; Mayfield 2016.

Bernardia myricifolia (Scheele) S. Watson, (with leaves like genus *Myrica*, WAX-MYRTLE, Myricaceae), MOUSE'S-EYE, PALO DE TARUGO, SOUTHWEST BERNARDIA, BRUSH MYRTLECROTON, OREJA-DE-RATÓN, MYRTLECROTON. Unarmed, dioecious, much-branched, semi-evergreen woody shrub, 1–3 m tall, sap clear; leaves alternate, simple, 1.5–4.5 cm long, broadly elliptic to orbiculate, crenate, with thickened teeth, reticulate-veined, lower surface dense stellate pubescent, usually with crateriform (= bowl-shaped) laminar glands on the proximal half of the blade, margins revolute; flowers small and inconspicuous, solitary (female) or in inconspicuous racemes (male); staminate flowers: sepals 3, ca. 2 mm long, petals absent; stamens usually 12–18; pistillate flowers: sepals 3–5, fruit a usually 3-celled capsule, 7–8 mm long, grayish brown, tomentose; seeds usually 3. Caliche slopes, canyons, brushy areas, limestone soils; Bexar, Comal, Hays, Travis (BRIT), Goliad, Jackson, and Lavaca (Turner et al. 2003) cos. in s (primarily sw margin) portion of East TX; also Edwards Plateau, Tran-Pecos, South TX Plains, and Gulf Prairies and Marshes; in the U.S. known only from TX; also Mexico. Spring–fall. According to Mayfield (2016), reports of *B. myricifolia* from AZ and CA are plants he identifies as *B. incana* C.V. Morton, and reports of *B. myricifolia* from NM are misidentified specimens of *B. obovata* I.M. Johnst. This species is extremely heat and drought tolerant.

CAPERONIA A. St.-Hil. FALSE CROTON

A mainly tropical genus of 32 species of the New World and Africa (Gillespie 2016a). (named "for the apothecary Caperoni"—Quattrocchi 2000)
REFERENCE: Gillespie 2016a.

Caperonia palustris (L.) A. St.-Hil., (of marshes), SACATRAPO, TEXASWEED, MEXICANWEED, BIRD-EYE, MULE'S-EAR. Monoecious annual herb ca. 0.2–1 m tall with glandular and nonglandular hairs; sap clear; leaves alternate, the blades lanceolate to elliptic-lanceolate or rarely ovate, usually 6–15 cm long, serrate; inflorescences spikes in upper leaf axils; staminate flowers distal with 5-lobed calyces, the lobes ca. 1 mm long, petals obovate, slightly exceeding the calyx, ca. 1.4 mm long; stamens in 2 whorls of ca. 5 each; pistillate flowers proximal, 2–2.3 mm long, becoming 3.5–5.5 mm in fruit; petals 5–6, 1–2(–2.4) mm long, whitish, soon falling; fruit a trilobed capsule, ca. 5–7 mm in diam., densely covered in gland-tipped trichomes. Jefferson, Liberty, Orange (BRIT), Brazos, Hardin, Harris, and Waller (Turner et al. 2003) cos. mainly in se portion of East TX; also n Gulf Prairies and Marshes; AL, AR, FL, LA, MS, and TX; also widespread in the Neotropics. Jul–Oct. Native of the Neotropics. This species can be an agricultural weed (e.g., in soybean and rice fields). According to Correll and Johnston (1970) it "arrived in Tex. about 1920."

CNIDOSCOLUS Pohl BULL-NETTLE, MALA MUJER

A mainly tropical American genus, often with stinging hairs. The number of species in the genus varies widely: ca. 50 (Levin 2016b) to 99 (Maya-Lastra & Steinmann 2019). The leaves of some species are eaten as greens. (Greek: *cnide*, nettle, and *scolopes*, prickle or sting, alluding to the stinging hairs)
REFERENCES: McVaugh 1944; Levin 2016b; Maya-Lastra & Steinmann 2019.

Cnidoscolus texanus (Müll. Arg.) Small, (of Texas), TEXAS BULL-NETTLE, MALA MUJER, TREAD-SOFTLY. Perennial with milky sap, to 80(–100) cm tall from a large (to 20 cm thick), deep, branching root; stems densely and leaves more sparsely hispid with pale, extremely painful, needle-like stinging hairs; leaves alternate, long-petioled, blades ca. 6–15 cm wide, cordate, palmately deeply 3–5-lobed, the lobes coarsely toothed or again lobed; stipules inconspicuous, narrow, toothed; flowers in terminal, cymose, pedunculate inflorescences slightly shorter to slightly longer than the leaves; male and female flowers in the same inflorescence; perianth large, white, showy, sweet-scented, of a single whorl of parts; perianth of male flowers with subcylindrical tube 15–20 mm long, slightly longer than the (4–)5 subrotate lobes; perianth of female flowers 10–17 mm long, 5-parted essentially to base; capsules 15–20 mm long, hispid, with 3 large seeds, 14–18 mm long. Sandy open woods, fields, disturbed areas; widespread throughout East TX; widespread in TX except Panhandle and Trans-Pecos; AR, LA, OK, and TX. Apr–Jul, less freely to Sep. [*Jatropha texana* Müll. Arg.] The seeds are reported to be edible; however, there are reports that they may contain some cyanide (Burrows & Tyrl 2001). If the foliage is touched, glass-like hairs break off in the skin and act like hypodermic needles; they release a toxin which causes an intense burning sensation; this type of effect is known as contact urticaria; the stinging hairs can penetrate even heavy clothing such as jeans; subsequent to the sting, the affected skin can be red, swollen, and irritated for a number of days. According to Lampe (1986), only four families (Euphorbiaceae, Hydrophyllaceae, Loasaceae, and Urticaceae) have stinging hairs—East TX has stinging representatives in the Euphorbiaceae and Urticaceae.

CROTON L. CROTON

Ours annual or perennial, often aromatic herbs or shrubs with stellate hairs or peltate scales on the epidermis of at least some parts; sap not milky; leaves alternate or subopposite, or those under the flowers apparently whorled, entire or toothed; stipules small, usually falling early; flowers solitary or in spike-like racemes or head-like clusters, unisexual; inflorescences unisexual (in dioecious plants) or bisexual (in monoecious plants), pistillate flowers proximal, staminate flowers distal; staminate flowers with (4–)5(–6) sepals united at base; petals as many or none; stamens 5–20; pistillate flowers with 5–6(–9) sepals united at or near base; petals none on most East TX species, or minute; pistil 1, usually 3-carpellate; fruit a 1- to usually 3-seeded, globose or subglobose capsule, or in *C. michauxii* an achene.

A huge genus (ca. 1,250 species) of tropical and warm areas (van Ee & Berry 2016) ranging from temperate herbs to tropical rain forest trees. Oil of croton, one of the most purgative substances known, is obtained from an Old World species, *C. tiglium* L. Many contain alkaloids and are used medicinally; an Ecuadorian species is currently being investigated for medicinal uses. *Croton* species are distasteful and rarely eaten by livestock; they thus increase under overgrazing; some species are toxic. The widely cultivated horticultural "crotons" with variegated foliage belong to the Euphorbiaceae genus *Codiaeum* Rumph. ex A. Juss., a group of 15 species native from Malesia to the Pacific. Key adapted from Weakley, A.S., and Southeastern Flora Team (2022). (Greek: *croton*, a tick; the Greek name of the CASTOR-OIL-PLANT of this family, from similarity of the seed to a tick)

REFERENCES: Ferguson 1901; Johnston 1959; Ginzbarg 1992; Webster 1992, 1993; Aplet et al. 1994; van Ee & Berry 2016.

1. Plants shrubs, (2–)5–30 dm tall; styles 3, unbranched or 2-branched, the terminal segments therefore 3 (or 6).
  2. Plants 2–10 dm tall; upper and lower surfaces of mature leaf blades with stellate pubescence, dense on lower surface, the upper surface green, the lower grayish green; leaf blade apex acute; on sw margin of East TX ____ **C. fruticulosus**
  2. Plants 5–35 dm tall; upper and lower surfaces of mature leaf blades strikingly different in appearance, the upper green and essentially glabrous (rarely a few peltate scales present), the lower silvery-white with reddish brown tinge due to it being covered with silvery-white peltate scales (some reddish brown), the actual leaf surface not visible; leaf blade apex obtuse: in East TX known only from Bell and Travis cos. ____ **C. alabamensis**
1. Plants annuals or perennial herbs or subshrubs to 20 dm tall; styles 2 or 3, 2-branched or several-branched, the terminal segments therefore 4, 6, or more.
  3. Leaves with coarsely serrate or crenate margins, blades sparsely to densely covered with stellate hairs; 2 cup-shaped yellowish glands present near the junction of the petiole and the leaf blade base (lower surface) ____ **C. glandulosus**
  3. Leaves with entire margins; glands absent at the junction of the petiole and the leaf blade base.
    4. Lower leaf surface densely lepidote (= covered with small scurfy scales) or stellate-lepidote, silvery; plants annual to perennial.
      5. Inflorescences unisexual; plants dioecious; staminate petals absent ____ **C. dioicus**
      5. Inflorescences bisexual; plants monoecious; staminate petals 5.
        6. Petioles 2–10(–20) mm long, at least some of them > 1/3 as long as the leaf blades; fruit 3-locular, dehiscent; seeds 3 per fruit, 2.5–5 mm long; petals of the staminate flowers lepidote ____ **C. argyranthemus**
        6. Petioles 1–5 mm long, < 1/5 the length of the leaf blades; fruit 1-locular, indehiscent; seeds 1 per fruit, 2–2.5 mm long; petals of the staminate flowers glabrous on the surfaces, villous on the margins ____ **C. michauxii**
    4. Lower leaf surface stellate-pubescent, not silvery; plants annual (perennial only in *C. pottsii*).
      7. Plants dioecious, the inflorescences therefore unisexual; staminate flowers lacking petals or petals (4–)5; capsules warty or smooth.
        8. Perennial herbs 1–6 dm tall; staminate flowers with (4–)5 petals; capsules smooth ____ **C. pottsii**
        8. Annual herbs 2–15 dm tall; staminate flowers lacking petals; capsules warty.
          9. Plants 5–15 dm tall; lower leaf surfaces densely stellate-tomentose; capsules 8–9 mm long ____ **C. parksii**
          9. Plants 2–9 dm tall; lower leaf surfaces densely appressed stellate; capsules 5–8 mm long ____ **C. texensis**
      7. Plants monoecious, the inflorescences either bisexual or unisexual; staminate flowers with 3–5 petals; capsules smooth.
        10. Lower leaf surfaces appearing brown-dotted, because some of the stellate hairs have dark brown centers; terminal style segments 4; ovaries 2-locular, only 1 fertile ____ **C. monanthogynus**
        10. Lower leaf surfaces not brown-dotted (none of the stellate hairs with brown centers); terminal style segments 6–18 (–24); ovaries 3 locular, all 3 usually fertile.
          11. Pistillate flowers with 5–6 sepals; terminal style segments 6; herbs 1–6 dm tall.
            12. Perennial herbs; inflorescence with staminate and pistillate flowers (12–)20–25 each; pistillate sepals 5; staminate flowers with 10–15 stamens ____ **C. pottsii**
            12. Annual herbs; inflorescence with 3–8 staminate flowers and 1–3 pistillate flowers; pistillate sepals 5–6; staminate flowers with 7–9(–12) stamens ____ **C. lindheimerianus**
          11. Pistillate flowers with 6–9 sepals; terminal style segments 12–18(–24); herbs 3–20 dm tall.
            13. Staminate flowers with sepals and petals 0.8–1 mm long; styles 2–3 mm long ____ **C. capitatus**
            13. Staminate flowers with sepals and petals 1–2 mm long; styles 3–4 mm long.

14. Pistillate flowers with sepals whitish appressed-tomentose; styles 3, terminal segments 12, with each style branching once into 4 segments ________ **C. heptalon**
14. Pistillate flowers with sepals yellowish woolly-tomentose; styles 3, terminal segments 12, with each style branching into 2, and then branching again to make 4 segments ________ **C. lindheimeri**

Croton alabamensis E.A. Sm. ex Chapm. var. **texensis** Ginzbarg, (sp.: of Alabama; var.: of Texas), TEXABAMA CROTON, ALABAMA CROTON. Monoecious shrub 5–35 dm tall; leaves petiolate, blades ovate or narrowly ovate, to 10 cm long, upper and lower leaf surfaces strikingly different in appearance, the upper green and glabrous, the lower silvery-white with reddish brown tinge due to it being covered with silvery-white peltate scales (some reddish brown), the actual leaf surface not visible, turning orange in fall; younger leaves and young twigs more reddish brown due to greater number of pigmented scales; inflorescences bisexual or unisexual, a terminal 6–14-flowered raceme. Limestone canyon slopes, in understory of forests or in full sun; Travis (BRIT) and Bell (Kartesz 2022) cos.; at w margin of East TX in Coryell Co. (BRIT) just to the nw; first discovered in 1989; described by Ginzbarg (1992). Aplet et al. (1994) gave additional information. Feb–Mar. The only other variety, var. *alabamensis*, is known only from 2 cos. in Alabama (van Ee & Berry 2016). NatureServe: T2, imperiled variety.

Croton argyranthemus Michx., (silver flowered) SILVER CROTON, HEALING CROTON. Plant herbaceous or subshrubby, monoecious annual or short-lived perennial, usually 3 dm or less tall; leaf blades 1–5 cm long, 0.5–2(–3) cm wide, oval to narrowly obovate proximally, oblong to broadly elliptic distally, lower leaf surface densely lepidote, upper leaf surface sparsely lepidote to glabrate; inflorescences bisexual, staminate flowers 15–35, sepals 5, petals 5; pistillate flowers 2–8, sepals 5–7, petals absent; capsules 3–4 mm in diam. Upland forests, roadsides; sandy soils; widespread in East TX except n and w margins; also South TX Plains and s Gulf Marshes and Prairies; AL, FL, GA, LA, and TX. Apr–Aug.

Croton capitatus Michx., (capitate, headed), WOOLLY CROTON, HOGWORT. Monoecious annual herb to 3–8 dm tall, densely stellate tomentose; leaf blades 3–8(–15) cm long, 1–4 cm wide, ovate to lanceolate-elliptic, apex usually acute to obtuse, stellate pubescent; flowers in terminal spike-like racemes or capitate; staminate flowers 5–15, sepals 5, petals 5; pistillate flowers 3–6, sepals 6–9, recurved, petals absent, styles 3, two or three times bifid. Sandy or calcareous prairies, pastures, roadsides; throughout East TX; widespread primarily in the e 1/2 of TX; widespread in the e 1/2 of the U.S. (except absent from FL and GA); also Ontario, Canada. Jun–Oct. While generally avoided, cattle can be poisoned by eating the plant fresh or in hay (Boughton 1931; Muenscher 1951). Similar to *C. lindheimeri* (see distinguishing characters under that description.

Croton dioicus Cav., (having pistils and stamens in different flowers and on different plants), GRASSLAND CROTON, HIERBA DEL GATO, ROSVAL, RUBALDO. Subshrub 2–5(–9) dm tall, stellate-lepidote; leaf blades with petiole usually < 1/2 blade length, blade 1–6.5 cm long, 0.6–2.2 cm wide, narrowly elliptic-ovate to lanceolate, base rounded, apex acute to rounded, lower leaf surface densely silvery lepidote or stellate-lepidote, upper leaf surface less so; inflorescences unisexual racemes; staminate flowers 4–16, sepals 5, petals absent; pistillate flowers 2–5, sepals 5, petals absent, style 2–3 times branched with 12–24 terminal segments; capsule 5–6 mm in diam. Limestone, canyons, disturbed areas; in East TX known only from Hays Co. (*Stanfield s.n.*, no date, but label states "New York Botanical Garden, Herbarium of Dr. John K. Small, purchased 1900 NY"); Edwards Plateau, Trans-Pecos and South TX Plains; NM and TX. Mar–Nov. [*C. bigbendinesis* B.L. Turner, *C. neomexicanus* Müll. Arg.]

Croton fruticulosus Engelm. ex Torr., (shrubby and dwarf), ENCINILLA, HIERBA LOCA. Monoecious shrub to ca. 10 dm tall, stellate pubescent; leaf blades 2–8 cm long, 2–4 cm wide, ovate to ovate-lanceolate, base cordate to truncate, apex attenuate to acute, stellate pubescent; inflorescences bisexual with 10–20 staminate flowers, sepals 5, petals 5; pistillate flowers 2–5, sepals 5, petals absent, styles bifid to base, capsules 5–6 mm in diam. In brush on limestone uplands; sw margin of East TX; also Edwards Plateau and Trans-Pecos; AZ, NM, and TX. May–fall.

Croton glandulosus L., (glandular), VENTE-CONMIGO, TROPIC CROTON. Monoecious annual, normally erect, to 12 dm tall, widely branched, stellate-pubescent; leaf blades 1–7 cm long, 0.3–3 cm wide, ovate to oblong-lanceolate, cup-shaped gland present on each side of the lower surface of the base of the leaf blade where the petiole attaches, margin coarsely crenate to serrate-dentate, sparsely to densely stellate pubescent; staminate flowers 10–20, sepals 5, petals 5; pistillate flowers 1–4, sepals 5, petals absent or 5, rudimentary, styles 3, deeply bifid; capsules 4–5 mm in diam. Stream bottoms or disturbed habitats; throughout East TX; throughout e 1/2 of TX; widespread in the e 1/2 of the U.S.; varieties are not distinguished on the county distribution map. May–Oct. Two avenues are provided below for keying out the varieties of *C. glandulosus*.

1. Leaf blades sparsely to moderately stellate pubescent.
   2. Leaf blades 2–7 cm long, moderately stellate pubescent; plants 2–12 dm tall; glands sessile, wavy wrinkle when dry, 0.5–0.8 mm wide ________ var. **septentrionalis** (in part)

Argythamnia simulans [CP1]

Bernardia myricifolia [VIN]

Caperonia palustris [CO1]

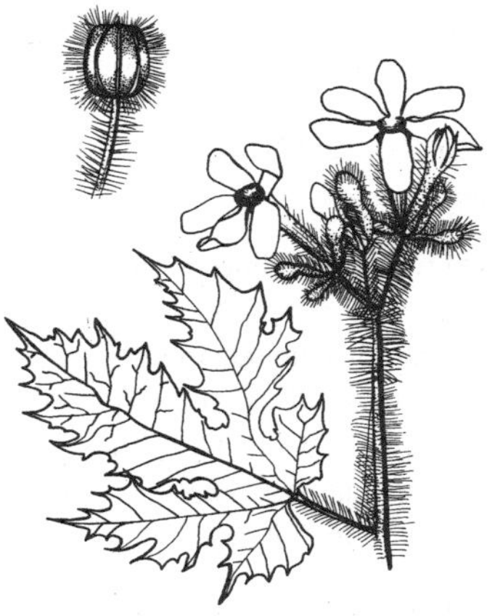

Cnidoscolus texanus [BT3]

Croton alabamensis var. texensis [SID]

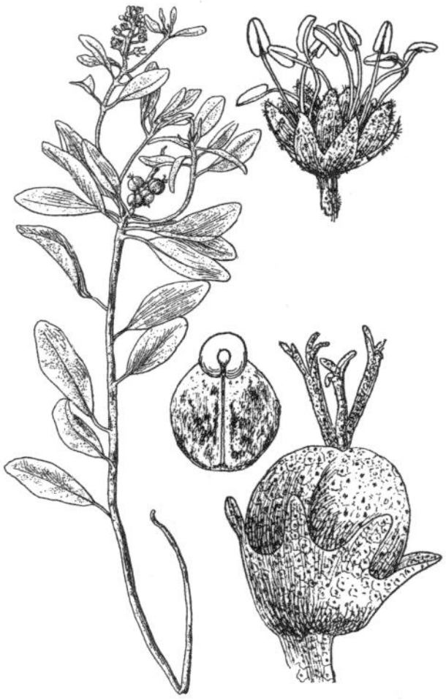

Croton argyranthemus [ARM]

Croton capitatus [ARM]

2. Leaf blades 1–2(–3) cm long, sparsely to moderately stellate pubescent; plants 1–2 dm tall; glands stipitate, circular when dry, 0.1–0.4 mm wide ____ var. **lindheimeri**
1. Leaf blades moderately to densely stellate pubescent.
   3. Leaf blades moderately stellate pubescent; radii of stellate hairs equal; plants 2–12 dm tall ____ var. **septentrionalis** (in part)
   3. Leaf blades densely stellate pubescent, radii of stellate hairs unequal; plants 1–2 dm tall ____ var. **pubentissimus**

1. Plants 1–2 dm tall; leaf blades 1–2(–3) cm long, 0.3–0.8(–1.3) cm wide; glands stipitate, circular when dry, 0.1–0.4 mm wide ____ var. **lindheimeri**
1. Plants 1–12 dm tall; leaf blades 2–7 cm long, 0.7–3 cm wide; glands sessile, wavy-wrinkled when dry, 0.5–0.8 mm wide.
   2. Plants 1–2 dm tall; stems densely stellate pubescent, hairs spreading, radii unequal (central branch often erect and much longer); leaf blades densely stellate-villous ____ var. **pubentissimus**
   2. Plants 2–12 dm tall; stems moderately stellate pubescent, hairs appressed, radii equal; leaf blades moderately stellate pubescent ____ var. **septentrionalis**

var. **lindheimeri** Müll. Arg., (for Ferdinand Jacob Lindheimer, 1801–1879, German born TX botanist), LINDHEIMER'S CROTON. Scattered in East TX; mostly in the e 1/2 of TX; KS, NM, OK, and TX.

var. **pubentissimus** Croizat, (very hairy). Mostly in the Gulf Prairies and Marshes; Colorado, Harris, and Victoria (Kartesz 2022) cos.

var. **septentrionalis** Müll. Arg., (northern), NORTHERN CROTON. Throughout East TX; mostly in the e 2/3 of TX; from NY s to FL and w to MN and TX.

Croton heptalon (Kuntze) B.W. van Ee & P.E. Berry, (with seven unequal parts), WOOLLY CROTON. Annual herb 5–15 dm tall, stellate pubescent; leaf blades 3–10 cm long, 1–5 cm wide, ovate-lanceolate, base cordate to rounded, apex acute; inflorescences bisexual racemes; staminate flowers 3–10, sepals (4–)5, petals 5; pistillate flowers 4–8, sepals 7–8, petals absent; capsules 6–7 mm in diam. Beaches, coastal dunes, roadsides; scattered in the s portion of East TX; Edwards Plateau, South TX Plains, and Gulf Prairies and Marshes; in the U.S. known only from TX; also Mexico. May–Dec. [*C. capitatus* var. *albinoides* (A.M. Ferguson) Shinners]

Croton lindheimeri (Engelm. & A. Gray) Alph. Wood, (for Ferdinand Jacob Lindheimer, 1801–1879, German born TX botanist), LINDHEIMER'S CROTON. Monoecious annual 5–20 dm tall; stems stellate hairy; leaf blades 3–7 cm long, ovate-lanceolate, lower surface not brown dotted; inflorescences bisexual racemes; staminate flowers 8–15, sepals (4–)5, petals 5; pistillate flowers 2–7, sepals 7–8, petals absent; capsules 8–9 mm in diam. Old pastures, forest openings, bottomlands, disturbed areas; throughout much of East TX; South TX Plains and Gulf Prairies and Marshes; mostly se U.S. from TN and SC s to FL and w to OK and TX. May–Dec. Similar to *C. capitatus*, but *C. lindheimeri* is distinguished by its more yellowish pubescence on young growth and consistently acute leaf tips.

Croton lindheimerianus Scheele, (for Ferdinand Jacob Lindheimer, 1801–1879, German born TX botanist), THREE-SEED CROTON. Monoecious annual 1–5 dm tall; leaf blades densely stellate pubescent, thick, usually obtuse to acute; the 3 styles bifid to base. Sandy ground; widespread in East TX; widespread but local over much of TX; mostly AR, AZ, NM, OK, and TX. Jul–Oct. Another variety, var. *tharpii* M.C. Johnst., THARP'S CROTON, occurs far to the w in the Tran-Pecos and adjacent areas.

Croton michauxii G.L. Webster, (for Andrè Michaux, 1746–1802, French botanist and explorer of North America. The date of his death sometimes misquoted as 1803, is 1802 (Pluchet 2004)). Erect, slender, loosely branched, monoecious annual to 6(–9) dm tall; stems stellate-lepidote, silvery; leaves alternate or subopposite, very short-petioled, blades 1.5–4 cm long, 0.1–0.4(–1.5) cm wide, linear-lanceolate to ovate-lanceolate; inflorescences bisexual spikes; staminate flowers 1–12, sepals 5, petals 5, 0.6–1 mm long, stamens 5–7; pistillate flowers 1–6, sepals (4–)5, petals absent, styles 3, minutely bifid; fruit an achene 1–1.2 mm in diam. Jun–Oct. This taxon previously treated in *Crotonopsis* Michx. is now lumped into the genus *Croton* (Webster 1992). According to Webster (1992), "…the main generic character of *Crotonopsis*—unicarpellate gynoecium and indehiscent fruit—represents merely the end-point in a reduction series from the 3-carpellate gynoecium of most *Croton* species through the 2-carpellate gynoecium of *Croton monanthogynus* Michx. to the 1-carpellate gynoecium of *Crotonopsis*." Consequently, there appears to be a much stronger case for treating *Crotonopsis* as a section of *Croton* than as an independent genus." [*Crotonopsis linearis* Michx.]

1. Inflorescences 1–3 cm long; achenes usually 3–6, densely stellate-lepidote, hairs ± spreading, radii separate most of the length, umbos (= raised central brownish portion) usually sharp, sometimes rounded; leaf blade upper surface hairs not overlapping ____ var. **michauxii**

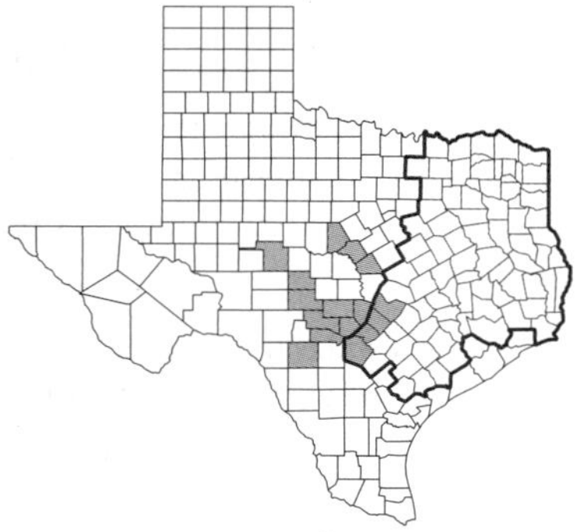

Argythamnia aphoroides

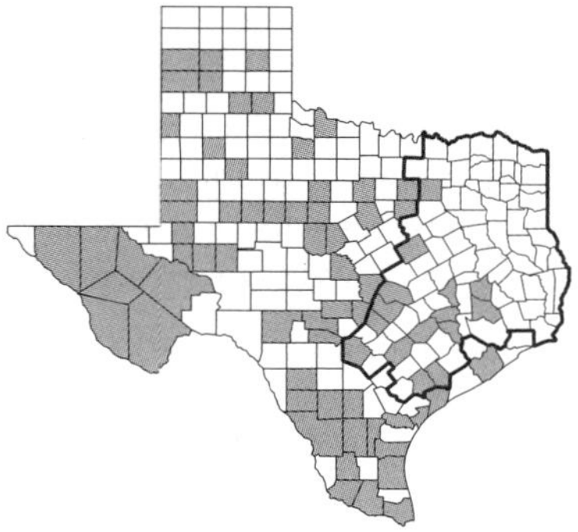

Argythamnia humilis

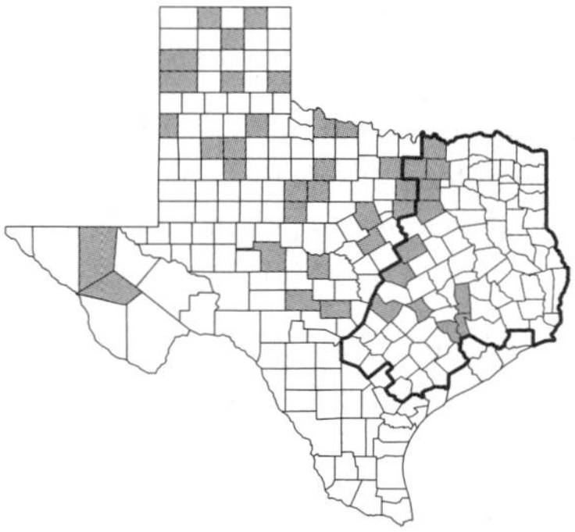

Argythamnia mercurialina var. mercurialina

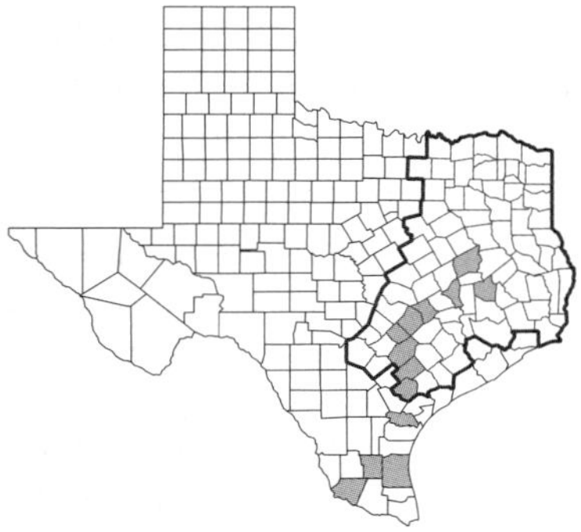

Argythamnia mercurialina var. pilosissima

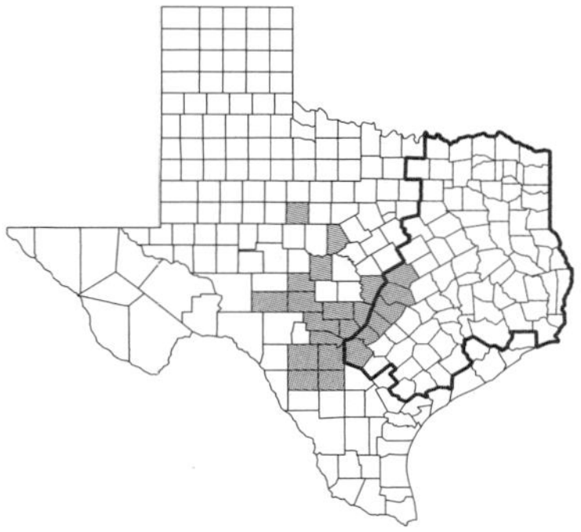

Argythamnia simulans

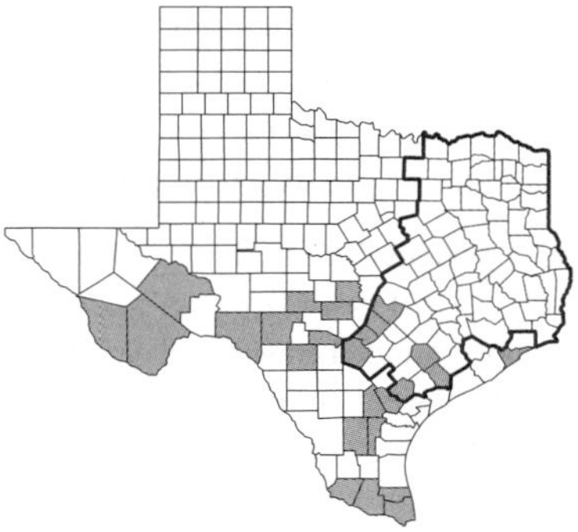

Bernardia myricifolia

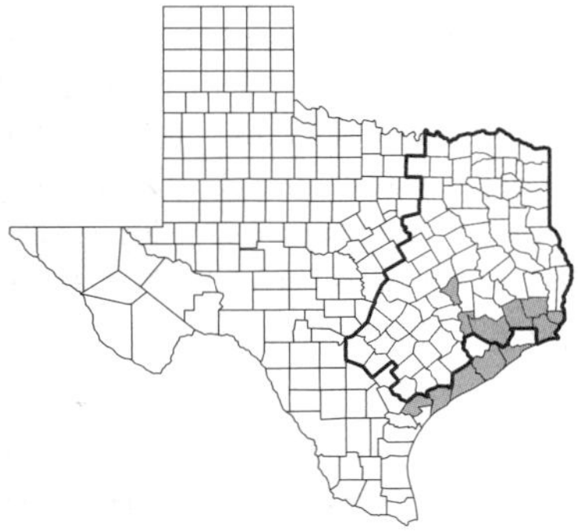

Caperonia palustris

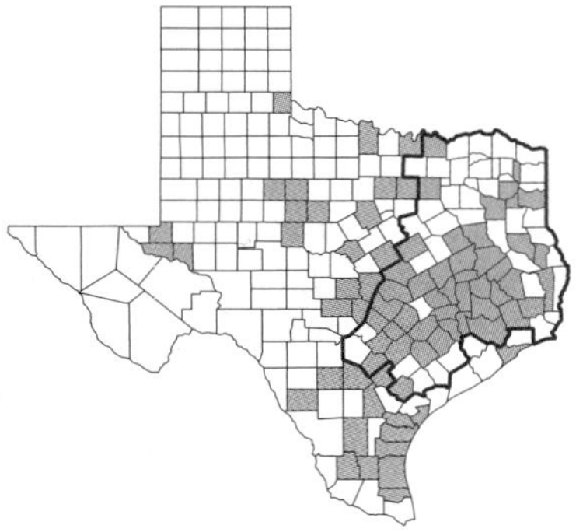

Cnidoscolus texanus

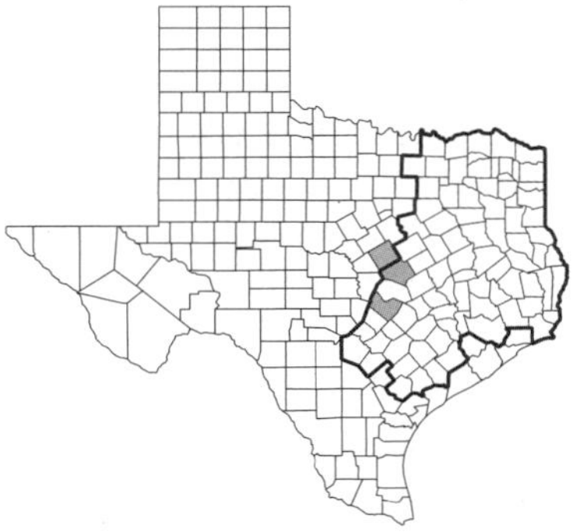

Croton alabamensis var. texensis

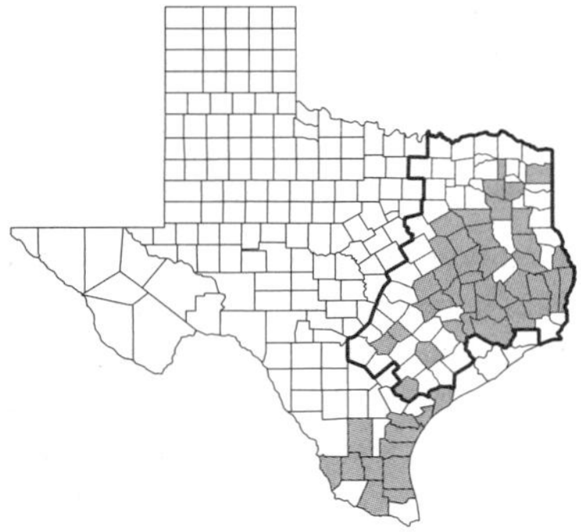

Croton argyranthemus

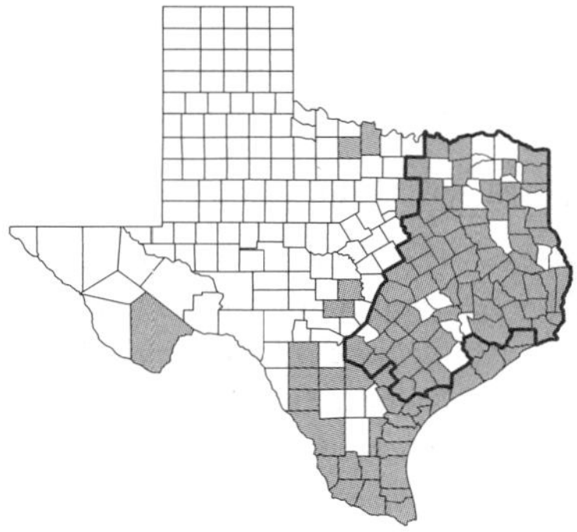

Croton capitatus

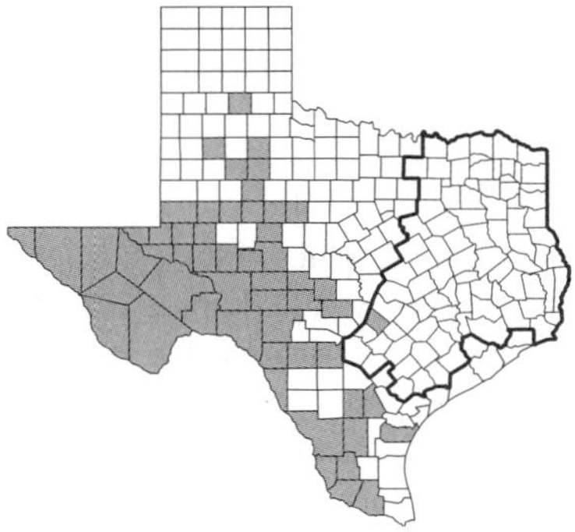

Croton dioicus

1. Inflorescences usually < 1 cm long; achenes 1–2, sparsely stellate-lepidote, hairs appressed, radii joined most or all of their length, umbos usually blunt, rarely sharp; leaf blade upper surface hairs overlapping near midvein _________ var. **elliptica**

var. **elliptica** (Willd.) B.W. van Ee & P.E. Berry, (shaped like an ellipse), ELLIPTIC-LEAF RUSHFOIL, WILLDENOW'S CROTON. Leaf blades 1.5–4 cm long, 0.1–0.4(–1.5) cm wide, lanceolate to ovate-lanceolate; radii usually 1–3; staminate flowers usually < 1 mm in diam. Sandstone and granite outcrops, disturbed sandy soils, forest openings; mostly in the e 2/3 of East TX; mostly e U.S. from VA s to FL and w to MO, OK, and TX. [*C. willdenowii* G.L. Webster, *Crotonopsis elliptica* Willd.]

var. **michauxii,** MICHAUX'S CROTON, NARROW-LEAF RUSHFOIL. Leaf blades 1.8–3.5 cm long, 0.1–0.3 cm wide, linear-lanceolate; radii usually 5–8; staminate flowers usually > 1 mm in diam. Deep sandy soils, forest openings, coasts; in the se portion of East TX; se U.S. from SC s to FL and w to AR and TX, also IA, IL, MO, and TN. [*Crotonopsis abnormis* Baill.]

Croton monanthogynus Michx., (one female flower), DOVEWEED, PRAIRIE-TEA, ONE-SEED CROTON. Annual monoecious herb to 5 dm tall, widely branched; leaf blades 1–3.5 cm long, 0.5–3 cm wide, rhombic-ovate to elliptic or lanceolate, lower leaf surface sparsely whitish stellate pubescent, some hairs with dark brown center; inflorescences bisexual, sometimes pistillate; staminate flowers 3–10, sepals 3–5, petals 3–5, 0.7–1 mm long; pistillate flowers 1–2(–5), sepals 5, petals absent, ovary 2-locular, 1 fertile, styles 2, bifid to base; capsules 1.8–2.2 mm in diam. Rocky or eroding ground, roadsides and waste places, particularly calcareous soils; throughout East TX; throughout most of TX, but rare in the Panhandle; scattered in e U.S. from PA s to GA; throughout from IN and KY s to AL and w to KS and TX; also n Mexico. Jun–Nov.

Croton parksii Croizat, (for H.B. Parks, 1879–1958, Texas botanist and coauthor (with V.L. Cory) of *Flora and Fauna of the Big Thicket* (1936)), PARKS'CROTON. Annual dioecious herb 5–15 dm tall; leaf blades 2.5–6 cm long, 1–2 cm wide, elliptic, base rounded, apex short rounded, lower surface densely stellate-tomentose; inflorescences unisexual, staminate inflorescence 3–20-flowered; staminate flowers (5–)6 sepals, 3 mm long, petals absent; pistillate inflorescence 1–3-flowererd; pistillate flowers 5 sepals, petals absent, styles 3, irregularly 2–3 times bifid, forming matted cluster; capsules 8–9 mm long, 9–10 mm in diam. Deep sandy soil; in East TX known only from Victoria (*Ajilvsgi 2148* BRIT) and Wilson (Croizat 1942) cos.; Gulf Prairies and Marshes and South TX Plains. Endemic to TX. Jun–Dec.

Croton pottsii (Klotzsch) Müll. Arg., (for John Potts, flourished 1840–1855, plant collector and manager of the Chihuahua Mint in Mexico), LEATHERWEED. Perennial monoecious or dioecious herb 1–6 dm tall, stellate pubescent; leaf blades 1.6–6 cm long, ovate-oblong, proximal rounder, distal more ovate-elliptic, base rounded, apex rounded to acute, both surfaces softly stellate pubescent, erect radii shorter than lateral radii; inflorescences bisexual or unisexual, racemes mostly 1.1–2.5 cm long; staminate flowers (12–)20–25, sepals (4–)5, 1–1.5 mm long, petals (4–)5, 1–1.3 mm long; pistillate flowers (12–)20–25, sepals 5, 1.5–3 mm long, petals absent, styles 3, bifid to base; capsules 4–5 mm in diam. Desert scrub, arid plains, rocky slopes; in East TX known only from Milam and Smith? (Kartesz 2022) cos.; Edwards Plateau, High Plains, and Trans-Pecos; AZ, NM, and TX, also Mexico. May–Oct. The East TX cos. are disjunct from most other TX localities and in habitats unusual for the species. *C. pottsii* may not be a part of the East TX flora.

Croton texensis (Klotzsch) Müll. Arg., (of Texas), TEXAS CROTON, SKUNKWEED. Dioecious annual herb 2–7(–9) dm tall, stellate pubescent; leaf blades 1–5 cm long, 0.5–2 cm wide, linear-lanceolate to nearly ovate-oblong, base truncate to rounded or subcordate, apex rounded to acute, densely stellate pubescent, more so below; inflorescences unisexual; staminate inflorescence 2–8 cm long, flowers 10–30, sepals 5, petals absent; pistillate inflorescence 1–2 cm long, flowers 1–6, sepals 5, petals absent, styles 3, many times divided, terminal segments 12–32+; capsule 4–5.5 mm wide, often slightly warty. Sand or sandy loam; mainly in the w 1/2 of East TX; widespread in TX except for South TX Plains and Trans-Pecos; widespread in the c U.S. Jun–Oct. While generally avoided, cattle can be poisoned by eating the plant fresh or in hay (Muenscher 1951).

DITRYSINIA Raf. GULF COAST SEBASTIAN-BUSH

A monotypic genus of the e U.S. (Mabberley 2017). We are following most recent authors (e.g., Webster 2014; Wurdack 2016a; Mabberley 2017; Kartesz 2022; Weakley, A.S., and Southeastern Flora Team 2022) in recognizing it as a distinct genus. It is apparently closely related to and was previously included in *Sebastiania* (e.g., Correll & Johnston 1970; Turner et al. 2003), SEBASTIAN-BUSH, ARROW-POISON. That genus, as currently defined (it was previously considerably larger), has ca. 25 species of tropical America. The seeds of some *Sebastiania* species, when occupied by a particular moth larva *Cydia* (or *Carpocapsa) saltitans* (the epithet from a root word meaning to jump), are played with as "jumping beans." *Sebastiania bilocularis* S. Watson, YERBA DE LA FLECHA, of sw North America, was the source of an arrow poison (Mabberley 2017). (Greek: *ditry*, two or three, and *syn*, together, alluding to number and union of stamens)

REFERENCES: Wurdack et al. 2005; Webster 2014; Wurdack 2016a; Mabberly 2017.

Croton dioicus [ARM]

Croton fruticulosus [ARM]

Croton glandulosus [ARM]

Croton glandulosus var. septentrionalis [ARM]

Croton heptalon [ARM]

Croton lindheimeri [ARM]

Croton lindheimerianus [ARM]

Croton michauxii var. elliptica [YA2]

Croton michauxii var. michauxii [MIC]

Ditrysinia fruticosa (Bartram) Govaerts & Frodin, (shrubby, bushy), GULF SEBASTIAN-BUSH. Shrub ca. 1–3 m tall, hairs unbranched; leaves alternate, simple, petioles reddish, 10 mm or less long, stipules 1–2 mm long, deciduous, blades 2–7(–10) cm long, ovate to lanceolate, elliptic to lanceolate-elliptic, basally cuneate to acute(–acuminate), apically acute to acuminate, entire, laminar glands absent or up to 5, abaxial; inflorescences narrow spike-like racemes to 4 cm long, greenish to greenish yellow; flowers small, unisexual, apetalous; staminate flowers on upper part of inflorescence, with 3 sepals and 3 exserted stamens; pistillate flowers on lower part of inflorescence, with 3 sepals (2 additional tiny sepals sometimes present) and 3 stigmas, branches recurved, capsules ca. 6–8 mm long, 3-locular, with 3 seeds; seeds 4.5–5 mm long. Stream banks; in the se portion of East TX; se U.S. from NC s to FL and w to AR and TX. Apr–Jun.

EUPHORBIA L. SPURGE

Annuals or perennials with milky sap; leaves alternate or opposite; stipules absent or present, or represented by minute glands; inflorescences dichotomously or trichotomously or more branched, subtended by opposite or whorled leafy bracts; except in the former genus *Chamaesyce* where inflorescences are terminal or axillary, solitary or in cymose clusters or dense, head-like glomerules; flowers greatly reduced, minute, unisexual (consisting only of a single pistil or a single stamen on a short pedicel), in small cup-like involucres (= cyathia); each cyathium containing one pistillate and several staminate flowers mixed with minute bracts; the cyathia also have fleshy glands (sometimes called involucral glands) on their rims; these glands in some species have small, petal-like, white or pinkish gland appendages—the cyathium is then termed a pseudanthium because of the resemblance to standard flowers; the glands can sometimes have tiny ± linear paired "horns"; pistillate pedicel elongating in age, the 3-locular capsule exserted from the involucre. In the past, there has not always been clarity in the terms used for the measurement of glands and gland appendages. We are following terminology suggested by M. Mayfield (pers. comm.)—in the key and descriptions the width of the gland signifies the measurement in the tangential dimension relative to the cyathium; the length of the gland signifies the measurement in the radial dimension relative to the cyathium; the width of the gland appendage signifies the measurement in the tangential dimension relative to the cyathium, and the length of the gland appendage signifies the measurement in the radial dimension relative to the cyathium—this means from attachment at the gland to the apex of the gland-appendage.

When considered broadly, *Euphorbia* is a huge, cosmopolitan, especially warm area genus of ca. 2,000 species (Berry et al. 2016)—the second largest genus of flowering plants (following only the legume genus *Astragalus*) (Euphorbia PBI Project 2008). Euphorbias are basically monoecious or dioecious herbs, succulents, shrubs, or trees with milky latex. A few years ago, there seemed to be no general agreement on whether *Euphorbia* should be recognized as one large genus or divided into a number of segregates. For example, in Texas, Turner et al. (2003) treated the segregates *Chamaesyce* and *Poinsettia* at the generic level, and Diggs et al. (1999) treated *Chamaesyce*, with 300+ species, as a distinct genus. However, molecular evidence (e.g., Steinmann & Porter 2002; Wurdack et al. 2005; Horn et al. 2012; Dorsey et al. 2013) has shown that the genus in the broad sense is monophyletic and there is now consensus that *Euphorbia* should be treated broadly to encompass a number of segregates including *Chamaesyce*, *Poinsettia* (Euphorbia PBI Project 2008), and *Tithymalus*. It is problematic for those wishing to split up the genus that a number of the previous subgenera and generic segregates do not appear supported by the new molecular evidence. For example, *Chamaesyce* seems clearly imbedded within other species of *Euphorbia* (Steinmann & Porter 2002). The genus is best defined morphologically by the presence of the cyathium. Many species are important ornamentals including *E. milii* Des Moul., (CROWN-OF-THORNS) and *E. pulcherrima* Willd. ex Klotzsch (POINSETTIA, named for Joel Poinsett, the first U.S. Minister to Mexico, who imported the plant into the U.S. in 1829—Spoerke & Smolinske 1990); the inflorescences are sometimes (e.g., in POINSETTIA) surrounded by colorful bracts (these often large and photosynthetic and possibly better referred to as leaves). Other species are considered serious invasive weeds (e.g., Eurasian *E. virgata* Waldst. & Kit., LEAFY SPURGE, in the n and w U.S. (sometimes confused with the Eurasian *E. esula* L. which is not part of the North American flora)). Some Old World species are ± leafless succulents which superficially resemble Cactaceae and function there as ecological equivalents of that family (note, however, that Euphorbias are easily distinguished by their very different flowers, milky sap, and paired spines). Physiologically, "*Euphorbia* is the only genus of plants known to include all three major photosynthetic systems (CAM, C3, and C4)" (Euphorbia PBI Project 2008). All species have a milky latex, which in many is apparently toxic and can cause dermatitis or other reactions in sensitive individuals; inflammation and large blisters can develop; the eyes are particularly sensitive. While livestock are sometimes accidentally poisoned, animals generally will not eat the plants; the toxins are complex terpenes such as euphorbol; some are suspected of acting as co-carcinogens (= promote the action of "sub"-carcinogenic doses of known carcinogens); while ingestion of euphorbias can produce gastric distress and irritation of the skin and mucus membranes, past reports of severe toxicity appear exaggerated (Wheeler 1941; Muenscher 1951; Kingsbury 1964, 1965; Kinghorn 1979; Lampe & McCann 1985; Lampe 1986; Burrows & Tryl 2001). *Euphorbia antisyphilitica* Zucc., CANDELILLA, of the TX Trans-Pecos, is the

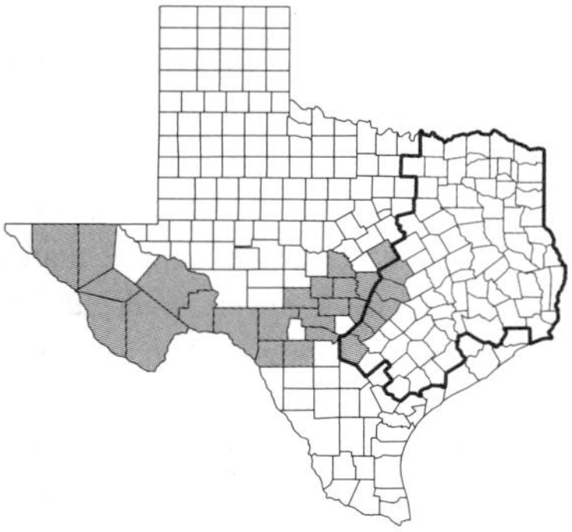

Croton fruticulosus

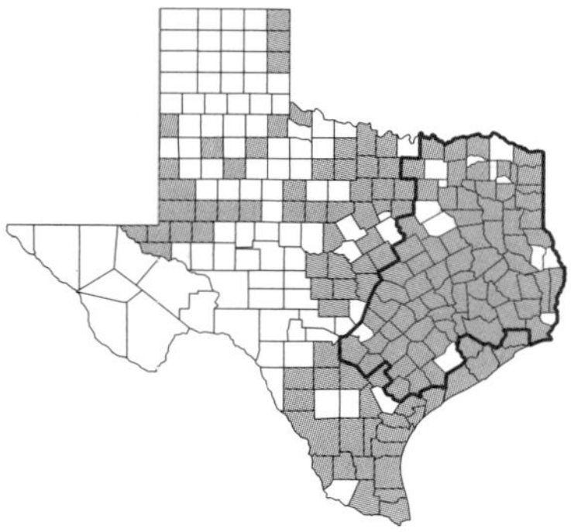

Croton glandulosus

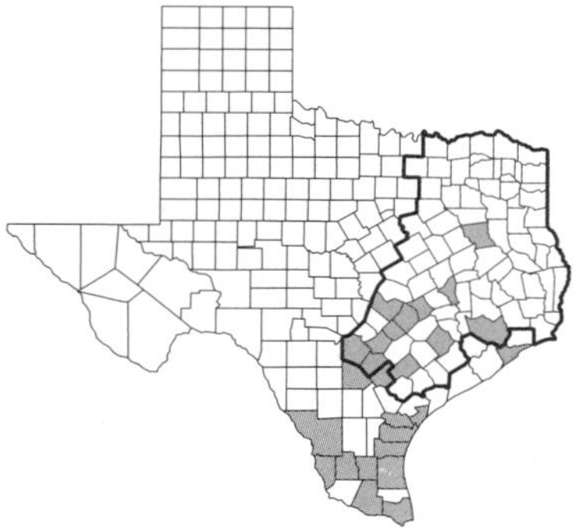

Croton heptalon

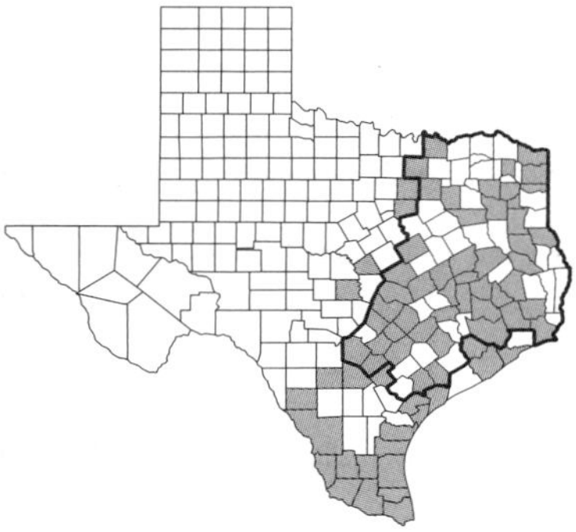

Croton lindheimerii

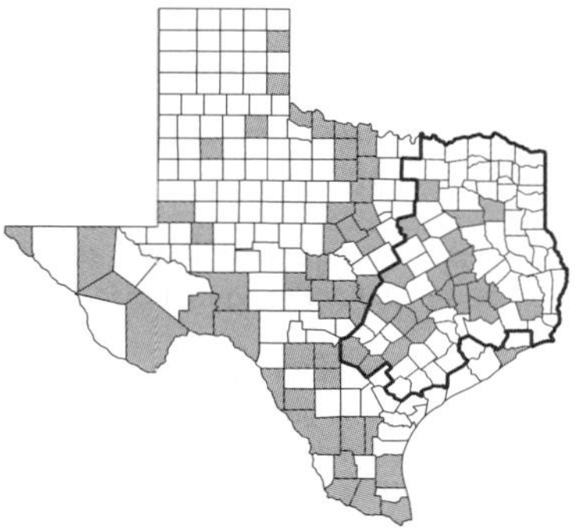

Croton lindheimerianus

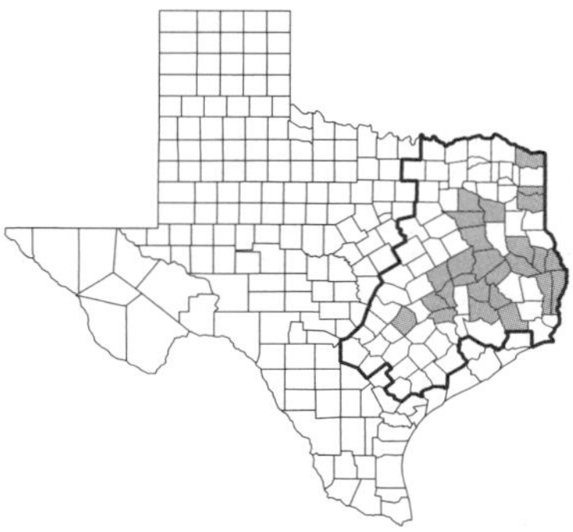

Croton michauxii var. elliptica

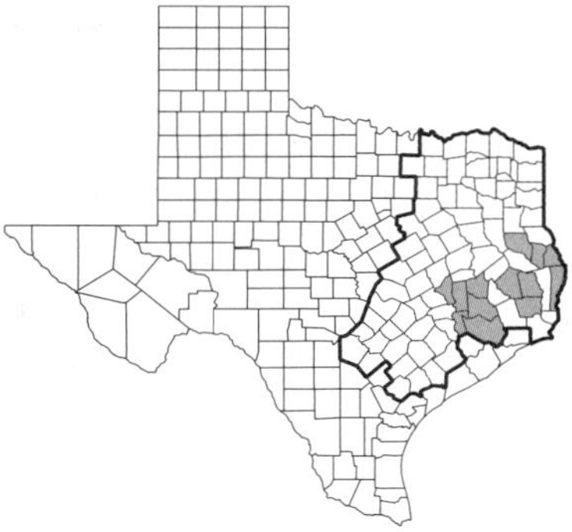

Croton michauxii var. michauxii

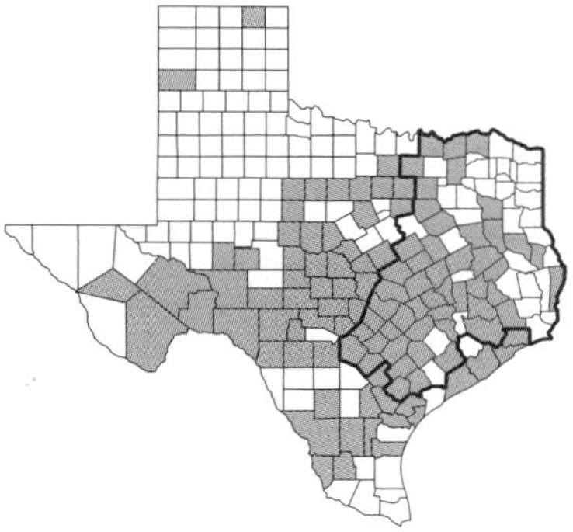

Croton monanthogynus

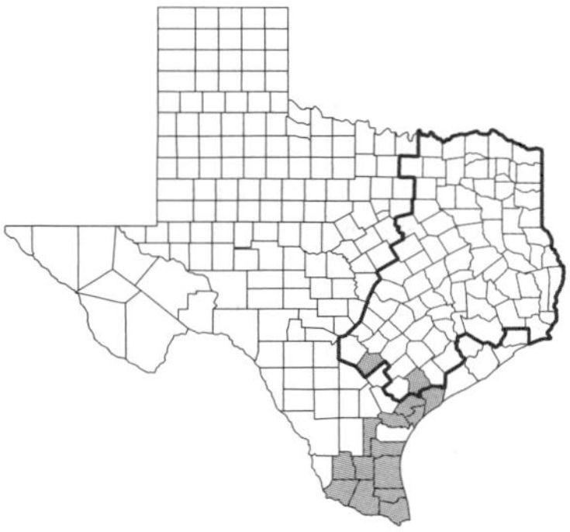

Croton parksii

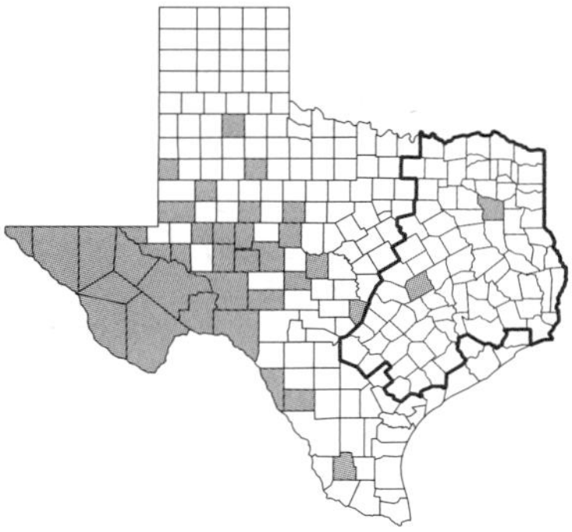

Croton pottsii

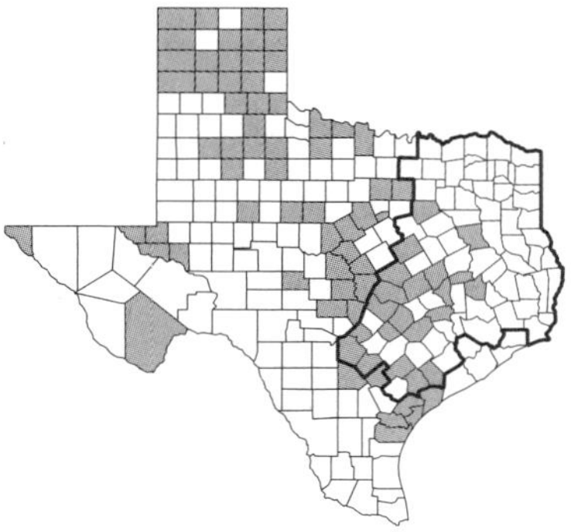

Croton texensis

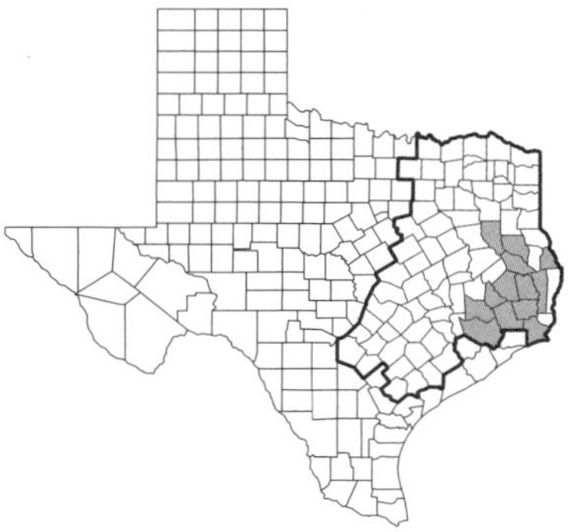

Ditrysinia fruticosa

source of Candelilla wax, one of the highest quality waxes known. The wax is obtained by boiling the stems in water and dilute acid and then scooping off the floating wax. Unfortunately, in some areas populations of this species have been greatly reduced (Powell 1988). (Named for Euphorbus, 1st century AD physician to King Juba of Mauritania, who used latex from some species for medicinal purposes)

Among the *Euphorbia* species occurring in East TX, there are several groups that are morphologically rather distinctive: 1) 19 species previously recognized in *Chamaesyce*—they are often prostrate in habit, have the main stem shorter than the branches, and have opposite leaves with blades basally asymmetrical; 2) 4 species (*E. cyathophora, E. davidii, E. dentata,* and *E. heterophylla*) that are sometimes segregated into the genus *Poinsettia*—they are characterized by having a solitary involucral gland without appendages; 3) 2 species (*E. bicolor* and *E. marginata*) that have very conspicuous, white-margined floral bracts and upper leaves and involucres with petal-like gland appendages; 4) 3 species (*E. corollata, E. discoidalis,* and *E. hexagona*) that have involucres with petal-like gland appendages, but lack colored floral bracts; 5) 8 species that have involucres with 4 glands which lack appendages but whose glands bear tiny, pointed, paired "horns"; and 6) 2 species (*E. spathulata* and *E. texana*) whose involucral glands bear neither appendages nor "horns."

References: Norton 1900; Wheeler 1936, 1941; Krochmal 1952; Burch 1966; Richardson 1968; Johnston 1975; Subils 1984; Park & Elisens 1997; Park 1998; Steinmann & Porter 2002; Euphorbia PBI Project 2008; Berry et al. 2016; Taylor 2016.

1. Stems often prostrate (some ascending and erect); the main stem short, basally branching, the branches much longer than the main stem; leaves opposite, the bases conspicuously to slightly asymmetrical (formerly segregated as *Chamaesyce*) ____ **Key A**
1. Stems erect to ascending, the main stem usually much longer than the branches along the main stem; leaves opposite or alternate, the bases symmetrical (all other *Euphorbia* segregates) ____ **Key B**

Key A

1. Capsules, ovaries, underside of leaf blades, and usually stems glabrous.
  2. Larger leaf blades linear or narrowly oblong, 4.5–10 times as long as wide ____ **E. missurica**
  2. Larger leaf blades oblong or elliptic to triangular or orbicular, < 4.5 times as long as wide.
    3. Larger leaf blades 13–35(–rarely more) mm long (including petioles); plants ascending or erect, 15–ca. 100 cm tall.
      4. Cyathia in dense head-like glomerules with reduced bract-like leaves subtending cyathia; gland appendages absent or 0.3–0.4 mm long; stems glabrous ____ **E. hypericifolia**
      4. Cyathia solitary or in small cymose clusters at distal nodes or on congested axillary branches (very short internodes); gland appendages 0.1–1 mm long; stems pubescent or glabrous.
        5. Stems glabrous or with straight hairs as long as those on leaves (when present); seed faces having 2–3 transverse ridges; capsules 1.5–1.6 mm long; gland appendages 0.1–0.3 mm long ____ **E. hyssopifolia**
        5. Stems with short crinkled hairs shorter than hairs on leaves (when present); seed faces with more than 3 transverse wrinkles (sometimes somewhat irregular); capsules 1.6–2.3 mm long; gland appendages 0.2–1 mm long ____ **E. nutans**
    3. Larger leaf blades 4–13(–15) mm long; plants usually prostrate or nearly so (except sometimes ascending in *C. fendleri* and *C. bombensis*), usually 1–10 cm in overall height.
      6. Stipules united into a single, triangular, entire or ragged-margined, white to pink scale on each side of the stem; stems often rooting at nodes.
        7. Cyathial glands 0.3–0.8 mm wide (tangential dimension; oblong in shape, the length or radial dimension much less); gland appendages usually conspicuous, 0.3–1 mm long (radial dimension); plants perennial; staminate flowers 12–30 per cyathium; capsules 1.1–2.3 mm long ____ **E. albomarginata**
        7. Cyathial glands ca. 0.2 mm wide (oblong in shape, the length or radial dimension much less); gland appendages usually inconspicuous, 0.1–0.2 mm long; plants annual; staminate flowers 3–8(–10) per cyathium; capsules 1.3–1.4 mm long ____ **E. serpens**
      6. Stipules cut into linear segments, or reduced to a linear-lanceolate, not noticeably white segment; stems not rooting at nodes.
        8. Plants annual from a taproot; stems usually prostrate or nearly so.
          9. Leaf blades oblong-ovate to orbicular, < twice as long as wide; petioles absent or very short, usually less than 1 mm long; gland appendages 1.1–1.5 mm long (radial dimension), showy; styles 0.5–0.8 mm long ____ **E. cordifolia**
          9. Leaf blades oblong or oblong-elliptic, twice as long as wide or longer; petioles of larger leaves up to 2 mm long; gland appendages absent or 0.1–1 mm long, not particularly showy; styles 0.1–0.6 mm long.
            10. Seeds transversely ridged; leaves usually minutely toothed at least near apex (under a handlens); gland appendages 0.1–0.3 mm long ____ **E. glyptosperma**
            10. Seeds smooth; leaves usually entire; gland appendages absent or 0.1–1 mm long.
              11. Leaf blades 2–3 mm wide; styles 0.2–0.3 mm long; gland appendages absent or 0.1–0.5 (–0.7) mm long; species scattered in East TX ____ **E. bombensis**

11. Leaf blades 2–6 mm wide; styles 0.2–0.6 mm long; gland appendages absent or 0.5–1 mm long; species rare in East TX ____ **E. geyeri**

8. Plants perennial from thick, woody taproot; stems prostrate to spreading or ascending ____ **E. fendleri**

1. Capsules, ovaries, underside of leaf blades, and stems sparsely to densely pubescent.

12. Upper leaf blades linear or nearly so, > 9 times as long as wide; plants erect perennials from woody taproots; species known in East TX only on sw margin ____ **E. angusta**

12. Upper leaf blades variously shaped but not linear, ca. 1–6 times as long as wide; plants annual or perennial, prostrate to erect; including species common and widespread in East TX.

13. Larger leaf blades broadly triangular-ovate; stems usually erect to ascending; stems villous to glabrous but also papillate (minutely so); mostly only on the sw margin of East TX ____ **E. villifera**

13. Larger leaf blades oblong to oblong-elliptic or oblong-obovate or ovate; stems prostrate to erect; including species common and widespread in East TX; stems variously pubescent to glabrate.

14. Leaf blades ca. 3 cm long or longer (< 3 cm in depauperate specimens); cyathia in large axillary and terminal head-like glomerules; some stem hairs spreading, to ca. 1 mm long, clear with several darkened septa ____ **E. hirta**

14. Leaf blades 2 cm long or less; cyathia variously distal and/or axillary, solitary and/or in small clusters, OR in dense head-like glomerules and terminal; stems hairs white, variously long, short, spreading, matted or appressed, some long, spreading clear hairs with darkened septa present.

15. Cyathia in dense, terminal, head-like glomerules; some stem hairs spreading, to ca. 1 mm long, clear with several darkened septa; in East TX known only from Guadalupe and Travis cos. ____ **E. ophthalmica**

15. Cyathia solitary and distal OR solitary, and in small clusters, distal and axillary.

16. Styles unbranched; leaf blades sparsely to densely pubescent above with hairs 0.5–1 mm long; seeds punctately pitted and mottled ____ **E. stictospora**

16. Styles bifid from only at apex to nearly their entire length; leaf blades glabrous or sparsely pubescent above with hairs up to 0.5 mm long; seeds transversely ridged (sometimes obscure), rugose or papillate, rarely smooth.

17. Capsules spreading-villous, the hairs mainly on the ridges of the capsules; leaf blades without reddish or purplish central splotch; pubescence of terminal internodes widely spreading or reflexed ____ **E. prostrata**

17. Capsules loosely appressed-pubescent, the hairs rather uniformly scattered over the surface; leaf blades sometimes with reddish or purplish central splotch or line; at least some hairs of terminal internodes loosely upwardly appressed, all of these white.

18. Stems rooting at the nodes or not; glands linear to narrowly oblong, 0.2–0.5 mm wide; appendages 0.2–1.5 mm wide; seeds 0.8–1.2 mm long; styles bifid only at apex to 1/2 their length; including species widespread in East TX.

19. Stems not rooting at the nodes; glands narrowly oblong to almost linear; styles 0.3–0.4 mm long, bifid only at apex, the branches clavate; widespread in East TX ____ **E. maculata**

19. Stems often rooting at the nodes; glands narrowly oblong; styles 0.5–0.8 mm long, bifid ca. 1/2 their length, the branches not clavate; only in se portion of East TX ____ **E. humistrata**

18. Stems not rooting at the nodes; glands slightly oblong to circular, 0.1–0.2 mm wide; appendages absent to 0.3 mm wide; seeds 0.7–0.9(–1.1) mm long; styles bifid 1/2 to nearly whole length; in East TX known only from Guadalupe and Travis cos. ____ **E. ophthalmica**

Key B

1. Involucral glands usually solitary (1 per involucre, (1–)2 in *E. dentata*), lacking petal-like appendages; larger leaf blades usually toothed or lobed, linear leaf blades (when present) often entire; leaves with petioles distinct, 2–50 mm long (species sometimes segregated into the genus *Poinsettia*).

2. Main stem leaves all or mostly alternate (sometimes lower leaves opposite), essentially glabrous above; leaf blade shape very variable, upper leaves sometimes fiddle-shaped.

3. Upper leaf blades usually with a striking rosy to reddish area at base; glands bilabiate, borne on a very short stalk or stalkless; seeds not angular; more common in East TX ____ **E. cyathophora**

3. Upper leaf blades usually green, occasionally pale white at base; glands with a circular opening, borne on a distinct stalk; seeds angular; rare in East TX ____ **E. heterophyla**

2. Main stem leaves all or mostly opposite (sometimes upper leaves alternate), pubescent above; upper leaves not fiddle-shaped, usually dentate (rarely entire) sometimes lighter towards base but without a striking rosy or reddish area.

4. Leaf blades usually widest at the middle, linear-elliptic to broadly elliptic; hairs of the lower leaf surfaces stiff, strongly tapered with a broad basal cell; seeds angular in cross-section, unevenly tuberculate, 2.4–2.9 mm long ____ **E. davidii**

4. Leaf blades usually widest below the middle, lanceolate to trullate (= trowel-shaped); hairs of the lower leaf surfaces weak, filiform, lacking a broad basal cell; seeds rounded in cross-section, evenly tuberculate, 1.7–2.0 (–2.2) mm long ____ **E. dentata**

1. Involucral glands (1–)2–5 per involucre, with OR without conspicuous or inconspicuous, white or pinkish, petal-like gland appendages; larger leaf blades usually entire OR very finely toothed (often distally); leaves often sessile or sometimes with distinct petioles.
  5. Involucres without conspicuous petal-like gland appendages (however, the glands can have tiny, ± linear, pointed, paired "horns"); stem leaves usually widest near apex.
    6. Leaf blades entire; capsules not warty; involucral glands each with 2 tiny pointed, paired horns; involucres 4-glanded.
      7. Uppermost leafy bracts partially fused, 1/4 to 1/2 (thus appearing perfoliate).
        8. Upper leafy bracts fused ca. 1/2 their length; capsules 2–2.5 mm long ____ **E. roemeriana**
        8. Upper leafy bracts fused ca. 1/4 their length; capsules 2.6–2.7 mm long ____ **E. ouachitana**
      7. Uppermost leafy bracts essentially free to their base.
        9. Seeds completely smooth; species of extreme s portion of the area (in East TX known only from Goliad & Gonzales cos. ____ **E. helleri**
        9. Seeds usually with at least some pitting, grooves, or tubercles (use lens).
          10. Uppermost leafy bracts from nearly as long as wide to much longer than wide, bluntly broadly triangular to very narrowly so; seeds with numerous small pits ± evenly distributed over the seed surface or relatively few larger pits in rows (at least adaxially or abaxially) to nearly smooth.
            11. Uppermost leafy bracts usually 5 mm or more wide, usually but not always broadly triangular or ovate to obovate; abaxial seed surface with 15–20 small pits to almost smooth OR 2 rows of distinct pits.
              12. Seeds bearing 2 rows of large distinct pits abaxially; capsules and ovaries winged ____ **E. peplus**
              12. Seeds bearing small indistinct pits or almost smooth abaxially; capsules and ovaries not winged ____ **E. tetrapora**
            11. Uppermost leafy bracts < 5 mm wide, usually very narrowly triangular; seeds with abaxial side bearing 4–6 large pits in rows ____ **E. peplidion**
          10. Uppermost leafy bracts mostly wider than long, suborbicular or reniform; seeds with numerous small pits ± evenly distributed over the seed surface at least on the abaxial side.
            13. Internodes of upper inflorescences very short resulting in overlapping leafy bracts, and a very dense inflorescence; seeds gray to purple-gray or sometimes nearly black ____ **E. longicruris**
            13. Internodes of inflorescences longer; upper leafy bracts not overlapping; inflorescences relatively less dense; seeds white to gray ____ **E. austotexana**
    6. Leaf blades finely toothed, at least around apex; capsules warty or smooth; involucral glands rounded, entire, without horns; involucres 4–5-glanded.
      14. Capsules and ovaries warty; plants 10–70 cm tall ____ **E. spathulata**
      14. Capsules and ovaries smooth; plants 7–20 cm tall ____ **E. texana**
  5. Involucres with white gland appendages, the gland appendages usually conspicuous and petal-like (but very small and whitish to greenish white in *E. hexagona*); stem leaves mostly widest near middle or of uniform width (sometimes ovate).
    15. Involucres with (1–)2–4 glands ____ **E. graminea**
    15. Involucres with 5 glands unless leafy bracts conspicuously white-margined, then glands 4–5.
      16. Stem leaves opposite, 1–6 mm wide; appendages white to green, tinged red, triangular to ovate; bracts never white-margined ____ **E. hexagona**
      16. Stem leaves alternate, in most species > 6 mm wide (< 6 mm wide in *E. discoidalis*); appendages white, orbiculate to oblong or spatulate; bracts white-margined or not so.
        17. Upper leafy bracts and upper leaves not white-margined (leaves sometimes absent from flowering plants); middle and upper stem leaves usually 12 mm wide or less (sometimes wider in *E. pubentissima*); capsules glabrous or with slight pubescence.
          18. Leaves 1.5–4 mm wide, > 8 times as long as wide, linear, conspicuously revolute; stems densely pubescent to glabrous; species known from se portion of East TX ____ **E. discoidalis**
          18. Leaves 5–18 mm wide, < 5.5 times as long as wide, linear-oblong to elliptic, elliptic-oblong, oblanceolate, or obovate, only sometimes slightly revolute in *E. corollata*; stems glabrous or slightly pubescent; including species widespread in East TX.
            19. Plants ca. 20–100 cm tall; white gland appendages 2.5–4.4 mm long, oblong, spatulate or rectangular; leaves sessile, leathery; lower inflorescence bracts smaller than stem leaves; species widespread and common in East TX ____ **E. corollata**
            19. Plants 30–65 cm tall; white gland appendages 1–2.2 mm long, oval or orbicular; leaves petiolate or subpetiolate, thin; lower inflorescence bracts similar in size to stem leaves; species possibly in East TX ____ *E. pubentissima*
        17. Upper leafy bracts and upper leaves prominently white-margined; middle and upper stem leaves usually > 12 mm wide; capsules pubescent.
          20. Uppermost (just below inflorescence), white-margined, leaf-like bracts linear, usually 5–10 times as long as broad; main stem leaves lanceolate to narrowly elliptic, usually 3–5 times as long as wide, pubescent ____ **E. bicolor**

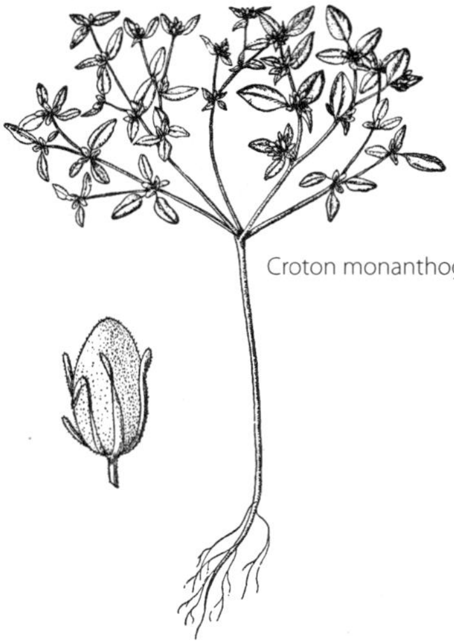

Croton monanthogynus [GLE]

Croton parksii [NAZ]

Croton pottsii [ARM]

Croton texensis [ARM]

Ditrysinia fruticosa [VIN]

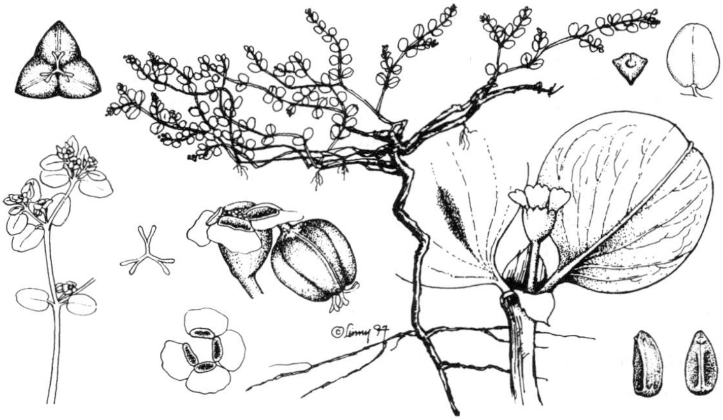

Euphorbia albomarginata [HEA, RHO]

20. Uppermost, white-margined, leaf-like bracts narrowly lanceolate to ovate or elliptic, < 4 times as long as broad; main stem leaves elliptic, oblong, or ovate, usually < 2.5 times as long as wide, often glabrous ____ **E. marginata**

Euphorbia albomarginata Torr. & A. Gray, (white-margined), WHITE-MARGIN EUPHORBIA. Prostrate, mat-forming perennial; leaf blades orbicular to oblong, entire; cyathia solitary at distal nodes; styles bifid nearly to base, gland appendages typically conspicuous, 0.3–1 mm long. Stream bottoms, roadsides, disturbed areas; mostly in sw portion of East Texas, also Rusk (*Beade 7129* IND) and Dallas (Kartesz 2024) cos.; mainly w 2/3 of TX; mostly sw U.S. from OK, LA and TX w to NV and CA; also Mexico. Late Apr–Oct. [*Chamaesyce albomarginata* (Torr. & A. Gray) Small]

Euphorbia angusta Engelm., (narrow), BLACK-FOOT EUPHORBIA. Perennial; stems erect, from a woody crown, 12–43 cm tall; leaf blades entire; lower leaf blades 7–25(–38) mm long, elliptic to linear-oblong, distal leaf blades > 9 times as long as wide, narrower, linear; stipules minute, distinct; cyathia at the upper nodes, solitary; glands narrowly oblong, 0.2–0.4 mm long; gland appendages white, 0.5–1.1 mm long, fan-shaped, distal margin irregularly toothed; styles bifid only at apex to almost 1/2 their length. Bexar, Comal, Hays, (BRIT), Travis, and Williamson (Turner et al. 2003) cos. on sw margin of East TX; mainly Edwards Plateau to Trans-Pecos; in the U.S. known only from TX; also Mexico. Spring–fall. [*Chamaesyce angusta* (Engelm.) Small]

Euphorbia austrotexana Mayfield var. **austrotexana,** (of South Texas), SOUTH TEXAS SPURGE. Erect annual, 10–22 cm tall, glabrous; leaf blades 5–18 mm long, 0.5–2.5 mm wide, linear to slightly lanceolate or linear-oblanceolate; upper leafy bracts distinct, reniform-ovate to subdeltate-ovate; glands 4; horns 0.5–0.7 mm long; ovary glabrous; styles bifid; capsules 1.8–2.2 mm long; seeds with deep irregular to rounded, shallow depressions over entire surface. Sandy soils; in East TX known only from Bexar and Wilson (Mayfield 2013) cos., also Atascosa Co. (Mayfield 2013). Endemic to TX. Mar–Apr.

Euphorbia bicolor Engelm. & A. Gray, (two-colored), SNOW-ON-THE-PRAIRIE. Conspicuous, erect, softly pilose annual; stems 40–100 cm tall; middle stem leaves 3–5 times as long as wide; leafy bracts and upper leaves prominently white-margined; involucral glands usually 4–5; gland appendages petal-like, white, 1.4–2.5 mm long. Prairies, roadsides, and waste ground; widespread in East TX except for ne portion; widespread in e 1/2 of TX; AR, LA, OK, and TX. Where the ranges of *E. bicolor* and *E. marginata* overlap, intermediate individuals can be found. Jul–Oct. The fruits are ballistic (= seeds thrown catapult-like from the capsule). The milky sap is caustic and can cause a skin rash in sensitive persons (M. Mayfield, pers. comm.).

Euphorbia bombensis Jacq., (of Isla de Tierra Bomba, near Cartagena, Colombia), DIXIE SANDMAT. Glabrous annual extremely similar to *E. geyeri*; stems prostrate or slightly ascending; gland appendages absent to 0.5(–0.7) mm long, semilunate, usually fringing edge of gland; styles bifid ca. 1/2 their length, the branches usually clavate. Sandy areas, roadsides; Anderson, Bastrop, Bexar, Burleson, Caldwell, Milam, and Wilson (Kartesz 2022) cos.; also South TX Plains and Gulf Prairies and Marshes; coastal U.S. from VA to TX; also Mexico and Central America. Feb–Sep. [*Chamaesyce ammannioides* (Kunth) Small, *C. bombensis* (Jacq.) Dugand, *E. ammannioides* Kunth] It has long been known that *E. ammannioides*, and *E. bombensis* are very similar, if not identical (Wheeler 1941). Texas plants have been treated as *E. ammannioides* by Correll and Johnston (1970) and *C. ammannioides* by Turner et al. (2003), but most recent authorities (e.g., Steinmann et al. 2016; Kartesz 2022; Weakley, A.S., and Southeastern Flora Team 2022) consider *E. ammannioides* conspecific with *E. bombensis* (*E. bombensis* has priority). *Euphorbia bombensis* is also similar to *E. geyeri* especially in Texas and can be difficult to distinguish. Steinmann et al. (2016) state, "*E. bombensis* differs from *E. geyeri* in its shorter, less conspicuous involucral gland appendages and its geographic restriction to the coastal plain." However, they also state, "Wheeler (1941) pointed out that plants of *E. bombensis* from Texas have more or less conspicuous involucral gland appendages." Steinmann et al. (2016) agree that there examination of specimens confirmed Wheeler's statement. Kartesz (2022) maps *E. bombensis* as strictly coastal from VA to TX; except TX is the only state with inland records (Kartesz 2022). We have seen no inland TX specimens.

Euphorbia cordifolia Elliott, (with heart-shaped leaves), HEART-LEAF SPURGE, HEART-LEAF EUPHORBIA. Usually prostrate annual; stems to 43 cm long; leaf blades entire, base rounded to cordate; cyathia solitary at distal nodes; gland appendages 1.1–1.5 mm long; styles bifid nearly to base. Loose sandy soils, open woods; widespread in East TX; also rarely scattered to the w of East TX and in the South TX Plains and s Gulf Prairies and Marshes; se U.S. from NC s to FL and w to OK and TX. Jun–Oct. [*Chamaesyce cordifolia* (Elliott) Small]

Euphorbia corollata L., (corolla-like), FLOWERING SPURGE, TRAMP'S SPURGE, BLOOMING SPURGE, WHITE PURSLEY, MILKWEED. Perennial with deep spreading rootstock, 20–100 cm tall; stems sometimes solitary or usually few, glabrous or slightly pubescent; leaves alternate, linear-oblong to elliptic-oblong or oblanceolate, leathery, sessile; inflorescences with 2–6 main rays; lower leafy bracts smaller than stem leaves; involucral glands 5, cupped, green; gland appendages petal-like, white, usually 2.5–4.4 mm long,

Euphorbia angusta [BOI, RHO]

Euphorbia austrotexana [CP1]

Euphorbia bicolor [BT3]

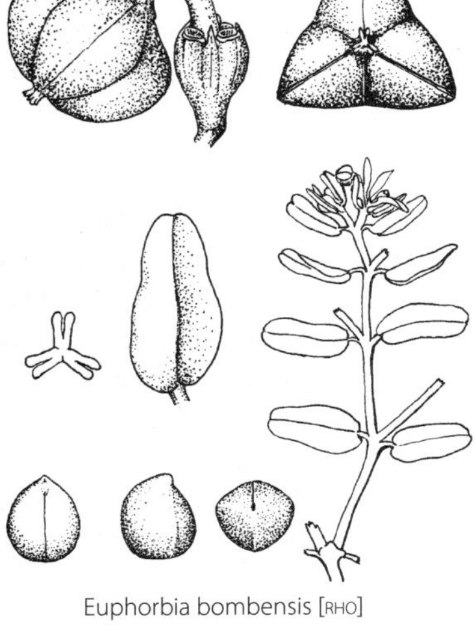

Euphorbia bombensis [RHO]

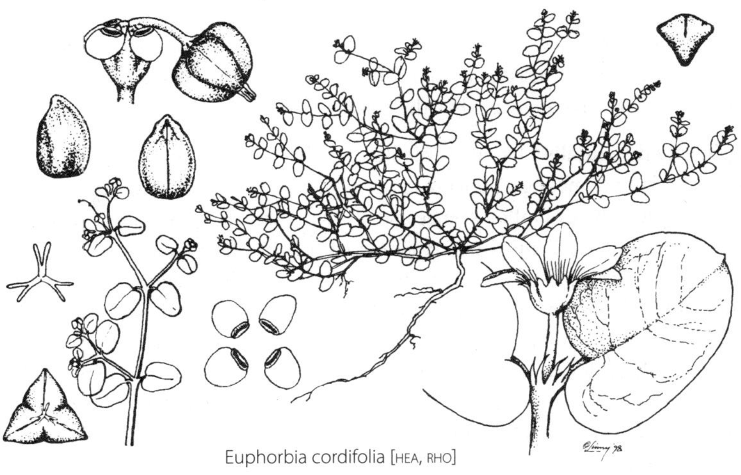

Euphorbia cordifolia [HEA, RHO]

oblong, spatulate or rectangular. Clay soils; throughout much of East TX; widespread in e 1/3 of TX w to Montague and Hays (BRIT) cos.; nearly throughout the e 1/2 of the U.S. and se Canada. May–Sep. [*Tithymalopsis corollata* (L.) Klotzsch & Garcke] The rootstock reportedly contains a poisonous substance, euphorbon (Burlage 1968). The Plains Indians used this plant in medicines as a laxative, as a treatment for rheumatism, and in vermifuge preparations. It was used in Anglo-American folk remedies as an emetic (Kindscher 1992.).

Euphorbia cyathophora Murray, (cup-bearing), WILD POINSETTIA, FIRE-ON-THE-MOUNTAIN, PAINTED EUPHORBIA, PAINTED SPURGE, PAINTED-LEAF. Nearly glabrous annual to 100 cm tall; leaf blades extremely variable in shape, elliptic to oblong-ovate or ovate, varying to linear-lanceolate or even linear, sometimes lobed and appearing fiddle-shaped, usually finely toothed to entire, the uppermost leaves and the floral bracts usually rosy to reddish or rosy to reddish and whitish at base, sometimes conspicuously colorful (thus reminding one of a very small cultivated POINSETTIA); involucral lobes with 3–5 triangular lobes; involucral gland 1, bilabiate, short-stalked or stalkless, without an annular rim; appendages absent; seed rounded in cross section, caruncle absent. Stream banks, open areas; scattered in East TX; scattered further w, particularly in the Edwards Plateau and Trans-Pecos; e U.S. from VA s to FL and w to MN and s to TX, widely scattered in sw U.S.; also Mexico and Central and South America. May–Sep. [*Poinsettia cyathophora* (Murray) Bartl.] This annual is sometimes cultivated in East TX native plant gardens, with quite colorful showy forms often selected. Once established, it can be weedy, with numerous individuals appearing year after year (G. Diggs, pers. obs.).

Euphorbia davidii Subils, (for David L. Anderson of Villa Mercedes, Argentina, who in 1984 collected one of the paratypes), DAVID'S SPURGE. Annual 20–70 cm tall; similar to *E. dentata* but more robust, with hairs on the upper stems and young growth coarse, slightly recurved, and stiff (not to the touch so much as to the eye); leaf blade lower surface with strigose, long-tapered hairs; involucral glands solitary, cupped, without appendages; capsules somewhat ovoid or broader at the base, tapering to the apex; seeds angular-ovoid with unevenly and roughly sculptured surface, with a suggestion of 2 transverse ridges in well-formed seeds on the abaxial surface, 2.4–2.9 mm long, with caruncle 1.0–1.4 mm wide; $n = 28$ (tetraploid). Disturbed sites, open slopes; in East TX known only from Travis (Kartesz 2022) and Milam (*Flinchbaugh EGF058* BRIT), and possibly Williamson (*Gibson 6886* TEX, image seen, not verifiable) cos; Trans-Pecos and the Panhandle; mostly c U.S. from IA s to TX and w to WY and AZ, scattered in ne U.S.; also Mexico. Summer–fall. Specimens of *E. davidii* are sometimes misidentified as the rather similar and more common *E. dentata*. Consequently, its range within TX is unclear but it is probably more common in w TX; however, the Flinchbaugh collection referenced above is well into the Blackland Prairie on the w edge of East TX; it is the most e record for East TX that we are aware of and should be expected in other areas of East TX; widespread in much of the U.S. except the far nw; also e Canada. Much of this description and the dichotomy in the key separating *E. davidii* and *E. dentata* are from M. Mayfield (pers. comm.).

Euphorbia dentata Michx., (toothed), TOOTHED SPURGE. Rather sparsely pubescent annual 15–60 cm tall, with long weak hairs on the young growth and abaxial leaf surfaces; leaf blades narrowly oblong-lanceolate to rhombic-ovate, rather coarsely and shallowly toothed; uppermost leaves and the floral bracts sometimes lighter towards base, but without a strikingly colored area; involucral glands (1–)2, cupped, oblong, without appendages; capsules smoothly rounded, depressed spherical; seeds very rounded to roundly ovoid, the abaxial surface evenly tuberculate with low sculpturing, 1.7–2.0(–2.2) mm long with caruncle ca. 0.8 mm or less wide; $n = 14$ (diploid). Stream bottoms, fields, roadsides and disturbed sites; widespread in East TX; widespread throughout most of TX; mostly widespread in e 1/2 of the U.S.; also ne Canada and Mexico. May–Oct, commonly two generations in one year. [*Poinsettia dentata* (Michx.) Klotzsch & Garcke] Specimens of *E. davidii* are sometimes misidentified as *E. dentata* but can be distinguished by the characters in the key.

Euphorbia discoidalis Chapm., (disk like, possibly in reference to the petaloid appendages), SUMMER SPURGE. Perennial 45–70 cm tall, somewhat resembling a miniature *E. corollata*; stems 1–2, densely pubescent, rarely glabrous; leaves alternate, linear, conspicuously revolute; base of inflorescence usually with three main rays; involucral glands 5, green or red, cupped; gland appendages petal-like, white, 1–1.7 mm long, semi-circular or orbicular. Open sandy woods; Hardin, Houston, Montgomery, Newton, Polk, Tyler (BRIT), Harris, San Augustine, Shelby (Kartesz 2024), Walker (Park 1998), and San Jacinto (Nesom & Brown 1998) cos.; Park (1998) also mapped two locations (counties not distinguished) in East TX.; Brazoria Co. in the n Coastal Prairies and Marshes, otherwise not known in TX; se U.S. from GA and FL w to TX. May–Sep. [*Tithymalopsis discoidalis* (Chapm.) Small] Park noted that this species "is recognized by the unique combination of revolute, linear leaves, a 3-rayed umbel, densely pubescent stems, and long primary internodes in dichasial branches." Park (1998) also suggested that it is of hybrid origin between *E. polyphylla* and *E. pubentissima*, two species that occur further e in the se U.S.

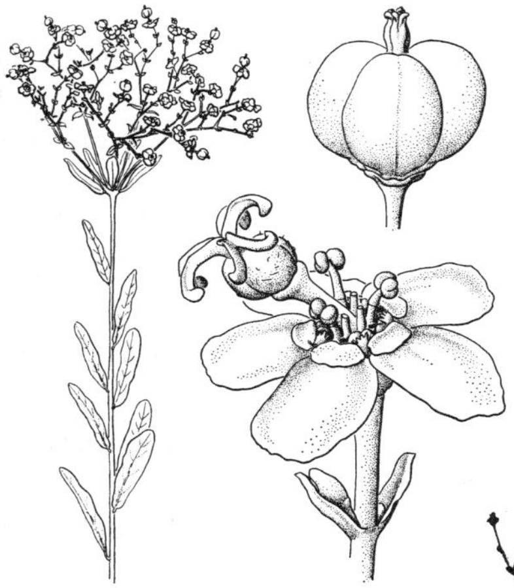

Euphorbia corollata [JAA]

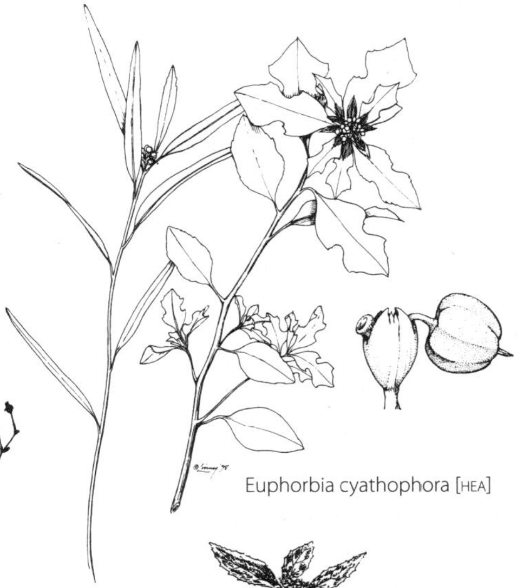

Euphorbia cyathophora [HEA]

Euphorbia davidii [KUR]

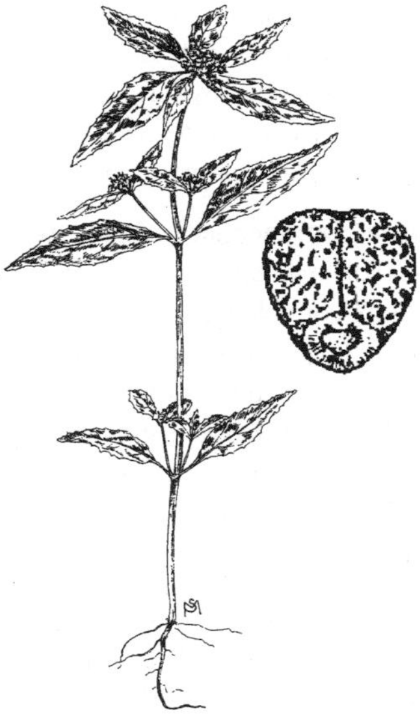

Euphorbia dentata [MAG]

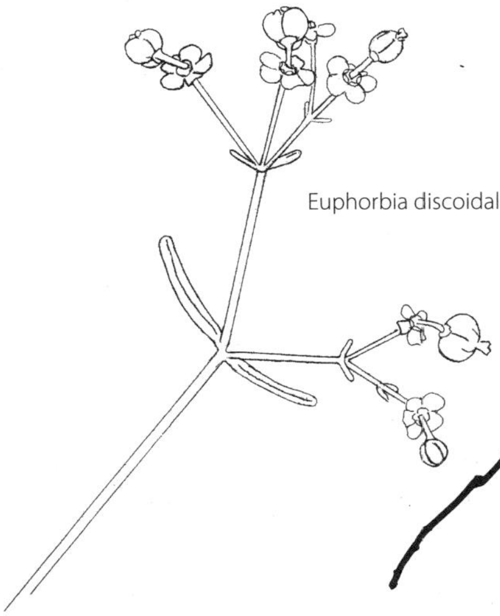

Euphorbia discoidalis [RJG]

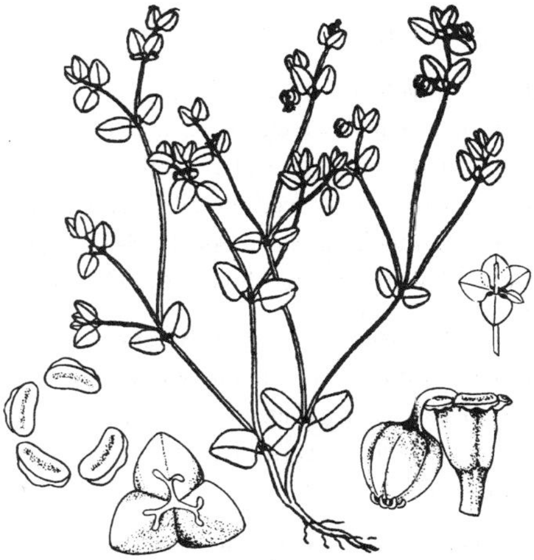

Euphorbia fendleri [BB2, RHO]

Euphorbia fendleri Torr. & A. Gray, (for August Fendler, 1813–1883, one of the first botanists to collect in New Mexico and Venezuela), FENDLER'S EUPHORBIA. Perennial with a woody, thickened taproot; stems rather wiry, prostrate to decumbent or ascending, to 12 cm long; leaf blades 3–8 mm long, 2.5–7 mm wide, orbicular-ovate to oblong, entire, apex rounded to obtuse; cyathia solitary at distal nodes; glands yellow-green to red, elliptic to oblong, 0.2–0.5 mm long, 0.4–0.9 mm wide; appendages absent or 0.1–0.6 mm long, white, rarely pink, distal margin entire or toothed; styles bifid at least 1/2 their length. Rocky or sandy soils; sw margin of East TX n to Bell Co. (BRIT); mainly w 2/3 of TX, except absent from South TX Plains and Cross Timbers and Prairies; w U.S. from SD s to TX and w to CA; also Mexico. Apr–Jun, sporadically later. [*Chamaesyce fendleri* (Torr. & A. Gray) Small]

Euphorbia geyeri Engelm., (for its discoverer Carl Andreas Geyer, 1809–1853, Austrian botanist), GEYER'S EUPHORBIA. Glabrous annual extremely similar to *E. bombensis*; leaf blades usually ca. 2 times as long as wide; stems prostrate; gland appendages absent or 0.5–1 mm long, usually rounded, sometimes pointed; styles bifid 1/3–1/2 their length, the branches ca. as wide as unbranched portion. Sandy open woods, disturbed areas; Bastrop (*Carr 7754* BRIT), Dallas (Wheeler 1941; a Reverchon collection from 1878), and Houston (*Corbin 12* SMU); Cross Timbers and Prairies just w of East TX; mainly Plains Country and Trans-Pecos; c U.S. from MI s to MO and TX and w to MT and NM; also Mexico. Jun–Oct. [*Chamaesyce geyeri* (Engelm.) Small] Discerning *E. geyeri* from *E. bombensis* can be difficult. See *E. bombensis* for discussion.

Euphorbia glyptosperma Engelm., (carved seed), RIDGE-SEED EUPHORBIA. Annual; stems ascending to decumbent, rarely prostrate, to 40 cm long; leaf blades finely toothed towards apex, sometimes reddish tinged or with reddish splotch; cyathia solitary or in small cymose clusters at distal nodes; gland appendages 0.1–0.3 mm long, distal margin irregularly lobed; styles bifid ca. 1/2 their length. Mostly sandy soils, stream bottoms, open woods, and disturbed areas; scattered on the w margin if East TX, Dallas s to Wilson cos.; mainly w 2/3 of TX; throughout much of the w 1/2 of the U.S., also the Great Lakes region; Canada and Mexico. Late May–Oct. [*Chamaesyce glyptosperma* (Engelm.) Small]

Euphorbia graminea Jacq., (grass-like), GRASS-LEAF SPURGE. Annual or rarely perennial; stems erect or ascending, branched, 30–80(–110) cm long; Leaves usually alternate, sometimes some opposite, blade 10–83 mm long, 3–39 mm wide, ovate, elliptic, linear-elliptic, or oblong, margins entire; upper leafy bracts often white; glands (1–)2–4, yellow to greenish, elliptic or oblong, 0.1–0.3 mm long; appendages white to tinged purple, ovate and often hood-like or forming narrow rim around distal margin of gland, 0.3–1.6 mm long, entire; ovary glabrous; styles 0.7–1 mm long, bifid from 1/2 to nearly entire length; capsules 2.5–3 mm long; seeds coarsely tuberculate with longitudinal rows of shallow pits. Disturbed, weedy, or urban areas; in East TX known only from Hardin and Harris (Brown et al. 2011) cos.; also Fort Bend and Hidalgo cos.; AL, AR, CA, FL, GA, MO, NM, SC; and TX; also Mexico and Central and South America. Year-round. [*Agaloma graminea* (Jacq.) D.B. Ward] Native of Mexico and South America.

Euphorbia helleri Millsp., (for Amos Arthur Heller, 1867–1944, who collected the type specimen, PA botanist and collector of plants of w North America), HELLER'S SPURGE. Annual to 30 cm tall; stem leaf blades spatulate, upper, sessile to the lowest with petioles to 3 mm long; uppermost leafy bracts orbicular-ovate to nearly reniform; involucral glands 4, elliptic, with a curved horn at each end, without appendages; seeds completely smooth. Open areas; Goliad and Gonzales (Kartesz 2022) cos. on sw margin of East TX; also South TX Plains and s Gulf Prairies and Marshes; LA and TX; also Mexico. Feb–Apr; also Nov. [*Tithymalus helleri* (Millsp.) Small]

Euphorbia heterophylla L., (bearing leaves of more than one kind), MEXICAN-FIREPLANT. Taprooted ascending to erect annual to 100 cm tall, stems sparsely pilose to villous; leaf blades narrowly lanceolate to elliptic or broadly obovate (then 4-lobed), margin sparsely glandular-serrate; involucral lobes divided into several smooth, linear lobes; involucral gland 1, circular with an annular rim, borne on a stalk; seeds broadly deltoid, angular in cross section, with a small caruncle. Disturbed areas, roadsides; in East TX known only from Harris and Travis (Kartesz 2022) cos.; South TX Plains and Gulf Prairies and Marshes; s most states from GA and FL w to CA; also Mexico; Central and South America. May–Sep. [*Poinsettia heterophylla* (L.) Klotzsch & Garcke]

Euphorbia hexagona Nutt. ex Spreng., (hexagonal, six-angled), GREEN SPURGE, SIX-ANGLE EUPHORBIA, SIX-ANGLE SPURGE. Minutely and sparsely pubescent annual, normally erect, to 80 cm tall; leaves opposite, entire, 1–6 mm wide; involucral glands 5, green to deep red; gland appendages white to green tinged with red, 0.7–1.7 mm long. Loose sandy open areas, stream banks or bottoms; scattered in s East TX; also scattered in Cross Timbers and Prairies and n Plains Country; widespread in the c U.S. from SD s to TX. Jun–Oct.

Euphorbia hirta L., (hairy), PILL-POD EUPHORBIA. Pubescent annual with flowering stems erect; stems 10–50(–70) cm long with long, spreading, clear hairs with darkened septa, appressed white hairs also present; leaves ca. 3 cm long or longer, elliptic-oblong; petioles 1–3 mm long; cyathia in dense axillary and terminal head-like glomerules; axillary glomerules either sessile or on tips

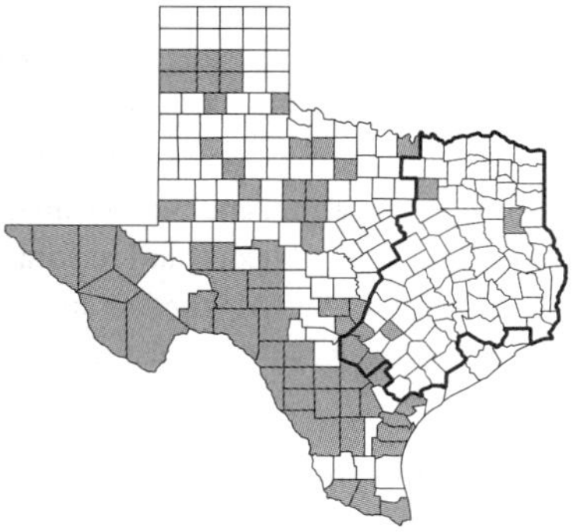
Euphorbia albomarginata

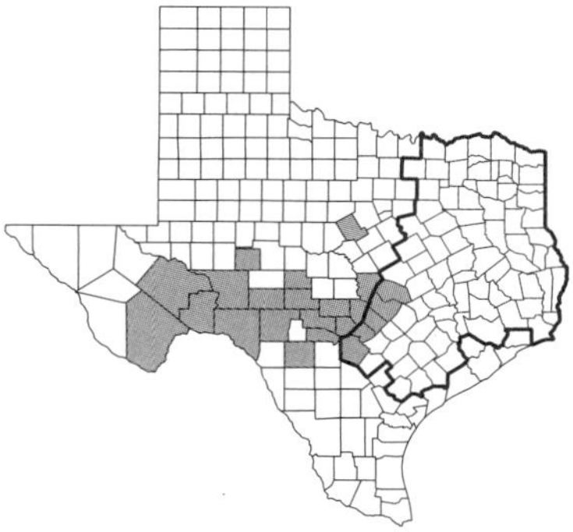
Euphorbia angusta

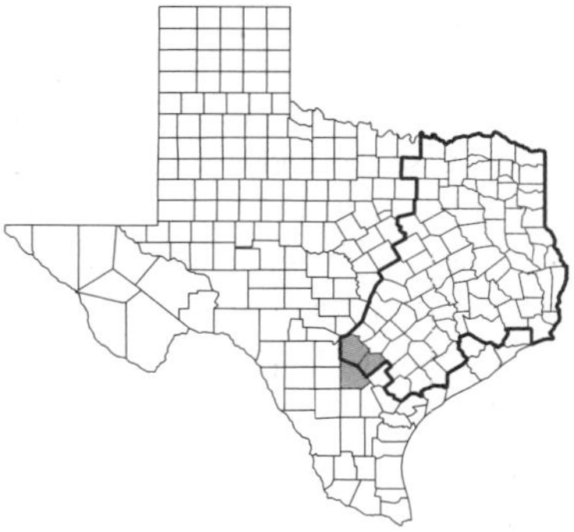
Euphorbia austrotexana var. austrotexana

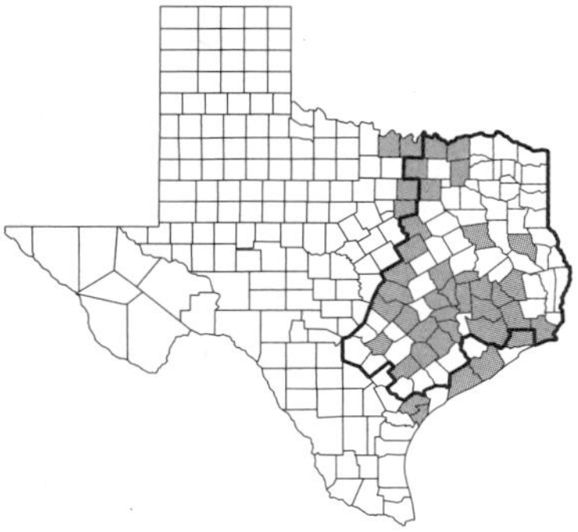
Euphorbia bicolor

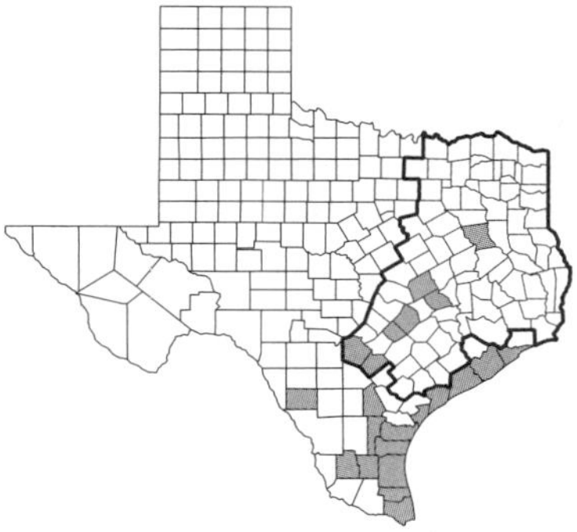
Euphorbia bombensis

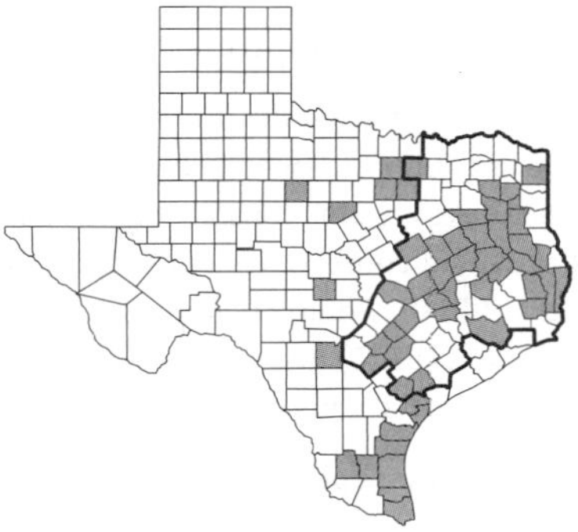
Euphorbia cordifolia

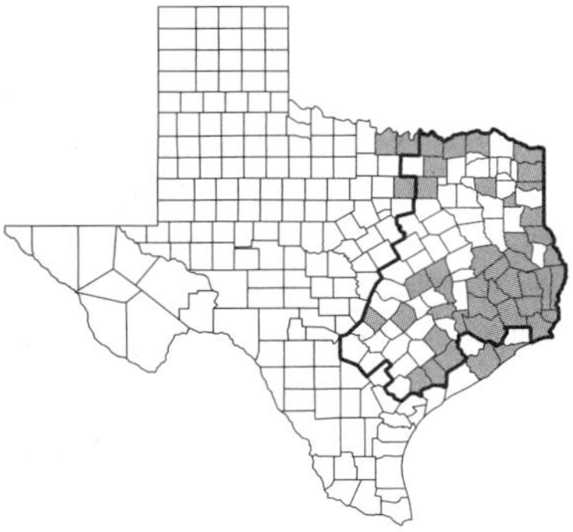
Euphorbia corollata

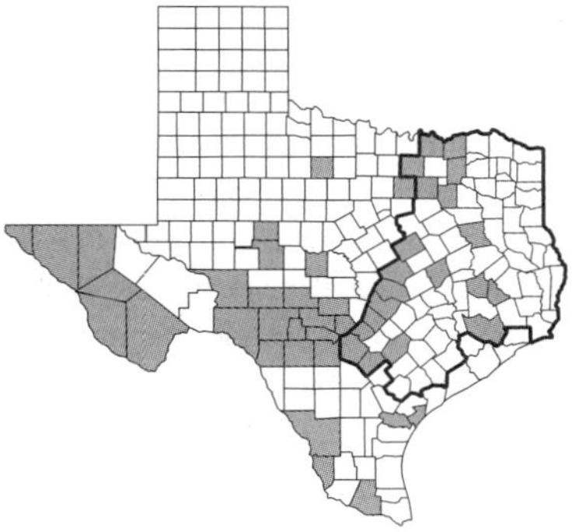
Euphorbia cyathophora

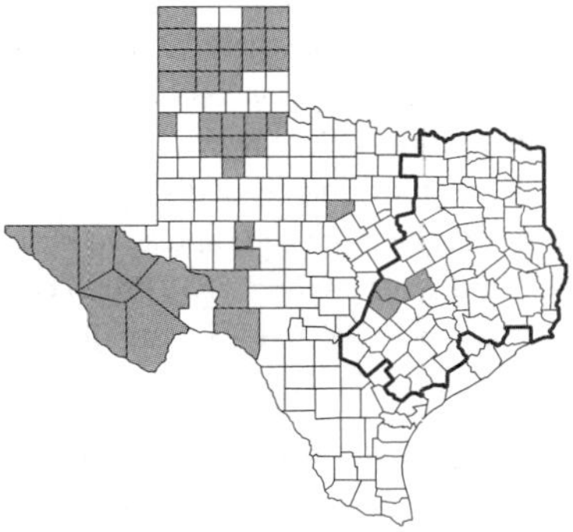
Euphorbia davidii

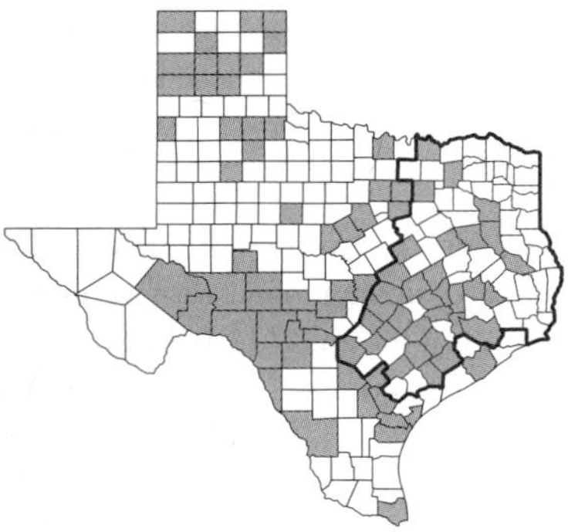
Euphorbia dentata

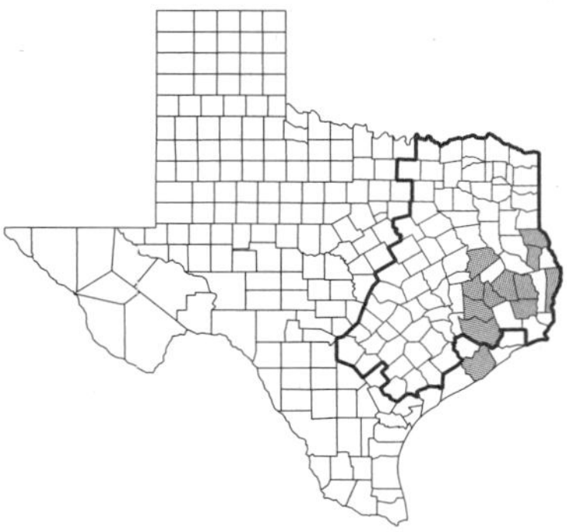
Euphorbia discoidalis

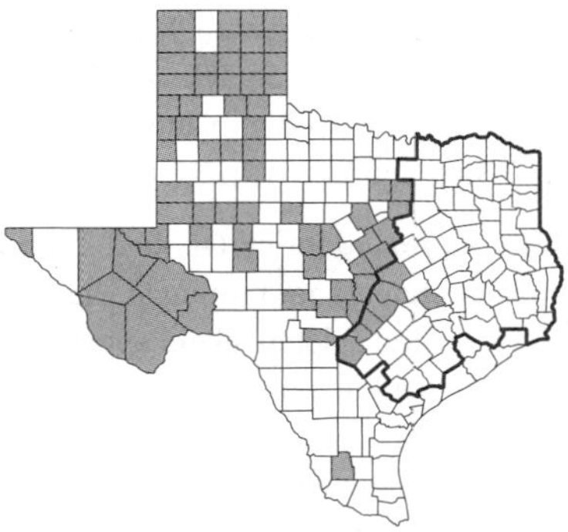
Euphorbia fendleri

of leafless, elongated stalks; glands circular, oblong, or reniform, 0.1–0.2 mm wide; gland appendages absent or 0.1–0.6 mm long; styles bifid 1/2 their length. Weeds in flower beds, nurseries, roadsides; Collin (*Lundell 18844* LL), Dallas (*Lipscomb 3385* BRIT), Harris (Kartesz 2024), Hunt (*White s.n.*, 2010, BAYLU), and Van Zandt (*Hill 314* BRIT) cos.; otherwise known in TX only from Cameron and Hidalgo (Kartesz 2022) cos. at extreme s tip of TX; widely scattered is se U.S. from VA s to FL and w to TX, except throughout FL, widely scattered in sw U.S., as a waif in several ne states; also Mexico; Central and South America. Reported as new for TX by Lipscomb (1984). Year-round. [*Chamaesyce hirta* (L.) Millsp.] Native of c Mexico and may range from South America to s U.S. Sometimes confused with *E. ophthalmica* (see discussion under that species). Used as an arrow poison ingredient in Zaire and medicinally in many countries in Africa (Neuwinger 1996); reportedly causes cardiac and respiratory depression (Burlage 1968).

Euphorbia humistrata Engelm., (prostrate), SPREADING SANDMAT. Prostrate to ascending, pubescent annual, often rooting at the nodes, usually mat-forming, very similar to *E. maculata*; styles bifid ca. 1/2 their length, 0.5–0.8 mm long, the branches not clavate; seeds smooth or papillate. Disturbed areas in sandy soils; mostly in the se portion of East TX, also Dallas (*Taylor 1845* BRIT) and Bowie (Wheeler 1941) cos.; otherwise not known in TX; e U.S. from MD s to FL and w to MN, KS, and TX; also Canada. Summer–fall. [*Chamaesyce humistrata* (Engelm.) Small]

Euphorbia hypericifolia L., (with leaves like *Hypericum*—ST. JOHN'S WORT), TROPICAL EUPHORBIA. Annual with taproot; stems to ca. 50 cm long; similar to *E. nutans*; branches erect to ascending; leaves usually without a central reddish or purplish splotch; glomerules of cyathia conspicuous; glands stipitate, subcircular; gland appendages absent or white to pinkish, 0.3–0.4 mm long, distal margin entire; styles bifid ca, 1/2 their length. Weed along sidewalks, and in flower gardens, nurseries, and landscapes; Brazos (BRIT), Collin (*Baucum s.n.*, 1977–1978, BRIT), Harris, Jefferson, Travis, and Walker (Turner et al. 2003) cos.; also Tarrant Co. (BRIT) just w of East TX and far s TX; s most states from SC s to FL and w to CA, very scattered except throughout FL.; also Central and South America. Summer. [*Chamaesyce hypericifolia* (L.) Millsp.]

Euphorbia hyssopifolia L., (having leaves resembling those of *Hyssop*, Lamiaceae, an aromatic herb used since at least Biblical times (e.g., John 19:29)), HYSSOP-LEAF SANDMAT. Plant to ca. 80 cm tall; main stem erect or ascending and resembling *E. nutans*, but distinguished by characters in the key; leaves uniformly green or suffused with purplish red; cyathia solitary or in small axillary leafy cymose clusters, glands 4(–5), fifth gland lacking appendage, gland appendages white to purplish red, 0.1–0.3 mm long; styles bifid more than 1/2 their length. Open areas, roadsides; according to Powell and Worthington (2018), Turner et al. (2003) lumped *E. hyssopifolia* and *E. nutans*, Kartesz (2022) maps *E. hyssopifolia* throughout most of East TX and cites many specimens from TEX; however, several East TX specimens at TEX have been variously annotated as *E. nutans* and *E. hyssopifolia*; considering the difficulty of distinguishing the two species, we are taking a conservative view and citing only specimens we have seen (it may be more wide spread in East TX); in East TX known only from Bell (*Keeney 3138* BRIT), Brazos (*Sperry s.n.*, 1967, SMU), and Titus (*Ajilvsgi 7038* BRIT) cos.; mostly Edwards Plateau and the Trans-Pecos; widespread across the s U.S. from SC s to FL and w to AZ and UT; also Mexico; Central and South America. Summer–fall. [*Chamaesyce brasiliensis* (Lam.) Small, *C. hyssopifolia* (L.) Small, *E. brasiliensis* Lam.] No county distribution map is provided.

Euphorbia longicruris Scheele, (long-legged), WEDGE-LEAF EUPHORBIA, WEDGE-LEAF SEDGE. Annual to 25 cm tall, at maturity tannish or coppery in color; leaf blades 5–15 mm long, 2–6 mm wide, obovate to spatulate with long-tapering base; branches of the inflorescence or their internodes mostly remaining short resulting in overlapping leafy bracts, usually not exceeding the very wide bracts until late fruiting stage; leafy bracts reniform or suborbicular, subconnate, upright or ascending and enclosing the involucres; involucral glands 4, crescent-shaped, with an erect horn at each end, without appendages. Limestone outcrops; Bell, Comal, Hays, McLennan, Navarro, Travis (BRIT), Bexar, and Williamson (Turner et al. 2003) cos. and disjunct e to Grimes Co. where it grows with other plants typical of the Edwards Plateau (Reed 2006); also Cross Timbers and Prairies, Edwards Plateau, and e Plains Country; AR, OK, and TX. Late Mar–early Jun.

Euphorbia maculata L., (spotted, from the spot on the leaf), SPOTTED EUPHORBIA, SPOTTED SPURGE. Prostrate or rarely suberect, pubescent annual, not rooting at the nodes; stems to 50 cm long; leaf blades finely toothed, oblong, often with a reddish or purplish central splotch; gland appendages white to red; styles 0.3–0.4 mm long, bifid only at the apex, the branches clavate; seeds with 3–4 low, transverse ridges. Sandy soils, stream banks, woods, disturbed sites; throughout East TX; throughout most of TX, but more common in the e 1/2 of the state; throughout the e 1/2 of the U.S., scattered in the w 1/2; also Canada; Mexico; Central and South America. [*Chamaesyce maculata* (L.) Small, *C. supina* (Raf.) Moldenke, *E. supina* Raf.] Reported to be toxic to livestock; when ingested by sheep, photosensitization, swelling of the head, and emaciation can result (Burlage 1968).

Euphorbia marginata Pursh, (margined), SNOW-ON-THE-MOUNTAIN. Conspicuous, pilose to glabrous annual to ca. 150 cm tall; leaf blades broadly ovate to elliptic; upper leaves and leafy bracts prominently white-margined; involucral glands 4–5, cupped;

Euphorbia geyeri [BB2, RHO]

Euphorbia glptospera [BB2, RHO]

Euphorbia graminea [CP1]

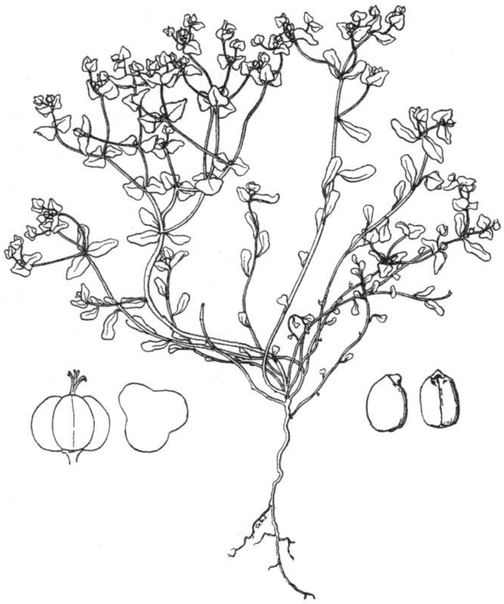
Euphorbia helleri [ARM]

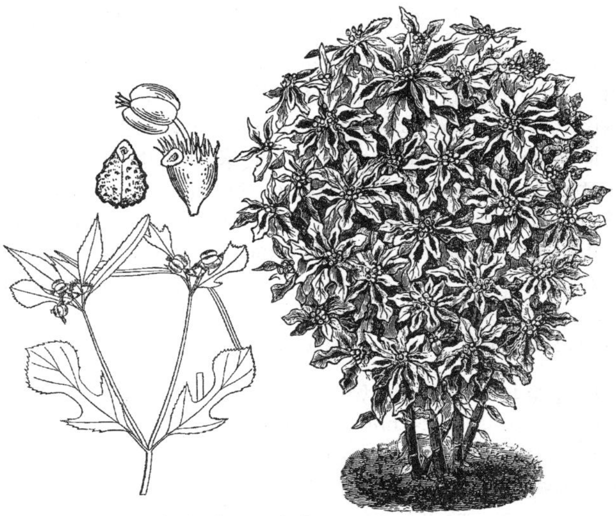
Euphorbia heterophylla [BB2, GAR]

gland appendages petal-like, orbiculate, white, 1.5–2.7 mm long. Calcareous uplands, stream bottoms, and low areas; mostly in the sw portion of East TX; throughout much of TX, except absent from South TX Plains; nearly throughout the U.S. (except the far nw) and se Canada; also Mexico. Jul–Oct. [*Agaloma marginata* (Pursh) Á. Löve & D. Löve] Reported to produce evil-tasting poisonous honey and to be toxic to livestock; the sap causes skin eruptions and has been used to brand cattle in preference to a hot iron (Kingsbury 1964, 1965; Burlage 1968); care should thus be taken to avoid the sap. This species is the western counterpart of the similar *E. bicolor*. It has been cultivated for its large, showy, white bracts.

Euphorbia missurica Raf. (of Missouri), PRAIRIE SPURGE, MISSOURI SPURGE. Annual with erect main stem; stems 10–60(–100) cm long and wide-spreading branches to 50 cm or more long; leaves with short petioles 1–3 mm long, blades linear to narrowly oblong, (4–)8–30 mm long, 3–7 mm wide, 4.5–10 times as long as wide, symmetrical or slightly asymmetrical at base, entire; gland appendages white (rarely pinkish), prominent, ca. 0.4–2.5 mm long; styles bifid 1/2 their length. Loose sandy soils, limestone outcrops; widespread in East TX; nearly throughout TX; widespread in c U.S. Late May–Oct. [*Chamaesyce missurica* (Raf.) Shinners, *C. missurica* (Raf.) Shinners var. *calcicola* Shinners]

Euphorbia nutans Lag., (nodding), EYEBANE. Annual with (1)–several low-spreading to erect stems to 80 cm long; sparsely to moderately pilose to villous or with short, crinkled hairs, hairs often concentrated at nodes and distally, occasionally in 2 bands along opposite sides of stem; leaf blades asymmetrically oblong, often with a central reddish or purplish splotch; gland appendages white to red, 0.2–1 mm long; styles bifid ca. 1/2 their length. Stream bottoms, roadsides, and disturbed soils; widespread in East TX; Gulf Prairies and Marshes, Edwards Plateau, and Trans-Pecos; widespread in the e 1/2 of the U.S.; also Canada; Mexico; Central and South America. May–Oct. The name *Chamaesyce maculata* has often been incorrectly applied to this species. [*Chamaesyce maculata* of authors, not (L.) Small, *C. nutans* (Lag.) Small, *E. maculata* of authors, not L.] Toxic and has caused death in lambs through photosensitization (Kingsbury 1964); the common name also suggests that care should be taken with this species.

Euphorbia ophthalmica Pers., (of or pertaining to the eye), FLORIDA SANDMAT, VAGRANT SANDMAT. Prostrate to rarely ascending annual; stems 6–22 cm long with short appressed hairs and longer (to 1 mm) clear, spreading hairs with darkened, sometimes red septa; leaf blades 4–13 mm long, 3–7 mm wide, ovate-oblong, sometimes with red line or spot on upper surface; cyathia in dense terminal head-like glomerules (often in pairs); glands circular to oblong, 0.1–0.2 mm wide; gland appendages absent to 0.2 mm long, forming a narrow rim around circular glands and distal in oblong glands; styles bifid 1/2 to nearly entire length. Disturbed areas in lawns and flower beds, roadsides; Guadalupe (*Lievens 7190* TEX) and Travis (*Taylor 1223* BRIT) cos.; also Presidio (*Fenstermacher s.n.*, 2015, SRSC) in w TX, Gillespie (*O'Kennon 7981p* BRIT), and Tarrant (*Taylor s.n.*, 2017, SRSC) just to the w of East TX; widely scattered in se U.S., except throughout much of FL; also single county records in CA, PA, and MO; also Mexico; Central and South America. Year-round. [*E. hirta* L. var. *procumbens* (Boiss.) N.E. Br., *Chameasyce ophthalmica* (Pers.) D.G. Burch] Possibly native to se U.S. Though widespread in FL, it is very scattered in the U.S. as noted above. Though *E. ophthalmica* was once a variety of *E. hirta*, it is now considered a separate species. *Euphorbia ophthalmica* (stems 6–22 cm long) is sometimes misidentified as a depauperate *E. hirta* (stems 10–50(–75) cm). Though their habits are very similar, *E. ophthalmica* has terminal dense head-like glomerules (often in pairs), and *E. hirta* has axillary and terminal dense head-like glomerules. ?

Euphorbia ouachitana Mayfield, (named for the region in which it is best represented: the Ouachita Mountains of se OK and AR), OUACHITA SPURGE. Plant annual with taproot; stems 12–28 cm long, erect-ascending, often basally decumbent, often branched near base, glabrous; leaf blades broadly oblanceolate to subspatulate, or lowest often orbiculate, 3–20 mm long, 3–9 mm wide; upper leafy bracts connate 1/4 of their length, not overlapping, broadly deltate to subreniform, base truncate to broadly obtuse, margins entire, apex obtuse to bluntly acuminate; involucral glands 4, crescent-shaped; 0.4–0.5 mm long; horns 0.2–0.4 mm long; styles 0.8 mm long, bifid; capsules 2.6–2.7 mm long; seeds dark brown, lustrous, oblong-ovoid, 1.5–2 mm long, with deep, rounded pits in regular vertical rows, 3–4 pits per vertical row on each adaxial facet. Semi-open forests, bluffs and ledges, stream banks; in TX known only from Red River Co. (Mayfield 2013); AR, MO, OK, TN, and TX. Spring. Sometimes confused with *E. commutata*, which is not known in TX. *Euphorbia ouachitana* can be distinguished by upper leafy bracts fused ca. 1/4 their length as opposed to distinct to subconnate in *E. commutata*.

Euphorbia peplidion Engelm., (possibly from Greek: *peplis*, ancient name used by Dioscorides for wild purslane, a species of *Euphorbia*), LOW SPURGE. Annual to 20 cm tall; stem leaves linear-oblanceolate to narrowly spatulate; upper leafy bracts very narrowly bluntly triangular; involucral glands 4, oblong or elliptic, with an erect horn at each end, without appendages. Limestone soils; sw portion of East TX; also adjacent South TX Plains, Edwards Plateau, and Trans-Pecos; endemic to TX. Spring. [*Tithymalus peplidion* (Engelm.) Small]

Euphorbia hexagona [BT3]

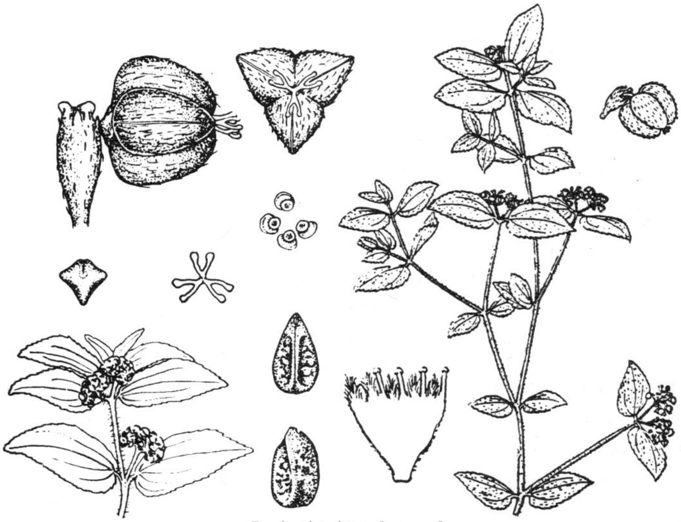
Euphorbia hirta [BR2, RHO]

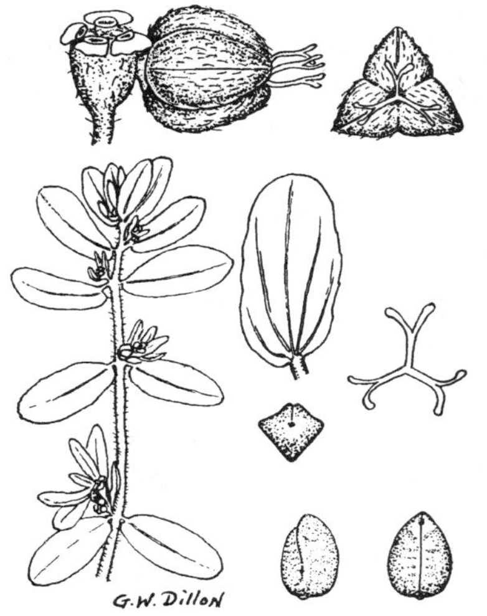

Euphorbia humistrata [RHO]

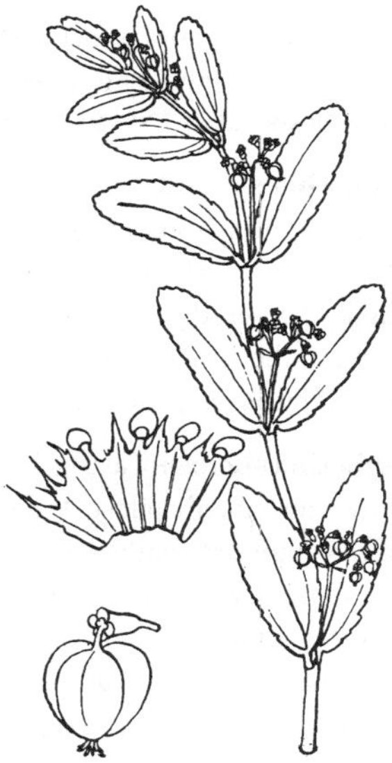
Euphorbia hypericifolia [BR2]

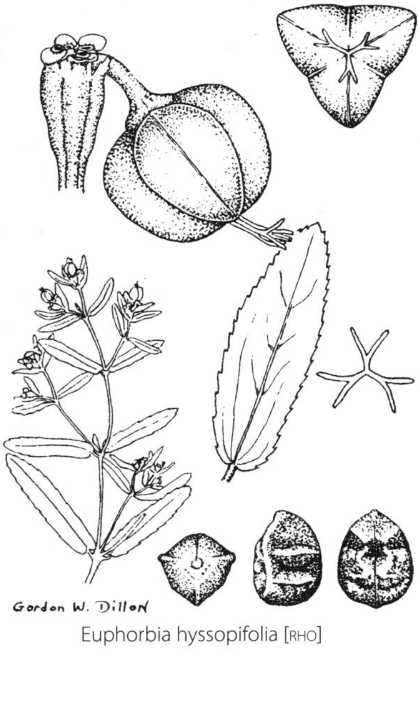

Euphorbia hyssopifolia [RHO]

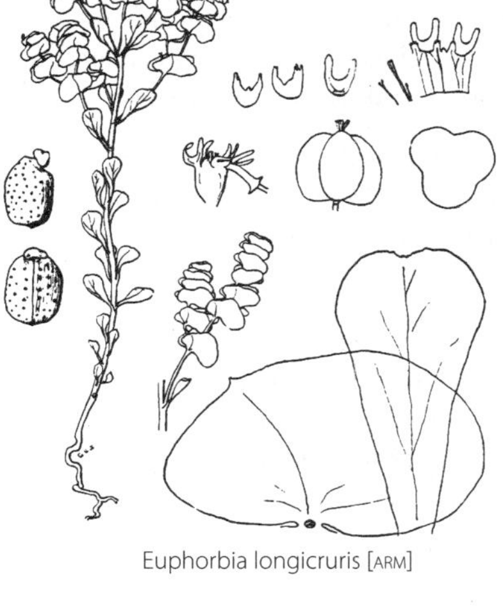
Euphorbia longicruris [ARM]

Euphorbia peplus L., (a shawl-like garment), PETTY SPURGE. Annual, branched or unbranched, 5–30 cm tall; leaf blades 5–25 mm long, 4–15 mm wide, obovate, oblong or suborbiculate; upper leafy bracts ovate to obovate; glands 4, elliptic; horns 0.4–0.6 mm long; capsules with 2 low longitudinal wings per lobe. Edges of gardens, weedy flower beds, roadsides; in East TX known only from Jefferson Co. (*Crockett 7126-A* LL), collected 24 April 1944 with a 1994 annotation: "introduced, but probably not persisting" (M. Mayfield); possibly also El Paso Co. (Kartesz 2022); scattered in ne and w U.S.; also Canada; Mexico; Central and South America. Year-round. [*Tithymalus peplus* (L.) Hill]

Euphorbia prostrata Aiton, (prostrate, flat to the ground), PROSTRATE EUPHORBIA. Annual similar to *E. maculata*; leaf blades varying to broadly elliptic, without central reddish or purplish splotch; styles bifid to base or nearly so. Stream banks, prairies, disturbed sites, chiefly clay soils; widespread in East TX; widespread in TX; most of the U.S. from PA s to FL and w to OR and CA; also Mexico; Central and South America. Jun–Oct, sporadically earlier. [*Chamaesyce prostrata* (Aiton) Small, *E. chamaesyce* of authors, not. L.] Can be toxic to cattle (Kingsbury 1964).

Euphorbia roemeriana Scheele, (for Ferdinand Roemer, 1818–1891, geologist, paleontologist, and explorer of TX), ROEMER'S SPURGE, ROEMER'S EUPHORBIA. Annual to 30 cm tall; stem leaves cuneate; leafy bracts of the floral branches paired, opposite, semicircular, connate for ca. 1/2 there length; involucral glands 4, with an erect horn at each end, without appendages. Calcareous soils; in the sw portion of East TX; otherwise mainly Edwards Plateau; endemic to TX. Spring.

Euphorbia serpens Kunth, (crawling), MAT EUPHORBIA, HIERBA DE GOLONDRINA. Annual; stems almost thread-like, frequently mat-forming, often rooting at the nodes; leaf blades ovate, oblong, or orbiculate; cyathia solitary at distal nodes; gland appendages forming narrow rim at edge of gland, 0.1–0.2 mm long, distal margin entire or with small rounded teeth; styles bifid ca. 1/2 or more of their length; seeds white to light pink. Stream bottoms, prairies, clay flats, disturbed sites; mostly s and w East TX; widespread in TX; mostly in c U.S.; also Mexico; Central and South America. Jul–Oct. [*Chamaesyce serpens* (Kunth) Small]

Euphorbia spathulata Lam., (spoon-shaped), WARTY EUPHORBIA, WARTY SPURGE. Glabrous, erect or ascending annual, 10–70 cm tall; leaves numerous, blades oblanceolate or spatulate, obtuse, toothed at least distally; upper leafy bracts ovate to deltoid ovate, sometimes clasping; involucral glands 4–5, minute, elliptic to oblong, without horns or appendages; styles bifid 1/3 to 1/2 their length. Prairies, disturbed areas; throughout East TX, except for ne portion; nearly throughout TX; nearly throughout the U.S., absent in the extreme ne U.S.; also s Canada (Ont.) Mexico; and South America. Late Mar–May. [*E. arkansana* Engelm. & A. Gray, *Tithymalus spathulatus* (Lam.) W.A. Weber]

Euphorbia stictospora Engelm., (straight seed), SLIM-SEED EUPHORBIA. Annual; stems prostrate, often mat-forming to 45 cm long, densely and evenly pilose; leaf blades usually oblong to oblong-obovate, occasionally nearly circular, without central red spot, apically minutely toothed; cyathia solitary at leaf nodes or in small cymose clusters on congested axillary branches (internodes very short); ovaries pilose to villous; gland appendages often unequal, sometimes 1 to all absent or to 0.3 mm long; styles unbranched. Limestone slopes, prairies, disturbed sites, clay soils; Collin, Dallas, Hunt (BRIT), Fannin, Lamar (Kartesz 2022), and McLennan (Hannick et al. 2013) cos.; mainly w1/2 of TX; c U.S. from ND s to TX and w to WY and AZ; also Mexico. Jun–Oct. [*Chamaesyce stictospora* (Engelm.) Small]

Euphorbia tetrapora Engelm., (four-pored), WEAK EUPHORBIA, WEAK SPURGE. Slender, erect, unbranched annual to 20 cm tall; stem leaves narrowly oblanceolate; branches of the inflorescence or their internodes longer than the leafy bracts; bracts bluntly triangular to narrowly so; involucral glands 4, with an erect horn at each end, without appendages. Sandy woods and fields; widespread in East TX; also w to West Cross Timbers and in the n Gulf Prairies and Marshes; LA, OK, and TX. Late Mar–May.

Euphorbia texana Boiss., (of Texas), TEXAS SPURGE. Erect or ascending annual 7–20 cm tall; leaf blades oblanceolate to oblong-oblanceolate, 8–15 mm long, 2–5 mm wide; upper leafy bracts elliptic to ovate, margin with rounded teeth; glands (4–)5; horns absent; ovary smooth, not warty, glabrous. Open ground, prairies; mostly in the s portion of East TX; LA and TX. Spring. [*Tithymalus texanus* (Boiss.) Millsp. ex A. Heller]

Euphorbia villifera Scheele, (bearing soft hairs), HAIRY EUPHORBIA. Annual or perennial with slender taproot or thickened woody rootstock; stems 10–30 cm long, ascending to erect, papillate, villous to glabrous; leaf blades 3–12 mm long, 2–10 mm wide, ovate, entire or minutely toothed, surfaces usually villous; cyathia solitary at distal nodes; gland appendages 0.2–0.4 mm long; ovaries glabrous, styles bifid 1/2 to their entire length. Limestone slopes, dry uplands; mostly on the sw margin of East TX; mainly Edwards Plateau and Trans-Pecos; in the U.S. known only from TX; also Mexico and Central America. Late May–Oct. [*Chamaesyce villifera* (Scheele) Small]

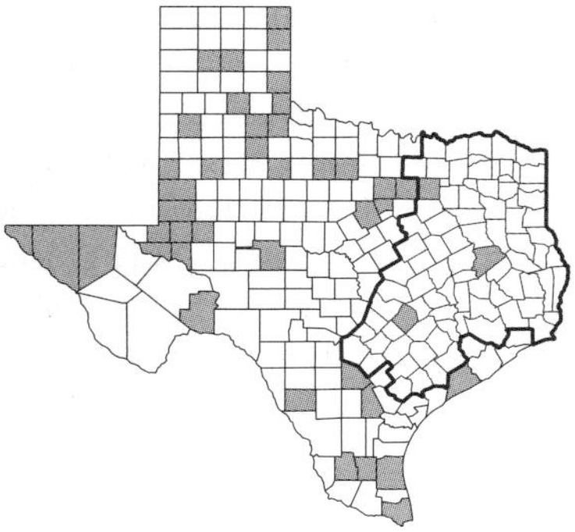
Euphorbia geyeri

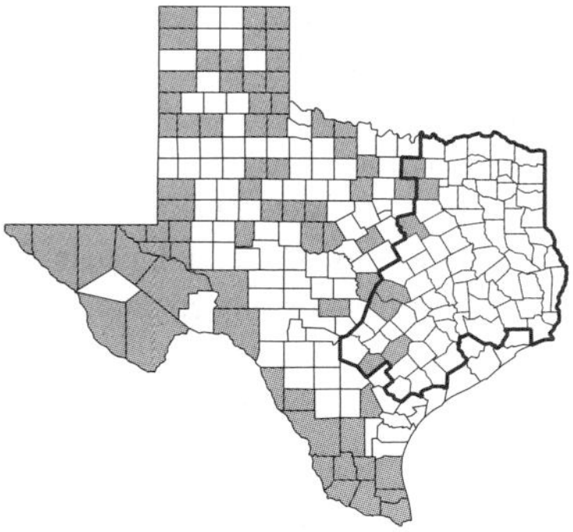
Euphorbia glyptosperma

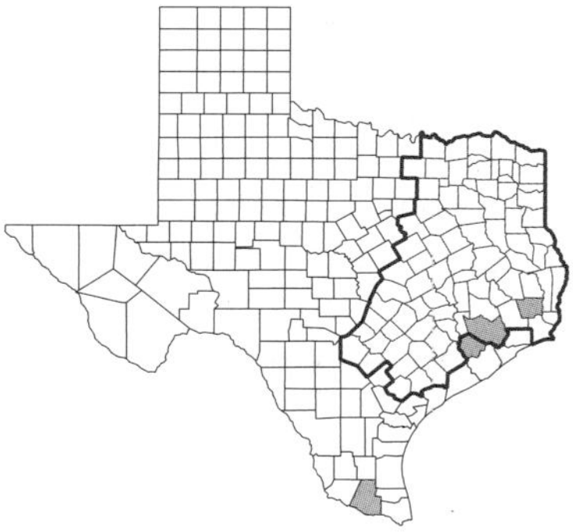
Euphorbia graminea

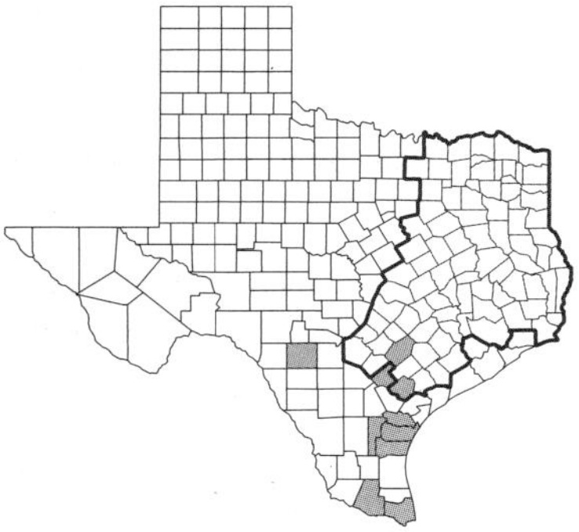
Euphorbia helleri

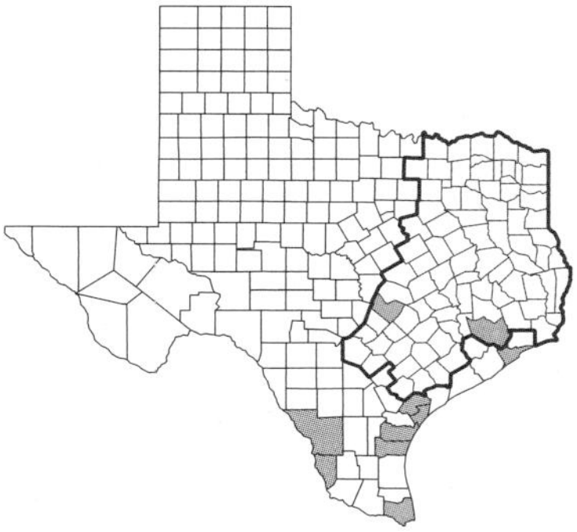
Euphorbia heterophylla

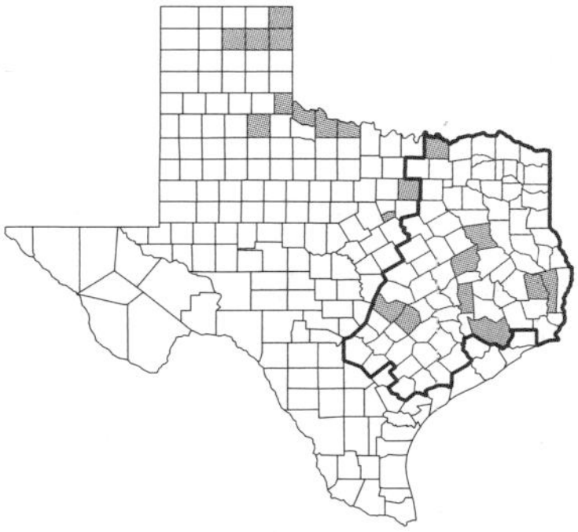
Euphorbia hexagona

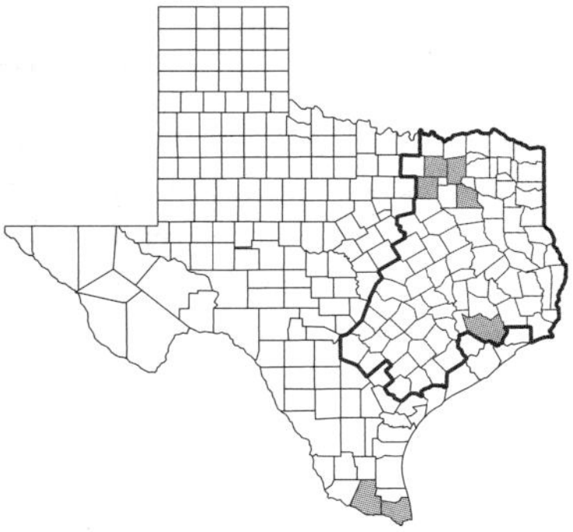
Euphorbia hirta

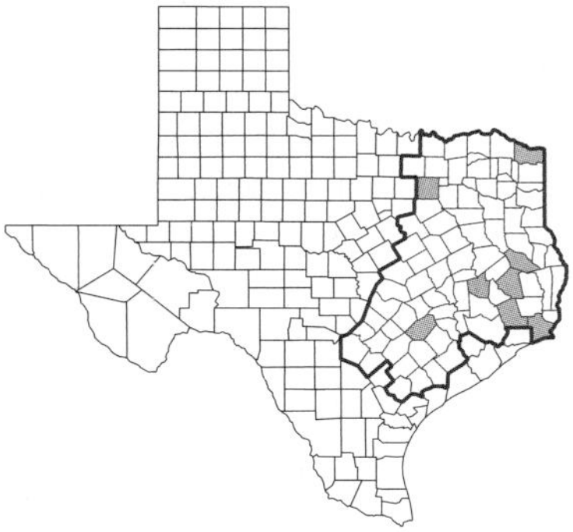
Euphorbia humistrata

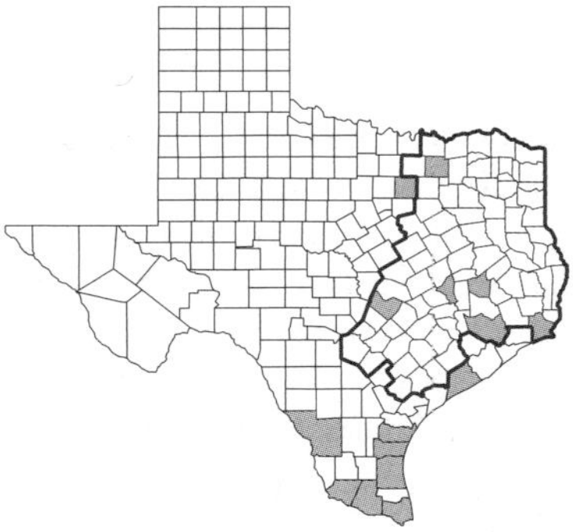
Euphorbia hypericifolia

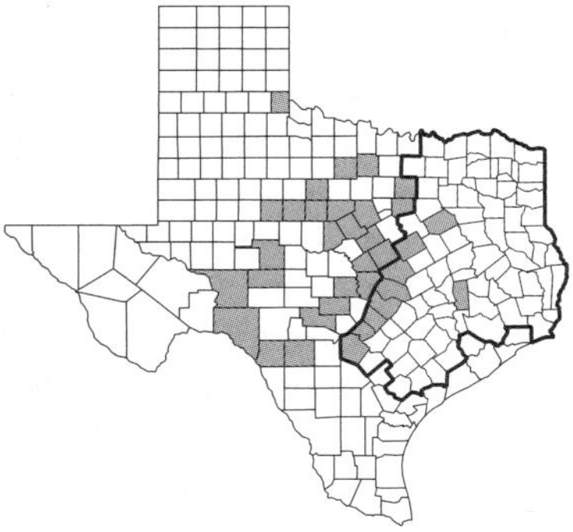
Euphorbia longicruris

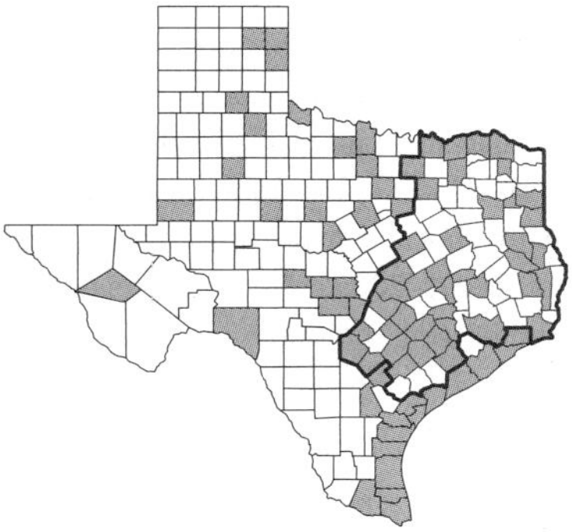
Euphorbia maculata

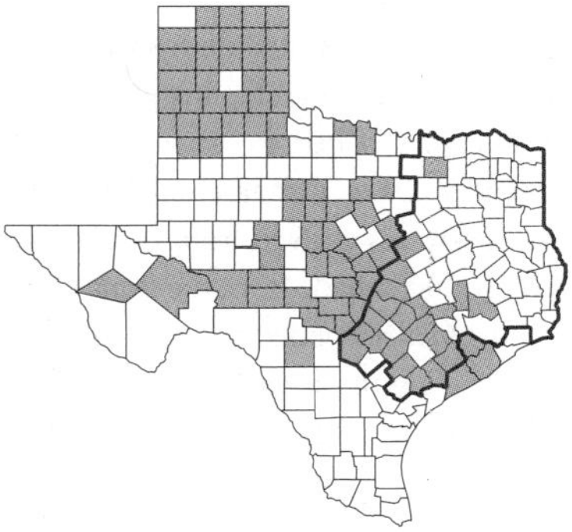
Euphorbia marginata

Euphorbia pubentissima Michx., (the most pubescent), FALSE SPURGE, a perennial closely related to *E. corollata* that has been reported for TX (Fernald 1950). Detailed treatments of this and related species (sect. *Tithymalopsis*—Park & Elisens 1997; Park 1998; Peirson et al. 2016; Weakley, A.S., and Southeastern Flora Team (2022) records it from the se U.S. only as far w as LA; we know of no definitive TX records. However, since it is known from LA (Park 1998) and has been reported from OK (as *E. corollata* var. *paniculata*—Oklahoma Vascular Plant Database 2008), it can reasonably be expected in East TX. It is therefore included here as a note and in the key to species to encourage collectors to look for it. It can be distinguished from *E. corollata* by its generally smaller size (plants 30–65 cm tall), solitary stems, petiolate or subpetiolate, thin leaves; lower inflorescence bracts similar in size to stem leaves, and gland appendages 1–2.2 mm long. [*E. corollata* L. var. *molle* Millsp., *E. corollata* var. *paniculata* Boiss.] Isozyme evidence (Park & Elisens 1997) indicates that *E. pubentissima* is more closely related to *E. discoidalis* than it is to *E. corollata*. Possibly in Bastrop Co. (*Ginzbarg 68* IND). Image on TORCH not verifiable.

MANIHOT Mill. MANIHOT

Perennial monoecious shrubs or small trees; latex whitish; roots tuberous or slender; leaves alternate, simple, palmately lobed, usually with 3–13 lobes, or a few (associated with the inflorescences) unlobed, long-petiolate, not peltate or only very slightly so (1–2 mm of tissue below insertion of petiole); stipules present; inflorescences panicles with the pistillate flowers proximal, the staminate distal; staminate flowers small, with perianth of 5 free or united tepals, green to yellowish green, reddish or purplish; pistillate flowers with ovary superior, 3 carpellate, subtended by a disk; staminate flowers with stamens usually 10, in 2 whorls, subtended by a disk; fruit a capsule; seeds caruncFRAGulate.

A genus of ca. 100 species of monoecious trees, shrubs, and herbs of tropical and warm areas of the Americas (Hayden 2016). (Name derived from a common name recorded for the plant when seen on the Brazilian coast in 1555, possibly from a Tupi (an indigenous Brazilian language) word, *maniot*—Gade 2002)

REFERENCES: Rogers & Appan 1973; Piperno & Pearsall 1998; Gade 2002; Hayden 2016.

1. Leaf blade lobes without secondary lobes; stem nodes distinctly enlarged; veins on lower surface of leaf pubescent, often reddish; roots enlarged, tuberous; capsules often winged; leaves with (2–)3–10(–12) lobes; petioles varying in color, but often with reddish pigmentation; inflorescences 2–10 cm long ____ **M. esculenta**
1. Median and adjacent leaf lobes with distal secondary lobes; stem nodes not or only slightly enlarged; veins on lower surface of leaf glabrous, yellow; roots slender, never enlarged; capsules not winged; leaves with 5–13 lobes; petioles yellow-green; inflorescences to 30 cm long ____ **M. grahamii**

Manihot esculenta Crantz, (edible), MANIOC, CASSAVA, TAPIOCA, YUCA. Shrub 1–4 m tall; roots tuberous; stems with nodes enlarged; leaf blades usually with 3–10 lobes, petiole 3–20 cm long; inflorescences axillary 2–10 cm long; capsules ca. 1.5 cm long, usually winged. Cultivated and perhaps escaping. Disturbed areas; reported from Hardin Co. (Turner et al. 2003), however, a specimen (*Johnston 12800* TEX) which maybe the origin of the Hardin Co. record was annotated to *M. grahamii* Hook (by W. John Hayden, University of Richmond). Additionally, Aplaca (2012) reported that many other vouchers from TX of *M. esculenta* have been annotated to *M. grahamii*. On the other hand, Hayden (2016) has *M. esculenta* mapped for TX. It may or may not be a part of the East TX flora. AL, FL, GA, and TX(?). Fall–winter [*Jatropha manihot* L.] The large tuberous roots are an important food crop in the tropics, especially on poor soils. It was brought into cultivation in the New World tropics and has long been cultivated, with a claim of 9,100 B.P. for domesticated forms (Piperno & Pearsall 1998; Gade 2002). Even today, it is an important part of the diet in a number of countries including Brazil, Mozambique, and Paraguay (Gade 2002). The plants are resistant to insect attack because of high levels of cyanogenic glycosides. However, special preparation is required to remove the cyanide and prevent a cyanide-related disease, konzo, which results from sustained eating of unprepared CASSAVA. There are numerous cultivars, with "sweet" ones having glycosides limited mainly to the bark of the roots. The starch-rich, potato-like roots are used in a variety of ways: baked, in soups, as a meal (Brazilian arrowroot), in breads, as tapioca (pudding), and in making alcoholic drinks (Mabberley 2017). Ornamental forms with unusual leaf coloration (variegated forms) are cultivated in East TX. The common name, CASSAVA, is derived from *casabe*, "originally an Arawak word from the Caribbean which referred quite specifically to the baked products made from the flour of the root" (Gade 2002). The common name, TAPIOCA, is derived from a Tupi word, *tipioca*, which originally applied to a product "made by heating the starch until the grains hydrolyze and aggregate into pearls" (Gade 2002). The common name, YUCA, is also from an indigenous (Taino Indian name from Hispaniola) word transferred through Spanish (Gade 2002). No county level distribution map is provided.

Manihot grahamii Hook., (for Robert Graham, 1786–1845, physician and a major figure in the creation of the Glasgow Botanic Gardens), GRAHAM'S MANIHOT, HARDY TAPIOCA. Shrub or low tree, in its native habitat to 7 m tall, with cyanide odor in all parts; roots slender, not tuberous; stem with nodes not or only slightly enlarged; leaf blades with 5–13 lobes, median and adjacent lobes with a pair of weakly defined rounded secondary lobes distal to middle, lateral lobes without secondary lobes, median lobe 5–24

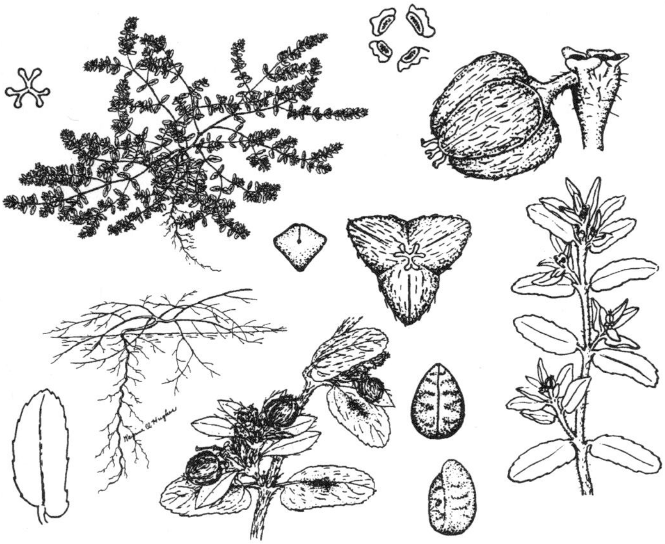

Euphorbia maculata [REE, RHO]

Euphorbia marginata [BT3]

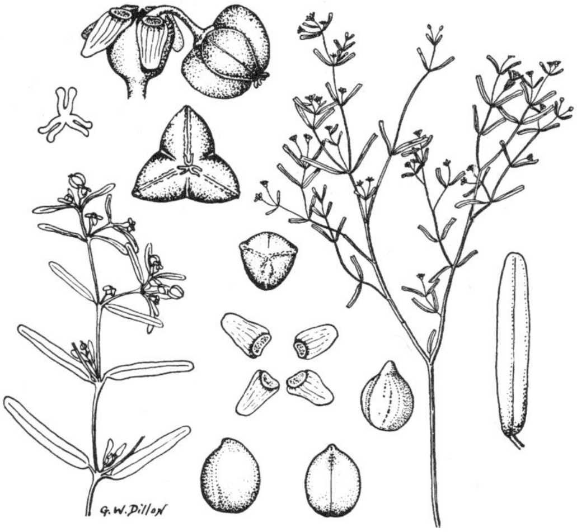

Euphorbia missurica [GLE, RHO]

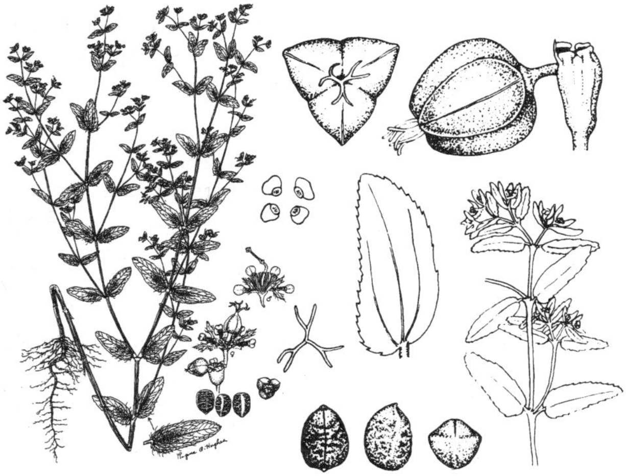

Euphorbia nutans [REE, RHO]

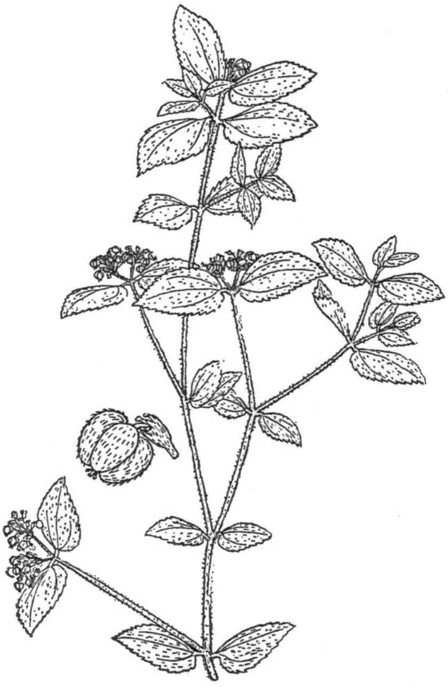

Euphorbia ophthalmica [BR2]

cm long, petiole 5–33 cm long; inflorescences to 30 cm long, capsule ca. 1.8 cm long, without wings, explosively dehiscent. Cultivated and apparently escaping; disturbed areas; known in TX only from Hardin (*Johnston 12800* TEX), Harris (Aplaca 2012), and Jasper (*Reynolds s.n.*, 2001, BAYLU; annotation: cultivated) cos.; AL, AR, FL, GA, LA, MS and TX. Apr–Aug. Native of tropical South America (n Argentina, se Brazil, Paraguay, Surinam, and Uruguay). This species is cultivated as an ornamental, and is also used in cassava breeding research.

RICINUS L. CASTOR-BEAN

A monotypic genus native from e and ne Africa (Gillespie 2016b), now naturalized throughout the tropics. (Latin: *ricinus*, tick, because of the resemblance of the seed to a tick)

Ricinus communis L., (common, general), CASTOR-BEAN, CASTOR-OIL-PLANT, PALMA CHRISTI, HIGUERILLA. Large, coarse, herbaceous, monoecious annual (in East TX) 1–5 m tall, glabrous or nearly so; sap watery; leaves alternate, with long petioles to 30+ cm long, blades peltate, with (5–)7–9(–11) palmate lobes, these serrate with the teeth gland-tipped; inflorescences 6–30 cm long, terminal, raceme- or panicle-like, the staminate flowers usually proximal, the pistillate flowers distal; staminate and pistillate sepals usually 5; petals 0; stamens numerous; styles 3, plumose, red; capsules round, 12–20 mm in diam., usually covered with soft, dark spines; seeds 8–11 mm long, glabrous, with mottled surface. Cultivated for its foliage; roadfill, waste areas, river banks; scattered in East TX; scattered in a few other TX localities, particularly in the far s; scattered in e U.S. from PA s to FL and w to KS and TX, also AZ and CA; however, throughout FL where it is considered invasive. Flowering throughout the growing season. Native to Africa and the Middle East. CASTOR-BEAN has a very long history of cultivation; it has been found in 6000 year-old Egyptian tombs (Zohary 1982). All parts of the plant, but especially the seeds, contain many alkaloids and toxic principles including a lectin (type of phytotoxin or toxic protein), ricin, one of the most toxic compounds known. If eaten, 1–8 seeds are fatally poisonous; the seeds are also the source of castor oil, used medicinally in the past; the toxins are not oil soluble. The foliage is also poisonous, but less so than the seeds. Castor oil is also used as an industrial lubricant. Livestock can be poisoned from eating the leaves or seeds (Muenscher 1951; Kingsbury 1964; Hardin & Arena 1974; Stephens 1980; Burrows & Tryl 2001; Mabberley 2017). Castor-bean is reported to be invasive in various parts of the world including Australia and Hawaii (Mabberley 2017).

STILLINGIA Garden QUEEN'S-DELIGHT, QUEEN'S-ROOT, TOOTHLEAF

Glabrous unarmed perennial herbs usually with several to many stems from a woody root; sap milky; leaves numerous, crowded, alternate, subsessile, evenly and finely glandular-toothed; flowers in terminal, sessile, spike-like inflorescences (which become over-topped by whorls of leafy branches developed from beneath them), unisexual with both sexes in the same inflorescence, borne in the axils of minute triangular-ovate bracts, the bracts with fleshy, saucer- or cup-shaped, gland-like stipules at each side, the glands larger than the bracts; staminate flowers above, on a fleshy axis which falls after flowering; pistillate flowers few, at base; gynobase (= lower portion of ovary) becoming thick and hard, persistent after the seeds fall, triangular; fruit a shallowly 3-lobed, 3-seeded capsule; seeds with a prominent caruncle (= outgrowth near the hilum).

A genus of ca. 33 species of tropical and warm America, Madagascar, and Fiji Islands (Huft 2016). Feeding experiments with *S. treculeana* (Müll. Arg.) I.M. Johnst., of s TX and the Edwards Plateau, indicated that the plant may be highly toxic due to hydrocyanic acid; small amounts of leaves and stems were lethal to sheep. (Named for Dr. Benjamin Stillingfleet, 1702–1771, English naturalist)

REFERENCE: Huft 2016.

1. Staminate flowers 3–7 in crowded bracted cymules; leaf blades finely dentate; inflorescences 3–9 cm long.
    2. Staminate flowers 4–7 per cymule; stem leaves lanceolate, elliptic to obovate, mostly 9–30 mm wide; capsules 6–12 mm in diam.; plants of sandy soil ____ **S. sylvatica**
    2. Staminate flowers 3–5 per cymule; stem leaves linear to linear-lanceolate, mostly 2–5 mm wide; capsules 6–8 mm in diam.; plants of limestone habitats ____ **S. texana**
1. Staminate flowers, bracted, 1 per node, not crowded; leaf blades coarsely dentate; inflorescences 2.5–5 cm long ____ **S. treculiana**

Stillingia sylvatica L., (forest-loving), QUEEN'S-DELIGHT. Stems ascending to erect, to ca. 1 m tall; inflorescences 3–9 cm long; pistillate flowers 3–4, sepals 3, persistent; gynobase lobes 4–6 mm long, columella not persistent; Sandy prairies, open woods, or open ground; nearly throughout East TX; widespread in TX; s U.S. from NC s to FL and w to NM. May–Jun. Used medicinally, but overdoses cause toxic symptoms; the latex is reported to have vesicant (= blister-causing) properties (Duke 1985; Lampe 1986).

Euphorbia ouatchitana [CP1]

Euphorbia peplidion [ARM]

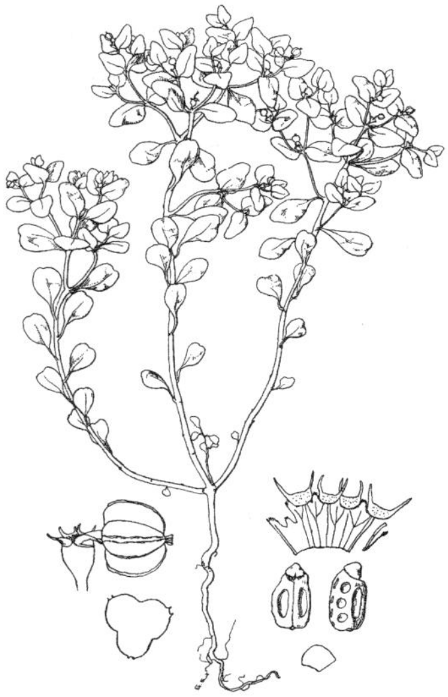
Euphorbia peplus [ARM]

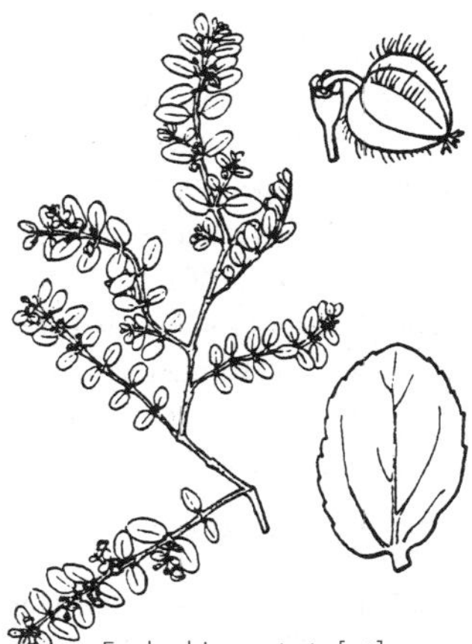
Euphorbia prostrata [BF2]

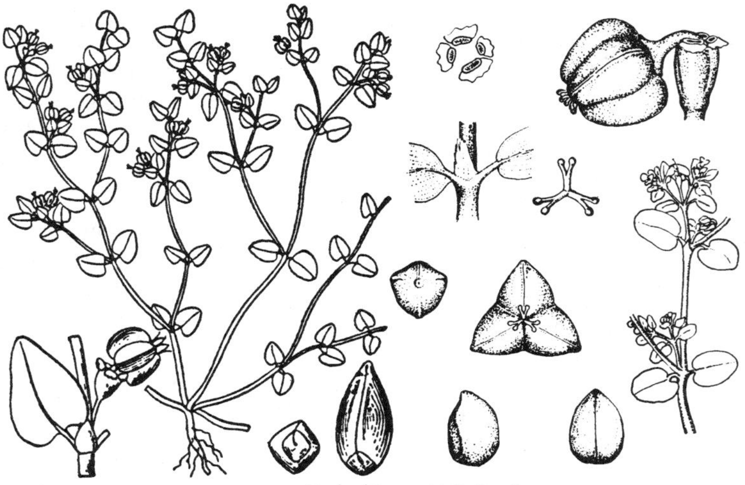
Euphorbia serpens [BB2, RHO]

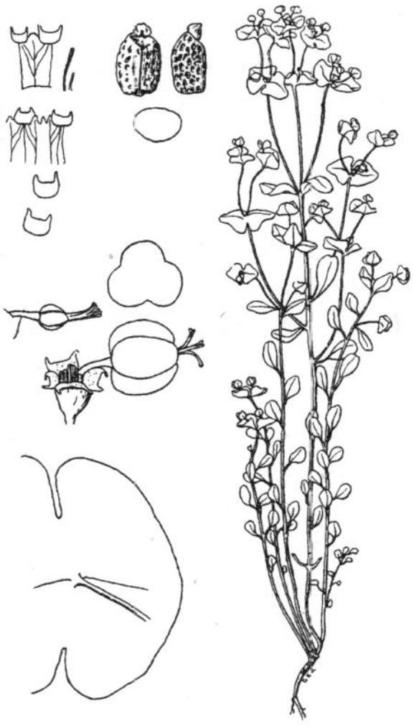
Euphorbia roemeriana [ARM]

Stillingia texana I.M. Johnst., (of Texas), TEXAS STILLINGIA, TEXAS TOOTHLEAF. Stems decumbent to erect, to ca. 6 dm long; inflorescences 3–9 cm long; pistillate flowers 3–4, sepals 3, persistent; gynobase lobes 2–3 mm long, columella often persistent. Gravelly or rocky, calcareous soils; w margin of East TX and Washington Co. (Kartesz 2022) further e; also Cross Timbers and Prairies and Edwards Plateau; OK and TX; also Mexico. Apr–Jun. The latex is reported to have vesicant (= blister-causing) properties (Lampe 1986). [*S. sylvatica* var. *linearifolia* (Torr.) Müll. Arg.] ☠

Stillingia treculiana (Müll. Arg.) I.M. Johnst., (for Auguste Trécul, 1818–1896, French botanist), TRECUL'S TOOTHLEAF, YERBA DE SAPO. Stems spreading, branching throughout, leaf blades 1–4 cm long, 0.5–1.5 cm wide, usually obovate-spatulate, rarely narrowly elliptic, coarsely serrate; inflorescences 2.5–5 cm long; pistillate flowers 3–4, sepals 3, ephemeral; gynobase lobes to 1 mm long, columella persistent. Calcareous sandy soils, gravelly soils; in East TX known only from Bexar Co. (Rogers 1951); not likely to be found in East TX proper; Edwards Plateau and South TX Plains; in the U.S. known only from TX; also Mexico. Mar–May.

TRAGIA L. NOSEBURN

Perennial herbs, erect to trailing or twining, usually with needle-like stinging hairs; sap not milky; leaves alternate, blades toothed; stipules foliaceous, lanceolate to ovate; flowers unisexual, in peduncled, axillary (at least appearing so), spike-like, minutely bracted racemes, apetalous, most often both sexes in the same inflorescence; staminate flowers distal, minute; pistillate proximal, solitary or few, larger; staminate flowers with 3–5(–7) sepals; stamens usually equaling the number of calyx lobes; pistillate flowers usually with 5–6 narrow sepals; styles 3; fruit an explosively dehiscent capsule; seeds without a caruncle.

A genus of ca. 175 species (Urtecho 2016) of tropical and warm areas of the world; usually with stiff, nettle-like stinging hairs. ☠ All East TX species have the stinging hairs and, therefore, this information has not been included in each species description. These hairs can be quite painful if allowed to come into contact with skin. The burning sensation is sometimes delayed for several minutes. It is apparently caused by the injection of a mixture of primarily proteins (Burrows & Tryl 2001). According to Lampe (1986), only four families (Euphorbiaceae, Hydrophyllaceae, Loasaceae, and Urticaceae) have stinging hairs—East TX has stinging representatives in Euphorbiaceae and Urticaceae. The following key is modified from those of Miller and Webster (1967) and Miller (in Correll & Johnston 1970). *Tragia* species are often difficult to distinguish and as Miller and Webster (1967) noted, their delineation, "is not an especially tidy one, as a considerable number (ca. 5–10%) of the specimens examined remain dubious or unassignable" [particularly fragmentary specimens]. (Name from *Tragus*, Latinized last name for early herbalist, Hieronymous Bock, 1498–1554)

REFERENCES: Miller & Webster 1967; Urtecho 2016.

1. Stamens usually 2 per flower (rarely 3); calyx lobes of staminate flowers usually 4–5(–6); species of the se portion of East TX (Big Thicket area).
    2. Leaf blades very variable, linear and entire to broadly oblanceolate (these usually distally toothed, sometimes elliptic, basally cuneate to acute; fruits 7–8 mm wide; seeds 3–4 mm long ____ **T. urens**
    2. Leaf blades suborbicular, oval or elliptic, coarsely toothed, basally rounded to obtuse, apically rounded to acute; fruits 9–13 mm wide; seeds 4–4.5 mm long ____ **T. smallii**
1. Stamens usually 3–4(–5 or even 10 in terminal flowers) per flower; calyx lobes of staminate flowers usually 3–4(–5) or even 7 in terminal flowers); including species widespread in East TX.
    3. Stems glabrous or nearly so; leaf blades with margins deeply and sharply toothed, basally cuneate; filaments united at least 1/3–1/2 their length forming a cylinder; species known in East TX only from the extreme sw corner (Comal Co.) ____ **T. nigricans**
    3. Stems pubescent; leaf blades with margins variously toothed, basally truncate to cordate or hastate; filaments united only at base; including species widespread in East TX.
        4. Persistent base of the staminate pedicel usually nearly as long as or longer than its subtending bract; stigmatic surfaces papillate; styles connate 1/3–1/2 their length, with a slight constriction where they join the ovary ____ **T. urticifolia**
        4. Persistent base of the staminate pedicel conspicuously shorter than its subtending bract; stigmatic surfaces smooth to papillate; styles free or connate, without a constriction where they join the ovary.
            5. Leaf blades broadly ovate, basally deeply cordate, apically acuminate, marginally coarsely serrate; capsules 11–13 mm in diam.; seeds 4.3–5.3 mm long ____ **T. cordata**
            5. Leaf blades variously linear-lanceolate to broadly triangular, ovate or rarely reniform but not broadly ovate (and apically acuminate), basally truncate to cordate, apically acute; marginally finely serrate to serrate; capsules 6–9 mm wide; seeds 2.6–4 mm long.
                6. Stigmas papillate; calyx lobes of pistillate flowers as long or longer than pistil; staminate flowers 14–75 per raceme, compactly arranged on axis ____ **T. betonicifolia**
                6. Stigmas smooth, undulate, or subpapillate; calyx lobes of pistillate flowers shorter than pistil; staminate flowers 2–20 per raceme, not compactly arranged.

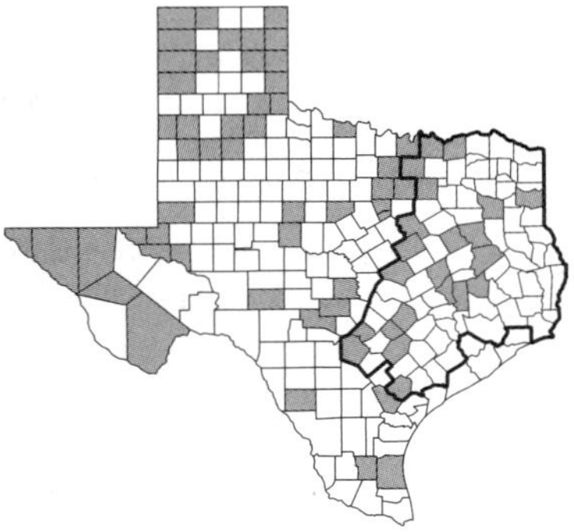

Euphorbia missurica

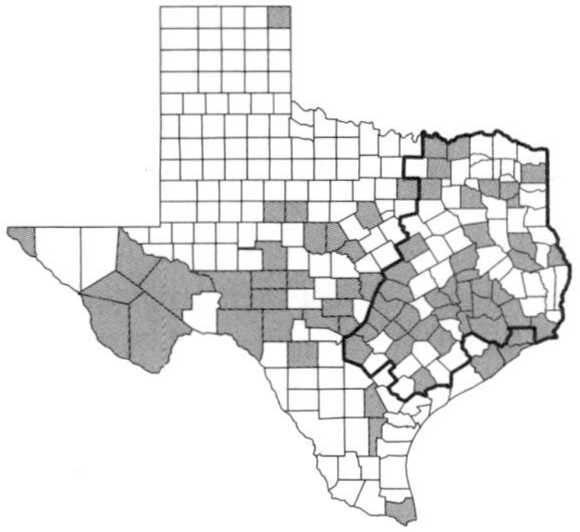

Euphorbia nutans

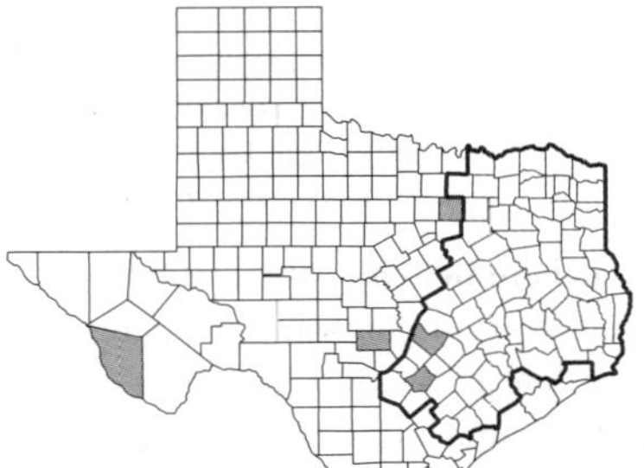

Euphorbia ophthalmica

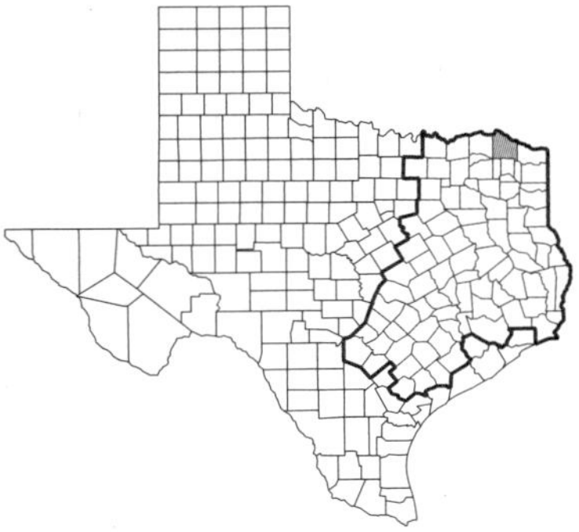

Euphorbia ouachitana

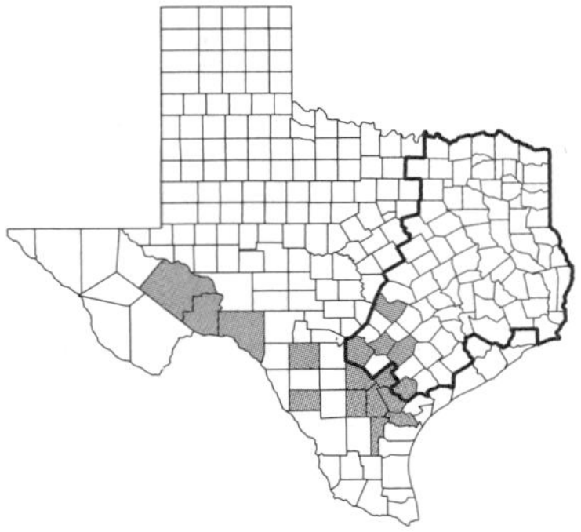

Euphorbia peplidion

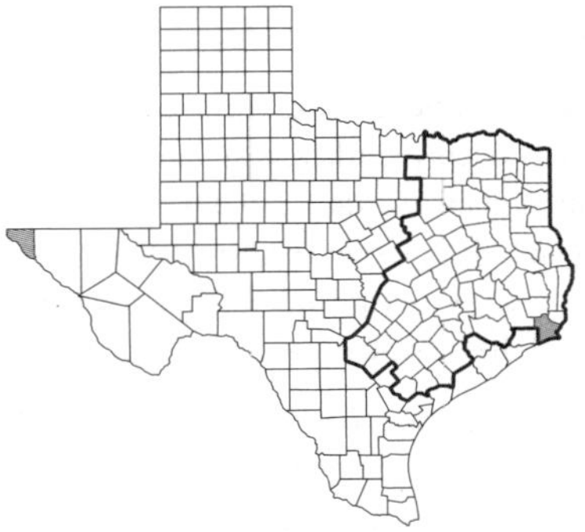

Euphorbia peplus

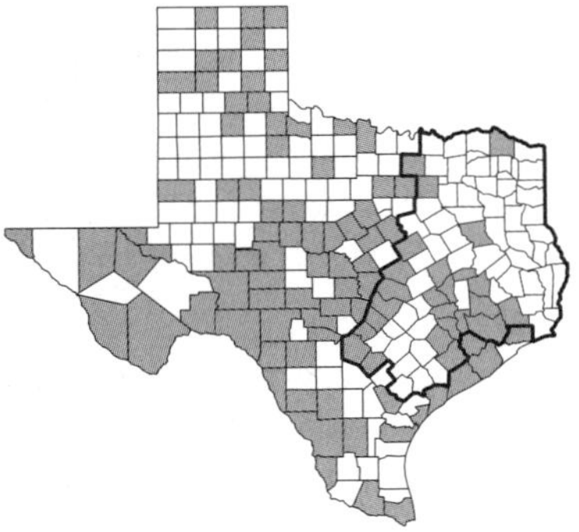

Euphorbia prostrata

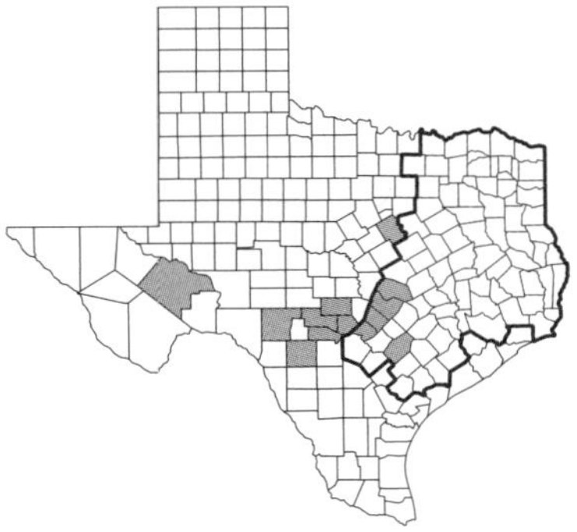

Euphorbia roemeriana

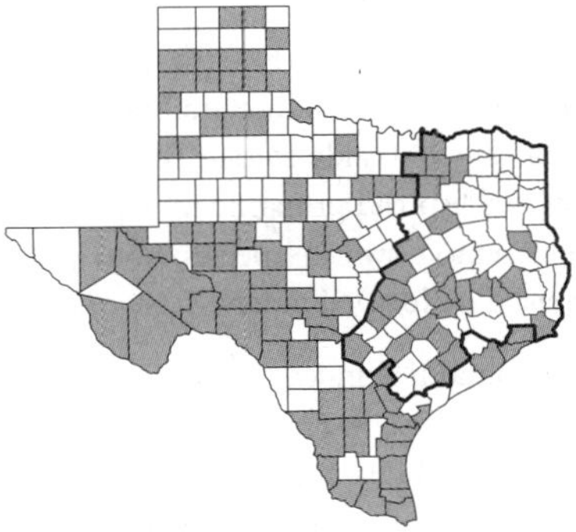

Euphorbia serpens

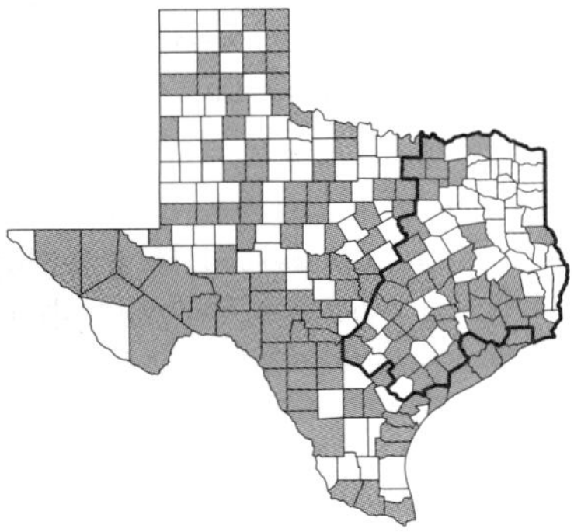

Euphorbia spathulata

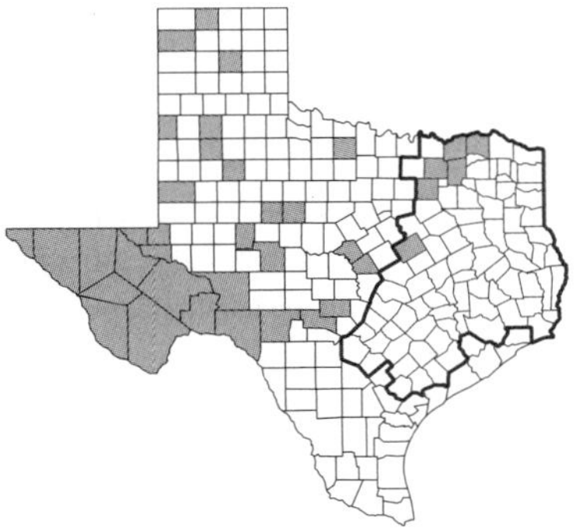

Euphorbia stictospora

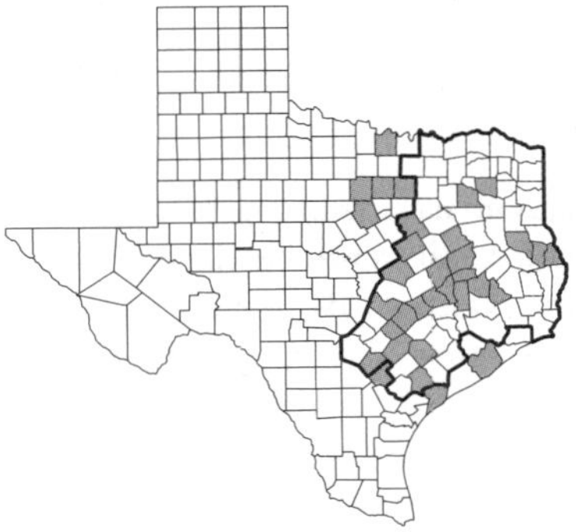

Euphorbia tetrapora

7. Larger leaf blades usually broadly triangular, usually 1–2.5 times as long as wide; stigmatic surfaces papillate; styles connate basally to 1/3 their length; some fruits with only 1 carpel maturing; these fruits with distinct wings **T. brevispica**
7. Larger leaf blades usually linear-lanceolate to ovate, usually 2–4 times as long as broad; stigmatic surfaces not papillate; styles connate 1/3–1/2 or more their length; fruits without wings **T. ramosa**

Tragia betonicifolia Nutt., (with leaves like *Betonica*—BETONY, now = *Stachys* in Lamiaceae), BETONY-LEAF NOSEBURN. Stems to 50 cm tall, erect, decumbent, or trailing, apex never flexuous; leaf blades triangular-ovate to triangular-lanceolate, usually cordate to truncate at base; pistillate calyx lobes 6, 1.4–3 mm long at anthesis, 3–5 mm long in fruit, lanceolate; styles connate basally; stigmatic surfaces papillate. Sandy soils; widespread in East TX; also w to Rolling Plains and Edwards Plateau and Gulf Prairies and Marshes; c U.S. from MO s to AL and w to KS and TX. May–Jun, Sep–Oct. [*T. urticifolia* Michx. var. *texana* Shinners]

Tragia brevispica Engelm. & A. Gray, (short-spiked), SHORT-SPIKE NOSEBURN. Stems to 100 cm long or longer, erect to trailing and twining, apex usually flexuous; pistillate calyx lobes 6, 1.3–2 mm long at anthesis, 1.8–3.5 long in fruit; some capsules with 1–3 distinct wings. Calcareous soils, open woods and prairies; widespread in w portion of East TX; also Edwards Plateau, South TX Plains, Gulf Prairies and Marshes, and Cross Timbers and Prairies; LA, OK, and TX; also Mexico. Apr–Oct.

Tragia cordata Michx., (heart-shaped), HEART-LEAF NOSEBURN. Stems to 150 cm long, trailing and twining; leaf blades broadly ovate, basally deeply cordate; pistillate calyx lobes 6(–7), 1.5–2 mm at anthesis, 2–3 mm long in fruit; capsules 5–7 mm long, 11–13 mm in diam.; seeds 4.3–5.3 mm long. Wooded areas; along streams, rich soils; Bowie (*Correll 15249* LL), Harrison (*Fleetwood 11808* SMU), Jasper (*Clark 3919* SMU), and Tyler (*Carr* (with *Paul Turner*) *18602* TEX) cos.; ec U.S. from IN s to FL and w to OK and TX. Spring–fall. [*T. macrocarpa* Willd.] Miller and Webster (1967) noted that the species is "not easily confused with any of the other U.S. taxa" and that it is "set apart by the vining habit, large deeply cordate leaves, and large capsules and seeds."

Tragia nigricans Bush ex Small, (blackish), DARK NOSEBURN. Stems to 55 cm tall, erect, apex never flexuous, blackening upon drying; leaf blades oblong to oblong-lanceolate, basally cuneate, coarsely serrate, tooth apices often recurved; pistillate calyx lobes (5–)6, 1–2.3 mm long at anthesis, 2.4–4 mm long in fruit. Open woods, fields; rocky soils; Comal Co. (Miller & Webster 1967) in extreme sw corner of East TX; also in adjacent area of Edwards Plateau and n South TX Plains; endemic to TX. Spring–fall. The filaments united at least 1/3–1/2 their length forming a cylinder is unique in the genus (Miller & Webster 1967).

Tragia ramosa Torr., (branched), BRANCHED NOSEBURN, CATNIP NOSEBURN. Subshrubs; stems to 50 cm tall, erect, decumbent, sometimes somewhat cespitose, trailing, with tendency toward twining; leaf blades usually linear-lanceolate to narrowly ovate, usually truncate to weakly cordate basally; pistillate calyx lobes 5–7, 0.8–2.5 mm long at anthesis, 1.5–3 mm long in fruit. Disturbed areas; widespread primarily in w portion of East TX; widespread in TX particularly in the w 2/3 of the state, except South TX Plains; a more sw U.S. species; from MO through AR to TX and w to NV and CA; also Mexico. Apr–Oct. [*T. nepetifolia* Cav. var. *leptophylla* (Torr.) Shinners, *T. nepetifolia* Cav. var. *ramosa* (Torr.) Müll. Arg.]

Tragia smallii Shinners, (for John Kunkel Small, 1869–1938, American botanist and author of numerous works including *Manual of the Southeastern Flora*), SMALL'S NOSEBURN. Stems to 25 cm tall, erect, apex never flexuous; leaf blades orbicular to elliptic, rounded to obtuse basally; pistillate calyx lobes 6, 1.3–1.7 mm long at anthesis, 1.5–2.3 in fruit. Open woods, clearings, roadsides; sandy soils; Hardin, Newton (Miller & Webster 1967), Angelina, Jasper, Jefferson, Sabine, Tyler (Turner et al. 2003), and Polk (Kartesz 2024) cos. in the Big Thicket area; se U.S. from GA and FL w to AR and TX. Spring–fall. Miller and Webster (1967) considered this species and *T. urens* to be in a different section (*Leptobotrys*) of the genus from the other TX species (which are in section *Tragia*); for example, note stamen number in first dichotomy of key to species. In fact, a molecular study by Cardinal-McTeague and Gillespie (2016) indicated that section *Leptobotrys* is distinct from all other North American *Tragia* and could be treated as a separate genus.

Tragia urens L., (stinging, burning), WAVY-LEAF NOSEBURN. Stems to 50 cm tall, erect, never flexuous; leaf blades broadly oblanceolate to linear, sometimes elliptic, basally acute to attenuate; pistillate calyx lobes 6, 1–1.5 mm long at anthesis, 1.5–1.8 mm long in fruit. Open woods, clearings, roadsides; sandy soils; Angelina, Hardin, Jasper, Polk, Sabine, San Augustine, Shelby, Trinity, and Tyler (Kartesz 2022) cos. in the Big Thicket area; se U.S. from VA s to FL and w to TX. Spring–fall. [*T. linearifolia* Elliott] Stinging hairs present, but according to Miller and Webster (1967), the "specific epithet is not very appropriate, since *T. urens* is the least painfully stinging of the U.S. species." Miller and Webster (1967) also thought that in "many ways *T. urens* is the most isolated of the U.S. species" and they hypothesized that it is "a phytogeographically isolated taxon in the United States, with a completely different origin from the noseburn complex in Texas and adjacent states."

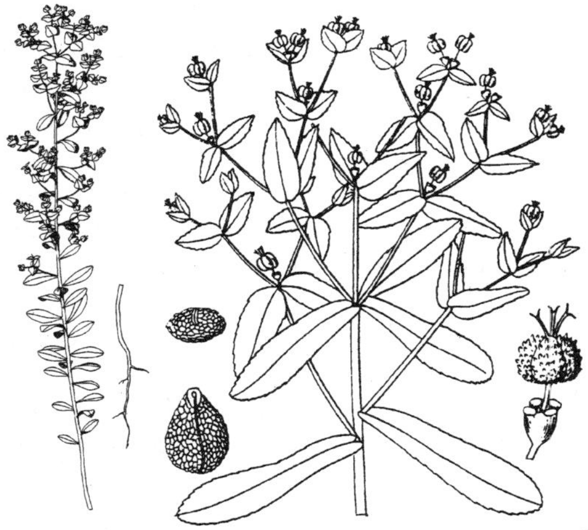

Euphorbia spathulata [BB1, STE]

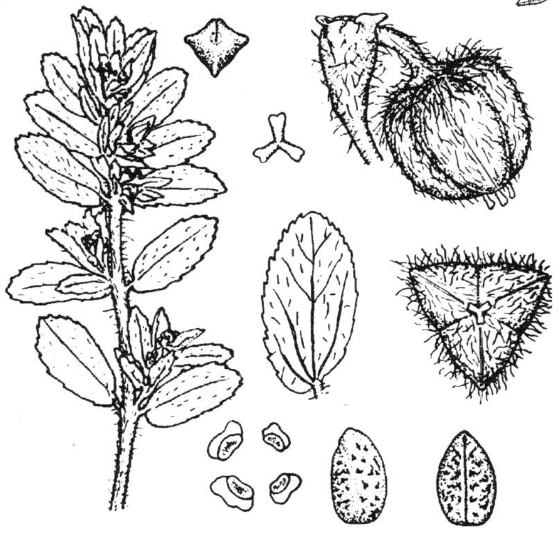

Euphorbia stictospora [BT3, RHO]

Euphorbia texana [CP1]

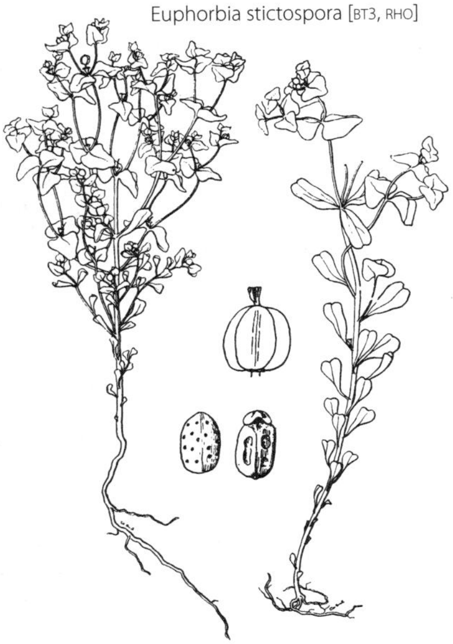

Euphorbia tetrapora [ARM]

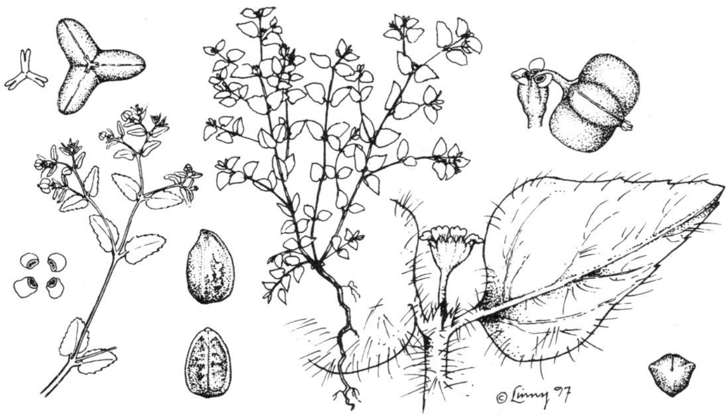

Euphorbia villifera [HEA]

Tragia urticifolia Michx., (with leaves like *Urtica*—NETTLE), NETTLE-LEAF NOSEBURN. Stems to 65 mm tall, erect to decumbent, apex never flexuous; leaf blades triangular-lanceolate, base truncate to cordate; pistillate calyx lobes (5–)6, 1.3–2.3 mm long at anthesis, 2–3 mm long in fruit, lanceolate to ovate. Sandy soils, fields, and open woods; widespread in East TX; also Brazoria and Galveston cos. in extreme n Gulf Prairies and Marshes; se U.S. from VA s to FL and w to AR and TX, also KY. Spring–fall. Miller and Webster (1967) noted that this is a "very distinct species" and that the "only species which is at all close to *T. urticifolia* is *T. betonicifolia*." Urtecho (2016) further notes, "Although similar to *T. betonicifolia*, *T. urticifolia* has fewer stems, shorter pistillate sepals, inflorescences with fewer staminate flowers that are not distally clustered, and leaf blades with very light green abaxial surfaces." According to M. Mayfield (pers. comm.), *T. urticifolia* occurs only in sandy habitats, these generally not found in the extreme w portion of East TX.

TRIADICA Lour. CHINESE TALLOW TREE

A genus of three species endemic to e and se Asia (Wurdack 2016b). It has often in the past been included in *Sapium* (e.g., Webster 1994b), MILKTREE, a genus of ca. 90–100 species (Webster 1994b) widespread in tropical and warm areas and in South America s to Patagonia. However, according to Webster (1994b), "*Triadica*, accepted by some Asian workers, is distinctively different from the neotropical species [of *Sapium*] in such features as its non-arillate seeds" and most authorities (e.g., Kartesz 2022) now recognize *Triadica* as a distinct genus. (Greek: *treis*, *tria*, three, in reference to the calyx, ovary, and fruit—Quattrocchi 2000)
REFERENCES: Esser 2002; Wurdack 2016b.

Triadica sebifera (L.) Small, (tallow-bearing), CHINESE TALLOW TREE, VEGETABLE TALLOW TREE, GRAY POPCORN TREE. Rapidly growing monoecious tree usually 3–10 m tall, unarmed, essentially glabrous, with milky sap; branches often slender and drooping; leaves alternate, the blades rhombic-ovate, 3–7(–10) cm long with acuminate apices, entire; petioles usually longer than blades, with 2 glands at juncture of blade and petiole; flowers in terminal spike-like inflorescences (3–)5–15 cm long; petals absent; staminate flowers in distal clusters; pistillate flowers solitary, proximal; fruit a 3-lobed, 3-seeded schizocarp ca. 1–1.3 cm long and about as broad; seeds 7–8 mm long, long persistent on the placenta after the fruit walls have fallen, chalky-white. Widely cultivated as ornamental shade tree and escaping in some areas; widespread in East TX, more commonly escaping in se TX where it is a problematic invader of native prairies and bottomland forests; Gulf Prairies and Marshes and Edwards Plateau; se U.S. from NC s to FL and w to OK and TX. Aug–Nov. Native of China and Japan. [*Sapium sebiferum* (L.) Dum. Cours., *Stillingia sebifera* (L.) Michx.] This species was introduced into the U.S. in South Carolina in the late 1770s and is now widespread; it displaces native vegetation and is considered one of the most serious invasive exotics in the s U.S. CHINESE TALLOW is one of the most problematic invasives (possibly the most problematic) in the se portion of East TX and the adjacent coastal prairie—large areas of the Big Thicket National Preserve are being invaded, as are remnant portions of the Coastal Prairie. The fatty covering (sarcotesta) of the seed yields oils and waxes and is used for candlewax and soap (Esser 2002; Mabberley 2017). Burlage (1968) reported the milky sap as poisonous.

VERNICIA Lour. VERNICIA

A genus of 3 closely related species of trees and shrubs distributed from Burma to Indochina, Thailand, S. China, and Japan (Stuppy et al. 1999). It was previously included in a broadly conceived *Aleurites* (5 species), but phylogenetic studies (van Welzen & Stuppy 1999) indicate that *Vernicia* is monophyletic. We are following most recent authorities (Gillespie 2016c; Mabberley 2017; Kartesz 2022; Weakley, A.S., and Southeastern Flora Team 2022) in recognizing *Vernicia*. The species are cultivated for shade and for oil from the seeds (including tung oil and candle-nut oil). (Latin: *vernix*, varnish, paint, in reference to the fact that it is possible to make varnish from the oily seeds—Quattrocchi 2000)
REFERENCES: Duke 1983; Stuppy et al. 1999, 2008; van Wenzel & Stuppy 1999; Gillespie 2016c.

Vernicia fordii (Hemsl.) Airy Shaw, (for Charles Ford, 1844–1927, plant collector in South China and Formosa), TUNG-OIL TREE. Small to medium-sized, monoecious, deciduous tree to ca. 20 m tall; bark smooth, thin; leaves alternate, simple, petioles 6–22 cm long, blades usually 10–25 cm long and 10–22 cm wide, with 3–5 major veins from base, cordate or with three shallow lobes, with 2 conspicuous, reddish, sessile glands at leaf base/petiole apex, a gland also in sinuses between lobes (if present); stipules 4–12 mm long; flowers appearing before or just with the leaves, in loose terminal clusters, the male and females flowers in the same inflorescence; staminate and pistillate petals 5, 20–35 mm long, usually pale pink suffused with reddish and with reddish veins in lower portion, yellow at very base, clawed; stamens (7–)8(–14); fruit capsular, ± subglobose, 4–6 cm long; seeds typically 4–5 per fruit, large, usually 20–25 mm long, warty and ridged. Cultivated and escaped; in TX known from Hardin (*Brown 24435* BRIT), Harris (Kartesz 2022), Newton (*Lewis & Lewis 123* TEX), and Walker (Turner et al. 2003) cos.; se U.S. from NC s to FL and w to AR and TX, also CA. Spring. Native from c Asia to Burma. [*Aleurites fordii* Hemsl.] This species has been cultivated in

Manihot esculenta [BA1]

Manihot grahamii [IPL]

Ricinus communis [BL1, LAM]

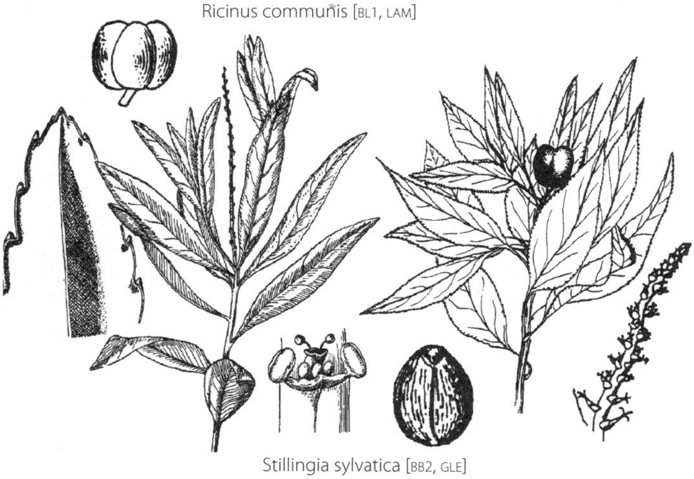
Stillingia sylvatica [BB2, GLE]

Stillingia texana [BB2]

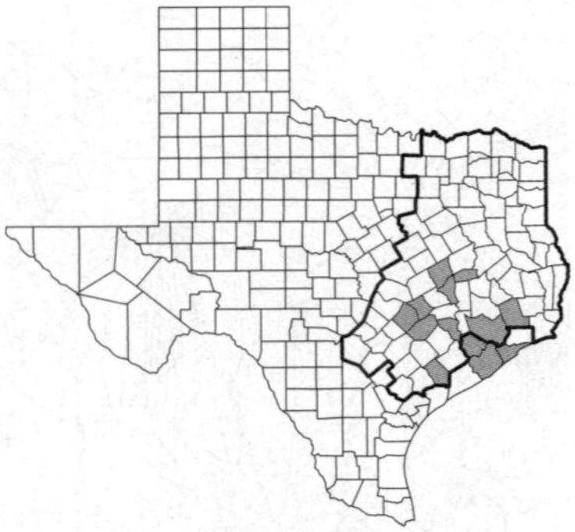

Euphorbia texana

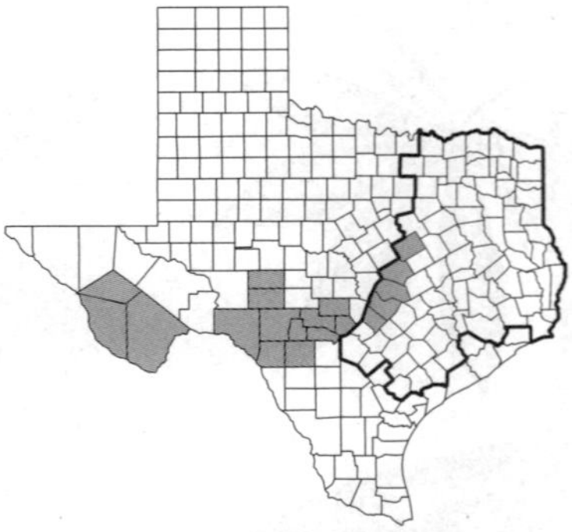

Euphorbia villifera

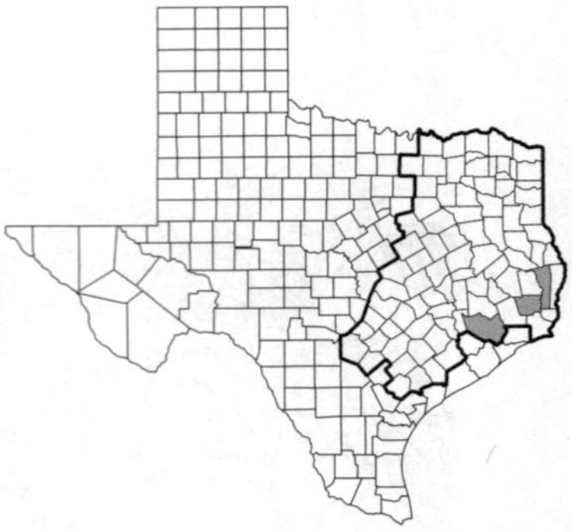

Manihot grahamii

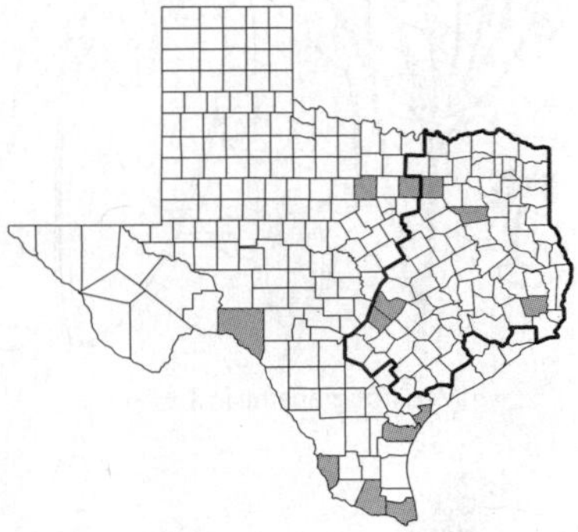

Ricinus communis

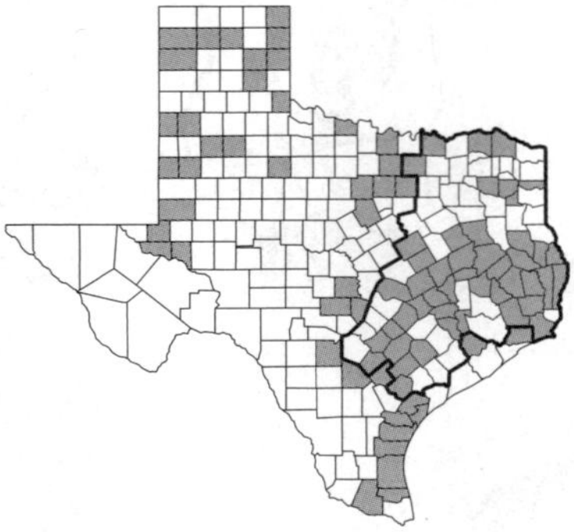

Stillingia sylvatica

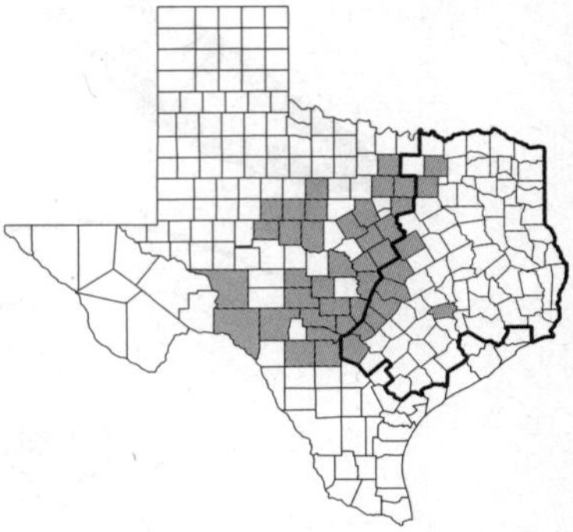

Stillingia texana

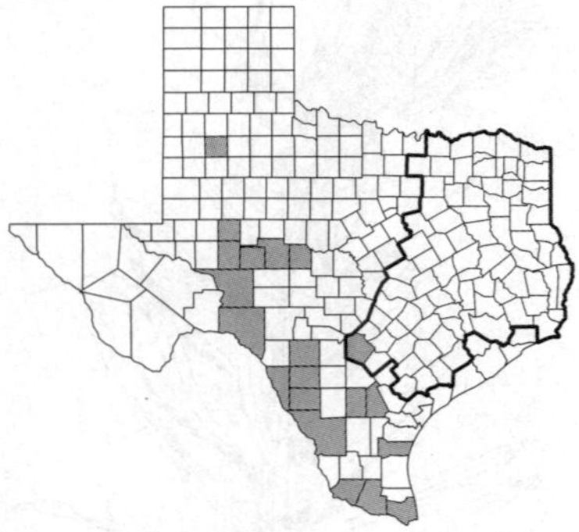

Stillingia treculiana

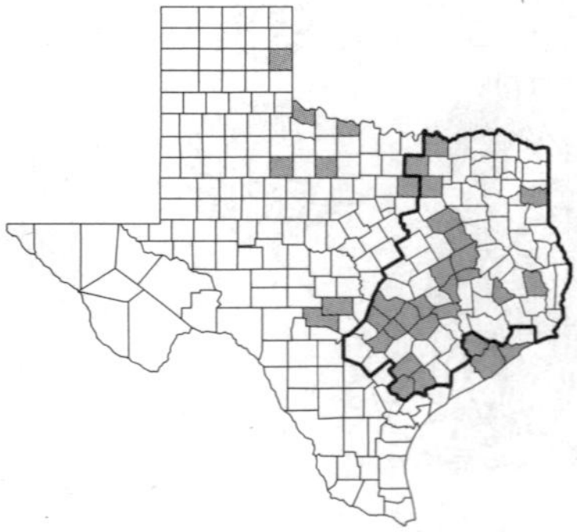

Tragia betonicifolia

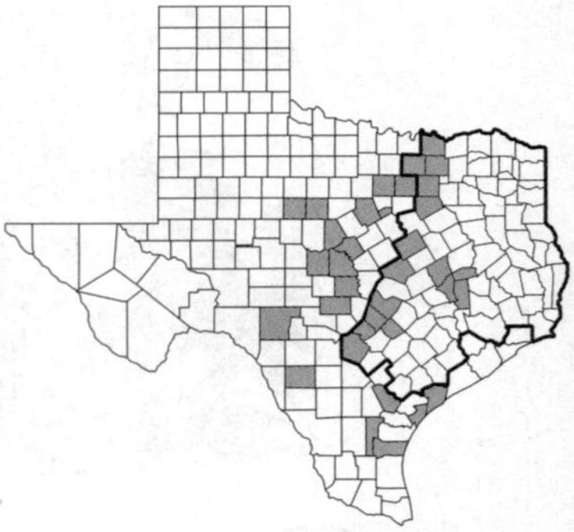

Tragia brevispica

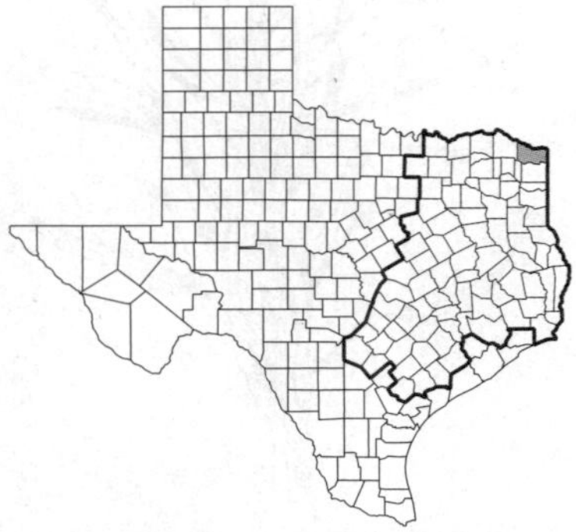

Tragia cordata

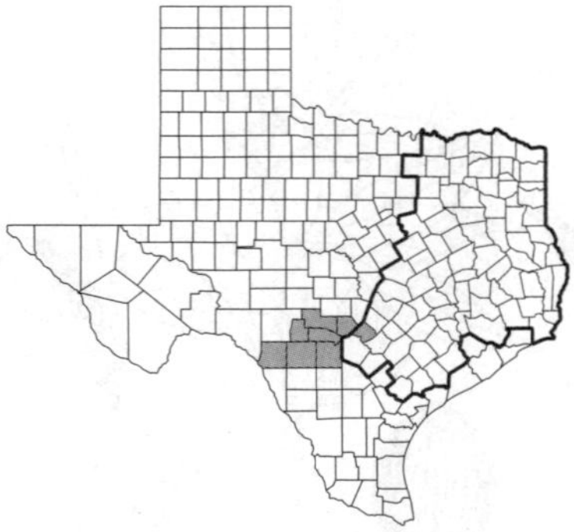

Tragia nigricans

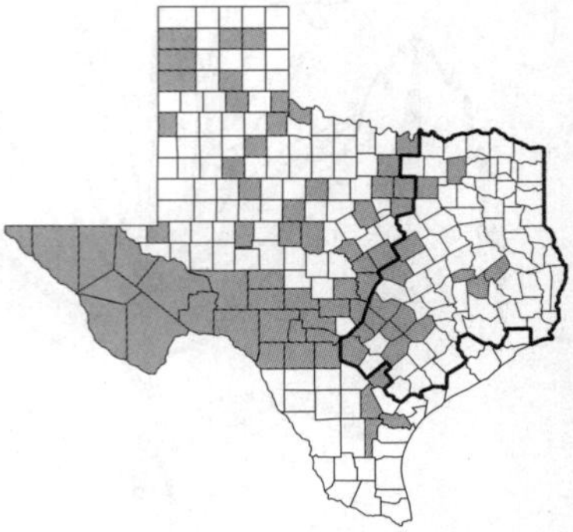

Tragia ramosa

Stillingia treculiana [EN1]

Tragia betonicifolia [RHO]

Tragia brevispica [RHO]

Tragia nigricans [RHO]

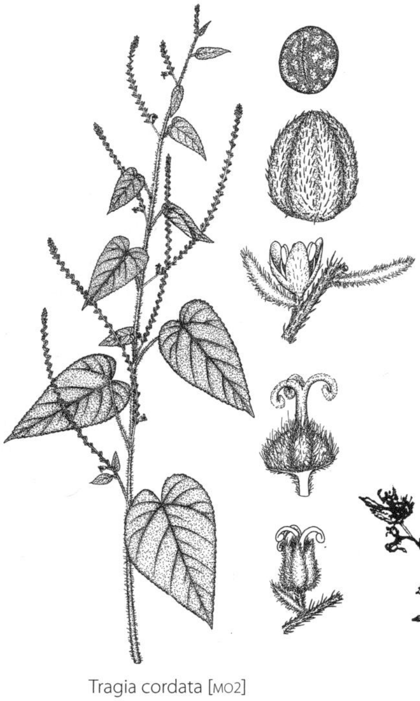

Tragia cordata [MO2]

Tragia ramosa [SHI]

Tragia smallii [RHO]

the Gulf states for its seeds, the source of tung oil (also known as China wood oil or nut oil), which is widely used in paints and quick-drying varnishes (Mabberley 2017); several layers of tung oil are often the final treatment for fine furniture. The seeds are reported to "have drastic purgative properties and can cause severe poisoning if eaten raw" (Stuppy et al. 2008).

FABACEAE Lindl. (LEGUMINOSAE Juss.)

LEGUME, BEAN, OR PULSE FAMILY

Contributed by **Eric Keith** (Raven Environmental Services, Inc.)

Plants herbaceous or woody, sometimes armed; leaves basal or alternate (sometimes crowded and appearing subopposite or whorled), compound or apparently simple (with only 1 leaflet); leaflets entire, or in a few genera toothed or lobed; stipules present (except *Acmispon*, *Lotus* and *Crotalaria*), falling early in many species; flowers solitary or in racemes, panicles, spikes, heads, or umbel-like clusters; sepals 5, separate or united; petals 1–5, equal or in most genera unequal; stamens 5, 10, or many; pistil 1; fruit a legume, developed from a 1-celled superior ovary with 1–many ovules and parietal placentation, in general opening along both sutures.

A huge (20,900 species in ca. 770 genera), cosmopolitan, vegetatively variable family ranging from herbs to shrubs, vines, and rain forest canopy trees (Vincent 2023). Nearly a third of all species are in 6 large genera: *Acacia*, *Astragalus*, *Cassia*, *Crotalaria*, *Indigofera*, and *Mimosa*. The Fabaceae is the third largest angiosperm family in terms of numbers (after Asteraceae and Orchidaceae), and in importance to humans is second only to Poaceae. It is also extremely important ecologically because of the symbiotic nitrogen-fixing *Rhizobium* bacteria associated with the roots of many species. The bacteria occur in nodules on the roots and these nodules are often easily seen when a plant is pulled up. The family includes many important timber trees, ornamentals, forage plants (e.g., *Trifolium*—CLOVERS and *Medicago*—ALFALFA), and particularly protein-rich food plants including *Arachis* (PEANUTS), *Cicer* (CHICK-PEAS), *Glycine* (SOY-BEANS), *Lens* (LENTILS), *Phaseolus* (BEANS), and *Pisum* (PEAS); in fact, seeds of legumes are the world's most important source of vegetable protein for man and animals (Isely 1990). However, in many species there are often toxic, defensive, non-protein amino acids in the seeds or vegetative tissues as well as alkaloids. For example, the tropical *Abrus precatorius* L. (PRECATORY-BEAN or ROSARY-PEA) has striking red and black seeds sometimes used in necklaces; they contain abrin, a protein so toxic that a single chewed seed is enough to kill a human (Kingsbury 1964). The wood of many legumes is hard and dense and a number of species, particularly in the tropics, are valued for their lumber (Yatskievych 2006).

While the Fabaceae is strongly supported as monophyletic (Stevens 2001 onwards), subfamilial classification has proven to be more complicated. Historically, Fabaceae were considered easily divided into three groups based on morphological features, especially of flowers: the caesalpinioids (Caesalpiniaceae R. Br., or Fabaceae subfamily Caesalpinioideae DC.), the mimosoids (Mimosaceae R. Br., or Fabaceae subfamily Mimosoideae DC.), and the papilionoids (Papilionaceae Giseke [Fabaceae in the strict sense], or Fabaceae subfamily Papilionoideae DC. [Faboideae Rudd]) (Cronquist 1981). Polyphyly of caesalpinioids has long been suspected, as have affinities to mimosoids, based on morphological data (Polhill & Vidal 1981). Phylogenetic studies have confirmed that mimosoids are embedded within a paraphyletic caesalpinioid clade (Bruneau et al. 2001, 2008). Most recently, the Legume Phylogeny Working Group (2017) has recognized six subfamilies: Caesalpinioideae (including a monophyletic "mimosoid clade"), Cercidoideae, Detarioideae, Dialioideae LPWG, Duparquetioideae LPWG, and Faboideae [as Papilionoideae]. Detariodeae, Dialioideae and Duparquetioideae do not occur in East TX. Arrangement of genera in this treatment follows the subfamilial classification of Legume Phylogeny Working Group, with the sequence of genera within the subfamilies adapted from Lewis at al. (2005).

At least 42 genera of the family have extrafloral nectaries (= nectar-producing organs in a location other than a flower). These glands, often located on the petiole, attract ants which remove eggs and larvae of insects that damage the plant—this is a type of defense mechanism known as animal guard defense—an animal (ant) is being paid (nectar) to protect the plant. Experiments have demonstrated increased damage to plants from which ants were excluded (Lersten & Brubaker 1987; Durkee et al. 1999; Rutter & Rausher 2004). Many species in the mimosoid clade display nighttime or "sleep" (nyctinastic) movements with leaflets folding together at night. This has been suggested as a mechanism to reduce water loss or alternatively as an adaptation to reduce damage by herbivores (Yatskievych 2006; LPWG 2017). The Fabaceae is the fourth largest family in the East TX flora (after Asteraceae, Poaceae, and Cyperaceae), with 278 distinct taxa, including fertile hybrids. Family name conserved from *Faba*, a genus of a single widely cultivated species, *F. vulgaris* Moench, (BROAD BEAN), of the Mediterranean region; it is now usually lumped into *Vicia* as *V. faba* L. (Latin: *faba*, bean) (subclass Rosidae)

TWINING BLUEHOOD (*Oxyrhynchus volubilis* Brandegee), a woody vine species native to Mexico, Bahamas, and Cuba, is mapped in TX by Delgado-Salinas (2023a) and in Travis Co. by Kartesz (2023). The inclusion of this species for TX is based on probable cultivated specimens from Austin collected in 1922 and 1923 since discussion of cultivation of the species is referenced on the herbarium sheets (*Britto s.n.*, 1922, NY; *Woodbridge s.n.*, 1922, US; *Piper s.n.*, 1923, NY, US). It was not mapped or treated in Turner (1959), Correll and Johnston (1970), Isely (1998), or Turner et al. (2003) for TX and no wild populations or additional collections are known from the state. (subclass Rosidae—Cronquist; order Fabales—APG IV)

FAMILY RECOGNITION IN THE FIELD: the characteristic most helpful in field recognition is the *legume* fruit—a 1-chambered, pod-like, often bean-like fruit; other helpful clues include leaves with 3 leaflets and pea-like flowers, pinnately compound leaves with flowers in heads with obviously exserted stamens, or stamens in a 9 (fused filaments) + 1 (separate filament) arrangement.

REFERENCES: Rydberg 1919–1920, 1923–1929; Britton & Rose 1928, 1930; Turner 1959; Isely 1973, 1975, 1981, 1982, 1986a, 1990, 1998; Elias 1974; Robertson & Lee 1976; Isely & Polhill 1980; Summerfield & Bunting 1980; Allen & Allen 1981; Duke 1981; Polhill & Raven 1981; Gunn 1983, 1984, 1991; Stirton 1987; Stubbendieck & Conard 1989; Ballenger et al. 1993; LPWG (2017); Weakley, A.S., and Southeastern Flora Team 2022; Vincent 2023.

KEY TO SUBFAMILIES OF FABACEAE

1. Corollas of 3 separate petals (larger standard, 2 wings) and 2 petals fused to form a keel (corollas rarely of only 1 petal); sepals at least partially fused; stamens usually enclosed by keel and usually not externally visible, their filaments usually fused or free in a few species; leaves simple, once pinnately compound (often with 3 leaflets OR with many leaflets), or palmately compound; standard outside of lateral petals in bud ____ **Faboideae**
1. Flowers either bilaterally or radially symmetrical, banner innermost or petals valvate (in mimosoid clade); sepals distinct or connate; stamens exerted and visible or included within flowers; leaves pinnate, bipinnate, or simple (in Cercidoideae).
   2. Flowers usually small and individually inconspicuous, arranged in dense heads or clusters, regular (= radially symmetrical); corollas usually sympetalous and so small as to be ± inevident; stamens 5–numerous, their filaments exserted beyond the petals and often showy; leaves twice pinnately compound ____ **Caesalpinioideae, mimosoid clade**
   2. Flowers whether small or large usually individually conspicuous, arranged in various types of inflorescences, usually zygomorphic (= bilaterally symmetrical); corollas usually easily seen, of separate petals, except 2 petals (= keel) fused along the margin in many; stamens 1–10, usually not exserted beyond the petals (except in *Erythrostemon*; leaves simple or palmately or once or twice pinnately compound (3–numerous leaflets).
      3. Leaves simple and entire, trees ____ **Cercidoideae**
      3. Leaves palmately or pinnately compound; trees, shrubs, or herbaceous perennials ____ **Caesalpinioideae, excluding mimosoid clade**

SUBFAMILY 1. CERCIDOIDEAE

Trees or shrubs, unarmed; leaves unifoliate; stipules lateral, free; sepals united or separate, spathaceous or 2–5 lobed; petals distinct (2–)5(or 6); stamens usually 10, rarely 1, 3 or 5, filaments partly connate or distinct. This subfamily is only represented by the genus *Cercis* in East TX.

SUBFAMILY 2. CAESALPINIOIDEAE (EXCLUDING MIMOSOID CLADE)

Trees, shrubs, or herbs; leaves once or twice compound, even- or odd-pinnate; stipules very small, often falling early; sepals united or separate; petals 5, well-developed, slightly or markedly unequal; stamens 5 or 10, separate.

KEY TO GENERA OF CAESALPINIOIDEAE (EXCLUDING MIMOSOID CLADE)

1. Trees or shrubs.
   2. Leaflets < 10 mm long.
      3. Plants armed with sharp spines; leaves with 2(–4) pinnae (each with leaflets), but each pinna appearing as a once pinnate leaf due to the absence of a common petiole; stamens small, inconspicuous ____ **Parkinsonia**
      3. Plants unarmed; leaves with 11–29 pinnae (each with leaflets); stamens very long, 70–90 mm, bright red ____ **Erythrostemon**
   2. Leaflets 15–50 mm long.
      4. Plants unarmed; leaves once pinnately compound with 4–24 leaflets; corollas conspicuous, 10–20 mm long, yellow to orangish; shrubs to 3.5 m tall ____ **Senna**
      4. Plants conspicuously armed with straight or branched thorns OR unarmed; leaves once or twice pinnately compound with 18–numerous leaflets; corollas inconspicuous, 4–10 mm long, greenish yellow or greenish white; trees to ca. 15(–35) m tall.
         5. Plants usually armed (cultivated forms sometimes unarmed); leaflets usually 4–15 mm wide, lanceolate to narrowly elliptic, rounded or ± blunt at apex, often crenulate, sometimes entire; native species widespread in TX ____ **Gleditsia**
         5. Plants unarmed; leaflets usually 15–30 mm wide, ovate, rather abruptly acuminate at apex, entire; cultivated species rarely escaping in TX ____ **Gymnocladus**

1. Herbaceous annuals or perennials.
    6. Leaves twice compound; leaflets 3–8 mm long ______ **Hoffmannseggia**
    6. Leaves once compound; leaflets usually much longer (7–170 mm long).
        7. Leaflets 2 ______ **Senna**
        7. Leaflets 12 to numerous.
            8. Leaflets 2.5 cm or more long; petiolar glands slender or stipitate or absent ______ **Senna**
            8. Leaflets 2 cm or less long; petiolar glands disc-shaped ______ **Chamaecrista**

MIMOSOID CLADE

Leaves twice even-pinnately compound; stipules thread-like or minute, often falling early; flowers in dense heads or spikes; sepals united; petals 4 or 5, equal, separate or united, so small as to be ± inconspicuous; stamens 5 to many, separate or the filaments all united toward base, exceeding the corolla, relatively conspicuous, typically being the primary visual attractant for pollinators.

KEY TO GENERA OF CAESALPINIOIDEAE (MIMOSOID CLADE)

1. Trees or woody shrubs or if herbaceous perennials, then stamens numerous.
    2. Stamens numerous.
        3. Stamens 23–36 mm long, pinkish to reddish; introduced species ______ **Albizia**
        3. Stamens < 20 mm long, white to yellow; native species.
            4. Stipular spines present, sometimes enlarged and inhabited by ants; ovaries sessile or subsessile; seeds sometimes surrounded by pulp ______ **Vachellia**
            4. Stipular spines absent; ovaries stipitate; seeds not surrounded by pulp.
                5. Filaments fused into a tube; low intricately branched shrubs or herbaceous perennials ______ **Calliandra**
                5. Filaments distinct; erect subshrubs, shrubs, or small trees.
                    6. Subshrubs; petiolar glands absent; prickles absent; inflorescences usually in globose heads; corollas greenish white, drying to pink-rose; stamens 175–250, creamy white, anther glands absent ______ **Acaciella**
                    6. Shrubs or small trees; petiolar glands usually present; prickles present; inflorescences spikes or heads; corollas white to creamy white or yellow; stamens 35–160, white, small anther glands sometimes present ______ **Senegalia**
    2. Stamens 10 or less.
        7. Branches armed with nodal spines; fruits (5–)7–20 cm long, about as thick as broad or 1.5–5.5 cm long and coiled, without prickles; flowers yellowish white or yellow ______ **Prosopis**
        7. Branches unarmed or armed with internodal prickles which are generally curved and flattened; fruits 1.5–5 cm long, flattened (sometimes contorted), with or without prickles; flowers white to pink or purple or bright yellow.
            8. Petiole nectar glands absent; legumes often segmented (craspedial), sometimes unsegmented; plants armed with recurved or straight prickles; corollas white or pink to purple ______ **Mimosa**
            8. Petiole nectar gland present (sometimes minute); legumes not segmented; plants without prickles; corollas yellow ______ **Leucaena**
1. Herbaceous perennials; stamens 10 or less.
    9. Plants armed with sharp recurved prickles capable of causing pain ______ **Mimosa**
    9. Plants unarmed or with numerous, weak, hair-like or bristle-like structures not painful to the touch.
        10. Flowers pink or purplish; fruits 15–20 mm long, with conspicuous appressed hairs ______ **Mimosa**
        10. Flowers whitish or yellow; fruits of variable lengths, sometimes much longer than 20 mm, usually glabrous or essentially so.
            11. Flowers whitish; fruits sessile; leaves with a petiolar gland between the lowest pair of pinnae (the gland sometimes minute) ______ **Desmanthus**
            11. Flowers yellow; fruits stipitate (= stalked); leaves without a petiolar gland ______ **Neptunia**

SUBFAMILY 3. FABOIDEAE

Leaves once compound with 3–many leaflets (simple or apparently so in species of *Baptisia* and *Crotalaria*), odd- or even-pinnate, or palmate; stipules various; sepals united (barely so in *Indigofera*); petals 5 (except in *Amorpha* and sometimes *Eysenhardtia*) and very unequal: uppermost petal (= standard or banner) usually largest, two lateral (= wings) smaller and separate, two lowest (= keel) smaller and united except at base; stamens 5–10; filaments separate or united.

KEY TO TRIBES AND GENERA OF FABOIDEAE

1. Stamens distinct.
    2. Leaves simple or palmately compound; herbs ______ **Baptisia**
    2. Leaves pinnately compound; trees or shrubs.

3. Leaflet blades not leathery; inflorescences racemes or panicles; calyces truncate; corollas usually white or yellow, rarely pink or purple; stamens 8; legumes fleshy, straight to curved, moniliform; seeds black _______ **Styphnolobium**
3. Leaflet blades leathery; inflorescences racemes; calyces with obvious lobes; corollas usually in shades of purple, rarely white; stamens 10; legumes papery, leathery, or woody, torose to torulose, straight to slightly curved or subglobose to cylindric; seeds usually red or dull red to reddish brown, rarely orange or yellow _______ **Dermatophyllum**

1. Stamens united, at least basally.
4. Leaves simple; petals yellow _______ **Crotalaria**
4. Leaves compound; petals variously colored including yellow.
5. Leaves palmately compound with 4–7 leaflets; stamens monadelphous; petals deep blue (rarely white) _______ **Lupinus**
5. Leaves pinnately compound with 3–many leaflets OR if palmately compound, then stamens diadelphous; stamens monadelphous or diadelphous; petals variously colored including blue.
6. Leaflets dentate, the teeth usually easily visible but very small and often visible only near the leaflet tips in *Trifolium arvense* and *T. pratense*; stipules adnate to petioles (Trifolieae).
7. Fruits coiled or curved, sometimes prickly; stems usually 4-angled _______ **Medicago**
7. Fruits straight, never prickly; stems terete (= round).
8. Inflorescences elongated slender racemes, loose enough that the axes are easily visible to the naked eye; petals deciduous, the mature fruit visible _______ **Melilotus**
8. Inflorescences globose to cylindrical, often head-like spikes or racemes, so dense that the axes are not easily visible; petals withering and persistent, concealing the mature fruits from view _______ **Trifolium**
6. Leaflets entire to dentate; stipules distinct or if united to petioles, then leaflets entire.
9. Tendrils present in place of terminal leaflets; leaflets entire to dentate (Vicieae).
10. Corollas yellow; leaflets absent, leaves consisting only of tendrils (however, leaf-like stipules present) _______ **Lathyrus**
10. Corollas white to blue, lavender, pink-purple, purple, or bicolored but not yellow; leaflets 2–numerous.
11. Leaflets 2; stems winged _______ **Lathyrus**
11. Leaflets 2–18; stems not winged (but can be angled or edged).
12. Stipules foliaceous, usually larger than leaflets; leaflets 2–6(–8) _______ **Pisum**
12. Stipules foliaceous or inconspicuous, not larger than leaflets; leaflets 4–18.
13. Leaflets usually 30–60 mm long, 10–30 mm wide; style flattened; rare introduced species _______ **Lathyrus**
13. Leaflets usually smaller, 5–35 mm long, 1.5–15 mm wide; style ± round in cross-section; widespread species common throughout East TX _______ **Vicia**
9. Tendrils absent; leaflets entire.
14. Leaves pinnately compound (3–numerous leaflets) or palmately compound with (3–)5–7 leaflets; leaflets usually glandular-punctate; fruits 1-seeded (Amorpheae and Psoraleeae).
15. Shrubs.
16. Petals white to pale yellow, 4–5; leaflets 3–13 mm long; in East TX only in extreme w part _______ **Eysenhardtia**
16. Petals bluish to purplish, 1 or 5; leaflets 1.5–80 mm long; widespread in East TX.
17. Petal 1 (standard only); leaflets (5–)15–80 mm long; small to large shrubs 0.3–3.5 m tall _______ **Amorpha**
17. Petals 5; leaflets 1.5–3.5(–5) mm long; small shrubs 0.3–1.2 m tall (only *D. frutescens*) _______ **Dalea**
15. Herbaceous plants.
18. Wing and keel petals attached laterally or apically on staminal tube; leaves pinnately compound with 3–numerous leaflets _______ **Dalea**
18. All petals basally attached; leaves palmately compound (3–7 leaflets) or pinnately compound with 3 leaflets.
19. Fruits conspicuously cross-wrinkled; leaves pinnately compound with 3 leaflets _______ **Orbexilum**
19. Fruits smooth, not cross-wrinkled; leaves pinnately compound with 3 leaflets (1 species) or palmately compound with 3–7 leaflets (sometimes reduced to 1 or 2 near top of stem) _______ **Pediomelum**
14. Leaves pinnately compound (3–numerous leaflets, often 3) OR palmately compound with 2 or 4 leaflets; leaflets not glandular-punctate; fruits 1- to several-seeded.
20. Fruits segmented, the segments separating at maturity or fruits 1-seeded (or in 1 species unsegmented and borne underground) (Aeschynomeneae, Coronilleae, and Desmodieae).
21. Leaves palmately or pinnately compound with 2 or 4 leaflets; flowers yellow.
22. Leaves palmately compound (all leaflets arising from a single point); leaflets 2–3 cm long; fruits borne above ground, breaking part into 2–8 segments; each flower enclosed by conspicuous bracts; native species _______ **Zornia**
22. Leaves pinnately compound (2 leaflets separated from the other 2 by a rachis); leaflets 2–6 cm long; fruits developing underground, not breaking apart into segments; flowers not enclosed in bracts; cultivated species rarely escaping _______ **Arachis**
21. Leaves odd-pinnately compound with 3–many leaflets; flowers variously colored including yellow.

23. Leaves pinnately compound with 3–70 leaflets, if 3 leaflets, to 12 mm long; fruit (loment) segments 4-angled to terete or quadrate to rounded.
24. Leaves with 7–25 leaflets; fruit (loment) not or only slightly constricted between seeds, 4-angled in cross-section or cylindric, indehiscent or tardily dehiscent, with weak longitudinal, anastomosing veins; glabrous erect or sprawling herbs to 1.2 m ____ **Securigera**
24. Leaves with 3–70 leaflets; fruit (loment) shallowly or deeply crenate or incised, with 2–18 segments, quadrate or rounded, the segments separating at maturity; low densely glandular herb or stout herb 0.5–2 m tall ____ **Aeschynomene**
23. Leaves pinnately compound with 3 leaflets, longer than 12 mm; fruit segments usually flattened.
25. Fruits of 1–10 segments, usually with hooked hairs causing them to stick to hair or clothing; leaflets usually stipellate.
26. Stipels present, persistent; calyx lobes longer than tube; loments with 1–10 segments, stipitate or sessile ____ **Desmodium**
26. Stipels absent or early-deciduous; calyx lobes shorter than tube; loments with 2–5 segments, distinctly stipitate ____ **Hylodesmum**
25. Fruits of 1(–2) segment, without hooked hairs; leaflets estipellate.
27. Petals orange or orange-yellow (very rarely white); stipules adnate to petioles; stamens monadelphous; anthers of 2 kinds ____ **Stylosanthes**
27. Petals purplish or pinkish to white; stipules free from petioles; stamens diadelphous; anthers of 1 kind.
28. Stipules nearly thread-like; leaflets not striate; plants perennial ____ **Lespedeza**
28. Stipules ovate or nearly so, striate; leaflets striate with conspicuous parallel lateral veins; plants annual ____ **Kummerowia**
20. Fruits non-segmented, several-seeded.
29. Inflorescences umbellate; leaves pinnately compound with 5 leaflets (the lower 2 stipular in position); petals yellow, marked with red ____ **Lotus**
29. Inflorescences racemose OR of solitary flowers; leaves pinnately compound with 3–numerous leaflets; petals variously colored.
30. Plants mostly herbaceous twining vines (except in *Acmispon*, *Glycine*, *Erythrina*, and 2 species in *Rhynchosia*); leaves mostly with 3 leaflets (except 5–7 leaflets in *Apios*, 1 species in *Clitoria*, and 1 species in *Galactia* and 1 leaflet for 2 species of *Rhynchosia*) (Phaseoleae).
31. Leaves with 5–7 leaflets.
32. Banner 40–55 mm long, much larger than other petals, rare introduced species ____ **Clitoria**
32. Banner 4–26 mm long, common native species.
33. Lower 2–3 pairs of leaflets attached at different places; leaflets 2–5(–10) cm long; corollas at least somewhat brownish red ____ **Apios**
33. Lower 2 pairs of leaflets attached at the same place (leaves almost palmate but with the terminal leaflet short-stalked); leaflets 0.6–2.4 cm long; corollas mostly lavender (standard often with some white) ____ **Galactia**
31. Leaves with 3 leaflets (1 leaflet for 2 species of *Rhynchosia*).
34. Plants erect herbs.
35. Corollas red, very long (30–53 mm long); flowers in a terminal raceme ____ **Erythrina**
35. Corollas white to pink, violet, or purple (rarely yellowish with pink tinge), 8 mm or less long; flowers axillary, either solitary or in racemes.
36. Leaflets 1–2 cm long; fruits not densely bristly, native species ____ **Acmispon**
36. Leaflets 3–15 cm long; fruits densely bristly, frequently cultivated but rarely escaped species ____ **Glycine**
34. Plants trailing or twining vines.
37. Standard (= banner) greatly exceeding keel and wings (ca. 2 times as long), 20–60 mm long.
38. Calyx tube short, ca. 4 mm long; calyx lobes linear, longer than the tube; fruits ca. 7–12(–14) cm long, sessile or nearly so ____ **Centrosema**
38. Calyx tube cylindrical, 10–20 mm long; calyx lobes ovate, much shorter than the tube; fruits 3–6(–8) cm long, stipitate, the stipe 10–20 mm long ____ **Clitoria**
37. Standard ± equaling or shorter than keel and wings, 4–25 mm long.
39. Corollas deep red, purple, or purple-black; one wing petal directed upward to adopt function of banner ____ **Macroptilium**
39. Corollas pink or yellow and wing petal not as described.
40. Keel of corollas strongly incurved.
41. Corollas pink or purple ____ **Strophostyles**
41. Corollas yellow ____ **Vigna**
40. Keel of corollas essentially straight.

42. Corollas yellow; fruits 1–2.5 cm long ____ **Rhynchosia**
42. Corollas white, pink, or purplish; fruits 1.5–30 cm long (except subterranean fruits).
    43. Corollas 25–30 mm long; fruits 10–30 cm long; rare cultivated species ____ **Vigna**
    43. Corollas 8–25 mm long; fruits 1.5–5.5 cm long; more widespread.
        44. Plants high climbing vines with stems often many meters long; corollas (10–)15–25 mm long; fruits densely and conspicuously long hairy ____ **Pueraria**
        44. Plants relatively low climbing vines, the stems to ca. 2 m long; corollas 8–15 mm long; fruits ± glabrous to pubescent, but not densely and conspicuously long hairy.
            45. Bracts subtending pedicels ± as wide as long (2–5 mm long), easily visible to the naked eye; calyx teeth nearly equal, ca. 1/2 as long as tube or less; leaflets gradually narrowed to ± acute apex, the larger leaflets 3–10.5 cm long, 1.6–8 cm wide; stipules 3.5–8.5 mm long, persistent; fruits 6–10 mm wide; corollas lilac or whitish ____ **Amphicarpaea**
            45. Pedicels without conspicuous bracts ± as wide as long; calyx teeth unequal, ca. as long as tube; leaflets usually abruptly narrowed to a rounded or obtuse apex, the larger leaflets 2–4(–6) cm long, 10–25(–35) mm wide; stipules 1–3 mm long, soon deciduous; fruits 3–6 mm wide; corollas pink or rose ____ **Galactia**

30. Plants erect, not viny (except in woody *Wisteria*); leaves pinnately compound with (3–)5–96 leaflets (Galegeae, Indigofereae, Loteae in part, Robinieae, and Tephrosieae).
    46. Trees, shrubs, or woody vines; flowers white or rose, reddish purple, or violet-blue; plants sometimes with stipular spines or conspicuously hispid young stems.
        47. Plants trees or shrubs (not vine-like); racemes axillary; young stems either hispid or with stipular spines ____ **Robinia**
        47. Plants woody vines or vine-like shrubs; racemes terminal; young stems neither hispid nor with stipular spines ____ **Wisteria**
    46. Herbaceous plants; flowers variously colored; plants without either stipular spines or hispid young stems.
        48. Leaves even-pinnately compound ____ **Sesbania**
        48. Leaves odd-pinnately compound.
            49. Leaflets mostly alternate along leaf rachis; petals reddish orange with tints of pink or salmon; pubescence of hairs attached at middle ____ **Indigofera**
            49. Leaflets opposite along leaf rachis; petals variously colored, but not reddish orange; pubescence of hairs attached basally or near middle.
                50. Main racemes terminal or if apparently lateral, then opposite a leaf; standard orbicular; fruits flattened, with neither suture intruding; style with fine stiff hairs ____ **Tephrosia**
                50. Main racemes axillary; standard obovate; fruits subcylindrical or somewhat triangular in cross-section to inflated, with 1 suture intruding; style glabrous.
                    51. Leaves all basal or crowded near base, the leafless peduncles more than twice as long as the leaf-bearing lower portion of the stem; flowers 15–26 mm long; keel tipped with a pointed slender beak ____ **Oxytropis**
                    51. Leaves all basal (1 species with flowers 8.5–14 mm long) or leaves usually extending well up the stem; keel acute or rounded, not beaked ____ **Astragalus**

ACACIELLA Britton & Rose ACACIA

A genus of 15 species including subshrubs (ours), shrubs, small trees, and perennial herbs, formerly included within the large genus *Acacia*, of s U.S. south to Argentina (Ebinger & Seigler 2023a) (Greek: *akis*, a sharp point for the genus *Acacia;* and Latin: *ella* meaning small alluding to small acacias) (subfamily Caesalpinioideae, Mimosoid clade, tribe Acacieae)
REFERENCES: Isely 1969, 1998; Rico-Arce & Bachman 2006; Ebinger & Seigler 2023a.

Acaciella angustissima (Mill.) Britton & Rose var. **angustissima,** (most narrow), FERN ACACIA, PRAIRIE ACACIA, WHITE-BALL ACACIA. Subshrub (in East TX) with creeping, woody root; stems sprawling to erect, to ca. 1 m tall, unarmed; leaves twice even-pinnately compound with 9–17 pairs of pinnae; leaflets ca. (20–)23–30 pair per pinna, slightly sensitive, very slowly folding together on being touched or in rain and displaying nighttime or "sleep" (nyctinastic) movements with leaflets folding together at night; inflorescences of many-flowered globose heads or spikes, the heads solitary or few in axils of upper leaves, also terminal, forming erect, ± leafy racemes of heads; flowers very small, whitish, petals inconspicuous, separate nearly to base; stamens numerous, 20–40 per flower, serving as the main attractant structure for the flower, filaments creamy white; fruits 3–7 cm long, ca. 6–10 per inflorescence, oblong, flat. Common in prairies on calcareous clay, occasionally on sandy soils in open areas; widespread in e 2/3 of TX; s U.S. from w LA, AR, and s MO w to AZ, KS, NM, OK, TX; disjunct in FL; Mexico s to Argentina. Late May–early Jul. [*Acacia angustissima* (Mill.) Kuntze var. *hirta* (Nutt.) B.L. Rob., *A. hirta* Nutt., *A. texensis* Torr. & A. Gray, *Acaciella angustissima* (Mill.) Britton & Rose, *A. hirta* (Nutt.) Britton & Rose] Rico-Arce and Bachman (2006), in a range-wide assessment of the genus, take a broad approach to the *Acaciella angustissima* complex, which they describe as "a very variable taxon" and include our material in *A. angustissima* var. *angustissima,* defined to include material from the s U.S., Mexico, Central America, and South America. Segregate species, subspecies, varieties, and forms have been recognized based on habit, leaflet shape, stem and leaflet indument, stem striations, number of pinnae, and leaflets per pinna (Ebinger & Seigler 2023a).

ACMISPON Raf. AMERICAN BIRD'S-FOOT-TREFOIL, PRAIRIE-TREFOIL

A genus of 20–30 species of temperate North America and South America. New World taxa often referred to *Lotus* are not closely related (Allan & Porter 2000; Degtjareva et al. 2006). (Greek: *acme*, point, apex, probably for the hooked-tipped fruit) (subfamily Faboideae, tribe Loteae)

References: Isely 1981, 1998; Grant & Small 1996; Allan & Porter 2000; Sokoloff 2000; Degtjareva et al. 2006; Brouillet 2023a.

Acmispon americanus (Nutt.) Rydb., (of America), WESTERN PRAIRIE-TREFOIL. Moderately short-pilose annual 5–60(–120) cm tall; leaves ± subsessile, the 3 leaflets 1–2 cm long, 2–3.5 times as long as wide; stipules apparently absent (or reduced to glands); flowers solitary in the leaf axils or in peduncled umbels; calyces nearly equaling the corollas; corollas 5–7(–9) mm long, initially often whitish or yellowish-white but veined or turning rosy or lavender-pink by late afternoon, lasting only one day; fruits dehiscent, 1.5–3 cm long. Sandy open ground; widespread in mostly c portion of TX; 3 disjunct distributions in the U.S. (see below); s Canada and Mexico. May–early Jul(–Oct). [*Lotus purshianus* Clem. & E.G. Clem., *L. unifoliolatus* (Hook.) Benth., *L. americanus* Vell.] The correct scientific name for this species has been somewhat confused. It has long gone under the names *Lotus unifoliolatus* or *L. purshianus* (Kartesz 1994; Jones et al. 1997; Diggs et al. 1999). Another named variety (*Acmispon americanus* var. *helleri* (Britton (Brouillet)) was the one considered to be in TX by Turner et al. (2003); however, that variety only occurs much to the e in GA, NC, SC, TN, and VA (Brouillet 2023a). It can be distinguished by the narrower leaflets (4–5 times as long as wide) and by the plants being glabrous or nearly so. This species has a unique geographic distribution occurring in three distinct groups: (1) Carolinas and adjacent states (var. *helleri*), (2) Central plains and prairies, e to Mississippi River (including TX), and (3) Pacific states, to AZ and w Idaho (var. *americanus*) (Isely 1998; Kartesz 2023).

AESCHYNOMENE L. SENSITIVE JOINT-VETCH, SHY LEAVES

Annual or perennial, unarmed herbs; pubescence usually including pustulate-based hairs (= swollen-based); stems erect or prostrate; leaves alternate, once even- or odd-pinnately compound, often sensitive to light and touch, stipulate; leaflets entire, finely punctate; inflorescences few-flowered, pedunculate, axillary, often sticky pubescent; calyx 2-lipped or with 5 subequal lobes, < tube; corollas whitish to yellow, infused or veined with reddish purple, pink, or orange; androecium monadelphous or diadelphous with two phalanges of 5 stamens each; style glabrous; fruits (loments) stipitate, shallowly or deeply crenate or incised, with 2–18 segments, quadrate or rounded, the segments separating at maturity; seeds 1 per segment.

A genus of ca. 150 species of mostly shrubs of tropical and warm areas of the world (Rudd & Raveill 2023a). Because of the lightness of the pith in several species, they are used in making hats, floats, and rafts (Rudd 1955). The common names indicate the sensitivity some species have to touch and light (from the ancient Greek name for a sensitive plant used by Pliny). (subfamily Faboideae, tribe Dalbergieae)

REFERENCES: Rudd 1955; Isely 1998; Cardoso et al. 2019; Cardoso et al. 2020; Rudd & Raveill 2023a.

1. Leaflets 3–7(–11) per leaf, obovate; petioles 1–1.5 cm long; corollas 4–6 mm long; perennial with stems trailing or sprawling, sometimes forming mats ____ **A. viscidula**
1. Leaflets 30–70 per leaf, linear-oblong or oblong; petioles 4–8 cm long; corollas 7–15 mm long; annual with stems usually erect.
    2. Leaflets (30–)50–70 per leaf, 2–10 mm long, 1–2.5 mm wide; corollas 7–10 mm long; calyces 4–6 mm long ____ **A. indica**
    2. Leaflets 30–40(–50) per leaf; 6–15 mm long, 2–3 mm wide; corollas (8–)10–15 mm long; calyces 5–8 mm long ____ **A. rudis**

Aeschynomene indica L., (of India—though this is apparently incorrect), INDIAN JOINT-VETCH, BUDDHA-PEA. Annual, usually erect, branched, glabrate to very sparsely pustulate; stems 0.5–2(–2.5) m tall; adventitious roots forming on stems in water; leaves once even-pinnately compound, with (30–)50–70 leaflets, reported to sometimes be sensitive; leaflets 2–10 mm long, 1–2.5 mm wide, linear-oblong or oblong, entire; inflorescences peduncled, few-flowered, axillary racemes; corollas pale yellow, partly

Tragia urens [FN2]

Tragia urticifolia [BT3]

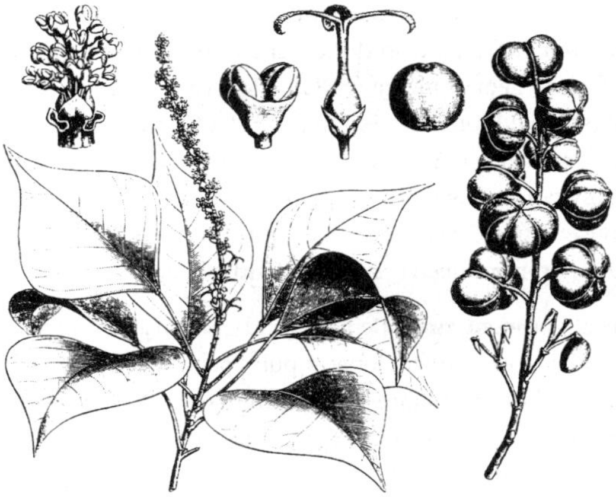
Triadica sebifera [EN1]

Vernicia fordii KGO]

Acaciella angustissima [VIN]

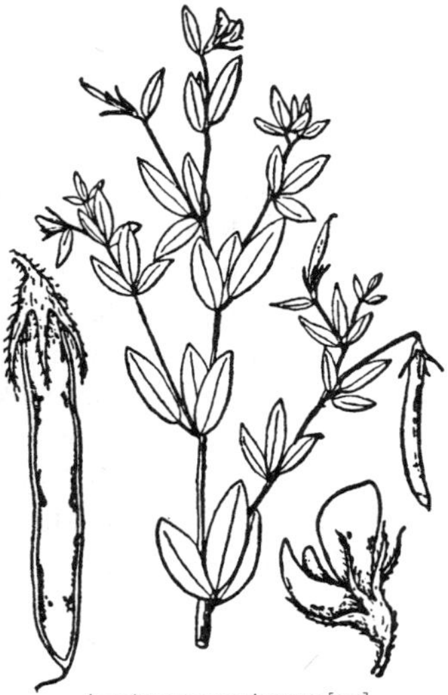
Acmispon americanus [BB2]

Aeschynomene indica [CO1]

Aeschynomene rudis [ARG]

infused with or the veins reddish purple; fruits (2.4–)3–5 cm long, with 5–8(–12) segments, the lower margin shallowly crenate, sometimes reddish-dotted, the result of where pustulate hairs had been. Moist to wet areas, weed in rice; Hardin, Harris, Jefferson, Leon, Orange (BRIT), Houston, Madison, and Trinity (Turner et al. 2003) cos.; otherwise mainly Gulf Prairies and Marshes with one location in San Saba County in the e Edwards Plateau; se U.S. from VA s to FL w to MI, OK, and TX; Puerto Rico; s Brazil; se Asia; Africa; Pacific Islands; and Australia. Jul–Nov. Toxic rotenoids contained in the seeds are reported to be fatally poisonous to mice and pigs (Latorre et al. 2011). It is considered a state noxious weed in HI (USDA Natural Resources Conservation Service 2008). Despite the scientific name, there is debate over whether this species is native to the U.S. (e.g., Isely 1998) or introduced (Carulli & Fairbrothers 1988), but the USDA Plants Database (2023) considers it native.

Aeschynomene rudis Benth., (wild), ROUGH JOINT-VETCH. Annual with stems erect, to 2 m tall, glabrous or moderately glandular-hispid; leaves 4–10 cm long; stipules peltate, 7–15 mm long, apex acute; leaflets 30–40(–50), blades nearly symmetric, oblong to linear-oblong, 6–15 mm long, 2–3 mm wide, margins usually entire, sometimes ciliate-denticulate, apex obtuse, surfaces glabrous; inflorescences 1–7-flowered, racemes and axis glabrous or glandular-hispidulous; flowers yellow usually with reddish center, often drying somewhat purplish, (8–)10–15 mm long, banner glabrous, glandular-ciliate; calyces 5–8 mm long, bilabiate; fruits 3–7.2 cm long with (3–)6–12 segments; abaxial margin crenate, adaxial margin ± straight, sutures between segments well-developed, margins not separating from valves; segments subquadrate, 4–6 mm long, 4–6 mm wide, reticulate venation incon spicuous to moderately conspicuous, center of each usually muricate or verrucose, faces subglabrous or lightly glandular-hispidulous, hairs bulbous based. Wet areas in standing water in rice fields or ditches adjacent to cultivated fields; Harris Co. (*Correll 31894* KANU, LL, NY); scattered in the U.S. in AL, AR, CA, FL, GA, LA, MO, NC, PA, SC, and TX; Mexico s to South America; Cuba. Aug–Sep. Native to South America. This species was initially reported in TX by Carulli et al. (1988) based on the Correll specimen and is mapped in TX by Rudd and Raveill (2023a). It is probably more common than currently reported based on its similarity to *A. indica* as the Correll specimen at KANU and LL are both annotated as that species.

Aeschynomene viscidula Michx., (minutely viscid, sticky), STICKY JOINT-VETCH. Prostrate perennial with stems to 1 m long, pustulate as well as with glandular hairs present on some structures, the plants with a viscid or sticky feel; leaves once odd-pinnately compound, with 3–7(–11) leaflets; leaflets 4–12 mm long, obovate, entire; inflorescences peduncled, axillary, 1–4-flowered; corollas off-white to yellow, sometimes with darker reddish purple or pink veins; fruits 0.8–1.5 cm long, with 2–3(–5) segments, the lower margin deeply crenate or incised. Sandy soils, grasslands, roadsides, and open wooded areas; Bexar, Harris (Brown et al. 2011), Guadalupe, and Wilson (BRIT) cos. in s portion of East TX; mainly s Gulf Prairies and Marshes; se U.S. in AL, FL, GA, MS, and TX. May–Nov. [*Ctenodon viscidulus* (Michx.) D.B.O.S. Cardoso & A. Delgado, *Secula viscidula* (Michx.) Small] Cardoso et al. (2020) segregated this species into the genus *Ctendodon* based on genetic and morphological data and indicated that the genus may be expanded to include taxa in the Old-World tropics. Until the taxonomy is settled, we follow Rudd and Raveill (2023a) who retained the species within *Aeschynomene*.

ALBIZIA Durazz. SILKTREE

A genus of ca. 140 species of trees, shrubs, or lianas of warm areas of Asia, Africa, and South America. Species are variously used for timber, gums, shade for coffee or tea, or as ornamentals (Rico-Arce 2023). (Named for Filippo delgi Albizzi, 18th-century Italian naturalist, who introduced the genus into European culture) (subfamily Caesalpinioideae, Mimosoid clade, tribe Ingeae)
REFERENCES: Isely 1970b, 1973; Elias 1974; Rico-Arce 2023.

Albizia julibrissin Durazz., (modification of Persian name), MIMOSA, SILKTREE. Broad-headed, fast growing, short-lived, small, unarmed tree with flat top, to ca. 8 m tall; leaves twice even-pinnately compound, with numerous small leaflets, the ultimate leaflets folding together at night (a nyctinastic or "sleep" movement) or in rain, or even slightly sensitive on being touched; petiole with a single, flat, elliptic gland near base; flowers in short-peduncled, hemispherical heads; heads clustered, conspicuous and very showy due to mass of colorful stamens; corollas narrowly funnelform, greenish to yellowish, inconspicuous, with united petals; stamens usually 20–40; filaments ca. 23–36 (rarely more) mm long, united and white in basal 1/4, separate and pink to rose-lavender or reddish above, serving as the main attractant structure for the flower; fruits flat, ca. 8–20 cm long. Commonly cultivated as a fast growing, showy yard tree, occasionally escaping; it can, however, aggressively invade native habitats; throughout East TX; mainly e 1/2 of TX but scattered locations westward; mainly e U.S. from s NY to FL and w to MO, OK, and TX, scattered locations westward including CA, NM, and UT; Mexico; Jamaica; South America; s Europe; Africa; and New Zealand. May–Jun. Native of s Asia.

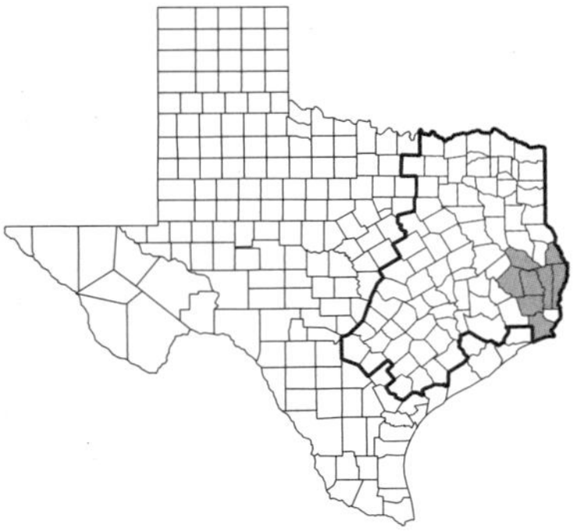

Tragia smallii

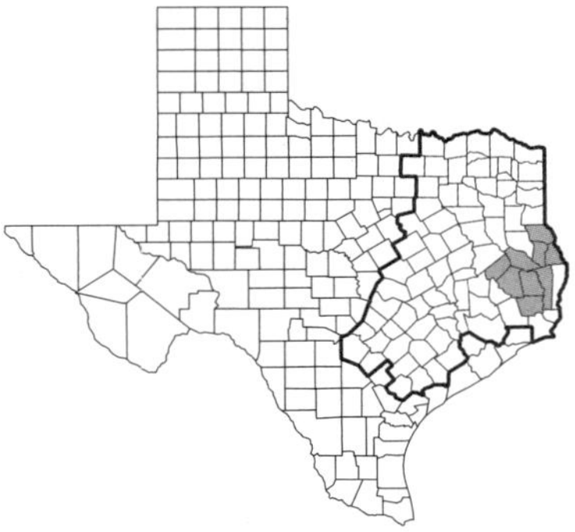

Tragia urens

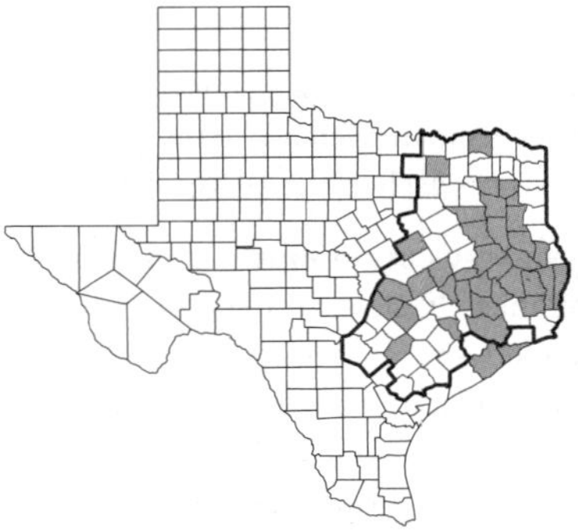

Tragia urticifolia

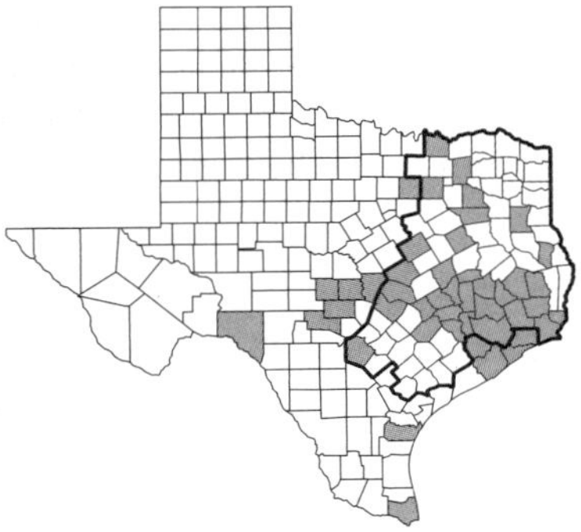

Triadica sebifera

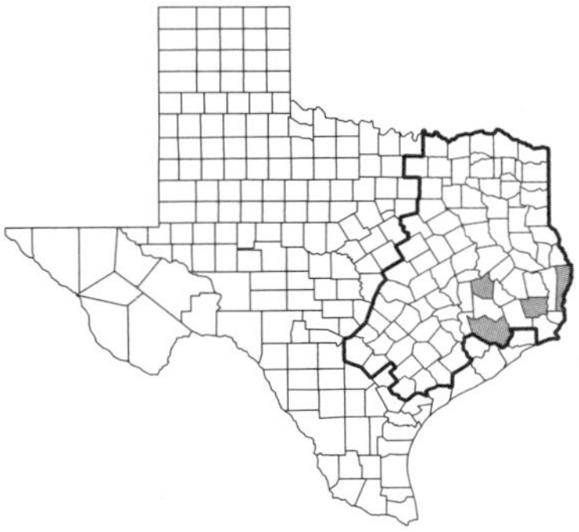

Vernicia fordii

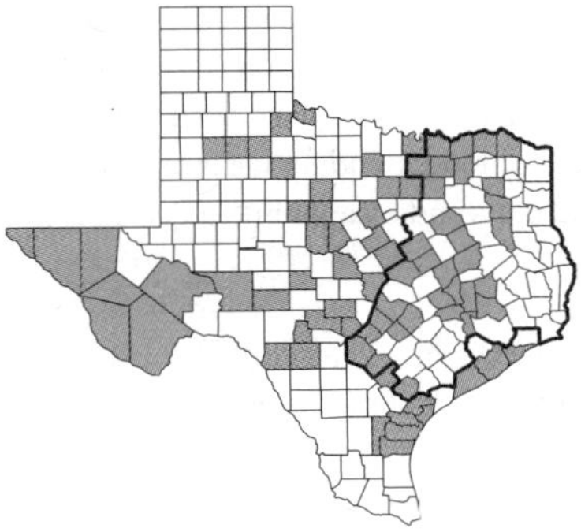

Acaciella angustissima

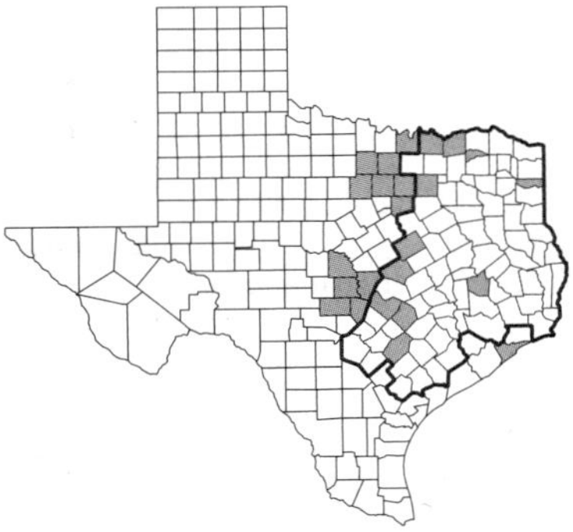

Acmispon americanus

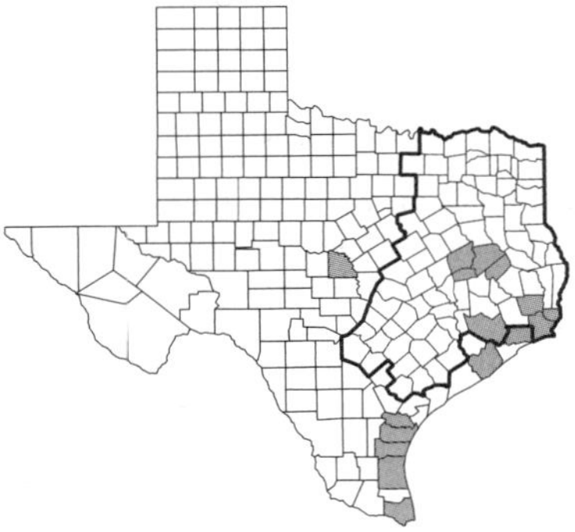

Aeschynomene indica

Aeschynomene rudis

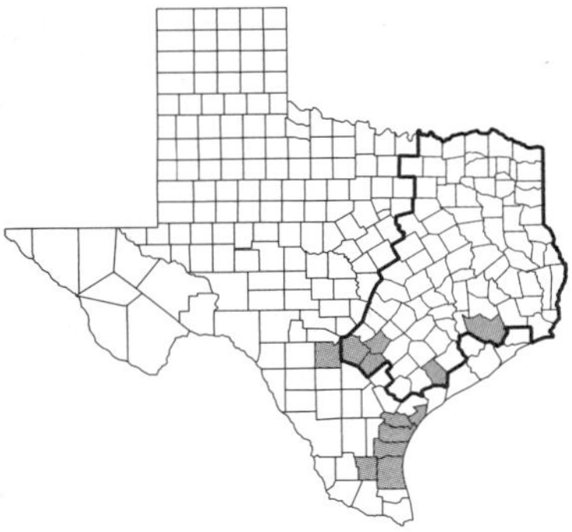

Aeschynomene viscidula

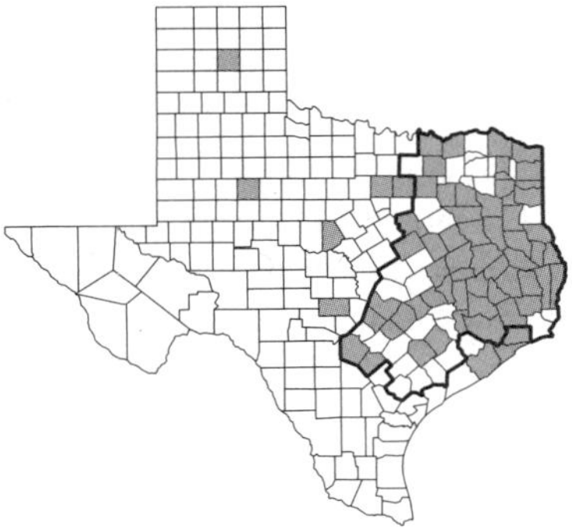

Albizia julibrissin

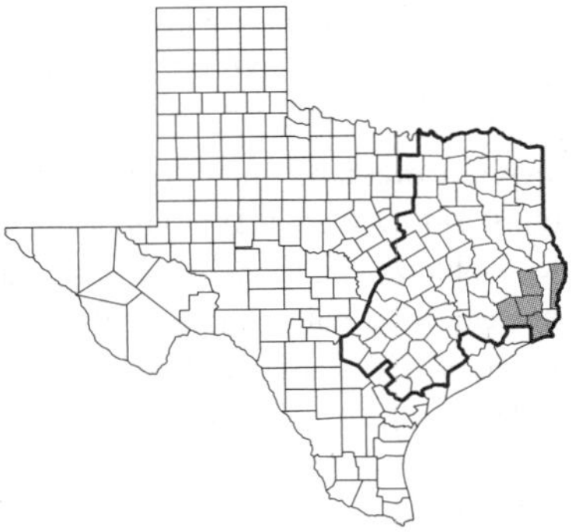

Alysicarpus ovalifolius

ALYSICARPUS Neck. ex Desv. ALYCE-CLOVER

An Old-World tropical genus of ca. 30 species (Ohashi 2023a); some are used for fodder or forage. (Greek: *halysis*, a chain, and *karpos*, fruit, in reference to the fruits—Quattrocchi 2000) (subfamily Faboideae, tribe Desmodieae)
REFERENCES: Isely 1990, 1998; Ohashi 2023a.

Alysicarpus ovalifolius (Schumach.) J. Léonard, (oval-leaved), ALYCE-CLOVER, ONE-LEAF-CLOVER, WHITE MONEYWORT. Annual or perennial (frost sensitive) herb, morphologically variable; stems single or numerous, sprawling to ascending, to 1 m long, sometimes rooting; leaves appearing simple, but actually reduced to only a single leaflet; leaflets (0.5–)1–5(–6.5) cm long, suborbicular to ovate or oblong-lanceolate, narrowing towards stem apex, entire, sometimes with an irregular lighter area near midrib; stipules amplexicaule, scarious, acuminate, conspicuous, 5–15 mm long; inflorescences raceme-like or branched of 6–12 flowers; calyx lobes 4, acuminate; corollas 5–6 mm long, variable in color, reddish pink to maroon or purplish, sometimes with a hint of yellowish; stamens 10, diadelphous; fruits 1–2 cm long, oblong or linear, ± smooth margined, but separating at maturity into 3–8 brick-shaped, reticulate segments. Roadsides, disturbed areas, and utility rights-of-way; known in TX only from Hardin (*McLeod s.n.* TEX), Jefferson (Turner et al. 2003), Liberty (*Sperry 3700* BRIT; *Brown & Marcus 19731* SBSC; Brown & Marcus 1998), Newton (iNaturalist Contributors 2023a), and Tyler (M. Reed, pers. comm.) cos.; se U.S. from VA to FL w to AR and TX; Mexico and Australia. Sep–Nov. [*Alysicarpus vaginalis* (L.) DC.] Native of the Old-World Tropics. This species is used for fodder and forage in the Gulf Coast states (Ohashi 2023a). *Alysicarpus ovalifolius* is occasionally still planted in TX in food plots and escapes into adjacent areas (E. Keith, pers. obs.).

AMORPHA L. FALSE INDIGO

Unarmed erect shrubs or subshrubs; leaves once odd-pinnately compound; leaflets many (ca. 9–31), oblong or elliptic, often gland-dotted; stipules slender, falling when the leaves expand; flowers many, in terminal, solitary or panicled, slender, dense, erect, spike-like racemes (young tips often nodding); calyces 5-toothed, often gland-dotted; corollas usually of one petal (standard), sometimes with very reduced additional ones; androecium monadelphous, stamens 10; style slender, pubescent, stigma terminal, capitate; fruits indehiscent and gland-dotted.

A North American genus of 16 species (Straub et al. 2023), characterized by corollas reduced to only a single petal (standard). Two species treated here are of conservation concern. (Greek: *amorphos*, shapeless deformed, from the absence of four of the petals) (subfamily Faboideae, tribe Amorpheae)
REFERENCES: Wilbur 1975; Isely 1998; Weakley, A.S., and Southeastern Flora Team 2022; Straub et al. 2023.

1. Leaves sessile or subsessile, the petioles up to 5(–8) mm long; low shrubs or subshrubs (0.3–)0.5–0.8(–1.2) m tall, usually (but not always) canescent (= whitish or grayish-white due to hairs); all calyx lobes acute ____ **A. canescens**
1. Leaves with petioles 10–70 mm long; shrubs, usually 1–3 m tall, not canescent; adaxial 2 calyx lobes often rounded.
  2. Leaflets 15–30(–50) mm wide.
    3. Leaflets (9–)11–19 per leaf, with conspicuous reticulate venation beneath; inflorescences usually 5–10 in a compound cluster; petioles 2–7 cm long; calyx lobes all acute or acuminate; species of e portion of East TX ____ **A. paniculata**
    3. Leaflets (7–)9–11(–15) per leaf, without conspicuous venation beneath; inflorescences solitary or few in a cluster; petioles 1–3 cm long; 2 of the calyx lobes blunt or rounded; species of sw margin of East TX ____ **A. roemeriana**
  2. Leaflets 4–15(–20) mm wide.
    4. Petiolules (= stalks of leaflets) with conspicuous raised glands; mature inflorescences 15–30 cm long; calyces with conspicuous glands; midnerve of leaflets terminating in glandular knob, 0.1–0.2 mm long ____ **A. laevigata**
    4. Petiolules not or only inconspicuously glandular; mature inflorescences 10–15(–20) cm long; calyx at most inconspicuously glandular; midnerve of leaflets terminating in exerted mucro, 0.5–1.5 mm long ____ **A. fruticosa**

Amorpha canescens Pursh, (gray-pubescent), LEADPLANT. Rhizomatous, usually conspicuously canescent, gray-green, low shrub (0.3–)0.5–0.8(–1.2) m tall; leaflets 11–41(–49), (5–)15–20 mm long, (2–)4–7(–12) mm wide, usually elliptic-oblong to ovate-oblong, the venation inconspicuous; racemes usually numerous near stem apices, often forming a dense compound cluster; calyces usually canescent, the lobes narrowly triangular-lanceolate; standard 4.5–6 mm long, bright violet; fruits 3–4.5 mm long. Sandy prairies and pine forests; scattered in East TX from the Pineywoods to the sw portion of East TX; also with several very scattered localities elsewhere in the state, particularly the e Panhandle; Correll and Johnston (1970) noted it has "a very peculiar disjunct distribution"; primarily c U.S. from s MI s to n AR, OK, disjunct s into TX, and w to ne WY and e NM; s Canada. May–Jul. [*A. brachycarpa* E.J. Palmer] Isely (1998) speculated that this species "was probably once cosmopolitan across the prairies and plains of the central United States. It is still abundant and conspicuous where some of the original vegetation yet persists. A shrublet, it is immediately discerned, even at a distance, against the usual prairie background by its gray aspect and

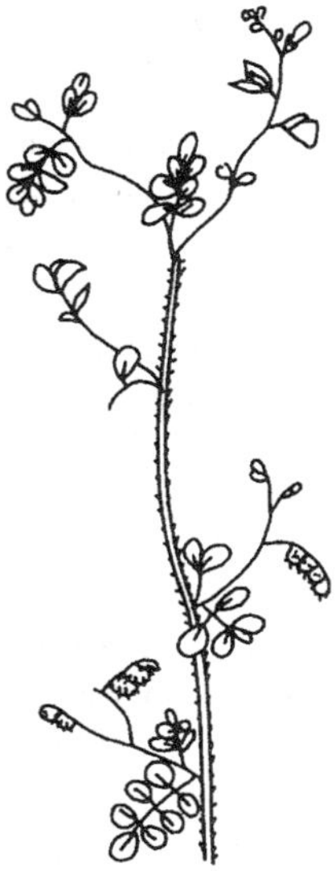
Aeschynomene viscidula [DN2]

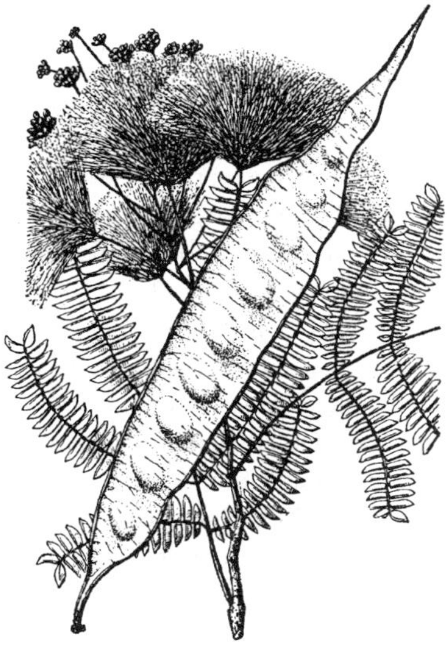
Albizia julibrissin [BR1]

Alysicarpus ovalifolius [FDC]

Amorpha canescens [GLE]

Amorpha fruticosa [CO1]

Amorpha laevigata [VIN]

Amorpha paniculata [VIN]

violet spikes." Despite its rarity in East TX, it is relatively common in longleaf pine forests in the southern Angelina National Forest in Angelina and Jasper counties (E. Keith, pers. obs.).

Amorpha fruticosa L., (shrubby, bushy), FALSE INDIGO, BASTARD INDIGO, INDIGO-BUSH AMORPHA, DESERT FALSE INDIGO. Much-branched shrub 1–3(–3.5) m tall, dark-green, sometimes forming thickets; leaflets 9–21(–31), 10–40 mm long, pubescent at least on veins on lower surface and sparingly gland-dotted, with an exerted mucro; racemes solitary or in clusters of 2–4; calyces ca. 3–5 mm long, 2 of the calyx lobes blunt often rounded; standard 5–6 mm long, dark blue to reddish purple or violet; anther orangish; fruits ca. 5–8 mm long, often curved, usually glabrous, but conspicuously gland-dotted. Low forests and woodlands and margins of river bottoms, swamps, ponds, and streams, occasionally in roadside ditches and moist urban waste areas, often on limestone in c TX westward; widespread in TX. Throughout the U.S. from ME s to FL, and w to ND s through OK and TX as a native and introduced w to OR and WA to s CA; Canada and Mexico (Baja California, Chihuahua); introduced also in Europe; Asia; and Africa. Apr–May. [*A. fruticosa* var. *angustifolia* Pursh, *A. fruticosa* var. *croceolanata* (P. Watson) Mouill., *A. fruticosa* var. *occidentalis* (Abrams) Kearney & Peebles] Reported as poisonous to livestock and to contain alkaloids (Burlage 1968). However, it has been found to possess anti-tumor, anti-inflammatory, and antimicrobial compounds, and has also been investigated for its insecticidal and insect repellant properties (Straub et al. 2023). This species has sometimes (e.g., Correll & Johnston 1970) been divided into several varieties. However, we are following most recent authors (e.g., Isely 1998, Straub et al. 2023) in not recognizing infraspecific taxa.

Amorpha laevigata Nutt. (smooth), SMOOTH FALSE INDIGO, SMOOTH INDIGO-BUSH. Shrub 1–2.5(–3) m tall, glabrate or with sparse hairs, but conspicuously glandular; leaflets (7–)13–21 per leaf, conspicuously glandular on lower surface, 15–30(–40) mm long, 7–15(–20) mm wide, elliptic-oblong or oblong; calyces glabrous, conspicuously gland-dotted, 2 of the calyx lobes rounded; standard blue to blue-violet or purplish; fruits ca. 5 mm long, conspicuously gland-dotted. Prairies, open woods, rocky ridges, and occasionally on roadsides; known in TX only from Angelina (iNaturalist Contributors 2023m), Cass, Harris (BRIT, TEX), Morris (Turner et al. 2003), and Van Zandt (Correll & Johnston 1970) cos. in East TX and Chambers Co. (BRIT) in the n Gulf Prairies and Marshes; sc U.S. from e OK s to TX. May–Jun. Correll and Johnston (1970) suggested that this species is "perhaps only a form of *A. texana* Buckley [= *A. roemeriana* Scheele]." Isely (1998) noted that *A. laevigata* is "usually quickly recognizable by its glabrate aspect, bluish-tinted flowers, and rather uniformly elliptic-oblong leaflets. It is apparently now very rare in TX having only been observed or collected twice since the 1960's in Angelina (iNaturalist Contributors 2023m) and Harris (*Rosen & Burnet 6587* BRIT, TEX) cos. NatureServe: G3, vulnerable; S1, critically imperiled.

Amorpha paniculata Torr. & A. Gray, (with flowers in panicles), PANICLED AMORPHA. Shrub 1–3 m tall; leaflets (9–)11–19, tomentose at least on lower surface, often densely so and with conspicuous raised veins on the lower surface, (20–)30–60(–80) mm long, oblong-elliptic to oblong; racemes clustered 5–10 together in a compound inflorescence exserted above the leaves; standard purple; fruits 6–8 mm long, conspicuously glandular. Low woods often at the ecotone of flood plains and uplands, but also wet areas such as sloughs and marshes, often locally abundant; in TX limited to, but widespread in e part of East TX w to Henderson (BRIT) and Robertson (Isely 1998) cos; sc U.S. in s AR s to n LA and TX, and w to se OK. May–Jun. Isely (1998) noted that this "distinctive species is easily recognized by its robust clusters of racemes and the villosulous, reticulate leaflets."

Amorpha roemeriana Scheele, (for Ferdinand Roemer, 1818–1891, geologist, paleontologist, and explorer of TX), TEXAS INDIGO-BUSH, TEXAS INDIGO-BUSH AMORPHA. Shrub 1–3(–3.5) m tall, similar to *A. paniculata*; leaflets 15–40(–50) mm long, broadly to narrowly ovate to elliptic-oblong, with pubescence at least along the inconspicuous veins on lower surface; petiolules conspicuously glandular; racemes solitary–few in a cluster; calyces pubescent or sometimes glabrous, gland-dotted; standards purplish to bluish; fruits 6–7 mm long, conspicuously glandular. Stream banks; Comal, Travis (BRIT), and Hays (Turner et al. 2003) cos. on sw margin of East TX; mainly e Edwards Plateau; endemic to TX. Apr–Jun. [*A. texana* Buckley, *A. texana* var. *glabrescens* E.J. Palmer] It is similar to *A. paniculata* but differs in having fewer leaflets with inconspicuous veins beneath, usually 15–40 mm long, and a smaller inflorescence 10–20 cm long (Isely 1998). NatureServe: G3, vulnerable.

AMPHICARPAEA Elliott ex Nutt. HOG-PEANUT

A genus of 3 species of e Asia, North America, and Africa (Delgado-Salinas & Parker 2023); all with cleistogamous flowers giving rise to subterranean fruits. The generic name was spelled *Amphicarpa* in past literature and orthographically corrected in Turner and Fearing (1964). (Greek: *amphi*, of both kinds, and *carpos*, fruit, in allusion to the two different fruit types) (subfamily Faboideae, tribe Phaseoleae)

REFERENCES: Turner & Fearing 1964; Isely 1998; Delgado-Salinas & Parker 2023.

Amphicarpaea bracteata (L.) Fernald, (bracteate, with bracts), SOUTHERN HOG-PEANUT. Annual with twining or sprawling stems 0.3–2 m long, sometimes forming a ground cover; leaves alternate, pinnately compound with 3 leaflets; leaflets 3–10.5 cm long, 1.6–8 cm wide, entire; petioles 2–10 cm long; stipules 3.5–8.5 mm long, persistent; flowers of two types: 1) axillary racemes of 1–17 chasmogamous flowers on peduncles 1–6 cm long; calyces 4–5 mm long, with 4 lobes; corollas whitish, lilac or purple; stamens 10, diadelphous; fruits 1.5–4 cm long, flattened, curved or often asymmetric or falcate; 2) racemes of cleistogamous, inconspicuous, often apetalous flowers near or underground producing fleshy, nut-like or tuber-like, often subterranean fruits, 6–12 mm in diam. Rich woods and often abundant adjacent to intermittent streams; scattered in East TX; c and e U.S. from ME s to n FL, w to ND, OK and TX; s Canada and Mexico (Nuevo León, Puebla). May–Sep, but typically late summer. [*A. bracteata* var. *comosa* (L.) Fernald] The underground fleshy fruits, when cooked, are edible and were used by Native Americans (Dore 1970; Delgado-Salinas & Parker 2023). This species often forms an extensive groundcover in rich woods along intermittent streams in East TX (Isely 1998; E. Keith, pers. obs.).

Apios Fabr. ROUNDNUT, POTATO-BEAN

A genus of 5 species of e Asia and North America (Woods 2023a). The habit (vine), characteristic leaves, and striking flower color make the single East TX member of this genus easily recognizable in the field. (Greek: *apios*, pear, from the somewhat pyriform tuberous enlargement of the rhizomes) (subfamily Faboideae, tribe Phaseoleae)
References: Seabrook & Dionne 1976; Isely 1998; Woods 2023a.

Apios americana Medik., (of America), GROUNDNUT, AMERICAN POTATO-BEAN. Perennial from rhizomes bearing a chain-like series of tuberous-enlarged areas to 6 cm in diam.; stems annual, twining and climbing, to 4 m long, sparsely to densely pubescent; leaves once odd-pinnately compound; leaflets 5–7, ovate- or elliptic-lanceolate, 2–5(–10) cm long, pubescent, especially beneath, rounded at base; petioles 15–70 mm long; stipules nearly thread-like, 4–7 mm long, soon deciduous; flowers slightly sweet-scented, in axillary, short-peduncled, spike-like racemes; pedicels 2–6 mm long; corollas 9–13 mm long; standard cupped or hooded, brown-red to purple-red with pale back; wings down-curved, brown-red or purple-red; keel sickle-shaped, brownish or dull red; stamens 10, diadelphous; fruits linear, 5–12 cm long, 4–7 mm wide, slightly flattened. Damp woods and thickets along streams and lakes and adjacent roadside ditches; widespread in East TX; rare in Edwards Plateau and one disjunct location in the Panhandle; e and c U.S. from ME s to FL and w to ND s to OK and TX, with a disjunct station in CO; Quebec, Nova Scotia, and New Brunswick in Canada. Jun–Aug. The strings of edible tubers were cooked and eaten by Native Americans and early explorers (Tull 1987; Heiser 1993). Elias and Dykeman (1982) report that roasted seeds are also edible.

Arachis L. PEANUT

A South American genus with the number of species not agreed upon, but most recent authorities suggest ca. 70 species (Ferguson et al. 2005; Krapovickas et al. 2007; Rudd & Raveill 2023b). The rich yellow to nearly orange flowers, conspicuous stipules, 4 leaflets, and readily identified subterranean fruits make the single species occasionally escaped in East TX easily recognizable in the field. (from an old Greek plant name, *arakos* or *arachos*, used by Theophrastus for another member of the family). (subfamily Faboideae, tribe Dalbergieae)
References: Hermann 1954; Gregory et al. 1980; Krapovickas et al. 2007; Rudd & Raveill 2023b.

Arachis hypogaea L., (underground), COMMON PEANUT, PEANUT, GOOBER, GROUNDNUT, MONKEYNUT, EARTHNUT. Annual with erect or decumbent, villous stems to 80 cm long; leaves pinnately compound, usually with 4 leaflets; leaflets 2–6 cm long; stipules prominent, linear-subulate, 10–50 mm long; flowers axillary; pedicel (actually sterile lower portion of fruit) eventually elongating and pushing the fruit underground where it matures; corollas yellow (standard can be nearly orange), 1–1.5 cm long; stamens 10, monadelphous; fruits subterranean, indehiscent, with 1–3 seeds, oblong, 1–5 cm long, 1–2 cm wide. Rarely escaping cultivation, but occasionally found in sandy areas along rivers and weedy areas adjacent to cultivated fields; sparsely scattered in East TX; also a few scattered locations further w; primarily escaped in e U.S. from CT s to FL, w to KS, OK, and TX, disjunct in UT; Asia; Africa; and Australia. Jun–Oct. Native of South America (pre-Columbian cultures have PEANUT images in their art; by the time of the Spanish conquest peanut had spread to Mesoamerica; the Spanish name *cacauhuate*, is derived from a Nauhtl (language of the Aztecs) word). It is cultivated worldwide in suitable climates (Isely 1998). It is an allotetraploid, having an A and a B genome (AABB), and is thought to have evolved relatively recently from a single hybridization event, either between the unreduced gametes of two diploid species pertaining to different genomes or two haploid gametes which subsequently underwent spontaneous doubling, thereby restoring fertility. Either way, this natural hybridization event reproductively isolated *Arachis monticola* Krapov. & Rigoni (wild progenitor or weedy form of cultivated peanut) and *A. hypogaea* from their original genome donors and other wild species (Ferguson et al. 2005). Seijo et al. (2004) and Seijo et al. (2007) provide evidence that *A. duranensis* Krapov. & W.C. Greg. and *A. ipaensis* Krapov. & W.C. Greg. are the wild diploid progenitors. Leaf lesions caused by the rust fungus *Puccinia arachidis* Speg. can sometimes be found (J. Hennen, pers. comm.). This is an important crop locally in areas of sandy soils (e.g., annual Peanut Festival in Whitesboro, in w Grayson Co.); it is cultivated for its edible seeds and for oil (obtained by crushing the seeds), and the cake/meal (left over after oil extraction) is used for animal food. Some individuals develop severe peanut allergies to such an extent that eating a single seed or even breathing dust from peanuts can cause a life-threatening anaphylactic reaction. Peanuts and peanut products (e.g., meal) can also be contaminated with aflatoxin, produced by the fungus *Aspergillus flavus* Link (Lewis & Elvin-Lewis 1977), which is potentially carcinogenic.

ASTRAGALUS L. LOCOWEED, MILK-VETCH

Ours herbaceous, unarmed, annuals or perennials, acaulescent or caulescent, prostrate or decumbent to wide-spreading or erect or nearly so; pubescence if present of hairs attached at base or attached above their bases so that they have a long and a short arm; leaves once odd-pinnately compound, estipellate, stipulate; flowers in peduncled, axillary heads or head-like or spike-like racemes; petals various shades of pink, lavender, or purple to white or yellowish; androecium diadelphous, stamens 10; pistils sessile or stipitate, style usually glabrous, stigma minute; fruits inflated and bladdery or linear to narrowly oblong and not inflated.

A huge genus of 2,300 to 3,000 species, *Astragalus* is the largest genus of plants in terms of described species (Frodin 2004; Weakley, A.S., and Southeastern Flora Team 2022; Welsh 2023a). It occurs mainly in the n temperate zone, especially w North America and c and w Asia and extends to Chile, n India, and the mountains of tropical Africa, and it is notoriously difficult taxonomically. Some species concentrate selenium in their tissues (POISON-VETCHES); such plants can cause a toxic, sometimes fatal response in livestock—when symptoms include emaciation, hoof deformity, and lameness, the condition has been referred to as alkali disease; when blindness, excitement, and other nervous symptoms occur, the condition has been referred to as blind staggers; other *Astragalus* species (LOCOWEEDS) are apparently toxic due to indolizidine alkaloids (e.g., swainsonine) and cause locoism; symptoms in livestock include intoxication, paralysis, respiratory problems, and sudden death; affected horses become excessively excited and wild and are dangerous to ride (Kingsbury 1964, 1965; Blackwell 1990; James & Welsh 1992). Other species, however, are cultivated for hay or forage. *Oxytropis*, sometimes included in *Astragalus*, is here treated as a separate genus. (Ancient Greek name of a leguminous plant; perhaps from *astragalus*, ankle bone or dice, possibly alluding to rattling of seeds within fruit) (subfamily Faboideae, tribe Galegeae)

REFERENCES: Barneby 1964; Isely 1983–1986, 1998; Welsh 2023a.

1. Leaves all basal or crowded near base of stem, the leafless peduncles more than twice as long as the leaf-bearing lower portion of the stem; leaflets, above and beneath, densely gray-pubescent; pubescence consisting of hairs attached above their bases so that they have a long and a short arm (= malpighian or dolabriform hairs); plants perennial.
  2. Leaflets linear or narrowly lanceolate, 4–15 times as long as wide; keel tipped with a pointed, slender beak; fruits 7–10(–17) mm long (without beak) ____ see **Oxytropis lambertii**
  2. Leaflets lanceolate to elliptic-orbicular or obovate, 1.5–4 times as long as wide; keel acute or rounded, not beaked; fruits 12–37 mm long (without beak).
    3. Corollas variable in color, greenish white to yellowish or tinged with red or purple or purple-tipped or -margined, 8.5–14 mm long; calyx tubes usually 3–4.5 mm long, campanulate; species scattered on n, s, and w margins of East TX ____ **A. lotiflorus**
    3. Corollas pink-purple, usually 14–25 mm long; calyx tubes usually, 6–9 mm long, usually cylindric; species only known nw of East TX, previously reported from Bell Co. ____ **A. missouriensis**
1. Leaves extending well up the stem, the peduncles about equaling or shorter than the leaf-bearing portion of stem; leaflets glabrous to sparsely pubescent above or else plant an annual; pubescence consisting of hairs attached at their bases (not malpighian or dolabriform—except in *A. canadensis*).
  4. Plants perennial, with rhizomes or woody taproot.
    5. Plants erect and tall (0.4–1.6 m); stipules dry, papery, deciduous; inflorescences densely spike-like, the flowers usually 20 or more; stems and leaves with hairs attached above their bases so that they have a long and a short arm ____ **A. canadensis**
    5. Plants prostrate to ascending-erect, usually < 0.4 m tall; stipules herbaceous, persistent; inflorescences with 1–20 flowers; stems and leaves with hairs attached at their bases.
      6. Leaflets acute or obtuse, pubescent beneath, at least when young; mature fruits 1.8–27 mm wide.
        7. Calyces 7–10+ mm long, the tube longer than lobes; corollas 12–20 mm long; mature fruits 10–27 mm wide.
          8. Ovaries and fruits glabrous; plants glabrate or strigulose and usually without slender rhizomes; mature fruits 11–27 mm wide; species widespread in East TX ____ **A. crassicarpus**
          8. Ovaries and fruits densely gray-pubescent; plants pubescent with flexuous villous hairs and older plants with slender rhizomes; mature fruits 10–13 mm wide; species known in East TX only from w margin of the area ____ **A. plattensis**

7. Calyces 2.5–5 mm long, the tube about equaling lobes; corollas 4–13 mm long; mature fruits 1.8–3.5 mm wide _______ **A. nuttallianus**

6. Leaflets notched or truncate, glabrous beneath; mature fruits 1.8–7 mm wide.

9. Corollas greenish-white to cream or yellowish cream in color, 12–18 mm long; mature fruits 5–8 mm wide; species of forested habitats in East TX _______ **A. soxmaniorum**

9. Corollas variously pink to purple, lilac or whitish, 4–13 mm long (can be longer in a variety of *A. distortus* not known in TX); mature fruits 1.8–7 mm wide; species widespread in East TX.

10. Mature fruits 3.5–7 mm wide, unilocular (or with an incipient septum), not reticulate; inflorescences with 8–20 flowers, spicate or racemose; calyces with short white or white and black hairs; corollas 8.2–15.5 mm long _______ **A. distortus**

10. Mature fruits 1.8–3.5 mm wide, bilocular, reticulate; inflorescences with 1–20 flowers (usually only 1–6), head-like, subumbellate or spicate or racemose; calyces varying from glabrous to villous, without black hairs; corollas 4–13 mm _______ **A. nuttallianus**

4. Plants annual, with slender taproot.

11. Leaflets of all leaves notched to truncate apically.

12. Fruits small and broad, 3–9 mm long; peduncles with wide-spreading hairs or strigose; w and sw portions of East TX.

13. Fruits peltate, 3.5–7 mm long and 4.5–7.5(–9) mm wide; peduncles strigose; keel shorter than the wings; corollas 6–8(–9) mm long; uncommon in sw portion of East TX _______ **A. brazoensis**

13. Fruits ovate-lunate in side view, ovate-deltoid from above, 5.5–9 mm long and 2.5–5.5 mm wide; peduncles with wide-spreading hairs; keel longer than the wings (= exserted); corollas 4–7 mm long; very rare in w portion of East TX _______ **A. reflexus**

12. Fruits elongated, 12–37 mm long; peduncles with appressed or ascending hairs or glabrous; more widespread.

14. Corollas (12–)13–18 mm long; wings wholly white in apical 2/3; ovules 8–12; style pubescent; mature fruits 3.5–6.5 mm wide, stipitate, the stipes 1–3 mm long _______ **A. lindheimeri**

14. Corollas 4–13 mm long; wings colored to apex along lower edge; ovules 10–26; style glabrous; mature fruits 1.8–3.5 mm wide, sessile or nearly so, the stipes < 1 mm long.

15. Inflorescences racemose, axis elongating to 1–3 cm in fruit; keel tip obtuse; flowers 4–20, 8–13 mm long; species rare in sw corner of East TX _______ **A. macilentus**

15. Inflorescences subumbellate, axis scarcely elongating to 1.5 cm in fruit; keel tip acute; flowers 2–10, 2–10 mm long, more widespread.

16. Mature fruits 20–37 mm long, straight or gently and evenly curved for their whole length, glabrous; ovules 20–26; standard 8.3–13 mm long; widespread in East TX primarily on sandy soils _______ **A. leptocarpus**

16. Mature fruits 12–26 mm long, usually curved near base, otherwise straight or nearly so, glabrous or with pubescence; ovules 10–18; standard 3.7–13 mm long (but usually only 4–9 mm); on the western and southern margins of East TX primarily on calcareous substrates _______ **A. nuttallianus**

11. Leaflets of middle and upper leaves acute to obtuse at apex.

17. Calyx teeth 1.5–2.5 times as long as calyx tube; mature fruits densely pilose-hairy, 7–13 mm long, straight or nearly so, erect; calyces over half as long as corollas to nearly as long as corollas; ovules 5–9; species known only from sw portion of East TX _______ **A. wrightii**

17. Calyx teeth about equal in length to calyx tube; mature fruits glabrous or sparsely hairy, 12–26 mm long, curved near base or straight, spreading, ascending, or deflexed, not erect; calyces half as long as corollas or less; ovules 10–18; including species widespread in TX.

18. Flowers relatively large, the standard mostly 7–9 mm long; racemes mostly with 4–10 flowers; fruits consistently glabrous _______ **A. pleianthus**

18. Flowers small, the standard mostly 5–7 mm long; racemes mostly with 1–5(–6) flowers; fruits glabrous or with pubescence _______ **A. nuttallianus**

Astragalus brazoensis Buckley, (Presumably named for the Brazos River, despite only being collected much farther south of the river), BRAZOS MILK-VETCH. Annual, with basally branched, spreading, ascending to erect stems; stems 5–35 cm tall; leaves (2–)2.5–7.5(–9) cm long, leaflets 13–17, cuneate and obovate to broadly oblong, emarginate, 3–10 mm long, glabrate, stipules free; inflorescences mostly exserted, initially subcapitate to racemose, slightly elongating in fruit with 8–20 ascending then deflexed flowers; calyx tubes 1.5–2.5 mm long, lobes 1–2 mm long; corollas 6–8(–9) mm long, lavender to bicolored, keel acute; ovules ca. 4; fruits distinctly stipitate, 1.5–2.5 mm long, deflexed, body strongly obcompressed, peltate, subcircular but wider than long in dorsal view; 3.5–7 mm long and 4.5–7.5(–9) mm wide. Grasslands, stream margins and disturbed areas; sw portion of East TX in Bexar, Goliad, and Wilson cos. (Turner et al. 2003); s TX; slightly into Tamaulipas, Mexico. Mar–May. This species superficially resembles *A. nuttallianus* in flower but is easily distinguished from all other TX species by the peltate fruits (Isely 1998).

Astragalus canadensis L., (of Canada), CANADA MILK-VETCH. Perennial, inconspicuously pubescent or glabrous, usually tall and robust; stems 0.4–1.6 m tall, erect, leafy, branched above; inflorescences densely spike-like, the flowers usually 20 or more; corollas (11.3–)13.2–17.5 mm long, greenish white, light yellowish green, yellowish white, or dull straw-colored; fruits 10–15 mm long, 4–5.2 mm wide. Low moist prairies, creek banks, and rich woods; scattered sparsely in East TX, primarily in ne quarter, disjunct to se TX; e U.S. from VT s to n GA w to WA and UT s to e OK and TX; s Canada. May–Jul. This species is the most widespread species in the genus in North America but is rare in TX with an unusual disjunct distribution in se TX, primarily in Brazoria Co. (Isely 1998; E. Keith, pers. obs.).

Astragalus crassicarpus Nutt., (thick-fruited), GROUND-PLUM, BUFFALO-PLUM, POMME DE PRAIRIE, GROUND-PLUM MILK-VETCH. Perennial; stems 10–50(–60) cm long, decumbent, ascending, or suberect; leaflets 13–29, broadly elliptic, oblong, or oblanceolate, 0.5–1.5 cm long, glabrate above; inflorescences usually not exceeding the leaves, with (5–)8–20 often showy flowers; calyx tube subcylindric, 5–9 mm long, strigulose to tomentose, lobes 2–4 mm long; corollas lilac, pink-purple, or purple or greenish white; fruits globose to ovoid or obovoid, sometimes superficially plum-like or cherry-like in appearance, up to 1.5 times as long as wide, 15–27 mm long, 11–27 mm wide (in our varieties), often reddish on upper surface. Native Americans ate the immature, succulent, plum-like or cherry-like fruits; care must be taken because there are several closely related poison MILK-VETCHES or LOCOWEEDS. If fruits are present, the species is "essentially unmistakable" (Isely 1998). The three varieties below can be separated by the key modified from Barneby (1964) and Isely (1998).

1. Flowers greenish white or cream (–slight lavender), the keel commonly maculate; peduncles at flowering 5–12 cm long; calyces with somewhat shorter (to only 1 mm) white to cream hairs; plants of open woodlands in ne TX ____ var. **trichocalyx**
1. Flowers purple, pink-purple, or lilac; peduncles at flowering 3–5(–8) cm long; calyces with relatively long (to. 1.5 mm) white and black or sometimes mostly black hairs; plants of prairies in nw and s portions of East TX.
   2. Stems arising singly or few together from slender, widely creeping, subterranean caudex-branches and often with rhizomes, forming loosely matted or colonial growths; ovules 34–50; plants primarily of s 1/2 of East TX ____ var. **berlandieri**
   2. Stems arising together from the crown of the taproot or shortly forking, determinate caudex at or just below soil-level; ovules 38–68; plants of nw part of East TX ____ var. **crassicarpus**

var. **berlandieri** Barneby, (for Jean Louis Berlandier, 1805–1851, French botanist who explored TX, NM, and Mexico, and one of the first botanists to make extensive collections in TX), BERLANDIER GROUND-PLUM. Similar to var. *crassicarpus*, but rhizomes often present and with fewer stems; corollas lilac or pink-purple. Soils derived from limestone or clay, gravelly areas, prairies; s 1/2, particularly sw portion of East TX; also adjacent Edwards Plateau; endemic to TX. Mar–Jun.

var. **crassicarpus,** GROUND-PLUM, MILK-VETCH, INDIAN-PEA. Several stems arising from central caudices; corollas purple or lilac. Clayey or rocky prairies, open woodlands; nw part of East TX; w across TX to Panhandle and Trans-Pecos; c U.S. from MN w to MT s to OK and TX and w to ne NM. (late Jan–)Mar–Jun.

var. **trichocalyx** (Nutt.) Barneby, (with a hairy calyx), LARGE GROUND-PLUM. Similar to var. *crassicarpus*, but distinguished by its calyx pubescence, "greater vigor" (Isely 1998) as seen in the longer inflorescences, and corolla color; corollas greenish white or cream (–slightly lavender). Open woodlands and remnant prairies; rare and known only from Anderson, Bowie, Henderson, Leon, Rains (BRIT), Van Zandt, and Wood (Kartesz 2023) cos. in ne part of East TX; sc U.S. from IL s to LA and w to se KS, c OK, and TX. Mar–Jun. [*A. trichocalyx* Nutt.] Isely (1998) noted that differentiation from var. *crassicarpus* is clinal, but Welsh (1960) recognized it as a distinct species.

Astragalus distortus Torr. & A. Gray, (twisted, from the fruit shape), OZARK MILK-VETCH. Low perennial; stems decumbent to prostrate, 10–40 cm long; inflorescences with 8–numerous flowers; calyces 3–6.1 mm long, with white or mixed black and white hairs; corollas 8.2–15.5 mm long, variable in color, pink-purple, lilac to whitish, sometimes with markings; fruits 13–25 mm long. The 2 varieties below can be distinguished by the following key modified from Barneby (1964) and Isely (1998).

1. Flowers relatively large, the standard 11–15.5 mm long, the keel (7–)7.4–9.5 mm long; fruits curved and deeply sulcate both dorsally and ventrally; ovules 26–37; mostly known n of Red River, but reported for Titus Co. in ne TX ____ var. **distortus**
1. Flowers smaller, the standard 8.2–12 mm long, the keel 5.5–7.2 mm long; fruits straighter and shallowly sulcate dorsally but not or only obscurely so ventrally; ovules 16–28; throughout East TX ____ var. **engelmannii**

var. **distortus,** OZARK MILK-VETCH, BENT-POD MILK-VETCH. Prairies and post oak woodlands; mapped in Titus Co. in East TX and Wise Co. in Cross Timbers and Prairies (Kartesz 2023); mainly sc U.S. from IL w to e IA and s to MS and w to c OK and possibly TX, disjunct and rare in WV and VA. Mar–Jul. The two TX specimens reported in Kartesz (2023) in Titus (*Alexander 13* TAES) and

Amorpha roemeriana [VIN]

Amphicarpaea bracteata [BB2]

Apios americana [BB2]

Arachis hypogaea [EN2]

Astragalus canadensis [DIL]

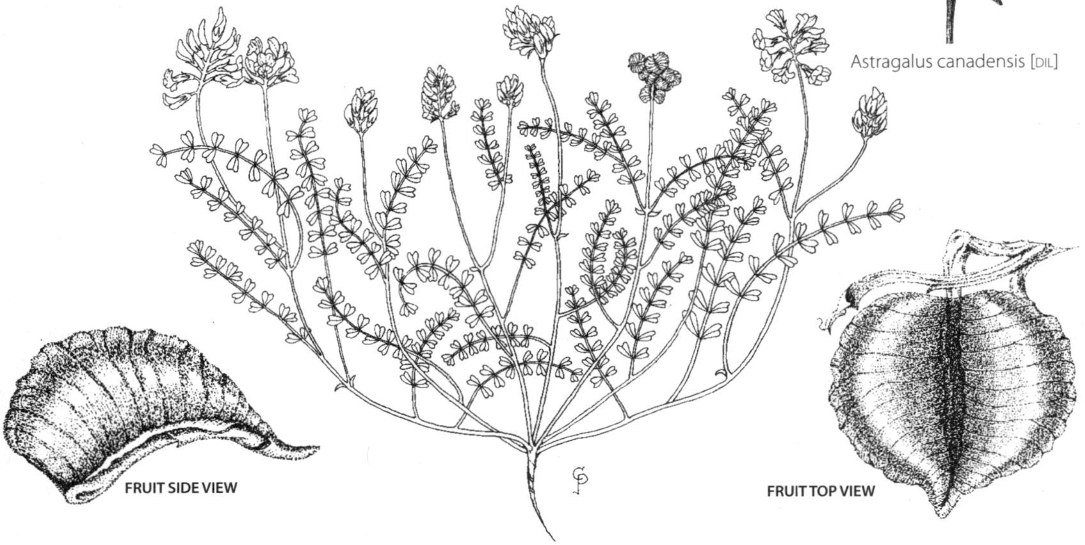

Astragalus brazoensis [CP1]

Wise (*O'Kennon 18290* BRIT) cos. may be misidentified *A. distortus* var. *engelmannii*, but it is included here because of proximity and its possibility of occurrence in East TX.

var. **engelmannii** (E. Sheld.) M.E. Jones, (for George Engelmann, 1809–1884, German-born botanist and physician of St. Louis), ENGELMANN'S MILK-VETCH. Open woodlands and prairies, on calcareous substrates, rocky outcroppings and occasionally on sandy soils; widespread in East TX; otherwise mainly in adjacent Gulf Prairies and Marshes; sc U.S. in AR, MS, LA, and TX. Mar–May. This variety was not previously reported n of the Red River (Isely 1998); however, Kartesz (2023) mapped this variety throughout AR.

Astragalus leptocarpus Torr. & A. Gray, (thin- or slender-fruited), SLIM-POD MILK-VETCH, BODKIN MILK-VETCH. Inconspicuously appressed-pubescent or glabrous annual; stems erect, ascending, or decumbent, 3–25(–35) cm long; racemes subcapitate, with 3–7(–12) flowers; corollas 8–13 mm long, purple-blue drying violet, the center of standard and upper sides of wings and keel often white; fruits 20–37 mm long, straight or slightly curved, linear, 2–3.1(–3.5) mm wide. Sandy open woods and sandy prairies adjacent to beaches; widespread in East TX, also adjacent Cross Timbers and Prairies and Gulf Prairies and Marshes; sc U.S. from AR and LA w to TX. Mar–Apr. *Astragalus leptocarpus* "is closely related to *A. nuttallianus* and might logically be treated as forming part of that polymorphic complex; for apart from the ordinarily longer pod enclosing from one to four extra pairs of ovules and seeds, it possesses no single attribute which cannot be matched somewhere in *A. nuttallianus*. In practice it is easily distinguished from sympatric varieties of *A. nuttallianus*: from var. *trichocarpus* by its glabrous ovary; from var. *austrinus* by its retuse leaflets; and from var. *nuttallianus*, which it most nearly resembles in its green, glabrescent foliage and in shape of the leaflets, by the nearly always longer and more amply proportioned flower" (Barneby 1964).

Astragalus lindheimeri Engelm. ex A. Gray, (for Ferdinand Jacob Lindheimer, 1801–1879, German-born TX collector), LINDHEIMER'S MILK-VETCH, BUFFALO-CLOVER, MILK-VETCH. Inconspicuously appressed-pubescent annual; stems decumbent to ascending, to ca. 35 cm long; inflorescences racemose, with 2–8 crowded flowers; flowers sweet-scented, large and very showy; corollas (12–)13–18 mm long, bicolored, purple-blue, with center of standard, upper side and tips of wings, and tip of keel white, turning violet upon drying; fruits 17–27 mm long, flattened, curved-oblong. Sandy, silty, or rocky ground; w margin of East TX; also Cross Timbers and Prairies, Rolling Plains, e Edwards Plateau; also w OK (Barneby 1964). Mar–early May. This species is recognized by its relatively large showy flowers with white-tipped wings (Isely 1998).

Astragalus lotiflorus Hook., (with flowers like *Lotus*—DEER-VETCH, TREFOIL), LOTUS MILK-VETCH. Perennial, low, tufted, ± scapose; stems spreading to erect, to 18 cm tall, usually densely pubescent with hairs attached above base so that they have a long and a short arm (= malpighian or dolabriform hairs); chasmogamous and cleistogamous inflorescences both sometimes produced, commonly on different plants (cleistogamous inflorescences only occasionally seen); chasmogamous inflorescences subcapitate, pedunculate, with (3–)5–17 flowers; cleistogamous inflorescences subsessile, with 2–4 flowers, the flowers scarcely petaloid; chasmogamous flowers with corollas 8.5–14 mm long, greenish white to yellowish or tinged with red or purple or purple-tipped or -margined; fruits usually 1.3–3 cm long, 6–8 mm wide, the fruits produced by cleistogamous flowers among or hidden by the basal leaves. Limestone and sandstone outcrops; w margin of East TX; w across TX to Panhandle and Trans-Pecos; c and w U.S. from MN s to OK and TX and w to MT and NM; sc Canada. Late Mar–Apr. [*A. lotiflorus* var. *reverchonii* (A. Gray) M.E. Jones] Barneby (1964) considered the species to be a polymorphic indivisible entity.

Astragalus macilentus (Small) Cory, (thin, lean), MEAGER MILK-VETCH. Annual, stems erect or ascending, to 30(–40) cm tall; inflorescences with 4–20 flowers, these usually rather loosely arranged, but occasionally dense and showy, the axis elongating, 1–3 cm long in fruit; corollas (7–)8–13 mm long, usually light purple; fruits 12–25 mm long. Rocky and gravelly open areas; Hays, Travis (BRIT), Bastrop, and Comal (Isely 1998) cos. in sw portion of East TX; mainly Edwards Plateau and Trans-Pecos; also reported in NM (Kartesz 2023). Mar–May. [*A. nuttallianus* DC. var. *macilentus* (Small) Barneby] This species is part of the *A. nuttallianus* "complex" and is often treated as a variety of that species (e.g., Barneby 1964; Correll & Johnston 1970; Diggs et al. 1999; Welsh 2023a). Isely (1983–1986, 1998), however considered it a separate species indicating that although partially sympatric with *A. nuttallianus*, it does not intergrade with that species. Isely (1998) noted that "*A. macilentus* differs from *A. nuttallianus* in its elongating inflorescence, larger flowers, substipitate pods, obtuse keel, and although partly sympatric with the other species is not intergradient with it." See further discussion under *A. nuttallianus*. In its grandest form, it is the showiest of the species/varieties in the complex and usually growing with or in proximity to *A. nuttallianus* without any intergrades (E. Keith, pers. obs.). Welsh (2023a) noted that species level recognition may be justified despite treating it as a variety of *A. nuttallianus*.

Astragalus missouriensis Nutt. (of Missouri), MISSOURI MILK-VETCH. Plant perennial, low tufted or prostrate, ± scapose, stems 1–10 cm long, silvery-or greenish-gray due to numerous hairs attached above base so that they have a long and a short arm (= malpighian or dolabriform hairs); inflorescences with 3–10(–13) flowers; corollas (7.5–)14–25(–30) mm long, pink-purple;

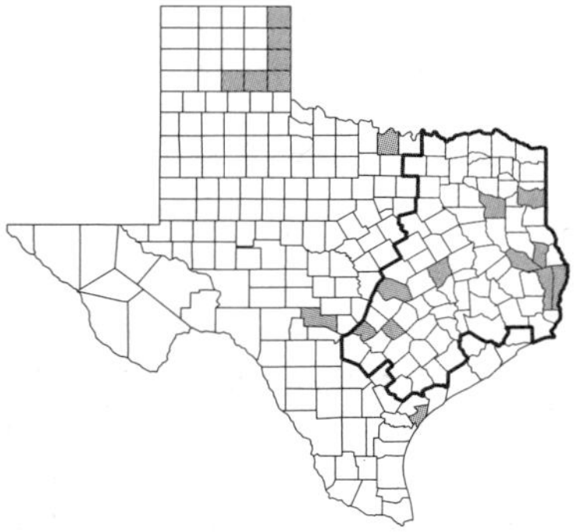

Amorpha canescens

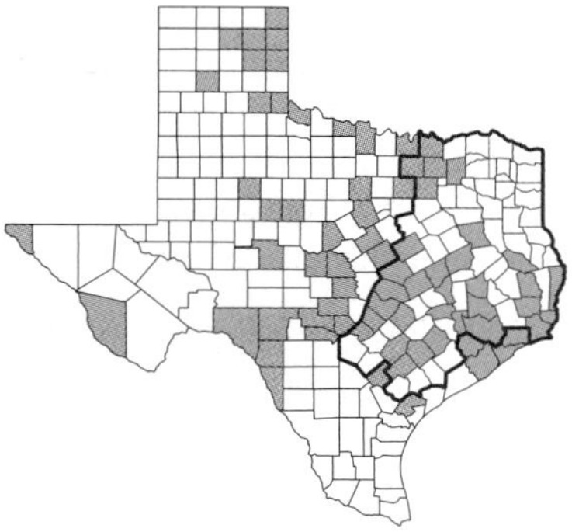

Amorpha fruticosa

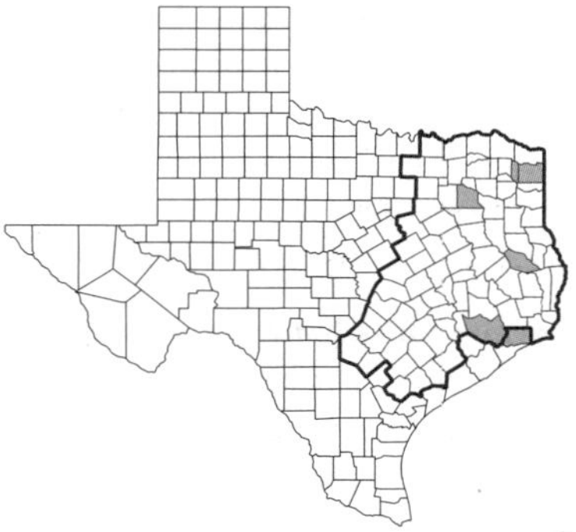

Amorpha laevigata

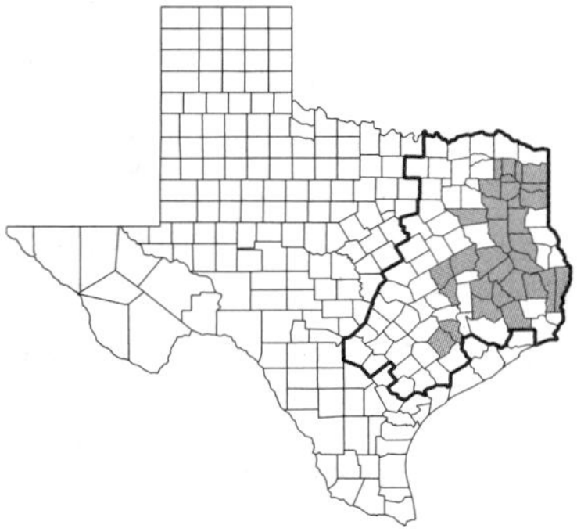

Amorpha paniculata

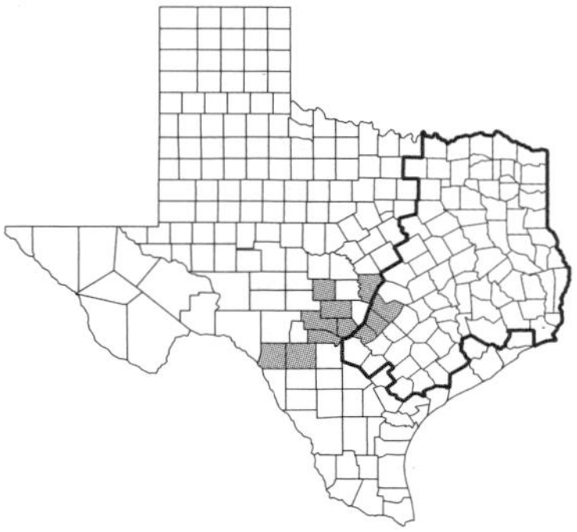

Amorpha roemeriana

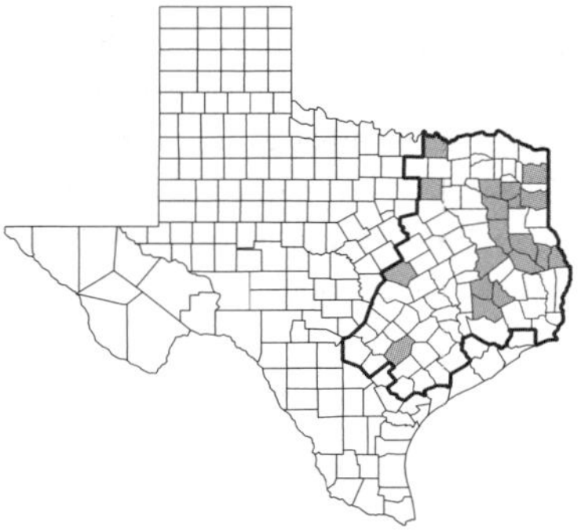

Amphicarpaea bracteata

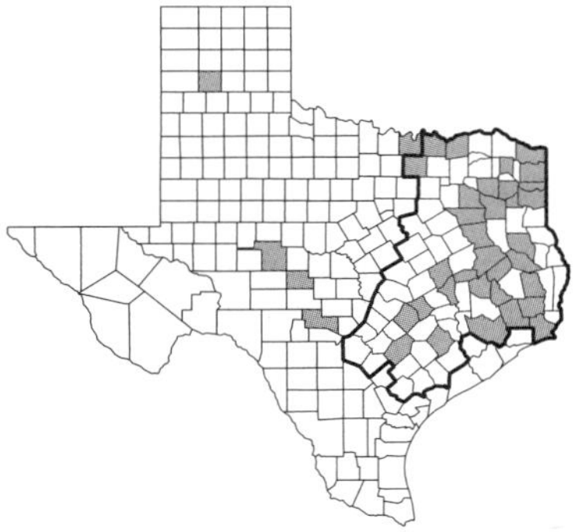

Apios americana

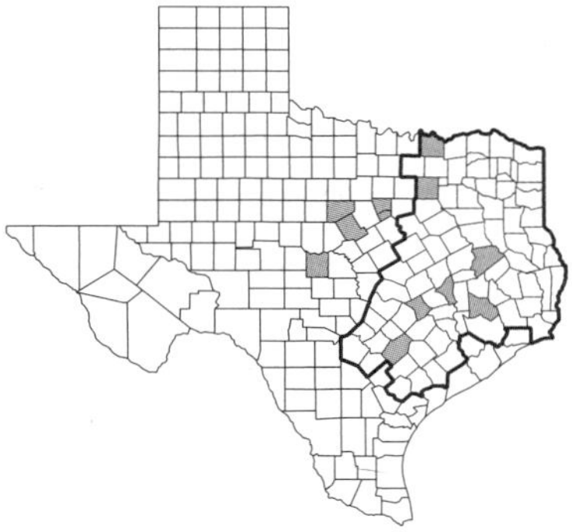

Arachis hypogaea

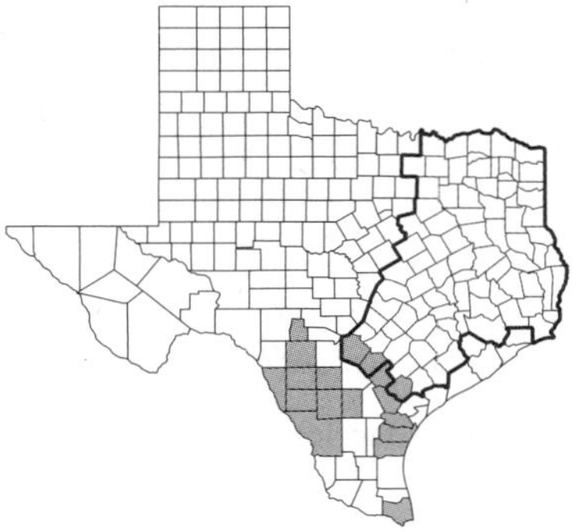

Astragalus brazoensis

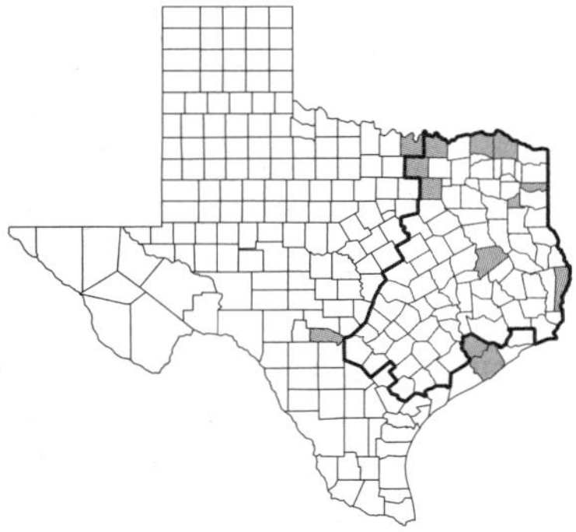

Astragalus canadensis

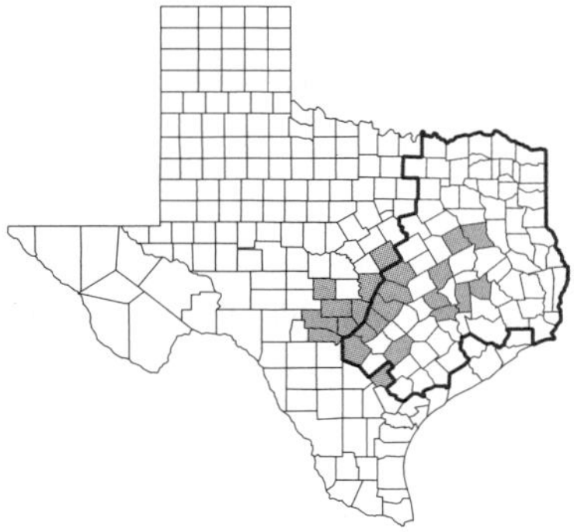

Astragalus crassicarpus var. berlandieri

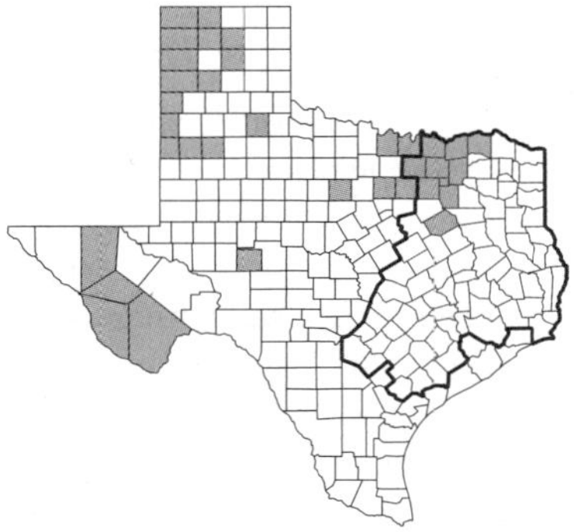

Astragalus crassicarpus var. crassicarpus

ovules numerous; fruits ascending, divergent, or humistrate (= laid flat on the soil); subterete or obcompressed and fleshy (10–)12–25(–28) mm long, (4–)6–9 mm wide. Diverse open, dry habitats; possibly Bell Co. (Turner et al. 2003) at w margin of East TX; primarily Panhandle and far w TX (Isely 1998); w U.S. from MN s to TX and w to MT, nw NM, and se UT; s Canada. Mar–Jun. The species is considered to have 4 varieties, with only var. *missouriensis* occurring in TX (Isely 1998). Turner et al. (2003) mapped a disjunct location in Bell Co. presumably from a specimen annotated by Dr. Turner (*Carr & Sanchez 17312* TEX). This specimen is *A. lotiflorus* with campanulate rather than cylindric calyx tubes.

Astragalus nuttallianus DC., (for Sir Thomas Nuttall, 1786–1859, English-American botanist), TURKEY-PEA, SMALL-FLOWER MILK-VETCH. Glabrous to pubescent annual (but can also be keyed as a perennial in the key); stems erect to decumbent, to 35(–45) cm long; leaflets 7–19, various shapes, truncate or retuse, but some varieties acute or obtuse; inflorescences with 1–8 flowers; corollas 3.7–13 mm long, varying in color from various shades of pinkish to purplish, often with variable white markings, to whitish; fruits 12–26 mm long, 1.8–3.5 mm wide. This is an extremely variable species including and closely related to a number of often difficult to distinguish taxa occurring in or near East TX. It should be noted that this taxonomically difficult group has had an unsettled history. Barneby (1964) and Correll and Johnston (1970) treated the species very broadly and recognized numerous varieties within a diverse *A. nuttallianus*, while Turner et al. (2003) recognized a number as distinct species and segregates including *A. austrinus* with two varieties. Isely (1998) on the other hand, treated *A. pleianthus* and *A. macilentus* as distinct species, but retained the other varieties within *A. nuttallianus*. He justified this by indicating that the morphological differentiation of *A. pleianthus* and *A. macilentus* from *A. nuttallianus* is "approximately equivalent to that ... of [other] conventionally accepted" species within the genus. We agree with Isely's justification and have noted the obvious morphological differences in the field without any intergrades. We are following Welsh (2023a) in treating the remainder of the taxa as varieties within *A. nuttallianus*.

1. Leaflets of all leaves notched or truncate-emarginate ____ **A nuttallianus** var. **nuttallianus**
1. Leaflets not all notched or truncate-emarginate, elliptic or ovate and obtuse to subacute in at least some upper or often all leaves.
   2. Fruits hirsutulous, the hairs 0.5–1 mm long; leaflets commonly 11–13 ____ **A. nuttallianus** var. **trichocarpus**
   2. Fruits glabrous or if pubescent the hairs shorter, appressed, 0.5 mm long or less; leaflets commonly 7–11 ____ **A nuttallianus** var. **austrinus**

var. **austrinus** (Small) Barneby, (southern). Abundant in a wide variety of dry habitats including roadsides and ruderal areas; western portions of East TX; c TX w to the Panhandle and Trans-Pecos; sw U.S. KS, w OK and TX w to s CA; Mexico. Mar–May. [*A. austrinus* (Small) E.D. Schulz] Isely (1998) noted the morphological similarity to the sympatric *A. nuttallianus* var. *nuttallianus* with which it often grows but is consistently separated by the acute or obtuse versus truncate or emarginate leaflets of var. *nuttallianus*. No reliable county level data available, therefore no distribution map provided.

var. **nuttallianus,** NUTTALL'S MILK-VETCH, TURKEY-PEA, PEAVINE. Prairies and open woodlands, common in disturbed and ruderal habitats, primarily in calcareous soils; widespread in the e 2/3 of TX; sc U.S. AL w to KS, OK, and TX. Mar–Jun. This variety is recognized by its truncate or emarginate leaflets, grayish pubescence, and relatively few purple and white flowers (1–8) (Isely 1998).

var. **trichocarpus** Torr. & A. Gray, (hairy-fruited), SOUTHWESTERN MILK-VETCH. Open rangeland, prairies, roadsides typically on limestone or sandy soil; widespread mainly in s 1/2 of East TX; also Gulf Prairies and Marshes, South TX Plains, and sparsely scattered to the w; s U.S. in NM, OK, and TX. Mar–May. [*A. austrinus* (Small) E.D. Schulz var. *trichocarpus* (Torr. & A. Gray) B.L. Turner] This variety is very similar to var. *austrinus* but is distinguished by its consistently longer spreading or curved pubescence 0.5–1 mm long on the pods (Isely 1998; E. Keith, pers. obs.).

Astragalus plattensis Nutt., (of the Platte River), PLATTE RIVER MILK-VETCH, GROUND-PLUM. Perennial resembling *A. crassicarpus*, usually largely decumbent, more pubescent throughout with flexuous, villous hairs; stems short, 10–20(–30) mm long; leaflets 13–25, obovate to oblong, 4–15 mm long; inflorescences with 6–12 flowers; flowers sweet-scented; calyx tube 2–3 mm long, generally white or occasionally black-strigulose, lobes 1.5–3 mm long; corollas 14–20 mm long; standard yellowish white to light or deep lavender or rosy lavender; wings similar or deeper-colored; keel lavender to purple or red-purple with dark tip; ovary and fruit densely pubescent; fruits ca. 10–13 mm long and wide, ovoid to ellipsoid or subglobose. Sandy or rocky prairies or open woods; Dallas, Hays, Hill, Rockwall (BRIT), and Travis (Turner et al. 2003) cos. on w margin of East TX; mainly e Edwards Plateau and Cross Timbers and Prairies nw to Panhandle; sc U.S. from ND s to s TX and w to CO, sw MT, and WY. Mar–Jul. *Astragalus plattensis* is included within the range of *A. crassicarpus*, and in flower the two species may be confused. When flowering, *A. plattensis* will have ovaries with hairs, whereas *A. crassicarpus* has glabrous ovaries (Welsh 2023a).

Astragalus pleianthus (Shinners) Isely, (many-flowered), EDWARDS PLATEAU MILK-VETCH. Annual; stems prostrate to erect, usually basally branched, to 40 cm long; leaflets (9–)13–19, lower leaves truncate or emarginate and upper leaves with at least some

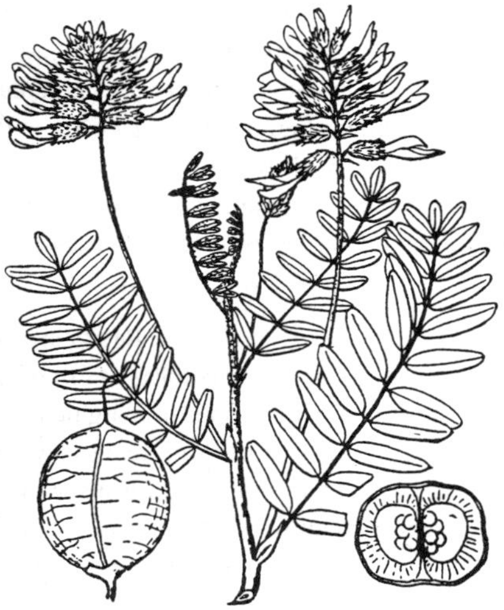
Astragalus crassicarpus var. berlandieri [BB2]

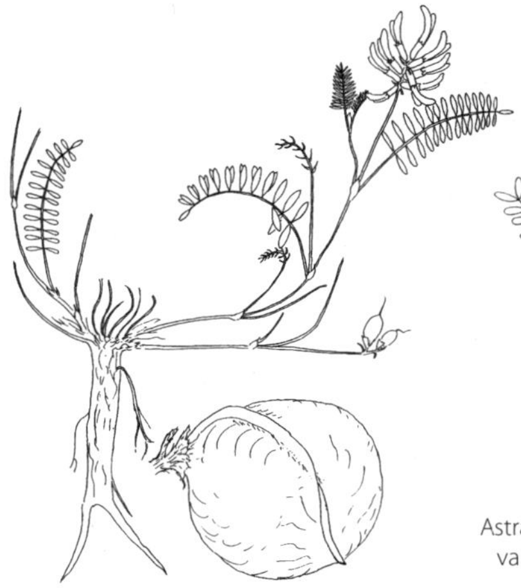
Astragalus crassicarpus var. crassicarpus [BT3]

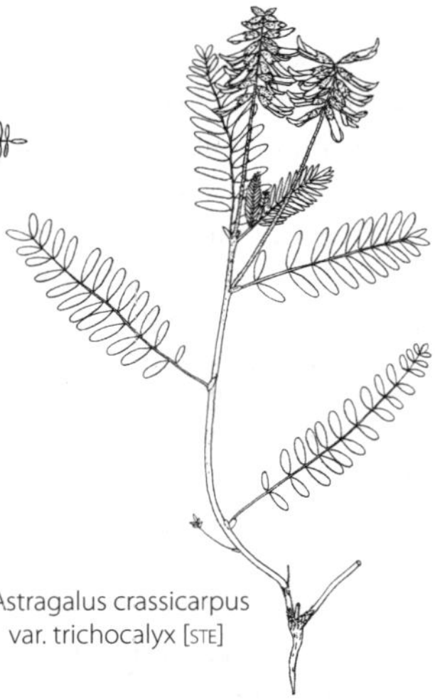
Astragalus crassicarpus var. trichocalyx [STE]

Astragalus distortus var. distortus [BB2]

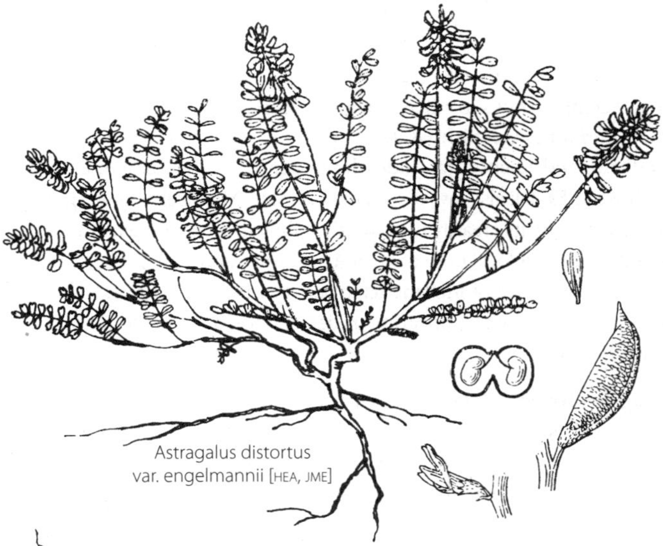
Astragalus distortus var. engelmannii [HEA, JME]

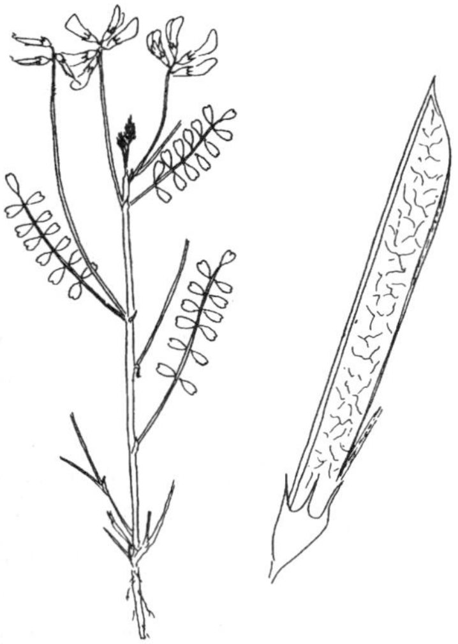
Astragalus leptocarpus [BT3]

Astragalus lindheimeri [BT3]

Astragalus lotiflorus [BT3]

acute; inflorescences with 4–10 flowers, the axis only ca. 1 cm long even in fruit; corollas 7–9(–9.5) mm long, purple; fruits 12–25 mm long, 2–3 mm wide. Calcareous soils; prairies, open woodlands; scattered, but mostly in the w portion of East TX; scattered in the e 2/3 of TX; endemic to TX. Mar–May. [*A. austrinus* (Small) E.D. Schulz var. *pleianthus* Shinners, *A. nuttallianus* var. *pleianthus* (Shinners) Barneby] This species is part of the *A. nuttallianus* "complex" and is often treated as a variety of that species (e.g., Barneby 1964; Correll & Johnston 1970; Diggs et al. 1999; Welsh 2023a). Isely (1983–1986, 1998), however, considered it a separate species indicating that "though sympatric with several of its congeners, it seemingly does not blend with any of them." Welsh (2023a) acknowledged that "justification exists for elevating this taxon to species." See further discussion under *A. nuttallianus*.

Astragalus reflexus Torr. & A. Gray, (reflexed, bent sharply backward), TEXAS MILK-VETCH, DROOPING MILK-VETCH. Rather sparsely pilose, prostrate to erect annual; stems to 35 cm long; racemes with 3–10 flowers, subcapitate to racemose; corollas 4–7 mm long, bicolored, bluish violet or reddish violet with white (wings white; standard and keel sometimes white-tipped or margined); keel exserted; fruits 5.5–9 mm long, 2.5–5.5 mm wide. Prairies, roadsides, edges of ponds and lakes in moist calcareous clay soils; scattered in East TX, but mainly on w margin; endemic to c TX. Feb–May. The exserted keel and short broad fruits are noteworthy for identification (Isely 1998). This species is now very rare with only one known extant population on Fort Cavazos Military Reservation in Bell Co. (*Hansen 4827* BAYLU; *5834* BAYLU, TEX; *6479* BAYLU, TEX). Eric Knight, environmental consultant and author, confirmed that this population was still extant in 2022 (iNaturalist contributors 2023b). Other than the Fort Cavazos population, it was last collected in 1965 in Zavala Co. in s TX (*Ramirez & Cardenas 12* TEX). NatureServe: G3, vulnerable; S3, vulnerable.

Astragalus soxmaniorum Lundell, (named for George M. and Florence Soxman, Dallas area plant collectors), SOXMAN'S MILK-VETCH. Perennial, but short-lived; stems numerous, decumbent, 10–30 cm long; inflorescences with 8–numerous flowers; corollas 12–18 mm long, greenish white to cream or yellowish cream; fruits 13–20(–25) mm long, 5–8 mm wide. Sandy forests and woodlands and adjacent roadsides; limited in TX to the e part of East TX w to Freestone Co. (BRIT) and s to Hardin Co. (Correll & Johnston 1970); also sw AR and nw LA. Mar–May. Isely (1990) noted that "*A. soxmaniorum* resembles *A. distortus* but is easily distinguished by its creamy to greenish yellow flowers and usually broader pods." Isely (1998) also pointed out that this species "is undoubtedly a local derivative of *A. distortus*, and it might be reasonable to regard it as no more than a sympatric variety. But there is no evidence of intermediacy between the two taxa ..." This species is rare throughout its range and often disappears for many years before reappearing under unknown favorable conditions (E. Keith, pers. obs.). NatureServe: G3, vulnerable; S3, vulnerable.

Astragalus wrightii A. Gray, (for Charles Wright, 1811–1885, TX collector), WRIGHT'S MILK-VETCH. Gray-pilose, erect annual to 35 cm tall; leaflets (7–)11(–13) acute or obtuse, 0.5–1.5 cm long; inflorescences head-like, 3–8-flowered, on stiffly erect peduncles; flowers sessile or nearly so; corollas 5.2–6.2 mm long, reddish violet, lilac, or whitish with lilac tinge, often barely longer than the villous calyx; fruits 7–13 mm long, 2.5–3.5 mm wide. Oak-juniper woodlands, grasslands, and outcrops on limestone soils; sw margin of East TX and disjunct to Montgomery (*McCleod s.n.*, 1946, TEX) and possibly Walker cos. (Turner et al. 2003) to the e, also adjacent Edwards Plateau; endemic to TX. Mar–May. Isely (1998) noted that this species is distinctive in the combination of its stiffly erect peduncles, head-like inflorescences, flowers with petals often barely longer than the villous calyx, and cluster of erect fruits.

BAPTISIA Vent. WILD INDIGO, FALSE INDIGO

Herbaceous perennials with woody crown and tough roots; foliage sometimes darkening upon drying; leaves very short-petioled, palmately compound with 3 leaflets (or uppermost with only 1 or 2); stipules large to small, persistent or falling early; flowers terminal, large, solitary or in erect to horizontal or drooping racemes or solitary in the leaf axils; flowers papilionaceous; calyx tube campanulate, lobes 4, subequal, < or > tube length; corollas variously colored, white, yellow, lavender, or blue; androecium of 10 stamens, separate; style longer than ovary, glabrous; fruits rounded or subcylindric, turgid or inflated, becoming woody, often with a beak; seeds numerous or few.

A genus of 17 species endemic to the e U.S. (Kartesz 2023; Turner 2023a); East TX species can usually be easily recognized to genus by the combination of the palmately 3-foliate leaves, free stamens, and inflated fruits. *Baptisia tinctoria* (L.) R. Br. was formerly used as a dye plant; others are used as ornamentals or medicinally; a number contain alkaloids. The plants are avoided by grazing animals and some species are reportedly toxic to livestock (Isely 1990). Hybridization is well known in *Baptisia* (Turner & Alston 1959; Alston & Turner 1962, 1963). Kosnik et al. (1996) described a *Baptisia* hybrid complex in the nw part of East TX including a new hybrid. In nearly every case where two or more parental species have been found growing together in that area, hybrid individuals have also been found (Kosnik et al. 1996). The key below includes the named hybrids known to

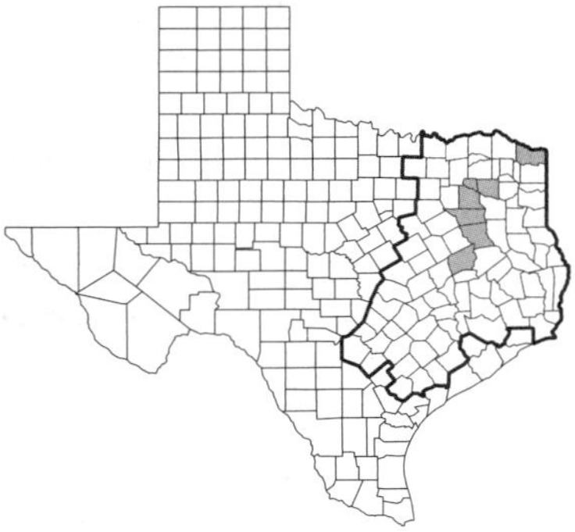
Astragalus crassicarpus var. trichocalyx

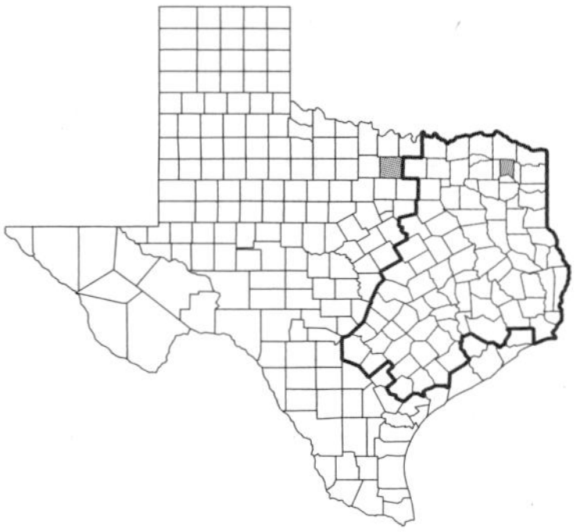
Astragalus distortus var. distortus

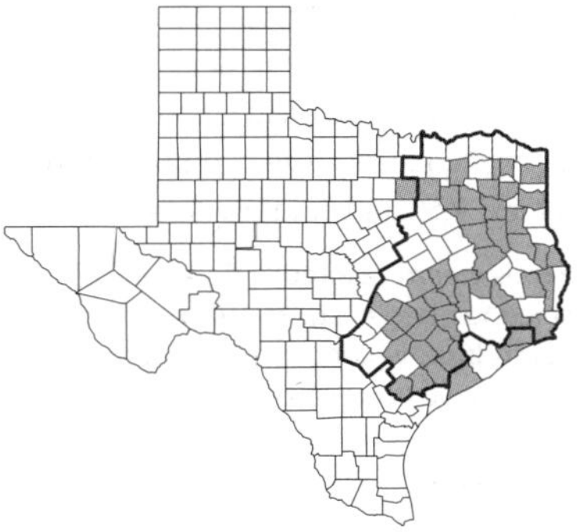
Astragalus distortus var. engelmannii

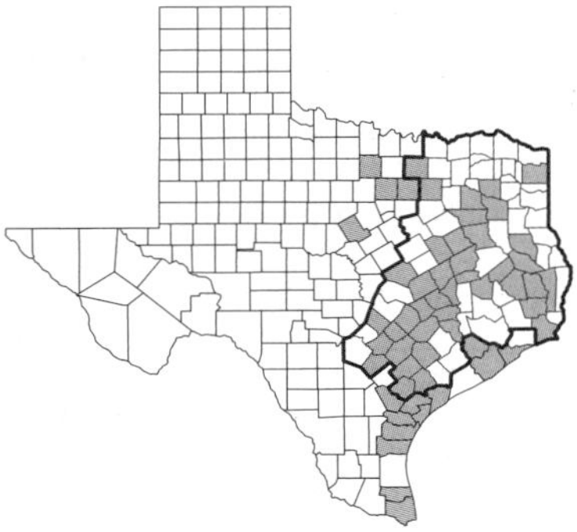
Astragalus leptocarpus

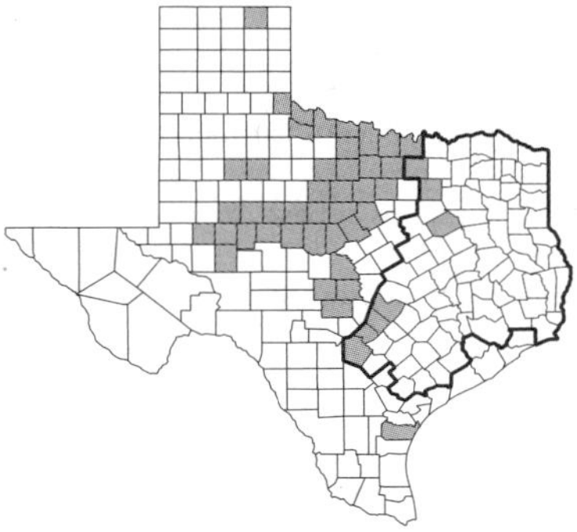
Astragalus lindheimeri

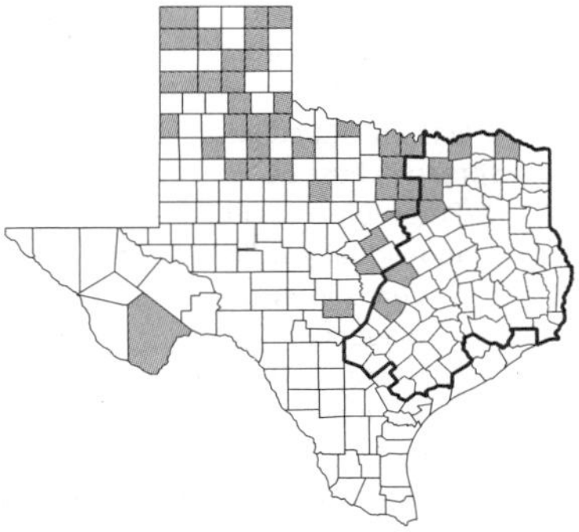
Astragalus lotiflorus

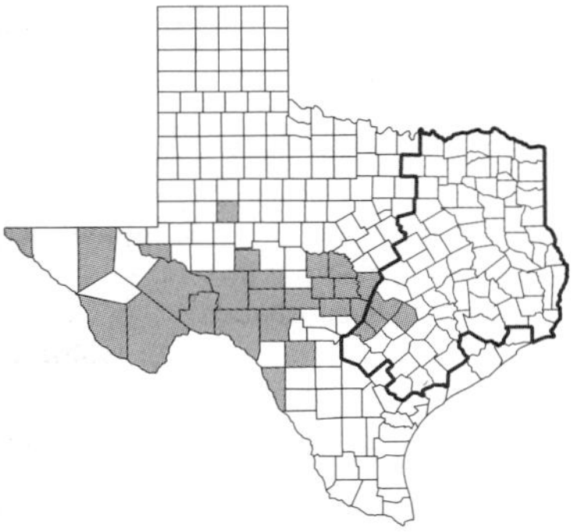
Astragalus macilentus

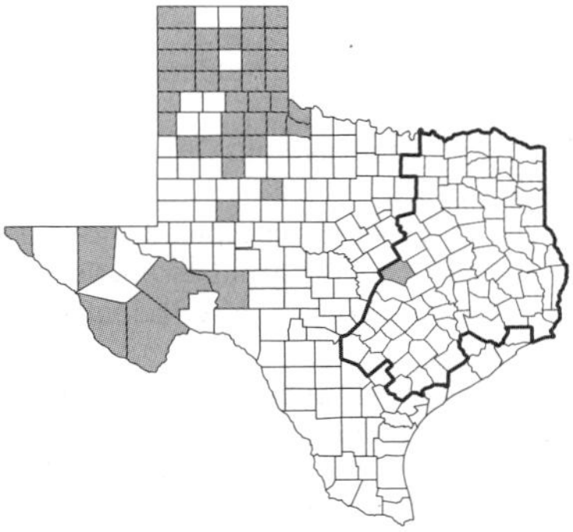
Astragalus missouriensis

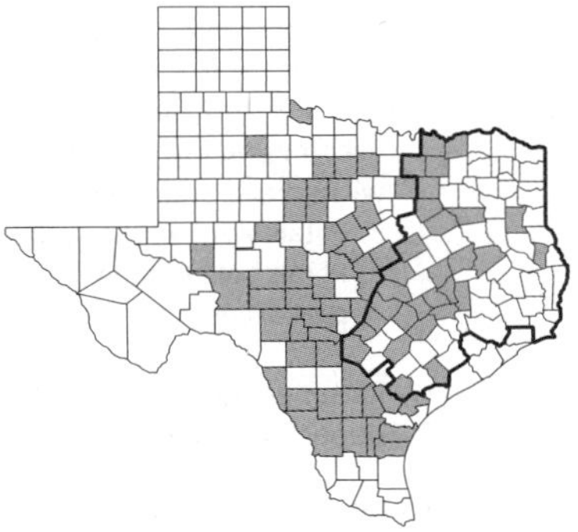
Astragalus nuttallianus var. nuttallianus

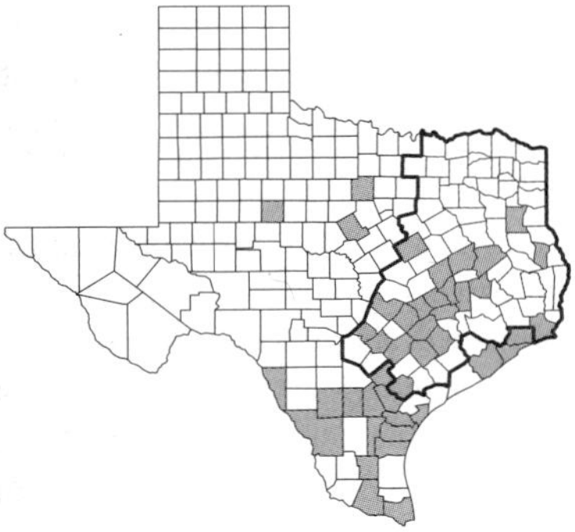
Astragalus nuttallianus var. trichocarpus

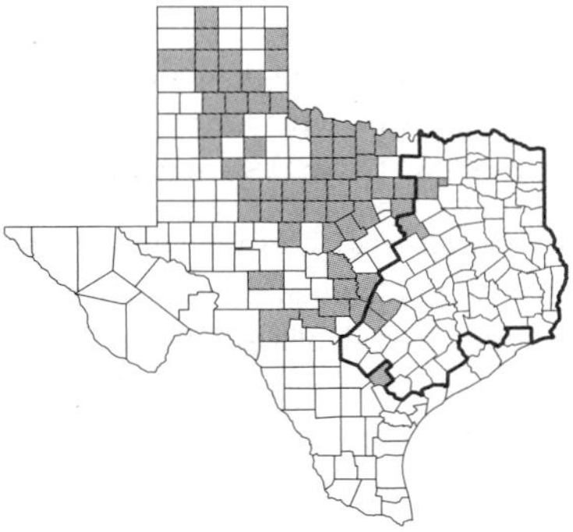
Astragalus plattensis

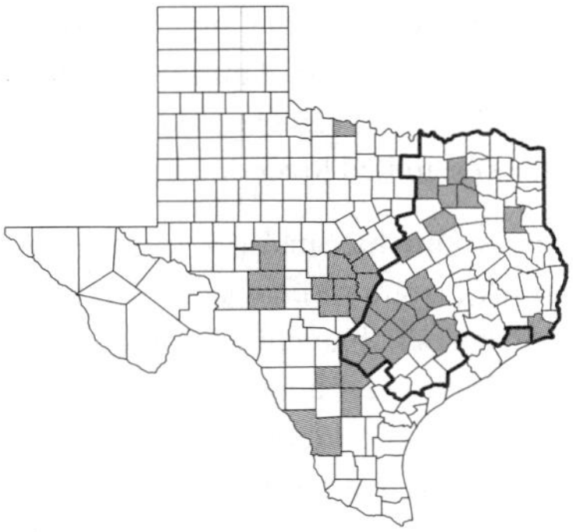
Astragalus pleianthus

occur in East TX. Other species are known to hybridize in TX but have not been formally named, including primarily hybrids of *B. leucantha* Torr. & A. Gray and *B. nuttalliana* (Alston & Turner 1963). (Greek: *baptis*, dye or dip, from the economic use of some species which yield a poor indigo dye) (subfamily Faboideae, tribe Thermopsideae).

REFERENCES: Larisey 1940; Turner & Alston 1959; Alston & Turner 1962, 1963; Isely 1981; Kosnik et al. 1996; Turner 2006a; Weakley, A.S., and Southeastern Flora Team 2022; Turner 2023a.

1. Petals cream-colored; inflorescences held distinctly below the level of the leaves; bracts and stipules usually conspicuous, large (e.g., stipules 4–10 mm long) and persistent ____ **B. leucophaea**
1. Petals blue-violet to yellow, orange, brick-red, or white, but not cream; inflorescences at least partially erect and held above the level of the leaves OR flowers solitary in the upper leaf axils; bracts and stipules often inconspicuous, small, quickly deciduous OR (in some hybrids) sometimes somewhat persistent.
   2. Flowers solitary in the upper leaf axils (at least some) or in short terminal racemes with 1–4 flowers; petals yellow ____ **B. nuttalliana**
   2. Flowers in racemes of 5–20 or more flowers; petals white, to yellow, orangish, brick-red, or blue-violet, sometimes multicolored.
      3. Petals white (standard can have some purplish splotches); plants glabrous and often glaucous; inflorescences vertical (strictly erect); species of e portion of East TX ____ **B. lactea**
      3. Petals blue-violet to yellow, orangish, or brick-red, sometimes multicolored; inflorescences vertical (strictly erect) OR angled about 45°; plants glabrous and glaucous to moderately pubescent; including species widespread in East TX.
         4. Petals blue-violet (keel sometimes whitish); stems at base thick (> 5 mm in diam.); plants usually growing individually; inflorescences one to a few per plant.
            5. Petals intensely to moderately blue-violet; inflorescences vertical (strictly erect); foliage glabrous and glaucous ____ **B. minor**
            5. Petals moderately to pale blue-violet; inflorescences angled about 45°; foliage slightly pubescent, not glaucous ____ **B. ×bicolor**
         4. Petals yellow to orangish or brick-red, sometimes partially blue-violet, but in these cases the corollas multicolored and at least partially some other color (not simply blue-violet); stems at base thin (< 5 mm diam.); plants usually growing in dense clusters; inflorescences usually numerous per plant.
            6. Petals brick-red and yellow to orangish to blue-violet and yellow, different petals of the same flower often of different colors; foliage glabrous; rare hybrid known only from Fannin and Grayson cos. on extreme nw margin of East TX ____ **B. ×variicolor**
            6. Petals yellow, all ± the same color; foliage glabrous to moderately pubescent; including species widespread and often abundant in East TX.
               7. Racemes nearly vertical (strictly erect); foliage glabrous OR moderately pubescent; including species widespread and often abundant in East TX.
                  8. Petals vividly yellow; foliage glabrous; species widespread and often abundant in East TX ____ **B. sphaerocarpa**
                  8. Petals pale to medium yellow; foliage moderately pubescent (in most of East TX) OR glabrous (in s part of East TX); rare hybrid, usually present only in mixed populations of *B. lactea* and *B. leucophaea* ____ **B. ×sulphurea**
               7. Racemes angled about 45°; foliage moderately pubescent; rare hybrid, usually present only in mixed populations of *B. sphaerocarpa* and *B. leucophaea* ____ **B. ×bushii**

Baptisia ×bicolor Greenm. & Larisey [*B. minor* × *B. leucophaea*], (two-colored), TWO-COLOR WILD INDIGO. Plant usually growing individually, intermediate to both parents in numerous characters, of medium height (30–60 cm tall), generally branching at or near the ground; stems and foliage moderately pubescent; leaves and stipules of moderate length (center leaflet 27–51 mm long; stipules 4–10 mm long); flowers borne on 1–several racemes angled about 45° above the ground; floral bracts quickly deciduous; pedicels 10–18 mm long; corollas large (27–33 mm long), moderately to pale blue-violet. Prairies, pastures, soils often intermediate between clay and sand or in areas where soil types are in proximity; known in TX only in Fannin and Grayson (BRIT) cos. in Red River drainage on nw margin of East TX; sc U.S. from MO w to KS s to OK and TX. Apr.

Baptisia ×bushii Small [*B. leucophaea* × *B. sphaerocarpa*], (for Benjamin Franklin Bush, 1858–1937, amateur botanist of Missouri), BUSH'S WILD INDIGO. Plant usually growing individually, intermediate to both parents in numerous characters, low (26–36 cm tall), branching at or near the ground; stems and foliage moderately pubescent; leaves and stipules moderate to large (center leaflet 42–60 mm long; stipules 7–34 mm long); flowers on several racemes angled about 45° above the ground; floral bracts quickly deciduous; pedicels 7–15 mm long; corollas medium-sized (22–29 mm long), pale to medium yellow. Usually on sandy soils; Fannin and Grayson cos. in Red River drainage and disjunct to se TX; sc U.S. in AR, OK, s LA and TX. Mar–Apr. [*B. ×intermedia* Larisey, *B. ×stricta* Larisey] This nothospecies was originally described as a non-hybrid species (Small 1903) and continued to be identified as such in Larisey (1940). Turner (1959) included it in synonymy under *B. laevicaulis* (A. Gray ex Canby) Small without explanation. Turner and Alston (1959) and Alston and Turner (1962, 1963) underwent detailed studies

Astragulus macilentus [BT3]

Astragalus missouriensis [BB2]

Astragalus nuttallianus var. austrinus [BT3]

Astragalus nuttallianus var. trichocarpus [COC]

Astragalus plattensis [BB1]

Astragalus pleianthus [CP1]

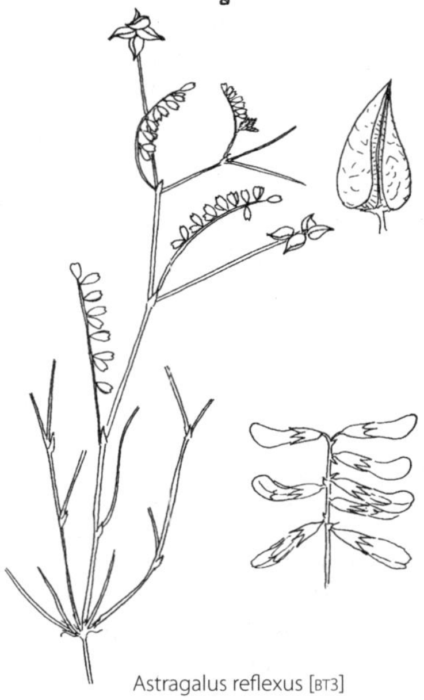
Astragalus reflexus [BT3]

of hybrids of *B. leucophaea* var. *laevicaulis* (as *B. laevicaulis*) and *B. sphaerocarpa* (as *B. viridis* Larisey in Turner & Alston (1959)), but never identify the hybrids as *B.* ×*bushii*. It was apparently first recognized as a hybrid by annotation of the type (*Bush* 68 NY) by Isely in 1978 and an isotype by unknown annotator and date (*Bush* 68 MO). It has two disjunct distributions presumably including the two varieties of *B. leucophaea* as parents in n TX, AR, and OK with var. *leucophaea* and in se TX and s LA with var. *laevicaulis* (Kartesz 2023).

Baptisia lactea (Raf.) Thieret, (feminine singular of *lacteus*, meaning milky, probably in reference to white flowers), WHITE WILD INDIGO. Plant robust, 0.5–1.2(–2 m) tall, usually erect or divaricate, glabrous and often glaucous; leaflets large, 20–60(–80) mm long; stipules small (ca. 5 mm or less long), quickly deciduous; flowers usually numerous, on 1–many vertical racemes; floral bracts quickly deciduous; pedicels 3–10 mm long; corollas (15–)20–25 mm long, white (standard can have some purplish splotches); fruits broadly ellipsoid to ellipsoid cylindric, apically truncate, abruptly beaked, 2–4(–5) cm long, 1–1.2(–1.5) cm wide, rigidly coriaceous, black at maturity, glabrous. Open woodlands, prairies, pastures, and roadsides usually on calcareous soils; mostly in e part of East TX w to Robertson Co. (Turner et al. 2003); also n Gulf Prairies and Marshes; c U.S. from OH s to MS and w to MN, SD, OK, and TX. Apr–Jun. [*Baptisia alba* (L.) R. Br. var. *macrophylla* (Larisey) Isely, *B. leucantha* Torr. & A. Gray, *B. pendula* Larisey var. *macrophylla* Larisey] This species was previously recognized as a variety of *B. alba*, which occurs in the se U.S. but only as far w as AL (Weakley, A.S., and Southeastern Flora Team 2022).

Baptisia leucophaea Nutt., (dusky-white), PLAINS WILD INDIGO, LONG-BRACT WILD INDIGO. Plant usually growing individually, typically low (21–36 cm tall), often wider than tall, branching at or near the ground, with all vegetative parts densely pubescent or glabrous; leaves and stipules large (center leaflet 30–60 mm long; stipules 10–40 cm long); flowers on reclining or hanging racemes; floral bracts persistent and large (10–30 mm long); pedicels long (15–35 mm long); corollas large (25–38 mm long); fruits (2–)3–4(–5) cm long, 1.5–2.5 cm wide, tapering to a slender beak (beak rather wide at base), black, pubescent or rarely later glabrate or glabrous. The plants turn blackish or silver-gray late in the summer, break off, and tumble in the wind (Ajilvsgi 1984). This species has had a diverse taxonomic and nomenclatural history and is often included as a variety of *B. bracteata* Muhl. ex Elliott (Isely 1990, 1998; Diggs et al. 1999), which is now considered a separate species occurring east of the Mississippi River (Weakley, A.S., and Southeastern Flora Team 2022; Turner 2023a). Two distinctive varieties occur in TX and mostly west of the Mississippi River and are described below.

1. Stem and leaves densely pubescent; leaflets, or at least those of lower leaves oblanceolate, relatively narrow, usually 2.5–5.5 times as long as wide; variety of the n part of East TX ____ var. **leucophaea**
1. Stem and leaves glabrous or nearly so; leaflets cuneate-rhombic or cuneate-obovate, relatively broader, 1.5–2.5 times as long as wide; variety of the s part of East TX ____ var. **laevicaulis**

var. **laevicaulis** A. Gray ex Canby, (smooth-stemmed). Plant essentially glabrous or nearly so; leaflets glabrous to 6 cm long. Open pine or mixed woodlands, coastal prairies, dunes, pastures, and roadsides; s part of East TX along the coast into s TX; in the U.S. otherwise known only from LA. Mar–May. [*B. bushii* of authors, not Small, *B. laevicaulis* (A. Gray ex Canby) Small, *B. bracteata* Muhl. ex Elliott var. *laevicaulis* (A. Gray ex Canby) Isely] This variety is distinctive in its ideal form by its glabrous stems and leaves and broadly rhombic leaves and racemes that often extend on to the ground (Isely 1998; E. Keith, pers. obs.). All examined specimens identified as var. *laevicaulis* from Arkansas and Oklahoma and mapped in Kartesz (2023) are var. *leucophaea*.

var. **leucophaea.** All vegetative parts densely pubescent; fruit pubescent or rarely later glabrate; leaflets villous to 8 cm long. Open oak or mixed woodlands, prairies, and pastures; n portion of East TX; c U.S. from WI and nw IN se to w KY and s to MS and w to e NE and TX. Mar–May. [*B. bracteata* Muhl. ex Elliott var. *leucophaea* (Nutt.) Kartesz & Gandhi, *B. leucophaea* var. *glabrescens* Larisey] This variety has often been treated as *B. bracteata* var. *leucophaea* (Diggs et al. 1999; Kartesz 2023). While some authorities (e.g., Turner et al. 2003) do not recognize varieties in this species, the morphological differences seem equivalent to or greater than those between varieties within other *Baptisia* species or other genera of the family. We are therefore following Isely (1998) and Turner 2023a, both treating *Baptisia* for the entire U.S., in recognizing the variation at the varietal level. This distinctive variety is recognized by its densely pubescent stems and more robust bushy aspect growing broader than high and with racemes that grow from the sides of the stem, but usually held above the ground (Isely 1998; E. Keith, pers. obs.).

Baptisia minor Lehm., (smaller), BLUE WILD INDIGO, WILD BLUE-INDIGO, BLUE-INDIGO. Plant usually growing individually, usually 46–74(–100) cm tall, of strictly erect posture, with a single thick (> 5 mm diam.), glaucous stem rising from the ground several cm before branching; branches generally few and rigid, spreading; leaves small (center leaflet 18–34 mm long) and glabrous; stipules small (4–11 mm long); flowers borne on a usually single, rigidly erect raceme; floral bracts quickly deciduous; pedicels 5–15 mm long; corollas large (27–36 mm long), blue-violet of variable intensity; fruits much inflated, usually much longer than wide,

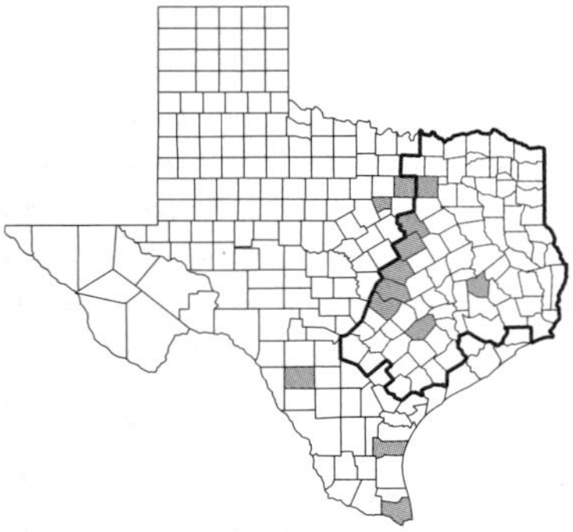

Astragalus reflexus

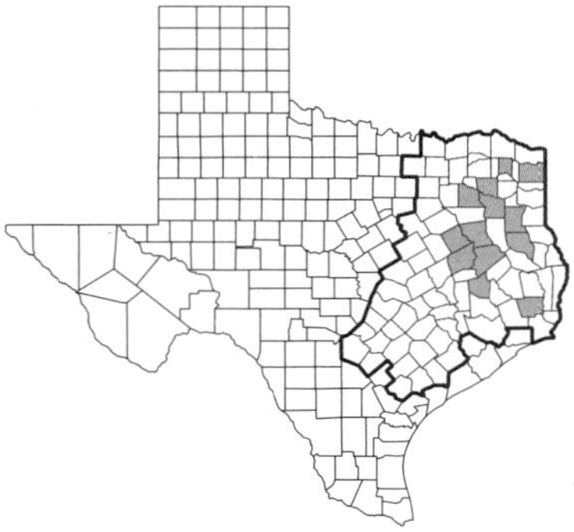

Astragalus soxmaniorum

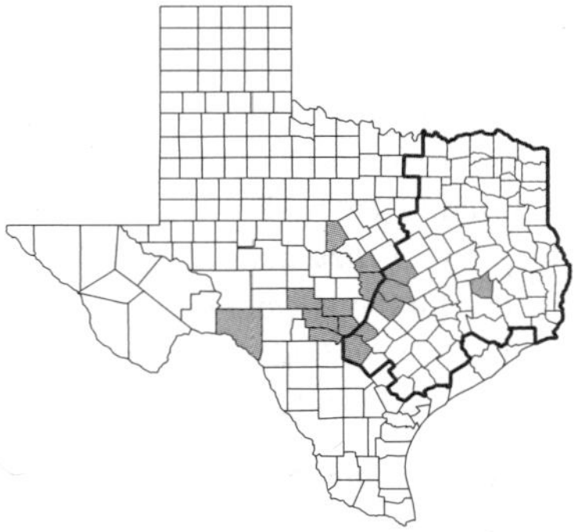

Astragalus wrightii

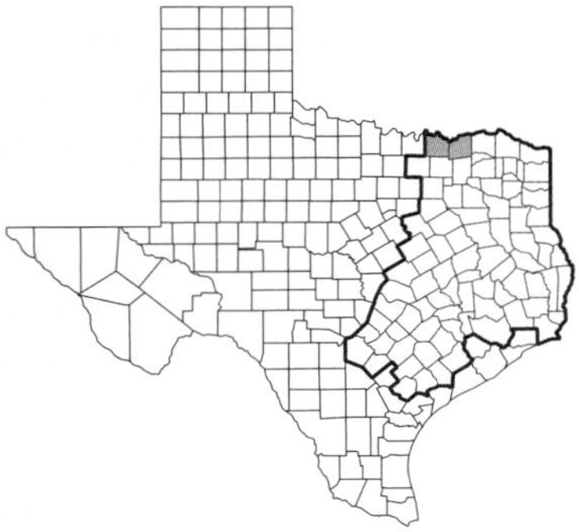

Baptisia ×bicolor

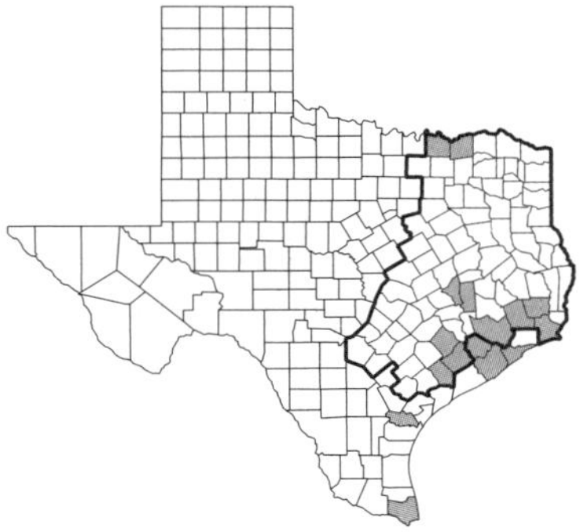

Baptisia ×bushii

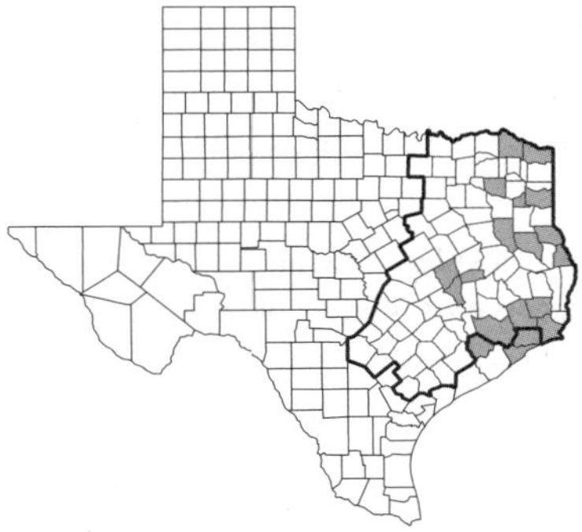

Baptisia lactea

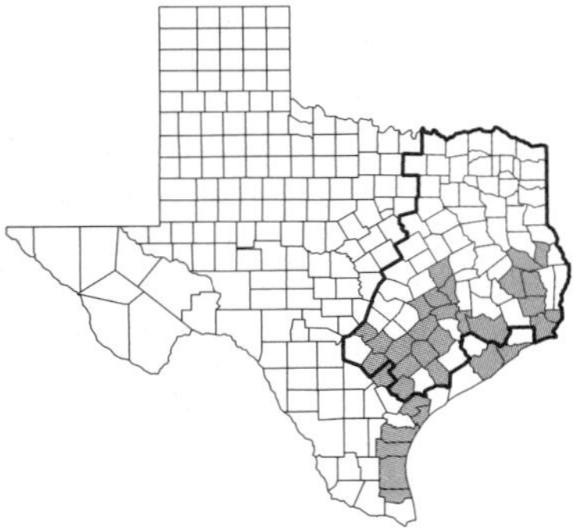

Baptisia leucophaea var. laevicaulis

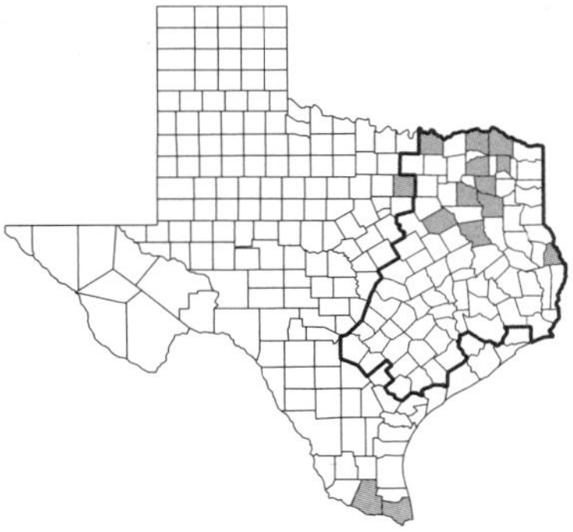

Baptisia leucophaea var. leucophaea

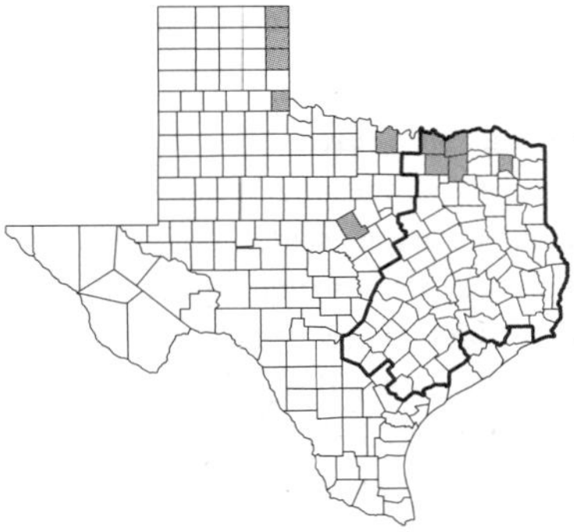

Baptisia minor

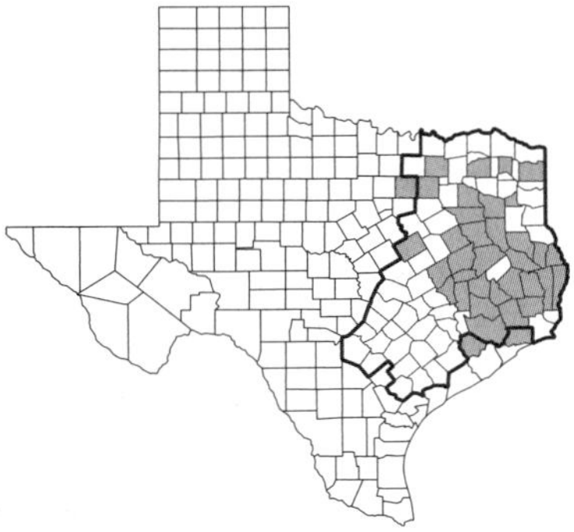

Baptisia nuttalliana

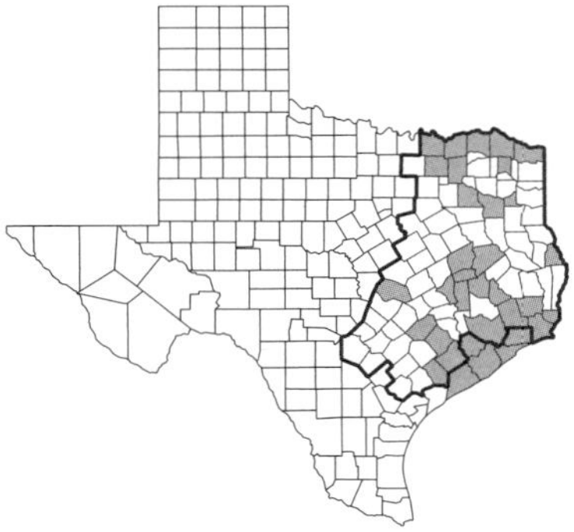

Baptisia sphaerocarpa

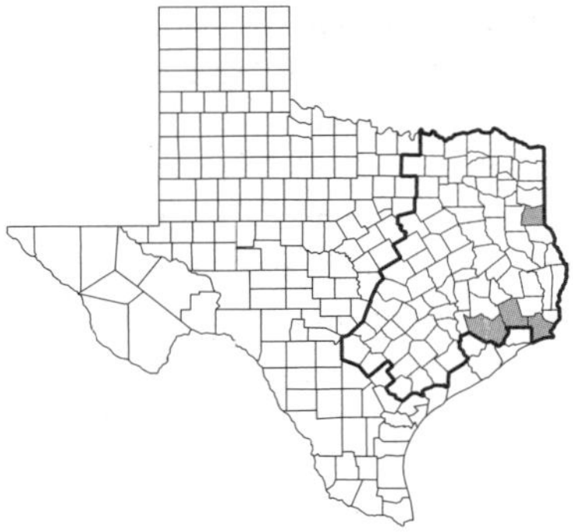

Baptisia ×sulphurea

30–60 mm long, 20–30 mm wide, often asymmetric, with a distinct persistent beak that is noticeably widened at base, black at maturity, glabrous. Clay soils, often on limestone, prairies, and pastures; Collin, Fannin, Grayson, Hunt, Titus (BRIT), and Dallas (Mahler 1988) cos. in n part of East TX, disjunct in Tyler Co. in se TX (J. Singhurst, pers. comm. and photograph); otherwise only known in TX from Comanche and Montague (BRIT) cos. in the Cross Timbers and Prairies and the Panhandle; c U.S. from IL and IA s to n AR w to se NE, OK, and TX. Apr–early Jun. [*B. australis* (L.) R. Br. var. *minor* (Lehm.) Fernald, *B. texana* Buckley, *B. vespertina* Small ex Rydb.] According to Ajilvsgi (1984), the sap turns purple when exposed to air; it is used in making a blue dye for dying wool for use in weaving. Weakley (2015, 2018) recognizes three species in the *B. australis* "complex" including *B. minor* (ours); *B. aberrans* (Larisey) Weakley, restricted to grassland/glade habitats in disjunct locations east of the Mississippi River; and *B. australis* (L.) R. Br., the more common and larger, blue-flowered species east of the Mississippi River.

Baptisia nuttalliana Small, (for Sir Thomas Nuttall, 1786–1859, English-American botanist), NUTTALL'S WILD INDIGO. Plant erect, to 120 cm tall, with well-developed individuals spherical and bush-like in form, puberulent, then glabrate; leaflets 25–60(–80) mm long; stipules very small, deciduous; pedicels 2–5 mm long; flowers solitary in upper axils or some in short, terminal, 1–4-flowered racemes, corollas 17–20 mm long, yellow; fruits ovoid to subspheroid, usually 8–13 mm long, (5–)8–13 mm wide, abruptly beaked, black, glabrate. Sandy open woods, pastures, and roadsides; widespread in East TX; also immediately adjacent Cross Timbers and Prairies and n Gulf Prairies and Marshes; sc U.S. in AR, LA, se OK, and TX. Apr–May. According to Isely (1990), it often forms extensive poorly fruiting stands. This is the only East TX *Baptisia* species with mostly axillary flowers.

Baptisia sphaerocarpa Nutt., (globe or spherical-fruited), GREEN WILD INDIGO, YELLOW BUSH-PEA. Plants often growing in a dense cluster with a number of apparently identical individuals in close proximity, suggesting vegetative spread, tall (40–70 cm) and of generally erect posture, with 1–many small, flexible, glabrous stems, branching close to the base; leaves large (center leaflet 36–80 mm long); upper leaves often reduced to 1–2 leaflets; stipules small (2–5(–6.1) mm long) to absent; flowers borne on 1–many flexible vertical racemes; floral bracts quickly deciduous; pedicels short (1–7 mm long); corollas small (17–25 mm long), intensely yellow; fruits subspheroid, 1.4–1.8 cm in diam., usually light brown at maturity (Isely (1990) also reported black), glabrous. Sandy or silty clay; widespread in East TX; also n Gulf Prairies and Marshes; sc U.S. from MO se to s MS and w to OK and TX. Mar–May. [*B. viridis* Larisey] In East TX, the bright yellow flowers in upright racemes and small, nearly round, light brown fruits immediately distinguish *B. sphaerocarpa* from the other three common species with larger, more elongate, black fruits (Isely 1990).

Baptisia ×sulphurea Engelm. [*B. lactea* × *B. sphaerocarpa*], (sulphur-yellow), SULPHUR WILD INDIGO. Plant erect, glabrous, intermediate to both parents in numerous characters; leaves on very short petioles, leaflets obovate, rhombic, obtuse, or occasionally emarginate; stipules and floral bracts small and quickly deciduous; corollas pale yellow; style much longer than ovary; fruit not seen, but Alston and Turner (1963) reported it as intermediate between the fruit morphology of the two parents. Prairies and pastures; Harris, Jefferson, and Liberty cos. in se TX and Panola Co. in ne TX; sc U.S. in AR, LA, OK, and TX. Apr–May. [*B. fragilis* Larisey, *B. macilenta* Small ex Larisey] Engelmann (1878) originally described the species and suggested that it may be of hybrid origin between *B. leucantha* (now *B. lactea*) and *B. sphaerocarpa*. Larisey (1940) listed it as a hybrid species in her monograph suggesting that it was a hybrid of *B. sphaerocarpa* and *B. australis*, however, listing *B. australis* as one of the putative parents was written in error since she annotated the holotype as *B. ×sulphurea* = *B. leucantha* × *B. sphaerocarpa* (*Butler s.n.*, 1877, MO).

Baptisia ×variicolor Kosnik, Diggs, Redshaw, & Lipscomb, [*B. minor* × *B. sphaerocarpa*], (variably colored), VARICOLORED WILD INDIGO. Plant often growing in a dense cluster with a number of apparently identical individuals in close proximity, suggesting vegetative spread, intermediate to both parents in numerous characters, but vegetatively often more similar to *B. sphaerocarpa* than *B. minor*, tall (40–76 cm) and of generally erect posture, with one to many small, flexible, glabrous stems, branching at or near the base and often; leaves medium-sized (center leaflet 30–50 mm long); stipules small (3–15 mm long) to absent; flowers borne on 1–many flexible vertical racemes; floral bracts quickly deciduous; corollas of intermediate size (22–27 mm long), brick-red and yellow to orangish to blue-violet and yellow. Prairies, pastures, soils often intermediate between clay and sand or in areas where soil types are in proximity; known in TX only in Fannin and Grayson (BRIT) cos. in the Red River drainage; also s OK (Kartesz 2023). Apr–May. This nothospecies was originally described in 1996 (Kosnik et al. 1996). Despite the original description including *B. australis* instead of *B. minor* as a parent, the ICBN, Code Article H.10: Note 2, states "Because the name of a nothotaxon at the rank of species or below has a type, statements of parentage play a secondary part in determining the application of the name" (K. Gandhi, pers. comm.).

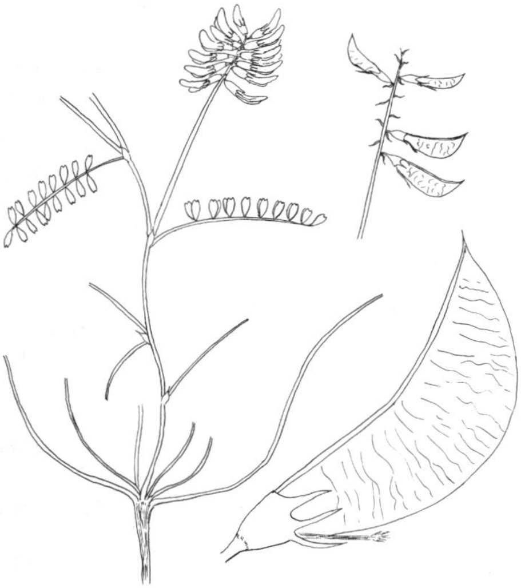
Astragalus soxmaniorum [BT3]

Astragalus wrightii [BT3]

Baptisia lactea [STE]

Baptisia leucophaea [STE]

Baptisia minor [STE]

Baptisia nuttalliana [COC]

Baptisia sphaerocarpa [BT3, GLE]

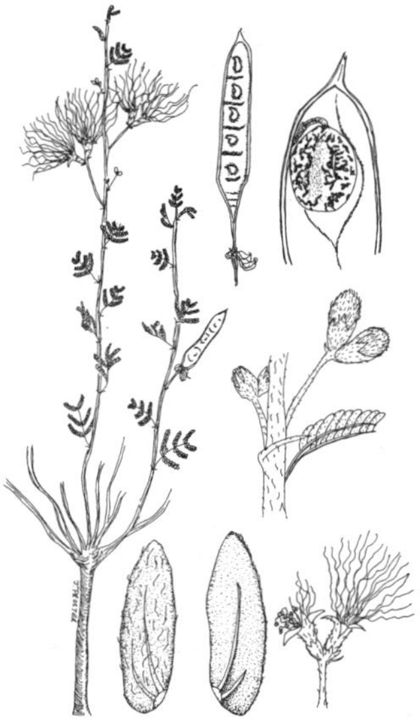
Calliandra biflora [SID]

CALLIANDRA Benth. STICK-PEA

Small unarmed shrubs or woody based herbs, 60 cm or less tall, somewhat similar to *Mimosa* species; leaves twice pinnately compound, with 1–3(–4) pairs of pinnae; leaflets 8–20(–28) per pinna, without glands; flowers aggregated, 2–12+ on short axillary peduncles, small and inconspicuous, yellowish white to cream or pink; stamens (14–)19–28, serving as the main attractant structure for the flower; fruits dehiscent (the two valves separating along both sutures and strongly curved outwards after dehiscence), usually with 1–6 seeds.

A genus of 130 species of mainly tropical America; some are cultivated as ornamentals, for their heads of flowers (showy stamens) (Hernández 2023). Four species are native to TX (2 in East TX), one of these, *C. iselyi* B.L. Turner, was described in 2000 (Turner 2000). (Greek: *kalos*, beautiful, and *andros*, male) (subfamily Caesalpinioideae, Mimosoid clade, tribe Ingeae)
REFERENCES: Isely 1998; Turner 2000; Souza et al. 2013; Hernández 2023.

1. Wand-like perennial or occasional subshrub, 30–60 cm tall; each peduncle bearing only 2(–3) flowers at its apex; fruits 7–10 mm wide at maturity; leaves with (1–)2–3 pairs of pinnae ______ **C. biflora**
1. Gnarled low-growing shrub with complex branching, 30 cm tall; each peduncle bearing 3–10(–15) flowers; fruits 5–7 mm wide at maturity; leaves with 1(–4) pair of pinnae (in East TX) ______ **C. eriophylla**

Calliandra biflora Tharp, (two-flowered), TWO-FLOWER STICK-PEA. Small subshrub growing from a woody caudex and rhizomes, usually 30–60 cm tall; leaves bipinnate with (1–)2–3 pairs of pinnae; leaflets 5–10 pairs per pinna, oblong-lanceolate, 5–9 mm long, sessile; stipules acicular ± 4 mm long; inflorescences peduncled, 8–10 mm long, usually 2-flowered or occasionally axillary and solitary, corolla 4–5 mm long, stamens reddish to 15 mm long; fruits 4–8 cm long, conspicuously pubescent on sutures. Grasslands and adjacent roadsides; Dewitt, Goliad (BRIT; Correll & Johnston 1970) and Victoria (BRIT) cos. at extreme s margin of East TX; also Karnes Co. (Isely 1998) in the adjacent South TX Plains; in the U.S. known only from TX; also adjacent Mexico. May–Jul. This species is rare and only known from the four counties listed above in the U.S. (Isely 1998). NatureServe: G3, vulnerable. ⚠

Calliandra eriophylla Benth., (Greek: *erion*, wool and *phyllon* in reference to hairy leaves), TEXAS FAIRY DUSTER, PINK MIMOSA, PINK-FLOWERED ACACIA, HUAJILLO, PLUMITA, GAVIA. Low contorted shrub usually to only 30 cm tall in TX, but to 60(–120) cm westward; pubescence of flexuous hairs; leaves alternate or 1–2 from spurs with 1–3(–4) pairs of pinnae, leaflets imbricate, 8–10 pairs per pinna, 3–6 mm long, 0.7–2 mm wide; inflorescences subsessile to 15 mm long; heads yellow-white to pink, 4–8 flowers; stamens ca. 10 mm long; fruits oblanceolate, 2.5–4(–10) cm long. Open limestone and caliche areas; Travis (BRIT) and Caldwell cos. near sw margin of East TX and Goliad Co. (Turner et al. 2003) at extreme s margin; mainly Edwards Plateau, Trans-Pecos, and South TX Plains; sw U.S. from TX w to CA; also adjacent Mexico. Mar–Jul, sometimes also in fall. Most TX plants have traditionally been referred to as *C. conferta* Benth. (Isely 1998; Turner 2000; Turner et al. 2003; Souza et al. 2013), however, we follow the most recent treatment in Hernández (2023) who wrote, "Specimens from across the range of *C. eriophylla* show that 1-paired leaves often occur intermixed with plants with 2 or 3 pairs of leaves. *Calliandra conferta* cannot be separated from *C. eriophylla* if the full morphological spectrum of the latter is considered."

CENTROSEMA (DC.) Benth. BUTTERFLY-PEA

A genus of about 35 species of warm areas of the Americas (Fantz 2023a); some are used as green manures under crops such as rubber and coconut. There are two native and one introduced species in the U.S. (Isely 1998). *Centrosema* and *Clitoria* are the only U.S. members of the family with large resupinate flowers (= turned upside down or inverted, with standard on lower side of the flower) (Isely 1990). *Clitoria*, however, has a longer calyx tube—ca. 10–20 mm long, oblong fruits—5–10 mm wide, and rarely twines in East TX (Isely 1998; E. Keith, pers. obs.). (Greek: *centron*, a spur, and *sema*, a standard) (subfamily Faboideae, tribe Phaseoleae)
REFERENCES: Isely 1998; Fantz 2002; Fantz 2023a.

Centrosema virginianum (L.) Benth., (of Virginia), BUTTERFLY-PEA. Perennial vine; stems glabrous or inconspicuously pubescent, trailing or usually twining, to 1.6 m long; leaves pinnately compound; leaflets 3, oblong-ovate to lanceolate, 2–5 cm long, scabrous-pubescent above, minutely pubescent or nearly glabrous and with prominent raised veins beneath; petioles not winged; flowers axillary, solitary (rarely in 2s); bracteoles calyx-like, 8–12 mm long, partly hiding calyces, calyces campanulate, 10–16 mm long, the tube ca. 4 mm long, the lobes longer than the tube; corollas 20–40 mm long; petals lavender to violet-blue, the standard with reddish central markings and whitish base, rarely all petals white; stamens 10, diadelphous; fruits linear, beaked, 7–12(–14) cm long, ca. 3–4 mm wide. Sandy woods, roadsides; widespread in the e 1/2 of TX; e and s U.S., NJ s to FL, w to e OK and TX; s to South America. Jun–Sep. Isely (1998) noted that this species extends s to Uruguay and Argentina and has the "widest distribution of any species of its genus." [*Clitoria virginiana* L.]

CERCIS L. REDBUD

A genus of 10 species disjunct in four n temperate areas in Asia, Europe, w North America, and e North America (extending s into ne Mexico) (Robertson & Lee 1976; Ballenger & Vincent 2023). These four widely separated occurrences represent Tertiary relict areas where remnants of a once widespread north temperate flora still survive (Wood 1972); for further discussion of this distribution see the treatment of *Campsis* (Bignoniaceae) and page 203 of Vol. 1 of The *Illustrated Flora of East Texas* (Diggs et al. 2006). Interestingly, based on molecular data, "*C. canadensis* from eastern North America is more closely related to *C. siliquastrum* from western Eurasia than to *C. occidentalis* from western North America" (Davis et al. 2002), reflecting Tertiary migration pathways when continental orientations, sea level, and climate were quite different than today. The species are deciduous trees with flowers appearing along the branches and trunk (= cauliflory) before the leaves have expanded. Some species are commonly cultivated as ornamentals. The newly circumscribed Cercidoideae is considered among the most basal elements in the family (Ballenger et al. 1993; Bruneau et al. 2001; LPWG 2017). (Ancient Greek name applied perhaps to a popular, but also to *C. siliquastrum* L., JUDASTREE; possibly from *kerkis*, a weaver's shuttle, alluding to the large woody fruits; the common name reflects the tradition that this was the tree from which Judas hanged himself) (subfamily Cercidoideae, tribe Cercideae)

REFERENCES: Hopkins 1942; Bruneau et al. 2001; Davis et al. 2002; Weakley, A.S., and Southeastern Flora Team 2022; Ballenger & Vincent 2023.

1. Plants usually single-trunked; blades of mature leaves thin (but firm), cordate, as long as wide or slightly longer, pointed, dull green on both surfaces; widespread in East TX ______ **C. canadensis** subsp. **canadensis**
1. Plants often multi-trunked; blades of mature leaves thick and leathery, cordate to cordate-reniform, wider than long, rounded or barely pointed, rich deep green, shiny, distinctly glaucous above; western portion of East TX ______ **C. canadensis** subsp. **texensis**

Cercis canadensis L., (of Canada), Shrubs or trees; leaves apparently simple (with 1 leaflet), palmately veined, cordate, ovate-acuminate or orbicular or subreniform with a deep but shallow sinus at base, thin to coriaceous, glabrous or pubescent, opening after the flowers; stipules deciduous; flowers pedicellate in clusters of 2–6 from spurs; calyces asymmetrically cup-shaped, 5-toothed; petals pinkish purple (very rarely white), ca. 8–13 mm long, very unequal, the flower simulating a papilionaceous form (but differing in having the banner smaller than and inside the wing petals and the keel petals not fused); stamens 10, all fertile, separate; fruits slightly stipitate, flattish, 4–10.9 cm long, 7–21 mm wide.

subsp. **canadensis,** REDBUD, JUDASTREE. **State tree of Oklahoma.** Small thornless, often understory tree, 3–10 m tall; leaves apparently simple (with 1 leaflet), opening after the flowers, cordate basally; flowers pedicellate in clusters of 2–6; calyces asymmetrically cup-shaped, 5-toothed; petals pinkish purple (very rarely white), ca. 8–11 mm long, very unequal, the flower simulating a papilionaceous form; fruits slightly stipitate, flattish, 4–10 cm long, 9–14(–18) mm wide. Stream bottoms and lower slopes, sandy or silty ground; widespread nearly throughout East TX except extreme w margin, also ne High Plains in TX; e U.S. from MA s to FL w to ne NE and s to w OK, and TX. Mar–Apr. Hybridization apparently occurs where it meets *C. canadensis* subsp. *texensis* in the western portion of East TX. The species is widely cultivated for its showy early spring flowers—our cultivated redbuds are mostly mongrel types derived from the two East TX subspecies (Mahler 1988). Yatskievych (2006) noted that the flowers produce nectar, and that pollination is by long- and short-tongued bees. Cox and Leslie (1991) reported that the buds, flowers, and young pods are edible—particularly sautéed; however, Burlage (1968) indicated that REDBUD is poisonous due to a saponin (no further details as to which organ, etc., were given). Ample quantities of flowers and young seedpods have been consumed by one of the authors with no ill effect (E. Keith, pers. obs.). Caution is, however, advised. In addition to the two subspecies included here, *Cercis canadensis* subsp. *mexicana* (Rose) A.E. Murray, occurs in the western Edwards Plateau and Trans-Pecos (Isely 1998; Turner et al. 2003). ☠

subsp. **texensis** (S. Watson) A.E. Murray, (of Texas), TEXAS REDBUD. Small thornless tree, 3–8 m tall; leaves apparently simple (with 1 leaflet), opening after the flowers, cordate or reniform basally, shiny and glaucous above; flowers pedicellate in clusters of 2–6; calyces asymmetrically cup-shaped, 5-toothed; petals pinkish purple, ca. 9–13 mm long, very unequal, the flower simulating a papilionaceous form; fruits slightly stipitate, flattish, 4–10 cm long, 7–17(–21) mm wide. Rocky limestone slopes; w margin of East TX; mostly in the middle 1/3 of TX; s OK and TX. [*C. texensis* (S. Watson) Sarg., *C. canadensis* var. *texensis* (S. Watson) M. Hopkins] This subspecies is treated as *C. texensis* in Weakley, A.S., and Southeastern Flora Team (2022).

CHAMAECRISTA (L.) Moench SENNA

Erect annuals or perennials, unarmed; leaves 2-ranked, once even-pinnately compound; petioles with a sessile to stalked, ± disc-shaped gland (extrafloral nectary); stipules well-developed, persistent; petals yellow, some of them sometimes with a reddish spot at base; stamens 5 or 10, all fertile, separate; anthers basifixed, apically dehiscent by pores, of 2 different lengths; fruits elastically (= explosively) dehiscent and capable of throwing seeds a meter or more.

A genus of ca. 330 species; pantropical but mainly South America to s U.S. (Marazzi 2023); sometimes lumped into the genus *Cassia* (e.g., Isely 1975), but differing in characters such as well-developed persistent stipules, anthers of 2 sizes apically dehiscent by pores, filaments not sigmoid, and fruits elastically (= explosively) dehiscent. Irwin and Barneby (1982) gave a detailed key separating *Cassia*, *Chamaecrista*, and *Senna* and more recent molecular work (Bruneau et al. 2001) supports this separation. The leaflets of *Chamaecrista* are slightly sensitive to touch (they do not respond instantly, rather over a period of time; they are not nearly as sensitive as those of the instantly responsive SENSITIVE-BRIARS in the genus *Mimosa*. Yatskievych (2006) noted that "Uprooting the plant also causes the leaflets to close, and herbarium specimens almost always have the leaves in the closed position. On their own, the leaves close and pull upward at night into a so-called sleeping position, a phenomenon known as nyctinasty, common to many Fabaceae. This movement is thought to control water loss or afford protection from herbivores." (Greek: *chamai*, on the ground or low, and *crista*, crest) (subfamily Caesalpinioideae, tribe Cassieae)

REFERENCES: Turner 1959; Isely 1975; Irwin & Barneby 1982; Rutter & Rausher 2004; Weakley, A.S., and Southeastern Flora Team 2022; Marazzi 2023.

1. Leaflets 2–8(–11) pairs per leaf, never more than 11; species of extreme sw margin of East TX **C. calycioides**
1. Leaflets 5–32 pairs per leaf, larger leaves with >11 pairs; species widespread in East TX (*C. flexuosa* only on sw margin).
   2. Perennial or annual (*C. chamaecristoides* only) with woody stems clustered from base, variously prostrate, decumbent, ascending or rarely erect; peduncles axillary or supra-axillary by adnation 0–15 mm above node; sw portion of East TX.
      3. Peduncles strictly axillary, 20–40 mm long; styles 4–9 mm long; inland **C. flexuosa**
      3. Peduncles supra-axillary, 1–10 mm above the node, to 15 mm long; styles 3.5–6 mm long; mostly coastal, but inland on dunes and saline soils **C. chamaecristoides**
   2. Annuals with single (often branched) stem from base, ascending or erect; peduncles adnate to stem 1–26 mm above the node (appearing to diverge from the stem in the internode); widespread and common in East TX.
      4. Flowers and fruits distinctly stalked (the pedicels 5–15 mm long); flowers 10–17 mm long, the largest petal only slightly larger than the other four, the petals yellow with some reddish at base; stamens 10 **C. fasciculata**
      4. Flowers and fruits essentially sessile (pedicels rarely to 3 mm long); flowers about 5(–8) mm long, the largest petal much larger than the other four, the petals yellow; stamens usually 5 **C. nictitans**

Chamaecrista calycioides (DC. ex Collad.) Greene, (presumably referring to the whorls of bracts below the calyx, resembling a calyx), WOODLAND SENSITIVE PEA. Perennial from woody taproot, decumbent to ascending to 0.5 m tall, pubescent with purple speckles, stems several, clustered; leaves 2-ranked, with 6–8(–11) pairs of leaflets; leaflets 4–10 mm long; petioles with a single inconspicuous gland below lowest pair of leaflets; stipules persistent, striate 3–6 mm long; flowers axillary, solitary, often partly submerged in bracts and appearing supra-axillary; sepals 9–11 mm long, densely glandular; petals mostly all yellow, or the upper 4 with a faded reddish spot at base, the lower slightly larger, 15–18 mm long; stamens 10, the anthers yellow; pistils pilose and conspicuous when elongating; fruits 3–4 cm long and 3–4 mm wide. Open areas on deep sandy soils; Bexar, Goliad, Dewitt, and Wilson (Kartesz 2023) cos. on extreme sw margin of East TX; in the U.S. known only from TX; s through Mexico and Central America to South America. Mar–Oct. [*Cassia aristella* (Pennell) Cory & H.B. Parks] If varieties are distinguished, *C. calycioides* var. *calycioides* is the only variety in TX (Isely 1998).

Chamaecrista chamaecristoides (Collad.) Greene var. **cruziana** (Britton & Rose) H.S. Irwin & Barneby, (sp.: like *Chamaecrista*, originally described in the genus *Cassia;* var.: presumably in reference to the type being collected in Veracruz). Annual, but often surviving the winter and forming woody stems, usually sprawling, 0.2–0.6 m long; stems several, clustered, often adventitiously rooting in shifting sand, usually densely pubescent with incumbent hairs; leaves 2-ranked, densely pubescent above and glabrate below with 8–18 pairs of leaflets; leaflets 3–12 mm long; petioles with a single (2–3 very small if present) gland situated near the middle of the petiole; sepals ovate or lanceolate acuminate 6–11 mm long; petals unequal and yellow, 3 shorter oblong or oblong obovate beyond the claw, 1 abaxial flabellate, 9–19 mm long; stamens 10, the anthers mostly red; fruits (25–)30–65 mm long, 4–5 mm wide. Coastal dunes, beaches, and sand flats; Colorado Co. (Kartesz 2023) on extreme s margin of East TX; mostly Gulf Prairies and Marshes and South TX Plains; in the U.S. known only from TX; also Mexico. Mar–Nov. [*Cassia fasciculata* Michx. var. *ferrisiae* (Britton) B.L. Turner, *Chamaecrista cruziana* Britton & Rose; *C. fasciculata* (Michx.) Greene var. *ferrisiae* (Britton ex Britton & Rose) B.L. Turner; *C. fasciculata* (Michx.) Greene var. *puberula*—valid combination not made but mapped as such

Calliandra eriophylla [MNY]

Centrosema virginianum [BOT]

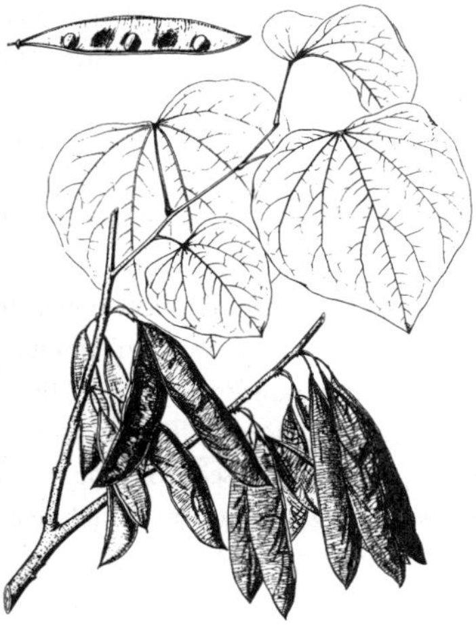
Cercis canadensis subsp. canadensis [SA3]

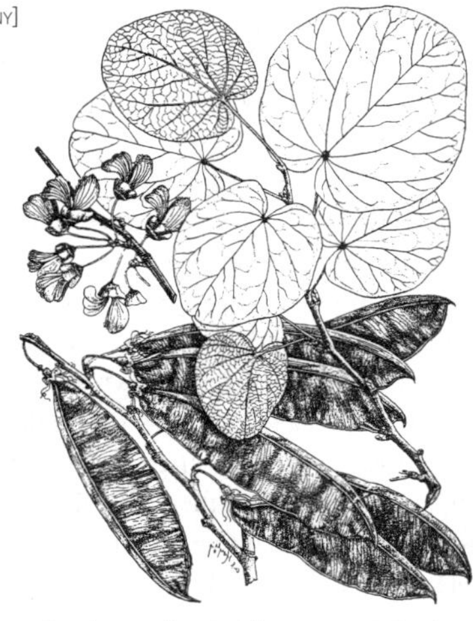
Cercis canadensis subsp. texensis [SUD]

Chamaecrista calycioides [MAR]

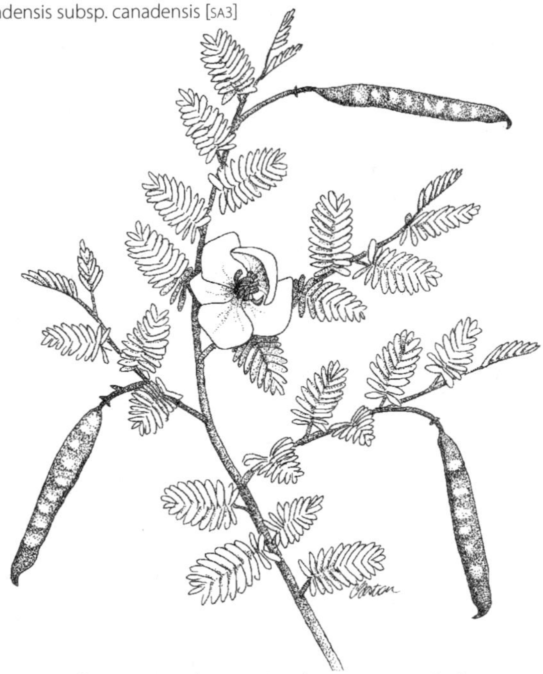
Chamaecrista chamaecristoides var. cruziana [CLA]

Chamaecrista fasciculata [JON]

in Isely (1998) and Turner et al. (2003)] Three varieties are recognized in *C. chamaecristoides* including var. *cruziana* an annual herb that often survives the winter and forms woody stems, and the other two varieties, var. *brandegeei* (Rose) H.S. Irwin & Barneby and var. *chamaecristoides*, perennials that occur only in Mexico and differ from each other and var. *cruziana* in the width of their pods, 3–4 mm and 5–6 mm, respectively (Marazzi 2023). *C. chamaecristoides* var. *cruziana* has long been considered a variety of *C. fasciculata* and occasionally intergrades with that erect growing, obligately annual species (Turner 1959; Correll & Johnston 1970; Hatch et al. 1990; Turner et al. 2003). Irwin and Barneby (1982) summarized the taxonomic problem posed by the occasional intergrades by comparing the problem as the "sort faced by a surgeon required to separate a pair of Siamese twins with one heart in common; irreparable damage to the integrity of one twin is inevitable." However, this variety is usually distinguished by its sprawling habit with a woody caudex, usually densely pubescent stems and leaves, and supra-axillary (versus inter-axillary) emerging peduncles (E. Keith, pers. obs.). Turner et al. (2003) maps this variety (as *C. fasciculata* var. *puberla*) more widely northward and eastward in the Gulf Prairies and Marshes and into the Pineywoods.

Chamaecrista fasciculata (Michx.) Greene, (fascicled, clustered), PARTRIDGE-PEA, SHOWY PARTRIDGE-PEA, BEEFLOWER, PRAIRIE SENNA. Annual 0.6–1 m tall, glabrate or sparsely hirsute; leaves 2-ranked, usually with 8–20 pairs of leaflets, very slightly touch sensitive; leaflets 5–15 mm long, glabrate or sparsely hirsute; petioles with a conspicuous (using hand lens) sessile or short-stalked, orangish, disk-like, nectar-producing gland ca. 1.5 mm across, the gland located from near the middle of the petiole to somewhat closer to the lower pair of leaflets; petals yellow, the upper 4 with a reddish spot at base, the lower slightly larger; anthers yellow and/or purple, 6–10 mm long, one larger than the other nine; fruits (3–)5–8 cm long. Open woods or prairies, disturbed areas, often on sandy soils; widespread in e 1/2 of TX; e U.S. from MA s to s FL and w to se SD s to TX. Jun–Oct. [*Cassia chamaecrista* L., *C. fasciculata* Michx., *C. fasciculata* var. *puberula* (Greene) J.F. Macbr., *C. fasciculata* var. *robusta* (Pollard) J.F. Macbr., *C. fasciculata* var. *rostrata* (Wooton & Standl.) B.L. Turner; *C. puberula* Kunth] This species is an important wildlife food (Yatskievych 2006), but in large quantities can cause digestive/toxicity problems apparently due to anthraquinone compounds (Kingsbury 1964; Burrows & Tryl 2001). The flowers produce no nectar, with pollination resulting from bumblebees that collect pollen from the anther pores by vibrating their thoracic/flight muscles—this is known as the "buzz," "vibrator," or "sonication" pollination syndrome. While pollen is the food reward used by the bees, at least some pollen is inadvertently transferred to other flowers, thereby allowing successful reproduction. The bees shake the anthers by vibrating/shivering their thoracic flight muscles at a certain frequency; this audible "buzz" sets up a resonance and the otherwise inaccessible pollen is released from the terminal pores of the anthers and collected by the insect. The porate anthers serve as an effective dispensing mechanism, which releases some pollen, but not so much that all available pollen is given to a single insect. This limitation maximizes the chance of successful pollen dispersal (Wolfe & Estes 1992). The flowers also exhibit the characteristic of enantiostyly—the presence of left- and right-handed flowers. On left-handed flowers, a bee receives pollen on its right side, and when it moves to a right-handed flower, the left side of the bee contacts the stigma (Gardner & Robertson 2000). In contrast to the flowers, there are extrafloral (= in a location other than the flower) nectarines on the leaf petioles. These nectar-producing glands attract ants which remove eggs and larvae of insects that damage the plant—this is a type of defense mechanism known as animal guard defense—an animal (ant) is being paid (nectar) to protect the plant. Experiments have demonstrated increased damage on plants from which ants were excluded (Ajilvsgi 1984; Durkee et al. 1999; Rutter & Rausher 2004). Population numbers of the species often vary widely from year to year with annual late winter burning increasing stem counts substantially and subsequently decreasing with less frequent fire and increased cover of perennial grasses (Gardner & Robertson 2000). ☠

Chamaecrista flexuosa (L.) Greene var. **texana** (Buckley) H.S. Irwin & Barneby, (sp.: tortuous, zigzag; var.: of Texas), TEXAS SENSITIVE PEA. Perennial 0.2–0.6 m tall; stems numerous, clustered, often zigzag, cinereous; leaves 2-ranked, with 11–18 pairs of leaflets; leaflets 3–9 mm long, glabrate; petioles with a single inconspicuous gland below lowest pair of leaflets; peduncles noticeably long 1.5–25 cm long in flower to 3.5 cm in fruit; petals yellow or orange-yellow, the upper 4 with a reddish spot at base, the lower slightly larger; 10–15 mm long; stamens 10, anthers yellow; fruits 3–5 cm long. Prairies, roadsides, sandy soils; Bexar, Dewitt, Goliad (BRIT), Victoria, and Wilson cos. (Isely 1998) on extreme s margin of East TX; in the U.S. known only from TX; also Mexico. Mar–Nov. [*Cassia itzana* Lundell, *C. texana* Buckley, *Chamaecrista texana* (Buckley) Pennell] The typical variety occurs from Mexico south to Argentina (Irwin & Barneby 1982). This species is readily distinguished by its shrubby habit with woody and zigzagged stems and long peduncles when in flower (Turner 1959; E. Keith, pers. obs.).

Chamaecrista nictitans (L.) Moench, (blinking, moving), SENSITIVE-PEA, SENSITIVE PARTRIDGE-PEA. Annual 0.1–0.6 m tall, puberulent or pilose; leaves 2-ranked, slightly touch sensitive; leaflets 5–12 mm long with 7–25 pairs of leaflets, glabrous or pubescent with whitish hairs; petiolar gland usually stalked or sometimes subsessile; flowers 1 or 2 in supra-axillary fascicles; petals yellow, one almost twice the size of the other four, 4–7 mm long; anthers pinkish to rose, 2–3 mm long; fruits 2–4(–5) cm long. Sandy open woods, disturbed areas, often a pioneer species in abandoned fields; scattered in East TX, w through w TX; e U.S. from NH s

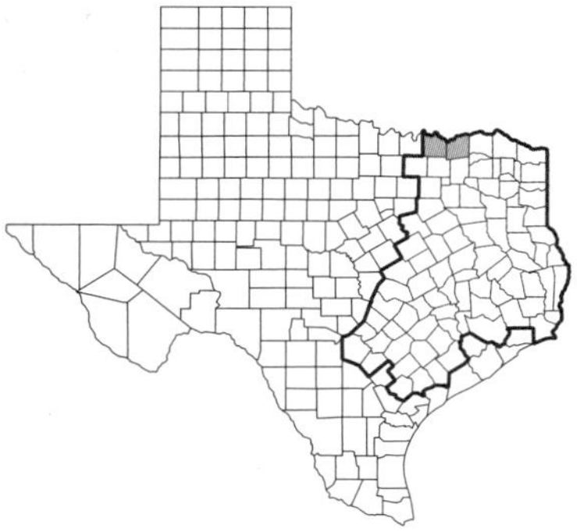

Baptisia ×variicolor

Calliandra biflora

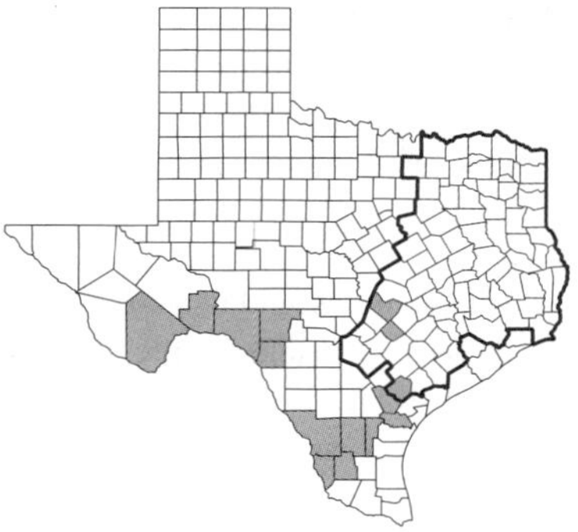

Calliandra eriophylla

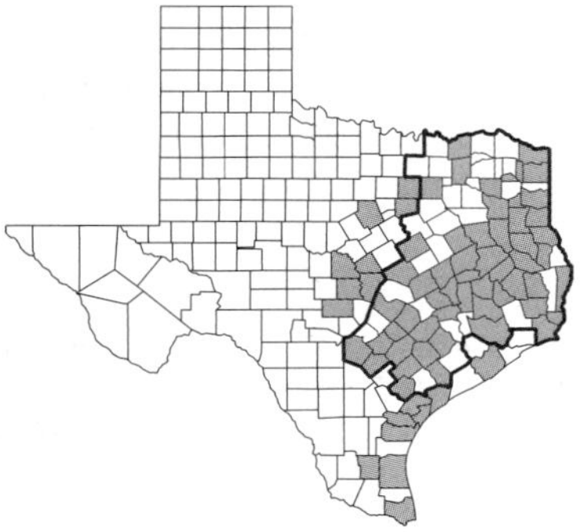

Centrosema virginianum

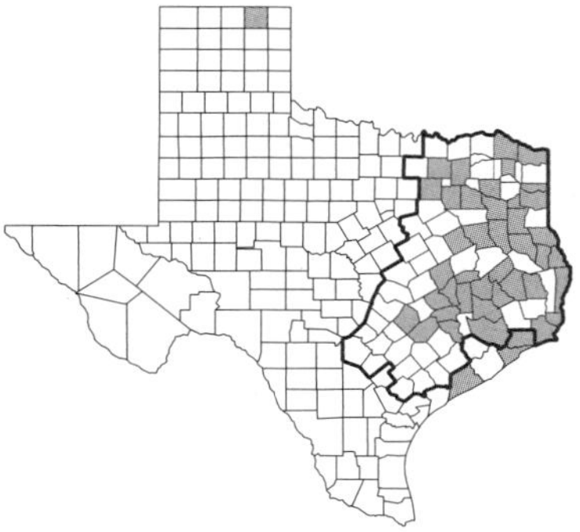

Cercis canadensis subsp. canadensis

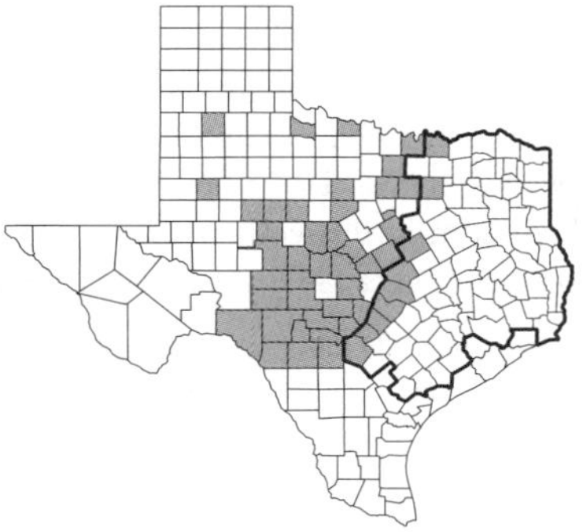

Cercis canadensis subsp. texensis

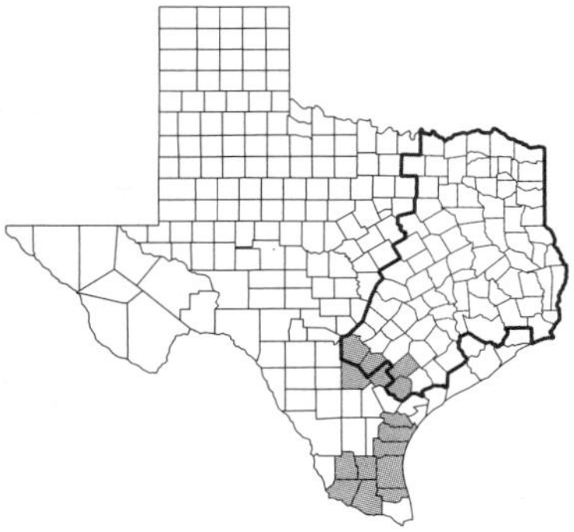

Chamaecrista calycioides

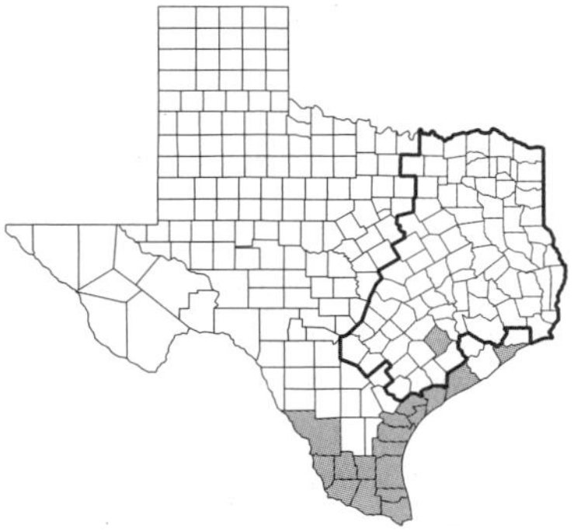

Chamaecrista chamaecristoides var. cuziana

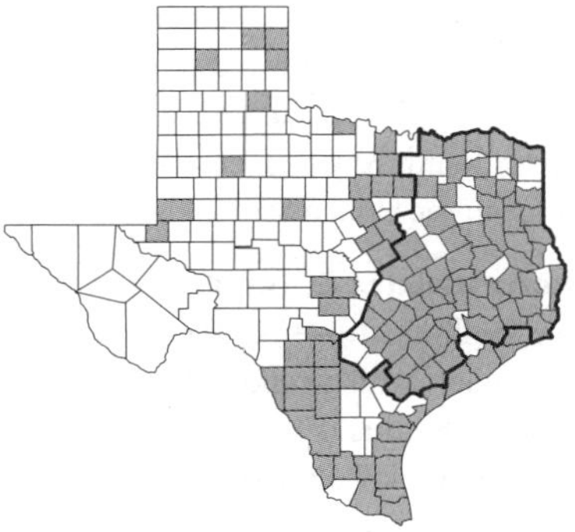

Chamaecrista fasciculata

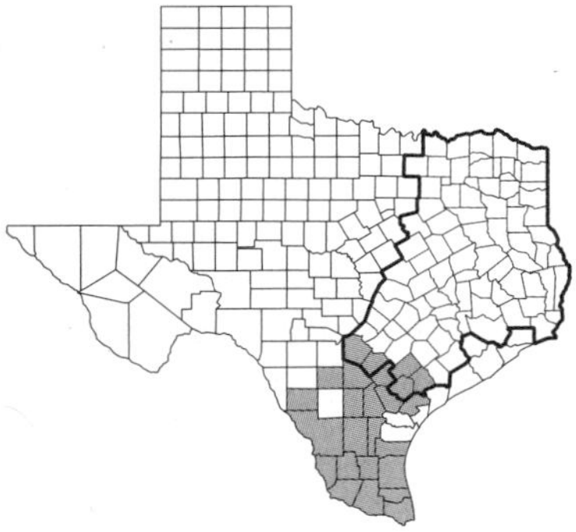

Chamaecrista flexuosa var. texana

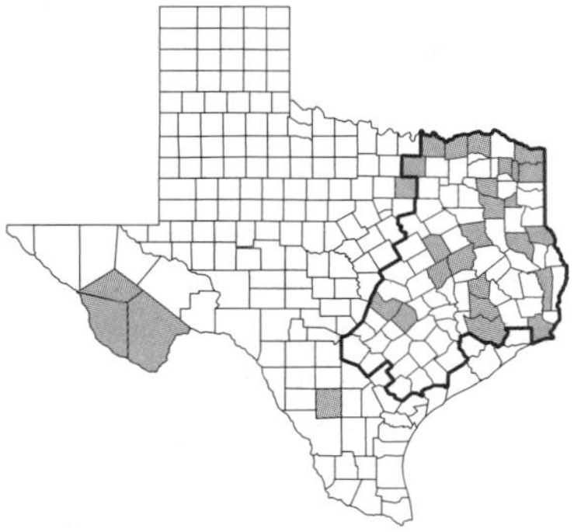

Chamaecrista nictitans

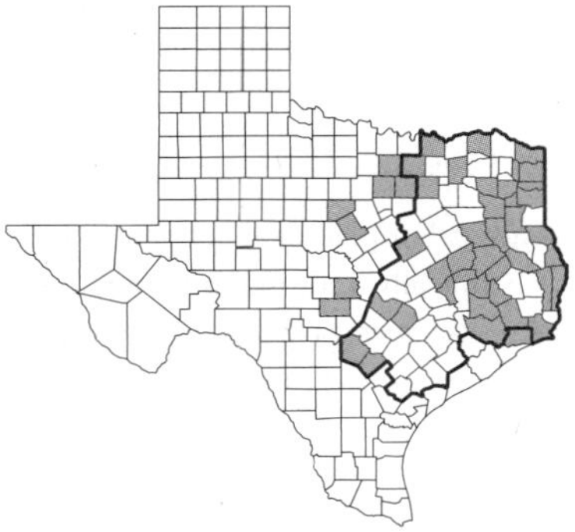

Clitoria mariana

to FL w to KS and s to OK and TX and sw to AZ; s to Argentina. Jul–Nov. [*Cassia nictitans* L.] This species has sometimes been divided into several subspecies and varieties (e.g., Irwin & Barneby 1982), but only one of these (var. *nictitans*) occurs in East TX (Isely 1998). The leaves are reported to act as cathartics (Burlage 1968). Isely (1998) refers to this species as the "ugly duckling" of the two common species in East TX when compared to the showy *C. fasciculata*.

CLITORIA L. PIGEON-WINGS, BUTTERFLY-PEA

Herbs with stems ascending and low-spreading, trailing or twining; leaves once odd-pinnately compound with 3–7 leaflets, evidently petioled; leaflets entire; flowers solitary or few on short axillary peduncles, inverted, with standard at base; corollas large, variously blue to lavender, the standard much larger than the other petals, with darker or reddish markings or yellowish to whitish area near center, the keel shorter than the wings; androecium diadelphous, anthers monomorphic; style laterally compressed, barbellate on the inner margin; fruits stipitate, dehiscent, oblong to linear-oblong, flattened, with few–several seeds.

A genus of about 60 species primarily of the tropics, especially of the Americas (Fantz 2023b). The flowers are inverted so that the anthers and pistil touch visiting insects on the back. Some are cultivated as climbing ornamentals and others used as aphrodisiacs, possibly based on the ancient idea known as the Doctrine of Signatures. *Centrosema* and *Clitoria* are the only U.S. members of the family with large resupinate flowers (= turned upside down or inverted, with standard on lower side of the flower) (Isely 1990). *Centrosema*, however, has a shorter calyx tube—ca. 4 mm, linear fruits (3–4 mm wide), and more frequently twining stems. (Derivation from the small keel relative to the large standard, suggesting mammalian clitoris) (subfamily Faboideae, tribe Phaseoleae)
REFERENCES: Fantz 1977, 1991; Isely 1998; Fantz 2023b.

1. Leaves with 3 leaflets; standard with darker or reddish markings toward center; common native species ____ **C. mariana**
1. Leaves with 5–7 leaflets; standard with a yellowish and whitish area toward center; rarely escaped ornamental species ____ **C. ternatea**

Clitoria mariana L., (of Maryland), SPOONFLOWER, ATLANTIC PIGEON-WINGS. Perennial herb; stems glabrous or minutely pubescent above, ascending to low-spreading or trailing, rarely shortly twining, to 1 m long; leaflets 3, ovate-elliptic to oblong-lanceolate, usually 2–4+ cm long, glabrous, much paler below; stipules linear-lanceolate; flowers solitary or few on short, axillary peduncles; corollas very large, the standard 3.5–6 cm long, 3–4 cm wide; petals violet-blue or lavender-blue, the standard with darker or reddish markings toward center and pale or whitish base; fruit ca. 3–5 cm long, linear-oblong. Sandy open woods; widespread in e 1/2 of TX except far s; e U.S. from s NJ s to FL and w to e NE and s to TX, disjunct w in AZ and disjunct and rare n in MN and WI; s to South America. Late May–Sep. Isely (1998) noted that this species "possibly has the largest flower of any herbaceous legume in the United States."

Clitoria ternatea L., (with parts in threes), BUTTERFLY-PEA. Perennial herb; stems trailing or twining, 1–3 m long; leaflets 5–7, ovate to elliptic, 1.5–4(–5.5) cm long; flowers solitary on short axillary peduncle, corollas 3.5–4 cm long, blue to violet(–white), the standard with a yellowish and whitish area toward center; fruit ca. 10 cm long, oblong. Cultivated and apparently rarely escaping; Bexar and Travis (Turner et al. 2003) cos. at sw margin of East TX; also Cameron Co. in South TX Plains (Turner et al. 2003), not known elsewhere in the state; s U.S. in FL and TX; West Indies; Central America; South America; Asia; Pacific Islands; and Australia. Flowering year-round in mild winters. Native to Old World Tropics. *Clitoria ternatea* is cultivated and naturalized pantropically (Fantz 2023b). Ornamental forms with extra petals are often cultivated (Isely 1990). The flowers have long been used to create a natural dye for edible dishes and drinks and may have health benefits and medicinal uses. Teas made from the flowers are called Anchan Tea, Color Changing Tea, Clitoria Tea, Thai Blue Tea, Blue Tea, Blue Chai (India) and Aparajita Tea in different parts of the world (Terahara et al. 1990; Fantz 1991).

CROTALARIA L. RATTLEPOD

Ours annual or perennial herbs with stems sometimes winged above; leaves sessile or subsessile, apparently simple in our species (technically a compound leaf reduced to one leaflet—referred to as leaves below); flowers in axillary or terminal, 2–many-flowered racemes, pedicelled; sepals united up to halfway; petals yellow; stamens 10, monadelphous, anthers dimorphic; ovary with numerous ovules, styles pubescent; fruits much inflated, with the numerous seeds rattling inside when dry.

A genus of ca. 600 species of the tropics and subtropics, mostly (511 species) in Africa and Madagascar; the genus is characterized by the inflated fruit (Nesom 2023a). Some are used as sources of fodder or fiber; others can cause poisoning in livestock or humans, apparently due to pyrrolizidine alkaloids (e.g., monocrotaline); such poisoning is known as crotalism (Kingsbury 1964; Fuller & McClintock 1986; Turner & Szczawinski 1991; Hardin & Brownie 1993). (Greek: *crotalon*, a rattle, from the loose seeds rattling in the coriaceous inflated pods) (subfamily Faboideae, tribe Crotalarieae)

REFERENCES: Senn 1939; Windler 1974; Isely 1998; Nesom 2023a.

1. Corollas 1.7–3.0 cm long; leaves 4–15 cm long; stipules not decurrent on the stem and not conspicuously foliose; escaped cultivated species.
   2. Bracts in inflorescence conspicuous, ovate, 5–8 mm long; stipules ovate to broadly lanceolate, 4–7 mm long; calyces glabrous or essentially so; leaves 5–15 cm long ____ **C. spectabilis**
   2. Bracts in inflorescence small (3 mm or less long) or absent, linear to awl-shaped if present; stipules bristle-like, very small, < 2 mm long or absent; calyces with short appressed hairs; leaves 4–8 cm long ____ **C. retusa**
1. Corollas 0.7–1.4 cm long; leaves 1–8 cm long; stipules of at least the upper leaves conspicuously decurrent on the stem, giving the impression of a downward pointing arrowhead; native species.
   3. Perennial; stems and sepals appressed-pubescent or nearly glabrous, the hairs 0.3–1 mm long; leaves usually 4–10 mm wide, 4–10 times as long as wide, glabrous above; long reported but not known to occur in East TX ____ *C. purshii*
   3. Annual; stems and sepals ± spreading-pilose or somewhat appressed pilose, the hairs usually 1–2 mm long; leaves 8–15 mm wide, 4–8 times as long as wide, pubescent or rarely glabrate above; species widespread in East TX ____ **C. sagittalis**

Crotalaria retusa L., (retuse, notched slightly at a rounded apex). Erect annual; stems 20–80 cm tall; leaves obovate to spatulate or oblanceolate, 4–8 cm long, strigose pubescence below; stipules inconspicuous, < 2 mm long or absent; racemes with 5–20 flowers, 10–30 cm long; sepals 12–15 mm long, strigose; petals yellow or yellow streaked with red, 20–25(–30) mm long; fruits 2.5–5 cm long, glabrous. Cultivated and escapes in disturbed areas, apparently not long persistent; Hunt and Victoria (Turner et al. 2003) cos.; also Tarrant Co. (Senn 1939) and Cameron and Hidalgo cos. in s TX (E. Keith, pers. obs.); introduced in s U.S. from FL w to TX; Mexico; West Indies; Central America; South America; Asia; Indian Ocean Islands; Pacific Islands; and Australia. Flowering year-round in mild winters. Probably native to Africa, now pantropical. Significant losses have occurred in fowl due to the alkaloid monocrotaline in the seeds (Kingsbury 1964); poisoning is similar to that caused by *C. spectabilis* (Burlage 1968).

Crotalaria sagittalis L., (sagittate, arrow-like), ARROW CROTALARIA. Erect annual or weak perennial; stems 10–40 cm tall, with usually conspicuous ± spreading hairs; leaves elliptic-oblong to linear-lanceolate, 1–8 cm long, usually with pilose pubescence above and below, but occasionally glabrate above, 4–8 times as long as wide; stipules of middle and upper leaves conspicuous, decurrent; racemes with 2–5 flowers; sepals 10–12 mm long, pilose; petals yellow, 9–11 mm long; fruits 1–2.5 cm long, glabrous. Sandy areas; widespread in East TX; also immediately adjacent Cross Timbers and Prairies and n Gulf Prairies and Marshes; e U.S. from NH s to FL and w to e NE, e OK, and TX, disjunct in s AZ; Mexico; West Indies; and Central America. Apr–Sep. According to McGregor (1986), there are reports of fatal poisoning in horses from eating fresh or dried plants; Burlage (1968) reported this species as the cause of "bottom disease" or "Missouri Bottom Disease." "The condition is characterized by slow emaciation, weakness, stupor, fatal hemorrhages and degenerative changes in the liver and spleen." Plants with glabrous relatively narrow leaves have been misidentified in TX as *C. purshii* DC. (Senn 1939) but can be distinguished from that species by the consistently long pilose (versus short strigose) pubescence on the sepals (E. Keith, pers. obs.).

Crotalaria spectabilis Roth, (spectacular, remarkable, showy), SHOWY CROTALARIA, RATTLE-BOX. Erect annual 50–100(–200) cm tall; leaves usually obovate, 5–15 cm long, 25–50 mm wide, glaucous and strigulose below; stipules 4–7 mm long, flap-like, ± persistent; racemes with up to 20–50 flowers, 15–50 cm long; sepals 10–15 mm long, glabrous; petals yellow, the standard can have veins darker with purple streaks, 15–25 mm long; fruits 3–5 cm long. Cultivated and escapes, weedy areas and roadsides; Brazos (Nesom 2023a), Dallas, Harris (Isely 1998), Liberty (Nesom 2023a) and Titus (BRIT) (Turner et al. 2003) cos.; also Hood and Montague (BRIT) cos. to the w and Karnes Co. (Isely 1998) just s of East TX; introduced in s U.S. from VA s to FL and w to s IL and MO and s to TX; Mexico; West Indies; Central America; South America; Reunion Island; and Pacific Islands. Feb–Oct. Native of s Asia, now pantropical. This species was introduced into the U.S. in 1921 for use as a cover crop and soil enricher; however, its use was abandoned with the advent of mechanized harvesting of crops and the associated danger of grain contamination by its seeds (Morton 1982). The plants, hay, and seeds are toxic to animals and humans due to the presence of pyrrolizidine alkaloids including monocrotaline; liver damage and heart failure can result, and animal fatalities are known; human poisoning has resulted from eating the seeds or drinking a "medicinal" tea; severe liver damage can occur (Muenscher 1951; Morton 1982).

Crotalaria purshii DC., (for Frederick Traugott Pursh, 1774–1820, German explorer, collector, horticulturist, and author who published on plant collections of the Lewis and Clark expedition and wrote *Flora Americae Septentrionalis*), PURSH'S CROTALARIA. Erect to ascending perennial to 50+ cm tall; leaves linear or oblong-lanceolate, 3–8 cm long, glabrous above, 4–10 times long as wide; stipules of middle and upper leaves conspicuous, decurrent; sepals 9–12 mm long, strigose; racemes with (2–)4–6(–12) flowers; petals yellow, 6–12 mm long; fruits 1.5–3 cm long. Pine flatwoods, savannahs, pocosins, and dry sand hills; previously reported in East TX from Harrison, Hunt (Nesom 2023a), Trinity (Senn 1939), Upshur, Van Zandt (BRIT), and Wood (BRIT) cos. (Turner et al. 2003; Kartesz 2023); se U.S. from s VA s to FL and w to se LA. Jul–Sep. Senn (1939) reported Trinity (*Tharp 808*, US) and Tarrant co. (*Killian 6986*, US) specimens and two unspecified locations (*Meyer s.n.* and *Wright s.n.* NY) in TX. Turner (1959) treated the specimens listed in Senn (1939) as *C. sagittalis*, and Windler (1974) and Isely (1998) mapped the species only as far west as se LA. Weakley, A.S., and Southeastern Flora Team (2022) mapped the species in TX with an "N" indicating that it is not actually present in the state. Nesom (2023a) indicated that the species "is known from Hunt County" but did not cite a specimen or provide additional information, and we could not locate a specimen from that county. After reviewing the digital specimens listed in Senn (1939) and those mapped in Kartesz (2023), all are misidentifications of *C. sagittalis*, including most that have subsequently been annotated as such. We include *C. purshii* here because of the previous reports and its potential, but unlikely, presence in TX. This report essentially eliminates the species from the flora of TX. No county level map is provided.

DALEA L. PRAIRIE-CLOVER

Ours unarmed herbaceous perennials (–annuals) or low shrubs; leaves once odd-pinnately compound; leaflets few to many, resin-dotted; stipules slender, falling early; flowers in terminal spikes, these often dense and cone-like; bracts often resin-dotted; petals usually 5, separate, papilionaceous or not; stamens 5–10, monadelphous, attached to staminal tube, either laterally or apically; ovules 2, collateral; fruits indehiscent. This genus was traditionally included in the genus *Petalostemon*; according to Correll and Johnston (1970) who recognized it, *Petalostemon* is "… not at all well-delimited from *Dalea* but technically keyed based on the insertion of four of the petals at the rim of the 'stamen tube' instead of along the length of it." The genus is distinguished from other members of the tribe Amorpheae by the possession of both of two basic characters: there are often only five stamens, and the four petals other than the standard are inserted at the apex of the staminal tube and alternate with the filaments, while the standard is inserted at the floral base (Wemple 1970). Detailed diagrams of these unique floral structures can be found in Wemple and Lersten (1966).

A New World genus of ca. 170 species native from Canada to Argentina, but especially in Mexico and the Andes. They mainly occur in dry or desert areas (Sutherland 2023). (Named for Samuel Dale, 1659–1739, English botanist and physician who practiced at Essex) (subfamily Faboideae, tribe Amorpheae)

REFERENCES: Shinners 1949b, 1949c, 1953b; Wemple & Lersten 1966; Wemple 1970; Barneby 1977; Turner 2013; Sutherland 2023.

1. Small shrubs with clearly woody branches; petals purple; species of extreme w margin of East TX ____ **D. frutescens**
1. Herbaceous perennials; petals purple to pink, white, or yellow; including species widespread in East TX.
   2. Calyx teeth triangular or lanceolate, broadly or narrowly acute, much shorter than or equaling the tube; stamens usually 5; 1 petal inserted near rim of floral cup, the other 4 at end of the "stamen tube."
      3. Petals white; calyces usually glabrous outside or nearly so (except pubescent on one variety of *D. phleoides*); spikes 6–10 mm wide in fruit (or in flower excluding corollas).
         4. Leaflets 13–41(–49) per leaf; calyces asymmetrical, the lobes displaced to one side ____ **D. phleoides**
         4. Leaflets 5–11 per leaf; calyces symmetrical.
            5. Spikes elongate and cylindric, 14–70(–100) mm long in flower, relatively few per plant; leaflets (6–)9–35 mm long; floral bracts below buds or very young flowers nearly equaling or exceeding the calyces (but bracts deciduous) ____ **D. candida**
            5. Spikes globose or subglobose, 7–16 mm long in flower, numerous per plant; leaflets 6–12(–15) mm long; floral bracts below buds or very young flowers 1/3–2/3 as long as the calyces ____ **D. multiflora**
      3. Petals pale yellow or pink to purple; calyces densely pubescent or pilose outside, at least on the teeth; spikes 6–20+ mm wide in fruit (or in flower excluding corollas).
         6. Petals pale yellow; leaves strikingly densely long pilose-villous; stems densely pilose; spikes (12–)15–20+ mm wide in fruit (or in flower excluding corollas). ____ **D. obovata**
         6. Petals pink to purple; leaves glabrous or with only sparse pubescence; stems glabrous or with short or sparse pubescent; spikes 6–15 mm wide in fruit (or in flower excluding corollas).
            7. Leaves with (9–)11–17(–21) leaflets, the leaflets flat or only loosely folded; spikes 6–10 mm wide in fruit (or in flower excluding corollas); species of the e part of East TX, primarily the Pineywoods ____ **D. grisea**
            7. Leaves with 3–9(–11) leaflets, the leaflets often tightly folded to involute or almost tubular; spikes 7–15 mm wide in fruit (or in flower excluding corollas); including species widespread in East TX.
               8. Calyx tube with velvety pubescence, at least some of the hairs retrorse (= the free end pointing down towards base of calyx tube); leaflets 3–5 per leaf, 5–12 mm long; species found in East TX only in extreme w tier of counties ____ **D. tenuis**

Chamaecrista flexuosa var. texana [CLA]

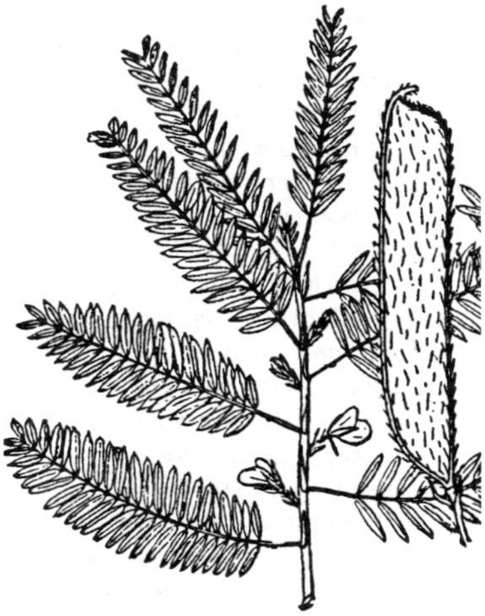
Chamaecrista nictitans [BB2]

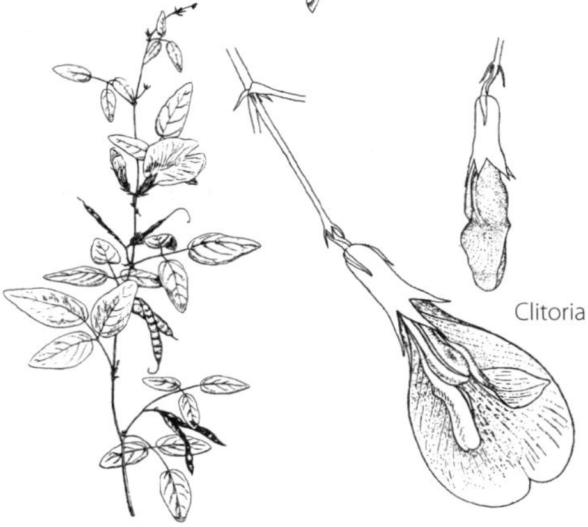
Clitoria mariana [BOT, WIL]

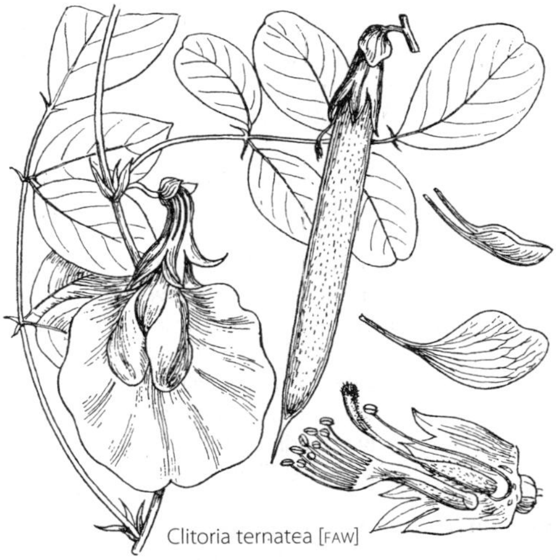
Clitoria ternatea [FAW]

Crotalaria retusa [WIL]

Crotalaria sagittalis [RHO]

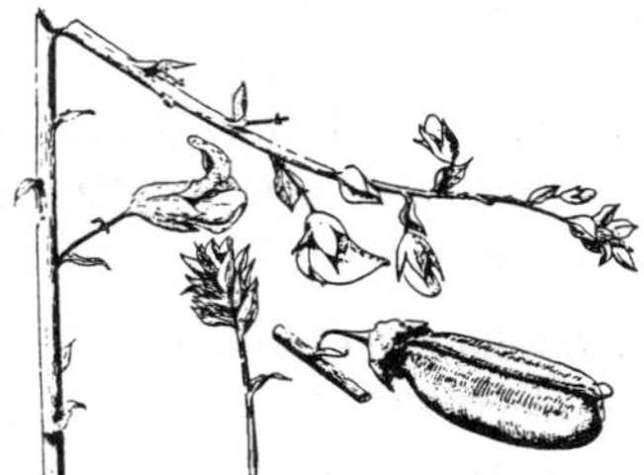
Crotalaria spectabilis [GRE]

8. Calyx tube not velvety, variously with fine hairs over the entire surface or limited to lines, the hairs antrorse (= the free end pointing upwards towards the summit); leaflets 3–9 per leaf, 8–25 mm long; including species widespread in East TX.
9. Leaflets (3–)–7(–9) per leaf; calyces (4.4–)5–6.2 mm long, pilosulous, with lines of antrorse, appressed hairs proximal to sinuses and on margins of lobes; spikes 10–15 mm wide in fruit (or in flower excluding corollas); widespread in East TX ________ **D. compacta**
9. Leaflets usually 3–5 per leaf; calyces 3–5(–5.2) mm long, pubescence not restricted to lines, or, if partly restricted, then hairs near base of calyx retrorse; spikes 7–12 mm wide in fruit (or in flower excluding corollas); species mainly in n part of East TX ________ **D. purpurea**
2. Calyx teeth with bristle-like tips, slightly shorter to much longer than the tube; stamens usually (5–)7–10; 1 petal attached near rim of floral cup, the other 4 at various places on the "stamen tube," but not at its end.
10. Petals white; flowers widely spaced in slender, drooping spikes; plants usually with a single main stem from the base (this can be quite branched); leaflets completely glabrous ________ **D. enneandra**
10. Petals yellow or purple; flowers crowded in rather slender to thick, spreading to erect spikes; plants usually with several main stems from the base; leaflets glabrous or pubescent.
11. Petals purple; leaflets glabrous or pubescent.
12. Leaflets usually with dense pubescence; stems prostrate; spikes narrow, 3–8 mm wide in fruit (or in flower excluding corollas); species limited to extreme nw margin of East TX ________ **D. lanata**
12. Leaflets completely glabrous; stems decumbent to ascending; spikes 9–13 mm wide in fruit (or in flower excluding corollas); species of s part of East TX.
13. Peduncles conspicuously elongate, often longer than leafy portion of stem, much > 5 cm long; leaflets (5–)11–17 per leaf, 2–7(–10) mm long; stamens 5 (staminodes also often present); on deep sandy soils ________ **D. emarginata**
13. Peduncles not elongate, usually 5 cm or less long; leaflets 5–11(–13) per leaf, 2–15 mm long; stamens 10; limestone or mostly calcareous substrates.
14. Calyx long pilose and plumose, hairs to 2 mm long, calyx tube 2.6–3.3 mm long; leaflets (3–) 5–7 per leaf; on mostly calcareous substrates in s portion of East TX________ **D. pogonathera**
14. Calyx sericeous-pilose, hairs to 1 mm long, calyx tube 3.3–3.9 mm long; leaflets (3–)–11 per leaf; on limestone on sw margin of East TX ________ **D. lasiathera**
11. Petals yellow; leaflets all pubescent.
15. Main leaves of the stem with 3 leaflets (some upper leaves with only 1 or 2); stems usually trailing________ **D. hallii**
15. Main leaves of the stem with 5–7(–9) leaflets (mostly 5); stems decumbent to erect.
16. Spikes (12–)14–21 mm wide in fruit (or in flower excluding corollas), 20–50 mm long; stems usually erect, (20–)30–75 cm long; standard (blade and claw) 6.3–8.6 mm long; yellow petals staying yellow even when old or dry; w portion of East TX ________ **D. aurea**
16. Spikes 7–13(–15) mm wide in fruit (or in flower excluding corollas), 10–30 mm long; stems usually spreading, (5–)10–35 cm long; standard 4.4–5.5 mm long; yellow petals often fading to pink or brown; primarily sw portion of East TX ________ **D. nana**

Dalea aurea Nutt. ex Pursh, (golden, in reference to the flowers), GOLDEN DALEA, GOLDEN PRAIRIE-CLOVER, SILK-TOP DALEA. Stems spreading to usually erect, (20–)30–75 cm long, silky-pilose, inconspicuously glandular; leaflets 3–7, generally 5, 4–20 mm long, pubescent; spikes 1–6 cm long and 14–21 mm wide, with a densely pilosulous axis; calyces pilose with hairs 1–2 mm long; petals yellow, 10–15 mm long. Prairies and dry hills on a variety of soils, but most common on gravelly limestone soils; mainly w portion of East TX, e to Grimes Co. (Turner et al. 2003); mainly w 2/3 of TX; w U.S. from SD s to TX and w to e CO and s to e AZ; s into Mexico. Late May–Jul(–Sep). *Dalea ceciliana* B.L. Turner, previously included within the concept of this species, is endemic to the South Texas Sandsheet and adjacent similar habitats and has only been recorded s of East TX (Turner 2010).

Dalea candida Willd., (shining or pure white, in reference to the flowers), WHITE PRAIRIE-CLOVER. Stems erect or ascending, 25–100 cm long; leaflets 5–9, (6–)9–35 mm long, linear-lanceolate or oblanceolate, acute to obtuse, glabrous; peduncles subsessile, but leaves at the nodes below the spike are typically reduced to stipuliform bracts, leaving the stem leafless and like a peduncle for 1–20 cm; spikes ovoid to cylindric 6–10 mm wide, glabrous or minutely pilosulous; calyces 3–4.2 mm long, glabrous except for a ring of small glands near top; petals white. Prairies and open woods; sandy, rocky, or clayey soils; scattered in East TX, occurring most commonly in n TX and se TX, rare elsewhere (Kartesz 2023); also adjacent Cross Timbers and Prairies and n Gulf Prairies and Marshes; c U.S. from e WI s to w AL and w to ND and s to c OK and TX; also s Canada. Late May–Aug. [*Petalostemon candidus* Michx.] *Dalea oligophylla* (Torr.) Shinners is often included as a variety of this species (Turner 1959; Correll & Johnston 1970; Diggs et al. 1999; Sutherland 2023), but it occurs to the west of East TX. It can be distinguished by the more loosely flowered spikes and finely pubescent calyces (Isely 1998) where he noted that the two "are essentially allopatric, differ in several features, and can usually be distinguished at a glance" and should thus be recognized at the species level.

Dalea compacta Spreng., (compact, dense), SHOWY PRAIRIE-CLOVER, COMPACT PRAIRIE-CLOVER, PLAINS PRAIRIE-CLOVER. Stems ascending or erect, 30–80 cm tall; leaflets (3–)–7(–9), 10–25 mm long, glabrous or with sparse pubescence and punctate-glandular, marginally involute; spikes usually 10–14 mm thick (slenderer on small, new shoots of mowed plants), pubescent; calyces (4.4–)5–6.2 mm long, gray-bald except villosulous in lines and primarily towards apex; corollas 8–10 mm long, purple or rose in various shades. The range map for this species in Turner et al. (2003) has the mapped localities for the two varieties reversed.

1. Spikes nearly sessile or on peduncles 3 cm or less long; bracts equal to or much longer than calyces at flowering, 4–8 mm long and evident to the naked eye; leaflets often > 3 mm wide; stems decumbent or weakly ascending; plants of se and nw parts of East TX ____ var. **compacta**
1. Spikes on peduncles 2–15 cm long; bracts shorter than or subequal to calyces at flowering, 3–5 mm long, hardly evident; leaflets < 3 mm wide; stems erect to decumbent; plants widespread in East TX ____ var. **pubescens**

var. **compacta,** PLAINS PRAIRIE-CLOVER. Stems commonly decumbent; leaflets > 3 mm wide; otherwise as given in key. Calcareous prairies; Hardin, Harris, Polk, San Augustine, and Tyler (Turner et al. 2003) cos. in se TX and disjunct in nw portion of East TX (Kartesz 2023); adjacent n Gulf Prairies and Marshes; sc U.S. in sw AR, se OK and e 1/2 of TX (Kartesz 2023). May–early Jul. [*Petalostemom decumbens* Nutt.]

var. **pubescens** (A. Gray) Barneby, (pubescent, downy), SHOWY PRAIRIE-CLOVER. Stems commonly erect or ascending; leaflets < 3 mm wide; otherwise as in key. Prairies, on clay and limestone; widespread in e 1/2 of TX except far s; also nearby portions of adjacent vegetational areas including Cross Timbers and Prairies, e Edwards Plateau, and Gulf Prairies and Marshes, disjunct to High Plains and Trans-Pecos; s U.S. in w LA and s OK w to w TX, introduced in NM (Kartesz 2023). May–Jul(–Aug). [*D. helleri* Shinners, *Petalostemon pulcherrimus* (A. Heller) A. Heller] This variety is similar to *D. purpurea*; see discussion under that species.

Dalea emarginata (Torr. & A. Gray) Shinners, (with a shallow notch at the end), WEDGELEAF PRAIRIE-CLOVER. Annual from a taproot; stems decumbent to ascending, 20–60 cm tall, glabrous; leaflets (5–)11–17 per leaf, cuneate-oblong to obovate, emarginate, bicolored, dark green above and glaucescent and dotted beneath; peduncles often longer than leafy portion of stem; calyces 3–4.2 mm long, densely pilosulous with pale yellow or almost colorless glands; petals rose-purple. Sandy soils; s part of East TX n to Travis (BRIT) and Brazos (Turner et al. 2003) cos.; widespread in s 1/3 of TX, primarily in South TX Plains and Gulf Prairies and Marshes; s U.S. in sw LA and se TX; e Mexico. Sep–May. [*Petalostemon emarginatus* Torr. & A. Gray] Isely (1998) noted that this "species is instantly known in its region by its excessively long, often curving peduncles." Under favorable conditions, plants typically begin flowering in late summer and continue with several flushes or without interruption throughout the winter into the following spring (Barneby 1977).

Dalea enneandra Nutt., (with nine stamens), NINE-ANTHER PRAIRIE-CLOVER, BIG-TOP DALEA. Stems 20–150 cm tall, usually solitary (–few), bushy-branched above, from a tough, woody, orange root; leaflets usually 5–11, narrowly linear to oblong, 5–12 mm long, glabrous; calyx teeth conspicuously white-pilose; petals essentially white (keel pale yellow; standard and wings white, but the standard can have a greenish "eye"); stamens 9 (hence the scientific name). Rocky, sandy, or silty soils; mainly w margin of East TX and Austin and Brazos (Turner et al. 2003) cos. further e; unusually distributed mainly in a wide band from the panhandle to the central Gulf Prairies and Marshes; c U.S. from w IA and MO s to OK and TX and w to e MT and s to NM. Jun–Sep. [*D. laxiflora* Pursh, *D. enneandra* var. *pumila* (Shinners) B.L. Turner] Isely (1998) noted that the "plumose calyx serves as a parachute for the wind-borne fruits." While lumped by most authorities (e.g., Turner et al. 2003), *D. enneandra* var. *pumila*, originally named by Shinners who observed the plants for many years in the field, may deserve varietal recognition. The plants are consistently short (20–30 cm tall) and compact, with shorter, denser inflorescences, and overall, in the field have a different aspect from typical *D. enneandra*. These plants are known from Ellis (BRIT) and Hill (BRIT) cos. in East TX and from Gillespie and Kerr cos. on the Edwards Plateau (Shinners 1949c). Barneby (1977) reported that similar plants also occur in Kansas.

Dalea frutescens A. Gray, (shrubby, bushy), BLACK DALEA, BLACK PRAIRIE-CLOVER. Small, intricately branched shrubs, 30–80(–120) cm tall, the branches clearly woody; leaflets (9–)13–17(–19), 1.5–3.5(–5) mm long, glabrous, bicolored, bright green above and paler below, punctate; flowers in short, few-flowered spikes; bracts conspicuously gland-dotted; calyx lobes shorter than tube, the tube glabrous with red or orange blister glands; petals purple. Rocky hillsides, washes, and stony clay flats, usually abundant; w margin of East TX; mainly w 2/3 of TX; sw U.S. from s OK and TX w to s NM; Mexico. Apr–Sep. [*D. frutescens* var. *laxa* (Rydb.) B.L. Turner] *Pilostyles thurberi* A. Gray (THURBER'S PILOSTYLES), a member of the Apodanthaceae, parasitizes *D. frutescens* in TX; its vegetative structures are ± entirely within the tissues of the host plant with the visible portions bud-like; only the small flowers and sometimes a few subtending scale-like leaves are externally visible.

Dalea grisea (Torr. & A. Gray) Shinners, (gray), SILKY PRAIRIE-CLOVER, OKLAHOMA PRAIRIE-CLOVER. Stems usually erect, 50–90 cm tall, stems reddish; leaflets (9–)11–17(–21), 6–12 mm long, with thin pubescence (appearing ± glabrate), greenish gray foliage; spikes short-peduncled; calyces 4–5 mm long, densely pilosulous; petals pink or pink-purple. Open woodlands and ruderal areas in deep sandy soils; mainly Pineywoods, but w to Milam (Turner 1959) and Falls (Turner et al. 2003) cos.; sc U.S. in AR, nw LA and TX (Kartesz 2023). Jun–Aug [*D. villosa* (Nutt.) Spreng. var. *grisea* (Torr. & A. Gray) Barneby, *Petalostemon griseus* Torr. & A. Gray] This taxon has been variously treated as a distinct species (Correll & Johnston 1970; Wemple 1970; Turner et al. 2003) or as a variety of *D. villosa* (Barneby 1977; Isely 1990, 1998; POWO 2023; Sutherland 2023), a species that occurs in w TX. The two are clearly related, but since there is no geographic overlap and they are usually easily distinguished in the field (and generally on herbarium sheets), we are following the recent treatment in Weakley, A.S., and Southeastern Flora Team (2022) in recognizing this taxon as a distinct species.

Dalea hallii A. Gray, (for its collector, Elihu Hall, 1822–1882, collected in TX and botanized with Charles C. Parry in the Rocky Mts.), HALL'S DALEA, HALL'S PRAIRIE-CLOVER. Stems usually trailing (the ends turning upward), purplish, to 35 cm long, appressed-pubescent; leaflets 3 (1 in distal portion of stem), 10–25 mm long, pubescent below; spikes sessile or short-pedunculate, few and loosely flowered, 1–5 cm long and 14–17 mm wide; calyces red-brown, the tube and teeth pilose; petals yellow. Gravelly prairies and woodlands, eroding limestone slopes; Collin, Grayson (BRIT), Dallas (Barneby 1977), and Fannin (Turner et al. 2003) cos. in nw part of East TX and Hays Co. (Turner et al. 2003) in sw corner of East TX; also Cross Timbers and Prairies and e Edwards Plateau; endemic to TX. Late May–Jun, repeating in Sep. Type is from Dallas, Texas (Barneby 1977). There are only three yellow-flowered *Dalea* species in East TX, the other two being *D. aurea* and *D. nana*; *D. hallii*, with 3 leaflets, is easily distinguished from the other two possessing 5–9 leaflets. NatureServe: G3, vulnerable; S2, imperiled. ⚠

Dalea lanata Spreng., (woolly), WOOLLY DALEA, WOOLLY PRAIRIE-CLOVER. Perennial from a tough orange root, prostrate, the stems trailing to somewhat ascending, often mat-forming to 20–70 cm long; stems and leaves gray-pubescent; leaflets 5–15, 4–12 mm long, obovate or elliptic; spikes initially compact, loosening post-anthesis, 2–7 cm long and 7–8 mm wide; calyces villosulous or glabrous, inconspicuously glandular; petals purple or red-purple. Deep sandy soils including dunes and river alluvium; Grayson Co. (BRIT; Isely 1998) in Red River drainage (exposed open sandy area near river); mainly w 2/3 of TX; c U.S. from AR and nw LA w through OK and TX to se CO and e NM. May–Sep.

Dalea lasiathera A. Gray, (with woolly awns, of the calyx teeth), PURPLE DALEA. Stems decumbent to ascending, 10–35 cm tall; leaflets (3–)7–11, 5–15 mm long, glabrous; bracts conspicuously gland-dotted to the naked eye; spikes moderately dense, flowers mostly 2–3 ranked and not concealing the glabrous axis; calyx teeth always shorter than tube, pilose but not plumose; petals bright purple or magenta, banner with a yellow gland-sprinkled eye. Limestone hillsides; Travis Co. (Turner et al. 2003) on sw margin of East TX; mainly Edwards Plateau and scattered in w 1/2 of TX; Nuevo León in Mexico. Mar–Jun. This species is very similar to *D. pogonathera* but can be distinguished by the characters and distribution listed in the key.

Dalea multiflora (Nutt.) Shinners, (many-flowered), ROUND-HEAD DALEA, ROUND-HEAD PRAIRIE CLOVER, WHITE PRAIRIE-CLOVER. Bushy-branched; stems erect or ascending, 30–80 cm long; leaflets (5–)7–11(–13), 6–12(–15) mm long, glabrous, linear or narrowly oblanceolate, deep or pallid green; spikes dense, subglobose to shortly oblong, 7–9 mm wide; calyces glabrous except for ciliolate teeth; petals white. Prairies, on limestone or calcareous clay; widely scattered in e 2/3 of TX, isolated blackland prairie locations in the Pineywoods; throughout c portion of TX; sc U.S. w MO w to NE s to TX and w to e CO, also n Mexico. May–Sep. [*Petalostemon multiflorus* Nutt.] This species is distinguished by its globose inflorescences and preference for calcareous prairies (Barneby 1977).

Dalea nana Torr. ex A. Gray, (dwarf), DWARF DALEA, DWARF PRAIRIE CLOVER. Stems decumbent to ascending, (5–)10–35 cm long; leaflets (3–)5(–7), 5–15 mm long, pubescent; spikes variable in density, moderately loose to very dense and cone-like, 5–40 mm long and 7–13(–15) mm wide; calyces 2–2.8 mm long, pilose and glandular; petals yellow, often fading to pink or brown. Sandy open areas; primarily sw part of East TX, disjunct eastward to Liberty and Waller cos. in se TX; mainly w 2/3 of TX; sw U.S. from sw KS, w OK, and TX w to se CO and AZ; Tamaulipas and Chihuahua in Mexico. Late Mar–Sep. *Dalea nana* is a variable species that was historically divided into two problematically identified varieties (var. *nana* and var. *carnescens* (Rydb.) Kearney & Peebles) (e.g., Correll & Johnston 1970; Barneby 1977; Isely 1998). Isely (1998) indicated that the two "probably are justified by ecological differentiation" (calciphile vs. arenophile). *Dalea nana* var. *carnescens* is sometimes recognized as a distinct species, *D. rubescens* S. Watson, and occurs west of East TX (Weakley, A.S., and Southeastern Flora Team 2022; Kartesz 2023).

Dalea obovata (Torr. & A. Gray) Shinners, (obovate, inversely ovate, egg-shaped with wider portion near the apex), PUSSYFOOT, STINKING PRAIRIE-CLOVER. Annual or short-lived perennial; stems decumbent to ascending, 20–55 cm long, with conspicuous lanate-pilose pubescence; leaflets 7–9(–11), 10–16 mm long, with conspicuous gray pubescence; peduncle stout, continuous

Dalea aurea [MNY]

Dalea candida [MIC, MNY]

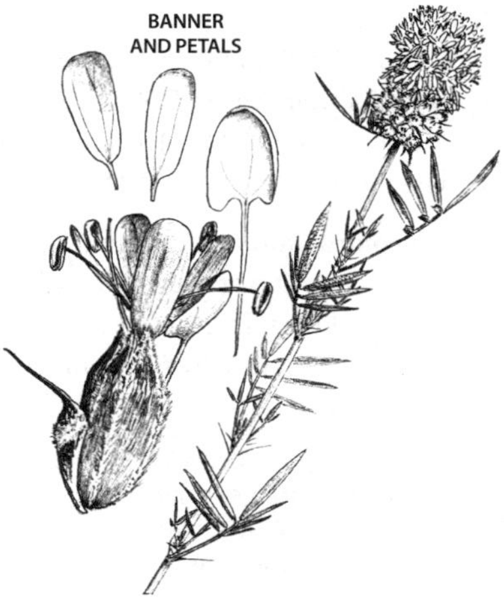

Dalea compacta var. compacta [MNY]

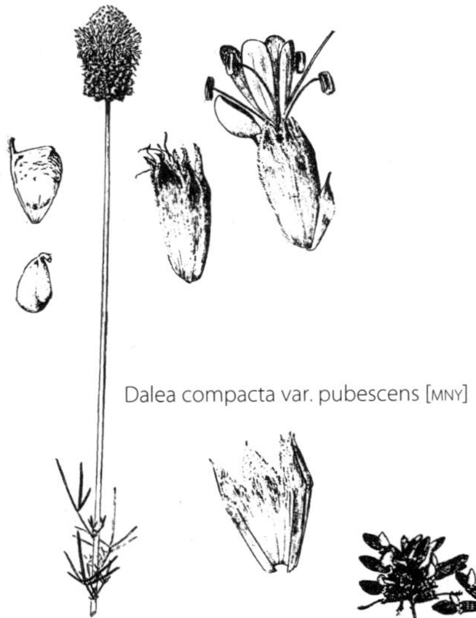

Dalea compacta var. pubescens [MNY]

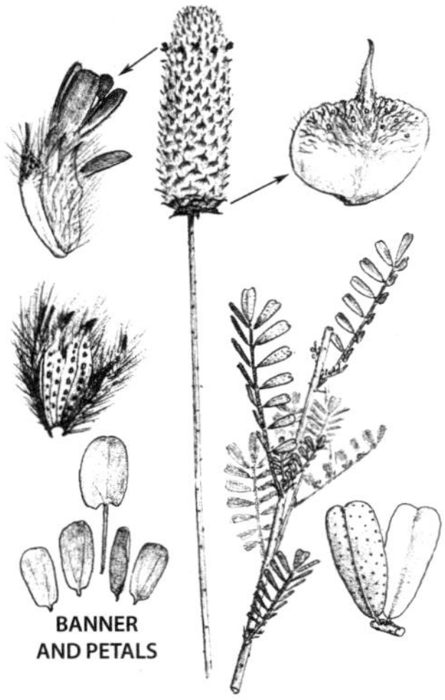

Delea emarginata [MNY]

Dalea enneandra [MNY]

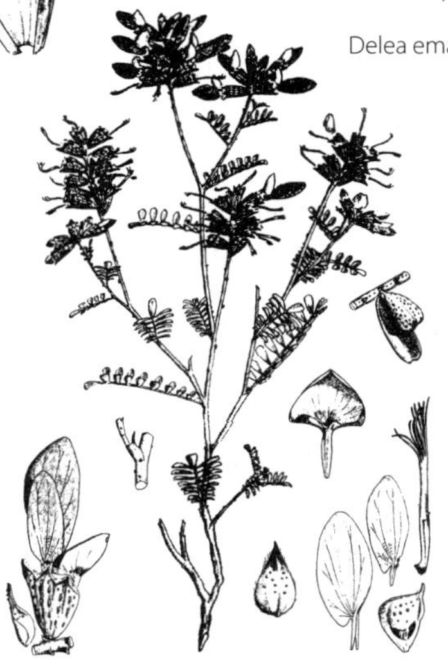

Dalea frutescens [MNY]

Dalea grisea [MNY]

with the stem, 0–3 cm long; spikes dense, 3–11 cm long, (12–)15–20 mm wide in fruit (or in flower excluding corollas), cylindric; calyces 3.3–5.8 mm long, densely pilose; petals whitish or pale yellow. Open sandy woodlands, grasslands, and dunes; s part of East TX from Austin (BRIT) and Travis (Turner et al. 2003) cos.; also South TX Plains and s Gulf Prairies and Marshes; into n Mexico. Apr–Jul (Oct–Jan). [*Petalostemon obovatus* Torr. & A. Gray] A very distinctive species with long shaggy pubescence, large subsessile spikes, and tiny fleeting flowers make this species unmistakable from any others in TX (Barneby 1977).

Dalea phleoides (Torr. & A. Gray) Shinners, (resembling *Phleum* of the Poaceae, in reference to the narrow spike), SLIMSPIKE PRAIRIE-CLOVER, LONG-BRACT PRAIRIE-CLOVER. Stems erect or ascending, 20–80 cm tall, variably pubescent and glandular, leafy in the lower 1/2–2/3 of the plant; leaflets green or grayish green (11–)13–37(–49) per leaf, 2–14 mm long; spikes becoming elongate, 2–10(–13) cm long, usually 6–8 mm wide in fruit; calyces glabrous to densely pubescent, 2.5–3.8 mm long with scattered glands; petals white.

1. Leaflets 13–23(–25), blades 5–14 mm long; spike axes and calyces usually glabrous; plants mostly of se portion of East TX, but scattered northward ____ var. **phleoides**
1. Leaflets (11–)23–37(–49); blades 2–7 mm long; spike axes and calyces usually densely pubescent, but occasionally less pubescent to glabrous in s and w portion of range; plants of w portion of East TX ____ var. **microphylla**

var. **microphylla** (Torr. & A. Gray) Barneby, (small-leaved), DRUMMOND'S PRAIRIE-CLOVER. Stems erect or ascending, variably pubescent and glandular, leafy in the lower 1/2–2/3 of the plant; leaflets green or grayish green (11–)23–41(–49) per leaf, 2–7 mm long, 0.5–5 mm wide; spikes becoming elongate, 2–10(–13) cm long, 6–8 mm wide in fruit; calyces usually pubescent, 2.5–3 mm long with scattered glands; petals white. Sandy open woodlands and rights-of-way; primarily in w portion of East TX; also Cross Timbers and Prairies and South TX Plains; sc U.S. in sw AR, s OK, and TX. May–Jul, repeating in September. [*D. drummondiana* Shinners, *D. carrizoana* B.L. Turner, *Petalostemon microphyllus* (Torr. & A. Gray) A. Heller, *P. phleoides* Torr. & A. Gray var. *microphyllus* Torr. & A. Gray] Historically treated as this variety (Barneby 1977; Isely 1998; Diggs et al. 1999), however, Turner et al. (2003), Turner (2013), Weakley,A.S., and Southeastern Flora Team (2022), and POWO (2023) recognize it as a distinct species, *D. drummondiana*. Since the ranges of the taxa overlap in the s portion of East TX and intermediate character states are occasionally observed, we are following the most recent treatment in Sutherland (2023) in retaining this taxon as a variety of *D. phleoides*. Turner (2013) described a new species, *D. carrizoana* B.L. Turner, for plants with glabrous leaves, calyces, and spike axes and generally with fewer leaflets (11–23) from Carrizo sands in DeWitt, Dimmitt, Karnes, and Wilson cos. in s portion of East TX and n portion of the South TX Plains. Several collections in the western portion of the range of *D. phleoides* var. *microphylla* are also glabrous and morphologically identical to the specimens that Turner (2013) described as *D. carrizoana*, so we are tentatively including it within this variety.

var. **phleoides,** SLIMSPIKE PRAIRIE CLOVER, LONG-BRACT PRAIRIE-CLOVER. Stems mostly erect to ascending; leaflets 11–23(–25); blades 5–14 mm long, 2–7 mm wide; spike axis glabrous; calyces glabrous or glabrate, 3–3.8 mm long; otherwise as given in main description and key. Open sandy areas; widespread mainly in se part of East TX; also Gulf Prairies and Marshes; sc U.S in nw LA, se OK, and TX (Kartesz 2023). May–Jul(–Sep) [*Petalostemon glandulosus* J.M. Coult. & Fisher, *Petalostemon phleoides* Torr. & A. Gray] This variety can usually be distinguished from the four other white-flowered taxa in East TX by the number of leaflets: *D. candida* has only 5–9 leaflets, *D. enneandra* has 6–11, *D. multiflora* has (5–)7–11, and *D. phleoides* var. *microphylla* rarely has fewer than 23 leaflets.

Dalea pogonathera A. Gray var. **walkerae** (Tharp & F.A. Barkley) B.L. Turner, (sp.: bearded awn of the plumose calyx teeth, var.: for its collector, Thelma Ratcliffe Walker, 1908–1999, s TX artist and plant collector), WALKER PRAIRIE-CLOVER, PRAIRIE-CLOVER. Stems decumbent or prostrate, 7–25 cm long, simple or sparingly branched above; leaflets usually (3–)5–7, 2–10 mm long, narrowly oblanceolate to narrowly oblong-elliptic, emarginate, glandular; spikes 1–8 cm long, 9–12 mm wide in fruit; calyces 5.2–10.5 mm long, densely villosulous; flowers pale or vivid violet, the banner pallid or greenish. Low hills and plains, on sandy clays; Bexar, Guadalupe, and Wilson cos. in s portion of East TX (Kartesz 2023), South TX Plains and w Edwards Plateau; s to Mexico. Flowering year-round. [*Parosela walkerae* Tharp & F.A. Barkley] *Dalea pogonathera* var. *pogonathera* occurs in the w half of TX and can be distinguished by its wider flowering spikes and longer calyx lobes (Isely 1998).

Dalea purpurea Vent., (purple), PURPLE PRAIRIE-CLOVER, PRAIRIE-CLOVER. Stems erect to ascending, 20–90 cm tall, simple or sparingly branched above; leaflets usually 3–5(–7), 8–24 mm long, linear, glabrous or villous, folded to involute; spikes 1–7 cm long, at least well-developed ones, usually 7–12 wide in fruit; bracts exceeding calyx in bud and ± conspicuous; calyces 3–5(–5.2) mm long, densely pilosulous (golden hairs when dry); petals various shade of purple. Prairies; rare and scattered throughout East TX, mainly in w 2/3 of TX; c U.S. from WI s to AL and w to e WA and s to NM, OK and TX; s Canada. May–Aug. [*Petalostemon purpureus* (Vent.) Rydb.] Correll and Johnston (1970) noted that "some populations seem to show the effect of introgression of

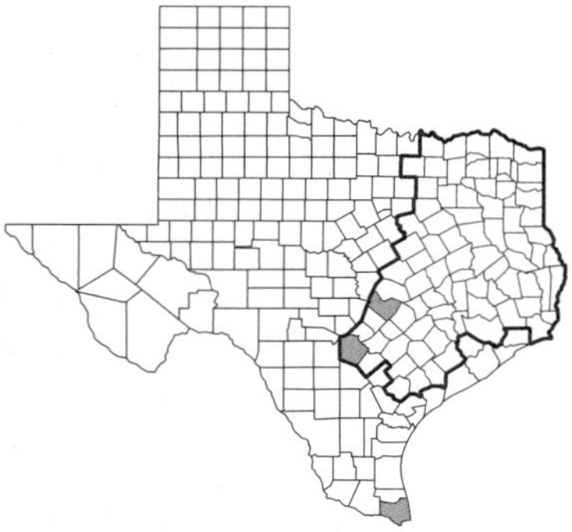

Clitoria ternatea

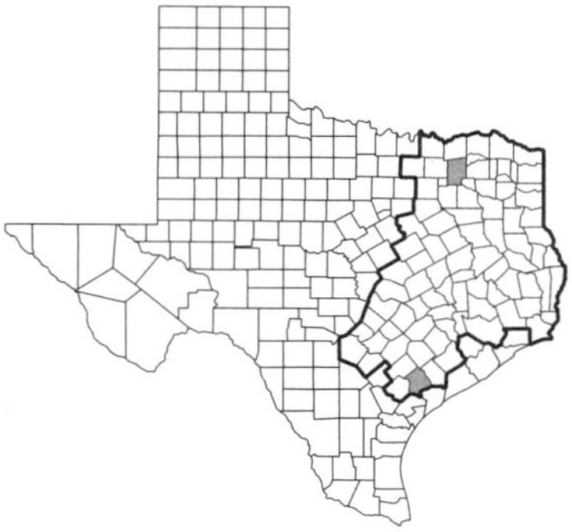

Crotalaria retusa

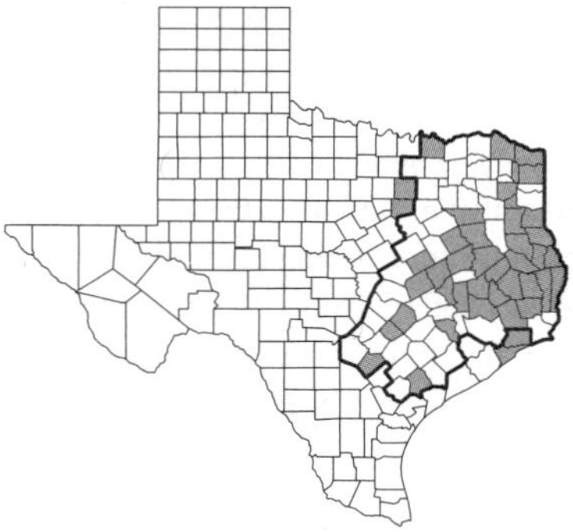

Crotalaria sagittalis

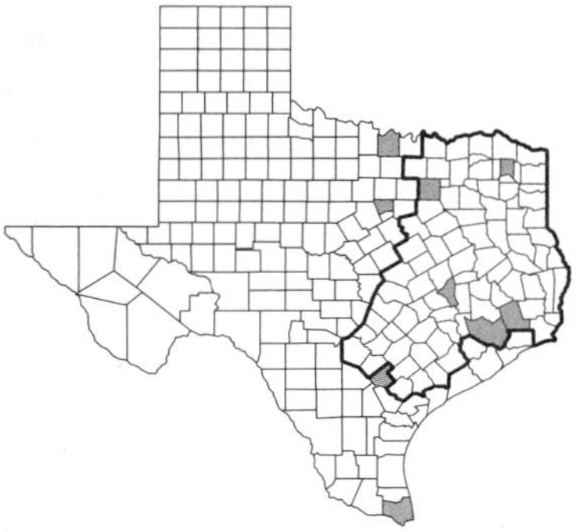

Crotalaria spectabilis

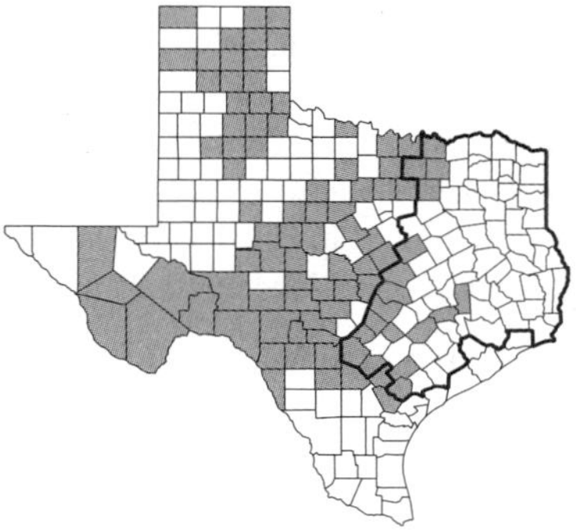

Dalea aurea

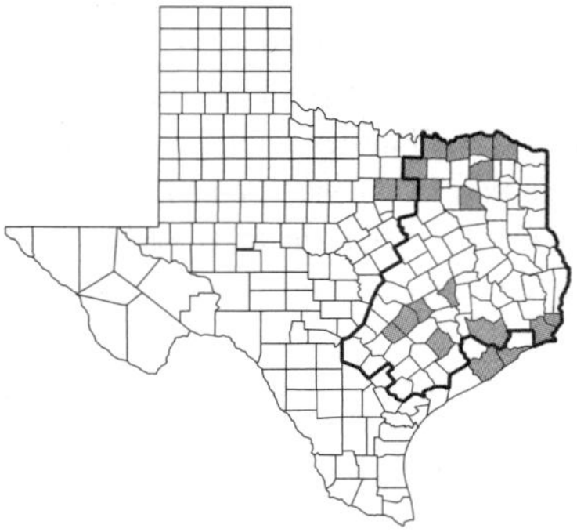

Dalea candida

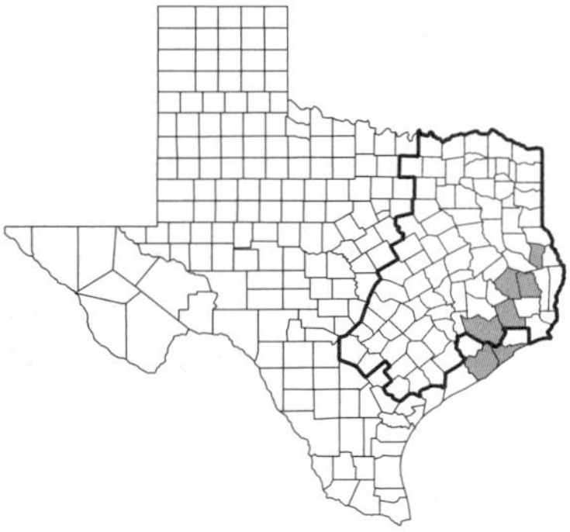

Dalea compacta var. compacta

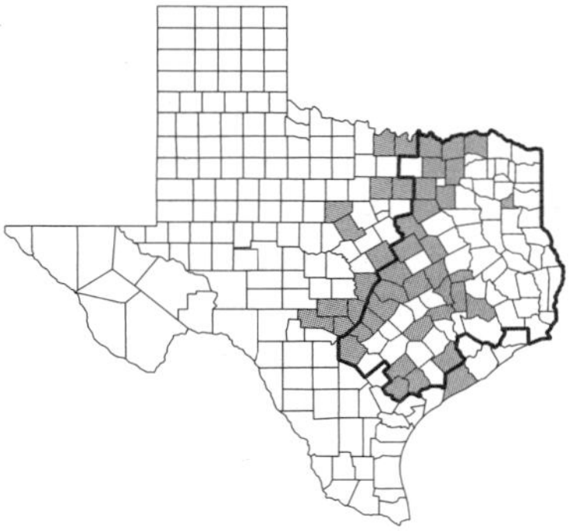

Dalea compacta var. pubescens

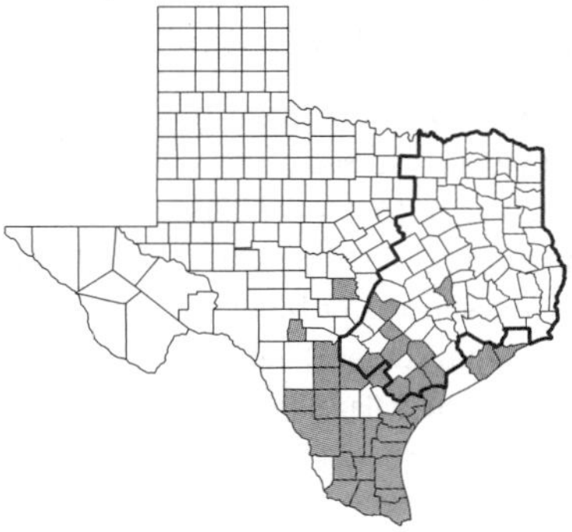

Dalea emarginata

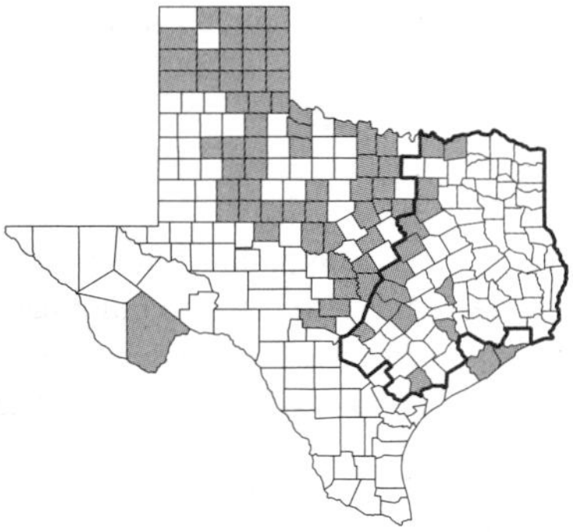

Dalea enneandra

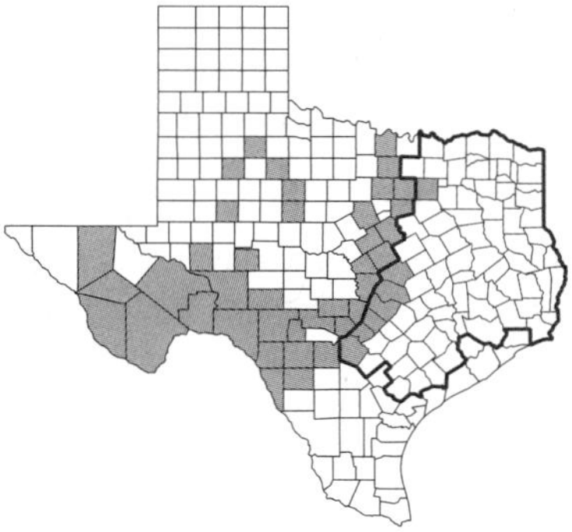

Dalea frutescens

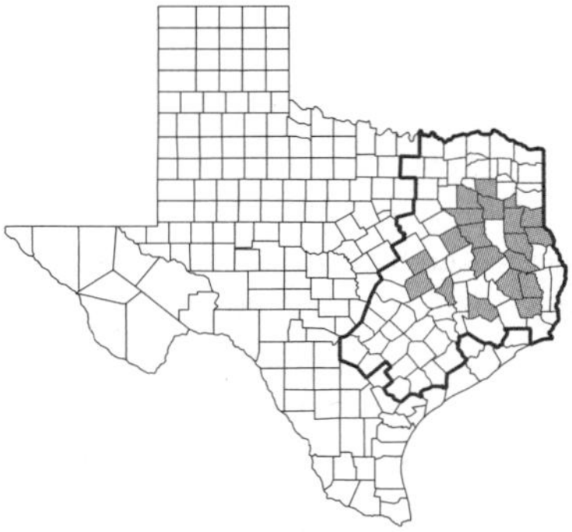

Dalea grisea

this species with ... *Petalostemon pulcherrimus* (A. Heller) A. Heller." Wemple (1970) also found evidence of hybridization of *D. purpurea* with both varieties of *D. compacta* (which he treated as 2 distinct species). *Dalea purpurea* is widespread throughout the c U.S. and adjacent Canada, but reaches its s limit in TX, primarily occurring only in the n part of East TX; here it seems to narrowly geographically overlap and be replaced to the s by the very similar *D. compacta* var. *pubescens*. These two taxa are so similar that in extreme n East TX they are very difficult, if not impossible to always distinguish. In fact, further research, probably molecular studies, will be needed to clarify the exact relationship between *D. purpurea* and *D. compacta* var. *pubescens*. In the narrow band of overlap, if other characters are intermediate or unclear, leaflet number can be helpful—Wemple (1970) noted that in his study of over 800 specimens of *D. purpurea*, almost all had 3–5 leaflets and only 10 exhibited leaves with 7 leaflets, while *D. compacta* had 5–9 (typically 7–9) leaflets; further, in *D. purpurea* the bracts are longer than the calyx in bud and ± conspicuous, while they are hardly evident in *D. compacta* var. *pubescens*.

Dalea tenuis (J.M. Coult.) Shinners, (slender, thin), PINK-GLOBE PRAIRIE-CLOVER. Stems erect or ascending, 15–50 cm tall, glabrous or inconspicuously pubescent, widely branched; leaflets 3–5, 5–12 mm long, glabrous or nearly so, involute; spikes compact 10–35 mm long, 7–10(–12) mm wide in fruit (or in flower excluding corollas), on slender often unusually elongate peduncles sometimes shorter but often 8–15 cm long; calyces 2–3 mm long, velvety with very short appressed and retrorse hairs, eglandular; petals pink-purple. Eroding limestone slopes; w margin of East TX, disjunct eastward in Grimes Co. (BRIT); mainly Cross Timbers and Prairies and Edwards Plateau, rare in Trans-Pecos (Turner et al. 2003); also c OK (Kartesz 2023). May–Jul(–Oct). [*D. purpurea* Vent. var. *tenuis* (J.M. Coult.) Shinners, *D. stanfieldii* (Small) Shinners, *Petalostemon stanfieldii* Small, *P. tenuis* (J.M. Coult.) A. Heller, *P. violaceus* Michx. var. *tenuis* J.M. Coult.] In East TX, the only other *Dalea* this species can be easily confused with is *D. compacta* var. *compacta*. That species, however, has longer (10–25 mm long) and more numerous leaflets (5–9).

DERMATOPHYLLUM Scheele NECKLACE-POD

A genus of six species of shrubs and small trees in sc and sw United States and n Mexico (Vincent & Kearns 2023). This genus was originally proposed in 1848 describing *D. speciosum* Sheele (now *D. secundiflora*) (Scheele 1848). It is now recognized as a segregate of the larger paraphyletic genus *Sophora* (Gandhi et al. 2011). This distinction is also supported by molecular data (Wojciechowski et al. 2004). (Greek: *dermato*, skin, and *phyllum*, leaf, presumably in reference to the thick coriaceous leaves) (subfamily Faboideae, tribe Sophoreae)
REFERENCES: Rudd 1972; Merrill 1977; Gandhi et al. 2011; Vincent & Kearns 2023.

Dermatophyllum secundiflorum (Ortega) Gandhi & Reveal, (with flowers on only one side of a stalk), MOUNTAIN-LAUREL, TEXAS MOUNTAIN-LAUREL, MESCAL-BEAN, FRIJOLITO. Plant usually a shrub, sometimes a small tree, often with multiple trunks; leaves evergreen, leaflets (5–)7–11, elliptic-obovate or oblong, 2–5(–8) cm long, 0.7–3.6 cm wide; flowers fragrant; corollas 15–16 mm long; fruits torulose, knobby with 1–4 fertile segments; seeds bright red, 10–15 mm in diam. Rocky limestone slopes; sw margin of East TX; widespread in s 1/2 of TX; sw U.S. in TX and s NM; Mexico; also frequently cultivated and escaping elsewhere. Mar. [*Calia secundiflora* (Ortega) Yakovlev; *Sophora secundiflora* (Ortega) Lag. ex. DC.] One of the most beautiful native plants of Texas; however, the foliage and seeds are poisonous to livestock (cause stiffness, trembling, collapse, and exercise induced seizures), and the flowers and bright red seeds are poisonous to humans. Since the seeds are used in ornamentation, this is a concern for children. Before the use of peyote, MESCAL-BEAN was used as a supposed hallucinogen for thousands of years in sw North America by Native Americans, causing dizziness, numbness, weakness, and stupor; however, it is very toxic (due to cytisine and related quinolizidine alkaloids) and fatal in excess (Standley 1922; Kingsbury 1964; Lampe & McCann 1985; Burrows & Tyrl 2001). The name MESCAL-BEAN is supposedly derived from "use (misuse) in adulterating weak whisky or mescal liquor, a small amount making the liquor so potent that over-indulgence brought on dizziness, disorientation and sometimes death" (Crosswhite 1980). Merrill (1977) describes extensive ethnographic and archeological information on this plant. ☠

DESMANTHUS Willd. BUNDLE-FLOWER

Unarmed herbaceous or sub-shrubby perennials; root often woody; leaves twice even-pinnately compound, with a petiolar gland between or near the lowest pair of pinnae (gland sometimes minute) or sometimes glands also between other pair of pinnae; flowers in peduncled, axillary heads; petals separate, linear; stamens 5 or 10; filaments white or yellowish white, separate, serving as the main attractant structure for the flower; ovary subsessile; fruits flattened, 1–several-seeded, dehiscent. Many species have nyctinastic (= nighttime or "sleep") leaf movements (Luckow 1993).

A genus of 25 species of warm areas of the Americas, introduced in Asia, Africa, Indian Ocean and Pacific Islands, and Australia (Luckow 2023). In TX, vegetative plants can be distinguished from similar genera such as *Neptunia* and *Mimosa* by possessing a petiolar gland between the lowest pair of pinnae and narrowly linear stipules (Turner 1950a) (Greek: *desme*, a

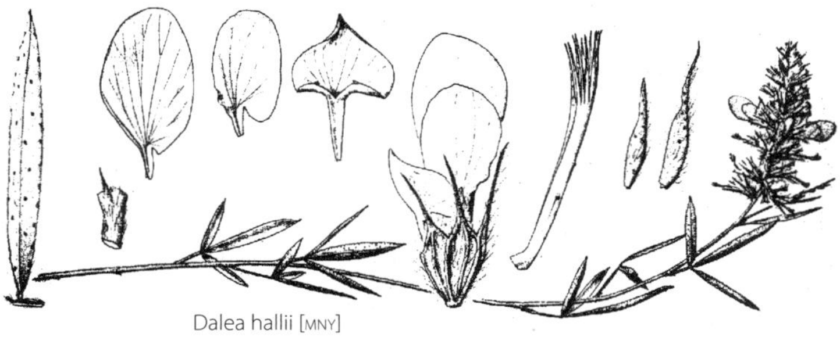
Dalea hallii [MNY]

Dalea lanata [MNY]

Dalea lasiathera [MNY]

Dalea mulitflora [MNY]

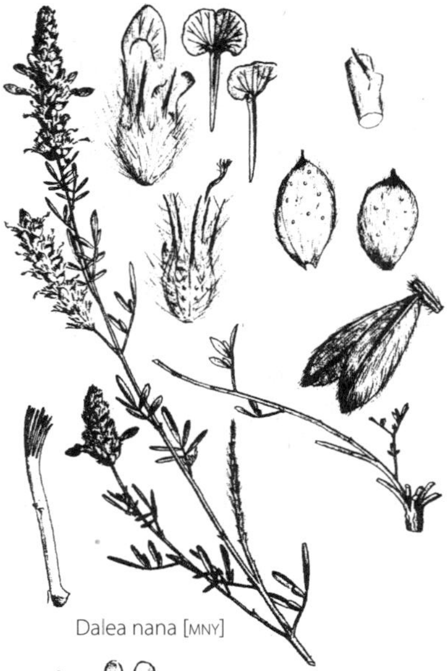
Dalea nana [MNY]

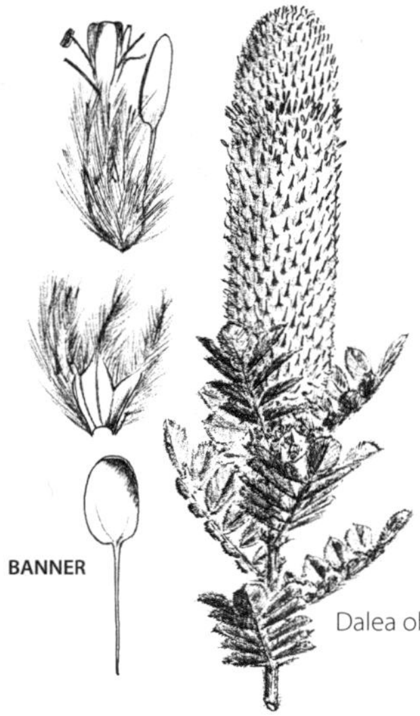

Dalea obovata [MNY]

Dalea phleoides
var. microphylla [MNY]

bundle, and *anthos*, a flower, presumably from the flowers clustered in heads) (subfamily Caesalpinioideae, Mimosoid clade, Mimoseae)

REFERENCES: Turner 1950a, 1950b; Latting 1961; Isely 1970a, 1973; Luckow 1993, 2023.

1. Species known only from extreme sw margin of East TX; leaflets with raised ± reticulate (= net-like) veins at least on the lower surface; mature fruit-bearing peduncles 3–12 cm long.
   2. Leaflets glabrous or with only sparse pubescence marginally, glaucous; mature fruit-bearing peduncles 7–14 cm long; fruits tapering to a point ____ **D. reticulatus**
   2. Leaflets obviously pubescent on lower surfaces; mature fruit-bearing peduncles 2–6.2 cm long; fruits obtuse or with an abrupt mucro, but not tapering to a point ____ **D. obtusus**
1. Including species widespread in East TX; leaflets without raised reticulate veins; mature fruit-bearing peduncles 6 cm or less long.
   3. Young stems minutely pubescent all around; peduncles ± equaling or exceeding the subtending leaves ____ **D. velutinus**
   3. Young stems glabrous except on the angles; peduncles shorter than the subtending leaves.
      4. Leaves with 1–6 pairs of pinnae (each with numerous leaflets); stamens 10.
         5. Stipules conspicuously pubescent; nectaries on petiole large, wider than petiole, elliptic, somewhat flattened; fruits falcate; taproot red, cylindrical or napiform (= turnip-shaped) ____ **D. acuminatus**
         5. Stipules usually glabrous; nectaries on petiole small, orbicular; fruits straight or slightly falcate; taproot brown, cylindrical ____ **D. virgatus**
      4. Leaves (except smallest) with 4–12(–15) pairs of pinnae; stamens 5.
         6. Plants 0.3–1 m tall; heads 4–12-flowered; fruits nearly straight, at maturity 2–3 mm wide, 35–60 mm long, usually ca. 7 times or more longer than wide, readily dehiscent along both sutures; leaves with 4–7(–8) pairs of pinnae ____ **D. leptolobus**
         6. Plants usually over 1 m tall; heads 20–80-flowered; fruits strongly sickle-shaped, at maturity 4.5–7 mm wide, 15–32 mm long, 3–4 times longer than wide, dehiscent along both sutures but tardily so on one side; leaves with 7–12(–15) pairs of pinnae ____ **D. illinoensis**

Desmanthus acuminatus Benth., (tapering to a long narrow point), SHARP-POD BUNDLE-FLOWER. Stems sprawling to decumbent or ascending, usually 20–60 cm long; leaves with 1–4 pairs of pinnae; leaflets 6–12 pairs per pinna, 2.5–4 mm long, linear-oblong, non-reticulate; petioles 3–6 mm long; heads with 6–13 flowers; fruits ca. 30–50 mm long, linear, straight to somewhat curved. Gravelly or sandy open areas; s and w portions of East TX; also just to the w in adjacent vegetational areas; endemic to TX. Apr–May(–Oct). [*D. virgatus* (L.) Willd. var. *acuminatus* (Benth.) Isely] While this taxon is often treated as a variety of *D. virgatus* (Jones et al. 1997; Isely 1998), we are following Luckow (1993, 2023) in treating it as a distinct species where numerous differences are cited.

Desmanthus illinoensis (Michx.) MacMill. ex B.L. Rob. & Fernald, (of Illinois), ILLINOIS BUNDLE-FLOWER, ILLINOIS DESMANTHUS, PRAIRIE-MIMOSA, PRICKLEWEED. Plant shrub-like, with usually 2–6, erect or spreading stems, 20–150 cm tall; leaves with 7–12(–15) pairs of pinnae; leaflets usually 15–25 pairs per pinna, 2–3 mm long, linear, non-reticulate; petioles 2–10 mm long; fruits 1.5–2.5 cm long, 3–4 times as long as broad (vs. at least 7 times in our other species), broadly oblong, strongly sickle-shaped, tightly clustered. Prairies, ditches, stream bottoms, fields, roadsides, and low areas, usually clay soils; nearly throughout TX except in the extreme s part of the state; e and c U.S. from PA s to n FL and w to ND s to NM, OK, and TX, disjunct in s NV and sw UT. May–Jun, sporadically to Sep. The strongly sickle-shaped, twisted fruits are distinctive (Luckow 1993). This is the most widespread species of the genus in the U.S. The leaves exhibit nyctinastic (nighttime or "sleep") movements with leaflets folding together at night (but they are not touch sensitive). The tiny petiolar glands, located between the lowest pair of pinnae or often also other pairs of pinnae, are worth looking at with a hand lens or scope; they are yellowish, raised, circular in outline, and sometimes glisten. They are presumably extrafloral nectaries, which are known in other mimosoid legumes to attract ants who remove eggs and larvae of insects that damage the plant—this is a type of defense mechanism known as animal guard defense—an animal (ant) is being paid (nectar) to protect the plant. It is being studied as a potential food and forage crop for temperate climates (Luckow 2023).

Desmanthus leptolobus Torr. & A. Gray, (thin-lobed), PRAIRIE BUNDLE-FLOWER, PRAIRIE-MIMOSA. Stems several, prostrate to suberect, 30–100 cm long; leaves with 4–7(–8) pairs of pinnae; leaflets 10–24 pairs per pinna, essentially linear or narrowly elliptic, non-reticulate, usually 2.5–5 mm long; petioles 2–5 mm long; stamens 5 per flower (10 in all other East TX species except also 5 in *D. illinoensis*); fruits linear, straight or essentially so, with undulate margins, 3.5–6 cm long. Prairies and open ground, clayey, rocky, or less often sandy soils, often a weed in lawns; widespread mainly in w 1/2 of East TX; also Cross Timbers and Prairies, e Rolling Plains and e Edwards Plateau; sc U.S. in KS, OK, and TX, introduced in MO and NM. May–Jul. The scalloped edges of the legume where it is constricted between the seeds and the longitudinal insertion of the seeds are diagnostic (Luckow 2023).

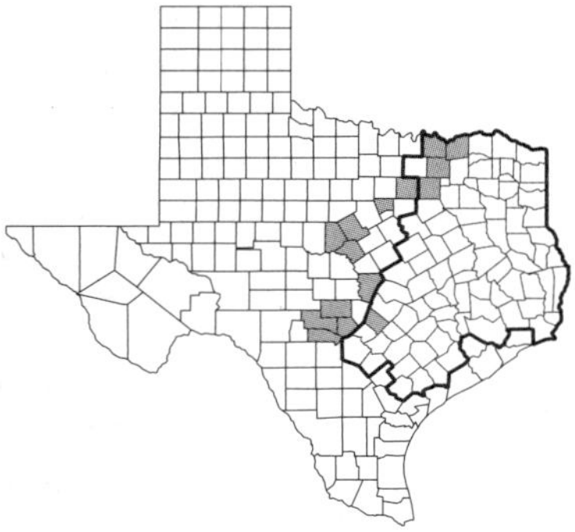

Dalea hallii

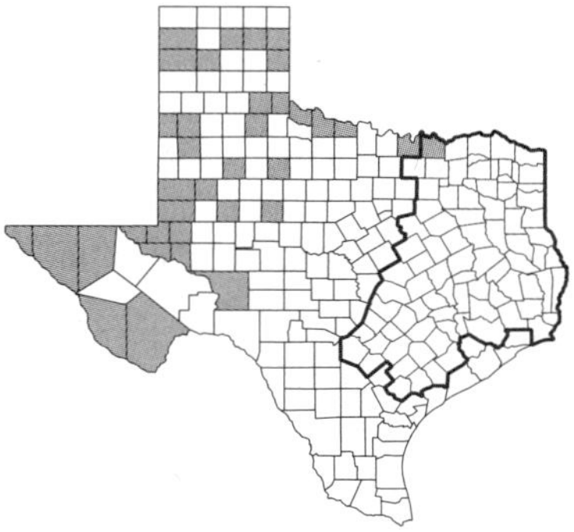

Dalea lanata

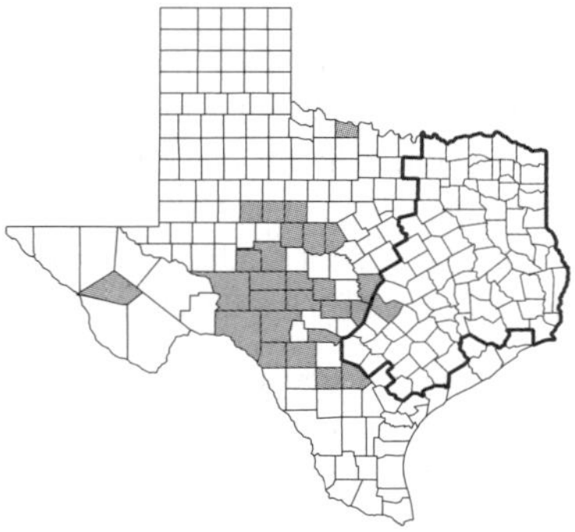

Dalea lasiathera

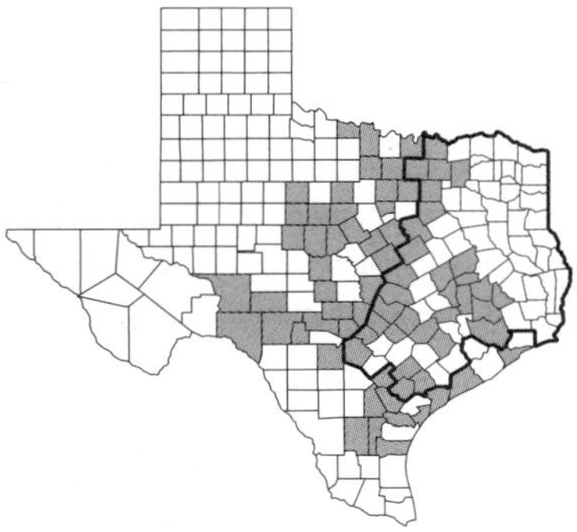

Dalea multiflora

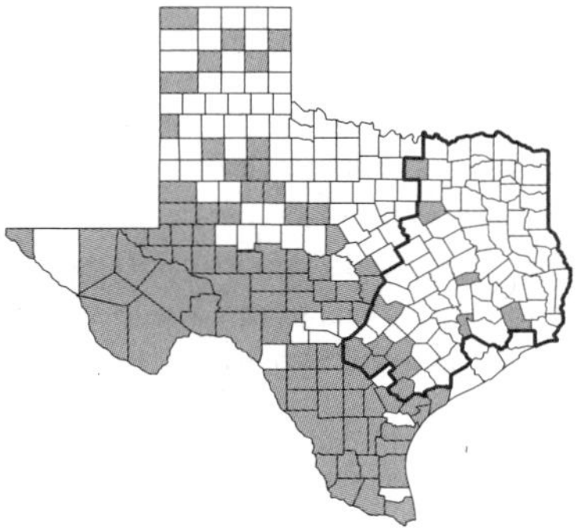

Dalea nana

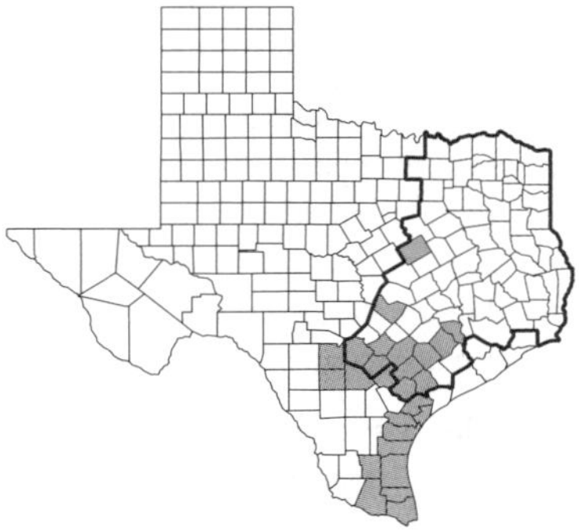

Dalea obovata

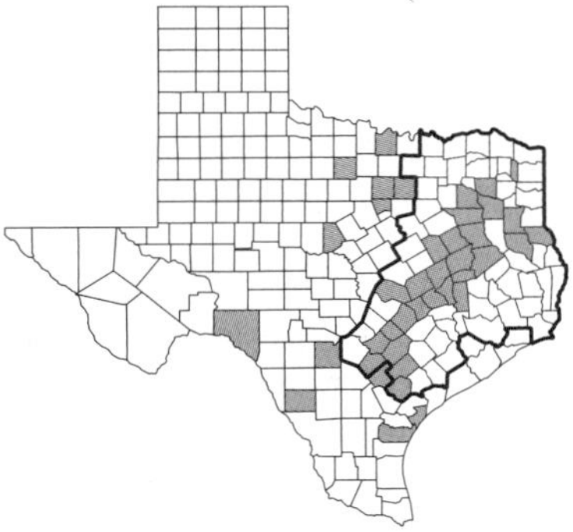

Dalea phleoides var. microphylla

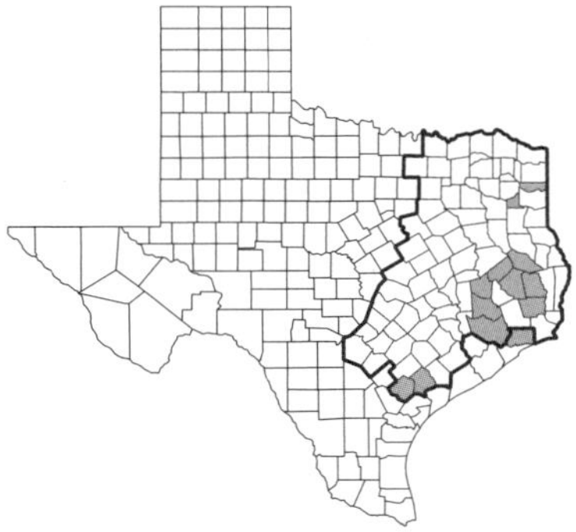

Dalea phleoides var. phleoides

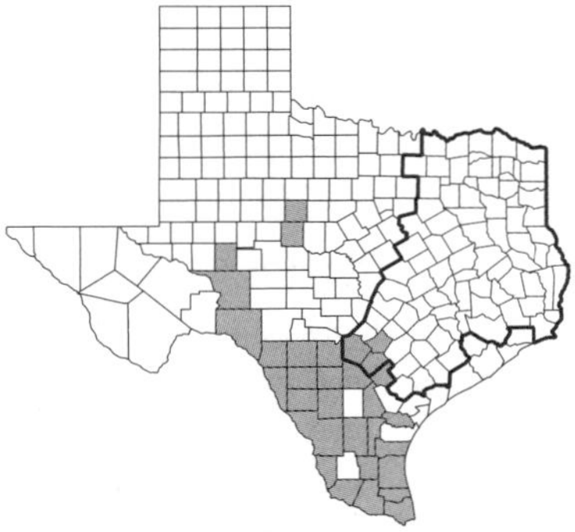

Dalea pogonathera var. walkerae

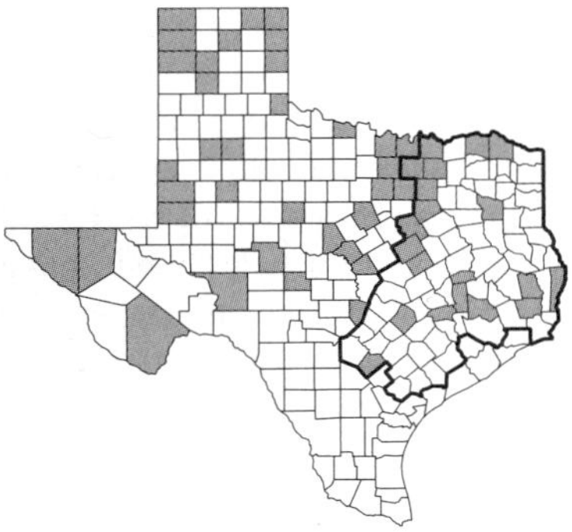

Dalea purpurea

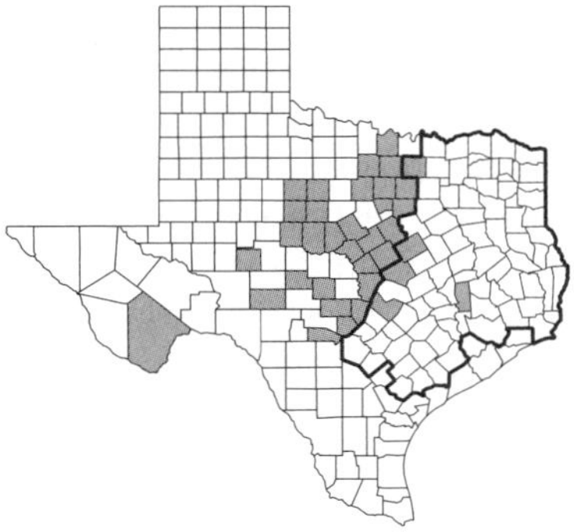

Dalea tenuis

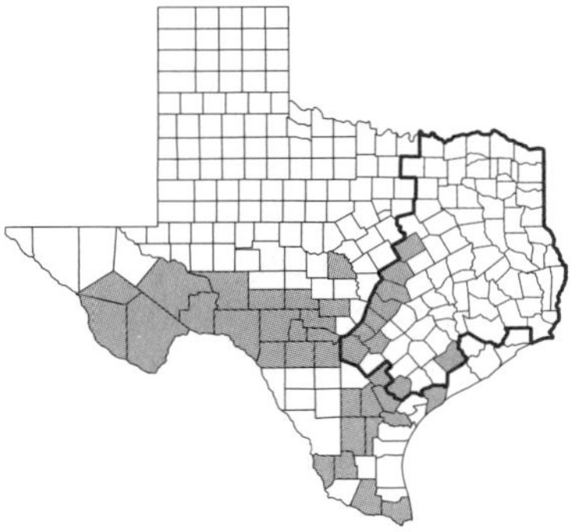

Dermatophyllum secundiflorum

Desmanthus obtusus S. Watson, (obtuse, blunt), BLUNT-POD BUNDLE-FLOWER. Plant somewhat similar to *D. reticulatus* (one of two East TX species with raised, reticulate veins on leaflets), much-branched basally; stems sprawling, at least near base, 60 cm or less long; leaves with 1–4 pairs of pinnae; leaflets 8–15 pairs per pinna, 3–6 mm long, oblong, pubescent below; larger leaves with petioles usually 10(–15) mm long or less; corolla 3 mm long, stamens 10; fruits 25–40 mm long, linear, obtuse or with an abrupt mucro, but not tapering to a point. Limestone based soils; Bexar Co. (Turner et al. 2003) in extreme sw corner of East TX; mainly w 1/2 of TX; adjacent Mexico. Apr–Aug. Isely (1998) noted that "this species presumed to be related to *Desmanthus velutinus* from which it is distinguished by its raised leaf nervation."

Desmanthus reticulatus Benth., (netted), NET-LEAF BUNDLE-FLOWER. Plant much-branched basally; stems sprawling, at least basally, 10–50(–100) cm long, pubescent all around; leaves with 2–4 pairs of pinnae; leaflets 4–10(–12) pairs per pinna, 3–6.5 mm long, oblong to elliptic or elliptic-obovate, glabrous or nearly so, usually glaucous below, reticulate; larger leaves with petioles 10 mm or more long; flowers pinkish white, 10–12 flowered; corolla 3–4 mm long, dark-tipped; fruits 25–50 mm long, linear. Rare in calcareous prairies and gravelly limestone grasslands; Bexar, Comal, Goliad, Gonzales, Guadalupe, Travis (BRIT), and Wilson (Kartesz 2023) cos. in sw part of East TX; endemic to c TX. Apr–Jun. The most distinctive features of this species are its broad, glaucous, and reticulate leaflets and long peduncles (Isely 1998). NatureServe: G3, vulnerable; S3, vulnerable.

Desmanthus velutinus Scheele, (velvety). Plant much-branched basally; stems spreading to erect, 20–50 cm long, villous with flexuous velvety pubescence; leaves with 3–7(–8) pairs of pinnae; leaflets 6–14 pairs per pinna, 2.5–4.5 mm long, linear-oblong, pubescent, non-reticulate; petioles 4–12 mm long; flowers 15–33 per head, corolla 3–4 mm long, stamens 10; fruits 30–80 mm long, linear, straight. Rocky limestone outcrops, streambeds, and roadsides; scattered mainly in the w part of East TX; widely scattered in the state except in the nw and e portions; sw U.S. from TX w to s NM; adjacent Mexico (Coahuila). Apr–Nov. This species can be recognized by its velvety pubescent, non-reticulate leaves (Isely 1998).

Desmanthus virgatus (L.) Willd., (twiggy). Plant branched, sometimes much so basally; stems prostrate to erect, to 1.5 m long; foliage on living plants blue-green, glaucous, subsessile or short petioled, leaves with 2–6 pairs of pinnae; leaflets 6–15 pairs per pinna, non-reticulate, foliar glands circular and not exceeding 1 mm in diameter, petioles usually 1–3(–5) mm long; heads with 3–10 flowers; fruits 22–88 mm long, linear, straight or slightly falcate. Disturbed areas including mud flats, scrubby shrublands, and shell mounds; s portion of East TX n to Brazos, Robertson, and Travis (Kartesz 2023) cos.; s 1/2 of TX; se U.S. in FL, nw LA, and s TX; s through Mexico and Central America to South America. Apr–Nov. [*D. depressus* Humb. & Bonpl. ex Willd., *D. virgatus* var. *depressus* (Humb. & Bonpl. ex Willd.) B.L. Turner] While TX material is sometimes treated as *D. virgatus* var. *depressus* (Kartesz 1994; Jones et al. 1997; Isely 1998), we are following Luckow (1993) in not recognizing this variety. She indicated that the holotype of *D. virgatus* (at LINN) corresponds to what has traditionally been called *D. virgatus* var. *depressus*.

DESMODIUM Desv. TICK-CLOVER, TICK TREFOIL, BEGGAR'S-LICE, BEGGAR'S-TICKS

Ours perennial herbs (annual in *D. tortuosum*); leaves pinnately compound; leaflets 3 (except *D. psilophyllum* with 1 leaflet); stipules broad or narrow, persistent or falling early; flowers 3.5–13 mm long, usually many, in erect, narrow racemes or panicles; petals pink, lavender, red-purple, purple, or whitish, often drying bluish or a striking blue-green; calyces campanulate with 5 unequal lobes, upper pair generally fused; fruits (loments) flat, constricted into ca. 2–10 one-seeded segments, usually with small hooked hairs. In East TX, the genus is easily identified in the field because it is one of two (closely related *Hylodesmum* being the other) native genera with flat fruits breaking into segments that readily adhere to fur and clothing.

A genus of ca. 120 species of warm areas of the world, especially e Asia, Brazil, and Mexico (Ohashi 2023b). The fruits fall into 1-seeded segments which stick to hair or clothing, hence the common names. Species are variously used as fodder, green manure, or medicinally. The descriptions and key to species are modified from Isely (1998). While the genus is easily recognized, determination to species is often challenging and many specimens are misidentified. Mature fruits are usually important in the identification of several species. A specimen misidentified as *D. illinoense* (*Gentry s.n.*, 1950, UT) and mapped in Kartesz (2023) in Grayson Co. is *D. tweedyi*. *Desmodium scorpiurus* (Sw.) Poir. in F. Cuvier was incorrectly mapped in TX in Isely (1998), a mistake which was perpetuated in Turner et al. (2003). That species is only known from Florida in the U.S. (Isely 1998, in text). Specimens identified and mapped as *D. strictum* in Kartesz (2023) from Brazos, Galveston, and Lamar cos. are all *D. paniculatum*. Specimens identified as *D. tenuifolium* from Newton Co. at BRIT and LSU and mapped in Kartesz (2023) are misidentified specimens of *D. paniculatum* and *D. obtusum*. *Desmodium strictum* and *D. tenuifolium* are only known from Louisiana eastward (Isely 1998; E. Keith, pers. obs.) (Greek: *desmos*, a chain, alluding to the connected segments of the fruit giving it a chain-like appearance) (subfamily Faboideae, tribe Desmodieae)

REFERENCES: Schubert 1950; Isely 1998; Thomas 2020; Weakley, A.S., and Southeastern Flora Team 2022; Ohashi 2023b.

Dalea phleoides var. phleoides [MNY]

Dalea pogonathera [MN2]

Dalea purpurea [MIC, MNY]

Dalea tenuis [MNY]

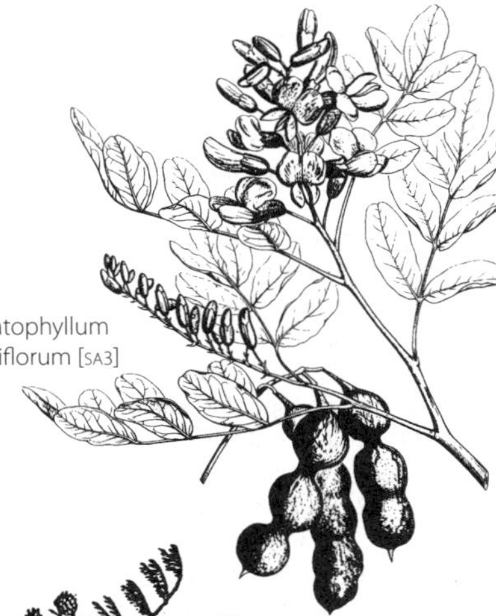
Dermatophyllum secundiflorum [SA3]

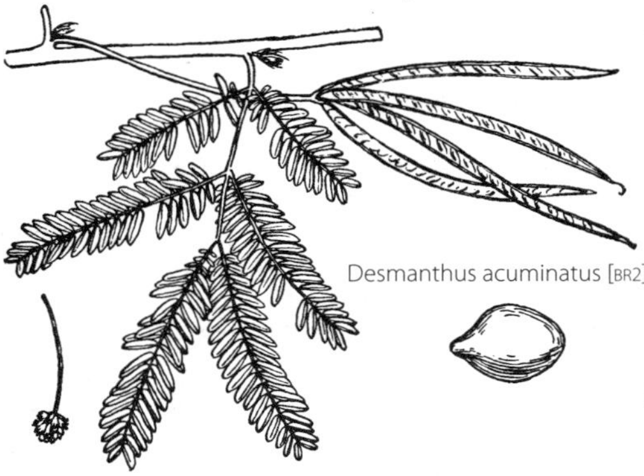
Desmanthus acuminatus [BR2]

Desmanthus illinoensis [BB2]

1. Leaves apparently simple (with 1 leaflet) ____ **D. psilophyllum**
1. Leaves compound with 3 leaflets.
   2. Fruits crenate below, the incisions/indentations between fruit segments shallow, usually cutting < 1/2 the distance through the fruit; leaflets usually 4 cm or less long; species rare in s portion of East TX ____ **D. incanum**
   2. Fruits notched or incised below, the incisions/indentations deep, cutting 1/2 to almost all the way through the fruit; leaflets of various lengths; including species widespread and common in East TX.
      3. Plants vine-like in appearance, prostrate; leaflets suborbicular or broadly ovate.
         4. Corollas 9–11 mm long; terminal leaflet (2.5–)3–5.5 cm long; stipules clasping, ovate, conspicuous, 8–12 mm long; stems without short, hooked hairs, pilose (= long straight hairs); fruit segments 3–6(–7) ____ **D. rotundifolium**
         4. Corollas 3.5–5 mm long; terminal leaflet 1.5–2.5(–4) cm long; stipules not clasping, lanceolate to ± linear, not conspicuous, 2–5 mm long; stems with short, hooked hairs, a few long straight hairs also sometimes present; fruit segments 2–3(–4) ____ **D. lineatum**
      3. Plants not vine-like, either erect, ascending, or sprawling; leaflets variously shaped.
         5. Leaflets with large conspicuous pale to whitish areas along midvein; petals white with greenish or yellowish tinge, sometimes with lavender base; w portion of East TX ____ **D. tweedyi**
         5. Leaflets without pale to whitish areas along midvein or less pronounced; petals variously colored, usually pink to purple; more widespread.
            6. Stipules (5–)7–20 mm long, > 1 mm wide at the base, clasping or truncate at the base, persistent (most or all of the stipules persisting through the year); leaflets 1.5--3 times as long as wide.
               7. Corollas ca. 6 mm long; calyces 2–2.2 mm long; fruits equally crenate above and below, the segments each 3–3.5 mm long, rounded above and below; plant annual ____ **D. tortuosum**
               7. Corollas 8–13 mm long; calyces 3–5 mm long; fruits sinuate above, deeply notched below, the segments each 6.5–11 mm long, convex above, obtusely angled below; plant perennial.
                  8. Inflorescence axis pilose (with long straight hairs) as well as having short, hooked hairs; pedicels long, in flower 8–15 mm long; stipules ovate and broadly clasping the stem ____ **D. canescens**
                  8. Inflorescence axis not pilose (glabrate or only with short, hooked hairs); pedicels shorter, in flower 4–8 mm long; stipules lanceolate or subulate ____ **D. cuspidatum**
            6. Stipules 2–6(–9.5) mm long, < 1 mm wide at the base, narrowed or parallel-sided to the base, caducous (most or all of the stipules falling soon after expansion of the leaves); leaflets 1–8(–10) times as long as wide.
               9. Stems with flanges of tissue; fruit segments 8–10 mm long, glabrous except for ciliate margins; corollas 6–7 mm long, blue-green; rare species known in East TX only from Comal Co. on sw margin of area ____ **D. lindheimeri**
               9. Stems without flanges of tissue; fruit segments not as above; corollas various; including widespread and common species.
                  10. Fruits mostly with 1–3 (less often 3) segments, these usually rounded below; flowers small, 3–7 mm long.
                      11. Leaflets 4–10 times as long as wide, narrow, narrowly oblong to linear; leaves sessile or subsessile (petioles usually 3 mm or less long); pedicels 5 mm or less long ____ **D. sessilifolium**
                      11. Leaflets < 3.5 times as long as wide (very rarely to 5 times as long as wide), usually broadly ovate to elliptic; leaves usually petiolate (petioles 1–25 mm long); pedicels 3–20 mm long.
                          12. Leaflets usually villous below, velvety to the touch and appearing ashy in color; corollas 6–7 mm long; fruit segments often 3 ____ **D. nuttallii** in part
                          12. Leaflets glabrate to variously pubescent below, but neither velvety nor appearing ashy; corollas 4–6 mm long; fruit segments usually 2, less often 3.
                              13. Lateral leaflets of middle and lower leaves nearly as long as petioles; petioles 10–25 mm long; stems and leaves essentially glabrous or sparsely puberulent with short, hooked hairs ____ **D. marilandicum**
                              13. Lateral leaflets of middle and lower leaves distinctly longer than petioles; petioles 1–12 mm long; stems and leaves pilose (= with long ± straight hairs) to villous (= with long curved or bent hairs) and/or sparsely to densely puberulent with short, hooked hairs.
                                  14. Terminal leaflets usually noticeably longer and narrower than lateral leaflets (25–)30–75 mm long, 2–3.5 times as long as wide; stems pubescent with short, hooked hairs, but without pilose hairs or these rare ____ **D. obtusum**
                                  14. Terminal leaflets similar to lateral leaflets, 9–30 mm long, usually 2.5 times as long as wide or less; stems sparsely to densely pilose (long ± straight hairs) ____ **D. ciliare**
                  10. Fruits mostly with 3–6 segments, these usually obtusely angled below (except can be rounded in *D. nuttallii*); flowers larger, usually 6–10 mm long.
                      15. Leaves densely villous and velvety on the lower surface; stem densely pubescent with uncinate (= hooked on the end) or non-uncinate hairs.
                          16. Fruits nearly straight, with (3–)4–5(–6) segments, 5–9 mm long and obtusely angled; terminal leaflet usually at least 2/3 as wide as long, rhombic to deltoid ____ **D. viridiflorum**

16. Fruits usually curved, with 2–4 segments, these 4–5 mm long and often rounded below; terminal leaflet ca. 1/2 as wide as long, elliptic-ovate ____ **D. nuttallii** (in part)

15. Leaves glabrous to moderately appressed-villous on the lower surface; stem glabrate, pilose or uncinate pubescent.
    17. Terminal leaflets narrow, usually lanceolate to narrowly oblong, usually (2.5–)3–10 times as long as wide, coriaceous; stems near middle usually glabrate; species widespread in East TX ____ **D. paniculatum**
    17. Terminal leaflets rather broad, very narrowly ovate to broadly ovate or rhombic, 1–3(–4) times as long as wide (if > 2.5 times as long as wide, then stems near middle with definite pubescence of short, hooked hairs), not coriaceous; species primarily of the e part of East TX.
        18. Plants glabrous or nearly so; leaves glaucous beneath; corollas 8–10 mm long ____ **D. laevigatum**
        18. Plants with pubescence on stems and leaves (particularly beneath); leaves green or light green beneath; corollas usually 6–8 mm long.
            19. Leaflet lower surface glabrous, except for the conspicuous uncinate pubescence on the veins, green beneath; stems and petioles glabrous or uncinate-puberulent; rare and scattered in e 1/4 of TX ____ **D. fernaldii**
            19. Leaflet lower surface strigose to conspicuously sub-appressed-villous, and sometimes also uncinate-puberulent, paler beneath; stems and petioles glabrate to conspicuously pilose or uncinate-puberulent; more widespread in East TX.
                20. Ventral margins of most articles (= segments) straight to slightly convex; leaves gradually but noticeably reduced in size and petiole length distally along the stem, often with small leaves extending well onto the flowering branches; terminal leaflets lanceolate to broadly ovate, broadest nearer the base than the middle, terminal leaflet emarginate; widespread in various habitats throughout East TX ____ **D. glabellum**
                20. Ventral margins of most articles concave; leaves mostly all the same size and not or little extending onto the flowering branches; terminal leaflets narrowly ovate to broadly elliptic-ovate, broadest near the middle, terminal leaflet acute; uncommon, but underreported in mesic to dry forests in e 1/4 of TX ____ **D. perplexum**

Desmodium canescens (L.) Poir., (gray pubescent), HOARY TICK-CLOVER, HOARY TICKTREFOIL. Stems erect or ascending, 0.5–1.2(–2) m tall, pilose and also usually with short hooked (uncinate) hairs; leaflets 5–10 cm long, 1.5–2 times as long as wide, with some short hooked hairs; petioles 19–100 mm long; inflorescence axis pilose; pedicels 8–13 mm long (the flowers often abort leaving the persistent and conspicuous pedicels); corollas 9–13 mm long, purplish or pinkish becoming greenish; fruits with 4–6 segments, each segment 6.5–10 mm long, convex above, obtusely angled below. Open woodlands and forests on sandy and calcareous soils; scattered in the n and e parts of East TX; c and e U.S. from MA s to FL and w to e NE and s to OK and TX. Jul–Oct. This species is distinguished from all others by the tall, pilose stems with large clasping stipules, and long pedicels that are conspicuous when the frequently aborted flowers fall (Isely 1998).

Desmodium ciliare (Muhl. ex Willd.) DC., (ciliate, fringed), LITTLE-LEAF TICK-CLOVER, HAIRY SMALL-LEAF TICKTREFOIL. Stems ascending to erect, 0.4–1.5 m tall, sparsely to densely pilose; leaflets small, 1.5–3 cm long, 1.5–2.3(–5) times as long as wide; petioles short, 1–5 mm long; pedicels 3–8 mm long; corollas 3.5–5 mm long, lavender-purple or pink; fruits with 1–2(–3) segments, each segment 4–5 mm long. Sandy woods and open areas; widespread in East TX; also adjacent n Gulf Prairies and Marshes; e and s U.S. from MA s to FL and w to MO, OK, and TX; s Canada and Cuba. Jul–Oct. [*D. marilandicum* (L.) DC. var. *ciliare* (Muhl. ex Willd.) H. Ohashi] Isely (1998) noted that some populations "are of, or include, plants with various features of *D. marilandicum*"; hybridization between these related species has been hypothesized. Ohashi (2023b) included this species within the concept of *D. marilandicum*, but the two entities often grow together in East TX and are quite distinctive without any evidence of introgression (E. Keith, pers. obs.). Plants occurring east of the Mississippi River with narrow leaflets have historically gone under the name *D. ciliare* var. *lancifolium* Fernald & B.G. Schub. (now *D. marilandicum* var. *lancifolium* (Fernald & B.G. Schub.) H. Ohashi (Ohashi 2013)) but are not known to occur in TX (Isely 1998).

Desmodium cuspidatum (Muhl. ex Willd.) DC. ex G. Don, (tipped with a firm point), LARGE-BRACT TICKTREFOIL. Plant erect or ascending; stems 0.5–1.5 m tall, glabrous or slightly pilose; leaflets 5–10 cm long, 1.5–3 times as long as wide, conspicuously acuminate, glabrate or somewhat villous below; stipules persistent; pedicels 4–8 mm long; calyces 3–4 mm long; corollas 8–12 mm long, purple; fruits with 4–6 segments, each segment 8–11 mm long. Rich woods and uncommon in ruderal areas; rare in ne TX only known from Bowie and Smith (Kartesz 2023) cos. in TX; e U.S. from NH s to n FL w to se NE s to OK and TX. Jul–Oct. This species is typically robust, generally glabrate with long petioled leaves, large acuminate leaflets, and large loments (Isely 1998).

Desmodium fernaldii B.G. Schub., (for Merritt Lyndon Fernald, 1873–1950, Harvard botany professor and prolific researcher of the flora of the eastern U.S.), FERNALD'S TICKTREFOIL. Plant erect or ascending; stems 0.5–1 m tall, glabrous or slightly uncinate-pubescent; leaflets 3–9 cm long, 1.5–2 times as long as wide, glabrate except vein uncinulate both above and below; stipules deltoid, caducous; pedicels 4–8(–10) mm long; calyces 3–4 mm long; corollas 6–8(–10) mm long, purplish; fruits with (2–)3–5 segments, each segment 4–8 mm long. Dry sandy pine-hardwood forests and woodlands; uncommon and scattered in East TX but probably more common than reports indicate due to its similarity to *D. glabellum*, confirmed in Montgomery (*Shinners 16566* BRIT) (iNaturalist Contributors 2023c), Newton (*Cory 10084* GH; *Keith 277* BRIT), and Smith (*Keith 895* BRIT; *Keith 1102* SHST) cos.; se U.S. from se VA s to n GA w to TX. Jul–Oct. [*D. paniculatum* (L.) DC. var. *fernaldii* (B.G. Schub.) H. Ohashi] This species is included within the concept of *D. paniculatum* in Ohashi (2023b), however, characteristics of the leaves described in the key are distinctive and obvious especially when similar species are found growing together (E. Keith, pers. obs.). This species is common at Tyler State Park in Smith Co. and grows with *D. paniculatum* and *D. glabellum* in upland sandy pine-hardwood forests without intergrades.

Desmodium glabellum (Michx.) DC., (somewhat glabrous), DILLENIUS' TICKTREFOIL. Stems erect or ascending, 0.3–1(–1.5) m tall, with pubescence of hooked hairs; leaflets 2–8(–10) cm long, 1.5–3(–4) times as long as wide, with variable pubescence particularly beneath; pedicels 5–10(–12) mm long; calyces 2–3 mm long; corollas 6–7(–8) mm long, lilac to purple; fruit segments (2–)3–5, each segment 4–8 mm long. Alluvial to upland woodlands and forests, streambanks, and old fields on diverse soil types; mainly e part of East TX and Grayson Co. (Turner et al. 2003) in the nw part of the area; also n Gulf Prairies and Marshes; c and e U.S. from MA s to FL and w to IA, e KS, OK, and TX. Jun–Oct. This species is included within the concept of *D. paniculatum* in Ohashi (2023b), but clear distinctions in the leaf width ratio and thinner texture of the leaves readily separate the species and remain constant when growing in proximity; see discussion under *D. paniculatum*. An excellent synopsis of how to differentiate this species from the similar, but substantively different *D. perplexum* (as outlined in key) is described in Thomas (2020).

Desmodium incanum (Sw.) DC., (grayish or hoary), ZARZABACOA COMUN, CREEPING BEGGERWEED, SPANISH-CLOVER, SPANISH TICKTREFOIL. Plant stoloniferous or rhizomatous; stems ascending to erect, 0.1–0.6(–3) m tall; leaflets 2–4(–9) cm long, elliptic to ovate; stipules persistent and clasping, ovate-deltate 5–10 mm long; pedicels 5–9 mm long; calyces 1.5–2 mm long; corollas 5–8 mm long, purple; fruits merely crenate below, with 4–7(–9) segments, each segment 3.5–5 mm long. Ruderal and disturbed habitats; Dewitt, and Harris (Kartesz 2023) cos. in extreme s part of East TX; also Brazoria, San Patricio, and Victoria (Turner et al. 2003) cos. in the Gulf Prairies and Marshes; se U.S. from s AL, FL, s GA, and TX; Mexico; Taiwan; Africa; Indian Ocean Islands; Pacific Islands including HI; and Australia. Jun–Oct. [*D. canum* (J.F. Gmel.) Schinz & Thell., *D. frutescens* Schindl.] Apparently native to the New World tropics, now widespread. This species is recognized by the combination of its stoloniferous habit and simple racemes with fruits having broad isthmi between the segments (Isely 1998). Nicolson (1978) discusses the nomenclatural history of this species.

Desmodium laevigatum (Nutt.) DC., (smooth or slippery, presumably in reference to the lack of pubescence), SMOOTH TICKTREFOIL. Plant glabrous or glabrate, both stems and leaves lacking hairs or essentially so; stems 0.3–1.2(–1.5) m tall; leaflets 3.5–8 cm long, ovate to elliptic-ovate, glaucous beneath; stipules ovate-lanceolate, caducous; pedicels 5–15(–20) mm long; calyces 3–4 mm long; corollas 8–10 mm long, lavender to purple; fruits with (2–)3–5 segments, each segment (4–)5–8 mm long. Sandy forests and woodlands; widespread mainly in Pineywoods in e part of East TX; also n Gulf Prairies and Marshes; e U.S. from NY s to n FL and w to s MO, OK, and TX. May–Oct. This species can sometimes be confused with broad-leaved individuals of the similar *D. paniculatum* and *D. glabellum* (nearly identical fruits); however, those species always have leaflets with some pubescence and smaller flowers (Isely 1998).

Desmodium lindheimeri Vail, (for Ferdinand Jacob Lindheimer, 1801–1879, German born TX botanist), LINDHEIMER'S TICKTREFOIL. Plant erect; stems 0.4–0.6 m tall, flanged, with short hooked hairs and sparsely pilose; leaflets 5–9 cm long, 2.2–2.5 times as long as wide, densely villous and velvety to the touch beneath; stipules broadly ovate and semi-clasping, caducous; pedicels ca. 5 mm long; corollas 6–7 mm long, blue-green; fruits with 3–5 segments, each 8–10 mm long. Forested limestone canyons; Comal Co. (Turner et al. 2003) in sw corner of East TX; also Uvalde Co. (*Carr & Wolfe 16051* TEX) to the w; in the U.S. known only from TX; also Mexico. Oct–Nov. Isely (1998) believed that the Lindheimer collection from Comal Co. in 1850 was the only one known for the U.S., but it has been collected in several locations in Comal County in recent years, primarily along the Guadalupe River.

Desmodium lineatum DC., (lined, in reference to the green-lined stems), SAND TICKTREFOL. Plant vine-like in appearance, prostrate; stems 1 m or less long, with short hooked hairs or also slightly pilose; leaflets suborbicular or broadly ovate, the largest 1.5–2.3 cm long, 1–1.2 times as long as wide, glabrate or inconspicuously villous beneath; stipules lanceolate and inconspicuous;

Desmanthus leptolobus [BBV]

Desmanthus obtusus [LEI]

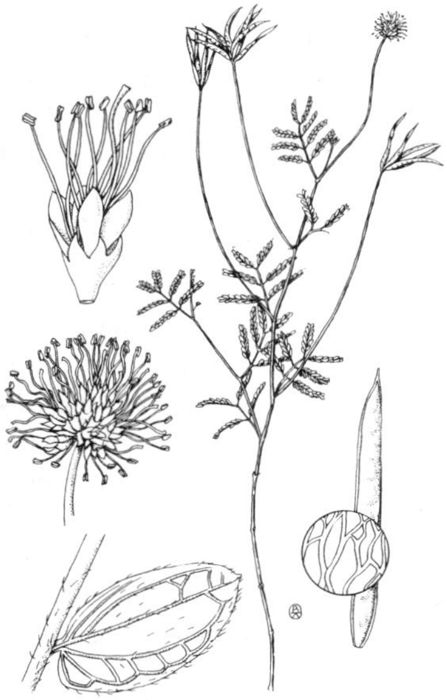

Desmanthus reticulatus [SBM]

Desmanthus velutinus [HEA]

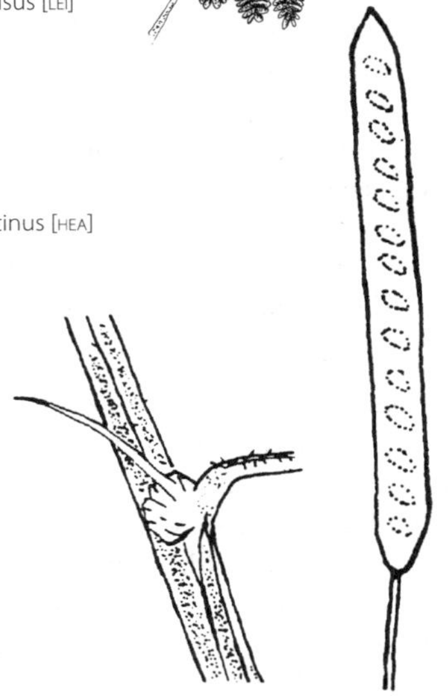

Desmanthus virgatus [F&L]

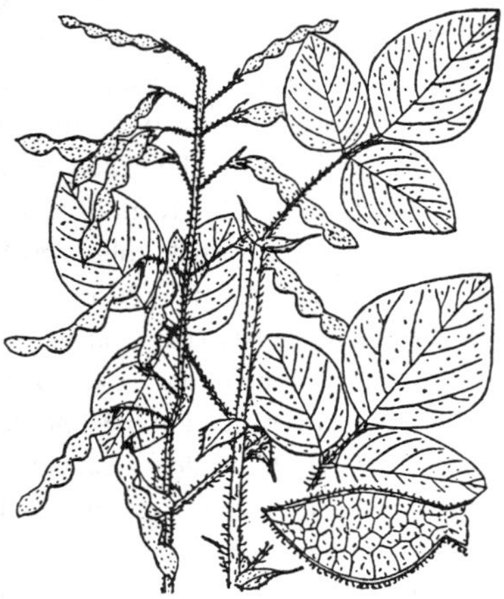

Desmodium canescens [BB2]

Desmodium ciliare [WIL]

Desmodium cuspidatum [STR]

inflorescences ascending; pedicels 6–12 mm long; calyces ca. 2 mm long; corollas 3.5–5 cm long, purple (–white); fruits with 2–3(–4) segments, each segment 4–5 mm long. Sandy wooded areas or occasionally open fields; uncommon in the Big Thicket area of se East TX; otherwise unknown in the state; se U.S. from MD s to n FL and w to TX. Aug–Oct. [*D. arenicola* (Vail) F.J. Herm.] Isely (1998) pointed out that this species is distinctive in its prostrate habit and small leaves with mostly suborbicular leaflets.

Desmodium marilandicum (L.) DC., (of Maryland), MARYLAND TICK-CLOVER, SMOOTH SMALL-LEAF TICKTREFOIL. Plant glabrous; stems ascending or erect, 0.3–1 m tall, glabrous or rarely with sparse pubescence; leaflets small, 1.5–2.5(–4) cm long, 1.2–1.5(–2) times as long as wide, pale below; petioles elongate, 10–25 mm long; stipules subulate and caducous; pedicels conspicuous, 8–15 mm long; calyces ca. 2 mm long; corollas ca. 5 mm long, lavender to red-violet, often drying bluish green; fruit segments 2(–3), each segment 3.5–4.5 mm long. Open woods; widely but sparsely scattered in East TX; e and sc U.S. from MA s to n FL and w to MO, OK, and TX. Aug–Oct. Isely (1998) noted the similarity of this species to *D. ciliare* and *D. obtusum* with some specimens difficult to determine, "possibly due to hybridization." Ohashi (2023b) included *D. ciliare* and *D. obtusum* within the concept of this species, however, we continue to follow the traditional species concepts detailed in Isely (1998) due to the consistent morphological differences outlined in the key even when growing in proximity (E. Keith, pers. obs.).

Desmodium nuttallii (Schindl.) B.G. Schub., (for Sir Thomas Nuttall, 1786–1859, English-American botanist), NUTTALL'S TICK-CLOVER, NUTTALL'S TICKTREFOIL. Plant usually with velvety pubescence and ashy gray in color; stems ascending to erect, 0.3–1 m tall; terminal leaflet often largest; leaflets 2–8 cm long, 1.5–2(–2.2) times as long as wide, beneath usually conspicuously villous and therefore ± velvety to the touch and ashy gray in color; petioles 10–30 mm long; inflorescence bracts usually villous; stipules subulate, caducous; pedicels 4–10 mm long; calyces 2–3 mm long; corollas 6–7 mm long, purple or pink; fruits curved, the segments 2–4, each segment 4–7 mm long. Sandy open woods; Jasper, Hardin, Harrison, Polk, Tyler, Upshur, Walker, and Wood (Kartesz 2023) cos. in the Pineywoods; e U.S. from NY (historical) s to n FL and w to MO, e OK, and TX. Jun–Oct. This species is similar to *D. viridiflorum* (velvety pubescence) but quite different in fruit—that species has nearly straight fruits with (3–)4–5(–6) segments. *Desmodium nuttallii* can also be confused with *D. glabellum*, but that species has glabrate inflorescence bracts and leaves that are not velvety to the touch (Isely 1998).

Desmodium obtusum (Muhl. ex Willd.) DC., (blunt), RIGID TICK-CLOVER, STIFF TICK-CLOVER, STIFF TICKTREFOIL. Stems ascending-erect, 0.5–1.5 m tall, with short hooked hairs; leaflets (2.5–)4–6(–7.5) cm long, 2.5–4 times as long as wide, the terminal leaflets noticeably longer and narrower; petioles 3–12 mm long; stipules deltate-subulate, caducous; pedicels 4–10 mm long; calyces 2–3 mm long; corollas 4.5–6 mm long, pink-purple (rarely white); fruit segments 2–3, each segment 3–5 mm long. Open woods often along streams; widely but sparsely scattered in East TX; also Galveston Co. (Turner et al. 2003) in the n Gulf Prairies and Marshes; e U.S. from MA s to FL and w to e MO, e OK, and TX. Aug–Oct. [*D. rigidum* (Elliott) DC.] This species is similar to *D. ciliare* and *D. marilandicum* with some specimens difficult to determine, "possibly due to hybridization" (Isely 1998). Ohashi (2023b) includes this species within the concept of *D. marilandicum* var. *lanceolatum*; however, that variety described from Virginia, only occurs east of the Mississippi River, and has glabrous or glabrate lanceolate and acuminate leaflets rather than oval or ovate obtuse leaflets that are consistently pubescent (Fernald 1938; E. Keith, pers. obs.).

Desmodium paniculatum (L.) DC., (with flowers in panicles), PANICLED TICK-CLOVER, PANICLED TICKTREFOIL. Stems 0.3–1(–1.5) m tall, ascending-spreading to erect, usually with few, long, wide-spreading branches, usually glabrate at mid-stem; leaflets 2–5(–6) cm long, narrow, usually (2.5–)3–8 times as long as wide, coriaceous, glossy and dark green above, with pubescence beneath, though this can be sparse; petioles (5–)20–50 mm long; stipules subulate, caducous; pedicels usually 6–12 mm long; calyces 2–3 mm long; corollas 6–7(–8) mm long, lilac to purple; fruit segments 3–5, each segment 3.5–7 mm long. Low woods, fencerows, and disturbed open areas; widespread mainly in e 1/2 of TX; e U.S. from MA s to FL and w to NE, OK, and TX; introduced in Japan. Jun–Oct. [*D. dichromum* Shinners is included in *D. paniculatum*; it is known only from a single collection n of Mineral Wells in Palo Pinto Co., *D. paniculatum* var. *epetiolatum* B.G. Schub.] This species is "probably the most cosmopolitan and abundant member of the genus in Eastern North America" (Isely 1998). The narrow coriaceous leaflets, sparse pubescence, and angled fruit segments usually make it easy to recognize. This species is part of the "*D. paniculatum* Group" including *D. glabellum* and *D. perplexum*. It "fades into" and apparently hybridizes with both, and "definitive determination of some specimens is difficult" (Isely 1998). While the three can often be easily recognized, "a good case could be made for considering them somewhat defined genetic phases of a single species" (Isely 1998). This is apparently an example where evolution has been "caught in action"; the three entities are distinct, but not completely so. There is no perfect taxonomic solution for such situations; therefore, we are following Isely (1998), who studied these species throughout their entire ranges, in recognizing the variation at the specific level. Plants with very short petioles have been named *D. paniculatum* var. *epetiolatum* (Schubert 1950),

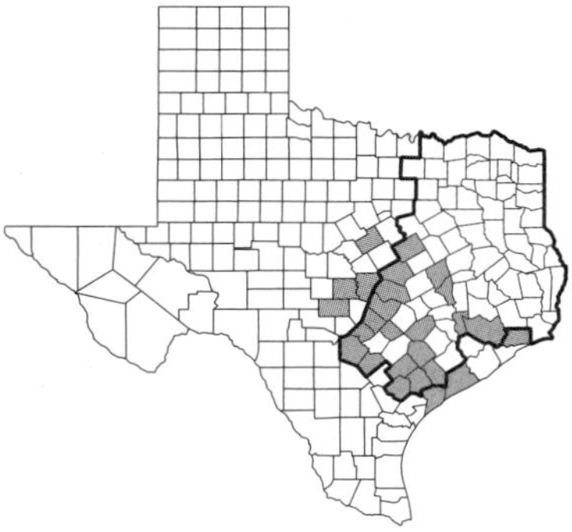

Desmanthus acuminatus

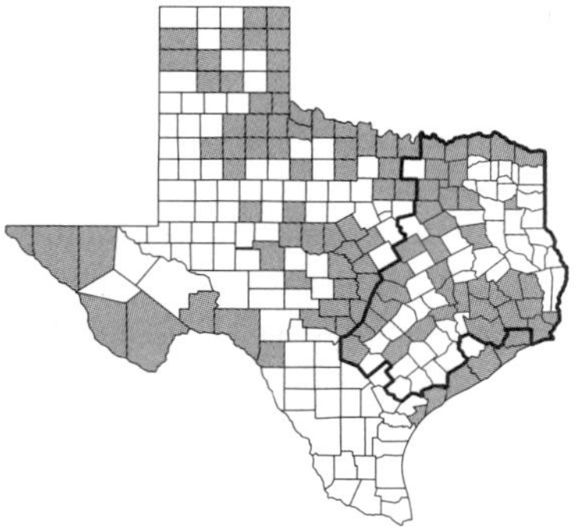

Desmanthus illinoensis

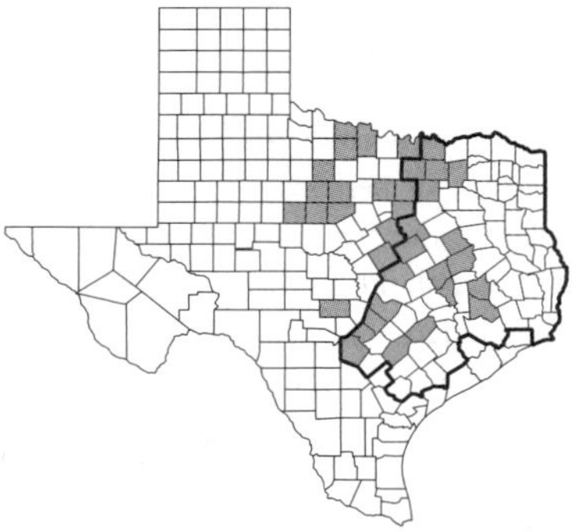

Desmanthus leptolobus

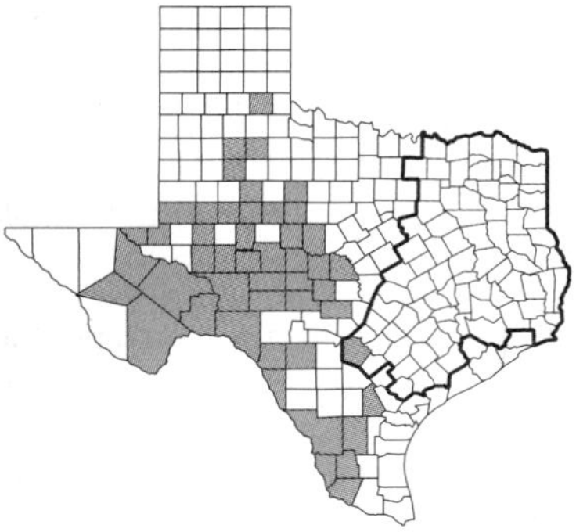

Desmanthus obtusus

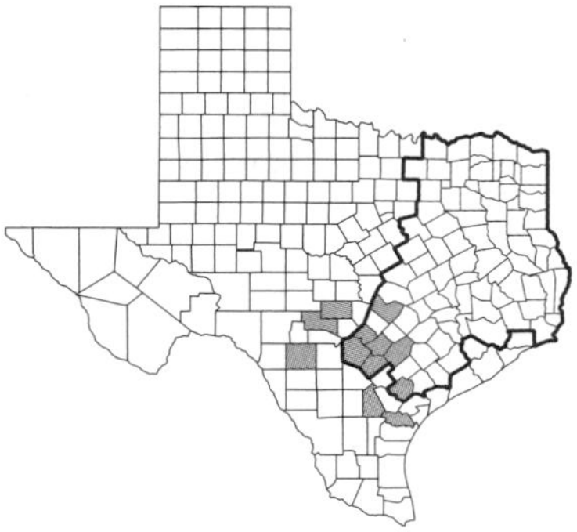

Desmanthus reticulatus

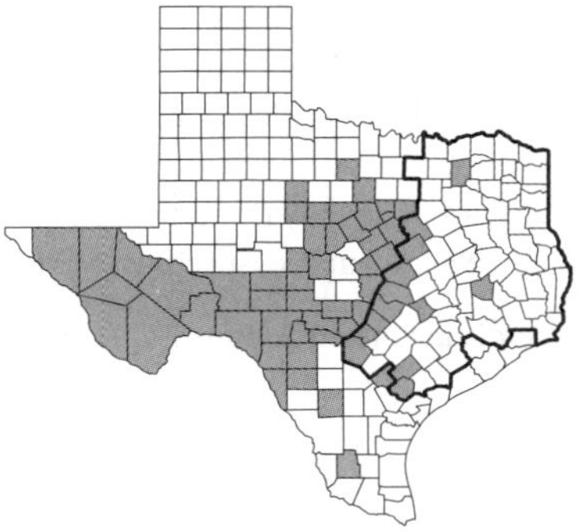

Desmanthus velutinus

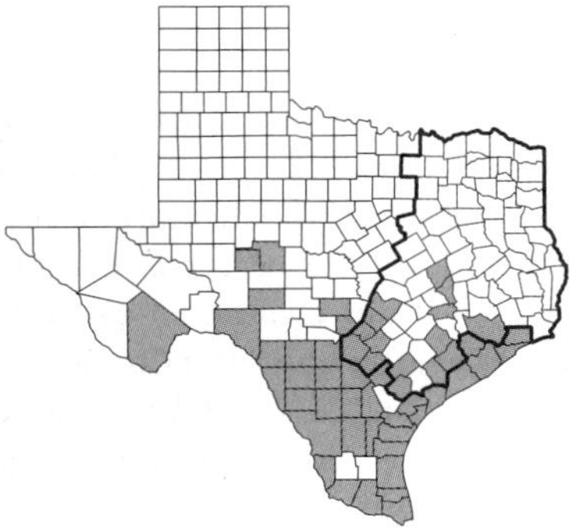

Desmanthus virgatus

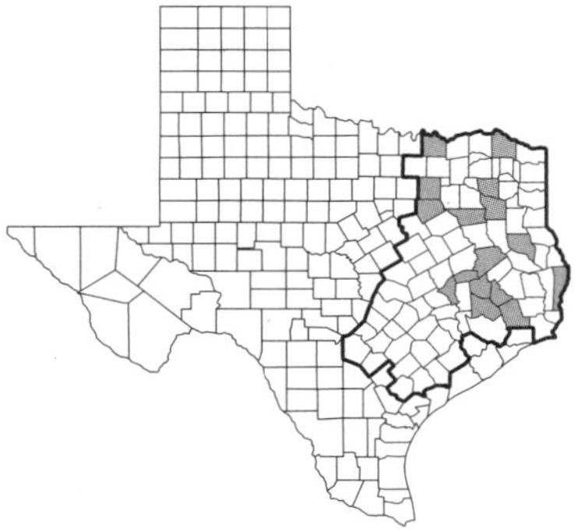

Desmodium canescens

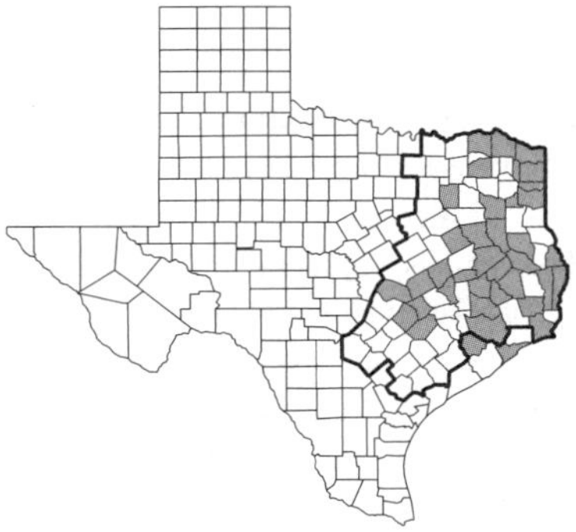

Desmodium ciliare

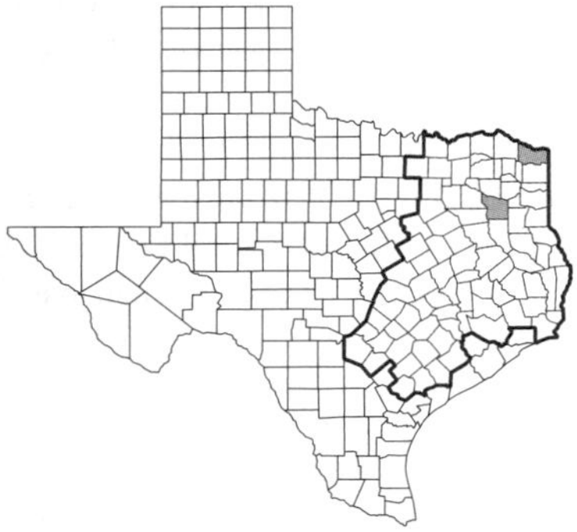

Desmodium cuspidatum

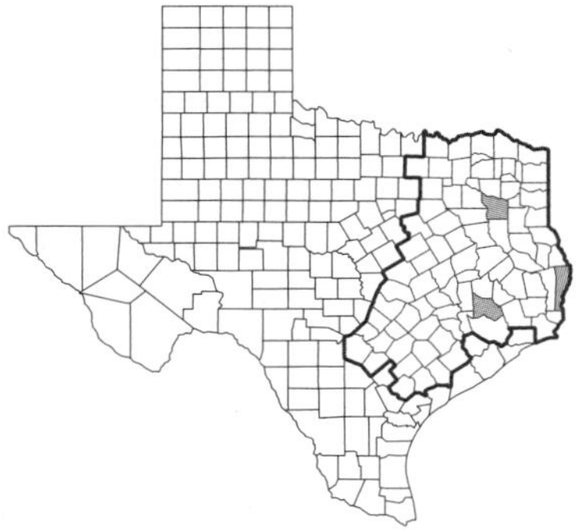

Desmodium fernaldii

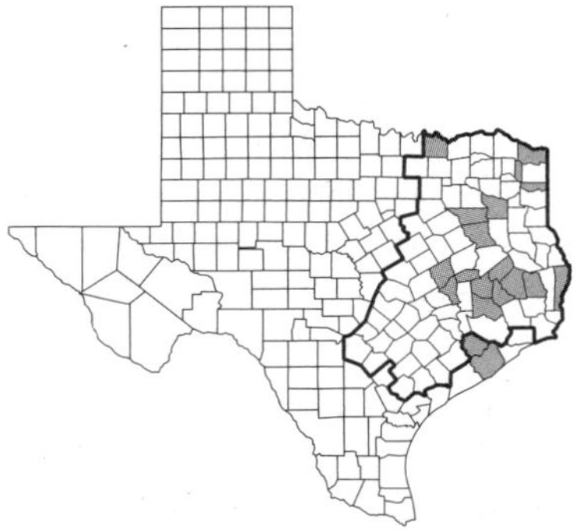

Desmodium glabellum

but occasionally occur intermixed with more typical plants or occasionally in pure populations (iNaturalist Contributors 2023g; E. Keith, pers. obs.).

Desmodium perplexum B.G. Schub., (perplexed, puzzling), PERPLEXED TICKTREFOIL. Plants similar to *D. glabellum*; stems near middle and/or petioles often pilose; leaflets 1.5–3(–3.5) times as long as wide, sparsely to abundantly villous below; stipules subulate, caducous; calyces 2–3 mm long; corollas 6–8(–9) mm long; fruits with (2–)3–5 segments, (4–)5–8.5 mm long. Mesic to dry woodlands and forests; Dallas (*Reverchon 22* GH; iNaturalist Contributors 2023d), Lamar (iNaturalist Contributors 2023e), Liberty (*Keith 325* BRIT), Tyler (iNaturalist Contributors 2023f), and Walker (*Cory 10393* GH) cos.; e U.S. from ME s to n FL and w to IA, e OK, and TX. Aug–Oct. [*D. dillenii* Darl.—a confused name (Isely 1998)] This species is part of the "*D. paniculatum* Group" and is similar to and easily confused with *D. paniculatum* (with narrower leaflets) and *D. glabellum*; see discussion under *D. paniculatum* (Isely 1998). Thomas (2020) provides an excellent synopsis for distinguishing this species from *D. glabellum*, from which our key is derived.

Desmodium psilophyllum Schltdl., (slender or naked-leaved), WRIGHT'S TICK-CLOVER, SIMPLE-LEAF TICK-CLOVER, SIMPLE-LEAF TICKTREFOIL. Stems ascending to erect, 0.3–0.8 m tall, glabrate or minutely and inconspicuously pubescent; leaves apparently simple, reduced to a single leaflet, the leaflet 3–8 cm long, 1–2 cm wide; petioles 6–9 mm long; pedicels 4–10 mm long; calyces 1.5–2 mm long; corollas 4–5 mm long, pink or lavender-pink; fruit with 3–5 segments, each segment 2.5–4.5 mm long. Stream banks and rocky slopes; Austin, Bell, Bexar, Comal, Hays, McLennan (Kartesz 2023), and Williamson (*Bodin s.n.*, 1891, MO) cos. in w part of East TX; mainly Edwards Plateau and Trans-Pecos; sw U.S. in sw AZ and TX; s to Central America. May–Oct. [*D. wrightii* A. Gray] The Austin Co. specimen (*Wright 142* MO) mapped in Kartesz (2023) is from Austin in Travis Co., TX.

Desmodium rotundifolium DC., (round-leaved), PROSTRATE TICKTREFOIL, DOLLAR-LEAF. Plant vine-like in appearance, prostrate; stems 1–2+ m long, pilose; leaflets suborbicular to broadly ovate, the larger ones (2.5–)3–5.5 cm long, 0.8–1.2 times as long as wide; stipules ovate-lanceolate, persistent and conspicuous; inflorescences erect; pedicels (5–)10–15 mm long; calyces 2.5–4 mm long; corollas 9–11 mm long, blue-purple; fruits with 3–6(–7) segments, each segment 4.5–6.5 mm long, but plants often not fruiting. Open woodlands and forests; Pineywoods of East TX, more common in se TX; e U.S. from NH s to n FL and w to MI sw to MO, e OK, and TX. May–Oct. Turner et al. (2003) and Kartesz (2023) only mapped counties in se TX, but this species does occur throughout ne TX being particularly common in Tyler State Park in Smith County (E. Keith, pers. obs.). Isely (1998) noted that this species is "usually easily recognized by its prostrate habit and large, suborbicular leaflets."

Desmodium sessilifolium Torr. & A. Gray, (sessile-leaved), SESSILE-LEAF TICK-CLOVER, SESSILE-LEAF TICKTREFOIL. Stems ascending-erect to erect, 0.5–1(–2) m tall, branched above, densely spreading-pubescent; leaves sessile or subsessile (petioles 3 mm long or less); leaflets (3–)4–8 cm long, 5–10 mm wide, 4–10 times as long as wide, densely soft-pubescent beneath, with prominent, raised veinlets; stipules subulate and awn-tipped, caducous; pedicels 2–5 mm long; calyces 1.5–2 mm long; corollas 4–5 mm long, pale lavender to reddish purple to nearly white, withering greenish; fruit segments 1–2(–3), each segment 4.5–6 mm long. Sandy open woods, roadsides, and fields; widespread in e 1/2 of TX, with a disjunct location far to the w in Potter Co. (Turner et al. 2003) in the Rolling Plains; e and c U.S. from MA (historical) s to n FL w to e KS, s NE, OK, and TX. May–Jul, rarely to Sep. This distinctive species "is usually easily known by its barely petioled leaves with narrow leaflets, and its 2-segmented loments" (Isely 1990, 1998). It is the first *Desmodium* species to bloom in the spring in East TX (E. Keith, pers. obs.).

Desmodium tortuosum (Sw.) DC., (winding, twisted), DIXIE TICKTREFOIL. Vigorous annual; stems erect, 0.5–2 m tall, pilose and with short hooked hairs; leaflets usually 4–10 cm long, with pubescence beneath; stipules persistent, the lower ones clasping; inflorescence axis pilose or not so; pedicels 10–15 mm long; calyces 2–2.2 mm long; corollas ca. 6 mm long, lavender; fruits with (3–)4–6(–7) segments, each segment 3–3.5 mm long, rounded above and below. Weedy habitats; in TX known only from Jasper, Newton, Tyler, and Walker (Turner et al. 2003) cos. in East TX, also Brazoria Co. from Gulf Prairies and Marshes (*Bush 327* GH); possibly no longer present in the state as the most recent collection was made in 1968 (*Correll 36529* KANU, NY, TEX); se U.S. from NC to FL and w to AR and TX; West Indies; Mexico s to South America. Sep–Oct. Native of tropical America. This species has been "cultivated to a limited extent for green manure and forage" (Isely 1998). The Caldwell Co. specimen mapped in Kartesz (2023) (*Lievens 524* LSU, SEU) is *D. glabellum*, which it superficially resembles vegetatively.

Desmodium tweedyi Britton, (for Frank Tweedy, 1854–1937, topographic engineer and botanical collector in the nw U.S.), TWEEDY'S TICK-CLOVER, TWEEDY'S TICKTREFOIL. Stems erect, to 1.25 m tall, short-pilose; leaflets 3–10 cm long, ovate to rhombic-elliptic, with raised veinlets beneath, forming a prominent network, densely spreading-pubescent beneath, with very distinctive large conspicuous pale to whitish areas along midvein; petioles 40–90 mm long; stipules lanceolate, broad-based, conspicuous, and persistent; inflorescence axis not pilose; pedicels 10–22 mm long; calyces 2.5–3.5 mm long; corollas white; fruits with 3–5 segments, each segment 6–8 mm long. Forests along streams on clay and limestone soils; Collin, Dallas, and Grayson (Turner et

Desmodium fernaldii [WIL]

Desmodium glabellum [BB1]

Desmodium lindheimeri [LEI]

Desmodium incanum [FAW]

Desmodium laevigatum [STR]

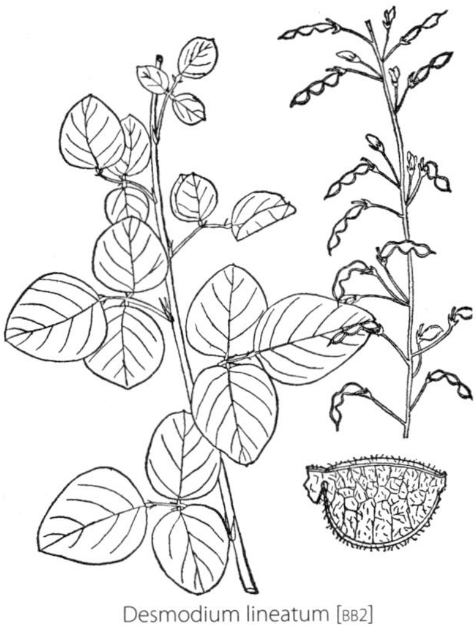
Desmodium lineatum [BB2]

Desmodium marilandicum [WIL]

Desmodium nuttallii [WIL]

al. 2003) cos. in nw part of East TX; sparsely scattered further w in TX; s U.S. in c OK and TX. Jun–Aug. Isely (1998) noted that the "irregular pale blotching along the midrib of leaflets is possibly a unique character," however, *D. glabellum* and *D. paniculatum* often also possess pale blotching but it is generally much less pronounced (E. Keith, pers. obs.). This species is somewhat similar to *D. canescens* (which has the inflorescence axis pilose and lacks the whitish areas along the midvein of the leaflets) (Isely 1998).

Desmodium viridiflorum (L.) DC., (green-flowered), VELVET-LEAF TICK-CLOVER, VELVET-LEAF TICKTREFOIL. Stems erect or ascending, 0.3–1 m tall, usually villous; leaflets broadly ovate to rhombic, 5–12(–15) cm long, 1–1.5(–1.9) times as long as wide, pubescent with short, hooked and strigose hairs above, beneath conspicuously villous and therefore usually ± velvety to the touch and ashy gray in color; petioles 20–50 mm long; stipules deltate, caducous; pedicels 3–9 mm long; calyces 2–3 mm long; corollas 7–8 mm long, purple to pink or light lavender; fruit segments (3–)4–5(–6), each segment 5–9 mm long. Open sandy woodlands and forests; widespread in East TX; also Denton Co. just to the w; e and c U.S. from NJ s to FL and w to se KS, e OK, and TX. Sep–Oct. Without fruits this species can sometimes be confused with *D. nuttallii* (very different fruits) (Isely 1998). The dense hooked hairs on the upper surfaces of the leaves often cling to clothing as readily as the fruits (E. Keith, pers. obs.).

ERYTHRINA L. CORAL-BEAN, COLORÍN

A genus of about 120 species of warm areas of the world. The species have red to orange flowers and are apparently all bird-pollinated. Species are variously used as shade trees, ornamentals with showy flowers (CORAL TREES), for their colorful seeds used as beads, and contain amino acids that provide a base for tortillas, desserts, pies, soups, and baby food (Nesom 2023b). Some have extrafloral nectaries to attract ants which guard the plant against herbivores (Lersten & Brubaker 1987). ☠ The seeds of all species are reported to be poisonous; some are used as fish poisons (Burlage 1968). *Erythrina crista-galli* is an ornamental small tree that is cultivated in East TX and has been documented as escaped in Harris Co. (*Brown 15516* NLU) but has not persisted as a member of the TX flora (Nesom 2015d). (Greek: *erythros*, red, referring to the color of the flowers) (subfamily Faboideae, tribe Phaseoleae)

REFERENCES: Lersten & Brubaker 1987; Nesom 2015d, 2016, 2023b.

Erythrina herbacea L., (herbaceous, not woody), CORAL-BEAN, EASTERN CORAL-BEAN, CHEROKEE-BEAN. Subshrub or shrub, in East TX dying back nearly or completely to the ground in winter (can become a small tree in warmer areas), with separate leaf-bearing and flowering stems; stems prickly, 0.5–2+ m tall; leaves alternate, pinnately compound with 3 leaflets; leaflets hastately 3-lobed to widely deltoid, (2–)4–13 cm long, glabrous or nearly so, stipellate, the small stipels bearing minute extrafloral nectaries; inflorescence a terminal, long pedunculate raceme; pedicels 3–9 mm long; calyces ± tubular, 5–11.5 mm long; corollas scarlet, very showy, narrow and elongate, standard 30–53 mm long, remaining folded; wings and keel 5.5–13 mm long; stamens 10, usually diadelphous, nearly as long as the standard; fruits 7–15(–21) cm long, 1.2–1.6 cm wide, constricted between the seeds, the stipe 1.5–4.5 cm long; seeds scarlet, 5–13 mm long. Sandy forests, woodlands, and old fields; common in se and East TX; cultivated as an ornamental, apparently escaped in Tarrant Co.; se U.S. from NC s to FL and w to s AR and TX. Apr–Jun. Austin (1975) considered this species adapted for pollination by ruby-throated hummingbirds (*Archilochus colubris*). The seeds have been used as beads; they contain numerous toxic alkaloids (e.g., erysodine, erysopine) and are used as a rat or dog poison in Mexico (Schmutz & Hamilton 1979; Morton 1982); according to Burlage (1968), the poison is similar in action to curare. Nesom (2016) described new species from Mexico that have historically been included within this species. ☠

ERYTHROSTEMON Klotzsch BIRD-OF-PARADISE

A genus of 31 species (34 taxa), shrubs, trees, and suffrutescent perennial herbs, of tropical and subtropical Americas. It includes many species previously placed in Central American and Mexican *Caesalpinia* or *Poincianella*. Twenty-two species are found across the southern U.S., Mexico and Central America, one occurs in the Caribbean (Cuba and Hispaniola), eight occur in South America, with one endemic in the Caatinga vegetation of Brazil, and the other seven in Argentina, Bolivia, Chile, and Paraguay (Gagnon et al.) 2016. (Greek: *erythron*, red and *stemon*, stamen) (subfamily Caesalpinioideae, tribe Caesalpinieae).

REFERENCES: Gagnon et al. 2016, Sotuyo 2023.

Erythrostemon gilliesii (Hook.) Klotzsch, (for John Gillies, 1792–1834, Scottish physician and botanist who collected in Argentina, Brazil, and Chile), BIRD-OF-PARADISE, POINCIANA, POP-BEAN BUSH. Unarmed shrub or tree to ca. 5 m tall; leaves twice pinnately compound with 11–29 pinnae (each with many leaflets), the pinnae paired or not so; leaflets 3–5(–8) mm long; inflorescences densely glandular-pubescent with yellow glands; sepals separate to base, the lower broader than others, deciduous in fruit; petals 5, yellow, 25–30 mm long; stamens 10, 70–90 mm long, much longer than petals, the filaments free, bright red; fruits flat, 6–12 cm long, 1.5–2 cm wide, glandular-pubescent. Cultivated for its showy flowers and escapes in dry ruderal habitats;

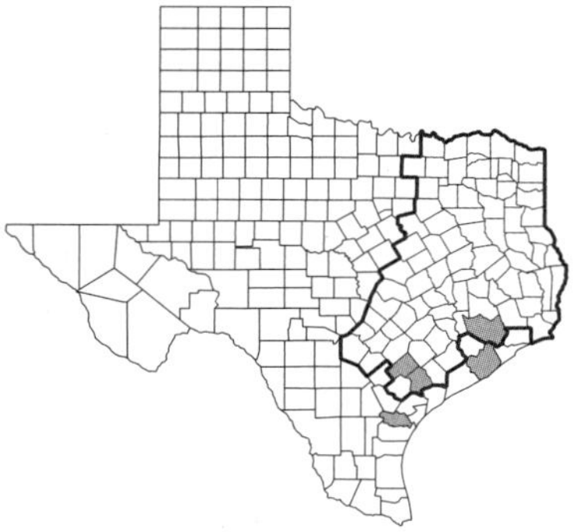

Desmodium incanum

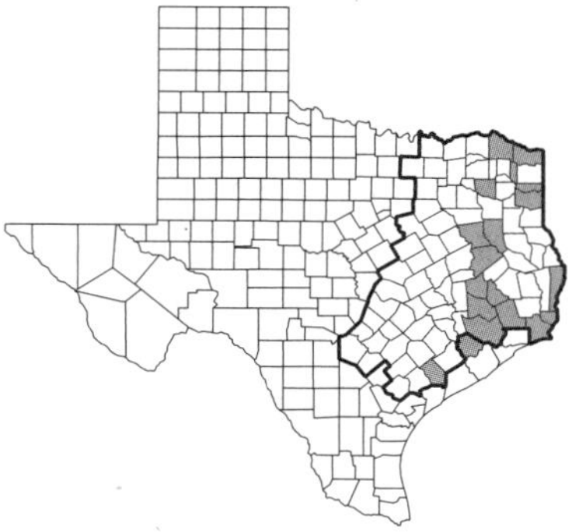

Desmodium laevigatum

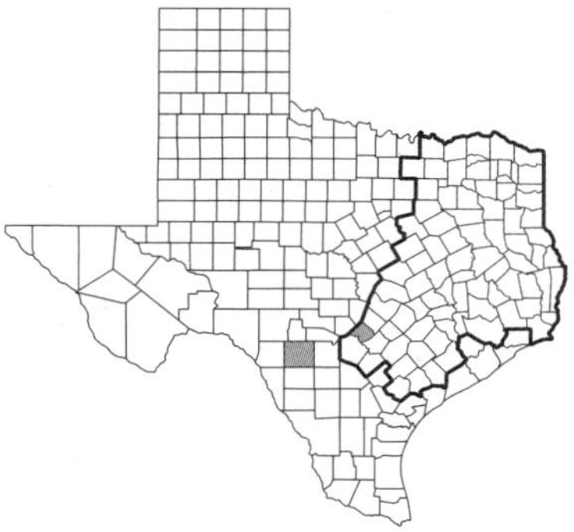

Desmodium lindheimeri

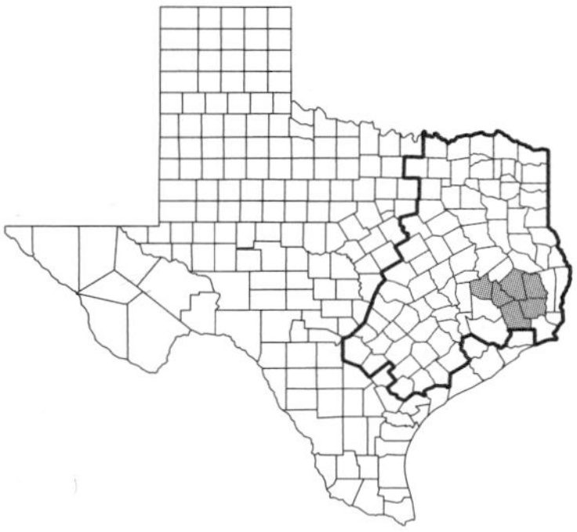

Desmodium lineatum

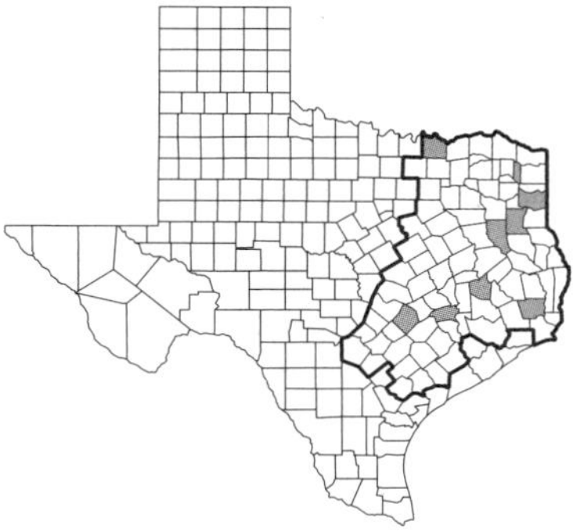

Desmodium marilandicum

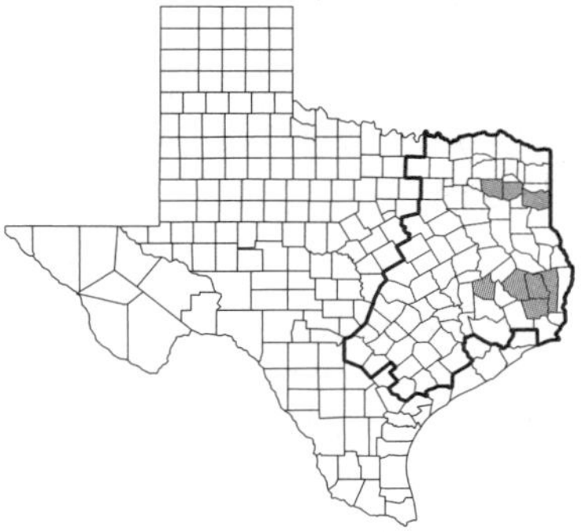

Desmodium nuttallii

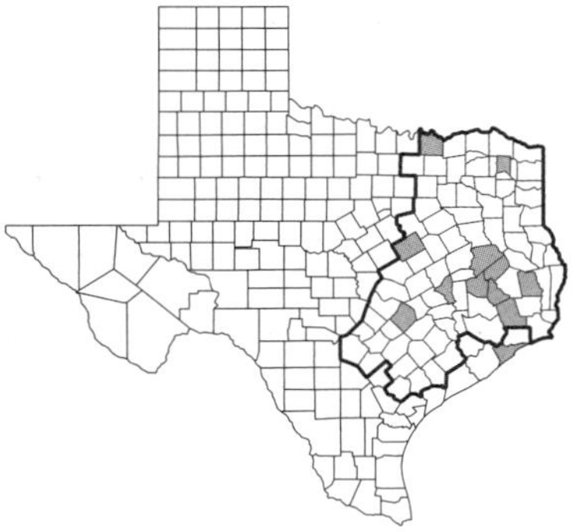

Desmodium obtusum

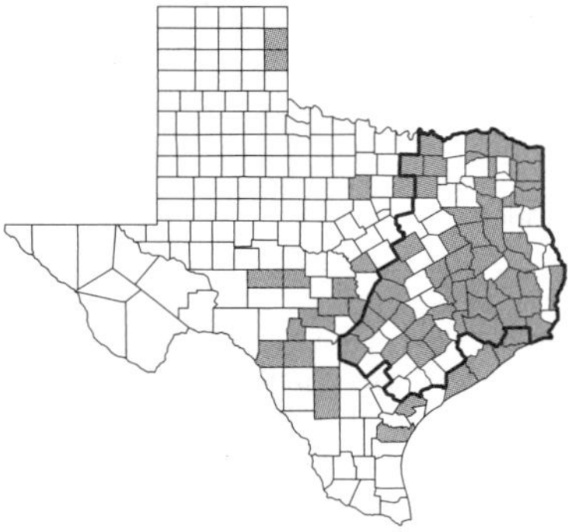

Desmodium paniculatum

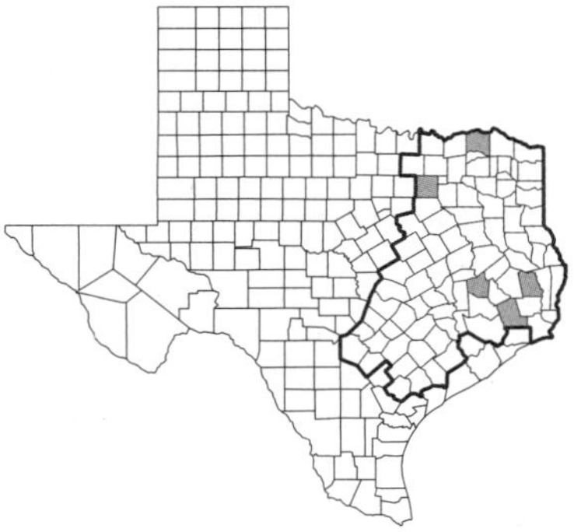

Desmodium perplexum

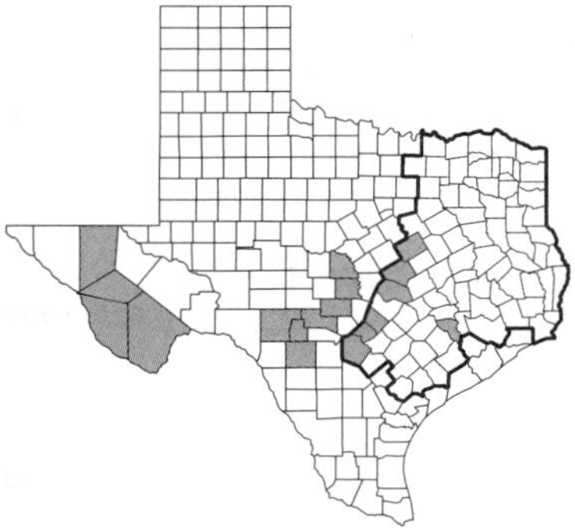

Desmodium psilophyllum

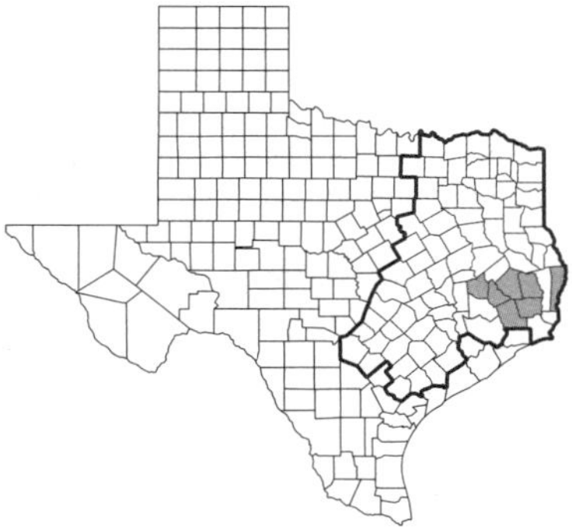

Desmodium rotundifolium

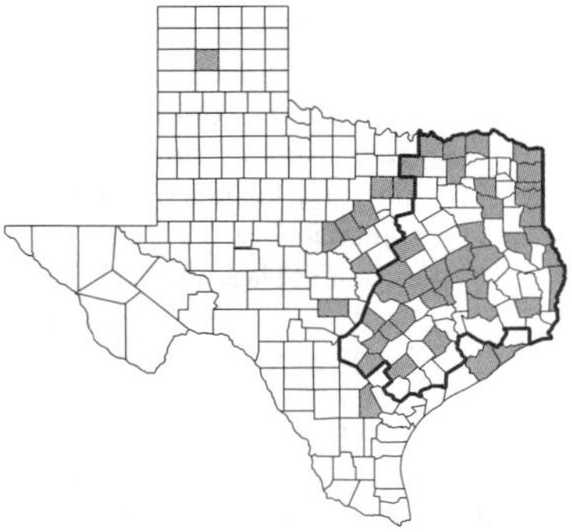

Desmodium sessilifolium

scattered in East TX (mainly to the w); w 2/3 of TX; s and sw U.S. from OK and TX w to CA; established in tropical regions of New World. Feb–Jul. [*Caesalpinia gilliesii* (Hook.) D. Dietr., *Poinciana gilliesii* Hook.] Native to Argentina and Uruguay. The green pods are reported to be severely irritating to the digestive tract (Kingsbury 1964). ☠

EYSENHARDTIA Kunth KIDNEYWOOD, BEEBRUSH

A genus of about 15 species (Goldman 2023) native from the sw U.S. to Guatemala, primarily a Mexican species. (Named for C.W. Eysenhardt, M.D., professor at Univ. of Konigsberg, Prussia) (subfamily Faboideae, tribe Amorpheae)
REFERENCES: Lang & Isely 1982, Goldman 2023.

Eysenhardtia texana Scheele, (of Texas), TEXAS KIDNEYWOOD, VARA DULCE. Unarmed shrub 1–3(–5) m tall; leaves crowded, once odd-pinnately compound; leaflets (7–)13–47, 3–13 mm long, oblong, gland-dotted, aromatic, with the aroma of tangerine rinds; stipules thread-like, persistent; flowers in terminal, erect, solitary or panicled, slender, spike-like racemes, sweet-scented, 3–11(–15) cm long; petals 4–5, white, separate, corollas 5–8 mm long; stamens 10, diadelphous, yellow; fruits indehiscent, laterally compressed, 6–13 mm long. Open and shrubby habitats on limestone and caliche; sw portion of East TX; widespread primarily in s 1/2 of TX; also Mexico. Mar–Nov. [*E. angustifolia* Pennell] Dyes have been obtained from the wood and the wood is fluorescent in water (Powell 1988); according to Crosswhite (1980), the wood "when leached in water turned the water orange, fluorescing blue against a black background. This was used as a diuretic [during frontier days]." Stems and leaves of this species apparently have antibacterial and antifungal chemical properties (Wächter et al. 1999).

GALACTIA P. Browne MILK-PEA

Prostrate perennials herbs with trailing or twining stems (erect in one species), with conspicuous pubescence to nearly glabrous; leaves alternate and pinnately compound or rarely simple; leaflets 1–3(–5), entire, 6–85 mm long; stipules thread-like, falling early; flowers solitary or in axillary, uncrowded, spike-like racemes with 2–16 flowers; calyces campanulate and 5-lobed (appearing 4); petals pink, roseate, pink-purple, lavender, or white; stamens 10, diadelphous; styles filiform, glabrous, stigma terminal, capitate; nectary at ovary base; fruits oblong-linear, several-seeded.

☛A genus of ca. 110 species of warm areas of the world, especially the Americas (Nesom 2023c). Many have latex (rare in Fabaceae), hence the common name (Diggs et al. 1999). (Greek: *gala*, milk; Patrick Browne originally stating that it has "milky branches") (subfamily Faboideae, tribe Phaseoleae)
REFERENCES: Vail 1895; Rogers 1949; Duncan 1979; Nesom 2015d, 2023c.

1. Leaves simple ____ **G. marginalis**
1. Leaves once pinnately compound with 3–5 leaflets.
  2. Leaflets 5, Goliad Co. in extreme s East TX ____ **G. heterophylla**
  2. Leaflets 3, more widespread.
    3. Stems erect, with only 4–6(–8) leaves; flowers white (–pale lavender) ____ **G. erecta**
    3. Stems trailing or twining, with numerous leaves; flowers pink to roseate, pink-purple, or lavender.
      4. Stems trailing but not twining; leaflets thick, leathery, with densely appressed pubescence on both surfaces but particularly beneath, prominently reticulate-veined beneath to the naked eye; plants with cleistogamous flowers and small, peanut-like, 1-seed, underground fruits; species known only from sw portion of East TX ____ **G. canescens**
      4. Stems usually twining (at least at the tips); leaflets (except *G. texana,* which possesses thick leaves) thin, with pubescence ± spreading to nearly glabrate, not prominently reticulate beneath; plants without cleistogamous flowers or underground fruits; including species widespread in East TX.
        5. Fruits sickle-shaped; flowers 1–5, solitary and subsessile or 2–5 on inflorescence axis 1–4(–40) mm long; leaflets thick, somewhat leathery; plants growing on rocky limestone soils in w part of East TX ____ **G. texana**
        5. Fruits ± straight; flowers 1–8(–16), solitary and pedicellate or 2–8(–16) on inflorescence axis 10–150(–280) mm long; leaflets thin, not leathery; plants growing on various, but often sandy or gravelly soils, widespread in East TX.
          6. Leaflet blades narrowly oblong to oblong-lanceolate, to 75 mm long, 6–14 mm wide; legumes 3.5–6.5 cm long; seeds 12–15; rare in sw portion of East TX ____ **G. longifolia**
          6. Leaflet blades elliptic to broadly elliptic or oblong to lanceolate-elliptic, to 50 mm long, 5–21(–25) mm wide; legumes 2–5.2 cm long; seeds 5–11; more widespread.
            7. Leaflet blades mostly elliptic to broadly elliptic, (5–)10–21(–25) mm wide; stems moderately to densely hirsute to hirsute-villous with spreading-deflexed hairs; corollas 7–10 mm long ____ **G. regularis**
            7. Leaflet blades oblong to lanceolate-elliptic, 5–15(–17) mm wide; stems sparsely to moderately strigose with tightly to loosely appressed, retrorse hairs, sometime glabrate; corollas 9–14 mm long ____ **G. volubilis**

Desmodium obtusum [WIL]

Desmodium paniculatum [WIL]

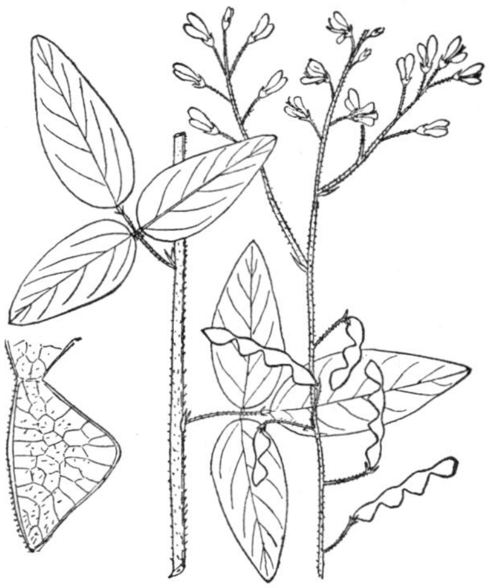

Desmodium perplexum [BB2]

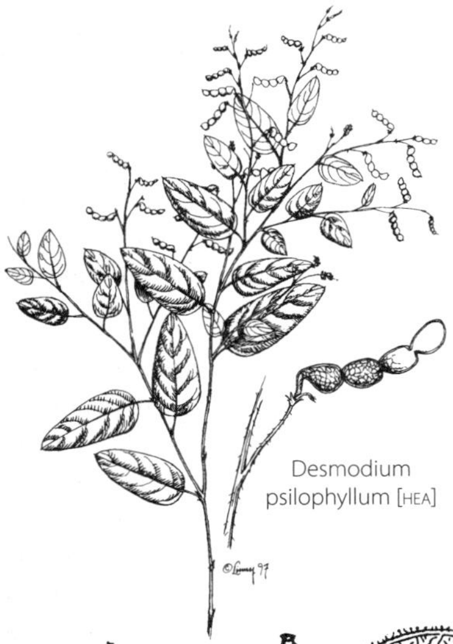

Desmodium psilophyllum [HEA]

Desmodium rotundifolium [STR]

Desmodium tortuosum [RAD]

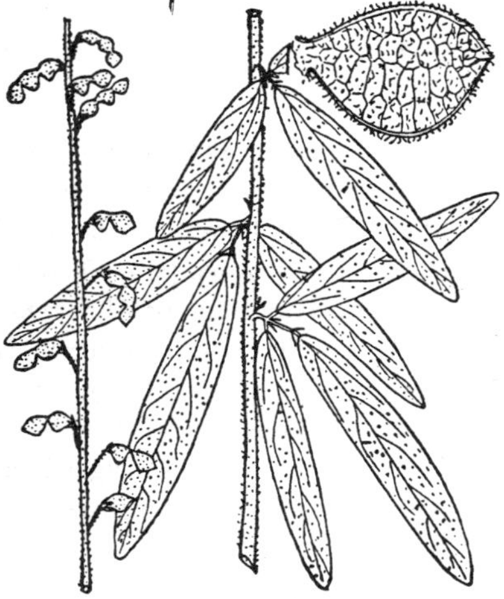

Desmodium sessilifolium [BB2]

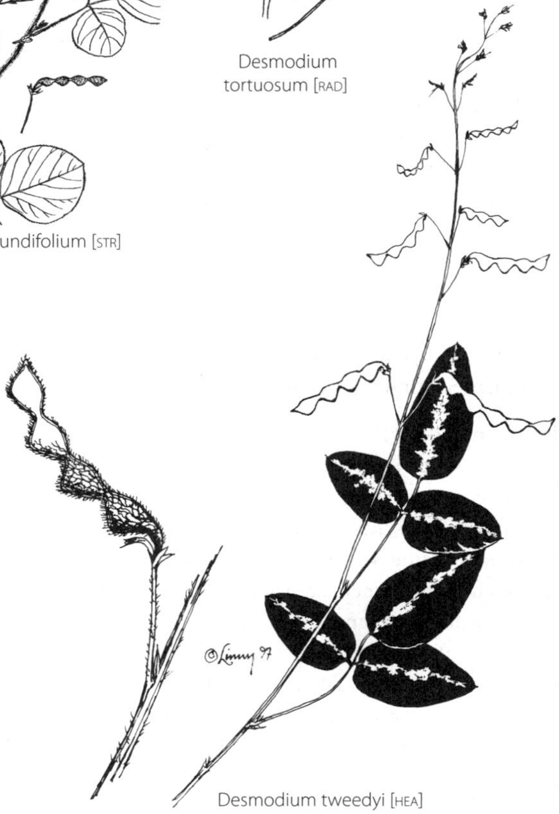

Desmodium tweedyi [HEA]

Galactia canescens Benth., (gray-pubescent), HOARY MILK-PEA. Stems trailing (not twining), appressed-pubescent, to 2 m long; leaflets 1–3 cm long, elliptic to orbicular-ovate; inflorescences with 2–several flowers; calyces 5–8 mm long, hirsute to strigose-hirsute; corollas 8–12 mm long, pinkish to pink-purple, developing into fruits above ground, the fruits 3–4 cm long; cleistogamous flowers and small, peanut-like, 1-seeded fruits also produced underground (this is the only East TX species that has underground fruits). Sandy soils; w margin of East TX; mainly s TX, e Edwards Plateau, and disjunct n in the Cross Timbers and Prairies; only in TX in U.S.; also Mexico. May–Jun. [*Heterocarpaea texana* Scheele] *Galactia canescens* is characterized by its prostrate, strigose stems rooting at the nodes, slender rhizomes often bearing subterranean flowers and fruits, and broadly oblong to suborbiculate leaflets with glaucous and densely strigose abaxial surfaces (Nesom 2023c).

Galactia erecta (Walter) Vail, (erect, upright), ERECT MILK-PEA. Stems erect, to only 40 cm tall; leaves few, widely spaced, palmately compound; leaflets 2–4(–5) cm long, usually linear-oblong to elliptic; inflorescences reduced, few-flowered (flowers 2–4(–8)), subsessile, the flowers appearing clustered in the leaf axils; calyces 5–6 mm long, short-strigose; corollas 7–8 mm long, usually white(–pale lavender); fruits straight or falcate, 2–4 cm long, densely hirsute. Dry upland areas, long-leaf pine forests; rare and known only from limited area in the Big Thicket n to Angelina Co. and w to Walker Co.; se U.S. from NC s to n FL and w to TX. May–Jun. *Galactia erecta* is distinct in its erect habit, slender rhizomes, subsessile leaflets, and flowers in axillary-sessile clusters (Nesom 2023c).

Galactia heterophylla A. Gray, (various-leaved), GRAY'S MILK-PEA. Herbs from a very slender, elongate, woody taproot, also sometimes with rhizomes or rhizome-like caudex branches from root apex; stems procumbent, not twining, densely strigose, hairs closely appressed, retrorse; leaflets 5, blades elliptic or broadly elliptic to elliptic-obovate or oblong-elliptic, 0.6–2.4 cm long, 0.2–1.2 cm wide, herbaceous, veins not raised, margins strigose, apex obtuse to rounded or shallowly retuse, surfaces smooth-glaucous abaxially and moderately strigose, hairs closely appressed, usually glabrous adaxially, rarely sparsely strigose; flowers solitary and axillary or 2–5 in reduced pseudoracemes; axis 1–10 mm long; calyces 7–9 mm long, short-strigose with closely appressed hairs; corolla pink, deep pink, cerise, rose-purple, lavender, purple, violet-red, 11–15 mm long; fruits straight, (20–)30–45 mm long, 5–7 mm wide, moderately strigose, hairs closely appressed. Blackbrush-cenizo brush, mesquite brush, openings in live oak-thorn scrub, granite slopes, limestone ridges, roadsides, caliche outcrops, sandy prairies, hard-packed sands, sandy silt, sandy clay, gravelly sandy loam; Goliad Co. in s portion of East TX; otherwise only in c and s TX in the U.S.; Mexico (Coahuila, Nuevo León, Tamaulipas). (Mar–)Apr–Sep. *Galactia heterophylla* is characterized by its prostrate, non-twining stems and particularly by its five leaflets with glabrous adaxial surfaces (Nesom 2023c).

Galactia longifolia (Jacq.) Benth., (long-leaved), LONGLEAF MILK-PEA. Stems climbing-twining, moderately strigose, hairs loosely appressed, retrorse; leaflets 3, blades narrowly oblong to oblong-lanceolate, 2.5–7.5 cm long, 0.6–1.4 cm wide, herbaceous, veins not raised, apex rounded to obtuse, surfaces moderately to densely minutely strigulose, with closely appressed hairs; flowers solitary and axillary or 2–8(–16) in pseudoracemes on distal 1/4 of axis, sometimes in fascicles; axis 30–120(–280) mm long; calyces 5–6 mm long, strigulose; corolla lavender, 10–11 mm long; fruits straight, 3.5–6.5 cm long, sparsely minutely strigulose, hairs closely appressed. Coastal prairies, clay, poorly drained sandy loam; Dewitt and Jackson cos. in s portion of East TX; otherwise only known from Aransas, Brazoria, Calhoun, and Victoria (Kartesz 2023) cos. in Gulf Prairies and Marshes in TX; also FL in the U.S.; West Indies (Hispaniola, Lesser Antilles, Puerto Rico); South America (Argentina, French Guiana, Paraguay). Apr–Aug (–Sep). [*Galega longifolia* Jacq.] The irregularly scattered distribution of this species (FL, TX, Caribbean, French Guiana, and southern South America) suggests that it may not be monophyletic (Nesom 2023c).

Galactia marginalis Benth., (in reference to the thickly margined leaves), EDIBLE MILK-PEA. Stems ascending, to only 30 cm long; leaves with a single leaflet, narrowly elliptic to narrowly lanceolate, 2.5–8.5 cm long, 0.4–1(–1.6) cm wide, short-petioled, the petioles only 3–5 mm long; inflorescences axillary, reduced giving the appearance of 1–3 subsessile flowers; corollas 13–16 mm long, purplish; fruits 2.5–3.5 cm long. Oak woodlands, sandy coastal prairies or caliche habitats; sw portion of East TX; also adjacent parts of other vegetational areas; Mexico (Coahuila and Tamaulipas); South America in Argentina, Brazil, Paraguay, and Uruguay. The South American plants appear essentially identical (Burkart 1971). Mar–Oct. [*G. heterophylla* (Gillies ex Hook.) Vail—an illegitimate homonym] Correll and Johnston (1970) noted that "the enlarged root is edible and formerly was used extensively" and that the "species tends to disappear in grazed ranges and is now quite rare." This species is characterized by prostrate stems, glabrous, 1-foliate leaves with closely reticulate, raised venation and a completely encircling marginal vein, and relatively large, solitary, axillary flowers (Nesom 2023c). It often forms extensive colonies of vegetative plants with relatively few plants flowering at one time (E. Keith, pers. obs.).

Galactia regularis (L.) Britton, Sterns, & Poggenb., (regular, Linnaeus thinking the species was a typical one in the genus *Dolichos*, in which this species was originally named), EASTERN MILK-PEA. Stems climbing and twining, 0.5–1.5(–3) m long; leaflets

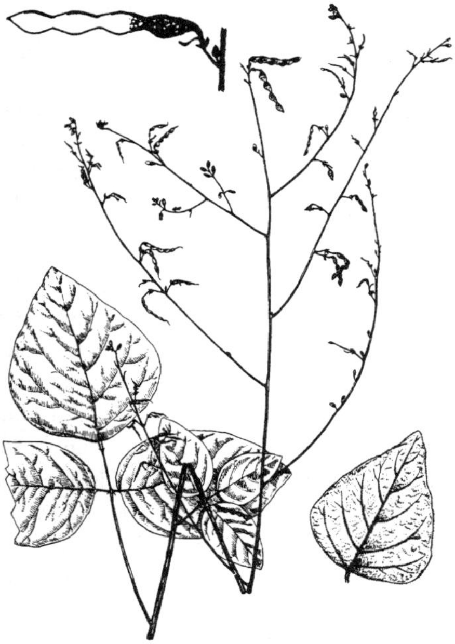

Desmodium viridiflorum [WIL]

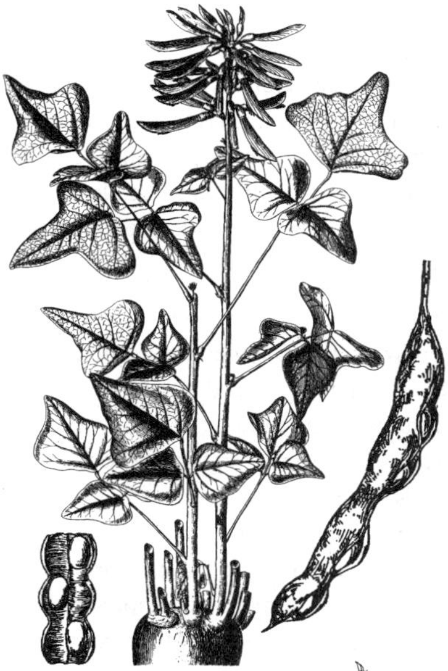

Erythrina herbacea [DIL]

Eysenhardtia texana [POW]

Erythrostemon gilliesii [BT3]

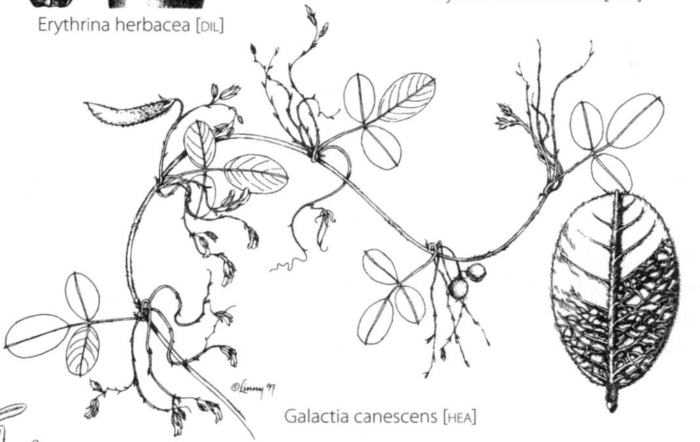

Galactia canescens [HEA]

Galactia erecta [WIL]

ovate to ovate-lanceolate, (1–)1.4–4(–5) cm long, (0.5–)1–2.1(–2.5) cm wide; inflorescences usually with flowers solitary and axillary or 2–7(–9) in reduced pseudoracemes, corollas 7–10 mm long, pink-purple to lavender; fruits 2–5 cm long. Dry sandy soils, in a wide variety of habitats, primarily open forests and woodlands; widespread in East TX; e U.S. from NY s to FL and w to se KS, OK, and TX. Jun–Aug. Turner et al. (2003) treated this species under the name of the somewhat similar *G. striata* (Jacq.) Urb. (FLORIDA HAMMOCK MILK-PEA), which has conspicuously striped standards and wide fruits (6–9 m wide—the other species have fruits 6 mm or less wide); however, most authorities (e.g., Isely 1990, 1998; Nesom 2023c) consider *G. striata* to occur in the U.S. only in s FL. *Galactia regularis* is the most widespread species of the genus in the e U.S. and is characterized by its relatively small flowers, elliptic leaflets, and twining, deflexed-hirsute-villous stems (Nesom 2023c).

Galactia texana (Scheele) A. Gray, (of Texas), TEXAS MILK-PEA. Stems twining, 0.5–1.5 m long; leaflets broadly ovate to elliptic, (1.5–)2–4.2 cm long, 1–2.5(–3.2) cm wide, thick, somewhat leathery; inflorescences with flowers solitary and axillary or 2–5 in reduced pseudoracemes; corollas 8–11 mm long, lavender to lavender-pink; fruits sickle-shaped, 3–6 cm long, minutely strigulose and appressed pubescent. Oak-juniper, ash-juniper, and oak woodlands, valley and canyon bottoms, road banks, gravelly limestone outcrops and slopes, streamsides, terraces on limestone and rocky clay; w margin of East TX and Fayette Co. (Turner et al. 2003) just to the e; also Edwards Plateau and Cameron Co. in s TX; Mexico (Coahuila, Nuevo Léon, Tamaulipas); South America (Argentina). Jun–Aug. *Galactia texana* is characterized by its twining stems, relatively short, few-flowered inflorescences, and falcate fruits (Nesom 2023c). The South American plants cannot be distinguished from those in the U.S. (Isely 1998). This sw U.S.–South America disjunction is an interesting biogeographic pattern shared with *G. marginalis* and *G. longifolia*. While it has been suggested that they were introduced from South America, Isely (1998) noted that "Perhaps this is plausible for a single species, but the unlikely event three times over suggests some biogeographic event or pattern of the past."

Galactia volubilis (L.) Britton, (twining around a support), DOWNY MILK-PEA. Stems twining, herbaceous, usually sparsely to moderately strigose with tightly to loosely appressed, retrorse hairs, sometimes glabrate, 1–2 m long; leaflets oblong to narrowly elliptic or narrowly lanceolate-elliptic, 1–4(–4.5) cm long, 0.5–1.5(–1.7) cm wide, veins not raised, apex rounded to truncate or shallowly retuse, surfaces short-strigose abaxially, glabrous or sparsely strigose adaxially; inflorescences with flowers solitary and axillary or 2–6(–10) in reduced pseudoracemes on distal 1/3–3/4 of axis, sometimes in fascicles; corollas 9–14 mm long, pink to roseate or pink-purple, fading pale; fruits (2–)2.5–5.2 cm long, ± straight. Swamp forests and borders, brackish marshes, ditches, shell mounds, hammock edges, live oak woods, pine and oak-pine woods, prairies, roadsides, cutover woods, disturbed sites; widespread in East TX and margins of immediately adjacent vegetational areas and Cameron Co. in s TX; se U.S. from NC s to FL and w to TX. Apr–Sep. [*G. mississippiensis* (Vail) Rydb., *G. macreei* M.A. Curtis] *Galactia volubilis* is characterized by its slightly larger flowers, characteristically narrow oblong leaflets, and twining, thin, sparsely, antrorsely strigose to glabrate stems, and it is distributed mostly on the coastal plain (Nesom 2023c). *Galactia volubilis* and *G. regularis* appear to be mostly distinct in their area of sympatry (*G. regularis* has a wider geographic range); but apparent intermediates are occasionally encountered. Flowering times coincide and there is at least some overlap in habitat, allowing opportunities for hybridization, but the relative stability of the two forms even where sympatric suggests that some kind of reproductive isolation is in effect, perhaps post-zygotic. The most prominent distinction between the two species is in leaf shape; leaves of *G. volubilis* also tend to be thinner and glaucous on the abaxial surface. Vestiture of *G. volubilis* is relatively sparse or surfaces may be glabrous or nearly so (Nesom 2023c).

GLEDITSIA J. Clayton HONEY-LOCUST

Polygamous or dioecious trees or shrubs, usually armed with straight or branched thorns; leaves alternate, deciduous, once or twice pinnately compound, sometimes a leaf partly once pinnate and partly bipinnate; leaflets entire or crenulate; flowers small, in lateral, catkin-like racemes appearing with the leaves; calyces bell-shaped with 3–5 lobes; petals 3–5, resembling the calyx lobes, inconspicuous, greenish white, greenish yellow or yellowish; stamens usually 5–8, separate; fruits stalked, flattened; seeds 1–25(–30).

A genus of ca. 16 species, with two in e North America, also e, se, w Asia, w Africa; warm temperate, subtropical, or tropical areas within northern temperate zone; introduced in South America, Europe, Australia (Poindexter et al. 2023a). Species are variously used as cultivated ornamentals, for shade, timber, or as hedges. Molecular research indicates that *G. aquatica* and *G. triacanthos* are sister species and these two are then most closely related to an Asian species, *G. japonica* (Schnabel & Wendel 1998). This is yet another example of the well-known floristic connection between North America and Asia (there is a detailed discussion of this biogeographic disjunction in the treatment of the genus *Campsis* (Bignoniaceae). (Simplified and Latinized from name of Johann Gottlieb Gleditsche, 1714–1786, a botanist contemporary with Linnaeus) (subfamily Caesalpinioideae, tribe Caesalpinieae)

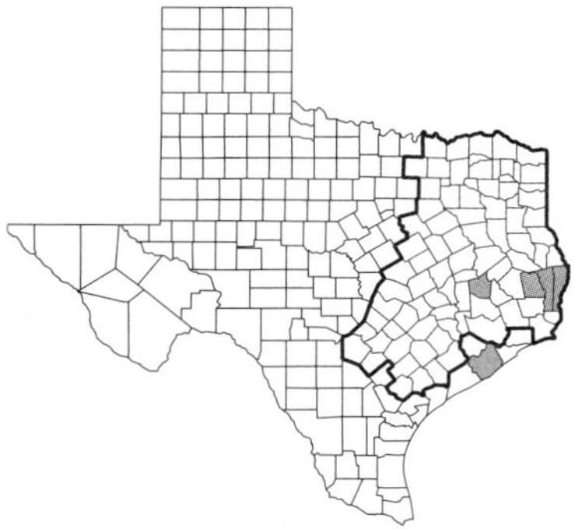

Desmodium tortuosum

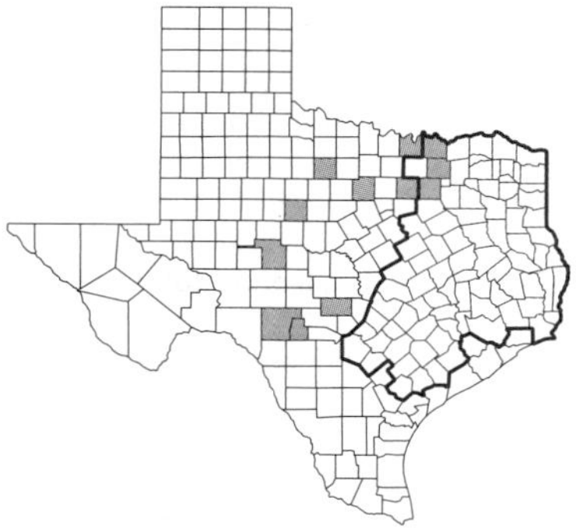

Desmodium tweedyi

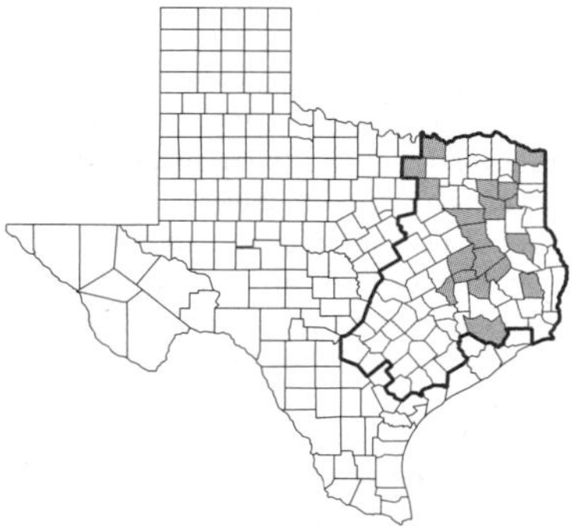

Desmodium viridiflorum

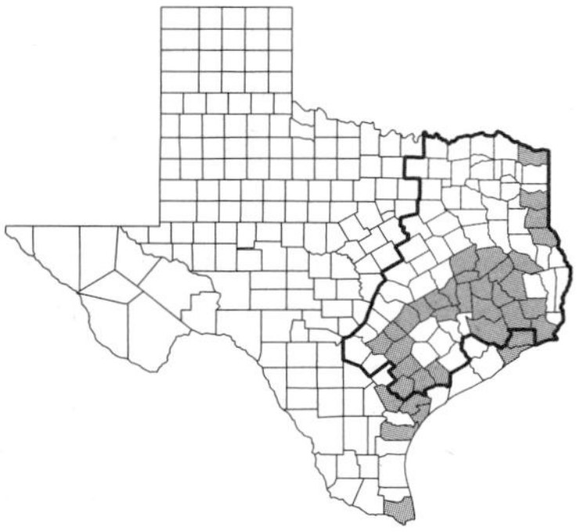

Erythrina herbacea

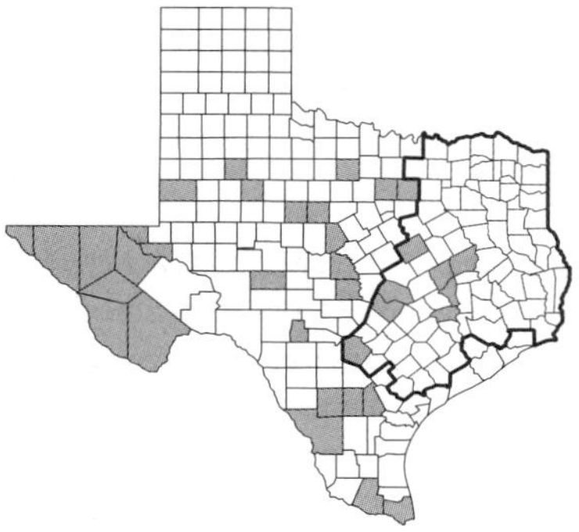

Erythrostemon gilliesii

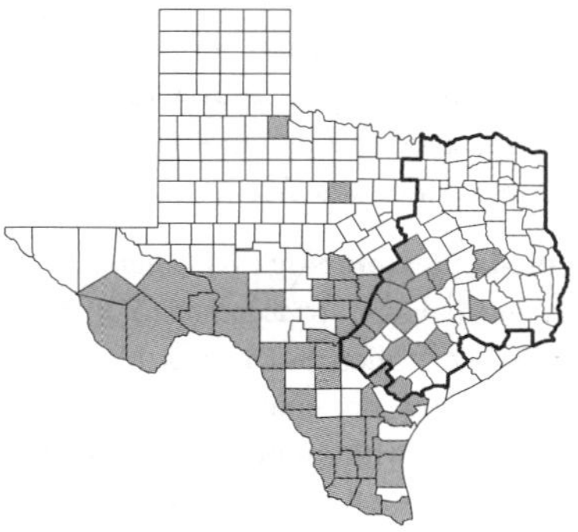

Eysenhardtia texana

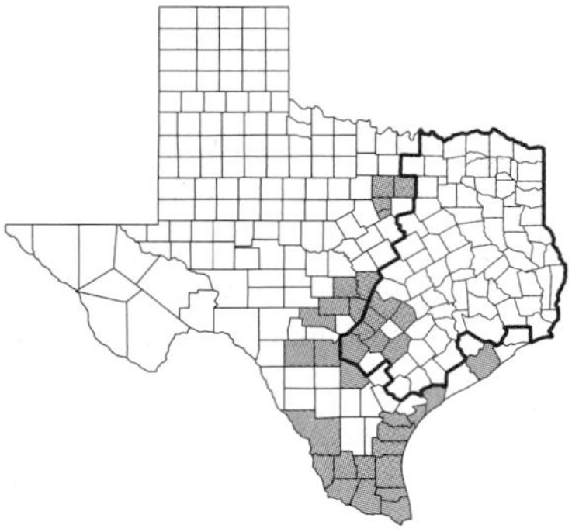

Galactia canescens

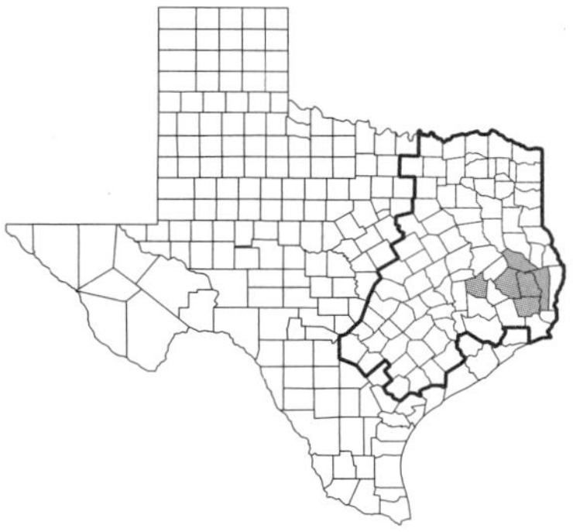

Galactia erecta

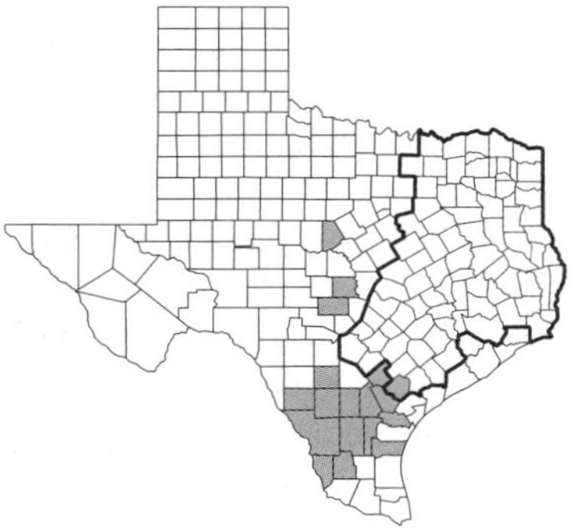

Galactia heterophylla

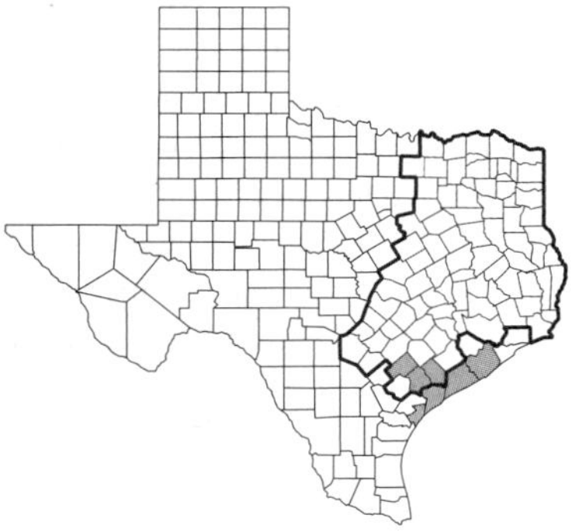

Galactia longifolia

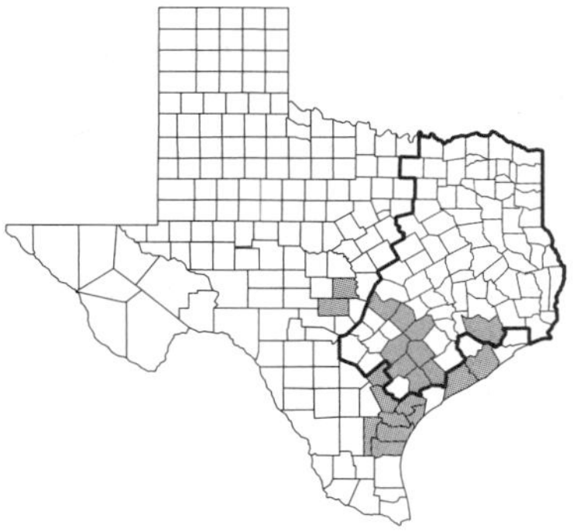

Galactia marginalis

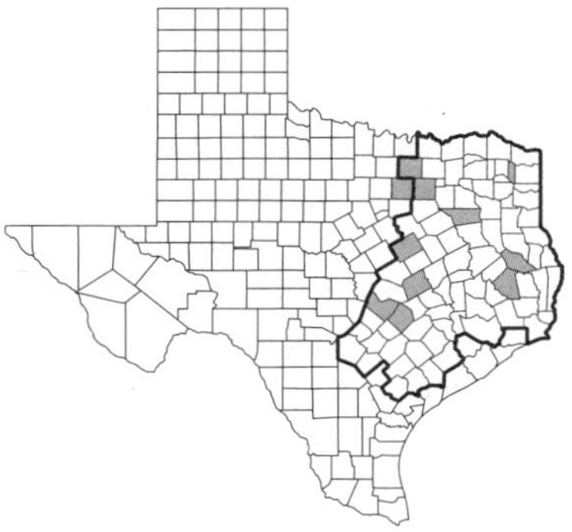

Galactia regularis

REFERENCES: Michener 1986; Schnabel & Wendel 1998; Bruneau et al. 2001; Poindexter et al. 2023a.

1. Fruits with only 1(–3) seeds, 3–5(–8) cm long, ± ovate, neither twisted nor contorted; mature leaf axes, stalks of leaflets, and leaflets glabrous or with a few hairs; seeds not surrounded by a sugary pulp; in swampy habitats ________ **G. aquatica**
1. Fruits many-seeded, 9–40 cm long, elongate, oblong, often twisted or contorted; mature leaf axes, stalks of leaflets, and often midribs on lower surfaces of leaflets usually definitely hairy; seeds usually surrounded by a sugary pulp; more diverse and often upland habitats.
   2. Fruits usually 20–40 cm long, with sugary pulp, often twisted or contorted; mature leaf axes pubescent; species widespread and common in East TX ________ **G. triacanthos**
   2. Fruits usually 10–15 cm long, without sugary pulp, flat; mature leaf axes often glabrous or nearly so; rare hybrid ________ **G. ×texana**

Gleditsia aquatica Marshall, (growing in or near water), WATER HONEY-LOCUST, WATER-LOCUST. Trees to 50 m tall, armed or unarmed; thorns if present simple or few-branched; leaves 10–20 cm long, resembling *G. triacanthos* except for pubescence and often slightly larger leaflets, bipinnate with 4 or 5 pairs of pinnae, leaflets 7–9 pairs, 1–2.5 cm long; racemes 5–15 cm long, perianth 2.5–3 mm long; legumes 3–5(–8) long, (0.5–)1.5–2 cm wide, capable of floating and presumably dispersed by water; seeds 10–12(–15) mm in diam. Swampy habitats, along streams, bottomland woods, often with BALD-CYPRESS; widespread in East TX; se U.S. from SC s to FL and w to Mississippi River Valley where it grows from IL and IN s to LA and w to TX. Apr–Jun. This species is flood tolerant and while in nature usually found in or near water, in cultivation it can grow in upland settings (e.g., yards and parks) (Yatskievych 2006).

Gleditsia ×texana Sarg., [*G. aquatica* × *G. triacanthos*] (of Texas), TEXAS HONEY-LOCUST, TEXAS-LOCUST. Trees intermediate between *G. aquatica* and *G. triacanthos* to 36 m tall; usually unarmed, but sometimes armed; leaves pinnate or bipinnate, 15 cm long, pinnae 3–7, leaflets 8–15 pairs, blades 1.5–2.5 cm long, surfaces pale green abaxially, dark green and shiny adaxially; racemes 8–10 cm long; staminate flowers with yellow corollas, stamens exserted, anthers green; legumes straight, 10–15 long, 2 cm wide, usually without pulp, with oblique and rounded base and blunt apex, surfaces puberulent; seeds many, compressed, and elliptic. Moist to dry woods; Brazos and Houston (Kartesz 2023) cos.; also n Gulf Prairies and Marshes; se U.S. in AR, FL, IN, LA, SC, and TX (all within the overlapping range of both "parents"). This taxon is of hybrid origin, based on its intermediate morphology (Poindexter et al. 2023a).

Gleditsia triacanthos L., (three-thorned), COMMON HONEY-LOCUST, HONEYSHUCK. Trees to 40 m tall, with long, stout, branched thorns on trunk and mostly simple ones on branches, rarely thornless; thorns sometimes very conspicuous, 6–15(–40) cm long; once pinnate leaves with 10–14 pairs of leaflets; twice pinnate leaves with 2–8 pairs of pinnae; leaflets 1.3–3(–3.5) mm long; racemes 5–9 cm long, petals 3–5 mm long; legumes 20–40 cm long, 2–3(–4) cm wide; seeds 10 mm or less in diam., with a sweet jelly-like pulp, this pulp drying and becoming almost raisin-like in taste (but with an unpleasant aftertaste). Bottomland forests and wet open riparian areas, also weedy invader of prairies and fields; widespread in East TX w to West Cross Timbers; also scattered in Rolling Plains and Edwards Plateau; throughout the U.S. and Ontario in Canada; the native distribution is obscured because of widespread cultivation (Poindexter et al. 2023a); introduced in Africa, Asia, Australia, Europe, and South America. Apr–Jun. [*G. triacanthos* var. *inermis* (L.) Castigl.] The very long thorns observed on some individuals are among the most striking examples of physical plant defense seen on any species in East TX; Native Americans used them for fishing-spear tips (Cox & Leslie 1991). The thorns are reduced branch systems, this derivation becoming obvious when abnormal thorns occasionally sprout from small leaves. The thorns are probably an anachronistic physical defense against large now extinct herbivores (e.g., mammoths, ground sloths) that inhabited this region during ice age times and only became extinct within the last 15,000 years (Yatskievych 2006). Likewise, the fruits are ideally adapted to be consumed by large herbivores and the seeds subsequently dispersed—the fruits have a sweet nutritious pulp, are indehiscent, and have seeds that can survive passage through a digestive tract. Other native East TX plants have similar adaptations—examples include BOIS D'ARC (*Maclura pomifera*, Moraceae), MESQUITE (*Prosopis glandulosa*, Fabaceae), PAWPAW (*Asimina triloba*, Annonaceae), and PRICKLY-ASH (*Zanthoxylum clava-herculis*, Rutaceae). Thornless trees are commonly found in cultivation; however, the thornless condition is genetically unstable and appears to vary with age (Michener 1986). The sweet fruits (containing up to 30 percent sugar) are eaten by cattle, and Native Americans ate the honey-like substance in the young pods (Cox & Leslie 1991). According to Crosswhite (1980), "Early settlers using them for food called them 'Honey-Shucks,' eating the young pods and considering the older ones too bitter." When older they have a somewhat raisin-like flavor but leave an unpleasant aftertaste. In portions of the Blackland Prairie under conditions of overgrazing and fire suppression, this species can become a problematic invader of native prairie remnants (E. Keith, pers. obs.).

Galactia heterophylla [HEA]

Galactia longifolia [WIG]

Galactia marginalis [DRW]

Galactia regularis [BB2]

Galactia texana [DRW]

Galactia volubilis [WIL]

Gleditsia aquatica [SA3]

GLYCINE Willd. SOY-BEAN

A genus of 9 species native from Asia to Australia including the important crop plant SOY-BEAN (Tucker & Alier 2023). (Greek, *glyco*, sweet, alluding to the sweet taste of the tubers of a species of *Apios* at one time included in *Glycine*) (subfamily Faboideae, tribe Phaseoleae)

REFERENCES: Hermann 1962; Hymowitz & Newell 1981; Tucker & Alier 2023.

Glycine max (L.) Merr., (old name), SOY-BEAN, SOYA-BEAN. More or less erect, bushy, annual herb, 0.2–1 m tall, densely long-hairy; leaves pinnately compound with 3 entire leaflets; leaflets 3–15 cm long, broadly ovate to elliptic-lanceolate; flowers in rather inconspicuous axillary racemes of (1–)5–8(–12) flowers; calyces 4–7 mm long, densely pubescent; petals 4–7 mm long, white to pink, violet, or purple; stamens diadelphous; fruits 2.5–8.5 cm long, with conspicuous, dense, bristly pubescence, pendant, with 2–4 seeds. Waif along railroads and disturbed areas; widely cultivated; Grayson and Hardin cos. (Kartesz 2023); probably widespread as a waif in TX; e 2/3 of U.S. from MA s to FL and w to KS and TX; s Canada in Ontario and Quebec; introduced in South America, Africa, Asia, and Australia. Jun–Sep. Native of e Asia. An old and very important crop cultivated primarily for its seeds which are extremely rich in protein and as a source of oil; SOYBEAN is also the source of soy sauce and bean curd or tofu. It was apparently derived in ne China ca. 11th century BC from *Glycine soja* Siebold & Zucc.; first cultivated in the U.S. in 1924; perhaps the world's most important pulse and currently one of the major agricultural crops of the U.S.; in this country grown mainly as an oil seed and meal crop (Isely 1990, 1998).

GYMNOCLADUS Lam. COFFEE TREE

A genus of ca. 4 species including one in e North America and the remaining species in e and se Asia, but introduced in South America, Europe, and Australia (Poindexter et al. 2023b); this disjunct distribution pattern is discussed in detail under the genus *Campsis* (Bignoniaceae). The species are trees with the fruits opening along a partial suture like a follicle (a seemingly ancestral characteristic). While morphologically quite distinct (e.g., spines absent, fruit differences), *Gymnocladus* shows many similarities to *Gleditsia* (e.g., plants polygamous or dioecious, seeds with copious endosperm, roots lacking nodule formation, and calyx similarities—Lee 1976) and molecular research shows that the two are closely related (Bruneau et al. 2001). (Greek: *gymnos*, naked, and *clados*, branch, referring to the branches being naked when the deciduous leaves are shed) (subfamily Caesalpinioideae, tribe Caesalpinieae)

REFERENCES: Lee 1976; Schnabel & Wendel 1998; Bruneau et al. 2001; Poindexter et al. 2023b.

Gymnocladus dioicus (L.) K. Koch, (dioecious, having the sexes on separate plants), KENTUCKY COFFEE TREE. Unarmed, dioecious or polygamous tree to 30 m tall, sometimes with root suckers; leaves alternate, irregularly twice pinnately compound, with 3–7 pairs of pinnae, the pinnae with 3–7 pairs of leaflets in addition to a terminal leaflet; leaflets 2–7 cm long, 1.5–3 cm wide, ovate, rather abruptly acuminate at apex, entire; petioles 20–40 cm long; inflorescences racemose to paniculate; flowers usually functionally imperfect, fragrant; hypanthium 6–12(–18) mm long; perianth radially symmetrical, of 2 similar 5-merous whorls 4–10 mm long, greenish white; stamens 10; fruits asymmetrically oblong, 12–25 cm long, 3–7 cm wide, woody at maturity; seeds 1–4, 15–20 mm in diam., turgid flattened, embedded in a jelly-like pulp. Cultivated and long persisting; apparently rarely escaping into native woodlands; Brazos (*Parks 145* TAES) and Walker cos. (*McCleod s.n.*, 1960, TEX) (Walker Co. population is no longer present, E. Keith, pers. obs.); also Tarrant Co. (S. Urshel, pers. comm.) in the Cross Timbers and Prairies; introduced and scattered in e U.S. outside its native range. May–Jun. Native to the e and c U.S. from NY s to GA and w ND and s to OK just n of East TX. The common name is probably derived from the use of roasted seeds as a (poor) coffee substitute by early settlers and during the Revolutionary War (Yatskievych 2006). However, poisoning in animals (potentially fatal) has been reported from eating the foliage or sprouts and in humans from eating the fruit pulp or seeds; the toxic substance is thought to be a quinolizidine alkaloid or cytisine (Kingsbury 1964; Turner & Szczawinski 1991; Hardin & Brownie 1993). The fruits are adapted for dispersal by large herbivores (see discussion under the related *Gleditsia triacanthos*). In the past there were problems with the generic name *Gymnocladus* Lam. (a possible earlier available name making *Gymnocladus* illegitimate) (N. Harriman, pers. comm.). Further, there were problems with the gender of *Gymnocladus*; Lamarck, who named the genus, treated it as feminine, but according to the *International Code of Botanical Nomenclature* (e.g., Greuter et al. 1994), the name should be masculine—hence disagreement over the spelling of the epithet ("*dioicus*" or "*dioica*"). However, the International Code of Botanical Nomenclature (McNeill et al. 2006) conserved the generic name as masculine making the correct spelling of the epithet "dioicus."

HOFFMANNSEGGIA Cav. RUSH-PEA

Ours unarmed subshrubs or perennial herbs; leaves alternate, twice pinnately compound; leaves with 1–6 pairs of pinnae plus a terminal pinna; leaflets 4–11 pairs per pinna, blade margins entire, glabrous or pubescent; stipules persistent; inflorescences terminal racemes with 3–27 flowers; sepals 5, fused at base for ca. 1 mm, persistent through fruit maturity; petals 5, yellow, 3–14 mm long, the standard with red markings; stamens 10, free; ovary elongate, laterally compressed; fruits dehiscent or indehiscent, if dehiscent, the valves curving outward, with 1–11 seeds, 2–7 mm long and 1.5–6 mm in diam.

A monophyletic genus of 22 species disjunctly distributed in North and South America, with 11 species in the sw U.S. and adjacent Mexico and 12 species in s South America (Simpson & Ulibarri 2006). In the past there has been confusion and taxonomic uncertainty between *Hoffmannseggia* and the related *Caesalpinia* and *Pomaria*; recent molecular work has resolved the issues (Simpson 1999; Simpson & Ulibarri 2006). Some sw U.S. species have tubers that are edible when roasted. Despite some variation in spelling, the correct form of the genus name is *Hoffmannseggia* (Simpson & Ulibarri 2006). (Named for J. Centurius, 1766–1849, Count of Hoffmannsegg, Germany, also referred to as John C. Hoffmannsegg) (subfamily Caesalpinioideae, tribe Caesalpinieae)

REFERENCES: Brummitt & Ross 1974; McVaugh 1975; Simpson & Miao 1997; Simpson 1998, 1999, 2013; Simpson & Ulibarri 2006.

1. Shrubs to 30 cm tall; leaves with 1 pair of pinnae plus a terminal pinna, each with 3–6 pairs of leaflets; fruit lunate or falcate in outline (wider in middle), dehiscent, valves simply flaring, slightly twisted, or each spirally twisted around itself ____ **H. drummondii**
1. Herbs or subshrubs, 5–30(–50) cm tall; leaves with 2–5(–6) pairs of pinnae, plus a terminal pinna; leaflets 4–11 pairs per pinna; legumes trapezoidal, arcuate, broadly oblong to suborbicular, falcate, or rectangular in outline, indehiscent or dehiscent and each valve spirally twisted around itself.
   2. Banner claw margins sparsely stipitate-glandular; legumes trapezoidal in outline, dehiscent, valves each tightly spirally twisted around itself after separating, pubescent to sparsely villous and stipitate-glandular ____ **H. oxycarpa**
   2. Banner claw margins notably stipitate-glandular; legumes rectangular or arcuate, indehiscent, valves flat, sparsely tomentose and stipitate-glandular ____ **H. glauca**

Hoffmannseggia drummondii Torr. & A. Gray, (for Thomas Drummond, 1780–1835, Scottish botanist and collector in North America), DRUMMOND'S RUSH-PEA. Shrub to 30 cm tall, often with scattered glandular hairs; leaves with 1 pair of pinnae plus a terminal pinna; leaflets 3–6 pairs per pinna, sometimes with scattered glandular hairs beneath; stipules ca. 1 mm long; inflorescences with glandular hairs, with 3–8 flowers; banner yellow with red markings, 3–7 mm long; lateral petals 3–7 mm long; fruits lunate (widest in middle), tapering to both ends, 16–20 mm long and 6–8 mm wide with 1–2 seeds. Sandy clay soils; Gonzales and Wilson (Simpson 1999) cos. at extreme s margin of East TX; mainly a small area of ne South TX Plains and adjacent Gulf Prairies and Marshes; in the U.S. known only for TX; also n Mexico. Usually Feb–Apr, sporadically year-round. [*Caesalpinia drummondii* (Torr. & A. Gray) Fisher, *C. texensis* (Fisher) Fisher, *Hoffmannseggia texensis* Fisher] While this species has often been treated in *Caesalpinia* (e.g., Correll & Johnston 1970; Isely 1975; Kartesz 1994; Jones et al. 1997), molecular evidence shows it should be included in *Hoffmannseggia* (Simpson 1999; Simpson & Ulibarri 2006). Simpson (1999) noted that this species "is among the rarest of *Hoffmannseggia* species probably because of the destruction of its Blackland Prairie habitat in south TX." NatureServe: G3, vulnerable; S3, vulnerable. ⚠

Hoffmannseggia glauca (Ortega) Eifert, (whitened with a coating or bloom), INDIAN RUSH-PEA, HOG-POTATO, SICKLE-POD RUSH-PEA, MESQUITEWEED, HOG-PEANUT, CAMOTE DE RATÓN. Perennial herb 5–30(–50) cm tall; tubers to ca. 3 cm in diam. sometimes produced on underground stems; stems finely pubescent at least when young, and with scattered multicellular hair; leaves with (2–)3–5 (–6) pairs of pinnae plus a terminal pinna; leaflets not glandular-pubescent beneath; stipules 1.5–4 mm long; inflorescences often with glandular hairs, with 4–15 flowers; banner yellow with red markings, 5–14 mm long; lateral petals ca. 13 mm long; fruits rectangular to arcuate or sometimes expanded apically, with ± parallel margins, 20–40 mm long and 5–8 mm wide. Roadsides and disturbed areas; Austin, Dallas, and Rusk (Simpson 1999) cos.; mainly w 2/3 of TX; sw U.S. from KS and CO s to TX and w to CA; also Mexico and disjunct to South America (Argentina, Bolivia, Chile, Peru—Simpson 1999). Mostly Mar–Apr. [*H. densiflora* Benth.] The underground tubers have been called "potatoes" in English and "camotes" in Spanish (Simpson 1999); according to Ajilvsgi (1984), they were roasted by Native Americans and have sometimes been fed to hogs; Moerman (1998) indicated that they were eaten either cooked or raw. This is the only North American species that forms tubers; however, this characteristic can be seen in many South American taxa. Simpson (1999) noted that this species can be a weed in agricultural fields.

Hoffmannseggia oxycarpa Benth., (bearing sharp, pointed fruit), SHARP-POD RUSH-PEA. Subshrub, 10–34(–40) cm tall; from woody taproot. Leaves (7–)11–70(–93) mm long, 16–38 mm wide; stipules lanceolate, 2–5 mm long, 1–1.5 mm wide, margins villous, glandular; leaves with 2–4 pairs of pinnae plus a terminal pinna; leaflets 4–9 pairs per pinna, blades ovate to oblong, 2–12 mm long and 1–6 mm wide, surfaces glabrous; inflorescences 4–18-flowered, terminal, 0.6–2.1 cm long; rachis and pedicels villous and stipitate-glandular; flowers turning downward, broadly turbinate, 8–10.8 mm long and ca. 8 mm wide; calyces persistent, edges red, distinct portion 5–8 mm long and 1.5–2 mm wide, villous and stipitate-glandular; banner yellow dotted with red, fading pinkish red, 6–13 mm long, 3–6 mm wide, glabrous abaxially, with few trichomes at base of claw adaxially or claw margins sparsely hairy; lateral petals yellow, 6–10 mm long, 3–4 mm wide, glabrous abaxially, eglandular; fruits trapezoidal in outline, elongate, 8–36 mm long, dehiscent, margins pubescent, apex acute; valves each spirally twisted around itself after separating, pubescent to sparsely villous, with mixed yellow or orange capitate, multicellular, glandular trichomes; seeds 4–6. Rocky limestone soils; Bexar Co. in extreme sw East TX; NM and TX; Mexico (Coahuila, Nuevo León, Tamaulipas). Mostly Feb–Apr, sporadically year-round. Subspecies *arida* (Rose) B.B. Simpson occurs in Hidalgo and Querétaro, Mexico, and differs from subsp. *oxycarpa* by being more robust and by legumes having black-tipped, multicellular trichomes (Simpson 2023).

HYLODESMUM H. Ohashi & R.R. Mill FOREST TICKTREFOIL

Herbs, perennial, unarmed; stems ascending to erect or spreading, terete, usually pubescent; leaves whorled or alternate, usually odd-pinnate; stipules present; leaflets usually 3, stipels present or absent, usually reduced, scarious, blade margins entire, ciliate, surfaces pubescent; inflorescences 2–26-flowered, terminal, sometimes also axillary, sometimes a fertile shoot separately arising from basal part of vegetative stem, pseudoracemes, sometimes in panicles, lax-flowered, branched or unbranched; primary bracts subtending flower cluster with secondary bracts each subtending 1 pedicel; flowers papilionaceous; calyces broadly campanulate, lobes 5, usually appearing 4-lobed, adaxial pair connate except apically, with 2 minute teeth, lobes shorter than tube; corollas pink, pink-purple, or white, banner clawed or tapering proximally, blade usually broadly obovate, often with pair of spots (false nectar guides) at base; wing and keel petals clawed, keel usually connate along abaxial margin of blade (distinct in *H. pauciflorum*); stamens 10, monadelphous; fruits loments, distinctly stipitate, glabrous or puberulent, compressed, very deeply incised abaxially, straight or shallowly undulate adaxially, 1–4-jointed, lateral faces densely uncinate-puberulent; segments obliquely depressed or very shallowly obovate or obtriangular; sutures glabrate, abaxial suture very deeply incised, adaxial distinctly thickened, connections between segments (isthmi) less than 1/5 as broad as pod; seeds 2–5, flat, obliquely depressed-obovate.

A genus of ca. 11 species from North America, n Mexico, Asia, n Africa. Plants of *Hylodesmum* and *Desmodium* are alike in having three-foliolate leaves and uncinate-puberulent loments that are easily separable into 1-seeded segments; *Hylodesmum* differs from *Desmodium* in having calyx lobes shorter than the tubes, monadelphous stamens, long-stipitate loments with abaxial sutures incised to the adaxial sutures, shallowly obtriangular segments, and seeds without a rim-aril around the hilum (Ohashi 2023c). Phylogenetic relationships between the three North American species were illustrated in phylogenetic trees in Ohashi et al. (2018) and discussed by Li et al. (2019). (subfamily Faboideae, tribe Desmodieae)

REFERENCES: Ohashi et al. 2018; Ohashi 2023c.

1. Stems of two types, one bearing leaves, the other a leafless inflorescence arising from ground level (rarely with 1 or 2 leaves); pedicels usually > 10 mm long ______ **H. nudiflorum**
1. Stems of one type, bearing both leaves and inflorescences; pedicels < 10 mm long.
   2. Stems usually branched; leaves alternate, dispersed over the stem; petals white; leaflets acute or only slightly acuminate; inflorescences usually 20 cm or less long ______ **H. pauciflorum**
   2. Stems unbranched below inflorescence; leaves clustered, almost in a whorl; petals usually pink to purple; leaflets conspicuously and often rather abruptly acuminate; inflorescences 30–60 cm long ______ **H. glutinosum**

Hylodesmum glutinosum (Muhl. ex Willd.) H. Ohashi & R.R. Mill, (glutinous, sticky), POINTED-LEAF TICKTREFOIL. Stems erect, 0.3–0.8 m tall; leaves usually 6 to 8 crowded just below the inflorescence in a whorl-like cluster; leaflets 5–13 cm long, usually 1–1.5 times as long as wide, conspicuously and often rather abruptly acuminate; petioles elongate 6–14.5 cm long; inflorescences 30–60 cm long; pedicels stout, 3–8 mm long; corollas 5–7 mm long, pink to pink-purple, drying bluish (rarely white); fruits long-stipitate, with 1–4 segments, each segment 7.5–11 mm long. Rich woods; uncommon and widely scattered in East TX; e U.S. from ME s to n FL and w to ND and s to OK and TX; se Canada; Mexico (Nuevo León, Puebla, San Luis Potosí). Jun–Jul. [*Desmodium glutinosum* (Muhl. ex Willd.) Alph. Wood] The combination of leaves in a whorl-like cluster and broadly acuminate leaflets are distinctive (Isely 1998). Related to *Desmodium nudiflorum* and *D. pauciflorum*, Isely (1998) noted that this is a group of three species "of Asiatic relationships that have exserted-stipitate, deeply incised loments and shallowly lobed calyces ... They usually are plants of well-shaded woodlands."

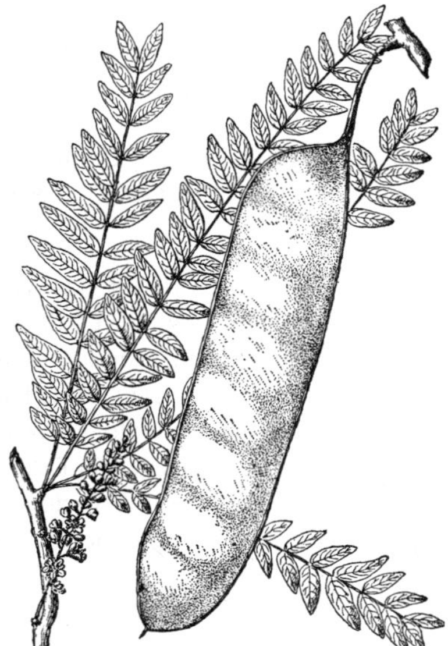

Gleditsia ×texana [BR1]

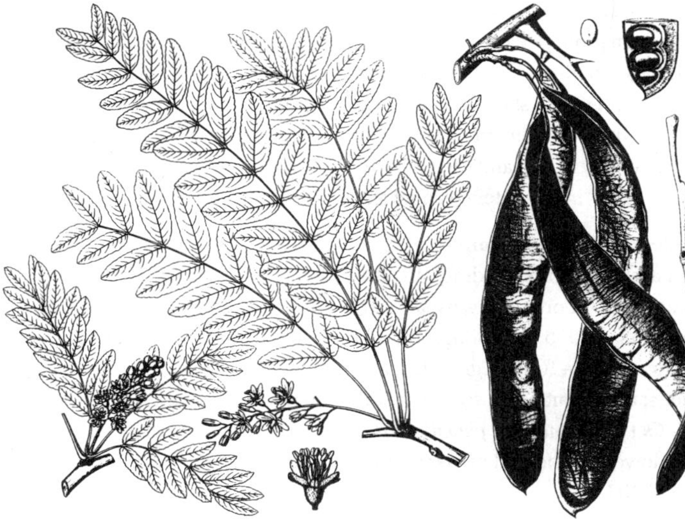

Gleditsia triacanthos [SA3]

Glycine max [HE2]

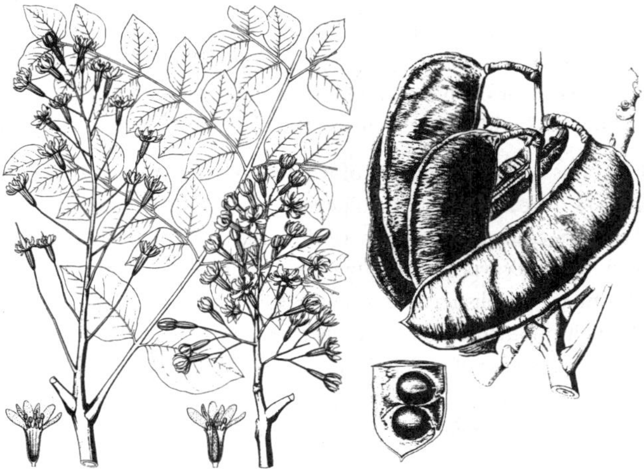

Gymnocaldus dioicus [SA3]

Hoffmannseggia drummondii [THO]

Hylodesmum nudiflorum (L.) H. Ohashi & R.R. Mill, (with naked flowers), NAKED-FLOWER TICKTREFOIL. Stems of two types arising from ground level: 1) sterile, leafy and 2) fertile, bearing inflorescence; sterile stems 10–50 cm tall, fertile stems to 70 cm tall; leaves usually 4–6(–7), these in a cluster at stem apex or slightly spread out; leaflets 4.5–12 cm long, 1.2–1.5 times as long as wide; fertile stem slightly to much exceeding the sterile; pedicels 10–25 cm long; corollas 6–9 mm long, pink; fruits long-stipitate, with 1–4 segments, each segment 7–12 mm long. Rich woods; scattered in the Pineywoods of East TX; e U.S. from ME s to n FL and w to MN and s to OK and TX; s Canada. May–Jul(–Sep). [*Desmodium nudiflorum* (L.) DC.] Isely (1998) noted that this species is "immediately recognized by its separate foliar and flowering stems."

Hylodesmum pauciflorum (Nutt.) H. Ohashi & R.R. Mill, (few-flowered), FEW-FLOWER TICK-CLOVER, FEW-FLOWER TICKTREFOIL. Stems ascending or spreading, 0.2–0.6 m long; leaves usually only 4–6, well-petioled; leaflets 3–9 cm long, 1.2–1.5 times as long as wide; inflorescences usually 20 cm or less long; rachis white-pilose and densely uncinate-puberulent, pedicels 2–7 mm long; corollas 4.5–6.5 mm long, white, keel distinct, not connate along abaxial margin (exposing reproductive organs); fruits long-stipitate, with 1–2(–3) segments, each segment 9–14 mm long. Rich, most woodlands, forested slopes, and dry calcareous uplands; uncommon and scattered in the Pineywoods of East TX; e U.S. from NY s to n FL and w to se KS, e OK, and TX. Jun–Aug(–Oct). [*Desmodium pauciflorum* (Nutt.) DC.] The alternate leaves (vs. clustered/almost whorled in related species) and white flowers are helpful in identifying this species (Isely 1998). The absence of fusion of the keel petals is unusual for the tribe (Ohashi 2023c).

INDIGOFERA L. INDIGO

Perennials, pubescent with hairs attached at the middle with the 2 ends free (= dolabriform pubescence); leaves alternate, once odd-pinnately compound, the leaflets usually 3–17 per leaf; stipules subulate to thread-like; flowers in axillary, peduncled, spike-like racemes; calyces with five teeth; corollas light brick red through purple-pink to orangish or greenish yellow; stamens diadelphous, 9 with filaments fused, the 10th free; fruit oblong, cylindric, or ovoid, straight or curved; seeds 1–12.

A genus of ca. 750 species of tropical and warm areas of the world. Several tropical species (e.g., *I. suffruticosa* Mill., *I. tinctoria* L.) are a source of the blue dye indigo, now largely replaced by synthetics. Some species are cultivated as ornamentals. *Indigofera* is easily recognized since all species have malpighian hairs that may be appressed or ascending and may have unequal branching; other types of hairs may also be present. In addition, the corollas range in color from shades of pink to red, salmon pink, orange, purplish red, or greenish yellow; petals are caducous; anthers are mucronate distally; seeds are cuboid or ellipsoid, and usually mottled. Another feature characteristic of *Indigofera* species is a floral-tripping mechanism in the flowers that results in a sudden release of pollen. If the base of the banner petal is touched, the claw splits and detaches from the calyx instantly, causing other petals to collapse and the pollen to be thrown from the anthers (Lievens & Vincent 2023). (Latin: *indigus*, indigo, and *fero*, to bear, in reference to the dye) (subfamily Faboideae, tribe Indigofereae)
REFERENCES: Isely 1998; Lievens & Vincent 2023.

1. Stems prostrate to ascending; leaflets borne alternate along the rachis ____ **I. miniata**
1. Stems erect; leaflets borne opposite or nearly so.
   2. Annual, biennial, or short-lived perennial herbs; stems, petioles, and legumes glandular, or brownish hirsute, or pilose, hairs long-spreading, or long-spreading intermixed with appressed ____ **I. hirsuta**
   2. Perennial herbs, shrubs, or subshrubs; glands and brownish hairs absent, hairs appressed or ascending, or hairs crisped, curling.
      3. Corollas reddish, 6–8 mm long; leaflets densely strigose on both upper and lower surfaces; fruits at maturity 18–25 mm long, with a swollen reddish or brownish knob at base; sw portion of East TX ____ **I. lindheimeriana**
      3. Corollas purple-pink to orangish or greenish yellow, 5–6 mm long; leaflets sparsely strigose on both surfaces or glabrate on upper surface; fruits at maturity 15–20 mm long, without a reddish knob; more widespread ____ **I. suffruticosa**

Indigofera hirsuta L., HAIRY INDIGO. Herbs, annual or biennial, brownish hirsute or pilose, hairs long-spreading; stems erect or sprawling, unbranched or branched, 3–10 dm long; leaves 4–12.5 cm long; stipules narrowly subulate, 10–13 mm long; petiole 0.8–1.2 cm long; stipels 1–3 mm long; petiolules 1.5–2 mm long; leaflets 3–7(or 9), opposite, blades elliptic to obovate, 15–60 mm long, 7–25 mm wide, base cuneate, apex rounded, short-mucronate, surfaces pubescent; peduncles 4–9 cm long; racemes 10–20+-flowered, dense, 6–20 cm long; pedicels 1–2 mm long; flowers 6–7 mm long; calyces 3.5–5 mm long, lobes setaceous, bristly-plumose; corolla salmon red to maroon; fruits dark brown, deflexed, cylindric, straight, 15–22 mm long, leathery, hispid; seeds 6–8, greenish, cuboid. Roadsides, old fields, disturbed woodlands, urban waste areas; Harris and Lee (Kartesz 2023) cos.; introduced in se U.S. in AL, FL, GA, MS, SC, and TX; introduced also in Mexico (Guerrero, Oaxaca, Tabasco); West Indies (Lesser Antilles, Puerto Rico); South America (Brazil, Ecuador, Guyana, Paraguay, Peru, Venezuela); Pacific Islands; and

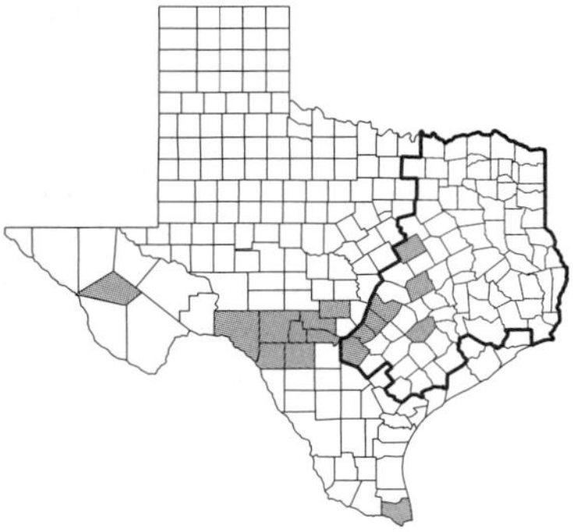

Galactia texana

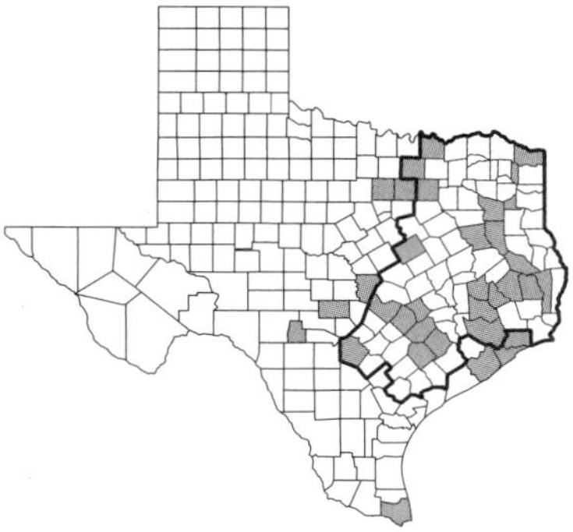

Galactia volubilis

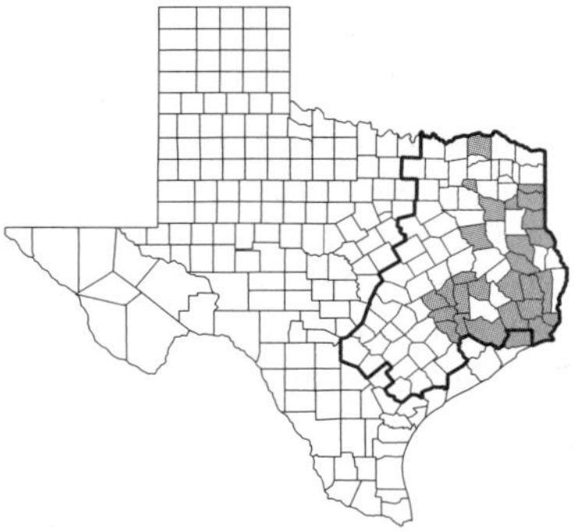

Gleditsia aquatica

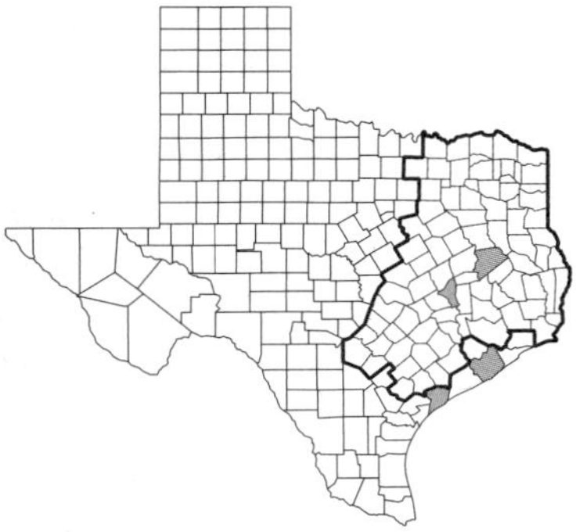

Gleditsia ×texana

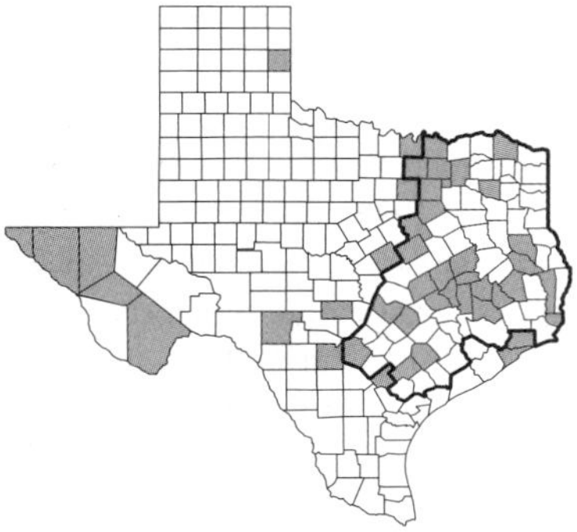

Gleditsia triacanthos

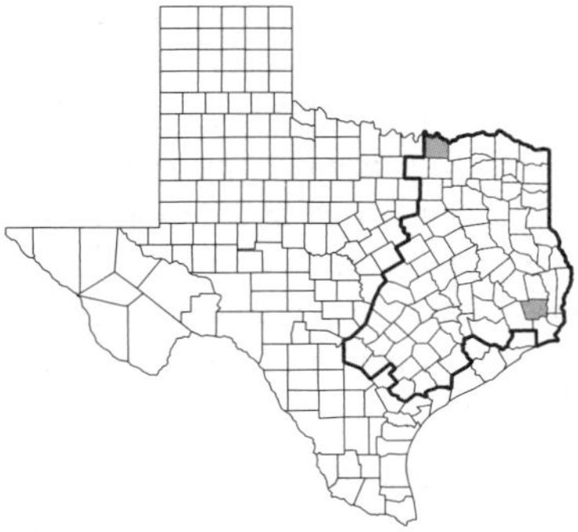

Glycine max

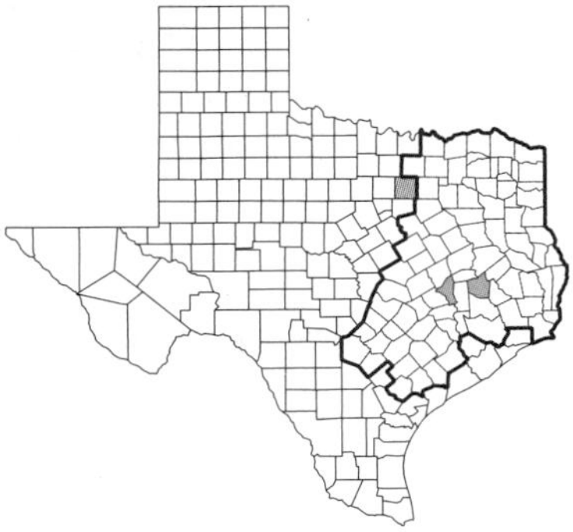

Gymnocladus dioicus

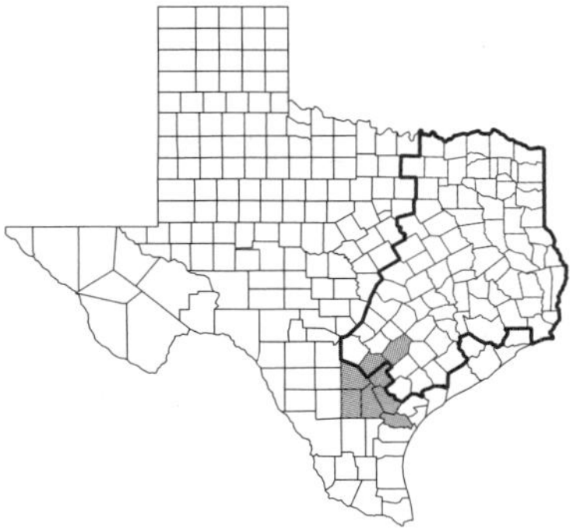

Hoffmannseggia drummondii

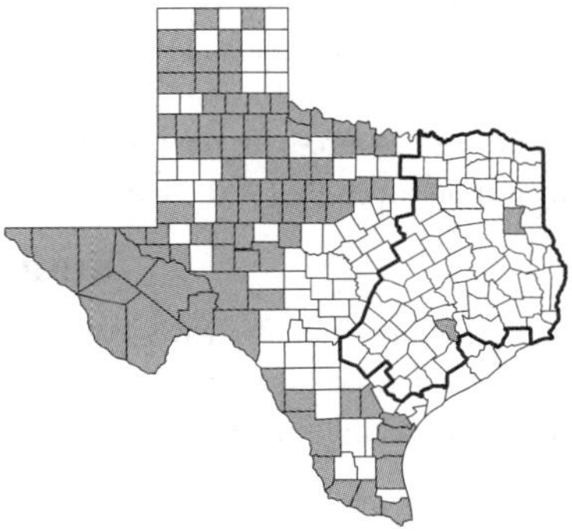

Hoffmannseggia glauca

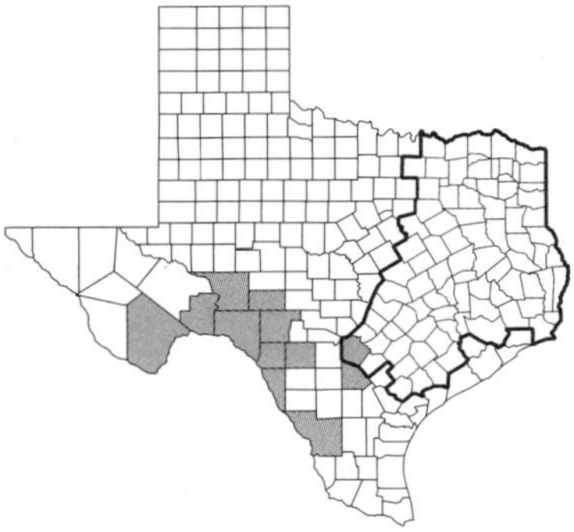

Hoffmannseggia oxycarpa

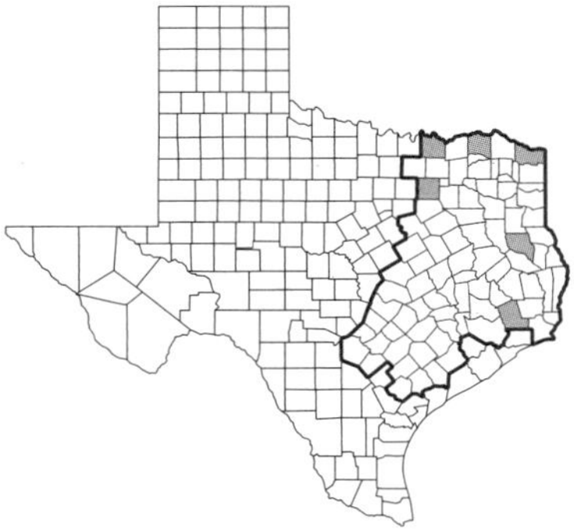

Hylodesmum glutinosum

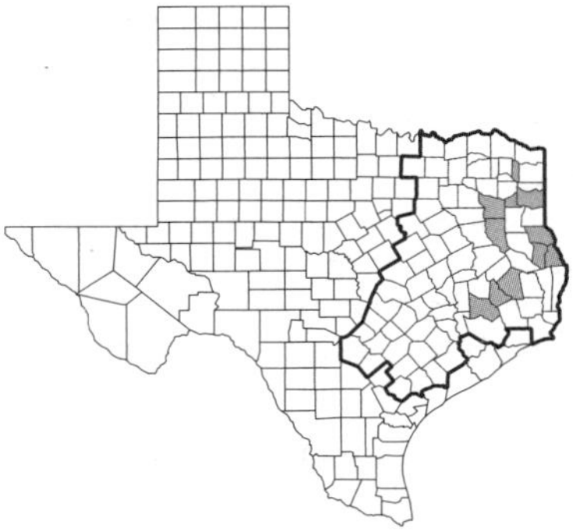

Hylodesmum nudiflorum

Australia. Flowering year-round. Native to se Asia and Africa. This introduced species was originally reported new to TX from Harris Co. in 2011 (Brown et al. 2011).

Indigofera lindheimeriana Scheele, (for Ferdinand Jacob Lindheimer, 1801–1879, German born TX botanist), LINDHEIMER'S INDIGO. Erect perennial, ashy gray due to strigose pubescence; stems 0.5–1.5 m tall; leaflets 7–15, opposite, 8–16 mm long and 4–8 mm wide, densely pubescent; racemes 8–30+ flowered, 5–12 cm long; calyx teeth not or only slightly longer than tube; corollas 6–8 mm long, reddish; fruits 18–25 mm long, with a distinctive swollen reddish or brownish knob at base, oblong, curved. River and creek bottoms, dry beds on limestone, occasionally adjacent roadsides; Bexar and Comal (Kartesz 2023) cos. in extreme sw corner of East TX; Edwards Plateau and eastern Trans-Pecos; in the U.S. known only from TX; also adjacent Mexico in Coahuila and Nuevo León. May–Sep. Isely (1998) noted that this species is "probably related to the more widespread *I. suffruticosa*."

Indigofera miniata Ortega, (cinnabar-red), SCARLET-PEA. Appressed-pubescent perennial from tough taproot; stems ascending to prostrate, to 50 cm long, usually freely branched; leaflets (3–)5–11(–17); racemes 2–20-flowered, dense or lax, 1.5–9 cm long; calyces 4–6 mm long, teeth twice as long as tube; corollas light brick-red to salmon-rose, the standard abruptly bent back from near base, 6–12 mm long; fruits 10–40 mm long, 2–3 mm in diam., straight; seeds 2–8. Prairies, open woods, and roadsides; widespread in the e 2/3 of TX; se and sc U.S. from FL w and nw to KS, OK, and TX; also Mexico; Cuba; and Central America. Jan–Oct. [*I. miniata* var. *leptosepala* (Nutt.) B.L. Turner] Varieties (intergrading) are sometimes recognized in this species (e.g., Turner 1959; Correll & Johnston 1970; Diggs et al. 1999); however, we are following Isely (1998) who studied the species throughout its range in North America and placed the varieties (apart from *I. miniata* var. *texana* (Buckley) B.L. Turner discussed below) in synonymy noting that they "are less than satisfactory." Turner et al. (2003) mapped *I. miniata* var. *texana* as endemic to the Llano Uplift in c TX. Lievens and Vincent (2023) concur with this distribution but consider it a species, *Indigofera texana* Buckley.

Indigofera suffruticosa Mill., (somewhat shrubby), UPRIGHT INDIGO, INDIGO. Erect perennial, clump-forming; stems 0.5–1(–2) m tall; leaflets 9–17, 15–20(–40) mm long, 5–10(–15) mm wide; racemes with 20–30 flowers or more, 3.5–5.5 mm long; calyces 1.5–2 mm long, calyx teeth shorter or only slightly longer than tube; corollas purple-pink to orangish or greenish yellow, 5–6 mm long; fruits 15–20 mm long, curved, oblong; seeds 4–6. Dry, sandy, open woodlands, along streams, abandoned fields, and ruderal areas; s and se parts of East TX; also Gulf Prairies and Marshes and South TX Plains; se U.S. from NC s to FL and w to TX; Central America; South America; Asia; Africa; Pacific Islands; and Australia. Jul–Nov. Native of tropical America. This species was one of the sources for the precursors of the blue dye indigo; it was planted in colonial America when indigo was commercially important (Isely 1990, 1998).

KUMMEROWIA Schindl. LESPEDEZA, BUSH-CLOVER

Low or prostrate annuals; leaves palmately compound with 3 leaflets; leaflets striate with parallel veins; stipules ovate, striate, 3–8 mm long; flowers axillary, solitary or few, shorter than the leaves, cleistogamous flowers sometimes present; fruits indehiscent, ca. 3 mm long, subtended by persistent calyx, 1-seeded.

An Asian genus of 2 species; in the past lumped into *Lespedeza*, but differing in its annual habit, striate leaflets, conspicuous stipules, and solitary flowers (Isely 1990). Turner et al. (2003) treated the species in *Lespedeza*, but we are following most recent authors (e.g., Isely 1998; Ohashi 2023d) in recognizing *Kummerowia* at the generic level. "Both species are widely planted in the southern states for forage and soil improvement" (Isely 1998). (For Heinrich Johannes Gotthilf Kummerow, 1860–1929, German professor in Poznań, now in Poland) (subfamily Faboideae, tribe Desmodieae)
REFERENCES: Isely 1948; Ohashi 2023d.

1. Stem hairs antrorse (= free end of hair pointing up toward apex of stem, the free end of the hair distal to the attached end); leaflet margins not conspicuously ciliate, notched or emarginate at apex; main stem leaves with petioles (2–)4–10 mm long; calyx glabrous ____ **K. stipulacea**
1. Stem hairs retrorse (= free end of hair pointing down toward base of stem); leaflet margins usually conspicuously spreading-ciliate, apex acute or obtuse; main stem leaves with petioles 1–2(–4) mm long; calyx lobe margins densely ciliate ____ **K. striata**

Kummerowia stipulacea (Maxim.) Makino, (having stipules), KOREAN BUSH-CLOVER, KOREAN LESPEDEZA. Stems erect or ascending, hairs antrorse, 10–50 cm long; leaflets 6–13 mm long, 3–9 mm wide, apex retuse or emarginate; inflorescences with 1 chasmogamous flower (5–6 mm long) and occasionally 1 cleistogamous flower (0.5 mm long) or none; calyces glabrous or nearly so; chasmogamous flower standard violet with dark base; wings white; keel white with red-black tip; fruits 2.5–3.5 mm long, covered by calyx for 1/2 the fruit length. Cultivated and naturalized in pastures, lawns, roadsides, and other ruderal areas;

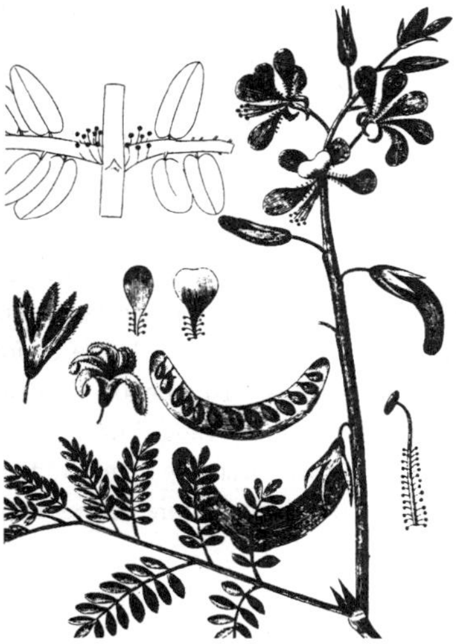

Hoffmannseggia glauca [LAM]

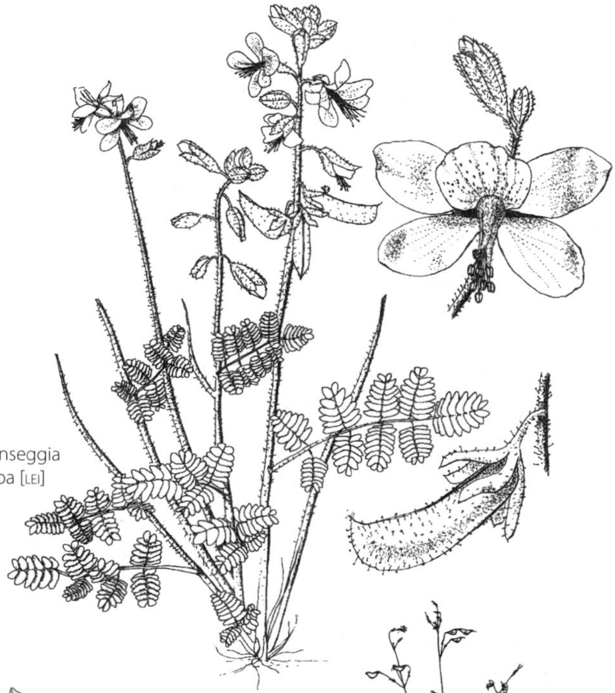

Hoffmannseggia oxycarpa [LEI]

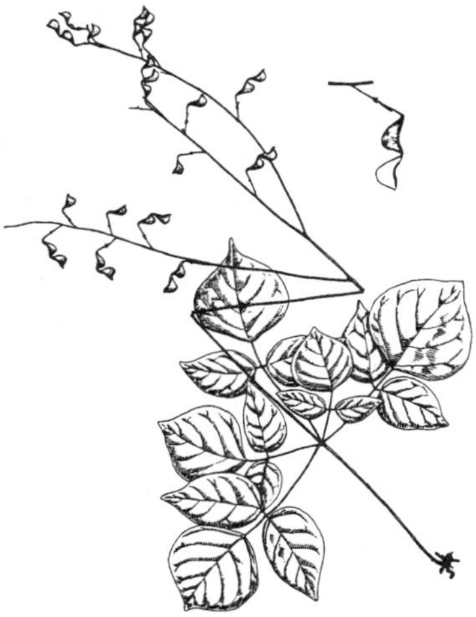

Hylodesmum glutinosum [WIL]

Hylodesmum nudiflorum [YA2]

Hylodesmum pauciflorum [WIL]

Indigofera hirsuta [FDC]

Indigofera lindheimeriana [VIN]

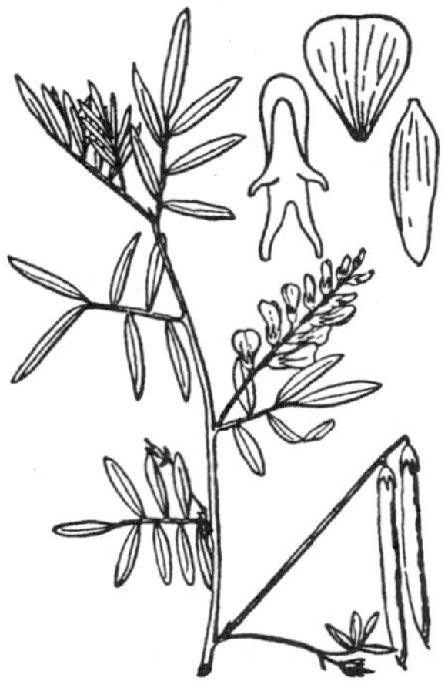

Indigofera miniata [BB2]

Angelina, Cherokee, Fannin, Grayson, Lamar, and Smith (Kartesz 2023) cos.; mainly n part of e TX, but also Lubbock and Lynn (Kartesz 2023) cos. in the High Plains; e U.S. from NY s to FL and w to e NE, e KS, e OK, and TX; introduced in Asia and Australia. Jul–Sep. Native of Korea. [*Lespedeza stipulacea* Maxim.] This species was introduced into the U.S. in 1919 and used for pasture, hay, and as a cover crop (Isely 1948); there are reports of death in cattle due to uncontrollable hemorrhaging (Isely 1948).

Kummerowia striata (Thunb.) Schindl., (striated, striped), JAPANESE BUSH-CLOVER, COMMON LESPEDEZA, JAPANESE LESPEDEZA. Stems freely branched, prostrate to erect, pubescent or in age nearly glabrous, (7–)20–50 cm long; main stem leaves appearing subsessile; leaflets 4–23 mm long, 2–8 mm wide, apex acute or obtuse; inflorescences with (1)2 chasmogamous flowers (5–6 mm long) or two cleistogamous flowers (2.5–3 mm long); calyx lobes usually ciliate; chasmogamous flower standard rose-violet; wings white; keel white, often with purple tip; fruits 3–3.5 mm long, covered by calyx for 1/2–3/4 the fruit length. Cultivated and naturalized, in roadsides, pastures, lawns, eroding and barren ruderal areas, and open woods; widespread primarily in the e part of East TX; e U.S. from VT s to FL and w to e IA, e KS, e OK, and TX; Ontario. Jun–Sep. [*Lespedeza striata* (Thunb.) Hook. & Arn.] Native of China and Japan. This species was introduced into the U.S. in 1846 and used for hay, pasture, and as a cover crop (Isely 1948).

LACKEYA Fortunato, L.P. Queiroz, & G.P. Lewis CLUSTER-PEAS

A monotypic genus of the se U.S. It was formerly included in *Dioclea* (e.g., Isely 1990, 1998) or *Galactia* (Maxwell 1979), but has more recently been recognized as a distinct genus (Fortunato et al. 1996). Phylogenetic studies suggest that its closest relationships are not with *Dioclea* (de Queiroz et al. 2003). (Named for James A. Lackey, b. 1948, specialist on the Leguminosae, particularly the Phaseoleae) (subfamily Faboideae, tribe Phaseoleae)
REFERENCES: Isely 1998; Keener 2023.

Lackeya multiflora (Torr. & A. Gray) Fortunato, L.P. Queiroz, & G.P. Lewis, (many-flowered), BOYKIN'S CLUSTER-PEA. High climbing, twining, perennial vine; stems 3–5 m long; leaves alternate, once pinnately trifoliolate, the leaflets ovate to often orbicular to subreniform, large to 13 cm long and 12 cm wide, 1–1.3 times as long as wide, entire; petioles well-developed; stipules lanceolate, 2.5–3 mm long; inflorescences axillary pedunculate racemes with numerous crowded flowers; calyxes 4.5–7 mm long, 4-lobed; corollas 14–20 mm long, partly pinkish-red, partly whitish, often with whitish striations (drying lavender); stamens 10, monadelphous; fruit flattened, oblong, 3–7 cm long, with few seeds. Wooded areas and alluvial soils; e part of East TX in the Pineywoods w to Trinity Co.; otherwise known in TX only from Hood Co. (Turner et al. 2003); se U.S. from KY se to GA and sw to TX. May–Aug. [*Dolichos multiflorus* Torr. & A. Gray, *Dioclea boykinii* A. Gray ex S. Watson, *D. multiflora* (Torr. & A. Gray) Mohr, *Galactia mohlenbrockii* R.H. Maxwell] While often in the past treated in *Dioclea* (e.g., Isely 1990, 1998), we are following most recent authors (Fortunato et al. 1996; Keener 2023) in recognizing this species in the genus *Lackeya*. Isely (1998) noted that it has the largest trifoliate leaves of any U.S member of the BEAN family except the introduced KUDZU.

LATHYRUS L. PEAVINE, VETCHLING

Annual or perennial, trailing, sprawling, or ascending to climbing herbs with winged stems (except in *L. venosus*); leaves pinnately compound, the rachis terminating in a tendril that is often branched; leaflets in East TX species 2 (except in *L. aphaca* and *L. venosus*), estipellate; stipules semi-sagittate, usually persistent and conspicuous; racemes few- to many-flowered, axillary, pedunculate; calyces essentially symmetrical basally; corollas white, pink-purple, lavender-blue, blue-purple, or bicolored; stamens 10, diadelphous; style dilated and flattened, pubescent on upper surface (= adaxial side); fruits elastically (= explosively) dehiscent, several-seeded, essentially linear, usually somewhat flattened.

A genus of ca.150 species of the n temperate zone, mountains of e Africa, and temperate South America; they are usually climbers with branched tendrils and winged stems (Broich 2023a). Many are used as ornamentals and a number as fodders. Eating the seeds of some species can lead to lathyrism (including spinal cord degeneration and paralysis of the legs); heavy losses of human life have occurred under conditions where other food was not available (e.g., Africa); poisonous due to the presence of toxic non-protein amino acids (Tampion 1977; Schmutz & Hamilton 1979; Fuller & McClintock 1986). According to Kupicha (1981), the genus is closely related to *Vicia*; with 2 rare exceptions (*L. aphaca* and *L. venosus*), East TX *Lathyrus* species can be readily distinguished by the combination of winged stems and leaves with only two leaflets (*Vicia* has stems unwinged and leaves with 4–18 leaflets). (Greek: *la*, very, and *thoures*, stimulant or passionate, the name due either to the seeds which were said to have irritant properties or from the reputation of the first described species as an aphrodisiac) (subfamily Faboideae, tribe Vicieae)
REFERENCES: Shinners 1948a; Hitchcock 1952; Kupicha 1981, 1983; Broich 2023a.

1. Corollas yellow; leaflets absent, the leaf consisting only of a tendril; stipules leaf-like, hastate, 5–40 mm wide; stems not winged ____ **L. aphaca**
1. Corollas various shades of blue to purple, pink, or white; leaflets 2–14; stipules various, usually < 10 mm wide; stems winged or not winged (in *L. venosus*).
   2. Leaflets 8–14; stems not winged ____ **L. venosus**
   2. Leaflets 2; stems winged.
      3. Corollas 18–25 mm long, usually pink-purple to white; racemes usually with 5 or more flowers; fruits 6–9(–12) cm long; petioles with conspicuous wings ca. 3–4 mm wide on each side; plants perennial ____ **L. latifolius**
      3. Corollas 7–14 mm long, lavender-blue, pink-purple, blue-purple, or bicolored; racemes with 1–4 flowers; fruits 2.5–4.5 cm long; petioles wingless or with wings to ca. 1 mm wide on each side; plants annual.
         4. Corollas 6–10 mm long, pale or light lavender-blue; fruits and enlarging ovaries glabrous; mature fruits 2–4 mm wide ____ **L. pusillus**
         4. Corollas 7–14 mm long, with standard pink-purple or blue-purple with white eye, the wings paler, and the keel nearly white; fruits and enlarging ovaries hirsute; mature fruits 5–8 mm wide ____ **L. hirsutus**

Lathyrus aphaca L., (name used by Pliny for a leguminous plant), YELLOW-PEA. Sprawling, suberect, or slightly scandent annual; stems glabrous, 0.3–0.6(–1) m long; leaflets absent (leaves of seedlings with 1 pair of leaflets); tendrils unbranched; stipules leaf-like, 1–3.5 cm long, ovate to broadly lanceolate, hastate; racemes with 1(–2) flowers; calyces 6–10 mm long; corollas yellow, 6–13 mm long; fruits 1.5–4 cm long, glabrous. Roadsides and other ruderal areas; currently only officially documented in Kaufman Co. in East TX and Wise (BRIT) Co. in Cross Timbers and Prairies Region (Kartesz 2023), however, it is more widespread than officially reported in ne TX (E. Keith, pers. obs.); widely scattered in the U.S. in AL, AR, CA, KY, LA, MD, NE, OR, PA, TN, TX, and WA; introduced also in n and e Africa. May-Jun. Native of Eurasia. This species was first reported in TX in 1997 from Kaufman Co. (Jones & Reznicek 1997) and appears to be spreading along roadsides in ne TX (iNaturalist 2023; E. Keith, pers. obs.). Beardsley and Browne (1972) initially reported it in the se U.S. from Tennessee. The leaf-like stipules are the primary photosynthetic structures of this species (which lacks leaflets) (Isely 1998).

Lathyrus hirsutus L., (hairy), ROUGH-PEA, SINGLETARY-PEA, SINGLETARY VETCHLING, CALEY-PEA. Ascending or sprawling, glabrous or sparsely pubescent annual; stems 0.2–1 m long; leaflets 2, 2–8 cm long; petioles and rachises winged; stipules 5–10 mm long, 1–3 mm wide, linear; racemes with 1–2(–4) flowers; calyces 5–7 mm long, lobes equal; corollas 7–14 mm long; ovary and fruits hirsute with pustulate hairs 1 mm or more long, legume 25–40 mm long, 5–8 mm wide. Occasionally cultivated for pasture or soil improvement, escaped on roadsides and other ruderal areas; widely distributed in East TX; also margins of adjacent vegetational areas; scattered mainly in e U.S. from NJ s to FL, n in Mississippi River Valley to MI and IL, and w to NM through OK and TX, also CA and OR; introduced in South America and n and e Africa. Mar–May. Native of the Mediterranean region. Reported as poisonous by Burlage (1968).

Lathyrus latifolius L., (broad-leaved), PERENNIAL SWEET-PEA, EVERLASTING-PEA, PERENNIAL-PEA. Glabrous perennial; stems sprawling or climbing, 0.8–2 m long, broadly winged; leaflets 2, 4–12(–15) cm long; tendrils branched; stipules large, 3–5 cm long, 4–15 mm wide, lanceolate to ovate; petioles conspicuously winged; racemes with 5–15 flowers; calyces 10–11 mm long, lobes unequal; corollas 18–25 mm long, usually pinkish purple but varying to white or even red or striped; fruits 50–110 mm long, 5–10 mm wide, glabrous. Cultivated in TX as an ornamental, persists and escapes along roadsides and other disturbed areas; Dallas (*Lipscomb 1409* BRIT, KANU; *Lipscomb 1929* BRIT), Fannin (*Holmes & Singhurst 11414* TEX), and Newton (*Keith 718* BAYLU) cos.; throughout U.S. except FL and ND (Kartesz 2023); also introduced in s Canada; Mexico; South America; Asia; Africa; Pacific Islands including HI; and Australia. May–Sep. Native of Europe. Seeds reported poisonous by Burlage (1968).

Lathyrus pusillus Elliott, (very small), LOW PEAVINE, TINY-PEA. Sprawling or low-climbing, glabrous or slightly pubescent annual; stems to 0.6 m long; leaflets 2, 1.5–6 cm long; stipules linear, 1–6 cm long, 1–5 mm wide, the upper lobe 2–3 times as long as the lower; racemes with 1–2 flowers; calyces 5–8 mm long; corollas 6–10 mm long; ovary glabrous; fruits 30–50 mm long, 2–4 mm wide. Roadsides, prairies, pastures, and open woodlands; widespread primarily in the e 1/2 of TX, less common in Edwards Plateau and n portion of South TX Plains; se and c U.S. in AL, AR, FL, KS, LA, MS, MO, NC, OK, TX, and VA, introduced in OR; also South America. Mar–May. *Lathyrus pusillus*, the only North American member of section *Notolathyrus*, a section centered in South America, is also the only annual species of *Lathyrus* native to North America. It is native to the southeastern U.S.; it has been cultivated as a forage crop in Oregon and occasionally escapes (Broich 2023a). Reported to cause livestock poisoning (Kingsbury 1964).

Lathyrus venosus Muhl. ex Willd., (veiny), VEINY-PEA. Perennial; stems 0.4–1 m long, nearly erect or scandent (= climbing); leaflets 8–14, 2.5–7 cm long, 1–3.5 cm wide; stipules linear-lanceolate to lanceolate, 1–2 cm long, usually < 10 mm wide, 1/6–1/2 as long as leaflets; racemes with 5–20 flowers; calyces 6–14 mm long, with unequal lobes; corollas 10–22 mm long, light lavender, blue purple, rose, or white with lavender lines; fruits 3–5 cm long, 3–7 mm wide, usually glabrous. Open woods; scattered mainly in n part of East TX; e U.S. from NY s to GA and w to ND se to MO and s to e OK and TX, introduced in NM; s Canada. Apr–Jun. [*L. venosus* var. *arkansanus* Fassett, *L. venosus* subsp. *arkansanus* (Fassett) C.L. Hitchc.] Broich (2023a) distinguished three varieties and mapped two of these in TX: *L. venosus* var. *intonsus* Butters & H. St. John and *L. venosus* var. *arkansanus* Fassett. Variety *intonsus* is distinguished from the other two varieties by having pubescence throughout while the other two are glabrous. Isely (1998) discusses the varieties originally proposed by Butters and St. John (1917), Fassett (1937), and Hitchcock (1952) but couldn't find consistent differences and determined the intermediate forms were more abundant than the pure forms described to distinguish the varieties. If varieties are distinguished, *L. venosus* var. *arkansanus* appears to be the only variety in East TX and is characterized by its glabrous foliage, stems, and calyces; smaller flowers (10–16 mm long); and calyx teeth that are oblong-lanceolate (versus linear-triangular) and often longer (versus shorter) than the calyx tubes (Hitchcock 1952).

LESPEDEZA Michx. BUSH-CLOVER

Perennial herbs or in one species (*L. bicolor*) a shrub; leaves alternate and pinnately compound with 3 leaflets; leaflets entire; stipules inconspicuous, slender, linear to nearly thread-like; inflorescences with (1–)2–40 flowers, axillary or appearing terminal due to reduction of subtending leaves; cleistogamous flowers often in proximal fascicles; flowers papilionaceous, chasmogamous or cleistogamous, chasmogamous pedicellate or subsessile, cleistogamous enclosed in calyx with reduced corolla; calyx campanulate, lobes 4 or 5, often longer than tube, adaxial 2 distinct or proximally ± connate (and thus calyx 4-lobed), apex 2-toothed; petals white to cream to purplish or pinkish (on open flowers; cleistogamous flowers with reduced perianth also produced by some species); stamens 10, diadelphous; ovary minutely stipitate, style adaxially incurved, slightly exserted from stamens; fruits flattened, 1-seeded, indehiscent, usually ovate or rounded; style elongate on chasmogamous fruits (though easily broken off), recurved tightly on cleistogamous fruits.

A genus of 47 species of temperate North and South America, tropical and e Asia, and Australia (Ohashi 2023e); some are cultivated for forage, fodder, or green manure. *Lespedeza* species are well known to hybridize (Turner 1959; Clewell 1966, 1968; McGregor 1986d) and because of this, some individuals are difficult to definitively identify. A population including *L. virginica*, *L. procumbens*, and numerous hybrid individuals is known from Grayson Co. (G. Diggs, pers. obs.). Thirty putative hybrids are currently recognized (Ohashi 2023e) and many have been described and mapped in Clewell (1966). *Lespedeza thunbergii* has been planted in East TX as a wildlife food and may have escaped in the past in Smith Co. (*Cory 25862* GH) but is not known to have persisted or subsequently been collected in TX (Ohashi 2023e). It can be differentiated from similar *L. bicolor* in Isely (1998) and Ohashi (2023e) (see discussion under *L. bicolor*). (Named for Vincente Manual de Céspedes, d. 1794, Spanish governor of East Florida, on account of the hospitality provided in the late 1700s to Michaux during his explorations there; later misspelled, probably by Michaux's editor as de Lespedez) (subfamily Faboideae, tribe Desmodieae)

REFERENCES: Clewell 1966, 1968; Isely 1998; Ohashi 2023e.

1. Plants basically prostrate (procumbent to trailing) for most of their length OR sometimes weakly ascending; inflorescences (of petaliferous flowers) usually long-pedunculate and exceeding subtending leaves.
   2. Stem hairs spreading — **L. procumbens**
   2. Stem hairs appressed or rarely short-pilose on *L. texana*.
      3. Leaflets gray or whitish beneath, subcoriaceous; plants ascending, then arching and spreading; flowers dark purple; species found on rocky limestone soil, primarily in w portion of East TX — **L. texana**
      3. Leaflets green beneath, not coriaceous; plants prostrate, trailing, or ascending; flowers pink to light purple; widespread or rare in ne TX.
         4. Keel petals usually longer than wings; leaflet blades apiculate and often of two sizes; stems ascending or sprawling; rare in ne TX — **L. frutescens**
         4. Keel petals nearly equal to wings; leaflet blades scarcely or minutely apiculate, of uniform size; stems procumbent or trailing; widespread in East TX — **L. repens**
1. Plants erect or strongly ascending; inflorescences (of petaliferous flowers) various, short-pedunculate and not greatly exceeding leaves OR long-pedunculate and exceeding leaves.
   5. Flowers white or cream-colored, sometimes with purplish markings; calyces ± equal to or longer than the mature fruits; flowers either 1–4 in axillary clusters OR in spike-like, ± globose or short-cylindric inflorescences of 10–45 flowers.
      6. Leaflets spatulate to cuneate (= wedge-shaped); flowers 1–4 in axillary clusters (these numerous); wing petals and keel ± equal in length — **L. cuneata**
      6. Leaflets not spatulate to cuneate; flowers in spike-like, globose or short-cylindric inflorescences of 10–45 flowers; wing petals longer than keel.
         7. Leaflets narrowly elliptic to oblong, the well-developed terminal leaflets > 2 times as long as wide; rachis (stalk of terminal leaflet) longer than petiole on well-developed leaves; calyx teeth 7–13 mm long; standard 7–12 mm long; inflorescences usually not or only slightly exserted beyond the leaves — **L. capitata**

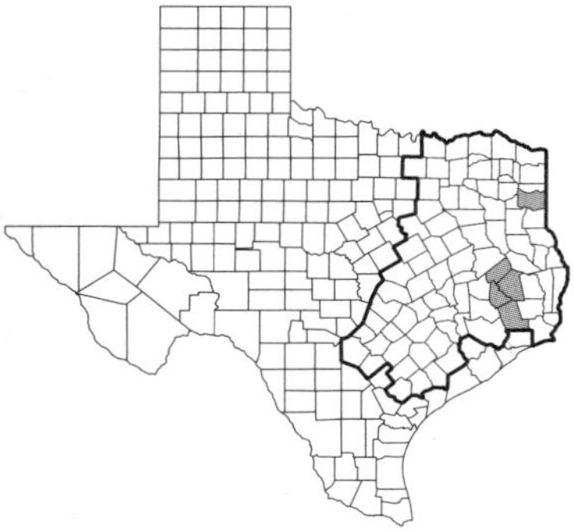

Hylodesmum pauciflorum

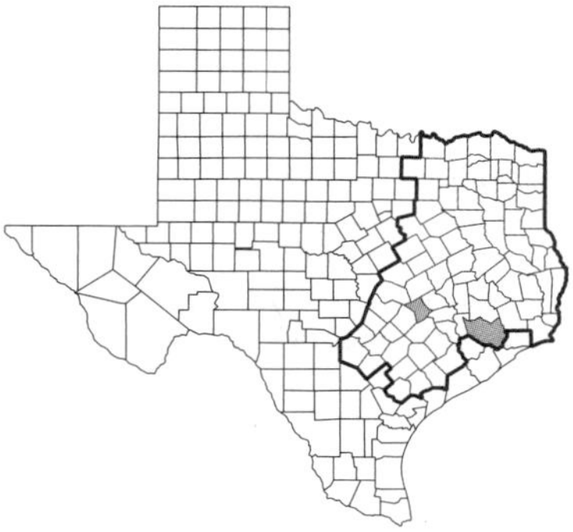

Indigofera hirsuta

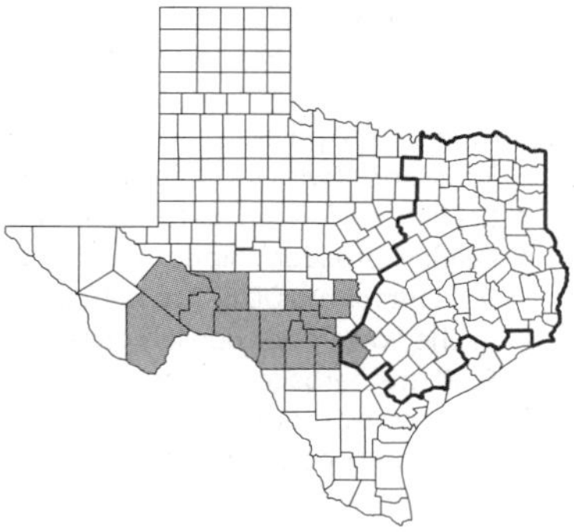

Indigofera lindheimeriana

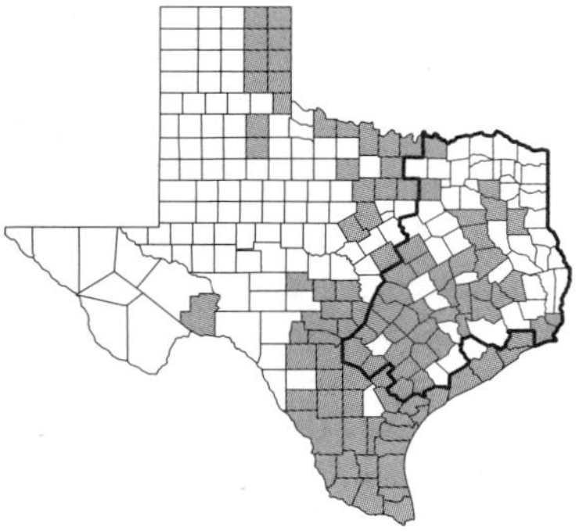

Indigofera miniata

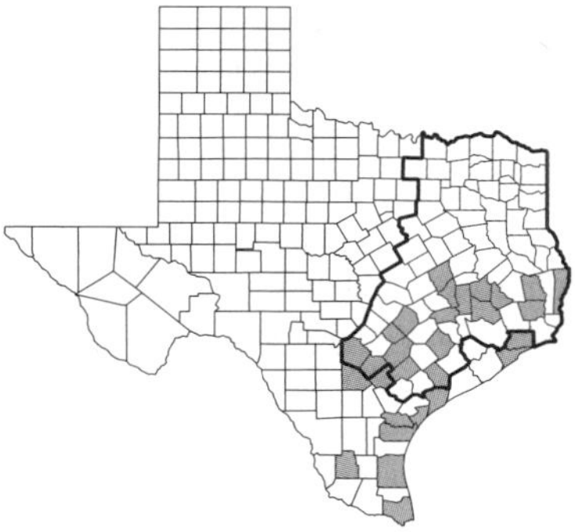

Indigofera suffruticosa

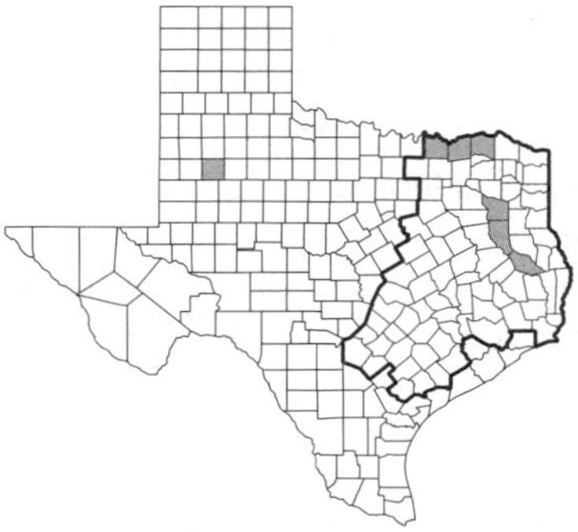

Kummerowia stipulacea

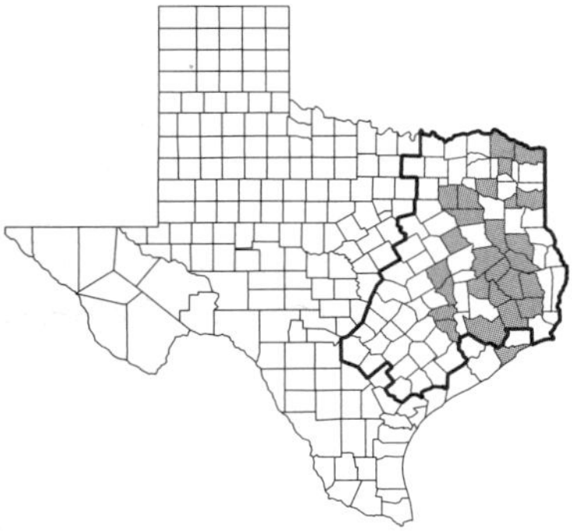

Kummerowia striata

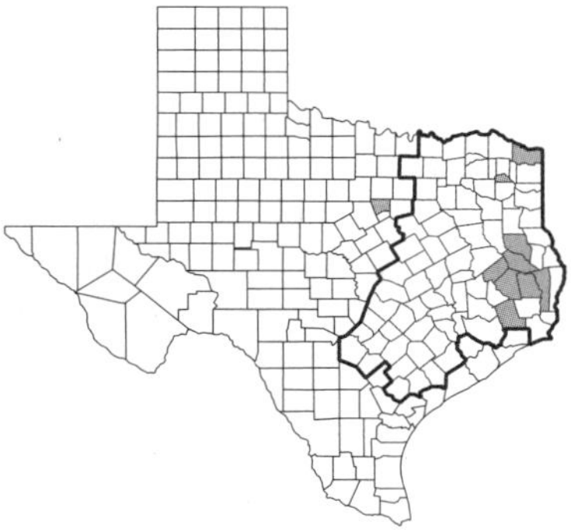

Lackeya multiflora

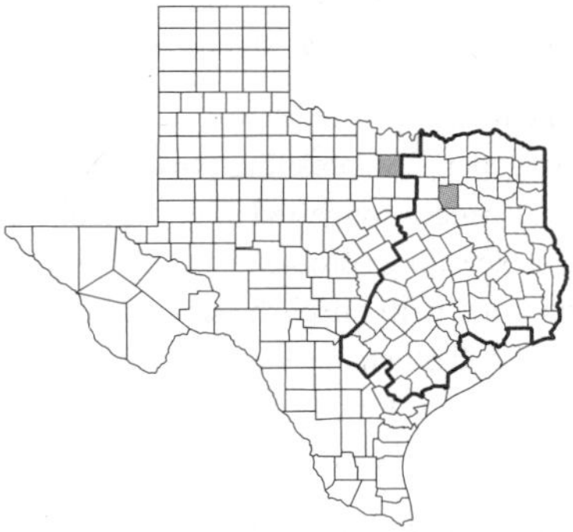

Lathyrus aphaca

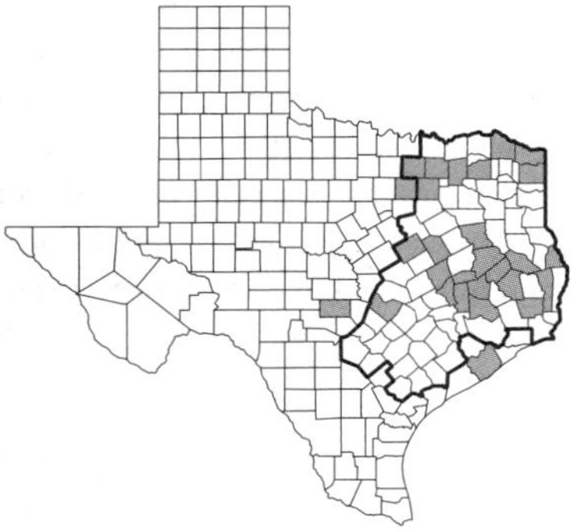

Lathyrus hirsutus

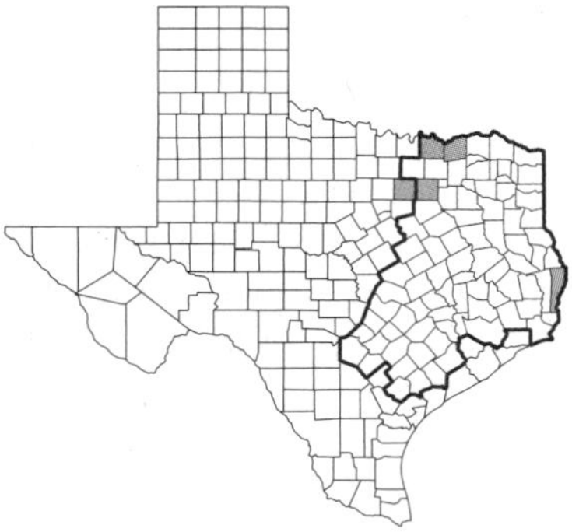

Lathyrus latifolius

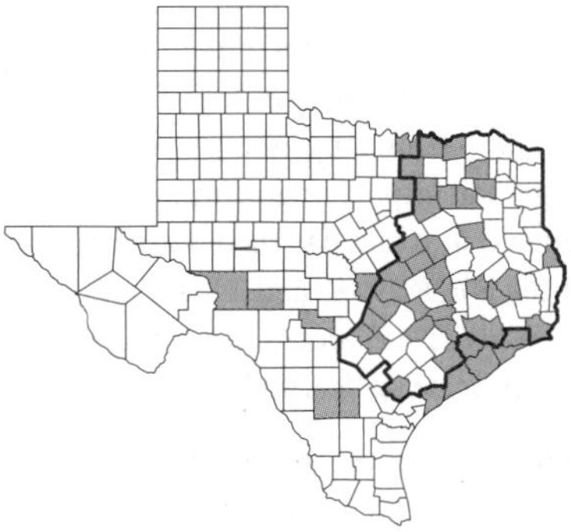

Lathyrus pusillus

7. Leaflets mostly ovate to elliptic, the well-developed terminal leaflets usually 1–2 times as long as wide; rachis equal to or shorter than petiole; calyx teeth 5–8 mm long; standard 6–8 mm long; inflorescences often conspicuously exserted beyond the leaves **L. hirta**

5. Flowers pink to purple, purple-red or violet; calyces 1/2 or less as long as mature fruits; flowers in axillary, racemose inflorescences with 4–14 flowers (usually not appearing spike-like).
8. Leaflets narrow, linear to linear-oblong, 3–8 times as long as wide **L. virginica**
8. Leaflets broader, elliptic or ovate, 1–3 times as long as wide.
9. Plants shrubs, 0.6–2 m tall; corollas 8–13 mm long; racemes stalked, much longer than subtending leaves, the flowers not densely crowded; introduced species rare in Pineywoods of East TX **L. bicolor**
9. Plants herbaceous perennials 0.2–1.5 m tall; corollas 5–8 mm long; plants EITHER 0.7 m or less tall OR with racemes shorter or only slightly longer than subtending leaves, the flowers densely crowded together; native species widespread and common in East TX.
10. Upper surface of leaflets and stems glabrous or with sparse pubescence; lower surface of leaflets with sparse pubescence; keel longer than wing petals, extending 1–2 mm past wings; racemes usually longer than subtending leaves, the flowers not densely crowded **L. violacea**
10. Upper surface of leaflets and stems conspicuously appressed-pubescent or pilose; lower surface of leaflets with dense pubescence; keel ± equal in length to wing petals, included within the wings; racemes shorter or only slightly longer than subtending leaves, the flowers densely crowded together **L. stuevei**

Lespedeza bicolor Turcz., (two-colored), SHRUB LESPEDEZA, BICOLOR LESPEDEZA. Plant shrubs, glabrate or inconspicuously pubescent; stems ascending, arched, 0.6–2 m tall; leaflets 2–6 cm long, 1–3.5 cm wide, elliptic; inflorescences with 5–15 flowers, definitely stalked, much longer than subtending leaves, the flowers not densely crowded; calyces 3–5 mm long, tube 1.5–2.4 mm long, lobes 4, 1.2–2.5 mm long; corollas 8–13 mm long, pink-purple or red-purple; fruits 5–7 mm long, 4–6 mm wide. Fields, waste places, roadsides; recently reported for the state and known in TX only from Jasper (E. Keith, pers. obs.), Newton (iNaturalist Contributors 2023h; E. Keith, pers. obs.), and San Jacinto (Nesom & Brown 1998) cos.; e U.S. from NH s to FL and w to IA, e KS, OK, and TX; Ontario. Jul–Sep. Native of temperate e Asia. This species has been planted in the e U.S. for erosion control, soil improvement, as a food for wildlife, and as an ornamental; however, it is considered by some to be an invasive exotic (Invasive.org 2006; Ohashi 2023e). Similar *L. thunbergia* has been planted in TX and can be distinguished by its larger flowers (12–16 mm long), long acuminate or acute calyx lobes (3.5–6 mm long) longer than the tubes, and relatively large loments (5–14 mm long) (Ohashi 2023e).

Lespedeza capitata Michx., (capitate, headed), ROUND-HEAD BUSH-CLOVER, ROUND-HEAD LESPEDEZA. Stems erect or arched, 0.5–1.5 m long, densely spreading-pubescent; leaflets 2–4 cm long, narrowly elliptic to oblong, often grayish below due to pubescence; inflorescences congested in upper axils, ± globose, of 15–30 crowded flowers, and not or only slightly exceeding leaves; calyces often exceeding corollas and fruits; calyces 8–12 mm long, longer than corollas; corollas 8–12 mm long, white or cream colored, sometimes with purplish markings; fruits 4–7 mm long. Sandy open woods or open habitats; uncommon and widely scattered in East TX; otherwise known in TX only from a few cos. in the Gulf Prairies and Marshes and High Plains; e U.S. from ME s to n FL and w to SD and s through OK into TX; s Canada. May–Sep. This species is recognized by its congested heads of flowers with deeply toothed calyces that exceed the flowers and fruit (Isely 1998).

Lespedeza cuneata (Dum. Cours.) G. Don, (wedge-shaped), SERICEA, SERICEA LESPEDEZA, CHINESE BUSH-CLOVER. Plant bushy-branched, often grayish green; stems 0.5–2 m tall, erect or nearly so, densely ascending or appressed white-pubescent on ridges; leaves conspicuously numerous; leaflets 0.7–3.2 cm long, spatulate to cuneate, the larger leaflets truncate or slightly notched apically; inflorescences numerous, axillary, individually small, of 1–4 flowers each, chasmogamous and cleistogamous; calyces often ± equal to fruits in length, 3–5 mm long; corollas 6–8 mm long, cream colored, with purplish marking; chasmogamous fruits 3.5–5 mm long, not or only faintly reticulate, cleistogamous fruits 3 mm long, suborbicular. Cultivated for erosion control, forage, soil improvement, and wildlife food (Isely 1998), but escaped and now a problematic invader, especially on sandy soils; scattered in n and e parts of East TX; also Gillespie Co. (Kartesz 2023) on the Edwards Plateau; e U.S. from MA s to FL and w to MN, NE, KS, OK, and TX; Ontario; introduced also in South Africa, Pacific Islands, and Australia. Jun–Oct. Native of e and c Asia. Isely (1998) noted that this species is sometimes confused with *L. virginica*, "but that species has [apically] tapering rather than truncate leaflets and strongly reticulate fruits."

Lespedeza frutescens (L.) Hornem., (shrubby, bushy), VIOLET BUSH-CLOVER, SHRUBBY LESPEDEZA. Stems ascending or sprawling, clustered, 10–50 cm long, branched much of length, sericeous (= silky) or glabrescent; leaves usually 2 sizes, axillary ones subtending racemes much smaller; stipules subulate, 2.5–6 mm long; petiole (5–)10–15(–20) mm long, longer than rachis; leaflet blades elliptic to narrowly elliptic, apex obtuse or retuse, apiculate, surfaces sericeous abaxially, glabrous adaxially; laterals similar to terminal, without oblique base; terminal blade 1–3(–4) cm long and 0.8–2.2 cm wide (0.6–1 cm long in axillary

Indigofera suffruticosa [BR2]

Kummerowia stipulacea [STE, WIL]

Kummerowia striata [RCA, STE, WIL]

Lackeya multiflora [GLE, HOL]

Lathyrus aphaca [BEN]

Lathyrus hirsutus [BEN]

Lathyrus latifolius [STE, WIL]

Lathyrus pusillus [WIL]

Lathyrus venosus [WIL]

leaves), length 1.4–2.5 times width; peduncles usually much longer than subtending leaves, sericeous; inflorescences slender, 4–7-flowered, flowers not clustered at apex, flowers chasmogamous and cleistogamous; pedicels 0.5–3 mm long, appressed-puberulent; bracteoles shorter than calyx tube; chasmogamous flowers 6.5–9 mm long; calyces 4–6 mm long, appressed-puberulent, tubes 1–1.2 mm long; lobes 4, lateral narrowly triangular, 2.5–3 mm long, adaxial connate proximally, apices acuminate; corollas purple; wings 5.5–6.5 mm long; keels 6–8 mm long; chasmogamous fruits as long as calyx, ovate to rounded, 5–7 mm long, cleistogamous fruits exserted from calyces, calyces 1/5 loment length, rounded, 4–5 mm long; stipes subsessile. Open deciduous, dry upland woodlands, prairie fragments, alluvial woodlands, ruderal areas; Cherokee (*Ajilvsgi 4957* BRIT), Fannin (*Correll 37937* KANU, LL, NY), Hunt (*Sanders 3449* BRIT, TEX) and Lamar (iNaturalist Contributors 2023i) cos.; e U.S. from ME s to n FL and w to MI, e OK, and TX. Jul–Oct. The name *Lespedeza violacea* was widely applied to *L. frutescens* in the past; however, the type specimen of the basionym (*Hedysarum violaceum*) represents the species previously called *L. intermedia*. The result is that the name *L. violacea* replaces what was called *L. intermedia*, and *L. frutescens* must be used for this species (Reveal & Barrie 1991). *Lespedeza frutescens* forms natural hybrids with *L. capitata, L. hirta, L. procumbens, L. repens, L. stuevei, L. violacea*, and *L. virginica*. Hybrids with *L. violacea* are common (Clewell 1966) and have been called *L.* ×*acuticarpa* Mackenzie & Bush (Ohashi 2023e). Isely (1998) noted that plants that grow in the shade often possess larger leaflets.

Lespedeza hirta (L.) Hornem., (hairy), HAIRY BUSH-CLOVER, HAIRY LESPEDEZA. Stems spreading to erect or over-arched, 0.5–2 m tall, densely spreading-pubescent; leaflets 1.5–4(–5) cm long, mostly ovate to elliptic, surfaces green, cinereous (= ash-colored), or silvery abaxially, sparsely to densely sericeous abaxially, glabrous, sericeous, pilose, or silvery with fine, dense, appressed-silky hairs adaxially; racemes 10–40-flowered, exserted from upper axils, often conspicuously pedunculate, short-cylindric; calyces typically exceeding corollas and fruit, 7–10 mm long; corollas 7–10 mm long, cream colored, with purplish markings; chasmogamous fruits 6–8 mm long, cleistogamous fruits ca. 6 mm long. Sandy woods, rocky woodlands and adjacent openings, common in frequently burned longleaf pine forests; widespread in East TX; otherwise in TX only in immediately adjacent n Cross Timbers and Prairies; e U.S. from s ME s to FL and w to se KS, OK, and TX; Ontario. Jul–Sep. *Lespedeza hirta* forms natural hybrids with several other species; *L. hirta* subsp. *curtissii* occurs on the coastal plain east of the Mississippi River (Ohashi 2023e).

Lespedeza procumbens Michx., (procumbent, prostrate), TRAILING BUSH-CLOVER, TRAILING LESPEDEZA. Stems prostrate or nearly so, to 1 m long, forming mats, usually freely branched, with pilose pubescence; leaflets oblong-elliptic to suborbicular 1–3.5 cm long, 1.2–1.8 cm wide, spreading-pubescent beneath; inflorescences exserted, with 4–14 flowers per raceme; calyces 3.5–4 mm long; corollas 5–6.8 mm long, rosy lavender to purple; fruits 4.5–5.5 mm long; axillary, ± sessile, cleistogamous inflorescences occasionally occur. Sandy or rocky woodlands and open habitats; widespread in East TX; otherwise in TX only in the immediately adjacent Edwards Plateau and n Cross Timbers and Prairies; e U.S. from NH s to n FL and w to s WI and s to se KS, OK, and TX. May–Sep. This species is similar to *L. repens* which has stems and peduncles sparsely short-appressed pubescent and typically only 4–8 flowers per inflorescence (Isely 1998).

Lespedeza repens (L.) Barton, (creeping), CREEPING BUSH-CLOVER, CREEPING LESPEDEZA. Stems usually completely trailing and freely branched, to 1 m long, often forming mats; leaves ± of one size, though reduced upwards, 1–2(–2.5) cm long, 0.5–1 cm wide; leaflets oblong-lanceolate or oblanceolate to obovate or short-elliptic, appressed-pubescent beneath; stipules 1.5–3(–4) mm long; racemes usually exserted, with 2–8(–10) uncrowded flowers; chasmogamous flowers purple, 6–8 mm long with calyces 3–4 mm long, 4 lobes; fruits 4–7 mm long, chasmogamous exerted from calyx for half its length, cleistogamous fruits rare and calyces only 1/4 to 1/3 loment length. Sandy soils, woods, fencerows, and adjacent roadsides; widespread in East TX; otherwise in TX only in immediately adjacent Edwards Plateau and East Cross Timbers; e U.S. from MA s to n FL and w to s WI and sw to KS, OK, and TX. Late May–Sep. This species is sometimes difficult to distinguish from *L. violacea* which has leaves of two distinct sizes due to smaller leaves in the axils of larger leaves and longer stipules, 2.5–6 mm long. This is the only species that initiates flowering in the late spring and continues through the growing season (Isely 1998). It forms natural hybrids with *L. frutescens* and *L. stuevei* (Ohashi 2023e).

Lespedeza stuevei Nutt., (for its discoverer and Nuttall's friend, Wilhelm Stüwe, 17??–18??, of Bremen, Prussia), TALL BUSH-CLOVER. Stems ascending to erect or arched, 0.5–1.5 m tall; stems and lower surface of leaflets densely spreading or occasionally appressed-pubescent; leaflets 0.5–4 cm long, 0.5–1.2 cm wide; stipules 3–5 mm long; racemes congested in upper axils with 5–15 flowers, shorter or only slightly longer than subtending leaves, the flowers densely crowded together; calyces 3–4 mm long; corollas 5–6 mm long, lavender-pink to rose-purple; fruits 4–6 mm long, sericeous or villous. Sandy open woods, fencerows, roadsides; widespread primarily in e 1/2 of TX; e U.S. from VT s to FL and w to KS, OK, and TX. Late May–Sep. This species apparently hybridizes with *L. virginica* (with narrower leaflets) and intermediates are difficult to categorize (Clewell 1966; Isely 1998).

Lespedeza bicolor [G&F]

Lespedeza capitata [BB1]

Lespedeza cuneata [WIL]

Lespedeza frutescens [GLE, STE]

Lespedeza hirta [MIC]

Lespedeza procumbens [MIC, WIL]

Lespedeza repens [WIL]

Lespedeza stuevei [WIL]

Lespedeza texana Britton ex Small, (of Texas), TEXAS BUSH-CLOVER, TEXAS LESPEDEZA. Stems initially erect then trailing to 1.5 m long, appressed pubescent or occasionally short-pilose; leaflets to 3.5 cm long, grayish or whitish green beneath, ± coriaceous, glabrous to short-pilose, (10–)15–25(–30) mm long, 6–10 mm wide; stipules 3–8 mm long; racemes exserted, with 3–8 uncrowded flowers; calyces 2–4 mm long, lobes 4; corollas 5.5–10 mm long, purple; fruits 4–7 mm long. Limestone hills and slopes; w margin of East TX; otherwise mainly Edwards Plateau and other widely scattered locations to the w; rare in s OK (Kartesz 2023); Mexico (Coahuila). Late May–Sep. Isely (1998) considered this taxon to be a "peripheral affiliate" of *L. repens* but indicated that "the inequalities seem of sufficient weight to accord it specific status." Correll and Johnston (1970) noted that the "firm, grayish leaflets and sprawling habit escape notice in pressed specimens." This has led some authors to consider it identifiable with *L. repens*. The habit, which is entirely unlike that of any other American species, is the most striking character in the field, yet it cannot be recognized from pressed specimens. The stems are erect up to about 15 cm high, the next 10 cm are arching, and the rest of each stem, which may extend over a meter in length, is trailing (Clewell 1966).

Lespedeza violacea (L.) Pers., (violet), VIOLET LESPEDEZA, WANDLIKE BUSH-CLOVER. Stems weakly ascending, wand-like, usually branched, 0.4–1.5 m long, pubescence appressed; leaves petioled, of two sizes, small leaves in axils of larger leaves (dimorphic leaves), glabrous or strigose abaxially; leaflets 1–4 cm long, 0.5–1.5 cm wide; stipules 2–4 mm long; inflorescences exserted, with 4–10 flowers, these not densely crowded, chasmogamous and cleistogamous; calyces 3–4.5 mm long; chasmogamous corollas 6–7 mm long, purple; fruits exserted from calyces, 1/2 loment length, chasmogamous ovate to rounded, 5–7 mm long, cleistogamous rounded, 4–6 mm long, appressed-puberulent. Open, upland rocky, deciduous woodlands, mixed or dry pine woods, mesic bottoms, prairies, old fields, roadsides, often on acidic sand; scattered in East TX; otherwise in TX mainly in e Cross Timbers and Prairies; e U.S. from ME s to n FL and w to s MN and s to se NE, e KS, OK, and TX; Ontario. Jul–Aug. The name *Lespedeza intermedia* has traditionally been used for plants here called *L. violacea* (Reveal & Barrie 1991). The species that previously was called *L. violacea* is now *L. frutescens* (Ohashi 2023e). It forms natural hybrids with *L. capitata, L. frutescens, L. hirta, L. procumbens, L. repens, L. stuevei,* and *L. virginica* (Ohashi 2023e).

Lespedeza virginica (L.) Britton, (of Virginia), SLENDER BUSH-CLOVER, SLENDER LESPEDEZA. Stems erect or nearly so, 0.3–1.6 m tall, wand-like; stems and leaflets appressed-pubescent to short pilose; leaflets narrow, linear to linear-oblong 1–3.5 cm long, 0.3–0.5 cm wide; stipules 2–6 mm long; inflorescences 4–14-flowered, from upper axils and often crowded, not or only slightly exserted, flowers chasmogamous and cleistogamous; chasmogamous calyces 3–4.5 mm long, lobes 4; corollas 4–8 mm long, pinkish to purple; chasmogamous fruits exserted from calyces, suborbicular, 4–7 mm long, cleistogamous fruits exserted from calyces, calyces 1/5–2/5 loment length, similar to chasmogamous, 3.5–5 mm long. Dry upland woodlands, clearings and borders, ruderal sites, old fields, eroded ridges, roadsides, pine barrens, moist savannas, and prairie fragments; widespread in e 1/2 of TX; e U.S. from NH s to n FL and w to sw MN s to e KS, OK, and TX; Ontario; Mexico (Nuevo León). Jul–Sep. This species is somewhat similar to *L. stuevei* (with wider leaflets), and intermediate apparent hybrids can be found (Isely 1998). *Lespedeza virginica* forms natural hybrids with, *L. capitata, L. frutescens, L. hirta, L. procumbens* (*L.* ×*brittonii* E.P. Bicknell), *L. repens, L. stuevei* (*L.* ×*neglecta* Mackenzie & Bush), and *L. violacea* (Ohashi 2023e).

LEUCAENA Benth. LEADTREE

Ours shrubs or trees, unarmed; stems erect, glabrous, sparsely hirsute, or tomentose; leaves alternate, even-bipinnate; stipules present, persistent or caducous, inconspicuous; solitary nectary gland present between or just below proximal-most pinnae, on rachis and rachillae; petiolate; pinnae 2–8 pairs; leaflets 8–32 per pinna, blade margins entire, surfaces pubescent or glabrous; inflorescences 45–190-flowered, axillary, globose heads, usually fascicled, rarely solitary; bracts present, connate, forming involucre at base of each inflorescence, 2–4-lobed; bracteoles peltate and marcescent (= withering but persistent), exceeding buds only when very young (except in *L. retusa*); flowers mimosoid; calyces obconic, tubular, or campanulate, lobes 5; corollas yellow, white, or greenish white; stamens 10, distinct; fruits legumes, shortly stipitate, compressed or flat, linear or oblong, dehiscent, valves sometimes curling, membranous to leathery, often succulent when immature, glabrous or pubescent; seeds 8–24, round, ovate, or weakly rhomboidal.

A genus of 22 species of evergreen shrubs or trees ranging from TX to Peru (Harriman 2023). A number have edible seeds or unripe fruits (e.g., eaten in Mexico) and various species are used as livestock forage and fodder, as a green manure, for firewood or charcoal, as pulp, or in soil conservation (Harriman 2023). Some have toxins in the seeds (e.g., mimosine) and ingestion can result in hair loss or infertility in humans, while others have toxins in the vegetative tissues and can cause toxicity (including hair loss) in non-ruminants (Tull 1987). (Greek: *leukos*, white, in reference to the flower color) (subfamily Caesalpinioideae, Mimosoid clade, tribe Mimoseae)

REFERENCES: Hughes 1998; Harriman 2023.

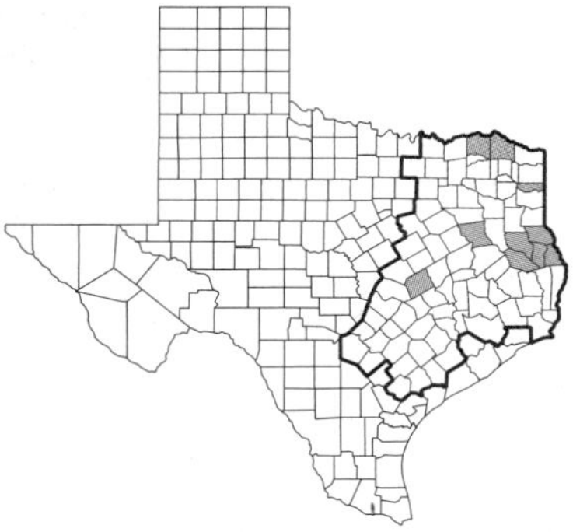

Lathyrus venosus

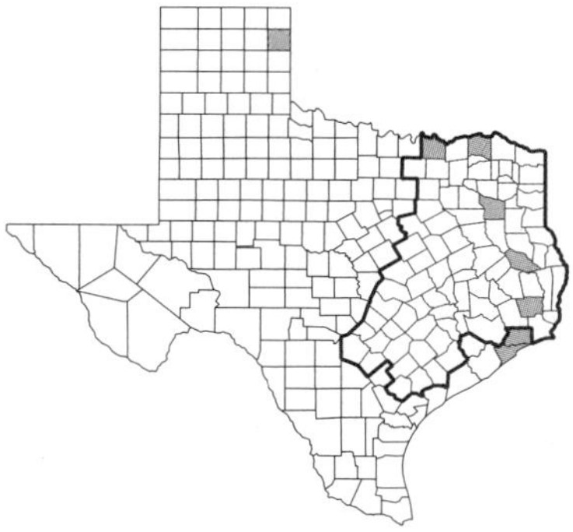

Lespedeza capitata

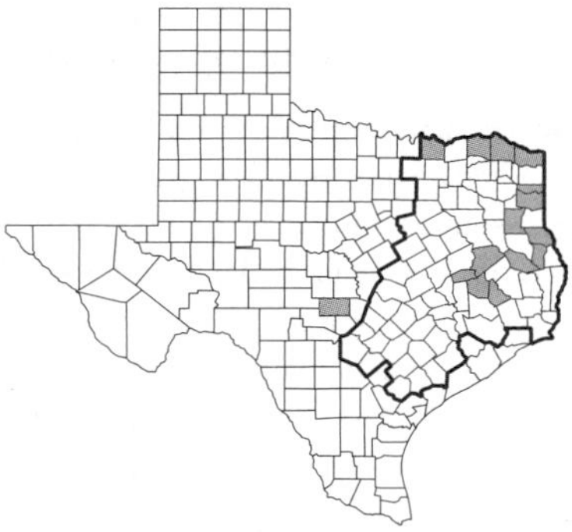

Lespedeza cuneata

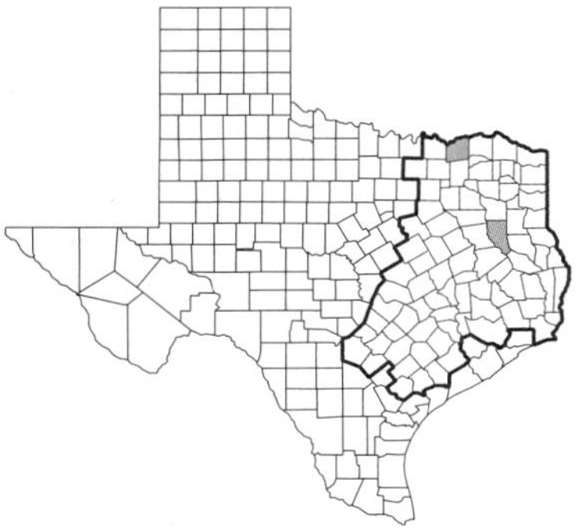

Lespedeza frutescens

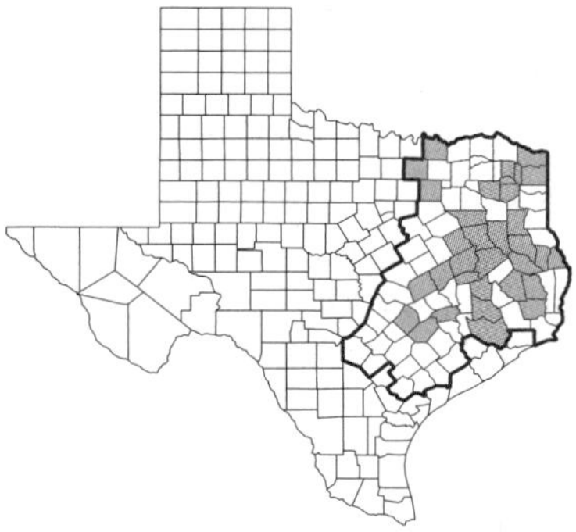

Lespedeza hirta

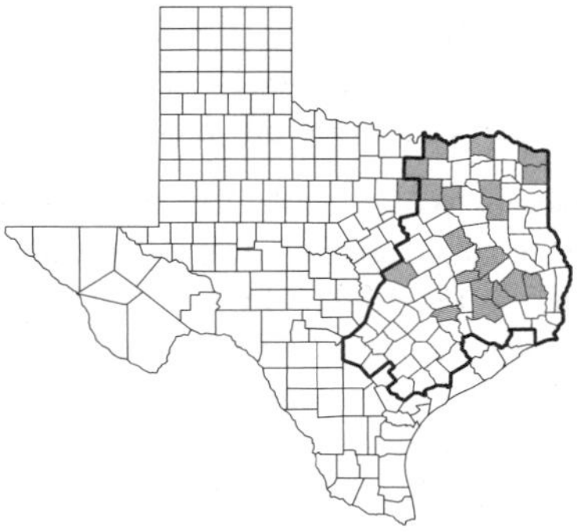

Lespedeza procumbens

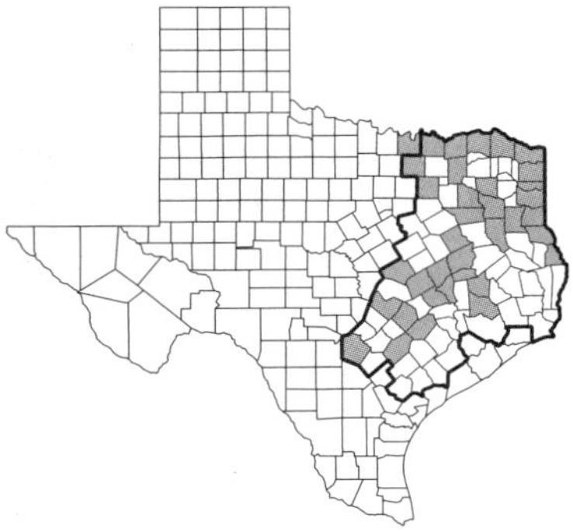

Lespedeza repens

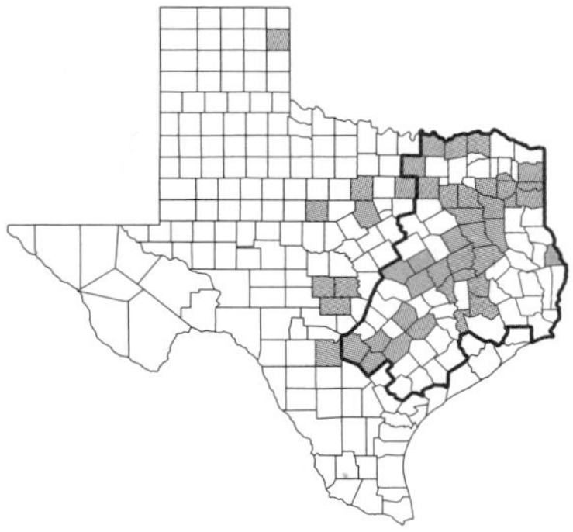

Lespedeza stuevei

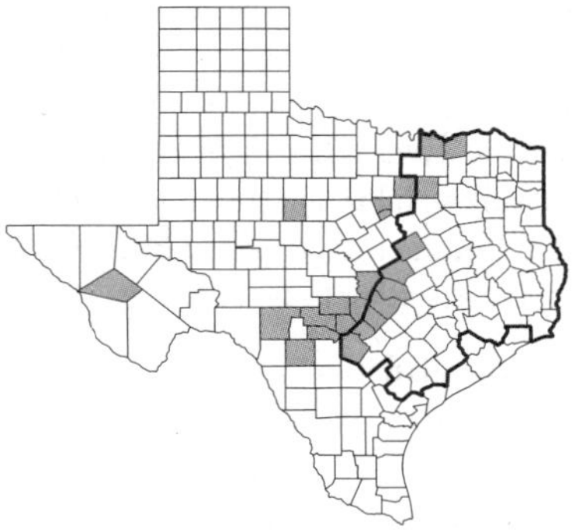

Lespedeza texana

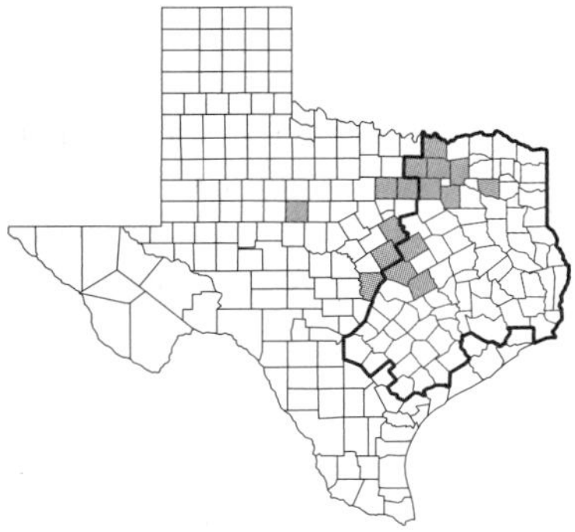

Lespedeza violacea

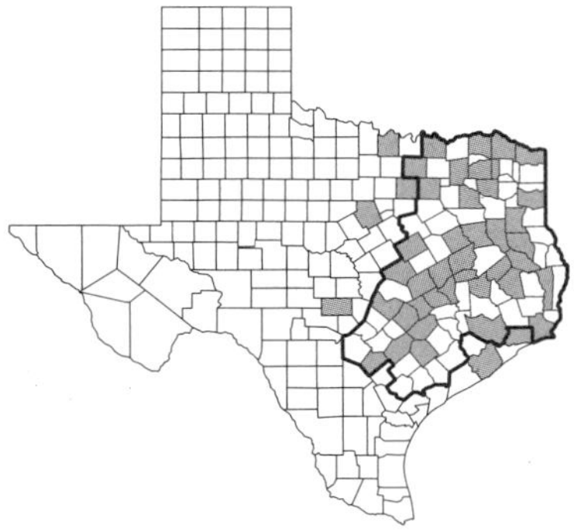

Lespedeza virginica

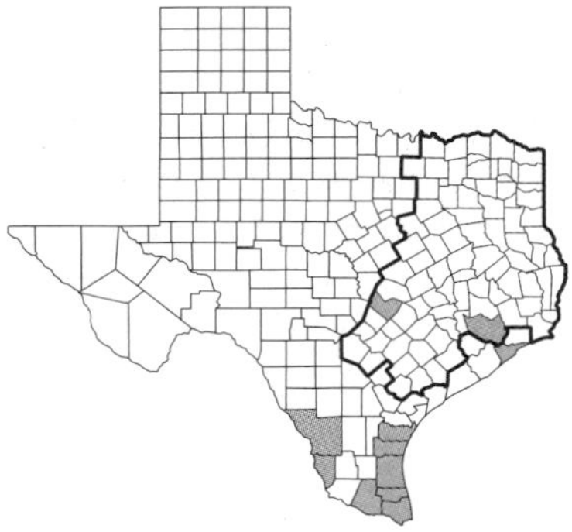

Leucaena leucocephala

1. Pinnae 2–4 pairs; leaflets 8–16 per pinna; flowers yellow ____ **L. retusa**
1. Pinnae 4–8 pairs; leaflets 26–32 per pinna; flowers white ____ **L. leucocephala**

Leucaena leucocephala (Lam.) de Wit, (Greek: *leuco*, white, and *cephala*, head, in reference to the round white flowers), JUMBAY, PEARL WATTLE, WHITE LEADTREE. Shrubs or trees 3–15(–20) m tall, twigs and petioles puberulent; leaves: petiole 2–3 cm long; gland saucer-shaped; pinnae 4–8 pairs; rachis 9–11 cm long; leaflets 26–32 per pinnae, rachilla 7–8 cm long, blade oblong, 8–14 mm long, 2–4.5 mm wide; peduncles 1–2.5 cm long, pubescent; inflorescences fascicled in distal axils, 1–2 cm diam.; bracts inconspicuous; flowers: calyx tube strigose; petals distinct, white, puberulent; anthers sparsely hairy, tip rounded; fruits green becoming red to brown, flat, compressed, 12–21 cm long, 1.4–2 cm wide. Ruderal areas, canal banks, fencerows, waste ground; introduced in Harris and Travis (Kartesz 2023) cos.; s half of TX but moving northward since it was only known from the Lower Rio Grande Valley in the mid-20th century (Turner 1959; Vines 1960); introduced in FL and TX; Mexico; West Indies; Central America; South America; and throughout tropical and subtropical areas worldwide. Flowering year-round. [*Mimosa leucocephala* Lam.] Probably native in the New World tropics. *Leucaena leucocephala* is planted for food and animal forage and is used as firewood (Harriman 2023). Hughes (1998) recognized three subspecies in *L. leucocephala*, two of which occur in TX: subsp. *glabrata* (Rose) Zárate (arborescent) and subsp. *leucocephala* (shrubby), with traits which are not determinable from herbarium specimens; their character states are otherwise overlapping (Harriman 2023). The seeds have historically been made into bracelets and necklaces and are sometimes cooked and eaten with rice (Vines 1960).

Leucaena retusa Benth., (notched at the tip), GOLDENBALL LEADTREE, LITTLELEAF LEADTREE, LITTLELEAF LEUCAENA. Shrub or tree 2–4 (–8) m tall, unarmed; leaves twice pinnately compound, the pinnae 2–4 pairs, the leaflets 8–16 per pinnae, (15–)20–26(–30) mm long, (8–)10–12(–15) mm wide; glands usually present between each pair of pinnae and often between each pair of leaflets; flowers in axillary, peduncled, globose heads ca. 2–2.5 cm in diam., bright yellow; stamens 10, separate; fruits linear, 12–25(–35) cm long. Limestone canyons and slopes; Brazos, Comal, and Travis (Kartesz 2023) cos. at sw margin of East TX; mainly Edwards Plateau and Trans-Pecos; also NM; Mexico (Chihuahua, Coahuila). Apr–Jul. [*Caudoleucaena retusa* (Benth.) Britton & Rose] It is occasionally cultivated as an ornamental outside its native range eastward (Isely 1998).

LOTUS L. DEER-VETCH, TREFOIL

A genus of ca. 125 species from Eurasia, Africa, Atlantic Islands (Azores), Pacific Islands (New Caledonia, Vanuatu), and Australia; introduced nearly worldwide (Brouillet 2023b). Morphological and molecular analyses (Allan et al. 2003; Degtjareva et al. 2008) have shown that the Eurasian members of *Lotus* are distinct from the North American *Acmispon* and *Hosackia*. Therefore, *Lotus* is here defined in its strict sense. *Lotus* species have been introduced to North America as forage crops and for road bank stabilization (Brouillet 2023b). (From the classical Greek name *lotus*, used for a number of different plants; restricted by Linnaeus to this genus) (subfamily Faboideae, tribe Loteae)
REFERENCES: Ottley 1944; Isely 1981; Brouillet 2023b.

Lotus corniculatus L., (horned), BIRD-FOOT TREFOIL, BIRD-FOOT DEER-VETCH, EGGS-AND-BACON. Herbs annual or perennial, 5–70 cm tall, glabrous or sparsely pilose to strigose; usually tap-rooted, rarely rhizomatous; stems ascending or prostrate, solid, not succulent; leaves 5–30 mm long; rachis (1–)2–10 mm long; leaflet blades sometimes asymmetric, obovate to oblanceolate, oblong, or ovate, 4–22 mm long, 2–11 mm wide, apex obtuse and mucronate or acuminate; inflorescences (1–)3–8(–10)-flowered; bracts (1–)3-foliolate; flowers 10–17 mm long; calyces 5–7.5 mm long, lobes not recurved in bud, usually triangular, rarely ovate, (1.5–)2.5–4.5 mm long, shorter to slightly longer than tube, tube villous; petals bright yellow, marked with red, turning orange, 8–14 mm long, wings shorter to longer than keels; fruits brown, narrowly oblong, 15–35 mm long, 2–3 mm wide, nonseptate; seeds 5–30, yellowish or light to dark brown, mottled or sometimes not, globose to round-oblong, 1–1.7 mm long, smooth. Cultivated and escapes into disturbed weedy areas and roadsides; Grayson Co. (Turner et al. 2003), a collection from Denton Co. (Kartesz 2023; *Cory 53249* GH) is a cultivated specimen; also Cross Timbers and Prairies and Trans-Pecos (Kartesz 2023); throughout the U.S. except FL and LA; s Canada; Mexico; Central America, South America; Iceland; and Australia. May–Aug. Native of Eurasia and n Africa. It is used for forage or fodder (Brouillet 2023b), but also known to be cyanogenic and poisonous (Scriber 1978).

LUPINUS L. BLUEBONNET, LUPINE, LUPIN

Annuals or in one of our species perennial; stems erect to decumbent or prostrate, branched or unbranched, usually pubescent; leaves alternate, usually palmately compound, usually cauline, sometimes crowded near base or basal; petiolate; leaflets 5–11, surfaces glabrous or pubescent, stipules adnate to petiole base; inflorescences with flowers few–numerous, terminal, racemes, erect; flowers spirally arranged or whorled, papilionaceous, chasmogamous; calyx bilabiate, lobes connate, entire or toothed;

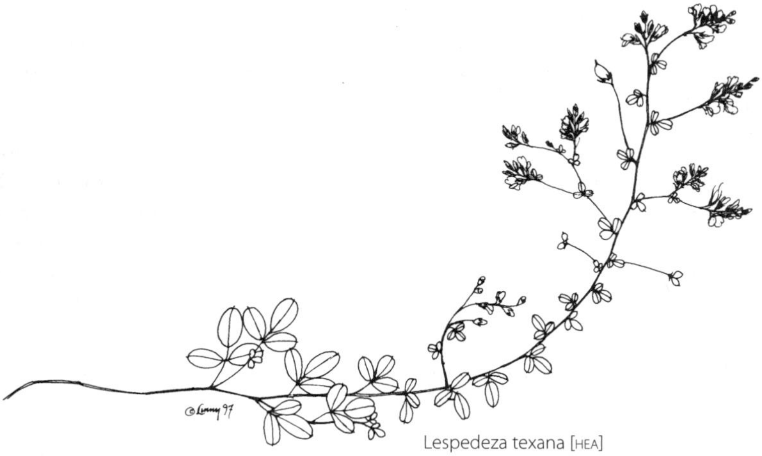

Lespedeza texana [HEA]

Lespedeza violacea [WIL]

Lespedeza virginica [WIL]

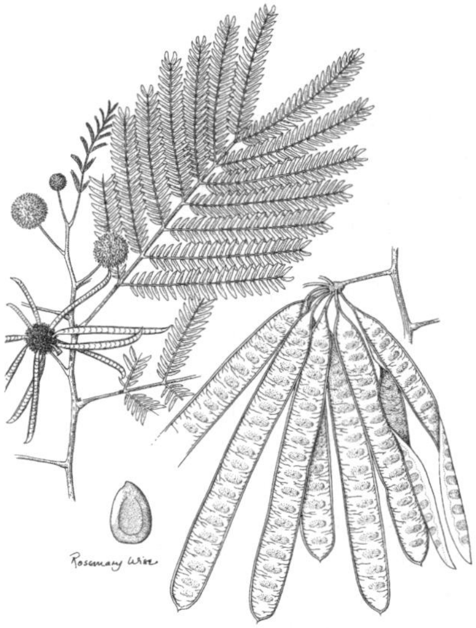

Leucaena leucocephala [SB2]

Leucaena retusa [SA4]

Lotus corniculatus [BEN, BL3]

corollas usually blue to purple, sometimes white, yellow, pink or rose; banners with central groove, sides reflexed; wings connivent at tips, corrugated; keels usually attenuate; stamens 10, monadelphous; styles brushy; fruits legumes, sessile, straight, laterally compressed, usually oblong, splitting along both margins, valves usually twisted after dehiscence, usually pubescent; seeds (1–)2–12, usually smooth, rarely ridged or tuberculate, spheric, lentiform, or angulate.

A genus of ca. 270 species widespread geographically including North America, Mexico, South America, Europe (Mediterranean), Africa; and introduced in China, s Africa, Iceland, New Zealand, and Australia (Sholars & Riggins 2023). While most species are herbaceous annuals or perennials, a few are shrubs, and one Mexican species is a small tree. A number are cultivated as ornamentals, as fodder, as green manure, and in land restoration (because of nitrogen fixation) (Isely 1998). However, many species have alkaloids (e.g., lupinine) and can cause poisoning in humans or livestock (Kingsbury 1964; Schmutz & Hamilton 1979).

In 1901, the Texas legislature was in the process of adopting a state flower. "… in the House, debates were flying fast and furious as one legislator launched his appeal for his favorite, to be followed by more eloquent protestations of the virtues of yet another. Phil Clement of Mills pleaded the case of the open cotton boll, which he likened to 'the white rose of commerce.' John Nance Garner, later the vice-president of the United States, jousted on behalf of the prickly-pear cactus flower. … Then up to the podium strode John M. Green of Cuero. As Green made his appeal for the beautiful bluebonnet, calls came from the floor asking, 'What the devil is a bluebonnet?' … 'You must mean *el conejo*.' 'The rabbit' was a name used by the Mexicans because the waving white tip reminded them of the bobbing tail of a cottontail rabbit. 'No, no, no,' roared another. 'He's referring to what some have called "buffalo clover".' … At this point a group of stalwart Texas women rose to the cause. … The National Society of the Colonial Dames of America in the State of Texas … had originated the idea of using the bluebonnet as the state flower and they were not going to let their favored blossom be left by the roadside for a cactus bloom or cotton boll just because a bunch of representatives didn't know what it was. … A bluebonnet painting was sent for, and one painted by Miss Mode Walker of Austin was carried into the chamber. We are told by Mary Daggett Lake that 'deep silence reigned for an instant. Then deafening applause fairly shook the old walls.' The bluebonnet had won hands down." (Andrews 1986). Unfortunately, due to confusion about common names and the fact that the legislators probably didn't know there were six bluebonnet species in the state, *Lupinus subcarnosus*, which some felt was the least attractive of the bluebonnet species, was officially designated as the state flower, rather than the more beautiful and widespread *L. texensis*. As a result, "For seventy years the argument kicked up dust in the halls of the state Capital until the politically astute representatives … decided to correct their oversight. In 1971, in order to make certain that they would not be caught in another botanical trap, they covered all their bases … by offering an additional resolution that would include 'any other variety of Bluebonnet not heretofore recorded'." (Andrews 1986). As a result, because there are seven *Lupinus* species in TX, there are seven state flowers. Other species in the state (from the Trans-Pecos and the High Plains) include *L. argenteus*, *L. concinnus*, *L. havardii*, and *L. plattensis*.(Derivation uncertain; Latin *lupinus*, wolf, or *lupe*, sadness, and *inus*, possession, perhaps alluding either to plants supposedly overrunning the ground as an animal might or to harsh taste of seeds causing facial contortion]) (subfamily Faboideae, tribe Genisteae)

REFERENCES: Shinners 1953c; Turner 1957; Andrews 1986; Turner & Andrews 1986; Sholars & Riggins 2023.

1. Leaves usually with 7–11 leaflets; plants rhizomatous, clump-forming perennial, known in East TX only from extreme se portion; racemes 8–20 cm long; mature fruits 3–5 cm long ____ **L. perennis**
1. Leaves usually with 5–6(–7) leaflets; plants annual, widespread in East TX; racemes 3–12 cm long; mature fruits 2.5–3.5(–4.2) cm long.
    2. Racemes without distinct white tips (the buds at the rounded raceme tips have yellowish gray or brownish hairs); leaflets rounded, very obtuse or indented at tip; wing petals inflated (cheek-like), light blue; plants typically growing on sandy soils ____ **L. subcarnosus**
    2. Racemes with conspicuous white tips (even from a distance; due to the incompletely expanded and silvery-white hairy buds at the pointed or acute raceme tips); leaflets obtuse to acute or abruptly short-pointed at tip; wing petals not inflated, deep blue (except for rare white albino forms); plants typically growing on limestone or clay soils and abundantly planted along roadsides ____ **L. texensis**

Lupinus perennis L., (perennial), SUNDIAL LUPINE, WILD LUPINE. Perennial, rhizomatous, forming clumps, to 80 cm tall, shaggy pubescent or glabrate; leaves with 7–11 leaflets, the leaflets narrowly oblanceolate, blades (15–)20–45(–50) mm long, 4–15 mm wide, abaxial surface appressed to spreading hairy, strigose, adaxial surface green, glabrous; racemes 8–20 cm long; calyces 5–6 mm long; petals blue or blue-violet (rarely white or pink), (8–)12–16 mm long; fruits 3–5 cm long, 8–10 mm wide. Sandy soils in pine barrens, openings in oak or conifer forests, bluffs, meadows, open rights-of-way; rare in Hardin, Jefferson, and Newton (Kartesz 2023) cos. in extreme se portion of East TX; e U.S. from ME s to n FL and w to e MN and ne IA in n portion of range and w to TX along Gulf Coast; Ontario; introduced in China. (Mar–)Apr–Jul(–Oct). [*L. perennis* subsp. *gracilis* (Chapm.) D.B. Dunn; *L. perennis* var. *gracilis* Chapm.] Shorter plants with more basal leaves have been recognized as subsp. *gracilis*;

intermediates between the named varieties are abundant (Sholars & Riggins 2023). This species is known to be variable in flower color, even within a colony (Isely 1990). It is toxic and potentially fatal to livestock, due to the alkaloid D-lupaine (Sholars & Riggins 2023).

Lupinus subcarnosus Hook., (somewhat fleshy), TEXAS BLUEBONNET, FLESHY-LEAF LUPINE, SANDYLAND BLUEBONNET. Winter annual to 40 cm tall, similar to *L. texensis* but distinguished as in the key above; the two apparently do not hybridize even when growing together (Isely 1998). Leaves cauline, often crowded near base; petiole 1–6 cm long; leaflets 5 or 6, blades 10–25 mm long, 4–15 mm wide, adaxial surface glabrate; racemes 6–12 cm long; calyces 5–6 mm long; petals blue, the wings lighter blue, the standard with white center (changing to purple to magenta red), 9–12 mm long; fruits 2.5–3.5 mm long, 6–8 mm wide, villous. Primarily sandy soils, open areas, conspicuous on roadsides; mainly in the se portion of East TX, n to Anderson (Turner et al. 2003) Co.; adjacent Gulf Prairies and Marshes and South TX Plains, planted and escaped in isolated locations westward and northward; native only in TX in the U.S.; Mexico (Coahuila, Nuevo Léon). Mar–May. This is the species that in 1901 was originally designated the state flower of Texas (see discussion under generic synopsis; Andrews 1986).

Lupinus texensis Hook., (of Texas), TEXAS BLUEBONNET, BUFFALO-CLOVER, TEXAS LUPINE. Winter annual to 40 cm tall; stems pilose with appressed or ascending hairs; leaves long-petioled, palmately compound; leaflets 5–6(–7), lanceolate or oblanceolate, glabrous above or nearly so, obtuse to acute or abruptly short-pointed apically, blades 10–25 mm long, 6–12 mm wide; racemes 3–12 cm long; calyces 6–8 mm long; petals deep blue (rarely white), 10–13 mm long; standard with white center (changing to purple to magenta red) bearing yellow-green dots; wing petals not inflated, nearly straight when viewed from front; fruits 2.5–3.5(–4.2) cm long. Rocky limestone soils, clay, open areas, conspicuous on roadsides; widespread in East TX, but primarily in the w portions of the area; widespread in TX except in the extreme e, sw, and nw; now widely cultivated and planted along highways in much of TX and adjacent OK; endemic to TX (can now be found planted and escaped in FL, LA and s OK) (Sholars & Riggins 2023). Mar–May. Several Brown Co. and Comanche Co. specimens (HPC) of *L. texensis* have inflated wing petals and are thus similar to *L. subcarnosus*. Apparently unpalatable to cattle. This species is widely planted by the TX Dept. of Transportation. Since it is a "winter" annual, it forms rosettes in the fall and overwinters as a small plant—therefore individuals trying to plant this species should do so in the fall (Sep–mid Nov).

MACROPTILIUM (Benth.) Urb. BUSH-BEAN

A western hemisphere species of 18 species native to s U.S., Mexico, West Indies, Central America, and South America. The greatest diversity in *Macroptilium* is found in South America (about 16 species); some species have been introduced for fodder and naturalized in the tropics and subtropics nearly worldwide (Torres-Colín & Delgado-Salinas 2023). (Greek: *makros*, large, and ptilon, wing, alluding to wind petals being larger than banner petal) (subfamily Faboideae, tribe Phaseoleae)
REFERENCES: Isely 1998; Torres-Colín & Delgado-Salinas 2023.

Macroptilium lathyroides (L.) Urb., (like the genus *Lathyrus*), PASEY BEAN, ONE-LEAF CLOVER. Herbs, biennial; root elongate; stems erect or semierect, with hollow pith, glabrescent; stipules retrorse, lanceolate, 42–60 mm long, 8–15 mm wide; leaflets 3, membranous, blades ovate, elliptic, or oblong, base attenuate, apex acute, surfaces strigulose abaxially, glabrescent adaxially, lateral blades 2–4.5 cm long, 0.6–2 cm wide, terminal blades 2–5.5 cm long, 0.6–4 cm wide; inflorescences 19–46 cm long, flowers 2–2.6 cm long; calyx tubular, 3–7 mm long, strigose or glabrescent, calyx lobes shorter than tube; corollas deep red, vivid purple, maroon, or blackish; banner oblong, 9.5–15 mm long, 8–9 mm wide; wings oblong, 11.5–26 mm long, 6.5(–12) mm wide, keel 9–16 mm long; fruits, linear, 9.5–10.5 cm long, 0.2–0.3 cm wide; seeds 15–20. Disturbed areas, often in landscaping; Harris Co. (Aplaca 2012; E. Keith, pers. obs.) in East TX; FL, GA, LA, SC, and TX; introduced also in Asia, Africa, and Australia. May-Nov. [*Phaseolus lathyroides* L.] Native of Mexico, West Indies, Central America, and South America. This species was first reported to TX from Harris County as an inadvertent introduction in landscaping plants (Aplaca 2012). It is now well established in the greater Houston area (iNaturalist 2023; E. Keith, pers. obs.). The leaves are very similar to *Strophostyles umbellata* but can readily be distinguished by the deep red, purple, or purplish black flowers.

MEDICAGO L. MEDIC, MEDICK, BUR-CLOVER, ALFALFA, BURWEED

Annual or perennial herbs; stems prostrate to erect; leaves pinnately compound; leaflets 3, with small, sharp teeth, at least in apical portion; flowers in axillary or terminal, peduncled, head-like or short cylindrical, spike-like racemes; petals yellow (in all East TX species except 1), blue, purple, or rarely white; stamens 10, diadelphous; fruits spirally coiled to curved.

An Old-World genus of ca. 85 species native to Europe, the Mediterranean, Asia, Ethiopia, and s Africa (Isely 1998; Small 2023a). A number are cultivated as ornamentals, for fodder or green manure (Small 2023a). All North American (and East TX) species are introduced from the Old-World. (Greek: *medice*, the name of alfalfa, because it came to the Greeks from Media)

(subfamily Faboideae, tribe Trifolieae)

REFERENCES: Wagner 1948; Lesins & Lesins 1979; Small & Jomphe 1989; Small 2023a.

1. Petals usually purple to pale blue, rarely white, yellow, or variegated, 5–15 mm long; ovaries and fruits without prickles; plants perennial ____ **M. sativa**
1. Petals yellow, 1.5–6 mm long; ovaries and fruits with or without prickles; plants annual.
  2. Stipules (2 at base of each leaf) entire to shallowly toothed; stems and leaflets pubescent.
    3. Ovaries and fruits without prickles, the fruits ± kidney-shaped, ± black at maturity, with 1 seed; flowers (5–) 15–50 per raceme; calyces 1–1.5 mm long ____ **M. lupulina**
    3. Ovaries and fruits prickly, the fruits coiled in 3–5 turns, brownish at maturity, with several seeds; flowers 2–5(–8) per raceme; calyces 2–2.3 mm long ____ **M. minima**
  2. Stipules deeply toothed to deeply lobed; stems sparsely pubescent or glabrous; leaflets glabrous.
    4. Leaflets with prominent, central, purple-red spot; stipules of upper leaves not divided more than half-way to base ____ **M. arabica**
    4. Leaflets without central spot; stipules of upper leaves divided more than half-way to base.
      5. Pedicels shorter than the calyx lobes, some flowers subsessile; mature fruits prickly or roughened with short points, loosely curled, barrel-shaped, 4–10 mm in diam. (excluding prickles) ____ **M. polymorpha**
      5. Pedicels equaling or exceeding the calyx lobes; mature fruits smooth, tightly coiled and flattened, disk-shaped, 8–20(–24) mm in diam. ____ **M. orbicularis**

Medicago arabica (L.) Huds., (of Arabia), SPOTTED BUR-CLOVER, SPOTTED MEDIC. Stems erect to decumbent, 10–60 cm long; stipules deeply dentate to lobed; leaflets 8–25 mm long, 7–20 mm wide with prominent, central, purple-red spot; terminal leaflet very short-stalked; calyces pubescent, 2–2.5 mm long; flowers (1–)2–5(–8) per head-like raceme; petals 4–5(–6) mm long; fruits prickly, with 3–7 coils, 4–8 mm in diam. (excluding prickles); prickles of fruits 2–3 mm long, recurved. Sandy pastures, lawns, and roadsides, chiefly in low ground; scattered in the e 1/2 of TX; first TX record in 1913 (Wagner 1948); scattered in the U.S. from ME s to n FL and nw to s IL then sw to e OK and TX, also CA, OR, and WA; British Columbia and New Brunswick in Canada; introduced in Central America; South America; New Zealand; and Australia. Mar–Apr. Native of s Europe, n Africa, and sw Asia. This species is easily recognized by its spotted leaflets and incised stipules (Isely 1998).

Medicago lupulina L., (resembling *Humulus lupulus*—HOPS, apparently based on a resemblance of the clusters of tiny fruits to those of hops), BLACK MEDICK, NONE-SUCH, HOP-CLOVER, YELLOW TREFOIL. Stems ascending or trailing, to 60 cm long; stipules entire or irregularly toothed; leaflets 10–20 mm long, 6–15 mm wide; racemes 7–10 mm long, with (5–)15–50 flowers; calyces 1–1.5 mm long; petals 2–4 mm long; fruits with 1 partial coil, curved, kidney-shaped, 2–3 mm in diam., without prickles, but with visible nerves. Planted for pasture and soil improvement (due to nitrogen fixation); established as a weed on roadsides, in lawns, and waste places; widely scattered in TX, but more commonly in the e 1/2 of the state; first TX record 1937 (Wagner 1948); throughout the U.S.; s half of Canada; Greenland; introduced nearly worldwide. Apr–May, less commonly to Jul. Native of Eurasia. Speculated by some to be the original Irish SHAMROCK; however, it is more likely that either *Trifolium dubium* Sibth. or *T. repens* L. is actually the SHAMROCK (Nelson 1991). It is often mistaken for the superficially similar species SMALL HOP CLOVER (*Trifolium dubium*), but can be distinguished by the angled stems and incised stipules (Isely 1998).

Medicago minima (L.) Bartal., (least, smallest), BUR-CLOVER, SMALL MEDIC, SMALL BUR-CLOVER. Similar to *M. lupulina* except for the fruits and usually short stems 10–40 cm long; stipules entire or minutely dentate at base; leaflets obovate or oblanceolate, 5–8(–12) mm long, 2–7 mm wide; inflorescences with 2–5(–8) flowers, calyces 2–2.3 mm long, pubescent; flowers 2–4.5(–6) mm long; fruits with 3–4 coils, 2.5–6 mm in diam. (excluding prickles); prickles of fruits 1.5–3 mm long, hooked at tip. Lawns, pastures, roadsides, grasslands, and open oak-juniper woodlands; widely scattered in TX except the far n, far w and far s, most common in c TX; first TX record in 1914 (Wagner 1948); scattered throughout most of U.S., primarily in s half; South America; Hawaii; New Zealand; and Australia. Mar–May, rarely again in Sep. Native of Eurasia.

Medicago orbicularis (L.) Bartal., (orbicular, round), BUTTON-CLOVER, BUTTON MEDIC. Similar to *M. polymorpha*; stems procumbent to ascending, 10–50 cm long; stipules laciniate; leaflets obovate to cuneate, 7–13 mm long, 5–10 mm wide; flowers 1–2(–5) per head-like raceme; calyces glabrous or sparsely pubescent, 1.5–3 mm long; petals 3–4(–6) mm long; fruits with (2–)3–7 coils, 8–20(–24) mm in diam., without prickles. Ruderal and fallow habitats including lawns and dried ponds; scattered and uncommon throughout TX; first TX record in 1915 (Wagner 1948); scattered locations in the U.S., IL, NJ, and PA northward and se U.S. from NC s to n FL and w to OK and TX, disjunct in CA; introduced also in Argentina; s Africa; Hawaii; and Australia. Apr–early Jun. Native of the Mediterranean region. This species is distinctive in its large, flat, spineless burs (Isely 1998).

Medicago polymorpha L., (of many forms), CALIFORNIA BUR-CLOVER, BUR-CLOVER. Plants prostrate to ascending, sometimes forming mats; stems 5–50 cm long; stipules laciniate; leaflets obovate, obcordate, or cuneate, 8–20 mm long, 7–18(–20) mm wide;

Lupinus perennis [WIL]

Lupinus subcarnosus [TRP]

Lupinus texensis [JON]

Macroptilium lathyroides [COZ]

Medicago arabica [BB2]

Medicago lupulina [REE]

terminal leaflet rather long-stalked; flowers (1–)2–6(–10) per head-like raceme; calyces sparsely pubescent, 2–3 mm long; petals 3.5–6 mm long; fruits coiled 2–7 times, 4–10 mm in diam. (excluding prickles); prickles of fruits usually well-developed, 2–3 mm long, hooked at tips, rarely absent. Fallow fields, ruderal habitats, lawns, and roadsides; widespread mainly in the e 1/2 of TX; first TX record in 1900 (Wagner 1948); throughout the U.S. except the nc portion, but including AK; s Canada; Mexico; Central America; South America; New Zealand; and Australia. Mar–early May. Native of Eurasia. [*M. polymorpha* var. *vulgaris* (Benth.) Shinners, *M. hispida* Gaertn., *M. denticulata* Willd., *M. nigra* (L.) Krock.] *Medicago polymorpha* is one of the more important annual medics that have been developed for use as pasture forage for dry, hot environments. About a dozen cultivars have been bred (Small 2023a).

Medicago sativa L., (cultivated), ALFALFA, LUCERNE. Glabrous or pubescent perennial, erect, 30–100 cm tall; stipules entire or basally toothed; leaflets 5–35 mm long, 2–15 mm wide; racemes 10–40 mm long, crowded, with 3–30(–50) flowers; calyces glabrous or pubescent with glandular or eglandular hairs; corollas 5–15 mm long, purple or rarely yellow, white, green, or variegated; fruits loosely coiled in 1–6 turns, 4–14 mm in diam., without prickles. Cultivated and persisting, locally established as an escape on roadsides and other ruderal areas; nearly throughout TX; first TX record in 1915 (Wagner 1948); throughout the U.S. including AK; introduced throughout Canada; West Indies; Central America; South America; Pacific Islands; and Australia. Late Apr–Sep. Native of sw Asia. This species is a tetraploid long cultivated (> 6,000 years) for fodder and silage, particularly for horses (Small 2023a); the Arabic name ALFALFA is derived from old Irani "*aspa-sti*" = horsefodder (Online Etymology Dictionary 2023) or from the Arabic word *al-fac-facah* meaning the 'father of foods' because of its very rich nutritional properties. Its health benefits were recognized thousands of years ago in various parts of Asia as alfalfa has been discovered in 6,000-year-old Persian ruins and it is mentioned in ancient Turkish literature written in 1300 B.C. In China, traditional practitioners have used the herb to treat ulcers and as an appetite stimulant (Gawel 2012). It is considered by some to be the most important forage plant in the world (Isely 1990). However, ALFALFA hay can sometimes contain blister beetles (*Epicauta* species) which, if ingested, are extremely dangerous and even potentially fatal to horses as well as other livestock (Hardin & Brownie 1993). Such beetles have been found in ALFALFA hay being fed to horses in TX. Humans also use various parts of the plant as sprouts, greens, and as tea which is reported to prevent hay fever and other ailments and as a mild laxative. The nectar of the flowers also yields a good honey (Tull 1987). Small (2023a) distinguished *M. sativa* subsp. *falcata* (L.) Arcang. and *M. sativa* subsp. ×*varia* (Martyn) Arcang., a hybrid of subsp. *sativa* and subsp. *falcata* by possessing yellow or variegated flowers, however, we are following Isely (1998) in not distinguishing subspecies and who wrote "alfalfa is a tetraploid, mostly hybrid complex that, as a consequence of human manipulation … is now subject to diverse classifications and nomenclature … the group is a continuum, and arbitrary segmentation is useful primarily for gene pool resource and plant breeding purposes." The yellow flowered plants are not known to occur in East TX (Isely 1998; Kartesz 2023).

MELILOTUS (L.) Mill. SWEET-CLOVER, MELILOT

Annuals or biennials with erect or spreading, branched stems; leaves pinnately compound; leaflets 3, with small, sharp teeth, at least in apical portion; flowers in axillary, erect, elongated, spike-like racemes; petals yellow or white; stamens 10, diadelphous; fruits ovoid to nearly globose, indehiscent, 1(–2) seeded, 5 mm or less long.

A genus ca. 18 species of fragrant herbs native to temperate and subtropical Eurasia and n Africa; cultivated for forage, hay, and green manure; also valued as a bee plant for honey production (Small 2023b); similar to *Medicago*. All East TX species are introduced. SWEET-CLOVERS can be recognized in the field by the combination of their slender racemes of small yellow or white flowers, indehiscent 1–2-seeded fruits, and often fragrant herbage. Sweet-clover poisoning, a hemorrhagic disease of cattle in which the animal bleeds to death, is caused by ingestion of improperly cured or molded sweet-clover hay (coumarin glycosides release coumarin which becomes converted to dicoumarin, a toxic substance—dicoumarin prevents blood clotting; it is now used medicinally as an anticoagulant) (Kingsbury 1964; Lewis & Elvin-Lewis 1977). (Greek: *meli*, honey, and *lotus*, some leguminous plant) (subfamily Faboideae, tribe Trifolieae)

REFERENCES: Hennen 1951; Stevenson 1969; Small 2023b.

1. Corollas 1.5–3.5 mm long, yellow; inflorescences 1–2(–3) cm long in flower; plants 0.6 m or less tall; mature fruits 1.5–3(–4) mm long, reticulate-veined or irregularly bumpy ____ **M. indicus**
1. Corollas 3.5–7 mm long, yellow or white; inflorescences 4–12 cm long in flower; plants 0.4–2 m tall; mature fruits 2.5–5 mm long, reticulate-veined or transverse rugose.
    2. Corollas yellow, usually 4–7 mm long; standard about as long as wings, the keel shorter; fruits light brown or tan when mature ____ **M. officinalis**
    2. Corollas white, 3.5–5(–6) mm long; standard markedly longer than the ± equal wings and keel; fruits brown or black when mature ____ **M. albus**

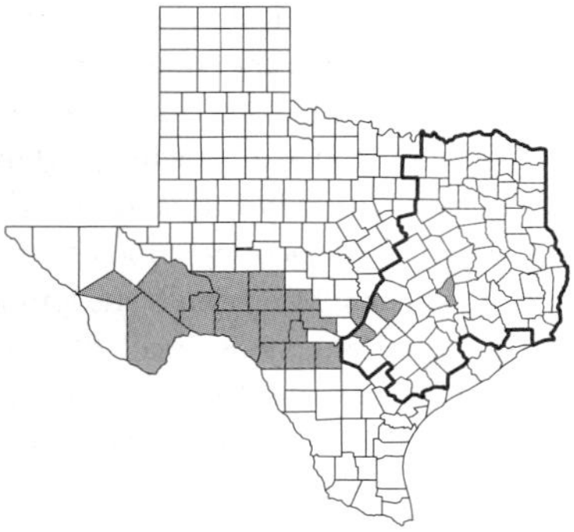

Leucaena retusa

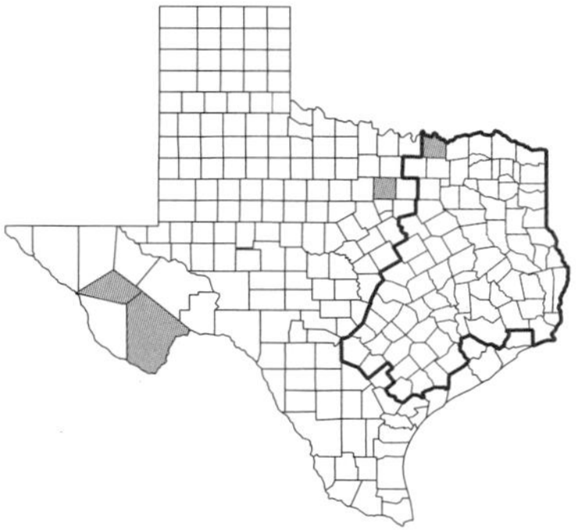

Lotus corniculatus

Lupinus perennis

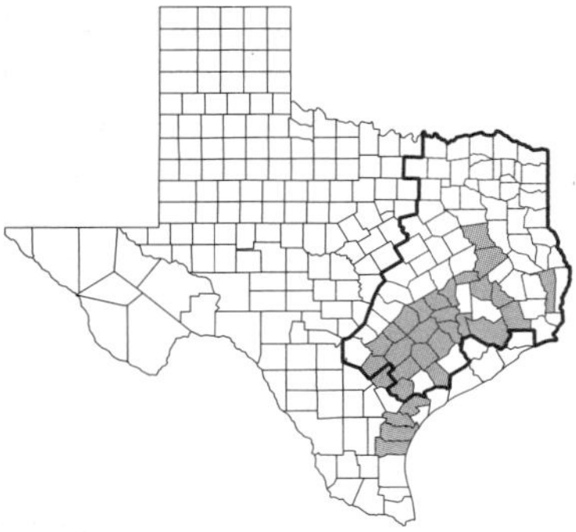

Lupinus subcarnosus

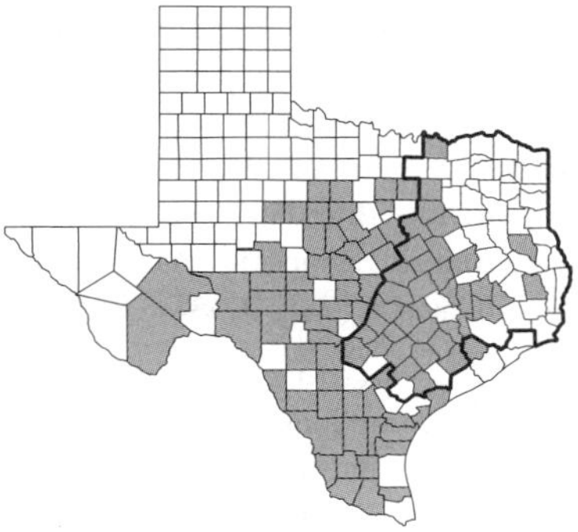

Lupinus texensis

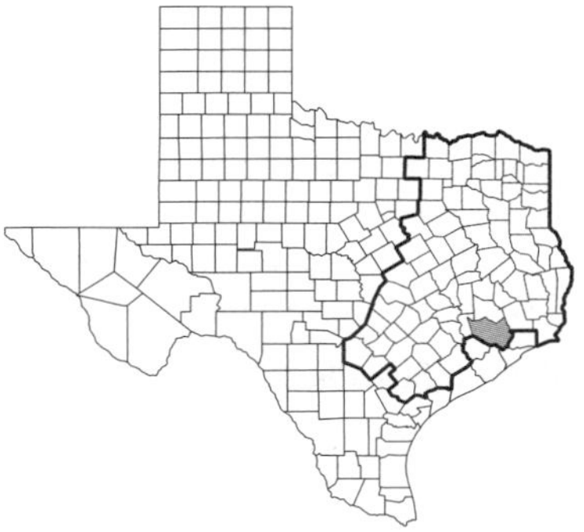

Macroptilium lathyroides

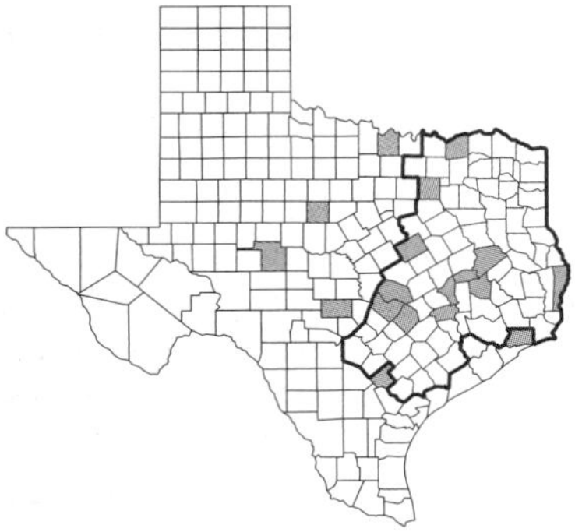

Medicago arabica

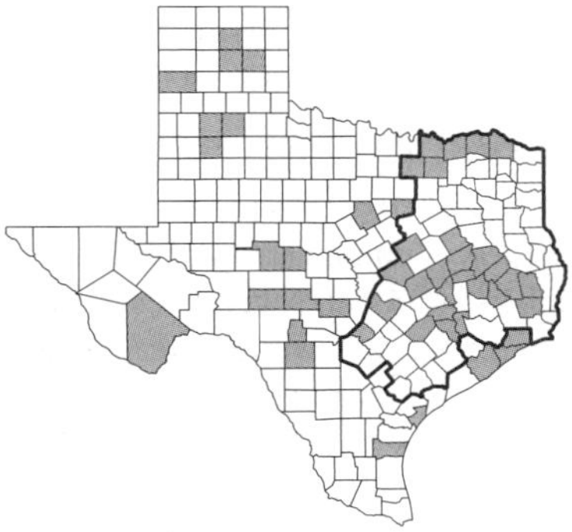

Medicago lupulina

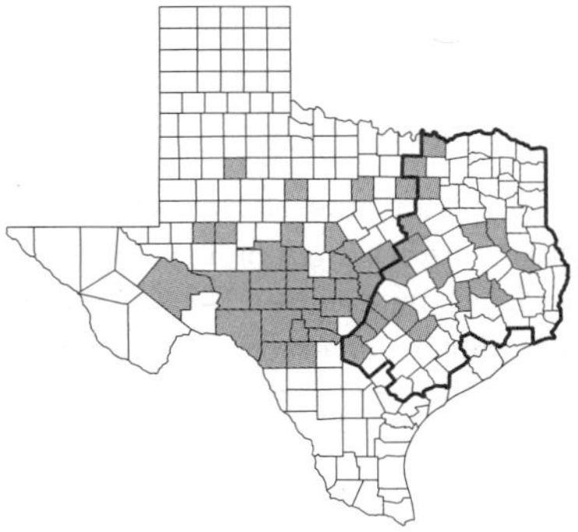

Medicago minima

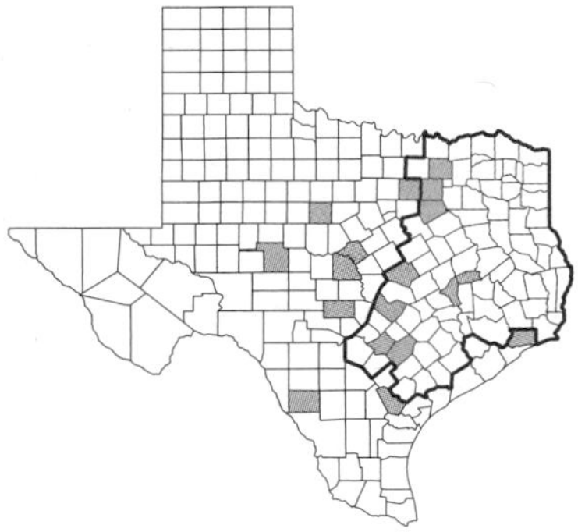

Medicago orbicularis

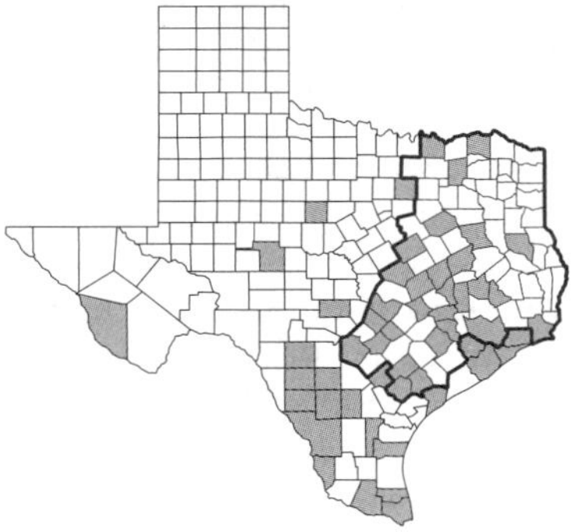

Medicago polymorpha

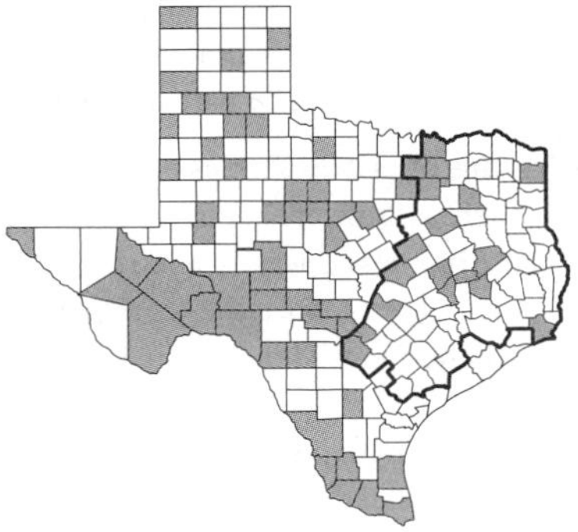

Medicago sativa

Melilotus albus Medik., (white), WHITE SWEET-CLOVER, WHITE MELILOT, HUBAM-CLOVER (an annual strain). Annual or biennial; stems erect or ascending, 0.5–2 m tall; stipules lanceolate- subulate to setaceous, 4–6(–10) mm long, margins usually entire, sometimes toothed at base; leaflet blades narrowly oblong-obovate to suborbiculate-oblong, 10–24(–50) mm long, 5–12(–15) mm wide, margins dentate, fragrant when crushed; racemes 40–80(–120)-flowered; flowers 3.5–5(–6) mm long; corolla white; legumes obovoid, subglobose, or globose-ovoid, 3–5 mm long, reticulate-veined, glabrous. Roadsides, fields, railroads, and disturbed sites; widespread in TX; throughout the U.S. including AK; also introduced in Canada; Greenland; Mexico; West Indies; South America; and Australia. May–Sep. Native of Europe and Asia. This species is vegetatively very similar to *M. officinalis* and lumped by some into that species (e.g., Kartesz 2023); we are following most recent authors (e.g., Isely 1990, 1998; Small 2023b) in recognizing it as a separate species. While often difficult to identify from herbarium specimens (flower color faded, mature fruits often absent), Isely (1998) noted that they sometimes occur together and "are evidently incompatible." This species is known to cause sweet-clover poisoning in cattle. This species is also an important bee plant for honey production (Small 2023b). No county distribution map provided.

Melilotus indicus (L.) All., (of India), SOUR-CLOVER, YELLOW SOUR-CLOVER, ALFALFILLA, SMALL MELILOT. Annual 0.1–0.6 m tall; stems erect or ascending; stipules lanceolate-subulate to setaceous, (3–)4–6(–8) mm long, margins subentire (entire or toothed at base); leaflet blades lanceolate-oblong, oblanceolate, or obovate, (8–)12–25 mm long, (2–)7–10 mm wide, margins dentate; racemes 10–50-flowered; flowers 1.5–3.5 mm long; corolla yellow; legumes subglobose, 1.5–3(–4) mm long, strongly reticulate-veined, glabrous. Roadsides, pastures, fields, and disturbed sites; widely scattered across TX, but rare in far n; widely scattered across the U.S., but more common southward; introduced in Nova Scotia; Greenland; Mexico; Honduras; South America; and New Zealand. Apr–Aug. Native of the Mediterranean region and s Asia. Though having the same flower color as *M. officinalis*, this species is more diminutive in numerous characters (Isely 1998).

Melilotus officinalis (L.) Lam., (medicinal), YELLOW SWEET-CLOVER, YELLOW MELILOT. Annual or biennial 0.4–2 m tall; stems decumbent to erect; stipules lanceolate or subulate, 3–6(–12) mm long, margins entire; leaflet blades obovate, obovate-oblong, ovate, or ovate-lanceolate, 8–25 mm long, 4–15 mm wide, margins dentate, fragrant when crushed; racemes 30–70(–80)–flowered; flowers 4–7 mm long; corolla yellow; legumes ovoid, 2.5–5 mm long, transversely rugose, areoles notably elongated, glabrous. Roadsides and railroads, disturbed areas; widely scattered across TX, but rare in far w and far s; throughout the U.S. including AK; introduced throughout Canada; also in Mexico; South America; Africa; and Australia. Apr–Sep. Native of Europe and Asia. The leaves have a characteristic, vanilla-like odor when crushed or dried and have been used in making sachets (Ajilvsgi 1984). This species is known to cause sweet-clover poisoning in cattle (Kingsbury 1964).

MIMOSA L. CATCLAW, SENSITIVE-BRIAR

Our species are prostrate, low spreading, or trailing, herbaceous perennials to 4 m long or erect shrubs, usually armed; leaves alternate, usually even-bipinnate, rarely some leaves unipinnate (*M. borealis*); stipules present; petiolate, petiole eglandular (glandular); pinnae (0 or)1–14(–33) pairs; leaflets 1–40(–95) pairs, opposite, blade margins entire, surfaces glabrous or pubescent; peduncles unarmed or prickly; inflorescences 5–180-flowered, usually axillary, sometimes also terminal, heads, spikes, or racemes; bracts present; flowers mimosoid, bisexual, or staminate present at base of inflorescence; calyces campanulate, lobes 4 or 5(–6), valvate; corollas white, pink, purple, or purple-pink, lobes 4 or 5(–6), valvate; stamens 8 or 10 (2 times corolla lobes) or 4(5) (as many as corolla lobes), distinct or connate at bases, exserted, white, pink or lilac; anthers dorsifixed, introrse, eglandular; ovaries sessile or stipitate; styles longer than stamens; stigmas poriform, cupuliform, or tubular; fruits legumes, sessile or stipitate, straight or curved, constricted or not between seeds, linear, oblong, or tetragonal, dehiscent, craspedial (= segmented into 1-seeded portions separating from persistent marginal sutures (replum)), sometimes unsegmented; sometimes armed with recurved or straight prickles, glabrous or pubescent, valves segmented with 1 seed each or entire, usually wider than margin, persistent margin usually narrow, sometimes as wide as or wider than valves, prickly or unarmed; seeds 1–30.

A genus of ca. 530 species of tropical and warm areas, especially the Americas. The group includes herbs, shrubs, lianas, and trees, often with stipular spines (Grether 2023). Some *Mimosa* species contain the toxic amino acid, mimosine, and if eaten can cause hair loss, gastritis, and cataracts in animals (Spoerke & Smolinske 1990). As treated here, *Mimosa* includes *Schrankia*, the SENSITIVE-BRIARS; since the segregate genus was named, virtually all the characters defining it have been found within other groups of the genus *Mimosa* (Yatskievych 2006). We are thus following most recent authors (e.g., Isely 1998; Turner et al. 2003; Yatskievych 2006; Grether 2023) in including *Schrankia* within a broadly conceived *Mimosa*. Further, we are following Turner (1994d) and Grether (2023) for nomenclature of the *M. quadrivalis* complex. Turner (1994d) argued that the former *Schrankia* species are isolated geographically, do not appreciably intergrade, and when occasionally growing together apparently do not hybridize. Barneby (1991) and Isely (1998) treated them all as varieties of *M. quadrivalis*. The vast number of taxonomic changes and disagreement over species limits in *Mimosa* has led to many misidentifications in herbaria, regional floras, and online forums.

The leaves of several *Mimosa* species (previously treated in *Schrankia*) are touch sensitive and display rapid movements upon being touched; these movements result from pressure changes in the pulvinules (= tiny swollen joints) at the base of each leaflet and the larger pulvini (singular: pulvinus) at the base of each leaf (Wills & Irwin 1961); this is possibly an adaptation to reduce water loss (via transpiration) or a defense against herbivores. The leaves also exhibit nyctinastic (nighttime or "sleep") movements with leaflets folding together and petioles drooping at night. (Greek: *mimos*, mimic, referring to the sensitive leaves) (subfamily Caesalpinioideae, Mimosoid clade, tribe Mimoseae)

REFERENCES: Isely 1971a, 1971b, 1973; Barneby & Isely 1986; Barneby 1991; Turner 1994d, 1994e; Grether 2023.

1. Plants prostrate, sprawling, or trailing, sometimes mat-forming; flowers pink; fruits with conspicuous hairs or distinctly prickly over the surface; stems herbaceous.
   2. Stems with numerous weak, hair-like or bristle-like structures not painful to the touch, recurved prickles absent; mature fruits 10–28 mm long, noticeably flattened, with numerous appressed hairs on the surface, but prickles absent ____ **M. strigillosa**
   2. Stems conspicuously armed with numerous recurved prickles capable of causing pain; mature fruits 20–120 mm long, not noticeably flattened OR flattened in *M. roemeriana*, usually covered with erect, recurved to straight prickles (*M. quadrivalvis* complex).
      3. Leaflets with evident venation above and raised midrib and side-veins beneath.
         4. Fruits 4–12 cm long with acute or beaked apex; peduncles in late flower or fruit 3.5–9.5 cm long; species widespread in East TX, but less so in se part ____ **M. nuttallii**
         4. Fruits 2–4.5 cm long with round apex; peduncles in late flower or fruit 7–13(–20) cm long; species of se part of East TX ____ **M. hystricina**
      3. Leaflets smooth or only midrib raised beneath.
         5. Pinnae (3–)4–9 pairs; leaflets 0.5–1.5 mm wide; legumes linear, to 140 mm long; deep sandy soils in se TX ____ **M. microphylla**
         5. Pinnae 1–3 or 2–5 pairs; leaflets 0.8–3 mm wide; legumes linear-oblong, to 70(–100) mm long; sandy or calcareous soils in w or s portions of East TX.
            6. Pinnae 2–5 pairs; petiole as long as or shorter than rachis; mature fruits flattened at maturity, 3–6 times as wide as thick, 3–5(–7) mm wide; species of w margin of East TX, usually on calcareous soils ____ **M. roemeriana**
            6. Pinnae 1–3 pairs; petiole usually much longer than rachis; mature fruits usually 4-sided to nearly terete, essentially unflattened, at most 2 times as wide as thick, 1–4 mm wide; species of s portion of East TX, usually on sandy soils ____ **M. latidens**
1. Plants erect shrubs OR (in the rare, white-flowered *M. malacophylla* plants clambering, forming a tangle or a sprawling shrub); flowers white or pink; fruits glabrous or with recurved prickles limited to the margins; stems woody.
   7. Plants clambering in trees and forming a tangle, 2–3 m high, or in some cases a sprawling shrub; leaflets 5–14 mm long; mature fruits 45–75 mm long, 6–11 mm wide, glabrous; filaments white or whitish ____ **M. malacophylla**
   7. Plants rounded much-branched shrubs to ca. 1(–2) m tall; leaflets usually 5 mm or less long; mature fruits 15–50(–60) mm long, 3–8(–10) mm wide, typically but not always with recurved prickles on the margins; filaments pink, white tinged with pink, or whitish.
      8. Corolla lobes free nearly to base, longer than tube; mature fruits strongly constricted between the seeds, jointed and twisted when mature; twigs usually ± straight, the prickles mostly between the nodes; filaments pink, fading pale ____ **M. borealis**
      8. Corolla lobes shorter than tube; mature fruits not or only moderately constricted between the seeds, neither jointed nor twisted; twigs often zig-zag, prickles usually on or near nodes; filaments white, tinged with pink.
         9. Prickles usually paired on stem rib leading to stipule; corolla pubescent; fruits linear, 3–4 mm wide ____ **M. biuncifera**
         9. Prickles usually solitary, occasionally paired, opposite stem rib leading to petiole; corolla glabrous or puberulent; fruits oblong, 5–8 mm wide ____ **M. texana**

Mimosa biuncifera Benth., (bearing two hooks), CATCLAW, WAIT-A-BIT, WAIT-A-MINUTE. Rounded or spindly, often much-branched shrub 1–2 m tall; stems usually zig-zag in appearance; prickles usually paired near the nodes; leaves with 1–8 pairs of pinnae; leaflets 5–9(–12) pairs per pinna; inflorescences with 17–40 flowers, axillary, globose heads; peduncles 0.7–1.5 cm long; corolla tube as longer or longer than lobes; filaments white, tinged with pink; fruits 25–45 mm long, 3–4 mm broad, often with numerous, conspicuous, recurved prickles on the margins; seeds 3–8. Calcareous or sandy rocky open areas, also alluvial soils; sw margin of East TX; mainly w 1/2 of TX; sw U.S. in AZ, NM, and TX; widespread in Mexico. Apr–Aug. [*M. aculeaticarpa* Ortega var. *biuncifera* (Benth.) Barneby, *M. lindheimeri* A. Gray] This species has long been included as a variety of *M. aculeaticarpa* (e.g., Isely 1998; Diggs et al. 1999; Turner et al. 2003; Kartesz 2023), but we are following Grether (2023) who chose to treat it as a species. Isely (1998) discussed variation in pinnae number in this species and mapped populations where pinnae were predominantly 1–3 pairs per leaf versus other populations where pinnae were 4–8 pairs; the populations on the sw margin of East TX have 1–3 pairs; however, Isely (1998) mapped also populations in counties just west of East TX with 4–8 pairs of pinnae, as well as populations with 1–3 pairs of pinnae.

Mimosa borealis A. Gray, (northern), CATCLAW, FRAGRANT MIMOSA, PINK MIMOSA. Rounded much-branched shrub 0.5–2 m tall; stems ± straight, prickles mostly between the nodes; leaves with 1–3(–5) pairs of pinnae per leaf, leaflets (1–)2–7(–8) pairs per pinna; inflorescences with (5–)10–30 flowers, axillary, globose heads; flowers very fragrant; corolla lobes free nearly to base, the lobes definitely longer than the tube; filaments pink, fading pale, conspicuous; fruits 15–45(–60) mm long, 6–7.5(–10) mm broad, stipitate, jointed and twisted, eventually breaking into segments, often without spine-like structures on the margins, sometimes with occasional recurved prickles, glabrous; seeds (1–)2–7(–8). Rocky open woodlands or grasslands, often on limestone; sw margin of East TX; mainly w 1/2 of TX; sw U.S. in CO, KS, NM, OK, and TX. Mar–May(–Oct). [*Mimosa fragrans* A. Gray] *Mimosa borealis* is recognized by its heavily armed straight branched stems with internodal prickles and fragrant bright pink flower clusters (Isely 1998).

Mimosa hystricina (Small ex Britton & Rose) B.L. Turner, (porcupine-like, bristly), BRISTLY SENSITIVE-BRIAR. Stems sprawling or trailing, 0.5–2 m long, ribbed with recurved prickles along the ribs; pinnae 3–5 pairs per leaf; leaflets 11–17 pairs per pinna, obviously veined above and below, 4–7 mm long, 1.5–2.5 mm wide, the ultimate leaflets folding together at night or on being touched; peduncles 7–13(–20) cm long; inflorescences with 100–180 flowers, axillary, globose heads; petals united nearly halfway; filaments conspicuous, pink; fruits tetragonal, densely prickly, 20–45 mm long, 8–10 mm wide; seeds 5–14. Sandy open woodlands, prairies, and adjacent rights-of-way; se portion of East TX w to Brazos Co (Kartesz 2023); sc U.S. in sw LA and se TX. (Jan–)Feb–May(–Oct). [*M. quadrivalvis* L. var. *hystricina* (Small ex Britton & Rose) Barneby, *Schrankia hystricina* (Small ex Britton & Rose) Standl.] Isely (1998) noted that specimens of *M. hystricina* and *M. nuttallii* "that lack pods are often difficult to differentiate."

Mimosa latidens (Small) B.L. Turner, (broad-toothed), KARNES SCHRANKIA. Stems sprawling or trailing, 0.2–0.5 m long, stems ribbed with relatively weak recurved prickles on the ribs; pinnae with 1–3 pairs per leaf; petiole 2–4 cm long and longer than the primary rachis (0.7–2.5 cm long), leaflets 6–9 pairs per pinna, 3–6 mm long, 0.8–2 mm wide, margins ciliate, central vein only evident abaxially, the ultimate leaflets folding together at night or on being touched; inflorescences with 20–70 flowers, axillary, globose heads; petals united nearly halfway; filaments conspicuous, pink; fruits 25–60 mm long, 2.5–4 mm wide; seeds 6–9. Sandy open areas; s portion of East TX n to Robertson Co. (Kartesz 2023); also Gulf Prairies and Marshes and South TX Plains; sc U.S. in LA (disjunct in Acadia Par.) and TX; Mexico (Coahuila, Nuevo León, San Luis Potosí, Tamaulipas). Mar–Oct. [*M. quadrivalvis* L. var. *latidens* (Small) Barneby, *Schrankia latidens* (Small) K. Schum., *S. microphylla* of TX authors, not (Dryand.) J.F. Macbr.] The precise range of this species is somewhat muddled because of its similarity to *M. microphylla*, which is now recognized as occurring in se TX (Grether 2023). In TX, it is only known as far ne in a line from Robertson Co. se to Fort Bend Co. where it is replaced by *M. microphylla* eastward. Most LA specimens are *M. microphylla* except for one collection from Acadia Par. (*Brown 5823* LSU).

Mimosa malacophylla A. Gray, (soft-leaved), RASPILLA, SOFT-LEAF MIMOSA. Plants clambering in trees, viny, forming a tangle, to 2–3 m high or in some cases a sprawling shrub; stems ribbed, tomentulose or glabrous; prickles along ribs, recurved; pinnae 4–6 pairs, leaflets 4–6 pairs per pinna, 5–14 mm long, 2.5–6.5 mm wide, margins ciliate or eciliate, revolute, reticulate veins prominent abaxially and adaxially, apex mucronate, surfaces glabrous or puberulent; peduncles 0.8–1 cm long; inflorescences with 40–70 flowers, axillary, globose heads; corolla white, tomentulose or glabrous, lobes (4 or)5, filaments connate at bases, white; fruits stipitate, curved, oblong, 4.5–7.5 cm long, 6–11 mm wide, reticulate-veined, constricted between seeds; seeds 6–8. Scrub woodlands, thornscrub, caliche ridges, shrub thickets; Hays Co. (Kartesz 2023) in East TX; mainly in South TX Plains; Mexico (Coahuila, Nuevo León, Tamaulipas). Apr–Nov. [*Mimosa malacophylla* var. *glabrata* Benth.; *M. wootonii* Standl.]

Mimosa microphylla Dryand., (small-leaved), PRICKLY RED MIMOSA, LITTLE-LEAF SENSITIVE-BRIAR. Stems sprawling or trailing to 1 m long, ribbed, glabrous or puberulent; weak prickles along ribs, recurved; petiole 3–5 cm long, primary rachis 6–8.5 cm long; pinnae (3–)5–8 pairs, leaflets 10–15(–17) pairs, 2.5–5 mm long, 0.5–1.5 mm wide, margins ciliate, 1 central vein evident abaxially, peduncles 2–5(–10) cm long; inflorescences with 60–120 flowers, axillary, globose heads; corolla purplish pink, glabrous, lobes 5; fruits sessile, straight, linear, tetragonal, 90–140 mm long, 1.5–4 mm wide; seeds ca. 18. Sandy open woodlands, fields, and roadsides; in TX only in se portion from Montgomery, Waller, and Walker cos. eastward; se U.S. from VA w to IL and s to East TX. Apr–Aug(–Jan). [*M. quadrivalvis* L. var. *angustata* (Torr. & A. Gray) Barneby, *Morongia uncinata* (Willd.) Britton, *Schrankia angustata* Torr. & A. Gray, *S. uncinata* Willd.] Grether (2023) listed only Waller Co. in TX, but this species is commonly found throughout se TX in sandy soils. Isely (1998), Weakley, A.S. and Southeastern Flora Team (2022), and Kartesz (2023) map the species from e LA eastward, but it is the most common small-flowered *Mimosa* in sw LA as well (see discussion under *M. latidens*).

Medicago minima [BEN]

Medicago obicularis [MOR]

Medicago polymorpha [GLE]

Medicago sativa [BB2, GLE]

Melilotus albus [BEN]

Melilotus indicus [MOR]

Melilotis officinalis [BL3]

Mimosa biuncifera [VIN]

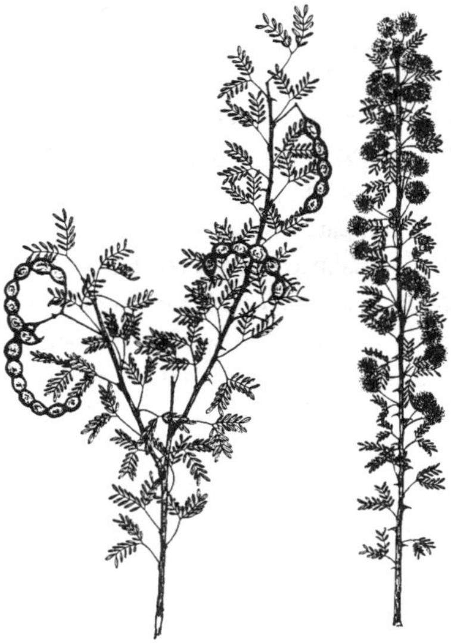

Mimosa borealis [VIN]

Mimosa nuttallii (DC. ex Torr. & A. Gray) B.L. Turner, (for Sir Thomas Nuttall, 1786–1859, English-American botanist), CATCLAW SENSITIVE-BRIAR, CATCLAW SCHRANKIA, SHAME-BOY, NUTTALL'S SENSITIVE-BRIAR. Stems sprawling or trailing, 0.5–1.2 m long; pinnae usually 4–8 pairs per leaf; leaflets 11–16 pairs per pinna, reticulate veins prominent, especially abaxially, the ultimate leaflets folding together at night or on being touched; peduncles 3.5–9.5 mm long; inflorescences with 90–140 flowers; petals united nearly halfway; filaments conspicuous, pink, the filaments of all flowers in a head forming a dense pink ball; fruits 4–12 cm long, 4-sided or nearly terete, densely or sparsely prickly; seeds 4–15. Open woods, prairies, and roadsides; widespread in TX except in far se, sw and s. Apr–Jun. [*M. quadrivalvis* L. var. *nuttallii* (DC. ex Torr. & A. Gray) Beard ex Barneby, *Schrankia nuttallii* (DC. ex Torr. & A. Gray) Standl., *Schrankia uncinata* of TX authors, not Willd.] Bernhardt (1990) studied the pollination ecology of this species. The flowers produce fragrance, but no nectar, with pollen serving as the nutritive reward for the bee pollinators. Isely (1998) noted that this is the "most widespread and distinctive" member of the *M. quadrivalis* complex. He also indicated that to the south this taxon possibly merges with *M. hystricina*.

Mimosa roemeriana Scheele, (for Ferdinand Roemer, 1818–1891, geologist, paleontologist, and explorer of TX), ROEMER'S SENSITIVE-BRIAR, ROEMER'S SCHRANKIA. Stems sprawling or trailing, 0.3–1 m long, ribbed, glabrous or puberulent, prickles along ribs, recurved; pinnae 2–5 pairs per leaf, leaflets 8–12 pairs per pinna, 4–9.5 mm long, 1–3 mm wide, central vein evident abaxially, the ultimate leaflets folding together at night or on being touched; inflorescences with 100–180 flowers, axillary, globose heads; petals united nearly halfway; filaments conspicuous, pink; fruits 40–70(–100) mm long, 3–5(–7) mm wide, flattened, densely prickly; seeds 12–18. Dry rocky prairies, limestone slopes, occasionally on sandstone; w margin of East TX; mainly Cross Timbers and Prairies and Edwards Plateau; s OK; Mexico (Nuevo León, Tamaulipas). Mar–Jul. [*M. quadrivalvis* L. var. *platycarpa* (A. Gray) Barneby, *Schrankia roemeriana* (Scheele) Blank.]

Mimosa strigillosa Torr. & A. Gray, (somewhat strigose, with stiff bristles), POWDERPUFF, VERGONZOSA, HERBACEOUS MIMOSA. Stems sprawling, often forming mats, 0.1–0.5 m long, with numerous stiff bristle-like structures, not painful to the touch; pinnae 3–7 pairs per leaf; leaflets (8–)10–15(–19) pairs per pinna, 3–5.5(–8) mm long, 0.5–1.5 mm wide; inflorescences with 80–140 flowers, axillary, subglobose heads; filaments pink to lavender-pink or purple-pink; fruits straight, obliquely oblong, 10–28 mm long, 6–9 mm wide, constricted between seeds; seeds 1–4. Open areas including lawns and roadsides; scattered mainly in e 1/2 of TX; se U.S. from FL w to e OK and TX; Mexico (Tamaulipas, Veracruz) and South America. Apr–Oct.

Mimosa texana (A. Gray) Small, (of Texas), TEXAS MIMOSA. Erect shrub 0.3–2 m tall; stems zig-zag; prickles solitary or occasionally paired, opposite rib leading to petiole; leaves with 3–7 pairs of pinnae; leaflets 3–14 pairs per pinna, 1.5–4.5 mm long, 0.5–2 mm wide; inflorescences with 10–30 flowers, axillary, globose heads; corolla with 5 lobes, filaments white; fruits 2–4 cm long, 5–8 mm wide; seeds 3–6. Hays and Travis (Kartesz 2023) cos. at extreme sw margin of East TX; in TX mainly in extreme s and sw parts; sw U.S. in NM and TX; Mexico (Coahuila, Durango, Nuevo León, San Luis Potosí, Sonora, Tamaulipas). Mar–Jul. [*M. borealis* A. Gray var. *texana* A. Gray] The position of the prickles opposite the rib leading to the petiole distinguishes this species from *M. biuncifera* when fruits are not evident (Isely 1998).

Mimosa monclovensis R. Grether & Marc. F. Simon is mapped in Weakley, A.S., and Southeastern Flora Team (2022) as occurring in East TX probably based on confusing language in Grether and Simon (2018) that seems to list the range of the species from sw LA sw to Coahuila, Mexico, however, that range description is for *M. latidens*. Grether (2023) describes the species as occurring in Coahuila and TX and "in southern TX southward from San Antonio according to Barneby (1991)." Barneby (1991) describes its range (treated as *M. quadrivalvis* var. *nelsonii*) as "apparently local from San Antonio sw into ne Coahuila." This range is apparently based on a specimen from Duval Co. (*Croft 154* NY) in s TX since no other TX specimens are listed. Turner (1994e) mapped the species (treated as *Mimosa subinermis* (S. Watson) B.L. Turner, an illegitimate name) as only occurring in n Mexico and indicated that it only grows on calcareous substrates. Isely (1998) mirrored the treatment and range of Barneby (1991). Based on review of this literature and examination of specimens, it appears that *M. monoclovensis* rarely occurs in s TX on calcareous substrates but has not been documented in and probably does not occur in East TX.

NEPTUNIA Lour. PUFF

Prostrate, unarmed, perennial herbs; taproot thickened, woody; stems annual; leaves twice even-pinnately compound, touch sensitive; stipules lanceolate; inflorescences peduncled axillary heads, 15–60-flowered, rounded to slightly cylindrical, in some species flowers all with anther-bearing stamens, in others flowers in upper part of head with stamens bearing anthers, the lower flowers smaller and with sterile, flattened, petal-like staminodes; sepals and petals 5 each; stamens usually 10, separate, yellow, the stamens and/or staminodes serving as the main attractant structures for the flower; fruits stipitate, oblong, flattened; seeds 1–20.

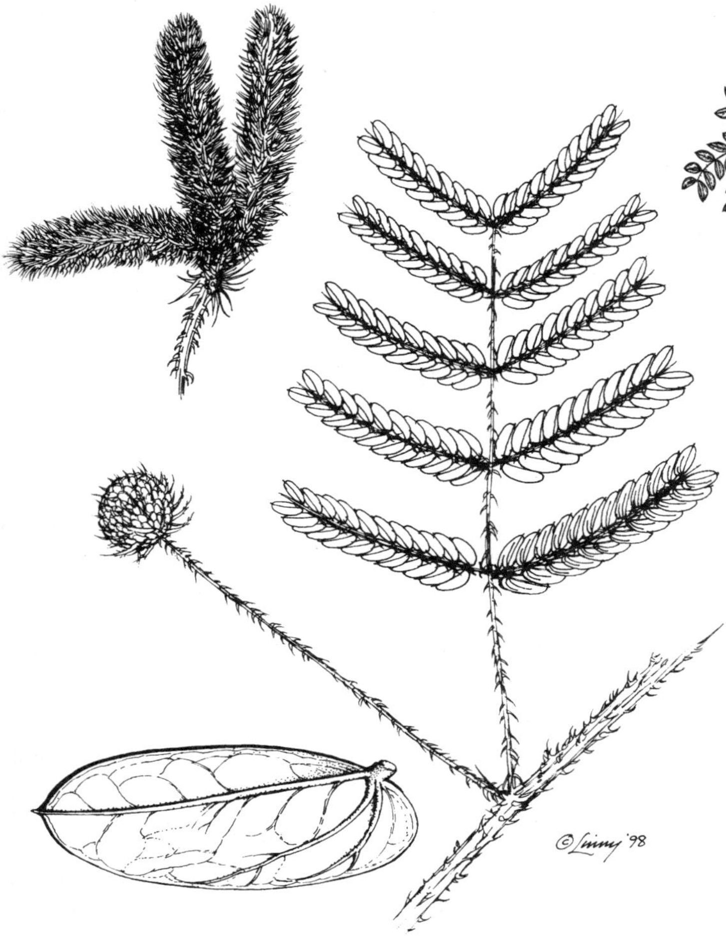

Mimosa hystricina [HEA]

Mimosa malacophylla [VIN]

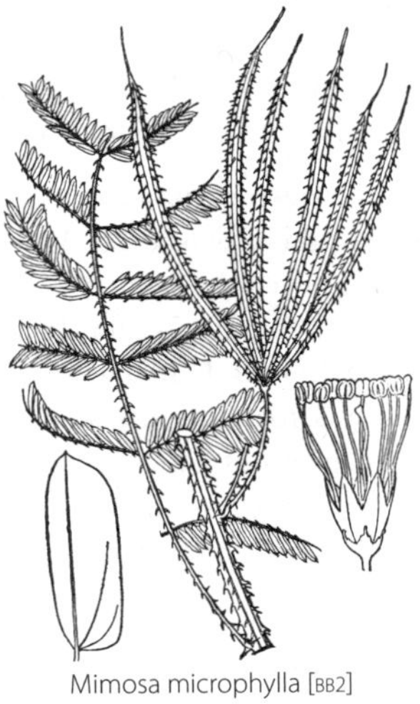

Mimosa microphylla [BB2]

Mimosa latidens [HEA]

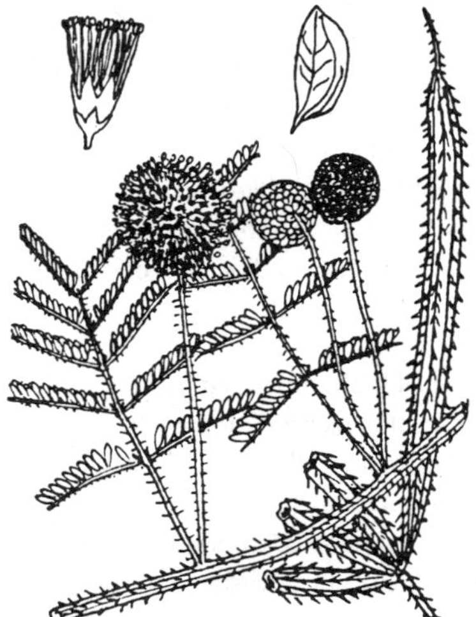

Mimosa nuttallii [RYD]

A genus of ca. 12 species native to tropical and warm areas of the world, especially Australia and the U.S. (Nesom 2023d). As in some *Mimosa*, the leaves of several *Neptunia* species are touch sensitive (leaflets instantly folding together on being touched); this is possibly an adaptation to reduce water loss (via transpiration) or a defense against herbivores. The leaves also exhibit nyctinastic (nighttime or "sleep") movements with leaflets folding together at night. (Named for Neptune, in Roman religion a god who had to do with perpetuity of springs and streams; sometimes referred to as god of the seas; so named because some species are aquatic) (subfamily Caesalpinioideae, Mimosoid clade, tribe Mimoseae)

REFERENCES: Turner 1951; Isely 1970a, 1973; Nesom 2023d.

1. Flowers all with stamens alike and bearing anthers; heads with 30–60 flowers; stipe of fruits 4–14 mm long; leaflets 8–18 pairs per pinna; calyces 1–2 mm long ______ **N. lutea**
1. Flowers in upper part of head with stamens bearing anthers, the lower flowers smaller and with sterile, flattened, petal-like staminodes; heads with 15–30 flowers; stipe of fruits 1–4 mm long; leaflets (8–)14–40(–43) pairs per pinna; calyces 2–2.7 mm long.
   2. Legumes 5–13 mm wide, base usually tapering to stipe; stipe 2–4 mm long, usually longer than calyx; pinnae (2–)3–6 pairs per leaf, blade margins ciliate ______ **N. pubescens**
   2. Legumes 8–17 mm wide, base usually rounded to stipe; stipe 1–2(–3) mm long, ± equal to calyx; pinnae 2 or 3(–4) pairs per leaf, blade margins eciliate ______ **N. microcarpa**

Neptunia lutea (Leavenw.) Benth., (yellow), YELLOW-PUFF, YELLOW NEPTUNIA. Perennial, unarmed (in contrast to the armed sensitive species of *Mimosa* previously treated in *Schrankia*); stems vine-like, usually densely and minutely pubescent, to 2 m long; foliage blue-green; leaves with 2–11 pairs of pinnae, leaflets 16–36 per pinna, blades 3.5–6 mm long, with raised-reticulate venation abaxially, margins ciliate, sensitive—folding together on being touched; flowers in ovoid or shortly cylindric heads ca. 2 cm long on erect, axillary peduncles; calyces 1–2 mm long, petals separate almost or quite to base, 2–3 mm long; stamens bright yellow; fruits 2.5–5 cm long, 10–15 mm wide; seeds 1–6. Diverse open habitats (some preference for sand or sandy clay); widespread in East TX (less in the Pineywoods); in much of the e 1/2 of TX; se U.S. from AL w to s KS, OK, and TX. Apr–Jun, sporadically later.

Neptunia microcarpa Rose, (small-fruited), RIO GRANDE PUFF. Herb, stems erect to decumbent or prostrate; petiole 1.5–2 cm long, eglandular, pinnae 2 or 3(–4) pairs; leaflets 30–80 per pinna, blades 4–6.5 mm long, with raised-reticulate venation abaxially, margins eciliate; peduncles 2.5–11 cm long, spikes 20–30-flowered, spheric to ovoid; bracts (0–)1 or 2, linear-subulate, 1–3 mm long, 1–2 mm wide; flowers dimorphic, proximal with flattened, petaloid staminodes, distal anther-bearing; calyces 2–2.7 mm long; fruits 8–17 mm wide, base usually rounded to stipe; stipe 1–2(–3) mm long, ± equal to calyx; seeds 5–9. Streamsides, floodplains, roadsides, often on limestone or silty gravel. Bexar and Comal cos. in sw portion of East TX (several collections; E. Keith, pers. obs.), mostly sw TX; also Mexico (Coahuila, Nuevo León, Tamaulipas). May–Aug(–Oct). [*N. palmeri* Britton & Rose, *N. pubescens* Benth. var. *microcarpa* (Rose) Windler] *Neptunia microcarpa* was treated as conspecific with *N. pubescens* by Windler (1966) and Isely (1998). Morphological differences are consistent, and the two taxa are disjunct in geographical range and different in ecology (Nesom 2023d). Turner (1951) and Turner et al. (2003) treated each at specific rank.

Neptunia pubescens Benth., (pubescent, downy), TROPICAL NEPTUNIA, TROPICAL-PUFF, PRAIRIE NEPTUNIA, SAND NEPTUNIA. Stems erect to decumbent, glabrate or inconspicuously pubescent; petiole 1.2–3 cm long, eglandular; pinnae (2–)3–6 pairs; leaflets (16–)28–60(–86) per pinna, blades 2.3–8.5 mm long, with raised-reticulate venation abaxially, margins ciliate, sensitive; heads round or ovoid, 15–30-flowered; stamens and strap-like or petal-like staminodes yellow, the staminodes ca. 3 mm long, often lost as inflorescence matures; fruits 1–4 cm long; 5–13 mm wide; seeds 4–11. Diverse open habitats including roadsides and lawns; scattered in s 1/2 of East TX n to Freestone Co. (Turner et al. 2003); mainly Gulf Prairies and Marshes and South TX Plains. Apr–Sep(–Oct). [*N. floridana* Small, *N. pubescens* var. *floridana* (Small) B.L. Turner, *N. pubescens* var. *lindheimeri* (B.L. Rob.) B.L. Turner] The heteromorphic flowers of *Neptunia pubescens*, *N. macrocarpa*, and other legume genera as well, result mostly from suppression and modification of structures after they are initiated rather than complete loss of structures (Tucker 1988).

ORBEXILUM Raf. SNAKEROOT

Perennial herbs; unarmed; with lignescent rhizome, tuber, or fusiform taproot; stems erect to ascending, usually gland-dotted, pubescent or glabrous; leaves alternate, unifoliolate, palmate, or odd-pinnate; stipules present, persistent or caducous; petiolate; stipels absent; leaflets 1–3(–7), blade margins entire, surfaces glandular-punctate or eglandular, pubescent or glabrous; inflorescences pedunculate, 5–50-flowered, terminal, spicate; bracts present; bracteoles absent; flowers papilionaceous; calyces tubular-campanulate, lobes 5, lobes 2–3 times tube length; corollas violet, purple, or purplish blue, banner, wings, and keel well-developed, keel connate apically; stamens 10, diadelphous or proximally monadelphous; anthers in 2 series, proximal dorsifixed, distal basifixed, introrse; fruits loments, 0.4–1.2 cm long, indehiscent, thickly leathery, rugose or papillose, glandular-punctate or eglandular, glabrous; seeds 1.

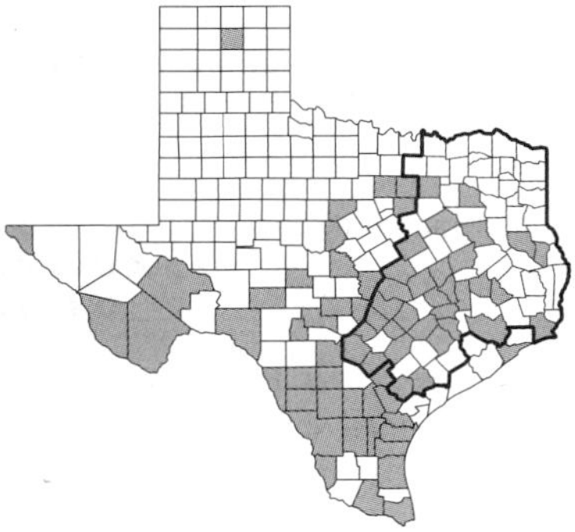
Melilotus indicus

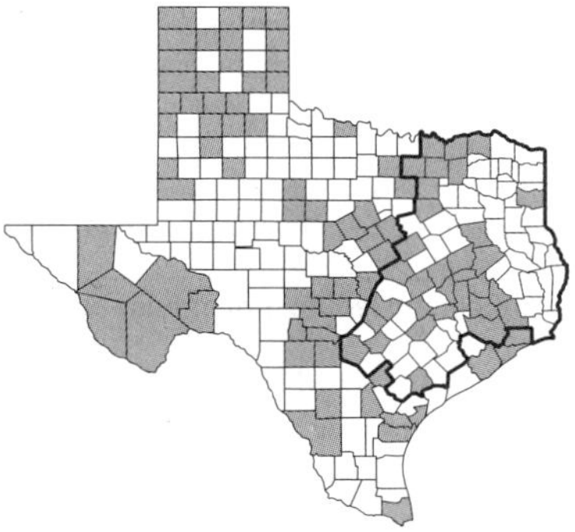
Melilotus officinalis

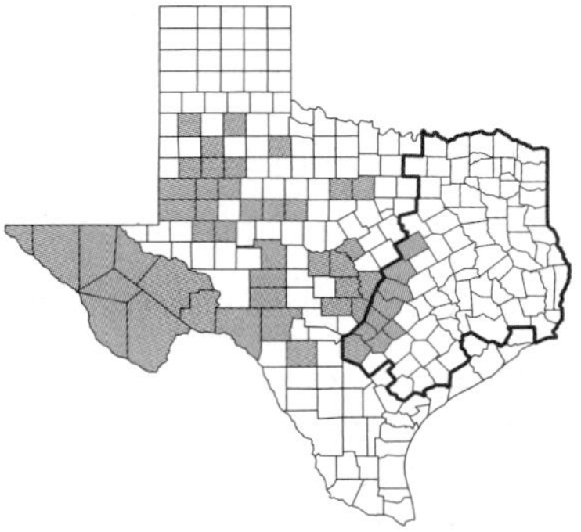
Mimosa biuncifera

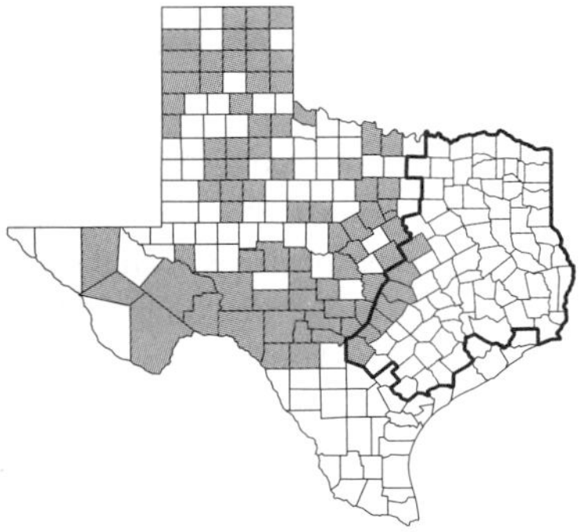
Mimosa borealis

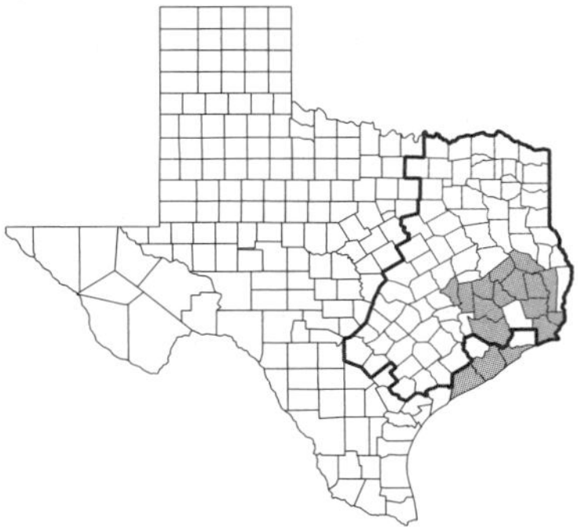
Mimosa hystricina

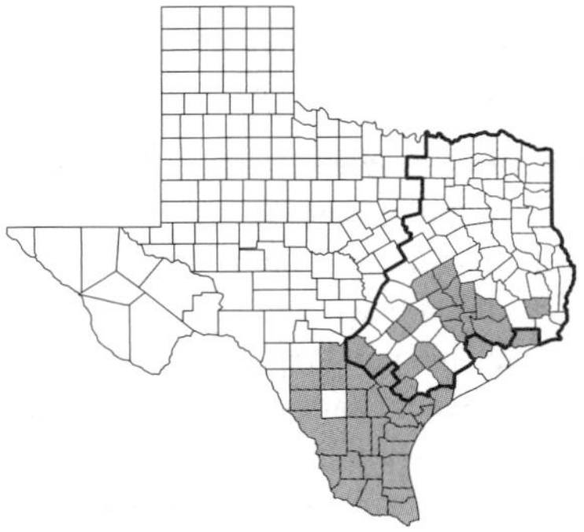
Mimosa latidens

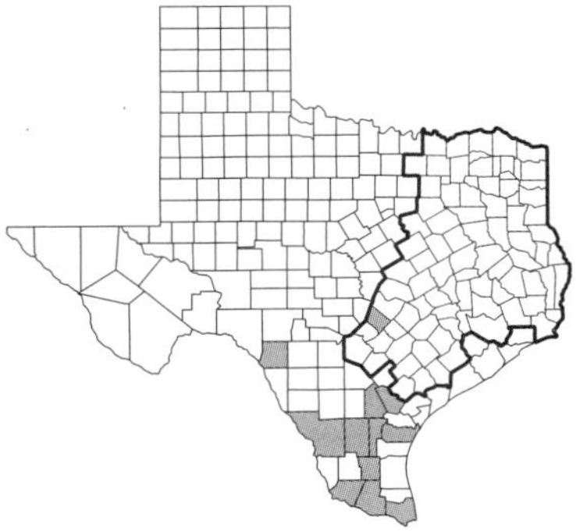
Mimosa malacophylla

Mimosa microphylla

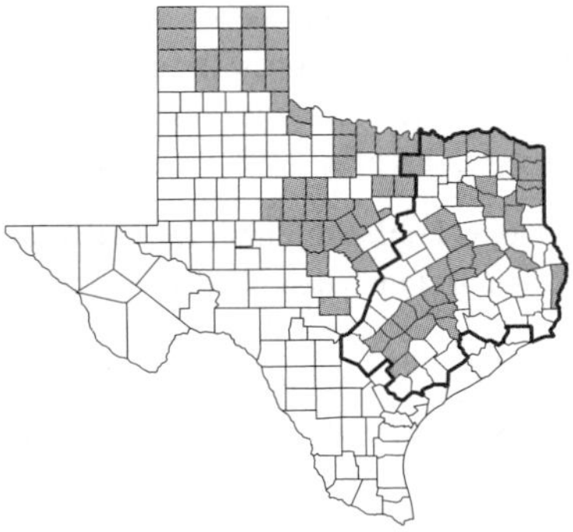
Mimosa nuttallii

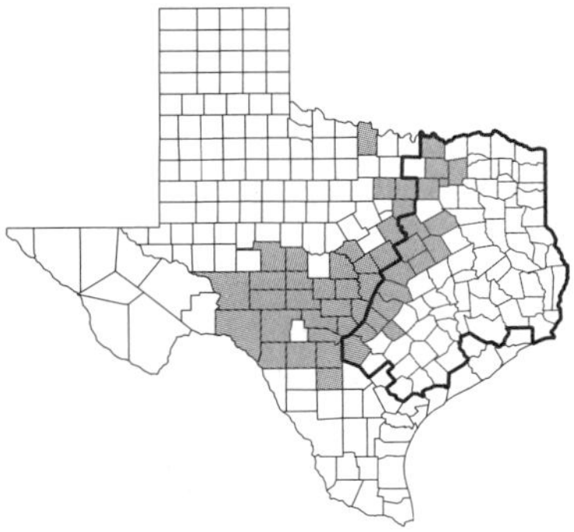
Mimosa roemeriana

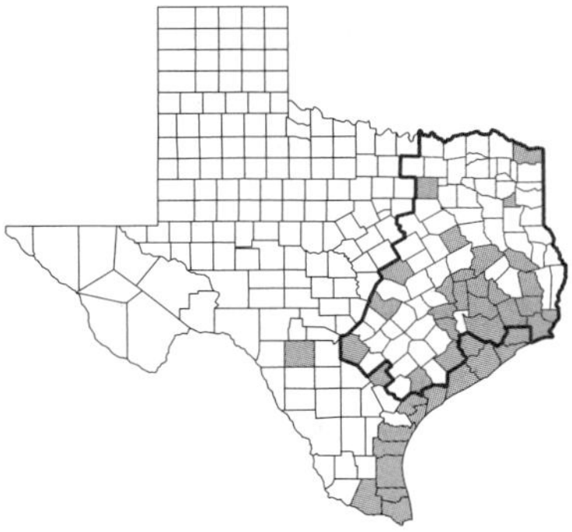
Mimosa strigillosa

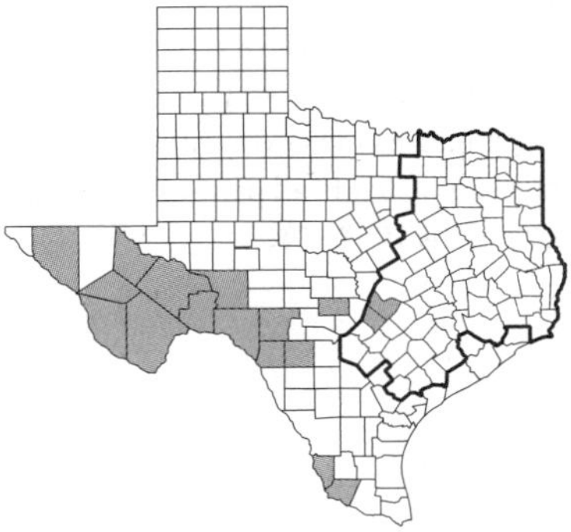
Mimosa texana

A genus of 11 species native to the U.S. and Mexico; it was previously placed in *Psoralea* (Turner 2023b). (Latin: *orbis*, circular, and *exilum*, spreading or vexillum, alluding to the shape of the banner in *O. latifolium*, the type species—Turner 2023b) (subfamily Faboideae, tribe Psoraleeae)
REFERENCES: Isely 1986a; Grimes 1990; Turner 2023b.

1. Corollas purple, grayish lavender or lilac (rarely white), 5–7 mm long; bracts, calyces, and legumes eglandular ______ **O. pedunculatum**
1. Corollas deep purple, 8–10 mm long; bracts, calyces, and legumes glandular-punctate ______ **O. simplex**

Orbexilum pedunculatum (Mill.) Rydb., (with a flower stalk), SAMPSON'S SNAKEROOT, BOBSROOT SNAKEROOT. Plant erect, 0.6–0.8 m tall; leaves trifoliate, leaflets 20–70 mm long, 6–22 mm wide, elliptic-lanceolate to lanceolate, the terminal leaflet usually not much larger than lateral leaflets, surfaces eglandular or sparsely glandular-punctate; peduncles 4–16 cm long; floral bracts and calyces eglandular; flowers various shades of lavender or rarely white, 5–7 mm long, calyces 4–7 mm long; fruits 3–5 mm long, 3–4 mm wide. Damp or dry sandy open ground and open woods; widespread in e part of East TX w to Lamar (BRIT), Austin, and Brazos (Kartesz 2023) cos.; in TX known only from East TX; e U.S. from OH se to n FL and w to se KS, e OK, and TX. Apr–Jun. [*O. pedunculatum* var. *eglandulosum* (Elliott) Isely, *Psoralea psoralioides* (Walter) Cory var. *eglandulosa* (Elliott) F.L. Freeman]

Orbexilum simplex (Nutt. ex Torr. & A. Gray) Rydb., (simple, unbranched), SINGLE-STEM SNAKEROOT. Plant erect, to 1 m tall, minutely and inconspicuously pubescent; leaves usually trifoliate, but occasionally unifoliate or palmately 5–7 foliate, leaflets 40–80 mm long, 6–15 mm wide, linear-lanceolate or lanceolate, the terminal leaflet often conspicuously larger than the lateral leaflets; floral bracts and calyces purple, gland-dotted; flowers deep purple, 8–10 mm long, calyces 5–7 mm long; fruits suborbicular, 4–6 mm long, 3–5 mm wide, glandular–punctate. Damp sandy woods or open ground; widespread in e part of East TX w to Brazos (Kartesz 2023) and Limestone (BRIT) cos.; in TX known only from East TX; sc U.S. in AL, AR, LA, MS, se OK, and TX. May–Jun. [*Psoralea simplex* Nutt. ex Torr. & A. Gray] This species is readily distinguished from *O. pedunculatum* by its showy deep purple flowers that often grow in large colonies in East TX (E. Keith, pers. obs.).

OXYTROPIS DC. LOCOWEED, PURPLE LOCO, CRAZYWEED

A genus of ca. 300 species of the n temperate and arctic, particularly c Asia; some are harmful to livestock and a number are cultivated as ornamentals; the genus is sometimes lumped into *Astragalus* (Welsh 2023b). (Greek: *oxys*, sharp, and *tropis*, keel, alluding to the keel of the flowers) (subfamily Faboideae, tribe Galegeae)
REFERENCES: Barneby 1952; Welsh 2023b.

Oxytropis lambertii Pursh var. **articulata** (Greene) Barneby, (sp.: for Aylmer Bourke Lambert, 1761–1842, from whose cultivated plants Pursh described it, var.: articulated, jointed), LOCOWEED, PURPLE LOCO, LAMBERT'S CRAZYWEED, WHITE LOCO, LAMBERT'S LOCO, CRAZYWEED, POINT LOCO. Perennial pubescent herb, often forming colonies; plant tufted, ± scapose, erect, to 35 cm tall; pubescence of hairs attached above base so that they have a long and a short arm (= malpighian or dolabriform hairs); leaves once odd-pinnately compound, often dimorphic; leaflets 7–19 per pinna, sometimes falcate; stipules 7–24 mm long; peduncles 10–24 cm long; flowers 6–28 per raceme, with sweet carnation scent; corollas 15–26 mm long, light to deep purple-red to almost white, withering to blue-violet; fruits 7–10(–17) mm long, the beak 3–7 mm long, strigose. Limestone outcrops and prairies; w margin of East TX; mostly in the High and Rolling Plains and scattered locations in the Cross Timbers and Prairies; sc U.S. in KS, NM, OK, and TX. Apr–Jun. [*Astragalus lambertii* (Pursh) Spreng. var. *abbreviatus* (Greene) Shinners] Two additional varieties: var. *lambertii* and var. *bigelovii* A. Gray occur to the n and w of TX (Welsh 2023b). As in some *Astragalus* species, toxins (apparently indolizidine alkaloids—e.g., swainsonine) can cause fatal loco poisoning or locoism in livestock; horses are particularly susceptible; symptoms include trembling, abortion, tendency to become excessively excited and wild, and eventually, inability to eat, paralysis, and death; all parts of the plant are toxic and the poison is cumulative (Kingsbury 1964; Burlage 1968; James & Welsh 1992; Ralphs 1992).

PARKINSONIA L. PALOVERDE

A genus of ca. 12 species of dryer parts of the Americas and Africa; it is sometimes treated as including the related genus *Cercidium* (Krings et al. 2023). (Named for John Parkinson, London apothecary and botanical author, 1567–1650) (subfamily Caesalpinioideae, tribe Caesalpinieae)

Parkinsonia aculeata L., (prickly), RETAMA, PALOVERDE, HORSE-BEAN, JERUSALEM-THORN, MEXICAN PALOVERDE. Shrub or small tree to 10 m tall; branches green, photosynthetic, well-armed with sharp spines; leaves twice pinnately compound with 2(–4) wispy pinnae (each with leaflets), but each pinna appearing as a once pinnate leaf due to the absence of a common petiole; pinnae with

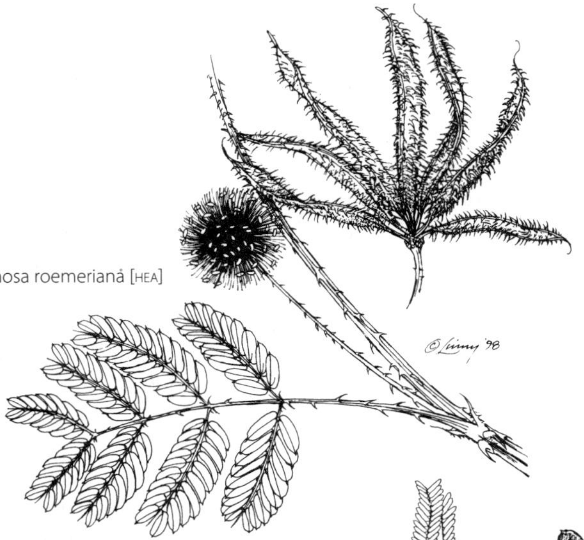

Mimosa roemeriana [HEA]

Mimosa strigillosa [COC]

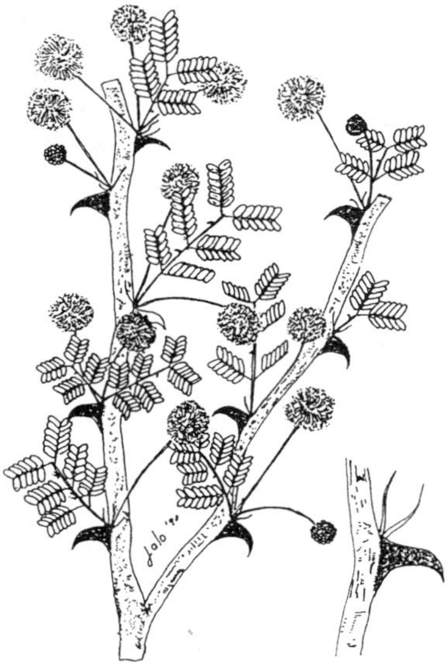

Mimosa texana [LNL]

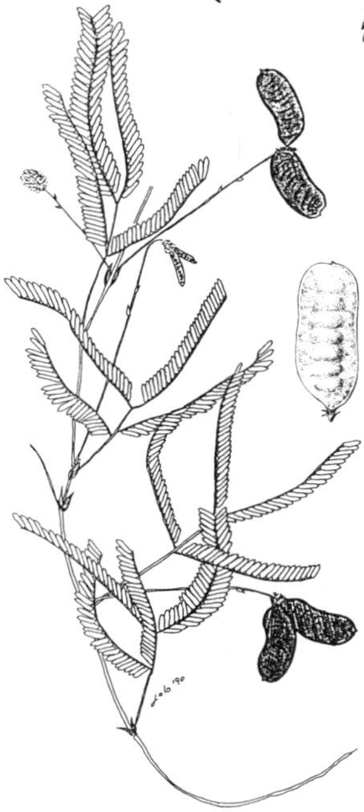

Neptunia microcarpa [AUJ]

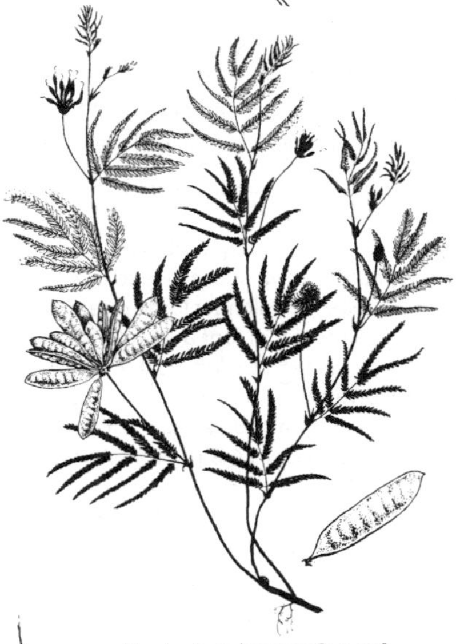

Neptunia pubescens [AUJ, LNL]

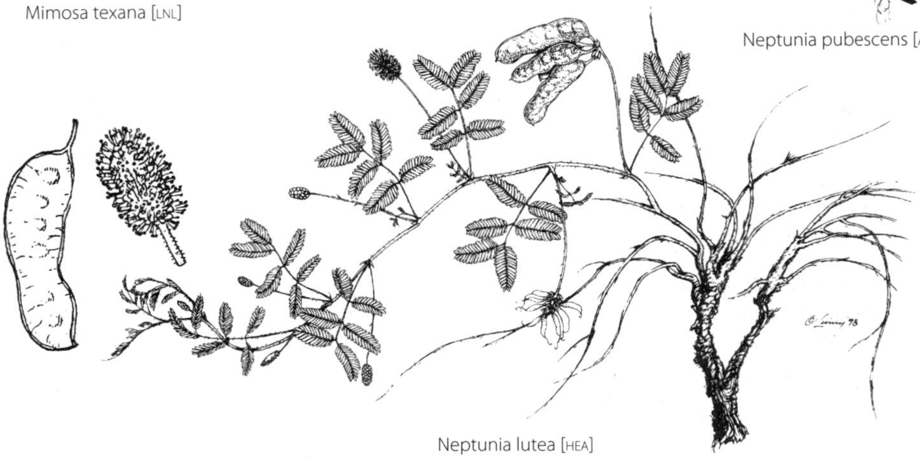

Neptunia lutea [HEA]

elongate, flattened, green rachis (persistent and photosynthetic under dry conditions after leaflets have fallen) and tiny leaflets; leaflets numerous, 2–8 mm long, drought deciduous; inflorescences racemes with 2–15 flowers; petals 8–13 mm long, yellow, the standard spotted with orange towards base, aging reddish orangish; stamens 10, separate, inconspicuous; fruits usually (2–)5–10 cm long, 4–6 mm wide, constricted between the seeds. Low poorly drained habitats, open woods, roadsides, and ruderal areas; s 1/2 of East TX, scattered further n; widespread mainly in s 1/2 of TX; also spreading from cultivation in Tarrant and Brown (Kartesz 2023) cos.; s U.S. from SC w to CA; Mexico; introduced in South America; Asia; Africa; and Australia. Mar–Sep. Native of tropical America n to TX; Correll and Johnston (1970) indicated this species is perhaps adventive in TX and possibly native to South America; Elias (1989) said it is native from s TX s through Mexico to South America. This species was used in a famous forensic botany case in 1993. A PALOVERDE tree was instrumental in linking a murder suspect to an Arizona crime site where the suspect allegedly dumped the body of a victim. Plant geneticist Tim Helentjaris of the University of Arizona demonstrated that two seed pods found in the back of suspect's truck came from a specific PALOVERDE tree scraped by the truck at the crime scene. This example is important because it was the first in which plant DNA was used in a criminal case (Mestel 1993; Yoon 1993). This species can be a problematic invasive exotic in other parts of the world (e.g., Australia).

PEDIOMELUM Rydb. SCURF-PEA

Perennials with tough or swollen (tuberous) root, caulescent or acaulescent; leaves once pinnately or palmately compound usually with 3–7 leaflets, often gland-dotted; inflorescences 3–many-flowered in axillary pseudoracemes; bracts present; calyces campanulate; corollas blue to lavender-purple, brick-red, or salmon-pink (rarely pink or white); stamens 10, monadelphous; fruits gland-dotted or eglandular, not cross-wrinkled, enclosed by enlarged, often gland-dotted calyces, dehiscent by transverse rupture and with the base of the fruit persistent on the receptacle; beak of fruit prominent, not cross-wrinkled; seed 1.

A genus of 25 species native from c Mexico n through the U.S. to sc Canada (Grimes 1990; Isely 1998); it was previously placed in *Psoralea* (Egan & Reveal 2023). Apart from *P. hypogaeum* and *P. rhombifolium*, East TX species usually have conspicuously gland-dotted leaves, fruits, calyces, and bracts. (Greek: *pedion*, plain, and *melo*, apple) (subfamily Faboideae, tribe Psoraleeae)
REFERENCES: Shinners 1951d; Ockendon 1965; Isely 1986a; Grimes 1990; Egan and Reveal 2023.

1. Leaves pinnately compound with 3 leaflets, not gland-dotted; stems normally prostrate; petals brick-red to salmon-pink (rarely pink) ____ **P. rhombifolium**
1. Leaves palmately compound with 3–7 leaflets, gland-dotted (except in *P. hypogaeum*); stems erect or ascending or plants acaulescent; petals blue to lavender or purple.
   2. Inflorescence a slender loose raceme or interrupted, spike-like raceme, < 2 cm thick; calyces 2–9 mm long; bracts up to 5 mm long; root thick, elongate.
      3. Flowers 5–6 mm long; calyx tube 1–1.5 mm long; pericarp (= fruit wall) thin, usually papery ____ **P. tenuiflorum**
      3. Flowers 8–11 mm long; calyx tube 2–4 mm long; pericarp thick, coriaceous (= leathery).
         4. Plants eglandular except for upper surfaces of leaflets; pedicels 1–3 mm long; petioles of middle and upper stem leaves 6 mm or more long (often much more); bracts 2–10 mm long; leaflets 4–8 times as long as wide ____ **P. digitatum**
         4. Plants gland-dotted on both leaf surfaces, stipules, bracts, and calyces; pedicels 3.5–10 mm long; petioles of middle and upper stem leaves 5 mm or less long; bracts 1.5–3.5 mm long; leaflets 10 times or more as long as wide ____ **P. linearifolium**
   2. Inflorescence a dense, spike-like raceme 2–4 cm thick; calyces 8–17 mm or more long; bracts 4–15 mm or more long; root thick, elongate or often tuberous-enlarged, sometimes conspicuously so.
      5. Leaves all basal or nearly so at flowering time; fruits long-beaked (beaks 7–19 mm long); foliage eglandular or inconspicuously gland-dotted ____ **P. hypogaeum**
      5. Leaves distributed well up the stem; fruits short-beaked (beaks 2–6 mm long); foliage usually conspicuously gland-dotted.
         6. Herbs 5–45 cm tall; peduncles shorter than or equal to subtending petioles; stipules 2.5–7 mm wide, elliptic-lanceolate to falcately oblanceolate or orbicular, the largest often in the middle of the plant ____ **P. latestipulatum**
         6. Herbs to 100 cm or taller; peduncles longer than subtending petioles; stipules (except quickly deciduous ones of very basal leaves) usually 2.5 mm or less wide, linear-lanceolate to lanceolate.
            7. Stems erect; leaflets (except those of quickly deciduous basal leaves) linear-lanceolate or narrowly oblong-lanceolate, 7–14 times as long as wide, 35–95 mm long, 3–9(–13) mm wide ____ **P. cyphocalyx**
            7. Stems decumbent to erect-ascending; leaflets broadly lanceolate to rhombic, elliptic, or obovate, 2–3(–5) times as long as wide, 20–50 mm long, (6–)10–24 mm wide ____ **P. cuspidatum**

Pediomelum cuspidatum (Pursh) Rydb., (with a sharp stiff point), TALL-BREAD SCURF-PEA. Stems decumbent or ascending, occasionally erect, to 100 cm long, freely branched, glandular throughout, strigose; leaves palmately 3–5-foliate, gland-dotted, leaflets broadly lanceolate to rhombic, elliptic, or obovate, 2–3(–5) times as long as wide, 20–50 mm long, (6–)10–24 mm wide; peduncles 0.6–15 cm long, longer than subtending petiole; flowers often with sweet-clover scent, 12–22 mm long; petals blue

Orbexilum pedunculatum [BB2]

Orbexilum simplex [HEA]

Oxytropis lambertii [REE]

Parkinsonia aculeata [SUD]

Pediomelum cuspidatum [BB2, BB3]

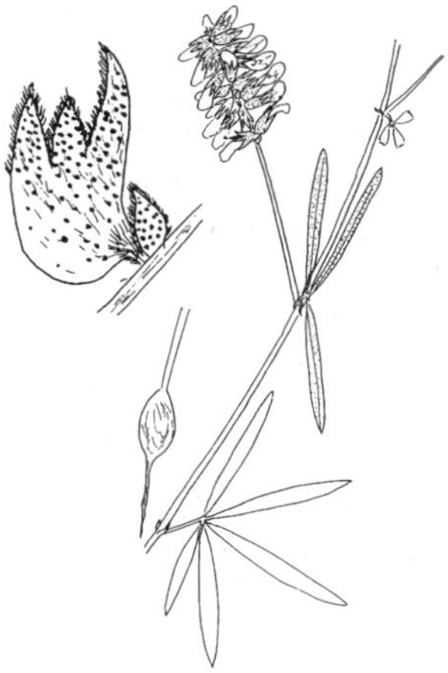
Pediomelum cyphocalyx [SHI]

Pediomelum digitatum [BB2]

Pediomelum hypogaeum var. scaposum [BT3]

Pediomelum hypogaeum var. subulatum [BT3]

or purplish; fruit body 6–8 mm long, gland-dotted, completely enclosed in calyces, the beak to 2 mm long. Clayey, rocky, or sandy prairies and open woodlands; w 1/2 of East TX; widespread in TX except far e, far s, and far w; c U.S in CO, KS, MT, NE, OK, SD, and TX. Apr–May. [*Psoralea cuspidata* Pursh] The horizontal stems are often bicolored, purple adaxially and light green abaxially (Egan & Reveal 2023).

Pediomelum cyphocalyx (A. Gray) Rydb., (bent or curved calyx), TURNIP-ROOT SCURF-PEA. Stems usually solitary, erect to 1 m tall, simple or sparingly branched above; leaves palmately 3–5-foliate, gland-dotted, leaflets linear-lanceolate or narrowly oblong-lanceolate, 7–14 times as long as wide, 35–95 mm long, 3–9(–13) mm wide; peduncles 2.5–7.5 cm long, longer than subtending petiole, strigose; flowers 12–18 mm long, calyces strongly gibbous (= inflated on one side) in fruit, 6–12 mm long; petals pale blue to lavender; fruit body 5–6 mm long, completely enclosed in calyx, glabrous, gland-dotted, the beak pubescent, 2–4 mm long. Limestone outcrops and open woodlands on limestone; w margin of East TX; also adjacent Cross Timbers and Prairies and e Edwards Plateau; endemic to a narrow n-s band in TX. Apr–Jun. [*Psoralea cyphocalyx* A. Gray] This narrowly endemic species is distinguished from similar species by its narrow leaflets and slender, erect habit (Egan & Reveal 2023).

Pediomelum digitatum (Nutt. ex Torr. & A. Gray) Isely, (finger- or hand-like), PALM-LEAF SCURF-PEA. Plant stiffly and openly branched from lower down, 30–90 cm tall; root not tuberous-enlarged; leaves palmately (3–)5(–7)-foliate, leaflets linear-oblanceolate, 0.9–5.5 cm long, 2–8 mm wide, gray, eglandular, and densely appressed-pubescent beneath, nearly glabrous but glandular above; peduncles 6.5–22 cm long, much longer than subtending petiole, spikes or spike-like racemes erect or ascending, interrupted; flowers 9–10.5 mm long; petals blue to purple; fruit body 5–6 mm long, glandular, beaks to 4 mm long. Open grasslands, shrublands, and adjacent roadsides; widespread in East TX w through Cross Timbers and Prairies to the Panhandle; c U.S. in AR, CO, KS, LA, NE, NM, OK, SD, TX, and WY. May–early Jul. [*Psoralea digitata* Nutt. ex Torr. & A. Gray, *P. digitata* var. *parvifolia* Shinners, *Pediomelum digitatum* var. *parvifolium* (Shinners) Gandhi & L.E. Br.] Variety *parvifolium*, with narrower linear leaflets, those of the middle leaves only 2–4 mm wide, is recognized by some authorities; we are following most recent authorities (e.g., Grimes 1990; Isely 1998; Weakley, A.S., and Southeastern Flora Team 2022; Egan & Reveal 2023), that describe no discrete break in leaflet width and do not recognize var. *parvifolium*.

Pediomelum hypogaeum (Nutt.) Rydb., (underground), EDIBLE SCURF-PEA, PRAIRIE-POTATO, POMME BLANCHE, POMME-DE-PRAIRIE. Plant to 25 cm tall, eglandular or with obscure glands; leaves all basal or nearly so; leaves palmately (3–)5-foliate, leaflets 1–7.5 cm long, 0.3–4 cm wide, elliptic to rhombic or oblanceolate; flowers in short, dense spikes, 11–18 mm long; petals lavender to purple; fruit body 5–6.5 mm long, the beak 7–19 mm long, projecting well past calyx. Mar–May.

1. Peduncles (= stalk of inflorescence) 1–3.5 cm long, < 1/2 as long as the petiole of the subtending leaf; w of East TX ____ var. *hypogaeum*
1. Peduncles (1.7–)4–14 cm long, > 1/2 as long as the petiole of the subtending leaf.
   2. Peduncles and petioles with appressed or closely ascending hairs; leaflets narrow, 4–11(–13) mm wide, 3–4 times as long as wide ____ var. **scaposum**
   2. Peduncles and petioles with wide-spreading hairs; leaflets broad, (8–)13–40 mm wide, 2 times as long as wide or less ____ var. **subulatum**

var. **scaposum** (A. Gray) Mahler, (with scapes). Rocky hills and grasslands primarily on limestone but sometimes sandy soils; sw margin of East TX; mainly e Cross Timbers and Prairies and e Edwards Plateau (type locality near Fredericksburg, Gillespie Co. just w of East TX); endemic to TX. [*Psoralea hypogaea* Nutt. var. *scaposa* A. Gray, *P. scaposa* (A. Gray) J.F. Macbr., *Pediomelum scaposum* (A. Gray) Rydb.]

var. **subulatum** (Bush) J.W. Grimes, (awl-shaped). Open woodlands and grasslands on dry sandy soils; widespread in East TX (type locality in Dallas Co.); also immediately adjacent Cross Timbers and Prairies and n Gulf Prairies and Marshes; sc U.S. in sw AR, se OK, and e 1/2 of TX. [*Psoralea subulata* Bush, *P. subulata* var. *minor* Shinners, *Pediomelum subulatum* (Bush) Rydb.]

var. *hypogaeum*. Sandy grasslands and open woodlands, not known from East TX, e to Parker Co. (Kartesz 2023) and several counties to the w of East TX; scattered in nw TX; e CO, w KS, MT, w NE, w OK, nw TX, and e WY. [*Psoralea hypogaea* Nutt., *P. scaposa* (A. Gray) J.F. Macbr. var. *breviscapa* Shinners] The tuber-like root was an important food source for Native Americans (Ajilvsgi 1984). This variety is included here and in the key for completeness and to encourage collectors to look for it on the w margin of East TX.

Pediomelum latestipulatum (Shinners) Mahler, (with broad stipules). Stems erect, 5–45 cm tall, glandular and pubescent throughout; rootstock usually turnip-shaped; stipules persistent, 8–15 mm long, 2.5–7 mm wide, leaves palmately 5–7-foliate, gland-dotted; leaflets usually elliptic to oblanceolate, 2–4.5 cm long, 5–15 mm wide; flowers 18–26 mm long; petals blue to lavender or purple; fruit body 5–6.5 mm long, gland-dotted, the beak narrow, 5–6 mm long. Mar–May.

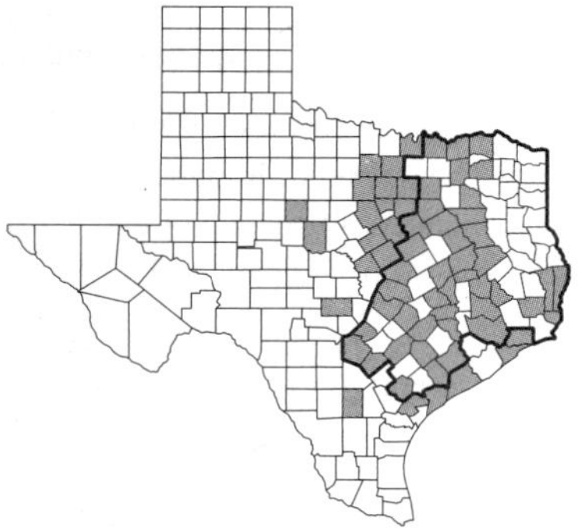
Neptunia lutea

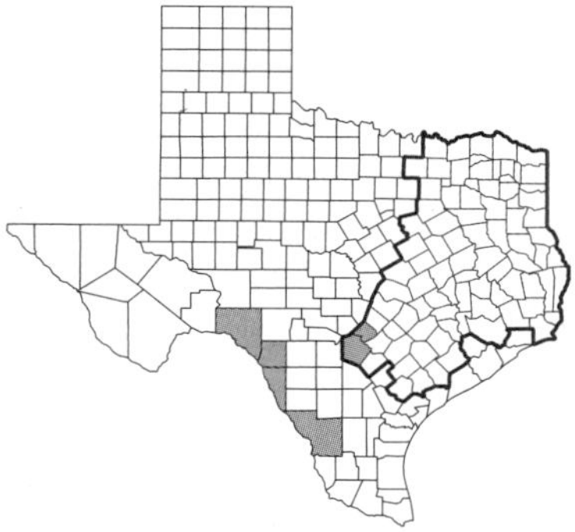
Neptunia microcarpa

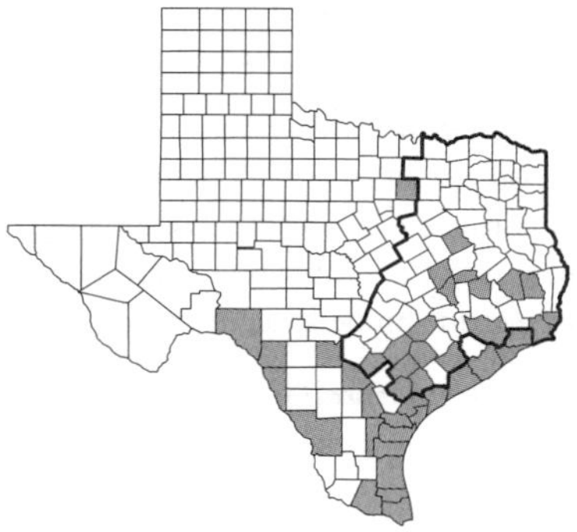
Neptunia pubescens

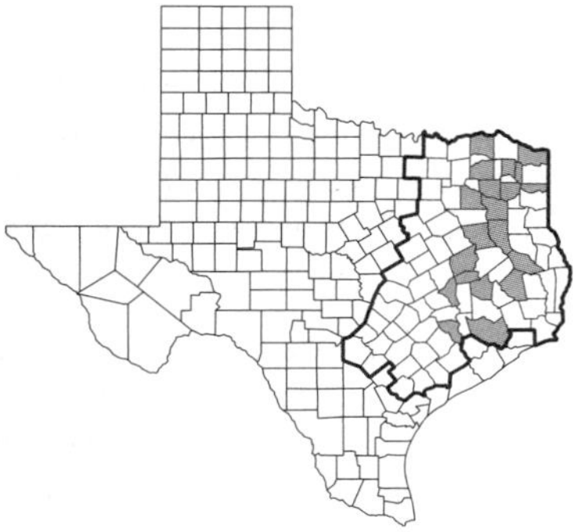
Orbexilum pedunculatum

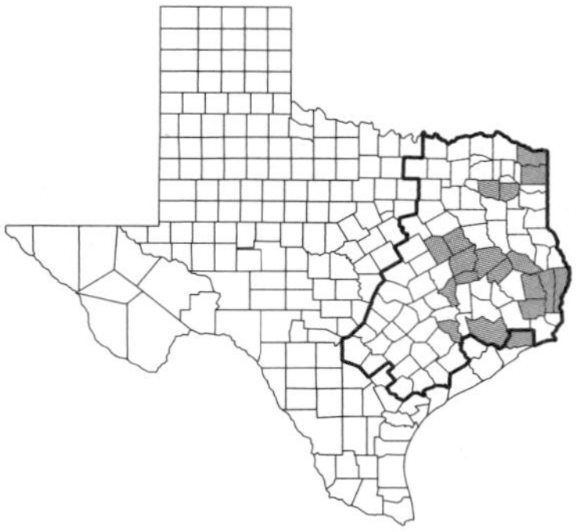
Orbexilum simplex

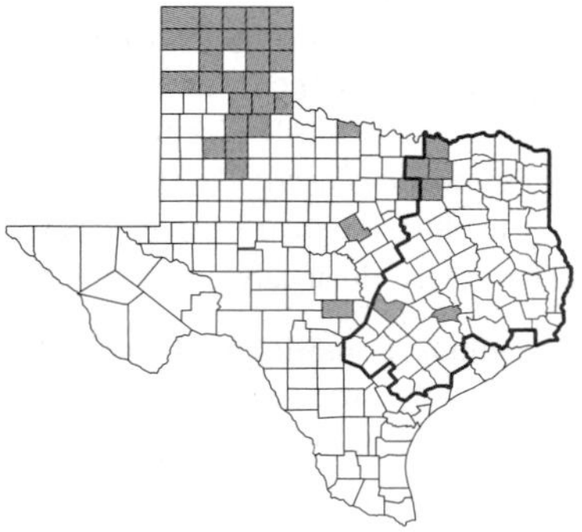
Oxytropis lambertii var. articulata

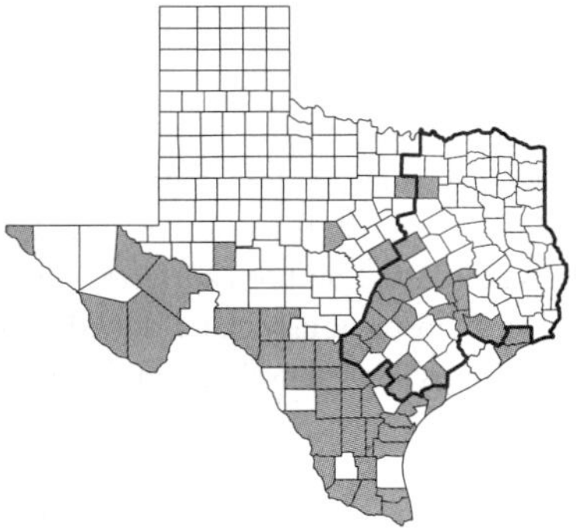
Parkinsonia aculeata

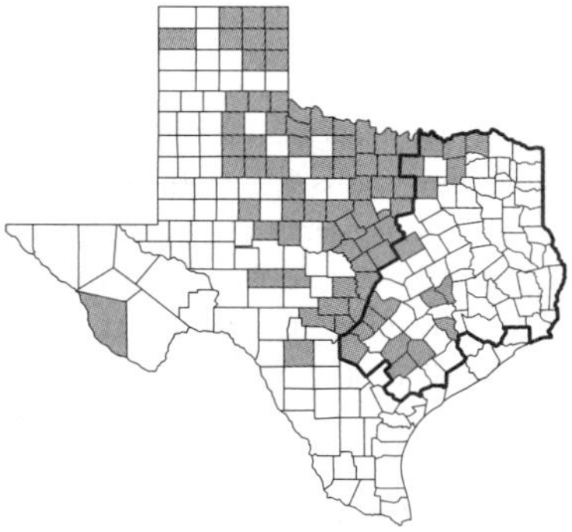
Pediomelum cuspidatum

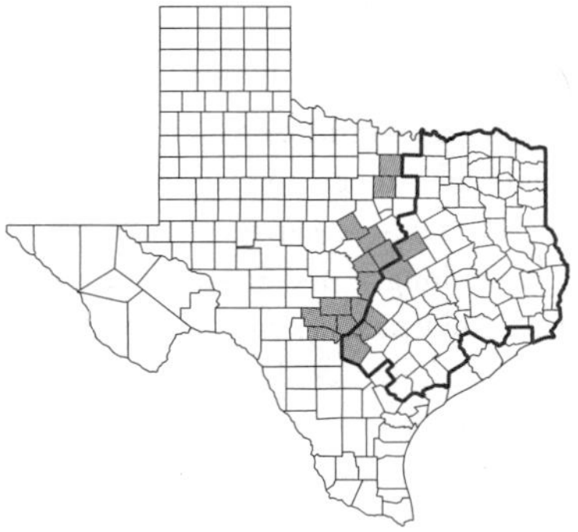
Pediomelum cyphocalyx

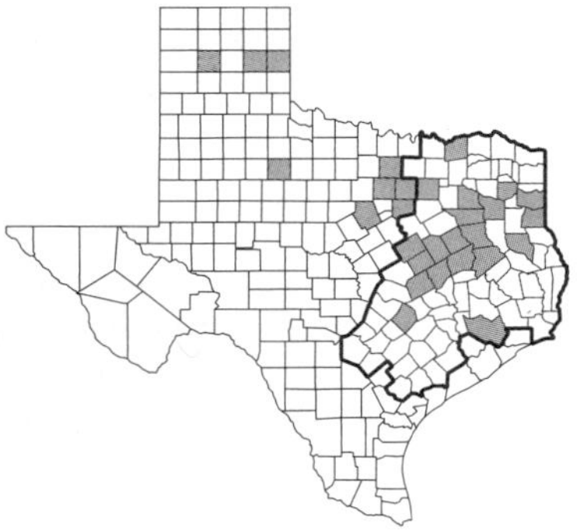
Pediomelum digitatum

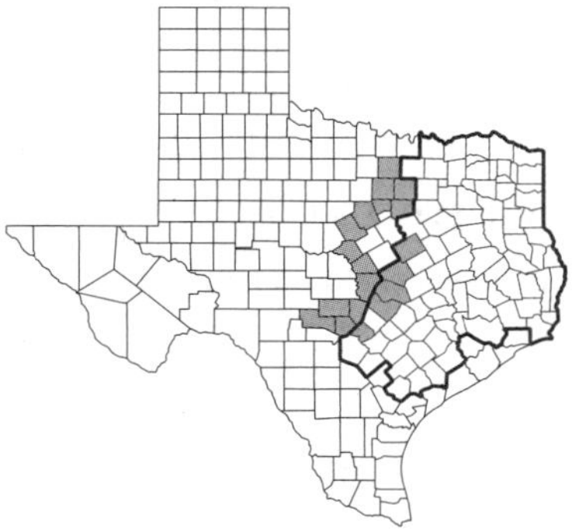
Pediomelum hypogaeum var. scaposum

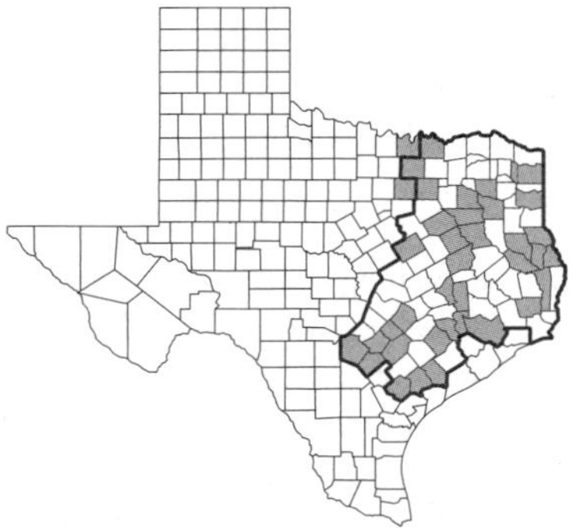
Pediomelum hypogaeum var. subulatum

1. Stems with appressed pubescence; variety of c TX ____ var. **appressum**
1. Stems with erect pubescence; variety of n 1/2 of TX ____ var. **latestipulatum**

var. **appressum** (Ockendon) Gandhi & L.E. Br., (pressed close to, lying flat against). According to Grimes (1990), while var. *appressum* is larger overall than var. *latestipulatum*, there is overlap in the size characters. The difference in pubescence is the only consistent character separating the varieties. Oak woodlands, ridges, and limestone outcrops; sw margin of East TX n to Williamson Co. otherwise mainly adjacent Edwards Plateau (Kartesz 2023); type is from Travis Co.; endemic to TX. [*Psoralea latestipulata* Shinners var. *appressa* Ockendon]

var. **latestipulatum.** Grasslands and open woodlands; McLennan, Smith, and Wood (Kartesz 2023) cos. in w and ne portions of East TX; mainly Grand Prairie w to Rolling Plains; endemic to TX. [*Psoralea latestipulata* Shinners]

Pediomelum linearifolium (Torr. & A. Gray) J.W. Grimes, (linear-leaved). Plant to 175 cm tall, branched above, glabrate to inconspicuously appressed-pubescent; leaves palmately 3(–4)-foliate, gland-dotted; leaflets linear or narrowly elliptic, 2–6 cm long, 3–6 mm wide; peduncles 0.8–10.5 cm long, usually much longer than subtending petiole; racemes slender, loose, drooping; flowers 8–11 mm long; petals blue to purplish; fruits 8–10.5 mm long including a broad beak to 3.5 mm long, gland-dotted. Rocky prairies and open woodlands, chiefly on limestone; w margin of East TX; also Cross Timbers and Prairies n and w to Panhandle; c U.S. in AR, CO, KS, NE, NM, OK, SD, and TX. Late May–Jun, sporadically later. [*Psoralea linearifolia* Torr. & A. Gray, *Psoralidium linearifolium* (Torr. & A. Gray) Rydb.] *Pediomelum linearifolium* is distinguished from its congeners by the tall, gangly habit (more than 1 m) and lax inflorescences that nod and by being glandular throughout (as opposed to *P. digitatum*, which has eglandular calyces) (Egan & Reveal 2023).

Pediomelum rhombifolium (Torr. & A. Gray) Rydb., (with rhomboid leaves), ROUND-LEAF SCURF-PEA. Plant sparsely to densely gray-pubescent; stems normally prostrate, viny, to 100(–200) cm long; root not tuberous-enlarged; leaflets and fruits eglandular; leaflets variable in shape, pinnately 3-foliate, leaflets 1–5 cm long and 0.6–3 cm wide; peduncles axillary, erect, 0.8–10 cm long; bracts inconspicuous; inflorescences subcapitate, globose to ovoid; flowers 5–11 mm long; petals brick-red to salmon-pink (rarely pink); fruit body 4.5–5 mm long, the beak 6–8 mm long. Grasslands, open woodlands, and adjacent roadsides usually on sandy soil; widespread in TX except High Plains; endemic to TX, s OK, w AR, and w LA; also isolated collections in Mexico. May–Jul, sporadically to Sep. [*Psoralea rhombifolia* Torr. & A. Gray] The salmon or brick red corollas are unique within the genus (Egan & Reveal 2023).

Pediomelum tenuiflorum (Pursh) A.N. Egan, (slender-leaved), SLIM-LEAF SCURF-PEA, WILD ALFALFA, SCURVY-PEA, SLENDER SCURFY-PEA. Much-branched perennial herb with erect or ascending stems 40–60(–130) cm tall; leaves palmately (1–)3–5-foliate, dispersed along the stems; lower surfaces of leaf blades and inflorescences strigose with conspicuous whitish hairs; leaflets (1–)1.4–3 (–4.1) cm long, 0.4–0.8(–1.5) cm wide, narrowly elliptic to oblong, conspicuously gland-dotted on both surfaces; stipules gland-dotted; racemes erect to drooping, with flowers numerous and often crowded; calyces conspicuously gland-dotted; corollas 5–6 mm long, dark blue, lavender to violet(–white); fruits 7–8 mm long, 3–4 mm wide, ellipsoidal, densely gland-dotted, not cross-wrinkled, exerted from the calyces, the calyx remnants basal only, indehiscent, with only 1 seed. Sandy or rocky prairies, open woods, and roadsides; widely scattered across TX except far s; c U.S. from IN and WI s to AR, OK, and TX and w to AZ and NV. Apr–Jul, sporadically later. [*Psoralea tenuiflora* Pursh, *Psoralidium tenuiflorum* (Pursh) Rydb.] *Pediomelum tenuiflorum* is widespread with morphological gradations across its distribution, especially in leaf and inflorescence size and shape and calyx pubescence. Rydberg (1919–1920) recognized as many as four separate species based on these differences, which seem to be environmentally influenced (Egan & Reveal 2023). While variable, breaking this species into infraspecific taxa does not seem justifiable (Isely 1998). Reported to be poisonous to livestock (Muenscher 1951), the toxicity unclear, but possibly due to furanocoumarins which can cause photodermatitis (Burrows & Tryl 2001).

PHASEOLUS L. BEAN, FRIJOLE

Unarmed, twining, perennial herbaceous vines; roots tuberous or elongated taproots or fibrous; leaves alternate, pinnately trifoliate, the leaflets ovate or broadly ovate to rhombic, sometimes lobed at base, acute to acuminate, entire, with uncinate hairs; stipules to 3 mm long, persistent; petioles well-developed; inflorescences axillary, long peduncled, exserted, pseudoracemes or panicles, inflorescences exserted, 5–20(–30) cm long with (1–)3–60 flowers, not crowded, papilionaceous; calyces with lobes shorter than the tube; corollas lavender-purple or pink to white, the keel contorted-incurved, the wings much longer than the keel, spreading; fruits elastically (= explosively) dehiscent, membranous, papery or leathery; seeds 1–20.

A genus of 70 species of tropical and warm areas of the Americas (Delgado-Salinas 2023b) characterized by uncinate (= apically hooked) hairs and flowers with curved/coiled keel; it is similar to and was in the past treated more broadly to include species of the related genus *Vigna*. Many are cultivated for their edible seeds (beans) and pods, including *P. acutifolius* A. Gray (TEPARY BEAN), *P. coccineus* L. (SCARLET RUNNER BEAN), *P. lunatus* L. (BUTTER or LIMA BEAN), and *P. vulgaris* L. (FRIJOLES, KIDNEY, STRING, SNAP, WAX, PINTO, or NAVY BEANS) (Isely 1998; Delgado-Salinas 2023b). The latter species is one of the most widely cultivated and important legumes and was very important in the CORN-BEAN-SQUASH agricultural system which provided the caloric underpinning for a number of pre-Colombian Native American civilizations in Mesoamerica (e.g., Aztecs). This importance is still reflected in the modern diets of most Mexicans and many people in the sw U.S. today. However, uncooked beans can be toxic to animals and humans due to lectins (have antinutritional effects) and hemagglutinins (damage digestive tract) (Burrows & Tyrl 2001). (Latin diminutive of Greek, *phaselos*, the name of another bean—Hyam & Pankhurst 1995) (subfamily Faboideae, tribe Phaseoleae)

REFERENCES: Delgado-Salinas 1985, 2023b; Freytag & Debouck 2002; Delgado-Salinas et al. 2006; Delgado-Salinas & Carr 2007.

1. Inflorescences usually contracted panicles with basal floral branches; corollas 10–12 mm long, pink to light purple or rarely white, flowering Jul–Oct; species of the Pineywoods ____ **P. polystachios**
1. Inflorescences pseudoracemes without basal floral branches; bracts 1–2 mm long; corollas 12–15 mm long, pink fading to dark pink, flowering Sep–Nov; species of the sw margin of East TX and further w ____ **P. texensis**

Phaseolus polystachios (L.) Britton, Sterns, & Poggenb., (many-spiked, the inflorescences sometimes being numerous), THICKET BEAN, WILD BEAN. Stems to 6 m long, puberulent; leaflets broadly ovate to rhombic, (1–)3–13 cm long, (1–)3–12 cm wide, acute to acuminate, with uncinate hairs; peduncles (1–)5–25 cm long; inflorescences 8–55 cm long; corollas pink to light purple or rarely white, 10–12 mm long, the keel contorted-incurved so as to touch the standard, the wings much longer than the keel, spreading; fruits 3–8.6 cm long, the valves twisting at maturity thus dispersing the seeds. Wooded areas; rare in TX, known only from Angelina, Cass, Harrison, Nacogdoches, Newton, and Tyler (Kartesz 2023) cos.; e U.S. from CT s to FL w to MO, e OK, and TX. Jul–Oct. *Phaseolus polystachios* is distinguished from other species of the genus by a combination of characters including its large inflorescences, with the presence of basal and lateral floral branches, floral nodes usually scattered along the axis, and banners without transverse thickenings (Delgado-Salinas 2023b). The uncinate (= apically hooked) hairs on the foliage can act as a mild "velcro"-like effect when pressed against cloth and removed quickly (Tenaglia 2007). The beans are reported to be edible (Steyermark 1963). NatureServe: G5, secure; S1, critically imperiled.

Phaseolus texensis A. Delgado & W.R. Carr, (of Texas, where it is endemic), TEXAS BEAN. Stems to 7 m long, with pubescence; leaflets ovate to broadly ovate, sometimes slightly lobed at base, 1.5–8.5 cm long, 2–7 cm wide, acute; peduncles 5–10 cm long; inflorescences 6–22 cm long; corollas pink fading to dark pink, 12–15 mm long, the keel contorted-incurved; fruits 3.5–5.2 cm long. Mixed woodlands, rocky canyons, limestone cliffs and outcrops, along streams; narrowly endemic to Travis Co. at extreme w edge of East Texas and central Edwards Plateau (Delgado-Salinas & Carr 2007; Kartesz 2023); endemic to TX. Sep–Nov. Correll and Johnston (1970) apparently included this taxon in *P. polystachios* noting that "Our Frio Canyon collection shows lobed leaflets and other characteristics which cast doubt on its referral here." Recently, Freytag and Debouck (2002) considered it to be conspecific with *P. polystachyos*; however, molecular evidence (Delgado-Salinas et al. 2006) clearly places it in a different section of the genus (sect. *Pedicellati*) from *P. polystachios* (sect. *Paniculati*). Delgado-Salinas and Carr (2007) described the species and noted that this TX endemic is a "rather rare species that deserves protection." NatureServe: G2, imperiled; S2, imperiled.

PISUM L. PEA

A genus of 3 species from s Europe, sw Asia, and n Africa (Broich 2023c). Numerous varieties and subspecies have been named for the various cultivars, which occasionally escape in North America (Isely 1998). (Latin *pisum* or Greek *pison*, pea) (subfamily Faboideae, tribe Vicieae)

REFERENCES: Isely 1998; Broich 2023c.

Pisum sativum L., (cultivated), COMMON PEA, GARDEN PEA, FIELD PEA. Herbaceous vine 10–200 cm long, branched basally; leaves alternate, even-pinnate, with branched tendrils, leaflets 2–8, opposite; inflorescences with 1–3(–4) flowers, 3–10 cm long; flowers papilionaceous; calyces campanulate with 5 lobes; corolla white or bicolored with lilac or reddish purple, 15–30 mm long; stamens 10, diadelphous; fruits 5–9 cm long, 1.5–2 cm wide; seeds 3–10, 4–12 mm in diam. Ruderal areas; occasionally escaped throughout East TX, also Denton and Montague cos. in Cross Timbers and Prairies (Kartesz 2023); throughout most of the U.S.; Canada and Greenland. Apr–Jul. [*P. sativum* L. subsp. *sativum*, *P. sativum* var. *arvense* (L.) Poir.] Native of s Europe, Asia, and n Africa. The GARDEN PEA is cultivated throughout North America and occasionally escapes in ruderal areas (Isely 1998). Broich (2023c) recognized *Pisum sativum* subsp. *sativum* including two varieties: *P. sativum* var. *sativum* and *P. sativum* var. *arvense* based on differences in flower color and size of the seeds. We are following Isely (1998) and Weakley, A.S., and Southeastern Flora

Team (2022) in not recognizing subspecies or varieties because of the overlapping characteristics of the varieties that intergrade even amongst populations. Isely (1998) also indicated that no wild prototype for the species is known.

PROSOPIS L. MESQUITE

Trees or shrubs, usually armed, rarely unarmed; taprooted; branches ascending or spreading; stems glabrous or pubescent; leaves alternate, even-bipinnate; stipules present, modified spines; petiolate; pinnae 1 or 2 pair, opposite; leaflets 12–48 per pinna, alternate or opposite, overlapping or not, blade margins entire, surfaces glabrous or pubescent; inflorescences 40–100+-flowered, axillary, spikes or heads, bracts absent; flowers mimosoid; calyx campanulate, lobes 5, connate proximally, sometimes striate; corolla yellow, cream-yellow, or purple-brown, petals connate or distinct, linear; stamens 10, distinct; fruits loments, stipitate or sessile, straight or spirally coiled, torulose, linear or cylindric, thickened, indehiscent.

A mainly American genus of ca. 48 species with some in sw Asia and Africa; mostly trees or shrubs, usually spiny (Trock 2023). The sweet, nutritious, often abundant fruits of *Prosopis* species were among the most important food plants for Native Americans in the sw U.S. (Felger 1977). Several species, including *P. glandulosa*, are known to contain alkaloids. (Derivation uncertain; perhaps Greek *prosopon*, mask, alluding to similarity with *Tamarindus* masking identity of *Prosopis*) (subfamily Caesalpinioideae, Mimosoid Clade, tribe Mimoseae)

REFERENCES: Isely 1972, 1973; Solbrig & Bawa 1975; Solbrig & Cantino 1975; Burkart 1976; Trock 2023.

1. Low shrubs; inflorescences globose heads; loments coiled; species rare in s portion of East TX ____ **P. reptans**
1. Shrubs or small trees; inflorescences in spikes; loments straight or curved; species abundant primarily in w portion of East TX ____ **P. glandulosa**

Prosopis glandulosa Torr., (glandular), HONEY MESQUITE, MESQUITE, ALGAROBA. Shrub or tree 2–10(–13) m tall, spreading by deep roots; branches armed with small, stout, straight spines at enlarged nodes; leaves twice even-pinnately compound with 1 pair of pinnae; leaflets 12–48 per pinna, glabrous, not crowded; flowers in drooping, short-peduncled, catkin- or spike-like racemes, yellowish white; petals separate nearly or quite to base; stamens 10, separate, serving as the main attractant structure for the flower; fruits 8–20 cm long, 7–13 mm wide, nearly as thick as broad, constricted between seeds. Various soils; throughout most of TX, common and locally abundant from Post Oak Savannah w, rare and local further e; frequently preserved as a semi-cultivated tree about houses. Apr–Jun, less freely to Jul or later. [*P. juliflora* (Sw.) DC. var. *glandulosa* (Torr.) Cockerell] MESQUITE is well known for its deep roots which can go to depths of 100 feet or more in search of water (Cox & Leslie 1991). Large, fist-sized, swollen galls are sometimes formed on the stems by the rust fungus *Ravenelia holwayi* Dietel; leaf lesions can also be present; the rust has a complex life cycle involving 5 different spore types (J. Hennen, pers. comm.). Smoke from the burning wood is widely used to flavor foods and the seeds of MESQUITE have long been eaten by livestock and humans; Native Americans made bread from the pods and an intoxicating beer by fermenting the meal (Standley 1922; Powell 1988); however, large amounts can cause poisoning in livestock (Kingsbury 1964). Furniture and flooring are made from the hard reddish brown wood. Isely (1998) noted that this is "one of the most successful and characteristic plants" of the sw U.S. While native here, it is a very problematic invader in other parts of the world—it is one of 32 plants on the World Conservation Union's (IUCN) list of "100 of the World's Worst Invasive Alien Species" (IUCN 2008). *Prosopis glandulosa* var. *torreyana* (L.D. Benson) M.C. Johnst. with smaller leaflets occurs west of the Pecos River (Isely 1998). In Texas this species was originally uncommon and restricted to stream banks and rocky slopes; it has greatly increased in abundance since settlement and is now a problematic invader of abused grasslands, particularly under abusive grazing regimes and fire suppression. However, early surveyor records (as far back as the 1830s) find mesquite as the most common tree in presettlement upland prairies in Navarro County suggesting "… the legendary spread of mesquite into North Texas by longhorn cattle may be an errant concept" (Jurney 1987). Roemer's (1849) mention of "extensive prairies covered with mesquite trees" also points to mesquite as a natural component of the vegetation. Likewise, early Spanish expedition diaries (e.g., Fray Isidro de Espinosa noted mesquite in Williamson County in 1716—Foster 1995), point to its widespread natural occurrence in the state. Nonetheless, mesquite undoubtedly has increased in many areas and the observations mentioned above are not so early as to preclude it having already been spread to some extent by land use changes. ⦸ ☠

Prosopis reptans Benth. var. **cinerascens** (A. Gray) Burkart, (sp.: creeping, var.: becoming ashy-gray), TORNILLO, DWARF SCREW-BEAN. Shrubs (in our area), 0.1–0.3(–0.4) m; branches distinctly flexuous; stems puberulent throughout; stipules modified paired spines, white, 4–9 cm long, sparsely short-pubescent; petiole/rachis 15–20 cm long; pinnae 2; leaflets 12–22 per pinna, alternate, touching or overlapping, blades oblong-obovate, 2.5–4 mm long, 1–2 mm wide, base oblique, margins distinctly ciliate, midvein displaced, apex rounded, surfaces cinereous (= ash-colored or gray) to glabrate; peduncles 3–4 cm long, tomentose to glabrescent; inflorescences globose heads, 1.5 cm long; sepals 2–2.5 mm long, glabrous; petals connate proximally, purple-

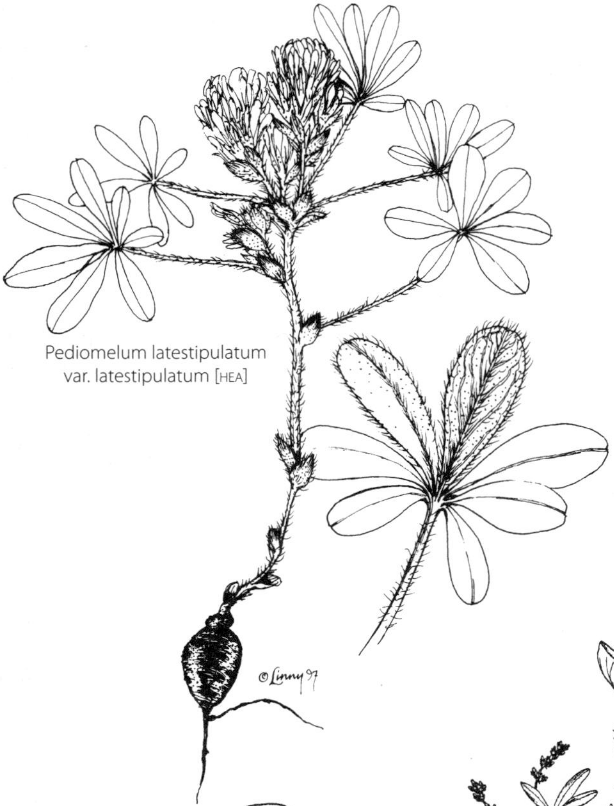

Pediomelum latestipulatum var. latestipulatum [HEA]

Pediomelum linearifolium [BB2]

Pediomelum tenuiflorum [GLE]

Pediomelum rhombifolium [COC]

Phaseolus polystachios [WIL]

brown with darker tips, ± 4 mm long, villous adaxially; stamens tan; filaments 1.5–2 mm long; anthers yellow; fruits yellow to light brown, cylindric, coiled (9–12 coils), 1.5–5.5 cm long, 0.6–0.8 cm wide, sparsely short-pubescent; seeds brown, oblong, 4–4.5 mm. Sandy bluffs, beaches, dunes, and grasslands; Wharton Co. in far s portion of East TX; s 1/3 of TX; Mexico (Nuevo León, San Luis Potosí, Tamaulipas). Mar–Jun. [*P. cinerascens* (A. Gray) A. Gray ex Benth.] Variety *reptans* occurs in Argentina and Peru (Trock 2023). Vegetative plants are superficially similar to species of *Mimosa* (Isely 1998).

PUERARIA DC. KUDZU, KUDZUVINE

A genus of 19 species of twiners with extra-floral nectaries native to tropical and e Asia and Pacific Islands (van der Maesen 2023). (Named for Marc Nicolas Puerari, 1766–1845, Swiss botanist) (subfamily Faboideae, tribe Phaseoleae)
REFERENCES: Frankel 1989; Ward 1998; van der Maesen 2023.

Pueraria montana (Lour.) Merr. var. **lobata** (Willd.) Maesen & S.M. Almeida ex Sanjappa & Predeep, (sp.: pertaining to mountains; var.: lobed), KUDZU, KUDSU, KUDZUVINE, JAPANESE ARROWROOT. Perennial vine with trailing or high-climbing, villous stems to 30 m long; foliage killed back to the ground during the winter in East TX; leaves pinnately compound with 3 leaflets; leaflets ovate-rhombic to ovate or nearly rotund, entire to often 2–3 lobed, pubescent below, 8–20(–26) cm long, 5–19(–22) cm wide; petioles very long, often as long as rest of the leaf; inflorescence an axillary raceme; pedicels 7–15 mm long; flowers fragrant; calyces gray to golden brown, corollas (10–)15–25 mm long, violet-purple or reddish purple; stamens monadelphous; fruits linear-oblong, 4–13 cm long, conspicuously tawny to reddish brown villous. Roadsides, waste places, forest edges; scattered in East TX; also Tarrant Co. (BRIT) just to the w; mostly e U.S. from MA s to FL and w to KS, e NE, e OK, and TX, disjunct in OR and WA; introduced in Mexico; Italy; Pacific Islands; and Australia. Jun–Sep. Native of China and Japan. [*Pueraria lobata* (Willd.) Ohwi] This species has been used for its edible root, as a green manure, for fodder, in textiles (fibers), and in misguided attempts to control erosion or serve as a ground cover. It was introduced to the U.S. in 1876 by the Japanese as a gift at the Philadelphia Centennial Exposition and was endorsed and encouraged by the U.S. Department of Agriculture as a soil binder and fertilizer (Frankel 1989). It grows extremely rapidly (a foot or more a day) and invades and literally covers native vegetation including large trees. In some areas of the se U.S. it is probably the most problematic noxious alien invader; KUDZU now covers more than one million acres in the southeastern U.S., is sometimes referred to as the "Scourge of the South," and is outlawed in several states (Frankel 1989). It is one of 32 plants on the World Conservation Union's (IUCN) list of "100 of the World's Worst Invasive Alien Species" (IUCN 2008). Recent studies with animals suggested that isoflavones and isoflavone glycosides from KUDZU may be useful in the treatment of alcohol abuse (Keung & Vallee 1993; Leung & Foster 1996). For an explanation of the nomenclature of this species, see Ward (1998).

RHYNCHOSIA Lour. SNOUT-BEAN

Trailing, twining or erect, usually obviously pubescent, perennial herbs with deep, often partly tuberous-thickened or branched roots; leaves alternate, simple or pinnately compound, petioled; leaflet(s) 1 or 3, with resin droplets (glands) beneath; stipules small; flowers axillary, solitary or in spike-like racemes; calyces with 5 lobes; petals yellow, sometimes tinged with brown or red; stamens 10, diadelphous; fruits sessile, flattened, gland-dotted; 1–2 seeds.

A mainly tropical genus of ca. 200 species (Woods 2023b); they can usually be recognized by the combination of ± yellow corollas, small typically 1 or 2-seeded fruits, and conspicuously glandular foliage (Isely 1990). (Greek: *rhynchos*, snout or beak, presumably in allusion to the somewhat beak-like form of the keel) (subfamily Faboideae, tribe Phaseoleae)
REFERENCES: Grear 1978; Woods 2023b.

1. Only 1 leaflet per leaf; inflorescences 3 cm or less long.
   2. Plants erect (never twining), 20 cm or less tall; stems commonly with only 4–6 leaves; species of se portion of East TX ____ **R. reniformis**
   2. Plants trailing or twining, to 1 m long; stems often with numerous leaves; species primarily of sw and extreme s portion of East TX ____ **R. americana**
1. Leaflets 3 per leaf; inflorescences variable, but often longer than 3 cm.
   3. Calyces 1.5–6 mm long, divided ca. 1/2 way to base, shorter than the corollas; corollas 4–7 mm long.
      4. Flowers 1–3 in the axils of the leaves; leaflets 2.2–5 times as long as wide; w portion of East TX ____ **R. texana**
      4. Flowers in conspicuous elongate racemes of 5–15 flowers; leaflets 0.8–1.2 times as long as wide; s portion of East TX ____ **R. minima**
   3. Calyces 6–15 mm long, divided nearly to base, ca. equal to or longer than the corollas; corollas 4.5–13 mm long.
      5. Axillary inflorescences 5–20 cm long in flower, with flowers distributed along a significant portion of the inflorescence axis, longer than leaves; leaf undersurfaces villous; species widespread in East TX ____ **R. latifolia**
      5. Axillary inflorescences (peduncle & raceme) 1–4 cm long in flower (can elongate in fruit and terminal raceme longer on *R. tomentosa*), with flowers usually clustered at tip, usually shorter than leaves; leaf undersurfaces

Phaseolus texensis [PRC]

Pisum sativum [BL3]

Prosopis glandulosa [SA2, SUD]

Prosopis reptans [VIN]

puberulent or strigose; species rare in East TX.

6. Stems ascending or erect; terminal leaflet elliptic; calyces adaxial lobes 2–2.5 mm long; known only from Jasper Co. in East TX ____ **R. tomentosa**
6. Stems usually prostrate or twining, rarely climbing or erect; terminal leaflet suborbicular; calyces adaxial lobes 3–5 mm long; rare in se TX ____ **R. difformis**

Rhynchosia americana (Mill.) Metz, (of America), AMERICAN SNOUT-BEAN, AMERICAN RHYNCHOSIA. Stems trailing or twining, to 1 m long; leaflet 1 (this is one of only 2 species of TX *Rhynchosia* with a single leaflet), suborbicular to reniform, 1.5–4 cm long, varying from wider than long to about as wide as long, 2–5.5 cm wide; racemes 1.5–5 cm long, with few flowers; calyces 7–13 mm long; corollas 6–10 mm long, yellow or orange-yellow, sometimes with a brownish tinge; fruits compressed 10–15 mm long, 5–8 mm wide, villosulous. Sandy or gravelly soils; prairies, open woodlands, sand dunes, and adjacent roadsides; sw and extreme s portion of East TX; Turner et al. (2003) mapped Trinity Co. further ne, but no specimens from that county could be located; otherwise mainly South TX Plains and Gulf Prairies and Marshes; in the U.S. known only from TX; also Mexico. Mar–Oct. [*Dolicholus americanus* (Mill.) Vail, *Rhynchosia menispermoidea* DC.]

Rhynchosia difformis (Elliott) DC., (of unusual or differing forms), DOUBLE FORM SNOUT-BEAN. Stems usually trailing or twining but can be suberect or ascending; leaflets 3, 2–6 cm long, 1–7 cm wide, usually villosulous beneath; racemes 1–4 cm long in flower (can elongate in fruit); calyces 9–12 mm long; corollas 6–10.5 mm long, yellow or orange-yellow; fruit compressed 14–20 mm long, 6–9 mm wide. Sandy soils in pinelands, fields, and clearings; species rare in se portion of East TX and Smith Co. (BRIT) southward; se U.S. from VA s to FL w to MO, OK and TX. May–Aug. [*Dolicholus tomentosus* (L.) Vail, *Rhynchosia lewtonii* (Vail) K. Schum., *R. tomentosa* (L.) Hook. & Arn. var. *volubilis* (Michx.) Torr. & A. Gray] Isely (1998) noted that this normally twining or trailing species can be suberect in the w extremity [TX] portion of its range, and that “Turner (1959) and Grear (1978) have suggested that Texas erect or suberect forms assigned to *R. difformis* may be the products of introgression between it and *R. tomentosa*.” However, the origin of this variation is still unclear and needs study.

Rhynchosia latifolia Nutt. ex Torr. & A. Gray, (broad-leaved), BROAD-LEAF SNOUT-BEAN, BROAD-LEAF RHYNCHOSIA. Stems trailing or twining, to 100 cm long; leaflets 3, 30–80 mm long, 10–60 mm wide, suborbicular to rhombic-ovate or elliptic-lanceolate; flowers in elongate racemes 5–20 cm long; calyces 8–15 mm long; corollas yellow (but can dry a dull pink), 8.5–14 mm long; fruits 15–20 mm long, 7–9 mm wide, shortly ovate-oblong. Sandy woods and roadsides; nearly throughout East TX w to West Cross Timbers (Parker Co. BRIT); also n Gulf Prairies and Marshes; sc U.S. in AR, LA, MO, MS, OK, and TX. Late May–Jun, sporadically Jul–Aug. [*Dolicholus latifolius* Kuntze] This is the most widespread and most common snout-bean in East TX.

Rhynchosia minima (L.) DC., (least, smallest), LEAST SNOUT-BEAN. Stems trailing to twining, to 2 m long; leaflets 3, 30–90 mm long, 5–70 mm wide, nearly or slightly longer than wide; racemes 2–20 cm long; calyces 3–6 mm long, puberulent or villosulous; corollas 4–6 mm long, yellow, sometimes with brownish tinge; fruits 9–20 mm long, 3–5 mm wide, scimitar-shaped (= like a curved sword). Open areas including prairies, old fields and roadsides, primarily on clay soils; widespread in s portion of East TX n to McLennan (Kartesz 2023) and Travis (BRIT) cos.; also adjacent Edwards Plateau and Gulf Prairies and Marshes; se US from FL and GA w to TX; West Indies; Bermuda; Central America; South America. Apr–Nov. [*Dolicholus minimus* (L.) Medik., *Rhynchosia minima* var. *diminifolia* Walraven]

Rhynchosia reniformis DC., (kidney-shaped), DOLLAR-LEAF, DOLLAR-LEAF RHYNCHOSIA, DOLLAR-LEAF SNOUT-BEAN. Plants ascending to erect, to only 20 cm tall; leaflet 1 (this is one of only 2 species of TX *Rhynchosia* with a single leaflet), usually reniform to subcordate, 20–50 mm long, 25–75 mm wide, varying from wider than long to slightly longer than wide; racemes 1–3.5 cm long; calyces usually 7–10 mm long; corollas 6–9 mm long, yellow; fruits 13–18 mm long, 5–7 mm wide, gland-dotted, hirtellous. Sandy soils, pine woods; se portion of East TX in Angelina, Hardin, Newton, Polk, Trinity, and Walker (Kartesz 2023) cos.; se U.S. from NC s to FL w to TX. Jun–Sep. [*Dolicholus simplicifolius* (Walt.) Vail, *Rhynchosia simplicifolia* DC.] *Rhynchosia reniformis* is unique in the combination of its short, erect habit and unifoliate leaves. Occasional plants that continue apical growth produce trifoliate leaves (Isely 1998).

Rhynchosia texana Torr. & A. Gray, (of Texas), TEXAS SNOUT-BEAN, TEXAS RHYNCHOSIA. Stems trailing or twining or ascending, to 70 cm long; leaflets 3, 5–45 mm long, 4–25 mm wide, ovate-lanceolate or elliptic to narrowly oblong; inflorescences reduced, 1–3 flowers in leaf axils; calyces 3–5 mm long, ca. half as long as corollas; corollas 3.5–7 mm long, yellow, often tinged with red or brown; fruits to 15–25 mm long, 4–6 mm wide, oblong, compressed. Gravelly or rocky limestone soils; w edge of East TX; widespread primarily in the s 1/2 of TX, but scattered n to Dallas Co. (Kartesz 2023); sw U.S. in AZ, NM, and TX; Mexico; South America. Jun–Sep. [*Dolicholus texensis* (Torr. & A. Gray) Vail, *Rhynchosia senna* Gillies ex Hook. & Arn. var. *angustifolia* (A. Gray) Grear, *R. senna* Gillies ex Hook. & Arn. var. *texana* (Torr. & A. Gray) M.C. Johnst.] Vegetative plants are superficially

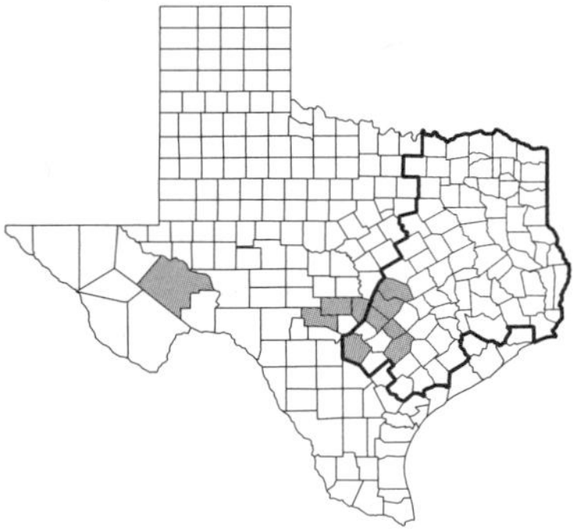
Pediomelum latestipulatum var. appressum

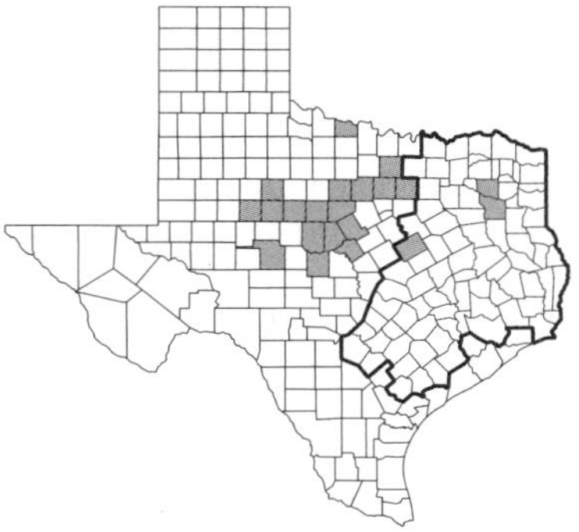
Pediomelum latestipulatum var. latestipulatum

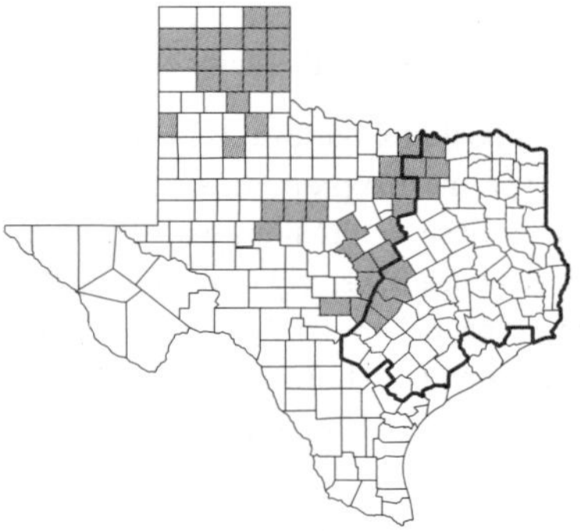
Pediomelum linearifolium

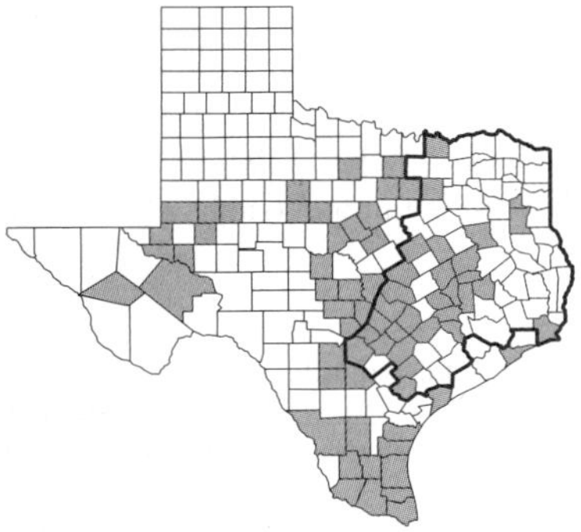
Pediomelum rhombifolium

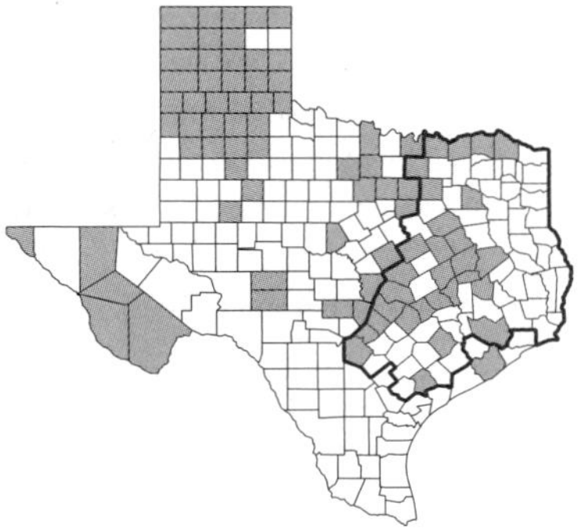
Pediomelum tenuiflorum

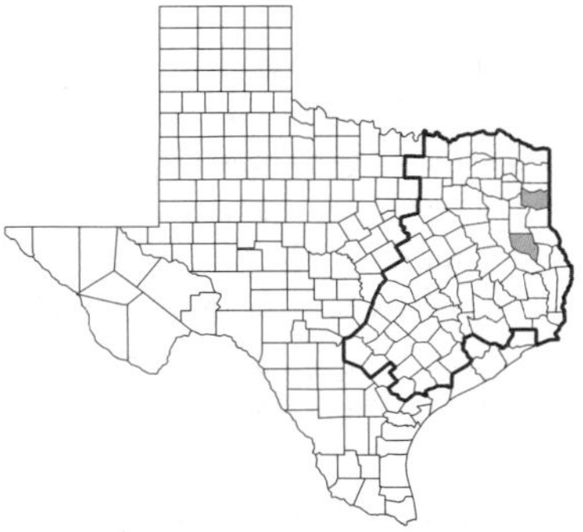
Phaseolus polystachios

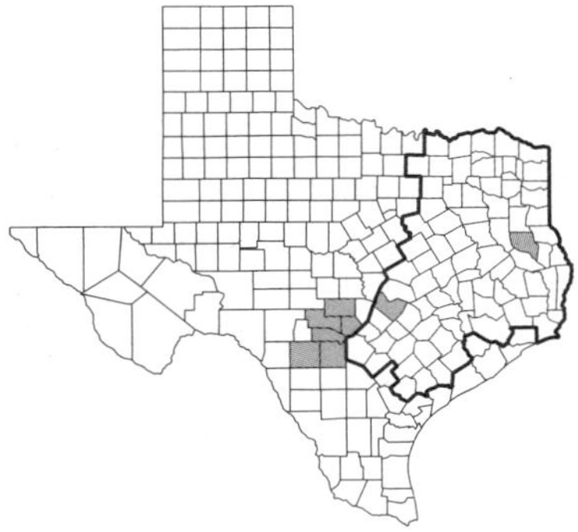
Phaseolus texensis

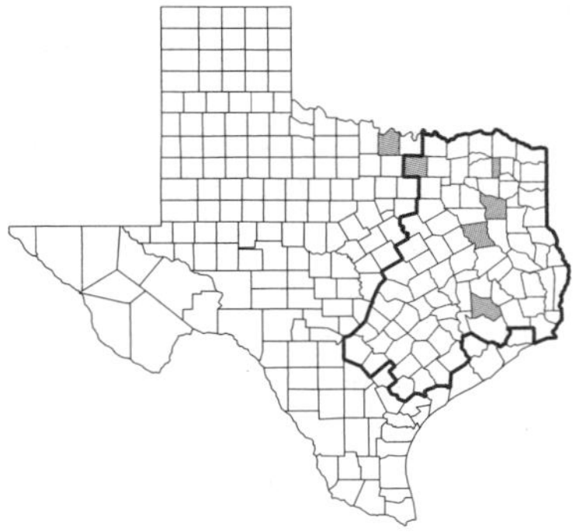
Pisum sativum

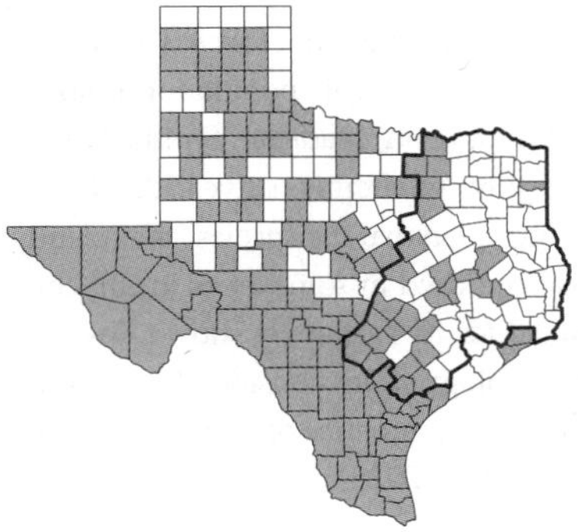
Prosopis glandulosa

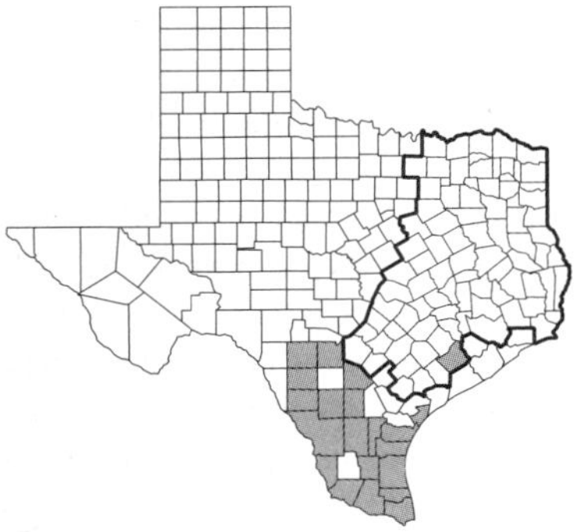
Prosopis reptans var. cinerascens

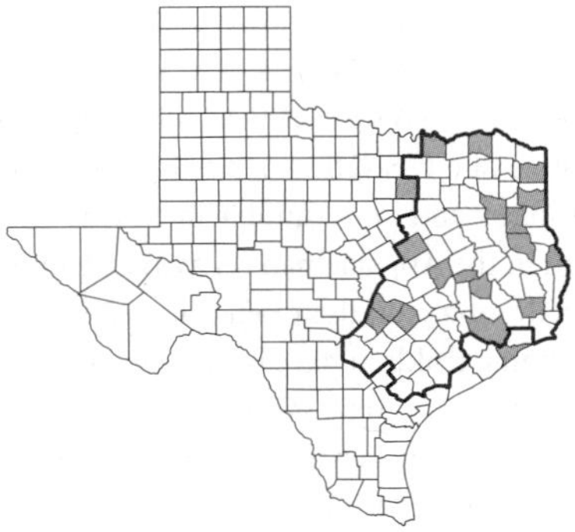
Pueraria montana var. lobata

similar to *Strophostyles leiosperma* (Torr. & A. Gray) Piper. Leaflets of the lower leaves tend to be broader than the narrower leaves at the tips of the stems (Isely 1998).

Rhynchosia tomentosa (L.) Hook. & Arn., (densely woolly, with matted hairs), SNOUT-BEAN, TWINING SNOUTBEAN—this common name seems inappropriate since the species is erect in contrast to many other twining species. Plant erect, to 70(–90) cm tall, tomentulose to hirtellous with tawny or silvery pubescence; leaflets 3, 20–50(–60) mm long, 10–40(–50) mm wide, tomentulose beneath, velvety to the touch, the terminal leaflet ovate to elliptic; racemes, shorter than leaves, axillary ones 1.5–4 cm long, terminal ones 6.5–23 cm long; calyces 6–9 mm long; corollas 4.5–9 mm long. Dry sandy open woods; in TX known only from Jasper Co. (Brown & Elsik 2002); se U.S. from DE s to FL and w to TX. May–Sep. [*Glycine tomentosa* L., *Rhynchosia erecta* (Walter) DC.] This species is still only known in TX from its original collection in Jasper Co. (Brown & Elsik 2002).

ROBINIA L. LOCUST

Trees or shrubs; leaves once odd-pinnately compound; leaflets 7–19, entire; stipules spiny or not so; flowers in drooping axillary racemes; calyces campanulate; fruits linear-oblong, compressed to flat; seeds 3–8(–16).

A North American genus of 4 species of se U.S. and Mexico. They are deciduous trees and shrubs usually with stipular spines and extrafloral nectaries (Isely 1998; Lavin & Lammers 2023); some are cultivated as ornamentals. Numerous species have been named in the past, but most are not usually recognized by modern taxonomists. (Named for Jean Robin, 1550–1629, herbalist to Henry IV of France, and his son Vespasian Robin, 1579–1662, who first cultivated the locust in Europe) (subfamily Faboideae, tribe Robinieae)

REFERENCES: Isely & Peabody 1984; Peabody 1984; Lavin & Lammers 2023.

1. Young stems and leaves densely and conspicuously hispid (= bristly); petals pink to rose or reddish purple; fruits densely hispid (but rarely developed); usually unarmed shrub ____ **R. hispida**
1. Young stems and leaves glabrous; petals whitish; fruits glabrous; tree usually armed with stipular spines ____ **R. pseudoacacia**

Robinia hispida L., (hispid, bristly), BRISTLY LOCUST, ROSE-ACACIA, MOSSY LOCUST. Shrub to 3 m tall (rarely small tree), root-suckering, with extremely conspicuous hairs 1–4 mm long; leaflets 7–13 per leaf; stipules not spiny; racemes with 3–11(–15) flowers; flowers inodorous, very showy; petals 2–2.5(–3) cm long, pink to rose or reddish purple, but can dry lavender to bluish; fruits 4–8 cm long, often not formed. Cultivated, persisting, and spreading around old homesites; known in TX only from Anderson, Grayson (BRIT, this population is spreading extensively from cultivation at an abandoned homesite—G. Diggs, pers. obs.), Franklin, Leon, Madison, Smith, Tyler, and Walker (Kartesz 2023) cos.; throughout the U.S. but mostly introduced; Eurasia. Mar–July. Apparently native to AL, GA, NC, SC, TN, and VA. Isely (1998) treated it as including several rather narrow endemics. He indicated that the species "is evidently mostly an aggregate of sterile clones that reproduce vegetatively." This variable species has been used in some areas in soil conservation plantings and in reclamation of strip mine spoils (Isely 1990). It is also cultivated for its extremely showy flowers. Hybrids with *R. pseudoacacia* are known from the e U.S. (Lavin & Lammers 2023). Morphology varies considerably within the species, particularly in habit and in the amount and distribution of pubescence; due to the predominance of asexual reproduction, populations in many cases are clones and quite uniform in these features (Lavin & Lammers 2023).

Robinia pseudoacacia L., (false *Acacia*), BLACK LOCUST, FALSE ACACIA, BASTARD ACACIA. Trees (1–)4–25 m tall, forming colonies by root sprouts; bark furrowed; leaflets usually 15–19 per leaf; blades elliptic, 25–45(–65) mm long, 10–20(–25) mm wide; racemes with 10–25 flowers; flowers fragrant, 1.5–2.5 cm long; fruits 4–10 cm long; seeds 4–8(–16). Open woodlands and disturbed areas; widespread in East TX; scattered further w; now widespread in the U.S., but common only in the e 1/2, but thought to be native from PA s to GA w to AR (Isely 1998); s Canada; South America (Argentina and Chile); Eurasia; and Australia. Mar–May. Native of e U.S. This species is very difficult to eradicate once established because of the root sprouts. The wood is one of the strongest and most durable in North America (Peattie 1948; Cox & Leslie 1991). This species is important in ecological succession in the Appalachians (Lavin & Lammers 2023). The leaves exhibit nyctinastic (nighttime or "sleep") movements, the leaflets drooping on their petiolules (Peattie 1948). The inner bark, young leaves, and seeds are toxic to all classes of livestock and humans; children have been poisoned by sucking on fresh twigs or eating inner bark or seeds; toxins include a lectin (= type of phytotoxin or toxic protein), known as robin, that agglutinates red blood cells, and robitin, a glycoside (Hardin & Arena 1974; Lewis & Elvin-Lewis 1977; Fuller & McClintock 1986). Fossil wood that is indistinguishable from that of *Robinia pseudoacacia* is encountered from the Late Eocene to Miocene in the south-central and western United States (Lavin & Lammers 2023).

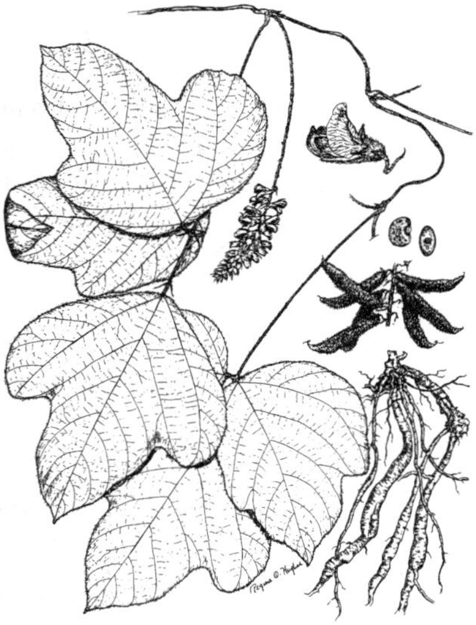

Pueraria montana var. lobata [REE]

Rhynchosia americana [WIL]

Rhynchosia difformis [WIL]

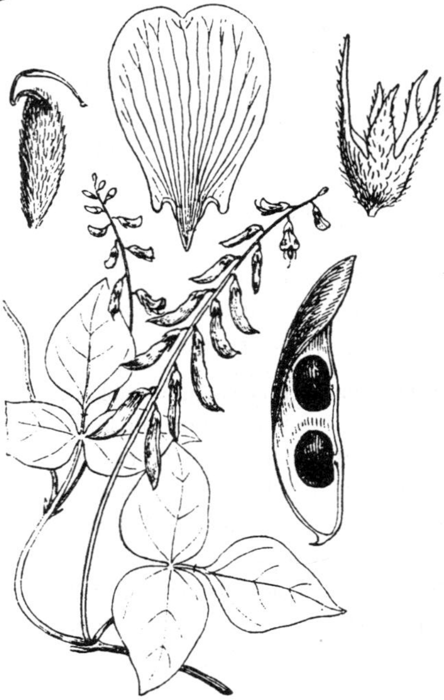

Rhynchosia minima [FAW]

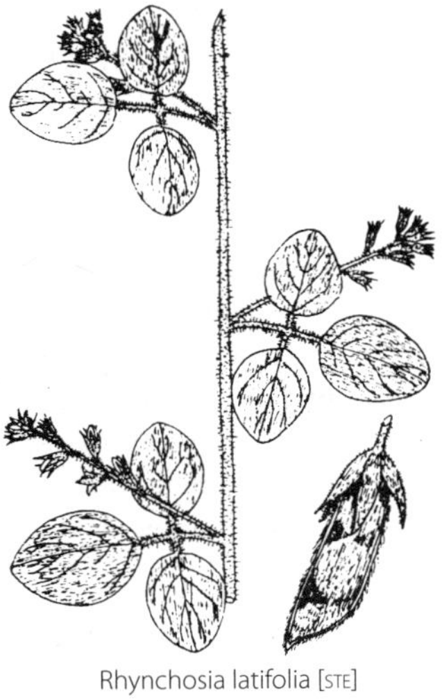

Rhynchosia latifolia [STE]

Rhynchosia reniformis [WIL]

Rhynchosia texana [HEA]

SECURIGERA DC. CROWN-VETCH

Herbs, annual or perennial, unarmed; stems spreading, furrowed or ridged, pubescent or glabrate; leaves alternate, odd-pinnate; stipules present, free, membranous or herbaceous, dark-tipped; petiolate; leaflets 7–25, blade margins entire, surfaces glabrous; inflorescences pedunculate, 3–15-flowered, axillary, umbels or heads; bracts absent; bracteoles present; flowers papilionaceous; calyces weakly zygomorphic, campanulate, lobes 5; corollas white, pink, purple, or bicolored, keel acute; stamens 10, diadelphous; fruits loments, stipitate, not or only slightly constricted between seeds, angled in cross-section or cylindric, indehiscent or tardily dehiscent, with weak longitudinal, anastomosing veins, glabrous; seeds 2–12, oblong, smooth.

An Old-World genus of 12 species ranging from the Mediterranean area including Europe to Somalia (Isely 1998; Ball 2023); it is similar to *Coronilla* (but has angled vs. rounded stems) and *Hippocrepis* (2 other Old-World genera) (Lassen 1989). Some are cultivated for erosion control. (Latin: *securis*, axe, and *-ger*, bearing, alluding to shape of fruit beak found in *Securigera securidaca*, the type species) (subfamily Faboideae, tribe Cornilleae)
REFERENCES: Tutin et al. 1968; Lassen 1989; Ball 2023.

1. Annual; leaflets 7–17; heads 3–6(–9)-flowered; corollas 4–7 mm ____ **S. cretica**
1. Perennial; leaflets (11–)15–25; heads (5–)10–20-flowered; corollas (8–)10–15 mm ____ **S. varia**

Securigera cretica (L.) Lassen, (of Crete, Greek island in the Mediterranean Sea), CRETAN CROWN-VETCH. Annual herb with erect stems to 90 cm tall; leaves once odd-pinnately compound with 7–17 leaflets, 5–20 mm long; stipules 1–3 mm long, free, linear, membranous; inflorescences axillary, with 3–6(–9) flowers per umbellate cluster; calyces 2–3 mm long; petals white to light pink, 4–7 mm long; stamens 10, diadelphous; ovary glabrous; fruits glabrous, linear-oblong with 4 angles, 3–8 cm long. Live oak woodlands and ruderal habitats; Bexar (Keith 2013b) and Comal cos. in sw portion of East TX; also Kendall Co. in Edwards Plateau (numerous obs., iNaturalist 2023); also MD. Mar–Jun. [*Coronilla cretica* L.] Native of Europe and the Mediterranean area. This species was not fully treated in Ball (2023) but was mentioned in genus description as being recorded from Maryland, where it was first reported for North America (Reed 1964). It was first reported as naturalized in Texas from Government Canyon State Natural Area in Bexar Co. in 2013 (Keith 2013b) and is now well established in the San Antonio area (iNaturalist 2023; E. Keith, pers. obs.).

Securigera varia (L.) Lassen, (variable), CROWN-VETCH, PURPLE CROWN-VETCH. Perennial herb with erect to ascending or sprawling stems 20–120 cm long, commonly forming a ground cover; leaves once odd-pinnately compound with (11–)15–25 leaflets, essentially sessile; stipules 1–2 mm long, dark tipped; inflorescences axillary, umbellate, on peduncles 4–14 cm long, of ca. 10–15 flowers; tiny bracts in an inconspicuous involucral ring; flowers (5–)10–20 per umbellate cluster; calyces 2.2–4 mm long; petals pink or partly whitish sometimes with purplish lines (can dry lavender) (wings typically lightest), (8–)10–15 mm long; stamens 10, diadelphous; ovary glabrous; fruits glabrous, linear, ± cylindrical with 4 angles, 2–6(–8) cm long, stipitate, with 2–10(–12) disarticulating segments. Roadsides and other ruderal areas, planted for erosion control; scattered in East TX (Kartesz 2023); also Montague Co. (Lipscomb 1984—first TX report) and Tarrant (BRIT) cos. just to the w of East TX and rare scattered occurrences further w; throughout the U.S.; Canada; ne Mexico; South America; New Zealand; and se Australia. May–Jul. [*Coronilla varia* L.] Native of Europe and the Mediterranean area. Diggs et al. (1999) treated this species in the genus *Coronilla*; however, a study by Lassen (1989) convincingly argued that *Securigera* is distinct from both *Coronilla* and *Hippocrepis* (two other Old-World genera). We are therefore recognizing this species in *Securigera*. According to McGregor (1986), the seeds are reported to be poisonous.

SENEGALIA Raf. LOCUST

Shrubs or small trees, armed with prickles, stipular spines absent; stems usually erect, rarely spreading or climbing; leaves alternate, even-bipinnate; stipules present, caducous, herbaceous; petiolate, gland present in channel; pinnae 1–14(–19) pairs, mostly opposite; leaflets 3–55 pairs per pinna, opposite, sessile or subsessile, blade margins entire, surfaces glabrous or pubescent; inflorescences 6–250-flowered, terminal or axillary, globose heads or spikes; bracts present; flowers mimosoid; calyces campanulate, lobes 5; corollas white, creamy white, or yellow, lobes 5; stamens 40–160, distinct, exserted, creamy or yellow, fading reddish brown; ovary usually stipitate with nectiferous disc at base, usually pubescent; style and stigma filiform; fruits legumes, stipitate or sessile, mostly flattened, straight to falcate, oblong, usually dehiscent along sutures, glabrous or pubescent; seeds 3–16(–25).

A genus of ca. 235 species from sw, sc United States, Mexico, West Indies, Central America, South America, Asia, Africa, Pacific Islands, Australia; nearly worldwide in tropical to warm temperate areas (Seigler & Ebinger 2023). *Senegalia* is separated from related genera by the presence of prickles, the absence of stipular spines, a nectiferous disk at the base of the stipitate ovary

Rhynchosia tomentosa [WIL]

Robinia hispida [LAM]

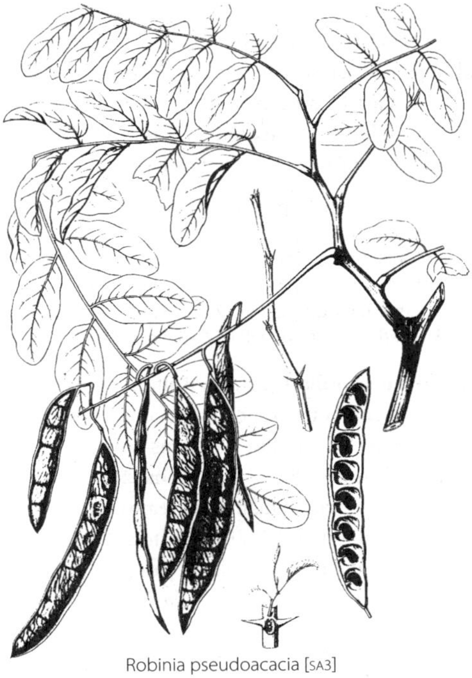

Robinia pseudoacacia [SA3]

Securigera cretica [FIO]

Securigera varia [WIL]

Senegalia berlandieri [VIN]

and pollen with pores instead of grooves. Based on morphological evidence, Pedley (1978, 1986) separated *Senegalia* from other members of *Acacia* in the broad sense. The separation was further supported by phylogenetic analysis of chloroplast matK and trnl sequence datasets, in which species of *Senegalia* formed a monophyletic clade (Seigler et al. 2006). The flowers of all four species in East TX are reported to produce an excellent honey (Vines 1960). (From species name *Mimosa Senegal* L., alluding to nativity) (subfamily Caesalpinioideae, Mimosoid clade, tribe Acacieae)
REFERENCES: Seigler et al. 2006; Seigler & Ebinger 2023.

1. Leaflets 25–55 pairs per pinna ______ **S. berlandieri**
1. Leaflets 3–10 pairs per pinna.
    2. Inflorescences globose or subglobose heads, lengths < 2 times widths ______ **S. roemeriana**
    2. Inflorescences cylindric spikes, lengths 3+ times widths.
        3. Leaflet blades usually 2.8–5.5 mm long; flowers sessile or short-pedicellate, pedicels 0–0.6 mm ______ **S. greggii**
        3. Leaflet blades usually 5.5–9.2 mm long; flowers pedicellate, pedicels 0.5–2.1 mm ______ **S. wrightii**

Senegalia berlandieri (Benth.) Britton & Rose, (for Jean Louis Berlandier, 1805–1851, French botanist who explored TX, NM, and Mexico, and one of the first botanists to make extensive collections in TX), GUAJILLO. Shrubs or small trees, 1–2(–6) m tall; stipular spines absent, but the stems armed with nearly straight prickles along the internodes; leaves with 7–14(–19) pairs of pinnae; leaflets 25–55 pairs per pinna, 2.8–7.2 mm long, 0.7–1.4(–1.8) mm wide; flowers in usually globose heads (sometimes twice as long as wide); filaments creamy white; fruits 7–16 cm long, 15–30 mm wide, flattened. Dry, mostly calcareous soils in thorn scrub; Bexar and Goliad (Kartesz 2023) cos. in sw portion of East TX; otherwise in s and w TX; only in TX in the U.S.; Mexico (Chihuahua, Coahuila, Durango, Guanajuato, Guerrero, Hidalgo, Nuevo León, Querétaro, San Luis Potosí, Tamaulipas, Zacatecas). Feb–May. [*Acacia berlandieri* Benth.] *Senegalia berlandieri* is abundant in sw and s Texas, where it commonly dominates the vegetation. The fertile named hybrids with *S. greggii* (= *S. ×emoryana*) and *S. wrightii* (= *S. ×turneri*) occur w and s of East TX and can be differentiated in Seigler and Ebinger (2023).

Senegalia greggii (A. Gray) Britton & Rose, (for Josiah Gregg, 1806–1850, who collected in Mexico and died in the wilderness in n CA), CATCLAW, WRIGHT'S ACACIA, JOINT-VETCH, UÑA DE GATO, HUISACHILLO. Shrub or small tree (1–)2–3(–6) m tall; stipular spines absent, but the stems armed with recurved prickles along the internodes or occasionally at a node; pinnae 1–3(–5) pairs per leaf; leaflets 3–6(–8) pairs per pinna, 2.8–5.5 mm long, 0.9–3.2 mm wide; flowers in elongate spikes 15–40 mm long, 2–6 times as long as wide, sweet-scented; filaments creamy white to yellowish; fruits 40–150 mm long, 13–21 mm wide, flattened. Rocky, gravelly, mostly calcareous soils; Bexar, Goliad, and Wilson (Kartesz 2023) cos. on sw margin of East TX; mainly to the s and w of East TX; sw U.S. from w OK and TX w to s CA; Mexico (Baja California, Chihuahua, Coahuila, Durango, Nuevo León, Sonora, Tamaulipas). Apr–Sep. [*Acacia greggii* A. Gray] Possibly poisonous to animals due to the high cyanide content of the foliage (Kingsbury 1964). The legumes were crushed into a meal and called "pinole" by Native Americans which was eaten in the form of a mush or cake (Vines 1960). This species is similar to *S. wrightii*, which is sometimes treated as a variety of this species (Turner et al. 2003). Isely (1998) treated them as distinct species but noted that "there is no hard line between them morphologically. Perhaps they should be regarded as expressions of one species." ☠

Senegalia roemeriana (Scheele) Britton & Rose, (for Ferdinand Roemer, 1818–1891, geologist, paleontologist, and explorer of TX), ROEMER'S ACACIA, CATCLAW. Shrub 1–2(–6) m tall with weak stems; stipular spines absent, but the stems armed with recurved prickles along the internodes; leaves with 1–4 pair of pinnae; leaflets 3–10 pairs per pinna, 5–14 mm long, 2–7 mm wide; flowers in globose heads 11–15 mm diam., fragrant; filaments creamy white to yellow; fruits 70–170 mm long, 15–30 mm wide, flattened. Dry, mostly calcareous soils, thorn scrub, oak-juniper forests, and disturbed areas; Bexar, Comal, Hays, and Travis (BRIT) cos. on sw margin of East TX; mainly w 1/2 of TX except Panhandle; sw U.S. in s NM and TX; Mexico (Chihuahua, Coahuila, Nuevo León, Tamaulipas). Mar–Jun. [*Acacia malacophylla* Benth., *A. roemeriana* Scheele] *Senegalia roemeriana* is uncommon and scattered and is rarely dominant (Seigler & Ebinger 2023).

Senegalia wrightii (Benth.) Britton & Rose, (for Charles Wright, 1811–1885, TX collector), CATCLAW ACACIA. Shrubs or trees to 12 m; bark dark gray or dark brown, shallowly furrowed; prickles recurved, scattered along twigs; stems not flexuous, usually glabrous; short shoots present; pinnae 1–3 pairs, 3–22 mm long; leaflets 3–7(–9) pairs per pinna, blades oblong or obovate to oblanceolate, 5.5–9.2 mm long, 2.2–4.5 mm wide; inflorescences loosely to densely flowered, cylindric spikes, 20–70 mm long, length 3+ times width, corollas creamy white to yellow, 1.8–2.8 mm long, glabrous; filaments 5–7 mm long; fruits 40–150 mm long, 15–25 mm wide. Dry, mostly calcareous soils, thorn scrub, desert grasslands, disturbed sites; Bexar, Goliad, and Wilson cos. (Turner et al. 2003); AZ and TX; Mexico (Baja California Sur, Chihuahua, Coahuila, Nuevo León, Tamaulipas). Apr–Jun. [*Acacia wrightii* Benth., *A. greggii* A. Gray var. *wrightii* (Benth.) Isely] *Senegalia wrightii* is closely related to *S. greggii*, under which *S. wrightii* is sometimes treated as a synonym or variety. *Senegalia wrightii* is distinguished from *S. greggii* in being usually

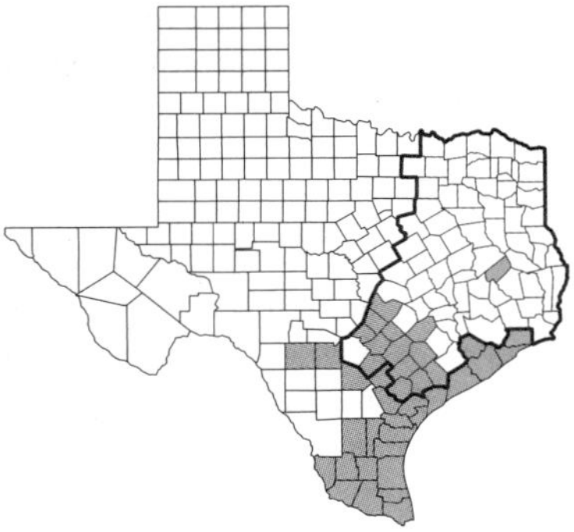

Rhynchosia americana

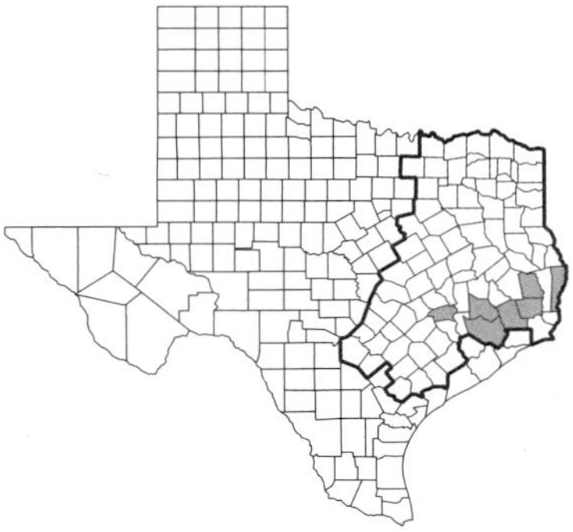

Rhynchosia difformis

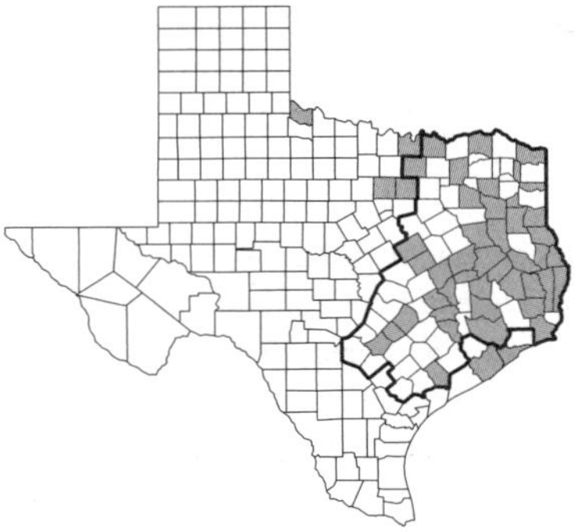

Rhynchosia latifolia

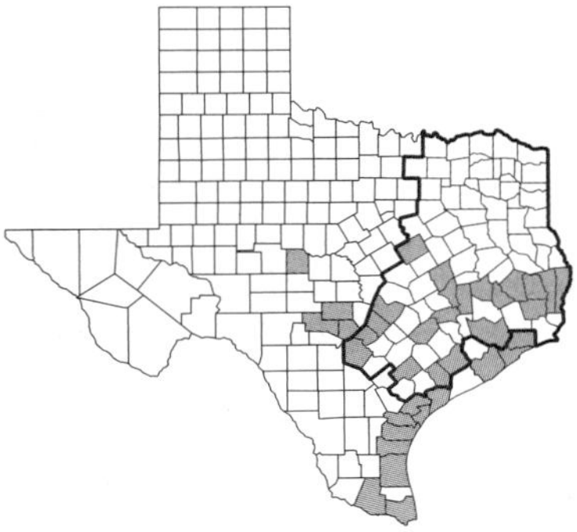

Rhynchosia minima

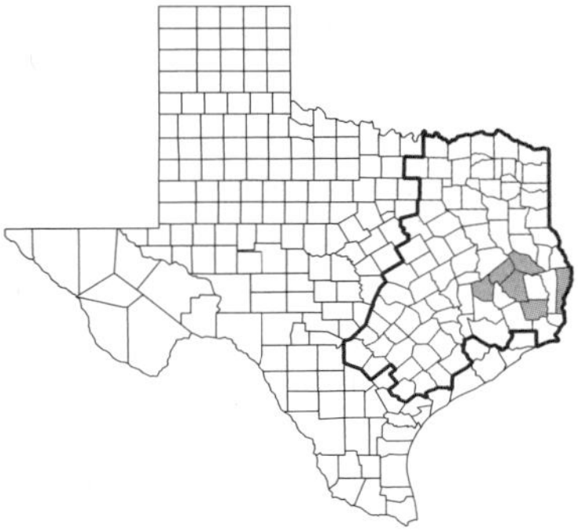

Rhynchosia reniformis

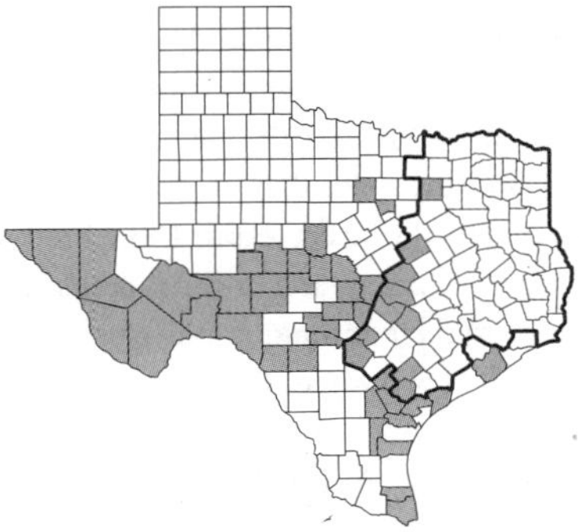

Rhynchosia texana

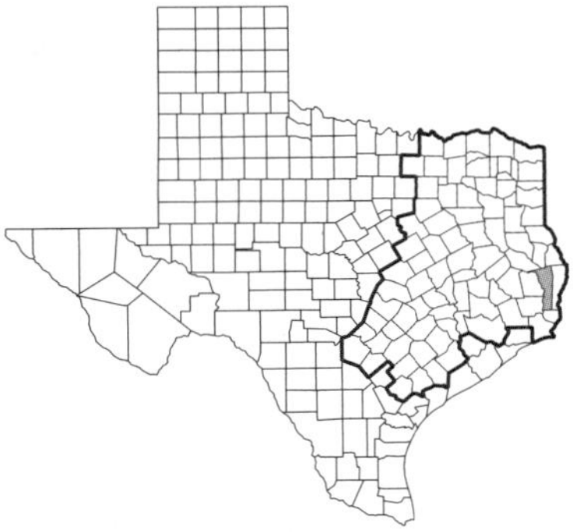

Rhynchosia tomentosa

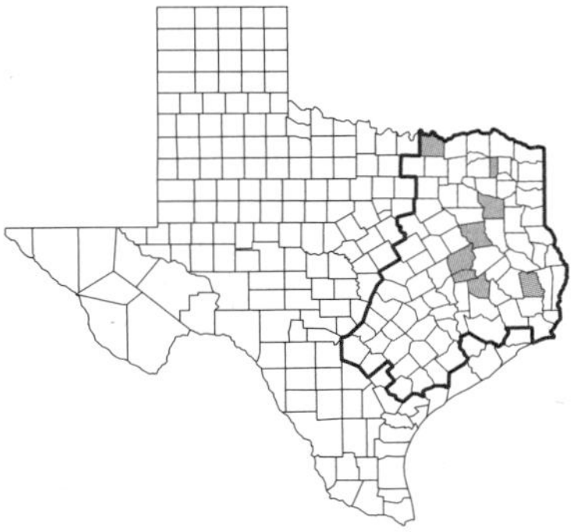

Robinia hispida

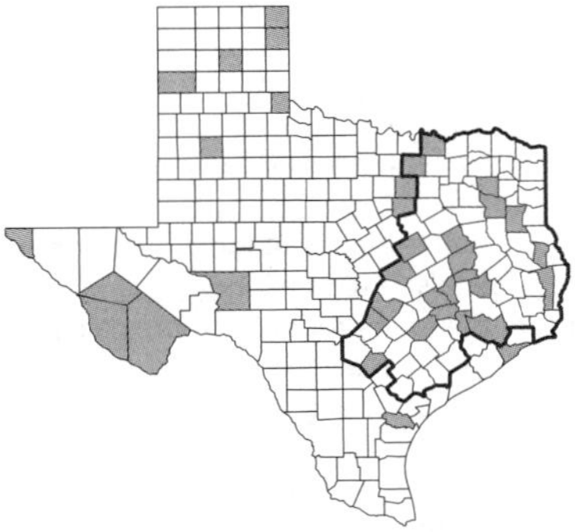

Robinia pseudoacacia

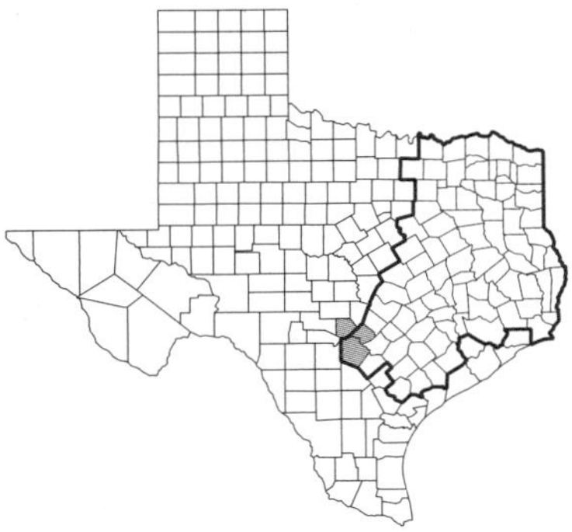

Securigera cretica

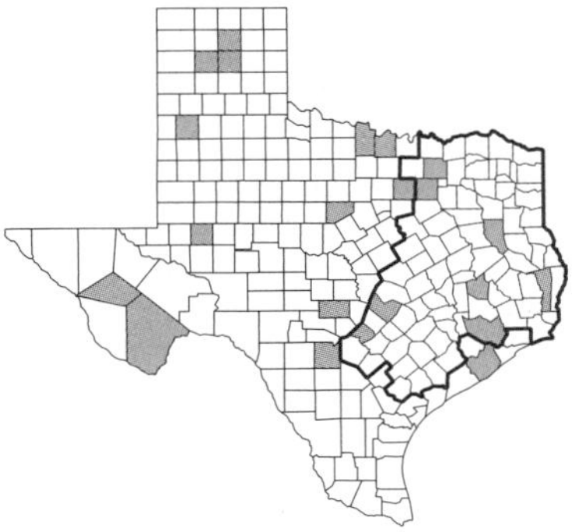

Securigera varia

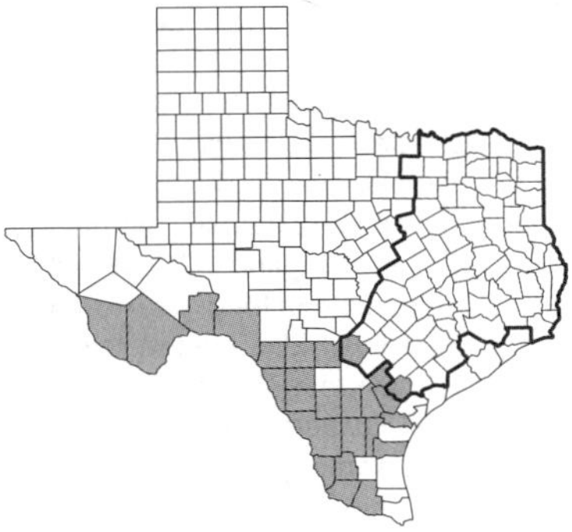

Senegalia berlandieri

more robust, with larger leaves, petioles, and petiolar glands, and leaflet blades are obovate to oblanceolate; *S. greggii* is smaller and leaflet blades are mostly oblong to only slightly obovate. Both hybridize with *S. berlandieri*, and probably with each other (Seigler & Ebinger 2023).

SENNA Mill. SENNA

Ours annual or perennial, unarmed herbs or shrubs; leaves once even-pinnately compound; petiolar glands (extrafloral nectaries) often present or less commonly absent; stipules small, often soon shed; inflorescences racemes with 1–40 flowers, these sometimes clustered; flower stalks without bracteoles; flowers perfect, caesalpinioid; calyces early or late caducous, corollas yellow or orange, petals subequal, stamens 10, unequal, usually not all fertile (often only ca. 7 functional), separate; anthers basifixed, curved, dehiscing by apical pores; fruits, legumes, not elastically (= explosively) dehiscent; seeds 10–20.

A genus of ca. 240 (Isely 1998) or up to ca. 350 species (Marrazzi & Vincent 2023) depending on circumscription; pantropical in distribution but mostly in the New World; extending northward in xeric regions (Isely 1990). *Senna* has sometimes been lumped into the genus *Cassia* (along with *Chamaecrista*), but *Senna* differs in characters such as anthers dehiscent by apical pores, filaments not sigmoid, and bracteoles absent (Irwin & Barneby 1982). Irwin and Barneby (1982) gave a detailed key separating *Cassia*, *Chamaecrista*, and *Senna*. Reported to have the "vibrator" or "buzz" pollination syndrome; pollinators (such as bumblebees) shake the anthers by vibrating their thoracic flight muscles at a certain frequency; this sets up a resonance in the anthers or the space they enclose, and the otherwise inaccessible pollen is released from the terminal pores of the anthers and collected by the insect (Barth 1985; Proctor et al. 1996). Extrafloral nectaries are known to protect some species from herbivore damage (see discussion under *S. occidentalis*). Some are poisonous or important medicinally (source of pharmaceutical senna) due to their anthraquinone glycosides; others are cultivated as ornamentals (Marazzi & Vincent 2023). *Senna bauhinioides* (A. Gray) H.S. Irwin & Barneby is mapped in Goliad County in Kartesz (2023) and an Oberlin College (OC) specimen is cited, but the specimen could not be found (Dr. Michael Moore, email comm.). It is included in the key for clarity but is not currently known in East TX. (Ancient Arabic name of these plants, *sana*) (subfamily Caesalpinioideae, tribe Cassieae)
REFERENCES: Irwin & Barneby 1982; Marrazzi & Vincent 2023.

1. Leaflets usually 2 (1 pair) per leaf; plants 0.7 m or less tall.
  2. Flowers solitary; calyces usually persistent into developing fruit; leaflets 2–5 mm wide; fruits 8–15 mm long; plants to 0.2 m tall ______ **S. pumilio**
  2. Flowers 1–5 per peduncle; calyces caducous; leaflets 4–16 mm wide; fruits 17–35(–50) mm long; plants 0.2–0.7 m tall.
    3. Leaflet blades lanceolate or lanceolate-oblong, length usually more than 4 times width; petals 12–17 mm long ______ **S. roemeriana**
    3. Leaflet blades oblong, ovate-oblong, oblong-elliptic, or obovate-elliptic, length usually < 4 times width; petals 5–10 mm long ______ *S. bauhinioides*
1. Leaflets 4–numerous (2 or more pairs) per leaf; plants variable in size, but often > 0.7 m tall.
  4. Leaflets 4–6 per leaf.
    5. Plants not malodorous; leaflets oblanceolate to lanceolate or lanceolate-elliptic, acute to acuminate at apex; petiolar gland (between lowest leaflets) erect; mature fruits terete, 4–12 cm long, 6–10 mm wide ______ **S. corymbosa**
    5. Plants malodorous; leaflets obovate, apex rounded and apiculate; petiolar gland appressed; mature fruits flat, (6–)16–20 cm long, 2.5–6 mm wide ______ **S. obtusifolia**
  4. Leaflets 8–numerous per leaf.
    6. Petioles without glands; mature fruits 10–15(–19) cm long, winged; showy cultivated shrub ______ **S. alata**
    6. Petioles with gland(s) between each pair of leaflets or below lowest pair of leaflets (near base of petiole); mature fruits 4–12(–15) cm long, not winged; perennial herbs or less showy non-native shrub.
      7. Petiolar glands more than one, occurring between each pair of leaflets; mature fruits 3–6.5 cm long; native species only in w portion of East TX ______ **S. lindheimeriana**
      7. Petiolar glands usually solitary, between only lowest pair of leaflets or well below lowest leaflets (usually near base of petiole); mature fruits 7–12(–16) cm long; native or non-native species widely distributed.
        8. Petiolar gland 1(2) between lowest pair of leaflets; fruits at maturity 9–16 mm wide, cylindric; petals 16–26 mm long; rare non-native shrub ______ **S. pendula**
        8. Petiolar gland below lowest pair of leaflets (usually near base of petiole); fruits at maturity 6–11 mm wide, laterally compressed; petals 8–17 mm long; native or widely naturalized perennial herbs.
          9. Leaflets 4–5(–6) pairs per leaf; flowers (1–)2–5 per raceme; mature fruits brown with lighter margins, somewhat raised over the seeds but without distinct septations between the seeds; naturalized non-native ______ **S. occidentalis**
          9. Leaflets 5–9 pairs per leaf; flowers 5–15 per raceme; mature fruits black without lighter margins, flat, with cross septations (visible externally as very distinct lines across the fruit); native species ______ **S. marilandica**

Senna alata (L.) Roxb., (winged), emperor's-candlesticks, candle-bush. Perennial shrub to 2–3(–4) m tall, usually dying back to the ground in winter in TX; leaflets 6–14 pairs per leaf, 7–21 cm long, 3–13.5 cm wide; inflorescence a terminal, showy, spike-like raceme 30–60(–100) cm long with 40 or more flowers, the inflorescences sometimes compound; corollas lemon yellow; upper 3 stamens much reduced, the lowest 3 stamens with larger anthers; our only *Senna* with winged fruit, 10–15(–19) cm long, 0.9–1.2 cm wide. Cultivated and escapes in disturbed habitats; Harris, Jefferson, (Kartesz 2023) and Walker (iNaturalist Contributors 2023j) cos.; s U.S. in AL, FL, LA, MS, and TX; Mexico; West Indies; Bermuda; Central America; Asia, Africa; Atlantic Islands; Indian Ocean Islands; Hawaii; and Australia. Jun–Oct. Native of tropical America. [*Cassia alata* L.] This species is cultivated worldwide for its showy inflorescences (Marrazzi & Vincent 2023).

Senna corymbosa (Lam.) H.S. Irwin & Barneby, (with corymbs), argentine senna. Glabrous perennial shrub or small tree to 3.5 m tall; evergreen where not killed back by cold; leaflets 2–3 pairs per leaf, usually 3, 2.5–6 cm long, 0.5–1.4 cm wide, oblanceolate to lanceolate or lanceolate-elliptic, acute to acuminate apically; flowers 4–18 in small racemes or clusters; corollas golden yellow; upper 3 stamens much reduced; lowest 3 stamens with long arching filaments; fruits pendulous, terete, cylindric, straight, 4–12 cm long, 0.6–1 cm wide, corrugated over seeds, indehiscent. Cultivated and escapes; Travis Co. on the w margin of East TX and Harris Co. (Kartesz 2023) on se corner of East TX; also Edwards Plateau and Gulf Prairies and Marshes (Kartesz 2023); se U.S. from SC s to FL and w to TX; w Europe and South Africa. Aug–Feb. Native of Argentina and Uruguay. [*Cassia corymbosa* Lam.] This species is reported to be able to tolerate light frosts (Isely 1998). It has been cultivated over two centuries and is a common ornamental in botanical gardens worldwide (Marrazzi & Vincent 2023).

Senna lindheimeriana (Scheele) H.S. Irwin & Barneby, (for Ferdinand Jacob Lindheimer, 1801–1879, German born TX botanist), lindheimer's senna. Erect perennial herb to 1.5 m tall; leaflets 4–8 pairs per leaf, 2–5 cm long, 0.8–2 cm wide, oblong-elliptic or ovate- to obovate-elliptic; inflorescences racemes with 5–25 flowers; corollas pale yellow; upper 3 stamens much reduced; fruits compressed, 3–6.5 cm long, 5–9 mm wide. Limestone soils in open woodlands and grasslands; Bexar, Comal, Hays, and Travis (Turner et al. 2003) cos.; mainly Edwards Plateau to Trans-Pecos. [*Cassia lindheimeriana* Scheele] The leaves are reported to be toxic but not fatal to grazing animals; they also act as a strong laxative (Kingsbury 1964; Ajilvsgi 1984). This attractive native species is now frequently cultivated in wildflower gardens outside its native range in East TX (E. Keith, pers. obs.).

Senna marilandica (L.) Link, (of Maryland), wild senna, maryland senna. Erect perennial herb from woody root, 0.6–2 m tall, essentially glabrous; leaflets 5–9 pairs per leaf; petiolar glands usually conical; inflorescences racemes with 5–15 flowers; corollas yellow; upper 3 stamens much reduced; lowest 3 stamens with somewhat larger anthers; fruits ascending, flat and curved downward, 6–11 cm long, 8–11 mm wide, flattened, the segments rectangular, tardily dehiscent. Open woods, fields, and creek banks; widespread in East TX; also Johnson (BRIT) and Tarrant (Kartesz 2023) cos. just to the w and Gulf Prairies and Marshes; e U.S. from NY s to FL and w to KS, e NE, OK, and TX, introduced in s CA. Jul–Sep. [*Cassia marilandica* L.] This species is sometimes cultivated as an ornamental. The leaves were used as a cathartic by Native Americans; glycosides are present and probably also saponins (Burlage 1968).

Senna obtusifolia (L.) H.S. Irwin & Barneby, (blunt-leaved), sickle-pod, coffeeweed. Erect annual to 1.2(–2.4) m tall, usually smaller (sometimes flowering when quite small), essentially glabrous; leaflets usually 3 pairs per leaf; petiolar glands slender; flowers axillary, 1–2(–3); corollas orangish yellow to pale yellow; upper 3 stamens much reduced; lowest 3 stamens with very large anthers; fruits terete or 4-angled, usually sickle-shaped, (6–)16–20 cm long, 2.5–6 mm wide, faintly corrugated, indehiscent. Ruderal and alluvial habitats in sandy soils; widespread in East TX; also adjacent Gulf Prairies and Marshes; e U.S. from MA s to FL and w to e KS, e NE, e OK, and East Texas, introduced in CA; also Asia; Africa; Atlantic Islands; Indian Ocean Islands; Pacific Islands; and Australia. Jul–Sep. [*Cassia obtusifolia* L.] Glycosides, alkaloids, and other toxins are present which can poison cattle and possibly other livestock if the leaves, stems, or seeds are eaten; fatalities can result (Hardin & Brownie 1993); Burrows and Tyrl (2001) noted that both this species and *S. occidentalis* are poisonous to livestock under certain conditions. The leaves have been used as a vegetable, a purgative drug, and poultice for sores, ulcers and insect bites; the seeds have been used as substitute for coffee and as a blue dye (Irwin & Barneby 1982).

Senna occidentalis (L.) Link, (western), coffee senna, stypicweed, bricho. Erect annual to 1–2.2 m tall, but sometimes flowering when much smaller; leaflets 4–5(–6) pairs per leaf, lanceolate to ovate, acute or acuminate at apex, gland at base of petiole; inflorescences racemes with (1–)2–5 flowers, the racemes sometimes congested; corollas yellow; upper 3 stamens much reduced; lowest 3 stamens with elongate filaments; fruits flat, slightly curved or straight, linear, 8–13.5 cm long, 6.5–9.5 mm wide. Roadsides and disturbed sites; widespread in s 1/2 of East TX; also Gulf Prairies and Marshes and South TX Plains; e U.S. from MA s to FL and w to OK and TX; Mexico; West Indies; Central America; South America; subtropical Eurasia; Africa; and Australia. Aug–Nov. Geographic origin unclear (Irwin & Barneby 1982), possibly pantropical, n to s U.S. [*Cassia occidentalis* L.]

Used as a coffee substitute in some areas; slightly toxic to livestock (Correll & Johnston 1970). The petiolar glands are known to be extrafloral (= in a location other than the flower). These nectar-producing structures have been shown experimentally to attract ants which remove eggs and larvae of insects that damage the plant—this is a type of defense mechanism known as animal guard defense—an animal (ant) is being paid (nectar) to protect the plant. Experiments have demonstrated increased damage to plants from which ants were excluded (Fleet & Young 2000).

Senna pendula (Humb. & Bonpl. ex Willd.) H.S. Irwin & Barneby var. **glabrata** (Vogel) H.S. Irwin & Barneby, (sp.: drooping, pendulous; var.: smooth, hairless), CLIMBING CASSIA, CHRISTMAS SENNA, PENDANT SENNA, GOLDEN SHOWER, CHRISTMAS CASSIA, BUTTERFLY SENNA, VALAMUERTO. Spreading or sprawling shrub to 3(–5) m; leaflets 4–5(–7) pairs, oblong to obovate, rounded apically, 2.5–6.5 cm long, 0.9–2.3 cm wide; inflorescences racemes of 4–35 flowers, the racemes sometimes clustered; corollas golden yellow; stamens unequal, 2 much longer than others; fruits 9–16 cm long, 9–16 mm wide. Cultivated and escaped in ruderal habitats; Bexar Co. (Turner et al. 2003) at extreme sw tip of East TX, Isely (1998) mapped 2 other cos. n and e; otherwise known in TX only from extreme s tip of state; also FL. Feb–Jun. Native of South America. [*Cassia indecora* Kunth var. *glabrata* Vogel, *C. coluteoides* Collad.] Somewhat similar to (but less hardy than—killed to the ground by frost) *S. corymbosa*, but with more leaflets that are rounded apically. Isely (1990) noted that most horticultural material of this species has gone under the name *Cassia bicapsularis* L. This is an extremely widespread variable species (17 varieties—Irwin & Barneby 1982) with material cultivated in TX being var. *glabrata*; however, some authorities (e.g., Turner et al. 2003) do not recognize infraspecific taxa. This species is on the FL invasive plant list (Florida Exotic Pest Plant Council 2007); however, given its rarity in East TX, at least at present it does not seem problematic in the state.

Senna pumilio (A. Gray) H.S. Irwin & Barneby, (dwarf), DWARF SENNA, PYGMY SENNA. Small perennial to 0.2 m tall, usually shorter, nearly glabrous; root woody and tuberous; leaflets 1 pair per leaf, 1–5 cm long, 2–5 mm wide, often folded and appearing linear, pubescent; flowers solitary, on elongate peduncles that hold the flowers beyond the leaves; corollas yellow; upper 3 stamens much reduced; fruits inflated, subglobose, indehiscent, 8–15 mm long, 6–8 mm wide. Gravelly, rocky, or sandy soils in open diverse upland habitats; Bexar Co. (Kartesz 2023) at extreme sw tip of East TX; widely scattered in s and w TX, sometimes abundant and showy; Mexico (Chihuahua, Coahuila, Durango, Nuevo León). Apr–early May. [*Cassia pumilio* A. Gray] Isely (1998) noted that flowering plants are superficially similar to buttercups (*Ranunculus* spp.).

Senna roemeriana (Scheele) H.S. Irwin & Barneby, (for Ferdinand Roemer, 1818–1891, geologist, paleontologist, and explorer of TX), TWO-LEAF SENNA. Perennial 0.2–0.7 m tall, from thick, woody root; stems and leaves pubescent, gray-green; leaflets 1 pair per leaf, 2–7 cm long, 4–14 mm wide, lanceolate or oblong-lanceolate; flowers 1–5 per peduncle; upper 3 stamens much reduced; fruits turgid, oblong, straight or curved, 20–35 mm long, 4.5–6.5 mm wide. Limestone outcrops, woodlands, and grasslands; w margin of East TX and Gonzales Co. (Kartesz 2023) slightly to the e; mainly w 2/3 of TX; sc U.S. in NM, OK, and TX; Mexico (Chihuahua, Coahuila, Nuevo León). Apr–May, sparingly Jun–Aug, and again in Sep. [*Cassia roemeriana* Scheele] *Senna roemeriana* is a common and conspicuous species in open upland habitats throughout the Edwards Plateau (Irwin & Barneby 1982).

SESBANIA Adans. RIVER-HEMP

Annual or perennial herbs or subshrubs unarmed; stems long, green, usually glabrous, unbranched below; leaves alternate, once even-pinnately compound; leaflets numerous, 10–96, 1–4 cm long, entire; flowers papilionaceous in axillary racemes with 1–18 flowers; calyces campanulate, the tube broader than long, the 5 lobes shorter than tube; corollas yellow, orange or red; keel petals auriculate; fruits legumes, oblong or linear, with few–numerous seeds, 1–40(–51).

A genus of ca. 70 species of tropical and subtropical regions nearly worldwide. *Sesbania* is distinctive among legumes and combines morphologies that are atypical among papilionoids. The most unusual morphology of *Sesbania* is the even- or odd-pinnate leaf in which the strictly opposite leaflets fold forward during night-time (nyctinasty), similar to the mimosoid clade of Caesalpinioideae (for example, the sensitive *Mimosa*). Root nodules of *Sesbania* are variable (produced on roots and laterally on stems) and differ from other legume groups, which generally have uniform nodule morphology. *Sesbania* fruits are also distinctive from other legume groups in possessing a well-developed, often spongy, mesocarp having four different variations: tardily dehiscent, linear, and many-seeded (sect. *Sesbania*); tardily dehiscent, bladdery-inflated, and 2-seeded (sect. *Glottidium* (Desv.) (Lavin), indehiscent, 4-winged, and several-seeded (sect. *Daubentonia* (DC.) Benth. & Hook. f.); and indehiscent, torulose, and several-seeded (Mexican sect. *Daubentoniopsis* (Rydb.) Lavin). These divergent fruit morphologies were used traditionally to separate the genus into these four sections, in addition to two others from Africa and the South Pacific. Several species are widely cultivated for multiple agronomic and ecological uses, especially for soil improvement as a green manure and is also used for shade, as windbreaks, cover crops, ornamentals, fish poisons (isoflavones), fiber sources, construction materials,

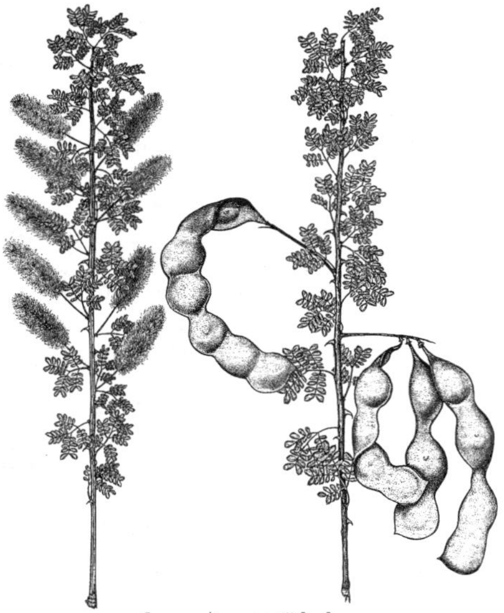

Senegalia greggii [VIN]

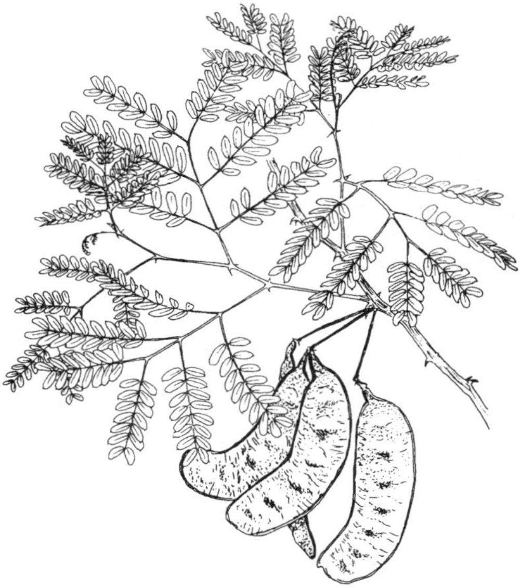

Senegalia roemeriana [LYN]

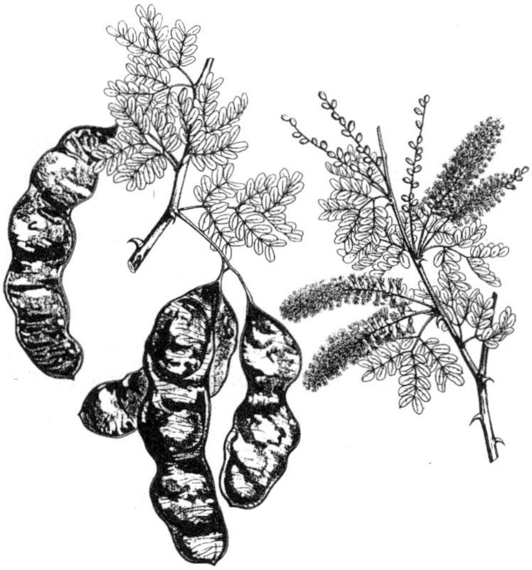

Senegalia wrightii [SA3]

Senna alata [WIG]

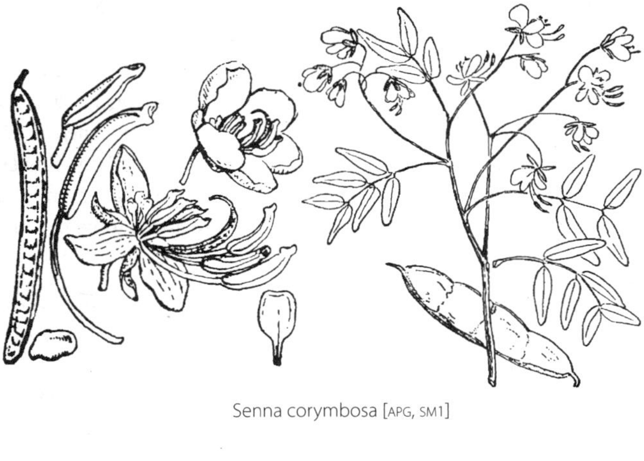

Senna corymbosa [APG, SM1]

Senna lindheimeriana [VIN]

and food for humans and livestock and for heavy-metal bioremediation (Farruggia 2023). (Either from the Arabic: *sesban*, name for *Sesbania sesban* (L.) Merr., or possibly from Persian, *sisabaan*, rope, fiber, or a kind of tree) (subfamily Faboideae, tribe Sesbanieae)

REFERENCES: Isely 1986a; McVaugh 1987; Farruggia 2023.

1. Fruits without wings, 2–4 mm wide, linear; corollas (7–)8–15(–19) mm long, yellow to greenish yellow or orangish yellow, keel blades with basal tooth; ovules and seeds (3–)12–40(–51).
   2. Plants glabrous or nearly so and often glaucous, the leaflets without silky pubescence; corollas yellow with purple or maroon at apex, (10–)14–15(–19) mm long; species widely scattered in e 1/2 of TX ____ **S. herbacea**
   2. Plants pubescent, the leaflets with silky pubescence on their undersurfaces; corollas yellow-orange throughout, (7–)8–9(–19) mm long; species extremely rare or possibly no longer extant in TX, known only from Harris Co. ____ **S. sericea**
1. Fruits flattened-inflated or with 4 conspicuous wings, 7–13(–16) mm wide, broadly oblong; corollas 8–25 mm long, yellow, yellow-orange OR scarlet to orange-red, keel blades without basal tooth; ovules and seeds 1–7(–10).
   3. Herbs, often woody at base; legumes flattened-inflated, flat to elliptic in cross section; seeds 1 or 2(–3); peduncles (3.7–)6.4–7.4(–12.2) cm long; banner yellow-orange with bright yellow spot at base, wings and keel darker ____ **S. vesicaria**
   3. Small trees, shrubs, or subshrubs, woody; legumes 4-angled or winged, square in cross section; seeds (1–)3–7(–10); peduncles (0.6–)0.9–2.4(–4.1) cm long; banner pale yellow, yellow, or orange-scarlet-red, sometimes with darker venation, wings and keel ± same color throughout.
      4. Corollas scarlet to orange-red, (1.8–)2–2.2(–2.5) mm long; legumes with pronounced, spreading wings, slightly torulose or not, acuminate or tapering to beak; species rare in East TX, known only from a few counties in the Pineywoods ____ **S. punicea**
      4. Corollas yellow or orange-yellow, (0.9–)1.3–1.5(–1.7) mm long; legumes with pronounced torulose wings, blunt or abruptly acuminate to beak; species widespread and common in East TX, particularly the s portion ____ **S. drummondii**

Sesbania drummondii (Rydb.) Cory, (for its discoverer, Thomas Drummond, 1780–1835, Scottish botanist and collector in North America), RATTLEBUSH, POISON-BEAN, COFFEE-BEAN, DRUMMOND'S SESBANIA. Subshrub, lower part becoming woody; plant to 3(–4) m tall; leaflets 16–50 per leaf, blades elliptic-ovate to oblong, surfaces glaucous and usually glabrous, sometimes with diffuse, appressed pubescence; flowers yellow or orange-yellow, often with red lines; fruits (3.2–)5.7–6.2(–9.3) cm long, broadly oblong, 4-sided and conspicuously 4-winged, apically short-beaked, the stipe ca. 1.5 cm long; seeds usually 3 or 4 per fruit. Moist riparian areas; mainly s 1/2 of East TX; scattered in e 1/2 of TX, but primarily the s part; se U.S. from SC s to FL and w to AR and TX. Jun–Oct. [*Daubentonia drummondii* Rydb.] The seeds become loose and rattle when moved and are poisonous to sheep, goats, and cattle when eaten, apparently due to the presence of saponins; affected animals develop diarrhea, weakness, lethargy, and sometimes die; the seeds are also poisonous to humans (Sperry et al. 1955; Kingsbury 1964; Ellis 1975; Lewis & Elvin-Lewis 1977; Crosswhite 1980).

Sesbania herbacea (Mill.) McVaugh, (herb-like, with succulent stems), COFFEE-BEAN, BEQUILLA, COLORADO RIVER-HEMP, BIG-POD SESBANIA, SESBANE, DANGLEPOD. Large erect annual to 4.5 m tall, openly branched; leaflets 20–80 per leaf, blades rectangular to oblong, surfaces glabrous; corollas yellow or yellow-orange, banner often with purple spots; fruits (8.3–)16.9–19.4(–23.5) cm long, narrowly linear, unwinged, glabrous, the sutures thickened, with beak usually 5–10 mm long, the stipe ca. 0.5 mm long; seeds (12–)29–36(–51)per fruit. Wet disturbed and riparian areas; scattered in e 1/2 of TX; throughout TX except High Plains; e and s U.S. from MA s to FL, nw to IL and w to TX, disjunct in AZ and s CA; Mexico (Baja California, Colima, Nayarit, Oaxaca, Sinaloa, Sonora); West Indies; Central and South America. Jul–Oct. [*Darwinia exaltata* Raf., *Sesbania exaltata* (Raf.) Rydb., *S. macrocarpa* Muhl. ex Raf.] This species has long gone under the name *S. exaltata* (e.g., Kartesz 1994; Jones et al. 1997; Isely 1998) and sometimes as *S. emerus* (Aubl.) Urb. (e.g., Turner et al. 2003). However, we are following Farruggia (2023) and McVaugh (1987) who made the combination, *S. herbacea*, based on an older name, [*Emerus herbacea* Mill.], published in 1768. Isely (1998) suggested that this species [as *S. exaltata*] is "probably an annual derivative of the tropical shrubby, large-flowered, spring blooming *S. emerus* that, in the United States, I find only in the Florida Keys." The species was used by Native Americans who made nets and fishlines from the strong fibers in the stems (Vines 1960). The seeds have been reported to be poisonous (Hardin & Brownie 1993).

Sesbania punicea (Cav.) Benth., (reddish purple), RATTLEBOX, RED SESBANIA, SCARLET-WISTERIA. Shrub or small tree, lower part becoming woody, to 4 m tall; leaflets 14–40 per leaf, blades elliptic to oblong-obovate, surfaces glabrous or sparsely pubescent; corollas scarlet or orange-red; fruits usually 8–10 cm long, broadly oblong, longitudinally 4-winged, usually with 6 or 7 seeds. Cultivated as an ornamental and escaping in wet and disturbed areas; in se part of Pineywoods; also Gulf Prairies and Marshes; se U.S. from VA s to FL and w to TX, disjunct in CA; also Mexico; West Indies; Central America; and South Africa. Jun–Oct.

Senna marilandica [WIL]

Senna obtusifolia [DIL]

Senna occidentalis [WIL]

Senna pendula var. glabrata [ABG]

Senna pumilio [HEA]

Senna roemeriana [HEA]

[*Daubentonia punicea* (Cav.) DC.] Native of South America. This species is known to hybridize with *S. drummondii* in s MS and there is evidence of introgression (Isely 1990).

Sesbania sericea (Willd.) Link, (silky), SILKY SESBAN, PAPAGAYO. Herb or subshrub, lower part becoming woody, to 6 m tall, stems golden tan with silky pubescence; leaflets 34–96, blades elliptic-ovate to broadly linear, densely golden tan pubescence on leaf undersides, glabrous above; corollas orangish yellow, sometimes with purple spots; fruits (1.7–)10–12.7(–18.8) cm long, linear, usually falcate, unwinged, with ca. 25–35 seeds. Disturbed bottomland areas; known in TX only from Harris Co. (*Kuntze 23815* NY; *Mears 728* TEX), last collected in TX in 1966; FL and TX; introduced in Mexico; Sri Lanka; West Indies; Central America; and South America. Jun–Oct. [*Coronilla sericea* Willd.] Probably native to Africa, but now in both the Old and New World tropics (Isely 1990). Isely (1990) noted that this species resembles *S. herbacea* but differs in having "distinctly pubescent leaflets." No other North American species shares this attribute (Farruggia 2023).

Sesbania vesicaria (Jacq.) Elliott, (bladder-like), BLADDERPOD, BAGPOD. Unarmed robust annual herb to 2.5 m tall, pith spongy; leaves once pinnately compound; leaflets 16–50, blades elliptic-ovate to oblong, glabrous; flowers 1–6(–10) in uncrowded axillary racemes; corollas usually 8–9 mm long, reddish brown to orange or yellowish or tinged with pink or red, banner with bright yellow spot, the wings and keel, often darker; fruits persistent and conspicuous, the body of fruits oblong, (4.5–)6.1–6.7(–8.2) cm long, 1.1–1.6 cm wide, acuminate at both ends, the valves of fruits separating at maturity into a thicker outside layer and a papery-membranous, thinner, inside layer; stipe of fruits 0.8–1.5 cm long; seeds 1 or 2(–3). Disturbed open areas and open wetland and riparian areas; abundant throughout East TX; e 1/2 of TX w to West Cross Timbers and Edwards Plateau, s to Gulf Prairies and Marshes and South TX Plains; se U.S. from NC s to FL and w to e OK and East TX. Jul–Sep. [*Glottidium vesicarium* (Jacq.) R.M. Harper] Without fruits this species resembles *Sesbania herbacea*, but that species has larger (11–19 mm long) yellow or yellow-orange flowers. The seeds and green leaves are reportedly poisonous to livestock due to the presence of saponins (Kingsbury 1964; Hardin & Brownie 1993).

STROPHOSTYLES Elliott WILD BEAN, FUZZY-BEAN

Annual or perennial twining or trailing vines, glabrous to pubescent; leaves pinnately compound with 3 leaflets, petiolate, stipulate; leaflets ovate to linear, entire or 3-lobed, stipellate; racemes axillary, long pedunculate, few-flowered; petals short-clawed, usually pinkish to purplish; keel strongly and conspicuously incurved or sickle-shaped; stamens 10, diadelphous; fruits linear, sessile, dehiscent; seeds 4–10.

A North American genus of 3 species (Delgado-Salinas et al. 2023). It is similar to *Phaseolus* ("very weakly distinguished"—Correll & Johnston 1970) but differs "in its long peduncles with few apical, subsessile flowers and its curved rather than spiral keel" (Isely 1990). East TX species are easily recognized in the field by the combination of these characters. Riley-Hulting et al. (2004) determined that *Strophostyles* is unequivocally monophyletic, sister to the South American genus *Dolichopsis* Hassl., and comprises the three taxa treated here with no detectable infraspecific taxa. (Greek: *strophe*, a turning, and *stylos*, a pillar or style, apparently in reference to the style which is curved to conform with the incurved or sickle-shaped keel surrounding it) (subfamily Faboideae, tribe Phaseoleae)

REFERENCES: Stace & Edye 1984; Riley-Hulting et al. 2004; Delgado-Salinas et al. 2023.

1. Banners 3.6–7(–8.3) mm long; leaflet blades lanceolate, margins entire, never deeply lobed, usually with persistent pubescence above; fruits pubescent at maturity, 1.2–4.7 cm long; peduncles slender, 1.2–11(–12.3) cm long ____ **S. leiosperma**
1. Banners 6.7–15 mm long; leaflet blades lanceolate to narrowly lanceolate or pandurate (= fiddle-shaped), margins entire or lobed, usually glabrate above; fruits glabrate at maturity, 3–9.6 cm long; peduncles stout, (2.4–)5–30 cm long.
   2. Leaflets typically lobed, the terminal one often > 2 cm wide, ca. 1.5–3 times as long as wide; keel petals with slender, curved beaks; bracteoles as long as or longer than calyx tube ____ **S. helvola**
   2. Leaflets rarely lobed, the terminal one usually < 2 cm wide, ca. 2–8 times as long as wide; keel petals with stout, erect, slightly curved beaks; bracteoles subtending individual flowers shorter than calyx tube ____ **S. umbellata**

Strophostyles helvola (L.) Elliott, (pale brownish yellow), AMBERIQUE-BEAN, TRAILING WILD BEAN. Perennial; stems to 3 m long, glabrous to pilose; leaflets ovate to ovate-lanceolate, sparsely strigose on both surfaces, usually glabrate above, lateral leaflets 1.3–6.3 cm long; 0.6–4.3 cm wide, terminal leaflet 1.8–7.2 cm long, 0.9–4.6 cm wide, often 3-lobed or with an indentation on one or both sides; peduncles (2.4–)5–21.5 cm long; bracteoles nearly equal in length to slightly longer than calyx tubes; calyces 2.3–6.1 mm long including longest lobe; banners 6.7–13.3 mm long, the keel incurved, darkened apically; fruits 3–9.6 cm long; seeds 5–10, covered with a cellular or waxy layer. Sandy open ground; common throughout East TX; widespread mainly in e 1/2 of TX, rare in Panhandle; e U.S. from ME s to FL and w to KS, NE, OK, SD, and TX. Jul–Oct. [*Phaseolus helvulus* L. (original

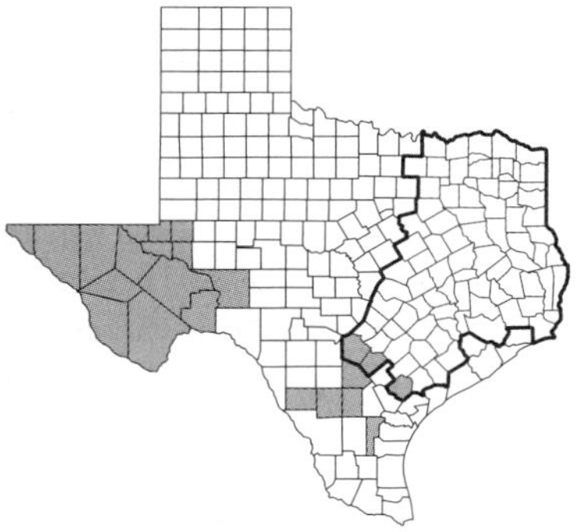
Senegalia greggii

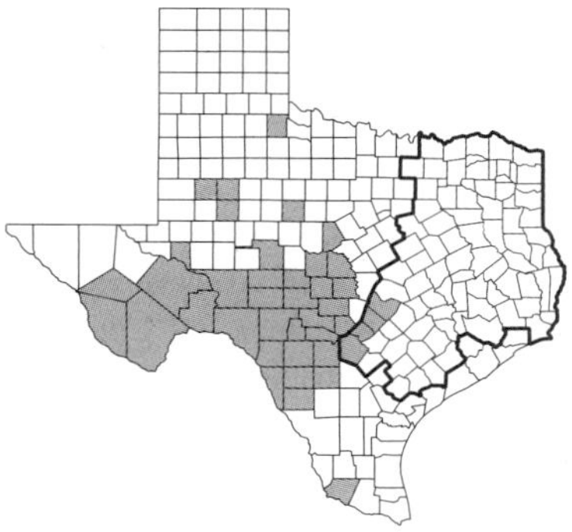
Senegalia roemeriana

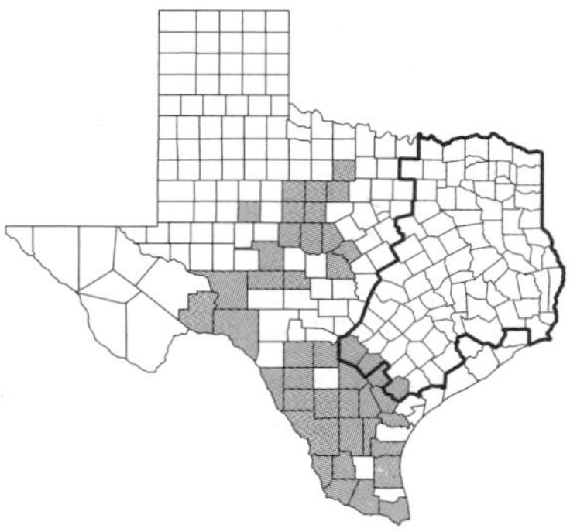
Senegalia wrightii

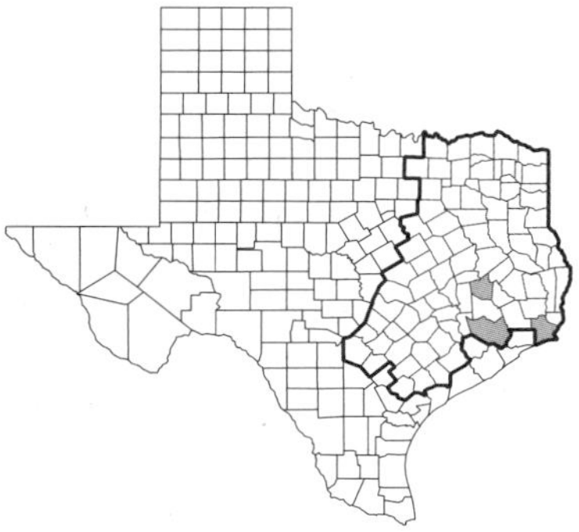
Senna alata

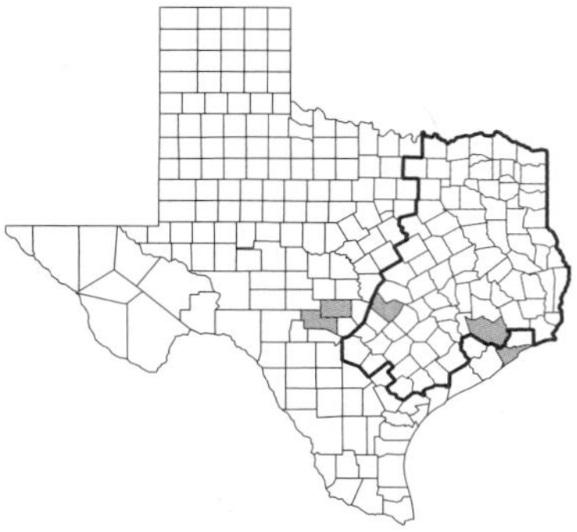
Senna corymbosa

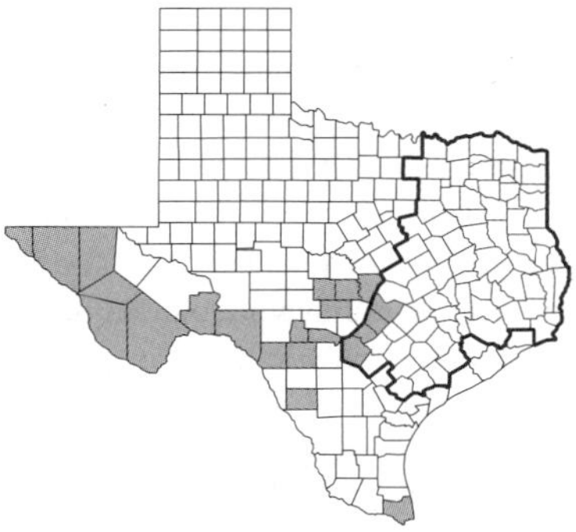
Senna lindheimeriana

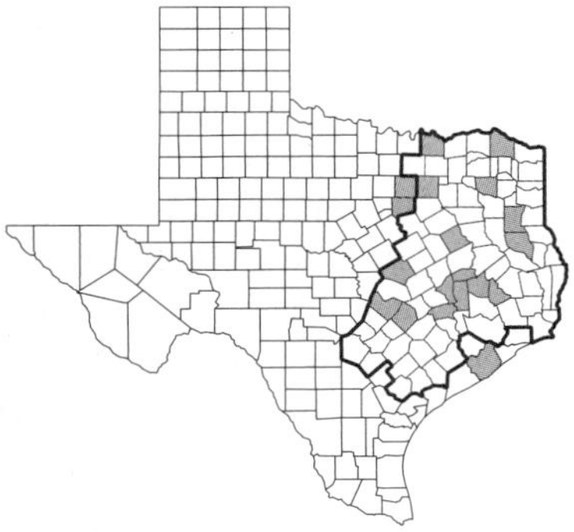
Senna marilandica

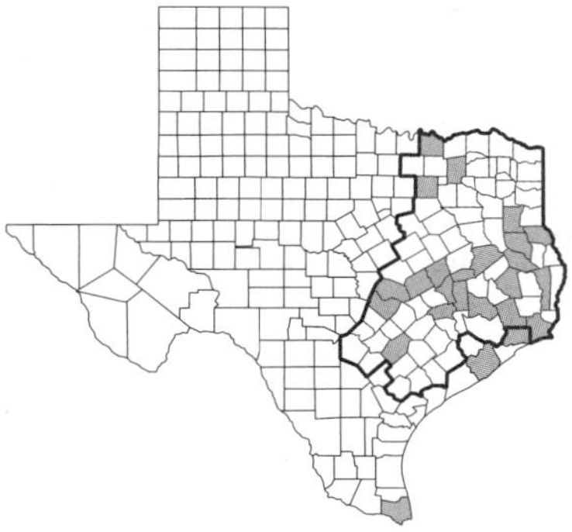
Senna obtusifolia

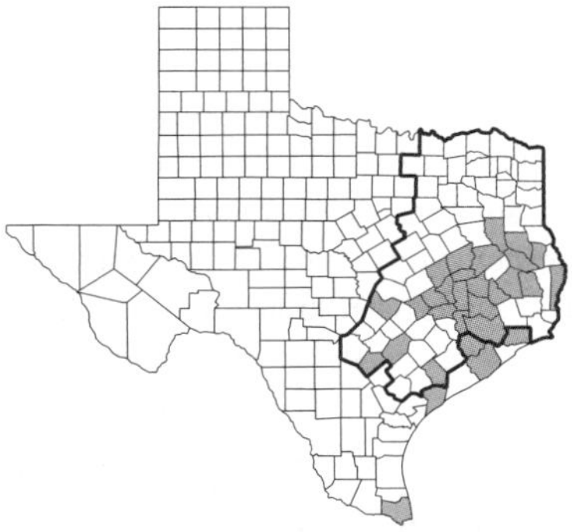
Senna occidentalis

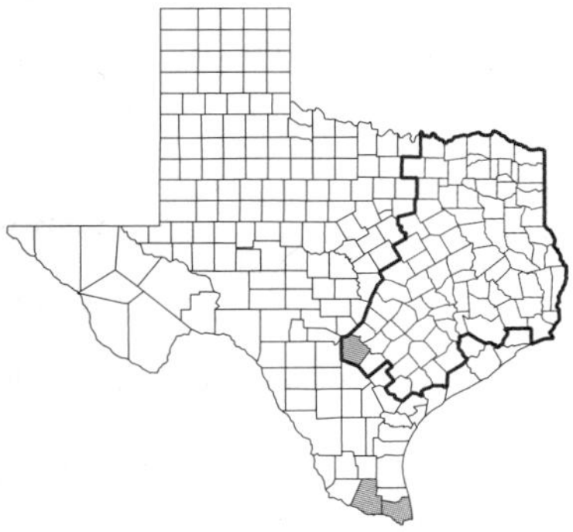
Senna pendula var. glabrata

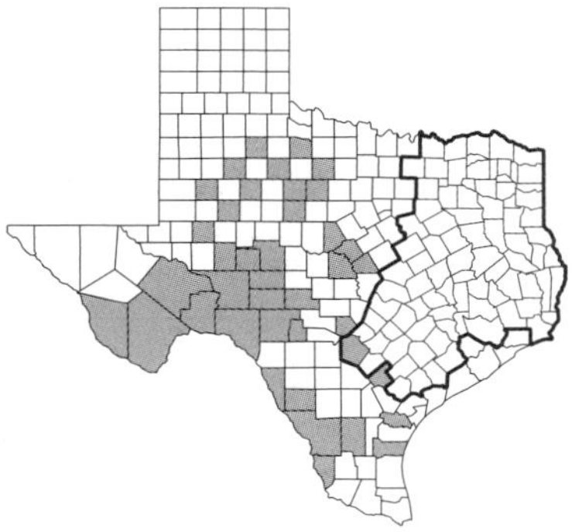
Senna pumilio

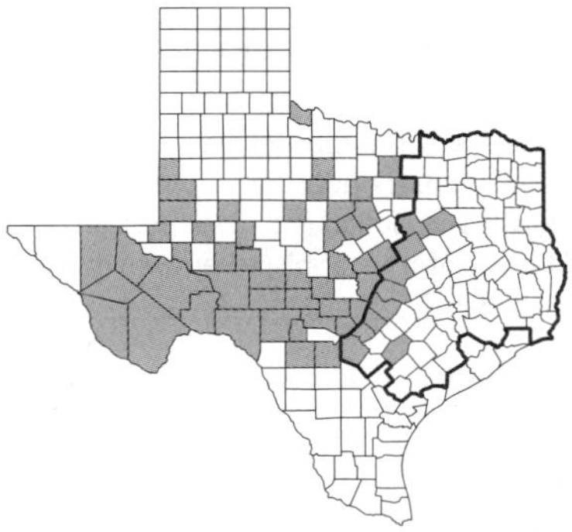
Senna roemeriana

spelling by Linnaeus of the specific epithet)] Verdcourt (1997) determined that the original epithet was probably written in error since "*helvulus*" does not have a meaning and that Willdenow (1802) corrected the spelling to "*helvolus*" which translates to "pale brownish yellow." This epithet was conserved by the General Committee of the ICBN (Nicolson 1999). The combination of lobed leaves, long glabrate fruits, and keel petals with slender curved beaks is distinctive.

Strophostyles leiosperma (Torr. & A. Gray) Piper, (smooth-seeded), SLICK-SEED FUZZY-BEAN. Annual or short-lived perennial vines to 1 m long, pilose; leaflets linear to narrowly ovate, pilose above and usually with pubescence below, lateral leaflets 1.5–5.6 cm long; 0.2–2.1 cm wide, terminal leaflet 1.7–5.4 cm long, 0.2–2.3 cm wide, entire, rarely with an indentation on one side or with very shallow lobing; peduncles 1.2–11(–12.3) cm long; bracteoles shorter than or nearly equal to calyx tube; calyces 1.3–3.9 mm long; banners 3.6–7(–8.3) mm long, the keel incurved; fruits 1.2–4.7 cm long; seeds 4–9, shiny and mottled. Open areas on sand or clay; scattered throughout East TX; widespread in TX, except Trans-Pecos; mostly c U.S. from IN s to n FL and w to CO and NM, including OK and TX, disjunct or introduced in PA; Ontario, Mexico (Chihuahua). Jun–Sep. [*Phaseolus leiospermus* Torr. & A. Gray] *Strophostyles leiosperma* is characterized by its narrower and sericeous leaflets and fruits and relatively small flowers making this species easily distinguishable from the other two species (Delgado-Salinas et al. 2023).

Strophostyles umbellata (Muhl. ex Willd.) Britton, (with umbels, the flowers arising from a central point), PINK FUZZY-BEAN. Perennial vine to 1.5 m long; leaflets lanceolate to narrowly elliptic to narrowly ovate (rarely linear), glabrate above and with only sparse pubescence below, rarely shallowly lobed, lateral leaflets 1.5–4.2 cm long; 0.3–2.7 cm wide, terminal leaflet (1.5–)2–4(–7) cm long, 0.2–2.2(–3) cm wide; peduncles (4.4–)6–30 cm long; bracteoles only 0.8–2.4 mm long, shorter than calyx tubes; calyces 2.4–7 mm long; banners 7.3–15 mm long, keel erect, slightly curved; fruits 3–7 cm long; seeds 5–10, with cellular or waxy covering. Sandy soils, pine woodlands and open fields; common in Pineywoods of East TX; widespread in e 1/2 of East TX w to Brazos, Henderson, and Lamar (Kartesz 2023) cos.; e U.S. from NY s to FL and w to KS, OK and TX. Jun–Sep. [*Phaseolus umbellatus* (Muhl. ex Willd.) Nutt.] The combination of usually unlobed leaves, short bracteoles, and relatively large flowers distinguishes this species (Isely 1990). The prominent keel beak with a thick base held close to the banner is also a distinguishing character when in flower (Delgado-Salinas et al. 2023).

STYLOSANTHES Sw. PENCIL-FLOWER

Herbs, perennial, or subshrubs, unarmed; roots thickened; stems prostrate to erect or ascending, sometimes viscid, pubescent or glabrous; leaves alternate, odd-pinnate; stipules present, foliaceous, amplexicaul, adnate to petiole for most of its length; petiolate; leaflets 3, blade margins entire, surfaces glandular-punctate or epunctate, glabrous or pubescent; inflorescences with 1–8 flowers, terminal or axillary, spikes or solitary flowers, erect; bracts present, persistent, foliaceous; bracteoles persistent, 1 outer, 1 or 2 inner; flowers papilionaceous; calyx tubular, lobes 5, unequal, adaxial 4-lobed; corolla yellow or orange-yellow, inserted near apex of calyx tube; petals evenly graduated; wings shorter and narrower than banner, not adnate to keel, clawed, auriculate; keel incurved or rostrate; stamens 10, monadelphous; anthers dimorphic, 5 versatile anthers alternating with 5 sub-basifixed; ovary subsessile, style curved, glabrous or pubescent; fruits loments, sessile or stipitate, compressed, articulate, ovate, individual segments indehiscent, pubescent, segments 2, breaking apart, densely pubescent, distal segment fertile, reticulate; style persistent as a beak, beak hooked or curled; seeds 1 or 2, black, compressed, ovoid, shiny.

A genus of 37 species of tropical and warm areas of the world (Mohlenbrock 2023a). Two South American species of *Stylosanthes* produce sticky secretions which immediately immobilize larvae of the cattle tick, *Boophilus microplus*, and the larvae are poisoned within 24 hours by an unidentified vapor from the secretions (Sutherst et al. 1982). The two East TX species are easily recognized in the field by the combination of conspicuous stipules and orange to orange-yellow flower color. (subfamily Faboideae, tribe Dalbergieae)

REFERENCES: Mohlenbrock 1957 [1958], 1958, 2023a.

1. Stems densely pubescent with short, viscid hairs; fruit beak to 1/5 as long as distal segment; known only from Dewitt and Goliad cos. in sw portion of East TX ____ **S. viscosa**
1. Stems glabrous or hispid but never glandular pubescent; fruit beak 1/2 as long as distal segment; widespread in East TX ____ **S. biflora**

Stylosanthes biflora (L.) Britton, Sterns, & Poggenb., (two-flowered), SIDE-BEAK PENCIL-FLOWER. Perennial herb from a tough, deep, vertically or obliquely branched root, glabrate or with bristly pubescence; stems decumbent to erect, 10–60 cm long; leaves pinnately compound; leaflets 3, narrowly lanceolate or oblanceolate to elliptic, 1–4 cm long, acute, with short, bristle-point; stipules fused and forming tube around stem, united with petiole ca. 2/3–3/4 its length, the apical 3–8 mm of stipules free, awl-like; free portion of petiole (part not united with stipules) short (1–4 mm), the leaves thus sometimes appearing subsessile; flowers solitary or up to 8 together, often only one open at a time, terminal, closely subtended by reduced leaves; petals orange

Sesbania drummondii [LYN]

Sesbania herbacea [GEO]

Sesbania punicea [REE]

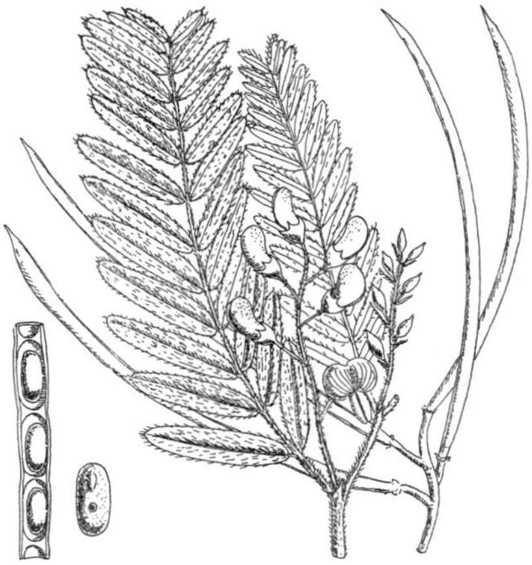

Sesbania sericea [KEW]

Sesbania vesicaria [GWO]

Strophostyles helvola [GLE]

Strophostyles leiosperma [GAT]

Strophostyles umbellata [BB2, WIL]

to orange-yellow, banner 4.5–7 mm long; wings and keel 3–4.5 mm long; fruits 2.5–5 mm long, with 2 segments, the lowest usually sterile, pedicel-like, and with white pubescence, the upper fertile, the beak hooked, 0.5–1 mm long. Sandy open woods and adjacent roadsides; widespread in East TX; also adjacent Cross Timbers and Prairies and n Gulf Prairies and Marshes; e and s U.S. from NY s to FL and w to WI then sw to KS, OK, and TX, disjunct to s AZ; Mexico. May–Oct. [*S. biflora* var. *hispidissima* (Michx.) Pollard & C.R. Bell]

Stylosanthes viscosa (L.) Sw., (sticky, clammy), VISCID PENCIL-FLOWER. Perennial herb, ascending and spreading or prostrate and matted, much branched, to 1 m tall, densely pubescent with short, viscid hairs; stipular sheaths 3–5-veined, teeth of distal stipules shorter than sheath; petiole 2.5–5 mm long; leaflet blades elliptic to lanceolate, 25 mm long, 5 mm wide, viscid, base cuneate, margins entire, apex acute or obtuse, surfaces punctate abaxially, short-hairy or hispidulous; inflorescences 2–5-flowered, ovoid, to 2 cm long, axis rudiment absent; outer bracts 3-foliolate, inner bracts 1-foliolate, sheath 5–7-veined, hispidulous and viscid abaxially; corolla yellow; banner suborbiculate, 4–7 mm long; wings 4–5 mm long; keel 3–4 mm long; fruits reticulate to 2.5 mm wide; fertile segment 1; proximal segment sterile; distal segment 2–4 mm long, short-hairy; beak strongly hooked, 0.5–1.5 mm long, often curled. Sandy prairies and rocky slopes; Dewitt (*Riedel s.n.*, 1941, TEX) and Goliad (Kartesz 2023) cos. in sw portion of East TX; e portion of s TX; only AZ and TX in the U.S.; Mexico; West Indies; Central America; and South America. Jun–Aug. [*Hedysarum hamatum* L. var. *viscosum* L.] Turner et al. (2003) mapped *S. viscosa* in Karnes Co. probably incorrectly based on the *Riedel s.n.* specimen from w Dewitt Co. and mapped in Turner (1959) on the boundary between Dewitt and Karnes cos. A few collections have since been made in Goliad Co. and are mapped in Kartesz (2023).

STYPHNOLOBIUM Schott PAGODA TREE

☛Species ca. 9, c, e United States, Mexico, Central America, nw South America (Colombia), Asia; introduced elsewhere in temperate and subtropical regions (Sugar & Krings 2023). [Greek *styphnos*, astringent, and *lobos*, pod, alluding to taste of fleshy pulp] (subfamily Faboideae, tribe Sophoreae)
REFERENCES: Sousa & Rudd 1993; Sugar & Krings 2023.

Styphnolobium affine (Torr. & A. Gray) Walp., (related), EVE'S-NECKLACE, TEXAS SOPHORA, EVE'S NECKLACE-POD. Plant usually a small tree to 10 m tall; petiole 1.8–2.5 cm long, leaflets (9–)13–19, petiolules 1–2 mm long, blades narrow to broadly elliptic or obovate, (0.6–)1–3.7(–4.5) cm long, (0.3–)0.6–1.7(–2) cm wide, apex obtuse to rounded, mucronate, surfaces evenly sparsely to densely pubescent abaxially, glabrous or glabrate adaxially; flowers showy, but obscured by new leaves, white to yellow or pink to purple; corollas 10–14 mm long; fruits with 1–4(–8) fertile segments, irregularly moniliform (= constricted at intervals and resembling a string of beads—hence the common name), initially gray hairy but maturing glossy black; seeds brown, 5.2–7 mm long. Rocky limestone slopes and ravines or calcareous soils in woods, floodplains, and stream margins; mostly w portion of East TX, uncommon and scattered eastward; mainly e 1/2 of TX; also Cross Timbers and Prairies and Edwards Plateau westward; sc U.S. in sw AR, w LA, s OK, and TX. Mar–May. [*Sophora affinis* Torr. & A. Gray] The seeds are reported to be poisonous (Correll & Johnston 1970). It is frequently cultivated as an ornamental and occasionally escapes outside its native range (Kartesz 2023; Sugar & Krings 2023). ☠

TEPHROSIA Pers. CATGUT, HOARY-PEA

Clump-forming perennials with tough roots; stems erect to sprawling, pubescent to glabrous; leaves once odd-pinnately compound; leaflets many, conspicuously pilose, often with prominent bristle-tip; stipules very slender; flowers in terminal or lateral, spike-like, short or elongate racemes with 8–45 flowers; petals large and showy; fruits legumes, laterally compressed, dehiscent; seeds 6–12.

☛A genus of ca. 400 species mainly in tropical and subtropical regions, especially Africa (Isely 1998; Nesom 2023e). ☠ Some *Tephrosia* species are the source of the alkaloid rotenone, used as a fish poison and insecticide (Jacobson 1982). Use of *Tephrosia* species as a fish poison developed independently in the Americas, Africa, Asia, and Australia; typically, the pounded leaves, branches, or roots are thrown into a body of water and the fish rise stunned to the surface (Wood 1949). (Greek: *tephros*, ash-colored or hoary, presumably from the dense pubescence of this color in some species) (subfamily Faboideae, tribe Millettieae)
REFERENCES: Wood 1949; Nesom 2023e.

1. Leaflets narrow, linear (narrowly to broadly), usually 3–6 times as long as wide; mature fruits 5 mm or less wide; species widespread in East TX.
   2. Corollas bicolored, the standard light yellow, the wings and keel partly or wholly pink or red; flowers crowded in thick, short racemes 2–8 cm long; stems ascending or erect, 30–70 cm tall; leaflets usually acute at apex (also bristle-tipped) ____ **T. virginiana**
   2. Corollas of 1 color, all petals white to yellowish white or whitish pink, changing to rosy or purple-red; flowers loosely spaced in long-peduncled, slender, elongate racemes (10–)20–40 cm long; stems spreading to largely

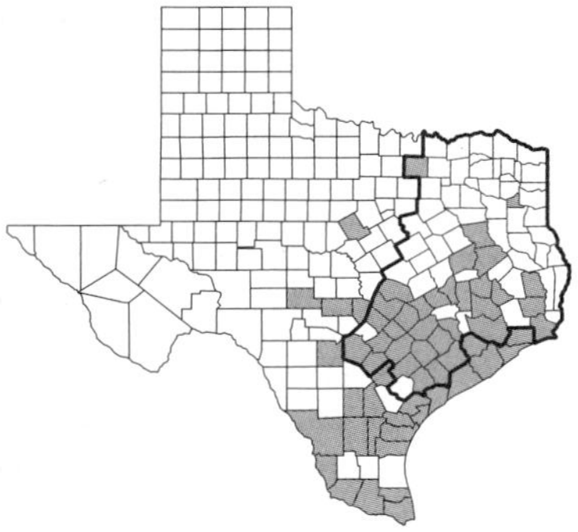
Sesbania drummondii

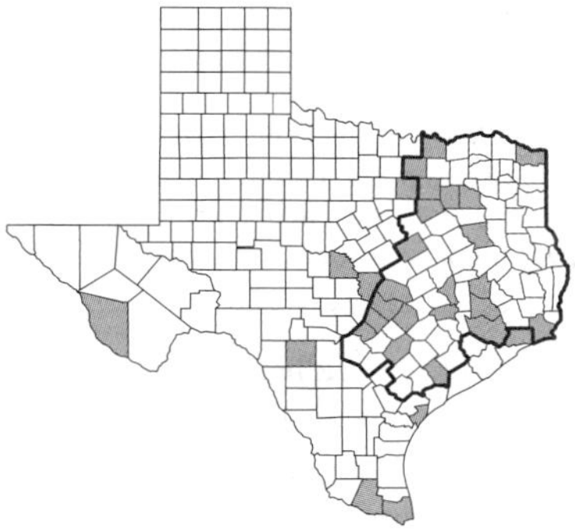
Sesbania herbacea

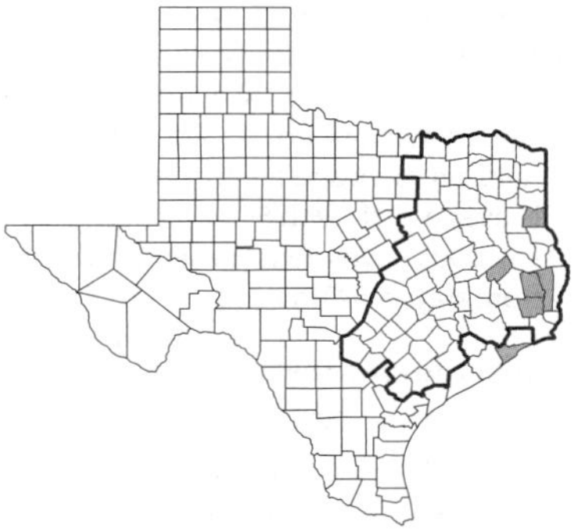
Sesbania punicea

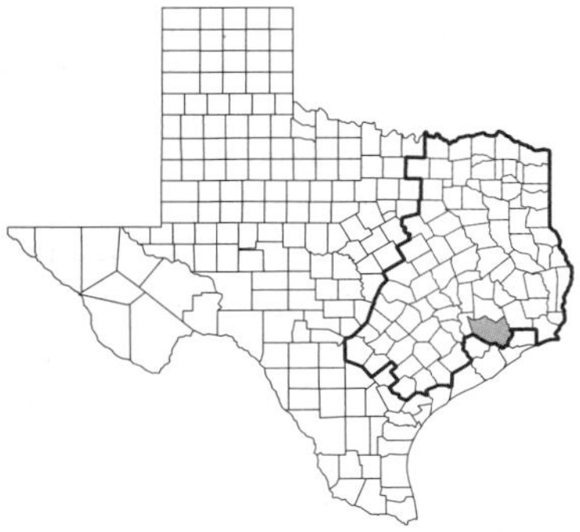
Sesbania sericea

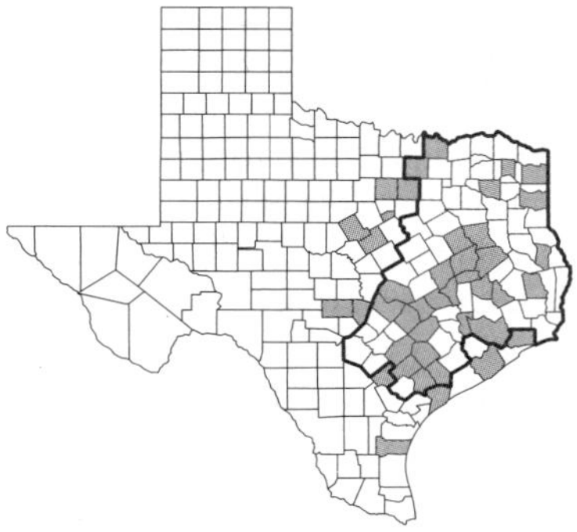
Sesbania vesicaria

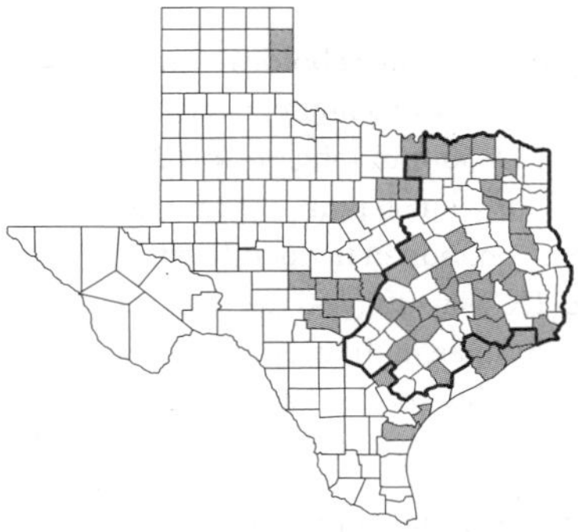
Strophostyles helvola

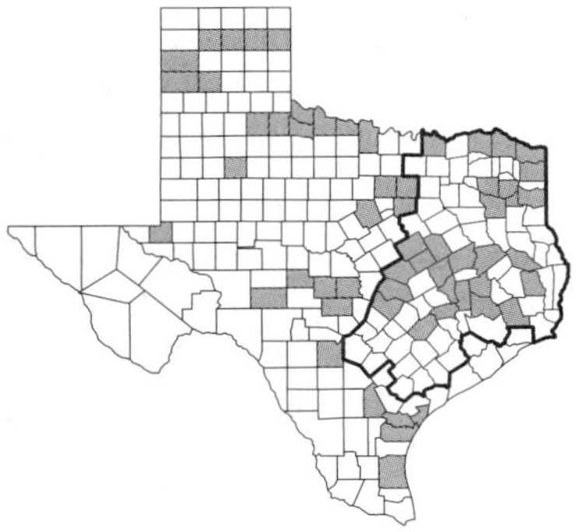
Strophostyles leiosperma

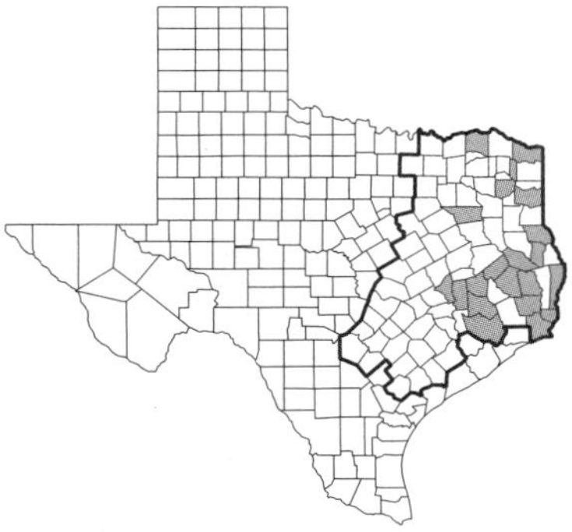
Strophostyles umbellata

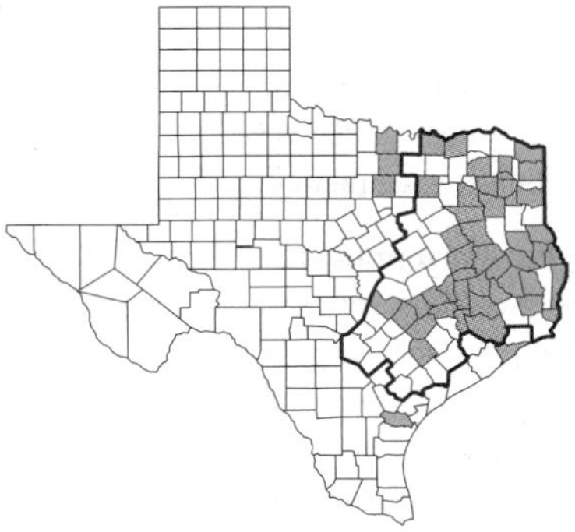
Stylosanthes biflora

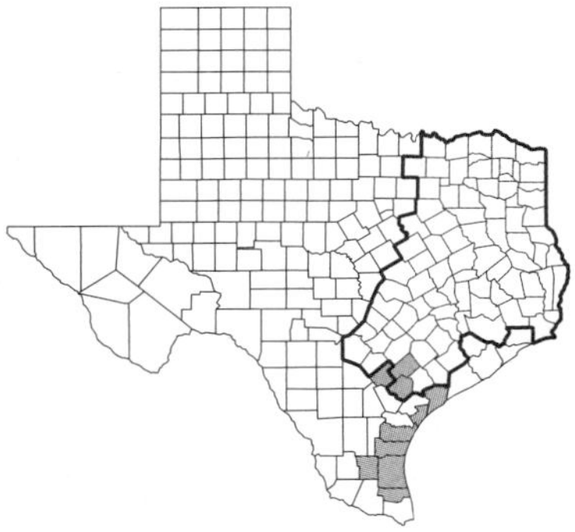
Stylosanthes viscosa

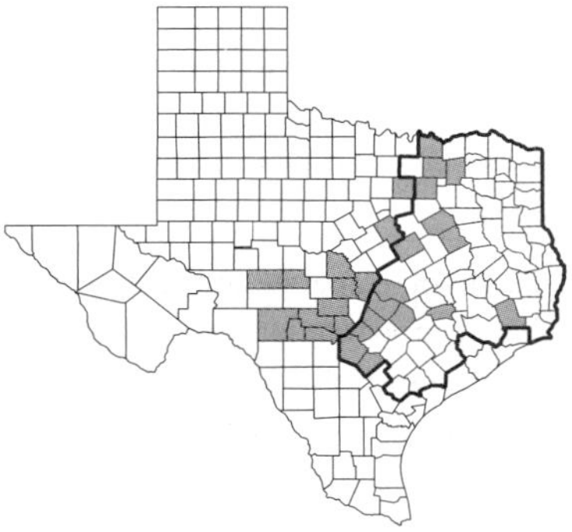
Styphnolobium affine

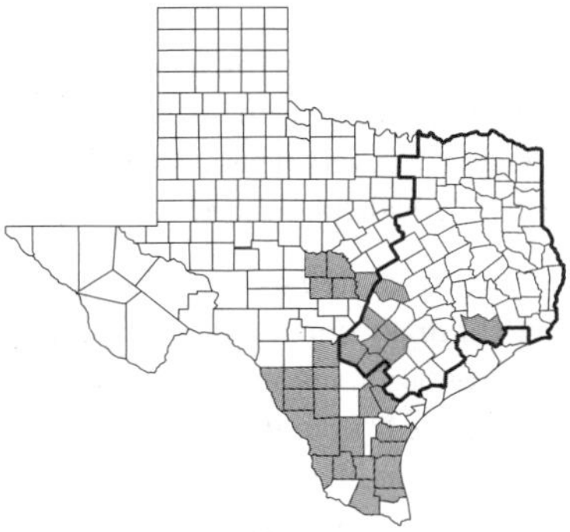
Tephrosia lindheimeri

decumbent, 30–100 cm long; leaflets usually obtuse to truncate (also bristle-tipped) ______ **T. onobrychoides**

1. Leaflets broad, obovate or elliptic to suborbicular, only 1–2 times as long as wide; mature fruits 6 mm or more wide; species limited to the sw margin of East TX.
   3. Leaves mostly with 9–13 leaflets (varying from 5 to 17), the leaflets with a conspicuous margin of whitish hairs; hairs on calyx and back of standard whitish; species of sw margin of East TX n to Williamson Co. ______ **T. lindheimeri**
   3. Leaves mostly with 5 or 7 leaflets (varying from 3 to 9), the leaflets without a conspicuous margin of hairs (inconspicuous tawny hairs present); hairs on calyx and back of standard golden to rusty; species known in East TX only from Hays and Travis cos. ______ **T. potosina**

Tephrosia lindheimeri A. Gray, (for Ferdinand Jacob Lindheimer, 1801–1879, German-born TX collector), LINDHEIMER'S TEPHROSIA, ROUND-LEAF TEPHROSIA. Stems prostrate to decumbent, 40–100 cm long, densely hirsute-pilose; leaflets (5–)9–13(–17), 12–30(–35) mm long, 12–23 mm wide; racemes with 18–25 flowers, corollas 12–15 mm long, pink, rose-purple, or violet-red, the standard with a white spot near base; fruits 40–50 mm long, 6–8 mm wide. Open areas with deep sand or limestone derived soils; sw margin of East TX n to Williamson Co. (Kartesz 2023), disjunct to Harris Co. in s part of East TX (*Fisher B15645* BRIT; *Small 11805* NY); also South TX Plains and e Edwards Plateau; endemic to TX. Apr–Jul.

Tephrosia onobrychoides Nutt., (resembling *Onobrychis*—another genus of Fabaceae), MULTI-BLOOM TEPHROSIA. Stems ascending to decumbent or reclining, rarely erect, 30–100 cm long; leaflets (9–)13–25(–29), blades obovate, elliptic to oblong-oblanceolate, (15–)20–50 mm long, 4–16(–20) mm wide; racemes with (12–)20–45 flowers, corollas 15–18 mm long, white to yellowish white or whitish pink, changing to rosy or purple-red; fruits 40–80 mm long, 4–5 mm wide. Diverse habitats on sands or clays, often common in frequently burned pine forests; throughout East TX, widespread in the e 1/2 of TX; s U.S. from w AL w to OK and TX. May–Jul, less freely to Sep. [*T. onobrychoides* var. *texana* (Rydb.) J.F. Macbr., *T. texana* (Rydb.) Cory] The light-colored flowers open in the evening and close the next morning; before closing, they change color becoming rosy or purple-red (Ajilvsgi 1984). This species has a surprisingly broad ecological amplitude, occurring in diverse habitats ranging from mima mound prairies in Grayson Co. in the extreme nw part of East TX to longleaf pine forests in the Big Thicket.

Tephrosia potosina Brandegee, (of San Luis Potosí, the Mexican state where the type was collected), EDWARDS PLATEAU HOARY-PEA. Species similar to *T. lindheimeri*; stems decumbent 25–50 cm long, hirsute to hirsute-pilose; leaflets mostly 5 or 7 per leaf (varying from 3 to 9), 12–30(–38) mm long, 12–25(–32) mm wide, obovate to orbicular; racemes with 8–20 flowers, corollas 12–15(–18) mm long, rose-purple, the standard with a greenish spot near base; fruits 32–50 mm long, 6–8 mm wide. Open areas on gravelly limestone or calcareous soils; species rare in TX, Comal and Hays (TEX) cos. at sw margin of East TX; also Uvalde Co. in South TX Plains (TEX); in the U.S. known only from TX; also n Mexico Apr–Aug. Turner et al. (2003) mapped Comal Co., but no specimens could be located. This species is very rare in TX with few known populations. ⚠

Tephrosia virginiana (L.) Pers., (of Virginia), GOAT'S-RUE, VIRGINIA TEPHROSIA, DEVIL'S-SHOESTRING, CATGUT. Stems erect, 30–70 cm tall, densely villous, sericeous, strigulose, or glabrate; leaflets (9–)13–23(–35), 10–25(–30) mm long, (3–)4–7 mm wide, elliptic to oblong, surfaces strigose to sericeous; racemes with 10–40 flowers, corollas 15–20 mm long, bicolored, the standard light yellow, the wings and keel partly or wholly pink or red; fruits 30–55 mm long, 4–5 mm wide. Sandy open woods or open ground; widespread in East TX; also scattered in TX to the w; e U.S. from NH s to n FL and w to KS, MN, NE, OK, and TX; s Ontario. May–Jun. Rotenone is present in the roots; the plant was used medicinally by Native Americans and early settlers (Ajilvsgi 1984) and by Native Americans as a vermifuge and fish poison (Burrows & Tyrl 2001).

TRIFOLIUM L. CLOVER

Annual or perennial herbs; leaves alternate, palmately or pinnately compound; leaflets 3, rounded or notched at apex, entire or with small, sharp teeth, usually with well-developed petioles; flowers sessile or pedicellate, in terminal or axillary, sessile to long-peduncled, very dense, head-like or spike-like inflorescences; petals usually withering and persistent long after flowering; stamens 10, diadelphous; fruits with 1–4(–9) seeds.

A genus of ca. 280 species of bee-pollinated herbs of temperate and subtropical areas (Vincent & Gillett 2023). Many species are agriculturally important as forage, fodder, hay, in crop rotation, and as bee plants; some species are cyanogenic or otherwise poisonous; a number are used as ornamentals, especially *T. incarnatum* in East TX. The common name CLOVER is said to have originated "from the great club that Hercules carried, known as 'clave trinodis'. It was said to have been a tree trunk with three large roots. The three leaves of the clover reminded Germanic tribes of this 'clave trinodis' and so in Danish it became 'klever' and in Anglo Saxon 'claefre' then in English 'claver' which eventually became 'clover'. The clover shape on the suit of playing cards we call 'clubs' is so called due to this relationship of clover to Hercules' club" (Wildflowers of the Southeastern U.S. 2008). (Latin: *tres*, three, and *folius*, leaf, from the three leaflets) (subfamily Faboideae, tribe Trifolieae)

Stylosanthes biflora [BB2]

Stylosanthes viscosa [MCR]

Styphnolobium affine [SA3]

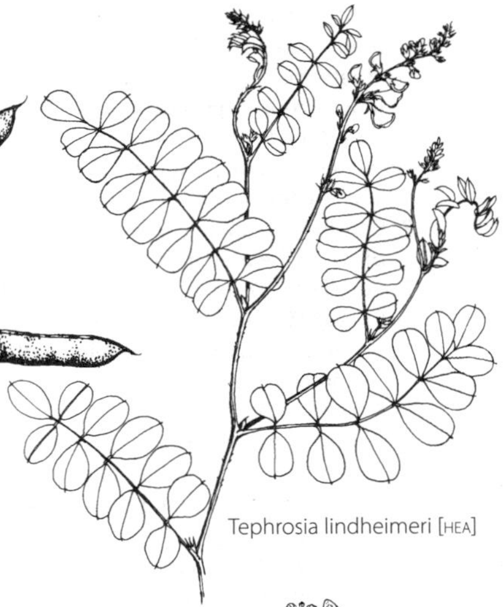

Tephrosia lindheimeri [HEA]

Tephrosia onobrychoides [HEA]

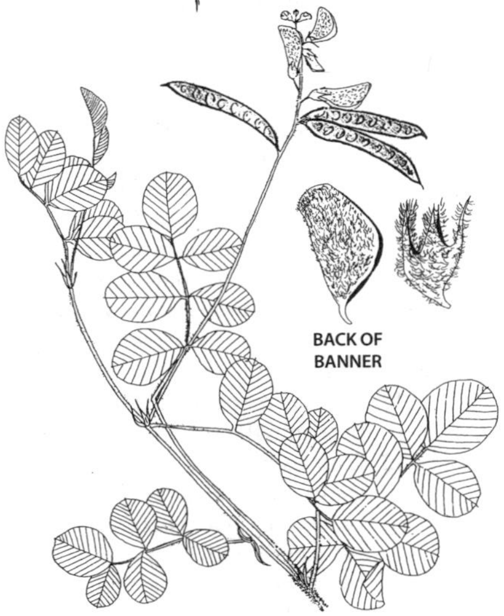

Tephrosia potosina [LNL]

REFERENCES: McDermott 1910; Hennen 1950; Isely 1951; Hermann 1953; Brown & Peterson 1984; Zohary & Heller 1984; Keith 2013a; Vincent & Gillett 2023.

1. Petals bright to pale yellow or greenish yellow; terminal leaflet on a stalk longer than those of the lateral leaflets (leaves pinnately compound).
   2. Petals striate-sulcate (with lines in shallow grooves—use lens); inflorescences 30–60-flowered; 7–10 mm wide; flowers 4.3–5 mm long, with banner 2.8–3.2 mm wide; petioles of middle stem leaves usually longer than leaflets ____ **T. campestre**
   2. Petals not striate-sulcate; inflorescences 5–20-flowered; ca. 6 mm wide; flowers 2.8–3.5 mm long, with banner 1–2 mm wide; petioles of middle stem leaves usually shorter than leaflets ____ **T. dubium**
1. Petals scarlet-red to pink-purple, purplish yellow, lavender-pink, pink, white, or yellowish white; terminal leaflet sessile or nearly so (leaves palmately compound).
   3. Petals a combination of yellow-white and purple; heads with flowers of two types: 2–6 outer fertile flowers and numerous, sterile, reduced inner flowers; peduncle in fruit elongating and pushing the bur-like inflorescence against or under the soil surface ____ *T. subterranean*
   3. Petals variously scarlet-red to pink-purple, lavender-pink, pink, white, or yellowish white; heads with only one type of flower—numerous fertile flowers, sterile reduced flowers lacking; peduncle not pushing inflorescence against or under the soil surface (except in rare *T. amphianthum* where solitary cleistogamous flowers near plant base push into the soil).
      4. Calyces pubescent.
         5. Flower clusters sessile or nearly so at the ends of the main branches (clusters immediately subtended by a pair of leaves; peduncles can be up to 10 mm long); corollas pink or rose-purple.
            6. Inflorescences oblong, 0.8–1.6 cm long, 0.6–1 cm wide; stipules subtending inflorescences distinctly striate; rare introduced species known only from Lamar Co. in East TX ____ **T. striatum**
            6. Inflorescences globose or ovoid, 1.2–7.7 cm long, 0.7–2.2 cm wide; stipules not distinctly striate, common introduced species ____ **T. pratense**
         5. Flower clusters on peduncles (5–)10 mm or longer, not immediately subtended by a pair of leaves; corollas variously colored.
            7. Corolla resupinate (corollas flipped or inverted so that standard is on the bottom) and bright pink; calyx bladdery inflated in fruit ____ **T. resupinatum**
            7. Corolla not resupinate, variously colored; calyces not inflated in fruit.
               8. Leaflets narrow, > 3 times as long as wide; stems, leaves, and calyces conspicuously silky pubescent; calyces usually longer than corollas (the calyx lobes are bristle-like and extend beyond the corollas); corollas 5–8 mm long ____ **T. arvense**
               8. Leaflets broader, < 3 times as long as wide (usually < 2 times as long as wide); stems, leaves, and calyces not all conspicuously silky pubescent; corollas usually as long as or longer than calyces (often much longer); corollas of various lengths, but often much longer than 8 mm.
                  9. Corollas a striking scarlet-red; flower clusters spike-like, ± cylindrical, at least 2 times as long as wide ____ **T. incarnatum**
                  9. Corollas not scarlet-red, varying from pink-purple approaching red to white, yellowish white, pink, lavender-pink; flower clusters globose or nearly so, as wide as long.
                     10. Individual flowers when fully opened sessile or nearly so and corollas white, turning pink; calyces ca. 20-ribbed, with bristle-like plumose lobes resulting in the heads becoming bur-like in fruit ____ **T. lappaceum**
                     10. Individual flowers when fully opened on pedicels 1–8 mm long and purple-pink, lavender, or dull red; calyces not as above; heads not bur-like.
                         11. Inflorescences 2–4 cm in diam. at maturity; calyces green; corollas 10–15 mm long, vivid dark pink (flowers often bicolored red and white or pink); species rare and declining ____ **T. reflexum**
                         11. Inflorescences 1–2 cm in diam. at maturity; calyces pink or magenta; corollas 4.5–7 mm long, whitish or pink fading to dull red; locally common ____ **T. carolinianum**
      4. Calyces mostly glabrous (some villous hairs may be present at orifice of tube; ciliate lobes on *T. bejariense*).
         12. Inflorescences axillary; peduncle absent or to 1 mm long; rare waif not established in East TX ____ *T. glomeratum*
         12. Inflorescences terminal or axillary with an obvious peduncle; native or naturalized species.
             13. Corollas magenta or pink; plants producing solitary cleistogamous flowers near plant base which push into the soil and develop into underground fruits; stems creeping; now very rare ____ **T. amphianthum**
             13. Corollas white to yellowish white, pinkish, or rose-pink; plants not producing underground fruits; stems creeping or decumbent to erect.
                 14. Inflorescences ovoid or oblong, 3–6 cm long; flowers sessile; calyces greatly inflated, contracted at mouth, with long (3–5 mm) awn-like teeth; corollas white, becoming pinkish ____ **T. vesiculosum**
                 14. Inflorescences globose, 1–3.5 cm long; pedicels 2–5 mm long; calyces not inflated; corollas variously colored.
                     15. Perennial; stems creeping and rooting at the nodes; peduncles growing from ground level; flower clusters short-racemose, the pedicels arising over a 2–5 mm distance at the end of the

peduncle (tear apart a flower cluster to see this character); lobes of calyx equal to or shorter than tube **T. repens**

15. Short-lived perennial or annual; stems decumbent, ascending, or erect, not rooting at the nodes; peduncles not growing from ground level; flower clusters umbellate, the pedicels all arising at the very end of the peduncle; lobes of calyx mostly longer than the tube.
    16. Calyx lobes broadly oblong, some of them nearly as wide as long, > 1 mm wide, reticulate nerved; calyces mostly glabrous but with ciliate lobes; pedicels villosulous; uncommon native species **T. bejariense**
    16. Calyx lobes < 1 mm wide, not reticulate-veined and without ciliate lobes; pedicels glabrous; uncommon naturalized species or not yet recorded in East TX.
        17. Inflorescences 1–1.8 cm in diam. with 15–50 flowers; calyx lobes proximally scarious-margined, commonly divergently twisted after antithesis, sinuses V-shaped; corolla faint pink, fading to white (or dirty white); widespread naturalized species **T. nigrescens**
        17. Inflorescence 1–2.5 cm in diam., with 10–70 flowers; calyx lobes neither scarious-margined nor bent outwards, sinuses broadly U-shaped; corolla distinctly rose-pink; not currently known from East TX but recorded near the borders with AR, LA, and OK *T. hybridum*

Trifolium amphianthum Torr. & A. Gray, (two types of flowers: showy chasmogamous flowers and ground-level cleistogamous flowers that grow into the ground and form a peanut, which is characteristic of the species), PEANUT CLOVER. Glabrous or sparsely pubescent creeping perennial, 10–25 cm long, rooting at the nodes; leaflet blades obcordate 0.5–1.8 cm long, 0.5–1.5 cm wide; chasmogamous flowers on long peduncle (6–8 cm) and in clusters of open flowers about twice as high as the leaves; corollas 7.5–11 mm long, 2–4 times as long as calyces, magenta or deep-pink; called PEANUT CLOVER because of fruits developing underground from cleistogamous flowers borne near plant base. Prairies, fields, open woods, and roadsides; historically widespread in East TX, but now very rare with most (if not all) historical populations no longer extant; also adjacent e Cross Timbers and Prairies and n Gulf Prairies and Marshes; LA and TX. Mar–Jun. [*T. roemerianum* Scheele] This native species has often been synonymized with the South American species *Trifolium polymorphum* Poir. (e.g., Jones et al. 1997; Diggs et al. 1999; Turner et al. 2003; Kartesz 2023), but Vincent and Gillett (2023) distinguish *T. amphianthum* by its much larger leaves and flowers and essentially glabrous leaves and peduncles. It has only been collected or observed twice since the 1980s in Fayette and Madison cos. Both locations have been searched on more than one occasion over three years, but no plants could be relocated (E. Keith, pers. obs.), and no extant populations are known. It is unknown why this once apparently common species can no longer be found. Campbell et. al. (1988) offered seven possible reasons for the sharp decline of native clover species in Kentucky including the endangered *Trifolium stoloniferum* Muhl. and rare *T. reflexum* Pourr. ex Willk. & Lange including initial habitat destruction; poor dispersal; loss of the natural grazing regime (including bison); increased grazing pressure from cattle, feral hogs, and white-tailed deer; competition from exotic plants; reduced fire frequency; and lack of rhizobial infection. It is probable that some or all of these factors have negatively impacted *T. amphianthum*. Keith (2013a) offered (and continues to offer) a reward to the first person finding a living population. No county distribution map provided. ⚠

Trifolium arvense L., (pertaining to cultivated fields), RABBIT-FOOT CLOVER, OLD-FIELD CLOVER. Annual herb, silky-villous, usually grayish green, 5–30 cm tall; leaflet blades 10–20 mm long, 2–5 mm wide, the margins with teeth small and visible only near the leaflet tips; flowers sessile in short, dense spikes; spikes 5–30 mm long, 8–15 mm in diam.; calyces prominently pilose-plumose, giving the whole inflorescence a silky-hairy appearance, the calyx lobes bristle-like; corollas rose, pinkish, or white, 5–8 mm long, usually exceeded by the calyx lobes. Roadsides, pastures, disturbed areas; spreading throughout East TX (E. Keith, pers. obs.) but most commonly in ne portion (Kartesz 2023); in TX only known from e 1/2 w to Grayson and McLennan cos. (Kartesz 2023); e, n, and w U.S. from ME s to FL and w across n U.S. and s to CA and OR; southward only as far w as IA, KS, OK, and East TX; Canada; s South America; s Africa; New Zealand; and Australia. Mar–Jun. Native of Europe, w Asia, and n Africa. Isely (1998) noted that this species can be an "aggressive primary invader" of sterile disturbed habitats. It is reported to produce dermatitis and photosensitization in domestic animals (Burlage 1968).

Trifolium bejariense Moric., (for San Antonio de Béxar—now San Antonio in Bexar County on se edge of the Edwards Plateau), BEJAR CLOVER. Annual; stems pilose or glabrate, 5–25 cm tall; leaflet blades 5–15 mm long, 3–8 mm wide, margins slightly denticulate or entire, apex acute to acuminate; inflorescences axillary or terminal with 5–15 flowers, globose or subglobose, 1–2 cm long, 1.5–2 cm wide; at least some calyx lobes as broad as long, very unequal; corollas 7–9 mm long, slightly longer than calyces, white to yellowish white and tinged with pink or purple. Open woods, prairies or adjacent roadsides, in sandy or sandy clay soils; widespread in East TX; also n Gulf Prairies and Marshes; mainly TX, but also AR and LA. Mar–May. This species is recognized for its subfoliaceous calyces (Isely 1998).

Trifolium campestre Schreb., (of the fields or plains), LARGE HOP CLOVER, LOW HOP CLOVER, HOP TREFOIL. Erect to decumbent annual 5–40 cm tall; leaflet blades rhombic to obovate, 4–15 mm long, 4–8 mm wide, base cuneate, lateral veins prominent, margins denticulate distally, apex truncate or retuse, surfaces strigose above; peduncles 1.5–3.5 cm long; inflorescences axillary with 30–60 flowers, globose or broadly ovoid, 0.8–1.5 cm long, 0.7–1 cm wide; corollas 5 mm or less long, bright yellow, conspicuously striate-sulcate. Waste areas and roadsides; widespread in East TX; scattered in TX except High Plains and Trans-Pecos; throughout the U.S.; Canada; South America; China; s Africa; Pacific Islands including HI; and Australia. Mar–May. Native of Europe; w Asia; and n Africa. This species along with its smaller relative, *Trifolium dubium*, are ubiquitous roadside plants in the spring in East TX.

Trifolium carolinianum Michx., (of Carolina), CAROLINA CLOVER, WILD WHITE CLOVER. Annual; stems glabrous or sparsely pilose; leaflet blades obovate, 4–14 mm long, 3–11 mm wide, margins denticulate, apex rounded or retuse; peduncles geniculate with inflorescences appearing sideways or inverted, 3.5–10 cm long; inflorescences with 5–40 flowers, globose, 1–2 cm in diam. at flowering; flowers quickly becoming reflexed; calyx lobes pink or magenta, 2–3 times as long as wide; corollas 1–2 times as long as calyces, 4.5–7 mm long, yellowish white or pinkish, changing to brown or dull red. Prairies, fields, lawns, and roadsides on mainly sandy soils; widespread in East TX; also n Gulf Prairies and Marshes; e and se U.S. from PA s to n FL, and w to se KS, MO, e OK, and East TX. Mar–May. This species is apparently related to *T. bejariense* but readily distinguished by the much narrower pink or magenta calyx lobes (Isely 1998).

Trifolium dubium Sibth., (doubtful), LEAST HOP CLOVER, LITTLE HOP CLOVER, SHAMROCK. Erect to decumbent annual 10–40 cm tall; leaflet blades obovate to elliptic-lanceolate, 10–20 mm long, 5–15 mm wide, margins dentate distally, apex rounded or retuse, surfaces glabrous or sparsely hairy; peduncles 1–1.5 cm long; inflorescences axillary or terminal with 5–20 flowers, ovoid or globose, 0.5–0.9 cm long; ca. 0.6 cm wide; flowers 2.8–3.5 mm long; calyces campanulate, 1.5–2 mm long, lobes unequal, longer than tube; corolla pale yellow becoming brown, 2.6–3.2 mm long. Waste areas, lawns, and roadsides; widely scattered in East TX; also a few other scattered localities in TX; throughout most of the U.S.; Canada; s South America; China; Africa; Pacific Islands including HI; and Australia. Mar–May. Native of Europe and w Asia. *Trifolium dubium* is superficially similar to *Medicago lupulina* L.; the latter may be distinguished by its toothed stipules, deciduous corollas, and shiny, black fruits (Vincent & Gillett 2023). This species and *T. repens* are the two species usually considered to be the Irish SHAMROCK (Nelson 1991).

Trifolium incarnatum L., (flesh-colored), CRIMSON CLOVER, ITALIAN CLOVER. Soft-pilose annual, 20–60(–80) cm tall; leaflet blades broadly ovate 10–30(–40) mm long, 10–15 mm wide; margins denticulate, apex emarginate or retuse, surfaces with spreading, pustulate-based hairs; inflorescences elongated, ± cylindrical, 1–2.5 cm in diam. with 25–100 flowers, tapering apically, showy; flowers sessile; calyx tubes densely long hairy; corollas 10–15 mm long, strikingly scarlet-red (white or pink mutants can occur). Sandy roadsides and fields; widespread in East TX, but more common in the Pineywoods; e 1/2 of TX; throughout most of the U.S. except desert sw; British Colombia; Chile; China; s Africa; Pacific Islands including HI; and Australia. Apr–May. Native of Europe. This species is often planted along roadsides for its showy scarlet-red corollas. It was introduced in 1818 and is now widely planted as a forage crop, for hay, or green manure (Isely 1998; Vincent & Gillett 2023). However, overripe crimson clover hay, with numerous stiff hairs on the pedicels and calyces, can be dangerous to horses; indigestible phytobezoars (= hairballs) can form in the stomach and death can result from impaction (Kingsbury 1964; Burlage 1968).

Trifolium lappaceum L., (perhaps from resemblance to bur or burdock from Latin, *lappa*, bur or burdock), LAPPA CLOVER, BURDOCK CLOVER, BUR CLOVER. Annual; stems decumbent to erect, 5–60 cm long, glabrous to thinly pubescent; leaflet blades ovate to obovate, 5–15 mm long, 3–20 mm wide, margins denticulate to dentate, apex rounded, truncate, or emarginate, hirsute; inflorescences terminal and globose with 40–60 flowers, 1–1.3 cm in diam. at flowering, in fruit bur-like (due to calyces), 1.5–2 cm in diam.; calyces ca. 20-ribbed, with bristle-like plumose lobes; corollas ± equal to calyces in length, 7–9 mm long, white turning pink. Weedy areas and roadsides; rapidly spreading in East TX and n Gulf Prairies and Marshes; otherwise, unknown from other regions in TX; widely scattered in the U.S. in AL, FL, LA, MS, NC, NJ, PA, SC, TN, and East TX; also Canary Islands; Uruguay; s Africa; and Australia. Mar–Jun. Native of Mediterranean region. According to Isely (1990), this species can be distinguished by its calyx tube which is 20-ribbed and glabrous, with pilose lobes. It was introduced to the U.S. in about 1903 and has been used as a winter annual pasture and hay crop as well as in roadside grass plantings as a nitrogen source (Vincent & Gillett 2023). First reported for TX by Brown and Peterson (1984), but now rapidly spreading along roadsides in the Pineywoods of East TX (Keith 2013a).

Trifolium nigrescens Viv., (growing dark in color), SMALL WHITE CLOVER. Annual herb, 10–60 cm tall, glabrous or glabrescent; stems usually erect or ascending, rarely prostrate, branched; leaflet blades obovate or obtriangular, 5–25 mm long, 3–25 mm

Tephrosia virginiana [GLE, WIL]

Trifolium amphianthum [ZO1]

Trifolium arvense [BB2]

Trifolium bejariense [ZO1]

Trifolium campestre [GLE]

Trifolium carolinianum [BB2]

Trifolium dubium [GLE]

Trifolium incarnatum [CO1]

wide, margins denticulate, apex rounded to emarginate, surfaces glabrous; peduncles 1.5–4 cm long; inflorescences axillary or terminal with 15–50 flowers, globose, 1–1.8 cm in diam.; flowers 6.5–9 mm long; calyces oblong, 3–5 mm long, glabrous or pubescent, veins 5–10, tube 1–2.5 mm long, lobes equal or unequal, triangular-lanceolate, recurved in fruit, margins green or purple, apex acute to acuminate, orifice open; corolla pink or white, 6–9 mm long. Roadsides, lawns, and fields; officially documented in Jefferson, Walker, and Williamson (Kartesz 2023) cos. in East TX, but much more common and spreading along roadsides throughout East TX (E. Keith, pers. obs.); not known in other regions of TX; se U.S. from GA and FL w to TX. Mar–May. Native of Mediterranean region. This species was first reported in TX in 2013 in Walker Co. but it is actually more common than currently reported (Keith 2013a).

Trifolium pratense L., (of meadows), RED CLOVER. Perennial or biennial; stems usually with spreading pubescence or glabrate, 20–70 cm tall; leaflet blades elliptic to obovate, 15–50 mm long, 7–15 mm wide, marginally with very small teeth to nearly entire, often with a light green or V-shaped marking or blotch; flowers sessile or nearly so, in ± sessile, subglobose to ovoid clusters, 1.2–7.7 cm long, the clusters above a pair of subsessile leaves; corollas 15–18 mm long, pinkish purple or reddish purple. Roadsides and fields; scattered in East TX; known only from e 1/2 of TX including Gulf Prairies and Marshes and Parker Co. to the w (Kartesz 2023); throughout the U.S. including AK; Canada; s South America; s Africa; Pacific Islands including HI; and Australia. May–Jul. Native of Europe. Said by some to be the Irish SHAMROCK; however, it is more likely that either *Trifolium dubium* or *T. repens* is actually the SHAMROCK (Nelson 1991). It is reported to cause slobbering, bloating, stiffness of gait, diarrhea, emaciation, and abortion in cattle; a saponin is present (Burlage 1968).

Trifolium reflexum L., (bent backwards), BUFFALO CLOVER. Annual or biennial, villous or glabrate; stems spreading to ascending or erect 10–60 cm long; leaflet blades ovate, obovate, oblong, or rhombic, margins entire or denticulate, 10–45 mm long, 7–20 mm wide; inflorescences 1.5–4 cm long, 2–4 cm wide, with 10–50 flowers, flowers pedicelled, ascending but eventually reflexed; calyx lobes 2 times or more as long as calyx tubes; corollas 10–15 mm long, varying from bright pink or white to bicolored, aging pink to bright red. Open woods and meadows; scattered mainly in ne part of East TX; otherwise unknown in the state; e U.S. from NJ s to n FL and w to e KS, se NE, e OK, and TX, but extirpated or declining in most states. Apr–Jun. This species has drastically declined throughout its range for unknown reasons; however, in areas frequently burned in AR, the species is still relatively common (E. Keith, pers. obs.). It occasionally reappears after many years following disturbance, then disappears until another disturbance such as fire creates suitable conditions for its reemergence (E. Keith, pers. obs.). Campbell et. al. (1988) hypothesized the possible reasons for this species decline in KY as further elaborated under *T. amphianthum*. Vincent (1991) also studied the species decline in OH. NatureServe: G3, vulnerable; SNR, no state status rank.

Trifolium repens L., (creeping), WHITE CLOVER, DUTCH CLOVER, LADINO CLOVER. Glabrous perennial, often mat-forming, 10–40 cm long; leaflet blades obovate, obcordate, or orbiculate, margins serrulate distally, 6–40 mm long, 4–25 mm wide; petioles long (5–20 cm); inflorescences axillary with 20–40 flowers, 1.5–3.5 cm long and wide; corollas 8–13 mm long, white or pinkish in age. Lawns, roadsides, and disturbed areas; widespread in East TX; scattered elsewhere in the state; throughout the U.S. including AK; Canada; Greenland; South America; Africa; and Pacific Islands. Mar–May, sporadically later. Native of Europe. This species and *T. dubium* are the two species usually considered to be the Irish SHAMROCK (Nelson 1991). It has long been cultivated in North America (probably since the early 1700s) and is regarded as the most important legume for grazing and is estimated to occupy 12 million acres of pasture in the U.S. (Isely 1998). However, in large quantities it can be toxic to livestock due to the presence of a cyanogenic glycoside (Lewis & Elvin-Lewis 1977).

Trifolium resupinatum L., (upside-down), PERSIAN CLOVER, REVERSED CLOVER. Glabrous annual with erect or partly decumbent stems 10–60 cm long; leaflet blades obovate, ovate, elliptic, or rhombic, margins spinulose-dentate, hairy below, glabrous above, 10–30 mm long, 5–30 mm wide; inflorescences hemispherical with 15–30 flowers, 5–10 mm in diam. at flowering, enlarging to 15–20 mm in fruit and forming stellate-spreading clusters, peduncle 1–6 cm long, but the individual flowers sessile or nearly so; flowers resupinate (flipped or inverted so that standard is on the bottom); calyces pilose, papery, prominently reticulate veined, with awn-like teeth < 2 mm long, inflated on one side in age; corollas 4–6 mm long, lavender-pink. Roadsides, fields, and lawns; widespread in East TX; also Denton Co. (BRIT) in the Eastern Cross Timbers and n Gulf Prairies and Marshes; mostly e U.S. from PA s to n FL and w to KS, ND, OK, SD, and TX but also CA, OR, and WA; Ontario; s South America; s Africa; Pacific Islands including HI; and Australia. Mar–May. Native of the Mediterranean region. Isely (1998) noted that this species and *T. incarnatum* are conspicuous on roadsides in East TX in the spring. PERSIAN CLOVER has been cultivated as a forage crop as early as the 1920s and began to spread in Louisiana after a flood of the Mississippi River (Vincent & Gillett 2023).

Trifolium striatum L., (striate or streaked, for the distinctive stipules subtending the inflorescences), KNOTTED CLOVER, PITTS' CLOVER. Herbs usually annual, rarely biennial, 10–50 cm tall, villous; stems erect, ascending, or decumbent, branched from base;

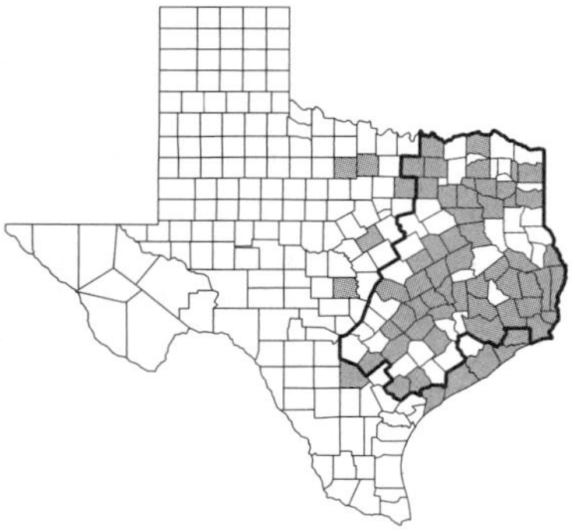

Tephrosia onobrychoides

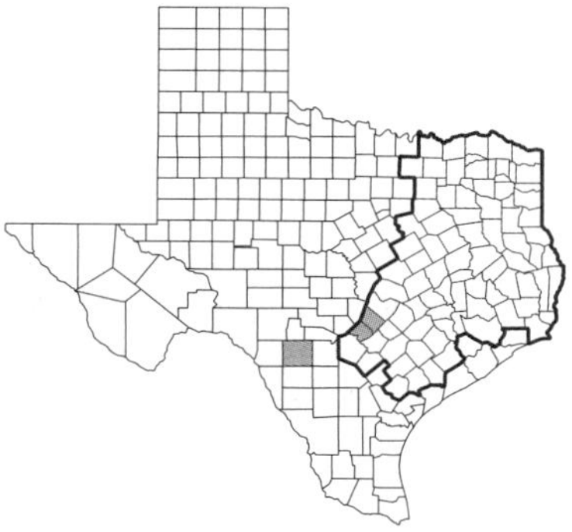

Tephrosia potosina

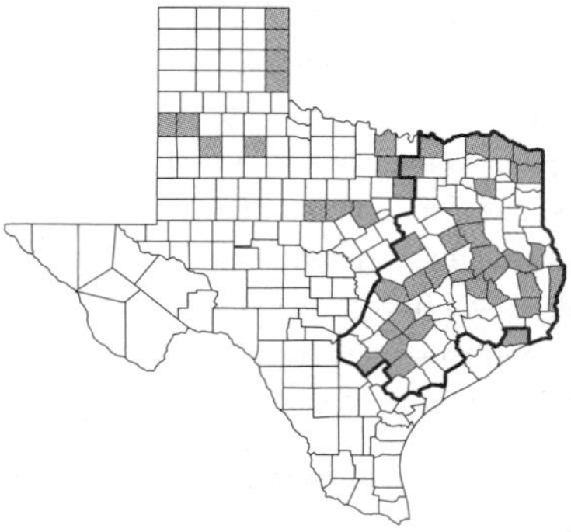

Tephrosia virginiana

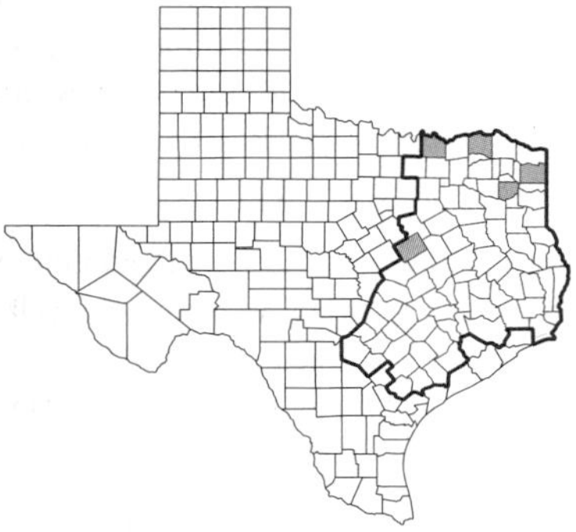

Trifolium arvense

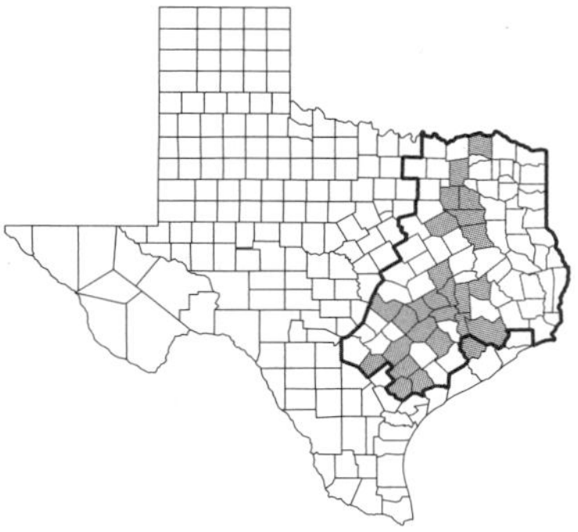

Trifolium bejariense

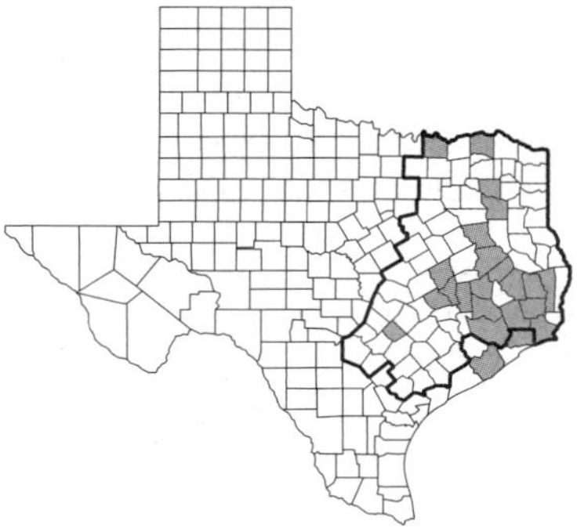

Trifolium campestre

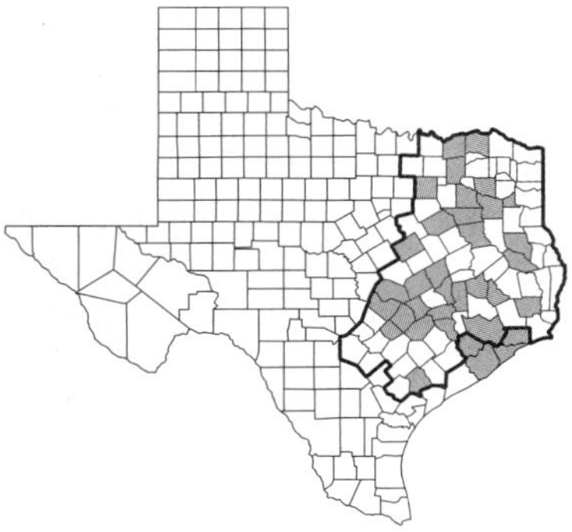

Trifolium carolinianum

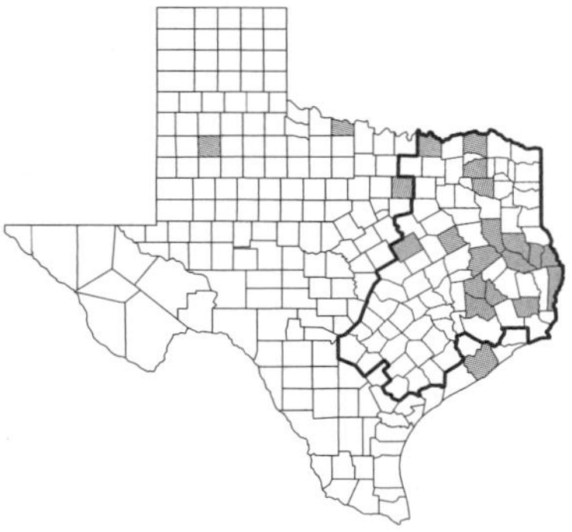

Trifolium dubium

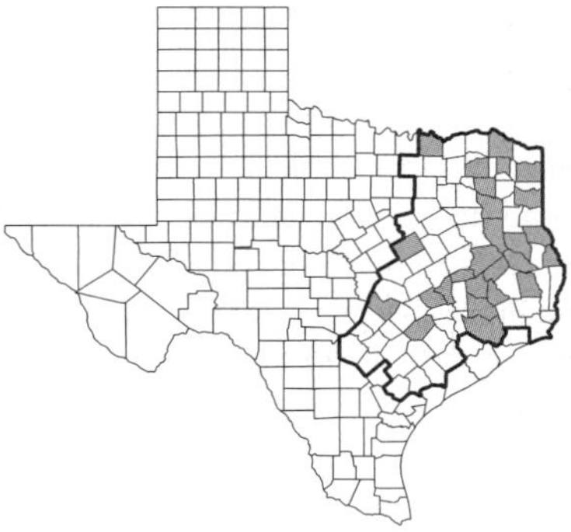

Trifolium incarnatum

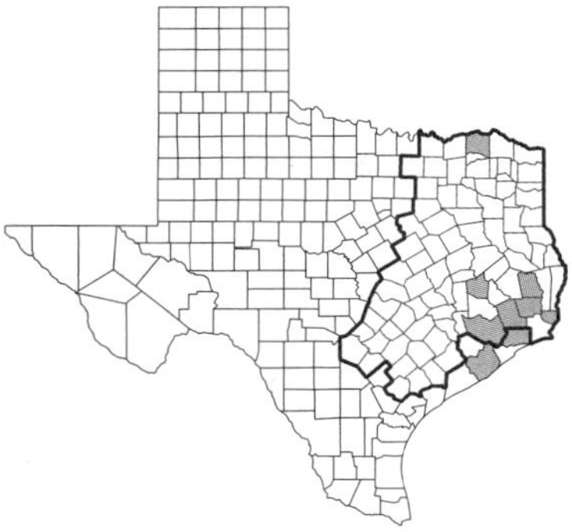

Trifolium lappaceum

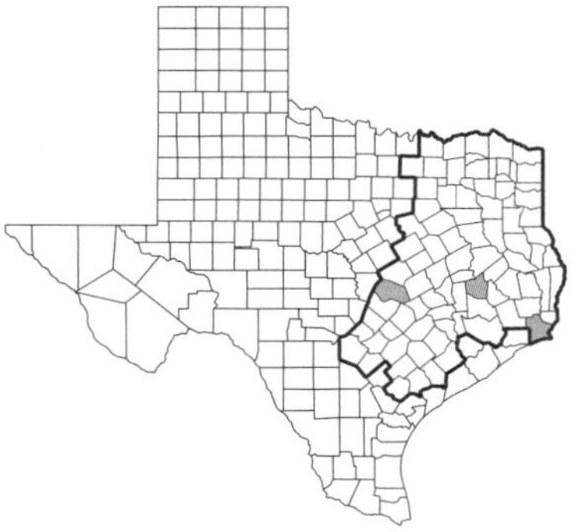

Trifolium nigrescens

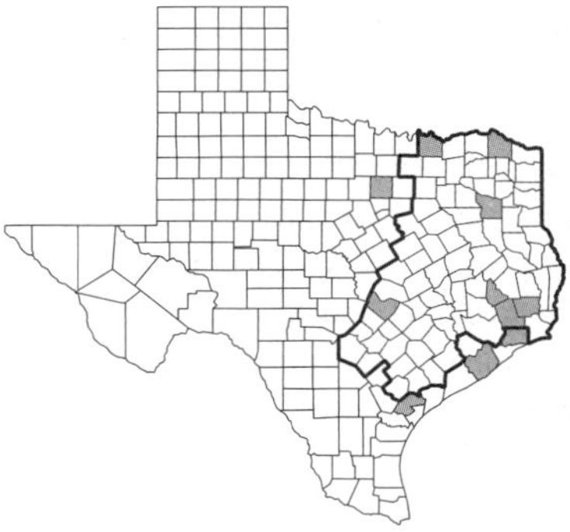

Trifolium pratense

leaflet blades obovate to oblong, 9–16 mm long, 4–10 mm wide, margins denticulate, apex obtuse, obcordate, or emarginate, surfaces hairy; peduncles 0–1 cm long; inflorescences axillary or terminal, solitary or paired, with 20–60 flowers, ovoid or oblong, 0.8–1.6 cm long, 0.6–1 cm wide, subtended by distinctly streaked stipular pseudo-involucre; flowers 5–7 mm long; calyces ellipsoid to urceolate, 3–4 mm long, hairy, distinctly 10-veined, lobes erect or spreading, unequal, abaxial longest and equal to tube, subulate; corollas pink. Disturbed areas, lawns, and cemeteries; known only from Lamar Co. in TX (*Edmiaston 35256* GH); sporadically scattered in the U.S.; British Columbia; Chile; New Zealand; and Australia. May–Aug. Native of s Eurasia. This species was first collected in the U.S. on ballast in New Jersey in 1880. It is now found sporadically as a weed of disturbed habitats and appears to be spreading rapidly. It has been called PITTS' CLOVER in the s U.S. because it was discovered by J.D. Pitts in a field of crimson clover as a weed, and he experimented with it as a forage plant (Vincent & Gillett 2023).

Trifolium vesiculosum Savi, (bladder-like), ARROW-LEAF CLOVER. Annual; stems 15–70 cm long, often dark in color; leaflet blades obovate, oblong, elliptic, or lanceolate, margins spinulose-denticulate, surfaces glabrous, 5–40 mm long, 5–15 mm wide; inflorescences globose, ovoid, or oblong with 50–100 flowers, 3–6 cm long, 2–3.5 cm wide, peduncled; flowers sessile; calyces greatly inflated, with long awn-like teeth; corollas white, becoming pink, 12–18 mm long, 1.5–2 times as long as calyces. Fields and roadsides; throughout East TX, but more common in n 1/2; also Tarrant (BRIT) and Wise (B. Lipscomb, pers. obs.) cos. just to the w; mostly se U.S. from VA s to FL and w to OK and TX; disjunct n to MA and w coast of U.S. in CA, OR, and WA. Apr–Aug. Native of s Europe. This species was first reported for the U.S. from LA and MS by Thieret (1969b) but was known to be introduced in cultivation in 1963 in w and s U.S. (Isely 1998; Vincent & Gillett 2023).

Trifolium glomeratum L., (club-shaped), CLUSTERED CLOVER, was collected twice in TX historically from cultivated and unintentionally introduced plants. A Freestone Co. specimen (*Smith s.n.*, 1953, SMU) was described as growing "in planting of yellow hop clover" and "may have been introduced with hop clover seed from Australia." A Jefferson Co. specimen (*Wolff s.n.*, 1947, BRIT) is described as "introduced on Rice Experiment Station, Beaumont." Vincent and Gillett (2023) mapped *T. glomeratum* in TX, probably based on these cultivated specimens, but it has apparently not naturalized in the state.

Trifolium hybridum L., (between two species), ALSIKE CLOVER, a Eurasian species, has been collected in every state In the U.S., except TX including very near the state borders with AR, LA, NM, and OK (Vincent & Gillett 2023). It is included in the key for clarity and because of the potential that it may be encountered in East TX.

Trifolium subterranean L., (below ground), SUBTERRANEAN CLOVER, SUB CLOVER. Pilose annual, stoloniferous, with procumbent stems to 50(–100) cm long; leaflets 10–15(–20) mm long; heads with 2 types of flowers: 1) 2–6 outer, petal-bearing fertile flowers; 2) numerous inner sterile flowers forming a mass of stiff, plumose, stalk-like modified calyces subsequently elongating, curving around the developing fruits, and forming a bur—this pushed against or under soil surface by elongation of the peduncle; corollas 8–14 mm long, yellow-white and purple; fruits inside of the compact bur. Weedy areas, roadsides; known in TX only from Brazos, Denton (Turner et al. 2003) and Jefferson (Kartesz 2024) cos. However, the Jefferson Co. specimen (*Unknown collector*, 1950, UTEP) is based on a cultivated specimen. The species was apparently only naturalized in TX temporarily from collections made in the same general area in 1946 and 1947 in Denton Co., (*Cory 53247* BRIT, NY; *Cory 53704* BRIT, NY; *Wagner 84* TEX). No specimens or observations could be located from Brazos Co.; Vincent and Gillett (2023) did not map it for TX. Very scattered in AR, FL, GA, LA, OK, MS, and SC. Spring–summer. Native of Europe and adjacent Africa and Asia. Isely (1998) noted that before flowering this species somewhat resembles *T. repens*—but is very different in flower color and especially in its unique fruit. It is cultivated as a forage in some parts of the world (e.g., Australia, w U.S.). However, it is reported to be toxic when eaten excessively; estrogen-like compounds can cause sterility in livestock (Isely 1993).

VACHELLIA Wight & Arn. ACACIA

Shrubs or trees, armed; stems erect to spreading, glabrous or pubescent; twigs terete, some flexuous, short shoots usually present; leaves alternate; stipules present, spinose, paired at nodes, straight, woody, in some species enlarged and inhabited by ants; petiolate, petiole channeled, glabrous or pubescent, petiolar gland 1; pinnae 1–6 pairs, mostly opposite; leaflets (4–)6–24 pairs per pinna, usually opposite, sessile or subsessile, blade margins entire, surfaces glabrous or pubescent; inflorescences with 50–numerous flowers, primarily axillary, globose heads or cylindrical spikes, solitary or clustered in leaf axis or on short shoots; bracts present; flowers mimosoid; calyces campanulate, lobes 4 or 5; corollas yellow, white, or cream, lobes 4 or 5; stamens 15–120, distinct, exserted, yellow or white, anthers dorsifixed, eglandular; ovary sessile or short-stipitate, usually glabrous, style and stigma filiform; fruits legumes, stipitate, flattened to terete, straight to falcate, linear to oblong, usually dehiscing along both sutures, sometimes indehiscent, glabrous or pubescent; seeds 6–10(–20).

A genus of ca. 160 species of the s and sw U.S., Mexico, West Indies, Central America, South America, Europe, Asia, Africa, and Australia and introduced nearly worldwide in tropical to warm temperate regions (Ebinger & Seigler 2023b). This genus has

Trifolium lappaceum [ZO1]

Trifolium nigrescens [FVM]

Trifolium reflexum [LU2]

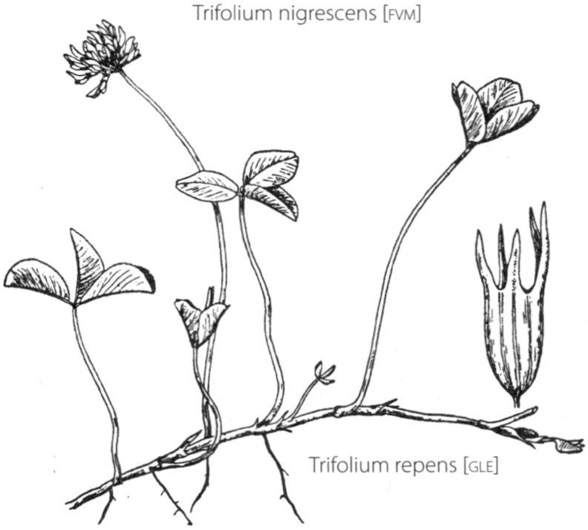

Trifolium repens [GLE]

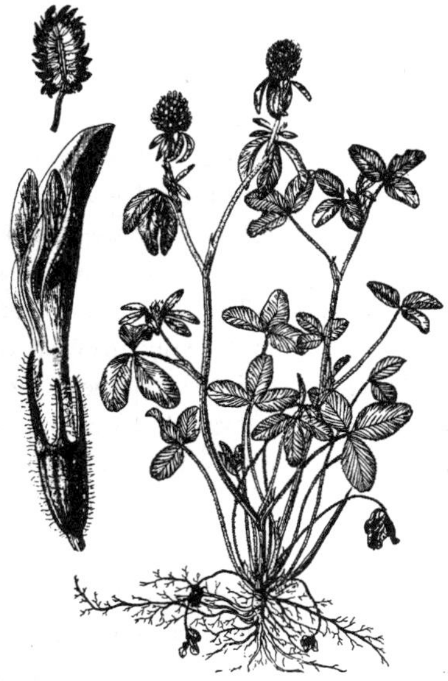

Trifolium pratense [BL3, USD]

Trifolium resupinatum [BEN, WIL]

traditionally been included in *Acacia* (Isely 1998; Diggs et al. 1999; Turner et al. 2003); however, there is considerable evidence that *Acacia* in the broad sense is not a natural or monophyletic group. Largely because the segregation of *Acacia* in the broad sense into at least five genera would result in requiring a change in generic epithet for about 1000 Australian species of subg. Phyllodineae (DC.) Seringe, at the 17th International Botanical Congress in 2005, the type of *Acacia* was changed from *A. scorpioides* (L.) W. Wight to the Australian species *A. penninervis* Seiber ex. DC. (Ebinger & Seigler 2023b). As a result of this change, plants of former *Acacia* subg. *Acacia* should be placed in *Vachellia* (Seigler & Ebinger 2005). (For Rev. George Harvey Vachell, 1799–1839, plant collector in China) (subfamily Caesalpinioideae, Mimosoid clade, tribe Acacieae)
REFERENCES: Isely 1969; Ebinger & Siegler 2023b.

1. Pinna pairs 1 per leaf; leaflet pairs 2–5 per pinna; leaflets (4–)6–13(–16) mm long; inflorescences cylindrical, white to cream ______ **V. rigidula**
1. Pinna pairs 1–6 per leaf; leaflet pairs 8–24 per pinna; leaflets 2–6.3 mm long; inflorescences spherical, yellow.
   2. Petiolar glands donut-shaped, at or just below lowermost pinnae; peduncles with tiny, red glands (use 10× magnification) (these deciduous and absent as the peduncle ages) ______ **V. bravoensis**
   2. Petiolar glands circular to slightly elongated, not donut-shaped, usually medial on petiole; peduncles lacking glands ______ **V. farnesiana**

Vachellia bravoensis (Isely) Seigler & Ebinger, (presumably in reference to Rio Bravo, a Mexican name for the Rio Grande), HUISACHILLO, TWISTED ACACIA, TENAZA PRIETO ACACIA, HUIZACHE CHINO, MEDUSA ACACIA, SCHAFFNER'S ACACIA, SCHAFFNER'S WATTLE. Shrub or small tree, 1–4(–6) m tall; stems slightly flexuous, glabrous or sparsely pubescent; stipular spines terete, straight, 1–15(–29) mm long, 0.5–1.5 mm wide near base; petiole 2–8 mm long, usually densely pubescent; petiolar gland 1, at or just below lowermost pinnae, sessile, circular, 0.2–1.1 mm diam., donut-shaped; leaves with 1–3(–4) pairs of pinnae; leaflets 10–24 pairs per pinna, 2–4.2(–5) mm long, 0.6–1.1 mm wide; heads globose; filaments yellow; fruits 7–16 cm long, 6–9 mm wide, linear, nearly terete, usually somewhat constricted between the seeds, covered with downy hairs, linear. Disturbed sites in thorn scrub woodlands and shrublands; Bexar, Dewitt, and Guadalupe (Kartesz 2023) cos. on s margin of East TX; mainly South TX Plains and s Gulf Prairies and Marshes; only in TX in U.S.; Mexico (Coahuila, Hidalgo, Nuevo León, San Luis Potosí, and Tamaulipas). Feb–Apr. [*Acacia schaffneri* (S. Wats.) F.J. Herm. var. *bravoensis* Isely]

Vachellia farnesiana (L.) Wight & Arn., (for Cardinal Odoardo Farnese, 1573–1626, of Rome, who grew the plant in his gardens), HUISACHE, SWEET ACACIA. Shrub or small tree 2–4(–8) m tall; stems usually slightly flexuous, usually glabrous, short shoots present; stipular spines terete, straight, 2–35(–55) mm long, 0.5–1.4 mm wide near base; petiole (3–)6–12(–17) mm long, usually sparsely to densely pubescent, hairs erect; petiolar gland 1, usually medial on petiole, sessile or short-stalked, circular to slightly elongated, 0.2–1 mm long, apex depressed; leaves with 2–6 pairs of pinnae; leaflets 8–19 pairs per pinna, 3–6.3 mm long, 0.5–1.7 mm wide; flowers in globose heads, sweet-scented; filaments bright yellow; fruits 3–17 cm long, 9–18 mm wide oblong, nearly terete, not constricted between the seeds, glabrous. Moist to seasonally dry sites in thickets, open fields and pastures, and disturbed areas; s portion of East TX disjunct n in Hopkins Co. (*Shinners 19009* SMU, TEX); mainly s 1/2 of TX; also Cross Timbers and Prairies (Kartesz 2023); s U.S. from SC s to FL and w to AZ; Mexico; West Indies; Central America; South America; introduced in Europe and Africa. Mar–Apr (in TX). [*Acacia minuta* (M.E. Jones) R.M. Beauch. subsp. *densiflora* (Alexander ex Small) R.M. Beauch., *A. smallii* Isely, *Vachellia densiflora* Alexander ex Small] Previously used as a source of fragrant oils for perfumes (Correll & Johnston 1970), valued as a honey plant in s TX (Wills & Irwin 1961), and used medicinally (Crosswhite 1980). The common name HUISACHE is derived from the Nahuatl (language of the Aztecs), *huitz-axin* (Crosswhite 1980). Texas populations of this species have sometimes been treated as *A. minuta* subsp. *densiflora* (Beauchamp 1980), *A. minuata* (orthographic error) (Jones et al. 1997), *A. smallii* (Powell 1988; Isely 1973, 1990, 1998) or *V. densiflora* (Weakley, A.S., and Southeastern Flora Team 2022) but we are following the treatment in Ebinger and Siegler (2023b). This species is migrating northward into previously unoccupied habitats in East TX but is negatively impacted by extended periods of freezing weather (E. Keith, pers. obs.). *Vachellia farnesiana* var. *minuta* (M.E. Jones) Seigler & Ebinger occurs in Mexico and San Diego County, CA, and var. *pinetorum* (F.J. Herm.) Seigler & Ebinger occurs in s FL (Ebinger & Siegler 2023b).

Vachellia rigidula (Benth.) Seigler & Ebinger, (somewhat rigid, stiff), BLACKBRUSH, CHAPARRO PRIETO. Shrub or small tree, 1–4 m tall, forming thickets; stems slightly flexuous, glabrous or sparsely puberulent, short shoots present; stipular spines terete, straight, 10–50(–80) mm long, 0.6–2.1 mm wide near base; petiole 2–5(–9) mm long, densely puberulent; petiolar gland 1, mostly medial on petiole, sessile, circular, 0.3–0.6(–1.1) mm diam., apex depressed; leaves usually with 1 pair of pinnae; leaflets 2–5 pairs per pinnae, (4–)6–13(–16) mm long, 2–6(–8) mm wide; flowers in elongate spikes 2–6 times as long as wide; filaments creamy white; fruits 4–10 cm long, 3–6 mm wide, linear, flattened. Open disturbed sites, rocky slopes, and arroyos, forming extensive thickets; Dewitt and Wilson cos. (Kartesz 2023) on s margin of East TX; abundant throughout s TX and Brewster

Co. in the Trans-Pecos; only in TX in the U.S.; Mexico (Coahuila, Jalisco, Michoacán, Nuevo León, Querétaro, San Luis Potosí, Tamaulipas, and Veracruz). Feb–Jul. [*Acacia rigidula* Benth.] The name *A. amentacea* DC. has sometimes been used for this species; however, that name was misapplied (Miller 2006). *Vachellia rigidula* rarely hybridizes with *V. bravoensis* (*V. ×ruthvenii* Seigler & Ebinger) and is only known from Dimmitt Co. in s TX (Ebinger & Siegler 2023b).

VICIA L. VETCH

Herbs annual, biennial, or perennial; decumbent, trailing to climbing herbs; leaves alternate, once pinnately compound, the rachis tip usually terminating in a simple or branched tendril, stipulate; leaflets estipellate, 2–15 pairs, entire, opposite to alternate on the rachis; flowers 1–2 in leaf axils or in few to many-flowered, axillary, spike-like racemes; calyces persistent, symmetrical or gibbous basally; corollas white to blue, lavender, pink-purple, purple, or bicolored; stamens 10, diadelphous; style nearly terete, bearded apically on side adjacent to standard or in a tuft about the apex; fruits elastically (= explosively) dehiscent, linear to oblong, flattened; seeds 2–15.

A mostly n temperate genus of ca. 140 species with some in South America and n Africa; they are mainly scrambling herbs often with alkaloids (Broich 2023b). Some are used as forage, fodder, or as green manure; others for their edible seeds including the Old World *V. faba* L. (BROAD-BEAN, FIELD-BEAN, HORSE-BEAN), important before introduction of New World *Phaseolus* L. species; some species, however, are toxic; the uncooked seeds of *V. faba* can cause hepatitis in some people of Italian or Jewish ancestry due to a genetic predisposition to a biochemical deficiency of red blood cells (La Vieille et al. 2019). Kartesz (2023) mapped *Vicia cracca* L. in TX based on a Navarro Co. specimen (*Worthington 5962*, COLO, UTEP); but it is a misidentified specimen of *V. villosa*. Broich (2023b) also did not map *V. cracca* in TX. In East TX, VETCHES are recognized in the field by the combination of their leaves with the rachis ending in a tendril, numerous small leaflets, and unwinged stems. (The classical Latin name, from *vincire*, to bind, alluding to the clasping tendrils) (subfamily Faboideae, tribe Vicieae)

REFERENCES: Shinners 1948a; Hermann 1960; Kupicha 1976; Lassetter 1978, 1984; Olwell 1982; Broich 2023b.

1. Flowers sessile or nearly so in the axils of the leaves, solitary or paired (rarely to 4); corollas yellow to variously white to light violet, light lavender, light blue or pinkish purple; fruits not stipitate.
   2. Corollas yellow (can have purplish tinge or streaks or can dry a drab violet), 18–35 mm long; fruits 7–14 mm wide.
      3. Corollas 18–25 mm long; fruits pilose with swollen-based hairs; one calyx lobe nearly as long as or longer than the calyx tube, the other calyx lobes much shorter ____ **V. lutea**
      3. Corollas 25–35 mm long; fruits puberulent becoming glabrous at maturity, without swollen-based hairs; calyx lobes subequal in length, all shorter than the calyx tube ____ **V. grandiflora**
   2. Corollas variously white to light violet, light lavender, light blue or pinkish purple, 5–30 mm long; fruits 3–11 mm wide.
      4. Corollas tiny, 5–8 mm long, light violet to whitish lavender or light blue; plant diminutive, the stems only 5–30 cm long; leaflets only 4–6(–8) per leaf ____ **V. lathyroides**
      4. Corollas usually conspicuous, 10–30 mm long, pink-purple to white; plant larger, the stems 30–100 cm long; leaflets 8–14 per leaf ____ **V. sativa**
1. Flowers 1–many in racemes with well-developed peduncles, the peduncles somewhat shorter to nearly equal to much longer than the leaves; corollas variously colored but never yellow; fruits sessile or stipitate 1–3 mm long.
   5. Corollas 8–18 mm long; calyces strongly swollen on one side basally OR not so.
      6. Calyces 5–6 mm long, strongly swollen on one side basally; fruits 6–11 mm wide; corollas 12–18 mm long, blue-purple or blue-purple and white (rarely white); flowers densely or loosely clustered in racemes; stems glabrate OR conspicuously villous with hairs 1–2 mm long; annual of open areas ____ **V. villosa**
      6. Calyces 2–2.5 mm long, symmetrically tapered basally; fruits 4–5 mm wide; corollas 8–12 mm long, white to lavender tinged or purple-tipped (can dry lilac or bluish); flowers loosely clustered in racemes; stems sparsely pubescent to glabrate; rhizomatous perennial typically of wooded areas and forest margins ____ **V. caroliniana**
   5. Corollas 2–8 mm long; calyces symmetrically tapered basally.
      7. Fruits hirsute, 6–10 mm long, sessile, with usually 2 seeds; corollas 2–4 mm long; non-native species rare in East TX ____ **V. hirsuta**
      7. Fruits usually glabrous or nearly so, 10–35 mm long, slightly stipitate (stipe 1–2 mm) OR sessile (in *V. minutiflora*), with 4–8 seeds; corollas 3.5–8 mm long; including species widespread and common in East TX.
         8. Racemes 3–20+-flowered ____ **V. ludoviciana**
         8. Racemes 1–3-flowered.
            9. Leaves with 2–8 leaflets; calyx lobes shorter than calyx tube; fruits sessile; ovules or seeds 6–8 per fruit ____ **V. minutiflora**
            9. Leaves with 4–14 leaflets; calyx lobes (at least one) nearly equaling or longer than the tube; fruits stipitate, the stipe ca. 1 mm long; ovules or seeds 4–8 per fruit.
               10. Fruits 1.6–3 cm long, 4–6 mm wide, asymmetrically acute at apex, with 4–8 seeds; corollas deep bluish purple to pale lavender or pinkish white; species widespread and common throughout East TX ____ **V. ludoviciana**

10. Fruits 1–1.5 cm long, 3–4 mm wide, symmetrically rounded at apex, usually with 4 seeds; corollas pale lavender often with purple lines; species uncommon and inconspicuous, mostly in Pineywoods ______ **V. tetrasperma**

Vicia caroliniana Walter, (of Carolina), CAROLINA VETCH, PALE VETCH, WOOD VETCH. Rhizomatous perennial; stems 30–150 cm long, sparsely pubescent to glabrate; leaflets 10–24, blades elliptic-oblong or oblanceolate to linear, 10–30 mm long, 2–6 mm wide, apex obtuse, surfaces glabrous or abaxial puberulent; inflorescences with 7–20 flowers, 3–10 cm long, shorter to longer than subtending leaf rachis; flowers 8–12 mm long; calyx base symmetric, lobes subequal, shorter than tube; corolla white with blue tip, banner pandurate, blade equal to or longer than claw, glabrous; fruits reddish brown, narrowly oblong, 15–30 mm long, 4–5 mm wide, obliquely long-tipped, glabrous; stipe to 2 mm long; seeds 5–8. Wooded areas, forest margins; in TX known only from, but sparsely scattered in East TX; e U.S. from NY s to n FL and w to MO, e OK, WI, and TX; Ontario. Mar–May. *Vicia caroliniana* is the only native and perennial species in East TX and the only species that grows in forested habitats.

Vicia grandiflora Scop., (large-flowered), LARGE YELLOW VETCH. Annual with inconspicuous pubescence; stems 30–60 cm long; leaflets 6–14 per leaf, those of the middle leaves oblong to elliptic-oblong or linear-oblong; flowers 1–3 in axils of upper leaves; corollas yellow, sometimes with purple streaks, but drying a drab violet, 25–35 mm long; fruits 25–45 mm long, 7–10 mm wide, sessile; seeds ca. 15. Roadsides, open woods, and cultivated fields; rare and scattered in ne TX in Cass, Fannin, Nacogdoches, and Rusk (Kartesz 2023) cos.; e U.S. from MA s to n FL and w to AR, MI, and TX. Apr–May. Native of Europe. *Vicia grandiflora* was first reported as naturalized in TX in Fannin Co. in 2002 (Singhurst et al. 2002).

Vicia hirsuta (L.) Gray, (hairy), TINY VETCH. Annual, glabrate or with inconspicuous pubescence; stems 20–80 cm long; leaflets 8–16 per leaf, those of the middle leaves narrowly elliptic to narrowly oblong or linear; flowers 2–7 on racemes 1–5 cm long; corollas dull white to pale blue or tinged with lavender, 2–4 mm long; fruits hirsute, 6–10 mm long, 3–4 mm wide, oblique at both ends; seeds usually 2. Roadsides, open woodlands, weedy areas; scattered in East TX in Anderson (iNaturalist Contributors 2023k), Hardin, Hill (TEX), Shelby, and Smith (Kartesz 2023) cos.; widely scattered in the U.S. from ME s to n FL and w to WI, OK, and TX, disjunct to CA, OR, WA, and AK; Greenland; Canada; South America. Apr–Jun. Native to Europe. Isely (1990) noted that among the tiny-flowered species, the combination of several flowers per inflorescence, calyx lobes often longer than the tube, and hirsute fruits, make this species recognizable. Fruits with two seeds are also helpful in identification.

Vicia lathyroides L., (like the genus *Lathyrus*), SPRING VETCH. Inconspicuous annual; stems 5–30 cm long; leaflets only 4–6(–8) per leaf, those of the middle leaves lanceolate to elliptic; flowers 1(–2), sessile or subsessile in the upper leaf axils; corollas light violet to whitish lavender or light blue, 5–8 mm long; fruits 15–25 mm long, 3–4 mm wide, sessile; seeds 6 or 7. Disturbed areas, fields, waste places, lawns; Smith Co.—this is the first and only report for the state (Hill et al. 2007); e U.S. from NH s to GA and w to MS, disjunct in CA and WA and TX; British Columbia. Apr–Jun. Native of Europe. Isely (1990) noted that this species resembles a miniature *V. sativa*.

Vicia ludoviciana Nutt. ex Torr. & A. Gray, (of Louisiana), DEER PEA VETCH. Glabrate or sparsely pubescent annual; sprawling or climbing, stems 10–200 cm long; leaflets 7–14, blades 5–25 mm long, 1–11 mm wide, ovate to linear; flowers 1–19, on a peduncle; corollas 4–7 mm long; fruits 16–30 mm long, 4–6 mm wide, stipitate to 1 mm long; seeds 4–8. The species is quite variable and has been divided into various species, subspecies, varieties, and races. Broich (2007) recognized seven races with overlapping characters within two distinctive varieties. Isely (1998) followed a similar treatment but distinguished the varieties as subspecies. We follow Broich (2007, 2023b) in recognizing *V. ludoviciana* var. *ludoviciana* and var. *leavenworthii*.

1. Flowers opening before elongation of peduncles; peduncles 1/8–1/2 as long as subtending leaves when all flowers have opened, elongating rapidly as fruits develop, ultimately equaling or slightly exceeding the leaves, racemes with 1–6 flowers; corollas pinkish white to light lavender; leaflets 11–14 ______ var. **leavenworthii**
1. Flowers opening after elongation of peduncles; peduncles more than 1/2 as long as to exceeding subtending leaves when all flowers have opened, elongating slightly in age (but tip of racemes often shriveled then), racemes with 1–19 flowers; corollas deep blue to bluish purple; leaflets 7–10(–14) ______ var. **ludoviciana**

var. **leavenworthii** (Torr. & A. Gray) Broich, (for its discoverer, Melines Conklin Leavenworth, 1796–1862, s U.S. botanist), LEAVENWORTH'S VETCH. Leaflets 11–14; inflorescences with 1–6 flowers; flowers opening before peduncle and internode elongation, usually containing young fruit when first open; corolla pinkish white to light lavender. Rocky, clayey, or occasionally sandy prairies, open woodlands, streamsides, and roadsides, often on limestone; mostly w portion of East TX; scattered throughout TX; sc U.S. in AR, LA, MO, MS, OK, and TX. Mar–May. [*V. leavenworthii* Torr. & A. Gray, *V. leavenworthii* var. *occidentalis* Shinners, *V. ludoviciana* subsp. *leavenworthii* (Torr. & A. Gray) Lassetter & C.R. Gunn]

Trifolium striatum [COS]

Trifolium vesiculosum [HEA]

Vachellia bravoensis [VIN]

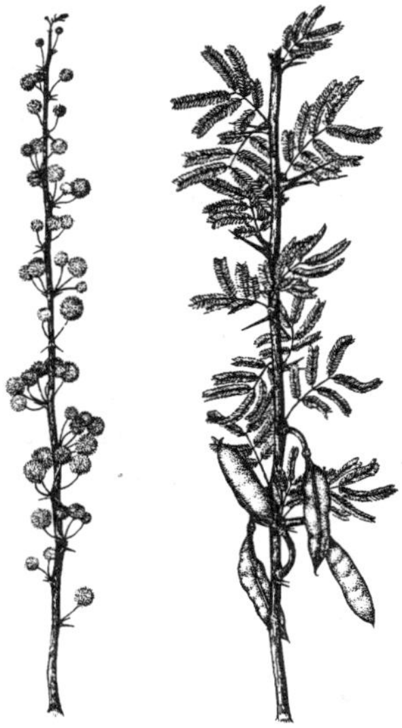

Vachellia farnesiana [VIN]

Vachellia rigidula [VIN]

Vicia caroliniana [WIL]

var. **ludoviciana.** Leaflets 7–10(–14); inflorescences with 1–19 flowers; flowers opening after peduncle and internode elongation, fruits not present when first open; corolla deep blue to bluish purple. Sandy soils in woodlands, grasslands, pastures, and roadsides; throughout East TX; throughout TX; mostly s U.S. from MS w to s CA and n to CO, UT, and WY; Mexico (Baja California, Baja California Sur, and Nuevo León). Mar–early May. [*V. exigua* Nutt, *V. ludoviciana* var. *laxiflora* Shinners, *V. ludoviciana* var. *texana* (Torr. & A. Gray) Shinners, *V. ludoviciana* var. *typica* Shinners [nom. illeg.]]

Vicia lutea L., (yellow), SMOOTH YELLOW VETCH. Annual, glabrate or sparsely villous; stems 20–60 cm long; leaflets of middle leaves 6–16 per leaf, 8–20 mm long, 1–5 mm wide, oblong to lanceolate-linear, sparsely pubescent with long pustulose-based hairs on undersurfaces, glabrous above; flowers 1–3 in leaf axils; corollas 18–25 mm long, yellow, sometimes with a purple tinge; fruits 25–35 mm long, 7–14 mm wide, pilose with swollen-based hairs, stipitate to 4 mm long; seeds 3–9. Fields and roadsides; only known from Brazos and Madison cos. in TX (Kartesz 2023); scattered in the U.S. in AL, CA, FL, LA, NC, OR, and TX. Apr–May. Native of Mediterranean Region. Neill (1999) first reported the species from TX in Madison Co.

Vicia minutiflora D. Dietr., (small-flowered), SMALL-FLOWER VETCH, PYGMY-FLOWER VETCH. Annual, glabrous or nearly so; stems 20–80 cm long; leaves with 2–8 leaflets, usually only 2 or 4 at the base, the leaflets of middle and upper leaves linear or narrowly oblong, ovate basally; inflorescences with only 1(–2) flowers; calyces 2.2–2.8 mm long; corollas 5–8 mm long, pale blue to lavender; fruits 15–30 mm long, 4–5 mm wide, glabrous or sparsely to densely pubescent, sessile; seeds 6–8. Sandy open woods, fencerows, and roadsides; widespread in East TX; also adjacent Edwards Plateau, n Gulf Prairies and Marshes, and ne Cross Timbers and Prairies; se U.S. from FL and GA w to OK and TX, n to KY and TN. Mar–May. [*V. micrantha* Lowe, *V. reverchonii* S. Watson]

Vicia sativa L., (cultivated), COMMON VETCH, SPRING VETCH, NARROW-LEAF VETCH. Annual, glabrous or (especially younger parts) ± pubescent; stems 30–100 cm long; leaflets 8–14, 15–30 mm long, 1.5–15 mm wide, ovate-oblong, narrowly elliptic, or linear; stipules foliaceous approaching leaflets in size, lobed or nearly entire; flowers solitary or paired in upper leaf axils, rarely to 4, 10–30 mm long; fruits 25–60 mm long, 3–11 mm wide, sessile; seeds 4–12. This is a complex group, often treated as made up of many infraspecific taxa; according to Isely (1990), determination of subspecies can sometimes be ambiguous. However, we are following Broich (2023b) who treated the two East TX entities at the varietal level.

1. Corollas 10–18(–20) mm long, pink-purple to white; calyces 7–11(–12) mm long; leaflets on upper leaves essentially linear, 1.5–4 mm wide, subacute to truncate or slightly notched at apex; fruits reddish brown to black, glabrous ____ var. **angustifolia**
1. Corollas 18–30 mm long, usually pink-purple; calyces 10–15 mm long; leaflets on upper leaves narrowly oblong, 5–15 mm wide, slightly or strongly notched at apex; fruits yellow to brown, pubescent ____ var. **sativa**

var. **angustifolia** L., (black), NARROW-LEAF VETCH. Leaflets 4–10 times as long as wide, blades narrowly elliptic to linear; flowers 10–18(–20) mm long; calyx lobes usually slightly shorter than tube; fruits reddish brown to black, glabrous. Roadsides, fields, and weedy areas; throughout East TX; scattered throughout TX except s TX; throughout the U.S. including AK; Greenland and Canada. [*V. angustifolia* L., *V. sativa* subsp. *nigra* (L.) Ehrh., *V. sativa* var. *nigra* L., *V. sativa* var. *segetalis* (Thuill.) Ser.] Native of Europe. Extrafloral nectaries are often present on the stipules; the nectaries are small, ± round, depressed, deep purple areas; under appropriate conditions glistening droplets of nectar can be seen in the nectaries; ants have been observed feeding at the nectaries in East TX (G. Diggs, pers. obs.), and Koptur (1979) determined that Argentine ants (*Iridomyrmex humilis* Mayr.) feeding on nectaries reduced herbivory. Turner et al. (2003) treated this taxon as a distinct species (*V. angustifolia*). Garrison et al. (2000) studied the ballistic seed projection of this variety with seeds travelling from 0.83–9.03 m/s (N = 78, mean = 4.64 m/s) and distances between 0.2–9 m (mean = 3.4 m).

var. **sativa,** COMMON VETCH. Leaflets 2–5(–7) times as long as wide, blades usually ovate-oblong to narrowly elliptic, sometimes linear; flowers18–30 mm long; calyx lobes usually equal to tube; fruits yellow to brown, pubescent. Rarely cultivated and escaping in fields, roadsides, and waste areas; throughout East TX; also adjacent Edwards Plateau; e U.S. from ME s to FL and w to IA, MN, and TX, disjunct in CA, OR, and WA; Mexico; South America; Asia; Africa; and Australia. Native to Europe. Extrafloral nectaries are often present on the stipules (Hermann 1960). Reported in some instances to be poisonous to livestock and humans due to lethal concentrations of a cyanogenic glycoside in the seeds (Kingsbury 1964).

Vicia tetrasperma (L.) Schreb., (four-seeded), LENTIL VETCH. Annual; glabrous or sparsely pubescent; stems 10–50 cm long; leaflets 4–12, 6–20 mm long, 1–5 mm wide, oblanceolate, elliptic, or linear; racemes 1–3 cm long with 1–3 flowers at apex; calyces 2–3 mm long; corollas 3–6 mm long, pale lavender often with purple lines; fruits apically rounded, glabrous, stipitate, usually with 4 seeds. Fields, roadsides, waste areas, and open woodlands; scattered throughout East TX; also Uvalde Co. (Kartesz 2023) in w Edwards Plateau; e U.S. from ME s to FL and w to MI, AR, and TX, disjunct w in CA, ID, OR, and WA; Canada and Chile.

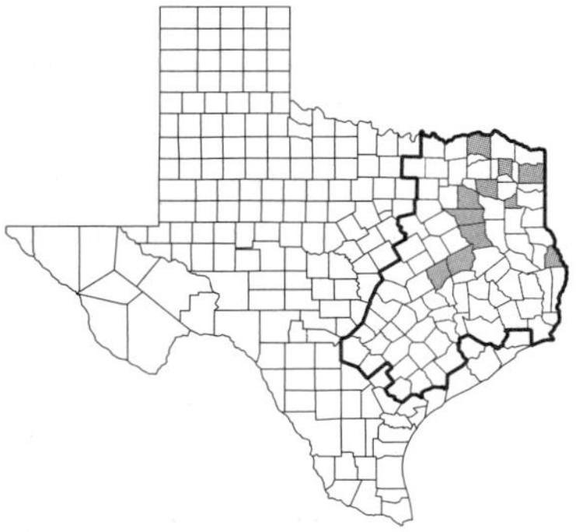

Trifolium reflexum

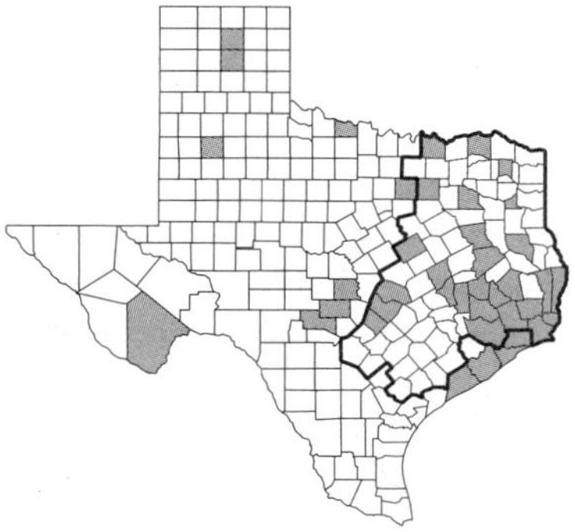

Trifolium repens

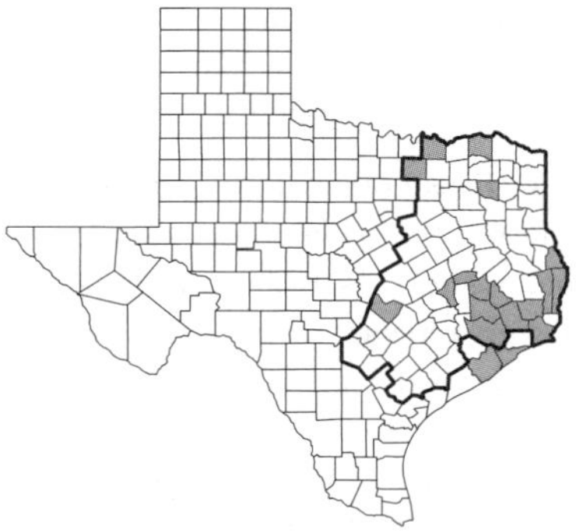

Trifolium resupinatum

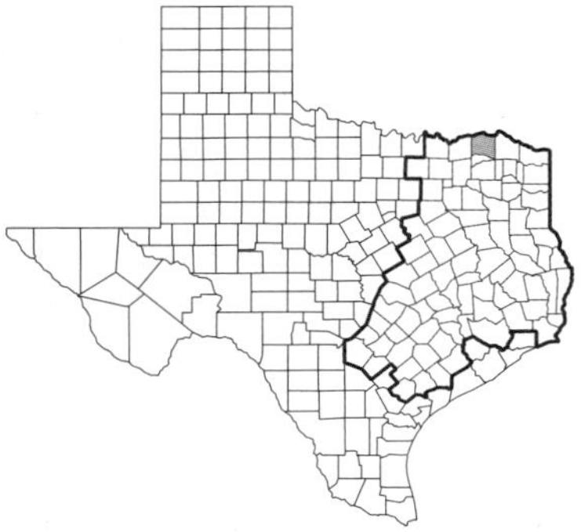

Trifolium striatum

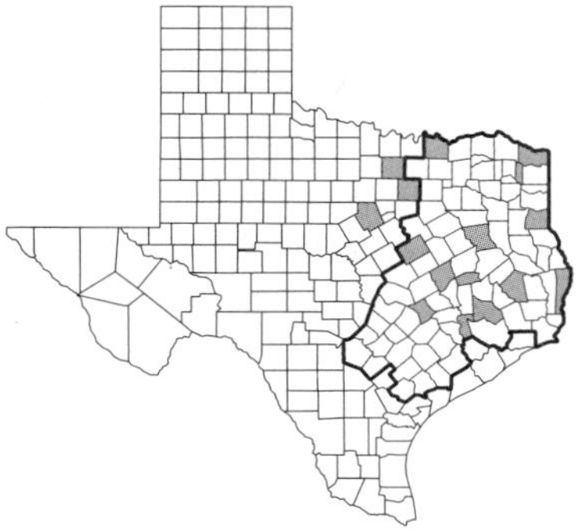

Trifolium vesiculosum

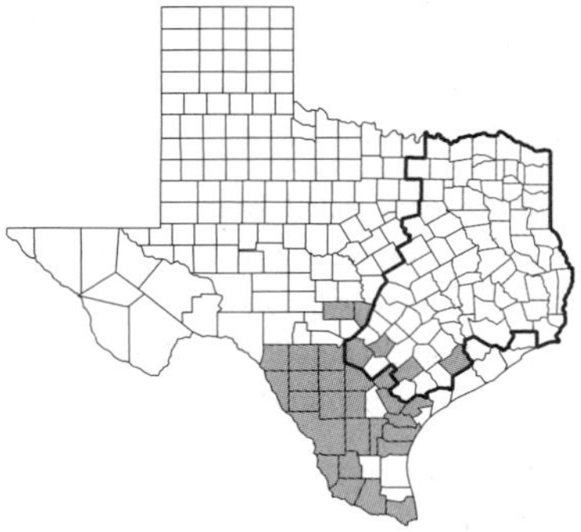

Vachellia bravoensis

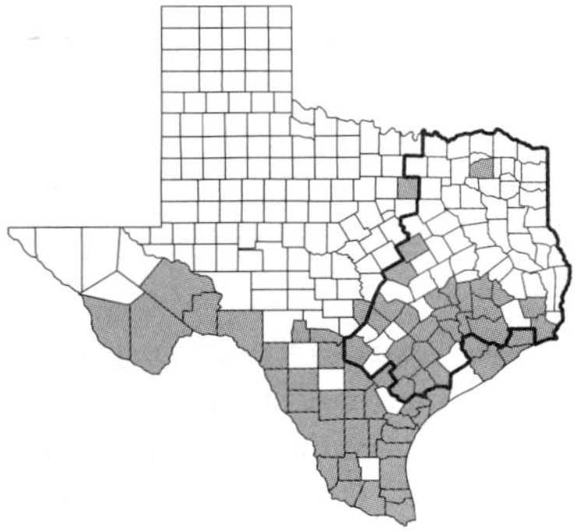

Vachellia farnesiana

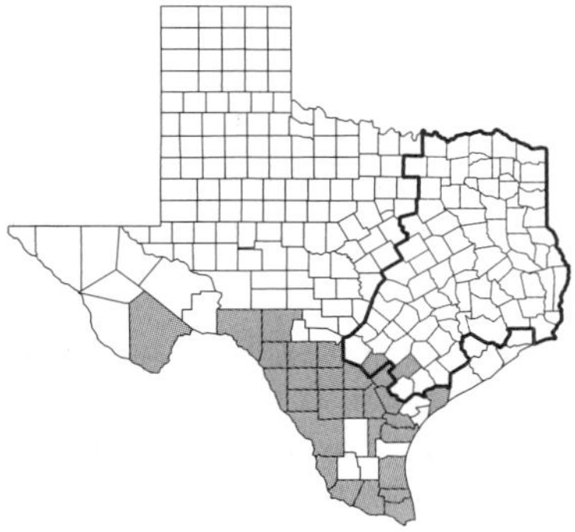

Vachellia rigidula

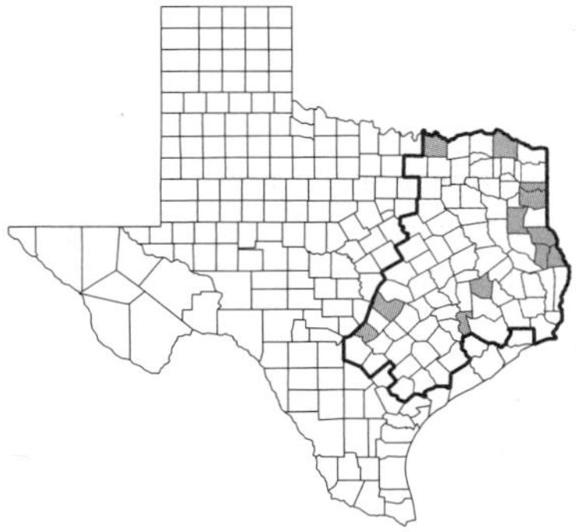

Vicia caroliniana

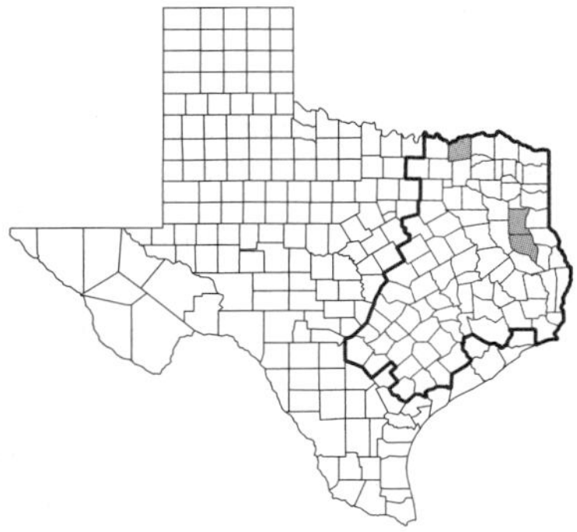

Vicia grandiflora

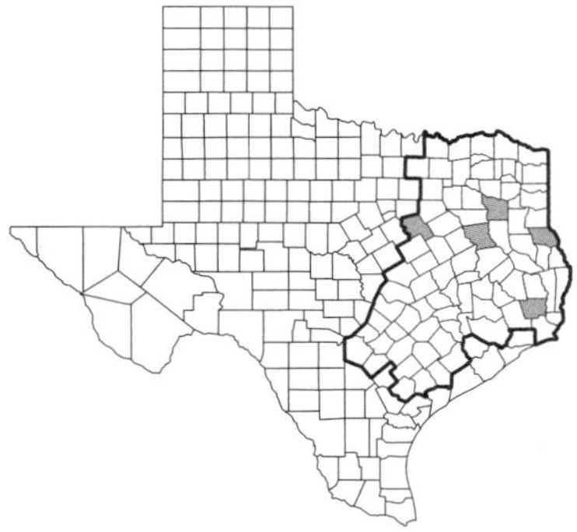

Vicia hirsuta

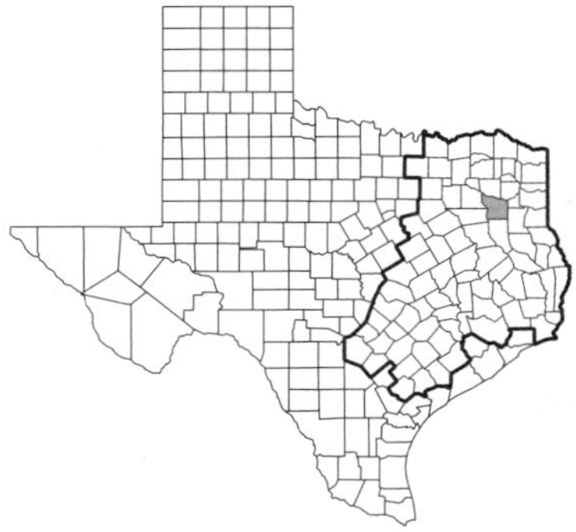

Vicia lathyroides

Apr–Jun. Native of Europe. The combination of tiny flowers, usually 1–2-flowered inflorescences, and apically rounded glabrous fruits with usually four seeds is helpful in distinguishing this species (Isely 1990). Nixon and Damuth (1987) first reported this species to TX in five East TX cos. indicating that it was rapidly spreading at the time.

Vicia villosa Roth, (villose, soft-hairy), Annual or biennial; stems 20–130 cm long, densely villous or glabrous; leaflets 10–18; racemes 3–19 cm long, usually 1-sided with 10–20 flowers; calyces 5–6 mm long; corollas blue-purple or blue-purple and white (rarely white), 12–18 mm long; fruits 15–40 mm long, 6–11 mm wide, glabrous, stipitate, 2–3 mm long; seeds 2–8. Both of the following varieties are cultivated for hay, forage, soil improvement, and erosion prevention; however, poisoning and death in cattle have been reported due to eating the seeds; neurological symptoms were similar to rabies, with bellowing, seizures, and even sudden death; sporadic dermatological effects also have been seen due to grazing the foliage ("vetch pasture disease") (Burrows & Tyrl 2001). Isely (1998) noted that numerous intermediates occur between the two subspecies; Turner et al. (2003) did not distinguish infraspecific taxa. We are following Broich (2023b) who treated the two overlapping entities at the varietal level. Specimens incorrectly identified as *V. cracca* in TX are usually this species. Varieties not distinguished on county distribution map.

1. Stems glabrous or sparsely and inconspicuously appressed-pubescent; flowers usually 10–20 per raceme, loosely arranged on rachis; upper calyx lobes lanceolate or narrowly lanceolate, 1–2(–2.4) mm long ________ var. **glabrescens**
1. Stems densely and conspicuously villous, the hairs spreading or ascending; flowers usually 20–40 per raceme, crowded on rachis; upper calyx lobes acicular, 2–4 mm long ________ var. **villosa**

var. **glabrescens** W.D.J. Koch, (nearly hairless), WINTER VETCH. Herbs glabrate, hairs inconspicuous, subappressed to spreading, 1 mm long; inflorescences with 10–20 flowers, loosely arranged on rachis, showy; flowers 12–16 mm long; upper calyx lobes lanceolate to narrowly lanceolate, 1–2(–2.4) mm long. Fields, roadsides, and waste areas; throughout East TX; most of TX, except High Plains and Trans-Pecos; throughout most of U.S., except interior w states; British Columbia and Nova Scotia. Apr–May. [*V. dasycarpa* Ten., *V. villosa* subsp. *varia* (Host) Corb.] Native of Europe.

var. **villosa.** HAIRY VETCH, RUSSIAN VETCH, WINTER VETCH. Herbs villous, hairs conspicuous, 1–2 mm long; inflorescences with 20–40 flowers, crowded on rachis; flowers 14–18 mm long; upper calyx lobes acicular, 2–4 mm long. Fields, roadsides, and waste areas; scattered in East TX in Brazos (Kartesz 2023), Dallas (Isely 1998), McLennan (Kartesz 2023), and Williamson (TEX) cos.; also Coleman and Randall (Isely 1998) cos. to the n and w; throughout the U.S.; Canada and South America. Apr–Jul. Native of Europe.

VIGNA Savi COWPEA

Erect or vining annual or perennial herbs; leaves alternate, pinnately compound, trifoliolate; leaflets entire; inflorescences axillary, racemes with 2–30 papilionaceous flowers, usually long-peduncled; fruit linear to linear-oblong, resupinate by twisting of pedicel, glabrous; seeds 5–25.

A mainly tropical, mostly Old-World genus of ca. 60 species of erect or twining herbs with extrafloral nectaries, fewer than ten species occur in the New World (Delgado-Salinas 2023c); in the past the species were sometimes included in the genus *Phaseolus* L. However, *Vigna* differs in such characters as having stipules often with appendages and the thickened part of the style less twisted. Several species are important pulses/beans/peas including *V. angularis* (Willd.) Ohwi & H. Ohashi (ADZUKI BEAN), *V. mungo* (L.) Hepper (URD), *V. radiata* (L.) R. Wilczek (MUNG BEAN, the "bean sprouts" sold commercially), and *V. unguiculata* (L.) Walp. (COWPEA, BLACK-EYED PEA, SOUTHERN PEA, CROWDER PEA) (Delgado-Salinas 2023c). (Named for Domenico Vigna, 17th century (d. 1647) Italian Professor of Botany at Pisa) (subfamily Faboideae, tribe Phaseoleae)

REFERENCES: Isely 1998; Delgado-Salinas 2023c.

1. Corollas yellow; leaflets fleshy or leathery; plant a trailing or twining perennial with stems to 1–3 m long; stipules 3–5 mm long, entire, without lobes; fruits with pubescence, 4–7 cm long; native species growing mostly in saline habitats ________ **V. luteola**
1. Corollas pink or purple to pale yellow; leaflets neither fleshy nor leathery; plant an erect bushy annual 1 m or less tall; stipules 5–15 mm long, with obvious lobes; fruits glabrous, 10–30 cm long; rarely escaped cultivated species ________ **V. unguiculata**

Vigna luteola (Jacq.) Benth., (yellowish), HAIRY COWPEA. Plant a vine, sometimes forming tangles to 3 m long; hirsute; leaflets variable in size and shape, ovate to lanceolate or narrowly lanceolate, 3–8 cm long, 0.5–5 cm wide; flowers usually 2–6 per inflorescence; bracteoles 1–1.5 mm long, not exceeding calyx tube; calyces 4.5–7 mm long; corollas 1.5–2 cm long; fruit linear-oblong, pendulous, 4–7 cm long; seeds 5–12. Mainly coastal locations on beaches, tidal flats, disturbed sites, roadsides, and rarely escaping cultivation outside native range; Brazos (escaped on Texas A&M campus, Kartesz 2023; E. Keith pers. obs.), Harris, and Orange (Kartesz 2023) cos. in s portion of East TX; mainly Gulf Prairies and Marshes and South TX Plains; e and se

Vicia grandiflora [WIL]

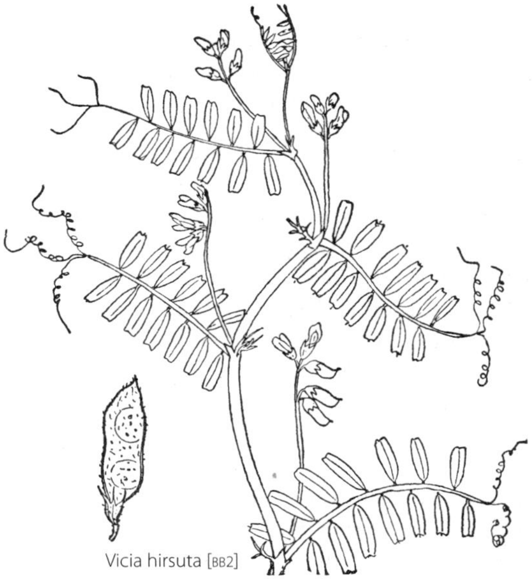

Vicia hirsuta [BB2]

Vicia lathyroides [FIO]

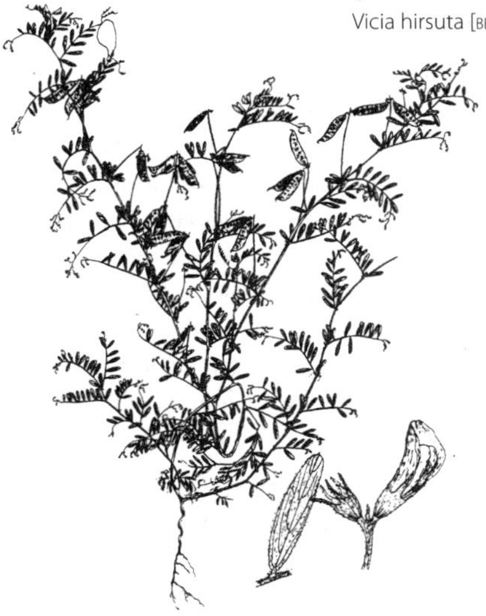

Vicia ludoviciana var. leavenworthia [HE1]

Vicia ludoviciana var. ludoviciana [HE1]

Vicia lutea [BEN]

Vicia minutiflora [HE1]

Vicia sativa var. angustifolia [HE1]

U.S. from PA s to FL and w to TX; Africa; South America; and Asia. Flowering year-round. [*Dolichos luteolus* Jacq., *D. mexicanus* Schltdl., *D. repens* L., not *Vigna repens* Baker 1876, *Orobus trifoliatus* Sessé & Moc., *Phaseolus hernandesii* Savi, *P. luteolus* (Jacq.) Gagnep., *Vigna villosa* Savi] *Vigna luteola* occurs in tropical coastal areas throughout the world and bears a complicated nomenclatural history (Pasquet 2001a,b; Delgado-Salinas 2023c).

Vigna unguiculata (L.) Walp., (clawed), COWPEA, BLACK-EYED PEA, SOUTHERN PEA, CROWDER PEA. Erect, trailing, or climbing annual; stems glabrate, often streaked red; stipules 5–15 mm long with basal appendages; leaflets 4–15 cm long, 3–8 cm wide, ovate, lanceolate, or rhombic-ovate; flowers 1–3 per inflorescence; bracteoles 4–6 mm long, exceeding calyx tubes; calyces ca. 1 cm long; corollas 2.5–3 cm long; fruits 10–30 cm long, ca. 1 cm wide, linear or falcate; seeds (8–)18–25. Rarely escaped in agricultural fields, roadsides, railroads, and disturbed areas; scattered in East TX in Brazos (NY), Collin (LL), Fayette (TEX), Madison (TAMU), and Tyler (TEX) cos.; in TX otherwise only known from Scurry Co. in the Rolling Plains and Galveston Co. on the coast (Kartesz 2023); e U.S. from PA s to FL and w to MI, MO, and TX; South America and Asia. May–Sep. [*Dolichos unguiculatus* L.] Native of w Africa. Isely (1990) noted that "*V. unguiculata* is one of the world's most important pulse and forage crops, particularly in Africa and Asia."

WISTERIA Nutt. WISTERIA

High climbing woody vines or vine-like shrubs with twining branches; plants unarmed; leaves once odd-pinnately compound; leaflets entire; racemes terminal, drooping, with numerous, large, showy, usually violet-blue to lavender, pinkish, or white flowers; calyces campanulate; fruits compressed, stipitate, seeds 1–5.

A genus of 8 species native to e Asia and e North America (Stritch 2023); this disjunct distribution pattern is discussed under the genus *Campsis* (Bignoniaceae). The Asiatic species are sometimes split as the segregate *Rehsonia* Stritch (Stritch 1984). Trusty et al. (2007) found that 24 of 25 individuals tested from scattered sites around the se U.S. showed genetic admixture (sometimes complicated) between *Wisteria floribunda* and *W. sinensis* and that the invasiveness of non-native species in the se U.S. may be attributable to the hybrids being more adaptable to naturalizing than the individual pure species. It is therefore probable that most of the material in East TX identified as either species is actually *W. ×formosa* ☠. All parts of *Wisteria* species should be considered poisonous, especially the fruits and seeds (Blackwell 1990) due to the presence of the glycoside wisterin (Schmutz & Hamilton 1979; Lampe & McCann 1985); as little as 2 seeds are enough to cause serious illness in children (Hardin & Arena 1974). Burrows and Tyrl (2001) discussed glycoprotein lectin which can cause gastrointestinal problems. (Named for Professor Caspar Wistar, 1761–1818, anatomist of Philadelphia) (subfamily Faboideae, tribe Millettieae)
REFERENCES: Wyman 1949; Valder 1995; Trusty et al. 2007; Stritch 2023.

1. Fruit and ovary glabrous, the fruits 12 mm or less wide; pedicels 5–10(–15) mm long; racemes usually 5–15 cm long; leaflet margins plane, apices acute or acuminate, pubescent; native species in woodland or forested habitats ______ **W. frutescens**
1. Fruit and ovary velvety pubescent (at least along sutures), the fruits 20–25 mm wide; pedicels 15–30 mm long; racemes often much longer; leaflet margins undulate, apices strongly acuminate, glabrate; introduced species in ruderal habitats or cultivated.
   2. Vines twining from left to right (counterclockwise).
      3. Racemes 10–33 cm long, flowers opening nearly simultaneously, before the leaves, unscented; banners 2.1–2.4 cm long ______ **W. sinensis**
      3. Racemes 15–36 cm long, flowers opening simultaneously or gradually opening from base to tip of raceme, with or after leaves, usually fragrant; banners 1.6–1.8 cm long ______ **W. ×formosa** (in part)
   2. Vines twining from right to left (clockwise).
      4. Leaves with 13–19 leaflets; racemes 20–132 cm long, flowers gradually opening from base to tip of raceme; auricles of banner's callosity 1.1–1.2 mm long, banner apex acuminate ______ **W. floribunda**
      4. Leaves with 7–15 leaflets; racemes 15–36 cm long, flowers opening simultaneously or gradually opening from base to tip of raceme; auricles of banner's callosity 0.7–0.8 mm long, banner apex rounded ______ **W. ×formosa** (in part)

Wisteria floribunda (Willd.) DC., JAPANESE WISTERIA. High climbing vines to 15 m long; stems twining clockwise; leaflets 13–19, blades elliptic to lanceolate, to 11.2 cm long, 2.6 cm wide; racemes to 132 cm long; pedicels to 2.8 cm long; flowers appearing with or after leaves, pleasantly scented; corollas azure, purple, rose, or white; banner 1.6–1.8 cm long, 1.6–1.7 cm wide, reflexed at base, apex acuminate, auricles of basal callosity 1.1–1.2 mm long; wings 1.4–1.5 cm long, 0.7–0.8 cm wide; keel 1.4–1.5 cm long, 0.7–0.8 cm wide; ovary pubescent; fruits flattened, oblong-elliptic or oblanceolate, 12–22(–30) cm long, tardily dehiscent, pubescent. Abandoned home sites, wooded edges of low fields, disturbed bottomland hardwood forests, and wet ground; no confirmed reports for TX and not mapped for TX in Stritch (2023), but probably escaped in Anderson Co. (iNaturalist Contributors 2023l), mapped in Newton Co. in Kartesz (2023), and as "rare" in East TX in Weakley, A.S., and Southeastern Flora Team (2022); e U.S. from ME s to FL and w to AR, LA, and probably TX; Argentina; Europe; and South Africa. Apr–Jun.

Vicia sativa var. sativa [HE1]

Vicia tetrasperma [BB2]

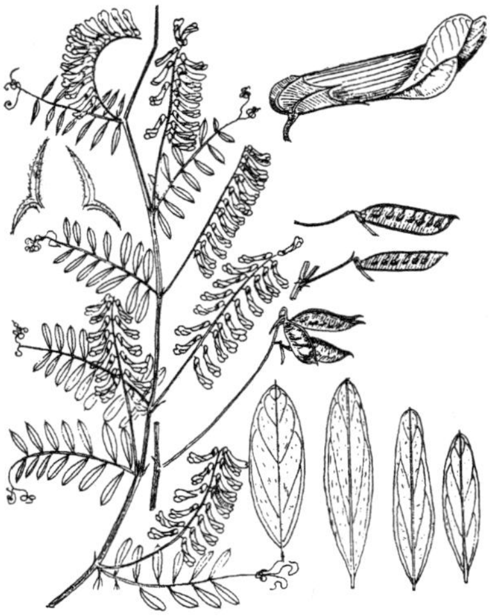

Vicia villosa var. glabrescens [HE1]

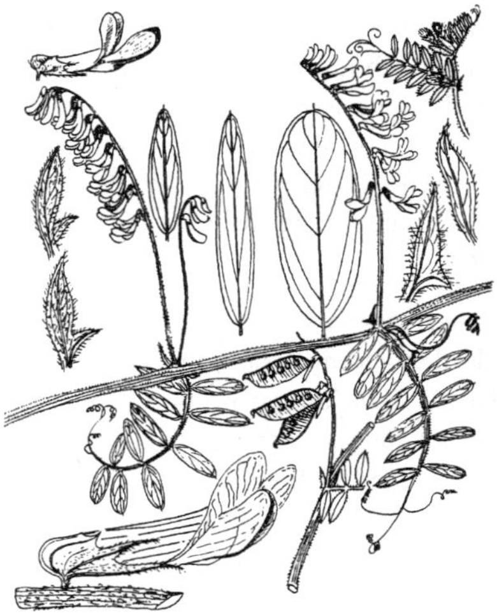

Vicia villosa var. villosa [HE1]

Vigna luteola [CO1]

Vigna unguiculata [WIL]

[*Glycine floribunda* Willd., *Kraunhia floribunda* (Willd.) Taub., *Rehsonia floribunda* (Willd.) Stritch] Native of Japan. This species is extensively cultivated in Japan and the U.S. where it occasionally escapes in similar habitats to those in Japan (Stritch 2023). It is most readily distinguished from *W. sinensis* by its sweetly fragrant flowers and clockwise twining and from their hybrid, *W. ×formosa* by the very long (to 132 cm) inflorescences with flowers always gradually opening from the base to the tip of inflorescences and longer, acuminate banner basal callosities. Reported as poisonous in Lampe and McCann (1985).

Wisteria ×formosa Rehder, (Chinese), FORMOSA WISTERIA. Stems twining clockwise or counterclockwise; leaflets 7–15, blades oblong-elliptic, to 8.2 cm long and 3.3 cm wide; racemes to 36 cm long; pedicels to 2.8 cm long; flowers appearing with or after leaves, pleasantly scented; corollas azure, purple, or white, 1.5–2 cm long; banner 1.6–1.8 cm long, 1.6–1.8 cm wide, reflexed at base, apex rounded, auricles of basal callosity 0.7–0.8 mm long; wings 1.7–1.8 cm long, 0.8–0.9 cm wide; keel 1.5–1.6 cm long, 0.7–0.8 cm wide; ovary pubescent; fruits flattened, oblong-elliptic or oblanceolate, 6–9.5 cm long, tardily dehiscent, pubescent. Abandoned home sites, wooded edges, and disturbed bottomlands; listed for TX in Stritch (2023) and probably much more common in East TX than currently documented (E. Keith, pers. obs.); e U.S. in AL, AR, FL, GA, KY, LA, MI, NC, SC, TX, and VA. Apr–Jun. [*Rehsonia ×formosa* (Rehder) Stritch] Parents native to Asia. *Wisteria ×formosa* originated spontaneously about 1905 at Holm Lea, the 150-acre estate owned by botanist Charles Sprague Sargent in Brookline, Massachusetts. The hybrid grew from seed borne by a white-petaled plant of *W. floribunda*; the pollen parent was assumed to have been *W. sinensis*, which was also cultivated on the estate. As an ornamental, it is regarded as superior to both parents, with large, fragrant flowers, and sequential anthesis combined with precocious blooming of *W. sinensis* (Rehder 1922). *Wisteria ×formosa* has been a source of confusion for botanists. Although Wyman (1949) indicated that it is more common in American horticulture than either parent, no major North American flora until Weakley, A.S., and Southeastern Flora Team (2022) and Stritch (2023) has included it in its keys and descriptions. As a result, its existence in the spontaneous flora of the continent has gone unrecognized. For example, Isely (1990) experienced difficulty in attempting to discriminate *W. floribunda* from *W. sinensis* in the se U.S. The lack of consistent character-state correlations he noted is exactly what would be seen if much of the studied material were actually *W. ×formosa*, as was borne out by molecular studies of weedy *Wisteria* in the se U.S. (Trusty et al. 2007). No county distribution map provided.

Wisteria frutescens (L.) Poir., (fruitful, fruit-bearing), AMERICAN WISTERIA. High climbing vine to 10(–15) m long; leaves with 9–15 leaflets; racemes appearing with leaves, 5–15(–22) cm long; rachises, pedicels, and calyces, often with club-shaped hairs; corollas 1.5–2 cm long, azure, purple, or rarely white; fruit 5–12(–15) cm long, 10–12 mm broad, oblong, explosively dehiscent. Bottomland forests, river and stream banks, mesic woodlands and forests; mostly in Pineywoods in East TX; also Gulf Prairies and Marshes; e U.S. from NY s to FL and w to KS, MO, and TX. Apr–Jun. [*W. frutescens* var. *macrostachya* Torr. & A. Gray, *W. macrostachya* (Torr. & A. Gray) Nutt. ex B.L. Rob. & Fernald] Isely (1990) noted that this species is "conspicuously variable in leaflet number and proportions, raceme length, and the evidence of "clavate glands" in the inflorescence. While it has often been divided into two separate species, we are following Isely (1990, 1998) and Stritch (2023) in treating the variation within one variable species. Presumably poisonous.

Wisteria sinensis (Sims) DC., (Chinese), CHINESE WISTERIA. High climbing vine or vine-like shrub with twining branches to 20 m long; stems twining counterclockwise; leaflets 7–13, blades ovate-elliptic to obovate, to 9 cm long and 3.5 cm wide; racemes to 33 cm long; pedicels to 3 cm long; flowers appearing before leaves, unscented; corolla azure, purple, rose, or rarely white, 2–2.5 cm long; banner 2.1–2.4 cm long, 2.1–2.3 cm wide, reflexed at base, apex rounded, auricles of basal callosity 0.7–0.8 mm long; wings 1.7–1.8 cm long, ca. 0.7 cm wide; keel 1.7–1.8 cm long, 0.7–0.8 cm wide; ovary pubescent; fruits flattened, oblong-elliptic or oblanceolate, 8–21 cm long, tardily dehiscent, pubescent. Widely cultivated and long persists around old home sites; scattered in s and e parts of East TX; also Edwards Plateau and Gulf Prairies and Marshes; e U.S from ME s to FL and w to AR, MI, MO, and TX, disjunct in AZ. Apr–Jun. Native of China. Poisonous. The fruits of this species are explosively dehiscent. In the fall of 2001, an Austin College botany student, Aire Bates, related that the *Wisteria* plant in her parent's yard was making considerable noise as the fruits "exploded" and threw out the seeds. She subsequently collected a specimen and numerous intact fruits. Over the course of several weeks, all of these fruits exploded, with fruit walls and seeds scattering over the faculty office where they were being kept. Post dehiscence, the fruit walls were significantly twisted. This species may not be as common as reported because of the possibility that naturalized plants may be *Wisteria ×formosa* (Trusty et. al 2007).

ZORNIA J.F. Gmel. ZORNIA

Herbaceous perennials; leaves alternate, palmately compound, with 2 or 4 leaflets; petioles ca. as long as leaflets; stipules not attached at base, with a lobe below attachment; inflorescences spikes, with several flowers; bracts very large, stipule-like, 2 per

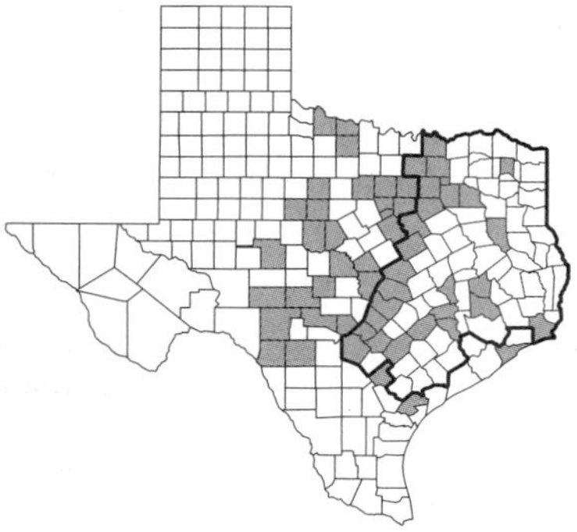

Vicia ludoviciana subsp. leavenworthii

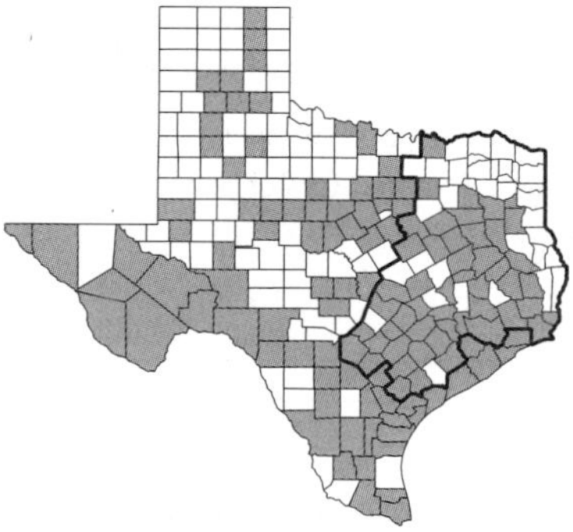

Vicia ludoviciana subsp. ludoviciana

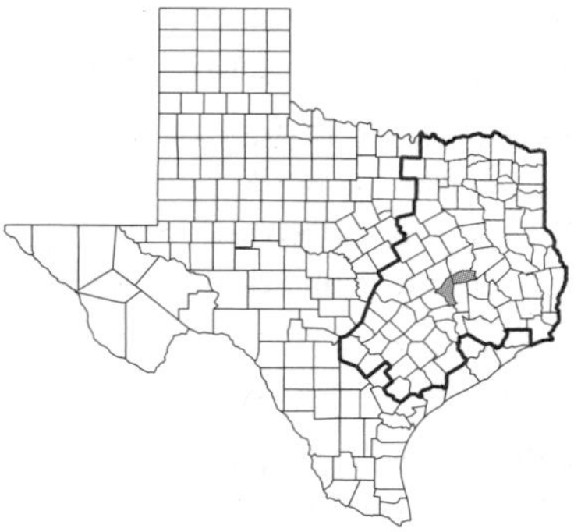

Vicia lutea

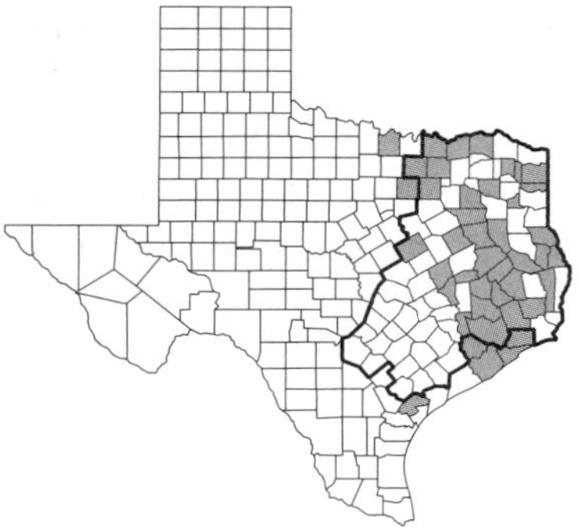

Vicia minutiflora

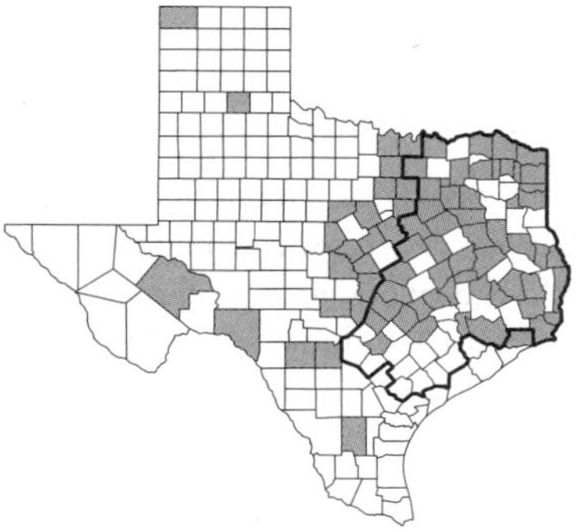

Vicia sativa var. angustifolia

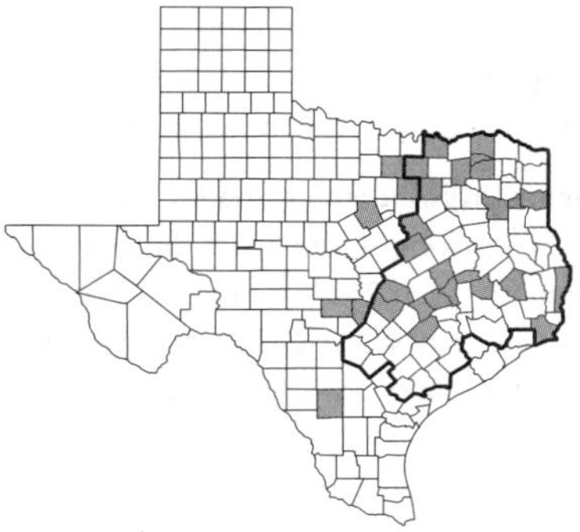

Vicia sativa var. sativa

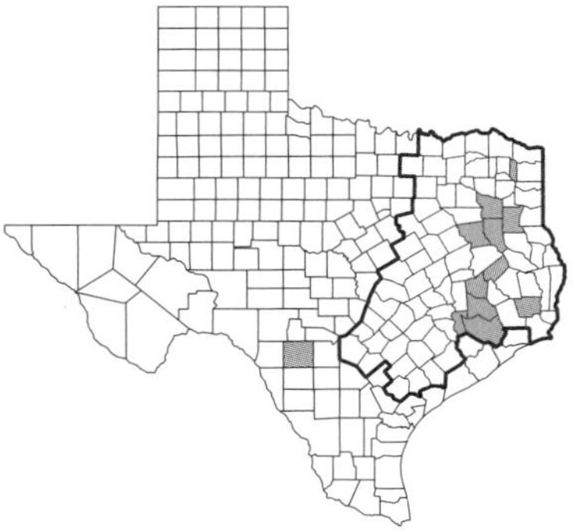

Vicia tetrasperma

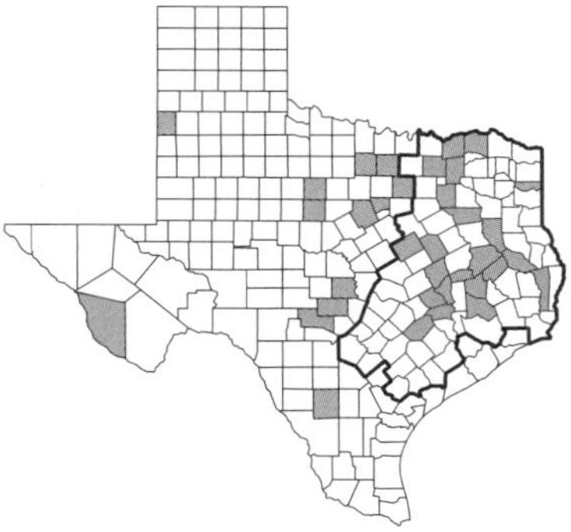

Vicia villosa

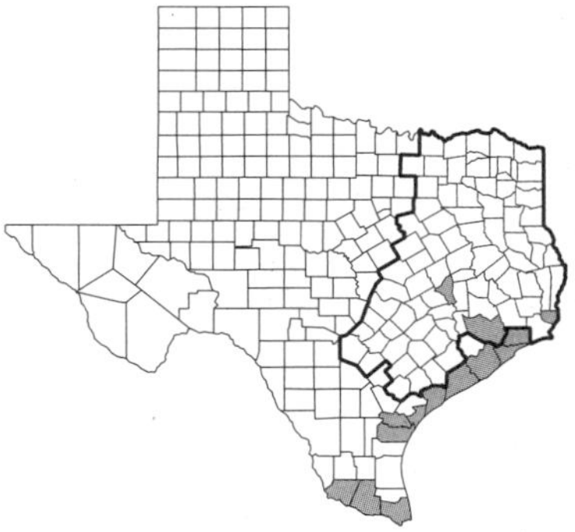

Vigna luteola

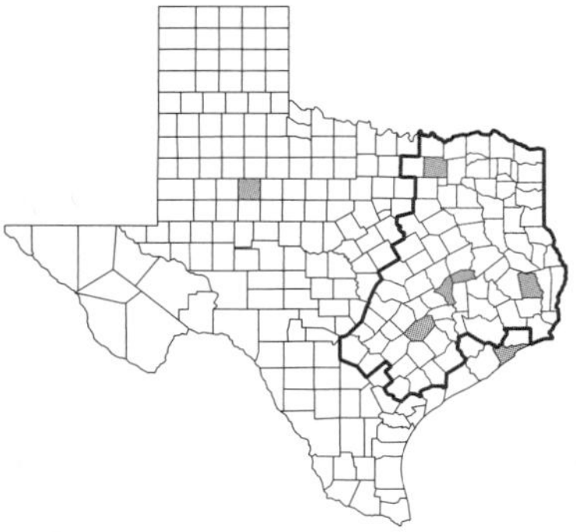

Vigna unguiculata

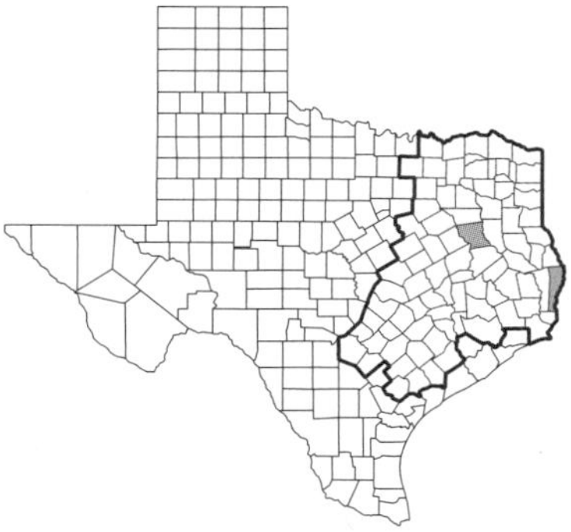

Wisteria floribunda

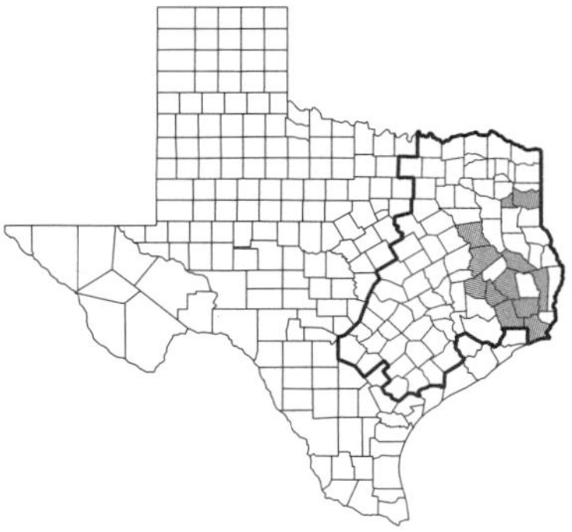

Wisteria frutescens

flower, partly obscuring flowers; corollas yellow to orangish yellow; stamens 10, monadelphous; styles curved; fruits flat, bristly, separating into 2–15 segments, often bristly with retrorse hairs; seeds 2–15.

A genus of ca. 50 (Isely 1998) to 75 (Mohlenbrock 2023b) species of warm areas of the world. The combination of large-bracted spikes and bristly flat fruits is distinctive. (Named for Johannes Zorn, 1739–1799, a German apothecary) (subfamily Faboideae, tribe Dalbergieae)
REFERENCES: Mohlenbrock 1961, 2023b.

1. Leaves with 4 leaflets ____ **Z. bracteata**
1. Leaves with 2 leaflets ____ **Z. latifolia**

Zornia bracteata J.F. Gmel., (bracteate, with bracts), BRACTED ZORNIA, VIPERINA. Glabrous perennial herb from a deep, tough, vertically or obliquely branched root; stems erect to prostrate, 20–80 cm long, sometimes mat-forming; leaflets typically 4, ca. 25 mm long, 8 mm wide, linear to narrowly elliptic, inconspicuously punctate; petioles as long as the leaflets; flowers widely spaced in long-peduncled, axillary spikes of 1–10 flowers, each flower with a pair of large, conspicuous, elliptic-ovate bracts 7–15 mm long; petals 10–14 mm long, yellow to orangish-yellow, with reddish veins; fruits with 2–6 segments, each 2.7–3.5 mm long, bristles to 1 mm long. Sandy open areas; widely scattered in the e half of TX; se U.S. from VA s to FL and w to TX. Apr–Oct. The combination of 4 leaflets, large, paired bracts, and bristly fruits make this species distinctive (Isely 1990).

Zornia latifolia Sm., (broad-leaved), ZAZABACOA DE DOS HOJAS. Erect, glabrous or pilose perennial herb; 20–50 cm tall; leaflets 2, ca. 25 mm long, 4 mm wide; spikes with 2–7 flowers; bracts partly hiding flowers; petals 13–15 mm long, yellow; fruits with 4–8 segments, each 2–2.4 mm long, reticulate, pilose to villous, with bristles 0.2–0.4 mm long. Sandy open areas; Dewitt, Goliad, and Lavaca (Kartesz 2023) cos. in sw portion of East TX; mainly s Gulf Prairies and Marshes and South TX Plains; FL and TX in the U.S.; Mexico; West Indies; Central America; and South America. Mar–Nov. [*Z. diphylla* (L.) Pers., *Z. diphylla* var. *vulgaris* Benth., *Z. gemella* Vogel] Mohlenbrock (2023b) treated this species as *Z. gemella*, but Fortuna-Perez and Azevedo Tozzi (2011) determined that *Z. latifolia* has priority.

Wisteria floribunda [FIO]

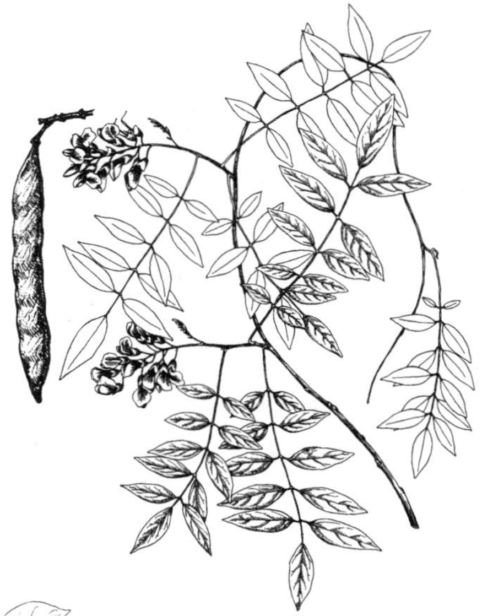

Wisteria frutescens [WIL]

Wisteria sinensis [WIL]

Zornia bracteata [WIL]

Zornia latifolia [FAW]

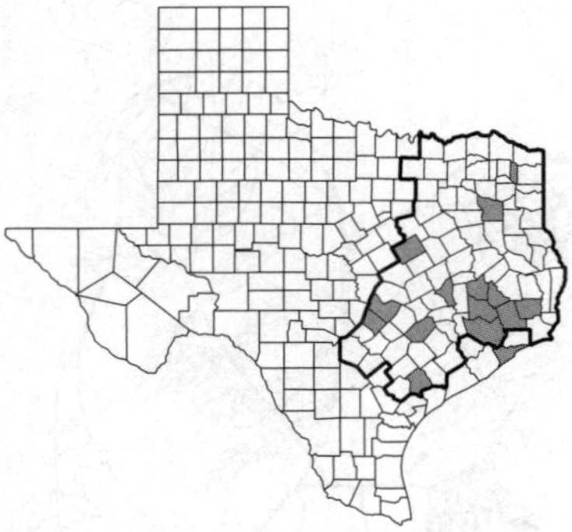
Wisteria sinensis

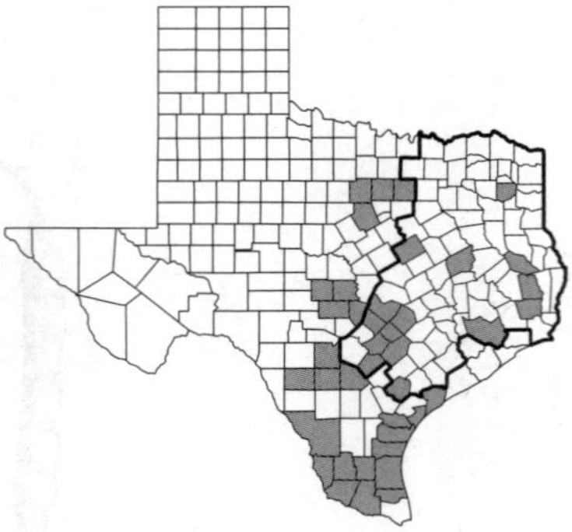
Zornia bracteata

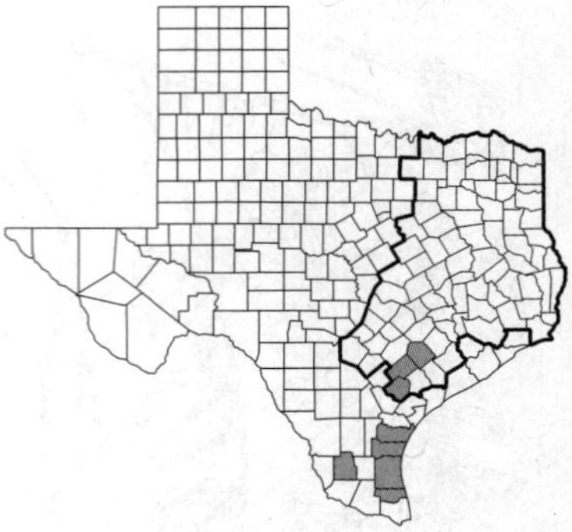
Zornia latifolia

ILLUSTRATION SOURCES

REF. CODE

ABG J. Adelaide Bot. Gard. 1975+. State Herbarium of South Australia. Adelaide. Reprinted with permission of the Board of the Botanic Gardens and State Herbarium. Adelaide.
ABR Abrams, L. 1923–1960. Illustrated flora of the Pacific states. Stanford University Press. Stanford, CA.
ADD Addisonia. 1916–1964. The New York Botanical Garden. Bronx, New York.
ALI Lounsberry, A. 1901. Southern wild flowers and trees. Fredrick Stokes Company. New York, NY.
ALL Allioni, C. 1785. Flora pedemontana. Joannes Michael Briolus. Torino, Italy.
ALS Aliso. 1948+. Rancho Santa Ana Botanic Garden. Claremont, CA. Reprinted with permission of Rancho Santa Ana Botanic Garden.
AMB Ann. Missouri Bot. Gard. 1914+. Missouri Botanical Garden. St. Louis. Reprinted with permission of the Missouri Botanical Garden Press.
ANS J. Acad. Nat. Sci. 1817–1842. Academy of Natural Sciences of Philadelphia. Philadelphia, PA.
APG Apgar, A.C. 1910. Ornamental shrubs of the United States. American Book Company. New York, NY.
ARF Parker, K.F. 1958. Arizona ranch, farm, and garden weeds. Agricultural Extension Service, Circ. 265. University of Arizona. Tucson.
ARG Anton, A.M. and Zuloaga, F.O., eds. 2012+. Flora Argentina: flora vascular de la República Argentina. Instituto de Botánica Darwinion, San Isidro, Argentina. Reprinted with permission.
ARM Annual Rep. Missouri Bot. Gard. 1889–1912. Missouri Botanical Garden. St. Louis. Reprinted with permission of the Missouri Botanical Garden Press.
AUJ Austral. J. Bot. 1953+. Commonwealth Scientific and Industrial Research Organization. Melbourne, Australia. Adapted from Windler, D.R. (1966). A revision of the genus *Neptunia* (Leguminosae). *Australian Journal of Botany* 14:379–420. Used with permission from CSIRO publishing.
BA1 Bailey, L.H. 1914–1917. The standard cyclopedia of horticulture. The Macmillan Company. New York, NY.
BA3 Bailey, L.H. 1900–1902. Cyclopedia of American horticulture. The Macmillan Company. New York, NY.
BAK Baker, J.J. 2026. Original illustrations. Published herein by The Botanical Research Institute of Texas. Fort Worth. Artist: Judith J. Baker. Copyright ©2026.
BB2 Britton, N.L. and Brown, A. 1913. An illustrated flora of the northern United States, Canada and the British possessions. Charles Scribner's Sons. New York, NY.
BBS Biltmore Bot. Stud. 1901–1902. Biltmore Herbarium. Biltmore, NC.
BCM Basset, I.J., Crompton, C.W., McNeill, J., and Taschereau, P.M. 1983. The genus *Atriplex* (Chenopodiaceae) in Canada (monograph no. 31). Agriculture Canada. Ottawa, Canada. Contains information licensed under the Open Government License-Canada.
BEL Benson, L. 1982. The cacti of the United States and Canada. Stanford University Press. Stanford, CA. Reprinted from *The Cacti of the United States and Canada* by Lyman Benson with permission of the publishers, Stanford University Press © 1982 by the Board of Trustees of the Leland Stanford Junior University.
BEN Bentham, G. 1865. Handbook of the British flora. (2nd ed.). Lovell Reeve and Company. London, England.
BL1 Baillon, H. 1871–1888. The natural history of plants. Lovell Reeve and Company. London, England.
BL2 Baillon, H. 1876–1892. Dictionnaire de botanique. Librairie Hachette. Paris, France.
BL3 Baillon, H. 1866–1895. Histoire des plantes. Librairie Hachette. Paris, France.
BOI Boissier, P.E. 1866. Icones Euphorbiarum. Victor Masson et fils. Paris, France.
BOT Bot. Gaz. (Crawfordsville). 1875+. University of Chicago Press. Chicago.
BR1 Britton, N.L. 1908. North American trees. Henry Holt and Company. New York, NY.
BR2 Britton, N.L. 1918. Flora of Bermuda. Charles Scribner's Sons. New York, NY.
BR3 Britton, N.L. and Rose, J.N. 1919–1923. The Cactaceae. Carnegie Institute. Washington, DC.
BRT J. Bot. Res. Inst. Texas. 2007+. Botanical Research Institute of Texas. Fort Worth.
BT3 Shinners, L.H., Whitehouse, E., and Mueller, P. 1999 and 2026. Previously published and unpublished original illustrations from the SMU herbarium. Published herein by The Botanical Research Institute of Texas. Fort Worth.
BTK Bot. Tidsskr. 1864–1981. Danish Botanical Association. Copenhagen. Used with permission.
BTT Brittonia. 1931/35+. The New York Botanical Garden. New York, NY. © The New York Botanical Garden. All rights reserved. Used by permission.
BUT Butcher, R.W. 1961. A new illustrated British flora: Volume 1-Lycopodiaceae to Salicaceae. Leonard Hill, Ltd. London, England.
CAV Cavanilles, A.J. 1791–1801 Icones et descriptiones plantarum. Regia Typographia. Matriti (Madrid), Spain.

CHA Chaudhri, M.N. 1968. Mededelingen. Botanisch Museum en Herbarium van de Rijksuniversiteit te Utrecht No. 285. Utrecht, Netherlands. Reprinted with permission of the Naturalis Biodiversity Center. Leiden, NL.

CLA Gaston, C. Original illustrations. Published herein by the Botanical Research Institute of Texas. Fort Worth. Artist: Clair Gaston. Copyright ©2026.

CO1 Correll, D.S. and Correll, H.B. 1972. Aquatic and wetland plants of southwestern United States. Environmental Protection Agency. Washington, DC.

CO3 Correll, D.S. and Correll, H.B. 1982. Flora of the Bahama archipelago. A. R. Gantner. Vaduz, Liechtenstein. Reprinted with permission of Koeltz Botanical Books.

COC Cocks, R.S. 1910. Leguminosae of Louisiana (Bull. no. 1). Louisiana State Museum. New Orleans.

COL Collett, Sir Henry. 1902. Flora Simlensis. Thacker, Spink, and Company. London, England.

COR Cornut, J.P. 1635. Canadensium plantarum, aliarumque nondum editarum historia. Simon le Moyne. Paris, France.

COS Coste, H. 1900–1906. Flore descriptive illustree de la France. Paul Klincksieck. Paris, France.

COZ Correll, D.S. and Correll, H.B. 1982. Flora of the Bahama archipelago. A.R. Gantner. Vaduz, Liechtenstein.

CP1 Patterson, C. 2026. Original illustrations. Published herein by The Botanical Research Institute of Texas. Fort Worth. Artist: Carolyn Patterson. Copyright ©2026.

CR2 Cronquist, A., Holmgren, A.H., Holmgren, N.H., Reveal, J.L., and Holmgren, P.K. 1972–2017 Intermountain flora: Vascular plants of the intermountain west, U.S.A. The New York Botanical Garden Press. Bronx, New York, NY. © The New York Botanical Garden Press. All rights reserved. Used by permission.

CRA Crawford, H.S., Kucera, C.L., and Ehrenreich, J.H. 1969. Ozark range and wildlife plants. United States Department of Agriculture. Washington, DC.

CRO Cronquist, A., Holmgren, A.H., Holmgren, N.H., Reveal, J.L., and Holmgren, P.K. 1972–2017. Intermountain flora: Vascular plants of the intermountain west, U.S.A. The New York Botanical Garden Press. Bronx, New York, NY. © The New York Botanical Garden Press. All rights reserved. Use by permission.

CSJ Cact. Succ. J. (Los Angeles). 1932+. Cactus and Succulent Society of America. Los Angeles, Pasadena, CA. Reprinted with permission of the Cactus and Succulent Society of America.

CTG Green, C.T. 1902. The flora of the Liverpool District. D. Marples and Company. Liverpool, England.

CUR Bot. Mag. 1787+. Royal Botanic Gardens, Kew. Richmond, England.

DAR Darlington, H.T. and Bessey, E.A. 1940. Some important Michigan weeds. Michigan State College Agricultural Experiment Station. East Lansing.

DAY Dayton, W.A. (Supervisor) (Prepared by Forest Service, United States Department of Agriculture) 1937. Range plant handbook. United States Government Printing Office. Washington, DC.

DEL Delessert, B. 1820–1846. Icones selectae plantarum. Fortin, Masson, and Company. Paris, France.

DEN Dentler, G. Original illustrations. Published herein by the Botanical Research Institute of Texas. Fort Worth. Artist: Gail Dentler. Copyright ©2026.

DHA Hall, D.W. 1993. Illustrated plants of Florida and the coastal plain. Maupin House. Gainesville, FL. Reprinted with permission of the author. Illustration by Edward H. Stehman.

DIL Dillenius J.J. 1732. Hortus elthamensis seu plantarum rariorum. Sumptibus Auctoris. London, England.

DN2 Hall, D.W. 1993. Illustrated plants of Florida and the Coastal Plain. Maupin House. Gainesville, FL.

DRW Darwiniana. 1922+. Instituto de Botánica Darwinion. Buenos Aires, Argentina. Reprinted with permission of Instituto de Botánica Darwinion.

EGG Eggers Ware, D.M. 1969. A revision of *Valerianella* in North America. Ph.D. dissertation. Vanderbilt University. Nashville, TN. Reprinted with permission of Donna Eggers Ware.

ELB LeBlanc, E. Original illustrations. Published herein by the Botanical Research Institute of Texas. Fort Worth. Artist: Ellen LeBlanc. Copyright ©2026.

EMO Emory, W.H. 1857–1859. United States and Mexican boundary survey. United States Government Printing Office. Washington, DC.

EN1 Engler, A. 1900–1953. Das pflanzenreich. Duncker and Humbolt. Berlin, Germany. Reproduced with permission of Dunker and Humbolt GmBh. Verlagsbuchhandlung. Berlin, Germany.

EN2 Engler, A. and Prantl, K. 1887–1915. Die naturlichen pflanzenfamilien. Gebrueder Borntraeger. Stuttgart. Originally published in "Die Naturlichen Pflanzenfamilien", reprint 1958/60 by J. Cramer in der Gebrueder Borntraeger Verlagsbuchhandlung, D–14129 Berlin D–70176 Stuttgart, Germany.

F&L Field & Lab. 1932–1970. Southern Methodist University Press. Dallas, TX.

FAW Fawcett, W. 1910–1936. Flora of Jamaica. British Museum (Natural History). London, England.

FDC Prepared for the Comité exécutif de la Flore du Congo belge et la Jardin Botanique de L'État. 1948–1963. Flore du Congo belge et du Ruanda-Urundi. L'Institut National pour L'Étude Agronommique du CongoBelge (The National Institute for the Agronomic Study of the of the Belgian Congo). Brussels.

FIE Fieldiana, Bot. 1895+. Field Museum of Natural History. Chicago, IL.

FIO Fiori, A. and Paoletti, G. 1923. Iconographia florae italicae. (2d ed.). Fiori Fratelli. San Casciano Val di Pesa, Florence, Italy.

FLH Degener, O. 1932–1986. Flora Hawaiiensis. Published by the author. Honolulu.

FMC Heller, A.A. 1895. Botanical explorations in southern Texas during the season of 1894. The New Era Printing House. Lancaster, PA.

FN2 Flora of North America Editorial Committee, eds. 1993+. Flora of North America north of Mexico. Oxford University Press, Inc. New York, NY. Courtesy of Flora of North America Association, Artists: *Tragia urens*, volume 12, Barbara Alongi (Plate 24 *Tragia ramosa*); *Soliva sessilis*, volume 19, Yevonn Wilson-Ramsey (Plate 64 *Cotula coronopifolia*); *Chrysactinia mexicana*, volume 21, Yevonn Wilson-Ramsey (Plate 36 *Chrysactinia mexicana*); *Gymnosperma glutinosum*, volume 20, Barbara Alongi (Plate 09 *Gutierrezia wrightii*); *Tetraneuris scaposa* var. *scaposa*, volume 21, Yevonn Wilson-Ramsey (Plate 68 *Plateilema plameri*); *Liatris elegans* var. *elegans*, volume 21, Linny Heagy (Plate 79 *Liatris elegans*); *Suaeda calceoliformis*, volume 4, Yevonn Wilson-Ramsey (Plate 62 *Suaeda linifolia*); *Lechea san-sabeana*, volume 6, Linny Heagy (Plate 45 *Cistus incanus*).

FNA Flora of North America Editorial Committee, eds. 1993+. Flora of North America north of Mexico. Oxford University Press, Inc. New York, NY. Couresy of Flora of North America Association. Artists: Barbara Alongi: *Aphanostephus ramosissimus* var. *ramosissimus*, volume 20; *Flaveria trinerva*, volume 21; *Gamochaeta calviceps* volume 19; *Gamochaeta coarctata*, volume 19; *Pterocaulon virgatum*, volume 19; *Rayjacksonia phyllocephala*, volume 20; *Schaefferia cuniefolia*, volume 12, Linny Heagy: *Erigeron procumbens*, volume 20; *Micropsis dasycarpa* volume 19, Marjorie C. Leggitt: *Heliopsis helianthoides* var. *scabra*, volume 21; *Solidago petioaris*, volume 20, John Myers: *Humulus lupulus*, volume 3; *Mammillaria prolifera* var. *texana*, volume 4, Yevonn Wilson-Ramsey: *Acourtia runcinata*, volume 19; *Cyclanthera naudiniana*, volume 6; *Celoisia nitida*, volume 4; *Dimorphotheca sinuata*, volume 19; *Gomphrena nealleyi*, volume 4; *Physaria gracilis* ssp. *nuttallii*, volume 7; *Pseudogynoxys chenopodioides*, volume 20.

FO2 Weakley, A.S., Ludwig, J.C., and Townsend, J.F. 2012. Flora of Virginia. Botanical Research Institute of Texas. Fort Worth. Illustrations by Lara Call Gastinger. © Flora of Virginia Project.

FOV Weakley, A.S., Ludwig, J.C., and Townsend, J.F. 2012. Flora of Virginia. Botanical Research Institute of Texas. Fort Worth. Illustrations by Lara Call Gastinger. © Flora of Virginia Project.

FSA Jessop, J.P. and Toelken, H.R., eds. 1986. Flora of South Australia. South Australian Government Printing Division. Adelaide. Reprinted with permission of Board of the Botanic Gardens and State Herbarium. Adelaide.

FSM Mathews, F.S. 1915. Field book of American trees and shrubs. G.P. Putnam's Sons. New York and London.

FVM Hegi, G. 1906–1931 Illustrierte flora von Mittel-Europa. A. Pichler's Witive and Sohn. Wien (Vienna), Austria.

G&F Gard. & Forest. 1888–1897. The Garden and Forest Publishing Company. New York, NY.

GAR Gartenflora. 1852–1938. Verlag von Ferdinand Enke. Stuttgart, Germany.

GAT Gates, F.C. 1941. Weeds in Kansas. Kansas State Board of Agriculture. Topeka.

GEN Genius, G. 2026. Original illustrations. Published herein by The Botanical Research Institute of Texas. Fort Worth. Artist: Gwen Genius. Copyright ©2026.

GEO Georgia, A.E. 1916. A manual of weeds. The Macmillan Company. New York, NY.

GL2 Gleason, H.A. 1952. The new Britton and Brown illustrated flora of the northeastern United States and adjacent Canada. The New York Botanical Garden. New York, NY. © The New York Botanical Garden. All rights reserved. Used by permission.

GLE Gleason, H.A. 1952. The new Britton and Brown illustrated flora of the northeastern United States and adjacent Canada. The New York Botanical Garden. New York, NY. © The New York Botanical Garden. All rights reserved. Used by permission.

GOR Gordon, J. Original illustrations. Published herein by the Botanical Research Institute of Texas. Fort Worth. Artist: Jeannice Gordon. Copyright ©2026.

GR1 Gray, A. 1848–1849. Genera florae Americae boreali-orientalis illustrata. James Munroe and Company. New York, NY.

GR2 Gray, A. 1852–1853. Plantae wrightianae. Smithsonian Institution Press. Washington, DC.

GRE Graham, E.H. 1941. Legumes for erosion control and wildlife (miscellaneous publication no. 412). United States Department of Agriculture. Washington, DC.

GWO Godfrey, R.K. and Wooten, J.W. 1979–1981. Aquatic and wetland plants of southeastern states. University of Georgia Press. Athens, GA. Reprinted with permission of the University of Georgia Press, Copyright © 1979, 1981.

HAL Hall, H.M. 1928. The genus *Haplopappus*. Carnegie Institute. Washington, DC.

HAR Quill, H. 2026. Original illustrations. Published herein by The Botanical Research Institute of Texas. Fort Worth. Artist: Harper Quill. Copyright ©2026.

HE1 Hermann, F.J. 1960. Vetches of the United States – native, naturalized, and cultivated. (Agricultural Handbook no. 168). United States Department of Agriculture. Washington, DC.

HE2 Hermann, F.J. 1962. A revision of the genus *Glycine* and its immediate allies (Technical Bulletin no. 1268). United States Department of Agriculture. Washington, DC.

HEA Diggs, G., Lipscomb, B., and O'Kennon, R.L. 1999. Shinners and Mahler's illustrated flora of North Central Texas. Original illustrations by Linda 'Linny' Heagy. Copyright ©1999. Botanical Research Institute of Texas. Fort Worth.

HO1 Hooker, W.J. 1829–1840. Flora boreali-americana. Treuttel and Wurtz. Paris, France.

HO5 Hooker's Icones Plantarum. 1867/71+. Williams and Norgate. London, England.

HOL Holmgren, N.H. 1998. Illustrated companion to Gleason and Cronquist's manual. The New York Botanical Garden Press. New York, NY. © The New York Botanical Garden Press. All rights reserved. Used by permission.

HOZ Holmgren, N.H. 1998. Illustrated companion to Gleason and Cronquist's manual. The New York Botanical Garden Press. New York, NY. © The New York Botanical Garden Press. All rights reserved. Used by permission.

HUM Humboldt, F.W.H.A. von. 1815–1825. Nova genera et species plantarum. Librairie Greque-Latine-Allemande. Paris, France.

HUT Hutchison, J. and Dalziel, J.M. 1927–1936. Flora of West Tropical Africa. Crown Agents for the Colonies. London, England. Reprinted with permission of the Crown Agents.

IPL Icon.Pl. 1837–1864. Longman, Orme, Brown, Green, and Longmans. London, England.

IV2 Ivey, R.D. 1995. Flowering plants of New Mexico (3rd ed.). Robert D. Ivey. Albuquerque, NM. Reprinted with permission of Vivian Ivey.

IVE Ivey, R.D. 1986. Flowering plants of New Mexico. (2nd ed.). Robert Dewitt Ivey. Albuquerque, NM. Reprinted with permission of Vivian Ivey.

JAA J. Arnold Arbor. 1920–1990. Arnold Arboretum of Harvard University. Cambridge, MA. Reprinted with permission of the Arnold Arboretum of Harvard University.

JAC Jacquin, N.J. 1763. Selectarum stirpium americanarum historia. Ex Oficina Krausiana. Vindobonae, Austria.

JEM J. Elisha Mitchell Sci. Soc. 1884+. North Carolina Academy of Science. Durham.

JEP Hickman, J.C., ed. The Jepson manual: Higher plants of California. University of California Press. Berkeley, CA. Reprinted with permission of the University of California Press.

JES Sowerby, J.E. 1860. British wild flowers. John E. Sowerby. London, England.

JLS J. Linn. Soc., Bot. 1865–1968. Williams and Norgate. London, England.

JME Jones, M.E. 1923. Revision of North American species of *Astragalus*. Marcus E. Jones. Salt Lake City, UT.

JMP Poole, J.M. and Riskind, D.H. 1987. Endangered, threatened, or protected native plants of Texas. Texas Parks and Wildlife Department. Austin. Reprinted with permission of Texas Parks and Wildlife.

JON Jones, F.B. 1982. Flora of the Texas coastal bend (3rd ed.). Welder Wildlife Foundation. Sinton, TX. Reprinted with permission of the Welder Wildlife Foundation.

JUJ Cabrera, A.L. 1978. Flora de la provencia de Jujuy: Part 10, Compositae. Instituto Nacional de Tecnologia Agropecuaria (INTA). Buenos Aires, Argentina. Reprinted with permission of Instituto Nacional de Tecnologia Agropecuaria (INTA).

KAR Karsten, H. 1891. Abbildungen zur deutschen flora. R. Friedlander and Sohn. Berlin, Germany.

KEB Kew Bull. 1946+. Royal Botanic Gardens, Kew. London, England.

KEW Bull. Misc. Inform. Kew. 1887–1942. Royal Botanic Gardens, Kew. London, England.

KGO Kurz, H. and Godfrey, R.K. 1962. Trees of northern Florida. University of Florida Press. Gainesville.

KIN Kinch, R.C. 1939. Nebraska weeds (Bulletin no. 101). Nebraska Department of Agriculture. Lincoln.

KOO Koorders, S.H. 1911–1937. Exkursionsflora von Java. Gustav Fischer. Jena, Germany. Reprinted with permission from Elsevier. Koorders, S.H. Exkursionsflora von Java, Volume 5. Part 2, Families 74–91, Page 644, Figure 930. Copyright 1926.

KUR Kurtziana. 1961+. Museo Botanico. Cordoba, Argentina. Reprinted with permission of Museo Botanico.

LAM Lamarck, J.B.A.P.M. 1791–1823.Tableau encyclopedique et methodique des trois regnes de la nature. Chez Pancouke. Paris, France.

LEI Ali-Sanchez, L. Original illustrations. Published herein by the Botanical Research Institute of Texas. Fort Worth. Artist: Leila Ali-Sanchez. Copyright ©2026.

LEM Le Maout, E. and Decaisne J. 1876. A general system of botany (2nd ed.). Longmans, Green, and Company. London, England.

LNL Estrada, A.E. and Marroquin de la Fuente, J.S. 1990. Leguminosas en el centro-sur de Nuevo Leon. Universidad Autonomo de Nuevo Leon. Linares, Nuevo Leon, Mexico.

LUN Lundell, C.L. 1961–1969. Flora of Texas. Texas Research Foundation. Renner.

LYN Lynch, D. 1981. Native and naturalized woody plants of Austin and the Hill Country. St. Edwards University. Austin, TX. Reprinted with permission of St. Edward's University and artist Nancy McGowan.

M&F Moore, R.J. and Frankton, C. 1974. The thistles of Canada. Research Branch, Canada Department of Agriculture. Ottawa, Canada. Reproduced under the Open Government License-Canada.

MAG Michigan Agric. Exp. Sta. Bull. 1885+. Michigan State University Press. East Lansing.

MAI Maire, R. 1952–1977. Flore du L'Afrique du nord. Paul Lechevalier. Paris, France.

MAR Martius, C.F.P. 1840–1906. Flora Brasiliensis. Monachii (Munich), Germany.

MAS Mason, H.L. 1957. A flora of the marshes of California. University of California Press. Berkeley, CA. Reprinted with permission of the University of California Press.

MCR Hammel, B.E., Grayum, M.H., Herrera, C., and Zamora, N., eds. 2010. Manual de plantas de Costa Rica, volume 5. Missouri Botanical Garden Press, St. Louis. Reprinted with permission.

MEP Parsons, M.E. 1909. The wild flowers of California. Cunningham, Curtiss, and Welch. San Francisco, CA.

MGH Mem. Gray Herb. 1917+. Harvard University. Cambridge, MA.

MIC Michaux, A. 1820. Flora boreali-americana. Bibliopola Jouanaux Junior. Paris, France.

MIT Mitt. Bot. Staatssamml. Munchen. 1950+. Botanische Staatssammlung. München (Munich), Germany. Reprinted with permission of Botanische Staatssammlung.

MN2 Mem. New York Bot. Gard. 1900+. New York Botanical Garden. New York, NY. © The New York Botanical Garden. All rights reserved. Used by permission.

MNY Mem. New York Bot. Gard. 1900+. New York Botanical Garden. New York, NY. © The New York Botanical Garden. All rights reserved. Used by permission.

MO2 Mohlenbrock, R.H. 1982. The illustrated flora of Illinois, flowering plants: Basswoods to spurges. Southern Illinois University Press. Carbondale. Reprinted with permission of Southern Illinois University Press.

MO4 Mohlenbrock, R.H. 2019. The illustrated flora of Illinois, flowering plants: Asteraceae part 2. Southern Illinois University Press. Carbondale. Reprinted with permission of the artist, Mark Mohlenbrock.

MOR Moris, J.H. 1837–1859. Flora Sardoa. Ex Regio Typographeo. Torino, Italy.

MOS Moss, C.E. 1920. The Cambridge British flora. Cambridge University Press. Cambridge, England.

MTB Mem. Torrey Bot. Club. 1889/90+. Torrey Botanical Society. New York, NY.

MUN Munz, P.A. 1935. A manual of southern California botany. Claremont Colleges. Claremont, CA.

NCS Editorial Subcommittee of the Regional Technical Committee of Project NC-10.1954. Weeds of the north central states. University of Illinois Agricultural Experiment Station. Urbana.

NEW Newhall, C.S. 1893. The shrubs of northeastern America. G.P. Putnam's Sons. New York, NY.

NIC Nicholson, G. 1885–1888. The illustrated dictionary of gardening. L. Upcott Gill. London, England.

NAZ Azure, N. 2026. Original illustrations. Published herein by The Botanical Research Institute of Texas. Fort Worth. Artist: Natch Azure. Copyright ©2026.

NSW Harden, G.J., ed. 1990. Flora of New South Wales. University of New South Wales Press. Sydney, Australia. Used with permission.

NZB New Zealand J. Bot. 1963+. Department of Scientific and Industrial Research. Wellington, Netherlands. Reprinted with permission of the Royal Society of New Zealand.

ORA Oregon Agric. Exp. Sta. Bull. 1914+. Oregon State University Agricultural Experiment Station. Corvallis.

PAM Pammel, L.H. 1913. The weed flora of Iowa. Iowa Geological Survey. Des Moines.

PCP Poole, J.M., Carr, W.R., Price, D.M., and Singhurst, J.R. 2007. Rare plants of Texas. Texas A&M University Press. College Station.

PFE Pfeifer, H.W. 1970. A taxonomic revision of the pentandrous species of *Aristolochia*. University of Connecticut. Storrs. Unable to locate copyright owner; author and artist deceased.

PHY Phytologia. 1933+. Texensis Publishing. Gruver, TX. Reprinted with permission of Phytologia and the authors.

PN2 Hitchcock, C.L., Cronquist, A., Ownbey, M., and Thompson, J.W. 1955–1969. Vascular plants of the Pacific northwest. University of Washington Press. Seattle. Reprinted with permission of the University of Washington Press.

PNW Hitchcock, C.L., Cronquist, A., Ownbey, M., and Thompson, J.W. 1955–1969. Vascular plants of the Pacific northwest. University of Washington Press. Seattle. Reprinted with permission of the University of Washington Press.

POP Pope, W.T. 1929. Manual of wayside plants of Hawaii. Advertiser Publishing Company, Ltd. Honolulu, HI.

POW Powell, A.M. 1988. Trees and shrubs of Trans-Pecos Texas. Big Bend Natural History Association, Inc. Alpine, TX. Reprinted with permission of Jim Henrickson, Michael Powell, and the Big Bend Natural History Assoc.

PRC Lundellia. 1998+. Billie L. Turner Plant Resources Center. Austin, TX. Reproduced from Delgado-Salinas, A., Carr, W.R., 2007. *Phaseolus texensis* (Leguminosae: Phaseolinae): A new species from the Edwards Plateau of central Texas, Lundellia 10:11-17, artist Albino Luna, with permission from the authors and the B.L. Turner Plant Resources Center, University of Texas.

RAD Radford, A.E., Ahles, H.E., and Bell, C.R. 1968. Manual of the vascular flora of the Carolinas. University of North Carolina Press. Chapel Hill, NC. From *Manual of the Vascular Flora of the Carolinas* by Albert E. Radford, Harry E. Ahles and C. Ritchie Bell, Copyright © 1968 by the University of North Carolina Press. Used by permission of the publisher. www.uncpress.org.

RBM Marcy, R.B. 1853. Exploration of the Red River of Louisiana. United States Government Printing Office. Washington, DC.

RCA Rep. Commiss. Agric. 1862–1893. United States Government Printing Office. Washington, DC.

REE Reed, C.F. 1970. Selected weeds of the United States (Agricultural Handbook no. 366). United States Department of Agriculture. Washington, DC.

RHO Rhodora. 1899+. New England Botanical Club. Cambridge, MA.

RIC Richardson, A. 1990. Plants of southernmost Texas. Gorgas Science Society. Brownsville, TX. Reprinted with permission of the author.

RJG George, R.J. 2026. Original illustrations. Published herein by the Botanical Research Institute of Texas. Fort Worth. Artist: Robert J. George. ©2026.

ROB Robbins, W.W., Bellue, M.K., and Ball, W.S. 1951. Weeds of California. California State Department of Agriculture. Sacramento.

ROE Roedner, B.J., Hamilton, D.A., and Evans, K.E. 1978. Rare plants of the Ozark Plateau. North Central Forest Experiment Station Forest Service-United States Department of Agriculture. St. Paul, MN.

ROS Rosendahl, C.O. and Butters, F.K. 1928. Trees and shrubs of Minnesota. The University of Minnesota Press. Minneapolis.

RUI Ruiz, H. and Pavon, J. 1794. Florae Peruvianae et chilensis prodromus. En la imprenta de Sancha. Madrid, Spain.

RYD Rydberg, P.A. 1932. Flora of the prairies and plains of central North America. Dover Publications Inc. New York, NY.

SA2 Sargent, C.S. 1905. Manual of the trees of North America (exclusive of Mexico). Houghton, Mifflin, and Company. Boston, MA.

SA3 Sargent, C.S. 1890–1902. The silva of North America. Houghton, Mifflin, and Company. Boston, MA.

SA4 Sargent, C.S. 1922. Manual of the trees of North America (exclusive of Mexico) (2nd ed.). Houghton Mifflin Company. Boston, MA and New York, NY.

SB2 Syst. Bot. Monogr. 1980+. American Society of Plant Taxonomists. Washington, D.C. From: Hughes, C. 1998. Monograph of *Leucaena* (Leguminosae-Mimosoideae). Systematic Botany Monographs, volume 55. Reprinted with permission of the American Society of Plant Taxonomists.

SBM Syst. Bot. Monogr. 1980+. American Society of Plant Taxonomists. Ann Arbor, MI. From: Tombs, S. 1980. Taxonomy of *Lygodesmia* (Asteraceae). Systematic Botany Monographs, volume 1. and Luckow, M. 1993. Monograph of *Desmanthus* (Leguminosae-Mimosoideae) Systematic Botany Monographs, volume 38. Reprinted with permission of the American Society of Plant Taxonomists and the authors.

SEM Semple, J. 2026. Original illustrations. Published herein by the Botanical Research Institute of Texas. Fort Worth. Artist: John Semple. Copyright ©2026. Used with permission of the artist.

SHI Shinners, L.H. 1958. Shinners' spring flora of the Dallas-Fort Worth area Texas. Lloyd Shinners. Dallas, TX.

SID Sida, Contributions to Botany. 1962+. Botanical Research Institute of Texas. Fort Worth.

SM1 Small, J.K. 1933. Manual of the southeastern flora. University of North Carolina Press. Chapel Hill, NC. From *Manual of the Southeastern Flora* by John Kunkel Small. Copyright ©1933 by John Kunkel Small, renewed 1961 by Kathryn Small Garber. Published by the University of North Carolina Press. Used by permission of the publisher. www.uncpress.org.

SMI Smith, J.E. and Sowerby, J. 1790–1814. English botany. Smith, J.E. London, England.

SNB Kirtikar, K.R and Basu, B.D. 1918. Indian medicinal plants. Sudhindra Nath Basu. Bahadurganj, India.

STE Steyermark, J.A. 1963. Flora of Missouri. Iowa State University Press. Ames.

STR Strausbaugh, P.D. and Core, E.L. 1978. Flora of West Virginia (2nd ed.). Seneca Books, Inc. Grantsville, WV.

SUD Sudworth G.B. 1908. Forest trees of the Pacific slope. United States Government Printing Office. Washington, DC.

SUL Sulman, F. 1913–1914. A popular guide to the wild flowers of New South Wales. Angus and Robertson, Ltd. Sydney, Australia.

SWF Biotic Consultants, Inc. 1990. Southern wetland flora. United States Department of Agriculture. Washington, DC. Used with permission of the artist: Mark Mohlenbrock.

SY2 Syst. Bot. 1976+. American Society of Plant Taxonomists. Washington, D.C. From: McCauley, R.A. 2004. New taxa and a new combination in the North American species of *Froelichia* (Amaranthaceae). Systematic Botany, volume 29, no. 1. Reprinted with permission of the American Society of Plant Taxonomists.

SYS Syst. Bot. 1976+. American Society of Plant Taxonomists. Washington, D.C. From: Eggers Ware, D.M. 1983. Genetic fruit polymorphism in North American *Valerianella* (Valerianaceae) and its taxonomic implications. Systematic Botany, volume 8, no. 1 and Semple, J.C. and Chinnappa, C.C. 1984. Observations on the cytology, morphology, and ecology of *Bradburia hirtella* (Compositae-Astereae) Systematic Botany. Volume 9, no. 1 Reprinted with permission of the American Society of Plant Taxonomists and the authors.

TAN Taylor, N.P. 1985. The genus *Echinocereus*. Royal Botanic Gardens, Kew. Richmond, England. Reprinted with permission of the Royal Botanic Gardens, Kew and the author. © Copyright The Board of Trustees of the Royal Botanic Gardens, Kew.

THO Thompson, M., 2026. Original illustrations. Published herein by the Botanical Research Institute of Texas. Fort Worth. Artist: Maria Thompson. Copyright ©2026.

TOR Bull. Torrey Bot. Club. 1870+. Torrey Botanical Society. Lancaster, PA.

TRP Hatch, S.L. and Pluhar, J. 1993.Texas range plants. Texas A&M University Press. College Station. Reprinted with the permission of the authors.

UCP Univ. Calif. Publ. Bot. 1902/03+. University of California Press. Berkeley.

UKS Univ. Kansas Sci. Bull. 1902–1996. University of Kansas. Lawrence.

USA Emory, W.H. Notes of a military reconnoissance. 1848. United State Army. Corps of Topographical Engineers. Washington, DC.

USC U.S.D.A. Circ. 1895–1901. United States Government Printing Office. Washington, DC.

USD Stefferud, A., ed. 1948. Grass: The yearbook of agriculture 1948. United States Department of Agriculture. Washington, DC.

USG Wheeler, G.M. 1878. Report upon United States geographical surveys west of the one hundredth meridian (Volume 6-Botany). United States Government Printing Office. Washington, DC.

USH Contr. U.S. Natl. Herb. 1890–1974. United States Government Printing Office. Washington, DC.

USW Engelmann, G. 1856. Reports of explorations and surveys, to ascertain the most practicable and economical route for a railroad from the Mississippi River to the Pacific Ocean. Volume 4. Government Printing Office. Washington, DC.

UWA Univ. Waterloo Biol. Ser. 1971+. University of Waterloo. Waterloo, Ontario, Canada. Reprinted with permission of J.C. Semple.

VIC Marie-Victorin, F. 1935. Flore laurentienne. Les Frères de Ècoles Chrètiennes. Montreal, Quebec, Canada.

VIN Vines, R.A. 1960. Trees, shrubs, and woody vines of the southwest. University of Texas Press. Austin.

VMU Mueller, F. von. 1889–1891. Iconography of Australian salsolaceous plants. R.S. Brain. Melbourne.

WEL Welch, A. 2026. Original illustrations. Published herein by the Botanical Research Institute of Texas. Fort Worth. Artist: Amy Welch. Copyright ©2026.

WF2 Radford, A.E., Ahles, H.E., and Bell, C.R. 1968. Manual of the vascular flora of the Carolinas. University of North Carolina Press. Chapel Hill, NC. From *Manual of the Vascular Flora of the Carolinas* by Albert E. Radford, Harry E. Ahles and C. Ritchie Bell, Copyright © 1968 by the University of North Carolina Press. Used by permission of the publisher. www.uncpress.org.

WFG Greene, W.F. and Blomquist, H.L. 1953. Flowers of the South: Native and exotic. University of North Carolina Press. Chapel Hill, NC. From *The Flowers of the South: Native and Exotic* by W.F. Greene and H.L. Blomquist. Copyright © 1953 by the University of North Carolina Press. Published by permission of the publisher. www.uncpress.org.

WIG Wight, R. 1838–1853. Icones plantarum indiae orientalis. J.B. Pharoah. Madras, Tamil Nadu, India.

WIL Wilbur, R.L. 1963. The leguminous plants of North Carolina (Technical Bulletin No. 151). North Carolina Agricultural Experiment Station. Raleigh.

WJB Bean, W.J. 1914. Trees and shrubs hardy in the British Isles, volume 2. John Murray. London, England.

WLD Willdenowia. 1953+. Botanical Garden and Botanical Museum. Berlin.

WLJ Jepson, W.L. 1923 and 1925. A manual of the flowering plants of California. Associated Students Store, University of California. Berkeley.

WU1 Wu, Z. and Raven, P.H., eds. 1994–2013. Flora of China. Science Press. Beijing, China and Missouri Botanical Garden Press. St. Louis, MO. Used with permission of the Missouri Botanical Garden Press, St. Louis and the Science Press, Beijing, China.

WYC Woon, Y.C. 1921. Chinese economic trees. Commercial Press, Ltd. Shanghai, China.

YA2 Yatskievych, G. 1999–2013. Steyermark's flora of Missouri. Missouri Department of Conservation. Jefferson City. Reprinted with permission of the Missouri Botanical Garden Press.

YAT Yatskievych, G. 1999–2013. Steyermark's flora of Missouri. Missouri Department of Conservation. Jefferson City. Reprinted with permission of the Missouri Botanical Garden Press.

ZO1 Zohary, M. and Heller, D. 1984. The genus *Trifolium*. Israel Academy of Sciences and Humanities. Jerusalem. Reprinted with permission of the Israel Academy of Sciences and Humanities.

DIVISION **MAGNOLIOPHYTA**

ANGIOSPERMS OR FLOWERING PLANTS

Worldwide, the Magnoliophyta (also known as the Angiospermae, Anthophyta, or Magnoliidae depending on author/taxonomic system) is composed of ca. (223,300–)352,000(–422,127) species in 13,405 genera (though there is substantial disagreement about the number of genera in some families) arranged into 413–416 families in 64 orders (Stevens 2001 onwards; APG IV 2016; Mabberley 2017); 190 of these families occur in East Texas. Depending on a variety of factors, including taxonomic philosophy (lumping versus splitting), more varied numbers of flowering plant families have been recognized, ranging from perhaps 387 to 685 (although a growing consensus views the 413–416 range to be a reasonable number). The rather broad range of the number of families mainly reflects differences in the rank at which groups are recognized (e.g., family versus subfamily) rather than differing views of evolutionary relationships (Cronquist 1988; Reveal 1993a, 1993b; Stevens et al. 2001 onwards; Thorne & Reveal 2007; APG IV 2016), although new revelations based on molecular research have also significantly impacted family recognition. The widely varying estimates of the number of angiosperm species is the result of such factors as our inadequate knowledge of tropical rain forest groups, the lack of modern taxonomic treatments for many genera, complications involving how to deal with specific boundaries including polyploidy and asexual reproduction, the implications of hybridization, cryptic species, and differences in taxonomic philosophy (see e.g., Stevens 2001 onwards). However, the number of angiosperm species approaching or more likely being somewhat over 300,000 now seems to be widely agreed upon (e.g., Stevens 2001 onwards; Wang 2023; Zhang & Ma 2024).

The Magnoliophyta has dominated the Earth's vegetation since the mid-Cretaceous (90 million years ago), and is the most important and most diverse group of plants on a worldwide basis, and is the cornerstone of most terrestrial ecosystems (Qiu et al. 1999; Zuntini et al. 2024). It is also the primary group upon which human civilization relies. For example, just three cereal grains, wheat, maize (corn), and rice, make up approximately 50% or more of the world's total human caloric intake (Sands et al. 2009; Singh 2022), and some authorities suggest that 15 crops make up 90% of the world's energy intake (Brewer 2022; Leite 2022).

The angiosperms are seed plants with flowers, seeds developing inside closed carpels, and double fertilization, a process by which cells in addition to the egg unite during fertilization to form a triploid endosperm (Mabberley 2017). "The term 'angiosperm' was coined in 1690 by the German botanist Paul Hermann (1646–1695) from Greek angeion (vessel) + spermos (seed), to name those plants that have their seeds contained [inside a protective carpel], in contrast to 'gymnosperms' that have their seeds exposed [or naked]" (Wang 2023). Recent, large scale molecular analyses have indicated that the angiosperms are a monophyletic group (Hamby & Zimmer 1992; Chase et al. 1993; Doyle et al. 1994; Stevens 2001 onwards; Gitzendanner et al. 2018; Zhang & Ma 2024; Lin et al. 2025) supporting the same conclusion based on morphological and anatomical characters. Current evidence favors the hypothesis that among living plants the angiosperms are sister to a clade that includes all extant gymnosperms (Stevens 2001 onward; Gitzendanner et al. 2018). In other words, the angiosperms are no more closely related to any group of living gymnosperm than to any other. This is in contrast to earlier conclusions linking the gymnosperm group Gnetophyta to the angiosperms (Chase et al. 1993; Doyle et al. 1994). However, since angiosperms are seed plants, botanists almost universally agree that they evolved from some now extinct group of gymnosperms.

The historical division of flowering plants into Class Monocotyledonae (monocots) and Class Dicotyledonae (dicots) is not supported by molecular data. Rather, the monocots appear to be a well-supported monophyletic group derived from *within* the dicots (e.g., Chase et al. 1993; Duvall et al. 1993; Qiu et al. 1993, 1999; Soltis et al. 2000; Savolainen et al. 2000; Stevens 2001 onwards; Zanis et al. 2003) The dicots are therefore paraphyletic and thus, from the cladistic standpoint, inappropriate for formal recognition (see Appendix 3). However, for practical reasons we are continuing to informally recognize these two traditional classes. Volume 1 of the *Illustrated Flora of East Texas* (Diggs et al. 2006) covered the monocots (along with ferns and similar plants and gymnosperms), while this current second volume and the future third volume treat the dicots (families listed alphabetically beginning with A through Fabaceae in Volume 2 and the remaining families in Volume 3).

REFERENCES: Cronquist 1981, 1988, 1993; Hamby & Zimmer 1992; Thorne 1992a, 1992b; Chase et al. 1993; Duvall et al. 1993; Qiu et al. 1993; Reveal 1993a, 1993b; Doyle et al. 1994; Takhtajan 1997; Qiu et al. 1999; Soltis et al. 1999, 2005, 2011; Stevens 2001 onwards; Wikström et al. 2001; Soltis & Soltis 2004; Thorne & Reveal 2007; APG IV 2016; Fay 2016; Gitzendanner et al. 2018; Singh 2022; Wang 2023; Zhang & Ma 2024; Zuntini et al. 2024; Lin et al. 2025.

DICOTS (PREVIOUSLY CLASS DICOTYLEDONAE)

Plants herbaceous or woody, often with secondary tissues derived from a vascular cambium; seedlings usually with 2 seed leaves (cotyledons); stems or branches elongating by apical growth; leaves not elongating once expanded; new leaves and branches developing from terminal or axillary buds; leaves when present alternate, opposite, whorled, or basal, commonly with a petiole and expanded blade; leaf blades usually net-veined, with midrib and less prominent, spreading branch-veins (these sometimes nearly parallel in very narrow leaves), but veins sometimes buried in thick or fleshy leaves; perianth herbaceous, with dissimilar inner and outer whorls (= petals and sepals), or all parts about alike (= tepals), the perianth parts separate or united, commonly in 2s or 5s or numerous, occasionally in 3s or perianth absent.

Worldwide, the dicots (previously Class Dicotyledonae), as a traditional group, is composed of ca. 215,200 species in 10,682 genera divided into 336 families and 52 orders (Mabberley 2017); 145 of these families occur in East TX. However, the historical division of flowering plants into Class Monocotyledonae (monocots) and Class Dicotyledonae (dicots), as mentioned above, is no longer supported by recent molecular studies. The dicots were first recognized in the 17th century by John Ray. From that time until the late 20th century, the combination of the presence of 2 seed leaves (cotyledons), usually net-veined leaves, and often 5-merous flowers (plus often some additional characteristics) distinguished the group. Only with the development of molecular methods did the actual evolutionary relationships within the angiosperms become clear. It is currently recognized that the monocots (now treated as Class Liliopsida) appear to be a well-supported monophyletic group derived from *within* the families traditionally treated as dicots (e.g., Chase et al. 1993; Duvall et al. 1993b; Qiu et al. 1993, 1999; Soltis et al. 2000; Savolainen et al. 2000; Stevens 2001 onwards; Zanis et al. 2003; Byng 2014; APG IV 2016; Mabberley 2017; Soltis et al. 2018; Guo et al. 2021; Li et al. 2021; Zuntini et al. 2024) (see Appendix 3). The dicots are therefore paraphyletic and thus, from the cladistic standpoint, inappropriate for formal recognition. In other words, it is now realized that a number of families with unusual characteristics formerly placed in the dicots (e.g., Aristolochiaceae, Magnoliaceae, Lauraceae, Nymphaeaceae) "are relicts of the group from which monocots and dicots in the narrow sense [eudicots] emerged ..." (Mabberley 2017). However, for practical reasons (e.g., volume one of the *Illustrated Flora of East Texas* treated the monocots), we are continuing to informally recognize the two traditional groups—monocots and dicots, and approximately half of the dicots are being treated in this second volume with the rest to be covered in a coming final third volume of the *Illustrated Flora of East Texas*. The traditional dicots are an extremely diverse group including a wide variety of different plants (ancient to modern lineages, herbs to trees, terrestrial to aquatic, 1-pored and 3-pored pollen, etc.). Within the dicots, the vast majority of species are in the monophyletic g roup known as the "eudicots" (these have triaperturate or 3-pored pollen, in contrast to the monocots and a few ancient dicot lineages which have uniaperturate or 1-pored pollen) (see Appendix 3). According to Herendeen and Crane (1995), "A useful model of angiosperm classification therefore recognizes the monocots and eudicots as two monophyletic clades (which together account for about 97% of extant angiosperm diversity) that are embedded in a systematically depauperate but morphologically diverse grade of magnoliid and (other primitive) dicotyledons." The exact relationships among the remaining 3% of angiosperms (non-monocots and non-eudicots—a group of ancient surviving lineages with 1-pored pollen sometimes referred to as the woody magnoliids and paleoherbs) was difficult to determine and only recently clarified. According to our present understanding, the earliest surviving branch of the angiosperms is the Amborellales, followed by the Nymphaeales, the Austrobaileyales, the "magnoliids" (comprised of Chloranthales, Magnoliales, Laurales, Canellales, and Piperales), and finally a branch leading to the monocots and eudicots (Stevens 2001 onwards; APG IV 2016; Guo et al. 2021). While our current understanding is still imperfect, the accomplishments and increased understanding of the past few years, made possible by molecular studies, are astonishing.

REFERENCES: Cronquist 1981, 1988, 1993; Wolf et al. 1989; Loconte & Stevenson 1991; Hamby & Zimmer 1992; Thorne 1992a, 1992b; Chase et al. 1993; Duvall et al. 1993; Qiu et al. 1993, 1999, 2000; Reveal 1993a, 1993b; Doyle et al. 1994; Herendeen & Crane 1995; Takhtajan 1997; Nandi et al. 1998; Soltis et al. 1999, 2000, 2018; Savolainen et al. 2000; Stevens 2001 onwards; Leitch & Hanson 2002; Reveal & Pires 2002; Zanis et al. 2002, 2003; Stuessy 2004; Byng 2014; APG IV 2016; Mabberley 2017; Guo et al. 2021; Li et al. 2021; Zuntini et al. 2024.

SUMMARY TREE OF ANGIOSPERM PHYLOGENY

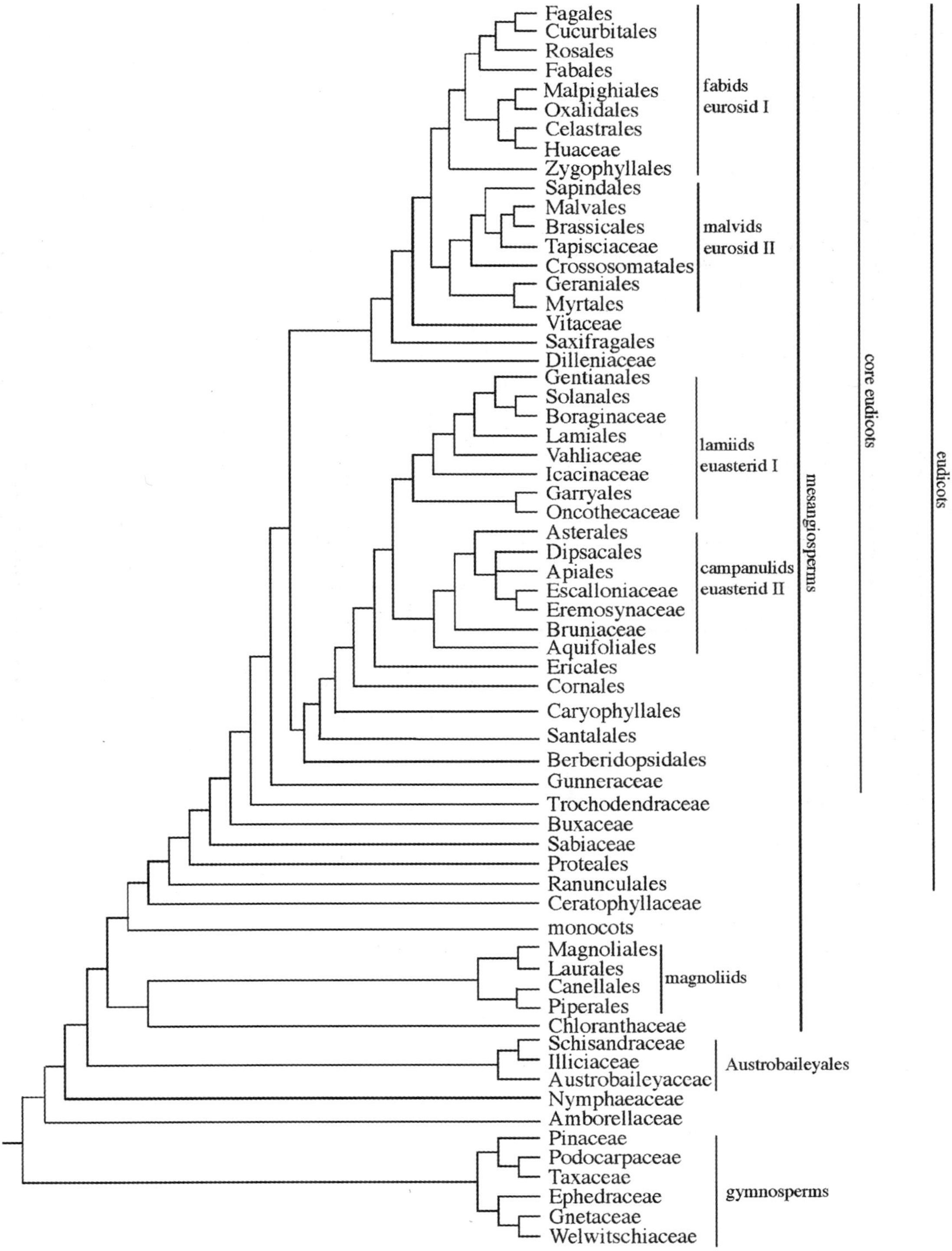

Schematic summary tree of angiosperm phylogeny (to the ordinal level). From Soltis, D.E, M. Moore, G. Burleigh, and P.S. Soltis. 2009. Molecular markers and concepts of plant evolutionary relationships: Progress, promise, and future prospects. Crit. Rev. Pl. Sci. 28(1):1–15. (Used with permission of D.E. Soltis)

Tree Diagram of the Magnoliid Clade

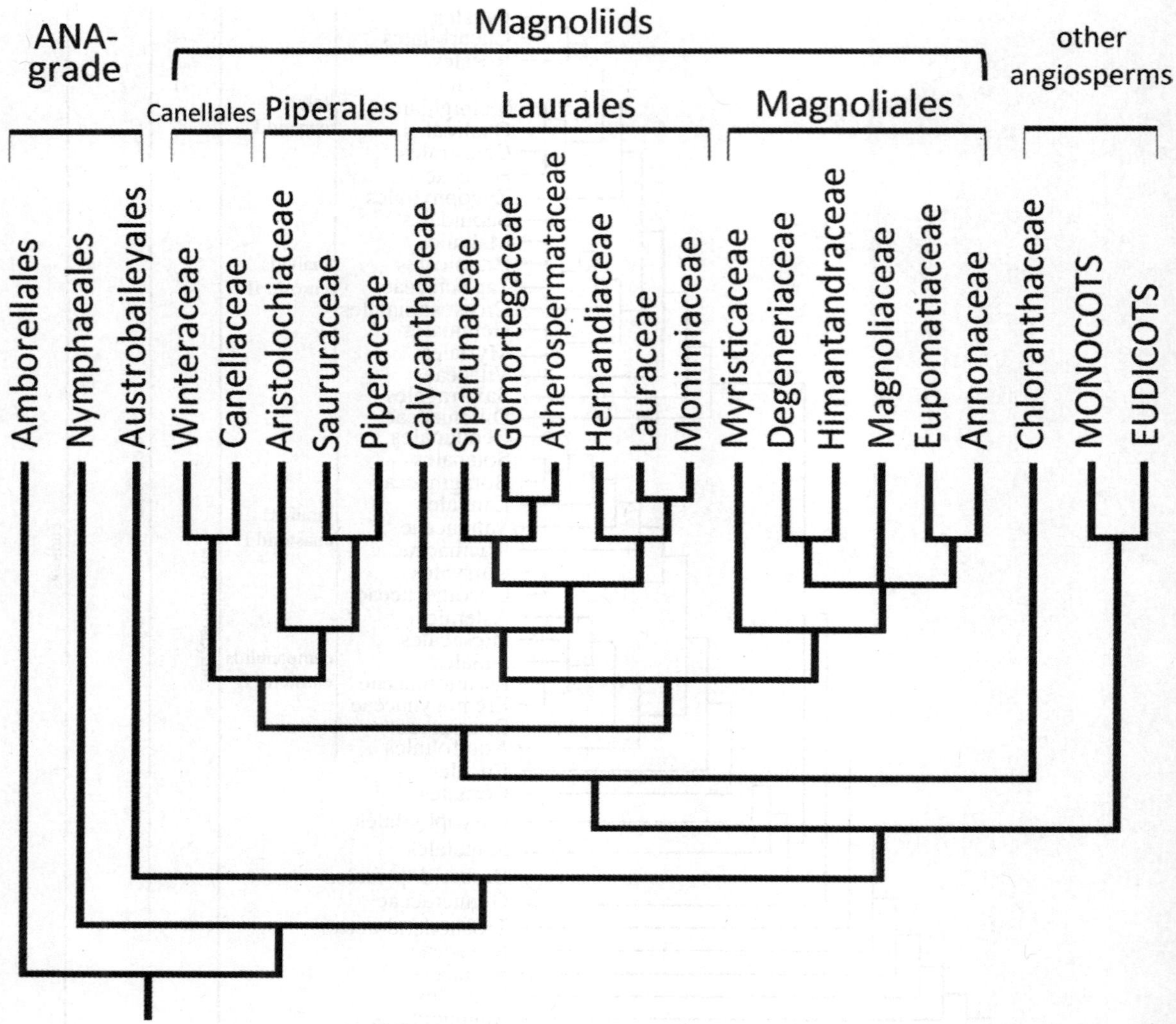

Tree diagram showing relationships within the magnoliid clade. From Rudall 2023. (Licensed under a Creative Commons Attribution 4.0 International License.)

Glossary

This glossary is modified from those of Shinners (1958) and Mahler (1988), with additional entries obtained or modified from a variety of sources including Lawrence (1951), Featherly (1954), Correll (1956), Gleason and Cronquist (1963), Radford et al. (1968), Correll and Johnston (1970), Gould (1975b), Lewis and Elvin-Lewis (1977), Benson (1979), Schmutz and Hamilton (1979), Fuller and McClintock (1986), Jones and Luchsinger (1986), Schofield (1986), Gandhi and Thomas (1989), Blackwell (1990), Isely (1990), Harris and Harris (1994), Spjut (1994), and Hickey and King (1997).

A

A- A prefix meaning without or not.

Abaxial Located on the side away from axis; e.g., lower leaf surface; contrasting with adaxial.

Aberrant Different from normal or typical condition.

Abortive Not developing or imperfectly developed; barren; defective.

Abscission Act or process of cutting off or shedding; e.g., the shedding or abscising of leaves.

Abscission layer Zone at base of petiole or other structure (e.g., pedicel) forming a layer of separation. This layer is important in the drop or shedding of leaves and fruits.

Acaulescent Stemless or apparently so; having leaves basal with stems not elongated.

Accessory fruit A fruit or assemblage of fruits with fleshy parts derived from organs other than the pistil; e.g., strawberry with fleshy receptacle with achenes (individual fruits) embedded in its surface.

Accrescent Enlarging after anthesis or with age, frequently in reference to the calyx.

Accumbent cotyledons Cotyledons lying face to face with the edges against the radicle.

Acerose (= Acicular) Needle-shaped or -like.

Achene Small, dry, indehiscent, one-seeded fruit with ovary wall free from seed.

Achlamydeous Lacking a perianth.

Achlorophyllus Lacking chlorophyll or apparently so; e.g., a number of non-green saprophytes or parasites.

Acicular (= Acerose) Needle-shaped or -like.

Acorn Fruit of a *Quercus* species (oak) composed of a nut and its cup or cupule made of fused bracts.

Acrid With sharp and harsh or bitterly pungent taste.

Acropetal Developing or maturing in succession from the base toward the apex.

Actinomorphic (= Regular) Radially symmetrical. The term usually refers to the arrangement of flower parts.

Acuminate Having a long, tapering point; longer tapering than acute.

Acute Forming a sharp angle of less than 90 degrees; less tapering than acuminate.

Ad- A prefix meaning to or toward.

Adaxial Located on side toward axis; e.g., upper leaf surface; contrasting with abaxial.

Adherent Touching or sticking together, when two organs or parts (typically dissimilar) touch each other but are not grown or fused together.

Adnate United or fused, when the fusion involves dissimilar structures; e.g., as in fusion of stamens and corolla.

Adpressed (= Appressed) Lying flat against a surface.

Adventitious Referring to structures or organs that develop in an unusual position; e.g., buds or roots that develop out of their usual place.

Adventive Not fully naturalized or established; of occasional occurrence.

Aerial Above ground level.

Aestival Appearing in or pertaining to the summer.

Aestivation Arrangement of young flower parts in the bud.

Agamospermy The production of seeds without fertilization.

Aggregate Crowded into a dense cluster or tuft.

Aggregate fruit A fruit formed by the clustering together of a number of separate pistils from a single flower; e.g., a blackberry is a cluster of druplets.

Aglycone The nonsugar component of a glycoside. Glycosides are composed of a sugar plus another compound (the aglycone); many aglycones are toxic.

Alate Winged.

Albidus White.

Albumen Nutritive material stored within the seed.

Alkaloid Any of a broad class (> 5, 000 known alkaloids) of bitter, usually basic (alkaline), organic compounds that contain nitrogen and typically have a ring in their structure. They are often physiologically active in animals; many are poisonous; many affect the nervous system; there are a number of general types based on chemical structure including indole, isoquinoline, piperidine, purine, pyrrolidine, quinoline, and tropane alkaloids; well known examples of alkaloids include atropine, caffeine, cocaine, quinine, morphine, nicotine, theobromine, and strychnine.

Allelopathy, Allelopathic Harmful or detrimental chemical effect by one species upon another; e.g., a plant producing phytotoxic compounds that inhibit the germination or growth of other plants.

Allergen Substance capable of inducing an allergic response.

Allergenic Causing an allergic response or an allergy to become manifest.

Allergy Hypersensitivity of the body cells to specific substances as antigens and allergens, resulting in various types of reactions (e.g., anaphylaxis, contact dermatitis, hay fever).

Alliaceous Onion-like.

Alluvial Of or pertaining to alluvium (= organic or inorganic materials, including soils, deposited by running water).

Alternate Bearing one leaf or other structure at a node; having only one attached at a given point; contrasting with opposite or whorled.

Alveolate Honeycombed.

Ament (= Catkin) A flexible often pendulous spike or spike-like raceme of small, inconspicuous, unisexual, apetalous, usually wind-pollinated flowers, the whole falling as one piece; e.g., male inflorescence of oaks or pecan.

Amentiferous Bearing aments.

Amethystine Violet-colored.

Amino acids Compounds containing both an amino group and a carboxyl group. They are the subunits (monomers) that are linked together by peptide bonds to form the polymers known as proteins; some nonprotein amino acids are found free in plants and are sometimes toxic; e.g., in *Lathyrus* (Fabaceae).

Amorphous Without regular or definite form; shapeless.

Amphitropous ovule Ovule that is half inverted so that the point of attachment is near the middle.

Amplexicaul (= Clasping) With base of leaf or other structure (e.g., stipule) wholly or partly surrounding the stem.

Ampliate Enlarged.

Anastomosing Net-like; with veins connecting by crossveins to form a network.

Anatropous ovule Ovule that is completely inverted, the micropylar end being essentially basal.

Androecium Collective term for the stamens or male structures of a flower.

Androgynophore A stalk bearing both androecium and gynoecium; e.g., in many Passifloraceae.

Androgynous Bearing staminate flowers above (= distal to) the pistillate in the same spike; e.g., in some Cyperaceae.

Androphore A support or column bearing stamens.

Anemophilous, Anemophily Wind-pollinated.

Angiosperm (= Flowering plant) Literally, "vessel seed"; a plant having its seeds enclosed in an ovary (= the proximal part of the carpel or "vessel"); a member of Division Magnoliophyta.

Annual Plant or root system living only one growing season (year); completing the growth cycle within one year.

Annular Arranged in a ring or circle.

Annulate With the appearance of rings; e.g., cross-ribbed or ringed spines of some Cactaceae.

Annulus A group or ring of thick-walled cells, on the sporangia of some ferns, that are involved in spore dehiscence.

Anterior Describing the position of an organ located toward the front in relation to the axis; e.g., in a flower the side away from the axis and toward the subtending bract.

Anther That part of a stamen producing the pollen.

Anther-cell (= Theca) One of the pollensacs or locules of an anther.

Antheridium Male sexual organ; structure forming male gametes, typically found in less derived plants (e.g., ferns) but so reduced evolutionarily as to not be present in flowering plants.

Anthesis (a) Time or process of flower expansion or opening; (b) also descriptive of period during which a flower is open and functional.

Anthocarp A structure that includes a fruit united with the perianth or the receptacle; e.g., Nyctaginaceae.

Anthocyanin A red, purplish, or blue water-soluble pigment found in most flowers. The color of these pigments is affected by pH (e.g., in *Hydrangea*); chemically, anthocyanins are phenolic.

Antipetalous Referring to stamens that are of the same number as, and borne in front of (= on the same radius as) the petals or corolla lobes.

Antisepalous Referring to stamens that are of the same number as, and borne in front of (= on the same radius as) the sepals or calyx lobes.

Antrorse Directed toward the summit, upward, or forward; e.g., pubescence directed up the stem, the free end of the hair above or distal to the attached end; contrasting with retrorse.

Antrorsely barbed With barbs (= points) pointing upward toward the summit or apex.

Apetalous Having flowers without petals.

Apetaly The condition of being without petals.

Aperture An opening.

Apex (pl. APICES) The tip or summit.

Aphyllopodic Lacking leaves at the base.

Aphyllous Leafless; e.g., Cuscutaceae.

Apical At the tip or apex; relating to the apex.

Apical bud (= Terminal bud) Bud at the end (= apex) of a stem or branch.

Apiculate Having a small sharp point formed by blade tissue (of a leaf, sepal, or petal) rather than by projection of a rib or vein; with an abrupt tip or projection.

Apocarpous With the carpels separate or free from one another.

Apogamous Forming a sporophyte without the union of gametes.

Apomixis A collective term for reproduction, including vegetative propagation, that does not involve sexual processes; any form of asexual reproduction.

Apopetalous (= Polypetalous) Referring to a corolla consisting of separate petals.

Apophysis Swelling or enlargement of the surface of an organ.

Aposepalous (= Polysepalous) Referring to a calyx consisting of separate sepals.

Appendage Any attached structure that is supplementary or secondary.

Appendiculate With an appendage.

Appressed (= Adpressed) Lying flat against a surface.

Approximate Close together.

Aquatic Living in water.

Arachnoid, Arachnose Cobwebby; cobweb-like, with entangled, slender, loose hairs; thinly pubescent with relatively long, usually appressed and interlaced hairs.

Arborescent Tree-like or becoming tree-like.

Archegonium Female sexual organ; structure forming female gametes, typically found in less derived plants (e.g., ferns) but so reduced evolutionarily as to not be present in flowering plants.

Arcuate Curved or bent like a bow, often used in reference to curving veins.

Arenaceous Sand-like or growing in sand.

Areolate Divided into small angular spaces; marked with areolae.

Areole, Areola (pl. AREOLAE) (a) Small space marked out on a surface, usually referring to the space bounded by veinlets on the surface of a leaf; (b) the small spine-bearing areas on a cactus stem.

Argillaceous Clayey; growing in clay or clay-like material.

Aril An appendage or covering on a seed, typically involved in dispersal by animals.

Arillate With an aril.

Aristate Bearing a stiff awn or bristle.

Aristulate Diminutive of aristate.

Armed Possessing sharp projections; e.g., prickles, spines, or thorns.

Aromatic (a) Generally, having a fragrant odor; (b) chemically, containing or patterned after benzene rings, with or without fragrance.

Article The individual unit of a constricted or jointed fruit; e.g., in some Fabaceae such as *Desmodium*.

Articulate Jointed; joined.

Articulation A separation place; joint.

Ascending, Ascendent Rising at an oblique angle.

Asepalous Without sepals.

Asexual Without sex; reproducing without sex.

Asperous Rough to the touch.

Assurgent Ascending, rising.
Astylous Without a style.
Asymmetrical Without symmetry.
Atomiferous Bearing very fine glands.
Atrocastaneous Very dark chestnut-colored.
Atropurpureous Dark purple; purple-black.
Attenuate Gradually tapering to a very slender tip, the taper more gradual than in acuminate.
Atypical Not typical; deviating from the norm.
Auricle Earlobe-like lobe or appendage; e.g., at the base of some leaves, sepals, etc.
Auriculate With an auricle.
Autotrophic Descriptive of an organism capable of making its own food, usually through photosynthesis; free living, not parasitic or saprophytic; e.g., green plants.
Autumnal Associated with or occurring in the fall of the year.
Awl-shaped (= Subulate) Tapering from the base to a slender or stiff point; narrow and sharp-pointed.
Awn Terminal slender bristle or hair-like extension or projection; e.g., in grasses, the prolongation of the midnerve of the glumes or lemmas.
Awn column In certain grasses, a prominent narrowed beak at the apex of the lemma. The awns arise from this structure.
Axial Relating to the axis.
Axil Angle between two organs; e.g., upper angle formed by a leaf and a stem.
Axile or **Axillary Placentation** Placentation with the ovules attached to the central axis of the ovary.
Axillary In an axil; e.g., in the angle between a leaf and a stem.
Axis (pl. AXES) (a) The central stem from which organs arise; (b) a portion of a plant from which a series of organs arises radially; e.g., the axis of an inflorescence.

B

Baccate Resembling or having the structure of a berry; berry-like.
Ballistic Referring to fruits that are forcibly or elastically dehiscent, whose seeds are thrown catapult-like; e.g., *Phyllanthus* (Euphorbiaceae).
Banner (= Standard) Adaxial and typically largest petal of a papilionaceous flower.
Barbed With short reflexed points like a multi-pronged fishhook.
Barbellate Diminutive of barbed; with short, fine, stiff hairs.
Bark Outer (= external to vascular cambium) protective tissues on the stems or roots of woody plants.
Basal Located at the base of a plant or of an organ.
Basal rosette Cluster of leaves on or near the ground.
Basal style Style projecting from among the lobes of a deeply lobed ovary.
Basifixed Attached basally, typically referring to attachment of an anther to a filament; contrasting with either dorsifixed or versatile.
Basilaminar At base of blade of leaf or other structure.
Basipetal Developing or maturing in succession from the apex toward the base.
Basiscopic Directed toward the base.
Basionym The original epithet assigned to a species (or other taxon of lower rank) by its author.
Beak A long, prominent, and relatively thickened point; a tapering projection; e.g., projection on a fruit resulting from a persistent style.
Beaked Ending in a beak.
Beard A group of long awns or bristle-like trichomes; a zone of pubescence; e.g., on some corollas.
Bearded Bearing long or stiff hairs, typically in a line or tuft.
Berry Indehiscent type of fruit with the entire pericarp fleshy and lacking a stone, usually with several to many seeds; e.g., tomato, grape.
Betalains Reddish, nitrogen-containing pigments (characteristic of most Caryophyllidae) that derive their name from the genus *Beta* (beets).
Bi-, Bis- Latin prefix signifying two, twice, or doubly.
BICOLORED Two-colored.
Biconvex Convex on both sides.
Bidentate Two-toothed.
Biennial Plant or root system living only two years (growing seasons), typically producing only leafy growth the first year, then flowering and dying the second.
Bifid Two-cleft, usually deeply so; with two lobes or segments.
Bifurcate Two-forked; e.g., some Y-shaped trichomes, stigmas, or styles.
Bilabiate Two-lipped, typically referring to corollas or calyces.
Bilateral Arranged on two sides; two-sided.
Bilaterally symmetricaL With only one plane of symmetry; divisible into halves in one plane only.
Bilocular Having two cavities.
Binomial The combination of a generic name and a specific epithet given to each species.
Binomial nomenclature System of nomenclature where each species has a two-part name composed of a generic name and a specific epithet.
Bipartite Two-parted; divided into two parts nearly to the base.
Bipinnate (= Twice-pinnate or 2-pinnate) Descriptive of a leaf with leaflets pinnately arranged on lateral axes that are themselves pinnately arranged on the main axis; with the primary divisions (pinnae) themselves pinnate.
Bisected Completely divided into two parts.
Biseriate In two whorls or cycles; e.g., a perianth with both calyx and corolla.
Bisexual flower Type of flower with both stamens and pistil(s) functional within the same flower.
Bivalvate Opening by two valves.
Bladder (a) A thin-walled, inflated structure; (b) a hollow, membranaceous appendage that traps insects; e.g., *Utricularia* (Lentibulariaceae).
Bladdery Thin-walled and inflated; like the bladder of an animal.
Blade Flat, expanded portion, as the main part of a leaf or petal.
Bloom (a) Flower or flowering; (b) coating of white wax or powder, as on plums or grapes.
Bole The main trunk or stem of a tree.
Boss A protrusion.
Brackish Somewhat salty.
Bract A modified reduced leaf typically subtending a flower or cluster of flowers. Bracts can vary from very leaf-like to scale-like or thread-like; in some cases they can be colorful and attract pollinators.

Bracteal Having the form or position of a bract.
Bracteate Having bracts.
Bracteolate Having bracteoles.
Bracteole, Bractlet A usually small bract borne on a secondary axis (e.g., on a pedicel).
Bracteose Having numerous or conspicuous bracts.
Branch A shoot or secondary stem growing from the main stem.
Branchlet The ultimate division of a branch.
Bristle Stiff, strong but slender hair or trichome.
Bristly Bearing bristles.
Broad (= Wide) Distance across a structure (equal to diameter if tubular); sometimes restricted to signify the width or diameter of three-dimensional structures.
Bryophyta Group containing the mosses, liverworts, and hornworts. The Bryophyta is not treated in this flora.
Bud (a) Undeveloped or unopened flower; (b) undeveloped, much-condensed shoots, containing embryonic (meristematic or growing) tissue, usually covered by scales or bracts. Such buds are usually found at the tips of stems or in the axils of leaves.
Bulb Underground structure composed of a short, disc-like stem and one or more buds surrounded by layers of thickened fleshy leaf bases or scales; e.g., an onion.
Bulbil Small bulbs produced in an inflorescence or in leaf axils; e.g., in *Allium* inflorescences.
Bulblet Small bulbs produced alongside a parent bulb; e.g., the numerous underground small bulbs produced by some garlics.
Bulbous, Bulbose Having bulbs or bulb-like structures.
Bullate Describing a surface with rounded elevations resembling blisters or puckers.
Bur, burr A structure with a rough or prickly envelope or covering; e.g., sandbur.
Bush (= Shrub) A woody perennial usually branching from the base with several main stems.

C

Ca. Latin, circa; abbreviation meaning about, around, approximately.
Caducous Falling off early, quickly, or prematurely; e.g., the sepals in some Papaveraceae.
Caespitose (= Cespitose) Growing in clumps or tufts.
Calcarate With a spur.
Calcareous Containing an excess of available calcium, usually in the form of the compound calcium carbonate; containing limestone or chalk.
Caliche A crust of calcium carbonate formed on stony soils in arid regions.
Callosity (= Callus). A hard protuberance or thickened, raised area.
Callous Having the texture of a callus.
Callus (= Callosity). A hard protuberance or thickened, raised area; e.g., thickened, hardened, basal portion of some lemmas in the Poaceae.
Calycine Resembling or pertaining to a calyx.
Calyculate Calyx-like; e.g., describing bracts that by their size or position are suggestive of a calyx.
Calyptra A lid, cap, cover, or hood; e.g., the lid of certain fruits and moss spore cases.
Calyx (pl. Calyces, Calyxes) Collective term for the sepals; outer series of floral "leaves", often enclosing the other flower parts in bud. The calyx is typically green but can be corolla-like and showy.
Calyx lobe One of the free projecting parts of a synsepalous calyx; also referred to as a calyx tooth.
Calyx tube The basal or tubular portion of a synsepalous calyx, as opposed to the free, distal calyx lobes.
Cambium The thin layer of delicate, rapidly dividing, meristematic cells that forms wood internally and bark externally; also known as vascular cambium.
Campanulate Bell-shaped; rounded at base with a broad flaring rim.
Campylotropus ovule Ovule curved in its development, so that the morphological apex lies near the base.
Canaliculate Longitudinally channeled or grooved.
Cancellate Latticed.
Cane Stem, specifically, (a) floricane, the flowering stem of *Rubus* species (blackberries and dewberries); (b) primocane, first-year leafy stem of the same; (c) persistent woody stems of *Arundinaria gigantea*, giant cane.
Canescent With whitish or grayish-white appearance due to abundance of soft short hairs.
Cap A convex, lid-like, removable covering; e.g., the apical portion of a circumscissile capsule. The term calyptra is used for the cap of some fruits and moss spore cases.
Capillary Hair-like; very slender.
Capitate (a) In heads, head-like, or head-shaped; aggregated into a dense or compact cluster; (b) referring to capitate hairs, like a pinhead on a stalk.
Capitellate Aggregated into a small, dense cluster; diminutive of capitate.
Capitulum (= Head) Dense cluster of sessile or nearly sessile flowers. This type of inflorescence is typical of the Asteraceae.
Capsular Having the structure of a capsule.
Capsule A dry dehiscent fruit developed from more than one carpel.
Carcinogen A substance potentially inducing cancer or malignancy.
Cardiac glycoside A glycoside (two-component molecule) that upon breakdown yields a heart stimulant as the aglycone (= non-sugar component). The aglycones are steroidal in structure and are typically poisonous; e.g., digoxin and digitoxin from *Digitalis*, used in treating heart trouble.
Carina (= Keel) The two lower (= abaxial) fused petals of a papilionaceous flower (Fabaceae); (b) prominent longitudinal ridge, shaped like the keel of a boat.
Carinate Keel-shaped; provided with a ridge or keel extending lengthwise along the middle. If more than one keel is present, the fact may be indicated by a numerical prefix.
Carnivorous Referring to plants that digest animal (primarily insect) tissue to obtain nutrients such as nitrogen.
Carnose Fleshy; succulent.
Carpel A modified leaf bearing ovules; a simple pistil or one unit of a compound pistil; female reproductive structure in flowering plants.
Carpellate Possessing carpels. The term is sometimes used with a numerical prefix to indicate the number of carpels.
Carpophore The slender prolongation of the floral axis between the carpels that supports the pendulous fruit segments (= mericarps) in the Apiaceae.

Cartilaginous Tough and hard but not bony; gristly; cartilage-like.
Caruncle Enlarged, somewhat spongy, seed appendage.
Caryopsis Achene-like, 1-seeded fruit with pericarp adnate to the seed coat; fruit typical of the Poaceae.
Castaneous Chestnut-colored; dark brown.
Catkin (= Ament) A flexible often pendulous spike or spike-like raceme of small, inconspicuous, unisexual, apetalous, usually wind-pollinated flowers, the whole falling as one piece; e.g., male inflorescence of oaks or pecan.
Caudate Having a tail or tail-like appendage.
Caudex (pl. Caudices) Woody stem base.
Caulescent With an evident leafy stem above ground.
Cauliflorous Having flowers borne along the stems or trunks.
Cauline Growing on or pertaining to the stem.
Cell (a) One of the living units of which a plant is composed; (b) (= locule) cavity or compartment containing the ovules in a carpel or the pollen in an anther.
Centrifugal Maturation of parts from the center toward the periphery.
Centripetal Maturation of parts from the periphery toward the center.
Ceraceous Waxy.
Ceriferous Wax-bearing; waxy.
Cernuous Nodding or drooping.
Cespitose (= Caespitose) Growing in clumps or tufts.
Chaff Thin, dry, or membranous scales or bracts, often used to refer to receptacular scales or bracts in many Asteraceae; see pale or palea.
Chaffy Thin, dry, or membranous.
Chalaza The basal part of the ovule where it attaches to the funiculus.
Channeled Deeply grooved.
Chartaceous Stiffly papery.
Chasmogamous Referring to flowers that open at anthesis; with pollination after opening of flowers; contrasting with cleistogamous.
Chiropterophily, Chiropterophilous Bat-pollination.
Chlorophyll The light-capturing pigment giving the green color to plants. Because chlorophyll absorbs less green than other wavelengths of light (and thus reflects and transmits relatively more green), leaves appear green to the human eye.
Choripetalous Composed of or characterized by separate petals.
Chorisepalous Composed of or characterized by separate sepals.
Chromosomes Thread-like "colored bodies" occurring in the nuclei of cells and containing the genetic material.
Ciliate With a marginal fringe of hairs similar to eye lashes.
Ciliolate Diminutive of ciliate.
Cilium (pl. Cilia) Marginal hair or trichome.
Cincinnus A curl; e.g., a helicoid (= curled or coiled) cyme, as in the Boraginaceae.
Cinereous Ash-colored; light-gray.
Circinate Coiled, with the apex innermost, as the young fronds in some ferns.
Circumscissile Dehiscing by a regular transverse line around the fruit or anther, the top coming off like a lid.
Cirrhous Tendril-like; e.g., a leaf with a slender coiled apex.
Cladophyll, Cladode (= Phylloclade) A portion of a stem having the general form and function of a leaf; a flattened photosynthetic stem.
Clambering Vine-like; growing over other plants often without the aid of tendrils or twining stems.
Clasping (= Amplexicaul) With base of leaf or other structure (e.g., stipule) wholly or partly surrounding the stem.
Class The unit, category, or rank in classification made up of one or more orders; ending in -ae or -opsida; sometimes divided into subclasses which in turn are made of orders.
Clathrate Latticed; with a series of crossed members.
Clavate Club-shaped; becoming gradually enlarged apically.
Clavellate Diminutive of clavate.
Claw Stalk-like basal portion of some petals or sepals.
Clawed Having a claw.
Cleft (a) Cut 1/2 or more the distance from the margin to midrib or from the apex to base; (b) generally, any deep cut.
Cleistogam A small flower that does not open and is necessarily self-pollinating.
Cleistogamous, Cleistogamic Referring to flowers not opening at anthesis and thus self-pollinating; with pollination prior to opening of flowers. Such flowers frequently have reduced or incompletely formed parts (e.g., petals).
Clesitogamy The self-pollination of flowers that do not open.
Clone A group of individuals of the same genotype, usually propagated vegetatively.
Clump A single plant with two to many, more or less crowded stems arising from a branched rootstock or short rhizome.
cm Centimeter; 10 mm; 1/100 of a meter; ca. 2/5 of an inch.
Coalescent Referring to organs of one kind that have grown together.
Coarctate Crowded together.
Cob Rachis or central stalk of the pistillate inflorescence of corn.
Coccus (pl.Cocci) (a) A berry; (b) one of the parts of a lobed or deeply divided fruit with 1-seeded sections.
Cochleate Coiled like a snail shell; spiral.
Coetaneous At the same time or of the same age; e.g., flowers and leaves appearing at the same time.
Coherent Descriptive of the close association of two similar structures without fusion.
Collar The outer side of a grass leaf at the junction of the blade and sheath.
Collateral Located side by side; e.g., ovules located side by side; e.g., accessory buds located on either side of a lateral bud.
Colonial Forming colonies usually by means of underground rhizomes, stolons, etc. The term is commonly used to describe groups of plants with asexual reproduction.
Colony A stand, group, or population of plants of one species, spreading vegetatively, or from seeds, or both.
-colpate A suffix referring to pollen grains having grooves (= colpi).
-colporate A suffix referring to pollen grains having grooves and pores.
Columella The persistent central axis around which the carpels of some fruits are arranged.
Column (a) United style and filaments in Orchidaceae; (b) united filaments in Malvaceae and Asclepiadaceae; (c) basal differentiated portion of the awn(s) in certain grasses.
Columnar Column-shaped.
Coma (a) A tuft of soft hairs or trichomes, as at the apices or bases of some seeds; (b) tuft of structures projecting from something (e.g., tuft of bracts projecting from heads of some *Eryngium* species).

Comate, Comose Resembling or provided with a coma.
Commisure The surface where organs are joined; e.g., the face by which one carpel joins another.
Complanate Flattened.
Complete With all of the usual parts; e.g., a flower with all four flower parts:sepals, petals, stamens, and pistils.
Complicate Folded together.
Composite (a) (= Compound) made up of several distinct parts; (b) common name for species of the Asteraceae.
Compound (= Composite) Made up of several distinct parts.
Compound inflorescence One having two or more degrees of branching; e.g., a compound umbel is one whose branches bear branchlets rather than ending directly in flowers.
Compound leaf A leaf that is cut completely to the base or midrib into segments (= leaflets) resembling miniature leaves; a leaf with two or more leaflets.
Compound ovary Ovary developed from two or more united carpels, as evidenced by the presence of two or more locules, valves, placentae, styles, or stigmas.
Compound pistil Pistil composed of two or more united carpels.
Compressed Flattened.
Concave Hollow; with a depression on the surface.
Concolor, Concolorous Of a uniform color.
Conduplicate Folded together lengthwise.
Cone (= Strobilus) A usually globose or cylindrical structure involved in reproduction and composed of an axis with a spiral, usually dense aggregation of sporophylls, bracts, or scales (these bearing spores, pollen, or seeds).
Confluent Blending of one part into another.
Congested Crowded together.
Conglomerate Densely clustered.
Conical, Conic Cone-shaped.
Coniferous Cone-bearing.
Conjugate Jointed in pairs.
Connate United or fused, when the fusion involves two or more similar structures; e.g., as in fusion of stamens into a tube.
Connate-perfoliate Both connate and perfoliate; e.g., two leaves grown together and completely encircling a stem; e.g., in *Eupatorium perfoliatum*.
Connective The tissue connecting the pollen-sacs of an anther. In certain plants the connective is prolonged at its base or apex.
Connivent Converging or nearly or quite in contact, but not fused; e.g., connivent stamens in *Solanum*.
Conserved Term applied to a scientific name whose use, even though illegitimate according to nomenclatural rules, is allowed by the International Code of Botanical Nomenclature; e.g., many family names long in use, such as Cactaceae and Caryophyllaceae, have been conserved to prevent confusion.
Conspecific Of the same species.
Constricted Tightened or drawn together or narrowed.
Contact dermatitis Inflammation of the skin due to contact with poisons, irritants, or sensitizers. In some individuals it can be caused by even the slightest contact; caused by a variety of plants including poison-ivy (*Toxicodendron radicans*).
Contiguous Touching but not fused.
Continuous Not interrupted; not articulated; not jointed.
Contorted Twisted or distorted.
Contracted Narrowed or shortened; reduced in size.
Contrary In an opposite direction or at right angles to.
Convergent Coming together or approaching.
Convex Rounded or bulged on the surface.
Convolute Rolled or twisted together when in an undeveloped stage.
Coralloid Coral-like.
Cordate, Cordiform (a) Heart-shaped; with a notch at the base and ovate in outline (the words apply specifically to flat surfaces and to solid shapes, respectively); (b) often referring only to the notched base of a structure; e.g., leaf base.
Coriaceous With texture like leather; tough; leathery.
Corm Bulb-like usually subterranean stem base, solid instead of with layers of modified leaves as in a true bulb.
Cormous Having a corm.
Corneous Horny in texture.
Corniculate Bearing a small horn or horns.
Cornute Horned or spurred.
Corolla Collective term for the petals; the inner series of floral "leaves". Corollas are typically but not always colorful and showy.
Corolla lobe A free projecting part of a synsepalous corolla; also referred to as a corolla tooth.
Corolla tube The basal or tubular portion of a synsepalous corolla, as opposed to the free, distal, corolla lobes.
Corona (= Crown) Projection or outgrowth of a corolla, perianth, or anthers, in the form of a fringe, cup, or tube, sometimes greatly resembling an extra corolla or perianth.
Coroniform scales Membranous scales broader than long.
Corpusculum In Asclepiadaceae, the gland-like clip connecting the two bands (= filament-like retinacula) attached to the pollinia; part of the translator.
Corrugate Having wrinkles or folds.
Cortex (a) The tissue between the stele and epidermis of a stem; (b) bark or rind.
Corymb A more or less flat-topped inflorescence (resulting from lower branches being longer than upper) that is indeterminate (i.e., with the outer flowers opening first); inflorescence superficially similar to an umbel but with the branches arising at different points rather than one.
Corymbiform Shaped like a corymb.
Corymbose In corymbs or corymb-like inflorescences (i.e., flat-topped).
Costa (pl. Costae) A rib; the midvein of a leaf.
Costate Longitudinally ribbed.
Costules Midveins of the pinnules.
Cotyledon Seed leaf; main leaf or leaves of the embryonic plant.
Coumarin glycoside A glycoside (two-component molecule) that upon breakdown yields coumarin as the aglycone (= non-sugar component). Coumarin can be converted to dicoumarin, a toxic compound which prevents blood clotting; e.g., in *Melilotus*.
Crateriform In the shape of a saucer; shallow and hemispherical.
Creeping Growing along the surface of the ground and emitting roots at intervals, usually from the nodes.
Crenate Scalloped with rounded teeth; shallowly round-toothed or with teeth obtuse.
Crenulate Diminutive of crenate; with small rounded teeth.
Crest An elevated ridge, process, or appendage on the surface of an organ or structure.

Crested Having a crest.

Crispate, Crisped Irregularly curled or twisted.

Cross-section A slice cut across an object; e.g., a slice of bread.

Crown (a) An irregular perennial or over-wintering stem or stem-root structure from which new growth arises; (b) (= Corona) projection or outgrowth of a corolla, perianth, or anthers, in the form of a fringe, cup, or tube, sometimes greatly resembling an extra corolla or perianth.

Crozier A young coiled leaf of some ferns.

Cruciate, Cruciform Cross-shaped.

Cruciferous (a) Cross-bearing; (b) specifically descriptive of cross-like arrangement of petals of members of the Brassicaceae (Cruciferae).

Crustaceous, Crustose With a brittle, hard texture.

Cryptogams An old term for plants that reproduce without flowers or seeds. Cryptogams typically reproduce by spores.

Crystalline Crystal-like.

Cucullate Hood-like.

Culm Stem of Poaceae and Cyperaceae.

Cultigen A plant known only in cultivation.

Cultivar A variety or race of a cultivated plant; abbreviated cv.

Cuneate, Cuneiform Wedge-shaped; triangular with tapering, straight-sided, narrow base.

Cup, Cupule The cup-like structure at the base of a fruit; e.g., acorn.

Cupuliform, Cupulate Cup-shaped.

Curvi- A prefix to denote curved or bent.

Cusp A sharp, abrupt, and often rigid point.

Cuspidate Bearing a cusp or strong sharp point.

Cut A general term for any dissection of a leaf or petal deeper than a lobe.

Cuticle The waxy, more or less waterproof coating secreted by the cells of the epidermis. The cuticle prevents water loss.

Cyanogenic glycoside A glycoside (two-component molecule) that upon breakdown yields hydrocyanic (prussic) acid (HCN) as the aglycone (= non-sugar component). Hydrocyanic acid is extremely dangerous, causing cyanide poisoning; e.g., amygdalin in cherry and peach leaves or apple seeds (Rosaceae).

Cyathiform Cup-shaped.

Cyathium (pl.Cyathia) (a) Cup-shaped structure producing unisexual flowers; (b) specifically, the units of the inflorescence in *Euphorbia*. In this case the cup contains a single pistillate flower and a number of staminate flowers, each consisting of a single stamen; on the rim of the cup there are glands and these often have a petal-like appendage; the whole structure superficially resembles a single flower.

-cyclic A suffix referring to the circles of different parts in a flower, commonly used with a numerical prefix; e.g., a Verbascum flower with sepals, a corolla, stamens, and an ovary is four-cyclic; compare with -merous.

Cylindrical, Cylindric Elongate, circular in cross-section; having the form of a cylinder.

Cymbiform Boat-like or boat-shaped.

Cyme A broad, flattish or convex, determinate inflorescence with the central flowers maturing first.

Cymose With the flowers in cymes; having an inflorescence type with the oldest flowers in the center.

Cymule A small or few-flowered cyme.

Cypsela Achene derived from an inferior ovary and adnate to the enclosing floral tube; e.g., in Asteraceae.

Cystolith A stone-like mineral concretion, usually of calcium carbonate.

D

DC. De Candolle, name of a distinguished family of Swiss botanists; specifically Augustin Pyramus, who sponsored early botanical exploration in Texas by Berlandier, and named many Texas species; A. DC.: Alphonse, son of the preceding.

Deciduous Falling away; not persistent over a long period of time.

Declinate, Declined Bent forward or downward.

Decompound More than once compound.

Decumbent Lying flat or reclining with terminal shoots or stem tips ascending.

Decurrent Extending down the stem and united with it, as in the continuation of leaf bases down the stem as wings.

Decurved Curved downward.

Decussate Arranged oppositely in pairs, each successive pair at right angles to the preceding one (resulting in the appearance of four rows), typically referring to arrangement of leaves.

Deflexed Bent downward.

Defoliate To shed or remove leaves.

Dehisce, Dehiscent To open at maturity to discharge the contents; e.g.; fruit releasing seeds or anther releasing pollen.

Dehiscence The process or act of opening.

Deliquescent Softening, dissolving, melting away, or wasting away; e.g., ephemeral petals of *Tradescantia*.

Deltate, Deltoid Shaped like an equilateral triangle, like the Greek letter delta (Δ).

Dendritic Tree-like, as in branching.

Dentate With sharp teeth not directed forward.

Denticulate Minutely dentate.

Depauperate Smaller than the usual natural size; short of the usual development; stunted; impoverished.

Depressed Low as if flattened.

Dermatitis Inflammation of the skin due to exposure to poisons, irritants, or sensitizers. It can be caused by a number of plants including poison ivy.

Descending With a gradual downward tendency.

Determinate Descriptive of an inflorescence whose flowers begin to open first at the top or center, progressing downward or outward; with the number of flowers fixed or limited in number; contrasting with indeterminate.

Di-, Dis- Greek prefix meaning two or double.

Diadelphous With filaments united so as to form two groups of stamens. The most common situation is nine in a group and a tenth separate; e.g., numerous Fabaceae.

Diandrous Having two stamens.

Diaphanous Transparent.

Diaphragm Dividing membrane or partition.

Dichasial With or in the form of a dichasium.

Dichasium A cyme with lateral branches on both sides of the main axis.

Dichotomous Forking regularly into two equal or nearly equal branches.

Dicotyledons (abbreviated Dicots) Flowering plants having two cotyledons, mostly net venation, and flower parts usually in 4s or 5s; one of the two classes of flowering plants, which, depending on the system of classification, is known as Dicotyledonae or Magnoliopsida. The dicots are now considered to be a paraphyletic group.

Didymous Developed in or occurring in pairs; twin.

Didynamous Having four stamens, two longer and two shorter.

Diffuse Loosely branching or spreading; of open growth; widely spread.

Digitate Fingered; with a number of structures attached at one point, like fingers on a hand.

Dilated Widened, flattened, broadened, or enlarged.

Dimidiate Halved, as if one half is missing.

Dimorphic, Dimorphism Occurring in two forms.

Dioecious With staminate flowers on one plant (staminate plant) and pistillate flowers on a different plant (pistillate plant); literally, "two houses."

Diploid Possessing two sets of chromosomes in each nucleus; twice the haploid number typical for gametes.

Diplostemonous The stamens in two series, those of the outer series alternating with the petals.

Dipterophily, Dipterophilous Pollination by dipterans (members of the insect order Diptera—flies, gnats, mosquitoes, and their relatives).

Disarticulating Separating; coming apart or falling apart.

Disc, Disk (a) A more or less fleshy or elevated development of the receptacle or of coalesced nectaries or staminodes about the pistil; (b) the central part of an Asteraceae head.

Disc floret (= Disk flower) In Asteraceae, small flower with tubular corolla, in disk (disc) portion of head; contrasting with ray (= ligulate) floret.

Disciform (a) Shaped like a disc; (b) in Asteraceae, in reference to a head with disk florets in center and marginal florets with ligule reduced or lacking.

Disclimax The condition where succession is indefinitely arrested or altered due to factors such as persistent disturbance (e.g., overgrazing).

Discoid (a) Shaped like a disc; (b) in reference to disc (disk) florets of an Asteraceae head; (c) in reference to the head of some Asteraceae with disk florets in center and marginal florets with ligule reduced or lacking; (d) without ligulate flowers.

Discoid head In Asteraceae, a head with only disk (disc) florets.

Discolorous Having the two surfaces of a structure dissimilar in color.

Discrete Separate.

Disjunct (a) Outside the main range of a species; (b) being divided into separate groups; disconnected.

Disk flower (= Disc floret) In Asteraceae, small flower with tubular corolla, in disk (disc) portion of head; contrasting with ray (= ligulate) florets.

Disparate Dissimilar; unequal.

Dissected Divided into numerous narrow or slender segments, the divisions usually deeper than lobes.

Distal Located at or toward the apex of a plant or organ; the terminal portion; the end opposite the attachment; contrasting with proximal.

Distant In reference to similar parts remote from one another; contrasting with approximate.

Distichous In two vertical rows or ranks, not spirally arranged; e.g., leaves occurring in two rows on opposite sides of a stem.

Distinct (= Free) Separate, not united or fused.

Disturbed Referring to a habitat that has been altered or modified but not completely destroyed.

Diurnal Occurring during the daytime.

Divaricate Very widely spreading.

Divergent Spreading, but less broadly than divaricate.

Divided (a) Cut 3/4–completely the distance from margin to midrib or from apex to base; (b) generally, cut deeply. DIVISION The highest rank, category, or taxon in the plant kingdom; made up of classes; ending in -phyta; equivalent to the rank of phylum in the animal kingdom.

dm Decimeter; 10 cm.

Doctrine of signatures Ancient belief that a plant structure that resembles a portion of the human body (a sign or signature) gives clues to its use. Some people believed that a creator had placed such signs to indicate the plant's value as a remedy for ailments of those body portions; e.g., Hepatica, named for its lobed, liver-shaped leaves, possibly useful for liver problems.

Dolabriform (a) Referring to pubescence where the hairs are attached near the middle or towards (but not at) one end and are thus 2-forked; (b) having the form of an ax or hatchet.

Dormant Not active.

Dorsal Referring to the back or outer surface of an organ; the side away from the axis; the lower or abaxial surface; contrasting with ventral.

Dorsifixed Attached by the back or dorsal edge, often in reference to the attachment of an anther to a filament; contrasting with either basifixed or versatile.

Dorsiventral Differentiated into upper and lower surfaces distinct in appearance or structure.

Double fertilization Process essentially unique to the angiosperms in which an egg unites with a sperm, forming a zygote, and a second sperm often unites with two nuclei resulting in a triploid endosperm. Double fertilization is also known in the small gymnosperm group the Gnetophyta.

Double flower (a) One with more than the normal number of petals (anywhere from a few more to many times the usual number); (b) in Asteraceae, double can refer to a head with more than the normal number of ray (= ligulate) florets.

Doubly serrate With coarse serrations bearing minute teeth on their margins, the teeth angled toward the apex of the structure.

Downy Closely covered with short, weak, soft hairs.

Drooping More or less erect at base but with upper part bending downward.

Drupaceous Pertaining to, or of the nature of a drupe.

Drupe One-seeded indehiscent fruit with a stony endocarp, the middle part fleshy or juicy, and an outer skin; e.g., plum or cherry.

Drupelet A small drupe; one drupe from an aggregate fruit composed of many drupes; e.g., a blackberry is an aggregate fruit composed of drupelets.

Duplex Double, as in pubescence composed of two kinds of hairs.

E

E East.

E-, Ex- Latin prefixes denoting without, that parts are missing.

Ebeneous Black, ebony-like.

Ebracteate Without bracts.

Echinate Prickly.

Eciliate Without cilia.

Ecological indicator An organism that is sensitive to pollution or some other environmental problem and can therefore be used as an indicator or gauge of the condition of an ecosystem.

Ecotone Transition zone between two biological communities; e.g., much of nc TX is an ecotone between the eastern deciduous forest and the central North American grassland.

Ecotype Those individuals adapted to only one of the kinds of environment occupied by a widespread species.

Edaphic Pertaining to soil conditions.

Edentate Without teeth.

Egg A female gamete or sex cell, in flowering plants contained in an ovule.

Eglandular Without glands.

Elaiosome An oily appendage on the seeds of some plants. These structures have apparently evolved to attract ants that act as dispersal agents.

Elimbate Referring to a corolla without a limb.

Ellipsoid A solid that is elliptic in outline.

Elliptic Shaped like an ellipse, with widest part at the middle; in the form of a flattened circle usually more than twice as long as wide.

Elongate Lengthened; stretched out.

Emarginate With a notch in the usually rounded apex.

Embryo The new plant in a seed.

Emersed, Emergent Raised above and out of the water.

Emetic A substance that causes vomiting.

Enation Outgrowth on the surface; epidermal outgrowth.

Endemic Confined geographically to or native to a single area.

Endocarp Innermost layer of pericarp or fruit wall.

Endogenous Produced deep within another body.

Endosperm The food reserve of many angiosperm seeds.

Ensiform Sword-shaped; e.g., leaves of an *Iris*.

Entire With smooth margins; without teeth.

Entomophilous, Entomophily Pollinated by insects.

Epappose Without pappus.

Ephemeral Lasting for a brief period; e.g., for only one day.

Epi- Greek prefix meaning upon or on.

Epicalyx A whorl or involucel of sepal-like bracts just below the true sepals; e.g., Malvaceae.

Epicarp (= Exocarp) The outermost layer of the pericarp or fruit wall.

Epicotyl The portion of the embryo just above the cotyledon(s); the young stem.

Epidermal Relating to the epidermis.

Epidermis The cellular covering of plant tissue below the cuticle.

Epigeous Growing upon or above the ground.

Epigynous Borne on the ovary; indicating a flower in which the hypanthium or the basal parts of the perianth are adnate to the ovary, the perianth and stamens thus appearing to rise from the summit of an inferior ovary.

Epigyny The state of being epigynous.

Epipetalous Descriptive of stamens in which the filaments are adnate to the corolla for all or part of their length; borne upon or arising from the petals or corolla.

Epiphyte A plant growing on another plant for physical support only and not parasitic; e.g., many Bromeliaceae including *Tillandsia* species.

Epiphytic Having the character of an epiphyte.

Equidistant Of equal distance.

Equitant Overlapping in two ranks; folded lengthwise and distichous; e.g., leaves folded around a stem in the manner of the legs of a rider around a horse.

Eradiate Lacking ray (= ligulate) florets; e.g., discoid heads in Asteraceae.

Erect (a) Growing essentially in a vertical position (e.g., whole plant); (b) a structure perpendicular to the object to which it is attached.

Ergot An alkaloid producing fungus, *Claviceps purpurea* (and related species).

Ergotism A sometimes lethal poisoning due to a number of alkaloids (e.g., lysergic acid hydroxyethylamide—LSD-like, egrotamine—vasoconstrictive) produced by ergot fungi, *Claviceps purpurea* (Fr.) Tul., *C. paspali* Stev. & Hall, and related species, which sometimes infect the inflorescences of members of the grass family (e.g., *Secale*—rye). Symptoms resulting from eating ergot-contaminated grain can include hallucinations, psychosis, convulsions, and gangrene of the extremities (due to vasoconstriction); referred to in ages past as Saint Anthony's Fire. The reference to fire resulted from the assumption that the burning sensations and blackened (gangrenous) limbs were retribution for sins. Saint Anthony, supposedly with special powers to protect against fire, infection, and epilepsy, was often prayed to for help by those with the condition. Large scale epidemics of ergotism in Europe prior to1800, from eating bread made with contaminated grain, resulted in 1, 000s of deaths; isolated instances still occur where grain purity is not controlled; ergotism is blamed by some for the hysteria that resulted in the Salem witch trials in 17th-century Massachusetts. See treatment of *Secale* (Poaceae) for references.

Erose With ragged margin, as if nibbled or chewed.

Escape A cultivated plant not purposely planted but found growing as though wild.

Escarpment A steep slope.

Essential oils Fragrant (often pleasantly so) substances that are typically rather simple terpenoid compounds (monoterpenes, sesquiterpenes). They are responsible for the fragrance of many flowers and other plant tissues and are currently used in aromatherapy; e.g., in mint family and citrus family.

Estipellate Without stipels.

Estipulate Without stipules.

Etiolate Lengthened and deprived of color by absence of light.

Evanescent Of short duration; quickly lost.

Even-pinnate Descriptive of a compound leaf with terminal leaflet absent; with an even number of leaflets.

Evergreen Remaining green through the winter.

Exalate Without wings.

Exalbuminous In reference to seeds without endosperm.

Excentric Not having the axis placed centrally.

Excurrent Extending beyond the tip or margin.

Excurved Curved outward or away from a central part.

Exfoliate To peel off or come off in scales, flakes, plates, layers, or shreds; e.g., some types of bark.
Exocarp (= Epicarp) The outermost layer of the pericarp or fruit wall.
Exotic Foreign; not native; from another geographic area.
Explanate Spread out flat.
Explosively dehiscent Descriptive of a fruit that suddenly and forcibly dehisces its seeds, with the seeds being thrown away from the plant; e.g., *Impatiens*.
Exserted Projecting out or beyond (e.g., stamens projecting beyond the corolla); contrasting with included.
Exstipulate Lacking stipules.
Extant Still existing; contrast with extinct.
Extinct No longer in existence; descriptive of a species for which living representatives no longer exist. Locally extinct refers to extinction in a given geographic region.
Extrafloral Outside the flower; e.g., extrafloral nectaries.
Extrorse Facing outward, away from the axis, typically used to refer to manner of anther dehiscence.
Exudate Material coming out slowly through small pores or openings.
Eye (a) The marked or contrastingly colored center of a flower; (b) a bud on a tuber; e.g., on potatoes.

F

f. (a) After an author's name: abbreviation of *filius*, the son, or "jr."; (b) abbreviation of forma or form (see next page).
Facet One of a set of small plane surfaces.
Facultative Not necessary or essential; optional; contrasting with obligate.
Falcate Sickle-shaped, with the tip curved to one side.
Fall Outer, spreading, often recurved, and commonly bearded perianth segment in *Iris*.
Family The unit, category, or rank in classification made up of one or more genera; ending in -aceae; sometimes divided into subfamilies, which in turn are made up of genera.
Farinaceous Mealy in texture.
Farinose Covered with a whitish mealy powder or mealiness.
Fasciated With an abnormal widening and flattening of the stem as if several stems had grown together.
Fascicle A condensed or close bundle or cluster.
Fasciculate Congested in bundles or clusters.
Fastigiate In reference to branches, close together and nearly parallel.
Fenestrate Perforated with holes, openings (windows), or translucent areas.
Feral Wild; not cultivated.
Ferrugineous Rust-colored.
Fertile Capable of normal reproductive functions, as a fertile stamen producing pollen, a fertile pistil producing ovules, or a fertile flower normally producing fruit (although it may lack stamens); e.g., used to describe Asteraceae flowers capable of maturing achenes, irrespective of ability to produce pollen.
Fertilization Union of two gametes (e.g., egg and a sperm) to form a zygote.
Fetid With a disagreeable odor.
Fibrillose Having small fibers or appearing finely lined.
Fibrous Resembling or having fibers.
Fibrous root system One with several roots about equal in size and arising from about the same place; contrasting with taproot.
-fid A suffix meaning deeply cut.
Filament (a) The thread-like stalk supporting an anther; (b) a thread or thread-like structure.
Filamentous, Filamentose Composed of filaments or threads; thread-like.
Filiferous With coarse marginal threads.
Filiform Slender; having the form of a thread; filamentous.
Fimbriate Fringed; with narrow or filiform appendages or segments along the margin.
Fimbrillate With a minute fringe.
First glume Lowermost of the two glumes in a grass spikelet.
Fissured Cracked or fractured.
Fistulose Hollow; lacking pith.
Flabellate, Flabelliform Fan-shaped; broadly wedge-shaped.
Flaccid Lax, weak, floppy, not rigid.
Flagelliform Whip-like.
Flange A rim-like structure.
Flavescent Yellowish or becoming yellow.
Fleshy Succulent, juicy, or pulpy.
Flexuous Zigzag; bending or curving alternately in opposite directions.
Floccose Covered with tufts of soft woolly hairs.
Flocculent Minutely floccose.
Flora (a) Collective term for the plants of an area; (b) a taxonomic work on the plants of an area.
Floral bract Reduced leaf subtending a flower in the inflorescence.
Floral envelope The calyx or corolla; the floral "leaves."
Floral tube, Floral cup Tube or cup formed by union of sepals, petals, and stamen bases. The structure can be either adnate to or free from the ovary; in some cases it is synonymous with hypanthium.
Floret (a) Small flower in a dense cluster; (b) in reference to individual flowers of Asteraceae and Poaceae; (c) in grasses referring to the lemma and palea together with the enclosed reproductive structures.
Floricane Flowering stem of the genus *Rubus* (dewberries and blackberries).
Floriferous Bearing or producing flowers.
Flower An axis bearing stamen(s), pistil(s), or both, and, in addition, often floral envelopes (= calyx and corolla); the reproductive structure of an angiosperm.
Fluted With alternating ridges and grooves.
Foliaceous Leaf-like.
Foliage Collective term for the leaves of a plant.
Foliate With leaves.
Foliolate With leaflets.
Foliose Leafy.
Follicle Dry, one-carpellate fruit dehiscing along one suture only.
Forb A herbaceous, non-grass-like plant.
Forked Dichotomous; divided into two equal or nearly equal branches.
Forma, Form (abbreviated f.) A taxon below the rank of variety used to refer to minor variations without distinctive geographic occurrence; e.g., occasional albinos or seasonal growth forms. This category is generally ignored in this book.

FORNICES Internal appendages in the upper throat of a corolla; e.g., in some Boraginaceae.
FOVEA (pl. FOVEAE). A pit or depression.
FOVEATE Pitted.
FOVEOLATE Minutely pitted.
FREE (= Distinct) Separate from one another.
FREE-CENTRAL PLACENTATION Placentation with the seeds attached to a central column and surrounded by a single continuous locule.
FROND The leaf of a fern, often compound or decompound.
FRUCTIFEROUS Producing or bearing fruit.
FRUGIVORE Animal that feeds on fruits.
FRUIT A mature, ripened pistil or ovary. In the case of accessory "fruits"other tissues may be involved.
FRUTICOSE Shrubby or bushy in sense of being woody.
FUGACIOUS Falling or disappearing early, usually in reference to parts of a flower.
FULVOUS Tawny, brownish yellow.
FUNICLE, FUNICULUS The stalk attaching an ovule or seed to the ovary wall or placenta.
FUNNELFORM (= Infundibuliform) Funnel-shaped; gradually widening upward.
FURROWED With longitudinal channels or grooves.
FUSCOUS Grayish-brown.
FUSED United by normal growth.
FUSIFORM Spindle-shaped; with broadest diameter at middle tapering to each end.

G

GALEA The helmet-like or hood-like upper lip of a bilabiate corolla or calyx, especially one that is strongly concave.
GALEATE Hooded, hood-like, or helmet-like.
GAMETE A sex cell; an egg or sperm.
GAMETOPHYTE The gamete-producing, typically haploid generation alternating with the sporophyte (= spore-producing, typically diploid); the stage in the life history of a plant that produces male or female cells (= gametes); the dominant generation in mosses and liverworts. In ferns and fern allies it is green and autotrophic, although small; in all flowering plants it is microscopic and develops within the tissues of the sporophyte.
GAMO- A prefix meaning united; e.g., gamopetalous or gamosepalous.
GAMOPETALOUS (= Sympetalous) With petals united, at least basally, forming a tube.
GAMOSEPALOUS (= Synsepalous) With sepals united, at least basally, forming a tube.
GASTROENTERITIS Inflammation of the stomach and intestines. It can be caused by a number of plant materials.
GEMINATE In pairs; twin.
GEMMA An asexual propagule sometimes appearing as, but not homologous with, a vegetative bud.
GENICULATE Bent abruptly, like a knee.
GENUS (pl. GENERA) The unit, category, or rank in classification between family and species; composed of one or more closely related species; sometimes divided into subgenera, which in turn are made up of species.
GIBBOUS Swollen basally on one side.
GLABRATE, GLABRESCENT Becoming hairless with age.
GLABROUS Without hairs.
GLADIATE Sword-shaped.
GLAND A secreting part or appendage, often protruding or wart-like.
GLANDULAR Having or bearing secreting organs, glands, or trichomes.
GLANDULAR-PUBESCENT With gland-tipped, pinhead-like hairs.
GLANDULAR-PUNCTATE With glands recessed in depressions.
GLAUCESCENT Becoming glaucous.
GLAUCOUS With waxy substances forming a whitish or gray-silvery covering or bloom.
GLOBOSE Nearly spherical or rounded.
GLOCHID (pl.GLOCHIDIA) An apically barbed bristle or hair; e.g., in many Cactaceae.
GLOMERATE In a dense cluster or glomerule.
GLOMERULATE Arranged in small dense clusters.
GLOMERULE A dense cluster of two or more structures.
GLUCOSIDE A glycoside with glucose as the sugar.
GLUMACEOUS With greenish bracts or petals similar in appearance to the glumes in grass spikelets.
GLUME One of a pair of bracts at the base of a grass spikelet.
GLUTINOUS Sticky, gluey, or resinous.
GLYCOSIDE Complex, two-component chemical compound that can break down or hydrolyze under certain conditions, yielding a sugar plus another compound (= aglycone) that can be physiologically active including poisonous. Types of glycosides include cardiac, coumarin, cyanogenic, mustard oil, steroidal, and saponic; the term glucoside refers to those in which the sugar molecule is glucose.
GLYCOSINOLATE (= Mustard oil glycoside) A complex molecule that upon breakdown yields a sugar, a sulfate fraction, and isothiocyanates (= mustard oils); e.g., in the Brassicaceae or mustard family.
-GONOUS A suffix meaning angled; e.g., trigonous means three-angled.
GRADUATED Referring to a sequence in shape or size; e.g., leaves becoming narrowed up a stem.
GRAIN The 1-seeded fruit typical of cereal crops; often used synonymously with caryopsis.
GRANULAR, GRANULOSE, GRANULATE Covered with minute, grain-like particles.
GRIT CELL (= Stone cell) A sclerotic or hardened cell, as in the flesh of pears.
GYMNOSPERMS Literally, "naked seed"; a polyphyletic assemblage of plants without flowers, the seeds "naked, "(= not enclosed in a special structure), often on the surface of thick or thin, sometimes woody cone scales.
GYNECANDROUS The pistillate flowers above (= distal to) the staminate of the same spike; e.g., in some Cyperaceae.
GYNOBASAL, GYNOBASIC Referring to or having a gynobase.
GYNOBASE An enlargement of the receptacle at the base of the ovary.
GYNODIOECIOUS Basically dioecious, but with some flowers perfect and others pistillate.
GYNOECIUM The pistil or pistils of a flower considered collectively; collective term for the female parts of a flower.
GYNOMONOECIOUS Having female and bisexual flowers on the same plant.
GYNOPHORE Prolonged stipe (= stalk) of a pistil.
GYNOSTEGIUM (a) Sheath or covering of the gynoecium; (b) in Asclepiadaceae, the columnar or disk structure made up of the connate stamens, style, and stigma.

Gypsiferous, Gypseous Containing gypsum (= calcium sulfate).

H

Habit Style or arrangement of growth; general appearance.

Habitat Type of locality in which a plant grows; e.g., prairie.

Hair An epidermal appendage that is usually slender, sometimes branched, not stiff enough to be called a spine, not flattened as a scale; often used synonymously with trichome.

Halberd-shaped (= Hastate). Arrowhead-shaped but with the two basal lobes turned outward.

Hallucinogen A material capable of causing the perception of imaginary sights, sounds, or objects through effects on the nervous systems. Various plant products, including certain alkaloids, are capable of such effects.

Halophyte A plant tolerant of salty or alkaline soils.

Hamate Hooked.

Haploid Having the reduced number of chromosomes typical of gametes; usually with a single set of chromosomes in each nucleus.

Hastate (= Halberd-shaped). Arrowhead-shaped but with the two basal lobes turned outward.

Haustorium (pl. Haustoria) Sucker-like attachment organ of parasitic plants by which they draw their food supply from the host-plant.; e.g., in Cuscutaceae and some Scrophulariaceae.

Hay fever Respiratory allergy, frequently due to plant substances or microstructures such as pollen; e.g., *Ambrosia* (ragweed) pollen is a well-known cause of hay fever.

Head (= Capitulum) Dense cluster of sessile or nearly sessile flowers; the type of inflorescence typical of the Asteraceae.

Heartwood The innermost and oldest wood, often with materials (e.g., toxins) giving it different characteristics from sapwood (e.g., more durability or resistance to rotting).

-hedral A suffix signifying surface, usually preceded by a number and then indicating the number of sides, as a tetrahedral spore.

Helicoid, Helical Coiled or spiraled, usually in reference to inflorescences.

Heliophyte A plant adapted to grow in or tolerate full sun.

Hemi- Greek prefix meaning half.

Hemiparasitic (= Semiparasitic) Descriptive of a plant that carries out photosynthesis but obtains some of its food, mineral nutrition, or water needs from another living organism (the host).

Herb A vascular plant lacking a persistent woody stem and typically dying back to the ground each season.

Herbaceous (a) Referring to the aerial shoot of a plant that does not become woody; typically dying back to the ground each year; (b) of a soft texture, as green leaves.

Herbage Collective term for the green or vegetative parts of a plant.

Herbarium (pl. Herbaria) A collection of dried pressed plants prepared for permanent preservation (see Appendix eight for further details).

Herbicidal Having the ability to kill plants.

Hermaphroditic With stamens and pistils in the same flower; bisexual.

Hesperidium A specific type of fruit usually associated with the citrus family; a berry developed from a pistil with numerous carpels, pulpy within, and externally covered with a hard rind; e.g., orange.

Hetero- Greek prefix meaning other, various, or having more than one kind.

Heterocarpous With more than one kind of fruit.

Heterochlamydeous With the perianth differentiated into a calyx and a corolla.

Heterogamous (a) With more than one kind of flower; (b) in Asteraceae, with each head composed of more than one kind of flower.

Heterogeneous Not uniform in kind.

Heterophyllous Having more than one form of leaf.

Heterosporous Having two spore types; e.g., Selaginella.

Heterostylous Having styles of different lengths.

Hexagonal Six-angled.

Hexaploid Having six sets of chromosomes.

Hilum Scar or mark on a seed indicating where the seed was attached by a funiculus (= stalk) to the ovary wall or placenta. The hilum is the "eye"of a bean or other large seeds.

Hip The "fruit" of a rose; actually a fleshy hypanthium or floral cup with the true fruits (= achenes) inside.

Hippocrepiform Horseshoe-shaped.

Hirsute With straight, moderately stiff hairs.

Hirsutulous Diminutive of hirsute.

Hirtellous Minutely hirsute.

Hispid Resembling hirsute but the hairs stiffer, ± bristly, feeling rough to the touch.

Hispidulose Minutely hispid.

Hoary Covered with a fine, white, whitish, or grayish white pubescence.

Holophyletic A term used to describe a group consisting of a common ancestor and all of its descendants. The term monophyletic is sometimes used in the same sense.

Holotype The one specimen used or designated by the author of a species or other taxon as the nomenclatural type in the original publication. The holotype is the specimen to which the scientific name is permanently attached; it is not necessarily the most typical or representative element of a taxon.

Homo- Greek prefix meaning all alike, very similar, same, or of one sort.

Homochlamydeous With a perianth of tepals undifferentiated into calyx and corolla.

Homogamous (a) With only one kind of flower; (b) with anthers and stigmas maturing simultaneously.

Homogeneous Of the same kind or nature; uniform; contrasting with heterogeneous.

Homosporous With spores all of one type.

Hood (a) A segment of the corona in Asclepiadaceae; (b) a hollow arched structure.

Hooded Descriptive of an organ with the lateral margins more or less inrolled and the apex more or less inflexed; helmet-like; shaped like a hood.

Honey-leaf Petal-like perianth part producing nectar; e.g., in some Ranunculaceae.

Horn An exserted tapering appendage resembling a cow's horn; e.g., appendage on the hood in some Asclepiadaceae.

Horny Hard or dense in texture.

Host Organism from which a parasite obtains nourishment.

Humic Consisting of or derived from humus (= organic portion of soil).

Humistrate Laid flat on the soil.

Humus Decomposing organic matter in the soil.
Husk The outer covering of some fruits, typically derived from the perianth or bracts.
Hyaline Thin, membranous, and transparent or translucent.
Hybrid (a) A cross between two unlike parents; (b) specifically, the offspring resulting from a cross between two species.
Hydathode An epidermal structure, usually marginal or terminal, that excretes water.
Hydrophily, Hydrophilous Water-pollination; water-mediated pollination; using water as the mechanism of transferring pollen; e.g., in some Callitrichaceae; see Philbrick (1991).
Hydrophyte A plant typically growing partially or wholly immersed in water; contrasting with mesophyte and xerophyte.
Hygroscopic Susceptible of expanding, shrinking, twisting, or untwisting on the application or removal of water or water vapor.
Hymenopterophily, Hymenopterophilous Pollination by hymenopterans (= members of the insect order Hymenoptera—bees, wasps, and their relatives).
Hypanthium (pl. Hypanthia)Cup-shaped or tubular structure formed by (a) fusion of the sepals, petals, and stamens; or (b) enlargement of the receptacle so that the perianth and androecium are attached above the gynoecium. A hypanthium may be adnate to ovary (resulting in epigyny) or free from ovary (perigyny); formerly referred to as calyx tube.
Hypocotyl Axis of an embryo below the cotyledons.
Hypogeous Below the ground.
Hypogynium A structure below the ovary, as in Scleria (Cyperaceae).
Hypogynous Referring to a flower having floral organs attached below the ovary. Hypogynous flowers have superior ovaries.

I

Illegitimate name Name unacceptable as the accepted scientific name because it is not the earliest one given to the plant in question, or published without description, or violating some other specific requirement of the International Code of Botanical Nomenclature.
Imbricate Overlapping like shingles on a roof.
Immaculate Not spotted.
Immersed (= Submerged, Submersed) Growing under water.
Imparipinnate Unequally or odd-pinnate, with a single terminal leaflet.
Imperfect In reference to a flower having either functional stamens or functional pistils, but not both; unisexual.
Incanous Gray or hoary.
Incipient Beginning to be; coming into being.
Incised Cut rather deeply and sharply; intermediate between toothed and lobed.
Included Not exserted; within; not projecting beyond the surrounding organ.
Incomplete Referring to a flower lacking one or more of the flower parts: sepals, petals, stamens, or pistils.
Incomplete veins In *Carex*, referring to veins that do not extend the entire length of the perigynium body.
Incurved Curved inward.
Indehiscent Referring to a fruit that does not open at maturity; contrasting with dehiscent.
Indeterminate Inflorescence whose flowers begin to open first at bottom or outside, progressing upward or inward with the number of flowers not predetermined at the beginning of flowering; growth of inflorescence not stopped by opening of the first flowers; contrasting with determinate.
Indigenous Native to an area; not introduced.
Indument, Indumentum Surface coating such as hairs, roughening, bloom, or glands.
Induplicate Folded or rolled inward.
Indurate Hardened.
Indusium (pl. Indusia) Epidermal outgrowth covering the sori or sporangia on fern fronds.
Inferior Descriptive of an ovary fused to the hypanthium or to the lower parts of the perianth and therefore appearing to be located below the rest of the flower; an ovary positioned below the point of attachment of the floral organs. This type of ovary is found in epigynous flowers.
Infertile (a) Incapable of normal reproductive functions; (b) specifically used to describe Asteraceae flowers incapable of maturing achenes, irrespective of presence of functional stamens.
Inflated With an internal air space; bladdery.
Inflexed Bent inward.
Inflorescence (a) Term commonly used to refer to the flowering structure of a plant; (b) a flower cluster; (c) arrangement of flowers on the floral axis; (d) manner of bearing flowers.
Infra- Latin prefix meaning below.
Infraspecific Within the species; referring to a unit of classification below the species; e.g., subspecies, variety, form.
Infructescence An inflorescence in the fruiting stage.
Infundibuliform (= Funnelform) Funnel-shaped; gradually widening upwards.
Innocuous Harmless, unarmed, spineless.
Innovation A basal shoot of a perennial grass.
Inrolled Rolled inward.
Insectivorous Consuming insects; referring to plants that digest insect tissue to obtain nutrients such as nitrogen.
Inserted Attached to another part or organ.
Insertion The place or mode of attachment of an organ.
Integument The covering of an organ; e.g., of the ovule.
Inter- Latin prefix meaning between.
Intercalary Medial in position.
Intercostal Located between the ribs or costae.
Interlacunar Between air spaces.
Internode Area of stem or other structure between two nodes.
Interrupted Not continuous or regular.
Interspecific Between different species.
Interstitial Referring to the space intervening between one thing and another.
Intra- Prefix used to denote within.
Introduced Brought from another geographic region; not native.
Introgression, Introgressive hybridization Successive crosses, first between plants of two species, then between the offspring of this cross and plants of one parent species, followed by further interbreeding between mongrels of varying percentage of impurity with purebreds of the parent line. This eventually leads to whole populations of one parent species being contaminated with genes derived from the other.

Introrse Facing inward, toward the axis, typically used to refer to manner of anther dehiscence.
Intruded Projecting inward or forward.
Invaginate To enclose in a sheath.
Involucel Diminutive of involucre; a secondary involucre; e.g., the bracts subtending the secondary umbels in Apiaceae or the whorl of bracts subtending a flower in Malvaceae.
Involucellate With an involucel.
Involucral, Involucrate Pertaining to or having an involucre.
Involucre The whorl of bracts subtending a flower cluster or flower; e.g., involucre of bracts (= phyllaries) subtending a head in Asteraceae or small involucre (= involucel) subtending a flower in many Malvaceae (specifically called an epicalyx).
Involute With margins or edges rolled inward toward the upper side.
Irregular (a) Structures not similar in size or shape; asymmetrical; (b) descriptive of a flower without any plane of symmetry; contrasting to regular and zygomorphic.
-ish Suffix meaning "slightly," often used with color terms.
Isodiametric Of equal dimensions.
Isolateral Equal-sided.
Isothiocyanate (= Mustard oil) Organic compound containing nitrogen and sulfur that has a pungent odor and taste and is irritating to the skin and mucous membranes. Isothiocyanates are toxic and can cause liver and kidney damage, as well as other problems; e.g., in the Brassicaceae or mustard family.
Isotype A specimen of the type collection other than the holotype; an extra or duplicate specimen made at the same time and place as the holotype.

J

Jointed With or apparently with nodes or points of articulation; e.g., jointed hairs of *Physalis* (Solanaceae).
Juvenile leaves In plants with more than one leaf type (e.g., leaf dimorphism), the leaves on new growth, often quite different in appearance from adult leaves; e.g., found in *Juniperus, Hedera*.

K

Keel (a) (= Carina) The two lower (= abaxial) fused petals of a papilionaceous flower (Fabaceae); (b) prominent longitudinal ridge, shaped like the keel of a boat.
Keeled With a ridge or keel.
Knees Erect woody projections; e.g., found in *Taxodium* (baldcypress).

L

L. Linnaeus, Swedish naturalist who established the binomial system of nomenclature; L.f., his son.
Labellum (= Lip) In Orchidaceae, the enlarged upper petal that appears to be the lowest petal because of twisting of the pedicel.
Labiate Lipped; differentiated into an upper and a lower portion.
Lacerate Irregularly cleft as if torn.
Laciniate Cut into long, narrow, ± equal divisions or segments.
Lactiferous, Lactescent With milky sap.
Lacuna (pl. Lacunae) A space, hole, cavity, or areole.
Lacunose Perforated, with holes, cavities, or depressions.
Laevigate Smooth as if polished.
Lamella A flat, thin plate.
Lamellate Made up of flat, thin plates.
Lamina The blade or expanded part of an organ.
Laminate In plates or layers.
Lanate, Lanose Woolly; covered with dense, long, entangled hairs resembling wool.
Lanceolate Lance-shaped; several times longer than wide, tapering at both ends, widest about a third above the base.
Lanuginose With a cottony or woolly appearance, the hairs shorter than in lanate.
Latent Dormant.
Lateral Belonging to or borne on the sides.
Lateral bud Bud in a leaf axil; contrasting with terminal or apical bud.
Latex A water insoluble mixture of organic compounds, predominantly hydrocarbons, produced in specialized cells called laticifers and often milky in color. The latex of some species has elastic properties (e.g., *Hevea brasiliensis*, rubber) while that of others contain compounds such as alkaloids (e.g., *Papaver somniferum*, opium poppy).
Laticiferous Latex-bearing.
Latisept With broad partitions in the fruits; e.g., Brassicaceae.
Latrorse Dehiscing laterally and longitudinally; e.g., some anthers.
Lax (a) Spread apart, loose, distant; (b) not rigid.
Leaf The primary photosynthetic organ of most plants, usually composed of an expanded blade and a stalk-like petiole.
Leaf scar The mark or scar left on the stem by the fall of a leaf.
Leaf trace A vascular bundle, one or more in number, extending from the stem into the leaf.
Leaflet A single, expanded segment or division of a compound leaf.
Lectins Certain plant proteins that cause linking or agglutination between cells. They can be toxic or mitogenic (= capable of stimulating mitosis); they apparently function in the binding of symbiotic nitrogen-fixing bacteria to roots and in protecting against pathogenic bacteria; often found in members of the Fabaceae.
Lectotype A specimen or other material selected by a later worker from the original material studied by the author of the species (or other taxon) to serve as the nomenclatural type when a holotype was not originally designated or was lost or destroyed.
Legume (a) Fruit type with a single carpel typically dehiscent along both sutures (= margins); (b) a member of the Fabaceae; (c) any fruit type within the Fabaceae.
Lemma The outer (= lowermost) of the two bracts enclosing the reproductive structures in the grass floret.
Lenticel A small corky pore or spot on the bark of young twigs found in many trees and shrubs and allowing gas exchange.
Lenticular Two-sided; lens-shaped.
Lepidote Covered with small scales; scurfy.
Lepidopterophily, Lepidopterophilous Pollination by lepidopterans (= members of the insect order Lepidoptera—butterflies and moths).
Liana, liane A woody climber (e.g., grape vine). Lianas are common in the tropics.
Ligneous Woody.
Lignescent Somewhat woody or becoming woody.
Ligulate Tongue-shaped; strap-shaped.
Ligulate floret (= Ray floret) Flower, with corolla expanded into a ligule, typical of many Asteraceae.

Ligulate head Head having only bisexual flowers with strap-shaped corollas.

Ligule (a) A strap-shaped limb or body; (b) strap-shaped part of ray (= ligulate) corolla in Asteraceae; (c) membranous or hairy appendage on adaxial surface of the leaf at junction of blade and sheath in Poaceae.

Liguliform Strap-shaped.

Limb (a) In a corolla of united petals, the main expanded portion, as distinguished from a basal tube; (b) expanded part of an organ.

Limbate With limb present.

Linear Resembling a line, long and narrow, with margins parallel to one another.

Lingulate Tongue-shaped.

Lip (a) Either of the principal lobes of a bilabiate or strongly zygomorphic corolla or calyx (e.g., Lamiaceae); (b) (= Labellum) the enlarged upper petal in Orchidaceae that appears to be the lowest petal because of twisting of the pedicel.

Lithophyte Plant that grows on rocks but derives its nourishment from the atmosphere and from accumulated humus.

Littoral Of a shore, particularly of the seashore.

Loam Soil consisting of a mixture of sand, clay, silt, and organic matter.

Lobate Having lobes.

Lobe A usually rounded segment or division of a leaf, petal, or other organ.

Lobed Having deep or coarse indentations of the margin, larger than mere teeth.(However, there is no sharp distinction between large teeth and small lobes.)

Lobulate Having small lobes.

Locular Having one or more locules.

Locule, Loculus The cavity, compartment, or cell containing the ovules in a carpel or the pollen in an anther.

Loculicidal Descriptive of a capsule dehiscing along the middle of the back of each locule or chamber (= along the midrib of each carpel).

Lodicule One of the two or three minute scales at the base of the ovary in most grasses, thought to be a rudiment of a perianth part. They swell and thus open the lemma and palea, allowing the reproductive parts to be exposed.

Loment, Lomentum An indehiscent fruit separating into one-seeded segments at maturity.

Longitudinal Lengthwise; along the long axis.

Lumper A taxonomist who in general has the tendency to lump segregates into larger groups; contrasting with splitter.

Lunate Crescent-shaped; half-moon-shaped.

Lurid Dirty, dingy.

Lustrous Shining.

Lutescent Becoming yellow.

Lyrate Lyre-shaped; pinnately lobed with the terminal lobe the largest.

M

m Meter; 10 decimeters; 39.37 inches.

Macro- Greek prefix denoting large or long.

Macrospore (= Megaspore) A large spore giving rise to the female gametophyte; the larger of two kinds of spores produced by heterosporous plants; a female spore.

Maculate With a spot or spots.

Malodorous Foul-smelling.

Malpighian Describing hairs lying parallel to a surface and attached by their middle; with two branches and almost no stalk; appearing to be an unbranched hair attached at the middle.

Mammiform Breast-shaped; conical with rounded apex.

Mammillate Having nipple-like structures.

Marbled With irregular streaks or blotches of color.

Marcescent Withering, but the remains persistent; e.g., the corollas of most *Trifolium*.

Margin Edge; the outer portion of a blade or other structure.

Marginal Attached to the edge or pertaining to the edge.

Marginate Distinctly margined; with a distinctly different margin.

Marsh Wet or periodically wet, treeless area.

Medial, Median Central, middle.

Medullary Pertaining to the pith.

Mega- Greek prefix meaning very large.

Megaphyll Leaf with branched veins.

Megasporangium Sporangium in which megaspores are formed.

Megaspore (= Macrospore) A large spore giving rise to the female gametophyte; the larger of two kinds of spores produced by heterosporous plants; a female spore.

Megasporophyll A sporophyll (= spore-bearing leaf) bearing one or more megaspores.

Membranaceous, Membranous Having the nature of a membrane, thin, somewhat flexible, translucent.

Mentum A projection near the base of some flowers in the Orchidaceae.

Mephitic Having an offensive odor.

Mericarp The individual, separated carpels of a schizocarpic fruit; e.g., one of the fruit segments in the Apiaceae or one of the "nutlets"in the Boraginaceae.

Meristem Embryonic or undifferentiated tissue, capable of developing into various organs.

Meristematic Pertaining to or with the nature of a meristem.

-merous Greek suffix used to refer to the number of parts (or multiples of such) in each circle of the floral organs, usually with a numerical prefix (e.g., a 3-merous perianth would mean there are three petals and three sepals or some multiple of three). The term often refers to the perianth only.

Mesa A flat-topped hill with abrupt or steeply sloping side or sides.

Mesocarp The middle layer of the pericarp or fruit wall.

Mesophyte Plant that grows under medium moisture conditions; contrasting with hydrophyte and xerophyte.

Micro- Greek prefix meaning small.

Microphyll A relatively small leaf with a single unbranched vein, typical of the Lycopodiophyta and Equisetophyta.

Microphyllous Having small leaves.

Micropyle A minute opening through the integuments into the ovule through which the pollen-tube usually enters and often distinguished in the mature seed as a slight depression.

Microsporangium Sporangium in which microspores are produced. In angiosperms, the microsporangium is equal to the pollen sac and there are typically four microsporangia per anther.

Microspore A small spore giving rise to the male gametophyte; the smaller of two kinds of spores produced by heterosporous plants; a male spore.

MICROSPOROPHYLL The sporophyll (= spore-bearing leaf) upon which microspores are produced.

MICROTUBERCULATE Minutely tuberculate.

MIDRIB The central or main rib or vein of a leaf or other similar structure.

MITRIFORM Shaped like a miter or bishop's hat.

MIXED INFLORESCENCE One in which the parts are not consistent in being all determinate or all indeterminate.

MM Millimeter; 1000 microns or 1/1000 of a meter.

MONADELPHOUS With all filaments united into a single tube surrounding the pistil.

MONANDROUS Having a single stamen.

MONILIFORM Like a string of beads.

MONO- Greek prefix meaning one or of one.

MONOCARPIC (a) Descriptive of a plant that flowers only once before dying; (b) having a single carpel.

MONOCEPHALOUS, MONOCEPHALIC Bearing only a single head.

MONOCHASIUM A cyme with lateral branching on only one side of the main axis.

MONOCHLAMYDEOUS Having only one set of floral envelopes; having perianth of a single series.

MONOCOLPATE With a single furrow; e.g., on a pollen grain.

MONOCOTYLEDONS (abbreviated MONOCOTS) Flowering plants having one cotyledon (= seed leaf), mostly parallel venation, and flower parts usually in threes; one of the two classes of flowering plants, which, depending on the system of classification, is known as Monocotyledonae or Liliopsida.

MONOECIOUS Plants with staminate flowers and pistillate flowers on the same plant but lacking perfect flowers.

MONOMORPHIC One form; contrasting with polymorphic.

MONOPHYLETIC A term previously used to describe a group of organisms with a common ancestor; more recently it has been used to describe a group consisting of a common ancestor and all of its descendants. Some authorities believe that a different term, holophyletic, should be used for a group consisting of a common ancestor and all of its descendants.

MONOPODIAL With an evident single and continuous axis.

MONOTYPIC Having a single type or representative; e.g., a genus with only one species.

MONTANE Pertaining to or living in mountains.

MOSTLY A quantitative term meaning "most of them."

MUCILAGINOUS Slimy; with mucilage.

MUCRO A short and small abrupt tip, as with the midrib extending as a short point.

MUCRONATE With a mucro.

MUCRONULATE Diminutive of mucronate.

MULTI- Latin prefix for many.

MULTICIPITAL Literally, "many-headed"; descriptive of a crown of roots or a caudex from which several stems arise.

MULTIFID Divided into many narrow segments or lobes.

MULTIFLOROUS Many-flowered.

MULTIPLE FRUIT (= Syncarp) A single "fruit" formed by the coalescence of several fruits from separate flowers; e.g., *Morus* (mulberry), *Maclura* (bois d'arc), *Ananas* (pineapple).

MURICATE With numerous minute short points; roughened.

MURICULATE Diminutive of muricate.

MUSTARD OIL (= Isothiocyanate) Organic compound containing nitrogen and sulfur, which has a pungent odor and taste and is irritating to the skin and mucous membranes. Mustard oils are toxic and can cause liver and kidney damage, as well as other problems; e.g., in the Brassicaceae or mustard family.

MUSTARD OIL GLYCOSIDE (= Glycosinolate) A complex molecule that upon breakdown yields a sugar, a sulfate fraction, and an irritating mustard oil or isothiocyanate; e.g., in the mustard family.

MUTICOUS Blunt, lacking a point.

MYCORRHIZA (pl. MYCORRIHIZAE) A mutually beneficial, symbiotic association of a fungus and the root of a plant. Mycorrihizal relationships are characteristic of most vascular plants.

MYCORRHIZAL Pertaining to mycorrhiza.

N

N North.

NAKED Lacking various coverings, organs, or appendages, almost always referring to organs or appendages present in other similar plants; e.g., a naked flower lacks perianth.

NAPIFORM Turnip-shaped; e.g., roots.

NASCENT In the act of being formed.

NATANT Floating underwater; immersed.

NATURALIZED Referring to an introduced foreign plant that has become part of the spontaneous, self-perpetuating flora of a region.

NAVICULATE, NAVICULAR Boat-shaped.

NECTAR A sugar-rich solution secreted by plants, typically produced in nectaries. Nectar production has apparently evolved to attract insects or other animals for pollination or other purposes.

NECTAR-GUIDE A line or other marking leading insects to the nectary; e.g., contrastingly colored lines on many corollas.

NECTARIFEROUS Having or producing nectar.

NECTARY A specialized nectar-secreting structure or area; there can be floral nectaries (in the flowers) or extra-floral nectaries (not associated with the flowers).

NEEDLE Linear, often stiff leaf as in *Pinus* (Pinaceae).

NEOTYPE A specimen selected by a later worker to serve as the nomenclatural type of a taxon when all material studied by the original author has been lost or destroyed.

NERVE A simple vein or slender rib of a leaf, bract, or other structure.

NERVED Having nerves.

NET-VEINED With veinlets branching irregularly and not uniformly angular, forming a net-like pattern.

NEUTER Sexless, as a flower that has neither functional stamens nor pistils.

NEUTRAL FLOWER A sterile flower; flower with a perianth but without functional sexual organs.

NIGRESCENT Turning black.

NITID Smooth and clear, lustrous, glittering.

NITRATES Nitrogen containing compounds that can be accumulated in plant tissues and cause toxicity in animals if ingested. During digestion, nitrates are converted to nitrites that are toxic due to their ability to render hemoglobin incapable of transporting oxygen.

NOCTURNAL Occurring at night or lasting for only one night.

NODAL Located at or pertaining to a node.

NODE Area of stem or axis at which branches, leaves, bracts, or flower stalks are attached; joint of a stem.

Nodding Hanging down.

Nodose Nodular, knotty, with semispheroid protuberances.

Nodulose Diminutive of nodose; having small, knobby nodes or knots.

Nut Hard-shelled and indehiscent fruit with a single seed.

Nutant Nodding.

Nutlet (a) Diminutive of nut; (b) used to refer to any small, dry, nut-like fruit or seed, thicker-walled than an achene; (c) seed-like sections into which the mature ovary breaks in Boraginaceae, Labiatae, and most Verbenaceae, each section consisting of one seed with extra coat formed from ovary wall or partitions, making it technically a fruit.

O

Ob- Latin prefix indicating the reverse or upside-down, as obcordate, meaning cordate or ovate with wider end at top or away from point of attachment.

Obconic, Obconical Inversely cone-shaped, with attachment at the small end.

Obdeltoid Inversely deltoid; triangle-shaped with base pointed.

Obdiplostemonous Describing a flower with two circles of stamens, in which those of the outer circle are borne in front of (= on the same radius as) the petals.

Oblanceolate Lanceolate with broadest part above the middle and tapering toward the base.

Oblate Nearly spherical but flattened at the poles.

Obligate Necessary or essential; contrasting with facultative.

Oblique Slanting; unequal-sided.

Oblong Longer than wide with sides nearly parallel.

Obovate Egg-shaped with attachment at narrow end; inversely ovate.

Obovoid Inversely ovoid; a solid that is obovate in outline.

Obpyriform Pear-shaped with attachment at narrow end.

Obsolescent Nearly or becoming obsolete; rudimentary; referring to structures that are not or are only slightly evident.

Obsolete Not apparent or evident; rudimentary; vestigial; extinct.

Obtuse Forming a blunt or rounded angle of more than 90 degrees; not pointed.

Ochraceous Ochre-colored, yellow with a tinge of red.

Ochroleucous Yellowish-white, buff.

Ocrea (pl. Ocreae) Tube formed by fused, sheathing stipules; e.g., in Polygonaceae.

Ocreate With an ocrea.

Ocreola (pl. Ocreolae) Secondary sheath; sheath surrounding a fascicle of flowers; e.g., in the Polygonaceae.

Odd-pinnate Compound leaf with terminal leaflet present; with an odd number of leaflets.

Offset A short lateral shoot, arising near the base of a plant, primarily propagative in function and thus can give rise to a new plant.

Oligo- Greek prefix signifying few.

Olivaceous Olive-green.

Ontogeny The developmental cycle of an individual organism.

Opaque Impervious to light.

Operculate Having an operculum.

Operculum Literally, "a small lid"; term applied to the terminal portion (= lid or cap) of a circumscissile fruit or other organ.

Opposite Arranged two at each node, on opposite sides of the axis.

Orbicular, Orbiculate With round, approximately circular outline.

Order The unit, category, or rank in classification made up of one or more families; ending in -ales; sometimes divided into suborders or superfamilies, which in turn are made up of families.

Orifice An opening.

Ornithophily, Ornithophilous Pollination by birds.

Ortho- Greek prefix signifying straight.

Orthotropous ovule An erect ovule, with the micropylar end at the summit.

Oval Broadly elliptic.

Ovary Basal, ovule-containing portion of the pistil in angiosperms, developing into the fruit.

Ovate Egg-shaped with widest part at the base.

Ovoid Solid oval or solid ovate.

Ovulate Bearing ovules.

Ovule An immature or unfertilized seed; the megasporangium of a seed plant; the egg-containing unit of the ovary.

Oxalates Salts of oxalic acid, a carboxylic acid that can occur in plant tissue either as the free acid (e.g., rhubarb leaves) or as salts in the form of calcium oxalate, potassium oxalate, or sodium oxalate. Ingestion of the soluble oxalates by animals results in the formation of calcium oxalate crystals, which when deposited in the kidneys and other organs can cause serious mechanical damage; calcium deficiencies can also result; calcium oxalate also occurs in some plant tissues in the form of raphides or needle-like crystals; e.g., in Araceae.

P

Palate The raised portion within the throat of a bilaterally symmetrical corolla.

Pale (= Palea) A chaffy scale or bract on the receptacle of many Asteraceae, often subtending the fruit; see chaff.

Palea (a) In Poaceae, the inner (= uppermost) bract of a floret, often partly enclosed by the lemma; (b) (= pale) the receptacular scale or bract in Asteraceae; see chaff.

Paleaceous Chaffy; with small membranaceous scales.

Pallid Pale in color.

Palmate Attached or radiating from one point, as leaflets in a palmately compound leaf, veins in some leaf blades, or fingers of a hand.

Palmately compound With the leaflets attached at one point at the apex of the petiole, like the fingers all attached to the palm of a hand.

Palmately trifoliolate Having three leaflets with the terminal leaflet unstalked, sessile.

Palmatifid Cut so as to appear nearly palmately compound.

Paludal Growing in marshes.

Pandurate, Panduriform Fiddle-shaped; obovate with a contraction on each side.

Panicle A branched usually elongate (not flat-topped) inflorescence with flowers on the branches of the primary axis; a compound racemose inflorescence. Panicles are indeterminate.

Paniculate Resembling a panicle.

Paniculiform Having the form and appearance but not necessarily the structure of a panicle.

Pannose Having the texture or appearance of woolen cloth or felt.

Papilionaceous, Papilionoid Descriptive of a flower having a bilaterally symmetrical corolla composed of a banner (= standard), two lateral wing petals, and a keel of two fused petals; descriptive of flowers characteristic of many Fabaceae in the subfamily Papilionoideae.

Papilla (pl. Papillae) Small pimple-like or nipple-like projection.

Papillate, Papilliform, Papillose Shaped like or bearing papillae.

Pappus The crown of bristles, awns, scales, etc.at the apex of the achene of most Asteraceae, thought to be the modified calyx.

Papyraceous Papery.

Parallel-veined With main veins parallel to each other. Such leaves are typical of many monocots.

Paraphyletic A term used to describe a taxonomic group consisting of an ancestral species and some but not all of its descendants. Paraphyletic groups can give an inaccurate view of phylogeny but are often useful in classification; see Appendix 6 for discussion.

Parasite A plant that derives its food, mineral nutrition, and/or water wholly or chiefly from another plant (the host) to which it is attached; contrasting with epiphyte, saprophyte, or autophyte.

Parasitic Deriving food, mineral nutrition, water, or any combination from another living organism (the host).

Parietal Located on the inner side of the exterior wall of the ovary.

Parietal placentation Placentation with ovules attached to the ovary wall instead of the axis.

Paripinnate Referring to a pinnately compound leaf with an even number of leaflets or pinnae.

Parted (a) Cut nearly but not quite the distance from the margin to midrib or from the apex to base; (b) more generally, any deep cut; (c) as a suffix can be combined with a numerical prefix to indicate the number of segments.

Parthenogenetic Developing without fertilization.

-partite A suffix synonymous with -parted.

Patelliform Disk-shaped; circular with narrow rim.

Patent Spreading.

Pauci- Latin prefix denoting few.

Pectinate With narrow closely set segments or units similar to the teeth of a comb.

Pedate Palmately divided with the lateral segments again divided.

Pedicel (a) Stalk supporting a single flower of an inflorescence; (b) in grasses the stalk supporting a single spikelet.

Pedicellate, Pediceled, Pedicelled Borne on a pedicel; contrasting with sessile.

Peduncle Stem or stalk of an inflorescence, flower cluster, or of the only flower in an inflorescence.

Peduncular Pertaining to or on a peduncle.

Pedunculate With a peduncle.

Pellicle A thin skin or filmy covering.

Pellucid Clear, almost transparent.

Peltate Shield-shaped, with stalk attached on the undersurface away from the margin or base (sometimes attached at the middle like the axis of an umbrella).

Pendent, Pendulous Hanging down or suspended.

Penicillate Like a brush, usually referring to a tuft of hairs.

Pentagonal Shaped like a pentagon; five-sided or -angled.

Pentamerous Having the parts in fives.

Pepo A specific type of fruit usually associated with the Cucurbitaceae; a many-seeded fruit with a hard rind; e.g., gourd.

Perennating Surviving from growing season to growing season.

Perennial Root system or plant living at least three growing seasons (years); lasting from year to year.

Perfect flower One with both functional stamens and pistils; a bisexual flower.

Perfoliate Referring to a sessile leaf or bract whose base completely surrounds the stem, the latter seemingly passing through the leaf or with two leaves basally united around the stem.

Perforate Having translucent dots that look like small holes, or pierced through.

Peri- Greek prefix denoting around, enclosing.

Perianth Collective term for the corolla and calyx, often used when corolla and calyx are not well differentiated.

Pericarp Fruit wall; structure enclosing the seeds in angiosperms. The pericarp can be subdivided into endocarp, mesocarp, and exocarp.

Perigynium Urn-shaped, sac-like or pouch-like structure surrounding the pistil in *Carex* (Cyperaceae).

Perigynous Having floral organs united forming a tube (= floral tube) surrounding but not adnate to the pistil, the floral organs thus borne around (versus above or below) the ovary.

Peripheral On or near the margin.

Perisperm The nutritional material of a seed surrounding the embryo and formed outside the embryo sac.

Perisporium, Perispore A thin envelope enclosing a spore.

Pernicious Highly hurtful or harmful.

Persistent Remaining attached; not falling off; contrasting with deciduous.

Personate Referring to a strongly two-lipped corolla with the throat closed by a palate.

Petal One of the next-to-outermost whorl of parts in a complete flower; a segment of the corolla; the inner series of floral "leaves." The petals are collectively referred to as the corolla.

Petaliferous Bearing petals.

Petal-like (= Petaloid) Having the character or appearance of a petal, usually showy.

Petaloid (= Petal-like) Having the character or appearance of a petal, usually showy.

Petiolar Pertaining to or located on the petiole.

Petiolate With a petiole.

Petiole Stalk of a leaf supporting the blade.

Petiolulate Having a petiolule.

Petiolule Stalk supporting a leaflet.

Phanerogam A seed plant or spermatophyte; contrasting with cryptogam.

Phenolic, Phenol An aromatic alcohol; more technically, a substance with an alcohol (= hydroxyl) group attached to an aromatic (= benzene or phenyl) ring structure. Plant phenolics are sometimes toxic; a number of plant resins or resinoids are phenolic in nature; e.g., urushiol from poison ivy, tetrahydrocannabinol (THC) from marijuana; other phenolic compounds include vitamin E and anthocyanins.

Phenology Study of the times at which various events occur in the life of a plant or a flower.

Phloem The food-conducting tissue in a plant.

Photodermatitis Skin inflammation resulting from increased sensitivity to sunlight, often caused by exposure to certain plants or plant-derived materials.

Photosynthesis The process by which plants convert carbon dioxide and water into carbohydrates in the presence of light.

Phyllary (= Involucral bract) One of the bracts subtending a head in the Asteraceae.

Phylloclade (= Cladophyll, Cladode) A portion of stem having the general form and function of a leaf.

Phyllode, Phyllodium An expanded bladeless petiole; a leaf reduced to a flattened petiole. These structures typically serve the photosynthetic function of a blade.

Phyllodial Having the character of a phyllode.

Phyllotaxy Arrangement of the leaves on their axis.

Phylogeny The evolutionary history of a group.

Phytophotodermatitis Plant-induced skin inflammation resulting from increased sensitivity to sunlight.

Phytotoxic Term used to describe materials that are toxic to plants.

Phytotoxin (a) General term for a plant-derived toxin; (b) more typically used to refer to toxic plant proteins or toxalbumins; e.g., ricin from *Ricinus*.

Pilose With long, soft, ± straight hairs; softer than hirsute, not flexuous or curved as in villous.

Pilosulous Diminutive of pilose.

Pinna (pl. Pinnae) A primary division of a compound leaf. A pinna can be simple (and thus equivalent to a leaflet) or compound and divided into leaflets. The term is frequently used in describing fern leaves—here the term is used for any primary division of a leaf (e.g., a leaflet or a leaf division divided into leaflets). In flowering plants the term is generally used to describe a primary division of a decompound leaf (a pinna is thus divided into leaflets).

Pinnate, Pinnatley compound Descriptive of a compound leaf with leaflets distributed along both sides of an elongate axis; feather-like. Bipinnate or 2-pinnate leaves have the leaflets distributed along a secondary axis; tripinnate or 3-pinnate leaves have the leaflets along a tertiary axis, etc.

Pinnately trifoliate Referring to trifoliate leaves in which the terminal leaflet is extended beyond the other two by having a distinct petiolule.

Pinnatifid Pinnately divided into stalkless segments, but the segments not distinct leaflets (i.e., not divided all the way to the midrib).

Pinnatisect Pinnately divided to the midrib.

Pinnipalmate Descriptive of a leaf exhibiting a combination of pinnate and palmate arrangements of the leaflets.

Pinnule A secondary pinna; division of a pinna or a leaflet.

Pisiform Pea-shaped.

Pistil The female reproductive organ of a flower consisting of the ovary, stigma, and style. Pistils can be either simple (of one carpel) or compound (with > one carpel); the pistils are collectively referred to as the gynoecium.

Pistillate With only the female reproductive structures being functional or with female reproductive structures only.

Pistillode, Pistillodium A rudimentary or vestigial pistil.

Pith The soft spongy center of a stem.

Pitted Having little depressions or cavities.

Placenta (pl. Placentae) Place of attachment of the ovule(s) within an ovary.

Placentation Arrangement of placentae and thus ovules in an ovary.

Plait A lengthwise fold or pleat.

Plaited (= Plicate) Folded like a fan.

Plane With a flat even surface.

Plano- A suffix denoting flat.

Pleated Folded.

Plicate (= Plaited) Folded like a fan.

-ploid A suffix used in genetics, prefixed by a term indicating number, to denote the number of sets of chromosomes in the nucleus.

Plumose Feathery; descriptive of a long hair with long, slender, lateral hairs.

Pod A legume or, more generally, a dry dehiscent fruit; sometimes loosely synonymous with capsule.

Pollen, Pollen grain The microscopic spheroidal structures that produce the male cells and borne within the anther of a flowering plant or on the microsporophyll of a gymnosperm; technically microgametophytes develop from microspores.

Pollen SAC A microsporangium containing pollen. In angiosperms there are typically four pollen sacs per anther.

Pollen tube The slender tube that develops from a pollen grain, penetrates the tissue of the ovary, and enters the ovule.

Pollinate To transfer pollen from a stamen to a stigma.

Pollination The act or process of pollinating.

Pollinium (pl. Pollinia) A mass of coherent pollen, as developed in Asclepiadaceae and Orchidaceae.

Poly- Greek prefix meaning many.

Polyadelphous With several groups of stamens.

Polygamo-dioecious Descriptive of dioecious plants having some perfect flowers; bearing on one plant flowers partly perfect and partly pistillate, on another plant flowers partly perfect and partly staminate.

Polygamo-monoecious Descriptive of monoecious plants having some perfect flowers; bearing partly perfect, partly unisexual flowers.

Polygamous Bearing unisexual and bisexual flowers on the same plant.

Polymorphic Of various forms; with three or more forms.

Polypetalous (= Apopetalous) With corolla of separate petals; contrasting with gamopetalous.

Polyphyletic A taxonomic group having species derived from more than one common ancestor, the species having been placed in the same group because of similarities due to convergent or parallel evolution. Polyphyletic taxa give an inaccurate view of phylogeny.

Polyploid A plant with three or more basic sets of chromosomes.

Polysepalous (= Aposepalous) Composed of or possessing separate sepals.

Polystichous Arranged in several longitudinal rows.

Pome A fleshy indehiscent fruit developing from a compound inferior ovary embedded in receptacular or floral cup tissue; e.g., apples or pears.

Pore A small aperature or opening; e.g., opening at end of anther in *Solanum* species.

Poricidal Opening by pores.

Porulus Somewhat porous (= pierced with small round holes).

Posterior Describing the position of an organ located on the side adjacent to the axis. In flowers, the upper lip is posterior, the lower anterior.

Praemorse Short and truncate at the base, as if eaten or chewed off; descriptive of the rhizomes in some species of *Viola* and *Hieracium*.

Precocious Developing or appearing very early; denoting flowers that appear in advance of the leaves.

Prehensile Clasping or grasping; e.g., tendrils.

Prickle A slender, sharp, epidermal outgrowth without vasculature; e.g., the armature of roses.

Primocane The first year's shoot of woody biennials, typically non-flowering; e.g., *Rubus*.

Prismatic Angulate with flat sides; shaped like a prism.

Procumbent Trailing or prostrate, not rooting.

Proliferous Bearing or developing offshoots or redundant parts; producing numerous new individuals, or parts, of the same kind; bearing offsets, bulbils, or other vegetative progeny.

Prophyll, Prophyllum (a) In the Poaceae, the first leaf of a lateral shoot or vegetative culm branch, consisting of a sheath without a blade; (b) bracteole subtending an individual flower; e.g., in some species of *Juncus*.

Prophyllate Furnished with prophylls.

Prostrate Lying flat.

Protandrous, Proterandrous Having the anthers mature before the pistils in the same flower.

Prothallus, Prothallium Gametophyte stage or generation in ferns and fernallies, bearing the sexual organs.

Protogynous Having the stigma receptive to pollen before pollen is shed from the anthers of the same flower.

Protuberance A protrusion, swelling, bump, or bulge.

Proximal Located nearest to the base or attachment point of a structure; contrasting with distal.

Pruinose Having a bloom on the surface; with a waxy or powdery secretion; with a surface coating more pronounced than glaucous.

Pseudo- Greek prefix meaning false, not genuine.

Pseudoantagonism Pollination system (e.g., in some Orchidaceae) in which a flower mimics a male insect, thereby attracting a male insect that "fights" with the flower in an effort to protect his territory. In the process of repeating this activity, pollen can be transferred from one plant to another.

Pseudobulb The thickened or bulb-like, above ground stems of certain orchids.

Pseudocarp A structure made up of the mature ovary combined with some other organ; e.g., the hip of a rose.

Pseudocopulation Pollination system (e.g., in some Orchidaceae) in which a flower mimics a female insect (e.g., wasp or fly), thereby attracting male insects who copulate with the flowers and in the process of repeating this activity transfer pollen from one plant to another.

Psychoactive Mind altering. A number of psychoactive materials are produced by plants.

Pteridology The study of ferns and similar plants.

Pteridophyte A fern or similar plant; any member of the Lycopodiophyta, Psilophyta, Polypodiophyta (formerly Pteridophyta), or Equisetophyta (formerly Sphenophyta).

Puberulent Minutely pubescent.

Pubescence The covering or indumentum of hairs on a plant without reference to specific type (e.g., pilose, hirsute).

Pubescent (a) General term for covering or indumentum of hairs; (b) sometimes used in a more restricted sense to refer to fine short hairs; downy.

Pulp The juicy or fleshy tissue of a fruit.

Pulvinate With a pulvinus; cushion-shaped.

Pulvinus The swollen base of a petiole, as in many Fabaceae. The structure is often involved in leaf movements.

Punctae Dots, depressions, or pits, commonly glandular in nature.

Punctate With translucent or colored dots, depressions, or pits.

Puncticulate Minutely punctate.

Pungent (a) Ending in a rigid sharp point or prickle; (b) acrid to the taste or smell.

Purpurascent Becoming or turning purple.

Pustulate With pustules, blisters, or blister-like swellings.

Putamen The bony endocarp or "pit" of some fruits.

Pyrene The "nutlet, ""pit, " or stone in a drupe or drupelet; a seed together with the bony endocarp.

Pyriform Pear-shaped.

Pyxis (pl. Pyxides) A capsule with circumscissile dehiscence, the top coming off as a lid; e.g., in *Plantago*.

Q

Quadr- Latin prefix meaning four.

Quadrangular Four-cornered; square.

Quadrate Nearly square.

Quasi- Prefix meaning "as if."

R

Raceme An unbranched (= simple) inflorescence with pedicelled flowers from an elongate main axis. Racemes are indeterminate.

Racemiform Having the form but not necessarily the technical character of a raceme.

Racemose In racemes or resembling a raceme.

Rachilla (a) A small rachis; (b) specifically the axis of a spikelet in the Poaceae.

Rachis (a) Leaflet-bearing central axis of a pinnately compound leaf; (b) axis of an inflorescence.

Radially symmetrical Descriptive of a structure that can be cut into halves from any marginal point through the center. When used in reference to flowers, the term is synonymous with actinomorphic or regular.

Radiate (a) Spreading from a common center; (b) in Asteraceae, describing a head with disk florets in center and a whorl of ray (= ligulate) florets around the periphery.

Radiate head In Asteraceae, a head with both peripheral ray (= ligulate) florets and central disk florets.

Radical leaves Leaves arising from a root crown; basal leaves.

Radicle Embryonic root of a germinating seed.

Rameal Pertaining to or located on a branch.

Ramification The arrangement of branching parts.

-ranked Suffix, when combined with a numerical prefix, indicating the number of longitudinal rows in which leaves or other structures are arranged along an axis or rachis.

Raphide Needle-shaped crystal in a plant cell, typically of calcium oxalate.

Raphidulous Resembling or having raphides.

Ray (a) Flat or strap-shaped type of corolla found in many Asteraceae; (b) primary branch of a compound umbel.

Ray floret (= Ligulate floret) Flower, with corolla expanded into a ligule, typical of many Asteraceae.

Receptacle (a) (= Torus) Tip of a stem or pedicel, usually more or less enlarged, on which flower parts are attached; (b) in Asteraceae, the structure on which the florets of a head are attached.
Receptacular Referring to the receptacle.
Reclined, Reclinate Bent or turned downward.
Recumbent Leaning or reclining.
Recurved Curved downward or backward.
Reduced Small but probably derived from larger forerunners.
Reflexed Abruptly bent downward.
Regular (= Actinomorphic or Radially symmetrical) Referring to a structure that can be cut into halves from any marginal point through the center; with the parts in each series alike or uniform. The term usually refers to flowers.
Relic (a) A long-surviving species whose relatives have become extinct; (b) a plant persisting in a relatively small portion or portions of its former range.
Remote Widely or distantly spaced.
Reniform Kidney-shaped.
Repand With a shallowly sinuate, undulating, or slightly wavy margin; less strongly wavy than sinuate.
Repent Creeping and rooting at the nodes.
Replicate Folded backward.
Replum The partition between the halves of a fruit; e.g., Brassicaceae.
Resin, Resinoid A miscellaneous or catchall term for a variety of amorphous, sometimes semisolid, perhaps gummy substances from plants. A number are toxic, sometimes extremely so; some are terpene derivatives while others are phenolic compounds; examples of phenolic resins are tetrahydrocannabinol (THC) from marijuana, urushiol from poison ivy, and hypericin from *Hypericum perforatum*.
Resinous, Resiniferous Producing or bearing resin.
Resupinate Upside down; inverted; turned 180 degrees; e.g., some members of the Fabaceae and Orchidaceae have resupinate flowers.
Reticulate Net-veined or with a net-like pattern.
Retinaculum (pl. Retinacula) (a) In Asclepiadaceae, the filament-like band connecting a pollinium to the gland-like corpusculum; (b) in Acanthaceae, the curved, hook-like, modified funiculus that retains the seed until maturity.
Retrorse Bent or turned backward or downward; e.g., hairs pointing down a stem, the free end of the hair below or proximal to the attached end; contrasting with antrorse.
Retrorsely barbed With barbs (= points) pointing downward toward base.
Retuse Shallowly notched at a rounded apex.
Revolute With margins rolled down and inward; e.g., revolute leaves.
Rhizoid A filamentous root-like structure (without the anatomy of a root) on the gametophyte of ferns or other non-seed-producing plants.
Rhizomatous Possessing a rhizome.
Rhizome Underground stem with nodes and scale leaves.
Rhizophore In certain Pteridophytes, a specialized leafless stem emitting roots.
Rhombic Somewhat diamond-shaped; shaped like two adjacent equilateral triangles.
Rhomboid, Rhomboidal A three-dimensional rhombic figure.
Rib (a) One of the principal longitudinal veins of a leaf or other organ; (b) ridge on a fruit.
Ribbed With prominent ribs or veins.
Ringent Gaping; open.
Root The portion of the main axis (or one of its subdivisions) of a plant usually found below ground and lacking nodes, internodes, or leaves.
Rootlet (a) A little root; (b) term often applied to the holdfast roots of certain climbing plants.
Rootstock According to Shinners (1958), a "weasel-word" indicating an elongate crown, rhizome, or rhizome-like structure; an old inaccurate term for rhizome.
Roseate Rosy or pinkish.
Rosette A cluster or whorl of leaves or other organs closely arranged in a radial pattern.
Rostellate Diminutive of rostrate.
Rostellum (a) A small beak; (b) in Orchidaceae, an extension from the upper edge of the stigma.
Rostrate Beaked.
Rostrum A beak or any beak-like extension.
Rosulate In rosettes.
Rosy The rich pink characteristic of the petals of some roses.
Rotate Wheel-shaped or saucer-shaped as in a sympetalous corolla with an obsolete or very short tube and a flat circular limb.
Rotund Essentially circular in outline.
Ruderal Weedy or growing in waste places.
Rudiment A structure very imperfectly developed, non-functional, or represented only by a vestige; e.g., rudimentary non-functional florets in some grass spikelets.
Rudimentary Having the character of a rudiment.
Rufescent Becoming reddish brown.
Rufous Reddish brown.
Rugose Wrinkled.
Rugulose Diminutive of rugose.
Ruminate Appearing as through chewed or wadded; roughly wrinkled.
Runcinate With margins that are coarsely serrate to sharply incised with the segments pointing toward the base, as in *Taraxacum*.
Runner An elongate, slender, prostrate stem taking root at the nodes or tip.
Rupturing Bursting open along irregular lines.
Russet Reddish brown.

S

s South.
Sac A pouch or bag.
Saccate Sac-like; pouch-like.
Sagittate With the form of an arrowhead; triangular with the basal lobes pointing downward or inward toward the petiole.
Salient Projecting forward.
Saline Of or pertaining to salt.
Salverform Descriptive of a corolla having a narrow, nearly straight basal tube, abruptly expanded at top into a flattened or saucer-shaped limb.
Samara A dry, indehiscent, winged fruit; a fruit that is achene-like but with a flat wing; e.g., in maple, *Acer* (Aceraceae).
Sap The juice of a plant.

Saponaceous Soapy, slippery to the touch.

Saponic or **Saponin glycoside** A glycoside (two-component molecule) that upon breakdown yields a saponin.

Saponin A soap-like molecule that lowers the surface tension of aqueous solutions. Saponins usually have a bitter acrid taste and are commonly irritating to mucous membranes; chemically saponins are either triterpenes or steroids; they can alter the permeability of cell membranes and may react with proteins and are thus potentially toxic; they are highly toxic to cold-blooded animals and have long been used to stun fish.

Saprophyte A non-photosynthetic plant without chlorophyll, deriving its food from dead organic material in the soil by mycorrhizal relationships or otherwise; e.g., *Monotropa* (Ericaceae).

Saprophytic Subsisting on dead organic matter.

Saxicolous Growing on rocks.

Scaberulous Minutely scabrous.

Scabridulous Slightly rough.

Scabrous Rough to the touch due to short stiff hairs or the structure of the epidermis.

Scalariform Ladder-like.

Scale Any small, thin, usually dry, appressed leaf or bract, often only vestigial.

Scalloped Crenate.

Scandent Climbing without aid of tendrils.

Scape A naked (leafless but scales or bracts may be present) flowering stem or stalk arising from the ground.

Scapiform Resembling a scape.

Scapose Bearing or resembling a scape.

Scarious Membranous, dry, papery, translucent or transparent, not green.

Scattered Distributed in an irregular manner; not clustered together.

Schizo- Greek preffix meaning split or divided.

Schizocarp A fruit that splits between carpels into one-seeded portions (= mericarps); e.g., in the Apiaceae or Boraginaceae.

Sciophyte A plant adapted to grow in or tolerate shade.

Scimitar-shaped With the shape of a curved sword or scimitar.

Sclerenchyma An internal tissue composed of hard, thick-walled cells.

Sclerenchymatous Having sclerenchyma.

Sclerotic Hardened; stony in texture.

Scorpioid Uncurling, like a scorpion's tail.

Scrobiculate Marked by minute or shallow depressions.

Scrotiform Pouch-like.

Scrub Vegetation of stunted or densely crowded bushes.

Scurfy With whitish, scaly, blister-like structures or scales on the surface.

Scutellum A shield-like protrusion on the calyx; e.g., in some Lamiaceae.

Second glume The uppermost of the two glumes of a grass spikelet.

Secondary compounds Naturally occurring plant materials not essential to the primary (= life-sustaining) metabolism of the plant; examples of categories include alkaloids and glycosides. Many are significant because of their physiological activity when given to animals; they are probably important to plants in defense against herbivores or microbes.

Secund Directed to one side; arranged on one side only; unilateral; one-sided.

Seed A fertilized ripened ovule, covered by a seed coat [developed from the integument(s)] and containing the embryo and in flowering plants the endosperm. The seed also contains the remnants of the nucellus (= sporangium) and megagametophyte.

Seed coat (= Testa) Outer covering of a seed, developed from the integument(s).

Seep A moist spot where underground water comes to or near the surface.

Segment One of the parts of a leaf, petal, or calyx that is divided but not truly compound.

Segregate Term used as a noun or adjective to refer to or describe a taxon that is sometimes recognized separately from a more inclusive group; e.g., depending on authority, the segregate, *Tovara* (or the segregate genus, *Tovara*), is either included in, or recognized separately from, the more inclusive genus *Polygonum*.

Seleniferous Containing selenium.

Selenium An element that is concentrated in the tissues of some plants (e.g., poison-vetches in genus *Astragalus*) and can cause a toxic, sometimes fatal response in livestock.

Self-fertile Capable of self-fertilization (= union of gametes from same plant).

Self-incompatible Incapable of self-fertilization.

Semi- Latin prefix meaning half.

Semiparasitic (= Hemiparasitic) Descriptive of a plant that carries out photosynthesis but obtains some of its food, mineral nutrition, or water needs from another living organism (the host).

Senescent Aging or aged.

Sensu lato "In a broad sense"; used to refer to the broad treatment of taxa; e.g., a genus sensu lato is one that has not been split into a number of segregates.

Sensu stricto "In a narrow sense"; used to refer to a restricted or narrow treatment of a taxonomic group; e.g., a genus sensu stricto is viewed in a more restricted sense than previously as the result of segregating or splitting out various taxa.

Sepal A single unit of the calyx; one of the outermost whorl of parts in a complete flower. Sepals typically cover the other flower parts during the bud stage; they are collectively referred to as the calyx.

Sepaloid Sepal-like, usually green and thicker in texture than a petal.

Separate Not joined; of individual units.

Septate With partitions or divisions.

Septicidal Descriptive of a capsule that dehisces along or through the septa or partitions separating its locules or seed chambers; contrasting with loculicidal.

Septifragal Breaking away at the partitions, as the valves of a capsule.

Septum A partition or cross wall within an organ; e.g., the septa of an ovary or of the leaf of a rush (*Juncus*).

Seriate, serial With series or distinct rows.

Sericeous Silky; with appressed, fine, and straight hairs.

Serotinous Literally, "late"; in *Salix* indicating those species in which the catkins develop later than the leaves; produced or occurring late in the season.

Serrate With pointed teeth sloping forward; saw-toothed.

Serrulate Finely serrate.

Sessile Without a pedicel, petiole, or stalk; inserted directly.

Seta (pl. Setae)A bristle.

Setaceous Bristle-like.

SETIFEROUS Bearing bristles.

SETIFORM Resembling a bristle.

SETOSE Bearing bristles.

SETULOSE Diminutive of setose.

SHEATH (a) ± tubular structure surrounding an organ; portion that clasps or encloses; (b) specifically, the basal tubular portion of the leaf in grasses and grass-like plants between the node and the blade.

SHOOT (a) A young stem or branch; (b) the ascending axis of a plant.

SHOULDER That part of an organ that rather abruptly curves inward.

SHRUB A woody perennial usually branching from the base with several main stems.

SIGMOID S-shaped; doubly curved.

SILICA Silicon dioxide; a white or colorless, very hard, crystalline mineral substance.

SILICEOUS Containing or composed of silica (the principal component of glass and sand); e.g., descriptive of certain hairs, with high silica content, that easily break off in the flesh.

SILICLE, SILICULA A short silique; short and broad capsular fruit of the mustard family. A silicle is not much longer than broad; sometimes defined as < 3 times as long as broad.

SILIQUE, SILIQUA Dry, dehiscent, variously shaped, many-seeded, 2-valved capsule with valves splitting from the bottom and leaving a false partition known as a replum; the fruit type typical of the mustard family. The term is sometimes restricted to long and narrowly cylindrical fruits 3 times as long as broad or longer; shorter fruits with the same structure are then referred to as silicles.

SILKY Sericeous; with appressed, fine, and straight hairs.

SILVERY With a whitish, metallic, more or less shining luster.

SIMPLE (a) Single, of one piece, not compound, as a pistil of one carpel; (b) descriptive of an unbranched stem, inflorescence, or other structure; (c) descriptive of a leaf that is not compound.

SIMPLE LEAF Single-bladed leaf, not divided into individual leaflets.

SIMPLE PISTIL Pistil composed of only one carpel (and with a single style branch).

SINUATE, SINUOUS Having the margin wavy with regular strong indentations.

SINUS The space or recess between two lobes, segments, or divisions of a leaf or other expanded organ.

SLOUGH A wet place or deep mud or mire; a sluggish channel.

SMOOTH Not rough to the touch; without vestiture or other special covering.

SOBOL A basal shoot.

SOBOLIFEROUS Producing basal shoots, clump-forming.

SOLITARY Borne singly.

SORDID Dirty white.

SORUS (pl. SORI) A cluster or heap of sporangia. The term is used mainly to refer to the sporangial clusters of ferns.

SPADIX (pl. SPADICES) Fleshy spike with inconspicuous flowers, usually surrounded or partially enclosed by a spathe; e.g., characteristic of Araceae.

SPATHACEOUS, SPATHIFORM Resembling a spathe.

SPATHE Enlarged leafy bract surrounding or partially enclosing an inflorescence.

SPATULATE Spatula-shaped; rounded above and gradually narrowed to base.

SPECIES Unit of classification below the rank of genus; a group of individuals that are actually or potentially capable of interbreeding in natural populations and are reproductively isolated from other such groups. Generally there are morphological characteristics that distinguish and can thus be used to separate such groups; the definition is complicated by instances of asexual reproduction.

SPECIFIC EPITHET The second half of the scientific name of a species, the scientific name being composed of the generic name and the specific epithet.

SPERMATOPHYTE A plant that produces seeds; all gymnosperms and angiosperms.

SPHERICAL Globular; orbicular.

SPICATE In spikes or resembling a spike.

SPICIFORM Having the form of a spike but not necessarily the technical structure.

SPIKE Indeterminate inflorescence with sessile flowers on an ± elongate floral axis.

SPIKELET The basic unit of the grass inflorescence, usually consisting of a short axis (= rachilla) bearing two empty bracts (= glumes) at the basal nodes and one or more florets above. Each floret usually consists of two bracts (the lemma = lower bract, and the palea = upper bract) and a flower. The flower usually includes two lodicules (= vestigial perianth segments that become turgid and open the bracts at flowering time), three stamens, and a pistil. The term spikelet is also used to refer to the basic unit of the inflorescence in the Cyperaceae.

SPINE A sharp-pointed structure, usually vascularized and thus ± stout or woody, generally modified from part or all of a leaf or stipule. A spine is sometimes distinguished from a thorn, which is a modified branch.

SPINESCENT Becoming spine-like; ending in a spine; having spines.

SPINIFEROUS, SPINOSE Having spines.

SPINULOSE With small spines over the surface, margin, or edge.

SPIRAL Describing the arrangement of like organs, such as leaves, at regular angular intervals along an axis.

SPIRAL THICKENING Thickening of the walls of a xylem cell laid down in the form of a spiral.

SPLITTER A taxonomist who in general has the tendency to split or divide larger taxa into a number of segregates; contrasting with lumper.

SPONGIOSE Soft, spongy.

SPORADIC Of irregular occurrence.

SPORANGIOPHORE An appendage holding a sporangium.

SPORANGIUM (pl. SPORANGIA) A spore case or spore sac.

SPORE An asexual, usually one-celled reproductive body; a cell resulting from meiotic cell division in a sporangium representing the first cell of the gametophyte generation.

SPOROCARP A specialized structure containing sporangia.

SPOROPHYLL Spore-bearing leaf.

SPOROPHYTE The spore-producing, typically diploid generation that alternates with the gametophyte (= gamete-producing, typically haploid); the dominant generation in most plants except mosses and liverworts.

SPREADING Diverging to the side, almost to the horizontal.

SPREADING HAIRS Hairs that are ± erect, not at all appressed.

SPUR (a) Basal, sac-like, hollow projection, short or long and narrow, from a corolla or calyx and often containing nectar; (b) a short lateral branch with little internode development resulting in closely clustered leaves or flowers.

Squama (pl. Squamae)A scale, usually reduced from and homologous to a leaf.

Squamella (pl. Squamellae) A tiny or secondary scale.

Squamose, Squamate Covered with scales; scaly.

Squamulose With small scales.

Squarrose Spreading rigidly at right angles or more; e.g., with leaves or bracts spreading and bending backward abruptly in the upper part.

Stalk The supporting structure of an organ; e.g., petiole, pedicel, peduncle.

Stamen The male reproductive organ of a flower consisting of a filament (= stalk) and anther (= pollen-bearing structure). The stamens of a flower are collectively referred to as the androecium.

Staminal Pertaining to a stamen.

Staminate Referring to a flower with only the stamens being functional or with stamens only.

Staminode, Staminodium (pl. Stamidodia) An abortive, sterile, or nonfunctional stamen with reduced anther or usually none. A staminode can be enlarged or widened, sometimes so much so as to be petal-like.

Standard (a) (= Banner) The usually large, upper (= adaxial) petal of a papilionaceous flower (Fabaceae); (b) the inner erect tepals or perianth segments in *Iris* (Iridaceae).

Stele The central vascular cylinder of a plant.

Stellate Star-shaped or star-like; when used in reference to hairs it means those branched hairs with a central stalk and branch hairs arising at the top of the stalk (like points of light coming out of a star).

Stem A major division of the plant-body in contrast to root and leaf, distinguished from both by certain anatomical features and commonly also by general aspect; the main axis or axes of a plant; the portion of the plant axis bearing nodes, leaves, and buds and usually found above ground.

Sterigma (pl. Sterigmata) A very short persistent stipe or stalk.

Sterile Unproductive; nonfruiting; without functional sex organs; without spores.

Sternotribal Descriptive of flowers in which anthers are positioned to dust pollen on underside of thorax of insects.

Steroid Any of a large group of lipid soluble organic compounds based on a complex framework of four interconnected rings of carbon atoms. A number of plant glycosides have steroids as the aglycone; these aglycones are often toxic.

Steroid glycoside A glycoside (two-component molecule) that upon breakdown yields a steroid as the aglycone (= non-sugar component). Cardiac glycosides and some saponic glycosides have steroidal aglycones; these are often poisonous.

Stigma Portion of pistil (usually terminal and small) modified (roughened, sticky, or branched) to catch or receive pollen; the receptive surface of the pistil.

Stigmatic Like or pertaining to a stigma.

Stipe (a) In general, a stalk; (b) specifically, the leaf stalk of a fern; (c) specifically, the narrow, stalk-like, basal portion of an ovary or fruit; e.g., in *Cleome*.

Stipel Appendage at the base of a leaflet, analogous to a stipule at base of a leaf.

Stipellate With stipels.

Stipitate With a stalk or stipe.

Stipular Pertaining to or located on a stipule.

Stipulate With stipules.

Stipule Paired appendages subtending the petiole of a leaf. Stipules can be conspicuous and persistent, small and rapidly lost, or totally lacking; the evolutionary origin and original function of stipules are unclear. They have become variously modified to serve as spines, additional photosynthetic area, etc.

Stolon A creeping horizontal stem that loops or runs along the surface of the ground and roots at the nodes.

Stoloniferous Producing stolons.

Stoma, Stomate (pl. Stomata) A minute opening (= "breathing" pore) between the epidermal cells of a leaf or stem through which gases and water-vapor enter and leave the plant.

Stomatiferous Bearing stomata.

Stone Bony endocarp of a drupe.

Stone cells (= Grit cells) Individual cells that have become hardened or sclerotic; e.g., in flesh of a pear.

Stramineous Straw-colored.

Striate With fine longitudinal lines, channels, or ridges.

Strict Very straight, unbranched, upright.

Strigillose, Strigulose Diminutive of strigose.

Strigose With sharply bent (at base) but otherwise straight, appressed hairs.

Strobilus (pl. Strobili) (= Cone) A usually cone-like, globose or cylindrical structure involved in reproduction and composed of an axis with a spiral, usually dense aggregation of sporophylls, bracts, or scales (these bearing spores, pollen, or seeds).

Strophiolate Having a strophiole.

Strophiole On certain seeds, an appendage to the hilum.

Strumose Descriptive of the surface of an organ bearing cushion-like swellings.

Style Portion of pistil between ovary and stigma, often elongated and narrow, frequently branched, with the number of branches often indicative of the number of carpels making up the pistil.

Stylopodium Enlarged base of style as in many Apiaceae.

Sub- Latin prefix meaning almost, somewhat, of inferior rank, beneath.

Subapical Almost at the apex.

Subbasal Almost at the base.

Subclass The unit, category, or rank in classification between class and order, composed of one or more orders; e.g., the Liliidae is a subclass of class Monocotyledonae.

Subcoriaceous Somewhat leathery in texture.

Suberose Corky in texture.

Subfamily The unit, category, or rank in classification between family and genus, composed of one or more genera.

Subgenus A unit, category, or rank in classification between genus and species, composed of one or more species.

Suborder The unit, category, or rank in classification between order and family, composed of one or more families.

Submerged, Submersed (= Immersed) Growing under water.

Subsessile Almost sessile.

Subspecies A unit, category, or rank in classification below the level of species and between species and variety; a geographically distinct variant. The categories of subspecies and variety are not used consistently by taxonomists.

Subtend To be present just below; e.g., an involucre of bracts immediately subtends the flowers of some Malvaceae.

Subterranean Below ground.

Subulate (= Awl-shaped) Tapering from the base to a slender or stiff point; narrow and sharp-pointed.

Succulent Fleshy, thickened.

Sucker A shoot of subterranean origin. Many tree species have sucker shoots with leaves atypical for the species (e.g., oaks).

Suffrutescent, Suffruticose Woody only at base with the upper parts herbaceous and annual; obscurely shrubby or fruticose.

Sulcate Grooved or furrowed longitudinally.

Sulcus A furrow or groove.

Superior ovary One that is separate from the other flower parts. The other flower parts may overtop it but are attached at or under its base; a superior ovary is found in either a hypogynous or a perigynous flower.

Supine Laying flat with face upward.

Supra-, Super- Latin prefix meaning above.

Supramedial Above the middle; when used to refer to the location of fern sori, it means somewhat beyond the middle of the distance between the leaf segment midvein and margin but not so much so as to be called submarginal.

Suppressed Failing to develop.

Surculose Producing suckers or shoots arising from underground parts.

Surcurrent Extending upward; said of a pinnule whose base extends upward and forms a wing along the rachis; opposite of decurrent.

Suspended Hanging downward.

Suture A line of dehiscence or a longitudinal seam.

Swale A moist meadowy area lower than the surrounding areas.

Swamp Wet or periodically wet area with some trees.

Sym- Greek prefix meaning with or together.

Symmetrical Possessing one or more planes of symmetry; regular in number and size of parts.

Sympatric Growing together with or having the same range as.

Sympetalous (= Gamopetalous) With petals united, at least basally, forming a tube.

Sympodial With the main axis or stem ceasing to elongate but growth being continued by the lateral branches.

Sympodium An apparent main axis formed of successive secondary axes, each of which represents one fork of a dichotomy, the other being much weaker or entirely suppressed.

Syn- Greek prefix meaning united.

Syncarp (= Multiple fruit) Used to refer to a structure composed of several more or less coalescent fruits from separate flowers; e.g., *Morus* (mulberry), *Maclura* (bois d'arc), *Ananas* (pineapple).

Syncarpous (a) Having carpels united; (b) of or pertaining to a syncarp.

Synconium The multiple, hollow "fruit" of a fig (*Ficus*), which is actually an enlarged fleshy branch or receptacle enclosing the inflorescence (with flowers borne inside). Much of the tissue of a fig is morphologically derived from the stem.

Syngenesious With anthers united into a tube surrounding the style. This condition is typical of Asteraceae.

Synonym A currently unaccepted scientific name for a taxon.

Synonymy Referring to the series of names no longer used for a taxon.

Synsepalous (= Gamosepalous) With sepals united, at least basally, forming a tube.

Systematics Scientific study of the kinds and diversity of living organisms and of the relationships between them. The term is often used synonymously with taxonomy.

T

Tannin A type of phenol (= aromatic alcohol) that can act as a digestion inhibitor by binding up proteins and thus stopping enzymatic action. They are widespread in plants apparently as a chemical defense against herbivores; used by humans to tan leather; when ingested in quantity they can be toxic to animals.

Taproot, Taprooted The primary descending root, giving off small laterals but not dividing; the one dominant root markedly larger than the others.

Tawny Dull brownish-yellow; fulvous.

Taxon (pl. Taxa) (a) General term referring to any unit of classification such as variety, subspecies, species, genus, family, etc.; (b) term used to refer to a specific variety, subspecies, etc.

Taxonomy The branch of science that deals with classification, identification, and nomenclature.

Teeth (plural of tooth) Marginal projections, protuberances, serrations, or dentations, usually sharply pointed.

Tendril A slender twining or coiling appendage or axis that enables plants to climb; often a modified leaf or stem.

Tepal Part of a perianth, usually of a perianth consisting of only one whorl, or of one not differentiated into sepals and petals; a part of the outermost whorl or whorls of flower parts.

Teratological Distinctly abnormal; malformed.

Terete Rounded or circular in cross-section, cylindrical.

Terminal At the tip or apex; distal.

Terminal bud (= Apical bud) Bud at the end (= apex) of a stem or branch.

Ternate, Ternary In threes.

Terpenes, Terpenoids Common organic compounds in plants that are products of acetate metabolism. Numerous kinds are known resulting from variations in the use of 5-carbon isoprene units in their structures; they are often combined with other substances to form complex molecules; essential oils, saponins, some resins, latex, cartenoid pigments, and steroids are examples of compounds that are terpene derivatives.

Terrestrial Growing in the ground; supported by soil; contrasting with aquatic.

Testa (= Seed coat) The outer covering of a seed; hardened mature integument(s).

Tetra- Greek prefix referring to four.

Tetrad A group of four similar objects; e.g., in Ericaceae, the four pollen grains remaining together.

Tetradynamous With four long stamens and two short stamens.

Tetragonal, Tetragonous Four-angled.

Tetrahedal Four-sided, as a three-sided pyramid and its base.

Tetramerous (= 4-merous) Having flower parts in fours or multiples of four.

Tetraploid With four sets of chromosomes; twice the normal diploid level.

Thalloid Consisting of a thallus; thallus-like.

Thallophytes Algae and fungi.

Thallus (a) A plant body not differentiated into true leaves, stems, or roots; (b) a flat, leaf-like organ.

Theca (= Anther cell) One of the pollen sacs or locules of an anther.

Thorn A sharp-pointed, stiff, woody structure derived from a modified branch.

Throat The area of juncture of limb and tube in a sympetalous corolla.

Thyrse A panicle-like inflorescence consisting of cymules, usually elongate and slender with main axis indeterminate and the lateral axes determinate.

Thyrsoid With the appearance of a thyrse.

Tomentose Covered with short, soft, curly, densely matted or entangled hairs.

Tomentulose Diminutive of tomentose.

Tomentum Densely matted wool.

Tooth (pl. Teeth) Any marginal projection, protuberance, serration, or dentation, usually sharp pointed.

Toothed With minor projections and indentations alternating along the margin.

Topotype A specimen from the original or type locality of that species or other taxon.

Torose Alternately contracted and expanded.

Tortuous Twisted or bent in different directions.

Torulose Cylindrical with slight constrictions at intervals; necklace-like.

Torus (= Receptacle) Tip of a stem or pedicel, usually more or less enlarged, on which flower parts are attached.

Toxalbumin Toxic plant proteins; e.g., ricin from *Ricinus*. The term phytotoxin is also often used in this context.

Trabecula A transverse partition or cross-bar.

Trabecular, Trabeculate Like or with a partition or cross-bar.

Trailing Prostrate but not rooting.

Translator In Asclepiadaceae, the wishbone-shaped combination of the clip (= gland-like corpusculum) and bands (= filament-like retinacula) connecting a pair of pollinia from adjacent anthers.

Translucent Allowing the passage of light rays but not transparent.

Transverse Lying or being across or in a cross direction.

Trapezoid A body with four unequal sides.

Tree A woody perennial with usually a solitary trunk or main stem.

Tri- Latin prefix indicating three, or three times.

Triad In threes.

Tribe The unit, category, or rank in classification between sub-family and genus, composed of one or more genera.

Trichome Any hair, hair-like projection, or bristle from the epidermal surface.

Trichotomous Forking into three equal parts.

Tricolpate Three-grooved; e.g., a three-grooved pollen grain.

Trifid Three-cleft.

Trifoliate Having three leaves. The term is often used synonymously with trifoliolate.

Trifoliolate Having three leaflets.

Trifurcate With three prongs; three-forked.

Trigonous, Trigonal Three-angled.

Trilobed, Trilobate Three-lobed.

Trimerous Having the parts in threes.

Trimorphic Of three forms.

Trinerved With three primary nerves; triple-nerved; usually with a midrib and two main lateral nerves arising from the base of the midrib.

Tripartite Three-parted.

Triquetrous Three-angled in cross-section.

Tristichous In three vertical rows.

Trullate Trowel-shaped, widest below the middle.

Trumpet-shaped Describing a gamopetalous corolla or gamosepalous calyx in which the tube gradually widens upward.

Truncate Ending abruptly as if cut off squarely at the end; appearing "chopped off."

Tube The narrow, cylindrical, basal portion of a sympetalous corolla or synsepalous calyx.

Tuber Modified underground stem; stem enlarged and subterranean with nodes, buds, and scale leaves, often serving to store food; e.g. in *Solanum* (potato).

Tubercle (a) A small, tuber-like structure; (b) small rounded protuberance or projection from a surface; (c) the persistent style base in some Cyperaceae.

Tuberculate Covered with tubercles or warty or nipple-like protuberances.

Tuberiferous Bearing tubers.

Tuberous Bearing tubers or resembling a tuber in appearance.

Tubular With the shape of a hollow cylinder.

Tuft, Tufted A cluster or fascicle of trichomes, leaves, or other elongate structures.

Tumid Swollen; inflated.

Tunic A loose, membranous, outer skin or coat.

Tunicate Describing a bulb in which the leaves are arranged in concentric circles; with coats or tunics.

Turbinate Top-shaped; inversely conical.

Turgid Swollen but solid or full; contrasting with inflated.

Turion A scaly swollen structure or offshoot, often serving to overwinter; e.g., in some *Myriophyllum* species the turions store carbohydrates and serve as propagules.

Twice-pinnately compound (= Bipinnate) Descriptive of a leaf with leaflets pinnately arranged on lateral axes that are themselves pinnately arranged on the main axis; with the primary divisions (= pinnae) themselves pinnate.

Twig A young woody stem; the growth of the current season.

Two-lipped (= Bilabiate) Descriptive of a corolla (or calyx) of united petals (or sepals) cut on the two sides forming an upper and a lower portion.

Type A plant specimen to which the name of a taxon is permanently attached. When any new taxon (e.g., species, variety) is named, the name has to be associated with a particular "type"specimen.

U

Ubiquitous Occurring everywhere.

Umbel Usually flat-topped or convex inflorescence with flower pedicels all attached at the same point, like the rays of an umbrella; inflorescence type typical of Apiaceae (Umbelliferae).

Umbellate Umbel-like, or in an umbel.

Umbellet A secondary or ultimate umbel; one of the smaller umbellate flower clusters in a compound umbel.

Umbelliform Resembling an umbel in appearance.

Umbo A rounded elevation or protuberance at the end of or on the side of a solid organ.
Umbonate With an umbo or projection.
Uncinate With apex hooked; e.g., hairs with a hook at tip.
Uncinulate Minutely uncinate.
Undulate Gently wavy, less pronounced than sinuate.
Unguiculate Narrowed at the base into a claw.
Uni- Latin prefix meaning one.
Unifoliate With only one leaf.
Unifoliolate Referring to a compound leaf reduced to a single leaflet; e.g., in some members of the Fabaceae.
Unilateral One-sided; developed or hanging on one side.
Unilocular Having only one locule or cell.
Uniseriate Arranged in a single horizontal row.
Unisexual Having only stamens or only pistils; of only one sex; having flowers either staminate or pistillate.
United Fused into one unit.
Urceolate Urn-shaped or pitcher-like, ovoid or subcylindrical in shape with narrowed top or opening; e.g., corollas in some Ericaceae.
Utricle (a) A small, bladder-like, one-seeded, usually indehiscent fruit; (b) a small bladder.

V

Valvate Meeting along the margins only and not overlapping; contrasting with imbricate.
Valve A separable part of a capsule; the units or pieces into which a capsule splits or divides in dehiscing.
Variegated Irregularly colored in patches; blotched.
Variety A unit, category, or rank in classification below the level of species, sometimes treated as a subdivision of subspecies; group of plants with minor characters or differences separating them from other similar plants. The terms variety and subspecies are used inconsistently by taxonomists.
Vascular Pertaining to the conducting tissues (xylem and phloem).
Vascular bundle Thread-like fiber of xylem and phloem in a stem or other organ.
Vascular cambium The thin layer of delicate rapidly dividing cells that form wood internally and bark externally; also known as cambium.
Vegetative organ Root, stem, leaf, or other non-reproductive organ of a plant.
Vein Strand or bundle of vascular tissue.
Veinlet A little or ultimate vein.
Velamen A thin sheath or covering; e.g., on orchid roots.
Velamentous With a thin sheath or covering.
Velum Thin flap of tissue.
Velutinous Velvety with numerous erect hairs.
Venation The pattern or arrangement of veins.
Ventral Situated on or pertaining to the adaxial side (= side toward axis) of an organ; typically the upper or inner surface; contrasting with dorsal.
Ventricose (= Gibbous) Swollen or inflated on one side.
Vermiform Worm-like.
Vermillion Scarlet; brillant red.
Vernal Appearing in spring.
Vernation The arrangement of leaves, sepals, or petals in the unopened bud.
Verrucose Covered with wart-like protuberances.
Verruculose Diminutive of verrucose.
Versatile Attached near the middle and often capable of swinging about the attachment point, typically referring to attachment of an anther to a filament; contrasting with either basifixed or dorsifixed.
Verticil (= Whorl) A whorl of three or more members or parts attached at the same node of the supporting axis.
Verticillaster A false whorl composed of pairs of opposite cymes; e.g., in some Lamiaceae.
Verticillate (= Whorled) With three or more leaves or flowers attached at the same node; in a circle or ring.
Vespertine Opening in the evening; e.g., night-blooming Cereus (Cactaceae).
Vesicle A small cavity or bladder.
Vessel Water conducting structure of the xylem, formed from the walls of a series of dead xylem cells stacked end to end.
Vestigial Reduced to a trace, rudiment, or vestige; degenerate; referring to a once more fully developed structure.
Vesture, Vestiture Any covering on a surface making it other than glabrous; e.g., hairs, scales.
Vexillum The standard or banner in papilionaceous flowers.
Villosulous Diminutive of villous.
Villous, Villose With long, soft, spreading, or ascending, unmatted hairs; shaggy.
Vine A plant that climbs by tendrils or other means, or that trails or creeps along the ground.
Virgate Wand-shaped; slender, straight, and erect.
Viscid, Viscous Sticky; with sticky surfaces formed by secretions; glutinous.
Viscidulous Slightly viscid.
Viviparous Germinating or sprouting from seed or bud while attached to the parent plant.
Voluble Twining.

W

W West.
Waif A species that is only fleetingly established and probably not permanently naturalized.
Weed (a) A plant growing where it is not wanted; (b) a plant with the genetic endowment to inhabit and thrive in places of continual disturbance, most especially in areas that are repeatedly affected by the activities of humankind.
Whorl, Whorled (= Verticillate) With three or more leaves or flowers attached at the same node; in a circle or ring.
Wide (= Broad) Distance across a structure (equal to diameter if tubular).
Wing (a) A thin, membranous or flat extension or projection; (b) the two lateral separate petals in some Fabaceae and Polygalaceae.
Winter annual An annual plant (the total life cycle taking one year or less) vegetatively persistent through the winter and flowering and fruiting in the late winter or early spring.
Woolly (= Lanate) With long, soft, and more or less matted or entangled hairs; wool-like.
Wort An old word of Anglo-Saxon origin meaning the equivalent of herbaceous plant.

X

Xeric Characterized by or pertaining to conditions of scanty moisture supply; dry.

Xero- Greek prefix signifying dry.

Xerophyte A plant that can subsist with a small amount of moisture, such as a desert plant; contrasting with hydrophyte and mesophyte.

Xerophytic Dry-adapted; drought resistant; contrasting with mesophytic and hydrophytic.

Xylem Water conducting tissue.

Z

Zygomorphic Referring to a flower or other structure with only one plane of symmetry; divisible into halves in one plane only; bilaterally symmetrical; e.g., with the left half a mirror image of the right; contrasting with both actinomorphic and irregular.

Zygote Cell produced from fertilization or the union of two gametes; a fertilized egg.

LEAF CHARACTERS

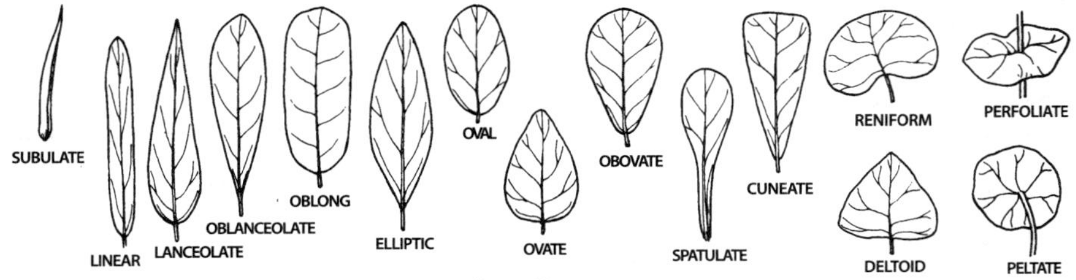

LEAF SHAPES

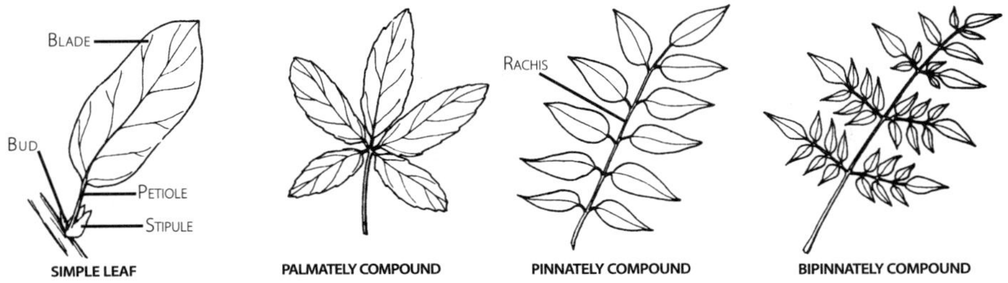

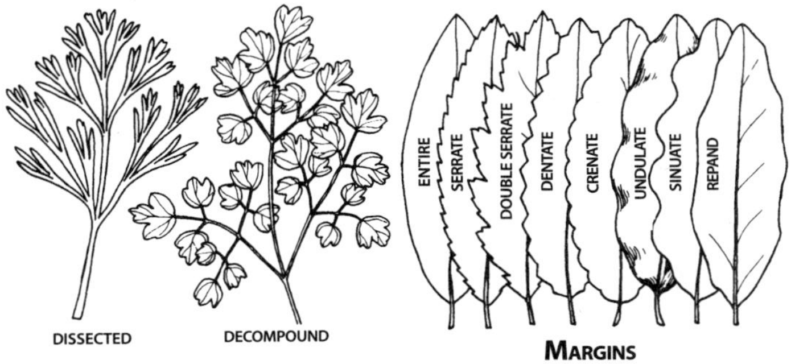

MARGINS

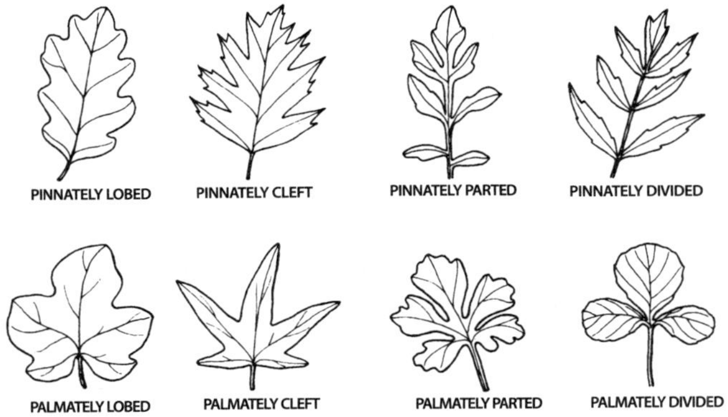

APEX OF BLADE

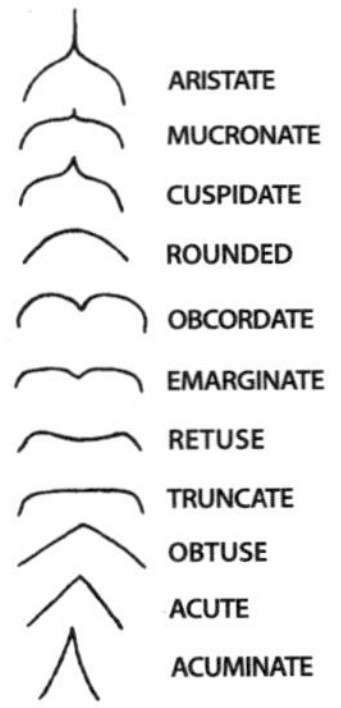

BASE OF BLADE

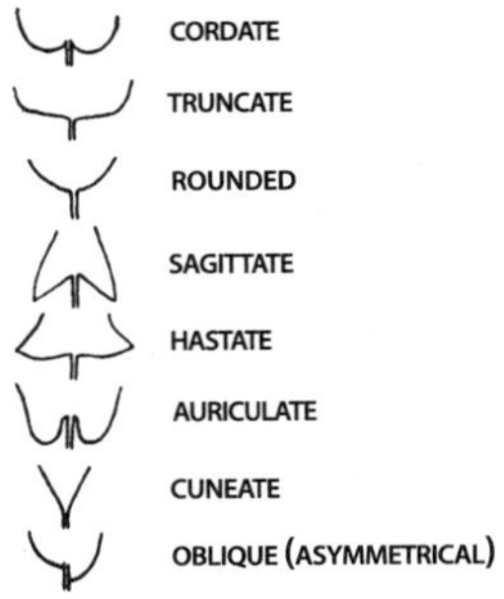

ADAPTED FROM MASON (1957), USED WITH PERMISSION OF UNIV. OF CALIFORNIA PRESS; ©1957

Leaf and Twig Structure and Arrangement

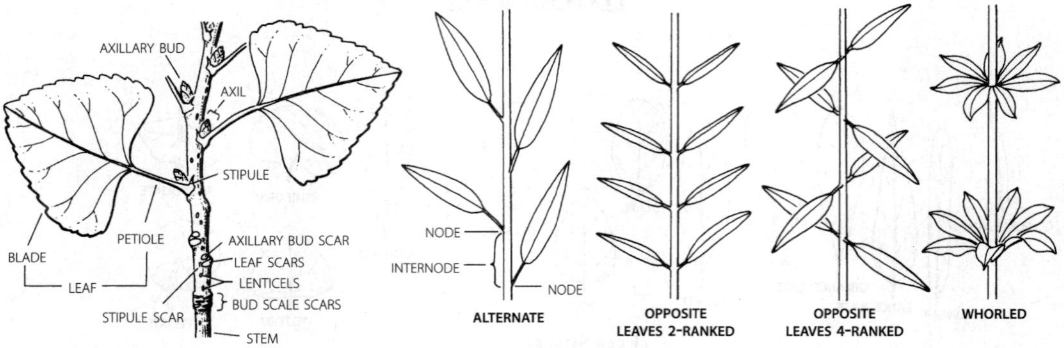

Flower Shape

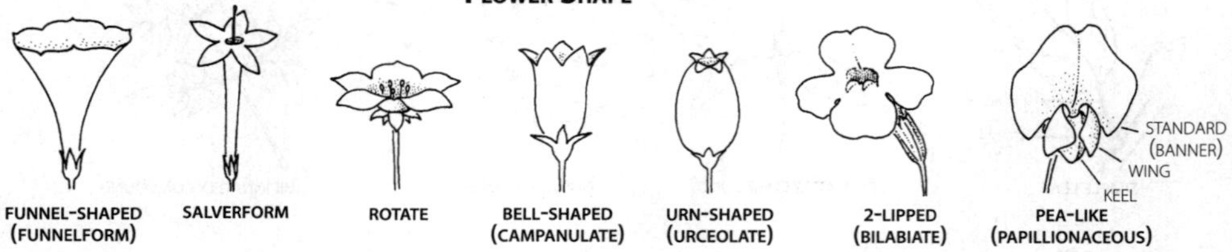

Flower Symmetry

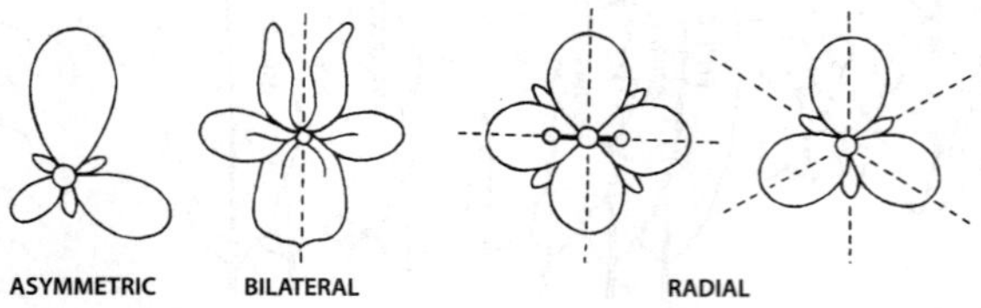

Flower Structure

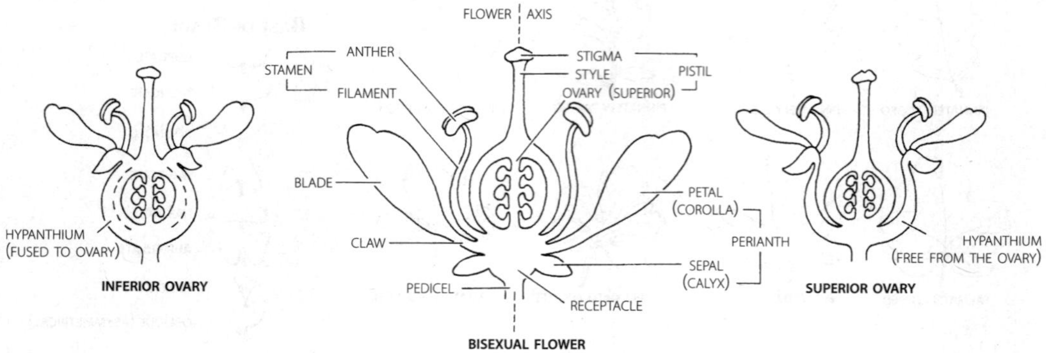

Adapted from Hickman (1993), used with permission of Univ. of California Press; ©1993

TYPES OF INFLORESCENCES

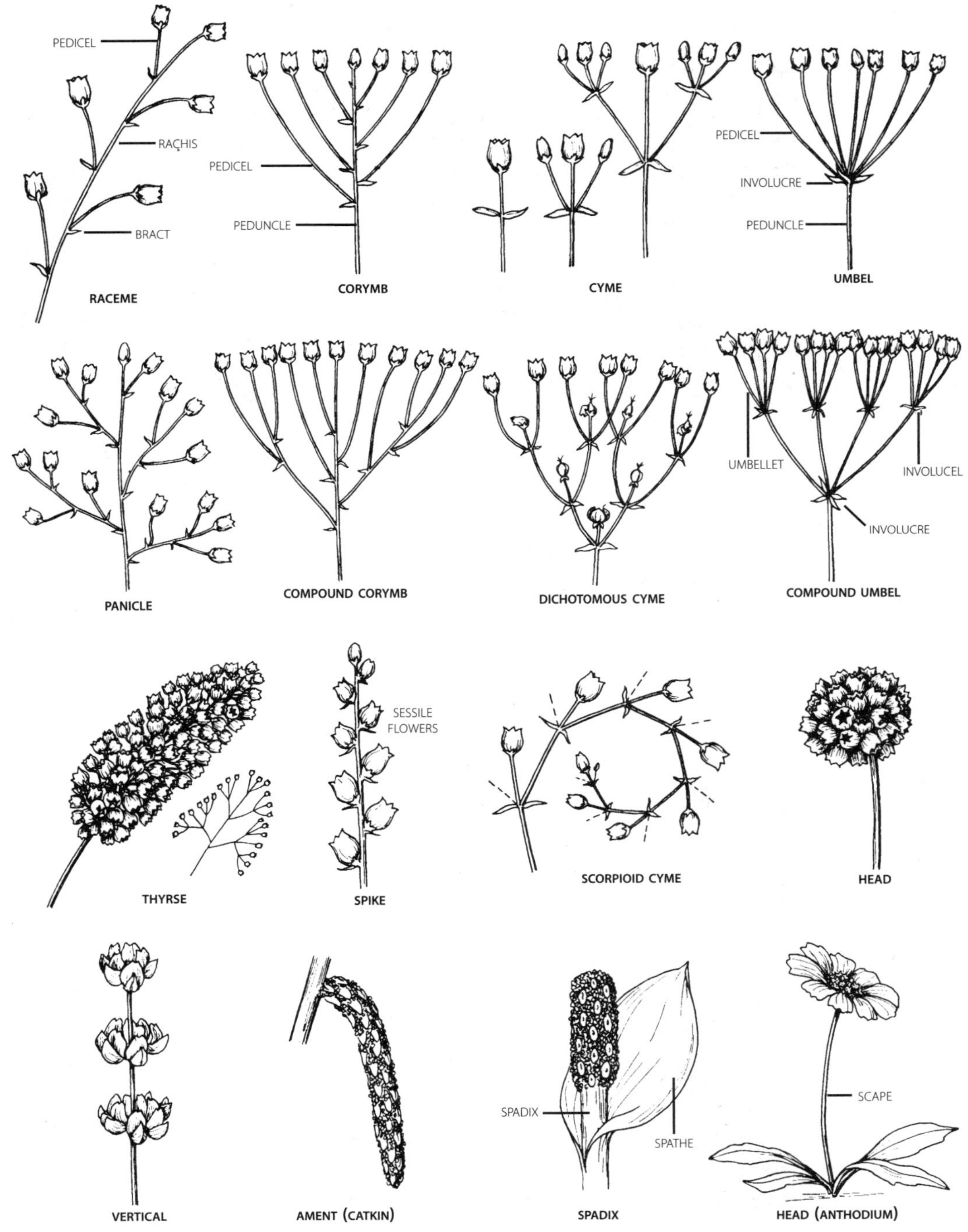

ADAPTED FROM MASON (1957), USED WITH PERMISSION OF UNIV. OF CALIFORNIA PRESS; ©1957

Types of Fruits and Root and Stem Variations

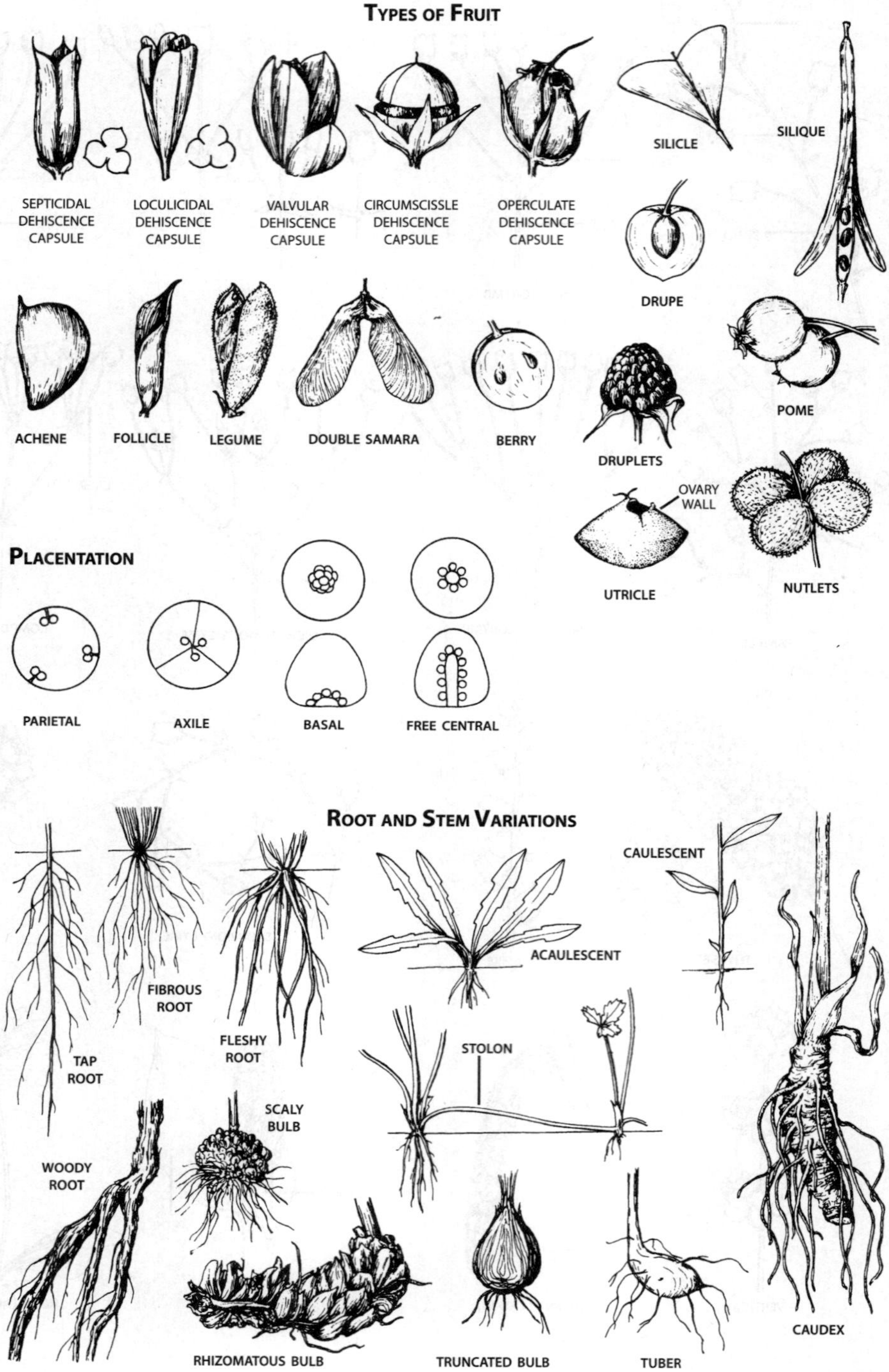

Adapted from Mason (1957), used with permission of Univ. of California Press; ©1957

Types of Pubescence

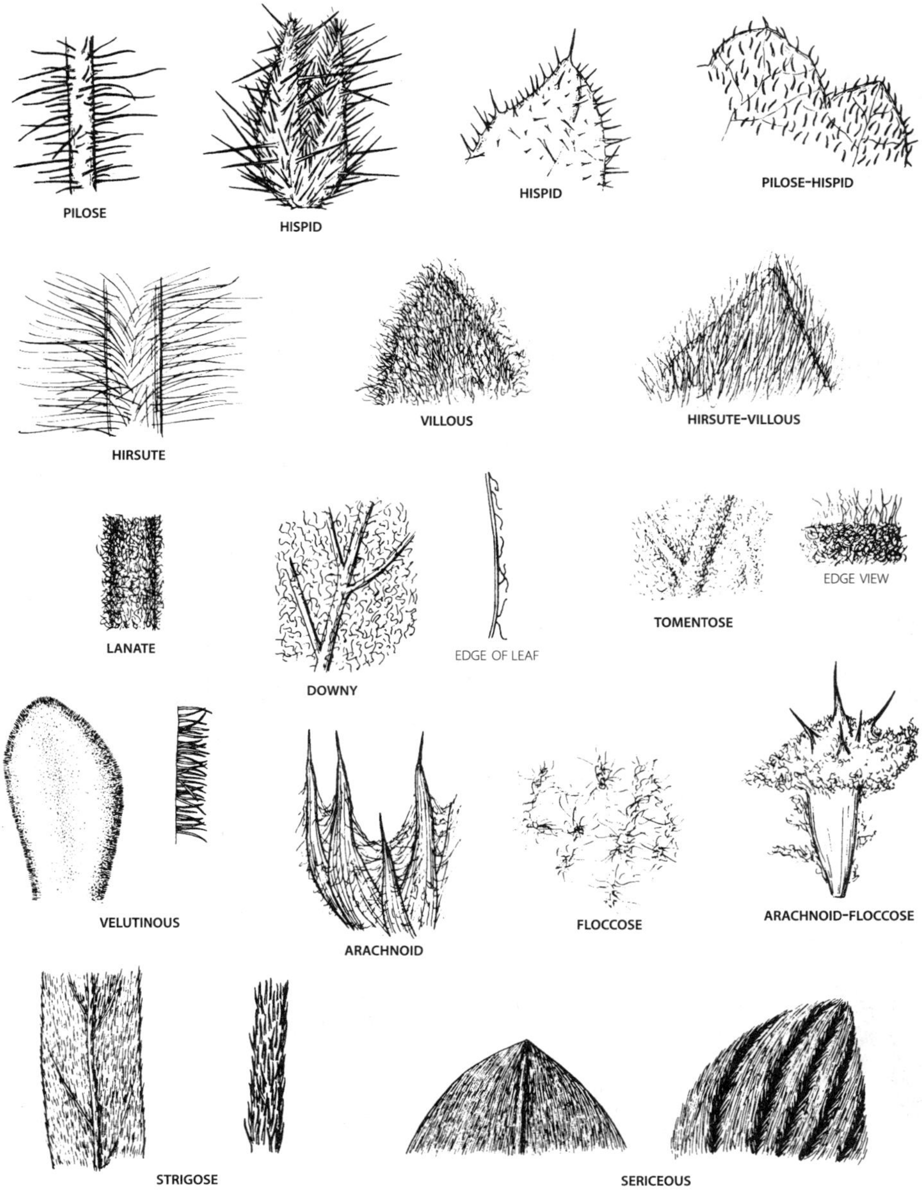

Adapted from Mason (1957), used with permission of Univ. of California Press; ©1957

Types of Hairs and Processes

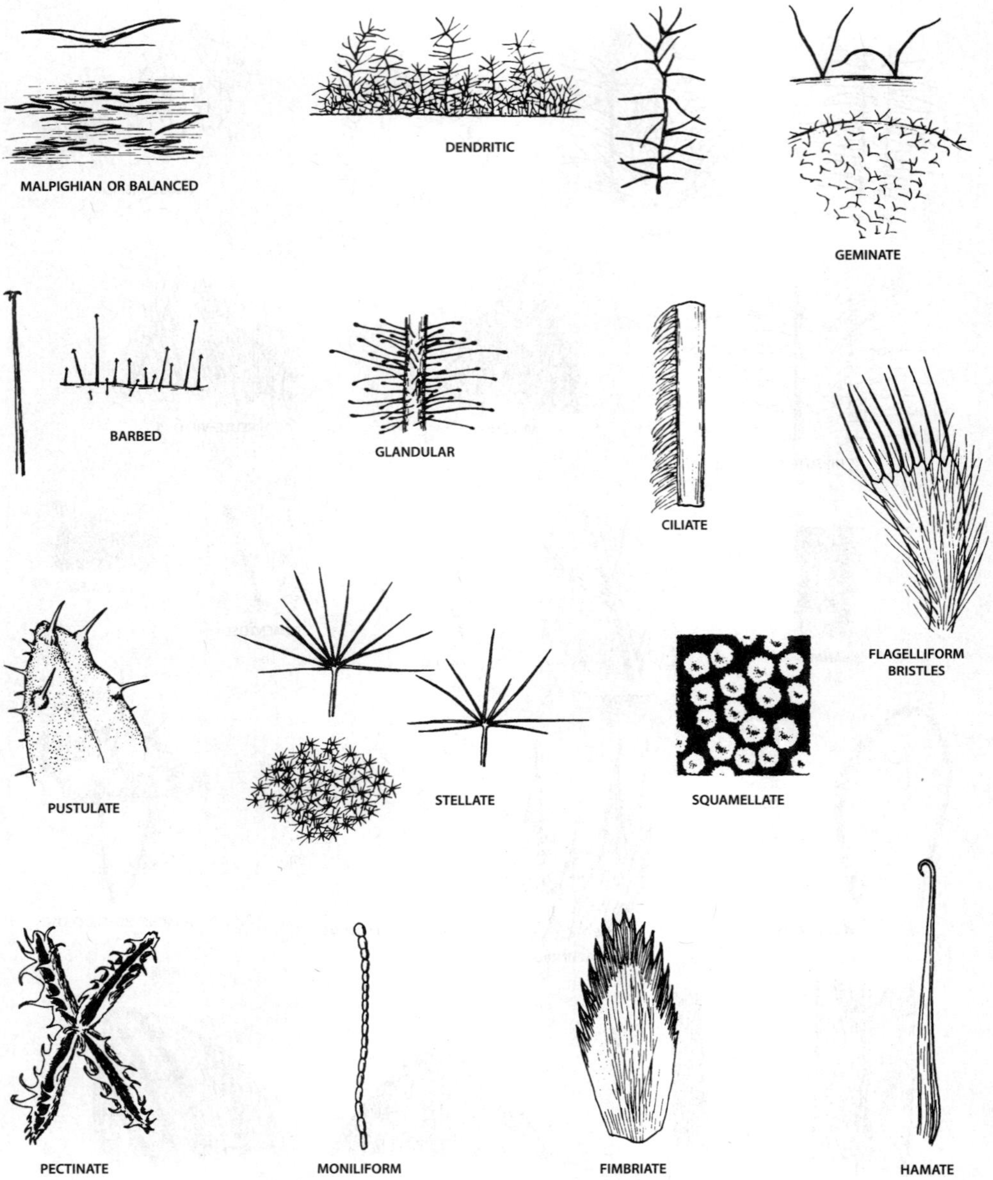

Adapted from Mason (1957), used with permission of Univ. of California Press; ©1957

General Characteristics of the Asteraceae (Compositae) Family [JEP]

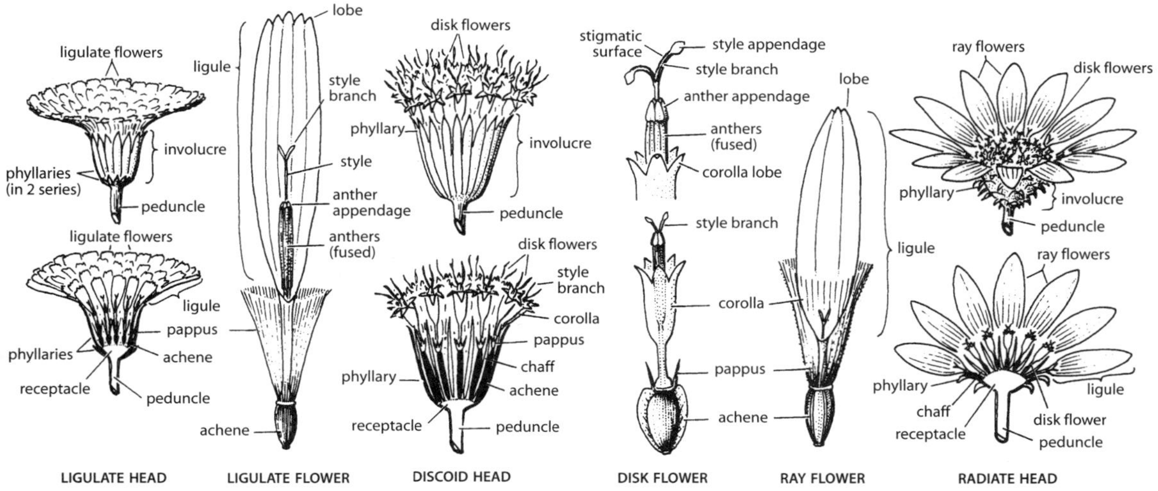

Asteraceae head types. Aa: Radiate head (e.g., *Gallardia*); **Ab:** Vertical section through radiate head; **a:** Receptacle; **b:** Disk flowers; **c:** Ray flowers; **d:** Phyllaries; **B:** Discoid head (e.g., *Eupatorium*); **C:** Ligulate head (e.g., *Cichorium*). [GAN]

ABROL, Y.P., E.E. CONN, and **J.R. STOKER.** 1966. Studies on the identification, biosynthesis and metabolism of a cyanogenic glucoside in *Nandina domestica* Thunb. Phytochemistry 5:1021–1027.

ACKERFIELD, J. and **J. WEN.** 2002. A morphometric analysis of *Hedera* L. (the ivy genus, Araliaceae) and its taxonomic implications. Adansonia 24:197–212.

ACKERFIELD, J. and **J. WEN.** 2003. Evolution of *Hedera* (the ivy genus, Araliaceae): Insights from chloroplast DNA data. Int. J. Pl. Sci. 164(4):593–602.

ACUÑA-CASTILLO, R., J.E. JIMÉNEZ, L.E. BEZERRA-SILVA, J. SAN GIL-LEÓN, and **M.A. BLANCO.** 2021. Note on the distribution of *Brasenia schreberi* (Cabombaceae) in the Neotropics, with a new country record for Costa Rica. Darwiniana Nueva Serie 9:364–374.

ADHIKARI, B., R. MILNE, T. PENNINGTON, T. SÄRKINEN, and **C.A. PENDRY.** 2015. Systematics and biogeography of *Berberis* s.l. inferred from nuclear ITS and chloroplast *ndhF* gene sequences. Taxon 64:39–48.

AELLEN, P. and **T. JUST.** 1943. Key and synopsis of the American species of the genus *Chenopodium* L. Amer. Midl. Naturalist 30: 47–76.

AFFOLTER, J.M. 1985. A monograph of the genus *Lilaeopsis* (Umbelliferae). Syst. Bot. Monogr. 6:1–140.

AGUILAR-ORTIGOZA, C.J. and **V. SOSA.** 2004. The evolution of toxic phenolic compounds in a group of Anacardiaceae genera. Taxon 53:357–364.

AGUILAR-ORTIGOZA, C.J., V. SOSA, and **M. AGUILAR-ORTIGOZA.** 2003. Toxic phenols in various Anacardiaceae species. Econ. Bot. 57: 354–364.

AHRENDT, L.W.A. 1961. *Berberis* and *Mahonia*: A taxonomic revision. J. Linn. Soc., Bot. 57:1–410.

AJILVSGI, G. 1984. Wildflowers of Texas. Shearer Publishing, Fredericksburg, TX.

AKHANI, H., G. EDWARDS, and **E.H. ROASLON.** 2007. Diversification of the Old World Salsoleae s.l. (Chenopodiaceae): Molecular and phylogenetic analysis of nuclear and chloroplast data sets and a revised classification. Int. J. Pl. Sci. 168:931–956.

AKHANI, H., W. GREUTER, and **E.H. ROALSON.** 2014. Notes on the typification and nomenclature of *Salsola* and *Kali* (Chenopodiaceae). Taxon 63:647–650.

ALEXANDER, P.J., M.D. WINDHAM, J.B. BECK, I.A. AL-SHEHBAZ, L. ALLPHIN, and **C.D. BAILEY.** 2013. Molecular phylogenetics and taxonomy of the genus *Boechera* and related genera (Brassicaceae: Boechereae). Syst. Bot. 38(1):192–209.

ALI, T., A. SCHMUKER, F. RUNGE, I. SOLOVYEVA, L. NIGRELLI, J. PAULE, A.K. BUCH, X. XIA, S. PLOCH, O. ORREN, and **V. KUMMER.** 2016. Morphology, phylogeny, and taxonomy of *Microthlaspi* (Brassicaceae: Coluteocarpeae) and related genera. Taxon 65(1):79–98.

ALLAN, G.J. and **J.M. PORTER.** 2000. Tribal delimitation and phylogenetic relationships of Loteae and Coronilleae (Faboideae: Fabaceae) with special reference to *Lotus*: Evidence from nuclear ribosomal ITS sequences. Amer. J. Bot. 87(12):1871–1881.

ALLAN, G.J., E.A. ZIMMER, W.L. WAGNER, and **D.D. SOKOLOFF.** 2003. Molecular phylogenetic analyses of tribe Loteae (Leguminosae): Implications for classification and biogeography. In: B.B. Klitgaard and A. Bruneau, eds. Advances in legume systematics. Part 10. Pp. 371–393. Royal Botanic Gardens, Kew, England, U.K.

ALLEN, C., P. LEWIS, and **D. LEWIS.** 2023. Vascular flora of Newton County, Texas, U.S.A. J. Bot. Res. Inst. Texas 17(1):369–387.

ALLEN, M.A. 2013. Synopsis of the genus *Stylisma* (Convolvulaceae) in Louisiana. J. Bot. Res. Inst. Texas 7:515–516.

ALLEN, M.A. 2023. *Stylisma*. In: Flora of North America Editorial Committee, eds. Fl. North Amer. 14:311–314. Oxford Univ. Press, New York and Oxford.

ALLEN, O.N. and **E.K. ALLEN.** 1981. The Leguminosae: A source book of characteristics, uses, and nodulation. Univ. of Wisconsin Press, Madison.

ALLRED, K.W. 2020. Flora neomexicana I: An annotated checklist, 3rd ed. Published by the author.

ALLRED, K.W., E.M. JERCINOVIC, and **R.D. IVEY.** 2020. Flora neomexicana III: An illustrated identification manual, Part 2: Dicotyledonous plants, 2nd ed. Allred and Ivey.

ALONSO, M.A. and **M.B. CRESPO.** 2008. Taxonomic and nomenclatural notes on South American taxa of *Sarcocornia* (Chenopodiaceae). Ann. Bot. Fenn. 45:241–254.

ALQAHTANI, A., J.-L. CHO, K.-H. WONG, K.-M. LI, V. RAZMOVSKI-NAUMOVSKI, and **G.Q. LI.** 2017. Differentiation of three *Centella* species in Australia as inferred from morphological characteristics, ISSR molecular fingerprinting, and phytochemical composition. Front. Pl. Sci. https://doi.org/10.3389/fpls.2017.01980.

AL-SHEHBAZ, I.A. 1985. The genera of Brassiceae (Cruciferae; Brassicaceae) in the southeastern United States. J. Arnold Arbor. 66(3):279–351.

AL-SHEHBAZ, I.A. 1986. The genera of Lepidieae (Cruciferae; Brassicaceae) in the southeastern United States. J. Arnold Arbor. 67(3):265–311.

AL-SHEHBAZ, I.A. 1991. The genera of Boraginaceae in the southeastern United States. J. Arnold Arbor. Supp. Ser. 1:1–169.

AL-SHEHBAZ, I.A. 2003. Transfer of most North American species of *Arabis* to *Boechera* (Brassicaceae). Novon 13(4):381–391.

AL-SHEHBAZ, I.A. 2010a. *Arabis mexicana* belongs to *Planodes* (Brassicaceae). Harvard Pap. Bot. 15(1):137–138.

AL-SHEHBAZ, I.A. 2010b. Brassicaceae. In: Flora of North America Editorial Committee, eds. Fl. North Amer. 7:224–746. Oxford Univ. Press, New York and Oxford.

AL-SHEHBAZ, I.A. 2010c. *Arabidopsis*. In: Flora of North America Editorial Committee, eds. Fl. North Amer. 7:447–450. Oxford Univ. Press, New York and Oxford.

AL-SHEHBAZ, I.A. 2010d. *Barbarea*. In: Flora of North America Editorial Committee, eds. Fl. North Amer. 7:460–463. Oxford Univ. Press, New York and Oxford.

AL-SHEHBAZ, I.A. 2010e. *Capsella*. In: Flora of North America Editorial Committee, eds. Fl. North Amer. 7:453–454. Oxford Univ. Press, New York and Oxford.

AL-SHEHBAZ, I.A. 2010f. *Chorispora*. In: Flora of North America Editorial Committee, eds. Fl. North Amer. 7:510–511. Oxford Univ. Press, New York and Oxford.

AL-SHEHBAZ, I.A. 2010g. *Erysimum*. In: Flora of North America Editorial Committee, eds. Fl. North Amer. 7:534–545. Oxford Univ. Press, New York and Oxford.

Al-Shehbaz, I.A. 2010h. *Iodanthus*. In: Flora of North America Editorial Committee, eds. Fl. North Amer. 7:484–485. Oxford Univ. Press, New York and Oxford.

Al-Shehbaz, I.A. 2010i. *Matthiola*. In: Flora of North America Editorial Committee, eds. Fl. North Amer. 7:253–255. Oxford Univ. Press, New York and Oxford.

Al-Shehbaz, I.A. 2010j. *Microthlaspi*. In: Flora of North America Editorial Committee, eds. Fl. North Amer. 7:599–600. Oxford Univ. Press, New York and Oxford.

Al-Shehbaz, I.A. 2010k. *Myagrum*. In: Flora of North America Editorial Committee, eds. Fl. North Amer. 7:568–569. Oxford Univ. Press, New York and Oxford.

Al-Shehbaz, I.A. 2010l. *Nasturtium*. In: Flora of North America Editorial Committee, eds. Fl. North Amer. 7:484–492. Oxford Univ. Press, New York and Oxford.

Al-Shehbaz, I.A. 2010m. *Planodes*. In: Flora of North America Editorial Committee, eds. Fl. North Amer. 7:492–493. Oxford Univ. Press, New York and Oxford.

Al-Shehbaz, I.A. 2010n. *Rorippa*. In: Flora of North America Editorial Committee, eds. Fl. North Amer. 7:493–506. Oxford Univ. Press, New York and Oxford.

Al-Shehbaz, I.A. 2010o. *Selenia*. In: Flora of North America Editorial Committee, eds. Fl. North Amer. 7:506–509. Oxford Univ. Press, New York and Oxford.

Al-Shehbaz, I.A. 2010p. *Sisymbrium*. In: Flora of North America Editorial Committee, eds. Fl. North Amer. 7:667–671. Oxford Univ. Press, New York and Oxford.

Al-Shehbaz, I.A. 2010q. *Streptanthus*. In: Flora of North America Editorial Committee, eds. Fl. North Amer. 7:700–723. Oxford Univ. Press, New York and Oxford.

Al-Shehbaz, I.A. 2010r. *Thlaspi*. In: Flora of North America Editorial Committee, eds. Fl. North Amer. 7:745–746. Oxford Univ. Press, New York and Oxford.

Al-Shehbaz, I.A. 2012. A generic and tribal synopsis of the Brassicaceae (Cruciferae). Taxon 61(5):931–954.

Al-Shehbaz, I.A. 2014. A synopsis of the genus *Noccaea* (Coluteocarpeae, Brassicaceae). Harvard Pap. Bot. 19(1):25–51.

Al-Shehbaz, I.A. and **V. Bates.** 1987. *Armoracia lacustris* (Brassicaceae), the correct name for the North American Lake Cress. J. Arnold Arbor. 68:357–359.

Al-Shehbaz, I.A. and **J.B. Beck.** 2010. *Leavenworthia*. In: Flora of North America Editorial Committee, eds. Fl. North Amer. 7:485–489. Oxford Univ. Press, New York and Oxford.

Al-Shehbaz, I.A. and **M.A. Beilstein.** 2010. *Camelina*. In: Flora of North America Editorial Committee, eds. Fl. North Amer. 7:451–453. Oxford Univ. Press, New York and Oxford.

Al-Shehbaz, I.A., M.A. Beilstein, and **E.A. Kellogg.** 2006. Systematics and phylogeny of the Brassicaceae (Cruciferae): An overview. Pl. Syst. Evol. 259:89–120.

Al-Shehbaz, I.A. and **J.F. Gaskin.** 2010. *Lepidium*. In: Flora of North America Editorial Committee, eds. Fl. North Amer. 7:570–595. Oxford Univ. Press, New York and Oxford.

Al-Shehbaz, I.A., K. Marhold, and **J. Lihová.** 2010a. *Cardamine*. In: Flora of North America Editorial Committee, eds. Fl. North Amer. 7:464–484. Oxford Univ. Press, New York and Oxford.

Al-Shehbaz, I.A. and **M.S. Mayer.** 2008. New or noteworthy *Streptanthus* (Brassicaceae) for the Flora of North America. Novon 18(3):279–282.

Al-Shehbaz, I.A, K. Mummenhoff, and **O. Appel.** 2002. *Cardaria*, *Coronopus*, and *Stroganowia* are united with *Lepidium* (Brassicaceae). Novon 12(1):5–11.

Al-Shehbaz, I.A. and **S.L. O'Kane.** 2002. *Lesquerella* is united with *Physaria* (Brassicaceae). Novon 12(3):319–329.

Al-Shehbaz, I.A. and **R.A. Price.** 1998. Delimitation of the genus *Nasturtium* (Brassicaceae). Novon 8:124–126.

Al-Shehbaz, I.A. and **M.D. Windham.** 2010. *Boechera*. In: Flora of North America Editorial Committee, eds. Fl. North Amer. 7:348–412. Oxford Univ. Press, New York and Oxford.

Al-Shehbaz, I.A., M.D. Windham, and **R. Elven.** 2010b. *Draba*. In: Flora of North America Editorial Committee, eds. Fl. North Amer. 7:269–347. Oxford Univ. Press, New York and Oxford.

Alston, R.E. and **R.E. Schultes.** 1951. Studies of early specimens and reports of *Ilex vomitoria*. Rhodora 53:274–280.

Alston, R.E. and **B.L. Turner.** 1962. New techniques in analysis of complex natural hybridization. Proc. Natl. Acad. Sci. U.S.A. 48: 130–137.

Alston, R.E. and **B.L. Turner.** 1963. Natural hybridization among four species of *Baptisia* (Leguminosae). Amer. J. Bot. 50:159–173.

Anderberg, A.A. 1991. Taxonomy and phylogeny of the tribe Gnaphalieae (Asteraceae). Opera Bot. 104:1–195.

Anderberg, A.A., B.G. Baldwin, R.G. Bayer, and **21 others.** 2006. Asteraceae. In: J.W. Kadereit and C. Jeffrey, eds. The families and genera of vascular plants, Volume VIII. Flowering plants. Eudicots: Asterales. Pp. 61–588. Springer, Berlin.

Anderberg, A.A., C. Rydin, and **M. Kallersjo.** 2002. Phylogenetic relationships in the order Ericales s. l.: Analyses of molecular data from five genes from the plastid and mitochondrial genomes. Amer. J. Bot. 89(4):677–687.

Anderberg, A.A. and **X. Zhang.** 2002. Phylogenetic relationships of Cyrillaceae and Clethraceae (Ericales) with special emphasis on the genus *Purdiaea* Planch. Organisms Diversity Evol. 2: 127–136.

Anderson, E.F. 2001. The cactus family. Timber Press, Portland, OR.

Anderson, L.C. 1970. Studies on *Bigelowia* (Astereae, Compositae). I. Morphology and taxonomy. Sida 3:451–465.

Anderson, L.C. 1977. Studies on *Bigelowia* (Asteraceae). III. Cytotaxonomy and biogeography. Syst. Bot. 2:209–218.

Anderson, L.C. 2006. *Arnoglossum*. In: Flora of North America Editorial Committee, eds. Fl. North Amer. 20:622–625. Oxford Univ. Press, New York and Oxford.

Anderson, T.E. 1995. The poison ivy, oak & sumac book: A short natural history and cautionary account. Acton Circle Publishing Co., Ukiah, CA.

Andersson, K. 1997. Poison oak! http://wildfirenews.com/fire/articles/oak.html. Wildland Firefighter Magazine.

Andres, T. 1987. *Cucurbita fraterna*, the closest wild relative and progenitor of *C. pepo*. Cucurbit Genet. Coop. Rep.10:69–71.

Andrews, J. 1986. The Texas bluebonnet. Univ. of Texas Press, Austin.

Aneja, K.R., S.A. Khan, and **S. Kaushal.** 2000. Management of horse purslane (*Trianthema portulacastrum* L.) with *Gibbago trianthemae* Simmons in India. In: R. Neal, ed. Proceedings of the X International Symposium on Biological Control of Weeds 4–14 July 1999. Montana State Univ., Bozeman, MT, Pp. 27–33.

Aneyefortexas. 2011. The beautiful Texas madrone. Posted 15 Oct 2011. https://aneyefortexas.wordpress.com/2011/10/15/the-beautiful-texas-madrone/. Accessed 25 Mar 2025.

Angerstein, M.B. and **D.E. Lemke.** 1994. First records of the aquatic weed *Hygrophila polysperma* (Acanthaceae) from Texas. Sida 16:365–371.

Antonelli, A. 2007. Higher level phylogeny and evolutionary trends in Campanulaceae subfam. Lobelioideae: Molecular signal overshadows morphology. Molec. Phylogen. Evol. 46:1–18.

Aoki, S., K. Uehara, M. Imafuku, M. Hasebe, and **M. Ito.** 2004. Phylogeny and divergence of basal angiosperms inferred from APETALA3- and PISTILLATA-like MADS-box genes. J. Pl. Res. 117:229–244.

APG. 1998. Angiosperm Phylogeny Group: K. Bremer, M.W. Chase, and P. Stevens, eds. An ordinal classification for the families of flowering plants. Ann. Missouri Bot. Gard. 85:531–553.

APG II. 2003. Angiosperm Phylogeny Group: B. Bremer, K. Bremer, M.W. Chase, J.L. Reveal, D.S. Soltis, P.S. Soltis, and P.F. Stevens, et al. An update of the Angiosperm Phylogeny Group classification for the orders and families of flowering plants. Bot. J. Linn. Soc. 141:399–436.

APG IV. 2016. Angiosperm Phylogeny Group: M.W Chase, M.J.M. Christenhusz, M.F. Fay, J.W. Byng, W.S. Judd, D.E. Soltis, D.J. Mabberley, A.N. Sennikov, P.S. Soltis, P.F. Stevens. An update of the Angiosperm Phylogeny Group classification for the orders and families of flowering plants: APG IV. Bot. J. Linn. Soc. 181: 1–20.

APHIS (Animal and Plant Health Inspection Service). 2019. Weed Risk Assessment for *Dolichandra unguis-cati* (Bignoniaceae) —Cat's claw. https://www.aphis.usda.gov/sites/default/files/dolichandra-unguis-cati.pdf. Accessed 24 Dec 2024.

Aplaca, J. 2010. The non-native flora of Texas. Thesis, Dept. of Biology, Texas State Univ., San Marcos. https://digital.library.txstate.edu/handle/10877/3149.

Aplaca, J. 2012. Non-native species new to Texas with comments on other species. Phytoneuron 2012-95:1–6.

Aplet, G.H., R.D. Laven, M.B. Falkner, and **R.B. Shaw.** 1994. Population and site characteristics of a recently discovered population of *Croton alabamensis* (Euphorbiaceae). Sida 16:37–51.

Applequist, W.L. and **P.P. Lowery.** 2024. *Osmorhiza*. In: Flora of North America Editorial Committee, eds. Fl. North Amer. 13:453–458. Oxford Univ. Press, New York and Oxford.

Applequist, W.L. and **R.S. Wallace.** 2001. Phylogeny of the portulacaceous cohort based on *ndhF* sequence data. Syst. Bot. 26:406–419.

Arias-Agudelo, L.M., F. González, J.P. Isaza, J.F. Alzate, and **N. Pabón-Mora.** 2019. Plastome reduction and gene content in New World *Pilostyles* (Apodanthaceae) unveils high similarities to African and Australian congeners. Molec. Phylogen. Evol. 135: 193–202.

Arriagada, J.E. and **N.G. Miller.** 1997. The genera of Anthemideae (Compositae; Asteraceae) in the southeastern United States. Harvard Pap. Bot. 2:1–46.

Arrington, J.M. and **K. Kubitzki.** 2003. Cistaceae. In: K. Kubitzki and C. Bayer, eds. The families and genera of vascular plants, Vol. V. Pp. 62–70. Springer-Verlag, Berlin.

Ashburner, K. and **H.A. McAllister.** 2013. The genus *Betula*: A taxonomic revision of birches. Royal Botanic Gardens, Kew, England, U.K.

Austin, D. 2002. Sundews: Discovering Florida's ethnobotany. Number 6 in the series. Palmetto 21:12–13.

Austin, D.F. 1975. Bird flowers in the eastern United States. Florida Sci. 38:1–12.

Austin, D.F. 1976. Varieties of *Ipomoea trichocarpa* (Convolvulaceae). Sida 6:216–220.

Austin, D.F. 1978. The *Ipomoea batatas* complex—I. Taxonomy. Bull. Torrey Bot. Club. 105:114–129.

Austin, D.F. 1979a. Plant without beginning or end. Fairchild Trop. Gard. Bull. 34:17–19.

Austin, D.F. 1979b. Comments on *Cuscuta*—For collectors and curators. Bull. Torrey Bot. Club 106:227–228.

Austin, D.F. 1980. Studies of the Florida Convolvulaceae—III. *Cuscuta*. Florida Sci. 43:294–302.

Austin, D.F. 1984. Studies of the Florida Convolvulaceae—IV *Ipomoea*. Florida Sci. 47:81–87.

Austin, D.F. 1986a. Convolvulaceae. In: Great Plains Flora Association. Flora of the Great Plains. Pp. 652–661. Univ. Press of Kansas, Lawrence.

Austin, D.F. 1986b. Cuscutaceae. In: Great Plains Flora Association Flora of the Great Plains. Pp. 661–666. Univ. Press of Kansas, Lawrence.

Austin, D.F. 1988. Taxonomy, evolution and genetic diversity of sweet potatoes and related wild species. In: Exploration, maintenance and utilization of sweet potato genetic resources, Proc. Planning Conf. Pp. 27–60. Centro Internacional de la Papa, Lima, Peru.

Austin, D.F. 1990. Annotated checklist of New Mexican Convolvulaceae. Sida 14:273–286.

Austin, D.F. 1992. Rare Convolvulaceae in the southwestern United States. Ann. Missouri Bot. Gard. 79:8–16.

Austin, D.F. 1998. Parallel and convergent evolution in the Convolvulaceae. In: P. Mathews, ed. Biodiversity and taxonomy of flowering plants, Calicut Univ., Calicut, India.

Austin, D.F. 2000. A revision of *Cressa* L. (Convolvulaceae). Bot. J. Linn. Soc. 133:27–39.

Austin, D.F. 2023. Convolvulaceae. In: Flora of North America Editorial Committee, eds. Fl. North Amer. 14:269–359. Oxford Univ. Press, New York and Oxford.

Austin, D.F. 2023a. *Convolvulus*. In: Flora of North America Editorial Committee, eds. Fl. North Amer. 14:336–338. Oxford Univ. Press, New York and Oxford.

Austin, D.F. 2023b. *Dichondra*. In: Flora of North America Editorial Committee, eds. Fl. North Amer. 14:306–308. Oxford Univ. Press, New York and Oxford.

Austin, D.F. 2023c. *Evolvulus*. In: Flora of North America Editorial Committee, eds. Fl. North Amer. 14:303–306. Oxford Univ. Press, New York and Oxford.

Austin, D.F. 2023d. *Ipomoea*. In: Flora of North America Editorial Committee, eds. Fl. North Amer. 14:342–359. Oxford Univ. Press, New York and Oxford.

Austin, D.F. 2023e. *Merremia*. In: Flora of North America Editorial Committee, eds. Fl. North Amer. 14:339–341. Oxford Univ. Press, New York and Oxford.

Austin, D.F. and **R.S. Bianchini.** 1998. Additions and corrections in American *Ipomoea* (Convolvulaceae). Taxon 47:833–838.

Austin, D.F., S. Demissew, and **J. young.** 1998. Studies of the Florida Convolvulaceae—VI. *Dichondra*. Florida Sci. 61(3/4):195–202.

Austin, D.F., G.M. Diggs, Jr., and **B.L. Lipscomb.** 1997. *Calystegia* (Convolvulaceae) in Texas. Sida 17:837–840.

Austin, D.F. and **Z. Huáman.** 1996. A synopsis of *Ipomoea* (Convolvulaceae) in the Americas. Taxon 45:3–38.

Ayala, F. and **J.W. O'Leary.** 1995. Growth and physiology of *Salicornia bigelovii* at suboptima salinity. Int. J. Pl. Sci. 156: 197–205.

Babcock, E.B. 1947. The genus *Crepis*. I, II. Univ. Calif. Publ. Bot. 21: 1–197; 22:199–1030.

Babcock, E.B. and **G.L. Stebbins.** 1938. The American species of *Crepis*. Publ. Carnegie Inst. Washington 504:1–199.

Babcock, E.B. and **G.L. Stebbins.** 1939. The genus *Youngia*. Publ. Carnegie Inst. Wash. 484:1–106.

Bacigalupi, R. 1931. A monograph of the genus *Perezia* section *Acourtia*, with a provisional key to the section *Euperezia*. Contr. Gray Herb. 97:1–81.

Badillo, V.M. 1971. Monografía de la família Caricaceae. Asociacion de Prefesores. Maracay, Venezuela.

Badillo, V.M. 2000. *Carica* L. vs. *Vasconcella* St.-Hil. (Caricaceae) con la rehabilitacion de este ultimo. Ernstia 10:74–79.

Baer, H. 1979. The poisonous Anacardiaceae. In: A.D. Kinghorn, ed. Toxic plants. Pp. 161–170. Columbia Univ. Press, New York.

Bailey, C.D., M.A. Koch, M. Mayer, K. Mummenhoff, S.L. O'Kane, Jr., S.I. Warwick, M.D. Windham, and **I.A. Al-Shehbaz.** 2006. Toward a global phylogeny of the Brassicaceae. Molec. Biol. Evol. 23(11):2142–2160.

Bailey, L.H. 1900. Cylopedia of American horticulture, 4 volumes. The Macmillan Co. London.

Bailey, L.H. 1929. The domesticated cucurbitas. Gentes Herbarum 2:63–115.

Bailey, L.H. 1930b. Three discussions in Cucurbitaceae. Gentes Herbarum 2:175–186.

Bailey, L.H. 1938. The garden of pinks. The MacMillan Co., New York.

Bailey, L.H. and **E.Z. Bailey.** 1976. Hortus third, a concise dictionary of plants cultivated in the United States and Canada. Revised and expanded by the staff of the Liberty Hyde Bailey Hortorium, Cornell Univ. MacMillan Publishing Co., New York.

Baker, A. 1993. Khat—Out of the bag: Deployment leads to questions about Somalian drug. Fort Worth Star-Telegram. 12 Jan 1993, Pp. 1,8.

Baker, M. and **R. Routson.** 2003. Progress on the taxonomy of the claret-cup cacti (*Echinocereus*, section Triglochidiatus) of the United States. Cact. Succ. J. (Los Angels) 75:217–223.

Ball, P.W. 2003a. *Salicornia*. In: Flora of North America Editorial Committee, eds. Fl. North Amer. 4:302. Oxford Univ. Press, New York and Oxford.

Ball, P.W. 2003b. *Sarcocornia*. In: Flora of North America Editorial Committee, eds. Fl. North Amer. 4:384–387. Oxford Univ. Press, New York and Oxford.

Ball, P.W. 2023. *Securigera*. In: Flora of North America Editorial Committee, eds. Fl. of North Amer. 11:501–502. Oxford Univ. Press, New York and Oxford.

Ball, P.W. 2024a. *Anethum*. In: Flora of North America Editorial Committee, eds. Fl. North Amer. 13:214–215. Oxford Univ. Press, New York and Oxford.

Ball, P.W. 2024b. *Coriandrum*. In: Flora of North America Editorial Committee, eds. Fl. North Amer. 13:264–265. Oxford Univ. Press, New York and Oxford.

Ball, P.W. 2024c. *Cyclospermum*. In: Flora of North America Editorial Committee, eds. Fl. North Amer. 13:263. Oxford Univ. Press, New York and Oxford.

Ball, P.W. 2024d. *Foeniculum*. In: Flora of North America Editorial Committee, eds. Fl. North Amer. 13:321–322. Oxford Univ. Press, New York and Oxford.

Ball, P.W. 2024e. *Sanicula*. In: Flora of North America Editorial Committee, eds. Fl. North Amer. 13:485–494. Oxford Univ. Press, New York and Oxford.

Ball, P.W. 2024f. *Scandix*. In: Flora of North America Editorial Committee, eds. Fl. North Amer. 13:495. Oxford Univ. Press, New York and Oxford.

Ball, P.W. 2024g. *Torilis*. In: Flora of North America Editorial Committee, eds. Fl. North Amer. 13:513–515. Oxford Univ. Press, New York and Oxford.

Ball, P.W. and **V.H. Heywood.** 1964. A revision of the genus *Petrorhagia*. Bull. Brit. Mus. (Nat. Hist.), Bot. 3:121–172.

Ballard, R. 1986. *Bidens pilosa* complex (Asteraceae) in North and Central America. Amer. J. Bot. 73:1452–1465.

Ballenger, J.A., E. Dickson, M. Meyer, and **J.J. Doyle.** 1993. DNA sequence data and the phylogeny of the Leguminosae. Amer. J. Bot. (Suppl.) 80(6):130–131.

Ballenger, J.A. and **M.A. Vincent.** 2023. *Cercis*. In: Flora of North America Editorial Committee, eds. Fl. of North Amer. 11:10–14. Oxford Univ. Press, New York and Oxford.

Baltzer, E.A. 1944. A monographic study of the genus *Palafoxia* and its immediate allies. Ann. Missouri Bot. Gard. 31:249–277.

Banasiak, L., A. Wojewódzka, J. Baczyński, J.P. Reduron, M. Piwczyński, R. Kurzyna-Młynik, R. Gutaker, A. Czarnocka-Cieciura, S. Kosmala-Grzechnik, and **K. Spalik.** 2016. Phylogeny of Apiaceae subtribe Daucinae and the taxonomic delineation of its genera. Taxon 65:563–585.

Barber, S.C. and **J.R. Estes.** 1978. Comparative pollination ecology of *Pyrrhopappus geiseri* and *P. carolinianus*. Amer. J. Bot. 65: 562–566.

Barkley, F.A. 1937. A monographic study of *Rhus* and its immediate allies in North and Central America including the West Indies. Ann. Missouri Bot. Gard. 24:265–498.

Barkley, F.A. 1961. Anacardiaceae. In: C.L. Lundell, ed. Flora of Texas 3:89–108. Texas Research Foundation, Renner.

Barkley, T.M. (Theodore) 1934–2004. Personal communication. Specialist in Asteraceae, especially *Senecio* and *Packera*, and flora of the Great Plains; spent career at Kansas State Univ., Manhattan; Flora of North America Editorial Committee and Research Associate at the Botanical Research Institute of Texas, Fort Worth.

Barkley, T.M. 1957. Generic key to the sumac family (Anacardiaceae). Lloydia 20:255–265.

Barkley, T.M. 1962. A revision of *Senecio aureus* Linn. and allied species. Trans. Kansas Acad. Sci. 65:318–408.

Barkley, T.M. 1978. *Senecio*. N. Amer. Fl. II. 10:50–139.

Barkley, T.M. 1980. Taxonomic notes on *Senecio tomentosus* and its allies (Asteraceae). Brittonia 32:291–308.

Barkley, T.M. 1981. *Senecio* and *Erechtites* (Compositae) in the North American flora: Supplementary notes. Brittonia 33:523–527.

Barkley, T.M. 1985a. Generic boundaries in the Senecioneae. Taxon 34:17–21.

Barkley, T.M. 1985b. Infrageneric groups in *Senecio*, s.l., and *Cacalia*, s.l. (Asteraceae: Senecioneae) in Mexico and Central America. Brittonia 37:211–218.

Barkley, T.M. 1986. Asteraceae. In: Great Plains Flora Association. Flora of the Great Plains. Pp. 838–1021. Univ. Press of Kansas, Lawrence.

Barkley, T.M. 1988. Variation among aureoid senecios of North America: A geohistorical interpretation. Bot. Rev. (Lancaster) 54:82–106.

Barkley, T.M. 2006a. *Emilia*. In: Flora of North America Editorial Committee, eds. Fl. North Amer. 20:605–607. Oxford Univ. Press, New York and Oxford.

Barkley, T.M. 2006b. *Erechtites*. In: Flora of North America Editorial Committee, eds. Fl. North Amer. 20:602–604. Oxford Univ. Press, New York and Oxford.

Barkley, T.M. 2006c. *Pseudogynoxys*. In: Flora of North America Editorial Committee, eds. Fl. North Amer. 20:608. Oxford Univ. Press, New York and Oxford.

Barkley, T.M. 2006d. *Senecio*. In: Flora of North America Editorial Committee, eds. Fl. North Amer. 20:544–570. Oxford Univ. Press, New York and Oxford.

Barkley, T.M., L. Brouillet, and **J.L. Strother.** 2006. Asteraceae. In: Flora of North America Editorial Committee, eds. Fl. North Amer. Vols. 19–21. Oxford Univ. Press, New York and Oxford.

Barkley, T.M., B.L. Clark, and **A.M. Funston.** 1996. The segregate genera of *Senecio sensu lato* and *Cacalia sensu lato* (Asteraceae: Senecioneae) in Mexico and Central America. In: D.J.N. Hind and H.J. Beentje, eds. Compositae: Systematics. Proceedings of the International Compositae Conference, Kew, 1994. (D.J.N. Hind, Editor-in-Chief), Vol. 1. Pp. 613–620. Royal Botanic Gardens, Kew, England, U.K.

Barkley, T.M. and **A. Cronquist.** 1978. *Erechtites*. In: N. Amer. Fl. II. 10:139–142.

Barlow, C. 2001. Anachronistic fruits and the ghosts who haunt them. Arnoldia (Jamaica Plain) 61(2):14–21.

Barlow, C. 2001b. The ghosts of evolution: Nonsensical fruits, missing partners, and other ecological anachronisms. Basic Books, New York.

Barnard-Kubow, K.B., C.L. Debban, and **L.F. Galloway.** 2015. Multiple glacial refugia lead to genetic structuring and the potential for reproductive isolation in a herbaceous plant. Amer. J. Bot. 102:1842–1853.

Barneby, R.C. 1952. A revision of the North American species of *Oxytropis*. Proc. Calif. Acad. Sci. 27:177–312.

Barneby, R.C. 1964. Atlas of North American *Astragalus*. Mem. New York Bot. Gard. 13. I:1–596, II:597–1188.

Barneby, R.C. 1977. Daleae imagines, an illustrated revision of *Errazurizia* Philippi, *Psorothamnus* Rydberg, *Marina* Liebmann, and *Dalea* Lucanus emend. Barneby, including all species of Leguminosae tribe Amorpheae Borissova ever referred to *Dalea*. Mem. New York Bot. Gard. 27:1–892.

Barneby, R.C. 1991. Sensitive *Censitae*, a description of the genus *Mimosa* Linnaeus in the New World. Mem. New York Bot. Gard. 65:1–835.

Barneby, R.C. and **D. Isely.** 1986. Reevaluation of *Mimosa biuncifera* and *M. texana* (Leguminosae: Mimosoideae). Brittonia 38: 119–122.

Barnes, G., D.G.O. Saunders, and **T. Williamson.** 2020. Banishing barberry: The history of *Berberis vulgaris* prevalence and wheat stem rust incidence across Britain. Plant Pathol. 69:1193–1202.

Barringer, K. 1993. New combinations in North American *Asarum* (Aristolochiaceae). Novon 3:225–227.

Barringer, K. 1997. *Aristolochia*. In: Flora of North America Editorial Committee, eds. Fl. North Amer. 3:45–50. Oxford Univ. Press, New York and Oxford.

Barringer, K. 2004. New Jersey pinweeds (*Lechea*, Cistaceae). J. Torrey Bot. Soc. 131:261–276.

Barringer, K. and **A.T. Whittemore.** 1997. Aristolochiaceae. In: Flora of North America Editorial Committee, eds. Fl. North Amer. 3:44. Oxford Univ. Press, New York and Oxford.

Barth, F.G. 1985. Insects and flowers: The biology of a partnership. Translated by M.A. Biederman-Thorson. Princeton Univ. Press, Princeton, NJ.

Barthlott, W. and **D.R. Hunt.** 1993. Cactaceae. In: K. Kubitzki, J.C. Rohwer, and V. Bittrich, eds. The families and genera of vascular plants, Vol. II. Pp. 161–197. Springer-Verlag, Berlin.

Bartish, I.V. and **U. Swenson.** 2004. Elaeagnaceae. In: K. Kubitzki, ed. The families and genera of vascular plants, Vol. VI. Pp. 131–134. Springer-Verlag, Berlin.

Bartoli, A. and **R.D. Tortosa.** 2012. Revision of the North American species of *Grindelia* (Asteraceae). Ann. Missouri Bot. Gard. 98(4): 447–513.

Baskin, C.C., J.M. Baskin, and **E.W. Chester.** 2003. Ecological life cycle of *Trepocarpus aethusae* (Nutt.) ex DC. and comparisons with two other winter annual Apiaceae native to eastern United States. Castanea 68:43–55.

Baskin, J.M. and **C. Baskin.** 1984. The ecological life cycle of *Campanula americana* in northcentral Kentucky. Bull. Torrey Bot. Club 111:329–337.

Baskin, J.M. and **C.C. Baskin.** 1989. The case of *Cotinus obovatus* Raf. (Anacardiaceae) in Kentucky. Castanea 54:119–120.

Baskin, J.M., K.M. Snyder, and **C.C. Baskin.** 1993. Nomenclatural history and taxonomic status of *Echinacea angustifolia*, *E. pallida*, and *E. tennesseensis* (Asteraceae). Sida 15:597–604.

Bassett, I.J. and **C.W. Crompton.** 1978. The genus *Suaeda* (Chenopodiaceae) in Canada. Canad. J. Bot. 56:581–591.

Bassett, I.J. and **C.W. Crompton.** 1983. The genus *Chenopodium* in Canada. Canad. J. Bot. 60:586–610.

Bassett, I.J., C.W. Crompton, J. McNeill, and **P.M. Taschereau.** 1983. The genus *Atriplex* (Chenopodiaceae) in Canada. Monogr. Res. Branch Canada Dept. Agric. 31:1–72.

Bates, D.M., R.W. Robinson, and **C. Jeffrey,** eds. 1990. Biology and utilization of the Cucurbitaceae. Cornell Univ. Press, Ithaca, NY.

Baumgardt, J.P. 1982. How to identify flowering plant families. Timber Press, Portland, OR.

Baxter, J.W. 1957. The genus *Cumminsiella*. Mycologia, 49:864–873.

Bayer, R.J. 1984. Chromosome numbers and taxonomic notes for North American species of *Antennaria* (Asteraceae: Inuleae). Syst. Bot. 9:74–83.

Bayer, R.J. 1985. Investigations into the evolutionary history of the polyploid complexes in *Antennaria* (Asteraceae: Inuleae). II. The *A. parlinii* complex. Rhodora 87:321–339.

Bayer, R.J. 2006. *Antennaria*. In: Flora of North America Editorial Committee, eds. Fl. North Amer. 19:388–415. Oxford Univ. Press, New York and Oxford.

Bayer, R.J. and **G.L. Stebbins.** 1982. A revised classification of *Antennaria* (Asteraceae: Inuleae) of the eastern United States. Syst. Bot. 7:300–313.

Beaman, J.H. 1957. The systematics and evolution of *Townsendia* (Compositae). Contr. Gray Herb. 183:1–151.

Beaman, J.H. 1990. Revision of *Hieracium* (Asteraceae) in Mexico and Central America. Syst. Bot. Monogr. 29:1–77.

Beardsley, R.L. and **E.T. Browne, Jr.** 1972. *Lathyrus aphaca* L. new to Tennessee and the southeast. Rhodora 74:155.

Beatley, J.C. 1973. Russian-thistle (*Salsola*) species in western United States. J. Range Managem. 26:225–226.

Beauchamp, R.M. 1980. New names in American acacias. Phytologia 46:5–9.

Beauverd, G. 1913. Contribution à l'étude des Composées, suite VIII. IV. Le genre *Micropsis* DC. (emend. Beauverd). Bull. Soc. Bot. Genève, sér. 2, 5:221–228.

Bebeau, G.D. 2015. *Cornus obliqua* Raf. Plants of the Eloise Butler Wildflower Garden. https://www.friendsofthewildflowergarden.org/pages/plants/paledogwood.html.

Becker, A.L., A.A. Crowl, J.L. Luteyn, and **11 others.** 2024. A global blueberry phylogeny: Evolution, diversification, and biogeography of Vaccinieae (Ericaceae). Molec. Phylogen. Evol. 201, 108202. https://doi.org/10.1016/j.ympev.2024.108202.

Behnke, H.-D. and **T.J. Mabry,** eds. 1994. Caryophyllales: Evolution and systematics. Springer-Verlag, New York.

Belcher, R.O. 1956. A revision of the genus *Erechtites* (Compositae) with inquiries into *Senecio* and *Arrhenechthites*. Ann. Missouri Bot. Gard. 43:1–85.

Bell, C.D., E.J. Edwards, S.T. Kim, and **M.J. Donoghue.** 2001. Dipsacales phylogeny based on chloroplast DNA sequences. Harvard Pap. Bot. 6:481–499.

Bell, C.R. 1963. The genus *Eryngium* in the southeastern United States. Castanea 28:73–79.

Bellot, S. and **S.S. Renner.** 2013. Pollination and mating systems of Apodanthaceae and the distribution of reproductive traits in parasitic angiosperms. Amer. J. Bot. 100:1083–1094.

Bellot, S. and **S.S. Renner.** 2014a. The systematics of the worldwide endoparasite family Apodanthaceae (Cucurbitales), with a key, a map, and color photos of most species. PhytoKeys 36:41–57.

Bellot, S. and **S.S. Renner.** 2014b. Exploring new dating approaches for parasites: The worldwide Apodanthaceae (Cucurbitales) as an example. Molec. Phylogen. Evol. 80:1–10.

Bellot, S. and **S.S. Renner.** 2015. The plastomes of two species in the endoparasite genus *Pilostyles* (Apodanthaceae) each retain just five or six possibly functional genes. Genome Biol. Evol. 8:189–201.

Bemis, W.P., A.M. Rhodes, T.W. Whitaker, and **S.G. Carmer.** 1970. Numerical taxonomy applied to *Cucurbita* relationships. Amer. J. Bot. 57:404–412.

Bemis, W.P., L.D. Curtis, C.W. Weber, and **J. Berry.** 1978. The feral buffalo gourd, *Cucurbita foetidissima*. Econ. Bot. 32:87–95.

Benet-Pierce, N. and **M.G. Simpson.** 2014. The taxonomy of *Chenopodium desiccatum* and *C. nitens*, sp. nov. J. Torrey Bot. Soc. 141:161–172.

Benet-Pierce, N. and **M.G. Simpson.** 2017. Taxonomic recovery of the species in the *Chenopodium neomexicanum* (Chenopodiaceae) complex and description of *Chenopodium sonorense* sp. nov. J. Torrey Bot. Soc. 144:339–356.

Benning, J.W. 2015. Odd for an ericad: Nocturnal pollination of *Lyonia lucida* (Ericaceae). Amer. Midl. Naturalist 174:204–217.

Benson, L. 1969. Cactaceae. In: C.L. Lundell, ed. Flora of Texas 2:221–317. Texas Research Foundation, Renner.

Benson, L. 1982. The cacti of the United States and Canada. Stanford Univ. Press, Stanford, CA.

Bernhardt, P. 1990. Anthecology of *Schrankia nuttallii* (Mimosaceae) on the tallgrass prairie. Pl. Syst. Evol. 170:247–255.

Berry, P.E., R. Riina, J.A. Peirson, Y. Yang, V.W. Steinmann, D.V. Geltman, J.J. Morawetz, and **N.I. Cacho.** 2016. *Euphorbia*. In: Flora of North America Editorial Committee, eds. Fl. North Amer. 12:237–324. Oxford Univ. Press, New York and Oxford.

Berry, P.E., V. Savolainen, K.J. Sytsma, J.C. Hall, and **M.W. Chase.** 2001. *Lissocarpa* is sister to *Diospyros* (Ebenaceae). Kew Bull. 56: 725–729.

Bicknell, E.P. 1895. The genus *Sanicula* in the eastern United States, with descriptions of two new species. Bull. Torrey Bot. Club 22:351–361.

Bicknell, E.P. 1901. A new horse gentian (Trisoteum) common in the eastern states. Torreya 1:25–28.

Bidartondo, M. 2005. The evolutionary ecology of myco-hertero-trophy. New Phytol. 167:335–352.

Bidartondo, M.I. and **T.D. Bruns.** 2001. Extreme specificity in epiparasitic Monotropoideae (Ericaceae): Widespread phylogenetic and geographical structure. Molec. Ecol. 10:2285–2295.

Bidartondo, M.I. and **T.D. Bruns.** 2002. Fine-level mycorrhizal specificity in the Monotropoideae (Ericaceae): Specificity for fungal species groups. Molec. Ecol. 11:557–569.

Biddulph, S.F. 1944. A revision of the genus *Gaillardia*. Res. Stud. State Coll. Wash. 13:195–256.

Bidlack, J.E. and **S.H. Jansky.** 2014. Stern's introductory plant biology, 13th ed. McGraw-Hill, New York, NY.

Bierner, M.W. 1972. Taxonomy of *Helenium* sect. *Tetrodus* and a conspectus of North American *Helenium* (Compositae). Brittonia 24:331–355.

Bierner, M.W. 1989. Taxonomy of *Helenium* sect. *Amarum* (Asteraceae). Sida 13:453–459.

Bierner, M.W. 1990. Present status of *Amblyolepis* (Asteraceae: Heliantheae). Madroño 37:133–140.

Bierner, M.W. 2006a. *Amblyolepis*. In: Flora of North America Editorial Committee, eds. Fl. North Amer. 21:420–421. Oxford Univ. Press, New York and Oxford.

Bierner, M.W. 2006b. *Helenium*. In: Flora of North America Editorial Committee, eds. Fl. North Amer. 21:426–435. Oxford Univ. Press, New York and Oxford.

Bierner, M.W. 2006c. *Hymenoxys*. In: Flora of North America Editorial Committee, eds. Fl. North Amer. 21:435–443. Oxford Univ. Press, New York and Oxford.

Bierner, M.W. and **R.K. Jansen.** 1998. Systematic implications of DNA restriction site variation in *Hymenoxys* and *Tetraneuris* (Asteraceae, Helenieae, Gaillardiinae). Lundellia 1:17–26.

Bierner, M.W. and **B.L. Turner.** 2003. Taxonomy of *Tetraneuris* (Asteraceae: Helenieae: Tetraneurinae). Lundellia 6:44–96.

Bierner, M.W. and **B.L. Turner.** 2006. *Tetraneuris*. In: Flora of North America Editorial Committee, eds. Fl. North Amer. 21:447–453. Oxford Univ. Press, New York and Oxford.

Bierzychudek, P. 1982. Life histories and demography of temperate forest herbs: a review. New Phytol. 90:757–776.

Bihrmann's Caudiciforms. 2025. https://bihrmann.com/caudiciforms/. Accessed 26 Mar 2025.

Binns, S.E., B.R. Baum, and **J.T. Arnason.** 2002. A taxonomic revision of *Echinacea* (Asteraceae: Heliantheae). Syst. Bot. 27(3):610–632.

Bittrich, V. 1993. Caryophyllaceae. In: K. Kubitzki, J.C. Rohwer, and V. Bittrich, eds. The families and genera of vascular plants, Vol. II. Pp. 206–236. Springer-Verlag, Berlin.

Bittrich, V. and **H.E.K. Hartmann.** 1988. The Aizoaceae—A new approach. Bot. J. Linn. Soc. 97:239–254.

Björkman, E. 1960. *Monotropa hypopitys* L.—An epiparasite on tree roots. Physiol. Pl. (Copenhagen) 13:308–327.

Blackwell, W.H. 1990. Poisonous and medicinal plants. Prentice Hall, Englewood Cliffs, NJ.

Blackwell, W.H., Jr., M.D. Baechle, and **G. Williamson.** 1978. Synopsis of *Kochea* (Chenopodiaceae) in North America. Sida 7:248–254.

Blake, S.F. 1918. A revision of the genus *Viguiera*. Contr. Gray Herb. 54:1–205.

Blake, S.F. 1921. Revision of the genus *Acanthospermum*. Contr. U.S. Natl. Herb. 20:383–392.

Blarer, A., D.L. Nickrent, and **P.K. Endress.** 2004. Comparative floral structure and systematics in Apodanthaceae (Rafflesiales). Pl. Syst. Evol. 245:119–142.

Blomquist, H.L. 1957. A revision of *Hexastylis* of North America. Brittonia 8:255–281.

Blum, W., M. Lange, W. Rischer, and **J. Rutow.** 1998. *Echinocereus* monographie. Published by the authors. Aachen, Germany.

Boetsch, J.R. 2002. The Aizoaceae and Molluginaceae of the southeastern United States. Castanea 67:42–53.

Bogle, A.L. 1969. The genera of Portulacaceae and Basellaceae in the southeastern United States. J. Arnold Arbor. 50:566–598.

Bogle, A.L. 1970. The genera of Molluginaceae and Aizoaceae in the southeastern United States. J. Arnold Arbor. 51:431–462.

Bogle, A.L.1986. The floral morphology and vascular anatomy of the Hamamelidaceae: Subfamily Liquidambaroideae. Ann. Missouri Bot. Gard. 73:325–347.

Bogle, A.L., T. Swain, R.D. Thomas, and **E.D. Kohn.** 1971. *Geocarpon*: Aizoaceae or Caryophylaceae? Taxon 20:473–477.

Bogler, D.J. 2006. Apocynaceae (Dogbane family). In: G. Yatskievych. Steyermark's flora of Missouri. Vol. 2. Pp. 115–126. Missouri Botanical Garden Press in cooperation with the Missouri Dept. of Conservation, St. Louis.

Bogler, D.J. 2006a. *Crepis*. In: Flora of North America Editorial Committee, eds. Fl. North Amer. 19:222–239. Oxford Univ. Press, New York and Oxford.

Bogler, D.J. 2006b. *Hypochaeris*. In: Flora of North America Editorial Committee, eds. Fl. North Amer. 19:297–299. Oxford Univ. Press, New York and Oxford.

Bogler, D.J. 2006c. *Leontodon*. In: Flora of North America Editorial Committee, eds. Fl. North Amer. 19:294–296. Oxford Univ. Press, New York and Oxford.

Bogler, D.J. 2006d. *Lygodesmia*. In: Flora of North America Editorial Committee, eds. Fl. North Amer. 19:369–373. Oxford Univ. Press, New York and Oxford.

Bogler, D.J. 2006e. *Pinaropappus*. In: Flora of North America Editorial Committee, eds. Fl. North Amer. 19:374–376. Oxford Univ. Press, New York and Oxford.

Bogler, D.J. 2006f. *Prenanthes*. In: Flora of North America Editorial Committee, eds. Fl. North Amer. 19:264–271. Oxford Univ. Press, New York and Oxford.

Bohac, J.R., P.D. Dukes, and **D.F. Austin.** 1995. Sweetpotato. In: J. Smartt and N.W. Simmonds, eds. Evolution of Crop Plants, ed. 2. Pp. 57–62. Longman Group, Ltd., London.

Bolick, M.R. 1985. Cladistic analysis of *Iva*: A case in point. Taxon 34:81–84.

Bone, T.S., S.R. Downie, J.M. Affolter, and **K. Spalik.** 2011. A phylogenetic and biogeographic study of the genus *Lilaeopsis* (Apiaceae tribe Oenantheae). Syst. Bot. 36:789–805.

Borgen, L. 2010. *Lobularia*. In: Flora of North America Editorial Committee, eds. Fl. North Amer. 7:597–598. Oxford Univ. Press, New York and Oxford.

Borhidi, A., E. Gondar, and **Z. Orosz-Kovacs.** 1989. The reconsideration of the genus *Cordia* (Boraginaceae). Acta Bot. Hung. 34:375–423.

Borja, G. 2013. Evolutionary History of *Paysonia* (Brassicaceae). PhD dissertation, Oklahoma State Univ., Norman.

Boufford, D.E. and **S.A. Spongberg.** 1981. *Calycanthus floridus* (Calycanthaceae)—A nomenclatural note. J. Arnold Arbor. 62: 265–266.

Boufford, D.E. and **S.A. Spongberg.** 1983. Eastern Asian-eastern North American phytogeographical relationships–A history from the time of Linnaeus to the twentieth century. Ann. Missouri Bot. Gard. 70:423–439.

Boughton, L.L. 1931. *Croton capitatus* as a poisonous forage plant. Trans. Kansas Acad. Sci. 34: no page number given.

Boulos, L. 1973. Révision systématique du genere *Sonchus* L. s.l.: IV. Sous-genre: 1. *Sonchus*. Bot. Not. 126:155–196.

Boulos, L. 1976. *Sonchus*. In: T.G. Tutin, V.H. Heywood, N.A. Burges, D.M. Moore, D.H. Valentine, S.M. Walters, and D.A. Webb, eds. Flora Europaea 4:327–328. Cambridge Univ. Press, Cambridge, England, UK.

Bousquet, J., S.H. Strauss, and **P. Li.** 1992. Complete congruence between morphological and rbcL-based molecular phylogenies in birches and related species (Betulaceae). Molec. Biol. Evol. 9:1076–1088.

Bowden, W.M. 1959a. Cytotaxonomy of *Lobelia* L. section *Lobelia*. I. Three diverse species and seven small-flowered species. Canad. J. Genet. Cytol. 1:49–64.

Bowden, W.M. 1959b. Phylogenetic relationships of twenty-one species of *Lobelia* L. section *Lobelia*. Bull. Torrey Bot. Club 86: 94–108.

Bowden, W.M. 1960a. Cytotaxonomy of *Lobelia* L. section *Lobelia*. II. Four narrow-leaved species and five medium-flowered species. Canad. J. Genet. Cytol. 1:11–27.

Bowden, W.M. 1960b. Cytotaxonomy of *Lobelia* L. Section *Lobelia*. III. *L. siphilitica* L. and *L. cardinalis* L. Canad. J. Genet. Cytol. 2: 234–251.

Bowden, W.M. 1961. Interspecific hybridization in *Lobelia* L. Section *Lobelia*. Canad. J. Bot. 39:1679–1693.

Bowden, W.M. 1964a. Cytogenetics of *Lobelia* ×*speciosa* Sweet (*L. cardinalis* L. × *L. siphilitica* L.). Canad. J. Genet. Cytol. 6:121–139.

Bowden, W.M. 1982. The taxonomy of *Lobelia* ×*speciosa* s.l. and its parental species, *L. siphilitica* and *L. cardinalis* s.l. (Lobeliaceae). Canad. J. Bot. 60:2054–2070.

Bradley, C.E. and **A.J. Haagen-Smit.** 1949. The essential oil of *Pectis papposa*. Econ. Bot. 3:407–412.

Bradley, T.R. 1975. Hybridization between *Triodanis perfoliata* and *Triodanis biflora* (Campanulaceae). Brittonia 27:110–114.

Braukmann, T. and **S. Stefanovic.** 2012. Plastid genome evolution in mycoheterotrophic Ericaceae. Pl. Molec. Biol. 79:5–20.

Breckenridge, F.G.I. 1983. *Berberis swaseyi*—Facing natural extinction? Texas Native Pl. Soc. News 1:1.

Breen, P. 2025. Oregon State University. College of Agricultural Sciences. Dept. of Horticulture. https://landscapeplants.oregonstate.edu/plants/leucothoe-axillaris#:~:text=Poisonous%3A%20Leaves%20and%20nectar%20from,and%20vomiting%2C%20diarrhea%2C%20paralysis.

Bremer, K. 1994. Asteraceae cladistics & classification. Timber Press, Portland, OR.

Bremer, K. 1996. Major clades and grades of the Asteraceae. In: D.J.N. Hind et al., eds. 1996. Proceedings of the International Compositae Conference, Kew, 1994. 2 vols. Kew. Vol. 1.

Bremer, K., A.A. Anderberg, P.O. Karis, and **J. Lundberg.** 1994. Tribe Eupatorieae (Chapter 23). In: Asteraceae cladistics & classification. Pp. 625–680. Timber Press, Portland, OR.

Breslin, P.B., M.F. Wojciechowski, and **L.C. Majure.** 2021. Molecular phylogeny of the Mammilloid clade (Cactaceae) resolves the monophyly of *Mammillaria*. Taxon 70(2):308–323.

Brewer, G. 2022. Back to the future: The green revolution. Royal Botanic Gardens Kew, England, U.K.

Brewer-Carias, C. and **J.A. Steyermark.** 1976. Hallucinagenic snuff drugs of the Yanomano Caburiwe-Teri in the Cauaburi River, Brazil. Econ. Bot. 30:57–66.

Briand, C.H. 2005. The common persimmon (*Diospyros virginiana* L.): The history of an underutilized fruit tree (16th–19th centuries). Huntia 12:71–89.

Britton, N.L. and **J.N. Rose.** 1905. Crassulaceae. N. Amer. Fl. 22: 7–74.

Britton, N.L. and **J.N. Rose.** 1919–1923. The Cactaceae: Descriptions and illustrations of plants of the cactus family, 4 vols. Publ. No. 248. Carnegie Institution of Washington, Washington, D.C.

Britton, N.L. and **J.N. Rose.** 1928. Mimosaceae. N. Amer. Fl. 23: 1–194.

Britton, N.L. and **J.N. Rose.** 1930. Caesalpiniaceae. N. Amer. Fl. 23: 201–349.

Britton, N.L. and **J.N. Rose.** 1937. The Cactaceae: Descriptions and illustrations of plants of the cactus family, 4 vols. Reprinted from the original Carnegie Institution of Washington printing, Publ. No. 248, 1919–1923. Scott E. Haselton, Abbey San Encino Press, Pasadena, CA.

Britton, N.L. and **J.N. Rose.** 1963. The Cactaceae: Descriptions and illustrations of plants of the cactus family. Republication of the 1937 edition in 4 vols. bound as 2 vols. by Dover Publications, Inc., New York.

Brizicky, G.K. 1962d. The genera of Anacardiaceae in the southeastern United States. J. Arnold Arbor. 43:359–375.

Brizicky, G.K. 1963a. Taxonomic and nomenclatural notes on the genus *Rhus* (Anacardiaceae). J. Arnold Arbor. 44:60–80.

Brizicky, G.K. 1964a. The genera of Celastrales in the southeastern United States. J. Arnold Arbor. 45:206–234.

Brizicky, G.K. 1964b. The genera of Cistaceae in the southeastern United States. J. Arnold Arbor. 45:346–357.

Brock, M. 2024. *Thaspium*. In: Flora of North America Editorial Committee, eds. Fl. North Amer. 13:507–510. Oxford Univ. Press, New York and Oxford.

Broe, M.B. 2014. Phylogenetics of the Monotropoideae (Ericaceae) with special focus on the genus *Hypopitys* Hill, together with a novel approach to phylogenetic inference using lattice theory. Ph.D. dissertation, Ohio State Univ., Columbus.

Broich, S.L. 2007. New combinations in North American *Lathyrus* and *Vicia* (Fabaceae: Faboideae: Fabeae). Madroño 54:63–71.

Broich, S.L. 2023a. *Lathyrus*. In: Flora of North America Editorial Committee, eds. Fl. of North Amer. 11:1008–1030. Oxford Univ. Press, New York and Oxford.

Broich, S.L. 2023b. *Vicia*. In: Flora of North America Editorial Committee, eds. Fl. of North Amer. 11:994–1005. Oxford Univ. Press, New York and Oxford.

Broich, S.L. 2023c. *Pisum*. In: Flora of North America Editorial Committee, eds. Fl. of North Amer. 11:1030–1031. Oxford Univ. Press, New York and Oxford.

Brooks, B. 1999. The Toxicodendrons: Poison ivy, poison oak and poison sumac. http://nac.tamu.edu/x075bb/caddo/poison.html. Accessed Jun 2007.

Brooks, R.E. 1977. Poison ivy and poison oak in Kansas. Bull. State. Bio. Sur. Kansas 4:1–38.

Brooks, R.E. 1986b. Campanulaceae. In: Great Plains Flora Association. Flora of the Great Plains. Pp. 808–816. Univ. Press of Kansas, Lawrence.

Brooks, R.E. 1986c. Caprifoliaceae. In: Great Plains Flora Association. Flora of the Great Plains. Pp. 823–832. Univ. Press of Kansas, Lawrence.

Brouillet, L. 2006a. *Eurybia*. In: Flora of North America Editorial Committee, eds. Fl. North Amer. 20:365–382. Oxford Univ. Press, New York and Oxford.

Brouillet, L. 2006b. *Matricaria*. In: Flora of North America Editorial Committee, eds. Fl. North Amer. 19:540–542. Oxford Univ. Press, New York and Oxford.

Brouillet, L. 2006c. *Taraxacum*. In: Flora of North America Editorial Committee, eds. Fl. North Amer. 19:239–252. Oxford Univ. Press, New York and Oxford.

Brouillet, L. 2023a. *Acmispon*. In: Flora of North America Editorial Committee, eds. Fl. of North Amer. 11:507–531. Oxford Univ. Press, New York and Oxford.

Brouillet, L. 2023b. *Lotus*. In: Flora of North America Editorial Committee, eds. Fl. of North Amer. 11:538–541. Oxford Univ. Press, New York and Oxford.

Brouillet, L., J.C. Semple, G.A. Allen, K.L. Chambers, and **S.D. Sundberg.** 2006. *Symphyotrichum*. In: Flora of North America Editorial Committee, eds. Fl. North Amer. 20:465–539. Oxford Univ. Press, New York and Oxford.

Brown, G.D. 1956. Taxonomy of American *Atriplex*. Amer. Midl. Naturalist 55:199–210.

Brown, G.K., L.A. Craven, F. Udovicic, and **P.Y. Ladiges.** 2006. Phylogenetic relationships of *Rhododendron* section *Vireya* (Ericaceae) inferred from the ITS nrDNA region. Austral. Syst. Bot. 19:329–342.

Brown, L.E. 2014. A checklist of the vascular plants of the Houston area. http://npsot.org/wp/houston/files/2014/08/A-Flora-2.pdf. Digital manuscript available through the Native Plant Society of Texas. Accessed 2020.

Brown, L.E. and **I.S. Elsik.** 2002. Notes on the flora of Texas with additions and other significant records. II. Sida 20(1):437–444.

Brown, LE, P.A. Harcombe, W.W. Pruess, I.S. Elsik, B.R. MacRoberts, M.H. MacRoberts, and **S.D. Jones.** 2006b. Annotated checklist of the vascular flora of the Lance Rosier Unit of the Big Thicket National Preserve, Hardin County, Texas. Sida 22(2):1175–1189.

Brown, L.E., K. Hillhouse, B.R. MacRoberts, and **M.H. MacRoberts.** 2002. The vascular flora of Windham Prairie, Polk County, East Texas. Texas J. Sci. 54(3):227–240.

Brown, L.E., E.L. Keith, D.J. Rosen, and **J. Liggio.** 2007. Notes on the flora of Texas with additions and other significant records. III. J. Bot. Res. Inst. Texas 1(2):1255–1264.

Brown, L.E., B.R. MacRoberts, M.H. MacRoberts, P.A. Harcombe, W.W. Pruess, I.S. Elsik, and **D. Johnson.** 2005. Annotated checklist of the vascular flora of the Turkey Creek Unit of the Big Thicket National Preserve, Tyler and Hardin counties, Texas. Sida 21(3):1807–1827.

Brown, L.E., B.R. MacRoberts, M.H. MacRoberts, P.A. Harcombe, W.W. Pruess, I.S. Elsik, and **S.D. Jones.** 2006a. Annotated checklist of the vascular flora of the Big Sandy Creek Unit, Big Thicket National Preserve, Texas. Sida 22(1):705–723.

Brown, L.E., B.R. MacRoberts, M.H. MacRoberts, P.A. Harcombe, W.W. Pruess, I.S. Elsik, and **S.B. Walker.** 2008a. Annotated checklist of the vascular flora of the Beech Creek Unit of the Big Thicket National Preserve, Tyler County, Texas. J. Bot. Res. Inst. Texas 2(1):651–660.

Brown, L.E., B.R. MacRoberts, M.H. MacRoberts, and **W.W. Pruess.** 2009. Annotated checklist of the vascular flora of the Menard Creek Corridor Unit of the Big Thicket National Preserve, Polk, Liberty, and Hardin counties, Texas. J. Bot. Res. Inst. Texas 3(1): 443–455.

Brown, L.E. and **S.J. Marcus.** 1998. Notes on the Flora of Texas with additions and other significant records. Sida 18(1):315–324.

Brown, L.E. and **C.D. Peterson.** 1984. *Carex rosea* (Cyperaceae), *Trifolium lappaceum* (Fabaceae), and *Aira caryophyllea* (Poaceae) new to Texas. Sida 10:263–264.

Brown, L.E., W.W. Pruess, I.S. Elsik, B.R. MacRoberts, M.H. MacRoberts, and **S.B. Walker.** 2008b. Annotated checklist of the vascular flora of the Loblolly Unit of the Big Thicket National Preserve, Liberty County, Texas. J. Bot. Res. Inst. Texas 2(2): 1481–1489.

Brown, L.E., P.V. Roling, J.L. Aplaca, and **E. Keith.** 2011. Notes on the flora of Texas, Arkansas, and Louisiana, with additions and other records. IV. Phytoneuron 2011-18:1–12.

Brown, M.L. and **R.G. Brown.** 1984. Herbaceous plants of Maryland. The Book Center, Univ. of Maryland, College Park., printed by Port City Press, Inc., Baltimore, MD.

Brown, R.C. 1978. Biosystematics of *Psilostrophe* (Compositae: Helenieae). II. Artificial hybridization and systematic treatment. Madroño 25:187–201.

Bruederle, L.P. and **N. Vorsa.** 1994. Genetic differentiation of diploid blueberry, *Vaccinium* sect. *Cyanococcus* (Ericaceae). Syst. Bot. 19:337–349.

Brummitt, R.K. 1965. New contributions in North American *Calystegia*. Ann. Missouri Bot. Gard. 52:214–216.

Brummitt, R.K. 1981. Further new names in the genus *Calystegia* (Convolvulaceae). Kew. Bull. 35:327–334.

Brummitt, R.K. 2023. *Calysetgia*. In: Flora of North America Editorial Committee, eds. Fl. North Amer. 14:321–336. Oxford Univ. Press, New York and Oxford.

Brummitt, R.K. and **J.H. Ross.** 1974. Proposal to conserve the generic name 3557 *Hoffmannseggia* Cav. nom. illegit. over *Larrea* Ortega (Leguminosae-Caesalpinioideae). Taxon 23(2–3):433–435.

Bruneau, A., F. Forest, P.S. Herendeen, B.B. Klitgaard, and **G.P. Lewis.** 2001. Phylogenetic relationships in the Caesalpinioideae (Leguminosae) as inferred from choroplast *trnL* intron sequences. Syst. Bot. 26:487–514.

Bruneau, A., M. Mercure, GL. Lewis, and **R.L. Herendeen.** 2008. Phylogenetic patterns and diversification in the caesalpinioid legumes. Botany (Ottawa) 86:697–718.

Buchmann, S.L. 1983. Buzz pollination in angiosperms. In: C.E. Jones and R.J. Little, eds. Handbook of experimental pollination biology, Pp. 73–113. Van Nostrand Reinhold Co. New York.

Buckingham, G.R. 1996. Biological control of alligatorweed, *Alternanthera philoxeroides*, the world's first aquatic weed success story. Castanea 61:232–243.

Burch, D. 1966. The application of the Linnaean names of some New World species of *Euphorbia* subgenus *Chamaesyce*. Rhodora 68:155–166.

Burckhalter, R.E. 1992. The genus *Nyssa* (Cornaceae) in North America: A revision. Sida 15:323–342.

Burckhardt, D. 2005. Biology, ecology, and evolution of gall-inducing psyllids (Hemiptera: Psylloidea). In: A. Raman, C.W. Schaefer, and T.M. Withers, eds. Biology, ecology and evolution of gall-inducing arthropods. Pp. 143–157. Science Publishers, Enfield, NH.

Burckhardt, D. and **Y. Basset.** 2000. The jumping plant lice (Hemiptera, Psylloidea) associated with *Schinus* (Anacardiaceae): Systematics, biogeography, and host plant relationships. J. Nat. Hist. 34:57–155.

Burkart, A. 1971. El genero *Galactia* (Legum.—Phaseoleae) en Sudamerica con especial referencia a la Argentina y paises vecinos. Darwiniana 16:663–796.

Burkart, A. 1976. A monograph of the genus *Prosopis* (Leguminosae subfam. Mimosoideae). J. Arnold Arbor. 57:219–249, 450–525.

Burlage, H.M. 1968. Index of plants of Texas with reputed medicinal and poisonous properties. Published by the author, Austin, TX.

Burleigh, J.G. and **T.P. Holtsford.** 2003. Molecular systematics of the eastern North American *Silene* (Caryophyllaceae): Evidence from nuclear ITS and chloroplast *trn*L intron sequences. Rhodora 105:76–90.

Burrows, G.E. and **R.J. Tyrl.** 2001. Toxic plants of North America. Iowa State Univ. Press, Ames.

Bush, C.M., W.S. Judd, L. Frazier, and **K.A. Kron.** 2010. The phylogeny of *Leucothoë* s.l. (Ericaceae: Vaccinioideae) based on morphological and molecular (*ndhF, matK*, and nrITS) data. Syst. Bot. 35:201–206.

Bush, C.M., L. Lu, P.W. Fritsch, D.-Z. Li, and **K.A. Kron.** 2009. Phylogeny of Gaultherieae (Ericaceae: Vaccinioideae) based on DNA sequence data from *matK, ndhF*, and nrITS. Int. J. Pl. Sci. 170:355–364.

Buthod, A. (Amy). Personal communication. Collections manager, Robert Bebb Herbarium, University of Oklahoma.

Buthod, A.K. and **L. Miller.** 2023. *Geocarpon minimum* (Caryophyllaceae), new to Oklahoma, U.S.A. J. Bot. Res. Inst. Texas 17(2):535–539.

Butters, F.K. and **H. St. John.** 1917. Studies in certain North American species of *Lathyrus*. Rhodora 19:156–163.

Butterworth, C.A., J.H. Cota-Sanchez, and **R.S. Wallace.** 2002. Molecular systematics of tribe Cacteae (Cactaceae: Cactoideae): A phylogeny based on rpl16 intron sequence variation. Syst. Bot. 27:257–270.

Butterworth, C.A. and **R.S. Wallace.** 2004. Phylogenetic studies of *Mammillaria* (Cactaceae)—Insights from chloroplast sequence variation and hypothesis testing using the parametric bootstrap. Amer. J. Bot. 91:1086–1098.

Bygrave, P. 1996. Allspices and winter-sweets: An account of Calycanthaceae in the wild and cultivation. New Plantsman 3: 40–49.

Byng, J.W. 2014. The Flowering Plants Handbook, Plant Gateway Ltd., Hertford, UK.

Bywater, M. and **G.E. Wickens.** 1984. New World species of the genus *Crassula*. Kew Bull. 39:699–728.

CABI. 2019. *Anredera cordifolia* [original text by Nick Pasiecznik]. In: Invasive Species Compendium. Wallingford, UK: CAB International. https://www.cabi.org/isc/datasheet/112290. Accessed Aug 2019.

Cabrera, A.L. 1949. Sinopsis del género *Soliva* (Compositae). Notas Mus. La Plata, Bot. 14:123–139.

Cabrera, A.L. and **A.M. Ragonese.** 1978. Revisión del género *Pterocaulon* (Compositae). Darwiniana 21:185–257.

Callaway, M.B. 1993. Pawpaw (*Asimina triloba*): A "tropical" fruit for temperate climates. In: J. Janick and J.E. Simon, eds., New Crops. Pp. 505–515. Wiley, New York.

Callaway, M.B. and **D.J.Callaway.** 1992. Our native pawpaw: the next new commercial fruit? Arnoldia 52(3):20–29.

Calviño, C.I. and **G.A. Levin.** 2024. *Eryngium*. In: Flora of North America Editorial Committee, eds. Fl. North Amer. 13:298–318. Oxford Univ. Press, New York and Oxford.

Calviño, C.I., S.G. Martínez, and **S.R. Downie.** 2008. The evolutionary history of *Eryngium* (Apiaceae, Saniculoideae): Rapid radiations, long distance dispersals, and hybridizations. Molec. Phylogen. Evol. 46:1129–1150.

Calviño, C.I., S.G. Martínez, and **S.R. Downie.** 2010. Unraveling the taxonomic complexity of *Eryngium* L. (Apiaceae, Saniculoideae): Phylogenetic analysis of 11 non-coding cpDNA loci corroborates rapid radiations. Pl. Diversity Evol. 128:137–149.

Camilli, K. 2003. Giant Asian dodder: A new invasive plant detected in Texas. Invasive Plants of the Eastern United States, unpaginated. http://www.invasive.org/eastern/other/texas/html. Accessed October 2021.

Camp, W.H. 1935. The genus *Gaylussacia* in North America north of Mexico. Bull. Torrey Bot. Club 62:129–132.

Camp, W.H. 1941. Studies in the Ericales: A review of the North American Gaylussacieae; with remarks on the origin and migration of the group. Bull. Torrey Bot. Club 68:531–551.

Camp, W.H. 1945. The North American blueberries with notes on other groups of Vacciniaceae. Brittonia 5:203–275.

Campbell, J.J.N., M. Evans, M.E. Medley, and **T.N. Taylor.** 1988. Buffalo clovers in Kentucky (*Trifolium stoloniferum* and *T. reflexum*): Historical records, presettlement environment, rediscovery, endangered status, cultivation, and chromosome number. Rhodora. 90:399–418.

Campbell, J.J.N. and **W.R. Seymour, Jr.** 2013. Towards a revision of the *Rudbeckia fulgida* complex (Asteraceae), with description of a new species from the blacklands of southern USA. Phytoneuron 2013-90:1–27.

Campbell, T.A. 1984. Agronomic and chemical evaluation of smooth sumac, *Rhus glabra*. Econ. Bot. 38:218–223.

Campos-Ríos, M.G. and **F. Chiang-Cabrera.** 2012. El género *Varronia* P. Browne (Cordiaceae, Boraginales) en México. Polibotánica 33: 1–32.

Cane, J.H. and **D. Schiffhauer.** 2003 Dose-response relationships between pollination and fruiting refine pollinator comparisons for cranberry (*Vaccinium macrocarpon* [Ericaceae]). Amer. J. Bot. 90(10):1425–1432.

Cane, J.H., G.C. Eickwort, F.R. Wesley, and **J. Spielholz.** 1985. Pollination ecology of *Vaccinium stamineum* (Ericaceae: Vaccinioideae). Amer. J. Bot. 72:135–142.

Caputo, P., S. Cozzolino, and **A. Moretti.** 2004. Molecular phylogenetics of Dipsacaceae reveals parallel trends in seed dispersal syndromes. Pl. Syst. Evol. 246:163–175.

Cardinal, S., S.L. Buchmann, and **A.L. Russell.** 2018. The evolution of floral sonication, a pollen foraging behavior used by bees (Anthophila). Evolution 72:590–600.

Cardinal-McTeague, W.M. and **L.J. Gillespie.** 2016. Molecular phylogeny and pollen evolution of Euphorbiaceae Tribe Plukenetieae. Sys. Bot. 41(2):329–347.

Cardoso, D.B.O.S., C.M.J. Mattos, F. Filardi, A. Delgado-Salinas, M. Lavin, P.L.R. de Moraes, F. Tapia-Pastrana, and **H.C. de Lima.** 2020. A molecular phylogeny of the pantropical papilionoid legume *Aeschynomene* supports reinstating the ecologically and morphologically coherent genus *Ctenodon*. Neodiversity 13:1–38.

Cardoso, D.B.O.S., G. Ramos, B. São-Mateus, W. Messias, and **L. Paganucci de Queiroz.** 2019. *Aeschynomene chicocesariana*, a striking new unifoliolate legume species from the Brazilian Chapada Diamantina and its phylogenetic placement in the Dalbergioid clade. Syst. Bot. 44:810–817.

Carlquist, S. 1976. Tribal interrelationships and phylogeny of the Asteraceae. Aliso 8:465–492.

Carlsen, T., R. Elven, W. Bleeker, H. Hurka, and **C. Brochmann.** 2009. Biogeography and phylogeny of *Cardamine* (Brassicaceae). Ann. Missouri Bot. Gard. 96:215–236.

Carr, W. 2008. An annotated checklist of the vascular plants of Travis County, Texas. Digitally published 2008 draft. https://fwbg.org/science-conservation/brit-press/

Carr, W. 2009. No place but Texas: an annotated list of plant taxa endemic to the lone star state. Digitally-published 2009 draft. The Nature Conservancy of Texas, Austin. http://www.guynesom.com/Texendemicsweb.htm. Accessed 2025.

Carulli, J.P. and **D.E. Fairbrothers.** 1988. Allozyme variation in three eastern United States species of *Aeschynomene* (Fabaceae) including the rare *A. virginica*. Syst. Bot. 13:559–566.

Carulli, J.P., A.O. Tucker, and **N.H. Dill.** 1988. *Aeschynomene rudis* Benth. (Fabaceae) in the United States. Bartonia 54:18–20.

Carvalho, F.A. 2013 onwards. e-Monograph of Caricaceae. Version 1, November 2013. [Database continuously updated]. https://herbaria.plants.ox.ac.uk/bol/caricaceae. Accessed Aug 2019.

Carvalho, F.A., D. Filer, and **S.S. Renner.** 2014. Taxonomy in the electronic age and an e-monograph of the papaya family (Caricaceae) as an example. Cladistics 31:321–329.

Carvalho, F.A. and **S.S. Renner.** 2012. A dated phylogeny of the papaya family (Caricaceae) reveals the crop's closest relatives and the family's biogeographic history. Molec. Phylogen. Evol. 65:46–53.

Carvalho, F.A. and **S.S. Renner.** 2013. Correct names for some of the closest relatives of *Carica papaya*: A review of the Mexican/Guatemalan genera *Jarilla* and *Horovitzia*. PhytoKeys 29:63–74.

Carvalho, F.A. and **S.S. Renner.** 2014. The Phylogeny of the Caricaceae. In: R. Ming and P.H. Moore, eds. Genetics and Genomics of Papaya. Plant Genetics and Genomics: Crops and Models 10. Springer Science and Business Media, New York.

Castetter, E.F., P. Pierce, and **K.H. Schwerin.** 1975. A reassessment of the genus *Escobaria*. Cact. Succ. J. (Los Angeles) 47:60–70.

Cathey, H.M. 1990. USDA plant hardiness zone map. Indicator plant examples. http://www.usna.usda.gov/Hardzone/hrdzon5.html. accessed Dec 2006.

Cellinese, N., S.A. Smith, E.J. Edwards, S.-T. Kim, R.C. Haberle, M. Avramakis, and **M.J. Donoghue.** 2009. Historical biogeography of the endemic Campanulaceae of Crete. J. Biogeogr. 36:1253–1269.

Cervantes, M.A. and **J.J. Ramirez.** 2002. *Bernardia fonsecae* (Euphorbiaceae), a new species from Guerrero, Mexico. Novon 12:38–41.

Chacón, J., F. Luebert, H.H. Hilger, S. Ovchinnikova, F. Selvi, L. Cecchi, C.M. Guilliams, K. Hasenstab-Lehman, K. Sutorý, M.G. Simpson, and **M. Weigend.** 2016. The borage family (Boraginaceae s.str.): A revised infrafamilial classification based on new phylogenetic evidence, with emphasis on the placement of some enigmatic genera. Taxon 65:523–546.

Chamberlain, D.F., R. Hyam, G. Argent, G. Fairweather, and **K.S. Walter.** 1996. The genus *Rhododendron*: Its classification and synonymy. Royal Botanic Garden, Edinburgh, Edinburgh, Scotland, U.K.

Chambers, K.L. 2004. Taxonomic notes on *Krigia* (Asteraceae). Sida 21:225–236.

Chambers, K.L. and **R.J. O'Kennon.** 2006. *Krigia*. In: Flora of North America Editorial Committee, eds. Fl. North Amer. 19:362–367. Oxford Univ. Press, New York and Oxford.

Champie, C. 2003. Discovery of a new variant of *Cylindropuntia leptocaulis* (DC.) F.M. Knuth. Cact. Succ. J. (Los Angeles) 75: 271–273.

Chandler, G.T. and **G.M. Plunkett.** 2004. Evolution in Apiales: Nuclear and chloroplast markers together in (almost) perfect harmony. Bot. J. Linn. Soc. 144:123–147.

Chandler, J.M., C.C. Jan, and **B.H. Beard.** 1986. Chromosomal differentiation among the annual *Helianthus* species. Syst. Bot. 11:354–371.

Channell, R.B. 1957. A revisional study of the genus *Marshallia* (Compositae). Contr. Gray Herb. 181:41–132.

Chapin, F.S., III. and **C. Körner.** 1995. Patterns, causes, changes and consequences of biodiversity in arctic and alpine ecosystems. In: F.S. Chapin, III and C. Körner, eds. Arctic and alpine biodiversity: Patterns, causes and ecosystem consequences. Pp. 313–320. Springer, Berlin, Germany.

Chapman, G.C. and **S.B. Jones.** 1978. Biosystematics of the Texanae Vernonias (Vernonieae: Compositae). Sida 7:264–281.

Charters, M.L. 2005. California plant names: Latin and Greek meanings and derivations: A dictionary of botanical etymology. http://www.calflora.net/botanicalnames/index.html. Accessed Jun 2007.

Charters, M.L. 2023. California plant names: Latin and Greek meanings and derivations; An annotated dictionary of botanical and biographical etymology. http://www.calflora.net/botanicalnames/. Accessed 24 Oct 2024.

Chase, M.W., D.E. Soltis, R. Olmstead, D. Morgan, and **38 others.** 1993. Phylogenetics of seed plants: An analysis of nucleotide sequences from the plastid gene *rbc*L. Ann. Missouri Bot. Gard. 80:528–580.

Chatrou, L.W., M.D. Pirie, R.H.J. Erkens, T.L.P. Couvreur, K.M. Neubig, J.R. Abbott, J.B. Mols, J.S. Maas, R.M.K. Saunders, and **M.W. Chase.** 2012. A new subfamilial and tribal classification of the pantropical flowering plant family Annonaceae informed by molecular phylogenetics. Bot. J. Linn. Soc. 169: 5–40.

Chaudhri, M.N. 1968. A revision of the Paronychiinae. Meded. Bot. Mus. Herb. Rijks Univ. Utrecht 285:306–397.

Chaurasiya, A., R.K. Pal, P.K. Verma, A. Katiyar, R. Kumar, and **N. Kumar.** 2021. An updated review on Malabar spinach (*Basella alba* and *Basella rubra*) and their importance. J. Pharmacognosy Phytochem. 10:1201–1207.

Cheatham, S. and **M.C. Johnston.** 1995. The useful wild plants of Texas, the southeastern and southwestern United States, the southern Plains, and northern Mexico. Vol. 1, *Abronia–Arundo*. Useful Wild Plants, Inc., Austin, TX.

Chen, Z. and **J. Li.** 2004. Phylogenetics and biogeography of *Alnus* (Betulaceae) inferred from sequences of nuclear ribosomal DNA ITS region. Int. J. Pl. Sci. 165:325–335.

Chen, X.-H., K.-L. Xiang, L. Lian, H.-W. Peng, A.S. Erst, X.-G. Xiang, Z.-D. Chen, and **W. Wang.** 2020. Biogeographic diversification of *Mahonia* (Berberidaceae): Implications for the origin and evolution of East Asian subtropical evergreen broadleaved forests. Molec. Phylogen. Evol. 151:e106910. https://doi: 10.1016/j.ympev.2020.106910.

Chen, Z.-D., S.R. Manchester, and **H.-Y. Sun.** 1999. Phylogeny and evolution of the Betulaceae as inferred from DNA sequences, morphology, and paleobotany. Amer. J. Bot. 86:1168–1181.

Chernetskyy, M. 2011. Problems in nomenclature and systematics in the subfamily Kalanchoideae (Crassulaceae) over the years. Acta Agrobot. 64:67–74.

Christenhusz, M.J.M., M.S. Vorontsova, M.F. Fay, and **M.W. Chase.** 2015. Results from an online survey of family delimitation in angiosperms and ferns: Recommendations to the Angiosperm Phylogeny Group for thorny problems in plant classification. Bot. J. Linn. Soc. 178:501–528.

Chung, K.-F. 2007. Inclusion of the South Pacific Alpine genus *Oreomyrrhis* (Apiaceae) in *Chaerophyllyum* based on nuclear and chloroplast DNA sequences. Sys. Bot. 32:671–681.

Chung, K.-F., C.-I. Peng, S.R. Downie, K. Spalik, and **B.A. Schaal.** 2005. Molecular Systematics of the Trans-Pacific Alpine genus *Oreomyrrhis* (Apiaceae): phylogenetic affinities and biogeographic implications. Amer. J. Bot. 92(12):2054–2071.

Clark, C. 1975. Ecogeographic races of *Lesquerella engelmannii* (Cruciferae); distribution, chromosome numbers, and taxonomy. Brittonia 27:263–278.

Clark, R.C. 2023. Hollies (Aquifoliaceae) of North America North of Mexico. Phytoneuron 2023-13:1–19.

Clark, T. and **K.C. Lund.** 2012. Presence of the indole alkaloid reserpine in *Bignonia capreolata* L. Inter. J. Pharmacog. Phytochem. Res. 4:89–91.

Clark, W. 1806. Journals of the Lewis and Clark Expedition, Thursday 18th of September 1806. https://lewisandclarkjournals.unl.edu/item/lc.jrn.1806-09-18.

Clausen, J. 1949. Evolutionary patterns in the genus *Crepis*. Evolution 3:185–188.

Clausen, R.T. 1975. *Sedum* of North America, north of the Mexican Plateau. Cornell Univ. Press, Ithaca, NY.

Clemants, S.E. 1992. Chenopodiaceae and Amaranthaceae of New York state. Bull. New York State Mus. Sci. Serv. Bull. No. 485:i–vi, 1–100.

Clemants, S.E. 2003a. *Alternanthera*. In: Flora of North America Editorial Committee, eds. Fl. North Amer. 4:447–451. Oxford Univ. Press, New York and Oxford.

Clemants, S.E. 2003b. *Blutaparon*. In: Flora of North America Editorial Committee, eds. Fl. North Amer. 4:56. Oxford Univ. Press, New York and Oxford.

Clemants, S.E. 2003c. *Gomphrena*. In: Flora of North America Editorial Committee, eds. Fl. North Amer. 4:451–454. Oxford Univ. Press, New York and Oxford.

Clemants, S.E. 2003d. *Gossypianthus*. In: Flora of North America Editorial Committee, eds. Fl. North Amer. 4:438–439. Oxford Univ. Press, New York and Oxford.

Clemants, S.E. 2003e. *Guilleminea*. In: Flora of North America Editorial Committee, eds. Fl. North Amer. 4:437–438. Oxford Univ. Press, New York and Oxford.

Clemants, S.E. 2003f. *Iresine*. In: Flora of North America Editorial Committee, eds. Fl. North Amer. 4:454–456. Oxford Univ. Press, New York and Oxford.

Clemants, S.E. and **S.L. Mosyakin.** 2003a. *Dysphania*. In: Flora of North America Editorial Committee, eds. Fl. North Amer. 4: 267–275. Oxford Univ. Press, New York and Oxford.

Clemants, S.E. and **S.L. Mosyakin.** 2003b. *Chenopodium*. In: Flora of North America Editorial Committee, eds. Fl. North Amer. 4: 275–299. Oxford Univ. Press, New York and Oxford.

Clevinger, J.A. 2006. *Silphium*. In: Flora of North America Editorial Committee, eds. Fl. North Amer. 21:77–82. Oxford Univ. Press, New York and Oxford.

Clewell, A.F. 1966. Native North American species of *Lespedeza* (Leguminosae). Rhodora 68:359–408.

Clewell, A.F. 1968. Some comments on American Lespedezas (Leguminosae). Sida 3:206–208.

Clewell, A.F. and **J.W. Wooten.** 1971. A revision of *Ageratina* (Compositae: Eupatorieae) from Eastern North America. Brittonia 23:123–143.

Clonts, J.A. and **S. McDaniel.** 1978. *Elephantopus*. N. Amer. Fl. II. 10:196–202.

Coder, K.D. 2016. Southern catalpa, *Catalpa bignonioides*. Univ. of Georgia Warnell School of Forestry & Natural Resources Outreach. Pub. 15:1–15.

Cody, W.J. 1954. A history of *Tillaea aquatica* (Crassulaceae) in Canada and Alaska. Rhodora 56:96–101.

Coffey, T. 1993. The history and folklore of North American wildflowers. Houghton Mifflin, Boston, MA.

Cohen, J.I. 2015. *Adelinia* and *Andersonglossum* (Boraginaceae), two new genera from New World species of *Cynoglossum*. Syst. Bot. 40:611–619.

Cohen, J.I. 2018. A revision of the Mexican species of *Lihospermum* (Boraginaceae). Ann. Missouri Bot. Gard. 103:200–257.

Cohen, J.I. and **J.I. Davis.** 2009. Nomenclatural changes in *Lithospermum* (Boraginaceae) and related taxa following a reassessment of phylogenetic relationships. Brittonia 61:101–111.

Coleman, J.R. 1966. A taxonomic revision of section *Ximenesia* of the genus *Verbesina* (Compositae). Amer. Midl. Naturalist 76: 475–481.

Coleman, J.R. 1968. *Verbesina*, a cytotaxonimic study. Rhodora 70: 95–102.

Constance, L. and **J.M. Affolter.** 1997 A pair of species of *Tauschia* Schltdl. (Umbelliferae/Apiaceae) from Mexico. Brittonia 49: 458–462.

Constance, L. and **R.H. Shan.** 1948. The genus *Osmorhiza* (Umbelliferae): A study in geographic affinities. Univ. Calif. Publ. Bot. 23:111–156.

Cooperrider, T.S. 1985. *Thaspium* and *Zizia* (Umbelliferae) in Ohio. Castanea 50:116–119.

Coquillat, M. 1951. Sur les plantes les plus communes a la surface du globe. Bull. Mens. Soc. Linn. Lyon 20(7):165–170.

Core, E.L. 1941. The North American species of *Paronychia*. Amer. Midl. Naturalist 26:396–397.

Correa, A., M.D. and **T.R. dos Santos Silva.** 2005. *Drosera* (Droseraceae). Fl. Neotrop. Monogr. 96:1–65.

Correll, D.S. 1965. Some additions and contributions to the flora of Texas. Wrightia 3:126–140.

Correll, D.S. and **H.B. Correll.** 1972. Aquatic and wetland plants of southwestern United States. U.S. Environmental Protection Agency. U.S. Government Printing Office, Washington, D.C.

Correll, D.S. and **M.C. Johnston.** 1970. Manual of the vascular plants of Texas. Texas Research Foundation, Renner.

Cory, V.L. 1940. Six thistles recently introduced into Texas. Madroño 5:200–202.

Cory, V.L. 1946. The genus *Palafoxia* in Texas. Rhodora 48:84–86.

Cory, V.L. 1947. Old World plants apparently recently introduced into Texas. Madroño 9:64.

Cory, V.L. 1948. Some first records of plant species collected in Texas. Field & Lab. 16:82–89.

Cory, V.L. 1950. Additional records of plants introduced into Texas. Field & Lab. 18:89–92.

Cosner, M.E., R.K. Jansen, and **T.G. Lammers.** 1994. Phylogenetic relationships in Campanulales based on *rbcL* sequences. Pl. Syst. Evol. 190:79–95.

Cosner, M.E, L.A. Raubeson, and **R.K. Jansen.** 2004. Chloroplast DNA rearrangements in Campanulaceae: Phylogenetic utility of highly rearranged genomes. B.M.C. Evol. Biol. https://doi.org/10.1186/1471-2148-4-27.

Costea, M., M.A. García, and **S. Stefanović.** 2015. A phylogenetically based infrageneric classification of the parasitic plant genus *Cuscuta* (Dodders, Convolvulaceae). Syst. Bot. 40:269–285.

Costea, M. and **G.L. Nesom.** 2023. *Cuscuta*. In: Flora of North America Editorial Committee, eds. Fl. North Amer. 14:271–302. Oxford Univ. Press, New York and Oxford.

Costea, M., G.L. Nesom, and **S. Stefanović.** 2006a. Taxonomy of the *Cuscuta pentagona* complex (Convolvulaceae) in North America. Sida 22:151–175.

Costea, M., G.L. Nesom, and **S. Stefanović.** 2006b. Taxonomy of *Cuscuta gronovii* and *Cuscuta umbrosa* (Convolvulaceae). Sida 22:197–207.

Costea, M., G.L. Nesom, and **S. Stefanović.** 2006c. Taxonomy of the *Cuscuta indecora* (Convolvulaceae) complex in North America. Sida 22:209–225.

Costea, M., A. Sanders, and **G. Waines.** 2001. Preliminary results toward a revision of the *Amaranthus* hybridus species complex (Amaranthaceae). Sida 19:975–992.

Cota, J.H. and **C.T. Philbrick.** 1994. Chromosome number variation and polyploidy in the genus *Echinocereus* (Cactaceae). Amer. J. Bot. 81:1054–1062.

Cotton, T. and **A.D. Nelson.** 2018. *Drosera (Droseraceae)*. Texas J. Sci. 70(1):30–55.

Coulter, J.M. and **J.N. Rose.** 1900. Monograph of the North American Umbelliferae. Contr. U.S. Natl. Herb. 7:1–256.

Coulter, J.M. and **W.H. Evans.** 1890. A revision of the North American Cornaceae. Bot. Gaz. 15:30–38, 86–97.

Couplan, F. 1998. The encyclopedia of edible plant of North America. Keats Publishing. New Canaan, CT.

Couvreur, T.L.P., A.J. Helmstetter, E.J.M. Koenen, K. Bethune, R.D. Brandão, S.A. Little, H. Sauquet, and **R.H.J. Erkens.** 2019. Phylogenomics of the major tropical plant family Annonaceae using targeted enrichment of nuclear genes. Front. Pl. Sci. https://doi.org/10.3389/fpls.2018.01941.

Cox, C.B. and **P.D. Moore.** 1993. Biogeography: An ecological and evolutionary approach, 5th ed. Blackwell Scientific Publications, Oxford, England, UK.

Cox, P.A. 1988. Hydrophilous pollination. Annual Rev. Ecol. Syst. 19:261–279.

Cox, P.B. and **L.E. Urbatsch.** 1990. A phylogenetic analysis of the coneflower genera (Asteraceae: Heliantheae). Syst. Bot. 15:394–402.

Cox, P.B. and **L.E. Urbatsch.** 1994. A taxonomic revision of *Rudbeckia* subg. *Macrocline* (Asteraceae: Heliantheae: Rudbeckiinae). Castanea 59:300–318.

Cox, P.B. and **L.E. Urbatsch.** 2006. *Ratibida*. In: Flora of North America Editorial Committee, eds. Fl. North Amer. 21:60–63. Oxford Univ. Press, New York and Oxford.

Cox, P.W. and **P. Leslie.** 1991. Texas trees. Corona Publishing Co., San Antonio, TX.

Craig, R.T. 1945. The *Mammillaria* handbook. Abbey Garden Press, Pasadena, CA.

Crawford, D.J. and **H.D. Wilson.** 1986. *Chenopodium*. In: Great Plains Flora Association. Flora of the Great Plains. Pp. 166–173. Univ. Press of Kansas, Lawrence.

Croizat, L. 1942. New and critical Euphorbiaceae chiefly from the southeastern United States. Bull. Torrey Bot. Club 69(6):445–460.

Cronquist, A. 1943. The separation of *Erigeron* from *Conyza*. Bull. Torrey Bot. Club 70:629–632.

Cronquist, A. 1947. A revision of the North American species of *Erigeron* north of Mexico. Brittonia 6:121–302.

Cronquist, A. 1977. The Compositae revisited. Brittonia 29:137–153.

Cronquist, A. 1980. Asteraceae. Vascular flora of the southeastern United States 1:1–261. The Univ. of North Carolina Press, Chapel Hill.

Cronquist, A. 1981. An integrated system of classification of flowering plants. Columbia Univ. Press, New York.

Cronquist, A. 1982. Reduction of *Pseudotaenidia* to *Taenidia* (Apiaceae). Brittonia 34:365–367.

Cronquist, A. 1985. History of generic concepts in the Compositae. Taxon 34:6–10.

Cronquist, A. 1988. The evolution and classification of flowering plants, 2nd ed. New York Bot. Garden, Bronx.

Cronquist, A. 1993. A commentary on the general system of classification of flowering plants. In: Flora of North America Editorial Committee, eds. Fl. North Amer. 1:272–293. Oxford Univ. Press, New York and Oxford.

Cronquist, A. and **R.F. Thorne.** 1994. Nomenclatural and taxonomic history. In: H.-D. Behnke and T.J. Mabry, eds. Caryophyllales: Evolution and systematics. Pp. 5–25. Springer-Verlag, New York.

Crook, R. and **R. Mottram.** 1997. *Opuntia* index: Part 3: Nomenclatural note and F. Bradleya 15:98–112.

Crosswhite, F.S. 1980. Dry country plants of the south Texas Plains. Desert Pl. 2:141–179.

Crosswhite, F.S. 1984. Crassulacean acid metabolism. Desert Pl. 5:192.

Crosswhite, F.S. and **C.D. Crosswhite.** 1985. The southwestern pipevine (*Aristolochia watsonii*) in relation to snakeroot oil, swallowtail butterflies, and Ceratopogonid flies. Desert Pl. 6:203–207.

Crosthwaite, S. (Sally). Personal communication. Botanist trained at Vanderbilt and Columbia, plant collector in north TX, and former instructor at Austin College, Sherman, TX; retired, Grayson Co., TX.

Crow, G.E. 1978. A taxonomic revision of *Sagina* (Caryophyllaceae) in North America. Rhodora 80:1–91.

Crow, G.E. 1979. The systematic significance of seed morphology in *Sagina* (Caryophyllaceae) under scanning electron microscopy. Brittonia 31:52–63.

Crow, G.E. 2005. *Sagina*. In: Flora of North America Editorial Committee, eds. Fl. North Amer. 5:140–147. Oxford Univ. Press, New York and Oxford.

Crow, G.E. and **C.B. Hellquist.** 2000a. Aquatic and wetland plants of northeaster North America. Vol. 1. Pteridophytes, gymnosperms, and angiosperms: Dicotyledons, Univ. of Wisconsin Press, Madison.

Crowl, A.A., P.W. Fritsch, G.P. Tiley, N.P. Lynch, T.G. Ranney, H. Ashrafi, and **P.S. Manos.** 2022. A first complete phylogenomic hypothesis for diploid blueberries (*Vaccinium* section *Cyanococcus*). Amer. J. Bot. 109:1596–1606.

Crowl, A.A., E. Mavrodiev, G. Mansion, R. Haberle, A. Pistarino, G. Kamari, D. Phitos, T. Borsch, and **N. Cellinese.** 2014. Phylogeny of Campanuloideae (Campanulaceae) with emphasis on the utility of nuclear pentatricopeptide repeat (PPR) genes. P.L.o.S. One 9, e94199.

Crowl, A.A., N.W. Miles, C.J. Visger, K. Hansen, T. Ayers, R. Haberle, and **N. Cellinese.** 2016. A global perspective on Campanulaceae: Biogeographic, genomic and floral evolution. Amer. J. Bot. 103:233–245.

Cuénoud, P. 2003. Introduction to expanded Caryophyllales. In: K. Kubitzki, ed. The families and genera of vascular plants, Vol. V. Pp. 1–4, Springer-Verlag, Berlin.

Cullings, K. 2000. Reassessment of phylogenetic relationships of some members of the Monotropoideae based on partial 28S ribosomal RNA gene sequencing. Canad. J. Bot. 78:1–2.

Cullings, K.W. and **L. Hileman.** 1997. The Monotropoideae is a monophyletic sister group to the Arbutoideae (Ericaceae): A molecular test of Copeland's hypothesis. Madroño 44: 297–299.

Cupido, C.N., J.M. Prebble, and **W.M.M. Eddie.** 2014. Phylogeny of southern African and Australasian Wahlenbergioides (Campanulaceae) based on ITS and *trn*L-F sequence data: Implications for a reclassification. Syst. Bot. 38:523–535.

Dabydeen, S. 1997. Natural hybridization in the genus *Cirsium*: *C. altissimum* × *C. discolor*: cytological and morphological evidence. Rhodora 99(898):152–160.

Dallimore, W. 1912. The Fullers Teasel. Kew Bull. 1912.345–350.

Daniel, T.F. 1980. Range extensions of *Carlowrightia* (Acanthaceae) and a key to the species of the United States. S. W. Naturalist 25:425–426.

Daniel, T.F. 1982. *Anisacanthus andersonii* (Acanthaceae), a new species from northwestern Mexico. Bull. Torrey Bot. Club 109: 148–151.

Daniel, T.F. 1983. *Carlowrightia* (Acanthaceae). Fl. Neotrop. Monogr. 34:1–116.

Daniel, T.F. 1984a. The Acanthaceae of the southwestern United States. Desert Pl. 5:162–179.

Daniel, T.F. 1984b. (738) Proposal to conserve 8014 *Carlowrightia* against *Cardiacanthus* (Acanthaceae). Taxon 33:122–123.

Daniel, T.F. 1988. Taxonomic, nomenclatural and reproductive notes on *Carlowrightia* (Acanthaceae). Brittonia 40:245–255.

Daniel, T.F. 1995. Acanthaceae. In: D.E. Breedlove, ed. Flora of Chiapas 4:1–158. California Acad. Sciences, San Francisco.

Daniel, T.F. 2000. Additional chromosome numbers of American Acanthaceae. Syst. Bot. 25:15–25.

Daniel, T.F. 2003. New and reconsidered Mexican Acanthaceae X. Flora del Bajío region. Novon 13:37–48.

Daniel, T.F. 2005. Catalog of Honduran Acanthaceae with taxonomic and phytogeographic notes. Contr. Univ. Michigan Herb. 24:51–108.

Daniel, T.F. 2011. *Justicia* (Acanthaceae) in Texas. J. Bot. Res. Inst. Texas 5:595–618.

Daniel, T.F. 2013. Taxonomic, distributional, and nomenclatural notes in North American Acanthaceae. In: W.R. Buck and R.F.C. Naczi, eds. Harmony and grit: paper celebrating the Holmgren's completion of Inermountain flora vol.108. Mem. New York Bot. Gard. 108:85–114.

Daniel, T.F. and **J. Hendrickson.** 1982. On the recognition of *Anisacanthus junceus* (Acanthaceae). Brittonia 34:177–180.

Daoud, H.S. and **R.L. Wilbur.** 1965. A revision of the North American species of *Helianthemum* (Cistaceae). Rhodora 67: 63–82, 201–216, 255–312.

D'Arcy, W.G. 1975. *Pterocaulon*, in Ann. Missouri Bot. Gard. 62: 1048–1049.

Daughtrey, M. 1993. Dogwood anthracnose disease: Native fungus or exotic invader? In: B.N. McNight, ed. Biological pollution: The control and impact of invasive exotic species. Indiana Academy of Science, Indianapolis.

Davis, A.M. and **O.W. Van Auken.** 2019. Ecological characteristics and distribution of *Arbutus xalapensis* (Texas madrone) populations in central Texas. Madroño 66:47–55.

Davis, C.C. and **M.W. Chase.** 2004. Elatinaceae are sister to Malpighiaceae; Peridiscaceae belong to Saxifragales. Amer. J. Bot. 91:262–273.

Davis, C.C., P.W. Fritsch, J. Li, and **M.J. Donoghue.** 2002. Phylogeny and biogeography of *Cercis* (Fabaceae): Evidence from nuclear ribosomal ITS and chloroplast *ndhF* sequence data. Syst. Bot. 27:289–302.

Decker, D.S. 1988. Origin(s), evolution, and systematics of *Cucurbita pepo* (Cucurbitaceae). Econ. Bot. 42:4–15.

Decker, D.S. and **H.D. Wilson.** 1986. Numerical analysis of seed morphology in *Cucurbita pepo*. Syst. Bot. 11:595–607.

Decker, D.S. and **H.D. Wilson.** 1987. Allozyme variation in the *Cucurbita pepo* complex: *C. pepo* var. *ovifera* vs. *C. texana*. Syst. Bot. 12:263–273.

Decker-Walters, D.S. 1990. Evidence for multiple domestication of *Cucurbita pepo*. In: D.M. Bates, R.W. Robinson, and C. Jeffrey, eds. Biology and utilization of the Cucurbitaceae. Pp. 96–101. Cornell Univ. Press, Ithaca, New York.

Decker-Walters, D.S., S.-M. Chung, J.E. Staub, H.D. Quemada, and **A.I. López-Sesé.** 2002. The origin and genetic affinities of wild populations of melon (*Cucumis melo*, Cucurbitaceae) in North America. Pl. Syst. Evol. 233:183–197.

Deegan, R.D. 2012. Finessing the fracture energy barrier in ballistic seed dispersal. Proc. Natl. Acad. Sci. U.S.A. 109:5166–5169.

DeFilipps, R.A. 1976. *Hypochoeris*. In: T.G. Tutin, V.H. Heywood, N.A. Burges, D.M. Moore, D.H. Valentine, S.M. Walters, and D.A. Webb, eds. Flora Europaea 4:308–310. Cambridge Univ. Press, Cambridge, England, UK.

Degtjareva, G.V., T.E. Kramina, D.D. Sokoloff, T.H. Samigullin, G. Sandral, and **C.M. Valiejo-Roman.** 2008. New data on nrITS phylogeny of *Lotus* (Leguminosae, Loteae). Wulfenia 15:35–49.

Degtjareva, G.V., T.E. Kramina, D.D. Sokoloff, T.H. Samigullin, C.M. Valiejo-Roman, and **A.S. Antonov.** 2006. Phylogeny of the genus *Lotus* (Leguminosae, Loteae): Evidence from nrITS sequences and morphology, Canad. J. Bot. 84:813–830.

DeJong, D.C.D. 1965. A systematic study of the genus *Astranthium* (Compositae, Astereae). Publ. Mus. Michigan State Univ., Biol. Ser. 2:429–528.

Delgado-Salinas, A. 1985. Systematics of the genus *Phaseolus* (Leguminosae) in North and Central America. Ph.D. dissertation (Botany). Univ. of Texas, Austin.

Delgado-Salinas, A. 2023a. *Oxyrhynchus*. In: Flora of North America Editorial Committee, eds. Fl. of North Amer. 11:412–413. Oxford Univ. Press, New York and Oxford.

Delgado-Salinas, A. 2023b. *Phaseolus*. In: Flora of North America Editorial Committee, eds. Fl. of North Amer. 11:414–423. Oxford Univ. Press, New York and Oxford.

Delgado-Salinas, A. 2023c. *Vigna*. In: Flora of North America Editorial Committee, eds. Fl. of North Amer. 11:405–408. Oxford Univ. Press, New York and Oxford.

Delgado-Salinas, A., R. Bibler, and **M. Lavin.** 2006. Phylogeny of the genus *Phaseolus* (Leguminosae): A recent diversification in an ancient landscape. Syst. Bot. 31:779–791.

Delgado-Salinas, A. and **W.R. Carr.** 2007. *Phaseolus texensis* (Leguminosae: Phaseolinae): A New Species from the Edwards Plateau of Central Texas. Lundellia 10:11–17.

Delgado-Salinas, A., M. Lavin, and **E.T. Riley.** 2023. *Strophostyles*. In: Flora of North America Editorial Committee, eds. Fl. of North Amer. 11:423–425. Oxford Univ. Press, New York and Oxford.

Delgado-Salinas, A. and **M.A. Parker.** 2023. *Amphicarpaea.* In: Flora of North America Editorial Committee, eds. Fl. of North Amer. 11:399–401. Oxford Univ. Press, New York and Oxford.

De Queiroz, L.P., R.H. Fortunato, and **A.M. Giulietti.** 2003. Phylogeny of the Diocleinae (Papilionoideae: Phaseoleae) based on morphological characters. In: B.B. Klitgaard and A. Bruneau, eds. Advances in legume systematics. Part 10. Pp. 303–324. Royal Botanic Gardens, Kew, England, U.K.

De Santis, S., F. Michelangeli, F. Spada, and **D. Magri.** 2024. Longitudinal population dynamics of Mediterranean-Atlantic Arbutus during the last 30 ka. Rev. Palaeobot. Palynol. https://doi.org/10.1016/j.revpalbo.2024.105099.

Desrochers, A.M., J.F. Bain, and **S.I. Warwick.** 1988. The biology of Canadian weeds. 89. *Carduus nutans* L. and *Carduus acanthoides* L. Canad. J. Pl. Sci. 68:1053–1068.

Diamond, A.R. 2023. New combinations in *Asarum* (Aristolochiaceae) for two recently described *Hexastylis* species. J. Bot. Res. Inst. Texas 17:427–429.

Diels, L. 1906. Droseraceae. In: A. Engler, ed. Das Pflanzenreich 4.112:1–136. Verlag von Wilhelm Engelmann, Leipzig.

Diggs, G.M., Jr. 1982. The Campanulaceae (including Lobelioideae) of Virginia. Virginia J. Sci. 33:206–221.

Diggs, G.M., Jr. and **B.L. Lipscomb.** 2005. *Hydrocotyle sibthorpioides* (Apiaceae) new for Texas and notes on introduced species. Sida 21:2449–2456.

Diggs, G.M. and **B.L. Lipscomb.** 2014. The ferns and Lycophytes of Texas. Sida Bot. Misc. 38.

Diggs, G.M., Jr., B.L. Lipscomb, and **R.J. George.** 2025. *Asarum arifolium* (Aristolochiaceae) new for Texas (U.S.A.). J. Bot. Res. Inst. Texas 19(2):147–153.

Diggs, G.M., Jr., B.L. Lipscomb, and **R.J. O'Kennon.** 1999. Shinners and Mahler's Illustrated Flora of North Central Texas. Sida Bot. Misc. 16.

Diggs, G.M., Jr., B.L. Lipscomb, M.D. Reed, and **R.J. O'Kennon.** 2006. Illustrated flora of East Texas. Vol. 1. Sida Bot. Misc. 26.

Diggs, G.M., Jr. and **R.J. O'Kennon.** 2003. A rockhouse microhabitat in the West Cross Timbers of North Central Texas. Sida 20: 1327–1333.

Diggs, G.M., Jr., R.J. O'Kennon, and **B.L. Lipscomb.** 1997b. *Hypochaeris glabra* (Asteraceae), a new record for Texas. Sida 17(3):633–634.

Dilcher, D. 2000. Geological history of the vegetation in southwest United States. In: B.L. Lipscomb, J.J. Pipoly, III, and R.W. Sanders, eds. Floristics in the new millennium: Proceeding of the flora of the southeast US symposium. Sida Bot. Misc. 18:1–21.

Dillenberger, M.S. and **J.W. Kadereit.** 2014. Maximum polyphyly: Multiple origins and delimitation with plesiomorphic characters require a new circumscription of *Minuartia* (Caryophyllaceae). Taxon 63:64–88.

Dixon, G.R. 2007. Vegetable brassicas and related crucifers. Crop Production Science in Horticulture, 14. CABI, Walllingford, Oxfordshire, UK.

Doffitt, C., C. Allen, P. Lewis, and **D. Lewis.** 2014. *Amsonia ludoviciana* (Apocynaceae) new to the flora of Texas. J. Bot. Res. Inst. Texas 8(2):663–664.

Dogterom, M.H., M.L. Winston, and **A. Mukai.** 2000. Effect of pollen load size and source (self, outcross) on seed and fruit production in highbush blueberry cv. 'Bluecrop' (Vaccinium corymbosum; Ericaceae). Amer. J. Bot. 87(11):1584–1591.

Donoghue, M.J., C.D. Bell, and **J. Li.** 2001. Phylogenetic patterns in northern hemisphere plant geography. Int. J. Pl. Sci. 162 (6 Suppl.):S41–S52.

Donoghue, M.J, T. Eriksson, P.A. Reeves, and **R.G. Olmstead.** 2001. Phylogeny and phylogenetic taxonomy of Dipsacales with special reference to *Sinadoxa* and *Tetradoxa* (Adoxaceae). Havard Pap. Bot. 6:459–479.

Donoghue, M.J. and **B.R. Moore.** 2003. Toward an integrative historical biogeography. Integr. Comp. Biol. 43:261–270.

Donoghue, M.J, R.G. Olmstead, J.F. Smith, and **J.D. Palmer.** 1992. Phylogenetic relationships of Dipsacales based on *rbc*L sequences. Ann. Missouri Bot. Gard. 79:333–345.

Donoghue, M.J. and **S.A. Smith.** 2004. Patterns in the assembly of temperate forests around the Northern Hemisphere. Philos. Trans., Ser. B 359:1633–1644.

Dore, W.G. 1970. A wild ground-bean, *Amphicarpaea*, for the garden. Greenh. Gard. Grass 9(2):7–11.

Dorsey, B.L., T. Haevermans, X. Aubriot, J.J. Morawetz, R. Riina, V.W. Steinmann, and **P.E. Berry.** 2013. Phylogenetics, morphological evolution, and classification of *Euphorbia* subgenus *Euphorbia.* Taxon 62(2):291–315.

Downie, S.R., R.L. Hartman, F.-J. Sun, and **D.S. Katz-Downie.** 2002 Polyphyly of the spring-parsleys (Cymopterus): Molecular and morphological evidence suggests complex relationships among the perennial endemic genera of western North American Apiaceae. Canad. J. Bot. 80(12):1295–1324.

Downie, S.R., D.S. Katz-Downie, and **K.J. Cho.** 1997. Relationships in the Caryophyllales as suggested by phylogenetic analyses of partial chloroplast DNA ORF2280 homolog sequences. Amer. J. Bot. 84:253–273.

Downie, S.R., D.S. Katz-Downie, F.-J. Sun, and **C.-S. Lee.** 2008. Phylogeny and biogeography of Apiaceae tribe Oenantheae inferred from nuclear rDNA ITS and cpDNA psbl-5'trnK$^{(UUU)}$ sequences, with emphasis on the North American Endemics clade. Botany (Ottawa) 86:1039–1064.

Downie, S.R., D.S. Katz-Downie, and **M.F. Watson.** 2000. A phylogeny of the flowering plant family Apiaceae based on chloroplast DNA *rpl16* and *rpoC1* intron sequences: Towards a suprageneric classification of subfamily Apioideae. Amer. J. Bot. 87:273–292.

Downie, S.R. and **J.D. Palmer.** 1994. A chloroplast DNA phylogeny of the Caryophyllales based on structural and inverted repeat restriction site variation. Syst. Bot. 19:236–252.

Downie, S.R., K. Spalik, D.S. Katz-Downie, and **J.-P. Reduron.** 2010. Major clades within Apiaceae subfamily Apioideae as inferred by phylogenetic analysis of nrDNA ITS sequences. Pl. Diversity Evol. 128:111–136.

Doyle, J.A., M.J. Donoghue, and **E.A. Zimmer.** 1994. Integration of morphological and ribosomal RNA data on the origin of angiosperms. Ann. Missouri Bot. Gard. 81:419–450.

Dress, W.J. 1966. Notes on the cultivated Compositae: 9. *Onopordum.* Baileya 14:74–86.

Du, Z.-Y., J. Cheng, and **Q.-Y. (J.) Xiang.** 2023b. RAD-seq data provide new insights into biogeography, diversity anomaly, and species delimitation in eastern Asian–North American disjunct clade *Benthamidia* of *Cornus* (Cornaceae). J. Syst. Evol. 62:1–19.

Du, Z.-Y., Q.-Y. (J.) Xiang, J. Cheng, W. Zhou, Q.-F. Wang, D.E. Soltis, and **P.S. Soltis.** 2023a. An updated phylogeny, biogeography, and PhyloCode-based classification of Cornaceae based on three sets of genomic data. Amer. J. Bot. 2023 Feb; 110(2):e16116. https://doi: 10.1002/ajb2.16116. Epub 2023 Feb 15. PMID: 36480351.

Duangjai, S., R. Samuel, F. Munzinger, F. Forest, B. Wallnöffer, M.H.J. Barfuss, G. Fischer, and **M.W. Chase.** 2009. A multilocus plastid phylogenetic analysis of the pantropical genus *Diospyros* (Ebenaceae), with an emphasis on the radiation and biogeographic origins on the New Caledonian endemic species. Molec. Phylogen. Evol. 52:602–620.

Duangjai, S., B. Wallnöfer, R. Samuel, J. Munzinger, and **M.W. Chase.** 2006. Generic delimitation and relationships in Ebenaceae sensu lato: Evidence from six plastid DNA regions. Amer. J. Bot. 93:1808–1827.

Duistermaat, H. 1996. Monograph of *Arctium* L. (Asteraceae). Gorteria, Suppl. 3:1–143.

Duke, J.A. 1981. Handbook of legumes of world economic importance. Plenum, New York.

Duke, J.A. 1983. *Aleurites fordii* Hemsl. (in Handbook of Energy Crops., unpublished). http://www.hort.purdue.edu/newcrop/duke_energy/Aleurites_fordii.html. Accessed Jul 2008.

Duke, J.A. 1985. CRC handbook of medicinal herbs. CRC Press, Inc., Boca Raton, FL.

Duncan, T.M. and **M.D. Rausher.** 2013. Morphological and genetic differentiation and reproductive isolation among closely related taxa in the *Ipomoea* series *Batatas*. Amer. J. Bot. 100(11): 2183–2193.

Duncan, W.H. 1959. Leaf variation in *Liquidambar styraciflua* L. Castanea 24:99–111.

Duncan, W.H. 1964. New *Elatine* (Elatinaceae) populations in the southeastern United States. Rhodora 66:47–53.

Duncan, W.H. 1979. Changes in *Galactia* (Fabaceae) of the southeastern United States. Sida 8:170–180.

Duncan, W.H., P.L. Piercy, S.D. Feurt, and **R. Starling.** 1957. Toxicological studies of southeastern plants. II. Compositae. Econ. Bot. 11:75–85.

Durand, H.K. 1972. Texas mahonia—A neglected economic plant. Econ. Bot. 26:319–325.

Durkee, L.H. 1999. Five new taxa and two new combinations of Acanthaceae from Central America. Novon 9:503–510.

Durkee, L.T., M.H. Harber, L. Dorn, and **A. Remington.** 1999. Morphology, ultrastructure, and function of extrafloral nectarines in three species of Caesalpiniaceae. J. Iowa Acad. Sci. 106:82–88.

Dute, R.R., D.R. Folkerts, J.E. Watkins, M.E. Saxon, and **R.S. Boyd.** 2004. Floral anatomy and pollination biology of *Cyrilla racemiflora* (Cyrillaceae). S. E. Naturalist (Steuben) 3:309–326.

Duvall, M.R., M.T. Clegg, M.W. Chase, W.D. Clark, and **7 others.** 1993. Phylogenetic hypotheses for the monocotyledons constructed from *rbc*L sequence data. Ann. Missouri Bot. Gard. 80: 607–619.

Dyal, S.C. 1938. *Valerianella* in North America. Rhodora 40:185–212.

Gilman, E.F. and **D.G. Watson.** 1994. *Pisticia chinensis*: Chinese pistache. U.S. Forest Service Fact Sheet. http://hort.ufl.edu/trees/PISCHIA.pdf. Accessed Jun 2007.

Easterly, N.W. 1957. A morphological study of *Ptilimnium*. Brittonia 9:136–145.

Ebinger, J.E. and **D.S. Siegler.** 2023a. *Acaciella*. In: Flora of North America Editorial Committee, eds. Fl. of North Amer. 11:121–123. Oxford Univ. Press, New York and Oxford.

Ebinger, J.E. and **D.S. Siegler.** 2023b. *Vachellia*. In: Flora of North America Editorial Committee, eds. Fl. of North Amer. 11:111–117. Oxford Univ. Press, New York and Oxford.

Eckenwalder, J.E. 2009. Ebenaceae. In: Flora of North America Editorial Committee, eds. Fl. North Amer. 8:247–250. Oxford Univ. Press, New York and Oxford.

Eddie, W.M.M., T. Shulkina, J. Gaskin, R.C. Haberle, and **R.K. Jansen.** 2003. Phylogeny of Campanulaceae s. str. Inferred from ITS sequences of nuclear ribosomal DNA. Ann. Missouri Bot. Gard. 90:554–575.

EDDMapS. 2016. Early Detection & Distribution Mapping System. The University of Georgia—Center for Invasive Species and Ecosystem Health. http://www.eddmaps.org/. Accessed May 2016.

EDDMapS. 2021. Early Detection & Distribution Mapping System. The University of Georgia—Center for Invasive Species and Ecosystem Health. http://www.eddmaps.org/. Accessed Febuary 2021.

EDDMapS. 2024. Early Detection & Distribution Mapping System. The University of Georgia—Center for Invasive Species and Ecosystem Health. http://www.eddmaps.org/. Accessed Febuary 2024.

Edwards, A.L. and **R. Wyatt.** 1994. Population genetics of the rare *Asclepias texana* and its widespread sister species, *A. perennis*. Syst. Bot. 19:291–307.

Edwards, E.J., R. Nyffeler, and **M.J. Donoghue.** 2005. Basal cactus phylogeny: Implications of *Pereskia* (Cactaceae) paraphyly for the transition to the cactus life form. Amer. J. Bot. 92: 1177–1188.

Egan, A.N. and **J.L. Reveal.** 2023. *Pediomelum*. In: Flora of North America Editorial Committee, eds. Fl. of North Amer. 11:476–494. Oxford Univ. Press, New York and Oxford.

Eggers, D.M. 1969. A revision of *Valerianella* in North America. Ph.D. dissertation, Vanderbilt Univ., Nashville, TN.

Eggers-Ware, D.M. 1983. Genetic fruit polymorphism in North American *Valerianella* (Valerianaceae) and its taxonomic implications. Syst. Bot. 8:33–44.

Elakovich, S.D. and **J.W. Wooten.** 1987. An examination of the phytotoxicity of the watershield, *Brasenia schreberi*. J. Chem. Ecol. 13:1935–1940.

Elias, T.S. 1974. The genera of Mimosoideae (Leguminosae) in the southeastern United States. J. Arnold Arbor. 55:67–118.

Elias, T.S. 1989. Field guide to North American trees. Grolier Book Clubs Inc., Danbury, CT.

Elias, T.S. and **P.A. Dykeman.** 1982. Field guide to North American edible wild plants. Outdoor Life Book, New York.

Eliasson, U.H. 1988 Floral morphology and taxonomic relations among the genera of Amaranthaceae in the New World and the Hawaiian Islands. Bot. J. Linn. Soc. 96:235–283.

Elisens, W.J., R.D. Boyd, and **A.D. Wolfe.** 1992. Genetic and morphological divergence among varieties of *Aphanostephus skirrhobasis* (Asteraceae-Astereae) and related species with different chromosome numbers. Syst. Bot. 17:380–394.

Ellis, M.D. 1975. Poisonous plants. In: M.D. Ellis, ed. Dangerous plants, snakes, arthropods & marine life of Texas. Pp. 3–120. U.S.

Dept. of Health, Education, and Welfare, Public Health Service. U.S. Government Printing Office, Washington, D.C.

Emboden, W.A. 1974. *Cannabis*—A polytypic genus. Econ. Bot. 28: 304–310.

Endress, M.E. and **P.V. Bruyns.** 2000. A revised classification of Apocynaceae. Bot. Rev. (Lancaster). 66(1):1–56.

Endress, M.E., U. Meve, D.J. Middleton, and **S. Liede-Schumann.** 2018. Apocynaceae. In: J. Kadereit and V. Bittrich, eds. The families and genera of flowering plants. XV. Flowering plants. Eudicots. Apiales, Gentianales (except Rubiaceae). Springer International Publishing, Cham, Switzerland. Pp. 207–411.

Endress, P.K. 1989. A suprageneric taxonomic classification of the Hamamelidaceae. Taxon 38:371–376.

Endress, P.K. 1993a. Calycanthaceae. In: K. Kubitzki, ed. The evolution of floral biology of basal angiosperms. Springer-Verlag, Berlin, Heidelberg, New York. p. 197–200.

Endress, P.K. 1993b. Hamamelidaceae. In: K. Kubitzki, J.C. Rohwer, and V. Bittrich, eds. The families and genera of vascular plants, Vol. II. Pp. 322–331. Springer-Verlag, Berlin.

Endress, P.K. 1994. Evolutionary aspects of the floral structure in *Ceratophyllum*. Pl. Syst. Evol. 8:175–183.

Eriksson, R. 2007. A synopsis of Basellaceae. Kew Bull. 62:297–320.

Ernst, W.R. 1963a. The genera of Capparaceae and Moringaceae in the southeastern United States. J. Arnold Arbor. 44:81–95.

Ernst, W.R. 1963b. The genera of Hamamelidaceae and Platanaceae in the southeastern United States. J. Arnold Arbor. 44:193–210.

Ernst, W.R. 1964a. The genera of Berberidaceae, Lardizabalaceae, and Menispermaceae in the southeastern United States. J. Arnold Arbor. 45:1–35.

Eschmann-Grupe, G., H. Hurka, and **B. Neuffer.** 2003. Species relationships within *Diplotaxis* (Brassicaceae) and the phylogenetic origin of *D. muralis*. Pl. Syst. Evol. 243:13–29.

Esmailbegi, S., I.A. Al-Shehbaz, M. Pouch, T. Mandáková, K. Mummenhoff, M.R. Rahiminejad, M. Mirtadzadini, and **M.A. Lysak.** 2018. Phylogeny and systematics of the tribe Thlaspideae (Brassicaceae) and the recognition of two new genera. Taxon 67(2):324–340.

Esser, H.J. 2002. A revision of *Triadica* Lour. (Euphorbiaceae). Harvard Pap. Bot. 7:17–21.

Euphorbia PBI Project. 2008. *Euphorbia* Planetary Biodiversity Inventory. http://www.euphorbiaceae.org/index.html. Accessed Aug 2008.

Eyde, R.H. 1987. The case for keeping *Cornus* in the broad Linnaean sense. Syst. Bot. 12:509–518.

Eyde, R.H. 1988. Comprehending *Cornus*: Puzzles and progress in the systematics of the dogwoods. Bot. Rev. (Lancaster) 233–351.

Eyde, R.H. and **C.C. Tseng.** 1971. What is the primitive floral structure of Araliaceae? J. Arnold Arbor. 52:205–239.

Ezcurra, C. and **T.F. Daniel.** 2007. *Ruellia simplex*, an older and overlooked name for *Ruellia tweediana* and *Ruellia coerulea* (Acanthaceae). Darwiniana 4:201–203.

Fan, C. and **Q.-Y. Xiang.** 2003. Phylogenetic analyses of Cornales based on 26S rRNA and combined 26S rDNA-matK-rbcL sequence data. Amer. J. Bot. 90:1357–1372.

Fantz, P.R. 1977. A monograph of the genus *Clitoria* (Leguminosae: Glycineae). Ph.D. dissertation, Univ. of Florida, Gainesville.

Fantz, P.R. 1991. Ethnobotany of *Clitoria* (Leguminosae). Econ. Bot. 45:511–520.

Fantz, P.R. 2002. Distribution of *Centrosema* (Leguminosae: Phaseoleae: Clitoriinae) for the Flora of North America project. Vulpia 1:41–81.

Fantz, P.R. 2023a. *Centrosema*. In: Flora of North America Editorial Committee, eds. Fl. of North Amer. 11:376–378. Oxford Univ. Press, New York and Oxford.

Fantz, P.R. 2023b. *Clitoria*. In: Flora of North America Editorial Committee, eds. Fl. of North Amer. 11:371–375. Oxford Univ. Press, New York and Oxford.

Farr, E.R., J.A. Leussink, and **F.A. Stafleu,** eds. 1979. Index nominum genericorum (plantarum), Vol. 1, Aa–Epochnium. Regnum Veg. 100.

Farruggia, F.T. 2023. *Sesbania*. In: Flora of North America Editorial Committee, eds. Fl. of North Amer. 11:496–501. Oxford Univ. Press, New York and Oxford.

Fassett, N.C. 1937. Notes from the herbarium of the University of Wisconsin–XV. Rhodora 39:377.

Fassett, N.C. 1953. A monograph of *Cabomba*. Castanea 18: 116–128.

Faust, W.Z. 1972. A biosystematic study of the Interiores species group of the genus *Vernonia* (Compositae). Brittonia 24: 363–378.

Faust, W.Z. and **S.B. Jones, Jr.** 1973. The systematic value of trichome complements in a North American group of *Vernonia* (Compositae). Rhodora 75:517–528.

Fay, M.F. 2016. APG–classification by consensus. Royal Botanic Gardens, Kew. https://www.kew.org/read-and-watch/apg-classification-consensus.

Feddema, C. 1972. *Sclerocarpus uniserialis* (Compositae) in Texas and Mexico. Phytologia 23:201–209.

Fehrenbach, W.E. 1984. The eastern hop hornbeam: its natural history and landscape potential. Arnoldia (Jamaica Plain) 44: 23–27.

Feinbrun, N. 1972. *Cuscuta* L. In: T.B. Tutin, V.H. Heywood, N.A. Burges, D.M. Moore, D.H. Valentine, S.M. Walters, and D.A. Webb, eds. Flora Europaea 3:74–77. Cambridge Univ. Press, Cambridge, England, UK.

Feist, M.A.E. 2010. The reinstatement of *Ptilimnium texense* (Apiaceae) and a new key to the genus. J. Bot. Res. Inst. Texas 4: 641–651.

Feist, M.A.E. 2024a. *Daucosma*. In: Flora of North America Editorial Committee, eds. Fl. North Amer. 13:294–295. Oxford Univ. Press, New York and Oxford.

Feist, M.A.E. 2024b. *Limnosciadium*. In: Flora of North America Editorial Committee, eds. Fl. North Amer. 13:343–344. Oxford Univ. Press, New York and Oxford.

Feist, M.A.E. 2024c. *Ptilimnium*. In: Flora of North America Editorial Committee, eds. Fl. North Amer. 13:482–485. Oxford Univ. Press, New York and Oxford.

Feist, M.A.E. 2024d. *Tiedemannia*. In: Flora of North America Editorial Committee, eds. Fl. North Amer. 13:511–513. Oxford Univ. Press, New York and Oxford.

Feist, M.A.E. and **S.R. Downie.** 2008. A phylogenetic study of *Oxypolis* and *Ptilimnium* (Apiaceae) based on nuclear rDNA ITS sequences. Sys. Bot. 33:447–458.

Feist, M.A.E., S.R. Downie, A.R. Magee, and **M. Liu.** 2012. Revised generic delimitations for *Oxypolis* and *Ptilimnium* (Apiaceae) based on leaf morphology, comparative fruit anatomy, and phylogenetic analysis of nuclear rDNA ITS and cpDNA trnQ-trnK intergenic spacer sequence data. Taxon 61:402–418.

Felger, R.S. 1977. Mesquite in Indian cultures of southwestern North America. In: B.B. Simpson, ed. Mesquite: Its biology in two desert ecosystems. Pp. 150–176. Dowden, Hutchinson & Ross, Inc., Stroudsburg, PA.

Felger, R.S., D.F. Austin, T.R. Van Devender, J.J. Sánchez-Escalante, and **M. Costea.** 2012. Convolvulaceae of Sonora, Mexico. I. *Covolvulus, Cressa, Dichondra, Evolvulus, Ipomoea, Jacquemontia, Merremia,* and *Operculina*. J. Bot. Res. Inst. Texas 6:459–527.

Feng, C.-M., S.R. Manchester, and **Q.-Y. Xiang.** 2009. Phylogeny and biogeography of Alangiaceae (Cornales) inferred from DNA sequences, morphology, and fossils. Molec. Phylogen. Evol. 51: 201–214.

Ferguson, A.M. 1901. Crotons of the United States. Rep. (Annual) Missouri Bot. Gard. 12:33–74.

Ferguson, D.J. 1989. Revision of the U.S. Members of the *Echinocereus triglochidiatus* group. Cact. Succ. J. (Los Angeles) 61:217–224.

Ferguson, I.K. 1966a. The genera of Caprifoliaceae in the southeastern United States. J. Arnold Arbor. 47:33–59.

Ferguson, I.K. 1966b. Notes on the nomenclature of *Cornus*. J. Arnold Arbor. 47:100–105.

Ferguson, I.K. 1966c. The Cornaceae of the southeastern United States. J. Arnold Arbor. 47:106–116.

Ferguson, I.K. and **G.K. Brizicky.** 1965. Nomenclatural note on *Dipsacus fullonum* and *Dipsacus sativus*. J. Arnold Arbor. 46: 362–365.

Ferguson, M.E., A. Jarvis, H.T. Stalker, D.E. Williams, L. Guarino, J.F.M. Valls, R.N. Pittman, and **P.J. Bramel.** 2005. Biogeography of wild *Arachis* (Leguminosae): Distribution and environmental characterization. Biodivers. & Conservation 14(7):1777–1798.

Fernald, M.L. 1897. A systematic study of the United States and Mexican species of *Pectis*. Proc. Amer. Acad. Arts 33:57–86.

Fernald, M.L. 1909. The North American species of *Barbarea*. Rhodora. 11:134–141.

Fernald, M.L. 1934. *Draba* in temperate northeastern America. Rhodora 36:241–261, 285–305, 314–344, 353–371, 392–404.

Fernald, M.L. 1938. Noteworthy plants of southeastern Virginia. Rhodora 40:437.

Fernald, M.L. 1940a. Some spermatophytes of eastern North America. Rhodora 42:281–302.

Fernald, M.L. 1940b. A synopsis of *Boltonia*. Rhodora 42:482–492.

Fernald, M.L. 1943. Notes on *Hieracium*. Rhodora 45:317–325.

Fernald, M.L. 1946b. Identifications and reidentifications of North American plants. Rhodora 48:137–162, 184–197, 207–216.

Fernald, M.L. 1950. Gray's manual of botany, 8th ed. Reprinted 1987. Dioscorides Press, Portland, OR.

Fernald, M.L. and **B.G. Schubert.** 1949. Some identities in *Breweria*. Rhodora 51:35–42.

Ferren, W.R., Jr. 2003a. *Sesuvium*. In: Flora of North America Editorial Committee, eds. Fl. North Amer. 4:80–81. Oxford Univ. Press, New York and Oxford.

Ferren, W.R., Jr. 2003b. *Trianthema*. In: Flora of North America Editorial Committee, eds. Fl. North Amer. 4:82–83. Oxford Univ. Press, New York and Oxford.

Ferren, W.R., Jr. and **H.J. Schenk.** 2003. *Suaeda*. In: Flora of North America Editorial Committee, eds. Fl. North Amer. 4:390–398. Oxford Univ. Press, New York and Oxford.

Ferry, R.J., Sr. and **R.J. Ferry, Jr.** 1987. *Calycanthus brockiana* (Calycanthaceae), a new spicebush from North Central Georgia. Sida 12:339–341.

Filipowicz, N. and **S.S. Renner.** 2010. The worldwide holoparasitic Apodanthaceae confidently placed in the Cucurbitales by nuclear and mitochondrial gene trees. B.M.C. Evol. Biol. 10:219. https://doi: 10.1186/1471-2148-10-219.

Fior, S., P.O. Karis, and **A.A. Anderberg.** 2003. Phylogeny, taxonomy, and systematic position of *Clethra* (Clethraceae, Ericales) with notes on biogeography: Evidence from plastid and nuclear DNA sequences. Int. J. Pl. Sci. 164:997–1006.

Firbank, L.G. 1988. Biological flora of the British Isles. 165. *Agrostemma githago* L. (*Lychnis githago* (L.) Scop.). J. Ecol. 76: 1232–1246.

Fischer, E. 2004. Celastraceae. In: K. Kubitzki, ed. The families and genera of vascular plants, Vol. VI. Pp. 29–64. Springer-Verlag, Berlin.

Fischer, E., I. Theisen, and **L.G. Lohmann.** 2004. Bignoniaceae. In: J.W. Kadereit, ed. The families and genera of vascular plants, Vol. VII. Pp. 9–38. Springer-Verlag, Berlin.

Fischer, P.C. 1980. The varieties of *Coryphantha vivipara*. Cact. Succ. J. (Los Angeles) 52:186–191.

Fishbein, M. 2023a. *Asclepias*. In: Flora of North America Editorial Committee, eds. Fl. North Amer. 14:156–227. Oxford Univ. Press, New York and Oxford.

Fishbein, M. 2023b. *Cynanchum*. In: Flora of North America Editorial Committee, eds. Fl. North Amer. 14:228–230. Oxford Univ. Press, New York and Oxford.

Fishbein, M. 2023c. *Funastrum*. In: Flora of North America Editorial Committee, eds. Fl. North Amer. 14:230–236. Oxford Univ. Press, New York and Oxford.

Fishbein, M. 2023d. *Metastelma*. In: Flora of North America Editorial Committee, eds. Fl. North Amer. 14:256–260. Oxford Univ. Press, New York and Oxford.

Fishbein, M. 2023e. *Pattalias*. In: Flora of North America Editorial Committee, eds. Fl. North Amer. 14:261–262. Oxford Univ. Press, New York and Oxford.

Fishbein, M., D.E. Lemke, and **A. Krings.** 2023. Apocynaceae. In: Flora of North America Editorial Committee, eds. Fl. North Amer. 14:103–268. Oxford Univ. Press, New York and Oxford.

Fishbein, M. and **A. McDonnell.** 2023. *Matelea*. In: Flora of North America Editorial Committee, eds. Fl. North Amer. 14:239–256. Oxford Univ. Press, New York and Oxford.

Fisher, D.D., H.J. Schenk, J.A. Thorsch, and **W.R. Ferren, Jr.** 1997. Leaf anatomy and subgeneric affiliations of C3 and C4 species of *Suaeda* (Chenopodiaceae) in North America. Amer. J. Bot. 84: 1198–1210.

Fisher, T.R. 1957. Taxonomy of the genus *Heliopsis* (Compositae). Ohio J. Sci. 57:171–191.

Fisher, T.R. 1958. Variation in *Heliopsis helianthoides* (L.) Sweet (Compositae). Ohio J. Sci. 58:97–107.

Flack, S. and **E. Furlow.** 1996. America's least wanted. Nat. Conservancy Mag. 46(6):17–23.

Fleet, R.R. and **B.L. Young.** 2000. Facultative mutualism between imported fire ants (*Solenopsis invicta*) and a legume (*Senna occidentalis*). S.W. Naturalist 45:289–298.

Fleming, K.M., J.R. Singhurst, and **W.C. Holmes.** 2002. Vascular flora of Big Lake Bottom Wildlife Management Area, Anderson County, Texas. Sida 20(1):355–371.

Floden, A. 2019. Distribution of *Thaspium chapmanii* and *T. barbinode* (Apiaceae) in Missouri. Missouriensis 37:45–49.

Flora of China Editorial Committee, eds. 2003. Flora of China, Vol. 8, Brassicaceae–Saxifragaceae. Science Press, Beijing, and Missouri Botanical Garden Press, St. Louis.

Flora of North America Editorial Committee, eds. 2006. Flora of North America north of Mexico, Vols. 19–21, Compositae. Oxford Univ. Press, New York and Oxford.

Flora of North America Editorial Committee, eds. 2010. Flora of North America north of Mexico, Vol. 7, Salicaceae–Brassicaceae. Oxford Univ. Press, New York and Oxford.

Flores Olvera, H. 2003. Classification of the North American species of Atriplex section Obione (Chenopodiaceae) based on numerical taxonomic analysis. Taxon 52(2):247–260.

Florida Exotic Pest Plant Council. 2007. 2007 list of invasive plant species. http://www.fleppc.org/list/07list_ctrfld.pdfl. Accessed Jun 2008.

Floyd, J.W. 2002. Phylogenetic and biogeographic patterns in *Gaylussacia* (Ericaceae) based on morphological, nuclear DNA, and chloroplast DNA variation. Syst. Bot. 27:99–115.

Fonseca, L.H.M., S.M. Cabral, M. De Fatima Agra, and **L.G. Lohmann.** 2017. Taxonomic revision of *Dolichandra* (Bignonieae, Bignoniaceae). Phytotaxa 301:1–70. Magnolia Press, Aukland, New Zealand.

Fonseca, L.H.M. and **L.G. Lohmann.** 2015a. Biogeography and evolution of *Dolichandra* (Bignonieae, Bignoniaceae). Bot. J. Linn. Soc. 179:403–420.

Fonseca, L.H.M. and **L.G. Lohmann.** 2015b. Taxonomic updates in *Dolichandra* Cham. (Bignonieae, Bignoniaceae). Phytokeys 46: 35–43.

Ford-Lloyd, B.V. and **J.T. Williams.** 1975. A revision of *Beta* section *Vulgares* (Chenopodiaceae), with new light on the origin of cultivated herbs. Bot. J. Linn. Soc. 71:89–102.

Forrester, M.B. 2018. Pediatric *Nandina domestica* ingestions reported to poison centers. Human Exp. Toxicol. 37:338–342.

Fortuna-Perez, A.P. and **A.M. Goulart de Azevedo Tozzi.** 2011. Nomenclatural changes to *Zornia* (Leguminosae, Papilionoideae, Dalbergieae) in Brazil. Novon 21:331–337.

Fortunato, R.H., L.P. de Queiroz, and **G.P. Lewis.** 1996. *Lackeya*, a new genus in tribe Phaseoleae subtribe Diocleinae (Leguminosae: Papilionoideae) from North America. Kew Bull. 51:365–370.

Fosberg, F.R. 1936. Varieties of the desert willow, *Chilopsis linearis*. Madroño 3:362–366.

Foster, K.E., M.M. Karpiscak, J.G. Taylor, and **N.G. Wright.** 1983. Guayule, jojoba, buffalo gourd and russian thistle: Plant characteristics, products and commercialization potential. Desert Pl. 5:112–117, 126.

Foster, S. 1991. *Echinacea*: Nature's immune enhancer. Healing Arts Press, Rochester, VT.

Foster, W.C. 1995. Spanish expeditions into Texas, 1689–1768. Univ. of Texas Press, Austin.

Fowler, N. 2014. Bracted twistflower (*Streptanthus bracteatus*), ecological characterization and data base construction. Section 6 final report, Texas Parks and Wildlife Department, Austin, and U.S. Fish and Wildlife Service, Albuquerque, NM.

Francis, A. and **S.I. Warwick.** 2009. The Biology of Canadian weeds. 142. *Camelina alyssum* (Mill.) Thell.; *C. microcarpa* Andrz. ex DC.; *C. sativa* (L.) Crantz. Canad. J. Pl. Sci. 89:91–810.

Franck, A.R. and **D. Salman.** 2024. Scientific note: Typification of *Vaccinium elliottii* (Ericaceae), a distinct species. Castanea 89: 182–189.

Frankel, E. 1989. Distribution of *Pueraria lobata* in and around New York City. Bull. Torrey Bot. Club 116:390–394.

Frankel, E. 1991. Poison ivy, poison oak, poison sumac and their relatives pistachios, mangoes, cashews. The Boxwood Press, Pacific Grove, CA.

Franzke, A., M.A. Lysak, I.A. Al-Shehbaz, M.A. Koch, and **K. Mummenhoff.** 2011. Cabbage family affairs: The evolutionary history of Brassicaceae. Trends Pl. Sci. 16(2):108–116.

Freeman, C.C. and **T.M. Barkley.** 1995. A synopsis of the genus *Packera* (Asteraceae: Senecioneae) in Mexico. Sida 16:699–709.

Freeman, D.C., E.D. McArthur, and **K.T. Harper.** 1984. The adaptive significance of sexual lability in plants using *Atriplex canescens* as a principal example. Ann. Missouri Bot. Gard. 71:265–277.

Freudenstein, J.V. and **M.B. Broe.** 2024. Phylogenetic relationships within Monotropoideae (Ericaceae): Implications for taxonomy and character evolution. Syst. Bot. 49:412–426.

Freudenstein, J.V., M.B. Broe, and **E.R. Feldenkris.** 2016. Phylogenetic relationships at the base of Ericaceae: Implications for vegetative and mycorrhizal evolution. Taxon 65: 794–804.

Freytag, G.F. and **D.G. Debouck.** 2002. Taxonomy, distribution, and ecology of the genus *Phaseolus* (Leguminosae–Papilionoideae) in North America, Mexico, and Central America. Sida Bot. Misc. 23.

Fritsch, P.W. (Peter). Personal communication. 2024. Expert on Ericales, especially the Vaccinioideae, focusing biogeography, taxonomy, floristics, and phylogenetics, using molecular and morphological data. Vice President of Research and Conservation, Botanical Research Institute of Texas, Fort Worth, U.S.A.

Fritsch, P.W., A.A. Crowl, H. Ashrafi, and **P.S. Manos.** 2022 [2024]. Systematics and evolution of *Vaccinium* sect. *Cyanococcus* (Ericaceae): Progress and prospects. Rhodora 124:301–332.

Frisch, P.W., A.W. Crowl, and **P.S. Manos.** 2024b. *Vaccinium virgatum*: A species to be recognized. J. Bot. Res. Inst. Texas 18:293–309.

Frisch, P.W., A.W. Crowl, and **P.S. Manos.** 2025. Evolution and taxonomic revision of "residual highbush" blueberries of *Vaccinium* sect. *Cyanococcus*. J. Bot. Res. Inst. Texas 19(4): 419–476.

Frodin, D.G. 2004. History and concepts of big plant genera. Taxon 53(3):753–776.

Frodin, D.G. and **R. Govaerts.** 2003. World checklist and bibliography of Araliaceae.. Botanic Gardens, Kew, England, UK.

Fu, C.-N., Z.-Q. Mo, J.-B. Yang, X.-J. Ge, D.-Z. Li, Q.-Y. (J.) Xiang, and **L.-M. Gao.** 2019. Plastid phylogenomics and biogeographic analysis support a trans-Tethyan origin and rapid early radiation of Cornales in the mid-Cretaceous. Molec. Phylogen. Evol. 140:106601.

Fualkner, P.L., C.S. Reid, B.R. MacRoberts, and **M.H. MacRoberts.** 2008. Noteworthy vascular plant collections from northwest Louisiana. J. Bot. Res. Inst. Texas 2:643–647.

Fuentes-Bazan, S., P. Uotila, and **T. Borsch.** 2012. A novel phylogeny-based generic classification for *Chenopodium* sensu lato, and a tribal rearrangement of Chenopodioideae (Chenopodiaceae). Willdenowia 42:6–24.

Fuller, T.C. and **E. McClintock.** 1986. Poisonous plants of California. Univ. of California Press, Berkeley.

Funk, V.A., A. Susanna, T.F. Stuessy, and **R.J. Bayer,** eds. 2009. Systematics, evolution and biogeography of Compositae. International Association for Plant Taxonomy. Vienna, Austria.

Furlow, J.J. 1987a. The *Carpinus caroliniana* complex in North America. I. A multivariate analysis of geographical variation. Syst. Bot. 12:21–40.

Furlow, J.J. 1987b. The *Carpinus caroliniana* complex in North America. II. Systematics. Syst. Bot. 12:416–434.

Furlow, J.J. 1990. The genera of Betulaceae in the southeastern United States. J. Arnold Arbor. 71:1–67.

Furlow, J.J. 1997. Betulaceae. In: Flora of North America Editorial Committee, eds. Fl. North Amer. 3:507–538. Oxford Univ. Press, New York and Oxford.

Gabel, J.D. and **D.H. Les.** 2001. *Neobeckia aquatica* (North American Lake Cress) Conservation and Research Plan. New England Plant Conservation Program, Framingham, Massachusetts, USA.

Gaddy, L. 1988. A review of the taxonomy and biogeography of *Hexastylis* (Aristolochiaceae). Castanea 52:186–196.

Gade, D.W. 2002. Names for *Manihot esculenta*: Geographical variations and lexical clarification. J. Latin Amer. Geogr. 1:55–74.

Gadella, T.W.J. 1966. Some notes on the delimitation of genera in the Campanulaceae. I, II. Proceed. Royal Netherlands Acad. Sci., series C 269:502–521.

Gagnon, E., A. Bruneau, C.E. Hughes, L. Paganucci de Quieroz, and **G.P. Lewis.** 2016. A new generic system for the pantropical Caesalpinia group (Leguminosae). PhytoKeys 71:1–160.

Gaiser, L.O. 1946. The genus *Liatris*. Rhodora 48:165–183, 216–263, 326, 331–382, 393–412.

Gale, J., S. Rachmilevitch, J. Reuveni, and **M. Volokita.** 2001. The high oxygen atmosphere toward the end-Cretaceous; a possible contributing factor to the KT boundary extinctions and to the emergence of C4 species. J. Exp. Bot. 52(357):801–809.

Gandhi, K.N. (Kanchi). Personal communication. Senior Nomenclature Registrar and Bibliographer in the Department of Botany Harvard University Herbaria & Libraries (HUH and HUL).

Gandhi, K.N. and **L.E. Brown.** 1993. *Crepis zacintha* (Asteraceae), adventive in Texas, and first record for North America. Sida 15: 547–549.

Gandhi, K.N. and **R.D. Thomas.** 1984. Observations on the floral structure of *Dyssodia tenuiloba* (DC.) B.L. Robinson and *Matricaria matricarioides* (Less.) T.C. Porter (Asteraceae). Phytologia 55: 253–255.

Gandhi, K.N. and **R.D. Thomas.** 1989. Asteraceae of Louisiana. Sida Bot. Misc. 4.

Gandhi, K.N., R.D. Thomas, and **S.L. Hatch.** 1987. Cuscutaceae of Louisiana. Sida 12:361–379.

Gandhi, K.N., M.A. Vincent, and **J.L. Reveal.** 2011. *Dermatophyllum*, the correct name for *Calia* (Fabaceae). Phytoneuron 2011-57:1–4.

Gara, B. and **G. Muenchow.** 1990. Chasmogamy/cleistogamy in *Triodanis perfoliata* (Campanulaceae): Some CH/CL comparisons in fitness parameters. Amer. J. Bot. 77:1–6.

Garcia-Jacas, N., A. Susanna, T. Garnatje, and **R. Vilatersana.** 2001. Generic delimitation and phylogeny of the subtribe Centaureinae (Asteraceae): A combined nuclear and chloroplast DNA analysis. Ann. Bot. (Oxford) 87(4):503–515.

Gardner, D.C. and **K.R. Robertson.** 2000. Effects of annual burning on populations of *Cassia fasciculata* (Fabaceae: Caesalpinioideae) with a review of its systematics and biology. Erigenia 18:22–29.

Garrison, W.J., G.L. Miller, and **R. Raspet.** 2000. Ballistic seed projection in two herbaceous species. Amer. J. Bot. 87(9): 1257–1264.

Gates, F.C. 1945. *Amphiachyris dracunculoides* as a poisonous plant. Trans. Kansas Acad. Sci. 48:87–89.

Gawel, E. 2012. Chemical composition of lucerne leaf extract (EFL) and its applications as a phytobiotic in human nutrition. Acta. Sci. Pol. Technol. Aliment. 11(3):303–310.

Geiser, S.W. 1948. Naturalists of the frontier, 2nd ed. Southern Methodist Univ. Press, Dallas, TX.

Gentry, A.H. 1992. Bignoniaceae, in Fl. Neotrop. Monogr. 25: 273–292.

George, R.J. and **E.S. Nixon.** 1990. The herbaceous flora of three Weches Formation outcrops in eastern Texas. Sida 14(1):117–127.

Gerard, J. 1597. The herball or generall historie of plantes. John Norton, London.

German, D.A., J.R. Grant, M.A. Lysak, and **I.A. Al-Shehbaz.** 2011. Molecular phylogeny and systematics of the tribe Chorisporeae (Brassicaceae). Pl. Syst. Evol. 294(1–2):65–86.

Giannasi, D.E., G. Zurawski, G. Learn, and **M.T. Clegg.** 1992. Evolutionary relationships of the Caryophyllidae based on comparative *rbc*L sequences. Syst. Bot. 17:1–15.

Gillespie, L.J. 2016a. *Caperonia*. In: Flora of North America Editorial Committee, eds. Fl. North Amer. 12:179–181. Oxford Univ. Press, New York and Oxford.

Gillespie, L.J. 2016b. *Ricinus*. In: Flora of North America Editorial Committee, eds. Fl. North Amer. 12:160–161. Oxford Univ. Press, New York and Oxford.

Gillespie, L.J. 2016c. *Vernicia*. In: Flora of North America Editorial Committee, eds. Fl. North Amer. 12:225–226. Oxford Univ. Press, New York and Oxford.

Gillis, W.T. 1971. The systematics and ecology of poison-ivy and the poison-oaks (*Toxicodendron*, Anacardiaceae). Rhodora 73: 72–159, 161–237, 370–443, 465–540.

Gillis, W.T. 1975. Poison-ivy and its kin. Arnoldia (Jamaica Plain) 35: 93–123.

Gillis, W.T. 1977. *Pluchea* revisited. Taxon 26:587–591.

Ginzbarg, S. 1992. A new disjunct variety of *Croton alabamensis* (Euphorbiaceae) from Texas. Sida 15:41–52.

GISD (Global Invasive Species Database). 2006. *Anredera cordifolia*. http://www.issg.org/database/species/ecology.asp?si=776&fr=1&sts=sss. Accessed October 2006.

Gitzendanner, M.A., P.S. Soltis, G.K.S. Wong, B.R. Ruhfel, and **D.E. Soltis.** 2018. Plastid phylogenomic analysis of green plants: A billion years of evolutionary history. Amer. J. Bot. 105: 291–301.

Gleason, H.A. 1906. A revision of the North American Vernonieae. Bull. New York Bot. Gard. 4:144–243.

Gleason, H.A. 1922. Vernonieae. N. Amer. Fl. 33:47–110.

Glenn, E.P., J.W. O'Leary, M. Watson, T.L. Thompson, and **R.O. Kuehl.** 1991. *Salicornia Bigelovii* Torr.: An oilseed halophyte for seawater irrigation. Science 251:1065–1067.

Godfrey, R.K. 1952. *Pluchea*, section *Stylimnus*, in North America. J. Elisha Mitchell Sci. Soc. 68:238–271.

Godfrey, R.K. and **J.W. Wooten.** 1981. Aquatic and wetland plants of the southeastern United States, dicotyledons. Vol. 2. Univ. of Georgia Press, Athens.

Goetsch, L.A., A.J. Eckert, and **B.D. Hall.** 2005. The molecular systematics of *Rhododendron* (Ericaceae): A phylogeny based upon RPB2 gene sequences. Syst. Bot. 30:616–626.

Goldblatt, P. and **P.K. Endress.** 1977. Cytology and evolution in Hamamelidaceae. J. Arnold Arbor. 58:67–71.

Goldman, D.H. 2023. *Eysenhardtia*. In: Flora of North America Editorial Committee, eds. Fl. of North Amer. 11:271–273. Oxford Univ. Press, New York and Oxford.

González, F.A. and **D.W. Stevenson.** 2002. A phylogenetic analysis of the subfamily Aristolochioideae (Aristolochiaceae). Revista Acad. Colomb. Ci. Exact. 26(98):25–58.

González, V.C. 1972. The ecology of *Hexastylis arifolia*, an evergreen herb in the North Carolina deciduous forest. Ph.D. dissertation, Duke Univ, Durham, NC.

Gonzalez-Elizondo, M.S., M. Gonzalez-Elizondo, and **P.D. Sorensen.** 2012. *Arbutus bicolor* (Ericaceae, Arbuteae), a new species from Mexico. Acta Bot. Mex. 99:55–72.

Gonzalez-Villarreal, L.M. 2009. Neotropical Clethraceae. In: W. Milliken, B. Klitaard, and A. Baracat. 2009 onwards. Neotropikey—Interactive key and information resources for flowering plants of the Neotropics. http://www.kew.org/science/tropamerica/neotropikey/families/Clethraceae.htm. Accessed Aug 2019.

Good, D.A. 1984. A revision of the Mexican and Central American species of *Cerastium* (Caryophyllaceae). Rhodora 86:339–379.

Goodman, G.J. and **C.A. Lawson.** 1992. Two new combinations and a name change from the Long Expedition of 1820. Rhodora 94:381–382.

Goodrich, K.R., M.L. Zjhra, C.A. Ley, and **R.A. Raguso.** 2006. When flowers smell fermented: The chemistry and ontogeny of yeasty floral scent in pawpaw (*Asimina triloba*: Annonaceae). Int. J. Pl. Sci. 167:33–46.

Goodson, B. and **I.A. Al-Shehbaz.** 2010. *Descurainia*. In: Flora of North America Editorial Committee, eds. Fl. North Amer. 7:518–529. Oxford Univ. Press, New York and Oxford.

Goodson, B., S. Rehman, and **R. Jansen.** 2011. Molecular systematics and biogeography of *Descurainia* (Brassicaceae) based on nuclear ITS and non-coding chloroplast DNA. Syst. Bot. 36: 957–980.

Goodwillie, C. and **E. Stewart.** 2013. Cleistogamy and hybridization in two subspecies of *Triodanis perfoliata* (Campanulaceae). Rhodora 115:42–60.

Gordon, R. 2023. Pawpaw: The story of America in one fruit. Library of Congress Blogs. https://blogs.loc.gov/families/2023/09/paw-paw-the-story-of-america-in-one-fruit/. Accessed 17 Oct 2024.

Gorelick, R. 2002. DNA sequences and cactus classification—a short review. Bradleya 20:1–4.

Gottschling, M. and **H.H. Hilger.** 2001. Phylogenetic analysis and character evolution of *Ehretia* and *Bourreria* (Ehretiaceae, Boraginales) and their allies based on ITS1 sequences. Bot. Jahrb. Syst. 123:249–268.

Gottschling, M., F. Luebert, H.H. Hilger, and **J.S. Miller.** 2014. Molecular delimitations in the Ehretiaceae (Boraginales). Molec. Phylogen. Evol. 72:1–6.

Gottschling, M., H.H. Hilger, M. Wolf, and **N. Diane.** 2001. Secondary structure of the ITS1 transcript and its application in a reconstruction of the phylogeny of Boraginales. Pl. Biol. (Stuttgart) 3:629–636.

Gottschling, M., J.S. Miller, M. Weigend, and **H.H. Hilger.** 2004. Congruence of a phylogeny of Cordiaceae (Boraginales) inferred from ITS1 sequence data with morphology, ecology, and biogeography. Ann. Missouri Bot. Gard. 92:425–437.

Gottschling, M., M. Weigend, and **H.H. Hilger.** 2016. Ehretiaceae. In: J.W. Kadereit and V. Bittrich, eds. The families and genera of vascular plants, Vol. XIV. Flowering plants. Eudicots: Aquifoliales, Boraginales, Bruniales, Dipsacales, Escalloniales, Garryales, Paracryphiales, Solanales (except Convolvulaceae), Icacinaceae, Metteniusaceae, Vahliaceae. Pp. 165–178. Springer, Cham, Switzerland.

Gould, K.R. and **M.J. Donoghue.** 2000. Phylogeny and biogeography of *Triosteum* (Caprifoliaceae). Harvard Pap. Bot. 5:157–166.

Govaerts, R., D.G. Frodin, and **A. Radcliff-Smith.** 2000. World Checklist and Bibliography of Euphorbiaceae (and Pandaceae). 4 vols. Royal Botanic Gardens, Kew, England, U.K.

Govoni, D. 1973. The taxonomy of the genus *Lithospermum* L. (Boraginaceae) in the western Great Plains. Ph.D. dissertation, Univ. of Nebraska, Lincoln.

Graham, A. 1972. Outline of the origin and historical recognition of floristic affinities between Asia and eastern North America. In: A. Graham, ed. Floristics and paleofloristics of Asia and eastern North America. Elsevier Publishing Co., Amsterdam, Netherlands.

Graham, A. 1993a. History of the vegetation: Cretaceous (Maastrichtian)–Tertiary. In: Flora of North America Editorial Committee, eds. Fl. North Amer. 1:57–70. Oxford Univ. Press, New York and Oxford.

Graham, A. 1999a. Late Cretaceous and Cenozoic history of North American vegetation north of Mexico. Oxford Univ. Press, New York and Oxford.

Graham, A. 1999b. The Tertiary history of the northern Temperate Element in the northern Latin American Biota. Amer. J. Bot. 86: 32–38.

Graham, S.A. 1964. The Elaeagnaceae in the southeastern United States. J. Arnold Arbor. 45:274–278.

Graham, S.A. 1966. The genera of Araliaceae in the southeastern United States. J. Arnold Arbor. 47:126–136.

Graham, V.A.W. 1988. Delimitation and infra-generic classification of *Justicia* (Acanthaceae). Kew Bull. 43:551–624.

Grant, V. 2003. Incongruence between cladistic and taxonomic systems. Amer. J. Bot. 90:1263–1270.

Grant, V. and **K.A. Grant.** 1979a. The pollination spectrum in the southwestern American cactus flora. Pl. Syst. Evol. 133:29–37.

Grant, V. and **K.A. Grant.** 1979b. Systematics of the *Opuntia phaeacantha* group in Texas. Bot. Gaz. 140:199–207.

Grant, V. and **K.A. Grant.** 1979c. Hybridization and variation in the *Opuntia phaeacantha* group in central Texas. Bot. Gaz. 140:208–215.

Grant, V. and **K.A. Grant.** 1982. Natural pentaploids in the *Opuntia lindheimeri-phaeacantha* group in Texas. Bot. Gaz. 143:117–120.

Grant, V., K.A. Grant, and **P.D. Hurd.** 1979. Pollination of *Opuntia lindheimeri* and related species. Pl. Syst. Evol. 132:313–320.

Grant, V. and **P.D. Hurd.** 1979. Pollination of the southwestern Opuntias. Pl. Syst. Evol. 133:15–28.

Grant, W.F. 1961. Speciation and basic chromosome number in the genus *Celosia*. Canad. J. Bot. 39 45–50.

Grant, W.F. 1962. Speciation and nomenclature in the genus *Celosia*. Canad. J. Bot. 40:1355–1363.

Grant, W.F. and **E. Small.** 1996. The origin of the *Lotus corniculatus* (Fabaceae) complex: A synthesis of diverse evidence. Canad. J. Bot. 74(7):975–989.

Gray, A. 1886. Synoptical flora of North America 2(1):396.

Grear, J.W. 1978. A revision of the New World species of *Rhynchosia* (Leguminosae-Faboideae). Mem. New York Bot. Gard. 31:1–168.

Green, P.S. 1962. Watercress in the New World. Rhodora 64:32–43.

Greenberg, A.K. and **M.J. Donoghue.** 2011. Molecular systematics and character evolution in Caryohyllaceae. Taxon 60:1637–1652.

Greene, E.L. 1894. *Triodanis biflora*. Man. Bot. San Francisco Bay, p. 230. Cubrey & Co., San Francisco, CA.

Greenman, J.M. 1915–1918. Monograph of the North and Central American species of the genus *Senecio*. Ann. Missouri Bot. Gard. 2:573–627; 3:85–195; 4:15–37; 5:37–109.

Greer, L.F. 1997. *Thelesperma curvicarpum* (Asteraceae), an achene form in populations of *T. simplicifolium* var. *simplicifolium* and *T. filifolium* var. *filifolium*. S. W. Naturalist 42:242–244.

Gregory, M.P. 1956. A phyletic rearrangement of the Aristolochiaceae. Amer. J. Bot. 43:110–122.

Gregory, W.C., A. Krapovickas, and **M.P. Gregory.** 1980. Structure, variation, evolution, and classification in *Arachis*. In: R.J. Summerfield and A.H. Bunting, eds. Advances in legume science. Pp. 469–481. Royal Botanic Gardens, Kew, England, U.K.

Grether, R. 2023. *Mimosa*. In: Flora of North America Editorial Committee, eds. Fl. of North Amer. 11:93–104. Oxford Univ. Press, New York and Oxford.

Grether, R. and **M.F. Simon.** 2018. *Mimosa monclovensis* (Leguminosae), a new name for a taxon in ser. Quadrivalves in the southern USA and northeastern Mexico. Phytoneuron 2018-39:1–3.

Greuter, W., F.R. Barrie, H.M. Burdet, W.G. Chaloner, V. Demoulin, D.L. Hawksworth, P.M. Jørgensen, D.H. Nicolson, P.C. Silva, P. Trehane, and **J. McNeill,** eds. 1994. International code of botanical nomenclature. Regnum Veg. 131.

Greuter, W., R.K. Brummitt, E. Farr, N. Kilian, P.M. Kirk, and **P.S. Silva.** 1993. Names in current use for extant plant genera. Regnum Veg. 129.

Grimes, J.W. 1990. A revision of the New World species of Psoraleeae (Leguminosae: Papilionoideae). Mem. New York Bot. Gard. 61: 1–114.

Grimm, G.W. and **S.S. Renner.** 2013. Harvesting Betulaceae sequences from GenBank to generate a new chronogram for the family. Bot. J. Linn. Soc. 172:465–477.

Gunn, C.R. 1983. A nomenclator of legume (Fabaceae) genera. U.S.D.A. Tech. Bull. 1680:1–224.

Gunn, C.R. 1984. Fruits and seeds of genera in the subfamily Mimosoideae (Fabaceae). U.S.D.A. Tech. Bull. 1681:1–194.

Gunn, C.R. 1991. Fruits and seeds of genera in the subfamily Caesalpiniodeae (Fabaceae). U.S.D.A. Tech. Bull. 1755:1–407.

Guo, X., D. Fang, S.K. Sahu, and **14 others.** 2021. *Chloranthus* genome provides insights into the early diversification of angiosperms. Nat. Commun. 12, article no. 6930. https://doi.org/10.1038/s41467-021-26922-4.

Guzmán, B. and **P. Vargas.** 2009. Historical biogeography and character evolution of Cistaceae (Malvales) based on analysis of plastid *rbc-trnF* sequences. Organisms Diversity Evol. 9:83–99.

Gustafsson, M.H.G. and **K. Bremer.** 1995. Morphology and phylogenetic interrelationships of the Asteraceae, Calyceraceae, Campanulaceae, Goodenianceae, and related families (Asterales). Amer. J. Bot. 82:250–265.

Guthrie, R.L. 1969. A biosystematic study of *Taenidia* and *Pseudotaenedia* (Umbelliferae). Ph.D. dissertation, West Virginia Univ., Morgantown.

Haberle, R.C., A. Dang, T. Lee, and **R.K. Jansen.** 2009. Taxonomic and biogeography implications of a phylogenetic analysis of the Campanulaceae based on three chloroplast genes. Taxon 58:715–734.

Hagen, S.H. 1941. A revision of the North American species of the genus *Anisacanthus*. Ann. Missouri Bot. Gard. 28:385–408.

Haines, A. 2006. *Euthamia*. In: Flora of North America Editorial Committee, eds. Fl. North Amer. 20:97–100. Oxford Univ. Press, New York and Oxford.

Haines, A. 2011. Flora Novae Angliae: A manual for the identification of native and naturalized higher vascular plants of New England. Yale Univ. Press, New Haven, CT.

Halkier, B.A. and **J. Gershenzon.** 2006. Biology and biochemistry of glucosinolates. Annual Rev. Pl. Biol. 57:303–333.

Hall, G.W. 1967. A biosystematic study of the North American complex of the genus *Bidens* (Compositae). Ph.D. dissertation, Indiana Univ., Bloomington.

Hall, H.M. 1928. The genus *Haplopappus*: A phylogenetic study in the Compositae. Publ. Carnegie Inst. Wash. 389:1–391.

Hall, H.M. and **F.E. Clements.** 1923. The phylogenetic method in taxonomy: The North American species of *Artemisia, Chrysothamnus* and *Atriplex*. Publ. Carnegie Inst. Wash. 326: 1–355.

Hall, J.C., K.J. Sytsma, and **H.H. Iltis.** 2002. Phylogeny of Capparaceae and Brassicaceae based on chloroplast sequence data. Amer. J. Bot. 89:1826–1842.

Hamby, R.K. and **E.A. Zimmer.** 1992. Ribosomal RNA as a phylogenetic tool in plant systematics. In: P.S. Soltis, D.E. Soltis, and J.J. Doyle, eds. Molecular systematics of plants. Pp. 50–91. Chapman and Hall, New York.

Hamilton, W. 1983. Cretaceous and Cenozoic history of the northern continents. Ann. Missouri Bot. Gard. 70:440–458.

Hannick, V.C., J.N. Mink, J.R. Singhurst, and **W.C. Holmes.** 2013. Annotated checklist of the vascular flora of McLennan County, Texas. Phytoneuron 2013-29:1–37.

Hansen, L.L. 2010. Annotated checklist of the vascular plants of Fort Hood, Texas. J. Bot. Res. Inst. Texas 4(1):523–558.

Hara, H. 1969a. The correct author's name of *Citrullus lanatus* (Cucurbitaceae). Taxon 18:346–347.

Harbaugh, D.T., M. Nepokroeff, R.K. Rabeler, J. McNeill, E.A. Zimmer, and **W.L. Wagner.** 2010. A new lineage-based tribal classification of the family Caryophyllaceae. Int. J. Pl. Sci.171: 185–198.

Harcombe, P.A., I.S. Elsik, W.W. Pruess, and **L.E. Brown.** 2007. The vascular flora of a woodland park site in east Harris County, Texas. J. Bot. Res. Inst. Texas 1(1):693–711.

Hardin, J.W. 1971. Studies of the southeastern United States flora: 1. Betulaceae. J. Elisha Mitchell Sci. Soc. 87:39–41.

Hardin, J.W. 1984. The Demoulin Rule in nomenclature. Sida 10:252.

Hardin, J.W. and **J.M. Arena.** 1974. Human poisoning from native and cultivated plants, 2nd ed. Duke Univ. Press, Durham, NC.

Hardin, J.W. and **C.F. Brownie.** 1993. Plants poisonous to livestock and pets in North Carolina. North Carolina Agricultural Research Service, North Carolina State Univ., Raleigh. Bull. No. 414 (revised).

Hardway, T.M., K. Spalik, M.F. Watson, D.S. Katz-Downie, and **S.R. Downie.** 2004. Circumscription of Apiaceae tribe Oenantheae. S. African J. Bot. 70:393–406.

Harms, R.T. 2007. A field study of hybridization between *Berberis swaseyi* and *B. trifoliolata* (Berberidaceae) in Hays County, Texas. Lundellia 10:18–31.

Harms, R.T. 2011. A study of *Chaptalia* (Asteraceae: Mutisieae) in central, southern, and western Texas and the resurrection of *Chaptalia carduacea*. Lundellia 14:8–26.

Harms, V.L. 1965. Biosystematic studies in the *Heterotheca subaxillaris* complex. Trans. Kansas Acad. Sci. 68:244–257.

Harms, V.L. 1968. Nomenclatural changes and taxonomic notes on *Heterotheca*, including *Chrysopsis*, in Texas and adjacent states. Wrightia 4:8–20.

Harms, V.L. 1974. A preliminary conspectus of *Heterotheca* sect. *Chrysopsis*. Castanea 39:155–165.

Harpke, D., A. Peterson, M.H. Hoffmann, and **M. Röser.** 2006. Phylogenetic evaluation of chloroplast *trn*L-*trn*F DNA sequence variation in the genus *Mammillaria* (Cactaceae). Schlechtendalia 14:7–16.

Harriman, N.A. (Neil) 1938–2018. Personal communication. Professor, retired from University of Wisconsin, Oshkosh.

Harriman, N.A. 2006. *Sclerocarpus*. In: Flora of North America Editorial Committee, eds. Fl. North Amer. 21:137. Oxford Univ. Press, New York and Oxford.

Harriman, N.A. 2023. *Leucaena*. In: Flora of North America Editorial Committee, eds. Fl. of North Amer. 11:82–84. Oxford Univ. Press, New York and Oxford.

Hartman, R.L. 1986a. Apocynaceae. In: Great Plains Flora Association. Flora of the Great Plains. Pp. 610–613. Univ. Press of Kansas, Lawrence.

Hartman, R.L. 1986b. Asclepiadaceae. In: Great Plains Flora Association. Flora of the Great Plains. Pp. 614–637. Univ. Press of Kansas, Lawrence.

Hartman, R.L. 1990. A conspectus of *Machaeranthera* (Asteraceae: Astereae). Phytololgia 68:439–465.

Hartman, R.L. 2006. *Xanthisma*. In: Flora of North America Editorial Committee, eds. Fl. North Amer. 20:383–393. Oxford Univ. Press, New York and Oxford.

Hartman, R.L. and **G.L. Nesom.** 2012. Taxonomy of the genus *Vesper* (Apiaceae). Phytoneuron 2012-94:1–9.

Hartman, R.L. and **R.K. Rabeler.** 2005a. *Spergula*. In: Flora of North America Editorial Committee, eds. Fl. North Amer. 5:14–16. Oxford Univ. Press, New York and Oxford.

Hartman, R.L. and **R.K. Rabeler.** 2005b. *Spergularia*. In: Flora of North America Editorial Committee, eds. Fl. North Amer. 5:16–23. Oxford Univ. Press, New York and Oxford.

Hartman, R.L. and **R.K. Rabeler.** 2005c. *Loeflingia*. In: Flora of North America Editorial Committee, eds. Fl. North Amer. 5:26–27. Oxford Univ. Press, New York and Oxford.

Hartman, R.L., R.K. Rabeler, and **F.H. Utech.** 2005b. *Arenaria*. In: Flora of North America Editorial Committee, eds. Fl. North Amer. 5:51–56. Oxford Univ. Press, New York and Oxford.

Hartman, R.L., J.W. Thieret, and **R.K. Rabeler.** 2005a *Paronychia*. In: Flora of North America Editorial Committee, eds. Fl. North Amer. 5:30–43. Oxford Univ. Press, New York and Oxford.

Hartmann, H.E.K. 1993. Aizoaceae. In: K. Kubitzki, J.C. Rohwer, and V. Bittrich, eds. The families and genera of vascular plants, Vol. II. Pp. 37–69. Springer-Verlag, Berlin.

Harvey, W.D. 1998. Harmful or potentially harmful fish, shellfish, and aquatic plants. In: Texas Administrative Code. Title 31, Part 2 (Texas Parks and Wildlife Department regulations). Chapter 57, subchapter A, section 57.111–112.

Hasenstab-Lehman, K.E. 2017. Phylogenetics of the borage family: Delimiting Boraginales and assessing closest relatives. Aliso 35(1):41–49.

Hassan, N.M.S., U. Meve, and **S. Liede-Schumann.** 2005. Seed coat morphology of Aizoaceae-Sesuvioideae, Gisekiaceae and Molluginaceae and its systematic significance. Bot. J. Linn. Soc. 148:189–206.

Hassan, N.S., J.Thiede, and **S. Liede.** 2005. Phylogenetic analysis of Sesuvioideae (Aizoaceae) inferred from nrDNA internal transcribed spacer (ITS) sequences and morphological data. Pl. Syst. Evol. 255:121–143.

Hatch, S.L., K.N. Gandhi, and **L.E. Brown.** 1990. Checklist of the vascular plants of Texas. Texas Agric. Exp. Sta. Misc. Publ. 1655: 1–158.

Hatch, S.L. and **A.T. Slack.** 2008. *Lilaeopsis carolinensis* (Apiaceae) a species new to Texas and a key to *Lilaeopsis* in Texas. J. Bot. Res. Inst. Texas 2:1497–1498.

Hausen, B.M., J. Bröhas, W.A. König, H. Faasch, H. Hahn, and **G. Bruhn.** 1987. Allergic and irritant contact dermatitis from falcarinol and didehydrofalcarinol in common ivy (*Hedera helix* L.). Contact Dermatitis 17:1–9.

Havard, V. 1896. Drink plants of the North American Indians. Bull. Torrey Bot. Club 23:33–46.

Hayashi, M., K.L. Feilich, and **D.J. Ellerby.** 2009. The mechanics of explosive seed dispersal in orange jewelweed (*Impatiens capensis*). J. Exp. Bot. 60:2046–2053.

Hayden, W.J. 2016. *Manihot*. In: Flora of North America Editorial Committee, eds. Fl. North Amer. 12:192–196. Oxford Univ. Press, New York and Oxford.

Hayne, F.G. 1830. Getreue darstellung und bescheribung der in der arzneykunde gebräuchlichen gewächse (Faithful representation and description of the plants commonly used in pharmacology). Vol. 11. Berlin.

Haynes, R.R. 1998. Noteworthy collections: Alabama [*Elatine triandra*]. Castanea 63:81–82.

He, J.M. and **Q. Mu.** 2015. The medicinal uses of the genus *Mahonia* in traditional Chinese medicine: An ethnopharmacological, phytochemical and pharmacological review. J. Ethnopharmacol. 175:668–683.

Heard, S.B. and **J.C. Semple.** 1988. The *Solidago rigida* complex (Compositae: Astereae): A multivariate morphometric analysis and chromosome numbers. Canad. J. Bot. 66:1800–1807.

Heiser, C.B., Jr. 1951. The sunflower among the North American Indians. Proc. Amer. Philos. Soc. 95:432–448.

Heiser, C.B., Jr. 1954. Variation and subspeciation in the common sunflower, *Helianthus annuus*. Amer. Midl. Naturalist 52:287–305.

Heiser, C.B., Jr. 1979. The gourd book. Univ. of Oklahoma Press, Norman.

Heiser, C.B., Jr. 1990. New perspectives on the origin and evolution of New World domesticated plants: Summary. Econ. Bot. 44 (Suppl.):111–116.

Heiser, C.B., Jr. 1993. Ethnobotany and economic botany. In: Flora of North America Editorial Committee, eds. Fl. North Amer. 1:199–206. Oxford Univ. Press, New York and Oxford.

Heiser, C.B., Jr., S.B. Clevenger, and **W.C. Martin, Jr.** 1969. The North American sunflowers (*Helianthus*). Mem. Torrey Bot. Club 22:1–218.

Hennen, J.F. (Joe) 1928–2024. Personal communication. Mycologist specializing in rusts; retired from career at Purdue and subsequently Research Associate at the Botanical Research Institute of Texas, Fort Worth.

Hennen, J.F. 1950. The true clovers (*Trifolium*) of Texas. Field & Lab. 18:159–164.

Hennen, J.F. 1951. The sweet clovers (*Melilotus*) of Texas. Field & Lab. 19:87–89.

Henrickson, J. 1985. A taxonomic revision of *Chilopsis* (Bignoniaceae). Aliso 11:179–197.

Henrickson, J. 1986. *Anisacanthus quadrifidus* sensu lato (Acanthaceae). Sida 11:286–299.

Henrickson, J. 1987. A taxonomic reevaluation of *Gossypianthus* and *Guilleminea* (Amaranthaceae). Sida 12:307–337.

Henrickson, J. 1993. Amaranthaceae. In: J.C. Hickman, ed. The Jepson manual: Higher plants of California. Pp. 130–134. Univ. of California Press, Berkeley.

Henrickson, J. 1996. Studies in *Macrosiphonia* (Apocynaceae): Generic recognition of *Telosiphonia*. Aliso 14:179–195.

Henrickson, J. 1999a. Studies in the genus *Dyschoriste* (Acanthaceae): 1. Plants of northern Mexico, Texas to Arizona. Lundellia 2: 72–99.

Henrickson, J. 1999b. Studies in New World *Amaranthus* (Amaranthaceae). Sida 18:783–807.

Henrickson, J. and **T.F. Daniel.** 1979. Three new species of *Carlowrightia* (Acanthaceae) from the Chihuahuan Desert region. Madroño 26:26–36.

Henrickson, J. and **S. Sundberg.** 1986. On the submersion of *Dicraurus* into *Iresine* (Amaranthaceae). Aliso 11:355–364.

Hensel, A., M. Maas, J. Sendker, M. Lechtenberg, F. Petereit, A. Deters, T. Schmidt, and **T. Stark.** 2011. *Eupatorium perfoliatum* L.: Phytochemistry, traditional use and current applications. J. Ethnopharmacol. 138(3):641–651.

Hepper, F.N. 1990. Pharaoh's flowers: The botanical treasures of Tutankhamun. HMSO, London, England, UK.

Herendeen, P.S. and **P.R. Crane.** 1995. The fossil history of the monocotyledons. In: P.J. Rudall, P.J. Cribb, D.F. Cutler, and C.J. Humphries, eds. Monocotyledons: systematics and evolution. Vol. 1:1–21. Royal Botanic Gadens, Kew, England, U.K.

Hermann, F.J. 1953. A botanical synopsis of the cultivated clovers (*Trifolium*). U.S.D.A. Monogr. 22:1–45.

Hermann, F.J. 1954. A synopsis of the genus *Arachis*. U.S.D.A. Monogr. 19:1–26.

Hermann, F.J. 1960. *Vicia*: Vetches of the United States–native, naturalized, and cultivated. U.S.D.A. Handb. No. 168.

Hermann, F.J. 1962. A revision of the genus *Glycine* and its immediate allies. U.S.D.A. Techn. Bull. 1268:1–82.

Hernández, H.M. 2023. *Calliandra*. In: Flora of North America Editorial Committee, eds. Fl. of North Amer. 11:126–129. Oxford Univ. Press, New York and Oxford.

Hernández-Ledesma, P., W.G. Berendsohn, T. Borsch, and **16 others.** 2015. A taxonomic backbone for the global synthesis of species diversity in the angiosperm order Caryophllales. Willdenowia 45:281–383.

Hershkovitz, M.A. and **E.A. Zimmer.** 1997. On the evolutionary origins of the cacti. Taxon 46:217–232.

Heywood, J.S. and **D.A. Levin.** 1984. Allozyme variation in *Gaillardia pulchella* and *G. amblyodon* (Compositae): Relation to morphological and chromosomal variation and to geographical isolation. Syst. Bot. 9:448–457.

Heywood, V.H., ed. 1993. Flowering plants of the world. Oxford Univ. Press, New York.

Hidalgo, O., T. Garnatje, A. Susanna, and **J. Mathez.** 2004. Phylogeny of Valerianaceae based on *matK* and ITS markers with reference to *matK* individual polymorphism. Ann. Bot. (Oxford). 93:283–293.

Hiern, W.P. 1873. A monograph of Ebenaceae. Trans. Cambridge Philos. Soc. 12:27–300.

Hileman, L.C., M.C. Vasey, and **V.T. Parker.** 2001. Phylogeny and biogeography of the Arbutoideae (Ericaceae): Implications for the Madrean-Tethyan hypothesis. Syst. Bot. 26:131–143.

Hill, S., R. Loper, J.R. Singhurst, and **W.C. Holmes.** 2007. *Vicia lathyroides* (Fabaceae): new to the flora of Texas. J. Bot. Res. Inst. Texas 1(2):1253–1254.

Hill, S., J.R. Singhurst, and **W.C. Holmes.** 2016. *Scleranthus annuua* (Caryophyllaceae) naturalized in Texas. Phytoneuron 2016-41:1–3.

Hillard, O.M. and **B.L. Burtt.** 1981. Some generic concepts in Compositae–Gnaphaliinae. Bot. J. Linn. Soc. 82:181–232.

Hilsenbeck, R.A. 1989a. Taxonomy of *Yeatesia* (Acanthaceae). Syst. Bot. 14:427–437.

Hilsenbeck, R.A. 1989b. A new species of *Siphonoglossa* (Acanthaceae) and some infrageneric transfers. Madroño 36:198–207.

Hilsenbeck, R.A. 1989c. Generic affinities and typification of eleven species excluded from *Siphonoglossa* Oerst. (Acanthaceae). Phytologia 67:227–234.

Hilsenbeck, R.A. 1990a. Pollen morphology and systematics of *Siphonoglossa* sensu lato (Acanthaceae). Amer. J. Bot. 77:27–40.

Hilsenbeck, R.A. 1990b. Systematics of *Justicia* section *Pentaloba* (Acanthaceae). Pl. Syst. Evol. 169:219–235.

Hiroe, M. 1979. Umbelliferae of world. Ariake Book Company, Tokyo, Japan.

Hitchcock, C.L. 1936. The genus *Lepidium* in the United States. Madroño 3:265–320.

Hitchcock, C.L. 1952. A revision of the North American species of *Lathyrus*. Univ. Wash. Publ. Biol. 15:1–104.

Hoagland, B.W. and **A. Buthod.** 2004. *Alternanthera paronichhyoides* (Amaranthaceae) and *Rumex maritimus* (Polygonaceae) new to Oklahoma. Sida 21:1199–1200.

Hodgdon, A.R. 1938. A taxonomic study of *Lechea*. Rhodora 40:29–131.

Hoey, M.T. and **C.R. Parks.** 1994. Genetic divergence in *Liquidambar styraciflua*, *L. formosana*, and *L. acalycina* (Hamamelidaceae). Syst. Bot. 19:308–316.

Hoffman, M.T. 1992. Functional dioecy in *Echinocereus coccineus* (Cactaceae): Breeding system, sex ratios, and geographic range of floral dimorphism. Amer. J. Bot. 79:1383–1388.

Hohmann, N., E.M. Wolf, M.A. Lysak, and **M.A. Koch.** 2015. A time-calibrated road map of Brassicaceae species radiation and evolutionary history. Pl. Cell 27(10):2770–2784.

Holloway, J.E. 2005. A dictionary of common wildflowers of Texas & the Southern Great Plains. Texas Christian Univ. Press, Fort Worth.

Holm, R.W. 1950. The American species of *Sarcostemma* R. Br. (Asclepiadaceae). Ann. Missouri Bot. Gard. 37:477–560.

Holmes, W.C. 1981. *Mikania* (Compositae) of the United States. Sida 9:147–158.

Holmes, W.C. 1990. The genus *Mikania* (Compositae-Eupatorieae) in Mexico. Sida Bot. Misc. 5.

Holmes, W.C. 2006. *Mikania*. In: Flora of North America Editorial Committee, eds. Fl. North Amer. 21:545–547. Oxford Univ. Press, New York and Oxford.

Holmes, W.C. 2010. Caricaceae. In: Flora of North America Editorial Committee, eds. Fl. North Amer. 7:170–171. Oxford Univ. Press, New York and Oxford.

Holmes, W.C., K.M. Fleming, R. Loper, R. Loper, and **J.S. Singhust.** 1999. *Triosteum* (Caprifoliaceae) in Texas. Sida 18:931–933.

Holmes, W.C. and **J.R. Singhurst.** 2012. *Pityopsis oligantha* (Asteraceae) new to Texas. Phytoneuron 2012-110:1–4.

Holmes, W.C., J.R. Singhurst, J.N. Mink, and **M. White.** 2012. *Selenia aurea* Nuttall (Brassicaceae) in Texas: A review. Phytoneuron 2012-62:1–8.

Holmgren, N.H. 2003. *Monolepis*. In: Flora of North America Editorial Committee, eds. Fl. North Amer. 4:302. Oxford Univ. Press, New York and Oxford.

Holmgren, N.H. and **P.K. Holmgren.** 2004. Lectotypification of ten names of North American taxa in Clusiaceae, Elatinaceae, Ericaceae, Primulaceae, and Violaceae. Brittonia 56:252–254.

Holtum, J.A.M., L.P. Hancock, E.J. Edwards, and **K. Winter.** 2018. Crassulacean acid metabolism in the Basellaceae (Caryophyllales). Pl. Biol. (New York) 20:409–414.

Hong, D.-Y. and **Q. Wang.** 2015. A new taxonomic system of the Campanulaceae s.s. J. Syst. Evol. 53:203–209.

Hopkins, C.O. and **W.H. Blackwell.** 1977. Synopsis of *Suaeda* (Chenopodiaceae) in North America. Sida 7:147–173.

Hopkins, M. 1942. *Cercis* in North America. Rhodora 44:193–211.

Horn, J.W., B. Van Ee, J.J. Morawetz, and **R. Riina.** 2012. Phylogenetics and the evolution of major structural characters in the giant genus *Euphorbia* L. (Euphorbiaceae). Molec. Phylogen. Evol. 63(2):305–326.

Howe, W.T. 1975. The butterflies of North America. Doubleday & Co., Inc., Garden City, NY.

Howell, J.T. 1959. Distribution data on weedy thistles in western North America. Leafl. W. Bot. 9:17–29.

Hsieh, C.-L., C.-C. Yu, Y.-L. Huang, and **K.-F. Chung.** 2022. *Mahonia* vs. *Berberis* unloaded: Generic delimitation and infrafamilial classification of Berberidaceae based on plastid phylogenomics. Frontiers Pl. Sci. (Lausanne) 12:720171. https://doi: 10.3389/fpls.2021.720171.

Hsü, J. 1983. Late Cretaceious and Cenozoic vegetation in China, emphasizing their connections with North America. Ann. Missouri Bot. Gard. 70:490–508.

Huber, H. 1993. Aristolochiaceae. In: K. Kubitzki, J.C. Rohwer, and V. Bittrich, eds. The families and genera of vascular plants, Vol. II. Pp. 129–137. Springer-Verlag, Berlin.

Huft, M.J. 2016. *Stillingia*. In: Flora of North America Editorial Committee, eds. Fl. North Amer. 12:233–237. Oxford Univ. Press, New York and Oxford.

Hughes, C. 1998. Monograph of *Leucaena* (Leguminosae-Mimosoideae). Syst. Bot. Monogr. 55:1–244.

Hunt, D.R. 1971. Schumann and Buxbaum reconciled. The Schumann system of *Mammillaria* classification brought provisionally up-to-date. Cact. Succ. J. Gr. Brit. 33:53–72.

Hunt, D.R. 1978. Amplification of the genus *Escobaria*. Cact. Succ. J. Gr. Brit. 40:13.

Hunt, D.R. 1999. CITES Cactaceae checklist, 2nd ed. Royal Botanic Gardens, Kew, England, U.K. and the International Organization for Succulent Plant Study (IOS).

Hunt, D.R. and **N.P. Taylor,** eds. 1990. The genera of Cactaceae: Progress towards consensus. Bradleya 8:85–107.

Hurka, H., N. Friesen, D.A. German, A. Franzke, and **B. Neuffer.** 2012. 'Missing link' species *Capsella orientalis* and *Capsella thracica* elucidate evolution of model plant genus *Capsella* (Brassicaceae). Molec. Ecol. 21:1223–1238.

Hurka, H. and **B. Neuffer.** 1997. Evolutionary processes in the genus *Capsella* (Brassicaceae). Pl. Syst. Evol. 206:295–316.

Hyam, R. and **R. Pankhurst.** 1995. Plants and their names: A concise dictionary. Oxford Univ. Press, New York.

Hyatt, P.E. 2006. *Sonchus*. In: Flora of North America Editorial Committee, eds. Fl. North Amer. 19:273–276. Oxford Univ. Press, New York and Oxford.

Hymowitz, T. and **C.A. Newell.** 1981. Taxonomy of the genus *Glycine*, domestication and uses of soybeans. Econ. Bot. 35:272–288.

Ickert-Bond, S.M., K.B. Pigg, and **J. Wen.** 2005. Comparative infructescence morphology in *Liquidambar* (Altingiaceae) and its evolutionary significance. Amer. J. Bot. 92:1234–1255.

Ickert-Bond, S.M., K.B. Pigg, and **J. Wen.** 2007. Comparative infructescence morphology in *Altingia* (Altingiaceae) and discordance between morphological and molecular phylogenies. Amer. J. Bot. 94:1094–1115.

Ickert-Bond, S.M. and **J. Wen.** 2006. Phylogeny and biogeography of Altingiaceae: Evidence from combined analysis of five non-coding chloroplast regions. Molec. Phylogen. Evol. 39:512–528.

Ickert-Bond, S.M. and **J. Wen.** 2013. A taxonomic synopsis of Altingiaceae with nine new combinations. Phytokeys 39: 512–528. https://phytokeys.pensoft.net/article/1500/.

Iltis, H.H. 1957. Studies in the Capparidaceae. III. Evolution and phylogeny of the western North American Cleomoideae. Ann. Missouri Bot. Gard. 44:77–119.

Iltis, H.H. 1958. Studies in the Capparidaceae—IV. *Polanisia* Raf. Brittonia 10:33–58.

Ilts, H.H. 1958 [1959]. Studies in the Capparidaceae—V. Capparidaceae of New Mexico. S. W. Naturalist 3:133–144.

Iltis, H.H. 1966. Studies in the Capparidaceae VIII. *Polanisia dodecandra* (L.) DC.: Further notes on its typification. Rhodora 68:41–47.

Iltis, H.H. and **G.C. Tucker.** 2010. *Tarenaya*. In: Flora of North America

Editorial Committee, eds. Fl. North Amer. 7:218–219. Oxford Univ. Press, New York and Oxford.

iNaturalist.org. 2016. http://www.inaturalist.org/. Accessed Sep 2016.

iNaturalist.org. 2021. The California Academy of Sciences. http://www.inaturalist.org/. Accessed 2019–2021. [Note: all cited iNaturalist observation numbers (from Asteraceae and Brassicaceae) form the final portion of the permanent URL, i.e., observation 71085243 can be viewed at https://www.inaturalist.org/observations/71085243]

iNaturalist. 2023. Available from https://www.inaturalist.org. Accessed April 15, 2023.

iNaturalist contributors, iNaturalist. 2023a. iNaturalist Research-grade Observations. iNaturalist.org. Occurrence dataset https://doi.org/10.15468/ab3s5x accessed via GBIF.org on 2023-01-16. https://www.gbif.org/occurrence/2238777823.

iNaturalist contributors, iNaturalist. 2023b. iNaturalist Research-grade Observations. iNaturalist.org. Occurrence dataset https://doi.org/10.15468/ab3s5x accessed via GBIF.org on 2023-01-27. https://www.gbif.org/occurrence/3764396775.

iNaturalist contributors, iNaturalist. 2023c. iNaturalist Research-grade Observations. iNaturalist.org. Occurrence dataset https://doi.org/10.15468/ab3s5x accessed via GBIF.org on 2023-03-04. https://www.gbif.org/occurrence/3772862031.

iNaturalist contributors, iNaturalist. 2023d. iNaturalist Research-grade Observations. iNaturalist.org. Occurrence dataset https://doi.org/10.15468/ab3s5x accessed via GBIF.org on 2023-03-11. https://www.gbif.org/occurrence/136522939.

iNaturalist contributors, iNaturalist. 2023e. iNaturalist Research-grade Observations. iNaturalist.org. Occurrence dataset https://doi.org/10.15468/ab3s5x accessed via GBIF.org on 2023-03-11. https://www.gbif.org/occurrence/2465090109.

iNaturalist contributors, iNaturalist. 2023f. iNaturalist Research-grade Observations. iNaturalist.org. Occurrence dataset https://doi.org/10.15468/ab3s5x accessed via GBIF.org on 2023-03-11. https://www.gbif.org/occurrence/3985753306.

iNaturalist contributors, iNaturalist. 2023g. iNaturalist Research-grade Observations. iNaturalist.org. Occurrence dataset https://doi.org/10.15468/ab3s5x accessed via GBIF.org on 2023-03-11. https://www.gbif.org/occurrence/3307129305.

iNaturalist contributors, iNaturalist. 2023h. iNaturalist Research-grade Observations. iNaturalist.org. Occurrence dataset https://doi.org/10.15468/ab3s5x accessed via GBIF.org on 2023-04-18. https://www.gbif.org/occurrence/2237413473.

iNaturalist contributors, iNaturalist. 2023i. iNaturalist Research-grade Observations. iNaturalist.org. Occurrence dataset https://doi.org/10.15468/ab3s5x accessed via GBIF.org 2023-07-29. https://www.gbif.org/occurrence/4091700308.

iNaturalist contributors, iNaturalist. 2023j. iNaturalist Research-grade Observations. iNaturalist.org. Occurrence dataset https://doi.org/10.15468/ab3s5x accessed via GBIF.org on 2023-06-16. https://www.gbif.org/occurrence/2517959204.

iNaturalist contributors, iNaturalist. 2023k. iNaturalist Research-grade Observations. iNaturalist.org. Occurrence dataset https://doi.org/10.15468/ab3s5x accessed via GBIF.org on 2023-07-13. https://www.gbif.org/occurrence/2236928349.

iNaturalist contributors, iNaturalist. 2023l. iNaturalist Research-grade Observations. iNaturalist.org. Occurrence dataset https://doi.org/10.15468/ab3s5x accessed via GBIF.org on 2023-07-17. https://www.gbif.org/occurrence/2595989272.

iNaturalist contributors, iNaturalist. 2023m. iNaturalist Research-grade Observations. iNaturalist.org. Occurrence dataset https://doi.org/10.15468/ab3s5x accessed via GBIF.org on 2023-07-29. https://www.gbif.org/occurrence/4129760302.

iNaturalist. 2024. https://www.inaturalist.org. Accessed 23 December 2024.

Inbar, M., M. Wink, and **D. Wool.** 2004. The evolution of host plant manipulation by insects: Molecular and ecological evidence from gall-forming aphids on *Pistacia*. Molec. Phylogen. Evol. 32(2):504–511.

Ingram, J. 1961. Studies in the cultivated Ericaceae, I. *Leucothoe*. Baileya 9:57–66.

Ingram, J. 1967. A revisional study of *Argythamnia* subgenus *Argythamnia* (Euphorbiaceae). Gentes Herbarum 10:1–38.

Ingram, J. 1979. *Leucothoe* revisited (Ericaceae). Baileya 20:141–145.

Ingram, J. 1980. The generic limits of *Argythamnia* (Euphorbiaceae) defined. Gentes Herbarum 11:427–436.

Invasive.org (Center for Invasive Species and Ecosystem Health). 2024. Carolina fanwort, fanwort, *Cabomba caroliniana* A. Gray. https://www.invasive.org/browse/subinfo.cfm?sub=5223. Accessed 29 Dec 2024.

Invasive.org (Center for Invasive Species and Ecosystem Health). 2006. Invasive and exotic weeds. http://www.invasive.org/weeds.cfm. Accessed Jun 2008.

Irving, E. and **R. Hebda.** 1993. Concerning the origin and distribution of Rhododendrons. J. Amer. Rhododendron Soc. 47(3): 139–146, 157–162.

Irwin, H.S. and **R.C. Barneby.** 1982. The American Cassiinae. A synoptical revision of Leguminosae tribe Cassieae subtribe Cassiinae in the New World. Mem. New York Bot. Gard. 35:1–918.

Isely, D. 1948. *Lespedeza striata* and *L. stipulacea*. Rhodora 50:21–27.

Isely, D. 1951. The Leguminosae of the north-central United States: I. Loteae and Trifolieae. Iowa State Coll. J. Sci. 25:439–482.

Isely, D. 1969. Legumes of the United States: I. Native *Acacia*. Sida 3:365–386.

Isely, D. 1970a. Legumes of the United States: II. *Desmanthus* and *Neptunia*. Iowa State J. Sci. 44:495–511.

Isely, D. 1970b. Legumes of the United States V. *Albizia*, *Lysiloma*, *Leucaena*, *Adenanthera*; and rejected genera of the Mimosoideae. Castanea 35:244–260.

Isely, D. 1971a. Legumes of the United States: III. *Schrankia*. Sida 4:232–245.

Isely, D. 1971b. Legumes of the United States. IV. *Mimosa*. Amer. Midl. Naturalist 85:410–424.

Isely, D. 1972. Legumes of the U.S. VI. *Calliandra*, *Pithecellobium*, and *Prosopis*. Madroño 21:273–298.

Isely, D. 1973. Leguminosae of the United States: I. Subfamily Mimosoideae. Mem. New York Bot. Gard. 25(1):1–152.

Isely, D. 1975. Leguminosae of the United States: II. Subfamily Caesalpinioideae. Mem. New York Bot. Gard. 25(2):1–228.

Isely, D. 1981. Leguminosae of the United States. III. Subfamily Papilionoideae: Tribes Sophoreae, Podalyrieae, Loteae. Mem. New York Bot. Gard. 25(3):1–264.

Isely, D. 1982. Leguminosae and *Homo sapiens*. Econ. Bot. 36:46–70.

Isely, D. 1983–1986. *Astragalus* L. (Leguminosae: Papilionoideae). Iowa State J. Res. 58:3–172; 59:99–209; 60:183–320; 61:157–289.

Isely, D. 1986a. Notes about *Psoralea* sensu auct., *Amorpha*, *Baptisia*, *Sesbania* and *Chamaecrista* (Leguminosae) in the southeastern United States. Sida 11:429–440.

Isely, D. 1990. Leguminosae (Fabaceae). Vascular flora of the southeastern United States 3(2):1–258. Univ. North Carolina Press, Chapel Hill.

Isely, D. 1998. Native and naturalized Leguminosae (Fabaceae) of the United States (exclusive of Alaska and Hawaii). Monte L. Bean Life Science Museum, Brigham Young Univ., Provo, UT.

Isely, D. and **F.J. Peabody.** 1984. *Robinia* (Leguminosae: Papilionoideae). Castanea 49:187–202.

Isely, D. and **R.M. Polhill.** 1980. Leguminosae: Subfamily Papilionoideae. Taxon 29:105–119.

IUCN. 2008. 100 of the world's worst invasive alien species. http://www.iucn.org/places/medoffice/invasive_species/docs/invasive_species_booklet.pdf. Accessed Jun 2008.

Jackson, R.C. 1960. A revision of the genus *Iva* L. Univ. Kansas Sci. Bull. 41:793–874.

Jackson, R.C. 1963. Cytotaxonomy of *Helianthus ciliaris* and related species of the southwestern U.S. and Mexico. Brittonia 15: 260–271.

Jackson, S.W. 1963. Hybridization among three species of *Ratibida*. Univ. Kansas Sci. Bull. 44:3–27.

Jacobson, H.A., J.B. Petersen, and **D.E. Putnam.** 1988. Evidence of pre-columbian *Brassica* in the northeastern United States. Rhodora 90(86):355–362.

Jacobson, M. 1982. Plants, insects, and man—Their interrelationships. Econ. Bot. 36:346–354.

James, L.F. and **S.L. Welsh.** 1992. Poisonous plants of North America. In: L.F. James, R.F. Keeler, E.M. Bailey, Jr., P.R. Cheeke, and M.P. Hegarty, eds. Poisonous plants: Proceedings of the third international symposium. Pp. 94–104. Iowa State Univ. Press, Ames.

James, R.L. 1948. Some hummingbird flowers east of the Mississippi. Castanea 12:97–109.

Jansen, R.K. 1981. Systematics of *Spilanthes* (Compositae: Heliantheae). Syst. Bot. 6:231–257.

Jansen, R.K. 1985a. The systematics of *Acmella* (Asteraceae–Heliantheae). Syst. Bot. Monogr. 8:1–115.

Jansen, R.K. 1985b. Systematic significance of chromosome number in *Acmella* (Asteraceae). Amer. J. Bot. 72:1835–1841.

Jansen, R.K. and **K.-J. Kim.**1996. Implications of chloroplast DNA data for the classification and phylogeny of the Asteraceae. In: D.J.N. Hind and H.J. Beentje, eds. Compositae: Systematics. Proceedings of the International Compositae Conference, Royal Botanic Gardens, Kew, 1994. Volume 1. Pp. 317–339. Royal Botanic Gardens Kew, England, U.K.

Jansen, R.K., E.B. Smith, and **D.J. Crawford.** 1987. A cladistic study of North American *Coreopsis* (Asteraceae: Heliantheae). Pl. Syst. Evol. 157:73–84.

Jansen, R.K., H.J. Michaels, and **J.D. Palmer.** 1991. Phylogeny and character evolution in the Asteraceae based on chloroplast DNA restriction site mapping. Syst. Bot. 16:98–115.

Jansen, R.K., H.J. Michaels, R.S. Wallace, K.-J. Kim, S.C. Keeley, L.E. Watson, and **J.D. Palmer.** 1992. Chloroplast DNA variation in the Asteraceae: Phylogenetic and evolutionary implications. In: P.S. Soltis, D.E. Soltis, and J.J. Doyle, eds. Molecular systematics of plants. Pp. 252–279. Chapman and Hall, New York.

Jansen, S.A., I. Kleerekooper, Z.L.M. Hofman, I.F.P.M. Kappen, A. Stary-Weinzinger, and **M.A.G. van der Heyden.** 2012. Grayanotoxin poisoning: 'Mad Honey Disease' and beyond. Cardiovasc. Toxicol. 12:208–215.

Janssens, S.B., K. Geuten, Y.M. Yuan, Y. Song, P. Küpfer, and **E.F. Smets.** 2006. Phylogenetics of *Impatiens* and *Hydrocera* (Balsaminaceae) using chloroplast atpB-rbcL spacer sequences. Syst. Bot. 31:171–180.

Janzen, D.H. and **P.S. Martin.** 1982. Neotropical anachronisms: The fruits the gomphotheres ate. Science 215:19–27.

Jarvinen, P., A. Palmé, J.O. Morales, M. Lännenpää, M. Keinänen, T. Sopanen, and **M. Lascoux.** 2004. Phylogenetic relationships of *Betula* species (Betulaceae) based on nuclear ADH and chloroplast MATK sequences. Amer. J. Bot. 91:1834–1845.

Jasiewicz, A. 1976. *Scabiosa*. In: T.G. Tutin, V.H. Heywood, N.A. Burges, D.H. Valentine, S.M. Walters, and D.A. Webb, eds. Flora Europaea. 4:68–74. Cambridge Univ. Press, Cambridge, England, UK.

Jeffrey, C. 1975. Further notes on Cucurbitaceae: IV. Some New-World taxa. Kew Bull. 33:347–380.

Jeffrey, C. 1990. Appendix: An outline classification of the Cucurbitaceae. In: D.M. Bates, R.W. Robinson, and C. Jeffrey, eds. Biology and utilization of the Cucurbitaceae. Pp. 449–463. Cornell Univ. Press, Ithaca, NY.

Jensen, U., I. Vogel-Bauer, and **M. Nitschke.** 1994. Leguminlike proteins and the systematics of the Euphorbiaceae. Ann. Missouri Bot. Gard. 81:160–179.

Johnson, D.M. 2003. Phylogenetic significance of spiral and distichous architecture in the Annonaceae. Syst. Bot. 28:503–511.

Johnson, G.P. 1997. Calycanthaceae. In: Flora of North America Editorial Committee, eds. Fl. North Amer. 3:23–25. Oxford Univ. Press, New York and Oxford.

Johnston, I.M. 1924. Studies in the Boraginaceae. II. 29. *Lappula*. Contr. Gray Herb. 70:47–51.

Johnston, I.M. 1925a. Studies in Boraginaceae. V. 1. Concerning the range and identity of certain American species. Contr. Gray Herb. 75:40–49.

Johnston, I.M. 1925b. Studies in the Boraginaceae. IV. The North American species of *Cryptantha*. Contr. Gray Herb. 74:1–114.

Johnston, I.M. 1952. Studies in the Boraginaceae, XXIII. A survey of the genus *Lithospermum*. J. Arnold Arbor. 33:299–366.

Johnston, I.M. 1954a. Studies in the Boraginaceae, XXVI. Further reevaluations of the genera of the Lithospermeae. J. Arnold Arbor. 35:1–81.

Johnston, I.M. 1954b. Studies in the Boraginaceae, XXVII. Some general observations concerning the Lithospermeae. J. Arnold Arbor. 35:158–166.

Johnston, I.M. 1967. Boraginaceae. In: C.L. Lundell, ed. Flora of Texas. 1:123–221. Texas Research Foundation, Renner.

Johnston, M.C. and **D.H. Riskind.** 1975. McKinney Falls State Park plant list. Unpublished report. Texas Parks and Wildlife Department, Austin.

Johnston, M.C. 1956. The Texas species of *Dyssodia* (Compositae). Field & Lab. 24:60–69.

Johnston, M.C. 1959. The Texas species of *Croton* (Euphorbiaceae). S. W. Naturalist 3:175–203.

Johnston, M.C. 1963a. The geography of the five Texas species of *Dichondra* (Convoluvlaceae). Wrightia 2:252–253.

Johnston, M.C. 1975. Studies of the *Euphorbia* species of the Chihuahuan Desert region and adjacent areas. Wrightia 5:120–143.

Johnston, M.C. and **B.L. Turner.** 1962. Chromosome numbers of *Dyssodia* (Compositae: Tagetinae) and phyletic interpretations. Rhodora 64:2–15.

Jones, A.G. 1978a. The taxonomy of *Aster* section *Multiflori* (Asteraceae)—I. Nomenclatural review and formal presentation of taxa. Rhodora 80:319–357.

Jones, A.G. 1978b. The taxonomy of *Aster* section *Multiflori* (Asteraceae)—II. Biosystematic investigations. Rhodora 80: 453–490.

Jones, A.G. 1980. A classification of the New World species of *Aster* (Asteraceae). Brittonia 32:230–239.

Jones, A.G. 1983. Nomenclatural changes in *Aster* (Asteraceae). Bull. Torrey Bot. Club 110:39–42.

Jones, A.G. 1984. Nomenclatural notes on *Aster* (Asteraceae)—II. New combinations and some transfers. Phytologia 55:373–388.

Jones, A.G. 1987. New combinations and status changes in *Aster* (Asteraceae). Phytologia 63:131–133.

Jones, A.G. 1992. *Aster* and *Brachyactis* (Asteraceae) in Oklahoma. Sida Bot. Misc. 8.

Jones, A.G. and **M.T. Deonier.** 1965. Interspecific crosses among *Ipomoea lacunosa*, *I. ramoni*, *I. trichoarpa*, and *I. triloba*. Bot. Gaz. 126:226–232.

Jones, A.G. and **D.A. Young.** 1983. Generic concepts of *Aster* (Asteraceae): A comparison of cladistic, phenetic, and cytological approaches. Syst. Bot. 8:71–84.

Jones, G.N. 1940. A monograph of the genus *Symphoricarpos*. J. Arnold Arbor. 21:201–252.

Jones, K.E., N. Korotkova, J. Petersen, T. Henning, T. Borsch, and **N. Kilian.** 2017. Dynamic diversification history with rate upshifts in Holarctic bell-flowers (*Campanula* and allies). Cladistics 33: 637–666.

Jones, R.L. 1983. A systematic study of *Aster* section *Patentes* (Asteraceae). Sida 10:41–81.

Jones, R.L., C.T. Witsell, and **G.L. Nesom.** 2008. Distribution and taxonomy of *Symphyotrichum sericeum* and *S. pratense* (Asteraceae: Astereae). J. Bot. Res. Inst. Texas 2(1):731–739.

Jones, S.B. and **W.Z. Faust.** 1978. Compositae tribe Vernonieae. N. Amer. Fl. II. 10:180–202.

Jones, S.D. and **A.A. Reznicek.** 1997 *Lathyrus aphaca* (Fabaceae), previously unreported for Texas. Phytologia 82(1):1–2.

Jones, S.D., J.K. Wipff, and **P.M. Montgomery.** 1997. Vascular plants of Texas: A comprehensive checklist including synonymy, bibliography, and index. Univ. of Texas Press, Austin.

Jordan, R. 1991. Putative natural hybrid of *Eupatorium capillifolium* × *E. glaucescens* from Hardin County, Texas. Phytologia 71(5): 360–361.

Jordon-Thaden, I.E., I.A. Al-Shehbaz, and **M.A. Koch.** 2013. Species richness of the globally distributed, arctic–alpine genus *Draba* L. (Brassicaceae). Alpine Bot. 123(2):97–106.

Jordon-Thaden, I.E., I. Hase, I.A. Al-Shehbaz, and **M.A. Koch.** 2010. Molecular phylogeny and systematics of the genus *Draba* (Brassicaceae) and identification of its most closely related genera. Molec. Phylogen. Evol. 55(2):524–540.

Judd, W.S. 1981. A monograph of *Lyonia* (Ericaceae). J. Arnold Arbor. 62(1):63–128.

Judd, W.S. 1995. *Lyonia*. In: J.L. Luteyn, ed. Ericaceae Part II, the superior-ovaried genera. Fl. Neotrop. Monogr. 66:222–294.

Judd, W.S. 2001. Phylogeny and biogeography of *Lyonia* sect. *Lyonia* (Ericaceae). In: C.A. Woods and F.E. Sergile, eds. Biogeography of the West Indies: Patterns and perspectives, 2nd ed. Pp.63–75. CRC Press, Boca Raton, FL

Judd, W.S. 2009. *Lyonia*. In: Flora of North America Editorial Committee, eds. Fl. North Amer. 8:500–503. Oxford Univ. Press, New York and Oxford.

Judd, W.S., C.S. Campbell, E.A. Kellogg, P.F. Stevens, and **M.J. Donoghue.** 2002. Plant systematics: A phylogenetic approach, 2nd ed. Sinauer Associates, Inc., Sunderland, MA.

Judd, W.S., C.S. Campbell, E.A. Kellogg, P.F. Stevens, and **M.J. Donoghue.** 2016. Plant systematics: A phylogenetic approach, 4th ed. Sinauer Associates, Inc., Sunderland, MA.

Judd, W.S. and **I.K. Ferguson.** 1999. The genera of Chenopodiaceae in the southeastern United States. Harvard Pap. Bot. 4:365–416.

Judd, W.S. and **K.A. Kron.** 1993. Circumscription of Ericaceae (Ericales) as determined by preliminary cladistic analyses based on morphological, anatomical, and embryological features. Brittonia 45:99–114.

Judd, W.S. and **K.A. Kron.** 2009. *Rhododendron*. In: Flora of North America Editorial Committee, eds. Fl. North Amer. 8:455–473. Oxford Univ. Press, New York and Oxford.

Judd, W.S., N.C. Melvin, III, K. Waselkov, and **K.A. Kron.** 2012. Taxonomic revision of *Eubotrys* (Ericaceae, Gaultherieae). Brittonia 64:–178.

Judd, W.S., N.C. Melvin, III, K. Waselkov, and **K.A. Kron.** 2013. A taxonomic revision of *Leucothoë* (Ericaceae; Tribe Gaultherieae). Brittonia 65(4):417–438.

Judd, W.S., R.W. Sanders, and **M.J. Donoghue.** 1994. Angiosperm family pairs: Preliminary phylogenetic analyses. Harvard Pap. Bot. 5:1–51.

Jurney, D.H. 1987. Presettlement vegetation recorded in land surveyors' notes. In: D.H. Jurney and R.W. Moir. Historic buildings, material culture, and people of the prairie margin. Richland Creek Technical Series Volume V. Pp. 211–227. Archaeology Research Program, Institute for the Study of Earth and Man, Southern Methodist Univ., Dallas, TX.

Jury, S.L. 2004. The Euro+Med treatment of *Hedera* (Araliaceae)—Recent studies and a new name. Willdenowia 34:351–352.

Kadereit, G., P. Ball, S. Beer, L. Mucina, D. Sokoloff, P. Teege, A.E. Yaprak, and **H. Freitag.** 2007. A taxonomic nightmare comes true: Phylogeny and biogeography of glassworts (*Salicornia* L., Chenopodiaceae). Taxon 56:1143–1170.

Kadereit, G., T. Borsch, K. Weising, and **H. Freitag.** 2003. Phylogeny of Amaranthaceae, and Chenopodiaceae and evolution of C4-photosynthesis. Int. J. Pl. Sci. 164(6):959–986.

Kadereit, G. and **H. Freitag.** 2011. Molecular phylogeny of Camphorosmeae (Camphorosmoideae, Chenopodiaceae): Implications for biogeography, evolution of C4-photosynthesis and taxonomy. Taxon 60:51–78.

Kadereit, J.W., D.C. Albach, F. Ehrendorfer, and **19 others.** 2016. Which changes are needed to render all genera of the German flora monophyletic? Willdenowia 46:39–91.

Kalisz, S. and **G.M. Wardle.** 1994. Life history variation in *Campanula americana* (Campanulaceae): Population differentiation. Amer. J. Bot. 81:521–527.

Kang, J.-S., V.N.L. Giang, H.-S. Park, Y.S. Park, W. Cho, V.B. Nguyen, H. Shim, N.E. Waminal, J.Y. Park, H.H. Kim, and **T.-J. Yang.** 2023. Evolution of the Araliaceae family involved rapid diversification of the Asian Palmate group and *Hydrocotyle* specific mutational pressure. Sci. Rep. 13. https://doi.org/10.1038/s41598-023-49830-7.

Kapil, R.N. and **A.K. Bhatnagar.** 1994. The contribution of embryology to the systematics of the Euphorbiaceae. Ann. Missouri Bot. Gard. 81:145–159.

Karaman-Castro, V. and **L.E. Urbatsch.** 2006. *Boltonia*. In: Flora of North America Editorial Committee, eds. Fl. North Amer. 20: 353–357. Oxford Univ. Press, New York and Oxford.

Karis, P.O. 1995. Cladistics of the subtribe Ambrosiinae (Asteraceae: Heliantheae). Syst. Bot. 20:40–54.

Karis, P.O. and **O. Ryding.** 1994. Tribe Helenieae. In: K. Bremer. Asteraceae cladistics & classification. Pp. 521–558. Timber Press, Portland, OR.

Kartesz, J.T. 1994. A synonymized checklist of the vascular flora of the United States, Canada, and Greenland, 2nd ed. (2 vols.). Timber Press, Portland, OR.

Kartesz, J.T. 1999. A synonymized checklist and atlas with biological attributes for the vascular flora of the United States, Canada, and Greenland, 1st ed. In: J.T. Kartesz and C.A. Meacham. Synthesis of the North American Flora, CD format, Version 1.0. North Carolina Botanical Garden, Chapel Hill.

Kartesz, J.T. 2005 (unpublished, but widely distributed). Synthesis of the North American flora, Version 2.0 (software program by C.A. Meacham). Missouri Botanical Garden Press, St. Louis.

Kartesz, J.T. 2015. The Biota of North America Program (BONAP) Taxonomic Data Center. Chapel Hill, N.C. [Maps generated from Kartesz, J.T. 2015–onward. Floristic Synthesis of North America, Biota of North America Program (BONAP)]. http://www.bonap.net/tdc. Accessed 2019–2021 (for Brassicaceae).

Kartesz, J.T. 2015. The Biota of North America Program (BONAP). Taxonomic Data Center. http://www.bonap.net/tdc. Chapel Hill, N.C. [Maps generated from Kartesz, J.T. 2015. Floristic Synthesis of North America, Version 1.0. Biota of North America Program (BONAP).]

Kartesz, J.T. 2020. The Biota of North America Program (BONAP). Synthesis of the North American Flora (CD version).

Kartesz, J.T. 2021. The Biota of North America Program (BONAP). Synthesis of the North American Flora (CD version).

Kartesz, J.T. 2022. The Biota of North America Program (BONAP). Synthesis of the North American Flora (CD version).

Kartesz, J.T. 2023. The Biota of North America Program (BONAP). Synthesis of the North American Flora (CD version).

Kartesz, J.T. 2024. The Biota of North America Program (BONAP). Synthesis of the North American Flora (CD version).

Kato, H, K. Oginuma, Z. Gu, B. Hammel, and **H. Tobe.** 1999. Phylogenetic relationships of Betulaceae based on matK sequences with particular reference to the position of *Ostryopsis*. Acta Phytotax. Geobot. 49:89–97.

Kaul, R.B.1986. Boraginaceae. In: Great Plains Flora Association. Flora of the Great Plains. Pp. 683–701. Univ. Press of Kansas, Lawrence.

Kaul, R.B. 1986. Elatinaceae. In: Great Plains Flora Association Flora of the Great Plains. Pp. 236. Univ. Press of Kansas, Lawrence.

Kawaguchi, S.T. 1996. Systematics of *Rayjacksonia* (Asteraceae: Astereae). M.S. thesis. University of Kansas, Lawrence.

Kearns, D.M. 1994. The genus *Ibervillea* (Cucurbitaceae): An enumeration of the species and two new combinations. Madroño 41:13–22.

Keddy, P.A. 2010. Wetland ecology: Principles and conservation, 2nd ed. Cambridge Univ. Press, Cambridge, UK.

Keeley, S.C. and **R.K. Jansen.** 1991. Evidence from chloroplast DNA for the recognition of a new tribe, the Tarchonantheae, and the tribal placement of *Pluchea* (Asteraceae). Syst. Bot. 16: 173–181.

Keener, B.R. 2020. A new *Hexastylis* (Aristolochiaceae) from Northeast Alabama (U.S.A.) with notes on the species "groups" within the genus. J. Bot. Res. Inst. Texas 14:161–166.

Keener, B.R. 2023. *Lackeya*. In: Flora of North America Editorial Committee, eds. Fl. of North Amer. 11:370–371. Oxford Univ. Press, New York and Oxford.

Keener, B.R. and **G. Todia.** 2021. A new *Hexastylis* (Aristolochiaceae) with uncertain origin in the southeastern U.S.A., discovered through social media. J. Bot. Res. Inst. Texas 15:319–325.

Keil, D.J. 1977. A revision of *Pectis* section *Pectothrix* (Compositae: Tageteae). Rhodora 79:32–78.

Keil, D.J. 2004. New taxa and new combinations in North American *Cirsium* (Asteraceae: Cardueae). Sida 21(1):207–219.

Keil, D.J. 2006a. *Arctium*. In: Flora of North America Editorial Committee, eds. Fl. North Amer.19:168–171. Oxford Univ. Press, New York and Oxford.

Keil, D.J. 2006b. *Carduus*. In: Flora of North America Editorial Committee, eds. Fl. North Amer.19:91–94. Oxford Univ. Press, New York and Oxford.

Keil, D.J. 2006c. *Carthamus*. In: Flora of North America Editorial Committee, eds. Fl. North Amer. 19:178–181. Oxford Univ. Press, New York and Oxford.

Keil, D.J. 2006d. *Cirsium*. In: Flora of North America Editorial Committee, eds. Fl. North Amer. 19:95–164. Oxford Univ. Press, New York and Oxford.

Keil, D.J. 2006e. *Engelmannia*. In: Flora of North America Editorial Committee, eds. Fl. North Amer. 21:87. Oxford Univ. Press, New York and Oxford.

Keil, D.J. 2006f. *Onopordum* In: Flora of North America Editorial Committee, eds. Fl. North Amer. 19:87–88. Oxford Univ. Press, New York and Oxford.

Keil, D.J. 2006g. *Pectis*. In: Flora of North America Editorial Committee, eds. Fl. North Amer. 21:222–230. Oxford Univ. Press, New York and Oxford.

Keil, D.J. 2006h. *Plectocephalus*. In: Flora of North America Editorial Committee, eds. Fl. North Amer. 19:175–177.Oxford Univ. Press, New York and Oxford.

Keil, D.J. 2006i. *Silybum*. In: Flora of North America Editorial Committee, eds. Fl. North Amer. 19:164. Oxford Univ. Press, New York and Oxford.

Keil, D.J. and **J. Oschmann.** 2006. *Centaurea*. In: Flora of North America Editorial Committee, eds. Fl. North Amer. 19:181–194. Oxford Univ. Press, New York and Oxford.

Keith, E.L. 2004a. *Dimorphotheca sinuata* and *Zinnia violacea* (Asteraceae), two escaped cultivars new to Texas. Phytologia 86(1):29–31.

Keith, E.L. 2004b. *Tagetes erecta* (Asteraceae), a new escape to the flora of Texas. Phytologia 86(2):104–106.

Keith, E.L. 2012. *Polytaenia albiflora* (Apiaceae), a new species from the Balcones Canyonlands in the Edwards Plateau of Texas. Phytoneuron 2012-84:1–8.

Keith, E.L. 2013a. *Trifolium nigrescens* (Fabaceae), new to the Texas flora. Phytoneuron 2013-32:1–6.

Keith, E.L. 2013b. *Securigera cretica* (Fabaceae), new to the Texas flora and second record for North America. Phytoneuron 2013-39:1–3.

Keith, E.L., J.R. Singhurst, and **S. Cook.** 2004 *Geocarpon minimum* (Caryophyllaceae), new to Texas. Sida 21:1165–1169.

Keith, E.L., J.R. Singhurst, R.J. Lewandowski, and **W.C. Holmes.** 2017. *Cornus alternifolia* (Cornaceae) in Texas. Phytoneuron 2017-8:1–4.

Keith, E.L., J. Wright, and **W. Godwin.** 2017. Naturalized occurrence of *Ruellia dipteracanthus* (Acanthaceae) in the USA. Phytoneuron 2017-57:1–3.

Kelch, D.G. and **B.G. Baldwin.** 2003. Phylogeny and ecological radiation of New World thistles (*Cirsium*, Cardueae–Compositae) based on ITS and ETS rDNA sequence data. Molec. Ecol. 12: 141–151.

Keller, H.W. 2014. The genus *Echinacea* (Asteraceae): flora, stem, and petiole morphology. J. Bot. Res. Inst. Texas 8(1):87–126.

Kelly, L.M. 1997. A cladistic analysis of *Asarum* (Aristolochiaceae) and implications for the evolution of herkogamy. Amer. J. Bot. 84:1752–1765.

Kelly, L.M. 1998. Phylogenetic relationships in *Asarum* (Aristolochiaceae) based on morphology and ITS sequences. Amer. J. Bot. 85:1454–1467.

Kelly, L.M. 2001. Taxonomy of *Asarum* section *Asarum* (Aristolochiaceae). Syst. Bot. 26:17–53.

Kelly, L.M. and **F. González.** 2003. Phylogenetic relationships in Aristolochiaceae. Syst. Bot. 28:236–249.

Kennedy, W. 1841. The rise, progress, and prospects of the Republic of Texas. London. Reprinted in one volume by Molyneaux Craftsmen, Inc., Fort Worth, Texas, 1925.

Kent, D.H. 1997. The correct authority for lesser chickweed, *Stellaria pallida* (Caryophyllaceae). Watsonia 21:364.

Kers, L.E. 2003. Capparaceae. In: K. Kubitzki and C. Bayer, eds. The families and genera of vascular plants, Vol. V. Pp. 36–56. Springer-Verlag, Berlin.

Kessler, E. 1987. *Carthamus lanatus* L. (Asteraceae: Cynareae)—A potentially serious plant pest in Oklahoma. Proc. Oklahoma Acad. Sci. 67:39–43.

Kessler, P.J.A. 1993a. Annonaceae. In: K. Kubitzki, J.C. Rohwer, and V. Bittrich, eds. The families and genera of vascular plants, Vol. II. Pp. 93–129. Springer-Verlag, Berlin.

Keung, W.-M. and **B.L. Vallee.** 1993. Daidzin and daidzein suppress free-choice ethanol intake by Syrian golden hamsters. Proc. Natl. Acad. Sci. U.S.A. 90:10008–10012.

Kiefer, C., C. Dobeš, and **M.A. Koch.** 2009. *Boechera* or not? Phylogeny and phylogeography of eastern North American *Boechera* species (Brassicaceae). Taxon 58(4):1109–1121.

Kiger, R.W. 2006. *Cosmos*. In: Flora of North America Editorial Committee, eds. Fl. North Amer. 21:203–205. Oxford Univ. Press, New York and Oxford.

Kim, K.-J., R.K. Jansen, and **B.L. Turner.** 1992b. Evolutionary implications of intraspecific chloroplast DNA variation in dwarf dandelions (*Krigia*; Asteraceae). Amer. J. Bot. 79:708–715.

Kim, K.-J. and **T.J. Mabry.** 1991. Phylogenetic and evolutionary implications of nuclear ribosomal DNA variation in dwarf dandelions (*Krigia*–Lactuceae–Asteraceae). Pl. Syst. Evol. 177:53–69.

Kim, K.-J. and **B.L. Turner.** 1992. Systematic overview of *Krigia* (Asteraceae-Lactuceae). Brittonia 44:173–198.

Kim, K.-J., B.L. Turner, and **R.K. Jansen.** 1992a. Phylogenetic and evolutionary implications of interspecific chloroplast DNA variation in *Krigia* (Asteraceae-Lactuceae). Syst. Bot. 17:449–469.

Kim, Y.-D. and **R.K. Jansen.** 1994. Characterization and phylogenetic distribution of a chloroplast DNA rearrangement in the Berberidaceae. Pl. Syst. Evol. 193:107–114.

Kim, Y.-D. and **R.K. Jansen.** 1996 [1997]. Phylogenetic implications of *rbcL* and ITS sequence variation in the Berberidaceae. Syst. Bot. 21:381–396.

Kim, Y.-D. and **R.K. Jansen.** 1998. Chloroplast DNA restriction site variation and phylogeny of the Berberidaceae. Amer. J. Bot. 85:1766–1778.

Kim, Y.-D., S.-H. Kim, C.H. Kim, and **R.K. Jansen.** 2004a. Phylogeny of Berberidaceae based on sequences of the chloroplast gene ndhF. Biochem. Syst. Ecol. 32:291–301.

Kim, Y.-D., S.-H. Kim, and **L.R. Landrum.** 2004b. Taxonomic and phytogeographic implications from ITS phylogeny in *Berberis* (Berberidaceae). J. Pl. Res. 117:175–182.

Kindscher, K. 1992. Medicinal wild plants of the prairie. Univ. Press of Kansas, Lawrence.

Kindscher, K. 2006. The conservation status of *Echinacea* species. Kansas Biological Survey, Univ. of Kansas. Lawrence, KS.

Kindscher, K., ed. 2016. *Echinacea*: Herbal medicine with a wild history. Springer International, Switzerland.

King, B.L. and **S.B. Jones.** 1975. The *Vernonia lindheimeri* complex (Compositae). Brittonia 27:74–86.

King, M.J. and **S.L. Buchmann.** 1995. Bumblebee-initiated vibration pollen release mechanism of *Rhododendron* pollen. Amer. J. Bot. 82:1407–1411.

King, R.M. and **H. Robinson.** 1970a. Studies in the Eupatorieae (Compositae). XVIII. New combinations in *Fleischmannia*. Phytologia 19:201–207.

King, R.M. and **H. Robinson.** 1970b. Studies in the Eupatorieae (Compositae). XIX. New combinations in *Ageratina*. Phytologia 19:208–229.

King, R.M. and **H. Robinson.** 1970c. Studies in the Eupatorieae (Compositae). XIII. The genus *Conoclinum*. Phytologia 19: 299–300.

King, R.M. and **H. Robinson.** 1970d. A monograph of the genus *Trichocoronis*. Phytologia 19:497–500.

King, R.M. and **H. Robinson.** 1970e. *Eupatorium*, a composite of arcto-tertiary distribution. Taxon 19:769–774.

King, R.M. and **H. Robinson.** 1980. Studies in the Eupatorieae (Compositae). XXIV. The genus *Chromolaena*. Phytologia 47(3):230–251.

King, R.M. and **H. Robinson.** 1987. The genera of the Eupatorieae (Asteraceae). Monogr. Syst. Bot. Missouri Bot. Gard. 22:1–581.

Kinghorn, A.D. 1979. Cocarcinogenic irritant Euphorbiaceae. In: A.D. Kinghorn, ed. Toxic plants. Pp. 137–159. Columbia Univ. Press, New York.

Kingsbury, J.M. 1964. Poisonous plants of the United States and Canada. Prentice-Hall, Englewood Cliffs, NJ.

Kingsbury, J.M. 1965. Deadly harvest: A guide to common poisonous plants. Holt, Rinehart and Winston, New York.

Kirkbride, J.H., Jr. 1993. Biosystematic monograph of the genus *Cucumis* (Cucurbitaceae). Parkway Publishers, Boone, NC.

Kirkbride, J.H., Jr. and **R.T. Olsen.** 2011. Neotypification of *Catalpa speciosa* (Bignoniaceae). Taxon 60:1760–1763.

Kirkpatrick, Z.M. 1992. Wildflowers of the western plains. Univ. of Texas Press, Austin.

Klak, C., P.V. Bruyns, and **T.A.J. Hedderson.** 2007. A phylogeny and new classification for Mesembryanthemoideae (Aizoaceae). Taxon 56:737–756.

Klak, C., A. Khunou, G. Reeves, and **T.A.J. Hedderson.** 2003. A phylogenetic hypothesis for the Aizoaceae (Caryophyllales) based on four plastid DNA regions. Amer. J. Bot. 90:1433–1445.

Klooster, M.R. and **T.M. Culley.** 2010. Population genetic structure of the mycoheterotroph *Monotropa hypopitys* L. (Ericaceae) and differentiation between red and yellow color forms. Int. J. Pl. Sci.171:167–174.

Kobuski, C.E. 1928. A monograph of the American species of the genus *Dyschoriste*. Ann. Missouri Bot. Gard. 15:9–90.

Koch, M. and **I.A. Al-Shehbaz.** 2004. Taxonomic and phylogenetic evaluation of the American "*Thlaspi*" species: Identity and relationship to the Eurasian genus *Noccaea* (Brassicaceae). Syst. Bot. 29(2):375–384.

Koch, M., I.A. Al-Shehbaz, and **K. Mummenhoff.** 2003. Molecular systematics, evolution, and population biology in the mustard family (Brassicaceae). Ann. Missouri Bot. Gard. 90(2):151–171.

Koch, M., M. Kiefer, D.A. German, I.A. Al-Shehbaz, A. Franzke, K. Mummenhoff, and **R. Schmickl.** 2020. BrassiBase: Tools and biological resources for Brassicaceae character and trait studies—Version 1.3. Univ. of Heidelberg, Germany. http://brassibase.cos.uni-heidelberg.de/. Accessed 2019–2022.

Kong, A., J. Cifuentes, A. Estrada-Torres, L. Guzmán-Dávalos, R. Garibay-Orijel, and **B. Buyck.** 2015. Russulaceae associated with mycoheterotroph *Monotropa uniflora* (Ericaceae) in Tlaxcala, Mexico: A phylogenetic approach. Cryptog. Mycol. 36(4):479–512.

Konstantinova, A.I. and **E.Y. Yembaturova.** 2010. Structural traits of some species of *Hydrocotyle* (Araliaceae) and their significance for constructing the generic system. Pl. Diverstiy Evol. 128:329–346.

Koptur, S. 1979. Facultative mutualism between weedy vetches bearing extrafloral nectaries and weedy ants in California. Amer. J. Bot. 66(9):1016–1020.

Kosnik, M.A., G.M. Diggs, Jr., P.A. Redshaw, and **B.L. Lipscomb.** 1996. Natural hybridization among three sympatric *Baptisia* (Fabaceae) species in North Central Texas. Sida 17:479–500.

Kral, R. 1960a. A revision of *Asimina* and *Deeringothamnus* (Annonaceae). Brittonia 12:233–278.

Kral, R. 1966. Observations on the flora of the southeastern United States with special reference to northern Louisiana. Sida 2: 395–408.

Kral, R. 1997. Annonaceae. In: Flora of North America Editorial Committee, eds. Fl. North Amer. 3:11–20. Oxford Univ. Press, New York and Oxford.

Krapovickas, A. and **W.C. Gregory.** 1994. Taxonomía del género *Arachis* (Leguminosae). Bonplandia (Corrientes) 8:1–186.

Krapovickas, A., W.C. Gregory, D.E. Williams, and **C.E. Simpson.** 2007. Taxonomy of the genus *Arachis* (Leguminosae). Bonplandia (Corrientes) 16 (supl.):1–205.

Kriebel, R., J.P. Rose, P. Bastide, D. Jolles, M. Reginato, and **K.J. Sytsma.** 2023. The evolution of Ericaceae flowers and their pollination syndromes at a global scale. Amer. J. Bot 110 (9): e16220. https://doi.org/10.1002/ajb2.16220.

Krings, A. 2003. Typification and nomenclatural history of *Trachelospermum difforme* (Apocynaceae). Sida 20:1641–1644.

Krings, A. 2023. *Gonolobus*. In: Flora of North America Editorial Committee, eds. Fl. North Amer. 14:236–238. Oxford Univ. Press, New York and Oxford.

Krings, A., W. Wall, and **A. Broadhead.** 2023. *Parkinsonia*. In: Flora of North America Editorial Committee, eds. Fl. of North Amer. 11:69–71. Oxford Univ. Press, New York and Oxford.

Krochmal, A. 1952. Seeds of weedy *Euphorbia* species and their identification. Weeds 1:243–255.

Krochmal, A., L. Wilken, and **M. Chien.** 1972. Lobeline content of four Appalachian lobelias. Lloydia 35:303–304.

Kron, K.A. 1993. A revision of *Rhododendron* section *Pentathera*. Edinburgh J. Bot. 50:249–364.

Kron, K.A. 1996. Phylogenetic relationships of Empetraceae, Epacridaceae, Ericaceae, Monotropaceae, and Pyrolaceae: Evidence from nuclear ribosomal 18s sequence data. Ann. Bot. (Oxford) 77:293–303.

Kron, K.A. and **M.W. Chase.** 1993. Systematics of the Ericaceae, Empetraceae, Epacridaceae and related taxa based upon *rbc*L sequence data. Ann. Missouri Bot. Gard. 80:735–741.

Kron, K.A., R. Fuller, D.M. Crayn, P.A. Gadek, and **C.J. Quinn.** 1999. Phylogenetic relationships of epacrids and vaccinioids (Ericaceae s. l.) based on matK sequence data. Pl. Syst. Evol. 218(1–2):55–65.

Kron, K.A. and **S.L. Johnson.** 1997. Phylogenetic analysis of the monotropoids and pyroloids (Ericaceae) using nrITS and 18S sequence data. (Meeting of the Botanical Society of America and the Canadian Botanical Association). Amer. J. Bot. 84(Suppl.):205–206.

Kron, K.A. and **W.S. Judd.** 1990. Phylogenetic relationships within the *Rhodoreae* (Ericaceae) with specific comments on the placement of *Ledum*. Syst. Bot. 15:57–68.

Kron, K.A. and **W.S. Judd.** 1997 [1998]. Systematics of the *Lyonia* group (Andromedeae, Ericaceae) and the use of species as terminals in higher-level cladistic analyses. Syst. Bot. 22(3): 479–492.

Kron, K.A., W.S. Judd, P.F. Stevens, D.M. Crayn, A.A. Anderberg, P.A. Gadek, C.J. Quinn, and **J.L. Luteyn.** 2002a. A phylogenetic classification of Ericaceae: Molecular and morphological evidence. Bot. Rev. (Lancaster) 68:335–423.

Kron, K.A., E.A. Powell, and **J.L. Luteyn.** 2002b. Phylogenetic relationships within the blueberry tribe (Vaccinieae, Ericaceae) based on sequence data from *matK* and nuclear ribosomal ITS regions, with comments on the placement of *Satyria*. Amer. J. Bot. 89(2):327–336.

Kruckeberg, A.R. and **R.D. Reeves.** 1995. Nickel accumulation by serpentine species of *Streptanthus* (Brassicaceae): Field and greenhouse studies. Madroño 42(4):458–469.

Kubitzki, K. 1993a. Betulaceae. In: K. Kubitzki, J.C. Rohwer, and V. Bittrich, eds. The families and genera of vascular plants, Vol. II. Pp. 152–157. Springer-Verlag, Berlin.

Kubitzki, K. 1993b. Cannabaceae. In K. Kubitzki, J.C. Rohwer, and V.

Bittrich, eds. The families and genera of vascular plants, Vol. II. Pp. 204–206. Springer-Verlag, Berlin.

Kubitzki, K. 2003a. Droseraceae. In: K. Kubitzki and C. Bayer, eds. The families and genera of vascular plants, Vol. V. Pp. 198–202. Springer-Verlag, Berlin.

Kubitzki, K. 2003b. Drosophyllaceae. In: K. Kubitzki and C. Bayer, eds. The families and genera of vascular plants, Vol. V. Pp. 203–205. Springer-Verlag, Berlin.

Kubitzki, K., ed. 2004a. The families and genera of vascular plants, Vol. VI. Springer-Verlag, Berlin.

Kubitzki, K. 2004b. Cyrillaceae. In: K. Kubitzki, ed. The families and genera of vascular plants, Vol. VI. Pp. 114–116. Springer-Verlag, Berlin.

Kubitzki, K. and **C. Bayer,** eds. 2003. Families and genera of vascular plants, Vol. V. Springer-Verlag, Berlin.

Kuby, J. 1997. Immunology, 3rd ed. W.H. Freeman and Co., New York.

Kuijt, J., D. Bray, and **A.R. Olson.** 1985. Anatomy and ultrastructure of the endophytic system of *Pilostyles thurberi* (Rafflesiaceae). Canad. J. Bot. 63:1231–1240.

Kuo, P.-C., Y.-C. Li, and **T.-S. Wu.** 2012. Chemical Constituents and Pharmacology of the *Aristolochia* species. J. Tradit. Complement. Med. 2:249–266. https://doi: 10.1016/s2225-4110(16)30111-0.

Kupicha, F.K. 1976. The infrageneric structure of *Vicia*. Notes Roy. Bot. Gard. Edinburgh 34:287–326.

Kupicha, F.K. 1981. Vicieae (Adans.) DC. In: R.M. Polhill and P.H. Raven, eds. Advances in legume systematics 1:377–381.

Kupicha, F.K. 1983. The infrageneric structure of *Lathyrus*. Notes Roy. Bot. Gard. Edinburgh 41:209–244.

Kurashige, Y., J.-I. Etoh, T. Handa, K. Takayanagi, and **T. Yukawa.** 2001. Sectional relationships in the genus *Rhododendron* (Ericaceae): Evidence from *mat*K and *trn*K intron sequences Pl. Syst. Evol. 228:1–14.

Lammers, T.G. In press. *Lobelia*. In: Flora of North America Editorial Committee, eds. Fl. North Amer. Vol. 18. Oxford Univ. Press, New York and Oxford.

Lammers, T.G. 1992. Circumscription and phylogeny of Campanulales. Ann. Missouri Bot. Gard. 79:388–413.

Lammers, T.G. 1993. Chromosome numbers of Campanulaceae. III. Review and integration of data for subfamily Lobelioideae. Amer. J. Bot. 80:660–675.

Lammers, T.G. 2006. Validation of names at subspecific rank in North American Campanulaceae. Novon 16:69–73.

Lammers, T.G. 2007. World checklist and bibliography of Campanulaceae. Royal Botanic Gardens, Kew, Richmond, Surrey, U.K.

Lammers, T.G. 2011. Revision of the infrageneric classification of *Lobelia* L. (Campanulaceae: Lobelioideae). Ann. Missouri Bot. Gard. 98:37–62.

Lamont, E.E. 1995. Taxonomy of *Eupatorium* section *Verticillata* (Asteraceae). Mem. New York Bot. Gard. 72:1–66.

Lamont, E.E. 2006. *Eutrochium*. In: Flora of North America Editorial Committee, eds. Fl. North Amer. 21:474–478. Oxford Univ. Press, New York and Oxford.

Lampe, K.E. 1986. Contact dermatitis from Sonoran Desert plants. Desert Pl. 8:32–37.

Lampe, K.E. and **M.A. McCann.** 1985. AMA handbook of poisonous and injurious plants. American Medical Association, Chicago, IL.

Lampert, E.C., L.A. Dyer, and **M.D. Bowers.** 2011. Chemical defense across three trophic levels: *Catalpa bignonioides*, the caterpillar *Ceratomia catalpae*, and its endoparasitoid *Cotesia congregata*. J. Chem. Ecol. 37:1063–1070.

Landrum, J.V. 2002. Four succulent families and 40 million years of evolution and adaptation to xeric environments: What can stem and leaf anatomical characters tell us about their phylogeny? Taxon 51:463–473.

Landry, G.P., W.D. Reese, and **C.M. Allen.** 1988. *Stellaria parva* Pederson new to North America. Phytologia 64:497.

Landry, G.P., W.D. Reese, and **K. Vincent.** 1989. *Stellaria parva* Pedersen (Caryophyllaceae) in North America. Sida 13(3): 373–375.

Lane, M.A. 1979. Taxonomy of the genus *Amphiachyris*. Syst. Bot. 4: 178–189.

Lane, M.A. 1980. New and re-instated combinations in *Gutierrezia* (Compositae: Astereae). Sida 8:313–314.

Lane, M.A. 1982. Generic limits of *Xanthocephalum*, *Gutierrezia*, *Amphiachyris*, *Gymnosperma*, *Greenella*, and *Thurovia*. Syst. Bot. 7:405–416.

Lane, M.A. 1985. Taxonomy of *Gutierrezia* (Compositae: Astereae) in North America. Syst. Bot. 10:7–28.

Lane, M.A. and **R.L. Hartman.** 1996. Reclassification of North American *Haplopappus* (Compositae: Astereae) completed. Amer. J. Bot. 83:356–370.

Lang, J.M. and **D. Isely.** 1982. *Eysenhardtia* (Leguminosae: Papilionoideae). Iowa State J. Res. 56:393–417.

Langer, R., I. Pein, and **B. Kopp.** 1995. Glandular hairs in the genus *Drosera* (Droseraceae). Pl. Syst. Evol. 194:163–172.

Larisey, M. 1940. A monograph of the genus *Baptisia*. Ann. Missouri Bot. Gard. 27:119–244.

Larsen, E.L. 1927. A revision of the genus *Townsendia*. Ann. Missouri Bot. Gard. 14:1–46.

Larsen, E.L. 1933. *Astranthium* and related genera. Ann. Missouri Bot. Gard. 20:23–44.

Larsen, E.L. 1940. *Lobelia* as a sure cure for venereal disease. Amer. J. Syphilis, Gonorrhea, Venereal Diseases 24:13–22.

Larsen, G.E. 1986. Caryophyllaceae. In: Great Plains Flora Association. Flora of the Great Plains. Pp.192–214. Univ. Press of Kansas, Lawrence.

Lassen, P. 1989. A new delimitation of the genera *Coronilla*, *Hippocrepis*, and *Securigera* (Fabaceae). Willdenowia 19:49–62.

Lassetter, J.S. 1978. Seed characters in some native American vetches. Sida 7:255–263.

Lassetter, J.S. 1984. Taxonomy of the *Vicia ludoviciana* complex (Leguminosae). Rhodora 86:475–505.

Latorre, A.O., G.A. Borghi, P.L. Lopes, K.C. Higa, L.M.X. Lopes, P.C. Maiorka, S.L. Gorniak, and **M. Haraguchi.** 2011. First report on rotenoids as neurotoxic principles of seeds from *Aeschynomene indica* (Leguminosae). J. Anim. Vet. Adv. 10(3): 291–294.

Latorre, D.L. and **F.A. Latorre.** 1977. Plants used by the Mexican Kickapoo Indians. Econ. Bot. 31:340–357.

Latting, J. 1961. The biology of *Desmanthus illinoensis*. Ecology 42:487–493.

La Vieille, S., D.E. Lefebvre, A.F. Khalid, M.R. Decan, and **S. Godefroy.** 2019. Dietary restrictions for people with glucose-6-phosphate dehydrogenase deficiency. Nutr. Rev. 77(2):96–106.

Lavin, M. and **T.G. Lammers.** 2023. *Robinia*. In: Flora of North America

Editorial Committee, eds. Fl. of North Amer. 11:545–549. Oxford Univ. Press, New York and Oxford.

Leake, J.R. 2005. Plants parasitic on fungi: Unearthing the fungi in myco-heterotrophs and debunking the 'saprophytic' plant myth. Mycologist 19:113–122.

Lee, C.-S. and **S.R. Downie.** 2006. Phylogenetic relationships within *Cicuta* (Apiaceae tribe Oenantheae) inferred from nuclear rDNA ITS and cpDNA sequence data. Canad. J. Bot. 84:453–468.

Lee, Y.S. and **D.B. Dickinson.** 1980. Field observations on hybrids between *Ambrosia bientata* and *A. trifida* (Compositae). Amer. Midl. Naturalist 103:180–184.

Lee, Y.S. 1981. Serological investigations in *Ambrosia* (Compositae: Ambrosieae) and relatives. Syst. Bot. 6(2)113–125.

Lee, Y.-T. 1976. The genus *Gymnocladus* and its tropical affinities. J. Arnold Arbor. 57:91–112.

Lehman, R.L. 2009. Plants of the Texas coastal bend. Texas A&M Univ. Press, College Station.

Leite, F.H.M., N. Khandpur, G. Calixto Andrade, K. Anastasiou, P. Baker, M. Lawrence, and **C.A. Monteiro.** 2022. Ultra-processed foods should be central to global food systems dialogue and action on biodiversity. BMJ Global Health. 2022;7:e008269. https://doi.org/10.1136/bmjgh-2021-008269.

Leitch, I.J. and **L. Hanson.** 2002. DNA C-values in seven families fill phylogenetic gaps in the basal angiosperms. Bot. J. Linn. Soc. 140:175–179.

Lemke, D.E. 1987. Recent collections and a redescription of *Amsonia ludoviciana* Vail (Apocynaceae). Sida 12(2):343–346.

Lemke, D.E. 2009. Cyrillaceae. In: Flora of North America Editorial Committee, eds. Fl. North Amer. 8:367–369. Oxford Univ. Press, New York and Oxford.

Lemke, D.E. 2015. *Lechea*. In: Flora of North America Editorial Committee, eds. Fl. North Amer. 6:389–397. Oxford Univ. Press, New York and Oxford.

Lemke, D.E. 2023a. *Apocynum*. In: Flora of North America Editorial Committee, eds. Fl. North Amer. 14:138–140. Oxford Univ. Press, New York and Oxford.

Lemke, D.E. 2023b. *Catharanthus*. In: Flora of North America Editorial Committee, eds. Fl. North Amer. 14:126–127. Oxford Univ. Press, New York and Oxford.

Lemke, D.E. 2023c. *Mandevilla*. In: Flora of North America Editorial Committee, eds. Fl. North Amer. 14:144–147. Oxford Univ. Press, New York and Oxford.

Lemke, D.E. 2023d. *Nerium*. In: Flora of North America Editorial Committee, eds. Fl. North Amer. 14:148. Oxford Univ. Press, New York and Oxford.

Lemke, D.E. 2023e. *Thyrsanthella*. In: Flora of North America Editorial Committee, eds. Fl. North Amer. 14:151. Oxford Univ. Press, New York and Oxford.

Lemke, D.E. 2023f. *Trachelospermum*. In: Flora of North America Editorial Committee, eds. Fl. North Amer. 14:151–152. Oxford Univ. Press, New York and Oxford.

Lemke, D.E. 2023g. *Vinca*. In: Flora of North America Editorial Committee, eds. Fl. North Amer. 14:135–136. Oxford Univ. Press, New York and Oxford.

Lemke, D.E. and **R.D. Worthington.** 1991. *Brassica* and *Rapistrum* (Brassicaceae) in Texas. S. W. Naturalist 36(2):194–197.

Lentink, D., W.B. Dickson, J.L. van Leeuwen, and **M.H. Dickinson.** 2009. Leading-edge vortices elevate lift of autorotating plant seeds. Science 324:1438–40.

Leonard, E.C. 1927. *Ruellia tuberosa* and a few of its close relatives. J. Wash. Acad. Sci. 17:509–520.

Lersten, N.R. and **C.L. Brubaker.** 1987. Extrafloral nectaries in Leguminosae: Review and original observations in *Erythrina* and *Mucuna* (Papilionoideae; Phaseoleae). Bull. Torrey Bot. Club 114:437–447.

Les, D.H. 1986a. The phytogeography of *Ceratophyllum demersum* and *C. echinatum* in glaciated North America. Canad. J. Bot. 64: 498–509.

Les, D.H. 1986b. The evolution of achene morphology in *Ceratophyllum* (Ceratophyllaceae), I. Fruit spine variation and relationships of *C. demersum*, *C. submersum*, and *C. apiculatum*. Syst. Bot. 11:549–558.

Les, D.H. 1988a. The origin and affinities of the Ceratophyllaceae. Taxon 37:326–345.

Les, D.H. 1988b. The evolution of achene morphology in *Ceratophyllum* (Ceratophyllaceae), II. Fruit variation and systematics of the "spiny-margined" group. Syst. Bot. 13:73–86.

Les, D.H. 1988c. The evolution of achene morphology in *Ceratophyllum* (Ceratophyllaceae), III. Relationships of the "facially-spined" group. Syst. Bot. 13:509–518.

Les, D.H. 1989. The evolution of achene morphology in *Ceratophyllum* (Ceratophyllaceae), IV. Summary of proposed relationships and evolutionary trends. Syst. Bot. 14:254–262.

Les, D.H. 1993. Ceratophyllaceae. In: K. Kubitzki, J.C. Rohwer, and V. Bittrich, eds. The families and genera of vascular plants, Vol. II. Pp. 246–250. Springer-Verlag, Berlin.

Les, D.H. 1994. Molecular systematics and taxonomy of lake cress (*Neobeckia aquatica*; Brassicaceae), an imperiled aquatic mustard. Aquatic Bot. 49:149–165.

Les, D.H. 1997. Ceratophyllaceae. In: Flora of North America Editorial Committee, eds. Fl. North Amer. 3:81–88. Oxford Univ. Press, New York and Oxford.

Les, D.H. 2017. Aquatic dicotyledons of North America: ecology, life history, and systematics. CRC press, Boca Raton, FL.

Lesica, P. 2022. Manual of Montana vascular plants, 2nd ed. BRIT Press, Fort Worth, TX.

Lesins, K.A. and **I. Lesins.** 1979. Genus *Medicago* (Leguminosae). W. Junk, The Hague, Netherlands.

Leuenberger, B.E. 1991. Interpretation and typification of *Cactus ficus-indica* L., and *Opuntia ficus-indica* (L.) Miller (Cactaceae). Taxon 40:621–627.

Leuenberger, B.E. 1993. Interpretation and typification of *Cactus opuntia* L., *Opuntia vulgaris* Mill., and *O. humifusa* (Rafin.) Rafin. (Cactaceae). Taxon 42:419–429.

Leung, A.Y. and **S. Foster.** 1996. Encyclopedia of common natural ingredients used in food, drugs, and cosmetics, 2nd ed. John Wiley and Sons, Inc., New York.

Levin, G.A. (Geoffrey). Personal communication. Specialist in *Acalypha* (Euphorbiaceae). Director of the Center for Biodiversity, Illinois Natural History Survey, Champaign, IL.

Levin, G.A. 1998. Evolution in the *Acalypha gracilens/monococca* complex (Euphorbiaceae): Morphological analysis. Syst. Bot. 23:269–287.

Levin, G.A. 1999. Notes on *Acalypha* (Euphorbiaceae) in North America. Rhodora 101:217–233.

Levin, G.A. 2016a. *Acalypha*. In: Flora of North America Editorial Committee, eds. Fl. North Amer. 12:162–172. Oxford Univ. Press, New York and Oxford.

Levin, G.A. 2016b. *Cnidoscolus*. In: Flora of North America Editorial Committee, eds. Fl. North Amer. 12:196–198. Oxford Univ. Press, New York and Oxford.

Levin, G.A. 2024a. *Lilaeopsis*. In: Flora of North America Editorial Committee, eds. Fl. North Amer. 13:340–343. Oxford Univ. Press, New York and Oxford.

Levin, G.A. 2024b. *Zizia*. In: Flora of North America Editorial Committee, eds. Fl. North Amer. 13:521–523. Oxford Univ. Press, New York and Oxford.

Levin, G.A. and **L.J. Gillespie.** 2016. Euphorbiaceae. In: Flora of North America Editorial Committee, eds. Fl. North Amer. 12:156–324. Oxford Univ. Press, New York and Oxford.

Lewis, G.P., B. Schrire, B. MacKinder, and **M. Lock,** eds. 2005. Legumes of the World. Royal Botanic Gardens, Kew, England, U.K.

Lewis, W.H. and **P.R. Fantz.** 1973. Tribal classification of *Triosteum* (Caprifoliaceae). Rhodora 75:120–121.

Lewis, W.H. 1971. Pollen differences between *Stylisma* and *Bonamia* (Convolvulaceae). Brittonia 23:331–334.

Lewis, W.H. and **M.P.F. Elvin-Lewis.** 1977. Medical botany: Plants affecting man's health. John Wiley & Sons, New York.

Lewis, W.H. and **P.R. Fantz.** 1973. Tribal classification of *Triosteum* (Caprifoliaceae). Rhodora 75:120–121.

Lewis, W.H. and **R.L. Oliver.** 1965. Realignment of *Calystegia* and *Convolvulus* (Convolvulaceae). Ann. Missouri Bot. Gard. 52:217–222.

Lewis, W.H., P. Vinay, and **V.E. Zenger.** 1983. Airborne and allergenic pollen of North America. Johns Hopkins Univ. Press, Baltimore, MD.

Li, B. and **B. Bartholomew.** 2008. Calycanthaceae. In: Z.-Y. Wu and P.H. Raven, eds. Flora of China (Menispermaceae through Capparaceae). 7:92–95. Science Press, Beijing, China and Missouri Botanical Garden, St. Louis.

Li, H.-C., X.-L. Zhao, X.-F. Gao, and **B. Xu.** 2019. Molecular phylogeny of the genus *Hylodesmum* (Fabaceae). Phytotaxa 403: 221–229.

Li, H.-L. 1952. Floristic relationships between eastern Asia and eastern North America. Trans. Amer. Philos. Soc. 42:371–409.

Li, H.T., Y. Luo, L. Gan, and **20 others.** 2021. Plastid phylogenomic insights into relationships of all flowering plant families. BMC Biol. 19, article no. 232. https://doi.org/10.1186/s12915-021-01166-2.

Li, J. 2008. Phylogeny of *Catalpa* (Bignoniaceae) inferred from sequences of chloroplast ndhF and nuclear ribosomal DNA. J. Syst. Evol. 46:341–348.

Li, J., J. Ledger, T. Ward, and **P. del Tredici.** 2004. Phylogentics of Calycanthaceae based on molecular and morphological data, with a special reference to divergent paralogues of the nrDNA ITS region. Harvard Pap. Bot. 9:69–82.

Li, J., S. Shoup, and **Z. Chen.** 2005. Phylogenetics of *Betula* (Betulaceae) inferred from sequences of nuclear ribosomal DNA. Rhodora 107:69–86.

Li, J., S. Shoup, and **T.S. Elias.** 2006. Molecular confirmation of intergeneric hybrid ×*Chitalpa tashkentensis* (Bignoniaceae). HortScience 41:1162–1164.

Li, P.T., A.J.M. Leeuwenberg, and **D.J. Middleton.** 1995. Apocynaceae. In: Z.-Y. Wu and P.H. Raven, eds. Flora of China (Gentianaceae through Boraginaceae). 16:143–188. Science Press, Beijing, China and Missouri Botanical Garden, St. Louis.

Li, Y. and **P.-T. Li.** 2000. Origin, evolution and distribution of the Calycanthaceae. Guihaia 20:295–300.

Li,Y, Y. Yang, L. Yu, X. Du, and **G. Ren.** 2018. Plastomes of nine hornbeams and phylogenetic implications. Ecol. Evol. 8:8770–8778.

Liede, S. 1996. *Sarcostemma* (Asclepiadaceae)—A controversial generic circumscription reconsidered: Morphological evidence. Syst. Bot. 21:31–44.

Liede, S. and **H. Kunze.** 2002. *Cynanchum* and the Cynanchinae (Apocynaceae-Asclepiadoideae): A molecular, anatomical and latex triterpenoid study. Organisms Diversity Evol. 2:239–269.

Liede, S. and **A. Täuber.** 2002. Circumscription of the genus *Cynanchum* (Apocynaceae-Asclepiadoideae). Syst. Bot. 27: 789–800.

Lievens, A.W. and **M.A. Vincent.** 2023. *Indigofera*. In: Flora of North America Editorial Committee, eds. Fl. of North Amer. 11:335–342. Oxford Univ. Press, New York and Oxford.

Lihova, J., K. Marhold, H. Kudoh, and **M.A. Koch.** 2006. Worldwide phylogeny and biogeography of *Cardamine flexuosa* (Brassicaceae) and its relatives. Amer. J. Bot. 93:1206–1231.

Lim, D. 1998. Microbiology, 2nd. ed. WCB/McGraw-Hill, Boston, MA.

Lin, D., B. Shao, Z. Gao, and **8 others.** 2025. Phylogenomics of angiosperms based on mitochondrial genes: Insights into deep node relationships. BMC Biol. 23, Article no. 45. https://doi.org/10.1186/s12915-025-02135-9.

Lin, H.-Y., M. Sun, Y.-J. Hao, D. Li, M.A. Gitzendanner, C.-X. Fu, D.E. Soltis, P.S. Soltis, and **Y.P. Zhao.** 2022. Phylogenetic diversity of eastern Asia–eastern North America disjunct plants is mainly associated with divergence time. Pl. Diversity 45:27–35.

Lindelof, K., J.A. Lindo, W. Zhou, X. Ji, and **Q.-Y. (J.) Xiang.** 2020. Phylogenomics, biogeography, and evolution of the blue- or white-fruited dogwoods (*Cornus*)—Insights into morphological and ecological niche divergence following intercontinental geographic isolation. J. Syst. Evol. 58:604–645.

Lindsey, A.H. 1975. Classification versus characterization in *Thaspium* and *Zizia* (Apiaceae). M.S. thesis, Univ. of North Carolina, Chapel Hill.

Lindsey, A.H. 1984. Reproductive biology of Apiaceae: 1. Floral visitors to *Thaspium* and *Zizia* and their importance in pollination. Amer. J. Bot. 71:375–387.

Lindsey, A.H. and **C.R. Bell.** 1985. Reproductive biology of Apiaceae. II. Cryptic specialization and floral evolution in *Thaspium* and *Zizia*. Amer. J. Bot. 72:231–247.

Lingbeck, J.M., C.A. O'Bryan, E.M. Martin, J.P. Adams, and **P.G. Crandall.** 2015. Sweetgum: An ancient source of beneficial compounds with modern benefits. Pharmacogn. Rev. 9(17): 1–11.

Linnaeus, C. 1753. Species Plantarum. 2 vols. Stockholm.

Linnaeus, C. 1754. Genera Plantarum, ed. 5. Stockholm.

Lipscomb, B.L. (Barney). Personal Communication. Botanist, Director of BRIT Press, Botanical Research Institute of Texas, Fort Worth.

Lipscomb, B.L. 1984. New additions or otherwise noteworthy plants of Texas. Sida 10:326–327.

Lipscomb, B.L. and **G.M. Diggs, Jr.** 1998. The use of animal-dispersed seeds and fruits in forensic botany. Sida 18:335–346.

Lipscomb, B.L. and **E.B. Smith.** 1977. Morphological intergradation of varieties of *Bidens aristosa* (Compositae) in northern Arkansas. Rhodora 79:203–213.

Lira, J. 1973. Breve diccionaria Kkechuwa-Español. Edición Popular. Cuzco, Peru.

Lira, R., J.L. Villaseñor, and **P.D. Davila.** 1997 [1998]. A cladistic analysis of the subtribe Sicyinae (Cucurbitaceae). Syst. Bot. 22:415–425.

Little, E.L. 1970. Endemic, disjunct and northern trees in the southern Appalachians. In: P.C. Holt and R.A. Paterson, eds. The distributional history of the biota of the southern Appalachians, Part II: Flora. Pp. 249–290. Research Division Monog. 2, Virginia Polytechnic Institute and State Univ., Blacksburg, VA.

Little, E.L. 1971. Atlas of United States trees, Vol. 1. Conifers and important hardwoods. U.S.D.A. Forest Serv. Misc. Publ. No. 1146.

Little, E.L. 1983. North American trees with relationships in eastern Asia. Ann. Missouri Bot. Gard. 70:605–615.

Liu, J., Z. Chen, and **A. Lu.** 2002. Molecular evidence for the sister relationship of the eastern Asia-North American intercontinental species pair in the *Podophyllum* group (Berberidaceae). Bot. Bull. Acad. Sin. 43:147–154.

Liu, Z.-W, Q.-R. Zhao, H.-P. Peng, and **J. Zhou.** 2017. Proposal to conserve the name *Hypopitys* (Ericaceae) with a conserved type. Taxon 66:987–988.

Lloyd, D.G. 1972. A revision of the New Zealand, Subantarctic, and South American species of *Cotula*, Section *Leptinella*. New Zealand J. Bot. 10:277–372.

Loconte, H. 1993. Berberidaceae. In: K. Kubitzki, J.C. Rohwer, and V. Bittrich, eds. The families and genera of vascular plants, Vol. II. Pp. 147–152. Springer-Verlag, Berlin.

Loconte, H. and **J.R. Estes.** 1989. Phylogenetic systematics of Berberidaceae and Ranunculales (Magnoliidae). Syst. Bot. 14: 565–579.

Loconte, H. and **D.W. Stevenson.** 1991. Cladistics of the Magnoliideae. Cladistics 7:267–296.

Lohmann, L.G. and **C.M. Taylor.** 2014. A new generic classification of tribe Bignonieae (Bignoniaceae). Ann. Missouri Bot. Gard. 99: 348–489.

Löhne, C., M.-J. Yoo, T. Borsch, J.H. Wiersema, V. Wilde, C.D. Bell, W. Barthlott, D.E. Soltis, and **P.S. Soltis.** 2008. Biogeography of Nymphaeales: Extant patterns and historical events. Taxon 57:1123–1146.

Loizeau, P.-A., G. Barriera, J.-F. Manen, and **O. Broennimann.** 2005. Towards an understanding of the distribution of Ilex L. (Aquifoliaceae) on a world-wide scale. Biol. Skr. 55:501–520.

Lonard, R.I. and **F.R. Waller.** 1971. *Helianthus simulans* E.E. Wats. (Compositae) in Robertson County, new to Texas. S. W. Naturalist 16(1):121.

Long, R.W. 1961. Convergent patterns in *Ruellia caroliniensis* and *R. humilis* (Acanthaceae). Bull. Torrey Bot. Club 88:387–396.

Long, R.W. 1966. Artificial interspecific hybridization in *Ruellia* (Acanthaceae). Amer. J. Bot. 53:917–927.

Long, R.W. 1970. The genera of Acanthaceae in the southeastern United States. J. Arnold Arbor. 51:257–309.

Long, R.W. 1971. Floral polymorphy and amphimictic breeding systems in *Ruellia caroliniensis* (Acanthaceae). Amer. J. Bot. 58:525–531.

Long, R.W. 1973. A biosystematic approach to generic delimitation in *Ruellia* (Acanthaceae). Taxon 22:543–555.

Long, R.W. 1974. Variation in natural populations of *Ruellia caroliniensis* (Acanthaceae). Bull. Torrey Bot. Club 101:1–6.

Long, R.W. 1975. Artificial interspecific hybridization in temperate and tropical species of *Ruellia* (Acanthaceae). Brittonia 27: 289–296.

Long, R.W. 1977. Artificial induction of obligate cleistogamy in species-hybrids in *Ruellia* (Acanthaceae). Bull. Torrey Bot. Club 104:53–56.

Long, R.W. and **L.J. Uttal.** 1962. Some observations on flowering in *Ruellia* (Acanthaceae). Rhodora 64:200–206.

Loos, P.M., T.C. Philipps, E.L. Keith, M. Buckingham, J.R. Singhurst, and **W.C. Holmes.** 2019. *Stokeisa laevis* (Asteraceae), a native of the Texas flora. Phytoneuron 2019-56:1–5.

Löve, D. and **P. Dansereau.** 1959. Biosystematic studies on *Xanthium*: Taxonomic appraisal and ecological status. Canad. J. Bot. 37(2): 173–208.

Lowden, R.M. 1978. Studies on the submerged genus *Ceratophyllum* L. in the Neotropics. Aquatic Bot. 4:127–142.

Lowry, P.P., II and **A.G. Jones.** 1984. Systematics of *Osmorhiza* Raf. (Apiaceae: Apioideae). Ann. Missouri Bot. Gard. 71:1128–1171.

Lowry, P.P., G.M. Plunkett, and **A.A. Oskolski.** 2001. Early lineages in Apiales: Insights from morphology, wood anatomy and molecular data. Edinburgh J. Bot. 58:207–220.

Lowry, P.P., G.M. Plunkett, and **J. Wen.** 2004. Generic relationships in Araliaceae: Looking into the crystal ball. S. African J. Bot. 70: 382–392.

LPWG (Legume Phylogeny Working Group). 2017. A new subfamily classification of the Leguminosae based on a taxonomically comprehensive phylogeny. Taxon 66(1):44–77.

Lu, L.-M., L.-F. Mao, T. Yang, and **20 others.** 2018. Evolutionary history of the angiosperm flora of China—Cradle vs. museum. Nature 554:234–238.

Luckow, M. 1993. Monograph of *Desmanthus* (Leguminosae-Mimosoideae). Syst. Bot. Monogr. 38:1–166.

Luckow, M. 2023. *Desmanthus* In: Flora of North America Editorial Committee, eds. Fl. of North Amer. 11:84–91. Oxford Univ. Press, New York and Oxford.

Luebert, F., L. Cecchi, M.W. Frohlich, and **17 others.** 2016. Familial classification of the Boraginales. Taxon 65:502–522.

Luken, J.O. and **J.W. Thieret.** 1995. Amur honeysuckle (*Lonicera maackii*; Caprifoliaceae): Its ascent, decline, and fall. Sida 16: 479–503.

Luken, J.O. and **J.W. Thieret.** 1996. Amur honeysuckle, its fall from grace. BioScience 46:18–24.

Luken, J.O., J.W. Thieret, and **J.T. Kartesz.** 1993. *Erucastrum gallicum* (Brassicaceae): Invasion and spread in North America. Sida 15(4):569–582.

Luken, J.O., T.C. Tholemeier, L.M. Kuddes, and **B.A. Kunkel.** 1995. Performance, plasticity, and acclimation potential of the non-indigenous shrub *Lonicera maackii* in contrasting light environments. Canad. J. Bot. 73:1953–1961.

Lundberg, J. and **K. Bremer.** 2003. A phylogenetic study of the order Asterales using one morphological and three molecular data sets. Int. J. Pl. Sci. 164. 553–578.

Lundell, C.L. 1941. Studies of American spermatophytes—I. Contr. Univ. Michigan Herb. 6:1–66.

Lundell, C.L. 1961a. Cyrillaceae. In: C.L. Lundell, ed. Flora of Texas 3:109–111. Texas Research Foundation, Renner.

Lundell, C.L. 1961b. Aquifoliaceae. In: C.L. Lundell, ed. Flora of Texas 3:112–122. Texas Research Foundation, Renner.

Lundell, C.L. 1969b. Celastraceae. In: C.L. Lundell, ed. Flora of Texas 2:339–355. Texas Research Foundation, Renner.

Luteyn, J.L., ed. 1995. Ericaceae, Part II, the superior-ovaried genera. Fl. Neotrop. Monogr. 66.

Luteyn, J.L., W.S. Judd, S.P. Vander Kloet, L.J. Dorr, G.D. Wallace, K.A. Kron, P.F. Stevens, and **S.E. Clemants.** 1996. Ericaceae of the southeastern United States. Castanea 61:101–144.

Lyrene, P. 1995. In defense of *Vaccinium elliottii*. Palmetto 15(Winter): 9–11.

Ma, J. 2016. *Schaefferia*. In: Flora of North America Editorial Committee, eds. Fl. North Amer. 12:131–132. Oxford Univ. Press, New York and Oxford.

Ma, J. and **G.A. Levin.** 2016. *Euonymus*. In: Flora of North America Editorial Committee, eds. Fl. North Amer. 12:122–125. Oxford Univ. Press, New York and Oxford.

Mabberley, D.J. 1987. The plant book, a portable dictionary of the higher plants. Reprinted with corrections in 1989. Cambridge Univ. Press, Cambridge, England, UK.

Mabberley, D.J. 1997. The plant book, a portable dictionary of the higher plants, 2nd ed. Cambridge Univ. Press, Cambridge, England, UK.

Mabberley, D.J. 2008. The plant book, a portable dictionary of the higher plants. Cambridge Univ. Press, Cambridge, England, UK.

Mabberley, D.J. 2017. The plant book, a portable dictionary of the higher plants. Cambridge Univ. Press, Cambridge, England, UK.

MacBride, J.F. 1917. A revision of the North American species of *Amsinckia*. Contr. Gray Herb. 49:1–16.

MacKenzie, K.K. 1906. *Onosmodium*. Bull. Torrey Bot. Club 32: 495–506.

MacRoberts, B.R., L.E. Brown, M.H. MacRoberts, W.W. Pruess, I.S. Elsik, and **S. Jones.** 2010. Annotated checklist of the vascular flora of the Jack Gore Baygall and Neches Bottom units of the Big Thicket National Preserve, Tyler, Jasper, and Hardin counties, Texas. J. Bot. Res. Inst. Texas 4(1):473–488.

MacRoberts, B.R. and **M.H. MacRoberts.** 1997. Floristics of beech-hardwood forests in East Texas. Phytologia 82:20–29.

Maddox, E. 1986. *Homalocephala texensis* "Texas horse crippler." Cact. Succ. J. (Los Angeles) 58:218–221.

Maguire, B. 1947. Studies in the Caryophyllaceae—III. A synopsis of the North American species of *Arenaria*, sect. *Eremogone* Fenzl. Bull. Torrey Bot. Club 74:38–56.

Maguire, B. 1950. Studies in the Caryophyllaceae—IV. A synopsis of the subfamily Silenoideae. Rhodora 52:233–245.

Maguire, B. 1951. Studies in the Caryophyllaceae—V. *Arenaria* in America north of Mexico. A conspectus. Amer. Midl. Naturalist 46:493–511.

Mahler, W.F. 1955. The genus *Baccharis* in the southwestern states of Oklahoma, Texas, and New Mexico. M.S. thesis, Oklahoma State Univ., Stillwater.

Mahler, W.F. 1981. Status report on *Brickellia dentata*. Report prepared for U.S. Fish & Wildlife Service, Albuquerque, NM.

Mahler, W.F. 1983. Rediscovery of *Hymenoxys texana* and notes on two other Texas endemics. Sida 10:87–91.

Mahler, W.F. 1987 *Leavenworthia texana* (Brassicaceae), a new species from Texas. Sida 12(1):239–242.

Mahler, W.F. 1988. Shinners' manual of the North Central Texas flora. Sida Bot. Misc. 3.

Mahler, W.F. and **U.T. Waterfall.** 1964. *Baccharis* (Compositae) in Oklahoma, Texas and New Mexico. S. W. Naturalist 9:189–202.

Majure, L.C. 2012. The evolution and systematics of the *Opuntia humifusa* complex. Ph.D. dissertation, Univ. of Florida.

Majure, L.C. 2022. *Opuntia*. In: A.S. Weakley. Flora of the southeastern United States (PDF). 1215–1220. University of North Carolina Herbarium, North Carolina Botanical Garden.

Majure, L.C., W.S. Judd, P.S. Soltis, and **D.E. Soltis.** 2017. Taxonomic revision of the *Opunita humifusa* complex (Opuntieae: Cactaceae) of the eastern United States. Phytotaxa 290(1):1–65.

Mammillaria Society. 2006. Mammillaria Society. http://www.cactus-mall.com/mammsoc/index.html. Accessed Dec. 2006.

Manchester, S.R., Z.-D. Chen, A.-M. Lu, and **K. Uemura.** 2009. Eastern Asian endemic seed plant genera and their paleogeographic history throughout the Northern Hemisphere. J. Syst. Evol. 47:1–42.

Manen, J.F., M.C. Boulter, and **Y. Naciri-Graven.** 2002. The complex history of the genus *Ilex* L. (Aquifoliaceae): Evidence from the comparison of plastid and nuclear DNA sequences and from fossil data. Pl. Syst. Evol. 235:79–98.

Manktelow, M. 2000. The filament curtain: A structure important to systematics and pollination biology in the Acanthaceae. Bot. J. Linn. Soc. 133:129–160.

Manning, S.D. 2000. The genera of Bignoniaceae in the southeastern United States. Harvard Pap. Bot. 5:1–77.

Manos, P.S. and **M.J. Donoghue.** 2001. Progress in northern hemisphere phytogeography: An introduction. Int. J. Pl. Sci. 162(6 Suppl.): S1–S2. https://donoghuelab.yale.edu/sites/default/files/110_manos_ijps01.pdf.

Manos, P.S. and **J.E. Meireles.** 2015. Biogeographic analysis of the woody plants of the southern Appalachians: Implications for the origins of a regional flora. Amer. J. Bot. 102:780–804.

Mansion, G., G. Parolly, A.A. Crowl, and **11 others.** 2012. How to handle speciose clades? Mass taxon-sampling as a strategy towards illuminating the natural history of *Campanula* (Campanuloideae). P.L.o.S. One 7(11):e50076. https://doi:10.1371/journal.pone.0050076.

Mapes, C., F. Basurto, and **R. Bye.** 1997. Ethnobotany of Quintonil: Knowledge, use and management of edible greens *Amaranthus* spp. (Amaranthaceae) in the Sierra Norte de Puebla, México. Econ. Bot. 51:293–306.

Marazzi, B. 2023. *Chamaecrista*. In: Flora of North America Editorial Committee, eds. Fl. of North Amer. 11:22–30. Oxford Univ. Press, New York and Oxford.

Marazzi, B. and **M.A. Vincent.** 2023. *Senna*. In: Flora of North America Editorial Committee, eds. Fl. of North Amer. 11:30–45. Oxford Univ. Press, New York and Oxford.

Marcus, J.A. 2018. *Eubotrys racemosus*. Lady Bird Johnson Wildflower Center. Plant database. https://www.wildflower.org/plants/result.php?id_plant=eura14#:~:text=Understory%20plant.,that%20is%20toxic%20to%20animals. Accessed 15 Mar 2025.

Marhold, K., M. Šlenker, H. Kudoh, and **J. Zozomová-Lihová.** 2016. *Cardamine occulta*, the correct species name for invasive Asian plants previously classified as *C. flexuosa*, and its occurrence in Europe. PhytoKeys 62:57–72.

Marlowe, K. and **L. Hufford.** 2007. Taxonomy and biogeography of *Gaillardia* (Asteraceae): A phylogenetic analysis. Syst. Bot. 32(1):208–226.

Marshall, W.T. and **T.M. Bock.** 1941. Cactaceae. Abbey Garden Press, Pasadena, CA.

Martinez-Laborde, J.B. 2010. *Diplotaxis*. In: Flora of North America Editorial Committee, eds. Fl. North Amer. 7:432–433. Oxford Univ. Press, New York and Oxford.

Massoni, J., F. Forest, and **H. Sauquet.** 2014. Increased sampling of both genes and taxa improves resolution of phylogenetic relationships within Magnoliidae, a large and early-diverging clade of angiosperms. Molec. Phylogen. Evol. 70:84–93.

Mat Salleh, K. 1991. *Rafflesia*, magnificent flower of Sabah. Borneo Publishing Co., Kota Kinabalu, Sabah, Malaysia.

Mathias, M.E. 1930. Studies in the Umbelliferae. III. A monograph of *Cymopterus* including a critical study of related genera. Ann. Missouri Bot. Gard. 17:213–476.

Mathias, M.E. 1965. Distribution patterns of certain Umbelliferae. Ann. Missouri Bot. Gard. 52:387–398.

Mathias, M.E. and **L. Constance.** 1941a. *Limnosciadium*, a new genus of Umbelliferae. Amer. J. Bot. 28:162–163.

Mathias, M.E. and **L. Constance.** 1941b. A synopsis of the North American species of *Eryngium*. Amer. Midl. Naturalist 25:361–387.

Mathias, M.E. and **L. Constance.** 1942. A synopsis of the American species of *Cicuta*. Madroño 6:145–151.

Mathias, M.E. and **L. Constance.** 1944–45. Umbelliferae. N. Amer. Fl. 28b:43–295.

Mathias, M.E. and **L. Constance.** 1961. Umbelliferae. In: C.L. Lundell, ed. Flora of Texas 3:263–329. Texas Research Foundation, Renner.

Mathias, M.E. and **L. Constance.** 1965. A revision of the genus *Bowlesia* Ruiz & Pav. (Umbelliferae-Hydrocotyloideae) and its relatives. Univ. Calif. Publ. Bot. 38:1–73.

Mavrodiev, E.V. 2015. 1001-A tool for binary representations of unordered multistate characters (with examples from genomic data). PeerJ PrePrints. https://dx.doi.org/10.7287/peerj.preprints.1153v1.

Maxwell, R.H. 1979. Transfer of *Dioclea multiflora* to *Galactia* (Leguminosae). Castanea 44:241–246.

May, R.A. (Ruth). 1919–2014. Personal communication. Wildflower artist and naturalist; instrumental in conservation activities at Harry S. Moss Park, Dallas; former Board Member of the Botanical Research Institute of Texas.

Maya-Lastra, C.A. and **V.W. Steinmann.** 2019. Evolution of the untouchables: Phylogenetics and classification of *Cnidoscolus* (Euphorbiaceae). Taxon 68(4):692–713.

Mayer, V. 1995. The epicalyx in fruits of *Scabiosa* and *Tremastelma* (Dipsacaceae): Anatomy and ecological significance. Bot. Jahrb. Syst. 117:333–357.

Mayer, V. and **F. Ehrendorfer.** 1999. Fruit differentiation, palynology, and systematics in the *Scabiosa* group of genera and *Pseudoscabiosa* (Dipsacaceae). Pl. Syst. Evol. 216:135–135.

Mayfield, M.H. (Mark). Personal communication. Specialist in Eurphobiaceae. Kansas State Univ., Manhattan.

Mayfield, M.H. 2013. Four new annual species of *Euphorbia* section *Tithymalus* (Euphorbiaceae) from North America. J. Bot. Res. Inst. Texas 7(2):633–647.

Mayfield, M.H. 2016. *Bernardia*. In: Flora of North America Editorial Committee, eds. Fl. North Amer. 12:182–184. Oxford Univ. Press, New York and Oxford.

McAllister, H. and **A. Rutherford.** 1990. *Hedera helix* L. and *Hedera hibernica* (Kirchner) Bean Araliaceae in the British Isles. Watsonia 18:7–15.

McArthur, E.D. and **B.L. Welch.** 1984. Proceedings—Symposium on the biology of *Artemisia* and *Chrysothamnus*. Intermountain Research Station, Ogden, UT.

McCain, J.W. and **J.F. Hennen.** 1982. Is the taxonomy of *Berberis* and *Mahonia* (Berberidaceae) supported by their rust pathogens *Cumminsiella santa* sp. nov. and other *Cumminsiella* species (Uredinales)? Syst. Bot. 7:48–59.

McCarty, M.K. 1978. The genus *Carduus* in the United States. In K.E. Frick, ed. Biological Control of Thistles in the genus *Carduus* in the United States. U.S. Department of Agriculture.

McCauley, R.A. 2004. New taxa and a new combination in the North American species of *Froelichia* (Amaranthaceae). Syst. Bot. 29:64–76.

McCauley, R.A. and **H.E. Ballard, Jr.** 2002. Inferring nativity and biogeographic affinities of central and marginal populations of *Froelichia floridana* (Amaranthaceae) from inter-simple sequence repeat (ISSR) markers. J. Torrey Bot. Soc. 129:311–325.

McCauley, R.A. and **H.E. Ballard, Jr.** 2007a. Systematics of North American *Froelichia* (Amaranthaceae subfam. Gomphrenoideae) I: Identification of consistent morphological variation and segregation of species complexes. Brittonia 59:255–274.

McCauley, R.A. and **H.E. Ballard, Jr.** 2007b. Systematics of North American *Froelichia* (Amaranthaceae Subfam. Gomphrenoideae) II: Phylogeny and biogeographic speciation patterns inferred from nrITS sequence data. Brittonia 59:275–289.

McClintock, E. 1993. Dipsacaceae. In: J.C. Hickman, ed. The Jepson manual: Higher plants of California. Pp. 540–541. Univ. of California Press, Berkeley.

McDade, L.A., T.F. Daniel, and **C.A. Kiel.** 2008. Toward a comprehensive understanding of the phylogenetic relationships among lineages of Acanthaceae s.l. (Lamiales). Amer. J. Bot. 95:1136–1152.

McDade, L.A., T.F. Daniel, C.A. Kiel, and **K.B. Vollesen.** 2005. Phylogenetic relationships among Acantheae (Acanthaceae): Major lineages present contrasting patterns of molecular evolution and morphological differentiation. Syst. Bot. 30:834–862.

McDade, L.A., T.F. Daniel, S.E. Masta, and **M.M. Riley.** 2000a. Phylogenetic relationships within the tribe Justicieae (Acanthaceae): Evidence from molecular sequences, morphology, and cytology. Ann. Missouri Bot. Gard. 87:435–458.

McDade, L.A., S.E. Masta, M.L. Moody, and **E. Waters.** 2000b. Phylogenetic relationships among Acanthaceae: Evidence from two genomes. Syst. Bot. 25:106–121.

McDermott, L.F. 1910. An illustrated key to the North American species of *Trifolium*. California Press, San Francisco.

McDonnell, M. and **M. Fishbein.** 2016. *Matelea hirtelliflora* (Apocynaceae), a new species of milkweed vine from Northeast Texas. Sys. Bot. 41(3):781–786.

McGee, M.D. 1973. *Rhododendron* ingestion. Natl. Clearinghouse Poison Control Center Bull. Sep–Oct 1973:1–2.

McGregor, R.L. 1968a. A new species and two new varieties of *Echinacea* (Compositae). Trans. Kansas Acad. Sci. 70:366–370.

McGregor, R.L. 1968b. The taxonomy of the genus *Echinacea* (Compositae). Univ. Kansas Sci. Bull. 48:113–142.

McGregor, R.L. 1984. *Camelina rumelica*, another weedy mustard established in North America. Phytologia 55(4):227–228.

McGregor, R.L. 1984b. Studies on the genus *Apocynum* (Apocynaceae) in Kansas. Contr. Univ. Kansas Herb. 9:1–12.

McGregor, R.L. 1985. Current status of the genus *Camelina* (Brassicaceae) in the prairies and plains of central North America. Contr. Univ. Kansas Herb. 15:1–13.

McGregor, R.L. 1985a. Musk thistle in Kansas: Observations from 1940–1985. Contr. Univ. Kansas Herb. 14:1–13.

McGregor, R.L. 1985d. Studies on the variability of *Lobelia spicata* infraspecific taxa in the prairies and plains of central North America with notes on *Lobelia appendiculata*. Contr. Univ. Kansas Herb. 16:1–10.

McGregor, R.L. 1986a. Amaranthaceae. In: Great Plains Flora Association. Flora of the Great Plains. Pp. 179–187. Univ. Press of Kansas, Lawrence.

McGregor, R.L. 1986b. Apiaceae. In: Great Plains Flora Association. Flora of the Great Plains. Pp. 584–604. Univ. Press of Kansas, Lawrence.

McGregor, R.L. 1986c. Balsaminaceae. In: Great Plains Flora Association. Flora of the Great Plains. Pp. 582–583. Univ. Press of Kansas, Lawrence.

McGregor, R.L. 1986d. Fabaceae. In: Great Plains Flora Association. Flora of the Great Plains. Pp. 416–490. Univ. Press of Kansas, Lawrence.

McGuffin, M., C. Hobbs, R. Upton, and **A. Goldberg,** eds. 1997. American Herbal Products Association's botanical safety handbook: Guidelines for the safe use and labeling for herbs in commerce. CRC Press, Boca Raton, FL.

McKay, D.L. and **J.B. Blumberg.** 2006. A review of the bioactivity and potential benefits of chamomile tea (*Matricaria recutita* L.). Phytotherapy Res. 20(7):519–530.

McLaughlin, S.P. 1982. A taxonomic revision of the southwestern species of *Amsonia* (Apocynaceae). Ann. Missouri. Bot. Gard. 69:336–350.

McNair, J.B. 1925. The taxonomy of poison ivy with a note on the origin of the generic name. Publ. Field Columbian Mus., Bot. Ser. 4:55–70.

McNeill, J. 1962. Taxonomic studies in the Alsinoideae: I. Generic and infra-generic groups. Notes Roy. Bot. Gard. Edinburgh 24:79–155.

McNeill, J. 1978. *Silene alba* and *S. dioica* in North America and the generic delimitation of *Lychnis*, *Melandrium* and *Silene* (Caryophyllaceae). Canad. J. Bot. 56:297–308.

McNeill, J. 1980. The delimitation of *Arenaria* (Caryophyllaceae) and related genera in North America, with 11 new combinations in *Minuartia*. Rhodora 82:495–502.

McNeill, J., F.R. Barrie, H.M. Burdet, V. Demoulin, D.L. Hawksworth, K. Marhold, D.H. Nicolson, J. Prado, P.C. Silva, J.E. Skog, J.H. Wiersema, and **N.J. Turland,** eds. 2006. International code of botanical nomenclature (Vienna Code). Regnum Veg. 146. Available on-line at: http://ibot.sav.sk/icbn/main.htm. Accessed Jun 2008.

McNeill, J., I.J. Bassett, and **C.W. Crompton.** 1977. *Suaeda calceoliformis*, the correct name for *Suaeda depressa* auct. Rhodora 79:133–137.

McVaugh, R. 1936. Studies in the taxonomy of eastern North American species of *Lobelia*. Rhodora 38:241–263, 273–329, 346–362.

McVaugh, R. 1940. Campanulaceae (Lobelioideae). In: R.E. Woodson and R.W. Shery. Contributions toward a flora of Panama. Ann. Missouri Bot. Gard. 27:347–353.

McVaugh, R. 1943. Campanulaceae (Lobelioideae). N. Amer. Fl. 32a:1–134.

McVaugh, R. 1944. The genus *Cnidoscolus*: Generic limits and intrageneric groups. Bull. Torrey Bot. Club 71:457–474.

McVaugh, R. 1945. The genus *Triodanis* Rafinesque, and its relationships to *Specularia* and *Campanula*. Wrightia 1:3–52.

McVaugh, R. 1948. Generic status of *Triodanis* and *Specularia*. Rhodora 50:38–49.

McVaugh, R. 1949. Botanical records of field work in the Southern United States in 1945 and 1947. Field & Lab. 17:123–142.

McVaugh, R. 1961. Campanulaceae. In: C.L. Lundell, ed. Flora of Texas 3:331–366. Texas Research Foundation, Renner.

McVaugh, R. 1972. The genus *Pinaropappus*. Contr. Univ. Michigan Herb. 9:371–378.

McVaugh, R. 1975. Committee for Spermatophyta: Conservation of generic names: 18. Proposal 377. (3557) *Hoffmannseggia* Cavanilles (1798) vs. *Larrea Ortega* (1797). Taxon 24(1):247.

McVaugh, R. 1982. The new synantherology vs. *Eupatorium* in Nueva Galicia. Contr. Univ. Michigan Herb. 15:181–190.

McVaugh, R. 1987. Flora Novo-Galiciana: A descriptive account of the vascular plants of western Mexico, Vol. 5, Leguminosae.

McVaugh, R. and **N.J. Smith.** 1967. *Calyptocarpus vialis* and *C. wendlandii* (Compositae). Brittonia 19:268–272.

McWilliams, E. (Edward). Personal communication. Horticulturist and specialist in Bromeliaceae; Texas A&M Univ., College Station.

McWilliams, E.L. and **M.A. Arnold.** 1998. Horticultural history repeating itself: Dispersal and the invasion lag phase of exotic plants on the TAMU Campus. http://www.ces.ncsu.edu/fletcher/programs/nursery/metria/metria10/exotics/exotics.htm. Accessed Jun 2006.

Mears, J.A. 1967. Revision of *Guilleminea* (*Brayulinea*) including *Gossypianthus* (Amaranthaceae). Sida 3:137–152.

Mears, J.A. 1975. The taxonomy of *Parthenium* section *Partheniastrum* DC. (Asteraceae-Ambrosiinae). Phytologia 31: 463–482.

Mears, J.A. 1977 [1978]. The nomenclature and type collections of the widespread taxa of *Alternanthera* (Amaranthaceae). Proc. Acad. Nat. Sci. Philadelphia 129:1–21.

Mears, J.A. 1980. The Linnaean species of *Gomphrena* L. (Amaranthaceae). Taxon 29:85–95.

Mears, J.A. 1982. A summary of *Blutaparon* Rafinesque including species earlier known as *Philoxerus* R. Brown. Taxon 31:111–117.

Meijer, W. 1993. Rafflesiaceae. In: K. Kubitzki, J.C. Rohwer, and V. Bittrich, eds. The families and genera of vascular plants, Vol. II. Pp. 557–565. Springer-Verlag, Berlin.

Melchert, T.E. 1975. New combinations in the Coreopsidinae. Phytologia 32:291–298.

Melchert, T.E. 1990. New taxa and new combinations in *Thelesperma*. Phytologia 69:216–226.

Mellichamp, T.L. 2015. Droseraceae. In: Flora of North America

Editorial Committee, eds. Fl. North Amer. 6:418–425. Oxford Univ. Press, New York and Oxford.

Melton, A.E., S. Chen, Y. Zhao, and **7 others.** 2020. Genetic insights into the evolution of genera with the eastern Asia–eastern North America floristic disjunction: a transcriptomics analysis. Amer. J. Bot. 107:1736–1748. https://doi: 10.1002/ajb2.1579. https://bsapubs.onlinelibrary.wiley.com/doi/full/10.1002/ajb2.1579.

Melville, R. 1958. Notes on *Alternanthera*. Kew Bull. 13:171–175.

Melvin, N.C. 1981. Additional comments on *Leucothoë* (Ericaceae). Baileya 21(3):127–130.

Melzer, B., R. Seidel, T. Steinbrecher, and **T. Speck.** 2012. Structure, attachment properties, and ecological importance of the attachment system of English ivy (*Hedera helix*). J. Exp. Bot. 63: 191–201.

Merrill, W.L. 1977. An investigation of ethnographic and archaeological specimens of mescalbeans (*Sophora secundiflora*) in American museums. Tech. Rep. No. 6. Museum of Anthropology, Univ. of Michigan, Ann Arbor.

Mestel, R. 1993. Murder trial features tree's genetic fingerprint. New Sci. 138:6.

Metcalfe, C.R. and **L. Chalk.** 1950. Anatomy of the dicotyledons. Vol. II. Clarendon Press, Oxford, England, UK.

Metcalfe, D.J. 2005. *Hedera helix* L. J. Ecol. 93:632–648.

Meyer, F.G. 1997. Hamamelidaceae. In: Flora of North America Editorial Committee, eds. Fl. North Amer. 2:362–367. Oxford Univ. Press, New York and Oxford.

Meyers, S.C., T. Jaster, K.E. Mitchell, T. Harvey, and **L.K. Hardison,** eds. 2020. Flora of Oregon. Vol. 2. BRIT Press, Fort Worth, TX.

Michaels, H.J., K.M. Scott, R.G. Olmstead, T. Szaro, R.K. Jansen, and **J.D. Palmer.** 1993. Interfamilial relationships of the Asteraceae: Insights from *rbc*L sequence variation. Ann. Missouri Bot. Gard. 80:742–751.

Michener, D.C. 1986. Phenotypic instability in *Gleditsia triacanthos* (Fabaceae). Brittonia 38:360–361.

Mielke, M.E. and **M.L. Daughtrey.** 1990. How to identify and control dogwood anthracnose. U.S.D.A., Forest Serv., Northeastern Area. Bull. NA-GR-18.

Miller, J. 2006. Acacia fact sheets. http://ccg.biology.uiowa.edu/jmiller/acaciaID.php. (Accessed Jun 2008).

Miller, J.M. 1988. Floral pigments and phylogeny in *Echinocereus* (Cactaceae). Syst. Bot. 13:173–183.

Miller, J.S. 1989. A revision of the New World species of *Ehretia* (Boraginaceae). Ann. Missouri Bot. Gard. 76:1050–1076.

Miller, J.S. and **M. Gottschling.** 2007. Generic classification in the Cordiaceae (Boraginales): resurrection of the genus *Varronia* P.Br. Taxon 56:163–169.

Miller, K.I. and **G.L. Webster.** 1967. A preliminary revision of *Tragia* (Euphorbiaceae) in the United States. Rhodora 69:241–305.

Miller, L.W. 1964. A taxonomic treatment of the species of *Acalypha* in the United States. Ph.D. dissertation, Purdue Univ., West Lafayette, IN.

Miller, L.W. 1969. *Acalypha gracilens* var. *delzii* L. Miller, var. nov. Sida 3:447.

Miller, L.W. 1970. *Acalypha*. In: D.S. Correll and M.C. Johnston. 1970. Manual of the vascular plants of Texas. Pp. 943–947. Texas Research Foundation, Renner.

Miller, N.G. 1970. The genera of the Cannabaceae in the southeastern United States. J. Arnold Arbor. 51:185–203.

Milliken, W. 2010. Neotropical Betulaceae. In: W. Milliken, B. Klitgård, and A. Baracat, eds. 2009 onwards. Neotropikey—Interactive key and information resources for flowering plants of the Neotropics. https://www.researchgate.net/publication/274347127_Neotropical_Betulaceae.

Millspaugh, C.F. and **E.E. Sherff.** 1919. Revision of the North American species of *Xanthium*. Publ. Field Columbian Mus., Bot. Ser. 4:9–49.

Milne, R.J. and **R.J. Abbott.** 2002. The origin and evolution of Tertiary relict floras. Advances Bot. Res. 38:281–314.

Milstead, W.L. 1964. A revision of the North American species of *Prenanthes*. Ph.D. dissertation, Purdue Univ., West Lafayette, IN.

Mink, J.N., T. Dickschat, J.R. Singhurst, and **W.C. Holmes.** 2017. *Carica papaya* (Caricaceae) as a probable waif in central Texas. Phytoneuron 2017-53:1–5.

Mink, J.N., J.R. Singhurst, and **W.C. Holmes.** 2012. A new species of *Hymenoxys* (Asteraceae, Helenieae, Tetraneuridinae) from Texas. Novon 22(1):56–59.

Miranda, F. and **A.J. Sharp.** 1950. Characteristics of the vegetation in certain temperate regions of eastern Mexico. Ecology 31: 313–333.

Mitchell, J.D. 1990. The poisonous Anacardiaceae genera of the world. In: G.T. Prance and M.J. Balick, eds. New directions in the study of plants and people: Research contributions from the Institute of Economic Botany. Bronx, New York Botanical Garden (Advances in Economic Botany, 8). Pp.103–129.

Miyake, T. and **T. Yahara.** 1998. Why does the flower of *Lonicera japonica* open at dusk? Canad. J. Bot. 76:1806–1811.

Moerman, D.E. 1998. Native American ethnobotany. Timber Press, Portland, OR.

Mohlenbrock, R.H. 1957 [1958]. A revision of the genus *Stylosanthes*. Ann. Missouri Bot. Gard. 44:299–355.

Mohlenbrock, R.H. 1958. The *Stylosanthes biflora* complex. Bull. Torrey Bot. Club 85:341–346.

Mohlenbrock, R.H. 1961. A monograph of the leguminous genus *Zornia*. Webbia 16:1–141.

Mohlenbrock, R.H. 2023a. *Stylosanthes*. In: Flora of North America Editorial Committee, eds. Fl. of North Amer. 11:322–324. Oxford Univ. Press, New York and Oxford.

Mohlenbrock, R.H. 2023b. *Zornia*. In: Flora of North America Editorial Committee, eds. Fl. of North Amer. 11:317–319. Oxford Univ. Press, New York and Oxford.

Mohr, B.A.R., M.E.C. Bernardes-de-Oliviera, and **D.W. Taylor.** 2008. *Pluricarpellatia*, a nymphaealean angiosperm from the Lower Cretaceous of northern Gondwana (Crato formation, Brazil). Taxon 57:1147–1158.

Molano-Flores, B. 2002. Pollination biology and flower visitors of the gynodioecious species *Lobelia spicata* Lam. (Campanulaceae). J. Torrey Bot. Soc. 129:187–193.

Monticello.org. 2004. The Thomas Jefferson Encyclopedia. Edward McMahon. https://www.monticello.org/research-education/thomas-jefferson-encyclopedia/bernard-mcmahon/. Accessed 27 December 2024.

Moore, D.M., T.G. Tutin, and **S.M. Walters.** 1976. Compositae. In: T.G. Tutin, V.H. Heywood, N.A. Burges, D.M. Moore, D.H. Valentine, S.M. Walters, and D.A. Webb, eds. Flora Europaea 4:103–410. Cambridge Univ. Press, Cambridge, England, UK.

Moore, M.J., C.D. Bell, P.S. Soltis, and **D.E. Soltis.** 2007. Using plastid genome-scale data to resolve enigmatic relationships among basal angiosperms. Proc. Natl. Acad. Sci. U.S.A. 104: 19363–19368.

Moore, M.J. (Michael). Email communication, June 15, 2023. College professor and herbarium curator (OC), Oberlin College, Oberlin, OH.

Moore, M.J., N. Hassan, M.A. Gitzendanner, R.A. Bruenn, and **13 others.** 2011. Phylogenetic analysis of the plastid inverted repeat for 244 species: Insights into deeper-level angiosperm relationships from a long, slowly evolving sequence region. Int. J. Pl. Sci. 172(4):541–558.

Moore, R.J. 1972. Distribution of native and introduced knapweeds in Canada and the United States. Rhodora 74:331–346.

Moore, R.J. and **C. Frankton.** 1969. Cytotaxonomy of some *Cirsium* species of the eastern United States. Canad. J. Bot. 47: 1257–1275.

Moraes, R.M., C. Burandt, Jr., M. Ganzera, X. Li, I. Khan, and **C. Canel.** 2000. The American mayapple revisited—*Podophyllum peltatum*—Still a potential cash crop? Econ. Bot. 54:471–476.

Moran, R.V. 2009a. *Crassula*. In: Flora of North America Editorial Committee, eds. Fl. North Amer. 8:150–155. Oxford Univ. Press, New York and Oxford.

Moran, R.V. 2009b. Crassulaceae. In: Flora of North America Editorial Committee, eds. Fl. North Amer. 8:147–229. Oxford Univ. Press, New York and Oxford.

Morefield, J.D. 1992. Resurrection and revision of *Hesperevax* (Asteraceae: Inuleae). Syst. Bot. 17:293–310.

Morefield, J.D. 2004. New taxa and names in North American *Ancistrocarphus*, *Diaperia*, and *Logfia* (Asteraceae: Gnaphalieae: Filagininae) and related taxa. Novon 14(4):463–475.

Morefield, J.D. 2006a. *Diaperia*. In: Flora of North America Editorial Committee, eds. Fl. North Amer. 19:460–463. Oxford Univ. Press, New York and Oxford.

Morefield, J.D. 2006b. *Micropsis*. In: Flora of North America Editorial Committee, eds. Fl. North Amer. 19:463–465. Oxford Univ. Press, New York and Oxford.

Morgan, D.R. 1993. A molecular systematic study and taxonomic revision of *Psilactis* (Asteraceae: Astereae). Syst. Bot. 18:290–308.

Morgan, D.R. 1997. Reticulate evolution in *Machaeranthera* (Asteraceae). Syst. Bot. 22:599–615.

Morgan, D.R. 2006. *Psilactis*. In: Flora of North America Editorial Committee, eds. Fl. North Amer. 20:462–464. Oxford Univ. Press, New York and Oxford.

Morgan, J.J. 1966. A taxonomic study of the genus *Boltonia* (Asteraceae). Ph.D. dissertation, Univ. of North Carolina, Chapel Hill.

Morin, N.R. In press. Campanulaceae. In: Flora of North America Editorial Committee, eds. Fl. North Amer. Vol. 18. Oxford Univ. Press, New York and Oxford.

Morin, N.R. In press. *Campanulastrum*. In: Flora of North America Editorial Committee, eds. Fl. North Amer. Vol. 18. Oxford Univ. Press, New York and Oxford.

Morin, N.R. In press. *Poolea*. In: Flora of North America Editorial Committee, eds. Fl. North Amer. Vol. 18. Oxford Univ. Press, New York and Oxford.

Morin, N.R. In press. *Triodanis*. In: Flora of North America Editorial Committee, eds. Fl. North Amer. Vol. 18. Oxford Univ. Press, New York and Oxford.

Morin, N.R. In press. *Wahlenbergia*. In: Flora of North America Editorial Committee, eds. Fl. North Amer. Vol. 18. Oxford Univ. Press, New York and Oxford.

Morin, N.R. 2020. Taxonomic changes in North American Campanuloideae (Campanulaceae). Phytoneuron 2020-49: 1–46.

Morris, A.B., S.M. Ickert-Bond, D.B. Brunson, D.E. Soltis, and **P.S. Soltis.** 2008. Phylogeographical structure and temporal complexity in American sweetgum (*Liquidambar styraciflua*; Altingiaceae). Molec. Ecol. 17:3889–3900.

Morton, C.M., M.W. Chase, K.A. Kron, and **S.M. Swensen.** 1996 [1997]. A molecular evaluation of the monophyly of the order Ebenales based upon *rbc*L sequence data. Syst. Bot. 21:567–586.

Morton, J.F. 1977. Major medicinal plants: Botany, culture and uses. Charles C. Thomas, Springfield, IL.

Morton, J.F. 1982. Plants poisonous to people in Florida and other warm areas, 2nd ed. Southeastern Printing Co., Stuart, FL.

Morton, J.K. 1972. On the occurrence of *Stellaria pallida* in North America. Bull. Torrey Bot. Club 99:95–97.

Morton, J.K. 2004. New combinations in North American Caryophyllaceae. Sida 21:887–888.

Morton, J.K. 2005a. *Cerastium*. In: Flora of North America Editorial Committee, eds. Fl. North Amer. 5:74–93. Oxford Univ. Press, New York and Oxford.

Morton, J.K. 2005b. *Stellaria*. In: Flora of North America Editorial Committee, eds. Fl. North Amer. 5:96–114. Oxford Univ. Press, New York and Oxford.

Morton, J.K. 2005c. *Silene*. In: Flora of North America Editorial Committee, eds. Fl. North Amer. 5:166–214. Oxford Univ. Press, New York and Oxford.

Mosyakin, S.L. 1996. A taxonomic synopsis of the genus *Salsola* (Chenopodiaceae) in North America. Ann. Missouri Bot. Gard. 83: 387–395.

Mosyakin, S.L. 2003a. *Cycloloma*. In: Flora of North America Editorial Committee, eds. Fl. North Amer. 4:264–265. Oxford Univ. Press, New York and Oxford.

Mosyakin, S.L. 2003b. *Kochia*. In: Flora of North America Editorial Committee, eds. Fl. North Amer. 4:310–312. Oxford Univ. Press, New York and Oxford.

Mosyakin, S.L. 2003c. *Salsola*. In: Flora of North America Editorial Committee, eds. Fl. North Amer. 4:302. Oxford Univ. Press, New York and Oxford.

Mosyakin, S.L. 2015. (2402) Proposal to conserve the name *Chenopodium* (Chenopodiaceae s.str.; Amaranthaceae sensu APG) with a conserved type. Taxon 64(6):1323–1325.

Mosyakin, S.L. and **S.E. Clemants.** 2002. New nomenclatural combinations in *Dysphania* R. Br. (Chenopodiaceae): Taxa occurring in North America. Ukrain. Bot. J. 59:380–385.

Mosyakin, S.L. and **S.E. Clemants.** 2008. Further tranfers of Glandular-pubescent species from *Chenopdium* subg. *Ambrosia* to *Dysphania* (Chenopodiaceae). J. Bot. Res. Inst. Texas 2(1): 425–431.

Mosyakin, S.L. and **K.R. Robertson.** 2003. *Amaranthus*. In: Flora of North America Editorial Committee, eds. Fl. North Amer. 4:410–435. Oxford Univ. Press, New York and Oxford.

Motz, V.A., C.P. Bowers, A.R. Kneubehl, E.C. Lendrum, L.M. Young, and **D.H. Kinder.** 2015. Efficacy of the saponin component of

Impatiens capensis Meerb. in preventing urushiol-induced contact dermatitis. J. Ethnopharmacol. 162:163–167.

Muenscher, W.C. 1951. Poisonous plants of the United States. Macmillan Co., New York.

Mulligan, G.A. 1980. The genus *Cicuta* in North America. Canad. J. Bot. 58:1755–1767.

Mulligan, G.A. 1995. Synopsis of the genus *Arabis* (Brassicaceae) in Canada, Alaska and Greenland. Rhodora 97:109–163.

Mulligan, G.A. and **L.G. Bailey.** 1975. The biology of Canadian weeds. 8. *Sinapis arvensis* L. Canad. J. Pl. Sci. 55:171–183.

Mulligan, G.A. and **I.J. Bassett.** 1959. *Achillea millefolium* complex in Canada and portions of the United States. Canad. J. Bot. 37: 73–79.

Mulligan, G.A. and **D.B. Munro.** 1990. Poisonous plants of Canada. Agri. Canada, Publ. 1842E. Canadian Government Publishing Centre, Ottawa.

Mummenhoff, K., H. Brüggemann, and **J.L. Bowman.** 2001. Chloroplast DNA phylogeny and biogeography of *Lepidium* (Brassicaceae). Amer. J. Bot. 88:2051–2063.

Murata, J., T. Ohi, S.G. Wu, D. Darnaedi, T. Sugawara, T. Nakanishi, and **H. Murata.** 2001. Molecular phylogeny of *Aristolochia* (Aristolochiaceae) inferred from matK sequences. Acta Phytotax. Geobot. 52:75–83.

Murrell, Z.E. 1993. Phylogenetic relationships in *Cornus* (Cornaceae). Syst. Bot. 18:469–495.

Murrell, Z.E. and **D.B. Poindexter.** 2016. Cornaceae. In: Flora of North America Editorial Committee, eds. Fl. North Amer. 12: 443–457. Oxford Univ. Press, New York and Oxford.

Murry, R.E., Jr. and **L.E. Urbatsch.** 1979. Preliminary reports on the Flora of Louisiana: 3. The families Droseraceae and Sarraceniaceae. Castanea 44:24–27.

Myint, T. 1966. Revision of the genus *Stylisma* (Convolvulaceae). Brittonia 18:97–116.

Nabhan, G.P. 1979. Southwestern Indian sunflowers. Desert Pl. 1:23–26.

Naczi, R.F.C. 2024. *Chaerophyllum*. In: Flora of North America Editorial Committee, eds. Fl. North Amer. 13:245–250. Oxford Univ. Press, New York and Oxford.

Nam, J-S., S.-Y. Park, H.-J. Lee, S.-O. Lee, H.-L. Jang, and **Y.-H. Rhee.** 2018. Correlation between acetogenin content and antiproliferative activity of pawpaw (*Asimina triloba* [L.] Dunal) fruit pulp grown in Korea. J. Food Sci. 83:1430–1435. https://doi: 10.1111/1750-3841.14144.

Nandi, O.I., M.W. Chase, and **P.K. Endress.** 1998. A combined cladistics analysis of angeiosperns using *rbc*L and non-molecular data sets. Ann. Missouri Bot. Gard. 85:137–212.

Nash, G.V. 1896. Revision of the genus *Asimina* in North America. Bull. Torrey Bot. Club 23:234–242.

National Park Service. 1993. Species in parks: Flora and fauna databases. http://www.ice.ucdavis.edu/nps/. Accessed Jul 2008.

NatureServe Explorer. 2019. NatureServe Explorer: an online encyclopedia of life. http://explorer.natureserve.org. Accessed 2019–2021.

Navarro, E., J. Bousquet, A. Moiroud, A. Munive, D. Piou, and **P. Normand.** 2003. Molecular phylogeny of *Alnus* (Betulaceae), inferred from nuclear ribosomal DNA ITS sequences. Pl. & Soil 254:207–217.

Nee, M. 1990. The domestication of *Cucurbita* (Cucurbitaceae). Econ. Bot. 44(Suppl. 3):56–68.

Neill, A.K. 1999. *Vicia lutea* (Fabaceae) new to Texas. Sida 18:(4): 1265–1266.

Neill, A.K. 2019. Asteraceae of East Texas. Published online. https://fwbgstg.wpengine.com/wp-content/uploads/2021/08/Asteraceae_East_Texas_download.pdf. Pp. 195.

Neill, A.K. and **H.D. Wilson.** 2001. The vascular flora of Madison County, Texas. Sida 19(4):1083–1121.

Neill, A.K. and **S.R. Kieschnick.** 2021. *Noccaea perfoliata* or *Microthlaspi perfoliatum* (Brassicaceae), new to the flora of Texas, U.S.A. J. Bot. Res. Inst. Texas 15(1):309–315.

Neinhuis, C., S. Wanke, K.W. Hilu, K. Müller, and **T. Borsch.** 2005. Phylogeny of Aristolochiaceae based on parsimony, likelihood, and Bayesian analyses of trnL-trnF sequences. Pl. Syst. Evol. 250: 7–26.

Nelson, E.C. 1991. Shamrock: Botany and history of an Irish myth. Boethius Press, Kilkenny, Ireland.

Nesom, G.L. (Guy). Personal communication. Specialist in Asteraceae; formerly of the Univ. of Texas, Austin, the Texas Regional Institute of Environmental Studies, Sam Houston State Univ., Huntsville, and the Botanical Research Institute of Texas.

Nesom, G.L. 1978. Chromosome numbers in *Erigeron* and *Conyza* (Compositae). Sida 7:375–381.

Nesom, G.L. 1988. Synopsis of *Chaetopappa* (Compositae-Astereae) with a new species and the inclusion of *Leucelene*. Phytologia 64:448–456.

Nesom, G.L. 1989a. Infrageneric taxonomy of New World *Erigeron* (Compositae: Astereae). Phytologia 67:67–93.

Nesom, G.L. 1989b. New species, new sections, and a taxonomic overview of American *Pluchea* (Compositae: Inuleae). Phytologia 67:158–167.

Nesom, G.L. 1990a. Taxonomic status of *Gamochaeta* (Asteraceae: Inuleae) and the species of the United States. Phytologia 68: 186–198.

Nesom, G.L. 1990b. Further definition of *Conyza* (Asteraceae: Astereae). Phytologia 68:229–233.

Nesom, G.L. 1990c. Studies in the systematics of Mexican and Texan *Grindelia* (Asteraceae: Astereae). Phytologia 68:303–332.

Nesom, G.L. 1990d. Infrageneric taxonomy of North and Central American *Baccharis* (Asteraceae: Astereae). Phytologia 69:40–46.

Nesom, G.L. 1990e. Taxonomy of *Solidago petiolaris* (Astereae: Asteraceae) and related Mexican species. Phytologia 69:445–456.

Nesom, G.L. 1991a. Taxonomy of *Isocoma* (Asteraceae: Astereae). Phytologia 70:69–114.

Nesom, G.L. 1991b. Union of *Bradburia* with *Chrysopsis* (Asteraceae: Astereae), with a phylogenetic hypothesis for *Chrysopsis*. Phytologia 71:109–121.

Nesom, G.L. 1991c. A phylogenetic hypothesis for the golden-asters (Asteraceae: Astereae). Phytologia 71:136–151.

Nesom, G.L. 1991d. Morphological definition of the *Gutierrezia* group (Asteraceae: Astereae). Phytologia 71:252–262.

Nesom, G.L. 1992a. Species rank for the varieties of *Grindelia microcephala* (Asteraceae: Astereae). Phytologia 73:326–329.

Nesom, G.L. 1992b. Transfer of *Chaetopappa elegans* to *Ionactis* (Asteraceae: Astereae). Phytologia 73:416–424.

Nesom, G.L. 1993a. Taxonomic infrastructure of *Solidago* and

Oligoneuron (Asteraceae: Astereae) and observations on their phylogenetic position. Phytologia 75:1–44.

Nesom, G.L. 1993b. Taxonomy of *Doellingeria* (Asteraceae: Astereae). Phytologia 75:358–365.

Nesom, G.L. 1994a. Subtribal classification of the Astereae (Asteraceae). Phytologia 76:193–274.

Nesom, G.L. 1994b. Review of the taxonomy of *Aster* sensu lato (Asteraceae: Astereae), emphasizing the New World species. Phytologia 77:141–297.

Nesom, G.L. 1995. Revision of *Chaptalia* (Asteraceae: Mutisieae) from North America and continental Central America. Phytologia 78: 153–188.

Nesom, G.L. 1997a. Taxonomic adjustments in North American *Aster* sensu latissimo (Asteraceae: Astereae). Phytologia 82:281–288.

Nesom, G.L. 1997b. The status of *Aster scabricaulis* (Asteraceae: Astereae), an endemic of the Gulf Coastal Plain. Phytologia 82(4): 300–315.

Nesom, G.L. 1997c. Review: "A revision of *Heterotheca* sect. *Phyllotheca* (Nutt.) Harms (Compositae: Astereae)" by J.C. Semple. Phytologia 83:7–21.

Nesom, G.L. 2001a. *Pseudognaphalium austrotexanum* (Asteraceae: Gnaphalieae), a new species from southeastern Texas and adjacent Mexico. Sida 19:507–511.

Nesom, G.L. 2001b. Notes on variation in *Pseudognaphalium obtusifolium* (Asteraceae: Gnaphalieae). Sida 19:615–619.

Nesom, G.L. 2001c. New records in *Pseudognaphalium* (Gnaphalieae: Asteraceae) for the United States. Sida 19:1185–1190.

Nesom, G.L. 2004a. Taxonomic reevaluations in North American *Erigeron* (Asteraceae: Astereae). Sida 21:19–40.

Nesom, G.L. 2004b. Notes on the distribution of *Pseudognaphalium luteoalbum* (Asteraceae: Gnaphalieae). Sida 21:463–464.

Nesom, G.L. 2005a. Infrageneric classification of *Liatris* (Asteraceae: Eupatorieae). Sida 21(3):1305–1321.

Nesom, G.L. 2005b. Taxonomic review of *Astranthium integrifolium* (Asteraceae: Astereae). Sida 21(4):2015–2021.

Nesom, G.L. 2005c. Taxonomy of the *Symphyotrichum* (*Aster*) *subulatum* group and *Symphyotrichum* (*Aster*) *tenuifolium* (Asteraceae: Astereae). Sida 21(4):2125–2140.

Nesom, G.L. 2006a. Taxonomic review of *Symphyotrichum patens* (Asteraceae: Astereae). Sida 22(2):1075–1080.

Nesom, G.L. 2006b. Taxonomic overview of the *Heterotheca villosa* complex (Asteraceae: Astereae). Sida 22(1):367–380.

Nesom, G.L. 2006c. *Ageratina*. In: Flora of North America Editorial Committee, eds. Fl. North Amer. 21:547–553. Oxford Univ. Press, New York and Oxford.

Nesom, G.L. 2006d. *Amphiachyris*. In: Flora of North America Editorial Committee, eds. Fl. North Amer. 20:87–88. Oxford Univ. Press, New York and Oxford.

Nesom, G.L. 2006e. *Aphanostephus*. In: Flora of North America Editorial Committee, eds. Fl. North Amer. 20:351–353. Oxford Univ. Press, New York and Oxford.

Nesom, G.L. 2006f. *Astranthium*. In: Flora of North America Editorial Committee, eds. Fl. North Amer. 20:203–204. Oxford Univ. Press, New York and Oxford.

Nesom, G.L. 2006g. *Bigelowia*. In: Flora of North America Editorial Committee, eds. Fl. North Amer. 20:95–96. Oxford Univ. Press, New York and Oxford.

Nesom, G.L. 2006h. *Chaetopappa*. In: Flora of North America Editorial Committee, eds. Fl. North Amer. 20:206–209. Oxford Univ. Press, New York and Oxford.

Nesom, G.L. 2006i. *Chaptalia*. In: Flora of North America Editorial Committee, eds. Fl. North Amer. 19:78–80. Oxford Univ. Press, New York and Oxford.

Nesom, G.L. 2006j. *Chromolaena*. In: Flora of North America Editorial Committee, eds. Fl. North Amer. 21:544–545. Oxford Univ. Press, New York and Oxford.

Nesom, G.L. 2006k. *Croptilon*. In: Flora of North America Editorial Committee, eds. Fl. North Amer. 20:228–230. Oxford Univ. Press, New York and Oxford.

Nesom, G.L. 2006l. *Erigeron*. In: Flora of North America Editorial Committee, eds. Fl. North Amer. 20:256–348. Oxford Univ. Press, New York and Oxford.

Nesom, G.L. 2006m. *Facelis*. In: Flora of North America Editorial Committee, eds. Fl. North Amer. 19:442–443. Oxford Univ. Press, New York and Oxford.

Nesom, G.L. 2006n. *Fleischmannia*. In: Flora of North America Editorial Committee, eds. Fl. North Amer. 21:540–541. Oxford Univ. Press, New York and Oxford.

Nesom, G.L. 2006o. *Gamochaeta*. In: Flora of North America Editorial Committee, eds. Fl. North Amer. 19:431–438. Oxford Univ. Press, New York and Oxford.

Nesom, G.L. 2006p. *Gutierrezia*. In: Flora of North America Editorial Committee, eds. Fl. North Amer. 20:88–94. Oxford Univ. Press, New York and Oxford.

Nesom, G.L. 2006q. *Gymnosperma*. In: Flora of North America Editorial Committee, eds. Fl. North Amer. 20:94–95. Oxford Univ. Press, New York and Oxford.

Nesom, G.L. 2006r. *Ionactis*. In: Flora of North America Editorial Committee, eds. Fl. North Amer. 20:82–84. Oxford Univ. Press, New York and Oxford.

Nesom, G.L. 2006s. *Isocoma*. In: Flora of North America Editorial Committee, eds. Fl. North Amer. 20:439–445. Oxford Univ. Press, New York and Oxford.

Nesom, G.L. 2006t. *Liatris*. In: Flora of North America Editorial Committee, eds. Fl. North Amer. 21:512–535. Oxford Univ. Press, New York and Oxford.

Nesom, G.L. 2006u. *Pluchea*. In: Flora of North America Editorial Committee, eds. Fl. North Amer. 19:478–484. Oxford Univ. Press, New York and Oxford.

Nesom, G.L. 2006v. *Pseudognaphalium*. In: Flora of North America Editorial Committee, eds. Fl. North Amer. 19:415–425. Oxford Univ. Press, New York and Oxford.

Nesom, G.L. 2006w. *Pterocaulon*. In: Flora of North America Editorial Committee, eds. Fl. North Amer. 19:476–477. Oxford Univ. Press, New York and Oxford.

Nesom, G.L. 2006x. *Rayjacksonia*. In: Flora of North America Editorial Committee, eds. Fl. North Amer. 20:437–439. Oxford Univ. Press, New York and Oxford.

Nesom, G.L. 2006y. *Thurovia*. In: Flora of North America Editorial Committee, eds. Fl. North Amer. 20:86–87. Oxford Univ. Press, New York and Oxford.

Nesom, G.L. 2006z. *Trichocoronis*. In: Flora of North America Editorial Committee, eds. Fl. North Amer. 21:487–488. Oxford Univ. Press, New York and Oxford.

Nesom, G.L. 2007a. Taxonomic overview of the *Heterotheca fulcrata* complex (Asteraceae: Astereae). J. Bot. Res. Inst. Texas 1(1):21–30.

Nesom, G.L. 2007b. Distribution of *Gamochaeta* (Asteraceae: Gnaphalieae) in Texas, Oklahoma, Arkansas, and Louisiana. J. Bot. Res. Inst. Texas 1(2):1125–1130.

Nesom, G.L. 2008. Taxonomic review of *Solidago petiolaris* and *S. wrightii* (Asteraceae: Astereae). Phytologia 90(1):21–35.

Nesom, G.L. 2009a. *Solidago dispersa* (Asteraceae: Astereae) replaces *Solidago ludoviciana* as the correct name. Phytologia 91:251–255.

Nesom, G.L. 2009b. *Thymophylla tenuiloba* and *T. wrightii* (Asteraceae: Tageteae). Phytologia 91:333–339.

Nesom, G.L. 2011a. New state records for *Citrullus*, *Cucumis*, and *Cucurbita* (Cucurbitaceae) outside of cultivation in the USA. Phytoneuron 2011-1:1–7.

Nesom, G.L. 2011b. Toward consistency of taxonomic rank in wild/domesticated Cucurbitaceae. Phytoneuron 2011-13:1–33.

Nesom, G.L. 2012a. Taxonomy of *Eurytaenia* (Apiaceae). Phytoneuron 2012-67:1–7.

Nesom, G.L. 2012b. Taxonomy of *Polytaenia* (Apiaceae): *P. nuttallii* and *P. texana*. Phytoneuron. 2012-66:1–12.

Nesom, G.L. 2012c. Taxonomy of *Apiastrum*, *Ammoselinum*, and *Spermolepis* (Apiaceae). Phytoneuron 2012-87:1–49.

Nesom, G.L. 2014. Taxonomy of *Cyclanthera* (Cucurbitaceae) in the USA. Phytoneuron 2014-11:1–17.

Nesom, G.L. 2015a. Key to native and cultivated species of *Erythrina* (Fabaceae) in the USA and comments on naturalization of *E. crista-galli*. Phytoneuron 2015-29:1–8.

Nesom, G.L. 2015b. Cucurbitaceae. In: Flora of North America Editorial Committee, eds. Fl. North Amer. 6:3–58. Oxford Univ. Press, New York and Oxford.

Nesom, G.L. 2015c. *Eurytaenia texana* (Apiaceae) native to Louisiana. Phytoneuron 2015-51:1–2.

Nesom, G.L. 2015d. Taxonomy of *Galactia* (Fabaceae) in the USA. Phytoneuron 2015-42:1–54.

Nesom, G.L. 2016. *Erythrina herbacea* (Fabaceae) and two close relatives from Mexico. Phytoneuron 2016-39:1–14.

Nesom, G.L. 2023a. *Crotalaria*. In: Flora of North America Editorial Committee, eds. Fl. of North Amer. 11:185–192. Oxford Univ. Press, New York and Oxford.

Nesom, G.L. 2023b. *Erythrina*. In: Flora of North America Editorial Committee, eds. Fl. of North Amer. 11:391–392. Oxford Univ. Press, New York and Oxford.

Nesom, G.L. 2023c *Galactia*. In: Flora of North America Editorial Committee, eds. Fl. of North Amer. 11:358–370. Oxford Univ. Press, New York and Oxford.

Nesom, G.L. 2023d. *Neptunia*. In: Flora of North America Editorial Committee, eds. Fl. of North Amer. 11:79–82. Oxford Univ. Press, New York and Oxford.

Nesom, G.L. 2023e. *Tephrosia*. In: Flora of North America Editorial Committee, eds. Fl. of North Amer. 11:347–355. Oxford Univ. Press, New York and Oxford.

Nesom, G.L. 2024a. *Ammoselinum*. In: Flora of North America Editorial Committee, eds. Fl. North Amer. 13:213–214. Oxford Univ. Press, New York and Oxford.

Nesom, G.L. 2024b. *Atrema*. In: Flora of North America Editorial Committee, eds. Fl. North Amer. 13:235–236. Oxford Univ. Press, New York and Oxford.

Nesom, G.L. 2024c. *Berula*. In: Flora of North America Editorial Committee, eds. Fl. North Amer. 13:236–237. Oxford Univ. Press, New York and Oxford.

Nesom, G.L. 2024d. *Centella*. In: Flora of North America Editorial Committee, eds. Fl. North Amer. 13:244–245. Oxford Univ. Press, New York and Oxford.

Nesom, G.L. 2024e. *Cryptotaenia*. In: Flora of North America Editorial Committee, eds. Fl. North Amer. 13:263–264. Oxford Univ. Press, New York and Oxford.

Nesom, G.L. 2024f. *Eurytaenia*. In: Flora of North America Editorial Committee, eds. Fl. North Amer. 13:319–320. Oxford Univ. Press, New York and Oxford.

Nesom, G.L. 2024g. *Petroselinum*. In: Flora of North America Editorial Committee, eds. Fl. North Amer. 13:471–472. Oxford Univ. Press, New York and Oxford.

Nesom, G.L. 2024h. *Polytaenia*. In: Flora of North America Editorial Committee, eds. Fl. North Amer. 13:476–477. Oxford Univ. Press, New York and Oxford.

Nesom, G.L. 2024i. *Sium*. In: Flora of North America Editorial Committee, eds. Fl. North Amer. 13:497. Oxford Univ. Press, New York and Oxford.

Nesom, G.L. 2024j. *Spermolepis*. In: Flora of North America Editorial Committee, eds. Fl. North Amer. 13:498–501. Oxford Univ. Press, New York and Oxford.

Nesom, G.L. 2024k. *Trepocarpus*. In: Flora of North America Editorial Committee, eds. Fl. North Amer. 13:515–516. Oxford Univ. Press, New York and Oxford.

Nesom, G.L. and **L.E. Brown.** 1998. Annotated checklist of the vascular plants of Walker, Montgomery, and San Jacinto counties, East Texas. Phytologia 84:107–153.

Nesom, G.L. and **P.L. Leary.** 1992. A new species of *Ionactis* (Asteraceae: Astereae) from southern Nevada and an overview of the genus. Brittonia 44:247–252.

Nesom, G.L. and **R.D. Noyes.** 1999. Notes on sectional delimitations in *Erigeron* (Asteraceae: Astereae). Sida 18:1161–1165.

Nesom, G.L. and **R.J. O'Kennon.** 2001. Two new species of *Liatris* series *Punctatae* (Asteraceae: Eupatorieae) centered in north central Texas. Sida (19):767–787.

Nesom, G.L., D.J. Rosen, and **S.K. Lawrence.** 2013. Notes on variation and geography in *Rayjacksonia phyllocephala* (Asteraceae: Astereae). Phytoneuron 2013-53:1–15.

Nesom, G.L, Y. Suh, D.R. Morgan, S.D. Sundberg, and **B.B. Simpson.** 1991. *Chloracantha*, a new genus of North American Astereae (Asteraceae). Phytologia 70:371–381.

Nesom, G.L., Y. Suh, and **B.B. Simpson.** 1993. *Prinopsis* (Asteraceae: Astereae) united with *Grindelia*. Phytologia 75:341–346.

Nesom, G.L. and **B.L. Turner.** 1995. Systematics of the *Sedum parvum* group (Crassulaceae) in northeastern Mexico and Texas. Phytologia 79:257–268.

Nesom, G.L. and **B.L. Turner.** 1998. Variation in the *Berlandiera pumila* (Asteraceae) complex. Sida 18:493–502.

Ness, J. 2003. *Catalpa bignonioides* alters extrafloral nectar production after herbivory and attracts ant bodyguards. Oecologia 134:210–218.

Neuwinger, H.D. 1996. African ethnobotany, poisons and drugs, chemistry, pharmacology, toxicology. Chapman & Hall, New York.

Neves, S.S. 2024. *Bupleurum*. In: Flora of North America Editorial Committee, eds. Fl. North Amer. 13:239–243. Oxford Univ. Press, New York and Oxford.

Neves, S.S. and **M.F. Watson.** 2004. Phylogenetic relationships in

Bupleurum (Apiaceae) based on nuclear ribosomal DNA ITS sequence data. Ann. Bot. (Oxford) 93:379–398.

Neyland, R. 2001. A phylogeny inferred from large ribosomal subunit (26S) rDNA sequences suggests that *Cuscuta* is a derived member of Convolvulaceae. Brittonia 53:108–115.

Neyland, R. 2004. The systematic significance of color variation in *Monotropa hypopithys* (Ericaceae) inferred from large ribosomal subunit (26S) r RNA gene sequences. Madroño 51(3):275–279.

Neyland, R. and **M.K. Hennigan.** 2004. A cladistic analysis of *Monotropa uniflora* (Ericaceae) inferred from large ribosomal subunit (26S) rRNA gene sequences. Castanea 69:265–271.

Nicely, K.A. 1965. A monographic study of the Calycanthaceae. Castanea 30:8–81.

Nickrent, D.L., A. Blarer, Y.-L. Qiu, V. Vidal-Russel, and **F.E. Anderson.** 2004. Phylogenetic inference in Rafflesiales: The influence of rate heterogeneity and horizontal gene transfer. B.M.C. Evol. Biol. 4:40. http://www.biomedcentral.com/1471-2148/4/40.

Nicolas, A.N. and **G.M. Plunkett.** 2009. The demise of subfamily Hydrocotyloideae (Apiaceae) and the re-alignment of its genera across the whole order Apiales. Molec. Phylogen. Evol. 53:134–151.

Nicolson, D.H. 1978. Illegitimate "basionyms," impact on priority and author citations, or the rise of *Desmodium incanum* and fall of *D. canum* (Fabaceae). Taxon 27:365–370.

Nicolson, D.H. 1980. Summary of cytological information on *Emilia* and the taxonomy of four Pacific taxa of *Emilia* (Asteraceae: Senecioneae). Syst. Bot. 5:391–407.

Nicolson, D.H. 1999. Report of the General Committee: 8. Taxon 48:377.

Niefoff, J. (Jerry). Personal communication. Soil Scientist, Idaho Panhandle National Forest.

Nienaber, M.A. 2005. *Geocarpon*. In: Flora of North America Editorial Committee, eds. Fl. North Amer. 5:148. Oxford Univ. Press, New York and Oxford.

Nieuwland, J.A. 1914. Critical notes on new and old genera of plants. Amer. Midl. Naturalist 3:170–197.

Nixon, E.S. and **S.C. Damuth.** 1987. Five additions to the Texas flora. Sida 12:421–423.

Nixon, E.S., J.R. Ward, and **B.L. Lipscomb.** 1983. Rediscovery of *Lesquerella pallida* (Cruciferae). Sida 10(2):167–175.

Norman, E.M., K. Rice, and **S. Cochran.** 1992. Reproductive biology of *Asimina parviflora* (Annonaceae). Bull. Torrey Bot. Club 119: 1–5.

North Carolina Cooperative Extension. 2024. *Campsis radicans*. https://plants.ces.ncsu.edu/plants/campsis-radicans/. Accessed 24 Dec 2024.

Northington, D.K. 1971. Taxonomy of *Pyrrhopappus*, a cytotaxonomic and chemotaxonomic study. Ph.D. dissertation, Univ. of Texas, Austin.

Northington, D.K. 1974. Systematic studies of the genus *Pyrrhopappus* (Compositae, Cichorieae). Special Publ. No. 6. The Museum, Texas Tech Univ., Lubbock.

Norton, J.B.S. 1900. A revision of the American species of *Euphorbia* of the section *Tithymalus* occurring north of Mexico. Rep. (Annual) Missouri Bot. Gard. 11:85–144.

Novak, S.J., D.E. Soltis, and **P.S. Soltis.** 1991. Ownbey's Tragopogons: 40 years later. Amer. J. Bot. 78:1586–1600.

Nowicke, J.W. 1994. A palynological study of Crotonoideae (Euphorbiaceae). Ann. Missouri Bot. Gard. 81:245–269.

Nowicke, J.W. 1996. Pollen morphology, exine structure and the relationships of Basellaceae and Didiereaceae to Portulacaceae. Syst. Bot. 21:187–208.

Nowicke, J.W., S.G. Shetler, and **N. Morin.** 1992. Exine structure of pantoporate *Campanula* (Campanulaceae) species. Ann. Missouri Bot. Gard. 79:65–80.

Nowicke, J.W. and **J.J. Skvarla.** 1981. Pollen morphology and phylogenetic relationships of the Berberidaceae. Smithsonian Contr. Bot. 50:1–83.

Nyffeler, R. 2002. Phylogenetic relationships in the cactus family (Cactaceae) based on evidence from *trnK/matK* and *trnL-trnF* sequences. Amer. J. Bot. 89:312–326.

Nyman, Y. 1993a. The pollen-collecting hairs of *Campanula* (Campanulaceae). I. Morphological variation and the retractive mechanism. Amer. J. Bot. 80:1427–1436.

Nyman, Y. 1993b. The pollen-collecting hairs of *Campanula* (Campanulaceae). II. Function and adaptive significance in relation to pollination. Amer. J. Bot. 80:1437–1443.

Ockendon, D.J. 1965. A taxonomic study of *Psoralea* subgenus *Pediomelum* (Leguminosae). S. W. Naturalist 10:81–124.

O'Donell, C.A. 1959. Las especies americanas de *Ipomoea* L. Sect. *Quamoclit* (Moench) Griseb. Lilloa 29:19–86.

Ohashi, H. 2013. New Combinations in North American *Desmodium* (Leguminosae: Tribe Desmodieae). J. Jap. Bot. 88:166–175.

Ohashi, H. 2023a. *Alysicarpus*. In: Flora of North America Editorial Committee, eds. Fl. of North Amer. 11:465–467. Oxford Univ. Press, New York and Oxford.

Ohashi, H. 2023b. *Desmodium*. In: Flora of North America Editorial Committee, eds. Fl. of North Amer. 11:442–462. Oxford Univ. Press, New York and Oxford.

Ohashi, H. 2023c. *Hylodesmum*. In: Flora of North America Editorial Committee, eds. Fl. of North Amer. 11:462–464. Oxford Univ. Press, New York and Oxford.

Ohashi, H. 2023d. *Kummerowia*. In: Flora of North America Editorial Committee, eds. Fl. of North Amer. 11:429–431. Oxford Univ. Press, New York and Oxford.

Ohashi, H. 2023e. *Lespedeza*. In: Flora of North America Editorial Committee, eds. Fl. of North Amer. 11:431–441. Oxford Univ. Press, New York and Oxford.

Ohashi, K., H. Ohashi, and **T. Nemoto.** 2018. Phylogenetic analyses of the *Desmodium* group of Leguminosae 2. Two new genera separated from *Desmodium* and two new combinations in *Grona* and *Sohmaea*. J. Jap. Bot. 93:293–306.

Ohba, H. 2009. *Sedum*. In: Flora of North America Editorial Committee, eds. Fl. of North Amer. 8:199. Oxford Univ. Press, New York and Oxford.

Ohi-Toma, T. and **J. Murata.** 2016. Nomenclature of *Isotrema*, *Siphisia*, and *Endodeca*, and their related infrageneric taxa of *Aristolochia* (Aristolochiaceae). Taxon 65:152–157.

Ohi-Toma, T., T. Sugawara, H. Murata, S. Wanke, C. Neinhuis, and **J. Murata.** 2006. Molecular phylogeny of *Aristolochia* sensu lato (Aristolochiaceae) based on sequences of rbcL, matK, and phyA genes, with special reference to differentiation in chromosome numbers. Syst. Bot. 31:481–492.

O'Kane, S.L. 2010a. *Paysonia*. In: Flora of North America Editorial

Committee, eds. Fl. North Amer. 7:611–616. Oxford Univ. Press, New York and Oxford.

O'Kane, S.L. 2010b. *Physaria*. In: Flora of North America Editorial Committee, eds. Fl. North Amer. 7:616–665. Oxford Univ. Press, New York and Oxford.

O'Kane, S.L. and **I.A. Al-Shehbaz.** 1997. A synopsis of *Arabidopsis* (Brassicaceae). Novon 7:323–327.

O'Kane, S.L. and **I.A. Al-Shehbaz.** 2002. *Paysonia*, a new genus segregated from *Lesquerella* (Brassicaceae). Novon 12(3):379–381.

O'Kane, S.L. and **I.A. Al-Shehbaz.** 2003. Phylogenetic position and generic limits of *Arabidopsis* (Brassicaceae) based on sequences of nuclear ribosomal DNA. Ann. Missouri Bot. Gard. 90:603–612.

Okeniyi, J.A., T.A. Ogunlesi, O.A. Oyelami, and **L.A. Adeyemi.** 2007. Effectiveness of dried *Carica papaya* seeds against human intestinal parasitosis: A pilot study. J. Med. Food 10:194–196.

O'Kennon, R.J. (Bob). Personal communication. Texas botanist and floristicist, Botanical Research Institute of Texas.

O'Kennon, R.J. 2019. *Ipomoea edwardsensis* (Convolvulaceae) at specific rank. J. Bot. Res. Inst. Texas 13(1):103–105.

O'Kennon, R.J., G.M. Diggs, Jr, and **B.L. Lipscomb.** 1998. *Lactuca saligna* (Asteraceae), a lettuce new for Texas. Sida 18(2):615–619.

O'Kennon, R.J. and **G. Nesom.** 1988. First report of *Cirsium vulgare* (Asteraceae) in Texas. Sida 13:115–116.

O'Kennon, R.J. and **G.L. Nesom.** 2002. A new variety of *Ipomoea costellata* (Convolvulaceae) from the Edwards Plateau region of Texas. Sida 20(1):39–45.

Oklahoma Vascular Plant Database. 2008. Oklahoma Vascular Plant Database, sponsored by the Oklahoma Biological Survey. http://www.biosurvey.ou.edu/atlasdesc.html. Accessed Aug 2008.

Ollerton, J., A. Killick, E. Lamborn, S. Watts, and **M. Whiston.** 2007. Multiple meanings and modes: On the many ways to be a generalist flower. Taxon 56:717–728.

Olmstead, R.G. 2013. Phylogeny and biogeography of Solanaceae, Verbenaceae, and Bignoniaceae: A comparison of continental and intercontinental diversification patterns. Bot. J. Linn. Soc. 171:80–102.

Olmstead, R.G., M.L. Zjhra, L.G. Lohmann, S.O. Grose, and **A.J. Eckert.** 2009. A molecular phylogeny and classification of Bignoniaceae. Amer. J. Bot. 96:1731–1743.

Olsen, J. 1979. Taxonomy of the *Verbesina virginica* complex (Asteraceae). Sida 8:128–134.

Olsen, J. 1985. Synopsis of *Verbesina* sect. *Ochratinia* (Asteraceae). Pl. Syst. Evol. 149:47–63.

Olsen, R.T. and **J.H. Kirkbride.** 2017. Taxonomic revision of the genus *Catalpa* (Bignoniaceae). Brittonia 69:387–421.

Olwell, M. 1982. A population study of the exomorphic variations in *Vicia minutiflora* Dietr. including *V. reverchonii* Wats. (Leguminosae). Sida 9:215–222.

Online Etymology Dictionary, Douglas Harper Inc. 2023. Retrieved 5 May 2023.

Ørgaard, M. 1991. The genus *Cabomba* (Cabombaceae)—a taxonomic study. Nordic J. Bot. 11:179–203. https://doi:10.1111/j.1756-1051.1991.tb01819.x.

O'Rourke George, L. 1997. *Podophyllum*. In: Flora of North America Editorial Committee, eds. Fl. North Amer. 3:287–288. Oxford Univ. Press, New York and Oxford.

Osborn, J.M. and **E.L. Schneider.** 1988. Morphological studies of the Nymphaeaceae sensu lato. XVI. The floral biology of *Brasenia schreberi*. Ann. Missouri Bot. Gard. 75:778–794.

Osborn, J.M., T.N. Taylor, and **E.L. Schneider.** 1991. Pollen morphology and ultrastructure of Cabombaceae: Correlations with pollination biology. Amer. J. Bot. 78:1367–1378.

Oswald, P.H. and **E.C. Nelson.** 2000. Jamaican plant genera named by Patrick Browne (ca. 1720–1790): A checklist with an attempt at an etymology. Huntia 11:17–30.

Ottley, A.M. 1944. The American Loti with special consideration of a proposed new section, *Simpeteria*. Brittonia 5:81–123.

Ownbey, M. 1950. Natural hybridization and amphiploidy in the genus *Tragopogon*. Amer. J. Bot. 37:487–499.

Oxelman, B., B. Ahlgren, and **M. Thulin.** 2002. Circumscription and phylogenetic relationships of *Gymnocarpos* (Carophyllaceae-Paronychioideae). Edinburgh J. Bot. 59:221–227.

Oxelman, B. and **M. Liden.** 1995. Generic boundaries in the tribe Sileneae (Caryophyllaceae) as inferred from nuclear rDNA sequences. Taxon 44:525–542.

Oxelman, B., M. Liden, and **D. Berglund.** 1997. Chloroplast rps16 intron phylogeny of the tribe Sileneae (Caryophyllaceae). Pl. Syst. Evol. 206:393–410.

Oxelman, B., M. Liden, R.K. Rabeler, and **M. Popp.** 2000. A revised generic classification of the tribe Sileneae (Caryophyllaceae). Nordic J. Bot. 20:743–748.

Özüdoğru, B., K. Özgişi, B. Tarikahya-Hacioğlu, A. Ocak, K. Mummenhoff, and **I.A. Al-Shehbaz.** 2019. Phylogeny of the genus *Noccaea* (Brassicaceae) and a critical review of its generic circumscription. Ann. Missouri Bot. Gard. 104(3):339–354.

Paclt, J. 1952. Synopsis of the genus *Catalpa* (Bignoniaceae) III, special part. Candollea 13:241–288.

Page, R.G. 2007. The *Cistus* and *Halimium* website. http://www.cistuspage.org.uk/. Accessed Jul 2007.

Palmer, E.J. and **J.A. Steyermark.** 1950. Notes on *Geocarpon minimum* Mackenzie. Bull. Torrey Bot. Club. 77(4):268–273.

Pammel, L.H. 1911. A manual of poisonous plants. The Torch Press, Cedar Rapids, IA.

Panero, J.L. and **V.A. Funk.** 2002. Toward a phylogenetic subfamilial classification for the Compositae (Asteraceae). Proc. Biol. Soc. Wash. 115:909–922.

Parfitt, B.D. and **A.C. Gibson.** 2003. Cactaceae (family description and key to genera). In: Flora of North America Editorial Committee, eds. Fl. North Amer. 4:92–99. Oxford Univ. Press, New York and Oxford.

Parfitt, B.D. and **D.J. Pinkava.** 1988. Nomenclatural and systematic reassessment of *Opuntia engelmannii* and *O. lindheimeri* (Cactaceae). Madroño 35:342–349.

Park, K. 1998. Monograph of *Euphorbia* sect. *Tithymalopsis* (Euphorbiaceae). Edinburgh J. Bot. 55:161–208.

Park, K.R. and **W.J. Elisens.** 1997. Isozyme and morphological divergence within *Euphorbia* section *Tithymalopsis* (Euphorbiaceae). Int. J. Pl. Sci. 158:465–475.

Parker, V.T. and **M.A. Stickrod.** 2022. Reproductive phenological shifts and other phylogenetic trait changes in the Arbutoideae (Ericaceae) in the context of drought, seed predation, and fire. Botany 100(4):387–399. https://doi: 10.1139/cjb-2021-0163. Accessed 25 Mar 2025.

Pasquet, R. 2001a. Notes on the genus *Vigna* (Leguminosae-Papilionoideae). Kew Bull. 56:223–227.

Pasquet, R. 2001b. *Vigna*. In: Flora Zambesiaca Managing Committee. Flora Zambesiaca. 3(5)121–156. Royal Botanic Gardens, Kew, England, U.K.

Patterson, T.F. 1994. A Taxonomic Revision of *Conoclinium* DC. (Asteraceae, Eupatorieae). M.S. thesis. Univ. of Texas, Austin.

Patterson, T.F. and **G.L. Nesom.** 2006. *Conoclinium*. In: Flora of North America Editorial Committee, eds. Fl. North Amer. 21:478–480. Oxford Univ. Press, New York and Oxford.

Paulsen, E., L.P. Christensen, and **K.E. Andersen.** 2010. Dermatitis from common ivy (*Hedera helix* L. subsp. *helix*) in Europe: Past, present, and future. Contact Dermatitis 62:201–209.

Payne, W.W. 1964. A re-evaluation of the genus *Ambrosia* (Compositae). J. Arnold Arbor. 45:401–430.

Payson, E.B. 1921. A monograph of the genus *Lesquerella*. Ann. Missouri Bot. Gard. 8:103–236.

Payson, E.B. 1922. Species of *Sisymbrium* native to America north of Mexico. Univ. Wyoming Publ. Bot. 1:1–27.

Payson, E.B. 1922b. A synoptical revision of the genus *Cleomella*. Univ. Wyoming Publ. Bot. 1:29–46.

Payson, E.B. 1927. A monograph of the section *Oreocarya* of *Cryptantha*. Ann. Missouri Bot. Gard. 14:211–358.

Peabody, F.J. 1984. Revision of the genus *Robinia* (Leguminosae: Papilionoideae). Ph.D. dissertation, Iowa State Univ., Ames.

Peattie, D.C. 1948. A natural history of trees of eastern and central North America. Houghton Mifflin Co., Boston, MA.

Pedersen, T.M. 2000. Studies in South American *Amaranthaceae* V. Bonplandia (Corrientes) 10:83–112.

Pedley, L. 1978. A revision of *Acacia* Mill. in Queensland. Austrobaileya 1:75–234.

Pedley, L. 1986. Derivation and dispersal of *Acacia* (Leguminosae), with particular reference to Australia, and the recognition of *Senegalia* and *Racosperma*. Bot. J. Linn. Soc. 92:219–254.

Pedrol, J. 1992. (1050) Proposal to amend 2261 *Suaeda*, nom. cons. (Chenopodiaceae). Taxon 41:337.

Peirson, J.A., V.W. Steinmann, and **J.J. Morawetz.** 2016. *Euphorbia* sect. *Alectoroctonum*. In: Flora of North America Editorial Committee, eds. Fl. North Amer. 12:240–251. Oxford Univ. Press, New York and Oxford.

Pell, S.K., J.D. Mitchell, and **A.J. Miller.** 2024a. Anacardiaceae. In: Flora of North America Editorial Committee, eds. Fl. North Amer. 13:37–63. Oxford Univ. Press, New York and Oxford.

Pell, S.K., J.D. Mitchell, and **A.J. Miller.** 2024b. *Rhus*. In: Flora of North America Editorial Committee, eds. Fl. North Amer. 13:45–53. Oxford Univ. Press, New York and Oxford.

Pell, S.K., J.D. Mitchell, and **A.J. Miller.** 2024c. *Toxicodendron*. In: Flora of North America Editorial Committee, eds. Fl. North Amer. 13:58–63. Oxford Univ. Press, New York and Oxford.

Pell, S.K., J.D. Mitchell, A.J. Miller, and **T.A. Lobova.** 2011. Anacardiaceae. In: K. Kubitzki, ed. The families and genera of vascular plants, Vol. X. Pp. 7–50. Springer-Verlag, Berlin, Heidelberg.

Pelton, J. 1964. A survey of the ecology of *Tecoma stans*. Butler Univ. Botanical Studies: Vol. 14, Article 11. Available at: http://digitalcommons.butler.edu/botanical/vol14/iss1/11.

Peng, Y., S.-B. Chen, L. Young, S.-L. Chen, and **P.-G. Xiao.** 2006. A pharmacophylogenetic study of the Berberidaceae (s.l.). Acta Phytotax. Sin. 44:241–257. https://doi: 10.1360/aps040149.

Perdue, R.E., Jr. 1957. Synopsis of *Rudbeckia* subgenus *Rudbeckia*. Rhodora 59:293–299.

Perry, L.M. 1935. *Evolvulus pilosus* an invalid name. Rhodora 37:63.

Perry, L.M. 1937. Notes on *Silphium*. Rhodora 39:281–297.

Persoon, C.H. 1805. Synopsis Plantarum 1:157.

Petersen, K.A., W.J. Elisens, and **J.R. Estes.** 1990. Allozyme variation in *Pyrrhopappus multicaulis* and *P. carolinianus* (Asteraceae): Relation to mating system and purported hybridization. Syst. Bot. 15:534–543.

Peterson, P.D. 2018. The barberry eradication program in Minnesota for stem rust control: A case study. Annual Rev. Phytopathol. 56:203–223.

Pfeifer, H.W. 1966. Revision of the North and Central American hexandrous species of *Aristolochia* (Aristolochiaceae). Ann. Missouri Bot. Gard. 53:115–196.

Philbrick, C.T. 1991. Hydrophily: Phylogenetic and evolutionary considerations. Rhodora 93:36–50.

Philbrick, C.T. 1993. Underwater cross-pollination in *Callitriche hermaphroditica* (Callitrichaceae): Evidence from randomly amplified polymorphic DNA markers. Amer. J. Bot. 80:391–394.

Philbrick, C.T. and **J.M. Osborn.** 1994. Exine reduction in underwater flowering *Callitriche* (Callitrichaceae): Implications for the evolution of hypohydrophily. Rhodora 96:370–381.

Phillippe, L.R. 1978. A biosystematic study of *Sanicula* section *Sanicula*. Ph.D. dissertation, Univ. of Tennessee, Knoxville.

Phillips, T.C., S.B. Walker, B.R. MacRoberts, and **M.H. MacRoberts.** 2007. Vascular flora of a longleaf pine upland in Sabine County, Texas. Phytologia 89(3):317–338.

Pietropaolo, J. and **P. Pietropaolo.** 1986. Carnivorous plants of the world. Timber Press, Portland, OR.

Pignone, D. and **C. Gómez-Campo.** 2011. *Eruca*. In K. Chittaranjan, ed. Wild crop relatives: Genomic and breeding resources: Oilseeds. Springer, Berlin, Heidelberg.

Piirainen, M., O. Liebisch, and **G. Kadereit.** 2017. Phylogeny, biogeography, systematics and taxonomy of Salicornioideae (Amaranthaceae/Chenopodiaceae)—A cosmopolitan, highly specialized hygrohalophyte lineage dating back to the Oligocene. Taxon 66(1):109–132.

Pilbeam, J. 1980. *Mammilaria*: A collectors's guide. Universe Books, New York, NY.

Pimenov, M.G. and **M.V. Leonov.** 1993. The genera of the Umbelliferae. Royal Botanic Gardens, Kew, England, U.K.

Pinkava, D.J. 1967. Biosystematic study of *Berlandiera* (Compositae). Brittonia 19:285–298.

Pinkava, D.J. 1999. Cactaceae cactus family: Part Three. *Cylindropunita* (Engelm.) Knuth chollas. J. Arizona-Nevada Acad. Sci. 32:32–47.

Pinkava, D.J. 1999. Vascular plants of Arizona: Cactaceae, cactus family: Part 3. *Cylindropuntia* (Engelm.) Kunth, chollas. J. Arizona-Nevada Acad. Sci. 32:32–47.

Pinkava, D.J. 2003a. *Cylindropuntia*. In: Flora of North America Editorial Committee, eds. Fl. North Amer. 4:103–118. Oxford Univ. Press, New York and Oxford.

Pinkava, D.J. 2003b. *Opuntia*. In: Flora of North America Editorial Committee, eds. Fl. North Amer. 4:123–148. Oxford Univ. Press, New York and Oxford.

Pinkava, D.J. 2003c. Cactaceae cactus family: Part 6. *Opuntia* Mill. J. Arizona-Nevada Acad. Sci. 35:137–150.

Pinkava, D.J. 2006. *Berlandiera*. In: Flora of North America Editorial Committee, eds. Fl. North Amer. 21:83–87. Oxford Univ. Press, New York and Oxford.

Piperno, D.R. and **D.M. Pearsall.** 1998. The origins of agriculture in the lowland Neotropics. Academic Press. San Diego.

Pippen, R.W. 1978. *Cacalia*. N. Amer. Fl. II. 10:151–159.

Pluchet, R. 2004. Michaux mysteries clarified. Castanea 228–232.

Plunkett, G.M. and **P.P. Lowry, Jr.** 2001. Relationships among "ancient araliads" and their significance for the systematics of Apiales. Molec. Phylogen. Evol. 19:259–276.

Plunkett, G.M., M.G. Pimenov, J.-P. Reduron, and **13 others.** 2018. Apiaceae. In: J. Kadereit and V. Bittrich, eds. The families and genera of flowering plants, Vol. XV. Flowering plants. Eudicots: Apiales, Gentianales (except Rubiaceae). Pp. 9–206. Springer, Cham, Switzerland.

Plunkett, G.M., D.E. Soltis, and **P.S. Soltis.** 1996. Higher level relationships of Apiales (Apiaceae and Araliaceae) based on phylogenetic analysis of rbcL sequences. Amer. J. Bot. 83(4):499–515.

Plunkett, G.M., D.E. Soltis, and **P.S. Soltis.** 1996 [1997]. Evolutionary patterns in Apiaceae: Inferences based on *mat*K sequence data. Syst. Bot. 21:477–495.

Plunkett, G.M., D.E. Soltis, and **P.S. Soltis.** 1997. Clarification of the relationship between Apiaceae and Araliaceae based on *mat*K and *rbc*L sequence data. Amer. J. Bot. 84:565–580.

Plunkett, G.M., J. Wen, and **P.P. Lowry, Jr.** 2004. Infrafamilial classifications and characters in Araliaceae: Insights from the phylogenetic analysis of nuclear (ITS) and plastid (trnL-trnF) sequence data. Pl. Syst. Evol. 245:1–39.

Poindexter, D.B., D. Adams, and **A.S. Weakley.** 2023a. *Gleditsia*. In: Flora of North America Editorial Committee, eds. Fl. of North Amer. 11:47–49. Oxford Univ. Press, New York and Oxford.

Poindexter, D.B., D. Adams, and **A.S. Weakley.** 2023b. *Gymnocladus*. In: Flora of North America Editorial Committee, eds. Fl. of North Amer. 11:46–47. Oxford Univ. Press, New York and Oxford.

Polhill, R.M. and **P.H. Raven,** eds. 1981. Advances in legume systematics. Vols. 1, 2. Proc. Intern. Legume Conference, Kew, England, UK.

Polhill, R.M. and **J.E. Vidal.** 1981. Caesalpinieae. In: R.M. Polhill and P.H. Raven, eds. Advances in Legume Systematics. 1:81–95. Royal Botanic Gardens, Kew, England, U.K.

Pollard, C.L. 1908. Calycanthaceae. N. Am. Fl. 22:237–238.

Pomper, K.W., S.B. Crabtree, D.R. Layne, and **R.N. Peterson.** 2008. Flowering and fruiting characteristics of eight pawpaw [*Asimina triloba* (L.)] Dunal selections in Kentucky. J. Amer. Pomol. Soc. 62:89–97.

Poole, J.M., W.R. Carr, D.M. Price, and **J.R. Singhurst.** 2007. Rare plants of Texas: A field guide. Texas A&M University Press, College Station.

Poole, J.M, J.R. Singhurst, and **W.R. Carr.** 2010. A list of the rare plants of Texas. Whitepaper produced jointly by the Wildlife Diversity Program of the Texas Parks and Wildlife Department and the Texas Conservation Data Center of The Nature Conservancy of Texas. December 2010 ed. PDF available online from the TPWD at: https://tpwd.texas.gov/publications/pwdpubs/media/pwd_rp_w7000_1142.pdf.

Post, A.R., R. Ali, A. Krings, J. Xiang, B.R. Sosinski, and **J.C. Neal.** 2011. On the identity of the weedy bittercresses (*Cardamine*: Brassicaceae) in United States nurseries: Evidence from molecules and morphology. Weed Sci. 59:123–135.

Post, A.R., A. Krings, J. Xiang, B.R. Sosinski, and **J.C. Neal.** 2009. Lectotypification of *Cardamine flexuosa*. J. Bot. Res. Inst. Texas 3(1):227–230.

Powell, A.M. 1973. Taxonomy of *Perityle* section *Laphamia* (Compositae-Helenieae-Peritylinae). Sida 5:61–128.

Powell, A.M. 1978. Systematics of *Flaveria* (Flaveriinae-Asteraceae). Ann. Missouri Bot. Gard. 65:590–636.

Powell, A.M. 1988. Trees and shrubs of Trans-Pecos Texas including Big Bend and Guadalupe Mountains National Parks. Big Bend Natural History Association, Inc., Big Bend National Park, TX.

Powell, E.A. and **K.A. Kron.** 2001. An Analysis of the phylogenetic relationships in the wintergreen group (*Diplycosia*, *Gaultheria*, *Pernettya*, *Tepuia*; Ericaceae). Syst. Bot. 26(4):808–817.

Powell, A.M. and **J.F. Weedin.** 2004. Cacti of the Trans-Pecos & adjacent areas. Texas Tech Univ. Press, Lubbock.

Powell, A.M., J.F. Weedin, and **S.A. Powell.** 2008. Cacti of Texas, a field guide: with emphasis on the Trans-Pecos species. Texas Tech Univ. Press, Lubbock.

Powell, A.M. and **R.D. Worthington.** 2018. Flowering plants of the Trans-Pecos Texas and adjacent areas. Sida Bot. Misc. 49.

POWO (2023). Plants of the World Online. Facilitated by the Royal Botanic Gardens, Kew. Published on the Internet; http://www.plantsoftheworldonline.org/ Retrieved 20 February 2023.

PPC (Parasitic Plant Connection). 2010. Apodanthaceae Takhtajan (Family description). Southern Illinois Univ. Carbondale, College of Science, Parasitic Plant Connection. http://www.parasiticplants.siu.edu/Apodanthaceae/description.html. Accessed Aug 2019.

Prasad, P., S. Savyasachi, L.P.A. Reddy, and **R.V. Sreedhar.** 2019. Physico-chemical characterization, profiling of total lipids and triacylglycerol molecular species of omega-3 fatty acid rich *B. arvensis* seed oil from India. J. Oleo Sci. 68:209–223.

Pratt, D.B. and **L.G. Clark.** 2001. *Amaranthus rudis* and *A. tuberculatus*: One species or two? J. Torrey Bot. Soc. 128:282–296.

Price, R.A. and **I.A. Al-Shehbaz.** 2001.A reconsideration of *Chaunanthus* (Brassicaceae). Novon 11:329–331.

Pringle, J.S. 1995. Possible eponymy of the generic name *Brasenia* Schreb. (Cabombaceae). Sida 16:597–600.

Pringle, J.S. 2005. *Gypsophila*. In: Flora of North America Editorial Committee, eds. Fl. North Amer. 5:153–156. Oxford Univ. Press, New York and Oxford.

Proctor, M., P. Yeo, and **A. Lack.** 1996. The natural history of pollination, Timber Press, Portland, OR.

Pruski, J.F. and **G. Sancho.** 2006. *Conyza sumatrensis* var. *leiotheca* (Compositae: Astereae), a new combination for a common neotropical weed. Novon 16(1):96–101.

Pryer, K.M. and **L.R. Phillippee.** 1989. A synopsis of the genus *Sanicula* (Apiaceae) in eastern Canada. Canad. J. Bot. 67:694–707.

Punt, W. 1962. Pollen morphology of the Euphorbiaceae with special reference to taxonomy. Wentia 7:1–116.

Pyankov, V.I., E.G. Artyusheva, G.E. Edwards, C.C. Black, Jr., and **P.S. Soltis.** 2001. Phylogenetic analysis of the tribe Salsoleae (Cenopodiaceae) based on ribosomal ITS sequences: Implications for the evolution of photosynthesis types. Amer. J. Bot. 88: 1189–1198.

Qin, H. and **M.G. Gilbert.** 2007. Elaeagnaceae (*Elaeagnus*). In: Z.-Y. Wu and P.H. Raven, eds. Flora of China 13:251–273. Science Press, Beijing, China and Missouri Botanical Garden, St. Louis.

Qiner, Y. and **S. Landrein.** 2011. Linnaeaceae. In: Z.-Y. Wu and P.H.

Raven, eds. Flora of China (Cucurbitaceae through Vallerianaceae, with Annonaceae and Berberidaceae). 19:642–648. Science Press, Beijing, China and Missouri Botanical Garden Press, St. Louis.

Qiu, Y-L., M.W. Chase, D.L. Les, and **C.R. Parks.** 1993. Molecular phylogenetics of the Magnoliidae: Cladistic analyses of nucleotide sequences of the plastid gene *rbc*L. Ann. Missouri Bot. Gard. 80:587–606.

Qiu, Y.-L., J. Lee, F. Bernasconi-Quadroni, D.E. Soltis, P.S. Soltis, M. Zanis, E.A. Zimmer, Z. Chen, V. Savolainen, and **M.W. Chase.** 1999. The earliest angiosperms: Evidence from Mitochondrial, plastid and nuclear genes. Nature 402:404–407.

Qiu, Y.-L., J. Lee, F. Bernasconi-Quadroni, D.E. Soltis, P.S. Soltis, M. Zanis, E.A. Zimmer, Z. Chen, V. Savolainen, and **M.W. Chase.** 2000. Phylogeny of basal angiosperms: Analysis of five genes from three genomes. Int. J. Pl. Sci. 16(6 Suppl.):53–S27.

Qiu, Y.-L.; L. Li, B. Wang, J.-Y. Xue, T.A. Hendry, R.-Q. Li, J.W. Brown, Y. Liu, G.T. Hudson, and **Z.-D. Chen.** 2010. Angiosperm phylogeny inferred from sequences of four mitochondrial genes. J. Syst. Evol. 48:391–425.

Quattrocchi, U. 2000. CRC world dictionary of plant names: Common names, scientific names, eponyms, synonyms, and etymology. 4 vols. CRC Press, Boca Raton, FL.

Rabeler, R.K. (Richard). Personal communication. Specialist in Caryophyllaceae, especially introduced species; Collections Manager at Univ. of Michigan Herbarium, Ann Arbor.

Rabeler, R.K. 1985. *Petrorhagia* (Caryophyllaceae) of North America. Sida 11:6–44.

Rabeler, R.K. 2004. Caryophyllaceae (pink family). In: N.P. Smith, S.A. Mori, A. Henderson, D.W. Stevenson, and S.V. Heald, eds. Flowering plants of the Neotropics. Pp. 88–90. Princeton Univ. Press, Princeton, NJ.

Rabeler, R.K. and **R.L. Hartman.** 2005a. Caryophyllaceae (family description and key to genera). In: Flora of North America Editorial Committee, eds. Fl. North Amer. 5:3–8. Oxford Univ. Press, New York and Oxford.

Rabeler, R.K. and **R.L. Hartman.** 2005b. *Dianthus*. In: Flora of North America Editorial Committee, eds. Fl. North Amer. 5:159–162. Oxford Univ. Press, New York and Oxford.

Rabeler, R.K. and **R.L. Hartman.** 2005c. *Petrorhagia*. In: Flora of North America Editorial Committee, eds. Fl. North Amer. 5: 162–165. Oxford Univ. Press, New York and Oxford.

Rabeler, R.K. and **R.L. Hartman.** 2006. Caryophylaceae. In: G. Yatskievych. Steyermark's flora of Missouri. Vol. 2. Pp. 794–849. Missouri Botanical Garden Press in cooperation with the Missouri Dept. of Conservation, St. Louis.

Rabeler, R.K., R.L. Hartman, and **F.H. Utech.** 2005. *Minuartia*. In: Flora of North America Editorial Committee, eds. Fl. North Amer. 5:116–136. Oxford Univ. Press, New York and Oxford.

Rabeler, R.K. and **A.A. Reznicek.** 1997. *Cerastium pumilum* and *Stellaria pallida* (Caryophyllaceae) new to Texas. Sida 17: 843–845.

Rabeler, R.K. and **J.W. Thieret.** 1988. Comments on the Caryophyllaceae of the southeastern United States. Sida 13:149–156.

Rabhi, M., A. Castagna, D. Remorini, C. Scattino, A. Smaoui, A. Ranieri, and **C. Abdelly.** 2012. Photosynthetic responses to salinity in two obligate halophytes: *Sesuvium portulacastrum* and *Tecticornia indica*. S. African J. Bot. 79:39–47.

Radcliffe-Smith, A. 2001. Genera Euphorbiacearum. Royal Botanic Gardens, Kew, England, U.K.

Radford, A.E., H.E. Ahles, and **C.R. Bell.** 1968. Manual of the vascular flora of the Carolinas. The Univ. of North Carolina Press, Chapel Hill.

Ralphs, M.H. 1992. Ecology, control, and grazing management of locoweeds in the western U.S. In: L.F. James, R.F. Keeler, E.M. Bailey, Jr., P.R. Cheeke, and M.P. Hegarty, eds. Poisonous plants: Proceedings of the third international symposium. Pp. 528–533. Iowa State Univ. Press, Ames.

Ramírez-Amezcua, Y. 2016. *Argythamnia*. In: Flora of North America Editorial Committee, eds. Fl. North Amer. 12:172–179. Oxford Univ. Press, New York and Oxford.

Randhawa, M.A., Z.A. Cheema, and **M.A. Ali.** 2002. Allelopathic effect of *Sorghum* water extract on the germination and seedling growth of *Trianthema portulacastrum*. Int. J. Agric. Biol. 4:383–384.

Ray, M.F. 1987. *Soliva* (Asteraceae: Anthemideae) in California. Madroño 34:228–239.

Ray, P.M. and **H.F. Chisaki.** 1957a. Studies on *Amsinckia*. I. A synopsis of the genus, with a study of heterostyly in it. Amer. J. Bot. 44:529–536.

Ray, P.M. and **H.F. Chisaki.**1957b. Studies on *Amsinckia*. II. Relationships among the primitive species. Amer. J. Bot. 44: 537–544.

Ray, P.M. and **H.F. Chisaki.** 1957c. Studies on *Amsinckia*. III. Aneuploid diversification in the Muricatae. Amer. J. Bot. 44: 545–554.

Razifard, H., G.C. Tucker, and **D.H. Les.** 2016. *Elatine*. In: Flora of North America Editorial Committee, eds. Fl. North Amer. 12: 349–353. Oxford Univ. Press, New York and Oxford.

Reed, C.F. 1969b. Amaranthaceae. In: C.L. Lundell, ed. Flora of Texas 2:89–150. Texas Research Foundation, Renner.

Reed, C.R. 1964. A flora of the chrome and manganese ore piles at Canton, in the Port of Baltimore, Maryland and at Newport News, Virginia, with descriptions. Phytologia 10:321–406.

Reed, M.D. (Monique) 1963–2025. Personal communication. Retired from Botanist at Tracy Herbarium (TAES), Texas A&M University, and Collections Manager of the Biology Department Herbarium (TAMU).

Reed, M.D. 2005. Manual of the dicot flora of Brazos and surrounding counties, 2nd ed. Publisher unknown.

Reed, M.D. 2004. *Silene conoidea* (Caryophyllaceae) new to Texas. Sida 21(1):493–494.

Reed, M.D. 2006. Key to the angiosperm flora of Brazos and surrounding counties. Published in xerox form by the Texas A&M Univ. Biology Dept., College Station.

Refulio-Rodriquez, N.F. and **R.G. Olmstead.** 2014. Phylogeny of Lamiidae. Amer. J. Bot. 101:287–299.

Rehder, A. 1903. Synopsis of the genus *Lonicera*. Rep. (Annual) Missouri Bot. Gard. 14:27–232.

Rehder, A.J. 1922. New species, varieties and combinations from the herbarium and the collections of the Arnold Arboretum. J. Arnold Arbor. 3:11–51.

Reid, C.S., P.L. Faulkner, B.R. MacRoberts, and **M.H. MacRoberts.** 2008. Noteworthy vascular plant collections from northwest Lousisana. J. Bot. Res. Inst. Texas 2(1):643–647.

Reid, C.S. and **D.X. Williams.** 2001. *Alternanthera sessilis* (Amaranthaceae) new to Arkansas. Sida 19:1197.

Ren, B.-Q., X.-G. Xiang, and **Z.-D. Chen.** 2010. Species identification of *Alnus* (Betulaceae) using nrDNA and cpDNA genetic markers. Molec. Ecol. Resources 10:594–605.

Ren, Z., A.J. Harris, R.B. Dikow, E. Ma, Y. Zhong, and **J. Wen.** 2017. Another look at the phylogenetic relationships and intercontinental biogeography of eastern Asian-North American *Rhus* gall aphids (Hemiptera: Aphididae: Eriosomatinae): Evidence from mitogenome sequences via genome skimming. Molec. Phylogen. Evol. 117:102–110.

Reveal, J.L. 1990a. (979) Proposal to conserve the name and type of *Carpinus virginiana* P.Miller (Betulaceae), the basionym of *Ostrya virginiana* (P.Miller) K.Koch. Taxon 39:357–359.

Reveal, J.L. 1990b. Minor new combinations in *Toxicodendron* (Anacardiaceae). Phytologia 69:275.

Reveal, J.L. 1991. Typification of six Philip Miller names of temperate North American *Toxicodendron* (Anacardiaceae) with proposals (999–1000) to *reject T. crenatum* and *T. volubile*. Taxon 40:333–335.

Reveal, J.L. 1993a. A splitter's guide to the higher taxa of the flowering plants (Magnoliophyta) generally arranged to follow the sequence proposed by Thorne (1992) with certain modifications. Phytologia 74:203–263.

Reveal, J.L. 1993b. Flowering plant families: An overview. In: Flora of North America Editorial Committee, eds. Fl. North Amer. 1:294–298. Oxford Univ. Press, New York and Oxford.

Reveal, J.L. 2000. (1452–1454). Three "superconservation" proposals for conserved family names in App. IIAB (Spermatophyta) of the *Code*. Taxon 49:291–294.

Reveal, J.L. 2012. An outline of a classification scheme for extant flowering plants. Phytoneuron 2012-37:1–221.

Reveal, J.L. and **F.R. Barrie.** 1991. On the identity of *Hedysarum violaceum* Linnaeus (Fabaceae). Phytologia 71:456–461.

Reveal, J.L. and **R.M. King.** 1973. Re-establishment of *Acourtia* D. Don (Asteraceae). Phytologia 15:228–232.

Reveal, J.L. and **J.C. Pires.** 2002. Phylogeny and classification of the Monocotyledons: An update. In: Flora of North America Editorial Committee, eds. Fl. North Amer. 26:3–36. Oxford Univ. Press, New York and Oxford.

Reverchon, J. 1880. Notes on some introduced plants in Dallas County, Texas. Bot. Gaz. 5:10.

Rhodes, A.M., W.P. Bemis, T.W. Whitaker, and **S.G. Carmer.** 1968. A numerical taxonomic study of *Cucurbita*. Brittonia 20:251–266.

Rice, B.A. 2011. The thread-leaf sundews *Drosera filiformis* and *Drosera tracyi* Carniv. Pl. Newslett. 40:4–16.

Richards, E.L. 1968. A monograph of the genus *Ratibida*. Rhodora 70:348–393.

Richardson, A.J. and **K. King.** 2011. Plants of deep South Texas: A field guide to the woody and flowering species. Texas A&M University Press, College Station, Texas.

Richardson, A.T. 1977. Monograph of the genus *Tiquilia* (*Coldenia*, sensu lato), Boraginaceae: Ehretioideae. Rhodora 79:467–572.

Richardson, J.W. 1968. The genus *Euphorbia* of the high plains and prairie plains of Kansas, Nebraska, South and North Dakota. Univ. Kansas Sci. Bull. 48:45–112.

Rickett, H.W. 1945a. Cornaceae. N. Amer. Fl. 28b:299–311.

Rico-Arce, M.L. 2023. *Albizia*. In: Flora of North America Editorial Committee, eds. Fl. of North Amer. 11:136–138. Oxford Univ. Press, New York and Oxford.

Rico-Arce, M.L. and **S. Bachman.** 2006. A taxonomic revision of *Acaciella* (Leguminosae, Mimosoideae). Anales Jard. Bot. Madrid 63:189–244.

Rieseberg, L.H. 1991. Homoploid reticulate evolution in *Helianthus* (Asteraceae): Evidence from ribosomal genes. Amer. J. Bot. 78:1218–1237.

Rieseberg, L.H. and **G.J. Seiler.** 1990. Molecular evidence and the origin and development of the domesticated sunflower (*Helianthus annus*, Asteraceae). Econ. Bot. 44(Suppl. 3):79–91.

Riess, E.M. 1958. The garden of chaparral: Plants of the lower Rio Grande Valley. Vantage Press, New York, NY.

Riley-Hulting, E.T., A. Delgado-Salinas, and **M. Lavin.** 2004. Phylogenetic systematics of *Strophostyles* (Fabaceae): A North American temperate genus within a neotropical diversification. Syst. Bot. 29(3):627–653.

Riskind, D.H. and **P.A. Amerson.** 1977. A second collection of *Monotropa hypopithys* L. for Texas and a new county record. S. W. Naturalist 22(1):136–137.

Rivadavia, F., K. Kondo, M. Kato, and **M. Hasebe.** 2003. Phylogeny of the sundews, *Drosera* (Droseraceae), based on chloroplast *rbcL* and nuclear 18S ribosomal DNA sequences. Amer. J. Bot. 90:123–130.

Roalson, E.H. and **K.W. Allred.** 1998. A clarification of *Centaurea americana* and *Centaurea rothrockii* (Compositae: Cardueae). New Mexico Botanist Newslett. 7:3–5.

Roberts, H.R, J.M. Warren, and **J. Provan.** 2018. Evidence for facultative protocarnivory in *Capsella bursa-pastoris* seeds. Sci. Rep. 8(1):10120. https://doi:10.1038/s41598-018-28564-x.

Robertson, K.R. 1981. The genera of Amaranthaceae in the southeastern United States. J. Arnold Arbor. 62:267–314.

Robertson, K.R. *Celosia*. 2003. In: Flora of North America Editorial Committee, eds. Fl. North Amer. 4:407–409. Oxford Univ. Press, New York and Oxford.

Robertson, K.R. 2023. *Jacquemontia*. In: Flora of North America Editorial Committee, eds. Fl. North Amer. 14:317–321. Oxford Univ. Press, New York and Oxford.

Robertson, K.R. and **S.E. Clemants** 2003. Amaranthaceae. In: Flora of North America Editorial Committee, eds. Fl. North Amer. 4:405–456. Oxford Univ. Press, New York and Oxford.

Robertson, K.R. and **Y.-T. Lee.** 1976. The genera of Caesalpinioideae (Leguminosae) in the southeastern United States. J. Arnold Arbor. 57:1–53.

Robinson, B.L. 1917. A monograph of the genus *Brickellia*. Mem. Gray Herb. 1:1–151.

Robinson, B.L. and **J.M. Greenman.** 1899. Synopsis of the genus *Verbesina*, with an analytical key to the species. Proc. Amer. Acad. Arts 34:534–566.

Robinson, H. 1974. Studies in the Senecioneae (Asteraceae). VI. the genus *Arnoglossum*. Phytologia 28:294–295.

Robinson, H. 1977. Studies in the Heliantheae (Asteraceae). VIII. Notes on genus and species limits in the genus *Viguiera*. Phytologia 36:201–215.

Robinson, H. 1978. Re-establishment of the genus *Smallanthus*. Phytologia 39:47–52.

Robinson, H. 1999. Generic and subtribal classification of American Vernonieae. Smithsonian Contr. Bot. 89:1–116.

Robinson, H. and **J. Cuatrecasas.** 1973. The generic limits of *Pluchea* and *Tessaria* (Inuleae, Asteraceae). Phytologia 27:277–285.

Robinson, H. and **J. Cuatrecasas.** 1977. Notes on the genus and species limits of *Pseudogynoxys* (Greenm.) Cabrera (Senecioneae, Asteraceae). Phytologia 36:177–192.

Robinson, H. and **R.M. King.** 1985. Comments on the generic concepts in the Eupatorieae. Taxon 34:11–16.

Rock, H.F.L. 1957. A revision of the vernal species of *Helenium* (Compositae). Rhodora 59:101–116, 128–158, 168–178, 203–216.

Rodman, J.E. 1974. Systematics and evolution of the genus *Cakile* (Cruciferae). Contr. Gray Herb. 205:3–146.

Rodman, J.E. 2010. *Cakile*. In: Flora of North America Editorial Committee, eds. Fl. North Amer. 7:424–428. Oxford Univ. Press, New York and Oxford.

Roemer, F. 1849. Texas with particular reference to German immigration and the physical appearance of the country. Translated from the German by O. Mueller. Standard Printing Company, San Antonio, TX. Reprinted 1983 by the German-Texan Heritage Society. Texian Press, Waco, TX.

Rogers, C.E., T.E. Thompson, and **G.J. Seiler.** 1982. Sunflower species of the United States. National Sunflower Association, Bismarck, ND.

Rogers, D.J. 1951. A revision of *Stillingia* in the New World. Ann. Missouri Bot. Gard. 38:207–259.

Rogers, D.J. and **S.G. Appan.** 1973. *Manihot*, Manihotoides (Euphorbiaceae). Fl. Neotropica Monogr. 13:1–272.

Rogers, H.J. 1949. The genus *Galactia* in the United States. Ph.D. dissertation, Duke Univ., Durham, NC.

Rogstad, S.H., K. Wolff, and **B.A. Schaal.** 1991. Geographical variation in *Asimina triloba* Dunal (Annonaceae) revealed by the M13 "DNA fingerprinting" probe. Amer. J. Bot. 78:1391–1396.

Rollins, R.C. 1941. A monographic study of *Arabis* in western North America. Rhodora 43:289–325, 348–411, 425–481.

Rollins, R.C. 1942. A systematic study of *Iodanthus*. Contr. Dudley Herb. 3:209–215.

Rollins, R.C. 1947. Generic revisions in the Cruciferae: *Sibara*. Contr. Gray Herb. 165:133–143.

Rollins, R.C. 1950. The guayule rubber plant and its relatives. Contr. Gray Herb. 172:1–73.

Rollins, R.C. 1955. The auriculate-leaved species of *Lesquerella* (Cruciferae). Rhodora 57:241–264.

Rollins, R.C. 1956. On the identity of *Lesquerella angustifolia*. Rhodora 58:199–202.

Rollins, R.C. 1961. Notes on American *Rorippa* (Cruciferae). Rhodora 63:1–10.

Rollins, R.C. 1981. Studies in the genus *Physaria* (Cruciferae). Brittonia 33:332–341.

Rollins, R.C. 1993. The Cruciferae of continental North America: Systematics of the mustard family from the Arctic to Panama. Stanford Univ. Press, Redwood City, CA.

Rollins, R.C. and **I.A. Al-Shehbaz.** 1986. Weeds of South-West Asia in North America with special reference to the Cruciferae. Proc. Roy. Soc. Edinburgh, B, Biol. Sci. 89:289–299.

Rollins, R.C. and **E.A. Shaw.** 1973. The genus *Lesquerella* (Cruciferae) in North America. Harvard Univ. Press, Cambridge, MA.

Roose, M.L. and **L.D. Gottlieb.** 1976. Genetic and biochemical consequences of polyploidy in *Tragopogon*. Evolution 30:818–830.

Roquet, C., I. Sanmartin, N. Garcia-Jacas, L. Sáez, A. Susana, N. Wikström, and **J.J. Aldasoro.** 2009. Reconstructing the history of Campanulaceae with a Bayesian approach to molecular dating and dispersal-vicariance analyses. Molec. Phylogen. Evol. 52:575–587.

Rosatti, T.J. 1986. The genera of Sphenocleaceae and Campanulaceae in the southeastern United States. J. Arnold Arbor. 67:1–64.

Rosatti, T.J. 1989. The genera of suborder Apocynineae (Apocynaceae and Asclepiadaceae) in the southeastern United States. J. Arnold Arbor. 70:307–401, 443–514.

Rose, J.P., T.J. Kleist, S.D. Löfstrand, B.T. Drew, J. Schönenberger, and **K.J. Sytsma.** 2018. Phylogeny, historical biogeography, and diversification of angiosperm order Ericales suggest ancient Neotropical and East Asian connections. Molec. Phylogen. Evol. 122:59–79.

Rossbach, G.B. 1958. The genus *Erysimum* (Cruciferae) in North America north of Mexico—A key to the species and varieties. Madroño 14:261–267.

Rostás, M., D. Maag, M. Ikegami, and **M. Inbar.** 2013. Gall volatiles defend aphids against a browsing mammal. B.M.C. Evol. Biol. 13:193.

Rudall, P.J. 2023. Phylogenetic, developmental and functional aspects of stomatal patterning: Lessons from magnoliids. Bot. Rev. (Lancaster) 89:1–18. https://doi: 10.1007/s12229-023-09287-9.

Rudd, V.E. 1955. The American species of *Aeschynomene*. Contr. U.S. Natl. Herb. 32:1–172.

Rudd, V.E. 1972. Leguminosae-Faboideae-Sophoreae. N. Amer. Fl. II 7:1–53.

Rudd, V.E. and **J.A. Raveill.** 2023a. *Aeschynomene*. In: Flora of North America Editorial Committee, eds. Fl. of North Amer. 11:329–335. Oxford Univ. Press, New York and Oxford.

Rudd, V.E. and **J.A. Raveill.** 2023b. *Arachis*. In: Flora of North America Editorial Committee, eds. Fl. of North Amer. 11:324–326. Oxford Univ. Press, New York and Oxford.

Ruffin, J. 1974. A taxonomic re-evaluation of the genera *Amphiachyris*, *Amphipappus*, *Greenella*, *Gutierrezia*, *Gymnosperma*, *Thurovia*, and *Xanthocephalum* (Compositae). Sida 5:301–333.

Ruiz, H. and **J. Pavon.** 1799. Prodromus et flora Peruviana, et Chilensis. 2:5 t. CXI. Fig. C. Typis Gabrieliscde de Sancha. Madrid.

Ruiz-Sanchez, E. and **J.F. Ornelas.** 2014. Phylogeography of *Liquidambar styraciflua* (Altingiaceae) in Mesoamerica: Survivors of a Neogene widespread temperate forest (or cloud forest) in North America? Ecol. Evol. 4:311–328. https://doi:10.1002/ece3.938.

Rust, R.W. 1977. Pollination in *Impatiens capensis* and *Impatiens pallida* (Balsaminaceae). Bull. Torrey Bot. Club. 104:361–367.

Rutherford, R.J. 1970. The anatomy and cytology of *Pilostyles thurberi* Gray (Rafflesiaceae). Aliso 7:263–288.

Rutter, M.T. and **M.D. Rausher.** 2004. Natural selection on extrafloral nectar production in *Chamaecrista fasciculata:* The costs and benefits of a mutualism trait. Evolution 58:2657–2668.

Ryan, F.J. and **D.R. Ayers.** 2000. Molecular markers indicate two cryptic, genetically different populations of Russian thistle (*Salsola tragus*) in California. Canad. J. Bot. 78:59–67.

Rydberg, P.A. 1910. Balsaminaceae. N. Amer. Fl. 25:93–96.

Rydberg, P.A. 1916. *Chamomilla* and *Matricaria*. N. Amer. Fl. 34: 229–234.

Rydberg, P.A. 1919–1920. Fabaceae: Psoraleae. N. Amer. Fl. 24: 1–136.

Rydberg, P.A. 1922. Ambrosiaceae. N. Amer. Fl. 22:3–44.

Rydberg, P.A. 1923–1929. Fabaceae: Indigofereae, Galegeae. N. Amer. Fl. 24:127–462.

Ryder, M.L. 1995. Is the Fuller's Teasel (*Dipsacus sativus*) really a distinct species? Linnean 11(4):21–27.

Rzedowski, J. and **G. Calderón de Rzedowski.** 1999. Hallazgo de plantas Mexicanas del género *Cotinus* (Anacardiaceae) con característicass de reliquias del Terciario. Acta Bot. Mex. 47:23–30.

Safford, W.E. 1918. *Cosmos sulphureus,* the Xochipalli or flower paint of the Aztecs. J. Wash. Acad. Sci. 8:613–619.

Sakai, S. 2002. *Aristolochia* spp. (Aristolochiaceae) pollinated by flies breeding on decomposing flowers in Panama. Amer. J. Bot. 89: 527–534.

Sánchez, D., B.Vázquez-Benítez, M. Vázquez-Sánchez, D. Aquino, and **L. Arias.** 2022. Phylogenetic relationships in *Coryphantha* and implications on *Pelecyphora* and *Escobaria* (Cacteae, Cactoideae, Cactaceae). PhytoKeys 188:115–165.

Sanchez, L.L. 1997. Vascular plant list of Fort Hood Military Reservation, Bell and Coryell counties, Texas. The Nature Conservancy, Unpublished March 1997 update, copy in the library of the Botanical Research Institute of Texas.

Sánchez del Pino, I. and **S.E. Clemants.** 2003. *Tidestromia*. In: Flora of North America Editorial Committee, eds. Fl. North Amer. 4: 439–443. Oxford Univ. Press, New York and Oxford.

Sánchez del Pino, I. and **H.F. Olvera.** 2002. New taxa and a new combination in *Tidestromia* (Amaranthaceae) from North America. Novon 12:399–407,

Sands, D.C., C.E. Morris, E.A. Dratz, and **A. Pilgeram.** 2009. Elevating optimal human nutrition to a central goal of plant breeding and production of plant-based foods. Pl. Sci. (Elsevier) 117:377–389.

Sauer, J. 1955. Revision of the dioecious amaranths. Madroño 13:5–46.

Sauer, J. 1967. The grain amaranths and their relatives: A revised taxonomic and geographic survey. Ann. Missouri Bot. Gard. 54:103–137.

Savolainen, V., M.W. Chase, S.B. Hoot, C.M. Morton, D.E. Soltis, C. Bayer, M.F. Fay, A.T. de Bruun, S. Sullivan, and **Y.-L. Qiu.** 2000. Phylogenetics of flowering plants based on combined analysis of plastid *atpB* and *rbc*L gene sequences. Syst. Biol. 49:306–362.

Savolainen, V., M.F. Fay, D.C. Albach, and **14 others.** 2000. Phylogeny of the eudicots: A nearly complete familial analysis based on *rbcL* gene sequences. Kew Bull. 55:257–309.

Schaneberg, B.T., W.L. Applequist, and **I.A. Khan.** 2002. Determination of aristolochic acid I and II in North American species of *Asarum* and *Aristolochia*. Pharmazie 57:686–689.

Scheele, A. 1848. Beiträge zur Flor von Texas. Linnaea 21:453–472.

Schenk, M.F., C.N. Thienpont, W.J.M. Koopman, L.J.W.J. Gilissen, and **M.J.M. Smulders.** 2008. Phylogenetic relationships in *Betula* (Betulaceae) based on AFLP markers. Tree Genet. Genomes 4:911. https://doi.org/10.1007/s11295-008-0162-0.

Schierenbeck, K.A. 2004. Japanese honeysuckle (*Lonicera japonica*) as an invasive species; history, ecology, and context. Crit. Rev. Pl. Sci. 23(5):391–400.

Schilling, E.E. 2006a. *Helianthus*. In: Flora of North America Editorial Committee, eds. Fl. North Amer. 21:141–169. Oxford Univ. Press, New York and Oxford.

Schilling, E.E. 2006b. *Viguiera*. In: Flora of North America Editorial Committee, eds. Fl. North Amer. 21:172–174. Oxford Univ. Press, New York and Oxford.

Schilling, E.E., A.J. Floden, A.S. Weakley, C. Winder, and **R.L. Small.** 2022. Molecular barcoding reveals unexpected diversity in eastern North American Stitchworts (Caryophyllaceae). Bot. J. Linn. Soc. 200(1):75–84.

Schilling, E.E., J.L. Panero, and **P.B. Cox.** 1999. Chloroplast DNA restriction site data support a narrowed interpretation of *Eupatorium* (Asteraceae). Pl. Syst. Evol. 219:209–223.

Schilling, E.E. and **D.M. Spooner.** 1988. Floral flavonoids and the systematics of *Simsia* (Asteraceae: Heliantheae). Syst. Bot. 13:572–575.

Schlauer, J. 1996. A dichotomous key to the genus *Drosera* L. (Droseraceae). Carniv. Pl. Newlett. 25:67–88.

Schlautman, B., G. Covarrubias-Pazaran, D. Fajardo, S. Steffan, and **J. Zalapa.** 2017. Discriminating power of microsatellites in cranberry organelles for taxonomic studies in *Vaccinium* and Ericaceae. Genet. Resources Crop Evol. 64:451–466.

Schlessman, M.A. 1984. Systematics of tuberous Lomatiums (Umbelliferae). Syst. Bot. Monogr. 4:1–55.

Schmidt, G.J. and **E.E. Schilling.** 2000. Phylogeny and biogeography of *Eupatorium* (Asteraceae: Eupatorieae) based on nuclear ITS sequence data. Amer. J. Bot. 87:716–726.

Schmutz, E.M. and **L.B. Hamilton.** 1979. Plants that poison. Northland Publishing, Flagstaff, AZ.

Schnabel, A. and **J.F. Wendel.** 1998. Cladistic biogeography of *Gleditisia* (Leguminosae) based on *ndhF* and *rpL16* chloroplast gene sequences. Amer. J. Bot. 85:1753–1765.

Schneider, E.L. and **S. Carlquist.** 1996. Conductive tissue in *Ceratophyllum demersum* (Ceratophyllaceae). Sida 17:437–443.

Schneider, E.L. and **J.M. Jeter.** 1982. Morphological studies of the Nymphaeaceae: 12. The floral biology of *Cabomba caroliniana.* Amer. J. Bot. 69:1410–1419.

Schneider, E.L., S.C. Tucker, and **P.S. Williamson.** 2003. Floral development in the Nymphaeales. Int. J. Pl. Sci. 164(Suppl. 5): S279–S292.

Schneider, J.V. and **C. Bayer.** 2004. Clethraceae. In: K. Kubitzki, ed. The families and genera of vascular plants, Vol. VI. Pp. 69–73. Springer-Verlag, Berlin.

Schönenberger, J. 1999. Floral structure, development and diversity in *Thunbergia* (Acanthaceae). Bot. J. Linn. Soc. 130:1–36.

Schönenberger, J., A.A. Anderberg, and **K.J. Sytsma.** 2005. Molecular phylogenetics and patterns of floral evolution in the Ericales. Int. J. Pl. Sci. 166:265–288.

Schönenberger, J. and **P.K. Endress.** 1998. Structure and development of the flowers in *Mendoncia*, *Pseudocalyx*, and *Thunbergia* (Acanthaceae) and their systematic implications. Int. J. Pl. Sci. 159:446–465.

Schubert, B.G. 1950. *Desmodium*: Preliminary studies–III. Rhodora 52:135–155.

Schubert, M.T.R. and **B.-E. van Wyk.** 1995. Two new species of *Centella* (Umbelliferae) with notes on infrageneric taxonomy. Nordic. J. Bot. 15:167–171.

Schulz, E.D. and **R. Runyon.** 1930. Texas cacti: A popular and scientific account of the cacti native of Texas. Trans. Texas Acad. Sci. 14:1–181.

Schwery, O., R.E. Onstein, Y. Bouchenak-Khelladi, Y. Xing, R.J.

Carter, and **H.P. Linder.** 2014. As old as the mountains: The radiations of the Ericaceae. New Phytol. 207:355–367.

Scotland, R.W., J.A. Sweere, P.A. Reeves, and **R.G. Olmstead.** 1995. Higher-level systematics of Acanthaceae determined by chloroplast DNA sequences. Amer. J. Bot. 82:266–275.

Scotland, R.W. and **K. Vollesen.** 2000. Classification of Acanthaceae. Kew Bull. 55:513–589.

Scott, A.J. 1977. Reinstatement and revision of Salicorniaceae J. Agardh. Bot. J. Linn. Soc. 75:357–374.

Scott, A.J. 1978. A review of the classification of *Chenopodium* L. and related genera (Chenopodiaceae). Bot. Jahrb. Syst. 100:205–220.

Scott, J.A. 1986. The butterflies of North America. Stanford Univ. Press, Stanford, CA.

Scott, R.W. 2006. *Brickellia*. In: Flora of North America Editorial Committee, eds. Fl. North Amer. 21:491–507. Oxford Univ. Press, New York and Oxford.

Scriber, J.M. 1978. Cyanogenic glycosides in *Lotus corniculatus*. Oecologia 34:143–155.

Seabrook, J.A. and **L.A. Dionne.** 1976. Studies on the genus *Apios*. I. Chromosome number and distribution of *Apios americana* and *A. priceana*. Canad. J. Bot. 54:2567–2572.

Seeligmann, P. and **R.E. Alston.** 1967. Complex chemical variation and the taxonomy of *Hymenoxys scaposa* (Compositae). Brittonia 19:205–211.

Seigler, D.S. 1994. Phytochemistry and systematics of the Euphorbiaceae. Ann. Missouri Bot. Gard. 81:380–401.

Seigler, D.S. and **J.E. Ebinger.** 2005. New combinations in the genus *Vachellia* (Fabaceae: Mimosoideae) from the New World. Phytologia 87:139–178.

Seigler, D.S. and **J.E. Ebinger.** 2023. *Senegalia*. In: Flora of North America Editorial Committee, eds. Fl. of North Amer. 11:117–121. Oxford Univ. Press, New York and Oxford.

Seigler, D.S., J.E. Ebinger, and **J.T. Miller.** 2006. The genus *Senegalia* (Fabaceae: Mimosoideae) from the New World. Phytologia 88:38–93.

Seijo, J.G., G.I. Lavia, A. Fernández, A. Krapovickas, D. Ducasse, and **E.A. Moscone.** 2004. Physical mapping of the 5S and 18S-25S rRNA genes by FISH as evidence that *Arachis duranensis* and *A. ipaensis* are wild diploid progenitors of *A. hypogaea* (Leguminosae). Amer. J. Bot. 91(9):1294–1303.

Seijo, J.G., G.I. Lavia, A. Fernández, A. Krapovickas, D. Ducasse, E.A. Moscone, and **D.J. Bertioli.** 2007. Genomic relationships between the cultivated peanut (*Arachis hypogaea*, Leguminosae) and its close relatives revealed by double GISH. Amer. J. Bot. 94(12):1963–1971.

Seine, R. and **W. Barthlott.** 1994. Some proposals on the infrageneric classification of *Drosera* L. Taxon 43:583–589.

Sell, P.D. 1976. *Hedypnois*. In: T.G. Tutin, V.H. Heywood, N.A. Burges, D.M. Moore, D.H. Valentine, S.M. Walters, and D.A. Webb, eds. Flora Europaea 4:151–152. Cambridge Univ. Press, Cambridge, England, UK.

Semple, J.C. 1974. The phytogeography and systematics of *Xanthisma texanum* DC. (Asteraceae); proper usage of infraspecific categories. Rhodora 76:1–19.

Semple, J.C. 1977. Cytotaxonomy of *Chrysopsis* and *Heterotheca* (Compositae-Astereae): A new interpretation of phylogeny. Canad. J. Bot. 55:2503–2513.

Semple, J.C. 1978. A revision of the genus *Borrichia* Adans. (Compositae). Ann. Missouri Bot. Gard. 65:681–693.

Semple, J.C. 1981. A revision of the goldenaster genus *Chrysopsis* (Nutt.) Ell. nom. cons. (Compositae-Astereae). Rhodora 83: 323–384.

Semple, J.C. 1992. Goldenrods of Ontario: *Solidago* and *Euthamia*. Univ. Waterloo Biol. Ser. 36:1–82.

Semple, J.C. 1996. A revision of *Heterotheca* sect. *Phyllotheca* (Nutt.) Harms (Compositae: Asteraceae): The prairie and montane goldenasters of North America. Univ. Waterloo Biol. Ser. 37: 1–164.

Semple, J.C. 2006a. *Borrichia*. In: Flora of North America Editorial Committee, eds. Fl. North Amer. 21:129–131. Oxford Univ. Press, New York and Oxford.

Semple, J.C. 2006b. *Bradburia*. In: Flora of North America Editorial Committee, eds. Fl. North Amer. 20:211–212. Oxford Univ. Press, New York and Oxford.

Semple, J.C. 2006c. *Chrysopsis*. In: Flora of North America Editorial Committee, eds. Fl. North Amer. 20:213–221. Oxford Univ. Press, New York and Oxford.

Semple, J.C. 2006d. *Heterotheca*. In: Flora of North America Editorial Committee, eds. Fl. North Amer. 20:230–256 Oxford Univ. Press, New York and Oxford.

Semple, J.C. 2006e. *Pityopsis*. In: Flora of North America Editorial Committee, eds. Fl. North Amer. 20:222–228. Oxford Univ. Press, New York and Oxford.

Semple, J.C. 2013. Application of the names *Solidago stricta* and *S. virgata* (Asteraceae: Astereae). Phytoneuron 2013-42:1–3.

Semple, J.C. 2016. Astereae Lab Research: Goldenrods. University of Waterloo, Canada. https://uwaterloo.ca/astereae-lab/research/goldenrods.

Semple, J.C., V. Blok, and **P. Heiman.** 1980. Morphological, anatomical, habit, and habitat differences among the goldenaster genera *Chrysopsis, Heterotheca, and Pityopsis* (Compositae–Astereae). Canad. J. Bot. 58:147–163.

Semple, J.C. and **F.D. Bowers.** 1985. A revision of the goldenaster genus *Pityopsis* Nutt. (Compositae: Asteraceae). Univ. Waterloo Biol. Ser. 29:1–34.

Semple, J.C. and **F.D. Bowers.** 1987. Cytogeography of *Pityopsis* Nutt., the grass-leaved goldenasters (Compositae: Astereae). Rhodora 89:381–389.

Semple, J.C. and **L. Brouillet.** 1980a. A synopsis of North American asters: The subgenera, sections and subsections of *Aster* and *Lasallea*. Amer. J. Bot. 67:1010–1026.

Semple, J.C. and **L. Brouillet.** 1980b. Chromosome numbers and satellite chromosome morphology in *Aster* and *Lasallea*. Amer. J. Bot. 67:1027–1039.

Semple, J.C. and **C.C. Chinnappa.** 1984. Observations on the cytology, morphology, and ecology of *Bradburia hirtella* (Compositae-Astereae). Syst. Bot. 9:95–101.

Semple, J.C. and **J.G. Chmielewski.** 1987. Revision of the *Aster lanceolatus* complex, including *A. simplex* and *A. hesperius* (Compositae: Astereae): A multivariate morphometric study. Canad. J. Bot. 65:1047–1062.

Semple, J.C. and **J.G. Chmielewski.** 2006. *Doellingeria*. In: Flora of North America Editorial Committee, eds. Fl. North Amer. 20: 43–46. Oxford Univ. Press, New York and Oxford.

Semple, J.C. and **R.E. Cook.** 2006. *Solidago*. In: Flora of North

America Editorial Committee, eds. Fl. North Amer. 20:107–166. Oxford Univ. Press, New York and Oxford.

Semple, J.C., S.B. Heard, and **C.S. Xiang.** 1996. The Asters of Ontario (Compositae: Astereae): *Diplactis* Raf., *Oclemena* Greene, *Doellingeria* Nees and *Aster* L. (including *Canaaanthus* Nesom, *Symphyotrichum* Nees and *Virgulus* Raf.). Univ. Waterloo Biol. Ser. 38:1–94.

Semple, J.C., H. Rahman, S. Bzovski, M.K. Sorour, K. Kornobis, R. Lopez Laphitz, and **L. Tong.** 2015. A multivariate morphometric study of the *Solidago altissima* complex and *S. canadensis* (Asteraceae: Astereae). Phytoneuron 2015-10:1–31.

Semple, J.C., T. Shea, H. Rahman, Y. Ma, and **K. Kornobis.** 2016. A multivariate study of the *Solidago sempervirens* complex (Asteraceae: Astereae: Solidago subsect. Maritimae). Phytoneuron 2016-73:1–31.

Semple, J.C., L. Tong, and **P. Pastolero.** 2012. Neotypification of *Solidago salicina* (Asteraceae: Astereae) and a multivariate comparison with *S. patula*. Phytoneuron 2012-56:1–6.–94.

Senn, H.A. 1939. The North American species of *Crotalaria*. Rhodora 41:317–367.

Sennblad, B. and **B. Bremer.** 1996. The familial and subfamilial relationships of Apocynaceae and Asclepiadaceae evaluated with *rbc*L data. Pl. Syst. Evol. 202:153–175.

Settle, W.J. and **T.R. Fisher.** 1970. The varieties of *Silphium integrifolium*. Rhodora 72:536–543.

Sexton, C. (Chuck). Personal communication. Wildlife Biologist, Balcones Canyonlands National Wildlife Refuge, U.S. Fish and Wildlife Service, Austin, TX.

Shan, R.H. and **L. Constance.** 1951. The genus *Sanicula* (Umbelliferae) in the Old World and the New. Univ. Calif. Publ. Bot. 25:1–78.

Sharma, O.P., R.K. Dawra, N.P. Kurade, and **P.D. Sharma.** 1998. A review of the toxicosis and biological properties of the genus *Eupatorium*. Nat. Tox. 6(1):1–14.

Sharples, M.T. and **E.A. Tripp.** 2019. Phylogenetic relationships within and delimitation of the cosmopolitan flowering plant genus *Stellaria* L. (Caryophyllaceae): Core stars and fallen stars. Syst. Bot. 44:857–876.

Shaw, J.M.H. 2000. A taxonomic revision of *Podophyllum* in the wild and in cultivation. New Plantsman 7:30–41.

Shaw, J.M.H. 2002. *Podophyllum*. In: W.T. Stern. The genus *Epimedium* and other herbaceous Berberidaceae. Pp. 239–314. Timber Press, Portland, OR.

Shen, Y., G. Yao, Y. Li, X. Tian, S. Li, N. Wang, C. Zhang, F. Wang, and **Y. Ma.** 2024. RAD-seq data reveals robust phylogeny and morphological evolutionary history of Rhododendron. Hort. Pl. J.10(3):866–878. https://www.sciencedirect.com/science/article/pii/S2468014123001425.

Shen, Z., X. Ding, J. Cheng, F. Wu, H. Yin, and **M. Wang.** 2022. Phylogenetic studies of magnoliids: Advances and perspectives. Front. Plant Sci. https://www.ncbi.nlm.nih.gov/pmc/articles/PMC9885158/.

Sherff, E.E. 1937. The genus *Bidens*. Publ. Field Mus. Nat. Hist., Bot. Ser. 16:1–709.

Sherff, E.E. 1955. *Cosmos,* in N. Amer. Fl. II. 2:130–146.

Sherff, E.E. and **E.J. Alexander.** 1955. Compositae-Heliantheae-Coreopsidinae. N. Amer. Fl. II. 2:1–190.

Shetler, S.G. 1963. A checklist and key to the species of *Campanula* native or commonly naturalized in North America. Rhodora 65:319–337.

Shetler, S.G. 1979. Pollen-collecting hairs of *Campanula* (Campanulaceae) I: Historical review. Taxon 28:205–215.

Shetler, S.G. and **N.R. Morin.** 1986. Seed morphology in North American Campanulaceae. Ann. Missouri Bot. Gard. 73:653–688.

Shi, S., Y.-L. Huang, Y. Zhong, Y. Du, Q. Zhang, H. Chang, and **D.E. Boufford.** 2001. Phylogeny of the Altingiaceae based on cpDNA matK, PY-IGS and nrDNA ITS sequences. Pl. Syst. Evol. 230: 13–24.

Shinners, L.H. 1946a. Revision of the genus *Chaetopappa*. Wrightia 1:63–81.

Shinners, L.H. 1946b. Revision of the genus *Leucelene* Greene. Wrightia 1:82–89.

Shinners, L.H. 1946c. Revision of the genus *Aphanostephus* DC. Wrightia 1:95–121.

Shinners, L.H. 1946d. Revision of the genus *Kuhnia* L. Wrightia 1:122–144.

Shinners, L.H. 1947a. Two anomalous new species of *Erigeron* L. from Texas. Wrightia 1:183–186.

Shinners, L.H. 1947b. Revision of the genus *Krigia* Schreber. Wrightia 1:187–206.

Shinners, L.H. 1948a. The vetches and pea vines (*Vicia* and *Lathyrus*) of Texas. Field & Lab. 16:18–29.

Shinners, L.H. 1949a. Nomenclature of species of dandelion and goatsbeard (*Taraxacum* and *Tragopogon*) introduced into Texas. Field & Lab. 17:13–19.

Shinners, L.H. 1949b. Transfer of Texas species of *Petalostemum* to *Dalea* (Leguminosae). Field & Lab. 17:81–85.

Shinners, L.H. 1949c. The genus *Dalea* (including *Petalostemum*) in North-Central Texas. Field & Lab. 17:85–89.

Shinners, L.H. 1949d. *Arenaria drummondii* Shinners, nom. nov. Field & Lab. 17:89.

Shinners, L.H. 1949e. The Texas species of *Conyza* (Compositae). Field & Lab. 17:142–144.

Shinners, L.H. 1950a. The Texas species of *Thelesperma* (Compositae). Field & Lab. 18:17–24.

Shinners, L.H. 1950c. The Texas species of *Cacalia* (Compositae). Field & Lab. 18:79–82.

Shinners, L.H. 1950d. The species of *Matelea* (including *Gonolobus*) in North Central Texas. Field & Lab. 18:73–78.

Shinners, L.H. 1951b. The north Texas species of *Heterotheca*, including *Chrysopsis* (Compositae). Field & Lab. 19:66–71.

Shinners, L.H. 1951c. Notes on Texas Compositae–VII. Field & Lab. 19:74–82.

Shinners, L.H. 1951d. The Texas species of *Psoralea* (Leguminosae). Field & Lab. 19:14–25.

Shinners, L.H. 1951e. The Texas species of *Evax* (Compositae). Field & Lab. 19:125–126.

Shinners, L.H. 1952. The Texas species of *Palafoxia* (Compositae). Field & Lab. 20:92–102.

Shinners, L.H. 1953a. Notes on Texas Compositae–IX. Field & Lab. 21:155–162.

Shinners, L.H. 1953b. Botanical notes. Field & Lab. 21:164–165.

Shinners, L.H. 1953c. The bluebonnets (*Lupinus*) of Texas. Field & Lab. 21:149–153.

Shinners, L.H. 1954. Notes on north Texas grasses. Rhodora 56(662): 25–38.

Shinners, L.H. 1956. *Argythamnia humilis* (Engelm. & Gray) Muell.-Arg. var. *laevis* (A. Gray) Shinners, comb. nov. Field & Lab. 24:38.

Shinners, L.H. 1957. *Wahlenbergia marginata* (Thunb.) A. DC. (Campanulaceae) in Louisiana. S. W. Naturalist 2:44.

Shinners, L.H. 1958. Spring flora of the Dallas–Fort Worth area Texas. Published by author, Dallas, TX.

Shinners, L.H. 1959. *Liatris tenuis*, n. sp. (Compositae), another endemic in southeastern Texas. S. W. Naturalist 4:207–208.

Shinners, L.H. 1961. Nomenclature of the Bignoniaceae of the southern United States. Castanea 26:109–118.

Shinners, L.H. 1962a. Synopsis of United States *Bonamia* including *Breweria* and *Stylisma* (Convolvulaceae). Castanea 27:65–77.

Shinners, L.H. 1962c. New names in *Arenaria* (Caryophyllaceae). Sida 1:49–52.

Shinners, L.H. 1962d. *Drosera* (Droseraceae) in the southeastern United States: An interim report. Sida 1:53–59.

Shinners, L.H. 1964. Texas *Evax* transferred to *Filago* (Compositae). Sida 1:252–253.

Shinners, L.H. 1965. *Ipomoea amnicola* (Convolvulaceae), a South American waif in Missouri. Rhodora 67:200.

Shinners, L.H. 1966. *Hypochoeris microcephala* var. *albiflora* (Compositae) in southeastern Texas: New to North America. Sida 2:393–394.

Shinners, L.H. 1967. *Bupleurum lancifolium* (*B. subovatum*) (Umbelliferae) in Texas. Sida 3:185.

Shinners, L.H. 1969. *Petrorhagia prolifera* (*Dianthus prolifer*, *Tunica prolifera*) (Caryophyllaceae) in Arkansas and Texas. Sida 3: 345–346.

Shinners, L.H. 1971. *Kuhnia* L. transferred to *Brickellia*. Sida 4:274.

Sholars, T. and **R. Riggens.** 2023. *Lupinus*. In: Flora of North America Editorial Committee, eds. Fl. of North Amer. 11:193–248. Oxford Univ. Press, New York and Oxford.

Shrestha, N., Z. Wang, X. Su, and **8 others.** 2018. Global patterns of Rhododendron diversity: The role of evolutionary time and diversification rates. Global Ecol. Biogeogr. 27:913–924. https://doi.org/10.1111/geb.12750.

Shulkina, T.V., J.F. Gaskin, and **W.M.M. Eddie.** 2003. Morphological studies toward an improved classification of Campanulaceae s. str. Ann. Missouri Bot. Gard. 90:576–591.

Shultz, L.M. 2003. *Beta*. In: Flora of North America Editorial Committee, eds. Fl. North Amer. 4:266–267. Oxford Univ. Press, New York and Oxford.

Shultz, L.M. 2003. *Spinacia*. In: Flora of North America Editorial Committee, eds. Fl. North Amer. 4:302. Oxford Univ. Press, New York and Oxford.

Shultz, L.M. 2006. *Artemisia*. In: Flora of North America Editorial Committee, eds. Fl. North Amer. 19:503–534. Oxford Univ. Press, New York and Oxford.

Shultz, L.M. and **W.A. Varga.** 2021. *Elaeagnus*. In: Flora of North America Editorial Committee, eds. Fl. North Amer. 10:416–419. Oxford Univ. Press, New York and Oxford.

Sieren, D.J. 1981. The taxonomy of the genus *Euthamia*. Rhodora 83:551–579.

Simmons, M.P. 2004. Balsaminaceae. In: K. Kubitzki, ed. The families and genera of vascular plants, Vol. VI. Pp. 20–25. Springer-Verlag, Berlin.

Simmons, M.P., J.A. Lombardi, and **L. Biral.** 2023. Classification of the Celastrales based on integration of genomic, morphological, and Sanger-sequence characters. Sys. Bot. 48(2):283–299.

Simões, A.O., M.E. Endress, T. van der Niet, L.S. Kinoshita, and **E. Conti.** 2004. Tribal and intergeneric relationships of Mesechiteae (Apocynoideae, Apocynaceae): Evidence from three noncoding plastid DNA regions and morphology. Amer. J. Bot. 91:1409–1418.

Simões, A.O., M.E. Endress, T. van der Niet, L.S. Kinoshita, and **E. Conti.** 2006. Is *Mandevilla* (Apocynaceae, Mesechiteae) monophyletic? Evidence from five plastid DNA loci and morphology. Ann. Missouri Bot. Gard. 93(4):565–591.

Simões, A.R. and **G. Staples.** 2017. Dissolution of Convolvulaceae tribe Merremieae and a new classification of the constituent genera. Bot. J. Linn. Soc. 183:561–586.

Simons, S.B. and **T.R. Seastedt.** 1999. Decomposition and nitrogen release from foliage of cottonwood (*Populus deltoides*) and Russian-olive (*Elaeagnus angustifolia*) in a riparian ecosystem. S. W. Naturalist 44:256–260.

Simpson, B.B. 1978. Mutisieae. N. Amer. Fl. II. 10:1–13.

Simpson, B.B. 1998. A revision of *Pomaria* (Fabaceae) in North America. Lundellia 1:46–71.

Simpson, B.B. 1999. A revision of *Hoffmansegia* (Fabaceae) in North America. Lundellia 2:14–54.

Simpson, B.B. 2004. The sumacs of Texas...they offer you a wide choice in attractive colors and shapes. Native Pl. Soc. Texas News 22(2):10–11.

Simpson, B.B. 2006. *Acourtia*. In: Flora of North America Editorial Committee, eds. Fl. North Amer. 19:72–74. Oxford Univ. Press, New York and Oxford.

Simpson, B.B. 2023. *Hoffmannseggia*. In: Flora of North America Editorial Committee, eds. Fl. of North Amer. 11:63–66. Oxford Univ. Press, New York and Oxford.

Simpson, B.B. and **B. Miao.** 1997. The circumscription of *Hoffmannseggia* (Fabaceae, Caesalpinioideae, Caesalpinieae) and its allies using morphological and cpDNA restriction site data. Pl. Syst. Evol. 205:157–178.

Simpson, B.B. and **E.A. Ulibarri.** 2006. A synopsis of the genus *Hoffmannseggia*. Lundellia 9:7–33.

Simpson, B.J. 1988. A field guide to Texas trees. Texas Monthly Press, Austin.

Sims, L.E. and **J. Price.** 1985. Nuclear DNA content variation in *Helianthus* (Asteraceae). Amer. J. Bot. 72:1213–1219.

Singh, V. 2022. Three crops rule the world: What it means for the planet's wildlife. Down to Earth (website). 2 Mar 2022.

Singhurst, J.R. (Jason). Personal communication. Botanist and Plant Community Ecologist with Texas Parks and Wildlife Department.

Singhurst, J.R., J.C. Cathy, D. Prochaska, H. Haucke, G.C. Kroh, and **W.C. Holmes.** 2003. The vascular flora of Gus Engeling Wildlife Management Area, Anderson County, Texas. S. E. Naturalist (Steuben) 2(3):347–368.

Singhurst, J.R., K.M. Fleming, and **W.C. Holmes.** 1998. *Eupatorium album* L. (Compositae): new to Texas. Phytologia 85(2):96–98.

Singhurst, J.R. and **W.C. Holmes.** 2001. *Dipsacus fullonum* (Dipsacaceae) and *Verbesina walteri* (Asteraceae), new to Texas. Sida 19:723–725.

Singhurst, J.R. and **W.C. Holmes.** 2010. *Prenanthes aspera*

(Asteraceae: Cichorieae): New to Texas. Phytoneuron 2010-52:1–4.

Singhurst, J.R. and **W.C. Holmes.** 2012. *Justicia brandegeeana* (Acanthaceae) new to the Texas flora. Phytoneuron 2012-114:1–3.

Singhurst, J.R. and **W.C. Holmes.** 2013. *Conyza floribunda* (Asteraceae) new to the flora of Texas. Phytoneuron 2013-13:1–4.

Singhurst, J.R. and **W.C. Holmes.** 2014. Rediscovery of *Selenia aurea* (Brassicaceae) in Texas. Phytoneuron 2014-49:1–3.

Singhurst, J.R., W.C. Holmes, and **W.R. Carr.** 2021. *Lonicera flava* (Caprifoliaceae) new to the Texas flora. Phytoneuron 2021-5:1–3.

Singhurst, J.R., J.N. Mink, K. Emde, L. Shen, D. Verser, and **W.C. Holmes.** 2018. The vascular flora and plant communities of Lawther-Deer Park Prairie, Harris County, Texas U.S.A. J. Bot. Res. Inst. Texas 12(2):721–733.

Singhurst, J.R., J.N. Mink, and **W.C. Holmes.** 2010. New and noteworthy plants of Texas. Phytologia 92(2):249–255.

Singhurst, J.R., R.J. O'Kennon, and **W.C. Holmes.** 2004. The genus *Prenanthes* (Asteraceae: Lactuceae) in Texas. Sida 21(1):181–191.

Singhurst, J.R., D.J. Rosen, and **W.C. Holmes.** 2009. Two additions to the vascular flora of Texas. Phytologia 91(1):69–72.

Singhurst, J.R., N. Shackelford, W. Newman, J.N. Mink, and **W.C. Holmes.** 2014. The ecology and abundance of *Hymenoxys texana* (Asteraceae). Phytoneuron 2014-19:1–19.

Singhurst, J.R., M. White, and **W.C. Holmes.** 2002. Noteworthy collections: Texas. Castanea 67(2):213–216.

Sinn, B.T., L.M. Kelly, and **J.V. Freudenstein.** 2015. Phylogenetic relationships in *Asarum*: Effect of data partitioning and a revised classification. Amer. J. Bot. 102:765–779.

Sinn, B.T., D.D. Sedmak, L.M. Kelly, and **J.V. Freudenstein.** 2018. Total duplication of the small single copy region in the angiosperm plastome: Rearrangement and inverted repeat instability in *Asarum*. Amer. J. Bot. 105:71–84.

Sipes, S., K.E. Huff Hartz, H. Amin, A. Anterola, and **D.L. Nickrent.** 2014. Floral scent and pollinators of the holoparasite *Pilostyles thurberi* (Apodanthaceae). J. Poll. Ecol. 12:31–39.

Siripun, K.C. and **E.E. Schilling.** 2006. *Eupatorium*. In: Flora of North America Editorial Committee, eds. Fl. North Amer. 21:462–474. Oxford Univ. Press, New York and Oxford.

Skallerup, H.R. 1953. The distribution of *Diospyros virginiana* L. Ann. Missouri Bot. Gard. 40:211–225.

Šlenker, M., J. Zozomová-Lihová, T. Mandáková, H. Kudoh, Y. Zhao, A. Soejima, T. Yahara, K. Skokanová, S. Španiel, and **K. Marhold.** 2018. Morphology and genome size of the widespread weed *Cardamine occulta*: How it differs from cleistogamic *C. kokaiensis* and other closely related taxa in Europe and Asia. Bot. J. Linn. Soc. 187(3):456–482.

Sleumer, H. 1967. Monographia Clethracearum. Bot. Jahrb. Syst. 87:36–175.

Sleumer, H. 1980. A system of the genus *Rhododendron* L. In: J.L. Luteyn and M.E. O'Brien, eds. Contributions toward a classification of *Rhododendron*: Proc. Inter. Rhododendron Conf., May 15–17, 1978. Bronx, NY. Pp. 1–18.

Slotte, T., A. Ceplitis, B. Neuffer, H. Hurka, and **M. Lascoux.** 2006. Intrageneric phylogeny of *Capsella* (Brassicaceae) and the origin of the tetraploid *C. bursa-pastoris* based on chloroplast and nuclear DNA sequences. Amer. J. Bot. 93(11):1714–1724.

Small, E. 1997. Cannabaceae. In: Flora of North America Editorial Committee, eds. Fl. North Amer. 3:381–387. Oxford Univ. Press, New York and Oxford.

Small, E. 2015. Evolution and classification of *Cannabis sativa* (Marijuana, Hemp) in relation to human utilization. Bot. Rev. (Lancaster) 81(3):189–294.

Small, E. 2023a. *Medicago*. In: Flora of North America Editorial Committee, eds. Fl. of North Amer. 11:983–994. Oxford Univ. Press, New York and Oxford.

Small, E. 2023b. *Melilotus*. In: Flora of North America Editorial Committee, eds. Fl. of North Amer. 11:977–980. Oxford Univ. Press, New York and Oxford.

Small, E. and **P.M. Catling.** 1999. Canadian medicinal crops. Ottawa, Canada: NRC Research Press.

Small, E. and **A. Cronquist.** 1976. A practical and natural taxonomy for *Cannabis*. Taxon 25:405–435.

Small, E. and **M. Jomphe.** 1989. A synopsis of the genus *Medicago* (Leguminosae). Canad. J. Bot. 67:3260–3294.

Small, E., J.Y. Perry, and **L.P. Lefkovitch.** 1976. A numerical taxonomic analysis of *Cannabis* with special reference to species delimitation. Syst. Bot. 1:67–84.

Small, J.K. 1896. Studies in the botany of the southeastern United States—VII. Bull. Torrey Bot. Club 23:405–410.

Small, J.K. 1897. The genus *Gaylussacia* in the Southern States. Bull. Torrey Bot. Club 24:441–445.

Small, J.K. 1903. Flora of the southeastern United States. Pp. 600–601, 1331. Published by the author.

Small, J.K. 1914a. Monotropaceae. N. Amer. Fl. 29:11–18.

Small, J.K. 1914b. Ericaceae (in part). N. Amer. Fl. 29:33–92.

Smith, A.C. 1944. Araliaceae. N. Amer. Fl. 28b:3–41.

Smith, A.R. 2006a. *Heliopsis*. In: Flora of North America Editorial Committee, eds. Fl. North Amer. 21:67–70. Oxford Univ. Press, New York and Oxford.

Smith, A.R. 2006b. *Zinnia*. In: Flora of North America Editorial Committee, eds. Fl. North Amer. 21:71–74. Oxford Univ. Press, New York and Oxford.

Smith, E.B. 1965. Taxonomy of *Haplopappus* section *Isopappus* (Compositae). Rhodora 67:217–238.

Smith, E.B. 1974. *Coreopsis nuecensis* (Compositae) and a related new species from southern Texas. Brittonia 26:161–171.

Smith, E.B. 1976. A biosystematic survey of *Coreopsis* in eastern United States and Canada. Sida 6:123–215.

Smith, E.B. 1981. New combinations in *Croptilon* (Compositae-Astereae). Sida 9:59–63.

Smith, E.B. 1982. Juvenile and adult leaflet phases in *Aralia spinosa* (Araliaceae). Sida 9:330–332.

Smith, E.B. and **H.M. Parker.** 1971. A biosystematic study of *Coreopsis tinctoria* and *C. cardaminefolia* (Compositae). Brittonia 23:161–170.

Smith, J.P., Jr. 1977. Vascular plant families. Mad River Press, Inc., Eureka, CA.

Smith, P.J. 1992. A revision of the genus *Wahlenbergia* (Campanulaceae) in Australia. Telopea 5:91–175.

Snogerup, S. and **B. Snogerup.** 2001. *Bupleurum* L. (Umbelliferae) in Europe—1. The annuals, B. sect. *Bupleurum* and sect. *Aristata*. Willdenowia. 31:205–308.

Sokoloff, D. 2000. New combinations in *Acmispon* (Leguminosae, Loteae). Ann. Bot. Fenn. 37(2):125–131.

Solbrig, O.T. 1960. The status of the genera *Amphipappus*,

Amphiachyris, Greenella, Gutierrezia, Gymnosperma and *Xanthocephalum*. Rhodora 62:43–54.

Solbrig, O.T. 1961. Synopsis of the genus *Xanthocephalum* (Compositae). Rhodora 63:151–164.

Solbrig, O.T. 1963. The tribes of Compositae in the southeastern United States. J. Arnold Arbor. 44:436–461.

Solbrig, O.T. and **K.S. Bawa.** 1975. Isosyme variation in species of *Prosopis* (Leguminosae). J. Arnold Arbor. 56:398–412.

Solbrig, O.T. and **P.D. Cantino.** 1975. Reproductive adaptations in *Prosopis* (Leguminosae, Mimosoideae). J. Arnold Arbor. 56:185–210.

Soltis, D.E. 1984. Karyotypes of species of *Asarum* and *Hexastylis* (Aristolochiaceae). Syst. Bot. 9:490–493.

Soltis, D.E., S.A. Smith, N. Cellenese, K.J. Wurdack, and **24 others.** 2011.Angiospermphylogeny:17genes,640taxa.Amer.J.Bot.98(4): 704–730.

Soltis, D.E. and **P.S. Soltis.** 1989. Allopolyploid speciation in *Tragopogon*: Insights from chloroplast DNA. Amer. J. Bot. 76: 1119–1124.

Soltis, D.E. and **P.S. Soltis.** 1997. Phylogenetic relationships in Saxifragaceae sensu lato: a comparison of topologies based on 18S rDNA and *rbc*L sequences. Amer. J. Bot. 84:504–522.

Soltis, D.E., P.S. Soltis, M.W. Chase, and **13 others.** 2000. Angiosperm phylogeny inferred from 18S rDNA, *rbcL*, and *atpB* sequences. Bot. J. Linn. Soc. 133:381–461.

Soltis, D.E., P.S. Soltis, P.K. Endress, and **M.W. Chase.** 2005. Phylogeny and evolution of angiosperms. Sinauer, Sunderland, MA.

Soltis, D.E., P.S. Soltis, P.K. Endress, M.W. Chase, S.R. Manchester, W.S. Judd, L.C. Majure, and **E.V. Mavrodiev.** 2018. Phylogeny and evolution of angiosperms, 2nd ed. Univ. Chicago Press, Chicago.

Soltis, D.E., P.S. Soltis, D.L. Nickrent, and **13 others.** 1997. Angiosperm phylogeny inferred from 18S ribosomal DNA sequences. Ann. Missouri Bot. Gard. 84:1–49.

Soltis, P.S. 2006. *Tragopogon*. In: Flora of North America Editorial Committee, eds. Fl. North Amer. 19:303–306. Oxford Univ. Press, New York and Oxford.

Soltis, P.S., G.M. Plunkett, S.J. Novak, and **D.E. Soltis.** 1995. Genetic variation in *Tragopogon* species: Additional origins of the allotetraploids *T. mirus* and *T. miscellus* (Compositae). Amer. J. Bot. 82:1329–1341.

Soltis, P.S. and **D.E. Soltis.** 1991. Multiple origins of the allotetraploid *Tragopogon mirus* (Compositae): rDNA evidence. Syst. Bot. 16:407–413.

Soltis, P.S. and **D.E. Soltis.** 2004. The origin and diversification of angiosperms. Amer. J. Bot. 91:1614–1626.

Soltis, P.S., D.E. Soltis, and **M.W. Chase.** 1999. Angiosperm phylogeny inferred from multiple genes as a tool for comparative biology. Nature 402:402–404.

Song, S., D. Zubov, H.P. Comes, and **6 others.** 2022. Plastid phylogenomics and plastome evolution of Nandinoideae (Berberidaceae). Frontiers Pl. Sci. (Lausanne) 13:913011. https:// doi: 10.3389/fpls.2022.913011.

Sørensen, P.D. 1995. *Arbutus*. In: J.L. Luteyn, ed. Ericaceae Part II, the superior-ovaried genera. Fl. Neotrop. Monogr. 66:194–221.

Sörensen, P.D. 2009. *Arbutus*. In: Flora of North America Editorial Committee, eds. Fl. North Amer. 8:398–400. Oxford Univ. Press, New York and Oxford.

Sorrells, J. and **R.J. Warren.** 2011. Ant-dispersed herb colonization lags behind forest re-establishment. J. Torrey Bot. Soc. 138:77–84.

Sorrie, B.A. 1998. Distribution of *Drosera filiformis* and *D. tracyi* (Droseraceae): Phytogeographic implications. Rhodora 100: 239–260.

Sorrie, B.A. 2015. *Crocanthemum*. In: Flora of North America Editorial Committee, eds. Fl. North Amer. 6:400–408. Oxford Univ. Press, New York and Oxford.

Sorrie, B.A., A.S. Weakley, and **G.C. Tucker.** 2009. *Gaylussacia*. In: Flora of North America Editorial Committee, eds. Fl. North Amer. 8:530–535. Oxford Univ. Press, New York and Oxford.

Sotuyo, S. 2023. *Erythrostemon*. In: Flora of North America Editorial Committee, eds. Fl. of North Amer. 11:56–59. Oxford Univ. Press, New York and Oxford.

Soule, J.A. 1993. Systematics of *Tagetes* (Asteraceae–Tageteae) (Mexico, Argentina). Ph.D. thesis. University of Texas.

Sousa, S.M. and **V.E. Rudd.** 1993. Revision del genero *Styphnolobium* (Leguminosae: Papilionoideae: Sophoreae). Ann. Missouri Bot. Gard. 80(1):270–283.

Souza, É.R. de, G.P. Lewis, F. Forest, A.S. Schnadelback, C. van den Berg, and **L.P. de Queiroz.** 2013. Phylogeny of *Calliandra* (Leguminosae: Mimosoideae) based on nuclear and plastid molecular markers. Taxon 62:1200–1219.

Spalik, K. and **S.R. Downie.** 2006. The evolutionary history of *Sium* sensu lato (Apiaceae): Dispersal, vicariance, and domestication as inferred from ITS rDNA phylogeny. Amer. J. Bot. 93: 747–761.

Spalik, K. and **S.R. Downie.** 2007. Intercontinental disjunctions in *Cryptotaenia* (Apiaceae, Oenantheae): An appraisal using molecular data. J. Biogeogr. 34:2039–2054.

Spalik, K., S.R. Downie, and **M.F. Watson.** 2009. Generic delimitations within the *Sium* alliance (Apiaceae tribe Oenantheae) inferred from cpDNA *rps*16-5´*trn*K$^{(UUU)}$ and nrDNA ITS sequences. Taxon 58:735–748.

Spangler, R.E. and **R.G. Olmstead.** 1999. Phylogenetic analysis of Bignoniaceae based on cpDNA gene sequences *rbc*L and *ndh*F. Ann. Missouri Bot. Gard. 86:33–46.

Spaulding, D.D. 2013. Key to the bindweeds (*Calystegia* and *Convolvulus*, Convolvulaceae) of Alabama and adjacent states. Phytoneuron 2013-83:1–12.

Spaulding, D.D. and **T.W. Barger.** 2016. Keys, distribution, and taxonomic notes for the lobelias (*Lobelia*, Campanulaceae) of Alabama and adjacent states. Phytoneuron 2016-76:1–60.

Spencer, W.P. and **J.K. Williams.** 2005. *Emilia fosbergii* (Asteraceae: Scenecionae) in Texas revisited. Sida 21(3):1937–1938.

Sperling, C.R. 1987. Systematics of the Basellaceae. Ph.D. dissertation. Harvard Univ., Cambridge, MA.

Sperling, C.R. and **V. Bittrich.** 1993. Basellaceae. In: K. Kubitzki, J.G. Rohwer, and V. Bittrich, eds. The Families and Genera of Vascular Plants. Vol. II. Flowering Plants. Dicotyledons: Magnoliid, Hamamelid, and Caryophyllid Families. Pp. 143–146. Spinger-Verlag, Berlin, Germany.

Sperry, O.E., J.W. Dollahite, J. Morrow, and **G.O. Hoffman.** 1955. Texas range plants poisonous to livestock. Texas Agric. Exp. Sta. Bull. No. 796.

Spethmann, W. 1987. A new infrageneric classification and phylogenetic trends in the genus Rhododendron (Ericaceae). Pl. Syst. Evol. 157:9–31.

Spoerke, D.G., Jr. and **S.C. Smolinske.** 1990. Toxicity of houseplants. CRC Press, Boca Raton, FL.

Spongberg, S.A. 1977. Ebenaceae hardy in temperate North America. J. Arnold Arbor. 58:146–160.

Spongberg, S.A. 1978. The genera of Crassulaceae in the southeastern United States. J. Arnold Arbor. 59:197–248.

Spooner, D., P. Rojas, M. Bonierbale, L.A. Mueller, M. Srivastov, D. Senalik, and **P. Simon.** 2013. Molecular phylogeny of *Daucus* (Apiaceae). Syst. Bot. 38:850–857.

Spooner, D.M. 1990. Systematics of *Simsia* (Compositae: Heliantheae). Syst. Bot. Monogr. 30:1–90.

Spooner, D.M. 2006. *Simsia*. In: Flora of North America Editorial Committee, eds. Fl. North Amer. 21:140–141. Oxford Univ. Press, New York and Oxford.

Spooner, D.M. 2024. *Daucus*. In: Flora of North America Editorial Committee, eds. Fl. North Amer. 13:295–296. Oxford Univ. Press, New York and Oxford.

Spurr, P.L. 2006. *Youngia*. In: Flora of North America Editorial Committee, eds. Fl. North Amer. 19:255–256. Oxford Univ. Press, New York and Oxford.

Srivastava, S., M. Srivastava, A. Misra, G. Pandey, and **A.K.S. Rawat.** 2015. A review on biological and chemical diversity in *Berberis* (Berberidaceae). EXCLI J. 14:247–267.

Stace, H.M. and **L.A. Edye,** eds. 1984. The biology of agronomy of *Stylosanthes*. Academic Press, San Francisco, CA.

Stalter, R. and **D.T. Kincaid.** 1994. Some observations on leaf form in *Ilex vomitoria* (Aquifoliaceae). Sida 16:79–92.

Standley, P.C. 1916. Chenopodiaceae. N. Amer. Fl. 21:3–93.

Standley, P.C. 1917. Amaranthaceae. N. Amer. Fl. 21:95–169.

Standley, P.C. 1922. Leguminosae. In: Trees and shrubs of Mexico. Contr. U.S. Natl. Herb. 23:348–515.

Standley, P.C. 1924. Trees and shrubs of Mexico. Contr. U.S. Natl. Herb. 23:1216–1224.

Stanford, J.W. (Jack). Personal communication. Specialist in plants of the Edwards Plateau and Fabaceae, especially *Mimosa*; Howard Payne Univ., Brownwood, TX.

Stanford, J.W. 1971. Vascular plants of the three central Texas counties of Brown, Comanche, and Hamilton. Ph.D. dissertation, Oklahoma State Univ., Stillwater.

Stanford, J.W. 1976. Keys to the vascular plants of the Texas Edwards Plateau and adjacent areas. Published by the author, Brownwood, TX.

Stanford, N. and **B.L. Turner.** 1988. The natural distribution and biological status of *Helenium amarum* and *H. badium* (Asteraceae, Heliantheae). Phytologia 65:141–146.

Stebbins, G.L. 1937. Critical notes on *Lactuca* and related genera. J. Bot. 75:12–18.

Stebbins, G.L. 1949. Speciation, evolutionary trends, and distributional patterns in *Crepis*. Evolution 3:188–193.

Stefanović, S.L., D.F. Austin, and **R.G. Olmstead.** 2003. Classification of Convolvulaceae: A phylogenetic approach. Sys. Bot. 28:791–806.

Stefanović, S., L. Krueger, and **R.G. Olmstead.** 2002. Monophyly of the Convolvulaceae and circumscription of their major lineages based on DNA sequences of multiple chloroplast loci. Amer. J. Bot. 89:1510–1522.

Stehlin, I.B. 1996. Outsmarting poison ivy and its cousins. U.S.D.A. Consumer (magazine), Sept. 1996. http://www.fda.gov/fdac/features/796_ivy.html. Accessed Jun 2007.

Steinmann, V.W. and **R.S. Felger.** 1997. The Euphorbiaceae of Sonora, Mexico. Aliso 16:1–71.

Steinmann, V.W., J.J. Morawetz, P.E. Berry, J.A. Peirson, and **Y. Yang.** 2016. *Euphorbia* sect. *Anisophyllum*. In: Flora of North America Editorial Committee, eds. Fl. North Amer. 12:251–293. Oxford Univ. Press, New York and Oxford.

Steinmann, V.W. and **J.M. Porter.** 2002. Phylogenetic relationships in Euphorbieae (Euphorbiaceae) based on ITS and *ndh*F sequence data. Ann. Missouri Bot. Gard. 89:453–490.

Stephens, H.A. 1980. Poisonous plants of the central United States. Univ. Press of Kansas, Lawrence.

Stephenson, A.G. 1982. The role of the extrafloral nectaries of *Catalpa speciosa* in limiting herbivory and increasing fruit production. Ecology 63:663–669.

Stephenson, A.G. and **W.W. Thomas.** 1977. Diurnal and nocturnal pollination of *Catalpa speciosa* (Bignoniaceae). Syst. Bot. 2:191–198.

Stuessy, T.F. 2004. A transitional-combinational theory for the origin of angiosperms. Taxon 53:3–16.

Stevens, P.F. 1995. Familial and infrafamilial relationships. In: J.L. Luteyn, ed. Ericaceae, Part II, the superior-ovaried genera. Fl. Neotrop. Monogr. 66:1–12.

Stevens, P.F. (2001 onwards). Angiosperm Phylogeny Website. Several Versions. [and more or less continuously updated since]. http://www.mobot.org/MOBOT/research/APweb/.

Stevens, P.F., J. Luteyn, E.G.H. Oliver, and **12 others.** 2004. Ericaceae. In: K. Kubitzki, ed. The families and genera of vascular plants, Vol. VI. Pp. 145–194. Springer-Verlag, Berlin.

Stevenson, G.A. 1969. An agronomic and taxonomic review of the genus *Melilotus* Mill. Canad. J. Pl. Sci. 49:1–20.

Steyermark, J.A. 1934. Studies in *Grindelia*. II. A monograph of the North American species of the genus *Grindelia*. Ann. Missouri Bot. Gard. 21:433–608.

Steyermark, J.A. 1963. Flora of Missouri. The Iowa State Univ. Press, Ames.

Steyermark, J.A., J.W. Voigt, and **R.H. Mohlenbrock.** 1959. Present biological status of *Geocarpon minimum* Mackenzie. Bull. Torrey Bot. Club 86:228–235.

Stiles, E.W. 1984. Fruit for all seasons. Nat. Hist. 93:42–53.

Stirton, C.H., ed. 1987. Advances in legume systematics. Part 3. Royal Botanic Gardens, Kew, England, U.K.

Stoyanov, K., T. Raycheva, and **I. Denev.** 2012. Molecular taxonomic analysis of *Monotropa hypopitys* and its relationship with Pyrolaceae. Biotechnology & Biotechnological Equipment 26:3116–3122. https://doi: 10.5504/BBEQ.2012.0040 W.

Stratton, J.M. 2024. *Conium*. In: Flora of North America Editorial Committee, eds. Fl. North Amer. 13:260–261. Oxford Univ. Press, New York and Oxford.

Straub, S.C.K., J.L. Reveal, and **A.S. Weakley.** 2023. *Amorpha*. In: Flora of North America Editorial Committee, eds. Fl. of North Amer. 11:258–270. Oxford Univ. Press, New York and Oxford.

Stritch, L.R. 1984. Nomenclatural contributions to a revision of the genus *Wisteria*. Phytologia 56:183–184.

Stritch, L.R. 2023. *Wisteria*. In: Flora of North America Editorial Committee, eds. Fl. of North Amer. 11:342–344. Oxford Univ. Press, New York and Oxford.

Strother, J.L. 1969. Systematics of *Dyssodia* Cavanilles (Compositae: Tageteae). Univ. Calif. Publ. Bot. 48:1–88.

Strother, J.L. 1977. Taxonomy of *Chrysactia, Harnackia*, and *Lescaillea* (Compositae: Tageteae). Madroño 24:129–192.

Strother, J.L. 1986. Renovation of *Dyssodia* (Compositae: Tageteae). Sida 11:371–378.

Strother, J.L. 1989. Chromosome numbers in *Thymophylla* (Compositae: Tageteae). Sida 13:351–358.

Strother, J.L. 1991. Taxonomy of *Complaya, Elaphandra, Iogeton, Jefea, Wamalchitamia, Wedelia, Zexmenia*, and *Zyzyxia* (Compositae-Heliantheae-Ecliptinae). Syst. Bot. Monogr. 33:1–111.

Strother, J.L. 2006a. *Acanthospermum*. In: Flora of North America Editorial Committee, eds. Fl. North Amer. 21:36–37. Oxford Univ. Press, New York and Oxford.

Strother, J.L. 2006aa. *Sphagneticola*. In: Flora of North America Editorial Committee, eds. Fl. North Amer. 21:126. Oxford Univ. Press, New York and Oxford.

Strother, J.L. 2006b. *Acmella*. In: Flora of North America Editorial Committee, eds. Fl. North Amer. 21:132–133. Oxford Univ. Press, New York and Oxford.

Strother, J.L. 2006bb. *Stokesia*. In: Flora of North America Editorial Committee, eds. Fl. North Amer. 19:201. Oxford Univ. Press, New York and Oxford.

Strother, J.L. 2006c. *Ambrosia*. In: Flora of North America Editorial Committee, eds. Fl. North Amer. 21:10–18. Oxford Univ. Press, New York and Oxford.

Strother, J.L. 2006cc. *Tagetes*. In: Flora of North America Editorial Committee, eds. Fl. North Amer. 21:235–236. Oxford Univ. Press, New York and Oxford.

Strother, J.L. 2006d. *Calyptocarpus*. In: Flora of North America Editorial Committee, eds. Fl. North Amer. 21:133. Oxford Univ. Press, New York and Oxford.

Strother, J.L. 2006dd. *Tetragonotheca*. In: Flora of North America Editorial Committee, eds. Fl. North Amer. 21:178–179. Oxford Univ. Press, New York and Oxford.

Strother, J.L. 2006e. *Chrysactinia*. In: Flora of North America Editorial Committee, eds. Fl. North Amer. 21:232. Oxford Univ. Press, New York and Oxford.

Strother, J.L. 2006ee. *Thelesperma*. In: Flora of North America Editorial Committee, eds. Fl. North Amer. 21:199–203. Oxford Univ. Press, New York and Oxford.

Strother, J.L. 2006f. *Cichorium*. In: Flora of North America Editorial Committee, eds. Fl. North Amer. 19:221–222. Oxford Univ. Press, New York and Oxford.

Strother, J.L. 2006ff. *Thymophylla*. In: Flora of North America Editorial Committee, eds. Fl. North Amer. 21:199–203. Oxford Univ. Press, New York and Oxford.

Strother, J.L. 2006g. *Conyza*. In: Flora of North America Editorial Committee, eds. Fl. North Amer. 20:348–350. Oxford Univ. Press, New York and Oxford.

Strother, J.L. 2006gg. *Townsendia*. In: Flora of North America Editorial Committee, eds. Fl. North Amer. 21:239–245. Oxford Univ. Press, New York and Oxford.

Strother, J.L. 2006h. *Coreopsis*. In: Flora of North America Editorial Committee, eds. Fl. North Amer. 21:185–198. Oxford Univ. Press, New York and Oxford.

Strother, J.L. 2006hh. *Verbesina*. In: Flora of North America Editorial Committee, eds. Fl. North Amer. 21:106–111. Oxford Univ. Press, New York and Oxford.

Strother, J.L. 2006i. *Dimorphotheca*. In: Flora of North America Editorial Committee, eds. Fl. North Amer. 19:380–381. Oxford Univ. Press, New York and Oxford.

Strother, J.L. 2006ii. *Vernonia*. In: Flora of North America Editorial Committee, eds. Fl. North Amer. 19:206–213. Oxford Univ. Press, New York and Oxford.

Strother, J.L. 2006j. *Dysodiopsis*. In: Flora of North America Editorial Committee, eds. Fl. North Amer. 21:337. Oxford Univ. Press, New York and Oxford.

Strother, J.L. 2006jj. *Wedelia*. In: Flora of North America Editorial Committee, eds. Fl. North Amer. 21:125–126. Oxford Univ. Press, New York and Oxford.

Strother, J.L. 2006k. *Eclipta*. In: Flora of North America Editorial Committee, eds. Fl. North Amer. 21:128–129. Oxford Univ. Press, New York and Oxford.

Strother, J.L. 2006kk. *Xanthium*. In: Flora of North America Editorial Committee, eds. Fl. North Amer. 21:19–20. Oxford Univ. Press, New York and Oxford.

Strother, J.L. 2006l. *Elephantopus*. In: Flora of North America Editorial Committee, eds. Fl. North Amer. 19:202–203. Oxford Univ. Press, New York and Oxford.

Strother, J.L. 2006m. *Gaillardia*. In: Flora of North America Editorial Committee, eds. Fl. North Amer. 21:421–426. Oxford Univ. Press, New York and Oxford.

Strother, J.L. 2006n. *Hedypnois*. In: Flora of North America Editorial Committee, eds. Fl. North Amer. 19:302. Oxford Univ. Press, New York and Oxford.

Strother, J.L. 2006o. *Hieracium*. In: Flora of North America Editorial Committee, eds. Fl. North Amer. 19:278–294. Oxford Univ. Press, New York and Oxford.

Strother, J.L. 2006p. *Hymenopappus*. In: Flora of North America Editorial Committee, eds. Fl. North Amer. 21:309–316. Oxford Univ. Press, New York and Oxford.

Strother, J.L. 2006q. *Iva*. In: Flora of North America Editorial Committee, eds. Fl. North Amer. 21:25–28. Oxford Univ. Press, New York and Oxford.

Strother, J.L. 2006r. *Lactuca*. In: Flora of North America Editorial Committee, eds. Fl. North Amer. 19:259–263. Oxford Univ. Press, New York and Oxford.

Strother, J.L. 2006s. *Leucanthemum*. In: Flora of North America Editorial Committee, eds. Fl. North Amer. 19:557–559. Oxford Univ. Press, New York and Oxford.

Strother, J.L. 2006t. *Lindheimera*. In: Flora of North America Editorial Committee, eds. Fl. North Amer. 21:75–76. Oxford Univ. Press, New York and Oxford.

Strother, J.L. 2006u. *Melampodium*. In: Flora of North America Editorial Committee, eds. Fl. North Amer. 21:34–35. Oxford Univ. Press, New York and Oxford.

Strother, J.L. 2006v. *Palafoxia*. In: Flora of North America Editorial Committee, eds. Fl. North Amer. 21:388–391. Oxford Univ. Press, New York and Oxford.

Strother, J.L. 2006w. *Parthenium*. In: Flora of North America Editorial Committee, eds. Fl. North Amer. 21:20–22. Oxford Univ. Press, New York and Oxford.

Strother, J.L. 2006x. *Psilostrophe*. In: Flora of North America Editorial Committee, eds. Fl. North Amer. 21:453–455. Oxford Univ. Press, New York and Oxford.

Strother, J.L. 2006y. *Pyrrhopappus*. In: Flora of North America

Editorial Committee, eds. Fl. North Amer. 19:376–378. Oxford Univ. Press, New York and Oxford.

Strother, J.L. 2006z. *Smallanthus*. In: Flora of North America Editorial Committee, eds. Fl. North Amer. 21:33–34. Oxford Univ. Press, New York and Oxford.

Strother, J.L. 2015. Cistaceae. In: Flora of North America Editorial Committee, eds. Fl. North Amer. 6:186–408. Oxford Univ. Press, New York and Oxford.

Strother, J.L. and **R.R. Weedon.** 2006. *Bidens*. In: Flora of North America Editorial Committee, eds. Fl. North Amer. 21:205–218. Oxford Univ. Press, New York and Oxford.

Strother, J.L. and **M.A. Wetter.** 2006. *Grindelia*. In: Flora of North America Editorial Committee, eds. Fl. North Amer. 21:424–436. Oxford Univ. Press, New York and Oxford.

Stubbendieck, J. and **E.C. Conard.** 1989. Common legumes of the Great Plains: An illustrated guide. Univ. of Nebraska Press, Lincoln.

Stubbendieck, J.L., G.Y. Friisoe, and **M.R. Bolick.** 1994. Weeds of Nebraska and the Great Plains. Nebraska Dept. Agric., Lincoln.

Stuckey, R.L. 1972. Taxonomy and distribution of the genus *Rorippa* (Cruciferae) in North America. Sida 4:279–430.

Stuessy, T.F. 1970. The genus *Acanthospermum*. Rhodora 72:106–109.

Stuessy, T.F. 1971. Systematic relationships in the white-rayed species of *Melampodium* (Compositae). Brittonia 23:177–190.

Stuessy, T.F. 1972. Revision of the genus *Melampodium* (Compositae: Heliantheae). Rhodora 74:1–70, 161–219.

Stuessy, T.F. 1979. Cladistics of *Melampodium* (Compositae). Taxon 28:179–195.

Stuessy, T.F. and **J.V. Crisci.** 1984. Phenetics of *Melampodium* (Compositae, Heliantheae). Madroño 31:8–19.

Stuessy, T.F. and **E. Hörandl.** 2014. The importance of comprehensive phylogenetic (evolutionary) classification—A response to Schmidt-Lebuhn's commentary on paraphyletic taxa. Cladistics 30:291–293.

Stuppy, W., P.C. van Welzen, P. Klinratana, and **M.C.T. Posa.** 1999. Revision of the genera *Aleurites*, *Reutealis* and *Vernicia* (Euphorbiaceae). Blumea 44:73–98.

Stuppy, W., P.C. van Welzen, P. Klinratana, and **M.C.T. Posa.** 2008. Taxonomic treatment No. 87. *Vernicia*. In: P.C. van Welzen and K. Chayamarit. Flora of Thailand. Euphorbiaceae. Nationaal Herbarium Nederland, Leiden; Forest Herbarium, National Park, Wildlife and Plant Conservation Department, Bangkok. http://www.nationaalherbarium.nl/Thaieuph/ThVspecies/ThVernicia.htm. Accessed Jul 2008.

Subils, R. 1984. Una nueva especie de *Euphorbia* sect. *Poinsettia* (Euphorbiaceae). Kurtziana 17:125–130.

Sugar, J.C. and **A. Krings.** 2023. *Styphnolobium*. In: Flora of North America Editorial Committee, eds. Fl. of North Amer. 11:160–161. Oxford Univ. Press, New York and Oxford.

Suh, Y. and **B.B. Simpson.** 1990. Phylogenetic analysis of chloroplast DNA in North American *Gutierrezia* and related genera (Asteraceae: Astereae). Syst. Bot. 15:660–670.

Sullivan, V.I. 1975. Pollen and pollination in the genus *Eupatorium* (Compositae). Canad. J. Bot. 53:582–589.

Summerfield, R.J. and **A.H. Bunting,** eds. 1980. Advances in legume science. Royal Botanic Gardens, Kew, England, U.K.

Sun, F.-J. and **S.R. Downie.** 2004. A molecular systematic investigation of *Cymopterus* and its allies (Apiaceae) based on phylogenetic analyses of nuclear (ITS) and plastid (*rps*16 intron) DNA sequences. S. African J. Bot. 70:407–416.

Sun, F.-J. and **S.R. Downie.** 2010. Phylogenetic analyses of morphological and molecular data reveal major clades within the perennial, endemic western North American Apiaceae subfamily Apioideae. J. Torrey Bot. Soc. 137:133–156.

Sun, F.-J., S.R. Downie, and **R.L. Hartman.** 2004. An ITS-based phylogenetic analysis of the perennial, endemic Apiaceae subfamily Apioideae of western North America. Sys. Bot. 29:419–431.

Sun, Y., M.J. Moore, J.B. Landis, N. Lin, L. Chen, T. Deng, J. Zhang, A. Meng, S. Zhang, K.S. Tojibaev, H. Sun, and **H. Wang.** 2018. Plastome phylogenomics of the early-diverging eudicot family Berberidaceae. Molec. Phylogen. Evol. 128:203–211.

Sundberg, S.D. 1991. Infraspecific classification of *Chloracantha spinosa* (Benth.) Nesom (Asteraceae, Astereae). Phytologia 70: 382–391.

Sundberg, S.D. and **D.J. Bogler.** 2006. *Baccharis*. In: Flora of North America Editorial Committee, eds. Fl. North Amer. 20:22–34. Oxford Univ. Press, New York and Oxford.

Sundberg, S.D. and **G.L. Nesom.** 2006. *Chloracantha*. In: Flora of North America Editorial Committee, eds. Fl. North Amer. 20:358. Oxford Univ. Press, New York and Oxford.

Sundell, E. 1981. The New World species of *Cynanchum* subgenus *Mellichampia* (Ascelpiadaceae). Evol. Monogr. 5:1–63.

Susanna, A., M. Galbany-Casals, K. Romaschenko, L. Barres, J. Martín, and **N. Garcia-Jacas.** 2011. Lessons from *Plectocephalus* (Compositae, Cardueae-Centaureinae): ITS disorientation in annuals and Beringian dispersal as revealed by molecular analyses. Ann. Bot. (Oxford) 108(2):263–277.

Sutherland, D.M. 2023. *Dalea*. In: Flora of North America Editorial Committee, eds. Fl. of North Amer. 11:283–315. Oxford Univ. Press, New York and Oxford.

Sutherst, R.W., R.J. Jones, and **H.J. Schnitzerling.** 1982. Tropical legumes of the genus *Stylosanthes* immobilize and kill cattle ticks. Nature 295:320–321.

Swanson, S.D. and **S.H. Sohmer.** 1976. The biology of *Podophyllum peltatum* L. (Berberidaceae), the may apple. II. The transfer of pollen and success of sexual reproduction. Bull. Torrey Bot. Club 103:223–226.

Swink, F. and **G. Wilhem.** 1994. Plants of the Chicago region, 4th ed. Indiana Acad. of Science, Indianapolis.

Taber, S.W. and **S.B. Fleenor.** 2005. Invertebrates of Central Texas Wetlands. Pp. 147–148. Texas Tech Univ. Press., Lubbock.

Tadesse, M., D.J. Crawford, and **S.C. Kim.** 2001. A cladistic analysis of morphological features in *Bidens* L. and *Coreopsis* L. (Asteraceae-Heliantheae) with notes on generic delimitation and systematics. Biol. Skr. 54:85–102.

Tadesse, M., D.J. Crawford, and **E.B. Smith.** 1995. Comparative capitular morphology and anatomy of *Coreopsis* L. and *Bidens* L. (Compositae), including a review of generic boundaries. Brittonia 47:61–91.

Takahashi, D. and **H. Setoguchi.** 2017. Molecular phylogeny and taxonomic implications of *Asarum* (Aristolochiaceae) based on ITS and *matK* sequences. Pl. Spec. Biol. 33:28–41.

Takahashi, M., J.W. Nowicke, and **G.L. Webster.** 1995. A note on remarkable exines in Acalyphoideae (Euphorbiaceae). Grana 34: 282–290.

Takhtajan, A.L. 1997. Diversity and classification of flowering plants. Columbia Univ. Press, New York, NY.

Tampion, J. 1977. Dangerous plants. Universe Books, New York.

Taylor, C.E.S. 1997. Keys to the Asteraceae of Oklahoma. Southeastern Oklahoma State Univ. Herbarium, Durant.

Taylor, C.E.S. and **R.J. Taylor.** 1983. New species, new combinations and notes on the goldenrods (*Euthamia* and *Solidago*–Asteraceae). Sida 10(2):176–183.

Taylor, C.E.S. and **R.J. Taylor.** 1984. *Solidago* (Asteraceae) in Oklahoma and Texas. Sida 10(3):223–251.

Taylor, M.L. and **H. Williams.** 2009. Consequences of pollination syndrome evolution for postpollination biology in an ancient angiosperm family. Int. J. Pl. Sci.170:584–598.

Taylor, N.C. 2016. Explorations into *Euphorbia* sect. *Anisophyllum* (Euphorbiaceae) in the trans-pecos region of Texas with a focus on the fendleri clade. M.S. thesis, Sul Ross State University.

Taylor, N.P. 1978. Review of the genus *Escobaria*. Cact. Succ. J. Gr. Brit. 40:31–37.

Taylor, N.P. 1986. The identification of Escobarias. Brit. Cactus Succ. J. 4:36–44.

Taylor, R.J. 1987. Populational variation and biosystematic interpretations in weedy dandelions. Bull. Torrey Bot. Club 114:109–120.

Taylor, R.J. and **C.E. Taylor.** 1981. Plants new to Arkansas, Oklahoma and Texas. Sida 9:25–28.

Taylor, R.J. and **C.E. Taylor.** 1994. An annotated list of the ferns, fern allies, gymnosperms and flowering plants of Oklahoma, 3rd ed. Southeastern Oklahoma State Univ. Herbarium, Durant, OK.

Tecnuextreme.com. 2007. The extreme poison ivy remedy. http://www.tecnuextreme.com/. Accessed Jun 2007.

Teixeira de Lima, C., F. de Assis Ribeiro dos Santos, and **A.M. Giulietti.** 2014. Morphological strategies of *Cabomba* (Cabombaceae), a genus of aquatic plants. Acta Bot. Brasil. 28: 327–338. https://www.researchgate.net/publication/278324100_Morphological_strategies_of_Cabomba_Cabombaceae_a_genus_of_aquatic_plants.

Tellman, B. 1997. Exotic pest plant introduction in the American southwest. Desert Pl. 13:3–9.

Tenaglia, D. 2007. Photographs and descriptions of the flowering and non-flowering plants of Missouri, USA. http://www.missouriplants.com. Accessed Jun 2008.

Tennessee Flora Committee. 2015. Guide to the vascular flora of Tennessee. The University Press of Tennessee, Knoxville.

Terahara, N., N. Saito, T. Honda, K. Toki, and **Y. Osajima.** 1990. Acylated anthocyanins of *Clitoria ternatea* flowers and their acyl moieties. Phytochemistry 29(3):949–953.

Texas Invasive Species Institute. 2014. English ivy, *Hedera helix*. http://www.tsusinvasives.org/home/database/hedera-helix. Accessed Jul 2019.

Texasinvasives.org. 2015. *Macfadyena unguis-cati*. Catclawvine. https://www.texasinvasives.org/plant_database/detail.php?symbol=MAUN3. Accessed 24 Dec 2024. also see the map at: https://www.texasinvasives.org/observations/mapping.php?species=MAUN3&search=go.

Tharp, B.C. and **F.A. Barkley.** 1949. Genus *Ruellia* in Texas. Amer. Midl. Naturalist 42:1–86.

Tharp, B.C. and **M.C. Johnston.** 1961. Recharacterization of *Dichondra* (Convolvulaceae) and a revision of the North American species. Brittonia 13:346–360.

Thiede, J., S.A. Schmidt, and **B. Rudolph.** 2007. Phylogenetic implication of the chloroplast rpoC1 intron loss in the Aizoaceae (Caryophyllales). Biochem. Syst. Ecol. 35:372–380.

Thieret, J.W. 1969b. *Trifolium vesiculosum* (Leguminosae) in Mississippi and Louisiana: New to North America. Sida 3: 446–447.

Thieret, J.W. 2005b. *Agrostemma*. In: Flora of North America Editorial Committee, eds. Fl. North Amer. 5:214–215. Oxford Univ. Press, New York and Oxford.

Thieret, J.W. and **R.K. Rabeler.** 2005a. *Vaccaria*. In: Flora of North America Editorial Committee, eds. Fl. North Amer. 5:156. Oxford Univ. Press, New York and Oxford.

Thieret, J.W. and **R.K. Rabeler.** 2005b. *Saponaria*. In: Flora of North America Editorial Committee, eds. Fl. North Amer. 5:157–158. Oxford Univ. Press, New York and Oxford.

Thieret, J.W. and **R.K. Rabeler.** 2005c. *Polycarpon*. In: Flora of North America Editorial Committee, eds. Fl. North Amer. 5:25–26. Oxford Univ. Press, New York and Oxford.

Thieret, J.W. and **R.K. Rabeler.** 2005d. *Scleranthus*. In: Flora of North America Editorial Committee, eds. Fl. North Amer. 5:149–151. Oxford Univ. Press, New York and Oxford.

Thomas, J.L. 1960. A monographic study of the Cyrillaceae. Contr. Gray Herb. 186:1–114.

Thomas, J.L. 1961. The genera of the Cyrillaceae and Clethraceae of the southeastern United States. J. Arnold Arbor. 42:96–106.

Thomas, J.R. 2020. *Desmodium glabellum* and *D. perplexum* (Fabaceae): a morphological reevaluation. Missouriensis 38:29–50.

Thomas, R.D. and **C.M. Allen.** 1996. Atlas of the vascular flora of Louisiana. Vol.2. Dicotyledons: Acanthaceae–Euphorbiaceae. Louisiana Dept. of Wildlife and Fisheries. Baton Rouge.

Thomas, S.K., X. Liu, Z.-Y. Du, Y. Dong, A. Cummings, L. Pokorny, Q.-Y. Xiang, and **J.H. Leebens-Mack.** 2021. Comprehending Cornales: Phylogenetic reconstruction of the order using the Angiosperms353 probe set. Amer. J. Bot. 108:1112–1121. https://www.researchgate.net/publication/353264212_Comprehending_Cornales_phylogenetic_reconstruction_of_the_order_using_the_Angiosperms353_probe_set.

Thompson, K.A., D.M. Sora, K.S. Cross, J.M. St. Germain, and **K. Cottenie.** 2014. Mucilage reduces leaf herbivory in Schreber's watershield, *Brasenia schreberi* J.F. Gmel. (Cabombaceae). Botany (Ottawa) 92:412–416. doi: https://doi.org/10.1139/cjb-2013-0296.

Thompson, S.W. and **T.G. Lammers.** 1997. Phenetic analysis of morphological variation in the *Lobelia cardinalis* complex (Campanulaceae: Lobelioideae). Syst. Bot. 22:315–331.

Thorne, R.F. 1992a. An updated classification of the flowering plants. Aliso 13:365–389.

Thorne, R.F. 1992b. Classification and geography of the flowering plants. Bot. Rev. (Lancaster) 58:225–348.

Thorne, R.F. 2010. Bataceae. In: Flora of North America Editorial Committee, eds. Fl. North Amer. 7:186–188. Oxford Univ. Press, New York and Oxford.

Thorne, R.F. and **J.L. Reveal.** 2007. An updated classification of the Class Magnoliopsida. Bot. Rev. (Lancaster) 73:67–181.

Tiffney, B.H. 1985a. Perspectives on the origin of the floristic similarity between eastern Asia and eastern North America. J. Arnold Arbor. 66:73–94.

Tiffney, B.H. 1985b. The Eocene North Atlantic land bridge: its importance in Tertiary and modern phytogeography of the Northern Hemisphere. J. Arnold Arbor. 66:243–273.

Tippo, O. and **W.L. Stern.** 1977. Humanistic botany. W.W. Norton & Co., Inc., New York.

Tölkin, H.R. 1977. A revision of the genus *Crassula* in southern Africa. Contr. Bolus Herb. 8:1–595.

Tomb, A.S. 1974. *Hypochoeris* in Texas. Sida 5:287–289.

Tomb, A.S. 1980. Taxonomy of *Lygodesmia* (Asteraceae). Syst. Bot. Monogr. 1:1–51.

Tô Ngô, L. and **W.L. Applequist.** 2023. *Amsonia*. In: Flora of North America Editorial Committee, eds. Fl. North Amer. 14:112–125. Oxford Univ. Press, New York and Oxford.

[TORCH] TORCH Portal. http//:portal.torcherbaria.org/portal/index.php.

Torres, A.M. 1963. Taxonomy of *Zinnia*. Brittonia 15:1–25.

Torres-Colín, L. and **A. Delgado-Salinas.** 2023. *Macroptilium*. In: Flora of North America Editorial Committee, eds. Fl. of North Amer. 11:426–429. Oxford Univ. Press, New York and Oxford.

Torrey, J. 1859. Botany. In: W.H. Emory, Report on the United States and Mexican boundary survey: Made under the direction of the secretary of the Interior. Wendell, Washington, DC.

Townsend, C.C. 1993. Amaranthaceae. In: K. Kubitzki, J.C. Rohwer, and V. Bittrich, eds. The families and genera of vascular plants, Vol. II. Pp. 70–91. Springer-Verlag, Berlin.

Toxic Plant Database. 2000. Toxic Plant Database. http://texnat.tamu.edu/cmplants/toxic/plants/oleander.html, accessed Dec. 2006.

Tripp, E.A. 2007. Evolutionary relationships within the species-rich genus *Ruellia* (Acanthaceae). Syst. Bot. 32:628–649.

Trock, D.K. 2006a. *Achillea*. In: Flora of North America Editorial Committee, eds. Fl. North Amer. 19:492–494. Oxford Univ. Press, New York and Oxford.

Trock, D.K. 2006b. *Packera*. In: Flora of North America Editorial Committee, eds. Fl. North Amer. 20:570–602. Oxford Univ. Press, New York and Oxford.

Trock, D.K. 2023. *Prosopis*. In: Flora of North America Editorial Committee, eds. Fl. of North Amer. 11:76–79. Oxford Univ. Press, New York and Oxford.

Trusty, J.L., B.G. Lockaby, W.C. Zipperer, and **L.R. Goertzen.** 2007. Identity of naturalized exotic *Wisteria* (Fabaceae) in the southeastern United States. Weed Res. 47:479–487.

Tucker, A.O., N.H. Dill, T.D. Pizzolato, and **R.D. Kral.** 1983. Nomenclature, distribution, chromosome numbers, and fruit morphology of *Oxypolis canbyi* and *O. filiformis* (Apiaceae). Syst. Bot. 8:299–304.

Tucker, G.C. 1986. The genera of Elatinaceae in the southeastern United States. J. Arnold Arbor. 67:471–483.

Tucker, G.C. 2009a. Ericaceae. In: Flora of North America Editorial Committee, eds. Fl. North Amer. 8:370–535. Oxford Univ. Press, New York and Oxford.

Tucker, G.C. 2009b. *Eubotrys*. In: Flora of North America Editorial Committee, eds. Fl. North Amer. 8:510–512. Oxford Univ. Press, New York and Oxford.

Tucker, G.C. 2009c. *Leucothoë*. In: Flora of North America Editorial Committee, eds. Fl. North Amer. Vol. 8:508–510. Oxford Univ. Press, New York and Oxford.

Tucker, G.C. 2010a. *Arivela*. In: Flora of North America Editorial Committee, eds. Fl. North Amer. 7:221–222. Oxford Univ. Press, New York and Oxford.

Tucker, G.C. 2010b. *Gynandropsis*. In: Flora of North America Editorial Committee, eds. Fl. North Amer. 7:222–223. Oxford Univ. Press, New York and Oxford.

Tucker, G.C. 2010c. *Polanisia*. In: Flora of North America Editorial Committee, eds. Fl. North Amer. 7:201–204. Oxford Univ. Press, New York and Oxford.

Tucker, G.C. 2016. *Bergia*. In: Flora of North America Editorial Committee, eds. Fl. North Amer. 12:349. Oxford Univ. Press, New York and Oxford.

Tucker, G.C. and **R.J. Alier.** 2023. *Glycine*. In: Flora of North America Editorial Committee, eds. Fl. of North Amer. 11:401–402. Oxford Univ. Press, New York and Oxford.

Tucker, G.C. and **S.C. Jones.** 2009. Clethraceae. In: Flora of North America Editorial Committee, eds. Fl. North Amer. Vol. 8:364–366. Oxford Univ. Press, New York and Oxford.

Tucker, G.C. and **S.S. Vanderpool.** 2010. Cleomaceae. In: Flora of North America Editorial Committee, eds. Fl. North Amer. 7:199–223. Oxford Univ. Press, New York and Oxford.

Tucker, S.C. 1988. Heteromorphic flower development in *Neptunia pubescens*, a mimosoid legume. Amer. J. Bot. 75:205–224.

Tull, D. 1987. Edible and useful plants of Texas and the Southwest. Univ. of Texas Press, Austin.

Tull, D. 2013. Edible and useful plants of the Southwest: Texas, New Mexico, and Arizona. Univ. of Texas Press, Austin.

Turland, N.J. and **M.W. Jackson.** 1997. Proposals to reject the names *Cerastium viscosum* and *C. vulgatum* (Caryophyllaceae). Taxon 46:775–778.

Turner, B.L. 1950. *Lobelia reverchoni* Turner, nom. nov. Field & Lab. 18:46.

Turner, B.L. 1950a. Vegetative key to Texas *Desmanthus* (Leguminosae) and similar genera. Field & Lab. 18:51–54.

Turner, B.L. 1950b. Texas species of *Desmanthus* (Leguminosae). Field & Lab. 18:54–65.

Turner, B.L. 1951. Revision of the United States species of *Neptunia* (Leguminosae). Amer. Midl. Naturalist 46:82–92.

Turner, B.L. 1956. A cytotaxonomic study of the genus *Hymenopappus* (Compositae). Rhodora 58:163–186, 208–242, 250–269, 295–308.

Turner, B.L. 1957. The chromosomal and distributional relationships of *Lupinus texensis* and *L. subcarnosus* (Leguminosae). Madroño 14:13–16.

Turner, B.L. 1959. The legumes of Texas. Univ. of Texas Press, Austin.

Turner, B.L. 1978. Taxonomic study of the scapiform species of *Acourtia* (Asteraceae-Mutisieae). Phytologia 38:456–468.

Turner, B.L. 1983. The Texas species of *Paronychia* (Caryophyllaceae). Phytologia 54:9–23.

Turner, B.L. 1984. Taxonomy of the genus *Aphanostephus* (Asteraceae-Astereae). Phytologia 56:81–101.

Turner, B.L. 1988a. Comments upon, and new combinations in *Heliopsis* (Asteraceae, Heliantheae). Phytologia 64:337–340.

Turner, B.L. 1988b. A new variety of *Sclerocarpus uniserialis* (Asteraceae, Heliantheae) from southernmost Texas. Phytologia 64:341–343.

Turner, B.L. 1988c. A new species of, and observations on, the genus *Smallanthus* (Asteraceae-Heliantheae). Phytologia 64: 405–407.

Turner, B.L. 1988d. New species and combinations in *Wedelia* (Asteraceae-Heliantheae). Phytologia 65:348–358.

Turner, B.L. 1989. An overview of the *Brickellia* (Kuhnia) *eupatorioides* (Asteraceae, Eupatorieae) complex. Phytologia 67:121–131.

Turner, B.L. 1991a. Texas species of *Ruellia* (Acanthaceae). Phytologia 71:281–299.

Turner, B.L. 1992. New species of *Wedelia* (Asteraceae, Heliantheae) from Mexico and critical assessment of previously described taxa. Phytologia 72:115–126.

Turner, B.L. 1993. New taxa, new combinations and nomenclatural notes in the genus *Acourtia* (Asteraceae, Mutisieae). Phytologia 74:385–413.

Turner, B.L. 1994d. Texas species of *Schrankia* (Mimosaceae) transferred to the genus *Mimosa*. Phytologia 76:412–420.

Turner, B.L. 1994e. Northern Mexican species of *Schrankia* (Mimosaceae) transferred to *Mimosa*. Phytologia 76:421–425.

Turner, B.L. 1995a. Synopsis of the genus *Onosmodium* (Boraginaceae). Phytologia 78:39–60.

Turner, B.L. 1995d [1996]. *Cerastium texanum* (Caryophyllaceae) does not occur in Texas. Phytologia 79:356–363.

Turner, B.L. 1996. Synoptical study of *Rhus virens* (Anacardiaceae) and closely related taxa. Phytologia 80:368–376.

Turner, B.L. 2000. The Texas species of *Calliandra* (Leguminosae, Mimosoideae). Lundellia 3:13–18.

Turner, B.L. 2003. New varietal and species combinations for taxa occurring in Texas. In: B.L. Turner, H. Nichols, G. Denny, and O. Doron. Atlas of the vascular plants of Texas, Vol. 1: Dicots. Pp. 4–8. Sida Bot. Misc. 24.

Turner, B.L. 2006a. Overview of the genus *Baptisia* (Leguminosae). Phytologia 88:253–268.

Turner, B.L. 2006b. Taxonomy and nomenclature of the *Erysimum asperum-E. capitatum* complex (Brassicaceae). Phytologia 88: 279–287.

Turner, B.L. 2007. Biological status and distribution of *Thelesperma flavodiscum* (Asteraceae: Coreopsideae). Phytologia 89(3): 290–292.

Turner, B.L. 2009. Taxonomy of *Asclepias hirtella* and *A. longifolia* (Apocynaceae). Phytologia 91(2):308–311,

Turner, B.L. 2009a. Taxonomy of *Iva angustifolia* and *I. asperifolia* (Asteraceae). Phytologia 91(1):76–83.

Turner, B.L. 2009b. *Iva corbinii* (Asteraceae): A remarkable new species from Travis County, Texas. Lundellia 12:5–7.

Turner, B.L. 2009c. Biological status of the varietal taxa of *Thymophylla pentachaeta* (Asteraceae: Tageteae). Phytologia 91(2):340–346.

Turner, B.L. 2010. A new species of *Dalea* (Fabaceae) from southernmost Texas. Phytoneuron 2010-50:1–5.

Turner, B.L. 2013. Taxonomy of *Tecoma stans* (Bignoniaceae) in North America. Phytologia 95:222–225. https://www.phytologia.org/uploads/2/3/4/2/23422706/953222-225turnertecoma.pdf.

Turner, B.L. 2013a. Taxonomy of the *Dalea phleoides* (Fabaceae) complex. Phytologia 95(4):274–278.

Turner, B.L. (with annotations by G. Nesom). 2016. Asteraceae of Texas. Unpublished Apr 2016 Draft. Copy in the library of the Botanical Research Institute of Texas.

Turner, B.L. 2023a. *Baptisia*. In: Flora of North America Editorial Committee, eds. Fl. of North Amer. 11:177–184. Oxford Univ. Press, New York and Oxford.

Turner, B.L. 2023b. *Orbexilum*. In: Flora of North America Editorial Committee, eds. Fl. of North Amer. 11:467–470. Oxford Univ. Press, New York and Oxford.

Turner, B.L. and **R. Alston.** 1959. Segregation and recombination of chemical constituents in a hybrid swarm of *Baptisia laevicaulis* × *B. viridis* and their taxonomic implications. Amer. J. Bot. 46: 678–686.

Turner, B.L. and **J. Andrews.** 1986. Lectotypification of *Lupinus subcarnosus and L. texensis* (Fabaceae). Sida 11:255–257.

Turner, B.L. and **D. Dawson.** 1980. Taxonomy of *Tetragonotheca* (Asteraceae-Heliantheae). Sida 8:296–303.

Turner, B.L. and **O.S. Fearing.** 1964. A taxonomic study of the genus *Amphicarpaea* (Leguminosae). S. W. Naturalist 9:207–218.

Turner, B.L. and **D.B. Horne.** 1964. Taxonomy of *Machaeranthera* sect. *Psilactis* (Compositae–Astereae). Brittonia 16:316–331.

Turner, B.L. and **M.C. Johnston.** 1956. Chromosome numbers and geographic distribution of *Lindheimera*, *Engelmannia*, and *Berlandiera* (Compositae-Heliantheae-Melampodinae). S. W. Naturalist 1:125–132.

Turner, B.L. and **K.-J. Kim.** 1990. An overview of the genus *Pyrrhopappus* (Asteraceae: Lactuceae) with emphasis on chloroplast DNA restriction site data. Amer. J. Bot. 77:845–850.

Turner, B.L. and **R.M. King.** 1962. A cytotaxonomic survey of *Melampodium* (Compositae-Heliantheae). Amer. J. Bot. 49: 263–269.

Turner, B.L. and **M.I. Morris.** 1976. Systematics of *Palafoxia* (Asteraceae: Helenieae). Rhodora 78:567–628.

Turner, B.L., H. Nichols, G. Denny, and **O. Doron.** 2003. Atlas of the vascular plants of Texas, Vol. 1: Dicots, Vol. 2: Ferns, gymnosperms, monocots. Sida Bot. Misc. 24.

Turner, B.L., M.W. Turner, and **J.C. Crutchfield.** 1988. Populational analyses and new combinations in *Psilostrophe tagetina* and *P. gnaphalodes* (Asteraceae, Heliantheae). Phytologia 65:231–240.

Turner, B.L. and **T. Watson.** 2007. Taxonomic revision of *Gaillardia* (Asteraceae) Phytologia Mem. 13.

Turner, B.L. and **M. Whalen.** 1975. Taxonomic study of *Gaillardia pulchella* (Asteraceae-Heliantheae). Wrightia 5:189–192.

Turner, B.L. and **L. Woodruff.** 1993. Annotated distribution of the monotypic genus *Lindheimera* (Asteraceae: Heliantheae). Sida 15:533–537.

Turner, N.W. and **A.F. Szczawinski.** 1991. Common poisonous plants and mushrooms of North America. Timber Press, Portland, OR.

Tutin, T.G. 1976. *Soliva* and *Gymnostyles*. In: T.G. Tutin, V.H. Heywood, N.A. Burges, D.M. Moore, D.H. Valentine, S.M. Walters, and D.A. Webb, eds. Flora Europaea 4:178. Cambridge Univ. Press, Cambridge, England, UK.

Tutin, T.G., V.H. Heywood, N.A. Burges, D.M. Moore, D.H. Valentine, S.M. Walters, and **D.A. Webb.** 1968. Flora Europaea. Vol. 2, Rosaceae to Umbelliferae. Cambridge Univ. Press, Cambridge, England.

Tveten, J.L. and **G.A. Tveten.** 1993. Wildflowers of Houston and southeast Texas. Univ. Texas Press, Austin.

Tyrl, R.J. 1975. Origin and distribution of polyploid *Achillea* (Compositae) in western North America. Britonnia 27:187–196.

Tyrl, R.J., S.C. Barber, P. Buck, J.R. Estes, P. Folley, L.K. Magrath, C.E.S. Taylor, and **R.A. Thompson.** (Oklahoma Flora Editorial Committee). 1994 (revised 1 Sep 1997). Key and descriptions for the vascular plant families of Oklahoma. Flora Oklahoma Incorporated, Noble, OK.

Uotila, P., A.P. Sukhorukov, N. Bobon, J. McDonald, A.A. Krinitsina, and **G. Kadereit.** 2021. Phylogeny, biogeography and systematics of Dysphanieae (Amaranthaceae). Taxon 70:526–551.

Urbatsch, L.E. 1972. Systematic study of the *Altissimae* and *Giganteae* species groups of the genus *Vernonia* (Compositae). Brittonia 24:229–238.

Urbatsch, L.E. and **P.B. Cox.** 2006. *Rudbeckia*. In: Flora of North America Editorial Committee, eds. Fl. North Amer. 21:44–60. Oxford Univ. Press, New York and Oxford.

Urbatsch, L.E. and **N.K. Fischer.** 1989. The subtribal affinities of the genus *Tetragonotheca* (Asteraceae: Heliantheae). Phytologia 67:405–416.

Urbatsch, L.E. and **R.K. Jansen.** 1995. Phylogenetic affinities among and within the coneflower genera (Asteraceae, Heliantheae), a chloroplast DNA analysis. Syst. Bot. 20:28–39.

Urbatsch, L.E, K.M. Neubig, and **P.B. Cox.** 2006. *Echinacea*. In: Flora of North America Editorial Committee, eds. Fl. North Amer. 21: 88–92. Oxford Univ. Press, New York and Oxford.

Urshel, S. (Susan). Personal communication. Professional gardener and horticulturalist, Fort Worth, TX.

Urtecho, R.J. 2016. *Tragia*. In: Flora of North America Editorial Committee, eds. Fl. North Amer. 12:184–191. Oxford Univ. Press, New York and Oxford.

USDA APHIS. 2017. Weed risk assessment for *Myagrum perfoliatum* L. (Brassicaceae)—Bird's-eye cress. USDA APHIS, Raleigh, NC. The PLANTS Database. National Plant Data Team, Greensboro, NC.

USDA Natural Resources Conservation Service. 2006. Federal noxious weeds. http://plants.nrcs.usda.gov/cgi_bin/topics.cgi?earl=noxious.cgi.

USDA Natural Resources Conservation Service. 2008. Invasive and noxious weeds. State noxious weeds. http://plants.usda.gov/java/noxComposite?stateRpt=yes.

USDA NRCS. 2021. The PLANTS Database. National Plant Data Team, Greensboro, NC. https://plants.sc.egov.usda.gov/. (Accessed 2019–2021 for Brassicaceae)

USDA Plants Database. 2006. USDA Natural Resources Conservation Service Plants Database. http://plants.usda.gov/.

USDA Plants Database. 2007. USDA Natural Resources Conservation Service Plants Database. http://plants.usda.gov/.

USDA Plants Database. 2008. USDA Natural Resources Conservation Service Plants Database. http://plants.usda.gov/.

USDA Plants Database. 2016. USDA Natural Resources Conservation Service Plants Database. http://plants.usda.gov/.

USDA Plants Database. 2019. USDA Natural Resources Conservation Service Plants Database. https://plants.usda.gov/.

USDA Plants Database. 2023. USDA Natural Resources Conservation Service Plants Database. http://plants.usda.gov/.

USDA Plants Database. 2024. *Mahonia* Nutt. https://plants.usda.gov/plant-profile/MAHON. Accessed 27 January 2024.

USDA Plants Database. 2025. USDA Natural Resources Conservation Service Plants Database. https://plants.usda.gov/.

Uttal, L.J. 1987. The genus *Vaccinium* L. (Ericaceae) in Virginia. Castanea 52:231–255.

Vail, A.M. 1895. A study of the genus *Galactia* in North America. Bull. Torrey Bot. Club 22:500–511.

Valcárcel, V., O. Fiz-Palacios, and **J. Wen.** 2014. The origin of the early differentiation of ivies (*Hedera* L.) and the radiation of the Asian palmate group (Araliaceae). Molec. Phylogen. Evol. 70:492–503.

Valder, P. 1995. Wisterias: A comprehensive guide. Timber Press, Portland, OR.

Vallejo-Marín, M. 2018. Buzz pollination: Studying bee vibrations on flowers. New Phytol. 224(3):1068–1074. https://doi: 10.1111/nph.15666.

Vander Kloet, S.P. 1988. The genus *Vaccinium* in North America. Research Branch Agric. Canada Publ. No. 1828.

Vander Kloet, S.P. 1992. On the etymology of *Vaccinium* L. Rhodora 94:371–373.

Vander Kloet, S.P. 2009. *Vaccinium*. In: Flora of North America Editorial Committee, eds. Fl. North Amer. 8:515–530. Oxford Univ. Press, New York and Oxford.

van der Maesen, L.J.G. 2023. *Pueraria*. In: Flora of North America Editorial Committee, eds. Fl. of North Amer. 11:397–398. Oxford Univ. Press, New York and Oxford.

Vanderpool, S.S. 2010. *Cleomella*. In Flora of North America Editorial Committee, eds. Fl. North Amer. 7:209–213. Oxford Univ. Press, New York and Oxford.

Van Ee, B.W. and **P.E. Berry.** 2016. *Croton*. In: Flora of North America Editorial Committee, eds. Fl. North Amer. 12:206–224. Oxford Univ. Press, New York and Oxford.

Van Horn, G.S. 1973. The taxonomic status of *Pentachaeta* and *Chaetopappa* with a revision of *Pentachaeta*. Univ. Calif. Publ. Bot. 65:1–41.

Van Ooststroom, S.J. 1934. A monograph of the genus *Evolvulus*. Meded. Bot. Mus. Herb. Rijks Univ. Utrecht 14:1–267.

Van Vleet, R.L. 1951. Phenotypic variations of *Erigeron strigosus* Muhl. (Compositae) in eastern Texas. Field & Lab. 19:161–163.

Van Welzen, P.C. and **W. Stuppy.** 1999. Phylogenetic considerations of Euphorbiaceae Tribe Aleuritideae. Ann. Missouri Bot. Gard. 86: 894–903.

Vargas, P., H.A. McAllister, C. Morton, S.L. Jury, and **M.J. Wilkinson.** 1999. Polyploid speciation in *Hedera* (Araliaceae): phylogenetic and biogeographic insights based on chromosome counts and ITS sequences. Pl. Syst. Evol. 219:165–179.

Venter, H.J.T. and **R.L. Verhoeven.** 1997. A tribal classification of the Periplocoideae (Apocynaceae). Taxon 46:705–720.

Verdcourt, B. 1997. (1298-1299) Proposals to conserve the names *Phaseolus helvolus* with a conserved type and spelling, and *Strophostyles* against *Phasellus* (Leguminosae). Taxon 46(2): 357–359.

Verkerk, R., M. Schreiner, A. Krumbein, E. Ciska, B. Holst, I. Rowland, R. de Schrijver, M. Hansen, C. Gerhäuser, R. Mithen, and **M. Dekker.** 2009. Glucosinolates in *Brassica* vegetables: the influence of the food supply chain on intake, bioavailability and human health. Molec. Nutr. Food Res. 53(S2): S219–S265.

Vernon, L. 1965. Biosystematic studies in the *Heterotheca subaxillaris* complex (Compositae: Astereae). Trans. Kansas Acad. Sci. 68:244–257.

Vincent, M.A. 2023. Fabaceae. In: Flora of North America Editorial Committee, eds. Fl. of North Amer. Vol. 11, Pp. 1108. Oxford Univ. Press, New York and Oxford.

Vincent, M.A. and **D.M. Kearns.** 2023. *Dermatophyllum*. In: Flora of North America Editorial Committee, eds. Fl. of North Amer. 11:162–164. Oxford Univ. Press, New York and Oxford.

Vincent, M.A. and **J.M. Gillett.** 2023. *Trifolium*. In: Flora of North America Editorial Committee, eds. Fl. of North Amer. 11:914–974. Oxford Univ. Press, New York and Oxford.

Vincent, M.A. Basellaceae. 2003. In: Flora of North America Editorial Committee, eds. Fl. North Amer. 4:505–508. Oxford Univ. Press, New York and Oxford.

Vines, R.A. 1960. Trees, shrubs and woody vines of the southwest. Univ. of Texas Press, Austin.

Visser, S., R.M. Danielson, and **D. Parkinson.** 1991. Field performance of *Elaeagnus commutata* and *Sheperdia canadensis* (Elaeagnaceae) inoculated with soil containing *Frankia* and vesicular-arbuscular mycorrhizal fungi. Canad. J. Bot. 69: 1321–1328.

Vivrette, N.J., J.E. Bleck, and **W.R. Ferren, Jr.** 2003. Aizoaceae. In: Flora of North America Editorial Committee, eds. Fl. North Amer. 4:75–91. Oxford Univ. Press, New York and Oxford.

Volponi, C.R. 1983. Sinopsis de las especies Argentinas de *Stellaria*. Lilloa 36:69–75.

Vuilleumier, B.S. 1969a. The genera of Senecioneae in the southeastern United States. J. Arnold Arb. 50:104–121.

Vuilleumier, B.S. 1969b. The tribe Mutisieae (Compositae) in the southeastern United States. J. Arnold Arbor. 50:620–625.

Vuilleumier, B.S. 1973. The genera of Lactuceae (Compositae) in the southeastern United States. J. Arnold Arbor. 54:42–93.

Wächter, G.A., J.J. Hoffmann, T. Furbacher, M.E. Blake, and **B.N. Timmermann.** 1999. Antibacterial and antifungal flavanones from *Eysenhardtia texana*. Phytochemistry 52:1469–1471.

Wagenknecht, B.L. 1960. Revision of *Heterotheca* section *Heterotheca*. Rhodora 62:61–76, 97–107.

Wagner, F.H. 1948. The bur clovers (*Medicago*) of Texas. Field & Lab. 16:3–7.

Walker, J.F., Y. Yang, T. Feng, and **18 others.** 2018. From cacti to carnivores: Improved phylotranscriptomic sampling and hierarchical homology inference provides further insight to the evolution of Caryophyllales. Amer. J. Bot. 105:446–462.

Wallace, G.D. 1975. Studies of the Monotropoideae (Ericaceae): Taxonomy and distribution. Wasmann J. Biol. 33:1–88.

Wallace, G.D. 1977. Studies of the Monotropoideae (Ericaceae). Floral nectaries: Anatomy and function in pollination ecology. Amer. J. Bot. 64:199–206.

Wallace, G.D. 1993. Ericaceae. In: J.C.Hickman, ed. The Jepson manual: Higher plants of California. Pp. 544–567. Univ. of California Press, Berkeley.

Wallace, G.D. 1995. Ericaceae subfamily Monotropoideae. In: J.L. Luteyn, ed. Ericaceae Part II, the superior-ovaried genera. Fl. Neotrop. Monogr. 66:13–27.

Wallace, G.D. 2009. *Monotropa*. In: Flora of North America Editorial Committee, eds. Fl. North Amer. 8:392–393. Oxford Univ. Press, New York and Oxford.

Wallace, R.S. 1995. Molecular systematic study of the Cactaceae: Using chloroplast DNA variation to elucidate cactus phylogeny. Bradleya 13:1–12.

Wallnöfer, B. 2001. The biology and systematics of Ebenaceae: A review. Ann. Naturhist. Mus. Wien, B 103:485–512.

Wallnöfer, B. 2004. Ebenaceae. In: K. Kubitzki, ed. The families and genera of vascular plants, Vol. VI. Pp. 125–130. Springer-Verlag, Berlin.

Walters, T.W. and **R. Wyatt.** 1982. The vascular flora of granite outcrops in the Central Mineral Region of Texas. Bull. Torrey Bot. Club 109:344–364.

Wang, N., H.A. McAllister, P.R. Bartlett, and **R.J.A. Buggs.** 2016. Molecular phylogeny and genome size evolution of the genus *Betula* (Betulaceae). Ann. Bot. (Oxford) 117:1023–1035.

Wang, W., Z.-D. Chen, Y. Liu, R.-Q. Li, and **J.-H. Li.** 2007. Phylogenetic and biogeographic diversification of Berberidaceae in the northern hemisphere. Syst. Bot. 32:731–742.

Wang, W., A.-M. Lu, Y. Ren, M.E. Endress, and **Z.-D. Chen.** 2009. Phylogeny and classification of Ranunculales: Evidence from four molecular loci and morphological data. Perspect. Pl. Ecol. Evol. Syst. 11:81–110.

Wang, X. 2023. Origin of angiosperms: Problems, challenges, and solutions. Life (Basel, Switzerland). 13(10):2029. https://doi: 10.3390/life13102029.

Wang, Y.-C., D.-C. Wu, J.-J. Liao, C.-H. Wu, W.-Y. Li, and **B.-C. Weng.** 2009. In vitro activity of *Impatiens balsamina* L. against multiple antibiotic-resistant *Helicobacter pylori*. Am. J. Chin. Med. 37:713–22. https://doi:10.1142/S0192415X09007181.

Wanke, S., F. González, and **C. Neinhuis.** 2006. Systematics of pipevines: Combining morphological and fast-evolving molecular characters to investigate the relationships within subfamily Aristolochioideae (Aristolochiaceae). Int. J. Pl. Sci.167:1215–1227. https://doi.org/10.1086/508024.

Wanke, S., M. Jaramillo, T. Borsch, T. Samain, M. Quandt, and **C. Neinhuis.** 2007. Evolution of Piperales—matK gene and trnK intron sequence data reveal lineage specific resolution contrast. Molec. Phylogen. Evol. 42:477–497. https://doi.org/10.1016/j.ympev.2006.07.007.

Wannan, B.S. 2006. Analysis of generic relationships in Anacardiaceae. Blumea 51:165–195.

Ward, D.B. 1974. Contributions to the flora of Florida: 6, *Vaccinium* (Ericaceae). Castanea 39:191–205.

Ward, D.B. 1978. Keys to the flora of Florida—7, Campanulaceae. Phytologia 39:1–12.

Ward, D.B. 1998. *Pueraria montana*: The correct scientific name of the kudzu. Castanea 63:76–77.

Warwick, S.I. 2010a. *Brassica*. In: Flora of North America Editorial Committee, eds. Fl. North Amer. 7:419–424. Oxford Univ. Press, New York and Oxford.

Warwick, S.I. 2010b. *Conringia*. In: Flora of North America Editorial Committee, eds. Fl. North Amer. 7:517. Oxford Univ. Press, New York and Oxford.

Warwick, S.I. 2010c. *Eruca*. In: Flora of North America Editorial Committee, eds. Fl. North Amer. 7:434–435. Oxford Univ. Press, New York and Oxford.

Warwick, S.I. 2010d. *Erucastrum*. In: Flora of North America Editorial Committee, eds. Fl. North Amer. 7:435. Oxford Univ. Press, New York and Oxford.

Warwick, S.I. 2010e. *Raphanus*. In: Flora of North America Editorial Committee, eds. Fl. North Amer. 7:438–440. Oxford Univ. Press, New York and Oxford.

Warwick, S.I. 2010f. *Rapistrum*. In: Flora of North America Editorial Committee, eds. Fl. North Amer. 7:440–441. Oxford Univ. Press, New York and Oxford.

Warwick, S.I. 2010g. *Sinapis*. In: Flora of North America Editorial Committee, eds. Fl. North Amer. 7:441–443. Oxford Univ. Press, New York and Oxford.

Warwick, S.I., A. Francis, and **D.J. Susko.** 2002. The biology of Canadian weeds. 9. *Thlaspi arvense* L.(updated). Canad. J. Pl. Sci. 82(4):803–823.

Warwick, S.I., K. Mummenhoff, C.A. Sauder, M.A. Koch, and **I.A. Al-Shehbaz.** 2010. Closing the gaps: Phylogenetic relationships in the Brassicaceae based on DNA sequence data of nuclear ribosomal ITS region. Pl. Syst. Evol. 285:209–232.

Waselkov, K. and **W.S. Judd.** 2008. A phylogenetic analysis of *Leucothoë* s.l. (Ericaceae; tribe Gaultherieae) based on phenotypic characters. Brittonia 60:382–394.

Wasshausen, D.C. 1966. Acanthaceae. In: C.L. Lundell, ed. Flora of Texas 1:223–282. Texas Research Foundation, Renner.

Wasshausen, D.C. 1998 Acanthaceae of the southeastern United States. Castanea 63:99–116.

Wasshausen, D.C. and **J.R.I. Wood.** 2003. The genus *Dyschoriste* (Acanthaceae) in Bolivia and Argentina. Brittonia 55:10–18.

Wasshausen, D.C. and **J.R.I. Wood.** 2004. Acanthaceae of Bolivia. Contr. U.S. Natl. Herb. 49:1–152.

Waterfall, U.T. 1971 [1972]. New taxa, combinations and distribution records for the Oklahoma flora. Rhodora 73:552–555.

Waterfall, U.T. 1972. Keys to the flora of Oklahoma, 5th ed. Published by the author for sale by the Student Union Bookstore, Oklahoma State Univ., Stillwater.

Watson, G. (Geraldine). 1925–2012. Personal communication. Botanist and artist who was instrumental in the formation of the Big Thicket National Preserve.

Watson, L.E. 2006a. *Anthemis*. In: Flora of North America Editorial Committee, eds. Fl. North Amer. 19:537–538. Oxford Univ. Press, New York and Oxford.

Watson, L.E. 2006b. *Cotula*. In: Flora of North America Editorial Committee, eds. Fl. North Amer. 19:543–544. Oxford Univ. Press, New York and Oxford.

Watson, L.E. 2006c. *Marshallia*. In: Flora of North America Editorial Committee, eds. Fl. North Amer. 21:456–458. Oxford Univ. Press, New York and Oxford.

Watson, L.E. 2006d. *Soliva*. In: Flora of North America Editorial Committee, eds. Fl. North Amer. 19:545–546. Oxford Univ. Press, New York and Oxford.

Watson, L.E. and **J.E. Estes.** 1990. Biosystematic and phenetic analysis of *Marshallia* (Asteraceae). Syst. Bot. 15:403–414.

Weakley, A.S. 2005. Flora of the Carolinas, Virginia, and Georgia. http://www.herbarium.unc.edu/FloraArchives/WeakleyFlora_2005-Jun.pdf. Accessed Jul 2008.

Weakley, A.S. 2015. Flora of the southern and mid-Atlantic states. Working draft of May 2015. University of North Carolina Herbarium, North Carolina Botanical Garden, Chapel Hill, U.S.A.

Weakley, A.S. 2018. Three southern blue baptisia species. In A. S. Weakley, D.B. Poindexter, R.J. LeBlond, B.A. Sorrie, E.L. Bridges, S.L. Orzell, A.R. Franck, M. Schori, B.R. Keener, A.R. Diamond, Jr., A.J. Floden, and R.D. Noyes. New combinations, rank changes, and nomenclatural and taxonomic comments in the vascular flora of the southeastern United States. III. J. Bot. Res. Inst. Texas 12:27–67.

[Weakley et al. 2015.] Weakley, A.S., and **Southeastern flora Team.** 2015. Flora of the southeastern United States. University of North Carolina Herbarium, North Carolina Botanical Garden, Chapel Hill, U.S.A.

Weakley, A.S., and **Southeastern flora Team.** 2020. Flora of the southeastern United States. University of North Carolina Herbarium, North Carolina Botanical Garden, Chapel Hill, U.S.A.

Weakley, A.S., and **Southeastern Flora Team.** 2022. Flora of the southeastern United States. University of North Carolina Herbarium, North Carolina Botanical Garden, Chapel Hill, U.S.A.

Weakley, A.S., and **Southeastern flora Team.** 2024. Flora of the southeastern United States. University of North Carolina Herbarium, North Carolina Botanical Garden, Chapel Hill, U.S.A.

Weakley, A.S., and **Southeastern Flora Team.** 2025. Flora of the southeastern United States Web App. University of North Carolina Herbarium, North Carolina Botanical Garden, Chapel Hill, U.S.A.

Weakley, A.S., R. LeBlond, B. Sorrie, C. Witsell, L. Estes, K. Gandhi, K. Mathews, and **A. Ebihara.** 2011. New combinations, rank changes, and nomenclatural and taxonomic comments in the vascular flora of the southeastern United States. J. Bot. Res. Inst. Texas 5:437–455.

Weakley, A.S., J.C. Ludwig, and **J.F. Townsend.** 2012. Flora of Virginia. Bland Crowder, ed. Foundation of the Flora of Virginia Project Inc., Richmond. BRIT Press, Fort Worth, TX.

Weatherby, C.A. 1927. The group of *Acalphya virginica* in eastern North America. Rhodora 29:193–204.

Weber, W.A. 1985. The genus *Teloxys* (Chenopodiaceae). Phytologia 58:477–478.

Webster, G.L. 1967. The genera of Euphorbiaceae in the southeastern United States. J. Arnold Arbor. 48:303–430.

Webster, G.L. 1992. Realignments in American *Croton* (Euphorbiaceae). Novon 2:269–273.

Webster, G.L. 1993. A provisional synopsis of the sections of the genus *Croton* (Euphorbiaceae). Taxon 42:793–823.

Webster, G.L. 1994a. Classification of the Euphorbiaceae. Ann. Missouri Bot. Gard. 81:3–32.

Webster, G.L. 1994b. Synopsis of the genera and suprageneric taxa of Euphorbiaceae. Ann. Missouri Bot. Gard. 81:33–144.

Webster, G.L. 2014. Euphorbiaceae. In: K. Kubitzki, ed. The families and genera of vascular plants, Vol. XI. Pp. 51–216. Springer-Verlag, Berlin, Heidelberg.

Weigend, M., F. Luebert, M. Gottschling, T.L.P. Couvreur, H.H. Hilger, and **J.S. Miller.** 2014. From capsules to nutlets—Phylogenetic relationships in the Boraginales. Taxon 30:508–518.

Wells, J.R. 1965. A taxonomic study of *Polymnia* (Compositae). Brittonia 17:144–159.

Welsh, S.L. 1960. Legumes of the north-central states: Galegeae. Iowa State J. Sci. 35:111–250.

Welsh, S.L. 1995. Names and types of perennial *Atriplex* Linnaeus (Chenopodiaceae) in North America selectively exclusive of Mexico. Great Basin Naturalist 55:322–334.

Welsh, S.L. 2003. *Atriplex*. In: Flora of North America Editorial Committee, eds. Fl. North Amer. 4:322–381. Oxford Univ. Press, New York and Oxford.

Welsh, S.L. 2023a. *Astragalus*. In: Flora of North America Editorial Committee, eds. Fl. of North Amer. 11:584–900. Oxford Univ. Press, New York and Oxford.

Welsh, S.L. 2023b. *Oxytropis*. In: Flora of North America Editorial

Committee, eds. Fl. of North Amer. 11:557–583. Oxford Univ. Press, New York and Oxford.

Welsh, S.L., C.W. Crompton, and **S.E. Clemants.** 2003. Chenopodiaceae (family description and key to genera). In: Flora of North America Editorial Committee, eds. Fl. North Amer. 4:258–261. Oxford Univ. Press, New York and Oxford.

Wemple, D.K. 1970. Revision of the genus *Petalostemon* (Leguminosae). Iowa State J. Sci. 45:1–102.

Wemple, D.K. and **N.R. Lersten.** 1966. An interpretation of the flower of *Petalostemum* (Leguminosae). Brittonia 18:117–126.

Wen, J. 1993. Generic delimitation of *Aralia* (Araliaceae). Brittonia 45:47–55.

Wen, J. 1999. Evolution of eastern Asian and eastern North American disjunct distributions in flowering plants. Annual Rev. Ecol. Syst. 30:421–455.

Wen, J. 2000. Internal transcribed spacer phylogeny of the Asian and eastern North American disjunct *Aralia* Sect. *Dimorphanthus* (Araliaceae) and its biogeographic implications. Int. J. Pl. Sci. 161: 959–966.

Wen, J. 2001a. Evolution of eastern Asian-North American biogeographic disjunctions: A few additional items. Int. J. Pl. Sci. 163(Suppl.):S117–S122.

Wen, J. 2001b. Evolution of the *Aralia-Panax* complex (Araliaceae) as inferred from nuclear ribosomal ITS sequences. Edinburgh J. Bot. 58:243–257.

Wen, J. 2011. Systematics and biogeography of *Aralia* L. (Araliaceae): Revision of *Aralia* sects. *Aralia, Humiles, Nanae*, and *Sciadodendron*. Contr. U.S. Natl. Herb. 57. Pp.172.

Wen, J. and **R.K. Jansen.** 1995. Morphological and molecular comparisons of *Campsis grandiflora* and *C. radicans* (Bignoniaceae), and eastern Asian and eastern North American vicariad pair. Pl. Syst. Evol. 196:173–183.

Wen, J., P.P. Lowry, II, J.L. Walck, and **K.-O. Yoo.** 2002. Phylogenetic and biogeographic diversification in *Osmorhiza* (Apiaceae). Ann. Missouri Bot. Gard. 89:414–428.

Wen, J., G.M. Plunkett, A.D. Mitchell, and **S.J. Wagstaff.** 2001. The evolution of Araliaceae: A phylogenetic analysis based on ITS sequences of nuclear ribosomal DNA. Syst. Bot. 26:144–167.

Wen, J. and **J.L. Reveal.** 1992. Typification of the Linnaean species of *Aralia* (Araliaceae). Taxon 41:69–75.

Wen, J., S. Shi, R.K. Jansen, and **E.A. Zimmer.** 1998. Phylogeny and biogeography of *Aralia* sect. *Aralia* (Araliaceae). Amer. J. Bot. 85: 866–875.

Wendling, B.M., K.E. Galbreath, and **E.G. DeChaine.** 2011. Resolving the evolutionary history of *Campanula* (Campanulaceae) in western North America. P.L.o.S. One. 6(9):e23559.

Weniger, D. 1970. Cacti of the southwest. Univ. of Texas Press, Austin.

Weniger, D. 1984. Cacti of Texas and neighboring states. Univ. of Texas Press, Austin.

Weniger, D. 1996. Catalpa (*Catalpa bignonioides*, Bignoniaceae) and bois d'arc (*Maclura pomifera*, Moraceae) in early Texas records. Sida 17:231–242.

Wheeler, L.C. 1936. Revision of the *Euphorbia polycarpa* group of the southwestern United States and adjacent Mexico; a preliminary treatment. Bull. Torrey Bot. Club 63:397–416.

Wheeler, L.C. 1941. *Euphorbia* subgenus *Chamaesyce* in Canada and the United States exclusive of southern Florida. Rhodora 43:97–154, 168–205, 223–286.

Wheeler, L.C. 1943. The genera of living Euphorbiaceae. Amer. Midl. Naturalist 30:456–503.

Whetstone, R.D., T.A. Atkinson, and **D.D. Spaulding.** 1997a. Berberidaceae (family description and key to genera). In: Flora of North America Editorial Committee, eds. Fl. North Amer. 3:272–273. Oxford Univ. Press, New York and Oxford.

Whetstone, R.D., T.A. Atkinson, and **D.D. Spaulding.** 1997b. *Nandina*. In: Flora of North America Editorial Committee, eds. Fl. North Amer. 3:273–274. Oxford Univ. Press, New York and Oxford.

Whitaker, T.W. and **W.P. Bemis.** 1964. Evolution in the genus *Cucurbita*. Evolution 18:553–559.

Whitaker, T.W. and **W.P. Bemis.** 1975. VIII. Origin and evolution of the cultivated *Cucurbita*. Bull. Torrey Bot. Club 102:362–368.

White, H.L., J.R. Branch, W.C. Holmes, and **J.R. Singhurst.** 1998. Comments on the distribution of *Sedum pulchellum* (Crassulaceae) in Texas. Sida 18:622–626.

White, M. 2008. *Asclepias hirtella* (Apocynaceae) newly documented for the flora of Texas. J. Bot. Res. Inst. Texas 2:1495–1496.

White, M. 2012. *Vernonia fasciculata* (Asteraceae): New to the flora of Texas. Phytoneuron 2012-90:1–6.

White, M. and **G.L. Nesom.** 2012. Two intersectional hybrids in *Liatris* (Asteraceae) from East Texas. Phytoneuron 2012-21:1–6.

White, M. and **J.R. Singhurst.** 2012. *Cerastium semidecandrum* (Caryophyllaceae) in Texas. Phytoneuron 2012-20:1–2.

White, P.S. 1984. The architecture of devil's walking stick, *Aralia spinosa* (Araliaceae). J. Arnold Arbor. 65:403–418.

Whittemore, A.T. 1997d. *Berberis*. In: Flora of North America Editorial Committee, eds. Fl. North Amer. 3:276–286. Oxford Univ. Press, New York and Oxford.

Whittemore, A.T. 2013. *Celtis*. In: G. Yatskievych. Steyermark's flora of Missouri. Vol 3. Pp. 1179–1184. Missouri Botanical Garden Press in cooperation with the Missouri Dept. of Conservation, St. Louis.

Whittemore, A.T. and **L.L. Gaddy.** 1997. *Hexastylis*. In: Flora of North America Editorial Committee, eds. Fl. North Amer. 3:45–50. Oxford Univ. Press, New York and Oxford.

Wiegand, K.M. 1923. Notes on *Triosteum perfoliatum* and related species. Rhodora 25:199–203.

Wiersema, J.H. 1997a. Nelumbonaceae. In: Flora of North America Editorial Committee, eds. Fl. North Amer. 3:64–65. Oxford Univ. Press, New York and Oxford.

Wiersema, J.H. 1997c. Cabombaceae. In: Flora of North America Editorial Committee, eds. Fl. North Amer. 3:78–80. Oxford Univ. Press, New York and Oxford.

Wikström, N., V. Savolainen, and **M.W. Chase.** 2001. Evolution of the angiosperms: Calibrating the family tree. Proc. Roy. Soc. Biol. Sci. Ser. B 268:2211–2220.

Wilbur, R.L. 1966. Notes on Rafinesque's species of *Lechea* (Cistaceae). Rhodora 68:192–208.

Wilbur, R.L. 1969. Cistaceae. In: C.L. Lundell, ed. Flora of Texas 2:1–17. Texas Research Foundation, Renner.

Wilbur, R.L. 1970. Taxonomic and nomenclatural observations on the eastern North American genus *Asimina* (Annonaceae). J. Elisha Mitchell Sci. Soc. 86:88–96.

Wilbur, R.L. 1975. A revision of the North American genus *Amorpha* (Leguminoseae–Psoraleae). Rhodora 77:337–409.

Wilbur, R.I. and **H.S. Daoud.** 1964. The genus *Helianthemum*

(Cistaceae) in the southeastern United States. J. Elisha Mitchell Sci. Soc. 80:38–43.

Wilbur, R.L. and **H.S. Daoud.** 1961. The genus *Lechea* (Cistaceae) in the southeastern United States. Rhodora 63:103–118.

Wildflowers of the southeastern U.S. 2008. Wildflowers of the southeastern U.S. Red clover (*Trifolium pratense*). http://www.2bnthewild.com/plants/H157.htm. Accessed Jul 2008.

Wilkinson, A.M. 1949. Floral anatomy and morphology of *Triosteum* and of the Caprifoliaceae in general. Amer. J. Bot. 36:481–489.

Willdenow, K.L. 1802. Species Plantarum. 3:1032.

Williams, B.R.M., T.C. Mitchell, J.R.I. Wood, D.J. Harris, R.W. Scotland, and **M.A. Carine.** 2014. Integrating barcode data in a monographic study of *Convolvulus*. Taxon 63:1287–1306.

Williams, C.R. 2019. New county records in Texas (U.S.A.) for the invasive aquatic plant *Hygrophila polysperma* (Acanthaceae). J. Bot. Res. Inst. Texas 13:349–353.

Williams, J.K. (Justin). Personal communication. Specialist on *Amsonia* (Apocynaceae) and floristics of the Texas panhandle; Professor of Biology, Sam Houston State Univ., Huntsville.

Williams, J.K. 1994. *Emilia fosbergii* (Asteraceae: Senecioneae), a new introduction to Texas. Sida 16(2):378–379.

Williams, J.K. 2019. Taxonomy of the *Amsonia tabernaemontana* complex (Apocynaceae: Rauvolfioideae). Phytoneuron 2019-13:1–7.

Williams, S.E., V.A. Albert, and **M.W. Chase.** 1994. Relationships of Droseraceae: A cladistic analysis of *rbc*L sequence and morphological data. Amer. J. Bot. 81:1027–1037.

Williamson, P.S. and **E.L. Schneider.** 1993a. Cabombaceae. In: K. Kubitzki, J.C. Rohwer, and V. Bittrich, eds. The families and genera of vascular plants, Vol. II. Pp. 157–161. Springer-Verlag, Berlin.

Wills, M.M. and **H.S. Irwin.** 1961. Roadside flowers of Texas. Univ. of Texas Press, Austin.

Willson, M.F. and **D.W. Schemske.** 1980. Pollinator limitation, fruit production, and floral display in pawpaw (*Asimina triloba*). Bull. Torrey Bot. Club 107:401–408.

Wilson, B.L. 2024. *Tauschia*. In: Flora of North America Editorial Committee, eds. Fl. North Amer. 13:503–507. Oxford Univ. Press, New York and Oxford.

Wilson, B.L., K.M. Mason, N. Otting, and **M.E. Darrach.** 2024. *Lomatium*. In: Flora of North America Editorial Committee, eds. Fl. North Amer. 13:345–441. Oxford Univ. Press, New York and Oxford.

Wilson, H.D. 1990. Quinua and relatives (*Chenopodium* sect. *Chenopodium* subsect. *Cellulata*). Econ. Bot. 44(Suppl.):92–110.

Wilson, J.S. 1964 [1965]. Variation of three taxonomic complexes of the genus *Cornus* in eastern United States. Trans. Kansas Acad. Sci. 67:747–817.

Wilson, K.A. 1960c. The genera of Convolvulaceae in the southeastern United States. J. Arnold Arbor. 41:298–317.

Wilson, K.L. 2017. Report of the General Committee: 20 [*Salsola* L. conserved]. Taxon 66(4):981.

Wilson, P. 1932. Basellaceae. N. Amer. Fl. 21:337–339.

Wilson, P.G. 1983. A taxonomic revision of the tribe Chenopodieae (Chenopodiaceae) in Australia. Nuytsia 4:135–262.

Wilson, P.G. 1984. Chenopodiaceae. In: Flora of Australia Editorial Committee, eds. Flora of Australia. 4:81–317. Australian Government Publishing Service. Canberra.

Wilson, P.G. 1987. Generic status in the Chenopodiaceae. Newslett. Austral. Syst. Bot. Soc. 53:78–85.

Windler, D.R. 1966. A revision of the genus *Neptunia* (Leguminosae). Austral. J. Bot. 14:379–420.

Windler, D.R. 1974. A systematic treatment of the native unifoliate crotalarias of North America (Leguminosae). Rhodora 76:151–204.

Wofford, B.E. 1981. External seed morphology of *Arenaria* (Caryophyllaceae) of the southeastern United States. Syst. Bot. 6:126–135.

Wojciechowski, M.F., M. Lavin, and **M.J. Sanderson.** 2004. A phylogeny of legumes (Leguminosae) based on analysis of the plastid matK gene resolves many well-supported subclades within the family. Amer. J. Bot. 91:1846–1862.

Woldemeskel, M. and **E.L. Styer.** 2010. Feeding behavior-related toxicity due to *Nandina domestica* in cedar waxwings (*Bombycilla cedrorum*). Vet. Med. Int. 2010 Dec 9; 2010:818159. https://doi:10.4061/2010/818159.

Wolf, K.H., M. Gouty, Y.-W. Yang, P.M. Sharp, and **W.-H. Li.** 1989. Date of the monocot-dicot divergence estimate from chloroplast DNA sequence data. Proc. Natl. Acad. Sci. U.S.A. 86:6201–6205.

Wolfe, A.D. and **J.R. Estes.** 1992. Pollination and the function of floral parts in *Chamaecrista fasciculata* (Fabaceae). Amer. J. Bot. 79:314–317.

Wood, C.E., Jr. 1949. The American barbistyled species of *Tephrosia* (Leguminosae). Rhodora 51:193–231, 233–302, 305–364, 369–384.

Wood, C.E., Jr. 1958. The genera of the woody Ranales in the southeastern United States. J. Arnold Arbor. 39:296–346.

Wood, C.E., Jr. 1959. The genera of the Nymphaeaceae and Ceratophyllaceae in the southeastern United States. J. Arnold Arbor. 40:94–112.

Wood, C.E., Jr. 1960. The genera of Sarraceniaceae and Droseraceae of the southeastern United States. J. Arnold Arbor. 41:152–163.

Wood, C.E., Jr. 1961. The genera of Ericaceae in the southeastern United States. J. Arnold Arbor. 42:10–80.

Wood, C.E., Jr. 1966. On the identity of *Drosera brevifolia*. J. Arnold Arbor. 47:89–99.

Wood, C.E., Jr. 1972. Morphology and phytogeography: The classical approach to the study of disjunctions. Ann. Missouri Bot. Gard. 59:107–124.

Wood, C.E., Jr. 1975. The Balsaminaceae in the southeastern United States. J. Arnold Arbor. 56:413–426.

Wood, C.E., Jr. and **R.B. Channell.** 1960. The genera of Ebenales in the southeastern United States. J. Arnold Arbor. 41:1–35.

Wood, J.R.I., T.C. Mitchell, M.A. Carine, D.J. Harris, and **R.W. Scotland.** 2015. A foundation monograph of *Convolvulus* L. (Convolvulceae). PhytoKeys 51:1–282.

Wood, J.R.I., P. Muñoz-Rodríquez, B.R.M. Williams, and **R.W. Scotland.** 2020. A foundation monograph of *Ipomoea* (Convolvulaceae) in the New World. PhytoKeys 143:1–823.

Woods, M. 2023a. *Apios*. In: Flora of North America Editorial Committee, eds. Fl. of North Amer. 11:378–380. Oxford Univ. Press, New York and Oxford.

Woods, M. 2023b. *Rhynchosia*. In: Flora of North America Editorial Committee, eds. Fl. of North Amer. 11:382–389. Oxford Univ. Press, New York and Oxford.

Woodson, R.E., Jr. 1928. Studies in the Apocynaceae. III. A monograph of the genus *Amsonia*. Ann. Missouri Bot. Gard. 15:379–434.

Woodson, R.E., Jr. 1929. Studies in the Apocynaceae. III A. A new species of *Amsonia* from the south-central states. Ann. Missouri Bot. Gard. 16:407–410.

Woodson, R.E., Jr. 1930. Studies in the Apocynaceae. I. A critical study of the Apocynoideae. Ann. Missouri Bot. Gard. 17:1–182.

Woodson, R.W., Jr. 1933. Studies in the Apocynaceae. IV: The American genera of Echitoideae. Ann. Missouri Bot. Gard. 20: 605–790.

Woodson, R.E., Jr. 1938. Apocynaceae. N. Amer. Fl. 29:103–192.

Woodson, R.E., Jr. 1941. The North American Asclepiadaceae. I. Perspective of the genera. Ann. Missouri Bot. Gard. 28: 193–244.

Woodson, R.E., Jr. 1947. Some dynamics of leaf variation in *Asclepias tuberosa*. Ann. Missouri Bot. Gard. 34:353–432.

Woodson, R.E., Jr. 1953. Biometric evidence of natural selection in *Asclepias tuberosa*. Proc. Natl. Acad. Sci. U.S.A. 39:74–79.

Woodson, R.E., Jr. 1954. The North American species of *Asclepias*. Ann. Missouri Bot. Gard. 41:1–211.

Woodson, R.E., Jr. 1962. Butterflyweed revisited. Evolution 16: 168–185.

Woolfe, J.A. 1992. Sweet potato: An untapped food resource. Cambridge Univ. Press, Cambridge, England, UK.

Wooten, J.W. and **A.F. Clewell.** 1971. *Fleischmannia* and *Conoclinum* (Compositae, Eupatorieae) in eastern North America. Rhodora 73:566–574.

Worobiec, G. 2003. New fossil floras from Neogene deposits in the Bełchatów lignite mine. Acta Palaeobot. (Suppl.) 3:3–133.

Wu, W., R. Zhou, Y. Huang, D.E. Boufford, and **S. Shi.** 2010. Molecular evidence for the natural intergeneric hybridization between *Liquidambar* and *Altingia*. J. Pl. Res. 123:231–239.

Wu, Z. 1983. On the significance of Pacific intercontinental discontinuity. Ann. Missouri Bot. Gard. 70:577–590.

Wunderlin, R.P. and **B.F. Hansen.** 2008. Atlas of Florida vascular plants. http://www.plantatlasbeta.usf.edu/Plant.aspx?id=223. Accessed Jun 2008.

Wunderlin, R.P. and **B.F. Hansen.** 2015. Flora of Florida. Volume 2. Dicotyledons: Cobombaceae through Geraniaceae. Univ. Press of Florida. Gainesville.

Wunderlin, R.P., B.F. Hansen, and **L.C. Anderson.** 2002. Plants new to the United States and Florida Sida 20:813–817.

Wunderlin, R.P., B.F. Hansen, and **A.R. Franck.** 2017. Flora of Florida. Volume 4. Dicotyledons: Combretaceae through Amaranthaceae. Univ.Press of Florida. Gainesville.

Wunderlin, R.P., B.F. Hansen, and **A.R. Franck.** 2018. Flora of Florida. Volume 5. Dicotyledons: Gisekiaceae through Boraginaceae. Univ.Press of Florida. Gainesville.

Wunderlin, R.P., B.F. Hansen, and **A.R. Franck.** 2019. Flora of Florida. Volume 6. Dicotyledons: Convolvulaceae through Paulowniaceae. Univ.Press of Florida. Gainesville.

Wunderlin, R.P., B.F. Hansen, and **A.R. Franck.** 2020. Flora of Florida. Volume 7. Dicotyledons: Orobanchaceae through Asteraceae. Univ. Press of Florida. Gainesville.

Wurdack, K.J. 2016a. *Ditrysinia*. Flora of North America Editorial Committee, eds. Fl. North Amer. 12:230. Oxford Univ. Press, New York and Oxford.

Wurdack, K.J. 2016b. *Triadica*. In: Flora of North America Editorial Committee, eds. Fl. North Amer. 12:226–227. Oxford Univ. Press, New York and Oxford.

Wurdack, K.J. and **C.C. Davis.** 2009. Malpighiales phylogenetics: Gaining ground on one of the most recalcitrant clades in the angiosperm Tree of Life. Amer. J. Bot. 96(8):1551–1570.

Wurdack, K.J., M.P. Hoffmann, and **M.W. Chase.** 2005. Molecular phylogenetic analysis of uniovulate Euphorbiaceae (Euphorbiaceae sensu stricto) using plastid rbcL and trnL-F DNA sequences. Amer. J. Bot. 92(8):1397–1420.

Wurdack, K.J., P. Hoffmann, R. Samuel, A. de Bruijn, M. van der Bank, and **M.W. Chase.** 2004. Molecular phylogenetic analysis of Phyllanthaceae (Phyllanthoideae pro parte, Euphorbiaceae sensu lato) using plastid rbcL DNA sequences. Amer. J. Bot. 91(11):1882–1900.

Wyman, D. 1949. The wisterias. Arnoldia (Jamaica Plain) 9:17–28.

Wynne, F.E. 1944. *Drosera* in eastern North America. Bull. Torrey Bot. Club 71:166–174.

Xi, Z., B.R. Ruhfel, H. Schaefer, A.M. Amorim, M. Sugumaran, K.J. Wurdack, P.K. Endress, M.L. Matthews, P.F. Stevens, S. Mathews, and **C.C. Davis.** 2012. Phylogenomics and a posteriori data partitioning resolve the Cretaceous angiosperm radiation Malpighiales. Proc. Natl. Acad. Sci. U.S.A. 109(43):17519–17524.

Xia, X.-M., M.-Q. Yang, C.-L. Cong, and **9 others.** 2022. Spatiotemporal evolution of the global species diversity of *Rhododendron*. Molec. Biol. Evol. 39(1):msab314. https://doi.org/10.1093/molbev/msab314.

Xiang, Q.-Y., S.J. Brunsfeld, D.E. Soltis, and **P.S. Soltis.** 1996 [1997]. Phylogenetic relationships in *Cornus* based on chloroplast DNA restriction sites: Implications for biogeography and character evolution. Syst. Bot. 21:515–534.

Xiang, Q.-Y., M. Moody, D.E. Soltis, C.Z. Fan, and **P.S. Soltis.** 2002. Relationships within Cornales and circumscription of Cornaceae–*matK* and *rbcL* sequence data and effects of outgroups and long branches. Molec. Phylogen. Evol. 24:35–47.

Xiang, Q.-Y. and **D.E. Soltis.** 2001. Dispersal-vicariance analysis on intercontinental disjuncts: Historical-biogeographic implications for angiosperms in the Northern Hemisphere. Int. J. Pl. Sci. 162(Suppl.):S29–S39.

Xiang, Q.-Y., D.E. Soltis, D.R. Morgan, and **P.S. Soltis.** 1993. Phylogenetic relationships of *Cornus* L. sensu lato and putative relatives inferred from *rbc*L sequence data. Ann. Missouri Bot. Gard. 80:723–734.

Xiang, Q.-Y., D.E. Soltis, and **P.S. Soltis.** 1998a. The eastern Asian and eastern and western North American floristic disjuction: Congruent phylogenetic patterns in seven diverse genera. Molec. Phylogen. Evol. 10:178–190.

Xiang, Q.-Y., D.E. Soltis, and **P.S. Soltis.** 1998b. Phylogenetic relationships of Cornaceae and close relatives inferred from *matK* and *rbcL* sequences. Amer. J. Bot. 85:285–297.

Xiang, Q.-Y. and **D.T. Thomas.** 2008. Phylogenomics and a posteriori data partitioning resolve the Cretaceous angiosperm radiation Malpighiales. J. Syst. Evol. 46:349–374.

Xiang, Q.-Y., D.T. Thomas, and **Q.P. Xiang.** 2011. Resolving and dating the phylogeny of Cornales—Effects of taxon sampling, data partitions, and fossil calibrations. Molec. Phylogen. Evol. 59: 123–138.

Xiang, Q.-Y., D.T. Thomas, W.-H. Zhang, S.R. Manchester, and **Z. Murrell.** 2006. Species level phylogeny of the genus *Cornus* (Cornaceae) based on molecular and morphological evidence—Implications for taxonomy and Tertiary intercontinental migration. Taxon 55:9–30.

Xiang, Q.-Y., J.L. Thorne, T. Seo, W. Zhang, D.T. Thomas, and **R.E. Ricklefs.** 2008. Rates of nucleotide substitution in Cornaceae (Cornales)-Pattern of variation and underlying causal factors. Molec. Phylogen. Evol. 49:327–342.

Xiaojie, L., J.M. Baskin, and **C.C. Baskin.** 1999. Contrasting dispersal phenologies in two fleshy-fruited congeneric shrubs, *Rhus aromatica* Ait. and *Rhus glabra* L. (Anacardiaceae). Canad. J. Bot. 77:976–988.

Xu, C. and **D.-Y. Hong.** 2021. Phylogenetic analyses confirm polyphyly of the genus *Campanula* (Campanulaceae s. str.), leading to a proposal for generic reappraisal. J. Syst. Evol. 59(3):475–489.

Xu, J.-B. and **K.-J. Cheng.** 2015. Studies on the alkaloids of the Calycanthaceae and their syntheses. Molecules 20(4): 6715–6738.

Yang, L., D. Su, X. Chang, and **7 others.** 2020. Phylogenomic insights into deep phylogeny of angiosperms based on broad nuclear gene sampling. Plant Comm. 1(2):100027. https://doi: 10.1016/j.xplc.2020.100027.

Yang, M.-Q., R. van Velzen, F.T. Bakker, A. Sattarian, D.-Z. Li, and **T.-S. Yi.** 2013. Molecular phylogenetics and character evolution of Cannabaceae. Taxon 62(3):473–485.

Yang, X.-Y., Z.-F. Wang, W.-C. Luo, X.-Y. Guo, C.-H. Zhang, J.-Q. Liu, and **G.-P. Ren.** 2018. Plastomes of Betulaceae and phylogenetic implications. J. Syst. Evol. https://doi.org/10.1111/jse.12479.

Yarborough, S.C. and **A.M. Powell.** 2006a. *Flaveria*. In: Flora of North America Editorial Committee, eds. Fl. North Amer. 21: 247–250. Oxford Univ. Press, New York and Oxford.

Yarborough, S.C. and **A.M. Powell.** 2006b. *Perityle*. In: Flora of North America Editorial Committee, eds. Fl. North Amer. 21:317–334. Oxford Univ. Press, New York and Oxford.

Yatskievych, G. 2004. How faster to master the aster disaster: A primer on the changing nomenclature of Missouri asters. Missouriensis 25:26–32.

Yatskievych, G. 2006. Steyermark's flora of Missouri. Vol. 2. Missouri Botanical Garden Press in cooperation with the Missouri Dept. of Conservation, St. Louis.

Yatskievych, G. 2013. Styermark's flora of Missouri. Vol. 3. Missouri Botanical Garden Press in cooperation with the Missouri Dept. of Conservation, St. Louis.

Yatskievych, G. 2015. Apodanthaceae. In: Flora of North America Editorial Committee, eds. Fl. North Amer. 6:183–184. Oxford Univ. Press, New York and Oxford.

Yatskievych, G. and **J. Turner.** 1990. Catalogue of the flora of Missouri. Monogr. Syst. Bot. Missouri Bot. Gard. 37:1–362.

Ye, W.-Q., Z.-Y. Yap, P. Li, H.P. Comes, and **Y.-X. Qiu.** 2018. Plastome organization, genome-based phylogeny and evolution of plastid genes in Podophylloideae (Berberidaceae). Molec. Phylogen. Evol. 127:978–987.

Ye, W.-Q., S.-S. Zhu, H.P. Comes, T. Yang, L. Lian, W. Wang, and **Y.-X. Qiu.** 2022. Phylogenomics and diversification drivers of the Eastern Asian–Eastern North American disjunct Podophylloideae. Molec. Phylogen. Evol. 169(5915):107427. https://doi: 10.1016/j.ympev.2022.107427.

Yesilyurt, J.C. 2010. Neotropical Cyrillaceae. In: W. Milliken, B. Klitgård., and A. Baracat. 2009 onwards. Neotropikey—Interactive key and information resources for flowering plants of the Neotropics. https://powo.science.kew.org/taxon/urn:lsid:ipni.org:names:30010307-2/general-information.

Yi, T.S., P.P. Lowry, G.M. Plunkett, and **J. Wen.** 2004. Chromosomal evolution in the Araliaceae and close relatives. Taxon 53: 987–1005.

Ying, J., D.E. Boufford, and **A.R. Brach.** 2011. Berbericaceae. In. Z.-Y. Wu and P.H. Raven, eds. Flora of China (Cucurbitaceae through Valerianaceae with Annonaceae and Berberidaceae). 19:714. Science Press, Beijing, China and Missouri Botanical Garden, St. Louis.

Ying, T.-S. 1983. The floristic relationships of the temperate forest regions of China and the United States. Ann. Missouri Bot. Gard. 70:597–604.

Yoo, M.-J., C.D. Bell, P.S. Soltis, and **D.E. Soltis.** 2005. Divergence times and historical biogeography of Nymphaeales. Syst. Bot. 30:693–704.

Yoon, C.K. 1993. Botanical witness for the prosecution. Science 260:894–895.

Young, D.A. 1978. Reevaluation of the sections of *Rhus* L. subgenus *Lobadium* (Raf.) T. & G. (Anacardiaceae). Brittonia 30:411–415.

Young, J.A. 1991. Tumbleweed. Sci. Amer. 264(3):82–87.

Young-Mathews, A. 2012. Plant guide for field mustard (*Brassica rapa* var. *rapa*). USDA-NRCS, Corvallis Plant Materials Center, OR.

Yu, C.-C. and **K.-F. Chung.** 2017. Why *Mahonia*? Molecular recircumscription of *Berberis* s.l. with the description of two new genera, *Alloberberis* and *Moranothamnus*. Taxon 66:1371–1392.

Yu, L. and **J. Lu.** 2011. Does landscape fragmentation influence sex ratio of dioecious plants? A case study of *Pistacia chinensis* in the Thousand Island Lake Region of China. P.L.o.S. One 6(8): e22903:22901–22908.

Yu, Y., Q.-Y. (J.) Xiang, P.S. Manos, D.E. Soltis, P.S. Soltis, B.-H. Song, S. Cheng, X. Liu, and **G. Wong.** 2017. Whole-genome duplication and molecular evolution in *Cornus* L. (Cornaceae): insights from transcriptome sequences. P.L.o.S. One https://doi:10.1371/journal.pone.0171361.

Yuncker, T.G. 1920. Revision of the North American and West Indian species of *Cuscuta*. Illinois Biol. Monogr. University of Illinois 6(2–3):91–231.

Yuncker, T.G. 1932. The genus *Cuscuta*. Mem. Torrey Bot. Club 18: 111–331.

Yuncker, T.G. 1961. Convolvulaceae: *Cuscuta*. In: C.L. Lundell, ed. Flora of Texas 3:123–150. Texas Research Foundation, Renner.

Zanis, M.J., D.E. Soltis, P.D. Soltis, S. Mathews, and **M.J. Donghue.** 2002. The root of the angiosperms revisited. Proc. Natl. Acad. Sci. U.S.A. 99:6848–6853.

Zanis, M.J., P.S. Soltis, Y.-L. Qiu, E. Zimmer, and **D.E. Soltis.** 2003. Phylogenetic analysis and perianth evolution in basal angiosperms. Ann. Missouri Bot. Gard. 90:129–150.

Zhang, G. and **H. Ma.** 2024. Nuclear phylogenomics of angiosperms and insights into their relationships and evolution. J. Integr. Pl. Biol. 66:546–578. https://doi.org/10.1111/jipb.13609.

Zhang, L.-B. and **M.P. Simmons.** 2006. Phylogeny and delimitation of the Celastrales inferred from nuclear and plastid genes. Syst. Bot. 31:122–137.

Zhang, X.-P. and **A.A. Anderberg.** 2002. Pollen morphology in the ericoid clade of the order Ericales, with special emphasis on Cyrillaceae. Grana 41:201–215.

Zhou, S., S.S. Renner, and **J. Wen.** 2006. Molecular phylogeny and intra- and intercontinental biogeography of Calycanthaceae. Molec. Phylogen. Evol. 39:1–15.

Zhu, J., L. Zhang, B. Ren, M. Chen, R. Li, Y. Zhou, Y. Liang, J. Li, and **Z. Chen.** 2018. Comparative flower and inflorescence organogenesis among genera of Betulaceae: Implications for phylogenetic relationships. Bot. Rev. (Lancaster) 84:79–98.

Zhu, X.-X., X.-Q. Li, S. Liao, C. Du, Y. Wang, Z.-H. Wang, J. Yan, Y.-J. Zuo, and **J.S. Ma.** 2019. Reinstatement of *Isotrema*, a new generic delimitation of *Aristolochia* subgen. *Siphisia* (Aristolochiaceae). Phytotaxa 401:1–23.

Zimmerman, A.D. 1985. Systematics of the genus *Coryphantha* (Cactaceae). Ph.D. dissertation, Univ. of Texas, Austin.

Zimmerman, A.D. and **B.D. Parfitt.** 2003a. *Echinocereus*. In: Flora of North America Editorial Committee, eds. Fl. North Amer. 4:157–174. Oxford Univ. Press, New York and Oxford.

Zimmerman, A.D. and **B.D. Parfitt.** 2003b. *Echinocactus*. In: Flora of North America Editorial Committee, eds. Fl. North Amer. 4: 188–191. Oxford Univ. Press, New York and Oxford.

Zimmerman, A.D. and **B.D. Parfitt.** 2003c. *Hamatocactus*. In: Flora of North America Editorial Committee, eds. Fl. North Amer. 4:218–220. Oxford Univ. Press, New York and Oxford.

Zimmerman, A.D. and **B.D. Parfitt.** 2003d. *Coryphantha*. In: Flora of North America Editorial Committee, eds. Fl. North Amer. 4:221–237. Oxford Univ. Press, New York and Oxford.

Zimmerman, A.D. and **B.D. Parfitt.** 2003e. *Mammillaria*. In: Flora of North America Editorial Committee, eds. Fl. North Amer. 4: 247–257. Oxford Univ. Press, New York and Oxford.

Zohary, D. and **M. Hopf.** 2000. Domestication of plants in the Old World: The origin and spread of cultivated plants in West Asia, Europe and the Nile Valley, 3rd ed. Oxford Univ. Press, New York and Oxford.

Zohary, M. 1982. Plants of the Bible. Cambridge Univ. Press, Cambridge, England, UK.

Zohary, M. and **D. Heller.** 1984. The genus *Trifolium*. Israel Academy of Sciences and Humanities, Jerusalem, Israel.

Zomlefer, W.B. 1994. Guide to flowering plant families. Univ. of North Carolina Press, Chapel Hill.

Zuntini, A.R., T. Carruthers, O. Maurin, and **many others.** 2024. Phylogenomics and the rise of the angiosperms. Nature 629: 843–850. https://doi.org/10.1038/s41586-024-07324-0.

The accepted scientific names of native or naturalized members of the East Texas flora (Vol. 2) are given in [**bold**]. Taxonomic synonyms and names of plants casually mentioned are in [*italics*]. Personal names, place names, and names of plant families not formally treated are given in [Roman type]. Common names are in [SMALL CAPS]. In order to help locate specific and infraspecific epithets, they are given in [Roman type] as separate entries followed on the next line by the genus or genera in which they are named.

A

B

F

G

H

I

M

O

P

Q

R

S

T

U

V

W

X

Y

Colophon

Five hundred copies were printed, using soy based inks, via offset lithography.

This book was printed and bound in Korea by OnGlobal Printing Co., LTD. The binding is flexbind using 350 gsm C1S.
Stock for the endpaper binding is 180 gsm woodfree/acid-free, White,
MooRim Paper, manufactured by Namhan Paper Company, LTD., South Korea.
Stock for the text pages is 70 gsm woodfree/acid-free, New classic paper, White,
manufactured by HanSol Paper Company, South Korea.
Archival permanence was kept in mind throughout the project.
Book layout, production, and print-ready file preparation was provided by Rebecca Horn, rhorngraphics, Plano, Texas.
Robert George, M.S., Project Research Assistant: Texas Floras Projects, was responsible for the selection,
screening, scanning, and proofing of all botanical illustrations and managed the layout
and production of the botanical illustration and map pages.
Fonts used throughout the book are ITC Berkeley Oldstyle Medium Std., Italic Std.,
and Black Std.; Myriad Pro Regular, Italic, Semibold, and Bold; and Symbol MT.
Visual icons (called dingbats) are used for ease of reference throughout the book.
To represent these specific "at-a-glance" icons, special digitized designs called Texicons were created,
while other icons were digitized using existing commercial typefaces Botanical MT and Warning Pi.
The majority of the book was electronically produced on the Macintosh platform
with Adobe InDesign®, Adobe Illustrator®, and Adobe Photoshop® programs.
The PC platform was used to produce the layouts of the botanical illustration pages,
with InDesign and Photoshop programs; map pages were made using QGIS (a GIS/mapping application),
Python scripting language, and a custom map generation script to make each county map.
Some map images and data (the Texas base map and counties) were from the U.S. Census Bureau.
These specifications are given in the spirit of sharing within the botanical community.
We hope our efforts are well-received and will contribute to furthering an interest
in the "art & science" found throughout the botanical world.

The Eri-Monotropa icon used on the front and back pages was created by Rebecca Horn,
using art produced by P.J.F. Turpin. The source of the art is
Dictionnaire des sciences naturelles (1816–1830).

Alphabetical List of Families

Volume Two: Acanthaceae through Fabaceae known to occur in East Texas

Alphabetical List of Counties

Wholly or Partially in East Texas

Anderson
Angelina
Austin
Bastrop
Bell
Bexar
Bowie
Brazos
Burleson
Caldwell
Camp
Cass
Cherokee
Collin
Colorado
Comal
Dallas
Delta
Denton
DeWitt
Ellis
Falls
Fannin
Fayette
Franklin
Freestone
Goliad
Gonzales
Grayson
Gregg
Grimes
Guadalupe
Hardin
Harris
Harrison
Hays
Henderson
Hill
Hopkins
Houston
Hunt
Jackson
Jasper
Jefferson
Kaufman
Lamar
Lavaca
Lee
Leon
Liberty
Limestone
Madison
Marion
McLennan
Milam
Montgomery
Morris
Nacogdoches
Navarro
Newton
Orange
Panola
Polk
Rains
Red River
Robertson
Rockwall
Rusk
Sabine
San Augustine
San Jacinto
Shelby
Smith
Titus
Travis
Trinity
Tyler
Upshur
Van Zandt
Victoria
Walker
Waller
Washington
Wharton
Williamson
Wilson
Wood